Atomic Numbers and Atomic Masses of the Elements

Based on ${}^{12}_{6}C$. Numbers in parentheses are the mass numbers of the most stable isotopes of radioactive elements.

Element	Symbol	Atomic Number	Atomic Mass	Element	Symbol	Atomic Number	Atomic Mass
Actinium	Ac	89	(227)	Meitnerium	Mt	109	(276)
Aluminum	Al	13	26.98	Mendelevium	Md	101	(258)
Americium	Am	95	(243)	Mercury	Hg	80	200.59
Antimony	Sb	51	121.76	Molybdenum	Mo	42	95.94
Argon	Ar	18	39.95	Neodymium	Nd	60	144.24
Arsenic	As	33	74.92	Neon	Ne	10	20.18
Astatine	At	85	(210)	Neptunium	Np	93	(237)
Barium	Ba	56	137.33	Nickel	Ni	28	58.69
Berkelium	Bk	97	(247)	Niobium	Nb	41	92.91
Beryllium	Be	4	9.01	Nitrogen	N	7	14.01
Bismuth	Bi	83	208.98	Nobelium	No	102	(259)
Bohrium	Bh	107	(264)	Osmium	Os	76	190.23
Boron	B	5	10.81	Oxygen	O	8	16.00
Bromine	Br	35	79.90	Palladium	Pd	46	106.42
Cadmium	Cd	48	112.41	Phosphorus	P	15	30.97
Calcium	Ca	20	40.08	Platinum	Pt	78	195.08
Californium	Cf	98	(251)	Plutonium	Pu	94	(244)
Carbon	C	6	12.01	Polonium	Po	84	(209)
Cerium	Ce	58	140.12	Potassium	K	19	39.10
Cesium	Cs	55	132.91	Praseodymium	Pr	59	140.91
Chlorine	Cl	17	35.45	Promethium	Pm	61	(145)
Chromium	Cr	24	52.00	Protactinium	Pa	91	(231)
Cobalt	Co	27	58.93	Radium	Ra	88	(226)
Copernicium	Cn	112	(285)	Radon	Rn	86	(222)
Copper	Cu	29	63.55	Rhenium	Re	75	186.21
Curium	Cm	96	(247)	Rhodium	Rh	45	102.91
Darmstadtium	Ds	110	(271)	Roentgenium	Rg	111	(280)
Dubnium	Db	105	(262)	Rubidium	Rb	37	85.47
Dysprosium	Dy	66	162.50	Ruthenium	Ru	44	101.07
Einsteinium	Es	99	(252)	Rutherfordium	Rf	104	(263)
Erbium	Er	68	167.26	Samarium	Sm	62	150.36
Europium	Eu	63	151.96	Scandium	Sc	21	44.96
Fermium	Fm	100	(257)	Seaborgium	Sg	106	(266)
Fluorine	F	9	19.00	Selenium	Se	34	78.96
Francium	Fr	87	(223)	Silicon	Si	14	28.09
Gadolinium	Gd	64	157.25	Silver	Ag	47	107.87
Gallium	Ga	31	69.72	Sodium	Na	11	22.99
Germanium	Ge	32	72.64	Strontium	Sr	38	87.62
Gold	Au	79	196.97	Sulfur	S	16	32.07
Hafnium	Hf	72	178.49	Tantalum	Ta	73	180.95
Hassium	Hs	108	(277)	Technetium	Tc	43	(98)
Helium	He	2	4.00	Tellurium	Te	52	127.60
Holmium	Ho	67	164.93	Terbium	Tb	65	158.93
Hydrogen	H	1	1.01	Thallium	Tl	81	204.38
Indium	In	49	114.82	Thorium	Th	90	(232)
Iodine	I	53	126.90	Thulium	Tm	69	168.93
Iridium	Ir	77	192.22	Tin	Sn	50	118.71
Iron	Fe	26	55.85	Titanium	Ti	22	47.87
Krypton	Kr	36	83.80	Tungsten	W	74	183.84
Lanthanum	La	57	138.91	Uranium	U	92	(238)
Lawrencium	Lr	103	(262)	Vanadium	V	23	50.94
Lead	Pb	82	207.19	Xenon	Xe	54	131.29
Lithium	Li	3	6.94	Ytterbium	Yb	70	173.04
Lutetium	Lu	71	174.97	Yttrium	Y	39	88.91
Magnesium	Mg	12	24.31	Zinc	Zn	30	65.41
Manganese	Mn	25	54.94	Zirconium	Zr	40	91.22

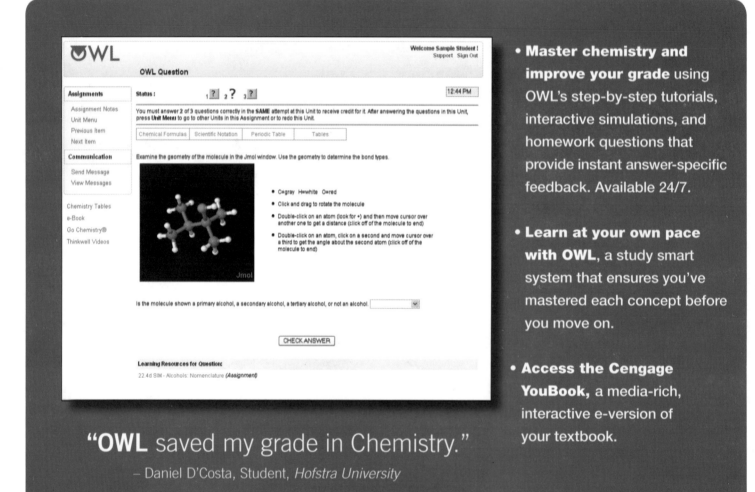

Sixth Edition

Organic and Biological CHEMISTRY

H. STEPHEN STOKER

Weber State University

BROOKS/COLE
CENGAGE Learning™

Australia • Brazil • Japan • Korea • Mexico • Singapore • Spain • United Kingdom • United States

BROOKS/COLE
CENGAGE Learning™

Organic and Biological Chemistry,
Sixth Edition

H. Stephen Stoker

Publisher: Mary Finch

Developmental Editor: Alyssa White

Editorial Assistant: Alicia Landsberg

Senior Media Editor: Lisa Weber

Marketing Manager: Nicole Hamm

Marketing Assistant: Julie Stefani

Marketing Communications Manager: Linda Yip

Content Project Manager: Teresa L. Trego

Design Director: Rob Hugel

Art Director: Maria Epes

Print Buyer: Judy Inouye

Rights Acquisitions Specialist: Dean Dauphinais

Production Service: PreMediaGlobal

Text Designer: tani hasegawa

Photo Researcher: Bill Smith Group

Text Researcher: Sue C. Howard

Copy Editor: PreMediaGlobal

OWL Producers: Stephen Battisti, Cindy Stein, David Hart (Center for Educational Software Development, University of Massachusetts, Amherst)

Cover Designer: tani hasegawa

Cover Image: All image Copyright Getty Images. From top to bottom: Jason Isley-Scubazoo; MIYAKO/a.collectionRF; Hola Images

Compositor: PreMediaGlobal

Library of Congress Control Number: 2011937395

ISBN-13: 978-1-133-10395-0

ISBN-10: 1-133-10395-2

Brooks/Cole
20 Davis Drive
Belmont, CA 94002-3098
USA

Cengage Learning is a leading provider of customized learning solutions with office locations around the globe, including Singapore, the United Kingdom, Australia, Mexico, Brazil, and Japan. Locate your local office at **www.cengage.com/global**

Cengage Learning products are represented in Canada by Nelson Education, Ltd.

To learn more about Brooks/Cole, visit **www.cengage.com/brookscole**

Purchase any of our products at your local college store or at our preferred online store **www.cengagebrain.com**

Printed in the United States of America
1 2 3 4 5 6 7 15 14 13 12 11

Brief Contents

Contents

PART II BIOLOGICAL CHEMISTRY

Preface

The positive responses of instructors and students who used the previous five editions of this text have been gratifying—and have led to the new sixth edition that you hold in your hands. This new edition represents a renewed commitment to the goals I initially set when writing the first edition. These goals have not changed with the passage of time. My initial and still ongoing goals are to write a text in which:

- The needs are simultaneously met for the many students in the fields of nursing, allied health, biological sciences, agricultural sciences, food sciences, and public health who are required to take such a course.
- The development of chemical topics always starts out at ground level. The students who will use this text often have little or no background in chemistry and hence approach the course with a good deal of trepidation. This "ground level" approach addresses this situation.
- The amount and level of mathematics is purposefully restricted. Clearly, some chemical principles cannot be divorced entirely from mathematics and, when this is the case, appropriate mathematical coverage is included.
- The early chapters focus on fundamental chemical principles, and the later chapters—built on these principles—develop the concepts and applications central to the fields of organic chemistry and biochemistry.

Focus on Biochemistry Most students taking this course have a greater interest in the biochemistry portion of the course than the preceding two parts. But biochemistry, of course, cannot be understood without a knowledge of the fundamentals of organic chemistry, and understanding organic chemistry in turn depends on knowing the key concepts of general chemistry. Thus, in writing this text, I essentially started from the back and worked forward. I began by determining what topics would be considered in the biochemistry chapters and then tailored the organic and then general sections to support that presentation. Users of the previous editions confirm that this approach ensures an efficient but thorough coverage of the principles needed to understand biochemistry.

Exciting New Art Program See the story of general, organic, and biological chemistry come alive on each page! In addition to the narrative, the new art and photography program helps tell a very important story—the story of ourselves and the world around us. Chemistry is everywhere! A new integrated talking label system in the art and photography program gives key figures a "voice" and helps students learn more effectively.

Emphasis on Visual Support I believe strongly in visual reinforcement of key concepts in a textbook; thus this book uses art and photos wherever possible to teach key concepts. Artwork is used to make connections and highlight what is important for the student to know. Reaction equations use color to emphasize the portions of a molecule that undergo change. Colors are likewise assigned to things like valence shells and classes of compounds to help students follow trends. Computer-generated, three-dimensional molecular models accompany many discussions in the organic and biochemistry sections of the text. Color photographs show applications of chemistry to help make concepts real and more readily remembered.

Visual summary features, called *Chemistry at a Glance,* pull together material from several sections of a chapter to help students see the larger picture. For example, Chapter 3 features a *Chemistry at a Glance* on the shell–subshell–orbital interrelationships; Chapter 10 presents buffer solutions; Chapter 2 includes IUPAC

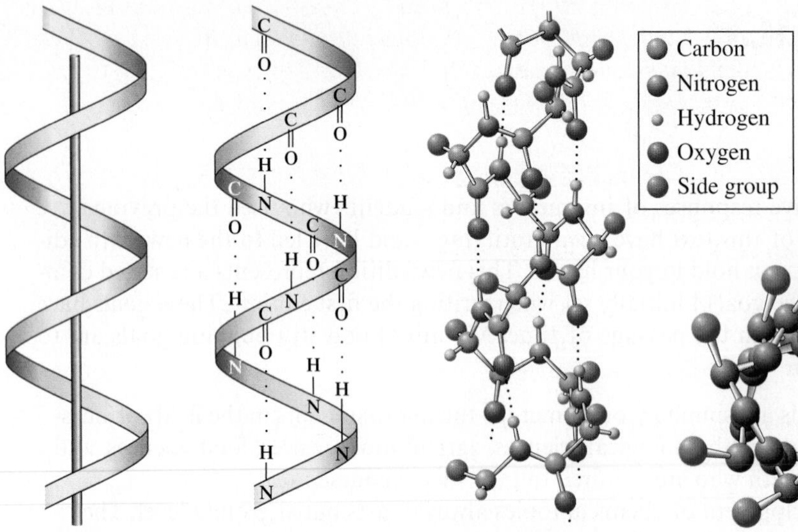

a Arrangement of protein backbone with no detail shown.

b Backbone arrangement with hydrogen-bonding interactions shown.

c Backbone atomic detail shown, as well as hydrogen-bonding interactions.

d Top view of an α helix showing that amino acid side chains (R groups) point away from the long axis of the helix.

A solid has a definite shape and a definite volume.

A liquid has an indefinite shape—it takes the shape of its container—and a definite volume.

A gas has an indefinite shape and an indefinite volume—it assumes the shape and volume of its container.

a b c

nomenclature for alkanes, alkenes, and alkynes; and Chapter 11 summarizes DNA replication. The *Chemistry at a Glance* feature serves both as an overview for the student reading the material for the first time and as a review tool for the student preparing for exams. Given the popularity of the *Chemistry at a Glance* summaries in the previous editions, several new ones have been added and several existing ones have been updated or expanded. New topics selected for *Chemistry at a Glance* boxes include:

- Metabolic reactions that involve nitrogen-containing compounds

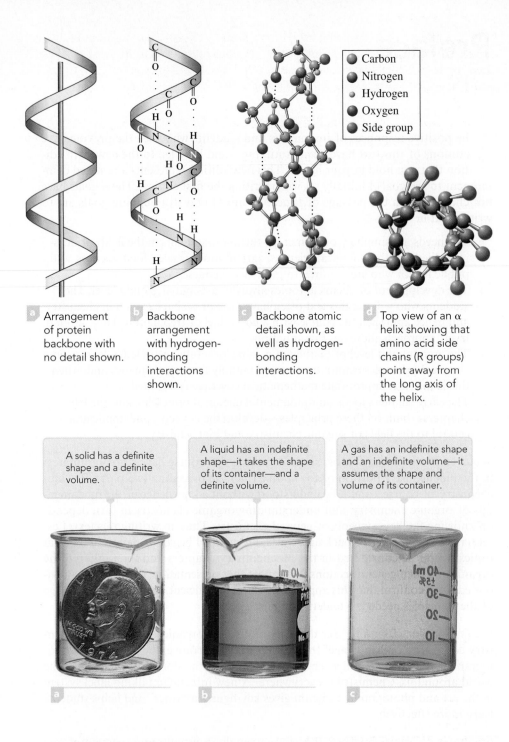

Emphasis on Chemical Relevancy In every chapter, *Chemical Connections* feature boxes show chemistry as it appears in everyday life. These boxes focus on topics that are relevant to a student's own life in terms of health issues, societal issues, and environmental issues. Many of the previous edition's feature "essays" have been updated to include the latest research findings. New topics selected for *Chemical Connections* emphasis in this edition are:

- Red wine and resveratrol
- Colostrum: immunoglobulins and much more
- Enzymes, prescription drugs, and the "grapefruit effect"
- Anticancer drugs that inhibit DNA synthesis

Commitment to Student Learning In addition to the study help *Chemistry at a Glance* offers, the text is built on a strong foundation of learning aids designed to help students master the course material.

- **Problem-solving pedagogy.** Because problem solving is often difficult for students in this course to master, I have taken special care to provide support to help students build their skills. Within the chapters, worked-out ***Examples*** follow the explanation of many concepts. These examples walk students through the thought processes involved in problem solving, carefully outlining all of the steps involved. Each is immediately followed by a ***Practice Exercise*** to reinforce the information just presented.

Diversity of Worked-out *Examples* Worked-out examples are a standard feature in the general chemistry portion of all textbooks for this market. This relates primarily to the mathematical nature of many general chemistry topics. In most texts, fewer worked-out examples appear in the organic chemistry chapters, and still fewer (almost none) are found in the biochemistry portion due to decreased dependence of the topical matter on mathematical concepts. Such is not the case in this textbook. All chapters in the latter portions of the text contain numerous worked-out examples. Several additional worked-out examples have been added to this new edition. Newly added worked-out examples involve the following topics:

- Predicting product identity in aldehyde/ketone redox reactions
- Changing a Fischer projection formula to a Haworth projection formula
- Drawing structural formulas for disaccharide hydrolysis products
- Determining relationships among DNA base sequences, mRNA base sequences, codons, anticodons, and amino acids
- Predicting the effect of a DNA point mutation

- **Margin notes.** Liberally distributed throughout the text, *margin notes* provide tips for remembering and distinguishing between concepts, highlight links across chapters, and describe interesting historical background information. An additional 70 margin notes, distributed throughout all chapters, have been added to the text in this revision.
- **Defined terms.** All definitions are highlighted in the text when they are first presented, using boldface and italic type. Each defined term appears as a complete sentence; students are never forced to deduce a definition from context. In addition, the definitions of all terms appear in the combined *Index/Glossary* found at the end of the text. A major emphasis in this new edition has been "refinements" of the defined terms. All defined terms were reexamined to see if they could be stated with greater clarity. The result was a "rewording" of many defined terms.
- **Concepts to Remember review.** A concise review of key concepts presented in each chapter appears at the end of the chapter, placed just before the end-of-chapter problems. This is a helpful aid for students as they prepare for exams.
- **End-of-chapter problems.** An extensive set of end-of-chapter problems complements the worked-out examples within the chapters. These end-of-chapter

problems are organized by topic and paired, with each pair testing similar material. The answer to the odd-numbered member of the pair is given at the back of the book. New to this edition are two problem-set features:

Problems denoted with a ▲ involve concepts found not only in the section under consideration but also concepts found in one or more earlier sections of the chapter.

Problems denoted with a ● cover concepts included in a *Chemical Connections* feature box found within the chapter.

Nearly 1100 (1092 to be exact) of the 3321 total end-of-chapter problems are new to this edition of the text. Although the number of end-of-chapter problems would have significantly exceeded that of most other texts even without these additions, the total number of such problems has been increased by 345.

Content Changes Coverage of a number of topics has been expanded in this edition. The two driving forces in expanded coverage considerations were (1) the requests of users and reviewers of the previous editions and (2) my desire to incorporate new research findings, particularly in the area of biochemistry, into the text. Topics with expanded coverage include:

- Halogenated methanes
- Ethanol uses
- Polyphenols
- Sunscreen and suntanning agents
- Differences between carbonyl and acyl compounds
- Decongestants and antihistamines
- Guidelines for identifying chiral centers
- Cyclic monosaccharide terminology
- Saponifiable and nonsaponifiable lipids
- Essential amino acids
- Extremozymes
- Prescription drugs that inhibit enzyme activity
- Individual B vitamins
- Nucleosides and nucleotides
- Recombinant DNA and genetic engineering
- Carboxylate ions in metabolic pathways
- B vitamins and the common metabolic pathway
- Lactate fermentation
- B vitamins and carbohydrate metabolism
- B vitamins and lipid metabolism
- Glutamate and aspartate production via transamination
- B vitamins and protein metabolism

Exciting New Media Options!

Chemistry CourseMate

CourseMate

Instant Access (two semester) ISBN: 978-1-133-35064-4

This book includes Chemistry CourseMate, a complement to your textbook. Chemistry CourseMate includes an interactive eBook, interactive teaching and learning tools such as quizzes, flashcards, videos, and more. Chemistry CourseMate also includes the Engagement Tracker, a first-of-its-kind tool that monitors student engagement in the course. Go to **login.cengage.com** to access these resources. Look for the CourseMate icon, which denotes a resource available within CourseMate.

GOB OWL Problems We've doubled the end-of-chapter problems that can now be assigned in OWL for GOB, the online homework and learning system available with this book.

***General, Organic, and Biological Chemistry,* 6th edition, Hybrid Version with OWL**
ISBN 13: 978-1-133-11064-4
ISBN 10: 1-133-11064-9

This briefer, paperbound version of *General, Organic, and Biological Chemistry* does not contain the end-of-chapter problems—these problems are available and assignable in OWL, the online homework and learning system for this book. Access to OWL and the Cengage YouBook is packaged with the hybrid version. The Cengage YouBook is the full version of the book with all end-of-chapter problem sets and questions.

Supporting Materials

OWL for General, Organic, and Biochemistry/Allied Health
Instant Access OWL with Cengage YouBook (6 months) ISBN: 978-1-133-17435-6
Instant Access OWL with Cengage YouBook (24 months) ISBN: 978-1-133-17429-5

OWL
Online Web Learning

By Roberta Day, Beatrice Botch, and David Gross of the University of Massachusetts, Amherst; William Vining of The State University of New York at Oneonta; and Susan Young of Hartwick College. **OWL** Online Web Learning offers more assignable, gradable content (including end-of-chapter questions specific to this textbook) and more reliability and flexibility than any other system. OWL's powerful course management tools allow instructors to control due dates, number of attempts to correctly answer questions, and whether students see answers or receive feedback on how to solve problems. OWL includes the **Cengage YouBook**, an interactive and customizable Flash-based eBook. Instructors can publish Web links, modify the textbook narrative as needed with the text edit tool, quickly reorder entire sections and chapters, and hide any content they don't teach to create an eBook that perfectly matches their syllabus. The Cengage YouBook includes animated figures, video clips, highlighting, notes, and more.

Developed by chemistry instructors for teaching chemistry, OWL is the only system specifically designed to support **mastery learning**, in which students work as long as required to master each chemical concept and skill. OWL has already helped hundreds of thousands of students master chemistry through a wide range of assignment types, including tutorials, interactive simulations, and algorithmically generated homework questions that provide instant, answer-specific feedback.

OWL is continually being enhanced with online learning tools to address the various learning styles of today's students, such as:

- **Quick Prep** review courses that help students learn essential skills to succeed in general and organic chemistry
- **Jmol** molecular visualization program for rotating molecules and measuring bond distances and angles
- **Go Chemistry**® mini video lectures on key concepts that students can play on their computers or download to their video iPods, smart phones, or personal video players

In addition, when you become an OWL user, you can expect service that goes far beyond the ordinary. To learn more or to see a demo, please contact your Cengage Learning representative or visit us at **www.cengage.com/owl**.

For Instructors

PowerLecture Instructor's CD/DVD Package with JoinIn® and ExamView®
ISBN: 978-1-133-10425-4

This digital library and presentation tool includes:

- **PowerPoint® lecture slides** written for this text by **Sreerama Lakshima** that instructors can customize by importing their own lecture slides or other materials.
- **Image libraries** that contain digital files for figures, photographs, and numbered tables from the text, as well as multimedia animations in a variety of digital formats. Instructors may use these files to print transparencies, create their own PowerPoint slides, and supplement their lectures.
- Digital files of the **Complete Solutions Manual** prepared by H. Stephen Stoker, Danny V. White, and Joanne A. White.
- Word files for the **Test Bank** prepared by Mark Erickson; Hartwick College.
- Digital files of the **Instructor's Resource Manual for the Lab Manual** prepared by G. Lynn Carlson.
- Sample chapters from the **Student Solutions Manual and Study Guide** written by Danny V. White and Joanne A. White.
- **ExamView testing software** that enables instructors to create, deliver, and customize tests using the more than 1500 test bank questions written specifically for this text.
- **JoinIn student response (clicker) questions** written for this book for use with the classroom response system of the instructor's choice.

Instructor Companion Site Supporting materials are available to qualified adopters. Please consult your local Cengage Learning sales representative for details. Go to **login.cengage.com**, find this textbook, and choose "Instructor Companion Site" to see samples of these materials, request a desk copy, locate your sales representative, download the WebCT or Blackboard versions of the Test Bank.

For Students

Visit CengageBrain.com To access these and additional course materials, please visit **www.cengagebrain.com**. At the CengageBrain.com home page, search for this textbook's ISBN (located on the back cover of your book). This will take you to the product page, where these resources can be found. (Instructors can log in at **login.cengage.com**.)

Instant Access Go Chemistry® for General Chemistry (27-video set)
ISBN: 978-0-495-38228-7
Pressed for time? Miss a lecture? Need more review? Go Chemistry for General Chemistry is a set of 27 downloadable mini video lectures. Developed by award-winning chemists, Go Chemistry helps you quickly review essential topics—whenever and wherever you want! Each video contains animations and problems and can be downloaded to your computer desktop or portable video player (like iPod or iPhone) for convenient self-study and exam review. Selected Go Chemistry videos have e-flashcards to briefly introduce a key concept and then test student understanding of the basics with a series of questions. OWL includes five Go Chemistry videos. Professors can package a printed access card for Go Chemistry with the textbook. Students can enter the ISBN above at **www.cengagebrain.com** to download two free videos or to purchase instant access to the 27-video set or to individual videos.

CengageBrain.com App Now, students can prepare for class anytime and anywhere using the CengageBrain.com application developed specifically for the Apple iPhone® and iPod touch®. This application allows students to access free study materials—book-specific quizzes, flash cards, related Cengage Learning materials, and more—so they can study the way they want to, when they want to . . . even on the go. To learn more about this complimentary application, please visit **www.cengagebrain.com**. Also available on iTunes.

Study Guide with Selected Solutions for General, Organic, and Biological Science, 6th edition By Danny V. White and Joanne A. White

The perfect way to build problem-solving skills, prepare for exams, and get the grade you want! This useful resource reinforces skills with activities and practice problems for each chapter. After completing the end-of-chapter exercises, you can check your answers for the odd-numbered questions. ISBN: 978-1-133-10423-0

Lab Manual for General, Organic, and Biological Science, 6th edition By G. Lynn Carlson

Each experiment in this manual was selected to match topics in the textbook and includes an introduction, a procedure, a page of pre-lab exercises about the concepts the lab illustrates, and a report form. Some experiments also include a scenario that places the experiment in a real-world context. In addition, each experiment has a link to a set of references and helpful online resources. ISBN: 978-1-133-10406-3

Survival Guide for General, Organic, and Biochemistry By Richard Morrison, Charles H. Atwood, and Joel Caughran (University of Georgia).

Available free in a package with any Cengage chemistry text or available for separate purchase at **www.cengagebrain.com**. Modeled after Atwood's widely popular *General Chemistry Survival Guide*, this straightforward, thorough guide helps students make the most of their study time for optimal exam results. The *Survival Guide* is packed with examples and exercises to help students master concepts and improve essential problem-solving skills through detailed step-by-step problem-solving sequences. This reader-friendly guide gives students the competency—and confidence—they need to survive and thrive in the GOB course. ISBN: 978-0-495-55469-1.

Chemistry CourseMate

Instant Access (two semester) ISBN: 978-1-133-35064-4

This book includes Chemistry CourseMate, which helps you make the grade. Chemistry CourseMate includes an interactive eBook with highlighting, note taking and search capabilities, as well as interactive learning tools such as quizzes, flashcards, videos, and more. Go to **login.cengage.com** to access these resources, and look for the CourseMate icon to find resources related to your text in Chemistry CourseMate.

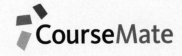

Acknowledgments

The publication of a book such as this requires the efforts of many more people than merely the author. Special thanks to the Editorial and Production Team at Cengage: Alyssa White, my Development Editor; Mary Finch, my Publisher; Teresa Trego, my Senior Content Production Manager; Lisa Weber and Stephanie Van Camp, who were in charge of the media program; and Nicole Hamm, my Marketing Manager. I would also like to thank Patrick Franzen, my Senior Project Manager at PreMediaGlobal, and my Photo Researcher Sarah Bonner (Bill Smith Group).

Apple, iPhone, iPod touch, and iTunes are trademarks of Apple Inc., registered in the U.S. and other countries.

I also appreciate the time and expertise of my reviewers, who read my manuscript and provided many helpful comments.

Special thanks to my accuracy reviewers:

David Shinn, *United States Merchant Marine Academy*

Keith Baessler, *United States Merchant Marine Academy*

Reviewers of the 5th edition:

Jennifer Adamski, *Old Dominion University*
M. Reza Asdjodi, *University of Wisconsin—Eau Claire*
Irene Gerow, *East Carolina University*
Ernest Kho, *University of Hawaii at Hilo*
Larry L. Land, *University of Florida*
Michael Myers, *California State University—Long Beach*
H. A. Peoples, *Las Positas College*
Shashi Rishi, *Greenville Technical College*
Steven M. Socol, *McHenry County College*

Reviewers of the 6th edition:

Maryfran Barber, *Wayne State University*
Keri Clemens, *Sierra College*
John Haseltine, *Kennesaw State University*
Maria Longas, *Purdue University*
Jennifer Powers, *Kennesaw State University*
Heather Sklenicka, *Rochester Community and Technical College/Science*
Angie Spencer, *Greenville Technical College*
David Tramontozzi, *Macomb CC/Science*

About the Cover

Learning Chemistry is like learning a new language—a language that will help you understand and communicate with the world around you in a new and exciting way. It reveals a world beyond what we can see and know with our eyes alone. It is also about you.

The sea as seen in this photo is swimming with fish, but it is also swimming with molecules—oxygen, hydrogen, carbon dioxide and much, much more.

Have you ever thought about the seasons and the chemical changes that occur in each one? Chemistry can explain how and why the world around you is changing. Like the leaves in this photograph—caretenoids are responsible for the brilliant reds, oranges, and yellows.

Chemistry is also a tool—a tool that can be used to help you and the world around you. For example chemistry makes it possible to produce and recycle plastics like the polyethylene water bottles pictured here. Plastic water bottles play an important role in distributing water across the world and recycling those plastics plays a role in preserving the environment.

Saturated Hydrocarbons

1

© Bill Ross/CORBIS

Crude oil (petroleum) constitutes the largest and most important natural source for saturated hydrocarbons, the simplest type of organic compound. Here is shown a pump and towers associated with obtaining crude oil from underground deposits.

CHAPTER OUTLINE

OWL

Sign in to OWL at **www.cengage.com/owl** to view tutorials and simulations, develop problem-solving skills, and complete online homework assigned by your professor.

This chapter is the first of six that deal with the subject of organic chemistry and organic compounds. Organic compounds are the chemical basis for life itself, as well as an important component of the current high standard of living enjoyed by people in many countries. Proteins, carbohydrates, enzymes, and hormones are organic molecules. Organic compounds also include natural gas, petroleum, coal, gasoline, and many synthetic materials such as dyes, plastics, and clothing fibers.

1.1 Organic and Inorganic Compounds

During the latter part of the eighteenth century and the early part of the nineteenth century, chemists began to categorize compounds into two types: organic and inorganic. Compounds obtained from living organisms were called *organic* compounds, and compounds obtained from mineral constituents of the Earth were called *inorganic* compounds.

During this early period, chemists believed that a special "vital force" supplied by a living organism was necessary for the formation of an organic compound. This concept was proved incorrect in 1828 by the German chemist Friedrick Wöhler. Wöhler heated an aqueous solution of two inorganic compounds, ammonium chloride and silver cyanate, and obtained urea (a component of urine).

$$NH_4Cl + AgNCO \longrightarrow \underset{\text{Urea}}{(NH_2)_2CO} + AgCl$$

The historical origins of the terms *organic* and *inorganic* involve the following conceptual pairings:

organic—living *organisms*
inorganic—*inanimate materials*

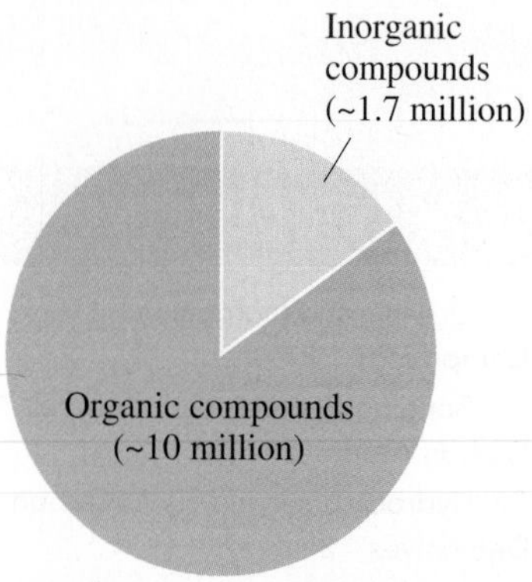

Figure 1.1 Sheer numbers is one reason why organic chemistry is a separate field of chemical study. Approximately 10 million organic compounds are known, compared to "just" 1.7 million inorganic compounds.

Soon other chemists had successfully synthesized organic compounds from inorganic starting materials. As a result, the vital-force theory was completely abandoned.

The terms *organic* and *inorganic* continue to be used in classifying compounds, but the definitions of these terms no longer reflect their historical origins. **Organic chemistry** *is the study of hydrocarbons (compounds of carbon and hydrogen) and their derivatives.* Nearly all compounds found in living organisms are still classified as organic compounds, as are many compounds that have been synthesized in the laboratory and have never been found in a living organism. **Inorganic chemistry** *is the study of all substances other than hydrocarbons and their derivatives.*

In essence, organic chemistry is the study of the compounds of one element (carbon), and inorganic chemistry is the study of the compounds of the other 117 elements. This unequal partitioning occurs because there are approximately 10 million organic compounds and only an estimated 1.7 million inorganic compounds (Figure 1.1). This is an approximately 6:1 ratio between organic and inorganic compounds.

1.2 Bonding Characteristics of the Carbon Atom

Some textbooks define organic chemistry as the study of carbon-containing compounds. Almost all carbon-containing compounds qualify as organic compounds. However, the oxides of carbon, carbonates, cyanides, and metallic carbides are classified as inorganic rather than organic compounds. Inorganic carbon compounds involve carbon atoms that are not bonded to hydrogen atoms (CO, CO_2, Na_2CO_3, and so on).

Why does the element carbon form six times as many compounds as all the other elements combined? The answer is that carbon atoms have the unique ability to bond to each other in a wide variety of ways that involve long chains of carbon atoms or cyclic arrangements (rings) of carbon atoms. Sometimes both chains and rings of carbon atoms are present in the same molecule.

The variety of covalent bonding "behaviors" possible for carbon atoms is related to carbon's electron configuration. Carbon is a member of Group IVA of the periodic table, so carbon atoms possess four valence electrons. In compound formation, four additional valence electrons are needed to give carbon atoms an octet of valence electrons. These additional electrons are obtained by electron sharing (covalent bond formation). The sharing of *four* valence electrons requires the formation of *four* covalent bonds.

Carbon can meet this four-bond requirement in three different ways:

1. *By bonding to four other atoms.* This situation requires the presence of four single bonds.

Four single bonds

2. *By bonding to three other atoms.* This situation requires the presence of two single bonds and one double bond.

Carbon atoms in organic compounds, in accordance with the octet rule, always form four covalent bonds.

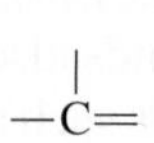

Two single bonds and one double bond

3. *By bonding to two other atoms.* This situation requires the presence of either two double bonds or a triple bond and a single bond.

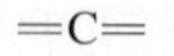

Two double bonds

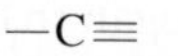

One triple bond and one single bond

1.3 Hydrocarbons and Hydrocarbon Derivatives

The field of organic chemistry encompasses the study of hydrocarbons and hydrocarbon derivatives (Section 1.1). A **hydrocarbon** *is a compound that contains only carbon atoms and hydrogen atoms.* Thousands of hydrocarbons are known. A **hydrocarbon derivative** *is a compound that contains carbon and hydrogen and one or more*

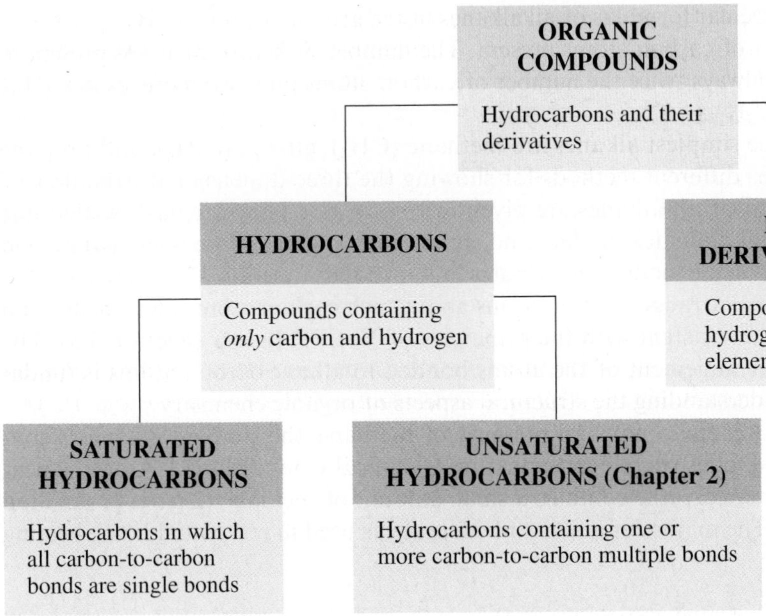

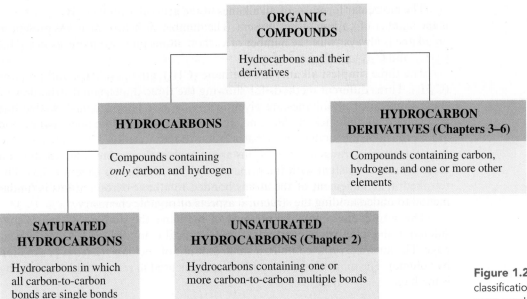

Figure 1.2 A summary of classification terms for organic compounds.

additional elements. Additional elements commonly found in hydrocarbon derivatives include O, N, S, P, F, Cl, and Br. Millions of hydrocarbon derivatives are known.

Hydrocarbons may be divided into two large classes: saturated and unsaturated. A **saturated hydrocarbon** *is a hydrocarbon in which all carbon–carbon bonds are single bonds.* Saturated hydrocarbons are the simplest type of organic compound. An **unsaturated hydrocarbon** *is a hydrocarbon in which one or more carbon–carbon multiple bonds (double bonds, triple bonds, or both) are present.* In general, saturated and unsaturated hydrocarbons undergo distinctly different chemical reactions.

Saturated hydrocarbons are the subject of this chapter. Unsaturated hydrocarbons are considered in the next chapter. Figure 1.2 summarizes the terminology presented in this section.

Two categories of saturated hydrocarbons exist, those with *acyclic* carbon atom arrangements and those with *cyclic* carbon atom arrangements. The term *acyclic* means "not cyclic." The following notations contrast simple acyclic and cyclic arrangements of six-carbon atoms.

The term *saturated* has the general meaning that there is no more room for something. Its use with hydrocarbons comes from early studies in which chemists tried to add hydrogen atoms to various hydrocarbon molecules. Compounds to which no more hydrogen atoms could be added (because they already contained the maximum number) were called saturated, and those to which hydrogen could be added were called unsaturated.

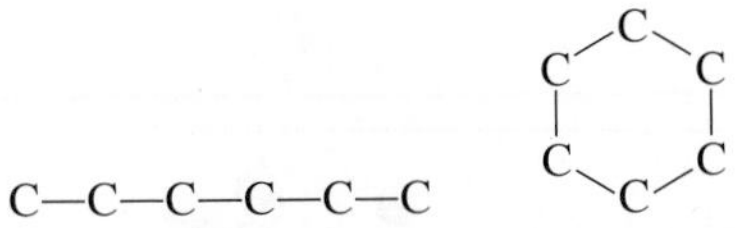

Acyclic Cyclic

Sections 1.4 through 1.11 of this chapter treat the subject of saturated hydrocarbons with *acyclic* carbon atom arrangements. A discussion of saturated hydrocarbons with *cyclic* carbon atom arrangements follows in Sections 1.12 through 1.14.

The prefix *a-* in acyclic means "not," so an acyclic hydrocarbon has an arrangement of carbon atoms within its structure that is *not* cyclic.

1.4 Alkanes: Acyclic Saturated Hydrocarbons

An **alkane** *is a saturated hydrocarbon in which the carbon atom arrangement is acyclic.* Thus an alkane is a hydrocarbon that contains only carbon–carbon single bonds (saturated) and has no rings of carbon atoms (acyclic).

The molecular formulas of all alkanes fit the general formula C_nH_{2n+2}, where *n* is the number of carbon atoms present. The number of hydrogen atoms present in an alkane is always twice the number of carbon atoms plus two more, as in C_4H_{10}, C_5H_{12}, and C_8H_{18}.

The three simplest alkanes are methane (CH_4), ethane (C_2H_6), and propane (C_3H_8). Three different methods for showing the three-dimensional structures of these simplest of all alkanes are given in Figure 1.3. They are dash-wedge-line structures, ball-and-stick models, and space-filling models. Note how each carbon atom in each of the models participates in four bonds (Section 1.2). Note also that the geometrical arrangement of atoms about each carbon atom is tetrahedral, an arrangement consistent with the principles of VSEPR theory (Section 5.8). The tetrahedral arrangement of the atoms bonded to alkane carbon atoms is fundamental to understanding the structural aspects of organic chemistry.

The natural environmental presence of methane, the simplest alkane, is considered in the focus on relevancy feature Chemical Connections 1-A on the next page. The air we breathe contains a small amount of methane (2 parts per million by volume). The major component of natural gas used in residential home heating is methane.

1.5 Structural Formulas

The structures of alkanes, as well as other types of organic compounds, are generally represented in two dimensions rather than three (Figure 1.3) because of the difficulty in drawing the latter. These two-dimensional structural representations make no attempt to portray accurately the bond angles or molecular geometry of molecules. Their purpose is to convey information about which atoms in a molecule are bonded to which other atoms.

Two-dimensional structural representations for organic molecules are called structural formulas. A **structural formula** *is a two-dimensional structural representation that shows how the various atoms in a molecule are bonded to each other.*

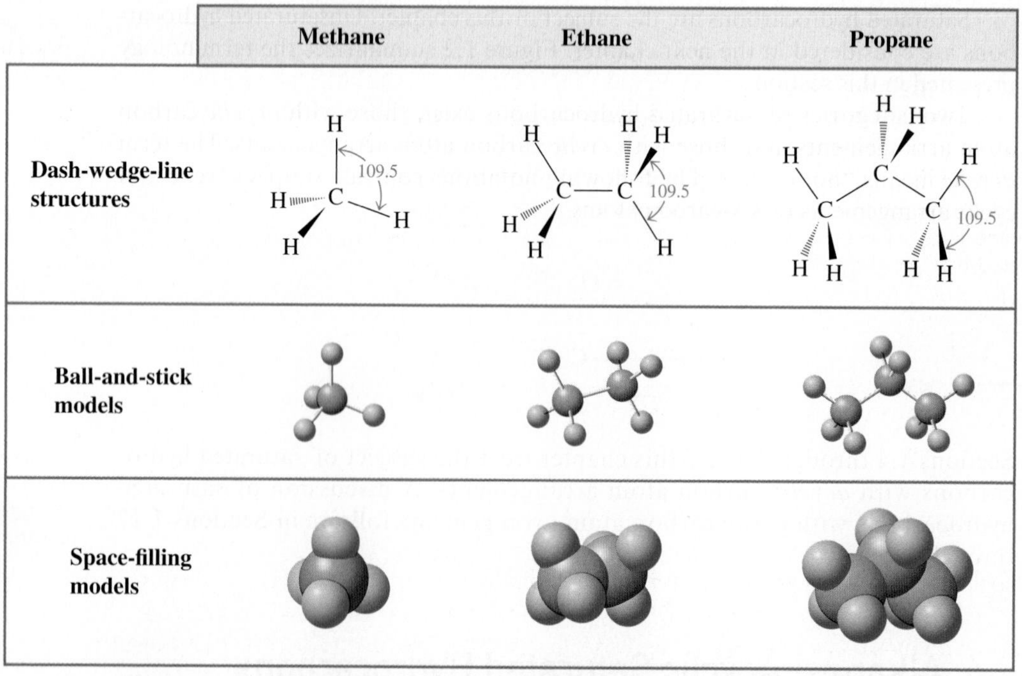

Figure 1.3 Three different three-dimensional ways of representing the structures of methane, ethane, and propane: dash-wedge-line structure, ball-and-stick model, and space-filling model.

CHEMICAL CONNECTIONS 1-A

The Occurrence of Methane

Methane (CH_4), the simplest of all hydrocarbons, is a major component of the atmospheres of Jupiter, Saturn, Uranus, and Neptune but only a minor component of Earth's atmosphere (see the accompanying table). Earth's gravitational field, being weaker than that of the large outer planets, cannot retain enough hydrogen (H_2) in its atmosphere to permit the formation of large amounts of methane; H_2 molecules (the smallest and fastest-moving of all molecules) escape from it into outer space.

The small amount of methane present in Earth's atmosphere comes from terrestrial sources. The decomposition of animal and plant matter in an oxygen-deficient environment—swamps, marshes, bogs, and the sediments of lakes—produces methane. A common name for methane, marsh gas, refers to the production of methane in this manner.

Composition of Earth's Atmosphere (in parts per million by volume)

Major Components		Minor Components	
nitrogen	780,800	argon	9340
oxygen	209,500	carbon dioxide	314
		neon	18
		helium	5
		methane	2
		krypton	1

Bacteria that live in termites and in the digestive tracts of plant-eating animals have the ability to produce methane from plant materials (cellulose). The methane output of a large cow (via belching and flatulence) can reach 20 liters per day. Livestock are the source of about 20% of methane emissions to the atmosphere each day.

Methane entering the atmosphere from terrestrial sources presents an environmental problem. Methane, like carbon dioxide, is a "greenhouse gas" that contributes to global warming. Methane is 15 to 30 times more efficient than carbon dioxide (the primary greenhouse in trapping reradiated heat from the Earth. Fortunately, atmospheric levels of methane (2.0 ppm by volume) are much lower than those of carbon dioxide (388 ppm by volume).

It should be noted that some greenhouse gas presence in the atmosphere is not only desirable but necessary. The average surface temperature of Earth is about 15ºC. In the absence of "normal amounts" of greenhouse gases, the Earth's average surface temperature would drop to about –18ºC, which would not be a good situation for humans. Of concern are the ever increasing amounts of greenhouse gases entering the atmosphere as the result of human activities and the associated increase, still small, in Earth's average surface temperature.

Methane gas is also found associated with coal and petroleum deposits. Methane associated with coal mines is considered a hazard. If left to accumulate, it can form pockets where air is not present, and asphyxiation of miners can occur. When mixed with air in certain ratios, it can also present an explosion hazard. Methane associated with petroleum deposits is most often recovered, processed, and marketed as *natural gas.* The processed natural gas used in the heating of homes is 85% to 95% methane by volume. Because methane is odorless, an odorant (smelly compound) must be added to the processed natural gas used in home heating. Otherwise, natural gas leaks could not be detected.

Doug Martin/Photo Researchers, Inc.

Decomposition of plant and animal matter in marshes is a source of methane gas.

Structural formulas are of two types: expanded structural formulas and condensed structural formulas. An **expanded structural formula** *is a structural formula that shows all atoms in a molecule and all bonds connecting the atoms.* When written out, expanded structural formulas generally occupy a lot of space, and condensed structural formulas represent a shorthand method for conveying the same information. A **condensed structural formula** *is a structural formula that uses groupings of atoms, in which central atoms and the atoms connected to them are written as a group, to* *convey molecular structural information.* The expanded and condensed structural formulas for methane, ethane, and propane follow.

Structural formulas, whether expanded or condensed, do not show the geometry (shape) of the molecule. That information can be conveyed only by 3-D drawings or models such as those in Figure 1.3.

	Methane	Ethane	Propane
Expanded structural formula	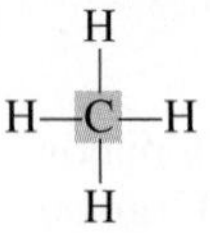	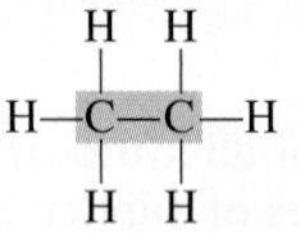	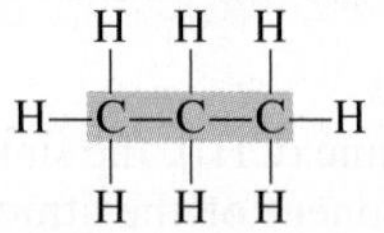
Condensed structural formula	CH_4	$CH_3—CH_3$	$CH_3—CH_2—CH_3$

The condensed structural formula for propane, $CH_3—CH_2—CH_3$, is interpreted in the following manner: The first carbon atom is bonded to three hydrogen atoms, and its fourth bond is to the middle carbon atom. The middle carbon atom, besides its bond to the first carbon atom, is also bonded to two hydrogen atoms and to the last carbon atom. The last carbon atom has bonds to three hydrogen atoms in addition to its bond to the middle carbon atom. As is always the case, each carbon atom has four bonds (Section 1.2).

The condensed structural formulas of hydrocarbons in which a long chain of carbon atoms is present are often condensed even more. The formula

$$CH_3—CH_2—CH_2—CH_2—CH_2—CH_2—CH_2—CH_3$$

can be further abbreviated as

$$CH_3—(CH_2)_6—CH_3$$

where parentheses and a subscript are used to denote the number of $—CH_2—$ groups in the chain.

It is important to note that expanded structural formulas show all bonds within a molecule and that condensed structural formulas show only certain bonds—the bonds between carbon atoms. Specifically, the bond line in the condensed structural formula

$$CH_3—CH_3$$

denotes the bond between the first carbon atom and the second carbon atom; it is not a bond between hydrogen atoms and the second carbon atom.

In situations where the focus is solely on the arrangement of carbon atoms in an alkane, *skeletal structural formulas* that omit the hydrogen atoms are often used. A **skeletal structural formula** *is a structural formula that shows the arrangement and bonding of carbon atoms present in an organic molecule but does not show the hydrogen atoms attached to the carbon atoms.*

C—C—C—C—C	means the same as	$CH_3—CH_2—CH_2—CH_2—CH_3$
Skeletal structural formula		Condensed structural formula

The skeletal structural formula still represents a unique alkane because we know that each carbon atom shown must have enough hydrogen atoms attached to it to give the carbon four bonds.

1.6 Alkane Isomerism

The molecular formulas CH_4, C_2H_6, and C_3H_8 represent the alkanes methane, ethane, and propane, respectively. Next in the alkane molecular formula sequence (C_nH_{2n+2}) is C_4H_{10}, which would be expected to be the molecular formula of the four-carbon alkane. A new phenomenon arises, however, when an alkane has four or more carbon atoms. There is more than one structural formula that is consistent with the molecular formula. Consequently, more than one compound exists with that molecular formula. The topic of *isomerism* addresses this situation.

Isomers *are compounds that have the same molecular formula (that is, the same numbers and kinds of atoms) but that differ in the way the atoms are arranged.* Isomers, even though they have the same molecular formula, are always different compounds with different properties.

The word *isomer* comes from the Greek *isos*, which means "the same," and *meros*, which means "parts." Isomers have the same parts put together in different ways.

There are two four-carbon alkane isomers, the compounds *butane* and *isobutane.* Both have the molecular formula C_4H_{10}.

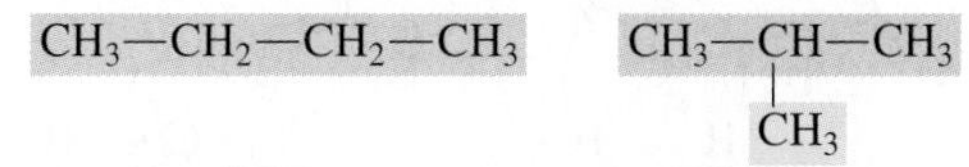

Butane Isobutane

Butane and isobutane are different compounds with different properties. Butane has a boiling point of -1°C and a melting point of -138°C, whereas the corresponding values for isobutane are -12°C and -159°C.

Contrasting the two C_4H_{10} isomers structurally, note that butane has a chain of four carbon atoms. It is an example of a continuous-chain alkane. A **continuous-chain alkane** *is an alkane in which all carbon atoms are connected in a continuous nonbranching chain.* The other C_4H_{10} isomer, isobutane, has a chain of three carbon atoms with the fourth carbon attached as a branch on the middle carbon of the three-carbon chain. It is an example of a branched-chain alkane. A **branched-chain alkane** *is an alkane in which one or more branches (of carbon atoms) are attached to a continuous chain of carbon atoms.*

The existence of isomers necessitates the use of structural formulas in organic chemistry. Isomers always have the same molecular formula and different structural formulas.

There are three isomers for alkanes with five carbon atoms (C_5H_{12}):

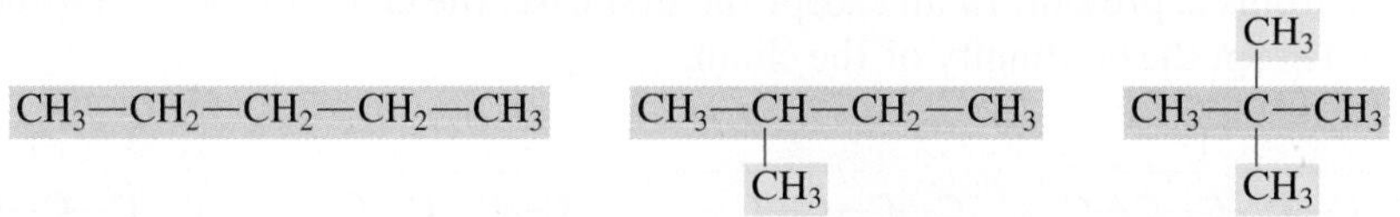

Pentane Isopentane Neopentane

Figure 1.4 shows space-filling models for the three isometric C_5 alkanes. Note how neopentane, the most branched isomer, has the most compact, most spherical three-dimensional shape.

The number of possible alkane isomers increases dramatically with increasing numbers of carbon atoms in the alkane, as shown in Table 1.1. Such isomerism is one of the major reasons for the existence of so many organic compounds.

Several different types of isomerism exist. The alkane isomerism examples discussed in this section are examples of *constitutional isomerism.* **Constitutional isomers** *are isomers that differ in the connectivity of atoms, that is, in the order in which atoms are attached to each other within molecules.* In Section 1.14 of this chapter and in later chapters of the text, additional types of isomerism besides *constitutional* isomerism will be encountered. When carbohydrates, lipids, and proteins are considered (Chapters 7 to 9), it will be common to find that different isomers elucidate different responses within the human body. Often, when many isomers are possible with the same molecular formula, only one isomer will be physiologically active.

Constitutional isomers are also frequently called *structural isomers.* The general characteristics of such isomers, independent of which name is used, are the same molecular formula and different structural formulas.

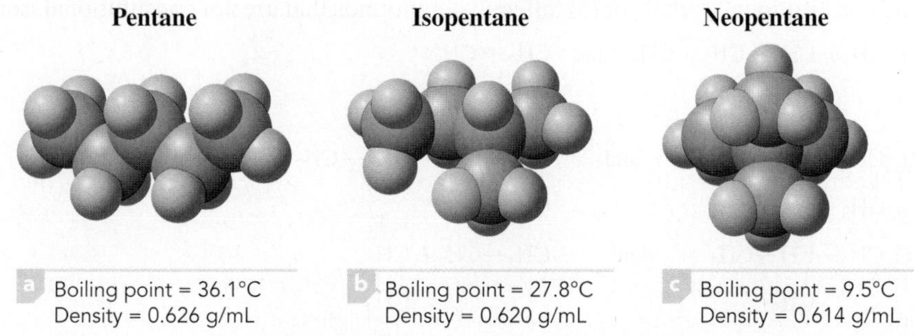

Figure 1.4 Space-filling models for the three isomeric C_5H_{12} alkanes.

Table 1.1 Number of Isomers Possible for Alkanes of Various Carbon Chain Lengths

Molecular Formula	Possible Number of Isomers
CH_4	1
C_2H_6	1
C_3H_8	1
C_4H_{10}	2
C_5H_{12}	3
C_6H_{14}	5
C_7H_{16}	9
C_8H_{18}	18
C_9H_{20}	35
$C_{10}H_{22}$	75
$C_{15}H_{32}$	4,347
$C_{20}H_{42}$	336,319
$C_{25}H_{52}$	36,797,588
$C_{30}H_{62}$	4,111,846,763

1.7 Conformations of Alkanes

Rotation about carbon–carbon single bonds is an important property of alkane molecules. Two groups of atoms in an alkane connected by a carbon–carbon single bond can rotate with respect to one another around that bond, much as a wheel rotates around an axle.

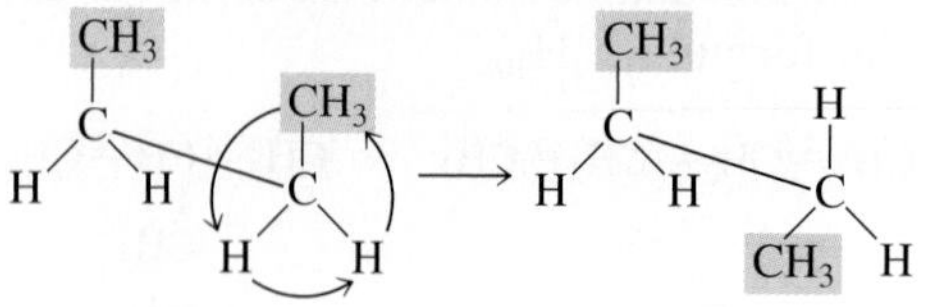

As a result of rotation around single bonds, alkane molecules (except for methane) can exist in infinite numbers of orientations, or conformations. A **conformation** *is the specific three-dimensional arrangement of atoms in an organic molecule at a given instant that results from rotations about carbon–carbon single bonds.*

The following skeletal formulas represent four different conformations for a continuous-chain six-carbon alkane molecule.

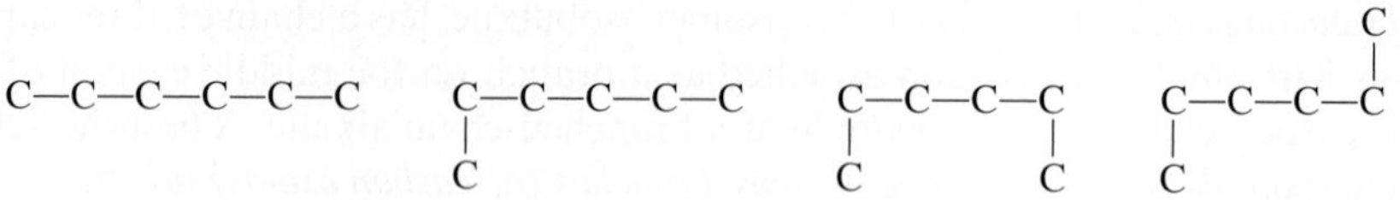

All four skeletal formulas represent the same molecule; that is, they are different conformations of the same molecule. In all four cases, a continuous chain of six carbon atoms is present. In all except the first case, the chain is "bent," but bends do not disrupt the continuity of the chain.

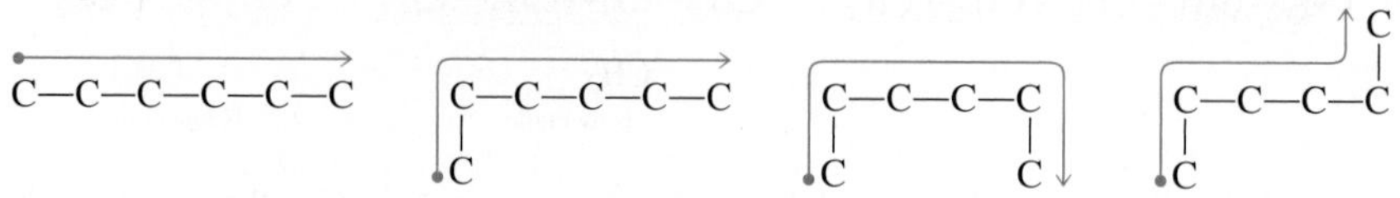

Note that the structures

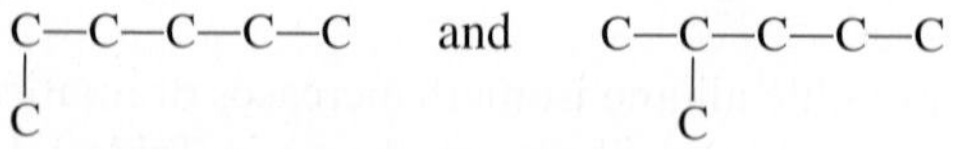

are not two conformations of the same alkane but, rather, represent two different alkanes. The first structure involves a continuous chain of six carbon atoms, and the second structure involves a continuous chain of five carbon atoms to which a branch is attached. There is no way that a continuous chain of six carbon atoms can be found in the second structure without "back-tracking," and "back-tracking" is not allowed.

Learning to recognize several different versions (conformations) of a molecule as being "the same" is an important skill. Like friends, molecules can be recognized independent of whether they are sitting, reclining, or standing.

EXAMPLE 1.1 Recognizing Different Conformations of a Molecule and Constitutional Isomers

Determine whether the members of each of the following pairs of structural formulas represent (1) different conformations of the same molecule (2) different compounds that are constitutional isomers or (3) different compounds that are not constitutional isomers.

a.

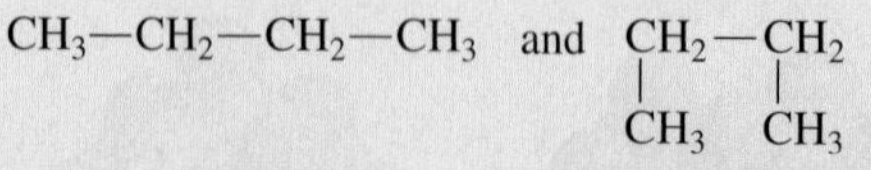

b.

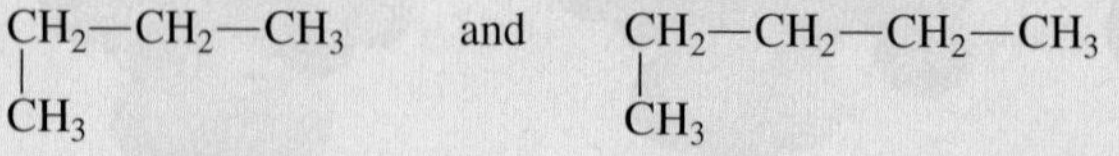

c.

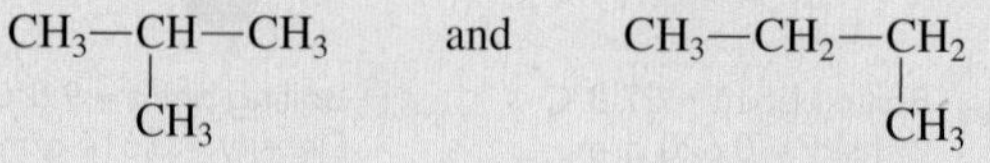

Solution

a. Both molecules have the molecular formula C_4H_{10}. The connectivity of carbon atoms is the same for both molecules: a continuous chain of four carbon atoms. For the second structural formula, the four-carbon-atom chain has two "bends" in it, which is fine because of the free rotation associated with single bonds in alkanes.

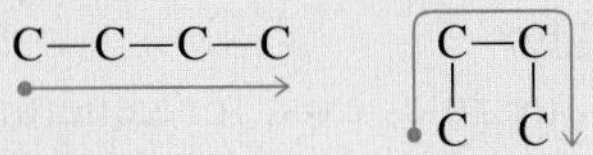

With the same molecular formula and the same connectivity of atoms, these two structural formulas are conformations of the same molecule.

b. The molecular formula of the first compound is C_4H_{10}, and that of the second compound is C_5H_{12}. Thus the two structural formulas represent different compounds that are not constitutional isomers. Constitutional isomers must have the same molecular formula.

c. Both molecules have the same molecular formula, C_4H_{10}. The connectivity of atoms is different. In the first case, a chain of three carbon atoms with a branch off the chain is present. In the second case, a continuous chain of four carbon atoms is present.

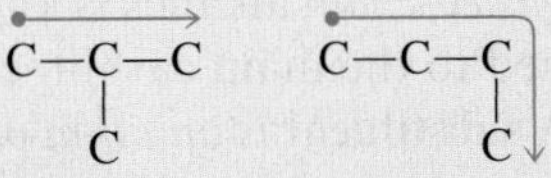

These two structural formulas are those of constitutional isomers.

Practice Exercise 1.1

Determine whether the members of each of the following pairs of structural formulas represent (1) different conformations of the same molecule (2) different compounds that are constitutional isomers or (3) different compounds that are not constitutional isomers.

a. $CH_3—CH_2—CH_2—CH_2—CH_3$ and

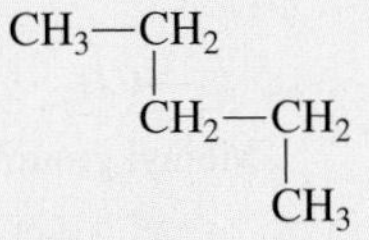

b.

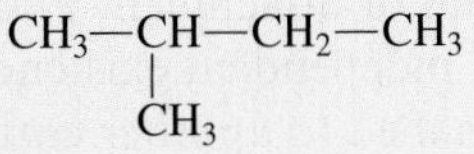

and

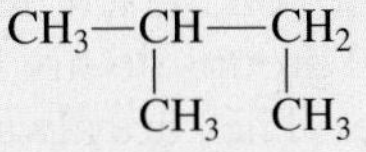

c.

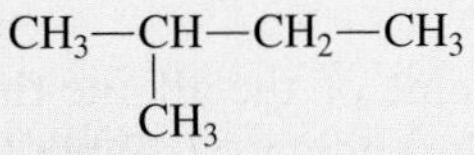

and

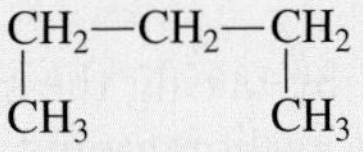

Answers: **a.** Different conformations; **b.** Different conformations; **c.** Constitutional isomers

The condensed structural formulas for branched-chain alkanes can be further condensed to give linear (straight-line) condensed structural formulas. The linear condensed structural formula for the alkane

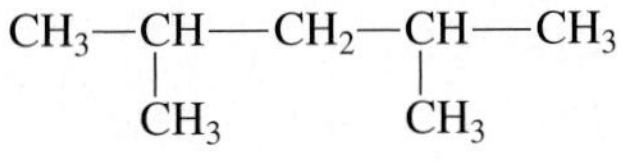

is

$CH_3—CH—(CH_3)—CH_2—CH—(CH_3)—CH_3$ or $(CH_3)_2—CH—CH_2—CH—(CH_3)_2$

Groups in parentheses in such formulas are understood to be attached to the carbon atom that *precedes* the group in the structural formula, unless the parenthesized group starts the formula. In that case, the group is attached to the carbon atom that *follows.* Writing structural formulas in this format is done primarily to reduce the vertical space that the structural formula takes.

1.8 IUPAC Nomenclature for Alkanes

IUPAC is pronounced "eye-you-pack."

Continuous-chain alkanes are also frequently called *straight-chain alkanes* and *normal-chain alkanes.*

It is important to know the prefixes given in the second column of Table 1.2. This is the way to count from 1 to 10 in "organic chemistry language."

When relatively few organic compounds were known, chemists arbitrarily named them using what today are called *common names.* These common names gave no information about the structures of the compounds they described. However, as more organic compounds became known, this nonsystematic approach to naming compounds became unwieldy.

Today, formal systematic rules exist for generating names for organic compounds. These rules, which were formulated and are updated periodically by the International Union of Pure and Applied Chemistry (IUPAC), are known as *IUPAC rules.* The advantage of the IUPAC naming system is that it assigns each compound a name that not only identifies it but also enables its structural formula to be drawn.

In Section 1.6, the existence of both continuous-chain alkanes and branched-chain alkanes was noted, with the former being the simpler type of compound from a structural viewpoint. IUPAC names for the first ten *continuous-chain* alkanes are given in Table 1.2. Note that all of these names end in *-ane,* the characteristic ending for all alkane names. Note also that beginning with the five-carbon alkane, Greek numerical prefixes are used to denote the actual number of carbon atoms in the continuous chain.

The key to naming *branched-chain* alkanes is knowing the name of the branch or branches that are attached to the main carbon chain. These branches are formally called substituents. A **substituent** *is an atom or group of atoms attached to a chain (or ring) of carbon atoms.* Note that *substituent* is a general term that applies to carbon-chain attachments in all organic molecules, not just alkanes.

For branched-chain alkanes, the substituents are specifically called *alkyl groups.* An **alkyl group** *is the group of atoms that would be obtained by removing a hydrogen atom from an alkane.*

The two most commonly encountered alkyl groups are the two simplest: the one-carbon and two-carbon alkyl groups. Their formulas and names are

$\wr$—CH_3	$\wr$—CH_2—CH_3
Methyl group	Ethyl group

In these formulas the "squiggle" on the left denotes the point of attachment to the carbon chain. Note that alkyl groups do not lead a stable, independent existence; that is, they are not molecules. They are always found attached to another entity (usually a carbon chain).

Alkyl groups are named by taking the stem of the name of the alkane that contains the same number of carbon atoms and adding the ending *-yl.* Table 1.3 gives the names for small continuous-chain alkyl groups.

Table 1.2 IUPAC Names for the First Ten Continuous-Chain Alkanes*

Molecular Formula	IUPAC Prefix	IUPAC Name	Condensed Structural Formula
CH_4	meth-	methane	CH_4
C_2H_6	eth-	ethane	CH_3—CH_3
C_3H_8	prop-	propane	CH_3—CH_2—CH_3
C_4H_{10}	but-	butane	CH_3—CH_2—CH_2—CH_3
C_5H_{12}	pent-	pentane	CH_3—CH_2—CH_2—CH_2—CH_3
C_6H_{14}	hex-	hexane	CH_3—CH_2—CH_2—CH_2—CH_2—CH_3
C_7H_{16}	hept-	heptane	CH_3—CH_2—CH_2—CH_2—CH_2—CH_2—CH_3
C_8H_{18}	oct-	octane	CH_3—CH_2—CH_2—CH_2—CH_2—CH_2—CH_2—CH_3
C_9H_{20}	non-	nonane	CH_3—CH_2—CH_2—CH_2—CH_2—CH_2—CH_2—CH_2—CH_3
$C_{10}H_{22}$	dec-	decane	CH_3—CH_2—CH_2—CH_2—CH_2—CH_2—CH_2—CH_2—CH_2—CH_3

* The IUPAC naming system also includes prefixes for naming continuous-chain alkanes that have more than 10 carbon atoms, but we will not consider them in this text.

Table 1.3 Names for the First Six Continuous-Chain Alkyl Groups

Number of Carbons	Structural Formula	Stem of Alkane Name	Suffix	Alkyl Group Name
1	$-CH_3$	meth-	–yl	methyl
2	$-CH_2-CH_3$	eth-	–yl	ethyl
3	$-CH_2-CH_2-CH_3$	prop-	–yl	propyl
4	$-CH_2-CH_2-CH_2-CH_3$	but-	–yl	butyl
5	$-CH_2-CH_2-CH_2-CH_2-CH_3$	pent-	–yl	pentyl
6	$-CH_2-CH_2-CH_2-CH_2-CH_2-CH_3$	hex-	–yl	hexyl

The ending *-yl,* as in meth*yl,* eth*yl,* prop*yl,* and but*yl,* appears in the names of all alkyl groups.

Formal IUPAC rules for naming branched-chain alkanes are as follows:

Rule 1: *Identify the longest continuous carbon chain (the parent chain), which may or may not be shown in a straight line, and name the chain.*

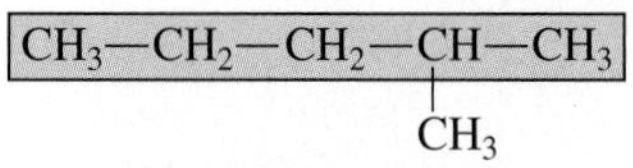

The parent chain name is *pentane,* because it has five carbon atoms.

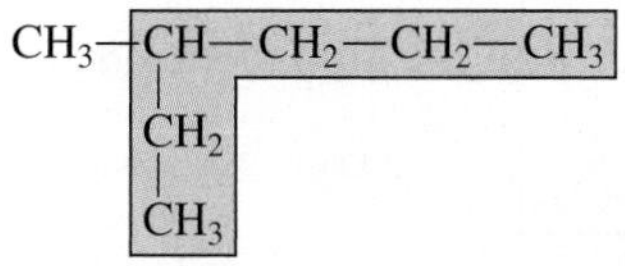

The parent chain name is *hexane,* because it has six carbon atoms.

An additional guideline for identifying the longest continuous carbon chain: If two different carbon chains in a molecule have the same largest number of carbon atoms, select as the parent chain the one with the larger number of substituents (alkyl groups) attached to the chain.

Rule 2: *Number the carbon atoms in the parent chain from the end of the chain nearest a substituent (alkyl group).*

There are always two ways to number the chain (either from left to right or from right to left). This rule gives the first-encountered alkyl group the lowest possible number.

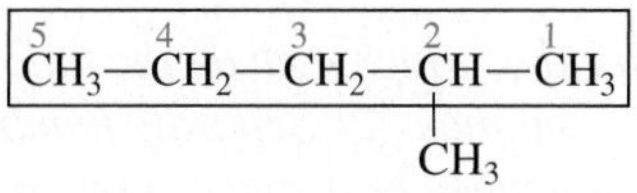

Right-to-left numbering system

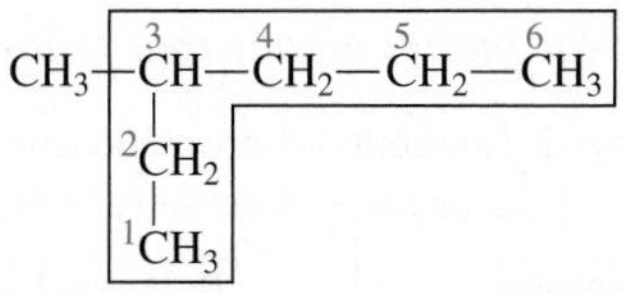

Left-to-right numbering system

Additional guidelines for numbering carbon-atom chains:

1. If both ends of the chain have a substituent the same distance in, number from the end closest to the second-encountered substituent.
2. If there are substituents equidistant from each end of the chain and there is no third substituent to use as the "tie-breaker," begin numbering at the end nearest the substituent that has alphabetical priority—that is, the substituent whose name occurs first in the alphabet.

Rule 3: *If only one alkyl group is present, name and locate it (by number), and prefix the number and name to that of the parent carbon chain.*

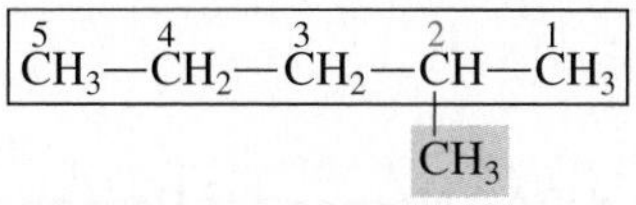

2-Methylpentane

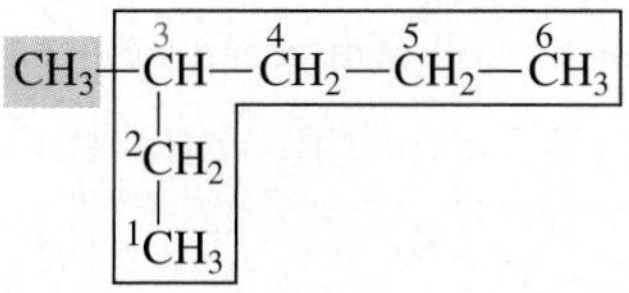

3-Methylhexane

Note that the name is written as one word, with a hyphen between the number (location) and the name of the alkyl group.

Rule 4: *If two or more of the same kind of alkyl group are present in a molecule, indicate the number with a Greek numerical prefix (di-, tri-, tetra-, penta-, and so forth). In addition, a number specifying the location of each identical group must be included. These position numbers, separated by commas, precede the numerical prefix. Numbers are separated from words by hyphens.*

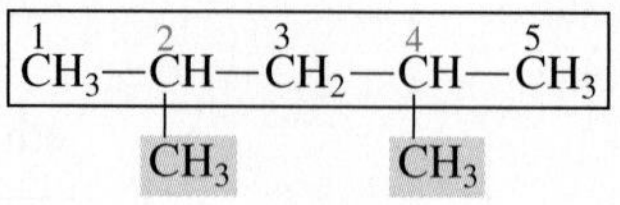

2,4-Dimethylpentane

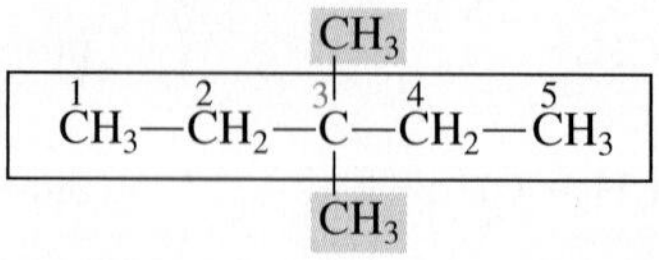

3,3-Dimethylpentane

There must be as many numbers as there are alkyl groups in the IUPAC name of a branched-chain alkane.

Note that the numerical prefix *di-* must always be accompanied by two numbers, *tri-* by three, and so on, even if the same number is written twice, as in 3,3-dimethylpentane.

Rule 5: *When two kinds of alkyl groups are present on the same carbon chain, number each group separately, and list the names of the alkyl groups in alphabetical order.*

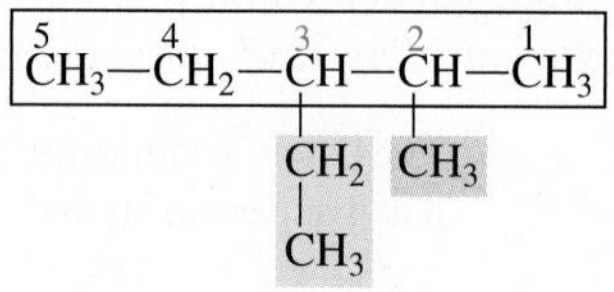

3-Ethyl-2-methylpentane

Note that ethyl is named first in accordance with the alphabetical rule.

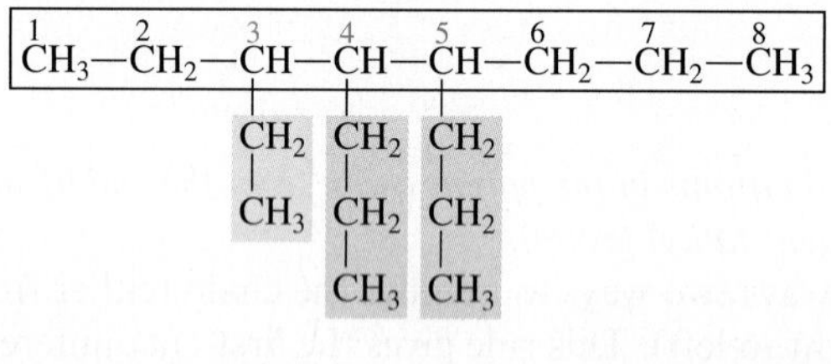

3-Ethyl-4,5-dipropyloctane

Numerical prefixes that designate numbers of alkyl groups, such as *di-*, *tri-*, and *tetra-*, are not considered when determining alphabetical priority for alkyl groups.

Note that the prefix *di-* does not affect the alphabetical order; *ethyl* precedes *propyl.*

Rule 6: *Follow IUPAC punctuation rules, which include the following: (1) Separate numbers from each other by commas. (2) Separate numbers from letters by hyphens. (3) Do not add a hyphen or a space between the last-named substituent and the name of the parent alkane that follows.*

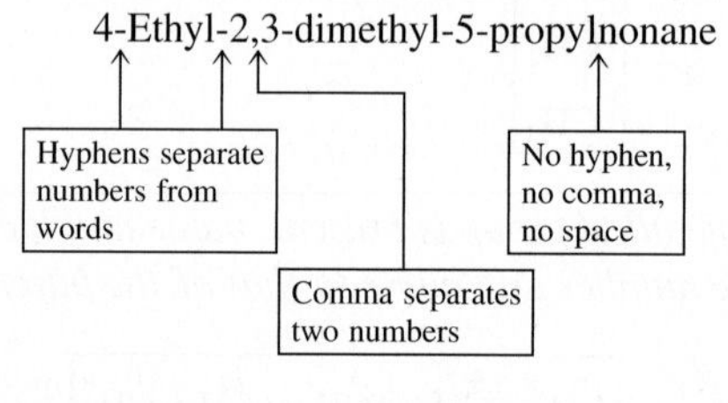

EXAMPLE 1.2 Determining IUPAC Names for Branched-Chain Alkanes

Give the IUPAC name for each of the following branched-chain alkanes.

a. 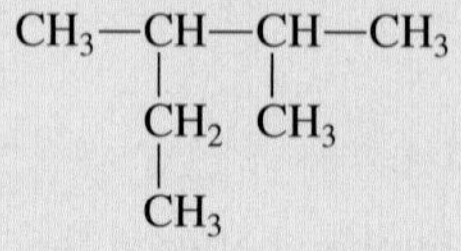

b. 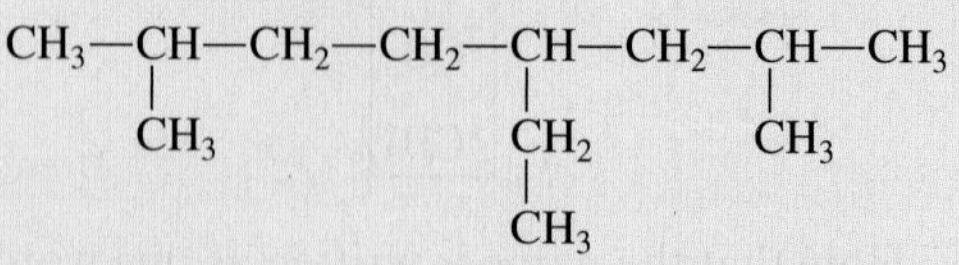

Solution

a. The longest carbon chain possesses five carbon atoms. Thus the parent-chain name is pentane.

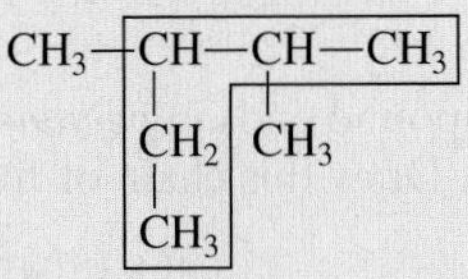

This parent chain is numbered from right to left because an alkyl substituent is closer to the right end of the chain than to the left end.

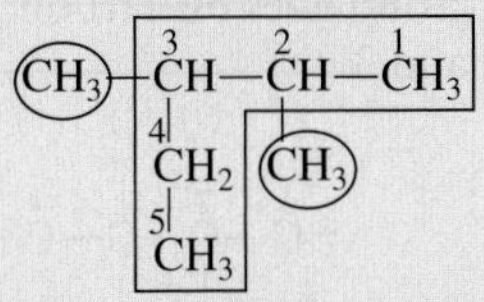

There are two methyl group substituents (circled). One methyl group is located on carbon 2 and the other on carbon 3. The IUPAC name for the compound is 2,3-dimethylpentane.

b. There are eight carbon atoms in the longest carbon chain, so the parent name is octane. There are three alkyl groups present (circled).

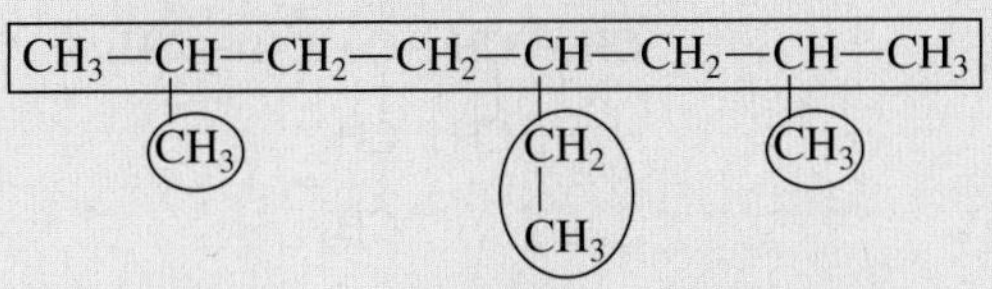

Selection of the numbering system to be used cannot be made based on the "first-encountered-alkyl-group rule" because an alkyl group is equidistant from each end of the chain. Thus the second-encountered alkyl group is used as the "tie-breaker." It is closer to the right end of the parent chain (carbon 4) than to the left end (carbon 5). Thus we use the right-to-left numbering system.

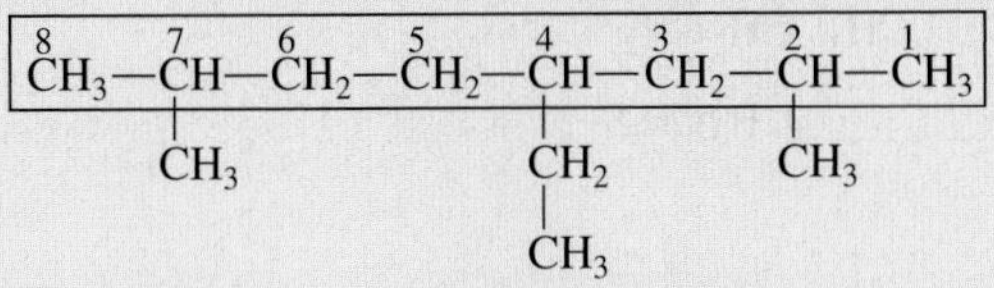

Always compare the total number of carbon atoms in the name with the number of carbon atoms in the structure to make sure they match. The name 4-ethyl-2,7-dimethyloctane indicates the presence of 2 + 2(1) + 8 = 12 carbon atoms. The structure does have 12 carbon atoms.

Two different kinds of alkyl groups are present: ethyl and methyl. Ethyl has alphabetical priority over methyl and precedes methyl in the IUPAC name. The IUPAC name is 4-ethyl-2,7-dimethyloctane.

Practice Exercise 1.2

Give the IUPAC name for each of the following alkanes.

a. 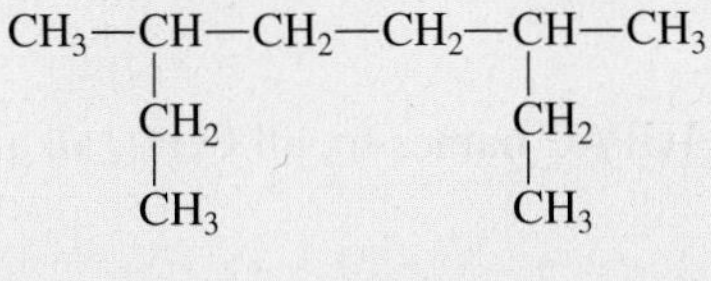 **b.** 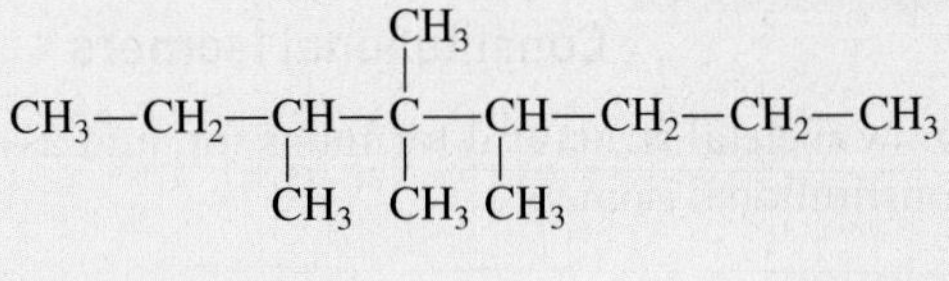

Answers: **a.** 3,6-Dimethyloctane; **b.** 3,4,4,5-Tetramethyloctane

Once the rules for naming alkanes are learned, it is relatively easy to reverse the procedure and translate the name of an alkane into a structural formula. Example 1.3 shows how this is done.

EXAMPLE 1.3 Generating the Structural Formula of an Alkane from Its IUPAC Name

Draw the condensed structural formula for 3-ethyl-2,3-dimethylpentane.

Solution

Step 1: The name of this compound ends in *pentane*, so the longest continuous chain has five carbon atoms. Draw this chain of five carbon atoms and number it.

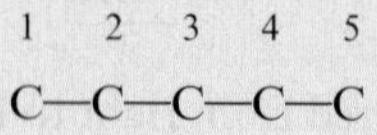

Step 2: Complete the carbon skeleton by attaching alkyl groups as they are specified in the name. An ethyl group goes on carbon 3, and methyl groups are attached to carbons 2 and 3.

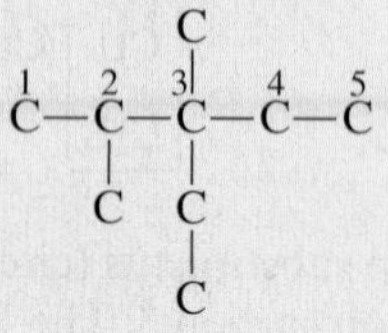

Step 3: Add hydrogen atoms to the carbon skeleton so that each carbon atom has four bonds.

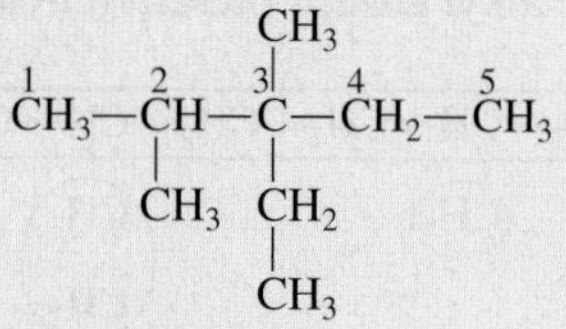

Practice Exercise 1.3

Draw the condensed structural formula for 4,5-diethyl-3,4,5-trimethyloctane.

Answer:

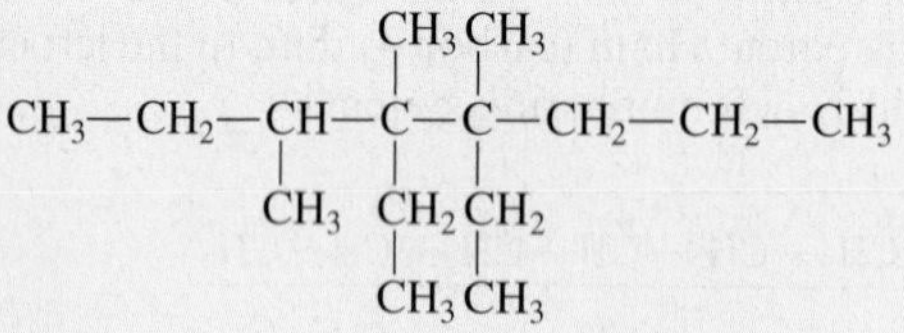

A few smaller branched alkanes have common names—that is, non-IUPAC names—that still have widespread use. They make use of the prefixes *iso* and *neo*, as in isobutane, isopentane, and neohexane. These prefixes denote particular end-of-chain carbon atom arrangements.

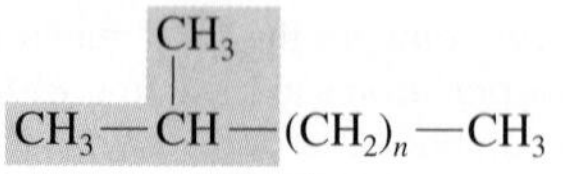

An isoalkane
(e.g., $n = 1$, Isopentane)

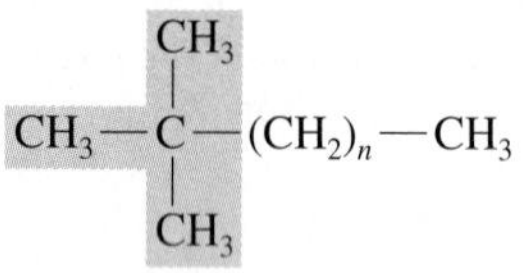

A neoalkane
(e.g., $n = 1$, Neohexane)

The following example, which involves determining the structural formulas for and naming of alkane constitutional isomers, serves as a good review of the structural and naming concepts for alkanes considered so far in this chapter.

EXAMPLE 1.4 Determining Structural Formulas for and Naming Alkane Constitutional Isomers

Draw skeletal structural formulas for, and assign IUPAC names to, all C_6H_{14} alkane constitutional isomers.

Solution

Table 1.1 indicates that there are five constitutional isomers with the chemical formula C_6H_{14}. Part of the purpose of this example is to consider the "thinking pattern" needed to identify these five isomers. There are two concepts embedded in the thinking pattern.

1. Carbon chains of varying length are examined for isomerism possibilities, starting with the chain of maximum length and then examining increasingly shorter chain lengths.
2. Substituents are added to the various carbon chains, with the number of added carbons determined by the chain length. Various location possibilities for the substituents are examined.

Step 1: A C_6 carbon chain is the longest chain possible; it contains all available carbon atoms.

C—C—C—C—C—C

This is the molecule hexane, the first of the five constitutional isomers. No substituents are added to this chain, as that would increase the carbon count beyond six.

Step 2: Decreasing the carbon-chain length by one gives a C_5 chain.

C—C—C—C—C

A methyl group must be added to the chain to bring the carbon count back up to six.

Theoretically, there are five possible positions for the methyl group:

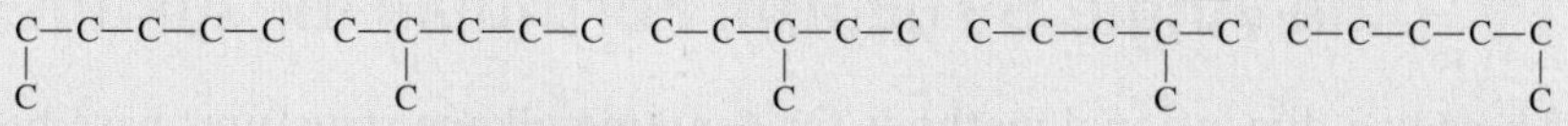

These five structures do not represent five new isomers. The first and last structures represent two alternate ways of drawing the molecule hexane, the first isomer. A methyl group (or any alkyl group) added to the end carbons of a carbon chain will always increase the chain length.

The second and third structures do represent new isomers:

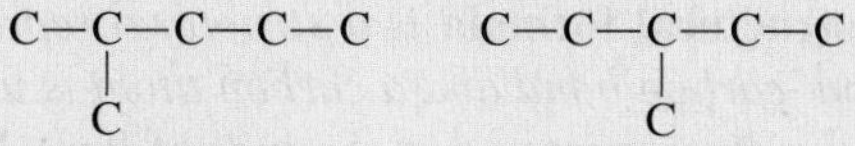

2-methylpentane 3-methylpentane

The fourth of the five structures is not a new isomer. Numbering its carbon chain from the right end shows that it is 2-methylpentane rather than 4-methylpentane. Thus the second and fourth structures are two representations of the same molecule.

Step 3: Decreasing the chain length to four carbon atoms is the next consideration. Two carbon atoms must now be added as attachments. This can be done in two ways—dimethyl and ethyl.

Examining dimethyl possibilities first, eliminating structures that have methyl groups on terminal carbon atoms gives the following possibilities.

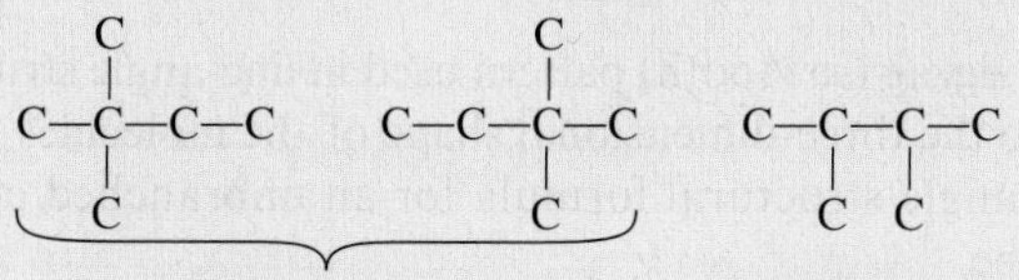

2,2-Dimethylbutane 2,3-Dimethylbutane

The first and second structures are the same; both represent the molecule 2,2-dimethylbutane, a fourth isomer.

The third structure, 2,3-dimethylbutane, is different from the other two. It is the fifth isomer.

What about ethyl butanes?

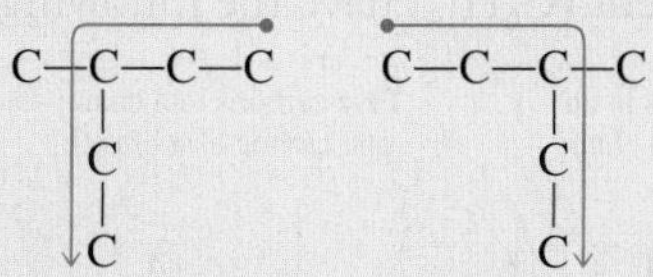

Neither of these structures is a new isomer because both have a five-carbon chain. Both structures are actually depictions of 3-methylpentane, one of the isomers previously identified.

Step 4: A chain length of three does not generate any new isomers. A trimethyl structure is impossible, as the middle carbon atom, the only carbon to which substituents can be attached, would have five bonds. An ethyl methyl structure extends the carbon chain length, as does a single three-carbon attachment.

Thus, there are five constitutional isomers: *hexane, 2-methylpentane, 3-methylpentane, 2,2-dimethylbutane, and 2,3-dimethylbutane.*

(continued)

Practice Exercise 1.4

Draw skeletal structural formulas for, and assign IUPAC names to, all C_5H_{12} alkane constitutional isomers.

Answer:

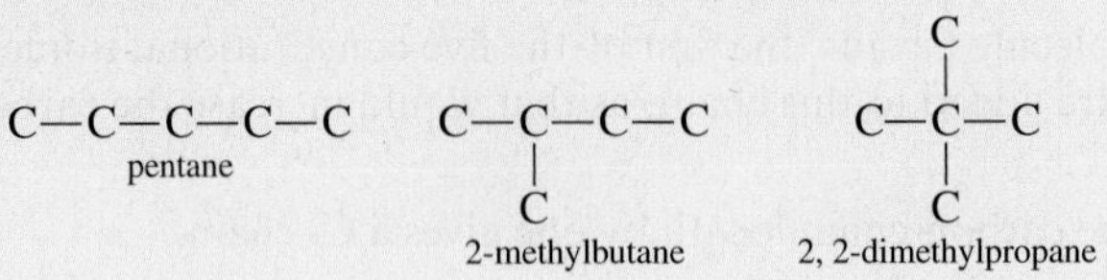

1.9 Line-Angle Structural Formulas for Alkanes

Three two-dimensional methods for denoting alkane structures have been used in previous sections of this chapter. They are expanded structural formulas, condensed structural formulas, and skeletal structural formulas. An even more concise method for denoting molecular structure of alkanes (and other hydrocarbons and their derivatives) exists. This method, *line-angle structural formulas,* is particularly useful for molecules in which higher numbers of carbon atoms are present.

A **line-angle structural formula** *is a structural representation in which a line represents a carbon–carbon bond and a carbon atom is understood to be present at every point where two lines meet and at the ends of lines.* Ball-and-stick models and line-angle structural formulas for the alkanes propane, butane, and pentane are as follows:

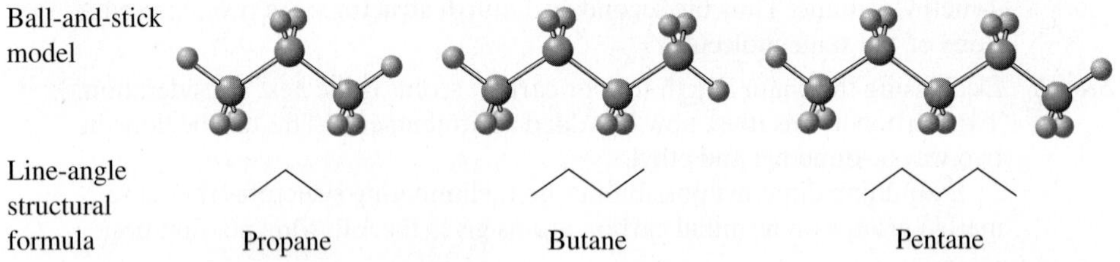

Note that the zigzag (sawtooth) pattern used in line-angle structural formulas has a relationship to the three-dimensional shape of the molecules that are represented.

The line-angle structural formula for an unbranched chain of eight carbon atoms would be

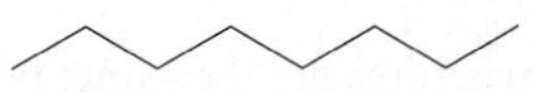

Octane

The structures of branched-chain alkanes can also be designated using line-angle structural formulas. The five constitutional alkane isomers in which six carbon atoms are present (C_6H_{14}) have the following line-angle formulas:

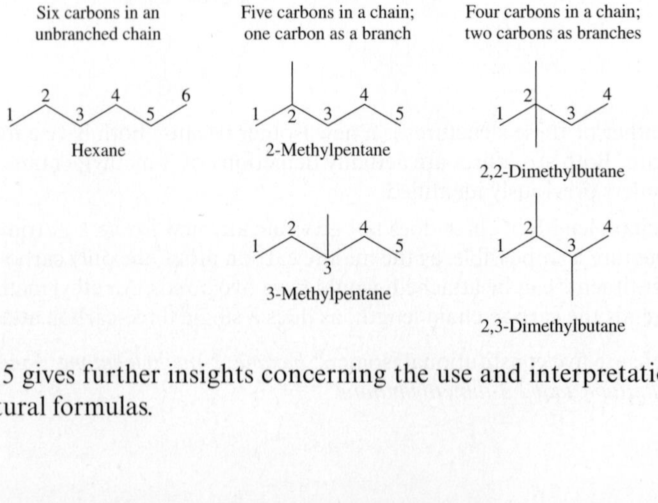

Example 1.5 gives further insights concerning the use and interpretation of line-angle structural formulas.

EXAMPLE 1.5 Generating Condensed Structural Formulas from Line-Angle Structural Formulas for Alkanes

For each of the following alkanes, determine the number of hydrogen atoms present on each carbon atom and then write the condensed structural formula for the alkane.

a. 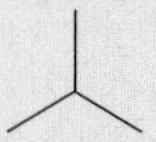 **b.** 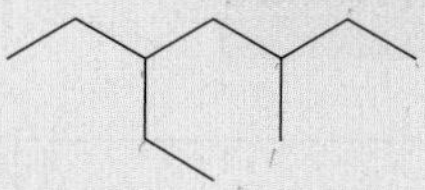

Solution

a. Each carbon atom in an alkane must be bonded to four atoms. Thus carbon atoms bonded to only one carbon atom have three hydrogen atoms attached; those bonded to two other carbon atoms have two hydrogen atoms attached; those bonded to three other carbon atoms have only one atom attached; and those bonded to four other carbon atoms bear no hydrogen atoms. For this alkane, each carbon atom's hydrogen content is indicated by circled numbers as follows.

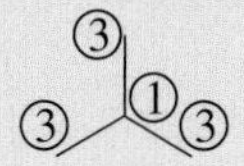

With this information on hydrogen content, the condensed structural formula is written as

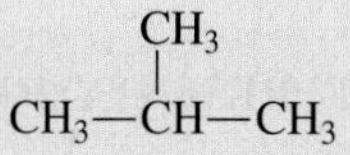

b. Using the methods of part **a,** the hydrogen content of this alkane is

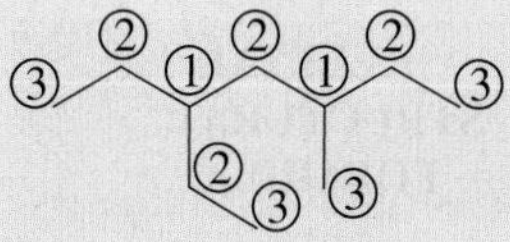

and the condensed structural formula becomes

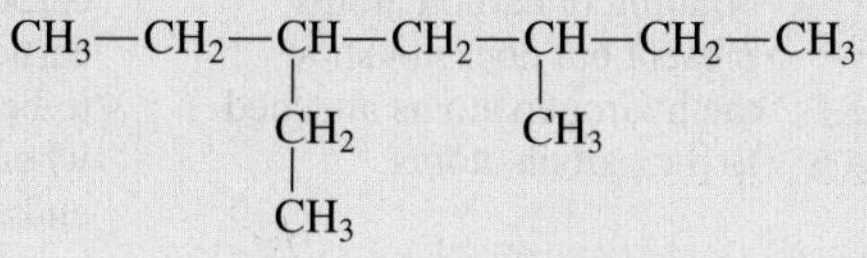

Practice Exercise 1.5

For each of the following alkanes, determine the number of hydrogen atoms present on each carbon atom and then write the condensed structural formula for the alkane.

a. 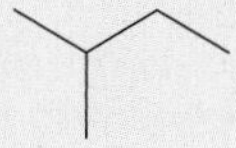 **b.** 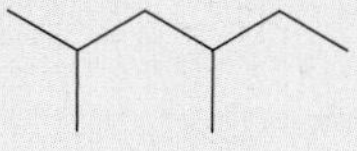

Answers: **a.** 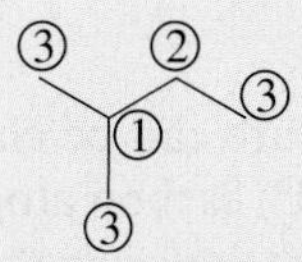 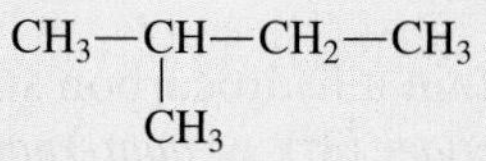

b. 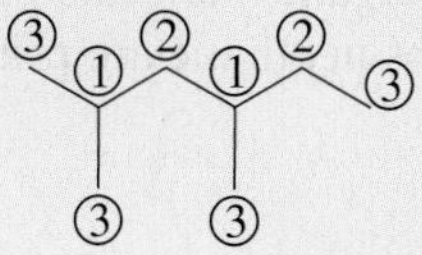 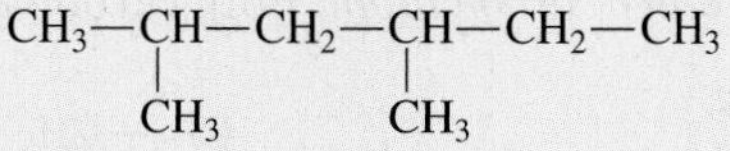

The Chemistry at a Glance feature on the next page contrasts the line-angle structural formula notation for alkanes with all other structural formula notations for alkanes encountered so far in this chapter.

CHEMISTRY AT A GLANCE Structural Representations for Alkane Molecules

THREE-DIMENSIONAL STRUCTURAL REPRESENTATIONS

DASH-WEDGE-LINE STRUCTURE

Dashes represent bonds receding behind the page, wedges bonds coming out of the page, and solid lines bonds in the plane of the page.

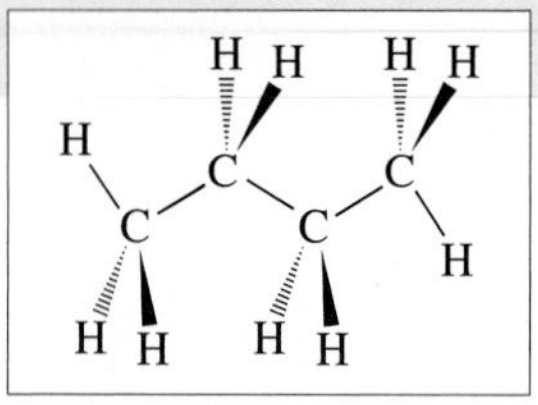

BALL-AND-STICK MODEL

This type of model emphasizes the connections (bonds) among the atoms and shows the tetrahedral arrangement of bonds about carbon atoms.

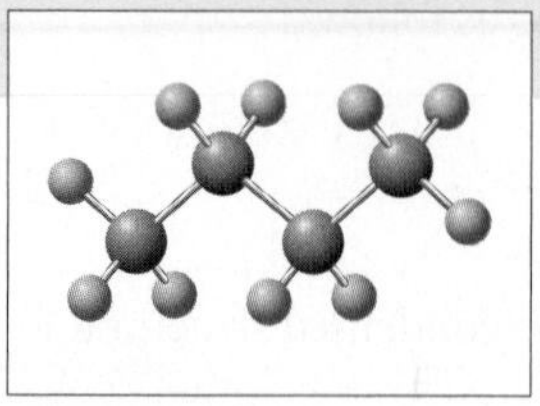

SPACE-FILLING MODEL

This type of model emphasizes the overall shape of the molecule and shows the tetrahedral arrangement of bonds about carbon atoms.

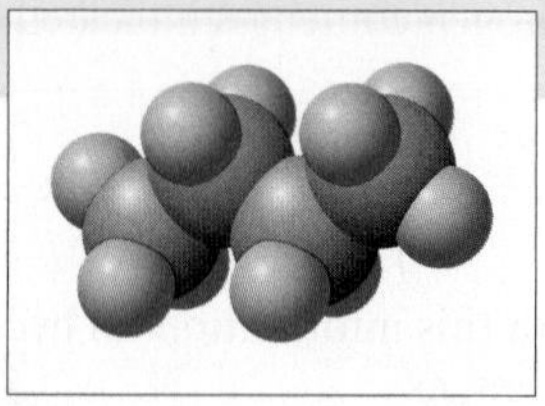

TWO-DIMENSIONAL STRUCTURAL REPRESENTATIONS

EXPANDED STRUCTURAL FORMULA

A structural formula that shows all atoms in a molecule and all bonds connecting the atoms.

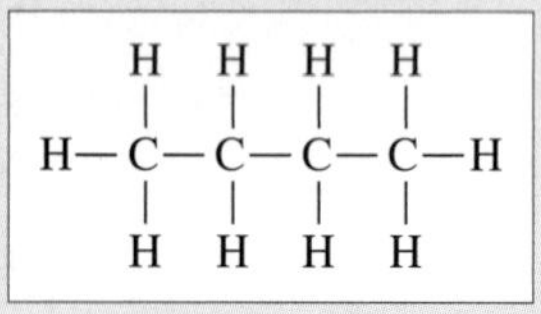

CONDENSED STRUCTURAL FORMULA

A structural formula that uses grouping of atoms, in which central atoms and the atoms connected to them are written as a group.

$CH_3—CH_2—CH_2—CH_3$

SKELETAL STRUCTURAL FORMULA

A structural formula that shows the arrangement and bonding of carbon atoms present but does not show the hydrogen atoms attached to the carbon atoms.

C—C—C—C

LINE-ANGLE STRUCTURAL FORMULA

A structural formula in which a line represents a carbon–carbon bond and a carbon atom is understood to be present at every point where lines meet and at the ends of lines.

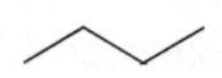

1.10 Classification of Carbon Atoms

Each of the carbon atoms within a hydrocarbon structure can be classified as a *primary* (1°), *secondary* (2°), *tertiary* (3°), or *quaternary* (4°) carbon atom.

A **primary carbon atom** *is a carbon atom in an organic molecule that is directly bonded to one other carbon atom.* Both carbon atoms in ethane are primary carbon atoms.

$$\underset{1°C}{CH_3}—\underset{1°C}{CH_3}$$

A **secondary carbon atom** *is a carbon atom in an organic molecule that is directly bonded to two other carbon atoms.* A propane molecule contains a secondary carbon atom as well as two primary carbon atoms.

$$\underset{1°C}{CH_3}—\underset{2°C}{CH_2}—\underset{1°C}{CH_3}$$

A **tertiary carbon atom** *is a carbon atom in an organic molecule that is directly bonded to three other carbon atoms.* The molecule 2-methylpropane contains a tertiary carbon atom.

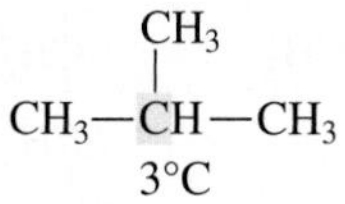

A **quaternary carbon atom** *is a carbon atom in an organic molecule that is directly bonded to four other carbon atoms.* The molecule 2,2-dimethylpropane contains a quaternary carbon atom.

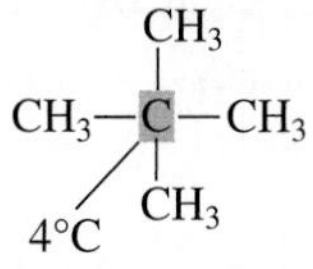

The notations 1°, 2°, 3°, and 4° are often used as designations for the terms *primary, secondary, tertiary,* and *quaternary.* Thus we can write

1° carbon atom
2° carbon atom
3° carbon atom
4° carbon atom

The alkane 2,2,3-trimethyl pentane is the simplest alkane in which all four types of carbon atoms (1°, 2°, 3°, and 4°) are present.

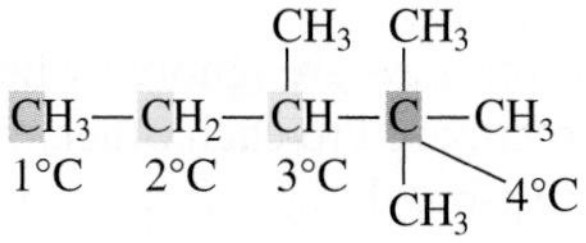

1.11 Branched-Chain Alkyl Groups

To this point in the chapter, all alkyl groups encountered in structures have been continuous-chain alkyl groups (Table 1.3), the simplest type of alkyl group. Just as there are continuous-chain and branched-chain alkanes, there are continuous-chain and branched-chain alkyl groups. There is one C_3 branched-chain alkyl group and three C_4 branched-chain alkyl groups. The names and structures of these four alkyl groups, the simplest of all branched-chain alkyl groups, are given in Figure 1.5.

For the two groups whose names contain the prefix *iso-*, the common structural feature is an end-of-chain arrangement that contains two methyl groups.

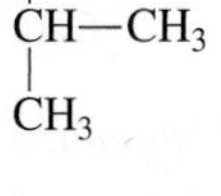

For the *secondary*-butyl group, the point of attachment of the group to the main carbon chain involves a *secondary* carbon atom. For the *tertiary*-butyl group, the point of attachment of the group to the main carbon chain involves a *tertiary* carbon atom. The name secondary-butyl is often shortened to sec-butyl or simply s-butyl. Similarly, tertiary-butyl is often written as tert-butyl or t-butyl.

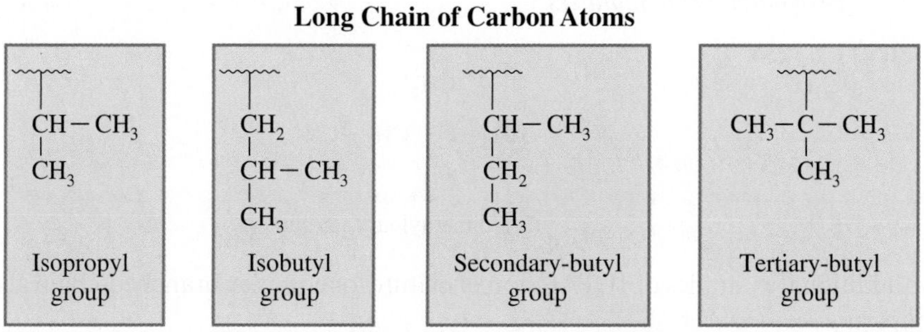

Figure 1.5 The four most common branched-chain alkyl groups and their IUPAC names.

It is important to be able to recognize various conformations of branched-chain alkyl groups. For example, these structures all represent an isopropyl group:

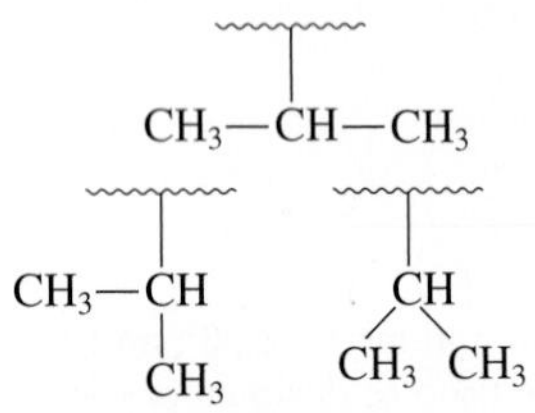

In each case, there is a chain of three carbon atoms with an attachment point (the long bond) involving the middle carbon atom of the chain.

Two examples of alkanes containing branched-chain alkyl groups follow.

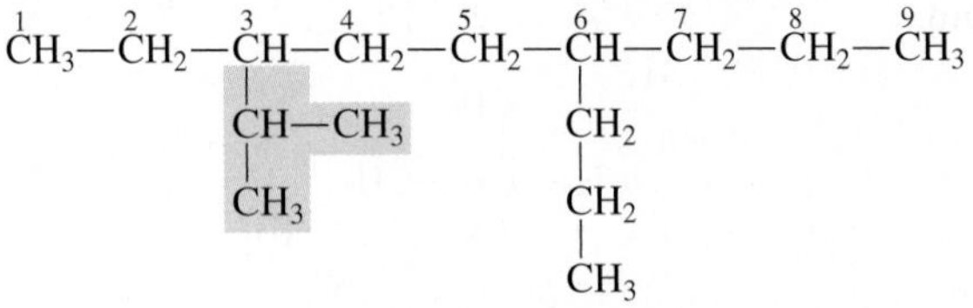

3-Isopropyl-6-propylnonane

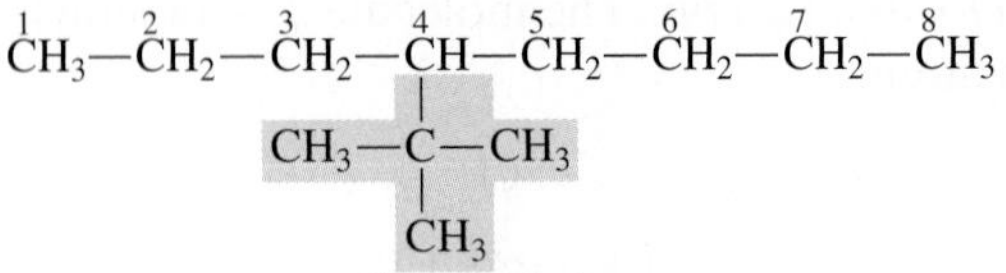

4-*tert*-Butyloctane

In IUPAC names, a hyphen always follows the designations *secondary* and *tertiary*, but no hyphen is used with the prefix *iso*.

secondary-butyl   tertiary-butyl   isobutyl

The hyphenated prefixes *sec*- and *tert*- are ignored when alphabetizing alkyl group names except when they are compared to each other. Thus, tert-butyl precedes isobutyl, and sec-butyl prcedes tert-butyl.

Complex Branched-Chain Alkyl Groups

"Simple" names, such as isobutyl and tert-butyl (Figure 1.5), do not exist for most branched-chain alkyl groups containing five or more carbon atoms. The IUPAC system provision for naming such larger groups involves naming them as if they were themselves compounds. The following rules are used.

Rule 1: *The longest continuous carbon chain that begins at the point of attachment of the alkyl group becomes the base name.*

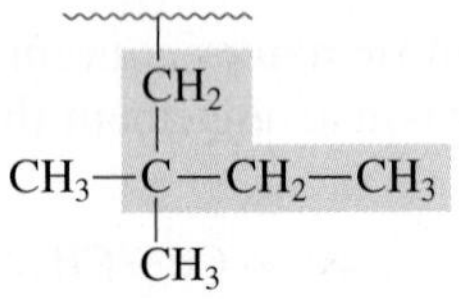

Rule 2: *The base chain is numbered beginning at the point of attachment.*

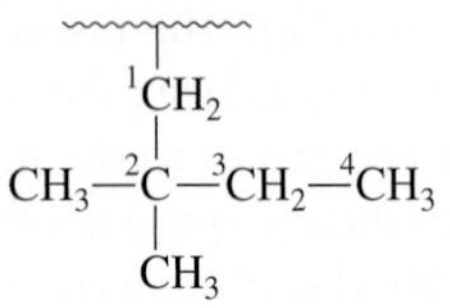

Rule 3: *Substituents on the base chain are listed in alphabetical order, using numerical prefixes when necessary, and substituent locations are designated using numbers.*

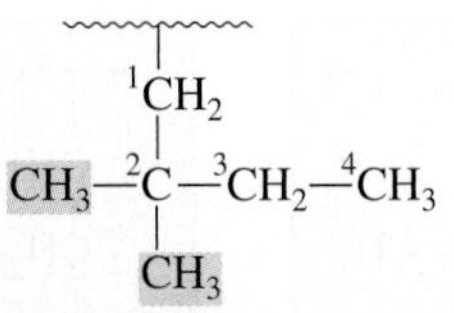

(2,2-dimethylbutyl) group

Two additional examples of IUPAC nomenclature for complex branched-chain alkyl groups are

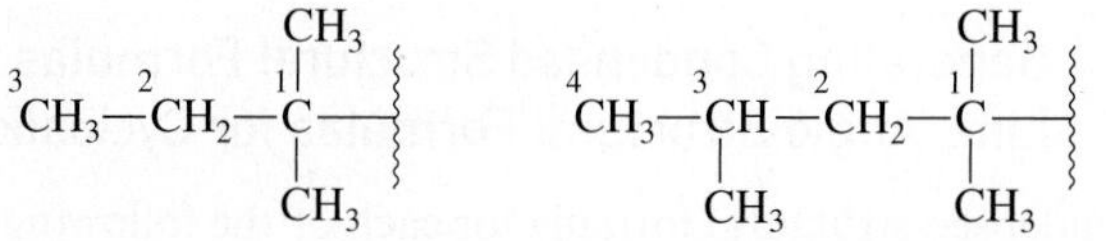

(1,1-dimethylpropyl) group (1,1, 3-trimethylbutyl) group

The C_3 and C_4 branched-chain alkyl groups (Figure 1.5) can also be named using the preceding rules. Thus there are two names for these groups. An alternate name for the sec-butyl group is the 1,1-dimethylethyl group.

A single compound or group can have several acceptable names, but no two compounds or groups can have the same name.

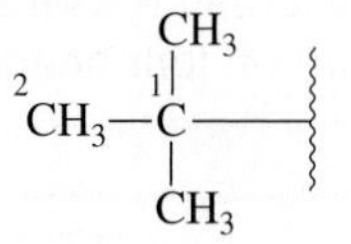

sec-butyl group
(1,1-dimethylethyl) group

1.12 Cycloalkanes

A **cycloalkane** *is a saturated hydrocarbon in which carbon atoms connected to one another in a cyclic (ring) arrangement are present.* The simplest cycloalkane is cyclopropane, which contains a cyclic arrangement of three carbon atoms. Figure 1.6 shows a three-dimensional model of cyclopropane's structure and those of the four-, five-, and six-carbon cycloalkanes.

It takes a minimum of three carbon atoms to form a cyclic arrangement of carbon atoms.

Cyclopropane's three carbon atoms lie in a flat ring. In all other cycloalkane molecules, some puckering of the ring occurs; that is, the ring systems are nonplanar, as shown in Figure 1.6.

The general formula for cycloalkanes is C_nH_{2n}. Thus a given cycloalkane contains two fewer hydrogen atoms than an alkane with the same number of hydrogen atoms (C_nH_{2n+2}). Butane (C_4H_{10}) and cyclobutane (C_4H_8) are not isomers; isomers must have the same molecular formula (Section 1.6).

Line-angle structural formulas are generally used to represent cycloalkane structures. The line-angle structural formula for cyclopropane is a triangle, that for cyclobutane a square, that for cyclopentane a pentagon, and that for cyclohexane a hexagon.

Cyclopropane Cyclobutane Cyclopentane Cyclohexane

In such structures, the intersection of two lines represents a CH_2 group. Three- and four-way intersections of lines are possible when substituents are present on a ring. A three-way intersection represents a CH group, and a four-way intersection is simply a carbon atom.

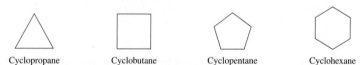

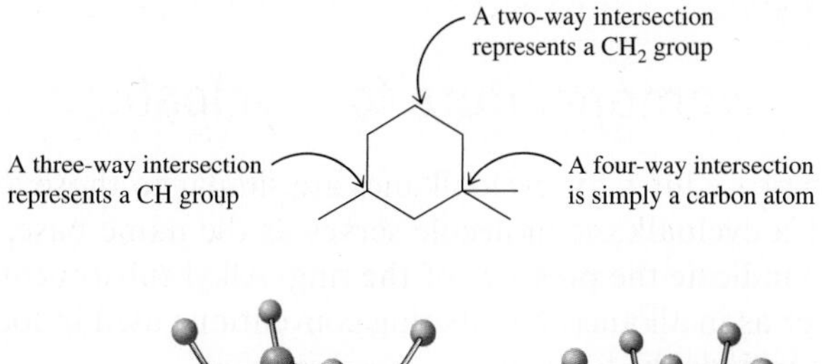

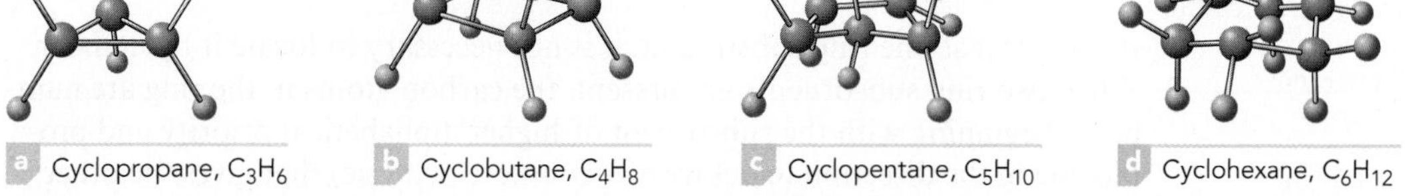

a Cyclopropane, C_3H_6 b Cyclobutane, C_4H_8 c Cyclopentane, C_5H_{10} d Cyclohexane, C_6H_{12}

Figure 1.6 Three-dimensional representations of the structures of simple cycloalkanes.

EXAMPLE 1.6 Generating Condensed Structural Formulas from Line-Angle Structural Formulas for Cycloalkanes

Generate the condensed structural formula for each of the following cycloalkanes.

Solution

a. First replace each angle and line terminus with a carbon atom, and then add hydrogens as necessary to give each carbon four bonds. The molecular formula of this compound is C_8H_{16}.

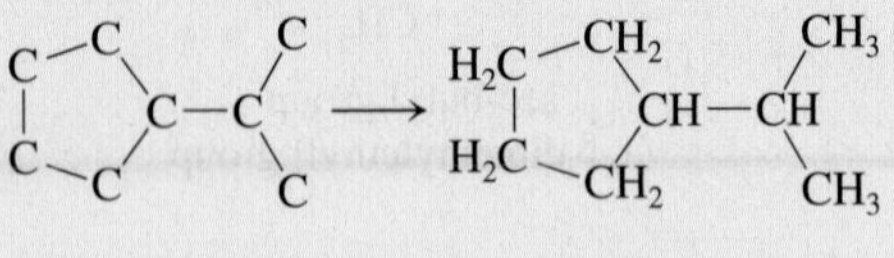

b. Similarly, we have

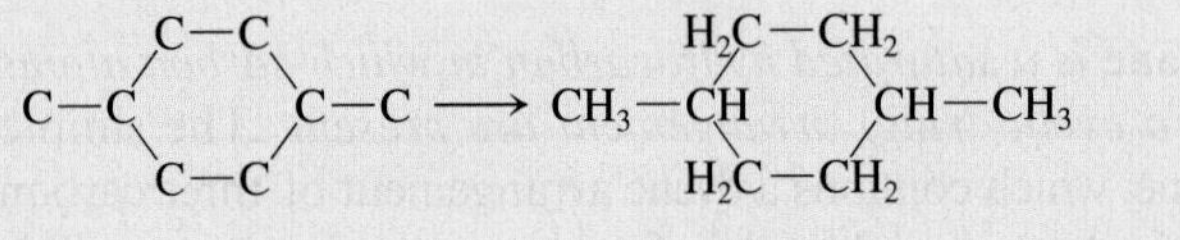

Practice Exercise 1.6

Generate the condensed structural formula for each of the following cycloalkanes.

Answers:

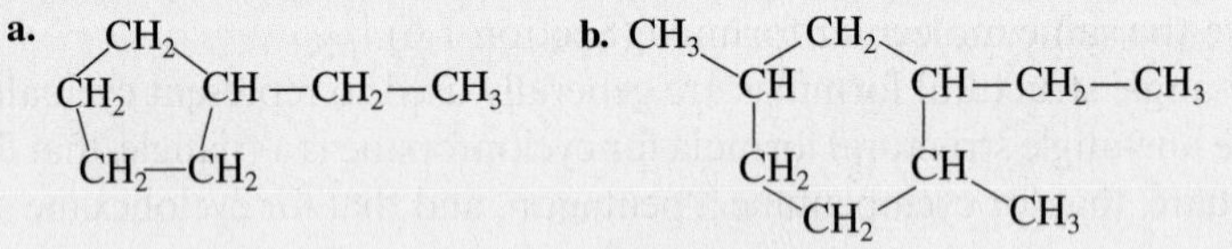

The observed C—C—C bond angles in cyclopropane are 60°, and those in cyclobutane are 90°, values that are considerably smaller than the 109° angle associated with a tetrahedral arrangement of bonds about a carbon atom. Consequently, cyclopropane and cyclobutane are relatively unstable compounds. Five- and six-membered cycloalkane structures are much more stable, and these structural entities are encountered in many organic molecules.

1.13 IUPAC Nomenclature for Cycloalkanes

IUPAC naming procedures for cycloalkanes are similar to those for alkanes. The ring portion of a cycloalkane molecule serves as the name base, and the prefix *cyclo-* is used to indicate the presence of the ring. Alkyl substituents are named in the same manner as in alkanes. Numbering conventions used in locating substituents on the ring include the following:

1. If there is just one ring substituent, it is not necessary to locate it by number.
2. When two ring substituents are present, the carbon atoms in the ring are numbered beginning with the substituent of higher alphabetical priority and proceeding in the direction (clockwise or counterclockwise) that gives the other substituent the lower number.

Cycloalkanes of ring sizes ranging from 3 to over 30 are found in nature, and, in principle, there is no limit to ring size. Five-membered rings (cyclopentanes) and six-membered rings (cyclohexanes) are especially abundant in nature.

3. When three or more ring substituents are present, ring numbering begins at the substituent that leads to the lowest set of location numbers. When two or more equivalent numbering sets exist, alphabetical priority among substituents determines the set used.

Example 1.7 illustrates the use of the ring-numbering guidelines.

EXAMPLE 1.7 Determining IUPAC Names for Cycloalkanes

Assign IUPAC names to each of the following cycloalkanes.

a. 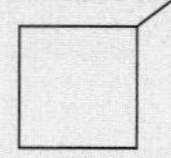 **b.** 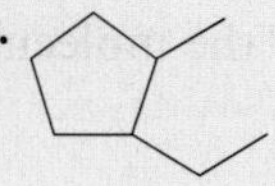 **c.** 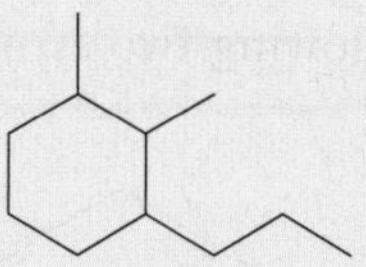

Solution

a. This molecule is a cyclobutane (four-carbon ring) with a methyl substituent. The IUPAC name is simply methylcyclobutane. No number is needed to locate the methyl group because all four ring positions are equivalent.

b. This molecule is a cyclopentane with ethyl and methyl substituents. The numbers for the carbon atoms that bear the substituents are 1 and 2. On the basis of alphabetical priority, the number 1 is assigned to the carbon atom that bears the ethyl group. The IUPAC name for the compound is 1-ethyl-2-methylcyclopentane.

c. This molecule is a dimethylpropylcyclohexane. Two different 1,2,3 numbering systems exist for locating the substituents. On the basis of alphabetical priority, the numbering system that has carbon 1 bearing a methyl group is used; methyl has alphabetical priority over propyl. Thus the compound name is 1,2-dimethyl-3-propylcyclohexane.

When a ring system contains fewer carbon atoms than an alkyl group attached to it, the compound is named as an alkane rather than as a cycloalkane; the ring is named as a cycloalkyl group.

$CH_3—CH—CH_2—CH_2—CH_2—CH_3$

2-cyclopentylhexane

Practice Exercise 1.7

Assign IUPAC names to each of the following cycloalkanes.

a. 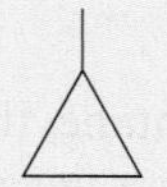 **b.** 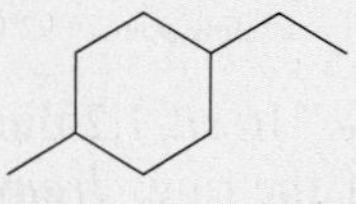 **c.** 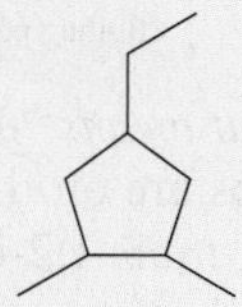

Answers: **a.** Methylcyclopropane; **b.** 1-Ethyl-4-methylcyclohexane; **c.** 4-Ethyl-1,2-dimethylcyclopentane

1.14 Isomerism in Cycloalkanes

Constitutional isomers are possible for cycloalkanes that contain four or more carbon atoms. For example, there are five cycloalkane constitutional isomers that have the formula C_5H_{10}: one based on a five-membered ring, one based on a four-membered ring, and three based on a three-membered ring. These isomers are

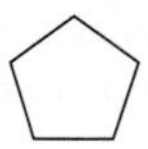
Cyclopentane

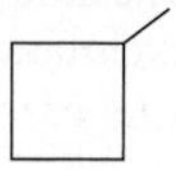
Methylcyclobutane

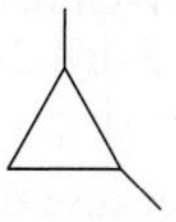
1,2-Dimethylcyclopropane

1,1-Dimethylcyclopropane

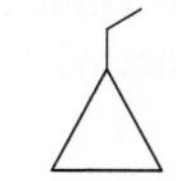
Ethylcyclopropane

A second type of isomerism, called *stereoisomerism,* is possible for some *substituted* cycloalkanes. Whereas constitutional isomerism results from differences in *connectivity,* stereoisomerism results from differences in *configuration.*

Cis–trans isomers have the same molecular formula and the same structural formula. The only difference between them is the orientation of atoms in space. *Constitutional isomers* have the same molecular formula but different structural formulas.

Stereoisomers *are isomers that have the same molecular and structural formulas but different orientations of atoms in space.* Several forms of stereoisomerism exist. The form associated with cycloalkanes is called *cis–trans isomerism.* ***Cis–trans*** **isomers** *are isomers that have the same molecular and structural formulas but different orientations of atoms in space because of restricted rotation about bonds.*

In alkanes, there is free rotation about all carbon–carbon bonds (Section 1.7). In cycloalkanes, the ring structure restricts rotation for the carbon atoms in the ring. The consequence of this lack of rotation in a cycloalkane is the creation of "top" and "bottom" positions for the two attachments on each of the ring carbon atoms. This "top–bottom" situation leads to *cis–trans* isomerism in cycloalkanes in which each of two ring carbon atoms bears two different attachments.

Consider the following two structures for the molecule 1,2-dimethylcyclopentane.

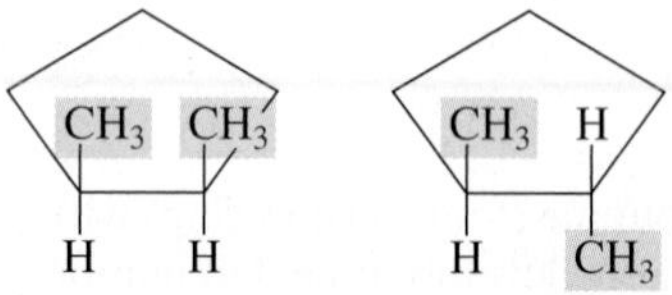

Structure A   Structure B

In structure A, both methyl groups are above the plane of the ring (the "top" side). In structure B, one methyl group is above the plane of the ring (the "top" side) and the other below it (the "bottom" side). Structure A cannot be converted into structure B without breaking bonds. Hence structures A and B are isomers; there are two 1,2-dimethylcyclopentanes. The first isomer is called *cis*-1,2-dimethylcyclopentane and the second *trans*-1,2-dimethylcyclopentane.

The Latin *cis* means "on the same side," and the Latin *trans* means "across from." Consider the use of the prefix *trans-* in the phrase "transatlantic voyage."

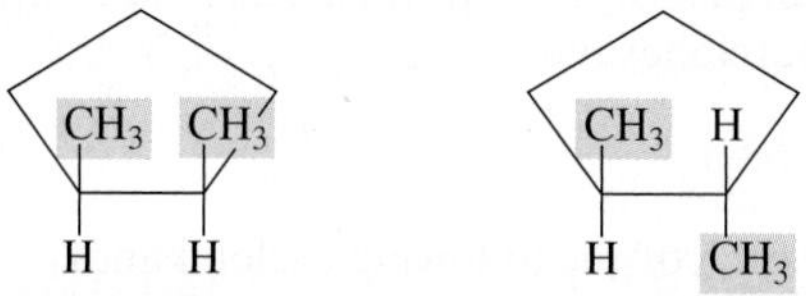

cis-1,2-Dimethylcyclopentane
Boiling point = 99°C

trans-1,2-Dimethylcyclopentane
Boiling point = 92°C

Cis–trans isomerism will also be encountered in the next chapter (Section 2.6), where the required restricted rotation barrier will be a carbon–carbon double bond rather than a ring of carbon atoms. Another type of stereoisomerism called enantiomerism (left- and right-handed forms of a molecule) will be considered in the discussion of carbohydrates in Chapter 7.

Cis- *is a prefix that means "on the same side."* In *cis*-1,2-dimethylcyclopentane, the two methyl groups are on the same side of the ring. ***Trans-*** *is a prefix that means "across from."* In *trans*-1,2-dimethylcyclopentane, the two methyl groups are on opposite sides of the ring.

Cis–trans isomerism can occur in rings of all sizes. The presence of a substituent on each of two carbon atoms in the ring is the requirement for its occurrence. In biochemistry, it will be found that the human body often selectively distinguishes between the *cis* and *trans* isomers of a compound. One isomer will be active in the body and the other inactive.

EXAMPLE 1.8 Identifying and Naming Cycloalkane *Cis–Trans* Isomers

Determine whether *cis–trans* isomerism is possible for each of the following cycloalkanes. If so, then draw structural formulas for the *cis* and *trans* isomers.

a. Methylcyclohexane
b. 1,1-Dimethylcyclohexane
c. 1,3-Dimethylcyclobutane
d. 1-Ethyl-2-methylcyclobutane

Solution

a. *Cis–trans* isomerism is not possible because there are not two substituents on the ring.

b. *Cis–trans* isomerism is not possible. There are two substituents on the ring, but they are on the same carbon atom. Each of two different carbons must bear substituents.

c. *Cis–trans* isomerism does exist.

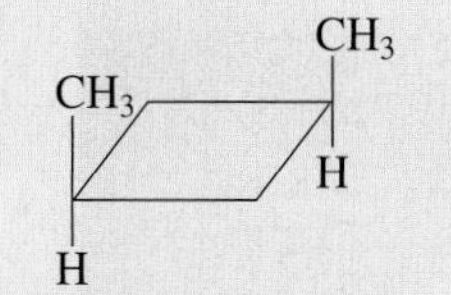

cis-1,3-Dimethylcyclobutane

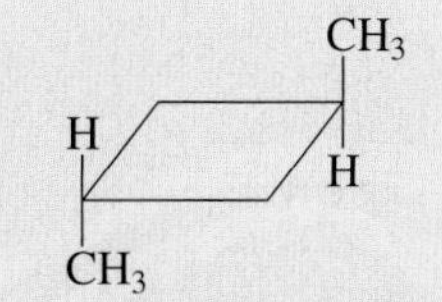

trans-1,3-Dimethylcyclobutane

d. *Cis–trans* isomerism does exist.

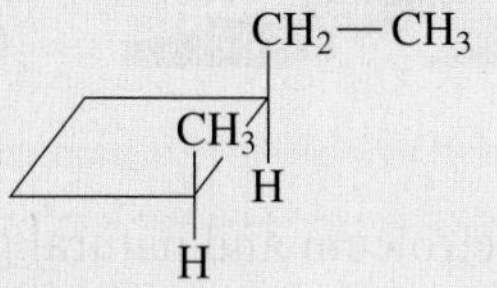

cis-1-Ethyl-2-methylcyclobutane

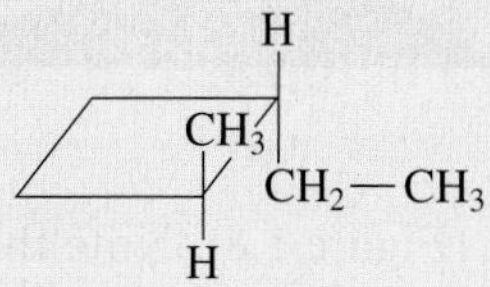

trans-1-Ethyl-2-methylcyclobutane

Practice Exercise 1.8

Determine whether *cis–trans* isomerism is possible for each of the following cycloalkanes. If so, then draw structural formulas for the *cis* and *trans* isomers.

a. 1-Ethyl-1-methylcyclopentane **b.** Ethylcyclohexane
c. 1,3-Dimethylcyclopentane **d.** 1,1-Dimethylcyclooctane

Answers: **a.** Not possible; **b.** Not possible; **c.**

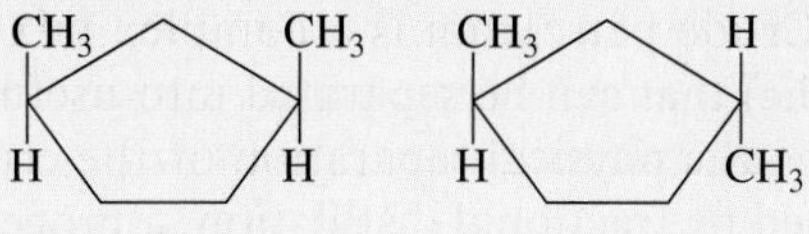

cis isomer *trans* isomer

d. Not possible

In cycloalkanes, *cis–trans* isomerism can also be denoted by using wedges and dotted lines. A heavy wedge-shaped bond to a ring structure indicates a bond *above* the plane of the ring, and a broken dotted line indicates a bond *below* the plane of the ring.

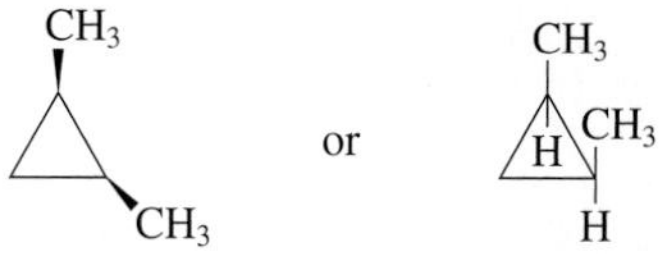

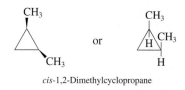

cis-1,2-Dimethylcyclopropane

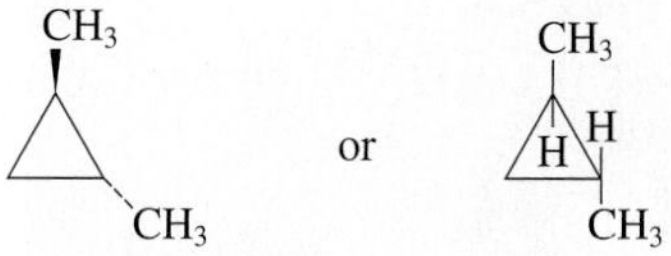

trans-1,2-Dimethylcyclopropane

Use of the terms *cis-* and *trans-* in designating stereoisomers in cycloalkanes is limited to substituted cycloalkanes in which the two substituted carbon atoms each have one hydrogen atom and one substituent other than hydrogen. The designations *cis-* and *trans-* become ambiguous in situations where either or both of the substituted carbons have two different substituents but no hydrogen atoms. Following is an example of such a situation in substituted cycloalkanes.

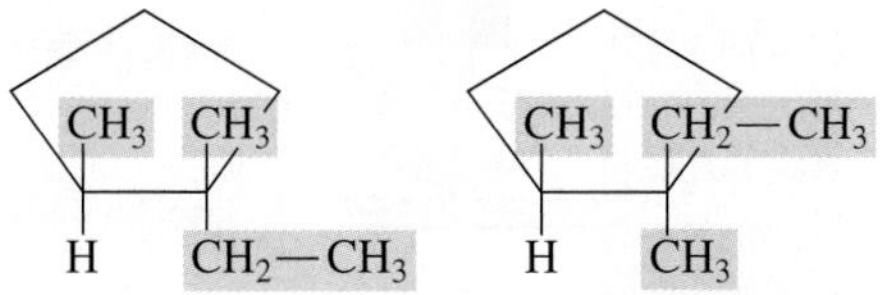

The first structure is the *cis-* isomer if the focus is on H and the ethyl group; the second structure is the *cis-* isomer if the focus is on H and the methyl group. A different nomenclature system, called the *E,Z* nomenclature system (which is not covered in this textbook), must be used to distinguish such isomerism.

1.15 Sources of Alkanes and Cycloalkanes

Alkanes and cycloalkanes are not "laboratory curiosities" but, rather, two families of extremely important naturally occurring compounds. Natural gas and petroleum (crude oil) constitute their largest and most important natural source. Deposits of these resources are usually associated with underground dome-shaped rock formations (Figure 1.7). When a hole is drilled into such a rock formation, it is possible to recover some of the trapped hydrocarbons—that is, the natural gas

The word *petroleum* comes from the Latin *petra,* which means "rock," and *oleum,* which means "oil."

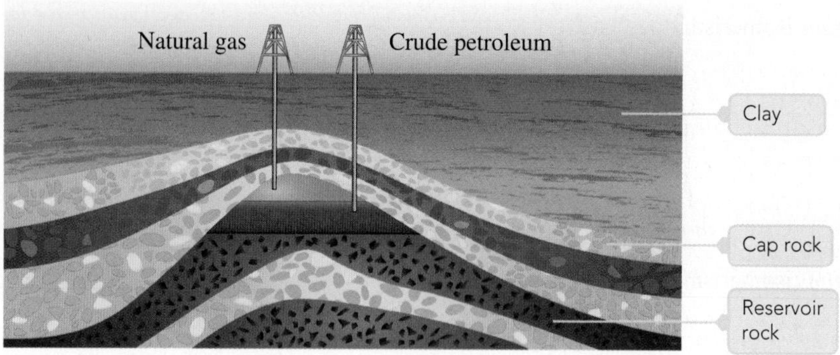

Figure 1.7 A rock formation such as this is necessary for the accumulation of petroleum and natural gas.

Figure 1.8 An oil rig pumping oil from an underground rock formation.

and/or petroleum (Figure 1.8). Note that petroleum and natural gas do not occur in the Earth in the form of "liquid pools" but, rather, are dispersed throughout a porous rock formation.

Unprocessed natural gas contains 50%–90% methane, 1%–10% ethane, and up to 8% higher-molecular-mass alkanes (predominantly propane and butanes). The higher alkanes found in crude natural gas are removed prior to release of the gas into the pipeline distribution systems. Because the removed alkanes can be liquefied by the use of moderate pressure, they are stored as liquids under pressure in steel cylinders and are marketed as bottled gas.

Crude petroleum is a complex mixture of hydrocarbons (both cyclic and acyclic) that can be separated into useful fractions through refining. During refining, the physical separation of the crude into component fractions is accomplished by fractional distillation, a process that takes advantage of boiling-point differences between the components of the crude petroleum. Each fraction contains hydrocarbons within a specific boiling-point range. The gasoline fraction consists primarily of alkanes and cycloalkanes with 5 to 12 carbon atoms present. The fractions obtained from a typical fractionation process are shown in Figure 1.9.

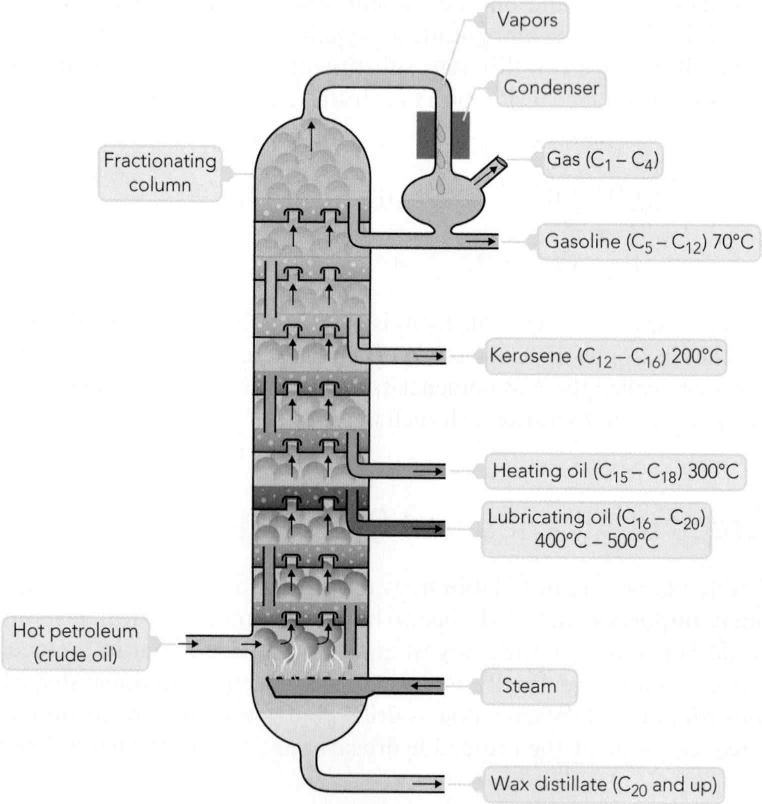

Figure 1.9 The complex hydrocarbon mixture present in petroleum is separated into simpler mixtures by means of a fractionating column.

1.16 Physical Properties of Alkanes and Cycloalkanes

In this section, a number of generalizations about the physical properties of alkanes and cycloalkanes will be considered.

1. *Alkanes and cycloalkanes are insoluble in water.* Water molecules are polar, and alkane and cycloalkane molecules are nonpolar. Molecules of unlike polarity have limited solubility in one another. The water insolubility of alkanes makes them good preservatives for metals. They prevent water from reaching the metal surface and causing corrosion. They also have biological functions as protective coatings (Figure 1.10).
2. *Alkanes and cycloalkanes have densities lower than that of water.* Alkane and cycloalkane densities fall in the range 0.6 g/mL to 0.8 g/mL, compared with water's density of 1.0 g/mL. When alkanes and cycloalkanes are mixed with water, two layers form (because of insolubility), with the hydrocarbon layer on top (because of its lower density). This density difference between alkanes/cycloalkanes and water explains why oil spills in aqueous environments spread so quickly. The *floating* oil follows the movement of the water.
3. *The boiling points of continuous-chain alkanes and cycloalkanes increase with an increase in carbon-chain length or ring size.* For continuous-chain alkanes, the boiling point increases roughly 30°C for every carbon atom added to the chain. This trend, shown in Figure 1.11, is the result of increasing London force strength. London forces become stronger as molecular surface area increases. Short, continuous-chain alkanes (1 to 4 carbon atoms) are gases at room temperature. Continuous-chain alkanes containing 5 to 17 carbon atoms are liquids, and alkanes that have carbon chains longer than this are solids at room temperature.

   Branching on a carbon chain lowers the boiling point of an alkane. A comparison of the boiling points of unbranched alkanes and their 2-methyl-branched isomers is given in Figure 1.11. Branched alkanes are more compact, with smaller surface areas than their straight-chain isomers.

   Cycloalkanes have higher boiling points than their noncyclic counterparts with the same number of carbon atoms (Figure 1.11). These differences are due in large part to cyclic systems having more rigid and more symmetrical structures.

   Cyclopropane and cyclobutane are gases at room temperature, and cyclopentane through cyclooctane are liquids at room temperature. Figure 1.12 is a physical-state summary for unbranched alkanes or unsubstituted cycloalkanes with 8 or fewer carbon atoms.

The alkanes and cycloalkanes whose boiling points are compared in Figure 1.11 constitute a *homologous series* of organic compounds. In a homologous series,

Figure 1.10 The insolubility of alkanes in water is used to advantage by many plants, which produce unbranched long-chain alkanes that serve as protective coatings on leaves and fruits. Such protective coatings minimize water loss for plants. Apples can be "polished" because of the long-chain alkane coating on their skin, which involves the unbranched alkanes $C_{27}H_{56}$ and $C_{29}H_{60}$. The leaf wax of cabbage and broccoli is mainly unbranched $C_{29}H_{60}$.

The variation of alkane and cycloalkane boiling point and melting point with carbon-chain length or ring size is used to advantage in gasoline production. Gasoline composition changes seasonally in locations that have very hot summers and very cold winters. "Summer" gasoline contains larger amounts of higher-boiling hydrocarbons so that it evaporates less readily. "Winter" gasoline contains larger amounts of lower-boiling hydrocarbons so that it freezes less readily.

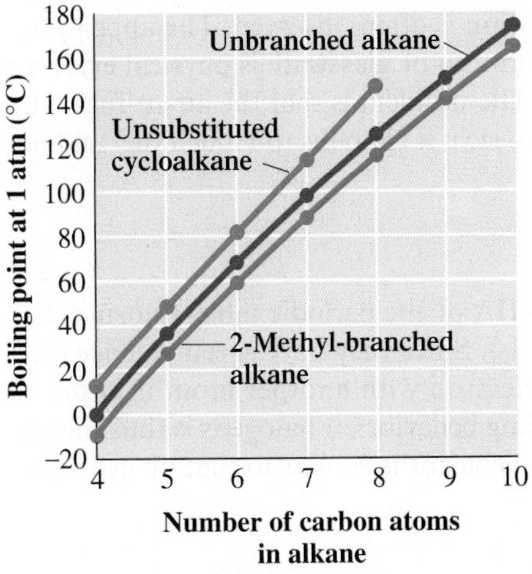

Figure 1.11 Trends in normal boiling points for continuous-chain alkanes, 2-methyl branched alkanes, and unsubstituted cycloalkanes as a function of the number of carbon atoms present. For a series of alkanes or cycloalkanes, melting point increases as carbon-chain length increases.

Unbranched Alkanes			
C_1	C_3	C_5	C_7
C_2	C_4	C_6	C_8

Unsubstituted Cycloalkanes			
	C_3	C_5	C_7
	C_4	C_6	C_8

☐ Gas ☐ Liquid

Figure 1.12 A physical-state summary for unbranched alkanes and unsubstituted cycloalkanes at room temperature and pressure.

the members differ structurally only in the number of $—CH_2—$ groups present. Members exhibit gradually changing physical properties and usually have very similar chemical properties.

The existence of homologous series of organic compounds gives organization to organic chemistry in the same way that the periodic table gives organization to the chemistry of the elements. Knowing something about a few members of a homologous series usually enables the properties of other members in the series to be deduced.

The physiological effects of alkanes on the human body depend heavily on alkane physical state (gas, liquid, or solid), which in turn depends heavily on carbon-chain length and degree of branching of the carbon chain. The focus on relevancy feature Chemical Connections 1-B on the next page discusses the physiological effects of alkanes. In general, liquid-state alkanes are skin "irritants" and solid-state alkanes "protect" the skin.

1.17 Chemical Properties of Alkanes and Cycloalkanes

The term *paraffins* is an older name for the alkane family of compounds. This name comes from the Latin *parum affinis*, which means "little activity." That is a good summary of the general chemical properties of alkanes.

Atmospheric levels of carbon dioxide are increasing annually as a result of the extensive use of alkanes and cycloalkanes as energy sources.

Alkanes are the least reactive type of organic compound. They can be heated for long periods of time in strong acids and bases with no appreciable reaction. Strong oxidizing agents and reducing agents have little effect on alkanes.

Alkanes are not absolutely unreactive. Two important reactions that they undergo are combustion, which is reaction with oxygen, and halogenation, which is reaction with halogens.

Combustion

A **combustion reaction** *is a chemical reaction between a substance and oxygen (usually from air) that proceeds with the evolution of heat and light (usually as a flame).* Alkanes readily undergo combustion when ignited. When sufficient oxygen is present to support total combustion, carbon dioxide and water are the products.

$$CH_4 + 2O_2 \longrightarrow CO_2 + 2H_2O + \text{heat energy}$$
$$2C_6H_{14} + 19O_2 \longrightarrow 12CO_2 + 14H_2O + \text{heat energy}$$

The exothermic nature of alkane combustion reactions explains the extensive use of alkanes as fuels. Natural gas, used in home heating, is predominantly methane. Propane is used in home heating in rural areas and in gas barbecue units (Figure 1.13). Butane fuels portable camping stoves. Gasoline is a complex mixture of many alkanes and other types of hydrocarbons.

Incomplete combustion can occur if insufficient oxygen is present during the combustion process. When this is the case, some carbon monoxide (CO) and/or elemental carbon are reaction products along with carbon dioxide (CO_2). In a chemical laboratory setting, incomplete combustion is often observed. The appearance of deposits of carbon black (soot) on the bottom of glassware is physical evidence that incomplete combustion is occurring. The problem is that the air-to-fuel ratio for the Bunsen burner is not correct. It is too rich; it contains too much fuel and not enough oxygen (air).

Figure 1.13 Propane fuel tank on a home barbecue unit.

Halogenation

The halogens are the elements of Group VIIA of the periodic table: fluorine (F_2), chlorine (Cl_2), bromine (Br_2), and iodine (I_2). Since they have seven valence electrons, halogen atoms need to share one electron with another atom in order to acquire an octet of electrons. Normal bonding behavior for halogens is thus the formation of one single bond. Their bonding behavior is similar to that of hydrogen, which also forms only one single bond.

CHEMICAL CONNECTIONS 1-B

The Physiological Effects of Alkanes

The simplest alkanes (methane, ethane, propane, and butane) are gases at room temperature and pressure. Methane and ethane are difficult to liquefy, so they are usually handled as compressed gases. Propane and butane are easily liquefied at room temperature under a moderate pressure. They are stored in low-pressure cylinders in a liquefied form. These four gases are colorless, odorless, and nontoxic, and they have limited physiological effects. The danger in inhaling them lies in potential suffocation due to lack of oxygen. The major immediate danger associated with a natural gas leak is the potential formation of an explosive air–alkane mixture rather than the formation of a toxic air–alkane mixture.

The C_5 to C_8 alkanes, of which there are many isomeric forms, are free-flowing, nonpolar, volatile liquids. They are the primary constituents of gasoline. These compounds are not particularly toxic, but gasoline should not be swallowed because (1) some of the additives present are harmful and (2) liquid alkanes can damage lung tissue because of physical rather than chemical effects. Physical effects include the dissolving of lipid molecules of cell membranes (see Chapter 8), causing pneumonia-like symptoms. Liquid alkanes can also affect the skin for related reasons. These alkanes dissolve natural body oils, causing the skin to dry out. (This "drying out" effect is easily noticed when paint thinner, a mixture of hydrocarbons, is used to remove paint from the hands.)

In direct contrast to liquid alkanes, solid alkanes are used to protect the skin. Pharmaceutical-grade *petrolatum* and *mineral oil* (also called liquid petrolatum), obtained as products from petroleum distillation, have such a function. Petrolatum is a mixture of C_{25} to C_{30} alkanes, and mineral oil involves alkanes in the C_{18} to C_{24} range.

Petrolatum (Vaseline is a well-known brand name) is a semi-solid hydrocarbon mixture that is useful both as a skin softener and as a skin protector. Many moisturizing hand lotions and some medicated salves contain petrolatum. Neither water nor water solutions (for example, urine) will penetrate protective petrolatum coatings. This explains why petrolatum products protect a baby's bottom from diaper rash.

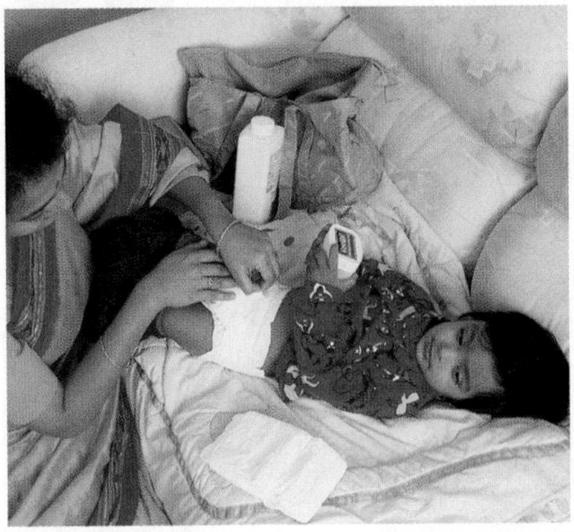

A semi-solid alkane mixture, such as Vaseline, is useful as a skin protector because neither water nor water solutions will penetrate a coating of it. Here, Vaseline is applied to a baby's bottom as a protection against diaper rash.

Mineral oil is often used to replace natural skin oils washed away by frequent bathing and swimming. Too much mineral oil, however, can be detrimental; it will dissolve nonpolar skin materials. Mineral oil has some use as a laxative; it effectively softens and lubricates hard stools. When taken by mouth, it passes through the gastrointestinal tract unchanged and is excreted chemically intact. Loss of fat-soluble vitamins (A, D, E, and K) can occur if mineral oil is consumed while these vitamins are in the digestive tract. Using a mineral-oil enema instead avoids this drawback.

A **halogenation reaction** *is a chemical reaction between a substance and a halogen in which one or more halogen atoms are incorporated into molecules of the substance.* Halogenation of an alkane produces a hydrocarbon derivative in which one or more halogen atoms have been substituted for hydrogen atoms. An example of an alkane halogenation reaction is

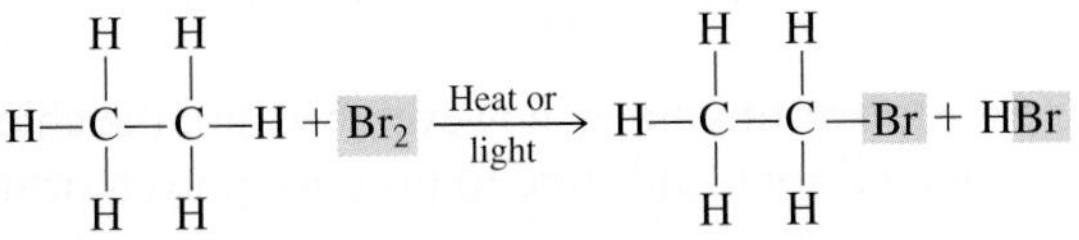

Alkane halogenation is an example of a substitution reaction, a type of reaction that occurs often in organic chemistry. A **substitution reaction** *is a chemical reaction in which part of a small reacting molecule replaces an atom or a group of atoms on a hydrocarbon or hydrocarbon derivative.* A diagrammatic representation of a substitution reaction is shown in Figure 1.14.

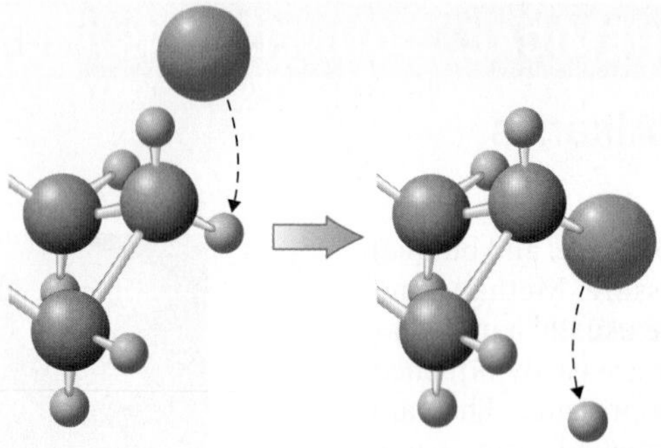

Figure 1.14 In an alkane substitution reaction, an incoming atom or group of atoms (represented by the orange sphere) replaces a hydrogen atom in the alkane molecule.

A *general* equation for the substitution of a single halogen atom for one of the hydrogen atoms of an alkane is

$$\underset{\text{Alkane}}{R{-}H} + \underset{\text{Halogen}}{X_2} \xrightarrow[\text{light}]{\text{Heat or}} \underset{\text{Halogenated alkane}}{R{-}X} + \underset{\text{Hydrogen halide}}{H{-}X}$$

Occasionally, it is useful to represent alkyl groups in a nonspecific way. The symbol R is used for this purpose. Just as *city* is a generic term for Chicago, New York, or San Francisco, the symbol R is a generic designation for any alkyl group. The symbol R comes from the German word *radikal,* which means, in a chemical context, "molecular fragment."

Note the following features of this general equation:

1. The notation R—H is a general formula for an alkane. R— in this case represents an alkyl group. Addition of a hydrogen atom to an alkyl group produces the parent hydrocarbon of the alkyl group.
2. The notation R—X on the product side is the general formula for a halogenated alkane. X is the general symbol for a halogen atom.
3. Reaction conditions are noted by placing these conditions on the equation arrow that separates reactants from products. Halogenation of an alkane requires the presence of heat or light.

(The symbol R is used frequently in organic chemistry and will be encountered in numerous generalized formulas in subsequent chapters; it always represents a generalized organic group in a structural formula. An R group can be an alkyl group—methyl, ethyl, propyl, etc.—or any number of other organic groups. Consider the symbol R to represent the Rest of an organic molecule, which is not specifically indicated because it is not the focal point of the discussion occurring at that time.)

The standard bonding behavior for the halogens is formation of one covalent bond. Thus in organic compounds, a carbon atom forms four bonds, a hydrogen atom forms one bond, and a halogen atom forms one bond.

In halogenation of an alkane, the alkane is said to undergo *fluorination, chlorination, bromination,* or *iodination,* depending on the identity of the halogen reactant. Chlorination and bromination are the two widely used alkane halogenation reactions. Fluorination reactions generally proceed too quickly to be useful, and iodination reactions go too slowly.

Halogenation usually results in the formation of a mixture of products rather than a single product. More than one product results because more than one hydrogen atom on an alkane can be replaced with halogen atoms. To illustrate this concept, let us consider the chlorination of methane, the simplest alkane.

Methane and chlorine, when heated to a high temperature or in the presence of light, react as follows:

$$CH_4 + Cl_2 \xrightarrow[\text{light}]{\text{Heat or}} CH_3Cl + HCl$$

The reaction does not stop at this stage, however, because the chlorinated methane product can react with additional chlorine to produce polychlorinated products.

$$CH_3Cl + Cl_2 \xrightarrow[\text{light}]{\text{Heat or}} CH_2Cl_2 + HCl$$

$$CH_2Cl_2 + Cl_2 \xrightarrow[\text{light}]{\text{Heat or}} CHCl_3 + HCl$$

$$CHCl_3 + Cl_2 \xrightarrow[\text{light}]{\text{Heat or}} CCl_4 + HCl$$

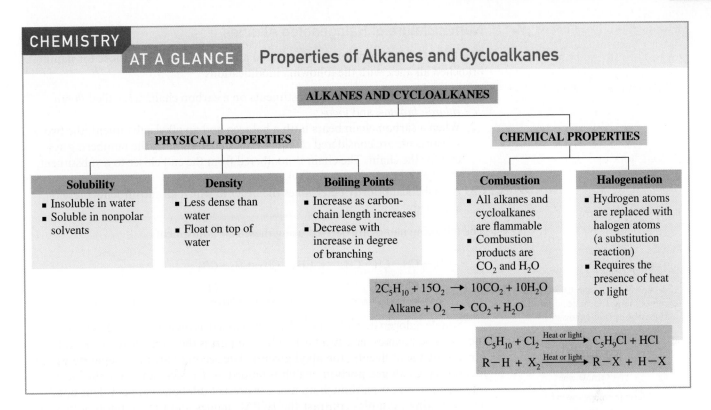

CHEMISTRY AT A GLANCE Properties of Alkanes and Cycloalkanes

ALKANES AND CYCLOALKANES

PHYSICAL PROPERTIES

Solubility
- Insoluble in water
- Soluble in nonpolar solvents

Density
- Less dense than water
- Float on top of water

Boiling Points
- Increase as carbon-chain length increases
- Decrease with increase in degree of branching

CHEMICAL PROPERTIES

Combustion
- All alkanes and cycloalkanes are flammable
- Combustion products are CO_2 and H_2O

$2C_5H_{10} + 15O_2 \rightarrow 10CO_2 + 10H_2O$

$\text{Alkane} + O_2 \rightarrow CO_2 + H_2O$

Halogenation
- Hydrogen atoms are replaced with halogen atoms (a substitution reaction)
- Requires the presence of heat or light

$C_5H_{10} + Cl_2 \xrightarrow{\text{Heat or light}} C_5H_9Cl + HCl$

$R{-}H + X_2 \xrightarrow{\text{Heat or light}} R{-}X + H{-}X$

By controlling the reaction conditions and the ratio of chlorine to methane, it is possible to *favor* formation of one or another of the possible chlorinated methane products.

The chemical properties of cycloalkanes are similar to those of alkanes. Cycloalkanes readily undergo combustion, as well as chlorination and bromination. With unsubstituted cycloalkanes, monohalogenation produces a single product because all hydrogen atoms present in the cycloalkane are equivalent to one another.

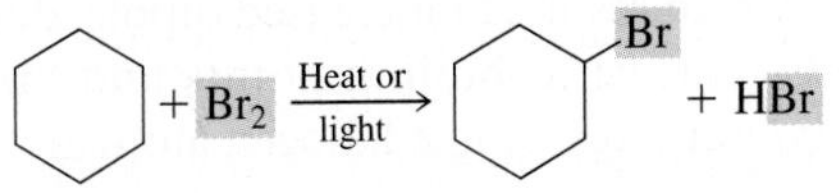

The Chemistry at a Glance feature above summarizes the physical properties and chemical reactions of alkanes and cycloalkanes.

1.18 Halogenated Alkanes and Cycloalkanes

A **halogenated alkane** *is an alkane derivative in which one or more halogen atoms are present.* Similarly, a **halogenated cycloalkane** *is a cycloalkane derivative in which one or more halogen atoms are present.* Produced by halogenation reactions (Section 1.17), these two types of compounds represent the first class of hydrocarbon derivatives (Section 1.3) formally considered in this text.

Alkanes have the general molecular formula C_nH_{2n+2} (Section 1.4). Halogenated alkanes containing one halogen atom have the general molecular formula $C_nH_{2n+1}X$; a halogen atom has replaced a hydrogen atom. If two halogen atoms are present in a halogenated alkane, the general molecular formula is $C_nH_{2n}X_2$. Since cycloalkanes have the general molecular formula C_nH_{2n} (Section 1.12), a halogenated cycloalkane with one halogen atom present will have a general molecular formula of $C_nH_{2n-1}X$.

Nomenclature of Halogenated Alkanes

The IUPAC rules for naming halogenated alkanes are similar to those for naming branched alkanes, with the following modifications:

1. Halogen atoms, treated as substituents on a carbon chain, are called *fluoro-*, *chloro-*, *bromo-*, and *iodo-*.
2. When a carbon chain bears both a halogen and an alkyl substituent, the two substituents are considered of equal rank in determining the numbering system for the chain. The chain is numbered from the end closer to a substituent, whether it be a *halo-* or an alkyl group.
3. Alphabetical priority determines the order in which all substituents present are listed.

The following names are derived using these rule adjustments.

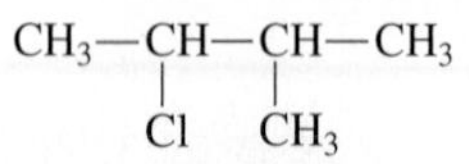

2-Chloro-3-methylbutane

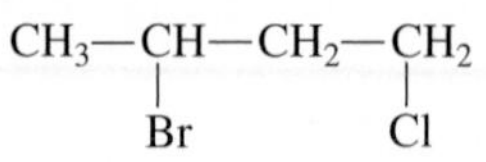

3-Bromo-1-chlorobutane

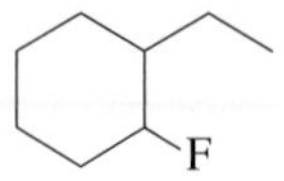

1-Ethyl-2-fluorocyclohexane

The contrast between IUPAC and common names for halogenated hydrocarbons is as follows:

IUPAC (one word)

haloalkane

chloromethane

Common (two words)

alkyl halide

methyl chloride

Simple halogenated alkanes can also be named as *alkyl halides.* These common (non-IUPAC) names have two parts. The first part is the name of the hydrocarbon portion of the molecule (the alkyl group). The second part (as a separate word) identifies the halogen portion, which is named as if it were an ion (chloride, bromide, and so on), even though no ions are present (all bonds are covalent bonds). The following examples contrast the IUPAC names and the common names (in parentheses) of selected halogenated alkanes.

CH_3-CH_2-Cl

Chloroethane
(ethyl chloride)

$CH_3-CH_2-CH_2-Br$

1-Bromopropane
(propyl bromide)

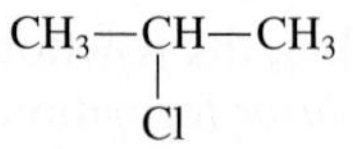

2-Chloropropane
(isopropyl chloride)

An alternative designation for a halogenated alkane is *alkyl halide*.

Physical Properties of Halogenated Alkanes

Halogenated alkane boiling points are generally higher than those of the corresponding alkane. An important factor contributing to this effect is the polarity of carbon–halogen bonds, which results in increased dipole–dipole interactions.

Two general trends relative to boiling points and melting points of halogenated hydrocarbons containing a single halogen atom are:

1. Boiling points and melting points increase as the size of the alkyl group present increases. This is due to increasing intermolecular forces associated with increased molecular surface area.
2. Boiling points and melting points increase as the size of the halogen atom increases from fluorine (F) to iodine (I).

Halogenated hydrocarbons do not have hydrogen-bonding capabilities since all hydrogen atoms are bonded to carbon atoms. Thus water solubility is very limited regardless of molecular size.

Some halogenated alkanes have densities that are greater than that of water, a situation not common for organic compounds. Chloroalkanes containing two or more chlorine atoms, bromoalkanes, and iodoalkanes are all more dense than water.

Halogenated Methanes

A diverse chemistry is associated with halogenated methane derivatives. Several derivatives are excellent nonflammable solvents for a variety of organic compounds. Several have good refrigerant properties. Some are useful as propellants for aerosol cans. On the downside, experience has shown that many derivatives exhibit varying degrees of toxicity.

CHEMICAL CONNECTIONS 1-C

Chlorofluorocarbons and the Ozone Layer

Chlorofluorocarbons (CFCs) are compounds composed of the elements chlorine, fluorine, and carbon. CFCs are synthetic compounds that have been developed primarily for use as refrigerants. The two most widely used of the CFCs are trichlorofluoromethane and dichlorodifluoromethane. Both of these compounds are marketed under the trade name Freon.

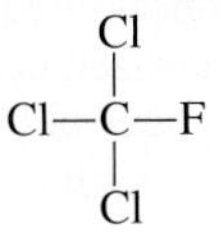

Trichlorofluoromethane
(Freon-11)

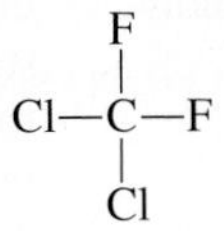

Dichlorodifluoromethane
(Freon-12)

Freon-11 and Freon-12 possess ideal properties for use as a refrigerant gas. Both are inert, nontoxic, and easily compressible. Prior to their development, ammonia was used in refrigeration. Ammonia is toxic, and leaking ammonia-based refrigeration units have been fatal.

It is now known that CFCs contribute to a serious environmental problem: destruction of the stratospheric (high-altitude) ozone that is commonly called the ozone layer. Once released into the atmosphere, CFCs persist for long periods without reaction. Consequently, they slowly drift upward in the atmosphere, finally reaching the stratosphere.

It is in the stratosphere, the location of the "ozone layer," that environmental problems occur. At these high altitudes, the CFCs are exposed to ultraviolet light (from the sun), which activates them. The ultraviolet light breaks carbon–chlorine bonds within the CFCs, releasing chlorine atoms.

$$CCl_2F_2 + \text{ultraviolet light} \longrightarrow CClF_2 + Cl$$

The Cl atoms so produced (called atomic chlorine) are extremely reactive species. One of the molecules with which they react is ozone (O_3).

$$Cl + O_3 \longrightarrow ClO + O_2$$

A reaction such as this upsets the O_3–O_2 equilibrium in the stratosphere.

The Montreal Protocol of 1987 (an international agreement on substances that deplete the ozone layer), and later amendments to this agreement, limit—and in some cases ban—future production and use of CFCs. The following graph shows the effects of the implementation of this agreement.

The most commonly used replacement compounds in refrigeration systems and as aerosol propellants for the phased-out CFCs are HFCs (hydrofluorocarbons), compounds that contain H and F (bonded to carbon atoms). The three most commonly used HFCs are

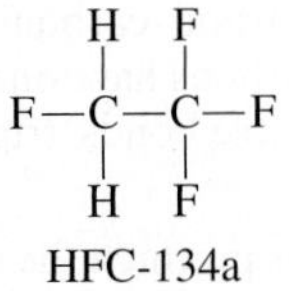

HFC-134a
1,1,1,2-tetrafluoroethane

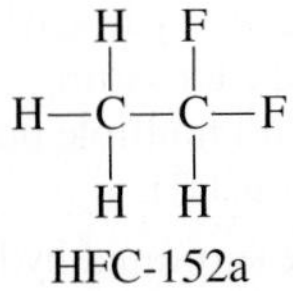

HFC-152a
1,1-difluoroethane

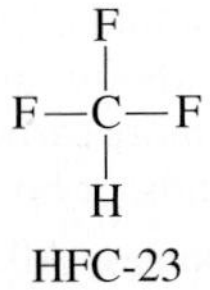

HFC-23
trifluoromethane

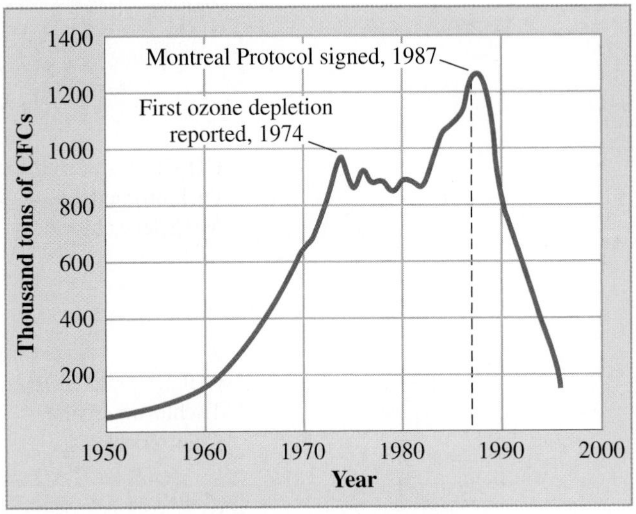

Worldwide Production of CFCs (1950–1996)

The presence of carbon-hydrogen bonds in these molecules makes them more reactive than CFCs, and they are generally destroyed at lower altitudes before they reach the stratosphere. Their refrigeration properties, although adequate, are not as good as those of the previously used CFCs.

HFCs, used heavily during the period 1990–2010, are now on the replacement (phase-out) list. They are now known to be potent greenhouse gases, and their concentrations in the lower atmosphere are increasing. HFC-134a, the most used of the HFCs, has a global warming potential that is 1340 times greater than that of an equivalent amount of carbon dioxide. HFC use in air-conditioning for new model cars is now restricted or banned in the European Union.

A new type of fluorinated hydrocarbon is being "hyped" as the "global replacement refrigerant." It is an HFO (hydrofluoroolefin). HFOs are unsaturated (a carbon-carbon double bond is present) rather than saturated hydrocarbon derivatives. The specific replacement compound coming to market is HFO-1234yf, a compound with the structure

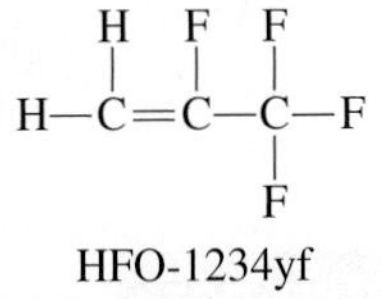

HFO-1234yf
2,3,3,3-tetrafluoropropene

The IUPAC name for this compound is 2,3,3,3-tetrafluoropropene (Section 2.3). The global warming potential of this HFO is only slightly greater than that of carbon dioxide. HFO-1234yf use began in Europe with 2011 model-year cars and will begin in the United States with 2013 model-year cars.

Table 1.4 Properties and Uses of the Chlorinated Methanes

Chemical Formula and Nomenclature	Physical Properties	Occurrence and Uses
CH_3Cl Chloromethane	Colorless, extremely flammable gas with a mildly sweet odor	present in volcanic gases; produced by algae and giant kelp; used as an industrial solvent; once used widely as a refrigerant; no longer found in consumer products because of toxicity concerns
CH_2Cl_2 Dichloromethane Methylene chloride	Colorless, volatile liquid with a mildly sweet odor	chemical intermediate in production of silicone polymers; used as a paint stripper and degreaser; once used to decaffeinate coffee but has been replaced by liquid carbon dioxide due to concern about trace amounts of CH_2Cl_2 remaining in the coffee
$CHCl_3$ Trichloromethane Chloroform	Colorless, very dense sweet-smelling liquid	used in manufacture of teflon; good industrial solvent; was an early popular anesthetic but used discontinued because of toxicity effects
CCl_4 Tetrachloromethane Carbon tetrachloride Carbon tet	Colorless, nonflammable sweet-smelling liquid	good solvent for fats, oils, and greases; once widely used in dry cleaning of clothing but has been replaced by tetrachloroethylene, which is more stable and less toxic

Two important categories of halogenated methanes are (1) the chlorinated methanes and (2) the chlorofluorocarbons called Freons.

Repeated chlorination of methane produces successively the compounds CH_3Cl, CH_2Cl_2, $CHCl_3$, and CCl_4. Table 1.4 gives nomenclature (which includes common names) and uses (both past and present) for these compounds.

Chlorofluorocarbons (CFCs) have the general molecular formula CCl_xF_{4-x}. They are manufactured under the trade name Freon. Representative of such compounds are CCl_3F (trichlorofluoromethane, or Freon-11) and CCl_2F_2 (dichlorodifluoromethane, or Freon-12). For many years, Freons were the refrigerant gas of choice for both industrial and home refrigeration and air-conditioning, as well as automobile air-conditioning. Indeed, it is CFCs that made refrigeration units available, at reasonable cost, to the general household after World War II.

Use of Freons in refrigeration and air-conditioning has now been discontinued. It was not a toxicity problem that caused the change but, rather, an atmospheric environmental problem. After many years of use, it became apparent that Freons were adversely affecting the ozone layer present in the upper atmosphere. The focus on relevancy feature Chemical Connections 1-C on the previous page gives details concerning the Freon-ozone layer problem and also dicusses replacement compounds for Freons.

Concepts to Remember

OWL Sign in at **www.cengage.com/owl** to view tutorials and simulations, develop problem-solving skills, and complete online homework assigned by your professor.

Carbon atom bonding characteristics. Carbon atoms in organic compounds must have four bonds (Section 1.2).

Types of hydrocarbons. Hydrocarbons, binary compounds of carbon and hydrogen, are of two types: saturated and unsaturated. In saturated hydrocarbons, all carbon–carbon bonds are single bonds. Unsaturated hydrocarbons have one or more carbon–carbon multiple bonds—double bonds, triple bonds, or both (Section 1.3).

Alkanes. Alkanes are saturated hydrocarbons in which the carbon atom arrangement is that of an unbranched or branched chain. The formulas of all alkanes can be represented by the general formula C_nH_{2n+2}, where n is the number of carbon atoms present (Section 1.4).

Structural formulas. Structural formulas are two-dimensional representations of the arrangement of the atoms in molecules. These formulas give complete information about the arrangement of the atoms in a molecule but not the spatial orientation of the atoms. Two types of structural formulas are commonly encountered: expanded and condensed (Section 1.5).

Isomers. Isomers are compounds that have the same molecular formula (that is, the same numbers and kinds of atoms), but that differ in the way the atoms are arranged (Section 1.6).

Constitutional isomers. Constitutional isomers are isomers that differ in the connectivity of atoms, that is, in the order in which atoms are attached to each other within molecules (Section 1.6).

Conformations. Conformations are differing orientations of the same molecule made possible by free rotation about single bonds in the molecule (Section 1.7).

Alkane nomenclature. The IUPAC name for an alkane is based on the longest continuous chain of carbon atoms in the molecule. A group of carbon atoms attached to the chain is an alkyl group. Both the position and the identity of the alkyl group are prefixed to the name of the longest carbon chain (Section 1.8).

Line-angle structural formulas. A line-angle structural formula is a structural representation in which a line represents a carbon–carbon bond and a carbon atom is understood to be present at every point where two lines meet and at the ends of the line. Line-angle structural formulas are the most concise method for representing the structure of a hydrocarbon or hydrocarbon derivative (Section 1.9).

Cycloalkanes. Cycloalkanes are saturated hydrocarbons in which at least one cyclic arrangement of carbon atoms is present. The formulas of all cycloalkanes can be represented by the general formula C_nH_{2n}, where *n* is the number of carbon atoms present (Section 1.12).

Cycloalkane nomenclature. The IUPAC name for a cycloalkane is obtained by placing the prefix *cyclo-* before the alkane name that corresponds to the number of carbon atoms in the ring. Alkyl groups attached to the ring are located by using a ring-numbering system (Section 1.13).

***Cis–trans* isomerism.** For certain disubstituted cycloalkanes, *cis–trans* isomers exist. *Cis–trans* isomers are compounds that have the same molecular and structural formulas but different arrangements of atoms in space because of restricted rotation about bonds (Section 1.14).

Natural sources of saturated hydrocarbons. Natural gas and petroleum are the largest and most important natural sources of both alkanes and cycloalkanes (Section 1.15).

Physical properties of saturated hydrocarbons. Saturated hydrocarbons are not soluble in water and have lower densities than water. Melting and boiling points increase with increasing carbon chain length or ring size (Section 1.16).

Chemical properties of saturated hydrocarbons. Two important reactions that saturated hydrocarbons undergo are combustion and halogenation. In combustion, saturated hydrocarbons burn in air to produce CO_2 and H_2O. Halogenation is a substitution reaction in which one or more hydrogen atoms of the hydrocarbon are replaced by halogen atoms (Section 1.17).

Halogenated alkanes. Halogenated alkanes are hydrocarbon derivatives in which one or more halogen atoms have replaced hydrogen atoms of the alkane (Section 1.18).

Halogenated alkane nomenclature. Halogenated alkanes are named by using the rules that apply to branched-chain alkanes, with halogen substituents being treated the same as alkyl groups (Section 1.18).

Exercises and Problems

OWL Interactive versions of these problems may be assigned in OWL.

Exercises and problems are arranged in matched pairs with the two members of a pair addressing the same concept(s). The answer to the odd-numbered member of a pair is given at the back of the book. Problems denoted with a ▲ involve concepts found not only in the section under consideration but also concepts found in one or more earlier sections of the chapter. Problems denoted with a ● cover concepts found in a Chemical Connections feature box.

Organic and Inorganic Compounds (Section 1.1)

1.1 Indicate whether each of the following statements is true or false.
  a. The number of organic compounds exceeds the number of inorganic compounds by a factor of about 2.
  b. Chemists now believe that a special "vital force" is needed to form an organic compound.
  c. Historically, the *org-* of the term *organic* was conceptually paired with the *org-* in the term *living organism.*
  d. Most but not all compounds found in living organisms are organic compounds.

1.2 Indicate whether each of the following statements is true or false.
  a. Approximately 10 million organic compounds have been characterized.
  b. The number of known organic compounds and the number of known inorganic compounds are approximately the same.
  c. In essence, organic chemistry is the study of the compounds of one element.
  d. Numerous organic compounds are known that do not occur in living organisms.

Bonding Characteristics of the Carbon Atom (Section 1.2)

1.3 Indicate whether each of the following situations meet or do not meet the "bonding requirement" for carbon atoms.
  a. Two single bonds and a double bond
  b. A single bond and two double bonds
  c. Three single bonds and a triple bond
  d. A double bond and a triple bond

1.4 Indicate whether each of the following situations meet or do not meet the "bonding requirement" for carbon atoms.
  a. Four single bonds
  b. Three single bonds and a double bond
  c. Two double bonds and two single bonds
  d. Two double bonds

Hydrocarbons and Hydrocarbon Derivatives (Section 1.3)

1.5 What is the difference between a *hydrocarbon* and a *hydrocarbon derivative*?

1.6 Contrast hydrocarbons and hydrocarbon derivatives in terms of number of compounds that are known.

1.7 What is the difference between a *saturated hydrocarbon* and an *unsaturated hydrocarbon*?

1.8 What structural feature is present in an unsaturated hydrocarbon that is not present in a saturated hydrocarbon?

1.9 Classify each of the following hydrocarbons as saturated or unsaturated.

a.
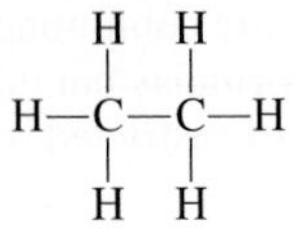
b.
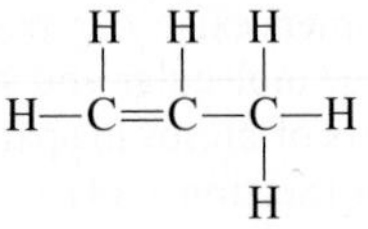
c.
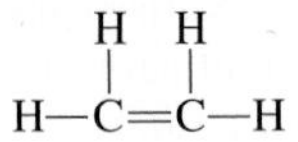
d.
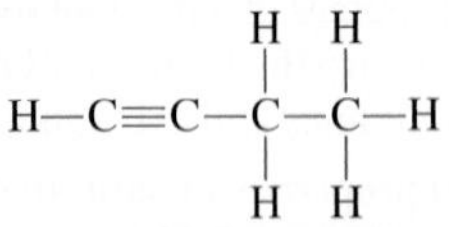

1.10 Classify each of the following hydrocarbons as saturated or unsaturated.

a. H—C≡C—H
b.
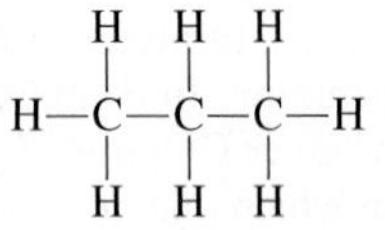
c.
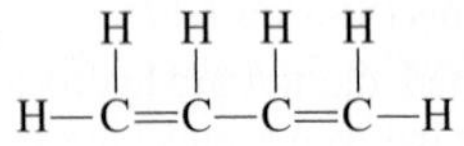
d.
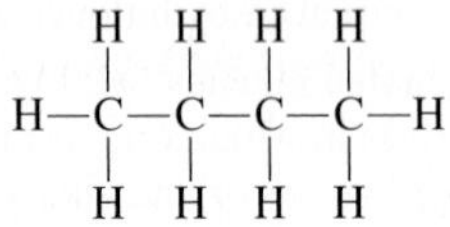

General Formulas for Alkanes (Section 1.4)

1.11 Using the general formula for an alkane, derive the following for specific alkanes.
a. Number of hydrogen atoms present when 8 carbon atoms are present
b. Number of carbon atoms present when 10 hydrogen atoms are present
c. Number of carbon atoms present when 41 total atoms are present
d. Total number of covalent bonds present in the molecule when 7 carbon atoms are present

1.12 Using the general formula for an alkane, derive the following for specific alkanes.
a. Number of carbon atoms present when 14 hydrogen atoms are present
b. Number of hydrogen atoms present when 6 carbon atoms are present
c. Number of hydrogen atoms present when 32 total atoms are present
d. Total number of covalent bonds present in the molecule when 16 hydrogen atoms are present

•1.13 (Chemical Connections 1-A) Indicate whether each of the following statements concerning the hydrocarbon methane is true or false.
a. Methane is a major component of the atmosphere of the planets Jupiter and Saturn.
b. Marsh gas is a common name for methane.
c. Methane production *within* the atmosphere is very limited because of the lack of H_2 gas in the atmosphere.
d. Methane differs from CO_2 in that it is not a greenhouse gas.

•1.14 (Chemical Connections 1-A) Indicate whether each of the following statements concerning the hydrocarbon methane is true or false.
a. Methane levels in the Earth's atmosphere are approximately 25 ppm.
b. Methane present in the Earth's atmosphere comes from terrestrial sources.
c. The digestive tracts of plant-eating animals contain bacteria that can produce methane.
d. Methane, when mixed with air in certain ratios, is an explosion hazard.

Structural Formulas (Section 1.5)

1.15 Convert the following expanded structural formulas for alkanes into condensed structural formulas where all carbon-carbon bonds are explicitly shown.

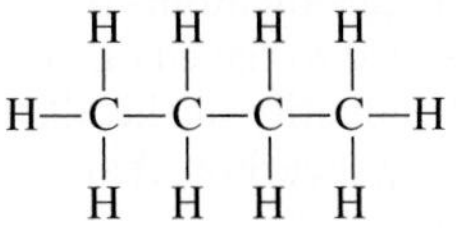
a.
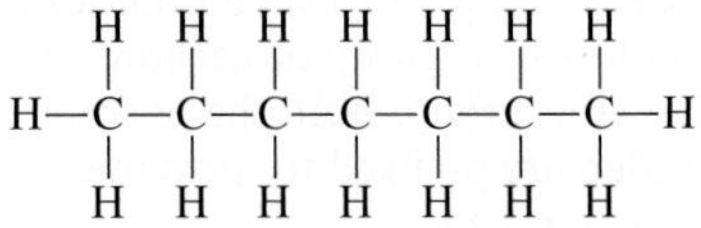
b.

1.16 Convert the following expanded structural formulas for alkanes into condensed structural formulas where all carbon-carbon bonds are explicitly shown.

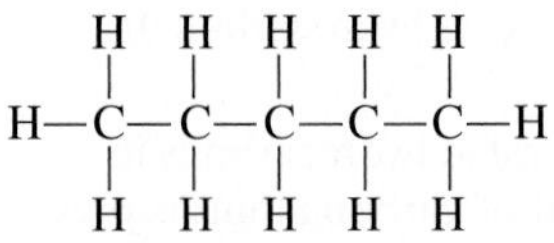
a.
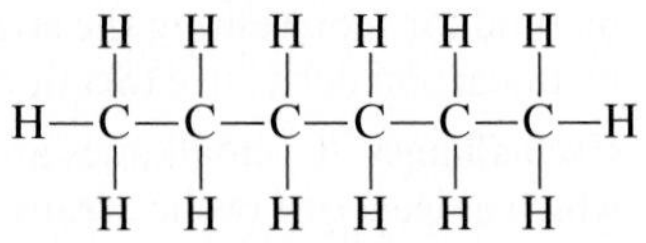
b.

1.17 Convert the expanded structural formulas in Problem 1.15 into condensed structural formulas where parentheses and a subscript are used to denote the number of CH_2 groups present.

1.18 Convert the expanded structural formulas in Problem 1.16 into condensed structural formulas where parentheses and a subscript are used to denote the number of CH_2 groups present.

1.19 Convert the expanded structural formulas in Problem 1.15 into skeletal structural formulas.

1.20 Convert the expanded structural formulas in Problem 1.16 into skeletal structural formulas.

1.21 Draw the indicated type of formula for the following alkanes.
a. The expanded structural formula for a continuous-chain alkane with the formula C_5H_{12}
b. The expanded structural formula for CH_3—$(CH_2)_6$—CH_3
c. The condensed structural formula using parentheses and a subscript to denote the number of CH_2 groups for the continuous-chain alkane $C_{10}H_{22}$
d. The molecular formula for the alkane CH_3—$(CH_2)_4$—CH_3

1.22 Draw the indicated type of formula for the following alkanes.
a. The expanded structural formula for a continuous-chain alkane with the molecular formula C_6H_{14}
b. The condensed structural formula using parentheses and a subscript to denote the number of CH_2 groups for the straight-chain alkane $C_{12}H_{26}$

c. The molecular formula for the alkane $CH_3{-}(CH_2)_6{-}CH_3$

d. The expanded structural formula for $CH_3{-}(CH_2)_3{-}CH_3$

1.23 Determine the following for the six-carbon alkane whose skeletal structural formula is

C—C—C—C—C—C

a. How many hydrogen atoms are present?

b. How many carbon-carbon bonds are present?

c. How many CH_2 groups are present?

d. How many total covalent bonds are present?

1.24 Determine the following for the five-carbon alkane whose skeletal structural formula is

C—C—C—C—C

a. How many hydrogen atoms are present?

b. How many carbon-hydrogen bonds are present?

c. How many CH_3 groups are present?

d. How many total covalent bonds are present?

Alkane Isomerism (Section 1.6)

1.25 What general requirement must be met before two compounds can be isomers?

1.26 Explain why two alkanes with the molecular formulas C_5H_{12} and C_6H_{14} could not be constitutional isomers.

1.27 Indicate whether each of the following would be expected to be the same or different for two alkane constitutional isomers.

a. Number of hydrogen atoms present in a molecule

b. Condensed structural formula

c. Boiling point

d. Melting point

1.28 Indicate whether each of the following would be expected to be the same or different for two alkane constitutional isomers.

a. Number of carbon atoms present in a molecule

b. Shape of molecule

c. Density

d. Molecular formula

1.29 What is the difference between a continuous-chain alkane and a branched-chain alkane?

1.30 The general formula for a continuous-chain alkane is C_nH_{2n+2}. What is the general formula for a branched-chain alkane?

1.31 With the help of Table 1.1, indicate how many constitutional isomers exist for each of the following.

a. Four-carbon alkanes b. Six-carbon alkanes

c. Eight-carbon alkanes d. Ten-carbon alkanes

1.32 With the help of Table 1.1, indicate how many constitutional isomers exist for each of the following.

a. Three-carbon alkanes b. Five-carbon alkanes

c. Seven-carbon alkanes d. Nine-carbon alkanes

1.33 How many of the numerous eight-carbon alkane constitutional isomers are continuous-chain alkanes?

1.34 How many of the numerous seven-carbon alkane constitutional isomers are continuous-chain alkanes?

Conformations of Alkanes (Section 1.7)

1.35 For each of the following pairs of structures, determine whether they are

1. Different conformations of the same molecule
2. Different compounds that are constitutional isomers
3. Different compounds that are not constitutional isomers

a.
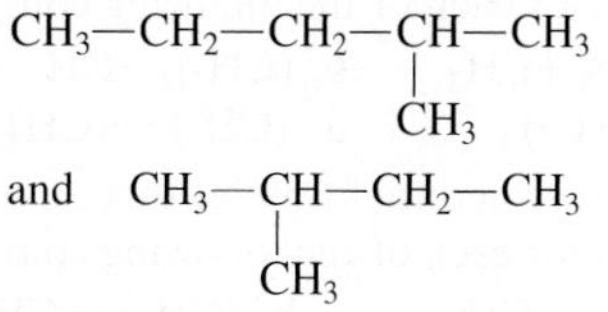

b.
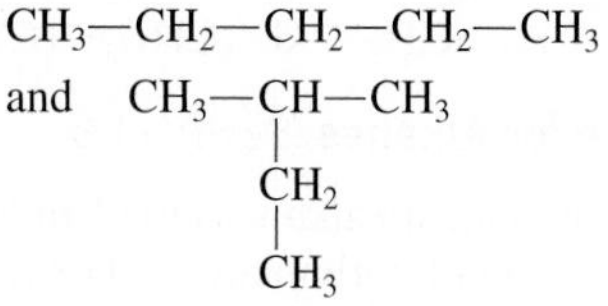

c.
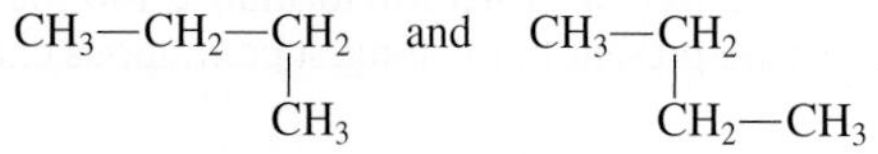

d.
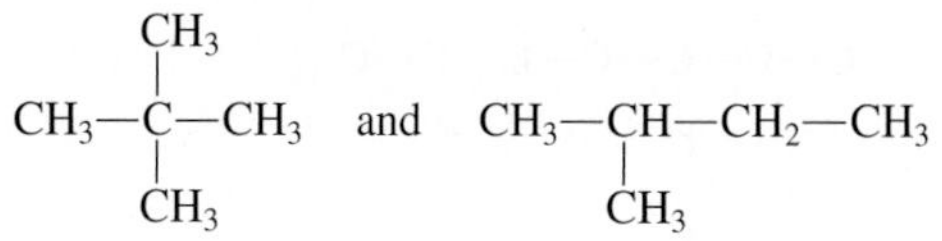

1.36 For each of the following pairs of structures, determine whether they are

1. Different conformations of the same molecule
2. Different compounds that are constitutional isomers
3. Different compounds that are not constitutional isomers

a.
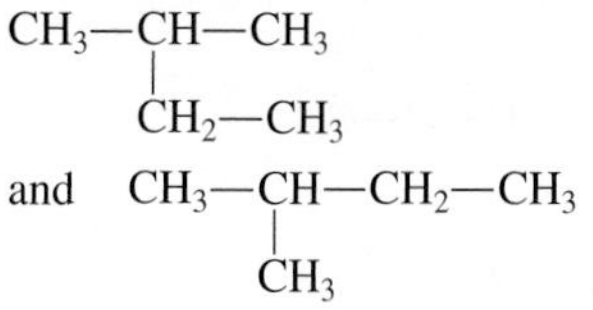

b.
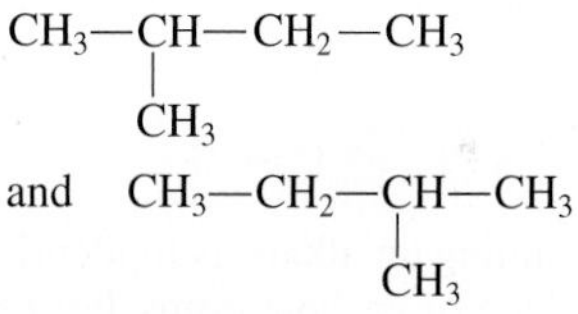

c.
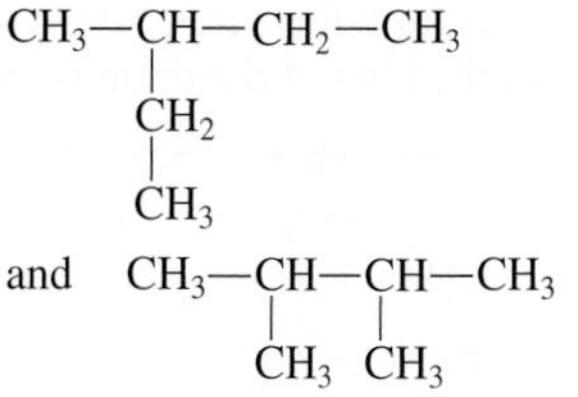

d.
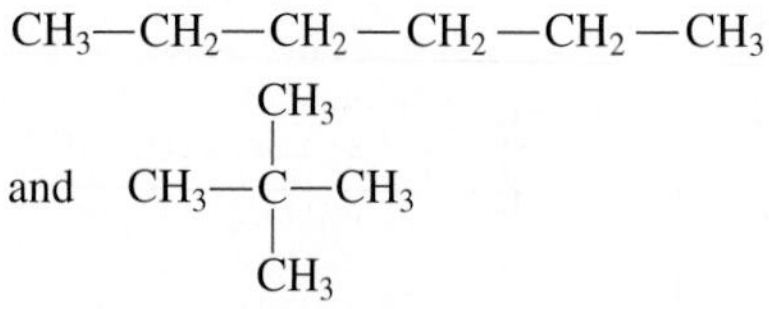

1.37 Convert each of the following linear condensed structural formulas into "regular" condensed structural formulas.

a. $CH_3{-}CH_2{-}CH{-}(CH_3){-}CH_2{-}CH_3$

b. $(CH_3)_2{-}CH{-}CH_2{-}CH{-}(CH_3)_2$

c. $CH_3{-}CH{-}(CH_3){-}CH_3$

d. $CH_3{-}CH_2{-}CH{-}(CH_2{-}CH_3){-}CH_2{-}CH_3$

1.38 Convert each of the following linear condensed structural formulas into "regular" condensed structural formulas.

a. $CH_3{-}CH{-}(CH_3){-}CH_2{-}CH_3$

b. $CH_3{-}C{-}(CH_3)_2{-}CH_3$

c. $(CH_3)_2{-}CH{-}CH_3$

d. $CH_3{-}CH_2{-}CH{-}(CH_3){-}CH{-}(CH_3){-}CH_3$

1.39 Draw condensed structural formulas that do not contain any parentheses for each of the following compounds.
a. $CH_3—(CH_2)_3—CH_3$ b. $(CH_3)_2—CH—CH_2—CH_3$
c. $(CH_3)_3—C—CH_3$ d. $(CH_3)_2—(CH)_2—(CH_3)_2$

1.40 Draw condensed structural formulas that do not contain any parentheses for each of the following compounds.
a. $CH_3—(CH_2)_2—CH_3$ b. $(CH_3)_2—CH—CH_3$
c. $(CH_3)_3—C—CH_2—CH_3$ d. $(CH_3)_3—(C)_2—(CH_3)_3$

IUPAC Nomenclature for Alkanes (Section 1.8)

1.41 The first step in naming an alkane is to identify the longest continuous chain of carbon atoms. For each of the following skeletal structural formulas, how many carbon atoms are present in the longest continuous chain?

a. 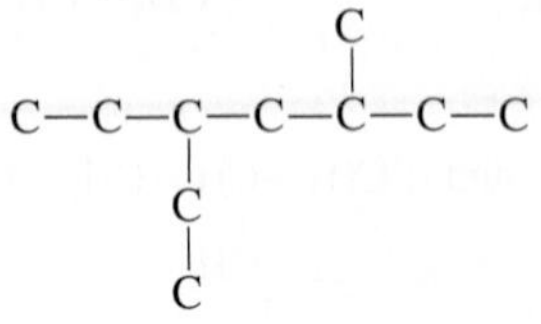

b. 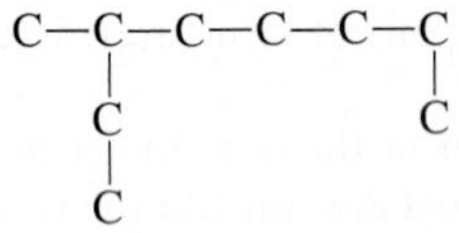

c. 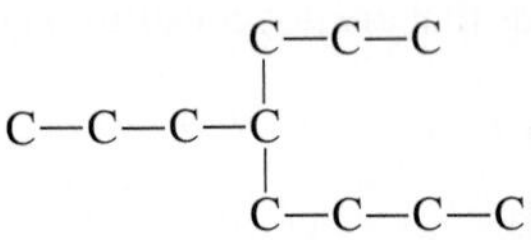

d. 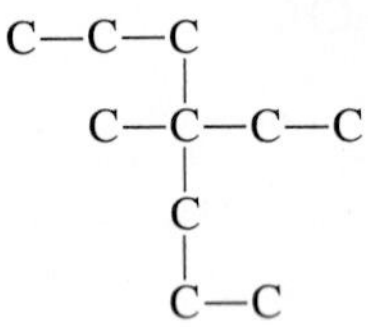

1.42 The first step in naming an alkane is to identify the longest continuous chain of carbon atoms. For each of the following skeletal structural formulas, how many carbon atoms are present in the longest continuous chain?

a. 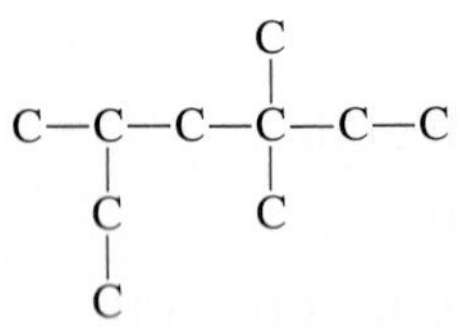

b. 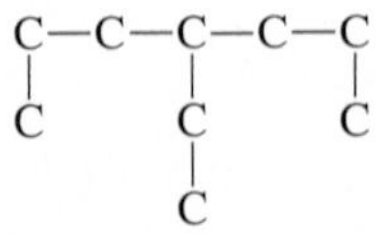

c. 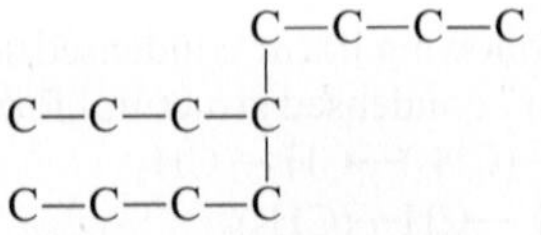

d. 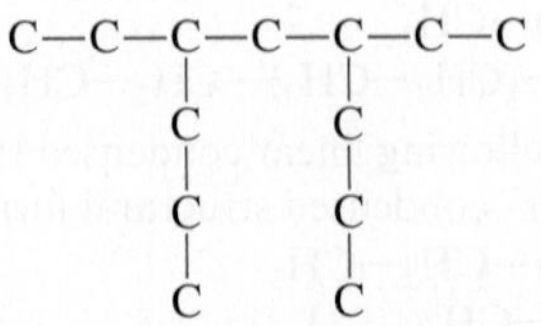

1.43 Give the IUPAC name for each of the following alkanes.

a. 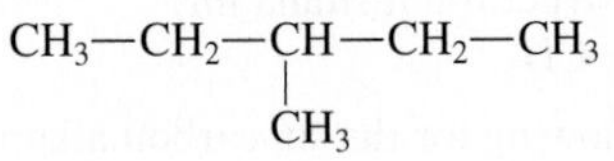

b. 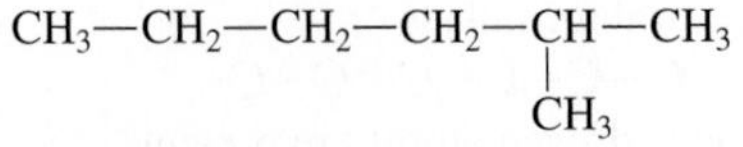

c. 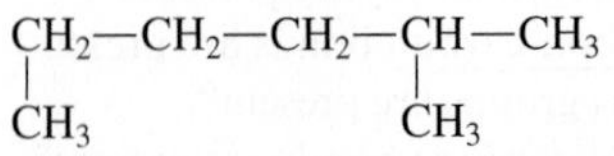

d. 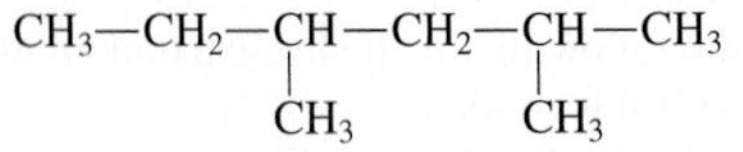

1.44 Give the IUPAC name for each of the following alkanes.

a. 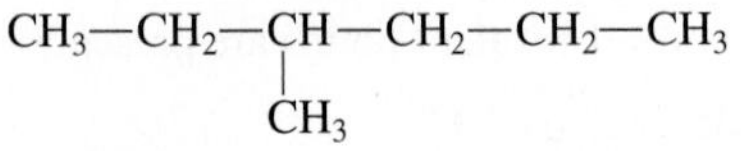

b. 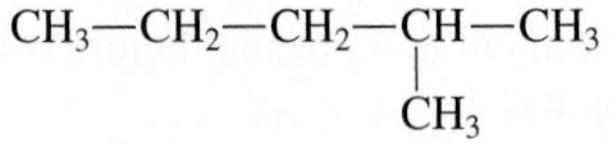

c. 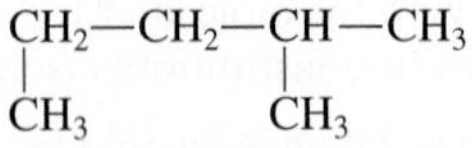

d. 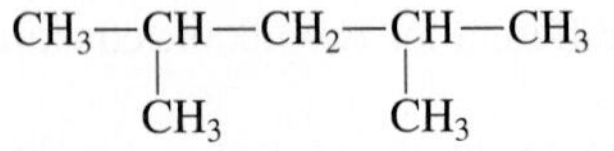

1.45 Give the IUPAC name for each of the following alkanes.

a. 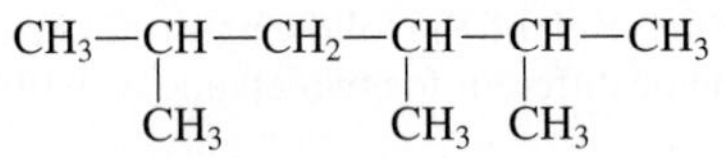

b. 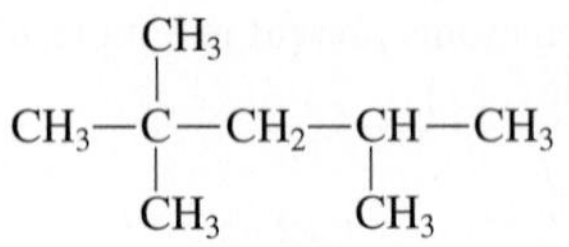

c. 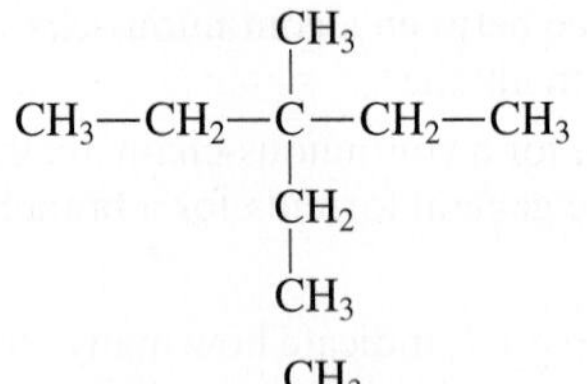

d. 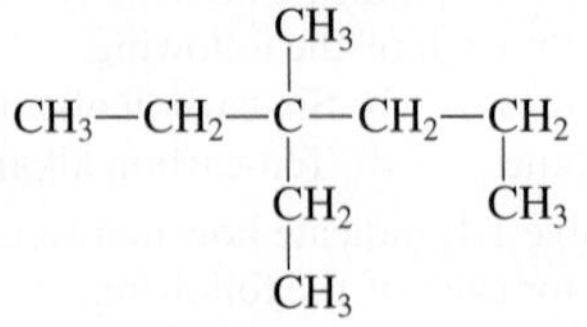

1.46 Give the IUPAC name for each of the following alkanes.

a. 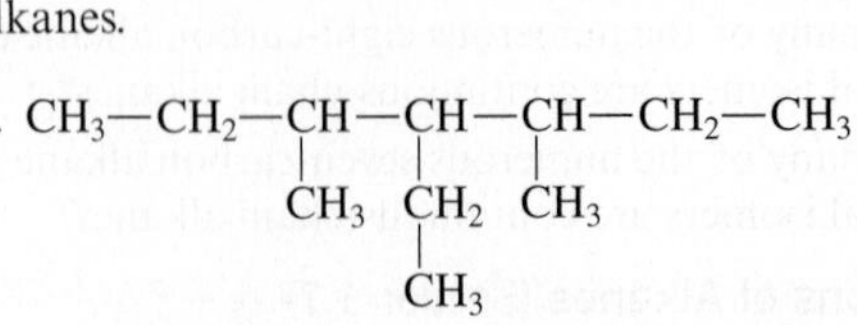

b. 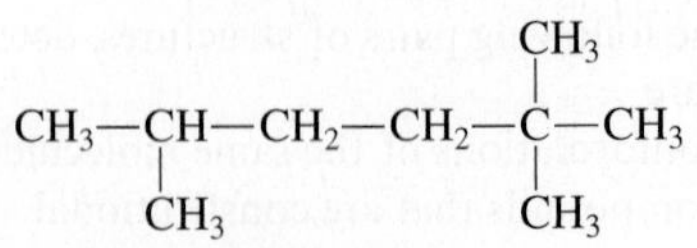

c. CH_3—CH—CH_2—CH_2—CH_3
   | |
   CH_3 CH_3

d. CH_3—CH—CH—CH_2—CH_3
   | | |
   CH_3 CH_3 CH_3

1.47 Draw a condensed structural formula for each of the following alkanes.
a. 3,4-Dimethylhexane b. 3-Ethyl-3-methylpentane
c. 3,5-Diethyloctane d. 4-Propylnonane

1.48 Draw a condensed structural formula for each of the following alkanes.
a. 2,4-Dimethylhexane b. 5-Propyldecane
c. 2,3,4-Trimethyloctane d. 3-Ethyl-3-methylheptane

•1.49 For each of the alkanes in Problem 1.47, determine (a) the number of alkyl groups present and (b) the number of substituents present.

•1.50 For each of the alkanes in Problem 1.48, determine (a) the number of alkyl groups present and (b) the number of substituents present.

1.51 Explain why the name given for each of the following alkanes is not the correct IUPAC name. Then give the correct IUPAC name for the compound.
a. 2-Ethyl-2-methylpropane
b. 2,3,3-Trimethylbutane
c. 3-Methyl-4-ethylhexane
d. 2-Methyl-4-methylhexane

1.52 Explain why the name given for each of the following alkanes is not the correct IUPAC name. Then give the correct IUPAC name for the compound.
a. 2-Ethylpentane b. 3,3,4-Trimethylpentane
c. 4-Ethyl-3-methylhexane d. 3-Ethyl-4-ethylhexane

1.53 How many of the 18 C_8 alkane constitutional isomers are named using the following base-chain names?
a. Octane b. Heptane c. Hexane d. Pentane

1.54 How many of the nine C_7 alkane constitutional isomers are named using the following base-chain names?
a. Heptane b. Hexane c. Pentane d. Butane

Line-Angle Structural Formulas for Alkanes
(Section 1.9)

1.55 Convert each of the following line-angle structural formulas to a skeletal structural formula.

a. 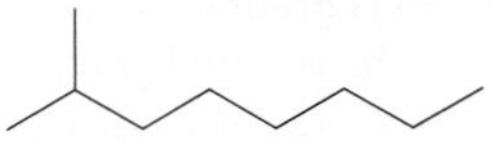
b. 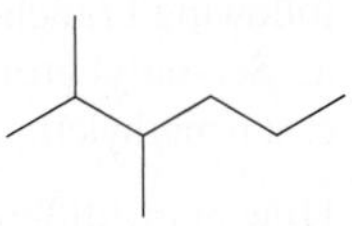
c. 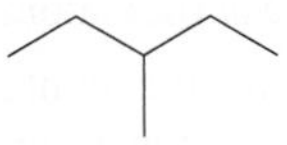
d. 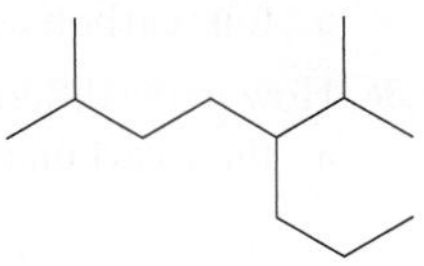

1.56 Convert each of the following line-angle structural formulas to a skeletal structural formula.

a. 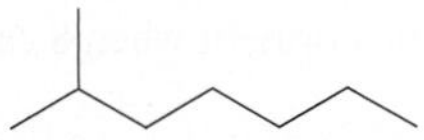
b. 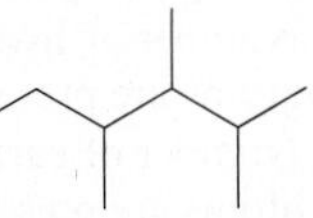
c. 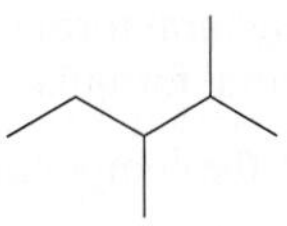
d. 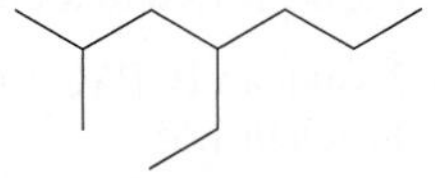

1.57 Convert each of the following line-angle structural formulas to a condensed structural formula.

a. 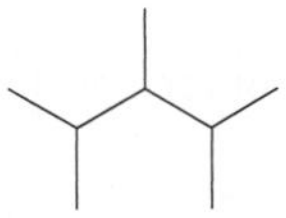
b. 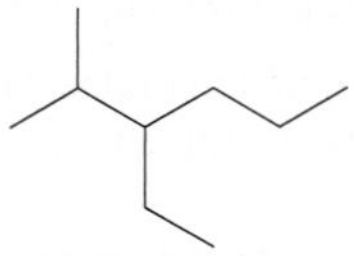
c. 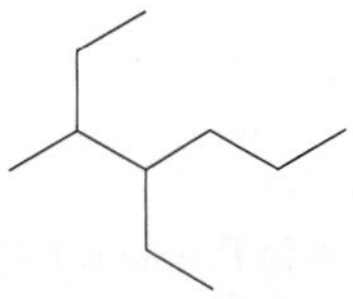
d. 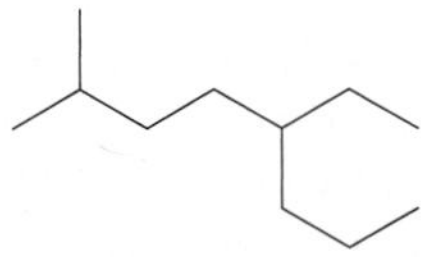

1.58 Convert each of the following line-angle structural formulas to a condensed structural formula.

a. 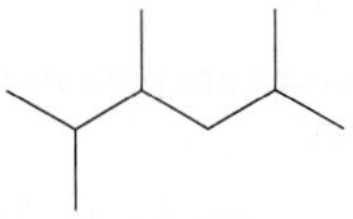
b. 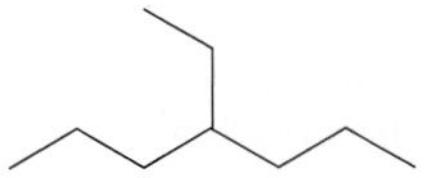
c. 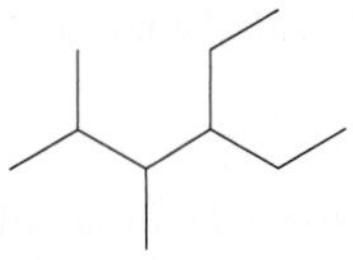
d. 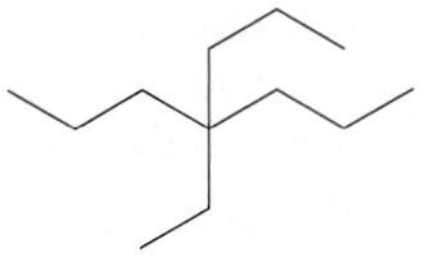

1.59 Do the line-angle structural formulas in each of the following sets represent (1) the same compound (2) constitutional isomers or (3) different compounds that are not constitutional isomers?

a. 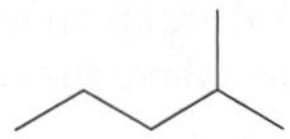 and 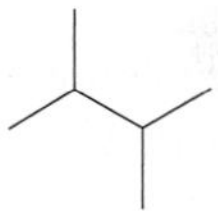

b. 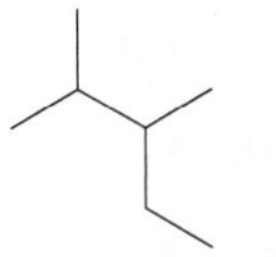 and 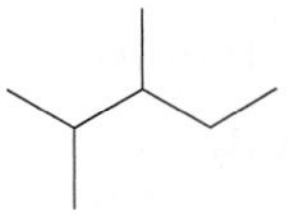

1.60 Do the line-angle structural formulas in each of the following sets represent (1) the same compound (2) constitutional isomers or (3) different compounds that are not constitutional isomers?

a. 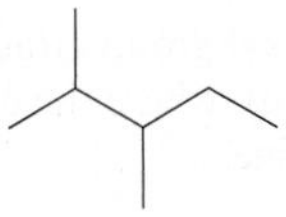 and 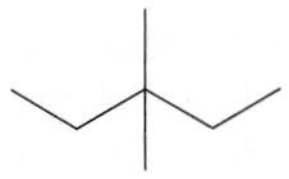

b. 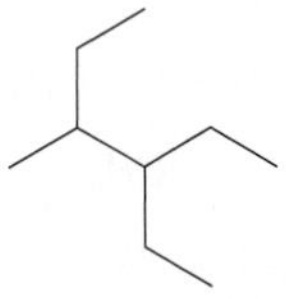 and 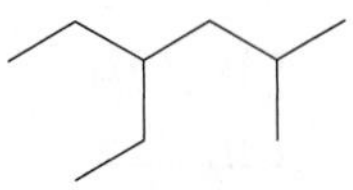

•1.61 Convert each of the condensed structural formulas in Problem 1.45 to a line-angle structural formula.

•**1.62** Convert each of the condensed structural formulas in Problem 1.46 to a line-angle structural formula.

•1.63 Assign an IUPAC name to each of the compounds in Problem 1.55.

•**1.64** Assign an IUPAC name to each of the compounds in Problem 1.56.

•1.65 Determine the molecular formula for each of the compounds in Problem 1.57.

▲**1.66** Determine the molecular formula for each of the compounds in Problem 1.58.

Classification of Carbon Atoms (Section 1.10)

1.67 For each of the alkane structures in Problem 1.45, give the number of (a) primary (b) secondary (c) tertiary and (d) quaternary carbon atoms present.

1.68 For each of the alkane structures in Problem 1.46, give the number of (a) primary (b) secondary (c) tertiary and (d) quaternary carbon atoms present.

•1.69 How many secondary carbon atoms are present in each of the structures in Problem 1.59?

•**1.70** How many secondary carbon atoms are present in each of the structures in Problem 1.60?

•1.71 Draw the condensed structural formula for an alkane with the molecular formula C_7H_{16} that contains
a. only 1° and 2° carbon atoms
b. both a 3° and 4° carbon atom

•**1.72** Draw the condensed structural formula for an alkane with the molecular formula C_7H_{16} that contains
a. five 1° carbon atoms
b. two adjacent 3° carbon atoms

Branched-Chain Alkyl Groups (Section 1.11)

1.73 Give the name of the branched alkyl group attached to each of the following carbon chains, where the carbon chain is denoted by a horizontal line.

a.
$$
\begin{array}{c}
\overline{} \\
| \\
CH-CH_3 \\
| \\
CH_3
\end{array}
$$

b.
$$
\begin{array}{c}
\overline{} \\
| \\
CH_2 \\
| \\
CH_3-CH-CH_3
\end{array}
$$

c.
$$
\begin{array}{c}
\overline{} \\
| \\
CH_3-CH-CH_3
\end{array}
$$

d.
$$
\begin{array}{c}
\overline{} \\
| \\
CH-CH_3 \\
| \\
CH_2 \\
| \\
CH_3
\end{array}
$$

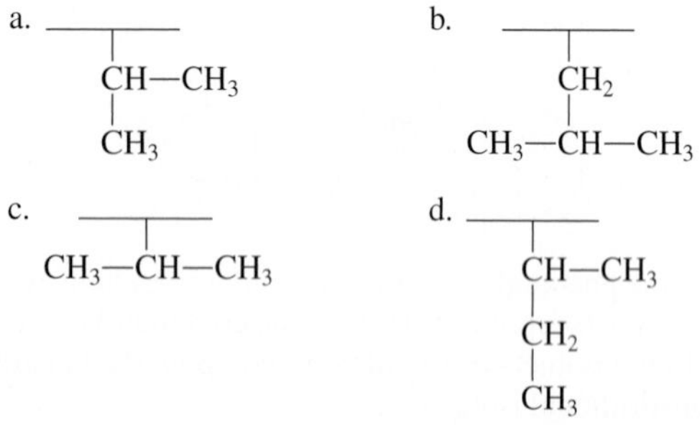

1.74 Give the name of the branched alkyl group attached to each of the following carbon chains, where the carbon chain is denoted by a horizontal line.

a.
$$
\begin{array}{c}
\overline{} \\
| \\
CH_2 \\
| \\
CH-CH_3 \\
| \\
CH_3
\end{array}
$$

b.
$$
\begin{array}{c}
\overline{} \\
| \\
CH_3-C-CH_3 \\
| \\
CH_3
\end{array}
$$

c.
$$
\begin{array}{c}
\overline{} \\
| \\
CH_3-CH_2-CH-CH_3
\end{array}
$$

d.
$$
\begin{array}{c}
\overline{} \\
| \\
CH \\
| \quad | \\
CH_3 \quad CH_3
\end{array}
$$

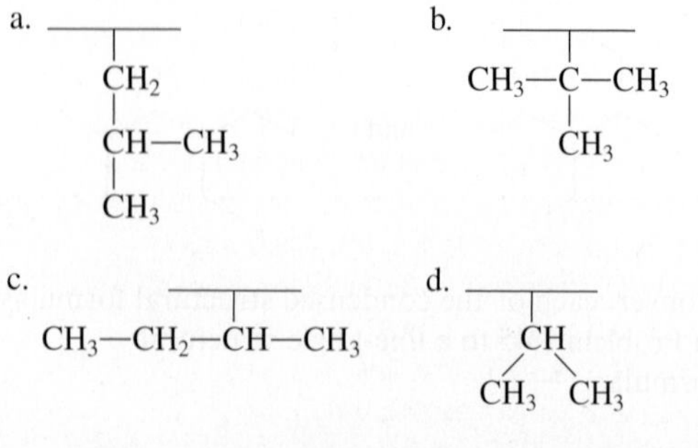

1.75 Draw condensed structural formulas for the following branched alkanes.
a. 5-(*sec*-Butyl)decane
b. 4,4-Diisopropyloctane
c. 5-Isobutyl-2,3-dimethylnonane
d. 4-(1,1-Dimethylethyl)octane

1.76 Draw condensed structural formulas for the following branched alkanes.
a. 5-Isobutylnonane
b. 4,4-Di(*sec*-butyl)decane
c. 4-(*tert*-Butyl)-3,3-diethylheptane
d. 5-(2-Methylpropyl)nonane

1.77 To which carbon atoms in a hexane molecule can each of the following alkyl groups be attached without extending the longest carbon chain beyond six carbons?
a. Ethyl b. Isopropyl c. Isobutyl d. *tert*-Butyl

1.78 To which carbon atoms in a heptane molecule can each of the following alkyl groups be attached without extending the longest carbon chain beyond seven carbons?
a. Ethyl b. Isopropyl c. *sec*-Butyl d. *tert*-Butyl

1.79 Using IUPAC rules, name the following "complex" five-carbon branched alkyl groups.
$$
\begin{array}{c}
CH_3 \\
| \\
\overline{}CH_2-CH-CH_2-CH_3
\end{array}
$$

b.
$$
\begin{array}{c}
CH_3 \\
| \\
\overline{}C-CH_2-CH_3 \\
| \\
CH_3
\end{array}
$$

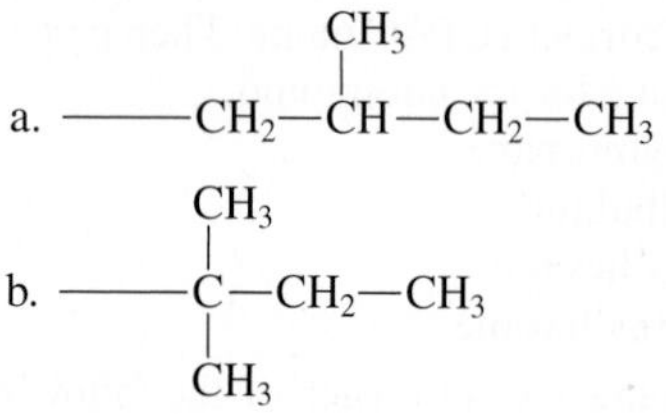

1.80 Using IUPAC rules, name the following "complex" five-carbon branched alkyl groups.
$$
\begin{array}{c}
CH_3 \\
| \\
\overline{}CH_2-CH-CH_3 \\
| \\
CH_3
\end{array}
$$

b.
$$
\begin{array}{c}
CH_3 \quad CH_3 \\
| \quad\quad | \\
\overline{}CH-CH-CH_3
\end{array}
$$

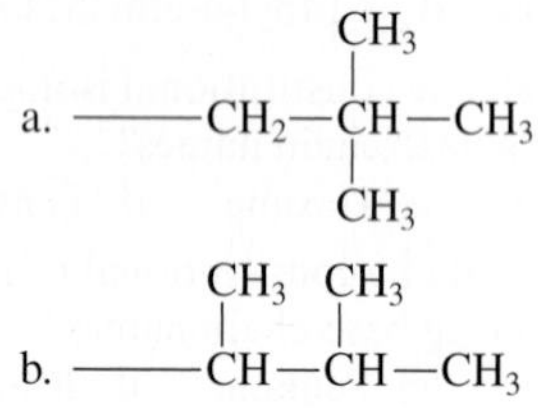

1.81 Give an acceptable alternate name for each of the following branched-chain alkyl groups.
a. Isopropyl group b. Tert-butyl group
c. (1-methylpropyl) group d. (2-methylpropyl) group

1.82 Give an acceptable alternate name for each of the following branched-chain alkyl groups.
a. Sec-butyl group b. Isobutyl group
c. (1-methylethyl) group d. (2,2-dimethylethyl) group

•1.83 How many different alkyl groups exist that contain
a. four carbon atoms b. five carbon atoms

•**1.84** How many different alkyl groups exist that contain
a. three carbon atoms b. six carbon atoms

Cycloalkanes (Section 1.12)

1.85 Using the general formula for a cycloalkane, derive the following for specific cycloalkanes.
a. Number of hydrogen atoms present when 8 carbon atoms are present
b. Number of carbon atoms present when 12 hydrogen atoms are present

c. Number of carbon atoms present when a total of 15 atoms are present
d. Number of covalent bonds present when 5 carbon atoms are present

1.86 Using the general formula for a cycloalkane, derive the following for specific cycloalkanes.
a. Number of hydrogen atoms present when 4 carbon atoms are present
b. Number of carbon atoms present when 6 hydrogen atoms are present
c. Number of hydrogen atoms present when a total of 18 atoms are present
d. Number of covalent bonds present when 8 hydrogen atoms are present

1.87 What is the molecular formula for each of the following cycloalkane molecules?

a. 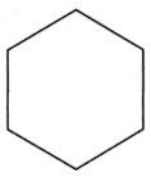 b. 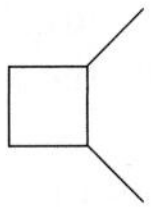 c.  d. 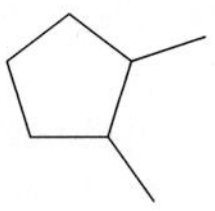

1.88 What is the molecular formula for each of the following cycloalkane molecules?

a. 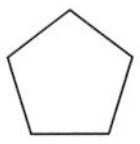 b. 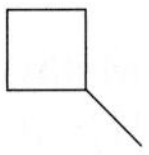 c.  d. 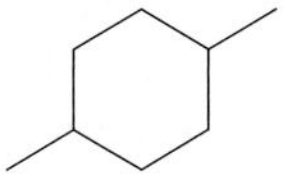

•1.89 How many secondary carbon atoms are present in each of the structures in Problem 1.87?

•**1.90** How many secondary carbon atoms are present in each of the structures in Problem 1.88?

IUPAC Nomenclature for Cycloalkanes (Section 1.13)

1.91 Assign an IUPAC name to each of the cycloalkanes in Problem 1.87.

1.92 Assign an IUPAC name to each of the cycloalkanes in Problem 1.88.

1.93 What is wrong with each of the following attempts to name a cycloalkane using IUPAC rules?
a. Dimethylcyclohexane
b. 3,4-Dimethylcyclohexane
c. 1-Ethylcyclobutane
d. 2-Ethyl-1-methylcyclopentane

1.94 What is wrong with each of the following attempts to name a cycloalkane using IUPAC rules?
a. Dimethylcyclopropane
b. 1-Methylcyclohexane
c. 2,5-Dimethylcyclobutane
d. 1-Propyl-2-ethylcyclohexane

1.95 Draw line-angle structural formulas for the following cycloalkanes.
a. Propylcyclobutane
b. Isopropylcyclobutane
c. 1,1-dimethylcyclobutane
d. (1-methylethyl)cyclobutane

1.96 Draw line-angle structural formulas for the following cycloalkanes.
a. Butylcyclopentane
b. Isobutylcyclopentane
c. 1,1-dimethylcyclopentane
d. (2-methylpropyl)cyclopentane

•1.97 What is the molecular formula for each of the following hydrocarbons?
a. 1,2-dimethylcyclohexane
b. 2,3-dimethylhexane
c. 1,2,3-trimethylcyclopentane
d. 2,2,3-trimethylpentane

•**1.98** What is the molecular formula for each of the following hydrocarbons?
a. 1,2-dimethylcyclopentane
b. 2,3-dimethylpentane
c. 1,2,3-trimethylcyclohexane
d. 2,3,4-trimethylhexane

•1.99 How many 3° carbon atoms are present in each of the compounds in Problem 1.97?

•**1.100** How many 3° carbon atoms are present in each of the compounds in Problem 1.98?

Isomerism in Cycloalkanes (Section 1.14)

1.101 Determine the number of constitutional isomers that are possible for each of the following situations.
a. Four-carbon cycloalkanes
b. Five-carbon cycloalkanes where the ring has three carbon atoms
c. Six-carbon cycloalkanes where the ring has five carbon atoms
d. Seven-carbon cycloalkanes where the ring has five carbon atoms

1.102 Determine the number of constitutional isomers that are possible for each of the following situations.
a. Five-carbon cycloalkanes
b. Six-carbon cycloalkanes where the ring has four carbon atoms
c. Six-carbon cycloalkanes where the ring has three carbon atoms
d. Seven-carbon cycloalkanes where the ring has four carbon atoms

1.103 Determine whether *cis–trans* isomerism is possible for each of the following cycloalkanes. If it is, then draw structural formulas for the *cis* and *trans* isomers.
a. Isopropylcyclobutane
b. 1,2-Diethylcyclopropane
c. 1-Ethyl-1-propylcyclopentane
d. 1,3-Dimethylcyclohexane

1.104 Determine whether *cis–trans* isomerism is possible for each of the following cycloalkanes. If it is, then draw structural formulas for the *cis* and *trans* isomers.
a. *sec*-Butylcyclohexane
b. 1-Ethyl-3-methylcyclobutane
c. 1,1-Dimethylcyclohexane
d. 1,3-Dipropylcyclopentane

•1.105 Indicate whether or not the members of each of the following pairs of hydrocarbons are constitutional isomers.
a. Hexane and cyclohexane
b. Hexane and methylcyclopentane
c. Cyclopentane and methylcyclobutane
d. cis-1,2-dimethylcyclobutane and trans-1,2-dimethylcyclobutane

•**1.106** Indicate whether or not the members of each of the following pairs of hydrocarbons are constitutional isomers.
a. 2-methylpentane and cyclopentane
b. 2-methylpentane and methylcyclopentane
c. Cyclohexane and methylcyclopentane
d. cis-1,3-dimethylcyclobutane and cis-1,2-dimethylcyclobutane

Sources of Alkanes and Cycloalkanes (Section 1.15)

1.107 In terms of the types of hydrocarbons present, what is the composition of unprocessed natural gas?

1.108 In terms of the types of hydrocarbons present, what is the composition of the gasoline fraction obtained by processing crude petroleum?

1.109 What physical property of hydrocarbons is the basis for the fractional distillation process for separating hydrocarbons?

1.110 Describe the process by which crude petroleum is separated into simpler mixtures (fractions).

Physical Properties of Alkanes and Cycloalkanes (Section 1.16)

1.111 Which member in each of the following pairs of compounds has the higher boiling point?
a. Hexane and octane
b. Cyclobutane and cyclopentane
c. Pentane and 1-methylbutane
d. Pentane and cyclopentane

1.112 Which member in each of the following pairs of compounds has the higher boiling point?
a. Methane and ethane
b. Cyclohexane and hexane
c. Butane and methylpropane
d. Pentane and 2,2-dimethylpropane

1.113 With the help of Figures 1.11 and 1.12, determine in which of the following pairs of compounds both members of the pair have the same physical state (solid, liquid, or gas) at room temperature and pressure.
a. Ethane and hexane
b. Cyclopropane and butane
c. Octane and 3-methyloctane
d. Pentane and decane

1.114 With the help of Figures 1.11 and 1.12, determine in which of the following pairs of compounds both members of the pair have the same physical state (solid, liquid, or gas) at room temperature and pressure.
a. Methane and butane
b. Cyclobutane and cyclopentane
c. Hexane and 2,3-dimethylbutane
d. Pentane and octane

1.115 Answer the following questions about the unbranched alkane that contains six carbon atoms.
a. Is it a solid, liquid, or gas at room temperature?
b. Is it less dense or more dense than water?
c. Is it soluble or insoluble in water?
d. Is it flammable or nonflammable in air?

1.116 Answer the following questions about the unsubstituted cycloalkane that contains six carbon atoms.
a. Is it a solid, liquid, or gas at room temperature?
b. Is it less dense or more dense than water?
c. Is it soluble or insoluble in water?
d. Is it flammable or nonflammable in air?

•1.117 (Chemical Connections 1-B) Indicate whether each of the following statements relating to the uses, physical properties, and physiological properties of alkanes is true or false.
a. Propane and butane are usually marketed in a liquefied form.
b. The C_5 to C_8 liquid alkanes are primary components of gasoline.
c. Solid-state alkanes are often used as protection against diaper rash.
d. Mineral oil contains longer carbon-chain alkanes than does petrolatum.

•1.118 (Chemical Connections 1-B) Indicate whether each of the following statements relating to the uses, physical properties, and physiological properties of alkanes is true or false.
a. Methane and ethane are gases that are easy to liquefy.
b. Gaseous alkanes have limited physiological effects.
c. Liquid-state alkanes can damage lung tissue by dissolving components of cellular membranes.
d. Petrolatum is a semi-solid C_{18}–C_{24} alkane mixture.

Chemical Properties of Alkanes and Cycloalkanes (Section 1.17)

1.119 Write the formulas of the products from the complete combustion of each of the following alkanes or cycloalkanes.
a. C_3H_8
b. Butane
c. Cyclobutane
d. $CH_3—(CH_2)_{15}—CH_3$

1.120 Write the formulas of the products from the complete combustion of each of the following alkanes or cycloalkanes.
a. C_4H_{10}
b. 2-Methylpentane
c. Cyclopentane
d. $CH_3—(CH_2)_7—CH_3$

1.121 Write molecular formulas for all the possible halogenated hydrocarbon products from the bromination of methane.

1.122 Write molecular formulas for all the possible halogenated hydrocarbon products from the fluorination of methane.

1.123 Write structural formulas for all the possible halogenated hydrocarbon products from the monochlorination of the following alkanes or cycloalkanes.
a. Ethane
b. Butane
c. 2-Methylpropane
d. Cyclopentane

1.124 Write structural formulas for all the possible halogenated hydrocarbon products from the monobromination of the following alkanes or cycloalkanes.
a. Propane
b. Pentane
c. 2-Methylbutane
d. Cyclohexane

Halogenated Alkanes and Cycloalkanes (Section 1.18)

1.125 Give both IUPAC and common names to each of the following halogenated hydrocarbons.
a. $CH_3—I$
b. $CH_3—CH_2—CH_2—Cl$
c. 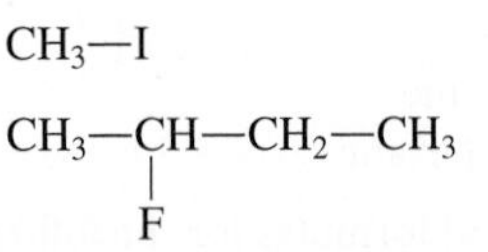
d. 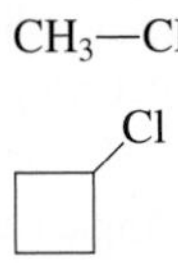

1.126 Give both IUPAC and common names to each of the following halogenated hydrocarbons.
a. $CH_3—CH_2—CH_2—CH_2—Br$
b. 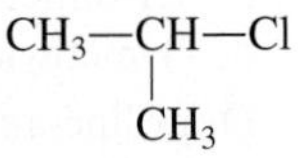
c. 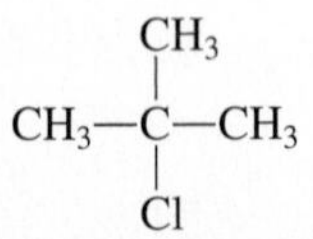
d. 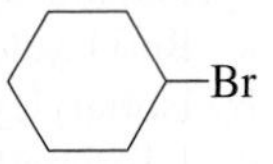

1.127 Draw structural formulas for the following halogenated hydrocarbons.
a. Trichloromethane
b. 1,2-Dichloro-1,1,2,2-tetrafluoroethane
c. Isopropyl bromide
d. trans-1-Bromo-3-chlorocyclopentane

1.128 Draw structural formulas for the following halogenated hydrocarbons.
a. Trifluorochloromethane
b. Pentafluoroethane
c. Isobutyl chloride
d. cis-1,2-Dichlorocyclohexane

1.129 Indicate whether each of the following molecular formulas is that for a halogenated alkane or that for a halogenated cycloalkane.
a. $C_4H_8Br_2$ b. $C_5H_9Cl_3$ c. $C_6H_{11}F$ d. $C_6H_{12}F_2$

1.130 Indicate whether each of the following molecular formulas is that for a halogenated alkane or that for a halogenated cycloalkane.
a. C_4H_7Cl b. $C_5H_{10}Cl_2$ c. $C_6H_{11}Br_3$ d. $C_6H_{10}F_2$

1.131 Which member of each of the following pairs of hydrocarbons and/or hydrocarbon derivatives would be expected to have the higher boiling point?
a. CH_3—Br or CH_3—I
b. CH_3—Cl or CH_3—CH_2—Cl
c. CH_3—Br or CH_3—CH_2—I
d. CH_4 or CH_3—Br

1.132 Which member of each of the following pairs of hydrocarbons and/or hydrocarbon derivatives would be expected to have the higher boiling point?
a. CH_3—CH_2—Br or CH_3—CH_2—Cl
b. CH_3—Cl or CH_3—CH_2—CH_2—Cl
c. CH_3—Cl or CH_3—CH_2—Br
d. CH_3—CH_3 or CH_3—CH_2—Cl

1.133 Indicate whether or not the following halogenated alkane chemical formulas are paired with a correct name.
a. CH_3Cl—chloroform
b. CCl_4—chloromethane
c. $CHCl_3$—methylene chloride
d. CH_2Cl_2—dichloromethane

1.134 Indicate whether or not the following halogenated alkane chemical formulas are paired with a correct name.
a. $CHCl_3$—chloromethane
b. CH_2Cl_2—carbon tet
c. CH_3Cl—methylene chloride
d. CCl_4—chloroform

1.135 Give the IUPAC names for the 8 isomeric halogenated hydrocarbons that have the molecular formula $C_5H_{11}Cl$.

1.136 Give the IUPAC names for the 9 isomeric halogenated hydrocarbons that have the molecular formula $C_4H_8Cl_2$.

•1.137 (Chemical Connections 1-C) Indicate whether each of the following statements concerning the atmospheric effects of "refrigerant" hydrocarbon derivatives is true or false.
a. The acronym HFC stands for hydrofluorocarbon.
b. Freon-11 and Freon-12 contain the elements hydrogen, carbon, and chlorine.
c. Freons react directly with ozone in the ozone layer.
d. HFOs are less potent greenhouse gases than HFCs.

•**1.138** (Chemical Connections 1-C) Indicate whether each of the following statements concerning the atmospheric effects of "refrigerant" hydrocarbon derivatives is true or false.
a. The acronym HFO stands for hydrofluorooxygen.
b. Freons interact with ultraviolet radiation to produce atomic Cl.
c. HFCs are less reactive than CFCs.
d. HFOs contain a carbon-carbon double bond.

Unsaturated Hydrocarbons

2

The unsaturated hydrocarbon ethene is used to stimulate the ripening process in fruit that has been picked while still green, such as bananas.

CHAPTER OUTLINE

Two general types of hydrocarbons exist, *saturated* and *unsaturated.* Saturated hydrocarbons, discussed in the previous chapter, include the alkanes and cycloalkanes. All bonds in saturated hydrocarbons are single bonds. Unsaturated hydrocarbons, the topic of this chapter, contain one or more carbon–carbon multiple bonds. There are three classes of unsaturated hydrocarbons: the *alkenes,* the *alkynes,* and the *aromatic hydrocarbons,* all of which are considered in this chapter.

2.1 Unsaturated Hydrocarbons

An **unsaturated hydrocarbon** *is a hydrocarbon in which one or more carbon–carbon multiple bonds (double bonds, triple bonds, or both) are present.* Unsaturated hydrocarbons have *physical* properties similar to those of saturated hydrocarbons. However, their *chemical* properties are much different. Unsaturated hydrocarbons are chemically more reactive than their saturated counterparts. The increased reactivity of unsaturated hydrocarbons is related to the presence of the carbon–carbon multiple bond(s) in such compounds. These multiple bonds serve as locations where chemical reactions can occur.

Whenever a specific portion of a molecule governs its chemical properties, that portion of the molecule is called a functional group. A **functional**

group *is the part of an organic molecule where most of its chemical reactions occur.* Carbon–carbon multiple bonds are the functional group for an unsaturated hydrocarbon.

The study of various functional groups and their respective reactions provides the organizational structure for organic chemistry. Each of the organic chemistry chapters that follow introduces new functional groups that characterize families of hydrocarbon derivatives.

Unsaturated hydrocarbons are subdivided into three groups on the basis of the type of multiple bond(s) present: (1) *alkenes,* which contain one or more carbon–carbon double bonds (2) *alkynes,* which contain one or more carbon–carbon triple bonds and (3) *aromatic hydrocarbons,* which exhibit a special type of "delocalized" bonding that involves a six-membered carbon ring (to be discussed in Section 2.12).

Alkanes and cycloalkanes (Chapter 1) lack functional groups; as a result, they are relatively unreactive.

Consideration of unsaturated hydrocarbons begins with a discussion of alkenes. Information about alkynes and aromatic hydrocarbons then follows.

2.2 Characteristics of Alkenes and Cycloalkenes

An **alkene** *is an acyclic unsaturated hydrocarbon that contains one or more carbon–carbon double bonds.* The alkene functional group is thus a C=C group. Note the close similarity between the family names *alkene* and *alkane* (Section 1.4); they differ only in their endings: *-ene* versus *-ane.* The *-ene* ending means a double bond is present.

The simplest type of alkene contains only one carbon–carbon double bond. Such compounds have the general molecular formula C_nH_{2n}. Thus alkenes with one double bond have two fewer hydrogen atoms than do alkanes (C_nH_{2n+2}).

The two simplest alkenes are ethene (C_2H_4) and propene (C_3H_6).

$CH_2{=}CH_2$ (Ethene) $\quad$ $CH_2{=}CH{-}CH_3$ (Propene)

The general molecular formula for an alkene with one double bond, C_nH_{2n}, is the same as that for a cycloalkane (Section 1.12). Thus such alkenes and cycloalkanes with the same number of carbon atoms are isomeric with one another.

Comparing the geometrical shape of ethene with that of methane (the simplest alkane) reveals a major difference. The arrangement of bonds about the carbon atom in methane is tetrahedral (Section 1.4), whereas the carbon atoms in ethene have a trigonal planar arrangement of bonds; that is, they form a flat, triangle-shaped arrangement (Figure 2.1). The two carbon atoms participating in a double bond and the four other atoms attached to these two carbon atoms always lie in a plane with a trigonal planar arrangement of atoms about each carbon atom of the double bond. Such an arrangement of atoms is consistent with the principles of VSEPR theory.

An older but still widely used name for alkenes is *olefins,* pronounced "oh-la-fins." The term *olefin* means "oil-forming." Many alkenes react with Cl_2 to form "oily" compounds.

A **cycloalkene** *is a cyclic unsaturated hydrocarbon that contains one or more carbon–carbon double bonds within the ring system.* Cycloalkenes in which there is only one double bond have the general molecular formula C_nH_{2n-2}. This general formula reflects the loss of four hydrogen atoms from that of an alkane (C_nH_{2n+2}). Note that two hydrogen atoms are lost because of the double bond and two because of the ring structure.

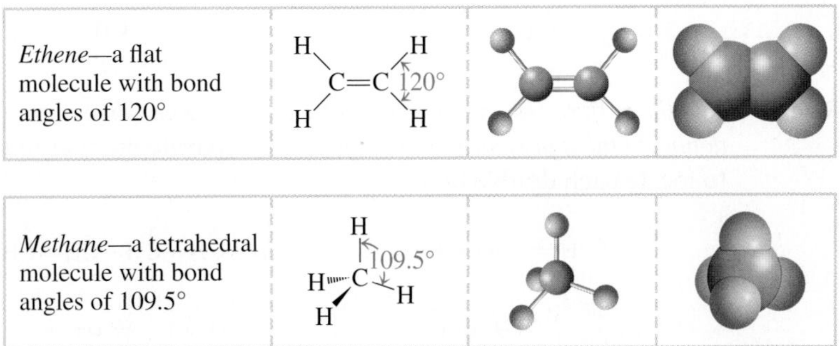

Figure 2.1 Three-dimensional representations of the structures of ethene and methane. In ethene, the atoms are in a flat (planar) rather than a tetrahedral arrangement. Bond angles are 120°.

The simplest cycloalkene is the compound cyclopropene (C_3H_4), a three-membered carbon ring system containing one double bond.

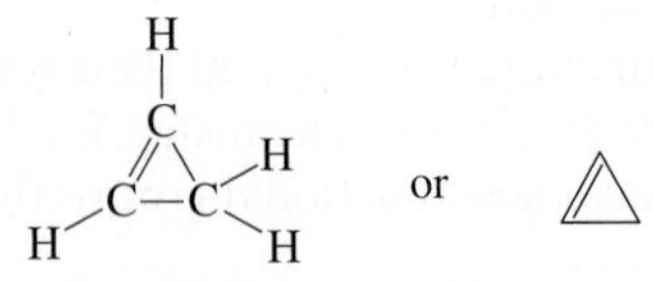

Cyclopropene

Alkenes with more than one carbon–carbon double bond are relatively common. When two double bonds are present, the compounds are often called *dienes;* for three double bonds the designation *trienes* is used. Cycloalkenes that contain more than one double bond are possible but are not common.

2.3 IUPAC Nomenclature for Alkenes and Cycloalkenes

IUPAC nomenclature rules are regularly revised. The most recent revisions occurred in 1979, 1993, and 2004. Issuance of an updated set of rules does not make earlier rule sets obsolete. Rather, with several sets of rules available and in use, there are often several acceptable ways to name a particular compound.

The IUPAC rules previously presented for naming alkanes and cycloalkanes (Sections 1.8 and 1.13) can be used, with some modification, to name alkenes and cycloalkenes.

Rule 1. *Replace the alkane suffix -ane with the suffix -ene, which is used to indicate the presence of a carbon–carbon double bond.*

Rule 2. *Select as the parent carbon chain the longest continuous chain of carbon atoms that contains both carbon atoms of the double bond.* For example, select

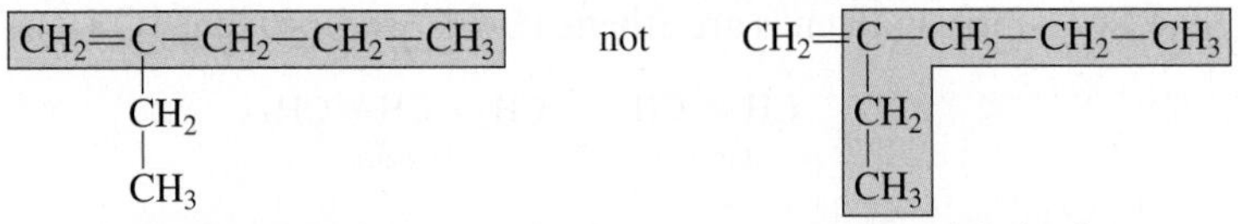

Longest carbon chain containing both carbon atoms of the double bond

Carbon chain that does not contain both carbon atoms of the double bond

Carbon–carbon double bonds take precedence over alkyl groups and halogen atoms in determining the direction in which the parent carbon chain is numbered.

Rule 3. *Number the parent carbon chain beginning at the end nearest the double bond.*

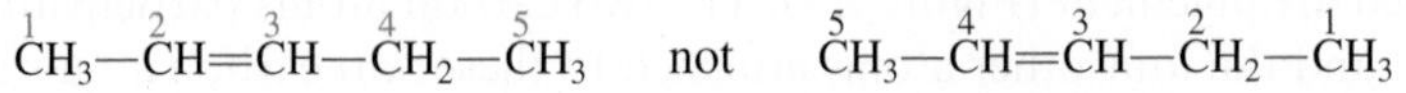

If the double bond is equidistant from both ends of the parent chain, begin numbering from the end closer to a substituent.

A number is not needed to specify double bond position in ethene and propene because there is only one way of positioning the double bond in these molecules.

Rule 4. *Give the position of the double bond in the chain as a single number, which is the lower-numbered carbon atom participating in the double bond.* This number is placed immediately before the name of the parent carbon chain.

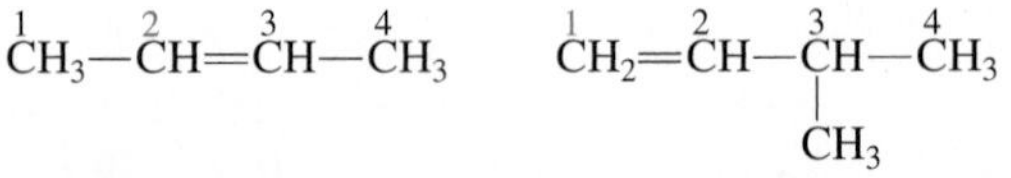

2-Butene

3-Methyl-1-butene

Rule 5. *Use the suffixes -diene, -triene, -tetrene, and so on when more than one double bond is present in the molecule.* A separate number must be used to locate each double bond.

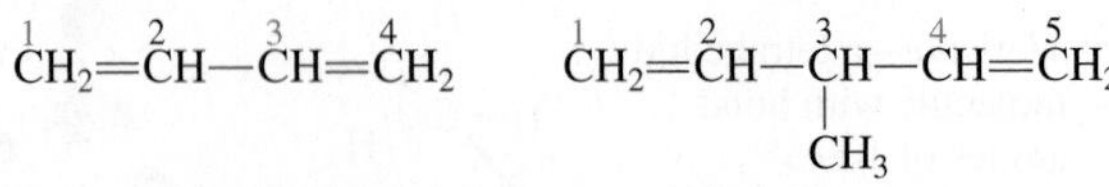

1,3-Butadiene

3-Methyl-1,4-pentadiene

Rule 6. *A number is not needed to locate the double bond in unsubstituted cycloalkenes with only one double bond because that bond is assumed to be between carbons 1 and 2.*

Rule 7. *In substituted cycloalkenes with only one double bond, the double-bonded carbon atoms are numbered 1 and 2 in the direction (clockwise or counterclockwise) that gives the first-encountered substituent the lower number.* Again, no number is used in the name to locate the double bond.

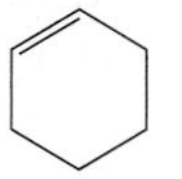

Cyclohexene

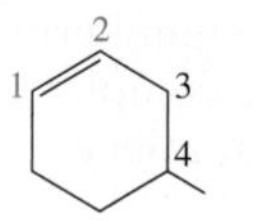

4-Methylcyclohexene

Rule 8. *In cycloalkenes with more than one double bond within the ring, assign one double bond the numbers 1 and 2 and the other double bonds the lowest numbers possible.*

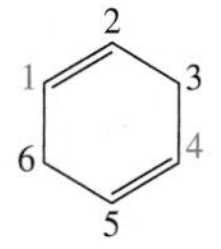

1,4-Cyclohexadiene

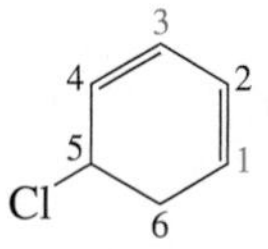

5-Chloro-1,3-cyclohexadiene

In 1993, IUPAC rules were revised relative to the positioning of numbers used in names. Instead of placing the number immediately in front of the base name (2-pentene), IUPAC recommended that the number be placed immediately before the part of the name that the number serves as a locator for (pent-2-ene). In this textbook, the newer system will be used only when it helps to clarify a name; otherwise the older system will be used. Two additional examples comparing the two systems are:

2-methyl-1-butene (old)
2-methylbut-1-ene (new)

1,3-butadiene (old)
buta-1,3-diene (new)

EXAMPLE 2.1 Assigning IUPAC Names to Alkenes and Cycloalkenes

Assign IUPAC names to the following alkenes and cycloalkenes.

a. $CH_3{-}CH{=}CH{-}CH_2{-}CH_2{-}CH_3$

b. $CH_3{-}CH_2{-}\underset{\displaystyle\underset{\displaystyle CH_3}{|}}{\underset{\displaystyle CH_2}{|}}{C}{=}CH_3$

c. 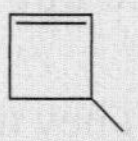

d. 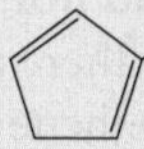

Solution

a. The carbon chain in this hexene is numbered from the end closest to the double bond.

$$\overset{1}{C}H_3-\overset{2}{C}H=\overset{3}{C}H-\overset{4}{C}H_2-\overset{5}{C}H_2-\overset{6}{C}H_3$$

The complete IUPAC name is 2-hexene.

b. The longest carbon chain containing *both* carbons of the double bond has four carbon atoms. Thus the base name is butene.

$$\boxed{CH_3{-}CH_2{-}C{=}CH_2}$$
$$\qquad\qquad\quad |$$
$$\qquad\qquad\quad CH_2$$
$$\qquad\qquad\quad |$$
$$\qquad\qquad\quad CH_3$$

The chain is numbered from the end closest to the double bond. The complete IUPAC name is 2-ethyl-1-butene.

c. This compound is a methylcyclobutene. The numbers 1 and 2 are assigned to the carbon atoms of the double bond, and the ring is numbered clockwise, which results in a carbon 3 location for the methyl group. (Counterclockwise numbering would have placed the methyl group on carbon 4.) The complete IUPAC name of the cycloalkene is 3-methylcyclobutene. The double bond is understood to involve carbons 1 and 2.

(continued)

d. A ring system containing five carbon atoms, two double bonds, and a methyl substituent on the ring is called a methylcyclopentadiene. Two different numbering systems produce the same locations (carbons 1 and 3) for the double bonds.

The counterclockwise numbering system assigns the lower number to the methyl group. The complete IUPAC name of the compound is 2-methyl-1,3-cyclopentadiene.

Practice Exercise 2.1

Assign IUPAC names to the following alkenes and cycloalkenes.

a. $CH_3-CH=CH-CH_2-CH(CH_3)-CH_3$

b. 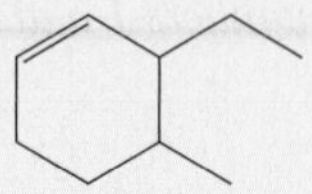

c. $CH_2=CH-CH=CH_2$

d. 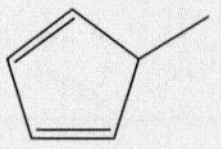

Answers: **a.** 5-Methyl-2-hexene; **b.** 3-Ethyl-4-methylcyclohexene; **c.** 1,3-Butadiene; **d.** 5-Methyl-1,3-pentadiene

Common Names (Non-IUPAC Names)

Despite the universal acceptance and precision of the IUPAC nomenclature system, some alkenes (those of low molecular mass) are known almost exclusively by common names.

The simpler members of most families of organic compounds, including alkenes, have common names in addition to IUPAC names. In many cases, these common (non-IUPAC) names are used almost exclusively for the compounds. It would be nice if such common names did not exist, but they do. There is no choice but to know these names; fortunately, there are not many of them.

The two simplest alkenes, ethene and propene, have common names. They are ethylene and propylene, respectively.

$CH_2=CH_2$	$CH_2=CH-CH_3$
Ethylene	Propylene

Ethylene (ethene) is a very important industrial chemical. In terms of amount, its industrial production exceeds that of any other organic compound. Propylene (propene) is also industrially produced on a relatively large scale. The focus on relevancy feature Chemical Connections 2-A on the next page gives further information about sources of and uses for ethylene.

Alkenes as Substituents

Just as there are *alkanes* and *alkyl groups* (Section 1.8), there are *alkenes* and *alkenyl groups.* An **alkenyl group** *is a noncyclic hydrocarbon substituent in which a carbon–carbon double bond is present.* The three most frequently encountered alkenyl groups are the one-, two-, and three-carbon entities, which may be named using IUPAC nomenclature (methylidene, ethenyl, and 2-propenyl) or common names (methylene, vinyl, and allyl).

$CH_2=$	$CH_2=CH-$	$CH_2=CH-CH_2-$
Methylene group (IUPAC name: methylidene group)	Vinyl group (IUPAC name: ethenyl group)	Allyl group (IUPAC name: 2-propenyl group)

CHEMICAL CONNECTIONS 2-A

Ethene: A Plant Hormone and High-Volume Industrial Chemical

Ethene (ethylene), the simplest unsaturated hydrocarbon (C_2H_4), is a colorless, flammable gas with a slightly sweet odor. It occurs naturally in *small* amounts in plants, where it functions as a plant hormone. A few parts per million ethene (less than 10 parts per million) stimulates the fruit-ripening process.

The commercial fruit industry uses ethene's ripening property to advantage. Bananas, tomatoes, and some citrus fruits are picked green to prevent spoiling and bruising during transportation to markets. At their destinations, the fruits are exposed to small amounts of ethene gas, which stimulates the ripening process.

© Alan Detrick/Photoresearchers

Ethene is the hormone that causes tomatoes to ripen.

Despite having no large natural source, ethene is an exceedingly important industrial chemical. Indeed, industrial production of ethene exceeds that of every other organic compound. Petrochemicals (substances found in natural gas and petroleum) are the starting materials for ethene production.

In one process, ethane (from natural gas) is *dehydrogenated* at a high temperature to produce ethene.

$$\underset{\text{Ethane}}{CH_3{-}CH_3} \xrightarrow{750C} \underset{\text{Ethene}}{CH_2{=}CH_2} + H_2$$

In another process, called *thermal cracking,* hydrocarbons from petroleum are heated to a high temperature in the absence of air (to prevent combustion), which causes the cleavage of carbon–carbon bonds. Ethene is one of the smaller molecules produced by this process.

Industrially produced ethene serves as a starting material for the production of many plastics and fibers. Almost one-half of ethene production is used in the production of the well-known plastic polyethylene (Section 2.10). Polyvinyl chloride (PVC) and polystyrene (Styrofoam) are two other important ethene-based materials. About one-sixth of ethene production is converted to ethylene glycol, the principal component of most brands of antifreeze for automobile radiators (Section 3.5).

The use of these alkenyl group names in actual compound nomenclature is illustrated in the following examples.

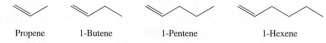

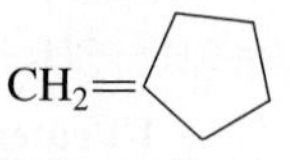

Methylene cyclopentane (IUPAC name: methylidenecyclopentane)

$CH_2{=}CH{-}Cl$

Vinyl chloride (IUPAC name: chloroethene)

$CH_2{=}CH{-}CH_2{-}Br$

Allyl bromide (IUPAC name: 3-bromopropene)

2.4 Line-Angle Structural Formulas for Alkenes

Line-angle formulas for the three- to six-carbon acyclic 1-alkenes are as follows.

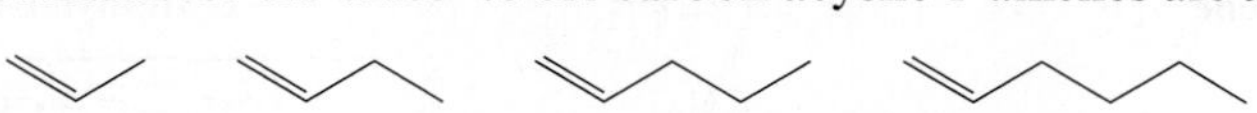

Propene 1-Butene 1-Pentene 1-Hexene

Representative line-angle structural formulas for substituent-bearing alkenes include

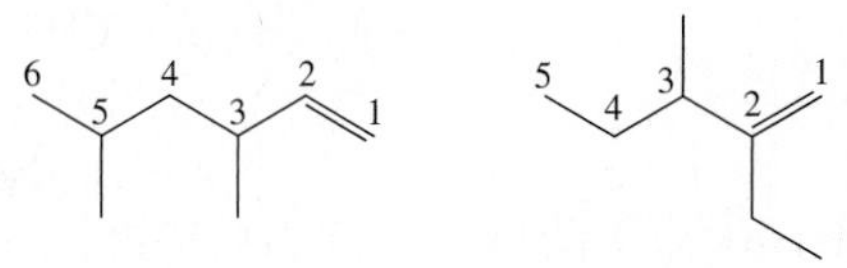

3,5-Dimethyl-1-hexene 2-Ethyl-3-methyl-1-pentene

Diene representations in terms of line-angle structural formulas include

1,4-Pentadiene 2-Methyl-1,3-butadiene

2.5 Constitutional Isomerism in Alkenes

Constitutional isomerism is possible for alkenes, just as it was for alkanes (Section 1.6). In general, there are more alkene isomers for a given number of carbon atoms than there are alkane isomers. This is because there is more than one location where a double bond can be placed in systems containing four or more carbon atoms. Figure 2.2 compares constitutional isomer possibilities for C_4 and C_5 alkanes and their counterpart alkenes with one double bond.

Two different subtypes of constitutional isomerism are represented among the alkene isomers shown in Figure 2.2: *positional* isomers and *skeletal* isomers. **Positional isomers** *are constitutional isomers with the same carbon-chain arrangement but different hydrogen atom arrangements as the result of differing location of the functional group present.* Positional isomer sets found in Figure 2.2 are:

1-butene and 2-butene

1-pentene and 2-pentene

2-methyl-1-butene, 3-methyl-1-butene, and 2-methyl-2-butene

Skeletal isomers *are constitutional isomers that have different carbon-chain arrangements as well as different hydrogen atom arrangements.* The C_4 alkenes 1-butene and 2-methylpropene are skeletal isomers. All alkane isomers discussed in the previous chapter were skeletal isomers; positional isomerism is not possible for alkanes because they lack a functional group.

FOUR-CARBON ALKANES (TWO ISOMERS)	FOUR-CARBON ALKENES (THREE ISOMERS)	FIVE-CARBON ALKANES (THREE ISOMERS)	FIVE-CARBON ALKENES (FIVE ISOMERS)
$CH_3-CH_2-CH_2-CH_3$ **Butane**	$CH_2=CH-CH_2-CH_3$ **1-Butene**	$CH_3-CH_2-CH_2-CH_2-CH_3$ **Pentane**	$CH_2=CH-CH_2-CH_2-CH_3$ **1-Pentene**
$CH_3-CH-CH_3$ | CH_3 **2-Methylpropane**	$CH_3-CH=CH-CH_3$ **2-Butene**	$CH_3-CH-CH_2-CH_3$ | CH_3 **2-Methylbutane**	$CH_3-CH=CH-CH_2-CH_3$ **2-Pentene**
	$CH_2=C-CH_3$ | CH_3 **2-Methylpropene**	CH_3 | CH_3-C-CH_3 | CH_3 **2,2-Dimethylpropane**	$CH_2=C-CH_2-CH_3$ | CH_3 **2-Methyl-1-butene**
			$CH_3-C=CH-CH_3$ | CH_3 **2-Methyl-2-butene**
			$CH_3-CH-CH=CH_2$ | CH_3 **3-Methyl-1-butene**

Figure 2.2 A comparison of structural isomerism possibilities for four- and five-carbon alkane and alkene systems.

EXAMPLE 2.2 Determining Structural Formulas for Alkene Constitutional Isomers

Draw condensed structural formulas for all alkene constitutional isomers that have the molecular formula C_5H_{10}.

Solution

The answers for this problem have already been considered. The structures of the five C_5H_{10} alkene constitutional isomers are given in Figure 2.2. The purpose of this example is to consider the "thinking pattern" used to obtain the given answers.

There are two concepts in the thinking pattern.

1. The different carbon skeletons (both unbranched and branched) that are possible using five carbon atoms are determined.
2. For each of the carbon skeletons determined, different positions for placement of the double bond are then considered.

Step 1: There are three possible arrangements for five carbon atoms:

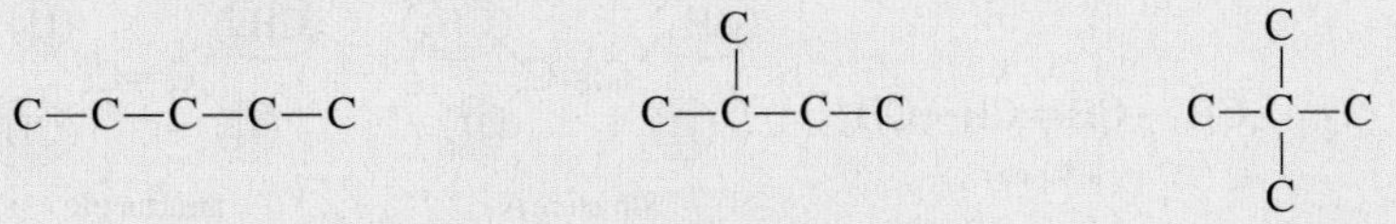

(These arrangements are the constitutional isomers for a 5-carbon alk*ane,* the situation considered in Section 1.6 of the previous chapter.)

Step 2: For the first carbon skeleton (the unbranched chain), there are two possible locations for the double bond; that is, there are two positional isomers:

$CH_2{=}CH{-}CH_2{-}CH_2{-}CH_3$ (1-Pentene) $CH_3{-}CH{=}CH{-}CH_2{-}CH_3$ (2-Pentene)

Moving the double bond farther to the right than in the second structure does not produce new isomers but, rather, duplicates of the two given structures. A double bond between carbons 3 and 4 (numbering from the left side) is the same as having the double bond between carbons 2 and 3 (numbering from the right side).

For the second carbon skeleton, there are three positional isomers—that is, three different positions for the double bond:

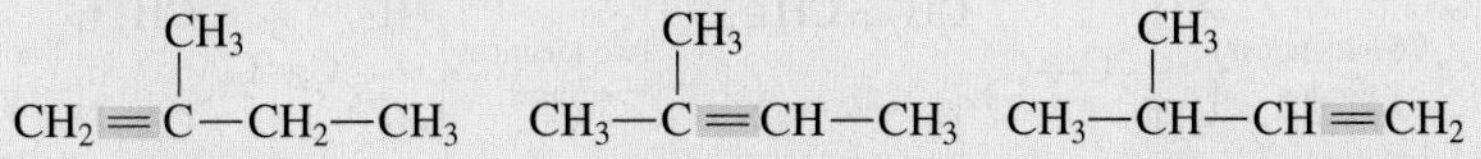

2-Methyl-1-butene   2-Methyl-2-butene   3-Methyl-1-butene

For the third carbon skeleton, alkene structures are not possible. Placing a double bond at any location within the structure creates a situation where the central carbon atom has five bonds. Thus there are five alkene constitutional isomers with the molecular formula C_5H_{10}.

Practice Exercise 2.2

Draw condensed structural formulas for all alkene constitutional isomers that have the molecular formula C_4H_8.

Answers: $CH_2{=}CH{-}CH_2{-}CH_3$; $CH_3{-}CH{=}CH{-}CH_3$;

1-Butene   2-Butene

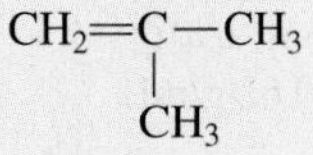

2-Methylpropene

2.6 *Cis–Trans* Isomerism in Alkenes

Cis–trans isomerism (Section 1.14) is possible for some alkenes. Such isomerism results from the structural rigidity associated with carbon–carbon double bonds: Unlike the situation in alkanes, where free rotation about carbon–carbon single bonds is possible (Section 1.7), no rotation about carbon–carbon double bonds (or carbon–carbon triple bonds) can occur.

To determine whether an alkene has *cis* and *trans* isomers, draw the alkene structure in a manner that emphasizes the four attachments to the double-bonded carbon atoms.

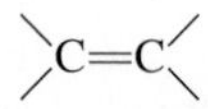

The double bond of alkenes, like the ring of cycloalkanes, imposes rotational restrictions.

If *each* of the two carbons of the double bond has two *different* groups attached to it, *cis* and *trans* isomers exist.

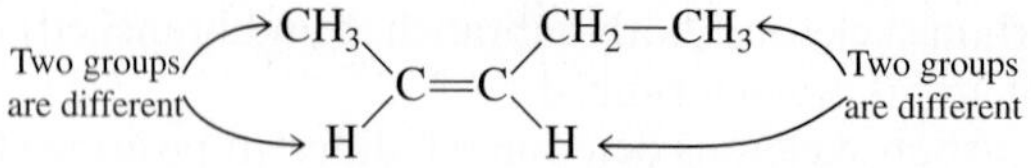

The simplest alkene for which *cis* and *trans* isomers exist is 2-butene.

$CH_3—CH{=}CH—CH_3$

2-butene

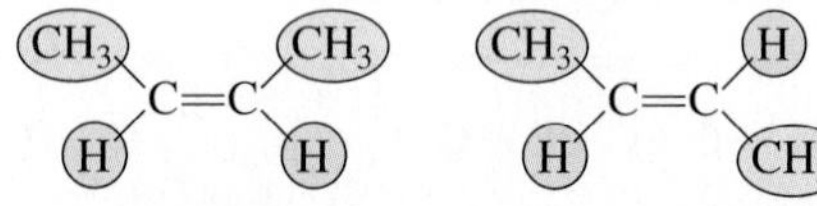

Structure A (*cis*-2-butene)

Structure B (*trans*-2-butene)

Recall from Section 1.14 that *cis* means "on the same side" and *trans* means "across from." Structure A is the *cis* isomer; both methyl groups are on the same side of the double bond. Structure B is the *trans* isomer; the methyl groups are on opposite sides of the double bond. The only way to convert structure A to structure B is to break the double bond. At room temperature, such bond breaking does not occur. Hence these two structures represent two different compounds (*cis–trans* isomers) that differ in boiling point, density, and so on. Figure 2.3 shows three-dimensional representations of the *cis* and *trans* isomers of 2-butene.

Cis–trans isomerism is not possible when one of the double-bonded carbons bears two identical groups. Thus neither 1-butene nor 2-methylpropene is capable of existing in *cis* and *trans* forms.

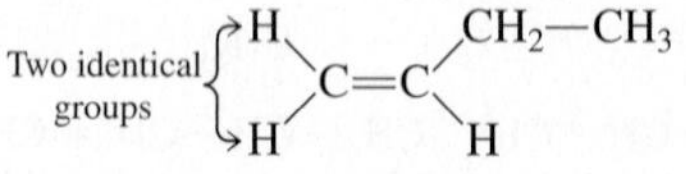

1-Butene

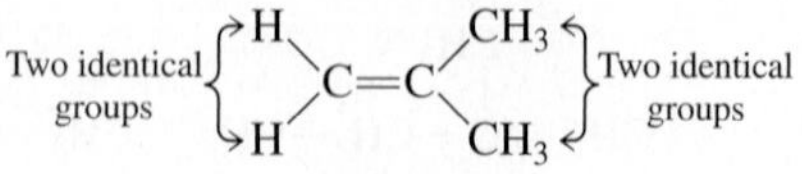

2-Methylpropene

When alkenes contain more than one double bond, *cis–trans* considerations are more complicated. Orientation about each double bond must be considered

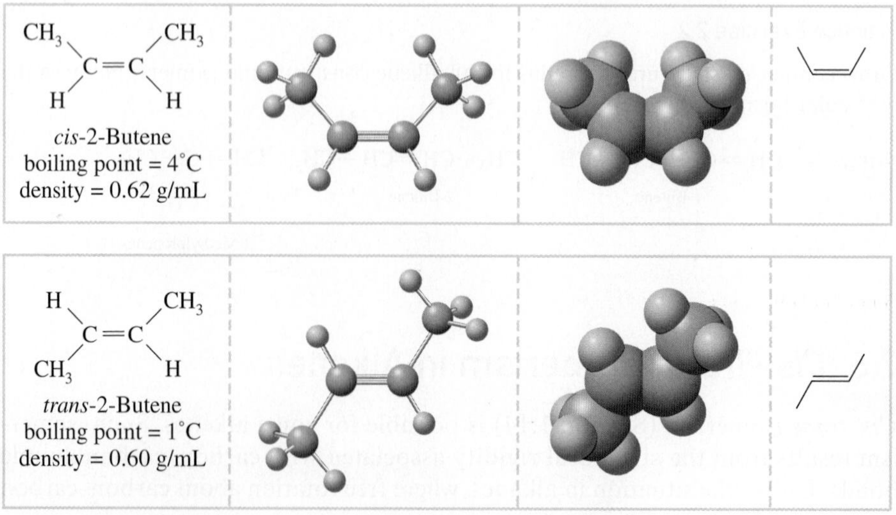

Figure 2.3 *Cis–trans* isomers: Different representations of the *cis* and *trans* isomers of 2-butene.

independently of that at other sites. For example, for the molecule 2,4-heptadiene (two double bonds) there are four different *cis–trans* isomers (*trans–trans, trans–cis, cis–trans,* and *cis–cis*). The structures of two of these isomers are

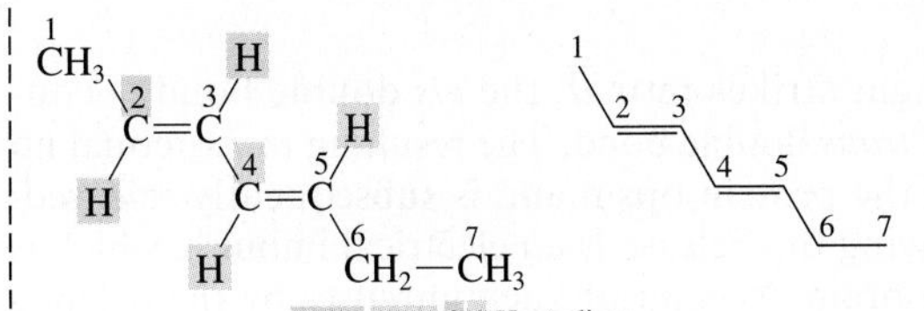

trans-trans-2,4-Heptadiene

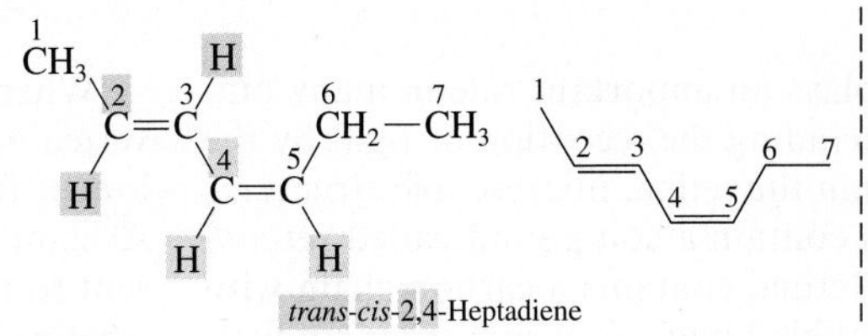

trans-cis-2,4-Heptadiene

EXAMPLE 2.3 Determining Whether *Cis–Trans* Isomerism Is Possible in Substituted Alkenes

Determine whether each of the following substituted alkenes can exist in *cis–trans* isomeric forms.

a. 1-Bromo-1-chloroethene **b.** 2-Chloro-2-butene

Solution

a. The condensed structural formula for this compound is

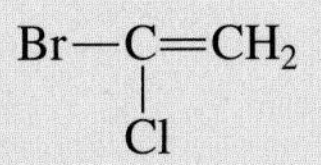

Redrawing this formula to emphasize the four attachments to the double-bonded carbon atoms gives

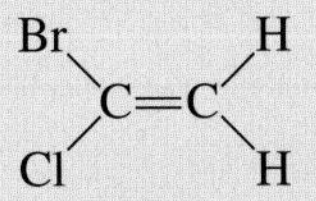

The carbon atom on the right has two identical attachments. Hence *cis–trans* isomerism is not possible.

b. The condensed structural formula for this compound is

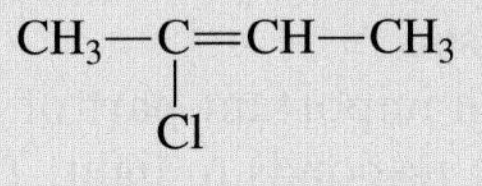

Redrawing this formula to emphasize the four attachments to the double-bonded carbon atoms gives

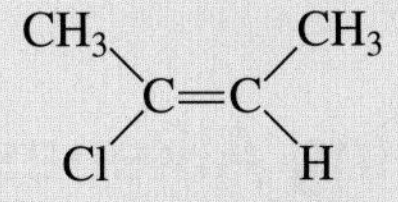

Because both carbon atoms of the double bond bear two different attachments, *cis–trans* isomers are possible.

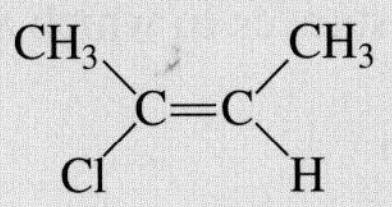

cis-2-Chloro-2-butene

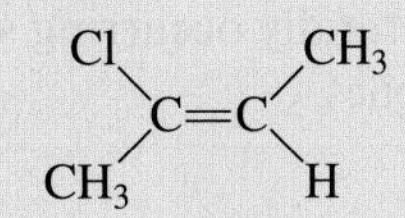

trans-2-Chloro-2-butene

Practice Exercise 2.3

Determine whether each of the following substituted alkenes can exist in *cis–trans* isomeric forms.

a. 1-Chloropropene **b.** 2-Chloropropene

Answers: **a.** Yes; **b.** No

CHEMICAL CONNECTIONS 2-B

Cis–Trans Isomerism and Vision

Cis–trans isomerism plays an important role in many biochemical processes, including the reception of light by the retina of the eye. Within the retina, microscopic structures called rods and cones contain a compound called *retinal,* which absorbs light. Retinal contains a carbon chain with five carbon–carbon double bonds, four in a *trans* configuration and one in a *cis* configuration. This arrangement of double bonds gives retinal a shape that fits the protein *opsin,* to which it is attached, as shown in the accompanying diagram.

When light strikes retinal, the *cis* double bond is converted to a *trans* double bond. The resulting *trans*-retinal no longer fits the protein opsin and is subsequently released. Accompanying this release is an electrical impulse, which is sent to the brain. Receipt of such impulses by the brain is what facilitates the vision process.

In order to trigger nerve impulses again, *trans*-retinal must be converted back to *cis*-retinal. This occurs in the membranes of the rods and cones, where enzymes change *trans*-retinal back into *cis*-retinal.

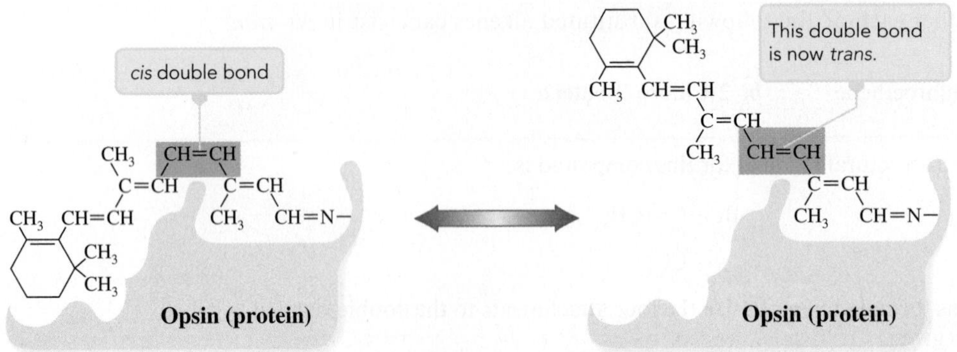

As noted previously (Section 1.14), *cis–trans* isomers are not constitutional isomers but, rather, stereoisomers. Each carbon atom in a pair of *cis–trans* isomers is bonded to the same atoms (groups); hence they cannot be constitutional isomers. The only difference, structurally, between the isomers is the orientation of the groups in space (stereoisomerism).

Cis–trans isomerism is an important part of many biochemical processes, including how the human eye responds to light. The focus on relevancy feature Chemical Connections 2-B above gives information on the specific role of *cis* and *trans* isomers in the process of vision.

2.7 Naturally Occurring Alkenes

Alkenes are abundant in nature. Many important biological molecules are characterized by the presence of carbon–carbon double bonds within their structure. Two important types of naturally occurring substances to which alkenes contribute are pheromones and terpenes.

Pheromones

A **pheromone** *is a compound used by insects (and some animals) to transmit a message to other members of the same species.* Pheromones are often alkenes or alkene derivatives. The biological activity of alkene-type pheromones is usually highly dependent on whether the double bonds present are in a *cis* or a *trans* arrangement (Section 2.6).

The sex attractant of the female silkworm is a 16-carbon alkene derivative containing an —OH group. Two double bonds are present, *trans* at carbon 10 and *cis* at carbon 12.

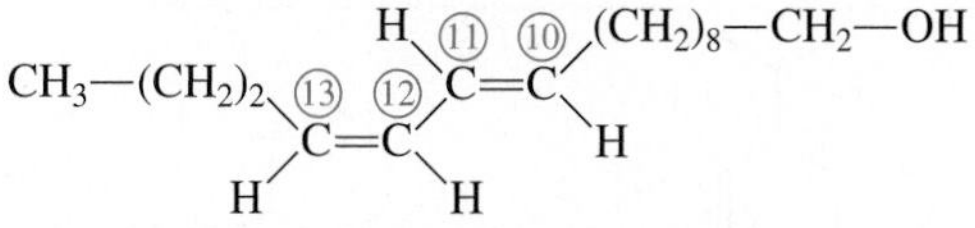

This compound is 10 billion times more effective in eliciting a response from the male silkworm than the 10-*cis*–12-*trans* isomer and 10 trillion times more effective than the isomer wherein both bonds are in a *trans* configuration.

Insect sex pheromones are useful in insect control. A small amount of synthetically produced sex pheromone is used to lure male insects of a single species into a trap (Figure 2.4). The trapped males are either killed or sterilized. Releasing sterilized males has proved effective in some situations. A sterile male can mate many times, preventing fertilization in many females, who usually mate only once. Sex attractant pheromones are now used to control the gypsy moth and the Mediterranean fruit fly.

© Steven Chadwick/Alamy

Figure 2.4 The application of sex pheromones in insect control involves using a small amount of synthetically produced pheromone to lure a particular insect into a trap. This is accomplished without harming other "beneficial" insects.

Terpenes

A **terpene** *is an organic compound whose carbon skeleton is composed of two or more 5-carbon isoprene structural units.* Isoprene (2-methyl-1,3-butadiene) is a five-carbon diene.

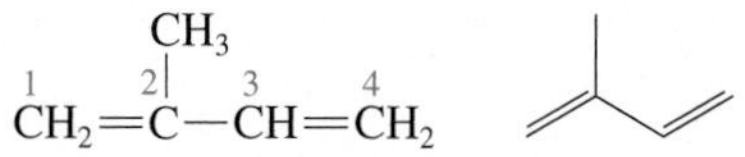

2-Methyl-1,3-butadiene (isoprene)

Terpenes are formed by joining the tail of one isoprene structural unit to the head of another unit.

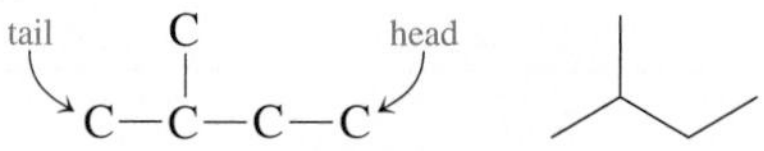

Isoprene structural unit

The isopentyl structural unit (Section 1.8) of isoprene maintains its identity when isoprene units are joined together; however, the positioning of the double bonds often changes.

Terpenes are among the most widely distributed compounds in the biological world, with more than 22,000 structures known. Such compounds are responsible for the odors of many trees and for many characteristic plant fragrances.

The number of carbon atoms present in a terpene is always a multiple of the number 5 (10, 15, and so on). Parts (a) and (b) of Figure 2.5 give the structures of selected 10- and 15-carbon terpenes found in plants. Beta-carotene is a terpene whose structure has 40 carbon atoms present in 8 isoprene units (Figure 2.5c).

In the human body, dietary beta-carotene (obtained by eating yellow-colored vegetables) serves as a precursor for vitamin A (Figure 2.6); splitting of a beta-carotene molecule produces two vitamin A molecules (Section 10.13). An additional role of beta-carotene in the body, independent of its vitamin A function, is that of antioxidant. An antioxidant is a substance that helps protect cells from damage from reactive oxygen-derived species called free radicals (Section 12.11). In later chapters, additional isoprene-based molecules important in the functioning of the human body will be encountered. They include vitamin K (Section 10.13), coenzyme Q (Section 12.7), and cholesterol (Section 8.9).

Carotenoids, a group of yellow, red, and orange pigments found in plants, are terpenes in which eight isoprene units are present. The focus on relevancy feature Chemical Connections 2-C on page 57 gives further information about the presence of these compounds and their color in various fruits, vegetables, and the leaves of many plants.

© Martha Cooper/Peter Arnold/Photo Library

Figure 2.6 The molecule *β*-carotene, whose structure is given in Figure 2.5, is responsible for the yellow-orange color of carrots, apricots, and yams.

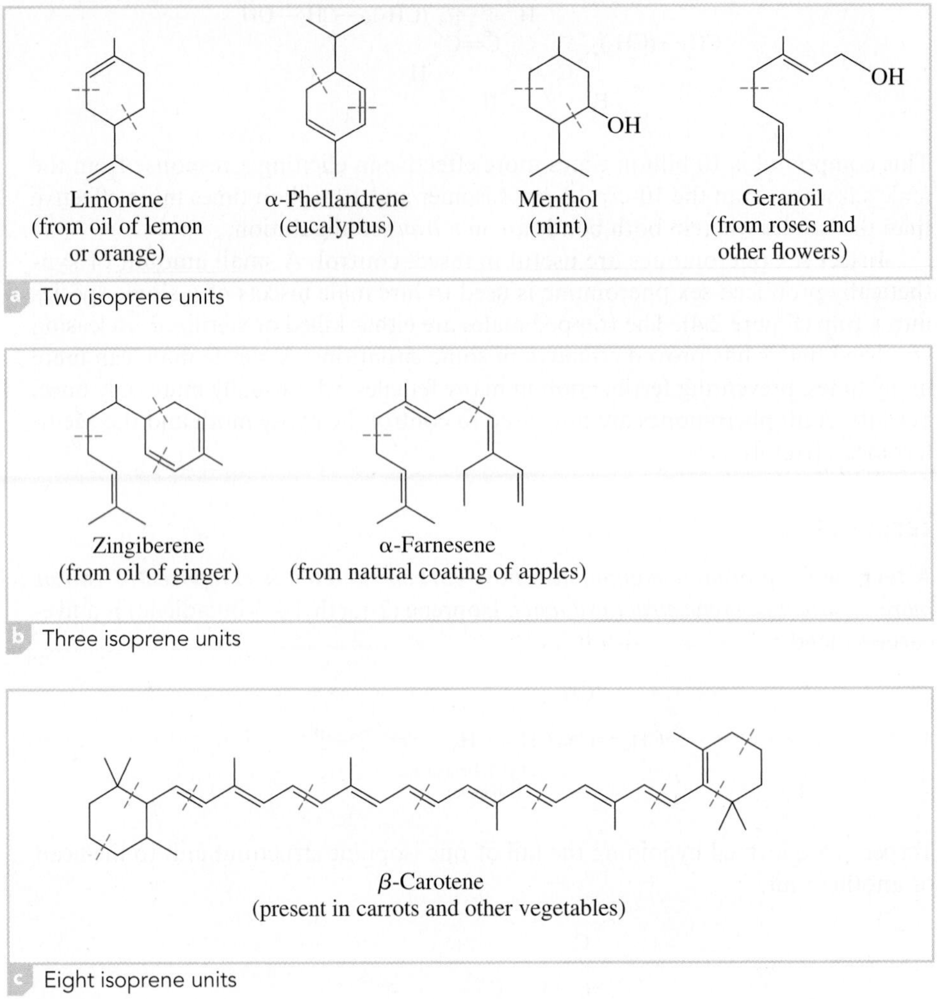

Figure 2.5 Selected terpenes containing two, three, and eight isoprene units. Dashed lines in the structures separate the individual isoprene units.

2.8 Physical Properties of Alkenes and Cycloalkenes

The general physical properties of alkenes and cycloalkenes include insolubility in water, solubility in nonpolar solvents, and densities lower than that of water. Thus they have physical properties similar to those of alkanes (Section 1.16). The melting point of an alkene is usually lower than that of the alkane with the same number of carbon atoms.

Alkenes with 2 to 4 carbon atoms are gases at room temperature. Unsubstituted alkenes with 5 to 17 carbon atoms and one double bond are liquids, and those with still more carbon atoms are solids. Figure 2.7 is a physical-state summary for unbranched 1-alkenes and unsubstituted cycloalkenes with 8 or fewer carbon atoms.

2.9 Chemical Reactions of Alkenes

Alkenes, like alkanes, are very flammable. The combustion products, as with any hydrocarbon, are carbon dioxide and water.

$$\underset{\text{Ethene}}{C_2H_4} + 3O_2 \longrightarrow 2CO_2 + 2H_2O$$

Pure alkenes are, however, too expensive to be used as fuel.

CHEMICAL CONNECTIONS 2-C

Carotenoids: A Source of Color

Carotenoids are the most widely distributed of the substances that give color to our world; they occur in flowers, fruits, plants, insects, and animals. Structurally, carotenoids are terpenes (Section 2.7) in which a tail-to-tail linkage between two C_{20} units occurs. Carotenoids are of two major types: the hydrocarbon class (the *carotenes*) and the oxygenated hydrocarbon (alcohol) class (the *xanthophylls*).

The two most familiar carotenoids are both carotenes—β-carotene and lycopene. These compounds are easily isolated from carrots and ripe tomatoes, respectively.

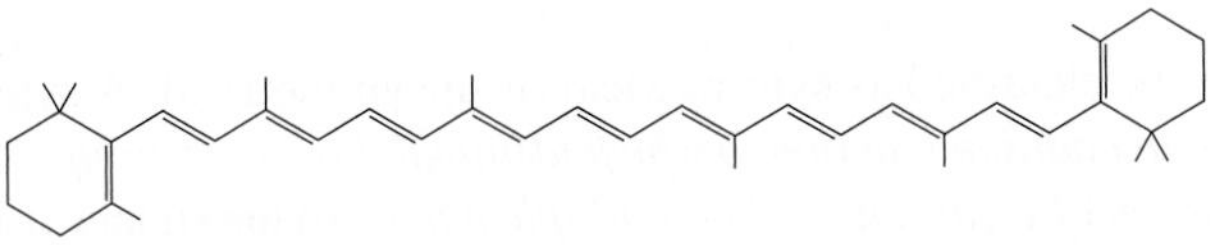

β-Carotene

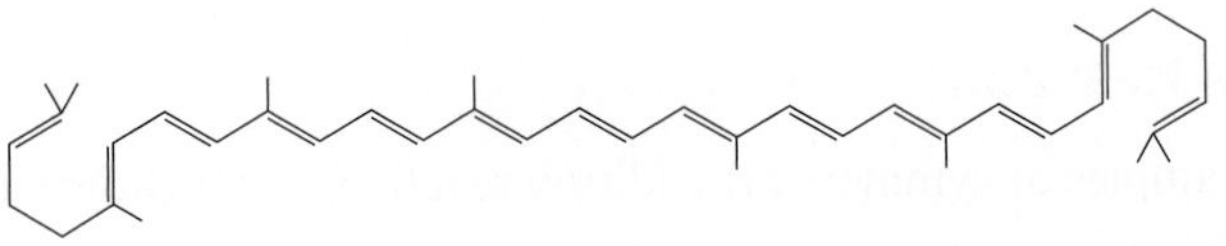

Lycopene

Present in both of these carotene structures is a *conjugated* system of 11 double bonds. (Conjugated double bonds are double bonds separated from each other by one single bond.) Color is frequently caused by the presence of compounds that contain extended conjugated-double-bond systems. When *visible* light strikes these compounds, certain wavelengths of the visible light are absorbed by the electrons in the conjugated-bond system. The unabsorbed wavelengths of visible light are reflected and are perceived as color.

The molecule β-carotene is responsible for the yellow-orange color in carrots, apricots, and yams. The yellow-orange color of autumn leaves also comes from β-carotene. Leaves contain chlorophyll (green pigment) and β-carotene (yellow-pigment) in a ratio of approximately 3 to 1. The yellow-orange β-carotene color is masked by the chlorophyll until autumn, when the chlorophyll molecules decompose as a result of lower temperatures and less sunlight and are not replaced.

The molecule lycopene is the red pigment in tomatoes and watermelon. Lycopene's structure differs from that of β-carotene in that the two rings in β-carotene have been opened. The ripening of a green tomato involves the gradual decomposition of chlorophyll with an associated unmasking of the red color of the lycopene present. A green pepper becomes red after ripening for the same reason.

Research studies indicate that lycopene has anticancer properties. One study comparing a group of men on a lycopene-rich diet with another group on a low-lycopene diet showed the incidence of prostate cancer was one-third lower in the lycopene-rich diet group.

Heat-processed tomatoes are a good source of dietary lycopene, with concentrated juice containing the highest levels of this substance. The lycopene in cooked tomatoes is

Yoshhio Sawaragi/Getty Images

Carotenoids, molecules that contain eight isoprene units, are responsible for the yellow-orange color of autumn leaves.

absorbed more readily during digestion than is the lycopene in raw tomatoes. Recent studies indicate that red seedless watermelon contains as much lycopene as cooked tomatoes.

Carotenoids such as β-carotene and lycopene are synthesized only by plants. They can, however, reach animal tissues via feed and can be modified and deposited therein. The yellowish tint of animal fat comes from β-carotene present in animal diets. The chicken egg yolk is another example of color imparted by dietary carotenoids.

Xanthophylls are oxygen-containing carotenoids in which two —OH groups (the alcohol functional group—Section 3.2) have replaced hydrogen atoms. The two best-known xanthophylls are zeaxanthin (pronounced *zee-uh-zan-thin*) and lutein (pronounced *lue-tin*). Both have structures very similar to β-carotene.

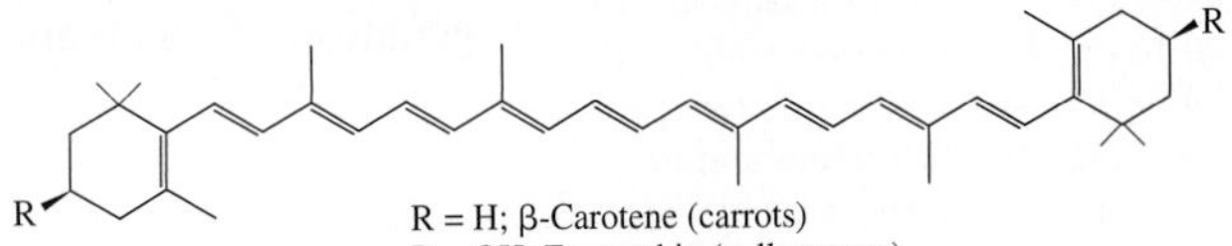

R = H; β-Carotene (carrots)
R = OH; Zeaxanthin (yellow corn)

Structurally, zeaxanthin and lutein are isomeric, with lutein differing from zeaxanthin only in the placement of a double bond in one of the end rings.

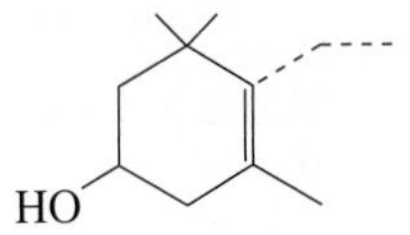

Zeaxanthin

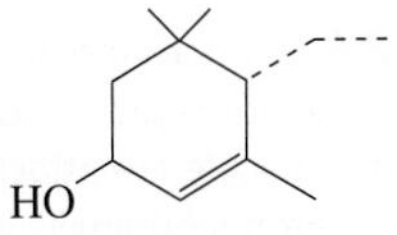

Lutein

The yellow color of corn involves zeaxanthin, and the red color of paprika relates to lutein's presence. At low concentrations these xanthophylls impart a yellow color, and at higher concentrations they impart an orange-red color. Zeaxanthin and lutein are also found in the retina of the eye, where they are believed to function as antioxidants, keeping the eyes safe from oxidative stress.

Unbranched 1-Alkenes			
☒	C_3	C_5	C_7
C_2	C_4	C_6	C_8

Unsubstituted Cycloalkenes			
☒	C_3*	C_5	C_7
☒	C_4*	C_6	C_8

□ Gas □ Liquid

*Cyclopropene and cyclobutene are relatively unstable compounds, readily converting to other hydrocarbons because of the severe bond angle strain associated with a small ring containing a double bond.

Figure 2.7 A physical-state summary for unbranched 1-alkenes and unsubstituted cycloalkenes with one double bond at room temperature and pressure.

Aside from combustion, nearly all other reactions of alkenes take place at the carbon–carbon double bond(s). These reactions are called *addition reactions* because a substance is *added* to the double bond. This behavior contrasts with that of alkanes, where the most common reaction type, aside from combustion, is *substitution* (Section 1.17).

An **addition reaction** *is a reaction in which atoms or groups of atoms are added to each carbon atom of a carbon–carbon multiple bond in a hydrocarbon or hydrocarbon derivative.* A general equation for an alkene addition reaction is

$$\rangle C{=}C\langle \;+\; A{-}B \longrightarrow -\underset{A}{\overset{|}{C}}-\underset{B}{\overset{|}{C}}-$$

In this reaction, the A part of the reactant A—B becomes attached to one carbon atom of the double bond, and the B part to the other carbon atom (Figure 2.8). As this occurs, the carbon–carbon double bond simultaneously becomes a carbon–carbon single bond.

Addition reactions can be classified as symmetrical or unsymmetrical. A **symmetrical addition reaction** *is an addition reaction in which identical atoms (or groups of atoms) are added to each carbon of a carbon–carbon multiple bond.* An **unsymmetrical addition reaction** *is an addition reaction in which different atoms (or groups of atoms) are added to the carbon atoms of a carbon–carbon multiple bond.*

Symmetrical Addition Reactions

The two most common examples of symmetrical addition reactions are hydrogenation and halogenation.

A **hydrogenation reaction** *is an addition reaction in which H_2 is incorporated into molecules of an organic compound.* In alkene hydrogenation, a hydrogen atom is added to each carbon atom of a double bond. This is accomplished by heating the alkene and H_2 in the presence of a catalyst (usually Ni or Pt).

$$\underset{\text{Propene}}{CH_2{=}CH{-}CH_3} + H_2 \xrightarrow[\substack{150^\circ C\\ 12\text{–}15\text{ atm}\\ \text{pressure}}]{\text{Ni or Pt}} \underset{\text{Propane}}{\overset{H\quad\;\; H}{CH_2{-}CH{-}CH_3}}$$

The identity of the catalyst used in hydrogenation is specified by writing it above the arrow in the chemical equation for the hydrogenation. In general terms, hydrogenation of an alkene can be written as

$$\underset{\text{Alkene}}{\rangle C{=}C\langle} + H_2 \xrightarrow[\substack{\text{Heat,}\\ \text{pressure}}]{\text{Ni or Pt}} \underset{\text{Alkane}}{-\overset{H}{\underset{|}{C}}-\overset{H}{\underset{|}{C}}-}$$

The following word associations are important to remember:

alkane—substitution reaction
alkene—addition reaction

An analogy can be drawn to a basketball team. When a *substitution* is made, one player leaves the game as another enters. The number of players on the court remains at five per team. If *addition* were allowed during a basketball game, two players could enter the game and no one would leave; there would be seven players per team on the court rather than five.

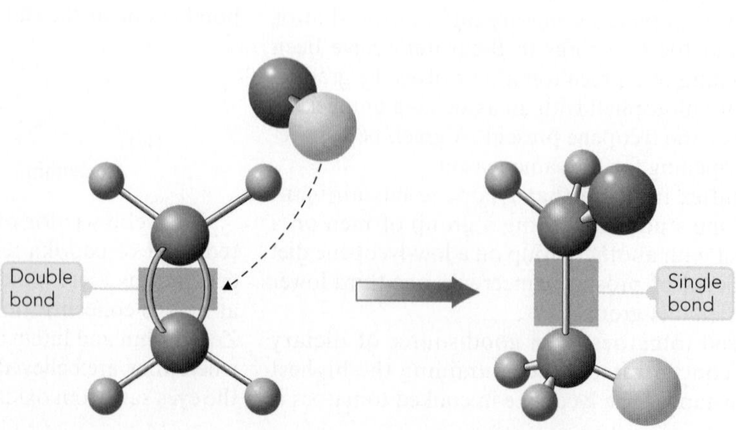

Figure 2.8 In an alkene addition reaction, the atoms provided by an incoming molecule are attached to the carbon atoms originally joined by a double bond. In the process, the double bond becomes a single bond.

The hydrogenation of vegetable oils is a very important commercial process today. Vegetable oils from sources such as soybeans and cottonseeds are composed of long-chain organic molecules that contain several double bonds. When these oils are hydrogenated, they are converted to low-melting solids that are used in margarines and shortenings (see Section 8.6).

A **halogenation reaction** *is an addition reaction in which a halogen is incorporated into molecules of an organic compound.* In alkene halogenation, a halogen atom is added to each carbon atom of a double bond. Chlorination (Cl_2) and bromination (Br_2) are the two halogenation processes most commonly encountered. No catalyst is needed.

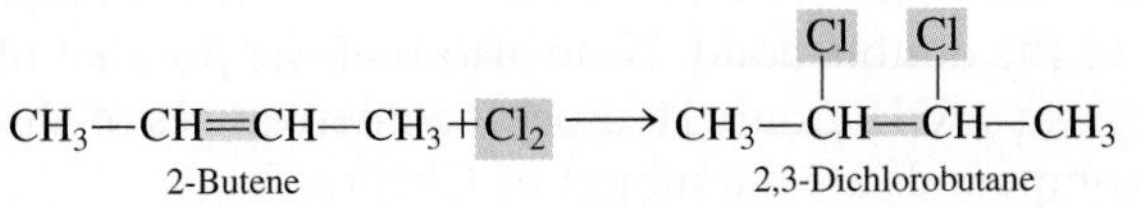

$$CH_3-CH{=}CH-CH_3 + Cl_2 \longrightarrow CH_3-\overset{\overset{\displaystyle Cl}{|}}{C}H-\overset{\overset{\displaystyle Cl}{|}}{C}H-CH_3$$
2-Butene 2,3-Dichlorobutane

In general terms, halogenation of an alkene can be written as

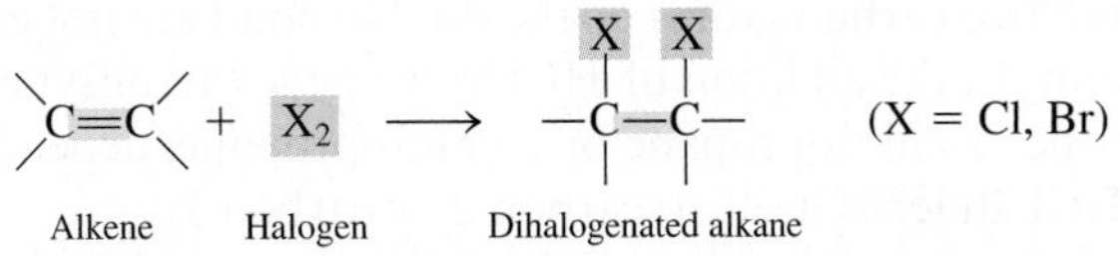

$$\overset{}{\underset{}{C}}{=}C + X_2 \longrightarrow -\overset{\overset{\displaystyle X}{|}}{C}-\overset{\overset{\displaystyle X}{|}}{C}- \quad (X = Cl, Br)$$
Alkene Halogen Dihalogenated alkane

Bromination is often used to test for the presence of carbon–carbon double bonds in organic substances. Bromine in water or carbon tetrachloride is reddish-brown. The dibromo compound(s) formed from the symmetrical addition of bromine to an organic compound is(are) colorless. Thus the decolorization of a Br_2 solution indicates the presence of carbon–carbon double bonds (Figure 2.9).

Unsymmetrical Addition Reactions

Two important types of unsymmetrical addition reactions are hydrohalogenation and hydration.

A **hydrohalogenation reaction** *is an addition reaction in which a hydrogen halide (HCl, HBr, or HI) is incorporated into molecules of an organic compound.* In alkene hydrohalogenation, one carbon atom of a double bond receives a halogen atom and the other carbon atom receives a hydrogen atom. Hydrohalogenation reactions require no catalyst. For *symmetrical* alkenes, such as ethene, only one product results from hydrohalogenation.

$$CH_2{=}CH_2 + H-Cl \longrightarrow \overset{\overset{\displaystyle H \quad\;\; Cl}{|\qquad|}}{CH_2{=}CH_2}$$
Ethene Chloroethane

Hydrogenation of an alkene requires a catalyst. No reaction occurs if the catalyst is not present.

The Chemical Connections feature "*Trans* Fatty Acids and Blood Cholesterol Levels" in Chapter 8 addresses health issues relative to consumption of partially hydrogenated products.

The addition of water to carbon–carbon double bonds occurs in many biochemical reactions that take place in the human body—for example, in the citric acid cycle (Section 12.6) and in the oxidation of fatty acids (Section 14.4).

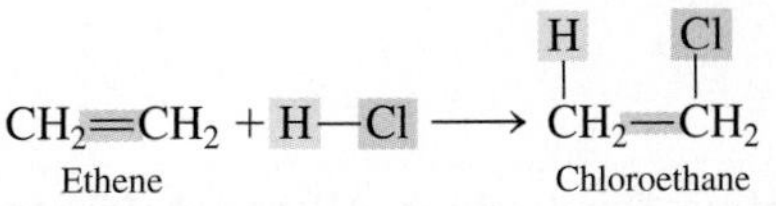

© Cengage Learning

Figure 2.9 A bromine in water solution is reddish-brown (left). When a small amount of such a solution is added to an unsaturated hydrocarbon, the added solution is decolorized as the bromine adds to the hydrocarbon to form colorless dibromo compounds (right).

A **hydration reaction** *is an addition reaction in which H_2O is incorporated into molecules of an organic compound.* In alkene hydration, one carbon atom of a double bond receives a hydrogen atom and the other carbon atom receives an —OH group. Alkene hydration requires a small amount of H_2SO_4 (sulfuric acid) as a catalyst. For *symmetrical* alkenes, only one product results from hydration.

$$\underset{\text{Ethene}}{CH_2{=}CH_2} + H{-}OH \xrightarrow{H_2SO_4} \underset{\text{An alcohol}}{\overset{H}{\overset{|}{C}}H_2{-}\overset{OH}{\overset{|}{C}}H_2}$$

In this equation, the water (H_2O) is written as H—OH to emphasize how this molecule adds to the double bond. Note also that the product of this hydration reaction contains an —OH group. Hydrocarbon derivatives of this type are called *alcohols.* Such compounds are the subject of Chapter 3.

When the alkene involved in a hydrohalogenation or hydration reaction is itself *unsymmetrical,* more than one product is possible. (An unsymmetrical alkene is one in which the two carbon atoms of the double bond are not equivalently substituted.) For example, the addition of HCl to propene (an unsymmetrical alkene) could produce either 1-chloropropane or 2-chloropropane, depending on whether the H from the HCl attaches itself to carbon 2 or carbon 1.

Edgar Fahs Smith Collection, University of Pennsylvania Library

Figure 2.10 Vladimir Vasilevich Markovnikov (1837–1904). A professor of chemistry at several Russian universities, Markovnikov synthesized rings containing four carbon atoms and seven carbon atoms, thereby disproving the notion of the day that carbon could form only five- and six-membered rings.

$$\underset{\text{Propene}}{CH_2{=}CH{-}CH_3} + HCl \longrightarrow \underset{\text{1-Chloropropane}}{\overset{Cl}{\overset{|}{C}}H_2{-}\overset{H}{\overset{|}{C}}H{-}CH_3}$$

or

$$\underset{\text{Propene}}{CH_2{=}CH{-}CH_3} + HCl \longrightarrow \underset{\text{2-Chloropropane}}{\overset{H}{\overset{|}{C}}H_2{-}\overset{Cl}{\overset{|}{C}}H{-}CH_3}$$

Two catchy summaries of Markovnikov's rule are "hydrogen goes where hydrogen is" and "the rich get richer" (in terms of hydrogen).

When two isomeric products are possible, one product usually predominates. The dominant product can be predicted by using Markovnikov's rule, named after the Russian chemist Vladimir V. Markovnikov (Figure 2.10). The surname Markovnikov is pronounced *Mar-cove-na-coff.* **Markovnikov's rule** *states that when an unsymmetrical molecule of the form* HQ *adds to an unsymmetrical alkene, the hydrogen atom from the* HQ *becomes attached to the unsaturated carbon atom that already has the most hydrogen atoms.* Thus the major product in our example involving propene is 2-chloropropane.

EXAMPLE 2.4 Predicting Products in Alkene Addition Reactions Using Markovnikov's Rule

Using Markovnikov's rule, predict the predominant product in each of the following addition reactions.

a. $CH_3{-}CH_2{-}CH_2{-}CH{=}CH_2 + HBr \rightarrow$

b. 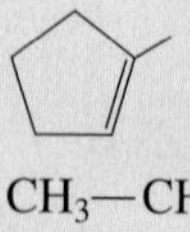 $+ HCl \rightarrow$

c. $CH_3{-}CH{=}CH{-}CH_2{-}CH_3 + HBr \rightarrow$

Solution

a. The hydrogen atom will add to carbon 1 because carbon 1 already contains more hydrogen atoms than carbon 2. The predominant product of the addition will be 2-bromopentane.

$$CH_3{-}CH_2{-}CH_2{-}\overset{②}{C}H{=}\overset{①}{C}H_2 + HBr \longrightarrow CH_3{-}CH_2{-}CH_2{-}\overset{Br}{\overset{|}{C}}H{-}\overset{H}{\overset{|}{C}}H_2$$

b. Carbon 1 of the double bond does not have any H atoms directly attached to it. Carbon 2 of the double bond has one H atom (H atoms are not shown in the structure but are implied) attached to it. The H atom from the HCl will add to carbon 2, giving 1-chloro-1-methylcyclopentane as the product.

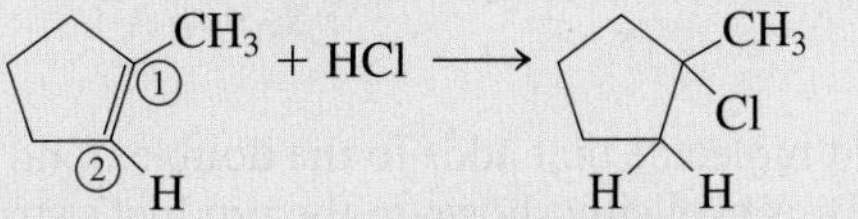

c. Each carbon atom of the double bond in this molecule has one hydrogen atom. Thus Markovnikov's rule does not favor either carbon atom. The result is two isomeric products that are formed in almost equal quantities.

$$CH_3-\underset{\displaystyle Br}{\underset{|}{CH}}-CH_2-CH_2-CH_3 \quad \text{and} \quad CH_3-CH_2-\underset{\displaystyle Br}{\underset{|}{CH}}-CH_2-CH_3$$

2-Bromopentane | 3-Bromopentane

Practice Exercise 2.4

Using Markovnikov's rule, predict the predominant product in each of the following addition reactions.

a. $CH_2{=}CH-CH_2-CH_3 + HCl \rightarrow$

b. 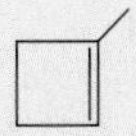 $+ HBr \rightarrow$

Answers: **a.** $CH_3-\underset{\displaystyle Cl}{\underset{|}{CH}}-CH_2-CH_3$ **b.** 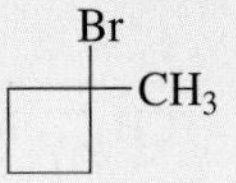

In compounds that contain more than one carbon–carbon double bond, such as dienes and trienes, addition can occur at each of the double bonds. In the complete hydrogenation of a diene and in that of a triene, the amounts of hydrogen needed are twice as much and three times as much, respectively, as that needed for the hydrogenation of an alkene with one double bond.

$$CH_2{=}CH-CH_2-CH_2-CH_2-CH_3 + H_2 \xrightarrow{Ni} CH_3-(CH_2)_4-CH_3$$
$$CH_2{=}CH-CH{=}CH-CH_2-CH_3 + 2H_2 \xrightarrow{Ni} CH_3-(CH_2)_4-CH_3$$
$$CH_2{=}CH-CH{=}CH-CH{=}CH_2 + 3H_2 \xrightarrow{Ni} CH_3-(CH_2)_4-CH_3$$

EXAMPLE 2.5 Predicting Reactants and Products in Alkene Addition Reactions

Supply the structural formula of the missing substance in each of the following addition reactions.

a. $CH_3-CH_2-CH{=}CH_2 + H_2O \xrightarrow{H_2SO_4} ?$

b. $? + Br_2 \rightarrow$ 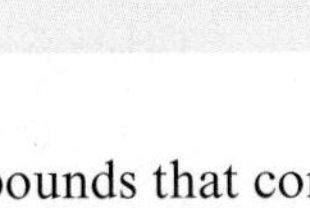

c. 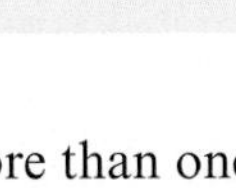 $+ ? \rightarrow$ 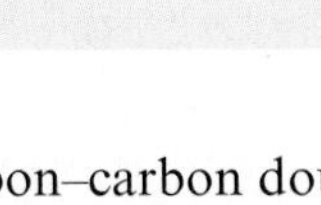

d. $CH_3-CH{=}CH-CH{=}CH_2 + 2H_2 \xrightarrow{Ni} ?$

Solution

a. This is a hydration reaction. Based on Markovnikov's rule, the H will become attached to carbon 1, which has more hydrogen atoms than carbon 2, and the —OH group will be attached to carbon 2.

$$CH_3-CH_2-\overset{\displaystyle OH}{\overset{|}{CH}}-CH_3$$

(continued)

b. The reactant alkene will have a double bond between the two carbon atoms that bromine atoms are attached to in the product.

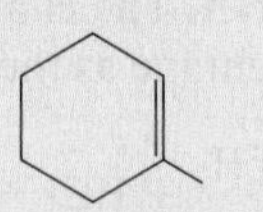

c. The small reactant molecule that adds to the double bond is HBr. The added Br atom from the HBr is explicitly shown in the product's structural formula, but the added H atom is not shown.

d. Hydrogen will add at each of the double bonds. The product hydrocarbon is pentane.

$$CH_3-CH_2-CH_2-CH_2-CH_3$$

Practice Exercise 2.5

Supply the structural formula of the missing substance in each of the following addition reactions.

a. $CH_3-CH_2-CH=CH_2 + HBr \rightarrow ?$

b. $? + H_2O \xrightarrow{H_2SO_4}$ 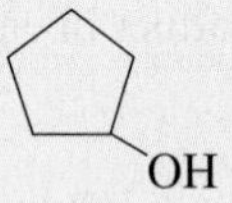

c. 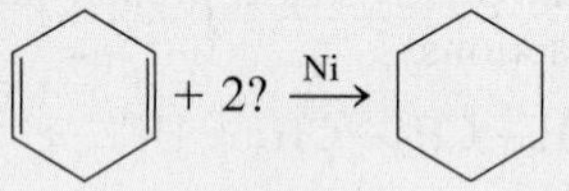

d. $CH_3-\underset{\displaystyle CH_3}{\underset{|}{C}}=CH-CH_3 + Cl_2 \longrightarrow ?$

Answers: **a.** $CH_3-CH_2-\underset{\displaystyle Br}{\underset{|}{CH}}-CH_3$; **b.** 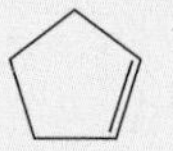 **c.** H_2; **d.** $CH_3-\overset{\displaystyle Cl}{\overset{|}{\underset{\displaystyle CH_3}{\underset{|}{C}}}}-\overset{\displaystyle Cl}{\overset{|}{CH}}-CH_3$

2.10 Polymerization of Alkenes: Addition Polymers

The word *polymer* comes from the Greek *poly*, which means "many," and *meros*, which means "parts."

A **polymer** *is a large molecule formed by the repetitive bonding together of many smaller molecules.* The smaller repeating units of a polymer are called *monomers.* A **monomer** *is the small molecule that is the structural repeating unit in a polymer.* The process by which a polymer is made is called *polymerization.* A **polymerization reaction** *is a chemical reaction in which the repetitious combining of many small molecules (monomers) produces a very large molecule (the polymer).* With appropriate catalysts, simple alkenes and simple substituted alkenes readily undergo polymerization.

The type of polymer that alkenes and substituted alkenes form is an *addition polymer.* An **addition polymer** *is a polymer in which the monomers simply "add together" with no other products formed besides the polymer.* Addition polymerization is similar to the addition reactions described in Section 2.9 except that there is no reactant other than the alkene or substituted alkene.

Polymer types other than addition polymers will be considered in Sections 4.12 and 5.18.

The simplest alkene addition polymer has ethylene (ethene) as the monomer. With appropriate catalysts, ethylene readily adds to itself to produce polyethylene.

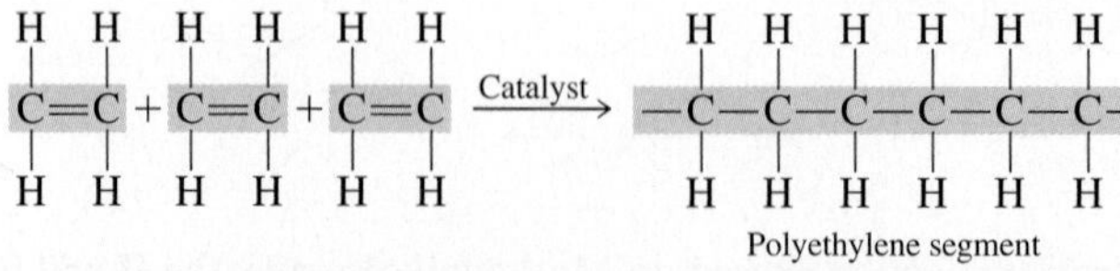

Polyethylene segment

An *exact* formula for a polymer such as polyethylene cannot be written because the length of the carbon chain varies from polymer molecule to polymer molecule. In recognition of this "inexactness" of formula, the notation used for denoting polymer formulas is independent of carbon-chain length. The formula of the simplest repeating unit (the monomer with the double bond changed to a single bond) is written in parentheses and then the subscript n is added after the parentheses, with n being understood to represent a very large number. Using this notation, the formula of polyethylene becomes

$$\left(\begin{matrix} H & H \\ | & | \\ -C- & C- \\ | & | \\ H & H \end{matrix}\right)_n$$

This notation clearly identifies the basic repeating unit found in the polymer.

Substituted-Ethene Addition Polymers

Many substituted alkenes undergo polymerization similar to that of ethene when they are treated with the proper catalyst. For a monosubstituted-ethene monomer, the general polymerization equation is

$$H_2C{=}\overset{\overset{Z}{|}}{C}H \xrightarrow{\text{Polymerization}} \left(CH_2-\overset{\overset{Z}{|}}{C}H\right)_n$$

Variation in the substituent group Z can change polymer properties dramatically, as is shown by the entries in Table 2.1, a listing of monomers for the five ethene-based polymers *polyethylene, polypropylene, poly(vinyl chloride) (PVC), Teflon,* and *polystyrene,* along with several uses of each. Figure 2.11 depicts the preparation of polystyrene. Figure 2.12 contrasts the structures of polyethylene, polypropylene, and poly(vinyl chloride) as depicted in space-filling models.

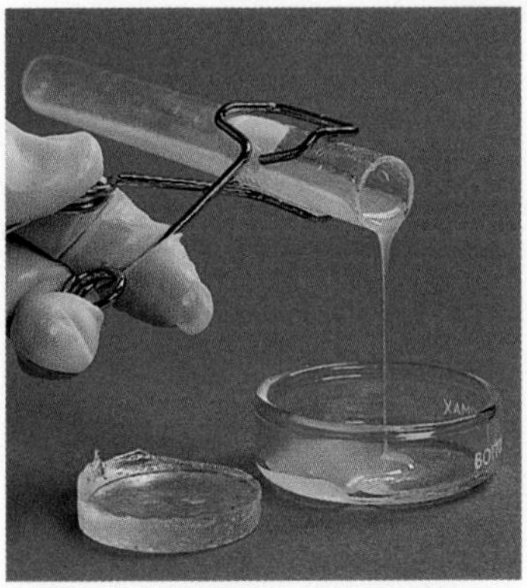

Figure 2.11 Preparation of polystyrene. When styrene, $C_6H_5{-}CH{=}CH_2$, is heated with a catalyst (benzoyl peroxide), it yields a viscous liquid. After some time, this liquid sets to a hard plastic (sample shown at left).

Table 2.1 Some Common Polymers Obtained from Ethene-Based Monomers

Polymer Formula and Name	Monomer Formula and Name	Uses of Polymer
$\left(\begin{matrix} H & H \\ \mid & \mid \\ -C- & C- \\ \mid & \mid \\ H & H \end{matrix}\right)_n$ polyethylene	$\begin{matrix} H & H \\ \mid & \mid \\ C & = C \\ \mid & \mid \\ H & H \end{matrix}$ ethylene	bottles, plastic bags, toys, electrical insulation
$\left(\begin{matrix} H & H \\ \mid & \mid \\ -C- & C- \\ \mid & \mid \\ H & CH_3 \end{matrix}\right)_n$ polypropylene	$\begin{matrix} H & H \\ \mid & \mid \\ C & = C \\ \mid & \mid \\ H & CH_3 \end{matrix}$ propylene	indoor–outdoor carpeting, bottles, molded parts (including heart valves)
$\left(\begin{matrix} H & H \\ \mid & \mid \\ -C- & C- \\ \mid & \mid \\ H & Cl \end{matrix}\right)_n$ poly(vinyl chloride) (PVC)	$\begin{matrix} H & H \\ \mid & \mid \\ C & = C \\ \mid & \mid \\ H & Cl \end{matrix}$ vinyl chloride	plastic wrap, bags for intravenous drugs, garden hose, plastic pipe, simulated leather (Naugahyde)
$\left(\begin{matrix} F & F \\ \mid & \mid \\ -C- & C- \\ \mid & \mid \\ F & F \end{matrix}\right)_n$ Teflon	$\begin{matrix} F & F \\ \mid & \mid \\ C & = C \\ \mid & \mid \\ F & F \end{matrix}$ tetrafluoroethylene	cooking utensil coverings, electrical insulation, component of artificial joints in body parts replacement
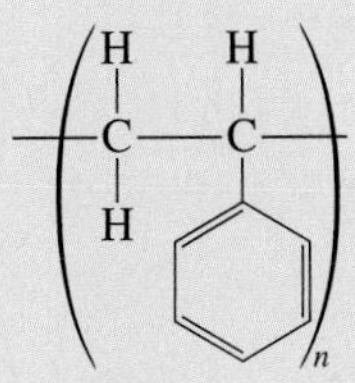 polystyrene	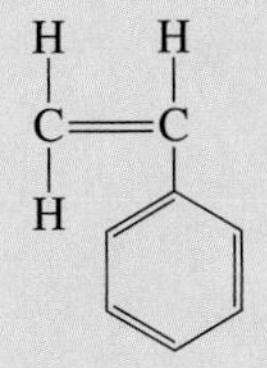 styrene	toys, Styrofoam packaging, cups, simulated wood furniture

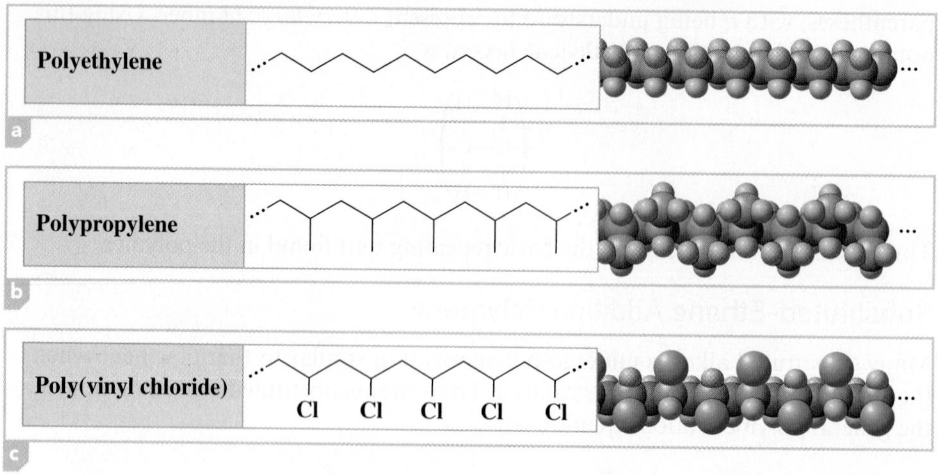

Figure 2.12 Line-angle structural formulas and space-filling models of segments of the ethene-based polymers (a) polyethylene, (b) polypropylene, and (c) poly(vinyl chloride).

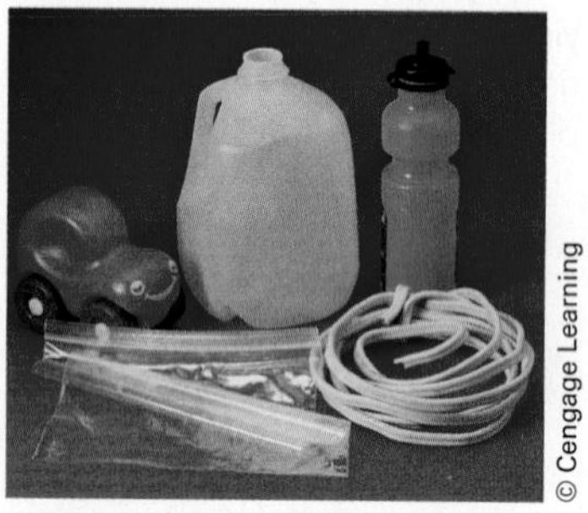

Figure 2.13 Examples of objects made of polyethylene. Polyethylene objects that are strong and rigid (bottles, toys, covering for wire) contain HDPE (high-density polyethylene). Polyethylene objects that are very flexible (plastic bags and packaging materials) most often contain LDPE (low-density polyethylene).

The properties of an ethene-based polymer depend not only on monomer identity but also on the average size (length) of polymer molecules and on the extent of polymer branching. For example, there are three major types of polyethylene: high-density polyethylene (HDPE), low-density polyethylene (LDPE), and linear low-density polyethylene (LLDPE). The major difference among these three materials is the degree of branching of the polymer chain. HDPE and LLDPE are composed of linear, unbranched carbon chains, whereas LDPE chains are branched. The strong and thick plastic bags from a shopping mall are made of LLDPE, the thin and flimsy grocery store plastic bags are made of HDPE, and the very wispy garment bags dry cleaners use are made of LDPE.

In general, HDPE materials are rigid or semi-rigid with uses such as threaded bottle caps, toys, bottles, and milk jugs, whereas LDPE materials are more flexible with uses such as plastic film and squeeze bottles (Figure 2.13). Objects made of HDPE hold their shape in boiling water, whereas objects made of LDPE become severely deformed at this temperature.

With their alkane-like structures, ethene-based addition polymers are very unreactive, as are alkanes. This unreactivity means that these polymers do not readily decompose when they are deposited into landfill sites.

Of the many hundreds of types of addition polymers in use, six account for approximately three-fourths by mass of polymer consumption. These six heavily used polymers are the focus of polymer recycling efforts. The six are assigned a numerical recycling code (the numbers 1 through 6) that relates to the ease of their recycling. The lower the recycling number, the easier it is to recycle the polymer. Table 2.2 lists the six most consumed addition polymers and their recycling codes.

After sorting by recycling code (see Figure 2.14), recycling involves shredding into small chips, removing components such as labels and adhesives, melting, and then remolding. Materials so recycled, because of possible residual contamination

Figure 2.14 Sorting of ethene-based polymers into various subtypes often occurs at recycling collection points.

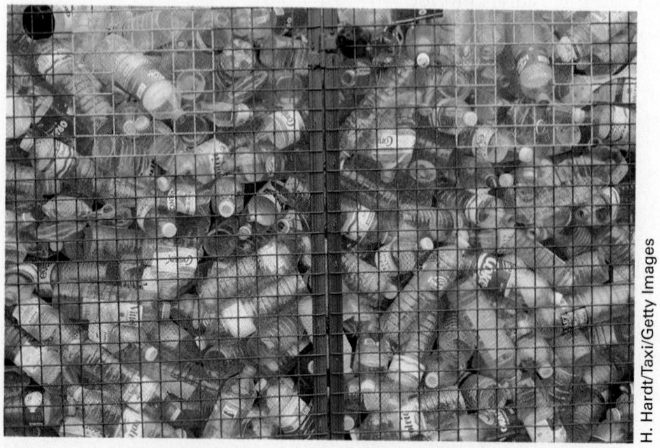

Table 2.2 Codes and Symbols Used in the Recycling of Ethene-Based Addition Polymers

Code	Type	Name	Examples
PET	♳ 1	Polyethylene terephthalate	soft drink bottles, peanut butter jars, vegetable oil bottles
HDPE	♴ 2	High-density polyethylene	milk, water, and juice containers; squeezable ketchup and syrup bottles
PVC	♵ 3	Polyvinyl chloride	shampoo bottles, plastic pipe, shower curtains
LDPE	♶ 4	Low-density polyethylene	six-pack rings, shrink-wrap, sandwich bags, grocery bags
PP	♷ 5	Polypropylene	margarine tubs, straws, diaper linings, toys
PS	♸ 6	Polystyrene	egg cartons, disposable utensils, packing peanuts, foam cups
Other	♹ 7	Multilayer plastics	various flexible items

with adhesives and the like, are not reused in products that store food and drink. End uses include insulating wrap for new home construction (such as Tyvek) and fibers for carpets.

Figure 2.15 Natural rubber being harvested in Malaysia.

Butadiene-Based Addition Polymers

When dienes such as 1,3-butadiene are used as the monomers in addition polymerization reactions, the resulting polymers contain double bonds and are thus still unsaturated.

$$\underset{\text{1,3-Butadiene}}{CH_2{=}CH{-}CH{=}CH_2} \xrightarrow{\text{Polymerization}} \underset{\text{Polybutadiene}}{+\!\!\left(CH_2{-}CH{=}CH{-}CH_2\right)\!\!+_n}$$

In general, unsaturated polymers are much more flexible than the ethene-based saturated polymers listed in Table 2.1. Natural rubber is a flexible addition polymer whose repeating unit is isoprene (Section 2.7)—that is, 2-methyl-1,3-butadiene (see Figure 2.15).

$$\underset{\substack{\text{Isoprene}\\ \text{(2-methyl-1,3-butadiene)}}}{CH_2{=}C(CH_3){-}CH{=}CH_2} \xrightarrow{\text{Polymerization}} \underset{\substack{\text{Polyisoprene}\\ \text{(natural rubber)}}}{\left(CH_2{-}C(CH_3){=}CH{-}CH_2\right)_n}$$

Addition Copolymers

Saran Wrap is a polymer in which two different monomers are present: chloroethene (vinyl chloride) and 1,1-dichloroethene.

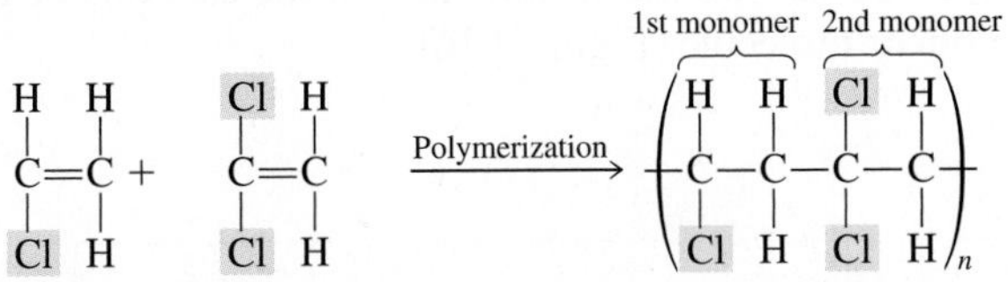

Vinyl chloride 1,1-Dichloroethene Saran Wrap

Such a polymer is an example of a *copolymer.* A **copolymer** *is a polymer in which two different monomers are present.* Another important copolymer is

CHEMISTRY AT A GLANCE Chemical Reactions of Alkenes

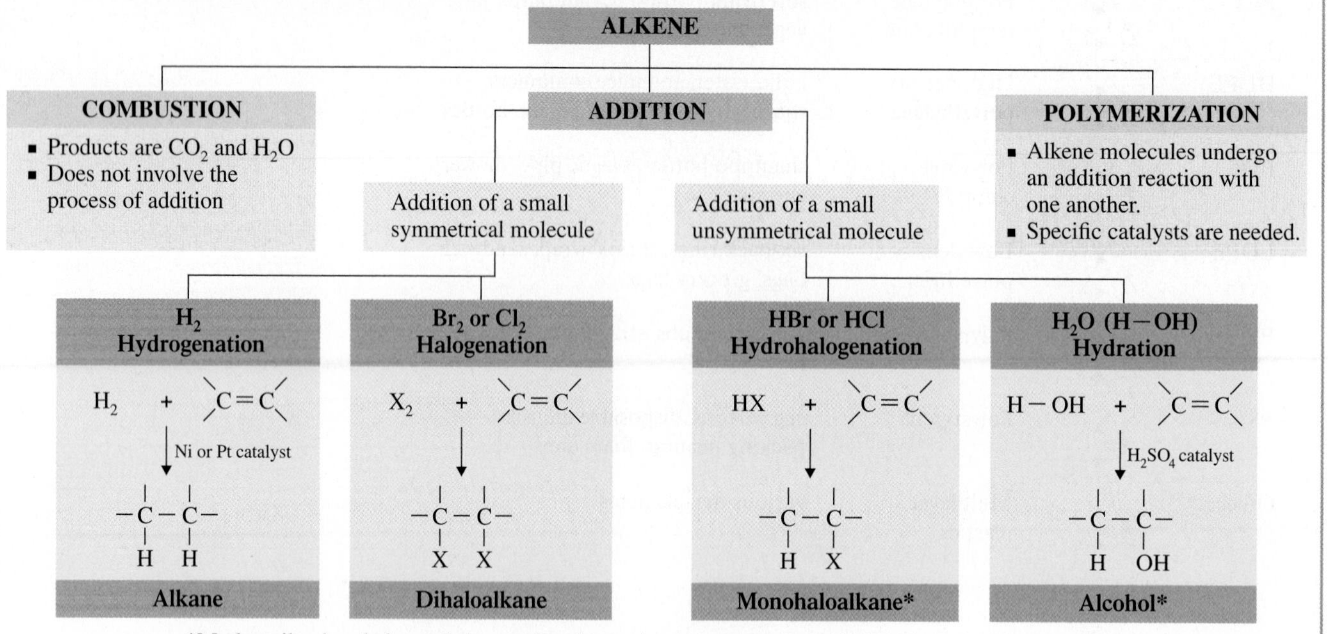

*Markovnikov's rule is needed to predict the product's exact structure if the alkene is unsymmetrical.

styrene–butadiene rubber, the leading synthetic rubber in use today. It contains the monomers 1,3-butadiene and styrene in a 3:1 ratio. It is a major ingredient in automobile tires.

The Chemistry at a Glance feature above summarizes the reaction chemistry of alkenes presented in this and the previous section.

2.11 Alkynes

Very few biological molecules are known that contain a carbon–carbon triple bond.

Alkynes are the second of the three classes of unsaturated hydrocarbons considered in this chapter; alkenes (previously considered) and aromatic hydrocarbons (yet to be considered) are the other two classes. An **alkyne** *is an acyclic unsaturated hydrocarbon that contains one or more carbon–carbon triple bonds.* The alkyne functional group is, thus, a C≡C group. As the family name *alkyne* indicates, the characteristic "ending" associated with a triple bond is *-yne.*

The general formula for an alkyne with one triple bond is C_nH_{2n-2}. Thus the simplest member of this type of alkyne has the formula C_2H_2, and the next member, with $n = 3$, has the formula C_3H_4.

$$\underset{\text{Ethyne}}{CH{\equiv}CH} \qquad \underset{\text{Propyne}}{CH{\equiv}C{-}CH_3}$$

The presence of a carbon–carbon triple bond in a molecule always results in a linear arrangement for the two atoms attached to the carbons of the triple bond. Thus ethyne is a linear molecule (Figure 2.16).

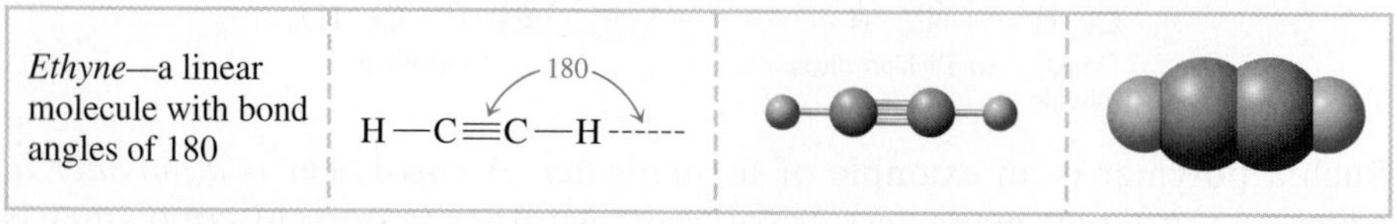

Figure 2.16 Structural representations of ethyne (acetylene), the simplest alkyne. The molecule is linear—that is, the bond angles are 180°.

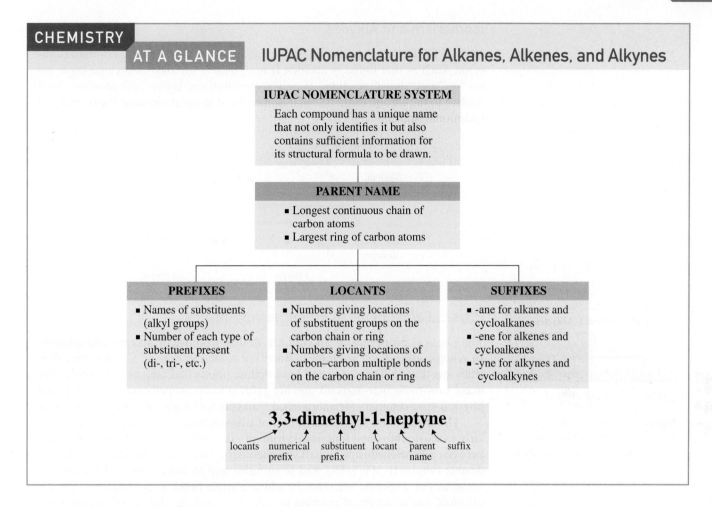

CHEMISTRY AT A GLANCE IUPAC Nomenclature for Alkanes, Alkenes, and Alkynes

IUPAC NOMENCLATURE SYSTEM

Each compound has a unique name that not only identifies it but also contains sufficient information for its structural formula to be drawn.

PARENT NAME

- Longest continuous chain of carbon atoms
- Largest ring of carbon atoms

PREFIXES

- Names of substituents (alkyl groups)
- Number of each type of substituent present (di-, tri-, etc.)

LOCANTS

- Numbers giving locations of substituent groups on the carbon chain or ring
- Numbers giving locations of carbon–carbon multiple bonds on the carbon chain or ring

SUFFIXES

- -ane for alkanes and cycloalkanes
- -ene for alkenes and cycloalkenes
- -yne for alkynes and cycloalkynes

3,3-dimethyl-1-heptyne

locants | numerical prefix | substituent prefix | locant | parent name | suffix

The simplest alkyne, ethyne (C_2H_2), is the most important alkyne from an industrial standpoint. A colorless gas, it goes by the common name *acetylene* and is used in oxyacetylene torches, high-temperature torches used for cutting and welding materials.

Cycloalkynes, molecules that contain a triple bond as part of a ring structure, are known, but they are not common. Because of the 180° angle associated with a triple bond, a ring system containing a triple bond has to be quite large. The smallest cycloalkyne that has been isolated is cyclooctyne.

IUPAC Nomenclature for Alkynes

The rules for naming alkynes are identical to those used to name alkenes (Section 2.3), except the ending *-yne* is used instead of *-ene*. Consider the following structures and their IUPAC names.

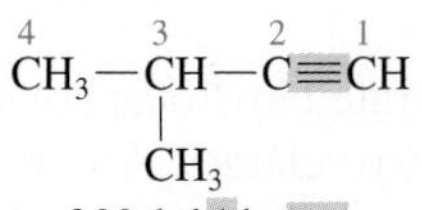
3-Methyl-1-butyne

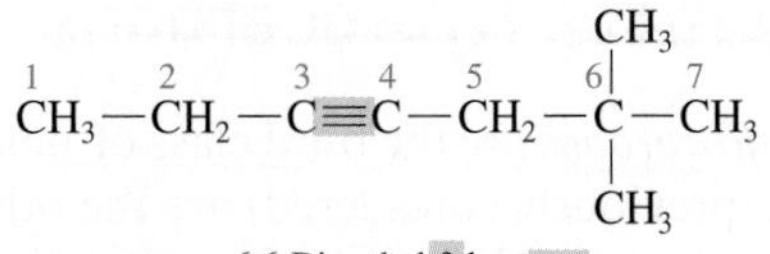
6,6-Dimethyl-3-heptyne

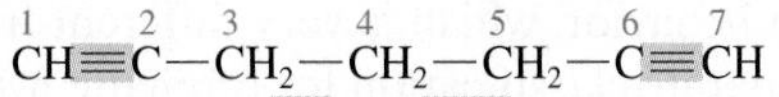
1,6-Heptadiyne

Common names for simple alkynes are based on the name *acetylene,* as shown in the following examples.

$CH\equiv CH$	$CH_3-C\equiv CH$	$CH_3-C\equiv C-CH_3$
Acetylene	Methylacetylene	Dimethylacetylene

Early cars had carbide headlights that produced acetylene by the action of slowly dripping water on calcium carbide. This same type of lamp, which was also used by miners, is still often used by spelunkers (cave explorers).

The Chemistry at a Glance feature above summarizes IUPAC nomenclature procedures for alkanes, alkenes, and alkynes.

Isomerism and Alkynes

Because of the linearity (180° angles) about an alkyne's triple bond, *cis–trans* isomerism, such as that found in alkenes, is not possible for alkynes because there are no "up" and "down" positions. However, constitutional isomers are possible—both relative to the carbon chain (skeletal isomers) and to the position of the triple bond (positional isomers).

Skeletal isomers:

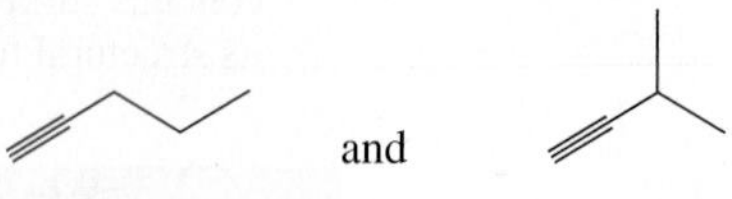

1-Pentyne and 3-Methyl-1-butyne

Positional isomers:

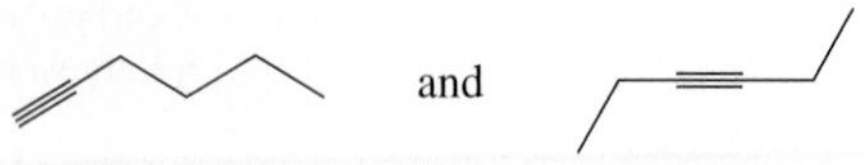

1-Hexyne and 3-Hexyne

Unbranched 1-Alkynes			
	C_3	C_5	C_7
C_2	C_4	C_6	C_8

Gas Liquid

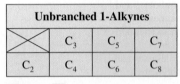

Figure 2.17 A physical-state summary for unbranched 1-alkynes at room temperature and pressure.

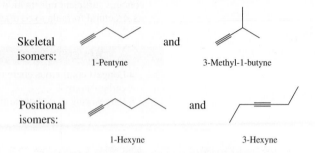

Physical and Chemical Properties of Alkynes

The physical properties of alkynes are similar to those of alkenes and alkanes. In general, alkynes are insoluble in water but soluble in organic solvents, have densities less than that of water, and have boiling points that increase with molecular mass. Low-molecular-mass alkynes are gases at room temperature. Figure 2.17 is a physical-state summary for unbranched 1-alkynes with eight or fewer carbon atoms.

The triple-bond functional group of alkynes behaves chemically quite similarly to the double-bond functional group of alkenes. Thus there are many parallels between alkene chemistry and alkyne chemistry. The same substances that add to double bonds (H_2, HCl, Cl_2, and so on) also add to triple bonds. However, two molecules of a specific reactant can add to a triple bond, as contrasted to the addition of one molecule of reactant to a double bond. In triple-bond addition, the first molecule converts the triple bond into a double bond, and the second molecule then converts the double bond into a single bond. For example, propyne reacts with H_2 to form propene first and then to form propane.

$$\underset{\text{(propyne)}}{\underset{\text{An alkyne}}{CH{\equiv}C{-}CH_3}} \xrightarrow[\text{Ni}]{H_2} \underset{\text{(propene)}}{\underset{\text{An alkene}}{CH_2{=}CH{-}CH_3}} \xrightarrow[\text{Ni}]{H_2} \underset{\text{(propane)}}{\underset{\text{An alkane}}{CH_3{-}CH_2{-}CH_3}}$$

Alkynes, like alkenes and alkanes, are flammable; that is, they readily undergo combustion reactions.

Students often ask whether it is possible to have hydrocarbons in which both double and triple bonds are present. The answer is yes. Immediately, another question is asked. How are such compounds named? Such compounds are called *alkenynes*. An example is

$$\underset{\text{1-Buten-3-yne}}{CH{\equiv}C{-}CH{=}CH_2}$$

A double bond has priority over a triple bond in numbering the chain when numbering systems are equivalent. Otherwise, the chain is numbered from the end closest to a multiple bond.

2.12 Aromatic Hydrocarbons

Aromatic hydrocarbons are the third class of unsaturated hydrocarbons; the alkenes and alkynes (previously considered) are the other two classes. An **aromatic hydrocarbon** *is an unsaturated cyclic hydrocarbon that does not readily undergo addition reactions.* This reaction behavior, which is very different from that of alkenes and alkynes, explains the separate classification for aromatic hydrocarbons.

The fact that, even though they are unsaturated compounds, aromatic hydrocarbons do not readily undergo addition reactions suggests that the bonding present in this type of compound must differ significantly from that in alkenes and alkynes. Such is indeed the case.

A consideration of the bonding present in *benzene,* the simplest aromatic hydrocarbon, is the key to understanding the "special type" of bonding that is characteristic of an aromatic hydrocarbon and to specifying the identity of the aromatic hydrocarbon functional group. Benzene, a flat, symmetrical molecule with

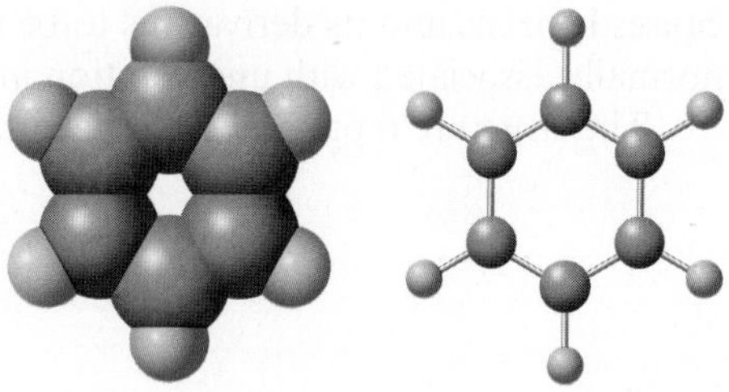

Figure 2.18 Space-filling and ball-and-stick models for the structure of benzene.

a molecular formula of C_6H_6 (Figure 2.18), has a structural formula that is often formalized as that of a cyclohexatriene—in other words, as a structural formula that involves a six-membered carbon ring in which three double bonds are present.

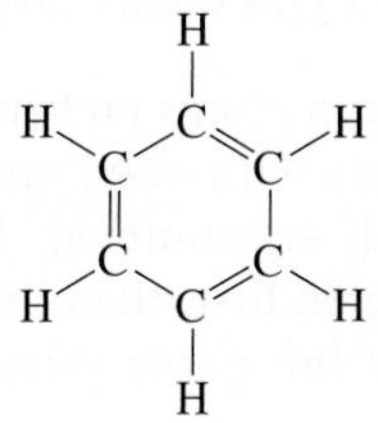

or

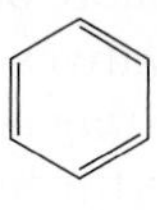

This structure is one of two equivalent structures that can be drawn for benzene that differ only in the locations of the double bonds (1,3,5 positions versus 2,4,6 positions):

and

Neither of these conventional structures, however, is totally correct. Experimental evidence indicates that all of the carbon–carbon bonds in benzene are equivalent (identical), and these preceding structures imply three bonds of one type (double bonds) and three bonds of a different type (single bonds).

The equivalent nature of the carbon–carbon bonds in benzene is addressed by considering the correct bonding structure for benzene to be an *average* of the two "triene" structures. Related to this "average"-structure situation is the concept that electrons associated with the ring double bonds are not held between specific carbon atoms; instead, they are free to move "around" the carbon ring. Thus the true structure for benzene, an intermediate between that represented by the two "triene" structures, is a situation in which all carbon–carbon bonds are equivalent; they are neither single nor double bonds but something in between. Placing a double-headed arrow between the conventional structures that are averaged to obtain the true structure is one way to denote the average structure.

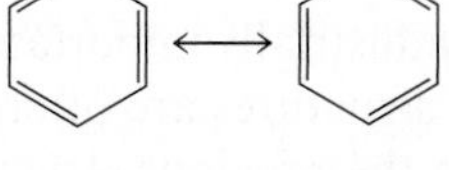

An alternative notation for denoting the bonding in benzene—a notation that involves a single structure—is

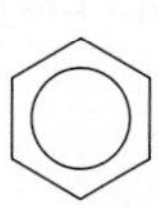

In this "circle-in-the-ring" structure for benzene, the circle denotes the electrons associated with the double bonds that move "around" the ring. Each carbon atom in the ring can be considered to participate in three conventional (localized) bonds (two C—C bonds and one C—H bond) and in one *delocalized bond* (the circle) that involves all six carbon atoms. A **delocalized bond** *is a covalent bond in which electrons are shared among more than two atoms.* This delocalized bond is what causes benzene and its derivatives to be resistant to addition reactions, a property normally associated with unsaturation in a molecule.

The structure represented by the notation

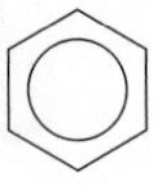

is called an *aromatic ring system,* and it is the functional group present in aromatic compounds. An **aromatic ring system** *is a highly unsaturated carbon ring system in which both localized and delocalized bonds are present.*

2.13 Names for Aromatic Hydrocarbons

Replacement of one or more of the hydrogen atoms on benzene with other groups produces benzene derivatives. Compounds with alkyl groups or halogen atoms attached to the benzene ring are commonly encountered. The naming of benzene derivatives with one substituent is considered first, then the naming of those with two substituents, and finally the naming of those with three or more substituents.

Benzene Derivatives with One Substituent

The IUPAC system of naming monosubstituted benzene derivatives uses the name of the substituent as a prefix to the name benzene. Examples of this type of nomenclature include

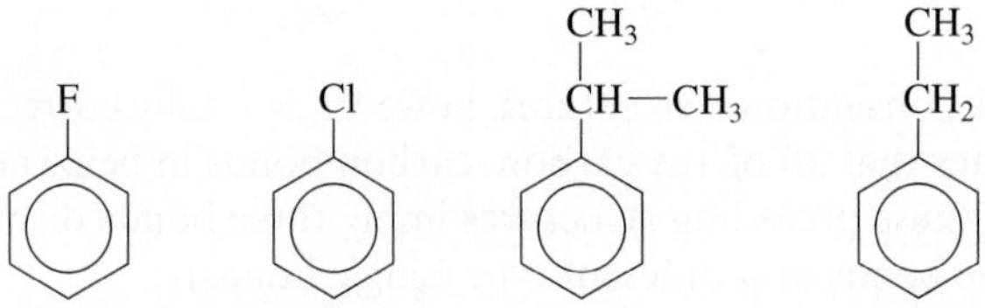

Fluorobenzene Chlorobenzene Isopropylbenzene Ethylbenzene

A few monosubstituted benzenes have names wherein the substituent and the benzene ring taken together constitute a new parent name. Two important examples of such nomenclature with hydrocarbon substituents are

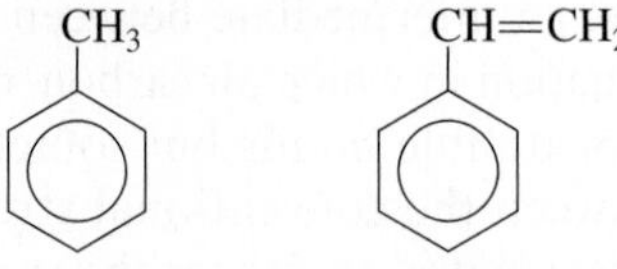

Toluene (not methylbenzene) Styrene (not vinylbenzene)

Both of these compounds are industrially important chemicals.

Monosubstituted benzene structures are often drawn with the substituent at the "12 o'clock" position, as in the previous structures. However, because all the hydrogen atoms in benzene are equivalent, it does not matter at which carbon of the ring the substituted group is located. Each of the following formulas represents chlorobenzene.

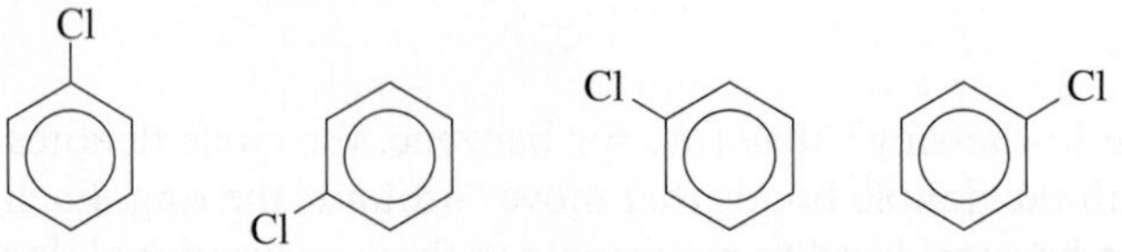

For monosubstituted benzene rings that have a group attached that is not easily named as a substituent, the benzene ring is often treated as a group attached to this substituent. In this reversed approach, the benzene ring attachment is called a *phenyl* group, and the compound is named according to the rules for naming alkanes, alkenes, and alkynes.

The word *phenyl* comes from "phene," a European term used during the 1800s for benzene. The word is pronounced *fen*-nil.

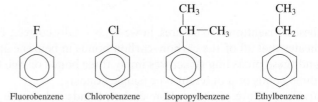

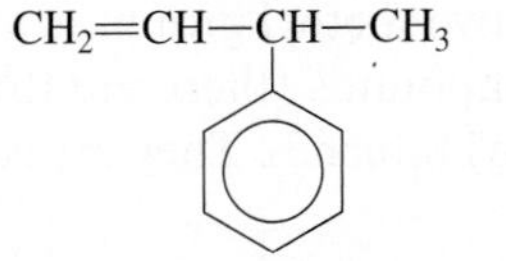

3-Phenyl-1-butene

Benzene Derivatives with Two Substituents

When two substituents, either the same or different, are attached to a benzene ring, three isomeric structures are possible.

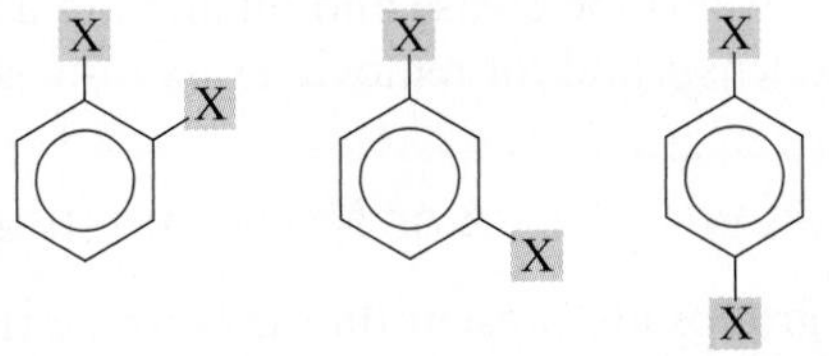

To distinguish among these three isomers, the positions of the substituents relative to one another must be specified. This can be done in either of two ways: by using numbers or by using nonnumerical prefixes.

When numbers are used, the three isomeric dichlorobenzenes have the first-listed set of names:

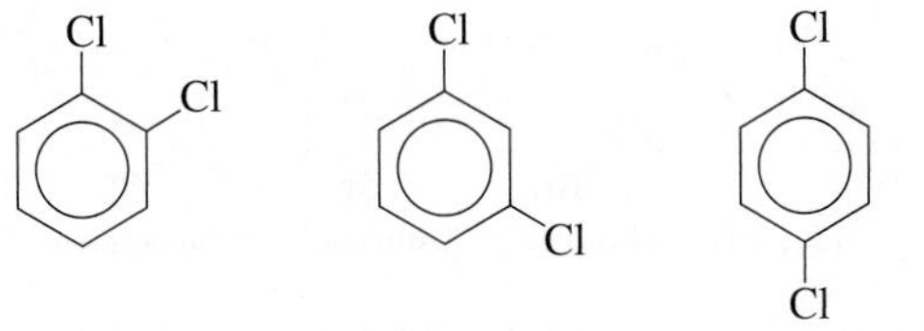

1,2-Dichlorobenzene (*ortho*-dichlorobenzene) 1,3-Dichlorobenzene (*meta*-dichlorobenzene) 1,4-Dichlorobenzene (*para*-dichlorobenzene)

Cis–trans isomerism is not possible for disubstituted benzenes. All 12 atoms of benzene are in the same plane—that is, benzene is a flat molecule. When a substituent group replaces an H atom, the atom that bonds the group to the ring is also in the plane of the ring.

The prefix system uses the prefixes *ortho-*, *meta-*, and *para-* (abbreviated *o-*, *m-*, and *p-*).

Ortho- means 1,2 disubstitution; the substituents are on adjacent carbon atoms.
Meta- means 1,3 disubstitution; the substituents are one carbon removed from each other.
Para- means 1,4 disubstitution; the substituents are two carbons removed from each other (on opposite sides of the ring).

When prefixes are used, the three isomeric dichlorobenzenes have the second-listed set of names above.

Learn the meaning of the prefixes *ortho-*, *meta-*, and *para-*. These prefixes are extensively used in naming disubstituted benzenes.

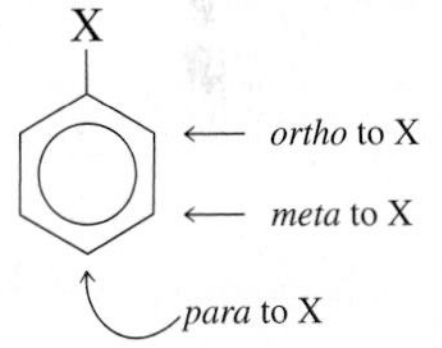

When one of the two substituents in a disubstituted benzene imparts a special name to the compound (as, for example, toluene), the compound is named as a derivative of that parent molecule. The special substituent is assumed to be at ring position 1.

The use of *ortho-*, *meta-*, and *para-* in place of position numbers is reserved for disubstituted benzenes. The system is not used with cyclohexanes or other ring systems.

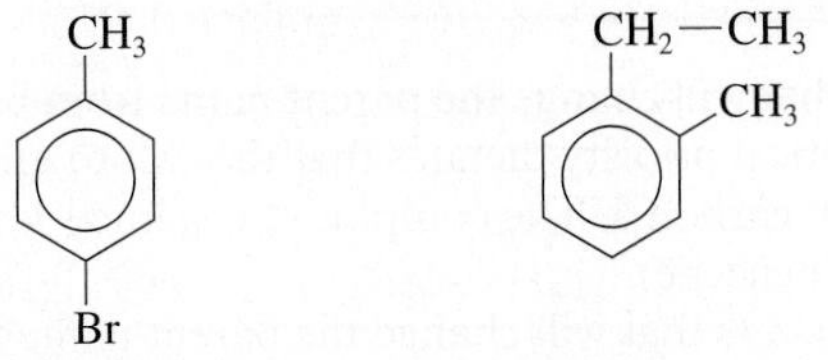

4-Bromotoluene (not 1-bromo-4-methylbenzene) 2-Ethyltoluene (not 1-ethyl-2-methylbenzene)

When neither substituent group imparts a special name, the substituents are cited in alphabetical order before the ending *-benzene.* The carbon of the benzene ring bearing the substituent with alphabetical priority becomes carbon 1.

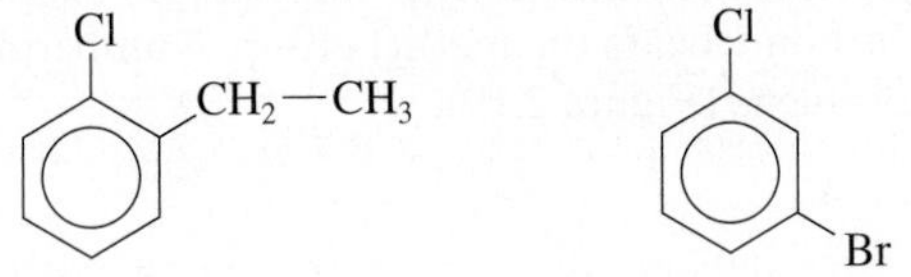

1-Chloro-2-ethylbenzene (not 2-chloro-1-ethylbenzene) 1-Bromo-3-chlorobenzene (not 3-bromo-1-chlorobenzene)

A benzene ring bearing two methyl groups is a situation that generates a new special base name. Such compounds (there are three isomers) are not named as dimethylbenzenes or as methyl toluenes. They are called xylenes.

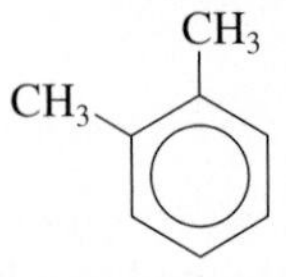
o-Xylene

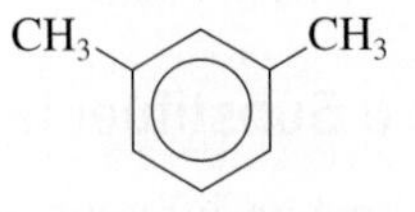
m-Xylene

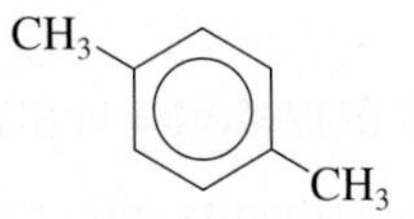
p-Xylene

The xylenes are good solvents for grease and oil and are used for cleaning microscope slides and optical lenses and for removing wax from skis.

When parent names such as *toluene* and *xylene* are used, additional substituents present cannot be the same as those included in the parent name. If such is the case, name the compound as a substituted benzene. The compound

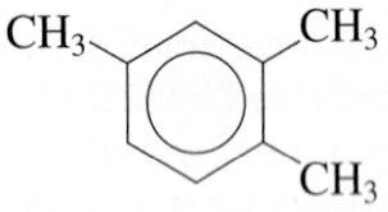

is named as a trimethylbenzene and not as a methylxylene or a dimethyltoluene.

Benzene Derivatives with Three or More Substituents

When more than two groups are present on the benzene ring, their positions are indicated with *numbers*. The ring is numbered in such a way as to obtain the lowest possible numbers for the carbon atoms that have substituents. If there is a choice of numbering systems (two systems give the same lowest set), then the group that comes first alphabetically is given the lower number.

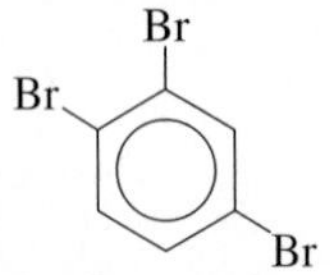
1,2,4-Tribromobenzene

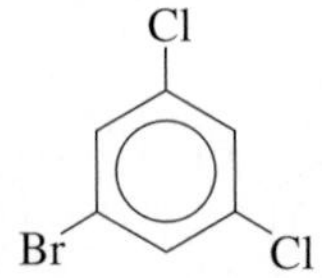
1-Bromo-3,5-dichlorobenzene

EXAMPLE 2.6 Assigning IUPAC Names to Benzene Derivatives

Assign IUPAC names to the following benzene derivatives.

a.
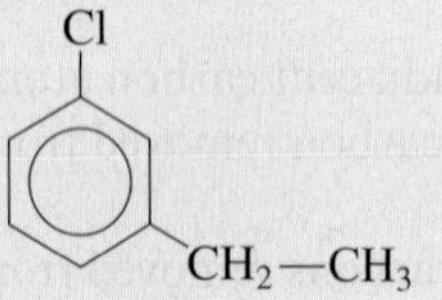

b.
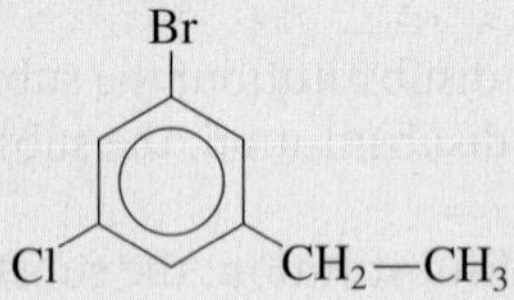

c.
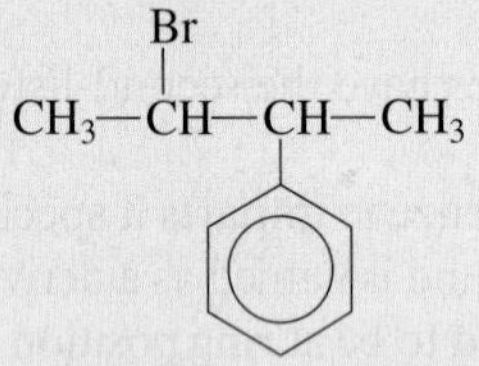

d.
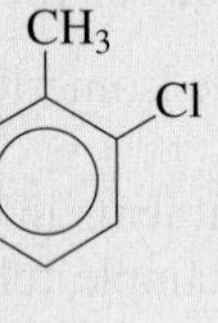

Solution

a. No substituents that will change the parent name from benzene are present on the ring. Alphabetical priority dictates that the chloro group is on carbon 1 and the ethyl group on carbon 3. The compound is named 1-chloro-3-ethylbenzene (or *m*-chloroethylbenzene).

b. Again, no substituents that will change the parent name from benzene are present on the ring. Alphabetical priority among substituents dictates that the bromo group is on carbon 1, the chloro group on carbon 3, and the ethyl group on carbon 5. The compound is named 1-bromo-3-chloro-5-ethylbenzene.

c. This compound is named with the benzene ring treated as a substituent—that is, as a phenyl group. The compound is named 2-bromo-3-phenylbutane.

d. The methyl group present on the benzene ring changes the parent name from benzene to toluene. Carbon 1 bears the methyl group. Numbering clockwise, we obtain the name 2-chlorotoluene (Figure 2.19).

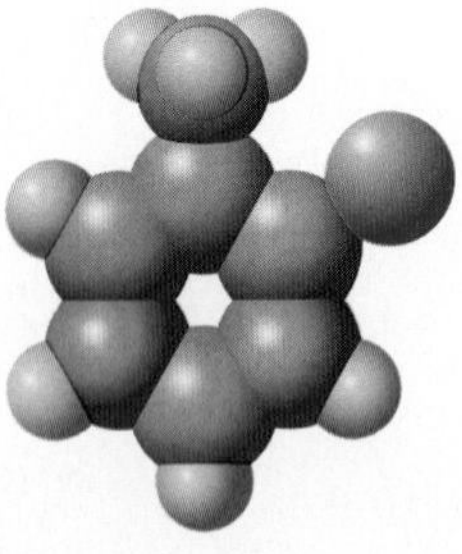
Figure 2.19 Space-filling model for the compound 2-chlorotoluene.

Practice Exercise 2.6

Assign IUPAC names to the following benzene derivatives.

a. 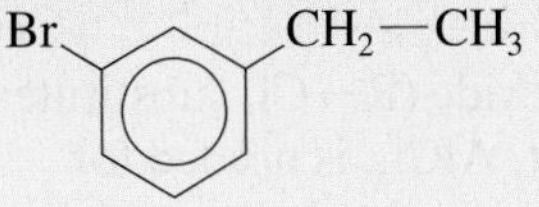

b. 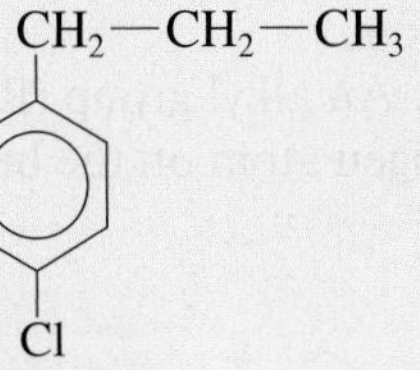

c. 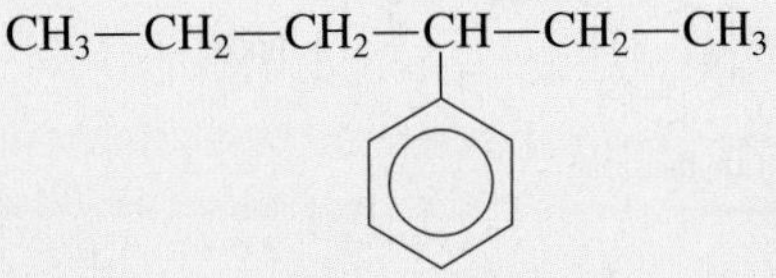

d. 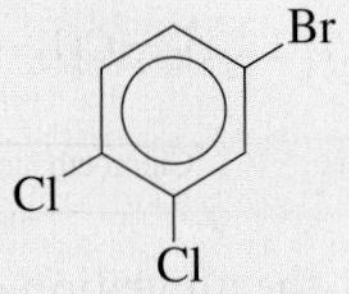

Answers: **a.** 1-Bromo-3-ethylbenzene or *m*-bromoethylbenzene; **b.** 1-Chloro-4-propylbenzene or *o*-chloropropylbenzene; **c.** 3-Phenylhexane; **d.** 4-Bromo-1,2-dichlorobenzene

2.14 Aromatic Hydrocarbons: Physical Properties and Sources

In general, aromatic hydrocarbons resemble other hydrocarbons in physical properties. They are insoluble in water, are good solvents for other nonpolar materials, and are less dense than water.

Benzene, monosubstituted benzenes, and many disubstituted benzenes are liquids at room temperature. Benzene itself is a colorless, flammable liquid that burns with a sooty flame because of incomplete combustion.

At one time, coal tar was the main source of aromatic hydrocarbons. Petroleum is now the primary source of such compounds. At high temperatures, with special catalysts, saturated hydrocarbons obtained from petroleum can be converted to aromatic hydrocarbons. The production of toluene from heptane is representative of such a conversion.

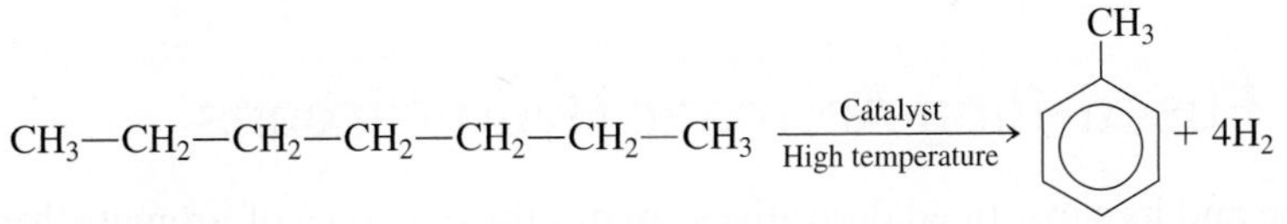

Benzene was once widely used as an organic solvent. Such use has been discontinued because benzene's short- and long-term toxic effects are now recognized. Benzene inhalation can cause nausea and respiratory problems.

Two common situations in which a person can be exposed to low-level benzene vapors are

1. Inhaling gasoline vapors while refueling an automobile. Gasoline contains about 2% (v/v) benzene.
2. Being around a cigarette smoker. Benzene is a combustion product present in cigarette smoke. For smokers themselves, inhaled cigarette smoke is a serious benzene-exposure source.

2.15 Chemical Reactions of Aromatic Hydrocarbons

It was previously noted that aromatic hydrocarbons do not readily undergo the addition reactions characteristic of other unsaturated hydrocarbons. An addition reaction would require breaking up the delocalized bonding (Section 2.12) present in the ring system.

If benzene is so unresponsive to addition reactions, what reactions does it undergo? Benzene undergoes *substitution* reactions. Recall from Section 1.17 that substitution reactions are characterized by different atoms or groups of atoms replacing hydrogen atoms in a hydrocarbon molecule. Two important types of substitution reactions for benzene and other aromatic hydrocarbons are alkylation and halogenation.

1. *Alkylation:* An alkyl group (R—) from an alkyl chloride (R—Cl) substitutes for a hydrogen atom on the benzene ring. A catalyst, $AlCl_3$, is needed for alkylation.

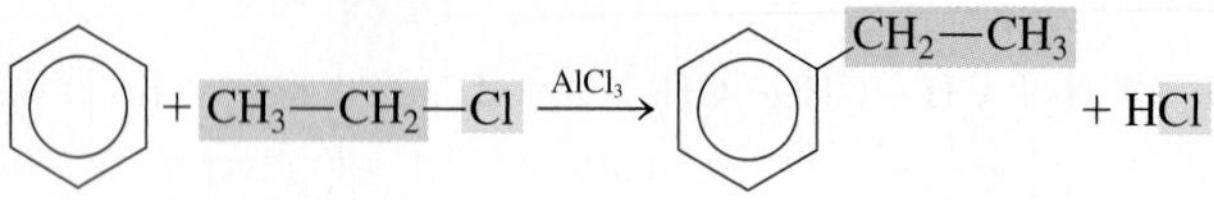

Benzene Chloroethane Ethylbenzene

Alkylation, the reaction that attaches an alkyl group to an aromatic ring, is also known as a *Friedel–Crafts reaction,* named after Charles Friedel and James Mason Crafts, the French and American chemists responsible for its discovery in 1877.

In general terms, the alkylation of benzene can be written as

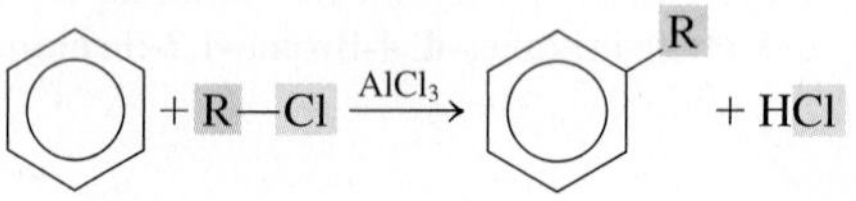

Alkylation is the most important industrial reaction of benzene.

2. *Halogenation* (bromination or chlorination): A hydrogen atom on a benzene ring can be replaced by bromine or chlorine if benzene is treated with Br_2 or Cl_2 in the presence of a catalyst. The catalyst is usually $FeBr_3$ for bromination and $FeCl_3$ for chlorination.

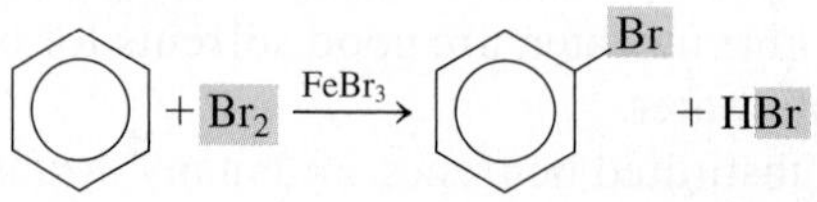

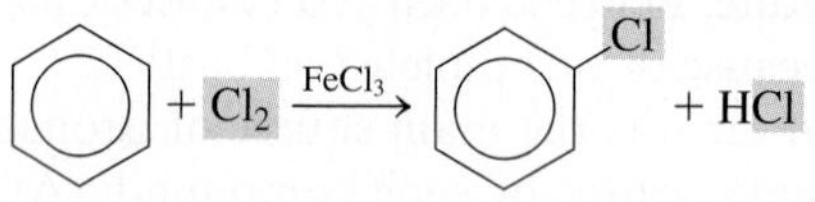

Aromatic halogenation differs from alkane halogenation (Section 1.17) in that light is not required to initiate aromatic halogenation.

2.16 Fused-Ring Aromatic Hydrocarbons

Benzene and its substituted derivatives are not the only type of aromatic hydrocarbon that exists. Another large class of aromatic hydrocarbons is the fused-ring aromatic hydrocarbons. A **fused-ring aromatic hydrocarbon** *is an aromatic hydrocarbon whose structure contains two or more carbon rings fused together.* Two carbon rings that share a pair of carbon atoms are said to be *fused.*

The three simplest fused-ring aromatic compounds are naphthalene, anthracene, and phenanthrene. All three are solids at room temperature.

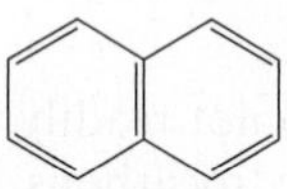

Naphthalene

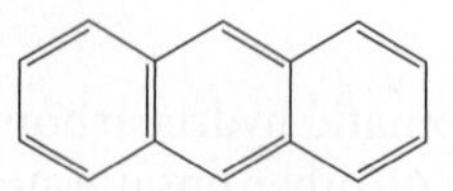

Anthracene

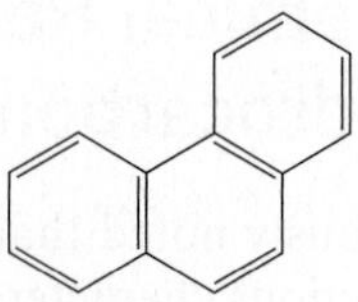

Phenanthrene

A number of fused-ring aromatic hydrocarbons are known to be carcinogens—that is, to cause cancer. Three of the most potent carcinogens are

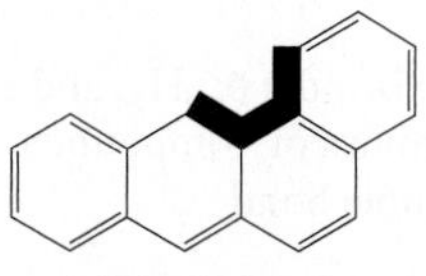

1,2-Benzanthracene

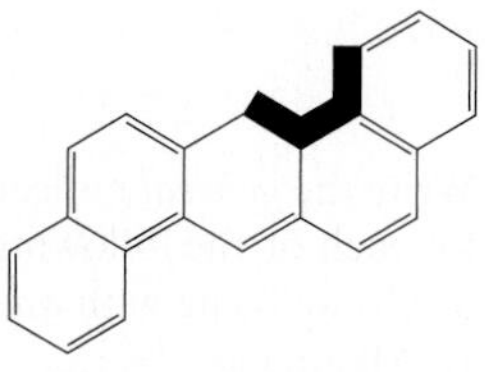

1,2,5,6-Dibenzanthracene

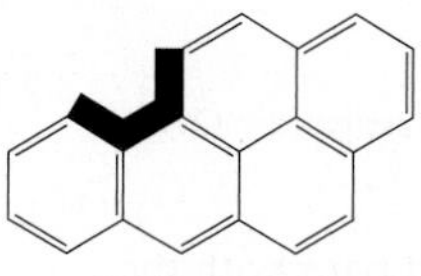

3,4-Benzpyrene

Very small amounts of these substances, when applied to the skin of mice, cause cancer.

Carcinogenic fused-ring aromatic hydrocarbons share some structural features. They all contain four or more fused rings, and they all have the same "angle" in the series of rings (the dark area in the structures shown).

Fused-ring aromatic hydrocarbons are often formed when hydrocarbon materials are heated to high temperatures. These resultant compounds are present in low concentrations in tobacco smoke, in automobile exhaust, and sometimes in burned (charred) food. The charred portions of a well-done steak cooked over charcoal are a likely source.

Angular, fused-ring hydrocarbon systems are believed to be partially responsible for the high incidence of lung and lip cancer among cigarette smokers because tobacco smoke contains 3,4-benzpyrene. The more a person smokes, the greater his or her risk of developing cancer.

It is now known that the high incidence of lung cancer in British chimney sweeps (documented over 200 years ago) was caused by fused-ring hydrocarbon compounds present in the chimney soot that the sweeps inhaled regularly.

Concepts to Remember

OWL Sign in at **www.cengage.com/owl** to view tutorials and simulations, develop problem-solving skills, and complete online homework assigned by your professor.

Unsaturated hydrocarbons. An unsaturated hydrocarbon is a hydrocarbon that contains one or more carbon–carbon multiple bonds. Three main classes of unsaturated hydrocarbons exist: alkenes, alkynes, and aromatic hydrocarbons (Section 2.1).

Alkenes and cycloalkenes. An alkene is an acyclic unsaturated hydrocarbon in which one or more carbon–carbon double bonds are present. A cycloalkene is a cyclic unsaturated hydrocarbon that contains one or more carbon–carbon double bonds within the ring system (Section 2.2).

Alkene nomenclature. Alkenes and cycloalkenes are given IUPAC names using rules similar to those for alkanes and cycloalkanes, except that the ending *-ene* is used. Also, the double bond takes precedence both in selecting and in numbering the main chain or ring (Section 2.3).

Isomerism in alkenes. Two subtypes of constitutional isomers are possible for alkenes: skeletal isomers and positional isomers. Positional isomers differ in the location of the functional group (double bond) present (Section 2.5).

***Cis–trans* isomerism in alkenes.** *Cis–trans* isomerism is possible for some alkenes because there is restricted rotation about a carbon–carbon double bond (Section 2.6).

Physical properties of alkenes. Alkenes and alkanes have similar physical properties. They are nonpolar, insoluble in water, less dense than water, and soluble in nonpolar solvents (Section 2.8).

Addition reactions of alkenes. Numerous substances, including H_2, Cl_2, Br_2, HCl, HBr, and H_2O, add to an alkene carbon–carbon double bond. When both the alkene and the substance to be added are unsymmetrical, the addition proceeds according to Markovnikov's rule: The carbon atom of the double bond that already has the greater number of H atoms gets one more (Section 2.9).

Addition polymers. Addition polymers are formed from alkene monomers that undergo repeated addition reactions with each other. Many familiar and widely used materials, such as fibers and plastics, are addition polymers (Section 2.10).

Alkynes and cycloalkynes. Alkynes and cycloalkynes are unsaturated hydrocarbons that contain one or more carbon–carbon triple bonds. They are named in the same way as alkenes and cycloalkenes, except that their parent names end in *-yne.* Like alkenes, alkynes undergo addition reactions. These occur in two steps, an alkene forming first and then an alkane (Section 2.11).

Aromatic hydrocarbons. Benzene, the simplest aromatic hydrocarbon, and other members of this family of compounds contain a six-membered ring with a cyclic, delocalized bond. This aromatic ring is often drawn as a hexagon containing a circle, which represents six electrons that move freely around the ring (Section 2.12).

Nomenclature of aromatic hydrocarbons. Monosubstituted benzene compounds are named by adding the substituent name to the word *benzene.* Positions of substituents in disubstituted benzenes are indicated by using a numbering system or the *ortho-* (1,2), *meta-* (1,3), and *para-* (1,4) prefix system (Section 2.13).

Chemical reactions of aromatic hydrocarbons. Aromatic hydrocarbons undergo substitution reactions rather than addition reactions. Important substitution reactions are alkylation and halogenation (Section 2.15).

Exercises and Problems

OWL Interactive versions of these problems may be assigned in OWL.

Exercises and problems are arranged in matched pairs with the two members of a pair addressing the same concept(s). The answer to the odd-numbered member of a pair is given at the back of the book. Problems denoted with a ▲ involve concepts found not only in the section under consideration but also concepts found in one or more earlier sections of the chapter. Problems denoted with a ● cover concepts found in a Chemical Connections feature box.

Unsaturated Hydrocarbons (Section 2.1)

2.1 How does an unsaturated hydrocarbon differ structurally from a saturated hydrocarbon?

2.2 What type of functional group is present in an unsaturated hydrocarbon?

2.3 What is the functional group present in an alkene?

2.4 What is the functional group present in an alkyne?

2.5 In general terms, compare the physical properties of unsaturated and saturated hydrocarbons.

2.6 In general terms, compare the chemical properties of unsaturated and saturated hydrocarbons.

2.7 Classify each of the following hydrocarbons as saturated or unsaturated.

a. $CH_3-CH_2-CH=CH-CH_3$

b. $CH_2=C(CH_3)-C(CH_3)_2-CH_3$

c. $CH_2=CH-CH_2-C(=CH_2)-CH_3$

d. $CH_2=CH-CH=CH-CH=CH_2$

2.8 Classify each of the following hydrocarbons as saturated or unsaturated.

a. $CH_3-CH=CH-CH=CH_2$

b. $CH_3-CH=CH-CH_3$

c. $CH_2=C(CH_3)-CH_2-CH_3$

d. $CH_2=C(CH_3)-CH=CH-CH=CH_2$

Characteristics of Alkenes and Cycloalkenes (Section 2.2)

2.9 Write the *molecular formula* for hydrocarbons with each of the following structural features.

a. Acyclic, four carbon atoms, no multiple bonds

b. Acyclic, five carbon atoms, one double bond

c. Cyclic, five carbon atoms, one double bond

d. Cyclic, seven carbon atoms, two double bonds

2.10 Write the *molecular formula* for hydrocarbons with each of the following structural features.

a. Acyclic, six carbon atoms, two double bonds

b. Acyclic, six carbon atoms, three double bonds

c. Cyclic, five carbon atoms, no multiple bonds

d. Cyclic, eight carbon atoms, four double bonds

2.11 Write the *general* molecular formula (C_nH_{2n} and so on) for each of the following families of compounds.

a. Cycloalkene with one double bond

b. Alkadiene

c. Diene

d. Cycloalkatriene

2.12 Write the *general* molecular formula (C_nH_{2n} and so on) for each of the following families of compounds.

a. Cycloalkadiene

b. Alkene with one double bond

c. Triene

d. Alkatriene

2.13 Classify each of the compounds in Problem 2.7 as an alkene with one double bond, as a diene, or as a triene.

2.14 Classify each of the compounds in Problem 2.8 as an alkene with one double bond, as a diene, or as a triene.

IUPAC Nomenclature for Alkenes and Cycloalkenes (Section 2.3)

2.15 Assign an IUPAC name to each of the following unsaturated hydrocarbons.

a. $CH_3-CH=CH-CH_3$

b. $CH_3-C(CH_3)=CH-CH(CH_3)-CH_3$

c.

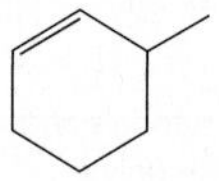

d.

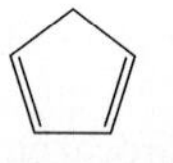

2.16 Assign an IUPAC name to each of the following unsaturated hydrocarbons.

a. $CH_3-CH_2-CH=CH-CH_3$

b. $CH_3-CH_2-C(CH_3)=CH-CH_3$

c.

d.

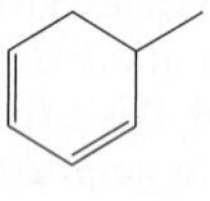

2.17 Assign an IUPAC name to each of the hydrocarbons in Problem 2.7.

2.18 Assign an IUPAC name to each of the hydrocarbons in Problem 2.8.

2.19 Draw a condensed structural formula for each of the following unsaturated hydrocarbons.

a. 3-Methyl-1-pentene

b. 3-Methylcyclopentene

c. 1,3-Butadiene

d. 3-Ethyl-1,4-pentadiene

2.20 Draw a condensed structural formula for each of the following unsaturated hydrocarbons.

a. 4-Methyl-1-hexene

b. 4-Methylcyclohexene

c. 1,3-Pentadiene

d. 2-Ethyl-1,4-pentadiene

2.21 The following names are *incorrect* by IUPAC rules. Determine the correct IUPAC name for each compound.
a. 2-Ethyl-2-pentene
b. 4,5-Dimethyl-4-hexene
c. 3,5-Cyclopentadiene
d. 1,2-Dimethyl-4-cyclohexene

2.22 The following names are *incorrect* by IUPAC rules. Determine the correct IUPAC name for each compound.
a. 2-Methyl-4-pentene
b. 3-Methyl-2,4-pentadiene
c. 3-Methyl-3-cyclopentene
d. 1,2-Dimethyl-3-cyclohexene

2.23 Draw a condensed structural formula for each of the following unsaturated hydrocarbons.
a. Ethylene b. Methylenecyclobutane
c. Vinyl bromide d. Allyl iodide

2.24 Draw a condensed structural formula for each of the following unsaturated hydrocarbons.
a. Propylene b. Methylenecyclopentane
c. Vinyl iodide d. Allyl chloride

▲2.25 Classify each of the following compounds as *saturated* or *unsaturated.*
a. Ethylcyclopentane b. Ethylcyclopentene
c. 1,3-butadiene d. 2-methyl-2-pentene

▲**2.26** Classify each of the following compounds as *saturated* or *unsaturated.*
a. 2-methylcyclopentene
b. 1,2-cyclopentadiene
c. 2,3-dimethylpentane
d. 1-ethyl-2-methylcyclohexene

▲2.27 How many hydrogen atoms are present in a molecule of each of the compounds in Problem 2.25?

▲**2.28** How many hydrogen atoms are present in a molecule of each of the compounds in Problem 2.26?

●2.29 (Chemical Connections 2-A) Indicate whether each of the following statements concerning ethylene is true or false.
a. Ethylene is a colorless, odorless, nonflammable gas at room temperature and pressure.
b. Exposing unripe fruit to small amounts of ethylene gas stimulates the ripening process.
c. Industrially, hydrogenation of ethane produces ethylene.
d. Almost one-half of industrially produced ethylene is used to make polyvinyl chloride.

●**2.30** (Chemical Connections 2-A) Indicate whether each of the following statements concerning ethylene is true or false.
a. Ethylene is a naturally occurring plant hormone that slows down the fruit-ripening process.
b. Large natural sources of ethylene do not exist.
c. Industrial production of ethylene exceeds that of every other organic chemical.
d. The antifreeze ingredient ethylene glycol is an ethylene-based chemical.

Line-Angle Structural Formulas for Alkenes (Section 2.4)

2.31 Draw a line-angle structural formula for each of the following unsaturated hydrocarbons.
a. 1-Butene b. 2-Butene
c. 1,3-Butadiene d. 3-Methyl-1-butene

2.32 Draw a line-angle structural formula for each of the following unsaturated hydrocarbons.
a. 1-Pentene b. 2-Pentene
c. 1,4-Pentadiene d. 4-Methyl-2-pentene

2.33 How many carbon atoms are present in each of the following unsaturated hydrocarbons?

a. 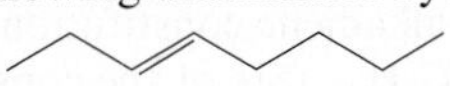 b. 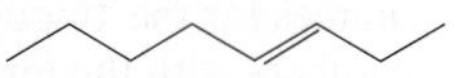

c. 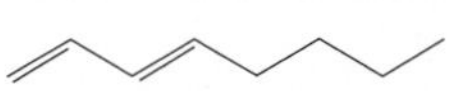 d. 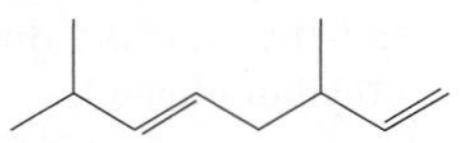

2.34 How many carbon atoms are present in each of the following unsaturated hydrocarbons?

a.  b. 

c. 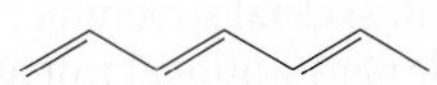 d. 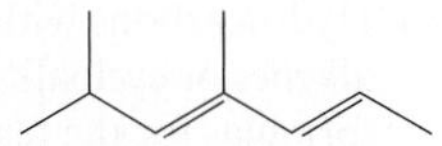

▲2.35 What is the molecular formula for each of the compounds in Problem 2.33?

▲**2.36** What is the molecular formula for each of the compounds in Problem 2.34?

▲2.37 What is the IUPAC name for each of the compounds in Problem 2.33?

▲**2.38** What is the IUPAC name for each of the compounds in Problem 2.34?

Constitutional Isomerism in Alkenes (Section 2.5)

2.39 For each of the following pairs of alkenes, indicate whether the members of each pair are *positional* constitutional isomers or *skeletal* constitutional isomers.
a. 1-Hexene and 2-hexene
b. 2-Methyl-1-pentene and 3-methyl-1-pentene
c. 3-Hexene and 2-ethyl-1-butene
d. 3-Methyl-2-pentene and 3-methyl-1-pentene

2.40 For each of the following pairs of alkenes, indicate whether the members of each pair are *positional* constitutional isomers or *skeletal* constitutional isomers.
a. 2-Hexene and 3-hexene
b. 4-Methyl-1-pentene and 4-methyl-2-pentene
c. 2,3-Dimethyl-2-butene and 3,3-dimethyl-1-butene
d. 2-Methyl-2-pentene and 2-methyl-1-pentene

2.41 How many constitutional isomers exist that fit each of the following specifications?
a. Unbranched chain of five carbon atoms, one carbon–carbon double bond
b. Unbranched chain of five carbon atoms, two carbon–carbon double bonds
c. Five carbon atoms, one methyl group, one carbon–carbon double bond
d. Five carbon atoms, two methyl groups, one carbon–carbon double bond

2.42 How many constitutional isomers exist that fit each of the following specifications?
a. Unbranched chain of six carbon atoms, one carbon–carbon double bond
b. Unbranched chain of six carbon atoms, two carbon–carbon double bonds
c. Six carbon atoms, one methyl group, one carbon–carbon double bond
d. Six carbon atoms, two methyl groups, one carbon–carbon double bond

2.43 Draw skeletal structural formulas and give the IUPAC names for the 13 possible alkene constitutional isomers with the formula C_6H_{12}. (Three of the constitutional isomers are hexenes, six are methylpentenes, three are dimethylbutenes, and one is an ethylbutene.)

2.44 Draw skeletal structural formulas and give the IUPAC names for the 16 possible alkadiene constitutional isomers with the formula C_6H_{10}. (Six of the constitutional isomers are hexadienes, eight are methylpentadienes, one is a dimethylbutadiene, and one is an ethylbutadiene.)

▲2.45 Hydrocarbons with the formula C_4H_8 can be either alkenes or cycloalkanes. Draw skeletal structural formulas for the five possible constitutional isomers that fit this formula; three are alkenes and two are cycloalkanes.

▲2.46 Hydrocarbons with the formula C_5H_{10} can be either alkenes or cycloalkanes. Draw skeletal structural formulas for the ten possible constitutional isomers that fit this formula; five are alkenes and five are cycloalkanes.

Cis–Trans Isomerism in Alkenes (Section 2.6)

2.47 For each molecule, indicate whether *cis–trans* isomers exist. If they do, draw the two isomers and label them as *cis* and *trans.*

a. $CH_2{=}CH{-}CH_3$ b. $CH_3{-}\underset{\displaystyle CH_3}{\underset{|}{C}}{=}CH{-}CH_3$

c. 3-Hexene d. 4-Methyl-2-pentene

2.48 For each molecule, indicate whether *cis–trans* isomers exist. If they do, draw the two isomers and label them as *cis* and *trans.*

a. $CH_3{-}CH_2{-}CH{=}CH_2$ b. $CH_3{-}CH_2{-}CH{=}\underset{\displaystyle Cl}{\underset{|}{CH}}$

c. 2-Pentene d. 1,2-Dichloroethene

2.49 Assign an IUPAC name to each of the following molecules. Include the prefix *cis-* or *trans-* when appropriate.

a. 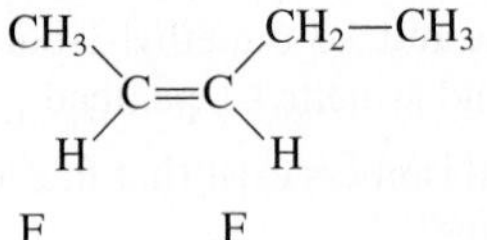

b. 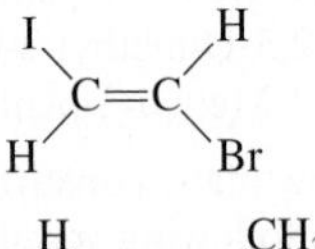

c. 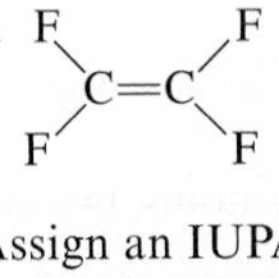

d. 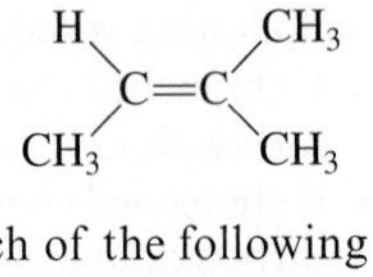

2.50 Assign an IUPAC name to each of the following molecules. Include the prefix *cis-* or *trans-* when appropriate.

a. 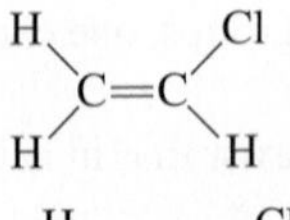

b. 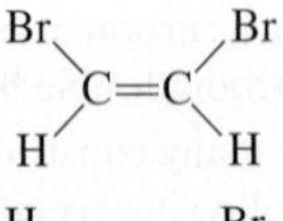

c. 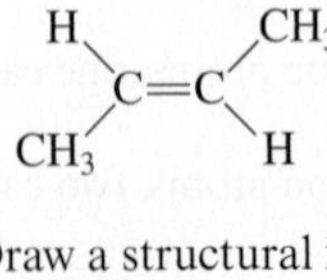

d. 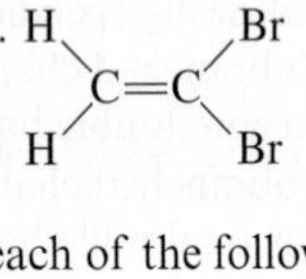

2.51 Draw a structural formula for each of the following compounds.

a. *trans*-3-Methyl-3-hexene b. *cis*-2-Pentene
c. *trans*-5-Methyl-2-heptene d. *trans*-1,3-Pentadiene

2.52 Draw a structural formula for each of the following compounds.

a. *trans*-2-Hexene
b. *cis*-4-Methyl-2-pentene
c. *cis*-1-Chloro-1-pentene
d. *cis*-1,3-Pentadiene

▲2.53 For each of the following molecules, indicate whether or not *cis–trans* isomerism is possible.

a. 2-methyl-1-pentene
b. 1-hexene
c. Methylcyclohexane
d. 1,2-diethylcyclopentane

▲2.54 For each of the following molecules, indicate whether or not *cis–trans* isomerism is possible.

a. 2-pentene
b. 1-chloro-2-pentene
c. Chlorocyclopentane
d. 1,2-dichlorocyclopentane

•2.55 (Chemical Connections 2-B) Indicate whether each of the following statements concerning the molecule retinal and its role in the vision process is true or false.

a. Retinal is a protein found in the retina of the eye.
b. The carbon chain in a retinal molecule contains five double bonds, all of which are in a *cis*-configuration.
c. When light strikes retinal, a *cis*-double bond is converted to a *trans*-double bond, with a resulting change in the shape of the molecule.
d. A change in double-bond configuration effects the release of retinal from the protein to which it is bound.

•2.56 (Chemical Connections 2-B) Indicate whether each of the following statements concerning the molecule retinal and its role in the vision process is true or false.

a. Opsin is a protein in the retina of the eye to which the molecule retinal is attached.
b. The carbon chain in a retinal molecule contains four double bonds, two in a *cis*-configuration and two in a *trans*-configuration.
c. Double-bond configuration determines the shape of the retinal molecule.
d. The release of a retinal molecule from its protein host triggers an electrical impulse that is sent to the brain.

Naturally Occurring Alkenes (Section 2.7)

2.57 What is the biochemical function of pheromones?

2.58 How are sex pheromones used in insect control?

2.59 What is the arrangement of carbon atoms in the basic structural unit of a terpene?

2.60 Why is the number of carbon atoms in a terpene always a multiple of the number 5?

2.61 What is the structural relationship between β-carotene and vitamin A?

2.62 What is an additional role for β-carotene that is not related to vitamin A?

•2.63 (Chemical Connections 2-C) Indicate whether each of the following statements concerning carotenoids is true or false.

a. Carotenoids are a group of color pigments found only in plants.
b. Carotenoids contain two C_{20} terpene units.
c. Carotenes are the oxygenated hydrocarbon subclass of carotenoids.
d. Lycopene is a carotene that gives tomatoes their red color.

•**2.64** (Chemical Connections 2-C) Indicate whether each of the following statements concerning carotenoids is true or false.
a. Xanthophylls and carotenes are two subclasses of carotenoids.
b. β-carotene is responsible for the yellow-orange color of carrots and yams.
c. The yellow-orange colors of autumn leaves come from lutein.
d. Xanthophylls function as antioxidants in the retina of the eye.

Physical Properties of Alkenes and Cycloalkenes (Section 2.8)

2.65 With the help of Figure 2.7, indicate whether each of the following alkenes would be expected to be a solid, a liquid, or a gas at room temperature and pressure.
a. Propene
b. 1-Pentene
c. 1-Octene
d. Cyclopentene

2.66 With the help of Figure 2.7, indicate whether each of the following statements is true or false.
a. 1-Butene has a density greater than that of water.
b. 1-Butene has a higher boiling point than 1-hexene.
c. 1-Butene is flammable, but 1-hexene is not.
d. Both 1-pentene and cyclopentene are gases at room temperature and pressure.

Chemical Reactions of Alkenes (Section 2.9)

2.67 Which of the following chemical reactions are addition reactions?
a. $C_4H_8 + Cl_2 \rightarrow C_4H_8Cl_2$
b. $C_6H_6 + Cl_2 \rightarrow C_6H_5Cl + HCl$
c. $C_3H_6 + HCl \rightarrow C_3H_7Cl$
d. $C_7H_{16} \rightarrow C_7H_8 + 4H_2$

2.68 Which of the following chemical reactions are addition reactions?
a. $C_3H_6 + Cl_2 \rightarrow C_3H_6Cl_2$
b. $C_8H_{10} \rightarrow C_8H_8 + H_2$
c. $C_6H_6 + C_2H_5Cl \rightarrow C_8H_{10} + HCl$
d. $C_4H_8 + HCl \rightarrow C_4H_9Cl$

2.69 Write a chemical equation showing reactants, products, and catalysts needed (if any) for the reaction of ethene with each of the following substances.
a. Cl_2 b. HCl c. H_2 d. HBr

2.70 Write a chemical equation showing reactants, products, and catalysts needed (if any) for the reaction of ethene with each of the following substances.
a. H_2O b. Br_2 c. HI d. I_2

2.71 Write a chemical equation showing reactants, products, and catalysts needed (if any) for the reaction of propene with each of the reactants in Problem 2.69. Use Markovnikov's rule as needed.

2.72 Write a chemical equation showing reactants, products, and catalysts needed (if any) for the reaction of propene with each of the reactants in Problem 2.70. Use Markovnikov's rule as needed.

2.73 Supply the structural formula of the product in each of the following alkene addition reactions.
a. $CH_3{-}CH{=}CH{-}CH_3 + Cl_2 \rightarrow ?$
b. $CH_3{-}CH_2{-}CH{=}CH_2 + HCl \rightarrow ?$
c. 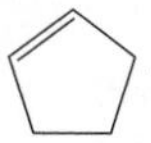 $+ H_2 \xrightarrow[\text{catalyst}]{\text{Ni}} ?$
d. 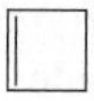 $+ H_2O \xrightarrow{H_2SO_4} ?$

2.74 Supply the structural formula of the product in each of the following alkene addition reactions.
a. $CH_3{-}CH_2{-}CH{=}CH_2 + Cl_2 \rightarrow ?$
b. $CH_3{-}\underset{\displaystyle CH_3}{\underset{|}{CH}}{-}CH{=}CH_2 + HBr \rightarrow ?$
c. 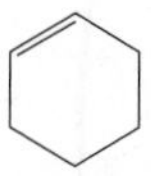 $+ H_2 \xrightarrow[\text{catalyst}]{\text{Ni}} ?$
d. $CH_3{-}CH{=}CH_2 + H_2O \xrightarrow{H_2SO_4} ?$

2.75 What reactant would you use to prepare each of the following compounds from cyclohexene?
a. 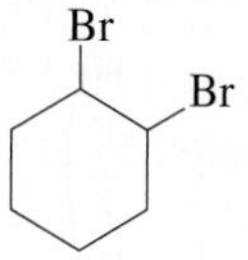

b. 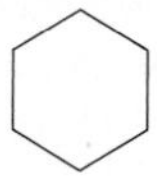
c. 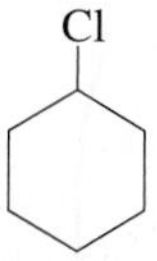

d. 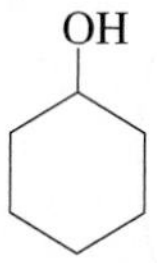

2.76 What reactant would you use to prepare each of the following compounds from cyclopentene?
a. 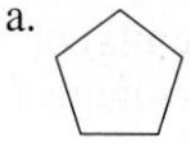
b. 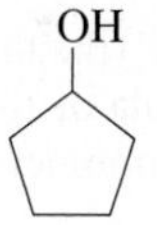

c. 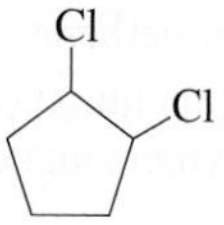

d. 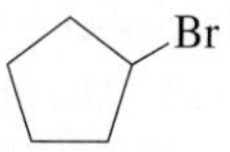

2.77 How many molecules of H_2 gas will react with 1 molecule of each of the following unsaturated hydrocarbons?
a. $CH_3{-}CH{=}CH{-}CH{=}CH{-}CH_3$
b. 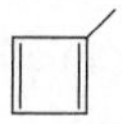
c. 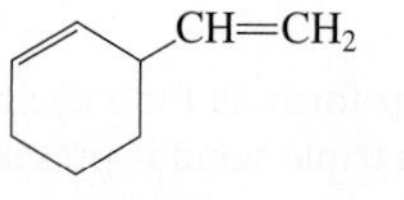

d. $CH_3{-}CH{=}C{=}\underset{\displaystyle CH_3}{\underset{|}{C}}{-}CH{=}CH_2$

2.78 How many molecules of H_2 gas will react with 1 molecule of each of the following unsaturated hydrocarbons?
a. $CH_3{-}CH{=}CH{-}CH_3$
b. 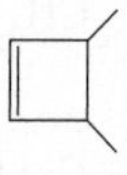
c. 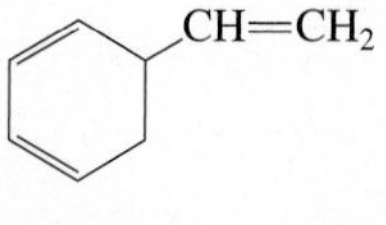

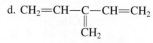

d. $CH_2{=}CH{-}\underset{\displaystyle CH_2}{\underset{\|}{C}}{-}CH{=}CH_2$

Polymerization of Alkenes (Section 2.10)

2.79 What is meant by the term *polymer*?

2.80 What is meant by the term *monomer*?

2.81 What is meant by the term *addition polymer*?

2.82 What is meant by the term *copolymer*?

2.83 Draw the structural formula of the monomer(s) from which each of the following polymers was made.

a. 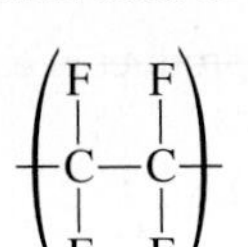

b. 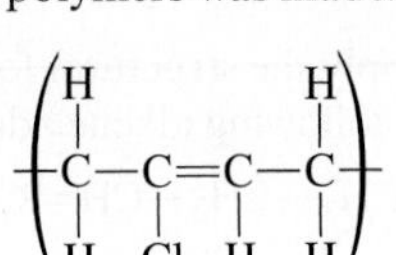

c. 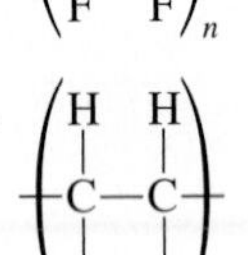

d. 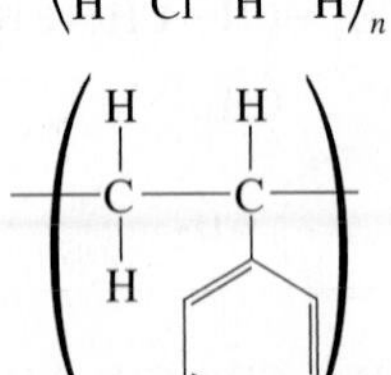

2.84 Draw the structural formula of the monomer(s) from which each of the following polymers was made.

a. 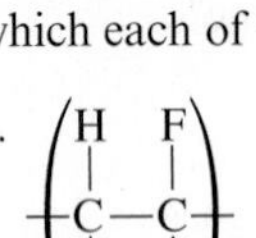

b. 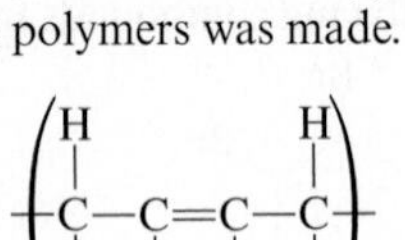

c. 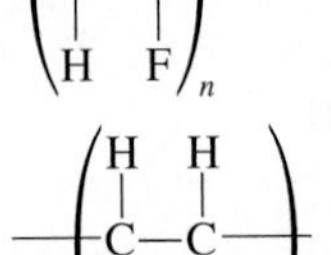

d. 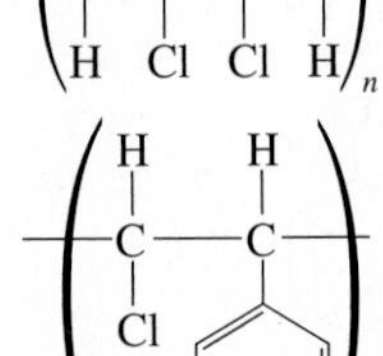

2.85 Draw the "start" (the first three repeating units) of the structural formula of the addition polymers made from the following monomers.
a. Ethylene
b. Vinyl chloride
c. 1,2-Dichloroethene
d. 1-Chloroethene

2.86 Draw the "start" (the first three repeating units) of the structural formula of the addition polymers made from the following monomers.
a. Propylene
b. 1,1,2,2-Tetrafluoroethene
c. 2-Methyl-1-propene
d. 1,2-Dichloroethylene

Alkynes (Section 2.11)

2.87 What is the general molecular formula for an alkyne in which two carbon–carbon triple bonds are present?

2.88 What is the general molecular formula for a cycloalkyne in which one carbon–carbon triple bond is present?

2.89 Assign an IUPAC name to each of the following unsaturated hydrocarbons.
a. $CH_3-CH_2-CH_2-CH_2-C\equiv CH$
b. 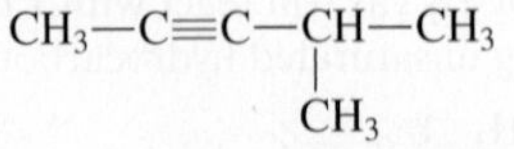
c. 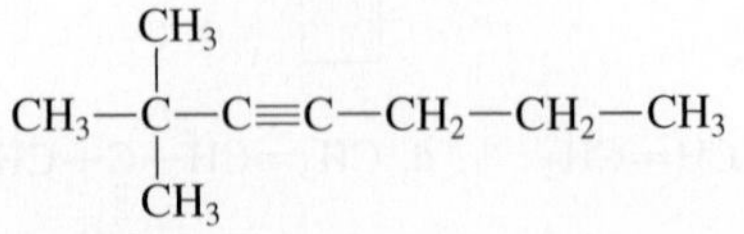
d.

2.90 Assign an IUPAC name to each of the following unsaturated hydrocarbons.
a. 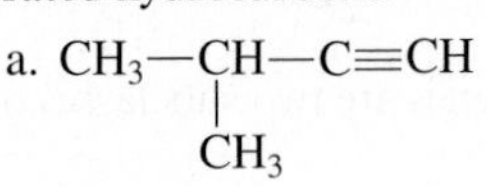
b. 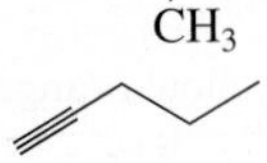
c. 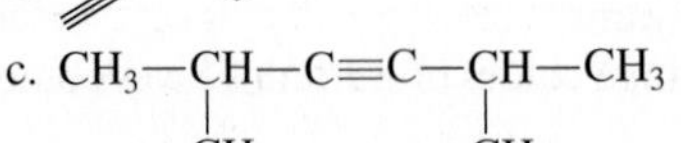
d. 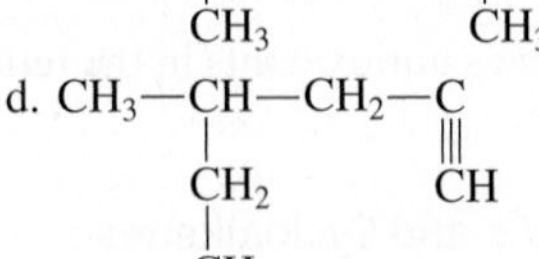

2.91 Draw skeletal structural formulas and give the IUPAC names for the three possible alkyne isomers with the molecular formula C_5H_8.

2.92 Draw skeletal structural formulas and give the IUPAC names for the seven possible alkyne isomers with the molecular formula C_6H_{10}. (Three of the constitutional isomers are hexynes, three are pentynes, and one is a butyne.)

2.93 Why is *cis–trans* isomerism not possible for an alkyne?

2.94 What are the bond angles about the triple bond in an alkyne?

2.95 Contrast alkynes and alkenes in terms of general physical properties.

2.96 Contrast alkynes and alkenes in terms of general chemical properties.

2.97 Supply the condensed structural formula of the product in each of the following alkyne addition reactions.
a. $CH\equiv CH + 2H_2 \xrightarrow{Ni} ?$
b. $CH_3-C\equiv CH + 2Br_2 \longrightarrow ?$
c. $CH_3-C\equiv CH + 2HBr \longrightarrow ?$
d. $CH\equiv CH + 1HCl \longrightarrow ?$

2.98 Supply the condensed structural formula of the product in each of the following alkyne addition reactions.
a. $CH_3-C\equiv C-CH_3 + 2Br_2 \longrightarrow ?$
b. $CH_3-C\equiv C-CH_3 + 2HBr \longrightarrow ?$
c. $CH\equiv C-CH_2-CH_3 + 1H_2 \xrightarrow{Ni} ?$
d. $CH\equiv C-CH_3 + 1HCl \longrightarrow ?$

▲2.99 Draw a condensed structural formula for each of the following hydrocarbons.
a. 5-methyl-2-hexyne
b. 2-methyl-2-butene
c. 1,6-heptadiyne
d. 3-penten-1-yne

▲2.100 Draw a condensed structural formula for each of the following hydrocarbons.
a. 2-butyne
b. 3,4-dimethyl-2-pentene
c. 1,6-heptadiene
d. 1-buten-3-yne

▲2.101 How many carbon atoms are present in a molecule of each of the following compounds?
a. Allyl chloride
b. Acetylene
c. Dimethylacetylene
d. 3-cyclobutyl-1-hexyne

▲2.102 How many carbon atoms are present in a molecule of each of the following compounds?
a. Vinyl bromide
b. Ethylene
c. Isopropylacetylene
d. 1-cyclobutyl-3-hexene

Aromatic Hydrocarbons (Section 2.12)

2.103 Draw a structural representation for the functional group present in an aromatic hydrocarbon.

2.104 A circle (ring) within a hexagon is often used to represent an aromatic carbon ring. What does the circle represent?

2.105 What is the shortcoming of representing the bonding in a benzene ring using alternating single and double carbon–carbon bonds?

2.106 What is a *delocalized* bond?

Nomenclature for Aromatic Hydrocarbons (Section 2.13)

2.107 Assign an IUPAC name to each of the following disubstituted benzenes. Use numbers rather than prefixes to locate the substituents on the benzene ring.

a. 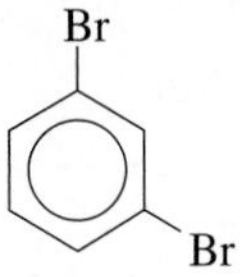

b. 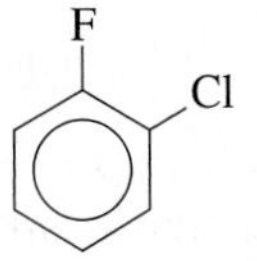

c. 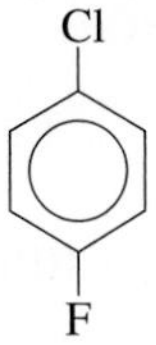

d. 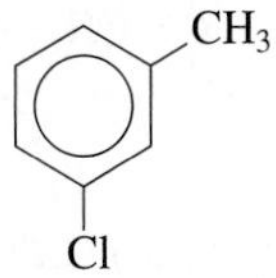

2.108 Assign an IUPAC name to each of the following disubstituted benzenes. Use numbers rather than prefixes to locate the substituents on the benzene ring.

a. 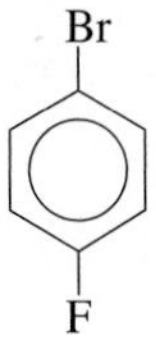

b. 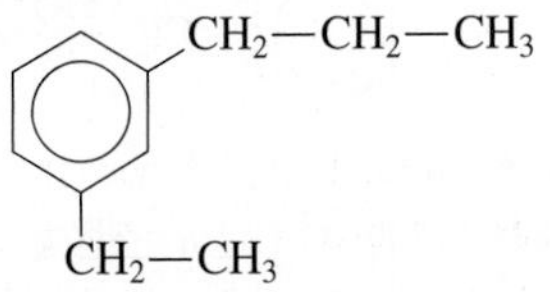

c. 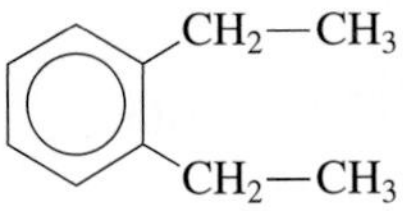

d. 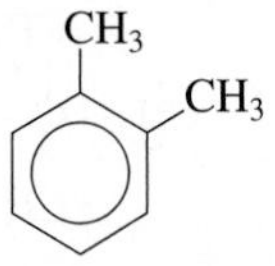

2.109 Assign each of the compounds in Problem 2.107 an IUPAC name in which the substituents on the benzene ring are located using the *ortho-*, *meta-*, *para-* prefix system.

2.110 Assign each of the compounds in Problem 2.108 an IUPAC name in which the substituents on the benzene ring are located using the *ortho-*, *meta-*, *para-* prefix system.

2.111 Assign an IUPAC name to each of the following substituted benzenes.

a. 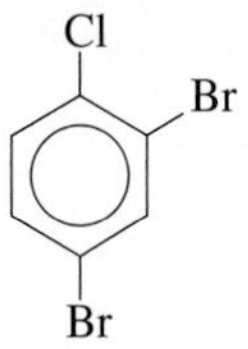

b. 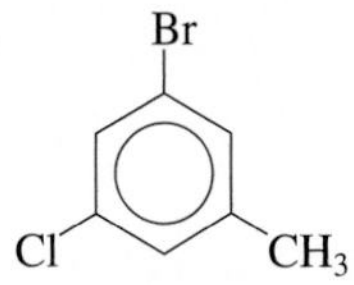

c. 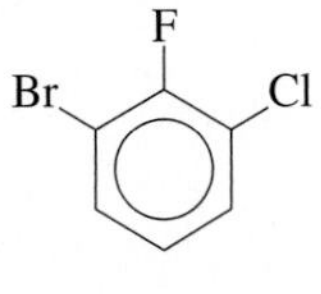

d. 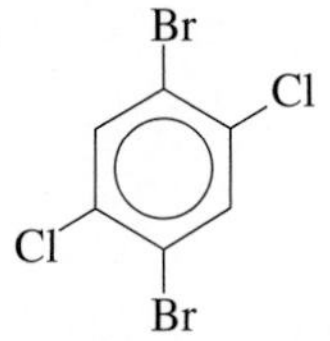

2.112 Assign an IUPAC name to each of the following substituted benzenes.

a. 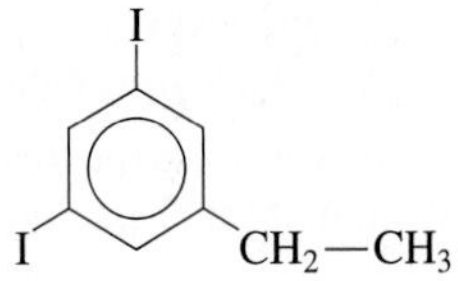

b. 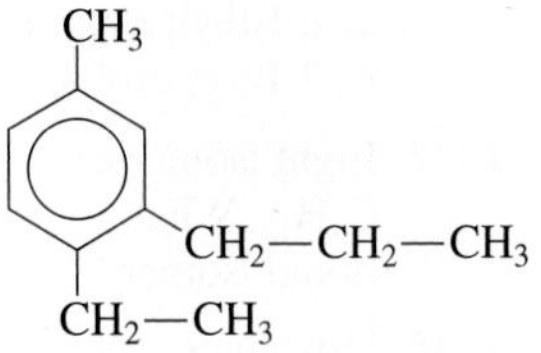

c. 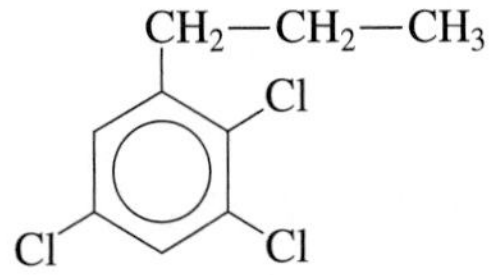

d. 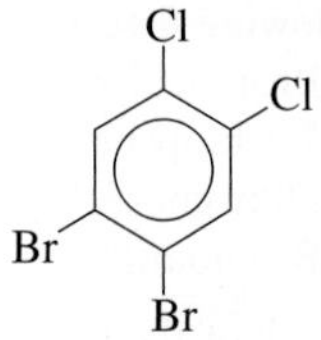

2.113 Assign an IUPAC name to each of the following compounds in which the benzene ring is treated as a substituent.

a. 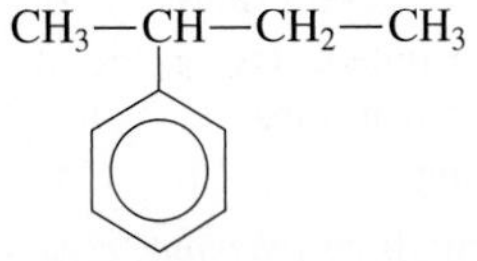

b. 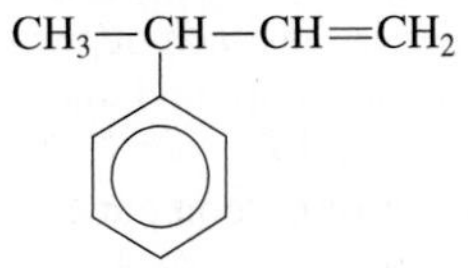

c. 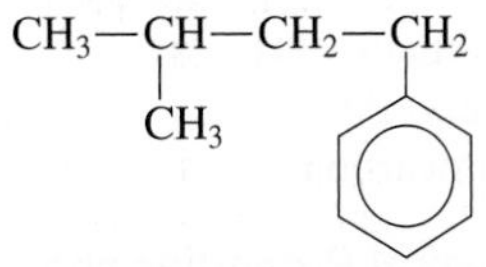

d. 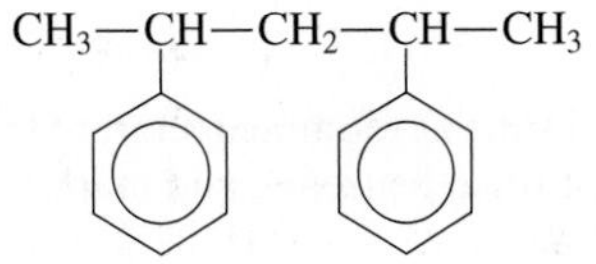

2.114 Assign an IUPAC name to each of the following compounds in which the benzene ring is treated as a substituent.

a. 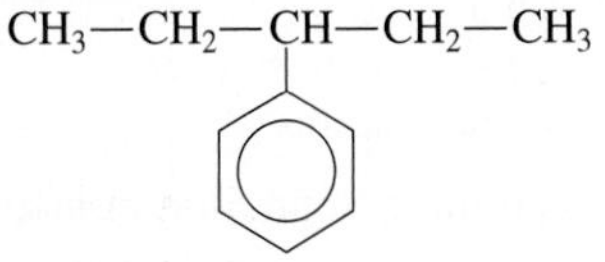

b. 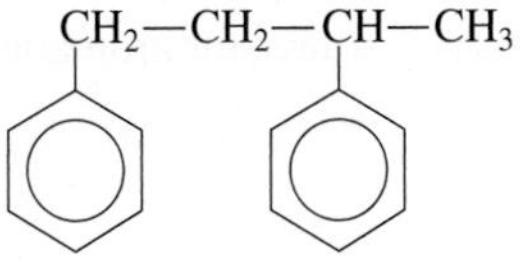

c. 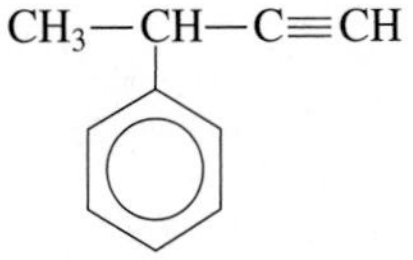

d. 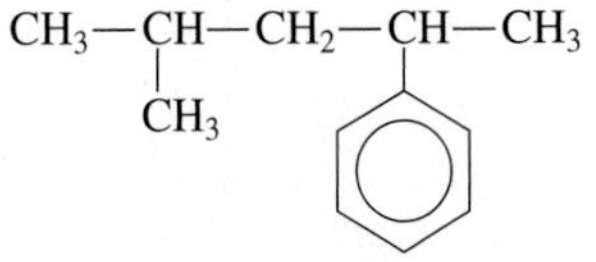

2.115 Write a structural formula for each of the following compounds.

a. 1,3-Diethylbenzene
b. *o*-Xylene
c. *p*-Ethyltoluene
d. Phenylbenzene

2.116 Write a structural formula for each of the following compounds.
a. *o*-Ethylpropylbenzene b. *m*-Xylene
c. 2-Bromotoluene d. 2-Phenylpropane

2.117 Eight isomeric substituted benzenes have the formula C_9H_{12}. What are the IUPAC names for these constitutional isomers?

2.118 How many constitutional isomers fit each of the following descriptions?
a. Bromochlorobenzenes
b. Trichlorobenzenes
c. Dibromodichlorobenzenes
d. Bromoanthracenes

▲2.119 Which of the four compounds *cyclohexane, cyclohexene, 1,3-cyclohexadiene,* and *benzene* has each of the following characteristics? More than one compound may be correct in a given situation.
a. Contains a 6-membered carbon ring
b. Has the generalized formula C_nH_{2n-2}
c. Undergoes substitution reactions
d. Has delocalized bonding

▲**2.120** Which of the four compounds *cyclohexane, cyclohexene, 1,3-cyclohexadiene,* and *benzene* has each of the following characteristics? More than one compound may be correct in a given situation.
a. Contains the same number of carbon and hydrogen atoms
b. Has the generalized formula C_nH_{2n-4}
c. Undergoes addition reactions
d. Is an unsaturated hydrocarbon

Aromatic Hydrocarbons: Physical Properties and Sources (Section 2.14)

2.121 What is the physical state at room temperature for benzene, monosubstituted benzenes, and many disubstituted benzenes?

2.122 Indicate whether or not each of the following is a general physical property of aromatic hydrocarbons.
a. Soluble in water
b. Less dense than water
c. Good solvent for nonpolar substances
d. All solids at room temperatures

2.123 What is currently the primary source for aromatic hydrocarbons?

2.124 What used to be the primary source for aromatic hydrocarbons?

Chemical Reactions of Aromatic Hydrocarbons (Section 2.15)

2.125 For each of the following classes of compounds, indicate whether addition or substitution is the most characteristic reaction.
a. Alkanes b. Dienes
c. Alkylbenzenes d. Cycloalkenes

2.126 For each of the following classes of compounds, indicate whether addition or substitution is the most characteristic reaction.
a. Alkynes b. Cycloalkanes
c. Aromatic hydrocarbons d. Saturated hydrocarbons

2.127 Complete the following reaction equations by supplying the formula of the missing reactant or product.

a.
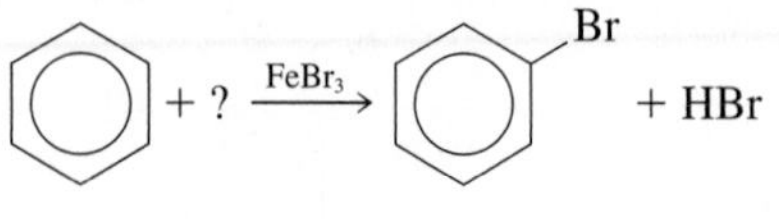

b.
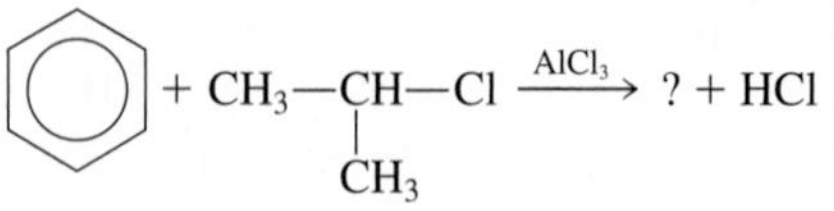

c.
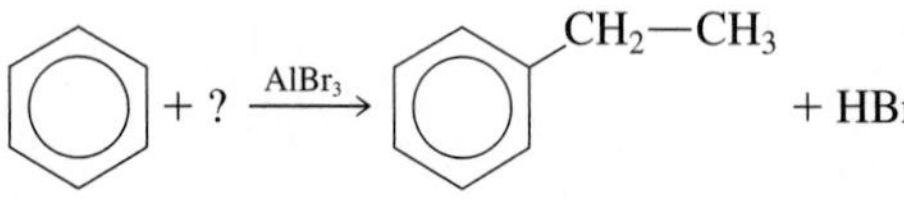

2.128 Complete the following reaction equations by supplying the formula of the missing reactant, product, or catalyst.

a.
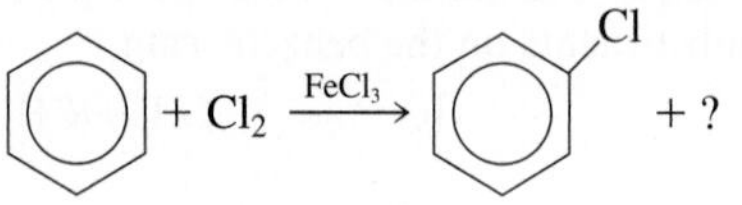

b.
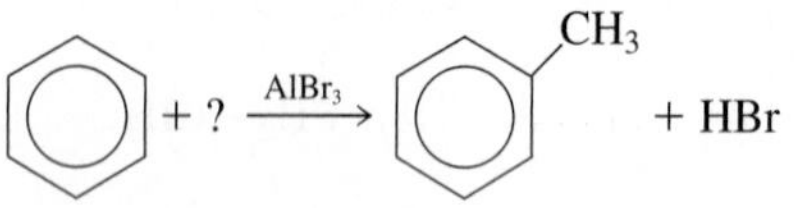

c.
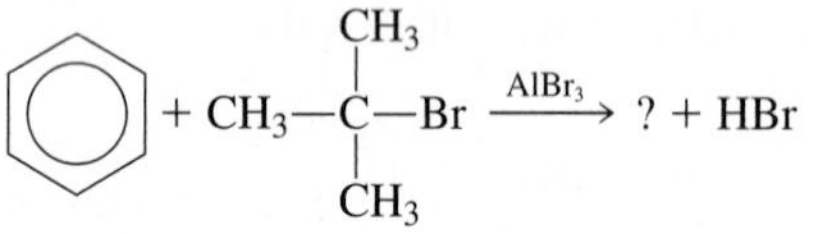

Fused-Ring Aromatic Hydrocarbons (Section 2.16)

2.129 What is the general characteristic associated with two carbon rings that are *fused* together?

2.130 What is the structural formula for naphthalene, the simplest fused-ring aromatic hydrocarbon?

Alcohols, Phenols, and Ethers

3

© Digoit Olivier/Peter Arnold/Photolibrary

Two alcohols, 1-octanol and 3-octanol, contribute to the distinctive flavor of mushrooms.

CHAPTER OUTLINE

OWL

Sign in to OWL at **www.cengage.com/owl** to view tutorials and simulations, develop problem-solving skills, and complete online homework assigned by your professor.

This chapter is the first of three that consider hydrocarbon derivatives with *oxygen-containing functional groups.* Many biochemically important molecules contain carbon atoms bonded to oxygen atoms.

In this chapter, hydrocarbon derivatives whose functional groups contain one oxygen atom participating in two single bonds (alcohols, phenols, and ethers) are considered. Chapter 4 focuses on derivatives whose functional groups have one oxygen atom participating in a double bond (aldehydes and ketones), and Chapter 5 examines functional groups that contain two oxygen atoms, one participating in single bonds and the other in a double bond (carboxylic acids, esters, and other acid derivatives).

3.1 Bonding Characteristics of Oxygen Atoms in Organic Compounds

An understanding of the bonding characteristics of the oxygen atom is a prerequisite for the study of compounds with oxygen-containing functional groups. Normal bonding behavior for oxygen atoms in such functional groups is the formation of two covalent bonds. Oxygen is a member of Group VIA of the periodic table and thus possesses six valence electrons. To complete its

$CH_3—OH$
One-carbon alcohol

$CH_3—CH_2—OH$
Two-carbon alcohol

$CH_3—CH_2—CH_2—OH$
Three-carbon alcohol

Figure 3.1 Space-filling models for the three simplest unbranched-chain alcohols: methyl alcohol, ethyl alcohol, and propyl alcohol.

octet by electron sharing, an oxygen atom can form either two single bonds or a double bond.

$$:\ddot{O}— \qquad :\ddot{O}=$$

Two single bonds — One double bond

Thus, in organic chemistry, carbon forms four bonds, hydrogen forms one bond, and oxygen forms two bonds.

$$—\overset{|}{\underset{|}{C}}— \qquad H— \qquad :\overset{|}{\ddot{O}}—$$

—C—	H—	:O—
4 valence electrons, 4 covalent bonds, no nonbonding electron pairs	1 valence electron, 1 covalent bond, no nonbonding electron pairs	6 valence electrons, 2 covalent bonds, 2 nonbonding electron pairs

3.2 Structural Characteristics of Alcohols

Alcohols are the first type of hydrocarbon derivative containing a single oxygen atom to be considered. They have the generalized formula

$$R—OH$$

An **alcohol** *is an organic compound in which an —OH group is bonded to a saturated carbon atom.* A *saturated* carbon atom is a carbon atom that is bonded to four other atoms.

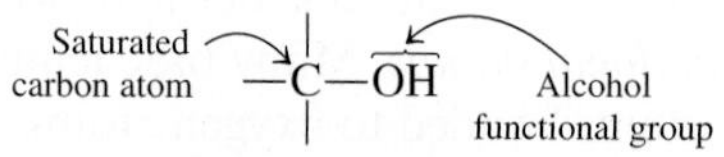

The hydroxyl group (—OH) should not be confused with the *hydroxide ion* (OH^-) that was repeatedly encountered in the general chemistry chapters of this text. Alcohols are not *hydroxides.* Hydroxides are ionic compounds that contain the OH^- polyatomic ion. Alcohols are not ionic compounds. In an alcohol, the —OH group, which is not an ion, is *covalently* bonded to a saturated carbon atom.

The —OH group, the functional group that is characteristic of an alcohol, is called a *hydroxyl group.* A **hydroxyl group** *is the —OH functional group.*

Examples of condensed structural formulas for alcohols include

$$CH_3—OH \qquad CH_3—CH_2—OH \qquad CH_3—CH_2—CH_2—OH$$

Space-filling models for these three alcohols, the simplest alcohols possible that have unbranched carbon chains, are given in Figure 3.1.

Alcohols may be viewed structurally as being alkyl derivatives of water in which a hydrogen atom has been replaced by an alkyl group.

$$H—\ddot{\underset{..}{O}}—H \qquad R—\ddot{\underset{..}{O}}—H$$

Water — An alcohol

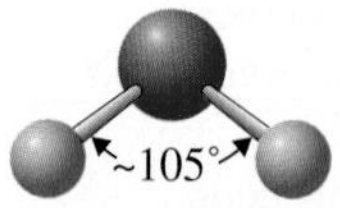

Water (HOH)

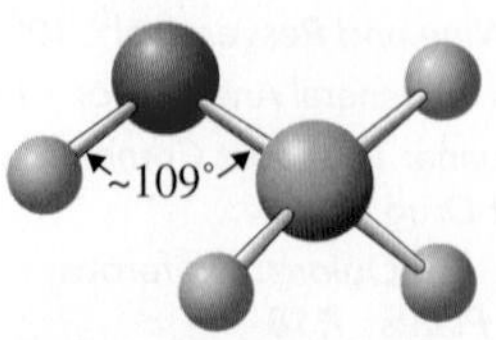

Methyl alcohol (CH_3OH)

Figure 3.2 The similar shapes of water and methanol. Methyl alcohol may be viewed structurally as an alkyl derivative of water.

Figure 3.2 shows the similarity in oxygen bond angles for water and $CH_3—OH$, the simplest alcohol.

Alcohols may also be viewed structurally as hydroxyl derivatives of alkanes in which a hydrogen atom has been replaced by a hydroxyl group.

$$R—H \qquad R—OH$$

An alkane — An alcohol

3.3 Nomenclature for Alcohols

Common names exist for alcohols with simple (generally C_1 through C_4) alkyl groups. A common name is assigned using the following rules:

Rule 1: *Name all of the carbon atoms of the molecule as a single* alkyl group.

Rule 2: *Add the word* alcohol, *separating the words with a space.*

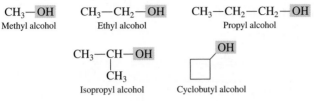

CH_3-OH Methyl alcohol

CH_3-CH_2-OH Ethyl alcohol

$CH_3-CH_2-CH_2-OH$ Propyl alcohol

$CH_3-\underset{\displaystyle CH_3}{\underset{|}{CH}}-OH$ Isopropyl alcohol

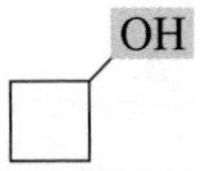

Cyclobutyl alcohol

IUPAC rules for naming alcohols that contain a single hydroxyl group follow.

Rule 1: *Name the longest carbon chain to which the hydroxyl group is attached.* The chain name is obtained by dropping the final *-e* from the alkane name and adding the suffix *-ol.*

Rule 2: *Number the chain starting at the end nearest the hydroxyl group, and use the appropriate number to indicate the position of the —OH group.* (In numbering of the longest carbon chain, the hydroxyl group has precedence over ("outranks") double and triple bonds, as well as over alkyl, cycloalkyl, and halogen substituents.)

Rule 3: Name and locate any other substituents present.

Rule 4: *In alcohols where the —OH group is attached to a carbon atom in a ring, the ring is numbered beginning with the —OH group.* Numbering then proceeds in a clockwise or counterclockwise direction to give the next substituent the lower number. The number "1" (for the hydroxyl group) is omitted from the name since by definition the hydroxyl-bearing carbon is carbon 1.

Table 3.1 gives both IUPAC and common names for monohydroxy alcohols that contain four or fewer carbon atoms.

Line-angle structural formulas for selected simple alcohols:

Propyl alcohol
(1-propanol)

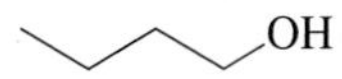

Butyl alcohol
(1-butanol)

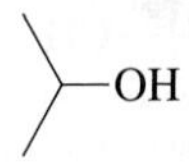

Isopropyl alcohol
(2-propanol)

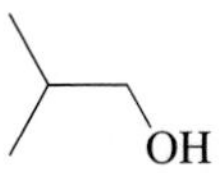

Isobutyl alcohol
(2-methyl-1-propanol)

Table 3.1 IUPAC and Common Names of Monohydroxy Alcohols That Contain Up to Four Carbon Atoms

Formula	IUPAC Name	Common Name
One carbon atom (CH_3OH)		
CH_3-OH	methanol	methyl alcohol
Two carbon atoms (C_2H_5OH)		
CH_3-CH_2-OH	ethanol	ethyl alcohol
Three carbon atoms (C_3H_7OH); two constitutional isomers exist		
$CH_3-CH_2-CH_2-OH$	1-propanol	propyl alcohol
$CH_3-\underset{\displaystyle CH_3}{\underset{\vert}{CH}}-OH$	2-propanol	isopropyl alcohol
Four carbon atoms (C_4H_9OH); four constitutional isomers exist		
$CH_3-CH_2-CH_2-CH_2-OH$	1-butanol	butyl alcohol
$CH_3-\underset{\displaystyle CH_3}{\underset{\vert}{CH}}-CH_2-OH$	2-methyl-1-propanol	isobutyl alcohol
$CH_3-CH_2-\underset{\displaystyle CH_3}{\underset{\vert}{CH}}-OH$	2-butanol	*sec*-butyl alcohol
$CH_3-\overset{\displaystyle CH_3}{\overset{\vert}{\underset{\displaystyle CH_3}{\underset{\vert}{C}}}}-OH$	2-methyl-2-propanol	*tert*-butyl alcohol

EXAMPLE 3.1 Determining IUPAC Names for Alcohols

Name the following alcohols, utilizing IUPAC nomenclature rules.

a.

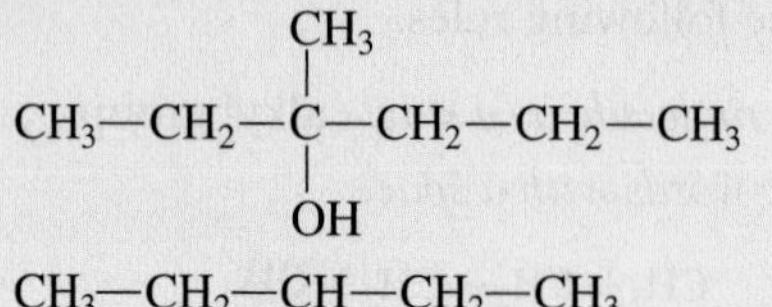

b.

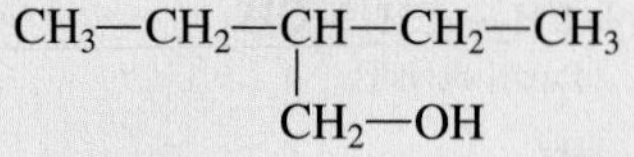

c.

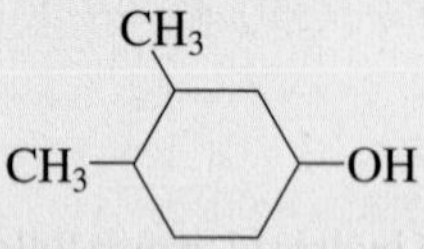

d.

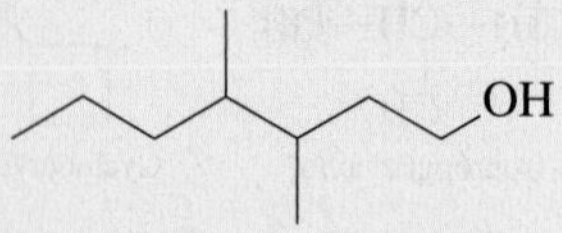

Solution

a. The longest carbon chain that contains the alcohol functional group has six carbons. Changing the *-e* to *-ol,* hexane becomes *hexanol.* Numbering the chain from the end nearest the —OH group identifies carbon number 3 as the location of both the —OH group and a methyl group. The complete name is 3-methyl-3-hexanol.

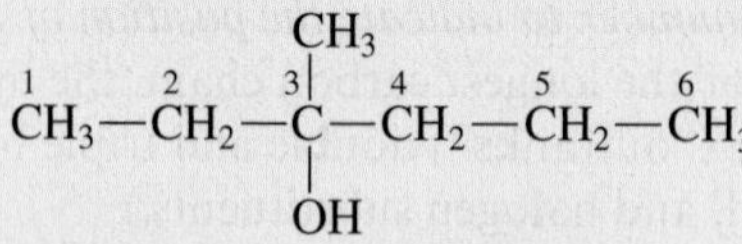

b. The longest carbon chain containing the —OH group has four carbon atoms. It is numbered from the end closest to the —OH group as follows:

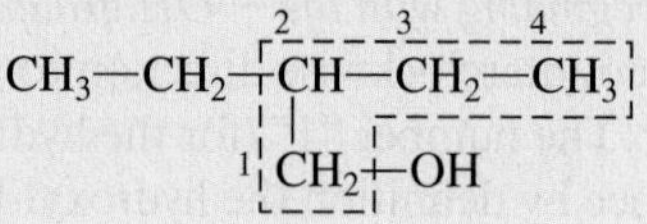

The base name is 1-butanol. The complete name is 2-ethyl-1-butanol.

c. This alcohol is a cyclohexanol. The carbon to which the —OH group is attached is assigned the number 1. The complete name for this alcohol is 3,4-dimethylcyclohexanol. Note that the number 1 is not part of the name.

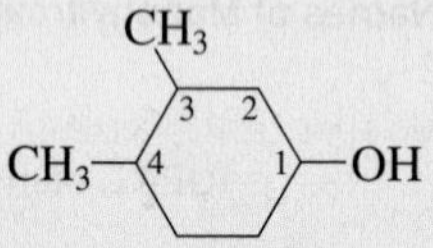

d. This alcohol is a dimethylheptanol. Numbering from right to left, the location of the hydroxyl group is 1, and locants for the methyl groups are 3 and 4. The complete IUPAC name is 3,4-dimethyl-1-heptanol.

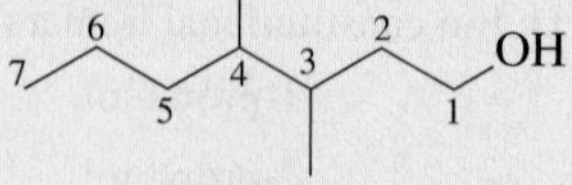

In the naming of alcohols with *unsaturated* carbon chains, two endings are needed: one for the double or triple bond and one for the hydroxyl group. The *-ol* suffix always comes last in the name; that is, unsaturated alcohols are named as *alkenols* or *alkynols.*

$$\overset{3}{CH_2}=\overset{2}{CH}-\overset{1}{CH_2}-OH$$

2-Propen-1-ol
(common name: allyl alcohol)

The contrast between IUPAC and common names for alcohols is as follows:

IUPAC (one word)
alkanol
ethanol

Common (two words)
alkyl alcohol
ethyl alcohol

Practice Exercise 3.1

Name the following alcohols utilizing IUPAC nomenclature rules.

a.

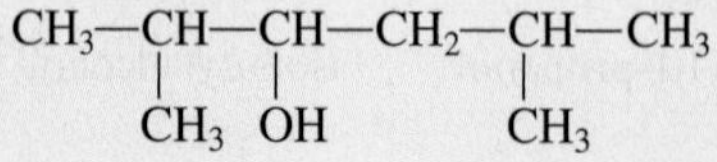

b.

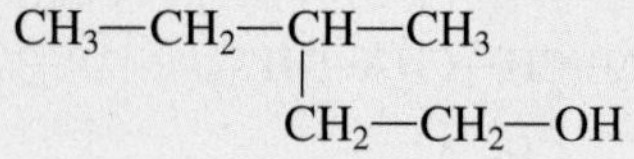

c.

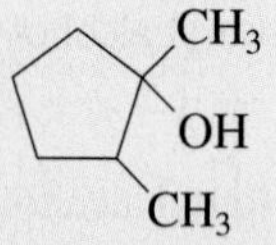

d.

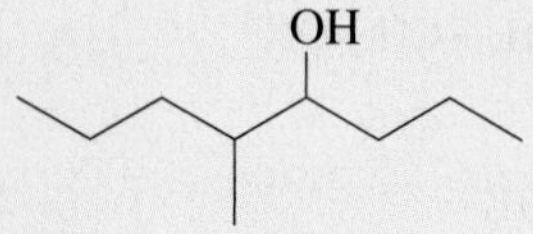

Answers: **a.** 2,5-Dimethyl-3-hexanol; **b.** 3-Methyl-1-pentanol; **c.** 1,2-Dimethylcyclopentanol; **d.** 5-Methyl-4-octanol

Alcohols with More Than One Hydroxyl Group

Polyhydroxy alcohols—alcohols that possess more than one hydroxyl group—can be named with only a slight modification of the preceding IUPAC rules. An alcohol in which two hydroxyl groups are present is named as a *diol,* one containing three hydroxyl groups is named as a *triol,* and so on. In these names for diols, triols, and so forth, the final *-e* of the parent alkane name is retained for pronunciation reasons.

A hydroxyl group as a substituent in a molecule is called a hydroxy group; an *-oxy* rather than an *-oxyl* ending is used.

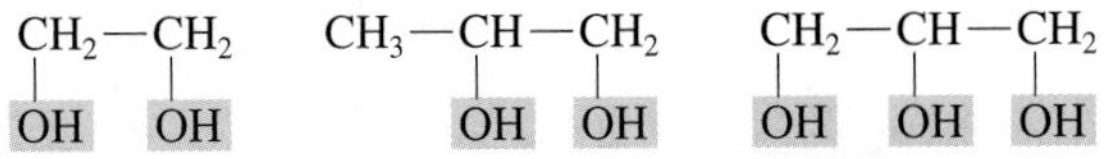

1,2-Ethanediol 1,2-Propanediol 1,2,3-Propanetriol

The first two of the preceding compounds have the common names *ethylene glycol* and *propylene glycol.* These two alcohols are synthesized, respectively, from the alkenes ethylene and propylene (Section 2.3); hence the common names.

3.4 Isomerism for Alcohols

Constitutional isomerism is possible for alcohols containing three or more carbon atoms. As with alkenes (Section 2.5), both *skeletal* isomers and *positional* isomers are possible. For monohydroxy saturated alcohols, there are two C_3 isomers, four C_4 isomers, and eight C_5 isomers. Structures for the C_3 and C_4 isomers are found in Table 3.1. The C_5 isomers are

Addition of a functional group greatly increases constitutional isomer possibilities. There are 75 alkane isomers with the formula $C_{10}H_{22}$ and 507 alcohol isomers with the formula $C_{10}H_{21}OH$.

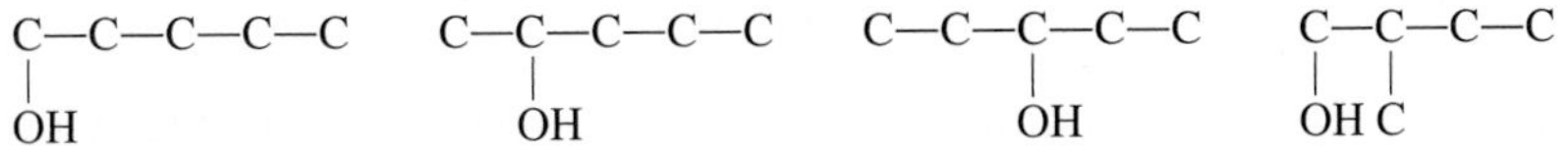

1-Pentanol 2-Pentanol 3-Pentanol 2-Methyl-1-butanol

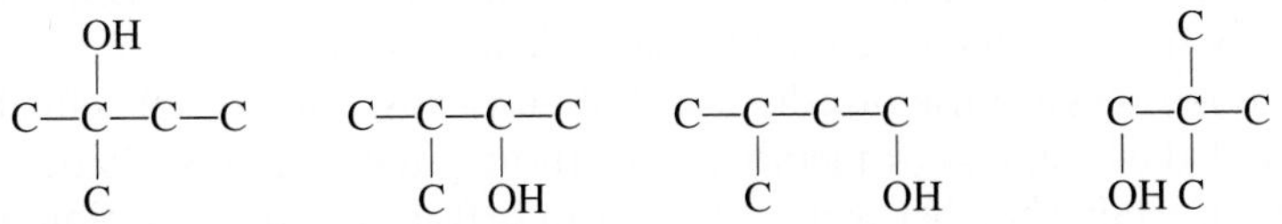

2-Methyl-2-butanol 3-Methyl-2-butanol 3-Methyl-1-butanol 2,2-Dimethyl-1-propanol

The three pentanols are positional isomers as are the four methylbutanols.

3.5 Important Commonly Encountered Alcohols

In this section, the properties and uses of six commonly encountered alcohols are considered: methyl, ethyl, and isopropyl alcohols (all monohydroxy alcohols), ethylene glycol and propylene glycol (both diols), and glycerol (a triol).

Methyl Alcohol (Methanol)

Methyl alcohol, with one carbon atom and one —OH group, is the simplest alcohol. It is a colorless liquid that has excellent solvent properties, and it is the solvent of choice for many shellacs and varnishes.

Specially designed internal combustion engines can operate using methyl alcohol as a fuel. For forty years, from 1965 to 2005, race cars at the Indianapolis Speedway were fueled with methyl alcohol (Figure 3.3). A major reason for the switch from gasoline to methyl alcohol relates to fires accompanying crashes. Methyl alcohol fires are easier to put out than gasoline fires because water mixes with and dilutes methyl alcohol. In 2006, a transition from methyl alcohol fuel use to ethyl alcohol race car fuel began, which is now complete. Reasons for the switch to ethyl alcohol are given in the discussion about ethyl alcohol later in this section.

Methyl alcohol is sometimes called *wood alcohol,* terminology that draws attention to an early method for its preparation—the heating of wood to a high

Jonathan Ferrey/Getty Images

Figure 3.3 Racing cars at the Indianapolis Speedway were fueled with methyl alcohol from 1965 to 2005.

temperature in the absence of air. Today, nearly all methyl alcohol is produced via the reaction between H_2 and CO.

$$CO + 2H_2 \xrightarrow[300°C - 400°C,\ 200\ atm]{ZnO-Cr_2O_3} CH_3-OH$$

Methyl alcohol poisoning is treated with ethyl alcohol, which ties up the enzyme that oxidizes methyl alcohol to its toxic metabolites. Ethyl alcohol has 10 times the affinity for the alcohol dehydrogenase enzyme that methyl alcohol has. This situation is considered further in Section 10.7.

Drinking methyl alcohol is very dangerous. Within the human body, methyl alcohol is oxidized by the liver enzyme *alcohol dehydrogenase* to the toxic metabolites formaldehyde and formic acid.

$$CH_3-OH \xrightarrow[dehydrogenase]{Alcohol} \underset{Formaldehyde}{H-\overset{O}{\overset{\|}{C}}-H} \xrightarrow[oxidation]{Further} \underset{Formic\ acid}{H-\overset{O}{\overset{\|}{C}}-OH}$$

Formaldehyde can cause blindness (temporary or permanent). Formic acid causes acidosis. Ingesting as little as 1 oz (30 mL) of methyl alcohol can cause optic nerve damage.

Ethyl Alcohol (Ethanol)

Many people imagine ethanol to be relatively nontoxic and methanol to be extremely toxic. Actually, their toxicities differ by a factor of only 2. Typical fatal doses for adults are about 100 mL for methanol and about 200 mL for ethanol, although smaller doses of methanol may damage the optic nerve.

The alcohol content of strong alcoholic beverages is often stated in terms of proof. *Proof* is twice the percentage of alcohol. This system dates back to the seventeenth century and is based on the fact that a 50% (v/v) alcohol–water mixture will burn. Its flammability was *proof* that a liquor had not been watered down.

Ethyl alcohol, the two-carbon monohydroxy alcohol, is the alcohol present in alcoholic beverages and is commonly referred to simply as alcohol or *drinking alcohol.* Like methyl alcohol, ethyl alcohol is oxidized in the human body by the liver enzyme *alcohol dehydrogenase.*

$$CH_3-CH_2-OH \xrightarrow[dehydrogenase]{Alcohol} \underset{Acetaldehyde}{CH_3-\overset{O}{\overset{\|}{C}}-H} \xrightarrow[oxidation]{Further} \underset{Acetic\ acid}{CH_3-\overset{O}{\overset{\|}{C}}-OH}$$

Acetaldehyde, the first oxidation product, is largely responsible for the symptoms of hangover. The odors of both acetaldehyde and acetic acid are detected on the breath of someone who has consumed a large amount of alcohol. Ethyl alcohol oxidation products are less toxic than those of methyl alcohol.

Long-term excessive use of ethyl alcohol may cause undesirable effects such as cirrhosis of the liver, loss of memory, and strong physiological addiction. Links have also been established between certain birth defects and the ingestion of ethyl alcohol by women during pregnancy (fetal alcohol syndrome).

Ethyl alcohol can be produced by yeast fermentation of sugars found in plant extracts (see Figure 3.4). The synthesis of ethyl alcohol in this manner, from grains

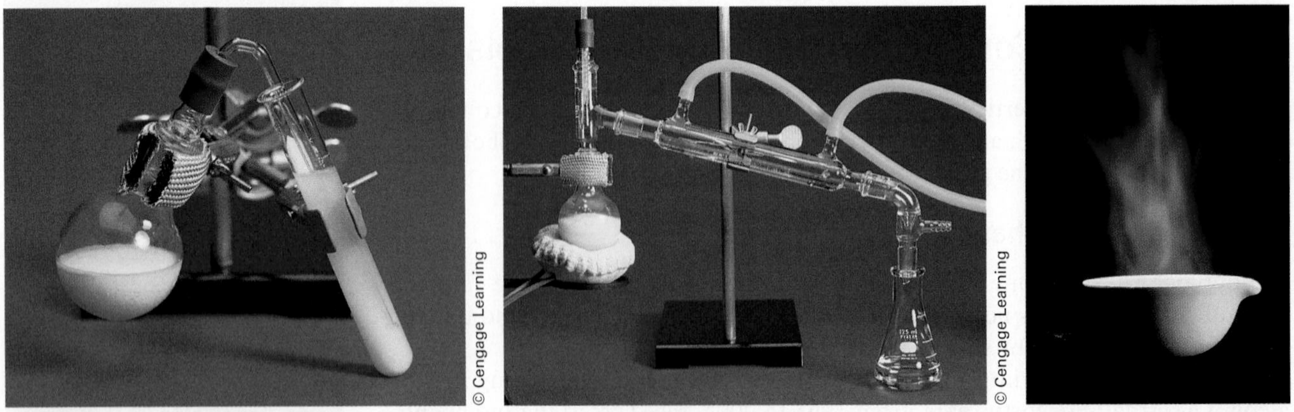

a A small amount of yeast has been added to the aqueous sugar solution in the flask. Yeast enzymes catalyze the decomposition of sugar to ethanol and carbon dioxide, CO_2. The CO_2 is bubbling through lime water, $Ca(OH)_2$, producing calcium carbonate, $CaCO_3$.

b More concentrated ethanol is produced from the solution in the flask by collecting the fraction that boils at about 78°C.

c Concentrated ethanol (50% v/v) burns when it is ignited.

Figure 3.4 An experimental setup for preparing ethyl alcohol by fermentation.

such as corn, rice, and barley, is the reason why ethyl alcohol is often called *grain alcohol.*

$$\underset{\text{(glucose)}}{\underset{\text{Sugar}}{C_6H_{12}O_6}} \xrightarrow[\text{Fermentation}]{\text{Yeast}} \underset{\text{Ethyl alcohol}}{2CH_3{-}CH_2{-}OH} + 2CO_2$$

Fermentation is the process by which ethyl alcohol for alcoholic beverages is produced. The maximum concentration of ethyl alcohol obtainable by fermentation is about 18% (v/v) because yeast enzymes cannot function in stronger alcohol solutions. Alcoholic beverages with a higher concentration of alcohol than this are prepared by either distillation or fortification with alcohol obtained by the distillation of another fermentation product. Table 3.2 lists the alcohol content of common alcoholic beverages and of selected common household products and over-the-counter drug products.

Table 3.2 Ethyl Alcohol Content (Volume Percent) of Common Alcoholic Beverages, Household Products, and Over-the-Counter Drugs

Product Type	Product	Volume Percent Ethyl Alcohol
Alcoholic Beverages	beer	3.2–9
	wine (unfortified)	12
	brandy	40–45
	whiskey	45–55
	rum	45
Flavorings	vanilla extract	35
	almond extract	50
Cough and Cold Remedies	Pertussin Plus	25
	Nyquil	25
	Dristan	12
	Vicks 44	10
	Robitussin, DM	1.4
Mouthwashes	Listerine	25
	Scope	18
	Colgate 100	17
	Cepacol	14
	Lavoris	5

Denatured alcohol is ethyl alcohol that has been rendered unfit to drink by the addition of small amounts of toxic substances (denaturing agents). Almost all of the ethyl alcohol used for industrial purposes is denatured alcohol.

Most ethyl alcohol used in industry is prepared from ethene via a hydration reaction (Section 2.9).

$$CH_2{=}CH_2 + H_2O \xrightarrow{\text{Catalyst}} CH_3{-}CH_2{-}OH$$

The reaction produces a product that is 95% alcohol and 5% water. In applications where water does interfere with its use, the mixture is treated with a dehydrating agent to produce 100% ethyl alcohol. Such alcohol, with all traces of water removed, is called *absolute alcohol.*

The largest single end use for ethyl alcohol is that of a motor vehicle fuel. Gasoline blends containing up to 10% ethyl alcohol (E10) are approved for use in all gasoline-powered vehicles manufactured in the United States. In 2011 the Environmental Protection Agency (EPA) approved the use of gasoline blends containing 15% ethyl alcohol (E15) in gasoline-power vehicles of model year 2006 or later. Blends with a greater alcohol content (up to E85) can be used in flexible-fuel vehicles that are sold as standard models by several automakers.

The ethyl alcohol fuel industry in the United States is based on the use of agricultural crops, predominantly corn. Sugars and starches present in the vegetation are fermented to produce the alcohol. Alcohol production in this manner supports local and regional economies while reducing the United States' reliance on imported petroleum. Blend use, particularly in cold weather with older and poorly maintained vehicles, also reduces carbon monoxide air pollution. On the downside, studies show that vehicle exhaust from *pure* ethanol use generates significantly greater amounts of compounds associated with smog production than does gasoline (1.7 versus 1.0 for gasoline on a relative scale). However, vehicle exhaust composition differences are minimal when *E10* fuel and gasoline are compared.

The previously mentioned switch from methyl alcohol to ethyl alcohol as a race car fuel was promoted and funded by a consortium of ethyl alcohol producers as a marketing effort to address public concerns that ethyl alcohol use in automobiles led to engine damage and poor performance. The transition from methyl to ethyl alcohol also boosts the fuel mileage of race cars. With the switch, the fuel-tank size of race cars has decreased, resulting in lower vehicle weight and decreased time required to refuel.

Isopropyl Alcohol (2-Propanol)

The "medicinal" odor associated with doctors' offices is usually that of isopropyl alcohol.

Isopropyl alcohol is one of two three-carbon monohydroxy alcohols; the other is propyl alcohol. A 70% isopropyl alcohol–30% water solution is marketed as *rubbing alcohol.* Isopropyl alcohol's rapid evaporation rate creates a dramatic cooling effect when it is applied to the skin, hence its use for alcohol rubs to combat high body temperature. It also finds use in cosmetics formulations such as after-shave lotion and hand lotions.

Isopropyl alcohol has a bitter taste. Its toxicity is twice that of ethyl alcohol, but it causes few fatalities because it often induces vomiting and thus doesn't stay down long enough to be fatal. In the body, it is oxidized to acetone.

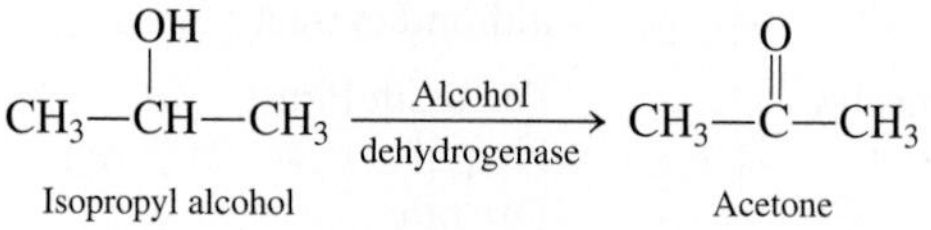

Large amounts (about 150 mL) of ingested isopropyl alcohol can be fatal; death occurs from paralysis of the central nervous system.

Ethylene Glycol (1,2-Ethanediol) and Propylene Glycol (1,2-Propanediol)

The ethylene glycol and propylene glycol used in antifreeze formulations are colorless and odorless; the color and odor of antifreezes come from additives for rust protection and the like.

Ethylene glycol and propylene glycol are synthesized from ethylene and propylene, respectively, hence their common names.

Ethylene glycol and propylene glycol are the two simplest alcohols possessing two —OH groups. Besides being diols, they are also classified as glycols. A **glycol** *is a diol in which the two —OH groups are on adjacent carbon atoms.*

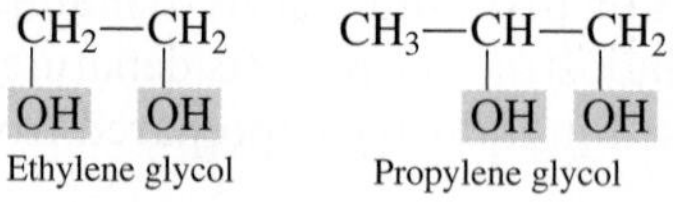

Both of these glycols are colorless, odorless, high-boiling liquids that are completely miscible with water. Their major uses are as the main ingredient in automobile "year-round" antifreeze and airplane "de-icers" (Figure 3.5) and as a starting material for the manufacture of polyester fibers (Section 5.18).

Ethylene glycol is extremely toxic when ingested. In the body, liver enzymes oxidize it to oxalic acid.

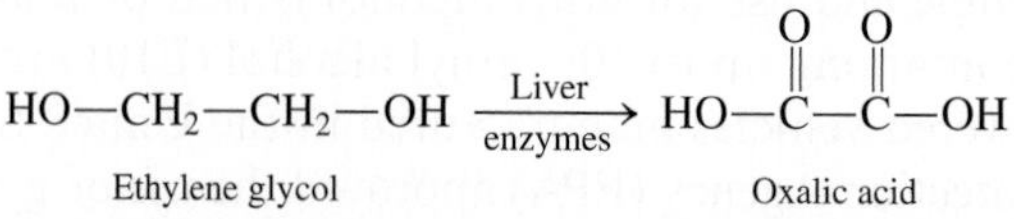

Oxalic acid, as a calcium salt, crystallizes in the kidneys, which leads to renal problems.

Propylene glycol, on the other hand, is essentially nontoxic and has been used as a solvent for drugs. Like ethylene glycol, it is oxidized by liver enzymes; however, pyruvic acid, its oxidation product, is a compound normally found in the human body, being an intermediate in carbohydrate metabolism (Chapter 13).

Figure 3.5 Ethylene glycol is the major ingredient in airplane "de-icers."

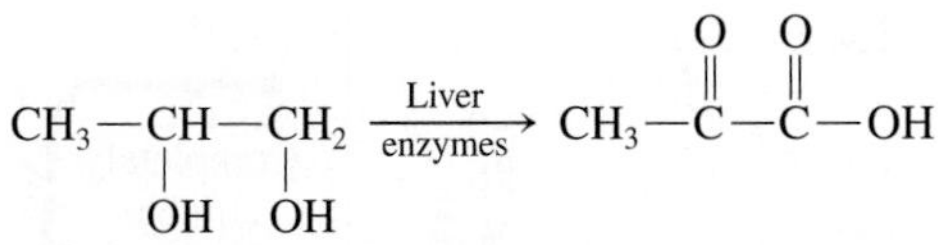

Propylene glycol — Pyruvic acid

Propylene glycol use as an antifreeze is increasing. Antifreeze brands marketed as "environmentally friendly" are usually propylene glycol formulations. Such antifreeze, however, costs more since propylene glycol production costs exceed those for ethylene glycol.

Accidental ethylene glycol poisoning is a problem that occurs much too frequently. Most often such poisonings occur when radiator fluid is changed and the spent fluid is not disposed of properly. Both cats and dogs as well as small children can be victims of such poisonings, which are often fatal.

Glycerol (1,2,3-Propanetriol)

Glycerol, which is often also called glycerin, is a clear, thick liquid that has the consistency of honey. Its molecular structure involves three —OH groups on three different carbon atoms.

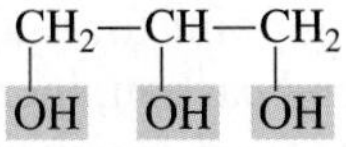

Glycerol is normally present in the human body because it is a product of fat metabolism. It is present, in combined form, in all animal fats and vegetable oils (Section 8.4). In some Arctic and northern species, glycerol functions as a "biological antifreeze" (Figure 3.6).

Because glycerol has a great affinity for water vapor (moisture), it is often added to pharmaceutical preparations such as skin lotions and soap. Florists sometimes use glycerol on cut flowers to help retain water and maintain freshness. Its lubricative properties also make it useful in shaving creams and in applications such as glycerol suppositories for rectal administration of medicines. It is used in candies and icings as a retardant for preventing sugar crystallization.

James Cotier/Stone/Getty Images

Figure 3.6 Glycerol is often called biological antifreeze. For survival in Arctic and northern winters, many fish and insects, including the common housefly, produce large amounts of glycerol that dissolve in their blood, thereby lowering the freezing point of the blood.

3.6 Physical Properties of Alcohols

Alcohol molecules have both polar and nonpolar character. The hydroxyl groups present are polar, and the alkyl (R) group present is nonpolar.

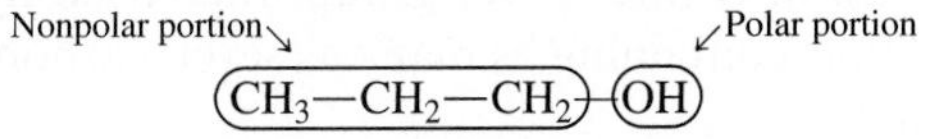

The physical properties of an alcohol depend on whether the polar or the nonpolar portion of its structure "dominates." Factors that determine this include the *length* of the nonpolar carbon chain present and the *number* of polar hydroxyl groups present (Figure 3.7).

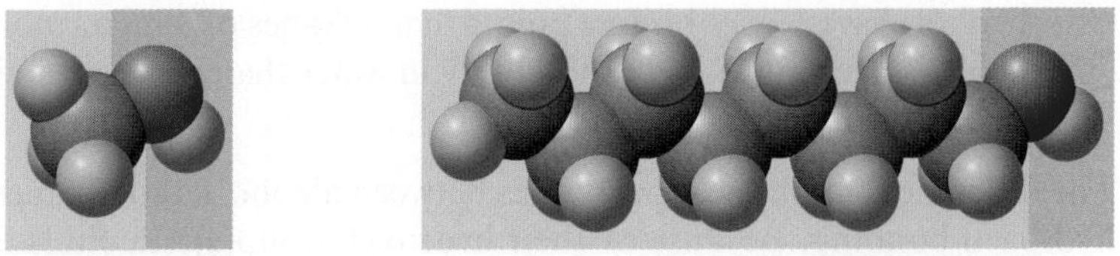

CH_3—OH
Nonpolar Polar

$CH_3CH_2CH_2CH_2CH_2CH_2CH_2CH_2$—OH
Nonpolar Polar

a **Methanol** The polar hydroxyl functional group dominates the physical properties of methanol. The molecule is completely soluble in water (polar) but has limited solubility in hexane (nonpolar).

b **1-Octanol** Conversely, the nonpolar portion of 1-octanol dominates its physical properties; it is infinitely soluble in hexane and has limited solubility in water.

Figure 3.7 Space-filling molecular models showing the nonpolar (green) and polar (pink) parts of methanol and 1-octanol.

Figure 3.8 (a) Boiling points and (b) solubilities in water of selected 1-alcohols.

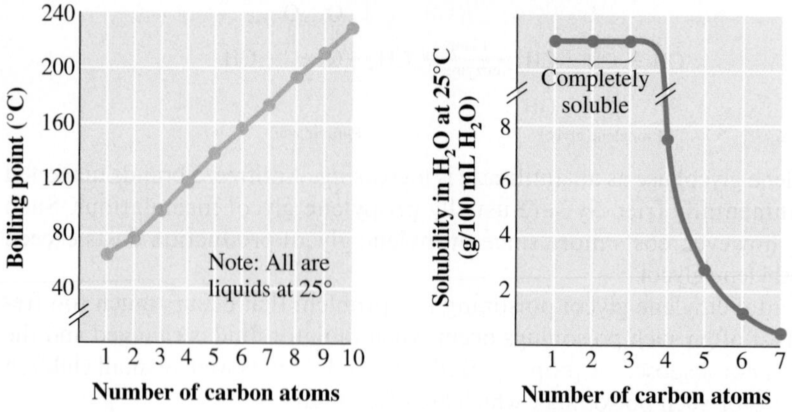

a Boiling points.

b Solubilities in water of selected 1-alcohols.

Boiling Points and Water Solubilities

Figure 3.8a shows that the boiling point for 1-alcohols, unbranched-chain alcohols with an —OH group on an end carbon, increases as the length of the carbon chain increases. This trend results from increasing London forces (Section 1.16) with increasing carbon-chain length. Alcohols with more than one hydroxyl group present have significantly higher boiling points (bp) than their monohydroxy counterparts.

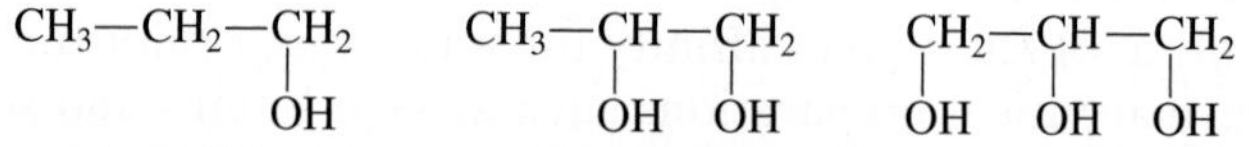

bp = 97°C   bp = 188°C   bp = 290°C

This boiling-point trend is related to increased hydrogen bonding between alcohol molecules (to be discussed shortly). Figure 3.9 is a physical-state summary for unbranched 1-alcohols and unsubstituted cycloalcohols with eight or fewer carbon atoms.

Small monohydroxy alcohols are soluble in water in all proportions. As carbon-chain length increases beyond three carbons, solubility in water rapidly decreases (Figure 3.8b) because of the increasingly nonpolar character of the alcohol. Alcohols with two —OH groups present are more soluble in water than their counterparts with only one —OH group. Increased hydrogen bonding is responsible for this. Diols containing as many as seven carbon atoms show appreciable solubility in water.

Unbranched 1-Alcohols			
C_1	C_3	C_5	C_7
C_2	C_4	C_6	C_8

Unsubstituted Cycloalcohols			
☒	C_3	C_5	C_7
☒	C_4	C_6	C_8

☐ Liquid

Figure 3.9 A physical-state summary for unbranched 1-alcohols and unsubstituted cycloalcohols at room temperature and pressure.

Alcohols and Hydrogen Bonding

A comparison of the properties of alcohols with their alkane counterparts (Table 3.3) shows that

1. Alcohols have *higher* boiling points than alkanes of similar molecular mass.
2. Alcohols have much *higher* solubility in water than alkanes of similar molecular mass.

The differences in physical properties between alcohols and alkanes are related to hydrogen bonding. Because of their hydroxyl group(s), alcohols can participate in hydrogen bonding, whereas alkanes cannot. Hydrogen bonding between alcohol molecules (Figure 3.10) is similar to that which occurs between water molecules.

Extra energy is needed to overcome alcohol–alcohol hydrogen bonds before alcohol molecules can enter the vapor phase. Hence alcohol boiling points are higher than those for the corresponding alkanes (where no hydrogen bonds are present).

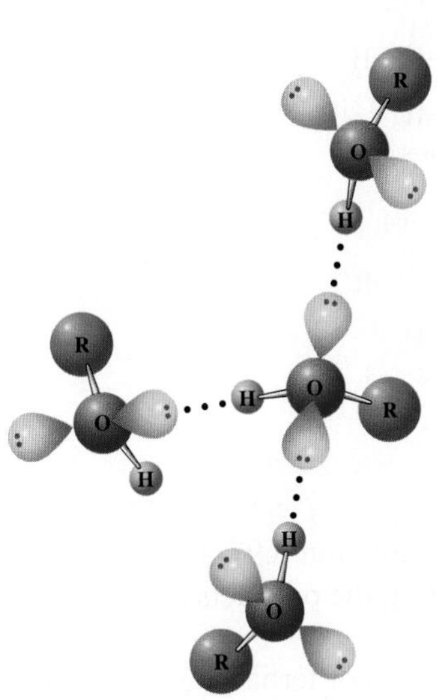

Figure 3.10 Alcohol boiling points are higher than those of the corresponding alkanes because of alcohol–alcohol hydrogen bonding.

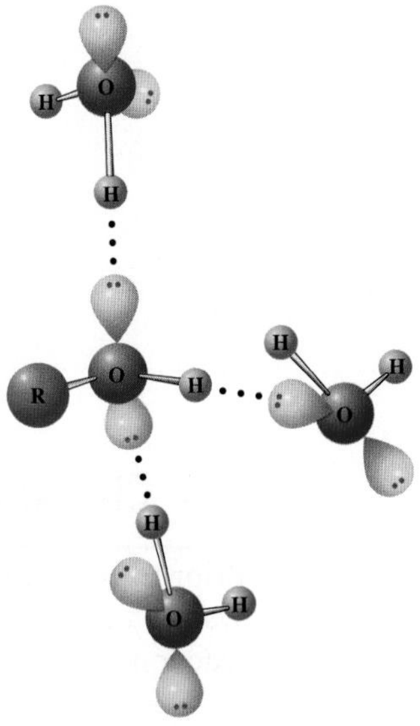

Figure 3.11 Because of hydrogen bonding between alcohol molecules and water molecules, alcohols of small molecular mass have unlimited solubility in water.

Alcohol molecules can also hydrogen-bond to water molecules (Figure 3.11). The formation of such hydrogen bonds explains the solubility of small alcohol molecules in water. As the alcohol chain length increases, alcohols become more alkane-like (nonpolar), and solubility decreases.

Table 3.3 A Comparison of Selected Physical Properties of Alcohols with Alkane Counterparts of Similar Molecular Mass

Type of Compound	Compound	Structure	Molecular Mass (amu)	Boiling Point (°C)	Solubility in Water
alkane	ethane	CH_3-CH_3	30	−89	slight solubility
alcohol	methanol	CH_3-OH	32	65	unlimited solubility
alkane	propane	$CH_3-CH_2-CH_3$	44	−42	slight solubility
alcohol	ethanol	CH_3-CH_2-OH	46	78	unlimited solubility
alkane	butane	$CH_3-CH_2-CH_2-CH_3$	58	−1	slight solubility
alcohol	1-propanol	$CH_3-CH_2-CH_2-OH$	60	97	unlimited solubility
alcohol	2-propanol	$CH_3-\underset{\underset{\displaystyle CH_3}{\vert}}{CH}-OH$	60	83	unlimited solubility

3.7 Preparation of Alcohols

A general method for preparing alcohols—the hydration of alkenes—was discussed in the previous chapter (Section 2.9). Alkenes react with water (an unsymmetrical addition agent) in the presence of sulfuric acid (the catalyst) to form an alcohol. Markovnikov's rule is used to determine the predominant alcohol product.

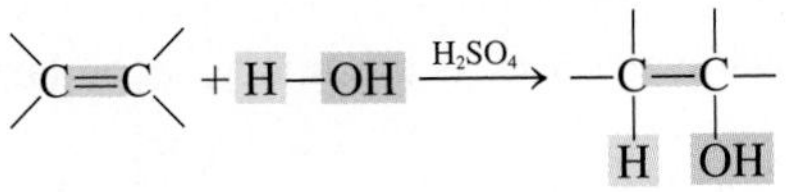

Another method of synthesizing alcohols involves the addition of H_2 to a carbon–oxygen double bond (a carbonyl group, C=O). (The carbonyl group is a functional group that will be discussed in detail in Chapter 4.) A carbonyl group behaves very much like a carbon–carbon double bond when it reacts with H_2 under the proper conditions. As a result of H_2 addition, the oxygen of the carbonyl group is converted to an —OH group.

Alcohols are intermediate products in the metabolism of both carbohydrates (Chapter 13) and fats (Chapter 14). In these metabolic processes, both addition of water to a carbon–carbon double bond and addition of hydrogen to a carbon–oxygen double bond lead to the introduction of the alcohol functional group into a biomolecule.

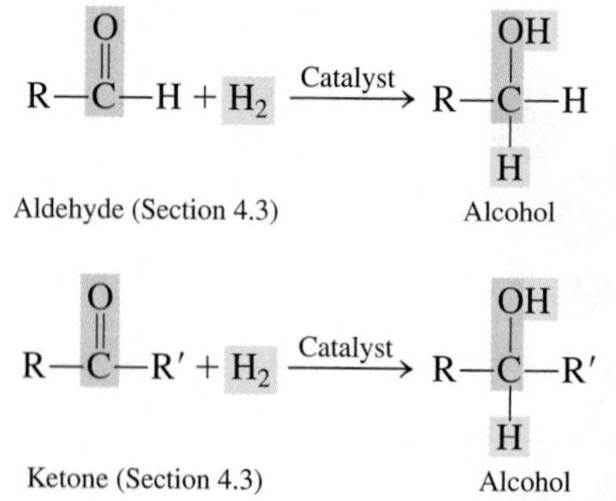

3.8 Classification of Alcohols

Prior to considering chemical reactions of alcohols (Section 3.9), a classification system for alcohols that is often needed when predicting the products in a chemical reaction involving an alcohol needs to be in place.

Pronounce 1° as "primary," 2° as "secondary," and 3° as "tertiary."

Alcohols are classified as primary (1°), secondary (2°), or tertiary (3°) depending on the number of carbon atoms bonded to the carbon atom that bears the hydroxyl group. A **primary alcohol** *is an alcohol in which the hydroxyl-bearing carbon atom is bonded to only one other carbon atom.* A **secondary alcohol** *is an alcohol in which the hydroxyl-bearing carbon atom is bonded to two other carbon atoms.* A **tertiary alcohol** *is an alcohol in which the hydroxyl-bearing carbon atom is bonded to three other carbon atoms.* Chemical reactions of alcohols often depend on alcohol class (1°, 2°, or 3°).

Methyl alcohol, CH_3—OH, an alcohol in which the hydroxyl-bearing carbon atom is attached to three hydrogen atoms, does not fit any of the alcohol classification definitions. It is usually grouped with the primary alcohols because its reactions are similar to theirs.

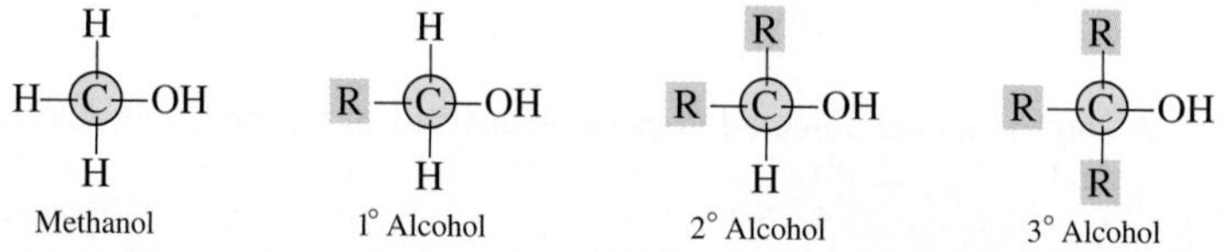

Although all alcohols are able to participate in hydrogen bonding (Section 3.6), increasing the number of R groups around the carbon atom bearing the OH group decreases the extent of hydrogen bonding. This effect, called stearic hindrance, becomes particularly important when the R groups are large. Thus, 1° alcohols are best able to hydrogen-bond and 3° alcohols are least able to hydrogen-bond.

For alcohols of similar molecular mass, 1° alcohols have higher boiling points than 2° alcohols, which in turn have higher boiling points than 3° alcohols, because of how stearic hindrance affects hydrogen bonding. The following data for three C_4 alcohols illustrates this situation.

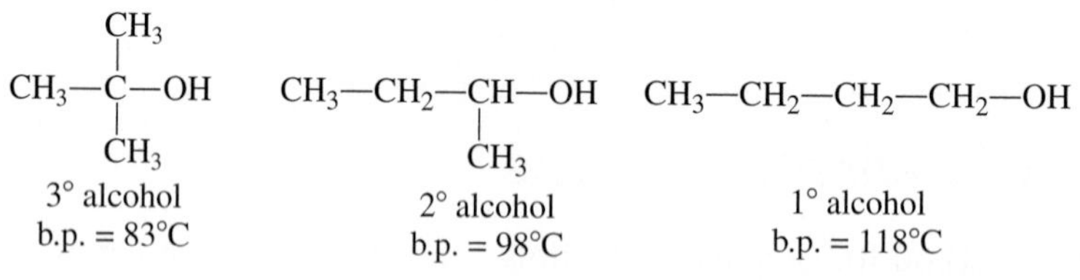

EXAMPLE 3.2 Classifying Alcohols as Primary, Secondary, or Tertiary Alcohols

Classify each of the following alcohols as a primary, secondary, or tertiary alcohol.

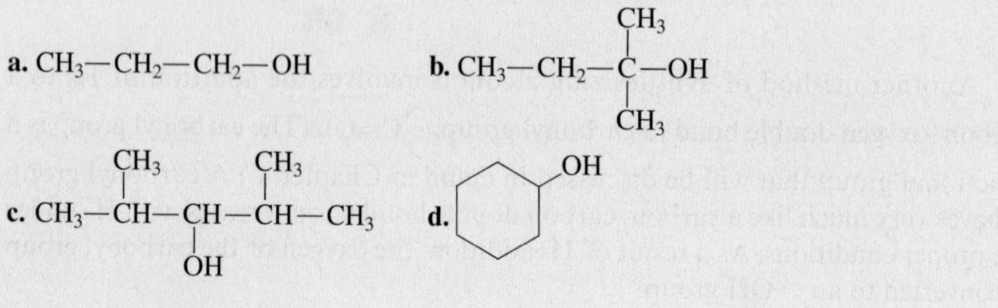

Solution

a. This is a primary alcohol. The carbon atom to which the —OH group is attached is bonded to only one other carbon atom.

b. This is a tertiary alcohol. The carbon atom bearing the —OH group is bonded to three other carbon atoms.

c. This is a secondary alcohol. The hydroxyl-bearing carbon atom is bonded to two other carbon atoms.

d. This is a secondary alcohol. The ring carbon atom to which the —OH group is attached is bonded to two other ring carbon atoms.

Practice Exercise 3.2

Classify each of the following alcohols as a primary, secondary, or tertiary alcohol.

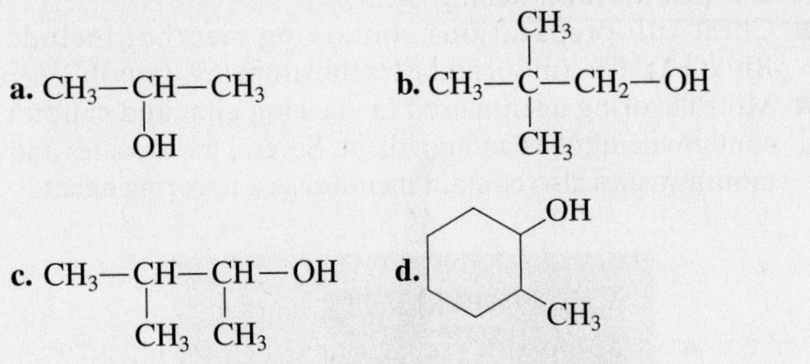

Answers: **a.** Secondary; **b.** Primary; **c.** Secondary; **d.** Secondary

Part d. of Example 3.2 illustrates that the 1°, 2°, and 3° classification system for alcohols applies not only to alcohols where the hydroxyl group is attached to an acyclic R group but also to alcohols where the R group is cyclic. All alcohols with cyclic R groups are either 2° or 3° alcohols; 1° alcohols with cyclic R groups are structurally not possible. A naturally occurring 2° alcohol with a cyclic R group whose name is familiar to most people is *menthol*. Many *mentholated* consumer products are available for use today. The focus on relevancy feature Chemical Connections 3-A on the next page gives information about this interesting secondary alcohol.

3.9 Chemical Reactions of Alcohols

Of the many chemical reactions that alcohols undergo, four will be considered in this section: (1) combustion (2) dehydration (3) oxidation and (4) halogenation.

Combustion

As has been seen in the previous two chapters, hydrocarbons of all types undergo combustion in air to produce carbon dioxide and water. Alcohols are also flammable; as with hydrocarbons, the combustion products are carbon dioxide and water. Both methanol and ethanol, as well as alcohol-gasoline mixtures such as E10 and E85 (Section 3.5), are used as automotive fuels.

Intramolecular Alcohol Dehydration

A **dehydration reaction** *is a chemical reaction in which the components of water (H and OH) are removed from a single reactant or from two reactants (H from one and OH from the other)*. In *intramolecular* dehydration, both water components are removed from the same molecule.

Reaction conditions for the intramolecular dehydration of an alcohol are a temperature of 180°C and the presence of sulfuric acid (H_2SO_4) as a catalyst. The dehydration product is an alkene.

Ease of alcohol dehydration depends on alcohol classification. Primary alcohols are the most difficult to dehydrate, requiring temperatures of around 180°C. Secondary alcohol dehydration occurs at lower temperatures, and tertiary alcohols dehydrate at temperatures slightly above room temperature.

CHEMICAL CONNECTIONS 3-A

Menthol: A Useful Naturally Occurring Terpene Alcohol

Menthol is a naturally occurring terpene (Section 2.7) alcohol with a pleasant, minty odor. Its IUPAC name is 2-isopropyl-5-methylcyclohexanol.

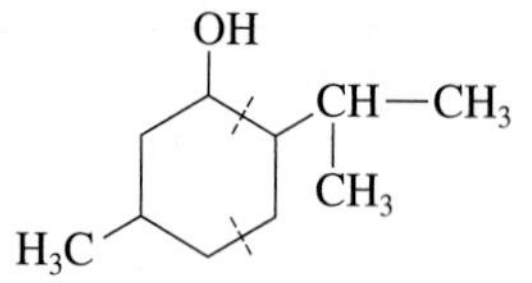

In the pure state, menthol is a white crystalline solid with a melting point of 41°C to 43°C. Menthol occurs naturally in peppermint oil. As is the case with many natural products, the demand for menthol exceeds its supply from natural sources. Methods now exist for the synthetic production of menthol.

Topical application of menthol to the skin causes a refreshing, cooling sensation followed by a slight burning-and-prickling sensation. Its mode of action is that of a *differential* anesthetic. It stimulates the receptor cells in the skin that normally respond to cold to give a sensation of coolness that is unrelated to body temperature. (This cooling sensation is particularly noticeable in the respiratory tract when low concentrations of menthol are inhaled.) At the same time as cooling is perceived, menthol can depress the nerves for pain reception.

Menthol's mode of action is opposite that of capsaicin (Section 6.14), the natural product responsible for the "spiciness" of hot peppers. Capsaicin stimulates heat sensors without causing an actual change in body temperature.

Numerous products contain menthol.

- Throat sprays and lozenges containing menthol temporarily soothe inflamed mucous surfaces of the nose and throat. Lozenges contain 2–20 milligrams of menthol per wafer.
- Cough drops and cigarettes of the "mentholated" type use menthol for its counterirritant effect.
- Pre-electric shave preparations and aftershave lotions often contain menthol. A concentration of only 0.1% (m/v) gives ample cooling to allay the irritation of a "close" shave.
- Many dermatologic preparations contain menthol as an anti-pruritic (anti-itching agent).
- Chest-rub preparations containing menthol include BENGAY [7% (m/v)] and Mentholatum [6% (m/v)].
- Mint flavoring agents used in chewing gum and candies contain menthol as an ingredient. Several toothpastes and mouthwashes also contain menthol as a flavoring agent.

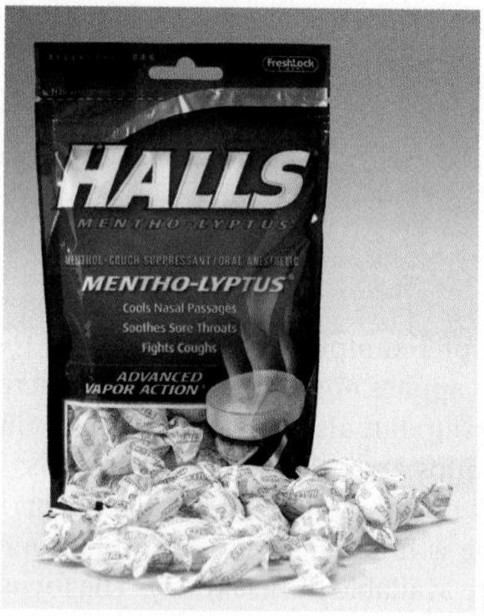

© Cengage Learning

Many kinds of cough drops contain menthol as a counterirritant.

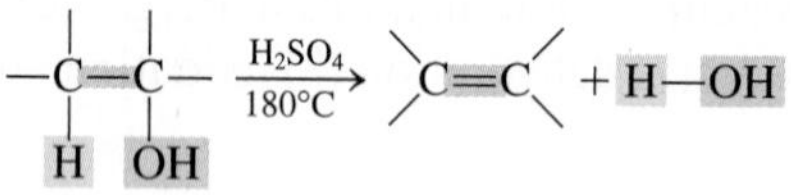

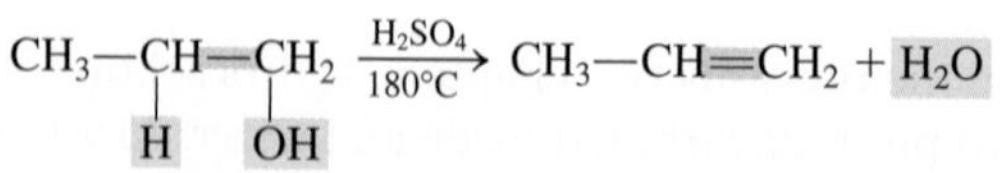

Intramolecular alcohol dehydration is an example of an *elimination reaction* (Figure 3.12), as contrasted to a substitution reaction (Section 1.17) and an addition reaction (Section 2.9). An **elimination reaction** *is a reaction in which two groups or two atoms on neighboring carbon atoms are removed, or eliminated, from a molecule, leaving a multiple bond between the carbon atoms.*

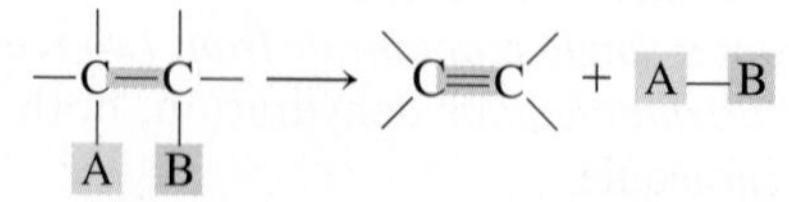

What occurs in an elimination reaction is the reverse of what occurs in an addition reaction.

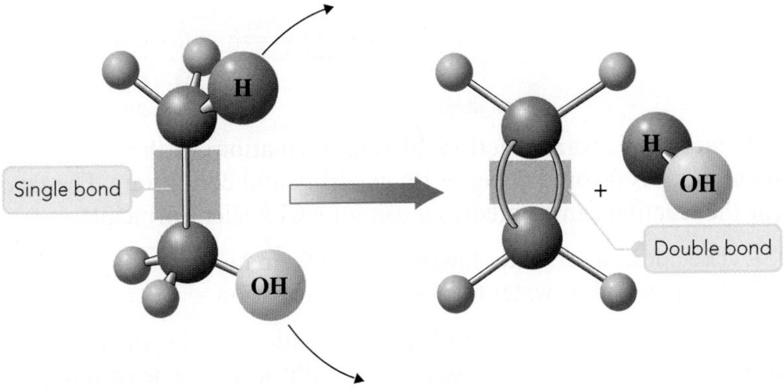

Figure 3.12 In an intramolecular alcohol dehydration, the components of water (H and OH) are removed from neighboring carbon atoms with the resultant introduction of a double bond into the molecule.

Dehydration of an alcohol can result in the production of more than one alkene product. This happens when there is more than one neighboring carbon atom from which hydrogen loss can occur. Dehydration of 2-butanol produces two alkenes.

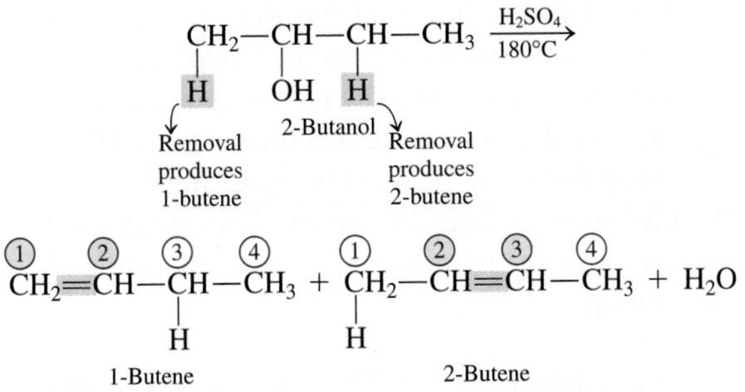

Dehydration of alcohols to form carbon–carbon double bonds occurs in several metabolic pathways in living systems, such as the citric acid cycle (Section 12.6) and the β oxidation pathway (Section 14.4). In these biochemical dehydrations, enzymes serve as catalysts instead of acids, and the reaction temperature is 37°C instead of the elevated temperatures required in the laboratory.

Although both products are formed, a large amount of one product is formed whereas only a small amount of the other product is formed. Prediction of which product is the major product and which is the minor product can be made using Zaitsev's (pronounced "*zait-zeff*") rule, a rule that carries the name of the Russian chemist Alexander Zaitsev. **Zaitsev's rule** states that *the major product in an intramolecular alcohol dehydration reaction is the alkene that has the greatest number of alkyl groups attached to the carbon atoms of the double bond.* In the preceding reaction, 2-butene (with two alkyl groups) is favored over 1-butene (with one alkyl group).

Alexander Zaitsev (1841–1910), a nineteenth-century Russian chemist, studied at the University of Paris and then returned to his native Russia to become a professor of chemistry at the University of Kazan.

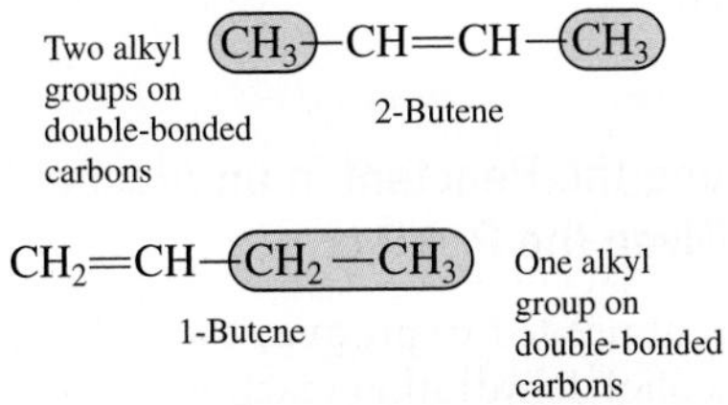

Alkene formation via intramolecular alcohol *dehydration* is the "reverse reaction" of the reaction for preparing an alcohol through *hydration* of an alkene (Section 2.9). This relationship can be diagrammed as follows:

An alternative way of expressing Zaitsev's rule is "Hydrogen atom loss, during intramolecular alcohol dehydration to form an alkene, will occur preferentially from the carbon atom (adjacent to the hydroxyl-bearing carbon) that already has the fewest hydrogen atoms."

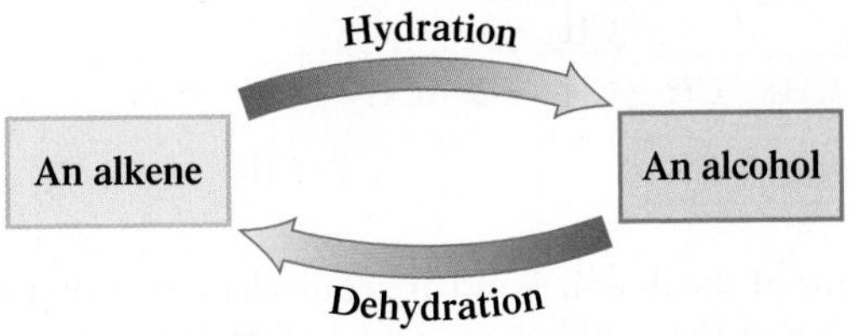

This "reverse reaction" situation illustrates the fact that many organic reactions can go both forward or backward, depending on reaction conditions. Noting relationships such as this helps in keeping track of the numerous reactions that hydrocarbon derivatives undergo. These two "reverse reactions" actually involve an equilibrium situation.

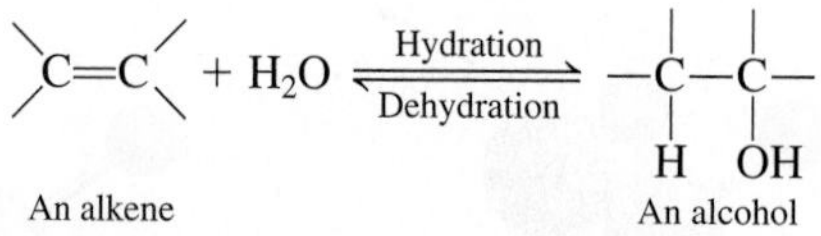

Whether the forward reaction (alcohol formation) or the reverse reaction (alkene formation) is favored depends on experimental conditions. The favored direction for the reaction can be predicted using Le Châtelier's principle.

1. The addition of water favors alcohol formation.
2. The removal of water favors alkene formation.

Experimental conditions for alcohol formation involve the use of a *dilute* sulfuric acid solution as a catalyst. *Concentrated* sulfuric acid (a dehydrating agent) as well as higher temperatures are used for alkene formation. Dilute acid solutions are mainly water; concentrated acid solutions have less water and heat also removes water.

Intermolecular Alcohol Dehydration

Ethers, like alcohols, constitute a many-membered important class of oxygen-containing hydrocarbon derivatives. An extended discussion of ethers, whose general properties are much different from those of alcohols, is found in Sections 3.15 through 3.19 of this chapter.

At a lower temperature (140°C) than that required for alkene formation (180°C), an *inter*molecular rather than an *intra*molecular alcohol dehydration process can occur to produce an ether—a compound with the general structure R—O—R. In such ether formation, two alcohol molecules interact, an H atom being lost from one and an —OH group from the other. The resulting "leftover" portions of the two alcohol molecules join to form the ether. This reaction, which gives useful yields only for primary alcohol reactants (2° and 3° alcohols yield predominantly alkenes), can be written as

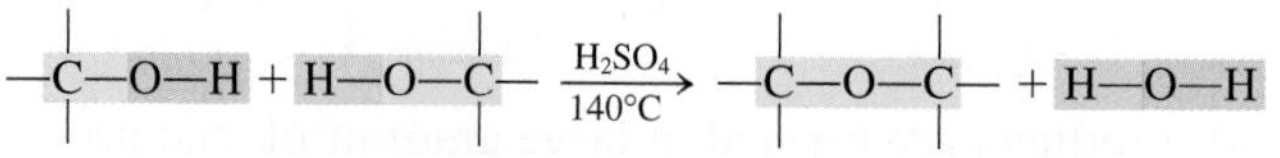

$$\underset{\text{Ethanol}}{CH_3—CH_2—O—H} + \underset{\text{Ethanol}}{H—O—CH_2—CH_3} \xrightarrow[140°C]{H_2SO_4}$$

$$CH_3—CH_2—O—CH_2—CH_3 + H_2O$$

The preceding reaction is an example of *condensation.* A **condensation reaction** *is a chemical reaction in which two molecules combine to form a larger one while liberating a small molecule, usually water.* In this case, two alcohol molecules combine to give an ether and water.

The following is a summary of products obtained from alcohol dehydration reactions using H_2SO_4 as a catalyst.

Primary alcohol	$\xrightarrow{180°C}$ alkene	$\xrightarrow{140°C}$ ether
Secondary alcohol	$\xrightarrow{180°C}$ alkene	$\xrightarrow{140°C}$ alkene
Tertiary alcohol	$\xrightarrow{180°C}$ alkene	$\xrightarrow{140°C}$ alkene

EXAMPLE 3.3 Predicting the Reactant in an Alcohol Dehydration Reaction When Given the Product

Identify the alcohol reactant needed to produce each of the following compounds as the *major product* of an alcohol dehydration reaction.

a. Alcohol $\xrightarrow[180°C]{H_2SO_4}$ $CH_3—CH{=}CH—CH_3$

b. Alcohol $\xrightarrow[180°C]{H_2SO_4}$

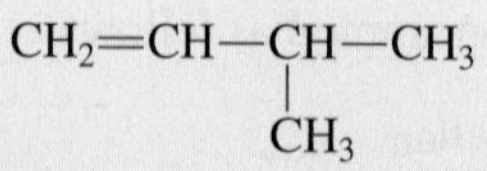

c. Alcohol $\xrightarrow[140°C]{H_2SO_4}$

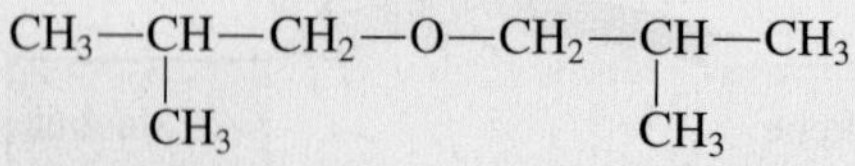

Solution

a. Both carbon atoms of the double bond are equivalent to each other. Add an H atom to one carbon atom of the double bond and an OH group to the other carbon atom of the double bond. It does not matter which goes where; you get the same molecule either way.

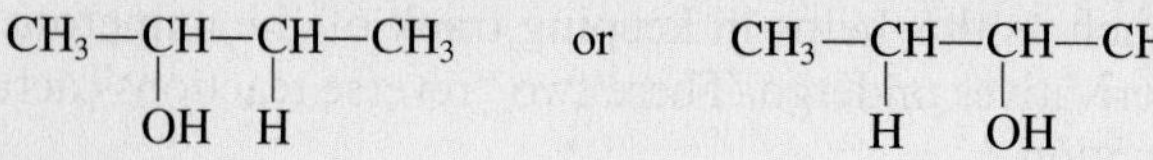

b. There are two possible parent alcohols: one with an —OH group on carbon 1 and the other with an —OH group on carbon 2.

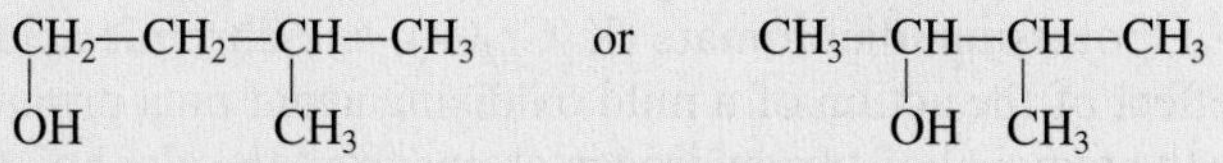

Based on the reverse of Zaitsev's rule, the hydrogen atom will go back on the double-bonded carbon that bears the most alkyl groups.

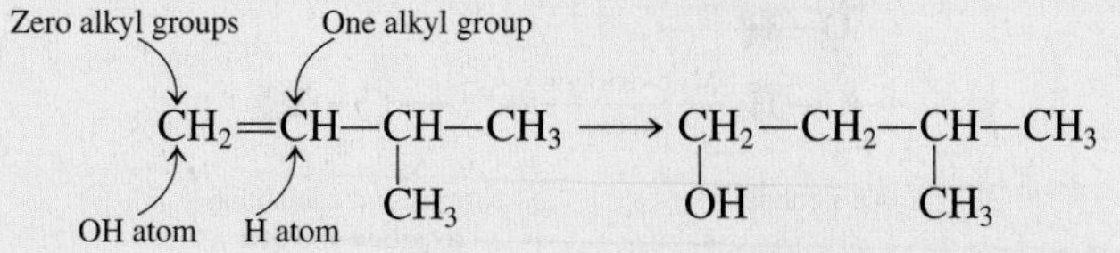

c. This is an ether. The primary alcohol from which the ether was formed will have the same alkyl group present as is in the ether. Thus the alcohol is

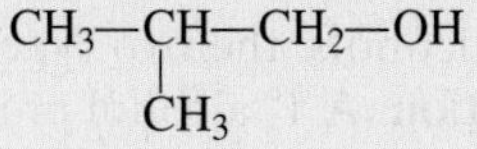

Practice Exercise 3.3

Identify the starting alcohol from which each of the following products was obtained by an alcohol dehydration reaction.

a. Alcohol $\xrightarrow[180°C]{H_2SO_4}$ $CH_2{=}CH{-}CH_2{-}CH_3$

b. Alcohol $\xrightarrow[180°C]{H_2SO_4}$

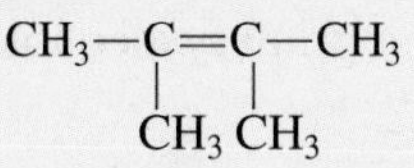

c. Alcohol $\xrightarrow[140°C]{H_2SO_4}$ $CH_3{-}CH_2{-}CH_2{-}O{-}CH_2{-}CH_2{-}CH_3$

Answers: **a.**

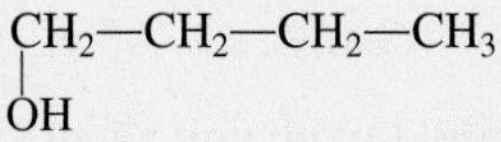

b.

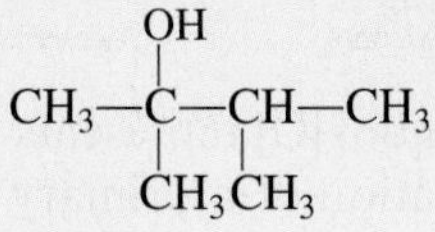

c. $CH_3{-}CH_2{-}CH_2{-}OH$

Oxidation

Before discussing alcohol oxidation reactions, a new method for recognizing when oxidation and reduction have occurred in a chemical reaction will be considered.

The processes of oxidation and reduction can be considered in the context of inorganic, rather than organic, reactions. Oxidation numbers are used to characterize oxidation–reduction processes. This same technique can be used in characterizing oxidation–reduction processes involving organic compounds, but it is not. Formal use of the oxidation number rules with organic compounds is usually cumbersome because of the many carbon and hydrogen atoms present; often, fractional oxidation numbers for carbon result.

A better approach for organic redox reactions is to use the following set of operational rules instead of oxidation numbers.

1. An *organic oxidation* is an oxidation that increases the number of C—O bonds and/or decreases the number of C—H bonds.
2. An *organic reduction* is a reduction that decreases the number of C—O bonds and/or increases the number of C—H bonds.

Note that these operational definitions for oxidation and reduction are "opposites." This is just as it should be; oxidation and reduction are "opposite" processes.

Some alcohols readily undergo oxidation with mild oxidizing agents; others are resistant to oxidation with these same oxidizing agents. Primary and secondary alcohols, but not tertiary alcohols, readily undergo oxidation in the presence of mild oxidizing agents to produce compounds that contain a carbon–oxygen double bond (aldehydes, ketones, and carboxylic acids). A number of different oxidizing agents can be used for the oxidation, including potassium permanganate ($KMnO_4$), potassium dichromate ($K_2Cr_2O_7$), and chromic acid (H_2CrO_4).

The net effect of the action of a mild oxidizing agent on a primary or secondary alcohol is the removal of two hydrogen atoms from the alcohol. One hydrogen comes from the —OH group, the other from the carbon atom to which the —OH group is attached. This H removal generates a carbon–oxygen double bond.

This chemical reaction is consistent with the operational definition given earlier in this section for an organic oxidation. A new C—O bond is formed and a C—H bond is broken.

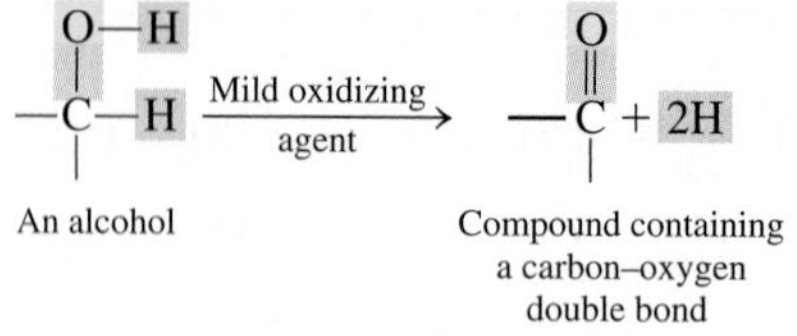

The two "removed" hydrogen atoms combine with oxygen supplied by the oxidizing agent to give H_2O.

Primary and secondary alcohols, the two types of oxidizable alcohols, yield different products upon oxidation. A 1° alcohol produces an *aldehyde* that is often then further oxidized to a *carboxylic acid,* and a 2° alcohol produces a *ketone.*

$$\text{Primary alcohol} \xrightarrow[\text{ox. agent}]{\text{Mild}} \text{aldehyde} \xrightarrow[\text{ox. agent}]{\text{Mild}} \text{carboxylic acid}$$

$$\text{Secondary alcohol} \xrightarrow[\text{ox. agent}]{\text{Mild}} \text{ketone}$$

$$\text{Tertiary alcohol} \xrightarrow[\text{ox. agent}]{\text{Mild}} \text{no reaction}$$

The general reaction for the oxidation of a primary alcohol is

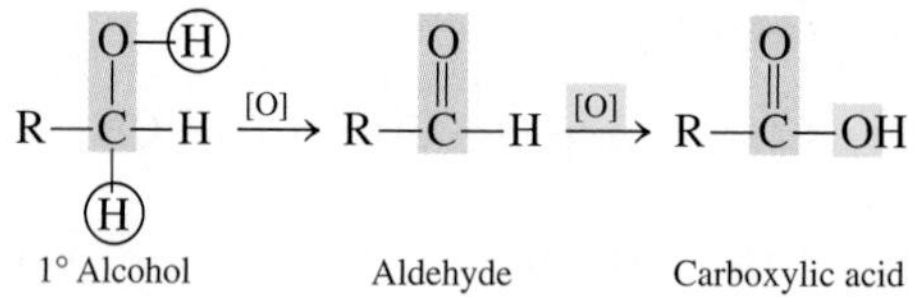

In this equation, the symbol [O] represents the mild oxidizing agent. The immediate product of the oxidation of a primary alcohol is an aldehyde. Because aldehydes themselves are readily oxidized by the same oxidizing agents that oxidize alcohols, aldehydes are further converted to carboxylic acids. A specific example of a primary alcohol oxidation reaction is

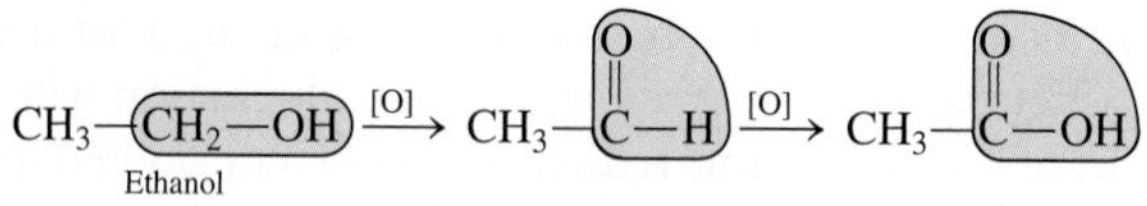

This specific oxidation reaction—that of ethanol—is the basis for the "breathalyzer test" used by law enforcement officers to determine whether an automobile driver is "drunk" (Figure 3.13).

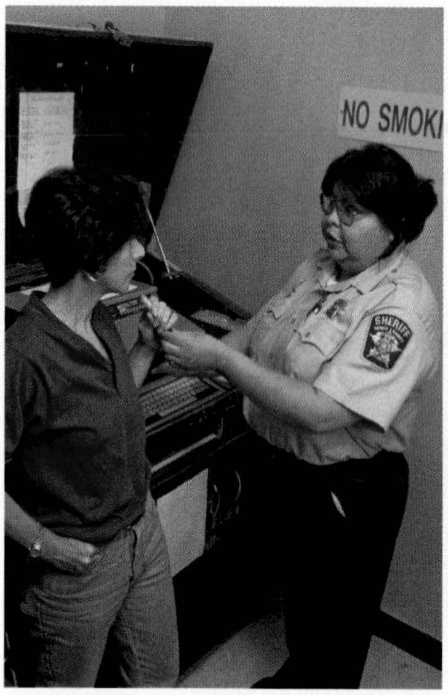

Bob Daemmrich/Stock Boston

Figure 3.13 The oxidation of ethanol is the basis for the "breathalyzer test" that law enforcement officers use to determine whether an individual suspected of driving under the influence (DUI) has a blood alcohol level exceeding legal limits

The DUI suspect is required to breathe into an apparatus containing a solution of potassium dichromate ($K_2Cr_2O_7$). The unmetabolized alcohol in the person's breath is oxidized by the dichromate ion ($Cr_2O_7^{2-}$), and the extent of the reaction gives a measure of the amount of alcohol present.

The dichromate ion is a yellow-orange color in solution. As oxidation of the alcohol proceeds, the dichromate ions are converted to Cr^{3+} ions, which have a green color in solution. The intensity of the green color that develops is measured and is proportional to the amount of ethanol in the suspect's breath, which in turn has been shown to be proportional to the person's blood alcohol level.

The general reaction for the oxidation of a secondary alcohol is

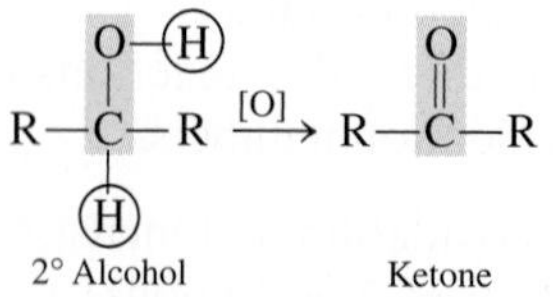

As with primary alcohols, oxidation involves the removal of two hydrogen atoms. Unlike aldehydes, ketones are resistant to further oxidation. A specific example of the oxidation of a secondary alcohol is

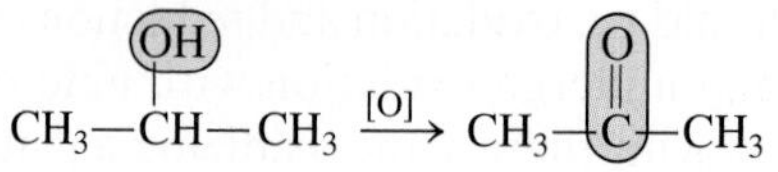

Tertiary alcohols do not undergo oxidation with mild oxidizing agents. This is because they do not have hydrogen on the —OH-bearing carbon atom.

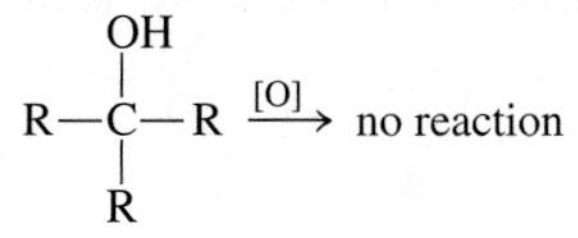

3° Alcohol

EXAMPLE 3.4 Predicting Products in Alcohol Oxidation Reactions

Draw the structural formula(s) for the product(s) formed by oxidation of the following alcohols with a mild oxidizing agent. If no reaction occurs, write "no reaction."

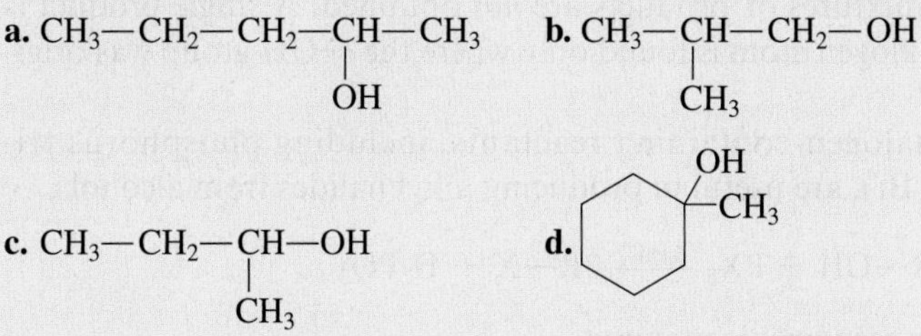

Solution

a. The oxidation product will be a ketone, as this is a 2° alcohol.

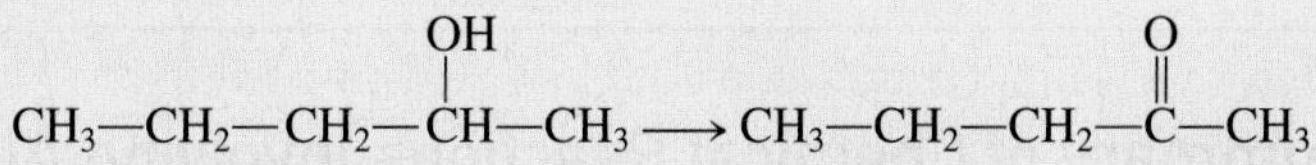

b. A 1° alcohol undergoes oxidation first to an aldehyde and then to a carboxylic acid.

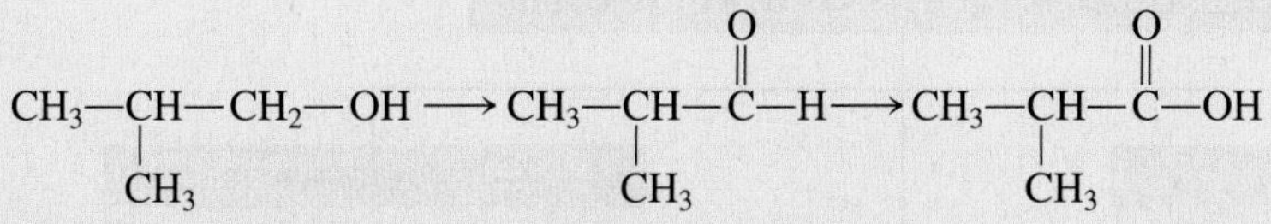

c. A ketone is the product from the oxidation of a 2° alcohol.

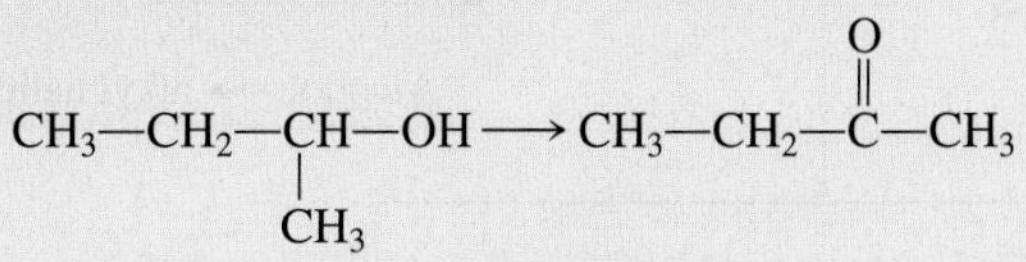

d. This cyclic alcohol is a tertiary alcohol. The hydroxyl-bearing carbon atom is attached to two ring carbon atoms and a methyl group. Tertiary alcohols do not undergo oxidation with mild oxidizing agents. Therefore, "no reaction."

Practice Exercise 3.4

Draw the structural formula(s) for the product(s) formed by oxidation of the following alcohols with a mild oxidizing agent. If no reaction occurs, write "no reaction."

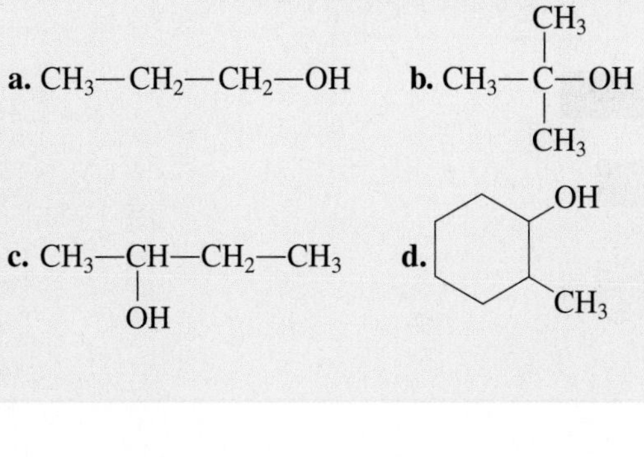

(continued)

Answers:

a. $CH_3—CH_2—\overset{O}{\overset{\|}{C}}—H$, $CH_3—CH_2—\overset{O}{\overset{\|}{C}}—OH$ **b.** No reaction

c. $CH_3—\overset{O}{\overset{\|}{C}}—CH_2—CH_3$ **d.** 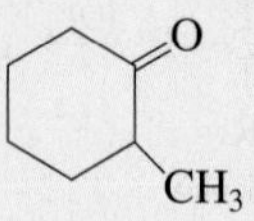

Halogenation

Alcohols undergo halogenation reactions in which a halogen atom is substituted for the hydroxyl group, producing an alkyl halide. Alkyl halide production in this manner is superior to alkyl halide production through halogenation of an alkane (Section 1.18) because mixtures of products are *not* obtained. A single product is produced in which the halogen atom is found only where the —OH group was originally located.

Several different halogen-containing reactants, including phosphorus trihalides (PX_3; X is Cl or Br), are useful in producing alkyl halides from alcohols.

$$3R—OH + PX_3 \xrightarrow{\text{heat}} 3R—X + H_3PO_3$$

Note that heating of the reactants is required.

The Chemistry at a Glance feature below summarizes the reaction chemistry of alcohols.

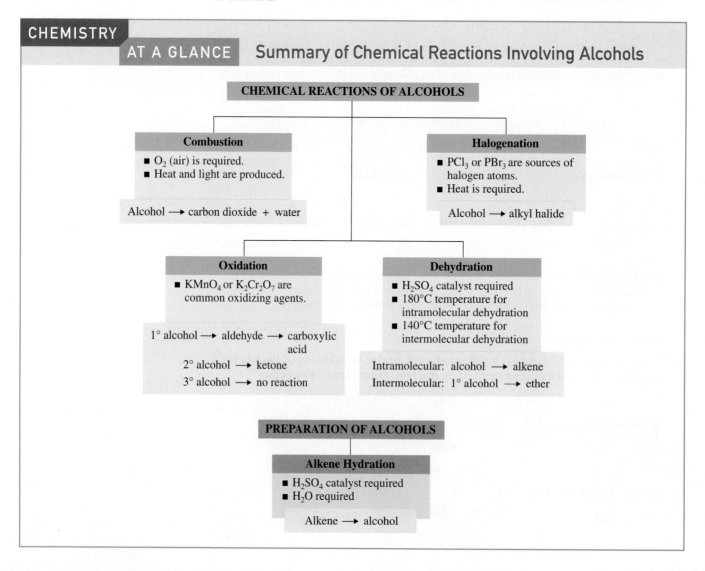

CHEMISTRY AT A GLANCE

Summary of Chemical Reactions Involving Alcohols

CHEMICAL REACTIONS OF ALCOHOLS

Combustion
- O_2 (air) is required.
- Heat and light are produced.

Alcohol → carbon dioxide + water

Halogenation
- PCl_3 or PBr_3 are sources of halogen atoms.
- Heat is required.

Alcohol → alkyl halide

Oxidation
- $KMnO_4$ or $K_2Cr_2O_7$ are common oxidizing agents.

1° alcohol → aldehyde → carboxylic acid
2° alcohol → ketone
3° alcohol → no reaction

Dehydration
- H_2SO_4 catalyst required
- 180°C temperature for intramolecular dehydration
- 140°C temperature for intermolecular dehydration

Intramolecular: alcohol → alkene
Intermolecular: 1° alcohol → ether

PREPARATION OF ALCOHOLS

Alkene Hydration
- H_2SO_4 catalyst required
- H_2O required

Alkene → alcohol

3.10 Polymeric Alcohols

It is possible to synthesize polymeric alcohols with structures similar to those of substituted polyethylenes (Section 2.9). One of the simplest such compounds is poly(vinyl alcohol) (PVA).

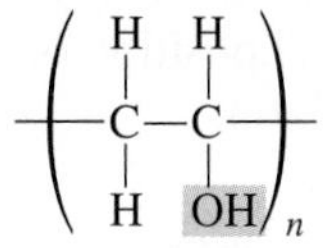

PVA

Poly(vinyl alcohol) is a tough, whitish polymer that can be formed into strong films, tubes, and fibers that are highly resistant to hydrocarbon solvents. Unlike most organic polymers, PVA is water-soluble. Water-soluble films and sheetings are important PVA products. PVA has oxygen-barrier properties under dry conditions that are superior to those of any other polymer. PVA can be rendered insoluble in water, if needed, by use of chemical agents that cross-link individual polymer strands.

3.11 Structural Characteristics of Phenols

A **phenol** *is an organic compound in which an —OH group is attached to a carbon atom that is part of an aromatic carbon ring system.*

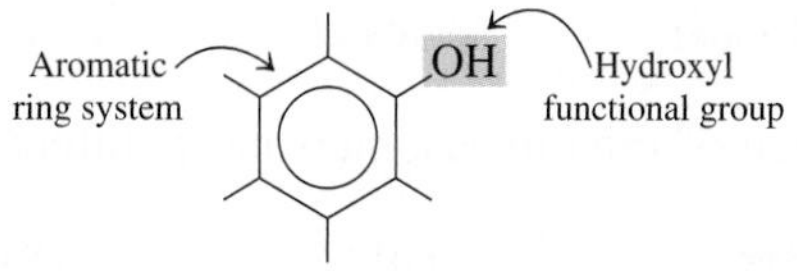

The general formula for phenols is Ar–OH, where Ar represents an *aryl group.* An **aryl group** *is an aromatic carbon ring system from which one hydrogen atom has been removed.*

The generic term *aryl group* (Ar) is the aromatic counterpart of the nonaromatic generic term *alkyl group* (R).

A hydroxyl group is thus the functional group for both phenols and alcohols. The reaction chemistry for phenols is sufficiently different from that for nonaromatic alcohols (Section 3.9) to justify discussing these compounds separately. Remember that phenols contain a "benzene ring" and that the chemistry of benzene is much different from that of other unsaturated hydrocarbons (Section 2.14).

The following are examples of compounds classified as phenols.

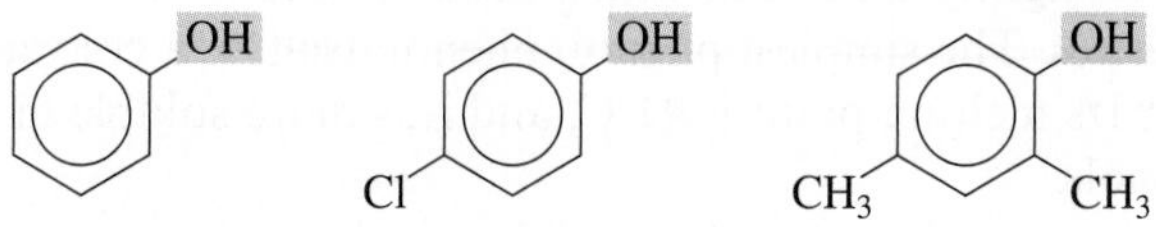

3.12 Nomenclature for Phenols

Besides being the name for a family of compounds, *phenol* is also the IUPAC-approved name for the simplest member of the phenol family of compounds.

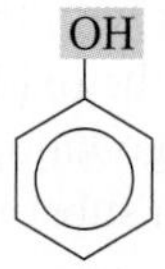

Phenol

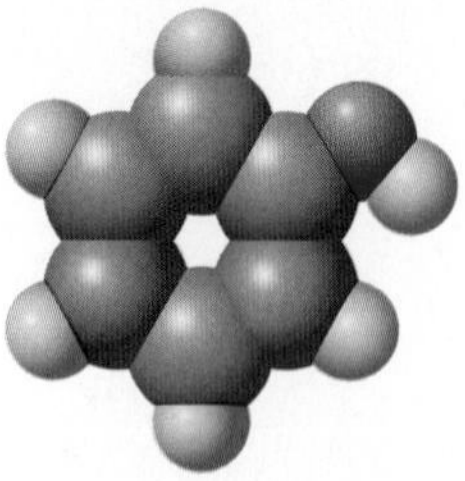

Figure 3.14 A space-filling model for *phenol*, a compound that has an —OH group bonded directly to a benzene (aromatic) ring.

A space-filling model for the compound *phenol* is shown in Figure 3.14. The name *phenol* is derived from a combination of the terms *phen*yl and alcoh*ol*.

The IUPAC rules for naming phenols are simply extensions of the rules used to name benzene derivatives with hydrocarbon or halogen substituents (Section 2.12). The parent name is phenol. Ring numbering always begins with the hydroxyl group and proceeds in the direction that gives the lower number to the next carbon atom bearing a substituent. The numerical position of the hydroxyl group is not specified in the name because it is 1 by definition.

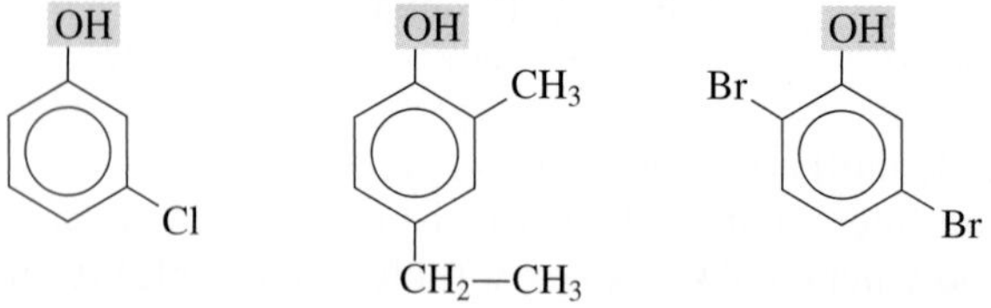

3-Chlorophenol (or *meta*-Chlorophenol) 4-Ethyl-2-methylphenol 2,5-Dibromophenol

Methyl and hydroxy derivatives of phenol have IUPAC-accepted common names. Methylphenols are called cresols. The name *cresol* applies to all three isomeric methylphenols.

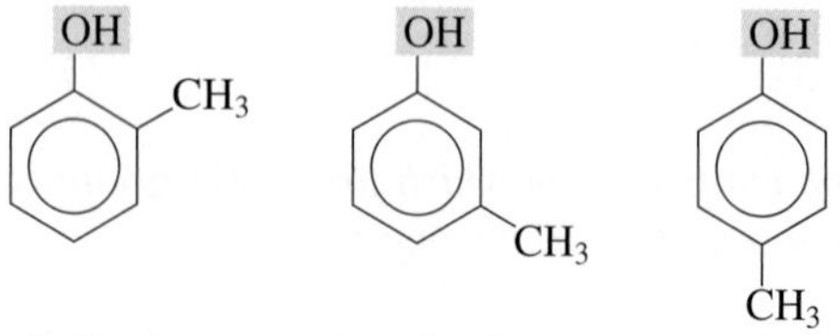

ortho-Cresol *meta*-Cresol *para*-Cresol

For hydroxyphenols, each of the three isomers has a different common name.

Several neurotransmitters in the human body (Section 6.10), including norepinephrine, epinephrine (adrenaline), and dopamine, are catechol derivatives.

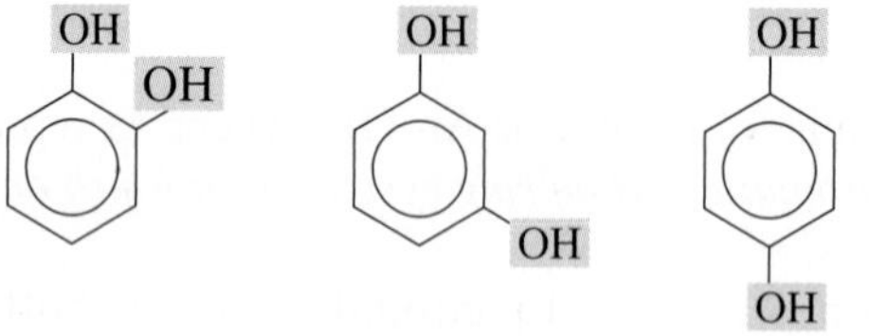

Catechol Resorcinol Hydroquinone

3.13 Physical and Chemical Properties of Phenols

Phenols are generally low-melting solids or oily liquids at room temperature. Most of them are only slightly soluble in water. Many phenols have antiseptic and disinfectant properties. The simplest phenol, phenol itself, is a colorless solid with a medicinal odor. Its melting point is 41°C, and it is more soluble in water than are most other phenols.

It has been previously noted that the chemical properties of phenols are significantly different from those of alcohols (Section 3.11). The similarities and differences between these two reaction chemistries are as follows:

1. Both alcohols and phenols are flammable.
2. Dehydration is a reaction of alcohols but not of phenols; phenols cannot be dehydrated.
3. Both 1° and 2° alcohols are oxidized by mild oxidizing agents. Tertiary (3°) alcohols and phenols do not react with the oxidizing agents that cause 1° and 2° alcohol oxidation. Phenols can be oxidized by stronger oxidizing agents.
4. Both alcohols and phenols undergo halogenation in which the hydroxyl group is replaced by a halogen atom in a substitution reaction.

Acidity of Phenols

One of the most important properties of phenols is their acidity. Unlike alcohols, phenols are weak acids in solution. As acids, phenols have K_a values of about 10^{-10}. Such K_a values are lower than those of most weak inorganic acids (10^{-5} to 10^{-10}). The acid ionization reaction for phenol itself is

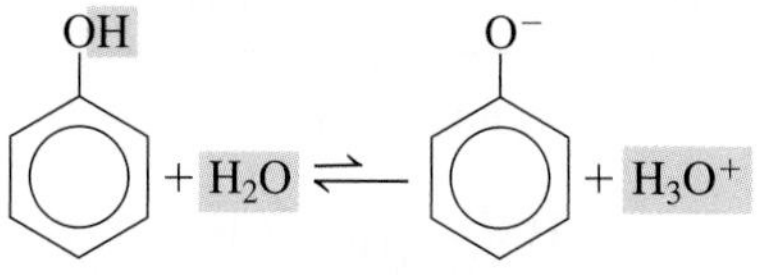

Phenol Phenoxide ion

Note that the negative ion produced from the ionization is called the phenoxide ion. When phenol itself is reacted with sodium hydroxide (a base), the salt sodium phenoxide is produced.

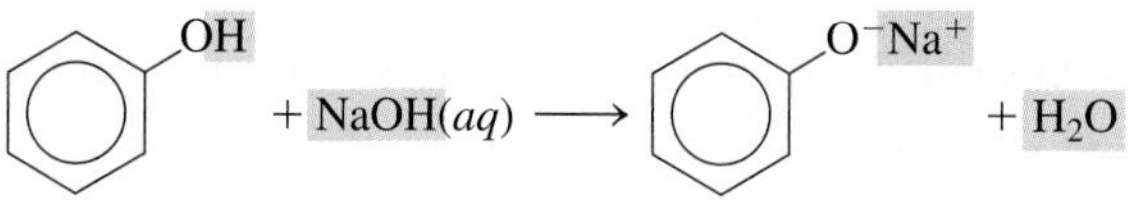

Phenol Sodium phenoxide

3.14 Occurrence of and Uses for Phenols

Dilute (2%) solutions of phenol have long been used as antiseptics. Concentrated phenol solutions, however, can cause severe skin burns. Today, phenol has been largely replaced by more effective phenol derivatives such as 4-hexylresorcinol. The compound 4-hexylresorcinol is an ingredient in many mouthwashes and throat lozenges.

An *antiseptic* is a substance that kills microorganisms on living tissue. A *disinfectant* is a substance that kills microorganisms on inanimate objects.

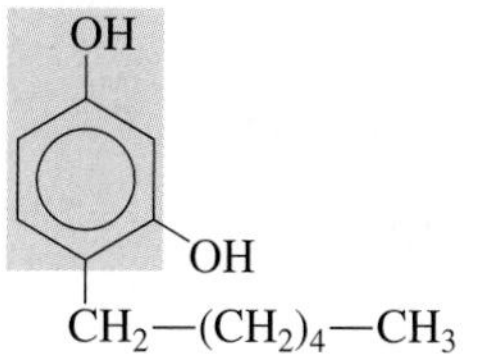

4-Hexylresorcinol

The "parent" name for a benzene ring bearing two hydroxyl groups "meta" to each other is resorcinol (Section 3.12).

The phenol derivatives *o*-phenylphenol and 2-benzyl-4-chlorophenol are the active ingredients in Lysol, a disinfectant for walls, floors, and furniture in homes and hospitals.

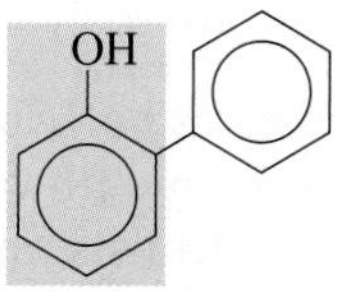

o-Phenylphenol

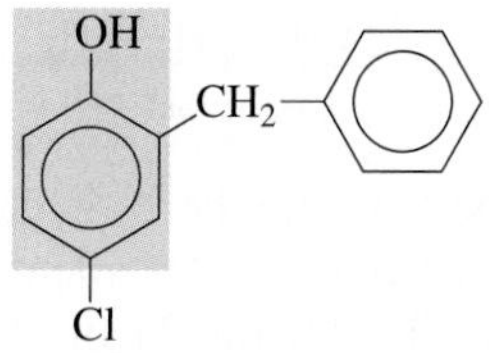

2-Benzyl-4-chlorophenol

A number of phenols possess antioxidant activity. An **antioxidant** *is a substance that protects other substances from being oxidized by being oxidized itself in preference to the other substances.* An antioxidant has a greater affinity for a particular oxidizing agent than do the substances the antioxidant is "protecting"; the antioxidant therefore reacts with the oxidizing agent first. Many foods sensitive to air are protected from oxidation through the use of phenolic antioxidants. Two commercial phenolic antioxidant food additives are BHA (butylated hydroxy anisole) and BHT (butylated hydroxy toluene) (Figure 3.15).

© Henry T. Kaiser/Envision

Figure 3.15 Many commercially baked goods contain the antioxidants BHA and BHT to help prevent spoilage.

Within the human body, natural dietary antioxidants also offer protection against undesirable oxidizing agents. They include vitamin C (Section 10.12), beta-carotene (Section 10.13), vitamin E (Section 10.14), and flavonoids (Section 1.11).

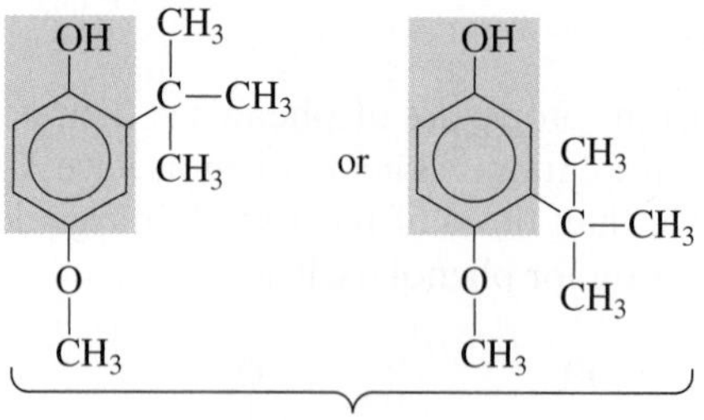

BHA (2 isomers)

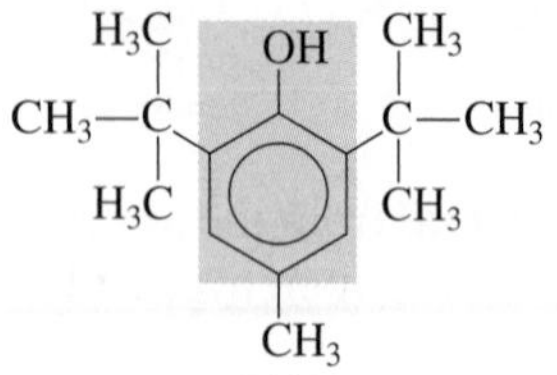

BHT

A naturally occurring phenolic antioxidant that is important in the functioning of the human body is vitamin E (Section 10.14).

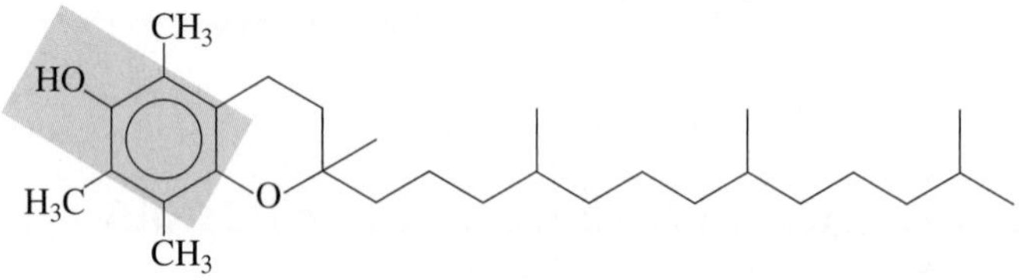

Vitamin E

© Michael Viard/Peter Arnold/Photolibrary

Figure 3.16 Nutmeg tree fruit. A phenolic compound, isoeugenol, is responsible for the odor associated with nutmeg.

A number of phenols found in plants are used as flavoring agents and/or antibacterials. Included among these phenols are

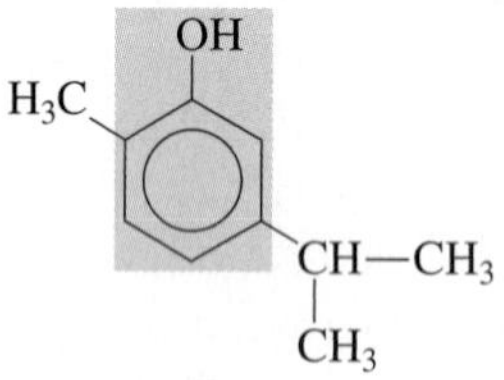

Thymol

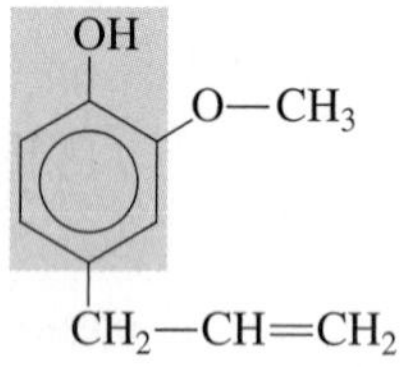

Eugenol

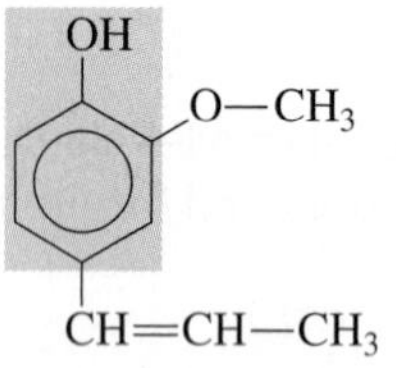

Isoeugenol

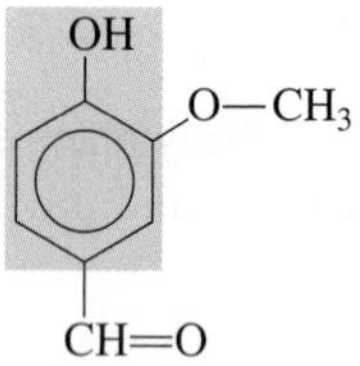

Vanillin

A diisopropyl phenol, with the medical name *Propofol*, is a short-acting intravenous agent used for the induction of general anesthesia and for sedation in several current medical contexts.

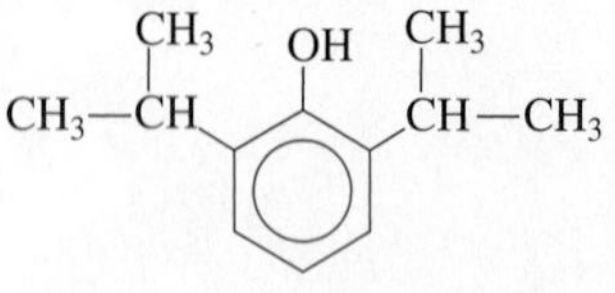

Propofol
(2,6-diisopropylphenol)

It is extensively used in medical contexts such as intensive care unit (ICU) sedation for intubated, mechanically ventilated adults and in procedures such as a colonoscopy. It provides no analgesia (pain relief).

Thymol, obtained from the herb thyme, possesses both flavorant and antibacterial properties. It is used as an ingredient in several mouthwash formulations.

Eugenol is responsible for the flavor of cloves. Dentists traditionally used clove oil as an antiseptic because of eugenol's presence; they use it to a limited extent even today.

Isoeugenol, which differs in structure from eugenol only in the location of the double bond in the hydrocarbon side chain, is responsible for the odor associated with nutmeg (Figure 3.16).

Vanillin, which gives vanilla its flavor, is extracted from the dried seed pods of the vanilla orchid. Natural supplies of vanillin are inadequate to meet demand for this flavoring agent. Synthetic vanillin is produced by oxidation of eugenol. Vanillin is an unusual substance in that even though its odor can be perceived at extremely low concentrations, the strength of its odor does not increase greatly as its concentration is increased.

Certain phenols exert profound physiological effects. For example, the irritating constituents of poison ivy and poison oak are derivatives of catechol (Section 3.12). These skin irritants have 15-carbon alkyl side chains with varying degrees of unsaturation (zero to three double bonds).

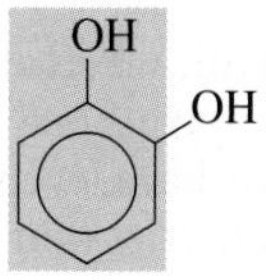
Catechol

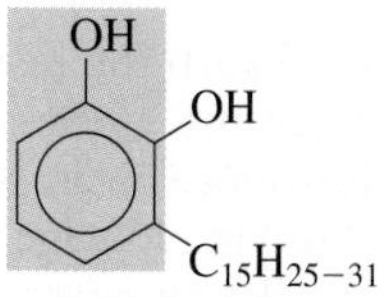

Poison ivy irritants

Polyphenols

A **polyphenol** *is a compound in which two or more phenol entities are present within the compound's structure.* One of the simplest types of polyphenols is that in which two phenol entities are connected via a short carbon chain.

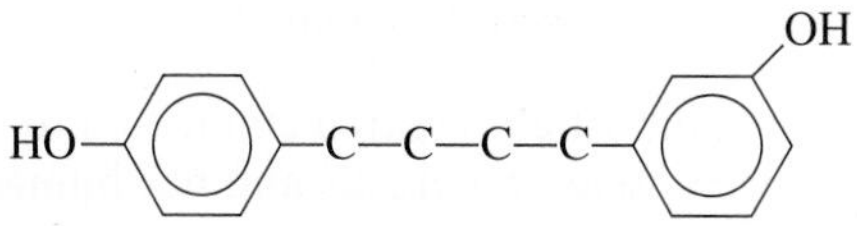

Large numbers of naturally-occurring polyphenols are found within the plant world, including those plants, fruits, and vegetables that are used by humans as a source of food. Generally these dietary polyphenols have more complicated structures than that of the preceding example and exhibit antioxidant properties to varying degrees.

For a typical adult diet in the United States, daily ingestion of dietary polyphenols from plant sources is estimated to be in the range of 200 mg to 1000 mg (1 gram). This is a larger dietary intake than is the intake of the well-studied vitamin dietary antioxidants (β-carotene, vitamin C, and vitamin E), which is about 100 mg a day.

The role that polyphenol antioxidants play in human body chemistry is an area of active research. It is a challenging area of research in which the answers to the questions being posed do not come easily. Any given plant food contains dozens, even hundreds, of different antioxidant compounds, not all of which are polyphenols.

One of the most studied of the many polyphenol antioxidants is the compound resveratrol, a substance that is found in grapes and wines made from the grapes. It is commonly linked to a situation called the "French Paradox." The focus on relevancy feature Chemical Connections 3-B on the next page gives current research findings relative to resveratrol and the associated "French Paradox."

3.15 Structural Characteristics of Ethers

An **ether** *is an organic compound in which an oxygen atom is bonded to two carbon atoms by single bonds.* In an ether, the carbon atoms that are attached to the oxygen atom can be part of alkyl, cycloalkyl, or aryl groups. Examples of ethers include

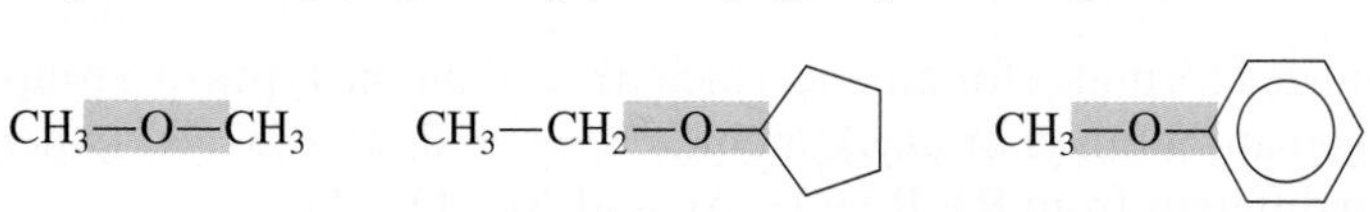

The two groups attached to the oxygen atom of an ether can be the same (first structure), but they need not be so (second and third structures).

All ethers contain a C—O—C unit, which is the ether functional group.

Ether functional group

---C—O—C---

CHEMICAL CONNECTIONS 3-B

Red Wine and Resveratrol

The "French Paradox" is a name associated with a study indicating that people in France were less likely to die of heart attacks than people living in the United States, despite both groups having similar high levels of saturated fats in their diets. A proposed explanation for this "paradox" was the idea that regular moderate consumption of red wine with meals (a common French, but not American, tradition) provides some type of added protection from cardiovascular disease.

This idea that red wine consumption has cardiovascular health benefits has spawned numerous investigational studies concerning what compounds might be present in red wine that produce such benefits. From such studies, one particular compound whose presence in red wine was discovered in 1992 has garnered much attention. It is the compound resveratrol.

Resveratrol (pronounced "ress-ver-A-trole") is an antioxidant compound produced by some plants, including grapes, to protect themselves against environmental stresses such as fungal diseases and sun damage. It is a polyphenol derivative of the aromatic hydrocarbon stilbene.

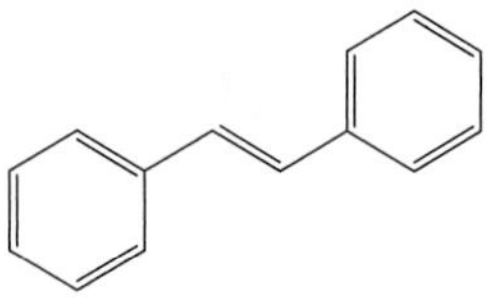

Stilbene

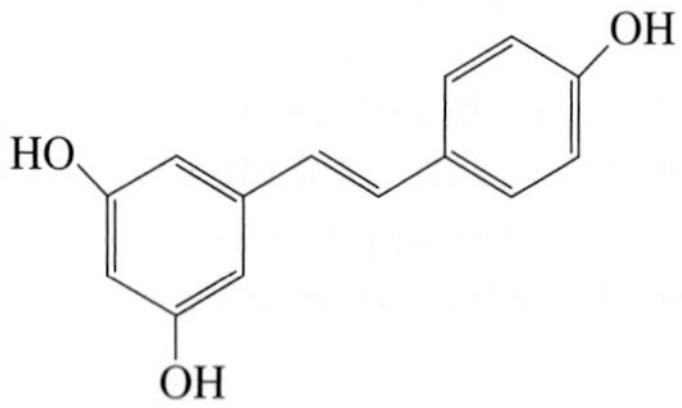

Resveratrol

Because of the double bond present in the carbon chain connecting the two benzene (phenol) centers, *cis*- and *trans*-isomers exist for both stilbene and resveratrol. It is the *trans*-isomer of resveratrol that is produced in plants.

Resveratrol is found in grapes, grape juice, berries of the *vaccinum* species (which includes blueberries and cranberries), and peanuts. Grapes contain the highest levels of resveratrol, where it is concentrated in the skin of the grapes. Red wines contain more resveratrol than white wines because red wine is fermented with grape skins, allowing the wine to absorb the resveratrol, whereas white wine is fermented after the grape skins have been removed. On an ounce-for-ounce basis, peanuts have levels of resveratrol about half those in red wine. Blueberries and cranberries contain much smaller amounts of the substance.

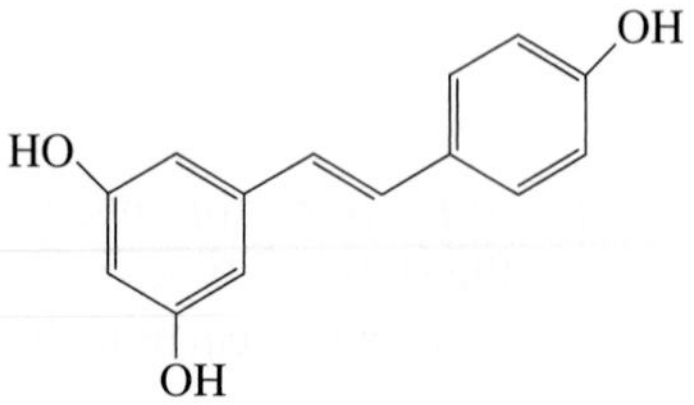

trans-Resveratrol

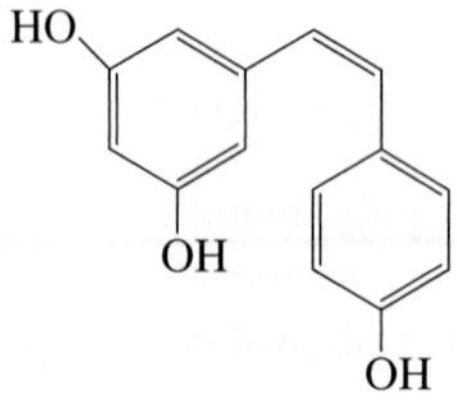

cis-Resveratrol

Almost all research conducted to date on resveratrol has been done on animals and not humans. Research involving mice given resveratrol shows a positive correlation with reduced risk of inflammation and blood clotting, as well as a protective effect against obesity and diabetes. Of importance, the dose of resveratrol used in the mice studies would be equivalent to people consuming more than 100 bottles of wine a day. Resveratrol administration has also increased the life spans of yeast, worms, fruit flies, fish, and mice that were fed a high-calorie diet. Whether such effects would be observed in humans is not yet known.

Resveratrol supplements are now available in the United States. Their source varies from extracts of the plant kojokon to red wine extracts and red grape extracts. The effectiveness and safety of the supplements is not well established. It is known that throat lozenges are the most effective way of administering resveratrol to humans. About 70% of an orally given resveratrol dose is absorbed; however, oral bioavailability is low because the absorbed resveratrol is rapidly metabolized in the intestine and liver. The oral availability of resveratrol from wine has been found to be no higher than that from a pill.

Resveratrol concentration levels found in the human body as the result of moderate drinking of red wine do not appear to be sufficiently high to explain the observations associated with the "French Paradox."

Generalized formulas for ethers, which depend on the types of groups attached to the oxygen atom (alkyl or aryl), include R—O—R, R—O—R′ (where R′ is an alkyl group different from R), R—O—Ar, and Ar—O—Ar.

Structurally, an ether can be visualized as a derivative of water in which both hydrogen atoms have been replaced by hydrocarbon groups (Figure 3.17). Note that unlike alcohols and phenols, ethers do not possess a hydroxyl (—OH) group.

$$\underset{\text{Water}}{H-\ddot{\underset{..}{O}}-H} \qquad \underset{\text{An ether}}{R-\ddot{\underset{..}{O}}-R}$$

3.16 Nomenclature for Ethers

Common names are almost always used for ethers whose alkyl groups contain four or fewer carbon atoms. There are two rules, one for unsymmetrical ethers (two different alkyl/aryl groups) and one for symmetrical ethers (two identical alkyl/aryl groups).

Rule 1: *For unsymmetrical ethers, name both hydrocarbon groups bonded to the oxygen atom in alphabetical order and add the word* ether, *separating the words with a space. Such ether names have three separate words within them.*

$CH_3—O—CH_2—CH_3$

Ethyl methyl ether

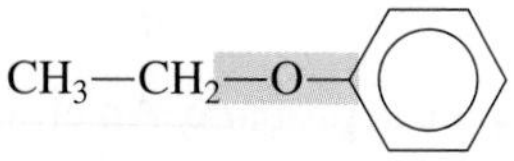

Ethyl phenyl ether

Rule 2: *For symmetrical ethers, name the alkyl group, add the prefix* di-, *and then add the word* ether, *separating the words with a space. Such ether names have two separate words within them.*

$CH_3—O—CH_3$ Dimethyl ether

$CH_3—CH_2—O—CH_2—CH_3$ Diethyl ether

Ethers with more complex alkyl/aryl groups are named using the IUPAC system. In this system, ethers are named as substituted hydrocarbons. The smaller hydrocarbon attachment and the oxygen atom are called an *alkoxy group,* and this group is considered a substituent on the larger hydrocarbon group. An **alkoxy group** *is an —OR group, an alkyl (or aryl) group attached to an oxygen atom.* Simple alkoxy groups include the following:

$CH_3—O—$ Methoxy group

$CH_3—CH_2—O—$ Ethoxy group

$CH_3—CH_2—CH_2—O—$ Propoxy group

The general symbol for an alkoxy group is R—O— (or RO—).

The rules for naming an ether using the IUPAC system are

Rule 1: *Select the longest carbon chain and use its name as the base name.*

Rule 2: *Change the* -yl *ending of the other hydrocarbon group to* -oxy *to obtain the alkoxy group name;* methyl *becomes* methoxy, ethyl *becomes* ethoxy, *etc.*

Rule 3: *Place the alkoxy name, with a locator number, in front of the base chain name.*

Two examples of IUPAC ether nomenclature, with the alkoxy groups present highlighted in each structure, are:

$(CH_3—O)—CH_2—CH_2—CH_2—CH_3$

1-Methoxybutane

$CH_3—CH(CH_3)—CH_2—(O—CH_2—CH_3)$

1-Ethoxy-2-methylpropane

The simplest aromatic ether involves a methoxy group attached to a benzene ring. This ether goes by the common name *anisole.*

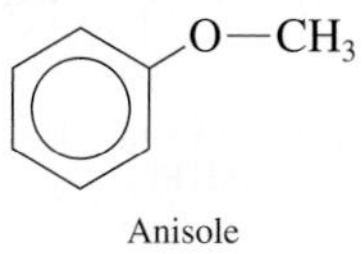

Anisole

Derivatives of anisole are named as substituted anisoles, in a manner similar to that for substituted phenols (Section 3.12). Anisole derivatives were encountered in Section 3.14 when considering antioxidant food additives: BHAs can be viewed as derivatives of either phenol or anisole.

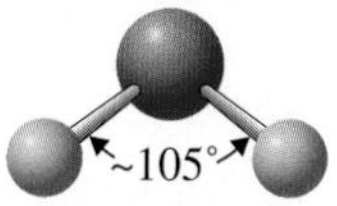

Water (HOH)

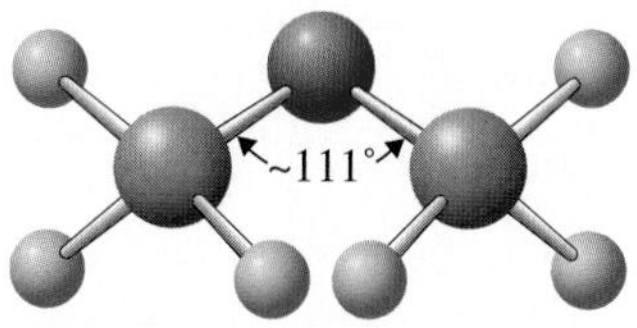

Dimethyl ether (CH_3OCH_3)

Figure 3.17 The similar shapes of water and dimethyl ether molecules. Dimethyl ether may be viewed structurally as a dialkyl derivative of water.

Line-angle structural formulas for selected simple ethers:

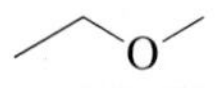

Ethyl methyl ether (methoxyethane)

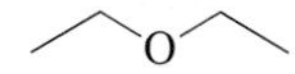

Diethyl ether (ethoxyethane)

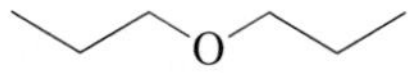

Dipropyl ether (1-propoxypropane)

It is possible to have compounds that contain both ether and alcohol functional groups such as

$CH_3—CH(OH)—CH_2—CH_2—O—CH_3$

4-Methoxy-2-butanol

(The alcohol functional group has higher priority in IUPAC nomenclature, so the compound is named as an alcohol rather than as an ether.)

The compound responsible for the characteristic odor of anise and fennel is *anethole,* an allyl derivative of anisole.

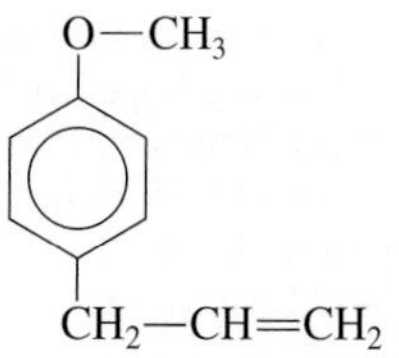

EXAMPLE 3.5 Determining IUPAC Names for Ethers

Name the following ethers utilizing IUPAC nomenclature rules.

a. $CH_3—CH_2—O—CH_2—CH_2—CH_3$

b. 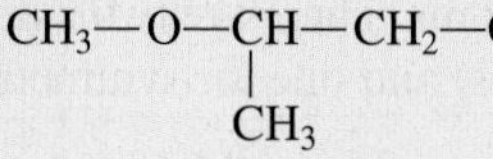

c. 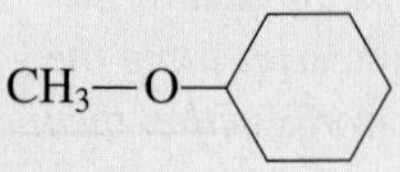

d. Ethyl methyl ether

Solution

a. The base name is propane. An ethoxy group is attached to carbon-1 of the propane chain.

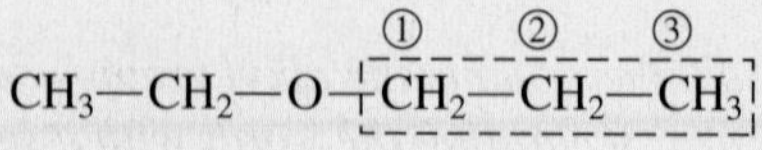

The IUPAC name is 1-ethoxypropane.

b. The base name is butane, as the longest carbon chain contains four carbon atoms.

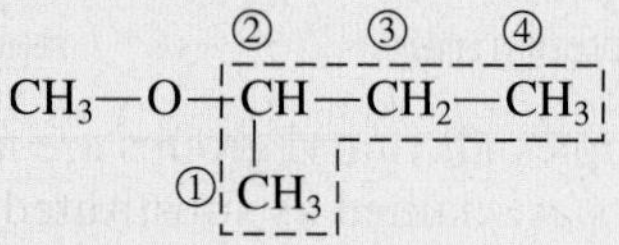

The IUPAC name is 2-methoxybutane.

c. The base name is cyclohexane. The complete IUPAC name is methoxycyclohexane. No number is needed to locate the methoxy group since all ring carbon atoms are equivalent to each other.

d. The ether structure is $CH_3—CH_2—O—CH_3$, and the IUPAC name is methoxyethane.

The contrast between IUPAC and common names for ethers is as follows:

IUPAC (one word)

alkoxyalkane

2-methoxybutane

Common (three or two words)

alkyl alkyl ether

ethyl methyl ether

or

dialkyl ether

dipropyl ether

Practice Exercise 3.5

Name the following ethers utilizing IUPAC nomenclature rules.

a. $CH_3—CH_2—CH_2—O—CH_2—CH_2—CH_3$

b. 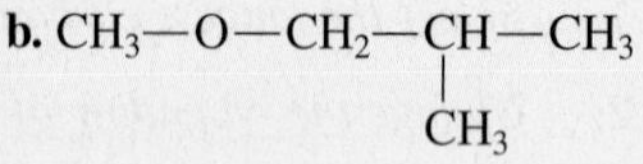

c. 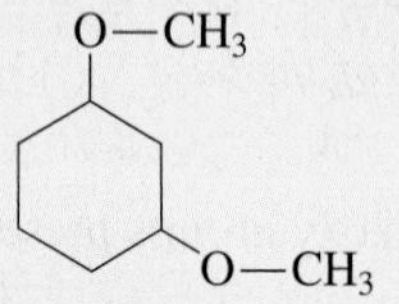

d. Dimethyl ether

Answers: **a.** 1-Propoxypropane; **b.** 1-Methoxy-2-methylpropane; **c.** 1, 3-Dimethoxycyclohexane; **d.** Methoxymethane

The ether MTBE (methyl *tert*-butyl ether) has been a widely used gasoline additive since the early 1980s.

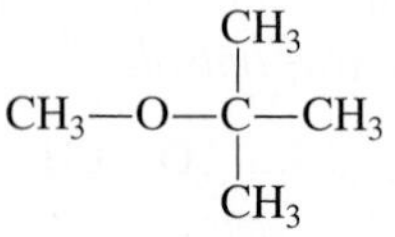

Methyl *tert*-butyl ether (MTBE)

Technically, the name methyl *tert*-butyl ether (MTBE) is incorrect because the convention for naming ethers dictates an alphabetical ordering of alkyl groups (*tert*-butyl methyl ether). However, the compound is called MTBE rather than TBME by those in the petroleum industry and by environmental scientists.

As an additive, MTBE not only raises octane levels but also functions as a clean-burning "oxygenate" in EPA-mandated reformulated gasolines used to improve air quality in polluted areas. The amount of MTBE used in gasoline is now decreasing in response to a growing problem: contamination of water supplies by small amounts of MTBE from leaking gasoline tanks and from spills. MTBE in

CHEMICAL CONNECTIONS 3-C

Ethers as General Anesthetics

For many people, the word *ether* evokes thoughts of hospital operating rooms and anesthesia. This response derives from the *former* large-scale use of diethyl ether as a general anesthetic. In 1846, the Boston dentist William Morton was the first to demonstrate publicly the use of diethyl ether as a surgical anesthetic.

In many ways, diethyl ether is an ideal general anesthetic. It is relatively easy to administer, it is readily made in pure form, and it causes excellent muscle relaxation. There is less danger of an overdose with diethyl ether than with almost any other anesthetic because there is a large gap between the effective level for anesthesia and the lethal dose.

Despite these ideal properties, diethyl ether is rarely used today because of two drawbacks: (1) It causes nausea and irritation of the respiratory passages and (2) it is a highly flammable substance, forming explosive mixtures with air, which can be set off by a spark.

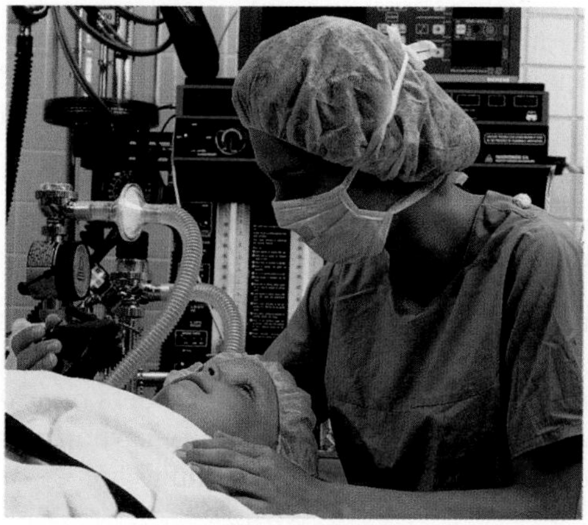

© SIU/Peter Arnold/Photolibrary

Preparing to administer an anesthetic to a child.

By the 1930s, nonether anesthetics had been developed that solved the problems of nausea and irritation. They also, however, were extremely flammable compounds. The simple hydrocarbon cyclopropane was the most widely used of these newer compounds.

It was not until the late 1950s and early 1960s that nonflammable general anesthetics became available. Anesthetic nonflammability was achieved by incorporating halogen atoms into anesthetic molecules. Three of the most used of these "halogenated" anesthetics are enflurane, isoflurane, and halothane.

Enflurane and isoflurane, which are constitutional isomers, are hexahalogenated ethers.

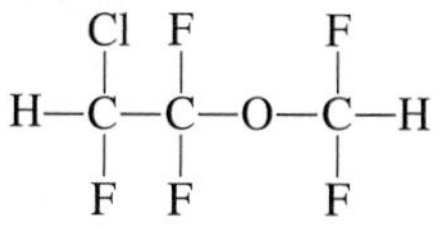

Enflurane

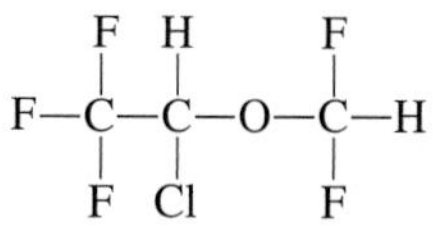

Isoflurane

With these compounds, induction of anesthesia can be achieved in less than 10 minutes with an inhaled concentration of 3% in oxygen.

Halothane, which is potent at relatively low doses and whose effects wear off quickly, is a pentahalogenated alkane derivative rather than an ether.

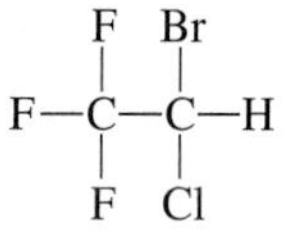

Halothane

It is the only inhalation anesthetic that contains a bromine atom.

Phase-out of the use of the preceding three compounds began in the late 1980s and early 1990s. They have largely been replaced by a "second generation" of similar halogenated ethers with even better anesthetic properties. In general, these new compounds have more fluorine atoms and fewer chlorine atoms. Two of the most prominent of these second-generation anesthetic agents now in use are Sevoflurane and Desflurane.

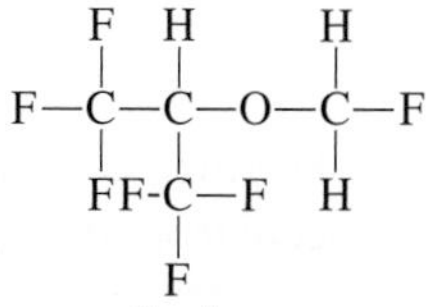

Sevoflurane

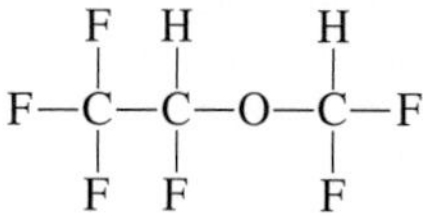

Desflurane

the water supplies is not a health-and-safety issue at this time, but its presence does affect taste and odor in contaminated supplies.

Compounds with ether functional groups occur in a variety of plants. The phenolic flavoring agents eugenol, isoeugenol, and vanillin (Section 3.14) are also ethers; each has a methoxy substituent on the ring.

Use of ethers as anesthetics is a vital part of current medical practice. Ethers presently in use are halogenated, rather than simpler nonhalogenated, ethers. It has been found that introduction of halogen atoms into an ether reduces or eliminates its flammability. Many polyhalogenated ethers are nonflammable. The focus on relevancy feature Chemical Connections 3-C above considers the use of ethers, from early nonhalogenated forms to current polyhalogenated forms, as anesthetics.

3.17 Isomerism for Ethers

Ethers contain two carbon chains (two alkyl groups), unlike the one carbon chain found in alcohols. Constitutional isomerism possibilities in ethers depend on (1) the partitioning of carbon atoms between the two alkyl groups and (2) isomerism possibilities for the individual alkyl groups present. Isomerism is not possible for a C_2 ether (two methyl groups) or a C_3 ether (a methyl and an ethyl group). For C_4 ethers, isomerism arises not only from carbon-atom partitioning between the alkyl groups (C_1—C_3 and C_2—C_2) but also from isomerism within a C_3 group (propyl and isopropyl). There are three C_4 ether constitutional isomers.

CH_3—CH_2—O—CH_2—CH_3
Diethyl ether
(C_2—C_2)

CH_3—O—CH_2—CH_2—CH_3
Methyl propyl ether
(C_1—C_3)

CH_3—O—CH—CH_3
|
CH_3
Isopropyl methyl ether
(C_1—C_3)

For C_5 ethers, carbon-partitioning possibilities are C_2—C_3 and C_1—C_4. For C_4 groups, there are four isomeric variations: butyl, isobutyl, *sec*-butyl, and *tert*-butyl (Section 1.11).

Functional Group Isomerism

Ethers and alcohols with the same number of carbon atoms and the same degree of saturation have the same molecular formula. The simplest manifestation of this phenomenon involves dimethyl ether, the C_2 ether, and ethyl alcohol, the C_2 alcohol. Both have the molecular formula C_2H_6O.

CH_3—O—CH_3
Dimethyl ether

CH_3—CH_2—OH
Ethyl alcohol

With the same molecular formula and different structural formulas, these two compounds are constitutional isomers. This type of constitutional isomerism is the subtype called *functional group isomerism.* **Functional group isomers** *are constitutional isomers that contain different functional groups.* When three carbon atoms are present, the ether–alcohol functional group isomerism possibilities are

Later in this chapter (Section 3.21) and in each of the next two chapters, other pairs of functional groups for which functional group isomerism is possible will be encountered.

CH_3—CH_2—O—CH_3
Ethyl methyl ether

CH_3—CH_2—CH_2—OH
Propyl alcohol

CH_3—CH—OH
|
CH_3
Isopropyl alcohol

All three compounds have the molecular formula C_3H_8O. Figure 3.18 shows molecular models for the isomeric propyl alcohol and ethyl methyl ether molecules.

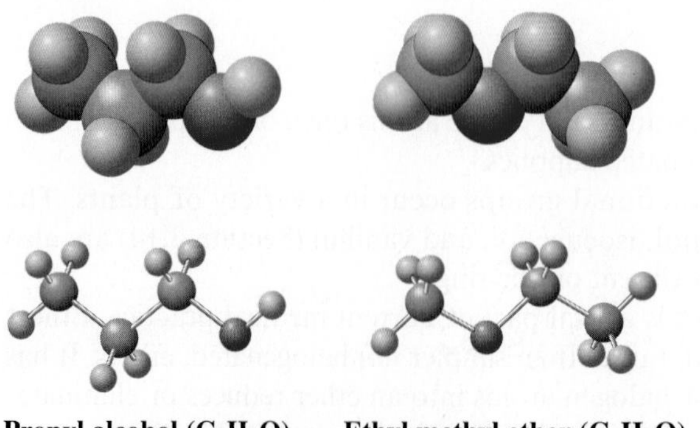

Propyl alcohol (C_3H_8O) **Ethyl methyl ether (C_3H_8O)**

Figure 3.18 Alcohols and ethers with the same number of carbon atoms and the same degree of saturation are functional group isomers, as is illustrated here for propyl alcohol and ethyl methyl ether.

3.18 Physical and Chemical Properties of Ethers

The boiling points of ethers are similar to those of alkanes of comparable molecular mass and are much lower than those of alcohols of comparable molecular mass.

Alkane	$CH_3—CH_2—CH_2—CH_2—CH_3$	Mol. mass = 72 amu bp = 36°C
Ether	$CH_3—CH_2—O—CH_2—CH_3$	Mol. mass = 74 amu bp = 35°C
Alcohol	$CH_3—CH_2—CH_2—CH_2—OH$	Mol. mass = 74 amu bp = 117°C

The much higher boiling point of the alcohol results from hydrogen bonding between alcohol molecules. Ether molecules, like alkanes, cannot hydrogen-bond to one another. Ether oxygen atoms have no hydrogen atom attached directly to them. Figure 3.19 is a physical-state summary for unbranched alkyl alkyl ethers where the alkyl groups range in size from C_1 to C_4.

Ethers, in general, are more soluble in water than are alkanes of similar molecular mass because ether molecules are able to form hydrogen bonds with water (Figure 3.20).

Ethers have water solubilities similar to those of alcohols of the same molecular mass. For example, diethyl ether and butyl alcohol have the same solubility in water. Because ethers can also hydrogen-bond to alcohols, alcohols and ethers tend to be mutually soluble. Nonpolar substances tend to be more soluble in ethers than in alcohols because ethers have no hydrogen-bonding network that has to be broken up for solubility to occur.

Two chemical properties of ethers are especially important.

1. *Ethers are flammable.* Special care must be exercised in laboratories where ethers are used. Diethyl ether, whose boiling point of 35°C is only a few degrees above room temperature, is a particular flash-fire hazard.
2. *Ethers react slowly with oxygen from the air to form unstable hydroperoxides and peroxides.*

R—O—O—H	R—O—O—R
Hydroperoxide	Peroxide

Such compounds, when concentrated, represent an explosion hazard and must be removed before *stored* ethers are used.

Like alkanes, ethers are unreactive toward acids, bases, and oxidizing agents. Like alkanes, they do undergo combustion and halogenation reactions.

The general chemical unreactivity of ethers, coupled with the fact that most organic compounds are ether-soluble, makes ethers excellent solvents in which to carry out organic reactions. Their relatively low boiling points simplify their separation from the reaction products.

A chemical reaction for the preparation of ethers has been previously considered. In Section 3.9 it was noted that the intermolecular dehydration of a primary alcohol will produce an ether.

$$\underset{\text{Alcohol}}{R_1—OH} + \underset{\text{Alcohol}}{R_2—OH} \xrightarrow[\text{Heat}]{H^+} \underset{\text{Ether}}{R_1—O—R_2} + \underset{\text{Water}}{H_2O}$$

Although additional methods exist for ether preparation, they will not be considered in this text.

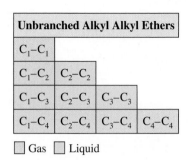

Unbranched Alkyl Alkyl Ethers			
C_1–C_1			
C_1–C_2	C_2–C_2		
C_1–C_3	C_2–C_3	C_3–C_3	
C_1–C_4	C_2–C_4	C_3–C_4	C_4–C_4

Gas   Liquid

Figure 3.19 A physical-state summary for unbranched alkyl alkyl ethers at room temperature and pressure.

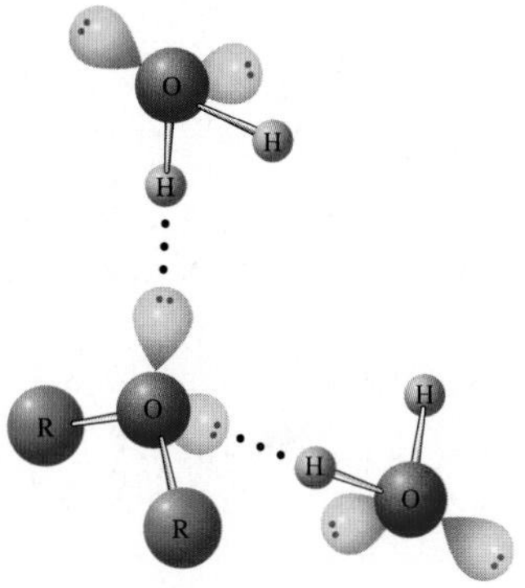

Figure 3.20 Although ether molecules cannot hydrogen-bond to one another, they can hydrogen-bond to water molecules. Such hydrogen bonding causes ethers to be more soluble in water than alkanes of similar molecular mass.

The term *ether* comes from the Latin *aether,* which means "to ignite." This name is given to these compounds because of their high vapor pressure at room temperature, which makes them very flammable.

3.19 Cyclic Ethers

Cyclic ethers contain ether functional groups as part of a ring system. Some examples of such cyclic ethers, along with their common names, follow.

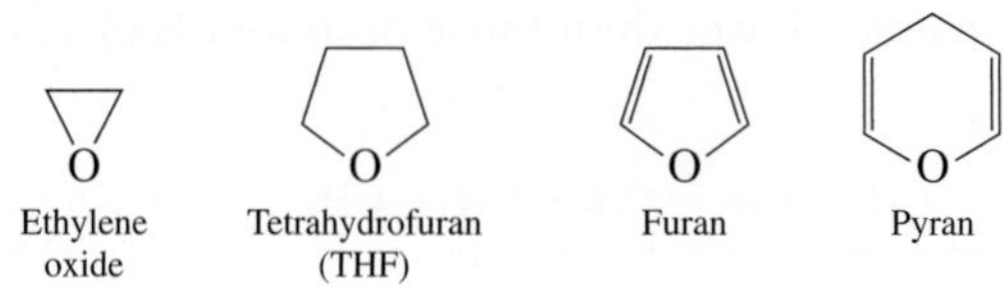

Ethylene oxide has few direct uses. Its importance is as a starting material for the production of ethylene glycol (Section 3.5), a major component of automobile antifreeze. THF is a particularly useful solvent in that it dissolves many organic compounds and yet is miscible with water. In the chapter on carbohydrate chemistry (Chapter 7), many cyclic structures that are polyhydroxy derivatives of the five-membered (furan) and six-membered (pyran) cyclic ether systems will be encountered. These carbohydrate derivatives are called *furanoses* and *pyranoses,* respectively (Section 7.10).

Cyclic ethers are our first encounter with heterocyclic organic compounds. A **heterocyclic organic compound** *is a cyclic organic compound in which one or more of the carbon atoms in the ring have been replaced with atoms of other elements.* The hetero atom is usually oxygen or nitrogen.

We have just seen that *cyclic ethers*—compounds in which the ether functional group is part of a ring system—exist. In contrast, cyclic alcohols—compounds in which the alcohol functional group is part of a ring system—do not exist. To incorporate an alcohol functional group into a ring system would require an oxygen atom with three bonds, and oxygen atoms form only two bonds.

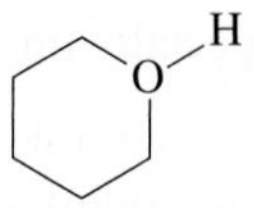

The oxygen atom in this structure has three bonds, which is not possible.

Compounds such as

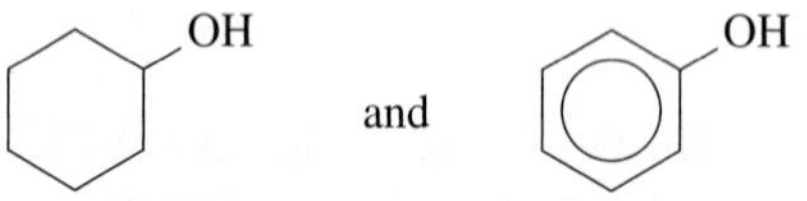

which do exist, are not cyclic alcohols in the sense in which the term is now being used because the alcohol functional group is attached to a ring system rather than being part of it.

Vitamin E, whose structure was given in Section 3.14, is both a phenol and a cyclic ether.

Cyclic ether systems are often encountered in "complex" molecules in which several functional groups are present. The molecule THC, a physiologically active ingredient present in marijuana, is such a molecule. THC's structure is based on a fused three-ring system in which one ring is a cycloalkene, one ring is a phenol, and one ring is a cyclic ether. The focus on relevancy feature Chemical Connections 3-D on the next page further considers the structure of THC and the physiological effects associated with the presence of THC in the human body.

3.20 Sulfur Analogs of Alcohols

Occasionally the term *thiol* is expanded to the term *thioalcohol*. In nomenclature, however, the correct term is *thiol*.

Many organic compounds containing oxygen have sulfur analogs, in which a sulfur atom has replaced an oxygen atom. Sulfur is in the same group of the periodic table as oxygen, so the two elements have similar electron configurations.

Thiols, the sulfur analogs of alcohols, contain —SH functional groups instead of —OH functional groups. The thiol functional group is called a *sulfhydryl group.* A **sulfhydryl group** *is the —SH functional group.* A **thiol** *is an organic compound in*

CHEMICAL CONNECTIONS 3-D

Marijuana: The Most Commonly Used Illicit Drug

Prepared from the leaves, flowers, seeds, and small stems of a hemp plant called *Cannabis sativa,* marijuana, which is also called pot or grass, is the most commonly used illicit drug in the United States. The most active ingredient of the many in marijuana is the molecule *tetrahydrocannabinol,* called THC for short. Three different functional groups are present in a THC molecule; it is a phenol, a cyclic ether, and a cycloalkene.

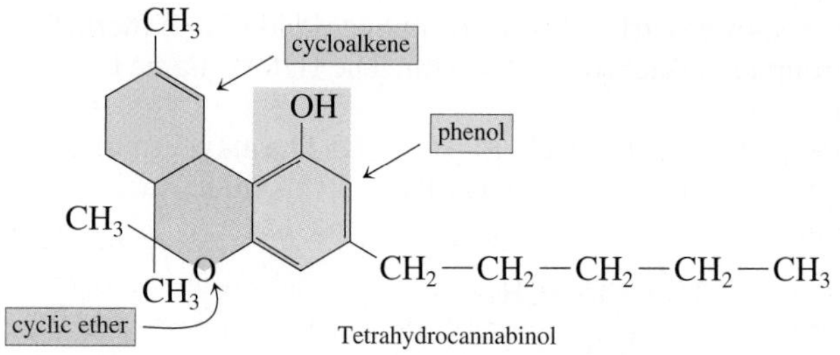

Tetrahydrocannabinol

The THC content of marijuana varies considerably. Most marijuana sold in the North American illegal drug market has a THC content of 1% to 2%.

Marijuana has a pharmacology unlike that of any other drug. A marijuana "high" is a combination of sedation, tranquilization, and mild hallucination. THC readily penetrates the brain. The portions of the brain that involve memory and motor control contain the receptor sites where THC molecules interact. Even moderate doses of marijuana cause short-term memory loss. Marijuana unquestionably impairs driving ability, even after ordinary social use. THC readily crosses the placental barrier and reaches the fetus. Heavy marijuana users experience inflammation of the bronchi, sore throat, and inflamed sinuses. Increased heart rate, to as high as a dangerous 160 beats per minute, can occur with marijuana use.

The onset of action of THC is usually within minutes after smoking begins, and peak concentration in plasma occurs in 10 to 30 minutes. Unless more is smoked, the effects seldom last longer than 2 to 3 hours. Because THC is only slightly soluble in water, it tends to be deposited in fatty tissues. Unlike alcohol, THC persists in the bloodstream for several days, and the products of its breakdown remain in the blood for as long as 8 days.

New research indicates that physical dependence on THC can develop. Drug withdrawal symptoms are seen in some individuals who have been exposed repeatedly to high doses.

which a sulfhydryl group is bonded to a saturated carbon atom. An older term used for thiols is *mercaptans.* Contrasting the general structures for alcohols and thiols, we have

Hydroxyl group → R—OH (An alcohol) and Sulfhydryl group → R—SH (A thiol)

Nomenclature for Thiols

Thiols are named in the same way as alcohols in the IUPAC system, except that the *-ol* becomes *-thiol.* The suffix *-thiol* indicates the substitution of a sulfur atom for an oxygen atom in a compound.

The root *thio-* indicates that a sulfur atom has replaced an oxygen atom in a compound. It originates from the Greek *theion,* meaning "brimstone," which is an older name for the element sulfur.

$CH_3—CH(OH)—CH_2—CH_3$ 2-Butanol

$CH_3—CH(SH)—CH_2—CH_3$ 2-Butanethiol

As in the case of diols and triols, the *-e* at the end of the alkane name is also retained for thiols.

Common names for thiols are based on use of the term *mercaptan,* the older name for thiols. The name of the alkyl group present (as a separate word) precedes the word *mercaptan.*

Even though thiols have a higher molecular mass than alcohols with the same number of carbon atoms, they have much lower boiling points because they do not exhibit hydrogen bonding as alcohols do.

$CH_3—CH_2—SH$ Ethyl mercaptan

$CH_3—CH(CH_3)—SH$ Isopropyl mercaptan

EXAMPLE 3.6 Determining IUPAC and Common Names for Thiols

Convert each of the following common names for thiols to IUPAC names or vice versa.

a. Propyl mercaptan **b.** Isobutyl mercaptan
c. 1-Butanethiol **d.** 2-Propanethiol

Solution

a. The structural formula for propyl mercaptan is $CH_3—CH_2—CH_2—SH$. In the IUPAC system, the name base is propane; the complete name is 1-propanethiol.

b. The structural formula for isobutyl mercaptan is

$$\begin{array}{l} CH_3—CH—CH_2—SH \\ \qquad\quad | \\ \qquad\; CH_3 \end{array}$$

The longest carbon chain has three carbon atoms (propane), and both a methyl group and a sulfhydryl group are attached to the chain. The IUPAC name is 2-methyl-1-propanethiol.

c. The structure of this thiol is $CH_3—CH_2—CH_2—CH_2—SH$. The alkyl group is a butyl group, giving a common name of butyl mercaptan for this thiol.

d. The thiol structural formula is

$$\begin{array}{l} CH_3—CH—CH_3 \\ \qquad\quad | \\ \qquad\; SH \end{array}$$

The sulfhydryl group is attached to an isopropyl group; the common name is isopropyl mercaptan.

Practice Exercise 3.6

Convert each of the following common names for thiols to IUPAC names or vice versa.

a. Methyl mercaptan **b.** *sec*-Butyl mercaptan
c. 2-Methyl-2-propanethiol **d.** 1-Pentanethiol

Answers: **a.** Methanethiol; **b.** 2-Butanethiol; **c.** *tert*-Butyl mercaptan; **d.** Pentyl mercaptan

Properties of Thiols

Two important properties of thiols are lower boiling points than their analogous alcohols and strong, usually disagreeable, odors. Their lower boiling points result from a lack of hydrogen bonding among thiol molecules; such molecules lack hydrogen atoms bonded to a very electronegative element (F, O, or N). The threshold of detection for thiol odors is exceptionally low (in the 1–3 parts per billion range); hence they are detectable long before concentration levels reach the toxic range.

Methanethiol (methyl mercaptan), with the formula $CH_3—SH$, is a colorless flammable gas, with an odor often described as that of rotten cabbage. It is the substance primarily responsible for "bad breath" and the smell of flatulence. Animals have the ability to produce methanethiol in their intestinal tract due to the action of bacteria on sulfur-containing proteins. Methanethiol production contributes to the odor associated with feedlots and barnyards.

In humans, methanethiol production is often a by-product of the metabolism of asparagus. When such metabolism occurs, a very strong odor is associated with urine in as few as 15 minutes after eating asparagus. This is not a universal problem because of genetic differences in how people metabolize sulfur-containing compounds. Studies show that a strong urine odor occurs in about 40% of the population; these individuals lack the ability to convert odiferous sulfur compounds present in asparagus to odor-free sulfate.

Ethanethiol (ethyl mercaptan), with the formula $CH_3—CH_2—SH$, is a very low-boiling flammable liquid with an odor described as very strong green onions.

When natural gas distributors began adding thiols to otherwise odorless natural gas (so that leaks could be detected), the original odorant was ethanethiol. Current natural gas odorants are usually mixtures of thiols and sulfides (Section 3.21) with *tert*-butylthiol as the major odiferous constituent.

The scent of skunks (Figure 3.21) is due primarily to two thiols.

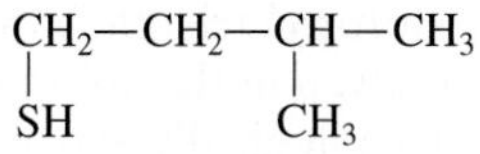

3-Methyl-1-butanethiol

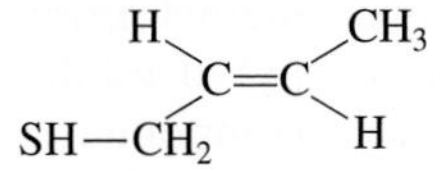

trans-2-Butene-1-thiol

Thiols are easily oxidized but yield different products than their alcohol analogs. Thiols form *disulfides.* Each of two thiol groups loses a hydrogen atom, thus linking the two sulfur atoms together via a disulfide group, —S—S—.

$$\text{R—SH} + \text{HS—R} \xrightarrow{\text{Oxidation}} \text{R—S—S—R} + 2\text{H}$$

A disulfide

Reversal of this reaction, a reduction process, is also readily accomplished. Breaking of the disulfide bond regenerates two thiol molecules.

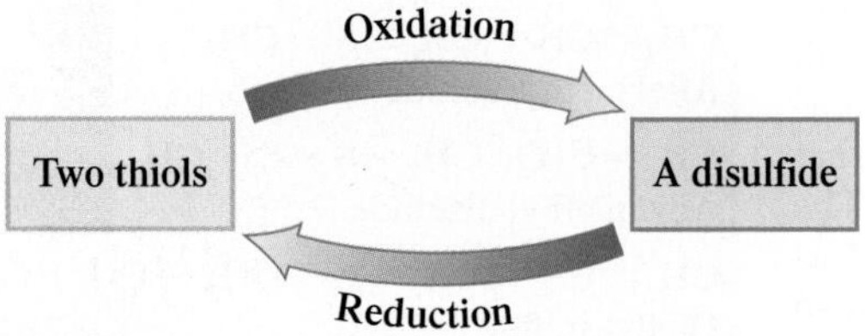

These two "opposite reactions" are of biological importance in the area of protein chemistry. Disulfide bonds formed from the interaction of two —SH groups contribute in a major way to protein structure (Chapter 9).

3.21 Sulfur Analogs of Ethers

Sulfur analogs of ethers are known as thioethers (or sulfides). A **thioether** *is an organic compound in which a sulfur atom is bonded to two carbon atoms by single bonds.* The generalized formula for a thioether is R—S—R. Like thiols, thioethers have strong characteristic odors.

Thioethers are named in the same way as ethers, with *sulfide* used in place of *ether* in common names and *alkylthio* used in place of *alkoxy* in IUPAC names.

$CH_3—S—CH_3$

Dimethyl sulfide (methylthiomethane)

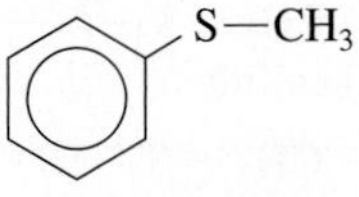

Methyl phenyl sulfide (methylthiobenzene)

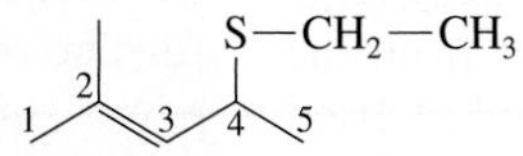

4-(ethylthio)-2-Methyl-2-pentene

In general, thiols are more reactive than their alcohol counterparts, and thioethers are more reactive than their corresponding ethers. The larger size of a sulfur atom compared to an oxygen atom (Figure 3.22) results in a carbon–sulfur

Figure 3.21 Thiols are responsible for the strong odor of "essence of skunk." Their odor is an effective defense mechanism.

A major contributor to the typical smell of the human armpit is a compound that contains both an alcohol and a thiol functional group.

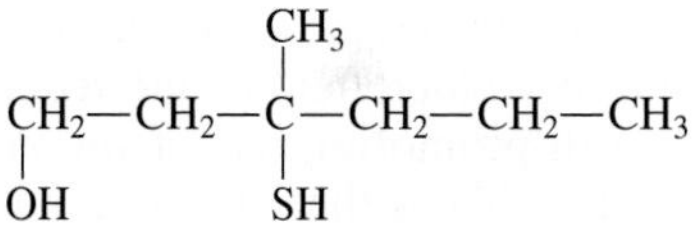

3-Methyl-3-sulfanyl-1-hexanol

Although hundreds of substances contribute to the aroma of freshly brewed coffee, the one most responsible for this characteristic odor is 2-(sulfhydrylmethyl)furan. Structurally, this compound is both a thiol and a cyclic ether.

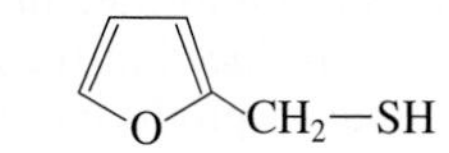

2-(Sulfhydrylmethyl)furan

Bacteria in the mouth interact with saliva and leftover food to produce such compounds as hydrogen sulfide, methanethiol (a thiol), and dimethyl sulfide (a thioether). These compounds, which have odors detectable in air at concentrations of parts per billion, are responsible for "morning breath."

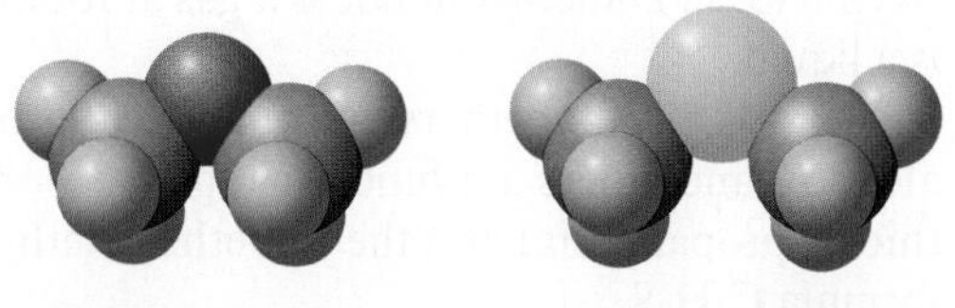

Dimethyl ether **Dimethyl sulfide**

Figure 3.22 A comparison involving space-filling models for dimethyl ether and dimethyl sulfide. A sulfur atom is much larger than an oxygen atom. This results in a carbon–sulfur bond being weaker than a carbon–oxygen bond.

CHEMICAL CONNECTIONS 3-E

Garlic and Onions: Odiferous Medicinal Plants

Garlic and onions, which botanically belong to the same plant genus, are vegetables known for the bad breath—and perspiration odors—associated with their consumption. These effects are caused by organic sulfur-containing compounds, produced when garlic and onions are cut, that reach the lungs and sweat glands via the bloodstream. The total sulfur content of garlic and onions amounts to about one percent of their dry weight.

Less well known about garlic and onions are the numerous studies showing that these same "bad breath" sulfur-containing compounds are health-promoting substances that have the capacity to prevent or at least ameliorate a host of ailments in humans and animals. The list of beneficial effects associated with garlic use is longer than that for any other medicinal plant. Only onions come close to having the same kind of efficacy. Garlic has been shown to function as an antibacterial, antiviral, antifungal, antiprotozal, and antiparasitic agent. In the area of heart and circulatory problems, garlic contains vasodilative compounds that improve blood fluidity and reduce platelet aggregation. The health-promoting role of onions has not been explored as thoroughly as that of garlic, but the studies undertaken so far seem to confirm that onions are second only to garlic in their "healing powers."

Whole garlic bulbs and whole onions that remain undisturbed and intact do not contain any strongly odiferous compounds and display virtually no physiological activity. The act of cutting or crushing these vegetables causes a cascade of reactions to occur in damaged plant cells. Exposure to oxygen in the air is an important facet of these reactions. More than one hundred sulfur-containing organic compounds are formed in garlic, and a similar number are probably produced in the less-studied onion. Many of the compounds so produced are common to both garlic and onions. The compounds associated with garlic ingestion that contribute to bad breath include allyl methyl sulfide, allyl methyl disulfide, diallyl sulfide, and diallyl disulfide. Their structures are given in the accompanying table.

Not all of the strongly odiferous compounds associated with garlic and onions elicit negative responses from the human olfactory system. For example, the smell of fried onions is considered a pleasant odor by most people. Compounds contributing to the "fried onion smell" include methyl propyl disulfide, methyl propyl trisulfide, allyl propyl disulfide, and dipropyl trisulfide. Structures for these compounds are also given in the accompanying table.

In addition to physiologically active sulfur compounds, garlic and onions also contain a variety of other healthful ingredients. Among these are the B vitamins thiamine and riboflavin and vitamin C. Almost all of the trace elements are also present, including manganese, iron, phosphorus, selenium, and chromium. The actual amount of a given trace element depends on the soil in which the garlic or onion was grown.

Getty Images

Garlic Breath

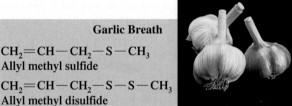

$CH_2{=}CH{-}CH_2{-}S{-}CH_3$
Allyl methyl sulfide

$CH_2{=}CH{-}CH_2{-}S{-}S{-}CH_3$
Allyl methyl disulfide

$CH_2{=}CH{-}CH_2{-}S{-}CH_2{-}CH{=}CH_2$
Diallyl sulfide

$CH_2{=}CH{-}CH_2{-}S{-}S{-}CH_2{-}CH{=}CH_2$
Diallyl disulfide

Getty Images

Fried Onions

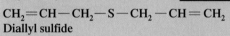

$CH_3{-}S{-}S{-}CH_2{-}CH_2{-}CH_3$
Methyl propyl disulfide

$CH_3{-}S{-}S{-}S{-}CH_2{-}CH_2{-}CH_3$
Methyl propyl trisulfide

$CH_2{=}CH{-}CH_2{-}S{-}S{-}CH_2{-}CH_2{-}CH_3$
Allyl propyl disulfide

$CH_3{-}CH_2{-}CH_2{-}S{-}S{-}S{-}CH_2{-}CH_2{-}CH_3$
Dipropyl trisulfide

covalent bond that is weaker than a carbon–oxygen covalent bond. An added factor is that sulfur's electronegativity (2.5) is significantly lower than that of oxygen (3.5). Dimethyl sulfide is a gas at room temperature, and ethyl methyl sulfide is a liquid.

Thiols and thioethers are functional group isomers in the same manner that alcohols and ethers are functional group isomers (Section 3.17). For example, the thiol 1-propanethiol and the thioether methylthioethane both have the molecular formula C_3H_8S.

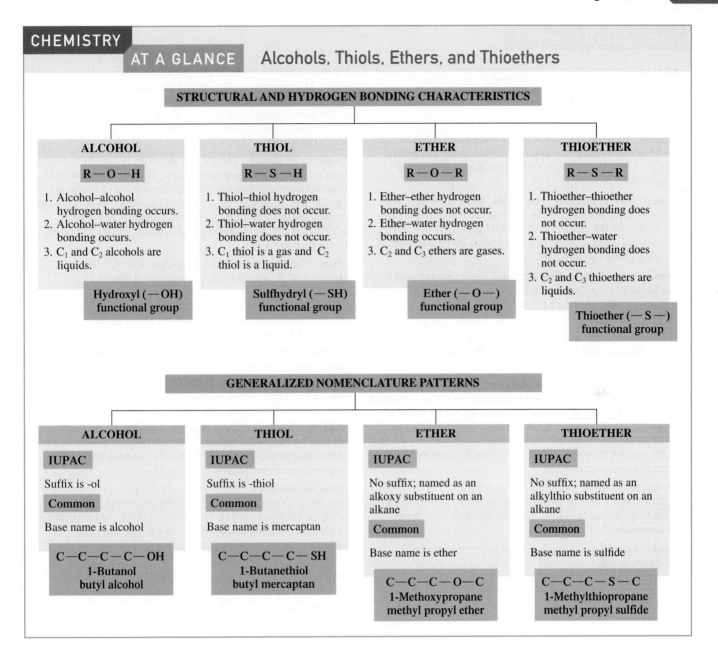

CHEMISTRY AT A GLANCE Alcohols, Thiols, Ethers, and Thioethers

STRUCTURAL AND HYDROGEN BONDING CHARACTERISTICS

ALCOHOL

R—O—H

1. Alcohol–alcohol hydrogen bonding occurs.
2. Alcohol–water hydrogen bonding occurs.
3. C_1 and C_2 alcohols are liquids.

Hydroxyl (—OH) functional group

THIOL

R—S—H

1. Thiol–thiol hydrogen bonding does not occur.
2. Thiol–water hydrogen bonding does not occur.
3. C_1 thiol is a gas and C_2 thiol is a liquid.

Sulfhydryl (—SH) functional group

ETHER

R—O—R

1. Ether–ether hydrogen bonding does not occur.
2. Ether–water hydrogen bonding occurs.
3. C_2 and C_3 ethers are gases.

Ether (—O—) functional group

THIOETHER

R—S—R

1. Thioether–thioether hydrogen bonding does not occur.
2. Thioether–water hydrogen bonding does not occur.
3. C_2 and C_3 thioethers are liquids.

Thioether (—S—) functional group

GENERALIZED NOMENCLATURE PATTERNS

ALCOHOL

IUPAC

Suffix is -ol

Common

Base name is alcohol

C—C—C—C—OH
1-Butanol
butyl alcohol

THIOL

IUPAC

Suffix is -thiol

Common

Base name is mercaptan

C—C—C—C—SH
1-Butanethiol
butyl mercaptan

ETHER

IUPAC

No suffix; named as an alkoxy substituent on an alkane

Common

Base name is ether

C—C—C—O—C
1-Methoxypropane
methyl propyl ether

THIOETHER

IUPAC

No suffix; named as an alkylthio substituent on an alkane

Common

Base name is sulfide

C—C—C—S—C
1-Methylthiopropane
methyl propyl sulfide

$CH_3—CH_2—CH_2—SH$ 1-Propanethiol

$CH_3—S—CH_2—CH_3$ Methylthioethane

Thioethers, like thiols, usually have strong odors. Dimethyl sulfide, $CH_3—S—CH_3$, the simplest thioether, is a water-soluble flammable liquid that has a cabbage-like smell described as pleasant (at low concentrations in air) and unpleasant (at higher concentrations in air).

Sulfides and disulfides contribute heavily to the odors of "garlic breath" and "fried onions." "Garlic breath" odorants are primarily sulfides, whereas the smell, usually considered pleasant, of fried onions is heavily influenced by the presence of disulfides. The focus on relevancy feature Chemical Connections 3-E on the previous page further considers studies about garlic and onions, plants which belong to the same plant genus.

The Chemistry at a Glance feature above contrasts the four major types of compounds considered in this chapter—alcohols, thiols, ethers, and thioethers—in terms of structure, hydrogen bonding characteristics, and nomenclature.

Concepts to Remember

OWL Sign in at **www.cengage.com/owl** to view tutorials and simulations, develop problem-solving skills, and complete online homework assigned by your professor.

Alcohols. Alcohols are organic compounds that contain an —OH group attached to a saturated carbon atom. The general formula for an alcohol is R—OH, where R is an alkyl group (Section 3.2).

Nomenclature of alcohols. The IUPAC names of simple alcohols end in *-ol,* and their carbon chains are numbered to give precedence to the location of the —OH group. Alcohol common names contain the word *alcohol* preceded by the name of the alkyl group (Section 3.3).

Isomerism for alcohols. Constitutional isomerism is possible for alcohols containing three or more carbon atoms. Both skeletal and positional isomers are possible (Section 3.4).

Physical properties of alcohols. Alcohol molecules hydrogen-bond to each other and to water molecules. They thus have higher-than-normal boiling points, and the low-molecular-mass alcohols are soluble in water (Section 3.6).

Classification of alcohols. Alcohols are classified on the basis of the number of carbon atoms bonded to the carbon attached to the —OH group. In primary alcohols, the —OH group is bonded to a carbon atom bonded to only one other C atom. In secondary alcohols, the —OH-containing C atom is attached to two other C atoms. In tertiary alcohols, it is attached to three other C atoms (Section 3.8).

Alcohol dehydration. Alcohols can be dehydrated in the presence of sulfuric acid to form alkenes or ethers. At 180°C, an alkene is produced; at 140°C, primary alcohols produce an ether (Section 3.9).

Alcohol oxidation. Oxidation of primary alcohols first produces an aldehyde, which is then further oxidized to a carboxylic acid. Secondary alcohols are oxidized to ketones, and tertiary alcohols are resistant to oxidation (Section 3.9).

Phenols. Phenols have the general formula Ar—OH, where Ar represents an aryl group derived from an aromatic compound. Phenols are named as derivatives of the parent compound phenol, using the conventions for aromatic hydrocarbon nomenclature (Sections 3.11 and 3.12).

Properties of phenols. Phenols are generally low-melting solids; most are only slightly soluble in water. The chemical reactions of phenols are significantly different from those of alcohols, even though both types of compounds possess hydroxyl groups. Phenols are more resistant to oxidation and do not undergo dehydration. Phenols have acidic properties, whereas alcohols do not (Section 3.13).

Ethers. The general formula for an ether is R—O—R′, where R and R′ are alkyl, cycloalkyl, or aryl groups. In the IUPAC system, ethers are named as alkoxy derivatives of alkanes. Common names are obtained by giving the R group names in alphabetical order and adding the word *ether* (Sections 3.15 and 3.16).

Functional group isomerism. Ethers and alcohols with the same number of carbon atoms and the same degree of saturation have the same molecular formula and are thus isomers of each other. This type of constitutional isomerism is known as functional group isomerism (Section 3.17).

Properties of ethers. Ethers have lower boiling points than alcohols because ether molecules do not hydrogen-bond to each other. Ethers are slightly soluble in water because water forms hydrogen bonds with ethers (Section 3.18).

Thiols and disulfides. Thiols are the sulfur analogs of alcohols. They have the general formula R—SH. The —SH group is called the sulfhydryl group. Oxidation of thiols forms disulfides, which have the general formula R—S—S—R. The most distinctive physical property of thiols is their foul odor (Section 3.20).

Thioethers (sulfides). Thioethers (sulfides) are the sulfur analogs of ethers. They have the general formula R—S—R (Section 3.21).

Exercises and Problems

OWL Interactive versions of these problems may be assigned in OWL.

Exercises and problems are arranged in matched pairs with the two members of a pair addressing the same concept(s). The answer to the odd-numbered member of a pair is given at the back of the book. Problems denoted with a ▲ involve concepts found not only in the section under consideration but also concepts found in one or more earlier sections of the chapter. Problems denoted with a ● cover concepts found in a Chemical Connections feature box.

Bonding Characteristics of Oxygen (Section 3.1)

3.1 In organic compounds, how many covalent bonds does each of the following types of atoms form?
a. Oxygen b. Hydrogen c. Carbon d. A halogen

3.2 Indicate whether or not each of the following covalent bonding behaviors are possible for an oxygen atom in an organic compound.
a. One single bond b. Two single bonds
c. One double bond d. Two double bonds

Structural Characteristics of Alcohols (Section 3.2)

3.3 What is the generalized formula for an alcohol?

3.4 What is the name of the functional group that characterizes an alcohol?

3.5 Contrast, in general terms, the structures of an alcohol and water.

3.6 Contrast, in general terms, the structures of an alcohol and an alkane.

Nomenclature for Alcohols (Section 3.3)

3.7 Assign an IUPAC name to each of the following alcohols.

a. 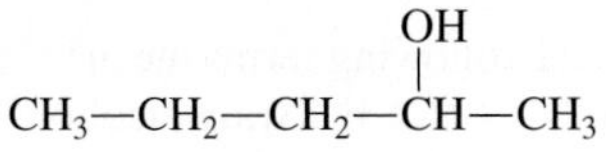

b. 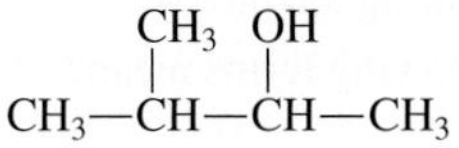

c. 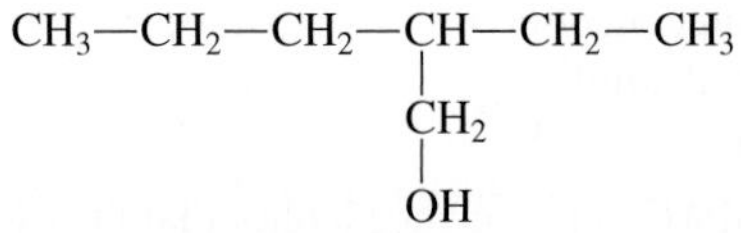

d. 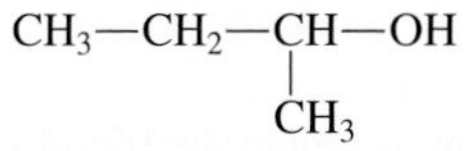

3.8 Assign an IUPAC name to each of the following alcohols.

a. 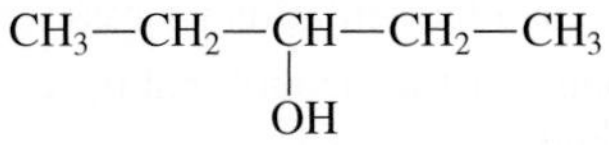

b. 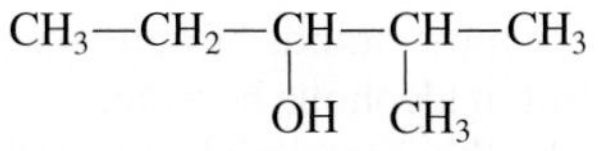

c. 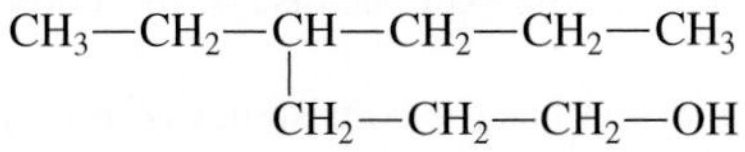

d. 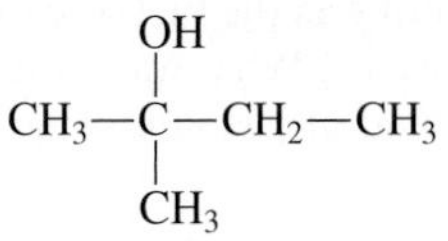

3.9 Assign an IUPAC name to each of the following alcohols.

a. 

b. 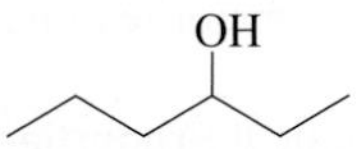

c. 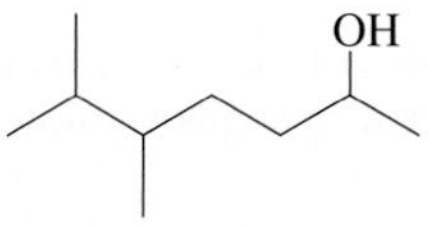

d. 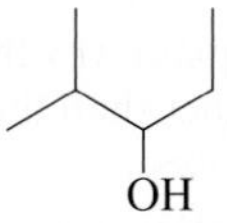

3.10 Assign an IUPAC name to each of the following alcohols.

a. 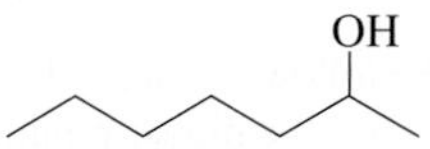

b. 

c. 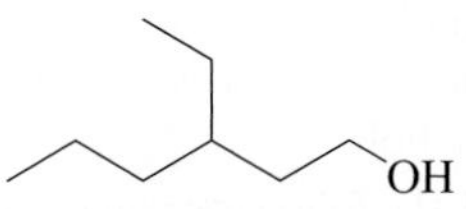

d. 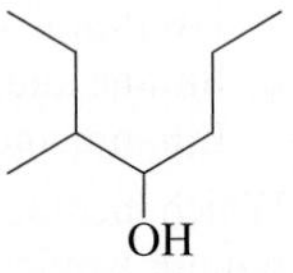

3.11 Write a condensed structural formula for each of the following alcohols.

a. 2-Methyl-1-propanol
b. 4-Methyl-2-pentanol
c. 2-Phenyl-2-propanol
d. 2-Methylcyclobutanol

3.12 Write a condensed structural formula for each of the following alcohols.

a. 2-Methyl-2-heptanol
b. 3-Ethyl-2-pentanol
c. 3-Phenyl-1-butanol
d. 3,5-Dimethylcyclohexanol

3.13 Write a condensed structural formula for, and assign an IUPAC name to, each of the following alcohols.

a. Pentyl alcohol
b. Propyl alcohol
c. Isobutyl alcohol
d. *sec*-Butyl alcohol

3.14 Write a condensed structural formula for, and assign an IUPAC name to, each of the following alcohols.

a. Butyl alcohol
b. Hexyl alcohol
c. Isopropyl alcohol
d. *tert*-Butyl alcohol

3.15 Assign an IUPAC name to each of the following polyhydroxy alcohols.

a. 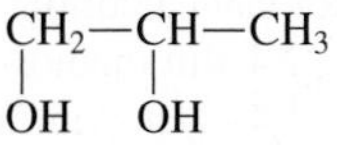

b. 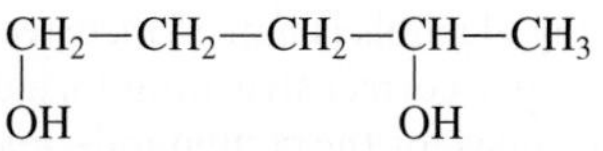

c. 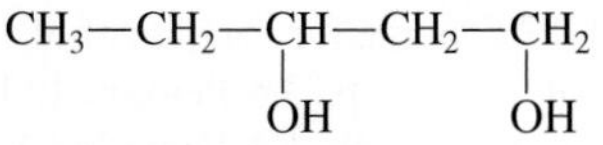

d. 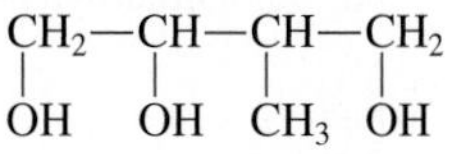

3.16 Assign an IUPAC name to each of the following polyhydroxy alcohols.

a. 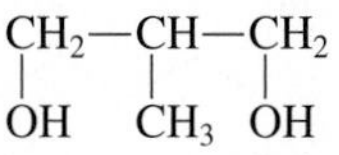

b. 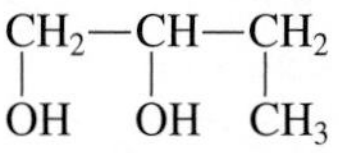

c. 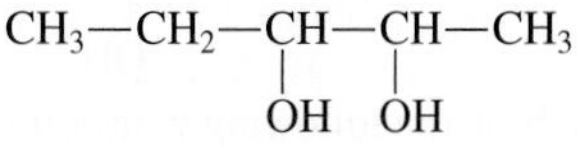

d. 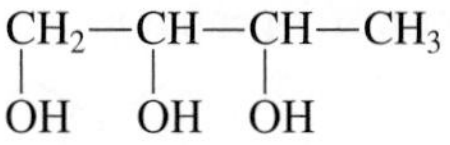

3.17 Utilizing IUPAC rules, name each of the following compounds. Don't forget to use *cis*- and *trans*- prefixes (Section 1.14) where needed.

a. 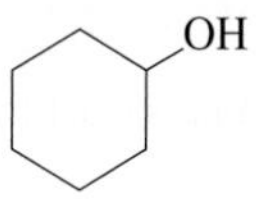

b. 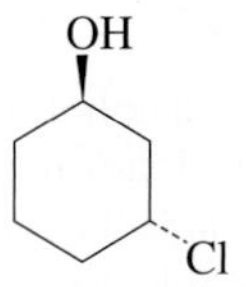

c. 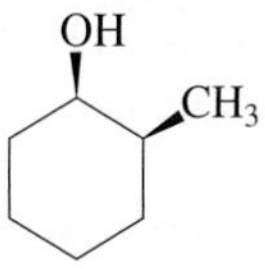

d. 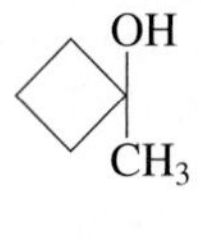

3.18 Utilizing IUPAC rules, name each of the following compounds. Don't forget to use *cis*- and *trans*- prefixes (Section 1.14) where needed.

a. 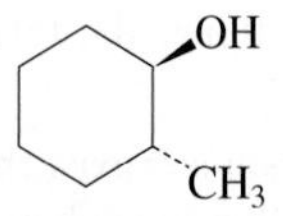

b. 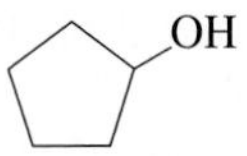

c. 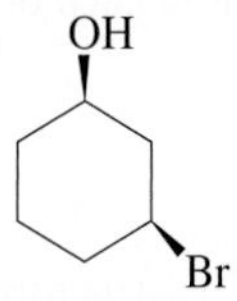

d. 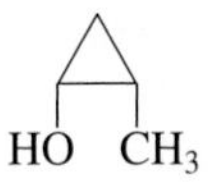

3.19 Write a condensed structural formula for each of the following *unsaturated* alcohols.

a. 4-Penten-2-ol
b. 1-Pentyn-3-ol
c. 3-Methyl-3-buten-2-ol
d. *cis*-2-Buten-1-ol

3.20 Write a condensed structural formula for each of the following *unsaturated* alcohols.
a. 1-Penten-3-ol
b. 3-Butyn-1-ol
c. 2-Methyl-3-buten-1-ol
d. *trans*-3-Penten-1-ol

3.21 Each of the following alcohols is named incorrectly. However, the names give correct structural formulas. Draw structural formulas for the compounds, and then write the correct IUPAC name for each alcohol.
a. 2-Ethyl-1-propanol
b. 2,4-Butanediol
c. 2-Methyl-3-butanol
d. 1,4-Cyclopentanediol

3.22 Each of the following alcohols is named incorrectly. However, the names give correct structural formulas. Draw structural formulas for the compounds, and then write the correct IUPAC name for each alcohol.
a. 3-Ethyl-2-butanol
b. 3,4-Pentanediol
c. 3-Methyl-3-butanol
d. 1,1-Dimethyl-1-butanol

Isomerism for Alcohols (Section 3.4)

3.23 Indicate whether each of the following compounds is or is not a constitutional isomer of 1-hexanol.

a. 

b. 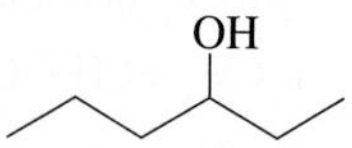

c. 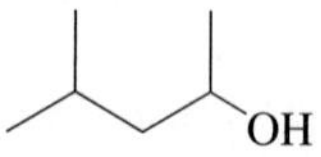

d. 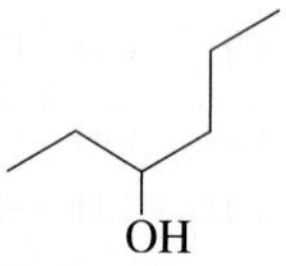

3.24 Indicate whether each of the following compounds is or is not a constitutional isomer of 2-pentanol.

a. 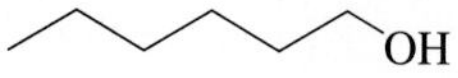

b. 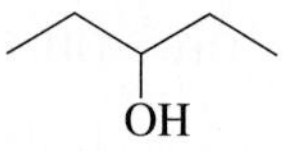

c. 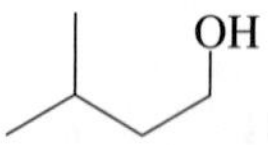

d. 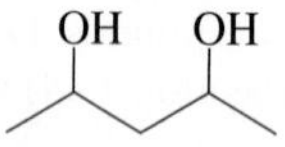

▲3.25 How many alcohol constitutional isomers exist that fit each of the following descriptions?
a. C_7 alcohols that are named as heptanols
b. C_7 alcohols that are named as dimethylcyclopentanols
c. Alcohols that have the molecular formula $C_5H_{12}O$
d. Saturated alcohols that have the molecular formula C_4H_8O

▲3.26 How many alcohol constitutional isomers exist that fit each of the following descriptions?
a. C_8 alcohols that are named as octanols
b. C_8 alcohols that are named as dimethylcyclohexanols
c. Alcohols that have the molecular formula $C_4H_{10}O$
d. Saturated alcohols that have the molecular formula $C_5H_{10}O$

▲3.27 How many alcohol constitutional isomers exist that have the chemical formula $C_5H_{12}O$ and that fit each of the following descriptions?
a. 1° alcohol
b. 2° alcohol
c. 3° alcohol
d. Has a carbon atom ring as part of its structure

▲3.28 How many alcohol constitutional isomers exist that have the chemical formula $C_4H_{10}O$ and that fit each of the following descriptions?
a. 1° alcohol
b. 2° alcohol
c. 3° alcohol
d. Has a carbon atom ring as part of its structure

Important Common Alcohols (Section 3.5)

3.29 What does each of the following terms mean?
a. Absolute alcohol
b. Grain alcohol
c. Rubbing alcohold.Drinking alcohol

3.30 What does each of the following terms mean?
a. Wood alcohol
b. Denaturated alcohol
c. 70-Proof alcohol
d. "Alcohol"

3.31 Give the IUPAC name of the alcohol that fits each of the following descriptions.
a. Moistening agent in many cosmetics
b. Major ingredient in "environmentally friendly" antifreeze formulations
c. Industrially produced from CO and H_2
d. Often produced via a fermentation process

3.32 Give the IUPAC name of the alcohol that fits each of the following descriptions.
a. Major ingredient in most brands of antifreeze
b. The alcohol present in alcoholic beverages
c. Often applied to the skin because of its "cooling effects"
d. Thick liquid that has the consistency of honey

3.33 From 1965 to 2005, methanol was the fuel of choice for Indianapolis Speedway race cars. What was a major reason for using methanol instead of gasoline, which was used previously?

3.34 Since 2007, Indianapolis Speedway race cars have used ethanol rather than methanol as a fuel. What was a major reason for this change?

Physical Properties of Alcohols (Section 3.6)

3.35 Explain why the boiling points of alcohols are much higher than those of alkanes with similar molecular masses.

3.36 Explain why the water solubilities of alcohols are much higher than those of alkanes with similar molecular masses.

3.37 Which member of each of the following pairs of compounds would you expect to have the higher boiling point?
a. 1-Butanol and 1-heptanol
b. Butane and 1-propanol
c. Ethanol and 1,2-ethanediol

3.38 Which member of each of the following pairs of compounds would you expect to have the higher boiling point?
a. 1-Octanol and 1-pentanol
b. Pentane and 1-butanol
c. 1,3-Propanediol and 1-propanol

3.39 Which member of each of the following pairs of compounds would you expect to be more soluble in water?
a. Butane and 1-butanol
b. 1-Octanol and 1-pentanol
c. 1,2-Butanediol and 1-butanol

3.40 Which member of each of the following pairs of compounds would you expect to be more soluble in water?
a. 1-Pentanol and 1-butanol
b. 1-Propanol and 1-hexanol
c. 1,2,3-Propanetriol and 1-hexanol

3.41 Determine the maximum number of hydrogen bonds that can form between an ethanol molecule and
a. other ethanol molecules
b. water molecules
c. methanol molecules
d. 1-propanol molecules

3.42 Determine the maximum number of hydrogen bonds that can form between a methanol molecule and
a. other methanol molecules
b. water molecules
c. 1-propanol molecules
d. 2-propanol molecules

Preparation of Alcohols (Section 3.7)

3.43 Write the structure of the expected predominant organic product formed in each of the following reactions.

a. $CH_2{=}CH_2 + H_2O \xrightarrow{H_2SO_4}$

b. $CH_3-CH_2-\overset{O}{\overset{\|}{C}}-H + H_2 \xrightarrow{\text{Catalyst}}$

c. $CH_3-CH_2-\underset{CH_3}{\underset{|}{C}}{=}CH_2 + H_2O \xrightarrow{H_2SO_4}$

d. $CH_3-CH_2-\overset{O}{\overset{\|}{C}}-CH_2-CH_3 + H_2 \xrightarrow{\text{Catalyst}}$

3.44 Write the structure of the expected predominant organic product formed in each of the following reactions.

a. $CH_3-CH{=}CH-CH_3 + H_2O \xrightarrow{H_2SO_4}$

b. $CH_3-CH_2-\overset{O}{\overset{\|}{C}}-CH_3 + H_2 \xrightarrow{\text{Catalyst}}$

c. $CH_3-\overset{O}{\overset{\|}{C}}-H + H_2 \xrightarrow{\text{Catalyst}}$

d. $CH_3-\underset{CH_3}{\underset{|}{CH}}-CH{=}CH-CH_3 + H_2O \xrightarrow{H_2SO_4}$

Classification of Alcohols (Section 3.8)

3.45 Classify each of the alcohols in Problem 3.7 as a primary, secondary, or tertiary alcohol.

3.46 Classify each of the alcohols in Problem 3.8 as a primary, secondary, or tertiary alcohol.

3.47 Classify each of the following alcohols as a primary, secondary, or tertiary alcohol.

a.
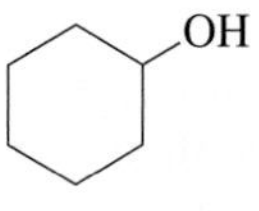
b.
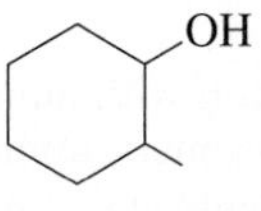
c.
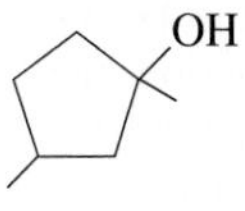
d.
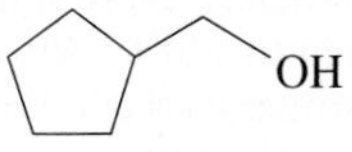

3.48 Classify each of the following alcohols as a primary, secondary, or tertiary alcohol.

a.
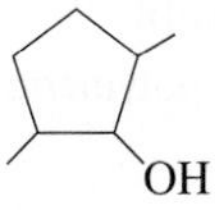
b.
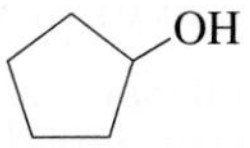
c.
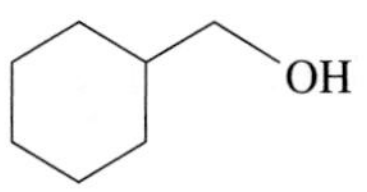
d.
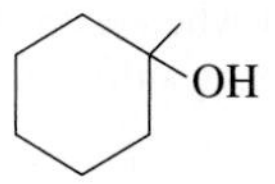

▲3.49 Classify each of the following alcohols as a primary, secondary, or tertiary alcohol.
a. 1-pentanol
b. 2-pentanol
c. 2-methyl-1-pentanol
d. 2-methyl-2-pentanol

▲3.50 Classify each of the following alcohols as a primary, secondary, or tertiary alcohol.
a. 1-butanol
b. 2-butanol
c. 2-methyl-1-butanol
d. 2-methyl-2-butanol

▲3.51 Draw a structural formula for the simplest alcohol (fewest carbon atoms) that fits each of the following descriptions.
a. tertiary alcohol with an acyclic R group
b. secondary alcohol with a cyclic R group

▲3.52 Draw a structural formula for the simplest alcohol (fewest carbon atoms) that fits each of the following descriptions.
a. secondary alcohol with an acyclic R group
b. tertiary alcohol with a cyclic R group

•3.53 (Chemical Connections 3-A) Indicate whether each of the following statements concerning the terpene alcohol menthol is true or false.
a. The R group present in menthol is a disubstituted cyclohexane ring.
b. Menthol is a naturally occurring terpene in which two isoprene units are present.
c. Menthol's mode of action is that of a differential anesthetic.
d. Menthol is used in toothpaste as a flavoring agent.

•3.54 (Chemical Connections 3-A) Indicate whether each of the following statements concerning the terpene alcohol menthol is true or false.
a. The two ring attachments in menthol's structure are methyl and ethyl.
b. In the pure state, at room temperature, menthol is a colorless liquid.
c. Menthol gives a sensation of coolness that is not related to body temperature.
d. Menthol is used in dermatologic preparations as an anti-itching agent.

Chemical Reactions of Alcohols (Section 3.9)

3.55 Draw the structure of the organic product expected to be predominant when each of the following alcohols is dehydrated using sulfuric acid at the temperature indicated.

a. $CH_3-\underset{OH}{\underset{|}{CH}}-CH_3 \xrightarrow[180°C]{H_2SO_4}$

b. $CH_3-CH_2-\underset{CH_3}{\underset{|}{CH}}-CH_2-OH \xrightarrow[180°C]{H_2SO_4}$

c. $CH_3-\underset{CH_3}{\underset{|}{CH}}-OH \xrightarrow[140°C]{H_2SO_4}$

d. $CH_3-CH_2-CH_2-OH \xrightarrow[140°C]{H_2SO_4}$

3.56 Draw the structure of the organic product expected to be predominant when each of the following alcohols is dehydrated using sulfuric acid at the temperature indicated.

a. $CH_3-CH(CH_3)-CH_2-OH \xrightarrow[140°C]{H_2SO_4}$

b. $CH_3-CH(CH_3)-CH_2-OH \xrightarrow[180°C]{H_2SO_4}$

c. $CH_3-CH(OH)-CH_2-CH_3 \xrightarrow[180°C]{H_2SO_4}$

d. $CH_3-CH(OH)-CH_2-CH_3 \xrightarrow[140°C]{H_2SO_4}$

3.57 Identify the alcohol reactant from which each of the following products was obtained by an alcohol dehydration reaction.

a. Alcohol $\xrightarrow[180°C]{H_2SO_4} CH_3-CH{=}C(CH_3)-CH_3$

b. Alcohol $\xrightarrow[180°C]{H_2SO_4} CH_3-CH{=}CH_2$

c. Alcohol $\xrightarrow[140°C]{H_2SO_4} CH_3-CH_2-O-CH_2-CH_3$

d. Alcohol $\xrightarrow[140°C]{H_2SO_4} CH_3-CH(CH_3)-CH_2-O-CH_2-CH(CH_3)-CH_3$

3.58 Identify the alcohol reactant from which each of the following products was obtained by an alcohol dehydration reaction.

a. Alcohol $\xrightarrow[180°C]{H_2SO_4} CH_2{=}C(CH_3)-CH_2-CH_3$

b. Alcohol $\xrightarrow[180°C]{H_2SO_4} CH_3-CH_2-CH{=}CH_2$

c. Alcohol $\xrightarrow[140°C]{H_2SO_4} CH_3-O-CH_3$

d. Alcohol $\xrightarrow[140°C]{H_2SO_4} CH_3-CH_2-O-CH_2-CH_3$

3.59 Draw the structure of the alcohol that could be used to prepare each of the following compounds in an oxidation reaction.

a. $CH_3-CH_2-C({=}O)-CH_3$

b. $CH_3-CH_2-C({=}O)-OH$

c. $CH_3-CH_2-C({=}O)-H$

d.

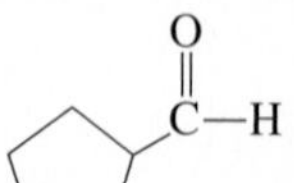

3.60 Draw the structure of the alcohol that could be used to prepare each of the following compounds in an oxidation reaction.

a. $CH_3-CH(CH_3)-C({=}O)-H$

b. $CH_3-CH(CH_3)-C({=}O)-CH_3$

c. $CH_3-CH(CH_3)-CH_2-C({=}O)-OH$

d.

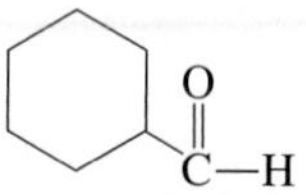

3.61 Draw the structure of the expected predominant organic product formed in each of the following reactions.

a. $CH_3-CH_2-CH_2-OH \xrightarrow[heat]{PCl_3}$

b. 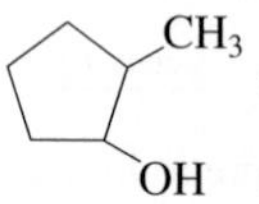 $\xrightarrow[180°C]{H_2SO_4}$

c. $CH_3-CH(OH)-CH_2-CH_3 \xrightarrow{K_2Cr_2O_7}$

d. $CH_3-CH_2-OH \xrightarrow[140°C]{H_2SO_4}$

3.62 Draw the structure of the expected predominant organic product formed in each of the following reactions.

a. 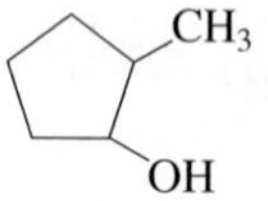 $\xrightarrow[heat]{PCl_3}$

b. $CH_2(OH)-CH_2-CH_2-CH_3 \xrightarrow[140°C]{H_2SO_4}$

c. $CH_3-CH_2-CH(OH)-CH_2-CH_3 \xrightarrow{K_2Cr_2O_7}$

d. $CH_3-CH(CH_3)-OH \xrightarrow[180°C]{H_2SO_4}$

▲**3.63** Three isomeric pentanols with unbranched carbon chains exist. Which of these isomers, upon dehydration at 180°C, yields only 1-pentene as a product?

▲**3.64** A mixture of methanol, 1-propanol, and H_2SO_4 (catalyst) is heated to 140°C. After reaction, the solution contains three isomeric ethers. Draw a condensed structural formula for each of these ethers.

Polymeric Alcohols (Section 3.10)

3.65 Describe the physical and chemical characteristics of the polymeric alcohol PVA [poly(vinyl alcohol)].

3.66 Draw a structural representation for the polymeric alcohol PVA [poly(vinyl alcohol)].

Structural Characteristics of Phenols (Section 3.11)

3.67 Explain why the first of the following two compounds is a phenol, and the second is not.

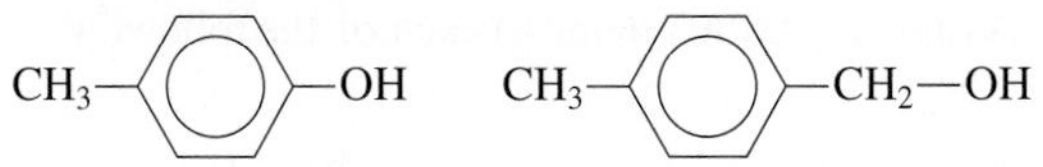

3.68 Explain why the first of the following two compounds is a phenol, and the second is not.

Nomenclature for Phenols (Section 3.12)

3.69 Name the following phenols.

a. 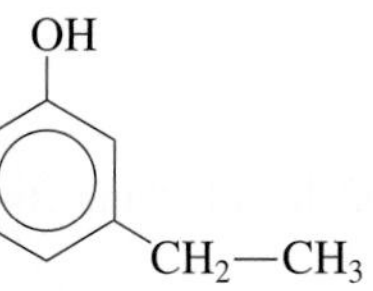

b. 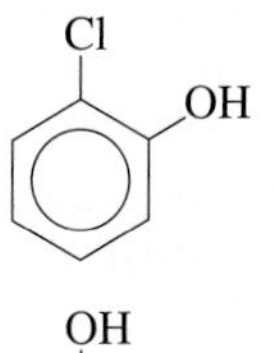

c. 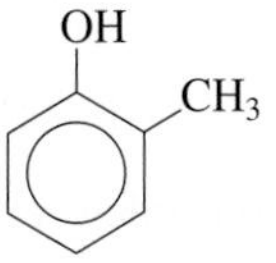

d. 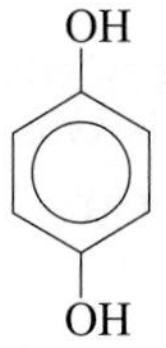

3.70 Name the following phenols.

a. 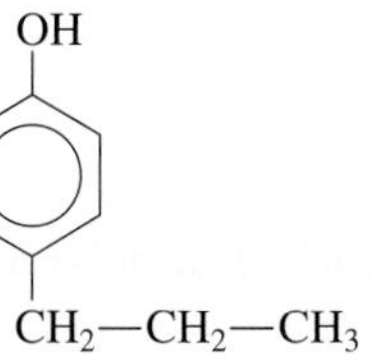

b. 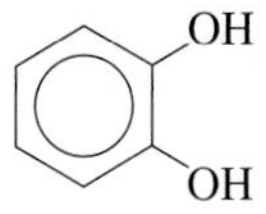

c. 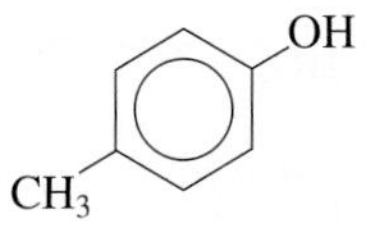

d. 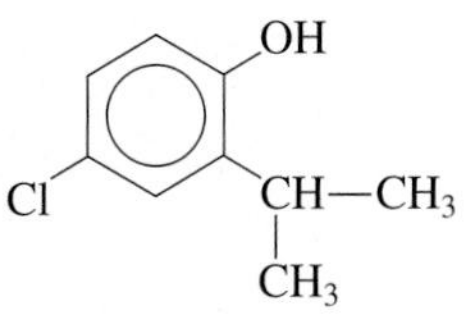

3.71 Draw a structural formula for each of the following phenols.
a. 2-Ethylphenol
b. 2,4-Dibromophenol
c. *m*-Cresol
d. Resorcinol

3.72 Draw a structural formula for each of the following phenols.
a. 3-Bromophenol
b. *o*-Cresol
c. Catechol
d. 2,6-Dichlorophenol

Physical and Chemical Properties of Phenols (Section 3.13)

3.73 Characterize phenols in terms of their physical state at room temperature.

3.74 Characterize phenols in terms of their solubility in water.

3.75 What is the difference, if any, between phenols and alcohols in terms of the following?
a. Flammability
b. Ability to undergo halogenation reactions

3.76 What is the difference, if any, between phenols and alcohols in terms of the following?
a. Reaction with weak oxidizing agents
b. Ability to undergo dehydration reactions

3.77 Phenols are weak acids. Write an equation for the acid ionization of the compound phenol.

3.78 How does the acidity of phenols compare with that of inorganic weak acids?

Occurrence and Uses for Phenols (Section 3.14)

3.79 Phenolic compounds are frequently used as antiseptics and disinfectants. What is the difference between an antiseptic and a disinfectant?

3.80 Phenolic compounds are frequently used as antioxidants. What is an antioxidant?

3.81 Describe the structures of the phenolic antioxidants BHA and BHT in terms of the group attached to the phenolic ring system.

3.82 Identity the flavor/odor and uses associated with each of the following phenolic flavoring agents.
a. Thymol b. Eugenol
c. Isoeugenol d. Vanillin

•3.83 (Chemical Connections 3-B) Indicate whether each of the following statements concerning the polyphenol compound resveratrol is true or false.
a. Resveratrol's structure contains two "phenol centers."
b. Plants produce the *cis*-isomer of resveratrol.
c. The resveratrol present in grapes is concentrated in their skin.
d. Red wines contain more resveratrol than white wines.

•3.84 (Chemical Connections 3-B) Indicate whether each of the following statements concerning the polyphenol compound resveratrol is true or false.
a. Resveratrol's structure contains three hydroxyl groups.
b. Resveratrol's biochemical function is that of an antioxidant.
c. Grapes are the only plant that produces resveratrol.
d. Resveratrol concentrations in moderate red-wine-drinking populations is sufficiently high to explain the "French Paradox."

Structural Characteristics of Ethers (Section 3.15)

3.85 Indicate whether each of the following structural notations denotes an ether.
a. R—O—R b. R—O—H
c. Ar—O—R d. Ar—O—Ar

3.86 What is the difference in meaning associated with each of the following pairs of notations?
a. R—O—R and R—O—H
b. Ar—O—R and Ar—O—Ar
c. Ar—O—H and Ar—O—R
d. R—O—R and Ar—O—Ar

▲3.87 Classify each of the following compounds as an *ether*, an *alcohol*, or *neither an ether or alcohol*.

a. 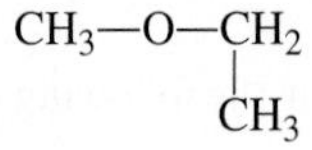

b. 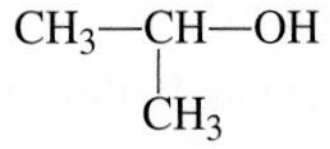

c. 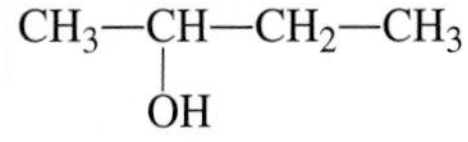

d. 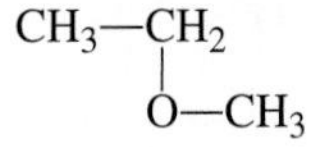

▲**3.88** Classify each of the following compounds as an *ether*, an *alcohol*, or as *neither an ether or alcohol.*

a. 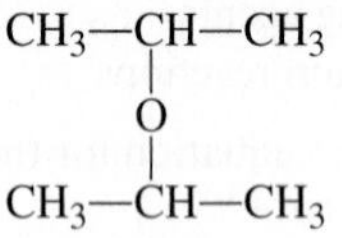 b. 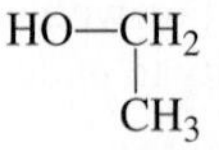

c. 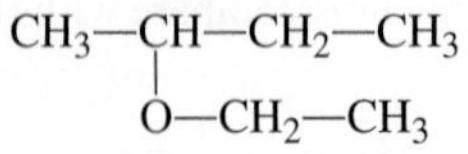 d. 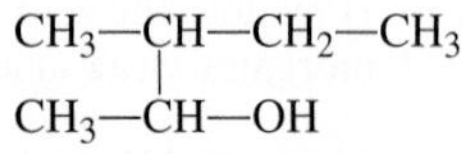

▲**3.89** Draw or write the following for the simplest ether (fewest carbon atoms) in which both R groups present are identical.

a. Expanded structural formula
b. Condensed structural formula
c. Skeletal structural formula
d. Molecular formula

▲**3.90** Draw or write the following for the simplest ether (fewest carbon atoms) in which the R groups present are different alkyl groups.

a. Expanded structural formula
b. Condensed structural formula
c. Skeletal structural formula
d. Molecular formula

Nomenclature for Ethers (Section 3.16)

3.91 Assign an IUPAC name to each of the following ethers.

a. $CH_3—CH_2—CH_2—O—CH_2—CH_3$

b. 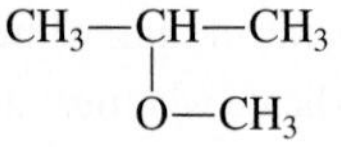

c. 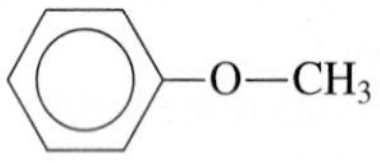

d. 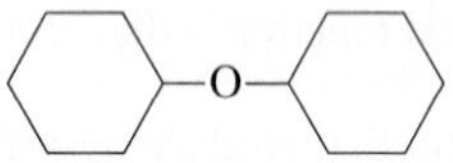

3.92 Assign an IUPAC name to each of the following ethers.

a. $CH_3—CH_2—O—CH_2—CH_3$

b. 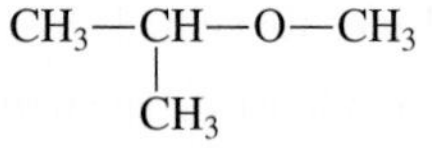

c. 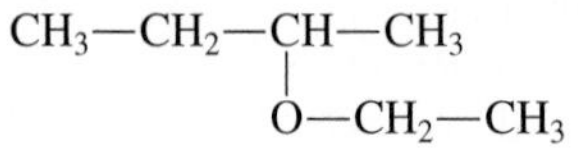

d. 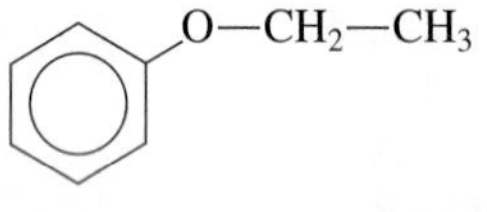

3.93 Assign a common name to each of the ethers in Problem 3.91.

3.94 Assign a common name to each of the ethers in Problem 3.92.

3.95 Assign an IUPAC name to each of the following ethers.

a. 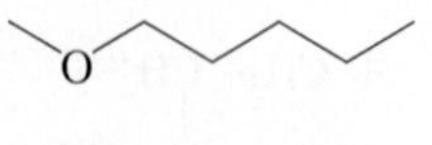 b. 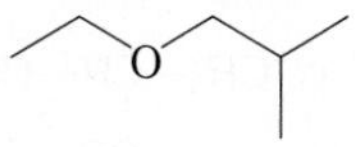

c. 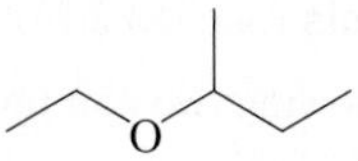 d. 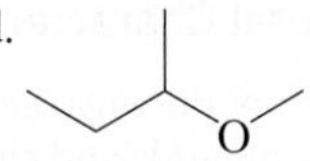

3.96 Assign an IUPAC name to each of the following ethers.

a. 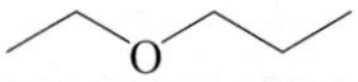 b. 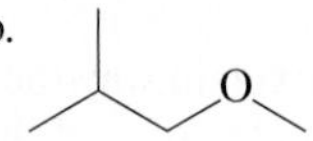

c. 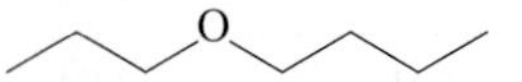 d. 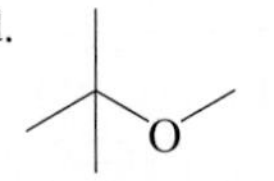

3.97 Draw the structure of each of the following ethers.

a. Isopropyl propyl ether
b. Ethyl phenyl ether
c. 3-Methylanisole
d. Ethoxycyclobutane

3.98 Draw the structure of each of the following ethers.

a. Butyl methyl ether
b. Anisole
c. Phenyl propyl ether
d. 1,3-Dimethoxybenzene

▲**3.99** Assign an IUPAC name to each of the following compounds.

a. 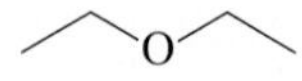 b. 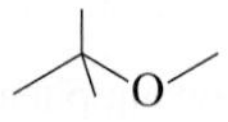

c. 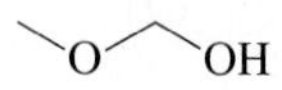 d. 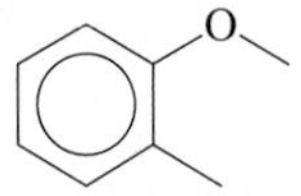

▲**3.100** Assign an IUPAC name to each of the following compounds.

a.  b. 

c. 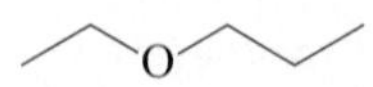 d. 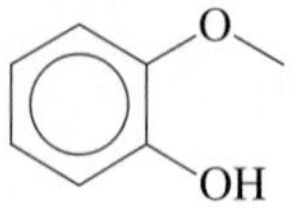

●**3.101** (Chemical Connections 3-C) Indicate whether each of the following statements concerning compounds used as general anesthetics is true or false.

a. Methyl ethyl ether was the first ether to be used as a surgical anesthetic.
b. Nonflammable-ether general anesthetics became available in the late 1970s.
c. Halothane was a widely used general anesthetic and is a halogenated alkane rather than a halogenated ether.
d. The "second generation" of nonflammable ethers contain more fluorine and less chlorine.

●**3.102** (Chemical Connections 3-C) Indicate whether each of the following statements concerning compounds used as general anesthetics is true or false.

a. Early use of ethers as anesthetics was hampered by flammability problems.
b. The "first generation" of nonflammable ethers were chlorofluoroethers.
c. Halothane is a pentahalogenated molecule with three different halogens present.
d. Sevoflurane and Isoflurane are both "second generation" nonflammable halogenated ethers.

Isomerism for Ethers (Section 3.17)

3.103 Indicate whether each of the following ethers is or is not a constitutional isomer of ethyl propyl ether.

a. 

b. 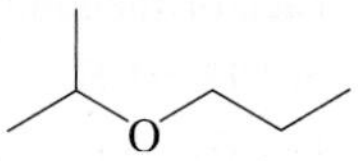

c. 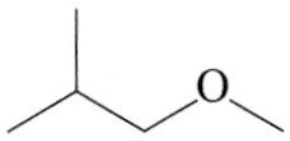

d. 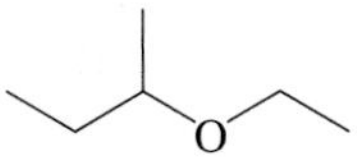

3.104 Indicate whether each of the following ethers is or is not a constitutional isomer of dipropyl ether.

a. 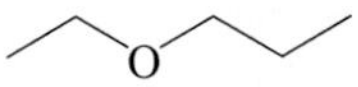

b. 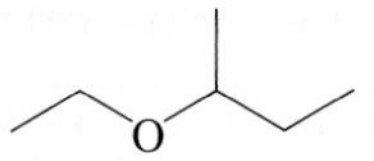

c. 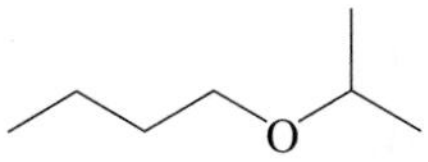

d. 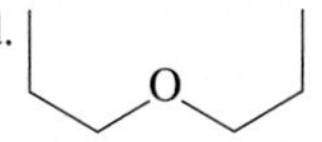

3.105 Give common names for all ethers that are constitutional isomers of ethyl propyl ether.

3.106 Give common names for all ethers that are constitutional isomers of butyl methyl ether.

3.107 How many different saturated ethers exist that contain
a. two carbon atoms
b. five carbon atoms
c. three carbon atoms plus a methoxy group
d. four carbon atoms plus a propoxy group

3.108 How many different saturated ethers exist that contain
a. three carbon atoms
b. four carbon atoms
c. three carbon atoms plus an ethoxy group
d. four carbon atoms plus a methoxy group

3.109 Draw condensed structural formulas for the following.
a. All ethers that are functional group isomers of 1-butanol
b. All alcohols that are functional group isomers of 2-methoxypropane

3.110 Draw condensed structural formulas for the following.
a. All ethers that are functional group isomers of 2-methyl-1-propanol
b. All alcohols that are functional group isomers of 1-ethoxyethane

▲3.111 Draw condensed structural formulas for the eight isomeric alcohols and six isomeric ethers that have the molecular formula $C_5H_{12}O$.

▲3.112 Draw condensed structural formulas for the four isomeric alcohols and three isomeric ethers that have the molecular formula $C_4H_{10}O$.

Physical and Chemical Properties of Ethers (Section 3.18)

3.113 Dimethyl ether and ethanol have the same molecular mass. Dimethyl ether is a gas at room temperature, and ethanol is a liquid at room temperature. Explain these observations.

3.114 Compare the solubility in water of ethers and alcohols that have similar molecular masses.

3.115 What are the two chemical hazards associated with ether use?

3.116 How do the chemical reactivities of ethers compare with those of
a. alkanes
b. alcohols

3.117 Explain why ether molecules cannot hydrogen-bond to each other.

3.118 How many hydrogen bonds can form between a single ether molecule and water molecules?

Cyclic Ethers (Section 3.19)

3.119 Classify each of the following molecular structures as that of a cyclic ether, a noncyclic ether, or a nonether.

a. 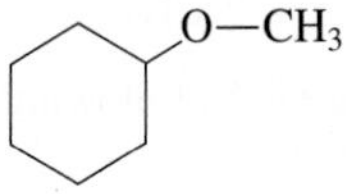

b. 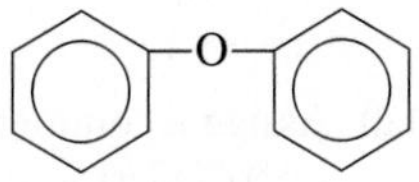

c. 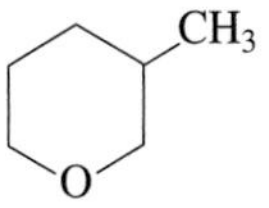

d. 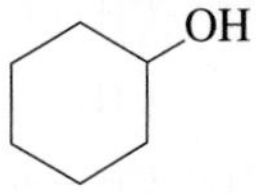

3.120 Classify each of the following molecular structures as that of a cyclic ether, a noncyclic ether, or a nonether.

a. 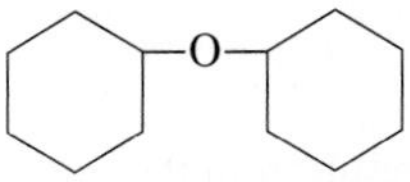

b. 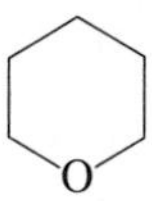

c. 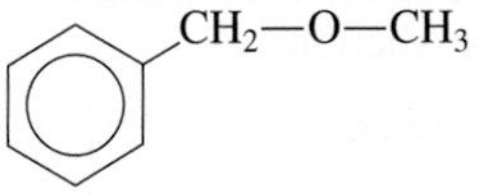

d. 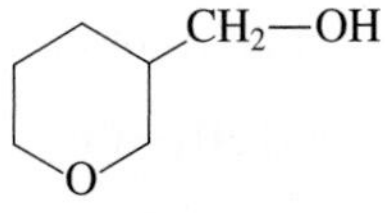

●3.121 (Chemical Connections 3-D) Indicate whether each of the following statements concerning THC, the primary active ingredient in marijuana, is true or false.
a. The acronym THC stands for tetrahalogencannabinol.
b. The biochemical effects of THC involve sedation, tranquilization, and mild hallucination.
c. The core of THC's structure is a fused set of three six-membered rings.
d. THC is only slightly soluble in water.

●3.122 (Chemical Connections 3-D) Indicate whether each of the following statements concerning THC, the primary active ingredient in marijuana, is true or false.
a. The THC content of marijuana sold in North America is 8%–10%.
b. Structurally, THC contains three different functional groups.
c. Four hydrocarbon ring attachments, including a butyl group, are present in THC's molecular structure.
d. THC breakdown products usually clear from the blood within 24 hours.

Sulfur Analogs of Alcohols (Section 3.20)

3.123 Contrast the general structural formulas for a thiol and an alcohol.

3.124 What is the generalized structure for and name of the functional group present in a thiol?

3.125 Draw a condensed structural formula for each of the following thiols.
a. 1-Butanethiol
b. 3-Methyl-1-pentanethiol
c. Cyclopentanethiol
d. 1,2-Ethanedithiol

3.126 Draw a condensed structural formula for each of the following thiols.
a. 1-Propanethiol
b. 1,3-Pentanedithiol
c. 3-Methyl-3-pentanethiol
d. 2-Methylcyclopentanethiol

3.127 Assign a common name to each of the following thiols.
a. $CH_3—SH$
b. $CH_3—CH_2—CH_2—SH$
c. $CH_3—CH_2—CH—SH$ (with CH_3 on the CH)
d. $CH_3—CH—CH_2—SH$ (with CH_3 on the CH)

3.128 Assign a common name to each of the following thiols.
a. $CH_3—CH_2—SH$
b. $CH_3—CH—SH$ (with CH_3 on the CH)
c. $CH_3—CH_2—CH_2—CH_2—SH$
d. $CH_3—C—SH$ (with CH_3 above and CH_3 below the C)

3.129 Contrast the products that result from the oxidation of an alcohol and the oxidation of a thiol.

3.130 Write the formulas for the sulfur-containing organic products of the following reactions.
a. $2CH_3—CH_2—SH \xrightarrow[\text{agent}]{\text{Oxidizing}}$
b. $CH_3—CH_2—S—S—CH_2—CH_3 \xrightarrow[\text{agent}]{\text{Reducing}}$

▲3.131 Assign an IUPAC name to each of the following compounds.
a. $CH_3—(CH_2)_4—SH$
b. $CH_3—(CH_2)_4—OH$
c. $CH_3—CH—CH—SH$ (with CH_3 on each CH)
d. $CH_3—CH—CH—OH$ (with CH_3 on each CH)

▲3.132 Assign an IUPAC name to each of the following compounds.
a. $CH_3—(CH_2)_3—CH—SH$ (with CH_3 on the CH)
b. $CH_3—(CH_2)_3—CH—OH$ (with CH_3 on the CH)
c. $CH_3—CH_2—CH—CH_2—OH$ (with CH_3 on the CH)
d. $CH_3—CH—CH_2—CH_2—SH$ (with CH_3 on the CH)

Sulfur Analogs of Ethers (Section 3.21)

3.133 Contrast the general structural formulas for a thioether and an ether.

3.134 What is the generalized structure for the functional group present in a thioether?

3.135 Assign both an IUPAC name and a common name to each of the following thioethers.
a. $CH_3—CH_2—S—CH_3$
b. $CH_3—CH—S—CH_3$ (with CH_3 on the CH)
c. 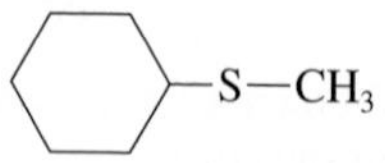
d. $CH_2{=}CH—CH_2—S—CH_3$

3.136 Assign both an IUPAC name and a common name to each of the following thioethers.
a. $CH_3—CH_2—S—CH_2—CH_3$
b. $CH_3—CH—S—CH_2—CH_3$ (with CH_3 on the CH)
c. 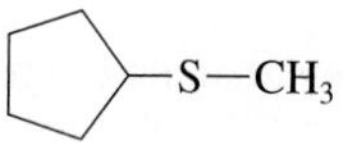
d. $CH_2{=}CH—CH_2—S—CH_2—CH_3$

▲3.137 Which of the terms *ether, alcohol, diol, thiol, thioether, thioalcohol, disulfide, sulfide,* and *peroxide* characterize(s) each of the following compounds? More than one term may apply in a given situation.
a. $CH_3—S—S—CH_3$
b. $CH_3—CH_2—O—O—CH_3$
c. $HO—CH_2—CH_2—OH$
d. $CH_3—O—CH_2—S—CH_3$

▲3.138 Which of the terms *ether, alcohol, diol, thiol, thioether, thioalcohol, disulfide, sulfide,* and *peroxide* characterize(s) each of the following compounds? More than one term may apply in a given situation.
a. $CH_3—S—CH_2—CH_3$
b. $HO—CH_2—CH_2—O—CH_3$
c. $HS—CH_2—CH_2—CH_2—OH$
d. $CH_3—O—CH_2—CH_2—SH$

●3.139 (Chemical Connections 3-E) Indicate whether each of the following statements concerning the active ingredients in garlic and onions is true or false.
a. The list of health benefits associated with garlic use is longer than that for any other plant.
b. Whole garlic and whole onions do not contain any strongly odiferous compounds.
c. Both sulfides and disulfides contribute to the odor of "garlic breath."
d. Propyl groups are prevalent in the odiferous compounds associated with the smell of fried onions.

●3.140 (Chemical Connections 3-E) Indicate whether each of the following statements concerning the active ingredients in garlic and onions is true or false.
a. Onions are second only to garlic in their health benefits.
b. The act of cutting or crushing starts the reactions that produce odiferous compounds in garlic and onions.
c. Both disulfides and trisulfides contribute to the odor of fried onions.
d. Allyl groups are prevalent in the odiferous compounds associated with the smell of ingested garlic.

Aldehydes and Ketones

4

Santokh Kochar/PhotoDisc/Getty Images

Benzaldehyde is the main flavor component in almonds. Aldehydes and ketones are responsible for the odor and taste of numerous nuts and spices.

CHAPTER OUTLINE

This is the second of three chapters that consider hydrocarbon derivatives *that contain the element oxygen.* The hydrocarbon derivatives considered in the previous chapter, alcohols and ethers, have the common structural feature of carbon–oxygen *single* bonds. In this chapter, aldehydes and ketones, the simplest types of hydrocarbon derivatives that contain a carbon–oxygen *double* bond, are considered.

4.1 The Carbonyl Group

Both aldehydes and ketones contain a carbonyl functional group. A **carbonyl group** *is a carbon atom double-bonded to an oxygen atom.* The structural representation for a carbonyl group is

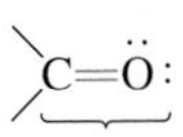

Carbonyl group

Carbon–oxygen and carbon–carbon double bonds differ in a major way. A carbon–oxygen double bond is *polar,* and a carbon–carbon double bond is *nonpolar.* The electronegativity of oxygen (3.5) is much greater than that of carbon (2.5). Hence the carbon–oxygen double bond is polarized, the

The word *carbonyl* is pronounced "carbon-EEL."

The difference in electronegativity between oxygen and carbon causes a carbon–oxygen double bond to be polar.

OWL

Sign in to OWL at **www.cengage.com/owl** to view tutorials and simulations, develop problem-solving skills, and complete online homework assigned by your professor.

oxygen atom acquiring a fractional negative charge (δ^-) and the carbon atom acquiring a fractional positive charge (δ^+).

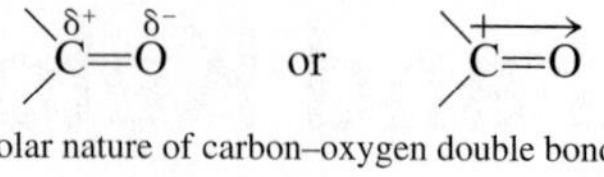

Polar nature of carbon–oxygen double bond

All carbonyl groups have a trigonal planar structure. The bond angles between the three atoms attached to the carbonyl carbon atom are 120°, as would be predicted using VSEPR theory.

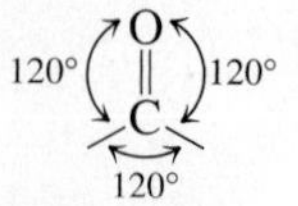

A trigonal planar structure

4.2 Compounds Containing a Carbonyl Group

The carbon atom of a carbonyl group must form two other bonds in addition to the carbon–oxygen double bond in order to have four bonds. The nature of these two additional bonds determines the type of carbonyl-containing compound it is. There are five major classes of carbonyl-containing hydrocarbon derivatives that are considered in this text.

1. **Aldehydes.** In an aldehyde, one of the two additional bonds that the carbonyl carbon atom forms must be to a hydrogen atom. The other may be to a hydrogen atom, an alkyl or cycloalkyl group, or an aromatic ring system.

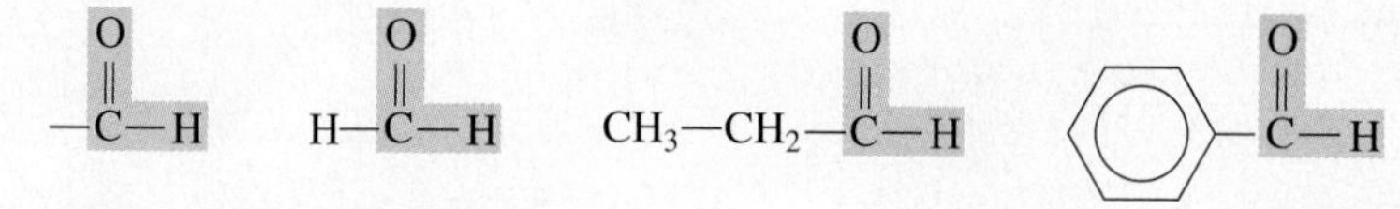

Aldehyde functional group — Simplest aldehyde — Other examples of aldehydes

2. **Ketones.** In a ketone, both of the additional bonds of the carbonyl carbon atom must be to another carbon atom that is part of an alkyl, cycloalkyl, or aromatic group.

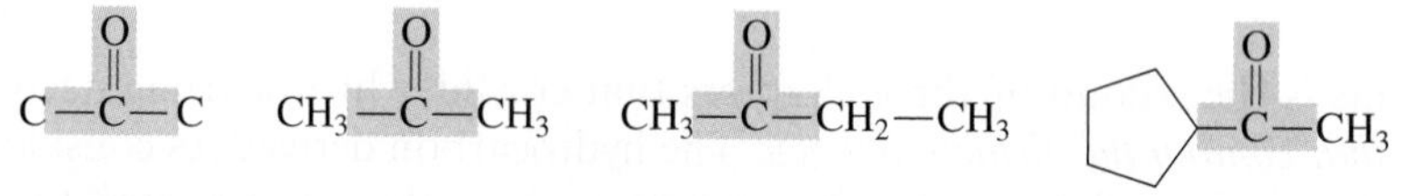

Ketone functional group — Simplest ketone — Other examples of ketones

3. **Carboxylic acids.** In a carboxylic acid, one of the two additional bonds of the carbonyl carbon atom must be to a hydroxyl group, and the other may be to a hydrogen atom, an alkyl or cycloalkyl group, or an aromatic ring system. The structural parameters for a carboxylic acid are the same as those for an aldehyde except that the mandatory hydroxyl group replaces the mandatory hydrogen atom of an aldehyde.

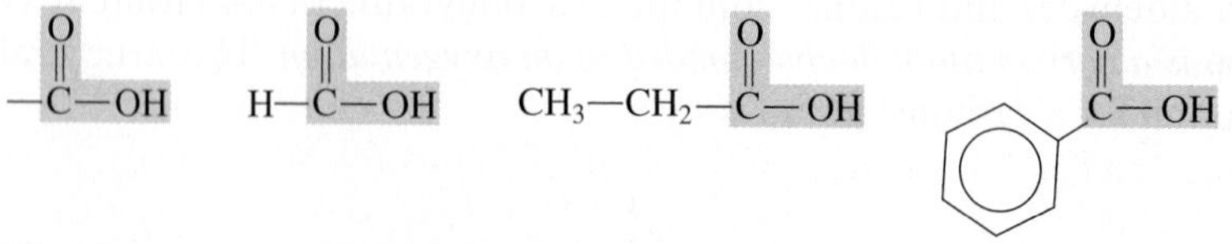

Carboxylic acid functional group — Simplest carboxylic acid — Other examples of carboxylic acids

4. **Esters.** In an ester, one of the two additional bonds of the carbonyl carbon atom must be to an oxygen atom, which in turn is bonded to an alkyl, cycloalkyl, or aromatic group. The other bond may be to a hydrogen atom, alkyl or cycloalkyl group, or an aromatic ring system. The structural parameters for

an ester differ from those for a carboxylic acid only in that an —OH group has become an —O—R or —O—Ar group.

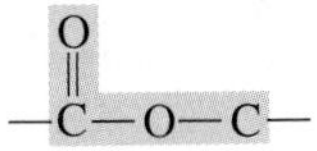

Ester
functional group

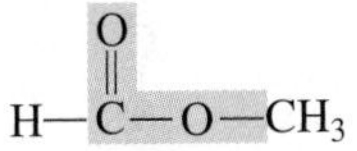

Simplest
ester

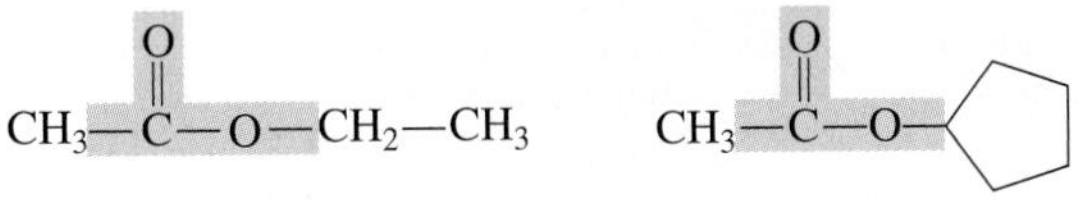

Other examples of esters

5. **Amides.** The previous four types of carbonyl compounds contain the elements carbon, hydrogen, and oxygen. Amides are different from these compounds in that the element nitrogen, in addition to carbon, hydrogen, and oxygen, is present. In an amide, an amino group (—NH_2) or substituted amino group replaces the —OH group of a carboxylic acid.

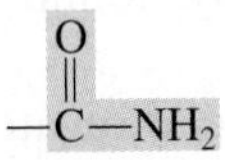

Amide
functional group

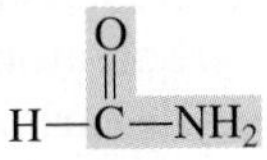

Simplest
amide

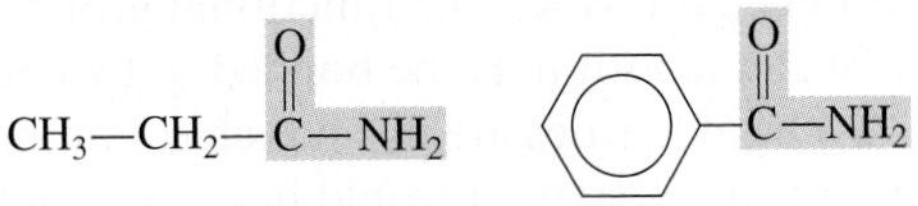

Other examples of amides

Aldehydes and ketones, the first two of the five major classes of carbonyl compounds, are the subject matter for the remainder of this chapter. They share the common structural feature of having *only hydrogen atoms or carbon atoms directly bonded to the carbonyl carbon atom.*

In Chapter 5, the third and fourth of the carbonyl classes, carboxylic acids and esters, are considered. The structural feature that links these two types of compounds together is that of having *an oxygen atom directly bonded to the carbonyl carbon atom.*

In Chapter 6, the fifth of the carbonyl classes, amides, is considered. In amides, *a nitrogen atom is directly bonded to the carbonyl carbon atom.* Thus, amides contain nitrogen, oxygen, carbon, and hydrogen.

4.3 The Aldehyde and Ketone Functional Groups

An **aldehyde** *is a carbonyl-containing organic compound in which the carbonyl carbon atom has at least one hydrogen atom directly attached to it.* The remaining group attached to the carbonyl carbon atom can be hydrogen, an alkyl group (R), a cycloalkyl group, or an aryl group (Ar). The aldehyde functional group is

The word *aldehyde* is pronounced "AL-da-hide."

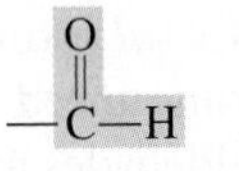

Linear notations for an aldehyde functional group and for an aldehyde itself are —CHO and RCHO, respectively. Note that the ordering of the symbols H and O in these notations is HO, not OH (which denotes a hydroxyl group). The C in the notation RCHO is the carbonyl carbon atom. It is bonded to the R group (single bond), the H atom (single bond), and the O atom (double bond).

A **ketone** *is a carbonyl-containing organic compound in which the carbonyl carbon atom has two other carbon atoms directly attached to it.* The groups containing these bonded carbon atoms may be alkyl, cycloalkyl, or aryl.

The word *ketone* is pronounced "KEY-tone."

The ketone functional group is

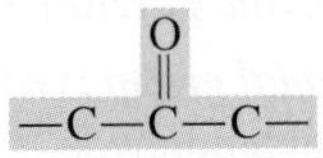

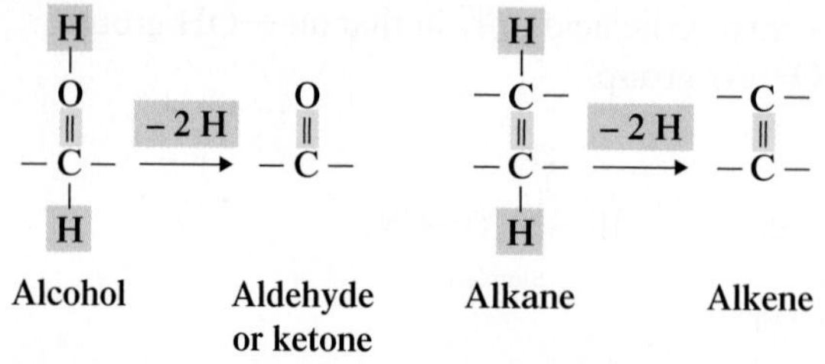

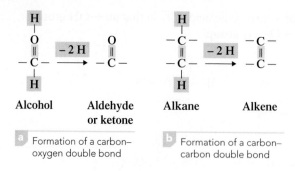

a Formation of a carbon–oxygen double bond

b Formation of a carbon–carbon double bond

Figure 4.1 Aldehydes and ketones are related to alcohols in the same manner that alkenes are related to alkanes; removal of two hydrogen atoms produces a double bond.

In an aldehyde, the carbonyl group is always located at the end of a hydrocarbon chain.

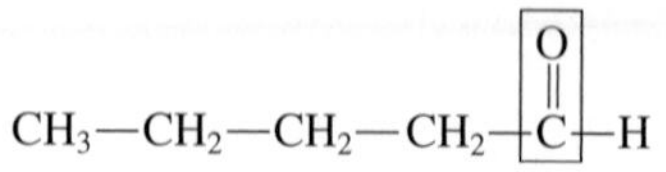

In a ketone, the carbonyl group is always at a nonterminal (interior) position on the hydrocarbon chain.

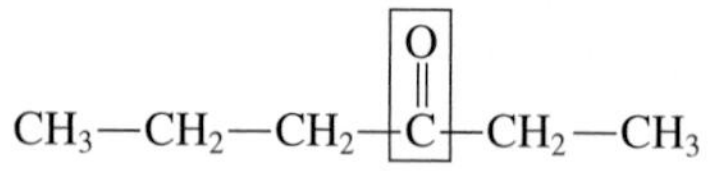

The general condensed formula for a ketone is RCOR, in which the oxygen atom is understood to be double-bonded to the carbonyl carbon at the left of it in the formula. If both R groups of the ketone are the same, the notation R_2CO is often used.

An aldehyde functional group can be bonded to only one carbon atom because three of the four bonds from an aldehyde carbonyl carbon must go to oxygen and hydrogen. Thus an aldehyde functional group is always found at the end of a carbon chain. A ketone functional group, by contrast, is always found within a carbon chain, as it must be bonded to two other carbon atoms.

Cyclic aldehydes are not possible. For an aldehyde carbonyl carbon atom to be part of a ring system, it would have to form two bonds to ring atoms, which would give it five bonds. Unlike aldehydes, ketones can form cyclic structures, such as

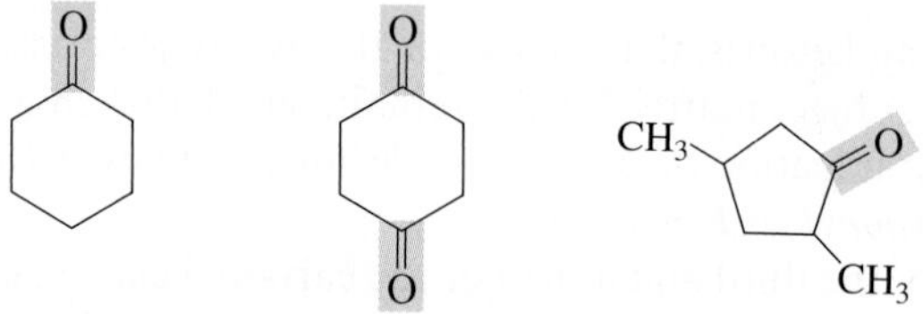

Six-membered ring, one ketone group

Six-membered ring, two ketone groups

Five-membered ring, one ketone group, two alkyl groups

Cyclic ketones are *not* heterocyclic ring systems as were cyclic ethers (Section 3.19).

Aldehydes and ketones are related to alcohols in the same manner that alkenes are related to alkanes. Removal of hydrogen atoms from each of two adjacent carbon atoms in an alkane produces an alkene. In a like manner, removal of a hydrogen atom from the —OH group of an alcohol and from the carbon atom to which the hydroxyl group is attached produces a carbonyl group (see Figure 4.1).

4.4 Nomenclature for Aldehydes

The carbonyl carbon atom in an aldehyde cannot have any number but 1, so the number does not have to included in the aldehyde's IUPAC name.

In IUPAC nomenclature, aldehydes are named using the suffix *-al*. Note the closeness of this suffix to that used for naming alcohols, *-ol*. The two suffixes are often confused with each other. The suffix *-al* (pronounced like the man's name Al) denotes an aldehyde; the suffix *-ol* (pronounced like the *ol* in old) denotes an alcohol.

The IUPAC rules for naming aldehydes are as follows:

Rule 1: *Select as the parent carbon chain the longest chain that* includes *the carbon atom of the carbonyl group.*

Rule 2: *Name the parent chain by changing the* -e *ending of the corresponding alkane name to* -al.

Rule 3: *Number the parent chain by assigning the number 1 to the carbonyl carbon atom of the aldehyde group. The number 1, however, does not become part of the name.*

Rule 4: *Determine the identity and location of any substituents, and append this information to the front of the parent chain name.*

Rule 5: *When an aldehyde functional group is attached to a carbon ring, name the ring and add the suffix* -carbaldehyde.

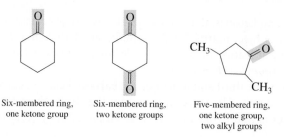

EXAMPLE 4.1 Determining IUPAC Names for Aldehydes

Assign IUPAC names to the following aldehydes.

a. $\mathrm{CH_3{-}CH_2{-}CH_2{-}CH_2{-}\overset{\displaystyle O}{\overset{\|}{C}}{-}H}$

b. 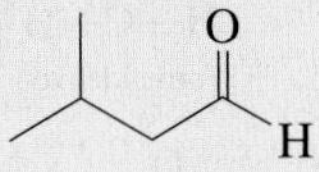

c. $\mathrm{CH_3{-}CH_2{-}CH_2{-}\underset{\substack{|\\ CH_2\\ |\\ CH_3}}{CH}{-}\overset{\displaystyle O}{\overset{\|}{C}}{-}H}$

d. $\mathrm{CH_3{-}CH_2{-}\underset{\substack{|\\ OH}}{CH}{-}CH_2{-}\overset{\displaystyle O}{\overset{\|}{C}}{-}H}$

Solution

a. The parent chain name comes from pentane. Remove the *-e* ending and add the aldehyde suffix *-al.* The name becomes *pentanal.* The location of the carbonyl carbon atom need not be specified because this carbon atom is always number 1. The complete name is simply *pentanal.*

b. The parent chain name is *butanal.* To locate the methyl group, we number the carbon chain beginning with the carbonyl carbon atom. The complete name of the aldehyde is *3-methylbutanal.*

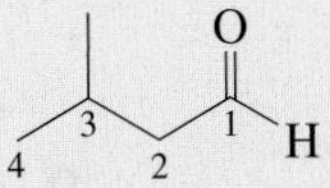

c. The longest chain containing the carbonyl atom is five carbons long, giving a parent chain name of *pentanal.* An ethyl group is present on carbon 2. Thus the complete name is *2-ethylpentanal.*

$$\mathrm{\overset{5}{C}H_3{-}\overset{4}{C}H_2{-}\overset{3}{C}H_2{-}\underset{\substack{|\\ CH_2\\ |\\ CH_3}}{\overset{2}{C}H}{-}\overset{\displaystyle O}{\overset{1\|}{C}}{-}H}$$

d. This is a hydroxyaldehyde, with the hydroxyl group located on carbon 3.

$$\mathrm{\overset{5}{C}H_3{-}\overset{4}{C}H_2{-}\underset{\substack{|\\ OH}}{\overset{3}{C}H}{-}\overset{2}{C}H_2{-}\overset{\displaystyle O}{\overset{1\|}{C}}{-}H}$$

The complete name of the compound is *3-hydroxypentanal.* An aldehyde functional group has priority over an alcohol functional group in IUPAC nomenclature. An alcohol group named as a substituent is a *hydroxy* group.

Practice Exercise 4.1

Assign IUPAC names to the following aldehydes.

a. $\mathrm{CH_3{-}\underset{\substack{|\\ CH_3}}{CH}{-}\overset{\displaystyle O}{\overset{\|}{C}}{-}H}$

b. $\mathrm{CH_3{-}CH_2{-}\underset{\substack{|\\ CH_3-CH_2-CH_2}}{CH}{-}\overset{\displaystyle O}{\overset{\|}{C}}{-}H}$

c. $\mathrm{CH_3{-}\underset{\substack{|\\ Cl}}{CH}{-}\underset{\substack{|\\ Cl}}{CH}{-}\overset{\displaystyle O}{\overset{\|}{C}}{-}H}$

d. 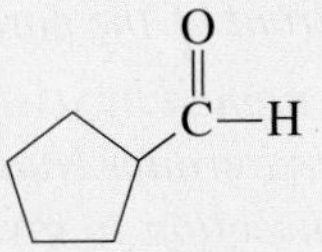

Answers: **a.** 2-Methylpropanal; **b.** 2-Ethylpentanal; **c.** 2,3-Dichlorobutanal; **d.** Cyclopentanecarbaldehyde

Line-angle structural formulas for the simpler unbranched-chain aldehydes:

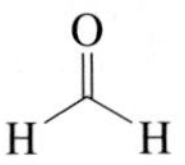

Methanal

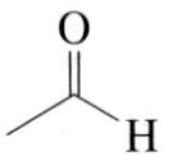

Ethanal

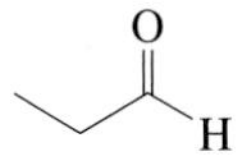

Propanal

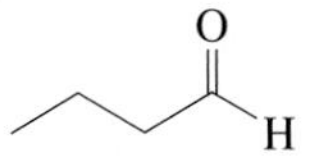

Butanal

When a compound contains more than one type of functional group, the suffix for only one of them can be used as the ending of the name. The IUPAC rules establish priorities that specify which suffix is used. For the functional groups discussed up to this point in the text, the IUPAC priority system is

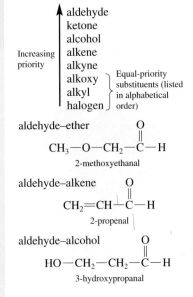

Increasing priority (↑):
- aldehyde
- ketone
- alcohol
- alkene
- alkyne
- alkoxy, alkyl, halogen — Equal-priority substituents (listed in alphabetical order)

aldehyde–ether

$$\mathrm{CH_3{-}O{-}CH_2{-}\overset{\displaystyle O}{\overset{\|}{C}}{-}H}$$

2-methoxyethanal

aldehyde–alkene

$$\mathrm{CH_2{=}CH{-}\overset{\displaystyle O}{\overset{\|}{C}}{-}H}$$

2-propenal

aldehyde–alcohol

$$\mathrm{HO{-}CH_2{-}CH_2{-}\overset{\displaystyle O}{\overset{\|}{C}}{-}H}$$

3-hydroxypropanal

Small unbranched aldehydes have common names:

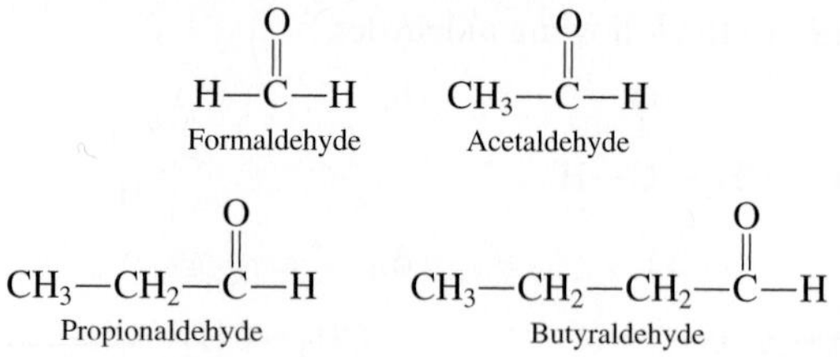

The contrast between IUPAC names and common names for aldehydes is as follows:

IUPAC (one word)

alkanal

pentanal

Common (one word)

(prefix) aldehyde*

butyraldehyde

*The common-name prefixes are related to natural sources for carboxylic acids with the same number of carbon atoms (Section 5.3).

Unlike the common names for alcohols and ethers, the common names for aldehydes are *one* word rather than two or three.

These aldehyde common names illustrate a second method for counting from one to four in organic chemistry language: *form-*, *acet-*, *propion-*, and *butyr-*. (The first method for counting from one to four is *meth-*, *eth-*, *prop-*, and *but-*, as in methane, ethane, propane, and butane.) This new counting method will be used again in the next chapter (Section 5.3) when common names for carboxylic acids are encountered.

Aromatic aldehydes are not cyclic aldehydes (which do not exist). The carbonyl carbon atom in an aromatic aldehyde is not part of the ring system.

In the IUPAC system, aromatic aldehydes—compounds in which an aldehyde group is attached to a benzene ring—are named as derivatives of benzaldehyde, the parent compound (see Figure 4.2).

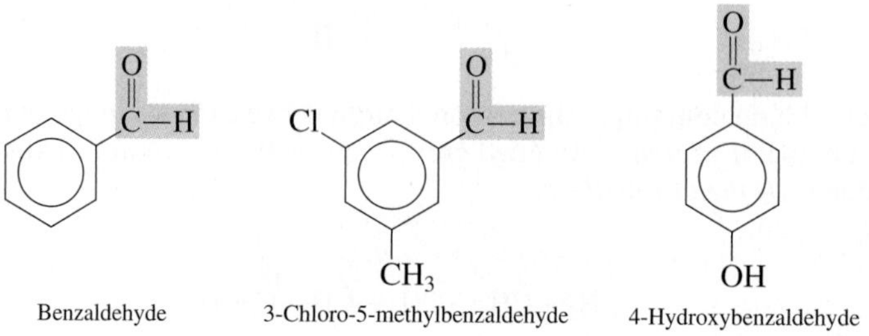

The last of these compounds is named as a benzaldehyde rather than as a phenol because the aldehyde group has priority over the hydroxyl group in the IUPAC naming system.

4.5 Nomenclature for Ketones

Assigning IUPAC names to ketones is similar to naming aldehydes except that the ending *-one* is used instead of *-al*. The rules for IUPAC ketone nomenclature follow:

Rule 1: *Select as the parent carbon chain the longest carbon chain that* includes *the carbon atom of the carbonyl group.*

Rule 2: *Name the parent chain by changing the* -e *ending of the corresponding alkane name to* -one. *This ending, -one, is pronounced "own."*

Rule 3: *Number the carbon chain such that the carbonyl carbon atom receives the lowest possible number. The position of the carbonyl carbon atom is noted by placing a number immediately before the name of the parent chain.*

Rule 4: *Determine the identity and location of any substituents, and append this information to the front of the parent chain name.*

Rule 5: *Cyclic ketones are named by assigning the number 1 to the carbon atom of the carbonyl group. Numbering then proceeds in a clockwise or counterclockwise direction to give the next encountered substituent the lower number. The "1," designating the carbonyl carbon atom location, is omitted from the name.*

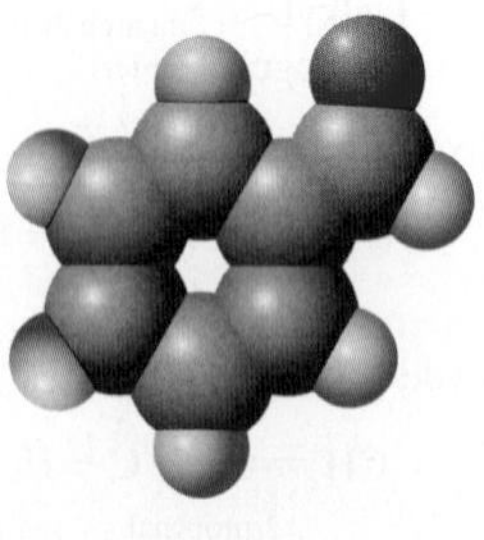

Figure 4.2 Space-filling model for benzaldehyde, the simplest aromatic aldehyde.

EXAMPLE 4.2 Determining IUPAC Names for Ketones

Assign IUPAC names to the following ketones.

a. 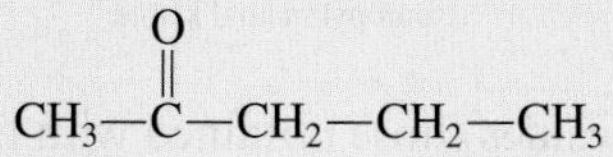

b. 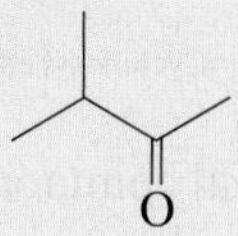

c. 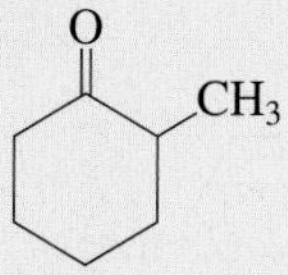

d. 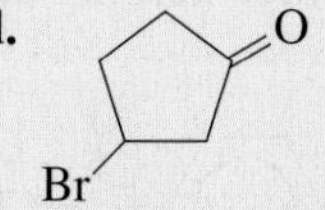

Solution

a. The parent chain name is *pentanone.* We number the chain from the end closest to the carbonyl carbon atom. Locating the carbonyl carbon at carbon 2 completes the name, *2-pentanone.*

b. The longest carbon chain of which the carbonyl carbon is a member is four carbons long. The parent chain name is *butanone.* There is one methyl group attached, and the numbering system is from right to left.

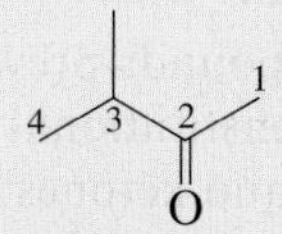

The complete name for the compound is *3-methyl-2-butanone.*

c. The base name is *cyclohexanone.* The methyl group is bonded to carbon 2 because numbering begins at the carbonyl carbon. The name is *2-methylcyclohexanone.*

d. This ketone has a base name of *cyclopentanone*. Numbering clockwise from the carbonyl carbon atom locates the bromo group on carbon 3. The complete name is *3-bromocyclopentanone.*

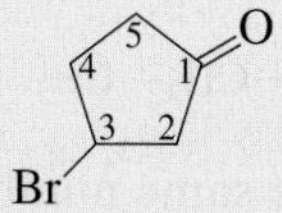

Practice Exercise 4.2

Assign IUPAC names to the following ketones.

a. 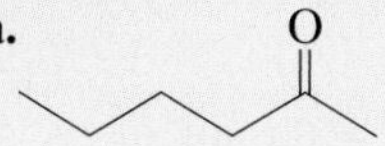

b. 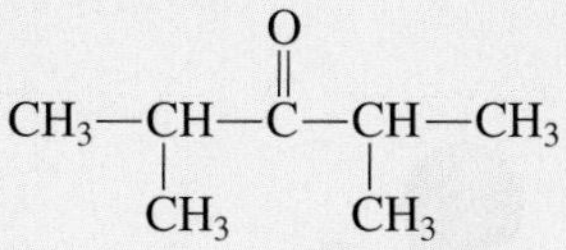

c. 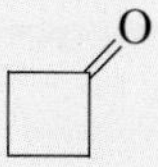

d. 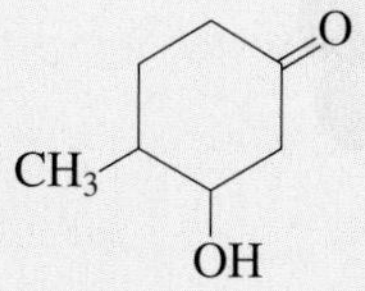

Answers: **a.** 2-Hexanone; **b.** 2,4-Dimethyl-3-pentanone; **c.** Cyclobutanone; **d.** 3-Hydroxy-4-methylcyclohexanone

In IUPAC nomenclature, the ketone functional group has precedence over all groups discussed so far except the aldehyde group. When both aldehyde and ketone groups are present in the same molecule, the ketone group is named as a substituent (the *oxo-* group).

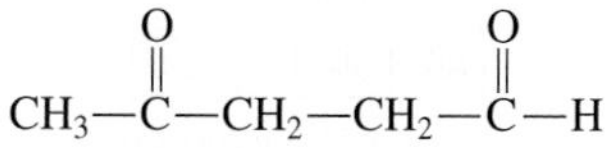

4-Oxopentanal

Line-angle structural formulas for the simpler unbranched-chain 2-ketones:

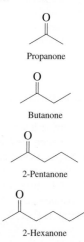

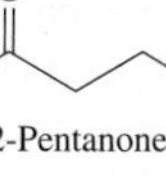

Propanone

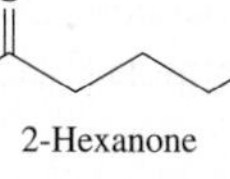

Butanone

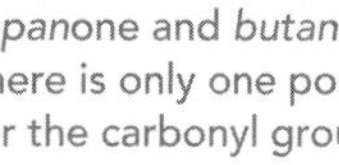

2-Pentanone

2-Hexanone

No locator number is needed in the names *propanone* and *butanone* because there is only one possible location for the carbonyl group.

The procedure for coining common names for ketones is the same as that used for ether common names (Section 3.16). They are constructed by giving, in alphabetical order, the names of the alkyl or aryl groups attached to the carbonyl functional group and then adding the word *ketone.* Unlike aldehyde common names, which are one word, those for ketones are two or three words.

The contrast between IUPAC names and common names for ketones is as follows:

IUPAC (one word)

alkanone

2-pentanone

Common (three or two words)

alkyl alkyl ketone

ethyl methyl ketone

or

dialkyl ketone

dipropyl ketone

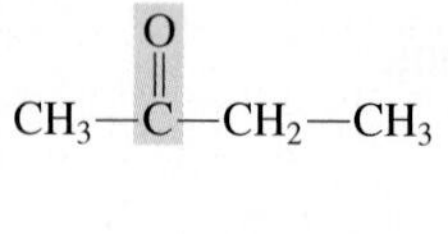
Ethyl methyl ketone

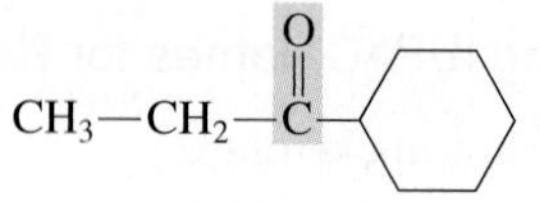
Cyclohexyl ethyl ketone

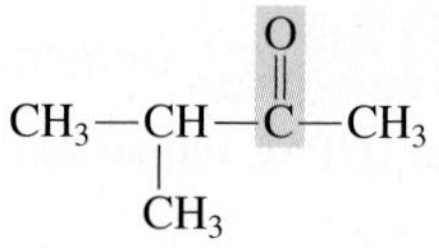
Isopropyl methyl ketone

Three ketones have additional common names besides those obtained with the preceding procedures. These three ketones are

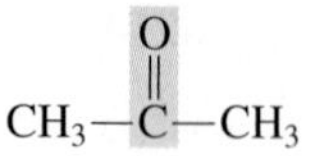
Acetone
(dimethyl ketone)

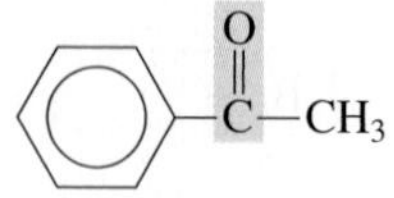
Acetophenone
(methyl phenyl ketone)

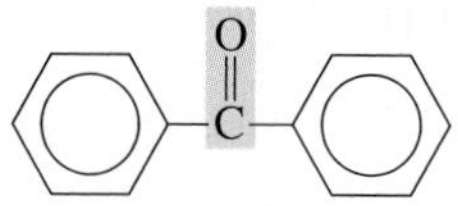
Benzophenone
(diphenyl ketone)

Acetophenone is the simplest aromatic ketone.

4.6 Isomerism for Aldehydes and Ketones

Like the classes of organic compounds previously discussed (alkanes, alkenes, alkynes, alcohols, ethers, etc.), constitutional isomers exist for aldehydes and for ketones, and *between* aldehydes and ketones (functional group isomerism). The compounds butanal and 2-methylpropanal are examples of skeletal aldehyde isomers; the compounds 2-pentanone and 3-pentanone are examples of positional ketone isomers.

Aldehyde skeletal isomers:
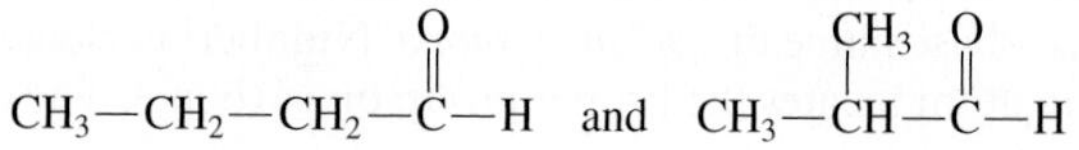

Ketone positional isomers:
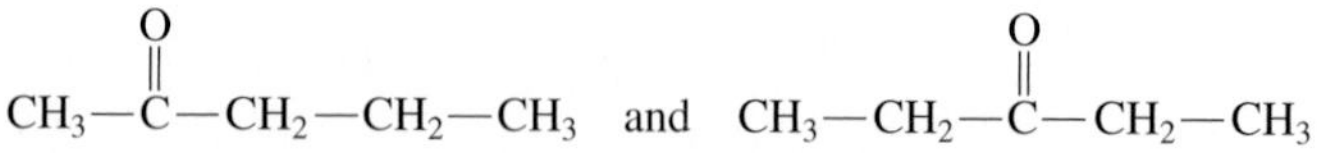

This is the third time functional group isomerism has been encountered. Alcohols and ethers (Section 3.17) and thiols and thioethers (Section 3.21) also exhibit this type of isomerism.

Aldehydes and ketones with the same number of carbon atoms and the same degree of saturation are functional group isomers. Molecular models for the isomeric C_3 compounds propanal and propanone, which both have the molecular formula C_3H_6O, are shown in Figure 4.3.

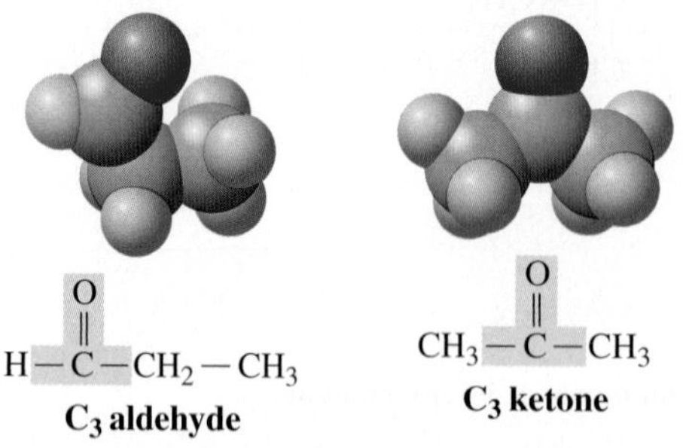

Figure 4.3 Aldehydes and ketones with the same number of carbon atoms and the same degree of saturation are functional group isomers, as is illustrated here for the three-carbon aldehyde (propanal) and the three-carbon ketone (propanone). Both have the molecular formula C_3H_6O.

4.7 Selected Common Aldehydes and Ketones

The simplest aldehyde is the one-carbon aldehyde formaldehyde (methanal). The simplest ketone is the three-carbon ketone acetone (dimethyl ketone, propanone). One- and two-carbon ketones cannot exist. A minimum of three carbon atoms is required for a ketone: one carbon atom for the carbonyl group and one carbon atom for each of the groups attached to the carbonyl carbon atom. These two simple substances, formaldehyde and acetone, are two of the most well known and most used of all carbonyl-containing compounds.

Formaldehyde

Formaldehyde is manufactured on a large scale by the oxidation of methanol.

$$CH_3{-}OH \xrightarrow[600°C-700°C]{\text{Ag catalyst}} H{-}\overset{\overset{\displaystyle O}{\|}}{C}{-}H + H_2$$

The oxidation of methanol to formaldehyde has been previously mentioned (Section 3.5). Ingested methanol is oxidized in the human body to formaldehyde, and it is the formaldehyde that causes blindness.

Its major use is in the manufacture of polymers (Section 4.12). At room temperature and pressure, formaldehyde is an irritating gas. Bubbling this gas through water produces *formalin*, an aqueous solution containing 37% formaldehyde by mass or 40% by volume. (This represents the solubility limit of formaldehyde gas in water.) Very little free formaldehyde gas is actually present in formalin; most of it reacts with water, producing methylene glycol.

$$H{-}\overset{\overset{\displaystyle O}{\|}}{C}{-}H + H{-}O{-}H \longrightarrow HO{-}CH_2{-}OH$$

Formalin is used for preserving biological specimens (Figure 4.4); anyone who has experience in a biology laboratory is familiar with the pungent odor of formalin. Formalin is also the most widely used preservative chemical in embalming fluids used by morticians. Its mode of action involves reaction with protein molecules in a manner that links the protein molecules together; the result is a "hardening" of the protein.

© 2005 Norbert Wu/ www.norbertwu.com

Figure 4.4 Formalin is used to preserve biological specimens. This coelacanth, a "prehistoric" fish, is preserved in formaldehyde.

Acetone

Acetone, a colorless, volatile liquid with a pleasant, mildly "sweet" odor, is the ketone used in largest volume in industry. Acetone is an excellent solvent because it is miscible with both water and nonpolar solvents. Acetone is the main ingredient in gasoline treatments that are designed to solubilize water in the gas tank and allow it to pass through the engine in miscible form. Acetone can also be used to remove water from glassware in the laboratory. And it is a major component of some nail-polish removers.

Small amounts of acetone are produced in the human body in reactions related to obtaining energy from fats. Normally, such acetone is degraded to CO_2 and H_2O. Diabetic people produce larger amounts of acetone, not all of which can be degraded. The presence of acetone in urine is a sign of diabetes. In severe diabetes, the odor of acetone can be detected on the person's breath.

Flavorants and Odorants

Aldehydes and ketones occur widely in nature. Naturally occurring compounds of these types, with higher molecular masses, usually have pleasant odors and flavors and are often used for these properties in consumer products (perfumes, air fresheners, and the like). Table 4.1 gives the structures and uses for selected naturally occurring aldehydes and ketones. The unmistakable odor of melted butter is largely due to the four-carbon diketone butanedione (Figure 4.5).

© Steven Needham/Envision

Figure 4.5 The delightful odor of melted butter is largely due to butanedione, whose structure is given in Table 4.1.

Steroid Hormones

Many important steroid hormones (Section 8.12) are ketones, including testosterone, the hormone that controls the development of male sex characteristics; progesterone, the hormone secreted at the time of ovulation in females; and

Table 4.1 Selected Aldehydes and Ketones Whose Uses Are Based on Their Odor or Flavor

Aldehydes

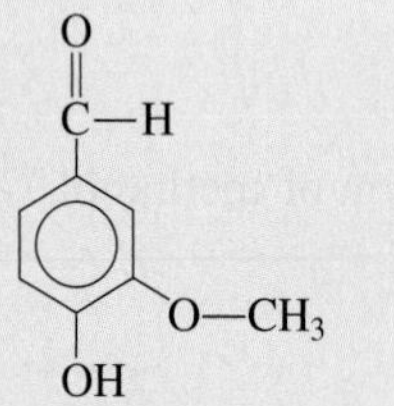

Vanillin
(vanilla flavoring)

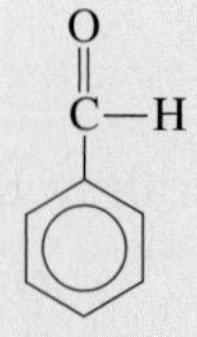

Benzaldehyde
(almond flavoring)

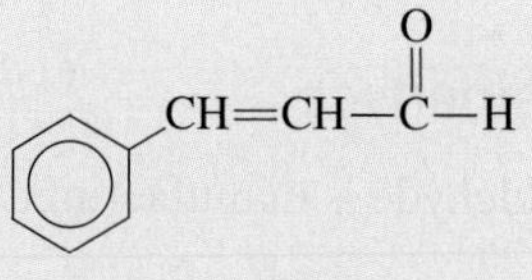

Cinnamaldehyde
(cinnamon flavoring)

Ketones

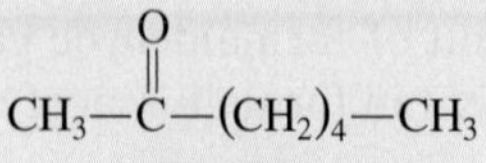

2-Heptanone
(clove flavoring)

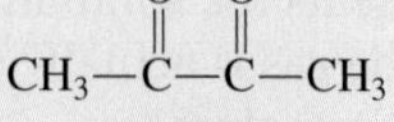

Butanedione
(butter flavoring)

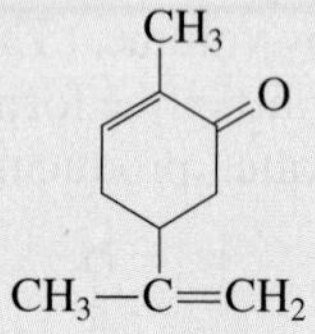

Carvone
(spearmint flavoring)

cortisone, a hormone from the adrenal glands. Synthetically produced cortisone is used as an anti-inflammatory medication.

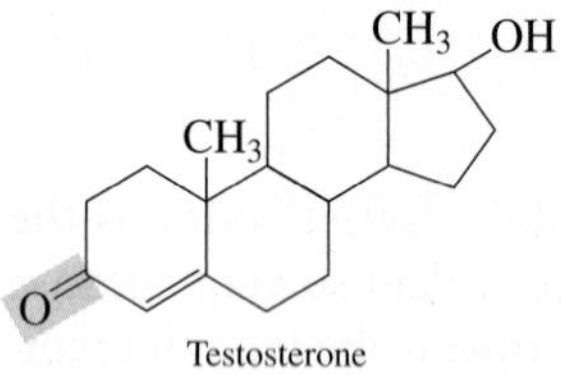

Testosterone

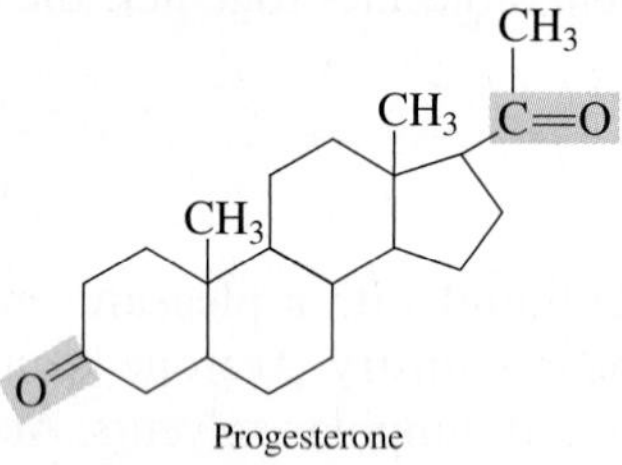

Progesterone

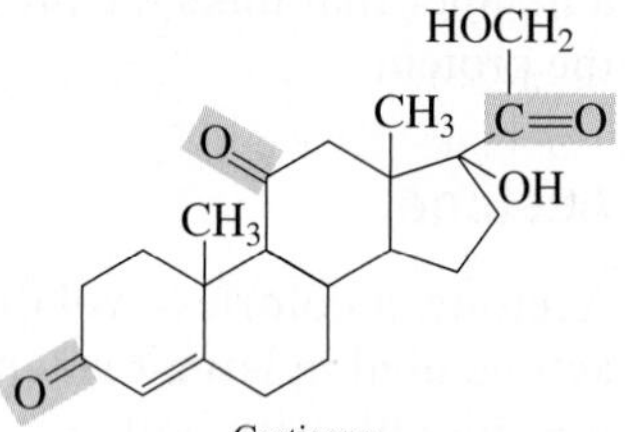

Cortisone

Sunscreens and Suntanning Agents

Many commercial sunscreen formulations have ketones as their active ingredients. Particularly common ingredients are aromatic ketones in which two non-fused benzene rings are present. Examples of such ingredients are avobenzone, oxybenzone, and dioxybenzone.

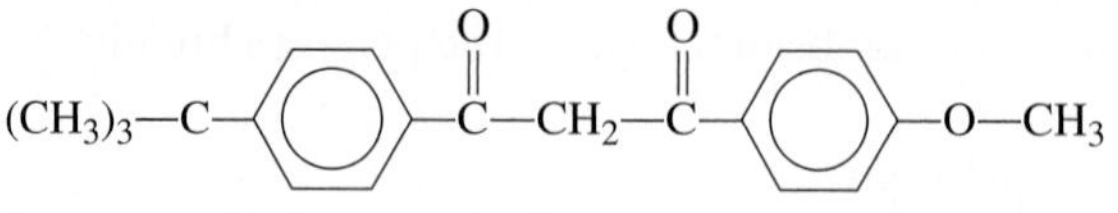

Avobenzone

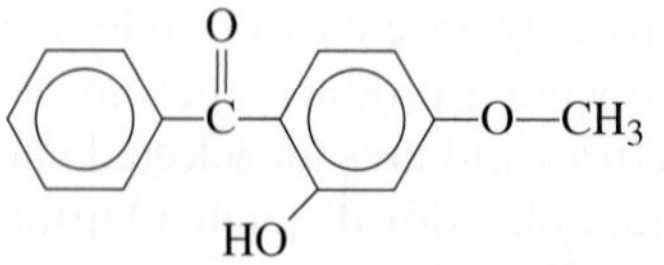

Oxybenzone

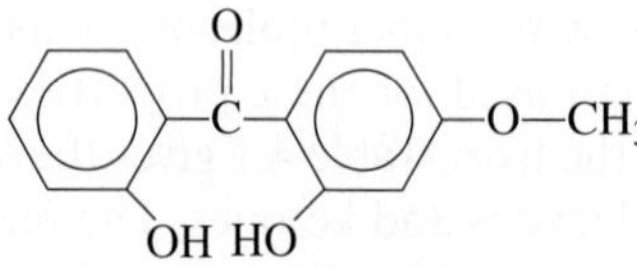

Dioxybenzone

Ketones of this type absorb ultraviolet (UV) light, which enables sunbathers to control UV radiation dosage and thus avoid sunburn (caused by too much UV radiation) and yet develop a suntan (caused by a small amount of UV radiation). Many people believe that a suntan is desirable. Studies show, however, that sunbathing as well as suntans, especially when sunburn results, ages the skin prematurely and increases the risk of skin cancer.

An interesting synthetic ketone is Novec 1230, a completely fluorinated compound with fire-suppressant properties.

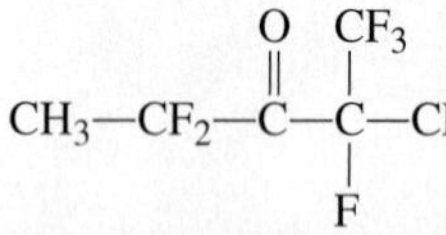

This ketone is a colorless liquid at room temperature that has the feel of water. When placed on a fire, it vaporizes and smothers the fire in a manner similar to that of CO_2. It is a nonconductor of electricity and thus can be used on electrical fires and in other contexts, such as hospitals, museums, and computer equipment rooms, where water cannot be used.

An important compound in the "sunless" tanning industry is the acetone derivative dihydroxyacetone (DHA).

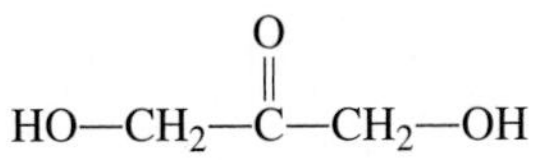

Dihydroxyacetone

DHA's mode of action is reaction with amino acids in skin protein, which produces different tones of coloration ranging from yellow to brown. The resulting pigments are called melanoidins. The coloration is similar to that produced by melanin, a pigment naturally present in the skin.

A "naturally-produced" suntan, obtained through continual gradual exposure to sunlight, involves the previously mentioned compound melanin, which has a polymeric cyclic ketone structure. The focus on relevancy feature Chemical Connections 4-A on the next page discusses the subject of "melanin pigmentation," a topic pertinent to both human skin and human hair.

4.8 Physical Properties of Aldehydes and Ketones

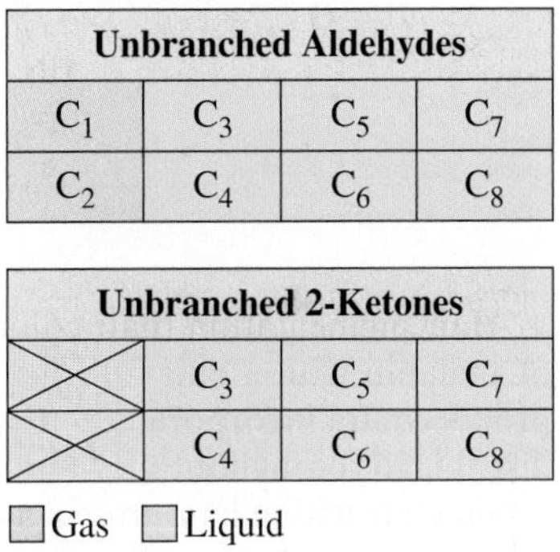

Unbranched Aldehydes
C1 C3 C5 C7
C2 C4 C6 C8
Unbranched 2-Ketones
C3 C5 C7
C4 C6 C8
Gas
Liquid

Figure 4.6 A physical-state summary for unbranched aldehydes and unbranched 2-ketones at room temperature and pressure.

The C_1 and C_2 aldehydes are gases at room temperature (Figure 4.6). The C_3 through C_{11} straight-chain saturated aldehydes are liquids, and the higher aldehydes are solids. The presence of alkyl groups tends to lower both boiling points and melting points, as does the presence of unsaturation in the carbon chain. Lower-molecular-mass ketones are colorless liquids at room temperature (Figure 4.6).

The boiling points of aldehydes and ketones are intermediate between those of alcohols and alkanes of similar molecular mass. Aldehydes and ketones have higher boiling points than alkanes because of dipole–dipole attractions between molecules. Carbonyl group polarity (Section 4.1) makes these dipole–dipole interactions possible.

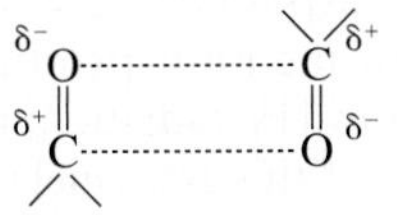

Dipole–dipole attraction

Aldehydes and ketones have lower boiling points than the corresponding alcohols because no hydrogen bonding occurs as it does with alcohols. Dipole–dipole attractions are weaker forces than hydrogen bonds. Table 4.2 provides boiling-point information for selected aldehydes, alcohols, and alkanes.

Aldehydes and ketones having fewer than six carbon atoms are soluble in both organic solvents and water. When six or more carbon atoms are present, such

The ordering of boiling points for carbonyl compounds (aldehydes and ketones), alcohols, and alkanes of similar molecular mass is

Alcohols > carbonyl compounds > alkanes

Table 4.2 Boiling Points of Some Alkanes, Aldehydes, and Alcohols of Similar Molecular Mass

Type of Compound	Compound	Structure	Molecular Mass	Boiling Point (8°C)
alkane	ethane	$CH_3—CH_3$	30	−89
aldehyde	methanal	H—CHO	30	−21
alcohol	methanol	$CH_3—OH$	32	65
alkane	propane	$CH_3—CH_2—CH_3$	44	−42
aldehyde	ethanal	$CH_3—CHO$	44	20
alcohol	ethanol	$CH_3—CH_2—OH$	46	78
alkane	butane	$CH_3—CH_2—CH_2—CH_3$	58	−1
aldehyde	propanal	$CH_3—CH_2—CHO$	58	49
alcohol	1-propanol	$CH_3—CH_3—CH_2—OH$	60	97

CHEMICAL CONNECTIONS 4-A

Melanin: A Hair and Skin Pigment

Human hair, as well as human skin, is colored naturally by the pigment melanin—a polymeric substance involving many interconnected *cyclic ketone* units. The more melanin a person produces, which is genetically controlled, the darker is his or her hair and skin. The number of melanin-producing cells is essentially the same in dark-skinned and light-skinned people, but they are more active in dark-skinned people. The following is a representation of a portion of the structure of polymeric melanin pigment.

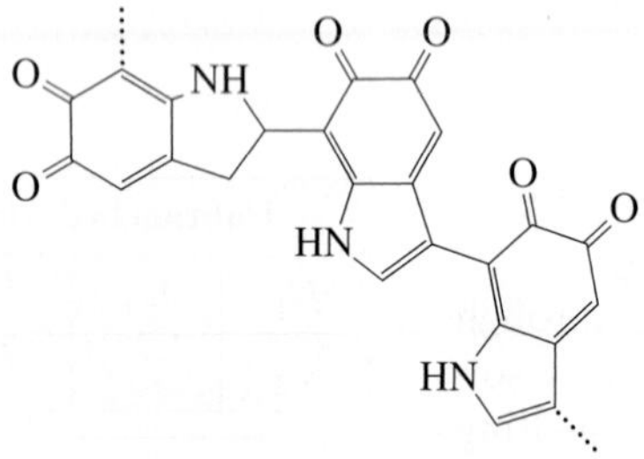

Hair pigmentation (hair color) results from biosynthesis of melanin within hair follicles. The melanin molecules so produced are incorporated into the growing hair shaft and distributed throughout the hair cortex. The melanin tends to accumulate within hair protein as granules. Hair, once it exits the scalp, is no longer alive and any damage to the melanin or the hair itself cannot be repaired by the body.

For most people, starting sometime in their thirties, the production of melanin in hair follicles begins to gradually decrease. Once a hair follicle stops producing melanin, the hair will be colorless but will appear white due to light scattering. A certain proportion of white hair in colored hair will make a head of hair appear gray.

Melanin molecules in the skin constitute a built-in defense system to protect the skin against UV radiation. The melanin molecules act as a protective barrier by absorbing and scattering the UV radiation. A dark-skinned person has more melanin molecules in the upper layers of the skin (and more protection against sunburn) than a light-skinned person.

Jim Reed/Getty Images

High level exposure to ultraviolet radiation causes the skin to burn.

Suntan and sunburn both involve melanin (as a skin pigment) and ultraviolet (UV) radiation. Sudden high-level exposure to UV radiation causes the skin to burn, and steady low-level exposure to UV radiation can cause tanning.

When melanin-producing cells deep in the skin are exposed to UV radiation, melanin production increases. The presence of this extra melanin in the skin gives the skin an appearance that is called a "tan." The larger the melanin molecules so produced, the deeper the tan. People who tan readily have skin that can produce a large amount of melanin.

When a person experiences a sunburn, the skin peels. When peeling occurs, any tan that has been built up (excess melanin) sloughs off with the dead skin. Thus the tanning process must begin anew.

compounds are soluble in organic solvents but not in water. The water solubility of the smaller aldehydes and ketones results from water being able to hydrogen bond to the lone pairs of electrons on the carbonyl oxygen atom (Figure 4.7). Table 4.3 gives water solubility data for selected aldehydes and ketones.

Figure 4.7 Low-molecular-mass aldehydes and ketones are soluble in water because of hydrogen bonding.

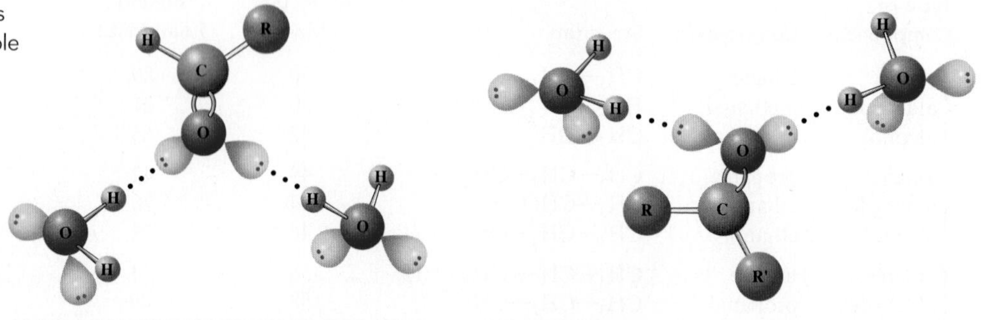

a Aldehyde–Water Hydrogen Bonding

b Ketone–Water Hydrogen Bonding

Table 4.3 Water Solubility (g/100 g H_2O) for Various Aldehydes and Ketones

Number of Carbon Atoms	Aldehyde	Water Solubility of Aldehyde	Ketone	Water Solubility of Ketone
1	methanal	infinite		
2	ethanal	infinite		
3	propanal	16	propanone	infinite
4	butanal	7	butanone	26
5	pentanal	4	2-pentanone	5
6	hexanal	1	2-hexanone	1.6
7	heptanal	0.1	2-heptanone	0.4
8	octanal	insoluble	2-octanone	insoluble

Although low-molecular-mass aldehydes have pungent, penetrating, unpleasant odors, higher-molecular-mass aldehydes (above C_8) are more fragrant, especially benzaldehyde derivatives. Ketones generally have pleasant odors, and several are used in perfumes and air fresheners.

4.9 Preparation of Aldehydes and Ketones

Aldehydes and ketones can be produced by the oxidation of primary and secondary alcohols, respectively, using mild oxidizing agents such as $KMnO_4$ or $K_2Cr_2O_7$ (Section 3.9).

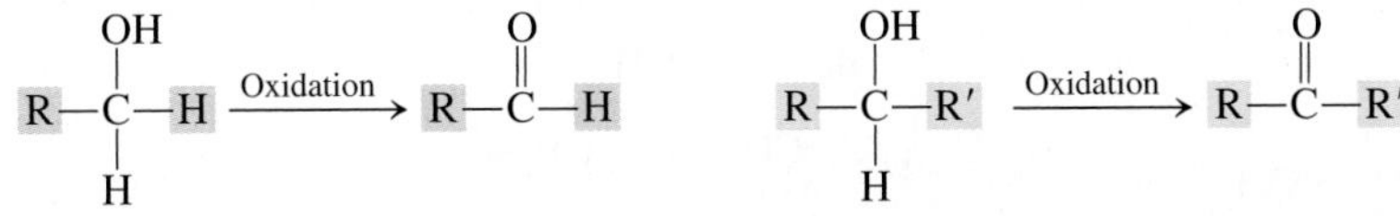

Primary alcohol Aldehyde Secondary alcohol Ketone

When this type of reaction is used for aldehyde preparation, reaction conditions must be sufficiently mild to avoid further oxidation of the aldehyde to a carboxylic acid (Section 3.9). Ketones do not undergo the further oxidation that aldehydes do.

The term *aldehyde* stems from *al*cohol *dehyd*rogenation, indicating that aldehydes are related to alcohols by the loss of hydrogen.

In the oxidation of an alcohol to an aldehyde or a ketone, the alcohol molecule loses H atoms. Recall that a *decrease in the number of C—H bonds* in an organic molecule is one of the operational definitions for the process of *oxidation* (Section 3.9).

EXAMPLE 4.3 Predicting Products in Alcohol Oxidation Reactions

Draw the structure of the aldehyde or ketone formed from the oxidation of each of the following alcohols. Assume that reaction conditions are sufficiently mild that any aldehydes produced are not oxidized further.

a. $CH_3—CH_2—CH_2—OH$

b.
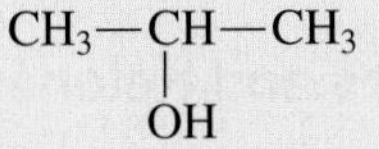

c.
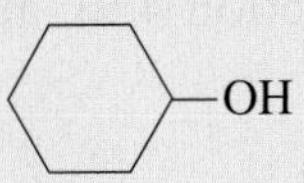

d.
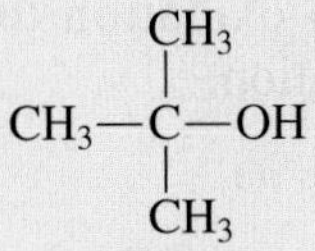

(continued)

Solution

a. This is a primary alcohol that will give the aldehyde *propanal* as the oxidation product.

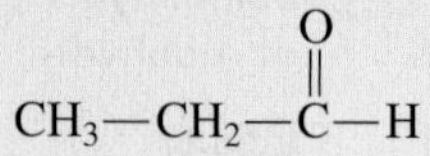

Propanal

b. This is a secondary alcohol. Upon oxidation, secondary alcohols are converted to ketones.

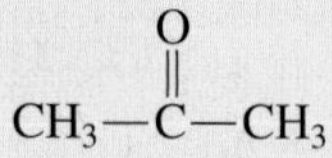

Propanone

c. This cyclic alcohol is a secondary alcohol; hence a ketone is the oxidation product.

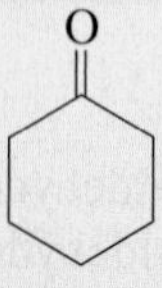

Cyclohexanone

d. This is a tertiary alcohol. Tertiary alcohols do not undergo oxidation (Section 3.9).

Practice Exercise 4.3

Draw the structure of the aldehyde or ketone formed from the oxidation of each of the following alcohols. Assume that reaction conditions are sufficiently mild that any aldehydes produced are not oxidized further to carboxylic acids.

a. 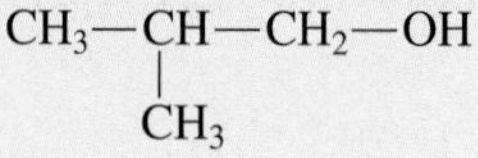

b. 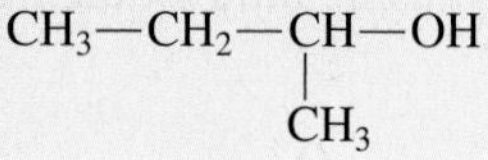

c. 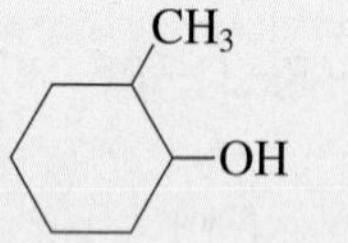

d. 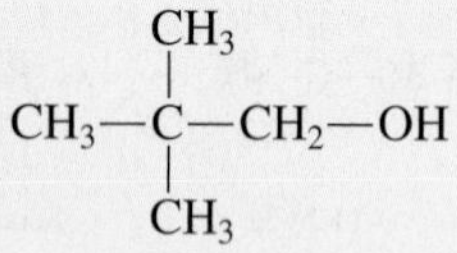

Answers: **a.** 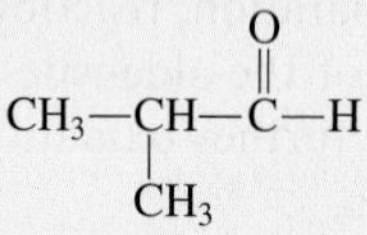

b. 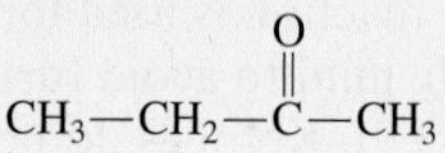

c. 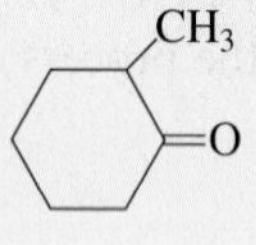

d. 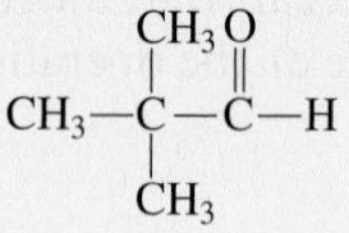

4.10 Oxidation and Reduction of Aldehydes and Ketones

Oxidation of Aldehydes and Ketones

Aldehydes readily undergo oxidation to carboxylic acids (Section 4.9), and ketones are resistant to oxidation.

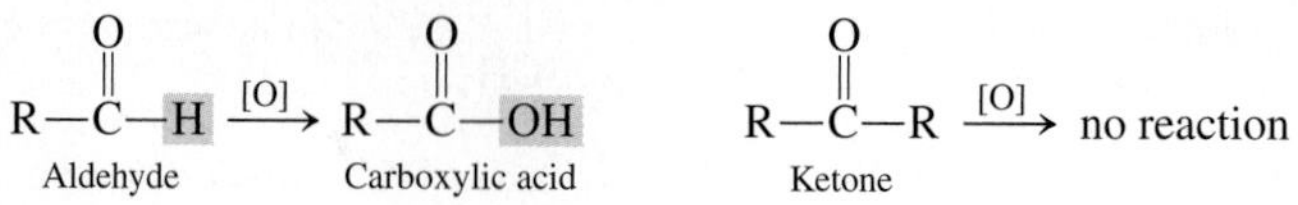

In aldehyde oxidation, the aldehyde gains an oxygen atom (supplied by the oxidizing agent). An *increase in the number of C—O bonds* is one of the operational definitions for the process of *oxidation* (Section 3.9).

Among the mild oxidizing agents that convert aldehydes into carboxylic acids is oxygen in air. Thus aldehydes must be protected from air. When an aldehyde is prepared from oxidation of a primary alcohol (Section 4.9), it is usually removed from the reaction mixture immediately to prevent it from being further oxidized to a carboxylic acid.

The oxidizing agent potassium dichromate ($K_2Cr_2O_7$) is often used for aldehyde oxidation reactions. This oxidizing agent was previously encountered in Section 3.7 as the agent used to oxidize primary and secondary alcohols.

Because both aldehydes and ketones contain carbonyl groups, we might expect similar oxidation reactions for the two types of compounds. Oxidation of an aldehyde involves breaking a carbon–hydrogen bond, and oxidation of a ketone involves breaking a carbon–carbon bond. The former is much easier to accomplish than the latter. For ketones to be oxidized, strenuous reaction conditions must be employed.

Several tests, based on the ease with which aldehydes are oxidized, have been developed for distinguishing between aldehydes and ketones, for detecting the presence of aldehyde groups in sugars (carbohydrates), and for measuring the amounts of sugars present in a solution. The most widely used of these tests are the Tollens test and Benedict's test.

The Tollens test, also called the silver mirror test, involves a solution that contains silver nitrate ($AgNO_3$) and ammonia (NH_3) in water. When Tollens solution is added to an aldehyde, Ag^+ ion (the oxidizing agent) is reduced to silver metal, which deposits on the inside of the test tube, forming a silver mirror. The appearance of this silver mirror (see Figure 4.8) is a positive test for the presence of the aldehyde group.

Reaction of Tollens Solution with aldehydes represents a case of *selective oxidation*; this solution will oxidize the aldehyde group present in a molecule but not any other functional groups present. Thus, if a molecule contains both an aldehyde group and an alcohol group, the former is oxidized but not the latter.

$$\underset{\text{Aldehyde}}{R{-}\overset{\overset{\displaystyle O}{\|}}{C}{-}H} + Ag^+ \xrightarrow[\text{heat}]{NH_3,\ H_2O} \underset{\text{Carboxylic acid}}{R{-}\overset{\overset{\displaystyle O}{\|}}{C}{-}OH} + \underset{\text{Silver metal}}{Ag}$$

The Ag^+ ion will not oxidize ketones.

Benedict's test is similar to the Tollens test in that a metal ion is the oxidizing agent. With this test, Cu^{2+} ion is reduced to Cu^+ ion, which precipitates from solution as Cu_2O (a brick-red solid; see Figure 4.9).

$$\underset{\text{Aldehyde}}{R{-}\overset{\overset{\displaystyle O}{\|}}{C}{-}H} + Cu^{2+} \longrightarrow \underset{\text{Carboxylic acid}}{R{-}\overset{\overset{\displaystyle O}{\|}}{C}{-}OH} + \underset{\text{Brick-red solid}}{Cu_2O}$$

Benedict's solution is made by dissolving copper sulfate, sodium citrate, and sodium carbonate in water.

a An aqueous solution of ethanal is added to a solution of silver nitrate in aqueous ammonia and stirred.

b The solution darkens as ethanal is oxidized to ethanoic acid, and Ag^+ ion is reduced to silver.

c The inside of the beaker becomes coated with metallic silver.

Figure 4.8 A positive Tollens test for aldehydes involves the formation of a silver mirror.

CHEMICAL CONNECTIONS 4-B

Diabetes, Aldehyde Oxidation, and Glucose Testing

Diabetes mellitus is a disease that involves the hormone insulin, a substance necessary to control blood-sugar (glucose) levels. There are two forms of diabetes. In one form, the pancreas does not produce insulin at all. Patients with this condition require injections of insulin to control glucose levels. In the second form, the body cannot make proper use of insulin. Patients with this form of diabetes can often control glucose levels through their diet but may require medication. If the blood-sugar level of a diabetic becomes too high, serious kidney damage can result.

Normal urine does not contain glucose. When the kidneys become overloaded with glucose (the blood-glucose level is too high), glucose is excreted in the urine. Benedict's test (Section 4.10) can be used to detect glucose in urine because glucose has an aldehyde group present in its structure.

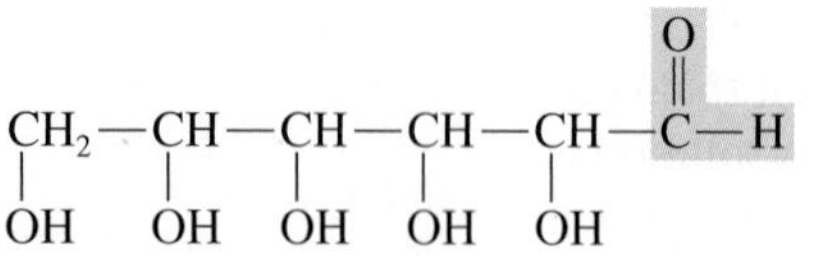

Glucose

For many years, diabetes monitoring involved testing urine for the presence of glucose with a Benedict's solution. The test was carried out using either plastic test strips coated with Benedict's solution or Clinitest tablets (a convenient solid form of Benedict's reagent). A few drops of urine were added to the plastic strip or tablet, and the degree of coloration was used to estimate the blood-glucose level. The solution turned greenish at a low glucose level, then yellow-orange, and finally a dark orange-red.

Today, use of Benedict's solution for urine glucose testing has been replaced largely by other chemical tests that provide better results. Blood-glucose testing has been found to detect rising glucose levels earlier than urine-glucose testing. Blood-glucose tests are now the preferred method for monitoring glucose levels.

Blood-glucose levels are now routinely monitored using electronic glucose meters. In the first generation of such meters, blood-glucose testing involved placing a drop of blood (from a finger prick) on a plastic strip containing a dye and an enzyme that would oxidize glucose's aldehyde group. A two-step reaction sequence then occurred. First, the enzyme caused glucose oxidation to a carboxylic acid, with hydrogen peroxide (H_2O_2) being another product of the reaction.

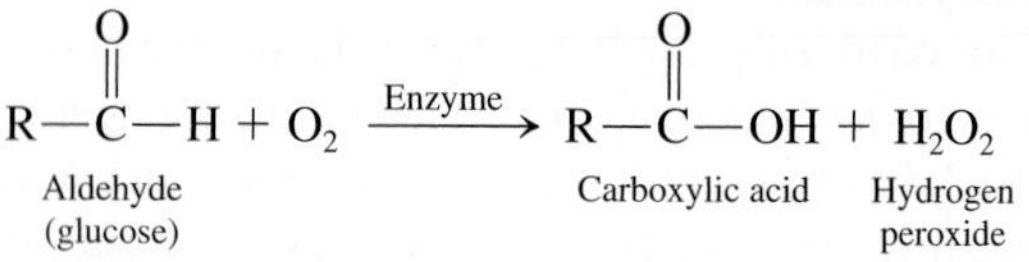

Then the H_2O_2 reacted with the dye to produce a colored product. The amount of color produced, measured by comparison with a color chart or by an electronic monitor, was proportional to the blood-glucose concentration.

Such glucose-monitoring systems have now been replaced by a second generation of meters that directly measure the amount of oxidizing agent that reacts with a known amount of blood. This value is electronically correlated with blood-glucose levels, and the concentration is displayed digitally.

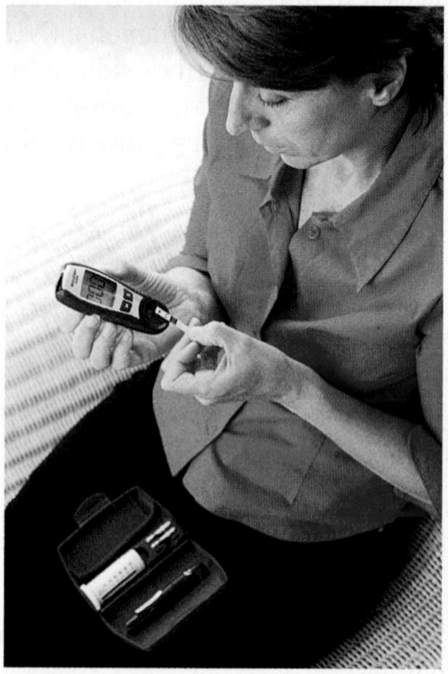

PHANIE/Photo Researchers, Inc.

In a blood-glucose test, a small drop of blood obtained from a finger prick is absorbed on a test strip present within the blood-glucose monitor. The results of the blood–strip interaction, which involves aldehyde oxidation, are processed electronically and displayed on an LCD readout.

Blood-glucose levels must be frequently measured when a person has diabetes. Glucose-monitoring systems now in use are based on aldehyde oxidation reactions. The focus on relevancy feature Chemical Connections 4-B above further considers the topic of diabetes, aldehyde oxidation, and glucose testing.

Reduction of Aldehydes and Ketones

Aldehydes and ketones are easily reduced by hydrogen gas (H_2), in the presence of a catalyst (Ni, Pt, or Cu), to form alcohols. The reduction of aldehydes produces primary alcohols, and the reduction of ketones yields secondary alcohols.

$$\text{Aldehyde} \xrightarrow{\text{Reduction}} \text{primary alcohol}$$

$$\text{Ketone} \xrightarrow{\text{Reduction}} \text{secondary alcohol}$$

Specific examples of such reactions follow.

Aldehyde reduction:

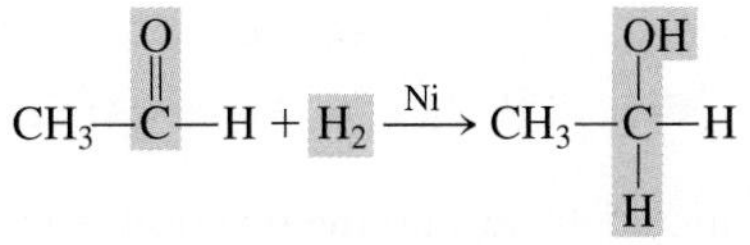

Ethanal Ethanol

Ketone reduction:

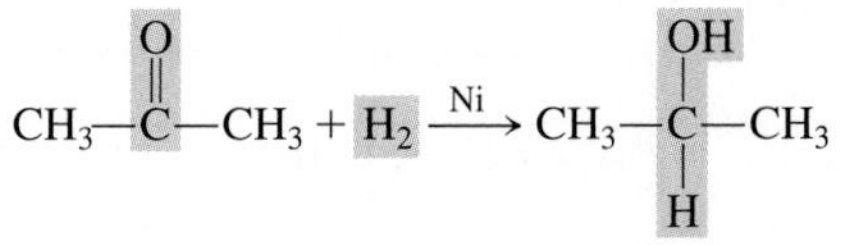

Propanone 2-Propanol

It is the addition of hydrogen atoms to the carbon–oxygen double bond that produces the alcohol in each of these reactions.

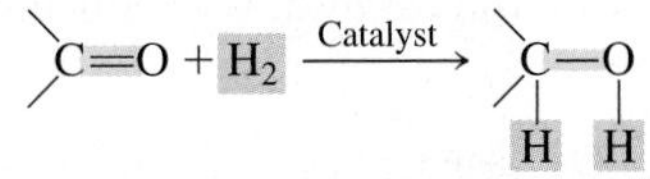

This hydrogen addition process is very similar to the addition of hydrogen to the carbon–carbon double bond of an alkene to produce an alkane, which was encountered in Section 2.9.

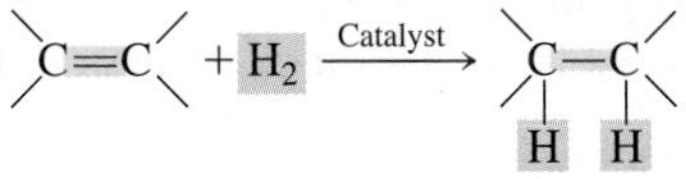

Aldehyde reduction and ketone reduction to produce alcohols are the "opposite" of the oxidation of alcohols to produce aldehydes and ketones (Section 4.9). These "opposite" relationships can be diagrammed as follows:

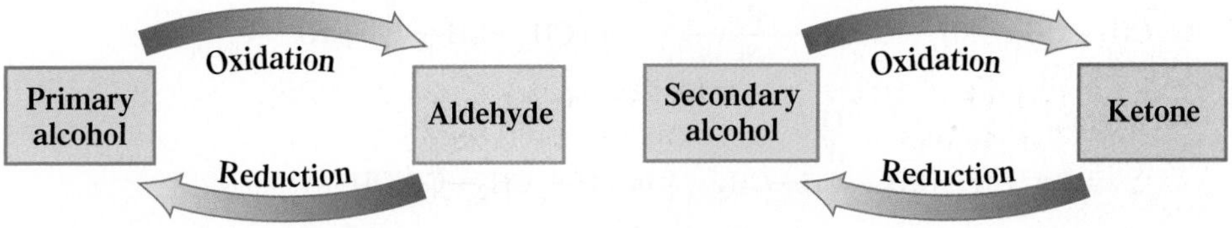

As was noted in Section 3.9, "keeping track" of such relationships is an aid in remembering organic chemistry reaction schemes.

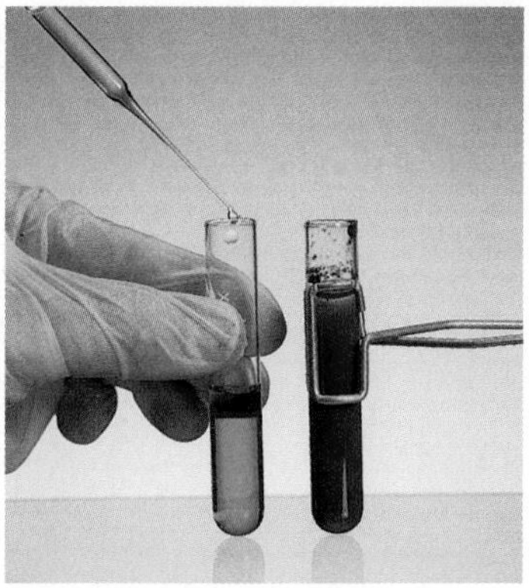

Figure 4.9 Benedict's solution, which is blue in color, turns brick-red when an aldehyde reacts with it.

EXAMPLE 4.4 Predicting Product Identity in Aldehyde/Ketone Oxidation/Reduction Reactions

Draw the structure of the organic product formed when each of the following aldehydes and ketones are subjected to oxidation or reduction. If no reaction occurs, state that such is the case.

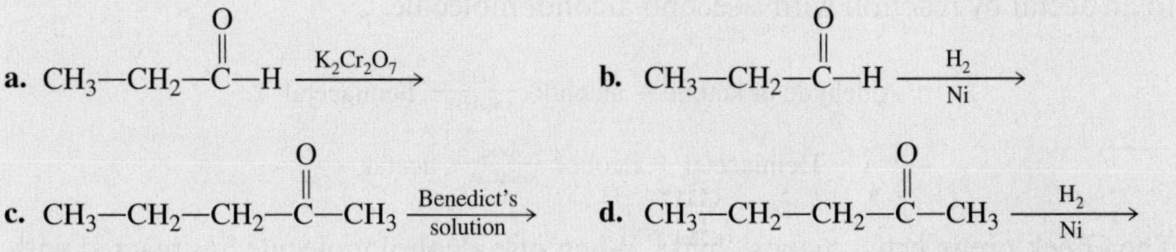

(continued)

Solution

a. The reactant is an aldehyde, and the reaction conditions are those of oxidation; $K_2Cr_2O_7$ is an oxidizing agent. Aldehyde oxidation produces a carboxylic acid.

$$CH_3-CH_2-\overset{\overset{O}{\|}}{C}-H \longrightarrow CH_3-CH_2-\overset{\overset{O}{\|}}{C}-OH$$

b. The reactant is an aldehyde, and the reaction conditions are those of reduction; H_2 gas, with Ni present as a catalyst, is a reducing agent. Aldehyde reduction produces a primary alcohol.

$$CH_3-CH_2-\overset{\overset{O}{\|}}{C}-H \longrightarrow CH_3-CH_2-\overset{\overset{OH}{|}}{CH_2}$$

c. The reactant is a ketone, and the reaction conditions are those of oxidation; Benedict's solution is an oxidizing agent. Ketones cannot be oxidized by mild oxidizing agents; therefore, no reaction occurs.

d. The reactant is a ketone, and the reaction conditions are those of reduction; as in Part b, H_2 gas, with Ni present as a catalyst, is a reducing agent. Ketone reduction produces a secondary alcohol.

$$CH_3-CH_2-CH_2-\overset{\overset{O}{\|}}{C}-CH_3 \longrightarrow CH_3-CH_2-CH_2-\overset{\overset{OH}{|}}{CH}-CH_3$$

Practice Exercise 4.4

Draw the structure of the organic product formed when each of the following aldehydes and ketones are subjected to oxidation or reduction. If no reaction occurs, state that such is the case.

a. $CH_3-CH_2-\overset{\overset{O}{\|}}{C}-CH_3 \xrightarrow[Ni]{H_2}$ **b.** $CH_3-CH_2-\overset{\overset{O}{\|}}{C}-H \xrightarrow[\text{solution}]{\text{Benedict's}}$

c. $CH_3-CH_2-CH_2-\overset{\overset{O}{\|}}{C}-H \xrightarrow[Ni]{H_2}$ **d.** $CH_3-CH_2-\overset{\overset{O}{\|}}{C}-CH_3 \xrightarrow[\text{solution}]{\text{Tollen's}}$

Answers:

a. $CH_3-CH_2-\overset{\overset{OH}{|}}{CH}-CH_3$ **b.** $CH_3-CH_2-\overset{\overset{O}{\|}}{C}-OH$

c. $CH_3-CH_2-CH_2-\overset{\overset{OH}{|}}{CH_2}$ **d.** no reaction occurs

4.11 Reaction of Aldehydes and Ketones with Alcohols

Aldehydes and ketones react with alcohols to form hemiacetals and acetals. Reaction with *one* molecule of alcohol produces a hemiacetal, which is then converted to an acetal by reaction with a second alcohol molecule.

Hemiacetal and acetal formation are very important biochemical reactions; they are crucial to understanding the chemistry of carbohydrates, which is considered in Chapter 7.

$$\text{Aldehyde or ketone} + \text{alcohol} \xrightarrow[\text{catalyst}]{\text{Acid}} \text{hemiacetal}$$

$$\text{Hemiacetal} + \text{alcohol} \xrightarrow[\text{catalyst}]{\text{Acid}} \text{acetal}$$

The Greek prefix *hemi-* means "half." When one alcohol molecule has reacted with the aldehyde or ketone, the compound is halfway to the final acetal.

Further information about these two reactions follows.

Hemiacetal Formation

Hemiacetal formation is an addition reaction in which a molecule of alcohol adds to the carbonyl group of an aldehyde or ketone. The H portion of the alcohol adds to the carbonyl oxygen atom, and the R—O portion of the alcohol adds to the carbonyl carbon atom.

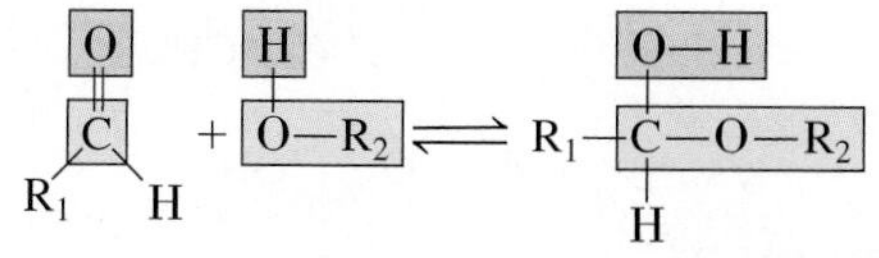

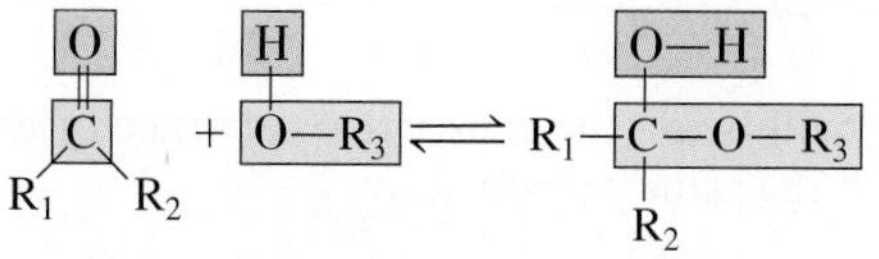

Formally defined, a **hemiacetal** *is an organic compound in which a carbon atom is bonded to both a hydroxyl group (—OH) and an alkoxy group (—OR).* The functional group for a hemiacetal is thus

Hemiacetals contain an alcohol group (hydroxyl group) and an ether group (alkoxy group) on the same carbon atom.

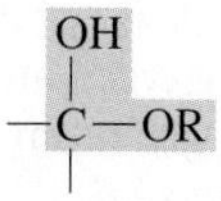

The carbon atom of the hemiacetal functional group is often referred to as the *hemiacetal carbon atom;* it was the carbonyl carbon atom of the aldehyde or ketone that reacted.

A reaction mixture containing a hemiacetal is always in equilibrium with the alcohol and carbonyl compound from which it was made, and the equilibrium lies to the carbonyl compound side of the reaction.

$$\text{Alcohol} + \text{aldehyde} \rightleftharpoons \text{hemiacetal}$$

$$\text{Alcohol} + \text{ketone} \rightleftharpoons \text{hemiacetal}$$

This situation makes isolation of the hemiacetal difficult; in practice, it usually cannot be done.

An important exception to this difficulty with isolation is the case where the —OH and $\rangle$C=O functional groups that react to form the hemiacetal come from the *same* molecule. This produces a *cyclic* hemiacetal rather than a noncyclic one, and cyclic acetals are more stable than the noncyclic ones and can be isolated.

Illustrative of intramolecular hemiacetal formation is the reaction

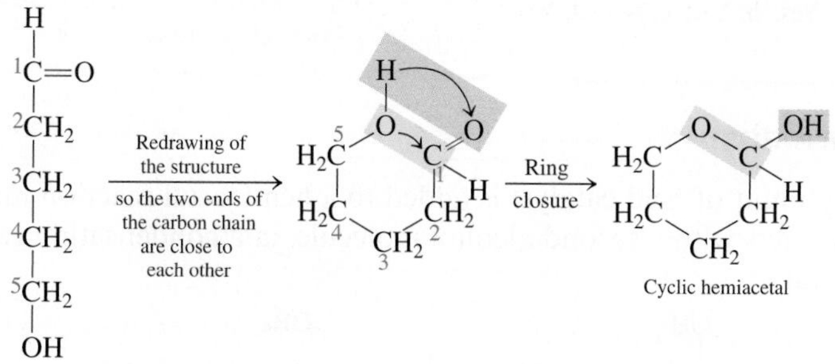

Cyclic hemiacetals are very important compounds in carbohydrate chemistry, the topic of Chapter 7. Glucose, the most common simple carbohydrate, exists primarily in the form of a cyclic hemiacetal (Section 7.10). This cyclic structure, which contains five —OH groups, is

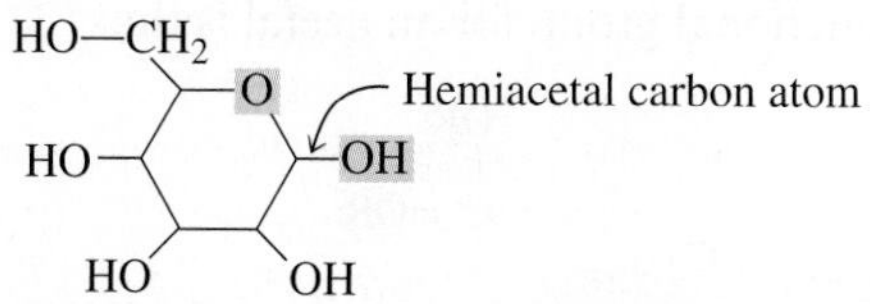

EXAMPLE 4.5 Recognizing Hemiacetal Structures

Indicate whether each of the following compounds is a hemiacetal.

a. 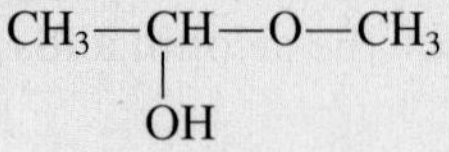

b. 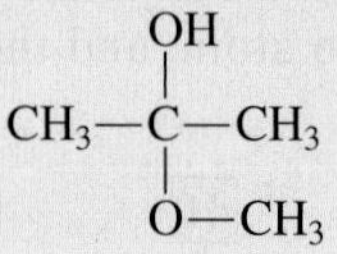

c. 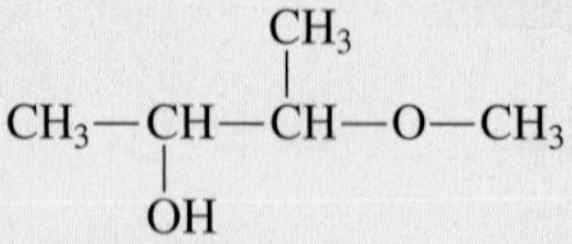

d. 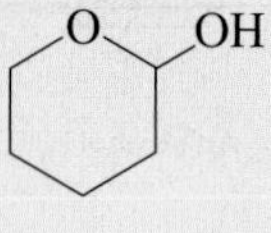

Solution

In each part, the focus will be on the presence or absence of an —OH group and an —OR group attached to the same carbon atom.

a. An —OH group and an —OR group are attached to the same carbon atom. The compound is a *hemiacetal.*

b. An —OH group and an —OR group are attached to the same carbon atom. The compound is a *hemiacetal.*

c. The —OH and —OR groups present in this molecule are attached to *different* carbon atoms. Therefore, the molecule is *not a hemiacetal.*

d. A ring carbon atom bonded to two oxygen atom is present: one oxygen atom in an —OH substituent and the other oxygen atom bonded to the rest of the ring (the same as an R group). This is a *hemiacetal.*

Practice Exercise 4.5

Indicate whether each of the following compounds is a hemiacetal.

a. 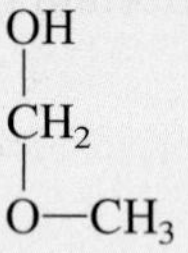

b. 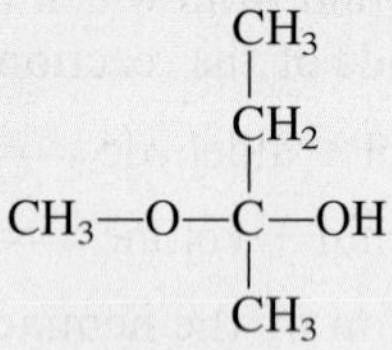

c. 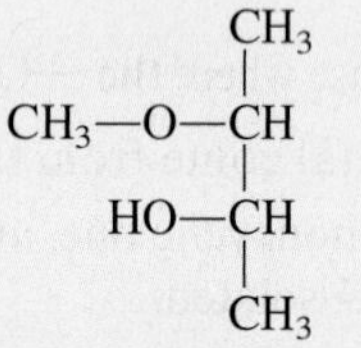

d. 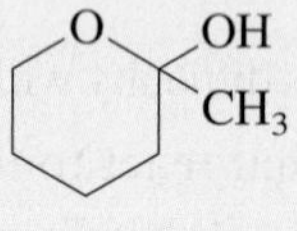

Answers: **a.** Yes; **b.** Yes; **c.** No; **d.** Yes

Acetal Formation

This is the second encounter with condensation reactions. The first encounter involved intermolecular alcohol dehydration (Section 3.9).

If a small amount of acid catalyst is added to a hemiacetal reaction mixture, the hemiacetal reacts with a second alcohol molecule, in a condensation reaction, to form an acetal.

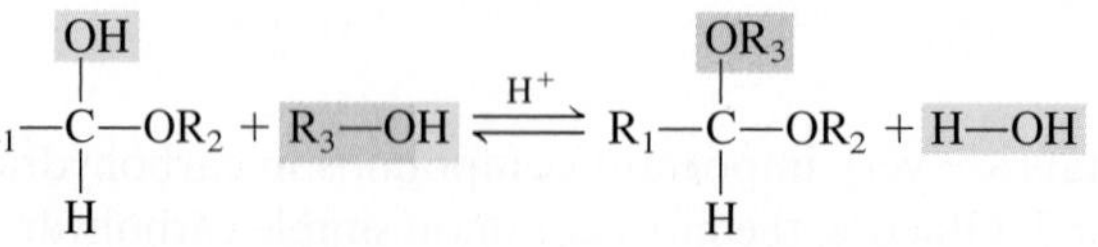

A hemiacetal An alcohol An acetal

Acetals have two alkoxy groups (—OR) attached to the same carbon atom.

An **acetal** *is an organic compound in which a carbon atom is bonded to two alkoxy groups (—OR).* The functional group for an acetal is thus

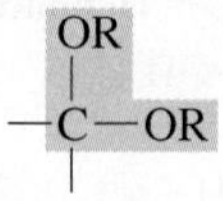

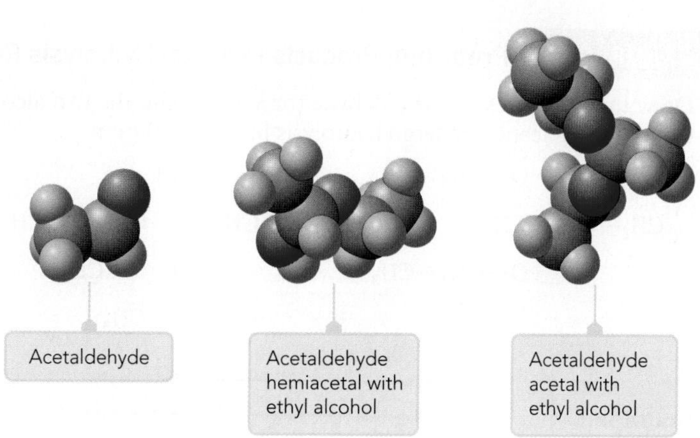

Figure 4.10 Molecular models for acetaldehyde and its hemiacetal and acetal formed by reaction with ethyl alcohol.

A specific example of acetal formation from a hemiacetal is

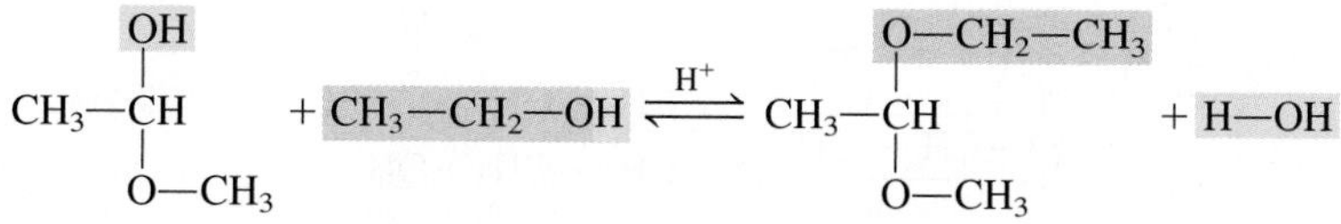

Note that acetal formation does not involve addition to a carbon–oxygen double bond as hemiacetal formation does; no double bond is present in either of the reactants involved in acetal formation. Acetal formation involves a substitution reaction; the —OR group of the alcohol replaces the —OH group on the hemiacetal.

Figure 4.10 shows molecular models for acetaldehyde (the two-carbon aldehyde) and the hemiacetal and acetal formed when this aldehyde reacts with ethyl alcohol.

An acetal is not an ether, even though both types of compounds have structures in which one or more alkoxy (—OR) groups are present.

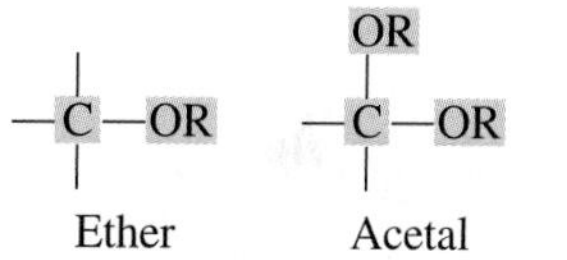

An acetal has two alkoxy groups bonded to the same carbon atom, whereas an ether has one alkoxy group bonded to a carbon atom.

Acetal Hydrolysis

Acetals, unlike hemiacetals, are easily isolated from reaction mixtures. They are stable in basic solution but undergo *hydrolysis* in acidic solution. A **hydrolysis reaction** *is the reaction of a compound with H_2O, in which the compound splits into two or more fragments as the elements of water (H— and —OH) are added to the compound.* The products of acetal hydrolysis are the aldehyde or ketone and alcohols that originally reacted to form the acetal.

In Section 13.1, enzyme-catalyzed hydrolysis of acetals will be discussed as an important step in the digestion of carbohydrates.

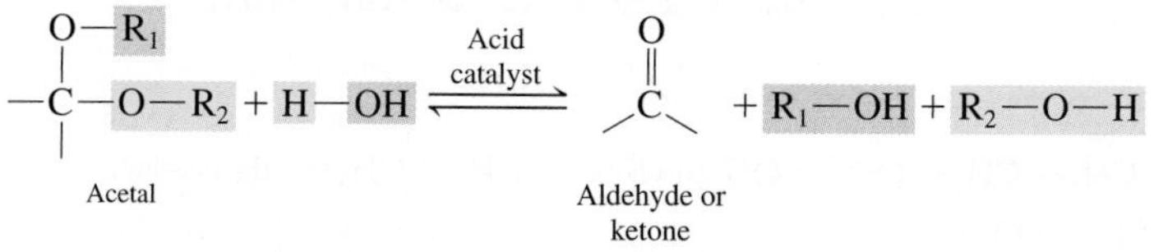

For example,

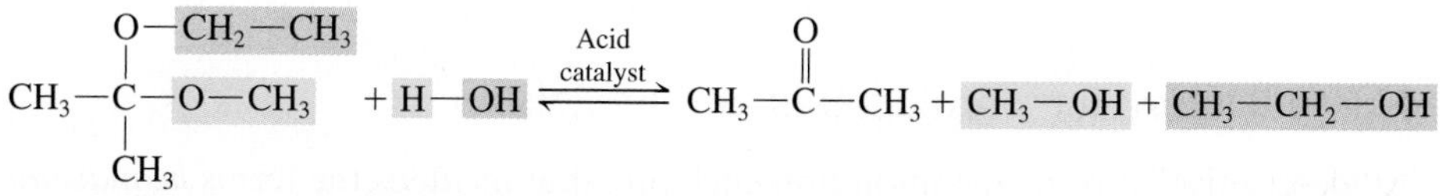

The carbonyl hydrolysis product is an aldehyde if the acetal carbon atom has a hydrogen atom attached directly to it, and it is a ketone if no hydrogen attachment is present. In the preceding example, the carbonyl product is a ketone because the two additional acetal carbon atom attachments are methyl groups.

EXAMPLE 4.6 Predicting Products in Acetal Hydrolysis Reactions

Draw the structures of the aldehyde (or ketone) and the two alcohols produced when the following acetals undergo hydrolysis in acidic solution.

a. 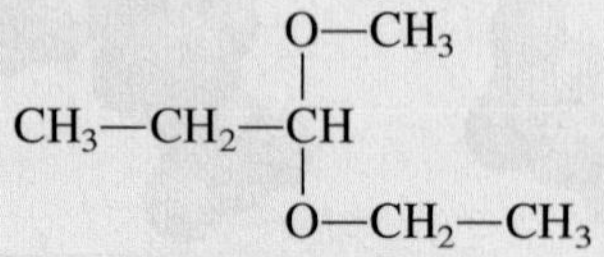

b. 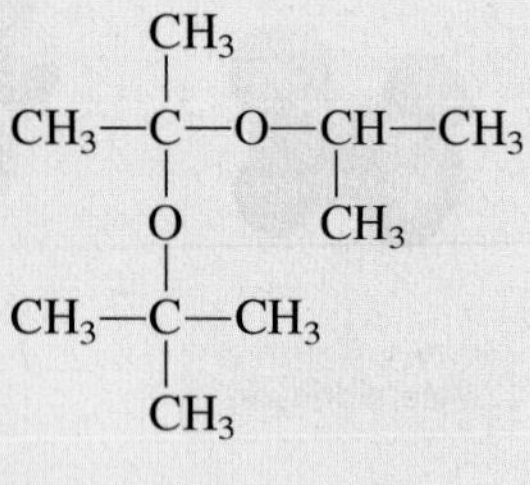

Solution

a. Each of the alkoxy (—OR) groups present will be converted into an alcohol during the hydrolysis. Because the acetal carbon atom has a H attachment, the remainder of the molecule becomes an aldehyde, with the carbon atom to which the alkoxy groups were attached becoming the carbonyl carbon atom.

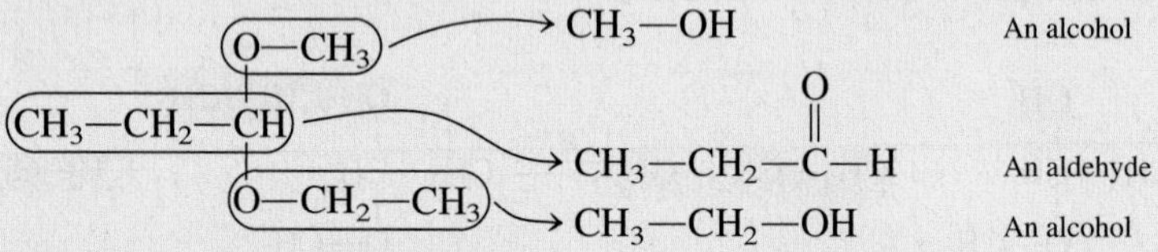

b. Again, each of the alkoxy groups present will be converted into an alcohol during the hydrolysis. Because the acetal carbon atoms lacks a H attachment, the remainder of the molecule becomes a ketone.

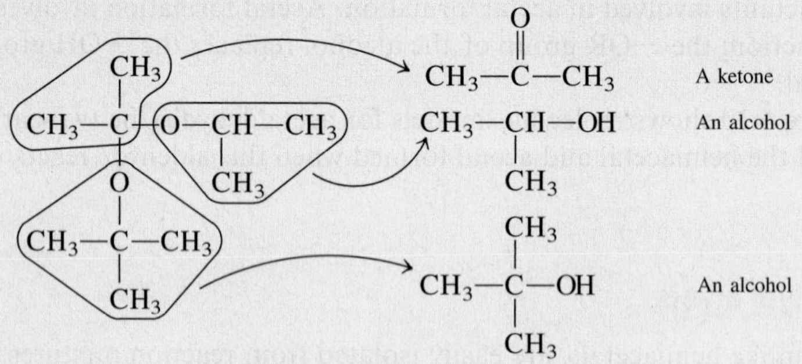

Practice Exercise 4.6

Draw the structures of the aldehyde (or ketone) and the two alcohols produced when the following acetal undergoes hydrolysis in acidic solution.

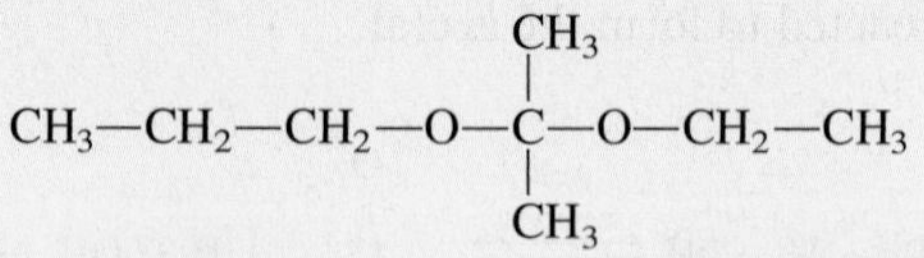

Answers: $CH_3—CH_2—CH_2—OH$ (alcohol), $CH_3—CH_2—OH$ (alcohol),

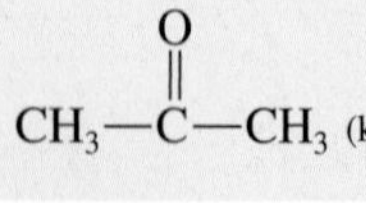 (ketone)

Nomenclature for Hemiacetals and Acetals

A "descriptive" type of common nomenclature that includes the terms *hemiacetal* and *acetal*, as well as the name of the carbonyl compound (aldehyde or ketone) produced in the hydrolysis of the hemiacetal or acetal, is commonly used in describing such compounds. Two examples of such nomenclature are

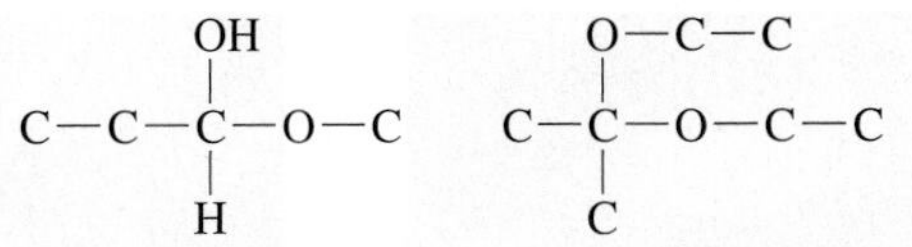

Methyl hemiacetal of propanal　　Diethyl acetal of propanone

The Chemistry at a Glance feature below summarizes reactions that involve aldehydes and ketones.

4.12 Formaldehyde-Based Polymers

Many types of organic compounds can serve as reactants (monomers) for polymerization reactions, including ethylenes (Section 2.10), alcohols (Section 3.10), and carbonyl compounds.

Formaldehyde, the simplest aldehyde, is a prolific "polymer former." As representative of its polymer reactions, the reaction between formaldehyde and phenol under acidic conditions to form a phenol–formaldehyde network polymer will be considered (Figure 4.11). A **network polymer** *is a polymer in which monomers are connected in a three-dimensional cross-linked network.*

When excess formaldehyde is present, the polymerization proceeds via mono-, di-, and trisubstituted phenols that are formed as intermediates in the reaction between phenol and formaldehyde.

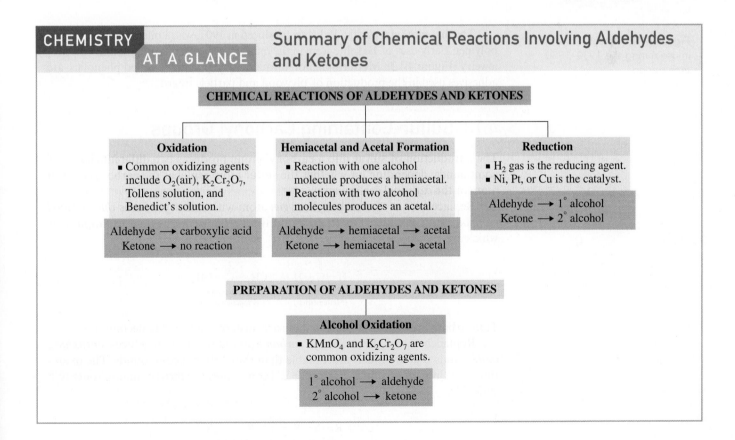

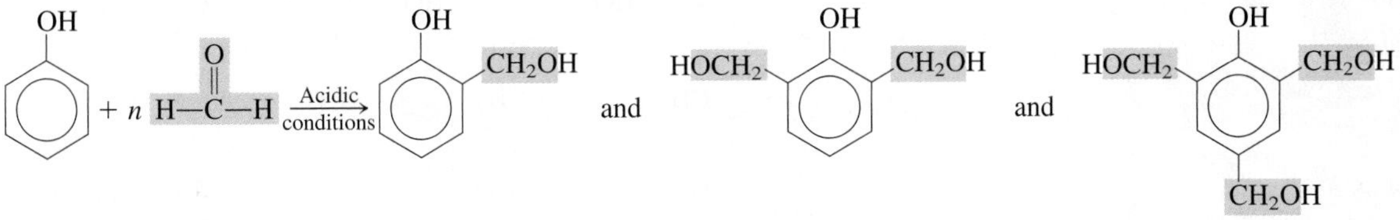

CHEMISTRY AT A GLANCE — Summary of Chemical Reactions Involving Aldehydes and Ketones

CHEMICAL REACTIONS OF ALDEHYDES AND KETONES

Oxidation

- Common oxidizing agents include O_2(air), $K_2Cr_2O_7$, Tollens solution, and Benedict's solution.

Aldehyde ⟶ carboxylic acid
Ketone ⟶ no reaction

Hemiacetal and Acetal Formation

- Reaction with one alcohol molecule produces a hemiacetal.
- Reaction with two alcohol molecules produces an acetal.

Aldehyde ⟶ hemiacetal ⟶ acetal
Ketone ⟶ hemiacetal ⟶ acetal

Reduction

- H_2 gas is the reducing agent.
- Ni, Pt, or Cu is the catalyst.

Aldehyde ⟶ 1° alcohol
Ketone ⟶ 2° alcohol

PREPARATION OF ALDEHYDES AND KETONES

Alcohol Oxidation

- $KMnO_4$ and $K_2Cr_2O_7$ are common oxidizing agents.

1° alcohol ⟶ aldehyde
2° alcohol ⟶ ketone

Figure 4.11 When a mixture of phenol and formaldehyde dissolved in acetic acid is treated with concentrated hydrochloric acid, a cross-linked phenol–formaldehyde network polymer is formed.

The substituted phenols then interact with each other by splitting out water molecules. The final product is a complex, large, three-dimensional network polymer in which monomer units are linked via methylene (—CH_2—) bridges.

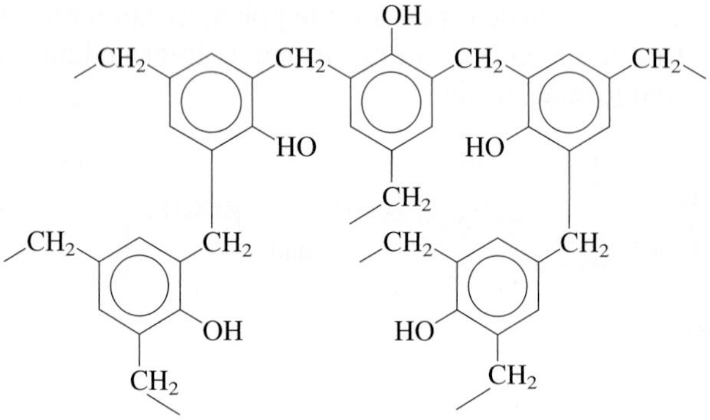

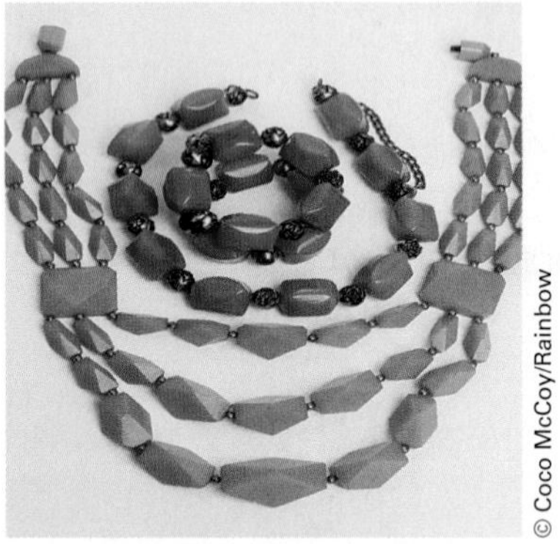

Figure 4.12 Bakelite jewelry in use during the 1930–1950 time period.

The first synthetic plastic, Bakelite, produced in 1907, was a phenol–formaldehyde polymer. Early uses of Bakelite were in the manufacture of billiard balls and "plastic" jewelry (Figure 4.12). Modern phenol–formaldehyde polymers, called phenolics, are adhesives used in the production of plywood and particle board.

4.13 Sulfur-Containing Carbonyl Groups

The introduction of sulfur into a carbonyl group produces two different classes of compounds, depending on whether the sulfur atom replaces the carbonyl oxygen atom or the carbonyl carbon atom.

Replacement of the carbonyl *oxygen* atom with sulfur produces *thiocarbonyl compounds*—thioaldehydes (thials) and thioketones (thiones)—the simplest of which are

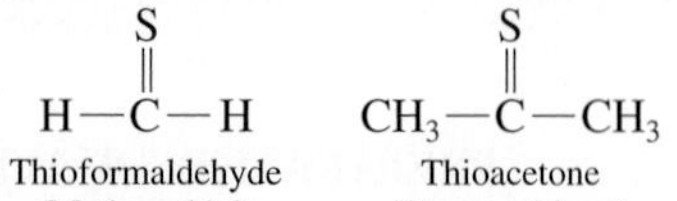

Thiocarbonyl compounds such as these are unstable and readily decompose.

Replacement of the carbonyl *carbon* atom with sulfur produces *sulfoxides,* compounds that are much more stable than thiocarbonyl compounds. The oxidation of a thioether (sulfide) [Section 3.21] constitutes the most common route to a sulfoxide.

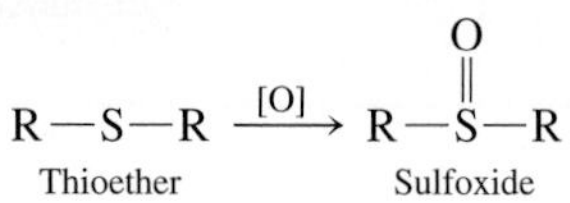

A highly interesting sulfoxide is DMSO (dimethyl sulfoxide), a sulfur analog of acetone, the simplest ketone.

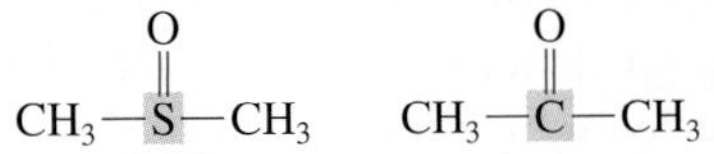

DMSO Acetone

DMSO is an odorless liquid with unusual properties. Because of the presence of the polar sulfur–oxide bond, DMSO is miscible with water and also quite soluble in less polar organic solvents. When rubbed on the skin, DMSO has remarkable penetrating power and is quickly absorbed into the body, where it relieves pain and inflammation. For many years, it has been heralded as a "miracle drug" for

CHEMICAL CONNECTIONS 4-C

Lachrymatory Aldehydes and Ketones

A lachrymator, pronounced "lack-ra-mater," is a compound that causes the production of tears. A number of aldehydes and ketones have lachrymatic properties.

Two lachrymal ketones are 2-chloroacetophenone and bromoacetone.

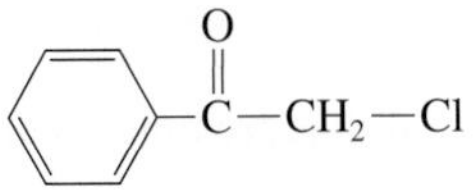

2-Chloroacetophenone (2-chloro-1-phenylethanone)

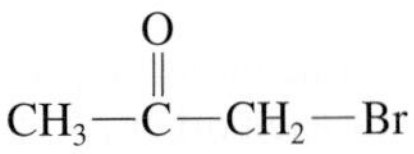

Bromoacetone (1-bromopropanone)

2-Chloroacetophenone is a component of the tear gas used by police and the military. It is also the active ingredient in mace canisters now marketed for use by individuals to protect themselves from attackers. The compound bromoacetone has been used as a chemical war gas.

Smoke contains compounds that cause the eyes to tear. A predominant lachrymator in wood smoke is formaldehyde, the one-carbon aldehyde. The smoke associated with an outdoor barbecue contains the unsaturated aldehyde *acrolein.*

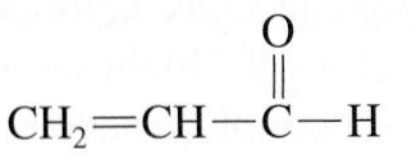

Acrolein (propenal)

Acrolein is formed as fats that are present in meat break down when heated. (Besides being a lachrymator, acrolein is responsible for the "pleasant" odor associated with the process of barbecuing meat.)

The lachrymatory compound associated with onions is a derivative of thiopropionaldehyde.

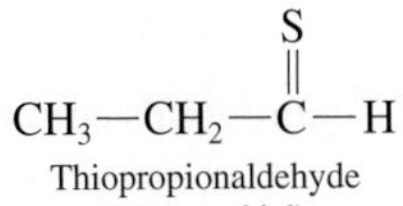

Thiopropionaldehyde (propanethial)

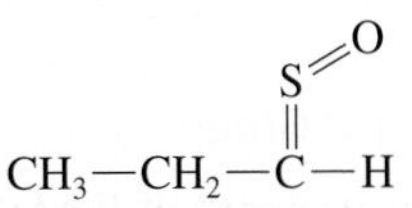

Thiopropionaldehyde-S-oxide (propanethial-S-oxide) [lachrymator in chopped onions]

Onions do not cause tear production until they are chopped or sliced. The onion cells damaged by these actions release the enzyme *allinase,* which converts an odorless compound naturally present in onions to the lachrymatic compound.

© Harvey Lloyd/Peter Arnold/Photolibrary

The smoke generated from outdoor barbecuing contains the lachrymatory aldehyde *acrolein.*

Scientists are not sure why thiopropionaldehyde-S-oxide causes tear production, but it is known that it is an unstable molecule that is readily broken up by water into propanal, hydrogen sulfide, and sulfuric acid.

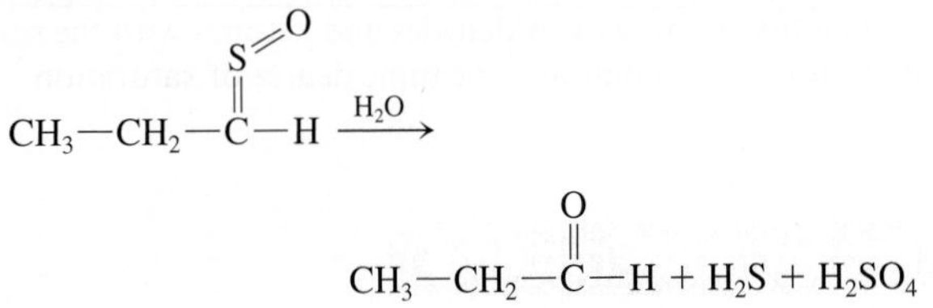

Sulfuric acid may be responsible for the eye irritation.

Many people can peel onions under water, or peel cold ones from the refrigerator, without crying. Water washes away the soluble lachrymator and also breaks it down chemically. If the onion is cold, the enzymatic reaction making the lachrymator is slower, so less is formed. Also the vapor pressure of the lachrymator is greatly reduced at the lower temperature, so its concentration in air is reduced.

arthritis, sprains, burns, herpes, infections, and high blood pressure. However, the FDA has for various reasons steadfastly refused to approve it for general medical use. For example, the FDA says that DMSO's powerful penetrating action could cause an insecticide on a gardener's skin to be accidentally carried into his or her bloodstream. Another complication is that DMSO is reduced in the body to dimethyl sulfide, a compound with a strong garlic-like odor that soon appears on the breath.

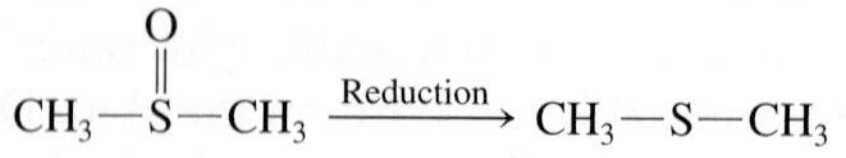

The FDA has approved DMSO for use in treating certain bladder conditions and as a veterinary drug for topical use in nonbreeding dogs and horses. For example, DMSO is used as an anti-inflammatory rub for race horses.

Several aldehydes, ketones, and thioaldehydes have eye-irritant properties that often cause tear production in humans. An oxygenated derivative of the three-carbon thioaldehyde propanethial is the compound that induces tear production when onions are chopped or sliced. The focus on relevancy feature Chemical Connections 4-C on the previous page presents information about this "onion compound," as well as about eye irritants associated with mace, tear gas, wood smoke, and barbecue smoke.

Concepts to Remember

OWL Sign in at **www.cengage.com/owl** to view tutorials and simulations, develop problem-solving skills, and complete online homework assigned by your professor.

The carbonyl group. A carbonyl group consists of a carbon atom bonded to an oxygen atom through a double bond. Aldehydes and ketones are compounds that contain a carbonyl functional group. The carbonyl carbon in an aldehyde has at least one hydrogen attached to it, and the carbonyl carbon in a ketone has no hydrogens attached to it (Sections 4.1 through 4.3).

Nomenclature of aldehydes and ketones. The IUPAC names of aldehydes and ketones are based on the longest carbon chain that contains the carbonyl group. The chain numbering is done from the end that results in the lowest number for the carbonyl group. The names of aldehydes end in *-al,* those of ketones in *-one* (Sections 4.4 and 4.5).

Isomerism for aldehydes and ketones. Constitutional isomerism is possible for aldehydes and for ketones when four or more carbon atoms are present. Aldehydes and ketones with the same number of carbon atoms and the same degree of saturation have the same molecular formula and thus are functional group isomers of each other (Section 4.6).

Physical properties of aldehydes and ketones. The boiling points of aldehydes and ketones are intermediate between those of alcohols and alkanes. The polarity of the carbonyl groups enables aldehyde and ketone molecules to interact with each other through dipole–dipole interactions. They cannot, however, hydrogen-bond to each other. Lower-molecular-mass aldehydes and ketones are soluble in water (Section 4.8).

Preparation of aldehydes and ketones. Oxidation of primary and secondary alcohols, using mild oxidizing agents, produces aldehydes and ketones, respectively (Section 4.9).

Oxidation and reduction of aldehydes and ketones. Aldehydes are easily oxidized to carboxylic acids; ketones do not readily undergo oxidation. Reduction of aldehydes and ketones produces primary and secondary alcohols, respectively (Section 4.10).

Hemiacetals and acetals. A characteristic reaction of aldehydes and ketones is the addition of an alcohol across the carbonyl double bond to produce hemiacetals. The reaction of a second alcohol molecule with a hemiacetal produces an acetal (Section 4.11).

Exercises and Problems

OWL Interactive versions of these problems may be assigned in OWL.

Exercises and problems are arranged in matched pairs with the two members of a pair addressing the same concept(s). The answer to the odd-numbered member of a pair is given at the back of the book. Problems denoted with a ▲ involve concepts found not only in the section under consideration but also concepts found in one or more earlier sections of the chapter. Problems denoted with a ● cover concepts found in a Chemical Connections feature box.

The Carbonyl Group (Section 4.1)

4.1 Indicate which of the following compounds contain a carbonyl group.

a. 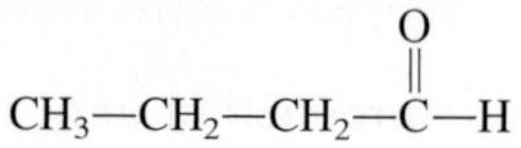

b. $CH_3—O—CH_3$

c. 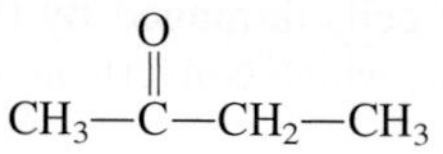

d. 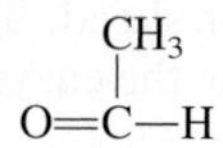

4.2 Indicate which of the following compounds contain a carbonyl group.

a. $CH_3-CH_2-\overset{O}{\overset{\|}{C}}-CH_2-CH_3$

b. $CH_3-CH_2-O-CH_2-CH_3$

c. $CH_3-CH_2-\overset{O}{\overset{\|}{C}}-H$

d. $CH_3-CH_2-\overset{CH_3}{\overset{|}{C}}=O$

4.3 What are the similarities and differences between the bonding in a carbon–oxygen double bond and that in a carbon–carbon double bond?

4.4 Use δ^+ and δ^- notation to show the polarity in a carbon–oxygen double bond.

4.5 What are the approximate bond angles between the atoms attached to the carbonyl carbon atom of a carbonyl group?

4.6 What is the geometrical arrangement for the atoms directly attached to the carbonyl carbon atom in a carbonyl compound?

Compounds Containing a Carbonyl Group (Section 4.2)

4.7 Indicate whether each of the following types of compounds contain a carbonyl group.
a. Aldehyde
b. Ester
c. Alcohol
d. Carboxylic acid

4.8 What elements are present in each of the following types of hydrocarbon derivatives?
a. Carboxylic acid
b. Amide
c. Aldehyde
d. Ketone

4.9 Identify the type of carbonyl-containing compound that each of the following generalized structural formulas represents.

a. $R-\overset{O}{\overset{\|}{C}}-H$ b. $R-\overset{O}{\overset{\|}{C}}-R$

c. $R-\overset{O}{\overset{\|}{C}}-NH_2$ d. $R-\overset{O}{\overset{\|}{C}}-OH$

4.10 Write the structural formula for the simplest member of each of the following types of hydrocarbon derivatives.
a. Aldehyde b. Ester
c. Ketone d. Carboxylic acid

The Aldehyde and Ketone Functional Groups (Section 4.3)

4.11 Classify each of the following structures as an aldehyde, a ketone, or neither.

a. $CH_3-CH_2-CH_2-\overset{O}{\overset{\|}{C}}-OH$

b. $CH_3-\overset{O}{\overset{\|}{C}}-CH_3$

c. $CH_3-O-CH_2-CH_3$

d. CH_3-CHO

4.12 Classify each of the following structures as an aldehyde, a ketone, or neither.

a. $CH_3-\overset{O}{\overset{\|}{C}}-CH_2-CH_3$

b. $CH_3-CH_2-CH_2-\overset{O}{\overset{\|}{C}}-O-CH_3$

c. $CH_3-CH_2-\underset{CH_3}{\underset{|}{CH}}-\overset{O}{\overset{\|}{C}}-H$

d. $CH_3-\overset{CH_3}{\overset{|}{\underset{CH_3}{\underset{|}{C}}}}-CH_2-CHO$

4.13 Draw the structures of the two simplest aldehydes and the two simplest ketones.

4.14 One- and two-carbon ketones do not exist. Explain why.

4.15 Classify each of the following structures as an aldehyde, a ketone, or neither.

a. 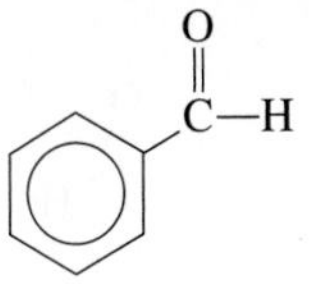

b. 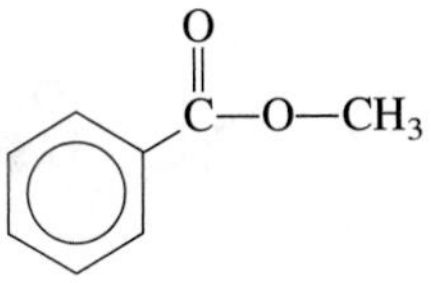

c. 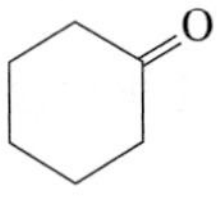

d. 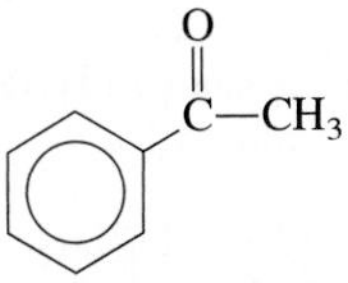

4.16 Classify each of the following structures as an aldehyde, a ketone, or neither.

a. 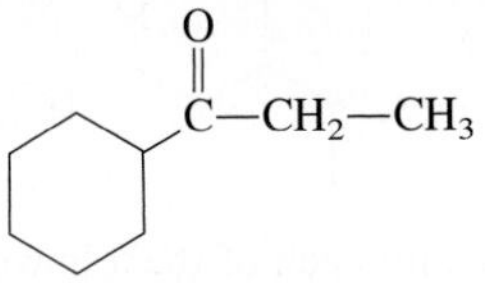

b. 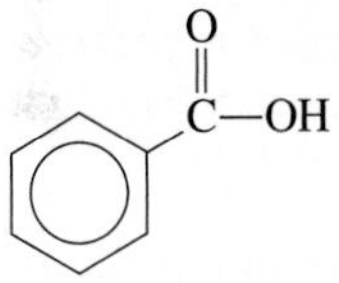

c. 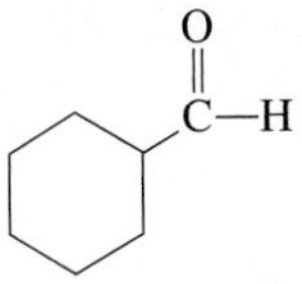

d. 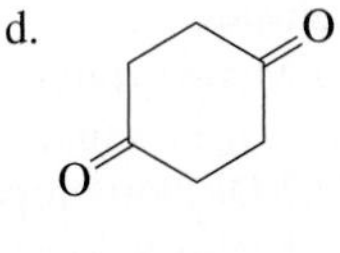

Nomenclature for Aldehydes (Section 4.4)

4.17 Assign an IUPAC name to each of the following aldehydes.

a. $CH_3-CH_2-\underset{CH_3}{\underset{|}{CH}}-\overset{O}{\overset{\|}{C}}-H$

b. $CH_3-\underset{CH_2-CH_2-CH_3}{\underset{|}{CH}}-CH_2-CH_2-\overset{O}{\overset{\|}{C}}-H$

c. 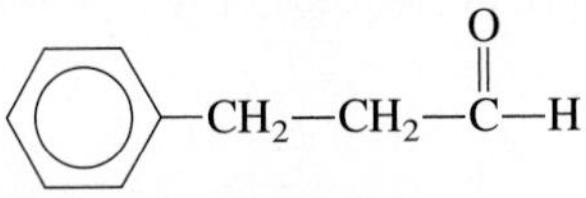

d. CH_3-CH_2-CHO

4.18 Assign an IUPAC name to each of the following aldehydes.

a.
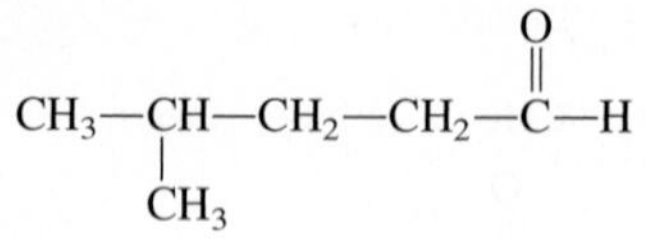

b.
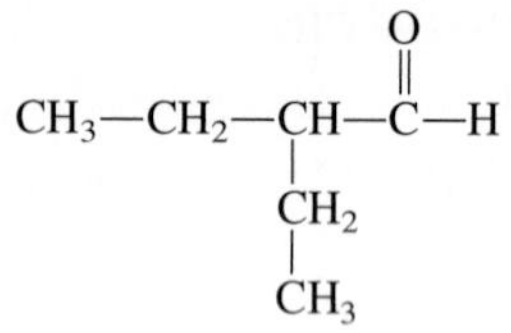

c.
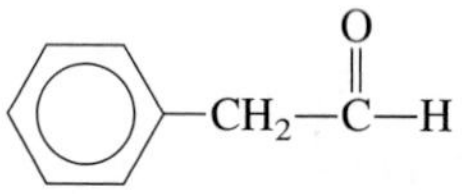

d. $CH_3—CH_2—CH_2—CHO$

4.19 Assign an IUPAC name to each of the following aldehydes.

a.
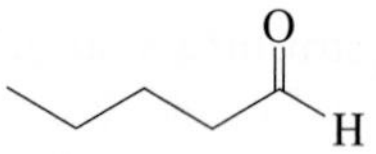

b.
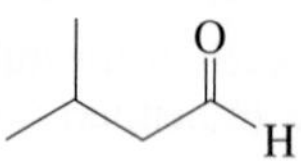

c.
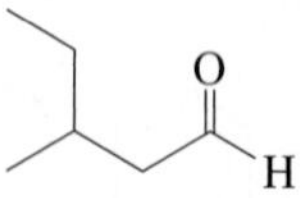

d.
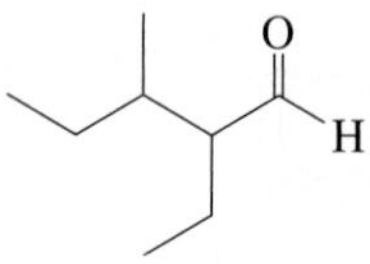

4.20 Assign an IUPAC name to each of the following aldehydes.

a.
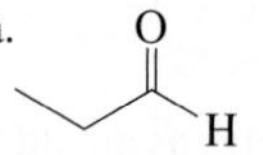

b.
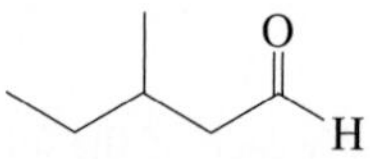

c.
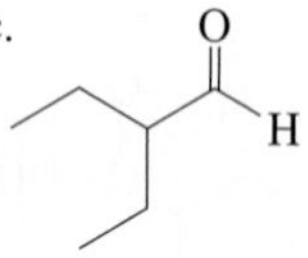

d.
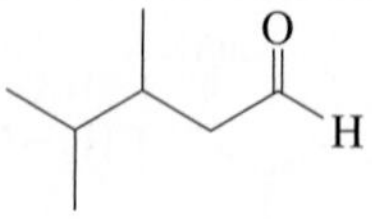

4.21 Draw a structural formula for each of the following aldehydes.

a. 3-Methylpentanal
b. 2-Ethylhexanal
c. 2,2-Dichloropropanal
d. 4-Hydroxy-2-methyloctanal

4.22 Draw a structural formula for each of the following aldehydes.

a. 2-Methylpentanal
b. 4-Ethylhexanal
c. 3,3-Dimethylhexanal
d. 2,3-Dibromopropanal

4.23 Draw a structural formula for each of the following aldehydes.

a. Formaldehyde
b. Propionaldehyde
c. 2-Chlorobenzaldehyde
d. 2,4-Dimethylbenzaldehyde

4.24 Draw a structural formula for each of the following aldehydes.

a. Acetaldehyde
b. Butyraldehyde
c. Dichloroacetaldehyde
d. 2-Methylbenzaldehyde

4.25 Assign a common name to each of the following aldehydes.

a.
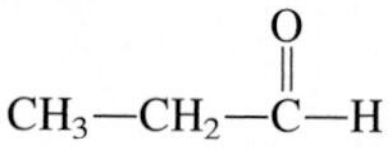

b. $CH_3—CH_2—CHO$

c.
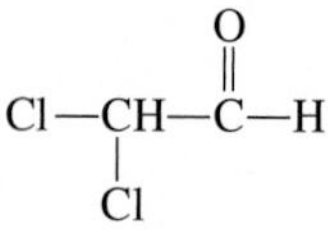

d.
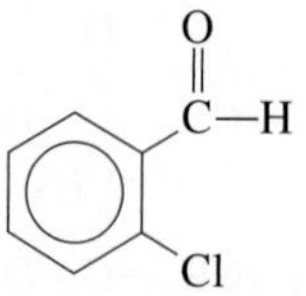

4.26 Assign a common name to each of the following aldehydes.

a.
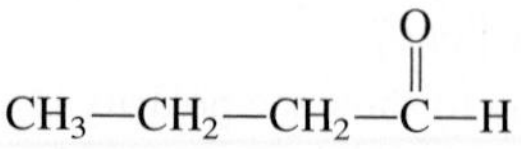

b. $CH_3—CH_2—CH_2—CHO$

c.
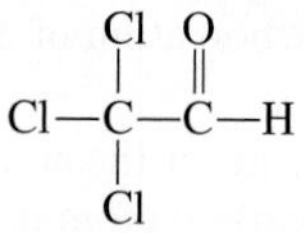

d.
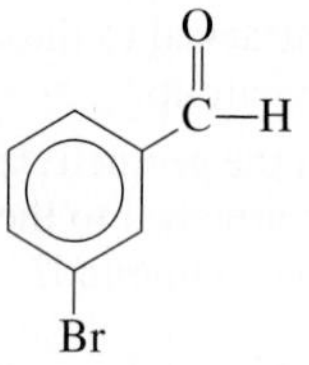

▲**4.27** Name the functional group(s) present in each of the following compounds.

a. Propanal
b. Propenal
c. 2-propanol
d. 3-hydroxypropanal

▲**4.28** Name the functional group(s) present in each of the following compounds.

a. Methanal
b. Methanol
c. 2-butenal
d. 3-methoxybutanal

Nomenclature for Ketones (Section 4.5)

4.29 Using IUPAC nomenclature, name each of the following ketones.

a.
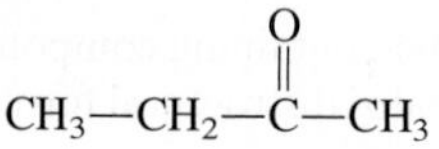

b.
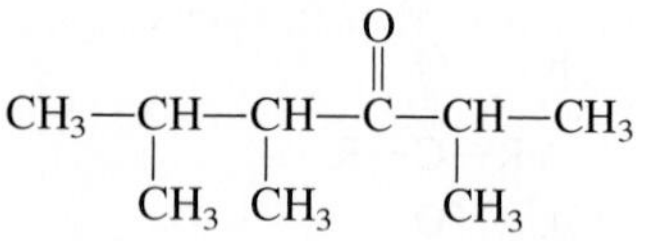

c.
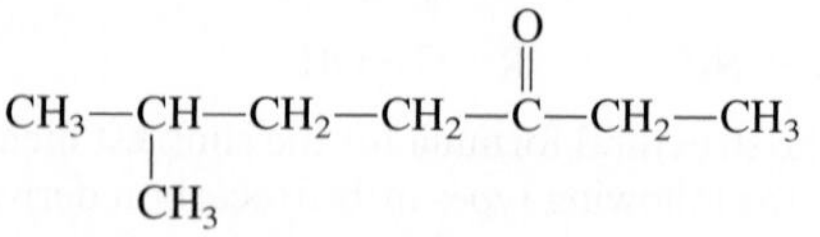

d.
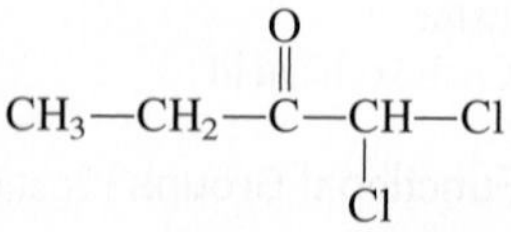

4.30 Using IUPAC nomenclature, name each of the following ketones.

a.
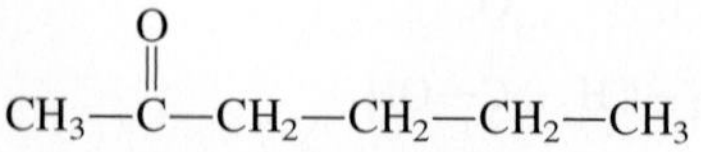

b.
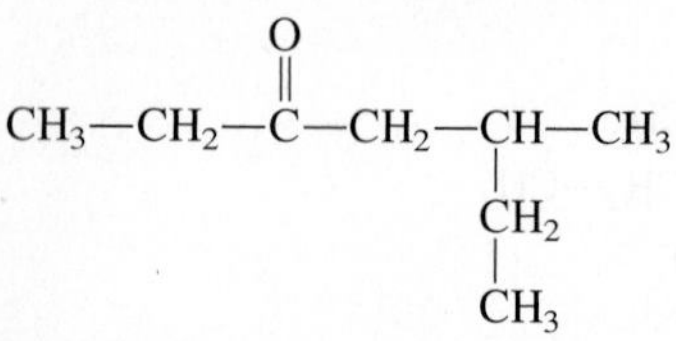

c.
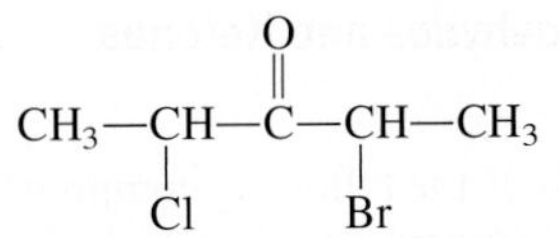

d.
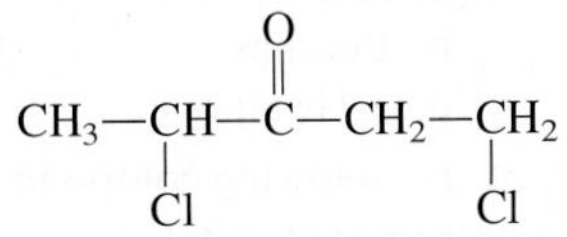

4.31 Assign an IUPAC name to each of the following ketones.

a.
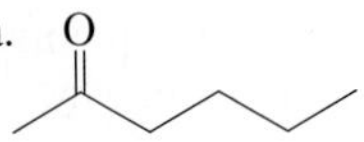
b.
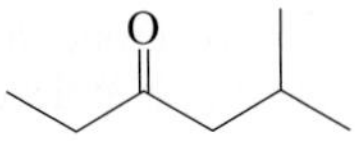

c.
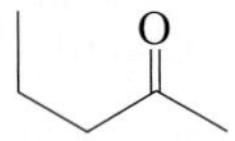
d.
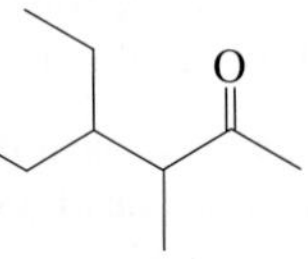

4.32 Assign an IUPAC name to each of the following ketones.

a.
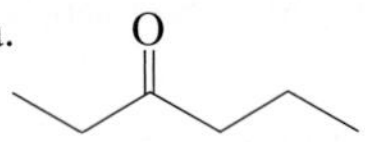
b.
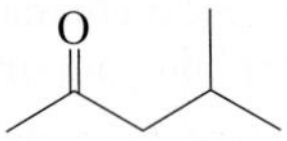

c.
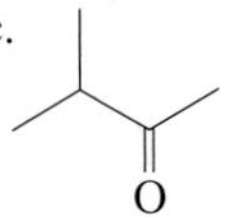
d.
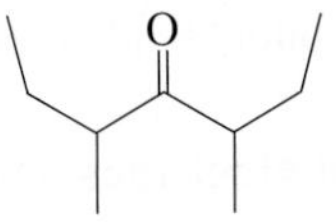

4.33 Using IUPAC nomenclature, name each of the following ketones.

a.
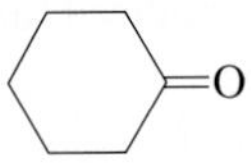
b.
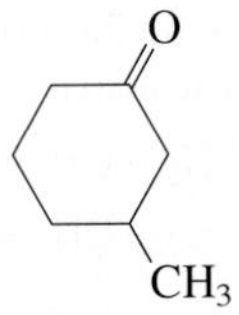

c.
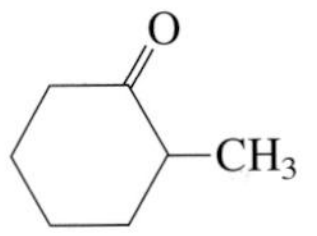
d.
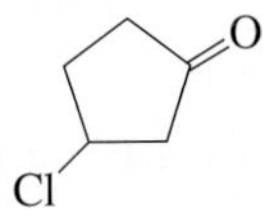

4.34 Using IUPAC nomenclature, name each of the following ketones.

a.
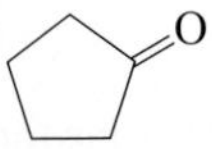
b.
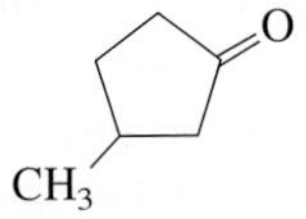

c.
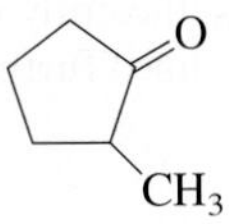
d.
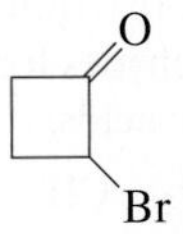

4.35 Draw a structural formula for each of the following ketones.
a. 3-Methyl-2-pentanone b. 3-Hexanone
c. Cyclobutanone d. Chloropropanone

4.36 Draw a structural formula for each of the following ketones.
a. 2-Methyl-3-pentanone b. 2-Pentanone
c. Bromopropanone d. Cyclopentanone

4.37 Draw a structural formula for each of the following ketones.
a. Isopropyl propyl ketone
b. Chloromethyl methyl ketone
c. Acetophenone
d. Methyl phenyl ketone

4.38 Draw a structural formula for each of the following ketones.
a. *tert*-Butyl methyl ketone
b. Dichloromethyl ethyl ketone
c. Benzophenone
d. Diphenyl ketone

▲4.39 Name the functional group(s) present in each of the following compounds.
a. Propanone b. Propanal
c. 4-Oxohexanal d. 4-Octen-2-one

▲4.40 Name the functional group(s) present in each of the following compounds.
a. Butanone b. Butanal
c. 2-Butanol d. 3-Oxobutanal

▲4.41 Indicate whether each of the following compounds would be named as an alcohol, an aldehyde, or a ketone.

a.
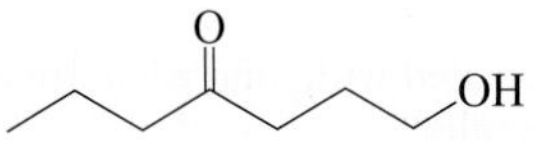

b.
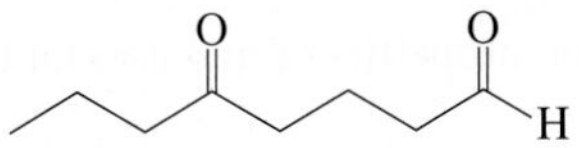

c.
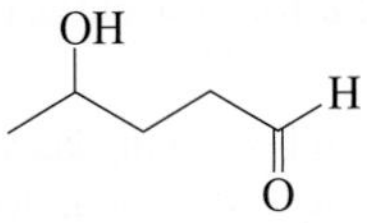

d.
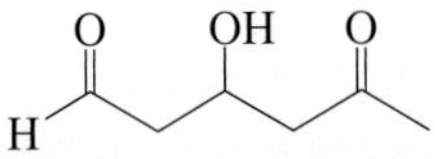

▲4.42 Indicate whether each of the following compounds would be named as an alcohol, an aldehyde, or a ketone.

a.
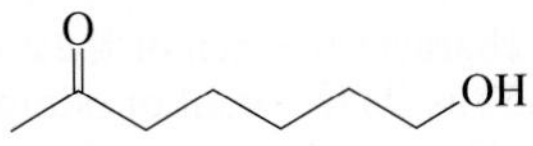

b.
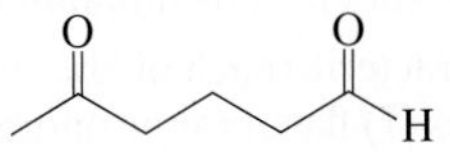

c.
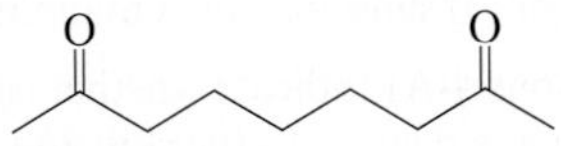

d.
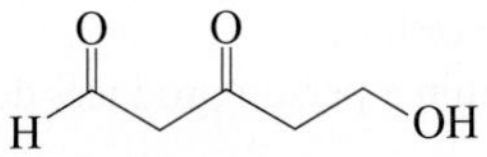

▲4.43 What is the molecular formula for each of the following compounds?
a. 2-Butanol b. Butanal
c. Butanone d. Butanedial

▲4.44 What is the molecular formula for each of the following compounds?
a. Propanal b. Propanedial
c. Propanone d. 2,4-Pentanedione

Isomerism for Aldehydes and Ketones (Section 4.6)

4.45 Give IUPAC names for all saturated unbranched-chain compounds that are named as the following.
a. Heptanals b. Heptanones

4.46 Give IUPAC names for all saturated unbranched-chain compounds that are named as the following.
a. Hexanals b. Hexanones

4.47 How many aldehydes and how many ketones exist with each of the following molecular formulas?
a. CH_2O b. C_3H_6O

4.48 How many aldehydes and how many ketones exist with each of the following molecular formulas?
a. C_2H_4O b. C_4H_8O

4.49 For which values of *x* is the ketone name *x*-methyl-3-hexanone a correct IUPAC name?

4.50 For which values of *x* is the ketone name *x*-methyl-3-pentanone a correct IUPAC name?

4.51 Draw skeletal structural formulas for the four aldehydes and three ketones that have the molecular formula $C_5H_{10}O$.

4.52 Draw skeletal structural formulas for the eight aldehydes and six ketones that have the molecular formula $C_6H_{12}O$.

Selected Common Aldehydes and Ketones (Section 4.7)

4.53 What is the difference between the substances formaldehyde and formalin?

4.54 What is the odor associated with, and what are the uses for, the substance formalin?

4.55 What are the general properties of and uses for the substance acetone?

4.56 What is the significance of the odor of acetone on the breath of a person?

4.57 How many aldehyde groups and how many ketone groups are present in each of the following molecules?
a. Vanillin b. Carvone
c. Cortisone d. Avobenzone

4.58 How many aldehyde groups and how many ketone groups are present in each of the following molecules?
a. Progesterone b. 2-Heptanone
c. Dihydroxyacetone d. Cinnamaldehyde

4.59 In terms of function, characterize each of the compounds in Problem 4.57 as (1) flavorant or odorant (2) steroid hormone or (3) sunscreen or suntanning agent.

4.60 In terms of function, characterize each of the compounds in Problem 4.58 as (1) flavorant or odorant (2) steroid hormone or (3) sunscreen or suntanning agent.

•4.61 (Chemical Connections 4-A) Indicate whether each of the following statements relating to the skin and hair pigment melanin is true or false.
a. The amount of melanin a person produces depends greatly on genetics.
b. Light-skinned people have more melanin-producing cells than do dark-skinned people.
c. Structurally, melanin is a polymeric material.
d. Three ketone groups are present in the repeating unit in melanin's structure.

•4.62 (Chemical Connections 4-A) Indicate whether each of the following statements relating to the skin and hair pigment melanin is true or false.
a. The more melanin a person produces, the darker the person's skin and hair.
b. A light-skinned person has more natural protection against sunburn than does a dark-skinned person.
c. Increased concentrations of melanin in the skin produce what is called a "suntan."
d. Structurally, melanin molecules are simple six-membered-ring cyclic ketones.

Physical Properties of Aldehydes and Ketones (Section 4.8)

4.63 Indicate whether each of the following compounds is a liquid or gas at room temperature (25°C).
a. Methanal b. Butanal
c. Propanone d. 2-Hexanone

4.64 Indicate whether each of the following compounds is a liquid or gas at room temperature (25°C).
a. Ethanal b. Hexanal
c. Butanone d. 2-Heptanone

4.65 Aldehydes and ketones have higher boiling points than alkanes of similar molecular mass. Explain why.

4.66 Aldehydes and ketones have lower boiling points than alcohols of similar molecular mass. Explain why.

4.67 How many hydrogen bonds can form between an acetone molecule and water molecules?

4.68 How many hydrogen bonds can form between an acetaldehyde molecule and water molecules?

4.69 Would you expect ethanal or octanal to be more soluble in water? Explain your answer.

4.70 Would you expect ethanal or octanal to have the more fragrant odor? Explain your answer.

Preparation of Aldehydes and Ketones (Section 4.9)

4.71 Draw the structure of the aldehyde or ketone formed from oxidation of each of the following alcohols. Assume that reaction conditions are sufficiently mild that any aldehydes produced are not oxidized further to carboxylic acids.

a. $CH_3—CH_2—CH_2—CH_2—CH_2—OH$

b. 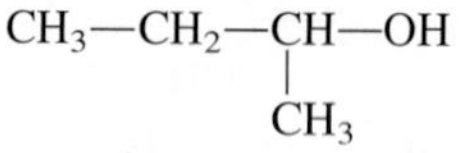

c. 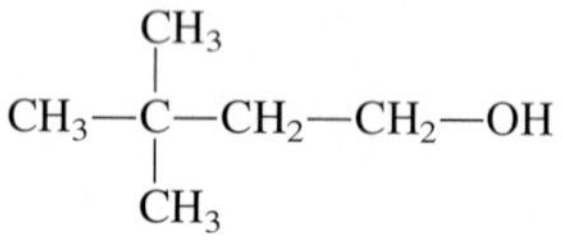

d. 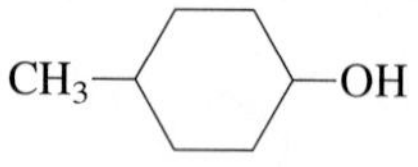

4.72 Draw the structure of the aldehyde or ketone formed from oxidation of each of the following alcohols. Assume that reaction conditions are sufficiently mild that any aldehydes formed are not oxidized further to carboxylic acids.

a. 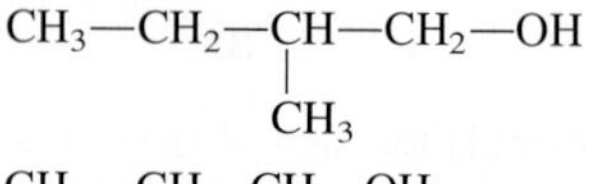

b. 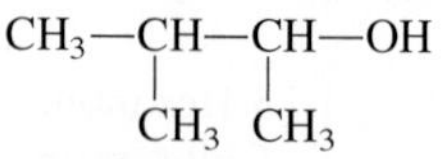

c. 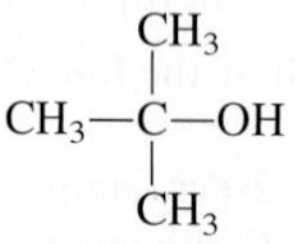

d. 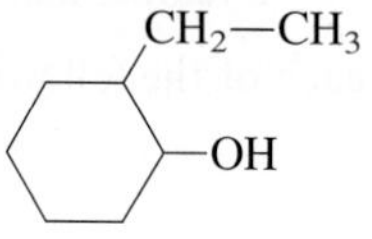

4.73 Draw the structure of the alcohol needed to prepare each of the following aldehydes or ketones by alcohol oxidation.
a. Diethyl ketone b. Phenylpropanone
c. Acetaldehyde d. 2-Ethylhexanal

4.74 Draw the structure of the alcohol needed to prepare each of the following aldehydes or ketones by alcohol oxidation.
a. Propanal b. Dipropyl ketone
c. 3-Phenyl-2-butanone d. Cyclohexanone

Oxidation and Reduction of Aldehydes and Ketones (Section 4.10)

4.75 Draw the structural formula of the organic product when each of the following aldehydes is oxidized to a carboxylic acid.
a. Ethanal b. Pentanal
c. Formaldehyde d. 3,4-Dichlorohexanal

4.76 Draw the structural formula of the organic product when each of the following aldehydes is oxidized to a carboxylic acid.
a. Butanal b. 2-Methylpentanal
c. Acetaldehyde d. Benzaldehyde

4.77 What are the characteristics of a positive Tollens test for aldehydes?

4.78 What are the characteristics of a positive Benedict's test for aldehydes?

4.79 What is the oxidizing agent in Benedict's solution?

4.80 What is the oxidizing agent in Tollens solution?

4.81 Which of the following compounds would react with Tollens solution?

a. $CH_3-CH_2-CH_2-C(=O)-CH_3$

b. $CH_3-CH_2-CH_2-C(=O)-H$

c. $CH_3-CH(OH)-CH_2-C(=O)-H$

d. $CH_3-CH(OH)-C(=O)-CH_3$

4.82 Which of the following compounds would react with Benedict's solution?

a. $CH_3-CH_2-C(=O)-H$

b. $CH_3-C(=O)-CH_3$

c. $CH_3-CH_2-CH(OH)-C(=O)-CH_2-CH_3$

d. $CH_3-CH(CH_3)-CH(CH_3)-CH_2-C(=O)-H$

4.83 Draw the structure of the major organic compound produced when each of the following compounds is reduced using molecular H_2 and a Ni catalyst.

a. $CH_3-CH_2-CH_2-C(=O)-H$

b. $CH_3-CH_2-C(=O)-CH_2-CH_3$

c. $CH_3-CH(CH_3)-CH_2-C(=O)-H$

d. $CH_3-CH(CH_3)-C(=O)-CH_2-CH_2-CH_3$

4.84 Draw the structure of the major organic compound produced when each of the following compounds is reduced using molecular H_2 and a Ni catalyst.

a. $CH_3-CH_2-CH_2-C(=O)-CH_3$

b. $CH_3-CH_2-CH_2-CH_2-C(=O)-H$

c. $CH_3-C(CH_3)_2-CH_2-CH_2-C(=O)-H$

d. $CH_3-CH_2-C(=O)-CH_3$

▲4.85 Which of the three compounds *pentanal*, 2-*pentanone*, and 2-*pentanol* will react with each of the following oxidizing or reducing agents? There may be more than one correct answer in a given situation.
a. $K_2Cr_2O_7$
b. Tollens solution
c. Benedict's solution
d. H_2, Ni catalyst

▲4.86 Which of the three compounds *hexanal*, 2-*butanone*, and 2-*propanol* will react with each of the following oxidizing or reducing agents? There may be more than one correct answer in a given situation.
a. $K_2Cr_2O_7$
b. Tollens solution
c. Benedict's solution
d. H_2, Ni catalyst

▲4.87 What is the IUPAC name of the aldehyde or ketone starting material needed to make each of the following compounds using an oxidation or reduction reaction?

a. CH_3-CH_2-OH b. $CH_3-CH(OH)-CH_2-CH_3$

c. $CH_3-CH_2-C(=O)-OH$ d. 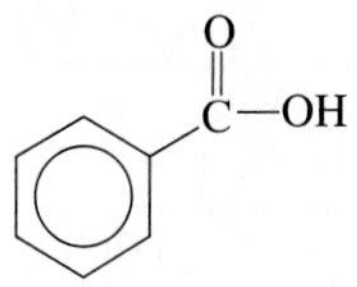

▲4.88 What is the IUPAC name of the aldehyde or ketone starting material needed to make each of the following compounds using an oxidation or reduction reaction?

a. $CH_3-CH_2-CH_2-OH$ b. $CH_3-CH(OH)-CH_3$

c. $CH_3-C(=O)-OH$ d. 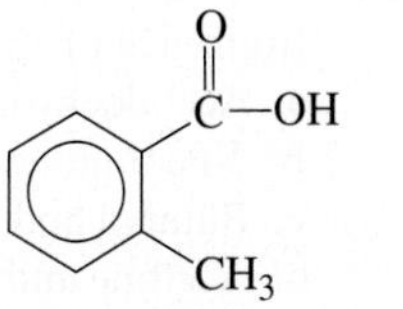

•4.89 (Chemical Connections 4-B) Indicate whether each of the following statements concerning diabetes and the chemical monitoring of this condition is true or false.
a. Diabetes results from having too much insulin in the blood.
b. The inability of the body to properly use insulin once it is produced is a cause of diabetes.
c. High levels of glucose in urine is a sign of diabetes.
d. Modern glucose-monitoring systems directly measure the amount of glucose in the blood.

•4.90 (Chemical Connections 4-B) Indicate whether each of the following statements concerning diabetes and the chemical monitoring of this condition is true or false.
a. The hormone insulin is involved in the regulation of blood-glucose levels.
b. Inadequate production of insulin by the pancreas is a cause of diabetes.
c. Blood-glucose testing has largely replaced urine-glucose testing in monitoring diabetes.
d. Glucose testing is based on the oxidation of the aldehyde group present in glucose molecules.

Hemiacetal Formation (Section 4.11)

4.91 When an alcohol molecule (R—O—H) adds across a carbon–oxygen double bond, into what "fragments" is the alcohol split?

4.92 When an alcohol molecule (R—O—H) adds across a carbon–oxygen double bond, which part of the alcohol molecule adds to the carbonyl oxygen atom?

4.93 Indicate whether each of the following compounds is a hemiacetal.

a. $CH_3—CH_2—O—CH_3$

b. 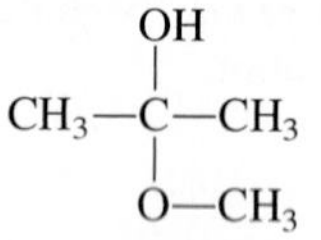

c. 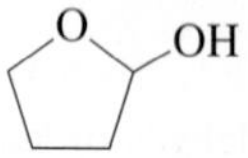

d. 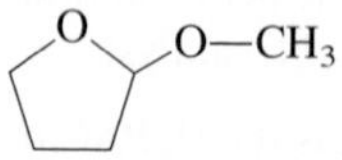

4.94 Indicate whether each of the following compounds is a hemiacetal.

a. 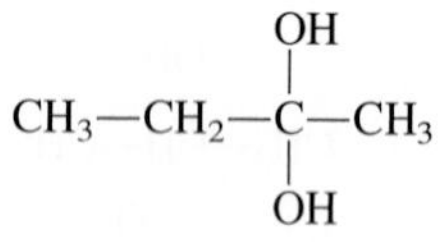

b. 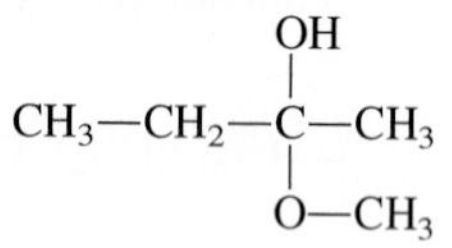

c. 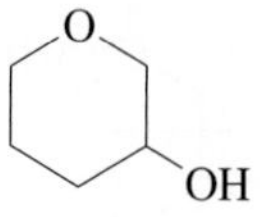

d. 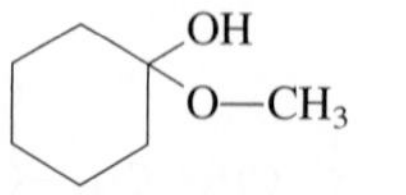

4.95 Draw the structural formula of the hemiacetal formed from each of the following pairs of reactants.
a. Acetaldehyde and ethyl alcohol
b. 2-Pentanone and methanol
c. Butanal and ethanol
d. Acetone and isopropyl alcohol

4.96 Draw the structural formula of the hemiacetal formed from each of the following pairs of reactants.
a. Acetaldehyde and methanol
b. 2-Pentanone and ethyl alcohol
c. Butanal and isopropyl alcohol
d. Acetone and ethanol

4.97 Draw the structural formula of the missing compound in each of the following reactions.

a. 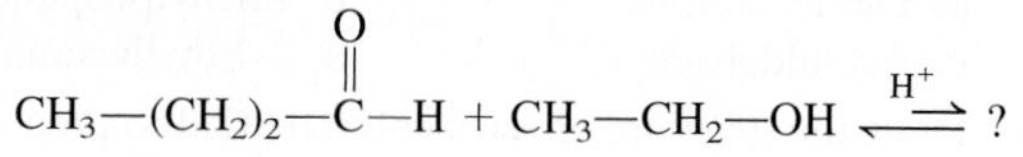

b. 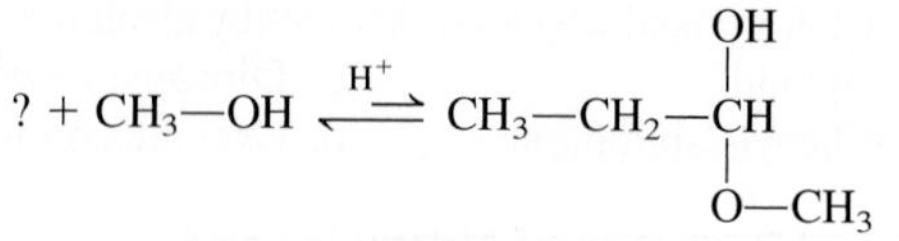

c. 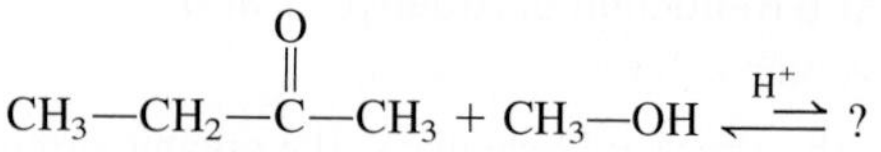

d. 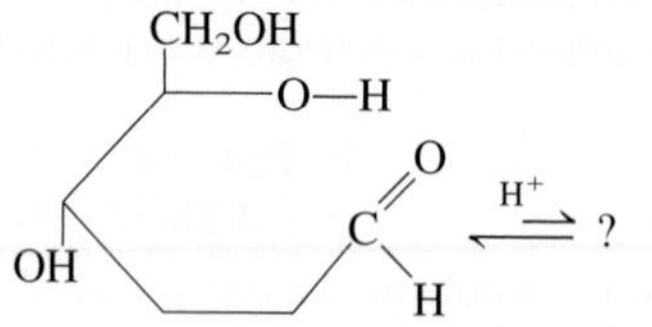

4.98 Draw the structural formula of the missing compound in each of the following reactions.

a. 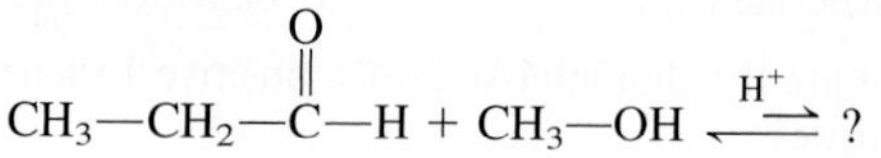

b. 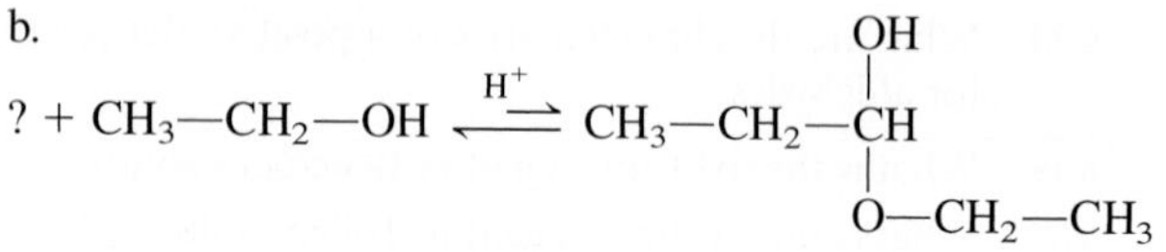

c. 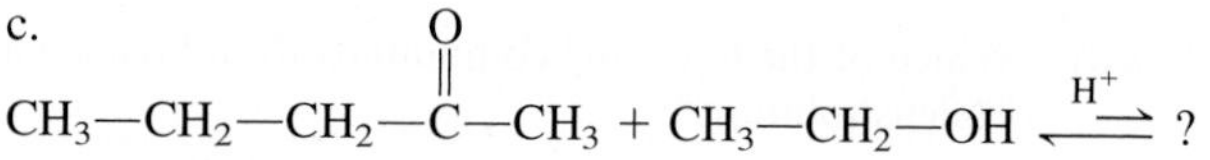

d. 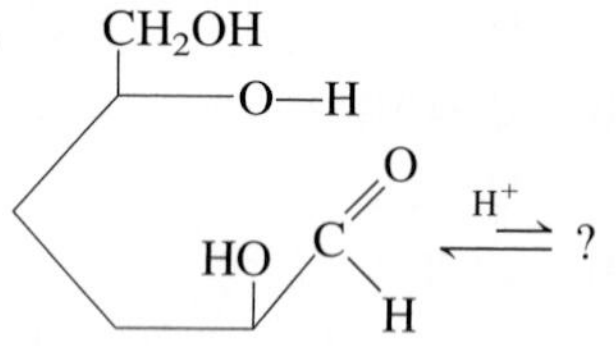

Acetal Formation (Section 4.11)

4.99 Indicate whether each of the following compounds is an acetal.

a. 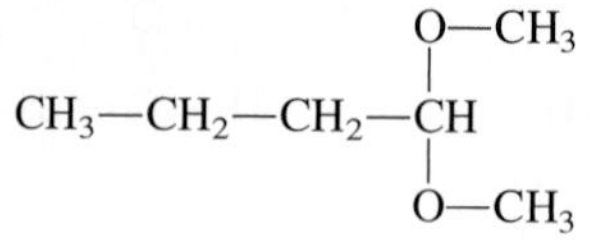

b. 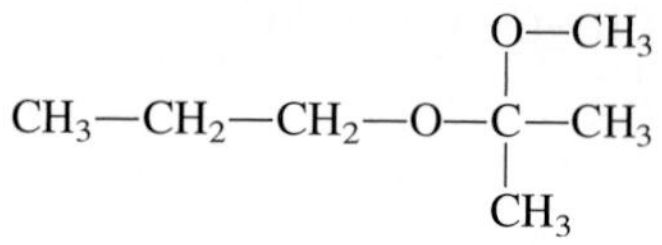

c. 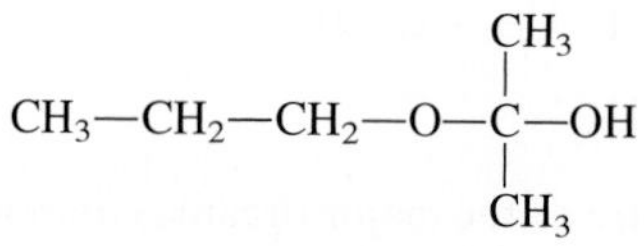

d. 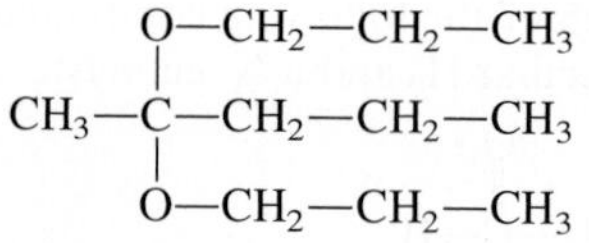

4.100 Indicate whether each of the following compounds is an acetal.

a. 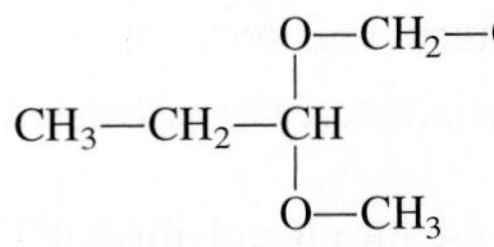

b. 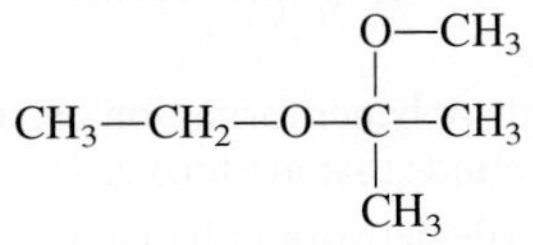

c. 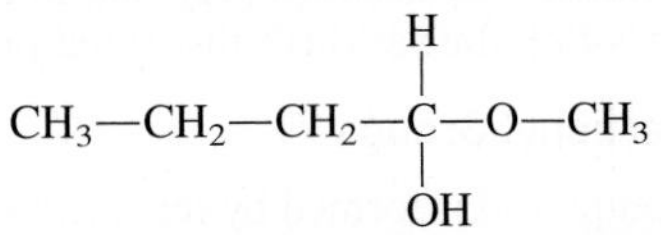

d. 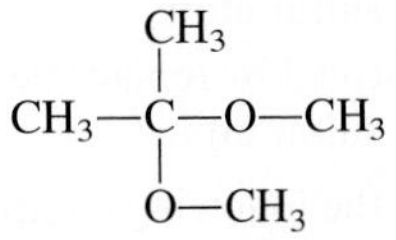

4.101 Draw the structural formula of the missing compound(s) in each of the following reactions.

a. 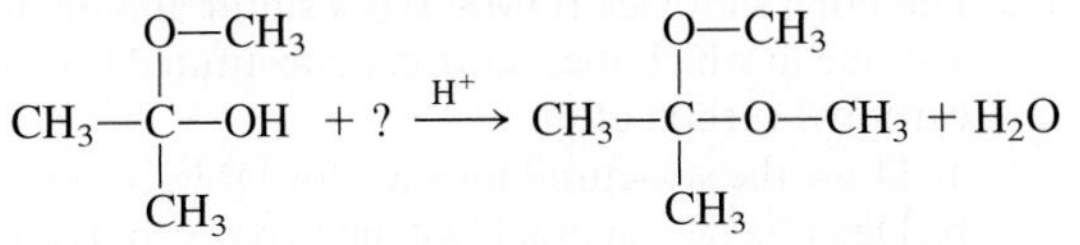

b. 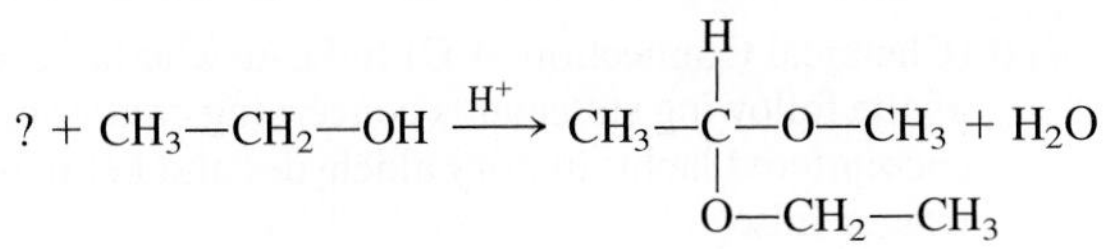

c. 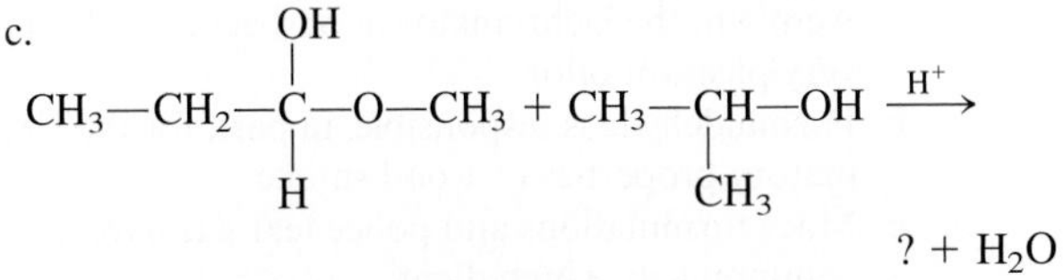

d. 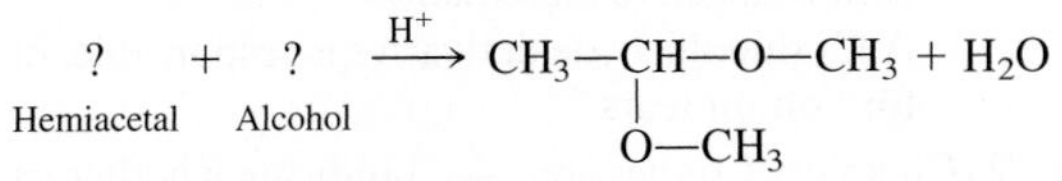

4.102 Draw the structural formula of the missing compound(s) in each of the following reactions.

a. 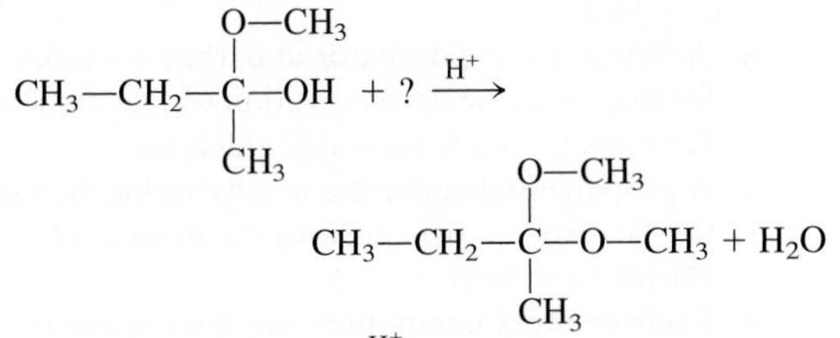

b. 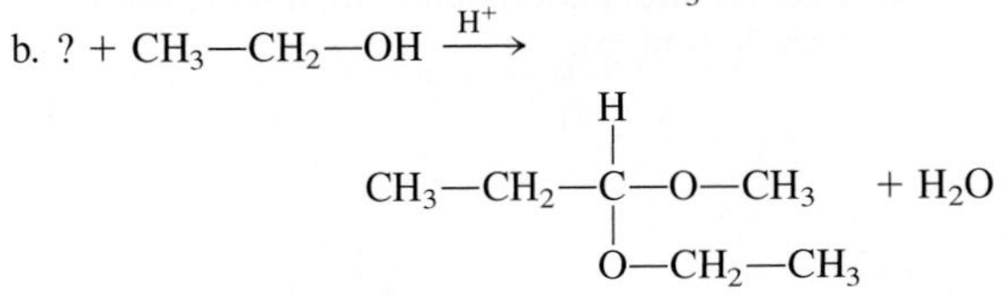

c. 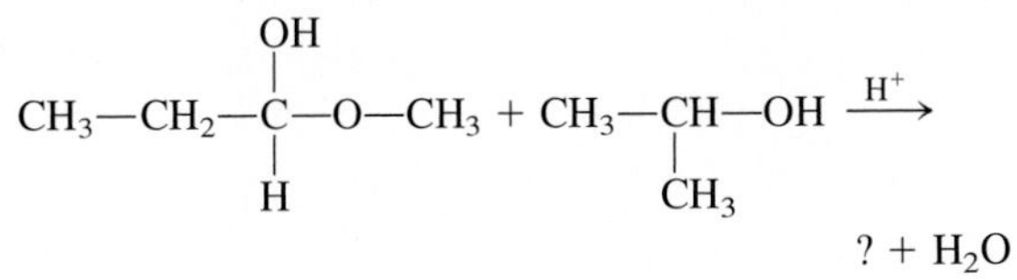

d. 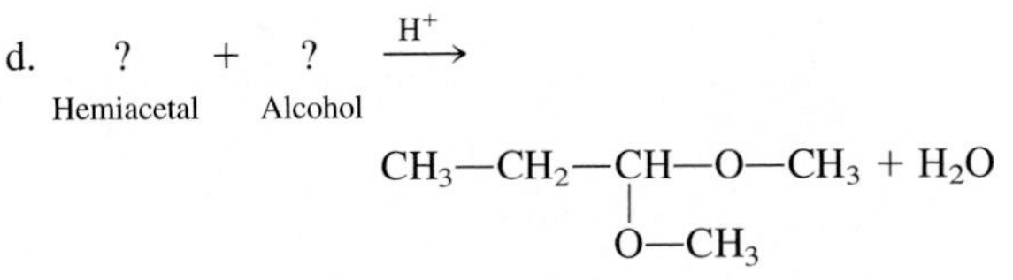

4.103 Draw the structural formulas of the aldehyde (or ketone) and the two alcohols produced when the following acetals undergo hydrolysis in acidic solution.

a. 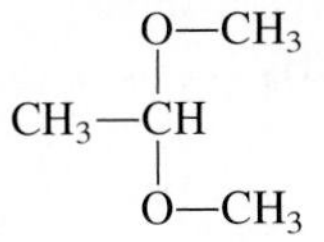

b. 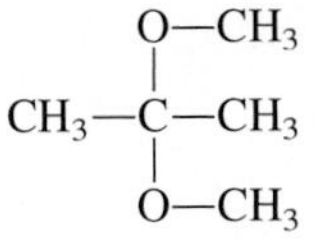

c. 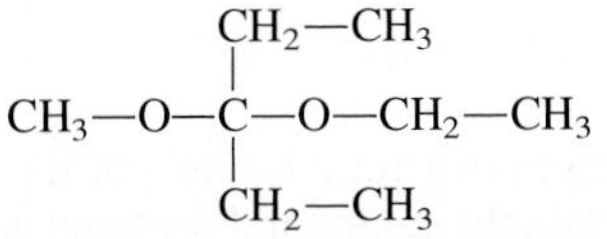

d. 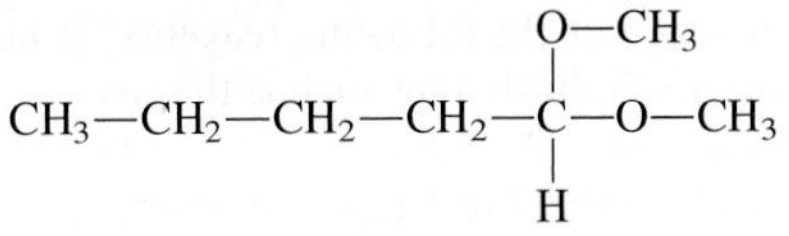

4.104 Draw the structural formulas of the aldehyde (or ketone) and the two alcohols produced when the following acetals undergo hydrolysis in acidic solution.

a. 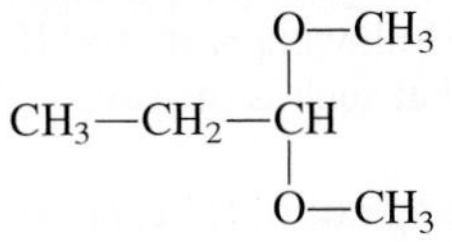

b. 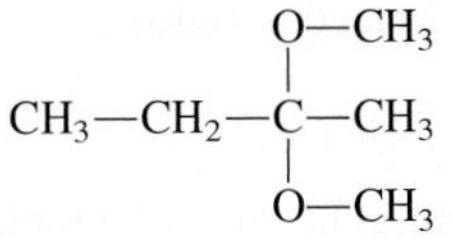

c. 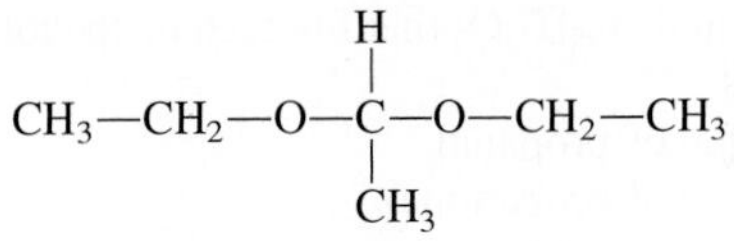

d. 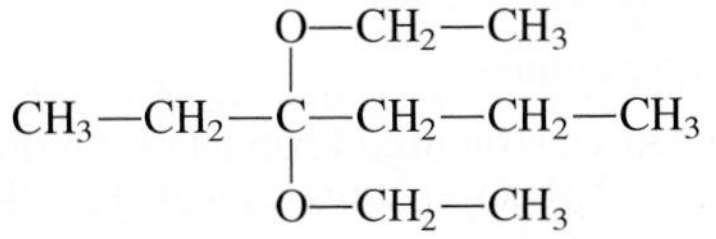

4.105 Name each of the compounds in Problem 4.103 in the manner described in Section 4.11.

4.106 Name each of the compounds in Problem 4.104 in the manner described in Section 4.11.

▲**4.107** What are the IUPAC names of the compounds formed when each of the following acetals or hemiacetals undergoes hydrolysis under acidic conditions?

a. 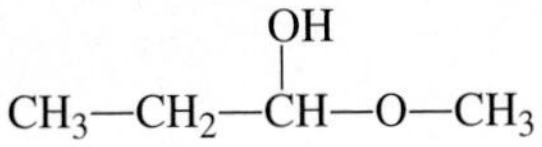

b. 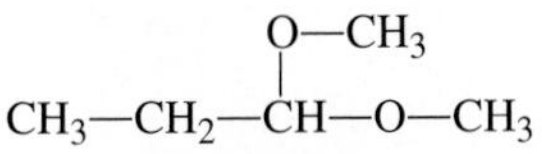

c. 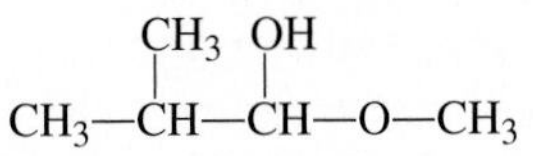

d. 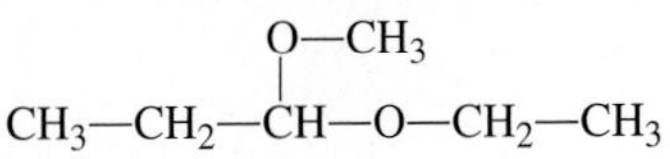

▲**4.108** What are the IUPAC names of the compounds formed when each of the following acetals or hemiacetals undergoes hydrolysis under acidic conditions?

a. 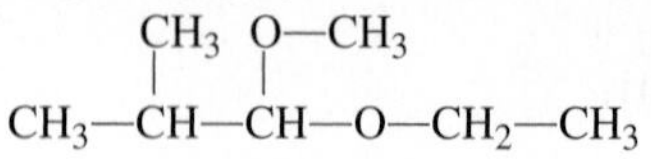

b. 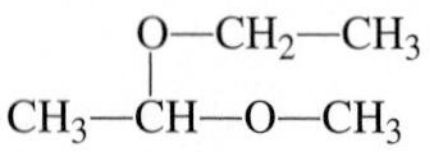

c. 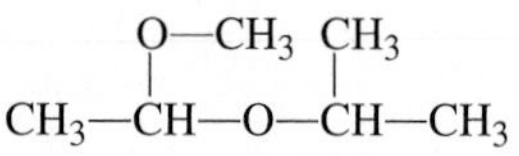

d. 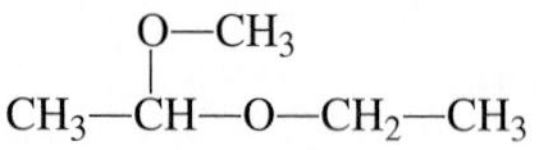

▲**4.109** What is the condensed structural formula of the organic product when the compound pentanal is treated with each of the following reagents? If no reaction occurs, indicate that such is the case.

a. H_2, Ni catalyst
b. $CH_3—CH_2—OH$ (1 to 1 reacting ratio), H^+ catalyst
c. $CH_3—OH$ (2 to 1 reacting ratio), H^+ catalyst
d. $K_2Cr_2O_7$

▲**4.110** What is the condensed structural formula of the organic product when the compound 3-pentanone is treated with each of the following reagents? If no reaction occurs, indicate that such is the case.

a. H_2, Ni catalyst
b. $CH_3—OH$ (1 to 1 reacting ratio), H^+ catalyst
c. $CH_3—CH_2—OH$ (2 to 1 reacting ratio), H^+ catalyst
d. $K_2Cr_2O_7$

▲**4.111** Draw the condensed structural formula of a molecule with the formula $C_5H_{12}O_2$ that fits each of the following descriptions.

a. Hemiacetal of propanal
b. Hemiacetal of propanone
c. Acetal of propanal
d. Acetal of propanone

▲**4.112** Draw the condensed structural formula of a molecule with the formula $C_6H_{14}O_2$ that fits each of the following descriptions.

a. Hemiacetal of butanol
b. Hemiacetal of butanone
c. Acetal of butanal
d. Acetal of butanone

Formaldehyde-Based Polymers (Section 4.12)

4.113 What are the structural characteristics associated with a network polymer?

4.114 What is a major current use for phenol–formaldehyde network polymers?

4.115 In phenol–formaldehyde polymer formation, what are the intermediate compounds that are formed?

4.116 In a phenol–formaldehyde network polymer, what type of "bridges" cross-link the various substituted phenols?

Sulfur-Containing Carbonyl Groups (Section 4.13)

4.117 What type of compound is formed by replacement of the carbonyl oxygen atom with a sulfur atom?

4.118 What type of compound is formed by replacement of the carbonyl carbon atom with a sulfur atom?

4.119 Draw structural formulas for the following compounds.

a. Thioformaldehyde
b. Methanethial
c. Thioacetone
d. Propanethione

4.120 Dimethyl sulfoxide (DMSO) is a sulfur analog of acetone in which the sulfur has substituted for the carbonyl carbon atom.

a. Draw the structural formula for DMSO.
b. Describe the "unusual" solubility properties of DMSO.

●**4.121** (Chemical Connections 4-C) Indicate whether each of the following statements concerning commonly encountered lachrymatory aldehydes and ketones is true or false.

a. Acrolein, the lachrymator in barbecue smoke, has a very pleasant odor.
b. Formaldehyde is responsible, in part, for the lachrymatory properties of wood smoke.
c. Mace formulations and police tear gas have a common active ingredient.
d. A C_4 thioaldehyde derivative is responsible, in part, for "onion tears."

●**4.122** (Chemical Connections 4-C) Indicate whether each of the following statements concerning commonly encountered lachrymatory aldehydes and ketones is true or false.

a. Acrolein, the lachrymator in barbecue smoke, is formed from the decomposition of fats in meat.
b. Acrolein is a C_4 unsaturated aldehyde.
c. A propanethial derivative is responsible, in part, for the tears associated with the process of chopping onions.
d. Lachrymatory compounds are not present in "whole" onions.

Carboxylic Acids, Esters, and Other Acid Derivatives

5

Esters, a type of carboxylic acid derivative, are largely responsible for the flavors and fragrances of ripe fruits such as red raspberries.

CHAPTER OUTLINE

In Chapter 4, the carbonyl group and two families of compounds—aldehydes and ketones—that contain this group were discussed. In this chapter, we discuss four more families of compounds in which the carbonyl group is present: carboxylic acids, esters, acid chlorides, and acid anhydrides.

5.1 Structure of Carboxylic Acids and Their Derivatives

A **carboxylic acid** *is an organic compound whose functional group is the carboxyl group.* What is a carboxyl group? A **carboxyl group** *is a carbonyl group (C=O) that has a hydroxyl group (—OH) bonded to the carbonyl carbon atom.* A general structural representation for a carboxyl group is

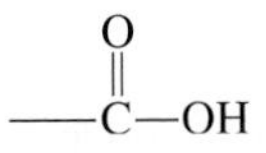

Abbreviated linear designations for the carboxyl group are

—COOH and —CO_2H

The term *carboxyl* is a contraction of the words *carbonyl* and *hydroxyl*.

Although a carboxyl group contains both a carbonyl group (C═O) and a hydroxyl group (—OH), the carboxyl group does not show characteristic behavior of either an alcohol or a carbonyl compound (aldehyde or ketone). Rather, it is a unique functional group with a set of characteristics different from those of its component parts.

The simplest carboxylic acid has a hydrogen atom attached to the carboxyl group carbon atom.

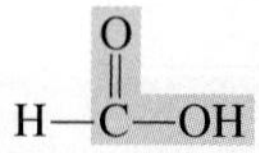

Structures for the next two simplest carboxylic acids, those with methyl and ethyl alkyl groups, are

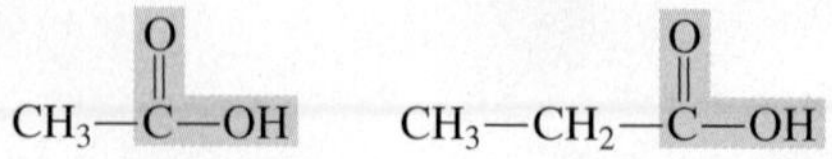

General formulas for carboxylic acids containing alkyl and aryl groups, respectively, are

R—COOH and Ar—COOH

The structure of the simplest aromatic carboxylic acid involves a benzene ring to which a carboxyl group is attached.

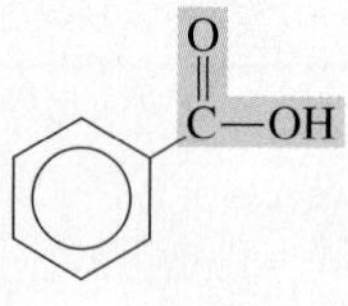

Cyclic carboxylic acids do not exist; having the carboxyl carbon atom as part of a ring system creates a situation where the carboxyl carbon atom would have five bonds. The nonexistence of cyclic carboxylic acids parallels the nonexistence of cyclic aldehydes (Section 4.3).

A **carboxylic acid derivative** *is an organic compound that can be synthesized from or converted into a carboxylic acid.* Four important families of carboxylic acid derivatives are esters, acid chlorides, acid anhydrides, and amides. The group attached to the carbonyl carbon atom distinguishes these derivative types from each other and also from carboxylic acids.

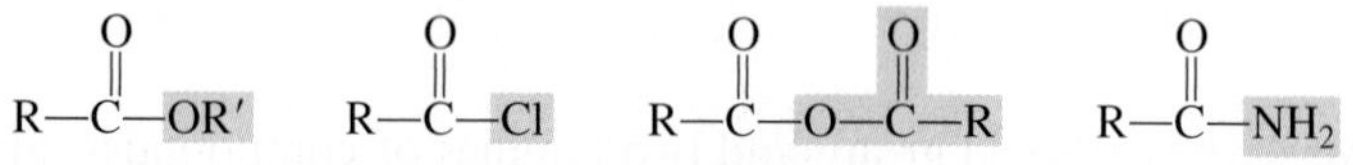

Ester Acid chloride Acid anhydride Amide

Carbonyl Compounds and Acyl Compounds

Generalized structures for carboxylic acids and carboxylic acid derivatives all have the form

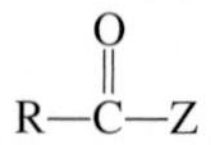

Structurally, these compound types differ from each other only in the identity of the Z entity present, as shown in the preceding set of carboxylic acid derivative structures.

Carboxylic acids and their derivatives have a common structural component, which is called an *acyl group*.

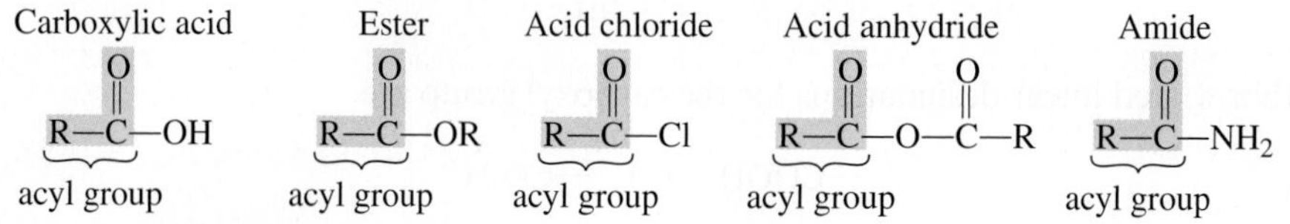

An **acyl group** *is the* R—C(=O)— *portion of a molecule with the general formula* R—C(=O)—Z. The acyl group is the portion of such a molecule that remains if the Z entity is removed. The linear designation for an acyl group is RCO—.

It is significant to note that aldehydes and ketones, the subject of the previous chapter, are also compounds that contain an acyl group.

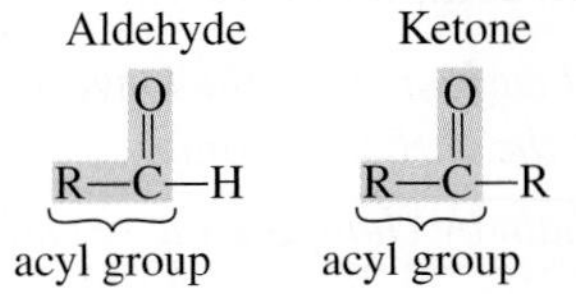

However, they are not carboxylic acid derivatives. Why is this so?

The chemical reactions that compounds containing an acyl group undergo are dependent on the nature of the bond between the carbonyl carbon atom of the acyl group and the "Z" group that is attached to it. Of particular concern is whether this bond is *nonpolar* or *polar*. On the basis of bond polarity, acyl group-containing compounds are classified into two types: carbonyl compounds (nonpolar bond) and acyl compounds (polar bond).

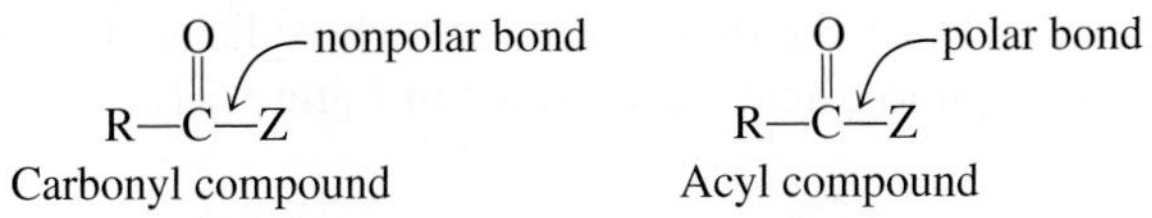

It is important to keep the terms *acyl group, carboxyl group,* and *carbonyl group* straight in the context of the structural characteristics of a carboxylic acid molecule. All three groups are considered to be present within such a molecule as shown in the following diagram.

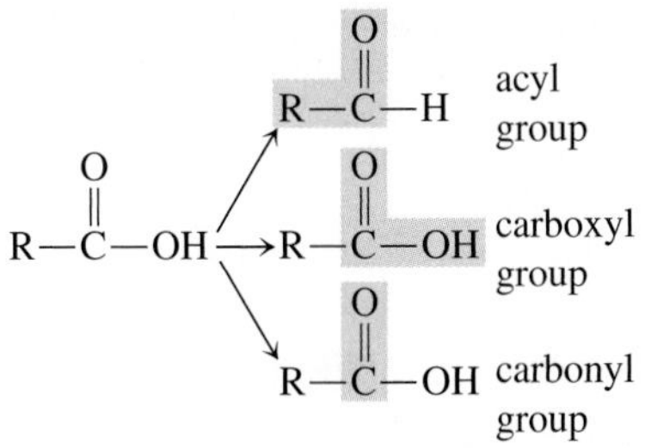

A **carbonyl compound** *contains an acyl group whose carbonyl carbon atom is bonded directly to a hydrogen atom or another carbon atom.* Carbon-carbon bonds and carbon-hydrogen bonds are nonpolar, as carbon and hydrogen electronegativities are almost the same. Aldehydes and ketones are examples of carbonyl compounds.

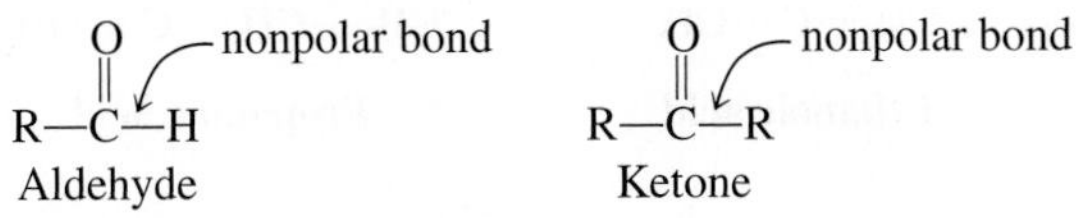

An **acyl compound** *contains an acyl group whose carbonyl carbon atom is bonded directly to an oxygen, nitrogen, or halogen atom.* Such bonds (C—O, C—N, and C—halogen) are polar because O, N, and the halogens have electronegativities significantly greater than that of carbon. Carboxylic acids and their derivatives (esters, acid chlorides, acid anhydrides, and amides) are examples of acyl compounds.

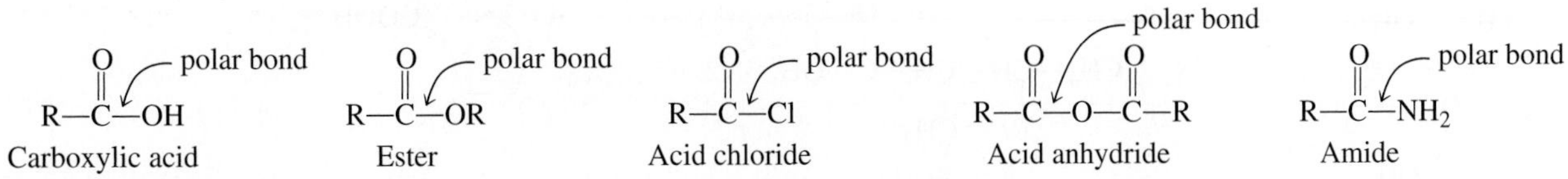

Structures, nomenclature, and chemical reactions for carbonyl compounds (aldehydes and ketones) were considered in the previous chapter. Structures, nomenclature, and chemical reactions for acyl compounds, except amides, are considered in this chapter. Amides are considered in the next chapter, which deals with nitrogen-containing hydrocarbon derivatives.

The reaction chemistry of acyl compounds is distinctly different from that of carbonyl compounds. For carbonyl compounds, reaction chemistry most often involves the carbon-oxygen double bond. For acyl compounds, as will be seen shortly, reaction chemistry most often involves the carbon-Z single bond, that is, the polar C—O, C—N, or C—halogen bond.

5.2 IUPAC Nomenclature for Carboxylic Acids

IUPAC rules for naming carboxylic acids resemble those for naming aldehydes (Section 4.4).

Monocarboxylic Acids

A **monocarboxylic acid** *is a carboxylic acid in which one carboxyl group is present.* IUPAC rules for naming such compounds are:

Rule 1: *Select as the parent carbon chain the longest carbon chain that* includes *the carbon atom of the carboxyl group.*

Rule 2: *Name the parent chain by changing the* -e *ending of the corresponding alkane to* -oic acid.

A carboxyl group must occupy a terminal (end) position in a carbon chain because there can be only one other bond to it.

Rule 3: *Number the parent chain by assigning the number 1 to the carboxyl carbon atom, but omit this number from the name.*

Rule 4: *Determine the identity and location of any substituents in the usual manner, and append this information to the front of the parent chain name.*

Rule 5: *If the carbonyl group is bonded to a carbon ring, name the ring and add the words* carboxylic acid. *The carbon bearing the carboxyl group is always carbon 1. Locate any other ring substituents in the usual manner.*

Space-filling models for the three simplest carboxylic acids—methanoic acid, ethanoic acid, and propanoic acid—are shown in Figure 5.1.

Figure 5.1 Space-filling models for the three simplest carboxylic acids: methanoic acid, ethanoic acid, and propanoic acid.

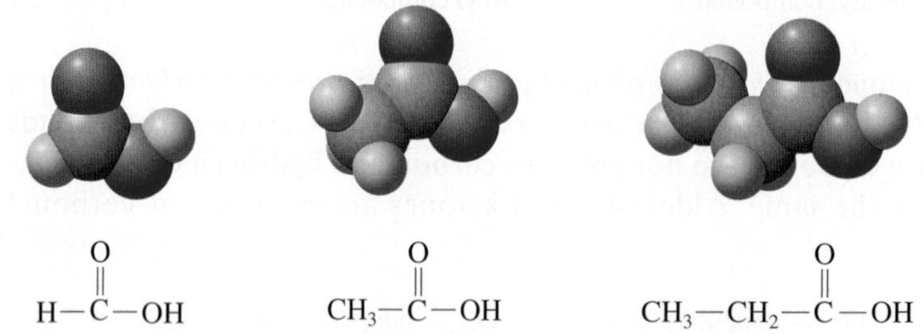

Methanoic acid **Ethanoic acid** **Propanoic acid**

Line-angle structural formulas for the simpler unbranched-chain carboxylic acids:

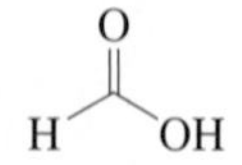

Methanoic acid

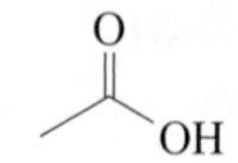

Ethanoic acid

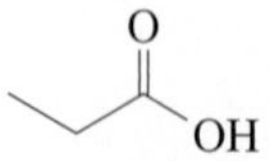

Propanoic acid

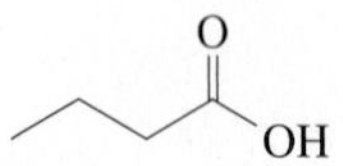

Butanoic acid

EXAMPLE 5.1 Determining IUPAC Names for Carboxylic Acids

Assign IUPAC names to the following carboxylic acids.

a. 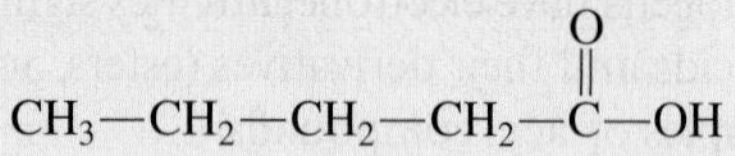

b. 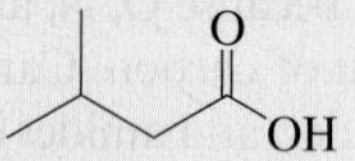

c. 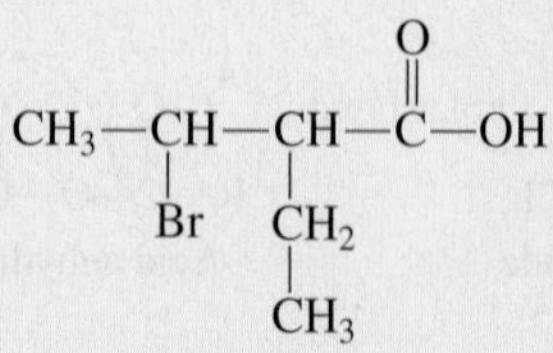

d. 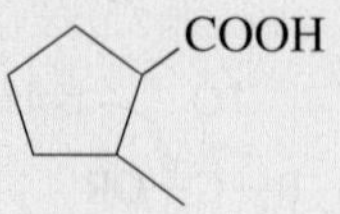

Solution

a. The parent chain name is based on pentane. Removing the *-e* ending from pentane and replacing it with the ending *-oic acid* gives *pentanoic acid.* The location of the carboxyl group need not be specified, because by definition the carboxyl carbon atom is always carbon 1.

b. The parent chain name is *butanoic acid.* To locate the methyl group substituent, we number the carbon chain beginning with the carboxyl carbon atom. The complete name of the acid is *3-methylbutanoic acid.*

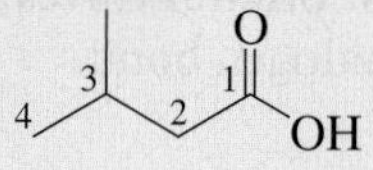

c. The longest carboxyl-carbon-containing chain has four carbon atoms. The parent chain name is thus *butanoic acid*. There are two substituents present, an ethyl group on carbon 2 and a bromo group on carbon 3. The complete name is *3-bromo-2-ethylbutanoic acid*.

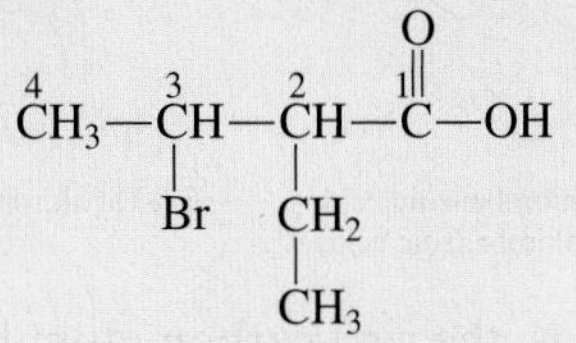

d. The ring contains five carbon atoms; it is cyclopentane. The ring carbon atom to which the carboxyl group is attached becomes carbon 1. Note that the carboxyl group is denoted using "linear" notation. Ring carbon 2 has a methyl attachment. The acid's name is *2-methylcyclopentanecarboxylic acid*.

The carboxyl functional group has the highest priority in the IUPAC naming system of all functional groups considered so far. When both a carboxyl group and a carbonyl group (aldehyde, ketone) are present in the same molecule, the prefix *oxo-* is used to denote the carbonyl group.

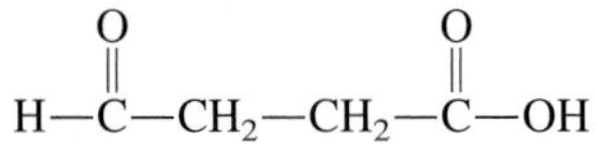

4-Oxobutanoic acid

Practice Exercise 5.1

Assign IUPAC names to the following carboxylic acids.

a.

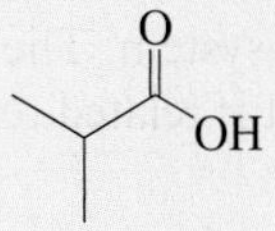

b.

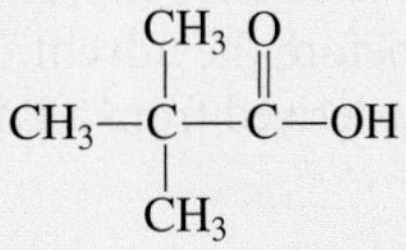

c.

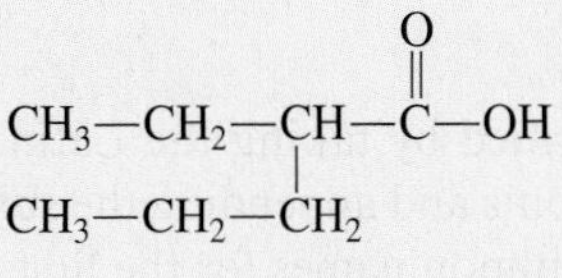

d. $CH_3—CH_2—CH_2—COOH$

Answers: **a.** 2-Methylpropanoic acid; **b.** 2,2-Dimethylpropanoic acid; **c.** 2-Ethylpentanoic acid; **d.** Butanoic acid

Dicarboxylic Acids

A **dicarboxylic acid** *is a carboxylic acid that contains two carboxyl groups, one at each end of a carbon chain.* Saturated acids of this type are named by appending the suffix *-dioic acid* to the corresponding alkane name (the *-e* is retained to facilitate pronunciation). Both carboxyl carbon atoms must be part of the parent carbon chain, and the carboxyl locations need not be specified with numbers because they will always be at the two ends of the chain.

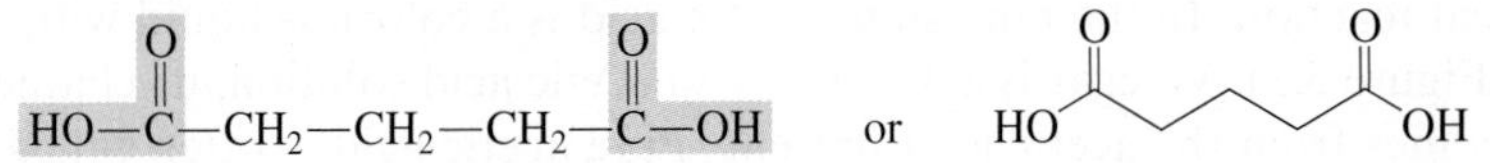

Pentanedioic acid

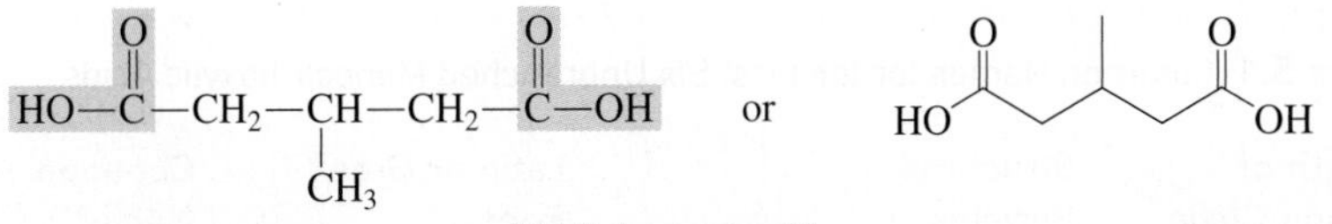

3-Methylpentanoic acid

Aromatic Carboxylic Acids

The simplest aromatic carboxylic acid is called benzoic acid (Figure 5.2).

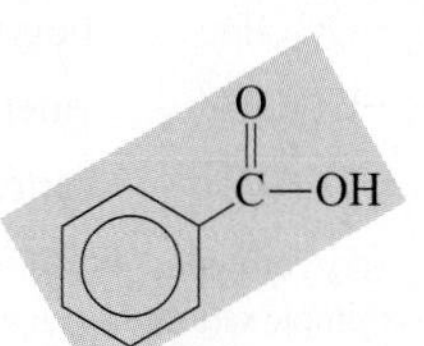

Benzoic acid

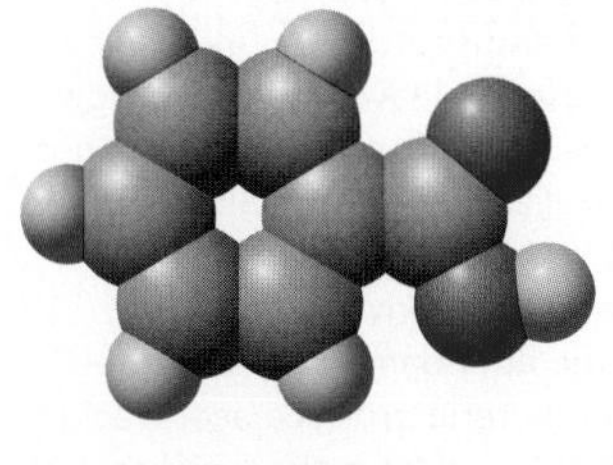

Figure 5.2 Space-filling model for benzoic acid, the simplest aromatic carboxylic acid.

Methyl benzoic acids go by the name *toluic acid.* (This situation parallels methyl benzene being called toluene.)

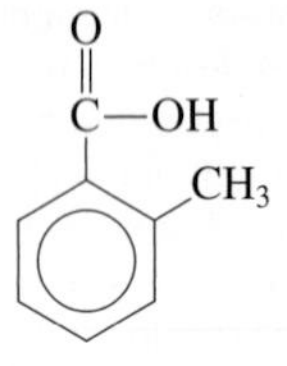

o-Toluic acid

Other simple aromatic acids are named as derivatives of benzoic acid.

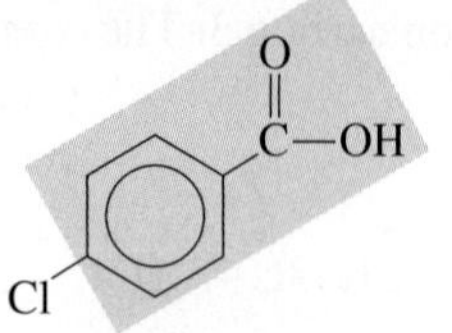

4-Chlorobenzoic acid (*p*-chlorobenzoic acid)

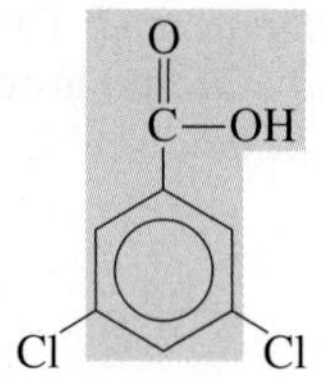

3,5-Dichlorobenzoic acid

In substituted benzoic acids, the ring carbon atom bearing the carboxyl group is always carbon 1.

5.3 Common Names for Carboxylic Acids

The common names of monocarboxylic acids are the basis for aldehyde common names (Section 4.4).

- C_1: formic acid and formaldehyde
- C_2: acetic acid and acetaldehyde
- C_3: propionic acid and propionaldehyde
- C_4: butyric acid and butyraldehyde

The use of common names is more prevalent for carboxylic acids than for any other family of organic compounds. Because of their abundance in nature, carboxylic acids were among the earliest classes of organic compounds to be studied, and they acquired names before the advent of the IUPAC naming system. These common names are usually derived from a Latin or Greek word that is related to a source for the acid.

Monocarboxylic Acids

The common name of a monocarboxylic acid is formed by taking the Latin or Greek root name for the specific number of carbon atoms and appending the suffix *–ic acid.* Table 5.1 gives the parent root names and common names for the first six unbranched monocarboxylic acids. The historical basis for the Latin-Greek root name system is as follows.

The stinging sensation associated with red ant bites is due in part to formic acid (Latin, *formica,* "ant"). Acetic acid gives vinegar its tartness (sour taste); vinegar contains small amounts of acetic acid (Latin, *acetum,* "sour"). Propionic acid is the smallest acid that can be obtained from fats (Greek, *protos,* "first," and *pion,* "fat"). Rancid butter contains butyric acid (Latin, *butyrum,* "butter"). Valeric acid, found in valerian root (an herb), has a strong odor (Latin, *valere,* "to be strong"). The skin secretions of goats contain caproic acid, which contributes to the odor associated with these animals (Latin, *caper,* "goats").

AP Photo/David Duprey

Figure 5.3 "Drug-sniffing" dogs used by narcotics agents can find hidden heroin by detecting the odor of acetic acid (vinegar odor). Acetic acid is a by-product of the final step in illicit heroin production, and trace amounts remain in the heroin.

Acetic acid is one of the most widely used of all carboxylic acids. Its primary use is as an *acidulant*—a substance that gives the proper acidic conditions for a chemical reaction. In the pure state, acetic acid is a colorless liquid with a sharp odor (Figure 5.3). Vinegar is a 4%–8% (v/v) acetic acid solution; its characteristic odor comes from the acetic acid present. Pure acetic acid is often called *glacial*

Table 5.1 Common Names for the First Six Unbranched Monocarboxylic Acids

Length of Carbon Chain	Structural Formula	Latin or Greek Root	Common Name*
C_1 monoacid	H—COOH	form-	formic acid
C_2 monoacid	CH_3—COOH	acet-	acetic acid
C_3 monoacid	CH_3—CH_2—COOH	propion-	propionic acid
C_4 monoacid	CH_3—$(CH_2)_2$—COOH	butyr-	butyric acid
C_5 monoacid	CH_3—$(CH_2)_3$—COOH	valer-	valeric acid
C_6 monoacid	CH_3—$(CH_2)_4$—COOH	capro-	caproic acid

*The mnemonic "*F*rogs *a*re *p*olite, *b*eing *v*ery *c*ourteous" is helpful in remembering, in order, the first letters of the common names of these six simple saturated monocarboxylic acids.

There is a connection between acetic acid and sourdough bread. The yeast used in leavening the dough for this bread is a type that cannot metabolize the sugar maltose as most yeasts do. Consequently, bacteria that thrive on maltose become abundant in the dough. These bacteria produce acetic acid and lactic acid from the maltose, and the dough becomes *sour* (acidic); hence the name *sourdough* bread.

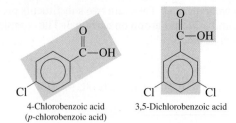

acetic acid because it freezes on a moderately cold day (f.p. = 17°C), producing icy-looking crystals.

When using common names for carboxylic acids, the positions (locations) of substituents are denoted by using letters of the Greek alphabet rather than numbers. The first four letters of the Greek alphabet are alpha (α), beta (β), gamma (γ), and delta (δ). The alpha-carbon atom is carbon 2, the beta-carbon atom is carbon 3, and so on.

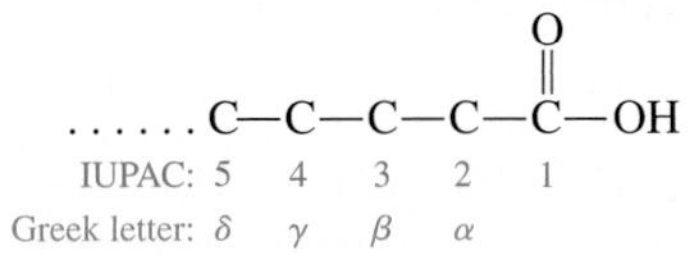

With the Greek-letter system, the compound

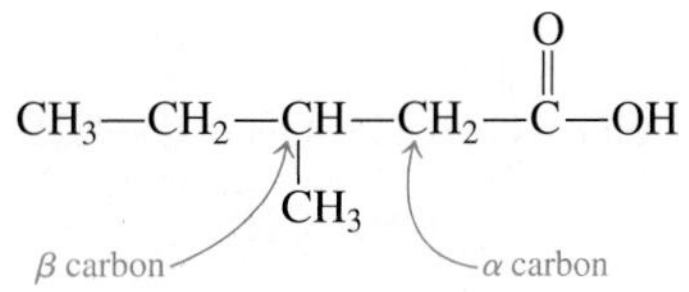

would be called *β-methylvaleric acid.*

Figure 5.4 contrasts the different carbon-atom numbering systems in IUPAC and common-name nomenclature for carboxylic acids.

Note that the common and IUPAC naming systems for carboxylic acids differ in three ways:

1. The base name for the carbon chain differs.
2. The suffix that ends the name differs, being *-ic acid* in the common system and *-oic acid* in the IUPAC system.
3. The carbon numbering system differs, involving Greek letters in the common system and Arabic numbers in the IUPAC system.

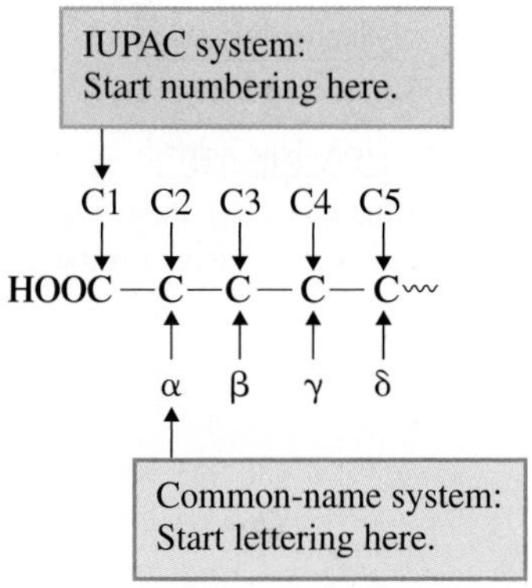

Figure 5.4 A Greek-letter numbering system is used in common-name nomenclature for carboxylic acids.

Dicarboxylic Acids

Common names for the first six dicarboxylic acids are given in Table 5.2. Oxalic acid, the simplest dicarboxylic acid, is found in plants of the genus *Oxalis,* which includes rhubarb and spinach, and in cabbage (see Figure 5.5). This acid and its salts are poisonous in *high* concentrations. The amount of oxalic acid present in spinach, cabbage, and rhubarb is not harmful. Oxalic acid is used to remove rust, bleach straw and leather, and remove ink stains. Succinic and glutaric acid and their derivatives play important roles in biochemical reactions that occur in the human body (Section 12.6).

Ingesting a fatal dose of oxalic acid from eating spinach would require that you eat nine pounds of spinach at one sitting, a slightly above-average serving of spinach!

Table 5.2 Common Names for the First Six Unbranched Dicarboxylic Acids

Length of Carbon Chain	Structural Formula	Latin or Greek Root	Common Name*
C_2 diacid	HOOC—COOH	oxal-	oxalic acid
C_3 diacid	HOOC—CH_2—COOH	malon-	malonic acid
C_4 diacid	HOOC—$(CH_2)_2$—COOH	succin-	succinic acid
C_5 diacid	HOOC—$(CH_2)_3$—COOH	glutar-	glutaric acid
C_6 diacid	HOOC—$(CH_2)_4$—COOH	adip-	adipic acid
C_7 diacid	HOOC—$(CH_2)_5$—COOH	pimel-	pimelic acid

*The mnemonic "*O*h *m*y, *s*uch *g*ood *a*pple *p*ie" is helpful in remembering, in order, the first letters of the common names of these six simple dicarboxylic acids.

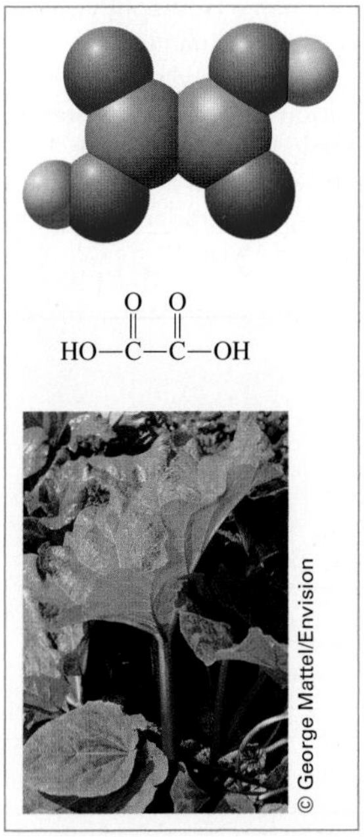

Figure 5.5 The C_2 dicarboxylic acid, oxalic acid, contributes to the tart taste of rhubarb stalks.

EXAMPLE 5.2 Generating the Structural Formulas of Carboxylic Acids from Their Common Names

Draw a structural formula for each of the following carboxylic acids.

a. Caproic acid
b. Glutaric acid
c. α,β-Dimethylsuccinic acid
d. β-Chlorobutyric acid

(continued)

The contrast between IUPAC names and common names for mono- and dicarboxylic acids is as follows:

Monocarboxylic Acids

IUPAC (two words)

alkanoic acid

Common (two words)

(prefix)ic acid*

Dicarboxylic Acids

IUPAC (two words)

alkanedioic acid

Common (two words)

(prefix)ic acid*

*The common-name prefixes are related to natural sources for the acids.

Solution

a. Caproic acid is the six-carbon unsubstituted monocarboxylic acid. Its structural formula is

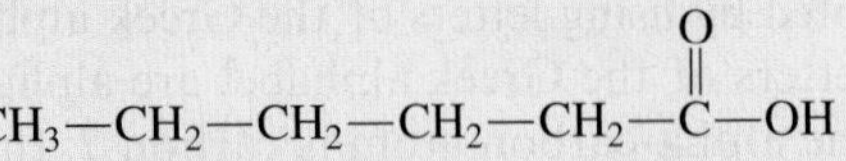

b. Glutaric acid is the five-carbon unsubstituted dicarboxylic acid, with a carboxyl group at each end of the carbon chain.

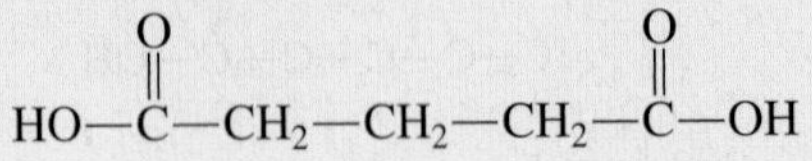

c. Succinic acid is the four-carbon unsubstituted dicarboxylic acid. Methyl groups are present on both the alpha- and beta-carbon atoms.

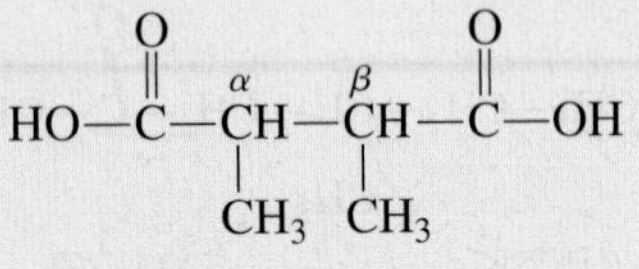

d. Butyric acid is the four-carbon unsubstituted monocarboxylic acid. A chloro group is attached to the beta-carbon atom (carbon 3).

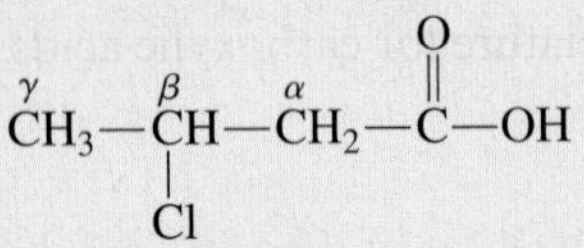

Practice Exercise 5.2

Draw a structural formula for each of the following carboxylic acids.

a. Adipic acid **b.** β-Chlorovaleric acid **c.** Malonic acid **d.** Phenylacetic acid

Answers:

a.

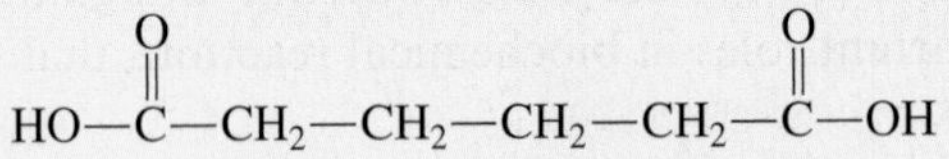

b.

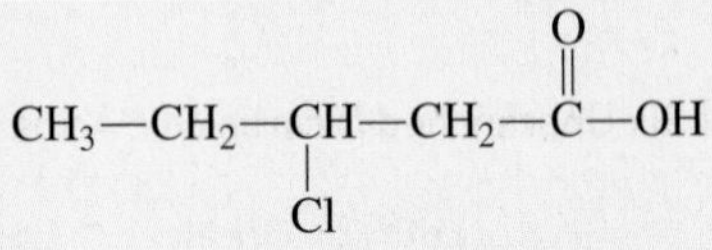

c.

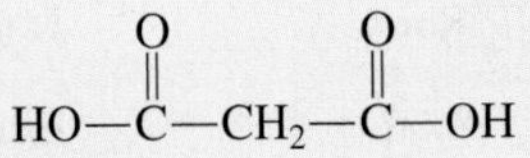

d.

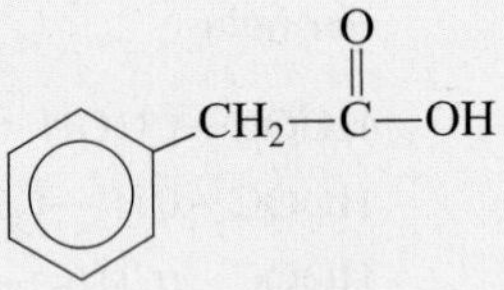

Industrially, the carboxylic acid produced in the greatest amount is a dicarboxylic acid with a benzene "core."

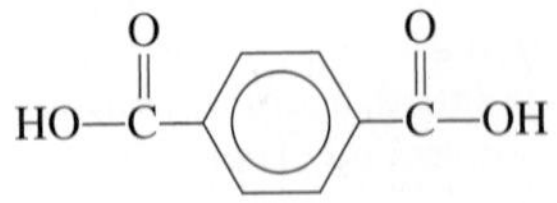

Its common name is terephthalic acid, and its IUPAC name is 1,4-benzenedicarboxylic acid. Terephthalic acid is the starting material for production of the polyester polymer PET (Section 5.17), which is marketed under tradenames such as Dacron and Mylar.

5.4 Polyfunctional Carboxylic Acids

A **polyfunctional carboxylic acid** *is a carboxylic acid that contains one or more additional functional groups besides one or more carboxyl groups.* Such acids occur naturally in many fruits, are important in the normal functioning of the human body (metabolism), and find use in over-the-counter skin-care products and in prescription drugs. Three commonly encountered types of polyfunctional carboxylic acids are *unsaturated* acids, *hydroxy* acids, and *keto* acids.

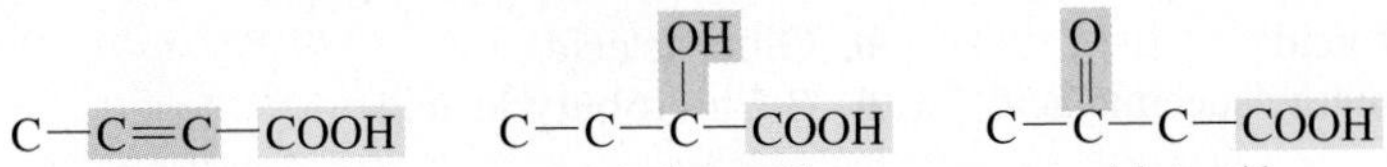

An unsaturated acid A hydroxy acid A keto acid

More information about these types of polyfunctional acids follows.

CHEMICAL CONNECTIONS 5-A

Nonprescription Pain Relievers Derived from Propanoic Acid

Consumers are faced with a shelf-full of choices when looking for an over-the-counter medicine to treat aches, pains, and fever. The vast majority of brands available, however, represent only four chemical formulations. Besides the long-available aspirin and acetaminophen, consumers can now purchase products that contain ibuprofen and naproxen.

Aspirin, acetaminophen, ibuprofen, and naproxen have antipyretic (fever-reducer) and mild analgesic (pain-reliever) properties. Aspirin, ibuprofen, and naproxen (but not acetaminophen) also have anti-inflammatory properties.

Ibuprofen and naproxen, the "newcomers" in the over-the-counter pain-reliever market, share a common structural feature. They are both derivatives of propanoic acid, the three-carbon monocarboxylic acid. Attached to the propanoic acid in both cases is an aromatic ring system with an attachment. In ibuprofen, the ring system is benzene and the ring attachment is an isobutyl group. In naproxen, the ring system is naphthalene (Section 2.16) with a methoxy attachment.

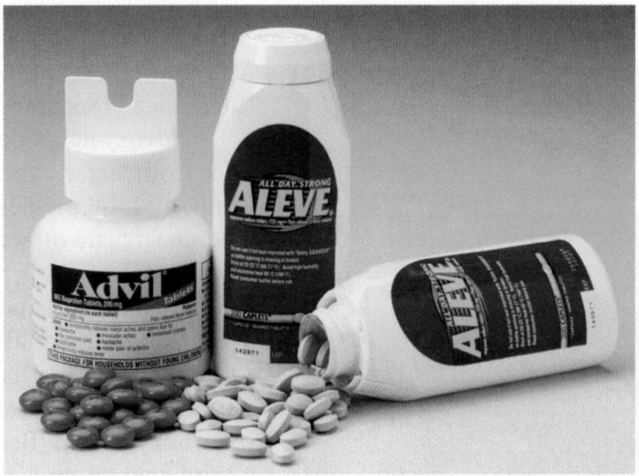

The active ingredients in Aleve and Advil are derivatives of propanoic acid.

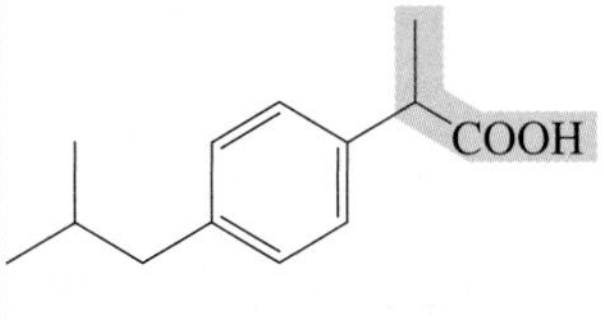

Ibuprofen

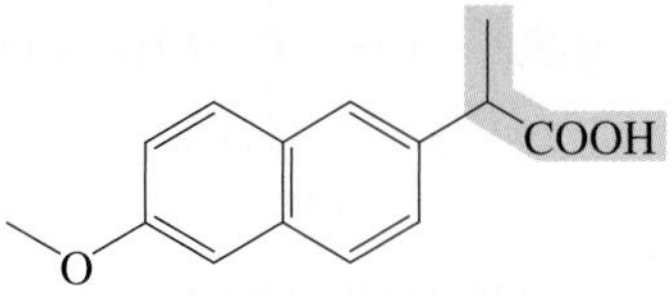

Naproxen

Ibuprofen, marketed under the brand names Advil, Motrin-IB, and Nuprin, was cleared by the FDA in 1984 for nonprescription sales. Numerous studies have shown that nonprescription-strength ibuprofen relieves minor pain and fever as well as does aspirin or acetaminophen. Like aspirin, ibuprofen reduces inflammation. (Prescription-strength ibuprofen has extensive use as an anti-inflammatory agent for the treatment of rheumatoid arthritis.) There is evidence that ibuprofen is more effective than either aspirin or acetaminophen in reducing dental pain and menstrual pain. Both aspirin and ibuprofen can cause stomach bleeding in some people, although ibuprofen seems to cause fewer problems. Ibuprofen is more expensive than either aspirin or acetaminophen.

Naproxen, marketed under the brand names Aleve and Anaprox, was cleared by the FDA in 1994 for nonprescription use. The effects of naproxen last longer in the body (8–12 hr per dose) than the effects of ibuprofen (4–6 hr per dose) and of aspirin and acetaminophen (4 hr per dose). Naproxen is more likely to cause slight intestinal bleeding and stomach upset than is ibuprofen. It is also not recommended for use by children under 12.

Unsaturated Acids

The simplest *unsaturated mono*carboxylic acid is propenoic acid (acrylic acid), a substance used in the manufacture of several polymeric materials. Two isomers exist for the simplest *unsaturated di*carboxylic acid, butenedioic acid. The two isomers have separate common names, fumaric acid (*trans*) and maleic acid (*cis*), a naming procedure seldom encountered.

$CH_2{=}CH{-}COOH$

Acrylic acid

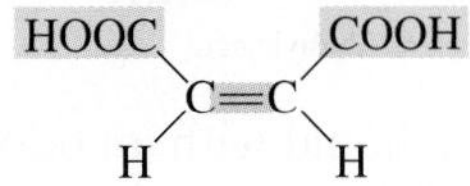

Maleic acid (*cis* isomer)

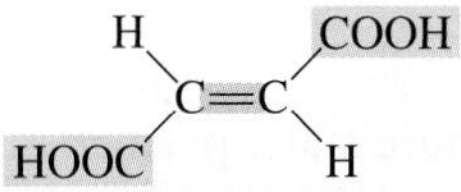

Fumaric acid (*trans* isomer)

Some antihistamines (Section 6.10) are salts of maleic acid. The addition of small amounts of maleic acid to fats and oils prevents them from becoming rancid. Fumaric acid is a *metabolic acid.* Metabolic acids are intermediate compounds in the metabolic reactions (Section 12.1) that occur in the human body.

The nonprescription pain relievers ibuprofen and naproxen are unsaturated carboxylic acids, with the unsaturation coming from the presence of an aromatic ring system within their structures. The focus on relevancy feature Chemical Connections 5-A above considers these compounds in further detail. They are two of the most used of the many "over-the-counter" medications now available.

An unsaturated monocarboxylic acid with the structure

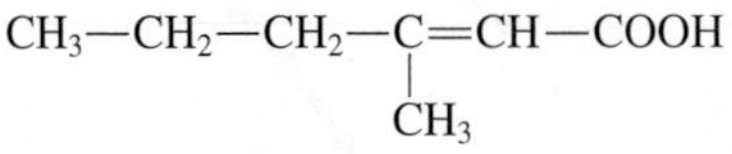

3-Methyl-2-hexenoic acid

has been found to be largely responsible for "body odor." It is produced by skin bacteria, particularly those found in armpits.

Hydroxy Acids

4-Hydroxybutanoic acid, which has the common name γ-hydroxybutyric acid (GHB), is an illegal recreational drug that depresses the central nervous system and causes intoxication. Its taste is masked when it is placed in alcoholic beverages; hence its illegal use as a "date rape" drug.

Four of the simpler *hydroxy* acids are

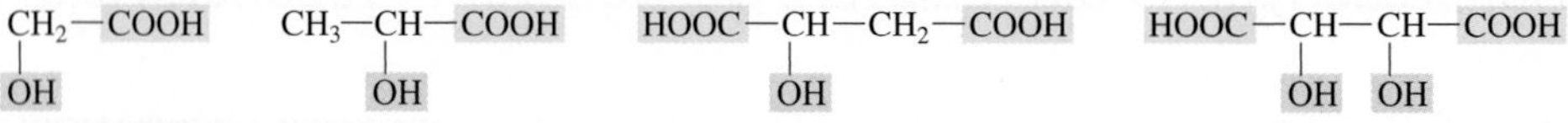

Glycolic acid   Lactic acid   Malic acid   Tartaric acid

Malic and tartaric acids are derivatives of succinic acid, the four-carbon unsubstituted diacid (Section 5.3).

Hydroxy acids occur naturally in many foods. Glycolic acid is present in the juice from sugar cane and sugar beets. Lactic acid is present in sour milk, sauerkraut, and dill pickles. Both malic acid and tartaric acid occur naturally in fruits. The sharp taste of some apples (fruit of trees of the genus *Malus*) is due to malic acid. Tartaric acid is particularly abundant in grapes (Figure 5.6). It is also a component of tartar sauce and an acidic ingredient in many baking powders. Lactic and malic acids are also *metabolic acids* (Section 12.4).

© F. Stuart Westmoreland/ Photo Researchers

Figure 5.6 Tartaric acid, the dihydroxy derivative of succinic acid, is particularly abundant in ripe grapes.

Citric acid, perhaps the best known of all carboxylic acids, is a hydroxy acid with a structural feature we have not previously encountered. It is a hydroxy *tri*carboxylic acid. Besides there being acid groups at both ends of a carbon chain, a third acid group is present as a substituent on the chain. An acid group as a substituent is called a *carboxy* group. Thus citric acid is a hydroxycarboxy diacid.

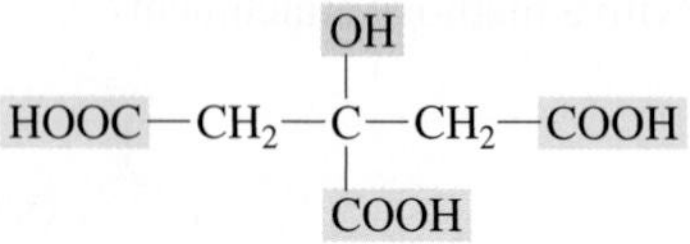

Citric acid

The IUPAC name for citric acid is 2-hydroxy-1,2,3-propanetrioic acid.

Citric acid gives citrus fruits their "sharp" taste; lemon juice contains 4%–8% citric acid, and orange juice is about 1% citric acid. Citric acid is used widely in beverages and in foods. In jams, jellies, and preserves, it produces tartness and pH adjustment to optimize conditions for gelation. In fresh salads, citric acid prevents enzymatic browning reactions, and in frozen fruits it prevents deterioration of color and flavor. Addition of citric acid to seafood retards microbial growth by lowering pH. Citric acid is also a metabolic acid (Section 12.4).

A number of skin-care products contain hydroxy acids. The focus on relevancy feature Chemical Connections 5-B on the next page considers such products, the advertising for which touts the presence of alpha-hydroxy acids as the active ingredients.

Keto Acids

Keto acids, as the designation implies, contain a carbonyl group within a carbon chain. Pyruvic acid, with three carbon atoms, is the simplest keto acid that can exist.

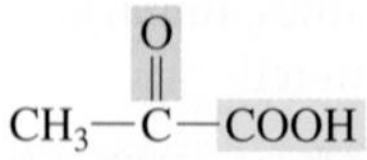

Pyruvic acid

In the pure state, pyruvic acid is a liquid with an odor resembling that of vinegar (acetic acid; Section 5.3). Pyruvic acid is a metabolic acid (Section 12.4).

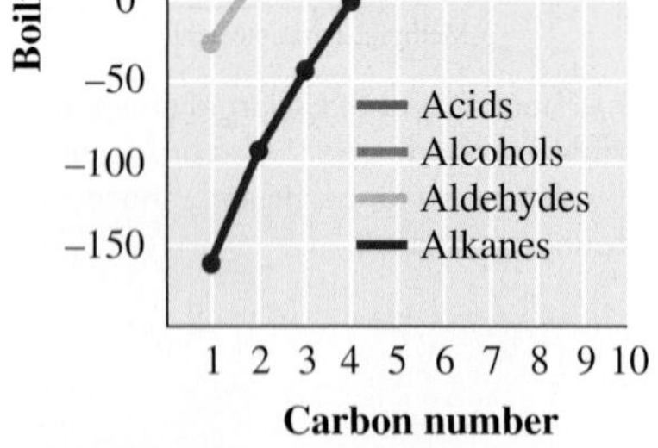

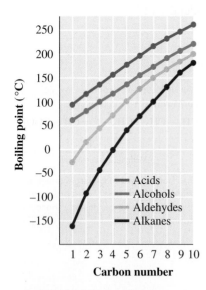

Figure 5.7 The boiling points of monocarboxylic acids compared to those of other types of compounds. All compounds in the comparison have unbranched carbon chains.

5.5 Physical Properties of Carboxylic Acids

Carboxylic acids are the most *polar* organic compounds we have discussed so far. Both the carbonyl part ($>C{=}O$) and the hydroxyl part (—OH) of the carboxyl functional group are polar. The result is very high melting and boiling points for carboxylic acids, the highest of any type of organic compound yet considered (Figure 5.7).

CHEMICAL CONNECTIONS 5-B

Carboxylic Acids and Skin Care

A number of carboxylic acids are used as "skin-care acids." Heavily advertised at present are cosmetic products that contain *alpha-hydroxy* acids, carboxylic acids in which a hydroxyl group is attached to the acid's alpha-carbon atom. Such cosmetic products address problems such as dryness, flaking, and itchiness of the skin and are highly promoted for removing wrinkles.

The alpha-hydroxy acids most commonly found in cosmetic products are glycolic acid and lactic acid, the two simplest alpha-hydroxy acids.

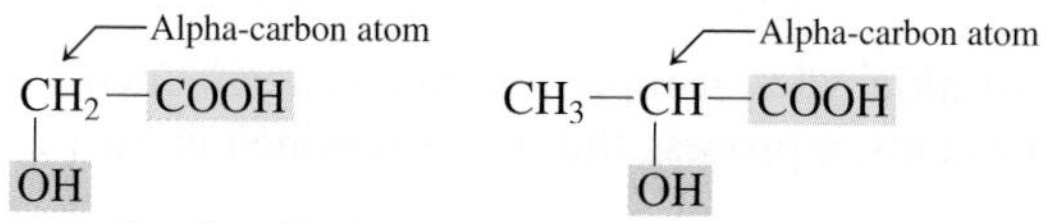

Glycolic acid   Lactic acid

Both acids are naturally occurring substances. Glycolic acid occurs in sugar cane and sugar beets, and lactic acid occurs in sour milk.

The use of alpha-hydroxy acids in cosmetics is considered safe at acid concentrations of less than 10%; higher concentrations can cause skin irritation, burning, and stinging. (Lactic acid becomes a prescription drug at concentrations of 12% or more.) One drawback of the cosmetic use of alpha-hydroxy products is that such use can increase the skin's sensitivity to the ultraviolet light component of sunlight; it is this component that causes sunburn. Individuals who are using "alpha-hydroxys" should apply a sunscreen whenever they go outside for an extended period of time.

Alpha-hydroxy acids are often touted as substances that reverse the aging process of skin. Such is not the case. Instead, these acids simply, react with outer-layer skin cells, causing them (and any blemishes they contain) to flake off. This exposes a new layer of skin cells that have not been subjected to sun exposure and which often temporarily have the appearance of "younger" skin.

Glycolic acid, at higher concentrations than that found in cosmetics, is used by dermatologists for the "spot" removal of *keratoses* (precancerous lesions and/or patches of darker, thickened skin).

Two skin-care products containing alpha-hydroxy acids.

Polyunsaturated carboxylic acids are used extensively in the treatment of severe acne. The prescription drugs Tretinoin and Accutane are such compounds.

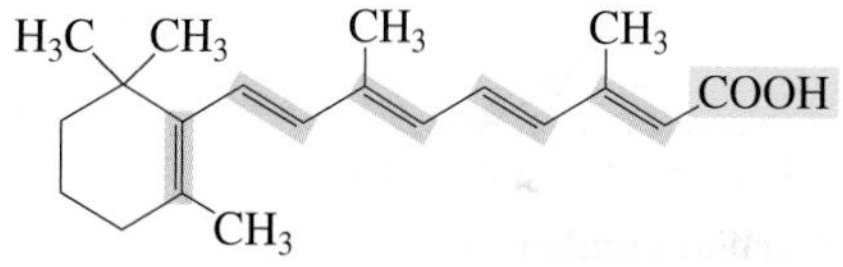

Tretinoin (5 *trans*-double bonds)

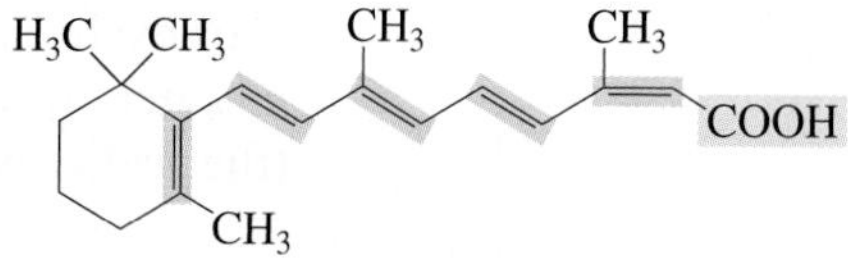

Accutane (4 *trans*- and 1 *cis*-double bonds)

Unsubstituted saturated monocarboxylic acids containing up to nine carbon atoms are liquids that have strong, sharp odors (Figure 5.8). Acids with 10 or more carbon atoms in an unbranched chain are waxy solids that are odorless (because of low volatility). Aromatic carboxylic acids, as well as dicarboxylic acids, are also odorless solids.

The high boiling points of carboxylic acids indicate the presence of strong intermolecular attractive forces. A unique hydrogen-bonding arrangement, shown in Figure 5.9, contributes to these attractive forces. A given carboxylic acid molecule forms two hydrogen bonds to another carboxylic acid molecule, producing a "complex" known as a *dimer.* Because dimers have twice the mass of a single molecule, a higher temperature is needed to boil a carboxylic acid than would be needed for similarly sized aldehyde and alcohol molecules in which dimerization does not occur.

Carboxylic acids readily hydrogen-bond to water molecules. Such hydrogen bonding contributes to water solubility for short-chain carboxylic acids. The unsubstituted C_1 to C_4 monocarboxylic acids are completely miscible with water.

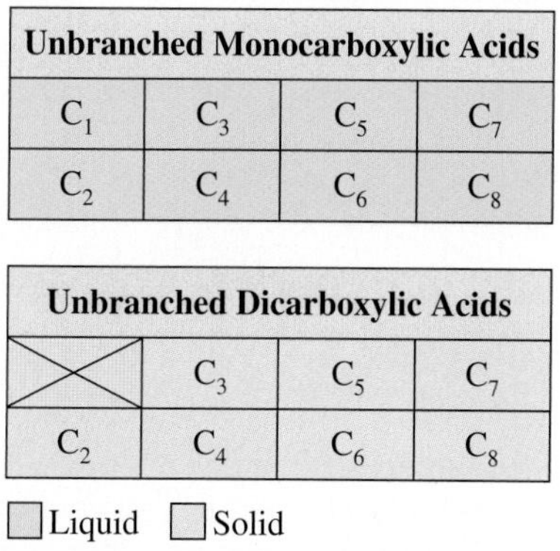

Unbranched Monocarboxylic Acids
C_1 C_3 C_5 C_7
C_2 C_4 C_6 C_8
Unbranched Dicarboxylic Acids
C_3 C_5 C_7
C_2 C_4 C_6 C_8
Liquid
Solid

Figure 5.8 A physical-state summary for unbranched mono- and dicarboxylic acids at room temperature and pressure.

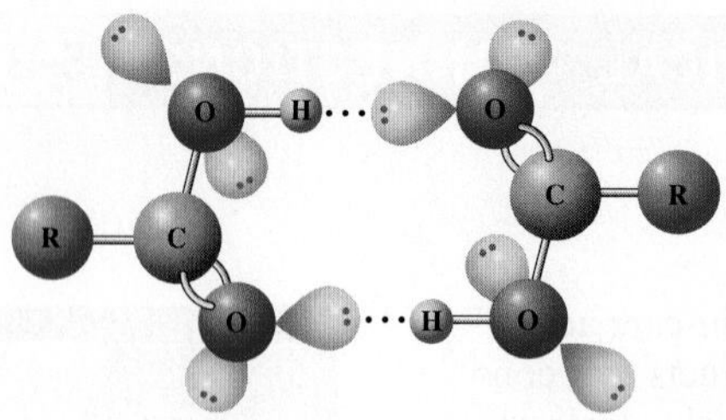

Figure 5.9 A given carboxylic acid molecule can form two hydrogen bonds to another carboxylic acid molecule, producing a "dimer," a complex with a mass twice that of a single molecule.

Solubility then rapidly decreases with carbon number, as shown in Figure 5.10. Short-chain dicarboxylic acids are also water-soluble. In general, aromatic acids are not water-soluble.

5.6 Preparation of Carboxylic Acids

Oxidation of primary alcohols or aldehydes, using an oxidizing agent such as CrO_3 or $K_2Cr_2O_7$, produces carboxylic acids, a process that was examined in Sections 3.9 and 4.10.

$$\text{Primary alcohol} \xrightarrow{[O]} \text{aldehyde} \xrightarrow{[O]} \text{carboxylic acid}$$

Aromatic acids can be prepared by oxidizing a carbon side chain (alkyl group) on a benzene derivative. In this process, all the carbon atoms of the alkyl group except the one attached to the ring are lost. The remaining carbon becomes part of a carboxyl group.

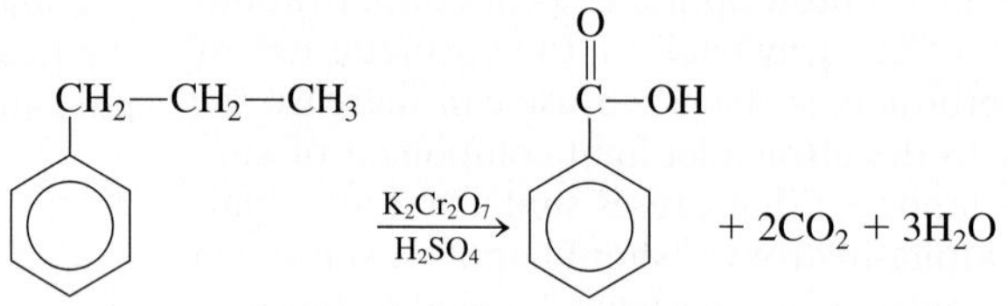

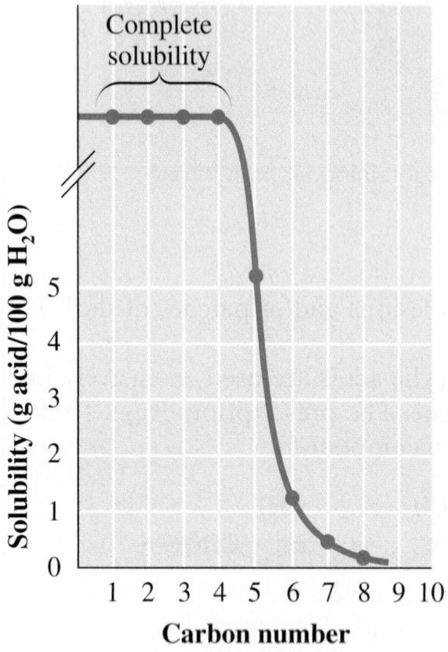

Figure 5.10 The solubility in water of saturated unbranched-chain carboxylic acids.

5.7 Acidity of Carboxylic Acids

Carboxylic acids, as the name implies, are *acidic.* When a carboxylic acid is placed in water, hydrogen ion transfer (proton transfer) occurs to produce hydronium ion (the acidic species in water) and carboxylate ion.

$$R{-}COOH + H_2O \longrightarrow \underset{\text{Hydronium ion}}{H_3O^+} + \underset{\text{Carboxylate ion}}{R{-}COO^-}$$

A **carboxylate ion** *is the negative ion produced when a carboxylic acid loses one or more acidic hydrogen atoms.*

Carboxylate ions formed from monocarboxylic acids always carry a −1 charge; only one acidic hydrogen atom is present in such molecules. Dicarboxylic acids, which possess two acidic hydrogen atoms (one in each carboxyl group), can produce carboxylate ions bearing a −2 charge.

Carboxylate ions are named by dropping the *-ic acid* ending from the name of the parent acid and replacing it with *-ate.*

$$\underset{\substack{\text{Acetic acid}\\\text{(ethanoic acid)}}}{CH_3{-}\overset{O}{\overset{\|}{C}}{-}OH} + H_2O \longrightarrow H_3O^+ + \underset{\substack{\text{Acetate ion}\\\text{(ethanoate ion)}}}{CH_3{-}\overset{O}{\overset{\|}{C}}{-}O^-}$$

$$\underset{\substack{\text{Oxalic acid}\\\text{(ethanedioic acid)}}}{HO{-}\overset{O}{\overset{\|}{C}}{-}\overset{O}{\overset{\|}{C}}{-}OH} + 2H_2O \longrightarrow 2H_3O^+ + \underset{\substack{\text{Oxalate ion}\\\text{(ethanedioate ion)}}}{{}^-O{-}\overset{O}{\overset{\|}{C}}{-}\overset{O}{\overset{\|}{C}}{-}O^-}$$

Table 5.3 Acid Strength for Selected Monocarboxylic Acids

Acid	K_a	Percent Ionization (0.100 M Solution)	pK_a
formic	1.8×10^{-4}	4.2%	3.75
acetic	1.8×10^{-5}	1.3%	4.75
propionic	1.3×10^{-5}	1.2%	4.89
butyric	1.5×10^{-5}	1.2%	4.82
valeric	1.5×10^{-5}	1.2%	4.82
caproic	1.4×10^{-5}	1.2%	4.85

Carboxylic acids are weak acids. The extent of proton transfer is usually less than 5%; that is, an equilibrium situation exists in which the equilibrium lies far to the left.

$$\underset{\text{More than 95\% of molecules in this form}}{R{-}COOH + H_2O} \rightleftharpoons \underset{\text{Less than 5\% of molecules in this form}}{H_3O^+ + R{-}COO^-}$$

Table 5.3 gives K_a values and percent ionization values in 0.100 M solution and pK_a values for selected monocarboxylic acids.

5.8 Carboxylic Acid Salts

In a manner similar to that of inorganic acids, carboxylic acids react completely with strong bases to produce water and a carboxylic acid salt.

$$\underset{\text{Carboxylic acid}}{CH_3{-}\overset{\overset{\large O}{\|}}{C}{-}OH} + \underset{\text{Strong base}}{NaOH} \longrightarrow \underset{\text{Carboxylic acid salt}}{CH_3{-}\overset{\overset{\large O}{\|}}{C}{-}O^-\ Na^+} + \underset{\text{Water}}{H_2O}$$

At normal human body pH values (pH = 7.35 to 7.45), most carboxylic acids exist as carboxylate ions. Acetic acid is in the form of acetate ion, pyruvic acid is in the form of pyruvate ion, lactic acid is in the form of lactate ion, and so on.

A **carboxylic acid salt** *is an ionic compound in which the negative ion is a carboxylate ion.*

Carboxylic acid salts are named similarly to other ionic compounds: *The positive ion is named first, followed by a separate word giving the name of the negative ion.* The salt formed in the preceding reaction contains sodium ions and acetate ions (from acetic acid); hence the salt's name is sodium acetate.

Carboxylic acid salt formation involves an acid–base neutralization reaction.

EXAMPLE 5.3 Writing Equations for the Formation of Carboxylic Acid Salts

Using an acid–base neutralization reaction, write a chemical equation for the formation of each of the following carboxylic acid salts.

a. Sodium propionate **b.** Potassium oxalate

Solution

a. This salt contains sodium ion (Na^+) and propionate ion, the three-carbon monocarboxylate ion.

$$CH_3{-}CH_2{-}\overset{\overset{\large O}{\|}}{C}{-}O^-Na^+$$

(continued)

From a neutralization standpoint, the sodium ion's source is the base sodium hydroxide, NaOH, and the negative ion's source is the acid propanoic acid. The acid–base neutralization equation is

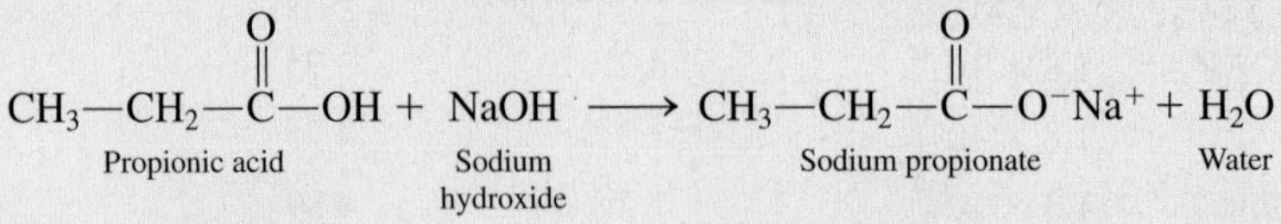

b. This salt contains potassium ions (K^+) whose source would be the base potassium hydroxide, KOH. The salt also contains oxalate ions, whose source would be the acid oxalic acid.

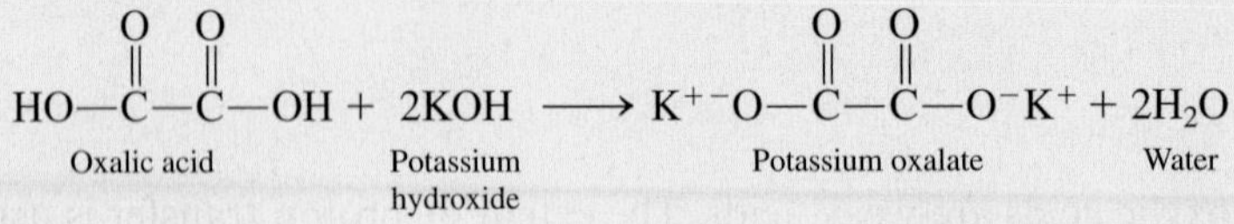

Note that two molecules of base are needed to react completely with one molecule of acid because the acid is a dicarboxylic acid.

Practice Exercise 5.3

Using an acid–base neutralization reaction, write a chemical equation for the formation of each of the following carboxylic acid salts.

a. Sodium formate **b.** Potassium malonate

Answers:

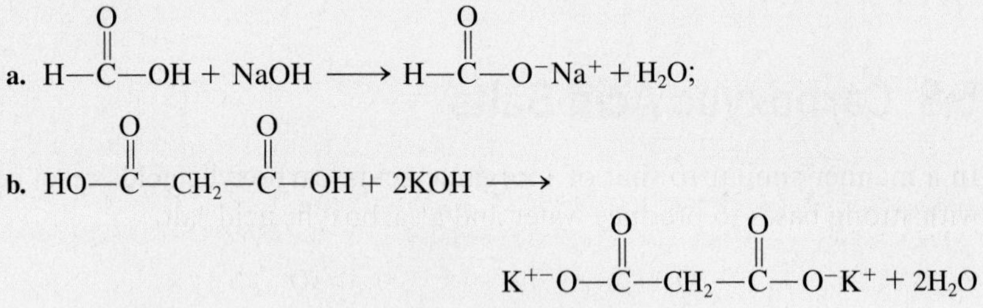

Converting a carboxylic acid salt back to a carboxylic acid is very simple. React the salt with a solution of a strong acid such as hydrochloric acid (HCl) or sulfuric acid (H_2SO_4).

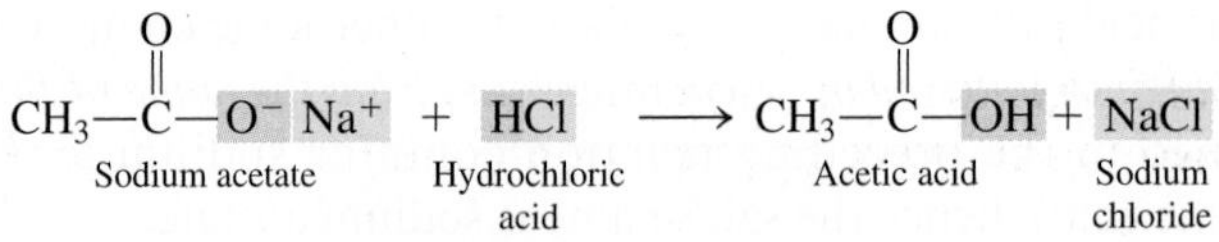

The interconversion reactions between carboxylic acid salts and their "parent" carboxylic acids are so easy to carry out that organic chemists consider these two types of compounds interchangeable.

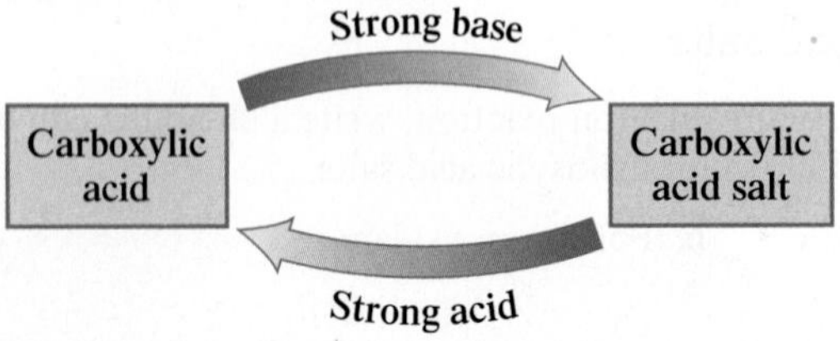

Uses for Carboxylic Acid Salts

The solubility of carboxylic acid salts in water is much greater than that of the carboxylic acids from which they are derived. Drugs and medicines that contain acid groups are usually marketed as the sodium or potassium salt of the acid. This greatly enhances the solubility of the medication, increasing the ease of its absorption by the body.

Many *antimicrobials,* compounds used as food preservatives, are carboxylic acid salts. Particularly important are the salts of benzoic, sorbic, and propionic acids.

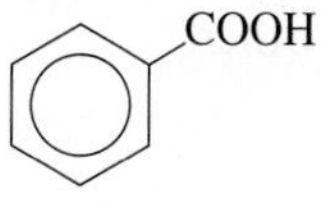

Benzoic acid

$CH_3-CH=CH-CH=CH-COOH$

Sorbic acid (2,4-hexadienoic acid)

CH_3-CH_2-COOH

Propionic acid

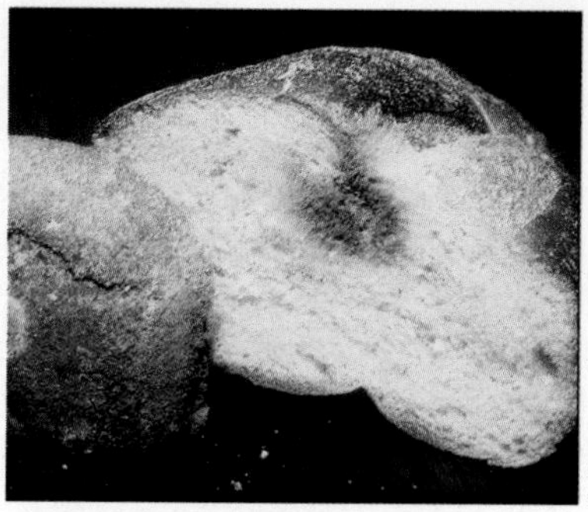

© Cengage Learning

Figure 5.11 Propionates, salts of propionic acid, extend the shelf life of bread by preventing the formation of mold.

The benzoate salts of sodium and potassium are effective against yeast and mold in beverages, jams and jellies, pie fillings, ketchup, and syrups. Concentrations of up to 0.1% (m/m) benzoate are found in such products.

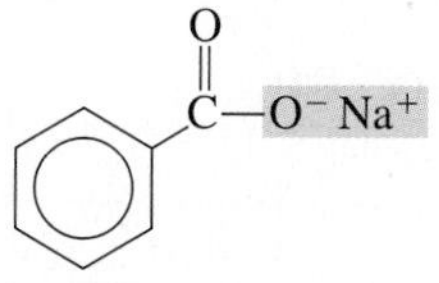

Sodium benzoate

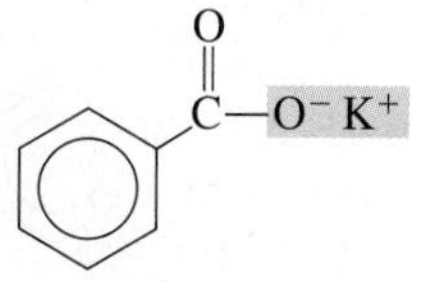

Potassium benzoate

The solubility of benzoic acid in water at 25°C is 3.4 g/L. The solubility of sodium benzoate, the sodium salt of benzoic acid, in water at 25°C is 550 g/L.

Sodium and potassium sorbates inhibit mold and yeast growth in dairy products, dried fruits, sauerkraut, and some meat and fish products. Sorbate preservative concentrations range from 0.02% to 0.2% (m/m).

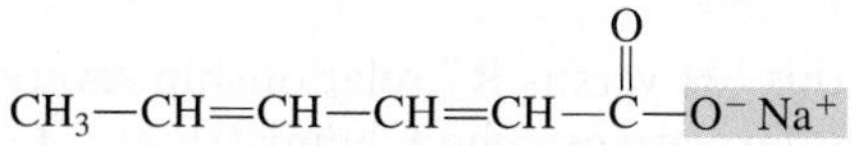

Sodium sorbate

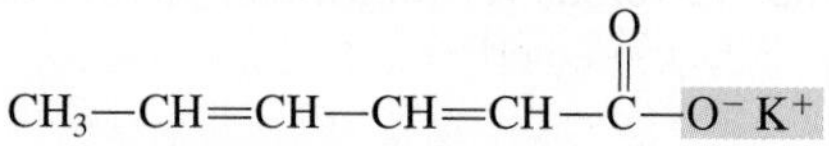

Potassium sorbate

Calcium and sodium propionates are used in baked products and also in cheese foods and spreads (Figure 5.11). Benzoates and sorbates cannot be used in yeast-leavened baked goods because they affect the activity of the yeast.

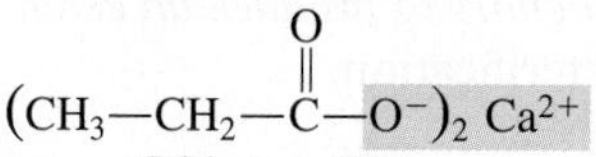

Calcium propionate

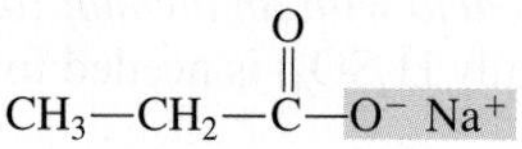

Sodium propionate

Carboxylate salts do not directly kill microorganisms present in food. Rather, they prevent further growth and proliferation of these organisms by increasing the pH of the foods in which they are used.

5.9 Structure of Esters

An **ester** *is a carboxylic acid derivative in which the —OH portion of the carboxyl group has been replaced with an —OR group.*

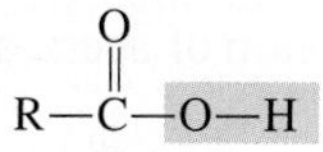

Carboxylic acid

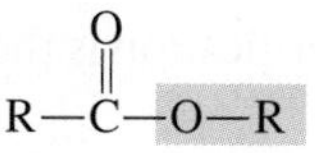

Ester

The ester functional group is thus

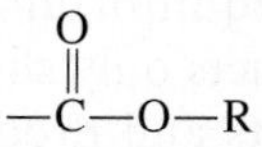

In linear form, the ester functional group can be represented as —COOR or $-CO_2R$.

The simplest ester, which has two carbon atoms, has a hydrogen atom attached to the ester functional group.

Note that the two carbon atoms present are not bonded to each other.

There are two three-carbon esters.

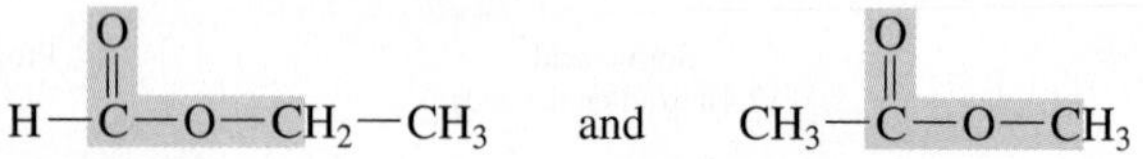

The structure of the simplest aromatic ester is derived from the structure of benzoic acid, the simplest aromatic carboxylic acid.

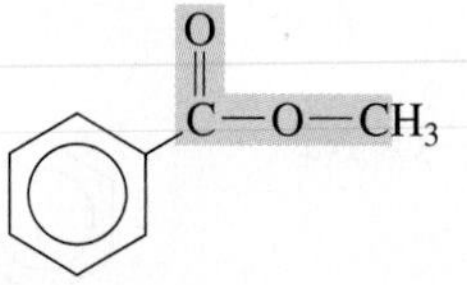

Note that the difference between a carboxylic acid and an ester is a "H versus R" relationship.

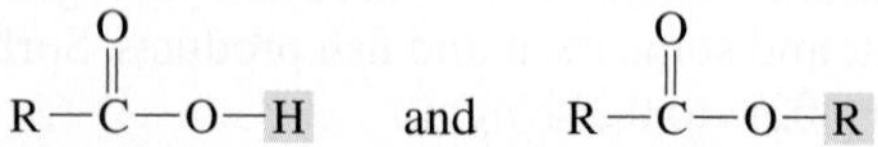

Acid Ester

We have encountered this "H versus R" relationship several times before in our study of hydrocarbon derivatives. The Chemistry at a Glance feature on the next page summarizes the "H versus R" relationships that have been encountered so far.

5.10 Preparation of Esters

Esterification is a *condensation reaction.* This is the third time this type of reaction has been encountered. The first encounter involved intermolecular alcohol dehydration (Section 3.9) and the second encounter involved the preparation of acetals (Section 4.11).

Esters are produced through *esterification.* An **esterification reaction** *is the reaction of a carboxylic acid with an alcohol (or phenol) to produce an ester.* A strong acid catalyst (generally H_2SO_4) is needed for esterification.

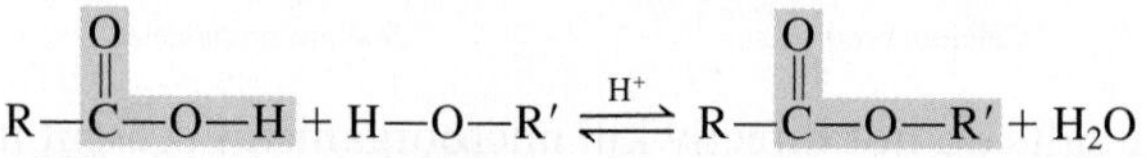

Carboxylic acid Alcohol Ester Water

In the esterification process, a —OH group is lost from the carboxylic acid, a —H atom is lost from the alcohol, and water is formed as a by-product. The net effect of this reaction is substitution of the —OR group of the alcohol for the —OH group of the acid.

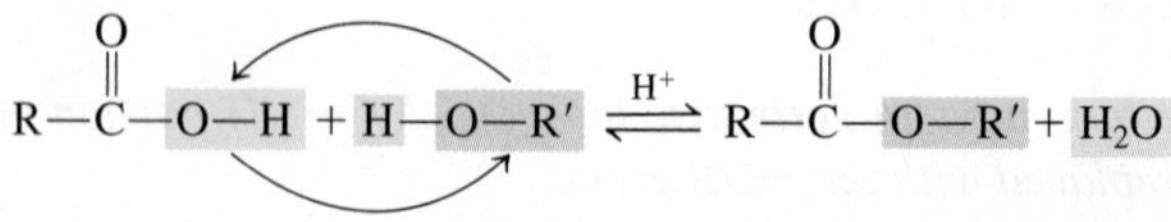

Studies show that in ester formation, the hydroxyl group of the acid (not of the alcohol) becomes part of the water molecule.

A specific example of esterification is the reaction of acetic acid with methyl alcohol.

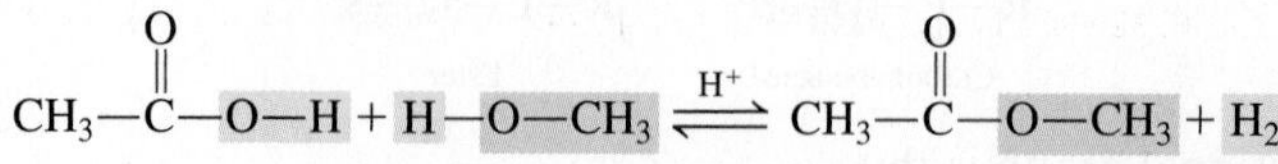

Esterification reactions are equilibrium processes, with the position of equilibrium usually favoring products only slightly. That is, at equilibrium, substantial amounts of both reactants and products are present. The amount of ester formed can be increased by using an excess of alcohol or by constantly

CHEMISTRY AT A GLANCE Summary of the "H Versus R" Relationship for Pairs of Hydrocarbon Derivatives

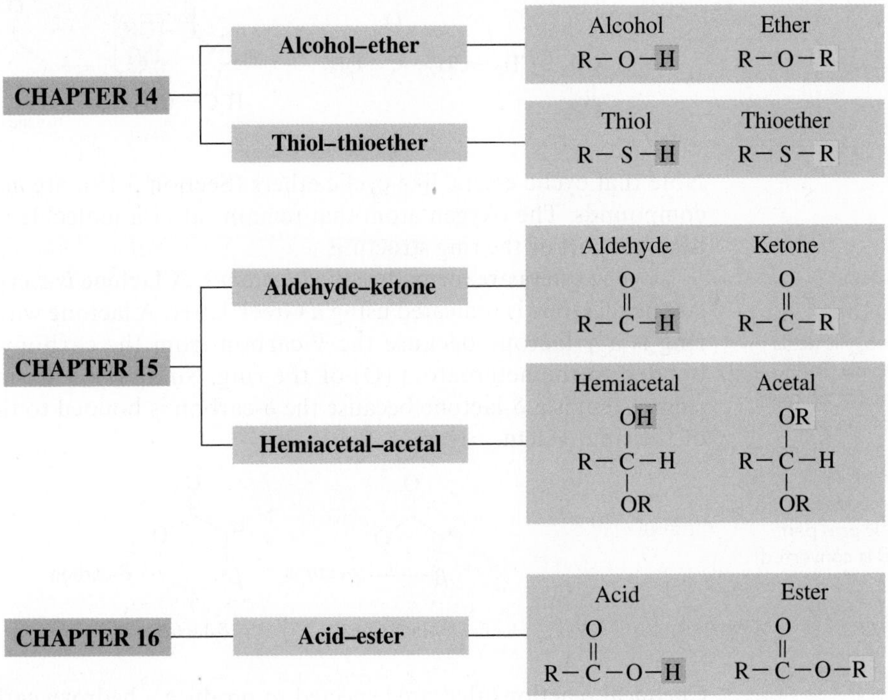

removing one of the products. According to Le Châtelier's principle, either of these techniques will shift the position of equilibrium to the right (the product side of the equation). This equilibrium problem explains the use of the "double-arrow notation" in all the esterification equations in this section.

It is often useful to think of the structure of an ester in terms of its "parent" alcohol and acid molecules; the ester has an acid part and an alcohol part.

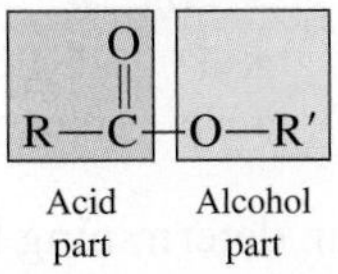

In this context, it is easy to identify the acid and alcohol from which a given ester can be produced; just add a —OH group to the acid part of the ester and a —H atom to the alcohol part to generate the parent molecules.

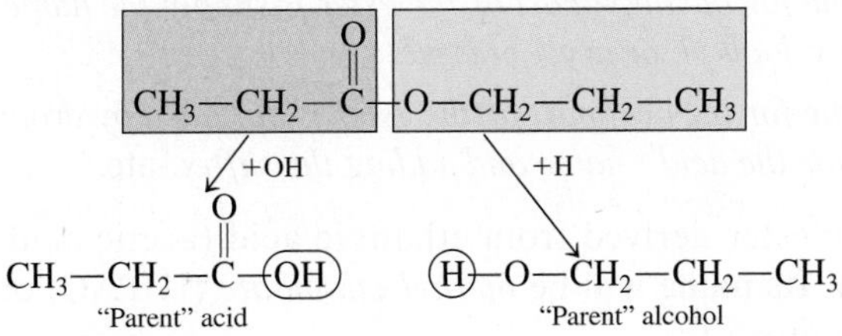

Cyclic Esters (Lactones)

Hydroxy acids—compounds which contain both a hydroxyl and a carboxyl group (Section 5.4)—have the capacity to undergo intermolecular esterification to form cyclic esters. Such internal esterification easily takes place in situations where a five- or six-membered ring can be formed.

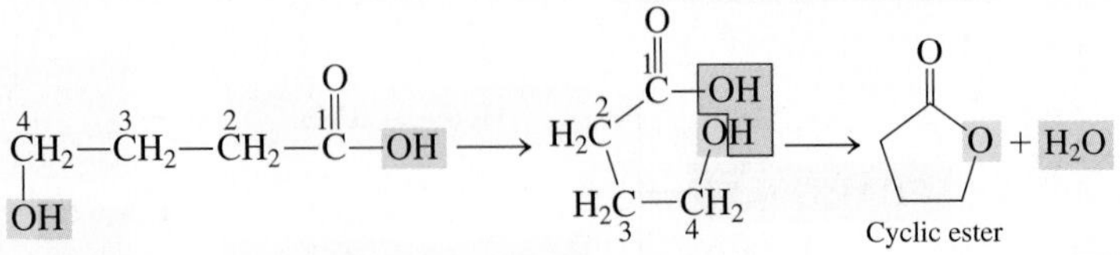

Note that cyclic esters, like cyclic ethers (Section 3.19), are *heterocyclic* organic compounds. The oxygen atom that remains after a molecule of water is formed becomes part of the ring structure.

Cyclic esters are formally called *lactones*. A **lactone** *is a cyclic ester.* The ring size in a lactone is indicated using a Greek letter. A lactone with a five-membered ring is a γ-lactone because the γ-carbon from the carbonyl carbon atom is bonded to the heteroatom (O) of the ring. Similarly, a six-membered lactone ring system is a δ-lactone because the δ-carbon is bonded to the heteroatom (O) of the ring system.

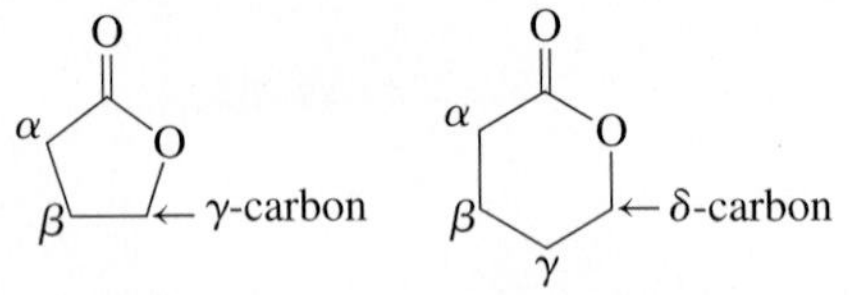

γ-lactone δ-lactone

Chemical reactions that are expected to produce a hydroxy carboxylic acid often yield a lactone instead if a five- or six-membered ring can be formed.

Salts and esters of carboxylic acids are named in the same way. The name of the positive ion (in the case of a salt) or the name of the organic group attached to the single-bonded oxygen of the carbonyl group (in the case of an ester) precedes the name of the acid. The *-ic acid* part of the name of the acid is converted to *-ate*.

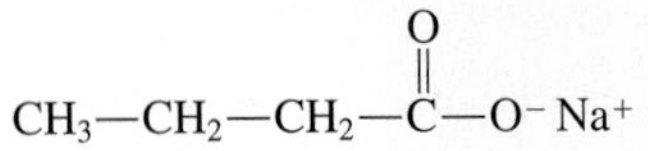

IUPAC: Sodium butanoate
Common: Sodium butyrate

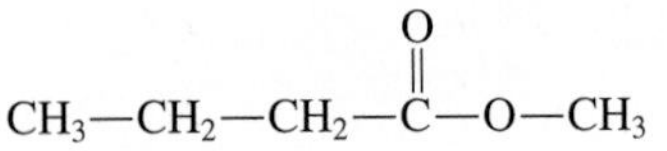

IUPAC: Methyl butanoate
Common: Methyl butyrate

5.11 Nomenclature for Esters

For naming purposes, an ester may be viewed structurally as containing an acyl group and an alkyl group.

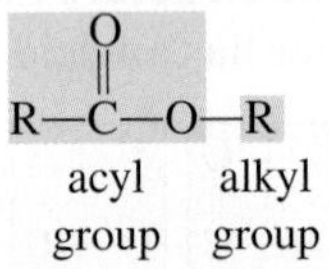

This is the basis for the rules used in determining both common and IUPAC names for esters. The rules are as follows:

Rule 1: *The name for the alkyl part of the ester appears first and is followed by a* separate word *giving the name for the acyl part of the ester.*

Rule 2: *The name for the alkyl part of the ester is simply the name of the R group (alkyl, cycloalkyl, or aryl) present.*

Rule 3: *The name for the acyl part of the ester is obtained by dropping the* -ic acid *ending for the acid's name and adding the suffix* -ate.

Consider the ester derived from ethanoic acid (acetic acid) and methanol (methyl alcohol). Its name will be *methyl ethanoate* (IUPAC) or *methyl acetate* (common); see Figure 5.12.

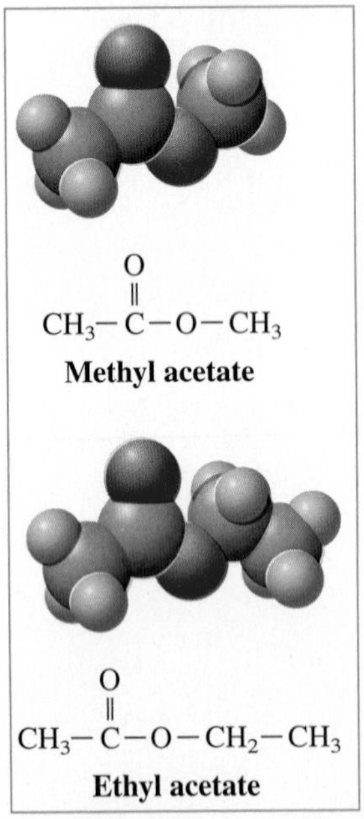

Figure 5.12 Space-filling models for the methyl and ethyl esters of acetic acid.

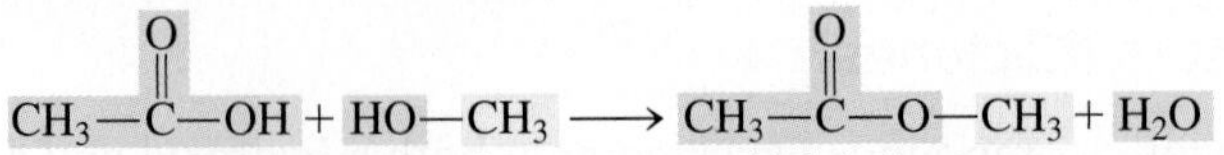

IUPAC:	Ethanoic acid	Methanol	Methyl ethanoate
Common:	Acetic acid	Methyl alcohol	Methyl acetate

Dicarboxylic acids can form diesters, with each of the carboxyl groups undergoing esterification. An example of such a molecule and how it is named is

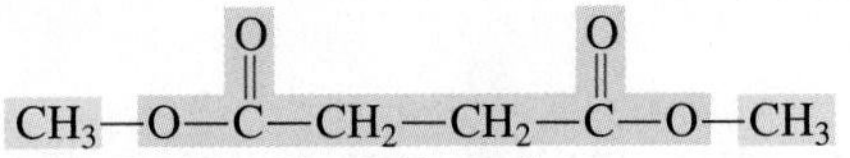

IUPAC: Dimethyl butanedioate
Common: Dimethyl succinate

Further examples of ester nomenclature, for compounds in which substituents are present, are

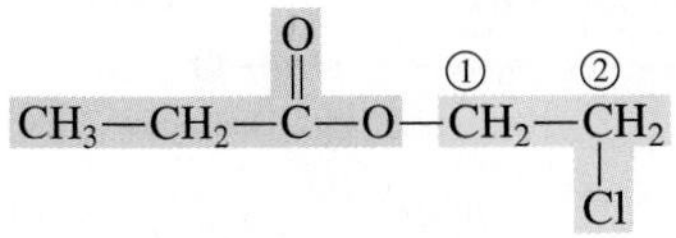

IUPAC: 2-Chloroethyl propanoate
Common: 2-Chloroethyl propionate

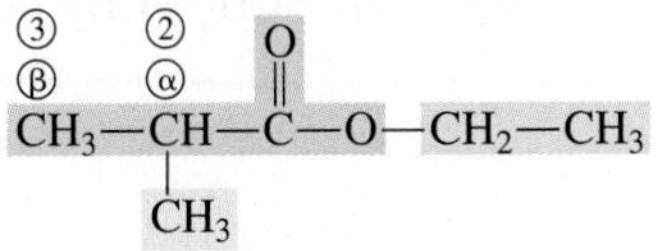

Ethyl 2-methylpropanoate
Ethyl α-methylpropionate

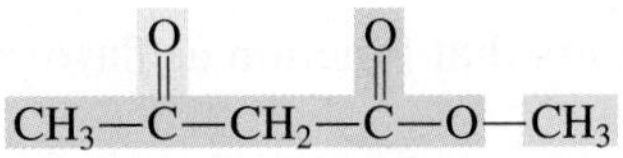

IUPAC: Methyl 3-oxobutanoate
Common: Methyl β-oxobutyrate

Line-angle structural formulas for the simpler unbranched-chain methyl esters:

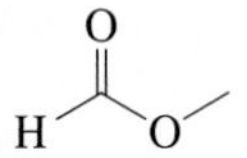

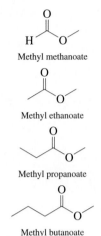

Methyl methanoate

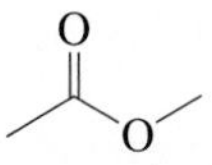

Methyl ethanoate

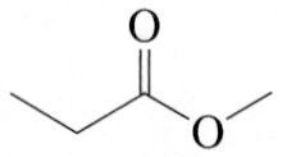

Methyl propanoate

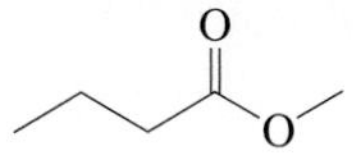

Methyl butanoate

EXAMPLE 5.4 Determining IUPAC and Common Names for Esters

Assign both IUPAC and common names to the following esters.

a. 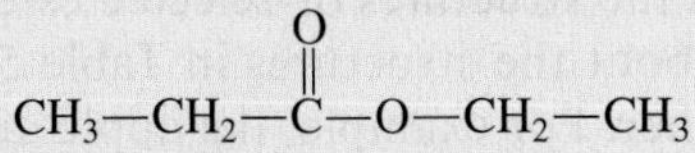

b. 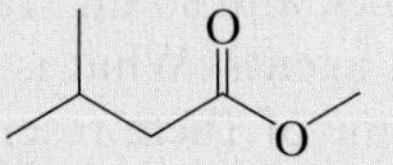

c. 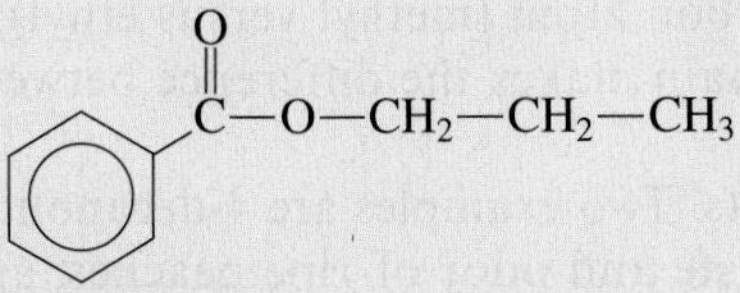

Solution

a. The name *ethyl* characterizes the alkyl group in the molecule. The name of the acid is propanoic acid (IUPAC) or propionic acid (common). Deleting the *-ic acid* ending and adding *-ate* gives the name *ethyl propanoate* (IUPAC) or *ethyl propionate* (common).

b. The name of the alkyl group is methyl (from methanol or methyl alcohol). The name of the five-carbon acid is 3-methylbutanoic acid or β-methylbutyric acid. Hence the ester name is *methyl 3-methylbutanoate* (IUPAC) or *methyl β-methylbutyrate* (common).

c. The name *propyl* characterizes the alkyl part of the molecule. The name of the acyl part of the molecule is derived from benzoic acid (both IUPAC and common name). Hence the ester name in both systems is *propyl benzoate.*

Practice Exercise 5.4

Assign both IUPAC and common names to the following esters.

a. 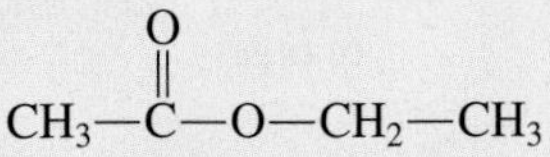

b. 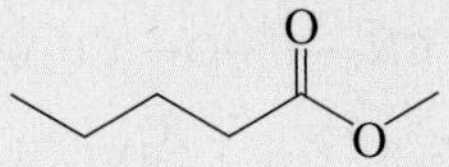

c. 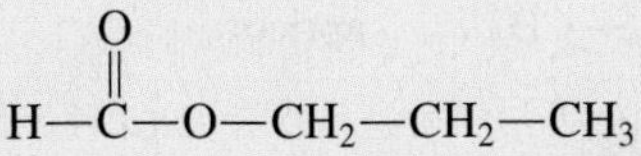

Answers: **a.** Ethyl ethanoate, ethyl acetate; **b.** Methyl pentanoate, methyl valerate; **c.** Propyl methanoate, propyl formate

The structural formulas of esters are usually written with the acyl group first (on the left) and the alkyl group last (on the right). When naming an ester, however, the situation is opposite. The alkyl group is named first, followed by the name of the acyl group.

The contrast between IUPAC names and common names for unbranched esters of carboxylic acids is as follows:

IUPAC (two words)

alkyl alkanoate

methyl propanoate

Common (two words)

alkyl (prefix)ate*

methyl acetate

*The common-name prefixes are related to natural sources for the "parent" carboxylic acids.

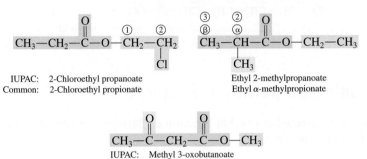

IUPAC names for lactones are generated by replacing the *-oic* ending of the parent hydroxycarboxylic acid name with *-olide* and identifying the hydroxyl-bearing carbon by number.

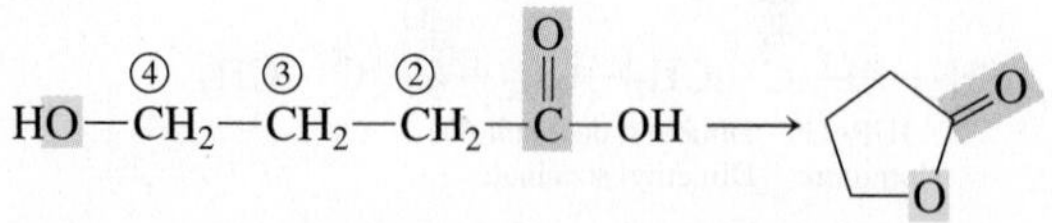

4-Hydroxybutanoic acid   4-Butanolide

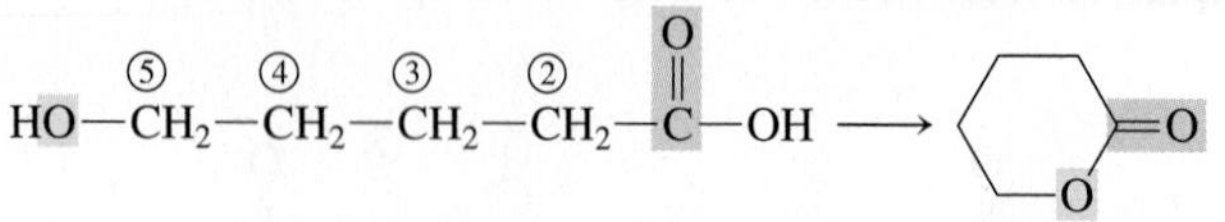

5-Hydroxypentanoic acid   5-Pentanolide

5.12 Selected Common Esters

In this section, selected esters that function as flavoring agents, pheromones, and medications are considered.

Flavor/Fragrance Agents

Esters are largely responsible for the flavor and fragrance of fruits and flowers. Generally, a natural flavor or odor is caused by a mixture of esters, with one particular compound being dominant. The synthetic production of these "dominant" compounds is the basis for the flavoring agents used in ice cream, gelatins, soft drinks, and so on. Table 5.4 gives the structures of selected esters used as flavoring agents. What is surprising about the structures in Table 5.4 is how closely some of them resemble each other. For example, the apple and pineapple flavoring agents differ by one carbon atom (methyl versus ethyl); a five-carbon chain versus an eight-carbon chain makes the difference between banana and orange flavor.

Numerous lactones are common in plants. Two examples are 4-decanolide, a compound partially responsible for the taste and odor of ripe peaches, and

Table 5.4 Selected Esters That Are Used as Flavoring Agents

IUPAC Name	Structural Formula	Characteristic Flavor and Odor
isobutyl methanoate	$H-\overset{\overset{O}{\|}}{C}-O-CH_2-\overset{\overset{CH_3}{\vert}}{CH}-CH_3$	raspberry
propyl ethanoate	$CH_3-\overset{\overset{O}{\|}}{C}-O-(CH_2)_2-CH_3$	pear
pentyl ethanoate	$CH_3-\overset{\overset{O}{\|}}{C}-O-(CH_2)_4-CH_3$	banana
octyl ethanoate	$CH_3-\overset{\overset{O}{\|}}{C}-O-(CH_2)_7-CH_3$	orange
pentyl propanoate	$CH_3-CH_2-\overset{\overset{O}{\|}}{C}-O-(CH_2)_4-CH_3$	apricot
methyl butanoate	$CH_3-(CH_2)_2-\overset{\overset{O}{\|}}{C}-O-CH_3$	apple
ethyl butanoate	$CH_3-(CH_2)_2-\overset{\overset{O}{\|}}{C}-O-CH_2-CH_3$	pineapple

coumarin (common name), the compound responsible for the pleasant odor of newly mown hay.

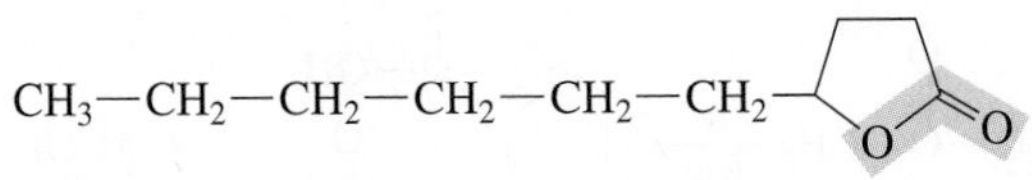

4-Decanolide
(peach odor)

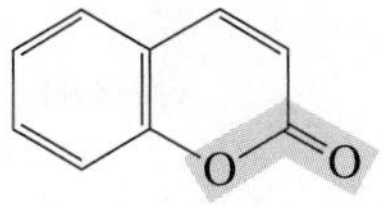

Coumarin
(newly mown hay odor)

Pheromones

A number of pheromones (Section 2.7) contain ester functional groups. The compound isoamyl acetate,

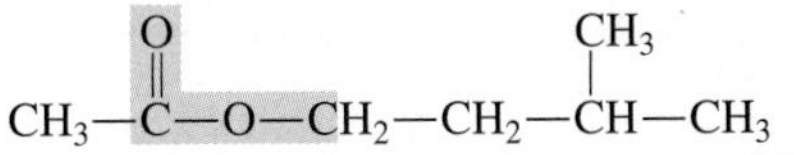

is an alarm pheromone for the honey bee. The compound methyl *p*-hydroxybenzoate,

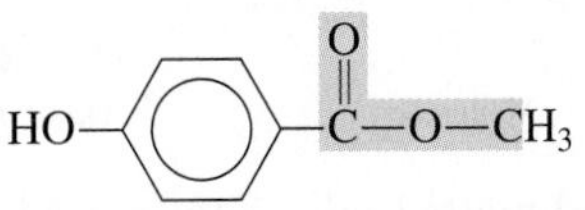

is a sexual attractant for canine species. It is secreted by female dogs in heat and evokes attraction and sexual arousal in male dogs.

The compound nepetalactone, a lactone present in the catnip plant, is an attractant for cats of all types. It is not considered a pheromone, however, because different species are involved (Figure 5.13).

Medications

Numerous esters have medicinal value, including benzocaine (a local anesthetic), aspirin, and oil of wintergreen (a counterirritant). The structure of benzocaine is

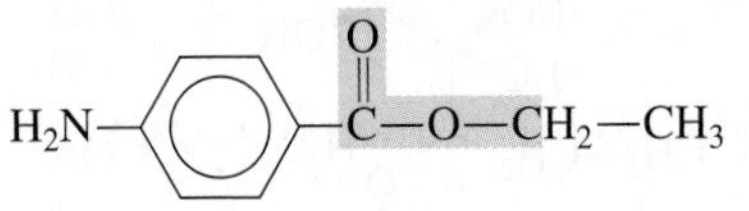

Both aspirin and oil of wintergreen are esters of salicylic acid, an aromatic hydroxyacid.

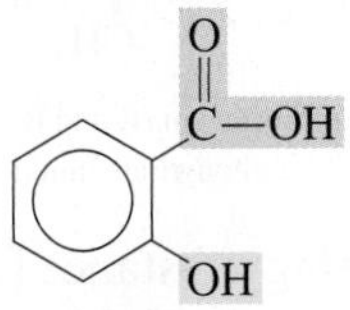

Salicylic acid

Because this acid has both an acid group and a hydroxyl group, it can form two different types of esters: one by reaction of its acid group with an alcohol, the other by reaction of its alcohol group with a carboxylic acid.

Figure 5.13 Cats of all types (from lions to house cats) are strongly attracted to the catnip plant. The attractant in the catnip plant is nepetalactone, a cyclic ester.

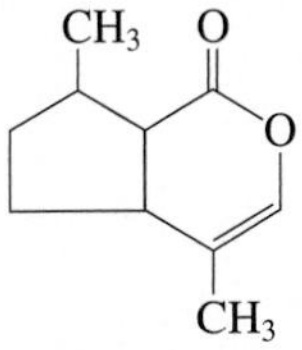

Nepetalactone

Reaction of acetic acid with the alcohol group of salicylic acid produces aspirin.

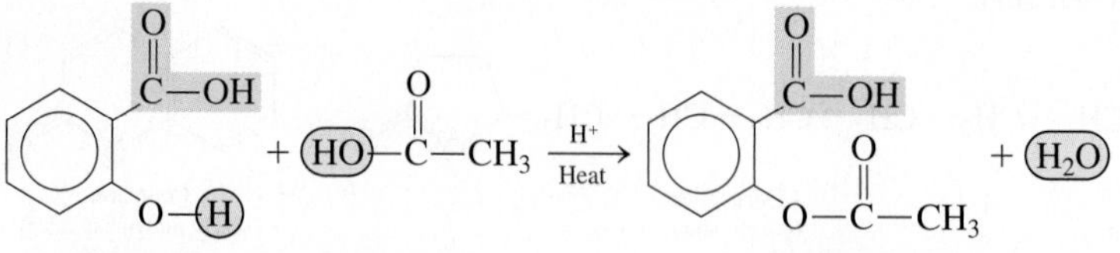

Salicylic acid   Acetic acid   Aspirin

Aspirin's mode of action in the human body is considered in the focus on relevancy feature Chemical Connections 5-C on the next page.

Reaction of methanol with the acid group of salicylic acid produces oil of wintergreen.

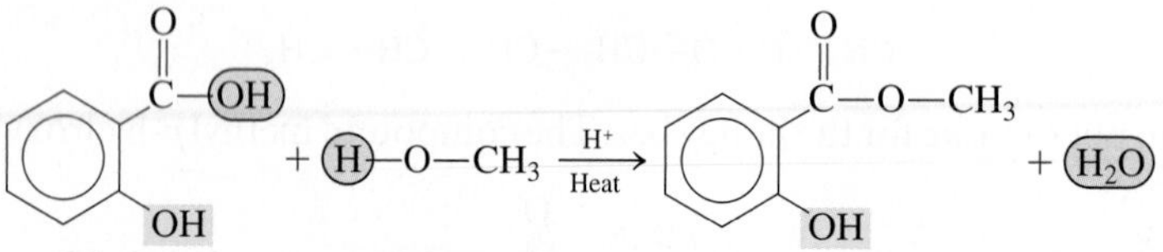

Salicylic acid   Methanol   Oil of wintergreen

Oil of wintergreen, also called methyl salicylate, is used in skin rubs and liniments to help decrease the pain of sore muscles. It is absorbed through the skin, where it is hydrolyzed to produce salicylic acid. Salicylic acid, as with aspirin, is the actual pain reliever.

The *macrolide antibiotics* are a family of large-ring lactones. Erythromycin, the best known member of this antibiotic family, has an antimicrobial spectrum similar to that of penicillins (Section 10.10) and is often used for people who have an allergy to penicillins. Structurally, this antibiotic contains a 14-membered lactone ring.

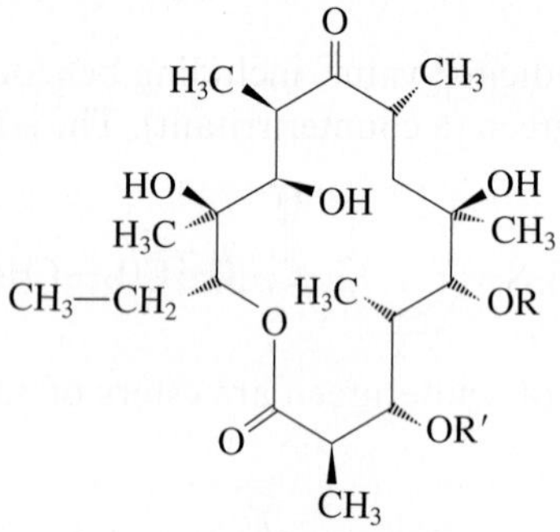

Erythromycin (R and R′ are carbohydrate units)

Erythromycin is a naturally occurring substance first isolated from a red-pigmented soil bacterium. The laboratory synthesis of this compound has now been achieved. Its chemical formula is $C_{37}H_{67}NO_{13}$.

5.13 Isomerism for Carboxylic Acids and Esters

As with the other families of organic compounds previously discussed, constitutional isomers based on different carbon skeletons and on different positions for the functional group are possible for carboxylic acids and esters as well as other types of carboxylic acid derivatives. The following two examples illustrate carboxylic acid skeletal isomerism and ester positional isomerism.

Carboxylic acid skeletal isomers:

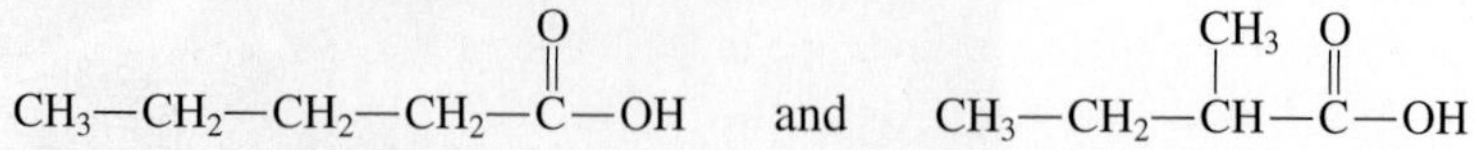

Pentanoic acid   and   2-Methylbutanoic acid

Ester positional isomers:

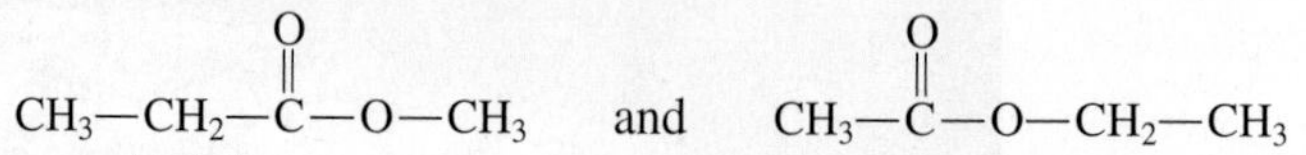

Methyl propanoate   and   Ethyl ethanoate

CHEMICAL CONNECTIONS 5-C

Aspirin

Aspirin, an ester of salicylic acid (Section 5.12), is a drug that has the ability to decrease pain (analgesic properties), to lower body temperature (antipyretic properties), and to reduce inflammation (anti-inflammatory properties). It is most frequently taken in tablet form, and the tablet usually contains 325 mg of aspirin held together with an inert starch binder.

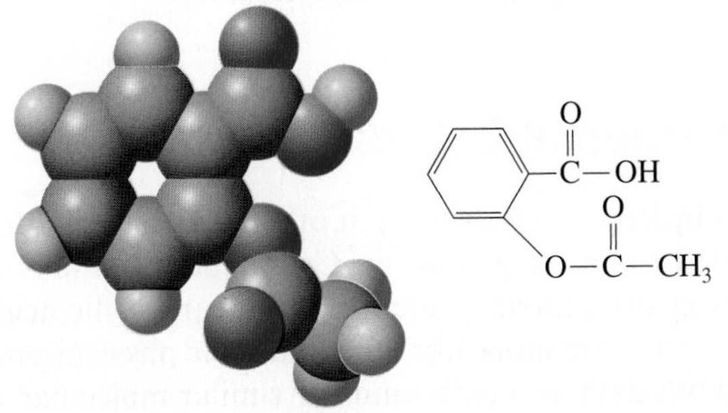

After ingestion, aspirin undergoes hydrolysis to produce salicylic acid and acetic acid. Salicylic acid is the active ingredient of aspirin—the substance that has analgesic, antipyretic, and anti-inflammatory effects.

Salicylic acid is capable of irritating the lining of the stomach, inducing a small amount of bleeding. Breaking (or chewing) an aspirin tablet, rather than taking it whole, reduces the chance of bleeding by eliminating drug concentration on one part of the stomach lining. Buffered aspirin products contain alkaline chemicals (such as aluminum glycinate or aluminum hydroxide) to neutralize the acidity of the aspirin when it contacts the stomach lining.

Aspirin—that is, salicylic acid—inhibits the synthesis of a class of hormones called prostaglandins (Section 8.13), molecules that cause pain, fever, and inflammation when present in the bloodstream in higher-than-normal levels. Salicylic acid's mode of action is irreversible inhibition (Section 10.7) of *cyclooxygenase,* an enzyme necessary for the production of prostaglandins.

Recent studies show that aspirin also increases the time it takes blood to coagulate (clot). For blood to coagulate, platelets must first be able to aggregate, and prostaglandins (which aspirin inhibits) appear to be necessary for platelet aggregation to occur. One study suggests that healthy men can cut their risk of heart attacks nearly in half by taking one baby aspirin per day (81 mg compared to the 325 mg in a regular tablet). Aspirin acts by making the blood less likely to clot. Heart attacks usually occur when clots form in the coronary arteries, cutting off blood supply to the heart.

Aspirin manufacturers indicate that "low dose" (81 mg) aspirin tablet use is rapidly increasing among adults in the United States. It is estimated that one-third of the adult population now regularly take "low-dose" aspirin for cardiovascular health reasons.

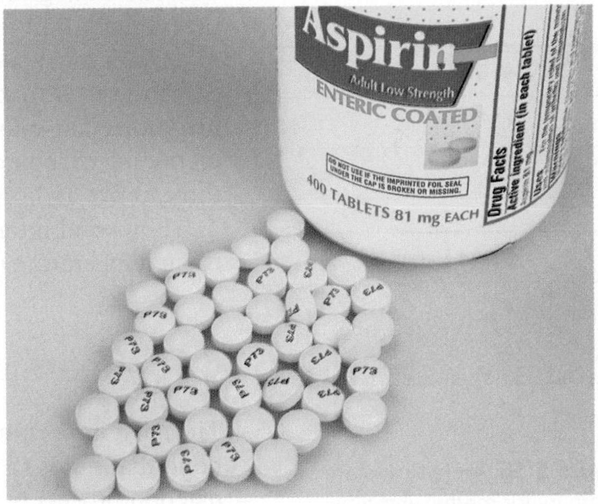

Newscom

Low-dose aspirin tablets contain 81 mg of aspirin.

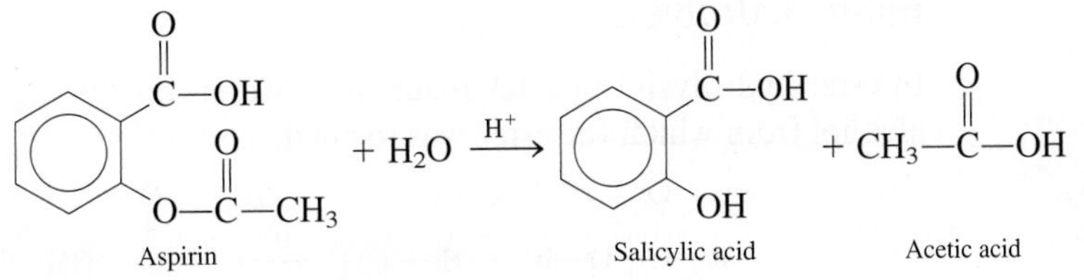

Carboxylic acids and esters with the same number of carbon atoms and the same degree of saturation are functional group isomers. The ester ethyl propanoate and the carboxylic acid pentanoic acid both have the molecular formula $C_5H_{10}O_2$ and are thus functional group isomers.

Carboxylic acid–ester functional group isomers:

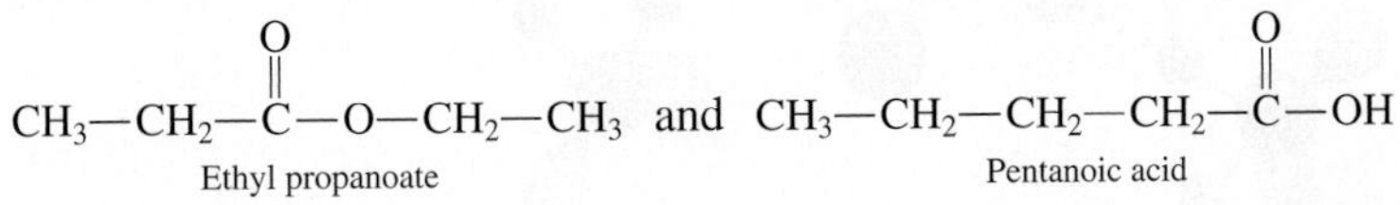

Esters and carboxylic acid functional group isomerism represents the fourth time we have encountered this type of isomerism. Previous examples are alcohol–ether, thiol–thioether, and aldehyde–ketone isomers.

Table 5.5 Boiling Points of Compounds of Similar Molecular Mass That Contain Different Functional Groups

Name	Functional-Group Class	Molecular Mass (amu)	Boiling Point (°C)
diethyl ether	ether	74	34
ethyl formate	ester	74	54
methyl acetate	ester	74	57
butanal	aldehyde	72	76
1-butanol	alcohol	74	118
propionic acid	acid	74	141

5.14 Physical Properties of Esters

Ester molecules cannot form hydrogen bonds to each other because they do not have a hydrogen atom bonded to an oxygen atom. Consequently, the boiling points of esters are much lower than those of alcohols and carboxylic acids of comparable molecular mass. Esters are more like ethers in their physical properties. Table 5.5 gives boiling-point data for compounds of similar molecular mass that contain different functional groups.

Water molecules can hydrogen-bond to esters through the oxygen atoms present in the ester functional group (Figure 5.14). Because of such hydrogen bonding, low-molecular-mass esters are soluble in water. Solubility rapidly decreases with increasing carbon chain length; borderline solubility situations are reached when three to five carbon atoms are in a chain.

Low- and intermediate-molecular-mass esters are usually colorless liquids at room temperature (see Figure 5.15). Most have pleasant odors (Section 5.12).

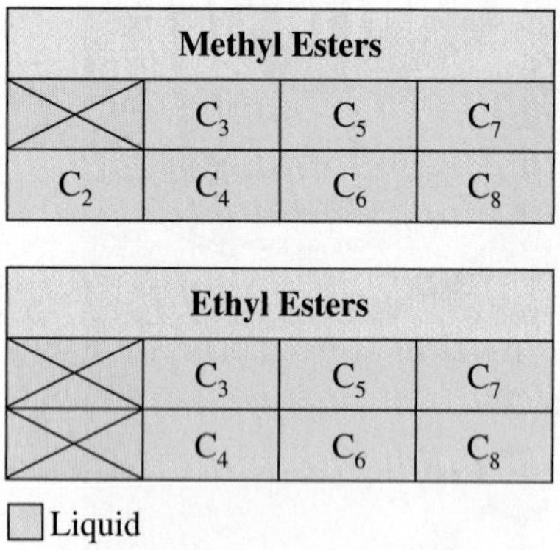

Methyl Esters
C3
C5
C7
C2
C4
C6
C8
Ethyl Esters
C3
C5
C7
C4
C6
C8
Liquid

Figure 5.15 A physical-state summary for methyl and ethyl esters of unbranched-chain carboxylic acids at room temperature and pressure.

5.15 Chemical Reactions of Esters

The most important reaction of esters involves breaking the carbon–oxygen single bond that holds the "alcohol part" and the "acid part" of the ester together. This reaction process is called either ester hydrolysis or ester saponification, depending on reaction conditions.

Ester Hydrolysis

In ester hydrolysis, an ester reacts with water, producing the carboxylic acid and alcohol from which the ester was formed.

This is the second time hydrolysis reactions have been encountered. The first encounter involved the hydrolysis of acetals (Section 4.11).

$$\mathrm{R{-}\overset{\overset{\large O}{\|}}{C}{-}O{-}R' + H{-}OH \xrightarrow{H^+} R{-}\overset{\overset{\large O}{\|}}{C}{-}OH + R'{-}O{-}H}$$

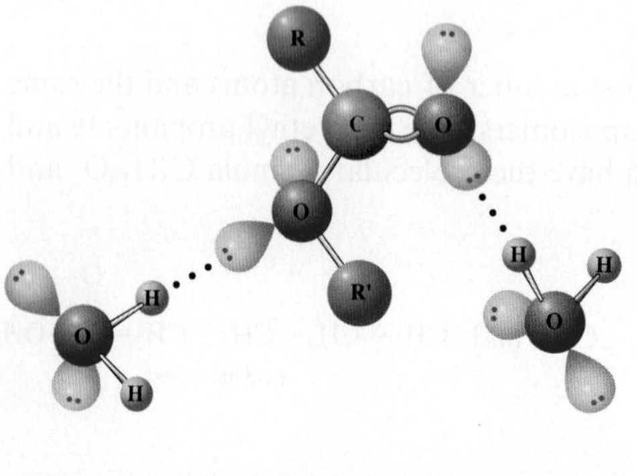

Figure 5.14 Low-molecular-mass esters are soluble in water because of ester–water hydrogen bonding.

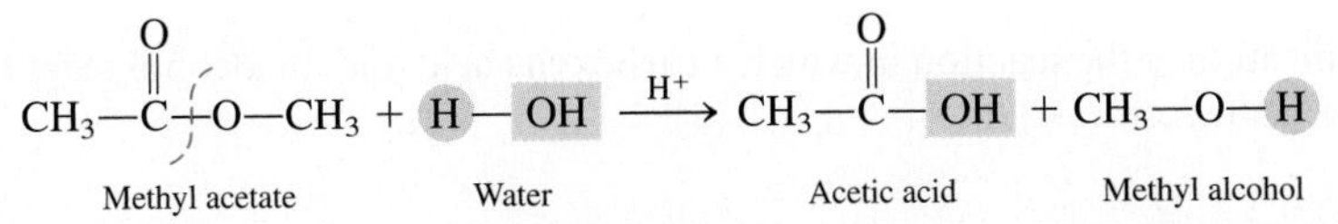

Ester hydrolysis requires the presence of a strong-acid catalyst or enzymes. Ester hydrolysis is the reverse of esterification (Section 5.10), the formation of an ester from a carboxylic acid and an alcohol.

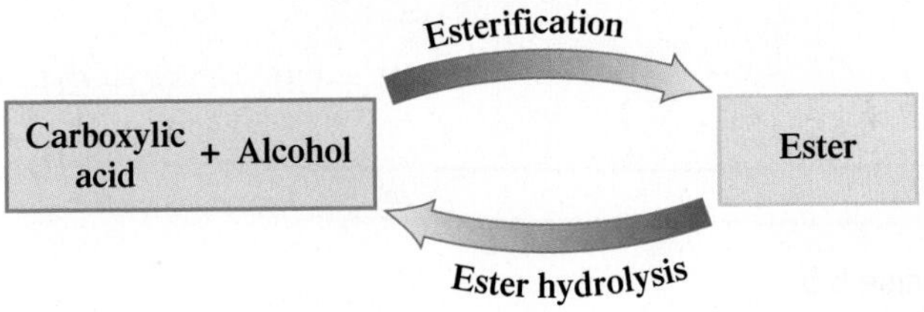

The breaking of a bond within a molecule and the attachment of the components of water to the fragments are characteristics of all hydrolysis reactions.

Ester Saponification

A **saponification reaction** *is the hydrolysis of an organic compound, under basic conditions, in which a carboxylic acid salt is one of the products.* Esters, amides (Section 6.17), and fats and oils (Section 8.6) all undergo saponification reactions.

In ester saponification, either NaOH or KOH is used as the base and the saponification products are an alcohol and a carboxylic acid salt. (Any carboxylic acid product formed is converted to its salt because of the basic reaction conditions.)

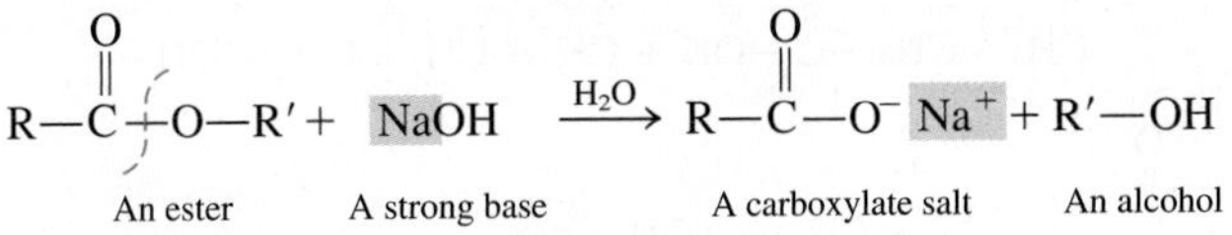

A specific example of ester saponification is

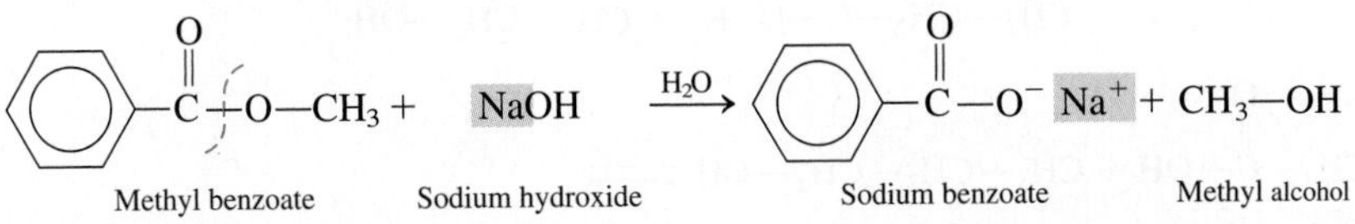

In both ester hydrolysis and ester saponification, an alcohol is produced. Under acidic conditions (ester hydrolysis), the other product is a carboxylic acid. Under basic conditions (ester saponification), the other product is a carboxylic acid salt.

EXAMPLE 5.5 Structural Equations for Reactions That Involve Esters

Write structural equations for each of the following reactions.

a. Hydrolysis, with an acidic catalyst, of ethyl acetate
b. Saponification, with NaOH, of methyl formate
c. Esterification of propionic acid using isopropyl alcohol

Solution

a. Hydrolysis, under acidic conditions, cleaves an ester to produce its "parent" carboxylic acid and alcohol.

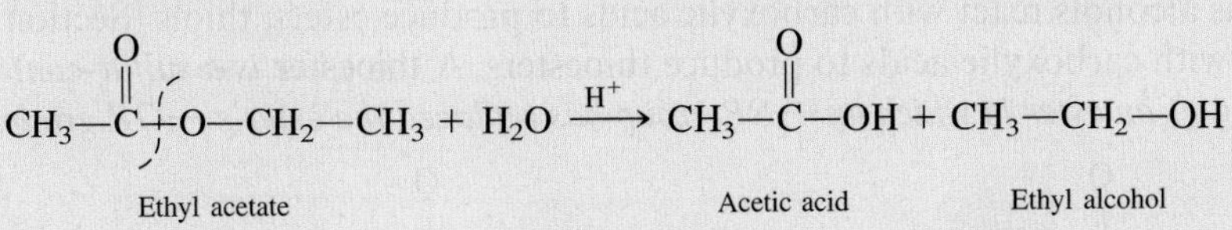

b. Saponification cleaves an ester to produce its "parent" alcohol and the *salt* of its "parent" carboxylic acid.

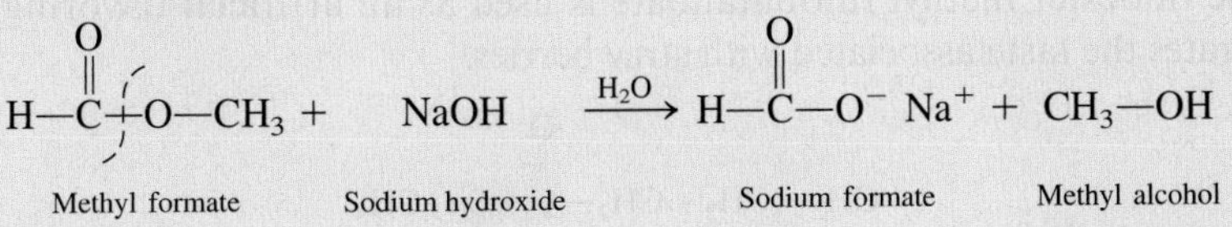

(continued)

c. Esterification is the reaction in which a carboxylic acid and an alcohol react to produce an ester.

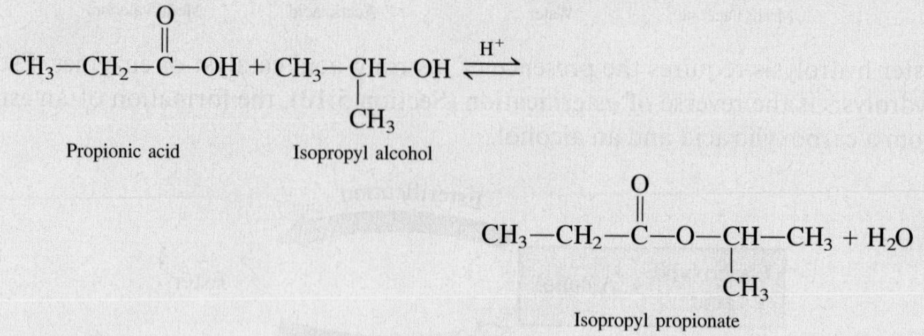

Practice Exercise 5.5

Write structural equations for each of the following reactions.

a. Hydrolysis, with an acidic catalyst, of propyl propanoate
b. Saponification, with KOH, of ethyl propanoate
c. Esterification of acetic acid with propyl alcohol

Answers:

a. 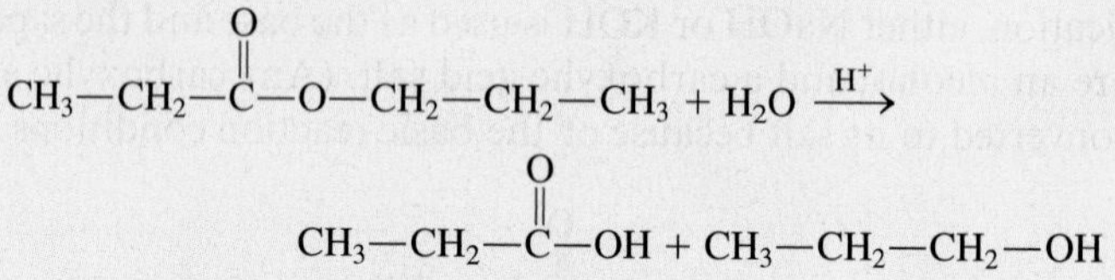

b. 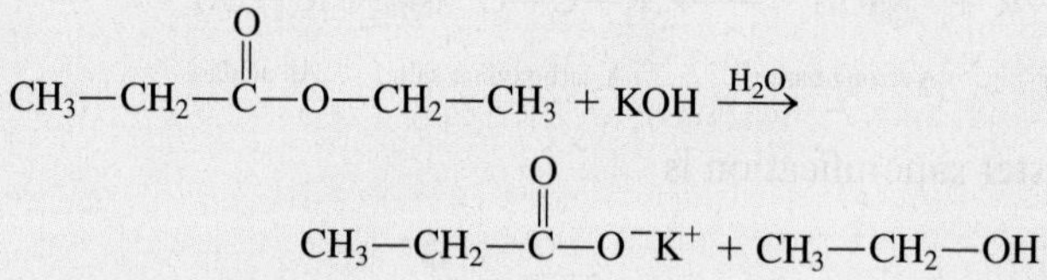

c. 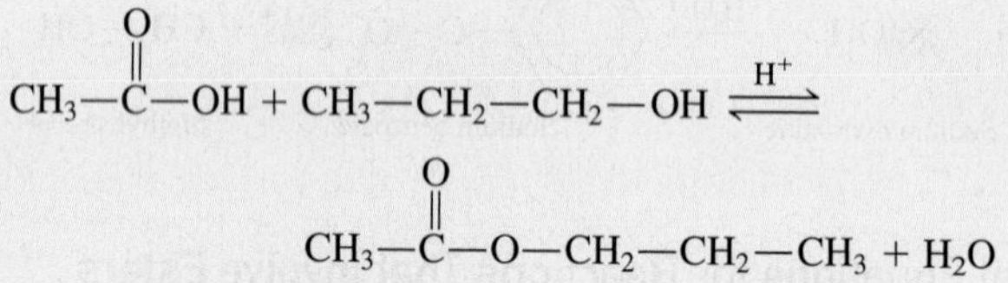

The Chemistry at a Glance feature on the next page summarizes reactions that involve carboxylic acids and esters.

5.16 Sulfur Analogs of Esters

Just as alcohols react with carboxylic acids to produce esters, thiols (Section 3.20) react with carboxylic acids to produce thioesters. A **thioester** *is a sulfur-containing analog of an ester in which an —SR group has replaced the ester's —OR group.*

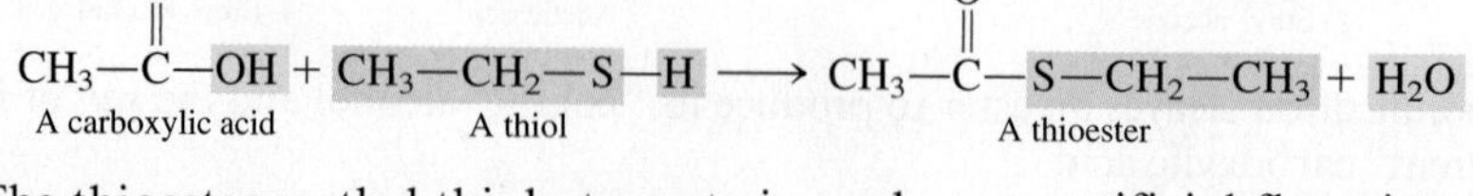

The ester functional group is

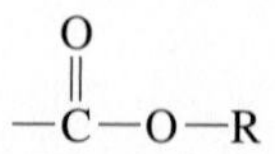

The thioester functional group is

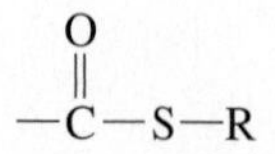

The thioester methyl thiobutanoate is used as an artificial flavoring agent. It generates the taste associated with strawberries.

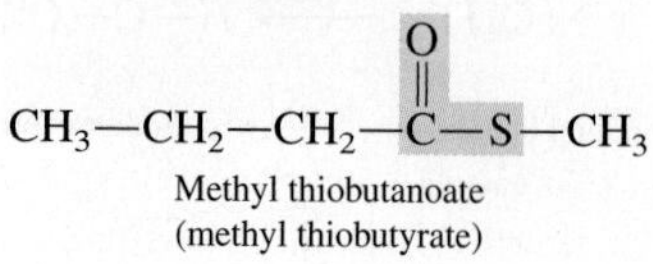

CHEMISTRY AT A GLANCE — Summary of Chemical Reactions Involving Carboxylic Acids and Esters

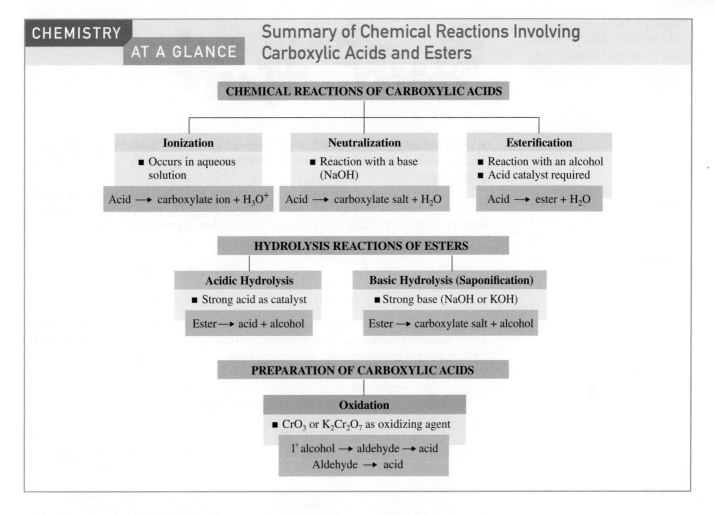

CHEMICAL REACTIONS OF CARBOXYLIC ACIDS

Ionization
- Occurs in aqueous solution

Acid ⟶ carboxylate ion + H_3O^+

Neutralization
- Reaction with a base (NaOH)

Acid ⟶ carboxylate salt + H_2O

Esterification
- Reaction with an alcohol
- Acid catalyst required

Acid ⟶ ester + H_2O

HYDROLYSIS REACTIONS OF ESTERS

Acidic Hydrolysis
- Strong acid as catalyst

Ester ⟶ acid + alcohol

Basic Hydrolysis (Saponification)
- Strong base (NaOH or KOH)

Ester ⟶ carboxylate salt + alcohol

PREPARATION OF CARBOXYLIC ACIDS

Oxidation
- CrO_3 or $K_2Cr_2O_7$ as oxidizing agent

1° alcohol ⟶ aldehyde ⟶ acid
Aldehyde ⟶ acid

Note, as seen in the IUPAC and common names of the preceding "strawberry" compound, that thioesters are named in a manner paralleling that for esters (Section 5.11) with the inclusion of the prefix *thio-* in the name.

The most important naturally occurring thioester is acetyl coenzyme A, whose abbreviated structure is

Acetyl coenzyme A

Coenzyme A, the parent molecule for acetyl coenzyme A, is a large, complex *thiol* whose structure, for simplicity, is usually abbreviated as CoA—S—H. The formation of acetyl coenzyme A (acetyl CoA) from coenzyme A can be envisioned as a thioesterification reaction between acetic acid and coenzyme A.

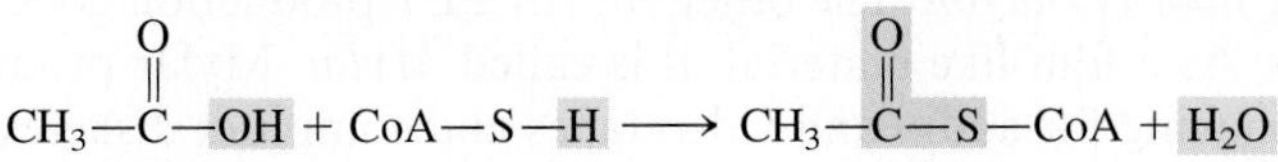

Acetic acid; Coenzyme A (a thiol); Acetyl coenzyme A (acetyl CoA)

Acetyl coenzyme A plays a central role in the metabolic cycles through which the body obtains energy to "run itself" (Section 12.6).

The complete structure of acetyl CoA is given in Section 12.3.

5.17 Polyesters

A **condensation polymer** *is a polymer formed by reacting difunctional monomers to give a polymer and some small molecule (such as water) as a by-product of the process. Polyesters* are an important type of condensation polymer. A **polyester**

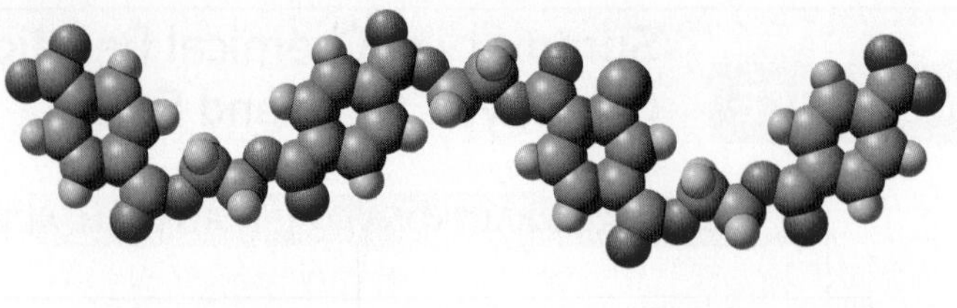

Figure 5.16 Space-filling model of a segment of the polyester condensation polymer known as poly(ethylene terephthalate), or PET.

is a condensation polymer in which the monomers are joined through ester linkages. Dicarboxylic acids and dialcohols are the monomers generally used in forming polyesters.

Condensation polymerization reactions produce two products: the polymer and a small molecule. This contrasts with addition polymerization reactions (Section 2.9) in which the polymer is the only product.

The best known of the many polyesters now marketed is *poly(ethylene terephthalate),* which is also known by the acronym *PET.* The monomers used to produce PET are terephthalic acid (a diacid) and ethylene glycol (a dialcohol).

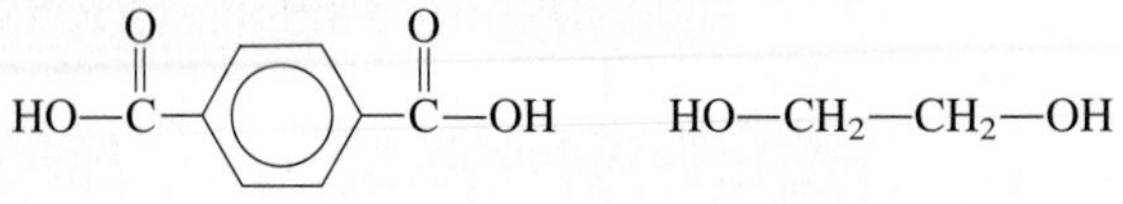

Terephthalic acid   Ethylene glycol

The reaction of one acid group of the diacid with one alcohol group of the dialcohol initially produces an ester molecule, with an acid group left over on one end and an alcohol group left over on the other end.

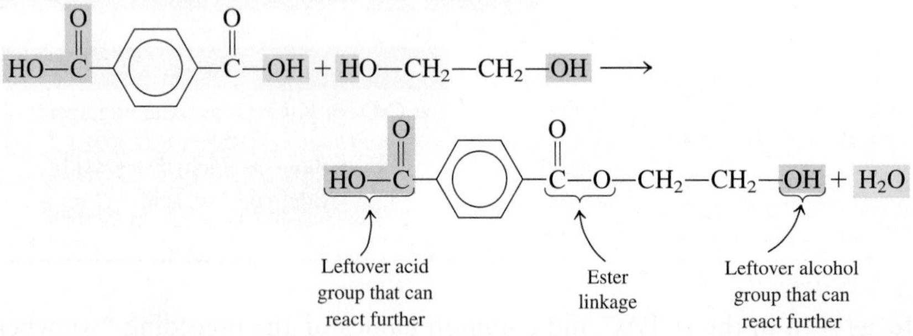

This species can react further. The remaining acid group can react with an alcohol group from another monomer, and the remaining alcohol group can react with an acid group from another monomer. This process continues until an extremely long polymer molecule called a *polyester* is produced (Figure 5.16).

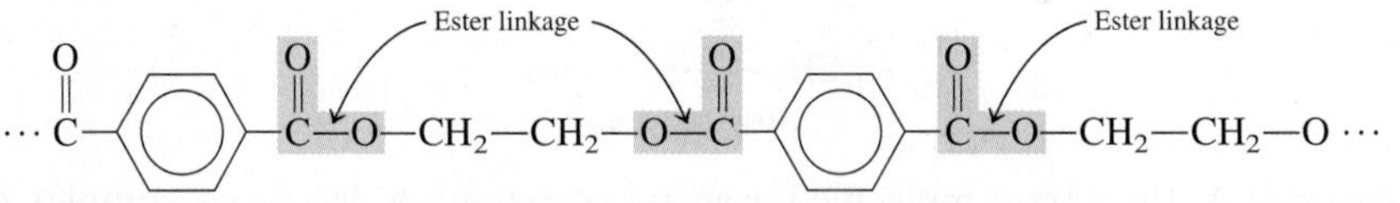

Poly(ethylene terephthalate), a polyester

About 50% of PET production goes into textile products, including clothing fibers, curtain and upholstery materials, and tire cord. The trade name for PET as a clothing fiber is *Dacron.* The other 50% of PET production goes into plastics applications. As a film-like material, it is called *Mylar.* Mylar products include the plastic backing for audio and video tapes and computer diskettes. Its chemical name PET is applied when this polyester is used in clear, flexible soft-drink bottles and as the wrapping material for frozen foods and boil-in-bag foods.

PET is also used in medicine. Because it is physiologically inert, PET is used in the form of a mesh to replace diseased sections of arteries. It has also been used in artificial heart valves (Figure 5.17).

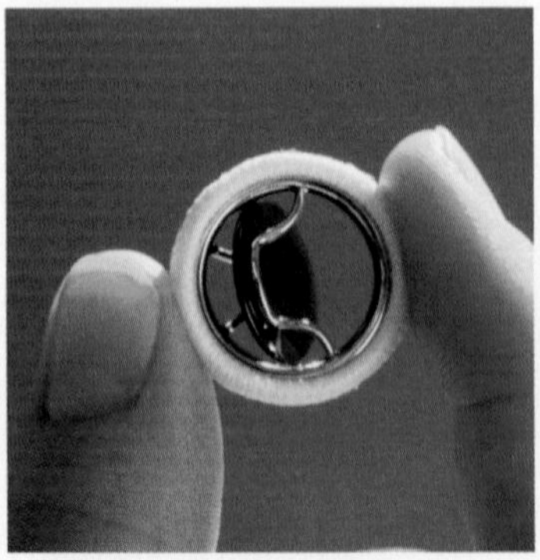

David Campione/Photo Researchers, Inc.

Figure 5.17 The polyester PET, as a fabric, is used in the construction of artificial heart valves.

Plastic bottles made of PET cannot be reused because they cannot withstand the high temperatures needed to sterilize them for reuse. Also, such bottles cannot be used for any food items that must be packaged at high temperatures, such as jams and jellies. For these uses, the polyester PEN (polyethylene naphthalate), a polymer that can withstand higher temperatures, is available.

The monomers for this polymer are ethylene glycol and one or more naphthalene dicarboxylic acids.

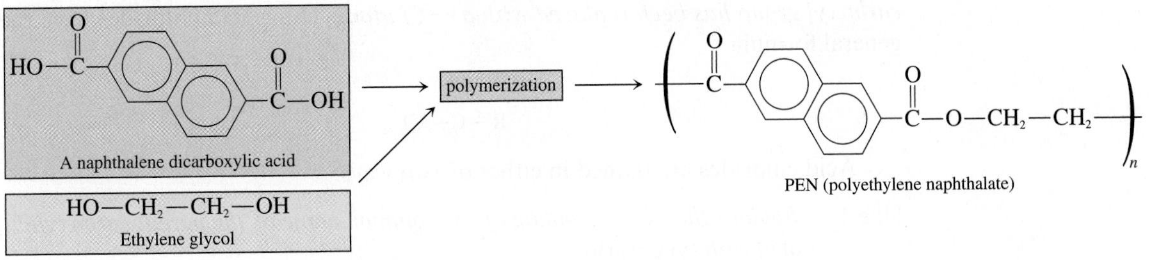

A variation of the diacid–dialcohol monomer formulation for polyesters involves using hydroxy acids as monomers. In this situation, both of the functional groups required are present in the same molecule.

A polymerization reaction in which lactic acid and glycolic acid (both hydroxy acids, Section 5.4) are monomers produces a biodegradable material (trade name *Lactomer*) that is used as surgical staples in several types of surgery. Traditional suture materials must be removed later on, after they have served their purpose. Lactomer staples start to dissolve (hydrolyze) after a period of several weeks. The hydrolysis products are the starting monomers, lactic acid and glycolic acid, both of which are readily metabolized by the human body. By the time the tissue has fully healed, the staples have fully degraded.

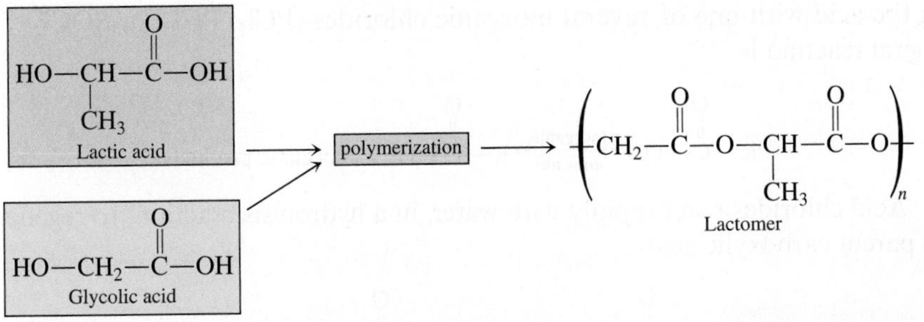

Another commercially available biodegradable polyester is PHBV, a substance that finds use in specialty packaging, orthopedic devices, and controlled drug-release formulations. Coating a drug formulation with PHBV, which degrades slowly over time, results in the drug being released slowly rather than all at once. The monomers for this polymer are 3-hydroxybutanoic acid (β-hydroxybutyrate) and 3-hydroxypentanoic acid (β-hydroxyvalerate). The properties of PHBV vary according to the reacting ratio for the two monomers. With more of the butanoic acid present, a stiffer polymer is produced, while more of the pentanoic acid imparts flexibility to the plastic. A nonconventional method for producing the polymer is used; the two monomers are produced by bacterial fermentation of mixtures containing acetic and propionic acids. The generalized formula for the polymer is written as

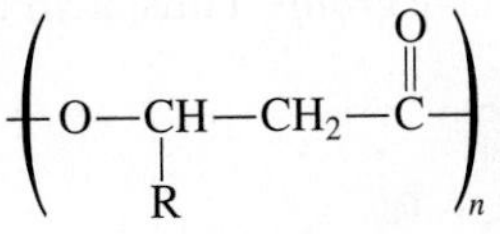
(R = $—CH_3$ and $—CH_2—CH_3$)

PHBV [poly(β-hydroxybutyrate-co-β-hydroxyvalerate)]

5.18 Acid Chlorides and Acid Anhydrides

Sections 5.9 through 5.17 have focused on the carboxylic acid derivatives called esters. This section considers acid chlorides and acid anhydrides, two of the other types of carboxylic acid derivatives mentioned in Section 5.1.

Acid Chlorides

An **acid chloride** *is a carboxylic acid derivative in which the —OH portion of the carboxyl group has been replaced with a —Cl atom.* Thus, acid chlorides have the general formula

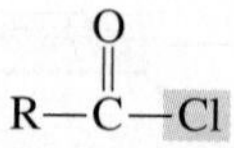

Acid chlorides are named in either of two ways:

Rule 1: *Replace the* -ic acid *ending of the common name of the parent carboxylic acid with* -yl chloride.

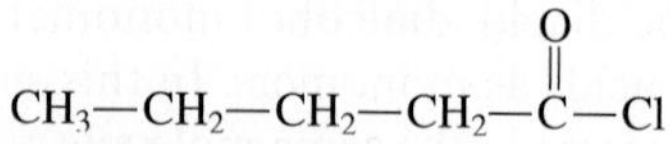

Butyric acid becomes butyryl chloride.

Rule 2: *Replace the* -oic acid *ending of the IUPAC name of the parent carboxylic acid with* -oyl chloride.

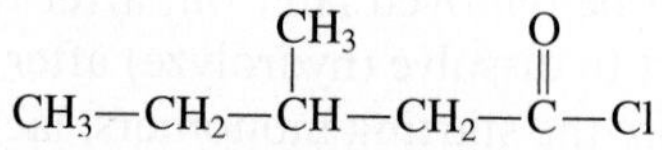

3-Methylpentanoic acid becomes 3-methylpentanoyl chloride.

Preparation of an acid chloride from its parent carboxylic acid involves reacting the acid with one of several inorganic chlorides (PCl_3, PCl_5, or $SOCl_2$). The general reaction is

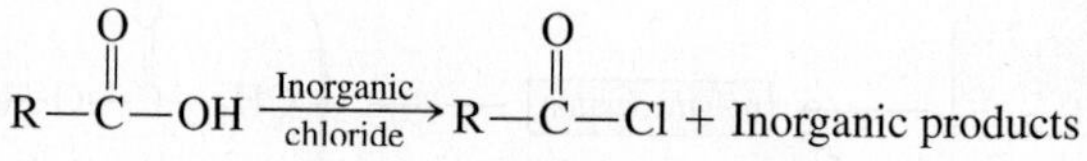

Acid chlorides react rapidly with water, in a hydrolysis reaction, to regenerate the parent carboxylic acid.

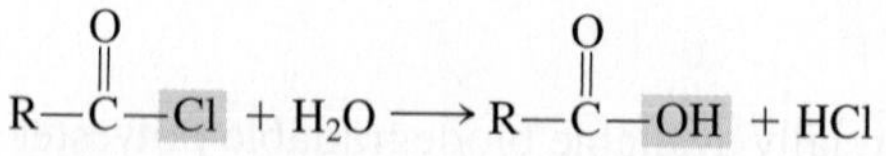

This reactivity with water means that acid chlorides cannot exist in biological systems.

Acid chlorides are useful starting materials for the synthesis of other carboxylic acid derivatives, particularly esters and amides. Synthesis of esters and amides using acid chlorides is a more efficient process than ester and amide synthesis using a carboxylic acid.

Acid Anhydrides

An **acid anhydride** *is a carboxylic acid derivative in which the —OH portion of the carboxyl group has been replaced with a* —O—C(=O)—R *group.* Thus, acid anhydrides have the general formula

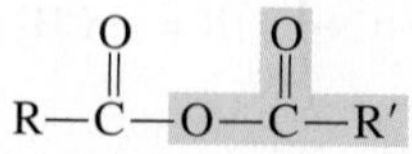

An acid anhydride's structure can be viewed as containing two carbonyl groups joined by a single oxygen atom or, alternatively, as two acyl groups joined by a single oxygen atom.

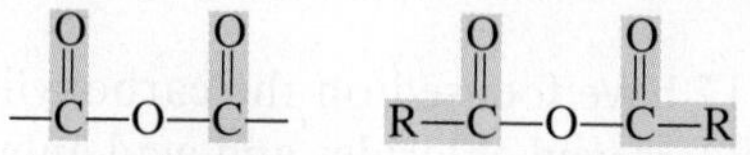

two carbonyl groups | two acyl groups

Symmetrical acid anhydrides (both R groups are the same) are named by replacing the *acid* ending of the parent carboxylic acid name with the word *anhydride.*

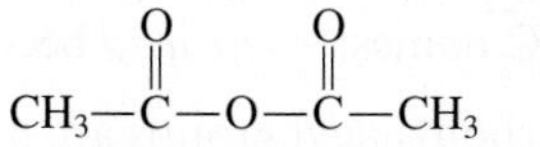

IUPAC name: Ethanoic anhydride
Common name: Acetic anhydride

Mixed acid anhydrides (different R groups present) are named by using the names of the individual parent carboxylic acids (in alphabetic order) followed by the word *anhydride.*

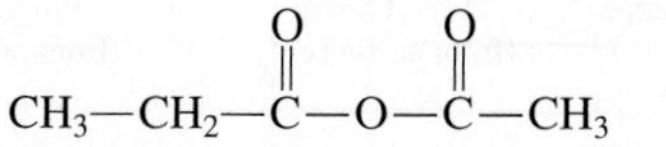

IUPAC name: Ethanoic propanoic anhydride
Common name: Acetic propionic anhydride

In general, acid anhydrides cannot be formed by directly reacting the parent carboxylic acids together. Instead, an acid chloride is reacted with a carboxylate ion to produce the acid anhydride.

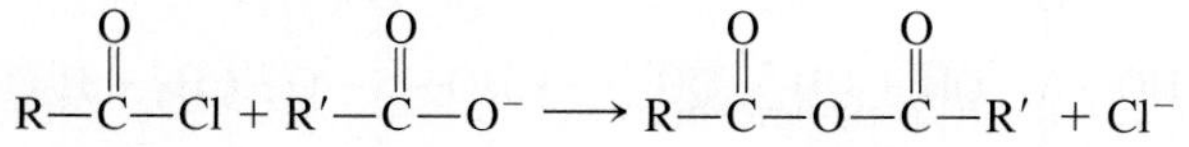

Acid chloride — Carboxylate ion — Acid anhydride

Acid anhydrides are very reactive compounds, although generally not as reactive as the acid chlorides. Like acid chlorides, they cannot exist in biological systems, as they undergo hydrolysis to regenerate the parent carboxylic acids.

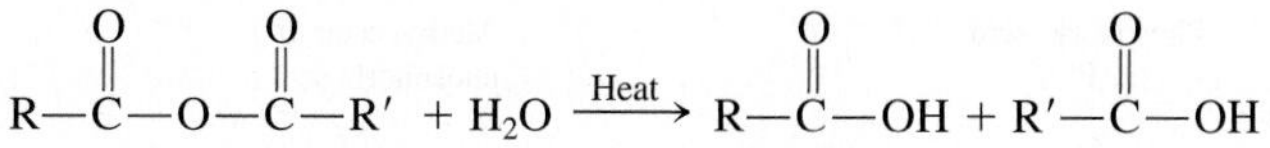

Acid anhydride — Acid — Acid

Acyl Transfer Reactions

When compounds containing acyl groups (Section 5.1) react with an alcohol or phenol, the acyl group is transferred to the oxygen atom of the alcohol or phenol; an ester is the product.

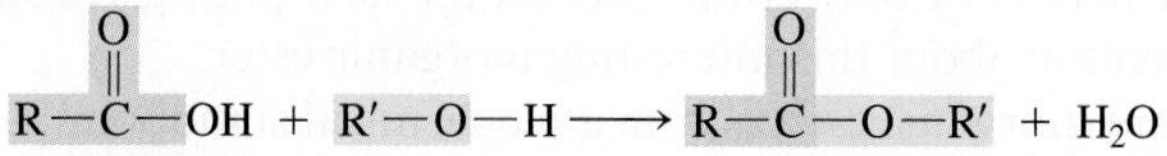

Carboxylic acid — Ester

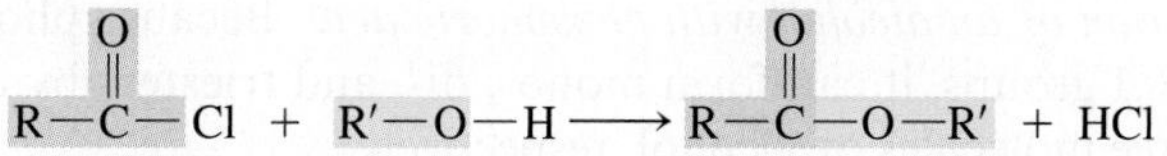

Acid chloride — Ester

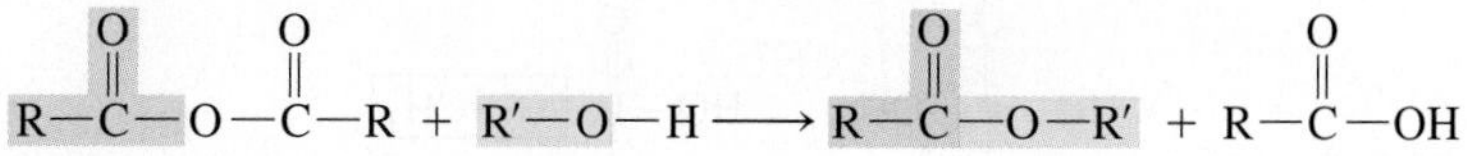

Acid anhydride — Ester

Chemical reactions such as these are called *acyl transfer reactions.* An **acyl transfer reaction** *is a chemical reaction in which an acyl group is transferred from one molecule to another.* Acyl transfer reactions occur frequently in biochemical systems. The process of protein synthesis (Section 11.11) is dependent upon acyl transfer reactions, as are many metabolic reactions. Often in metabolic reactions the thioester acetyl coenzyme A serves as an acyl transfer agent (Section 12.6).

An *acyl transfer reaction* is also called an *acylation reaction.*

The acyl group present in carboxylic acids and carboxylic acid derivatives is named by replacing the *-ic acid* ending of the acid name with the suffix *-yl.*

Common names: *-ic acid* become *-yl*
IUPAC names: *-oic acid* becomes *-oyl*

Thus, the two- and three-carbonyl acyl groups are named as follows:

IUPAC name:	Ethanoyl group (from ethanoic acid)	Propanoyl group (from propanoic acid)
Common name:	Acetyl group (from acetic acid)	Propionyl group (from propionic acid)

5.19 Esters and Anhydrides of Inorganic Acids

Inorganic acids such as sulfuric, phosphoric, and nitric acids react with alcohols to form esters in a manner similar to that for carboxylic acids and their derivatives.

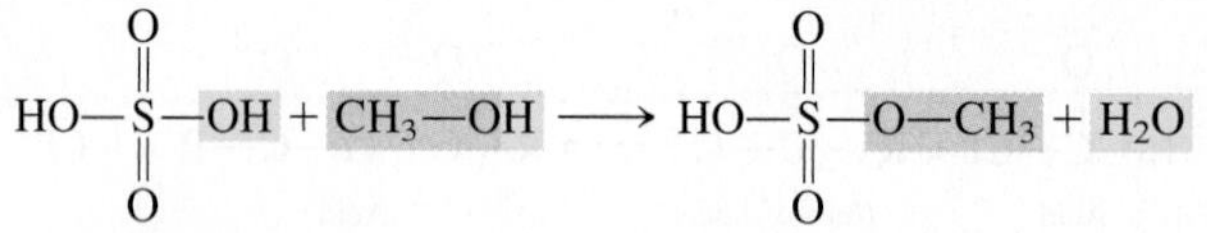

Sulfuric acid (H_2SO_4) — Methyl ester of sulfuric acid

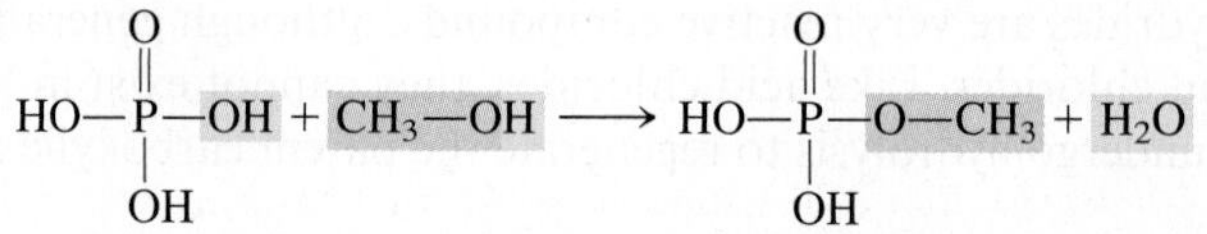

Phosphoric acid (H_3PO_4) — Methyl ester of phosphoric acid

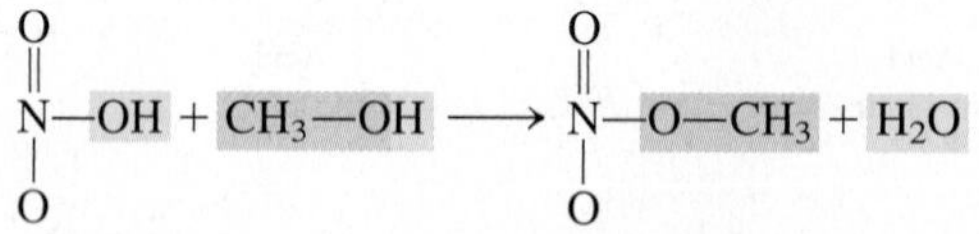

Nitric acid (HNO_3) — Methyl ester of nitric acid

The substance nitroglycerin, which, somewhat surprisingly, is used both as an explosive and as a heart medication, is an inorganic ester formed from three molecules of nitric acid and one molecule of the trihydroxy-alcohol glycerol. The focus on relevancy feature Chemical Connections 5-D on the next page gives structural as well as other information about this interesting inorganic ester.

Esters of inorganic acids undergo hydrolysis reactions in a manner similar to that for esters of carboxylic acids (Section 5.15).

The most important inorganic esters, from a biochemical standpoint, are those of phosphoric acid—that is, phosphate esters. A **phosphate ester** *is an organic compound formed by reaction of an alcohol with phosphoric acid.* Because phosphoric acid has three hydroxyl groups, it can form mono-, di-, and triesters by reaction with one, two, and three molecules of alcohol, respectively.

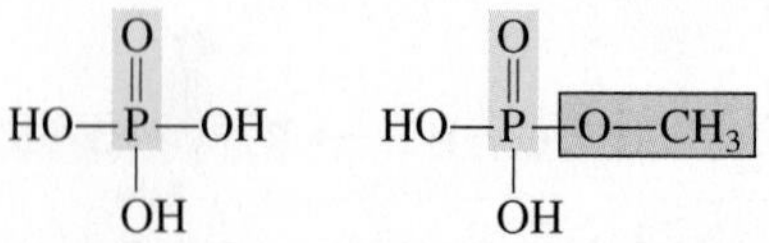

Phosphoric acid — Monoester (one —OR group)

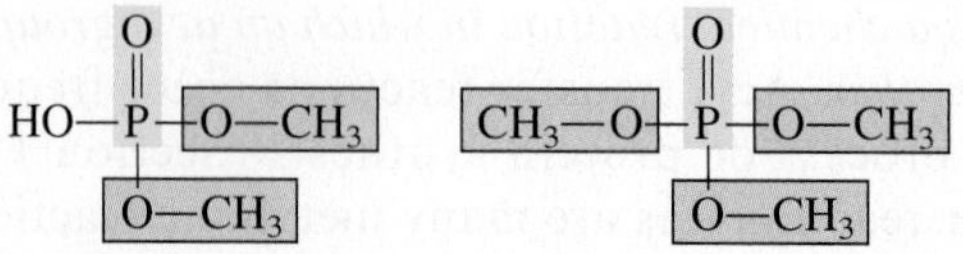

Diester (two —OR groups) — Triester (three —OR groups)

CHEMICAL CONNECTIONS 5-D

Nitroglycerin: An Inorganic Triester

The reaction of one molecule of glycerol (a trihydroxy-alcohol) with three molecules of nitric acid produces the trinitrate ester called nitroglycerin; it is a component of dynamite.

$$\begin{array}{l} CH_2{-}OH \\ | \\ CH{-}OH \\ | \\ CH_2{-}OH \end{array} + 3HO{-}NO_2 \longrightarrow \begin{array}{l} CH_2{-}O{-}NO_2 \\ | \\ CH{-}O{-}NO_2 \\ | \\ CH_2{-}O{-}NO_2 \end{array} + 3H_2O$$

In the pure state, nitroglycerin is a shock-sensitive liquid that can decompose to produce large volumes of gases (N_2, CO_2, H_2O, and O_2). When used in dynamite, it is adsorbed on clay-like materials, giving products that will not explode without a formal ignition system.

Besides being a component of dynamite explosives, nitroglycerin has medicinal value. It is used in treating patients with angina pectoris—sharp chest pains caused by an insufficient supply of oxygen reaching heart muscle. Its effect on the human body is that of a vasodilator, a substance that increases blood flow by relaxing constricted muscles around blood vessels.

Nitroglycerin medication is available in several forms: (1) as a liquid diluted with alcohol to render it nonexplosive (2) as a liquid adsorbed to a tablet for convenience of sublingual (under the tongue) administration (3) in ointments for topical use and (4) as "skin patches" that release the drug continuously through the skin over a 24-hr period. Nitroglycerin is rapidly absorbed through the skin, enters the bloodstream, and finds its way to heart muscle within seconds.

It is not nitroglycerin itself that relieves the heart muscle pain associated with angina, but, rather, a compound produced when the nitroglycerin is metabolized. That compound is the simple diatomic molecule NO (nitric oxide).

Nitroglycerin use for treatment of angina pain originated in the 1860s, more than 150 years ago. It came about from serendipitous observations. Factory workers involved in nitroglycerin production who suffered from angina pain noticed that their condition improved during the work week but then deteriorated on weekends. This regularly reported occurrence led physicians of the day to make the cause-effect connection that led to nitroglycerin use as a medication.

Alfred Nobel, who originally formulated what is now called dynamite, became a very rich man because of his discovery, amassing a fortune. It is his name and money that are associated with the prestigious Nobel Prizes awarded each year in the fall. In his later years, Alfred Nobel suffered from angina, and his doctor suggested that he use nitroglycerin to relieve the pain. He wrote to a friend, "Isn't it the irony of fate that I have been prescribed nitroglycerin? They call it Trinitrin so as not to scare the chemist and the public." He refused to take the medication. Further irony occurred in 1998 when the Nobel Prize in Physiology and Medicine was awarded to individuals involved in research on the body's generation of, as well as the biochemical effects of, NO in the human body.

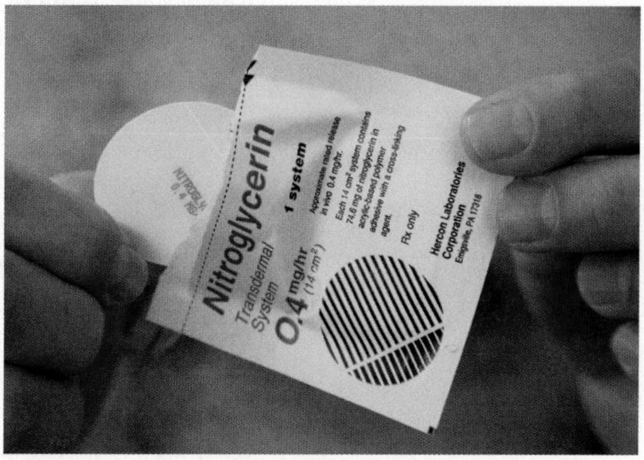

Custom Medical Stock Photo

Nitroglycerin is used in treating chest pains related to angina pectoris.

Phosphoric Acid Anhydrides

Three biologically important phosphoric acids exist: phosphoric acid, diphosphoric acid, and triphosphoric acid. Phosphoric acid, the simplest of the three acids, undergoes intermolecular dehydration to produce diphosphoric acid.

$$\underset{\text{Phosphoric acid}}{HO{-}\overset{\overset{\displaystyle O}{\|}}{\underset{\underset{\displaystyle OH}{|}}{P}}{-}OH} + \underset{\text{Phosphoric acid}}{HO{-}\overset{\overset{\displaystyle O}{\|}}{\underset{\underset{\displaystyle OH}{|}}{P}}{-}OH} \longrightarrow \underset{\text{Diphosphoric acid}}{HO{-}\overset{\overset{\displaystyle O}{\|}}{\underset{\underset{\displaystyle OH}{|}}{P}}{-}O{-}\overset{\overset{\displaystyle O}{\|}}{\underset{\underset{\displaystyle OH}{|}}{P}}{-}OH} + H_2O$$

Another intermolecular dehydration, involving diphosphoric acid and phosphoric acid, produces triphosphoric acid.

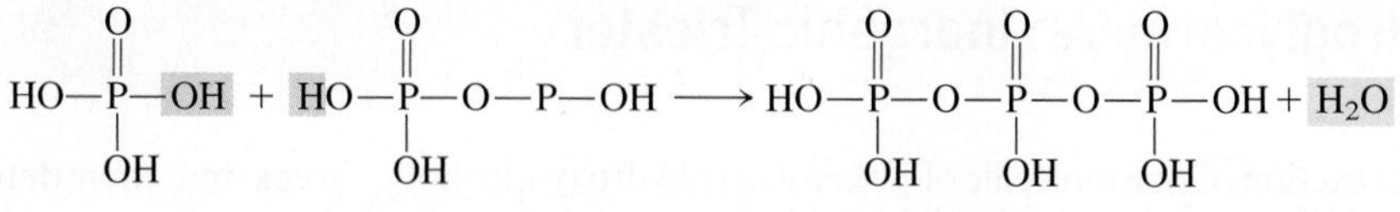

Triphosphoric acid

In the same manner that carboxylic acids are acidic (Section 5.7), phosphoric acid, diphosphoric acid, and triphosphoric acid are also acidic. The phosphoric acids are, however, polyprotic rather than monoprotic acids. The hydrogen atom in each of the —OH groups possesses acidic properties. All three phosphoric acids undergo esterification reactions with alcohols, producing species such as

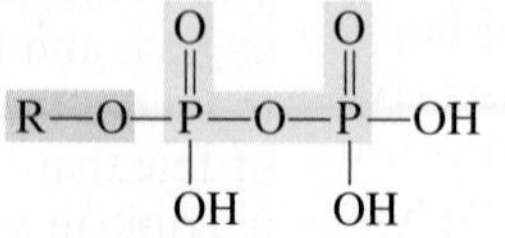

Diphosphate monoester

and

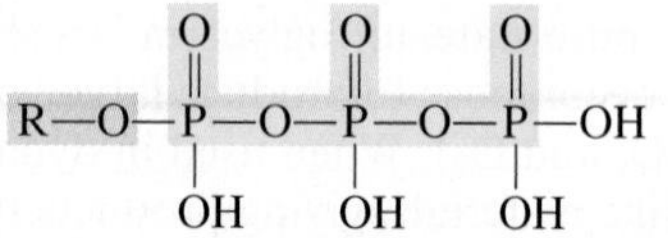

Triphosphate monoester

Esters of these types will be encountered in Chapter 12 when the biochemical production of energy in the human body is considered. Adenosine diphosphate (ADP) and adenosine triphosphate (ATP) are important examples of such compounds.

Diphosphoric acid and triphosphoric acid are phosphoric acid anhydrides as well as acids. Note the structural similarities between a carboxylic acid anhydride and diphosphoric acid.

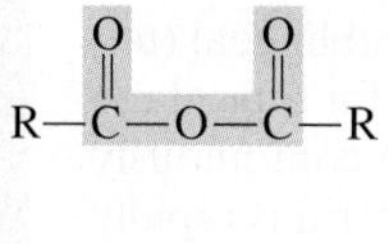

Carboxylic acid anhydride

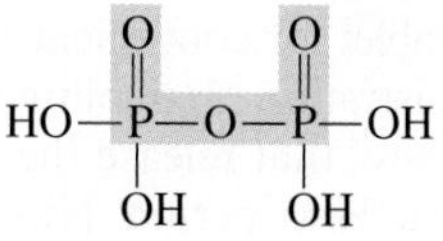

Diphosphoric acid

Phosphoric acid anhydride systems play important roles in cellular processes through which biochemical energy is produced. The presence of phosphoric anhydride systems in biological settings contrasts markedly with carboxylic acid anhydride systems (Section 5.18), which are not found in biological settings because of their reactivity with water.

Phosphorylation Reactions

In neutral solution and in alkaline solution (including most body fluids), phosphoric acid esters and anhydrides (as well as the esters and anhydrides of other inorganic acids) exist as ions rather than neutral molecules. Ion formation occurs because such molecules possess acidic hydrogen atoms, which are lost through ionization. The hydrogen atom present in each —OH group present in an inorganic ester or anhydride is acidic. The neutral and ionic forms of a monoester of a phosphoric acid monoester are

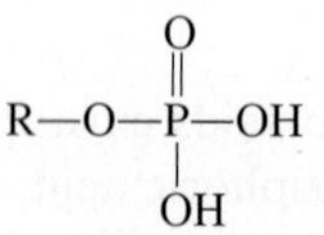

Neutral form

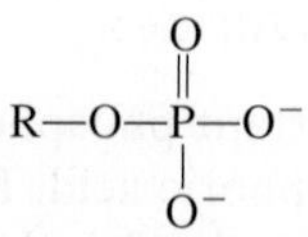

Ionic form

For a triphosphate monoester, the neutral and ionic forms are

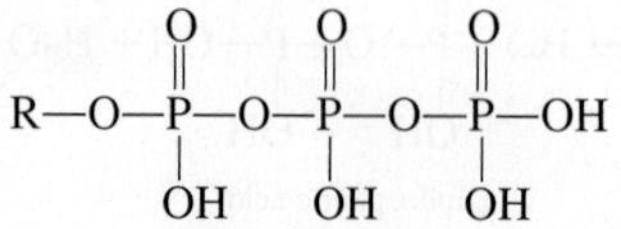

Neutral form

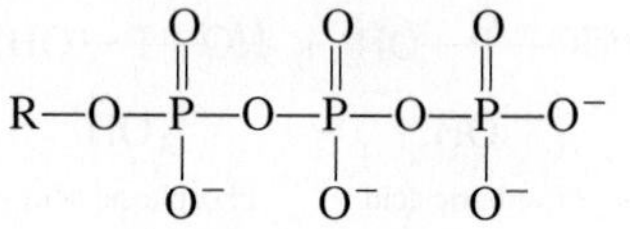

Ionic form

The terminal phosphate group in a phosphate-containing ester or anhydride, which has the formula PO_3^{2-}, is called a *phosphoryl group*. Note the location of the phosphoryl group in the following two structures.

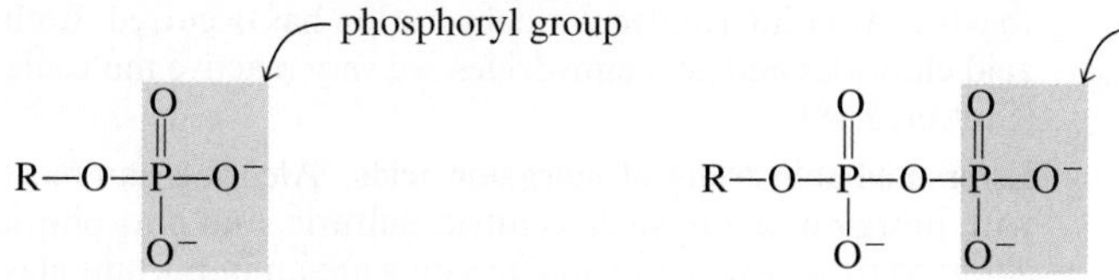

A **phosphoryl group** *is a* —PO_3^{2-} *group present in an organic molecule.*

A commonly encountered type of biochemical reaction is one that involves the transfer of a phosphoryl group from one organic molecule to another. This process is called *phosphorylation*. A **phosphorylation reaction** *is a chemical reaction in which a phosphoryl group is transferred from one molecule to another*. (Phosphorylation and acylation reactions are similar in that a group, phosphoryl or acyl, is transferred from one molecule to another.) The following structural equation typifies a phosphorylation reaction.

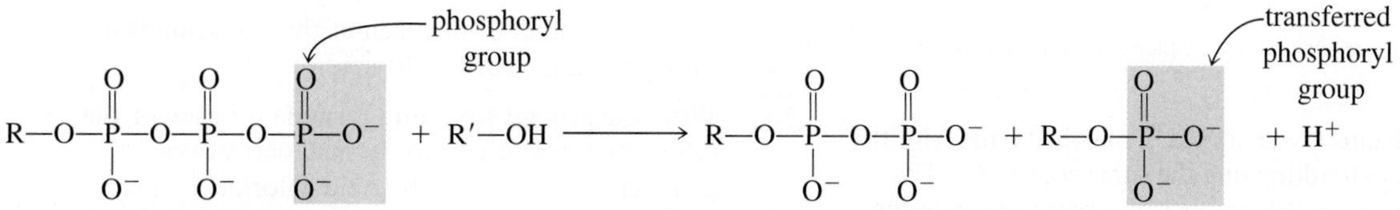

Triphosphate Diphosphate

The preceding kind of phosphorylation reaction, the conversion of a triphosphate to a diphosphate, will be encountered repeatedly in the biochemical chapters that deal with metabolic reactions (Chapters 12–15).

Concepts to Remember

OWL Sign in at **www.cengage.com/owl** to view tutorials and simulations, develop problem-solving skills, and complete online homework assigned by your professor.

The carboxyl group. The functional group present in carboxylic acids is the carboxyl group. A carboxyl group is composed of a hydroxyl group bonded to a carbonyl carbon atom. It thus contains two oxygen atoms directly bonded to the same carbon atom (Section 5.1).

Carboxylic acid derivatives. Four important families of carboxylic acid derivatives are esters, acid chlorides, acid anhydrides, and amides. The group attached to the carbonyl carbon atom distinguishes these derivatives from each other and also from carboxylic acids (Section 5.1).

Nomenclature of carboxylic acids. The IUPAC name for a monocarboxylic acid is formed by replacing the final *-e* of the hydrocarbon parent name with *-oic acid.* As with previous IUPAC nomenclature, the longest carbon chain containing the functional group is identified, and it is numbered starting with the carboxyl carbon atom. Common-name usage is more prevalent for carboxylic acids than for any other type of organic compound (Sections 5.2 and 5.3).

Types of carboxylic acids. Carboxylic acids are classified by the number of carboxyl groups present (monocarboxylic, dicarboxylic, etc.), by the degree of saturation (saturated, unsaturated, aromatic), and by additional functional groups present (hydroxy, keto, etc.) (Sections 5.4 and 5.5).

Physical properties of carboxylic acids. Low-molecular-mass carboxylic acids are liquids at room temperature and have sharp or unpleasant odors. Long-chain acids are waxlike solids. The carboxyl group is polar and forms hydrogen bonds to other carboxyl groups or other molecules. Thus carboxylic acids have relatively high boiling points, and those with lower molecular masses are soluble in water (Section 5.5).

Preparation of carboxylic acids. Carboxylic acids are synthesized through oxidation of primary alcohols or aldehydes using strong oxidizing agents. Aromatic carboxylic acids can be prepared by oxidizing a carbon side chain on a benzene derivative using a strong oxidizing agent (Section 5.6).

Acidity of carboxylic acids. Soluble carboxylic acids behave as weak acids, donating protons to water molecules. The portion of the acid molecule left after proton loss is called a carboxylate ion (Section 5.7).

Carboxylic acid salts. Carboxylic acids are neutralized by bases to produce carboxylic acid salts. Such salts are usually more soluble in water than are the acids from which they were derived. Carboxylic acid salts are named by changing the *-ic* ending of the acid to *-ate* (Section 5.8).

Esters. Esters are formed by the reaction of an acid with an alcohol. In such reactions, the —OR group from the alcohol replaces the —OH group in the carboxylic acid. Esters are polar compounds, but they cannot form hydrogen bonds to each other. Therefore, their boiling points are lower than those of alcohols and acids of similar molecular mass (Sections 5.9 and 5.10).

Nomenclature of esters. An ester is named as an alkyl (from the name of the alcohol reactant) carboxylate (from the name of the acid reactant) (Section 5.11).

Chemical reactions of esters. Esters can be converted back to carboxylic acids and alcohols under either acidic or basic conditions. Under acidic conditions, the process is called hydrolysis, and the products are the acid and alcohol. Under basic conditions, the process is called saponification, and the products are the acid salt and alcohol (Section 5.15).

Thioesters. Thioesters are sulfur-containing analogs of esters in which a —SR group has replaced the —OR group (Section 5.16).

Polyesters. Polyesters are polymers in which the monomers (diacids and dialcohols) are joined through ester linkages (Section 5.17).

Acid chlorides and acid anhydrides. An acid chloride is a carboxylic acid derivative in which the —OH portion of the carboxyl group has been replaced with a —Cl atom. An acid anhydride involves two carboxylic acid molecules bonded together after intermolecular dehydration has occurred. Both acid chlorides and acid anhydrides are very reactive molecules (Section 5.18).

Esters and anhydrides of inorganic acids. Alcohols can react with inorganic acids, such as nitric, sulfuric, and phosphoric acids, to form esters. Phosphate esters are an important class of biochemical compounds. Anhydrides of phosphoric acid (diphosphoric acid and triphosphoric acid) and their esters are also important types of biochemical molecules (Section 5.19).

Exercises and Problems

OWL Interactive versions of these problems may be assigned in OWL.

Exercises and problems are arranged in matched pairs with the two members of a pair addressing the same concept(s). The answer to the odd-numbered member of a pair is given at the back of the book. Problems denoted with a ▲ involve concepts found not only in the section under consideration but also concepts found in one or more earlier sections of the chapter. Problems denoted with a ● cover concepts found in a Chemical Connections feature box.

Structure of Carboxylic Acids and Their Derivatives (Section 5.1)

5.1 In which of the following compounds is a carboxyl group present?

a. $CH_3{-}CH_2{-}\overset{\overset{O}{\|}}{C}{-}OH$

b. $CH_3{-}CH_2{-}CH_2{-}\overset{\overset{O}{\|}}{C}{-}O{-}CH_3$

c. $CH_3{-}\overset{\overset{OH}{|}}{CH}{-}\overset{\overset{O}{\|}}{C}{-}CH_3$

d. $CH_3{-}CH_2{-}COOH$

5.2 In which of the following compounds is a carboxyl group present?

a. $CH_3{-}CH_2{-}\overset{\overset{O}{\|}}{C}{-}O{-}CH_3$

b. $CH_3{-}CH_2{-}\overset{\overset{O}{\|}}{C}{-}OH$

c. 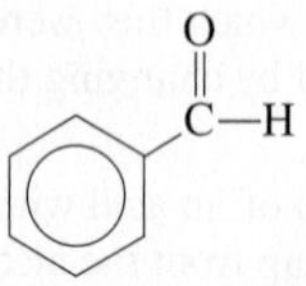

d. $CH_2{-}\underset{\underset{CH_3}{|}}{CH}{-}CH_2{-}CH_2{-}CO_2H$

5.3 Indicate whether or not each of the compounds in Problem 5.1 is a carboxylic acid.

5.4 Indicate whether or not each of the compounds in Problem 5.2 is a carboxylic acid.

5.5 Write the general structural formula for each of the following types of carboxylic acid derivatives.

a. Ester b. Acid chloride
c. Amide d. Acid anhydride

5.6 Write the general structural formula for the functional group present in each of the following carboxylic acid derivatives.

a. Amide b. Ester
c. Acid anhydride d. Acid chloride

5.7 For each of the following types of compounds, all of which have the general formula $R{-}\overset{\overset{O}{\|}}{C}{-}Z$, indicate whether or not the "C—Z bond" is polar or nonpolar.

a. $R{-}\overset{\overset{O}{\|}}{C}{-}NH_2$ **b.** $R{-}\overset{\overset{O}{\|}}{C}{-}CH_3$

c. $R{-}\overset{\overset{O}{\|}}{C}{-}O{-}\overset{\overset{O}{\|}}{C}{-}CH_3$ **d.** $R{-}\overset{\overset{O}{\|}}{C}{-}O{-}CH_3$

5.8 For each of the following types of compounds, all of which have the general formula $R{-}\overset{\overset{O}{\|}}{C}{-}Z$, indicate whether or not the "C—Z bond" is polar or nonpolar.

a. $R{-}\overset{\overset{O}{\|}}{C}{-}OH$ **b.** $R{-}\overset{\overset{O}{\|}}{C}{-}H$

c. $R{-}\overset{\overset{O}{\|}}{C}{-}Cl$ **d.** $R{-}\overset{\overset{O}{\|}}{C}{-}O{-}CH_2{-}CH_3$

5.9 Classify each of the compounds in Problem 5.7 as a *carbonyl compound* or an *acyl compound.*

5.10 Classify each of the compounds in Problem 5.8 as a *carbonyl compound* or an *acyl compound.*

5.11 Which of the designations *carboxyl group present, carbonyl group present*, and *acyl group present* applies to each of the following types of compounds? More than one designation may be correct in a given situation, or only one or none of the designations may apply.

a. Amide b. Carboxylic acid
c. Aldehyde d. Ester

5.12 Which of the designations *carboxyl group present*, *carbonyl group present*, and *acyl group present* applies to each of the following types of compounds? More than one designation may be correct in a given situation, or only one or none of the designations may apply.
a. Acid chloride b. Ketone
c. Acid anhydride d. Ether

IUPAC Nomenclature for Carboxylic Acids (Section 5.2)

5.13 Give the IUPAC name for each of the following carboxylic acids.

a. 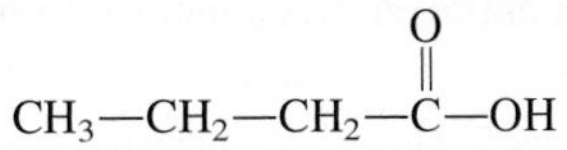

b. 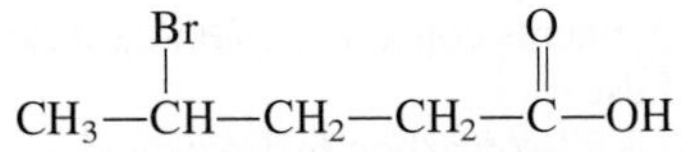

c. 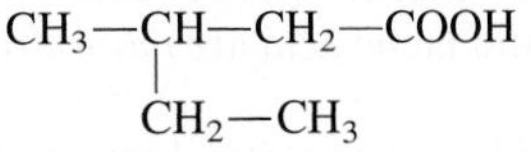

d. 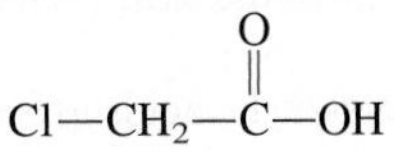

5.14 Give the IUPAC name for each of the following carboxylic acids.

a. 

b. 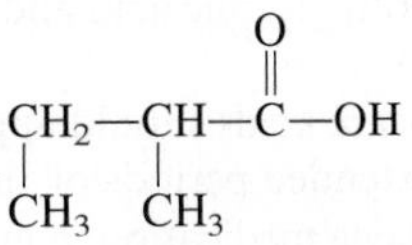

c. 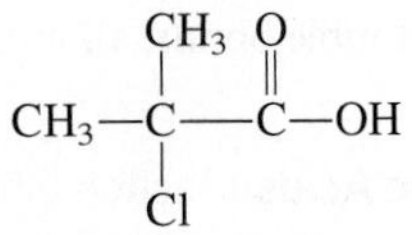

d. $HOOC—CH_3$

5.15 Assign an IUPAC name to each of the following carboxylic acids.

a. 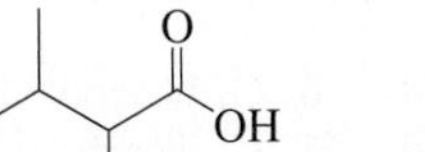
b. 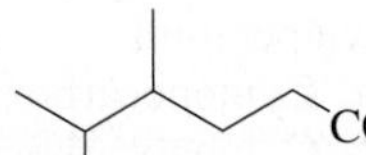
c. 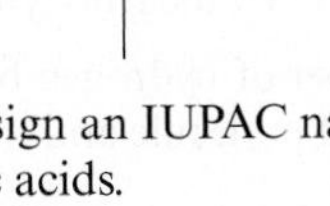
d. 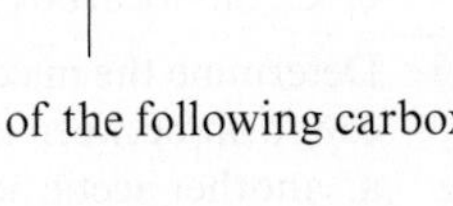

5.16 Assign an IUPAC name to each of the following carboxylic acids.

a. 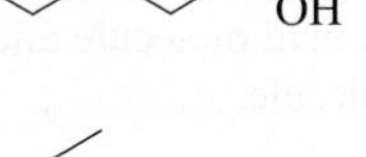
b. 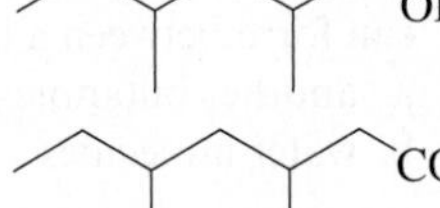
c. 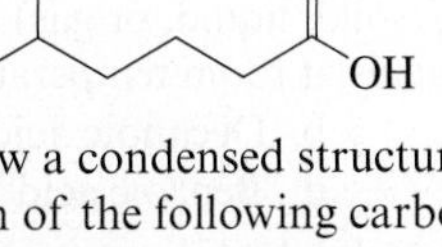
d. 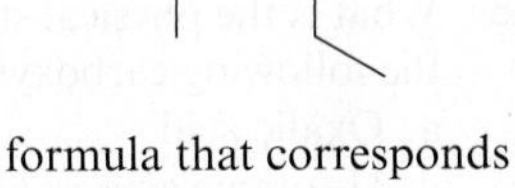

5.17 Draw a condensed structural formula that corresponds to each of the following carboxylic acids.
a. 2-Ethylbutanoic acid
b. 2,5-Dimethylhexanoic acid
c. Methylpropanoic acid
d. Dichloroethanoic acid

5.18 Draw a condensed structural formula that corresponds to each of the following carboxylic acids.
a. 3,3-Dimethylheptanoic acid
b. 4-Methylpentanoic acid
c. 3-Chloropropanoic acid
d. Trichloroethanoic acid

5.19 Give the IUPAC name for each of the following carboxylic acids.

a. 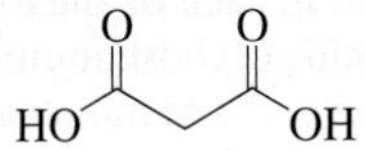

b. 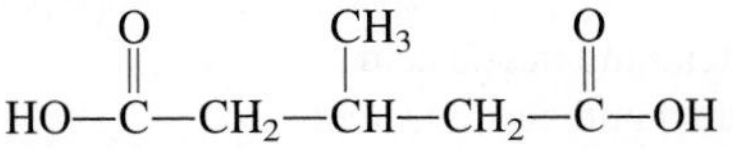

c. 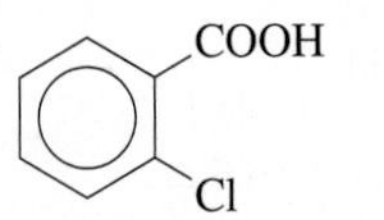
d. 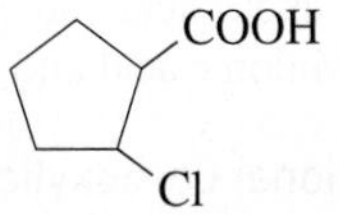

5.20 Give the IUPAC name for each of the following carboxylic acids.

a. 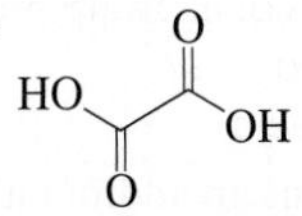

b. 

c. 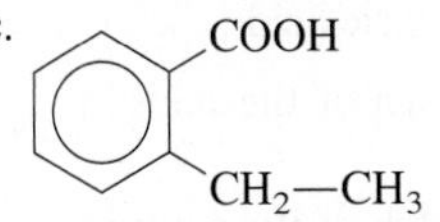
d. 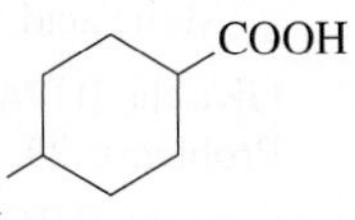

5.21 Draw a condensed structural formula that corresponds to each of the following carboxylic acids.
a. 2,2-Dimethylbutanoic acid
b. 2,2-Dimethylbutanedioic acid
c. 2,2-Dimethylcyclohexanecarboxylic acid
d. 2,4-Dichlorobenzoic acid

5.22 Draw a condensed structural formula that corresponds to each of the following carboxylic acids.
a. 2,3-Dichlorohexanoic acid
b. 2,3-Dichlorohexanedioic acid
c. 2,3-Dichlorocyclopentanecarboxylic acid
d. 3,5-Dichlorobenzoic acid

Common Names for Carboxylic Acids (Section 5.3)

5.23 Draw a condensed structural formula that corresponds to each of the following monocarboxylic acids.
a. Valeric acid b. Acetic acid
c. α-Chlorobutyric acid d. β-Bromocaproic acid

5.24 Draw a condensed structural formula that corresponds to each of the following monocarboxylic acids.
a. Butyric acid
b. Caproic acid
c. α-Methylpropionic acid
d. β-Chloro-β-iodocaproic acid

5.25 Draw a condensed structural formula that corresponds to each of the following dicarboxylic acids.
a. Malonic acid b. Succinic acid
c. γ-Bromopimelic acid d. α-Methylglutaric acid

5.26 Draw a condensed structural formula that corresponds to each of the following dicarboxylic acids.
a. Glutaric acid b. Pimelic acid
c. α-Methyladipic acid d. α,β-Dichlorosuccinic acid

5.27 Classify the two carboxylic acids in each of the following pairs as (1) both dicarboxylic acids (2) both monocarboxylic acids or (3) one dicarboxylic and one monocarboxylic acid.
a. Glutaric acid and valeric acid
b. Adipic acid and oxalic acid
c. Caproic acid and formic acid
d. Succinic acid and malonic acid

5.28 Classify the two carboxylic acids in each of the following pairs as (1) both dicarboxylic acids (2) both monocarboxylic acids or (3) one dicarboxylic and one monocarboxylic acid.
a. Formic acid and acetic acid
b. Butyric acid and succinic acid
c. Pimelic acid and caproic acid
d. Malonic acid and adipic acid

Polyfunctional Carboxylic Acids (Section 5.4)

5.29 Each of the following acids contains an additional type of functional group besides the carboxyl group. For each acid, specify the noncarboxyl functional group present.
a. Acrylic acid b. Lactic acid
c. Maleic acid d. Glycolic acid

5.30 Each of the following acids contains an additional type of functional group besides the carboxyl group. For each acid, specify the noncarboxyl functional group present.
a. Fumaric acid b. Pyruvic acid
c. Malic acid d. Tartaric acid

5.31 Give the IUPAC name for each of the acids in Problem 5.29.

5.32 Give the IUPAC name for each of the acids in Problem 5.30.

5.33 Draw a structural formula for each of the following acids.
a. 3-Oxopentanoic acid
b. 2-Hydroxybutanoic acid
c. *trans*-4-Hexenoic acid
d. α,β-Dihydroxyglutaric acid

5.34 Draw a structural formula for each of the following acids.
a. 3-Hydroxypentanoic acid
b. α,γ-Dihydroxyvaleric acid
c. 2-Oxobutanoic acid
d. *cis*-3-Heptenoic acid

▲5.35 Classify each of the following carboxylic acids as (1) an unsaturated acid (2) a hydroxy acid (3) a keto acid (4) a monocarboxylic acid or (5) a dicarboxylic acid. More than one response may apply for a given acid.
a. Acrylic acid b. Maleic acid
c. Succinic acid d. Caproic acid

▲5.36 Classify each of the following carboxylic acids as (1) an unsaturated acid (2) a hydroxy acid (3) a keto acid (4) a monocarboxylic acid or (5) a dicarboxylic acid. More than one response may apply for a given acid.
a. Lactic acid b. Glutaric acid
c. Valeric acid d. Pyruvic acid

●5.37 (Chemical Connections 5-A) Indicate whether each of the following statements relating to the over-the-counter medications ibuprofen and naproxen is true or false.
a. Both ibuprofen and naproxen have antipyretic properties.
b. Advil and Aleve are two brand names for ibuprofen.
c. A benzene ring system is part of the structure of naproxen.
d. The ring system in ibuprofen has an isobutyl group attachment.

●5.38 (Chemical Connections 5-A) Indicate whether each of the following statements relating to the over-the-counter medications ibuprofen and naproxen is true or false.
a. Both ibuprofen and naproxen have anti-inflammatory properties.
b. Motrin and Anaprox are two brand names for naproxen.
c. A naphthalene ring system is part of the structure of ibuprofen.
d. The ring system in naproxen has a methoxy group attachment.

●5.39 (Chemical Connections 5-B) Indicate whether each of the following statements concerning skin-care carboxylic acids is true or false.
a. Glycolic acid is a two-carbon hydroxy acid.
b. Both glycolic acid and lactic acid are naturally occurring substances.
c. Use of skin-care acids increases the skin's sensitivity to ultraviolet light.
d. Polyunsaturated hydroxycarboxylic acids are used in the treatment of severe acne.

●5.40 (Chemical Connections 5-B) Indicate whether each of the following statements concerning skin-care carboxylic acids is true or false.
a. Lactic acid is a four-carbon hydroxy acid.
b. At high concentrations, both glycolic acid and lactic acid cause skin irritation.
c. Individuals who use skin-care acids should apply sunscreen when outside for extended periods of time.
d. Accutane, a prescription acne medication, contains five carbon-carbon double bonds, all in the *trans*-configuration.

Physical Properties of Carboxylic Acids (Section 5.5)

5.41 Indicate whether each of the following unbranched saturated carboxylic acids is a solid, liquid, or gas at room temperature.
a. C_4 monocarboxylic acid b. C_4 dicarboxylic acid
c. C_5 monocarboxylic acid d. C_6 dicarboxylic acid

5.42 Indicate whether each of the following unbranched saturated carboxylic acids is a solid, liquid, or gas at room temperature.
a. C_3 monocarboxylic acid b. C_3 dicarboxylic acid
c. C_6 monocarboxylic acid d. C_5 dicarboxylic acid

5.43 Determine the maximum number of hydrogen bonds that can form between an acetic acid molecule and
a. another acetic acid molecule
b. water molecules

5.44 Determine the maximum number of hydrogen bonds that can form between a butanoic acid molecule and
a. another butanoic acid molecule
b. water molecules

5.45 What is the physical state (solid, liquid, or gas) of each of the following carboxylic acids at room temperature?
a. Oxalic acid b. Decanoic acid
c. Hexanoic acid d. Benzoic acid

5.46 What is the physical state (solid, liquid, or gas) of each of the following carboxylic acids at room temperature?
a. Succinic acid b. Octanoic acid
c. Pentanoic acid d. *p*-Chlorobenzoic acid

Preparation of Carboxylic Acids (Section 5.6)

5.47 Draw a structural formula for the carboxylic acid expected to be formed when each of the following substances is oxidized using a strong oxidizing agent.

a. CH_3-CH_2-OH

b. $CH_3-\overset{O}{\overset{\|}{C}}-H$

c. $CH_3-CH_2-\underset{|}{\overset{CH_3}{CH}}-CH_2-\overset{O}{\overset{\|}{C}}-H$

d. 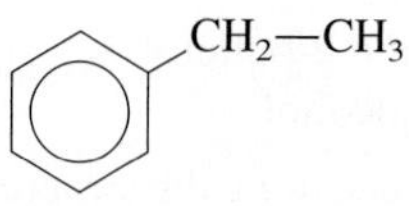

5.48 Draw a structural formula for the carboxylic acid expected to be formed when each of the following substances is oxidized using a strong oxidizing agent.

a. $CH_3-CH_2-\overset{O}{\overset{\|}{C}}-H$

b. $CH_3-CH_2-CH_2-OH$

c. $CH_3-\underset{CH_3}{\underset{|}{CH}}-\underset{CH_3}{\underset{|}{CH}}-\overset{O}{\overset{\|}{C}}-H$

d. 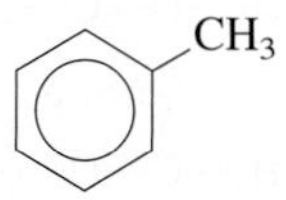

Acidity of Carboxylic Acids (Section 5.7)

5.49 How many *acidic* hydrogen atoms are present in each of the following carboxylic acids?

a. Pentanoic acid b. Citric acid
c. Succinic acid d. Oxalic acid

5.50 How many *acidic* hydrogen atoms are present in each of the following carboxylic acids?

a. Acetic acid b. Benzoic acid
c. Propanoic acid d. Glutaric acid

5.51 What is the charge on the carboxylate ion formed when each of the acids in Problem 5.49 ionizes in water?

5.52 What is the charge on the carboxylate ion formed when each of the acids in Problem 5.50 ionizes in water?

5.53 What is the name of the carboxylate ion that forms when each of the acids in Problem 5.49 ionizes in water? (Use an IUPAC carboxylate name if the acid name is IUPAC; use a common name if the acid name is common.)

5.54 What is the name of the carboxylate ion that forms when each of the acids in Problem 5.50 ionizes in water? (Use an IUPAC carboxylate name if the acid name is IUPAC; use a common name if the acid name is common.)

5.55 Write a chemical equation for the formation of each of the following carboxylate ions, in aqueous solution, from its parent acid.

a. Acetate b. Citrate
c. Ethanoate d. 2-Methylbutanoate

5.56 Write a chemical equation for the formation of each of the following carboxylate ions, in aqueous solution, from its parent acid.

a. Butanoate b. Succinate
c. Benzoate d. α-Methylbutyrate

Carboxylic Acid Salts (Section 5.8)

5.57 Give the IUPAC name for each of the following carboxylic acid salts.

a. $CH_3-\overset{O}{\overset{\|}{C}}-O^-\ K^+$

b. $\left(CH_3-CH_2-\overset{O}{\overset{\|}{C}}-O^-\right)_2 Ca^{2+}$

c. $K^{+\,-}O-\overset{O}{\overset{\|}{C}}-CH_2-CH_2-\overset{O}{\overset{\|}{C}}-O^-\ K^+$

d. 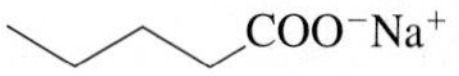

5.58 Give the IUPAC name for each of the following carboxylic acid salts.

a. $\left(H-\overset{O}{\overset{\|}{C}}-O^-\right)_2 Ca^{2+}$

b. $CH_3-CH_2-\overset{O}{\overset{\|}{C}}-O^-\ Na^+$

c. 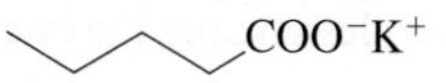

d. 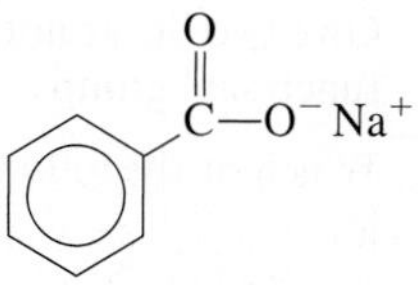

5.59 Give the common name for each of the carboxylic acid salts in Problem 5.57.

5.60 Give the common name for each of the carboxylic acid salts in Problem 5.58.

5.61 Write a chemical equation for the preparation of each of the salts in Problem 5.57 using an acid–base neutralization reaction.

5.62 Write a chemical equation for the preparation of each of the salts in Problem 5.58 using an acid–base neutralization reaction.

5.63 Write a chemical equation for the conversion of each of the following carboxylic acid salts to its parent carboxylic acid. Use hydrochloric acid (HCl) as the source of the needed hydronium ions.

a. Sodium butanoate b. Potassium oxalate
c. Calcium malonate d. Sodium benzoate

5.64 Write a chemical equation for the conversion of each of the following carboxylic acid salts to its parent carboxylic acid. Use hydrochloric acid (HCl) as the source of the needed hydronium ions.

a. Calcium propanoate b. Sodium lactate
c. Magnesium succinate d. Potassium benzoate

5.65 What is an antimicrobial?

5.66 Which three carboxylic acids have salts that are used extensively as food preservatives?

5.67 Which carboxylic acid has salts that are used to inhibit yeast and mold growth in the following?

a. Jams and jellies b. Dairy products
c. Dried fruit d. Cheese spreads

5.68 Which carboxylic acid has salts that are used to inhibit yeast and mold growth in the following?

a. Ketchup and syrup b. Sauerkraut
c. Pie filling d. Baked products

▲5.69 Classify each of the following substances as (1) a salt of a monocarboxylic acid or (2) a salt of a dicarboxylic acid.

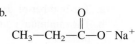

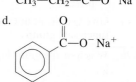

a. Sodium formate b. Potassium pentanedioate
c. Calcium oxalate d. Magnesium propanoate

▲5.70 Classify each of the following substances as (1) a salt of a monocarboxylic acid or (2) a salt of a dicarboxylic acid.
a. Potassium ethanoate
b. Magnesium acetate
c. Sodium succinate
d. Calcium butanedioate

▲5.71 How many ions are present in one formula unit of each of the compounds in Problem 5.69?

▲5.72 How many ions are present in one formula unit of each of the compounds in Problem 5.70?

Structure of Esters (Section 5.9)

5.73 Characterize an ester functional group in terms of the following.
a. The number of oxygen atoms that are present
b. The minimum number of carbon atoms that must be present

5.74 Give two linear notations for designating the ester functional group.

5.75 Which of the following structures represent esters?

a. $CH_3-CH_2-CH_2-\overset{O}{\overset{\|}{C}}-O-CH_3$

b. $CH_3-O-\overset{O}{\overset{\|}{C}}-CH_3$

c. $CH_3-O-CH_2-\overset{O}{\overset{\|}{C}}-CH_3$

d. 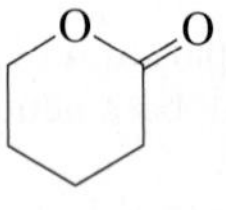

5.76 Which of the following structures represent esters?

a. $CH_3-\overset{CH_3}{\overset{|}{CH}}-\overset{O}{\overset{\|}{C}}-O-CH_3$

b. $CH_3-CH_2-O-\overset{O}{\overset{\|}{C}}-CH_2-CH_3$

c. 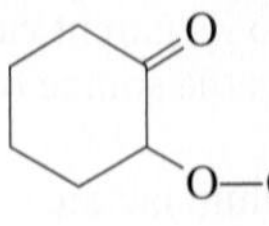

d. 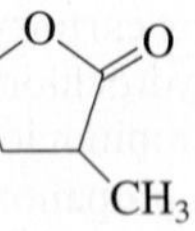

▲5.77 Classify each of the following substances as (1) an ester (2) a carboxylic acid or (3) a carboxylic acid salt.

a. $CH_3-\overset{O}{\overset{\|}{C}}-O-H$ b. $CH_3-\overset{O}{\overset{\|}{C}}-O-CH_3$

c. $CH_3-\overset{O}{\overset{\|}{C}}-O^-Na^+$ d. CH_3-COOH

▲5.78 Classify each of the following substances as (1) an ester (2) a carboxylic acid or (3) a carboxylic acid salt.

a. $H-\overset{O}{\overset{\|}{C}}-O-CH_3$ b. $H-\overset{O}{\overset{\|}{C}}-O^-K^+$

c. $H-COOH$ d. $H-\overset{O}{\overset{\|}{C}}-O-H$

Preparation of Esters (Section 5.10)

5.79 Draw the structure of the ester produced when each of the following pairs of carboxylic acid and alcohol react.
a. Propanoic acid and methanol
b. Acetic acid and 1-propanol
c. 2-Methylbutanoic acid and 2-propanol
d. Valeric acid and *sec*-butyl alcohol

5.80 Draw the structure of the ester produced when each of the following pairs of carboxylic acid and alcohol react.
a. Methanoic acid and 1-propanol
b. Propanoic acid and ethanol
c. 2-Methylpropanoic acid and 2-butanol
d. Valeric acid and isobutyl alcohol

5.81 For each of the following esters, draw the structural formula of the "parent" acid and the "parent" alcohol.

a. $CH_3-CH_2-\overset{O}{\overset{\|}{C}}-O-CH_2-CH_3$

b. $CH_3-O-\overset{O}{\overset{\|}{C}}-CH_2-CH_2-CH_3$

c. 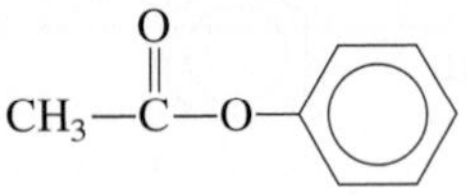

d. 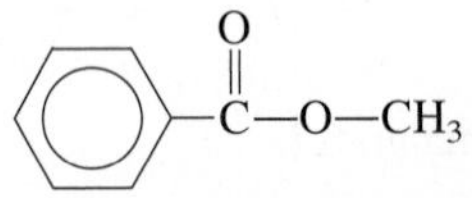

5.82 For each of the following esters, draw the structural formula of the "parent" acid and the "parent" alcohol.

a. $CH_3-\overset{O}{\overset{\|}{C}}-O-CH_2-CH_3$

b. $CH_3-O-\overset{O}{\overset{\|}{C}}-CH_2-CH_3$

c. 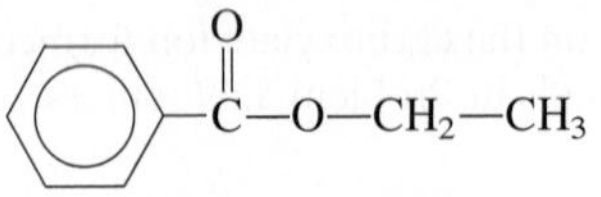

d. 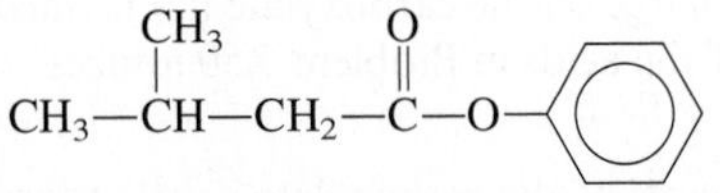

5.83 What are lactones, and by what chemical reaction are they produced?

5.84 What is the difference between a γ-lactone and a δ-lactone?

Nomenclature for Esters (Section 5.11)

5.85 Assign an IUPAC name to each of the following esters.

a. $CH_3-CH_2-\overset{O}{\overset{\|}{C}}-O-CH_3$

b. $H-\overset{O}{\overset{\|}{C}}-O-CH_3$

c. $CH_3-CH_2-CH_2-O-\overset{O}{\overset{\|}{C}}-CH_3$

d. $CH_3-CH_2-\overset{O}{\overset{\|}{C}}-O-\overset{CH_3}{\overset{|}{CH}}-CH_3$

5.86 Assign an IUPAC name to each of the following esters.

a.
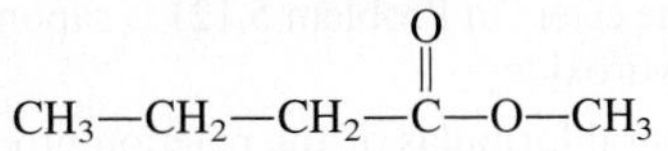

b.
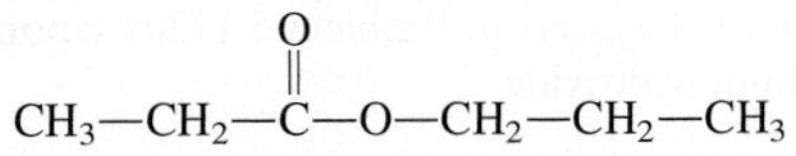

c.
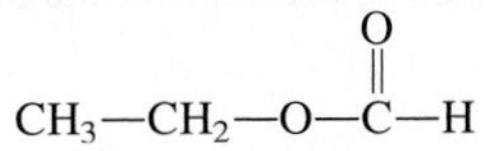

d.
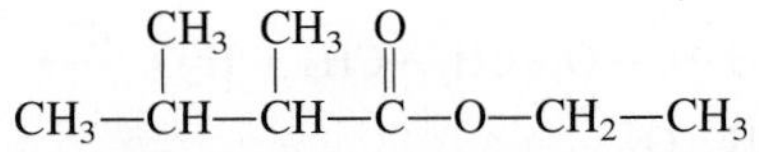

5.87 Assign a common name to each of the esters in Problem 5.85.

5.88 Assign a common name to each of the esters in Problem 5.86.

5.89 Assign an IUPAC name to each of the following esters.

a.
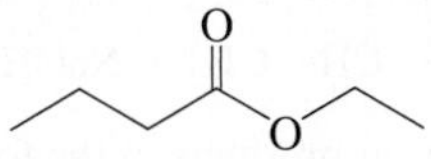

b.
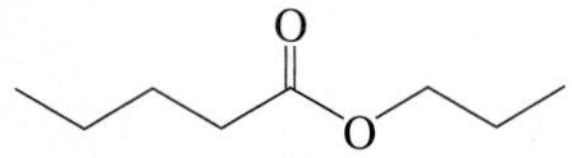

c.
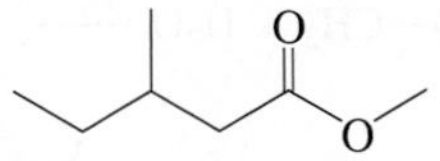

d.
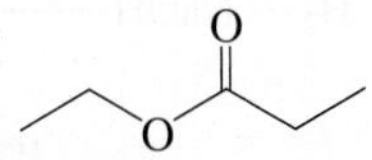

5.90 Assign an IUPAC name to each of the following esters.

a.
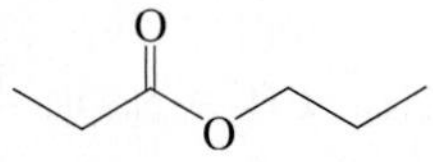

b.
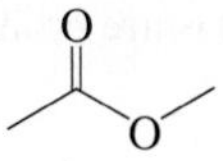

c.
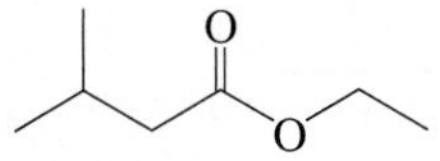

d.
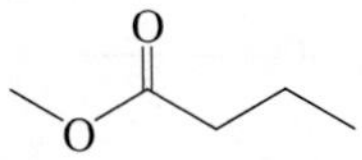

5.91 Draw a structural formula for each of the following esters.
a. Methyl formate
b. Ethyl phenylacetate
c. Isopropyl acetate
d. 2-Bromopropyl ethanoate

5.92 Draw a structural formula for each of the following esters.
a. Ethyl butyrate
b. Butyl ethanoate
c. 2-Methylpropyl formate
d. Ethyl α-methylpropionate

5.93 Assign IUPAC names to the esters that are produced from the reaction of the following carboxylic acids and alcohols.
a. Acetic acid and ethanol
b. Ethanoic acid and methanol
c. Butyric acid and ethyl alcohol
d. 2-Butanol and caproic acid

5.94 Assign IUPAC names to the esters that are produced from the reaction of the following carboxylic acids and alcohols.
a. Ethanoic acid and propyl alcohol
b. Acetic acid and 1-pentanol
c. Acetic acid and 2-pentanol
d. Ethanol and benzoic acid

▲5.95 Classify each of the following compounds as (1) an ester (2) a carboxylic acid or (3) a carboxylic acid salt.
a. Methyl succinate
b. Sodium succinate
c. Methyl butanoate
d. Methyl 2-methylbutanoate

▲5.96 Classify each of the following compounds as (1) an ester (2) a carboxylic acid or (3) a carboxylic acid salt.
a. Ethyl formate
b. Ethyl ethanoate
c. Potassium ethanoate
d. Potassium ethanedioate

▲5.97 How many carbon atoms are present in a molecule of each of the compounds in Problem 5.95?

▲5.98 How many carbon atoms are present in a molecule of each of the compounds in Problem 5.96?

Selected Common Esters (Section 5.12)

5.99 Structurally contrast the ester flavoring agents that generate apple and pineapple flavors in terms of their "acid part" and "alcohol part."

5.100 Structurally contrast the ester flavoring agents that generate pear and banana flavors in terms of their "acid part" and "alcohol part."

5.101 How does the structure of aspirin differ from that of salicylic acid?

5.102 How does the structure of oil of wintergreen differ from that of salicylic acid?

●5.103 Why is the lactone nepetalactone, present in the catnip plant, not considered a pheromone for cats?

●5.104 What is the original source of the large-ring lactone antibiotic erythromycin?

●5.105 (Chemical Connections 5-C) Indicate whether each of the following statements relating to aspirin and its effects on the human body is true or false.
a. A "regular" aspirin tablet contains 500 mg of aspirin held together with an inert starch binder.
b. Within the human body, aspirin undergoes hydrolysis to produce two different carboxylic acids.
c. Salicylic acid, rather than aspirin itself, is the "active ingredient" associated with aspirin use.
d. Aspirin increases the time it takes for blood to coagulate.

●5.106 (Chemical Connections 5-C) Indicate whether each of the following statements relating to aspirin and its effects on the human body is true or false.
a. A "baby" aspirin tablet contains 81 mg of aspirin held together with an inert starch binder.
b. Structurally, aspirin is both an ester and a carboxylic acid.
c. Salicylic acid, produced through aspirin hydrolysis, can cause stomach irritation.
d. Aspirin exerts its pain-relieving effects by inhibiting the synthesis of hormones called prostaglandins.

Isomerism for Carboxylic Acids and Esters (Section 5.13)

5.107 Give IUPAC names for the four isomeric C_5 monocarboxylic acids with saturated carbon chains.

5.108 Give IUPAC names for the eight isomeric C_6 monocarboxylic acids with saturated carbon chains.

5.109 Give IUPAC names for the four isomeric methyl esters that contain six carbon atoms and saturated carbon chains.

5.110 Give IUPAC names for the two isomeric ethyl esters that contain six carbon atoms and saturated carbon chains.

5.111 How many esters exist that are isomeric with 2-methylbutanoic acid?

5.112 How many esters exist that are isomeric with 2-methylpropanoic acid?

5.113 Draw condensed structural formulas for all carboxylic acids and all esters that have the molecular formula $C_3H_6O_2$.

5.114 Draw condensed structural formulas for all carboxylic acids and all esters that have the molecular formula $C_4H_8O_2$.

▲5.115 The general molecular formula for an alkane is C_nH_{2n}. What is the general molecular formula for an ester in which all carbon-carbon bonds are single bonds?

▲5.116 How does the general molecular formula for an ester compare to that for a carboxylic acid, when all carbon-carbon bonds in both types of compounds are single bonds?

Physical Properties of Esters (Section 5.14)

5.117 Explain why ester molecules cannot form hydrogen bonds to each other.

5.118 How many hydrogen bonds can form between a methyl acetate molecule and two water molecules?

5.119 Explain why esters have lower boiling points than carboxylic acids of comparable molecular mass.

5.120 Explain why esters are less soluble in water than carboxylic acids of comparable molecular mass.

Chemical Reactions of Esters (Section 5.15)

5.121 Write the structural formulas of the reaction products when each of the following esters is hydrolyzed under acidic conditions.

a. $CH_3-CH_2-C(=O)-O-CH_2-CH_3$

b. 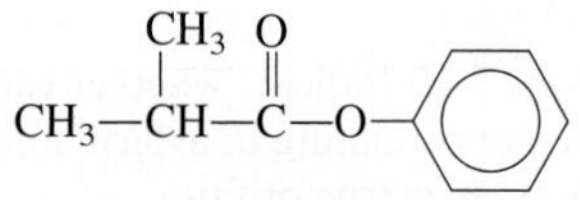

c. Methyl butanoate d. Isopropyl benzoate

5.122 Write the structural formulas of the reaction products when each of the following esters is hydrolyzed under acidic conditions.

a. $CH_3-CH(CH_3)-C(=O)-O-CH(CH_3)-CH_3$

b. 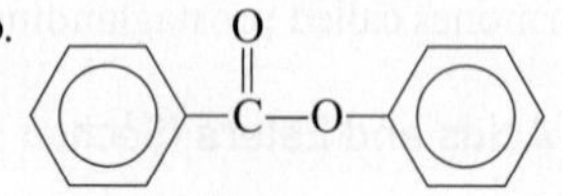

c. Ethyl valerate d. Pentyl benzoate

5.123 Write the structural formulas of the reaction products when each of the esters in Problem 5.121 is saponified using sodium hydroxide.

5.124 Write the structural formulas of the reaction products when each of the esters in Problem 5.122 is saponified using sodium hydroxide.

5.125 Draw structures of the reaction products in the following chemical reactions.

a. $CH_3-CH(CH_3)-C(=O)-O-CH_2-CH_3 + H_2O \xrightarrow{H^+}$

b. $CH_3-CH(CH_3)-C(=O)-O-CH_2-CH_3 + NaOH \xrightarrow{H_2O}$

c. $H-C(=O)-O-CH_2-CH_2-CH_2-CH_3 + H_2O \xrightarrow{H^+}$

d. $CH_3-C(=O)-O-CH_2-CH(CH_3)-CH(CH_3)-CH_3 + NaOH \xrightarrow{H_2O}$

5.126 Draw structures of the reaction products in the following chemical reactions.

a. $CH_3-CH(CH_3)-CH_2-C(=O)-O-CH_3 + H_2O \xrightarrow{H^+}$

b. $CH_3-CH(CH_3)-CH_2-C(=O)-O-CH_3 + NaOH \xrightarrow{H_2O}$

c. $CH_3-CH_2-C(=O)-O-(CH_2)_5-CH_3 + H_2O \xrightarrow{H^+}$

d. $CH_3-(CH_2)_5-C(=O)-O-CH_2-CH_3 + NaOH \xrightarrow{H_2O}$

▲5.127 What type(s) of organic compound(s) is/are produced in the following chemical reactions?

a. $CH_3-C(=O)-OH + NaOH \longrightarrow$

b. $CH_3-CH_2-C(=O)-OH + CH_3-OH \xrightarrow{H^+}$

c. $CH_3-C(=O)-O-CH_3 + H_2O \xrightarrow{H^+}$

d. $CH_3-CH_2-C(=O)-O-CH_3 + NaOH \xrightarrow{H_2O}$

▲5.128 What type(s) of organic compound(s) is/are produced in the following chemical reactions?

a. $CH_3-C(=O)-OH + CH_3-OH \xrightarrow{H^+}$

b. $CH_3-CH_2-C(=O)-OH + NaOH \longrightarrow$

c. $CH_3-C(=O)-O-CH_3 + NaOH \xrightarrow{H_2O}$

d. $CH_3-CH_2-C(=O)-O-CH_3 + H_2O \xrightarrow{H^+}$

Sulfur Analogs of Esters (Section 5.17)

5.129 Draw the structures of the thioesters formed as a result of each of the following reactions between carboxylic acids and thiols.

a.
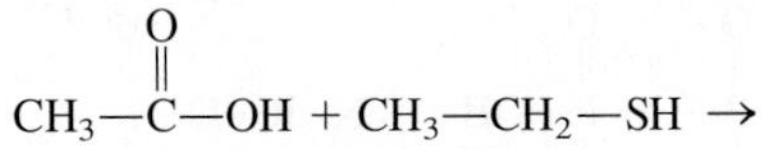

b.
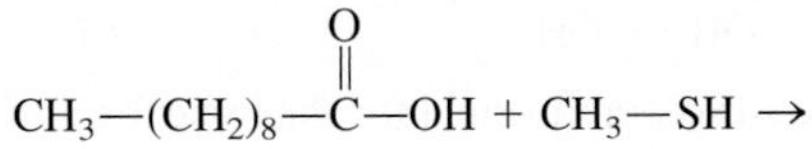

c.
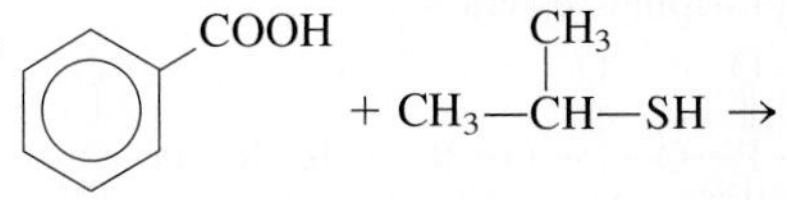

d.
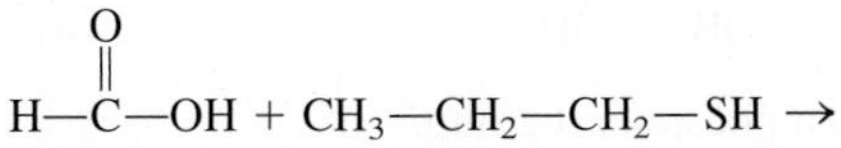

5.130 Draw the structures of the thioesters formed as a result of each of the following reactions between carboxylic acids and thiols.

a.
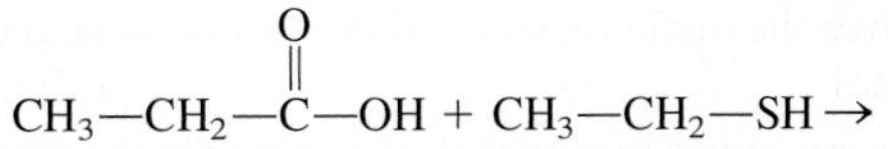

b. $CH_3—CH_2—CH_2—COOH + CH_3—SH \rightarrow$

c.
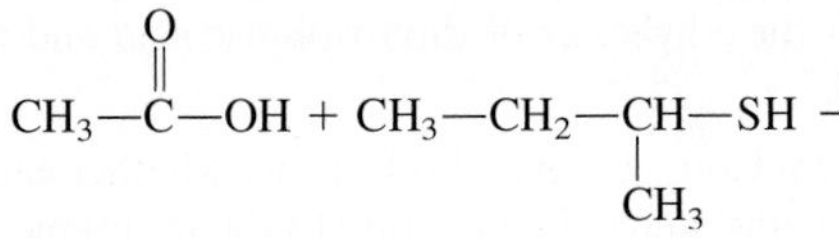

d.
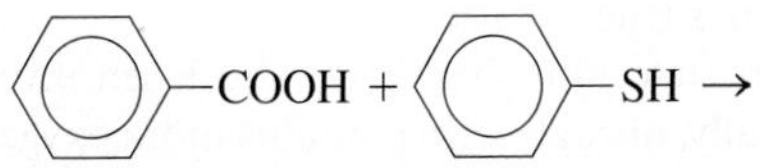

5.131 Draw the structural formula for each of the following thioesters.
a. Methyl thioethanoate
b. Methyl thioacetate
c. Ethyl thioformate
d. Ethyl thiomethanoate

5.132 Draw the structural formula for each of the following thioesters.
a. Methyl thiopropanoate
b. Methyl thiopropionate
c. Ethyl thioacetate
d. Ethyl thioethanoate

5.133 Formation of the thioester acetyl coenzyme A can be envisioned as a thioesterification reaction involving which two substances?

5.134 What is the generalized abbreviated structural formula for the thioester acetyl CoA?

Polyesters (Section 5.17)

5.135 Write the structure (two repeating units) of the polyester polymer formed from oxalic acid and 1,3-propanediol.

5.136 Write the structure (two repeating units) of the polyester polymer formed from malonic acid and ethylene glycol.

5.137 Draw the structural formulas of the monomers needed to form the following polyester.

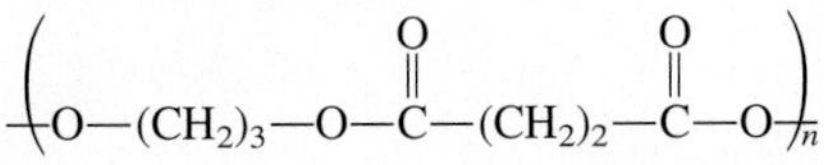

5.138 Draw the structural formulas of the monomers needed to form the following polyester.

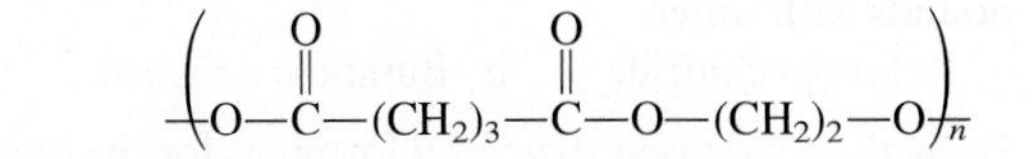

5.139 What are the names of the monomers used to produce the polyester whose acronym is PET?

5.140 What are the names of the monomers used to produce the polyester whose acronym is PEN?

5.141 What are some uses for the biodegradable polyester PHBV?

5.142 What is the major use for the biodegradable polyester known as *Lactomer*?

Acid Chlorides and Acid Anhydrides (Section 5.18)

5.143 Draw the condensed structural formula for each of the following compounds.
a. Propionyl chloride
b. 3-Methylbutanoyl chloride
c. Butyric anhydride
d. Butanoic ethanoic anhydride

5.144 Draw the condensed structural formula for each of the following compounds.
a. Acetyl chloride
b. 2-Methylbutanoyl chloride
c. Propionic anhydride
d. Ethanoic methanoic anhydride

5.145 Assign an IUPAC name to each of the following compounds.

a.
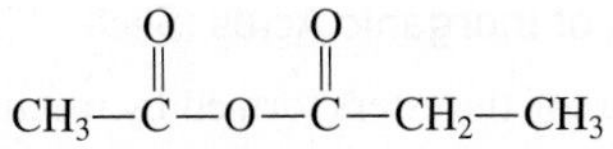

b.
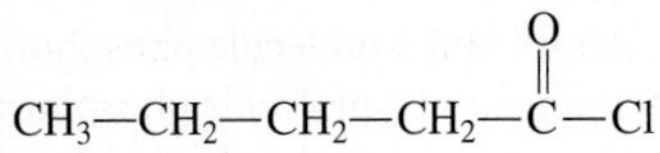

c.
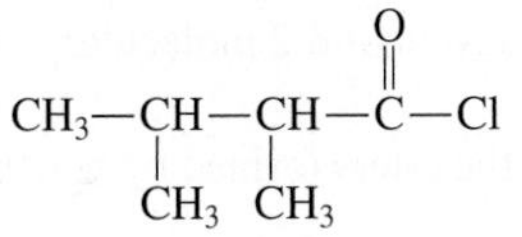

d.
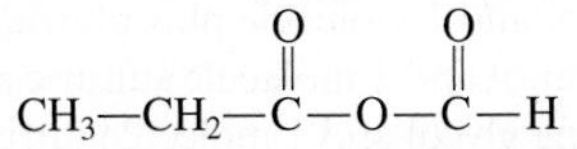

5.146 Assign an IUPAC name to each of the following compounds.

a.
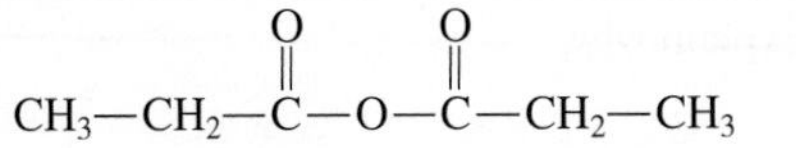

b.
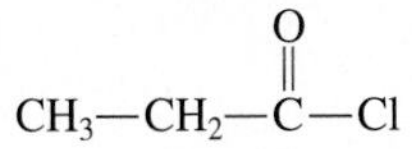

c.
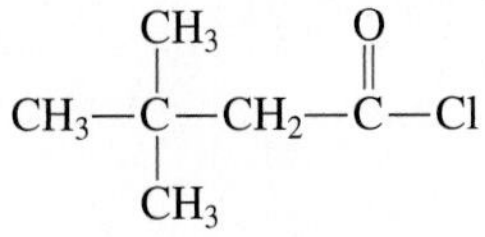

d.
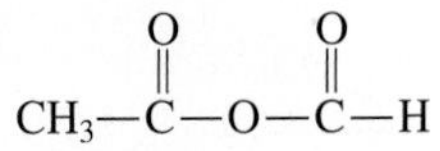

5.147 Draw a condensed structural formula for the organic product of the reaction of each of the following compounds with water.
a. Pentanoyl chloride b. Pentanoic anhydride

5.148 Draw a condensed structural formula for the organic product of the reaction of each of the following compounds with water.
a. Butanoyl chloride b. Butanoic anhydride

5.149 Draw the condensed structural formulas for the ester formed and the carboxylic acid formed when acetic anhydride reacts with the following alcohols.
a. Ethyl alcohol b. 1-Butanol

5.150 Draw the condensed structural formulas for the ester formed and the carboxylic acid formed when ethanoic anhydride reacts with the following alcohols.
a. Propyl alcohol b. 2-Butanol

5.151 What is the IUPAC name for the acyl group present in each of the following hydrocarbon derivatives?
a. Propanoic acid
b. Butanoic anhydride
c. Butanoyl chloride
d. Acetic acid

5.152 What is the IUPAC name for the acyl group present in each of the following hydrocarbon derivatives?
a. Butanoic acid b. Propanoic anhydride
c. Propanoyl chloride d. Formic acid

5.153 In general terms, what are the products obtained in an acyl transfer reaction involving an acid chloride and an alcohol?

5.154 In general terms, what are the products obtained in an acyl transfer reaction involving an acid anhydride and an alcohol?

Esters and Anhydrides of Inorganic Acids (Section 5.19)

5.155 Draw the structures of the esters formed by reacting the following substances.
a. 1 molecule methanol and 1 molecule phosphoric acid
b. 2 molecules methanol and 1 molecule phosphoric acid
c. 1 molecule methanol and 1 molecule nitric acid
d. 1 molecule ethylene glycol and 2 molecules nitric acid

5.156 Draw the structures of the esters formed by reacting the following substances.
a. 1 molecule ethanol and 1 molecule phosphoric acid
b. 2 molecules methanol and 1 molecule sulfuric acid
c. 1 molecule ethylene glycol and 1 molecule nitric acid
d. 1 molecule glycerol and 3 molecules nitric acid

5.157 Phosphoric acid can form triesters, but sulfuric acid cannot. Explain why.

5.158 Sulfuric acid can form diesters, but nitric acid cannot. Explain why.

5.159 Draw the structural formula for the ionic form of the following phosphate esters.

a. 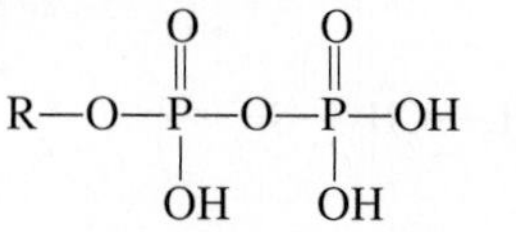

b. 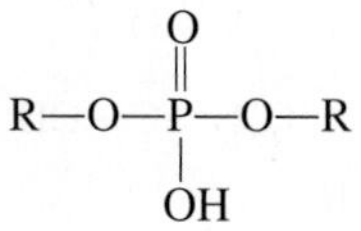

5.160 Draw the structural formula for the ionic form of the following phosphate esters.

a. 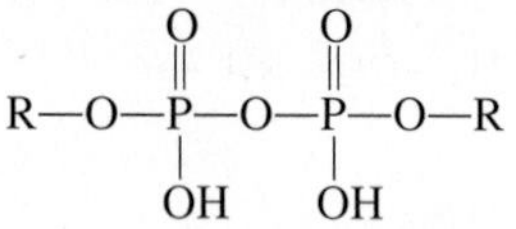

b. 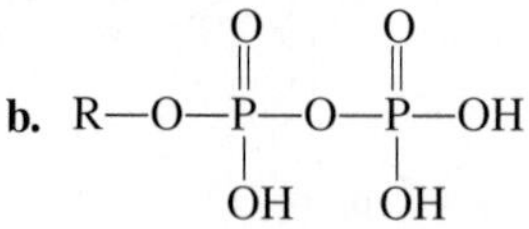

5.161 What is a phosphoryl group?

5.162 What is a phosphorylation reaction?

5.163 Draw structural formulas (ionic form) for the organic products produced from a phosphorylation reaction that involves the methyl ester of diphosphoric acid and ethyl alcohol.

5.164 Draw structural formulas (ionic form) for the organic products produced from a phosphorylation reaction that involves the ethyl ester of diphosphoric acid and methyl alcohol.

•5.165 (Chemical Connections 5-D) Indicate whether each of the following statements relating to the inorganic ester nitroglycerin is true or false.
a. In its pure state, nitroglycerin is a shock-sensitive solid.
b. Structurally, nitroglycerin contains more oxygen atoms than carbon atoms.
c. As a medication, nitroglycerin is a vasodilator that increases blood flow.
d. "Skin patch" nitroglycerin is readily absorbed through the skin into the bloodstream.

•5.166 (Chemical Connections 5-D) Indicate whether each of the following statements relating to the inorganic ester nitroglycerin is true or false.
a. When used in dynamite, nitroglycerin is adsorbed onto clay-like material.
b. Structurally, nitroglycerin contains the same number of carbon and nitrogen atoms.
c. It is not nitroglycerin itself that relieves heart muscle pain but, rather, one of its metabolic products.
d. Medicinal nitroglycerin is available in tablet form or as "skin patches."

Amines and Amides

6

Parachutist with a parachute made of the polyamide nylon.

CHAPTER OUTLINE

The four most abundant elements in living organisms are carbon, hydrogen, oxygen, and nitrogen. In previous organic chemistry chapters, compounds containing the first three of these elements have been considered. Alkanes, alkenes, alkynes, and aromatic compounds are all carbon–hydrogen compounds. Carbon–hydrogen–oxygen compounds include alcohols, phenols, ethers, aldehydes, ketones, carboxylic acids, and esters. This chapter extends the discussion of organic compounds by focusing on those that contain the element nitrogen.

Two important types of organic nitrogen-containing compounds will be considered: *amines* and *amides.* Amines are carbon–hydrogen–nitrogen compounds, and amides contain oxygen in addition to these elements. Amines and amides occur widely in living organisms. Many of these naturally occurring compounds are very active physiologically. In addition, numerous drugs used for the treatment of mental illness, hay fever, heart problems, and other physical disorders are amines or amides.

6.1 Bonding Characteristics of Nitrogen Atoms in Organic Compounds

An understanding of the bonding characteristics of the nitrogen atom is a prerequisite to the study of amines and amides. Nitrogen is a member of Group VA of the periodic table; it has five valence electrons and will form three covalent bonds to complete its octet of electrons. Thus, in organic chemistry, carbon forms four bonds (Section 1.2), nitrogen forms three bonds, and oxygen forms two bonds (Section 3.1).

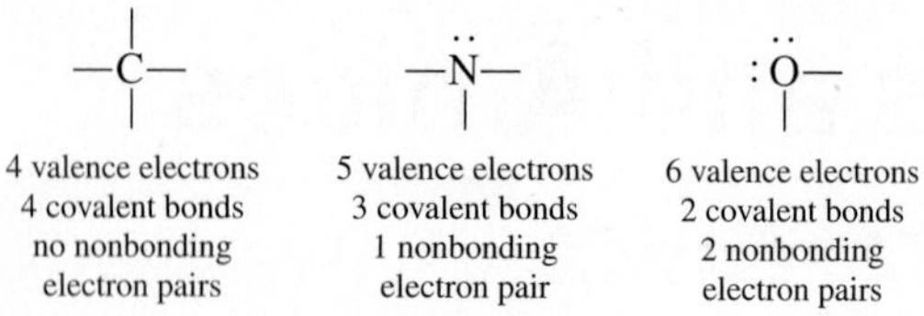

6.2 Structure and Classification of Amines

Amines bear the same relationship to ammonia that alcohols and ethers bear to water (Sections 3.2 and 3.15).

An **amine** *is an organic derivative of ammonia (NH_3) in which one or more alkyl, cycloalkyl, or aryl groups have replaced ammonia hydrogen atoms.* Amines are classified as primary (1°), secondary (2°), or tertiary (3°) on the basis of how many hydrocarbon groups have replaced ammonia hydrogen atoms (Figure 6.1). A **primary amine** *is an amine in which the nitrogen atom is bonded to one hydrocarbon group and two hydrogen atoms.* The generalized formula for a primary amine is RNH_2; one carbon–nitrogen bond is present. A **secondary amine** *is an amine in which the nitrogen atom is bonded to two hydrocarbon groups and one hydrogen atom.* The generalized formula for a secondary amine is R_2NH; two carbon–nitrogen bonds are present. A **tertiary amine** *is an amine in which the nitrogen atom is bonded to three hydrocarbon groups and no hydrogen atoms.* The generalized formula for a tertiary amine is R_3N; three carbon–nitrogen bonds are present.

The basis for the amine primary-secondary-tertiary classification system differs from that for alcohols (Section 3.8).

1. For alcohols how many R groups are on a *carbon* atom, the hydroxyl-bearing carbon atom, is the determining factor.
2. For amines how many R groups are on the *nitrogen* atom is the determining factor.

Tert-butyl alcohol is a *tertiary* alcohol, whereas *tert*-butylamine is a *primary* amine.

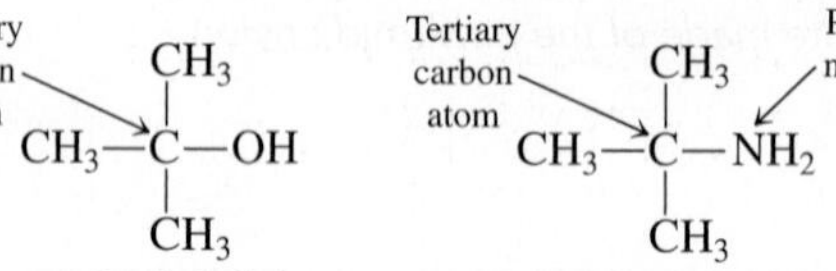

tert-Butyl alcohol (a tertiary alcohol)

tert-Butylamine (a primary amine)

The functional group present in a primary amine, the $—NH_2$ group, is called an *amino* group. An **amino group** *is the $—NH_2$ functional group.* Secondary and tertiary amines possess substituted amino groups.

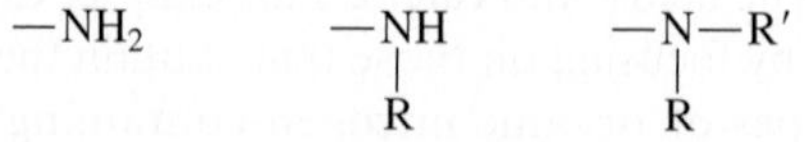

Amino group · Monosubstituted amino group · Disubstituted amino group

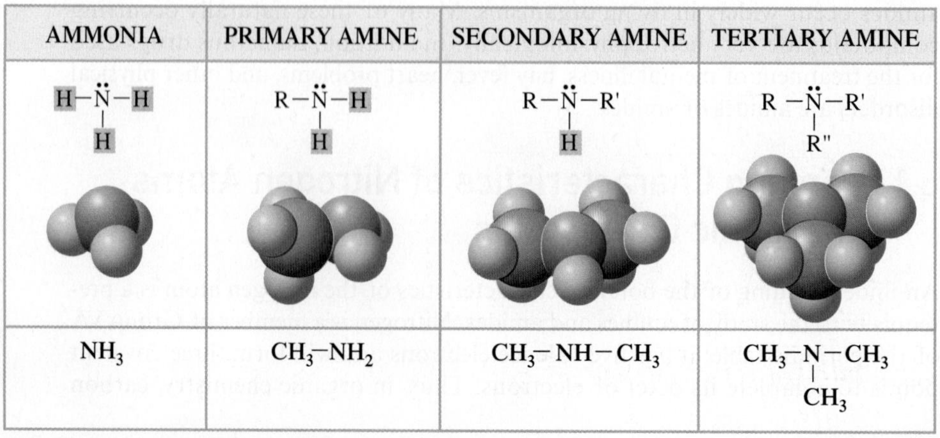

AMMONIA
PRIMARY AMINE
SECONDARY AMINE
TERTIARY AMINE
NH3
CH3–NH2
CH3–NH–CH3
CH3–N–CH3
CH3

Figure 6.1 Classification of amines is related to the number of R groups attached to the nitrogen atom.

EXAMPLE 6.1 Classifying Amines as Primary, Secondary, or Tertiary

Classify each of the following amines as a primary, secondary, or tertiary amine.

a. 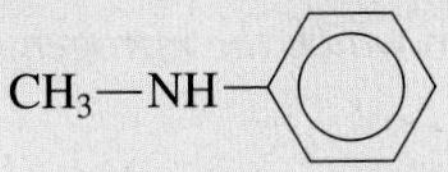

b. 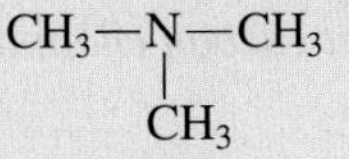

c. 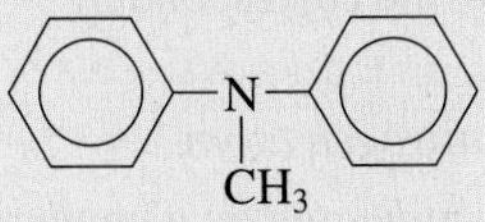

d. 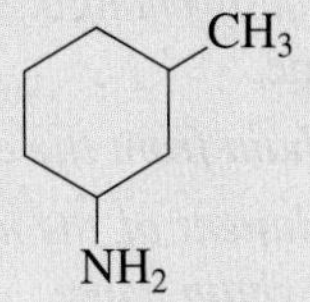

Solution

The number of carbon atoms directly bonded to the nitrogen atom determines the amine classification.

a. This is a secondary amine because the nitrogen is bonded to both a methyl group and a phenyl group.
b. This is a tertiary amine because the nitrogen atom is bonded to three methyl groups.
c. This is also a tertiary amine; the nitrogen atom is bonded to two phenyl groups and a methyl group.
d. This is a primary amine. The nitrogen atom is bonded to only one carbon atom, a carbon atom in the ring structure.

Practice Exercise 6.1

Classify each of the following amines as a primary, secondary, or tertiary amine.

a. $CH_3—CH_2—CH_2—NH_2$ **b.** $CH_3—NH—CH_2—CH_3$

c. 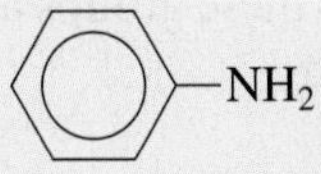

d. 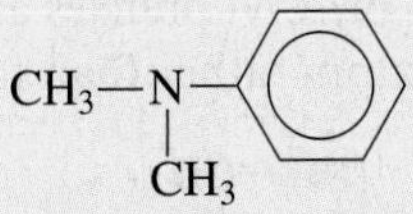

Answers: **a.** Primary; **b.** Secondary; **c.** Primary; **d.** Tertiary

Line-angle structural formulas for selected primary, secondary, and tertiary amines.

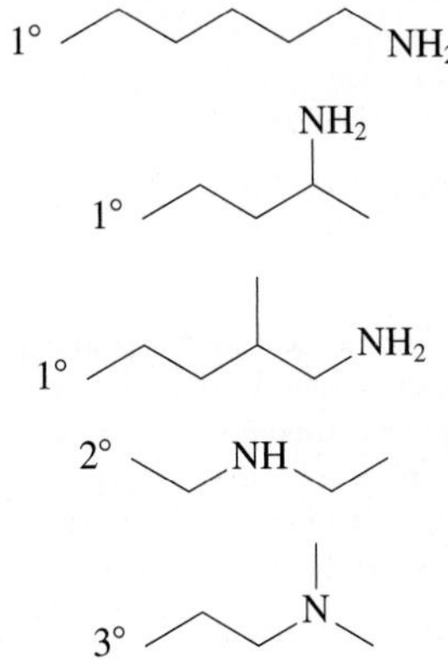

Cyclic amines exist. Such compounds, which are heterocyclic compounds (Section 3.19), are always either secondary or tertiary amines.

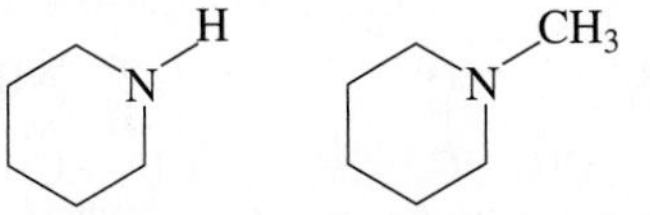

2° Cyclic amine 3° Cyclic amine

Numerous cyclic amine compounds are found in biochemical systems (Section 6.9).

Cyclic amines are the fourth type of heterocyclic compound to be considered; cyclic ethers (Section 3.19), cyclic forms of hemiacetals and acetals (Section 4.11), and cyclic esters (Section 5.9) were previously discussed. In cyclic amines, the heteroatom is nitrogen; in the other heterocyclic compounds, oxygen is the heteroatom.

6.3 Nomenclature for Amines

Both common and IUPAC names are extensively used for amines. In the common system of nomenclature, amines are named by listing the alkyl group or groups attached to the nitrogen atom in alphabetical order and adding the suffix *-amine;* all of this appears as one word. Prefixes such as *di-* and *tri-* are added when identical groups are bonded to the nitrogen atom.

$CH_3—CH_2—NH_2$ $CH_3—NH—CH_3$ 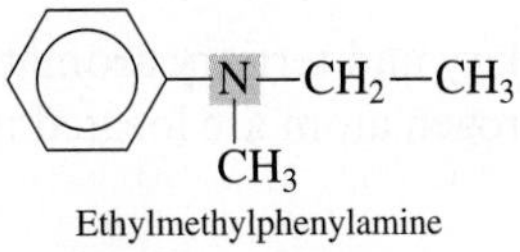

Ethylamine Dimethylamine Ethylmethylphenylamine

The common names of amines, like those of aldehydes, are written as a single word, which is different from the common names of alcohols (two words), ethers (two or three words), ketones (two or three words), acids (two words), and esters (two words).

The IUPAC rules for naming amines are similar to those for alcohols (Section 3.3). Alcohols are named as *alkanols* and amines are named as *alkanamines.* IUPAC rules for naming *primary* amines are as follows:

Rule 1: *Select as the parent carbon chain the longest chain to which the nitrogen atom is attached.*

Rule 2: *Name the parent chain by changing the -e ending of the corresponding alkane name to -amine.*

Rule 3: *Number the parent chain from the end nearest the nitrogen atom.*

Rule 4: *The position of attachment of the nitrogen atom is indicated by a number in front of the parent chain name.*

Rule 5: *The identity and location of any substituents are appended to the front of the parent chain name.*

IUPAC nomenclature for primary amines is similar to that for alcohols, except that the suffix is *-amine* rather than *-ol*. An —NH_2 group, like an —OH group, has priority in numbering the parent carbon chain.

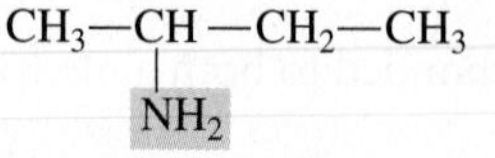

2-Butanamine

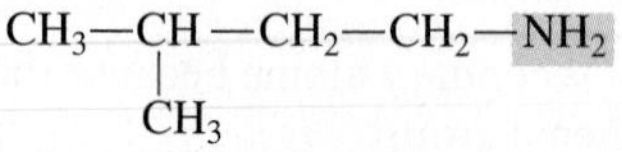

3-Methyl-1-butanamine

In diamines, the final *-e* of the carbon chain name is retained for ease of pronunciation. Thus the base name for a four-carbon chain bearing two amino groups is butan*e*diamine.

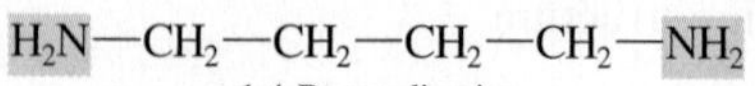

1,4-Butanediamine

Secondary and tertiary amines are named as *N*-substituted primary amines. The largest carbon group bonded to the nitrogen is used as the parent amine name. The names of the other groups attached to the nitrogen are appended to the front of the base name, and *N*- or *N,N*- prefixes are used to indicate that these groups are attached to the nitrogen atom rather than to the base carbon chain.

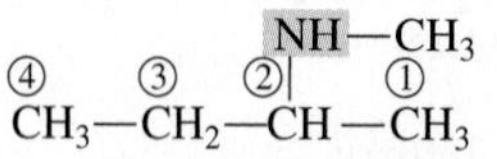

N-Methyl-2-butanamine

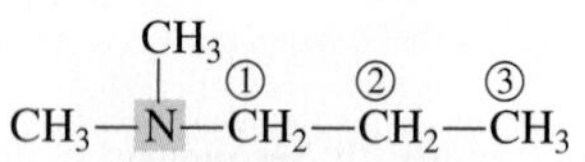

N,N-Dimethyl-1-propanamine

In IUPAC nomenclature, the amino group has a priority just below that of an alcohol. The priority list for functional groups is

- carboxylic acid
- aldehyde
- ketone
- alcohol
- amine

(↑ Increasing priority)

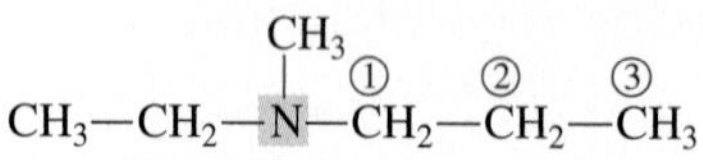

N-Ethyl-*N*-methyl-1-propanamine

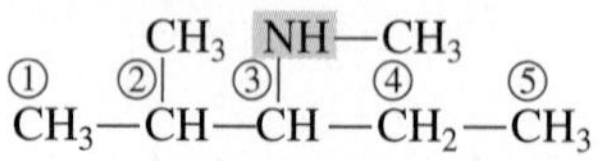

2,*N*-Dimethyl-3-pentanamine

In amines where additional functional groups are present, the amine group is treated as a substituent. As a substituent, an —NH_2 group is called an *amino* group.

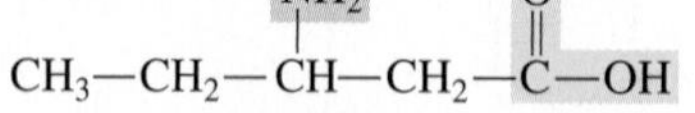

3-Aminopentanoic acid

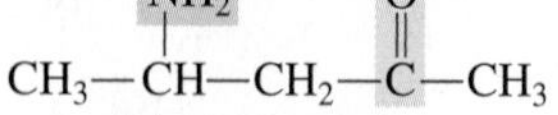

4-Amino-2-pentanone

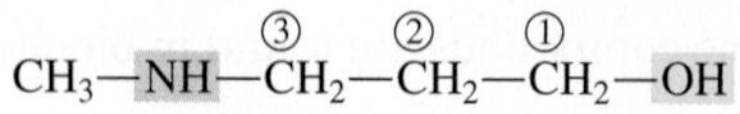

3-(*N*-Methylamino)-1-propanol

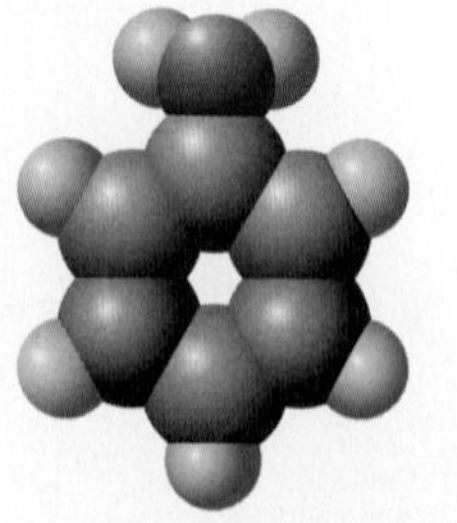

Figure 6.2 Space-filling model of aniline, the simplest aromatic amine. Aromatic amines, including aniline, are generally toxic; they are readily absorbed through the skin.

The simplest aromatic amine, a benzene ring bearing an amino group, is called *aniline* (Figure 6.2). Other simple aromatic amines are named as derivatives of aniline.

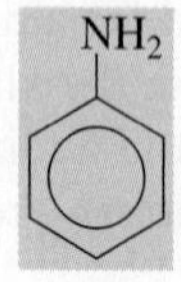

Aniline

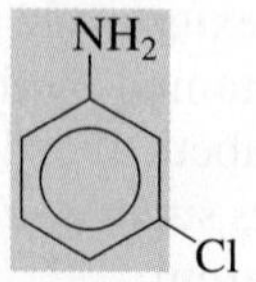

m-Chloroaniline

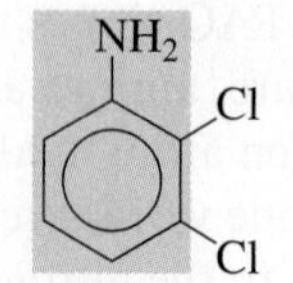

2,3-Dichloroaniline

In secondary and tertiary aromatic amines, the additional group or groups attached to the nitrogen atom are located using a capital *N*-.

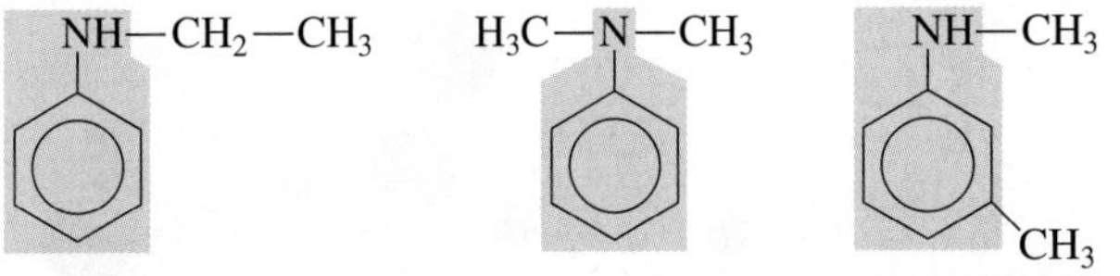

N-Ethylaniline *N,N*-Dimethylaniline 3,*N*-Dimethylaniline

EXAMPLE 6.2 Determining IUPAC Names for Amines

Assign IUPAC names to each of the following amines.

a. $CH_3—CH_2—NH—(CH_2)_4—CH_3$

b.

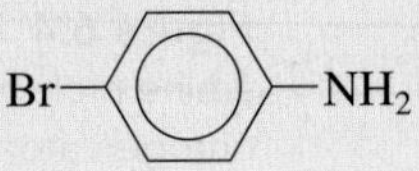

c. $H_2N—CH_2—CH_2—NH_2$

d.

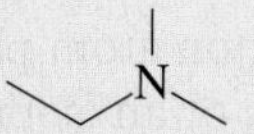

Solution

a. The longest carbon chain has five carbons. The name of the compound is *N-ethyl-1-pentanamine.*

b. This compound is named as a derivative of aniline: *4-bromoaniline* (or *p*-bromoaniline). The carbon in the ring to which the $—NH_2$ is attached is carbon 1.

c. Two $—NH_2$ groups are present in this molecule. The name is *1,2-ethanediamine.*

d. This is a tertiary amine in which the longest carbon chain has two carbons (ethane). The base name is thus *ethanamine.* Also two methyl groups are attached to the nitrogen atom. The name of the compound is *N,N-dimethylethanamine.*

Practice Exercise 6.2

Assign IUPAC names to each of the following amines.

a.

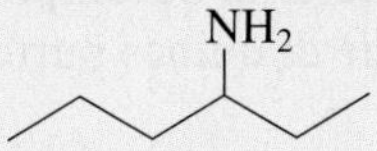

b. $CH_3—CH_2—CH_2—NH—CH_2—CH_2—CH_3$

c. $CH_3—N(CH_3)—CH_3$

d.

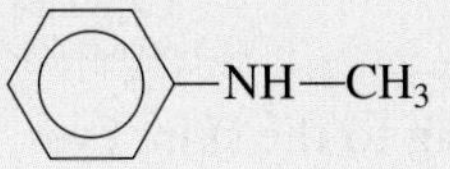

Answers: **a.** 3-hexanamine; **b.** *N*-propyl-1-propanamine; **c.** *N,N*-dimethylmethanamine; **d.** *N*-methylaniline

A benzene ring with both an amino group and a methyl group as substituents is called *toluidine.* This name is a combination of the names *toluene* and *aniline.*

The contrast between IUPAC names and common names for primary, secondary, and tertiary amines is as follows:

Primary Amines

IUPAC (one word)

alkanamine

Common (one word)

alkylamine

Secondary Amines

IUPAC (one word)

N-alkylalkanamine

Common (one word)

alkylalkylamine

Tertiary Amines

IUPAC (one word)

N-alkyl-*N*-alkylalkanamine

Common (one word)

alkylalkylalkylamine

6.4 Isomerism for Amines

Constitutional isomerism in amines can arise from several causes. Different carbon atom arrangements produce isomers, as in

$CH_3—CH_2—CH_2—CH_2—CH_2—NH_2$ (1-Pentanamine) and $CH_3—CH_2—CH(CH_3)—CH_2—NH_2$ (2-Methyl-1-butanamine)

Different positioning of the nitrogen atom on a carbon chain is another cause for isomerism, illustrated in the following compounds.

$CH_2(NH_2)—CH_2—CH_2—CH_3$ (1-Butanamine) and $CH_3—CH(NH_2)—CH_2—CH_3$ (2-Butanamine)

For secondary and tertiary amines, different partitioning of carbon atoms among the carbon chains present produces constitutional isomers. There are three

Unbranched Primary Amines			
C_1	C_3	C_5	C_7
C_2	C_4	C_6	C_8

Gas Liquid

Figure 6.3 A physical-state summary for unbranched primary amines at room temperature and room pressure.

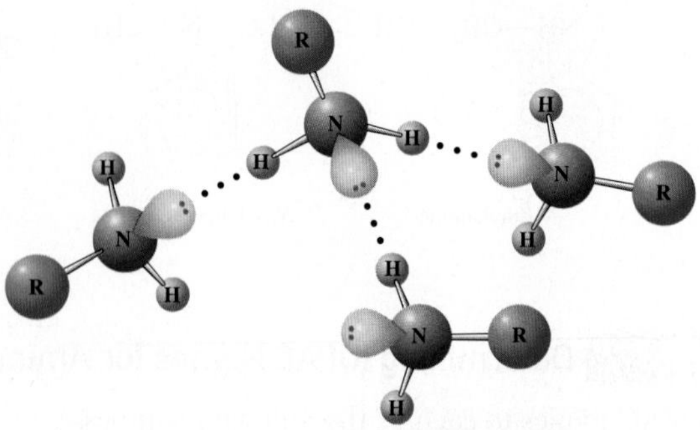

Figure 6.4 Hydrogen bonding interactions among amine molecules involves the hydrogen atoms and nitrogen atoms of amino groups.

C_4 secondary amines; carbon atom partitioning can be two ethyl groups, a propyl group and a methyl group, or an isopropyl group and a methyl group.

$$\underset{N\text{-Ethylethanamine}}{CH_3{-}CH_2{-}NH{-}CH_2{-}CH_3} \quad \text{and} \quad \underset{N\text{-Methyl-1-propanamine}}{CH_3{-}CH_2{-}CH_2{-}NH{-}CH_3} \quad \text{and} \quad \underset{N\text{-Methyl-2-propanamine}}{CH_3{-}\underset{\displaystyle CH_3}{\underset{|}{CH}}{-}NH{-}CH_3}$$

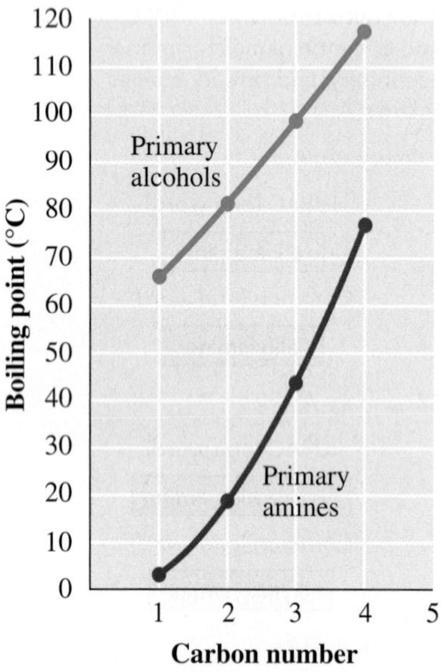

Figure 6.5 A comparison of boiling points of unbranched primary amines and unbranched primary alcohols.

6.5 Physical Properties of Amines

The methylamines (mono-, di-, and tri-) and ethylamine are gases at room temperature and have ammonia-like odors. Most other amines are liquids (Figure 6.3), and many have odors resembling that of raw fish. A few amines, particularly diamines, have strong, disagreeable odors. The foul odor arising from dead fish and decaying flesh is due to amines released by the bacterial decomposition of protein. Two of these "odoriferous" compounds are the diamines putrescine and cadaverine.

$$\underset{\substack{\text{Putrescine}\\ \text{(1,4-butanediamine)}}}{H_2N{-}(CH_2)_4{-}NH_2} \qquad \underset{\substack{\text{Cadaverine}\\ \text{(1,5-pentanediamine)}}}{H_2N{-}(CH_2)_5{-}NH_2}$$

The simpler amines are irritating to the skin, eyes, and mucous membranes and are toxic by ingestion. Aromatic amines are generally toxic. Many are readily absorbed through the skin and affect both the blood and the nervous system.

The boiling points of amines are intermediate between those of alkanes and alcohols of similar molecular mass. They are higher than alkane boiling points, because primary and secondary amines are capable of intermolecular hydrogen bonding, whereas alkane molecules lack such ability. Such amine intermolecular hydrogen bonding is depicted in Figure 6.4. Tertiary amines have boiling points that are much lower than primary and secondary amines of similar molecular mass; hydrogen bonding is not possible between tertiary amine molecules, as such amines have no hydrogen atoms directly bonded to the nitrogen atom.

The boiling points of amines are lower than those of corresponding alcohols (Figure 6.5), because N···H hydrogen bonds are weaker than O···H hydrogen bonds. [The difference in hydrogen-bond strength results from electronegativity differences; nitrogen is less electronegative than oxygen.]

Amines with fewer than six carbon atoms are infinitely soluble in water. This solubility results from hydrogen bonding between the amines and water. Even tertiary amines are water-soluble, because the amine nitrogen atom has a nonbonding electron pair that can form a hydrogen bond with a hydrogen atom of water (Figure 6.6).

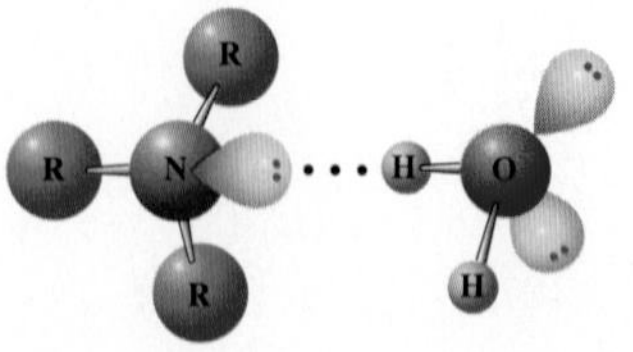

Figure 6.6 Low-molecular-mass amines are soluble in water because of amine–water hydrogen bonding interactions.

6.6 Basicity of Amines

Like the nitrogen atom in ammonia, an amine nitrogen atom possesses a lone pair of electrons.

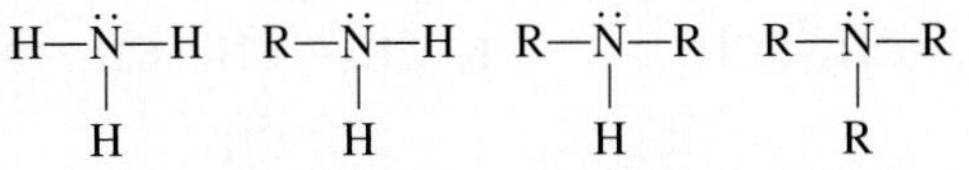

Ammonia 1° amine 2° amine 3° amine

The lone pair of electrons present on an amine nitrogen atom is generally not shown when drawing structural formulas for amines. However, in acid-base considerations, the presence of this lone pair of electrons is of vital importance.

This means that amines, like ammonia, are capable of exhibiting Brønsted-Lowry base behavior; that is, they can function as proton acceptors. Paralleling the behavior of ammonia, a basic solution is produced when an amine dissolves in water. A proton is transferred from a water molecule to the lone-pair site on the nitrogen atom. The following two equations show the similarities in the behaviors of ammonia and an amine in aqueous solution.

$$\ddot{N}H_3 + HOH \rightleftharpoons NH_4^+ + OH^-$$

Ammonia; Ammonium ion; Hydroxide ion

$$CH_3{-}\ddot{N}H_2 + HOH \rightleftharpoons CH_3{-}\overset{+}{N}H_3 + OH^-$$

Methylamine; Methylammonium ion; Hydroxide ion

The result of the interaction of an amine with water is a basic solution containing substituted ammonium ions and hydroxide ions. A **substituted ammonium ion** *is an ammonium ion in which one or more alkyl, cycloalkyl, or aryl groups have been substituted for hydrogen atoms.*

Amines, like ammonia, have a pair of unshared electrons on the nitrogen atom present. These unshared electrons can accept a hydrogen ion from water. Thus both amines and ammonia produce basic aqueous solutions.

Two important generalizations apply to substituted ammonium ions.

1. *Substituted ammonium ions are charged species rather than neutral molecules.*
2. *The nitrogen atom in an ammonium ion or a substituted ammonium ion participates in four bonds.* In a neutral compound, nitrogen atoms form only three bonds. Four bonds about a nitrogen atom are possible, however, when the species is a positive ion because the fourth bond is a coordinate covalent bond.

Naming the positive ion that results from the interaction of an amine with water is based on the following two rules:

Rule 1: *For alkylamines, the ending of the name of the amine is changed from* amine *to* ammonium ion.

Substituted ammonium ions always contain one more hydrogen atom than their "parent" amine. They also always carry a +1 charge, whereas the "parent" amine is a neutral molecule.

$$CH_3{-}CH_2{-}NH_2 \xrightarrow{H_2O} CH_3{-}CH_2{-}\overset{+}{N}H_3 + OH^-$$

Ethylamine; Ethylammonium ion

$$CH_3{-}N(CH_3){-}CH_2{-}CH_3 \xrightarrow{H_2O} CH_3{-}\overset{+}{N}H(CH_3){-}CH_2{-}CH_3 + OH^-$$

Ethyldimethylamine; Ethyldimethylammonium ion

Rule 2: *For aromatic amines, the final* -e *in the name aniline is replaced by* -ium ion.

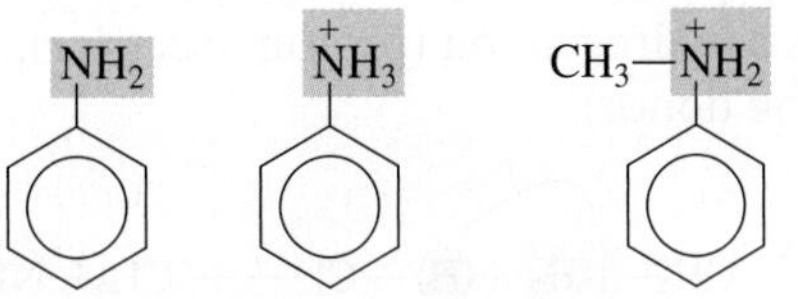

Aniline Anilinium ion *N*-methylanilinium ion

EXAMPLE 6.3 Determining Names for Substituted Ammonium and Substituted Anilinium Ions

Name the following substituted ammonium or substituted anilinium ions.

a. $CH_3{-}CH_2{-}\overset{+}{N}H_2{-}CH_2{-}CH_3$

b. $CH_3{-}CH(CH_3){-}CH_2{-}\overset{+}{N}H_3$

c. $CH_3{-}\overset{+}{N}H(CH_3){-}CH_3$

d. 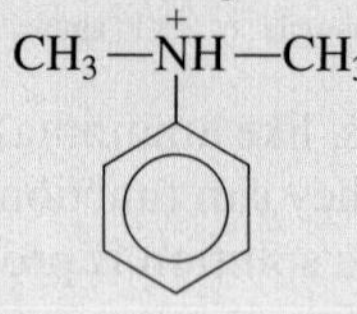

Solution

a. The parent amine is diethylamine. Replacing the word *amine* in the parent name with *ammonium ion* generates the name of the ion, *diethylammonium ion.*

b. The parent amine is isobutylamine. The name of the ion is *isobutylammonium ion.*

c. The parent amine is trimethylamine. The name of the ion is *trimethylammonium ion.*

d. The parent name is *N,N*-dimethylaniline. Replacing the word *aniline* in the parent name with *anilinium ion* generates the name of the ion, *N,N-dimethylanilinium ion.*

Practice Exercise 6.3

Name the following substituted ammonium or substituted anilinium ions.

a. $CH_3{-}CH_2{-}\overset{+}{N}H_2{-}CH_3$

b. $CH_3{-}CH(CH_3){-}\overset{+}{N}H_3$

c. $CH_3{-}CH_2{-}\overset{+}{N}H(CH_3){-}CH_2{-}CH_3$

d. 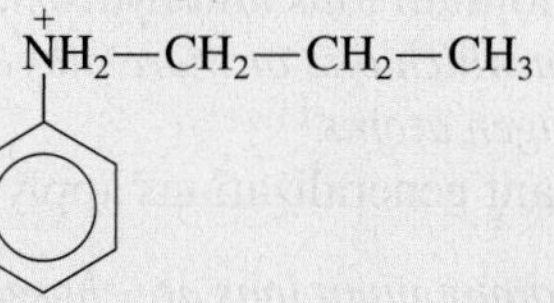

Answers: **a.** Ethylmethylammonium ion; **b.** Isopropylammonium ion; **c.** Diethylmethylammonium ion; **d.** *N*-propylanilinium ion

Amines, the most basic type of organic compound, are very weak bases when compared to strong inorganic bases like NaOH and KOH.

Amines are stronger proton acceptors than oxygen-containing organic compounds such as alcohols and ethers; that is, they are stronger bases than these compounds. A 0.1 M aqueous solution of methylamine has a pH of 11.8, and a 0.1 M aqueous solution of aniline has a pH of 8.6. These solutions are sufficiently basic to turn red litmus paper blue. Carboxylic acid salts (Section 5.8) are the only other type of organic compound sufficiently basic to turn red litmus paper blue.

6.7 Reaction of Amines with Acids

The reaction of an acid with a base (neutralization) produces a salt. Because amines are bases (Section 6.6), their reaction with an acid—either inorganic or carboxylic—produces a salt, an amine salt. In such a reaction, the proton (H^+) from the acid becomes bonded to the nitrogen atom of the amine by using the lone pair of electrons present on the nitrogen atom. In an amine-acid reaction, the amine always gains a hydrogen ion (proton acceptor), and the acid always loses a hydrogen ion (proton donor).

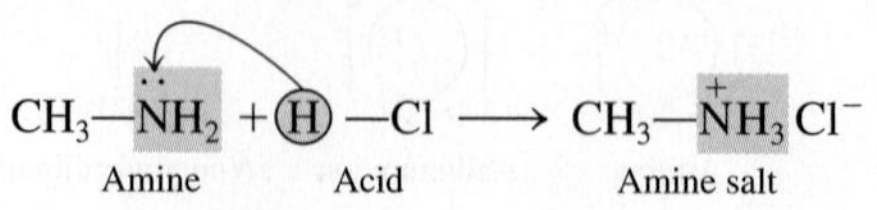

Aromatic amines react with acids in a similar manner.

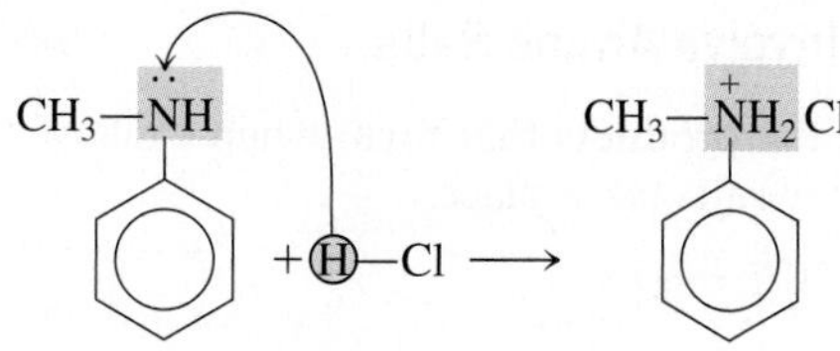

Amine Acid Amine salt

An **amine salt** *is an ionic compound in which the positive ion is a mono-, di-, or trisubstituted ammonium ion (RNH_3^+, $R_2NH_2^+$, or R_3NH^+) and the negative ion comes from an acid.* Amine salts can be obtained in crystalline form (odorless, white crystals) by evaporating the water from the acidic solutions in which amine salts are prepared.

Amine salts are named using standard nomenclature procedures for ionic compounds. The name of the positive ion, the substituted ammonium or anilinium ion, is given first and is followed by a separate word for the name of the negative ion.

$$CH_3-CH_2-\overset{+}{N}H_3\ Cl^- \qquad CH_3-\overset{+}{N}H_2-CH_3\ Br^-$$

Ethylammonium chloride Dimethylammonium bromide

An older naming system for amine salts, still used in the pharmaceutical industry, treats amine salts as amine–acid complexes rather than as ionic compounds. In this system, the amine salt made from dimethylamine and hydrochloric acid is named and represented as

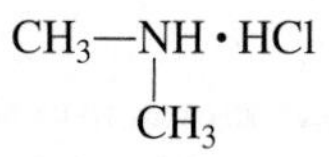

Dimethylamine hydrochloride

rather than as

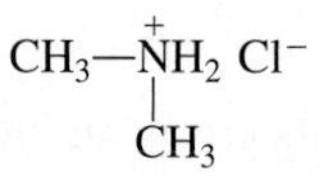

Dimethylammonium chloride

Many medication labels refer to hydrochlorides or hydrogen sulfates (from sulfuric acid), indicating that the medications are in a water-soluble ionic (salt) form.

Many higher-molecular-mass amines are water-insoluble; however, virtually all amine salts are water-soluble. Thus amine salt formation, like carboxylic acid salt formation (Section 5.8), provides a means for converting water-insoluble compounds into water-soluble compounds. Many drugs that contain amine functional groups are administered to patients in the form of amine salts because of their increased solubility in water in this form.

Many people unknowingly use acids to form amine salts when they put vinegar or lemon juice on fish. Such action converts amines in fish (often smelly compounds) to salts, which are odorless.

The process of forming amine salts with acids is an easily reversed process. Treating an amine salt with a strong base such as NaOH regenerates the "parent" amine.

$$\underset{\text{Amine salt}}{CH_3-\overset{+}{N}H_3\ Cl^-} + \underset{\text{Base}}{NaOH} \longrightarrow \underset{\text{Amine}}{CH_3-NH_2} + NaCl + H_2O$$

The "opposite nature" of the processes of amine salt formation from an amine and the regeneration of the amine from its amine salt can be diagrammed as follows:

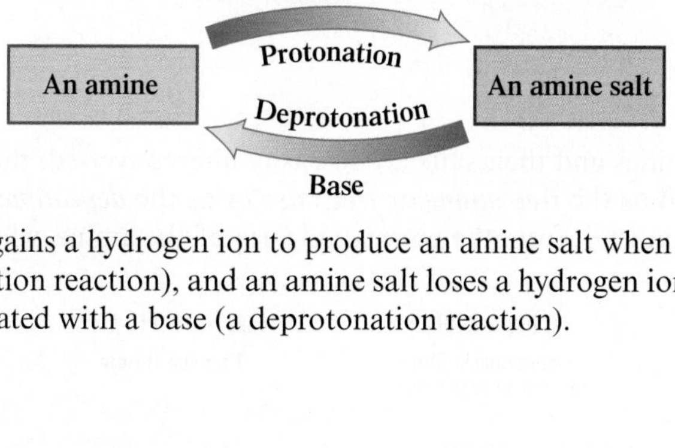

An amine gains a hydrogen ion to produce an amine salt when treated with an acid (a protonation reaction), and an amine salt loses a hydrogen ion to produce an amine when treated with a base (a deprotonation reaction).

An amine is neutral. When it accepts a H^+ from an acid, it becomes an ion with a +1 charge (the charge coming from the hydrogen ion); a substituted ammonium ion is formed.

The cocaine molecule is both an amine and an ester; one amine and two ester functional groups are present. As an illegal street drug, cocaine is consumed as a water-soluble amine salt and in a water-insoluble, free-base form (nonsalt form—freed of the base required to make the salt). Cocaine hydrochloride, the amine salt, is a white powder that is snorted or injected intravenously. Free-base cocaine is heated and its vapors are inhaled. Cocaine users and dealers call the water-insoluble form of the drug "crack." "Snow" and "coke" are street names for the water-soluble form of the drug.

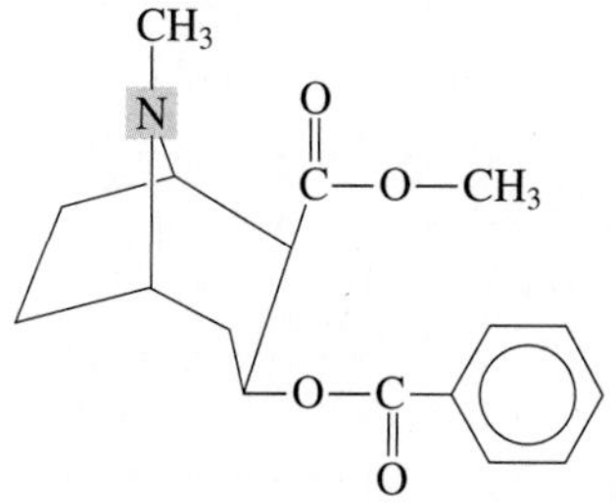

Cocaine
(an amine)

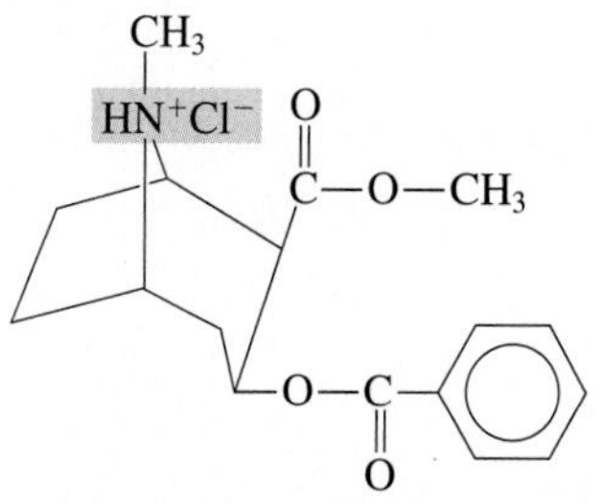

Cocaine hydrochloride
(an amine salt)

EXAMPLE 6.4 Writing Chemical Equations for Reactions That Involve Amine Salts

Write the structures of the products that form when each of the following reactions involving amines or amine salts takes place.

a. $CH_3—NH—CH_3 + HCl \longrightarrow$

b. 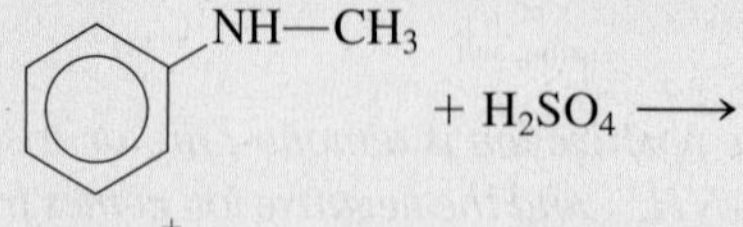

c. $CH_3—\overset{+}{N}H_2—CH_3\ Cl^- + NaOH \longrightarrow$

Solution

a. The reactants are an amine and a strong acid. Their interaction produces an amine salt.

$$CH_3—\ddot{N}H—CH_3 + (H)Cl \longrightarrow CH_3—\overset{+}{N}H_2—CH_3\ Cl^-$$

Note that the H^+ ion from the acid always bonds to the amine nitrogen atom at the location where the lone pair of electrons is found.

b. Again, the reaction is that of an amine with a strong acid. A hydrogen ion is transferred from the acid to the amine.

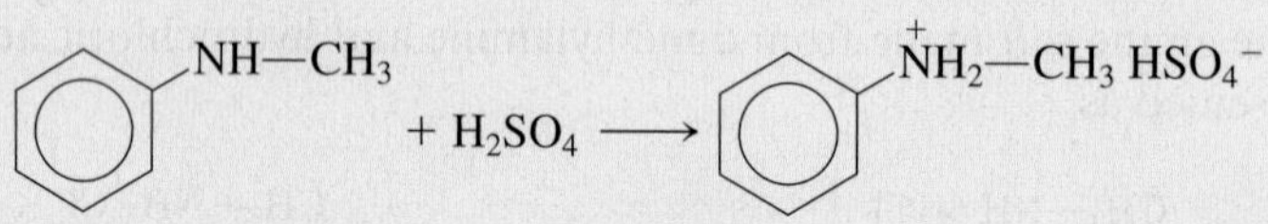

c. The reactants are an amine salt and a strong base. Their interaction regenerates the "parent" amine.

$$CH_3—\overset{+}{N}H_2—CH_3\ Cl^- + NaOH \longrightarrow CH_3—NH—CH_3 + NaCl + H_2O$$

Several physiological active amines are dispensed to consumers in the form of amine salts. An example is phenylephrine hydrochloride, the decongestant present in Sudafed SE, Theraflu, and DayQuil. The structure of this amine salt is

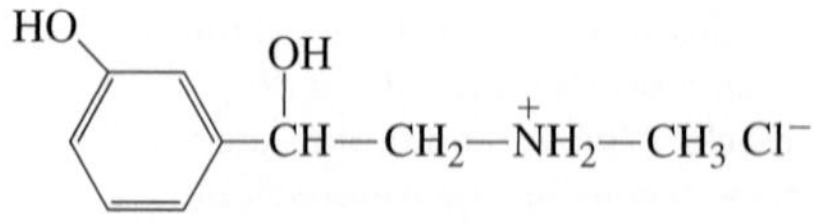

Practice Exercise 6.4

Write the structures of the products formed in the following reactions.

a. $CH_3—CH_2—N(CH_3)—CH_3 + HCl \longrightarrow$

b. $CH_3—CH_2—NH_2 + H_2SO_4 \longrightarrow$

c. $CH_3—CH_2—\overset{+}{N}H(CH_3)—CH_3\ Br^- + NaOH \longrightarrow$

Answers:

a. $CH_3—CH_2—\overset{+}{N}H(CH_3)—CH_3\ Cl^-$

b. $CH_3—CH_2—\overset{+}{N}H_3\ HSO_4^-$

c. $CH_3—CH_2—N(CH_3)—CH_3 + NaBr + H_2O$

Because amines and their salts are so easily interconverted, the amine itself is often designated as the *free amine* or *free base* or as the *deprotonated form* of the amine, to distinguish it from the *protonated form* of the amine, which is present in the amine salt.

$CH_3—NH_2$	$CH_3—\overset{+}{N}H_3\ Cl^-$
Deprotonated form (free amine or free base)	Protonated form

Whether an amine in solution exists in its deprotonated form or its protonated form is dependent on the pH of the solution. An aqueous solution of a simple amine has a pH of about 12 and at this pH the amine is almost entirely (99%) present in its deprotonated form. In biochemical solutions such as body fluids, which are buffered solutions with a pH near 7, amines exist predominantly in protonated form.

6.8 Alkylation of Ammonia and Amines

Several methods exist for preparing amines, one of which is *alkylation* in the presence of base. An **alkylation reaction** *is a chemical reaction in which an alkyl group is transferred from one molecule to another molecule.* In preparing amines, the alkylating agent is an alkyl halide (Section 1.18). General equations for the alkylation process, as it pertains to amines, are

$$\text{Ammonia} + \text{alkyl halide} \xrightarrow{\text{Base}} 1^\circ\ \text{amine}$$

$$1^\circ\ \text{Amine} + \text{alkyl halide} \xrightarrow{\text{Base}} 2^\circ\ \text{amine}$$

$$2^\circ\ \text{Amine} + \text{alkyl halide} \xrightarrow{\text{Base}} 3^\circ\ \text{amine}$$

$$3^\circ\ \text{Amine} + \text{alkyl halide} \xrightarrow{\text{Base}} \text{quaternary ammonium salt}$$

Alkylation under basic conditions is actually a two-step process. In the first step, using a primary amine preparation as an example, an amine salt is produced.

$$NH_3 + R{-}X \longrightarrow R{-}\overset{+}{N}H_3\ X^-$$

The second step, which involves the base present (NaOH), converts the amine salt to free amine.

$$R{-}\overset{+}{N}H_3\ X^- + NaOH \longrightarrow RNH_2 + NaX + H_2O$$

A specific example of the production of a primary amine from ammonia is the reaction of ethyl bromide with ammonia to produce ethylamine. The chemical equation (with both steps combined) is

$$NH_3 + CH_3{-}CH_2{-}Br + NaOH \longrightarrow CH_3{-}CH_2{-}NH_2 + NaBr + H_2O$$

If the newly formed primary amine produced in an ammonia alkylation reaction is not quickly removed from the reaction mixture, the nitrogen atom of the amine may react with further alkyl halide molecules, giving, in succession, secondary and tertiary amines.

$$NH_3 \xrightarrow[OH^-]{RX} \underset{\text{Primary amine}}{RNH_2} \xrightarrow[OH^-]{RX} \underset{\text{Secondary amine}}{R_2NH} \xrightarrow[OH^-]{RX} \underset{\text{Tertiary amine}}{R_3N}$$

Examples of the production of a 2° amine and a 3° amine via alkylation are

$$\underset{\text{Primary amine}}{CH_3{-}CH_2{-}NH_2} + \underset{\text{Alkyl halide}}{CH_3{-}Br} + \underset{\text{Base}}{NaOH} \longrightarrow \underset{\text{Secondary amine}}{CH_3{-}CH_2{-}NH{-}CH_3} + NaBr + H_2O$$

$$\underset{\text{Secondary amine}}{CH_3{-}CH_2{-}NH{-}CH_3} + \underset{\text{Alkyl halide}}{CH_3{-}Br} + \underset{\text{Base}}{NaOH} \longrightarrow \underset{\text{Tertiary amine}}{CH_3{-}CH_2{-}\underset{\underset{CH_3}{|}}{N}{-}CH_3} + NaBr + H_2O$$

Tertiary amines react with alkyl halides in the presence of a strong base to produce a quaternary ammonium salt. A **quaternary ammonium salt** *is an ammonium salt in which all four groups attached to the nitrogen atom of the ammonium ion are hydrocarbon groups.*

$$R{-}\underset{\underset{R}{|}}{N}{-}R + R{-}X \xrightarrow{OH^-} R{-}\overset{\overset{R}{|}}{\underset{\underset{R}{|}}{N^+}}{-}R\ X^-$$

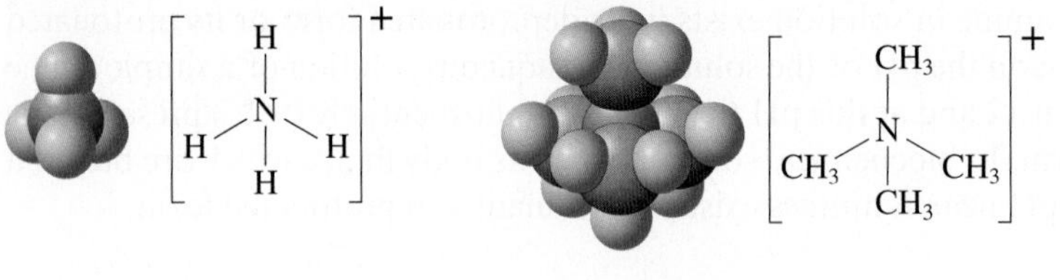

Figure 6.7 Space-filling models showing that the ammonium ion (NH_4^+) has a tetrahedral structure, as does the quaternary ammonium ion, in which four methyl groups are present [$(CH_3)_4N^+$].

Quaternary ammonium salts differ from amine salts in that the addition of a strong base does not convert quaternary ammonium salts back to their "parent" amines; there is no hydrogen atom on the nitrogen with which the OH^- can react. Quaternary ammonium salts are colorless, odorless, crystalline solids that have high melting points and are usually water-soluble.

Quaternary ammonium salts are named in the same way as amine salts (Section 6.7), taking into account that four organic groups are attached to the nitrogen atom rather than a lesser number of groups. Figure 6.7 contrasts the structures of an ammonium ion and a tetramethyl ammonium ion.

Compounds that contain quaternary ammonium ions are important in biochemical systems. Choline and acetylcholine are two important quaternary ammonium ions present in the human body. Choline has important roles in both fat transport and growth regulation. Acetylcholine is involved in the transmission of nerve impulses.

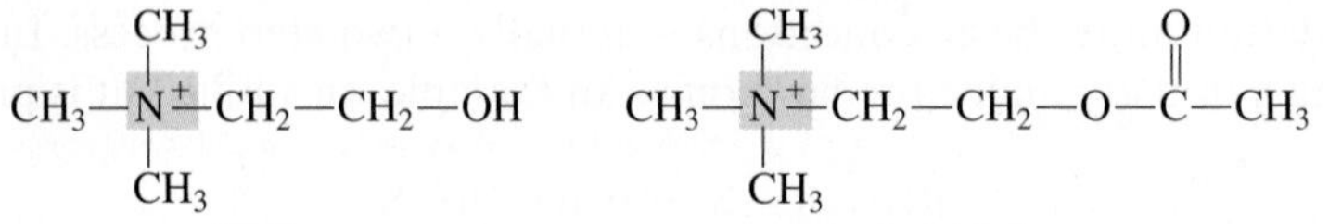

Choline Acetylcholine

6.9 Heterocyclic Amines

Heterocyclic amines are the first heterocyclic compounds we have encountered that have nitrogen heteroatoms. In previous chapters, we have discussed heterocyclic compounds with oxygen as the heteroatom: cyclic ethers (Section 3.19); the cyclic forms of hemiacetals and acetals (Section 4.11); and cyclic esters (Section 5.10).

Heterocyclic amines often have strong odors, some agreeable and others disagreeable. The "pleasant" aroma of many heat-treated foods is caused by heterocyclic amines formed during the heat treatment. The compounds responsible for the pervasive odors of popped popcorn and hot roasted peanuts are heterocyclic amines.

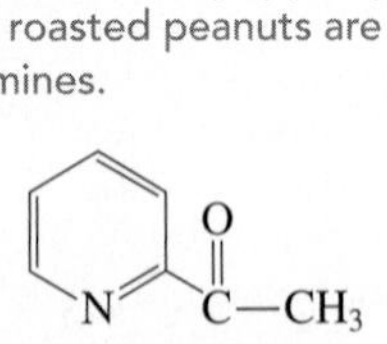

Methyl-2-pyridyl ketone (odor of popcorn)

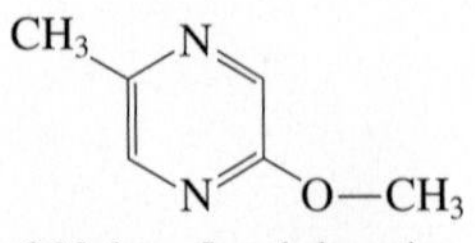

2-Methoxy-5-methylpyrazine (odor of peanuts)

A **heterocyclic amine** *is an organic compound in which nitrogen atoms of amine groups are part of either an aromatic or a nonaromatic ring system.* Heterocyclic amines are the most common type of heterocyclic organic compound (Section 3.19). Figure 6.8 gives structures for a number of "key" unsubstituted heterocyclic amines. These compounds are the "parent" compounds for numerous derivatives that are important in medicinal, agricultural, food, and industrial chemistry, as well as in the functioning of the human body.

Study of the heterocyclic amine structures in Figure 6.8 shows that (1) ring systems may be saturated, unsaturated, or aromatic (2) more than one nitrogen atom may be present in a given ring and (3) fused ring systems often occur.

The two most widely used central nervous system stimulants in our society, caffeine and nicotine, are heterocyclic amine derivatives. Caffeine's structure is

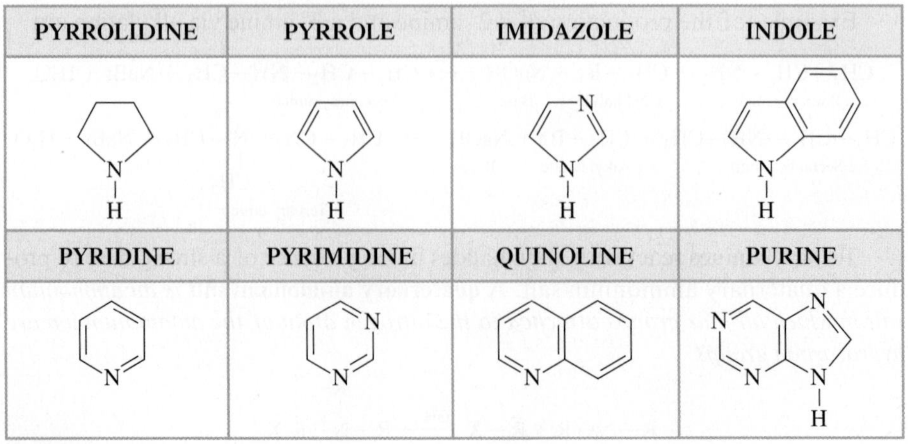

Figure 6.8 Structural formulas for selected heterocyclic amines that serve as "parent" molecules for more complex amine derivatives.

based on a purine ring system. Nicotine's structure contains one pyridine ring and one pyrrolidine ring.

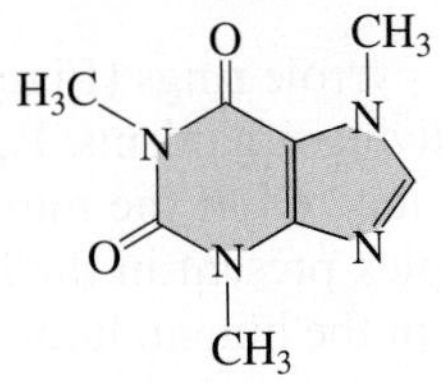

Caffeine

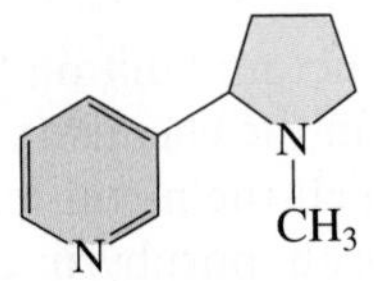

Nicotine

The focus on relevancy feature Chemical Connections 6-A below considers the biochemical effects and mode of action of caffeine in the human body, and the

CHEMICAL CONNECTIONS 6-A

Caffeine: The Most Widely Used Central Nervous System Stimulant

Caffeine, which is naturally present in coffee beans and tea leaves and is added to many soft drinks, is both the most widely used and the most frequently used central nervous system (CNS) stimulant in human society. Its mild temporary stimulant effects are often described as an increased feeling of alertness and/or a decreased feeling of drowsiness.

Studies indicate that 82% of the adult U.S. population drink coffee and 52% drink tea. More than one-half of the coffee drinkers consume three or more cups daily. In the soft drink market, the use of caffeine has now spread beyond cola drinks to fruit-flavored drinks.

Caffeine belongs to a family of compounds called *xanthines*. Its formal chemical name is 1,3,7-trimethylxanthine, and its structure is

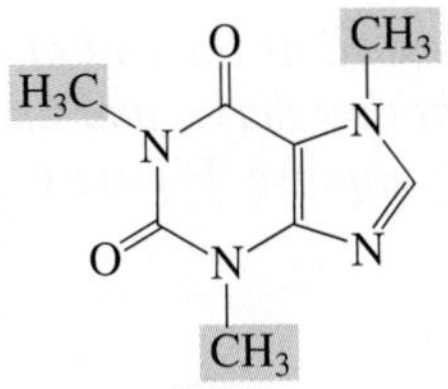

Caffeine
(1,3,7-trimethylxanthine)

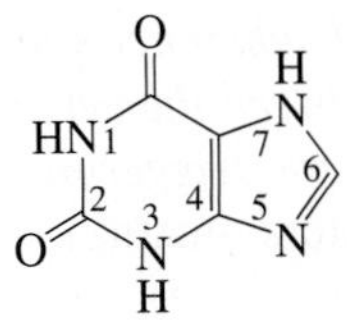

Xanthine

Besides its CNS system effects, caffeine also increases basal metabolic rate, increases heart rate by stimulating heart muscles, promotes secretion of stomach acid, functions as a bronchial tube dilator, and increases urine production because of its diuretic properties. The overall effect that most individuals experience from caffeine consumption is interpreted as a "lift."

Used in small quantities, caffeine's effects are temporary; hence it must be consumed on a regular basis throughout the day. Its half-life in the body is 3.0–7.5 hours. Caffeine tolerance develops in regular users of the substance. Over time, larger amounts of caffeine are needed in order for an individual to achieve his or her "lift."

Caffeine is mildly addicting. Most people who ordinarily consume substantial amounts of caffeine-containing beverages or drugs experience withdrawal symptoms if caffeine is eliminated. Such symptoms may include headache and depression for a period of several days. As a result of caffeine dependence, many people need a cup of coffee before they feel good each morning.

Michael S. Lewis/Getty Images

Caffeine is naturally present in coffee beans. Over 60 plants and trees cultivated by humans contain caffeine.

In large quantities, caffeine has been shown to cause significant undesirable effects including anxiety, sleeplessness, headaches, and dehydration. The latter occurs because of caffeine's diuretic effects.

Several studies indicate that caffeine has an effect on the child bearing abilities of women. One study indicates a decreased rate of conception for women who consume more than three cups of coffee daily. Another study indicates that the consumption of this amount of caffeine increases the risk of miscarriage by 30% compared to women who do not drink coffee. Lactating mothers probably should limit their caffeine intake because it appears in breast milk, thus affecting newborn babies.

Current scientific thought holds that caffeine's mode of action in the body is exerted through a chemical substance called cyclic adenosine monophosphate (cyclic AMP). Caffeine inhibits an enzyme that ordinarily breaks down cyclic AMP to its inactive end product. The resulting increase in cyclic AMP leads to increased glucose production within cells and thus makes available more energy to allow higher rates of cellular activity.

Coffee is the major source of caffeine for most Americans. However, substantial amounts of caffeine may be consumed in soft drinks, tea, and numerous nonprescription medications including combination pain relievers (Anacin, Midol, Empirin), cold remedies (Dristan, Triaminicin), and antisleep agents (NO-DOZ, Vivarin).

focus on relevancy feature Chemical Connections 6-B on the next page details the biochemical effects of nicotine and the effect that pH change has on the action of this molecule.

A large cyclic structure built on four pyrrole rings (Figure 6.8), called a porphyrin, is important in the chemistry of living organisms. Porphyrins form metal ion complexes in which the metal ion is located in the middle of the large ring structure. *Heme,* an iron–porphyrin complex present in the hemoglobin found in blood, is responsible for oxygen transport in the human body.

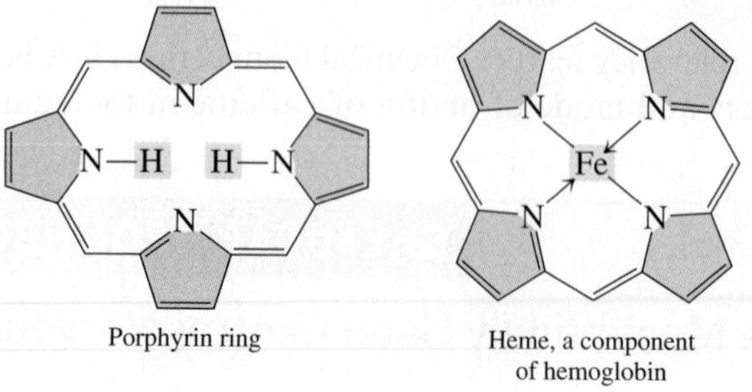

6.10 Selected Biochemically Important Amines

Many simple amines are physiologically active compounds that affect the functioning of the human body. Some are naturally present biosynthesized compounds, others are substances isolated from plants, and still others are laboratory-synthesized products that find use in both prescription and nonprescription drugs. In this section, amines that exert the following four different types of effects are considered: (1) neurotransmitters (2) central nervous system stimulants (3) decongestants and (4) antihistamines.

Neurotransmitters

A **neurotransmitter** *is a chemical substance that is released at the end of a nerve, travels across the synaptic gap between the nerve and another nerve, and then bonds to a receptor site on the other nerve, triggering a nerve impulse.* Figure 6.9 shows schematically how neurotransmitters function.

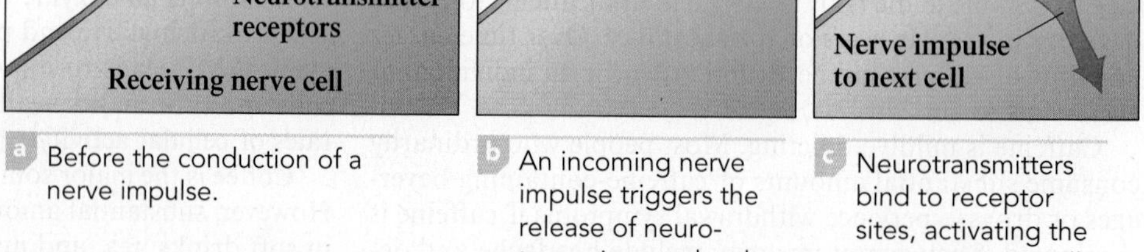

a Before the conduction of a nerve impulse.

b An incoming nerve impulse triggers the release of neuro-transmitter molecules.

c Neurotransmitters bind to receptor sites, activating the receptor nerve cell.

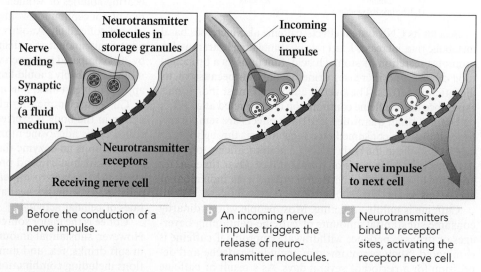

Figure 6.9 Neurotransmitters are chemical messengers between nerve cells. Neurotransmitters released from one nerve cell stimulate (activate) an adjacent nerve cell.

CHEMICAL CONNECTIONS 6-B

Nicotine Addiction: A Widespread Example of Drug Dependence

Next to caffeine, the heterocyclic amine nicotine is the most widely used central nervous system stimulant in our society. It is obtained primarily from the tobacco plant, where it constitutes 0.3% to 0.5% by dry mass of the plant. Biosynthesis of nicotine occurs in the roots of the plant and it then is translocated to the leaves, where it accumulates. In the pure state, in its free base form, nicotine is an oily liquid that is miscible with water. It readily reacts with water to form salts that are usually solids and water soluble, behavior associated with the amine nature of this substance (Section 6.7).

Structurally, nicotine contains both a pyrrolidine and a pyridine ring system (see Figure 6.8); these rings are connected through a carbon–carbon bond rather than fused. Two amine functional groups are present in a nicotine molecule.

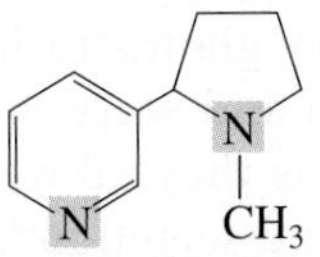

Nicotine

Cigarettes contain 8 to 20 milligrams of nicotine (depending on the brand), but only approximately 1 mg is actually absorbed by the body when a cigarette is smoked. This absorbed nicotine is responsible for the addictive process that develops with continued cigarette smoking. In its free-base form, nicotine will burn and its vapors will combust at 95°C in air. Thus, most of the nicotine is burned when a cigarette is smoked; however, enough is inhaled to provide the desired stimulatory effects.

Two biochemical effects of nicotine are

1. Nicotine acts on acetylcholine receptors, increasing their activity. This leads to an increased flow of adrenaline (Section 6.10), which in turn causes an increase in heart rate, blood pressure, and respiration rate, as well as higher blood-glucose levels.
2. Nicotine increases the release of the brain neurotransmitter dopamine (Section 6.10), which makes a person feel "good." Getting the dopamine boost is part of the addiction process.

Inhaled nicotine is rapidly distributed throughout the body. On average, it takes 7 seconds for nicotine to reach the brain once it is inhaled. The overall stimulatory effect of nicotine is, however, rather transient. After an initial response, depression follows. The half-life of nicotine in the body is between 1 and 2 hours.

Nicotine induces both physiological and psychological dependence. Withdrawal from nicotine is accompanied by headache, stomach pain, irritability, and insomnia. Full withdrawal can take six months or longer.

In large doses, nicotine is a potent poison. Nicotine poisoning causes vomiting, diarrhea, nausea, and abdominal pain. Death occurs from respiratory paralysis. The lethal dose of the drug is considered to be approximately 60 mg, a level that is never reached with cigarette smoking.

Nicotine is formed in the roots of tobacco plants and then translocates to the leaves, where it accumulates.

Nicotine can influence the action of prescription drugs that the smoker is taking concurrently. For example, Propoxyphene (Darvon) can lose its painkilling effect in a smoker, and tranquilizers such as Diazepam (Valium) exert a diminished effect on the central nervous system.

The carcinogenic effects associated with cigarette smoking are caused not by nicotine but by other substances present in tobacco smoke, including fused-ring aromatic hydrocarbons (Section 2.16).

There are actually three forms for a nicotine molecule, with the form adopted determined by the pH of the molecule's environment. At a low pH (less than 3.0) nicotine exists in a diprotonated form; at pH values between 3.0 and 8.0 it is monoprotonated; and at higher pH values it is in a nonprotonated (free-base) form (see Section 6.7).

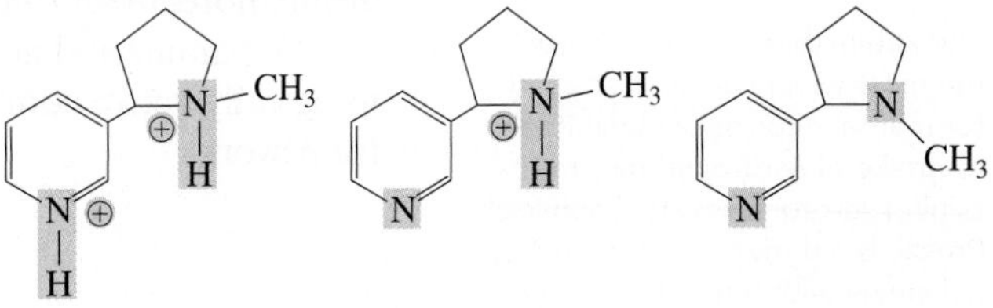

Diprotonated form Monoprotonated form Nonprotonated form (free-base form)

Note that the protonation sites in the molecules are the nitrogen atoms of the amine functional groups (Section 6.7). The nonprotonated (free-base) form of nicotine is very volatile and evaporates (as well as combusts) at the temperatures associated with a burning cigarette.

Cigarette manufacturers purposely add ammonia to tobacco leaves to raise the pH and make the free-base form more available to the smoker. Such pH adjustment also occurs for smokeless tobacco products. Products of lower pH are marketed for new users, and higher pH products—which supply more nicotine to the user—are available for "experienced" users.

The most important neurotransmitters in the human body are acetylcholine (Section 6.8) and the amines serotonin, dopamine, and norepinephrine.

Serotonin, naturally present in the brain, is involved in sleep, sensory perception, and the regulation of body temperature. Serotonin deficiency has been implicated in mental illness. Treatment of mental depression can involve the use of drugs that help maintain serotonin at normals levels by preventing its breakdown in the brain.

The names of the amine neurotransmitters are pronounced SER-oh-tone-in, DOPE-a-mean, and nor-ep-in-NEFF-rin.

Biosynthesis of serotonin depends on the presence of tryptophan, an amino acid (Section 9.2) found in proteins. Turkey meat contains higher levels of tryptophan than most proteins; hence a Thanksgiving dinner is often followed by a case of drowsiness.

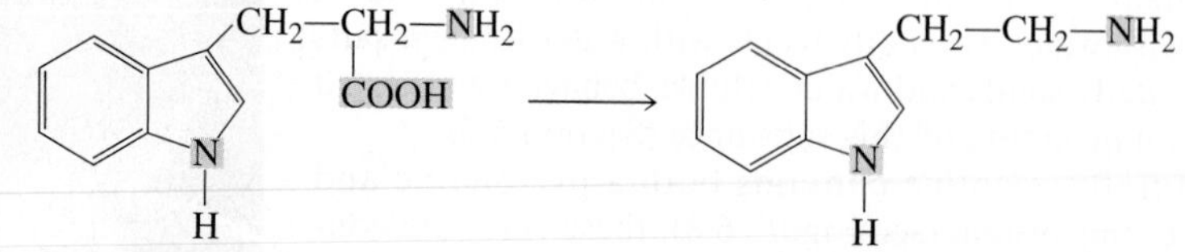

Tryptophan Serotonin

Recent research shows that serotonin also regulates lactation in women. Such serotonin is produced in the mammary glands rather than in the brain. At this location, it controls milk production and secretion. The concentration of serotonin builds up in mammary glands as they fill with milk. The increase in serotonin inhibits further milk synthesis and secretion by suppressing the expression of milk protein genes. After nursing, the cycle of milk and serotonin production begins again.

Dopamine and **norepinephrine** are structurally related compounds whose biosynthesis depends on the presence of tyrosine, an amino acid (Section 9.2) present in proteins such as meats, nuts, and eggs.

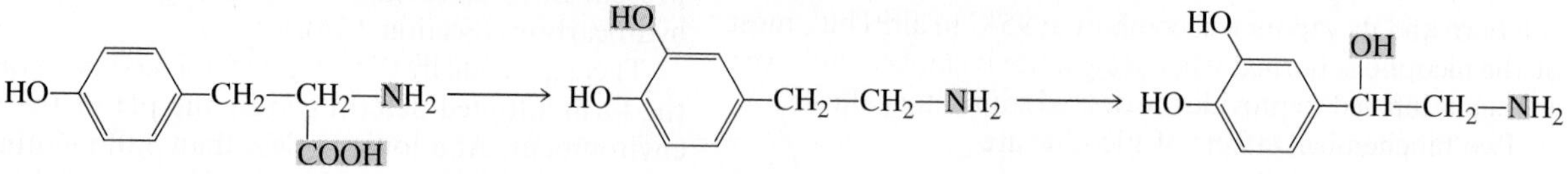

Tyrosine Dopamine Norepinephrine

Note that the structures of dopamine and norepinephrine differ only by a hydroxy group; norepinephrine is a hydroxydopamine.

Dopamine and norepinephrine are two members of an extended family of physiologically active compounds whose structures are based on a 2-phenylethylamine framework.

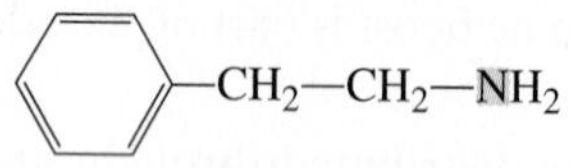

2-Phenylethylamine

Additional 2-phenylmethylamine derivatives will be encountered in later parts of this text section.

Fluoxetine (brand name: Prozac), the most widely prescribed drug for mental depression, inhibits the reuptake of serotonin, thus maintaining serotonin levels. Chemically, Prozac is a derivative of methyl propyl amine with three fluorine atoms present in the structure. The element fluorine is rarely encountered in biochemical molecules.

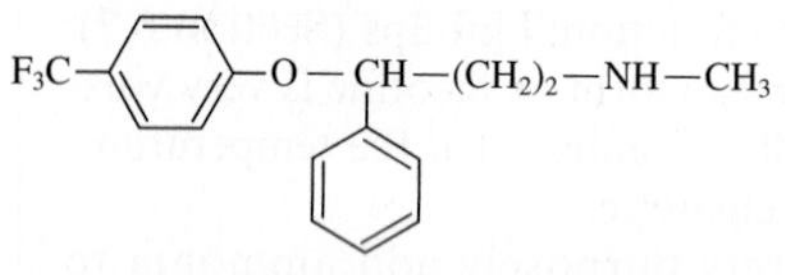

Dopamine is found in the brain. A deficiency of this neurotransmitter results in Parkinson's disease, a degenerative neurological disease. Administration of dopamine to a patient does not relieve the symptoms of this disease because dopamine in the blood cannot cross the blood-brain barrier. The drug L-dopa, which can pass through the blood-brain barrier, does give relief from Parkinson's disease. Inside brain cells, enzymes catalyze the conversion of L-dopa to dopamine.

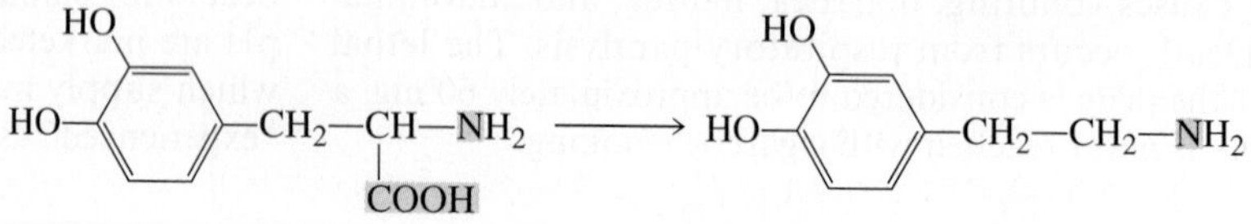

L-Dopa Dopamine

Norepinephrine is also needed at appropriate levels by a properly functioning brain. Higher than normal levels of this neurotransmitter cause a person to feel elated; however, extremely high levels can cause manic behavior.

Besides functioning as a neurotransmitter, norepinephrine can also function as a central nervous system stimulant. Norepinephrine produced by the adrenal glands exhibits this stimulant function. The normal mode of action for adrenal norepinephrine is to constrict (narrow) blood vessels, which increases blood pressure, and also to increase blood-glucose levels. Administered as a drug, norepinephrine is used to treat life-threatening low blood pressure that occurs when a person goes into septic shock caused by severe infection. It is also often used during CPR (cardiopulmonary resuscitation) procedures by medical personnel. Normally, norepinephrine chemistry is, however, associated with maintaining normal body chemistry rather than treating emergency conditions.

Shock is a medical condition that can be caused by trauma, heatstroke, blood loss, an allergic reaction, severe infection, poisoning, and severe burns. The organs of a person in shock are not getting enough blood and/or oxygen, which, if left untreated, can lead to permanent organ damage or death.

Central Nervous System Stimulants

Epinephrine, also known as adrenaline, is the most well-known central nervous system stimulant found in the human body. It is produced by the adrenal glands from norepinephrine.

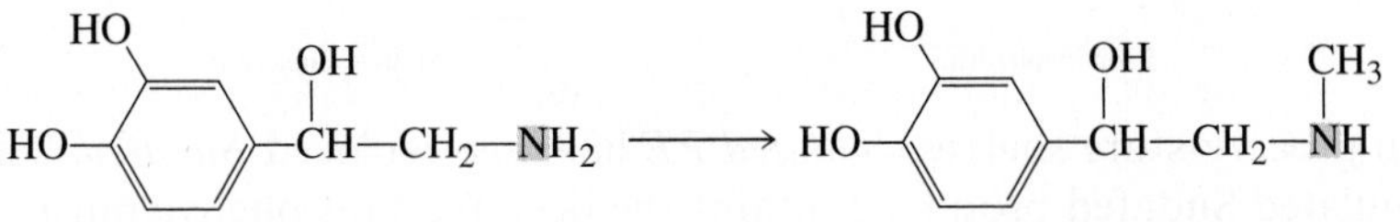

Norepinephrine — Epinephrine (adrenaline)

Note from these structures that epinephrine and norepinephrine differ in structure only in that the former has a methyl group substituent present on the nitrogen atom.

Pain, excitement, and fear trigger the release of large amounts of epinephrine into the bloodstream. The effect is increased blood-glucose levels, which in turn increase blood pressure, rate and force of heart contraction, and muscular strength. These changes cause the body to function at a "higher" level. Epinephrine is often called the "fight or flight" hormone.

The use of epinephrine as a drug is often the immediate treatment for anaphylactic shock resulting from a severe allergic reaction to a bee sting, a drug, or even peanuts. People known to be susceptible to such allergic events often carry a dose of epinephrine for emergency use (Figure 6.10).

Amphetamine and **methamphetamine,** both of which have 2-phenylethylamine derivative structures, are powerful *synthetic* central nervous system stimulants.

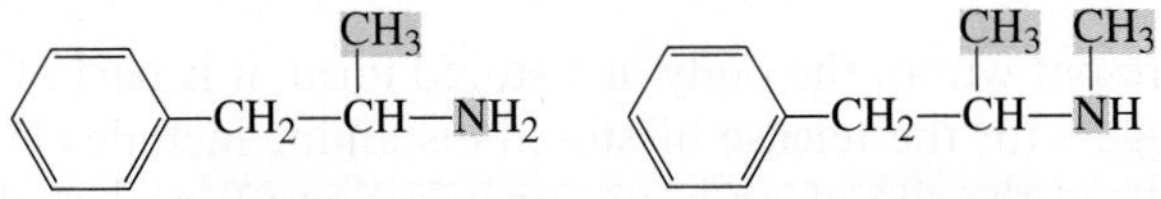

Amphetamine — Methamphetamine

Amphetamine, under the trade name Adderall, is approved by the U.S. Food and Drug Administration (FDA) for treatment of ADHD (Attention Deficit Hyperactivity Disorder). It increases alertness, concentration, and overall cognitive performance while decreasing fatigue. Methamphetamine is also approved by the FDA for treatment of ADHD under the tradename Desoxyn.

Both amphetamine and methamphetamine have a high potential for abuse and addiction. In the illegal drug market, methamphetamine is colloquially referred to simply as "meth." Pharmacologically, the physiological reward system accompanying the use of these synthetic central nervous system stimulants results from increased levels of serotonin, dopamine, and norepinephrine in the brain. These stimulants were once approved as appetite suppressants to help with weight control, but such use is now strongly discouraged.

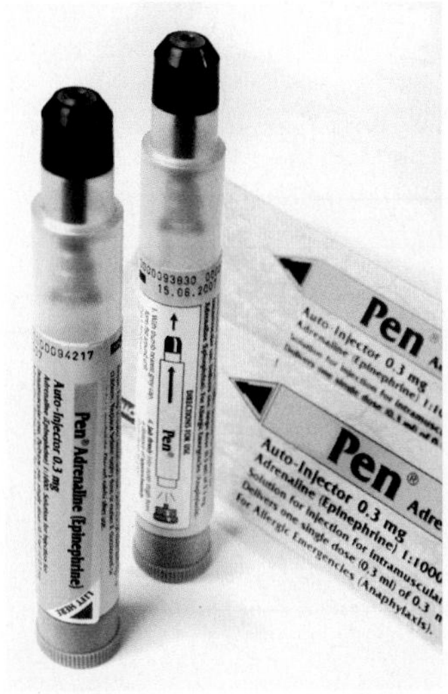

Medic Image/Universal Images Group/Getty Images

Figure 6.10 Individuals at known risk for anaphylactic shock caused by an allergen often carry with them at all times an epinephrine autoinjection pen.

Decongestants

Pseudoephedrine and **phenylephrine** are synthetic 2-phenylethylamine derivatives used as decongestants in several over-the-counter medications.

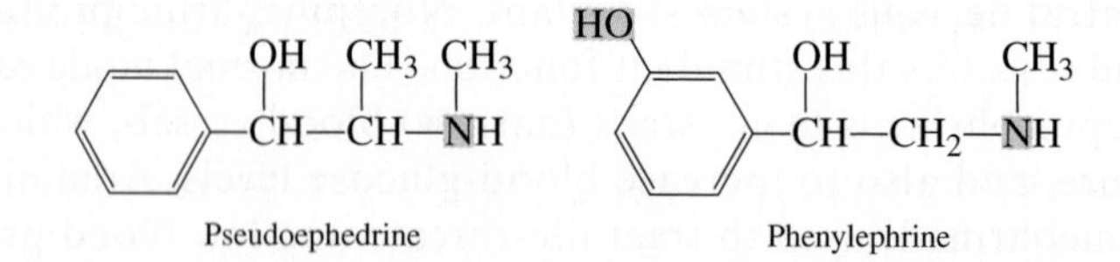

Pseudoephedrine is the decongestant present in the formulation marketed under the brand name *Sudafed.* Although still available for purchase without a prescription, Sudafed and other products containing this decongestant must now be kept behind the counter in a pharmacy so that their sale can be monitored. The reason for this recent change is that individuals within the illegal "meth" culture have discovered that pseudoephedrine can be easily converted to methamphetamine ("meth"). The structures of the two compounds differ only by a hydroxyl group.

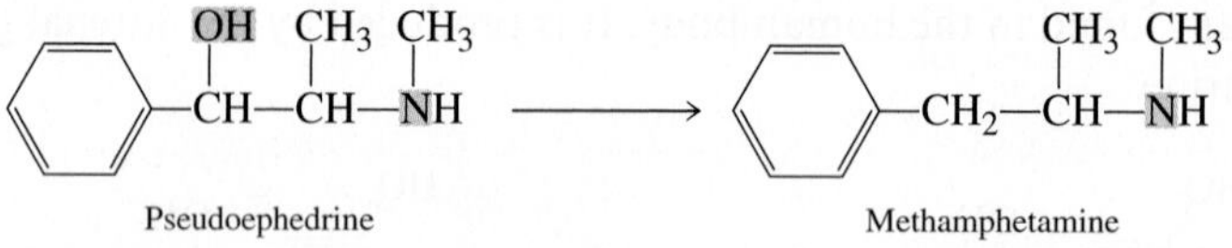

On grocery-store shelves, *Sudafed PE* has now replaced *Sudafed.* This new reformulated Sudafed product contains the decongestant phenylephrine rather than pseudoephedrine. The addition of a —OH group (on the benzene ring) and the removal of a methyl group (from the carbon chain) is a sufficient structural change to prevent this compound from being converted to methamphetamine.

Antihistamines

Antihistamines are medications that counteract the physiological effects of a naturally occurring substance within the body called histamine.

Histamine, a heterocyclic amine, is found in tissues throughout the human body. It is biosynthesized from the amino acid histidine, a component of protein (Section 9.2).

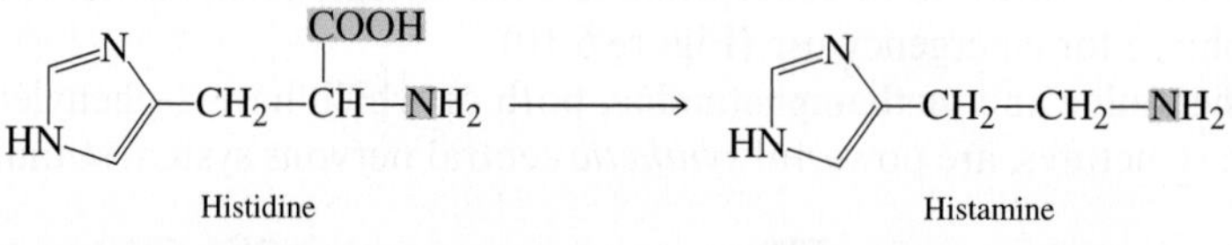

Histamine is present within the body in a stored form; it is part of more complex molecules. Triggers for the release of stored histamine include (1) substances released from cells at the site of an injury or infection (2) pollen, dust, and other allergens (3) viruses, such as that of the common cold and (4) chemicals to which an individual has become sensitized.

A major, though not the sole, function of histamine release is immunological in nature. Histamine release is part of the body's immune system response to invasion by a "foreign body." The "free" histamine increases the permeability, through dilation, of blood vessels, making them abnormally permeable to white blood cells and proteins needed to combat the invading foreign entities.

However, this increased vascular permeability also allows fluids to escape from the blood vessels to surrounding tissues, causing the classic symptoms of allergic reactions such as hay fever, as well as the common-cold symptoms of a runny nose and watery eyes.

Taken as medication, antihistamines counteract to some extent the "runny-nose, watery-eyes" problem. The molecular structures of antihistamines are such that they can occupy the receptor sites in nerves normally occupied by histamine molecules, thus blocking histamine from occupying the sites and exerting its effects.

A side effect of many antihistamine formulations is sedation (drowsiness); antihistamines cross the blood-brain barrier and affect receptors in the brain. Newer antihistamine formulations are now available that are nonsedating; they have molecular structures that prevent them from crossing the blood-brain barrier. An example of a nonsedating antihistamine is fexofenodine (brand name Allegra), which has the following complex heterocyclic amine structure.

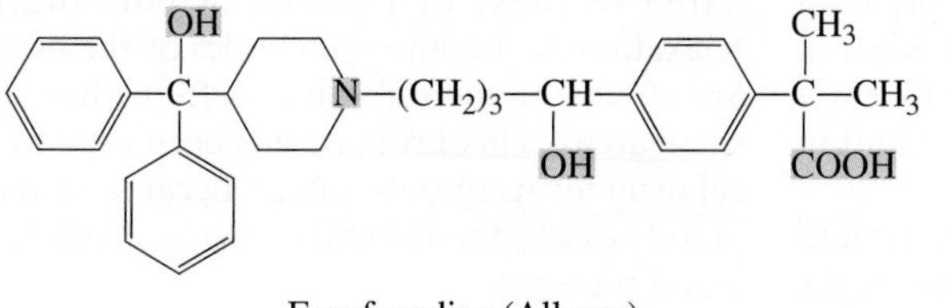

Fexofenodine (Allegra)

6.11 Alkaloids

People in various parts of the world have known for centuries that physiological effects can be obtained by eating or chewing the leaves, roots, or bark of certain plants. More than 5000 different compounds that are physiologically active have been isolated from such plants. Nearly all of these compounds, which are collectively called alkaloids, contain amine functional groups. An **alkaloid** *is a nitrogen-containing organic compound extracted from plant material.*

The name *alkaloid,* which means "like a base," reflects the fact that alkaloids react with acids. Such behavior is expected for substances with amine functional groups because amines are weak bases.

Three well-known substances previously considered in this chapter are alkaloids, although they were not called such at the time they were discussed. They are nicotine (obtained from the tobacco plant), caffeine (obtained from coffee beans and tea leaves), and cocaine (obtained from the coca plant). The active ingredient in chocolate, a substance called theobromine, is also an alkaloid that comes from beans. Theobromine's source is the cocoa tree, and its structure is related to that of caffeine. Theobromine's properties are, however, very different from those of caffeine. The focus on relevancy feature Chemical Connections 6-C on the next page considers further the alkaloid content of chocolate products.

A number of alkaloids are currently used in medicine. Quinine, which occurs in cinchona bark, is used to treat malaria; it is a powerful antipyretic (fever-reducer). Atropine, which is isolated from the belladonna plant, is used to dilate the pupil of the eye in patients undergoing eye examinations (Figure 6.11). Atropine is also used as a preoperative drug to relax muscles and reduce the secretion of saliva in surgical patients.

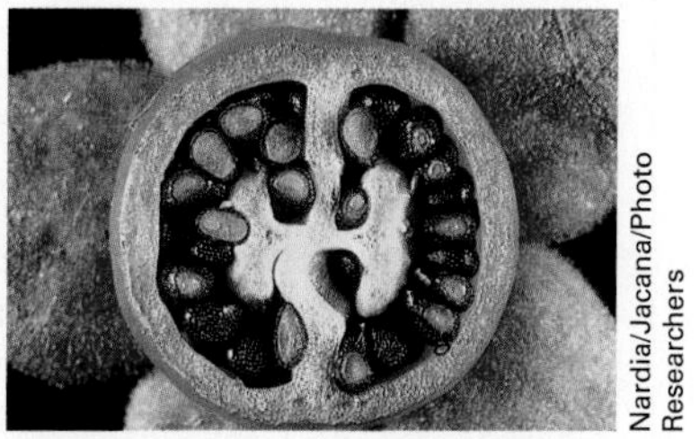

Figure 6.11 Fruit of the belladonna plant; the alkaloid atropine is obtained from this plant.

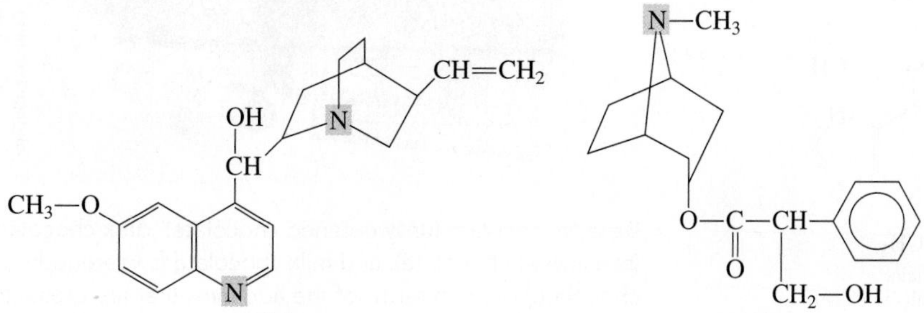

Quinine Atropine

Morphine, Codeine, and Heroin

An extremely important family of alkaloids is the narcotic painkillers, a class of drugs derived from the resin (opium) of the oriental poppy plant (see Figure 6.12). The most important drugs obtained from opium are morphine and codeine. Synthetic modification of morphine produces the illegal drug heroin.

CHEMICAL CONNECTIONS 6-C

Alkaloids Present in Chocolate

Chocolate is a food preparation made from the beans (seeds) of the tropical cacao tree. Growth conditions for cacao trees require a warm, moist climate like that found near the equator. The majority of the world's supply of cacao beans comes from the west coast of Africa—Ivory Coast, Ghana, and Nigeria. (Because of a mistake in spelling, probably made by early English importers, cacao beans are known as cocoa beans in English-speaking countries.)

All chocolate products are manufactured from ground cocoa beans. The heat from the grinding process causes the cocoa bean mixture to melt, forming a free-flowing mixture called *chocolate liquor.* Baker's chocolate (unsweetened chocolate) is simply cooled, hardened chocolate liquor. Dark chocolate (semisweet chocolate) has small amounts of sugar added. Milk chocolate (sweetened chocolate) has larger amounts of sugar, as well as milk solids and vanilla flavoring, added.

Because of their plant origins, chocolate products contain alkaloids. The dominant alkaloid present is theobromine, with caffeine being present in a smaller amount. The name theobromine comes from the Greek term *theobroma* meaning "food of the Gods." The concentrations of these two alkaloids in cocoa beans varies depending on the origin of the beans. The following table gives the theobromine and caffeine content of several types of chocolate.

Theobromine and Caffeine Content of Various Types of Chocolate

Product	Theobromine %	Caffeine %	Theobromine/ Caffeine Ratio
baker's chocolate	1.386	0.164	8.45 to 1
dark chocolate	0.474	0.076	6.3 to 1
milk chocolate	0.197	0.022	9.0 to 1

The caffeine content of a typical chocolate bar is 30 mg and that of a slice of chocolate cake 20–30 mg. By contrast, a cup of coffee contains 100–150 mg of caffeine and a twelve-ounce cola drink contains 33–52 mg.

Structurally, theobromine and caffeine differ only by a methyl group.

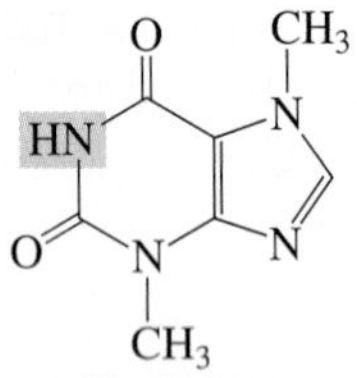

Theobromine
(3,7-dimethyxanthine)

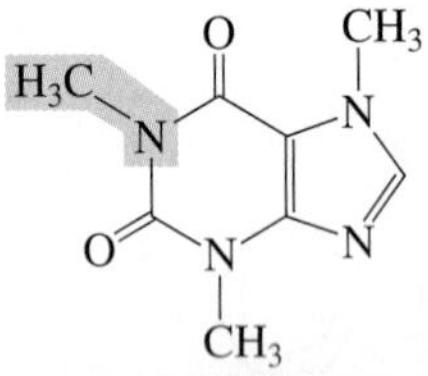

Caffeine
(1,3,7-trimethylxanthine)

This close structural similarity does not, however, translate into close pharmacological properties. Theobromine's central nervous system stimulant effects are minimal compared to those of caffeine. A mild diuretic effect and relaxation of the smooth muscles of the bronchi in the lungs are effects of theobromine; caffeine has similar effects in these areas. Theobromine has been used as a pharmaceutical drug for its diuretic effect. Because of its ability to dilate blood vessels, theobromine also has been used to treat high blood pressure.

Research shows that pets, especially dogs, are sensitive to theobromine because the animals metabolize theobromine more slowly than humans. A chocolate bar, when ingested, is poisonous to dogs and can even be lethal. The same holds true for cats.

Occasionally, chocolate is touted as a "health" food because cocoa beans have relatively high levels of several kinds of antioxidant flavonoids (see Chemical Connections 12-C on page 539). Studies show that people with high blood levels of flavonoids are at lower risk of developing heart disease, asthma, and type 2 diabetes. Dark chocolate contains the most cocoa and thus the most flavonoids. As a "health" food, however, chocolate should be consumed only occasionally because the downside of consumption is the high number of calories associated with chocolate.

Baker's chocolate (unsweetened chocolate), dark chocolate (semisweet chocolate), and milk chocolate (sweetened chocolate) differ in terms of the additives that are present.

These three compounds have similar chemical structures.

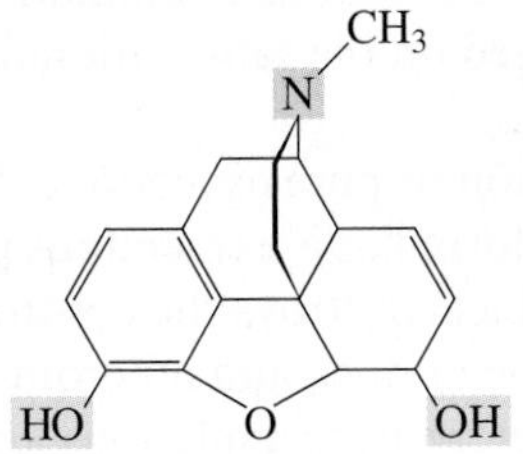

Morphine

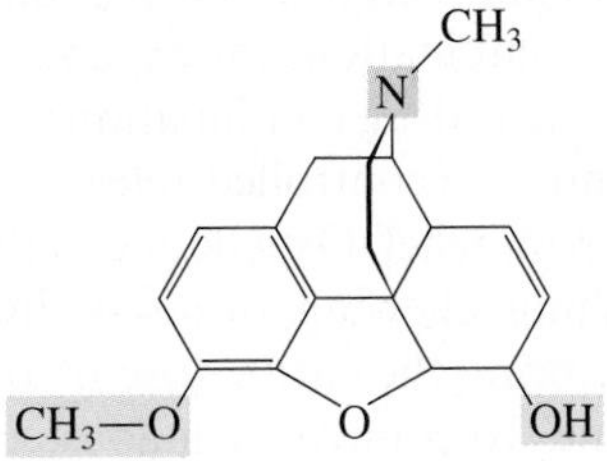

Codeine

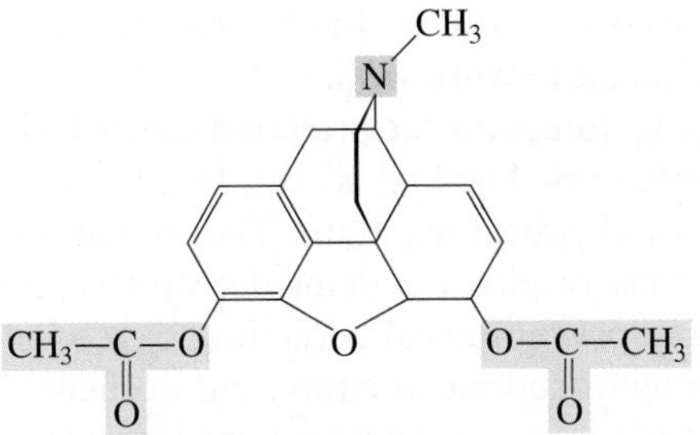

Heroin

Figure 6.12 Oriental poppy plants, the source of several narcotic painkillers, including morphine.

Morphine is one of the most effective painkillers known; its painkilling properties are about a hundred times greater than those of aspirin. Morphine acts by blocking the process in the brain that interprets pain signals coming from the peripheral nervous system. The major drawback to the use of morphine is that it is addictive.

Codeine is a methylmorphine. Almost all codeine used in modern medicine is produced by methylating the more abundant morphine. Codeine is less potent than morphine, having a painkilling effect about one-sixth that of morphine.

Heroin is a semi-synthetic compound, the diacetyl ester of morphine; it is produced from morphine. This chemical modification increases painkilling potency; heroin has more than three times the painkilling effect of morphine. Heroin's greater painkilling ability results from its molecular structure being less polar than that of morphine; hence it is more soluble in the body's fatty tissues. However, heroin is so addictive that it has no accepted medical use in the United States.

The demand for a painkiller as strong as morphine is very limited, and natural supply far exceeds demand. Demand for the much weaker codeine is, however, many times greater than its natural supply. Approximately 95% of the morphine obtained from legally grown poppy plants is converted to codeine, which is then converted to codeine derivatives, primarily hydrocodone and oxycodone.

Hydrocodone, Oxycodone, and OxyContin

The most widely prescribed painkillers in the United States, at present, for the relief of moderate to heavy pain contain the codeine derivatives hydrocodone or oxycodone as an active ingredient. Usually a second painkilling agent is present in addition to the codeine derivative.

The total alkaloid content of a poppy plant is about 20% codeine and 70% morphine. Biosynthetically, a poppy plant first produces codeine, which is then modified to generate morphine. Plant biochemists recently discovered the gene needed for the poppy plant to convert codeine to morphine. Using genetic engineering procedures to modify poppy plants so that the expression of this gene was reduced, it was found that codeine levels increased from 20% to 65% of the total alkaloid content whereas morphine levels were reduced from 70% to 15% of the total alkaloid content. This discovery may eventually lead to better methods for obtaining increased amounts of codeine directly from poppy plants.

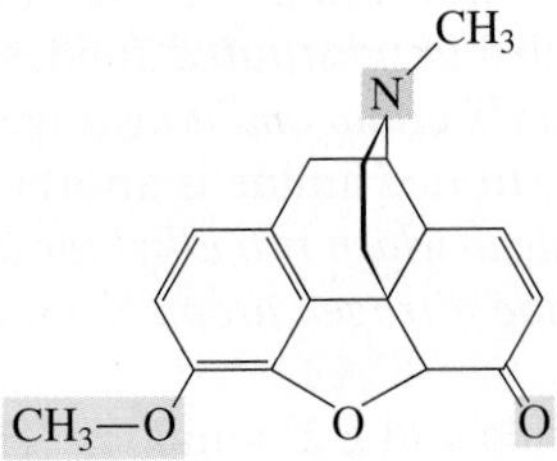

Hydrocodone

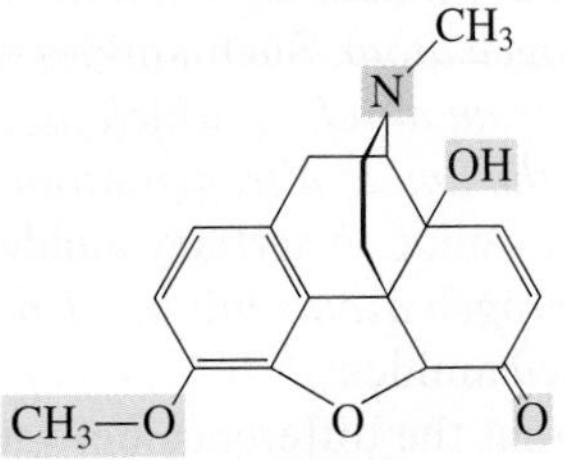

Oxycodone

Lortab and Norco are examples of hydrocodone-containing formulations in which acetaminophen is the second painkilling agent present. The painkilling effect of this two-drug combination is actually greater than that of the sum of the individual effects. The presence of the acetaminophen also limits the potential for the unsafe addictive side effects associated with higher than prescribed doses of hydrocodone because of the liver-toxicity problem associated with higher doses of the accompanying acetaminophen.

Poppy seeds contain small amounts of morphine. Eating poppy-seed cake, bagels, or muffins can often introduce enough morphine into a person's body to produce a positive test in random drug screening procedures.

Oxycodone-containing formulations include Percodan (an aspirin-oxycodone combination) and Percocet (an acetaminophen-oxycodone combination). The use of two-drug combinations for oxycodone is based on the same rationale as that for hydrocodone two-drug combinations.

OxyContin is a controlled-release formulation of pure oxycodone that provides 12 hours of pain relief. Oxycodone in this pure form has a tremendous potential for abuse. Crushing, chewing, or dissolving the tablets destroys the controlled-release feature, delivering the entire dose of oxycodone at once and a heroin-like high at the same time. To combat such abuse, abuse-deterrent formulations are in various stages of development. These new formulations are designed to be more resistant to mechanical forces such as crushing. In the presence of water, the formulations form a "goo" that is difficult to work with.

There are two driving forces for the preferred use of hydrocodone formulations over oxycodone formulations. The first is cost. Hydrocodone production requires a less-costly sequence of chemical reactions. The second is less potential for abuse of the drug. While the use of either codeine derivative can be habit-forming, and can lead to physical and psychological addiction, oxycodone effects are generally greater than those for hydrocodone at equivalent dosages. A driving force for the use of oxycodone formulations over hydrocodone formulations is that the onset of pain relief occurs faster for oxycodone.

6.12 Structure and Classification of Amides

An **amide** *is a carboxylic acid derivative in which the carboxyl —OH group has been replaced with an amino or a substituted amino group.* The amide functional group is thus

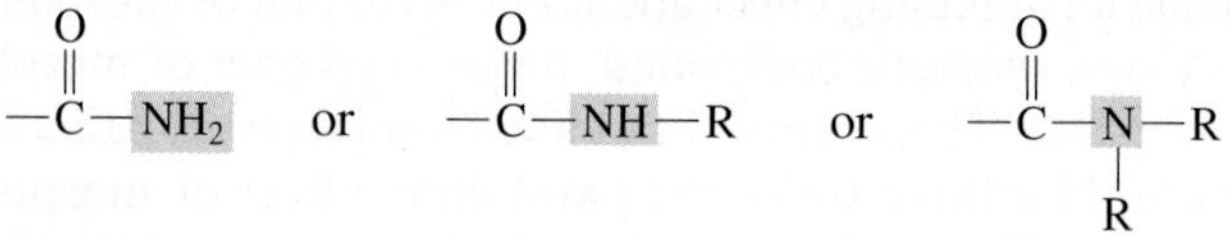

depending on the degree of substitution.

Amides, like amines, can be classified as primary (1°), secondary (2°), or tertiary (3°), depending on how many hydrogen atoms are attached to the nitrogen atom.

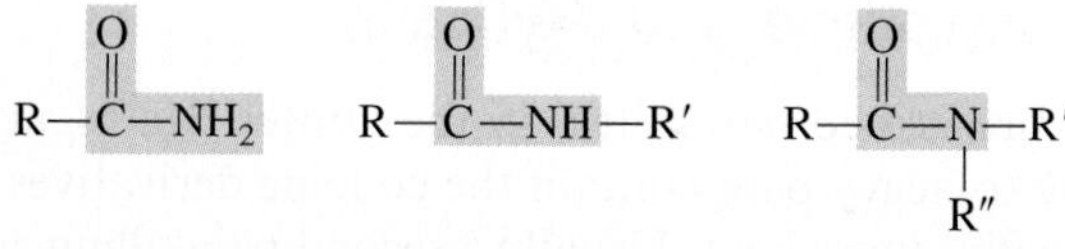

Primary amide   Secondary amide   Tertiary amide

Primary, secondary, and tertiary amides are also called unsubstituted, monosubstituted, and disubstituted amides, respectively.

A **primary amide** *is an amide in which two hydrogen atoms are bonded to the amide nitrogen atom.* Such amides are also called *unsubstituted* amides. A **secondary amide** *is an amide in which an alkyl (or aryl) group and a hydrogen atom are bonded to the amide nitrogen atom. Monosubstituted* amide is another name for this type of amide. A **tertiary amide** *is an amide in which two alkyl (or aryl) groups and no hydrogen atoms are bonded to the amide nitrogen atom.* Such amides are *disubstituted* amides.

Note that the difference between a 1° amide and a 2° amide is "H versus R" and that the difference between a 2° amide and a 3° amide is again "H versus R." These "H versus R" relationships are the same relationships that exist between 1° and 2° amines and 2° and 3° amines (Section 6.2), as is summarized in Figure 6.13.

The simplest amide has a hydrogen atom attached to an unsubstituted amide functional group.

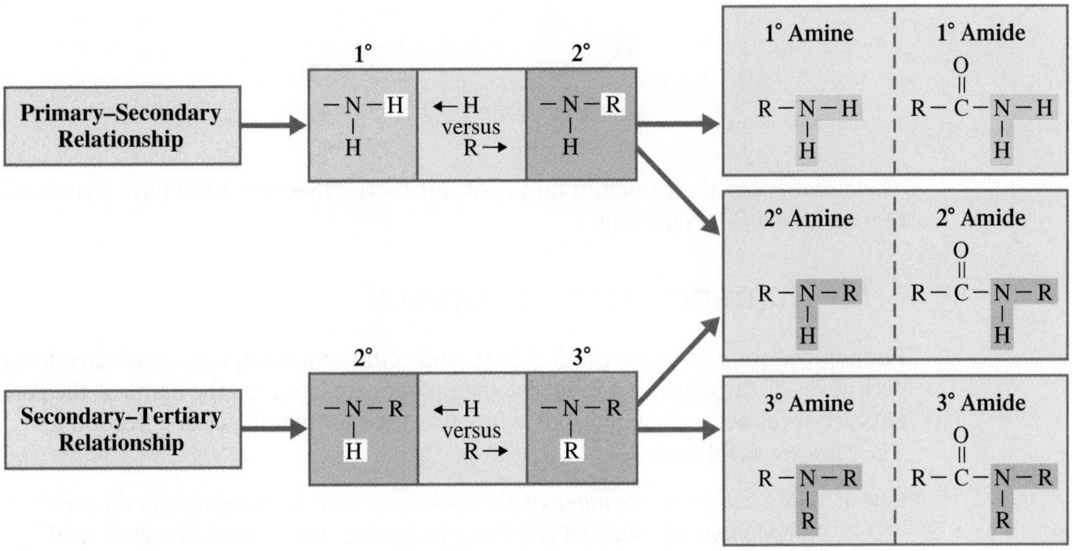

Figure 6.13 Primary, secondary, and tertiary amines and amides and the "H versus R" relationship.

Next in complexity are amides in which a methyl group is present. There are two of them, one with the methyl group attached to the carbon atom and the other with the methyl group attached to the nitrogen atom.

$$CH_3{-}\overset{\overset{\displaystyle O}{\|}}{C}{-}NH_2 \quad \text{and} \quad H{-}\overset{\overset{\displaystyle O}{\|}}{C}{-}NH{-}CH_3$$

Line-angle structural formulas or selected primary, secondary, and tertiary amides:

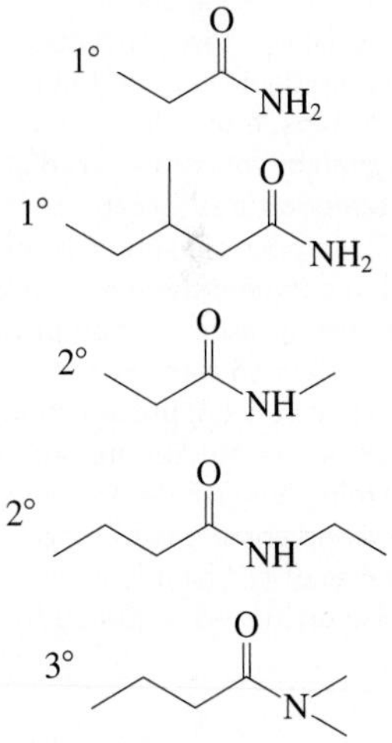

The first of these structures is a 1° amide, and the second structure is a 2° amide. The structure of the simplest aromatic amide involves a benzene ring to which an unsubstituted amide functional group is attached.

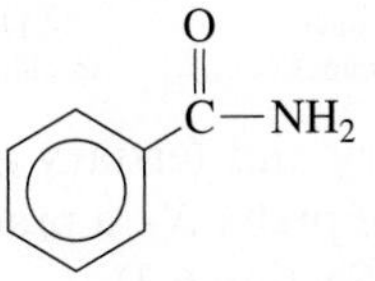

Cyclic amide structures are possible. Examples of such structures include

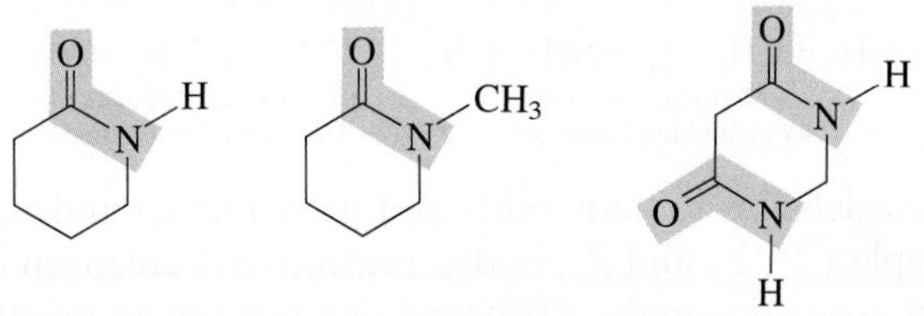

Cyclic amides are called *lactams,* a term that parallels the use of the term *lactones* for cyclic esters (Section 5.11).

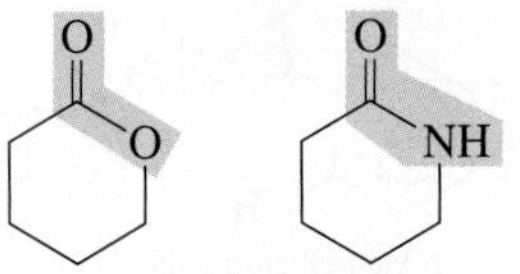

A lactone (a cyclic ester)　　A lactam (a cyclic amide)

A **lactam** *is a cyclic amide.* As with lactones (Section 5.11), the ring size in a lactam is indicated using a Greek letter. A lactam with a four-membered ring is a β-lactam because the β carbon from the carbonyl group is bonded to the heteroatom (O) of the ring. A lactam with a five-membered ring is a γ-lactam because the γ-carbon is bonded to the heteroatom (O) of the ring system.

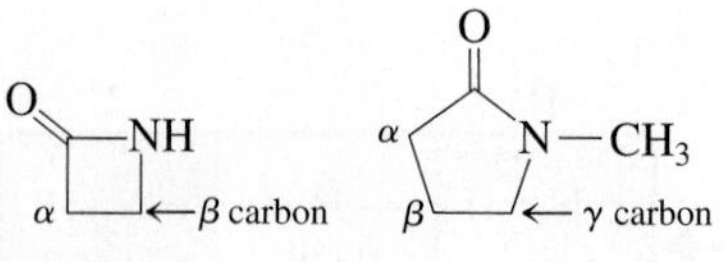

β-Lactam γ-Lactam

The members of the penicillin family of antibiotics (Section 10.9) have structures that contain a β-lactam ring.

6.13 Nomenclature for Amides

For nomenclature purposes (both IUPAC and common), amides are considered to be derivatives of carboxylic acids. Hence their names are based on the name of the parent carboxylic acid. (A similar procedure was used for naming esters; Section 5.11.) The rules are as follows:

Rule 1: *The ending of the name of the carboxylic acid is changed from* -ic acid *(common) or* -oic acid *(IUPAC) to* -amide. *For example,* benzoic acid *becomes* benzamide.

Rule 2: *The names of groups attached to the nitrogen (2° and 3° amides) are appended to the front of the base name, using an* N- *prefix as a locator.*

Acrylamide (2-propenamide), the simplest unsaturated amide, has the structure

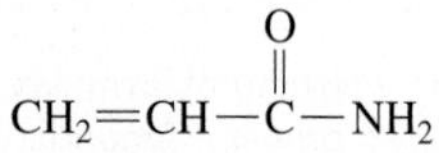

It is a neurotoxic agent and a possible human carcinogen.

Surprisingly, low concentrations of acrylamide have been found in potato chips, french fries, and other starchy foods prepared at high temperatures (greater than 120°C). Its possible source is the reaction between the amino acid asparagine (present in food proteins; Section 9.2) and carbohydrate sugars (Section 7.8) present in food.

Human risk studies are underway concerning acrylamide presence in fried and some baked foods. No traces of acrylamide have been found in uncooked or boiled foods.

Selected primary amide IUPAC names (with the common name in parentheses) are

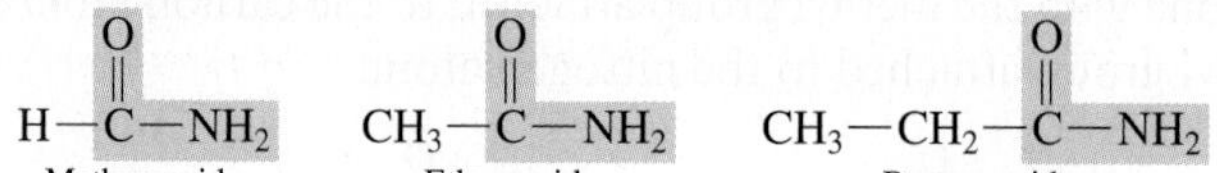

Methanamide (formamide) Ethanamide (acetamide) Propanamide (propionamide)

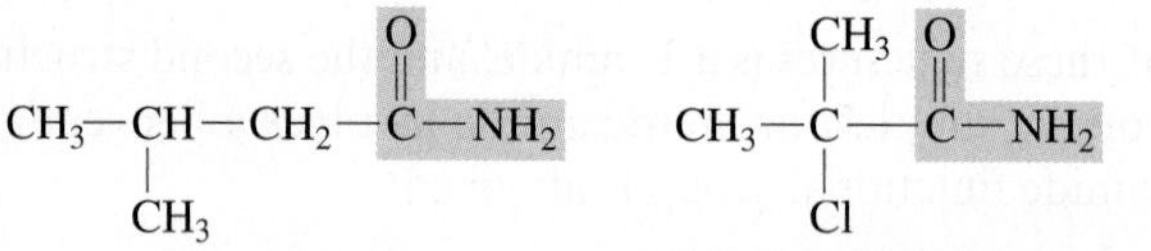

3-Methylbutanamide (β-methylbutyramide) 2-Chloro-2-methylpropanamide (α-chloro-α-methylpropionamide)

Nomenclature for secondary and tertiary amides, amides with substituted amino groups, involves use of the prefix *N*-, a practice that was previously encountered with amine nomenclature (Section 6.3).

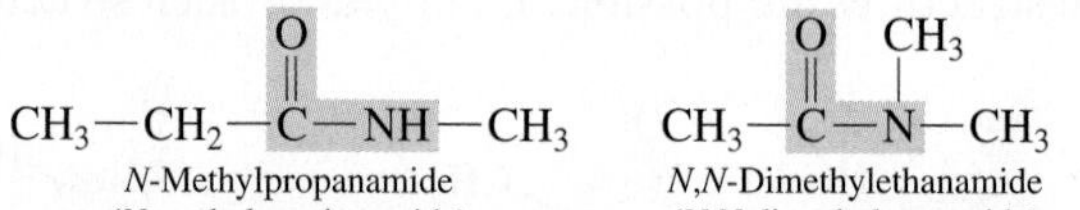

N-Methylpropanamide (*N*-methylpropionamide) *N,N*-Dimethylethanamide (*N,N*-dimethylacetamide)

Molecular models for methanamide and its *N*-methyl and *N,N*-dimethyl derivatives (the simplest 1°, 2°, and 3° amides, respectively) are given in Figure 6.14.

The simplest aromatic amide, a benzene ring bearing an unsubstituted amide group, is called *benzamide*. Other aromatic amides are named as benzamide derivatives.

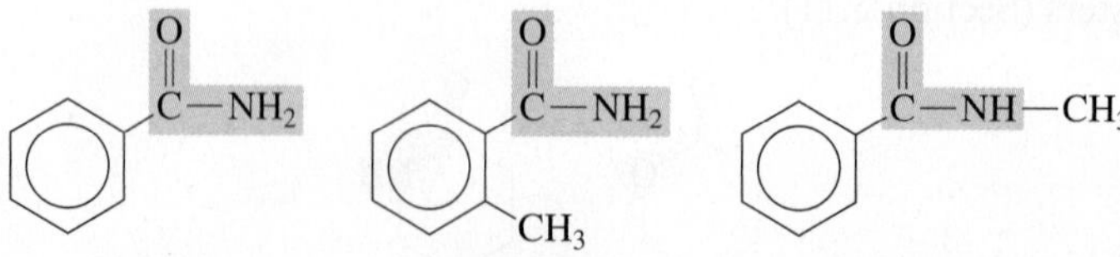

Benzamide 2-Methylbenzamide *N*-Methylbenzamide

When an amide functional group is attached to a nonaromatic ring, the suffix *carboxamide* is used in the name.

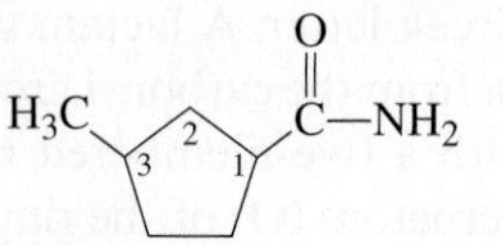

3-methylcyclopentanecarboxamide

EXAMPLE 6.5 Determining IUPAC and Common Names for Amides

Assign both common and IUPAC names to each of the following amides.

a. $CH_3{-}CH_2{-}CH_2{-}\overset{\overset{\displaystyle O}{\|}}{C}{-}NH_2$

b. $CH_3{-}\overset{\overset{\displaystyle Br}{|}}{CH}{-}\overset{\overset{\displaystyle O}{\|}}{C}{-}NH{-}CH_3$

c.

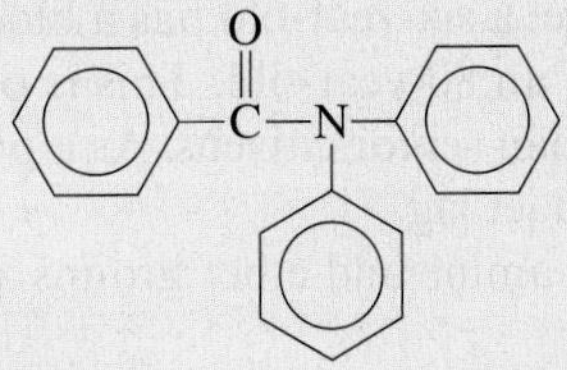

d.

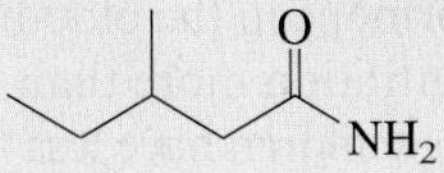

Solution

a. The parent acid for this amide is butyric acid (common) or butanoic acid (IUPAC). The common name for this amide is *butyramide,* and the IUPAC name is *butanamide.*

b. The common and IUPAC names of the acid are very similar; they are propionic acid and propanoic acid, respectively. The common name for this amide is *α-bromo-N-methylpropionamide,* and the IUPAC name is *2-bromo-N-methylpropanamide.* The prefix *N-* must be used with the methyl group to indicate that it is attached to the nitrogen atom.

c. In both the common and IUPAC systems of nomenclature, the name of the parent acid is the same: benzoic acid. The name of the amide is *N,N-diphenylbenzamide.*

d. The amide is derived from valeric acid (common name) or pentanoic acid (IUPAC name). The complete name must take into account the presence of the methyl group on the carbon chain. The amide's common name is *β-methylvaleramide* and its IUPAC name is *3-methylpentanamide.*

Practice Exercise 6.5

Assign both common and IUPAC names to each of the following amides.

a. $CH_3{-}CH_2{-}\underset{\underset{\displaystyle Br}{|}}{CH}{-}\overset{\overset{\displaystyle O}{\|}}{C}{-}NH_2$

b. $CH_3{-}\overset{\overset{\displaystyle O}{\|}}{C}{-}NH{-}CH_3$

c.

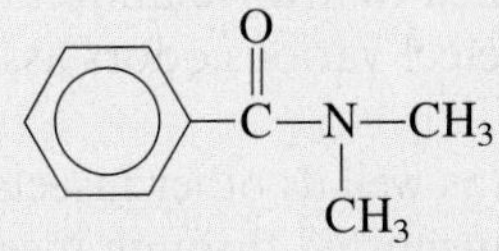

d.

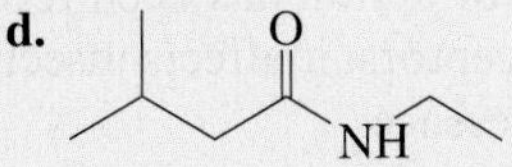

Answers: **a.** *α*-bromobutyramide, 2-bromobutanamide; **b.** *N*-methylacetamide, *N*-methylethanamide; **c.** *N,N*-dimethylbenzamide (both common and IUPAC name); **d.** *N*-ethyl-*β*-methylbutyramide, *N*-ethyl-3-methylbutanamide

The contrast between IUPAC names and common names for unbranched unsubstituted amides is as follows:

IUPAC (one word)

alkanamide

ethanamide

Common (one word)

(prefix)amide*

acetamide

*The common-name prefixes are related to natural sources for the acids.

$H{-}\overset{\overset{\displaystyle O}{\|}}{C}{-}NH_2$

Methanamide (a primary amide)

$H{-}\overset{\overset{\displaystyle O}{\|}}{C}{-}NH{-}CH_3$

***N*-Methyl methanamide (a secondary amide)**

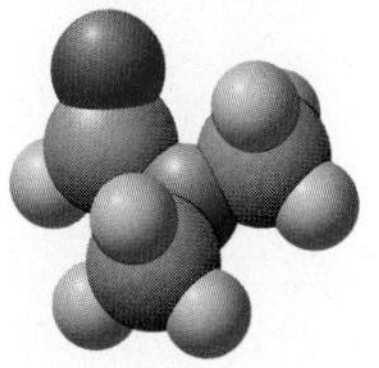

$H{-}\overset{\overset{\displaystyle O}{\|}}{C}{-}\underset{\underset{\displaystyle CH_3}{|}}{N}{-}CH_3$

***N,N*-Dimethyl methanamide (a tertiary amide)**

Figure 6.14 Space-filling models for the simplest primary, secondary, and tertiary amides.

6.14 Selected Amides and Their Uses

The simplest naturally occurring amide is urea, a water-soluble white solid produced in the human body from carbon dioxide and ammonia through a complex series of metabolic reactions (Section 15.4).

$$CO_2 + 2NH_3 \longrightarrow (H_2N)_2CO + H_2O$$

Urea is a one-carbon diamide. Its molecular structure is

$$H_2N{-}\overset{\overset{\displaystyle O}{\|}}{C}{-}NH_2$$

Urea formation is the human body's primary method for eliminating "waste" nitrogen. The kidneys remove urea from the blood and provide for its excretion in

Synthetically produced melatonin is under investigation as a drug for treating jet lag. Jet lag is a condition caused by desynchronization of the biological clock. It is usually caused by drastically changing the sleep–wake cycle, as when crossing several time zones during an airline flight or when performing shift work. Symptoms of jet lag include fatigue, early awakening or inability to sleep, and headaches. Studies indicate that melatonin taken in the evening in a new time zone will usually reset a person's biological clock and almost totally alleviate (or prevent) the symptoms of jet lag.

urine. With malfunctioning kidneys, urea concentrations in the body can build to toxic levels—a condition called *uremia.*

The complex amide melatonin is a hormone that is synthesized by the pineal gland and that regulates the sleep–wake cycle in humans. Melatonin levels within the body increase in evening hours and then decrease as morning approaches. High melatonin levels are associated with longer and more sound sleeping. The concentration of this hormone in the blood decreases with age; a six-year-old has a blood melatonin concentration more than five times that of an 80-year-old. This is one reason why young children have less trouble sleeping than senior citizens. As a prescription drug, melatonin is used to treat insomnia and jet lag.

Structurally, melatonin is a polyfunctional amide; amine and ether groups are also present as well as unsaturation.

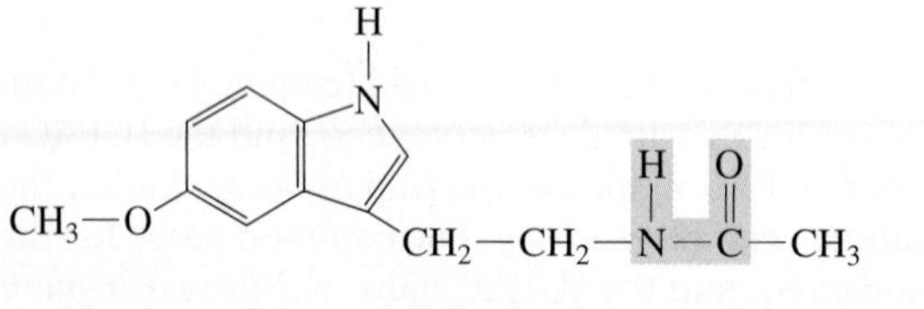

A number of synthetic amides exhibit physiological activity and are used as drugs in the human body. Foremost among them, in terms of use, is acetaminophen, which is the top-selling over-the-counter pain reliever. The focus on relevancy feature Chemical Connections 6-D on the next page considers the pharmacological chemistry of acetaminophen.

The ortho- and para-isomers (Section 2.13) of *N,N*-diethyltoluamide are less effective insect repellents than is the meta-isomer.

The active ingredient in most insect repellents currently available for purchase is the tertiary amide *N,N*-diethyl-m-toluamide, a substance better known by the acronym DEET.

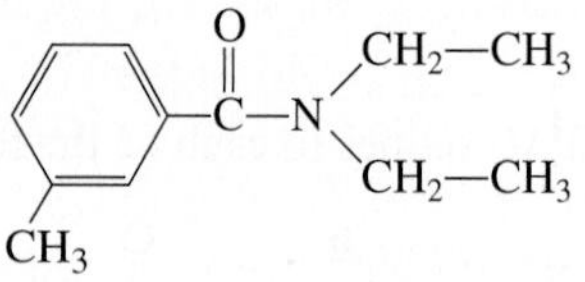

It is a compound approved for direct application to skin and clothing.

DEET is an insect *repellent;* it does not kill insects but, rather, "repels" them. DEET's actual mode of action has been recently determined to involve inhibition of insect olfactory receptors; it affects insects' ability to detect various odors associated with human presence.

DEET is effective against ticks, fleas, and mosquitos (as well as other insects), thus offering protection against insect-borne diseases transmitted through insect bites, such as Lyme disease (ticks) and West Nile virus (mosquitos).

6.15 Basicity of Amides

Both primary amines and primary amides have a $—NH_2$ group present in their structures. There is an important difference between these two $—NH_2$ groups. In an amine, the N atom is bonded to a carbon atom associated with a hydrocarbon group. In an amide, the N atom is bonded to the carbon atom of a carbonyl group. A carbonyl group is a polar entity, with the C atom possessing a partial positive charge and the O atom possessing a partial negative charge (Section 4.1).

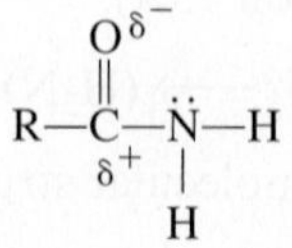

The partial positive charge on the carbonyl carbon atom is sufficient to exert an influence (attractive force) on the lone pair of electrons on the nitrogen atom, causing these electrons to be more tightly held by the nitrogen atom. The net result of this situation is that the nitrogen atom of the amide is prevented from acting as

CHEMICAL CONNECTIONS 6-D

Acetaminophen: A Substituted Amide

Often called the aspirin substitute, acetaminophen is the most widely used of all nonprescription pain relievers, accounting for over half of that market. Acetaminophen is a derivative of acetamide in which a hydroxyphenyl group has replaced one of the amide hydrogens.

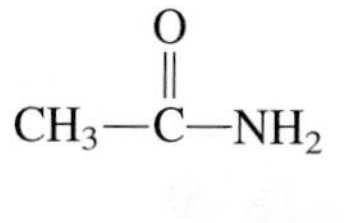

Acetamide

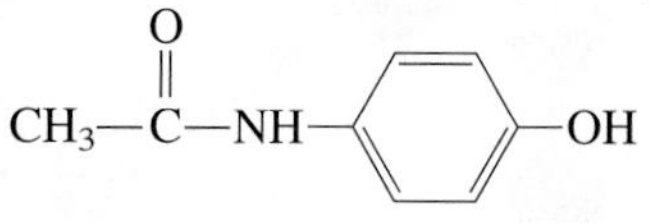

Acetaminophen

The pharmaceutical designation APAP for this compound comes from its IUPAC name, which is *N*-*a*cetyl-*p*-*a*mino*p*henol.

Acetaminophen is the active ingredient in Tylenol, Datril, Tempra, Equate, and Anacin-3. Excedrin, which contains both acetaminophen and aspirin, is a combination pain reliever.

Because acetaminophen does not have a carboxyl functional group, as do aspirin, ibuprofen, and naproxen, it does not have irritating effects on the intestinal tract. Unlike the other three pain relievers, however, acetaminophen is not effective against inflammation and is of limited use for the aches and pains associated with arthritis. Also, acetaminophen, unlike aspirin, does not inhibit platelet aggregation and is therefore not useful for the prevention of blood clots.

Acetaminophen is available in a liquid form that is used extensively for small children and other patients who have difficulty taking solid tablets. The wide use of acetaminophen for children has a drawback; it is the drug most often involved in childhood poisonings.

In *large* doses, acetaminophen can cause liver and kidney damage. Such effects are not found when acetaminophen is taken as directed. For this reason, the maximum adult daily dosage of 4 g should not be exceeded (eight 500 mg tablets), and extra-strength formulations should be used with great caution. Analgesic abuse is a real potential with the heavily advertised extra-strength formulations.

Acetaminophen's mode of action in the body is similar to that of aspirin—inhibition of prostaglandin synthesis.

Currently under research development is an acetaminophen derivative that is touted as being safer and more versatile than its parent compound. It is an acetaminophen molecule to which a saccharin molecule (Section 7.13) has been attached.

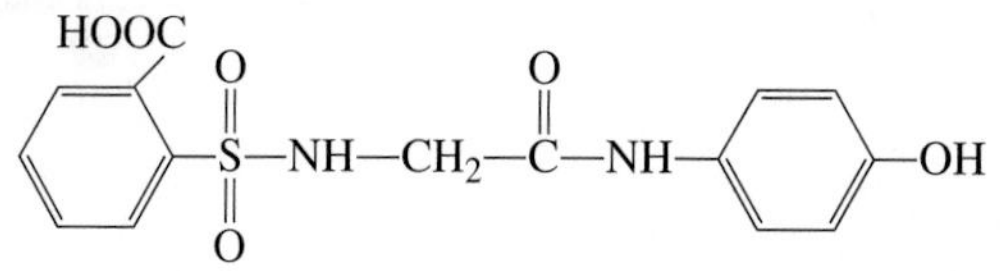

This saccharin derivative is just as potent as acetaminophen at relieving pain but has better water solubility and decreased liver toxicity. It may also be able to diminish pain that arises from nervous system damage, something that acetaminophen does not do.

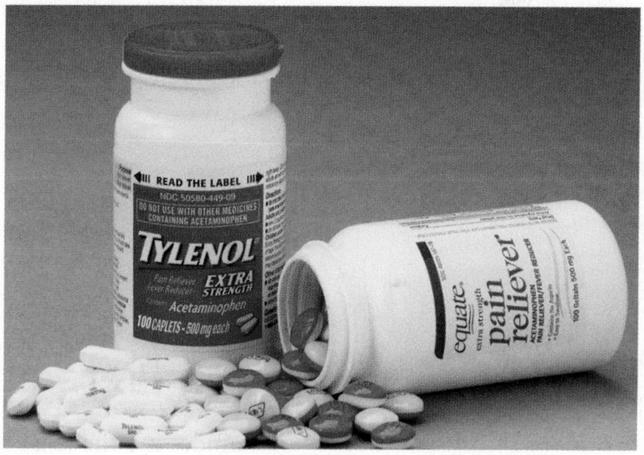

© Cengage Learning

Over-the-counter pain relievers such as Tylenol and Equate contain acetaminophen as the active ingredient.

a Brønsted-Lowry base, that is, as a proton acceptor. Thus amides do not exhibit basic behavior as do amines, even though a lone pair of electrons is present on the amide N atom. This also means that the N atom of an amide does not readily serve as a hydrogen bonding site.

6.16 Physical Properties of Amides

Methanamide and its *N*-methyl and *N,N*-dimethyl derivatives (the simplest 1°, 2°, and 3° amides, respectively), are all liquids at room temperature. All unbranched primary amides, except methanamide, are solids at room temperature (Figure 6.15), as are most other amides. In many cases, the amide melting point is even higher than that of the corresponding carboxylic acid. The high melting points result from the numerous intermolecular hydrogen-bonding possibilities that exist between amide H atoms and carbonyl O atoms. Figure 6.16 shows selected hydrogen-bonding interactions that are possible among several primary amide molecules.

Unbranched Primary Amides			
C_1	C_3	C_5	C_7
C_2	C_4	C_6	C_8

☐ Liquid ☐ Solid

Figure 6.15 A physical-state summary for unbranched primary amides at room temperature and pressure.

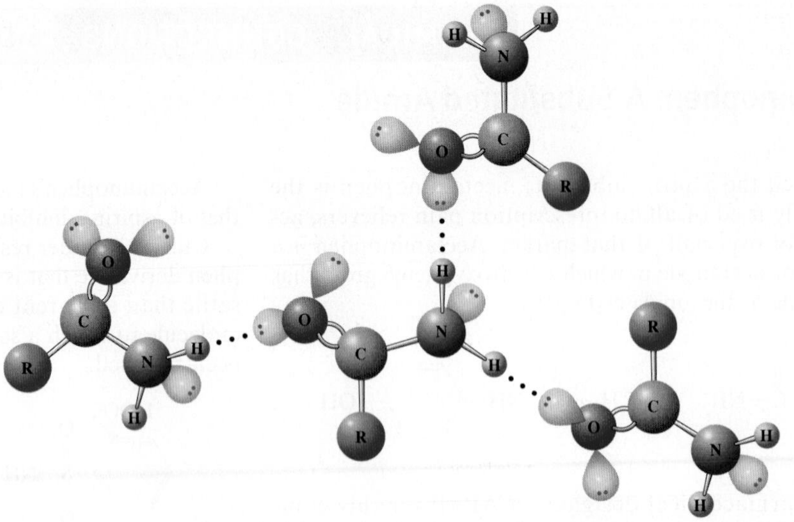

Figure 6.16 The high boiling points of amides are related to the numerous amide–amide hydrogen-bonding possibilities that exist.

Lidocaine (xylocaine), a substance commonly administered by injection as a dental anesthetic, is a synthetic molecule that contains both amide and amine functional groups.

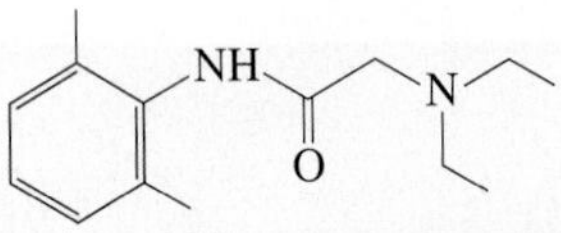

Another well-known local anesthetic is procaine (novocaine). Its structure contains two amine groups and an ester group but no amide group.

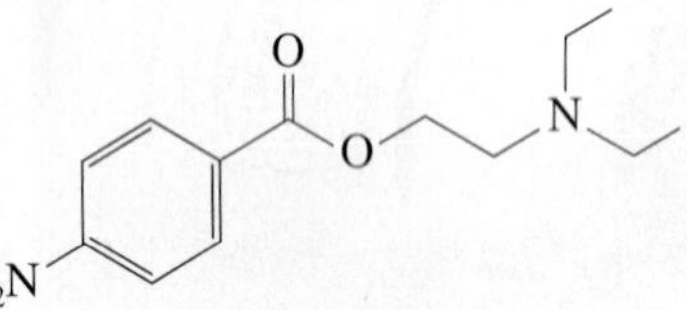

Both lidocaine and procaine share a common structural feature—the presence of a diethyl amino group (on the right side of each structure).

Fewer hydrogen-bonding possibilities exist for 2° amides because the nitrogen atom now has only one hydrogen atom; hence lower melting points are the rule for such amides. Still lower melting points are observed for 3° amides because no hydrogen bonding is possible. The disubstituted *N,N*-dimethylacetamide has a melting point of −20°C, which is about 100°C lower than that of the unsubstituted acetamide.

Amides of low molecular mass, up to five or six carbon atoms, are soluble in water. Again, numerous hydrogen-bonding possibilities exist between water and the amide. Even disubstituted amides can participate in such hydrogen bonding.

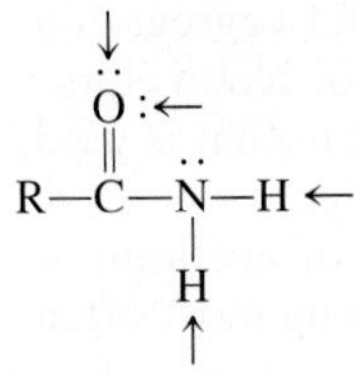

Arrows denote sites where hydrogen bonding to water can occur.

6.17 Preparation of Amides

Amides are the least reactive of the common carboxylic acid derivatives and they can be synthesized from an acid chloride, an acid anhydride, an ester, or the carboxylic acid itself.

The reaction of a carboxylic acid with ammonia or a 1° or 2° amine produces an amide, provided that the reaction is carried out at an elevated temperature (greater than 100°C) and a dehydrating agent is present.

$$\text{Ammonia} + \text{carboxylic acid} \xrightarrow[\text{Catalyst}]{100^\circ\text{C}} 1^\circ \text{ amide}$$

$$1^\circ \text{ Amine} + \text{carboxylic acid} \xrightarrow[\text{Catalyst}]{100^\circ\text{C}} 2^\circ \text{ amide}$$

$$2^\circ \text{ Amine} + \text{carboxylic acid} \xrightarrow[\text{Catalyst}]{100^\circ\text{C}} 3^\circ \text{ amide}$$

If the preceding reactions are run at room temperature (25°C), no amide formation occurs; instead an acid–base reaction occurs in which a carboxylic acid salt is produced. This acid–base reaction when a 1° amine is the reactant is

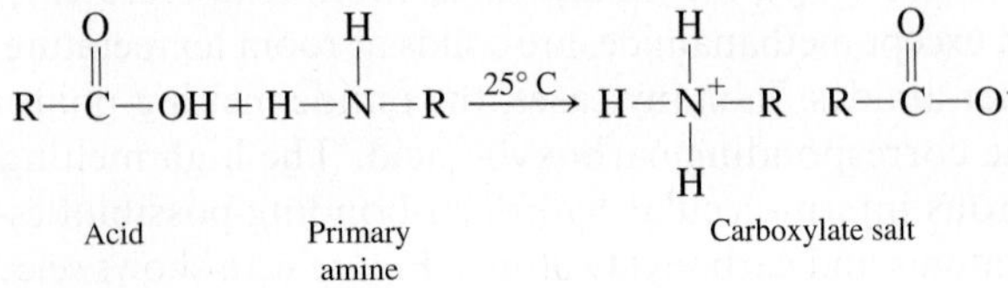

General structural equations for 1°, 2°, and 3° amide production from carboxylic acids are

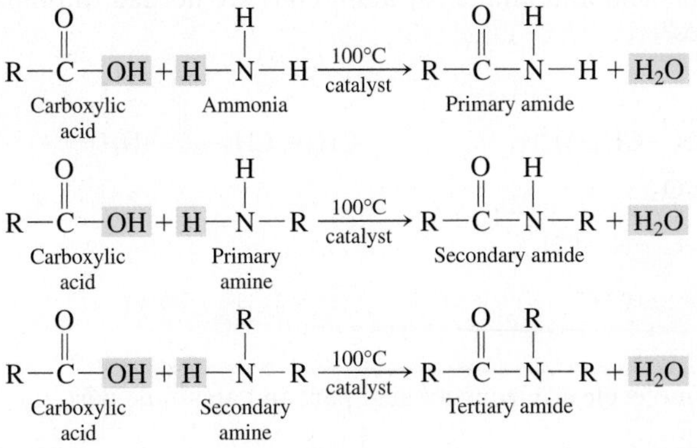

The reaction of a carboxylic acid with ammonia or an amine to produce an amide is similar to the reaction of a carboxylic acid with an alcohol to produce an ester. In both cases, water is formed as a by-product as the —OH part of the carboxylic acid is replaced.

These reactions are called *amidification* reactions. An **amidification reaction** *is the reaction of a carboxylic acid with an amine (or ammonia) to produce an amide.* In amidification, an —OH group is lost from the carboxylic acid, a —H atom is lost from the ammonia or amine, and water is formed as a by-product. Amidification reactions are thus *condensation* reactions.

This is the fourth time that *condensation* reactions have been encountered. Esterification (Section 5.10), acetal formation (Section 4.11), and intermolecular alcohol dehydration (Section 3.9) were the other three condensation reactions.

Two specific amidification reactions, in which a 2° amide and a 3° amide are produced, respectively, are

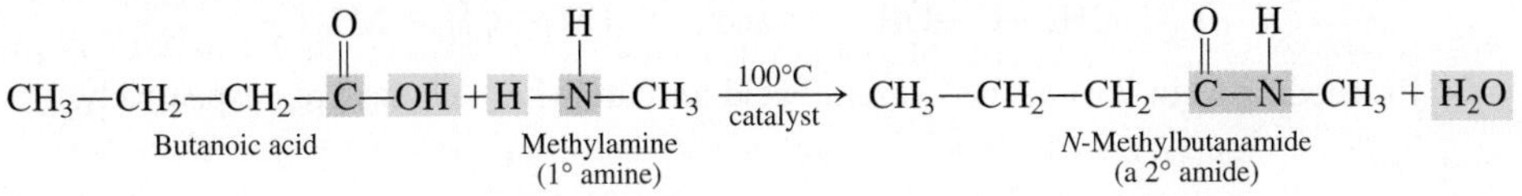

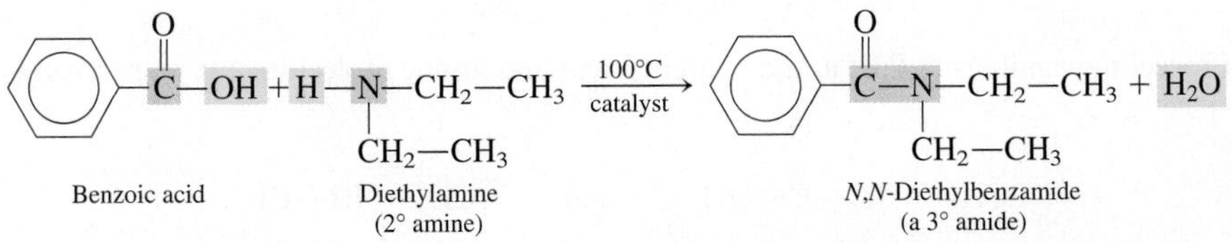

Just as it is useful to think of the structure of an ester (Section 5.9) in terms of an "acid part" and an "alcohol part," it is useful to think of an amide in terms of an "acid part" and an "amine (or ammonia) part."

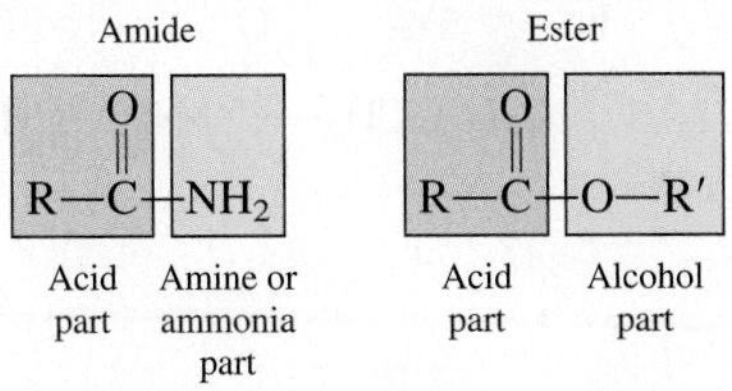

In this context, it is easy to identify the parent acid and amine from which a given amide can be produced; to generate the parent molecules, just add an —OH group to the acid part of the amide and an H atom to the amine part.

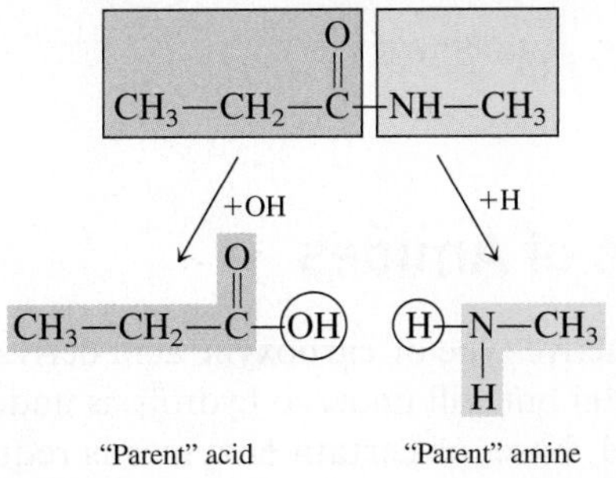

EXAMPLE 6.6 Predicting Reactants Needed to Prepare Specific Amides

What carboxylic acid and amine (or ammonia) are needed to prepare each of the following amides?

a. $CH_3-\overset{O}{\overset{\|}{C}}-NH-CH_2-CH_3$

b. $CH_3-CH_2-\overset{O}{\overset{\|}{C}}-NH_2$

c. $CH_3-CH_2-\overset{O}{\overset{\|}{C}}-\underset{CH_3}{\underset{|}{N}}-CH_3$

Solution

a. Viewing the molecule as having an acid part and an amine part, the following are obtained

$$\underbrace{\boxed{CH_3-\overset{O}{\overset{\|}{C}}}}_{\text{Acid part}}\underbrace{\boxed{-NH-CH_2-CH_3}}_{\text{Amine part}}$$

Adding an —OH group to the acid part and a H atom to the amine part, the "parent" molecules are obtained, which are

$$CH_3-\overset{O}{\overset{\|}{C}}-OH \quad \text{and} \quad CH_3-CH_2-NH_2$$

b. Proceeding as in Part **a,** the "parent" acid and amine molecules are, respectively,

$$CH_3-CH_2-\overset{O}{\overset{\|}{C}}-OH \quad \text{and} \quad NH_3$$

c. Proceeding again as in Part **a,** the "parent" acid and amine molecules are, respectively,

$$CH_3-CH_2-\overset{O}{\overset{\|}{C}}-OH \quad \text{and} \quad CH_3-NH-CH_3$$

Practice Exercise 6.6

What carboxylic acid and amine (or ammonia) are needed to prepare each of the following amides?

a. $CH_3-\overset{O}{\overset{\|}{C}}-\underset{CH_3}{\underset{|}{N}}-CH_2-CH_3$

b. $CH_3-CH_2-\overset{O}{\overset{\|}{C}}-NH-CH_3$

c. 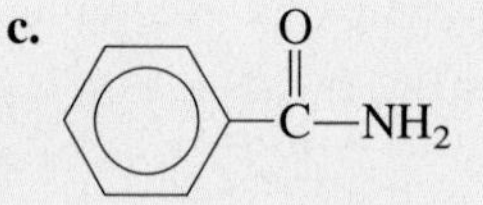

Answers:

a. $CH_3-\overset{O}{\overset{\|}{C}}-OH$, $CH_3-CH_2-NH-CH_3$

b. $CH_3-CH_2-\overset{O}{\overset{\|}{C}}-OH$, CH_3-NH_2

c. 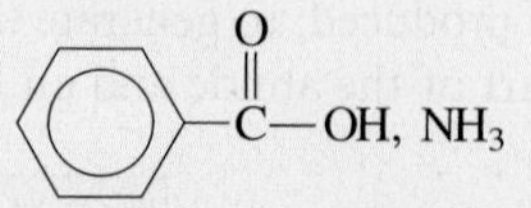

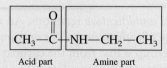

The relative lack of reactivity of amides becomes very important in protein chemistry considerations (Chapter 9). Proteins are polymers in which the monomers are connected to each other via amide bonds (Section 9.7). Proteins are relatively stable in aqueous solution in the absence of acid or base; hence they can exert their various biochemical effects in aqueous solution without hydrolysis occurring.

6.18 Hydrolysis of Amides

Amides are the least reactive type of carboxylic acid derivative. They are relatively stable in aqueous solution but will undergo hydrolysis under strenuous conditions; the presence of an acid, base, or certain enzymes is required as a catalyst, and sustained heating is also often required. In such hydrolysis, the bond between the carbonyl carbon atom and the nitrogen is broken, and free acid and free amine are produced.

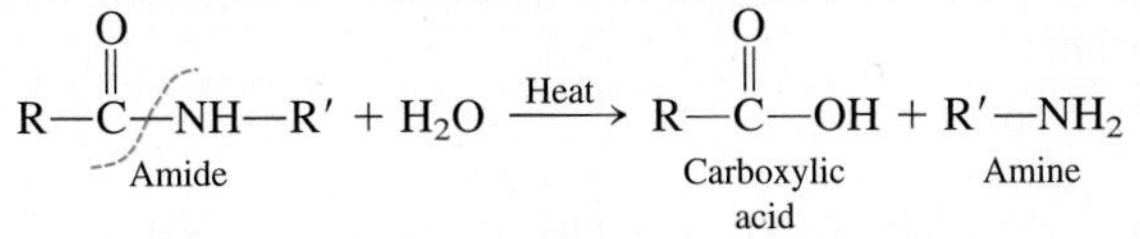

Acidic or basic hydrolysis conditions have an effect on the products. *Acidic* conditions convert the product amine to an amine salt (Section 6.7). *Basic* conditions convert the product carboxylic acid to a carboxylic acid salt (Section 5.8).

Amide hydrolysis under basic conditions is also called amide saponification, just as ester hydrolysis under basic conditions is called ester saponification (Section 5.15).

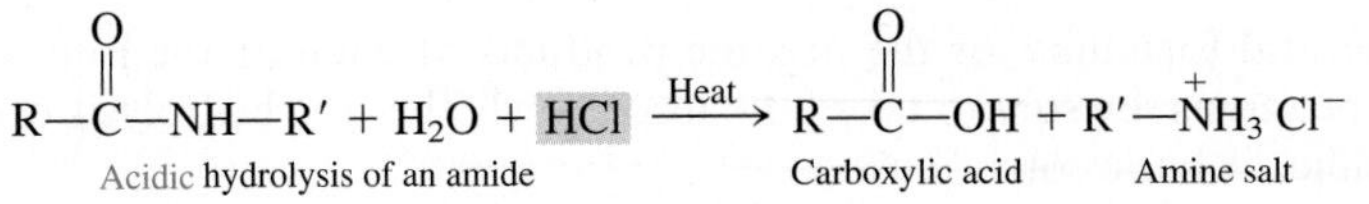

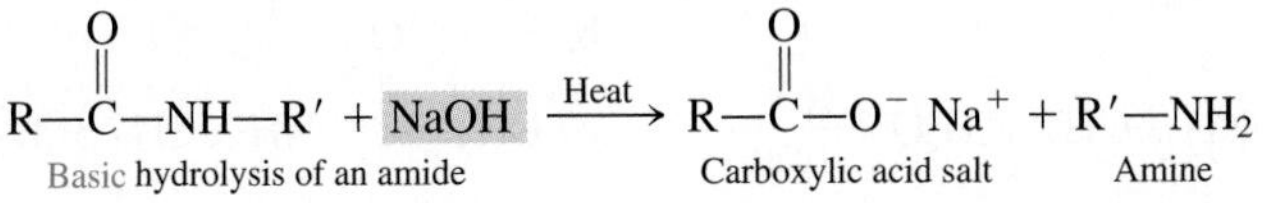

EXAMPLE 6.7 Predicting the Products of Amide Hydrolysis Reactions

Draw structural formulas for the organic products of each of the following amide hydrolysis reactions. Be sure to take into account whether the hydrolysis occurs under neutral, acidic, or basic conditions.

a.

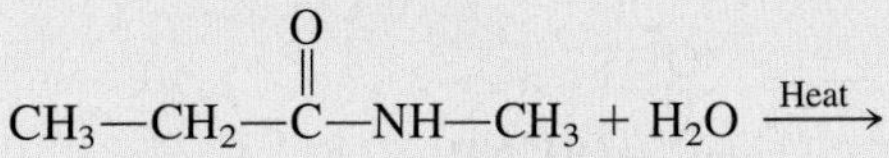

b.

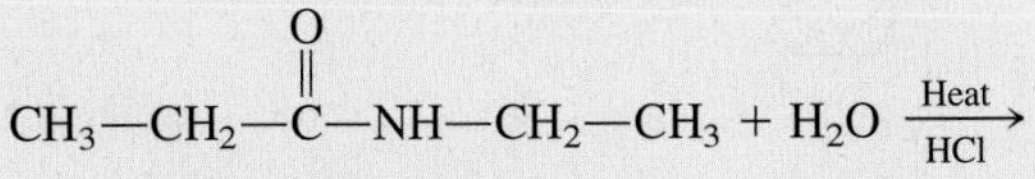

c.

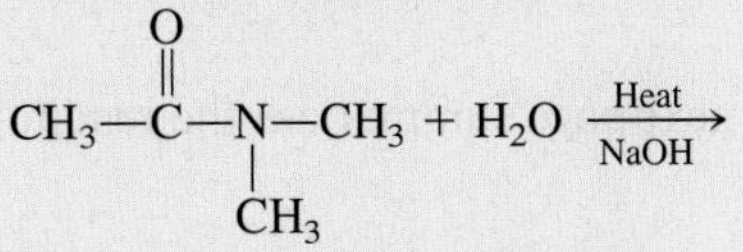

d.

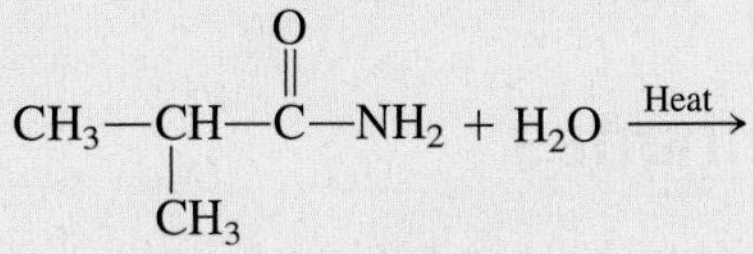

Solution

a. This reaction is hydrolysis under neutral conditions. The products will be the "parent" acid and amine for the amide. These "parents" are

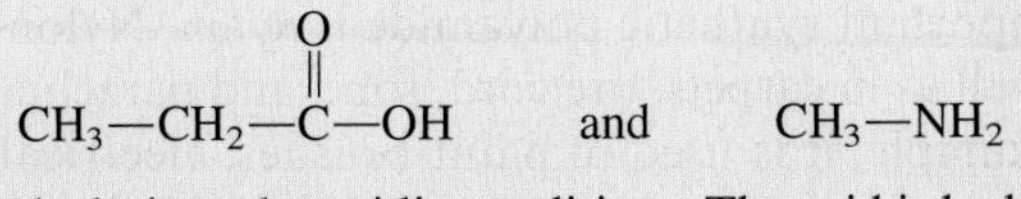

b. This reaction is hydrolysis under acidic conditions. The acid is hydrochloric acid (HCl). The products will be the "parent" carboxylic acid and the chloride salt of the amine. The HCl converts the amine to its chloride salt.

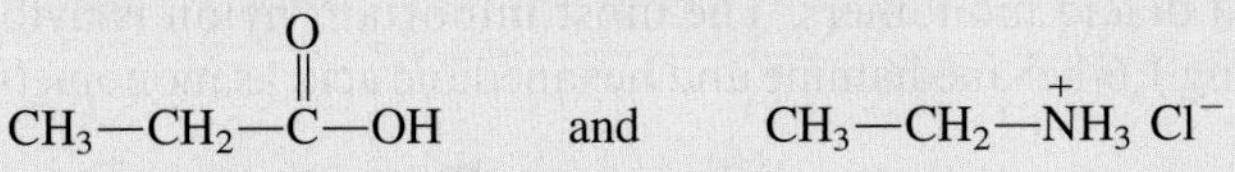

c. This reaction is hydrolysis under basic conditions. The base is sodium hydroxide (NaOH). The products will be the "parent" amine and the salt of the carboxylic acid. The NaOH converts the carboxylic acid to its sodium salt.

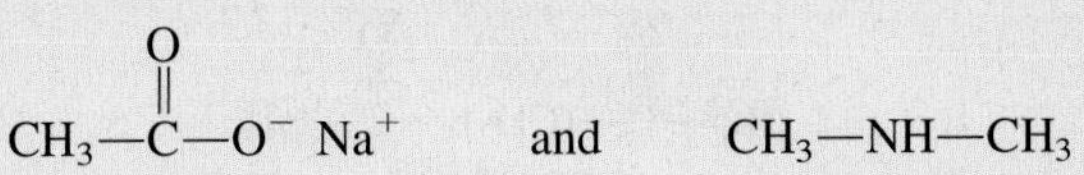

(continued)

d. This reaction is hydrolysis under neutral conditions. The products will be the "parent" acid and amine of the amide. Because the amide is unsubstituted, the parent amine is actually ammonia.

$$CH_3-CH(CH_3)-\overset{O}{\overset{\|}{C}}-OH \quad \text{and} \quad NH_3$$

Practice Exercise 6.7

Draw structural formulas for the organic products of each of the following amide hydrolysis reactions. Be sure to take into account whether the hydrolysis occurs under neutral, acidic, or basic conditions.

a. $CH_3-\overset{O}{\overset{\|}{C}}-NH-CH_3 + H_2O \xrightarrow[\text{NaOH}]{\text{Heat}}$

b. $CH_3-\overset{O}{\overset{\|}{C}}-NH-CH_3 + H_2O \xrightarrow[\text{HCl}]{\text{Heat}}$

c. $CH_3-\overset{O}{\overset{\|}{C}}-NH-CH_3 + H_2O \xrightarrow{\text{Heat}}$

d. 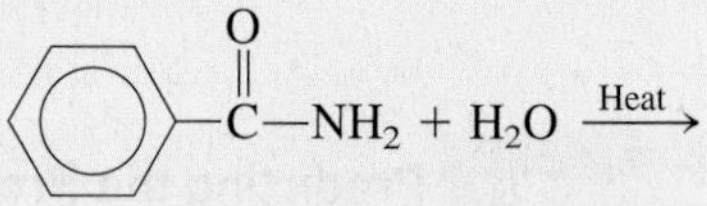

Answers:

a. $CH_3-\overset{O}{\overset{\|}{C}}-O^- Na^+, CH_3-NH_2$

b. $CH_3-\overset{O}{\overset{\|}{C}}-OH, CH_3-\overset{+}{N}H_3\ Cl^-$

c. $CH_3-\overset{O}{\overset{\|}{C}}-OH, CH_3-NH_2$

d. 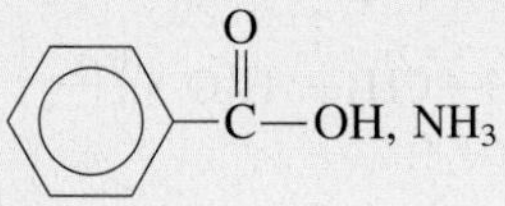

The Chemistry at a Glance feature on the next page summarizes the reactions that involve amines and amides.

6.19 Polyamides and Polyurethanes

Amide polymers—polyamides—are synthesized by combining diamines and dicarboxylic acids in a condensation polymerization reaction (Section 5.17). A **polyamide** *is a condensation polymer in which the monomers are joined through amide linkages.*

The most important synthetic polyamide is *nylon.* Nylon is used in clothing and hosiery, as well as in carpets, tire cord, rope, and parachutes. It also has non-fiber uses; for example, it is used in paint brushes, electrical parts, valves, and fasteners. It is a tough, strong, nontoxic, nonflammable material that is resistant to chemicals. Surgical suture is made of nylon because it is such a strong fiber.

There are actually many different types of nylon, all of which are based on diamine and diacid monomers. The most important nylon is nylon 66, which is made by using 1,6-hexanediamine and hexanedioic acid as monomers (Figure 6.17).

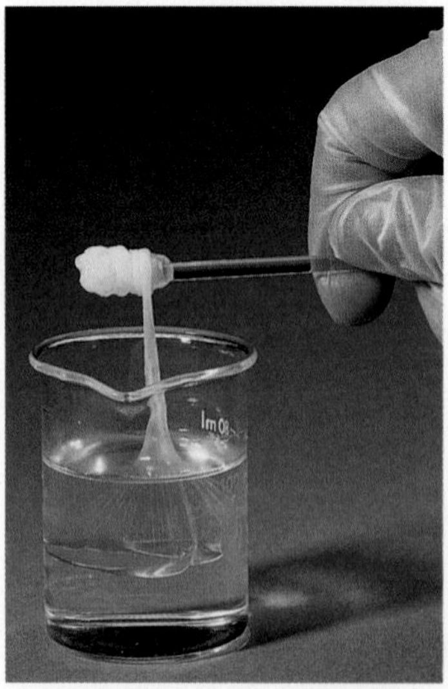

Figure 6.17 A white strand of a nylon polymer forms between the two layers of a solution containing a diacid (bottom layer) and a diamine (top layer).

$$H-\underset{H}{N}-(CH_2)_6-\underset{H}{N}-H$$

1,6-Hexanediamine

$$HO-\overset{O}{\overset{\|}{C}}-(CH_2)_4-\overset{O}{\overset{\|}{C}}-OH$$

Hexanedioic acid

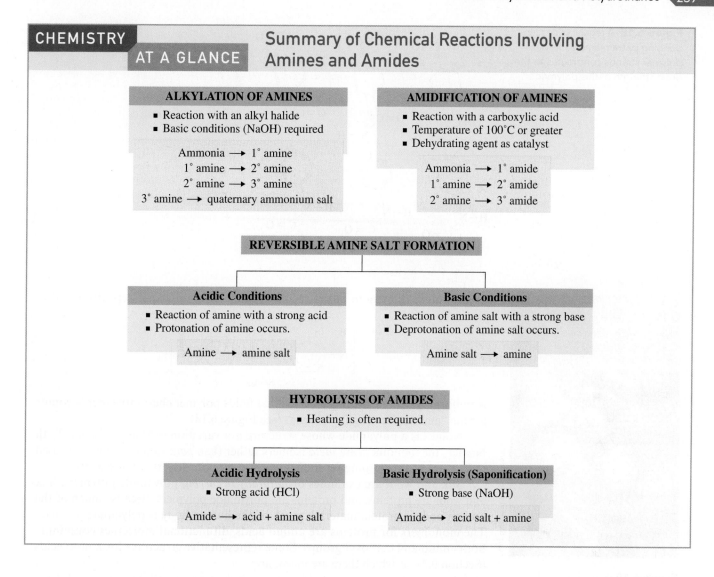

CHEMISTRY AT A GLANCE — Summary of Chemical Reactions Involving Amines and Amides

ALKYLATION OF AMINES

- Reaction with an alkyl halide
- Basic conditions (NaOH) required

Ammonia → 1° amine
1° amine → 2° amine
2° amine → 3° amine
3° amine → quaternary ammonium salt

AMIDIFICATION OF AMINES

- Reaction with a carboxylic acid
- Temperature of 100°C or greater
- Dehydrating agent as catalyst

Ammonia → 1° amide
1° amine → 2° amide
2° amine → 3° amide

REVERSIBLE AMINE SALT FORMATION

Acidic Conditions

- Reaction of amine with a strong acid
- Protonation of amine occurs.

Amine → amine salt

Basic Conditions

- Reaction of amine salt with a strong base
- Deprotonation of amine salt occurs.

Amine salt → amine

HYDROLYSIS OF AMIDES

- Heating is often required.

Acidic Hydrolysis

- Strong acid (HCl)

Amide → acid + amine salt

Basic Hydrolysis (Saponification)

- Strong base (NaOH)

Amide → acid salt + amine

The reaction of one acid group of the diacid with one amine group of the diamine initially produces an amide molecule; an acid group is left over on one end, and an amine group is left over on the other end.

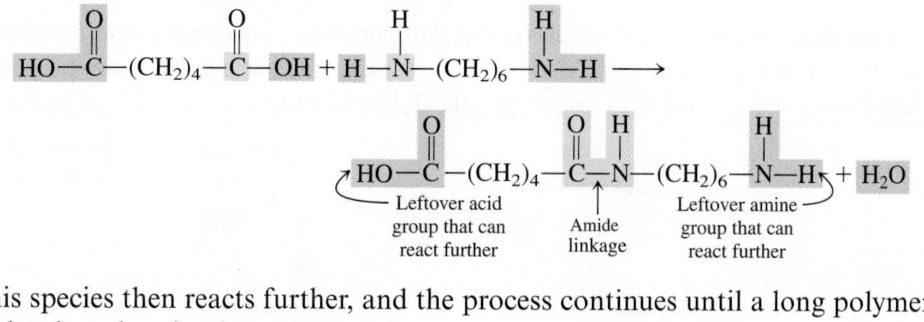

This species then reacts further, and the process continues until a long polymeric molecule, nylon, has been produced.

The name *nylon 66* comes from the fact that each of the monomers has six carbon atoms.

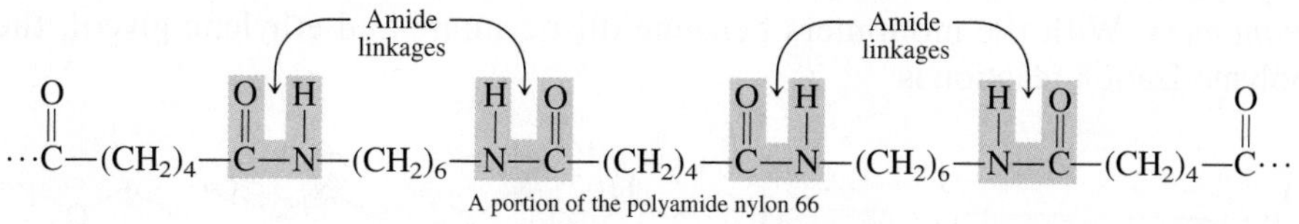

A portion of the polyamide nylon 66

Additional stiffness and toughness are imparted to polyamides if aromatic rings are present in the polymer "backbone." The polyamide Kevlar is now

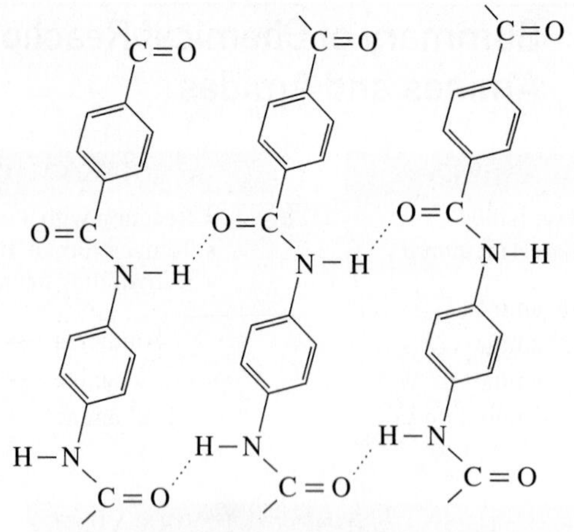

Figure 6.18 A regular hydrogen-bonding pattern among Kevlar polymer strands contributes to the great strength of this polymer.

Peter Skinner/Photo Researchers

Figure 6.19 Firefighters with flame-resistant clothing containing Nomex.

used in place of steel in bullet-resistant vests. The polymeric repeating unit in Kevlar is

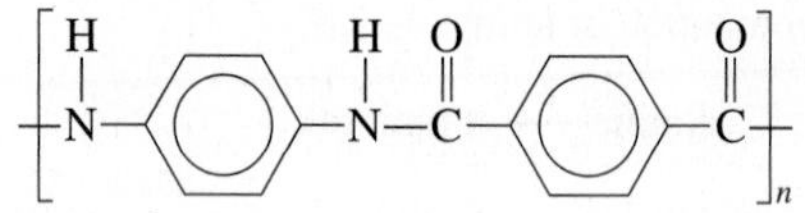

Kevlar

A uniform system of hydrogen bonds that holds polymer chains together accounts for the "amazing" strength of Kevlar (see Figure 6.18).

Nomex is a polyamide whose structure is a variation of that of Kevlar. With Nomex, the monomers are *meta* isomers rather than *para* isomers. Nomex is used in flame-resistant clothing for firefighters and race car drivers (Figure 6.19).

Silk and wool are examples of *naturally occurring* polyamide polymers. Silk and wool are proteins, and proteins are polyamide polymers. Because much of the human body is protein material, much of the human body is polyamide polymer. The monomers for proteins are amino acids, difunctional molecules containing both amino and carboxyl groups. Some representative structures for amino acids (Section 9.2), of which there are many, are:

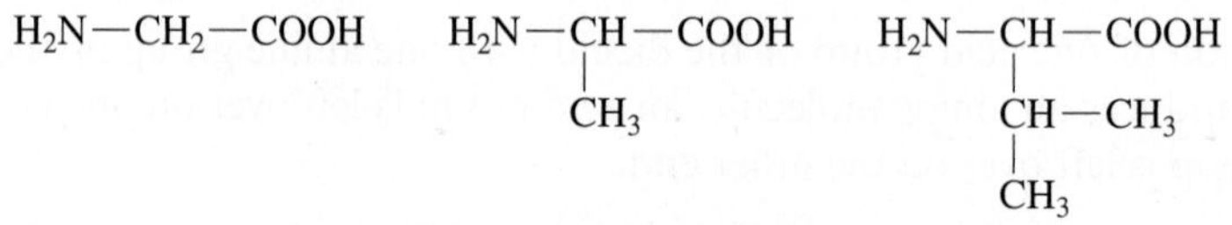

A **urethane** *is a hydrocarbon derivative that contains a carbonyl group bonded to both an —OR group and a —NHR (or —NR$_2$) group.* Such compounds are prepared by reaction of an alcohol with an isocyanate (RN=C=O).

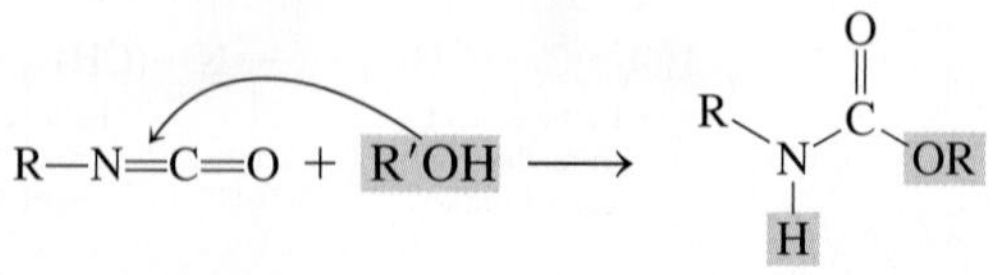

Isocyanate — Urethane

A **polyurethane** *is a polymer formed from the reaction of dialcohol and diisocyanate monomers.* With the monomers benzene diisocyanate and ethylene glycol, the polymerization reaction is

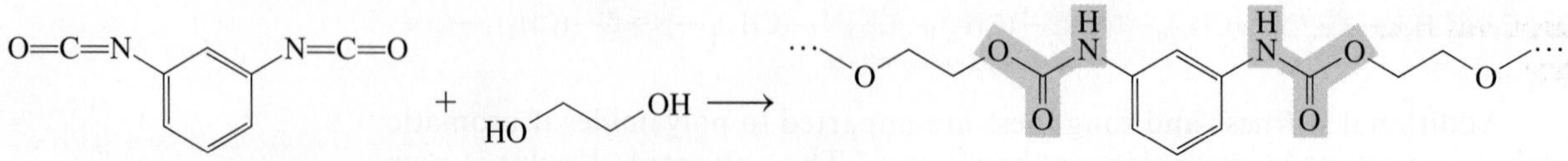

Benzene 2,6-diisocyanate — Ethylene glycol — A polyurethane

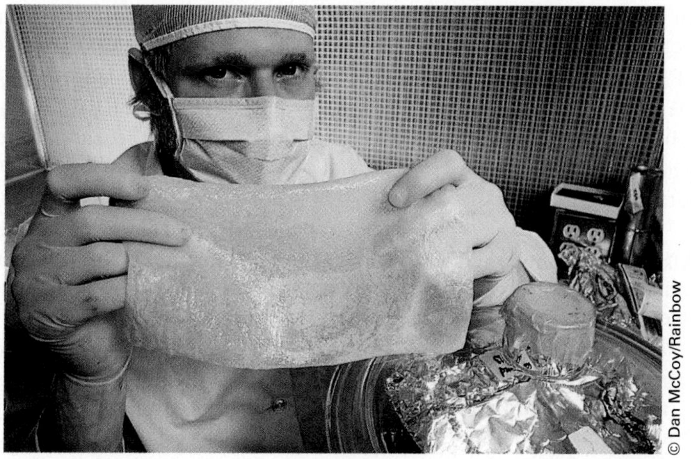

Figure 6.20 Polyurethanes have medical applications. For example, polyurethane membranes are used as skin substitutes for severe burn victims. Because they allow the passage of only oxygen and water, these membranes help patients recover more rapidly.

Structurally, polyurethanes have aspects of the structures of both polyesters and polyamides as shown in the following segment of a polyurethane structure.

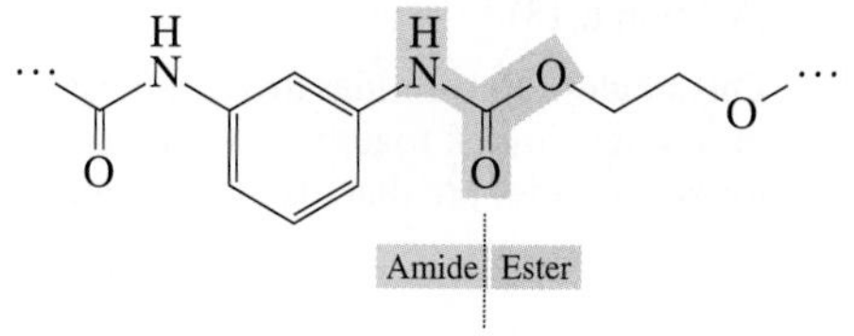

Foam rubber in furniture upholstery, packaging materials, life preservers, elastic fibers, and many other products contain polyurethane polymers (see Figure 6.20).

One of the best-known polyurethanes is Spandex, a strong yet flexible polymer used in both men's and women's athletic wear. It has also been used in support hosiery. On the molecular level, it has rigid regions (for strength) that are joined together by flexible segments.

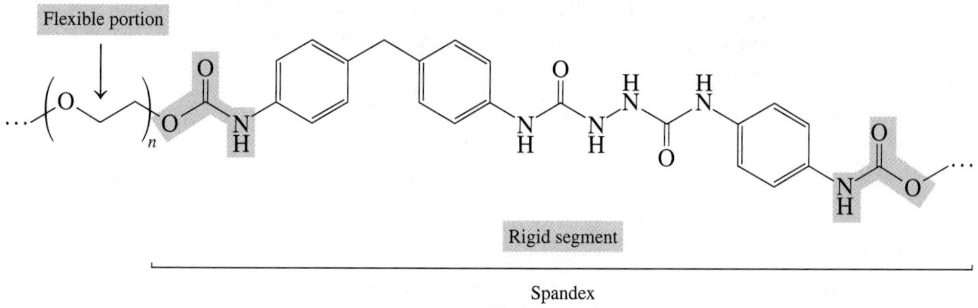

Spandex
Trade name: Lycra

Concepts to Remember

OWL Sign in at **www.cengage.com/owl** to view tutorials and simulations, develop problem-solving skills, and complete online homework assigned by your professor.

Structural characteristics of amines. Amines are derivatives of ammonia (NH_3) in which one or more hydrogen atoms have been replaced by an alkyl, a cycloalkyl, or an aryl group (Section 6.2).

Classification of amines. Amines are classified as primary, secondary, or tertiary, depending on the number of hydrocarbon groups (one, two, or three) directly attached to the nitrogen atom. The functional group present in a primary amine, the —NH_2 group, is called an *amino* group. Secondary and tertiary amines contain substituted amino groups (Section 6.2).

Nomenclature for amines. Common names for amines are formed by listing the hydrocarbon groups attached to the nitrogen atom in alphabetical order, followed by the suffix *-amine.* In the IUPAC system, the *-e* ending of the name of the longest carbon chain present is changed to *-amine,* and a number is used to locate the position of the amino group. Carbon-chain substituents are given numbers to designate their locations (Section 6.3).

Properties of amines. The methylamines and ethylamine are gases at room temperature; amines of higher molecular mass are usually liquids and smell like raw fish. Primary and secondary, but not tertiary, amines can participate in hydrogen bonding to other amine molecules (Section 6.5).

Basicity of amines. Amines are weak bases because of the ability of the unshared electron pair on the amine nitrogen atom to accept a proton in acidic solution (Section 6.6).

Amine salts. The reaction of a strong acid with an amine produces an amine salt. Such salts are more soluble in water than are the parent amines (Section 6.7).

Alkylation of ammonia and amines. Alkylation of ammonia, primary amines, secondary amines, and tertiary amines produces primary amines, secondary amines, tertiary amines, and quaternary ammonium salts, respectively (Section 6.8).

Heterocyclic amines. In a heterocyclic amine, the nitrogen atoms of amino groups present are part of either an aromatic or a nonaromatic ring system. Numerous heterocyclic amines are important biochemical compounds (Section 6.9).

Structural characteristics of amides. An amide is derived from a carboxylic acid by replacing the hydroxyl group with an amino or a substituted amino group (Section 6.12).

Classification of amides. Amides, like amines, can be classified as primary, secondary, or tertiary, depending on how many nonhydrogen atoms are attached to the nitrogen atom (Section 6.12).

Nomenclature for amides. The nomenclature for amides is derived from that for carboxylic acids by changing the *-oic acid* ending to *-amide*. Groups attached to the nitrogen atom of the amide are located using the prefix *N-* (Section 6.13).

Basicity of amides. Amides, unlike amines, do not exhibit basic properties. The nitrogen lone pair of electrons does not function as a proton acceptor because of the partial positive charge present on the nitrogen's neighboring carbonyl carbon atom (Section 6.15).

Physical properties of amides. Most unbranched amides are solids at room temperature and have correspondingly high boiling points because of strong hydrogen bonds between molecules (Section 6.16).

Preparation of amides. Reaction, at elevated temperature, of carboxylic acids with ammonia, primary amines, and secondary amines produces primary, secondary, and tertiary amides, respectively (Section 6.17).

Hydrolysis of amides. In amide hydrolysis, the bond between the carbonyl carbon atom and the nitrogen is broken, and free acid and free amine are produced. Acidic hydrolysis conditions convert the product amine to an amine salt. Basic hydrolysis conditions convert the product acid to an acid salt (Section 6.18).

Polyamides. Polyamides are condensation polymers with monomers joined together by amide linkages. The monomers for polyamides are diacids and diamines (Section 6.19).

Exercises and Problems

OWL Interactive versions of these problems may be assigned in OWL.

Exercises and problems are arranged in matched pairs with the two members of a pair addressing the same concept(s). The answer to the odd-numbered member of a pair is given at the back of the book. Problems denoted with a ▲ involve concepts found not only in the section under consideration but also concepts found in one or more earlier sections of the chapter. Problems denoted with a ● cover concepts found in a Chemical Connections feature box.

Bonding Characteristics of Nitrogen Atoms (Section 6.1)

6.1 Contrast N, O, and C atoms in terms of the number of covalent bonds they usually form in organic compounds.

6.2 Contrast N, O, and C atoms in terms of the number of nonbonding electron pairs they normally possess when they are present in organic compounds.

Structure and Classification of Amines (Section 6.2)

6.3 What is the generalized molecular formula for the following?
a. Primary amine b. Secondary amine c. Tertiary amine

6.4 Draw a generalized structural formula for the functional group present in the following.
a. Primary amine b. Secondary amine c. Tertiary amine

6.5 How many carbon–nitrogen bonds are present in each of the following types of amines?
a. 1° amine b. 2° amine c. 3° amine

6.6 How many nitrogen-hydrogen bonds are present in each of the following types of amines?
a. 1° amine b. 2° amine c. 3° amine

6.7 In which of the following compounds is an amine functional group present?

a. 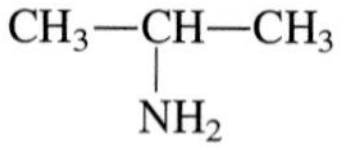

b. $CH_3—NH—CH_3$

c. 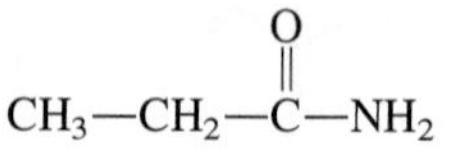

d. 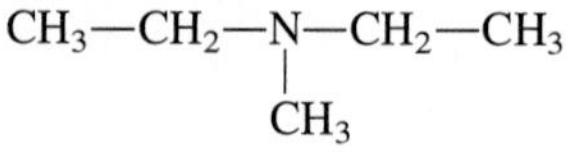

6.8 In which of the following compounds is an amine functional group present?

a. $CH_3—CH_2—CH_2—NH_2$

b. 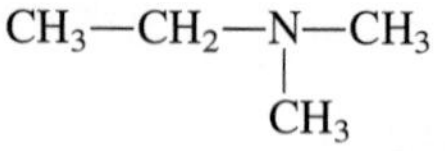

c. 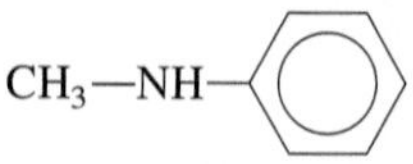

d. 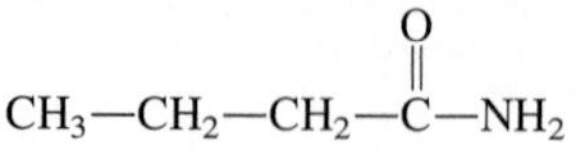

6.9 Classify each of the following amines as a primary, secondary, or tertiary amine.

a. $CH_3—NH_2$

b. 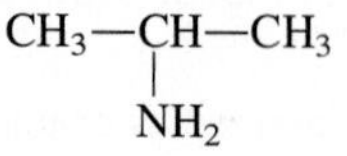

c. $CH_3—NH—CH_3$

d. 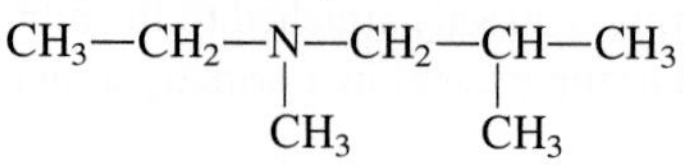

6.10 Classify each of the following amines as a primary, secondary, or tertiary amine.

a. 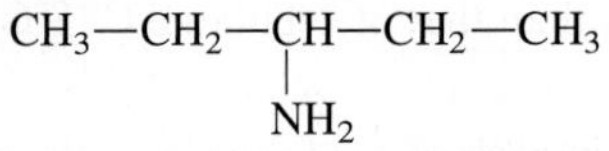

b. 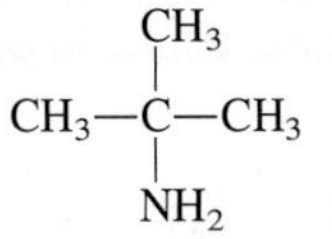

c. 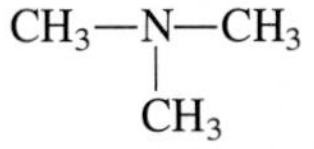

d. 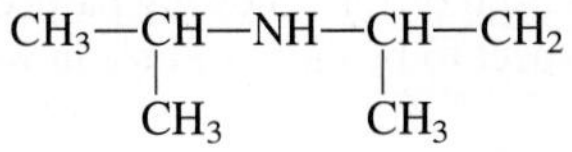

6.11 Classify each of the following amines as a primary, secondary, or tertiary amine.

a. 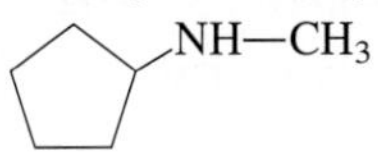

b. 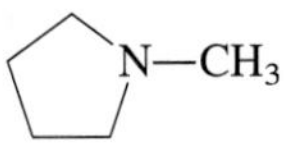

c. 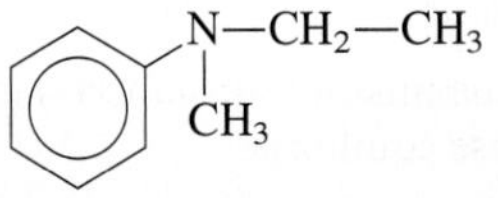

d. 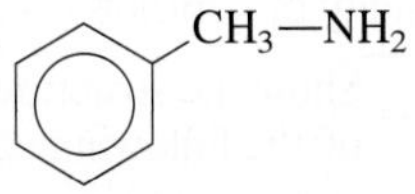

6.12 Classify each of the following amines as a primary, secondary, or tertiary amine.

a. 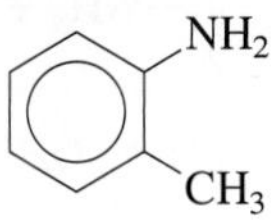

b. 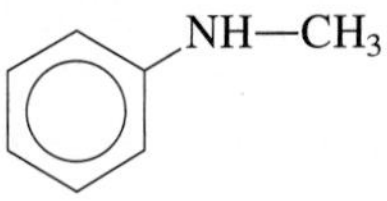

c. 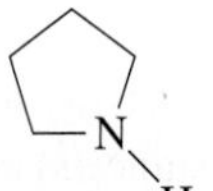

d. 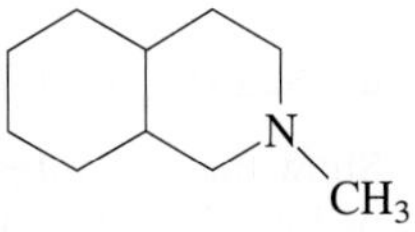

Nomenclature for Amines (Section 6.3)

6.13 Assign a common name to each of the following amines.

a. $CH_3—NH—CH_2—CH_3$

b. $CH_3—CH_2—CH_2—NH_2$

c. 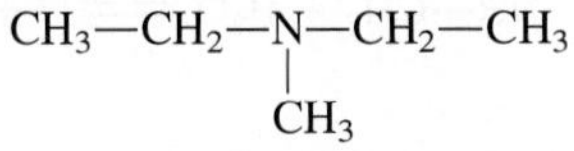

d. 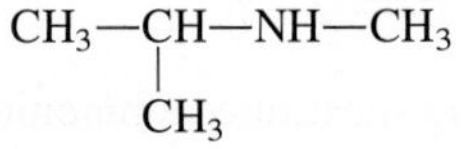

6.14 Assign a common name to each of the following amines.

a. 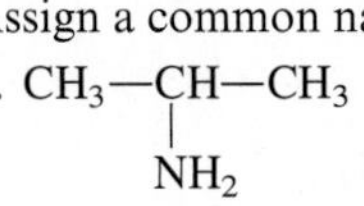

b. $H_2N—CH_2—CH_2—CH_2—CH_3$

c. 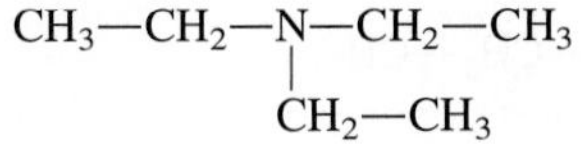

d. 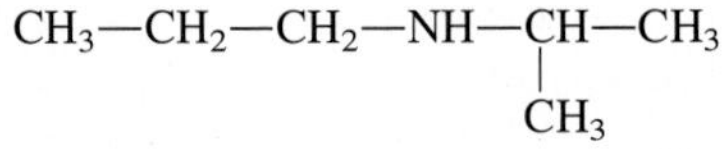

6.15 Assign an IUPAC name to each of the following amines.

a. 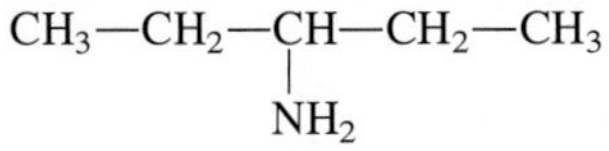

b. 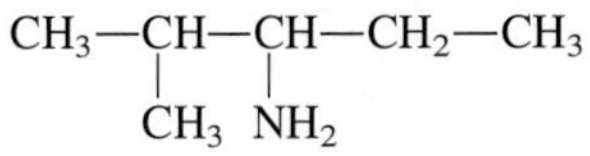

c. 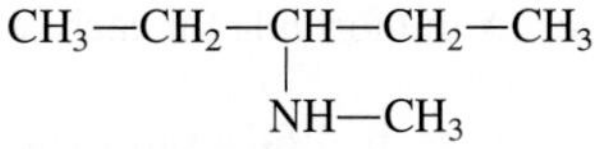

d. 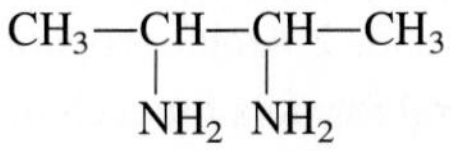

6.16 Assign an IUPAC name to each of the following amines.

a. $CH_3—CH_2—CH_2—NH_2$

b. 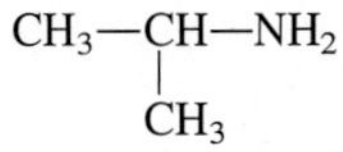

c. 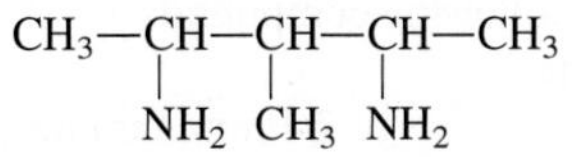

d. $CH_3—CH_2—CH_3—NH—CH_3$

6.17 Assign both a common name and an IUPAC name to each of the following amines.

a. 

b. 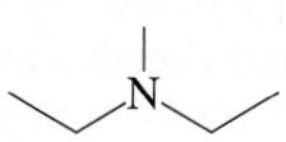

c. 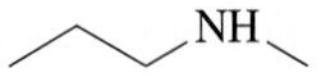

d. 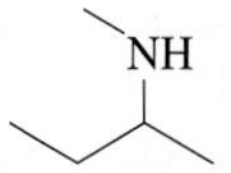

6.18 Assign both a common name and an IUPAC name to each of the following amines.

a. 

b. 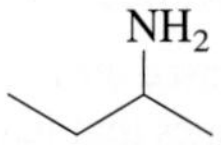

c. 

d. 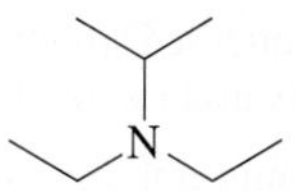

6.19 Name each of the following aromatic amines as a derivative of aniline.

a. 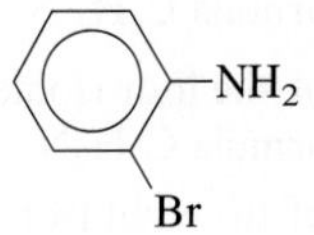

b. 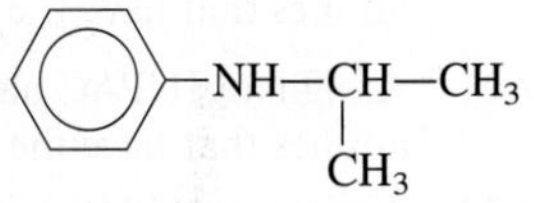

c. 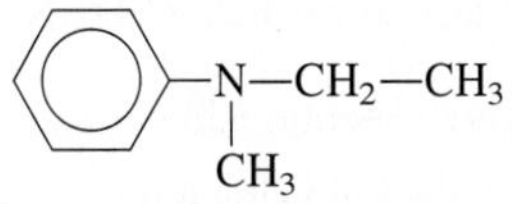

d. 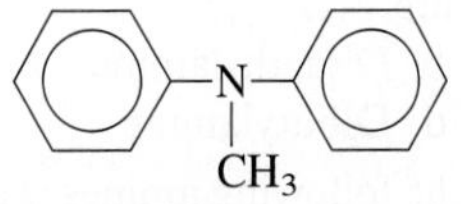

6.20 Name each of the following aromatic amines as a derivative of aniline.

a. 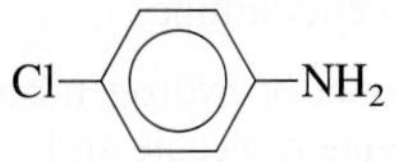

b. 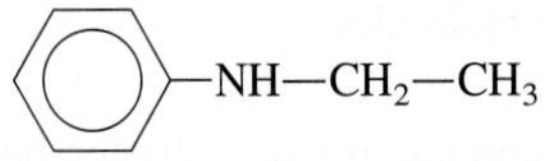

c. 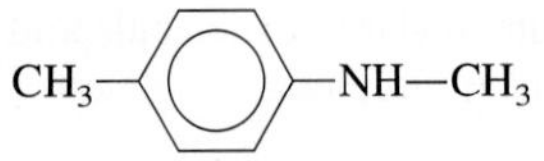

d. 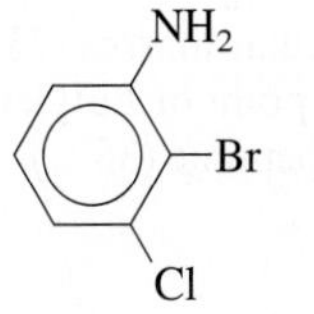

6.21 Draw a condensed structural formula for each of the following amines.
a. 2-Methyl-2-butanamine b. 1,6-Hexanediamine
c. 2-Amino-3-pentanone d. 2-Aminopropanoic acid

6.22 Draw a condensed structural formula for each of the following amines.
a. 2-Methyl-3-ethyl-1-hexanamine
b. 1,3-Pentanediamine
c. 3-Amino-2-pentanol
d. *N,N*-dimethyl-1-butanamine

▲6.23 Classify each of the following compounds as a 1° amine, 2° amine, or 3° amine.
a. Ethylpropylamine b. 2-Propanamine
c. N-Phenylaniline d. 2-Cyclohexylethanamine

▲**6.24** Classify each of the following compounds as a 1° amine, 2° amine, or 3° amine.
a. Diisopropylamine b. 3-Methylaniline
c. *N,N*-Dimethylmethanamine
d. *N*-Ethyl-2-methylcyclopentylamine

▲6.25 What is the molecular formula for each of the compounds in Problem 6.23?

▲**6.26** What is the molecular formula for each of the compounds in Problem 6.24?

Isomerism for Amines (Section 6.4)

6.27 Draw condensed structural formulas for the eight isomeric primary amines that have the molecular formula $C_5H_{13}N$.

6.28 Draw condensed structural formulas for the six isomeric secondary amines that have the molecular formula $C_5H_{13}N$.

6.29 Give common names for the three isomeric tertiary amines that have the molecular formula $C_5H_{13}N$.

6.30 Give common names for the seven isomeric tertiary amines that have the molecular formula $C_6H_{15}N$.

6.31 Assign an IUPAC name to each of the four isomeric amines that have the molecular formula C_3H_9N.

6.32 Assign an IUPAC name to each of the eight isomeric amines that have the molecular formula $C_4H_{11}N$.

Physical Properties of Amines (Section 6.5)

6.33 Indicate whether each of the following amines is a liquid or a gas at room temperature.
a. Butylamine b. Dimethylamine
c. Ethylamine d. Dibutylamine

6.34 Indicate whether each of the following amines is a liquid or a gas at room temperature.
a. Methylamine b. Propylamine
c. Trimethylamine d. Pentylamine

6.35 Determine the maximum number of hydrogen bonds that can form between a methylamine molecule and
a. other methylamine molecules
b. water molecules

6.36 Determine the maximum number of hydrogen bonds that can form between a dimethylamine molecule and
a. other dimethylamine molecules
b. water molecules

6.37 Although they have similar molecular masses (73 and 72 amu, respectively), the boiling point of butylamine is much higher (78°C) than that of pentane (36°C). Explain why.

6.38 Although they have similar molecular masses (73 and 74 amu, respectively), the boiling point of 1-butanamine is much lower (78°C) than that of 1-butanol (118°C). Explain why.

6.39 Which compound in each of the following pairs of amines would you expect to be more soluble in water? Justify each answer.
a. $CH_3—CH_2—NH_2$ and $CH_3—CH_2—CH_2—CH_2—CH_2—NH_2$
b. $CH_3—CH_2—CH_2—NH_2$ and $H_2N—CH_2—CH_2—CH_2—NH_2$

6.40 Which compound in each of the following pairs of amines would you expect to be more soluble in water? Justify each answer.
a. $CH_3—CH_2—CH_2—NH_2$ and $CH_3—CH_2—CH_2—CH_2—NH_2$
b. $CH_3—CH_2—NH—CH_3$ and $CH_3—N(CH_3)—CH_3$

Basicity of Amines (Section 6.6)

6.41 Show the structures of the missing substance(s) in each of the following acid–base equilibria.
a. $CH_3—CH_2—NH_2 + H_2O \rightleftharpoons ? + OH^-$
b. 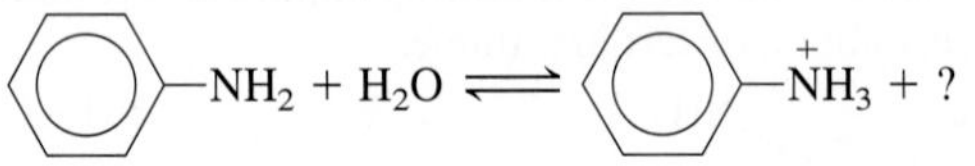
c. $? + H_2O \rightleftharpoons CH_3—CH(CH_3)—\overset{+}{N}H_2—CH_3 + OH^-$
d. Diethylamine $+ H_2O \rightleftharpoons ? + ?$

6.42 Show the structures of the missing substance(s) in each of the following acid–base equilibria.
a. $CH_3—CH_2—CH_2—NH_2 + H_2O \rightleftharpoons CH_3—CH_2—CH_2—\overset{+}{N}H_3 + ?$
b. 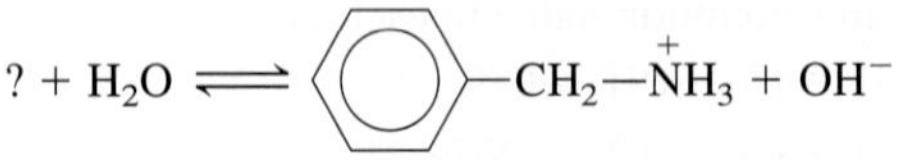
c. $CH_3—CH(CH_3)—CH_2—NH—CH_3 + H_2O \rightleftharpoons ? + OH^-$
d. Trimethylamine $+ H_2O \rightleftharpoons ? + ?$

6.43 Name each of the following substituted ammonium and substituted anilinium ions.
a. $CH_3—\overset{+}{N}H_2—CH_3$
b. $CH_3—CH_2—\overset{+}{N}H(CH_2—CH_3)—CH_2—CH_3$
c. 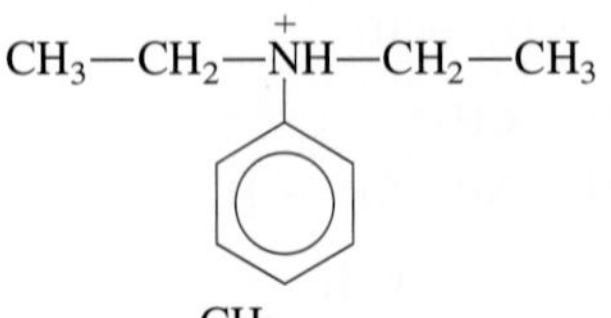
d. 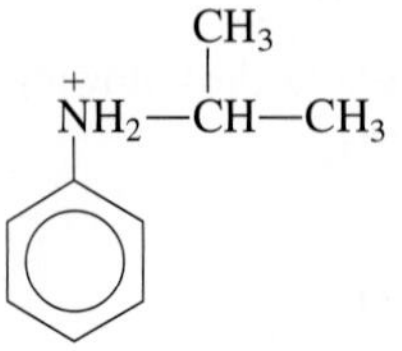

6.44 Name each of the following substituted ammonium and substituted anilinium ions.

a. $CH_3-\overset{+}{N}H_3$

b. $CH_3-CH_2-CH_2-\overset{+}{N}H_2-CH_3$

c. 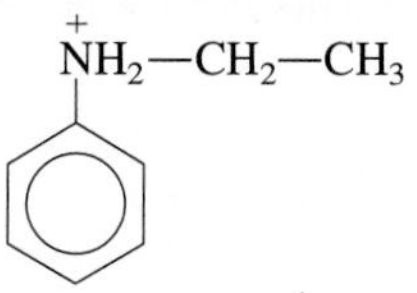

d. 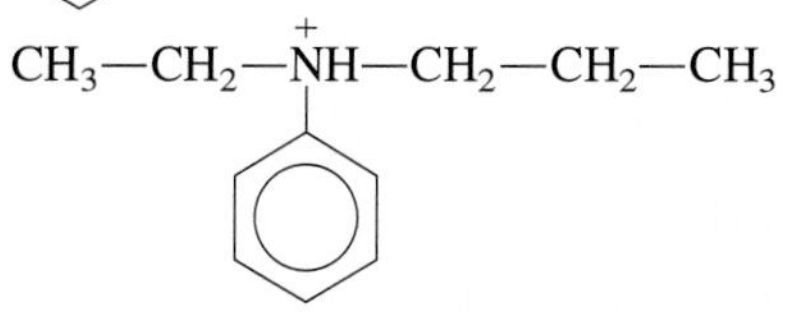

6.45 Draw a structural formula for the "parent" amine of each of the substituted ammonium and substituted anilinium ions in Problem 6.43.

6.46 Draw a structural formula for the "parent" amine of each of the substituted ammonium and substituted anilinium ions in Problem 6.44.

▲6.47 Draw skeletal structural formulas for the three isomeric disubstituted ammonium ions that have the chemical formula $C_4H_{12}N^+$.

▲**6.48** Draw skeletal structural formulas for the three isomeric trisubstituted ammonium ions that have the chemical formula $C_5H_{14}N^+$.

Amine Salts (Section 6.7)

6.49 Draw the structure of the missing substance in each of the following reactions involving amine salts.

a. $CH_3-CH_2-NH_2 + HCl \longrightarrow ?$

b. 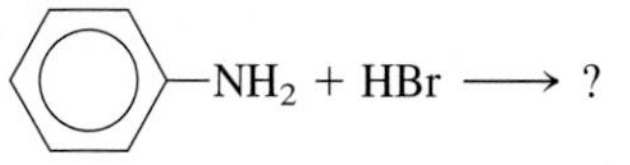

c. 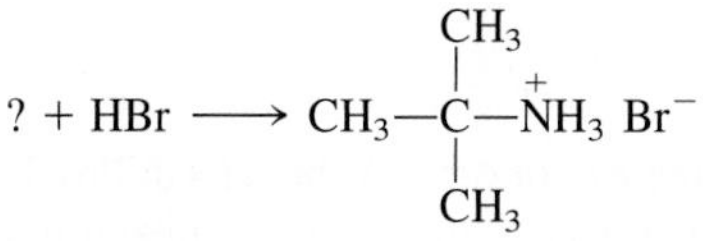

d. $CH_3-CH_2-NH-CH_3 + ? \longrightarrow CH_3-CH_2-\overset{+}{N}H_2-CH_3\ Cl^-$

6.50 Draw the structure of the missing substance in each of the following reactions involving amine salts.

a. $CH_3-CH_2-NH-CH_2-CH_3 + HBr \longrightarrow ?$

b. $CH_3-NH_2 + ? \longrightarrow CH_3-\overset{+}{N}H_3\ Cl^-$

c. 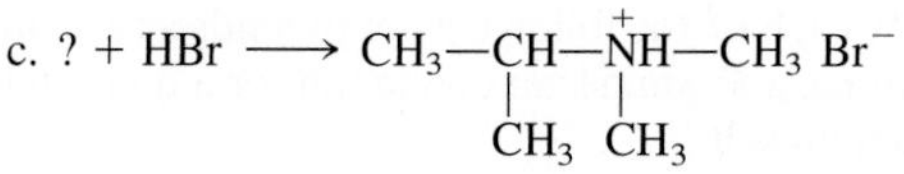

d. 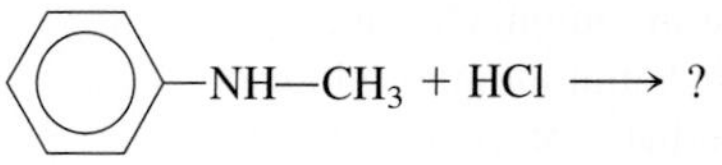

6.51 Draw the structure(s) of the missing substance(s) in each of the following reactions involving amine salts.

a. 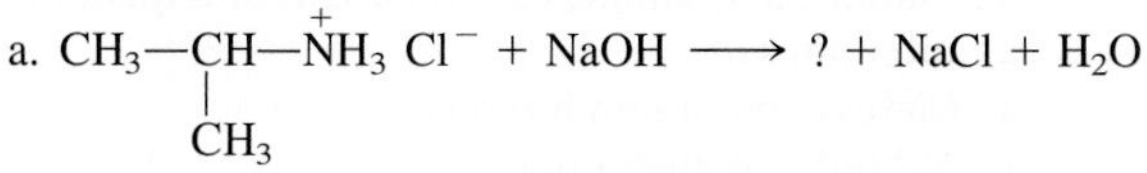

b. $? + NaOH \longrightarrow CH_3-NH-CH_3 + NaCl + H_2O$

c. 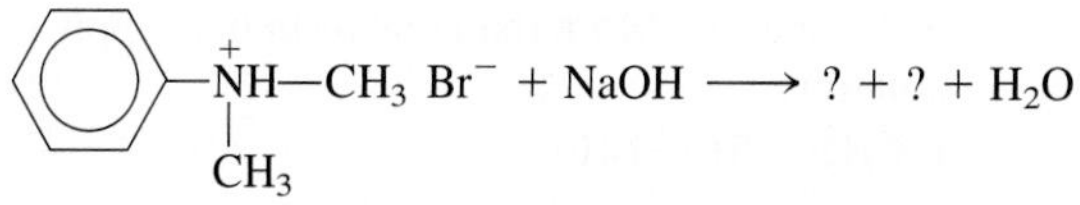

d. $CH_3-\overset{+}{N}H_2-CH_3\ Cl^- + NaOH \longrightarrow ? + NaCl + H_2O$

6.52 Draw the structure(s) of the missing substance(s) in each of the following reactions involving amine salts.

a. $CH_3-CH_2-CH_2-\overset{+}{N}H_3\ Br^- + NaOH \longrightarrow ? + NaBr + H_2O$

b. 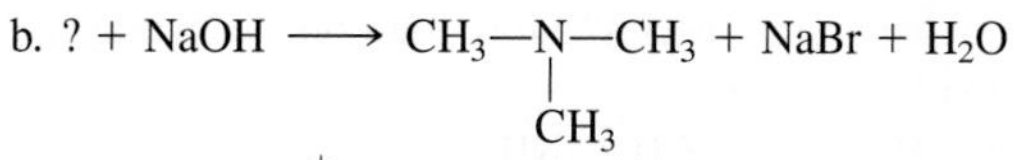

c. 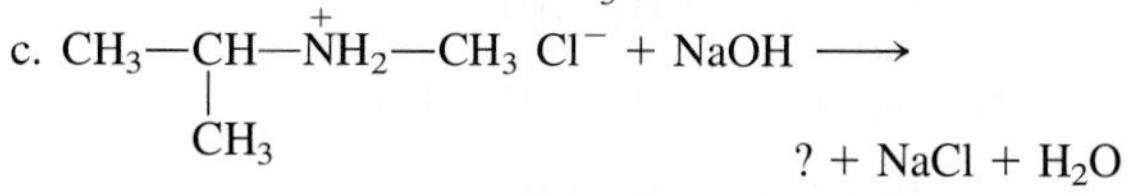

d. 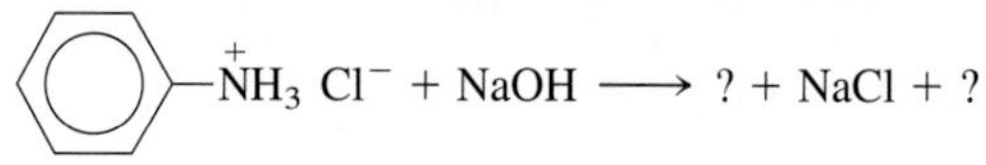

6.53 Name each of the following amine salts.

a. $CH_3-CH_2-CH_2-\overset{+}{N}H_3\ Cl^-$

b. 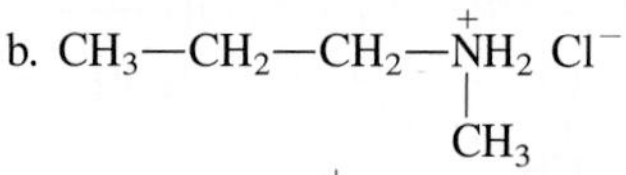

c. 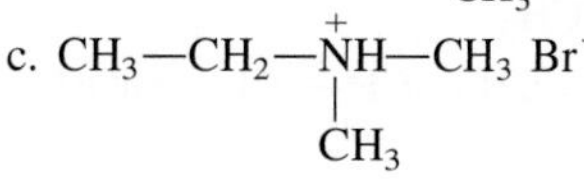

d. 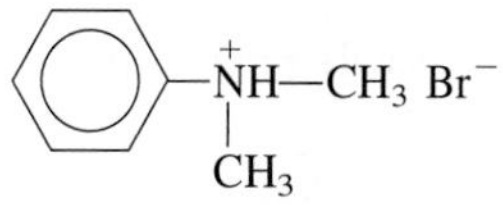

6.54 Name each of the following amine salts.

a. 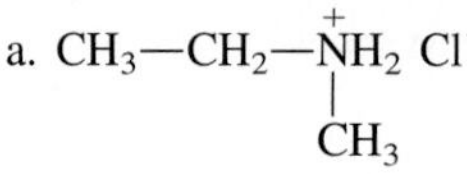

b. $CH_3-CH_2-CH_2-CH_2-\overset{+}{N}H_3\ Cl^-$

c. 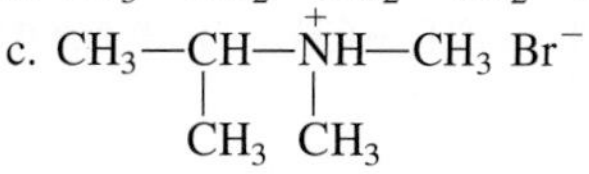

d. 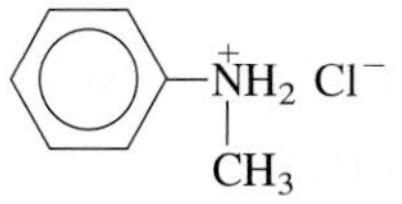

▲6.55 What is the IUPAC name of the "parent" amine for each of the amine salts in Problem 6.53?

▲**6.56** What is the IUPAC name of the "parent" amine for each of the amine salts in Problem 6.54?

6.57 Why are drugs that contain the amine functional group most often administered to patients in the form of amine chloride or hydrogen sulfate salts?

6.58 A student looking in an old chemistry book found the following name and structure for a compound.

$CH_3-CH_2-NH_2 \cdot HBr$

Ethylamine hydrobromide

What are the modern name and structural representation for this compound?

6.59 Which of the four terms *free amine, free base, deprotonated base,* and *protonated base* apply to each of the following amine species? More than one term may apply in a given situation.

a. $CH_3—CH_2—NH_2$

b. $CH_3—CH_2—CH_2—NH—CH_3$

c. $CH_3—\overset{+}{N}H_3$

d. $CH_3—CH_2—\overset{+}{N}H_2—CH_3$

6.60 Which of the four terms *free amine, free base, deprotonated base,* and *protonated base* apply to each of the following amine species? More than one term may apply in a given situation.

a. $CH_3—CH_2—CH_2—NH_2$

b. $CH_3—CH_2—CH_2—\overset{+}{N}H_2—CH_3$

c. $CH_3—NH—CH_2—CH_3$

d. $CH_3—CH_2—\overset{+}{N}H_3$

Alkylation of Ammonia and Amines (Section 6.8)

6.61 Identify the three products in each of the following reactions.

a. $NH_3 + CH_3—CH_2—CH_2—Cl + NaOH \longrightarrow$

b. $CH_3—Br + CH_3—\underset{\underset{CH_3}{|}}{CH}—NH—CH_3 + NaOH \longrightarrow$

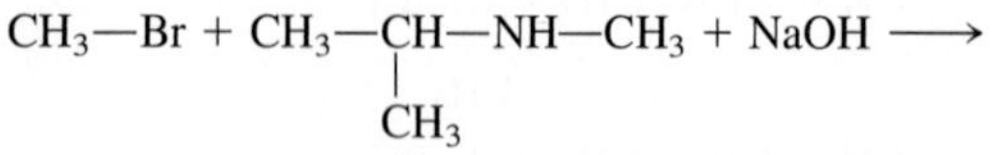

c. $CH_3—CH_2—NH_2 + CH_3—CH_2—Cl + NaOH \longrightarrow$

d. $CH_3—\underset{\underset{CH_3}{|}}{\overset{\overset{CH_3}{|}}{C}}—Br + NH_3 + NaOH \longrightarrow$

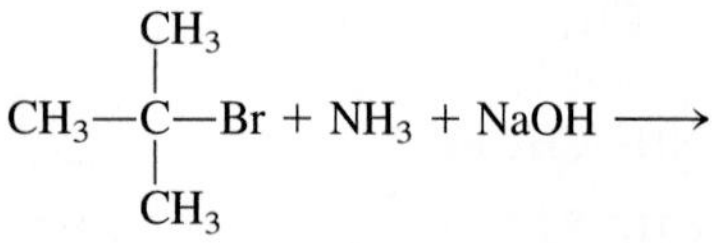

6.62 Identify the three products in each of the following reactions.

a. $CH_3—\underset{\underset{CH_3}{|}}{CH}—Cl + NH_3 + NaOH \longrightarrow$

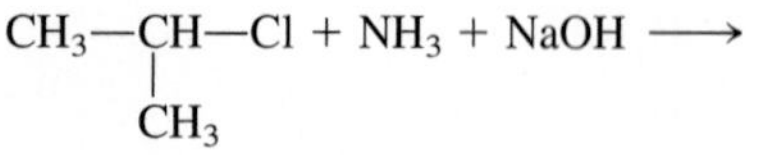

b. $CH_3—NH—CH_3 + CH_3—Br + NaOH \longrightarrow$

c. $CH_3—CH_2—CH_2—NH_2 + CH_3—CH_2—Br + NaOH \longrightarrow$

d. $CH_3—CH_2—\underset{\underset{CH_3}{|}}{CH}—Cl +$

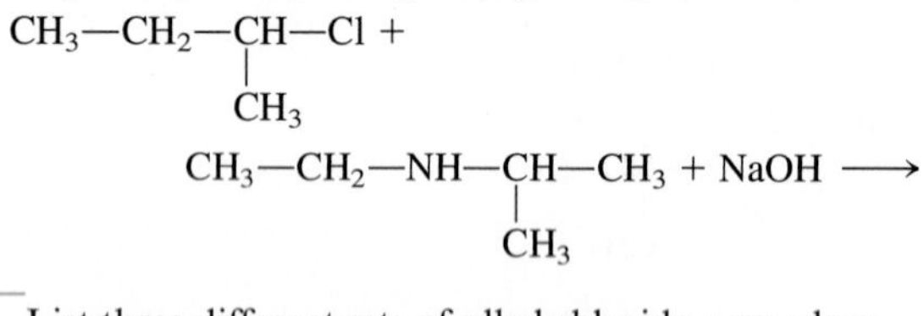

$CH_3—CH_2—NH—\underset{\underset{CH_3}{|}}{CH}—CH_3 + NaOH \longrightarrow$

6.63 List three different sets of alkyl chloride–secondary amine reactants that could be used to prepare the tertiary amine ethylmethylpropylamine.

6.64 List three different sets of alkyl chloride–secondary amine reactants that could be used to prepare the tertiary amine butylethylpropylamine.

6.65 Draw the structure of the amine or quaternary ammonium salt produced when each of the following pairs of compounds reacts in the presence of a strong base.

a. Trimethylamine and ethyl bromide

b. Diisopropylamine and methyl bromide

c. Ethylmethylpropylamine and methyl chloride

d. Ethylamine and ethyl chloride

6.66 Draw the structure of the amine or quaternary ammonium salt produced when each of the following pairs of compounds reacts in the presence of a strong base.

a. Dimethylamine and propyl bromide

b. Diethylmethylamine and isopropyl chloride

c. Methylpropylamine and ethyl chloride

d. Tripropylamine and propyl chloride

6.67 Classify each of the following salts as an amine salt or a quaternary ammonium salt.

a. $CH_3—\underset{\underset{CH_3}{|}}{\overset{+}{N}H}—CH_3\ Br^-$

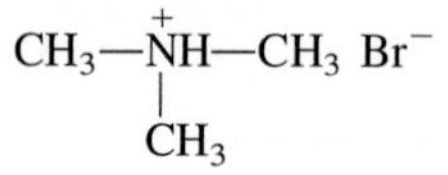

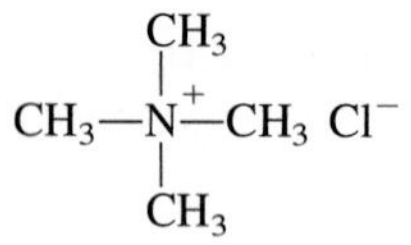

b. $CH_3—\underset{\underset{CH_3}{|}}{\overset{\overset{CH_3}{|}}{\overset{+}{N}}}—CH_3\ Cl^-$

c. $CH_3—CH_2—\overset{+}{N}H_2—CH_3\ Br^-$

d. $CH_3—CH_2—\underset{\underset{CH_3}{|}}{\overset{\overset{CH_3}{|}}{\overset{+}{N}}}—CH_2—CH_3\ Cl^-$

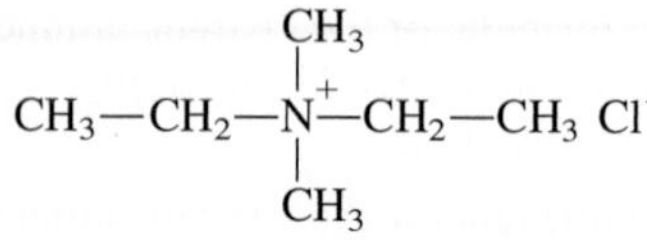

6.68 Classify each of the following salts as an amine salt or a quaternary ammonium salt.

a. $CH_3—\underset{\underset{CH_3}{|}}{\overset{\overset{CH_3}{|}}{\overset{+}{N}}}—CH_2—CH_3\ Cl^-$

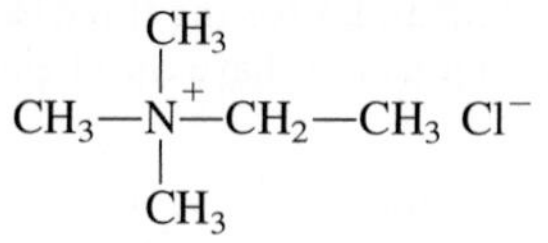

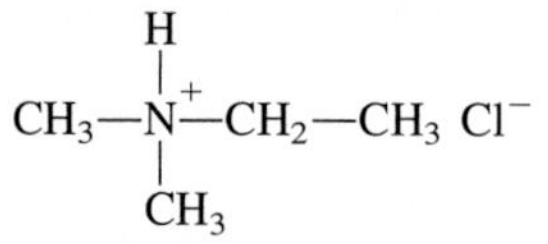

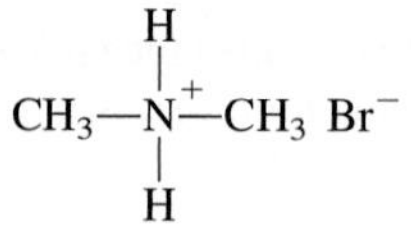

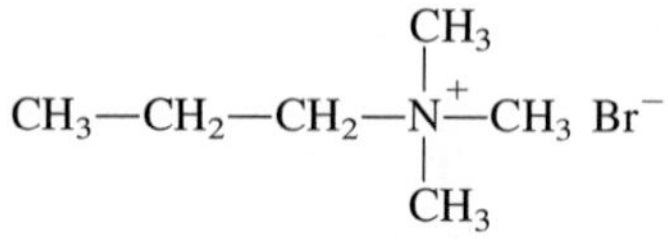

b. $CH_3—\underset{\underset{CH_3}{|}}{\overset{\overset{H}{|}}{\overset{+}{N}}}—CH_2—CH_3\ Cl^-$

c. $CH_3—\underset{\underset{H}{|}}{\overset{\overset{H}{|}}{\overset{+}{N}}}—CH_3\ Br^-$

d. $CH_3—CH_2—CH_2—\underset{\underset{CH_3}{|}}{\overset{\overset{CH_3}{|}}{\overset{+}{N}}}—CH_3\ Br^-$

▲**6.69** Indicate whether or not each of the salts in Problem 6.67 can be converted back to its "parent" amine through reaction with a strong base.

▲**6.70** Indicate whether or not each of the salts in Problem 6.68 can be converted back to its "parent" amine through reaction with a strong base.

6.71 Name each of the salts in Problem 6.67.

6.72 Name each of the salts in Problem 6.68.

▲**6.73** Classify each of the following compounds as a 1° amine, a 2° amine, a 3° amine, an amine salt, or a quaternary ammonium salt.

a. Dimethylammonium chloride

b. Tetramethylammonium bromide

c. *N,N*-Dimethyl-1-propanamine

d. 2,2-Dimethyl-1-propanamine

▲**6.74** Classify each of the following compounds as a 1° amine, a 2° amine, a 3° amine, an amine salt, or a quaternary ammonium salt.

a. Methylammonium bromide

b. *N*-Methylmethanamine

c. Diethyldimethylammonium chloride

d. Isopropylpropylamine

Heterocyclic Amines (Section 6.9)

6.75 Indicate whether or not each of the following compounds is a heterocyclic amine.

a. 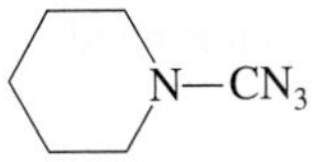

b. 

c. 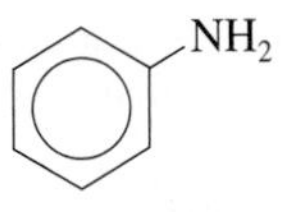

d. 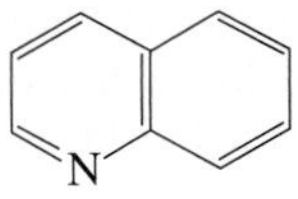

6.76 Indicate whether or not each of the following compounds is a heterocyclic amine.

a. 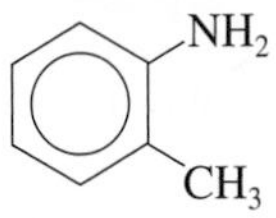

b. 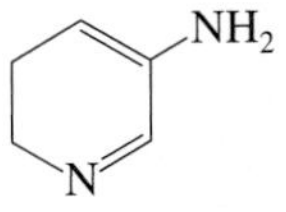

c. 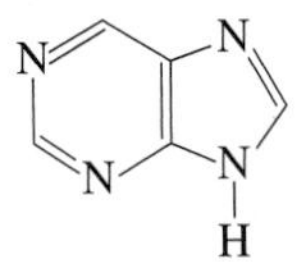

d. 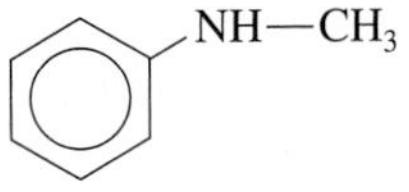

6.77 With the help of Figure 6.8, indicate whether the ring system in each of the following heterocyclic amines is (1) saturated (2) unsaturated or (3) fused. More than one choice may be correct for a given heterocyclic amine.
a. Pyrrolidine   b. Imidazole
c. Pyridine   d. Quinoline

6.78 With the help of Figure 6.8, indicate whether the ring system in each of the following heterocyclic amines is (1) saturated (2) unsaturated or (3) fused. More than one choice may be correct for a given heterocyclic amine.
a. Pyrrole   b. Indole   c. Pyrimidine   d. Purine

6.79 What is the difference, if any, between a porphyrin ring and a heme unit in terms of the number of pyrrole rings present?

6.80 What is the difference, if any, between a porphyrin ring and a heme unit in terms of iron content?

6.81 With the help of Figure 6.8, identify the heterocyclic amine ring system or systems present in each of the following substances.
a. Caffeine   b. Heme   c. Histamine   d. Serotonin

6.82 With the help of Figure 6.8, identify the heterocyclic amine ring system or systems present in each of the following.
a. Nicotine   b. Quinine
c. Odor of popcorn   d. Porphyrin ring

•6.83 (Chemical Connections 6-A) Indicate whether each of the following statements relating directly or indirectly to the substance caffeine is true or false.
a. Caffeine is a dimethyl derivative of the molecule xanthine.
b. Caffeine's half-life in the body is 3.0–7.5 hours.
c. Caffeine is mildly addictive, with headache being a major withdrawal symptom.
d. Increased glucose production caused by inhibition of the enzyme cyclic AMP is one of the effects of caffeine in the human body.

•6.84 (Chemical Connections 6-A) Indicate whether each of the following statements relating directly or indirectly to the substance caffeine is true or false.
a. Both a six-membered and four-membered ring system are present in caffeine's molecular structure.
b. Regular users of caffeine develop an increasing tolerance for the substance.
c. Caffeine is present in the milk of lactating mothers who drink caffeine-containing beverages.
d. Diuretic action (increased urine output) is one of the effects of caffeine within the human body.

•6.85 (Chemical Connections 6-B) Indicate whether each of the following statements relating directly or indirectly to the substance nicotine is true or false.
a. Nicotine's molecular structure involves two non-fused rings, each of which contains a nitrogen atom.
b. Because of nicotine's flammability, most of the nicotine in a cigarette is burned when it is smoked.
c. Within the human body, nicotine increases the production of the central nervous system stimulant adrenaline.
d. The carcinogenic effects associated with cigarette smoking are caused in part by nicotine.

•6.86 (Chemical Connections 6-B) Indicate whether each of the following statements relating directly or indirectly to the substance nicotine is true or false.
a. Nicotine's molecular structure contains two amine functional groups.
b. Within the human body, nicotine increases the release of the neurotransmitter dopamine.
c. Inhaled nicotine from a burning cigarette reaches the brain in approximately two minutes.
d. High-pH cigarette products supply more nicotine to the user than do low-pH cigarettes.

Biochemically Important Amines (Section 6.10)

6.87 Indicate whether each of the following statements about biochemically important amines is true or false.
a. Serotonin regulates lactation in women.
b. Structurally, dopamine and norepinephrine differ only by an —OH group.
c. L-Dopa and dopamine are two names for the same compound.
d. Prozac is a drug that helps maintain adrenaline levels in the brain.

6.88 Indicate whether each of the following statements about biochemically important amines is true or false.
a. Serotonin and dopamine are both 2-phenylethylamine derivatives.
b. A deficiency of dopamine is associated with Parkinson's disease.
c. Structurally, norepinephrine and epinephrine differ only by a $—CH_3$ group.
d. Epinephrine and adrenaline are two names for the same compound.

6.89 Indicate whether or not each of the following amines is a 2-phenylethylamine derivative.
a. Methamphetamine   b. Amphetamine
c. Histamine   d. Pseudoephedrine

6.90 Indicate whether each of the following amines is (1) a decongestant (2) an antihistamine (3) a central nervous system stimulant or (4) a protein component.
a. Phenylephrine   b. Histidine
c. Amphetamine   d. Fexofenodine

Alkaloids (Section 6.11)

6.91 Indicate whether or not each of the following substances is an alkaloid.
a. Nicotine   b. Quinine   c. Morphine   d. Cocaine

6.92 Indicate whether or not each of the following substances is an alkaloid.
a. Caffeine b. Theobromine
c. Atropine d. Codeine

6.93 Indicate whether each of the following statements about particular alkaloids is true or false.
a. Medicinally, atropine is used to dilate the pupil of the eye.
b. Quinine is a powerful antipyretic.
c. Morphine is 100 times more powerful as a painkiller than codeine.
d. The opium of the oriental poppy plant contains both morphine and heroin.

6.94 Indicate whether each of the following statements about particular alkaloids is true or false.
a. Atropine's molecular structure contains two amine groups.
b. Quinine is used in the treatment of malaria.
c. The demand for morphine as a painkiller exceeds that for codeine.
d. Structurally, codeine is a methyl morphine.

6.95 Describe each of the following molecules, which have the same central core, in terms of functional groups attached to the common central core.
a. Codeine b. Morphine
c. Hydrocodone d. Oxycodone

6.96 Indicate whether each of the following prescription painkillers contains *oxycodone* or *hydrocodone.*
a.Lortab b. Percocet c. OxyContin d. Percodan

•6.97 (Chemical Connections 6-C) Indicate whether each of the following statements relating directly or indirectly to chocolate is true or false.
a. The starting material for all types of chocolate is ground cacao beans.
b. Baker's chocolate is simply hardened chocolate liquor.
c. Two important alkaloids present in chocolate are theobromine and caffeine.
d. The alkaloid theobromine is a CNS system stimulant that is several times stronger than caffeine.

•6.98 (Chemical Connections 6-C) Indicate whether each of the following statements relating directly or indirectly to chocolate is true or false.
a. Dark chocolate does not contain added sugar or milk solids.
b. The theobromine/caffeine ratio in milk chocolate is approximately 3-to-1.
c. Structurally, theobromine is a methylcaffeine molecule.
d. Theobromine has the ability to dilate blood vessels.

Structure of and Classification of Amides (Section 6.12)

6.99 What is the generalized molecular formula for each of the following?
a. Primary amide b. Secondary amide
c. Tertiary amide

6.100 Draw a generalized structural formula for the functional group present in each of the following.
a. Primary amide b. Secondary amide
c. Tertiary amide

6.101 Indicate whether each of the following is a general structural characteristic of (1) a 1° amide (2) a 2° amide or (3) a 3° amide. More than one response may be correct for a given characteristic.
a. Contains one carbon–nitrogen bond
b. Contains two hydrogen–nitrogen bonds
c. Contains one carbon–oxygen double bond
d. Contains three R groups

6.102 Indicate whether each of the following is a general structural characteristic of (1) a 1° amide (2) a 2° amide or (3) a 3° amide. More than one response may be correct for a given characteristic.
a. Contains two carbon–nitrogen bonds
b. Contains zero hydrogen–nitrogen bonds
c. Contains two R groups
d. Contains a carbonyl group

6.103 Indicate whether or not each of the following compounds contains an amide functional group.

a.
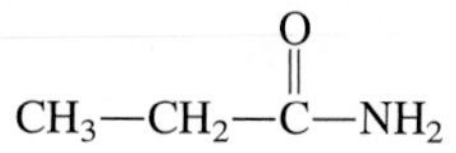

b.
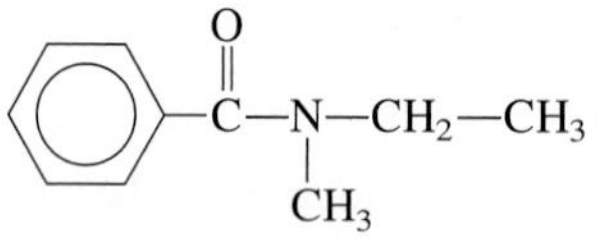

c.
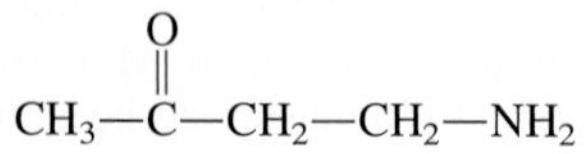

d.
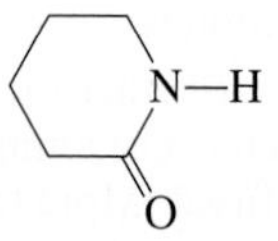

6.104 Indicate whether or not each of the following compounds contains an amide functional group.

a.
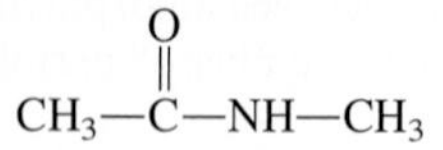

b.
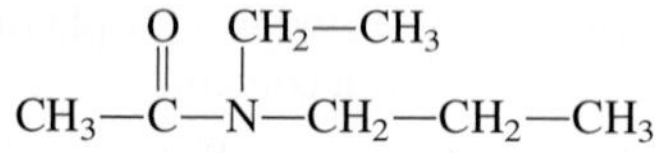

c.
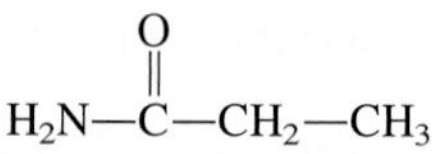

d.
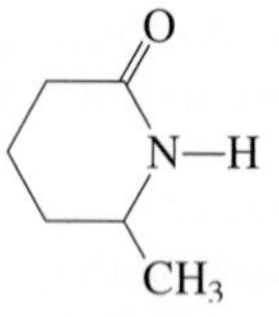

6.105 Classify each of the following amides as unsubstituted, monosubstituted, or disubstituted.

a.
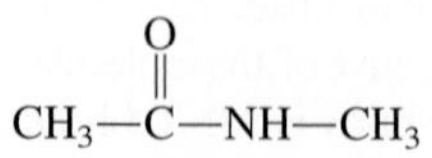

b.
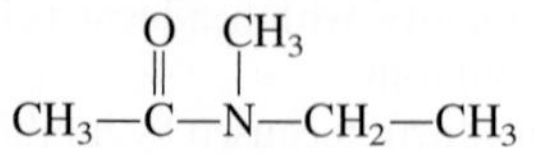

c.
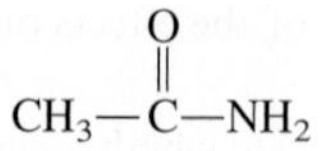

d.
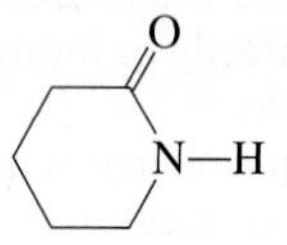

6.106 Classify each of the following amides as unsubstituted, monosubstituted, or disubstituted.

a. 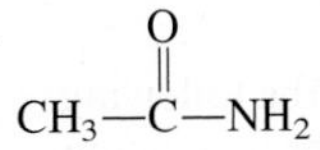

b. 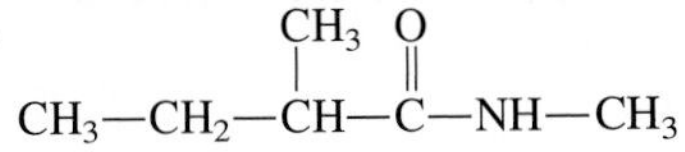

c. 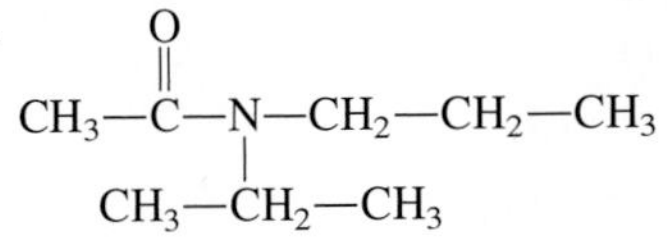

d. 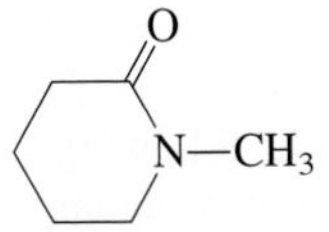

6.107 Classify each of the amides in Problem 6.105 as a primary, secondary, or tertiary amide.

6.108 Classify each of the amides in Problem 6.106 as a primary, secondary, or tertiary amide.

▲6.109 Classify each of the following compounds as (1) an amine (2) an amide or (3) both an amine and an amide.

a. 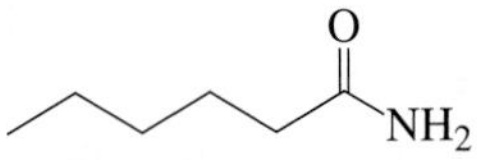 b. 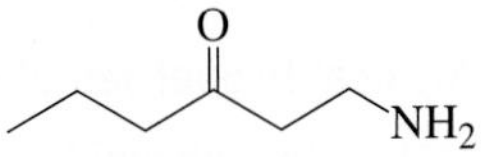

c. 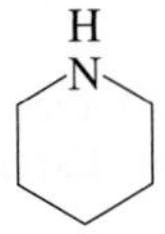 d. 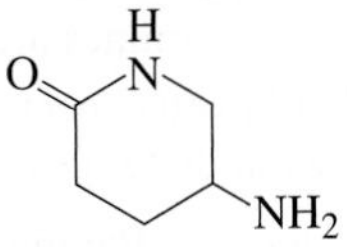

▲**6.110** Classify each of the following compounds as (1) an amine (2) an amide or (3) both an amine and an amide.

a. 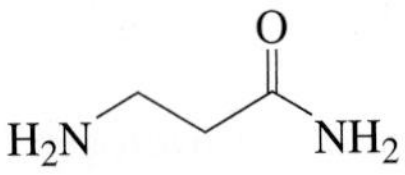 b. 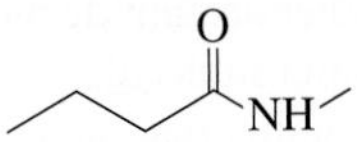

c. 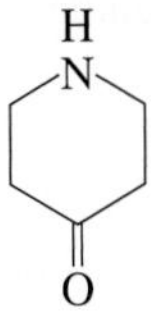 d. 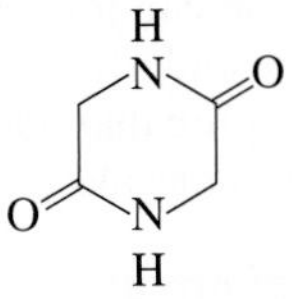

Nomenclature for Amides (Section 6.13)

6.111 Assign an IUPAC name to each of the following amides.

a. 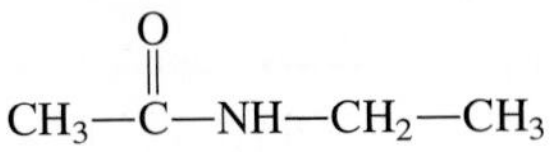

b. 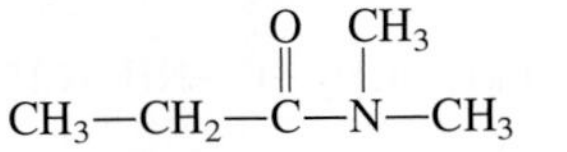

c. 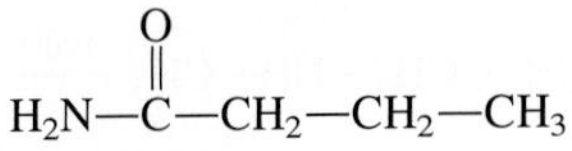

d. 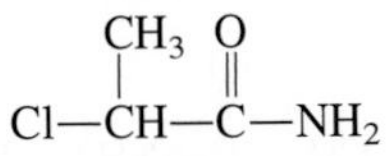

6.112 Assign an IUPAC name to each of the following amides.

a. 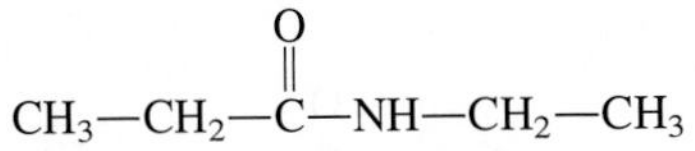

b. 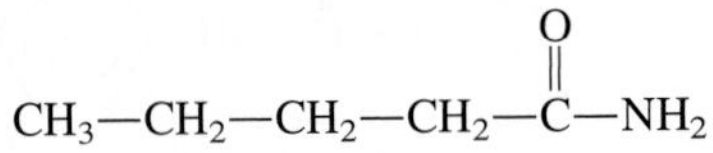

c. 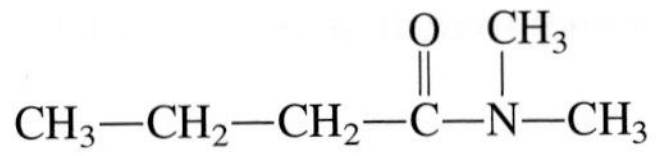

d. 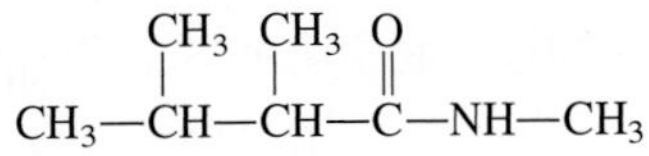

6.113 Assign a common name to each of the amides in Problem 6.111.

6.114 Assign a common name to each of the amides in Problem 6.112.

6.115 Assign an IUPAC name to each of the following amides.

a. 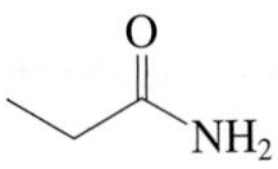

b. 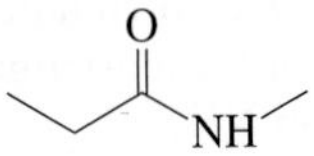

c. 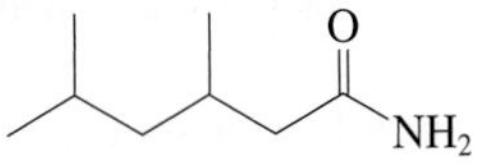

d. 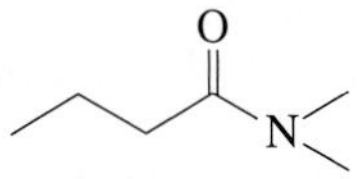

6.116 Assign an IUPAC name to each of the following amides.

a. 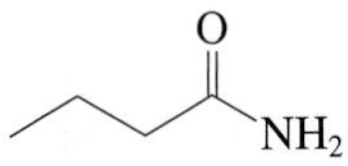

b. 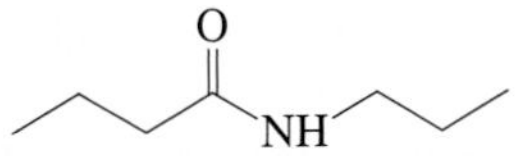

c. 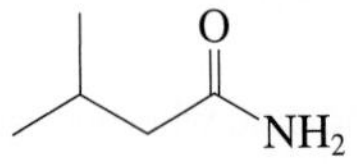

d. 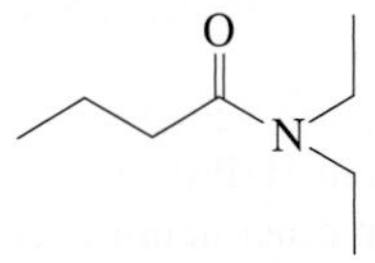

6.117 Write a structural formula for each of the following amides.

a. *N,N*-Dimethylacetamide b. α-Methylbutyramide
c. 3,*N*-Dimethylbutanamide d. Formamide

6.118 Write a structural formula for each of the following amides.

a. *N,N*-Diethylpropanamide
b. β-Methylbutyramide
c. *N*-Methylbenzamide
d. β,β,*N*-Trimethylbutyramide

▲6.119 Assign IUPAC names to each of the following compounds.

a. 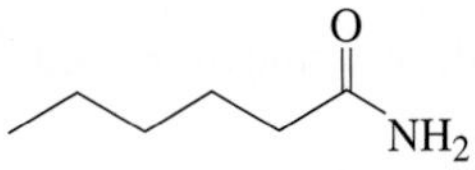 b. 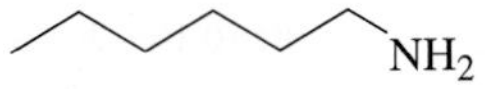

c. 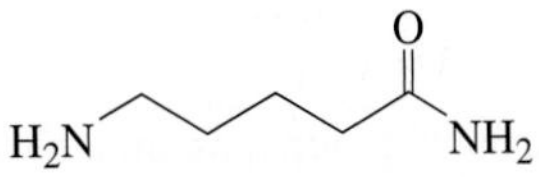

d. 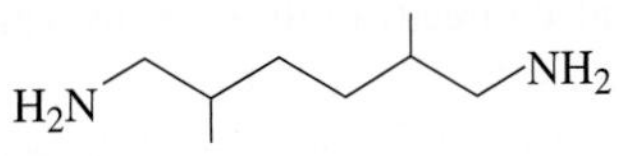

▲6.120 Assign IUPAC names to each of the following compounds.

a. 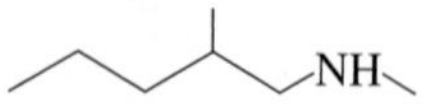 b. 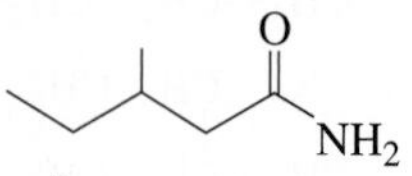

c. 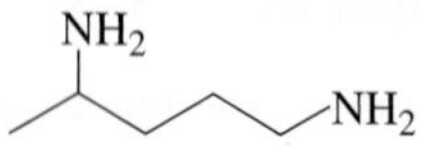 d. 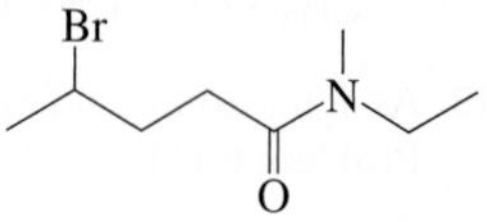

Selected Amides and Their Uses (Section 6.14)

6.121 Indicate whether each of the following statements about the compound urea is true or false.
a. Only one carbon atom is present in the structure of urea.
b. In the pure state, urea is a yellow-colored liquid.
c. Two of the starting materials for the biosynthetic production of urea are CO_2 and NH_3.
d. There are more nitrogen atoms than carbon atoms in the structure of urea.

6.122 Indicate whether each of the following statements about the compound melatonin is true or false.
a. Melatonin concentrations in the body decrease with age.
b. Low levels of melatonin are associated with a "good night's sleep."
c. As a prescription drug, melatonin is used to treat "jet lag."
d. There is an equal number of nitrogen atoms and carbon atoms in the structure of melatonin.

6.123 What is the mode of action for the insect repellent known as DEET?

6.124 What is the relationship between the acronym DEET and DEET's chemical name?

●6.125 (Chemical Connections 6-D) Indicate whether each of the following statements relating to the substance acetaminophen is true or false.
a. The pharmaceutical designation APAP for acetaminophen is related to its IUPAC name.
b. Acetaminophen differs from other major over-the-counter pain relievers in that it does not have anti-inflammatory properties.
c. Acetaminophen is an amide rather than an amine.
d. In large doses, acetaminophen can cause liver and kidney damage.

●**6.126** (Chemical Connections 6-D) Indicate whether each of the following statements relating to the substance acetaminophen is true or false.
a. In the designation APAP for acetaminophen, the two P's stand for propyl and phenol.
b. A major brand name for acetaminophen is Tylenol.
c. Acetaminophen is the pain reliever of choice for young children because of its availability in liquid form.
d. Acetaminophen's mode of action in the body is similar to that for aspirin.

Basicity of Amides (Section 6.15)

6.127 What effect does a carbonyl group have on a nitrogen atom that is bonded directly to the carbonyl carbon atom?

6.128 Explain why the nitrogen atom in an amide functional group does not readily participate in hydrogen bonding.

6.129 Indicate whether or not each of the following substances exhibits basic behavior in aqueous solution.
a. Propanamide
b. 1-Propanamine
c. $CH_3-CH_2-\overset{O}{\overset{\|}{C}}-NH_2$
d. $H_2N-CH_2-CH_2-\overset{O}{\overset{\|}{C}}-NH_2$

6.130 Indicate whether or not each of the following substances exhibits basic behavior in aqueous solution.
a. Methylethylamine
b. Butanamide
c. $H_2N-CH_2-CH_2-NH_2$
d. $H_2N-\overset{O}{\overset{\|}{C}}-CH_2-CH_2-\overset{O}{\overset{\|}{C}}-NH_2$

Physical Properties of Amides (Section 6.16)

6.131 Although amides contain a nitrogen atom, they are not bases as amines are. Explain why.

6.132 Would you expect *N*-ethylacetamide or *N,N*-diethylacetamide to have the higher boiling point? Explain.

6.133 Determine the maximum number of hydrogen bonds that can form between an acetamide molecule and
a. other acetamide molecules
b. water molecules

6.134 Determine the maximum number of hydrogen bonds that can form between a propanamide molecule and
a. other propanamide molecules
b. water molecules

Preparation of Amides (Section 6.17)

6.135 Draw the structures of the missing substances in each of the following reactions involving amides.

a. $CH_3-CH_2-\overset{O}{\overset{\|}{C}}-OH + ? \xrightarrow{100°C} CH_3-CH_2-\overset{O}{\overset{\|}{C}}-NH-CH_3 + H_2O$

b. $CH_3-\underset{CH_3}{\underset{|}{\overset{CH_3}{\overset{|}{C}}}}-\overset{O}{\overset{\|}{C}}-OH + CH_3-NH-CH_3 \xrightarrow{100°C} ? + H_2O$

c. $CH_3-\overset{O}{\overset{\|}{C}}-OH + ? \xrightarrow{100°C} CH_3-\overset{O}{\overset{\|}{C}}-NH_2 + H_2O$

d. 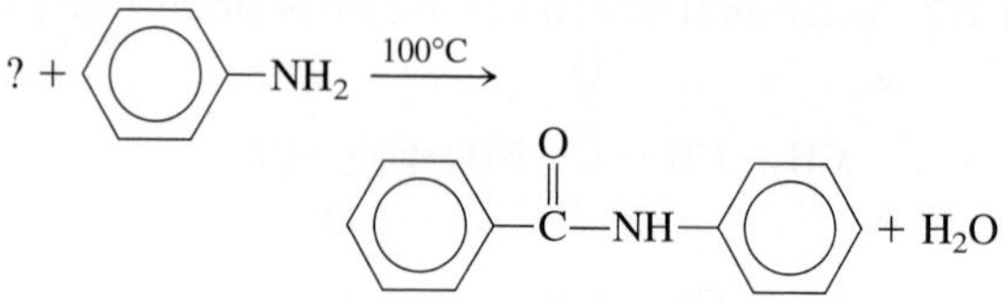

6.136 Draw the structures of the missing substances in each of the following reactions involving amides.

a.
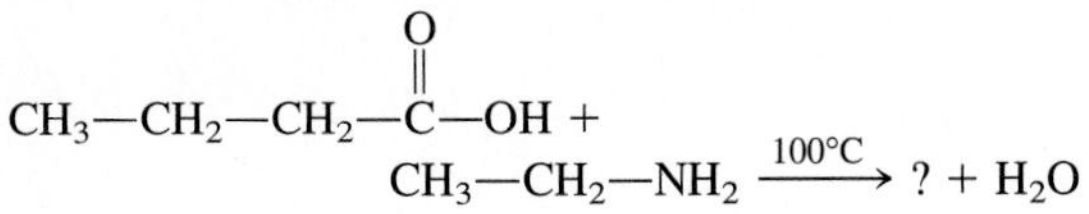

b.
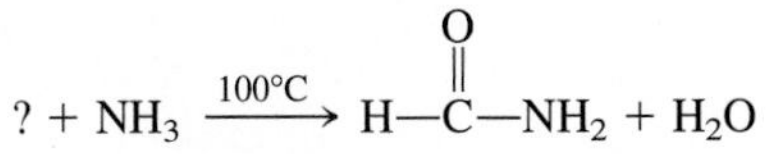

c.
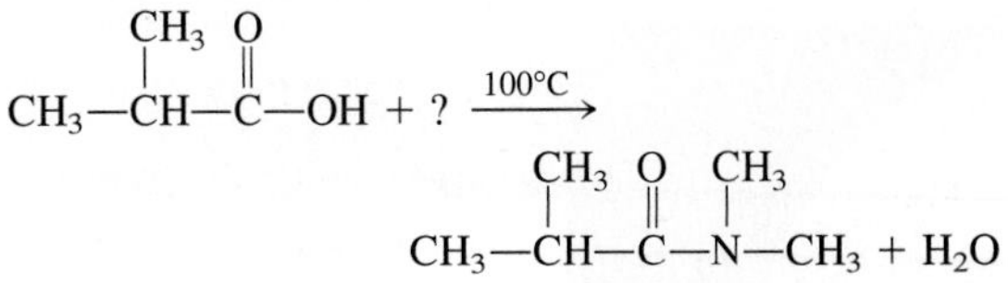

d.
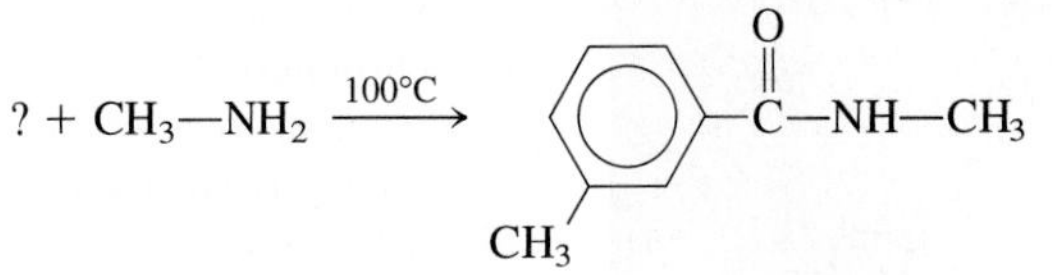

6.137 Draw the structures of the carboxylic acid and the amine from which each of the following amides could be formed.

a.
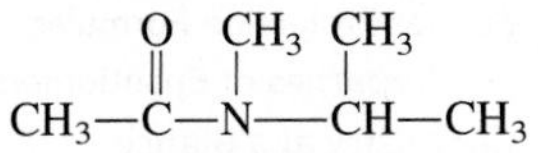

b. *N*-Methylpentanamide

c.
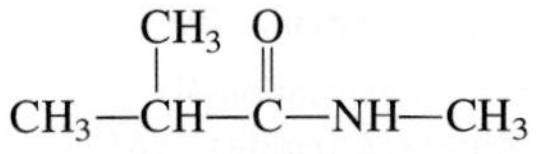

d. 2,3,*N*-Trimethylbutanamide

6.138 Draw the structures of the carboxylic acid and the amine from which each of the following amides could be formed.

a.
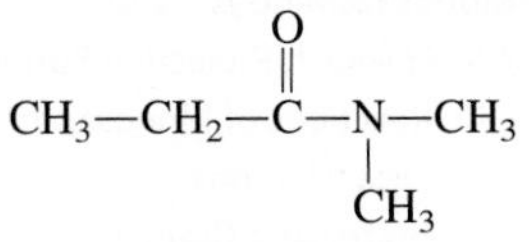

b. 2-Methylpentanamide

c.
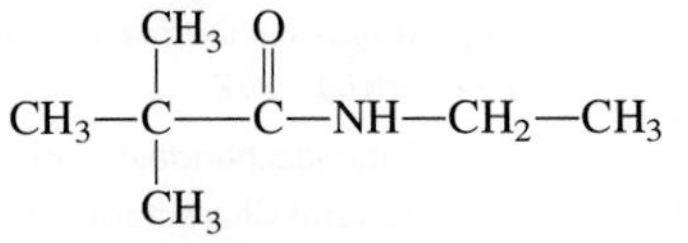

d. *N,N*-Diethylacetamide

Hydrolysis of Amides (Section 6.18)

6.139 Draw the structures of the organic products in each of the following hydrolysis reactions.

a.
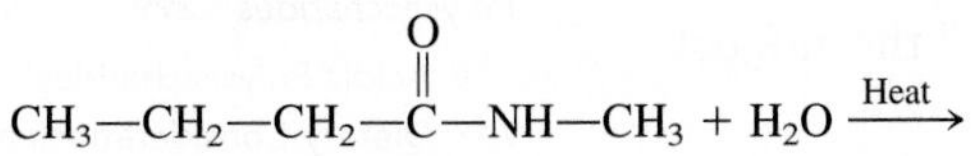

b.
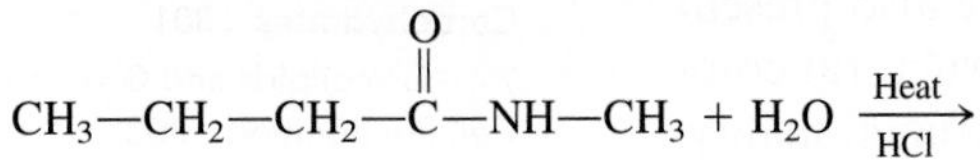

c.
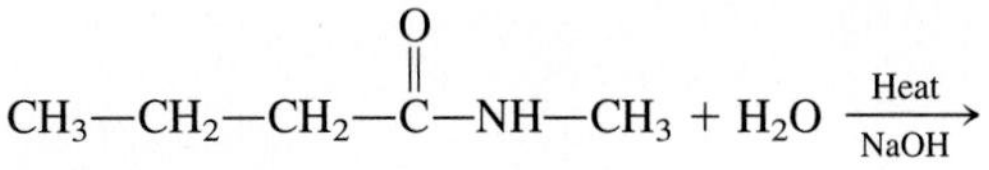

d.
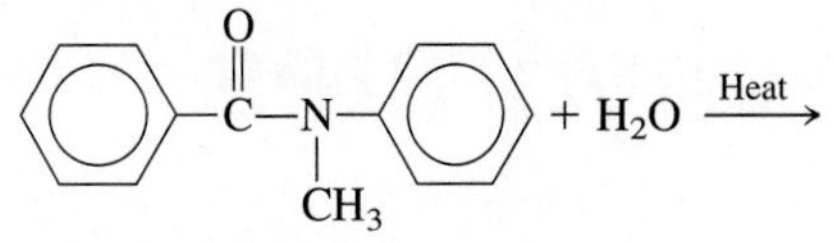

6.140 Draw the structures of the organic products in each of the following hydrolysis reactions.

a.
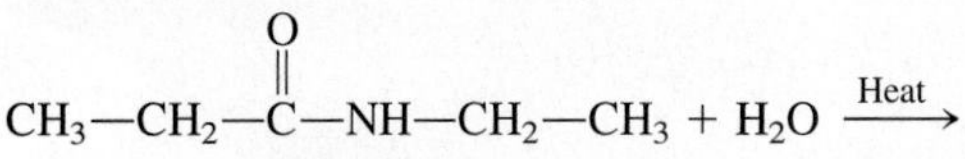

b.
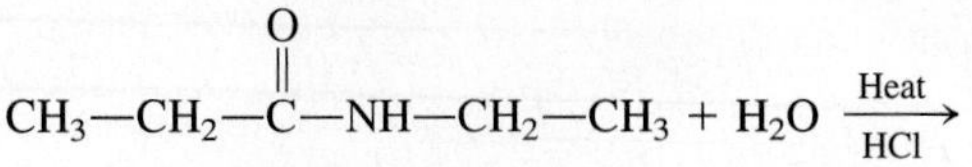

c.
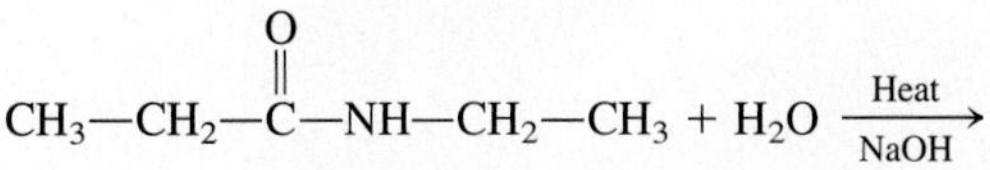

d.
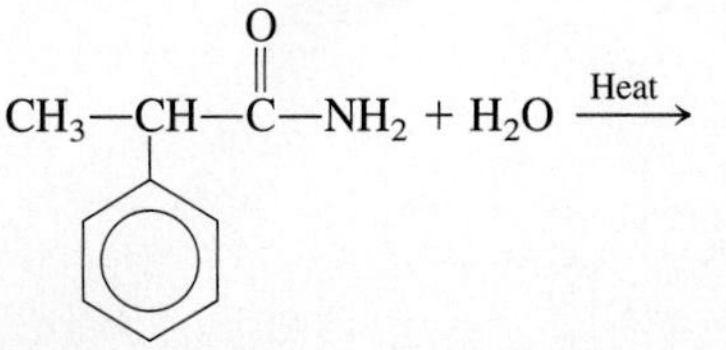

▲6.141 What type of organic product (or products) is (or are) produced in each of the following reactions?

a. 1-Butanamine + H_2O + HCl
b. *N*-Methylbutanamide + H_2O + HCl
c. *N*-Methylbutanamide + H_2O + NaOH
d. Chloride salt of 1-butanamine + H_2O + NaOH

▲**6.142** What type of organic product (or products) is (or are) produced in each of the following reactions?

a. 2-Butanamine + H_2O + HCl
b. *N,N*-Dimethylbutanamide + H_2O + HCl
c. *N,N*-Dimethylbutanamide + H_2O + NaOH
d. Chloride salt of 2-butanamine + H_2O + NaOH

Polyamides and Polyurethanes (Section 6.19)

6.143 List the general characteristics of the monomers needed to produce a polyamide.

6.144 Contrast the monomers needed to produce a polyamide with those needed to produce a polyester.

6.145 Draw a structural representation for the polyamide formed from the reaction of succinic acid and 1,4-butanediamine.

6.146 Draw a structural representation for the polyamide formed from the reaction of adipic acid and 1,2-ethanediamine.

6.147 Draw the generalized structural formula for a urethane.

6.148 What are the two types of monomers used to form a polyurethane polymer?

Carbohydrates

7

Inga Spence/Getty Images

The naturally present sugars fructose, glucose, and sucrose all contribute to the sweetness of ripe peaches.

CHAPTER OUTLINE

With this chapter, focus shifts from "organic chemistry" to "biochemistry." This chapter, and all remaining chapters in the textbook, deal with biochemical topics, that is, subjects associated with the chemistry of living systems. Only a few of the many facets of this vast subject can be considered. The approach for these considerations will be similar to that previously used for the organic subject matter. Initial chapters are devoted to each of the major classes of biochemical compounds (carbohydrates, lipids, proteins, and nucleic acids), followed by chapters addressing the metabolism of carbohydrates, lipids, and proteins. In this first "biochapter," the subject is carbohydrates.

The same functional groups found in organic compounds are also present in biochemical compounds. Usually, however, there is greater structural complexity associated with biochemical compounds as a result of polyfunctionality; several different functional groups are present. Often biochemical compounds interact with each other, within cells, to form larger structures. But the same chemical principles and chemical reactions associated with the various organic functional groups that were previously considered apply to these larger biochemical structures as well.

OWL

Sign in to OWL at **www.cengage.com/owl** to view tutorials and simulations, develop problem-solving skills, and complete online homework assigned by your professor.

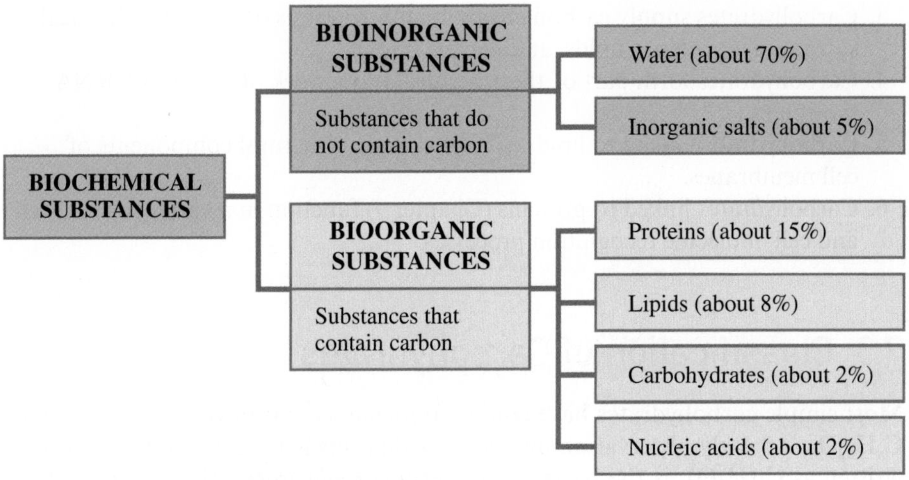

Figure 7.1 Mass composition data for the human body in terms of major types of biochemical substances.

7.1 Biochemistry—An Overview

Biochemistry *is the study of the chemical substances found in living organisms and the chemical interactions of these substances with each other.* Biochemistry is a field in which new discoveries are made almost daily about how cells manufacture the molecules needed for life and how the chemical reactions by which life is maintained occur. The knowledge explosion that has occurred in the field of biochemistry during the last decades of the twentieth century and the beginning of the twenty-first is truly phenomenal.

As isolated compounds, bioinorganic and bioorganic substances have no life in and of themselves. Yet when these substances are gathered together in a cell, their chemical interactions are able to sustain life.

A **biochemical substance** *is a chemical substance found within a living organism.* Biochemical substances are divided into two groups: bioinorganic substances and bioorganic substances. *Bioinorganic substances* include water and inorganic salts. *Bioorganic substances* include carbohydrates, lipids, proteins, and nucleic acids. Figure 7.1 gives an approximate mass composition for the human body in terms of types of biochemical substances present.

Although the human body is usually thought of as containing mainly organic (biochemical) substances, such substances make up only about one-fourth of total body mass. The bioinorganic substance water constitutes more than two-thirds of the mass of the human body, and another 4%–5% of body mass comes from inorganic salts.

7.2 Occurrence and Functions of Carbohydrates

Carbohydrates are the most abundant class of bioorganic molecules on planet Earth. Although their abundance in the human body is relatively low (Section 7.1), carbohydrates constitute about 75% by mass of dry plant materials (Figure 7.2).

Green (chlorophyll-containing) plants produce carbohydrates via *photosynthesis.* In this process, carbon dioxide from the air and water from the soil are the reactants, and sunlight absorbed by chlorophyll is the energy source.

$$CO_2 + H_2O + \text{solar energy} \xrightarrow[\text{Plant enzymes}]{\text{Chlorophyll}} \text{carbohydrates} + O_2$$

Plants have two main uses for the carbohydrates they produce. In the form of *cellulose,* carbohydrates serve as structural elements, and in the form of *starch,* they provide energy reserves for the plants.

Dietary intake of plant materials is the major carbohydrate source for humans and animals. The average human diet should ideally be about two-thirds carbohydrate by mass. Carbohydrates have the following functions in humans:

1. Carbohydrate oxidation provides energy.
2. Carbohydrate storage, in the form of glycogen, provides a short-term energy reserve.

Figure 7.2 Most of the matter in plants, except water, is carbohydrate material. Photosynthesis, the process by which carbohydrates are made, requires sunlight.

It is estimated that more than half of all organic carbon atoms are found in the carbohydrate materials of plants.

Human uses for carbohydrates of the plant kingdom extend beyond food. Carbohydrates in the form of cotton and linen are used as clothing. Carbohydrates in the form of wood are used for shelter and heating and in making paper.

3. Carbohydrates supply carbon atoms for the synthesis of other biochemical substances (proteins, lipids, and nucleic acids).
4. Carbohydrates form part of the structural framework of DNA and RNA molecules.
5. Carbohydrates linked to lipids (Chapter 8) are structural components of cell membranes.
6. Carbohydrates linked to proteins (Chapter 9) function in a variety of cell–cell and cell–molecule recognition processes.

7.3 Classification of Carbohydrates

Most simple carbohydrates have empirical formulas that fit the general formula $C_nH_{2n}O_n$. An early observation by scientists that this general formula can also be written as $C_n(H_2O)_n$ is the basis for the term *carbohydrate*—that is, "hydrate of carbon." It is now known that this hydrate viewpoint is not correct, but the term *carbohydrate* still persists. Today the term is used to refer to an entire family of compounds, only some of which have the formula $C_nH_{2n}O_n$.

A **carbohydrate** *is a polyhydroxy aldehyde, a polyhydroxy ketone, or a compound that yields polyhydroxy aldehydes or polyhydroxy ketones upon hydrolysis.* The carbohydrate glucose is a polyhydroxy aldehyde, and the carbohydrate fructose is a polyhydroxy ketone.

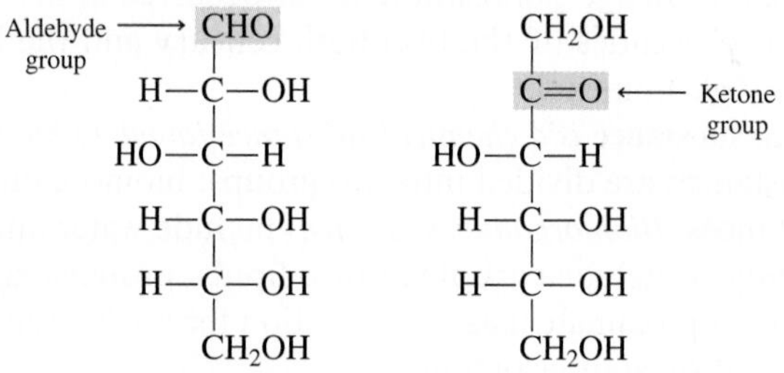

Glucose (a polyhydroxy aldehyde) Fructose (a polyhydroxy ketone)

A striking structural feature of carbohydrates is the large number of functional groups present. In glucose and fructose, a functional group is attached to each carbon atom.

Carbohydrates are classified on the basis of molecular size as monosaccharides, disaccharides, oligosaccharides, and polysaccharides.

The term *monosaccharide* is pronounced "mon-oh-SACK-uh-ride."

A **monosaccharide** *is a carbohydrate that contains a single polyhydroxy aldehyde or polyhydroxy ketone unit.* Monosaccharides cannot be broken down into simpler units by hydrolysis reactions. Both glucose and fructose are monosaccharides. Naturally occurring monosaccharides have from three to seven carbon atoms; five- and six-carbon species are especially common. Pure monosaccharides are water-soluble, white, crystalline solids.

A **disaccharide** *is a carbohydrate that contains two monosaccharide units covalently bonded to each other.* Like monosaccharides, disaccharides are crystalline, water-soluble substances. Sucrose (table sugar) and lactose (milk sugar) are disaccharides. Hydrolysis of a disaccharide produces two monosaccharide units.

The *oligo* in the term *oligosaccharide* comes from the Greek *oligos*, which means "small" or "few." The term *oligosaccharide* is pronounced "OL-ee-go-SACK-uh-ride."

An **oligosaccharide** *is a carbohydrate that contains three to ten monosaccharide units covalently bonded to each other.* "Free" oligosaccharides are seldom encountered in biochemical systems. They are usually found associated with proteins and lipids in complex molecules that have both structural and regulatory functions. Complete hydrolysis of an oligosaccharide produces several monosaccharide molecules; a trisaccharide produces three monosaccharide units, a hexasaccharide produces six monosaccharide units, and so on.

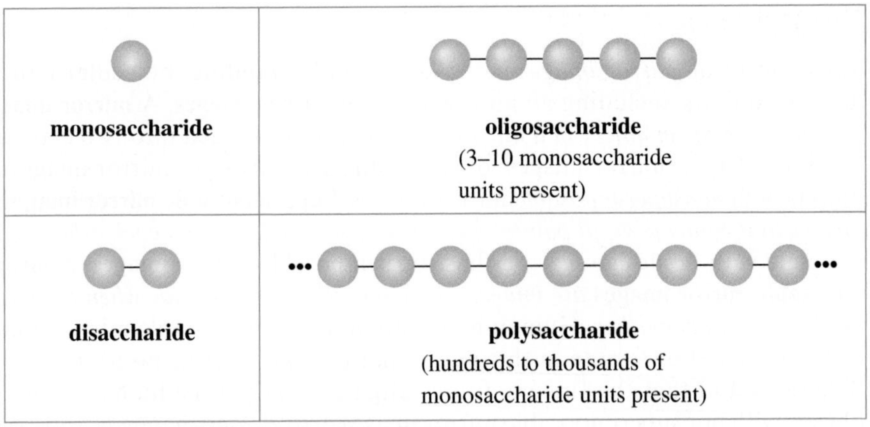

Figure 7.3 Monosaccharides can be bonded together to give disaccharides (two units), oligosaccharides (a few units), or polysaccharides (many units).

A **polysaccharide** *is a polymeric carbohydrate that contains many monosaccharide units covalently bonded to each other.* The number of monosaccharide units present in a polysaccharide varies from a few hundred units to more than a million units. Polysaccharides, like disaccharides and oligosaccharides, undergo hydrolysis under appropriate conditions to produce monosaccharides.

Types of carbohydrates are related to each other through hydrolysis.

Polysaccharides
↓ Hydrolysis
Oligosaccharides
↓ Hydrolysis
Disaccharides
↓ Hydrolysis
Monosaccharides

Both cellulose and starch are naturally occurring polysaccharides that are very prevalent in the world of plants. The paper on which this book is printed is mainly cellulose, as is the cotton in clothing fabrics and the wood used in home construction. Starch is a component of many types of foods, including bread, pasta, potatoes, rice, corn, beans, and peas.

Figure 7.3 shows diagrammatically the relationships among the various classes of saccharides.

7.4 Chirality: Handedness in Molecules

An important property of many molecules, including most carbohydrates, that has not been previously addressed in this text is the property of "handedness," which is a form of isomerism. Because of the importance of this property in carbohydrate chemistry considerations, details concerning it will now be presented, prior to discussing specifics about the various classes of carbohydrates. It will take several sections of text to formalize this "handedness concept" to the depth that is needed to understand carbohydrate chemistry.

Molecules that possess "handedness" exist in two forms: a "left-handed" form and a "right-handed" form. These two forms are related to each other in the same way that a pair of hands are related to each other. The relationship is that of *mirror images.* A left hand and a right hand are mirror images of each other, as shown in Figure 7.4.

The property of handedness is not restricted to carbohydrates. It is a general phenomenon found in all classes of organic compounds.

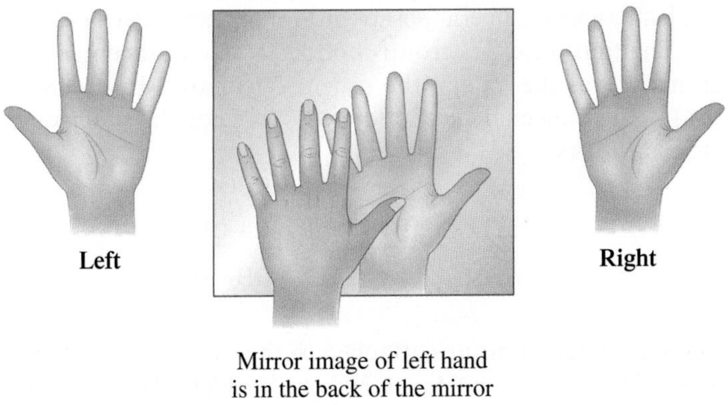

Figure 7.4 The mirror image of the right hand is the left hand. Conversely, the mirror image of the left hand is the right hand.

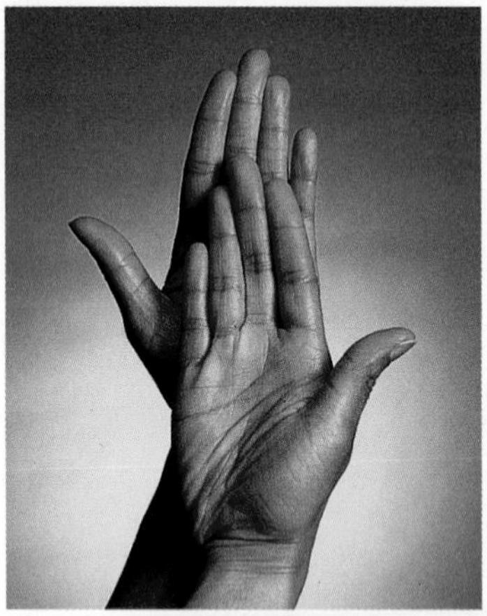

© Cengage Learning

Figure 7.5 A person's left and right hands are not superimposable upon each other.

Mirror Images

The concept of *mirror images* is the key to understanding molecular handedness. All objects, including all molecules, have mirror images. A **mirror image** *is the reflection of an object in a mirror.* Objects can be divided into two classes on the basis of their mirror images: objects with *superimposable* mirror images and objects with *nonsuperimposable* mirror images. **Superimposable mirror images** *are images that coincide at all points when the images are laid upon each other.* A dinner plate with no design features has superimposable mirror images. **Nonsuperimposable mirror images** *are images where not all points coincide when the images are laid upon each other.* Human hands are nonsuperimposable mirror images, as Figure 7.5 shows; note in this figure that the two thumbs point in opposite directions and that the fingers do not align correctly. Like human hands, all objects with nonsuperimposable mirror images exist in "left-handed" and "right-handed" forms.

Chirality

Every object has a mirror image. The question is, "Is the mirror image the same (superimposable) or different (nonsuperimposable)?"

Some, but not all, molecules possess handedness. What determines whether or not a molecule possesses handedness? For organic and bioorganic compounds, the structural requirement for handedness is the presence of a carbon atom that has four *different* groups bonded to it in a tetrahedral orientation. The tetrahedral orientation requirement is met only if the bonds to the four different groups are all single bonds.

Any molecule that contains a carbon atom with four different groups bonded to it in a tetrahedral orientation possesses handedness. The handedness-generating carbon atom is called a *chiral center.* A **chiral center** *is an atom in a molecule that has four different groups bonded to it in a tetrahedral orientation.*

A molecule that contains a chiral center is said to be *chiral.* A **chiral molecule** *is a molecule whose mirror images are not superimposable.* Chiral molecules have handedness. An **achiral molecule** *is a molecule whose mirror images are superimposable.* Achiral molecules do not possess handedness.

The term *chiral* (rhymes with *spiral*) comes from the Greek word *cheir,* which means "hand." Chiral objects are said to possess "handedness."

The simplest example of a chiral organic molecule, that is, an organic molecule that has left-handed and right-handed forms, is a trisubstituted methane molecule such as bromochloroiodomethane.

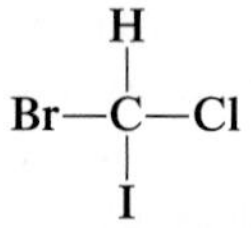

Bromochloroiodomethane

Note the four different groups attached to the carbon atom: —H, —Br, —Cl, and —I. The fact that this molecule has left-hand and right-hand forms, that is handedness, can be demonstrated using molecular models, as shown in Figure 7.6a. As shown in this figure, the two forms are nonsuperimposable mirror images of each other, just as two hands are nonsuperimposable mirror images of each other.

The four different groups bonded to a chiral center need not be just single atoms as was the case in the previous example. In the following chiral-center-containing molecule, glyceraldehyde, three of the four bonded groups are polyatomic entities.

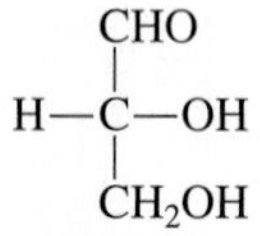

Glyceraldehyde

The four different groups attached to the carbon atom at the chiral center in this molecule are —H, —OH, —CHO, and —CH_2OH. The nonsuperimposability of the two mirror-image forms of glyceraldehyde is shown in Figure 7.6b.

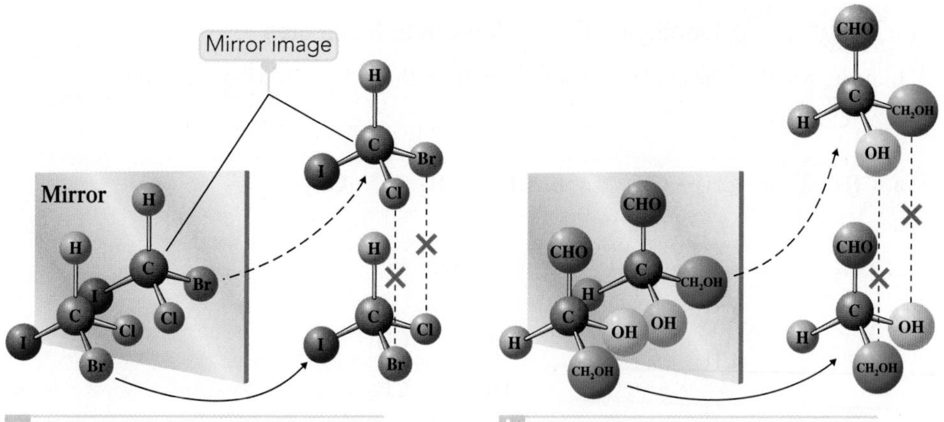

a The mirror image forms of the molecule bromochloroiodomethane are nonsuperimposable.

b The mirror image forms of the molecule glyceraldehyde are nonsuperimpoable.

Figure 7.6 Examples of simple molecules that are chiral.

Chiral centers within molecules are often denoted by a small asterisk. Note the chiral centers in the following molecules.

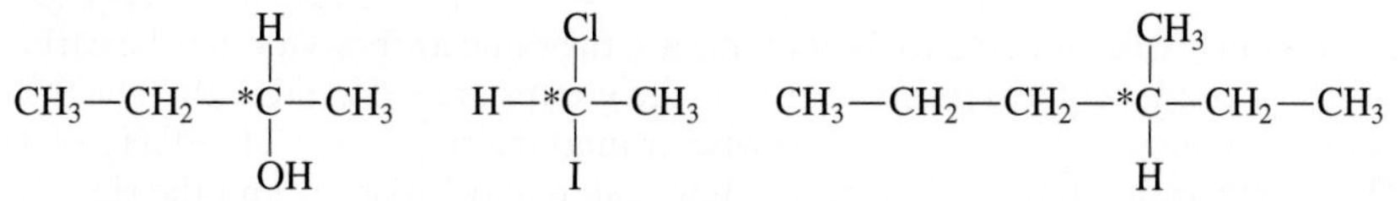

2-Butanol   1-Chloro-1-iodoethane   3-Methylhexane

A few chiral molecules are known that do not have a chiral center. Such exceptions are not important for the applications of the chirality concept that will be made in this text.

Guidelines for Identifying Chiral Centers

The following guidelines are helpful in determining the presence or lack of chiral centers in a molecule.

1. A carbon atom involved in a multiple bond (double or triple bond) cannot be a chiral center since it has fewer than four groups bonded to it. To have *four* groups present, all bonds about the chiral center must be single bonds.
2. A carbon atom that has two like groups bonded to it cannot be a chiral center since it does not meet the requirement of four *different* groups. The commonly encountered entities —CH_3 and —CH_2— in a structural formula never involve chiral centers because of the presence of two or more like hydrogen atoms.
3. Carbon atoms in a ring system, if not involved in multiple bonding, can be chiral centers. Such carbon atoms have four bonds—two to neighboring atoms in the ring and two to substituents on the ring. Chirality occurs when both (1) the two substituents are different and (2) the two "halves" of the ring emanating from the chiral center are different. This "difference in ring halves" concept is illustrated in Figure 7.7, where four different cyclic compounds are shown, one of which is chiral and three of which are achiral.

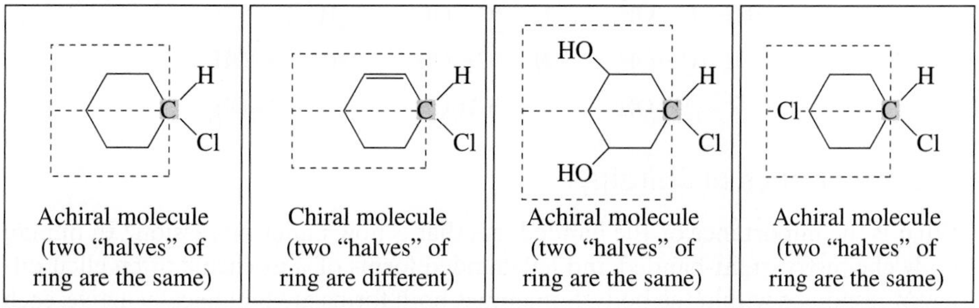

Figure 7.7 Examples of both chiral and achiral cyclic organic compounds. In achiral compounds, the two "halves" of the ring system are equivalent to each other; in chiral compounds, the two "halves" of the ring system are not equivalent.

EXAMPLE 7.1 Identifying Chiral Centers in Molecules

Indicate whether the circled carbon atom in each of the following molecules is a chiral center.

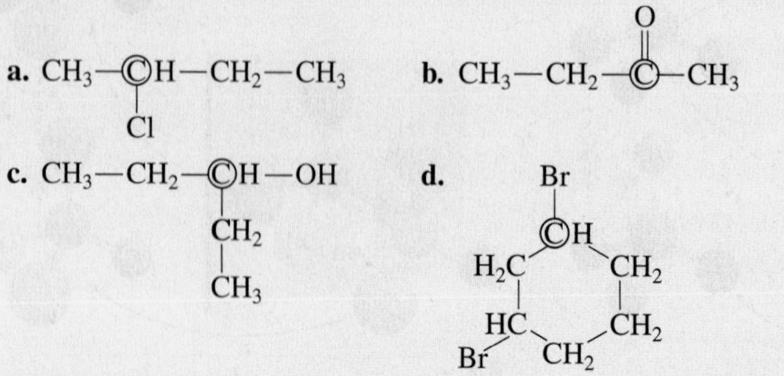

Solution

a. This is a chiral center. The four different groups attached to the carbon atom are —CH_3, —Cl, —CH_2—CH_3, and —H.

b. No chiral center is present. The carbon atom is attached to only *three* groups because it is involved in a double bond.

c. No chiral center is present. Two of the groups attached to the carbon atom are identical.

d. The chirality rules for ring carbon atoms are the same as those for acyclic carbon atoms. A chiral center is present. Two of the groups are —H and —Br. The third group, obtained by proceeding clockwise around the ring, is —CH_2—CH_2—CH_2. The fourth group, obtained by proceeding counterclockwise around the ring, is —CH_2—CHBr—CH_2.

Practice Exercise 7.1

Indicate whether the circled carbon atom in each of the following molecules is a chiral center.

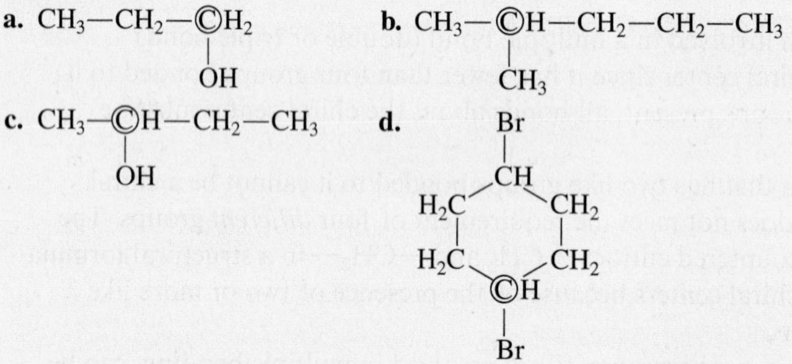

Answers: **a.** Not a chiral center; **b.** Not a chiral center; **c.** Chiral center; **d.** Not a chiral center

Remember the meaning of the structural notations —CHO and —CH_2OH.

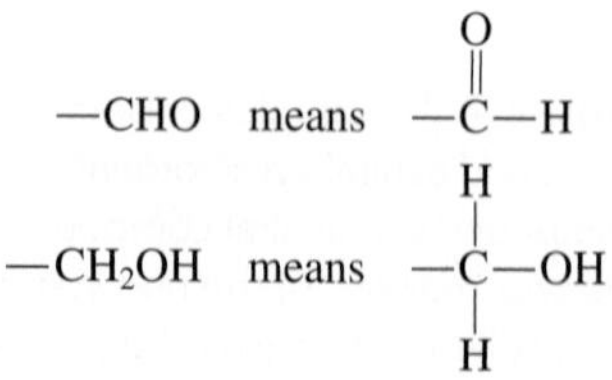

Organic molecules may contain more than one chiral center. For example, the following compound has two chiral centers.

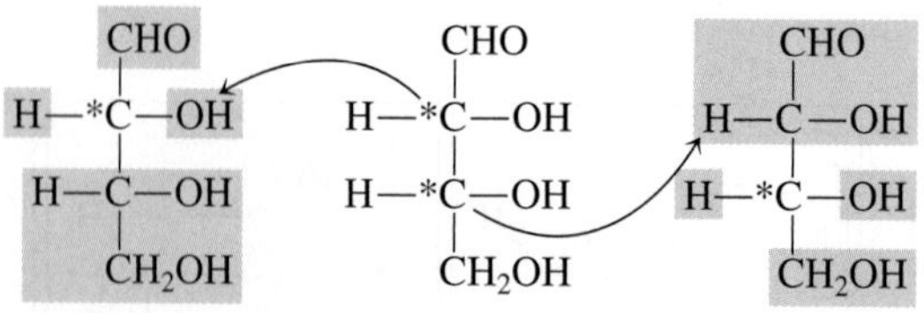

The Importance of Chirality

What is the importance of the handedness that is now under discussion? In human body chemistry, right-handed and left-handed forms of a molecule often elicit different responses within the body. Sometimes both forms are biologically active, each form giving a different response; sometimes both elicit the same response, but one form's response is many times greater than that of the other; and sometimes only one of the two forms is biochemically active. For example, studies show that the body's response to the right-handed form of the hormone epinephrine (Section 6.10) is 20 times greater than its response to the left-handed form.

Monosaccharides, the simplest type of carbohydrate and the building block for more complex types of carbohydrates (Section 7.3), are almost always "right-handed." Plants, the main dietary source of carbohydrates, produce only right-handed monosaccharides. Interestingly, the building blocks for proteins, amino acids (Section 9.4), are always left-handed molecules. The procedure for determining whether a given molecular structure represents a left-handed or right-handed molecule is considered in Section 7.6.

7.5 Stereoisomerism: Enantiomers and Diastereomers

The left- and right-handed forms of a chiral molecule are isomers. They are not *constitutional isomers,* the type of isomerism that has been encountered repeatedly in the organic chemistry chapters of the text, but rather are *stereoisomers.* **Stereoisomers** *are isomers that have the same molecular and structural formulas but differ in the orientation of atoms in space.* By contrast, atoms are connected to each other in different ways in constitutional isomers (Section 1.6).

There are two major structural features that generate *stereoisomerism:* (1) the presence of a chiral center in a molecule and (2) the presence of "structural rigidity" in a molecule. Structural rigidity is caused by restricted rotation about chemical bonds. It is the basis for *cis–trans* isomerism, a phenomenon found in some substituted cycloalkanes (Section 1.14) and some alkenes (Section 2.5). Thus handedness is this text's second encounter with stereoisomerism. (The discussion of *cis–trans* isomerism did not mention that it is a form of stereoisomerism.)

Stereoisomers can be subdivided into two types: *enantiomers* and *diastereomers.* **Enantiomers** *are stereoisomers whose molecules are nonsuperimposable mirror images of each other.* Left- and right-handed forms of a molecule with a single chiral center are enantiomers.

The term *enantiomer* comes from the Greek *enantios,* which means "opposite." It is pronounced "en-AN-tee-o-mer."

Diastereomers *are stereoisomers whose molecules are not mirror images of each other. Cis–trans* isomers (of both the alkene and the cycloalkane types) are diastereomers. Molecules that contain more than one chiral center can also exist in diastereomeric as well as enantiomeric forms, as will be shown in Section 7.6.

Some textbooks use the term *diastereoisomers* instead of *diastereomers.* The pronunciation for *diastereomer* is "dye-a-STEER-ee-o-mer."

Figure 7.8 shows the "thinking pattern" involved in using the terms *stereoisomers, enantiomers,* and *diastereomers.*

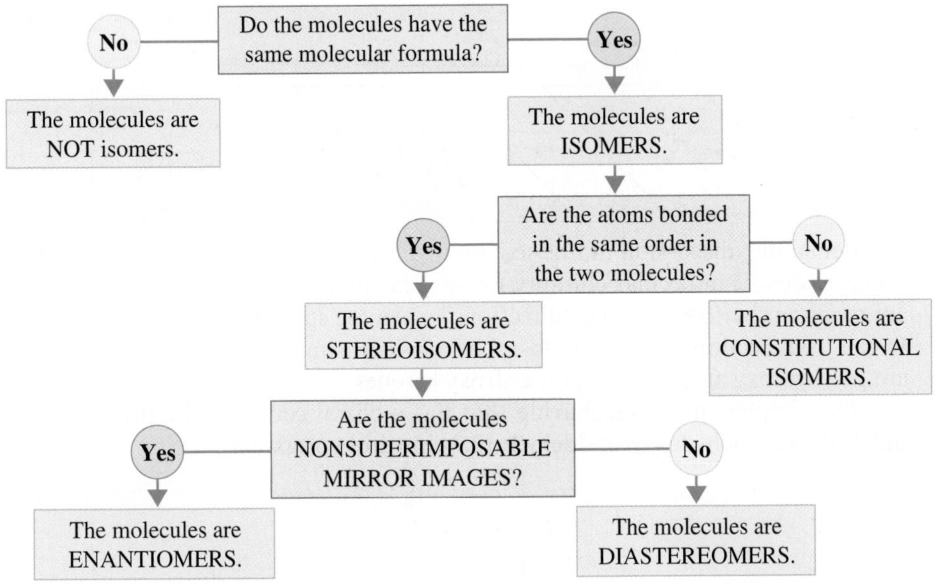

Figure 7.8 A summary of the "thought process" used in classifying molecules as enantiomers or diastereomers.

7.6 Designating Handedness Using Fischer Projection Formulas

Fischer projection formulas carry the name of their originator, the German chemist Hermann Emil Fischer (see Figure 7.9).

Courtesy of the Edgar Fahs Smith Collection, University of Pennsylvania Library

Figure 7.9 The German chemist Hermann Emil Fischer (1852–1919), the developer of the two-dimensional system for specifying chirality, was one of the early greats in organic chemistry. He made many fundamental discoveries about carbohydrates, proteins, and other natural products. In 1902 he was awarded the second Nobel Prize in chemistry.

Drawing *three*-dimensional representations of chiral molecules to specify handedness can be both time-consuming and awkward. Fischer projection formulas represent a method for giving molecular chirality specifications in *two* dimensions. A **Fischer projection formula** *is a two-dimensional structural notation for showing the spatial arrangement of groups about chiral centers in molecules.*

In a Fischer projection formula, a chiral center is represented as the intersection of vertical and horizontal lines. The atom at the chiral center, which is almost always carbon, is not explicitly shown.

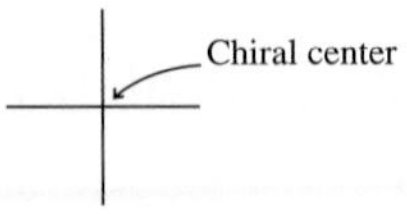

The arrangement of the four groups attached to the atom at the chiral center is specified using conventions based on the following interpretation for a *printed* three-dimensional model of the tetrahedral bond orientations about the chiral center.

1. The central atom (the chiral center) is considered to be in the plane of the paper.
2. Two of the bonds are considered to be directed into the printed page (*w* and *z* in the illustration that follows).
3. Two of the bonds are considered to be directed out of the printed page (*x* and *y* in the illustration that follows).

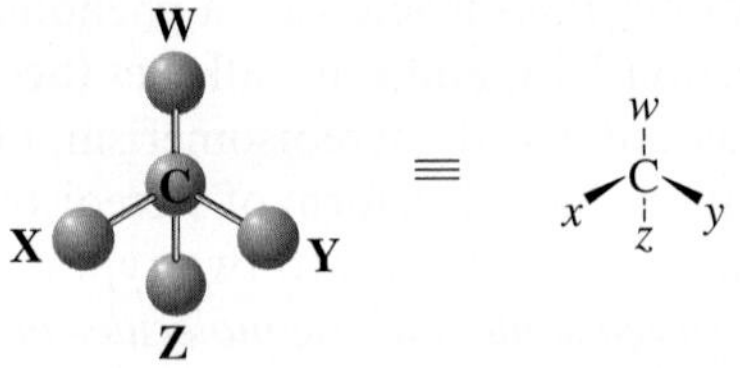

The conventions that connect these model specifications with a Fischer projection formula are

1. Vertical lines from the chiral center in a Fischer projection formula represent bonds to groups directed into the printed page (*w* and *z*).
2. Horizontal lines from the chiral center in a Fischer projection formula represent bonds to groups directed out of the printed page (*x* and *y*).

The Fischer projection formula thus becomes

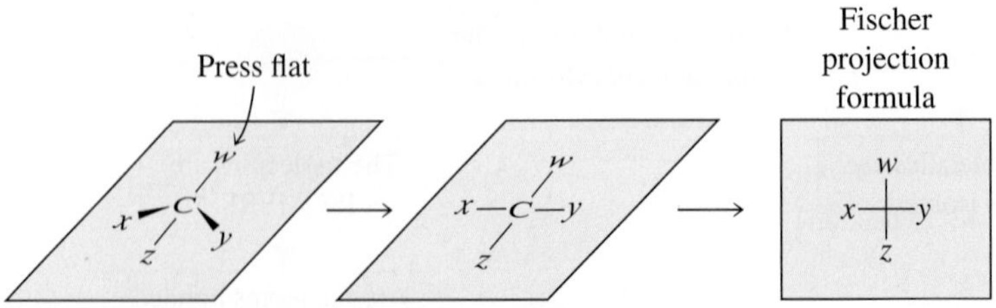

Remaining discussion in this section involves drawing Fischer projection formulas to denote molecular chirality for specific molecules. All example molecules are monosaccharides, and the "directions" given are specific for monosaccharides. Recall from Section 7.3 that monosaccharides (the simplest type of carbohydrate) are polyhydroxy aldehydes or polyhydroxy ketones.

The simplest monosaccharide that has a chiral center is the three-carbon polyhydroxy aldehyde glyceraldehyde (2,3-dihydroxypropanal).

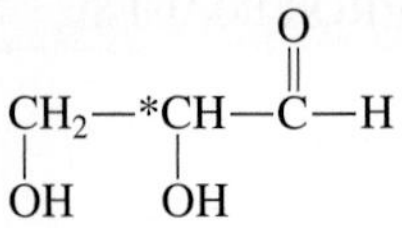

There are two stereoisomers for this compound—a pair of enantiomers. The Fischer projection formulas for these two "mirror image" molecules are

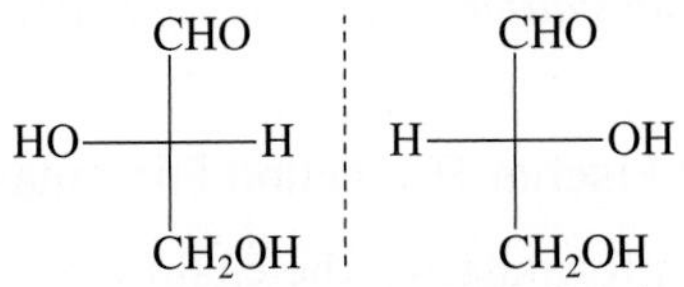

L-Glyceraldehyde D-Glyceraldehyde

By convention, in a Fischer projection formula for a monosaccharide, the *carbon chain* is always positioned *vertically* with the carbonyl group (aldehyde or ketone) at or near the top. An aldehyde group is denoted using the notation CHO (Section 4.3).

For this particular molecule, the handedness (right-handed or left-handed) of the two enantiomers is specified by using the designations D and L. The enantiomer with the chiral center —OH group on the right in the Fischer projection formula is by definition the right-handed isomer (D-glyceraldehyde), and the enantiomer with the chiral center —OH group on the left in the Fischer projection formula is by definition the left-handed isomer (L-glyceraldehyde).

The D and L designations for the handedness of the two members of an enantiomeric pair come from the Latin words *dextro,* which means "right," and *levo,* which means "left."

The compound 2,3,4-trihydroxybutanal, a monosaccharide with four carbons and *two* chiral centers, has the structural formula

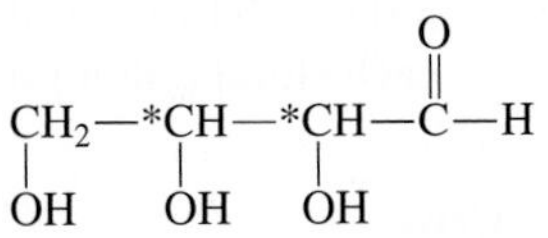

There are four stereoisomers for this compound—two pairs of enantiomers. The Fischer projection formulas for these stereoisomers are

To draw the mirror image of a Fischer projection structure, keep up-and-down and front-and-back aspects of the structure the same and reverse the left-and-right aspects.

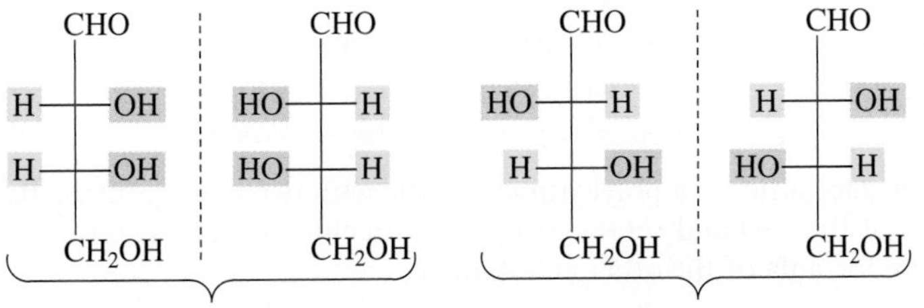

First enantiomeric pair Second enantiomeric pair

In the first enantiomeric pair, both chiral-center —OH groups are on the same side of the Fischer projection formula, and in the second enantiomeric pair, the chiral-center —OH groups are on opposite sides of the Fischer projection formula. These are the only —OH group arrangements possible.

Any given molecular structure can have only one mirror image. Hence enantiomers always come in pairs; there can never be more than two.

The D,L system used to designate the handedness of glyceraldehyde enantiomers is extended to monosaccharides with more than one chiral center in the following manner. The carbon chain is numbered starting at the carbonyl group end of the molecule, and the highest-numbered chiral center is used to determine D or L configuration.

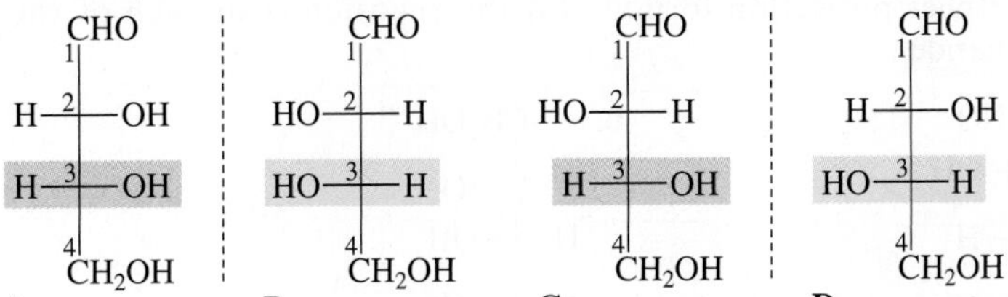

A D isomer **B** L isomer **C** D isomer **D** L isomer

The highest numbered chiral center in a monosaccharide is always the chiral center that is farthest from the monosaccharide's carbonyl group.

The D,L nomenclature gives the configuration (handedness) only at the highest-numbered chiral center. The configuration at other chiral centers in a molecule is accounted for by assigning a different common name to each pair of D,L enantiomers. In the present example, compounds A and B (the first enantiomeric pair) are D-erythrose and L-erythrose; compounds C and D (the second enantiomeric pair) are D-threose and L-threose.

What is the relationship between compounds A and C in the present example? They are diastereomers (Section 7.5), stereoisomers that are not mirror images of

Diastereomers that have two chiral centers must have the same handedness (both left or both right) at one chiral center and opposite handedness (one left and one right) at the other chiral center.

each other. Other diastereomeric pairs in the present example are A and D, B and C, and B and D. The members of each of these four pairs are epimers. **Epimers** *are diastereomers whose molecules differ only in the configuration at one chiral center.*

EXAMPLE 7.2 Drawing Fischer Projection Formulas for Monosaccharides

Draw a Fischer projection formula for the enantiomer of each of the following monosaccharides.

a.

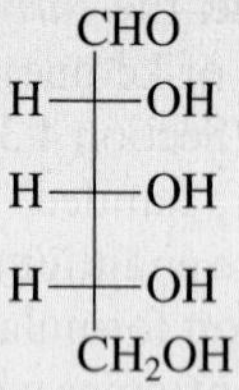

b.

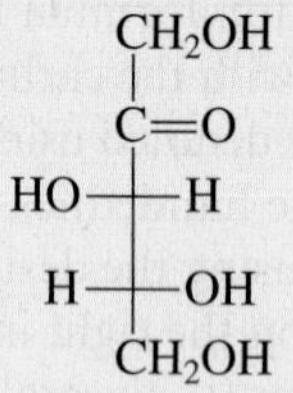

Solution

Knowing the Fischer projection formula of one member of an enantiomeric pair, the other enantiomer's Fischer projection formula is drawn by reversing the substituents that are in *horizontal* positions *at each chiral center.*

a. Three chiral centers are present in this polyhydroxy aldehyde. Reversing the positions of the —H and —OH groups at each chiral center produces the Fischer projection formula of the other enantiomer.

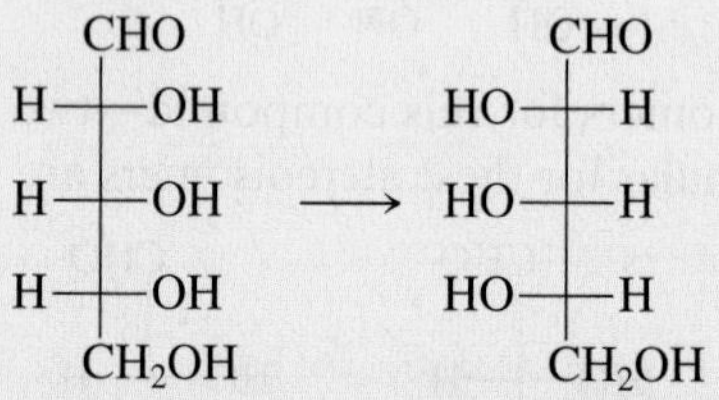

The given enantiomer The other enantiomer

b. This monosaccharide is a polyhydroxy ketone with two chiral centers. Reversing the positions of the —H and —OH groups at both chiral centers generates the Fischer projection formula of the other enantiomer.

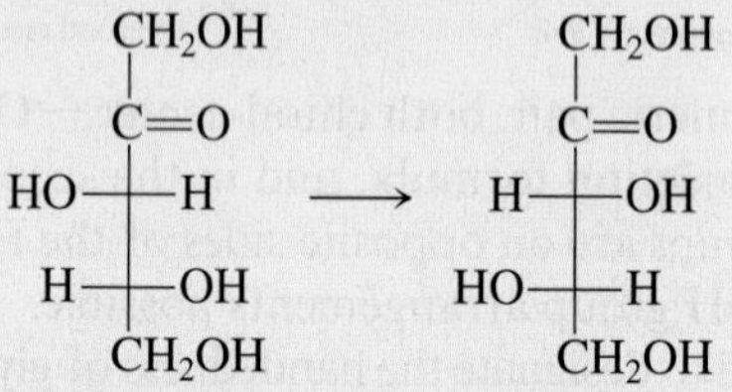

The given enantiomer The other enantiomer

Practice Exercise 7.2

Draw a Fischer projection formula for the enantiomer of each of the following monosaccharides.

a.

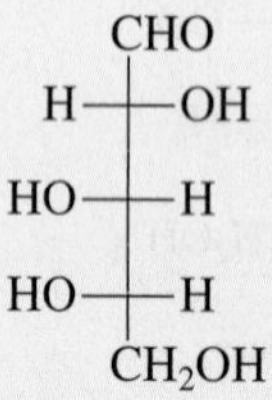

b.

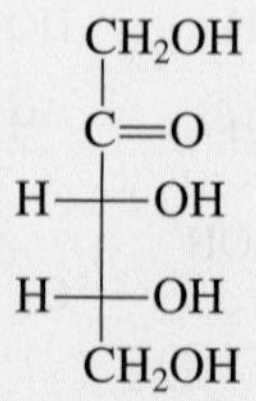

Answers: **a.**

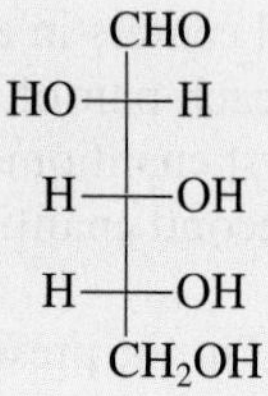

b.

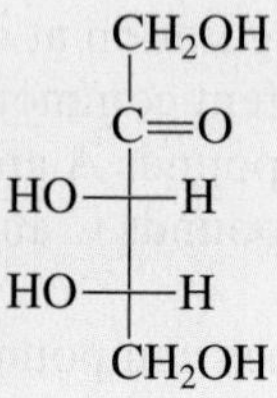

EXAMPLE 7.3 Classifying Monosaccharides as D or L Enantiomers

Classify each of the following monosaccharides as a D enantiomer or an L enantiomer.

a. 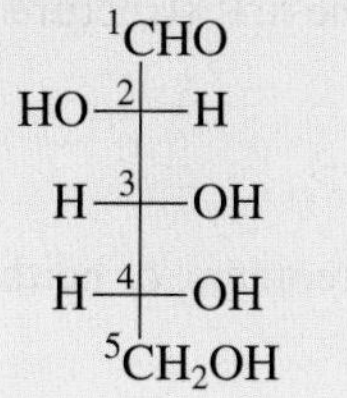

b. 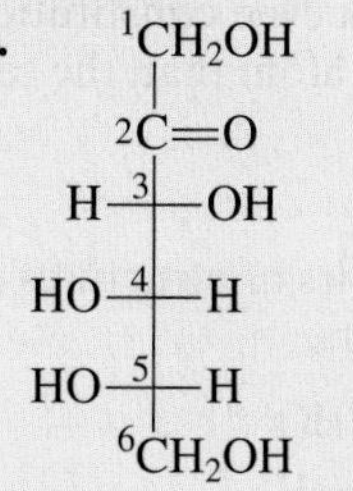

Solution

D or L configuration for a monosaccharide is determined by the highest-numbered chiral center, the one farthest from the carbonyl carbon atom.

a. The highest-numbered chiral center, which involves carbon 4, has the —OH group on the right. Thus this monosaccharide is a D enantiomer.
b. The highest-numbered chiral center, which involves carbon 5, has the —OH group on the left. Thus this monosaccharide is an L enantiomer.

Practice Exercise 7.3

Classify each of the following monosaccharides as a D enantiomer or an L enantiomer.

a. 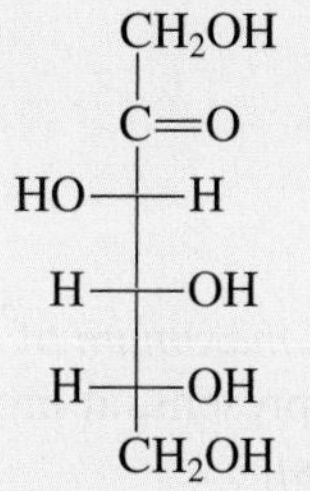

b. 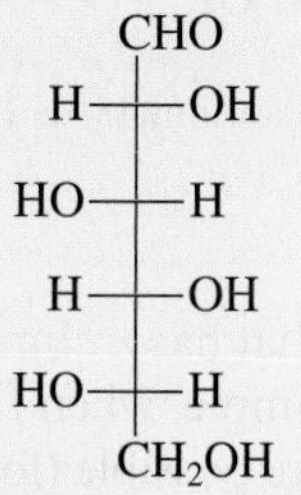

Answers: **a.** D enantiomer; **b.** L enantiomer

EXAMPLE 7.4 Recognizing Enantiomers and Diastereomers

Characterize each of the following pairs of structures as enantiomers, diastereomers, or neither enantiomers nor diastereomers.

a. 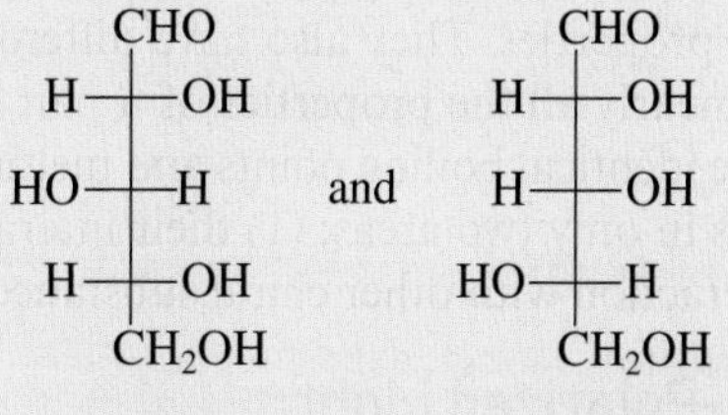

b. 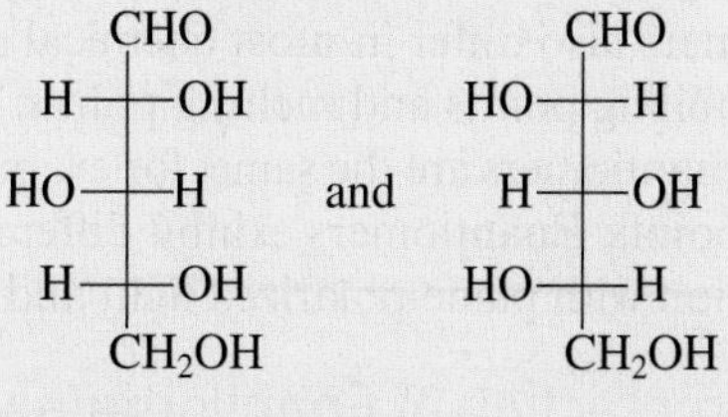

c. 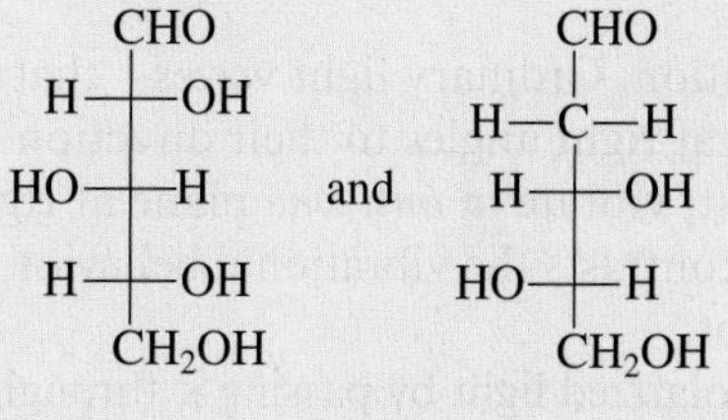

Solution

a. These two structures are *diastereomers.* All chiral centers have both a —H and —OH attachment. However, a mirror image relationship between these two attachments is present at only two of the three chiral centers. At the third chiral center, carbon 2, the —H and —OH orientation is the same.
b. These two structures represent *enantiomers*—a mirror-image substituent relationship exists between the two isomers at *each* chiral center.

(continued)

c. These two structures are *neither enantiomers nor diastereomers.* The connectivity of atoms differs in the two structures at carbon 2. Stereoisomers (enantiomers and diastereomers) must have the same connectivity throughout both structures. (The two structures are not even constitutional isomers because the first structure contains one more oxygen atom than the second.)

Practice Exercise 7.4

Characterize the following pairs of structures as enantiomers, diastereomers, or neither enantiomers nor diastereomers.

a.

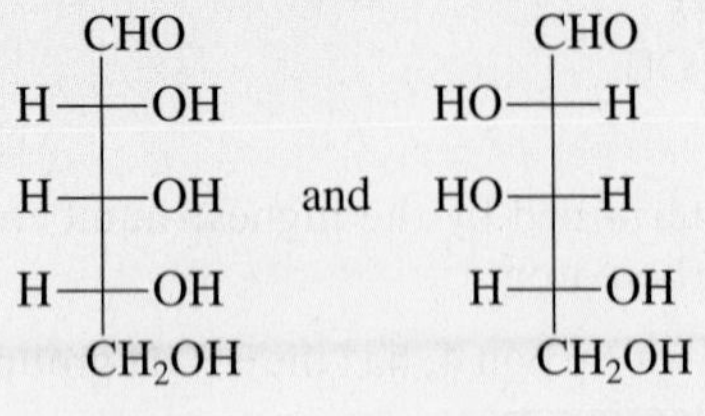

b.

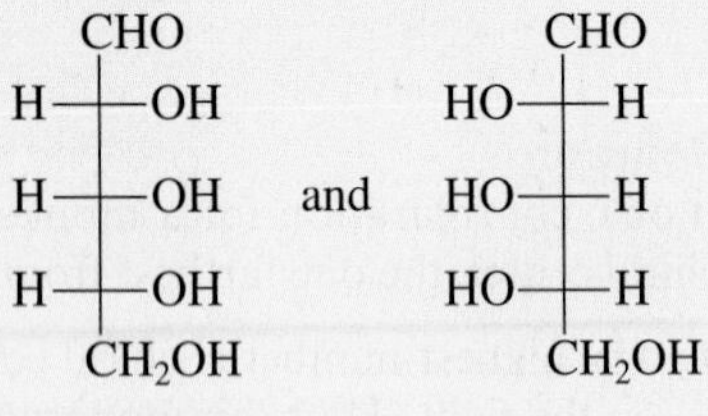

c.

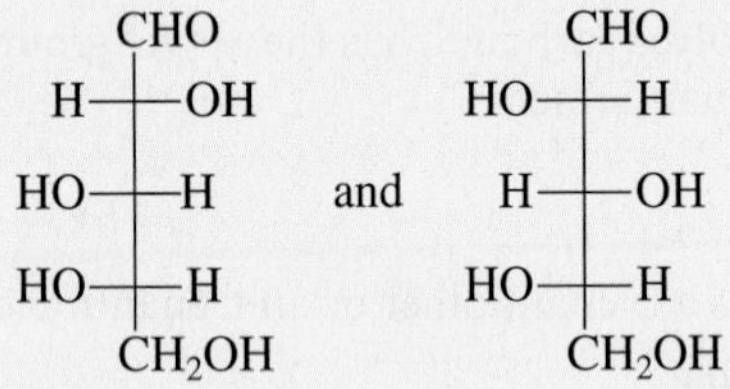

Answers: **a.** Diastereomers; **b.** Enantiomers; **c.** Diastereomers

The formula 2^n gives the maximum possible number of stereoisomers for a molecule with *n* chiral atoms. In a few cases, the actual number of stereoisomers is less than the maximum because of symmetry considerations that make some mirror images superimposable.

In general, a compound that has *n* chiral centers may exist in a *maximum* of 2^n stereoisomeric forms. For example, when three chiral centers are present, at most eight stereoisomers ($2^3 = 8$) are possible (four pairs of enantiomers).

The Chemistry at a Glance feature on the next page summarizes information about the various types of isomers we have encountered so far in the text—the various subtypes of constitutional isomers and the various subtypes of stereoisomers.

7.7 Properties of Enantiomers

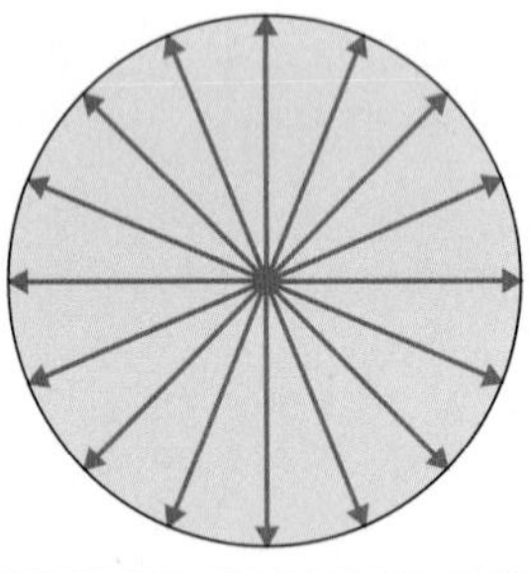

a Ordinary (unpolarized) light

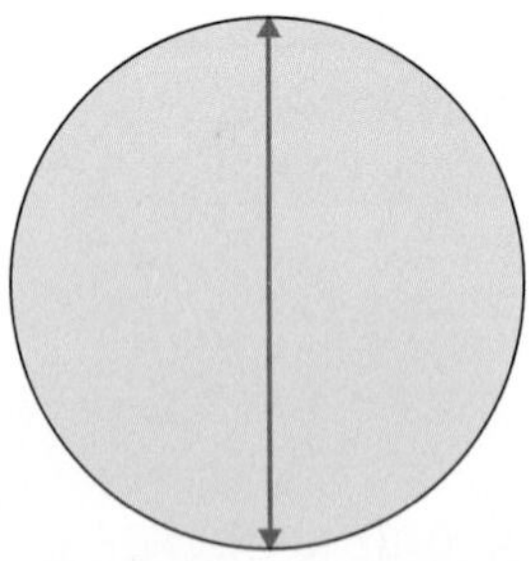

b Plane-polarized light

Figure 7.10 Vibrational characteristics of ordinary (unpolarized) light (a), and polarized light (b). The direction of travel of the light is toward the reader.

Constitutional isomers differ in most chemical and physical properties. For example, constitutional isomers have different boiling points and melting points. Diastereomers also differ in most chemical and physical properties. They also have different boiling points and melting points. In contrast, nearly all the properties of a pair of enantiomers are the same; for example, they have identical boiling points and melting points. Enantiomers exhibit different properties in only two areas: (1) their interaction with plane-polarized light and (2) their interaction with other chiral substances.

Interaction of Enantiomers with Plane-Polarized Light

All light moves through space with a wave motion. Ordinary light waves—that is, unpolarized light waves—vibrate in *all* planes at right angles to their direction of travel. Plane-polarized light waves, by contrast, vibrate in *only one* plane at right angles to their direction of travel. Figure 7.10 contrasts the vibrational behavior of ordinary light with that of plane-polarized light.

Ordinary light can be converted to plane-polarized light by passing it through a *polarizer,* an instrument with lenses or filters containing special types of crystals. When plane-polarized light is passed through a solution containing a *single* enantiomer, the plane of the polarized light is rotated counterclockwise (to the left) or clockwise (to the right), depending on the enantiomer. The extent of rotation depends on the concentration of the enantiomer as well as on its identity. Furthermore, the two enantiomers of a pair rotate the plane-polarized light the same number of degrees, but in opposite directions. If a 0.50 M solution of one enantiomer rotates the light 30° to the right, then a 0.50 M solution of the other enantiomer rotates the light 30° to the left.

CHEMISTRY AT A GLANCE Constitutional Isomers and Stereoisomers

CONSTITUTIONAL ISOMERS

Isomers in which the atoms have different connectivity

SKELETAL ISOMERS

Isomers with different carbon atom arrangements and different hydrogen atom arrangements

$CH_3-CH_2-CH_2-CH_3$

Butane (C_4H_{10})

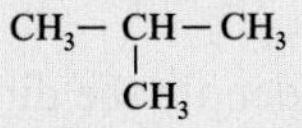

2-Methylpropane (C_4H_{10})

POSITIONAL ISOMERS

Isomers that differ in the location of the functional group

$CH_2=CH-CH_2-CH_3$

1-Butene (C_4H_8)

$CH_3-CH=CH-CH_3$

2-Butene (C_4H_8)

FUNCTIONAL GROUP ISOMERS

Isomers that contain different functional groups

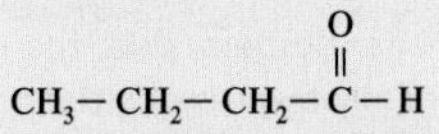

Butanal (aldehyde, C_4H_8O)

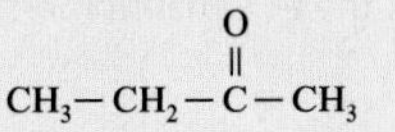

2-Butanone (ketone, C_4H_8O)

STEREOISOMERS

Isomers with atoms of the same connectivity that differ only in the orientation of the atoms in space

ENANTIOMERS

- Stereoisomers that are nonsuperimposable mirror images of each other
- Handedness (D and L forms) is determined by the configuration at the high-numbered chiral center

D and L Enantiomers

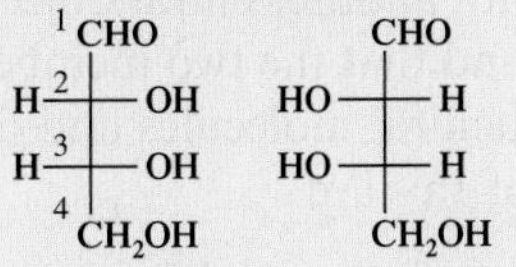

D-Erythrose L-Erythrose

DIASTEREOMERS

Stereoisomers that are not mirror images of each other

***CIS–TRANS* ISOMERS**

Stereoisomerism that results from restricted rotation about chemical bonds

- Is sometimes possible when a ring is present
- Is sometimes possible when a double bond is present

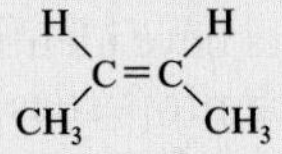

cis-2-butene

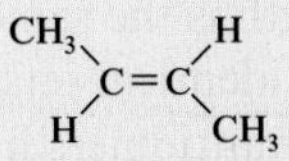

trans-2-butene

MOST OTHER DIASTEREOMERS (two or more chiral centers)

Stereoisomerism that results from

- A mirror image relationship at one (or more) chiral centers, and
- The same configuration at one (or more) chiral centers

Three chiral centers

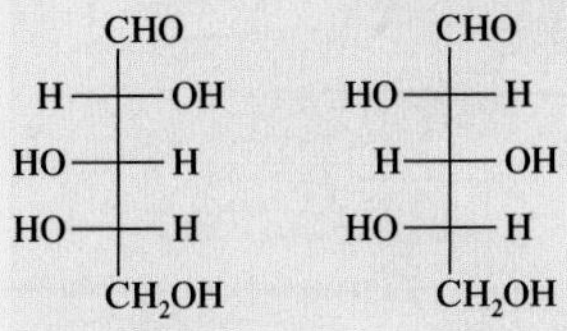

L-Arabinose L-Xylose

Instruments used to measure the degree of rotation of plane-polarized light by enantiomeric compounds are called *polarimeters.* The schematic diagram in Figure 7.11 shows the basis for these instruments.

Dextrorotatory and Levorotatory Compounds

Enantiomers are said to be optically active because of the way they interact with plane-polarized light. An **optically active compound** *is a compound that rotates the plane of polarized light.*

An enantiomer that rotates plane-polarized light in a clockwise direction (to the *right*) is said to be dextrorotatory (the Latin *dextro* means "right"). A **dextrorotatory compound** *is a chiral compound that rotates the plane of polarized light in a clockwise direction.* An enantiomer that rotates plane-polarized light in a counterclockwise

Achiral molecules are optically *inactive*. Chiral molecules are optically *active*.

Because of their ability to rotate the plane of polarized light, enantiomers are sometimes referred to as *optical isomers.*

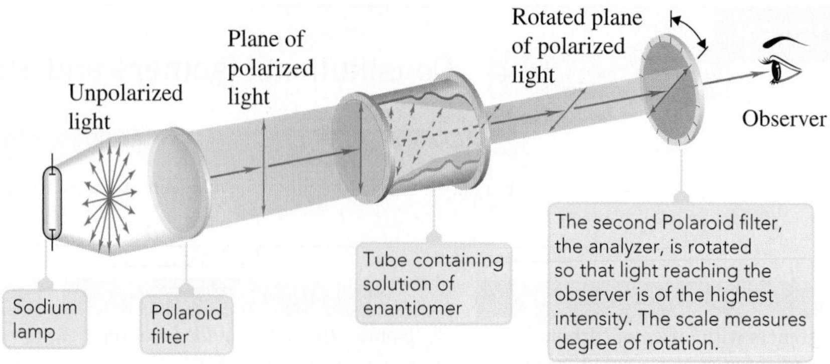

Figure 7.11 Schematic depiction of how a polarimeter works.

In any pair of enantiomers, one, the (+)-enantiomer, always rotates the plane of polarized light to the right, and the other, the (−)-enantiomer, to the left.

direction (to the *left*) is said to be levorotatory (the Latin *levo* means "left"). A **levorotatory compound** *is a chiral compound that rotates the plane of polarized light in a counterclockwise direction.* If one member of an enantiomeric pair is dextrorotatory, then the other member must be levorotatory.

A plus or minus sign inside parentheses is used to denote the direction of rotation of plane-polarized light by a chiral compound. The notation (+) means rotation to the right (clockwise), and (−) means rotation to the left (counterclockwise). Thus the dextrorotatory enantiomer of glucose is (+)-glucose.

Both handedness and direction of rotation of plane-polarized light can be incorporated into the name of an enantiomer. For example, the notation D-(+)-mannose specifies that the right-handed isomer of the monosaccharide mannose rotates plane-polarized light in a clockwise direction (to the right).

The handedness of enantiomers (D or L, Section 7.6) and the direction of rotation of plane-polarized light by enantiomers [(+) or (−)] are not connected entities. There is no way of knowing which way an enantiomer will rotate light until it is examined with a polarimeter. Not all D enantiomers rotate plane-polarized light in the same direction, nor do all L enantiomers rotate plane-polarized light in the same direction. Some D enantiomers are dextrorotatory; others are levorotatory.

Interactions Between Chiral Compounds

A left-handed baseball player (chiral) and a right-handed baseball player (chiral) can use the same baseball bat (achiral) or wear the same baseball hat (achiral). However, left- and right-handed baseball players (chiral) cannot use the same baseball glove (chiral). This nonchemical example illustrates that the chirality of an object becomes important when the object interacts with another chiral object.

Applying this generalization to molecules, it is found that the two members of an enantiomeric pair have the same interaction with achiral molecules and different interactions with chiral molecules. Ramifications of this are

1. Enantiomers have identical boiling points, melting points, and densities because such properties depend on the strength of intermolecular forces, and intermolecular force strength does not depend on chirality. Intermolecular force strength is the same for both forms of a chiral molecule because both forms have identical sets of functional groups.
2. A pair of enantiomers have the same solubility in an achiral solvent, such as ethanol, but differing solubilities in a chiral solvent, such as D-2-butanol.
3. The rate and extent of reaction of enantiomers with another reactant are the same if the reactant is achiral but differ if the reactant is chiral.
4. Receptor sites for molecules within the body have chirality associated with them. Thus enantiomers always generate different responses within the human body as they interact at such sites. Sometimes the responses are only slightly different, and at other times they are very different.

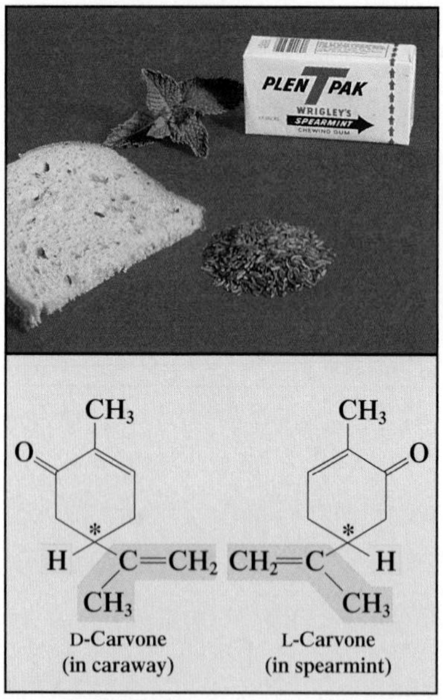

Figure 7.12 The distinctly different natural flavors of spearmint and caraway are caused by enantiomeric molecules. Spearmint leaves contain L-carvone, and caraway seeds contain D-carvone.

Two specific examples of differing chiral–chiral interactions involving enantiomers that occur within the human body are now considered. The first example involves taste perceptions. The distinctly different natural flavors "spearmint" and "caraway" are generated by molecules that are enantiomers interacting with chiral "taste receptors" (Figure 7.12).

The second example involves the body's response to the enantiomeric forms of the hormone epinephrine (adrenaline). The response of the body to the D isomer of the hormone is 20 times greater than its response to the L isomer of the hormone. Epinephrine binds to its cellular receptor site by means of a three-point contact,

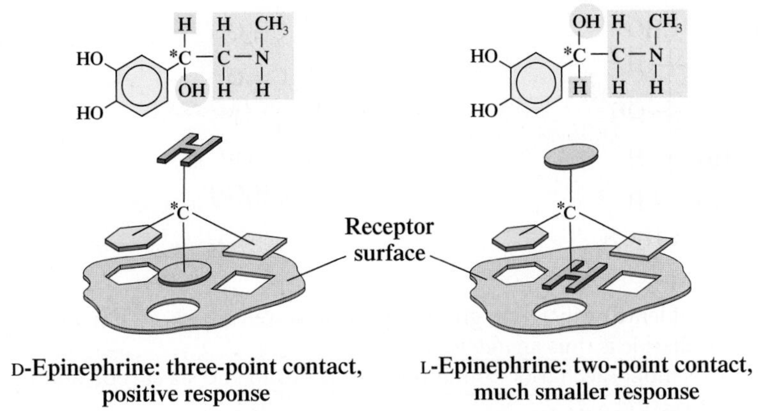

Figure 7.13 D-Epinephrine binds to the receptor at three points, whereas the biochemically weaker L-epinephrine binds at only two sites.

as is shown in Figure 7.13. D-Epinephrine makes a perfect three-point contact with the receptor surface, but the biochemically weaker L-epinephrine can make only a two-point contact. Because of the poorer fit, the binding of the L isomer is weaker, and less physiological response is observed.

7.8 Classification of Monosaccharides

With discussion about molecular chirality and its consequences (Sections 7.4 through 7.7) complete, additional details about carbohydrate chemistry will now be presented. The starting point for further carbohydrate considerations is monosaccharide classifications. It should be remembered that chirality is an underlying governing principle in all considerations.

Although there is no limit to the number of carbon atoms that can be present in a monosaccharide, only monosaccharides with three to seven carbon atoms are commonly found in nature. A three-carbon monosaccharide is called a *triose,* and those that contain four, five, and six carbon atoms are called *tetroses, pentoses,* and *hexoses,* respectively.

Monosaccharides are classified as *aldoses* or *ketoses* on the basis of type of carbonyl group (Section 4.1) present. An **aldose** *is a monosaccharide that contains an aldehyde functional group.* Aldoses are polyhydroxy aldehydes. A **ketose** *is a monosaccharide that contains a ketone functional group.* Ketoses are polyhydroxy ketones.

Structurally, monosaccharides have a chain of three to seven carbon atoms, with a carbonyl carbon group at either the terminal carbon (C1) or the carbon adjacent to it (C2). If the carbonyl group involves C1, the monosaccharide is an aldose; if it involves C2, the monosaccharide is a ketose.

Monosaccharides are often classified by both their number of carbon atoms and their functional group. A six-carbon monosaccharide with an aldehyde functional group is an *aldohexose;* a five-carbon monosaccharide with a ketone functional group is a *ketopentose.*

Monosaccharides are also often called sugars. Hexoses are six-carbon sugars, pentoses five-carbon sugars, and so on. The word *sugar* is associated with "sweetness," and most (but not all) monosaccharides have a sweet taste. The designation *sugar* is also applied to disaccharides, many of which also have a sweet taste. Thus **sugar** *is a general designation for either a monosaccharide or a disaccharide.*

The term *saccharide* comes from the Latin word for "sugar," which is *saccharum.*

EXAMPLE 7.5 Classifying Monosaccharides on the Basis of Structural Characteristics

Classify each of the following monosaccharides according to both the number of carbon atoms and the type of carbonyl group present.

a.

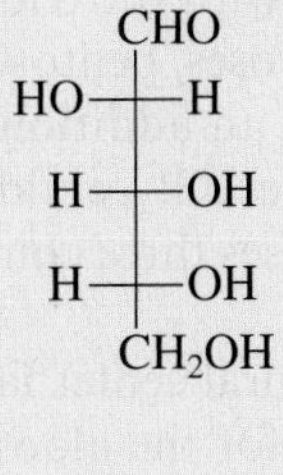

b.

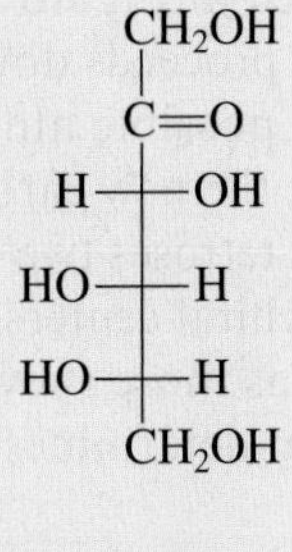

(continued)

c.

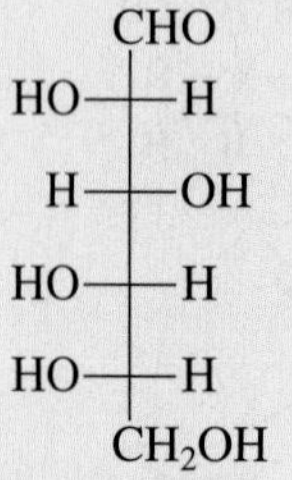

d.

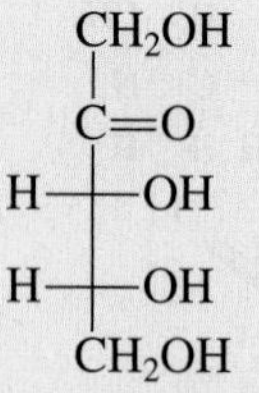

Solution

a. An aldehyde functional group is present as well as five carbon atoms. This monosaccharide is thus an *aldopentose.*

b. This monosaccharide contains a ketone group and six carbon atoms, so it is a *ketohexose.*

c. Six carbon atoms and an aldehyde group in a monosaccharide are characteristic of an *aldohexose.*

d. This monosaccharide is a *ketopentose.*

Practice Exercise 7.5

Classify each of the following monosaccharides according to both the number of carbon atoms and the type of carbonyl group present.

a.

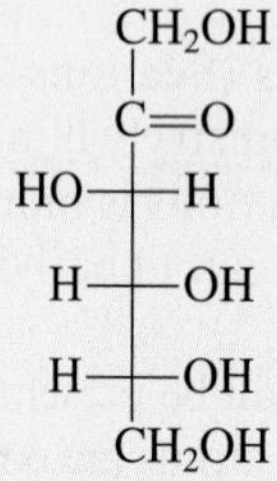

b.

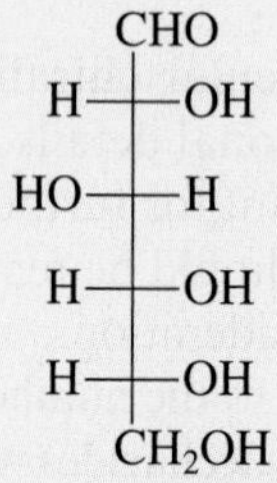

c.

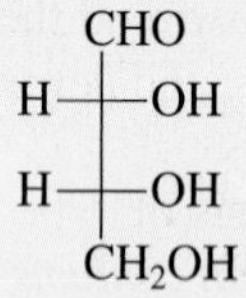

d.

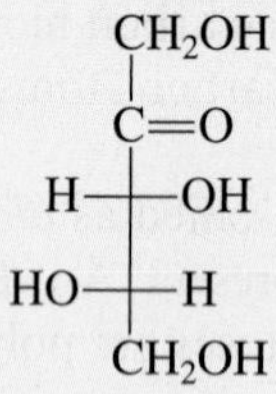

Answers: **a.** Ketohexose; **b.** Aldohexose; **c.** Aldotetrose; **d.** Ketopentose

Nearly all naturally occurring monosaccharides are D isomers. These D monosaccharides are important energy sources for the human body. L monosaccharides, which can be produced in the laboratory, cannot be used by the body as energy sources. Body enzymes are specific for D isomers.

In terms of carbon atoms, trioses are the smallest monosaccharides that can exist. There are two such compounds, one an aldose (glyceraldehyde) and the other a ketose (dihydroxyacetone).

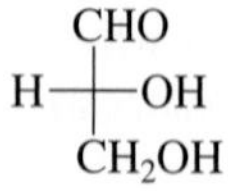

D-Glyceraldehyde

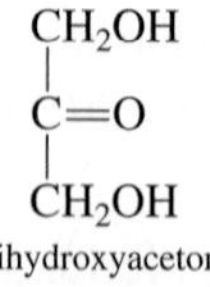

Dihydroxyacetone

These two triose structures serve as the reference points for consideration of the structures of aldoses and ketoses that contain more carbon atoms.

The Fischer projection formulas of all D aldoses containing three, four, five, and six carbon atoms are given in Figure 7.14. Figure 7.14 starts with the triose glyceraldehyde at the top and proceeds downward through the tetroses, pentoses, and hexoses. The number of possible aldoses doubles each time an additional carbon atom is added because the new carbon atom is a chiral center. Glyceraldehyde has one chiral center, the tetroses two chiral centers, the pentoses three chiral centers, and the hexoses four chiral centers.

In aldose structures such as those shown in Figure 7.14, the chiral center farthest from the aldehyde group determines the D or L designation for the aldose.

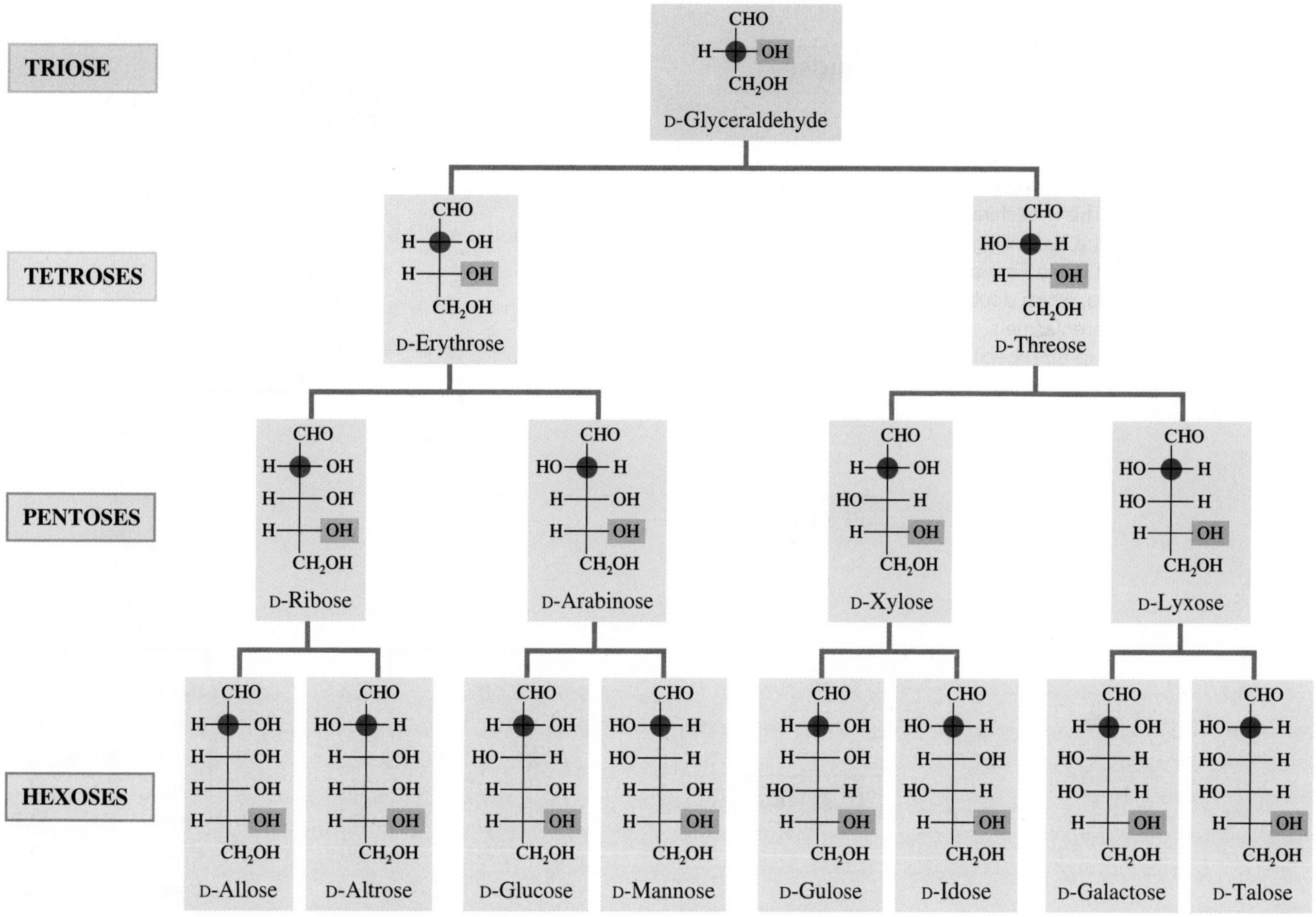

Figure 7.14 Fischer projection formulas and common names for D aldoses containing three, four, five, and six carbon atoms. The new chiral-center carbon atom added in going from triose to tetrose to pentose to hexose is marked in color. This new chiral center can have the hydroxyl group at the right or left in the Fischer projection formula, which doubles the number of stereoisomers. The hydroxyl group that specifies the D configuration is highlighted in dark pink.

The configurations about the other chiral centers present are accounted for by assigning a different common name to each set of D and L enantiomers. (Only the D isomer is shown in Figure 7.14; the L isomer is the mirror image of the structure shown.)

A major difference between glyceraldehyde and dihydroxyacetone is that the latter does not possess a chiral carbon atom. Thus D and L forms are not possible for dihydroxyacetone. This reduces by half (compared with aldoses) the number of stereoisomers possible for ketotetroses, ketopentoses, and ketohexoses. An aldohexose has four chiral carbon atoms, but a ketohexose has only three. Figure 7.15 gives the Fischer projection formulas and common names for the D forms of ketoses containing three, four, five, and six carbon atoms.

All monosaccharides have names that end in *-ose* except the trioses glyceraldehyde and dihydroxyacetone.

7.9 Biochemically Important Monosaccharides

Of the many monosaccharides, six that are particularly important in the functioning of the human body are the trioses D-glyceraldehyde and dihydroxyacetone and the D forms of glucose, galactose, fructose, and ribose. Glucose and galactose are aldohexoses, fructose is a ketohexose, and ribose is an aldopentose. All six of these monosaccharides are water-soluble, white, crystalline solids.

It is often useful to have the structures of the six monosaccharides considered in this section memorized.

Figure 7.15 Fischer projection formulas and common names for D ketoses containing three, four, five, and six carbon atoms. The new chiral-center carbon atom added in going from triose to tetrose to pentose to hexose is marked in color. This new chiral center can have the hydroxyl group at the right or left in the Fischer projection formula, which doubles the number of stereoisomers. The hydroxyl group that specifies the D configuration is highlighted in dark pink.

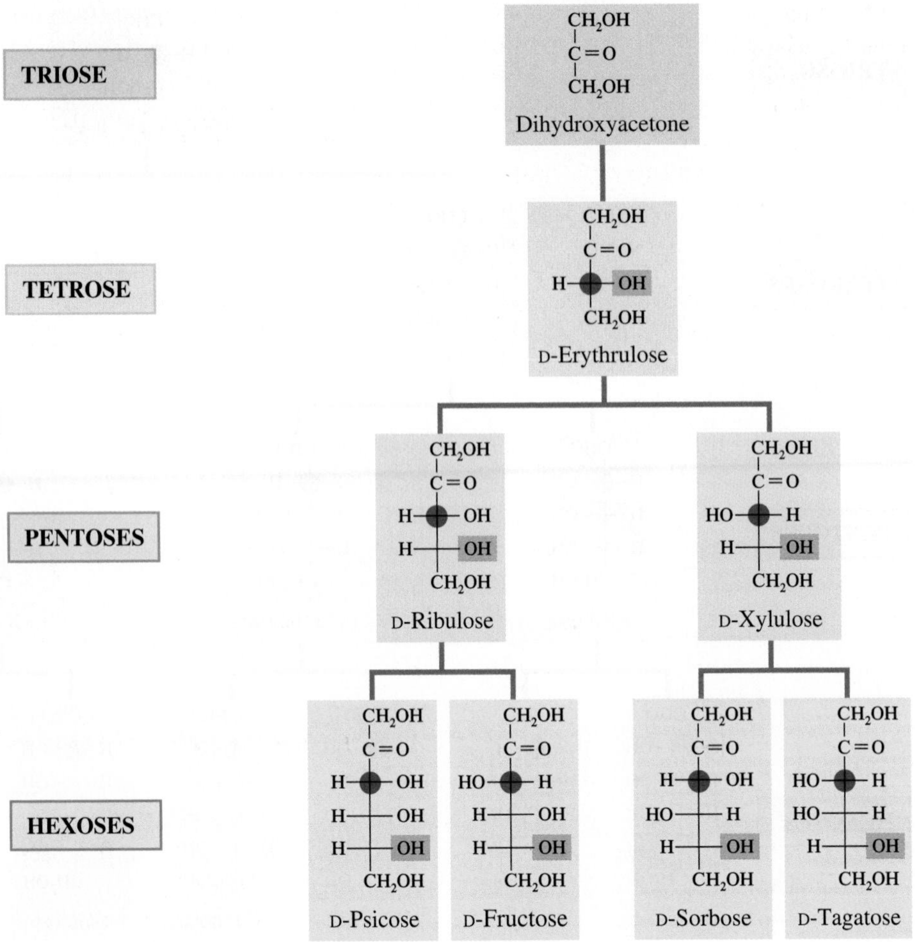

D-Glyceraldehyde and Dihydroxyacetone

The simplest of the monosaccharides, these two trioses are important intermediates in the process of glycolysis (Section 13.2), a series of reactions whereby glucose is converted into two molecules of pyruvate. D-Glyceraldehyde is a chiral molecule, but dihydroxyacetone is not.

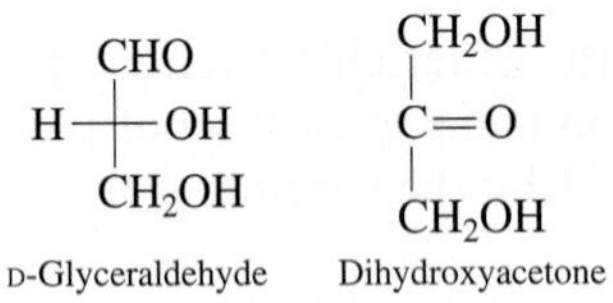

D-Glucose

D-Glucose tastes sweet, is nutritious, and is an important component of the human diet. L-Glucose, on the other hand, is tasteless, and the body cannot use it.

Of all monosaccharides, D-glucose is the most abundant in nature and the most important from a human nutritional standpoint. Its Fischer projection formula is

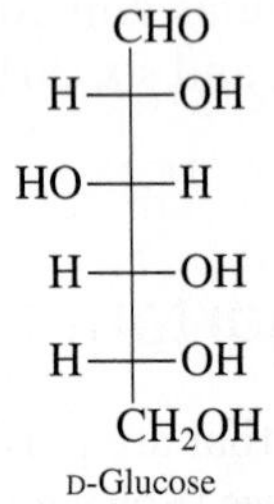

Ripe fruits, particularly ripe grapes (20%–30% glucose by mass), are a good source of glucose, which is often referred to as *grape sugar.* Two other names for D-glucose are *dextrose* and *blood sugar.* The name *dextrose* draws attention to the fact that the optically active D-glucose, in aqueous solution, rotates plane-polarized light to the right.

The term *blood sugar* draws attention to the fact that blood contains dissolved glucose. The normal concentration of glucose in human blood is in the range of 70–100 mg/dL (1 dL = 100 mL). The actual glucose concentration in blood is dependent on the time that has elapsed since the last meal was eaten. A concentration of about 130 mg/dL occurs in the first hour after eating, and then the concentration decreases over the next 2–3 hours back to the normal range. Cells use glucose as a primary source of energy (Figure 7.16).

Two hormones, insulin and glucagon (Section 13.9), have important roles in keeping glucose blood concentrations within the normal range, which is required for normal body function. Abnormal functioning of the hormonal control process for blood glucose levels leads to the condition known as diabetes (see Section 13.9).

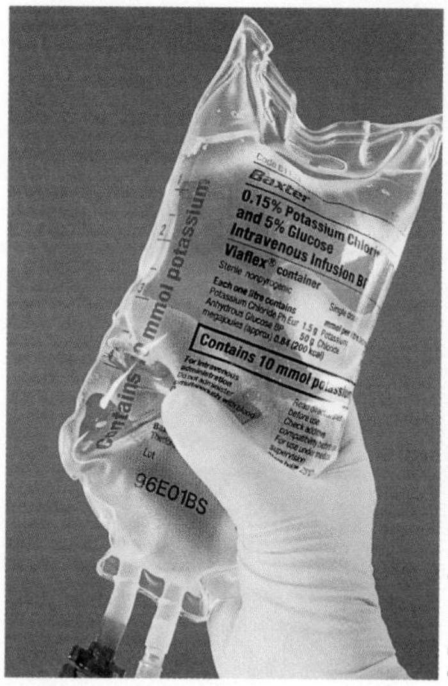

C. Paxton & J. Farrow/Photo Researchers

Figure 7.16 A 5% (m/v) glucose solution is often used in hospitals as an intravenous source of nourishment for patients who cannot take food by mouth. The body can use it as an energy source without digesting it.

D-Galactose

A comparison of the Fischer projection formulas for D-galactose and D-glucose shows that these two compounds differ only in the configuration of the —OH group and —H group on carbon 4.

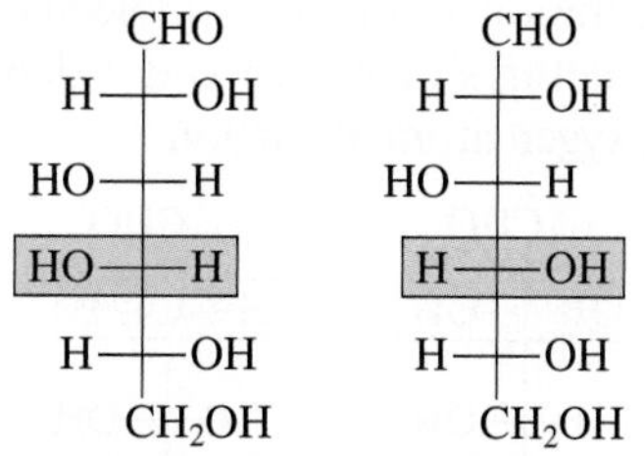

D-Galactose D-Glucose

D-Galactose and D-glucose are epimers (diastereomers that differ only in the configuration at one chiral center; Section 7.6).

D-Galactose is seldom encountered as a free monosaccharide. It is, however, a component of numerous important biochemical substances. In the human body, galactose is synthesized from glucose in the mammary glands for use in lactose (milk sugar), a disaccharide consisting of a glucose unit and a galactose unit (Section 7.13). D-Galactose is sometimes called *brain sugar* because it is a component of glycoproteins (protein–carbohydrate compounds; Section 7.18) found in brain and nerve tissue. D-Galactose is also present in the chemical markers that distinguish various types of blood—A, B, AB, and O (see Chemical Connections 7-D on page 293).

D-Fructose

D-Fructose is biochemically the most important ketohexose. It is also known as *levulose* and *fruit sugar.* Aqueous solutions of naturally occurring D-fructose rotate plane-polarized light to the left; hence the name *levulose.* The sweetest-tasting of all sugars, D-fructose is found in many fruits and is present in honey in equal amounts with glucose. It is sometimes used as a dietary sugar, not because it has fewer calories per gram than other sugars but because less is needed for the same amount of sweetness. The Chemical Connections 7-B feature on page 286 provides additional information about fructose use as a sweetener in the form of high fructose corn syrup (HFCS).

From the third to the sixth carbon, the structure of D-fructose is identical to that of D-glucose. Differences at carbons 1 and 2 are related to the presence of a ketone group in fructose and of an aldehyde group in glucose.

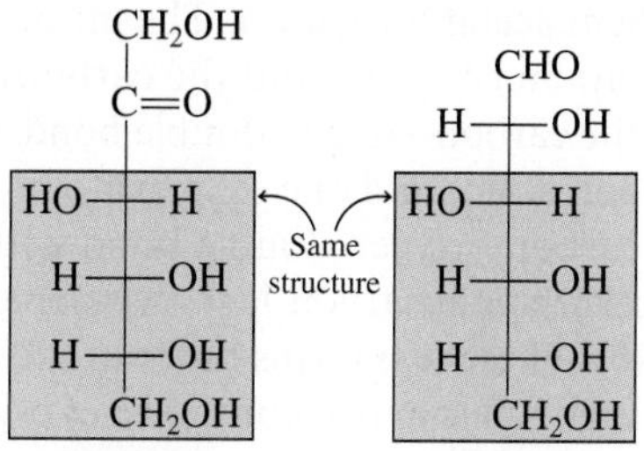

D-Fructose D-Glucose

D-Ribose

D-glucose, D-galactose, and D-fructose are all hexoses. D-Ribose is a pentose. If carbon 3 and its accompanying —H and —OH groups were eliminated from the structure of D-glucose, the remaining structure would be that of D-ribose.

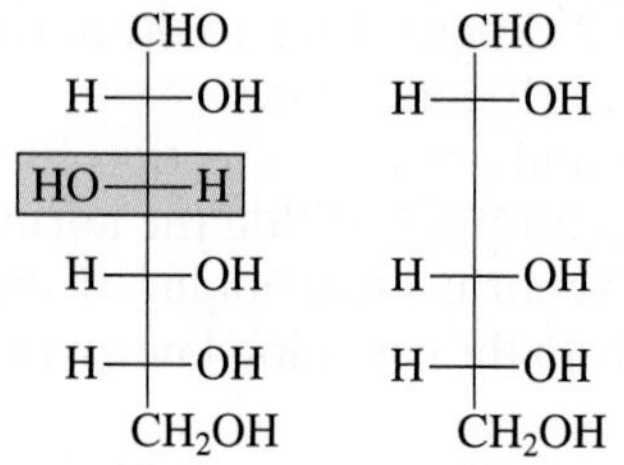

D-Glucose D-Ribose

D-Ribose is a component of a variety of complex molecules, including ribonucleic acids (RNAs) and energy-rich compounds such as adenosine triphosphate (ATP). The compound 2-deoxy-D-ribose is also important in nucleic acid chemistry. This monosaccharide is a component of DNA molecules. The prefix *deoxy-* means "minus an oxygen"; the structures of ribose and 2-deoxyribose differ in that the latter compound lacks an oxygen atom at carbon 2.

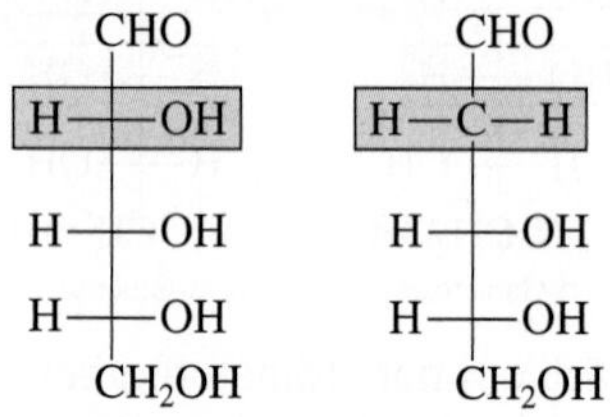

D-Ribose 2-Deoxy-D-ribose

7.10 Cyclic Forms of Monosaccharides

Intramolecular cyclic hemiacetal formation was previously discussed in the chapter on aldehydes and ketones (Section 4.11). Recall that hemiacetals have both an —OH group and an —OR group attached to the same carbon atom. In the cyclic hemiacetals that monosaccharides form, it is the carbonyl carbon atom that bears the —OH and —OR groups.

So far in this chapter, the structures of monosaccharides have been depicted as open-chain polyhydroxy aldehydes or ketones. However, experimental evidence indicates that for monosaccharides containing five or more carbon atoms, such open-chain structures are actually in equilibrium with two cyclic structures, and the cyclic structures are the dominant forms at equilibrium. The cyclic forms of monosaccharides result from the ability of their carbonyl group to react intramolecularly with a hydroxyl group. Structurally, the resulting cyclic compounds are cyclic hemiacetals.

Cyclic Forms of D-Glucose

Further details about the cyclic forms of monosaccharides are obtained by considering in depth the formation of the cyclic forms of D-glucose. The platform for this in-depth consideration is Figure 7.17.

In Figure 7.17, structure 2 is a rearrangement of the projection formula for D-glucose in which the carbon atoms have locations similar to those found for carbon atoms in a six-membered ring. All hydroxyl groups drawn to the right in the original Fischer projection formula appear below the ring. Those to the left in the Fischer projection formula appear above the ring.

Structure 3 in Figure 7.17 is obtained by rotating the groups attached to carbon 5 in a counterclockwise direction so that they are in the positions where it is easiest to visualize intramolecular hemiacetal formation. The intramolecular reaction occurs between the hydroxyl group on carbon 5 and the carbonyl group (carbon 1). The —OH group adds across the carbon–oxygen double bond, producing a heterocyclic ring that contains five carbon atoms and one oxygen atom.

Cyclization of glucose (hemiacetal formation) creates a new chiral center at carbon 1, and the presence of this new chiral center produces two stereoisomers, called α and β isomers.

Addition across the carbon–oxygen double bond with its accompanying ring formation produces a chiral center at carbon 1, so two stereoisomers are possible (see Figure 7.17, structures 4–6). These two forms differ in the orientation of the —OH group on the hemiacetal carbon atom (carbon 1). In α-D-glucose, the —OH group

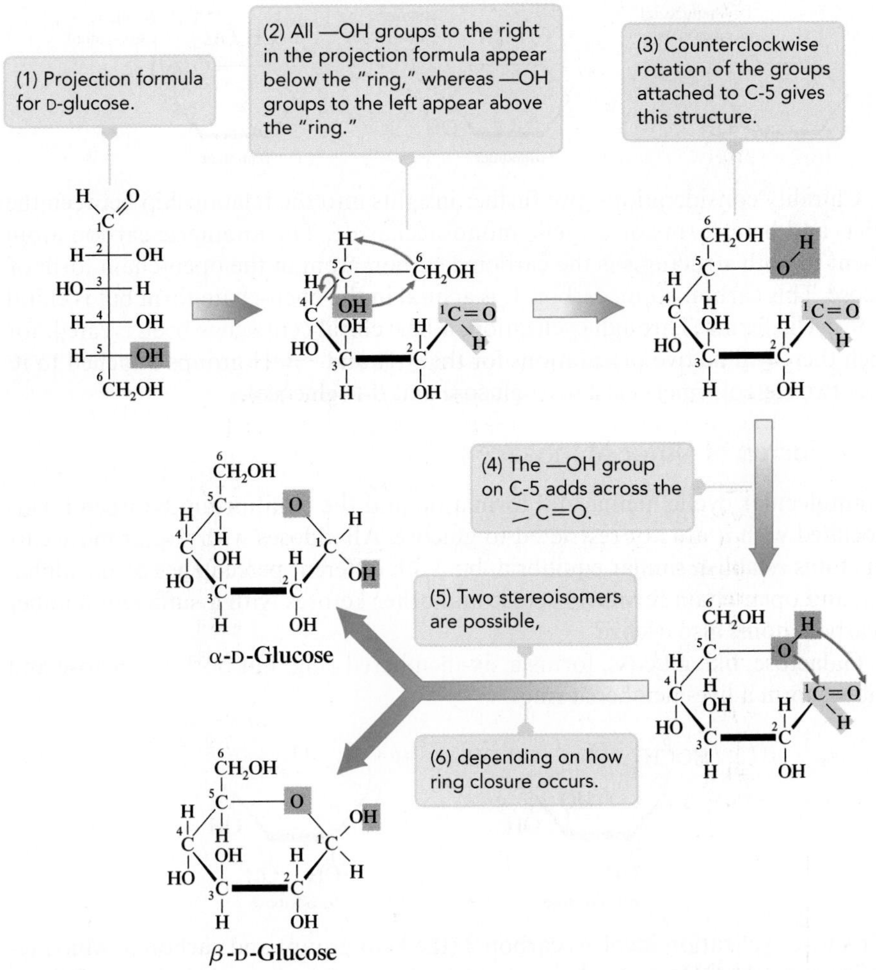

(1) Projection formula for D-glucose.
(2) All —OH groups to the right in the projection formula appear below the "ring," whereas —OH groups to the left appear above the "ring."
(3) Counterclockwise rotation of the groups attached to C-5 gives this structure.
(4) The —OH group on C-5 adds across the C=O.
(5) Two stereoisomers are possible,
(6) depending on how ring closure occurs.
α-D-Glucose
β-D-Glucose

Figure 7.17 The cyclic hemiacetal forms of D-glucose result from the intramolecular reaction between the carbonyl group and the hydroxyl group on carbon 5.

is on the opposite side of the ring from the CH_2OH group attached to carbon 5. In β-D-glucose, the CH_2OH group on carbon 5 and the —OH group on carbon 1 are on the same side of the ring.

In an aqueous solution of D-glucose, a dynamic equilibrium exists among the α, β, and open-chain forms, and there is continual interconversion among them. For example, a freshly mixed solution of pure α-D-glucose slowly converts to a mixture of both α- and β-D-glucose by an opening and a closing of the cyclic structure. When equilibrium is established, 63% of the molecules are β-D-glucose, 37% are α-D-glucose, and less than 0.01% are in the open-chain form.

$$\underset{(37\%)}{\alpha\text{-D-Glucose}} \rightleftarrows \underset{(\text{less than } 0.01\%)}{\text{Open-chain D-Glucose}} \rightleftarrows \underset{(63\%)}{\beta\text{-D-Glucose}}$$

Special Terminology for Cyclic Monosaccharide Structures

Special terminology exists for the hemiacetal carbon atom present in a cyclic monosaccharide structure. This carbon atom is called the *anomeric carbon atom.* An **anomeric carbon atom** *is the hemiacetal carbon atom present in a cyclic monosaccharide structure.* It is the carbon atom that is bonded to an —OH group and to the oxygen atom in the heterocyclic ring.

Cyclic monosaccharide formation always produces two stereoisomers—an alpha form and a beta form. These two isomers are called *anomers.* **Anomers** *are cyclic monosaccharides that differ only in the positions of the substituents on the anomeric (hemiacetal) carbon atom.* The α-stereoisomer has the —OH group on the opposite side of the ring from the —CH_2OH group, and the β-stereoisomer has the —OH group on the same side of the ring as the —CH_2OH group.

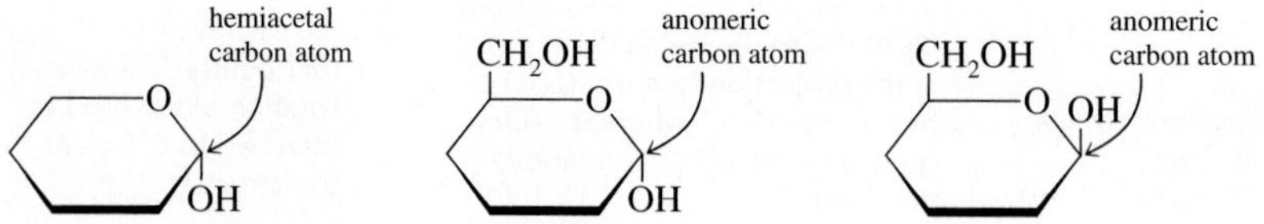

α anomer β anomer

Chirality considerations give further insights into the relationship between the alpha- and beta-forms of a cyclic monosaccharide. The anomeric carbon atom present in both structures is the carbonyl carbon atom in the open-chain form of glucose. This carbon atom, carbon 1, is achiral in the open-chain form but is chiral in the cyclic forms. Through cyclization, a new chiral center has been created, for which there will be two orientations for the —H and —OH groups attached to it; hence two stereoisomers exist (α-D-glucose and β-D-glucose).

Cyclic Forms of Other Monosaccharides

Intramolecular cyclic hemiacetal formation and the equilibrium between forms associated with it are not restricted to glucose. All aldoses with five or more carbon atoms establish similar equilibria, but with different percentages of the alpha, beta, and open-chain forms. Fructose and other ketoses with a sufficient number of carbon atoms also cyclize.

Galactose, like glucose, forms a six-membered ring, but both D-fructose and D-ribose form a five-membered ring.

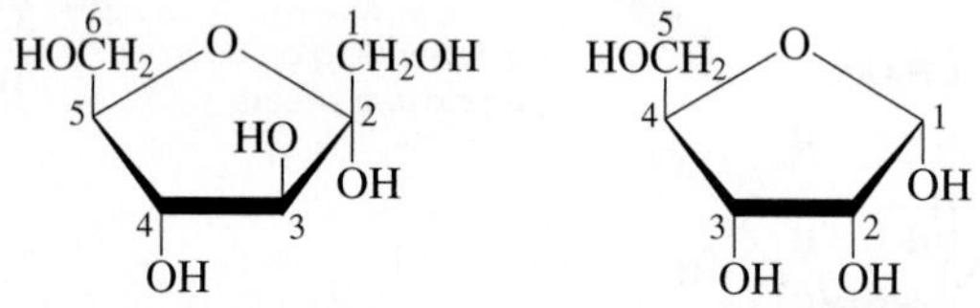

α-D-Fructose α-D-Ribose

D-Fructose cyclization involves carbon 2 (the keto group) and carbon 5, which results in two CH_2OH groups being outside the ring (carbons 1 and 6). D-Ribose cyclization involves carbon 1 (the aldehyde group) and carbon 4.

A cyclic monosaccharide containing a six-atom ring is called a *pyranose,* and one containing a five-atom ring is called *furanose* because their ring structures resemble the ring structures in the cyclic ethers *pyran* and *furan* (Section 3.19), respectively.

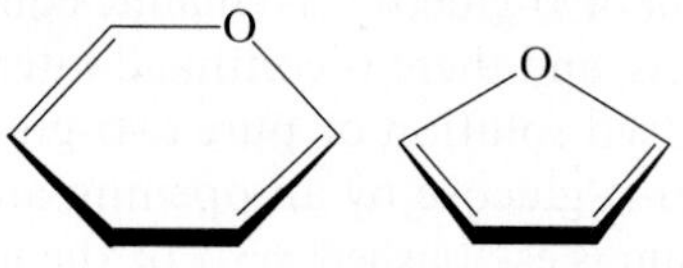

Pyran Furan

Such nomenclature leads to more specific names for the cyclic forms of monosaccharides—names that specify ring size. The more specific name for α-D-glucose is α-D-glucopyranose, and the more specific name for α-D-fructose is α-D-fructofuranose. The last part of each of these names specifies ring size.

EXAMPLE 7.6 Distinguishing Common Monosaccharides from Each Other on the Basis of Structural Characteristics

Which of the monosaccharides *glucose, fructose, galactose,* and *ribose* has each of the following structural characteristics? There may be more than one correct answer for a given characteristic.

a. It is a hexose.
b. It is an aldose.
c. Its cyclic form has a 5-membered ring.
d. Its cyclic form exists in alpha and beta forms.

Solution

a. *Glucose, fructose,* and *galactose.* The only monosaccharide of the four listed that is not a hexose is ribose, which is a pentose.

b. *Glucose, galactose,* and *ribose.* The only monosaccharide of the four listed that is not an aldose is fructose, which is a ketose.

c. *Fructose* and *ribose.* Fructose is a ketohexose and when ketoses cyclize, the number of atoms in the ring is always one less than the number of carbon atoms in the open chain form; the ring contains 4 carbon atoms and 1 oxygen atom and there are 2 carbon atoms outside the ring. Ribose is an aldopentose and when aldoses cyclize, the number of atoms in the ring is the same as the number of carbon atoms in the open chain form; the ring contains 4 carbon atoms and 1 oxygen atom and there is 1 carbon atom outside the ring.

d. *Glucose, fructose, galactose,* and *ribose.* All monosaccharides have cyclic structures that have alpha and beta forms. As ring closure occurs, the random "twisting" of the two ends of the chain can produce two orientations for the —OH group on the hemiacetal carbon atom, giving rise to alpha and beta forms.

Practice Exercise 7.6

Which of the monosaccharides *glucose, fructose, galactose,* and *ribose* has each of the following structural characteristics? There may be more than one correct answer for a given characteristic.

a. It is a pentose.
b. It is a ketose.
c. Its cyclic form has a 6-membered ring.
d. Its cyclic form has two carbon atoms outside the ring.

Answers: **a.** Ribose; **b.** Fructose; **c.** Glucose, galactose; **d.** Fructose

7.11 Haworth Projection Formulas

The structural representations of the cyclic forms of monosaccharides found in the previous section are examples of Haworth projection formulas. A **Haworth projection formula** *is a two-dimensional structural notation that specifies the three-dimensional structure of a cyclic form of a monosaccharide.* Such projections carry the name of their originator, the British chemist Walter Norman Haworth (Figure 7.18).

In a Haworth projection, the hemiacetal ring system is viewed "edge on" with the oxygen ring atom at the upper right (six-membered ring) or at the top (five-membered ring).

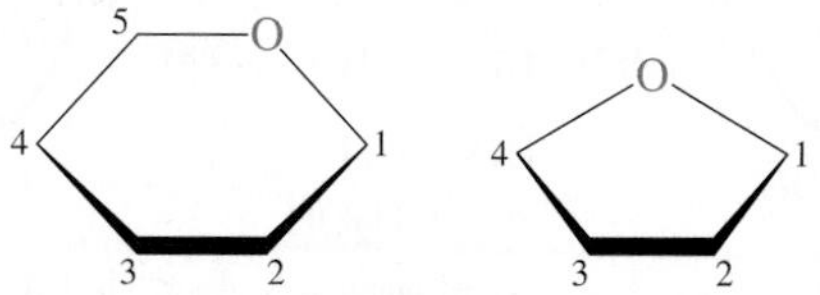

The D or L form of a monosaccharide is determined by the position of the terminal CH_2OH group on the highest-numbered ring carbon atom. In the D form, this group is positioned above the ring. In the L form, which is not usually encountered in biochemical systems, the terminal CH_2OH group is positioned below the ring.

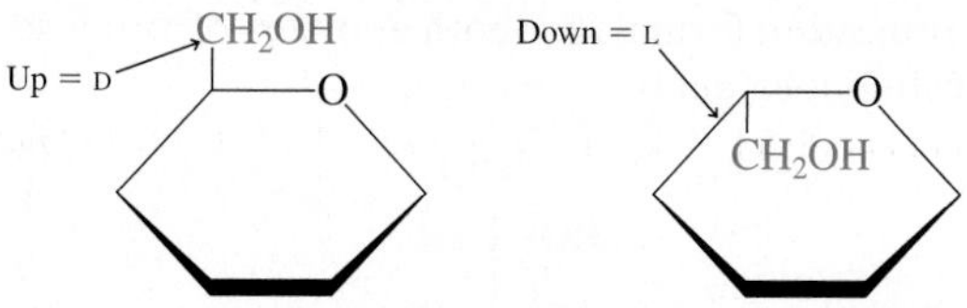

α or β configuration is determined by the position of the —OH group on carbon 1 relative to the CH_2OH group that determines D or L series. In a β configuration, both of these groups point in the same direction; in an α configuration, the two groups point in opposite directions.

© Hulton-Deutsch Collection/CORBIS

Figure 7.18 Walter Norman Haworth (1883–1950), the developer of Haworth projection formulas, was a British carbohydrate chemist. He helped determine the structures of the cyclic forms of glucose, was the first to synthesize vitamin C, and was a corecipient of the 1937 Nobel Prize in chemistry.

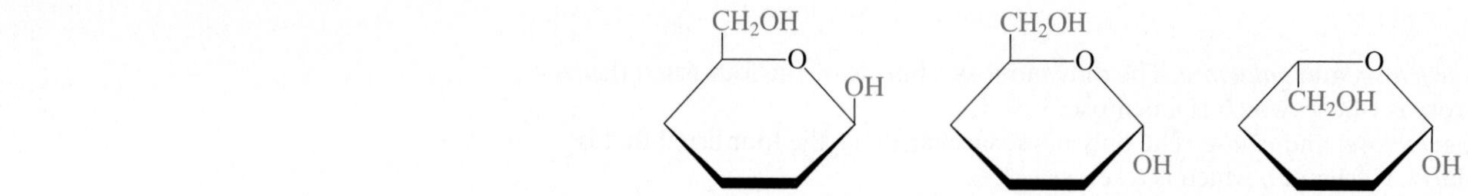

β-D-Monosaccharide α-D-Monosaccharide β-L-Monosaccharide

In situations where α or β configuration does not matter, the —OH group on carbon 1 is placed in a horizontal position, and a wavy line is used as the bond that connects it to the ring.

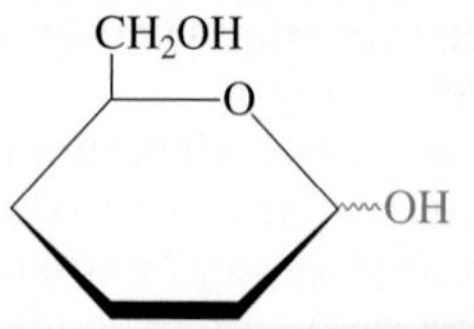

The specific identity of a monosaccharide is determined by the positioning of the other —OH groups in the Haworth projection formula. Any —OH group at a chiral center that is to the right in a Fischer projection formula points down in the Haworth projection formula. Any group to the left in a Fischer projection formula points up in the Haworth projection formula. The following is a matchup between Haworth projection formulas and a Fischer projection formula.

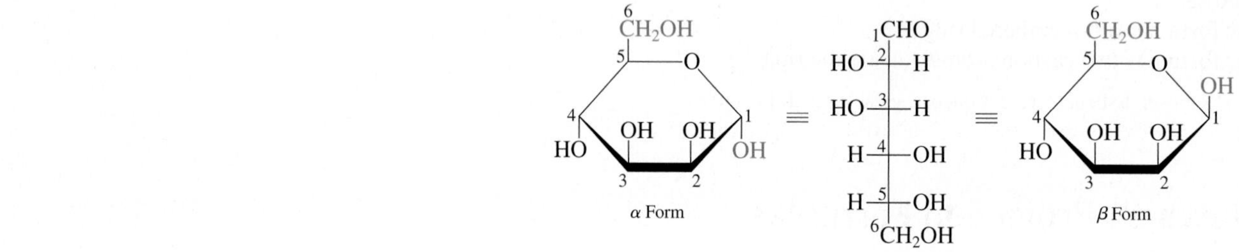

α Form β Form

Comparison of this Fischer projection formula with those given in Figure 7.14 reveals that the monosaccharide is D-mannose.

A similar matchup diagram for L-mannose shows that all up and down positions are inverted relative to the D-mannose diagram.

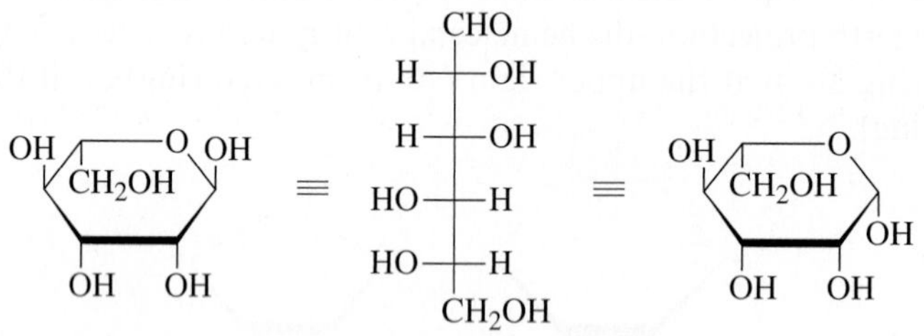

α-L-Mannose L-Mannose β-L-Mannose

EXAMPLE 7.7 Changing a Fischer Projection Formula to a Haworth Projection Formula

Draw the Haworth projection formula for both anomers of D-talose, a monosaccharide whose Fischer projection formula is

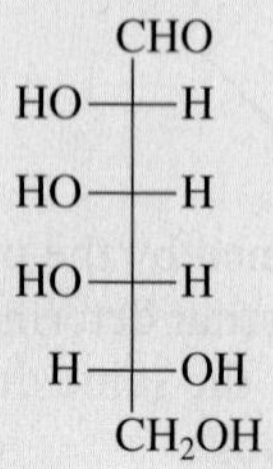

Solution

D-Talose is an aldohexose. An aldohexose cyclization produces a six-membered ring with one carbon atom (C6) outside the ring (Section 7.10).

Step 1: Draw a planar hexagon ring with the oxygen atom in the upper right corner of the ring.

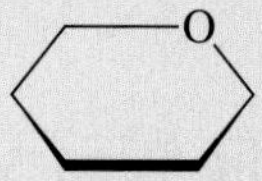

Step 2: Add a —CH_2OH group to the first carbon to the left of the ring oxygen atom. This attachment is drawn up since the monosaccharide is D-talose rather than L-talose. For L-talose, the —CH_2OH group would have been drawn down.

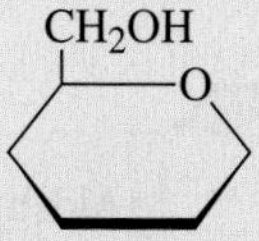

Step 3: Add to the anomeric carbon, the first carbon to the right of the ring oxygen atom, an —OH group that is drawn down for the α-anomer and drawn up for the β-anomer.

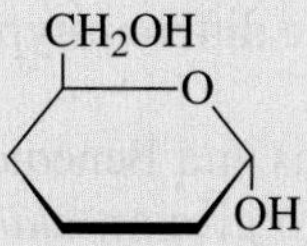

α-Anomer

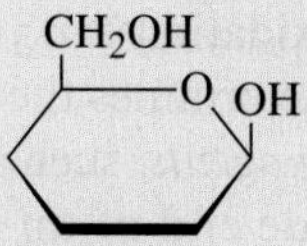

β-Anomer

Step 4: Add an —OH group to each of the remaining three carbons atoms (carbons 2–4). The —OH group is drawn down if it was to the right in the Fischer projection formula and drawn up if it was to the left in the Fischer projection formula.

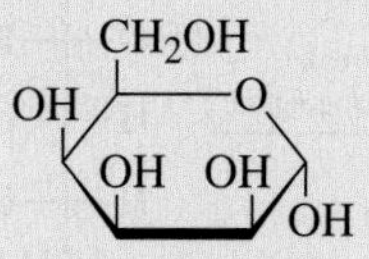

α-D-Talose

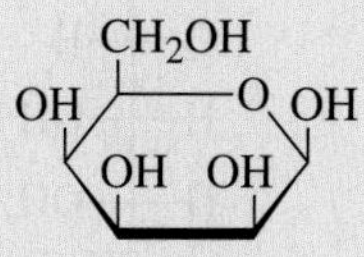

β-D-Talose

Practice Exercise 7.7

Draw the Haworth projection formula for both anomers of D-idose, a monosaccharide whose Fischer projection formula is

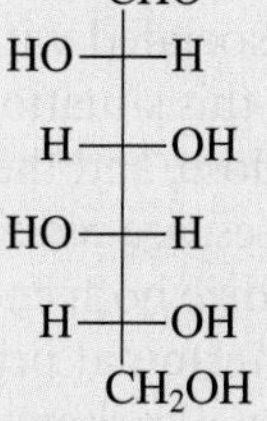

Answers:

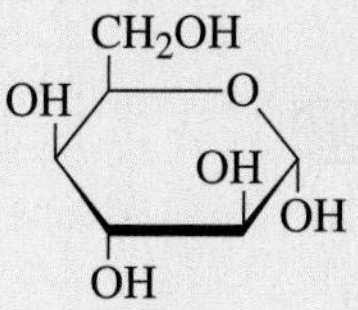

α-D-Iodose

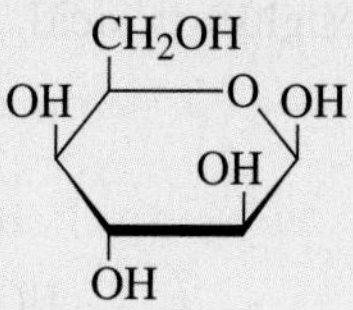

β-D-Iodose

7.12 Reactions of Monosaccharides

Five important reactions of monosaccharides are oxidation to acidic sugars, reduction to sugar alcohols, glycoside formation, phosphate ester formation, and amino sugar formation. In considering these reactions, glucose will be used as the monosaccharide reactant. Remember, however, that other aldoses, as well as ketoses, undergo similar reactions.

Oxidation to Produce Acidic Sugars

The redox chemistry of monosaccharides is closely linked to that of the alcohol and aldehyde functional groups. This latter redox chemistry, which was considered in Chapters 3 and 4, is summarized in the following diagram.

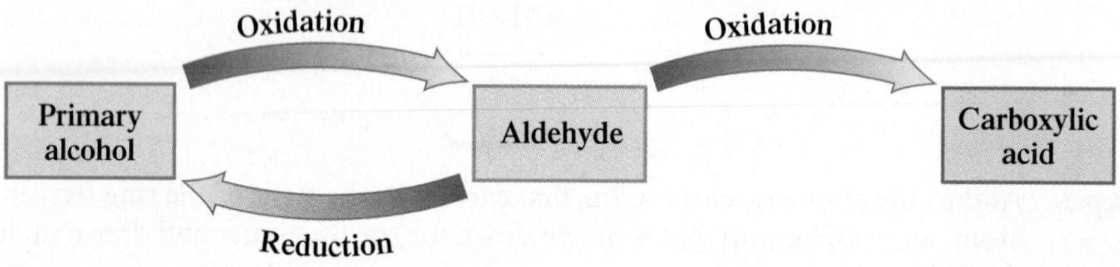

Glucose testing associated with diabetes (Chemical Connections 4-C on page 153) involves oxidation of glucose to produce an acidic sugar.

Monosaccharide oxidation can yield three different types of *acidic sugars.* The oxidizing agent used determines the product.

Weak oxidizing agents, such as Tollens and Benedict's solutions (Section 4.10), oxidize the aldehyde end of an aldose to give an *aldonic acid.* Oxidation of the aldehyde end of glucose produces gluconic acid, and oxidation of the aldehyde end of galactose produces galactonic acid. The structures involved in the glucose reaction are

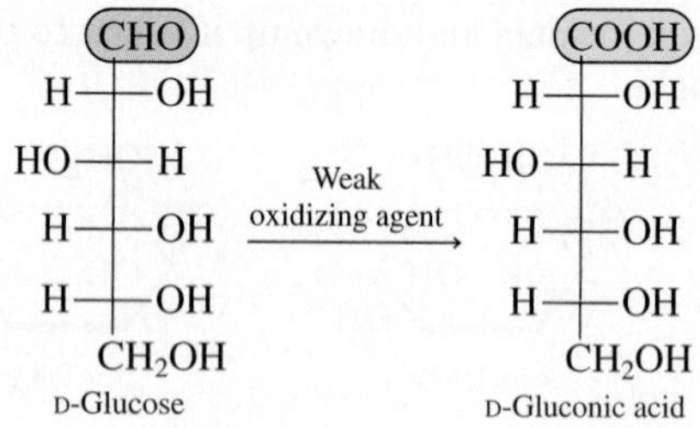

Because aldoses act as reducing agents in such reactions, they are called *reducing sugars.* With Tollens solution, glucose reduces Ag^+ ion to Ag, and with Benedict's solution, glucose reduces Cu^{2+} ion to Cu^+ ion (see Section 4.10). A **reducing sugar** *is a carbohydrate that gives a positive test with Tollens and Benedict's solutions.*

Under the basic conditions associated with Tollens and Benedict's solutions, ketoses are also reducing sugars. In this situation, the ketose undergoes a structural rearrangement that produces an aldose, and the aldose then reacts. Thus all monosaccharides, both aldoses and ketoses, are reducing sugars.

Strong oxidizing agents can oxidize both ends of a monosaccharide at the same time (the carbonyl group and the terminal primary alcohol group) to produce a dicarboxylic acid. Such polyhydroxy dicarboxylic acids are known as *aldaric acids.* For glucose, this oxidation produces glucaric acid.

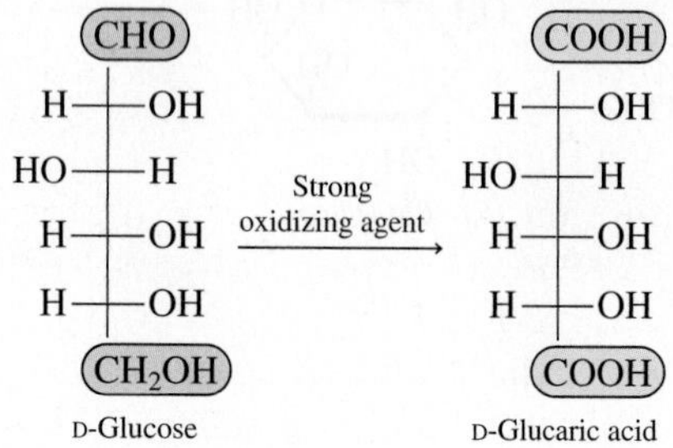

Although it is difficult to do in the laboratory, in biochemical systems *enzymes* can oxidize the primary alcohol end of an aldose such as glucose, without oxidation of the aldehyde group, to produce an *alduronic acid.* For glucose, such an oxidation produces D-glucuronic acid.

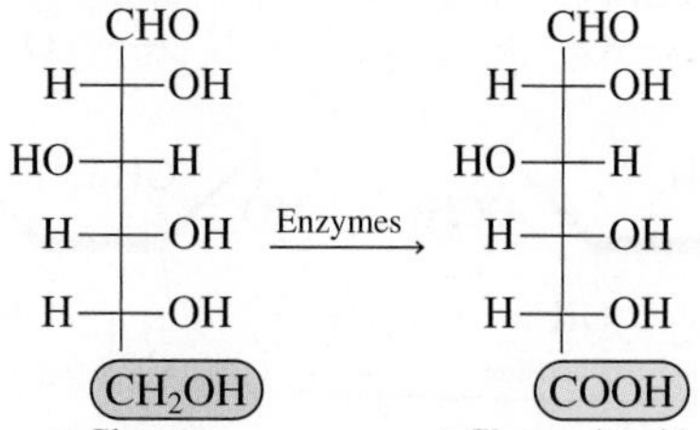

D-Glucose D-Glucuronic acid

The acidic polysaccharides hyaluronic acid and chitin (Section 7.18) both have structural components that are derived from glucuronic acid.

Reduction to Produce Sugar Alcohols

The carbonyl group present in a monosaccharide (either an aldose or a ketose) can be reduced to a hydroxyl group, using hydrogen as the reducing agent. For aldoses and ketoses, the product of the reduction is the corresponding polyhydroxy alcohol, which is sometimes called a *sugar alcohol.* For example, the reduction of D-glucose gives D-glucitol.

Some people are particularly sensitive to D-sorbitol. They absorb it poorly and it thus passes undigested into the large intestine, where it is metabolized by bacteria. A by-product of such bacterial action is large amounts of intestinal gas, which can cause major discomfort.

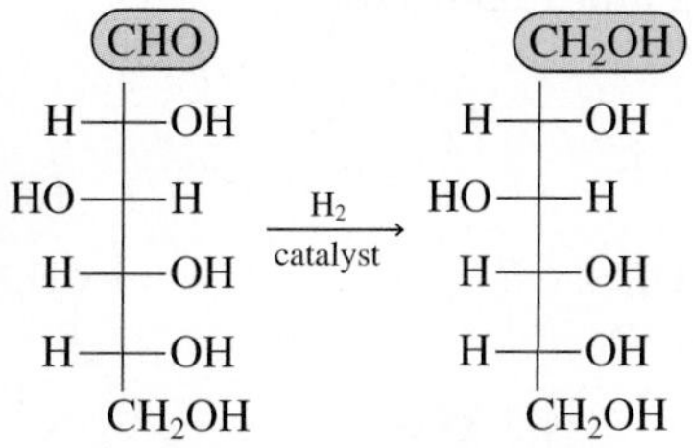

D-Glucose D-Glucitol

D-Glucitol is also known by the common name D-sorbitol. Hexahydroxy alcohols such as D-sorbitol have properties similar to those of the trihydroxy alcohol *glycerol* (Section 3.5). These alcohols are used as moisturizing agents in foods and cosmetics because of their affinity for water. D-Sorbitol is also used as a sweetening agent in chewing gum; bacteria that cause tooth decay cannot use polyalcohols as food sources, as they can glucose and many other monosaccharides.

D-Sorbitol accumulation in the eye is a major factor in the formation of cataracts due to diabetes.

Glycoside Formation

In Section 4.11, hemiacetals were shown to react with alcohols in acid solution to produce acetals. Because the cyclic forms of monosaccharides are hemiacetals, they react with alcohols to form acetals, as is illustrated here for the reaction of β-D-glucose with methyl alcohol.

Remember, from Section 4.11, that acetals have two —OR groups attached to the same carbon atom.

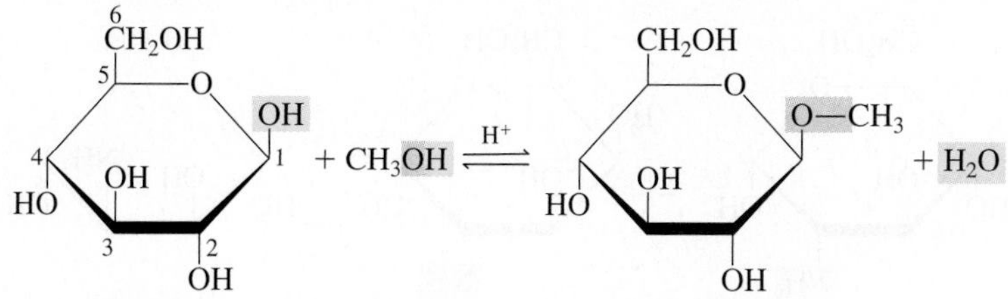

β-D-Glucose Methyl-β-D-glucoside

The general name for monosaccharide acetals is *glycoside.* A **glycoside** *is an acetal formed from a cyclic monosaccharide by replacement of the hemiacetal carbon —OH group with an —OR group.* More specifically, a glycoside produced from glucose is called a glucoside, that from galactose is called a galactoside, and so on. Glycosides, like the hemiacetals from which they are formed, can exist in both

α and β forms. Glycosides are named by listing the alkyl or aryl group attached to the oxygen, followed by the name of the monosaccharide involved, with the suffix *-ide* appended to it.

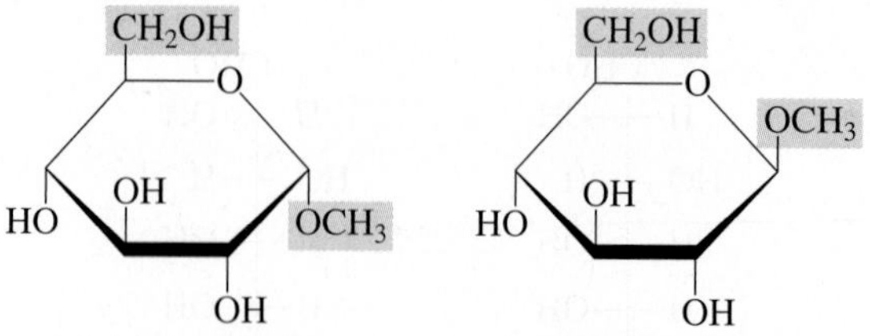

Methyl-α-D-glucoside Methyl-β-D-glucoside

Phosphate Ester Formation

The hydroxyl groups of a monosaccharide can react with inorganic oxyacids to form inorganic esters (Section 5.19). Phosphate esters, formed from phosphoric acid and various monosaccharides, are commonly encountered in biochemical systems. For example, specific enzymes in the human body catalyze the esterification of the hemiacetal group (carbon 1) and the primary alcohol group (carbon 6) in glucose to produce the compounds glucose 1-phosphate and glucose 6-phosphate, respectively (Sections 13.2 and 13.5–13.6).

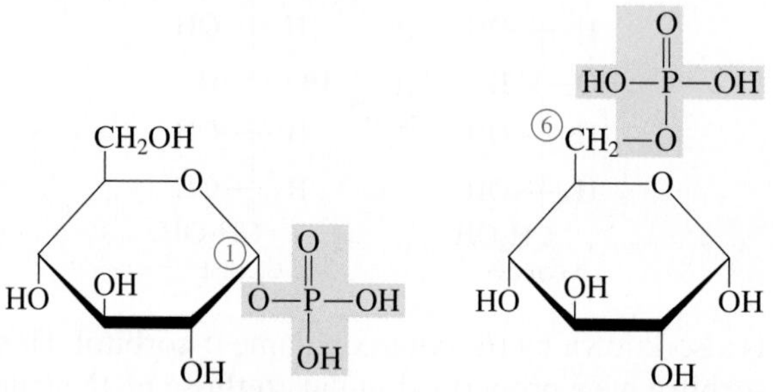

α-D-Glucose 1-phosphate α-D-Glucose 6-phosphate

These phosphate esters of glucose are stable in aqueous solution and play important roles in the metabolism of carbohydrates.

Amino Sugar Formation

If one of the hydroxyl groups of a monosaccharide is replaced with an amino group, an amino sugar is produced. In naturally occurring amino sugars, of which there are three common ones, the amino group replaces the carbon 2 hydroxyl group. The three common natural amino sugars are

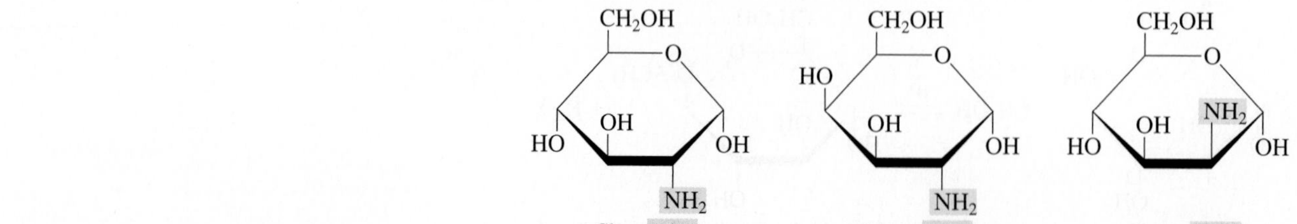

D-Glucosamine D-Galactosamine D-Mannosamine

Amino sugars and their *N*-acetyl derivatives are important building blocks of polysaccharides found in chitin (Section 7.16) and hyaluronic acid (Section 7.17). The *N*-acetyl derivatives of D-glucosamine and D-galactosamine are present in the biochemical markers on red blood cells, which distinguish the various blood types. (See Chemical Connections 7-D on page 293.)

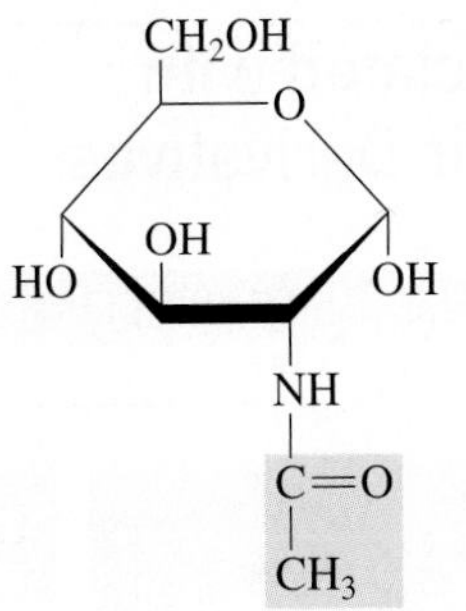

N-Acetyl-α-D-glucosamine

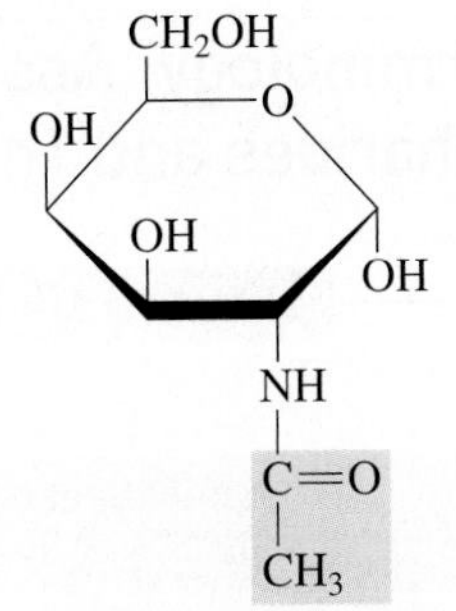

N-Acetyl-α-D-galactosamine

An *acetyl group* has the structure

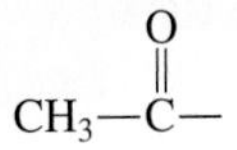

It can be considered to be derived from acetic acid by removal of the —OH portion of that structure.

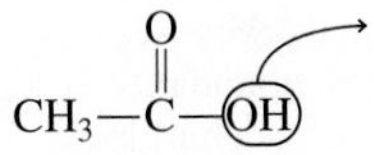

Acetic acid

The Chemistry at a Glance feature on the next page summarizes the "sugar terminology" associated with the common types of monosaccharides and monosaccharide derivatives considered so far in this chapter.

7.13 Disaccharides

A monosaccharide that has cyclic forms (hemiacetal forms) can react with an alcohol to form a glycoside (acetal), as noted in Section 7.12. This same type of reaction can be used to produce a *disaccharide*, a carbohydrate in which two monosaccharides are bonded together (Section 7.3). In disaccharide formation, one of the monosaccharide reactants functions as a hemiacetal, and the other functions as an alcohol.

$$\underset{\text{(Functioning as a hemiacetal)}}{\text{Monosaccharide}} + \underset{\text{(Functioning as an alcohol)}}{\text{monosaccharide}} \longrightarrow \underset{\text{(Glycoside)}}{\text{disaccharide}} + H_2O$$

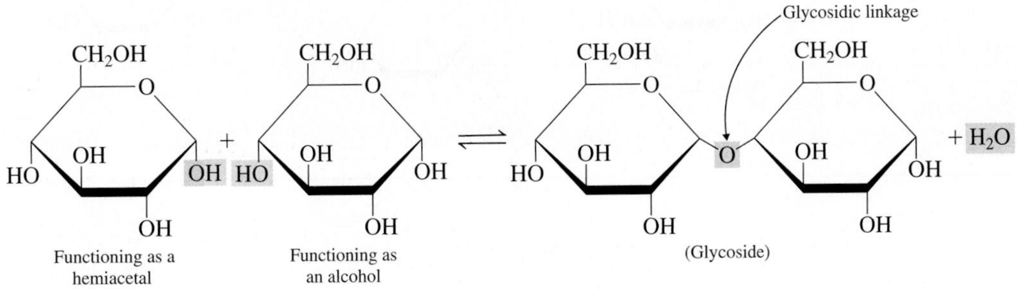

The bond that links the two monosaccharides of a disaccharide (glycoside) together is called a glycosidic linkage. A **glycosidic linkage** *is the bond in a disaccharide resulting from the reaction between the hemiacetal carbon atom —OH group of one monosaccharide and an —OH group on the other monosaccharide.* It is always a carbon–oxygen–carbon bond in a disaccharide.

In Section 7.10, it was noted that a cyclic monosaccharide contains a hemiacetal (anomeric) carbon atom. Many disaccharides contain both a hemiacetal carbon atom and an acetal carbon atom, as is the case for the preceding disaccharide structure. Hemiacetal and acetal locations within disaccharides play an important role in the chemistry of these substances.

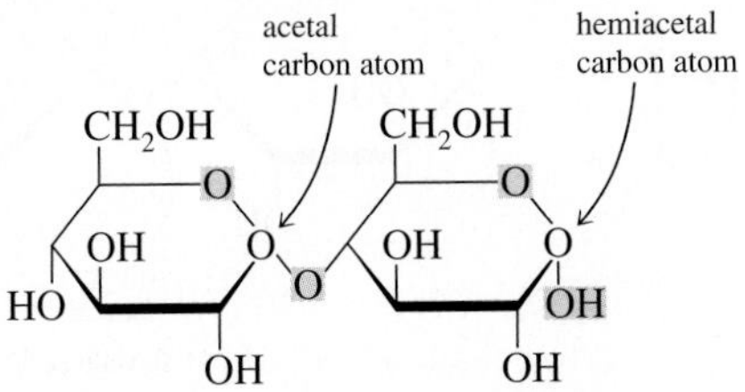

The remainder of this section deals with the specific structures and properties of four important disaccharides: maltose, cellobiose, lactose, and sucrose.

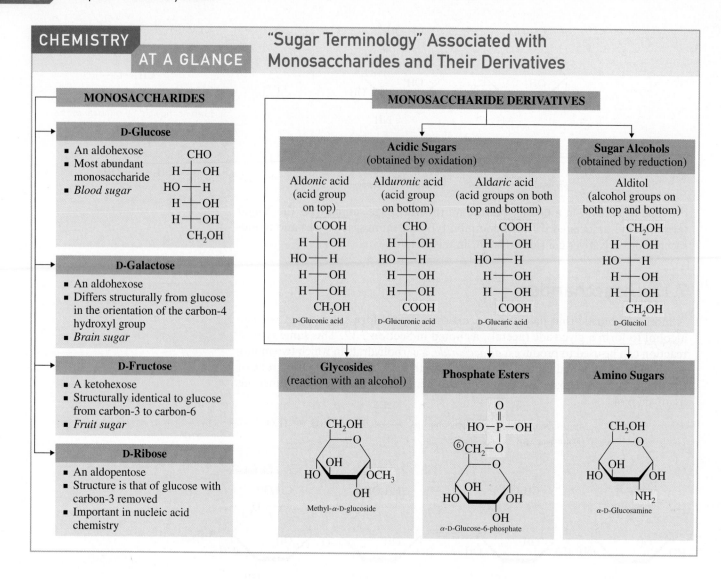

CHEMISTRY AT A GLANCE
"Sugar Terminology" Associated with Monosaccharides and Their Derivatives

MONOSACCHARIDES

D-Glucose

- An aldohexose
- Most abundant monosaccharide
- *Blood sugar*

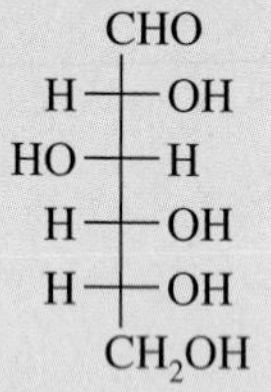

D-Galactose

- An aldohexose
- Differs structurally from glucose in the orientation of the carbon-4 hydroxyl group
- *Brain sugar*

D-Fructose

- A ketohexose
- Structurally identical to glucose from carbon-3 to carbon-6
- *Fruit sugar*

D-Ribose

- An aldopentose
- Structure is that of glucose with carbon-3 removed
- Important in nucleic acid chemistry

MONOSACCHARIDE DERIVATIVES

Acidic Sugars (obtained by oxidation)

Ald*onic* acid (acid group on top)

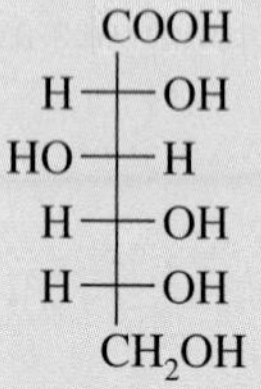

D-Gluconic acid

Ald*uronic* acid (acid group on bottom)

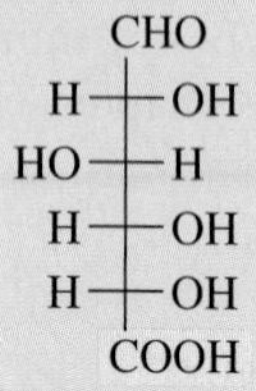

D-Glucuronic acid

Ald*aric* acid (acid groups on both top and bottom)

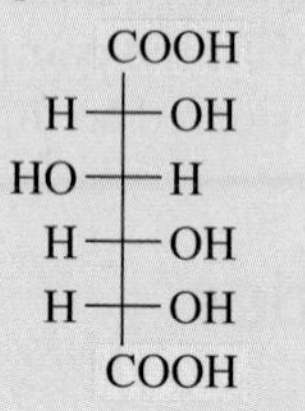

D-Glucaric acid

Sugar Alcohols (obtained by reduction)

Alditol (alcohol groups on both top and bottom)

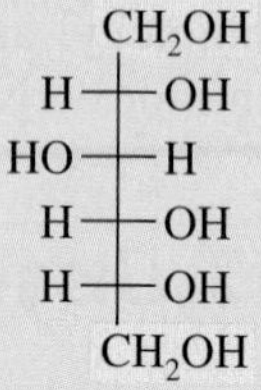

D-Glucitol

Glycosides (reaction with an alcohol)

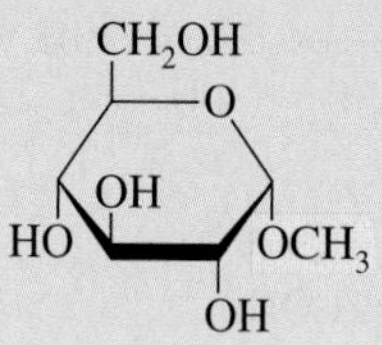

Methyl-α-D-glucoside

Phosphate Esters

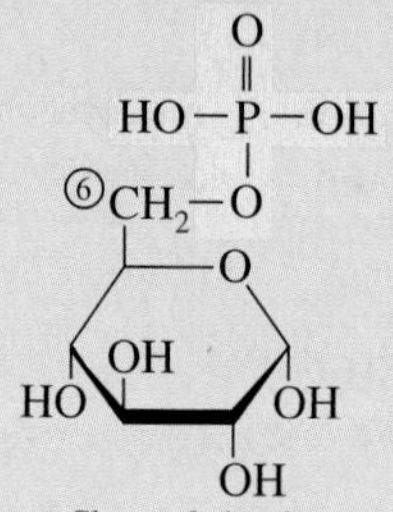

α-D-Glucose-6-phosphate

Amino Sugars

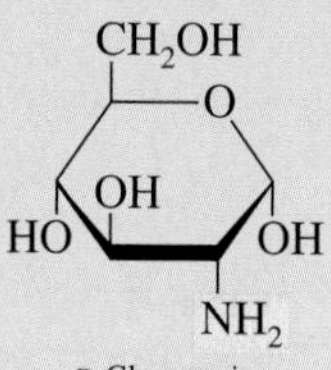

α-D-Glucosamine

Maltose

Maltose, often called *malt sugar,* is produced whenever the polysaccharide starch (Section 7.15) breaks down, as happens in plants when seeds germinate and in human beings during starch digestion. It is a common ingredient in baby foods and is found in malted milk. Malt (germinated barley that has been baked and ground) contains maltose; hence the name *malt sugar.*

Structurally, maltose is made up of two D-glucose units, one of which must be α-D-glucose. The formation of maltose from two glucose molecules is as follows:

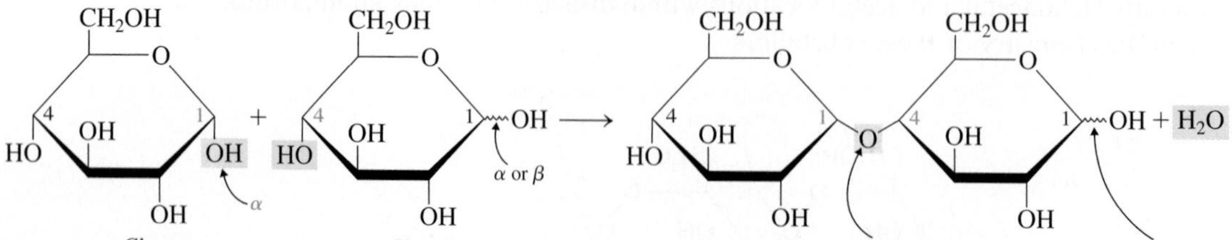

α-D-Glucose D-Glucose α(1→4) Linkage D-Maltose

The glycosidic linkage between the two glucose units is called an α(1 → 4) linkage. The two —OH groups that form the linkage are attached, respectively, to

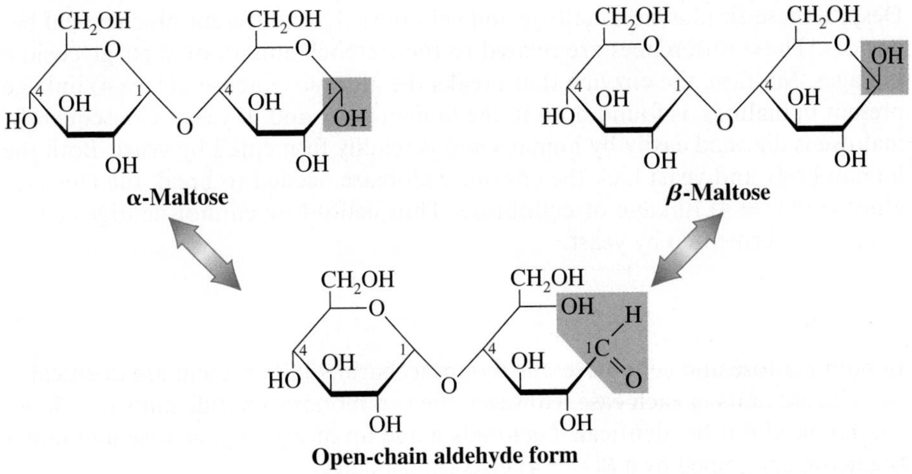

Figure 7.19 The three forms of maltose present in aqueous solution.

carbon 1 of the first glucose unit (in an α configuration) and to carbon 4 of the second.

Maltose is a reducing sugar (Section 7.12) because the glucose unit on the right has a hemiacetal carbon atom (C-1). Thus this glucose unit can open and close; it is in equilibrium with its open-chain aldehyde form (Section 7.10). This means there are actually three forms of the maltose molecule: α-maltose, β-maltose, and the open-chain form. Structures for these three maltose forms are shown in Figure 7.19. In the solid state, the β form is dominant.

The most important chemical reaction of maltose is that of hydrolysis. Hydrolysis of D-maltose, whether in a laboratory flask or in a living organism, produces two molecules of D-glucose. Acidic conditions or the enzyme *maltase* is needed for the hydrolysis to occur.

$$\text{D-Maltose} + H_2O \xrightarrow{H^+ \text{ or maltase}} 2\ \text{D-glucose}$$

It is important to distinguish between the structural notation used for an $\alpha(1 \rightarrow 4)$ glycosidic linkage and that used for a $\beta(1 \rightarrow 4)$ glycosidic linkage.

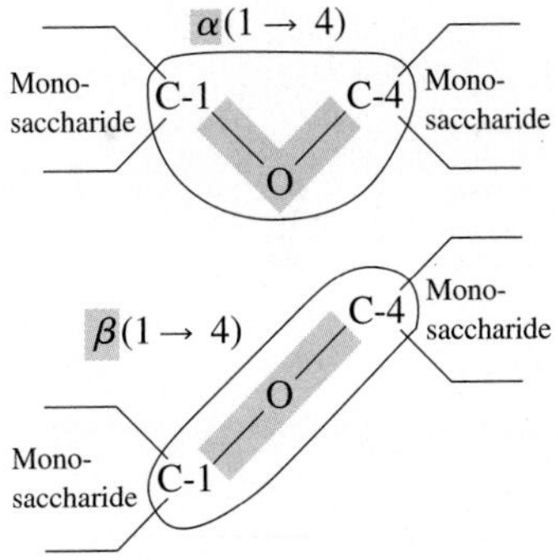

Cellobiose

Cellobiose is produced as an intermediate in the hydrolysis of the polysaccharide cellulose (Section 7.16). Like maltose, cellobiose contains two D-glucose monosaccharide units. It differs from maltose in that one of the D-glucose units—the one functioning as a hemiacetal—must have a β configuration instead of the α configuration for maltose. This change in configuration results in a $\beta(1 \rightarrow 4)$ glycosidic linkage.

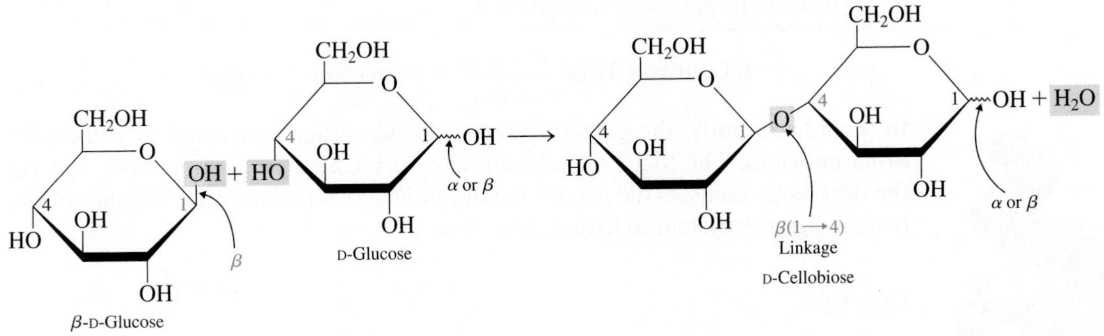

Like maltose, cellobiose is a reducing sugar, has three isomeric forms in aqueous solution, and upon hydrolysis produces two D-glucose molecules.

$$\text{D-Cellobiose} + H_2O \xrightarrow{H^+ \text{ or cellobiase}} 2\ \text{D-glucose}$$

Despite these similarities, maltose and cellobiose have different biochemical behaviors. These differences are related to the stereochemistry of their glycosidic linkages. Maltase, the enzyme that breaks the glucose–glucose $\alpha(1 \rightarrow 4)$ linkage present in maltose, is found both in the human body and in yeast. Consequently, maltose is digested easily by humans and is readily fermented by yeast. Both the human body and yeast lack the enzyme cellobiase needed to break the glucose–glucose $\beta(1 \rightarrow 4)$ linkage of cellobiose. Thus cellobiose cannot be digested by humans or fermented by yeast.

The α form of lactose is sweeter to the taste and more soluble in water than the β form. The β form can be found in ice cream that has been stored for a long time; it crystallizes and gives the ice cream a gritty texture.

Lactose

In both maltose and cellobiose, the monosaccharide units present are identical—two glucose units in each case. However, the two monosaccharide units in a disaccharide need not be identical. *Lactose* is made up of a β-D-galactose unit and a D-glucose unit joined by a $\beta(1 \rightarrow 4)$ glycosidic linkage.

The enzyme needed to break the $\beta(1 \rightarrow 4)$ linkage in lactose is different from the one needed to break the $\beta(1 \rightarrow 4)$ linkage in cellobiose. Because the two disaccharides have slightly different structures, different enzymes are required—*lactase* for lactose and *cellobiase* for cellobiose.

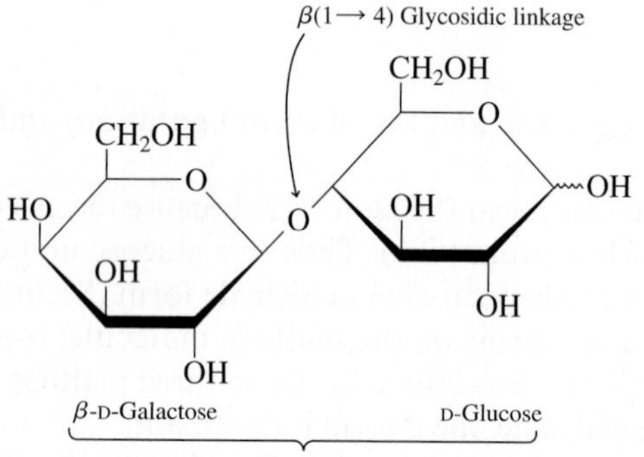

The glucose hemiacetal center is unaffected when galactose bonds to glucose in the formation of lactose, so lactose is a reducing sugar (the glucose ring can open to give an aldehyde).

Lactose is the major sugar found in milk. This accounts for its common name, *milk sugar.* Enzymes in mammalian mammary glands take glucose from the bloodstream and synthesize lactose in a four-step process. Epimerization (Section 7.6) of glucose yields galactose, and then the $\beta(1 \rightarrow 4)$ linkage forms between a galactose and a glucose unit. Lactose is an important ingredient in commercially produced infant formulas that are designed to simulate mother's milk. Souring of milk is caused by the conversion of lactose to lactic acid by bacteria in the milk. Pasteurization of milk is a quick-heating process that kills most of the bacteria and retards the souring process.

Lactose can be hydrolyzed by acid or by the enzyme *lactase,* forming an equimolar mixture of galactose and glucose.

$$\text{D-Lactose} + H_2O \xrightarrow{H^+ \text{ or lactase}} \text{D-galactose} + \text{D-glucose}$$

In the human body, the galactose so produced is then converted to glucose by other enzymes. The focus on relevancy feature Chemical Connections 7-A on the next page discusses the genetic condition *lactose intolerance,* an inability of the human digestive system to hydrolyze lactose.

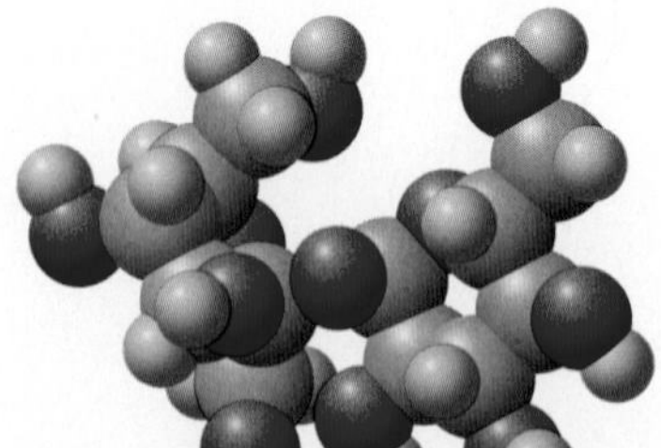

Figure 7.20 Space-filling model of the disaccharide sucrose.

Sucrose

Sucrose, common *table sugar,* is the most abundant of all disaccharides and occurs throughout the plant kingdom. It is produced commercially from the juice of sugar cane and sugar beets. Sugar cane contains up to 20% by mass sucrose, and sugar beets contain up to 17% by mass sucrose. Figure 7.20 shows a molecular model for sucrose.

CHEMICAL CONNECTIONS 7-A

Lactose Intolerance or Lactase Persistence

Lactose is the principal carbohydrate in milk. Human mother's milk obtained by nursing infants contains 7%–8% lactose, almost double the 4%–5% lactose found in cow's milk.

For many people, the digestion and absorption of lactose are a problem. This problem, called *lactose intolerance,* is a condition in which people lack the enzyme *lactase,* which is needed to hydrolyze lactose to galactose and glucose.

$$\text{Lactose} + H_2O \xrightarrow{\text{Lactase}} \text{glucose} + \text{galactose}$$

Many adults cannot drink milk without discomfort because of inadequate amounts of the enzyme *lactase* within their body.

Deficiency of lactase can be caused by a genetic defect, by physiological decline with age, or by injuries to the mucosa lining the intestines. When lactose molecules remain in the intestine undigested, they attract water to themselves, causing fullness, discomfort, cramping, nausea, and diarrhea. Bacterial fermentation of the lactose further along the intestinal tract produces acid (lactic acid) and gas, adding to the discomfort.

The level of the enzyme lactase in humans varies with age. Most children have sufficient lactase during the early years of their life when milk is a much-needed source of calcium in their diet. In adulthood, the enzyme level decreases, and lactose intolerance develops. This explains the change in milk-drinking habits of many adults. Some researchers estimate that as many as one in three adult Americans exhibits a degree of lactose intolerance.

The level of the enzyme lactase in humans varies widely among ethnic groups, indicating that the trait is genetically determined (inherited). The occurrence of lactose intolerance is lowest among Scandinavians and other northern Europeans and highest among native North Americans, Southeast Asians, Africans, and Greeks. The estimated prevalence of lactose intolerance is as follows:

80% Asian Americans	60% Inuits
80% Native Americans	50% Hispanics
75% African Americans	20% Caucasians
70% Mediterranean peoples	10% Northern Europeans

An alternate analysis of milk-drinking versus non-milk-drinking populations exists. Being able to drink milk is considered to be the "abnormal" situation, as less than 40% of the world's total adult population retain the ability to drink milk (digest lactose) after childhood. Thus more than 60% of adults (the majority) are non-milk-drinkers. The term *lactase persistence* is used to describe the condition where milk-drinking ability continues into adulthood.

Lactase persistence is considered to be a genetic adaptation possessed by many Americans and northern Europeans. Most lactase-persistent people do not realize their minority status relative to milk-drinking. No other species but humans, as adults, can drink milk, and then only some of them. There are no milk-drinking adult animals; only baby animals drink milk.

Note that the "normal" condition of lactose intolerance is not considered to be a food allergy situation. Food allergies arise when a person's immune system responds to an invading allergen. Lactose intolerance does not involve the body's immune system. Rather, it relates to an enzymatic condition—lack of the enzyme lactase.

For lactase-persistent people, after lactose has been metabolized into glucose and galactose, the galactose has to be converted into glucose before it can be used by cells. In humans, the genetic condition called *galactosemia* is caused by the absence of one or more of the enzymes needed for this conversion. In people with this condition, galactose and its toxic metabolic derivative galactitol (dulcitol) accumulate in the blood.

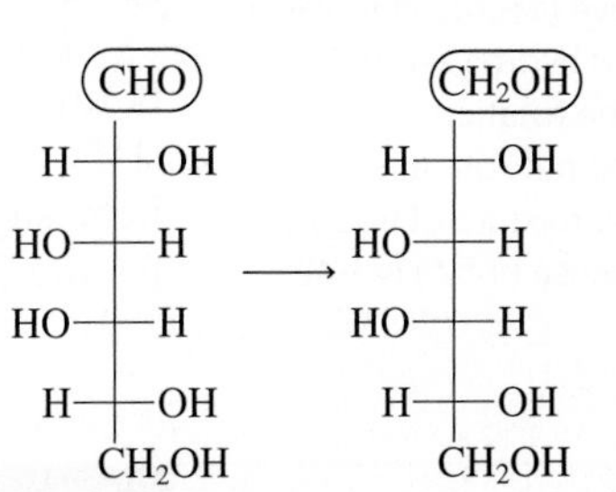

D-Galactose D-Galactitol (D-Dulcitol)

If not treated, galactosemia can cause mental retardation in infants and even death. Treatment involves exclusion of milk and milk products from the diet.

The glycosidic linkage in sucrose is very different from that in maltose, cellobiose, and lactose. The linkages in the latter three compounds can be characterized as "head-to-tail" linkages—that is, the front end (carbon 1) of one monosaccharide is linked to the back end (carbon 4) of the other monosaccharide. Sucrose has a "head-to-head" glycosidic linkage; the front ends of the two monosaccharides (carbon 1 for glucose and carbon 2 for fructose) are linked.

Figure 7.21 Honeybees and many other insects possess an enzyme called *invertase* that hydrolyzes sucrose to invert sugar. Thus honey is predominantly a mixture of D-glucose and D-fructose with some unhydrolyzed sucrose. The term *invert* sugar comes from the observation that the direction of rotation of plane-polarized light (Section 7.7) changes from positive (clockwise) to negative (counterclockwise) when sucrose is hydrolyzed to invert sugar. The rotation is +66° for sucrose. The net rotation for the invert sugar mixture of fructose (−92°) and glucose (+52°) is −40°.

The two monosaccharide units present in a D-sucrose molecule are α-D-glucose and β-D-fructose. The glycosidic linkage is not a $(1 \rightarrow 4)$ linkage, as was the case for maltose, cellobiose, and lactose. It is instead an $\alpha, \beta(1 \rightarrow 2)$ glycosidic linkage. The —OH group on carbon 2 of D-fructose (the hemiacetal carbon) reacts with the —OH group on carbon 1 of D-glucose (the hemiacetal carbon).

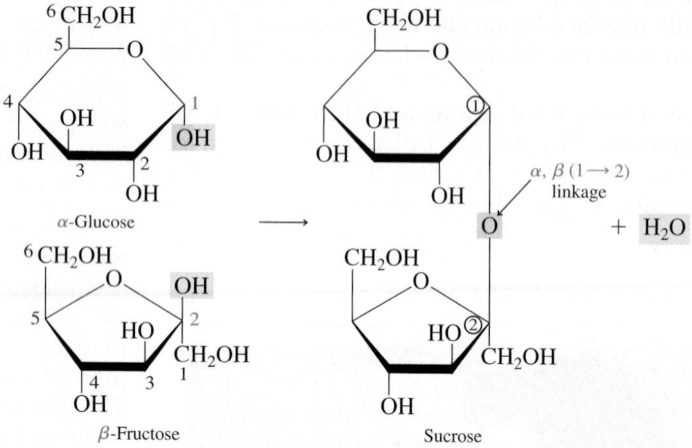

Sucrose, unlike maltose, cellobiose, and lactose, is a *nonreducing sugar.* In sucrose, the hemiacetal center (anomeric carbon atom) of each monosaccharide is involved in the glycosidic linkage. The result is a molecule that contains two acetal centers. Sucrose, in the solid state and in solution, exists in only one form—there are no α and β isomers, and an open-chain form is not possible.

Sucrase, the enzyme needed to break the $\alpha, \beta(1 \rightarrow 2)$ linkage in sucrose, is present in the human body. Hence sucrose is an easily digested substance. Sucrose hydrolysis (digestion) produces an equimolar mixture of glucose and fructose called *invert sugar* (see Figure 7.21).

$$\text{D-Sucrose} + H_2O \xrightarrow{H^+ \text{ or sucrase}} \underbrace{\text{D-glucose} + \text{D-fructose}}_{\text{Invert sugar}}$$

When sucrose is cooked with acid-containing foods such as fruits or berries, partial hydrolysis takes place, forming some invert sugar. Jams and jellies prepared in this manner are actually sweeter than the pure sucrose added to the original mixture because one-to-one mixtures of glucose and fructose taste sweeter than sucrose.

The focus on relevancy feature Chemical Connections 7-B, which starts at the bottom of this page, discusses the increasing trend of using high-fructose corn syrup (HFCS) instead of sucrose as a sweetener in beverages and processed foods. A related item, sugar substitutes (artificial sweeteners), is the topic of the focus on relevancy feature Chemical Connections 7-C that then follows.

CHEMICAL CONNECTIONS 7-B

Changing Sugar Patterns: Decreased Sucrose, Increased Fructose

Prior to 1970, the sweetener added to foods and beverages consumed by humans was almost always the disaccharide sucrose (table sugar). Since that time, sucrose use has steadily decreased, with the monosaccharide fructose progressively displacing sucrose as a sweetener. There is an almost 1-to-1 correlation between decrease in sucrose use and increase in fructose use, as shown in the graph on the top of the next page. Fructose use, per capita, now slightly exceeds that of sucrose.

Fructose, the most abundant sugar in most fruits and some vegetables, has a "sweetness factor" that is 73% greater than that of sucrose (see Chemical Connections 7-C on sweeteners found on page 288). It is the presence of fructose that gives ripe

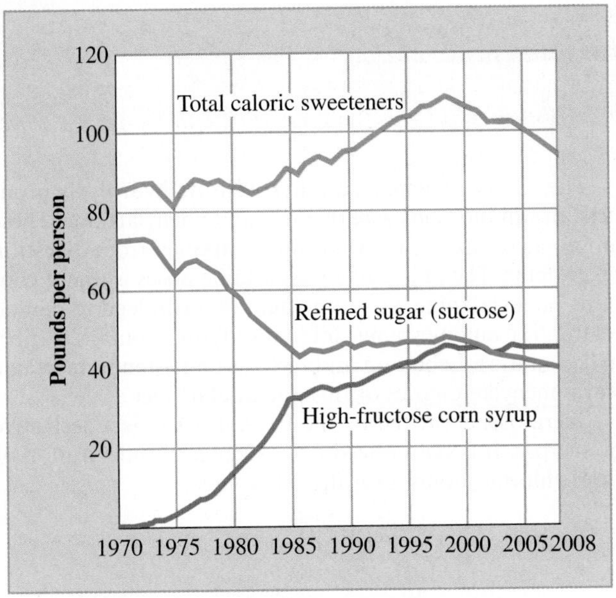

Per capita consumption in the United States, in pounds, of selected sweeteners (1970–2005).

fruit its "extra" sweetness. Fruits are, however, not the source of the fructose that is replacing sucrose in prepared foods and beverages. Rather, fructose in the form of high-fructose corn syrup (HFCS) made from milled corn is the source.

The replacement of sucrose with HFCS is mainly an economically driven decision. It costs less to sweeten products with HFCS than with sucrose for three reasons.

1. Fructose (obtained from corn) is cheaper than sucrose (obtained from sugar beets and sugar cane) in the United States due to the relative abundance of corn, farm subsidies for corn, and sugar import tariffs.
2. HFCS is easier to blend and transport because it is a liquid.
3. HFCS usage leads to products with a much longer shelf life.

HFCS production involves milling corn to produce corn starch and then hydrolyzing the starch (a glucose polymer, Section 7.16) to glucose using the enzymes alpha-amylase and glucoamylase (Section 10.3). The glucose so produced is then treated with the enzyme glucose isomerase to produce a mixture of glucose and fructose. This latter enzymatic process produces a mixture whose composition is 42% fructose, 50% glucose, and 8% other sugars (HFCS-42). Concentration procedures are then used to produce a syrup that is 90% fructose (HFCS-90). HFCS-42 and HFCS-90 are then blended to produce HFCS-55, which is 55% fructose, 41% glucose, and 4% other sugars. HFCS-55 is the sweetener of preference in the soft drink industry. Its properties, including sweetness, are very similar to those of sucrose, which is 50% fructose and 50% glucose. HFCS-42 is commonly used as a sweetener in many processed foods, and HFCS-90 is often used in baked goods.

Fructose considerations also play a major role in the formulation of most fruit juice beverages, which are not 100% one juice but rather fruit juice blends. Fructose from fruit, rather than HFCS, gives these blends their sweetness. The "base" juice for a fruit juice blend is almost always apple juice or pear juice or a mix of the two. To this is added a small amount of a stronger flavored juice such as blueberry or pomegranate juice. Why are apple juice and/or pear juice the "base" juice? Apples and pears are the two fruits with the highest fructose/glucose sugar ratio and therefore the greatest sweetness. The fructose/glucose sugar ratio is almost two times higher for these fruits than for most other fruits, as shown in the graphical information.

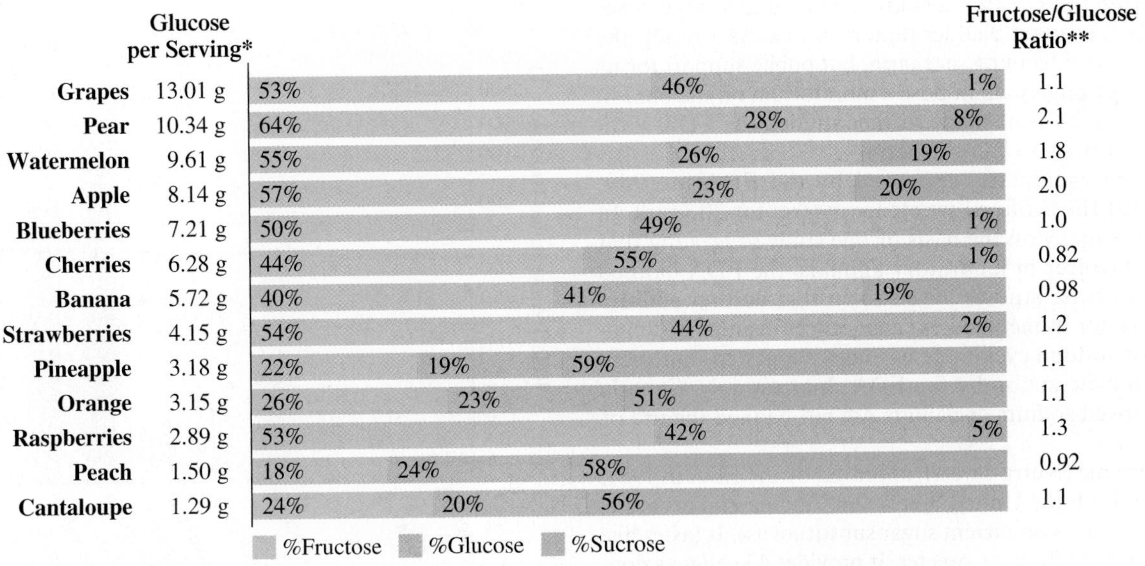

* Serving size: 1 medium-sized fruit (apple, banana, orange, peach, pear); 1 cup (blueberries, cherries, grapes, pineapple, raspberries, strawberries); 1/8 melon (cantaloupe); 1/16 melon (watermelon)

** Fructose/glucose ratio: Calculated treating the sucrose present as one-half fructose and one-half glucose.

CHEMICAL CONNECTIONS 7-C

Sugar Substitutes

Because of the high caloric value of sucrose, it is often difficult to satisfy a demanding "sweet tooth" with sucrose without adding pounds to the body frame or inches to the waistline. Sugar substitutes, which provide virtually no calories, are now used extensively as a solution to the "sucrose problem."

Three sugar substitutes that have been widely used are saccharin, sodium cyclamate, and aspartame.

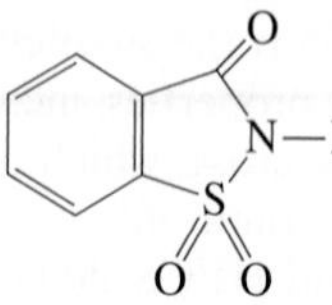

Saccharin

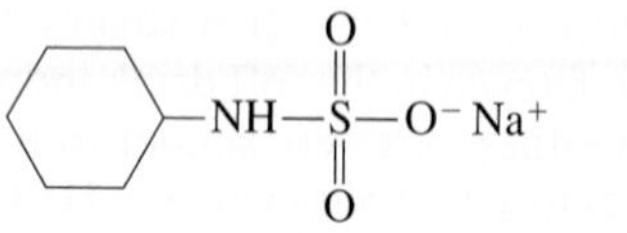

Sodium cyclamate

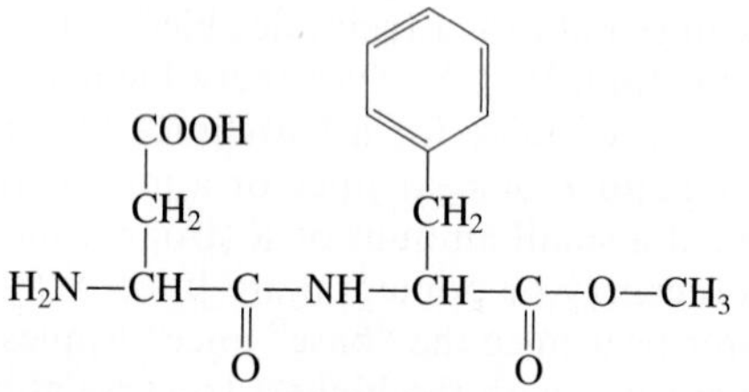

Aspartame

All three of these sugar substitutes have received much publicity because of concern about their safety.

Saccharin is the oldest of the artificial sweeteners, having been in use for more than 100 years. Questions about its safety arose in 1977 after a study suggested that large doses of saccharin caused bladder tumors in rats. As a result, the FDA proposed banning saccharin, but public support for its use caused Congress to impose a moratorium on the ban. In 1991, on the basis of many further studies, the FDA withdrew its proposal to ban saccharin.

Sodium cyclamate, approved by the FDA in 1949, dominated the artificial sweetener market for 20 years. In 1969, principally on the basis of one study suggesting that it caused cancer in laboratory animals, the FDA banned its use. Further studies have shown that neither sodium cyclamate nor its metabolites cause cancer in animals. Reapproval of sodium cyclamate has been suggested, but there has been little action by the FDA. Interestingly, Canada has approved sodium cyclamate use but banned the use of saccharin.

Aspartame (Nutra-Sweet), approved by the FDA in 1981, is used in both the United States and Canada and accounts for three-fourths of current sugar substitute use. It tastes like sucrose but is 150 times sweeter. It provides 4 kcal/g, as does sucrose, but because so little is used, its calorie contribution is negligible. Aspartame has quickly found its way into almost every diet food on the market today. Aspartame as well as saccharin are not heat-stable and therefore cannot be used in products that require cooking.

The safety of aspartame lies with its hydrolysis products: the amino acids aspartic acid and phenylalanine. These amino acids are identical to those obtained from digestion of proteins. The only danger aspartame poses is that it contains phenylalanine, an amino acid that can lead to mental retardation among young children suffering from PKU (phenylketonuria). Labels on all products containing aspartame warn phenylketonurics of this potential danger.

Sucralose, approved by the FDA in 1990, is a derivative of sucrose. It is synthesized from sucrose by substitution of three chlorine atoms for hydroxyl groups.

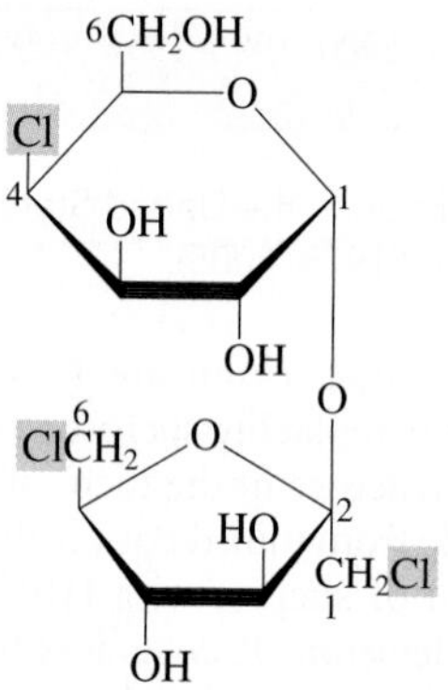

An advantage of sucralose over aspartame is that it is heat-stable and can therefore be used in cooked food. Aspartame loses its sweetness when heated. Sucralose is 600 times

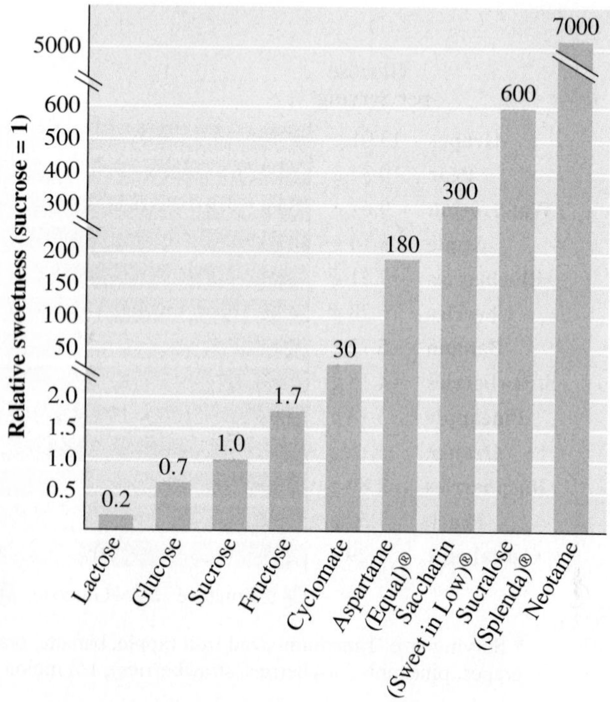

Sweetness of common sugars and sugar substitutes based on sucrose (table sugar) having an assigned value of 1.

sweeter than sucrose and has a similar taste. It is calorie-free because it cannot be hydrolyzed as it passes through the digestive tract.

A newer sugar substitute working its way into the marketplace is neotame, approved by the FDA in 2002 as a general-purpose sweetener. Heat-stable, this sweetener can be used in a wide variety of products including baked goods, frostings, frozen desserts, puddings, and fruit juices. With a sweetness 7000 times greater than that of sucrose, the small amounts needed to sweeten products means its caloric impact is negligible. Economic savings also occur because of the small amounts required.

Structurally, neotame is an aspartame derivative. The same two amino acids are present as in aspartame. It differs structurally from aspartame in that a 3,3-dimethylbutyl group is attached to the terminal $—NH_2$ group of aspartame. This "bulky" attachment prevents the breakdown of neotame into its component amino acids, as occurs for aspartame. Hence individuals with PKU can use neotame without concern.

Other Types of Glycosidic Linkages

Three of the four disaccharides considered in this section—maltose, cellobiose, and lactose—have $(1 \rightarrow 4)$ glycosidic linkages. The other disaccharide considered, sucrose, has a $(1 \rightarrow 2)$ glycosidic linkage. Other carbon atoms besides 1, 2, and 4 can also participate in glycosidic linkages. An additional common type of glycosidic linkage is one that involves carbons 1 and 6. Several disaccharides not detailed in this section have α-$(1 \rightarrow 6)$ glycosidic linkages. The following structure is that for two α-D-glucose molecules connected via an α-$(1 \rightarrow 6)$ glycosidic linkage.

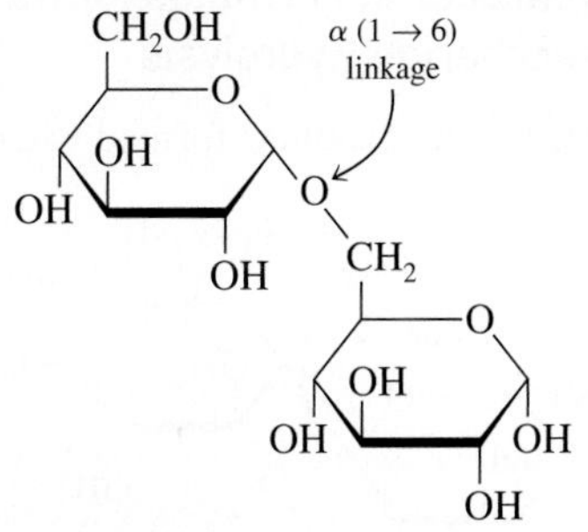

This disaccharide is an entirely different compound than maltose or cellobiose, both of which involve two glucose molecules connected, respectively, via $\alpha(1 \rightarrow 4)$ and $\beta(1 \rightarrow 4)$ glycosidic linkages.

EXAMPLE 7.8 Distinguishing Common Disaccharides from Each Other on the Basis of Structural and Reaction Characteristics

Which of the disaccharides *maltose, cellobiose, lactose,* and *sucrose* has each of the following structural or reaction characteristics? There may be more than one correct answer for a given characteristic.

a. Both monosaccharide units present are glucose.
b. It is a reducing sugar.
c. It can exist in alpha and beta forms.
d. Its glycoside linkage is an $\alpha(1 \rightarrow 4)$ linkage.

Solution

a. *Maltose* and *cellobiose.* Both of these disaccharides contain two glucose units. They differ from each other in the type of glycosidic linkage present; maltose contains an $\alpha(1 \rightarrow 4)$ linkage and cellobiose contains a $\beta(1 \rightarrow 4)$ glycosidic linkage.
b. *Maltose, cellobiose,* and *lactose.* A reducing sugar has one "free" hemiacetal carbon atom. Such is the case any time two monosaccharides are bonded through a $(1 \rightarrow 4)$ glycosidic linkage. Sucrose is the only one of the four disaccharides that does not have a $(1 \rightarrow 4)$ glycosidic linkage; its glycosidic linkage is of the $(1 \rightarrow 2)$ variety.

(continued)

c. *Maltose, cellobiose,* and *lactose.* A "free" hemiacetal carbon atom is a prerequisite for alpha and beta cyclic forms. Sucrose is the only one of the four disaccharides for which this is not the case. Reducing sugars (part **b**) and alpha-beta forms are based on the same structural feature.

d. *Maltose.* Maltose, cellobiose, and lactose all have (1 → 4) glycosidic linkages. However, only maltose has an $\alpha(1 \rightarrow 4)$ linkage; the other two monosaccharides have $\beta(1 \rightarrow 4)$ linkages.

Practice Exercise 7.8

Which of the disaccharides *maltose, cellobiose, lactose,* and *sucrose* has each of the following structural or reaction characteristics? There may be more than one correct answer for a given characteristic.

a. Two different monosaccharide units are present.
b. Hydrolysis produces only monosaccharides.
c. Its glycosidic linkage is a "head-to-head" linkage.
d. It is not a reducing sugar.

Answers: **a.** Lactose, sucrose; **b.** Maltose, cellobiose, lactose, sucrose; **c.** Sucrose; **d.** Sucrose

EXAMPLE 7.9 Drawing Structural Formulas for the Products from Disaccharide Hydrolysis

Draw the structural formulas for the products formed when the disaccharide lactose is hydrolyzed.

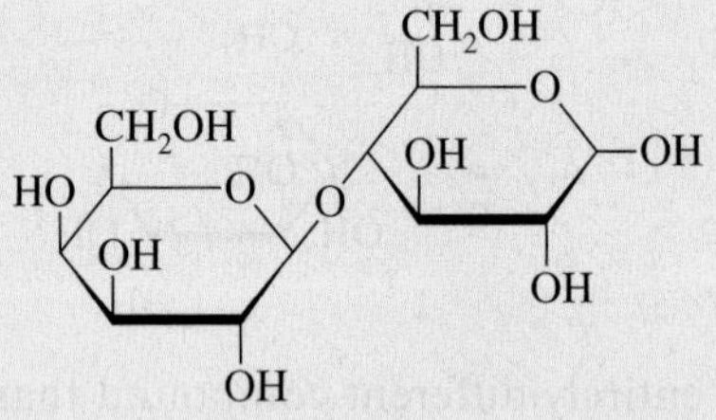

Solution

The glycosidic linkage that connects the two monosaccharide units present in a disaccharide structure is the "point of attack" during a hydrolysis reaction.

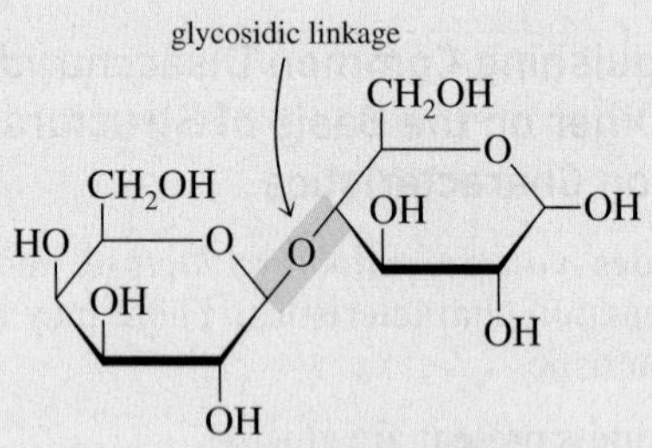

Cleavage of the glycoside linkage releases the two monosaccharides. Water participates in the reaction by supplying a H atom to one monosaccharide (the one that retains the O atom of the glycosidic linkage) and supplying a —OH group entity to the other monosaccharide.

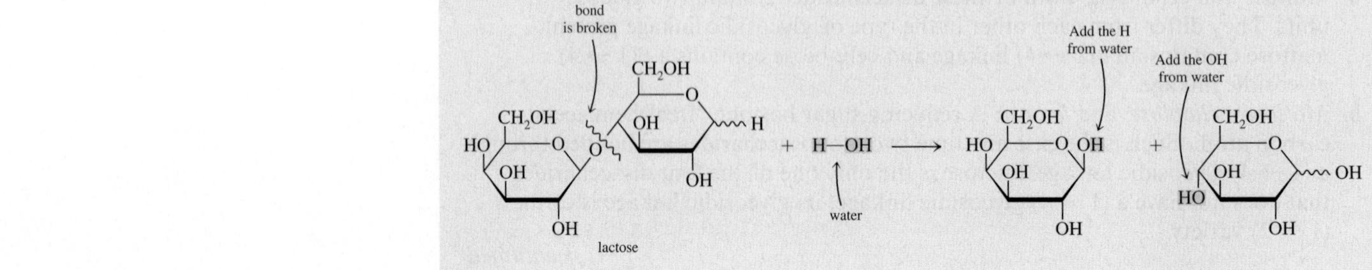

Practice Exercise 7.9

Draw the structural formulas for the products formed when the following disaccharide is hydrolyzed.

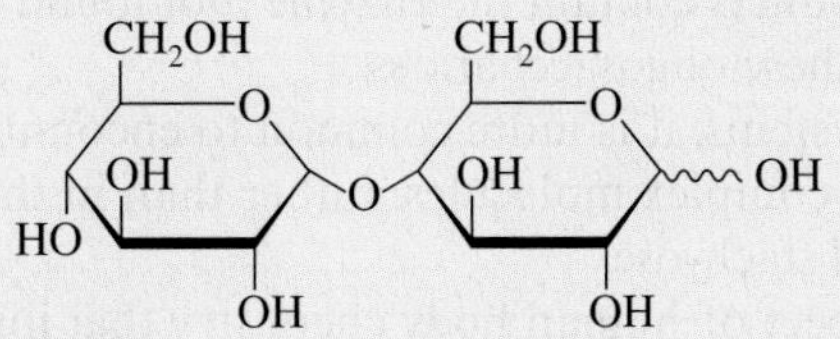

Answers:

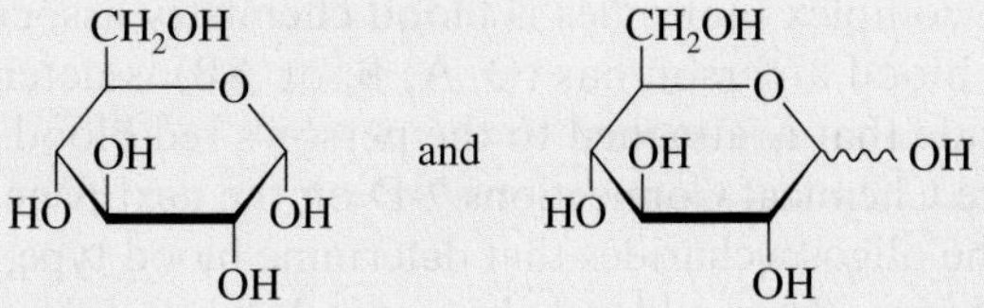

7.14 Oligosaccharides

Oligosaccharides are carbohydrates that contain three to ten monosaccharide units bonded to each other via glycoside linkages (Section 7.3). Two naturally occurring oligosaccharides found in onions, cabbage, broccoli, brussel sprouts, whole wheat, and all types of beans are the trisaccharide raffinose and the tetrasaccharide stachyose. Raffinose's monosaccharide components are galactose, glucose, and fructose. Stachyose's structure differs from that of raffinose in that an additional galactose unit is present. Detailed structures for these two oligosaccharides are given in Figure 7.22. Note the presence in both structures of two different types of glycosidic linkages—$\alpha(1 \rightarrow 6)$ and $\alpha,\beta(1 \rightarrow 2)$ linkages.

Humans lack the digestive enzymes necessary to metabolize either raffinose or stachyose. Hence these oligosaccharides, when ingested in food, pass undigested

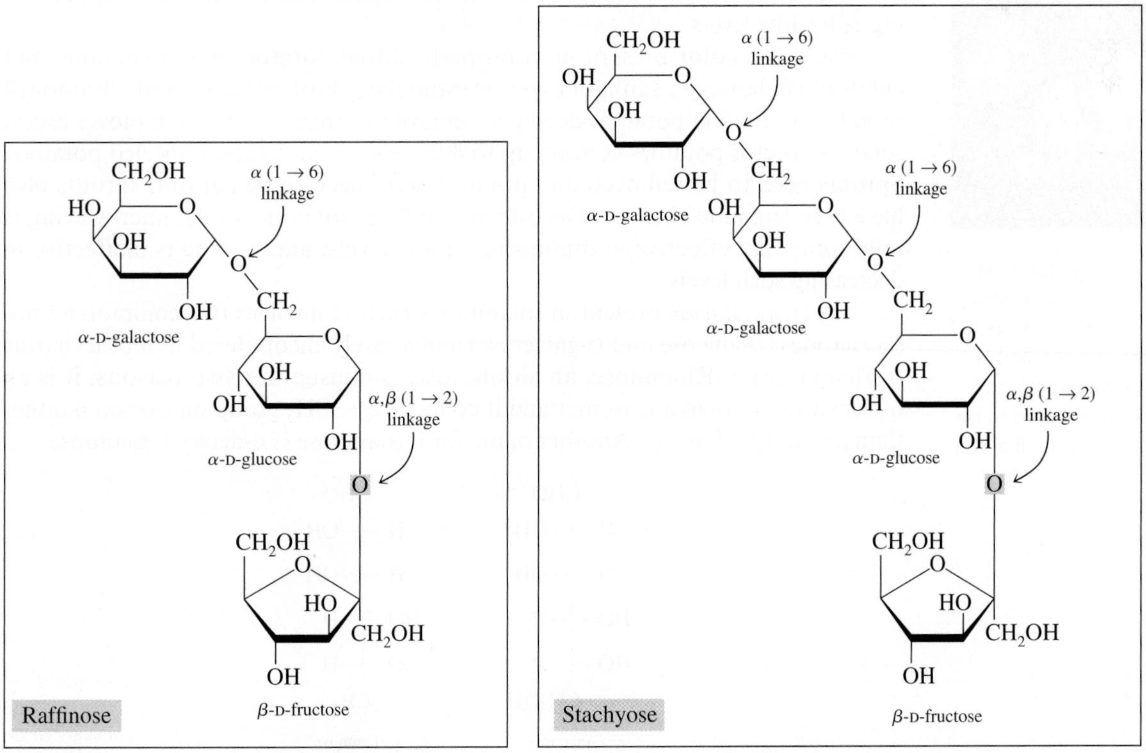

Figure 7.22 Structural formulas for the trisaccharide raffinose and the tetrasaccharide stachyose.

© iStockphoto.com/FotografiaBasica

Figure 7.23 In humans, intestinal bacteria action on the undigestable raffinose and stachyose present in beans produces gaseous products that can cause discomfort and flatulence.

into the large intestine, where bacteria act upon them. This bacterial action usually produces discomfort and flatulence (gas). Taking a preparation such as Beano with a meal heavy in beans (Figure 7.23) can reduce the amount of flatulence produced. Beano and related products contain the enzyme (not found in humans) that facilitates the digestion of these oligosaccharides.

In biochemical systems, it is more common to encounter oligosaccharides as components of more complex molecules, rather than in the "free" state as is the case with raffinose and stachyose.

An important aspect of human body chemistry that involves an oligosaccharide attachment to more complex molecules is blood chemistry associated with blood type. The type of blood a person has (O, A, B, or AB) is determined by the type of oligosaccharide that is attached to the person's red blood cells. The focus on relevancy feature Chemical Connections 7-D on the next page gives further information about the oligosaccharides that determine blood type, as well as about the role that blood type plays in blood chemistry. Interestingly, one of the monosaccharides present in the oligosaccharides associated with blood type is a 6-deoxy-L-monosaccharide.

Solanine, a compound found in the potato plant, is another example of an oligosaccharide-containing "complex" molecule. The framework for solanine's structure is a large multi-ring amine system (an alkaloid; Section 6.11) to which a trisaccharide is attached. A major function of the trisaccharide entity, with its many hydroxy groups, is to impart solubility to the compound.

Many plants, including the potato plant, produce toxins as a defense against insects and predators. Solanine is the potato plant's toxin. The small amount of solanine present in properly stored potatoes is not dangerous and actually contributes to the flavor of potatoes. (When excessive amounts of solanine are present, potatoes taste bitter.)

Solanine amounts in potatoes increase when potatoes sprout and when they are exposed to sunlight. (Potato farmers store their product in the dark to minimize solanine production.) Improperly stored potatoes "green"; that is, a green color develops in their skin and the flesh just under the skin (Figure 7.24). Such greening is a warning that solanine levels are higher than normal, even approaching unhealthy levels.

© Nigel Cattlin/Alamy

Figure 7.24 The presence of an underlying green color in the skin of potatoes denotes the presence of chlorophyll (the green color), as well as the toxin solanine (which is unseen). Such potatoes need to be deeply peeled (to remove all of the green color) before use.

The green color present in improperly stored potatoes is not solanine, but chlorophyll; however, sunlight exposure stimulates both solanine and chlorophyll production. Peeling potatoes deeply to remove the green color also removes excess solanine; peeled potatoes contain up to 80% less solanine than unpeeled potatoes. Sprouts need to be removed and potato "eyes" need to be cut out; sprouts also have high solanine content. Deep-frying reduces solanine levels; microwaving is only somewhat effective in diminishing these levels; and boiling is ineffective in decreasing such levels.

The trisaccharide present in solanine's structure involves two common monosaccharides (D-glucose and D-galactose) and a rarely encountered monosaccharide (L-rhamnose). L-Rhamnose, an aldohexose, is unusual for two reasons: it is an L-isomer rather than a D-isomer and it contains a $—CH_3$ group on carbon 6 rather than a $—CH_2OH$ group. Another name for L-rhamnose is 6-deoxy-L-mannose.

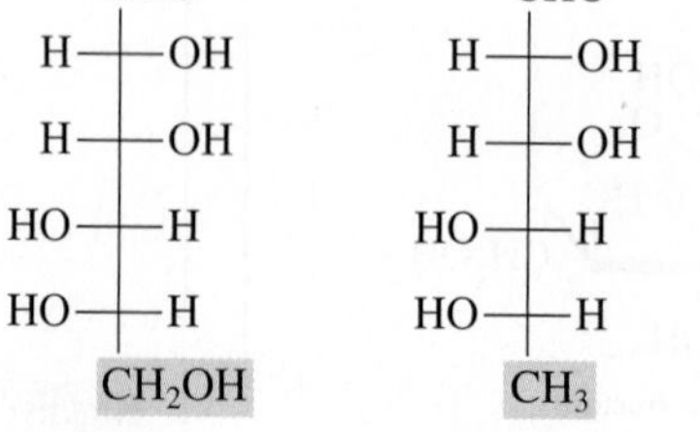

L-mannose

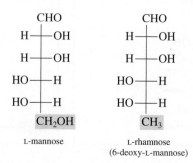

L-rhamnose (6-deoxy-L-mannose)

A complete structural representative for solanine is given in Figure 7.25.

CHEMICAL CONNECTIONS 7-D

Blood Types and Oligosaccharides

Human blood is classified into four types: A, B, AB, and O. If a blood transfusion is necessary and the patient's own blood is not available, the donor's blood must be matched to that of the patient. Blood of one type cannot be given to a recipient with blood of another type unless the two types are compatible. A transfusion of the wrong blood type can cause the blood cells to form clumps, a potentially fatal reaction. The following table shows compatibility relationships. People with type O blood are universal donors, and those with type AB blood are universal recipients.

Human Blood Group Compatibilities

	Recipient Blood Type			
Donor blood type	**A**	**B**	**AB**	**O**
A	+	−	+	−
B	−	+	+	−
AB	−	−	+	−
O	+	+	+	+

+ = compatible; − = incompatible

In the United States, sampling studies show that type O is the most common type of blood, with type A the second most common. There is a definite correlation between ethnicity and blood type, as shown in the following table.

Blood Types in the United States

	Caucasian	**African**	**Hispanic**	**Asian**
O	45%	51%	57%	40%
A	40%	26%	33%	27%
B	11%	19%	11%	26%
AB	4%	4%	2%	7%

Percentages may not add to 100 because of rounding.

The biochemical basis for the various types of blood involves oligosaccharide molecules that are attached to the plasma membrane of red blood cells. These oligosaccharide attachments, which are called biochemical markers, are of three different formulations: one is a tetrasaccharide and the other two are pentasaccharides.

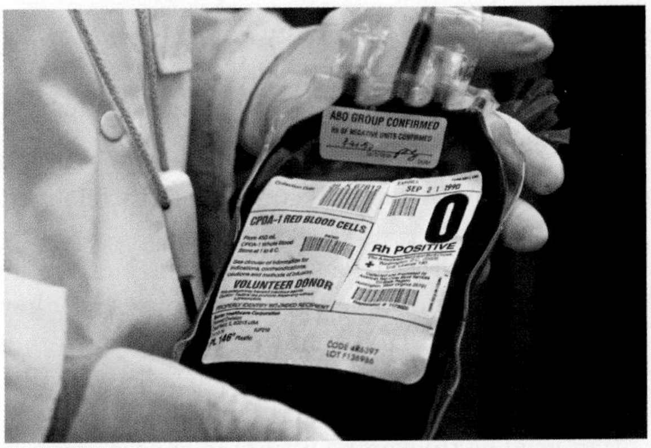

Matt Meadows/Peter Arnold Inc./Photolibrary

A unit of blood obtained from a blood bank.

Four monosaccharides contribute to the make-up of the oligosaccharide "marking system." One is the simple monosaccharide D-galactose and the other three are monosaccharide derivatives. Two of these are *N*-acetyl amino derivatives (Section 7.12), those of D-glucose and D-galactose. The third is L-fucose (6-deoxy-L-galactose), an L-galactose derivative in which the oxygen atom at carbon 6 has been removed (converting the —CH_2OH group to a —CH_3 group). The L configuration of this derivative is unusual in that L-monosaccharides are seldom found in the human body.

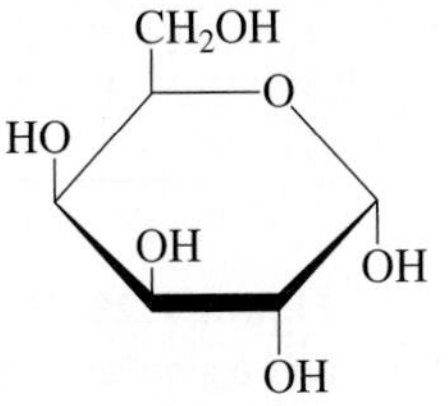

α-D-Galactose

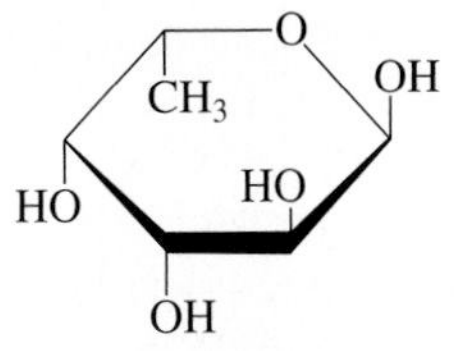

α-L-Fucose (α-6-Deoxy-L-galactose)

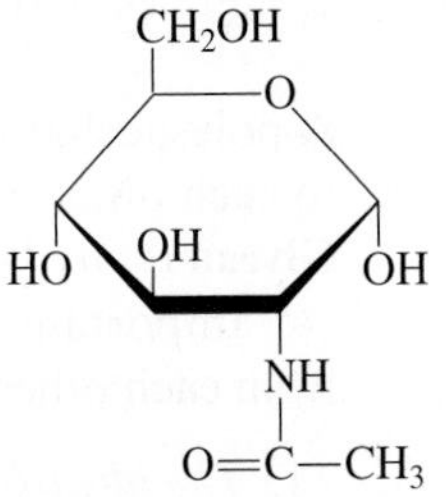

α-*N*-Acetyl-D-glucosamine

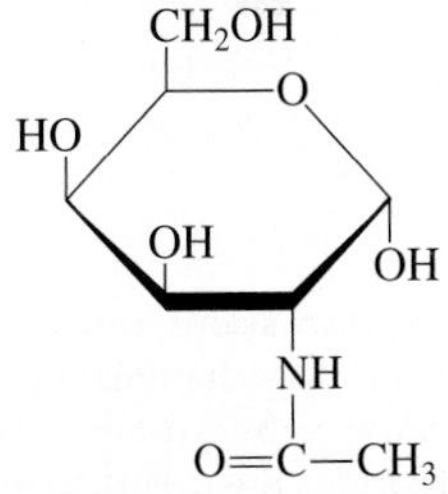

α-*N*-Acetyl-D-galactosamine

The arrangement of these monosaccharides in the biochemical marker determines blood type.

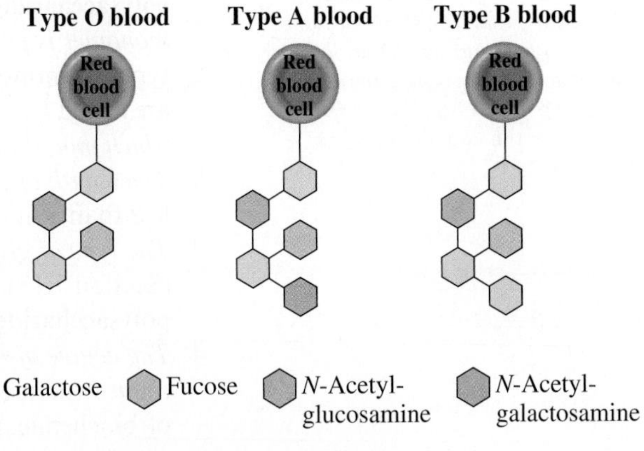

Note that all three of the oligosaccharide markers have a common four-monosaccharide sequence in their structure.

Galactose — *N*-acetylglucosamine — Galactose — Fucose

The absence or presence of a fifth monosaccharide (attached to the second galactose) determines blood type. Type O blood lacks a fifth monosaccharide unit. Type A blood has *N*-acetylgalactosamine as a fifth unit. Type B blood has galactose as a fifth unit. Type AB blood contains both type A and type B markers.

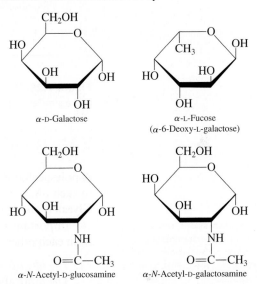

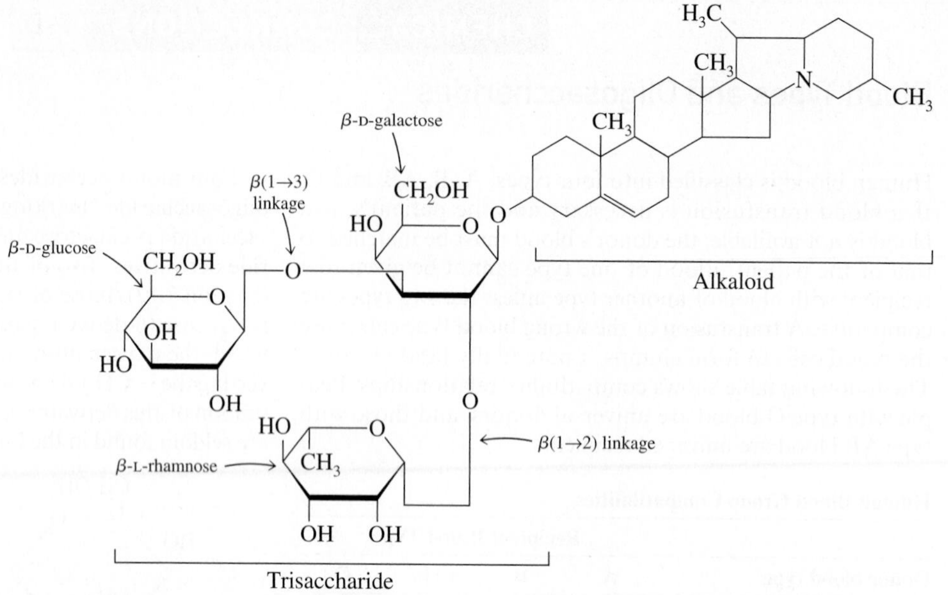

Figure 7.25 The structure of the potato toxin solanine has a large multi-ring alkaloid part and a smaller oligosaccharide part.

7.15 General Characteristics of Polysaccharides

A polysaccharide is a polymer that contains many monosaccharide units bonded to each other by glycosidic linkages. Polysaccharides are often also called *glycans.* **Glycan** *is an alternate name for a polysaccharide.*

In nutrition discussions, monosaccharides and disaccharides are called *simple carbohydrates,* and polysaccharides are called *complex carbohydrates.*

When some vegetables, such as peas or corn, age too much (become overly mature), some of their sugars (mono- and disaccharides) are converted to polysaccharides (starch; Section 7.16). This conversion makes the vegetables taste less sweet.

Important parameters that distinguish various polysaccharides (or glycans) from each other are:

1. *The identity of the monosaccharide repeating unit(s) in the polymer chain.* The more abundant polysaccharides in nature contain only one type of monosaccharide repeating unit. Such polysaccharides, including starch, glycogen, cellulose, and chitin, are examples of *homopolysaccharides.* A **homopolysaccharide** *is a polysaccharide in which only one type of monosaccharide monomer is present.* Polysaccharides whose structures contain two or more types of monosaccharide monomers, including hyaluronic acid and heparin, are called *heteropolysaccharides.* A **heteropolysaccharide** *is a polysaccharide in which more than one (usually two) type of monosaccharide monomer is present.*
2. *The length of the polymer chain.* Polysaccharide chain length can vary from less than a hundred monomer units to up to a million monomer units.
3. *The type of glycosidic linkage between monomer units.* As with disaccharides (Section 7.13), several different types of glycosidic linkages are encountered in polysaccharide structures.
4. *The degree of branching of the polymer chain.* The ability to form *branched-chain* structures distinguishes polysaccharides from the other two major types of biochemical polymers: proteins (Chapter 9) and nucleic acids (Chapter 11), which occur only as linear (unbranched) polymers.

Figure 7.26 illustrates important general structural considerations relative to polysaccharides.

Unlike monosaccharides and most disaccharides, polysaccharides are not sweet and do not test positive in Tollens and Benedict's solutions. They have limited water solubility because of their size. However, the —OH groups present can individually become hydrated by water molecules. The result is usually a thick colloidal suspension of the polysaccharide in water. Polysaccharides, such as flour and cornstarch, are often used as thickening agents in sauces, desserts, and gravy.

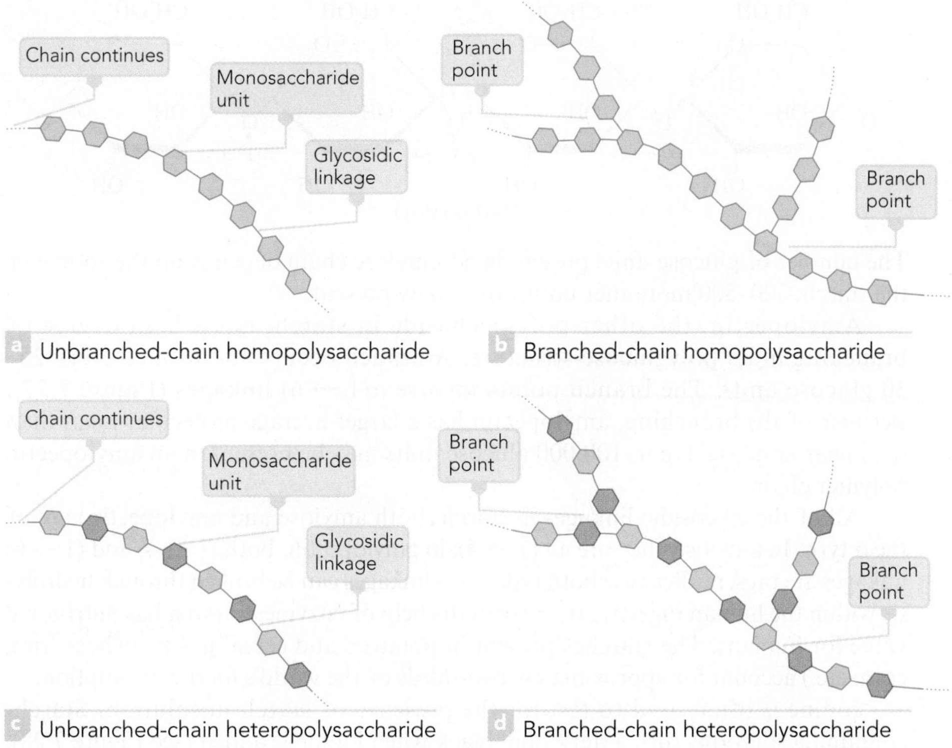

Figure 7.26 The polymer chain of a polysaccharide may be unbranched or branched. The monosaccharide monomers in the polymer chain may all be identical, or two or more kinds of monomers may be present.

Although there are many naturally occurring polysaccharides of biochemical importance, further polysaccharide discussion will be limited to six of them: starch, glycogen, cellulose, chitin, hyaluronic acid, and heparin. Starch and glycogen are examples of *storage polysaccharides* (Section 7.17); cellulose and chitin are *structural polysaccharides* (Section 7.18); and hyaluronic acid and heparin are *acidic polysaccharides* (Section 7.19).

7.16 Storage Polysaccharides

A **storage polysaccharide** *is a polysaccharide that is a storage form for monosaccharides and is used as an energy source in cells.* In cells, monosaccharides are stored in the form of polysaccharides rather than as individual monosaccharides in order to lower the osmotic pressure within cells. Osmotic pressure depends on the number of *individual* molecules present. Incorporating many monosaccharide molecules into a single polysaccharide molecule results in a dramatic reduction in molecular numbers. The most important storage polysaccharides are starch (in plant cells) and glycogen (in animal and human cells).

Starch

Starch is a homopolysaccharide containing only glucose monosaccharide units. It is the energy-storage polysaccharide in plants. If excess glucose enters a plant cell, it is converted to starch and stored for later use. When the cell cannot get enough glucose from outside the cell, it hydrolyzes starch to release glucose.

Two different polyglucose polysaccharides can be isolated from most starches: amylose and amylopectin. *Amylose,* a straight-chain glucose polymer, usually accounts for 15%–20% of the starch; *amylopectin,* a branched glucose polymer, accounts for the remaining 80%–85% of the starch.

In amylose's non-branched structure, the glucose units are connected by $\alpha(1 \rightarrow 4)$ glycosidic linkages.

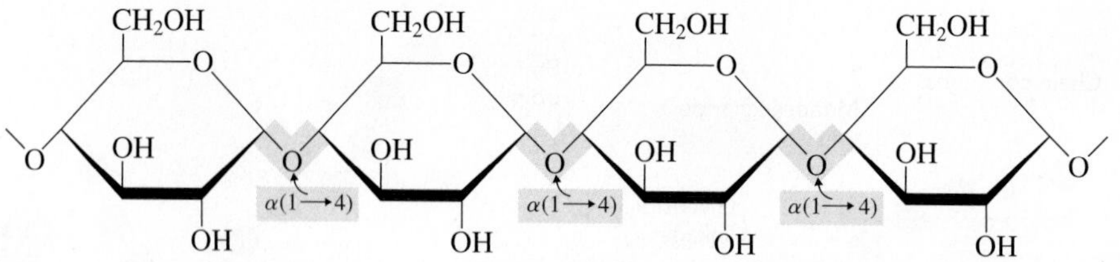

Starch (amylose)

Amylopectin is digested more rapidly than amylose. Digestive enzymes exert their effect primarily at the end (terminal) glucose units in a glucose chain. A branched-chain structure (amylopectin) has more "ends" than a straight-chain structure (amylose).

The number of glucose units present in an amylose chain depends on the source of the starch; 300–500 monomer units are usually present.

Amylopectin, the other polysaccharide in starch, has a high degree of branching in its polyglucose structure. A branch occurs about once every 25–30 glucose units. The branch points involve $\alpha(1 \rightarrow 6)$ linkages (Figure 7.27). Because of the branching, amylopectin has a larger average molecular mass than the linear amylose. Up to 100,000 glucose units may be present in an amylopectin polymer chain.

All of the glycosidic linkages in starch (both amylose and amylopectin) are of the α type. In amylose, they are all $(1 \rightarrow 4)$; in amylopectin, both $(1 \rightarrow 4)$ and $(1 \rightarrow 6)$ linkages are present. Because both types of α linkages can be broken through hydrolysis within the human digestive tract (with the help of enzymes), starch has nutritional value for humans. The starches present in potatoes and cereal grains (wheat, rice, corn, etc.) account for approximately two-thirds of the world's food consumption.

Iodine is often used to test for the presence of starch in solution. Starch-containing solutions turn a dark blue-black when iodine is added (see Figure 7.28). As starch is broken down through acid or enzymatic hydrolysis to glucose monomers, the blue-black color disappears.

Glycogen

Glycogen, like starch, is a polysaccharide containing only glucose units. It is the glucose storage polysaccharide in humans and animals. Its function is thus similar to that of starch in plants, and it is sometimes referred to as *animal starch.* Liver cells and muscle cells are the storage sites for glycogen in humans.

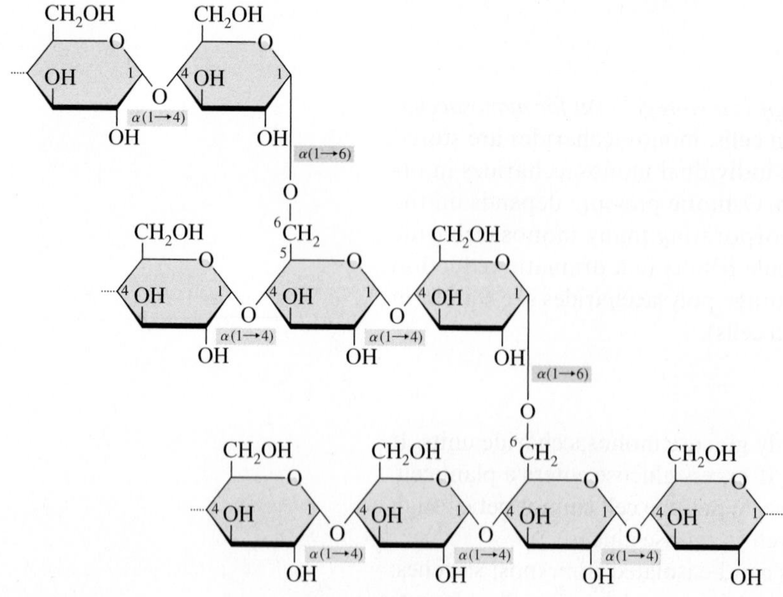

a Molecular structure of amylopectin.

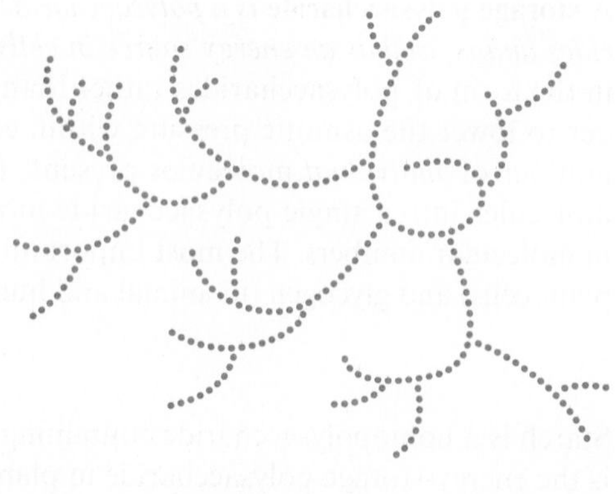

An $\alpha(1 \rightarrow 6)$ linkage is present in the amylopectin structure at each branch point.

b An overview of the branching that occurs in the amylopectin structure. Each dot is a glucose unit

Figure 7.27 Two perspectives on the structure of the polysaccharide amylopectin.

© Cengage Learning
© Cengage Learning

Figure 7.28 Use of iodine to test for starch. Starch-containing solutions and foods turn dark blue-black when iodine is added.

The glucose polymers amylose, amylopectin, and glycogen compare as follows in molecular size and degree of branching.

Amylose:	Up to 1000 glucose units; no branching
Amylopectin:	Up to 100,000 glucose units; branch points every 25–30 glucose units
Glycogen:	Up to 1,000,000 glucose units; branch points every 8–12 glucose units

Glycogen has a structure similar to that of amylopectin; all glycosidic linkages are of the α type, and both $(1 \rightarrow 4)$ and $(1 \rightarrow 6)$ linkages are present. Glycogen and amylopectin differ in the number of glucose units between branches and in the total number of glucose units present in a molecule. Glycogen is about three times more highly branched than amylopectin, and it is much larger, with up to 1,000,000 glucose units present.

When excess glucose is present in the blood (normally from eating too much starch), the liver and muscle tissue convert the excess glucose to glycogen, which is then stored in these tissues. Whenever the glucose blood level drops (from exercise, fasting, or normal activities), some stored glycogen is hydrolyzed back to glucose. These two opposing processes are called *glycogenesis* and *glycogenolysis,* the formation and decomposition of glycogen, respectively.

$$\text{Glucose} \underset{\text{Glycogenolysis}}{\overset{\text{Glycogenesis}}{\rightleftharpoons}} \text{glycogen}$$

The amount of stored glycogen in the human body is relatively small. Muscle tissue is approximately 1% glycogen, liver tissue 2%–3%. However, this amount is sufficient to take care of normal-activity glucose demands for about 15 hours. During strenuous exercise, glycogen supplies can be exhausted rapidly. At this point, the body begins to oxidize fat as a source of energy.

Many marathon runners eat large quantities of starch foods the day before a race. This practice, called *carbohydrate loading,* maximizes body glycogen reserves.

Glycogen is an ideal storage form for glucose. The large size of these macromolecules prevents them from diffusing out of cells. Also, conversion of glucose to glycogen reduces osmotic pressure. Cells would burst because of increased osmotic pressure if all of the glucose in glycogen were present in cells in free form. High concentrations of glycogen in a cell sometimes precipitate or crystallize into *glycogen granules.* These granules are discernible in photographs of cells under electron microscope magnification (Figure 7.29).

7.17 Structural Polysaccharides

A **structural polysaccharide** *is a polysaccharide that serves as a structural element in plant cell walls and animal exoskeletons.* Two of the most important structural polysaccharides are cellulose and chitin. Both are homopolysaccharides.

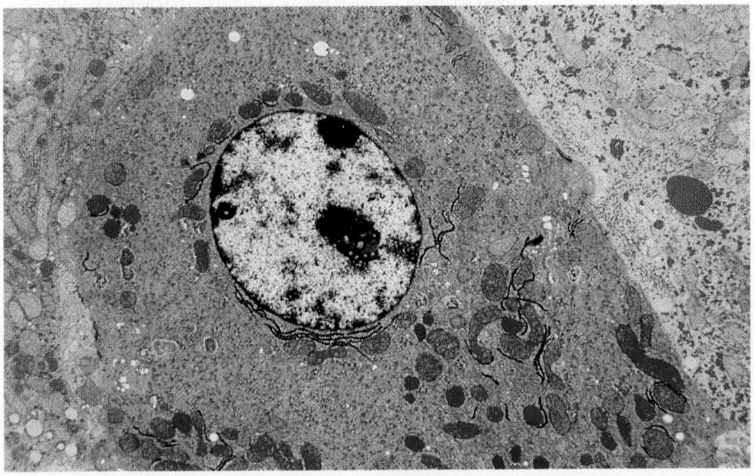
Dr. Dennis Kunkel/Phototake

Figure 7.29 The small, dense particles within this electron micrograph of a liver cell are glycogen granules.

Cellulose

Cellulose, the structural component of plant cell walls, is the most abundant naturally occurring polysaccharide. The "woody" portions of plants—stems, stalks, and trunks—have particularly high concentrations of this fibrous, water-insoluble substance.

Like amylose, cellulose is an unbranched glucose polymer. The structural difference between cellulose and amylose, which gives them completely different properties, is that the glucose residues present in cellulose have a beta-configuration whereas the glucose residues in amylose have an alpha-configuration. The glycosidic linkages in cellulose are therefore $\beta(1 \rightarrow 4)$ linkages rather than $\alpha(1 \rightarrow 4)$ linkages.

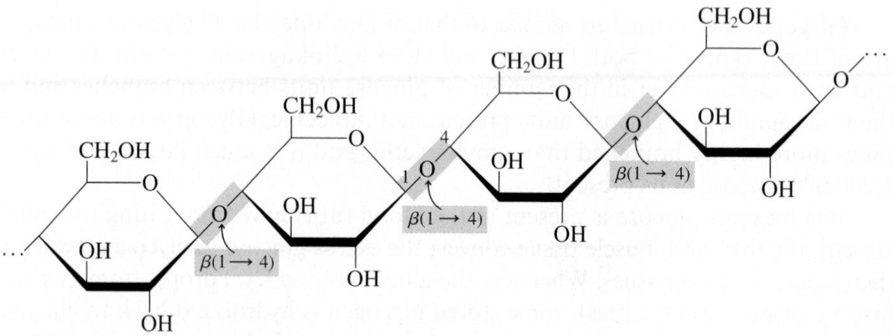

This difference in glycosidic linkage type causes cellulose and amylose to have different molecular shapes. Amylose molecules tend to have spiral-like structures, whereas cellulose molecules tend to have linear structures. The linear (straight-chain) cellulose molecules, when aligned side by side, become water-insoluble fibers because of inter-chain hydrogen bonding involving the numerous hydroxyl groups present.

Typically, cellulose chains contain about 5000 glucose units, which gives macromolecules with molecular masses of about 900,000 amu. Cotton is almost pure cellulose (95%), and wood is about 50% cellulose.

Even though it is a glucose polymer, cellulose is not a source of nutrition for human beings. Humans lack the enzymes capable of catalyzing the hydrolysis of $\beta(1 \rightarrow 4)$ linkages in cellulose. Even grazing animals lack the enzymes necessary for cellulose digestion. However, the intestinal tracts of animals such as horses, cows, and sheep contain bacteria that produce *cellulase,* an enzyme that can hydrolyze cellulose $\beta(1 \rightarrow 4)$ linkages and produce free glucose from cellulose. Thus grasses and other plant materials are a source of nutrition for grazing animals. The intestinal tracts of termites contain the same microorganisms, which enable termites to use wood as their source of food. Microorganisms in the soil can also metabolize cellulose, which makes possible the biodegradation of dead plants.

Despite its nondigestibility, cellulose is still an important component of a balanced diet. It serves as dietary fiber. Dietary fiber provides the digestive tract with "bulk" that helps move food through the intestinal tract and facilitates the excretion of solid wastes. Cellulose readily absorbs water, leading to softer stools and frequent bowel action. Links have been found between the length of time stools spend in the colon and possible colon problems.

High-fiber food may also play a role in weight control. Obesity is not seen in parts of the world where people eat large amounts of fiber-rich foods (Figure 7.30). Many of the weight-loss products on the market are composed of bulk-inducing fibers such as methylcellulose.

Some dietary fibers bind lipids such as cholesterol (Section 8.9) and carry them out of the body with the feces. This lowers blood lipid concentrations and, possibly, the risk of heart and artery disease.

About 25–35 grams of dietary fiber daily is a desirable intake. This is two to three times higher than the average intake of Americans.

Figure 7.30 A sandwich such as this is high in dietary fiber; that is, it is a cellulose-rich "meal."

The word *chitin* is pronounced "kye-ten"; it rhymes with *Titan.*

CHEMISTRY AT A GLANCE: Types of Glycosidic Linkages for Common Glucose-Containing Di- and Polysaccharides

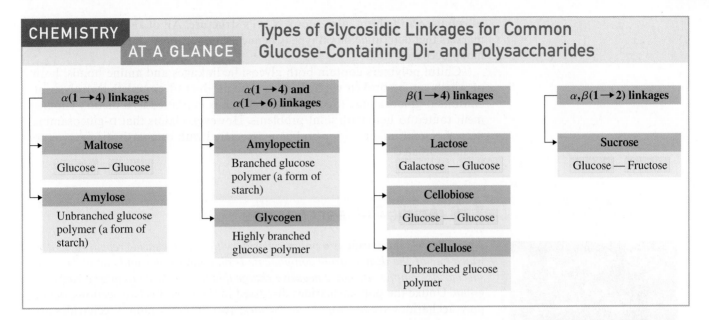

- **α(1→4) linkages**
  - **Maltose**: Glucose — Glucose
  - **Amylose**: Unbranched glucose polymer (a form of starch)
- **α(1→4) and α(1→6) linkages**
  - **Amylopectin**: Branched glucose polymer (a form of starch)
  - **Glycogen**: Highly branched glucose polymer
- **β(1→4) linkages**
  - **Lactose**: Galactose — Glucose
  - **Cellobiose**: Glucose — Glucose
  - **Cellulose**: Unbranched glucose polymer
- **α,β(1→2) linkages**
  - **Sucrose**: Glucose — Fructose

The Chemistry at a Glance feature above summarizes the types of glycosidic linkages present in commonly encountered glucose-containing di- and polysaccharides. Included in the summary is the $\beta(1 \rightarrow 4)$ linkage present in cellulose as well as in the disaccharides cellobiose and lactose (Section 7.13).

Chitin

Cellulose is the most abundant naturally occurring polysaccharide. Chitin is the second most abundant naturally occurring polysaccharide. Its function is to give rigidity to the exoskeletons of crabs, lobsters, shrimp, insects, and other arthropods (Figure 7.31). It also has been found in the cell walls of fungi.

Structurally, chitin is identical to cellulose, except the monosaccharide present is the glucose derivative *N*-acetyl-D-glucosamine (NAG) rather than D-glucose itself. The structure of NAG was previously encountered in Section 7.12 when amino sugar formation was discussed. Figure 7.32, which contrasts the structures

Figure 7.31 Chitin, a linear $\beta(1 \rightarrow 4)$ polysaccharide, produces the rigidity in the exoskeletons of crabs and other arthropods.

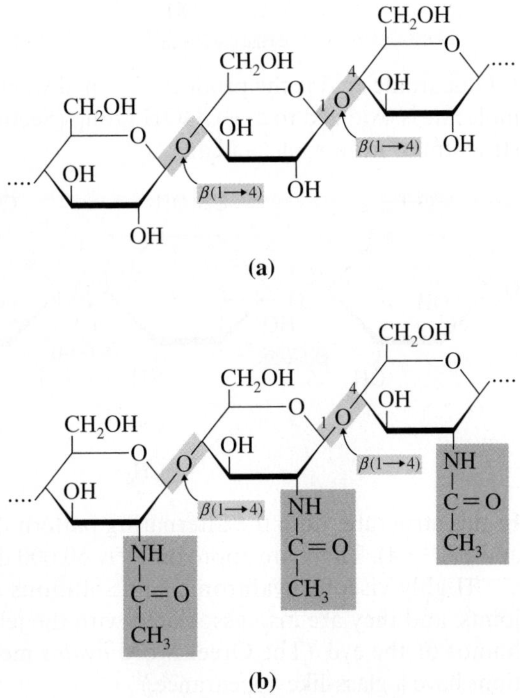

Figure 7.32 The structures of cellulose (a) and chitin (b). In both substances, all glycosidic linkages are of the $\beta(1 \rightarrow 4)$ type.

of cellulose and chitin, also shows NAG's structure. All of the oligosaccharide markers that determine blood type (Chemical Connections 7-D) contain an NAG monosaccharide unit.

Chitin polymers contain both glycosidic linkages and amine bonds, both of which can be broken via hydrolysis. The product of complete hydrolysis of chitin is D-glucosamine (Section 7.12), which is marketed as a dietary supplement touted to help with joint problems. However, claims that D-glucosamine decreases joint inflammation and pain associated with osteoarthritis have yet to be substantiated.

7.18 Acidic Polysaccharides

J.M Barey/Vandystadt/Photo Researchers

Figure 7.33 Acidic polysaccharides associated with the connective tissue of joints give hurdlers such as these the flexibility needed to accomplish their task.

An **acidic polysaccharide** *is a polysaccharide with a disaccharide repeating unit in which one of the disaccharide components is an amino sugar and one or both disaccharide components has a negative charge due to a sulfate group or a carboxyl group.* Unlike the polysaccharides discussed in the previous two sections, acidic polysaccharides are *heteropolysaccharides;* two different monosaccharides are present in an alternating pattern. Acidic polysaccharides are involved in a variety of cellular functions and tissues (Figure 7.33). Two of the most well-known acidic polysaccharides are hyaluronic acid and heparin, both of which have unbranched-chain structures.

Hyaluronic Acid

The structure of hyaluronic acid contains alternating residues of *N*-acetyl-β-D-glucosamine (NAG) and D-Glucuronate. NAG was previously encountered in the structure of chitin. D-glucuronate is the carboxylate ion (Section 5.7) formed when D-glucuronic acid loses its acidic hydrogen atom.

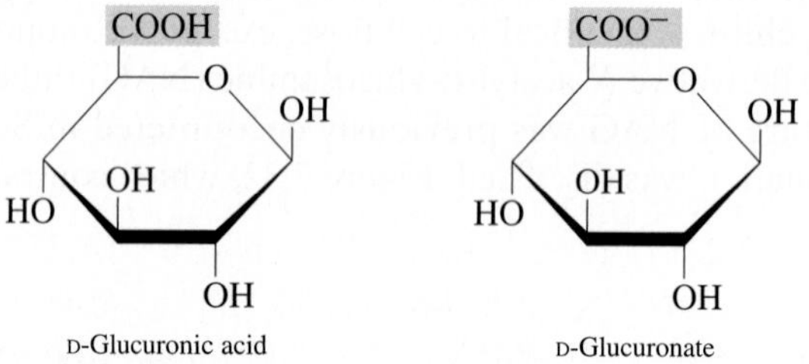

D-Glucuronic acid is the product obtained when the —CH_2OH group of a glucose molecule is oxidized to a —COOH group (Section 7.12). Two repeating units in the structure of hyaluronic acid are

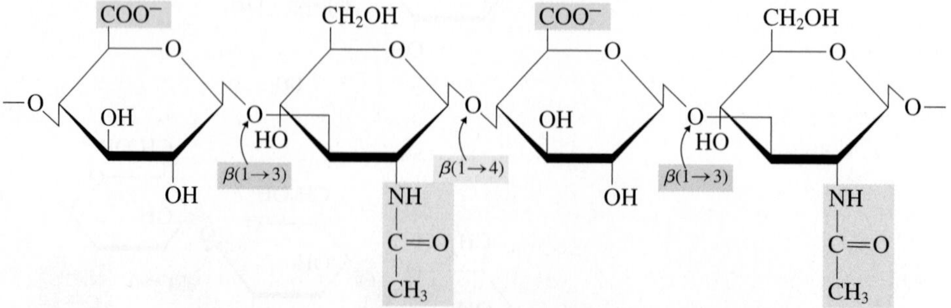

In this structure, note the alternating pattern of glycosidic bond types, $\beta(1 \rightarrow 3)$ and $\beta(1 \rightarrow 4)$. There are approximately 50,000 disaccharide units per chain.

Highly viscous hyaluronic acid solutions serve as lubricants in the fluid of joints, and they are also associated with the jelly-like consistency of the vitreous humor of the eye. (The Greek word *hyalos* means "glass"; hyaluronic acid solutions have a glass-like appearance.)

Heparin

Heparin is a small highly-sulfated polysaccharide with only 15–90 disaccharide residues per chain. The monosaccharides present in heparin's disaccharide repeating unit are a sulfate derivative of D-glucuronate (D-glucuronate-2-sulfate) and a doubly sulfated derivative of D-glucosamine (*N*-sulfo-D-glucosamine-6-sulfate). Both of these monosaccharide derivatives contain two negatively charged acidic groups.

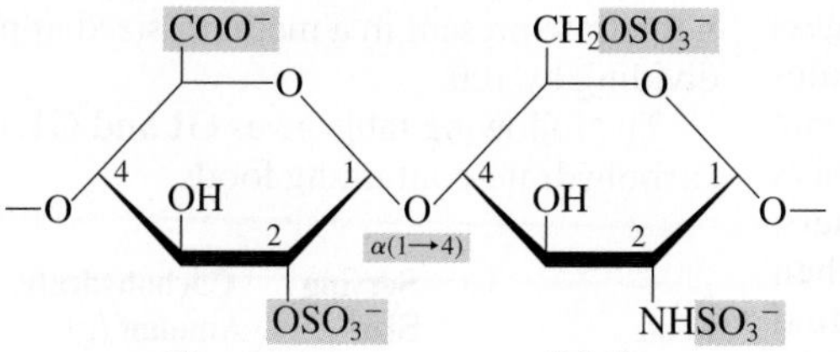

D-Glucuronate-2-sulfate
N-Sulfo-D-glucosamine-6-sulfate

Heparin is a blood anticoagulant. It is naturally present in mast cells and is released at the site of tissue injury. It prevents the formation of clots in the blood and retards the growth of existing clots within the blood. It does not, however, break down clots that have already formed.

Pharmaceutical-grade heparin is applied as an anticoagulant to the interior/exterior surface of external objects that come in contact with blood (test tubes, kidney dialysis machine surfaces, prosthetic implant materials) to prevent the blood from clotting. The source for pharmaceutical heparin is intestinal or lung tissue of slaughter-house animals (pigs and cows).

Mast cells are part of the body's immune system. They play an important role in wound healing and defense against pathogens. They are more abundant in tissues that commonly come into contact with the outside world (the skin, mucosa of the lungs, digestive tract, mouth, and nose) than in other tissues.

7.19 Dietary Considerations and Carbohydrates

Foods high in carbohydrate content constitute over 50% of the diet of most people of the world—rice in Asia, corn in South America, cassava (a starchy root vegetable) in parts of Africa, the potato and wheat in North America, and so on. Current nutritional recommendations support such a situation; a balanced diet should ideally be about 60% carbohydrate.

Nutritionists usually subdivide dietary carbohydrates into the categories *simple* and *complex*. A **simple carbohydrate** *is a dietary monosaccharide or dietary disaccharide.* Simple carbohydrates are usually sweet to the taste and are commonly referred to as sugars (Section 7.8). A **complex carbohydrate** *is a dietary polysaccharide.* The main complex carbohydrates are starch and cellulose, substances not generally sweet to the taste.

Simple carbohydrates provide 20% of the energy in the U.S. diet. Half of this energy content comes from *natural sugars* and the other half from *refined sugars* added to foods. A **natural sugar** *is a sugar naturally present in whole foods.* Milk and fresh fruit are two important sources of natural sugars. A **refined sugar** *is a sugar that has been separated from its plant source.* Sugar beets and sugar cane are major sources of refined sugars. Despite claims to the contrary, refined sugars are chemically and structurally no different from the sugars naturally present in foods. The only difference is that the refined sugar is in a pure form, whereas natural sugars are part of mixtures of substances obtained from a plant source.

Refined sugars are often said to provide *empty calories* because they provide energy but few other nutrients. Natural sugars, on the other hand, are accompanied by nutrients. A tablespoon of sucrose (table sugar) provides 50 calories of energy just as a small orange does. The small orange, however, also supplies vitamin C, potassium, calcium, and fiber; table sugar provides no other nutrients.

The major dietary source for complex carbohydrates in the U.S. diet is grains, a source of both starch and fiber as well as of protein, vitamins, and minerals. The pulp of a potato provides starch, and the skin provides fiber. Vegetables such as broccoli and green beans are low in starch but high in fiber.

CHEMICAL CONNECTIONS 7-E

Glycemic Response, Glycemic Index, and Glycemic Load

Shortly after ingestion of carbohydrate-containing foods, the body's blood-glucose level increases. This change in blood-glucose level that comes from carbohydrate ingestion is called the *glycemic response.* Different carbohydrate-containing foods elicit different glycemic responses; some foods are more easily and more rapidly digested than others and more glucose is obtained from some foods than others. For some foods blood-glucose levels rise quickly and then decrease quickly, whereas for other foods a more gradual increase and decrease in blood-glucose levels occurs. The accompanying graph shows such variation in glycemic response to food.

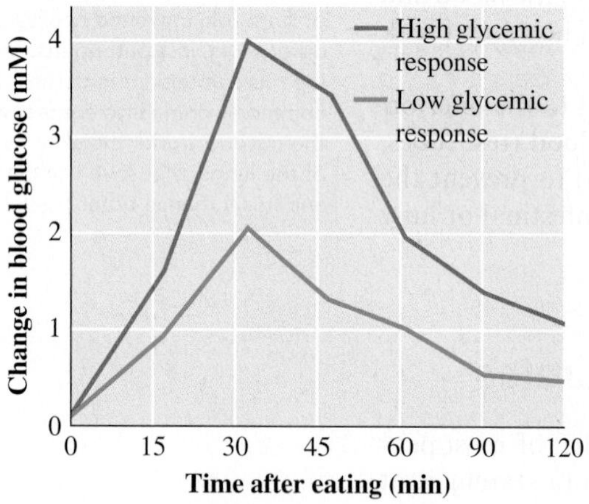

Blood-glucose levels maximize at higher values for some carbohydrate-containing foods than for others. The time required for maximization also varies from food to food.

Concern about how blood-glucose levels fluctuate with time after ingestion of carbohydrates is an important consideration, with both short- and long-term consequences, for many individuals, particularly those with diabetes. A rating system called the *glycemic index* (GI) categorizes foods according to glycemic response. The key parameter in this rating system is the ratio of blood-glucose change for a given food compared to a standard (which is usually pure glucose or white bread). Representative GI values for carbohydrate-containing foods, where glucose is the standard with an assigned value of 100, are 46 for orange juice, 48 for baked beans, 72 for short-grain white rice, and 85 for a baked potato.

A major problem associated with GI values is that they are all based on a serving size that provides 50 grams of carbohydrate. A 50-gram base size does not, however, reflect the amount of carbohydrate typically present in a serving of food. The *glycemic load* (GL) is an alternate rating system that takes into account both the glycemic index for and the actual amount of carbohydrate present in a serving of a specific food. A GL value is calculated by multiplying the GI for a specific food by the grams of carbohydrate in one serving and dividing by 100 (to give a percentage).

The GL value for an apple is 8. This value is obtained by multiplying the GI value for an apple, 38, by the 22 g of carbohydrate present in a medium-sized apple (one serving) and dividing by 100.

The following table gives GI and GL data for ten selected carbohydrate-containing foods.

Food	Serving Size	Carbohydrate Amount (g)	Glycemic Index (GI)*	Glycemic Load (GL)
Baked potato	1 cup	57	85	48
White rice	1 cup	53	72	38
White bread	1 slice	10	70	7
Brown rice	1 cup	46	55	25
Banana	1 medium	29	55	16
Potato chips	1 oz	15	54	8
Cooked carrots	1 cup	16	49	8
Chocolate	1 oz	18	49	9
Orange juice	1 cup	26	46	13
Apple	1 medium	22	38	8

* GI values are based on a value of 100 for glucose (the standard).

** GI food categories: low (below 55), intermediate (between 55 and 69), high (over 69)

*** GL food categories: low (below 10), intermediate (between 10 and 19), high (over 19)

Correlation between GI and GL values for various foods varies from good to poor. Note, for example, that white bread has a high GI value and a low GL value, and a banana has a borderline low GI value and an intermediate GL value.

Recent research suggests that using GI/GL values as the basis for weight management or disease prevention/control (diabetes) is not as "simple" as once thought; the actual effects of high GI/GL foods are not always that which might be predicted. Several reasons explain why this is the case.

1. High GI/GL foods are usually consumed in combination with low GI/GL foods as part of a meal; macaroni and cheese, cereal and milk, etc. The combined effect is much lower than the individual effects.
2. The presence of protein and fats in a meal affect glycemic response. Carbohydrates eaten with protein evoke a lower glycemic response than the same carbohydrates eaten without protein.
3. The "state" of a carbohydrate-containing food affects glycemic response. The carbohydrate makeup of overripe fruit is different than that of underripe fruit; overripe fruit is more easily digested and thus has a higher GI/GL. Cooking and food processing also make food more digestible, increasing the GI/GL value relative to raw unprocessed food. Larger bites of food are more slowly digested (due to less surface area) and have lower GI values.
4. Glycemic response variance from individual to individual occurs. Because of genetic differences, individuals do not always process a given type of food at exactly the same rate nor in the same way; for example, milk digestion can vary dramatically among individuals.

A developing concern about dietary intake of carbohydrates involves *how fast* a given dietary carbohydrate is broken down to generate glucose within the human body. The term *glycemic response* refers to how quickly carbohydrates are digested (broken down into glucose), how high blood glucose levels rise, and how quickly blood glucose levels return to normal. Two measurement systems called the *glycemic index (GI)* and the *glycemic load (GL)* have been developed for rating carbohydrate-containing foods in terms of their glycemic response. The focus on relevancy feature Chemical Connections 7-E on the previous page discusses this topic further.

7.20 Glycolipids and Glycoproteins: Cell Recognition

Prior to 1960, the biochemistry of carbohydrates was thought to be rather simple. These compounds served (1) as energy sources for plants, humans, and animals and (2) as structural materials for plants and arthropods.

It is now known that mono-, di-, and oligosaccharides attached through glycosidic linkages to lipid molecules (Chapter 8) and protein molecules (Chapter 9) have a wide range of biochemical functions, including allowing cells to interact with invading bacteria and viruses and enabling cells of differing function to recognize each other. Such carbohydrate-lipid and carbohydrate-protein molecules are called, respectively, *glycolipids* and *glycoproteins*. A **glycolipid** *is a lipid molecule that has one or more carbohydrate (or carbohydrate derivative) units covalently bonded to it.* Similarly, a **glycoprotein** *is a protein molecule that has one or more carbohydrate (or carbohydrate derivative) units covalently bonded to it.*

The prefix *glyco-*, used in the terms *glycolipid* and *glycoprotein*, is derived from the Greek word *glykys*, which means "sweet." Most monosaccharides and disaccharides have a sweet taste.

Glycolipids called cerebrosides and gangliosides occur extensively in brain tissue (Section 8.8). Glycoproteins called immunoglobins are key components of the body's immune system response to invading foreign materials (Section 9.18).

In glycolipids and glycoproteins associated with cell membrane structure, the lipid or protein part of the glycolipid or glycoprotein is incorporated into the cell membrane structure and the carbohydrate (oligosaccharide) part functions as a marker on the outer cell membrane surface. Cell recognition generally involves the interaction between the carbohydrate marker of one cell and a protein imbedded into the cell membrane of another cell.

In the human reproductive process, fertilization involves a binding interaction between oligosaccharide markers on the outer membrane surface of an ovulated egg and protein receptor sites on a sperm cell membrane. This binding process is followed by release of enzymes by the sperm cell which dissolve the egg cell membrane allowing for entry of the sperm and the ensuing egg–sperm fertilization process.

The cell recognition system associated with red blood cells, which is the basis for the various blood types, involves glycoproteins. This recognition system was previously discussed in Chemical Connections 7-D on page 293, although the term *glycoprotein* was not mentioned at that time.

Concepts to Remember

OWL Sign in at **www.cengage.com/owl** to view tutorials and simulations, develop problem-solving skills, and complete online homework assigned by your professor.

Biochemistry. Biochemistry is the study of the chemical substances found in living systems and the chemical interactions of these substances with each other (Section 7.1).

Carbohydrates. Carbohydrates are polyhydroxy aldehydes, polyhydroxy ketones, or compounds that yield such substances upon hydrolysis. Plants contain large quantities of carbohydrates produced via photosynthesis (Section 7.2).

Carbohydrate classification. Carbohydrates are classified into four groups: monosaccharides, disaccharides, oligosaccharides, and polysaccharides (Section 7.3).

Chirality and achirality. A chiral object is not identical to its mirror image. An achiral object is identical to its mirror image (Section 7.4).

Chiral center. A chiral center is an atom in a molecule that has four different groups tetrahedrally bonded to it. Molecules that contain a single chiral center exist in a left-handed and a right-handed form (Section 7.4).

Stereoisomerism. The atoms of stereoisomers are connected in the same way but are arranged differently in space. The major causes of stereoisomerism in molecules are structural rigidity and the presence of a chiral center (Section 7.5).

Enantiomers and diastereomers. Two types of stereoisomers exist: enantiomers and diastereomers. Enantiomers have structures that are nonsuperimposable mirror images of each other.

Enantiomers have identical achiral properties but different chiral properties. Diastereomers have structures that are not mirror images of each other (Section 7.5).

Fischer projection formulas. Fischer projection formulas are two-dimensional structural formulas used to depict the three-dimensional shapes of molecules with chiral centers (Section 7.6).

Chirality of monosaccharides. Monosaccharides are classified as D or L stereoisomers on the basis of the configuration of the chiral center farthest from the carbonyl group (Section 7.6).

Optical activity. Chiral compounds are optically active—that is, they rotate the plane of polarized light. Enantiomers rotate the plane of polarized light in opposite directions. The prefix (+) indicates that the compound rotates the plane of polarized light in a clockwise direction, whereas compounds that rotate the plane of polarized light in a counterclockwise direction have the prefix (−) (Section 7.7).

Classification of monosaccharides. Monosaccharides are classified as aldoses or ketoses on the basis of the type of carbonyl group present. They are further classified as trioses, tetroses, pentoses, etc. on the basis of the number of carbon atoms present (Section 7.8).

Important monosaccharides. Important monosaccharides include glucose, galactose, fructose, and ribose. Glucose and galactose are aldohexoses, fructose is a ketohexose, and ribose is an aldopentose (Section 7.9).

Cyclic monosaccharides. Cyclic monosaccharides form through an intramolecular reaction between the carbonyl group and an alcohol group of an open-chain monosaccharide. These cyclic forms predominate in solution (Section 7.10).

Haworth projection formulas. Haworth projection formulas are two-dimensional structural representations used to depict the three-dimensional structure of a cyclic form of a monosaccharide (Section 7.11).

Reactions of monosaccharides. Five important reactions of monosaccharides are (1) oxidation to an acidic sugar (2) reduction to a sugar alcohol (3) glycoside formation (4) phosphate ester formation and (5) amino sugar formation (Section 7.12).

Disaccharides. Disaccharides are glycosides formed from the linkage of two monosaccharides. The most important disaccharides are maltose, cellobiose, lactose, and sucrose. Each of these has at least one glucose unit in its structure (Section 7.13).

Oligosaccharides. Oligosaccharides are carbohydrates that contain three to ten monosaccharide units covalently bonded to each other. Two important naturally occurring oligosaccharides are raffinose and stachyose (Section 7.14).

Polysaccharides. Polysaccharides are polymers in which monosaccharides are the monomers. In homopolysaccharides, only one type of monomer is present. Two or more monosaccharide monomers are present in heteropolysaccharides. Storage polysaccharides (starch, glycogen) are storage molecules for monosaccharides. Structural polysaccharides (cellulose, chitin) serve as structural elements in plant cell walls and animal exoskeletons (Sections 7.15 to 7.18).

Glycolipids and glycoproteins. Glycolipids and glycoproteins are molecules in which oligosaccharides are attached through glycosidic linkages to lipids and proteins, respectively. Such molecules often govern how cells of differing function interact with each other (Section 7.20).

Exercises and Problems

OWL Interactive versions of these problems may be assigned in OWL.

Exercises and problems are arranged in matched pairs with the two members of a pair addressing the same concept(s). The answer to the odd-numbered member of a pair is given at the back of the book. Problems denoted with a ▲ involve concepts found not only in the section under consideration but also concepts found in one or more earlier sections of the chapter. Problems denoted with a ● cover concepts found in a Chemical Connections feature box.

Biochemical Substances (Section 7.1)

7.1 Define each of the following terms.
a. Biochemistry
b. Biochemical substance

7.2 Contrast the relative amounts, by mass, of bioorganic and bioinorganic substances present in the human body.

7.3 What are the four major types of bioorganic substances?

7.4 For each of the following pairs of bioorganic substances, indicate which member of the pair is more abundant in the human body.
a. Proteins and nucleic acids
b. Proteins and carbohydrates
c. Lipids and carbohydrates
d. Lipids and nucleic acids

Occurrence of Carbohydrates (Section 7.2)

7.5 Write a general chemical equation for photosynthesis.

7.6 What role does chlorophyll play in photosynthesis?

7.7 What are the two major functions of carbohydrates in the plant kingdom?

7.8 What are the six major functions of carbohydrates in the human body?

Classification of Carbohydrates (Section 7.3)

7.9 Define the term *carbohydrate.*

7.10 What functional group is present in all carbohydrates?

7.11 Indicate how many monosaccharide units are present in each of the following.
a. Disaccharide
b. Tetrasaccharide
c. Oligosaccharide
d. Polysaccharide

7.12 Identify, in general terms, the product produced from the complete hydrolysis of each of the following types of carbohydrates.
a. Disaccharide
b. Tetrasaccharide
c. Oligosaccharide
d. Polysaccharide

Chirality (Section 7.4)

7.13 Explain what the term *superimposable* means.

7.14 Explain what the term *nonsuperimposable* means.

7.15 In each of the following lists of objects, identify those objects that are chiral.
a. Nail, hammer, screwdriver, drill bit
b. Your hand, your foot, your ear, your nose
c. The words TOT, TOOT, POP, PEEP

7.16 In each of the following lists of objects, identify those objects that are chiral.
a. Baseball cap, glove, shoe, scarf
b. Pliers, scissors, spoon, fork
c. The words MOM, DAD, AHA, WAX

7.17 Indicate whether the circled carbon atom in each of the following molecules is a chiral center.

a. CH_3—ⒸH_2—OH

b. 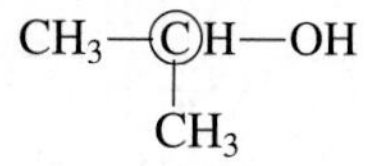

c. 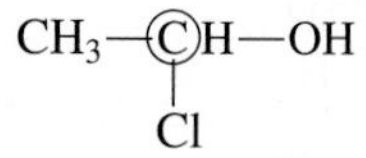

d. 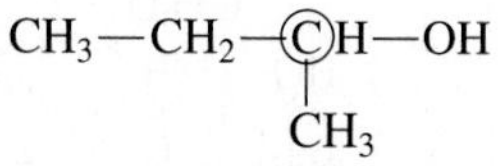

7.18 Indicate whether the circled carbon atom in each of the following molecules is a chiral center.

a. CH_3—ⒸH_2—NH_2

b. 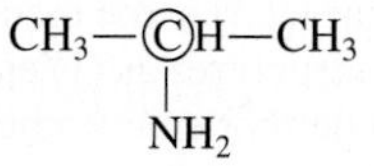

c. 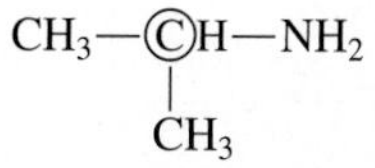

d. 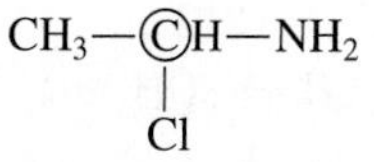

7.19 Use asterisks to show the chiral center(s), if any, in the following structures.

a. 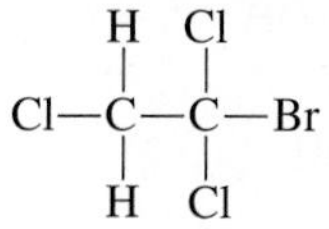

b. 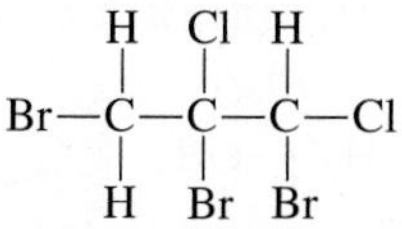

c. 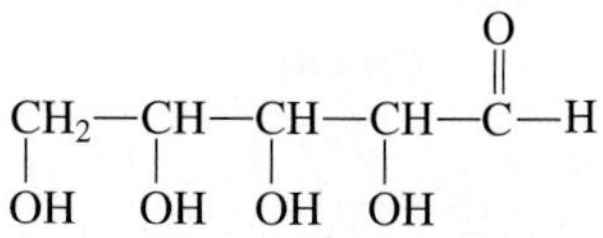

d. 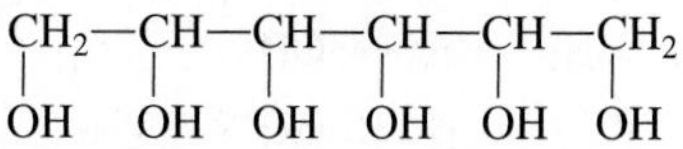

7.20 Use asterisks to show the chiral center(s), if any, in the following structures.

a. 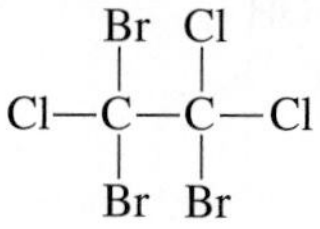

b. 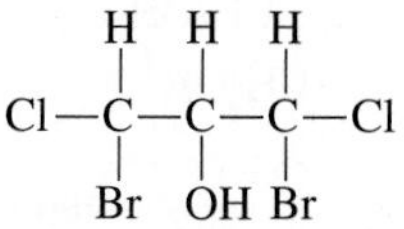

c. 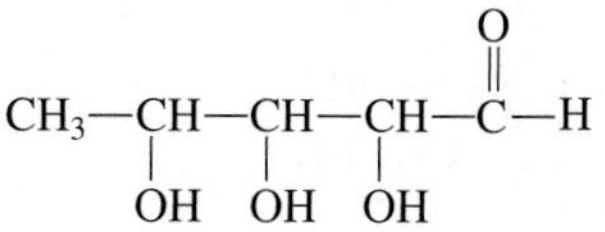

d. 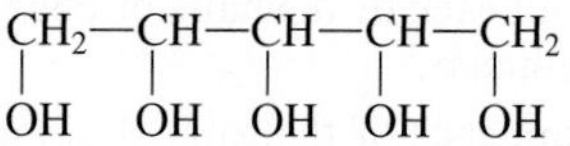

7.21 How many chiral centers are present in each of the following molecular structures?

a. 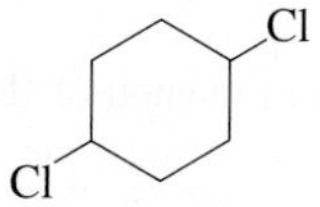

b. 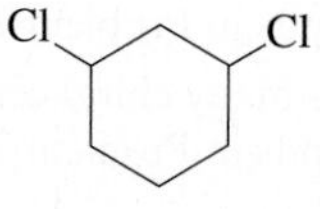

c. 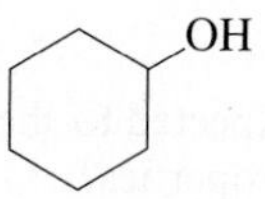

d. 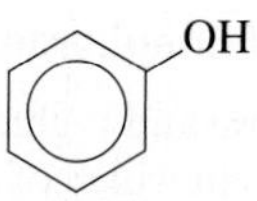

7.22 How many chiral centers are present in each of the following molecular structures?

a. 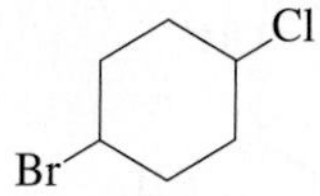

b. 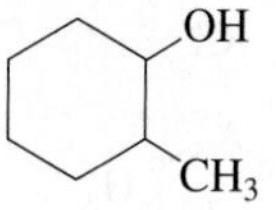

c. 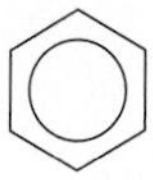

d. 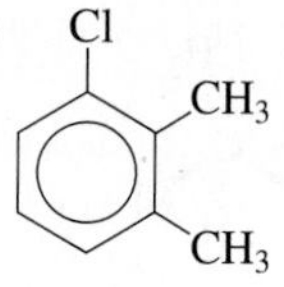

7.23 Classify each of the molecules in Problem 7.19 as chiral or achiral.

7.24 Classify each of the molecules in Problem 7.20 as chiral or achiral.

▲7.25 The alkane of lowest molecular mass that has a chiral center has the molecular formula C_7H_{16}. Identify the "groups" attached to the chiral center in a molecule of this compound.

▲7.26 The saturated alcohol of lowest molecular mass that has a chiral center has the molecular formula $C_4H_{10}O$. Identify the "groups" attached to the chiral center in a molecule of this compound.

Stereoisomerism: Enantiomers and Diastereomers (Section 7.5)

7.27 What is the difference between constitutional isomers and stereoisomers?

7.28 Both enantiomers and diastereomers are stereoisomers. How do they differ?

7.29 What are two major structural features that can generate stereoisomerism?

7.30 Explain why *cis–trans* isomers are diastereomers rather than enantiomers.

7.31 Indicate whether each of the following statements about *stereoisomers* is true or false.
a. Stereoisomers always have the same molecular formula.
b. Stereoisomers always have the same structural formula.
c. Stereoisomers are always nonsuperimposable mirror images of each other.
d. Stereoisomers always possess handedness.

7.32 Indicate whether each of the following statements about *enantiomers* is true or false.
a. Enantiomers always have the same molecular formula.
b. Enantiomers always have the same structural formula.
c. Enantiomers are always nonsuperimposable mirror images of each other.
d. Enantiomers always differ in handedness.

Fischer Projection Formulas (Section 7.6)

7.33 Draw the Fischer projection formula for each of the following molecules.

a. 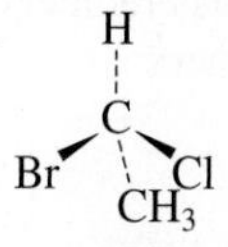

b. 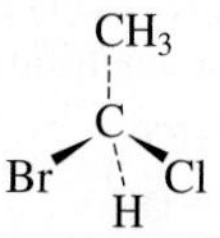

c. 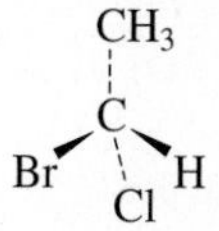

d. 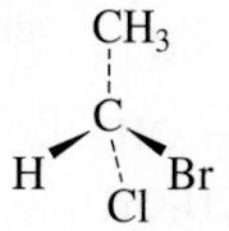

7.34 Draw the Fischer projection formula for each of the following molecules.

a.
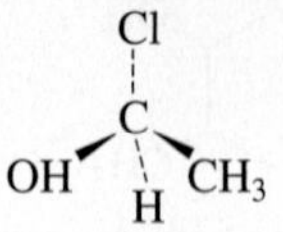

b.
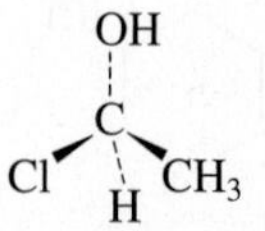

c.
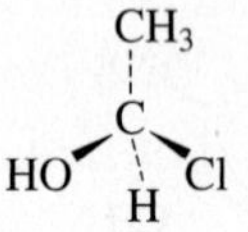

d.
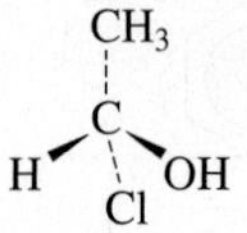

7.35 Draw a Fischer projection formula for the enantiomer of each of the following monosaccharides.

a.
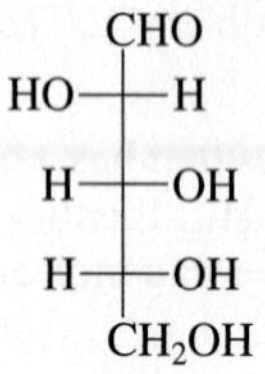

b.
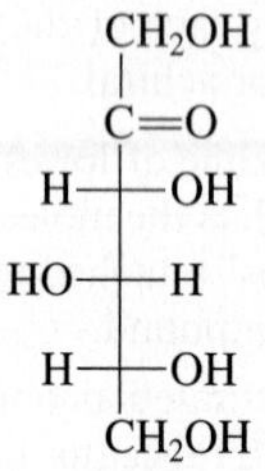

c.
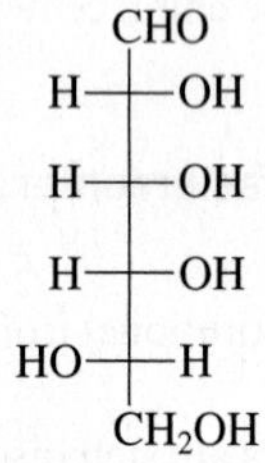

d.
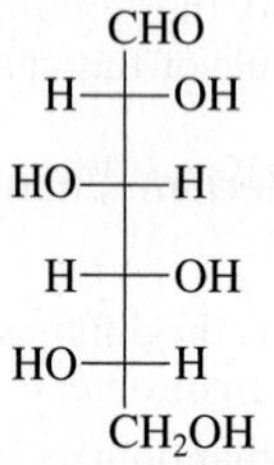

7.36 Draw a Fischer projection formula for the enantiomer of each of the following monosaccharides.

a.
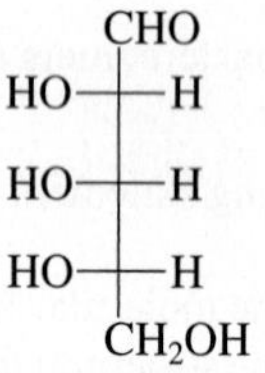

b.
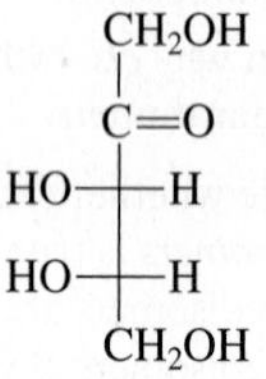

c.
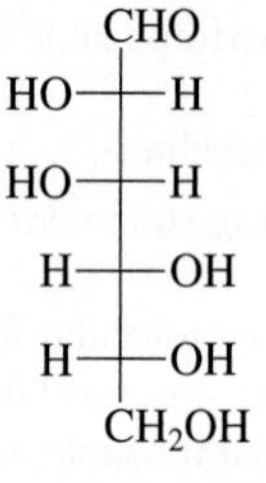

d.
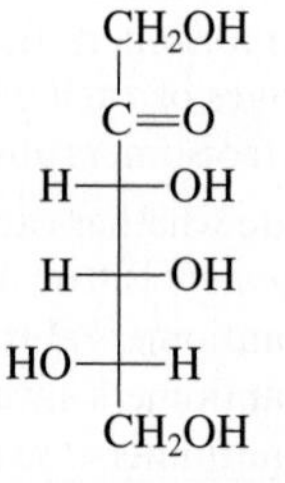

7.37 Classify each of the molecules in Problem 7.35 as a D enantiomer or an L enantiomer.

7.38 Classify each of the molecules in Problem 7.36 as a D enantiomer or an L enantiomer.

7.39 Characterize the members of each of the following pairs of structures as (1) enantiomers (2) diastereomers or (3) neither enantiomers nor diastereomers.

a.
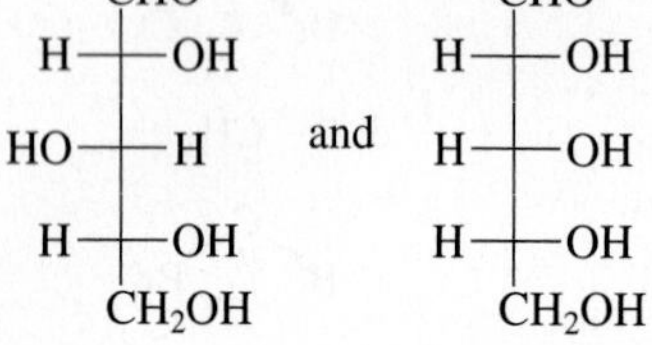

b.
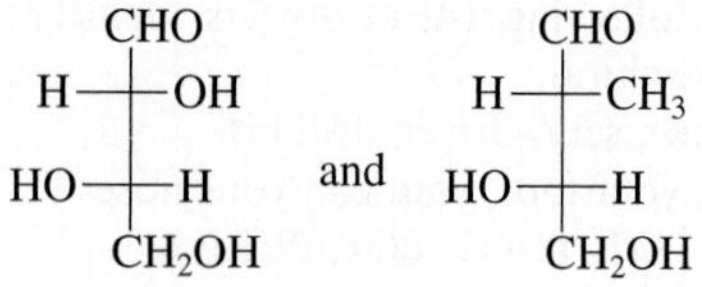

c.
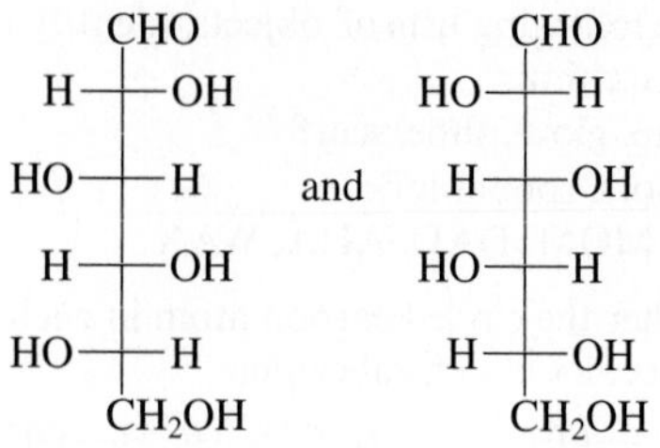

d.
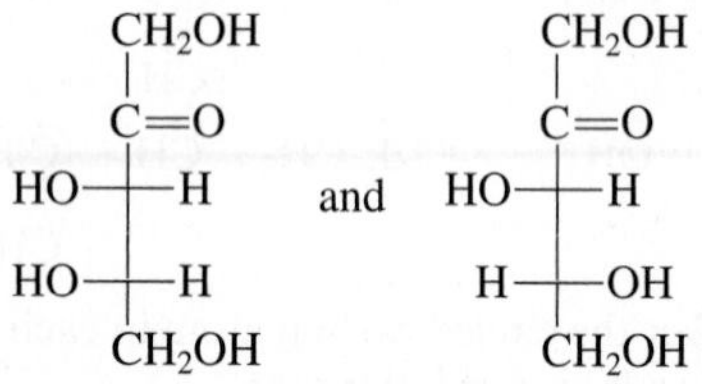

7.40 Characterize the members of each of the following pairs of structures as (1) enantiomers (2) diastereomers or (3) neither enantiomers nor diastereomers.

a.
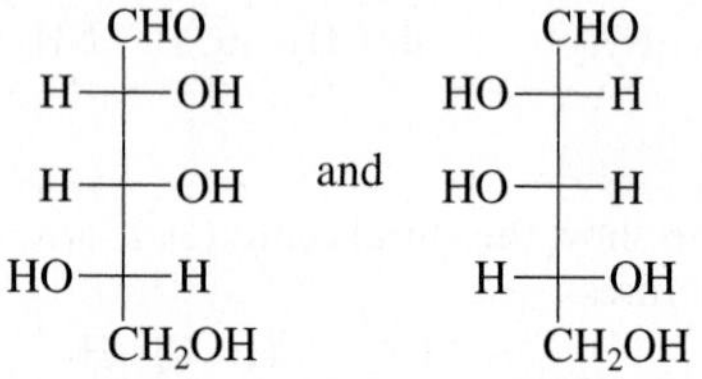

b.
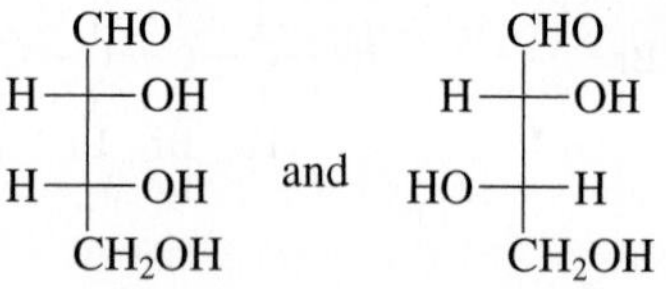

c.
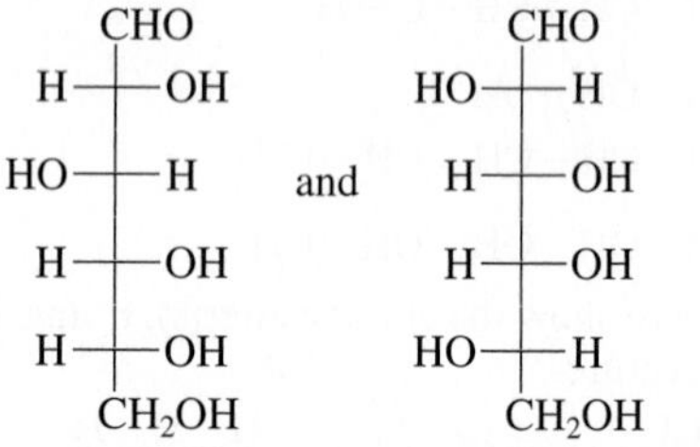

d.
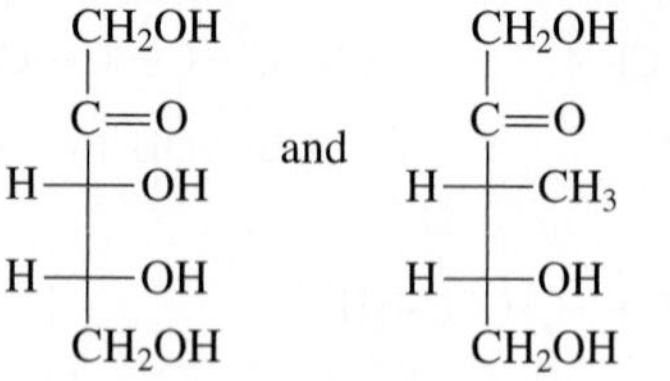

7.41 Indicate whether or not each of the pairs of compounds in Problem 7.39 are epimers.

7.42 Indicate whether or not each of the pairs of compounds in Problem 7.40 are epimers.

▲**7.43** How many chiral centers are present in each of the compounds in Problem 7.39?

▲**7.44** How many chiral centers are present in each of the compounds in Problem 7.40?

Properties of Enantiomers (Section 7.7)

7.45 D-glucose and L-glucose would be expected to show differences in which of the following properties?

a. Solubility in an achiral solvent
b. Density
c. Melting point
d. Effect on plane-polarized light

7.46 D-glucose and L-glucose would be expected to show differences in which of the following properties?
a. Solubility in a chiral solvent
b. Freezing point
c. Reaction with ethanol
d. Reaction with (+)-lactic acid

7.47 Compare (+)-lactic acid and (−)-lactic acid with respect to each of the following properties.
a. Boiling point
b. Optical activity
c. Solubility in water
d. Reaction with (+)-2,3-butanediol

7.48 Compare (+)-glyceraldehyde and (−)-glyceraldehyde with respect to each of the following properties.
a. Freezing point
b. Rotation of plane-polarized light
c. Reaction with ethanol
d. Reaction with (−)-2,3-butanediol

▲7.49 Indicate whether or not each of the following compounds is optically active.

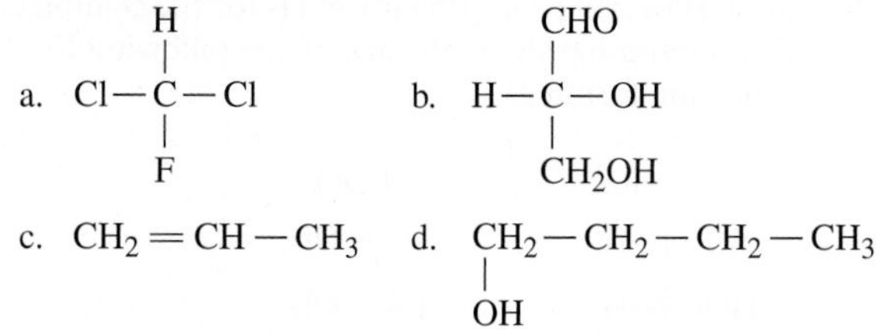

▲7.50 Indicate whether or not each of the following compounds is optically active.

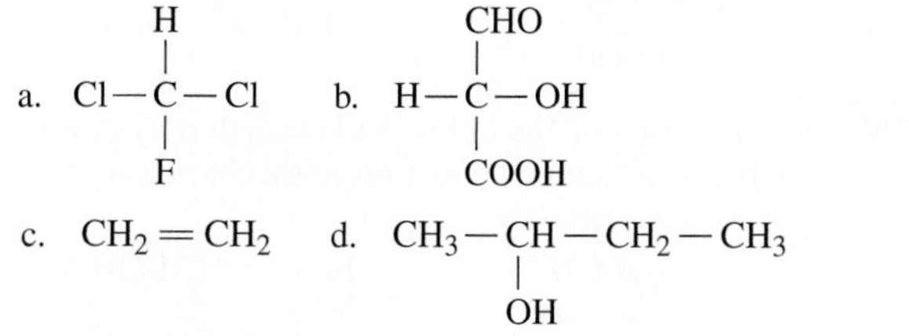

Classification of Monosaccharides (Section 7.8)

7.51 Classify each of the following monosaccharides as an aldose or a ketose.

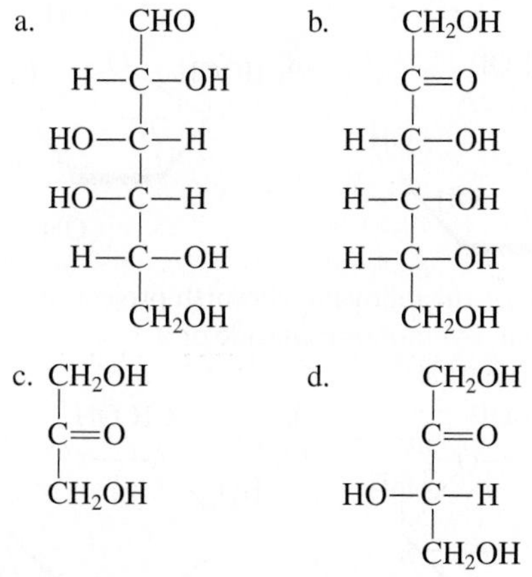

7.52 Classify each of the following monosaccharides as an aldose or a ketose.

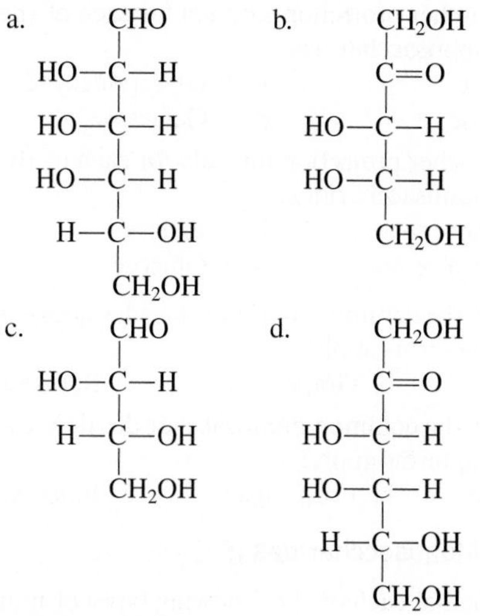

7.53 Classify each monosaccharide in Problem 7.51 by its number of carbon atoms and its type of carbonyl group.

7.54 Classify each monosaccharide in Problem 7.52 by its number of carbon atoms and its type of carbonyl group.

7.55 Using the information in Figures 7.14 and 7.15, assign a name to each of the monosaccharides in Problem 7.51.

7.56 Using the information in Figures 7.14 and 7.15, assign a name to each of the monosaccharides in Problem 7.52.

▲7.57 How many chiral centers are present in the structure of each of the monosaccharides in Problem 7.51?

▲7.58 How many chiral centers are present in the structure of each of the monosaccharides in Problem 7.52?

Biochemically Important Monosaccharides (Section 7.9)

7.59 Indicate at what carbon atom(s) the structures of each of the following pairs of monosaccharides differ.
a. D-Glucose and D-galactose
b. D-Glucose and D-fructose
c. D-Glyceraldehyde and dihydroxyacetone
d. D-Ribose and 2-deoxy-D-ribose

7.60 Indicate whether the members of each of the following pairs of monosaccharides have the same molecular formula.
a. D-Glucose and D-galactose
b. D-Glucose and D-fructose
c. D-Glyceraldehyde and dihydroxyacetone
d. D-Ribose and 2-deoxy-D-ribose

7.61 Indicate which of the terms *aldoses, ketoses, hexoses,* and *aldohexoses* apply to both members of each of the following pairs of monosaccharides. More than one term may apply in a given situation.
a. D-Glucose and D-galactose
b. D-Glucose and D-fructose
c. D-Galactose and D-fructose
d. D-Glyceraldehyde and D-ribose

7.62 Indicate which of the terms *aldoses, ketoses, trioses,* and *aldohexoses* apply to both members of each of the following pairs of monosaccharides. More than one term may apply in a given situation.
a. D-Glucose and D-ribose
b. D-Fructose and dihydroxyacetone
c. D-Glyceraldehyde and dihydroxyacetone
d. D-Galactose and D-ribose

7.63 Draw the Fischer projection formula for each of the following monosaccharides.
a. D-Glucose b. D-Glyceraldehyde
c. D-Fructose d. L-Galactose

7.64 Draw the Fischer projection formula for each of the following monosaccharides.
a. D-Galactose b. D-Ribose
c. Dihydroxyacetone d. L-Glucose

7.65 To which of the common monosaccharides does each of the following terms apply?
a. Levulose b. Grape sugar c. Brain sugar

7.66 To which of the common monosaccharides does each of the following terms apply?
a. Dextrose b. Fruit sugar c. Blood sugar

Cyclic Forms of Monosaccharides (Section 7.10)

7.67 Indicate whether each of the following types of monosaccharides, upon intramolecular cyclization, forms a six-membered ring, a five-membered ring, or a four-membered ring.
a. Aldohexose b. Ketohexose
c. Aldopentose d. Ketopentose

7.68 Identify the two carbon atoms, using numbers (C1, C2, C3, etc.), that bear the functional groups that interact during the intramolecular cyclization of each of the following types of monosaccharides.
a. Aldohexose b. Aldopentose
c. Ketohexose d. Ketopentose

7.69 How many carbon atoms end up outside the ring in the cyclization of the open-chain form of the following monosaccharides?
a. D-Glucose b. D-Galactose
c. D-Fructose d. D-Ribose

7.70 After cyclization of each of the monosaccharides in Problem 7.69, how many of the carbon atoms do not have a hydroxyl group directly attached to it?

7.71 One of the cyclic forms of D-glucose has the structure

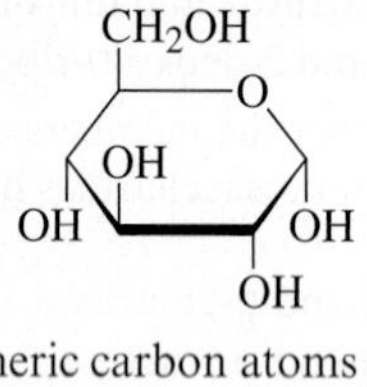

a. How many anomeric carbon atoms are present in this structure?
b. How many hemiacetal carbon atoms are present in this structure?

7.72 One of the cyclic forms of D-galactose has the structure

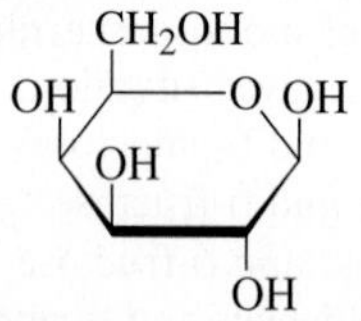

a. How many anomeric carbon atoms are present in this structure?
b. How many hemiacetal carbon atoms are present in this structure?

7.73 Draw the structure for the anomer of the monosaccharide in Problem 7.71.

7.74 Draw the structure for the anomer of the monosaccharide in Problem 7.72.

7.75 Classify the monosaccharide structure in Problem 7.71 as an α-anomer or a β-anomer.

7.76 Classify the monosaccharide structure in Problem 7.72 as an α-anomer or a β-anomer.

7.77 The structure of D-glucose is sometimes written in an open-chain form and sometimes in a cyclic form. Explain why either form is acceptable.

7.78 When pure α-D-glucose is dissolved in water, both α-D-glucose and β-D-glucose are soon present. Explain why this is so.

Haworth Projection Formulas (Section 7.11)

7.79 Draw Haworth projection formulas for the α-anomer of monosaccharides with each of the following Fischer projection formulas.

a.
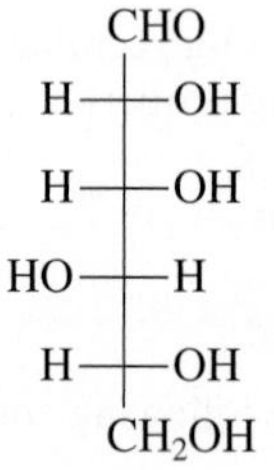

b.
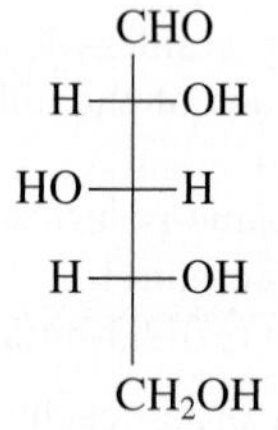

7.80 Draw Haworth projection formulas for the β-anomer of monosaccharides with each of the following Fischer projection formulas.

a.
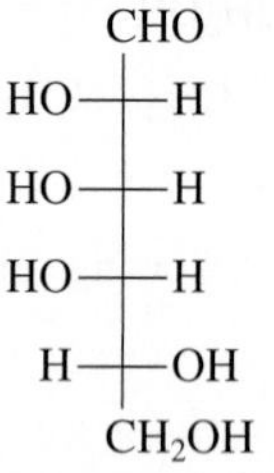

b.
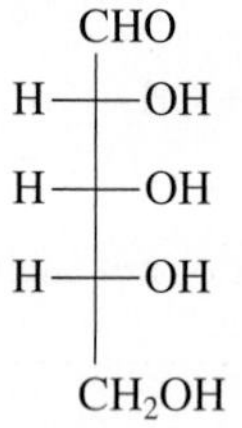

7.81 Identify each of the following Haworth projection formulas as that of an α-D-monosaccharide or a β-D-monosaccharide.

a.
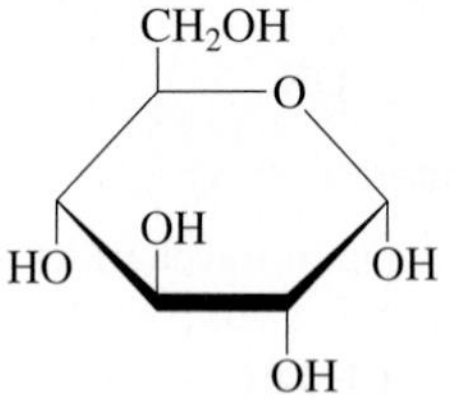

b.
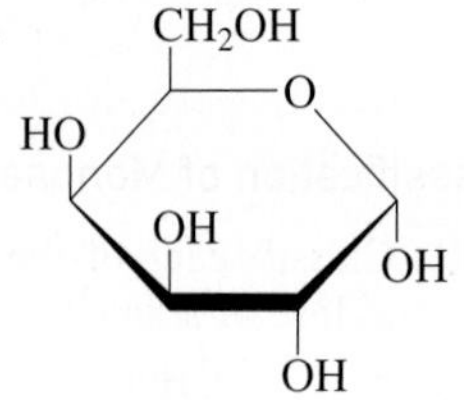

c.
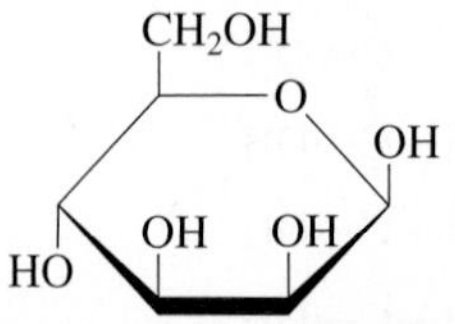

d.
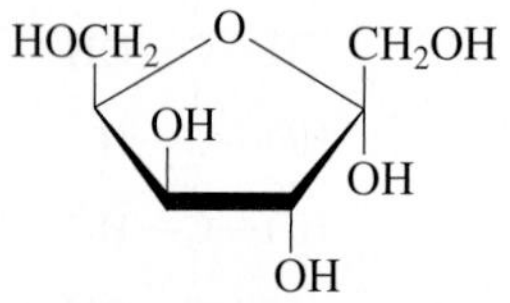

7.82 Identify each of the following Haworth projection formulas as an α-D-monosaccharide or a β-D-monosaccharide.

a.
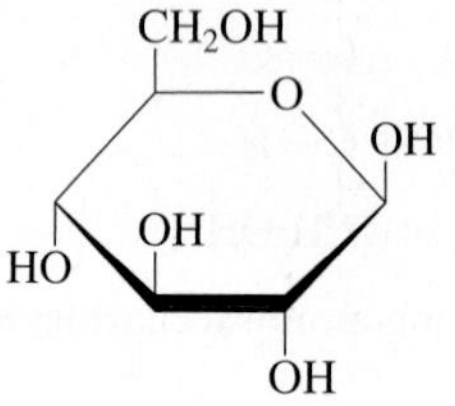

b.
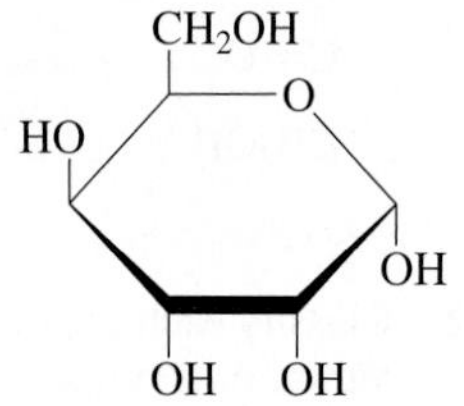

c.
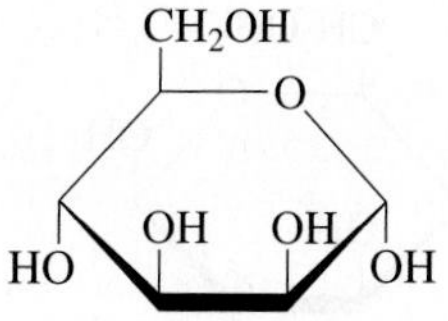

d.
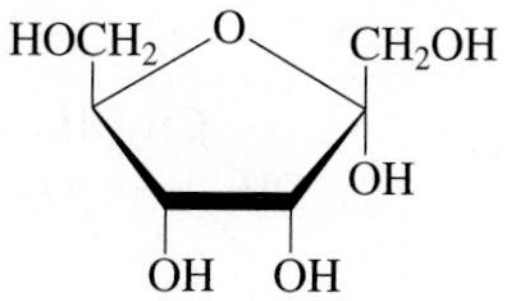

7.83 Draw the open-chain form for each of the monosaccharides in Problem 7.81.

7.84 Draw the open-chain form for each of the monosaccharides in Problem 7.82.

7.85 Using the information in Figures 7.14 and 7.15, assign a name to each of the monosaccharides in Problem 7.81.

7.86 Using the information in Figures 7.14 and 7.15, assign a name to each of the monosaccharides in Problem 7.82.

7.87 Draw the Haworth projection formula for each of the following monosaccharides.
a. α-D-Galactose b. β-D-Galactose
c. α-L-Galactose d. β-L-Galactose

7.88 Draw the Haworth projection formula for each of the following monosaccharides.
a. α-D-Mannose b. β-D-Mannose
c. α-L-Mannose d. β-L-Mannose

Reactions of Monosaccharides (Section 7.12)

7.89 Draw the Fischer projection formula for the galactose derivative formed when galactose undergoes each of the following changes.
a. The —CHO group is oxidized.
b. The —CH_2OH group is oxidized.
c. Both the —CHO group and —CH_2OH group are oxidized.
d. The —CHO group is reduced.

7.90 Draw the Fischer projection formula for the glucose derivative formed when glucose undergoes each of the following changes.
a. The —CHO group is oxidized.
b. Both the —CHO group and —CH_2OH group are oxidized.
c. The —CHO group is reduced.
d. The —CH_2OH group is oxidized.

7.91 Classify each of the galactose derivatives in Problem 7.89 as an acidic sugar or a sugar alcohol.

7.92 Classify each of the glucose derivatives in Problem 7.90 as an acidic sugar or a sugar alcohol.

7.93 Name each of the galactose derivatives in Problem 7.89.

7.94 Name each of the glucose derivatives in Problem 7.90.

7.95 Which of the following monosaccharides is a *reducing sugar*?
a. D-Glucose b. D-Galactose
c. D-Fructose d. D-Ribose

7.96 Which of the following monosaccharides will give a positive test with Benedict's solution?
a. D-Glucose b. D-Galactose
c. D-Fructose d. D-Ribose

7.97 Indicate whether each of the following structures is that of a glycoside.

a.
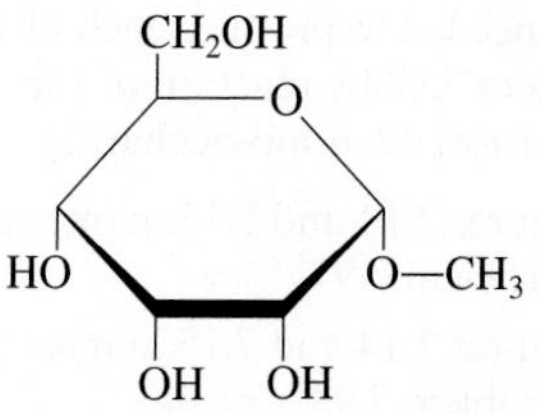

b.
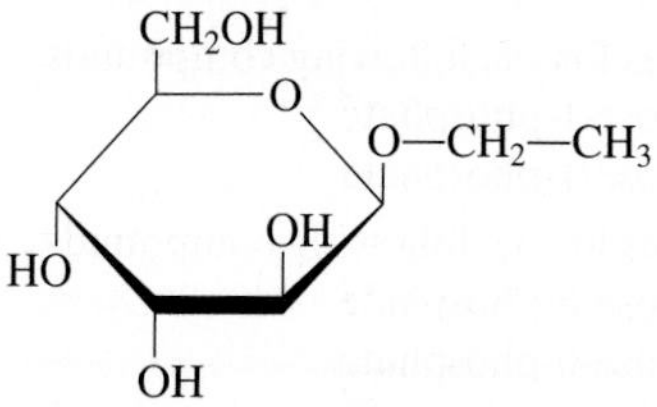

c.
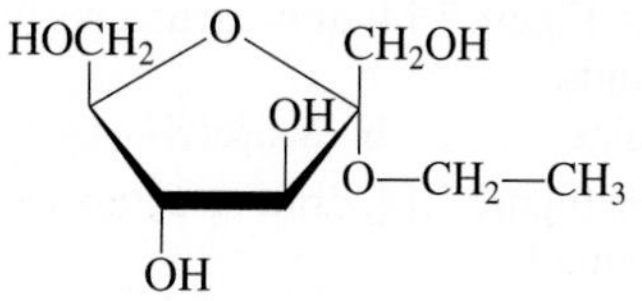

d.
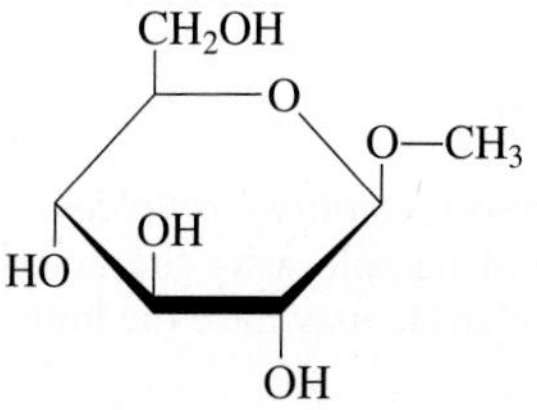

7.98 Indicate whether each of the following structures is that of a glycoside.

a.
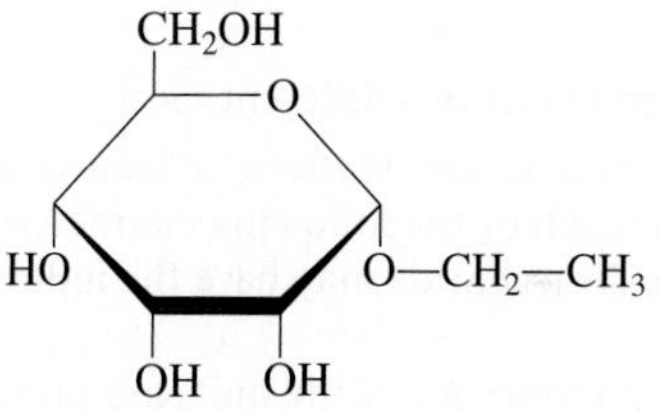

b.
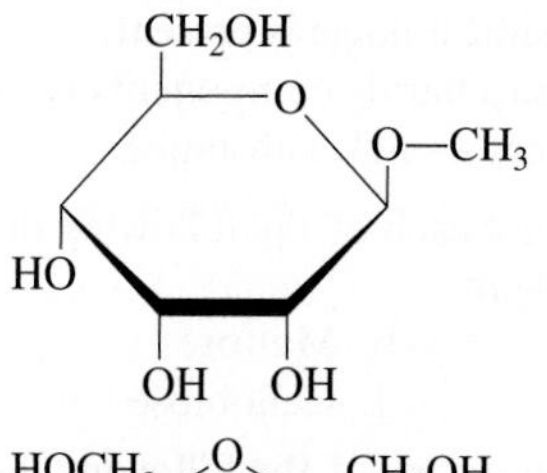

c.
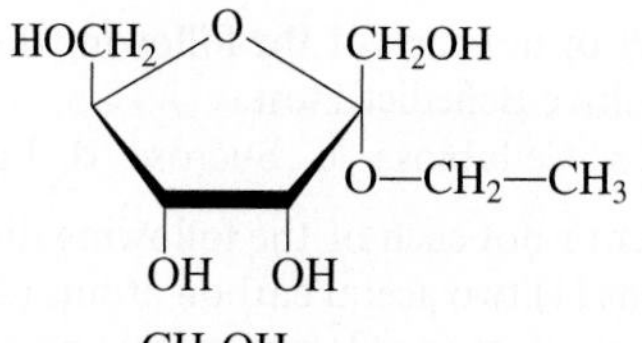

d.
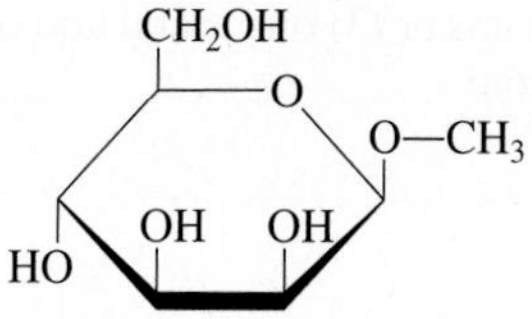

7.99 For each structure in Problem 7.97, identify the configuration at the acetal carbon atom as α or β.

7.100 For each structure in Problem 7.98, identify the configuration at the acetal carbon atom as α or β.

7.101 Identify the alcohol needed to produce each of the compounds in Problem 7.97 by reaction of the alcohol with the appropriate monosaccharide.

7.102 Identify the alcohol needed to produce each of the compounds in Problem 7.98 by reaction of the alcohol with the appropriate monosaccharide.

▲**7.103** With the help of Figures 7.14 and 7.15, name each of the compounds in Problem 7.97.

▲**7.104** With the help of Figures 7.14 and 7.15, name each of the compounds in Problem 7.98.

7.105 Draw structures for the following compounds.
a. α-D-Galactose-1-phosphate
b. β-D-Galactose-1-phosphate

7.106 Draw structures for the following compounds.
a. α-D-Galactose-6-phosphate
b. β-D-Galactose-6-phosphate

▲**7.107** With the help of Figure 7.14, draw structures for the following compounds.
a. α-D-gulosamine   b. *N*-acetyl-α-D-gulosamine

▲**7.108** With the help of Figure 7.14, draw structures for the following compounds.
a. α-D-allosamine   b. *N*-acetyl-α-D-allosamine

Disaccharides (Section 7.13)

7.109 Which of the disaccharides *maltose, cellobiose, lactose,* and *sucrose* has each of the following characteristics? More than one disaccharide may have the indicated characteristic.
a. Both monosaccharide units are the same.
b. An $\alpha(1 \rightarrow 4)$ glycosidic linkage is present.
c. One of the monosaccharide components is galactose.
d. Hydrolysis produces two different monosaccharides.

7.110 Which of the disaccharides *maltose, cellobiose, lactose,* and *sucrose* has each of the following characteristics? More than one disaccharide may have the indicated characteristic.
a. Two different monosaccharide units are present.
b. A $\beta(1 \rightarrow 4)$ glycosidic linkage is present.
c. One of the monosaccharide components is fructose.
d. Hydrolysis produces a single substance.

7.111 Indicate whether or not each of the following disaccharides is a reducing sugar.
a. Sucrose   b. Maltose
c. Lactose   d. Cellobiose

7.112 Indicate whether or not each of the following disaccharides gives a positive Benedict's test.
a. Maltose b. Cellobiose c. Sucrose d. Lactose

7.113 Indicate whether or not each of the following disaccharides contains (1) two acetal carbon atoms (2) two hemiacetal carbon atoms or (3) one acetal and one hemiacetal carbon atom.

a.
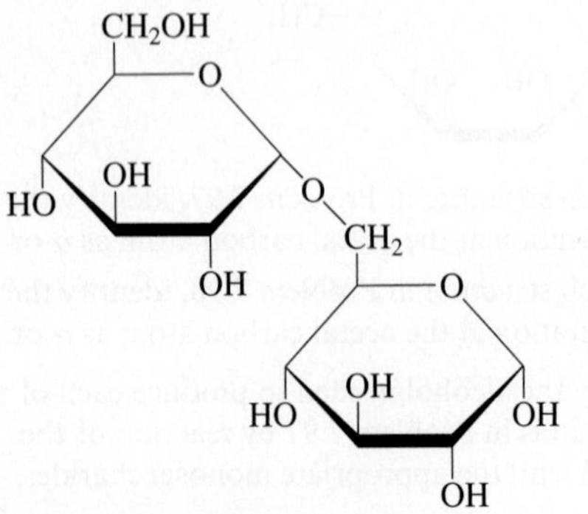

b.
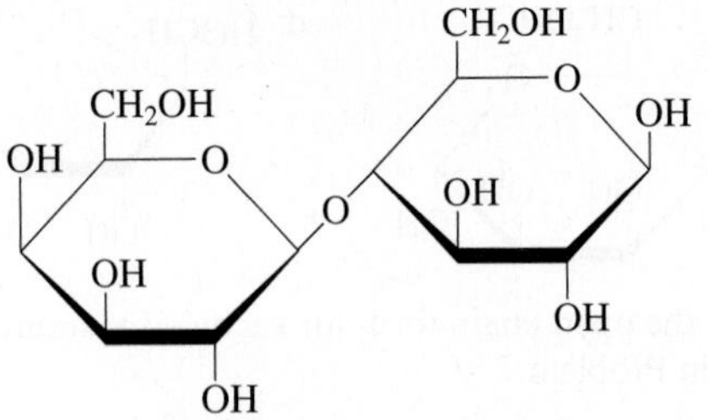

c.
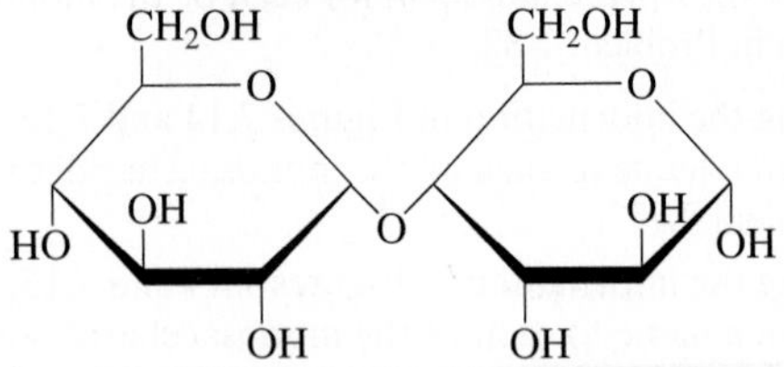

d.
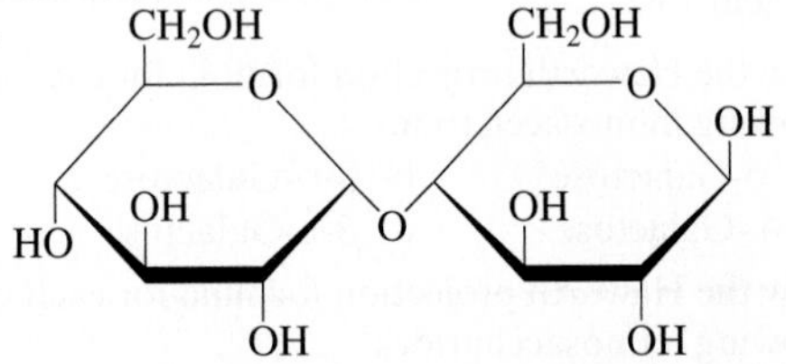

7.114 Indicate whether or not each of the following disaccharides contains (1) two acetal carbon atoms (2) two hemiacetal carbon atoms or (3) one acetal and one hemiacetal carbon atom.

a.
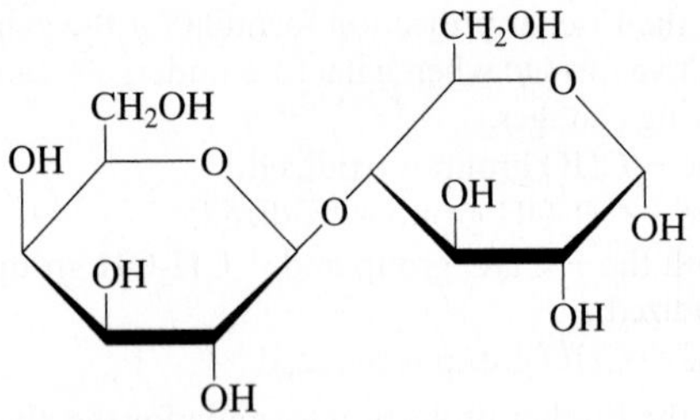

b.
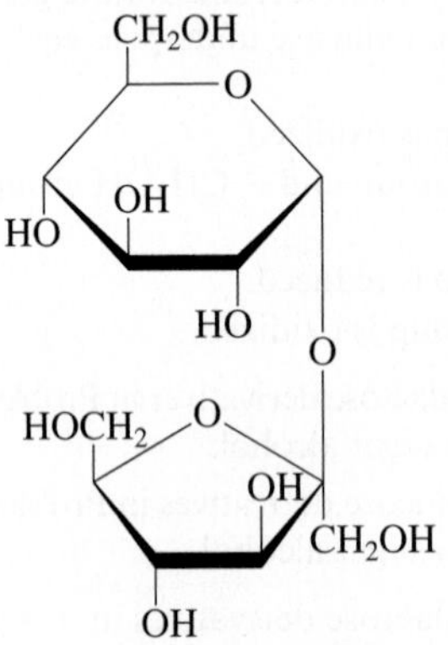

c.
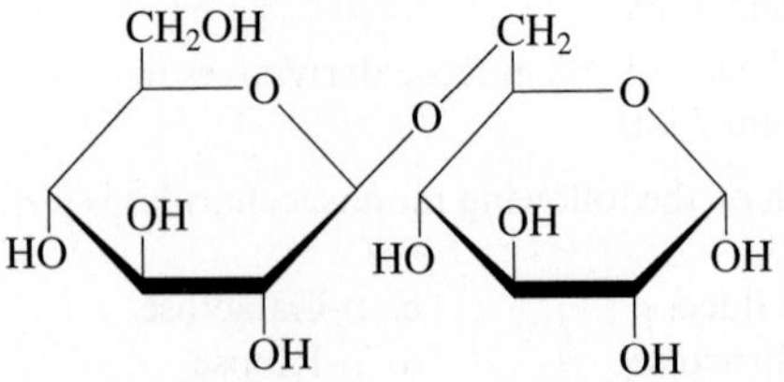

d.
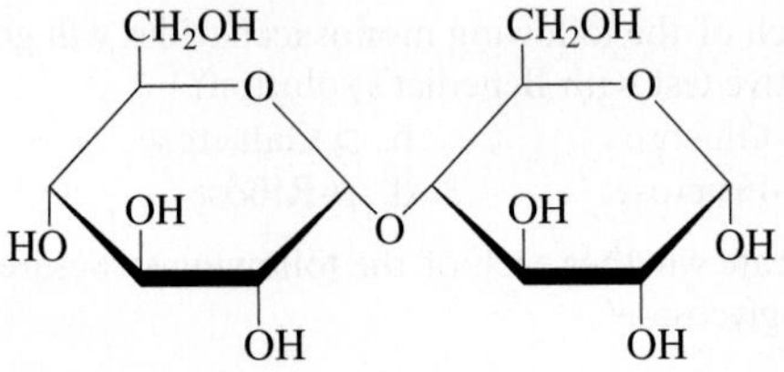

7.115 For each of the structures in Problem 7.113, specify whether the disaccharide is in an α configuration or a β configuration, or neither.

7.116 For each of the structures in Problem 7.114, specify whether the disaccharide is in an α configuration or a β configuration, or neither.

7.117 Identify each of the structures in Problem 7.113 as a reducing sugar or a nonreducing sugar.

7.118 Identify each of the structures in Problem 7.114 as a reducing sugar or a nonreducing sugar.

7.119 Draw the structures of the substances produced when each of the disaccharides in Problem 7.113 undergoes hydrolysis.

7.120 Draw the structures of the substances produced when each of the disaccharides in Problem 7.114 undergoes hydrolysis.

7.121 What type of glycosidic linkage [$\alpha(1 \rightarrow 4)$, etc.] is present in each of the disaccharides in Problem 7.113?

7.122 What type of glycosidic linkage [$\alpha(1 \rightarrow 4)$, etc.] is present in each of the disaccharides in Problem 7.114?

7.123 Draw the structure of the disaccharide sophorose, given that it contains an α-D-glucose unit, a β-D-glucose unit, and a $\beta(1 \rightarrow 2)$ glycosidic linkage.

7.124 Draw the structure of the disaccharide isomalatose, given that it contains an α-D-glucose unit, a β-D-glucose unit, and a $\alpha(1 \rightarrow 6)$ glycosidic linkage.

▲7.125 Indicate which of the terms *monosaccharide*, *disaccharide*, *reducing sugar*, *anomers*, *enantiomers*, and *aldohexose* applies to both members of each of the following pairs of substances. More than one term may apply to a given pair of substances.
a. α-D-Glucose and β-D-glucose
b. Sucrose and maltose
c. D-Fructose and L-fructose
d. Lactose and galactose

▲**7.126** Indicate which of the terms *monosaccharide*, *disaccharide*, *reducing sugar*, *anomers*, *enantiomers*, and *aldohexose* applies to both members of each of the following pairs of substances. More than one term may apply to a given pair of substances.
a. α-D-Glucose and α-D-galactose
b. Sucrose and cellobiose
c. Glyceraldehyde and dihydroxyacetone
d. D-Ribose and L-ribose

●7.127 (Chemical Connections 7-A) Indicate whether each of the following statements concerning lactose, lactase, and lactose intolerance is true or false.
a. The level of the enzyme lactase in humans decreases with age.
b. Lactase-persistent people cannot drink milk after childhood.
c. Lactose intolerance is lowest among Mediterranean people.
d. Lactase is the principal carbohydrate in milk.

●**7.128** (Chemical Connections 7-A) Indicate whether each of the following statements concerning lactose, lactase, and lactose intolerance is true or false.
a. The enzyme lactose is required for the hydrolysis of milk sugar.
b. Lactose-intolerant people have an allergy to the lactose present in milk.
c. Lactase persistence is a condition in which people cannot hydrolyze lactose in their digestive tract.
d. The level of lactase in humans varies by ethnic group.

●7.129 (Chemical Connections 7-B) Indicate whether each of the following statements concerning use of sucrose and fructose as sweeteners is true or false.
a. Fructose has a "sweetness factor" that is 7 times greater than that of fructose.
b. HFCS-42 contains 42% sucrose.
c. The switch from sucrose to HFCS was economically driven.
d. Oranges and grapes are the two fruits with the highest fructose/glucose ratio.

●**7.130** (Chemical Connections 7-B) Indicate whether each of the following statements concerning use of sucrose and fructose as sweeteners is true or false.
a. The source for HFCS is milled barley.
b. HFCS-90 is the sweetener used in the soft drink industry.
c. The acronym HFCS stands for "high-frequency concentrated sucrose."
d. Most fruits have a fructose/glucose ratio between 3 and 5.

●7.131 (Chemical Connections 7-C) Indicate whether each of the following statements concerning sugar substitutes is true or false.
a. Aspartame is the most widely used sugar substitute.
b. Sucralose is a sucrose molecule in which three of the hydroxyl groups have been replaced with methyl groups.
c. A requirement for FDA approval of a sugar substitute is that it must be heat-stable.
d. Sodium cyclamate is no longer used as a sugar substitute because it causes cancer in animals.

●**7.132** (Chemical Connections 7-C) Indicate whether each of the following statements concerning sugar substitutes is true or false.
a. Aspartame is a zero-calorie sugar substitute.
b. Saccharin is a sugar substitute that is banned in Canada but approved for use in the United States.
c. Sucralose has a sweetness factor greater than that of other FDA-approved sugar substitutes.
d. Neotame and aspartame have structures based on the same two amino acid building blocks.

Oligosaccharides (Section 7.14)

7.133 Characterize the oligosaccharide raffinose in terms of
a. total number of monosaccharide units present.
b. total number of different kinds of monosaccharide units present.
c. total number of glycosidic linkages present.
d. total number of different kinds of glycosidic linkages present.

7.134 Characterize the oligosaccharide stachyose in terms of
a. total number of monosaccharide units present.
b. total number of different kinds of monosaccharide units present.
c. total number of glycosidic linkages present.
d. total number of different kinds of glycosidic linkages present.

▲7.135 Indicate whether or not one or more galactose monosaccharide units is/are present in the structure of the following carbohydrates.
a. Sucrose b. Ribose c. Stachyose d. Lactose

▲7.136 Indicate whether or not one or more galactose monosaccharide units is/are present in the structure of the following carbohydrates.
a. Cellobiose b. Fructose
c. Raffinose d. Maltose

▲7.137 Identify the type(s) of glycosidic linkage(s) [$\alpha(1 \rightarrow 4)$, etc.] present in each of the following carbohydrates, or indicate that none are present.
a. Maltose b. Galactose
c. Stachyose d. Fructose

▲7.138 Identify the type(s) of glycosidic linkage(s) [$\alpha(1 \rightarrow 4)$, etc.] present in each of the following carbohydrates, or indicate that none are present.
a. Sucrose b. Raffinose
c. Cellobiose d. Lactose

•7.139 (Chemical Connections 7-D) Indicate whether each of the following statements concerning blood-type chemistry is true or false.
a. There are four types of oligosaccharide markers for red blood cells.
b. The oligosaccharide marker for type O blood is a tetrasaccharide.
c. Blood-type distribution varies by ethnic group.
d. The monosaccharide derivative *N*-acetylglucosamine is a component of all red blood cell biochemical markers.

•7.140 (Chemical Connections 7-D) Indicate whether each of the following statements concerning blood-type chemistry is true or false.
a. Oligosaccharide markers are attached to red blood cells via a galactose monosaccharide unit.
b. The oligosaccharide marker for type A blood is a hexasaccharide.
c. Fucose is a galactose derivative in which a $—CH_3$ group has replaced a $—CH_2OH$ group.
d. Type B blood is more prevalent in Asian people than in Hispanic people.

General Characteristics of Polysaccharides (Section 7.15)

7.141 What is the difference, if any, between a polysaccharide and a glycan?

7.142 What is the difference, if any, between a homopolysaccharide and a heteropolysaccharide?

7.143 What is the range for the polymer chain length in a polysaccharide?

7.144 Contrast polysaccharides with mono- and disaccharides in terms of general property differences.

Storage Polysaccharides (Section 7.16)

7.145 Indicate whether or not each of the following is a correct characterization for the amylose form of starch.
a. It is a homopolysaccharide.
b. It contains two different types of monosaccharide molecules.
c. It is a branched-chain glucose polymer.
d. All glycosidic linkages present are $\alpha(1 \rightarrow 4)$.

7.146 Indicate whether or not each of the following is a correct characterization for glycogen.
a. It is a homopolysaccharide.
b. It contains two different types of monosaccharide molecules.
c. It is a branched-chain glucose polymer.
d. All glycosidic linkages present are $\alpha(1 \rightarrow 4)$.

7.147 What is the difference, if any, between the amylose and amylopectin forms of starch in terms of the following?
a. Relative abundance
b. Length of polymer chain
c. Type of glycosidic linkages present
d. Type of monosaccharide monomers present

7.148 Which of the characterizations *homopolysaccharide, heteropolysaccharide, straight-chain polysaccharide,* and *storage polysaccharide* applies to both members of each of the following pairs of substances? More than one characterization may apply in a given situation.
a. Glycogen and starch
b. Amylose and amylopectin
c. Glycogen and amylose
d. Starch and amylopectin

Structural Polysaccharides (Section 7.17)

7.149 Indicate whether or not each of the following is a correct characterization for cellulose.
a. It is an unbranched glucose polymer.
b. Its glycosidic linkages are of the same type as those in starch.
c. It is a source of nutrition for humans.
d. One of its biochemical functions is that of dietary fiber.

7.150 Indicate whether or not each of the following is a correct characterization for chitin.
a. It is an unbranched polymer.
b. Two different types of monomers are present.
c. Glycosidic linkages present are the same as those in cellulose.
d. The monomers present are glucose derivatives rather than glucose itself.

7.151 Indicate whether or not each of the following characterizations applies to (1) both cellulose and chitin (2) to cellulose only (3) to chitin only or (4) to neither cellulose nor chitin.
a. Storage polysaccharide
b. Monomers are glucose units
c. Glycosidic linkages are all $\alpha(1 \rightarrow 4)$
d. An unbranched polymer

7.152 Indicate whether or not each of the following characterizations applies to (1) both cellulose and chitin (2) to cellulose only (3) to chitin only or (4) to neither cellulose nor chitin.
a. Structural polysaccharide
b. Monomers are glucose derivatives
c. Glycosidic linkages are all $(1 \rightarrow 4)$
d. Homopolysaccharide

▲7.153 Match each of the following structural characteristics to the polysaccharides *amylopectin, amylose, glycogen, cellulose,* and *chitin*. A specific characteristic may apply to more than one of the polysaccharides.
a. $\alpha(1 \rightarrow 4)$ glycosidic linkages are present.
b. All of the glycosidic linkages present are of the same type.

c. The polymer chain is unbranched.
d. The monosaccharide repeating unit is not glucose.

▲7.154 Match each of the following structural characteristics to the polysaccharides *amylopectin*, *amylose*, *glycogen*, *cellulose*, and *chitin*. A specific characteristic may apply to more than one of the polysaccharides.
a. $\beta(1 \rightarrow 4)$ glycosidic linkages are present.
b. Two different kinds of glycosidic linkages are present.
c. The polymer chain is branched.
d. The monosaccharide repeating unit is a glucose derivative.

Acidic Polysaccharides (Section 7.18)

7.155 Indicate whether each of the following statements about the polysaccharide hyaluronic acid is true or false.
a. One of its monosaccharide building blocks is NAG.
b. One of its monosaccharide building blocks has a −2 charge.
c. Two types of glycosidic linkages are present.
d. One of its biochemical functions is as a lubricant for joints.

7.156 Indicate whether each of the following statements about the polysaccharide heparin is true or false.
a. Its biochemical function is to dissolve blood clots.
b. Both of its monosaccharide building blocks contain the element sulfur.
c. $\beta(1 \rightarrow 3)$ glycosidic linkages are present.
d. It has one of the longest chain lengths of any polysaccharide.

▲7.157 Which of the characterizations *homopolysaccharide*, *heteropolysaccharide*, *branched polysaccharide*, and *unbranched polysaccharide* applies to both members of each of the following pairs of carbohydrates? More than one characterization may apply in a given situation.
a. Starch and cellulose
b. Glycogen and amylopectin
c. Amylose and chitin
d. Heparin and hyaluronic acid

▲7.158 Which of the characterizations *homopolysaccharide*, *heteropolysaccharide*, *branched polysaccharide*, and *unbranched polysaccharide* applies to both members of each of the following pairs of carbohydrates? More than one characterization may apply in a given situation.
a. Glycogen and starch
b. Amylose and amylopectin
c. Chitin and hyaluronic acid
d. Heparin and cellulose

▲7.159 Indicate whether each of the following is a *storage polysaccharide*, a *structural polysaccharide*, an *acidic polysaccharide*, or a *non-polysaccharide*.
a. Amylose b. Stachyose
c. Hyaluronic acid d. Cellulose

▲7.160 Indicate whether each of the following is a *storage polysaccharide*, a *structural polysaccharide*, an *acidic polysaccharide*, or a *non-polysaccharide*.
a. Heparin b. Chitin c. Glycogen d. Raffinose

▲7.161 Match the polysaccharides *amylopectin*, *amylose*, *cellulose*, *chitin*, *glycogen*, *heparin*, and *hyaluronic acid* to the following glycosidic linkage characterizations. More than one of the polysaccharides may be correct in a given situation.
a. All glycosidic linkages present are the same.
b. Some, but not all, glycosidic linkages are $\alpha(1 \rightarrow 4)$ linkages.
c. Both $\beta(1 \rightarrow 3)$ and $\beta(1 \rightarrow 4)$ glycosidic linkages are present.
d. All glycosidic linkages are $\alpha(1 \rightarrow 4)$ linkages.

▲7.162 Match the polysaccharides *amylopectin*, *amylose*, *cellulose*, *chitin*, *glycogen*, *heparin*, and *hyaluronic acid* to the following glycosidic linkage characterizations. More than one of the polysaccharides may be correct in a given situation.
a. Two different types of glycosidic linkages are present.
b. Some, but not all, glycosidic linkages are $\alpha(1 \rightarrow 6)$ linkages.
c. All glycosidic linkages are $(1 \rightarrow 4)$ linkages.
d. All glycosidic linkages are $\beta(1 \rightarrow 4)$ linkages.

Dietary Considerations and Carbohydrates (Section 7.19)

7.163 In a dietary context, what is the difference between a *simple* carbohydrate and a *complex* carbohydrate?

7.164 In a dietary context, what is the difference between a *natural* sugar and a *refined* sugar?

7.165 In a dietary context, what are *empty* calories?

7.166 In a dietary context, what is the *glycemic effect*?

●7.167 (Chemical Connections 7-E) Indicate whether each of the following statements concerning glycemic response measurement is true or false.
a. A GL value is usually 2 to 3 times larger than the GI value for the same food.
b. The standard for GI values is usually whole-wheat bread.
c. GI values are based on a food portion size that contains a specific amount of carbohydrate.
d. Larger bites of food evoke a different glycemic response than smaller bites of the same food.

●7.168 (Chemical Connections 7-E) Indicate whether each of the following statements concerning glycemic response measurement is true or false.
a. A GI value is a ratio of blood-glucose change compared to a standard.
b. Overripe fruit produces a different glycemic response than the same fruit when it is underripe.
c. The serving-size standard for GI values is 10 g of contained carbohydrate.
d. Some carbohydrate-containing foods have a high GI value and a low GL value.

Glycolipids and Glycoproteins (Section 7.20)

7.169 In terms of general structure, what is a glycolipid?

7.170 In terms of general structure, what is a glycoprotein?

7.171 Describe the general features of the cell recognition process in which glycoproteins participate.

7.172 Describe the general features of the cell recognition process in which glycolipids participate.

Lipids

8

Dan Guravich/Photo Researchers

Fats and oils are the most widely occurring types of lipids. Thick layers of fat help insulate polar bears against the effects of low temperatures.

CHAPTER OUTLINE

There are four major classes of bioorganic substances: carbohydrates, lipids, proteins, and nucleic acids (Section 7.1). In the previous chapter the first of these classes, carbohydrates, was considered. Attention now turns to the second of the bioorganic classes, the compounds called lipids.

Lipids known as fats provide a major way of storing chemical energy and carbon atoms in the body. Fats also surround and insulate vital body organs, providing protection from mechanical shock and preventing excessive loss of heat energy. Phospholipids, glycolipids, and cholesterol (a lipid) are the basic components of cell membranes. Several cholesterol derivatives function as chemical messengers (hormones) within the body.

8.1 Structure and Classification of Lipids

Unlike carbohydrates and most other classes of compounds, lipids do not have a common structural feature that serves as the basis for defining such compounds. Instead, their characterization is based on solubility characteristics. A **lipid** *is an organic compound found in living organisms that is insoluble (or only sparingly*

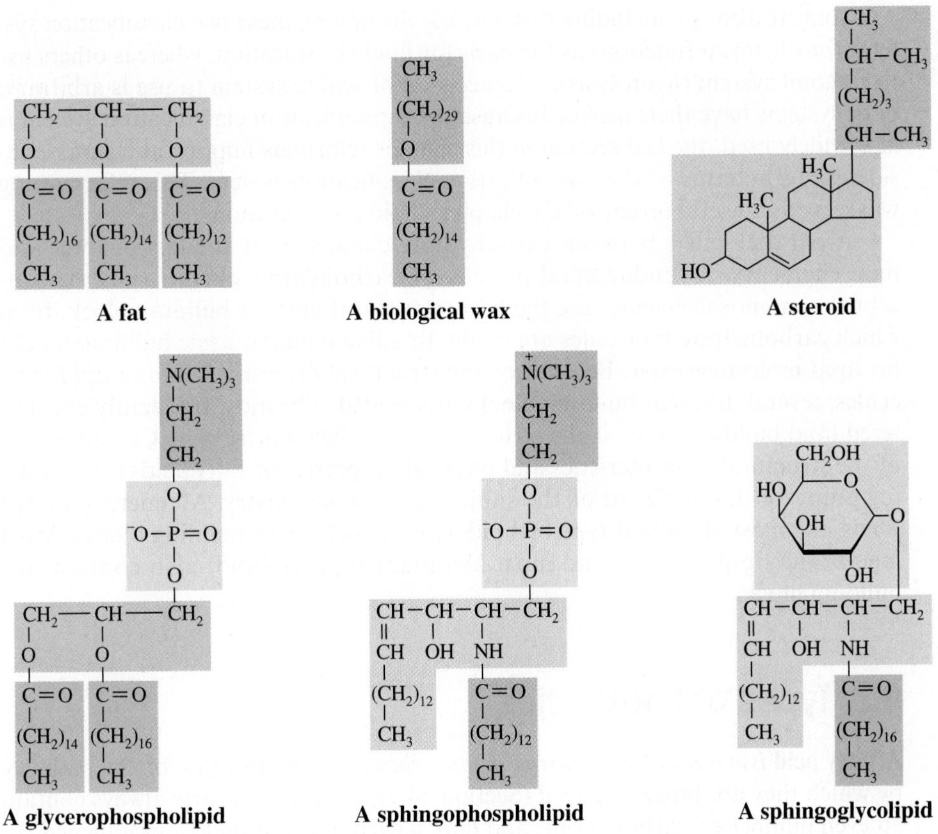

Figure 8.1 The structural formulas of these types of lipids illustrate the great structural diversity among lipids. The defining parameter for lipids is solubility rather than structure.

soluble) in water but soluble in nonpolar organic solvents. When a biochemical material (human, animal, or plant tissue) is homogenized in a blender and mixed with a nonpolar organic solvent, the substances that dissolve in the solvent are the lipids.

Figure 8.1 shows the structural diversity that is associated with lipid molecules. Some are esters, some are amides, and some are alcohols; some are acyclic, some are cyclic, and some are polycyclic. The common thread that ties all of the compounds of Figure 8.1 together is solubility rather than structure. All are insoluble in water.

Two common methods exist for subclassifying lipids into families for the purpose of study. One method uses the biochemical function of a lipid as the basis for classification, and the other method is based on whether or not a lipid can be broken down into smaller units through basic hydrolysis, that is, reaction with water under basic conditions. A hydrolysis reaction that occurs in basic solution is called a *saponification reaction* (Section 5.16).

Based on biochemical function, lipids are divided into five categories:

1. **Energy-storage lipids** (triacylglycerols)
2. **Membrane lipids** (phospholipids, sphingoglycolipids, and cholesterol)
3. **Emulsification lipids** (bile acids)
4. **Messenger lipids** (steroid hormones and eicosanoids)
5. **Protective-coating lipids** (biological waxes)

Based upon whether or not saponification occurs when a lipid is placed in basic aqueous solution, lipids are divided into two categories:

1. **Saponifiable lipids** (triacylglycerols, phospholipids, sphingoglycolipids, and biological waxes)
2. **Nonsaponifiable lipids** (cholesterol, steroid hormones, bile acids, and eicosanoids)

Saponifiable lipids are converted into smaller molecules when hydrolysis occurs. Nonsaponifiable lipids cannot be broken up into smaller units since they do not react with water.

Some textbooks, including this one, use the first of these two classification systems (biochemical function) as the basis for lipid classification, whereas others use the second system (hydrolysis). The decision of which system to use is arbitrary; both systems have their merits. Because the saponification classification system is also widely used, the last section in this chapter reformats important chapter considerations in terms of the saponification classification system. This reformatting will serve as a useful review of the chapter's lipid considerations.

A parallel exists between carbohydrate chemistry of the last chapter and lipid chemistry. A fundamental premise of carbohydrate chemistry is the concept that monosaccharides are the basic structural unit, or building block, from which carbohydrate molecules are made. In a like manner, basic building blocks for lipid molecules exist. Because of the structural diversity found in lipid molecules, several different building blocks are needed. The most frequently encountered lipid building block is the structural unit called a *fatty acid.* Consideration of the structural characteristics and physical properties of fatty acids is the starting point for development of the subject of lipid chemistry. All energy-storage lipids, the most abundant type of lipid, contain fatty acid building blocks. Most membrane lipids, the second most abundant type of lipid, also contain this building block.

8.2 Types of Fatty Acids

Fatty acids were first isolated from naturally occurring fats; hence the designation *fatty acids.*

A **fatty acid** *is a naturally occurring monocarboxylic acid.* Because of the pathway by which they are biosynthesized (Section 14.7), fatty acids nearly always contain an even number of carbon atoms and have a carbon chain that is unbranched. In terms of carbon chain length, fatty acids are characterized as *long-chain fatty acids* (C_{12} to C_{26}), *medium-chain fatty acids* (C_8 and C_{10}), or *short-chain fatty acids* (C_4 and C_6). Fatty acids are rarely found free in nature but rather occur as part of the structure of more complex lipid molecules.

Saturated and Unsaturated Fatty Acids

The carbon chain of a fatty acid may or may not contain carbon–carbon double bonds. On the basis of this consideration, fatty acids are classified as saturated fatty acids (SFAs), monounsaturated fatty acids (MUFAs), or polyunsaturated fatty acids (PUFAs).

A **saturated fatty acid** *is a fatty acid with a carbon chain in which all carbon–carbon bonds are single bonds.* The structural formula for the 16-carbon SFA is

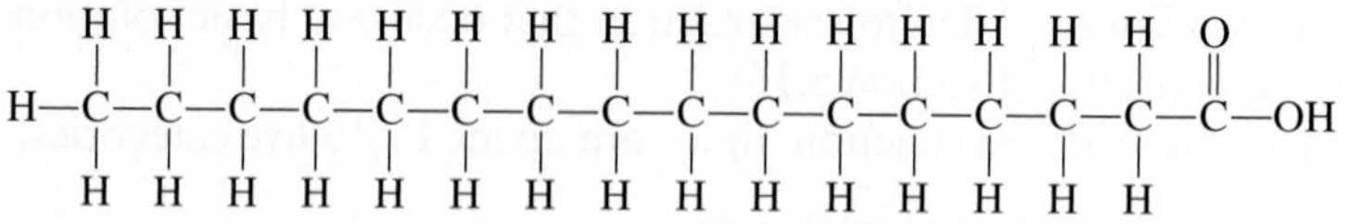

IUPAC name: hexadecanoic acid
Common name: palmitic acid

The structural formula for a fatty acid is usually written in a more condensed form than the preceding structural formula. Two alternative structural notations for palmitic acid are

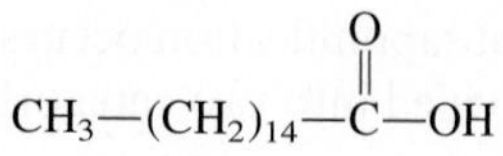

and

(Line-angle structural formulas were first encountered in Section 1.9.)

A **monounsaturated fatty acid** *is a fatty acid with a carbon chain in which one carbon–carbon double bond is present.* In biochemically important MUFAs, the configuration about the double bond is nearly always *cis* (Section 2.5). Different ways of depicting the structure of a MUFA follow.

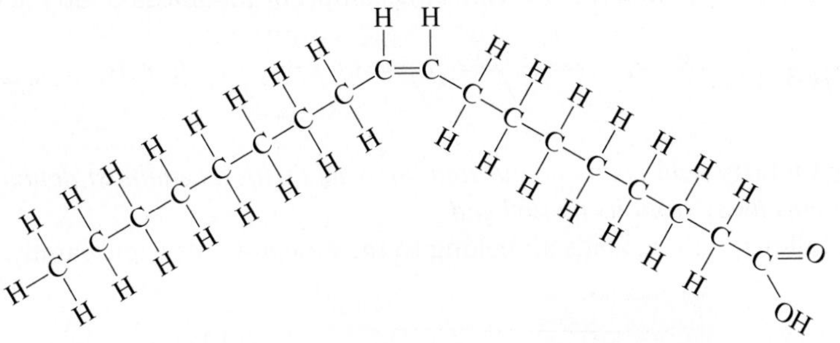

$$CH_3{-}(CH_2)_7{-}CH{=}CH{-}(CH_2)_7{-}\overset{\displaystyle O}{\overset{\|}{C}}{-}OH$$

IUPAC name: *cis*-9-octadecenoic acid
Common name: oleic acid

More than 500 different fatty acids have been isolated from the lipids of microorganisms, plants, animals, and humans. These fatty acids differ from one another in the length of their carbon chains, their degree of unsaturation (number of double bonds), and the positions of the double bonds in the chains.

The first of these structures correctly emphasizes that the presence of a *cis* double bond in the carbon chain puts a rigid 30° bend in the chain. Such a bend affects the physical properties of a fatty acid, as discussed in Section 8.3.

A **polyunsaturated fatty acid** *is a fatty acid with a carbon chain in which two or more carbon–carbon double bonds are present.* Up to six double bonds are found in biochemically important PUFAs.

Fatty acids are nearly always referred to using their common names. IUPAC names for fatty acids, although easily constructed, are usually quite long. These two types of names for an 18-carbon PUFA containing *cis* double bonds in the 9 and 12 positions are as follows:

IUPAC name: *cis,cis*-9,12-octadecadienoic acid
Common name: linoleic acid

The fatty acids present in naturally occurring lipids almost always have the following three characteristics:

1. An unbranched carbon chain
2. An even number of carbon atoms in the carbon chain
3. Double bonds, when present in the carbon chain, in a *cis* configuration

Unsaturated Fatty Acids and Double-Bond Position

A numerically based shorthand system exists for specifying key structural parameters for fatty acids. In this system, two numbers separated by a colon are used to specify the number of carbon atoms and the number of carbon–carbon double bonds present. The notation 18:0 denotes a C_{18} fatty acid with no double bonds, whereas the notation 18:2 signifies a C_{18} fatty acid in which two double bonds are present.

To specify double-bond positioning within the carbon chain of an unsaturated fatty acid, the preceding notation is expanded by adding the Greek capital letter delta (Δ) followed by one or more superscript numbers. The notation 18:3($\Delta^{9,12,15}$) denotes a C_{18} PUFA with three double bonds at locations between carbons 9 and 10, 12 and 13, and 15 and 16.

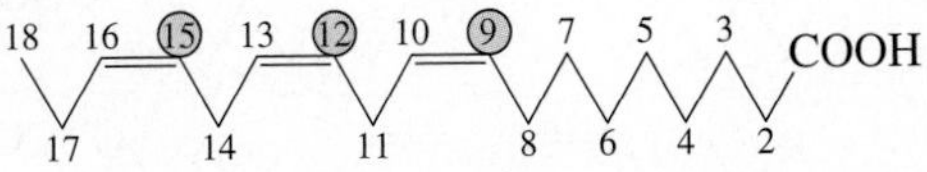

MUFAs are usually Δ^9 acids, and the first two additional double bonds in PUFAs are generally at the Δ^{12} and Δ^{15} locations. [A notable exception to this generalization is the biochemically important arachidonic acid, a PUFA with the structural parameters 20:4($\Delta^{5,8,11,14}$).] Denoting double-bond locations using this "delta notation" always assumes a numbering system in which the carboxyl carbon atom is C-1.

Several different "families" of unsaturated fatty acids exist. These family relationships become apparent when double-bond position is specified relative to the methyl (noncarboxyl) end of the fatty acid carbon chain. Double-bond positioning determined in this manner is denoted by using the Greek lowercase letter omega (ω). An **omega-3 fatty acid** *is an unsaturated fatty acid with its endmost double bond three carbon atoms away from its methyl end.* An example of an omega-3 fatty acid is

ω-3 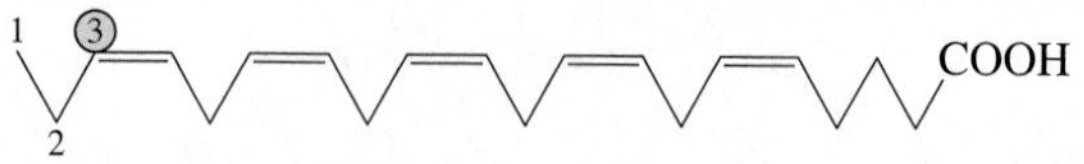

(20:5)

An **omega-6 fatty acid** *is an unsaturated fatty acid with its endmost double bond six carbon atoms away from its methyl end.*

The following three acids all belong to the omega-6 fatty acid family.

ω-6 

(14:1)

ω-6 

(18:2)

ω-6 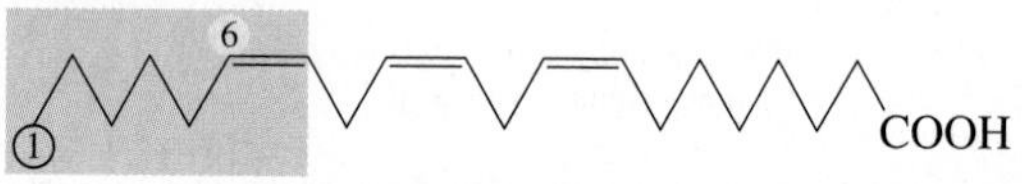

(20:3)

The structural feature common to these omega-6 fatty acids is highlighted with color in the preceding structural formulas. All the members of an omega family of fatty acids have structures in which the same "methyl end" is present.

Table 8.1 gives the names and structures of the fatty acids most commonly encountered as building blocks in biochemically important lipid structures, as well as the "delta" and "omega" notations for the acids.

Table 8.1 Selected Fatty Acids of Biological Importance

Structural Notation	Common Name	Structure
Saturated Fatty Acids		
12:0	lauric acid	
14:0	myristic acid	
16:0	palmitic acid	
18:0	stearic acid	
20:0	arachidic acid	
Monounsaturated Fatty Acids		
16:1 Δ^{9} ω-7	palmitoleic acid	
18:1 Δ^{9} ω-9	oleic acid	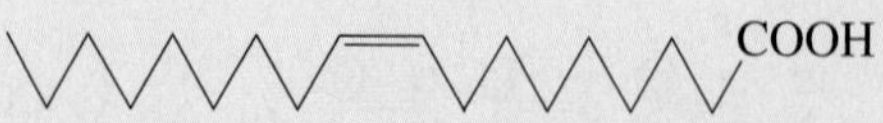
Polyunsaturated Fatty Acids		
18:2 $\Delta^{9,12}$ ω-6	linoleic acid	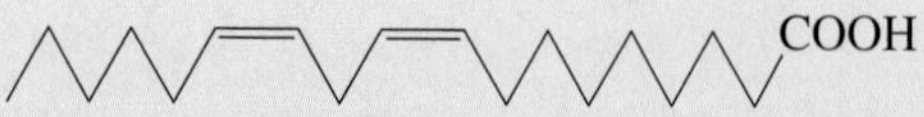
18:3 $\Delta^{9,12,15}$ ω-3	linolenic acid	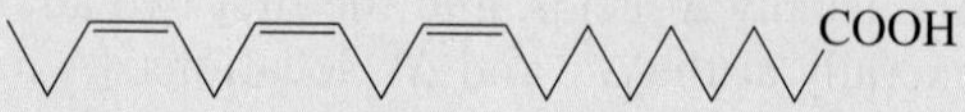
20:4 $\Delta^{5,8,11,14}$ ω-6	arachidonic acid	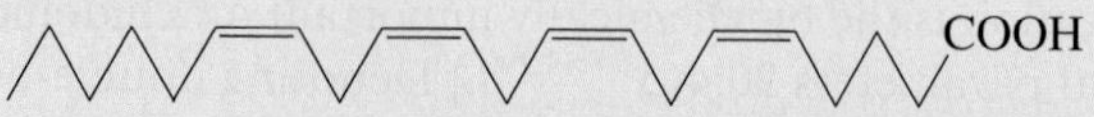
20:5 $\Delta^{5,8,11,14,17}$ ω-3	EPA (eicosapentaenoic acid)	
22:6 $\Delta^{4,7,10,13,16,19}$ ω-3	DHA (docosahexaenoic acid)	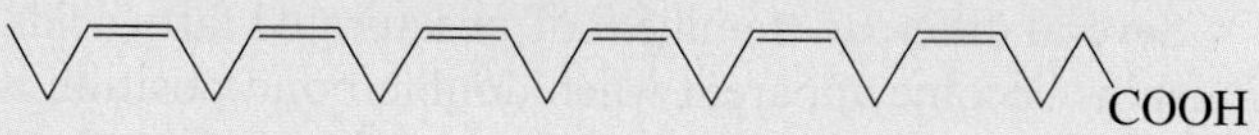

EXAMPLE 8.1 Classifying Fatty Acids on the Basis of Structural Characteristics

Classify the fatty acid with the following structural formula in the ways indicated.

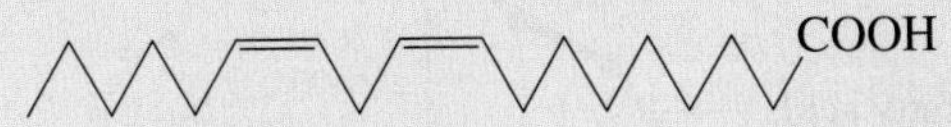

a. What is the type designation (SFA, MUFA, or PUFA) for this fatty acid?
b. On the basis of carbon chain length and degree of unsaturation, what is the numerical shorthand designation for this fatty acid?
c. To which "omega" family of fatty acids does this fatty acid belong?
d. What is the "delta" designation for the carbon chain double-bond locations for this fatty acid?

Solution

a. Two carbon–carbon double bonds are present in this molecule, which makes it a *polyunsaturated fatty acid (PUFA)*.
b. Eighteen carbon atoms and two carbon–carbon double bonds are present. The shorthand numerical designation for this fatty acid is thus *18:2*.
c. Counting from the methyl end of the carbon chain, the first double bond encountered involves carbons 6 and 7. This fatty acid belongs to the *omega-6* family of fatty acids.
d. Counting from the carboxyl end of the carbon chain, with C-1 being the carboxyl group, the double-bond locations are 9 and 12. This is a $\Delta^{9,12}$ *fatty acid.*

Practice Exercise 8.1

Classify the fatty acid with the following structural formula in the ways indicated.

a. What is the type designation (SFA, MUFA, or PUFA) for this fatty acid?
b. On the basis of carbon chain length and degree of unsaturation, what is the numerical shorthand designation for this fatty acid?
c. To which "omega" family of fatty acids does this fatty acid belong?
d. What is the "delta" designation for the carbon chain double-bond location for this fatty acid?

Answers: **a.** MUFA (monounsaturated fatty acid); **b.** 12:1 fatty acid; **c.** omega-3 fatty acid (ω-3); **d.** delta-9 fatty acid (Δ^9)

8.3 Physical Properties of Fatty Acids

The physical properties of fatty acids, and of lipids that contain them, are largely determined by the length and degree of unsaturation of the fatty acid carbon chain.

Water solubility for fatty acids is a direct function of carbon chain length; solubility decreases as carbon chain length increases. Short-chain fatty acids have a slight solubility in water. Long-chain fatty acids are essentially insoluble in water. The slight solubility of short-chain fatty acids is related to the polarity of the carboxyl group present. In longer-chain fatty acids, the nonpolar nature of the hydrocarbon chain completely dominates solubility considerations.

Fatty acids have low water solubilities, which decrease with increasing carbon chain length; at 30°C, lauric acid (12:0) has a water solubility of 0.063 g/L and stearic acid (18:0) a solubility of 0.0034 g/L. Contrast this with glucose's solubility in water at the same temperature, 1100 g/L.

Melting points for fatty acids are strongly influenced by both carbon chain length and degree of unsaturation (number of double bonds present). Figure 8.2 shows melting-point variation as a function of both of these variables. As carbon chain length increases, melting point increases. This trend is related to the greater surface area associated with a longer carbon chain and to the increased opportunities that this greater surface area affords for intermolecular attractions between fatty acid molecules.

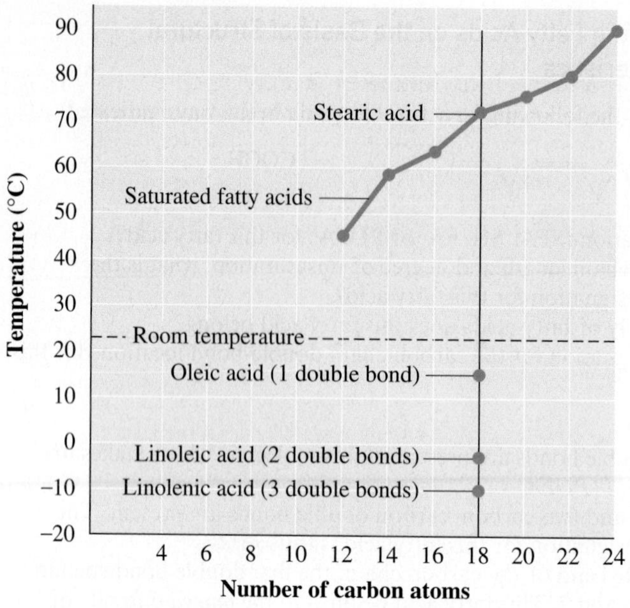

Figure 8.2 The melting point of a fatty acid depends on the length of the carbon chain and on the number of double bonds present in the carbon chain.

A trend of particular significance is that saturated fatty acids have higher melting points than unsaturated fatty acids with the same number of carbon atoms. The greater the degree of unsaturation, the greater the reduction in melting points. Figure 8.2 shows this effect for the 18-carbon acids with zero, one, two, and three double bonds. Long-chain saturated fatty acids tend to be solids at room temperature, whereas long-chain unsaturated fatty acids tend to be liquids at room temperature.

The decreasing melting point associated with increasing degree of unsaturation in fatty acids is explained by decreased molecular attractions between carbon chains. The double bonds in unsaturated fatty acids, which generally have the *cis* configuration, produce "bends" in the carbon chains of these molecules (Figure 8.3). These "bends" prevent unsaturated fatty acids from packing together as tightly as saturated fatty acids. The greater the number of double bonds, the less efficient the packing. As a result, unsaturated fatty acids always have fewer intermolecular attractions, and therefore lower melting points, than their saturated counterparts.

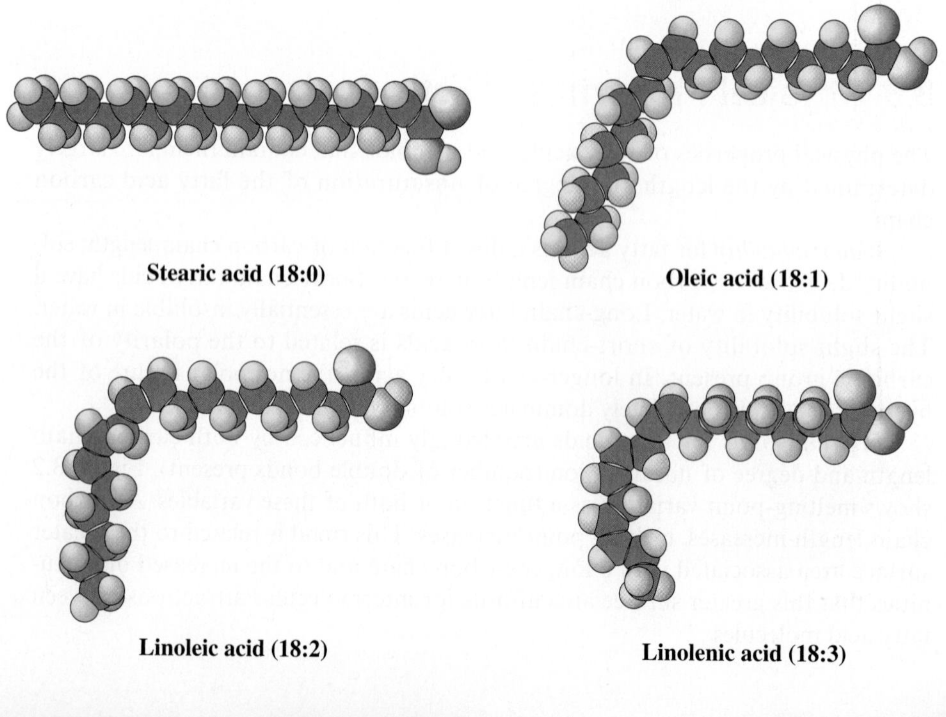

Figure 8.3 Space-filling models of four 18-carbon fatty acids, which differ in the number of double bonds present. Note how the presence of double bonds changes the shape of the molecule.

8.4 Energy-Storage Lipids: Triacylglycerols

With the notable exception of nerve cells, human cells store small amounts of energy-providing materials for use when energy demand is high. The most widespread energy-storage material within cells is the carbohydrate glycogen (Section 7.15); it is present in small amounts in most cells.

Lipids known as triacylglycerols also function within the body as energy-storage materials. Rather than being widespread, triacylglycerols are concentrated primarily in special cells (adipocytes) that are nearly filled with the material. Adipose tissue containing these cells is found in various parts of the body: under the skin, in the abdominal cavity, in the mammary glands, and around various organs (Figure 8.4). Triacylglycerols are much more efficient at storing energy than is glycogen because large quantities of them can be packed into a very small volume. These energy-storage lipids are the most abundant type of lipid present in the human body.

Manfred Kage/Peter Arnold, Inc.

Figure 8.4 An electron micrograph of adipocytes, the body's triacylglycerol-storing cells. Note the bulging spherical shape.

In terms of functional groups present, triacylglycerols are triesters; three ester functional groups are present. Recall from Section 5.10 that an ester is a compound produced from the reaction of an alcohol with a carboxylic acid. The alcohol involved in triacylglycerol formation is always glycerol, a three-carbon alcohol with three hydroxyl groups.

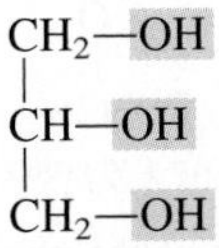

Glycerol

Fatty acids are the carboxylic acids involved in triacylglycerol formation. In the esterification reaction producing a triacylglycerol, a single molecule of glycerol reacts with three fatty acid molecules; each of the three hydroxyl groups present is esterified. Figure 8.5 shows the triple esterification reaction that occurs between glycerol and three molecules of stearic acid (18:0); note the production of three molecules of water as a by-product of the reaction.

Two general ways to represent the structure of a triacylglycerol are

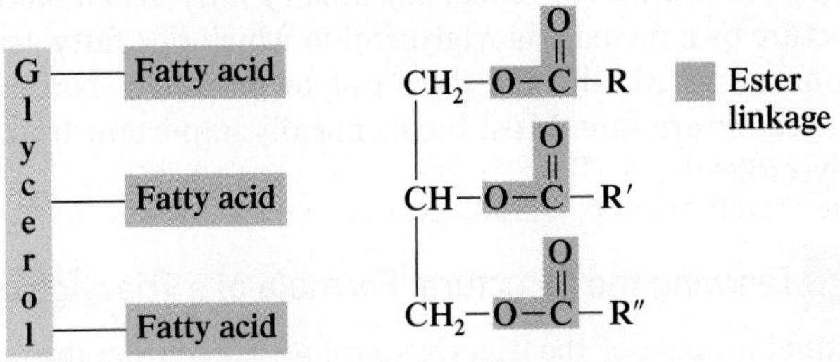

Triacylglycerols do not actually contain glycerol and three fatty acids, as the block diagram for a triacylglycerol implies. They actually contain a glycerol *residue* and three fatty acid *residues*. In the formation of the triacylglycerol, three molecules of water have been removed from the structural components of the triacylglycerol, leaving residues of the reacting molecules.

The first representation, a block diagram, shows the four subunits (building blocks) present in the structure: glycerol and three fatty acids. The second representation, a general structural formula, shows the three ester linkages present in a triacylglycerol. Each of the fatty acids is attached to glycerol through an ester linkage.

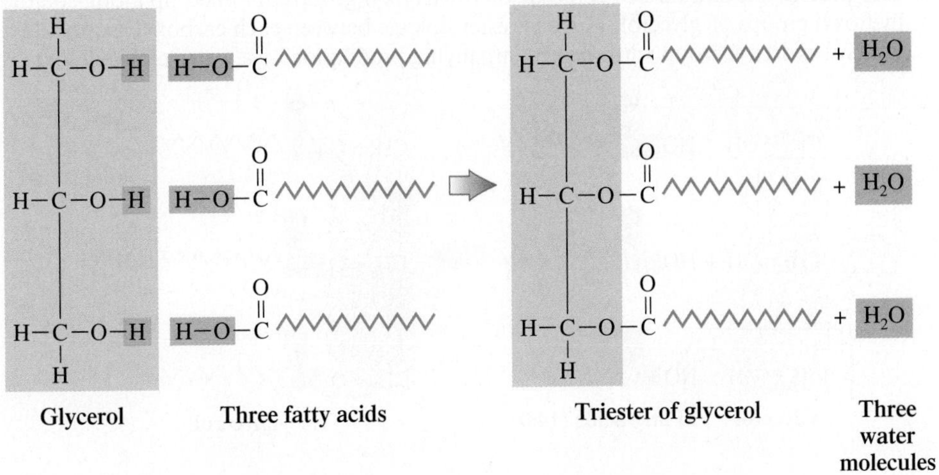

Figure 8.5 Structure of the simple triacylglycerol produced from the triple esterification reaction between glycerol and three molecules of stearic acid (18:0 acid). Three molecules of water are a by-product of this reaction.

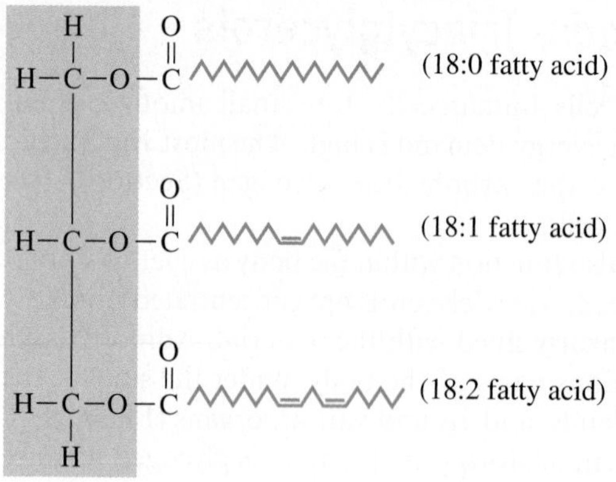

Figure 8.6 Structure of a mixed triacylglycerol in which three different fatty acid residues are present.

Formally defined, a **triacylglycerol** *is a lipid formed by esterification of three fatty acids to a glycerol molecule.* Within the name *triacylglycerol* is the term *acyl.* An *acyl group,* previously defined and considered in Section 5.1, is the portion of a carboxylic acid that remains after the —OH group is removed from the carboxyl carbon atom. The structural representation for an acyl group is

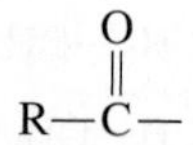

An acyl group

Thus, as the name implies, triacylglycerol molecules contain three fatty acid residues (three acyl groups) attached to a glycerol residue. An older name that is still frequently used for a triacylglycerol is *triglyceride.*

The triacylglycerol produced from glycerol and three molecules of stearic acid (shown in Figure 8.5) is an example of a simple triacylglycerol. A **simple triacylglycerol** *is a triester formed from the esterification of glycerol with three identical fatty acid molecules.* If the reacting fatty acid molecules are not all identical, then the result is a mixed triacylglycerol. A **mixed triacylglycerol** *is a triester formed from the esterification of glycerol with more than one kind of fatty acid molecule.* Figure 8.6 shows the structure of a mixed triacylglycerol in which one fatty acid is saturated, another monounsaturated, and the third polyunsaturated. Naturally occurring *simple* triacylglycerols are rare. Most biochemically important triacylglycerols are *mixed* triacylglycerols.

EXAMPLE 8.2 Drawing the Structural Formula of a Triacylglycerol

Draw the structural formula of the triacylglycerol produced from the reaction between glycerol and three molecules of myristic acid.

Solution

Table 8.1 shows that myristic acid is the 14:0 fatty acid. Draw the structure of glycerol and then place three molecules of myristic acid alongside the glycerol. The fatty acid placements should be such that their carboxyl groups are lined up alongside the hydroxyl groups of glycerol. Form an ester linkage between each carboxyl group and a glycerol hydroxyl group with the accompanying production of a water molecule.

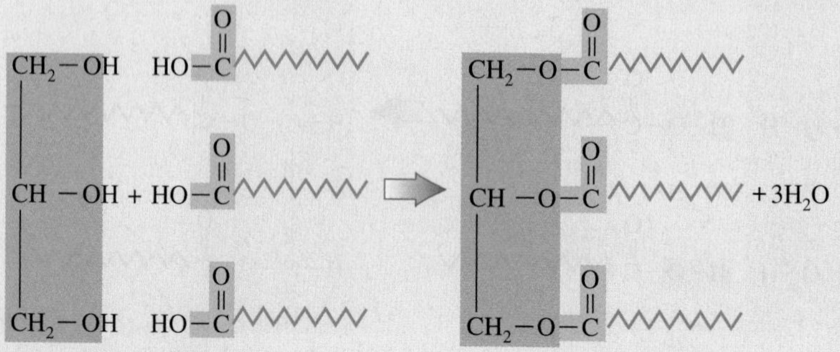

Glycerol   Fatty acids (14:0)   Triacylglycerol

Practice Exercise 8.2

Draw the structural formula of the triacylglycerol produced from the reaction between glycerol and three molecules of lauric acid.

Answer:

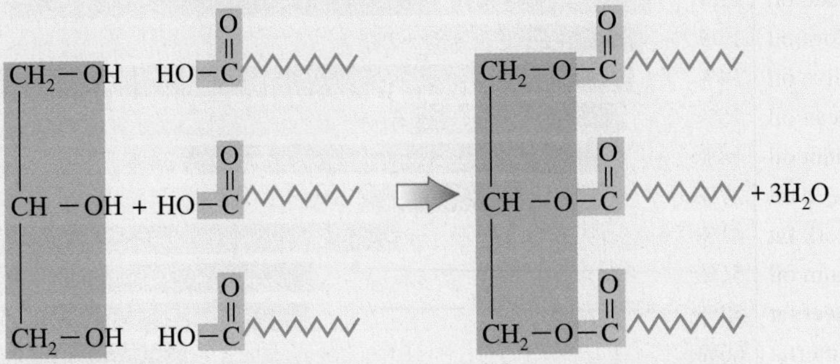

Glycerol   Fatty acids (12:0)   Triacylglycerol

Fats and Oils

Fats are naturally occurring mixtures of triacylglycerol molecules in which many different kinds of triacylglycerol molecules are present. *Oils* are also naturally occurring mixtures of triacylglycerol molecules in which there are many different kinds of triacylglycerol molecules present. Given that both are triacylglycerol mixtures, what distinguishes a fat from an oil? The answer is physical state at room temperature. A **fat** *is a triacylglycerol mixture that is a solid or a semi-solid at room temperature (25°C)*. Generally, fats are obtained from animal sources. An **oil** *is a triacylglycerol mixture that is a liquid at room temperature (25°C)*. Generally, oils are obtained from plant sources. Because they are mixtures, no fat or oil can be represented by a single specific chemical formula. Many different fatty acids are present in the triacyl-glycerol molecules found in the mixture. The actual composition of a fat or oil varies even for the species from which it is obtained. Composition depends on both dietary and climatic factors. For example, fat obtained from corn-fed hogs has a different overall composition than fat obtained from peanut-fed hogs. Flaxseed grown in warm climates gives oil with a different composition from that obtained from flaxseed grown in colder climates.

Additional generalizations and comparisons between fats and oils follow.

1. Fats are composed largely of triacylglycerols in which saturated fatty acids predominate, although some unsaturated fatty acids are present. Such triacylglycerols can pack closely together because of the "linearity" of their fatty acid chains (Figure 8.7a), thus causing the higher melting points associated with fats. Oils contain triacylglycerols with larger amounts of mono- and polyunsaturated fatty acids than those in fats. Such triacylglycerols cannot pack as tightly together because of "bends" in their fatty acid chains (Figure 8.7b). The result is lower melting points.

Petroleum oils (Section 1.15) are structurally different from *lipid oils*. The former are mixtures of alkanes and cycloalkanes. The latter are mixtures of triesters of glycerol.

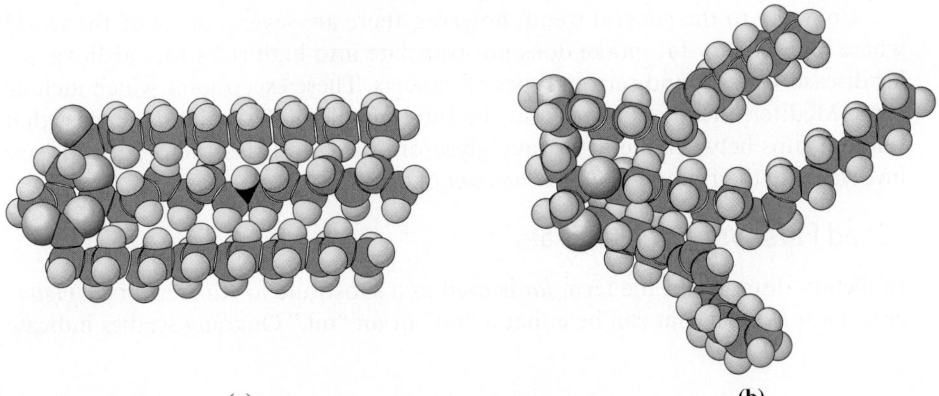

(a)   (b)

Figure 8.7 Representative triacylglycerols from (a) a fat and (b) an oil.

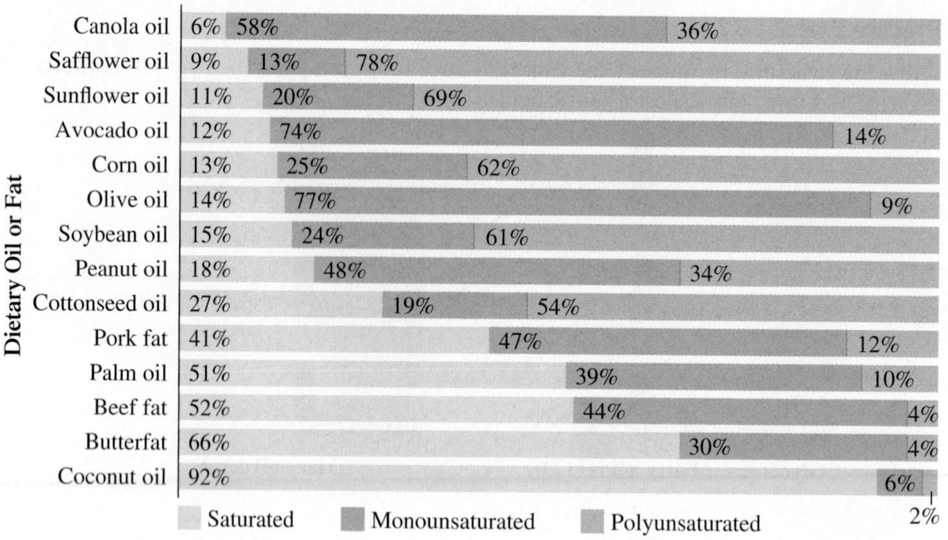

Figure 8.8 Percentages of saturated, monounsaturated, and polyunsaturated fatty acids in the triacylglycerols of various dietary fats and oils.

2. Fats are generally obtained from animals; hence the term *animal fat.* Although fats are solids at room temperature, the warmer body temperature of the living animal keeps the fat somewhat liquid (semi-solid) and thus allows for movement. Oils typically come from plants, although there are also fish oils. A fish would have some serious problems if its triacylglycerols "solidified" when it encountered cold water.
3. Pure fats and pure oils are colorless, odorless, and tasteless. The tastes, odors, and colors associated with dietary plant oils are caused by small amounts of other naturally occurring substances present in the plant that have been carried along during processing. The presence of these "other" compounds is usually considered desirable.

Fats contain both saturated and unsaturated fatty acids. Oils also contain both saturated and unsaturated fatty acids. The difference between a fat and an oil lies in which type of fatty acid is more prevalent. In fats, saturated fatty acids are more prevalent; in oils, unsaturated fatty acids are more prevalent.

Figure 8.8 gives the percentages of saturated, monounsaturated, and polyunsaturated fatty acids found in common dietary oils and fats. In general, a higher degree of fatty acid unsaturation is associated with oils than with fats. A notable exception to this generalization is coconut oil, which is highly saturated. This oil is a liquid not because it contains many double bonds within the fatty acids but because it is rich in *shorter-chain* fatty acids, particularly lauric acid (12:0).

8.5 Dietary Considerations and Triacylglycerols

In recent years, considerable research has been carried out concerning the role of dietary factors as a cause of disease (obesity, diabetes, cancer, hypertension, and atherosclerosis). Numerous studies have shown that, *in general,* nations whose citizens have high dietary intakes of triacylglycerols (fats and oils) tend to have higher incidences of heart disease and certain types of cancers. This is the reason for concern that the typical American diet contains too much fat and the call for Americans to reduce their total dietary fat intake.

Contrary to the general trend, however, there are several areas of the world where high dietary fat intake does not translate into high risks for cardiovascular disease, obesity, and certain types of cancers. These exceptions, which include some Mediterranean countries and the Inuit people of Greenland, suggest that relationships between dietary triacylglycerol intake and risk factors for disease involve more than simply the *total amount* of triacylglycerols consumed.

A grain- and vegetable-rich diet that contains small amounts of extra-virgin olive oil (three to four teaspoons daily) has been found to help people with high blood pressure reduce the amount of blood pressure medication they require, on average, by 48%. Substitution of sunflower oil for the olive oil resulted in only a 4% reduction in medication dosage.

The blood-pressure-reduction benefits of olive oil do not relate to the triacylglycerols present but rather come from other compounds naturally present, namely from antioxidant polyphenols olive oil contains. These antioxidants help promote the relaxation of blood vessels.

"Good Fats" Versus "Bad Fats"

In dietary discussions, the term *fat* is used as a substitute for the term *triacylglycerol.* Thus a dietary fat can be either a "fat" or an "oil." Ongoing studies indicate that both the *type of dietary fat* consumed and the *amount of dietary fat* consumed are important factors in determining human body responses to dietary fat. Current dietary fat recommendations are that people limit their total fat intake to 30% of total calories—with up to 15% coming from monounsaturated fat, up to 10% from polyunsaturated fat, and less than 10% from saturated fats.

These recommendations imply correctly that different types of dietary fat have different effects. In simplified terms, research studies indicate that saturated fats are "bad fat," monounsaturated fats are "good fat," and polyunsaturated fats can be both "good fat" and "bad fat." In the latter case, fatty acid omega classification (Section 8.2) becomes important, a situation addressed later in this section. Studies indicate that saturated fat can increase heart disease risk, that monounsaturated fat can decrease both heart disease and breast cancer risk, and that polyunsaturated fat can reduce heart disease risk but promote the risk of certain types of cancers.

Referring to Figure 8.8, note the wide variance in the three general types of fatty acids (SFAs, MUFAs, and PUFAs) present in various kinds of dietary fats. Dietary fats high in "good" monounsaturated fatty acids include olive, avocado, and canola oils. Monounsaturated fatty acids help reduce the stickiness of blood platelets. This helps prevent the formation of blood clots and may also dissolve clots once they form.

Many people do not realize that most tree nuts and peanuts are good sources of MUFAs. The focus on relevancy feature Chemical Connections 8-A on the next page looks at recent research on the fat content of nuts.

Freshly pressed extra-virgin olive oil contains a compound that has the same pharmacological activity as the over-the-counter pain reliever ibuprofen (Section 5.4). This finding suggests a possible explanation for some of the various health benefits attributed to a Mediterranean diet that typically is rich in olive oil. It is estimated that this olive oil compound, called oleocanthal, is present in a typical Mediterranean diet in an amount equivalent to about 10% of the ibuprofen dose recommended for headache relief.

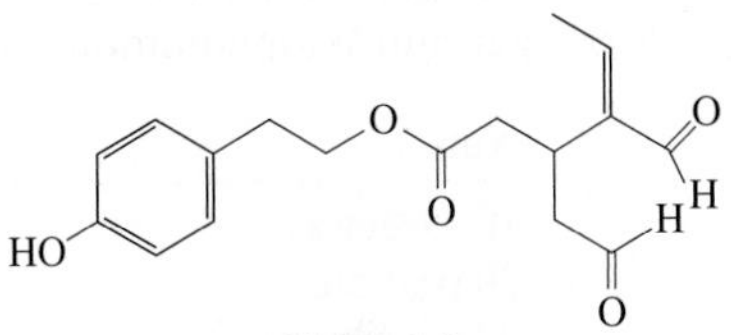

Oleocanthal

Omega-3 and Omega-6 Fatty Acids

In the 1980s, researchers found that the Inuit people of Greenland exhibit a low incidence of heart disease despite having a diet very high in fat. This contrasts markedly with studies on the U.S. population, which show a correlation between a high-fat diet and a high incidence of heart disease. What accounts for the difference between the two peoples? The Inuit diet is high in omega-3 fatty acids (from fish), and the U.S. diet is high in omega-6 fatty acids (from plant oils). An American consumes about double the amount of omega-6 fatty acids and half the amount of omega-3 fatty acids that an Inuit consumes.

Several large studies now confirm that benefits can be derived from eating several servings of fish each week. The choice of fish is important, however. Not all fish are equal in omega-3 fatty acid content. Cold-water fish, also called fatty fish because of the extra amounts of fat they have for insulation against the cold, contain more omega-3 acids than leaner, warm-water fish. Fatty fish include albacore tuna, salmon, and mackerel (Figure 8.9). Leaner, warm-water fish, which include cod, catfish, halibut, sole, and snapper, do not appear to offer as great a positive effect on heart health as do their "fatter" counterparts. (Note that most of the fish used in fish and chips (e.g., cod, halibut) is on the low end of the omega-3 scale.) Table 8.2 gives the actual omega-3 fatty acid concentrations associated with various kinds of cold-water fish.

Recommendations that the U.S. population increase their consumption of cold-water fish has sparked a demand for omega-3 fish oil supplements. Wild salmon populations are the primary source for such oil. With plunging salmon populations due to overfishing, disease, and pollution, concern is rising about a lack of adequate fish oil supplies. Research is well advanced in getting "fish oil without any fish." Fish do not make the omega-3 fatty acids they have within themselves. Rather, they obtain these fatty acids from the algae that they feed on. Genetic engineering experiments (Section 11.14) are underway in which the genes that allow algae to synthesize omega-3 fatty acids are incorporated into plants. In the future, land-based plants may become sources for omega-3 fatty acids.

Figure 8.9 Fish that live in deep, cold water—mackerel, herring, tuna, and salmon—are better sources of omega-3 fatty acids than other fish.

CHEMICAL CONNECTIONS 8-A

The Fat Content of Tree Nuts and Peanuts

People who bypass the nut tray at holiday parties usually believe a myth—that nuts are *unhealthful* high-fat foods. Indeed, nuts are high-fat food. However, the fat is "good fat" rather than "bad fat" (Section 8.5); that is, the fatty acids present are MUFAs and PUFAs rather than SFAs. In most cases, a handful of nuts is better for you than a cookie or bagel.

Numerous studies now indicate that eating nuts can have a strong protective effect against coronary heart disease. The most improvement comes from adding small amounts of nuts—an ounce (3–4 teaspoons)—to the diet five or more times a week. Raw, dry-roasted, or lightly salted varieties are best.

The recommendation of only one ounce of nuts per day relates to the high calorie content of nuts, which is 160 to 200 calories per ounce. The number of nuts and number of calories per ounce for common types of nuts is as follows:

Nuts	Calories
18 cashews	160
20 peanuts	160
47 pistachios	160
24 almonds	166
14 walnut halves	180
8 Brazil nuts	186
12 hazelnuts	188
15 pecan halves	190
12 macadamias	200

The amount of fat present in nuts ranges from 74% in the macadamia nut, 68% in pecans, and 63% in hazelnuts to around 50% in nuts such as the almond, cashew, peanut, and pistachio, as is shown in the table below.

The different fatty acid fractions (SFAs, MUFAs, and PUFAs) present in nuts also vary, but with definite trends. Unsaturated fatty acids always significantly dominate saturated fatty acids. The unsaturation/saturation ratio is highest for hazelnuts (11.9), pecans (10.9), walnuts (9.0), and almonds (9.0) and is lowest for cashews (3.9).

© Harris Shiffman/Shutterstock.com

The fat in nuts is "good fat"; the unsaturated/saturated fatty acid ratio is higher than that in most foods.

Their low amounts of saturated fatty acids are not the only reason why nuts help reduce the risk of coronary heart disease. Nuts also offer valuable antioxidant vitamins, minerals, and plant fiber protein. The protein content is highest (18%–26%) in the cashew, pistachio, almond, and peanut; here the amount of protein is about the same as in meat, fish, and cheese. The carbohydrate content of nuts is relatively low, less than 10% in most cases.

An unexpected discovery involving the anticancer drug Taxol and hazelnuts was made in the year 2000. The active chemical component in this drug, paclitaxel, was found in hazelnuts. It was the first report of this potent chemical being found in a plant other than in the bark of the Pacific yew tree, a slow-growing plant found in limited quantities in the Pacific Northwest. Although the amount of the chemical found in a hazelnut tree is about one-tenth that found in yew bark, the effort required to extract paclitaxel from these sources is comparable. Because hazelnut trees are more common, this finding could reduce the cost of the commercial drug and make it more readily available.

Fat and Fatty Acid Composition of Selected Nuts

	Total Fat (percentage of weight)	SFA (percentage of total fat)	MUFA (percentage of total fat)	PUFA (percentage of total fat)	UFA/SFA Ratio
almonds	52	10	68	22	9.0
cashews	46	20	62	18	3.9
hazelnuts	63	8	82	10	11.9
macadamias	74	16	82	2	5.4
peanuts	49	15	51	34	5.7
pecans	68	8	66	26	10.9
pistachios	48	13	72	15	6.6
walnuts	62	10	24	66	9.0

Table 8.2 **Omega-3 Fatty Acid Amounts Associated with Various Kinds of Cold-Water Fish**

Per 3.5-oz. Serving (raw)	Omega-3s (grams)*
mackerel	2.3
albacore tuna	2.1
herring, Atlantic	1.6
anchovy	1.5
salmon, wild king (Chinook)	1.4
salmon, wild sockeye (red)	1.2
tuna, bluefin	1.2
salmon, wild pink	1.0
salmon, wild Coho (silver)	0.8
oysters, Pacific	0.7
salmon, farm-raised Atlantic	0.6
swordfish	0.6
trout, rainbow	0.6

*Omega-3 content of fish can vary depending on harvest location and time of year.

Essential Fatty Acids

An **essential fatty acid** *is a fatty acid needed in the human body that must be obtained from dietary sources because it cannot be synthesized within the body, in adequate amounts, from other substances.* There are two essential fatty acids: *linoleic acid* and *linolenic acid.* Linoleic acid (18:2) is the primary member of the omega-6 acid family, and linolenic acid (18:3) is the primary member of the omega-3 acid family. Their structures are given in Table 8.1.

These two acids (1) are needed for proper membrane structure and (2) serve as starting materials for the production of several nutritionally important longer-chain omega-6 and omega-3 acids. When these two acids are missing from the diet, the skin reddens and becomes irritated, infections and dehydration are likely to occur, and the liver may develop abnormalities. If the fatty acids are restored, then the conditions reverse themselves. Infants are especially in need of these acids for their growth. Human breast milk has a much higher percentage of the essential fatty acids than cow's milk.

In 2001, the FDA gave approval for manufacturers of baby formula to add the fatty acids DHA (docosahexaenoic acid) and AA (arachidonic acid) to infant formulas. Human breast milk naturally contains these acids, which are important in brain and vision development. Because not all mothers can breast-feed, health officials regulate the ingredients in infant formula so that formula-fed babies get the next best thing to mother's milk.

Linoleic acid is the starting material for the biosynthesis of arachidonic acid.

$$\text{Linoleic acid (18:2)} \longrightarrow \text{arachidonic acid (20:4)}$$
Omega-6 fatty acids

Arachidonic acid is the major starting material for eicosanoids (Section 8.13), substances that help regulate blood pressure, clotting, and several other important body functions.

Linolenic acid is the starting material for the biosynthesis of two additional omega-3 fatty acids.

$$\text{Linolenic acid (18:3)} \longrightarrow \text{EPA (20:5)} \longrightarrow \text{DHA (22:6)}$$
Omega-3 fatty acids

EPA (eicosapentaenoic acid) and DHA (docosahexaenoic acid) are important constituents of the communication membranes of the brain and are necessary for normal brain development. EPA and DHA are also active in the retina of the eye.

CHEMICAL CONNECTIONS 8-B

Fat Substitutes

Sugar substitutes (artificial sweeteners) have been an accepted part of the diet of most people for many years. New since the 1990s are fat substitutes—substances that create the sensations of "richness" of taste and "creaminess" of texture in food without the negative effects associated with dietary fats (heart disease and obesity). Today, in most grocery stores, sitting next to almost all high-fat foods on the shelf are lower-fat counterparts (see accompanying photo). In most cases, the lower-fat products contain fat substitutes.

Foods that contain fat substitutes have less fat but not necessarily fewer calories.

Food scientists have been trying to develop fat substitutes since the 1960s. Now available for consumer use are two types of fat substitutes: *calorie-reduced* fat substitutes and *calorie-free* substitutes. They differ in their chemical structures and therefore in how the body handles them.

Simplesse, the best-known calorie-reduced fat substitute, received FDA marketing approval in 1990. It is made from the protein of fresh egg whites and milk by a procedure called microparticulation. This procedure produces tiny, round protein particles so fine that the tongue perceives them as a fluid rather than as the solid they are. Their fineness creates a sensation of smoothness, richness, and creaminess on the tongue.

In the body, Simplesse is digested and absorbed, contributing to energy intake. But 1 g of Simplesse provides 1.3 cal, compared with the 9 cal provided by 1 g of fat. Simplesse is used only to replace fats in *formulated* foods such as salad dressings, cheeses, sour creams, and other dairy products. Simplesse is unsuitable for frying or baking because it turns rubbery or rigid (gels) when heated. Consequently, it is not available for home use.

Olestra, the best-known calorie-free fat substitute, received FDA marketing approval in 1996. It is produced by heating cottonseed and/or soybean oil with sucrose in the presence of methyl alcohol. Chemically, olestra has a structure somewhat similar to that of a triacylglycerol; sucrose takes the place of the glycerol molecule, and six to eight fatty acids are attached by ester linkages to it rather than the three fatty acids in a triacylglycerol.

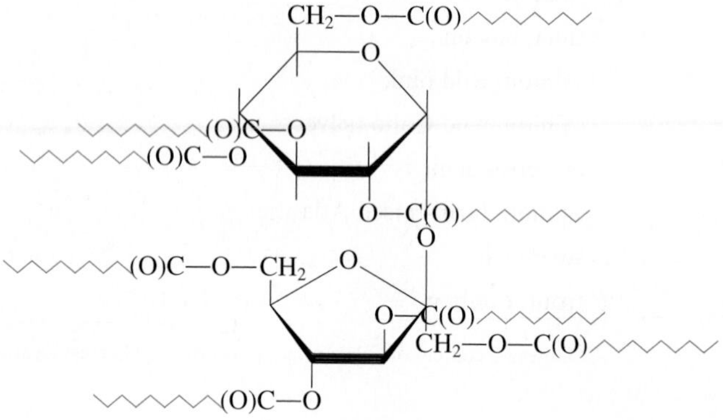

Olestra

Unlike triacylglycerols, however, olestra cannot be hydrolyzed by the body's digestive enzymes nor processed by colonic bacteria, and therefore passes through the digestive tract undigested.

Olestra looks, feels, and tastes like dietary fat and can substitute for fats and oils in foods such as shortenings, oils, margarines, snacks, ice creams, and other desserts. It has the same cooking properties as fats and oils.

In the digestive tract, Olestra interferes with the absorption of both dietary and body-produced cholesterol; thus it may lower total cholesterol levels. A problem with its use is that it also reduces the absorption of the fat-soluble vitamins A, D, E, and K. To avoid such depletion, Olestra is fortified with these vitamins. Another problem with Olestra use is that in some individuals it can cause gastrointestinal irritation and/or diarrhea. All products containing Olestra must carry the following label: "Olestra may cause abdominal cramping and loose stools. Olestra inhibits the absorption of some vitamins and other nutrients. Vitamins A, D, E, and K have been added."

The terminology used to describe products that contain fat substitutes can be confusing. Fat-free means less than 0.5 g of fat per serving. Low-fat means 3 g or less fat per 50 g serving. Reduced-fat or less-fat means at least 25% less fat per serving than the "regular" food. Calorie-free means less than 0.5 kilocalories per serving.

Table 8.3 gives pronunciation guidelines for the names of the two essential fatty acids and of the other acids mentioned that are biosynthesized from them.

Fat Substitutes (Artificial Fats)

In response to consumer demand for low-fat, low-calorie foods, food scientists have developed several types of "artificial fats." Such substances replicate the taste, texture, and cooking properties of fats but are themselves not lipids. The focus on relevancy feature Chemical Connections 8-B above considers further the topic of fat substitutes ("artificial fats").

Table 8.3 Biochemically Important Omega-3 and Omega-6 Fatty Acids

Omega-3 Acids	Omega-6-Acids
linolenic acid (18:3) (lin-oh-LEN-ic)	linoleic acid (18:2) (lin-oh-LAY-ic)
eicosapentaenoic acid (20:5) (EYE-cossa-PENTA-ee-NO-ic)	arachidonic acid (20:4) (a-RACK-ih-DON-ic)
docosahexaenoic acid (20:6) (DOE-cossa-HEXA-ee-NO-ic)	

8.6 Chemical Reactions of Triacylglycerols

The chemical properties of triacylglycerols (fats and oils) are typical of esters and alkenes because these are the two functional groups present in triacylglycerols. Four important triacylglycerol reactions are hydrolysis, saponification, hydrogenation, and oxidation.

Hydrolysis

Hydrolysis of a triacylglycerol is the reverse of the esterification reaction by which it was formed (see Figure 8.5). Triacylglyercol hydrolysis, when carried out in a laboratory setting, requires the presence of an acid or a base. Under acidic conditions, the hydrolysis products are glycerol and fatty acids. Under basic conditions, the hydrolysis products are glycerols and fatty acid salts.

Within the human body, triacylglycerol hydrolysis occurs during the process of digestion. Such hydrolysis requires the help of enzymes (protein catalysts; Section 10.1) produced by the pancreas. These enzymes cause the triacylglycerol to be hydrolyzed in a stepwise fashion. First, one of the outer fatty acids is removed, then the other outer one, leaving a monoacylglycerol. In most cases, this is the end product of the initial digestion (hydrolysis) of the triacylglycerol. Sometimes, enzymes remove all three fatty acids, leaving a free molecule of glycerol.

In situations where all three fatty acids are removed, the hydrolysis process is referred to as *complete hydrolysis,* which is depicted in Figure 8.10a. If one or more of the fatty acid residues remains attached to the glycerol, the hydrolysis process is called *partial hydrolysis* (Figure 8.10b).

Naturally occurring mono- and diacylglycerols are seldom encountered. *Synthetic* mono- and diacylglycerols are used as emulsifiers in many food products. Emulsifiers prevent suspended particles in colloidal solutions from coalescing and settling. Emulsifiers are usually present in so-called fat-free cakes and other fat-free products.

EXAMPLE 8.3 Writing a Structural Equation for the Hydrolysis of a Triacylglycerol

Write an equation for the acid-catalyzed hydrolysis of the following triacylglycerol.

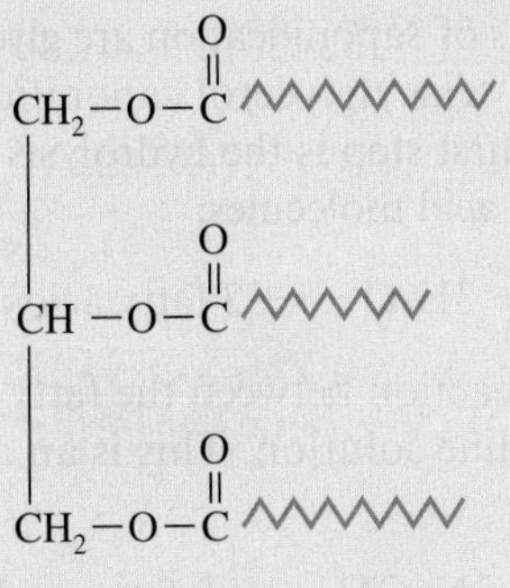

(continued)

Solution

Three water molecules are required for the hydrolysis, one to interact with each of the ester linkages present in the triacylglycerol. Breaking of the three ester linkages produces four product molecules: glycerol and three fatty acids.

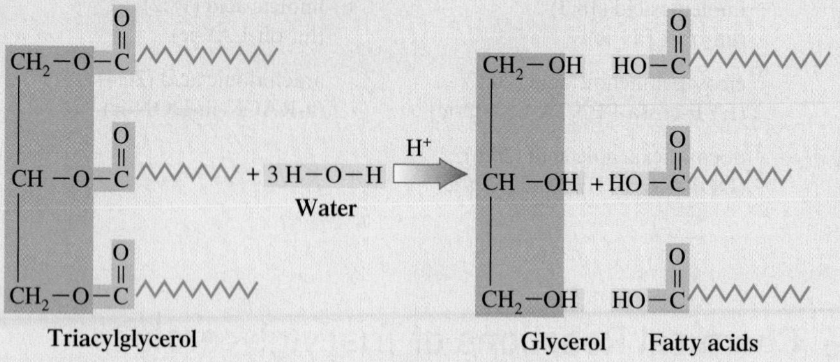

Practice Exercise 8.3

Write a structural equation for the acid-catalyzed hydrolysis of the following triacylglycerol.

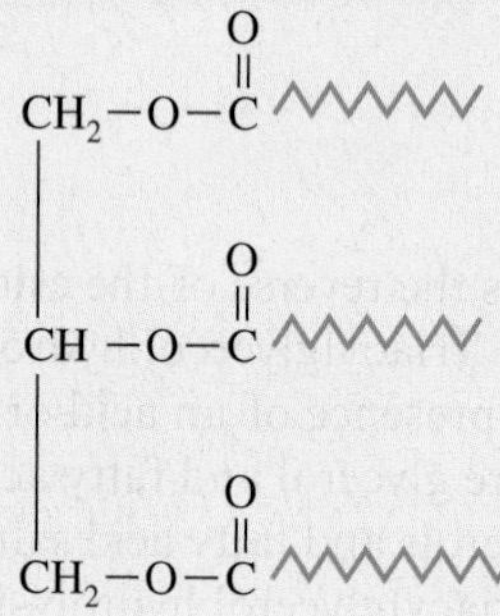

Answer:

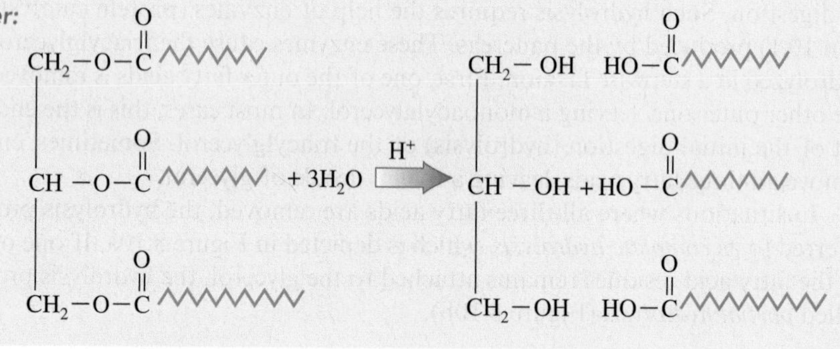

Saponification

Recall from Section 5.8 the structural difference between a carboxylic acid and a carboxylic acid salt.

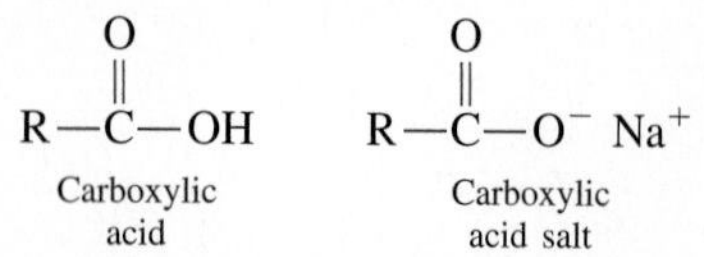

Saponification (Section 5.15) is a reaction carried out in an alkaline (basic) solution. For fats and oils, the products of saponification are glycerol and fatty acid *salts.*

The overall reaction of triacylglycerol saponification can be thought of as occurring in two steps. The first step is the hydrolysis of the ester linkages to produce glycerol and three fatty acid molecules:

$$\text{Fat or oil} + 3H_2O \longrightarrow 3 \text{ fatty acids} + \text{glycerol}$$

The second step involves a reaction between the fatty acid molecules and the base (usually NaOH) in the alkaline solution. This is an acid–base reaction that produces water plus salts:

$$3 \text{ fatty acids} + 3NaOH \longrightarrow 3 \text{ fatty acid salts} + 3H_2O$$

Figure 8.10 Complete and partial hydrolysis of a triacylglycerol.

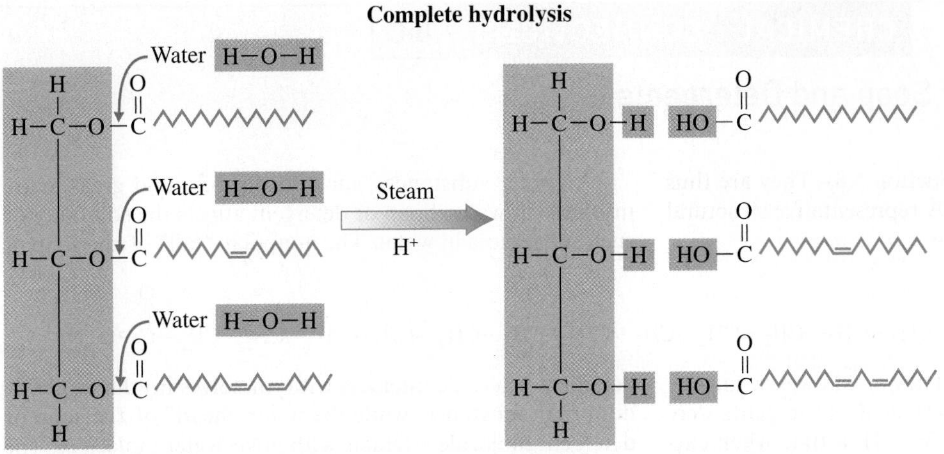

a Complete hydrolysis of a triacylglycerol produces glycerol and three fatty acid molecules.

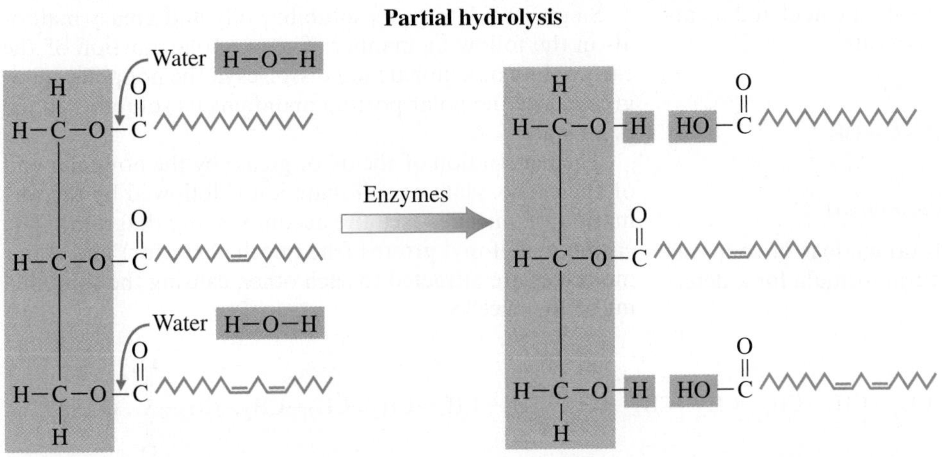

b Partial hydrolysis (during digestion) of a triacylglycerol produces a monoacylglycerol and two fatty acid molecules.

Saponification of animal fat is the process by which soap was made in pioneer times. Soap making involved heating lard (fat) with lye (ashes of wood, an impure form of KOH). Today most soap is prepared by hydrolyzing fats and oils (animal fat and coconut oil) under high pressure and high temperature. Sodium carbonate is used as the base.

The cleansing action of soap is related to the structure of the carboxylate ions present in the fatty acid salts of soap and the fact that these ions readily participate in micelle formation. A **micelle** *is a spherical cluster of molecules in which the polar portions of the molecules are on the surface, and the nonpolar portions are located in the interior.* The focus on relevancy feature Chemical Connections 8-C on the next page further discusses micelle formation as it relates to the cleansing action of soap.

Hydrogenation

Hydrogenation is a chemical reaction first encountered in Section 2.9. It involves hydrogen addition across carbon–carbon multiple bonds, which increases the degree of saturation as some double bonds are converted to single bonds. With this change, there is a corresponding increase in the melting point of the substance.

Hydrogenation involving just one carbon–carbon bond within a fatty acid residue of a triacylglycerol can be diagrammed as follows:

$$\text{---}CH_2{-}CH_2{-}CH{=}CH{-}CH_2{-}CH_2\text{---} + H_2 \longrightarrow \text{---}CH_2{-}CH_2{-}CH_2{-}CH_2{-}CH_2{-}CH_2\text{---}$$

Portion of an unsaturated fatty acid residue in a triacylglycerol containing one double bond

The double bond has been converted to a single bond; the degree of saturation has increased

CHEMICAL CONNECTIONS 8-C

The Cleansing Action of Soap and Detergents

Soaps are carboxylic acid salts (Section 5.8). They are thus ionic compounds, as are all salts. A representative structural formula for a carboxylic acid salt is

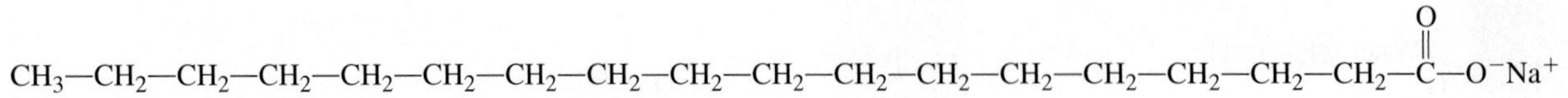

Detergents are also acid salts. They are, however, salts of sulfonic acids rather than carboxylic acids. Detergents were initially developed during World War II, a time when carboxylic acid supplies available for soap making were very limited. Sulfonic acids were used as substitutes for carboxylic acids. The general structures for a sulfonic acid and a carboxylic acid (for comparision purposes) are

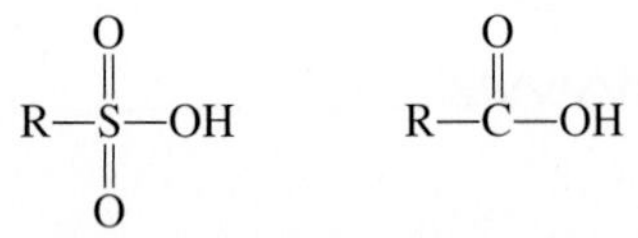

Sulfonic acid   Carboxylic acid

Detergents are synthetic structural analogs of soaps, as the following representative structural formula for a detergent molecule shows.

Nonpolar substances, such as fats, oils, and greases, are insoluble in water. Soap or detergent affects the solubility of such substances in water. The nonpolar "tail" of the soap or detergent molecule interacts with (dissolves in) the insoluble nonpolar substance, while the polar "head" of the soap or detergent molecule interacts with polar water molecules. The soap or detergent thus overcomes the nonpolar-polar solubility barrier.

Soaps and detergents solubilize oily and greasy materials in the following manner: The nonpolar portion of the carboxylate or sulfonate ion dissolves in the nonpolar oil or grease, and the polar portion maintains its solubility in the polar water.

The penetration of the oil or grease by the nonpolar end of the carboxylate or sulfonate ion is followed by the formation of micelles (see the accompanying diagram). The carboxyl sulfonyl groups (the micelle exterior) and water molecules are attracted to each other, causing the solubilizing of the micelle.

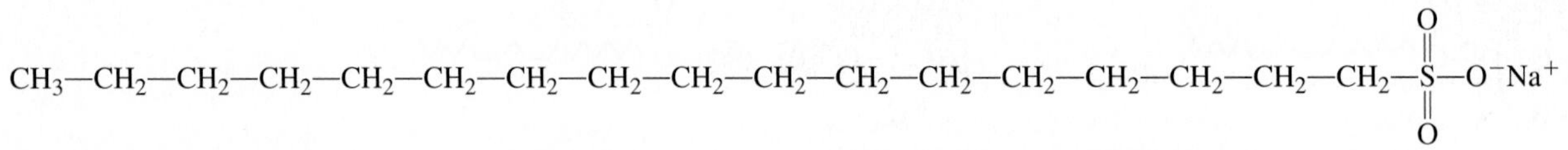

Structurally, both soaps and detergents contain a very small positive ion (usually Na^+ or K^+) and a negative ion that contains a very long carbon chain. The negative ion is the "active ingredient" in both soaps and detergents. In aqueous solution, salt dissociation occurs, which releases the salt's constituent ions. This allows the carboxylate ions (soaps) and sulfonate ions (detergents) present to exert their effects.

The cleansing action of soaps and detergents relates to the "dual polarity" that carboxylate and sulfonate ions possess. The long carbon chain present, which is called the "tail" of the ion, is nonpolar, whereas the small oxygen-containing group present, which is called the "head" of the ion, is polar.

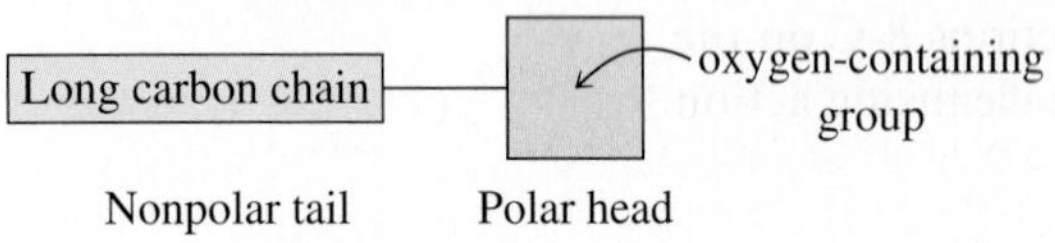

The micelles do not combine into larger drops because their surfaces are all negatively charged, and like charges repel each other. The water-soluble micelles are subsequently rinsed away, leaving a material devoid of oil and grease.

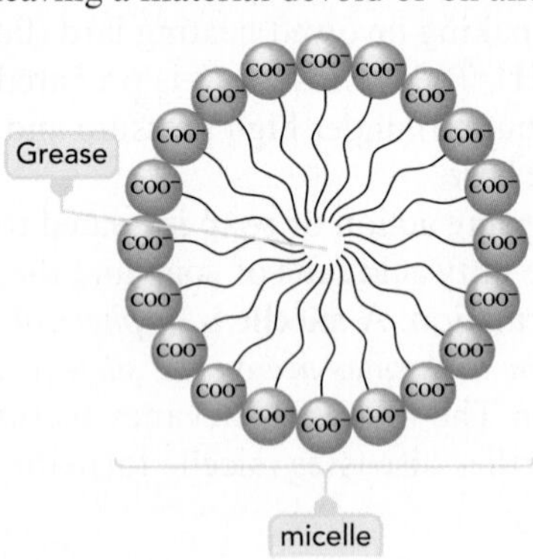

The structural equation for the complete hydrogenation of a triacylglycerol in which all three fatty acid residues are oleic acid (18:1) is shown in Figure 8.11.

Many food products are produced via partial hydrogenation. In partial hydrogenation, some, but not all, of the double bonds present are converted into single bonds. In this manner, liquids (usually plant oils) are converted into semi-solid materials.

Peanut butter is produced from peanut oil through partial hydrogenation. Solid cooking shortenings and stick margarine are produced from liquid plant oils through partial hydrogenation. Soft-spread margarines are also partial-hydrogenation products. Here, the extent of hydrogenation is carefully controlled to make the

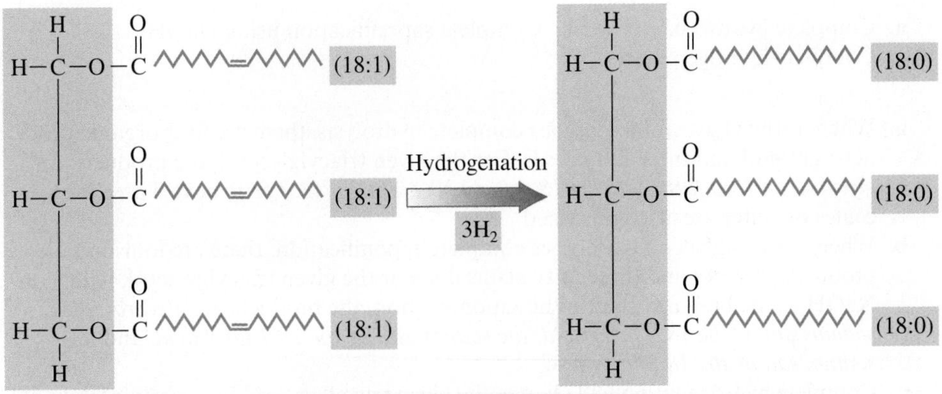

Figure 8.11 Structural equation for the complete hydrogenation of a triacylglycerol with oleic acid (18:1) fatty acid residues.

margarine soft at refrigerator temperatures (4°C). Concern has arisen about food products obtained from hydrogenation processes because the hydrogenation process itself converts some *cis* double bonds within fatty acid residues into *trans* double bonds, producing *trans* unsaturated fatty acids. The focus on relevancy feature Chemical Connections 8-D on page 335 further explores the subject of *trans*-fatty acids in food.

Oxidation

The carbon–carbon double bonds present in the fatty acid residues of a triacylglycerol are subject to oxidation with molecular oxygen (from air) as the oxidizing agent. Such oxidation breaks these bonds, producing both aldehyde and carboxylic acid products.

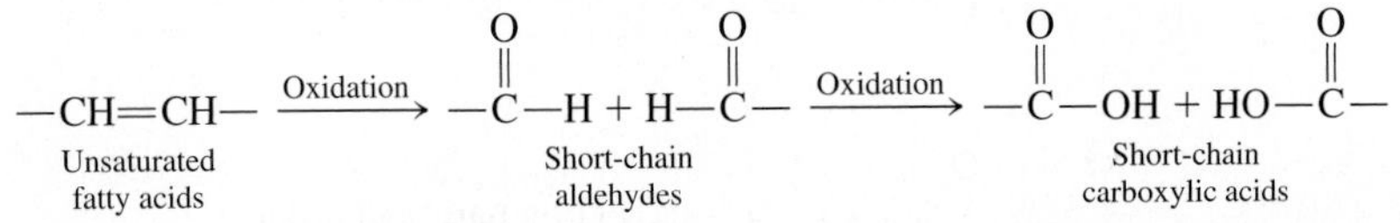

Antioxidants are compounds that are easily oxidized. When added to foods, they are more easily oxidized than the food. Thus they prevent the food from being oxidized (see Section 3.14).

The short-chain aldehydes and carboxylic acids so produced often have objectionable odors, and fats and oils containing them are said to have become *rancid.* To avoid this unwanted oxidation process, commercially prepared foods containing fats and oils nearly always contain *antioxidants*—substances that are more easily oxidized than the food. Two naturally occurring antioxidants are vitamin C (Section 10.12) and vitamin E (Section 10.13). Two synthetic oxidation inhibitors are BHA and BHT (Section 3.14). In the presence of air, antioxidants, rather than food, are oxidized.

EXAMPLE 8.4 Determining the Products for Reactions that Triacylglycerols Undergo

Using words rather than structural formulas, characterize the products formed when the following triacylglycerol undergoes the reactions listed.

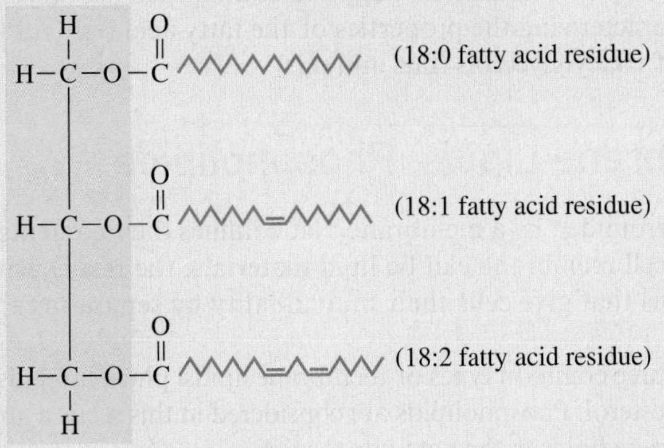

(continued)

a. Complete hydrolysis **b.** Complete saponification using NaOH
c. Complete hydrogenation

Solution

a. When a triacylglycerol undergoes complete hydrolysis, there are four organic products: glycerol and three fatty acids. For the given triacylglycerol, the products are *glycerol, an 18:0 fatty acid, an 18:1 fatty acid,* and *an 18:2 fatty acid.* Three molecules of water are also consumed.

b. When a triacylglycerol undergoes complete saponification, there are four organic products: glycerol and three fatty acid salts. For the given triacylglycerol, with NaOH as the base involved in the saponification, the products are *glycerol, the sodium salt of the 18:0 fatty acid, the sodium salt of the 18:1 fatty acid,* and *the sodium salt of the 18:2 fatty acid.*

c. Complete hydrogenation will change the given triacylglycerol into a *triacylglycerol in which all three fatty acid residues are 18:0 fatty acid residues.* That is, all of the fatty acid residues are completely saturated (there are no carbon–carbon double bonds).

Practice Exercise 8.4

Using words rather than structural formulas, characterize the organic products formed when the following triacylglycerol undergoes the reactions listed.

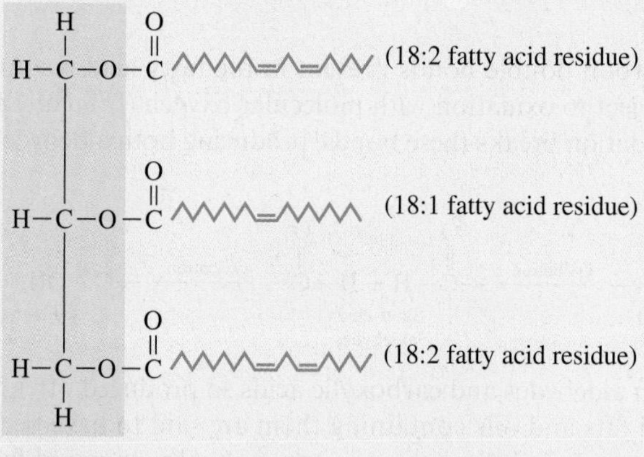

a. Complete hydrolysis **b.** Complete saponification using NaOH
c. Complete hydrogenation

Answers: **a.** Four products: glycerol and three fatty acids; **b.** Four products: glycerol and three fatty acid salts; **c.** One product: a triacylglycerol in which all fatty acid residues are saturated (18:0) residues

BRI/Vision/Photo Researchers

Figure 8.12 The oils (triacylglycerols) present in skin perspiration rapidly undergo oxidation. The oxidation products, short-chain aldehydes and short-chain carboxylic acids, often have strong odors.

Perspiration generated by strenuous exercise or by "hot and muggy" climatic conditions contains numerous triacylglycerols (oils). Rapid oxidation of these oils, promoted by microorganisms on the skin, generates the body odor that accompanies most "sweaty" people (Figure 8.12).

The Chemistry at a Glance on page 336 contains a summary of the terminology used in characterizing the properties of the fatty acid residues that are part of the structure of triacylglycerols (fats and oils).

8.7 Membrane Lipids: Phospholipids

All cells are surrounded by a membrane that confines their contents. Up to 80% of the mass of a cell membrane can be lipid materials; the rest is primarily protein. It is membranes that give cells their individuality by separating them from their environment.

There are three common types of membrane lipids: phospholipids, sphingoglycolipids, and cholesterol. Phospholipids are considered in this section and the other two types of membrane lipids in the next two sections.

CHEMICAL CONNECTIONS 8-D

Trans Fatty Acid Content of Foods

All current dietary recommendations stress reducing saturated fat intake because of its association with elevated blood cholesterol levels. In accordance with such recommendations, many people have switched from butter to margarine and now use partially hydrogenated vegetable oils rather than animal fat for cooking. Research now shows that partially hydrogenated vegetable oils also play a role in raising blood cholesterol levels. Why is this?

When the triacylglycerols in vegetable oils are subjected to partial hydrogenation (Section 8.6), two types of changes, rather than just one (as was previously thought), occur in the fatty acid residues present: (1) some of the *cis* double bonds present are converted to single bonds (the objective of the process) and (2) some of the remaining *cis* double bonds are converted to *trans* double bonds (an unanticipated result of the process). These latter *cis–trans* conversions affect the general shape of the fatty acid residues present in triacylglycerols, which in turn affects the biochemical behavior of the triacylglycerols. In the following diagram, note how conversion of a *cis,cis*-18:2 fatty acid to a *trans,trans*-18:2 fatty acid affects molecular shape. The *trans,trans*-18:2 fatty acid has a shape very much like that of an 18:0 saturated fatty acid (the structure on the right).

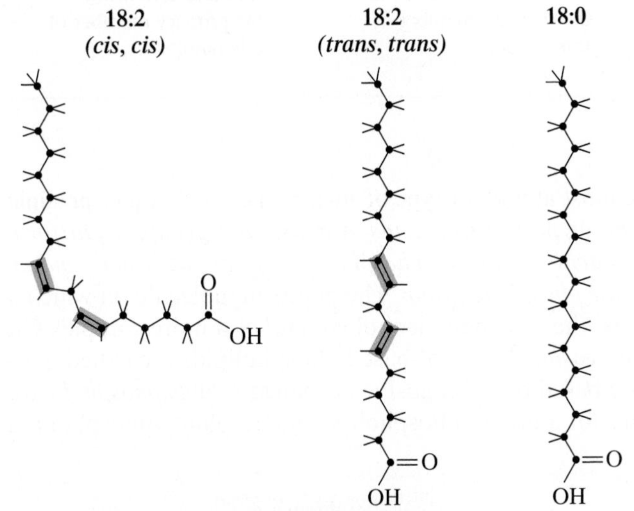

Studies show that fatty acids with *trans* double bonds affect blood cholesterol levels in a manner similar to saturated fatty acids.

Trans fatty acids (TFAs) are found naturally in meat and dairy products. Such TFAs, which constitute about 15%–20% of dietary TFA intake, are not thought to cause health problems. The remaining 80%–85% of dietary TFAs are the commercially produced TFAs associated with partial hydrogenation processes. A major focus, which even includes legislation, is currently on the reduction of such TFA levels in foods. Reduction efforts include new food preparation and processing methods. For example, fast-food chains have switched from using deep-frying oils with high TFA content to those with lower TFA levels. A new process called *transesterification*, which rearranges (reorders) the fatty acids within triacylglycerol molecules present in plant oils, is being investigated. This "rearranging" can change plant oil properties, in a positive manner, without introducing *trans* bonds into the triacylglcerols. Long-range reduction efforts even involve genetic engineering procedures to produce seeds that yield plants whose oils have more desirable triacylglycerol saturation/unsaturation ratios.

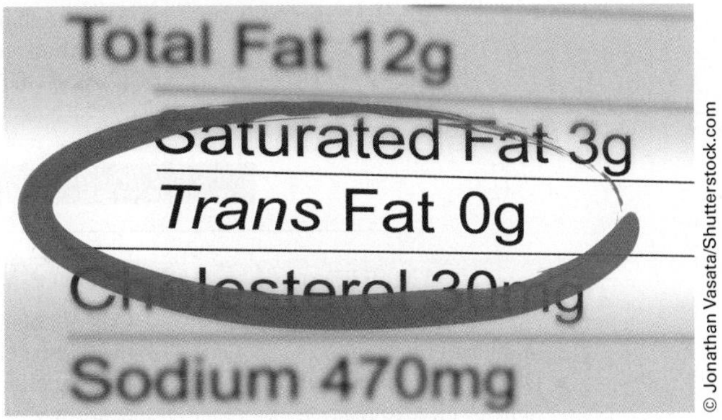

FDA regulations now require that *trans* fat content be listed as a separate item on nutrition facts labels.

Since 2006, the U.S. Food and Drug Administration (FDA) has required that the *trans* fat content of a food be included in the nutrition facts panel found on all food products (see accompanying photo). Prior to this rule change, the only way consumers could determine whether a food included *trans* fat was to look for the word *hydrogenated* on the list of ingredients. A food that listed partially hydrogenated oils among its first three ingredients usually contained substantial amounts of *trans* fatty acids as well as some saturated fat.

A label of "zero grams *trans* fat per serving" on a product does not mean the product is absolutely *trans* fat free. FDA regulations allow *trans* fat levels of less than 0.5 gram per serving to be labeled as 0 grams per serving. Also, naturally occurring *trans* fats present are not required to be included in *trans* fat values.

The health implications of *trans* fatty acids is still an area of active research; many answers are yet to be found. Preliminary studies indicate that *trans* fat raises bad (LDL) cholesterol, but it does not raise good (HDL) cholesterol. Saturated fat, on the other hand, raises both bad and good cholesterol. Thus, just as too much saturated fat isn't healthy, too much *trans* fat is also not healthy. Recommendations are that total fat intake be limited to 30% of daily calories and that combined saturated fat and *trans* fat intake should be limited to 10% or less of daily calories.

CHEMISTRY AT A GLANCE — Classification Schemes for Fatty Acid Residues Present in Triacylglycerols

FATTY ACIDS

- Most contain an even number of carbon atoms.
- Carbon chain length is up to 24 carbon atoms.

Classification Based on Degree of Unsaturation

SATURATED

- No double bonds are present in the carbon chain.
- Dietary effect is an increase in heart disease risk.

MONOUNSATURATED

- One double bond is present in the carbon chain.
- Dietary effect is a decrease in heart disease risk.

POLYUNSATURATED

- Two or more double bonds are present in the carbon chain.
- Dietary effect is "mixed"; there have been several conflicting studies relative to heart disease risk.

Classification Based on Configuration of Double Bond

CIS

- Naturally occurring fatty acids generally contain *cis* double bonds.

TRANS

- Hydrogenation converts some *cis* double bonds to *trans* double bonds.
- *Trans* fatty acids have effects on blood chemistry similar to those of saturated fatty acids.

Classification Based on Location of Double Bond

OMEGA-3

- First double bond is three carbons away from the CH_3 end of the carbon chain.
- Linolenic acid (18:3) is the primary member of this family.

OMEGA-6

- First double bond is six carbons away from the CH_3 end of the carbon chain.
- Linoleic acid (18:2) is the primary member of this family.

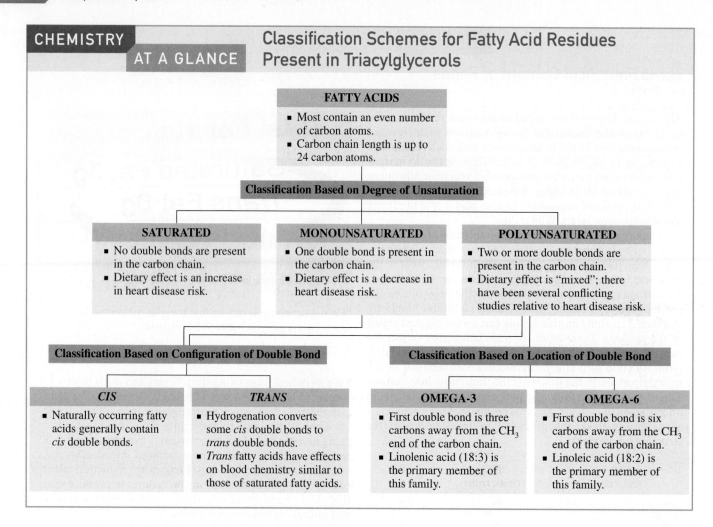

Phospholipids are the most abundant type of membrane lipid. A **phospholipid** *is a lipid that contains one or more fatty acids, a phosphate group, a platform molecule to which the fatty acid(s) and the phosphate group are attached, and an alcohol that is attached to the phosphate group.* The platform molecule on which a phospholipid is built may be the 3-carbon alcohol *glycerol* or a more complex C_{18} aminodialcohol called *sphingosine.* Glycerol-based phospholipids are called *glycerophospholipids,* and those based on sphingosine are called *sphingophospholipids.* The general block diagrams for a glycerophospholipid and a sphingophospholipid are as follows:

An aminodialcohol contains two hydroxyl groups, —OH, and an amino group, $—NH_2$.

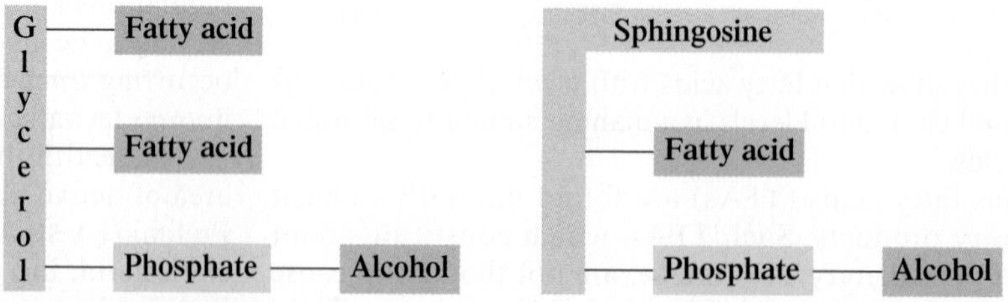

Glycerophospholipid   Sphingophospholipid

Glycerophospholipids

A **glycerophospholipid** *is a lipid that contains two fatty acids and a phosphate group esterified to a glycerol molecule and an alcohol esterified to the phosphate group.* All attachments (bonds) between groups in a glycerophospholipid are ester linkages, a situation similar to that in triacylglycerols (Section 8.4). However, glycerophospholipids have four ester linkages as contrasted to three ester linkages in triacylglycerols.

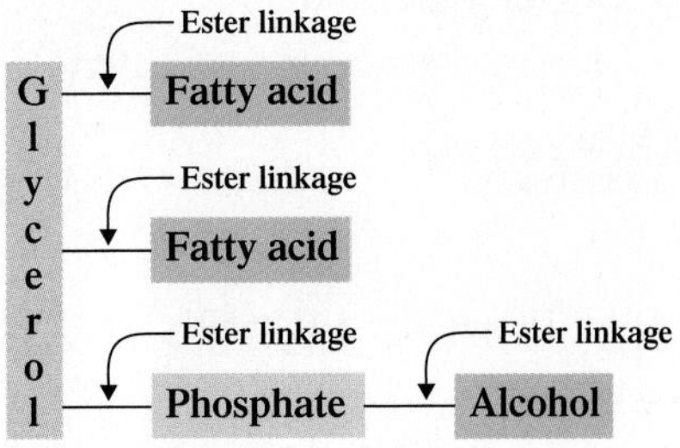

Glycerophospholipid
(four ester linkages)

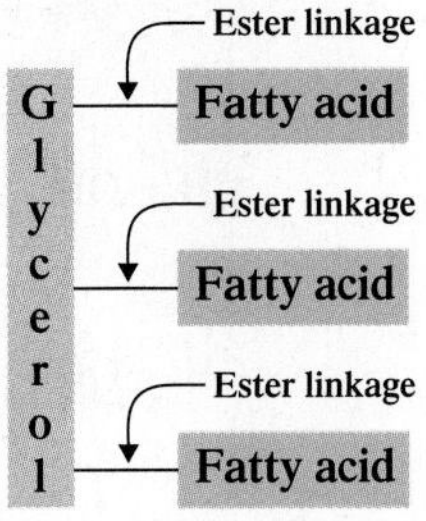

Triacylglycerol
(three ester linkages)

Because of the ester linkages present, glycerophospholipids undergo hydrolysis and saponification reactions in a manner similar to that for triacylglycerols (Section 8.6). There will be five reaction products, however, instead of the four for triacylglycerols.

Phosphoric acid is the parent source for the minus one charged phosphate group used in the formation of glycerophospholipids. The structures of these two entities are

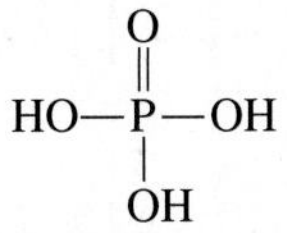

Phosphoric acid

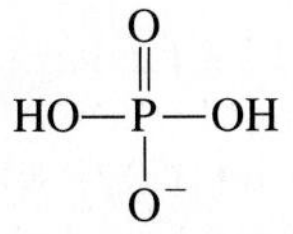

Minus one charged
phosphate group

Phosphoric acid structures were previously considered in Section 5.19 when esters of inorganic acids were considered.

The alcohol attached to the phosphate group in a glycophospholipid is usually one of three amino alcohols: choline, ethanolamine, or serine. The structures of these three amino alcohols, given in terms of the charged forms (Sections 6.8 and 9.4) that they adopt in neutral solution, are

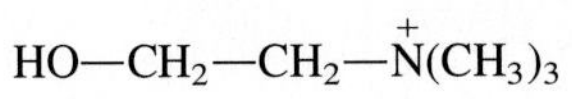

Choline
(a quaternary ammonium ion)

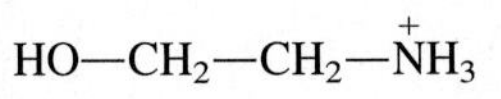

Ethanolamine
(positive-ion form)

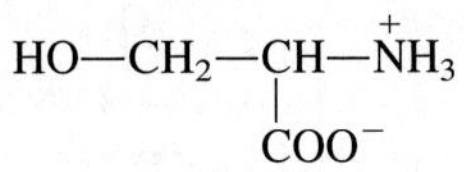

Serine
(two ionic groups present)

Glycerophospholipids containing these three amino alcohols are respectively known as phosphatidylcholines, phosphatidylethanolamines, and phosphatidylserines. The fatty acid, glycerol, and phosphate portions of a glycerophospholipid structure constitute a *phosphatidyl* group.

EXAMPLE 8.5 Drawing the Structural Formula for a Glycerophospholipid

Draw the structural formula for the glycerophospholipid that produces, upon hydrolysis, equimolar amounts of glycerol, phosphoric acid, and ethanolamine, and twice that molar amount of stearic acid, the 18:0 fatty acid.

Solution

The fact that the molar amount of stearic acid produced is twice that of the other products indicates that both fatty acid residues present in the glycerophospholipid are stearic acid residues.

To draw the structural formula for this lipid, arrange the component parts in the following manner. Draw on the left the structure of glycerol, the "backbone" of the structure. Then place alongside it the two stearic acid and one phosphoric acid molecules. The acid placements should be such that their acid groups are lined up alongside the hydroxyl groups of glycerol. Then place the ion form of ethanolamine alongside the phosphoric acid molecule.

(continued)

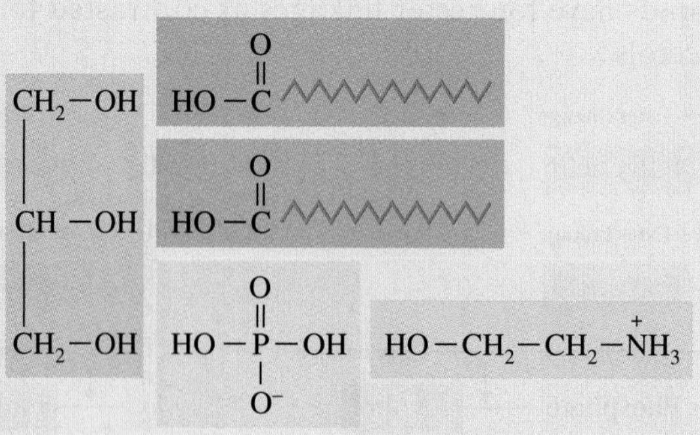

The five components of the overall structure are then bonded together via ester linkages. The product of the formation of each ester linkage is a molecule of water. The atoms involved in this water formation are highlighted in color in the structures at the left and the ester linkages formed are highlighted in color in the structure at the right.

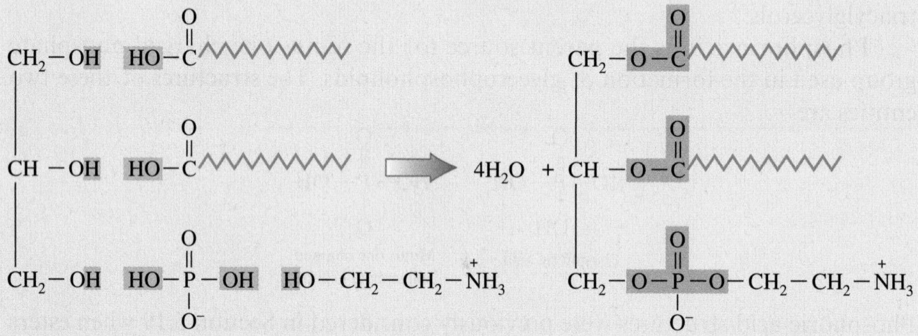

Practice Exercise 8.5

Draw the structure formula for the glycerophospholipid that produces, upon hydrolysis, equimolar amounts of glycerol, phosphoric acid, and choline, and twice that molar amount of lauric acid, the 12:0 fatty acid.

Answer:

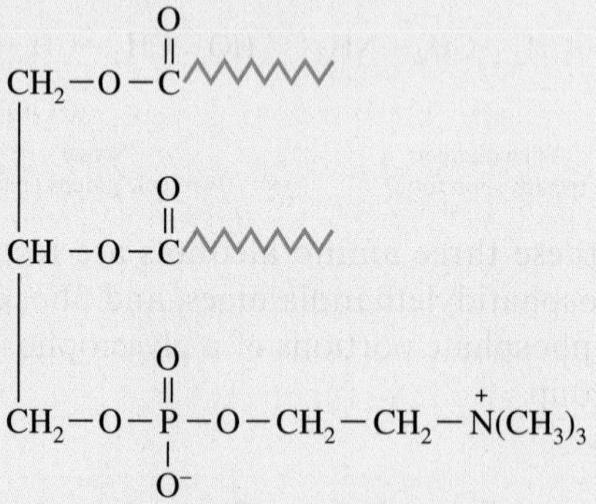

Although the general structural features of glycerophospholipids are similar in many respects to those of triacylglycerols, these two types of lipids have quite different biochemical functions. Triacylglycerols serve as storage molecules for metabolic fuel. Glycerophospholipids function almost exclusively as components of cell membranes (Section 8.10) and are not stored. A major structural difference between the two types of lipids, that of polarity, is related to their differing biochemical functions. Triacylglycerols are a nonpolar class of lipids, whereas glycerophospholipids are polar. In general, membrane lipids have polarity associated with their structures.

Further consideration of general glycerophospholipid structure reveals an additional structural characteristic of most membrane lipids. A phosphatidylcholine containing stearic and oleic acids will be used to illustrate this additional feature. The chemical structure of this molecule is shown in Figure 8.13a.

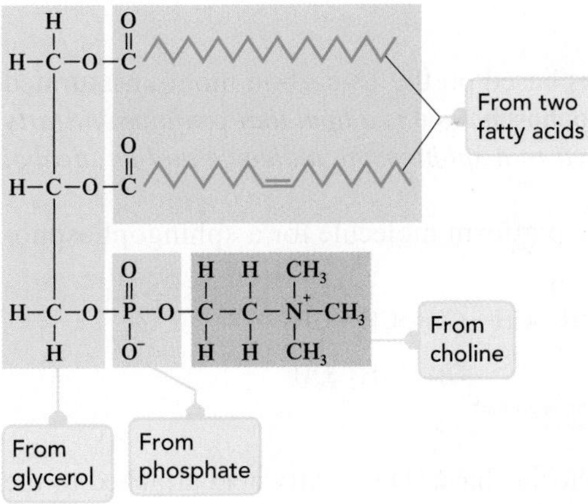

a Structural formula

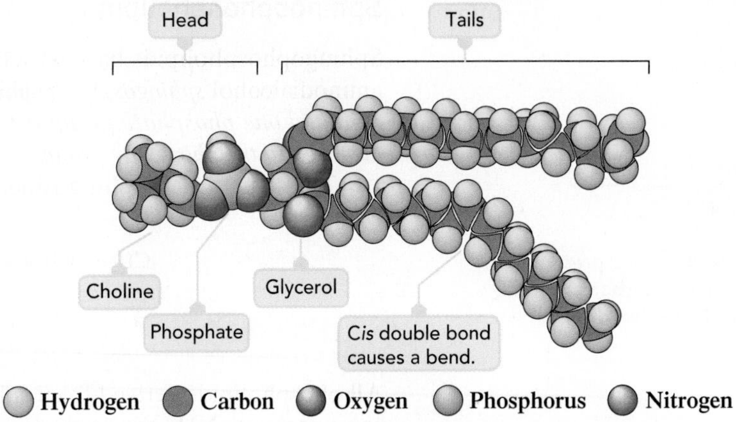

b Molecular model showing the "head and two tails" structure of a phosphatidylcholine molecule containing stearic acid (18:0) and oleic acid (18:1)

Figure 8.13

A molecular model for this compound, which gives the orientation of groups in space, is illustrated in Figure 8.13b. There are two important things to notice about this model: (1) There is a "head" part, the choline and phosphate and (2) there are two "tails," the two fatty acid carbon chains. The head part is polar. The two tails, the carbon chains, are nonpolar.

All glycerophospholipids have structures similar to that shown in Figure 8.13. All have a "head" and two "tails." A simplified representation for this structure uses a circle to represent the polar head and two wavy lines to represent the nonpolar tails.

Glycerophospholipids have a *hydrophobic* ("water-hating") portion, the nonpolar fatty acid groups, and a *hydrophilic* ("water-loving") portion, the polar head group.

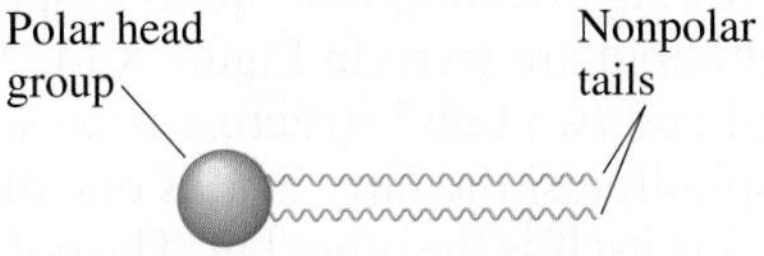

The polar head group of a glycerophospholipid is soluble in water. The nonpolar tail chains are insoluble in water but soluble in nonpolar substances. This dual polarity, which was previously encountered when soaps were discussed (see Chemical Connections 8-C on page 332), is a structural characteristic of most membrane lipids.

Phosphatidylcholines are also known as *lecithins.* There are a number of different phosphatidylcholines because different fatty acids may be bonded to the glycerol portion of the phosphatidylcholine structure. In general, phosphatidylcholines are waxy solids that form colloidal suspensions in water. Egg yolks and soybeans are good dietary sources of these lipids. Within the body, phosphatidylcholines are prevalent in cell membranes.

The amino alcohol in phosphatidylcholines (pronounced fahs-fuh-TIDE-ul-KOH-leens) is choline.

Periodically, claims arise that phosphatidylcholine should be taken as a nutritive supplement; some even maintain it will improve memory. There is no evidence that these supplements are useful. The enzyme *lecithinase* in the intestine hydrolyzes most of the phosphatidylcholine taken orally before it passes into body fluids, so it does not reach body tissues. The phosphatidylcholine present in cell membranes is made by the liver; thus phosphatidylcholines are not essential nutrients.

The food industry uses phosphatidylcholines as emulsifiers to promote the mixing of otherwise immiscible materials. Mayonnaise, ice cream, and custards are some of the products they are found in. It is the polar–nonpolar (head–tail) structure of phosphatidylcholines that enables them to function as emulsifiers.

Phosphatidylethanolamines and phosphatidylserines are also known as *cephalins.* These compounds are found in heart and liver tissue and in high concentrations in the brain. They are important in blood clotting. Much is yet to be learned about how these compounds function within the human body.

Sphingophospholipids

Sphingophospholipids have structures based on the 18-carbon monounsaturated aminodialcohol *sphingosine.* A **sphingophospholipid** *is a lipid that contains one fatty acid and one phosphate group attached to a sphingosine molecule and an alcohol attached to the phosphate group.*

The structure of sphingosine, the platform molecule for a sphingophospholipid, is

The double bond present in the sphingosine carbon chain is a *trans* double bond.

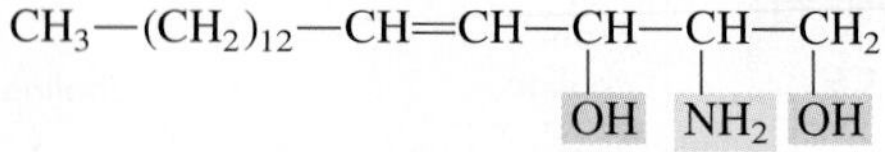

Sphingosine

All phospholipids derived from sphingosine have (1) the fatty acid attached to the sphingosine $—NH_2$ group via an *amide linkage* (2) the phosphate group attached to the sphingosine terminal —OH group via an *ester linkage* and (3) an additional alcohol esterified to the phosphate group. The general block diagram for a sphingophospholipid is

When sphingolipids were discovered over a century ago by the physician–chemist Johann Thudichum (1829–1901), their biochemical role seemed as enigmatic as the Sphinx, for which he named them.

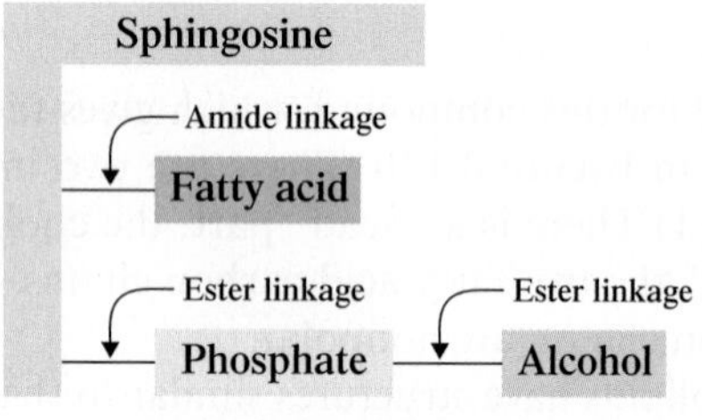

Sphingophospholipid
(two ester linkages and one amide linkage)

In sphingophospholipids, the first three carbon atoms at the polar end of sphingosine are analogous to the three carbon atoms of glycerol in glycerophospholipids.

Molecular models showing orientation of atoms in space for sphingosine itself and for a sphingophospholipid are given in Figure 8.14. Note that, as in glycerophospholipids, the "head and two tails" structure is present in sphingophospholipids. For sphingophospholipids, the fatty acid is one of the tails, and the long carbon chain of sphingosine itself is the other tail. The polar head is the phosphate group with its esterified alcohol.

Like glycerophospholipids, sphingophospholipids participate in hydrolysis and saponification reactions. Amide linkages behave much as ester linkages do in this type of reaction.

Sphingophospholipids in which the alcohol esterified to the phosphate group is *choline* are called *sphingomyelins.* Sphingomyelins are found in all cell membranes and are important structural components of the myelin sheath, the protective and insulating coating that surrounds nerves. The molecule depicted in Figure 8.14b is a sphingomyelin. The structural formula for a sphingomyelin in which stearic acid (18:0) is the fatty acid is

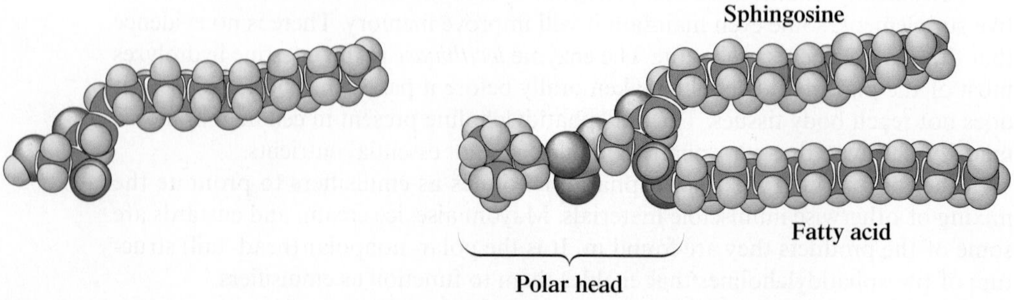

a Sphingosine
b A sphingophospholipid

Figure 8.14 Molecular models for (a) sphingosine and (b) a sphingophospholipid. The particular sphingophospholipid shown has choline as the alcohol esterified to the phosphate group. Note the "head and two tails" structure for the sphingophospholipid.

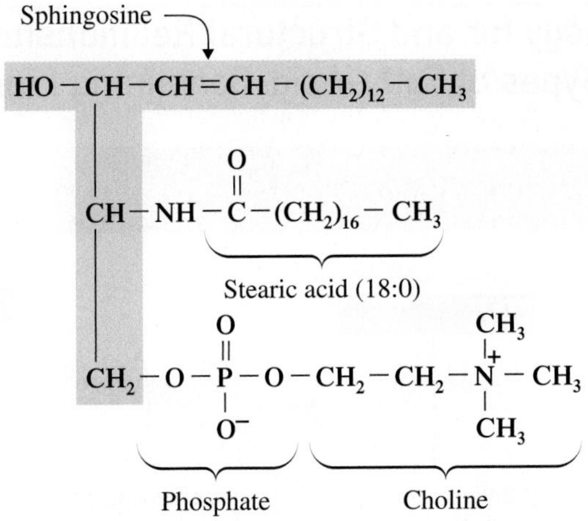

8.8 Membrane Lipids: Sphingoglycolipids

The second of the three major types of membrane lipids is *sphingoglycolipids.* A **sphingoglycolipid** *is a lipid that contains both a fatty acid and a carbohydrate component attached to a sphingosine molecule.* A fatty acid is attached to the sphingosine through an amide linkage, and a monosaccharide or oligosaccharide (Section 7.3) is attached to the sphingosine at the terminal —OH carbon atom through a glycosidic linkage (Section 7.13). The generalized block diagram for a sphingoglycolipid is

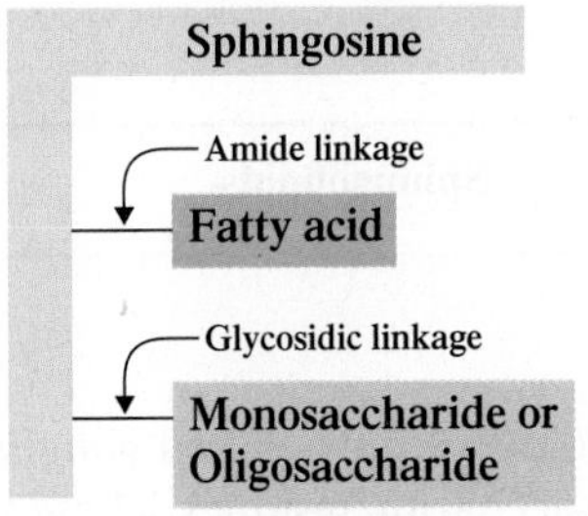

Sphingoglycolipid
(one amide linkage and one or more glycosidic linkages)

Like glycerophospholipids and sphingophospholipids (Section 8.7), sphingoglycolipids have a "head and two tail" structure. Sphingoglycolipids and sphingophospholipids have similar "tails," but their polar "heads" differ in the constituents present (mono- or oligosaccharide versus phosphate-alcohol).

Sphingoglycolipids undergo hydrolysis and saponification reactions; both the amide and the glycosidic linkages can be hydrolyzed.

The simplest sphingoglycolipids, which are called *cerebrosides,* contain a single monosaccharide unit—either glucose or galactose. As the name suggests, cerebrosides occur primarily in the brain (7% of dry mass). They are also present in the myelin sheath of nerves. The specific structure for a cerebroside in which stearic acid (18:0) is the fatty acid and galactose is the monosaccharide is

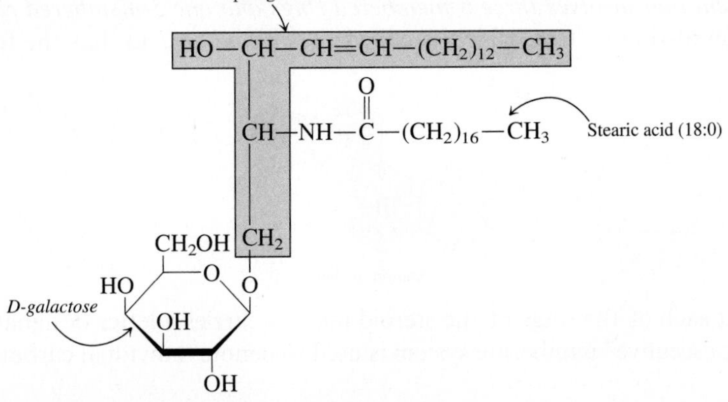

CHEMISTRY AT A GLANCE Terminology for and Structural Relationships Among Various Types of Fatty-Acid-Containing Lipids

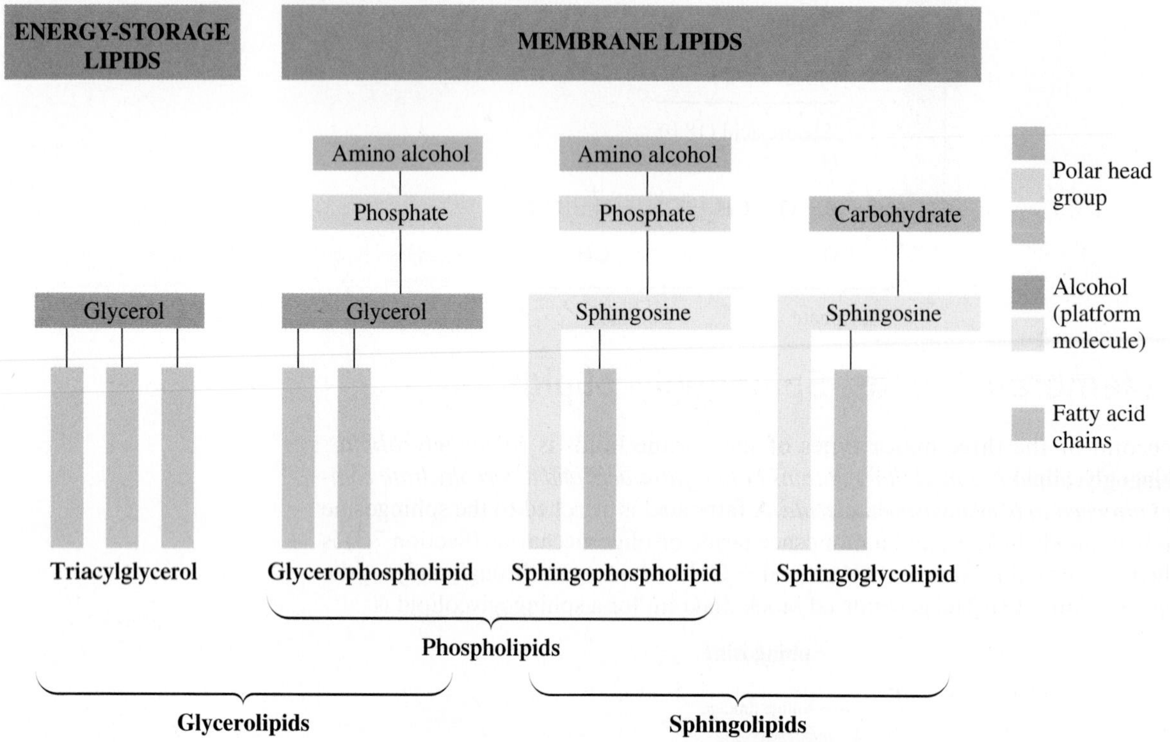

More complex sphingoglycolipids, called *gangliosides,* contain a branched chain of up to seven monosaccharide residues. These substances occur in the gray matter of the brain as well as in the myelin sheath.

The Chemistry at a Glance feature above summarizes terminology and structural relationships among the types of lipids that have been considered up to this point. The common thread among all of the structures is the presence of at least one fatty acid residue.

8.9 Membrane Lipids: Cholesterol

Cholesterol, the third of the three major types of membrane lipids, is a specific compound rather than a family of compounds like the phospholipids (Section 8.7) and sphingoglycolipids (Section 8.8). Cholesterol's structure differs markedly from that of other membrane lipids in that (1) there are no fatty acid residues present and (2) neither glycerol nor sphingosine is present as the platform molecule.

Cholesterol is a *steroid.* A **steroid** *is a lipid whose structure is based on a fused-ring system that involves three 6-membered rings and one 5-membered ring.* This steroid fused-ring system, which is called the *steroid nucleus,* has the following structure:

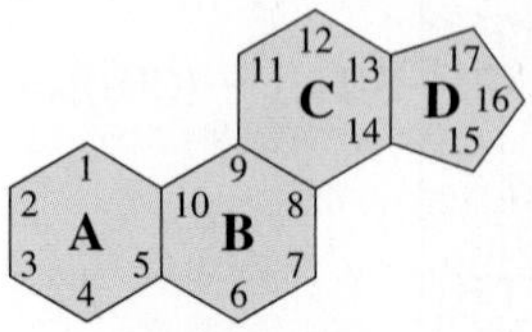

Steroid nucleus

Note that each of the rings of the steroid nucleus carries a letter designation and that a "consecutive" numbering system is used to denote individual carbon atoms.

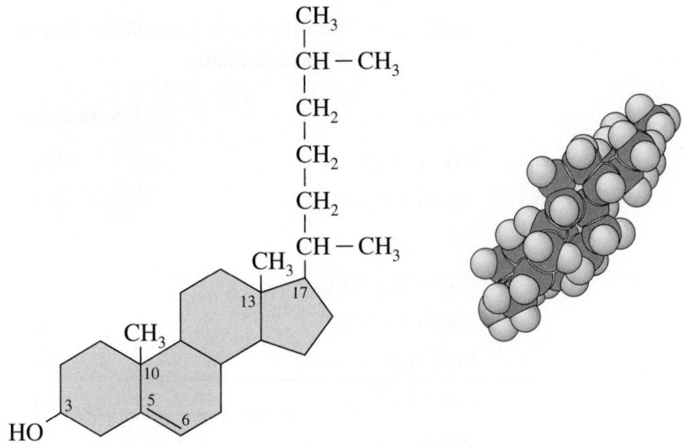

Figure 8.15 Structural formula and molecular model for the cholesterol molecule.

Numerous steroids have been isolated from plants, animals, and human beings. Location of double bonds within the fused-ring system and the nature and location of substituents distinguish one steroid from another. Most steroids have an oxygen functional group (=O or —OH) at carbon 3 and some kind of side chain at carbon 17. Many also have a double bond from carbon 5 to either carbon 4 or carbon 6.

Cholesterol *is a* C_{27} *steroid molecule that is a component of cell membranes and a precursor for other steroid-based lipids.* It is the most abundant steroid in the human body. The *-ol* ending in the name cholester*ol* conveys the information that an alcohol functional group is present in this molecule; it is located on carbon 3 of the steroid nucleus. In addition, cholesterol has methyl group attachments at carbons 10 and 13, a carbon–carbon double bond between carbons 5 and 6, and an eight-carbon branched side chain at carbon 17. Figure 8.15 gives both the structural formula and a molecular model for cholesterol. The molecular model shows the rather compact nature of the cholesterol molecule. The "head and two tails" arrangement found in other membrane lipids is not present. The lack of a *large* polar head group causes cholesterol to have limited water solubility. The —OH group on carbon 3 is considered the head of the molecule.

Besides being an important molecule in and of itself, cholesterol serves as a precursor for several other important steroid molecules including bile acids (Section 8.11), steroid hormones (Section 8.12), and vitamin D (Section 10.13).

Within the human body, cholesterol is found in cell membranes (up to 25% by mass), in nerve tissue, in brain tissue (about 10% by dry mass), and in virtually all fluids. Every 100 mL of human blood plasma contains about 50 mg of free cholesterol and about 170 mg of cholesterol esterified with various fatty acids.

Although a portion of the body's cholesterol is obtained from dietary intake, most of it is biosynthesized by the liver and (to a lesser extent) the intestine. Typically, 800–1000 mg are biosynthesized each day. Ingested cholesterol decreases biosynthetic cholesterol production. However, the reduction is less than the amount ingested. Therefore, total body cholesterol levels increase with increased dietary intake of cholesterol.

Biosynthetic cholesterol is distributed to cells throughout the body for various uses via the bloodstream. Because cholesterol is only sparingly soluble in water (blood), a protein carrier system is used for its distribution. These cholesterol–protein combinations are called *lipoproteins.*

The lipoproteins that carry cholesterol *from the liver* to various tissues are called LDLs (low-density lipoproteins), and those that carry excess cholesterol from tissues *back to the liver* are called HDLs (high-density lipoproteins). If too much cholesterol is being transported by LDLs or too little by HDLs, the imbalance results in an increase in blood cholesterol levels. High blood cholesterol levels contribute to atherosclerosis, a form of cardiovascular disease characterized by the buildup of plaque along the inner walls of arteries. Plaque is a mound of lipid material mixed with smooth muscle cells and calcium. Much of the lipid material in plaque is cholesterol. Plaque deposits in the arteries that serve the heart reduce blood flow to the heart muscle and can lead to a heart attack. Figure 8.16 shows the occlusion that can occur in an artery as a result of plaque buildup.

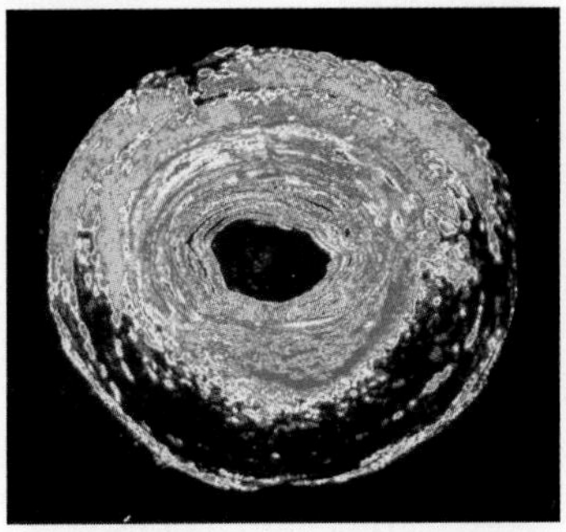

Figure 8.16 A severely occluded artery—the result of the buildup of cholesterol-containing plaque deposits.

Table 8.4 The Amount of Cholesterol Found in Various Foods

Food	Cholesterol (mg)
liver (3 oz)	410
egg (1 large)	213
shrimp (3 oz)	166
pork chop (3 oz)	83
chicken (3 oz)	75
beef steak (3 oz)	70
fish fillet (3 oz)	54
whole milk (1 cup)	33
cheddar cheese (1 oz)	30
Swiss cheese (1 oz)	26
low-fat milk (1 cup)	22

The cholesterol associated with LDLs is often called "bad cholesterol" because it contributes to increased blood cholesterol levels, and the cholesterol associated with HDLs is often called "good cholesterol" because it contributes to reduced blood cholesterol levels. The focus on relevancy feature Chemical Connections 9-F titled "Lipoproteins and Heart Disease Risk" on page 406 in the next chapter considers this topic in further detail.

Much still needs to be learned concerning the actual role played by serum cholesterol in plaque buildup within arteries. Current knowledge suggests that it makes good sense to reduce the amount of cholesterol (as well as saturated fats) taken into the body through dietary intake. People who want to reduce dietary cholesterol intake should reduce the amount of animal products they eat (meat, dairy products, etc.) and eat more fruit and vegetables. Plant foods contain negligible amounts of cholesterol; cholesterol is found primarily in foods of animal origin. Table 8.4 gives cholesterol amounts associated with selected foods.

8.10 Cell Membranes

Cell membranes are also commonly called *plasma membranes* because they separate the cytoplasm (aqueous contents) of a cell from its surroundings.

The percentage of lipid and protein components in a cell membrane is related to the function of the cell. The lipid/protein ratio ranges from about 80% lipid/20% protein by mass in the myelin sheath of nerve cells to the unique 20% lipid/80% protein ratio for the inner mitochondrial membrane (Section 12.2). Red blood cell membranes contain approximately equal amounts of lipid and protein. A typical membrane also has a carbohydrate content that varies between 2% and 10% by mass.

Prior to discussing additional types of lipid molecules—emulsification lipids (Section 8.11), messenger lipids (Sections 8.12 and 8.13), and protective-coating lipids (Section 8.14)—the discussion of membrane lipids will be extended to include how these types of lipids interact with each other to form cell membranes.

Living cells contain an estimated 10,000 different kinds of molecules in an aqueous environment confined by a *cell membrane.* A **cell membrane** *is a lipid-based structure that separates a cell's aqueous-based interior from the aqueous environment surrounding the cell.* Besides its "separation" function, a cell membrane also controls the movement of substances into and out of the cell. Up to 80% of the mass of a cell membrane is lipid material consisting primarily of the three types of membrane lipids just discussed: phospholipids, glycolipids, and cholesterol.

The keys to understanding the structural basis for a cell membrane are (1) the virtually insoluble nature of membrane lipids in water and (2) the "head and two tails" structure (Section 8.7) of phospholipids and sphingoglycolipids. When these lipids are placed in water, the polar heads of phospholipids and sphingoglycolipids favor contact with water (they are *hydrophilic;* Section 8.7 margin note), whereas their nonpolar tails interact with one another rather than with water (they are *hydrophobic*). The result is a remarkable bit of molecular architecture called a *lipid bilayer.* A **lipid bilayer** *is a two-layer-thick structure of phospholipids and glycolipids in which the nonpolar tails of the lipids are in the middle of the structure and the polar heads are on the outside surfaces of the structure.* Such a bilayer is six-billionths

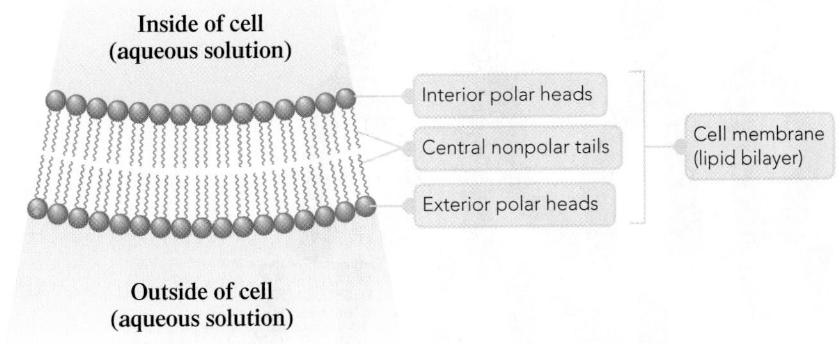

Figure 8.17 Cross-section of a lipid bilayer. The circles represent the polar heads of the lipid components, and the wavy lines represent the nonpolar tails of the lipid components. The heads occupy "surface" positions, and the tails occupy "internal" positions.

to nine-billionths of a meter thick—that is, 6 to 9 nanometers thick. There are three distinct parts to the bilayer: the exterior polar "heads," the interior polar "heads," and the central nonpolar "tails," as shown in Figure 8.17.

Figure 8.18, which is based on space-filling models for phospholipids, gives a "close-up" view of the arrangement of lipid molecules in a section of a lipid bilayer. Note the "exterior" nature of the polar heads of these membrane lipids.

A lipid bilayer is held together by intermolecular interactions, not by covalent bonds. This means each phospholipid or sphingolipid is free to diffuse laterally within the lipid bilayer. Most lipid molecules in the bilayer contain at least one unsaturated fatty acid. The presence of such acids, with the kinks in their carbon chains (Section 8.3), prevents tight packing of fatty acid chains (Figure 8.19). The open packing imparts a flexible or fluid character to the membrane—a necessity because numerous types of biochemicals must pass into and out of a cell.

The glycerophospholipids and sphingoglycolipids found in a lipid bilayer are chiral molecules with the chiral center (Section 7.4) at carbon 2 of the glycerol or sphingosine components of the molecules. The stereoconfiguration of these chiral molecules is always "left-handed," that is, L isomers. The fact that they all have the same configuration enhances their ability to aggregate together in the lipid bilayer.

Cholesterol molecules are also components of cell membranes. They regulate membrane rigidity. Because of their compact shape (Section 8.9; Figure 8.15), cholesterol molecules fit between the fatty acid chains of the lipid bilayer (Figure 8.20), restricting movement of the fatty acid chains. Within the membrane, the cholesterol molecule orientation is "head" to the outside (the hydroxyl group) and "tail" to the inside (the steroid ring structure with its attached alkyl groups).

Proteins are also components of lipid bilayers. The proteins are responsible for moving substances such as nutrients and electrolytes across the membrane, and they also act as receptors that bind hormones and neurotransmitters.

There are two general types of membrane proteins: *integral* and *peripheral.* An **integral membrane protein** *is a membrane protein that penetrates the cell membrane.* Some membrane proteins penetrate only partially through the lipid bilayer, whereas others go completely from one side to the other side of the lipid bilayer. A **peripheral membrane protein** *is a nonpenetrating membrane protein located on the surface of the cell membrane.* Intermolecular forces rather than chemical bonds govern the interactions between membrane proteins and the lipid bilayer. Figure 8.21 shows diagrammatically the relationship between membrane proteins and the overall structure of a cell membrane.

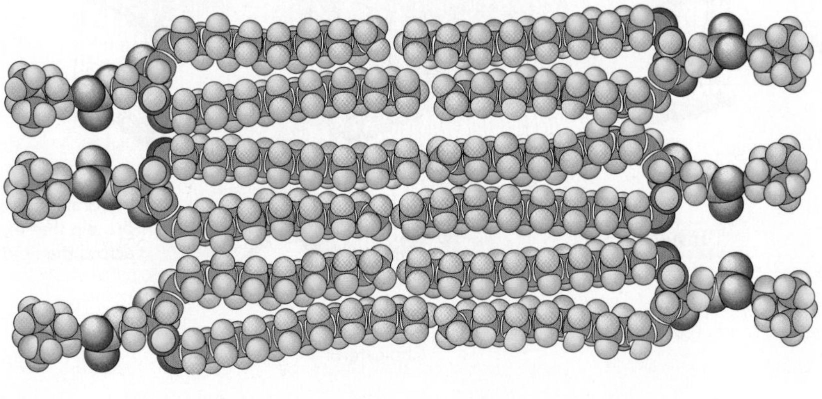

Figure 8.18 Space-filling model of a section of a lipid bilayer. The key to the structure is the "head and two tails" structure of the membrane lipids that constitute the bilayer.

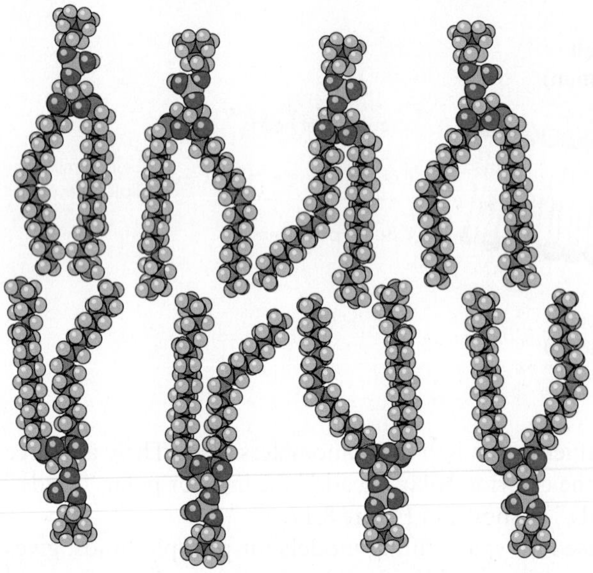

Figure 8.19 The kinks associated with *cis* double bonds in fatty acid chains prevent tight packing of the lipid molecules in a lipid bilayer.

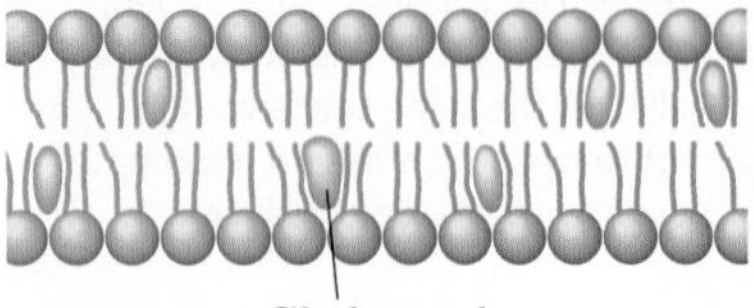

Figure 8.20 Cholesterol molecules fit between fatty acid chains in a lipid bilayer.

Small carbohydrate molecules, usually oligosaccharides, are also components of cell membranes. They are found on the *outer* membrane and are covalently bonded to protein molecules or lipid molecules. The carbohydrate chains extend outward from the membrane into the surrounding aqueous environment. Such carbohydrate-protein and carbohydrate-lipid combinations are called, respectively, glycoproteins and glycolipids (Section 7.20). Functionally, the carbohydrate portions of these glycoproteins and glycolipids are *markers*, substances that play key roles in the process by which different cells recognize each other. The red-blood-cell marker system previously discussed in Chemical Connections 7-D on page 293 involves glycoproteins.

Transport Across Cell Membranes

In order for cellular processes to be maintained, molecules of various types must be able to cross cell membranes. Three common transport mechanisms exist by which molecules can enter and leave cells. They are *passive* transport, *facilitated* transport, and *active* transport.

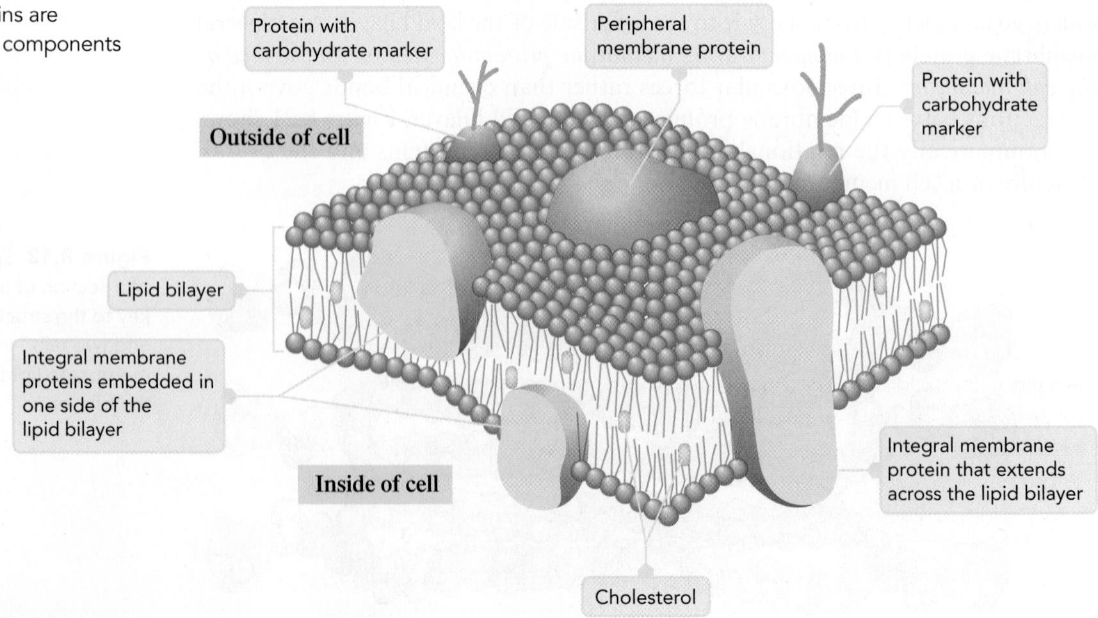

Figure 8.21 Proteins are important structural components of cell membranes.

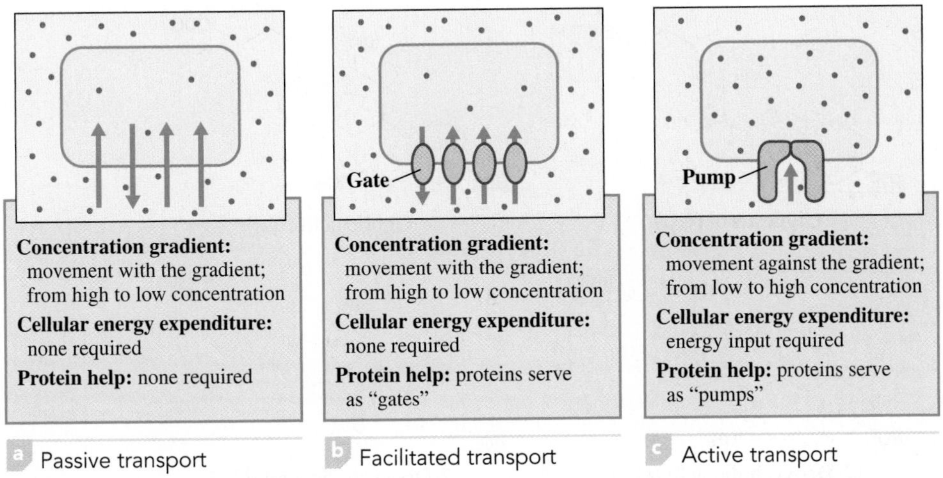

Gate
Pump
Concentration gradient: movement with the gradient; from high to low concentration
Cellular energy expenditure: none required
Protein help: none required
Concentration gradient: movement with the gradient; from high to low concentration
Cellular energy expenditure: none required
Protein help: proteins serve as "gates"
Concentration gradient: movement against the gradient; from low to high concentration
Cellular energy expenditure: energy input required
Protein help: proteins serve as "pumps"
a Passive transport
b Facilitated transport
c Active transport

Figure 8.22 Three processes by which substances can cross plasma membranes.

Passive transport *is the transport process in which a substance moves across a cell membrane by diffusion from a region of higher concentration to a region of lower concentration without the expenditure of any cellular energy.* Only a few types of molecules, including O_2, N_2, H_2O, urea, and ethanol, can cross membranes in this manner. Passive transport is closely related to the process of osmosis.

Facilitated transport *is the transport process in which a substance moves across a cell membrane, with the aid of membrane proteins, from a region of higher concentration to a region of lower concentration without the expenditure of cellular energy.* The specific protein molecules involved in the process are called *carriers* or *transporters.* A carrier protein forms a complex with a specific molecule at one surface of the membrane. Formation of the complex induces a conformational change in the protein that allows the molecule to move through a "gate" to the other side of the membrane. Once the molecule is released, the protein returns to its original conformation. Glucose, chloride ion, and bicarbonate ion cross membranes in this manner.

Active transport *is the transport process in which a substance moves across a cell membrane, with the aid of membrane proteins, against a concentration gradient with the expenditure of cellular energy.* Proteins involved in active transport are called "pumps," because they require energy much as a water pump requires energy in order to function. The needed energy is supplied by molecules such as ATP (Section 12.3). The need for energy expenditure is related to the molecules moving against a concentration gradient—from lower to higher concentration. It is essential to life processes to have some solutes "permanently" at different concentrations on the two sides of a membrane, a situation contrary to the natural tendency (osmosis) to establish equal concentrations on both sides of a membrane. Hence the need for active transport. Sodium, potassium, and hydronium ions cross membranes through active transport.

Figure 8.22 contrasts the processes of passive transport, facilitated transport, and active transport.

8.11 Emulsification Lipids: Bile Acids

An **emulsifier** *is a substance that can disperse and stabilize water-insoluble substances as colloidal particles in an aqueous solution.* Cholesterol derivatives called *bile acids* function as emulsifying agents that facilitate the absorption of dietary lipids in the intestine. Their mode of action is much like that of soap during washing (see Chemical Connections 8-C on page 332).

A **bile acid** *is a cholesterol derivative that functions as a lipid-emulsifying agent in the aqueous environment of the digestive tract.* From one-third to one-half of the daily production of cholesterol by the liver is used to replenish bile acid stores.

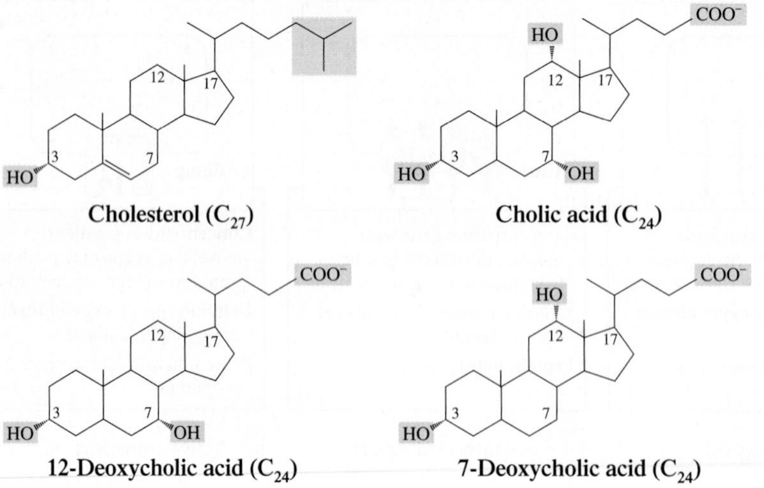

Figure 8.23 Structural formulas for cholesterol, cholic acid, and two deoxycholic acids.

Obtained by oxidation of cholesterol, bile acids differ structurally from cholesterol in three respects:

1. They are tri- or dihydroxy cholesterol derivatives.
2. The carbon 17 side chain of cholesterol has been oxidized to a carboxylic acid.
3. The oxidized acid side chain is bonded to an amino acid (either glycine or taurine) through an amide linkage.

The average bile acid composition in normal human adult bile is 38% cholic acid derivatives, 34% 7-deoxycholic acid derivatives, and 28% 12-deoxycholic acid derivatives. Glycine-containing derivatives predominate over taurine-containing derivatives by a 3:1 to 4:1 ratio. Uncomplexed (free) bile acids are not present in bile.

Figure 8.23 gives structural formulas for the three major types of bile acids produced from cholesterol by biochemical oxidation: cholic acid, 7-deoxycholic acid, and 12-deoxycholic acid. The structural formulas are those for these bile acids prior to the attachment of the amino acid to the carbon 17 side chain.

Bile acids always carry an amino acid (either glycine or taurine) attached to the side-chain carboxyl group via an amide linkage. The presence of this amino acid attachment increases both the polarity of the bile acid and its water solubility. Figure 8.24 shows the structures of glycocholic acid (glycine is the amino acid) and taurocholic acid (taurine is the amino acid).

The medium through which bile acids are supplied to the small intestine is *bile*. **Bile** *is a fluid containing emulsifying agents that is secreted by the liver, stored in the gallbladder, and released into the small intestine during digestion.* Besides

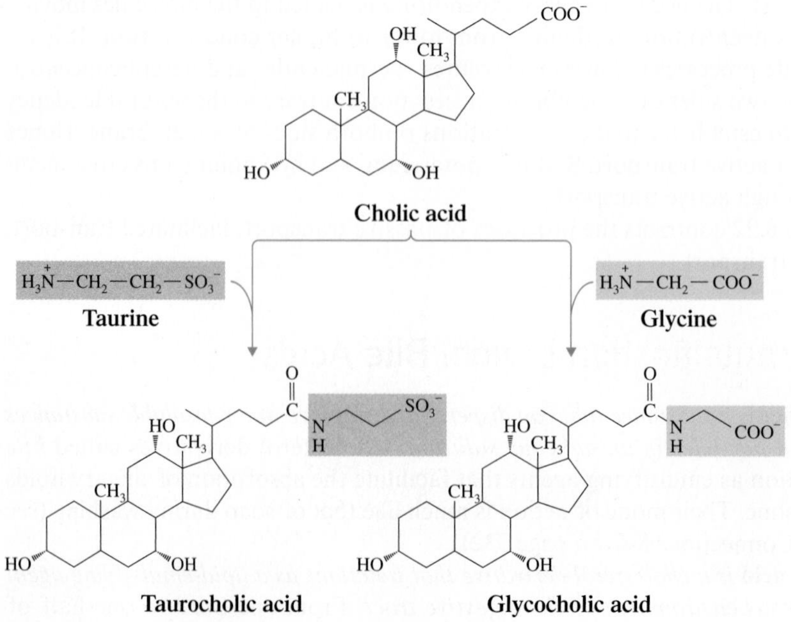

Figure 8.24 The structures of glycocholic acid and taurocholic acid.

bile acids, bile also contains bile pigments (breakdown products of hemoglobin; Section 15.7), cholesterol itself, and electrolytes such as bicarbonate ion. The bile acids that are present increase the solubility of the cholesterol in the bile fluid.

A number of factors, including increased secretion of cholesterol and a decrease in the size of the bile pool, can upset the balance between the cholesterol present in bile and the bile acid derivatives needed to maintain cholesterol's solubility in the bile. The result is the precipitation of crystallized cholesterol from the bile and the resulting formation of gallstones in the gallbladder. In Western countries, approximately 80% of gallstones are almost pure cholesterol (Figure 8.25).

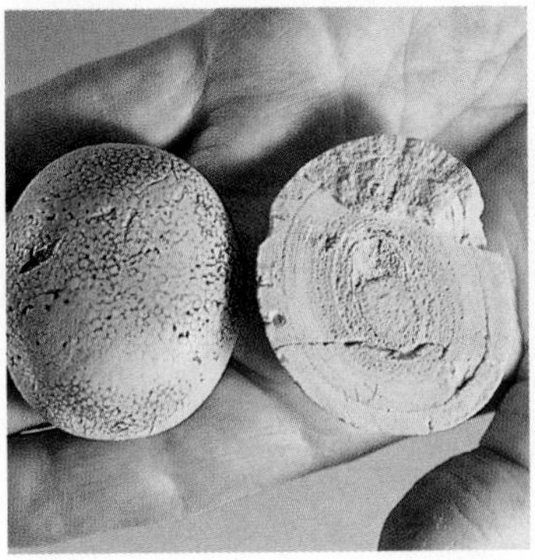

C. James Webb/Phototake

Figure 8.25 A large percentage of gallstones, the causative agent for many "gallbladder attacks," are almost pure crystallized cholesterol that has precipitated from bile solution.

8.12 Messenger Lipids: Steroid Hormones

Previously considered were lipids that function as energy-storage molecules (triacylglycerols; Section 8.4), as components of cell membranes (phospholipids, sphingoglycolipids, and cholesterol; Sections 8.7 through 8.9), and as emulsifying agents (bile acids; Section 8.11). An additional role played by lipids is that of "chemical messenger." *Steroid hormones* and *eicosanoids* are two large families of lipids that have messenger functions. In this section, steroid hormones, which are cholesterol derivatives, will be considered. In Section 8.13, eicosanoids, which are fatty acid derivatives, will be considered.

A **hormone** *is a biochemical substance, produced by a ductless gland, that has a messenger function.* Hormones serve as a means of communication *between* various tissues. Some hormones, though not all, are lipids.

A **steroid hormone** *is a hormone that is a cholesterol derivative.* There are two major classes of steroid hormones: (1) sex hormones, which control reproduction and secondary sex characteristics and (2) adrenocorticoid hormones, which regulate numerous biochemical processes in the body.

Sex Hormones

The sex hormones can be classified into three major groups:

1. Estrogens—the female sex hormones
2. Androgens—the male sex hormones
3. Progestins—the pregnancy hormones

Estrogens are synthesized in the ovaries and adrenal cortex and are responsible for the development of female secondary sex characteristics at the onset of puberty and for regulation of the menstrual cycle. They also stimulate the development of the mammary glands during pregnancy and induce estrus (heat) in animals.

Estrogens are a class of molecules rather than a single molecule. Statements like "the estrogen level is high" should be rephrased as "there is a high level of estrogens."

Androgens are synthesized in the testes and adrenal cortex and promote the development of male secondary sex characteristics. They also promote muscle growth.

Progestins are synthesized in the ovaries and the placenta and prepare the lining of the uterus for implantation of the fertilized ovum. They also suppress ovulation.

Figure 8.26a gives the structure of the primary hormone in each of the three subclasses of sex hormones. Other members of these hormone families are metabolized forms of the primary hormone.

Note in Figure 8.26a how similar the structures are for these principal hormones, and yet how different their functions are. The fact that seemingly minor changes in structure effect great changes in biofunction points out, again, the extreme specificity (Section 10.5) of the enzymes that control biochemical reactions.

Increased knowledge of the structures and functions of sex hormones has led to the development of a number of *synthetic* steroids whose actions often mimic those of the natural steroid hormones. The best known types of synthetic steroids are oral contraceptives and anabolic steroids.

Figure 8.26 Structures of selected sex hormones and synthetic compounds that have similar actions.

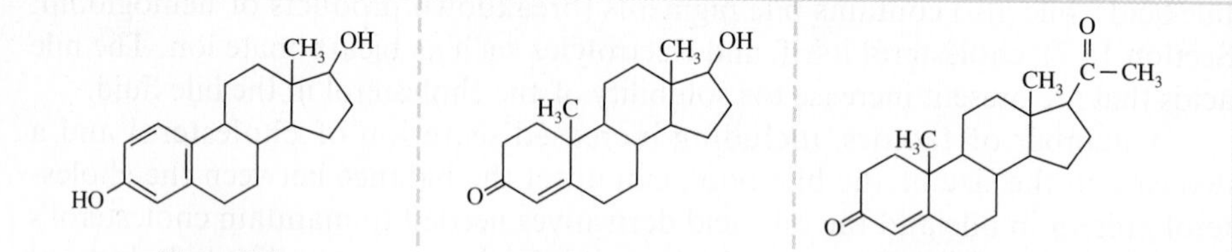

Estradiol (the primary estrogen; responsible for secondary female characteristics)

Testosterone (the primary androgen; responsible for secondary male characteristics)

Progesterone (the primary progestin; prepares the uterus for pregnancy)

a Natural hormones

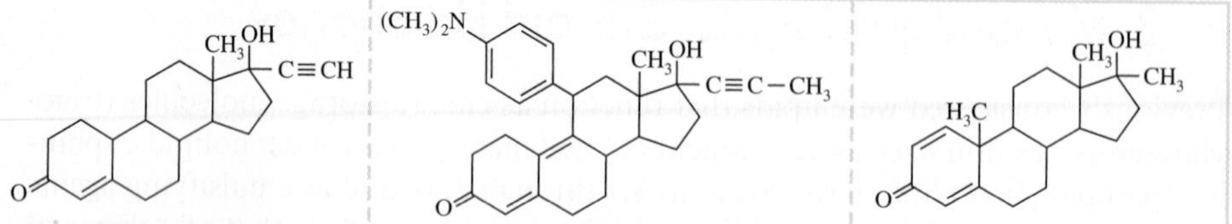

Norethynodrel (a synthetic progestin)

RU-486 (mifepristone; a synthetic abortion drug)

Methandrostenolone (a synthetic tissue-building steroid)

b Synthetic steroids

The C≡C functional group, which occurs in both norethynodrel (Enovid) and RU-486, is rarely found in biomolecules.

Oral contraceptives are used to suppress ovulation as a method of birth control. Generally, a mixture of a synthetic estrogen and a synthetic progestin is used. The synthetic estrogen regulates the menstrual cycle, and the synthetic progestin prevents ovulation, thus creating a false state of pregnancy. The structure of norethynodrel (Enovid), a synthetic progestin, is given in Figure 8.26b. Compare its structure to that of progesterone (the real hormone); the structures are very similar.

Interestingly, the controversial "morning after" pill developed in France and known as RU-486, is also similar in structure to progesterone. RU-486 interferes with gestation of a fertilized egg and terminates a pregnancy within the first nine weeks of gestation more effectively and safely than surgical methods. The structure of RU-486 appears next to that of norethynodrel in Figure 8.26b.

Anabolic steroids include the illegal steroid drugs used by some athletes to build up muscle strength and enhance endurance. Such steroids are now known to have serious side effects in the user. The structure of one of the more commonly used anabolic steroids, methandrostenolone, is given in Figure 8.26b. Note the similarities between its structure and that of the naturally occurring testosterone. The focus on relevancy feature Chemical Connections 8-E on the next page further explores the use, usually illegal, of anabolic steroids by athletes.

Adrenocorticoid Hormones

The second major group of steroid hormones consists of the adrenocorticoid hormones. Produced by the adrenal glands, small organs located on top of each kidney, at least 28 different hormones have been isolated from the adrenal cortex (the outer part of the glands).

There are two types of adrenocorticoid hormones.

1. *Mineralocorticoids* control the balance of Na^+ and K^+ ions in cells and body fluids.
2. *Glucocorticoids* control glucose metabolism and counteract inflammation.

The major mineralocorticoid is aldosterone, and the major glucocorticoid is cortisol (hydrocortisone). Cortisol is the hormone synthesized in the largest amount by the adrenal glands. Cortisol and its synthetic ketone derivative cortisone exert powerful anti-inflammatory effects in the body. Both cortisone and prednisolone,

CHEMICAL CONNECTIONS 8-E

Anabolic Steroid Use in Competitive Sports

The steroid hormone testosterone is the principal male sex hormone. It has masculinizing (androgenic) effects and muscle-building (anabolic) effects. Masculinizing effects of testosterone include the growth of facial and body hair, deepening of the voice, and maturation of the male sex organs. Testosterone's anabolic effects are responsible for the muscle development that boys experience at puberty.

Many synthetic derivatives of testosterone are now known. Some of these exert primarily androgenic effects, whereas others exert primarily anabolic effects. Androgenic steroids can be used to correct hormonal imbalances in the body. Anabolic steroids can be used to prevent the withering of muscle in persons recovering from major surgery or serious injuries.

Anabolic steroids have also been "discovered" by athletes, who have found that these compounds can be used to help build muscle mass and reduce the healing time for muscle injuries. Often the net result of anabolic hormone use by an athlete is enhanced athletic performance.

Anabolic steroid use by athletes in competitive sports is now referred to as "doping" and is banned by all major sporting organizations, including the International Olympic Committee (IOC), National Football League (NFL), National Basketball Association (NBA), and Major League Baseball (MLB). To enforce the ban on anabolic steroid use, athletes are subjected to random drug screening. In major events, such as the Olympics, testing is no longer random but mandatory for all competitors.

One major athletic event that has been plagued with allegations of doping in recent years is the Tour de France. This annual bicycle race, which is held in France and nearby countries, is one of the most physically grueling tests of endurance available to athletes (see accompanying photo). Lasting three weeks, the race covers 2200 miles, including major "mountain climbs." Numerous riders have been disqualified for testing positive for banned performance-enhancing drugs.

Major League Baseball is another sport that has encountered anabolic steroid problems. During the 1990s and the early part of the next decade, many players, including some of the League's most prominent ones, were involved in anabolic steroid use.

In response to the growing problem of illegal use of anabolic steroids, legislation was passed in the United States that supplemented (amended) the Controlled Substance Act. Called the Anabolic Steroid Control Act, which went into effect in 2005, this legislation makes possession of a banned anabolic steroid a federal crime. Still, some athletes continue to use these substances illegally.

© Dcslim/Dreamstime.com

Tour de France participants face one of the most physically grueling endurance tests in all of competitive athletics.

Two major reasons exist for prohibiting steroid use in athletes: (1) their use is considered a form of cheating because it confers an unfair advantage and (2) their "beneficial effects" are far outweighed by serious negative side effects.

Regarding the latter of these two reasons, medical evidence now indicates that anabolic steroid use has many side effects that range from some that are physically unattractive, such as acne and breast development in men, to others that are life-threatening, such as heart attacks and liver problems. Most of these alarming effects are reversible if the abuser stops taking the drugs, but some are permanent.

Steroid abuse disrupts the normal production of hormones in the body, causing both reversible and irreversible change. The male reproductive system is altered, causing testicular shrinkage and decreased sperm production. Both of these effects are reversible, but breast development is an irreversible change.

In the female body, steroid abuse causes masculinization. Breast size and body fat decrease, the skin becomes coarse, and the voice deepens. Some women may experience excessive growth of body hair but lose hair from the scalp.

The structures of synthetic anabolic steroids closely resemble that of the naturally occuring male sex hormone testosterone. Chemically, most anabolic steroids are testosterone derivatives. Testosterone can be viewed as the "template" structure from which many synthetic variations (analogs) have been "designed." Several of these testosterone derivatives increase the body's ability to metabolize ingested protein and facilitate the synthesis of new skeletal muscle proteins.

Two of the most publicized anabolic steroids that athletes have used illegally are "andro" and "THG." The structures of these two compounds, as well as that of testosterone for comparison purposes, are shown in the accompanying diagram.

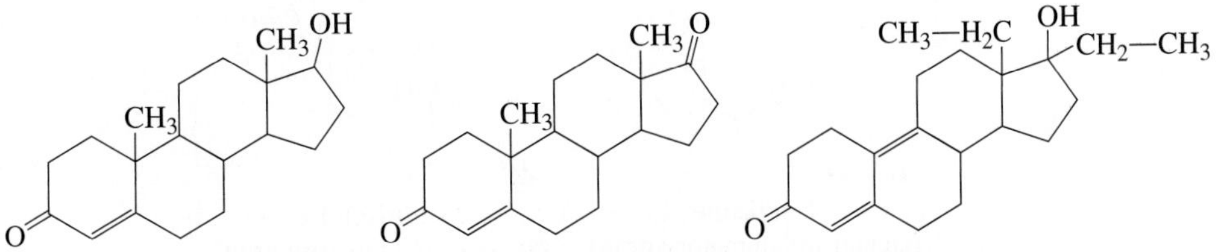

Testosterone | "Andro" Androstenedione | THG Tetrahydrogestrinone

(continued)

"Andro," which is short for androstenedione, was the steroid used during the 1990s and early part of the following decade in Major League Baseball. It was initially developed for use as a dietary supplement. At the time of its use in MLB, it was a legal supplement that could be purchased over the counter. Now, its possession without a prescription is illegal.

THG, short for tetrahydrogestrinone, is a "designer" steroid. Its molecular structure was designed so that drug testing procedures of the time could not detect the drug. For several years, it went undetected in drug testing. In 2003, the illegal use of this drug was discovered only after a spent syringe was sent anonymously to a testing agency with the comment that the syringe might contain a new performance-enhancing drug. Once the drug was identified, tests were rapidly developed for its detection. Some Olympic sprinters have been disqualified for using THG.

a similar synthetic derivative, are used as prescription drugs to control inflammatory diseases such as rheumatoid arthritis. Figure 8.27 gives the structures of these adrenocorticoid hormones.

8.13 Messenger Lipids: Eicosanoids

An **eicosanoid** *is an oxygenated* C_{20} *fatty acid derivative that functions as a messenger lipid.* The term *eicosanoid* is derived from the Greek word *eikos,* which means "twenty." The metabolic precursor for most eicosanoids is arachidonic acid, the 20:4 fatty acid.

Eicosanoids exert their effects at very low concentrations, sometimes less than one part in a billion (10^9).

Almost all cells, except red blood cells, produce eicosanoids. These substances, like hormones, have profound physiological effects at extremely low concentrations. Eicosanoids are hormonelike molecules rather than true hormones because they are not transported in the bloodstream to their site of action as true hormones are. Instead, they exert their effects in the tissues where they are synthesized. Eicosanoids usually have a very short "life," being broken down, often within seconds of their synthesis, to inactive residues (which are eliminated in urine). For this reason, they are difficult to study and monitor within cells.

The physiological effects of eicosanoids include mediation of

1. The inflammatory response, a normal response to tissue damage
2. The production of pain and fever
3. The regulation of blood pressure
4. The induction of blood clotting
5. The control of reproductive functions, such as induction of labor
6. The regulation of the sleep/wake cycle

Figure 8.27 Structures of selected adrenocorticoid hormones and related synthetic compounds.

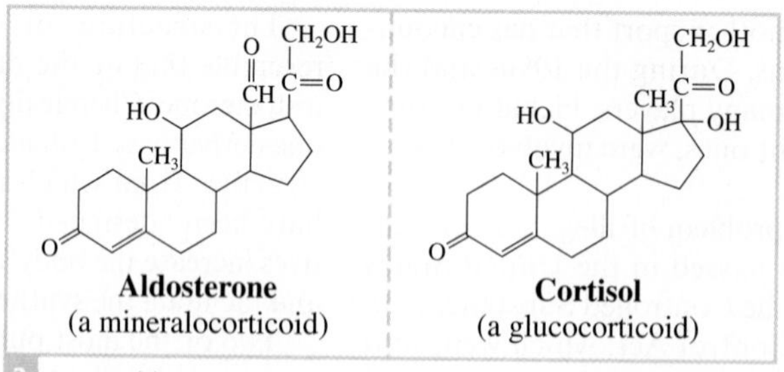

a Natural hormones

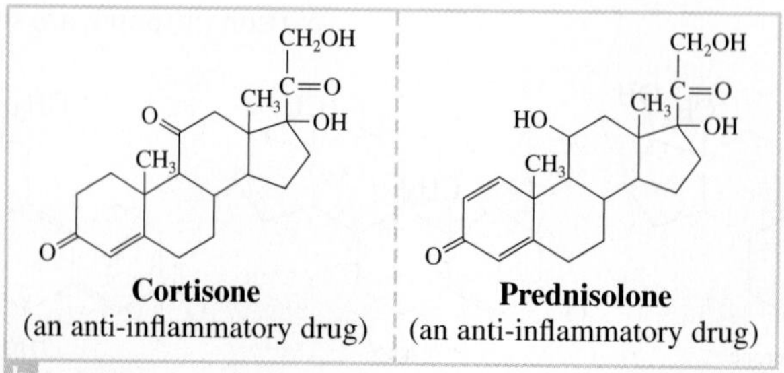

b Synthetic steroids

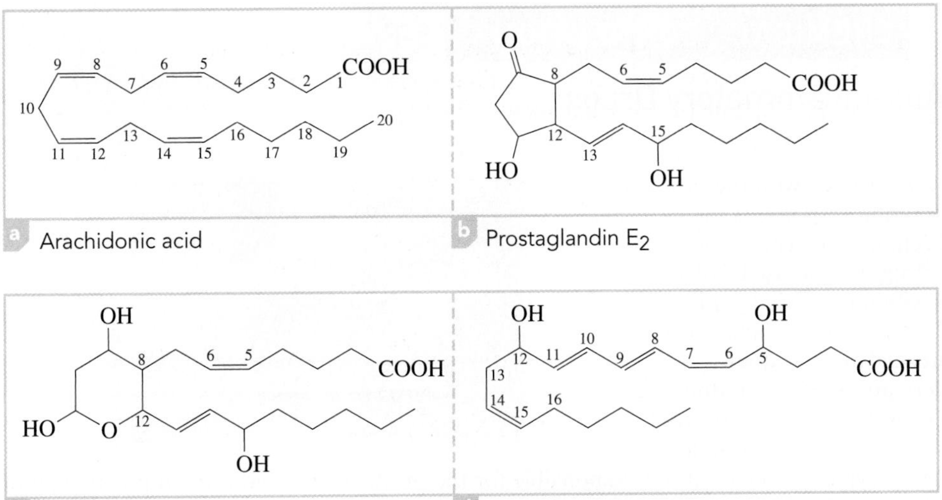

a Arachidonic acid

b Prostaglandin E_2

c Thromboxane B_2

d Leukotriene B_4

Figure 8.28 Relationship of the structures of various eicosanoids to their precursor, arachidonic acid.

The capital letter–numerical subscript designations for individual eicosanoids is based on selected structural characteristics of the molecules. The numerical subscript indicates the number of carbon–carbon double bonds present. The letters denote subgroups of molecules. The prostaglandin E group, for example, has a carbonyl group on carbon 9.

There are three principal types of eicosanoids: prostaglandins, thromboxanes, and leukotrienes.

Prostaglandins

A **prostaglandin** *is a messenger lipid that is a C_{20}-fatty-acid derivative that contains a cyclopentane ring and oxygen-containing functional groups.* Twenty-carbon fatty acids are converted into a prostaglandin structure when the eighth and twelfth carbon atoms of the fatty acid become connected to form a five-membered ring (Figure 8.28b).

Prostaglandins are named after the prostate gland, which was first thought to be their only source. Today, more than 20 prostaglandins have been discovered in a variety of tissues in both males and females.

Within the human body, prostaglandins are involved in many regulatory functions, including raising body temperature, inhibiting the secretion of gastric juices, increasing the secretion of a protective mucus layer into the stomach, relaxing and contracting smooth muscle, directing water and electrolyte balance, intensifying pain, and enhancing inflammation responses. Aspirin reduces inflammation and fever because it inactivates enzymes needed for prostaglandin synthesis.

The focus on relevancy feature Chemical Connections 8-F on the next page considers further the relationship between prostaglandin synthesis and inflammation in terms of how anti-inflammatory drugs exert their effects in the human body.

Thromboxanes

Naturally occurring fatty acids are normally found in the form of fatty acid residues in such molecules as triacylglycerols, phospholipids, and sphingolipids, substances previously considered in this chapter. In eicosanoids, the fatty acids are not esterified and carry out their biochemical roles in the form of the free acids, that is, free fatty acid derivatives.

A **thromboxane** *is a messenger lipid that is a C_{20}-fatty-acid derivative that contains a cyclic ether ring and oxygen-containing functional groups.* As with prostaglandins, the cyclic structure involves a bond between carbons 8 and 12 (Figure 8.28c). An important function of thromboxanes is to promote the formation of blood clots. Thromboxanes are produced by blood platelets and promote platelet aggregation.

Leukotrienes

A **leukotriene** *is a messenger lipid that is a C_{20}-fatty-acid derivative that contains three conjugated double bonds and hydroxy groups.* Fatty acids and their derivatives do not normally contain *conjugated* double bonds (see Chemical Connections 2-C on page 57), as do leukotrienes (Figure 8.28d). Leukotrienes are found in leukocytes (white blood cells). Their source and the presence of the three conjugated double bonds account for their name. Various inflammatory and hypersensitivity (allergy) responses are associated with elevated levels of leukotrienes. The development of drugs that inhibit leukotriene synthesis has been an active area of research.

CHEMICAL CONNECTIONS 8-F

The Mode of Action for Anti-Inflammatory Drugs

Injury or damage to bodily tissue is associated with the process of inflammation. This inflammation response is mediated by prostaglandin molecules (Section 8.13). The mode of action for most anti-inflammatory drugs now in use involves decreasing prostaglandin synthesis within the body by inhibiting the action of one or more of the enzymes (biochemical catalysts; Chapter 10) needed for prostaglandin synthesis.

Prostaglandin molecules are derivatives of arachidonic acid, a 20:4 fatty acid (Section 8.13). Anti-inflammatory *steroid* drugs such as cortisone (Section 8.12) inhibit the action of the enzyme *phospholipase A_2*, the enzyme that facilitates the breakdown of complex arachidonic acid-containing lipids to produce free arachidonic acid. Inhibiting arachidonic acid release stops the prostaglandin synthesis process, which in turn prevents (or diminishes) inflammation.

Besides anti-inflammatory *steroid* drugs, many *nonsteroidal* anti-inflammatory drugs (NSAIDs) are also available for inflammation control. Aspirin (see Chemical Connections 5-C on page 185) is the oldest and best known NSAID. Ibuprofen (Advil) and naproxen (Aleve) (see Chemical Connections 5-A on page 171), which were initially prescription medications, are now commonly used over-the-counter anti-inflammatory drugs. The most often prescribed prescription NSAID is *diclofenac* (see accompanying photo), a compound whose chemical structure contains chlorine atoms.

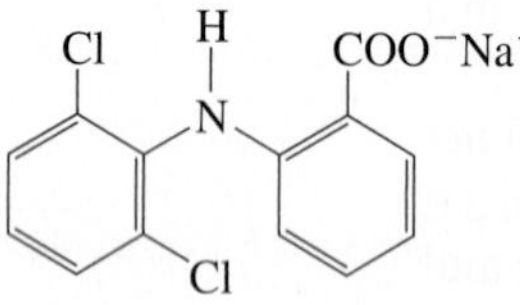

These substances are not only anti-inflammatories, but also have antipyretic (fever reducer) and mild analgesic (pain-reliever) properties. They prevent prostaglandin synthesis by inhibiting the enzyme needed for the ring closure reaction at carbons 8 and 12 in arachidonic acid, a necessary step in prostaglandin synthesis (Section 8.13). The enzyme that NSAIDs inhibit is called *cyclooxygenase,* an enzyme known by the acronym *COX.*

There are actually two forms of the COX enzyme: COX-1 and COX-2. COX-1 also produces prostaglandins that induce the production of a protective mucous coating for the stomach lining (a desirable situation). The COX-2 enzyme is

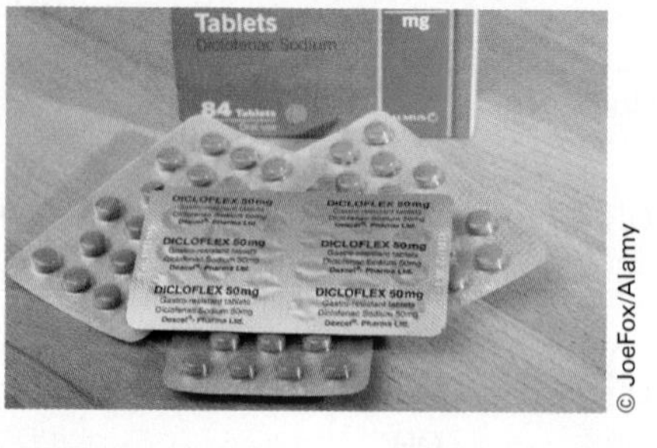

© JoeFox/Alamy

responsible for the production of additional prostaglandin molecules involved in inflammatory diseases such as arthritis (an undesirable situation).

Aspirin, ibuprofen, naproxen, and diclofenac inhibit both COX-1 and COX-2 enzymes. Use of these NSAIDs thus decreases inflammation but also makes an individual more susceptible to stomach irritation and ulcer formation, as protective mucous secretions are decreased when COX-1 is inhibited.

A new generation of prescription anti-inflammatory agents, developed in the 1990s, is now in use. These drugs are COX-2 inhibitors but not COX-1 inhibitors. The best known of these COX-2 inhibitors are Vioxx and Celebrex, whose chemical structures are

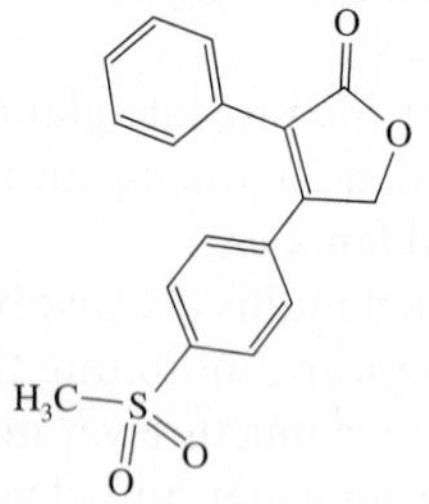

Vioxx

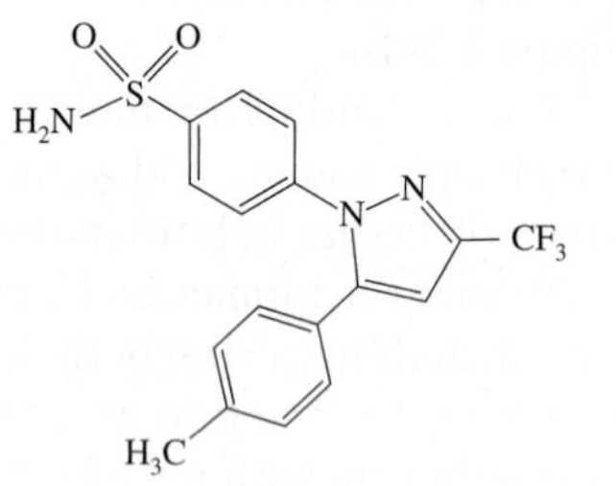

Celebrex

In 2004, Vioxx was withdrawn from the market because of concerns relating to heart attacks and strokes. A study indicated that patients taking Vioxx were twice as likely to suffer a heart attack or stroke as a control group involved in the study who were taking a placebo. The actual risk was 3.5% in the Vioxx group, compared with 1.9% in the control group, according to the FDA. The difference became apparent after 18 months of Vioxx use. Celebrex has fewer side effects than Vioxx, so it is still available and is widely used as a prescription medication.

8.14 Protective-Coating Lipids: Biological Waxes

A **biological wax** *is a lipid that is a monoester of a long-chain fatty acid and a long-chain alcohol.* Biological waxes are *monoesters,* unlike fats and oils (Section 8.4), which are *triesters.* The fatty acids found in biological waxes generally are saturated and contain from 14 to 36 carbon atoms. The alcohols found in biological waxes may be saturated or unsaturated and may contain from 16 to 30 carbon atoms.

The block diagram for a biological wax is

Long-chain fatty acid — Long-chain alcohol

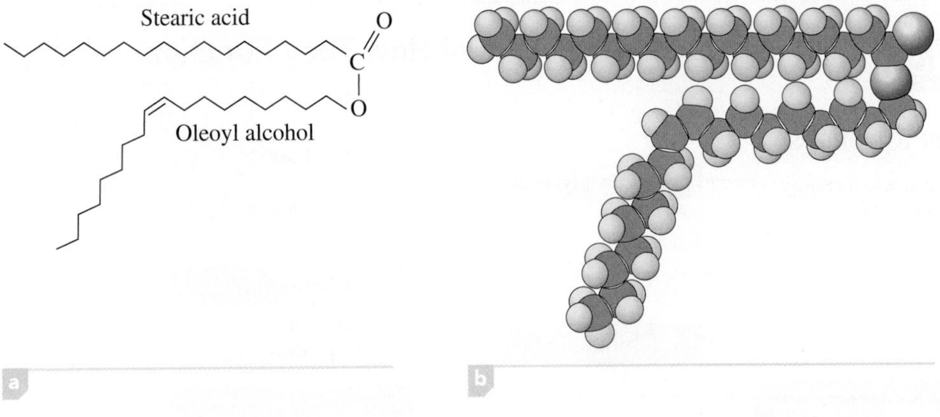

Figure 8.29 A biological wax has a structure with a small, weakly polar "head" and two long, nonpolar "tails." The polarity of the small "head" is not sufficient to impart any degree of water solubility to the molecule.

with the fatty acid and alcohol linked through an ester linkage. An actual structural formula for a biological wax that bees secrete and use as a structural material is

Fatty acid residue — Ester linkage — Alcohol residue

$$CH_3-(CH_2)_{14}-\overset{\displaystyle O}{\overset{\|}{C}}-O-(CH_2)_{29}-CH_3$$

A component of beeswax

The term wax derives from the old English word *weax*, which means "the material of the honeycomb."

Note that the general structural formula for a biological wax is the same as that for a simple ester (Section 5.9).

$$R-\overset{\displaystyle O}{\overset{\|}{C}}-O-R'$$

However, for waxes, both R and R′ must be long carbon chains (usually 20–30 carbon atoms).

The water-insoluble, water-repellent properties of biological waxes result from the complete dominance of the nonpolar nature of the long hydrocarbon chains present (from the alcohol and the fatty acid) over the weakly polar nature of the ester functional group that links the two carbon chains together (Figure 8.29).

In living organisms, biological waxes have numerous functions, all of which are related directly or indirectly to their water-repellent properties. Both humans and animals possess skin glands that secrete biological waxes to protect hair and skin and to keep it pliable and lubricated. With animal fur, waxes impart water repellency to the fur. Birds, particularly aquatic birds, rely on waxes secreted from preen glands to keep their feathers water-repellent. Such wax coatings also help minimize loss of body heat when the bird is in cold water. Many plants, particularly those that grow in arid regions, have leaves that are coated with a thin layer of biological waxes, which serve to prevent excessive evaporation of water and to protect against parasite attack (Figure 8.30). Similarly, insects with a high surface-area-to-volume ratio are often coated with a protective biological wax.

© Kevin Schaefer/Peter Arnold, Inc./Photolibrary

Figure 8.30 Plant leaves often have a biological wax coating to prevent excessive loss of water.

Biological waxes find use in the pharmaceutical, cosmetics, and "polishing" industries. Carnauba wax (obtained from a species of Brazilian palm tree) is a particularly hard wax whose uses involve high-gloss finishes: automobile wax, boat wax, floor wax, and shoe wax.

$$CH_3-(CH_2)_{28}-\overset{\displaystyle O}{\overset{\|}{C}}-O-(CH_2)_{31}-CH_3$$

A component of carnauba wax

When aquatic birds are caught in an oil spill, the oil dissolves the wax coating on their feathers. This causes the birds to lose their buoyancy (they cannot swim properly) and compromises their protection against the effects of cold water.

Lanolin, a mixture of waxes obtained from sheep wool, is used as a base for skin creams and ointments intended to enhance retention of water (which softens the skin).

Many synthetic materials are now available with properties that closely match—and even improve on—the properties of biological waxes. Such synthetic materials, which are generally polymers, have now replaced biological waxes in many cosmetics, ointments, and the like. The synthetic *carbowax,* for example, is a polyether.

Naturally occuring waxes, such as beeswax, are usually mixtures of several monoesters rather than being a single monoester. This parallels the situation for fats and oils, which are mixtures of numerous triesters (triacylglycerols).

CHEMISTRY AT A GLANCE

Types of Lipids in Terms of How They Function

ENERGY-STORAGE LIPIDS
Lipids stored for use when energy demand is high

- Triacylglycerols
  - Fats
  - Oils

MEMBRANE LIPIDS
Lipids that are structural components of cell membranes

- Phospholipids
  - Glycerophospholipids
    - Lecithins
    - Cephalins
  - Sphingophospholipids
    - Sphingomyelins
- Cholesterol
- Sphingoglycolipids
  - Cerebrosides
  - Gangliosides

EMULSIFICATION LIPIDS
Lipids that stabilize and disperse water-insoluble materials in aqueous solution

- Bile acids
  - Cholic acid
  - Deoxycholic acids

MESSENGER LIPIDS
Regulatory lipids that act in the tissue where they are synthesized or at other locations after transport via the bloodstream

- Steroid hormones
  - Sex hormones
    - Estrogens
    - Androgens
    - Progestins
  - Adrenocorticoids
    - Mineralocorticoids
    - Glucocorticoids
- Eicosanoids
  - Prostaglandins
  - Thromboxanes
  - Leukotrienes

PROTECTIVE-COATING LIPIDS
Water-insoluble, water-repellent lipids with protective-coating and lubricant functions

- Biological waxes

Human ear wax, which acts as a protective barrier against infection by capturing airborne particles, is not a true biological wax—that is, it is not a mixture of simple esters. Human ear wax is a yellow waxy secretion that is a mixture of triacylglycerols, phospholipids, and esters of cholesterol. Its medical name is *cerumen*.

Throughout this discussion, the term *biological wax,* rather than just *wax,* has been used. This is because the everyday meaning of the term *wax* is broader in scope than the chemical definition of the term *biological wax.* In general discussions, a **wax** *is a pliable, water-repelling substance used particularly in protecting surfaces and producing polished surfaces.* This broadened definition for waxes includes not only *biological waxes* but also *mineral waxes.* A **mineral wax** *is a mixture of long-chain alkanes obtained from the processing of petroleum.* (How mineral waxes are obtained from petroleum was considered in Section 1.15.) Mineral waxes, which are also called *paraffin waxes,* resist moisture and chemicals and have no odor or taste. They serve as a waterproof coating for such paper products as milk cartons and waxed paper. Most candles are made from mineral waxes. Some "wax products" are a blend of biological and mineral waxes. For example, beeswax is sometimes a component of candle wax.

The Chemistry at a Glance feature above summarizes the function-based lipid classifications that have been considered in this chapter. Subclassifications within each function classification are also given in this summary.

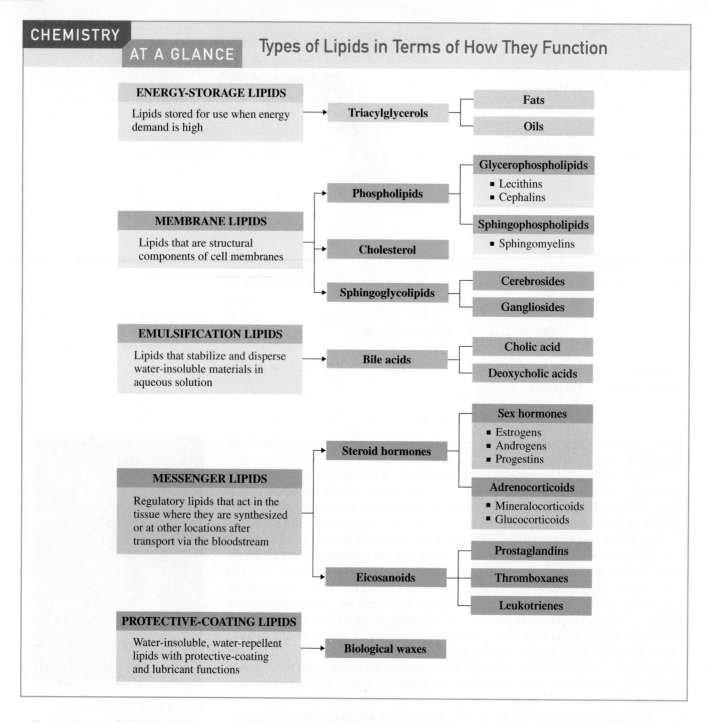

8.15 Saponifiable and Nonsaponifiable Lipids

At the beginning of this chapter, in Section 8.1, the following items were noted:

1. Two classification systems for lipids exist: a system based on biochemical function and a system based on the chemical reaction called saponification.
2. The biochemical function classification system would be used as the basis for this chapter's organization.
3. The saponification reaction classification system would be considered in the last section of this chapter, serving as a chapter summary.

Consideration of the saponification classification system for lipids is now in order.

A saponification reaction is a hydrolysis reaction that is carried out in basic solution (Section 5.16). Some lipids undergo saponification, whereas others do not. Such is the basis for this classification system. There are only two lipid categories in this system: saponifiable lipids and nonsaponifiable lipids. A **saponifiable lipid** *is a lipid that undergoes hydrolysis in basic solution to yield two or more smaller product molecules.* As a result of hydrolysis, a saponifiable lipid is broken up into smaller component parts. A **nonsaponifiable lipid** *does not undergo hydrolysis in basic solution.* Such lipids cannot be broken up into smaller component parts using hydrolysis.

There are five types of saponifiable lipids and four types of nonsaponifiable lipids. The saponifiable lipids are triacylglycerols, glycerophospholipids, sphingophospholipids, sphingoglycolipids, and waxes. The nonsaponifiable lipids are cholesterol, bile acids, steroid hormones, and eicosanoids (Figure 8.31).

What determines whether or not a lipid is saponifiable? It is the types of linkages (bonds) that hold its component parts (building blocks) together. Three types of linkages, all of which have been previously considered, can be hydrolyzed (broken) by reaction with water. These linkage types are (1) ester linkages (2) amide linkages and (3) glycosidic linkages. All three of these linkage types can be found in lipids.

Ester saponification (Section 5.16) and amide saponification (Section 6.17) are topics that have been previously considered.

Saponifiable lipids always contain at least one of these three types of linkages. The linkage "makeup" for the five kinds of saponifiable lipids is as follows:

1. Triacylglycerols—three ester linkages
2. Glycerophospholipids—four ester linkages
3. Sphingophospholipids—one amide and two ester linkages
4. Sphingoglycolipids—one amide, one ester, and one glycosidic linkage
5. Biological waxes—one ester linkage

Figure 8.32 shows pictorially the location of the various linkages in these five types of lipids.

Using the block diagrams in Figure 8.32, it is easy to visualize the products that result from saponification of the various types of saponifiable lipids. The products are simply the "building blocks" from which each lipid can be considered to have been made. The ester, amide, and glycosidic linkages present are broken through reaction with water, which releases (frees up) the lipid's component parts. Any fatty acid molecules released will be converted to fatty acid salts because of the basic conditions associated with saponification (Sections 5.16 and 6.17).

Hydrolysis and saponification reactions involving lipids were previously considered in Section 8.6 of this chapter. Examples 8.3 and 8.4, the worked-out

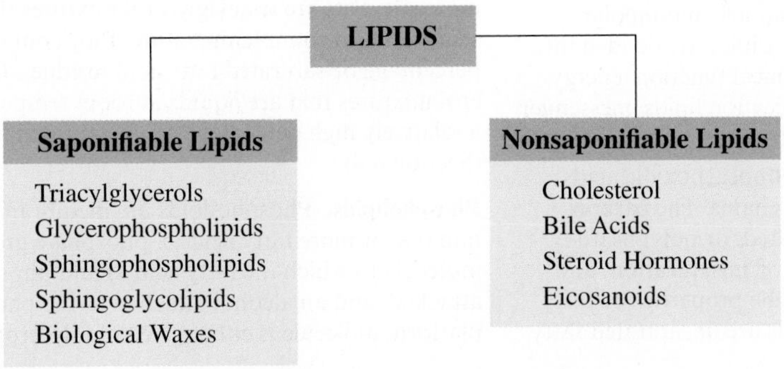

Figure 8.31 Classification of lipids into the categories *saponifiable* and *nonsaponifiable*.

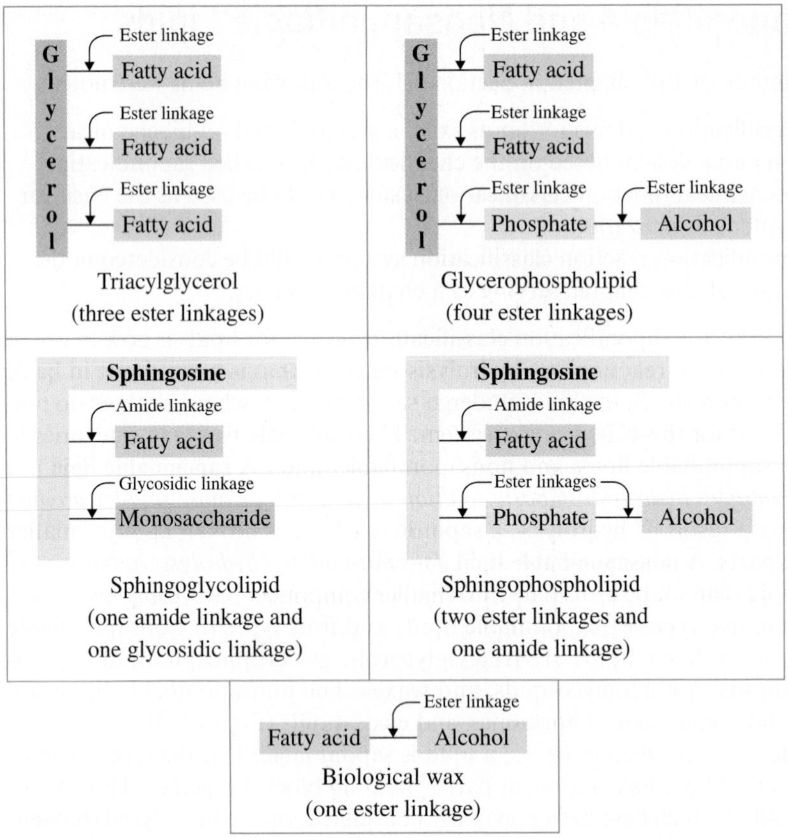

Figure 8.32 Types of linkages present in the five kinds of saponifiable lipids.

Hydrolysis reactions involving the ester, amide, and glycosidic linkages present in saponifiable lipids will be an important topic in Chapter 14, which considers the metabolic reactions that lipids undergo within the human body.

examples in that section, both deal with the concepts that have just been reviewed. In light of this review, it would be instructive to revisit these examples.

The nonsaponifiable lipids are cholesterol, bile acids, steroid hormones, and eicosanoids. Cholesterol, bile acids, and steroid hormones have structures containing the four-ring steroid nucleus as a platform. Eicosanoids have structures based on a single fatty acid molecule. Structurally, these nonsaponifiable lipids have two things in common: (1) only one building block is present and (2) there is no need for ester, amide, or glycosidic linkages to be present to link building blocks together since there is only one building block. There is no site (linkage) at which hydrolysis can occur, which occasions the nonsaponifiable classification for these substances.

Concepts to Remember

OWL Sign in at **www.cengage.com/owl** to view tutorials and simulations, develop problem-solving skills, and complete online homework assigned by your professor.

Lipids. Lipids are a structurally heterogeneous group of compounds of biochemical origin that are soluble in nonpolar organic solvents and insoluble in water. Lipids are divided into five major types on the basis of biochemical function: energy-storage lipids, membrane lipids, emulsification lipids, messenger lipids, and protective-coating lipids (Section 8.1).

Types of fatty acids. Fatty acids are monocarboxylic acids that contain long, unbranched carbon chains. The carbon chain may be saturated, monounsaturated, or polyunsaturated. Length of carbon chain, degree of unsaturation, and location of the unsaturation influence the properties of fatty acids. Omega-3 and omega-6 fatty acids are unsaturated fatty acids with the endmost double bond three and six carbons, respectively, away from the methyl end of the carbon chain (Section 8.2).

Triacylglycerols. Triacylglycerols are energy-storage lipids formed by esterification of three fatty acids to a glycerol molecule. Fats are triacylglycerol mixtures that are solids or semi-solids at room temperature; they contain a relatively high percentage of saturated fatty acid residues. Oils are triacylglycerol mixtures that are liquids at room temperature; they contain a relatively high percentage of unsaturated fatty acid residues (Section 8.4).

Phospholipids. Phospholipids are membrane lipids that contain one or more fatty acids, a phosphate group, a platform molecule to which the fatty acid(s) and phosphate group are attached, and an alcohol attached to the phosphate group. The platform molecule is either glycerol (glycerophospholipids) or sphingosine (sphingophospholipids). Phospholipids have a "head and two tails" structure. Lecithins, cephalins, and sphingomyelins are types of phospholipids (Section 8.7).

Sphingoglycolipids. Sphingoglycolipids are membrane lipids in which a fatty acid and a mono- or oligosaccharide are attached to the platform molecule sphingosine. Cerebrosides and gangliosides are types of sphingoglycolipids (Section 8.8).

Cholesterol. Cholesterol is a membrane lipid whose structure contains a steroid nucleus. It is the most abundant type of steroid. Besides its membrane functions, it also serves as a precursor for several other types of lipids (Section 8.9).

Lipid bilayer. A lipid bilayer is the fundamental structure associated with a cell membrane. It is a two-layer structure of lipid molecules (mostly phospholipids and glycolipids) in which the nonpolar tails of the lipids are in the interior and the polar heads are on the outside surfaces (Section 8.10).

Membrane transport mechanisms. The transport mechanisms by which molecules enter and leave cells include *passive* transport, *facilitated* transport, and *active* transport. Passive and facilitated transport follow a concentration gradient and do not involve cellular energy expenditure. Active transport involves movement against a concentration gradient and requires the expenditure of cellular energy (Section 8.10).

Bile acids. Bile acids are cholesterol derivatives that function as emsulsification lipids. They cause dietary lipids to be soluble in the aqueous environment of the digestive tract. Cholic acid and deoxycholic acids are the major types of bile acids (Section 8.11).

Steroid hormones. Steroid hormones are cholesterol derivatives that function as messenger lipids. The two major types of steroid hormones are sex hormones and adrenocorticoid hormones (Section 8.12).

Eicosanoids. Eicosanoids are fatty acid derivatives that function as messenger lipids. The major classes of eicosanoids are prostaglandins, thromboxanes, and leukotrienes (Section 8.13).

Biological waxes. Biological waxes are protective-coating lipids formed through the esterification of a long-chain fatty acid to a long-chain alcohol (Section 8.14).

Saponifiable and nonsaponifiable lipids. Saponifiable lipids are lipids that undergo hydrolysis in basic solution to yield two or more smaller product molecules. Nonsaponifiable lipids are lipids that do not undergo hydrolysis in basic solution (Section 8.15).

Exercises and Problems

OWL Interactive versions of these problems may be assigned in OWL.

Exercises and problems are arranged in matched pairs with the two members of a pair addressing the same concept(s). The answer to the odd-numbered member of a pair is given at the back of the book. Problems denoted with a ▲ involve concepts found not only in the section under consideration but also concepts found in one or more earlier sections of the chapter. Problems denoted with a ● cover concepts found in a Chemical Connections feature box.

Structure and Classification of Lipids (Section 8.1)

8.1 What characteristic do all lipids have in common?

8.2 What structural feature, if any, do all lipid molecules have in common? Explain your answer.

8.3 Would you expect lipids to be soluble or insoluble in each of the following solvents?
a. H_2O (polar)
b. $CH_3—CH_2—O—CH_2—CH_3$ (nonpolar)
c. $CH_3—OH$ (polar)
d. $CH_3—CH_2—CH_2—CH_2—CH_3$ (nonpolar)

8.4 Would you expect lipids to be soluble or insoluble in each of the following solvents?
a. $CH_3—(CH_2)_7—CH_3$ (nonpolar)
b. $CH_3—Cl$ (polar)
c. CCl_4 (nonpolar)
d. $CH_3—CH_2—OH$ (polar)

8.5 In terms of biochemical function, what are the five major categories of lipids?

8.6 What is the biochemical function of each of the following types of lipids?
a. Triacylglycerols b. Bile acids
c. Sphingoglycolipids d. Eicosanoids

Types of Fatty Acids (Section 8.2)

8.7 Classify each of the following fatty acids as long-chain, medium-chain, or short-chain.
a. Myristic (14:0) b. Caproic (6:0)
c. Arachidic (20:0) d. Capric (10:0)

8.8 Classify each of the following fatty acids as long-chain, medium-chain, or short-chain.
a. Lauric (12:0) b. Oleic (18:1)
c. Butyric (4:0) d. Stearic (18:0)

8.9 Classify each of the following fatty acids as saturated, monounsaturated, or polyunsaturated.
a. Stearic (18:0) b. Linolenic (18:3)
c. Docosahexaenoic (22:6) d. Oleic (18:1)

8.10 Classify each of the following fatty acids as saturated, monounsaturated, or polyunsaturated.
a. Palmitic (16:0) b. Linoleic (18:2)
c. Arachidonic (20:4) d. Palmitoleic (16:1)

8.11 Structurally, what is the difference between a SFA and a MUFA?

8.12 Structurally, what is the difference between a MUFA and a PUFA?

8.13 With the help of Table 8.1, classify each of the acids in Problem 8.9 as an omega-3 acid, an omega-6 acid, or neither an omega-3 nor an omega-6 acid.

8.14 With the help of Table 8.1, classify each of the acids in Problem 8.10 as an omega-3 acid, an omega-6 acid, or neither an omega-3 nor an omega-6 acid.

8.15 Draw the condensed structural formula for the fatty acid whose numerical shorthand designation is 18:2 ($\Delta^{9,12}$).

8.16 Draw the condensed structural formula for the fatty acid whose numerical shorthand designation is 20:4 ($\Delta^{5,8,11,14}$).

8.17 Using the structural information given in Table 8.1, assign an IUPAC name to each of the following fatty acids.
a. Myristic acid b. Palmitoleic acid

8.18 Using the structural information given in Table 8.1, assign an IUPAC name to each of the following fatty acids.
a. Stearic acid b. Linolenic acid

Physical Properties of Fatty Acids (Section 8.3)

8.19 What is the relationship between carbon chain length and melting point for fatty acids?

8.20 What is the relationship between degree of unsaturation and melting point for fatty acids?

8.21 Why does the introduction of a *cis* double bond into a fatty acid lower its melting point?

8.22 Why does increasing carbon chain length decrease water solubility for fatty acids?

8.23 In each of the following pairs of fatty acids, select the fatty acid that has the lower melting point.
a. 18:0 acid and 18:1 acid b. 18:2 acid and 18:3 acid
c. 14:0 acid and 16:0 acid d. 18:1 acid and 20:0 acid

8.24 In each of the following pairs of fatty acids, select the fatty acid that has the higher melting point.
a. 14:0 acid and 18:0 acid b. 20:4 acid and 20:5 acid
c. 18:3 acid and 20:3 acid d. 16:0 acid and 16:1 acid

Triacylglycerols (Section 8.4)

8.25 What are the four structural subunits that contribute to the structure of a triacylglycerol?

8.26 Draw the general block diagram for a triacylglycerol.

8.27 How many different kinds of functional groups are present in a triacylglycerol in which all three fatty acid residues come from saturated fatty acids?

8.28 How many different kinds of functional groups are present in a triacylglycerol in which all three fatty acid residues come from unsaturated fatty acids?

8.29 Draw the condensed structural formula of a triacylglycerol formed from glycerol and three molecules of palmitic acid.

8.30 Draw the condensed structural formula of a triacylglycerol formed from glycerol and three molecules of stearic acid.

8.31 Draw block diagram structures for the four different triacylglycerols that can be produced from glycerol, stearic acid, and linolenic acid.

8.32 Draw block diagram structures for the three different triacylglycerols that can be produced from glycerol, palmitic acid, stearic acid, and linolenic acid.

8.33 Identify, by common name, fatty acids present in each of the following triacylglycerols.

a.
$CH_2—O—\overset{O}{\overset{\|}{C}}—(CH_2)_{14}—CH_3$
$CH—O—\overset{O}{\overset{\|}{C}}—(CH_2)_{12}—CH_3$
$CH_2—O—\overset{O}{\overset{\|}{C}}—(CH_2)_7—CH{=}CH—(CH_2)_7—CH_3$

b.
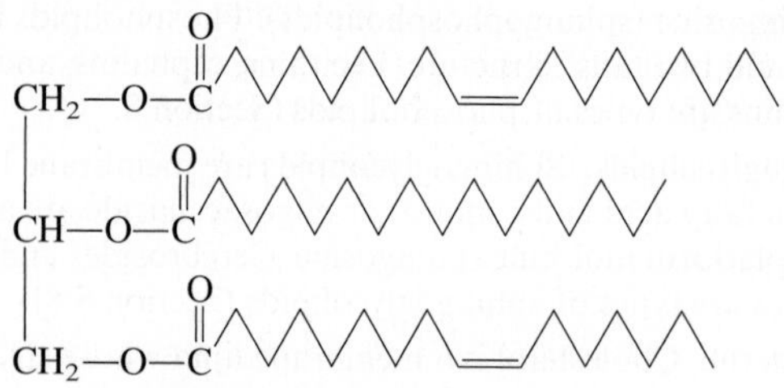

8.34 Identify, by common name, fatty acids present in each of the following triacylglycerols.

a.
$CH_2—O—\overset{O}{\overset{\|}{C}}—(CH_2)_{16}—CH_3$
$CH—O—\overset{O}{\overset{\|}{C}}—(CH_2)_7—CH{=}CH—(CH_2)_7—CH_3$
$CH_2—O—\overset{O}{\overset{\|}{C}}—(CH_2)_{12}—CH_3$

b.
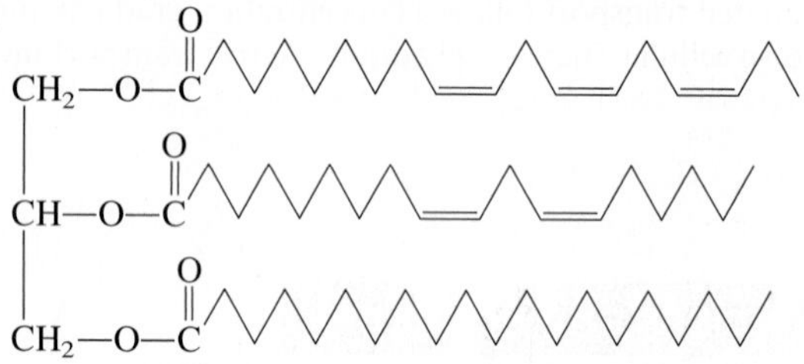

8.35 What is the difference in meaning, if any, between the members of each of the following pairs of terms?
a. Triacylglycerol and triglyceride
b. Triacylglycerol and fat
c. Triacylglycerol and mixed triacylglycerol
d. Fat and oil

8.36 What is the difference in meaning, if any, between the members of each of the following pairs of terms?
a. Triacylglycerol and oil
b. Triacylglycerol and simple triacylglycerol
c. Simple triacylglycerol and mixed triacylglycerol
d. Triglyceride and fat

▲8.37 For each of the triacylglycerol molecules in Problem 8.33, indicate how many of each of the following entities are present in the molecule. In certain cases, the answer may be zero.
a. Acyl groups
b. 18:2 fatty acid residues
c. SFA residues
d. Linolenic acid residues

▲8.38 For each of the triacylglycerol molecules in Problem 8.34, indicate how many of each of the following entities are present in the molecule. In certain cases, the answer may be zero.
a. Ester linkages
b. 18:1 fatty acid residues
c. PUFA residues
d. Stearic acid residues

Dietary Considerations and Triacylglycerols (Section 8.5)

8.39 In a dietary context, indicate whether each of the following pairings of concepts is correct.
a. "Saturated fat" and "good fat"
b. "Polyunsaturated fat" and "bad fat"

8.40 In a dietary context, indicate whether each of the following pairings of concepts is correct.
a. "Monounsaturated fat" and "good fat"
b. "Saturated fat" and "good and bad fat"

8.41 In a dietary context, which of the following pairings of concepts is correct?
a. "Cold-water fish" and "high in omega-3 fatty acids"
b. "Fatty fish" and "low in omega-3 fatty acids"

8.42 In a dietary context, which of the following pairings of concepts is correct?
a. "Warm-water fish" and "low in omega-3 fatty acids"
b. "Fish and chips" and "high in omega-3 fatty acids"

8.43 In a dietary context, classify each of the following fatty acids as an essential fatty acid or as a nonessential fatty acid.
a. Lauric acid (12:0)
b. Linoleic acid (18:2)
c. Myristic acid (14:0)
d. Palmitoleic acid (16:1)

8.44 In a dietary context, classify each of the following fatty acids as an essential fatty acid or as a nonessential fatty acid.
a. Stearic acid (18:0)
b. Linolenic acid (18:3)
c. Oleic acid (18:1)
d. Arachidic acid (20:0)

▲8.45 For each of the triacylglycerol molecules in Problem 8.33, indicate how many of each of the following entities are present in the molecule. In certain cases, the answer may be zero.
a. Omega-3 fatty acid residues
b. Omega-6 fatty acid residues
c. "Good" fatty acid residues
d. Δ^9 fatty acid residues

▲8.46 For each of the triacylglycerol molecules in Problem 8.34, indicate how many of each of the following entities are present in the molecule. In certain cases, the answer may be zero.
a. Omega-3 fatty acid residues
b. Omega-6 fatty acid residues
c. "Bad" fatty acid residues
d. $\Delta^{9,12}$ fatty acid residues

●8.47 (Chemical Connections 8-A) Indicate whether each of the following statements concerning commonly consumed nuts is true or false.
a. The carbohydrate content of common nuts ranges from 30% to 50%.
b. The SFA content of common nuts is almost double the MUFA content.
c. Among common nuts, cashews have the lowest percent fat.
d. The fatty acid unsaturation/saturation ratio for some nuts exceeds 10.

●8.48 (Chemical Connections 8-A) Indicate whether each of the following statements concerning commonly consumed nuts is true or false.
a. The protein content of common nuts ranges from 30% to 40%.
b. The PUFA content of common nuts almost always exceeds the MUFA content.
c. Common nuts, except for macadamia nuts, have a low percent of fat.
d. The fatty acid unsaturation/saturation ratio for some nuts is less than 1.

●8.49 (Chemical Connections 8-B) Indicate whether each of the following statements concerning fat substitutes is true or false.
a. Simplesse is a protein-based fat substitute.
b. Olestra passes through the human intestinal tract in an undigested form.
c. Simplesse has the same cooking properties as fats and oils.
d. For FDA approval, a fat substitute must be calorie-free.

●8.50 (Chemical Connections 8-B) Indicate whether each of the following statements concerning fat substitutes is true or false.
a. Olestra is a carbohydrate-based fat substitute that contains sucrose.
b. Simplesse has a structure that contains fatty acid residues.
c. Olestra is a "calorie-free" fat substitute.
d. For FDA approval, a fat substitute must be heat-stable.

Chemical Reactions of Triacylglycerols (Section 8.6)

8.51 Draw condensed structural formulas for all products obtained from the complete hydrolysis of the following triacylglycerol.

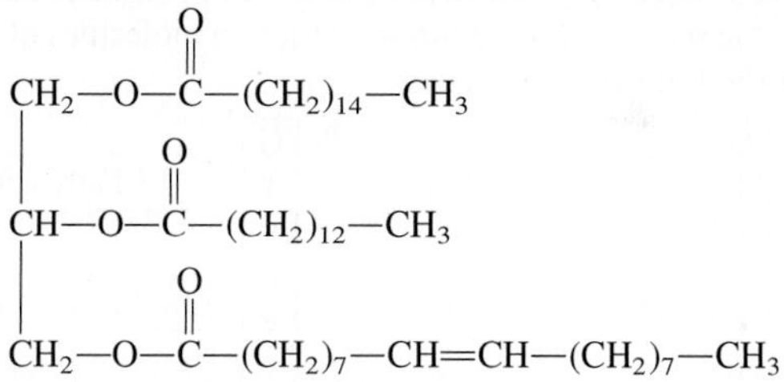

8.52 Draw condensed structural formulas for all products obtained from the complete hydrolysis of the following triacylglycerol.

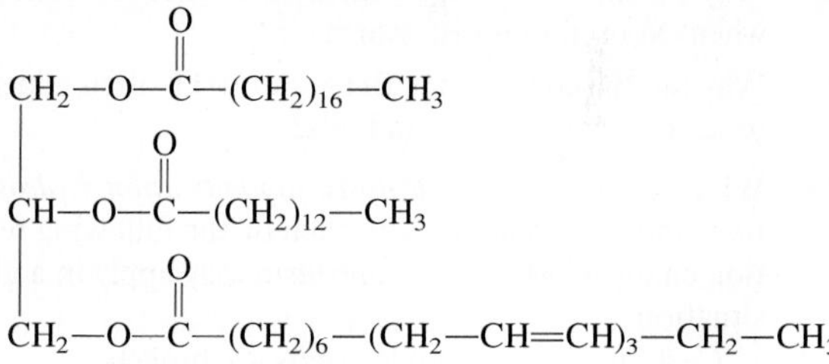

8.53 Draw condensed structural formulas for all products you would obtain from the saponification with NaOH of the triacylglycerol in Problem 8.51.

8.54 Draw condensed structural formulas for all products you would obtain from the saponification with KOH of the triacylglycerol in Problem 8.52.

8.55 Why can only unsaturated triacylglycerols undergo hydrogenation?

8.56 A food package label lists an oil as "partially hydrogenated." What does this mean?

8.57 How many molecules of H_2 will react with one molecule of the following triacylglycerol?

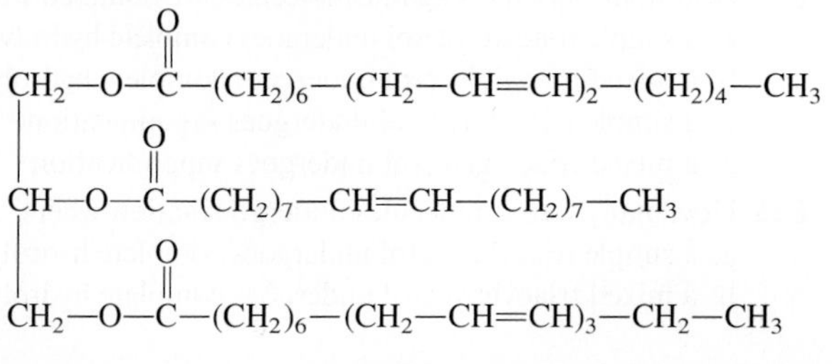

8.58 How many molecules of H_2 will react with one molecule of the following triacylglycerol?

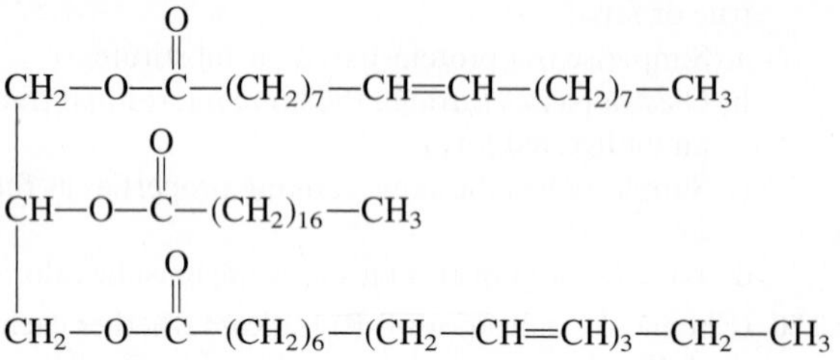

8.59 Draw block diagram structures for all possible products of the partial hydrogenation, with two molecules of H_2, of the following molecules.

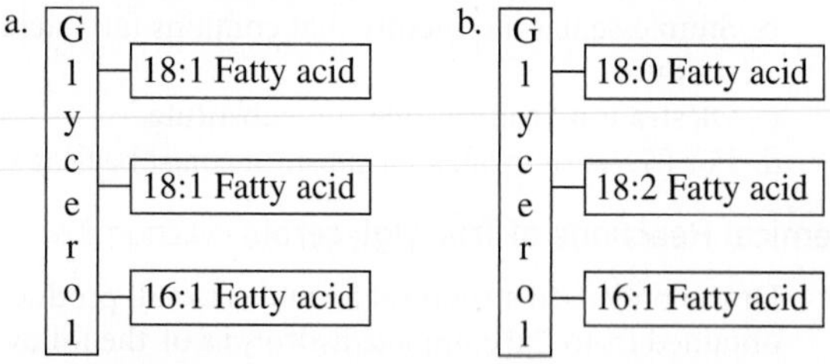

8.60 Draw block diagram structures for all possible products of the partial hydrogenation, with two molecules of H_2, of the following molecules.

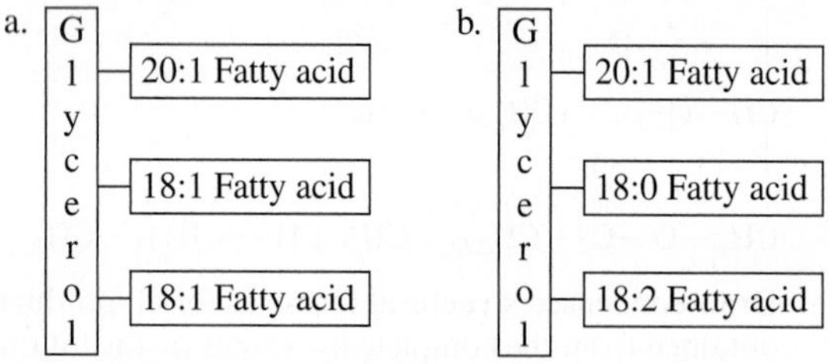

8.61 Why do animal fats and vegetable oils become rancid when exposed to moist, warm air?

8.62 Why are the compounds BHA and BHT often added to foods that contain fats and oils?

8.63 Which of the terms *hydrolysis, saponification, hydrogenation,* and *oxidation* apply to each of the following reaction changes? More than one term may apply in a given situation.
a. Carbon–carbon double bonds are broken.
b. Fatty acids are among the products.
c. *Cis* double bonds are converted to *trans* double bonds.
d. Water is a reactant.

8.64 Which of the terms *hydrolysis, saponification, hydrogenation,* and *oxidation* apply to each of the following reaction changes? More than one term may apply in a given situation.
a. Carbon–oxygen single bonds are broken.
b. Glycerol is among the products.
c. Fatty acid salts are among the products.
d. Carbon–carbon double bonds are changed to carbon–carbon single bonds.

▲8.65 How many *different* organic molecules are obtained when
a. a simple triacylglycerol undergoes complete hydrolysis
b. a mixed triacylglycerol undergoes complete hydrolysis
c. a simple triacylglycerol undergoes saponification
d. a mixed triacylglycerol undergoes saponification

▲**8.66** How many water molecules undergo reaction when
a. a simple triacylglycerol undergoes complete hydrolysis
b. a mixed triacylglycerol undergoes complete hydrolysis
c. a simple triacylglycerol undergoes saponification
d. a mixed triacylglycerol undergoes saponification

▲8.67 What type(s) of organic compound(s) are products of the following reactions? If no reaction occurs, indicate that such is the case.
a. An oil undergoes hydrolysis under acidic conditions.
b. A fat undergoes saponification.
c. A single saturated fatty acid undergoes hydrolysis.
d. A single unsaturated fatty acid undergoes saponification.

▲**8.68** What type(s) of organic compound(s) are products of the following reactions? If no reaction occurs, indicate that such is the case.
a. A fat undergoes hydrolysis under acidic conditions.
b. An oil undergoes saponification.
c. A single unsaturated fatty acid undergoes hydrolysis.
d. A single saturated fatty acid undergoes saponification.

●8.69 (Chemical Connections 8-C) Indicate whether each of the following statements concerning soaps and detergents is true or false.
a. Carboxylic acid salts are the primary ingredient in both soaps and detergents.
b. The active ingredient in a soap is the positive ion present, and in a detergent it is the negative ion present.
c. The cleansing action of both soaps and detergents involves micelle formation.
d. Detergents contain short-carbon-chain acid salts, and soaps contain long-carbon-chain acid salts.

●**8.70** (Chemical Connections 8-C) Indicate whether each of the following statements concerning soaps and detergents is true or false.
a. The acid salts present in a detergent contain the element sulfur.
b. The cleansing action of soaps, but not detergents, is based on polarity considerations.
c. Soaps, but not detergents, are salts of carboxylic acid.
d. The negative ion present in both soaps and detergents has "dual polarity."

●8.71 (Chemical Connections 8-D) Indicate whether each of the following statements relating to *trans* fatty acids is true or false.
a. Partial hydrogenation processes involving triacylglycerols effect the conversion of all *cis* double bonds to *trans* double bonds.
b. Partial hydrogenation processes involving triacylglycerols effect the conversion of some *cis* double bonds to *trans* double bonds.
c. *Trans* fatty acids occur naturally in meat and dairy products.
d. *Trans* fatty acids and saturated fatty acids have similar molecular shapes.

●**8.72** (Chemical Connections 8-D) Indicate whether each of the following statements relating to *trans* fatty acids is true or false.
a. Partial hydrogenation processes involving triacylglycerols effect the conversion of some *cis* double bonds to single bonds.
b. Partial hydrogenation processes involving triacylglycerols effect the conversion of some single bonds to *trans* double bonds.
c. Biochemically, *trans* fatty acids and saturated fatty acids have similar effects relative to cardiovascular disease.
d. The degree of carbon-chain bending is greater for *trans* fatty acids than *cis* fatty acids.

Phospholipids (Section 8.7)

8.73 The following is a block diagram for a glycerophospholipid where the building blocks are labeled with letters and the linkages between building blocks are labeled with numbers.

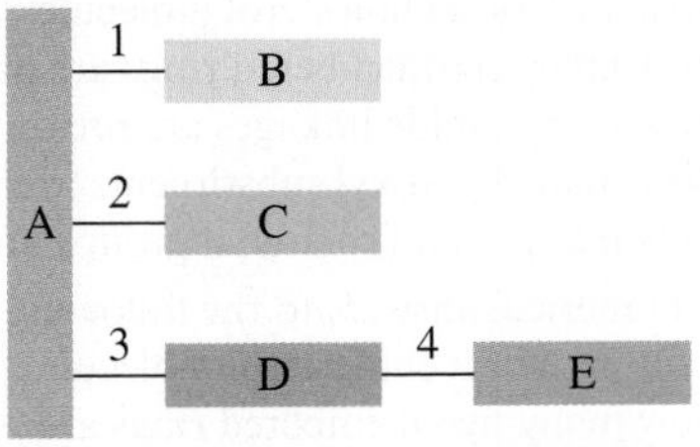

a. Which building blocks are fatty acid residues?
b. Which building blocks are alcohol residues?
c. Which linkages are ester linkages?
d. Which linkages involve a phosphate residue?

8.74 Using the block diagram in Problem 8.73 for a glycerophospholipid where the building blocks are labeled with letters and the linkages between building blocks are labeled with numbers,
a. which building blocks are glycerol residues?
b. which building blocks are phosphate residues?
c. which linkages involve an alcohol residue?
d. which linkages involve both glycerol and fatty acid residues?

8.75 A glycerophospholipid has a "head and two tails" structure. Classify each of the following components of a glycerophospholipid as being part of its "head" or part of its "tails."
a. Fatty acid b. Glycerol
c. Alcohol d. Phosphate

8.76 Indicate whether or not each of the following is a component of a phosphatidyl group.
a. Fatty acid b. Glycerol
c. Phosphate d. Ethanolamine

8.77 Indicate whether each of the following descriptions concerning the molecule sphingosine is true or false.
a. Four functional groups are present.
b. The carbon chain is unsaturated.
c. It has a head-and-two-tails structure.
d. An amino functional group is present.

8.78 Indicate whether each of the following descriptions concerning the molecule sphingosine is true or false.
a. Carbons 1, 2, and 3 all bear the same kind of functional group.
b. Three different kinds of functional groups are present.
c. It is an unsaturated diaminoalcohol.
d. The carbon chain contains 20 carbon atoms.

8.79 The following is a block diagram for a sphingophospholipid where the building blocks are labeled with letters and the linkages between building blocks are labeled with numbers.

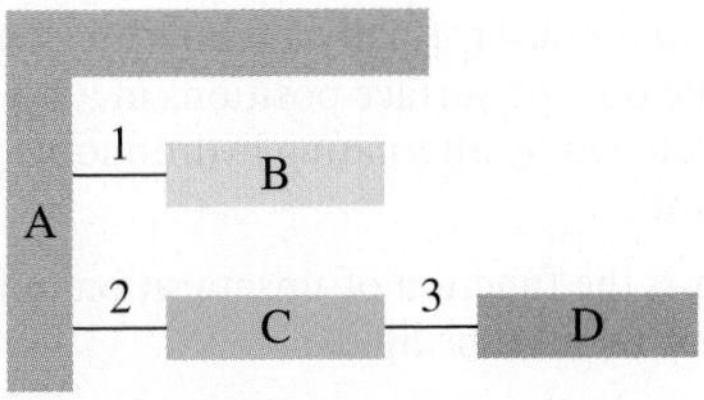

a. Which building blocks are fatty acid residues?
b. Which building blocks are phosphate residues?
c. Which linkages are amide linkages?
d. Which linkages involve a sphingosine residue?

8.80 Using the block diagram in Problem 8.79 for a sphingophospholipid where the building blocks are labeled with letters and the linkages between building blocks are labeled with numbers,
a. which building blocks are sphingosine residues?
b. which building blocks are alcohol residues?
c. which linkages are ester linkages?
d. which linkages involve a phosphate residue?

8.81 Indicate whether or not each of the following must contain both a phosphate and a choline building block.
a. Phosphatidylcholine b. Sphingomyelin
c. Sphingophospholipid d. Glycerophospholipid

8.82 Indicate whether or not a sphingomyelin contains each of the following building blocks.
a. Fatty acid b. Phosphate
c. Ethanolamine d. Choline

8.83 Which portion of the structure of a phospholipid has hydrophobic characteristics?

8.84 Which portion of the structure of a phospholipid has hydrophilic characteristics?

▲8.85 Which of the terms *triacylglycerol, glycerophospholipid,* and *sphingophospholipid* applies to each of the following characterizations? More than one term may apply in a given situation.
a. It contains four building blocks.
b. All linkages are ester linkages.
c. An alcohol building block is present.
d. It is an energy-storage lipid.

▲8.86 Which of the terms *triacylglycerol, glycerophospholipid,* and *sphingophospholipid* applies to each of the following characterizations? More than one term may apply in a given situation.
a. It contains five building blocks.
b. It contains two kinds of linkages.
c. A phosphate building block is present.
d. It is a membrane lipid.

▲8.87 Which of the terms *triacylglycerol, glycerophospholipid,* and *sphingophospholipid* applies to each of the following characterizations? More than one term may apply, or none of the terms may apply in a given situation.
a. Must contain ethanolamine
b. Phosphatidlycholine
c. Must contain an amide linkage
d. Nonsaponifiable

▲8.88 Which of the terms *triacylglycerol, glycerophospholipid,* and *sphingophospholipid* applies to each of the following characterizations? More than one term may apply, or none of the terms may apply in a given situation.
a. Must contain choline
b. Lecithin
c. Must contain an ester linkage
d. Saponifiable

Sphingoglycolipids (Section 8.8)

8.89 The following is a block diagram for a sphingoglycolipid where the building blocks are labeled with letters and the linkages between building blocks are labeled with numbers.

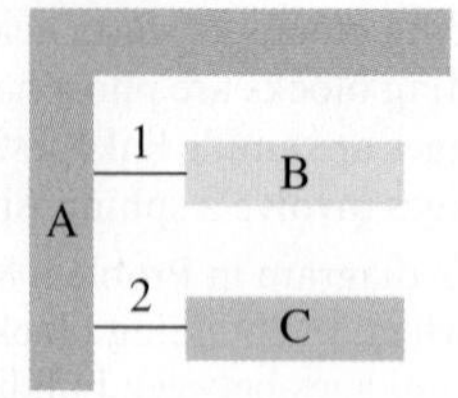

a. Which building blocks are fatty acid residues?
b. Which building blocks are carbohydrate residues?
c. Which linkages are amide linkages?
d. Which linkages could involve a monosaccharide?

8.90 Using the block diagram in Problem 8.89 for a sphingoglycolipid where the building blocks are labeled with letters and the linkages between building blocks are labeled with numbers,
a. which building blocks are sphingosine residues?
b. which building blocks are mono- or oligosaccharide residues?
c. which linkages are glycosidic linkages?
d. which linkages involve both sphingosine and fatty acid residues?

8.91 Indicate whether each of the following is a component of (1) a cerebroside but not a ganglioside (2) a ganglioside but not a cerebroside or (3) both a cerebroside and a ganglioside.
a. Monosaccharide
b. Oligosaccharide
c. Fatty acid
d. Sphingosine

8.92 A sphingoglycolipid has a "head and two tails" structure. Classify each of the following components of a sphingoglycolipid as being part of its "head" or part of its "tails."
a. Sphingosine
b. Fatty acid
c. Monosaccharide
d. Oligosaccharide

▲8.93 Which of the terms *glycerophospholipid, sphingophospholipid,* and *sphingoglycolipid* applies to each of the following characterizations? More than one term may apply, or none of the terms may apply in a given situation.
a. Membrane lipid
b. Saponifiable lipid
c. Amide linkage present
d. Has a "head and two tails" structure

▲**8.94** Which of the terms *glycerophospholipid, sphingophospholipid,* and *sphingoglycolipid* applies to each of the following characterizations? More than one term may apply, or none of the terms may apply in a given situation.
a. Energy-storage lipid
b. Nonsaponifiable lipid
c. Glycosidic linkage present
d. Does not have a "head and two tails" structure

Cholesterol (Section 8.9)

8.95 Indicate whether each of the following statements about a steroid nucleus is true or false.
a. Rings A, B, and C are identical.
b. Rings A and C are linked via a glycosidic linkage.
c. Rings A and B share a common side.
d. Twenty-seven carbon atoms are present in a steroid nucleus.

8.96 Indicate whether each of the following statements about a steroid nucleus is true or false.
a. Rings B, C, and D are identical.
b. Rings B and C are linked via an ester linkage.
c. Ring B is fused to two other rings.
d. Twenty carbon atoms are present in a steroid nucleus.

8.97 Give numerical answers to the following questions about the structure of a cholesterol molecule.
a. How many six-membered rings are present?
b. How many amide linkages are present?
c. How many hydroxyl substituents are present?
d. How many total functional groups are present?

8.98 Give numerical answers to the following questions about the structure of a cholesterol molecule.
a. How many five-membered rings are present?
b. How many ester linkages are present?
c. How many methyl substituents are present?
d. How many different kinds of functional groups are present?

8.99 Using the information in Table 8.4, rank a serving of the following four foods in order of increasing amount of cholesterol present: Swiss cheese, fish fillet, liver, chicken.

8.100 Using the information in Table 8.4, rank a serving of the following four foods in order of increasing amount of cholesterol present: egg, beef steak, cheddar cheese, whole milk.

8.101 In a dietary context, what is the difference between "good cholesterol" and "bad cholesterol"?

8.102 In a dietary context, how do HDL and LDL differ in function?

Cell Membranes (Section 8.10)

8.103 What is the general structural characteristic associated with the majority of lipids present in a lipid bilayer?

8.104 What constitutes the two outside surfaces of a lipid bilayer?

8.105 Indicate whether each of the following statements about a lipid bilayer is true or false.
a. The outside surface positions in a lipid bilayer are occupied by hydrophobic entities.
b. The interaction between the fluid *inside* a cell and the surface of a lipid bilayer is primarily an interaction between polar entities.
c. The outside surface positions in a lipid bilayer are occupied by the nonpolar tails of phospholipids and glycolipids.
d. The interactions between adjacent lipids in a lipid bilayer usually involve covalent bonding.

8.106 Indicate whether each of the following statements about a lipid bilayer is true or false.
a. The outside surface positions in a lipid bilayer are occupied by the polar heads of phospholipids and glycolipids.
b. The interaction between the fluid *outside* a cell and the surface of a lipid bilayer is primarily an interaction between nonpolar entities.
c. The interactions between the tails of adjacent lipids in a lipid bilayer usually are intermolecular forces.
d. The outside surface positions in a lipid bilayer are occupied by alternating hydrophobic and hydrophilic entities.

8.107 What is the function of unsaturation in the hydrocarbon tails of membrane lipids?

8.108 What function does cholesterol serve when it is present in cell membranes?

8.109 What is an integral membrane protein?

8.110 What is a peripheral membrane protein?

8.111 What is the difference between *passive transport* and *facilitated transport*?

8.112 What is the difference between *facilitated transport* and *active transport*?

8.113 Match each of the following statements related to membrane transport processes to the appropriate term: *passive transport, facilitated transport, active transport.* More than one term may apply in a given situation.
a. Movement across the membrane is against the concentration gradient.
b. Proteins serve as "gates."
c. Expenditure of cellular energy is required.
d. Movement across the membrane is from a high to a low concentration.

8.114 Match each of the following statements related to membrane transport processes to the appropriate term: *passive transport, facilitated transport, active transport.* More than one term may apply in a given situation.
a. Movement across the membrane is with the concentration gradient.
b. Proteins serve as "pumps."
c. Expenditure of cellular energy is not required.
d. Movement across the membrane is from a low to a high concentration.

▲8.115 A cell membrane contains higher than average levels of lipids with linolenic acid and oleic acid building blocks. Another cell membrane contains higher than average levels of lipids with stearic acid and palmitic acid building blocks. Property-wise, how will the two membranes differ?

▲8.116 A cell membrane contains higher than average amounts of cholesterol. Another cell membrane contains lower than average amounts of cholesterol. Property-wise, how will the two membranes differ?

Bile Acids (Section 8.11)

8.117 What is an emulsifier?

8.118 How do bile acids aid in the digestion of lipids?

8.119 Describe the structural differences between a bile acid and cholesterol.

8.120 Describe the structural differences between cholic acid and a deoxycholic acid.

8.121 Describe the structural differences between glycocholic acid and taurocholic acid.

8.122 Describe the structural differences between glycocholic acid and 7-deoxycholic acid.

8.123 What is the medium through which bile acids are supplied to the small intestine?

8.124 What is the chemical composition of bile?

8.125 At what location in the body are bile acids stored until needed?

8.126 What is the chemical composition of the majority of gallstones?

▲8.127 Classify each of the following as (1) an energy-storage lipid (2) a membrane lipid or (3) an emulsification lipid.
a. Fats b. Sphingoglycolipids
c. Triacylglycerols d. Cholesterol

▲8.128 Classify each of the following as (1) an energy-storage lipid (2) a membrane lipid or (3) an emulsification lipid.
a. Oils b. Bile acids
c. Glycerophospholipids d. Sphingophospholipids

▲8.129 Classify each of the lipids in Problem 8.127 as a saponifiable lipid or a nonsaponifiable lipid.

▲8.130 Classify each of the lipids in Problem 8.128 as a saponifiable lipid or a nonsaponifiable lipid.

Steroid Hormones (Section 8.12)

8.131 What are the two major classes of steroid hormones?

8.132 Describe the general function of each of the following types of steroid hormones.
a. Estrogens b. Androgens
c. Progestins d. Mineralocorticoids

8.133 How do the sex hormones estradiol and testosterone differ in structure?

8.134 What functional groups are present in each of the following steroid hormones or synthetic steroids?
a. Estradiol b. Testosterone
c. Progesterone d. Cortisone

8.135 Indicate whether or not each of the following is an adrenal corticoid hormone or a sex hormone.
a. Aldosterone b. Testosterone
c. Cortisol d. Estradiol

8.136 Indicate whether or not each of the following is a natural steroid hormone or a synthetic steroid.
a. Cortisone b. Progesterone
c. Prednisolone d. Norethynodrel

8.137 What is the biochemical function for each of the steroids listed in Problem 8.135?

8.138 What is the biochemical function for each of the steroids listed in Problem 8.136?

▲8.139 Draw the structure of an anabolic steroid with the following structural characteristics: C1 and C2 are involved in a double bond, C3 is part of a carbonyl group, C4 and C5 are involved in a double bond, C10 and C13 have methyl group attachments, and C17 has both a hydroxyl group and a methyl group as substituents.

▲8.140 Draw the structure of an anabolic steroid with the following structural characteristics: C3 is part of a carbonyl group; carbon–carbon double bonds involve C4 and C5, C9 and C10, and C11 and C12; C13 has a methyl group attachment; and C17 has a hydroxyl group attachment.

●8.141 (Chemical Connections 8-E) Indicate whether each of the following statements concerning anabolic steroid use is true or false.
a. Medically, anabolic steroids are used to prevent withering of muscles in long-term bed ridden patients.
b. Some, but not all, of the negative effects of illegal anabolic steroid use are reversible.
c. Structurally, testosterone contains methyl group ring attachments and THG contains ethyl group ring attachments.
d. Structurally, testosterone and "Andro" differ in that a ketone group has replaced an alcohol group.

●8.142 (Chemical Connections 8-E) Indicate whether each of the following statements concerning anabolic steroid use is true or false.

a. Illegal anabolic steroid use has very little effect on the performance of female athletes.
b. Use of illegal anabolic steroids affects the male reproductive system.
c. Structurally, testosterone and "Andro" contain the same number of carbon atoms.
d. Structurally, testosterone and THG differ in that all steroid ring attachments are different.

Eicosanoids (Section 8.13)

8.143 What is the major structural difference between a prostaglandin and its parent fatty acid?

8.144 What is the major structural difference between a leukotriene and its parent fatty acid?

8.145 What structural feature distinguishes a prostaglandin from a leukotriene?

8.146 What structural feature distinguishes a thromboxane from a leukotriene?

▲8.147 Classify each of the following types of lipids as (1) glycerol-based (2) sphingosine-based (3) steroid nucleus-based or (4) fatty acid-based.
a. Bile acids b. Oils
c. Prostaglandins d. Thromboxanes

▲8.148 Classify each of the following types of lipids as (1) glycerol-based (2) sphingosine-based (3) steroid nucleus-based or (4) fatty acid-based.
a. Fats b. Cholesterol
c. Eicosanoids d. Leukotrienes

▲8.149 Classify each of the following types of lipids as (1) an energy-storage lipid (2) a membrane lipid (3) an emulsification lipid or (4) a messenger lipid.
a. Bile acids b. Cholesterol
c. Eicosanoids d. Sphingophospholipids

▲8.150 Classify each of the following types of lipids as (1) an energy-storage lipid (2) a membrane lipid (3) an emulsification lipid or (4) a messenger lipid.
a. Triacylglycerols b. Glycerophospholipids
c. Prostaglandins d. Estrogens

●8.151 (Chemical Connections 8-F) Indicate whether each of the following statements concerning anti-inflammatory medications is true or false.
a. Both cortisone and ibuprofen are NSAIDs.
b. Both aspirin and Celebrex are COX-1 and COX-2 inhibitors.
c. COX-1 is involved in the production of a protective mucous coating for the stomach lining.
d. SAIDs inhibit the release of arachidonic acid from more complex lipids.

●8.152 (Chemical Connections 8-F) Indicate whether each of the following statements concerning anti-inflammatory medications is true or false.
a. Both diclofenac and naproxen are prescription NSAIDs.
b. Both aspirin and ibuprofen inhibit COX-2 but not COX-1.
c. COX-2 is involved in the production of prostaglandins associated with arthritis inflammation.
d. Prostaglandin synthesis is dependent on a supply of arachidonic acid.

Biological Waxes (Section 8.14)

8.153 Draw the general block diagram for a biological wax.

8.154 Draw the condensed structural formula of a wax formed from palmitic acid (Table 8.1) and cetyl alcohol, $CH_3—(CH_2)_{14}—CH_2—OH$.

8.155 What is the difference between a biological wax and a mineral wax?

8.156 Biological waxes have a "head and two tails" structure. Give the chemical identity of the head and of the two tails.

▲8.157 Which of the substance categories *biological wax, mineral wax, bile acid,* and *sphingophospholipid* has each of the following characteristics? More than one substance may have a given characteristic.
a. Structure is based on a steroid nucleus
b. A saponifiable lipid
c. Contains at least one fatty acid building block
d. Contains at least one amide linkage

▲8.158 Which of the substance categories *biological wax, mineral wax, bile acid,* and *sphingoglycolipid* has each of the following characteristics? More than one substance may have a given characteristic.
a. Contains at least one ester linkage
b. Contains at least one alcohol building block
c. Classified as a protective-coating lipid
d. Classified as a emulsification lipid

▲8.159 Indicate whether or not each of the following types of substances has a "head and two tails" structure.
a. Biological wax b. Cholesterol
c. Sphingoglycolipid d. Fat

▲8.160 Indicate whether or not each of the following types of substances has a "head and two tails" structure.
a. Mineral wax b. Sphingosine
c. Sphingophospholipid d. Oil

▲8.161 Indicate whether or not each of the following substances contains (1) an ester linkage (2) an amide linkage or (3) a glycosidic linkage. More than one characterization may apply, or none of the characterizations may apply.
a. Biological wax b. Triacylglycerol
c. Bile acid d. Sphingosine

▲8.162 Indicate whether or not each of the following substances contains (1) an ester linkage (2) an amide linkage or (3) a glycosidic linkage. More than one characterization may apply, or none of the characterizations may apply.
a. Mineral wax b. Prostaglandin
c. Oil d. Cholesterol

8.163 Indicate whether or not both members of each of the following pairs of substances are saponifiable lipids.
a. Sphingoglycolipids and sphingophospholipids
b. Biological waxes and mineral waxes
c. Triacylglycerols and steroid hormones
d. Eicosanoids and cholesterol

8.164 Indicate whether or not both members of each of the following pairs of substances are saponifiable lipids.
a. Glycerophospholipids and triacylglycerols
b. Cholesterol and sphingoglycolipids
c. Mineral waxes and bile acids
d. Eicosanoids and biological waxes

8.165 How many saponifiable linkages are present in each of the following types of substances?
a. Cholesterol b. Triacylglycerols
c. Bile acids d. Biological waxes

8.166 How many saponifiable linkages are present in each of the following types of substances?
a. Eicosanoids b. Sphingoglycolipids
c. Mineral waxes d. Sphingophospholipids

Proteins

9

© Jochem Wijnands/Alamy

The fibrous protein alpha-keratin is the major structural element present in sheep's wool.

CHAPTER OUTLINE

OWL

Sign in to OWL at **www.cengage.com/owl** to view tutorials and simulations, develop problem-solving skills, and complete online homework assigned by your professor.

In this chapter, the third of the bioorganic classes of molecules (Section 7.1) is considered, the compounds called proteins. An extraordinary number of different proteins, each with a different function, exist in the human body. A typical human cell contains about 9000 different kinds of proteins, and the human body contains about 100,000 different proteins. Proteins are needed for the synthesis of enzymes, certain hormones, and some blood components; for the maintenance and repair of existing tissues; for the synthesis of new tissue; and sometimes for energy.

9.1 Characteristics of Proteins

Next to water, proteins are the most abundant substances in nearly all cells—they account for about 15% of a cell's overall mass (Section 7.1) and for almost half of a cell's dry mass. All proteins contain the elements carbon, hydrogen, oxygen, and nitrogen; most also contain sulfur. The presence of nitrogen in proteins sets them apart from carbohydrates and lipids, which most often do not contain nitrogen. The average nitrogen content of proteins is 4.4% by mass. Other elements, such as phosphorus and iron, are essential constituents of certain specialized proteins. Casein, the main protein of milk, contains phosphorus, an element very important in the diet of infants and children. Hemoglobin, the oxygen-transporting protein of blood, contains iron.

The word *protein* comes from the Greek *proteios*, which means "of first importance." This reflects the key role that proteins play in life processes.

A **protein** *is a naturally occurring, unbranched polymer in which the monomer units are amino acids.* Thus the starting point for a discussion of proteins is an understanding of the structures and chemical properties of amino acids.

9.2 Amino Acids: The Building Blocks for Proteins

In an α-amino acid, the carboxyl group and the amino group are attached to the same carbon atom.

The nature of the side chain (R group) distinguishes α-amino acids from each other, both physically and chemically.

An **amino acid** *is an organic compound that contains both an amino* (—NH_2) *group and a carboxyl* (—COOH) *group.* The amino acids found in proteins are always α-amino acids. An **α-amino acid** *is an amino acid in which the amino group and the carboxyl group are attached to the α-carbon atom.* The general structural formula for an α-amino acid is

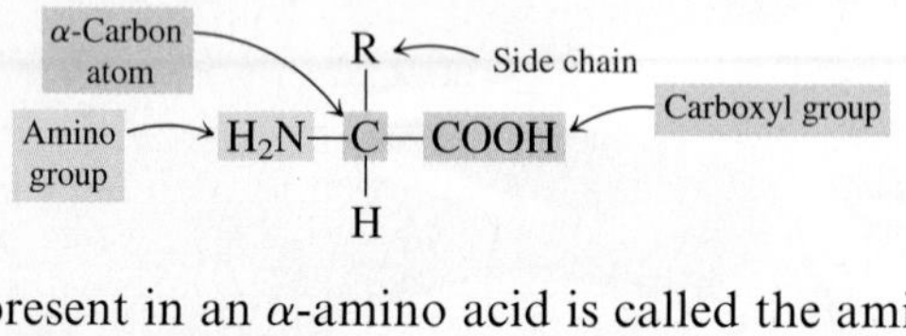

The R group present in an α-amino acid is called the amino acid *side chain.* The nature of this side chain distinguishes α-amino acids from each other. Side chains vary in size, shape, charge, acidity, functional groups present, hydrogen-bonding ability, and chemical reactivity.

More than 700 different naturally occurring amino acids are known, but only 20 of them, called standard amino acids, are normally present in proteins. A **standard amino acid** *is one of the 20 α-amino acids normally found in proteins.* The structures of the 20 standard amino acids are given in Table 9.1. Within Table 9.1, amino acids are grouped according to side-chain polarity. In this system, there are four categories: (1) nonpolar amino acids (2) polar neutral amino acids (3) polar acidic amino acids and (4) polar basic amino acids. This classification system gives insights into how various types of amino acid side chains help determine the properties of proteins (Section 9.12).

The nonpolar amino acid *proline* has a structural feature not found in any other standard amino acid. Its side chain, a propyl group, is bonded to both the α-carbon atom and the amino nitrogen atom, giving a cyclic side chain.

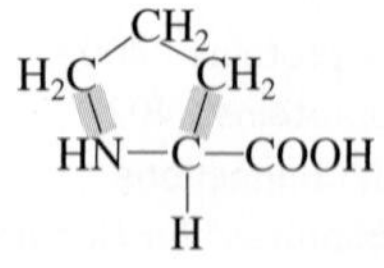

Proline

A **nonpolar amino acid** *is an amino acid that contains one amino group, one carboxyl group, and a nonpolar side chain.* When incorporated into a protein, such amino acids are *hydrophobic* ("water-fearing"); that is, they are not attracted to water molecules. They are generally found in the interior of proteins, where there is limited contact with water. There are nine nonpolar amino acids. Tryptophan is a borderline member of this group because water can weakly interact through hydrogen bonding with the NH ring location on tryptophan's side-chain ring structure. Thus, some textbooks list tryptophan as a polar neutral amino acid.

The three types of polar amino acids have varying degrees of affinity for water. Within a protein, such amino acids are said to be *hydrophilic* ("water-loving"). Hydrophilic amino acids are often found on the surfaces of proteins.

A **polar neutral amino acid** *is an amino acid that contains one amino group, one carboxyl group, and a side chain that is polar but neutral.* In solution at physiological pH, the side chain of a polar neutral amino acid is neither acidic nor basic. There are six polar neutral amino acids. These amino acids are more soluble in water than the nonpolar amino acids as, in each case, the R group present can hydrogen bond to water.

A variety of functional groups are present in the side chains of the 20 standard amino acids: six have alkyl groups (Section 1.8), three have aromatic groups (Section 2.11), two have sulfur-containing groups (Section 3.20), two have hydroxyl (alcohol) groups (Section 3.2), three have amino groups (Section 6.2), two have carboxyl groups (Section 5.1), and two have amide groups (Section 6.12).

A **polar acidic amino acid** *is an amino acid that contains one amino group and two carboxyl groups, the second carboxyl group being part of the side chain.* In solution at physiological pH, the side chain of a polar acidic amino acid bears a negative charge; the side-chain carboxyl group has lost its acidic hydrogen atom. There are two polar acidic amino acids: aspartic acid and glutamic acid.

A **polar basic amino acid** *is an amino acid that contains two amino groups and one carboxyl group, the second amino group being part of the side chain.* In solution at physiological pH, the side chain of a polar basic amino acid bears a positive charge; the nitrogen atom of the amino group has accepted a proton

Table 9.1 The 20 Standard Amino Acids, Grouped According to Side-Chain Polarity

Below each amino acid's structure are its name (with pronunciation), its three-letter abbreviation, and its one-letter abbreviation.

Nonpolar Amino Acids

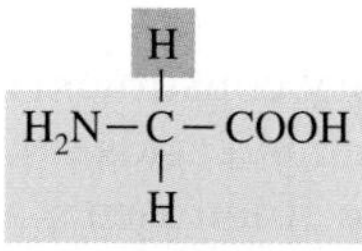

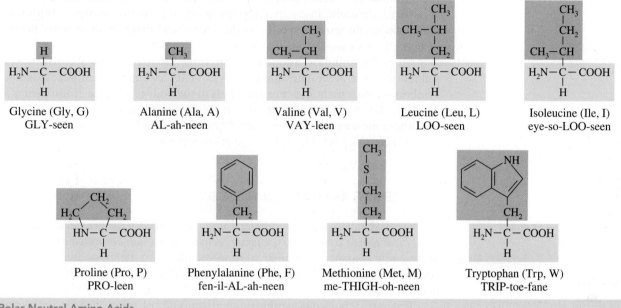

Glycine (Gly, G)
GLY-seen

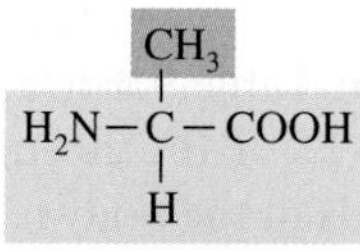

Alanine (Ala, A)
AL-ah-neen

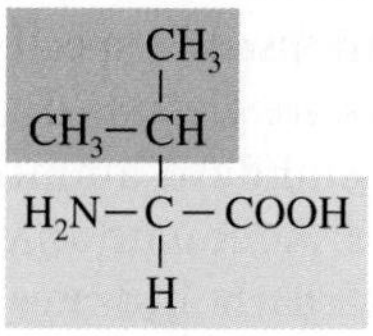

Valine (Val, V)
VAY-leen

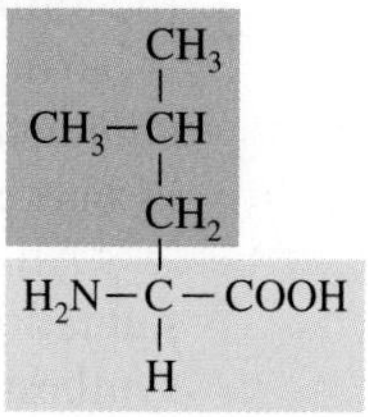

Leucine (Leu, L)
LOO-seen

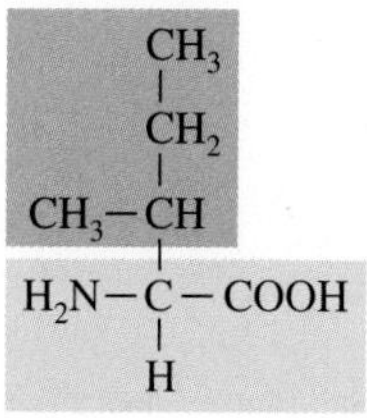

Isoleucine (Ile, I)
eye-so-LOO-seen

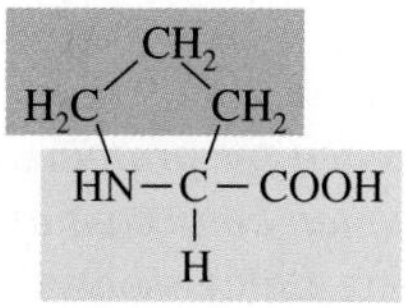

Proline (Pro, P)
PRO-leen

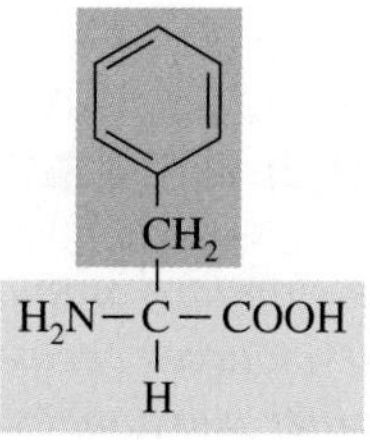

Phenylalanine (Phe, F)
fen-il-AL-ah-neen

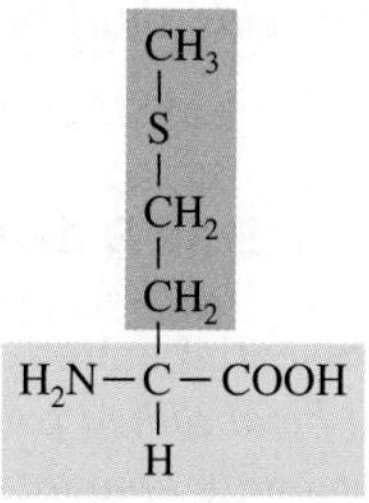

Methionine (Met, M)
me-THIGH-oh-neen

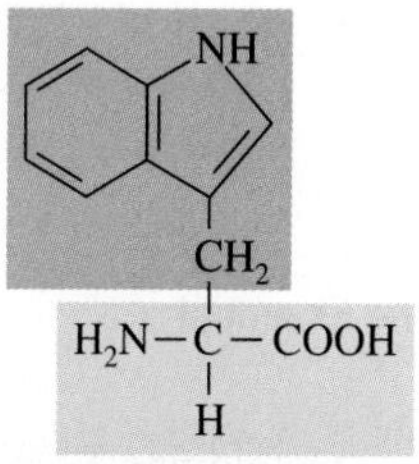

Tryptophan (Trp, W)
TRIP-toe-fane

Polar Neutral Amino Acids

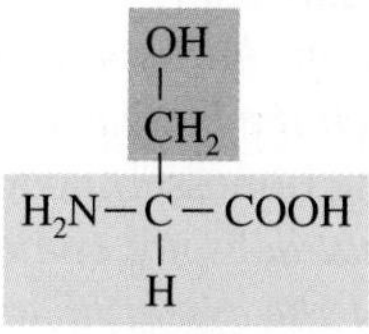

Serine (Ser, S)
SEER-een

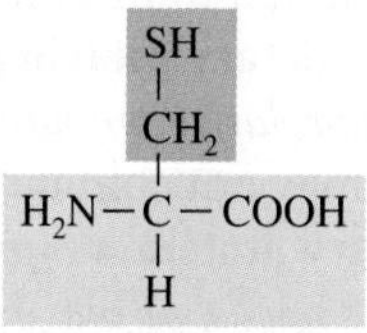

Cysteine (Cys, C)
SIS-teh-een

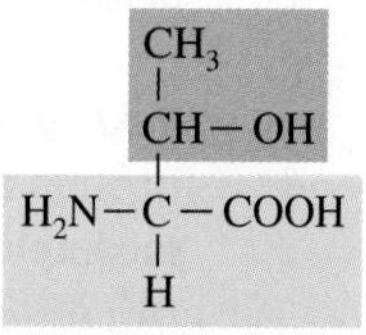

Threonine (Thr, T)
THREE-oh-neen

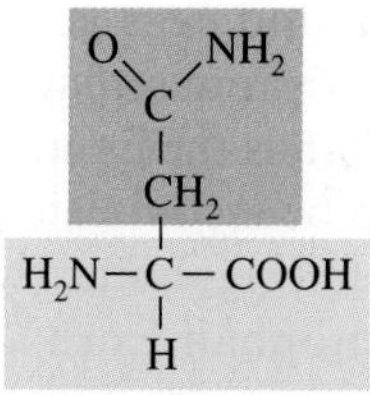

Asparagine (Asn, N)
ah-SPAR-ah-geen

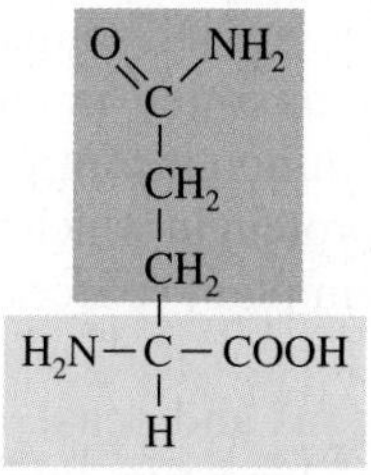

Glutamine (Gln, Q)
GLU-tah-meen

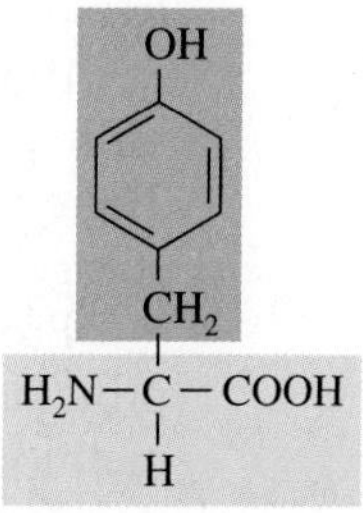

Tyrosine (Tyr, Y)
(TIE-roe-seen)

Polar Acidic Amino Acids

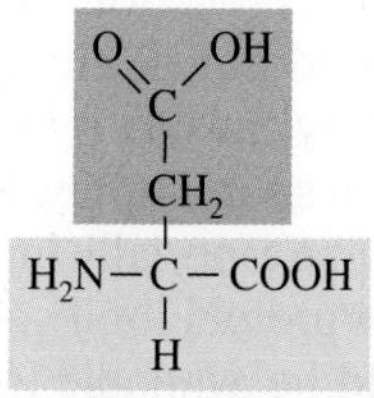

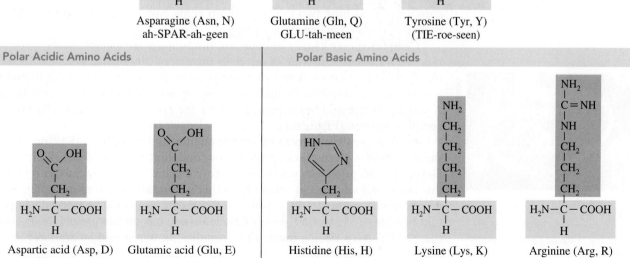

Aspartic acid (Asp, D)
ah-SPAR-tic acid

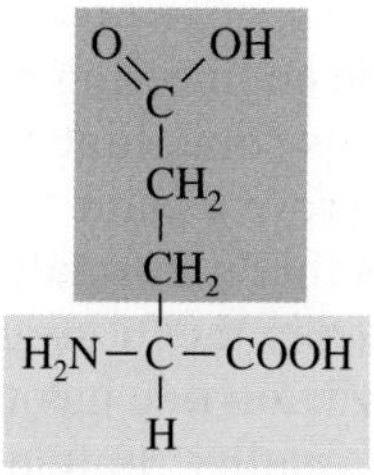

Glutamic acid (Glu, E)
glu-TAM-ic acid

Polar Basic Amino Acids

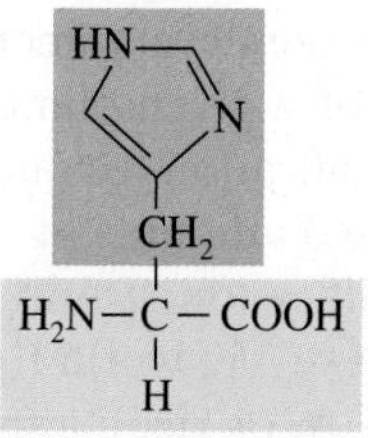

Histidine (His, H)
HISS-tuh-deen

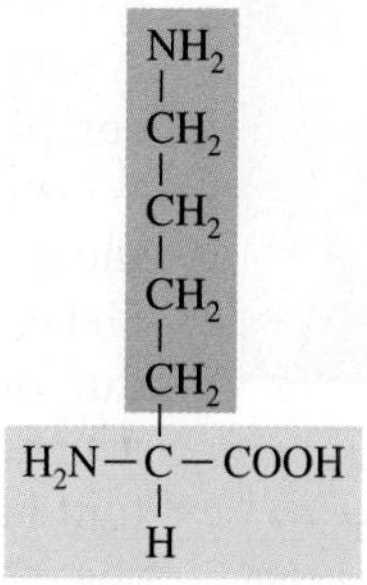

Lysine (Lys, K)
LYE-seen

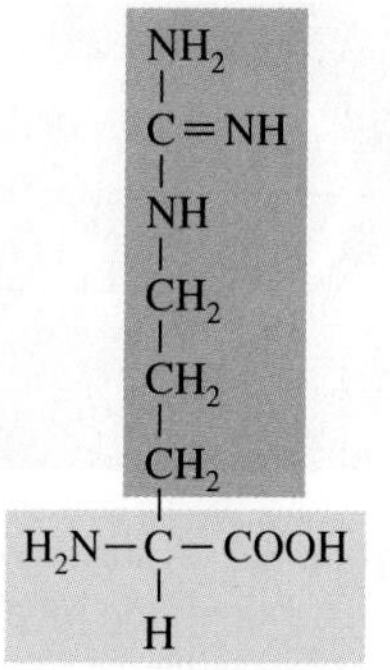

Arginine (Arg, R)
ARG-ih-neen

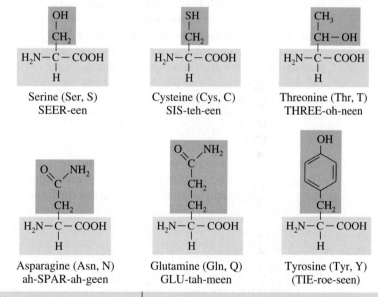

(basic behavior; Section 6.6). There are three polar basic amino acids: lysine, arginine, and histidine.

The names of the standard amino acids are often abbreviated using three-letter codes. Except in four cases, these abbreviations are the first three letters of the amino acid's name. Also, a one-letter code for amino acid names exists that is particularly useful in computer applications that involve proteins. Both sets of abbreviations are used in specifying the amino acid make-up of protein. Both sets of abbreviations are given in Table 9.1.

All of the standard amino acids in Table 9.1 are necessary constituents of human proteins. Adequate amounts of 11 of the 20 standard amino acids can be synthesized from carbohydrates and lipids in the body if a source of nitrogen is also available. However, the adult human body cannot produce adequate amounts of the other nine standard amino acids. These amino acids, which are called *essential amino acids*, must be obtained from dietary protein.

9.3 Essential Amino Acids

Table 9.2 The Essential Amino Acids

arginine*	methionine
histidine	phenylalanine
isoleucine	threonine
leucine	tryptophan
lysine	valine

* Arginine is required for growth in children but is not an essential amino acid for adults.

An **essential amino acid** *is a standard amino acid needed for protein synthesis that must be obtained from dietary sources because the human body cannot synthesize it in adequate amounts from other substances.* There are nine essential amino acids for adults, and a tenth one is needed for growth in children. Table 9.2 lists the essential amino acids.

The human body can synthesize small amounts of some of the essential amino acids, but not enough to meet its needs, especially in the case of growing children.

A **complete dietary protein** *is a protein that contains all of the essential amino acids in the same relative amounts in which the body needs them.* A complete dietary protein may or may not contain all of the nonessential amino acids. Conversely, an **incomplete dietary protein** *is a protein that does not contain adequate amounts, relative to the body's needs, of one or more of the essential amino acids.* Associated with the term *incomplete dietary protein* is the term *limiting amino acid.* A **limiting amino acid** *is an essential amino acid that is missing, or present in inadequate amounts, in an incomplete dietary protein.*

Protein from animal sources is usually complete dietary protein. Casein from milk and proteins found in meat, fish, and eggs are complete dietary proteins. There is one common incomplete dietary protein that comes from animal sources. It is gelatin, a protein in which tryptophan is the limiting amino acid.

Protein from plant sources tends to be incomplete dietary protein. With plant proteins, three amino acids are often limiting: lysine (wheat, rice, oats, and corn), methionine (beans and peas), and tryptophan (corn and beans). Note that both corn and beans have two limiting amino acids. Soy is the only common plant protein that is a complete dietary protein.

Complementary dietary proteins *are two or more incomplete dietary proteins that, when combined, provide an adequate amount of all essential amino acids relative to the body's needs.* A mix of plant proteins generally provides high-quality (complete) protein. Rice by itself is an incomplete dietary protein, as is beans. A serving of rice and beans provides all of the essential amino acids by protein complementation (Figure 9.1).

Figure 9.1 A serving of rice and beans, a common meal in many countries, involves complementary dietary proteins.

Prior to the development of genetic engineering procedures (Section 11.14), the quality of a given plant's protein was something that could not be changed. Genetic modification techniques can improve a plant's protein by causing it to produce increased amounts of amino acids that it normally has in short supply. Such genetic modification, if fully implemented in the future, would be especially important in areas of the world that rely heavily on one incomplete protein food source (beans or corn or rice); a much higher quality protein would be available to the people for consumption.

9.4 Chirality and Amino Acids

Four different groups are attached to the α-carbon atom in all of the standard amino acids except glycine, where the R group is a hydrogen atom.

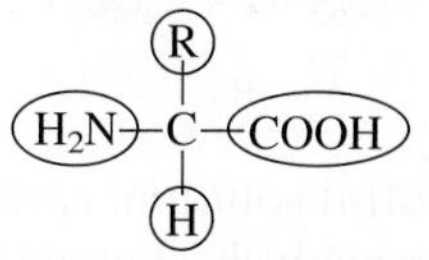

This means that the structures of 19 of the 20 standard amino acids possess a chiral center (Section 7.4) at this location, so enantiomeric forms (left- and right-handed forms; Section 7.5) exist for each of these amino acids.

With few exceptions (in some bacteria), the amino acids found in nature and in proteins are L isomers. Thus, as is the case with monosaccharides (Section 7.8), nature favors one mirror-image form over the other. Interestingly, for amino acids the L isomer is the preferred form, whereas for monosaccharides the D isomer is preferred.

The rules for drawing Fischer projection formulas (Section 7.6) for amino acid structures follow.

1. The —COOH group is put at the top of the projection formula, the R group at the bottom. This positions the carbon chain vertically.
2. The $—NH_2$ group is in a horizontal position. Positioning it on the left denotes the L isomer, and positioning it on the right denotes the D isomer.

Figure 9.2 shows molecular models that illustrate the use of these rules. Fischer projection formulas for both enantiomers of the amino acids alanine and serine are:

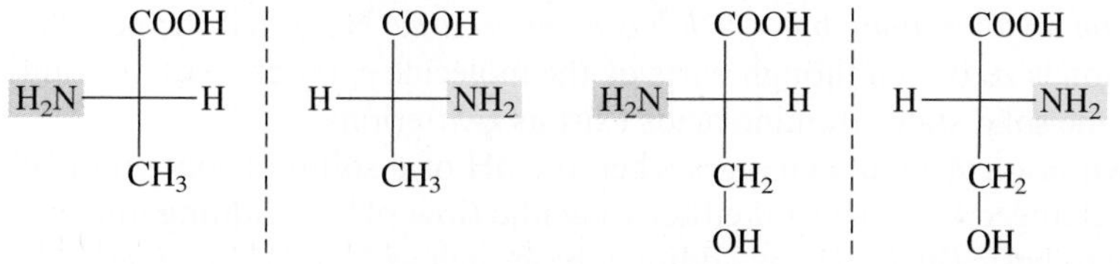

L-Alanine D-Alanine L-Serine D-Serine

Glycine, the simplest of the standard amino acids, is achiral. All of the other standard amino acids are chiral.

Because only L amino acids are constituents of proteins, the enantiomer designation of L or D will be omitted in subsequent amino acid and protein discussions. It is understood that it is the L isomer that is always present.

Two of the 19 chiral standard amino acids, isoleucine and threonine, possess two chiral centers (see Table 9.1). With two chiral centers present, four stereoisomers are possible for these amino acids. However, only one of the L isomers is found in proteins.

9.5 Acid–Base Properties of Amino Acids

In pure form, amino acids are white crystalline solids with relatively high decomposition points. (Most amino acids decompose before they melt.) Also, most amino acids are *not* very soluble in water because of strong intermolecular forces within their crystal structures. Such properties are those often exhibited by compounds in which charged species are present. Studies of amino acids confirm that they are charged species both in the solid state and in solution. Why is this so?

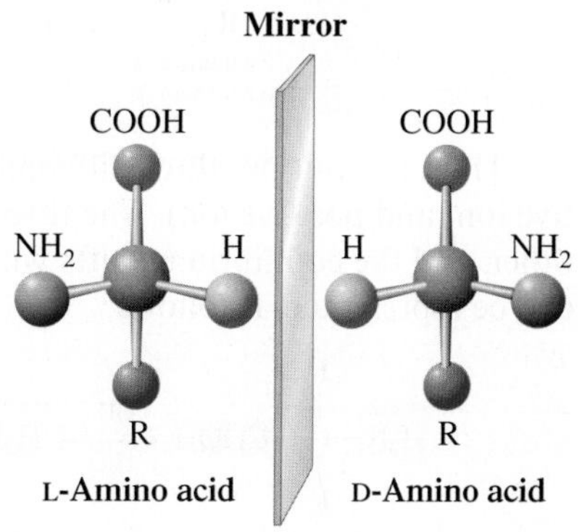

Figure 9.2 Designation of handedness in standard amino acid structures involves aligning the carbon chain vertically and looking at the position of the horizontally aligned $—NH_2$ group. The L form has the $—NH_2$ group on the left, and the D form has the $—NH_2$ group on the right.

Both an acidic group (—COOH) and a basic group (—NH$_2$) are present on the same carbon in an α-amino acid.

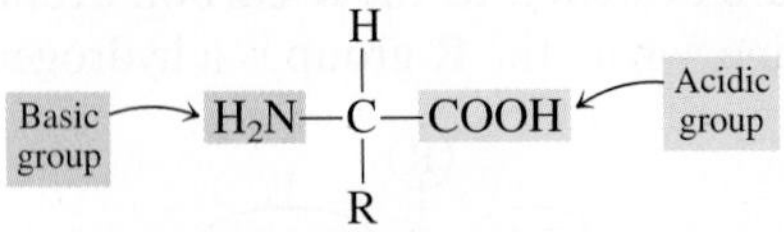

$$\text{Basic group} \longrightarrow H_2N-\underset{\underset{R}{|}}{\overset{\overset{H}{|}}{C}}-COOH \longleftarrow \text{Acidic group}$$

Section 5.8 indicated that in neutral solution, carboxyl groups have a tendency to lose protons (H$^+$), producing a negatively charged species:

$$-COOH \longrightarrow -COO^- + H^+$$

Section 6.6 indicated that in neutral solution, amino groups have a tendency to accept protons (H$^+$), producing a positively charged species:

$$-NH_2 + H^+ \longrightarrow -\overset{+}{N}H_3$$

Consistent with the behavior of these groups, in neutral solution, the —COOH group of an amino acid donates a proton to the —NH$_2$ of the same amino acid. We can characterize this behavior as an *internal* acid–base reaction. The net result is that in neutral solution, amino acid molecules have the structure

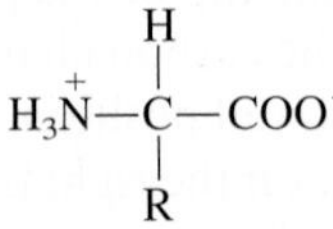

$$\overset{+}{H_3}N-\underset{\underset{R}{|}}{\overset{\overset{H}{|}}{C}}-COO^-$$

Such a molecule is known as a zwitterion, from the German term meaning "double ion." A **zwitterion** *is a molecule that has a positive charge on one atom and a negative charge on another atom, but which has no net charge.* Note that the net charge on a zwitterion is zero even though parts of the molecule carry charges. In solution and also in the solid state, α-amino acids exist as zwitterions.

Zwitterion structure changes when the pH of a solution containing an amino acid is changed from neutral either to acidic (low pH) by adding an acid such as HCl or to basic (high pH) by adding a base such as NaOH. In an acidic solution, the zwitterion accepts a proton (H$^+$) to form a positively charged ion.

$$\overset{+}{H_3}N-\underset{\underset{R}{|}}{\overset{\overset{H}{|}}{C}}-COO^- + H_3O^+ \longrightarrow \overset{+}{H_3}N-\underset{\underset{R}{|}}{\overset{\overset{H}{|}}{C}}-COOH + H_2O$$

Zwitterion (no net charge)　　　　　　Positively charged ion

In basic solution, the —$\overset{+}{N}H_3$ of the zwitterion loses a proton, and a negatively charged species is formed.

$$\overset{+}{H_3}N-\underset{\underset{R}{|}}{\overset{\overset{H}{|}}{C}}-COO^- + OH^- \longrightarrow H_2N-\underset{\underset{R}{|}}{\overset{\overset{H}{|}}{C}}-COO^- + H_2O$$

Zwitterion (no net charge)　　　　　　Negatively charged ion

Thus, in solution, three different amino acid forms can exist (zwitterion, negative ion, and positive ion). The three species are actually in equilibrium with each other, and the equilibrium shifts with pH change. The overall equilibrium process can be represented as follows:

$$\overset{+}{H_3}N-\underset{\underset{R}{|}}{\overset{\overset{H}{|}}{C}}-COOH \underset{H_3O^+}{\overset{OH^-}{\rightleftarrows}} \overset{+}{H_3}N-\underset{\underset{R}{|}}{\overset{\overset{H}{|}}{C}}-COO^- \underset{H_3O^+}{\overset{OH^-}{\rightleftarrows}} H_2N-\underset{\underset{R}{|}}{\overset{\overset{H}{|}}{C}}-COO^-$$

Acidic solution (low pH)　　　　Neutral solution (pH = 7.0)　　　　Basic solution (high pH)

In drawing amino acid structures, where handedness designation is not required, the placement of the four groups about the α-carbon atom is arbitrary. From this point on in the text, we will draw amino acid structures such that the —COOH group is on the right, the —NH$_2$ group is on the left, the R group points down, and the H atom points up. Drawing amino acids in this "arrangement" makes it easier to draw structures where amino acids are linked together to form longer amino acid chains.

Strong intermolecular forces between the positive and negative centers of zwitterions are the cause of the high melting points of amino acids.

From this point on in the text, the structures of amino acids will be drawn in their zwitterion form unless information given about the pH of the solution indicates otherwise.

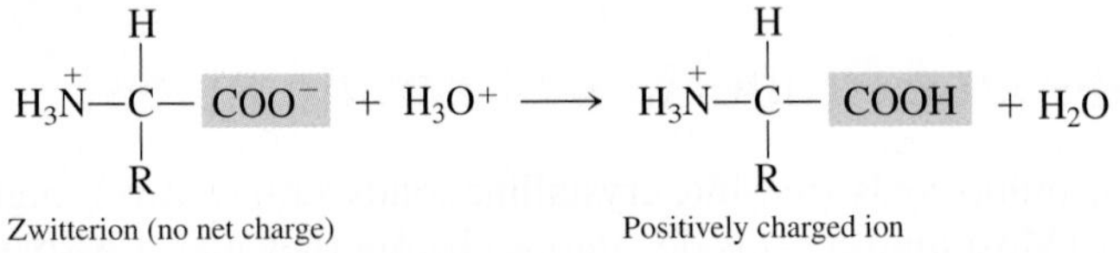

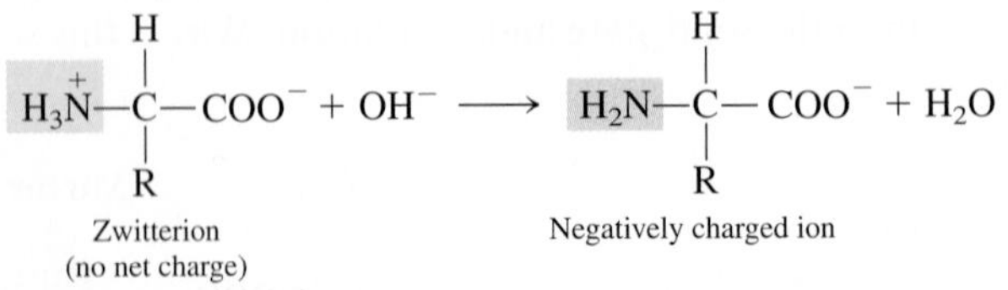

The ability of amino acids to react with both H$_3$O$^+$ and OH$^-$ ions means that amino acid solutions can function as buffers. The same is true for proteins, which are amino acid polymers (Section 9.1). The buffering action of proteins present in blood is a major function of such proteins.

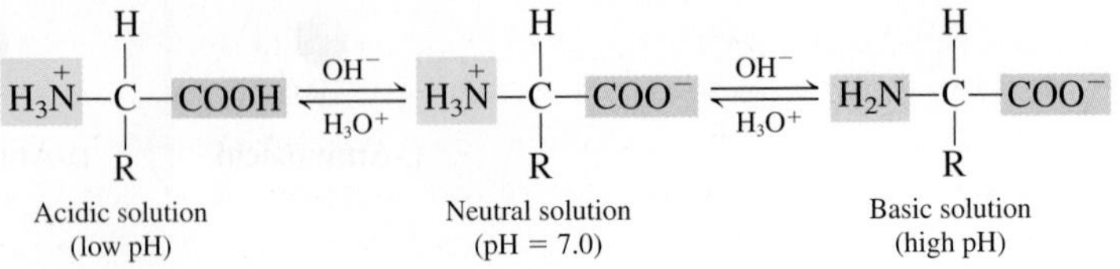

In acidic solution, the positively charged species on the left predominates; nearly neutral solutions have the middle species (the zwitterion) as the dominant species; in basic solution, the negatively charged species on the right predominates.

EXAMPLE 9.1 Determining Amino Acid Form in Solutions of Various pH

Draw the structural form of the amino acid alanine that predominates in solution at each of the following pH values.

a. pH = 1.0 **b.** pH = 7.0 **c.** pH = 11.0

Solution

At low pH, both amino and carboxyl groups are protonated. At high pH, both groups have lost their protons. At neutral pH, the zwitterion is present.

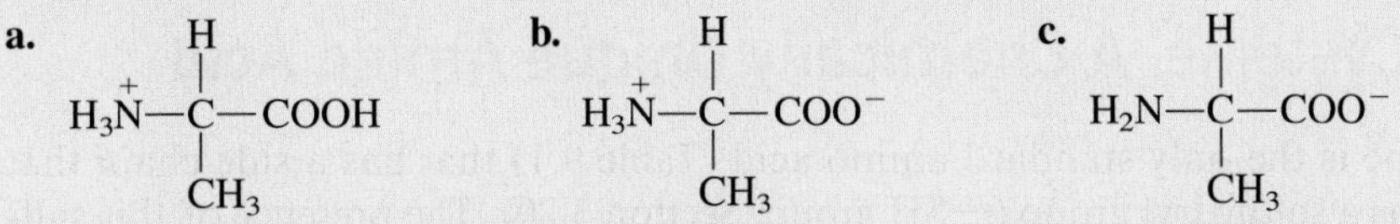

pH = 1.0 (net charge of +1) | pH = 7.0 (no net charge) | pH = 11.0 (net charge of −1)

Practice Exercise 9.1

Draw the structural form of the amino acid valine that predominates in solution at each of the following pH values.

a. pH = 7.0 **b.** pH = 12.0 **c.** pH = 2.0

Answers:

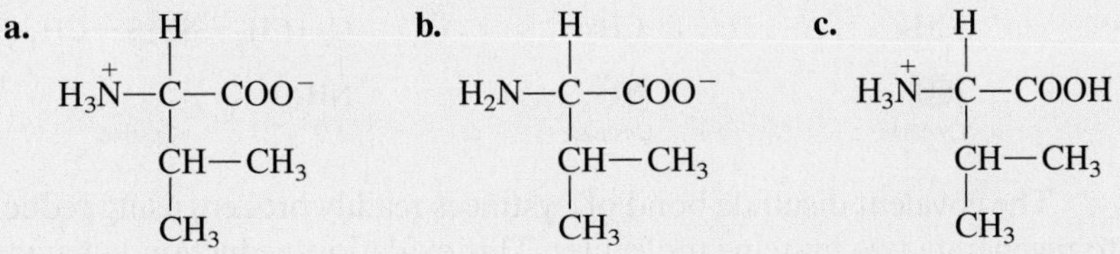

Guidelines for amino acid form as a function of solution pH follow.

Low pH: All acid groups are protonated (—COOH). All amino groups are protonated ($—\overset{+}{N}H_3$).

High pH: All acid groups are deprotonated ($—COO^-$). All amino groups are deprotonated ($—NH_2$).

Neutral pH: All acid groups are deprotonated ($—COO^-$). All amino groups are protonated ($—\overset{+}{N}H_3$).

The previous discussion assumed that the side chain (R group) of an amino acid remains unchanged in solution as the pH is varied. This is the case for neutral amino acids but not for acidic or basic ones. For these latter compounds, the side chain can also acquire a charge because it contains an amino or a carboxyl group that can, respectively, gain or lose a proton.

Because of the extra site that can be protonated or deprotonated, acidic and basic amino acids have four charged forms in solution. These four forms for aspartic acid, one of the acidic amino acids, are

The term *protonated* denotes gain of a H^+ ion, and the term *deprotonated* denotes loss of a H^+ ion.

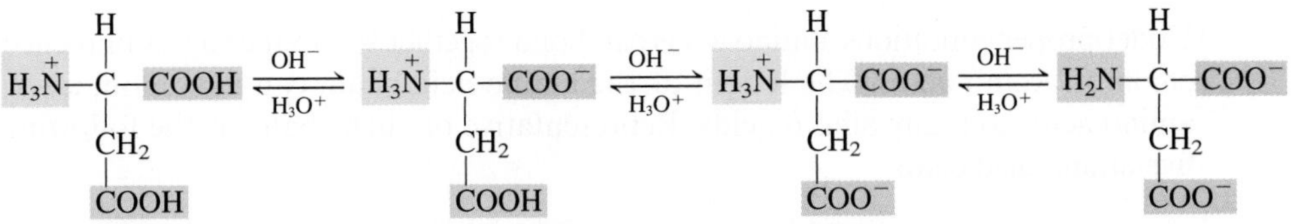

Low-pH form (+1 charge) | Moderately-low-pH form (no net charge) (zwitterion) | Intermediate-pH form (−1 net charge) | High-pH form (−2 net charge)

The existence of two low-pH forms for aspartic acid results from the two carboxyl groups being deprotonated at different pH values. For basic amino acids, two high-pH forms exist because deprotonation of the amino groups does not occur simultaneously. The side-chain amino group deprotonates before the α-amino group for histidine, but the opposite is true for lysine and arginine.

Side-chain carboxyl groups are weaker acids than α-carbon carboxyl groups.

Table 9.3 Isoelectric Points for the 20 Amino Acids Commonly Found in Proteins

Name	Isoelectric Point
alanine	6.01
arginine	10.76
asparagine	5.41
aspartic acid	2.77
cysteine	5.07
glutamic acid	3.22
glutamine	5.65
glycine	5.97
histidine	7.59
isoleucine	6.02
leucine	5.98
lysine	9.74
methionine	5.74
phenylalanine	5.48
proline	6.48
serine	5.68
threonine	5.87
tryptophan	5.88
tyrosine	5.66
valine	5.97

Isoelectric Points

An important pH value, relative to the various forms an amino acid can have in solution, is the pH at which it exists primarily in its zwitterion form, that is, its neutral form (no net charge). This pH value is known as the *isoelectric point* for the amino acid. An **isoelectric point** *is the pH at which an amino acid exists primarily in its zwitterion form.* At the isoelectric point, almost all amino acid molecules in a solution (more than 99%) are present in their zwitterion form.

Every amino acid has a different isoelectric point. Fifteen of the 20 amino acids, those with nonpolar or polar neutral side chains (Table 9.1), have isoelectric points in the range of 4.8–6.3. The three basic amino acids have higher isoelectric points, and the two acidic amino acids have lower ones. Table 9.3 lists the isoelectric points for the 20 standard amino acids.

9.6 Cysteine: A Chemically Unique Amino Acid

Cysteine is the only standard amino acid (Table 9.1) that has a side chain that contains a sulfhydryl group (—SH group; Section 3.20). The presence of this sulfhydryl group imparts to cysteine a chemical property that is unique among the standard amino acids. Cysteine, in the presence of mild oxidizing agents, readily *dimerizes,* that is, reacts with another cysteine molecule to form a cystine molecule. (A *dimer* is a molecule that is made up of two like subunits.) In cystine, the two cysteine residues are linked via a covalent disulfide bond.

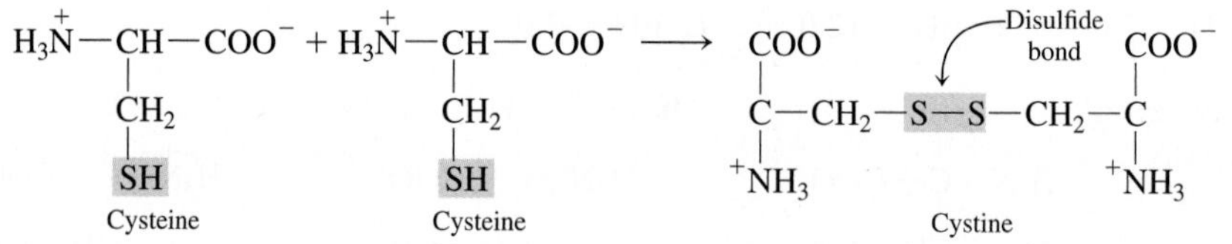

Cystine contains two *cysteine* residues linked by a disulfide bond.

The covalent disulfide bond of cystine is readily broken, using reducing agents, to regenerate two cysteine molecules. This oxidation–reduction behavior involving sulfhydryl groups and disulfide bonds was previously encountered in Section 3.20 when the reactions of thiols were considered.

$$-SH + HS- \underset{\text{Reduction}}{\overset{\text{Oxidation}}{\rightleftharpoons}} -S-S- + 2H$$

As will be shown in Section 9.11, the formation of disulfide bonds between cysteine residues present in protein molecules has important consequences relative to protein structure and protein shape.

9.7 Peptides

Under proper conditions, amino acids can bond together to produce an unbranched chain of amino acids. The length of the amino acid chain can vary from a few amino acids to many amino acids. Representative of such chains is the following five-amino-acid chain.

Amino acid—Amino acid—Amino acid—Amino acid—Amino acid

Such a chain of covalently linked amino acids is called a *peptide.* A **peptide** *is an unbranched chain of amino acids.* Peptides are further classified by the number of amino acids present in the chain. A compound containing two amino acids is specifically called a *dipeptide;* three amino acids joined together in a chain constitute a *tripeptide;* and so on. The name *oligopeptide* is loosely used to refer to peptides with 10 to 20 amino acid residues, and the name *polypeptide* is used to refer to longer peptides. A **polypeptide** *is a long unbranched chain of amino acids.*

Nature of the Peptide Bond

The bonds that link amino acids together in a peptide chain are called peptide bonds. There are four peptide bonds present in a pentapeptide.

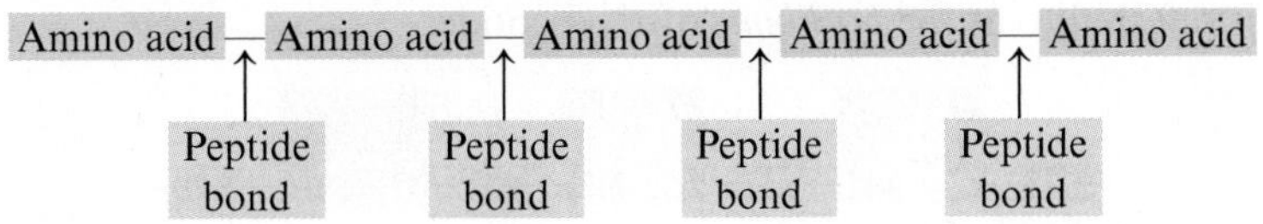

The nature of the peptide bond becomes apparent by reconsidering a chemical reaction previously encountered. In Section 6.16, the reaction between a carboxylic acid and an amine to produce an amide was considered. The general equation for this reaction was

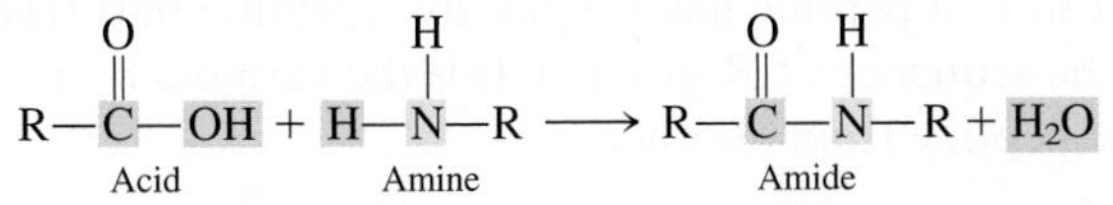

Two amino acids can combine in a similar way—the carboxyl group of one amino acid interacts with the amino group of the other amino acid. The products are a molecule of water and a molecule containing the two amino acids linked by an amide bond.

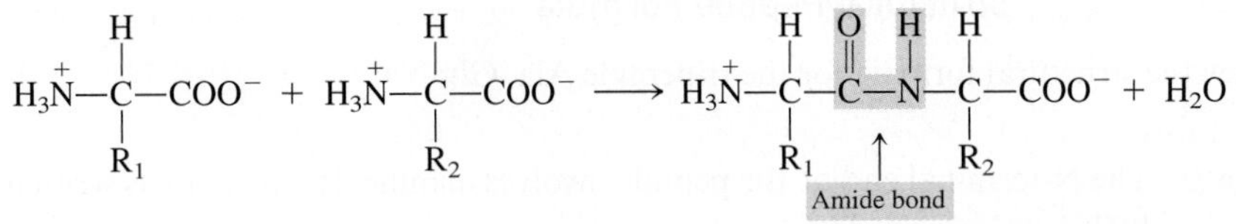

Amide bond formation is an example of a condensation reaction.

Removal of the elements of water from the reacting carboxyl and amino groups and the ensuing formation of the amide bond are better visualized when expanded structural formulas for the reacting groups are used.

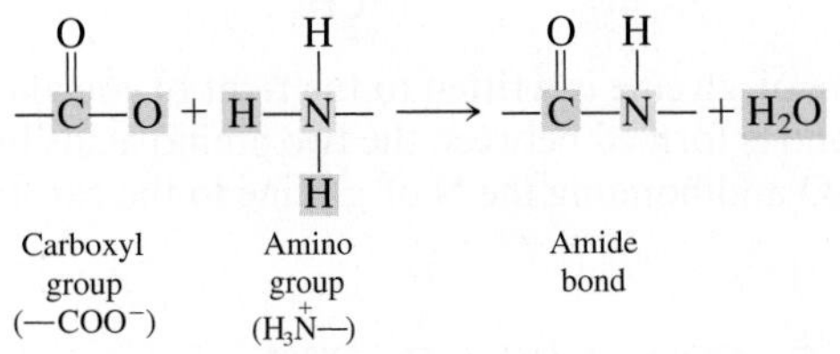

In amino acid chemistry, amide bonds that link amino acids together are given the specific name of *peptide bond.* A **peptide bond** *is a covalent bond between the carboxyl group of one amino acid and the amino group of another amino acid.*

In all peptides, long or short, the amino acid at one end of the amino acid sequence has a free $H_3\overset{+}{N}$ group, and the amino acid at the other end of the sequence has a free COO^- group. The end with the free $H_3\overset{+}{N}$ group is called the *N-terminal end,* and the end with the free COO^- group is called the *C-terminal end.* By convention, the sequence of amino acids in a peptide is written with the N-terminal end amino acid on the left. The individual amino acids within a peptide chain are called *amino acid residues.* An **amino acid residue** *is the portion of an amino acid structure that remains, after the release of H_2O, when an amino acid participates in peptide bond formation as it becomes part of a peptide chain.*

The structural formula for a peptide may be written out in full, or the sequence of amino acids present may be indicated by using the standard three-letter amino acid abbreviations. The abbreviated formula for the tripeptide

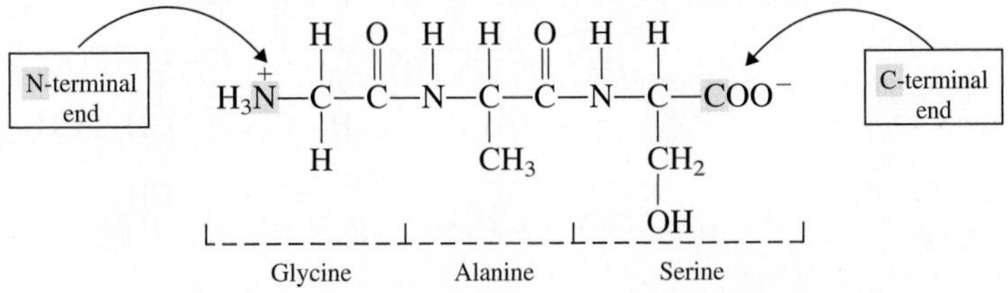

A peptide chain has *directionality* because its two ends are different. There is an N-terminal end and a C-terminal end. By convention, the direction of the peptide chain is always

N-terminal end → C-terminal end

The N-terminal end is always on the left, and the C-terminal end is always on the right.

which contains the amino acids glycine, alanine, and serine, is Gly–Ala–Ser. When we use this abbreviated notation, by convention, the amino acid at the N-terminal end of the peptide is always written on the left.

The repeating sequence of peptide bonds and α-carbon —CH groups in a peptide is referred to as the *backbone* of the peptide.

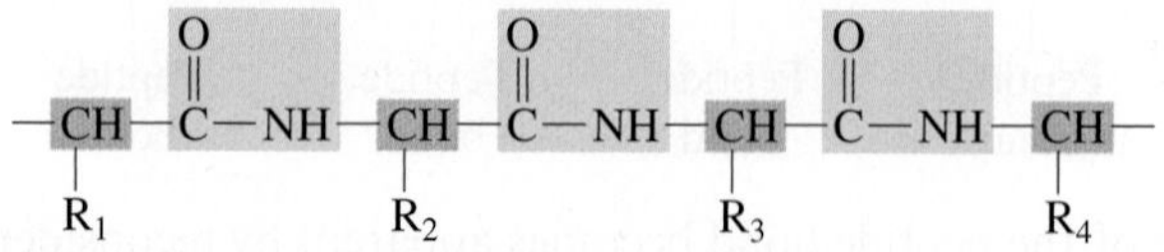

Backbone of peptide (in color)

The R group side chains are considered substituents on the backbone rather than part of the backbone.

Thus, structurally, a peptide has a regularly repeating part (the backbone) and a variable part (the sequence of R groups). It is the variable R group sequence that distinguishes one peptide from another.

EXAMPLE 9.2 Converting an Abbreviated Peptide Formula to a Structural Peptide Formula

Draw the structural formula for the tripeptide Ala–Gly–Val.

Solution

Step 1: The N-terminal end of the peptide involves alanine. Its structure is written first.

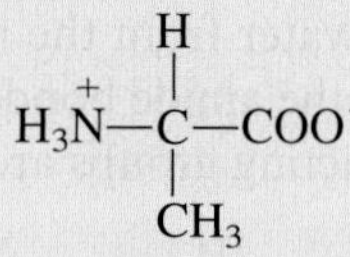

Step 2: The structure of glycine is written to the right of the alanine structure, and a peptide bond is formed between the two amino acids by removing the elements of H_2O and bonding the N of glycine to the carboxyl C of alanine.

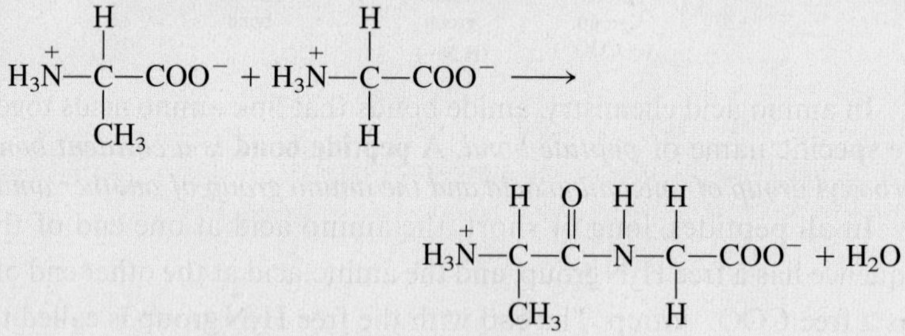

Step 3: To the right of the just-formed dipeptide, draw the structure of valine. Then repeat Step 2 to form the desired tripeptide.

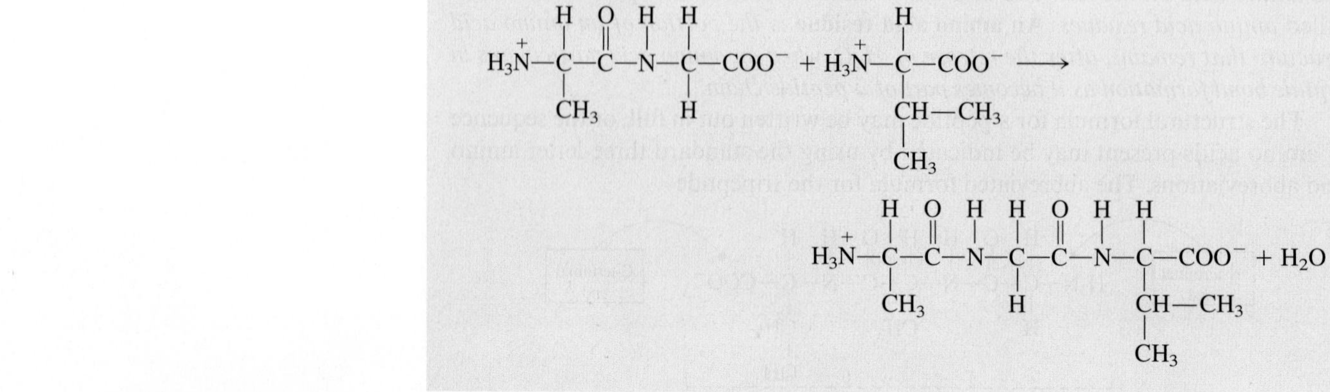

Practice Exercise 9.2

Draw the structural formula for the tripeptide Cys–Ala–Gly.

Answer:

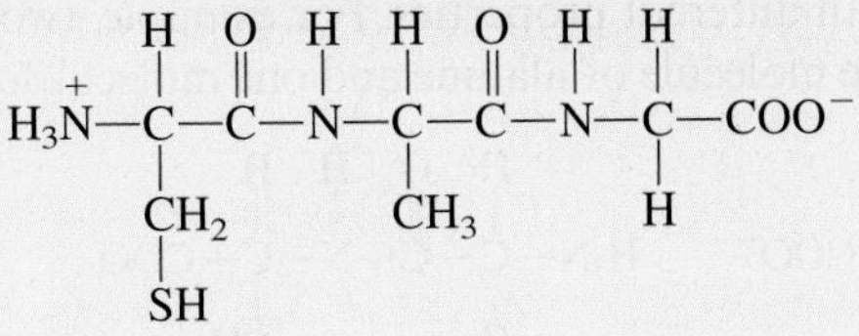

Peptide Nomenclature

Small peptides are named as derivatives of the C-terminal amino acid that is present. The IUPAC rules for doing this are:

Rule 1: *The C-terminal amino acid residue (located at the far right of the structure) keeps its full amino acid name.*

Rule 2: *All of the other amino acid residues have names that end in* -yl. *The* -yl *suffix replaces the* -ine *or* -ic acid *ending of the amino acid name, except for tryptophan (tryptophyl), cysteine (cysteinyl), glutamine (glutaminyl), and asparagine (asparaginyl).*

Rule 3: *The amino acid naming sequence begins at the N-terminal amino acid residue.*

EXAMPLE 9.3 Determining IUPAC Names for Small Peptides

Assign IUPAC names to each of the following small peptides.

a. Glu–Ser–Ala
b. Gly–Tyr–Leu–Val

Solution

a. The three amino acids present are glutamic acid, serine, and alanine. Alanine, the C-terminal residue (on the far right), keeps its full name. The other amino acid residues in the peptide receive "shortened" names that end in *-yl*. The *-yl* replaces the *-ine* or *-ic acid* ending of the amino acid name. Thus

glutamic acid becomes glutamyl
serine becomes seryl
alanine remains alanine

The IUPAC name, which lists the amino acids in the sequence from N-terminal residue to C-terminal residue, becomes *glutamylserylalanine.*

b. The four amino acids present are glycine, tyrosine, leucine, and valine. Proceeding as in part **a**, we note that

glycine becomes glycyl
tyrosine becomes tyrosyl
leucine becomes leucyl
valine remains valine

Combining these individual names gives the IUPAC name *glycyltyrosylleucylvaline.*

The complete name for a small peptide should actually include a handedness designation (L-designation; Section 9.4) before the name of each residue. For example, the dipeptide serylglycine (Ser–Gly) should actually be L-seryl–L-glycine (L-Ser–L-Gly). The L-handedness designation is, however, usually not included because it is understood that all amino acids present in a peptide, unless otherwise noted, are L enantiomers.

Practice Exercise 9.3

Assign IUPAC names to each of the following small peptides.

a. Gly–Ala–Leu **b.** Gly–Tyr–Ser–Ser

Answers: **a.** Glycylalanylleucine; **b.** Glycyltyrosylserylserine

Isomeric Peptides

Peptides that contain the same amino acids but in different order are different molecules (constitutional isomers) with different properties. For example, two different dipeptides can be formed from one molecule of alanine and one molecule of glycine.

Amino acid sequence in a peptide has biochemical importance. Isomeric peptides give different biochemical responses; that is, they have different biochemical specificities.

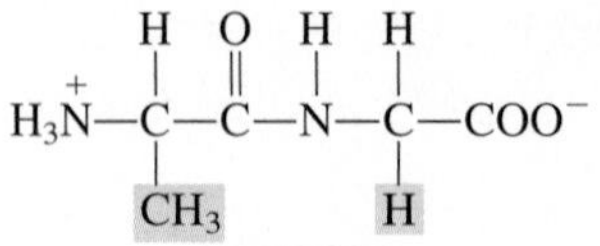

Ala–Gly

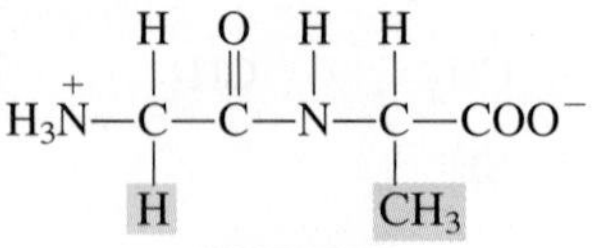

Gly–Ala

In the first dipeptide, the alanine is the N-terminal residue, and in the second molecule, it is the C-terminal residue. These two compounds are isomers with different chemical and physical properties.

The number of isomeric peptides possible increases rapidly as the length of the peptide chain increases. Let us consider the tripeptide Ala–Ser–Cys as another example. In addition to this sequence, five other arrangements of these three components are possible, each representing another isomeric tripeptide: Ala–Cys–Ser, Ser–Ala–Cys, Ser–Cys–Ala, Cys–Ala–Ser, and Cys–Ser–Ala. For a pentapeptide containing five different amino acids, 120 isomers are possible.

For a peptide containing one each of *n* different kinds of amino acids, the number of constitutional isomers is given by *n*! (*n* factorial).

$5! = 5 \times 4 \times 3 \times 2 \times 1 = 120$

9.8 Biochemically Important Small Peptides

Many relatively small peptides have been shown to be biochemically active. Functions for them include hormonal action, neurotransmission, and antioxidant activity.

Small Peptide Hormones

The two best-known peptide hormones, both produced by the pituitary gland, are *oxytocin* and *vasopressin.* Each hormone is a nonapeptide (nine amino acid residues) with six of the residues held in the form of a loop by a disulfide bond formed from the interaction of two cysteine residues (Section 9.6). Structurally, these nonapeptides differ in the amino acid present in positions 3 and 8 of the peptide chain. In both structures, an amine group replaces the C-terminal single-bonded oxygen atom.

Oxytocin plays a role in stimulating the flow of milk in a nursing mother. The baby's suckling action sends nerve signals to the mother's brain, triggering the release of oxytocin, via the blood, to the mammary glands. The oxytocin causes muscle contraction in the mammary gland, forcing out milk. As suckling continues, more oxytocin is released and more milk is available for the baby.

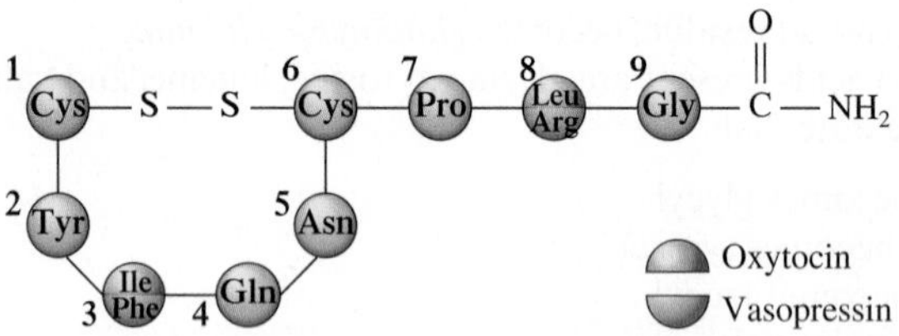

Oxytocin regulates uterine contractions and lactation. Vasopressin regulates the excretion of water by the kidneys; it also affects blood pressure. Another name for vasopressin is *antidiuretic hormone (ADH).* This name relates to vasopressin's function in the kidneys, which is to decrease urine output in order to decrease water elimination from the body. Such action is necessary when the body becomes dehydrated.

Small Peptide Neurotransmitters

Enkephalins are pentapeptide neurotransmitters produced by the brain itself that bind at receptor sites in the brain to reduce pain. The two best-known enkephalins are Met-enkephalin and Leu-enkephalin, whose structures differ only in the amino acid residue present at the C-terminal end (Section 9.6) of the peptide; this amino acid difference is incorporated into their names.

Met-enkephalin: Tyr–Gly–Gly–Phe–Met

Leu-enkephalin: Tyr–Gly–Gly–Phe–Leu

The pain-reducing effects of enkephalin action play a role in the "high" reported by long-distance runners, in the competitive athlete's managing to finish the game despite being injured, and in the pain-relieving effects of acupuncture.

The action of the prescription painkillers morphine and codeine is based on their binding at the same receptor sites in the brain as the naturally occurring enkephalins. Enkephalin pain relief is short-term, whereas morphine-codeine pain relief lasts much longer. Enzymes present in the brain readily hydrolyze the peptide linkages in enkephalins; morphine and codeine do not have such linkages and are unaffected by the hydrolysis enzymes.

The action of another important peptide hormone, the octapeptide angiotensin, is considered in the next chapter (Section 10.10).

Small Peptide Antioxidants

The tripeptide *glutathione* (Glu–Cys–Gly) is present in significant concentrations in most cells and is of considerable physiological importance as a regulator of oxidation–reduction reactions. Specifically, glutathione functions as an antioxidant (Section 3.13), protecting cellular contents from oxidizing agents such as peroxides and superoxides (highly reactive forms of oxygen often generated within the cell in response to bacterial invasion) (Section 12.11).

The tripeptide structure of glutathione has an unusual feature. The amino acid Glu, an acidic amino acid, is bonded to Cys through the side-chain carboxyl group rather than through its α-carbon carboxyl group.

Other antioxidants previously considered are BHA and BHT (Section 3.13) and β-carotene (Section 2.7).

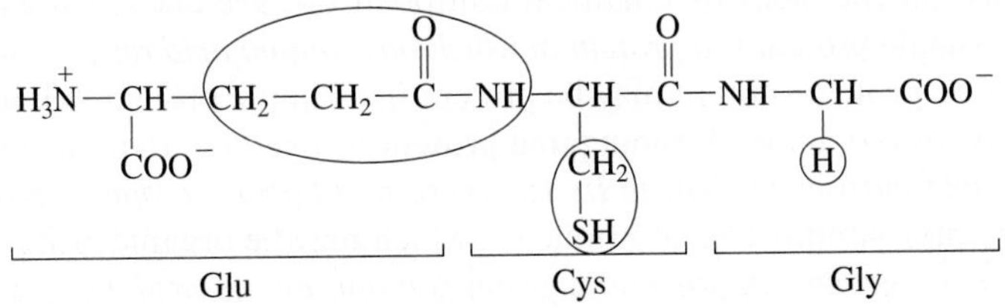

The side chains are circled in the above structure.

9.9 General Structural Characteristics of Proteins

In Section 9.1, a protein was defined simply as a naturally occurring, unbranched polymer in which the monomer units are amino acids. A more specific protein definition is now in order. A **protein** *is a peptide in which at least 40 amino acid residues are present.* The defining line governing the use of the term *protein*—40 amino acid residues—is an arbitrary line. The terms *polypeptide* and *protein* are often used interchangeably; a protein is a relatively long polypeptide. The key point is that the term *protein* is reserved for peptides with a large number of amino acids; it is not correct to call a tripeptide a protein. More than 10,000 amino acid residues are present in several proteins; 400–500 amino acid residues are common in proteins; small proteins contain 40–100 amino acid residues.

Proteins are the second type of biochemical polymer encountered in this text; the other was polysaccharides (Section 7.14). Protein monomers are amino acids, whereas polysaccharide monomers are monosaccharides.

More than one peptide chain may be present in a protein. On this basis, proteins are classified as *monomeric* or *multimeric.* A **monomeric protein** *is a protein in which only one peptide chain is present.* Large proteins, those with many amino acid residues, usually are multimeric. A **multimeric protein** *is a protein in which more than one peptide chain is present.* The peptide chains present in multimeric proteins are

Table 9.4 Types of Conjugated Proteins

Class	Prosthetic Group	Specific Example	Function of Example
hemoproteins	heme unit	hemoglobin myoglobin	carrier of O_2 in blood oxygen binder in muscles
lipoproteins	lipid	low-density lipoprotein (LDL) high-density lipoprotein (HDL)	lipid carrier lipid carrier
glycoproteins	carbohydrate	gamma globulin mucin interferon	antibody lubricant in mucous secretions antiviral protection
phosphoproteins	phosphate group	glycogen phosphorylase	enzyme in glycogen phosphorylation
nucleoproteins	nucleic acid	ribosomes viruses	site for protein synthesis in cells self-replicating, infectious complex
metalloproteins	metal ion	iron–ferritin zinc–alcohol dehydrogenase	storage complex for iron enzyme in alcohol oxidation

called *protein subunits.* The protein subunits within a multimeric protein may all be identical to each other or different kinds of subunits may be present. Proteins with up to 12 subunits are known. The small protein insulin, which functions as a hormone in the human body, is a multimeric protein with two protein subunits; one subunit contains 21 amino acid residues and the other 30 amino acid residues. The structure of insulin is considered in more detail in Section 9.12.

Proteins, on the basis of chemical composition, are classified as *simple* or *complex.* A **simple protein** *is a protein in which only amino acid residues are present.* More than one protein subunit may be present in a simple protein, but all subunits contain only amino acids. A **conjugated protein** *is a protein that has one or more non-amino acid entities present in its structure in addition to one or more peptide chains.* These non-amino acid components, which may be organic or inorganic, are called *prosthetic groups.* A **prosthetic group** *is a non-amino acid group present in a conjugated protein.*

Conjugated proteins may be further classified according to the nature of the prosthetic group(s) present. *Lipoproteins* contain lipid prosthetic groups, *glycoproteins* contain carbohydrate groups, *metalloproteins* contain a specific metal, and so on (Table 9.4). Some proteins contain more than one type of prosthetic group. In general, prosthetic groups have important roles in the biochemical functions for conjugated proteins. Several examples of glycoproteins and lipoproteins are discussed in Sections 9.18 and 9.19, respectively.

In general, the three-dimensional structures of proteins, even those with just a single peptide chain, are more complex than those of carbohydrates and lipids—the biomolecules discussed in the two previous chapters. Describing and understanding this complexity in protein structure involves considering four levels of protein structure. These four protein structural levels, listed in order of increasing complexity, are *primary* structure, *secondary* structure, *tertiary* structure, and *quaternary* structure. They are the subject matter for the next four sections of this chapter.

9.10 Primary Structure of Proteins

The primary structure of a protein is the *sequence* of amino acids in a protein chain—that is, the order in which the amino acids are connected to each other.

Primary protein structure *is the order in which amino acids are linked together in a protein.* Every protein has its own unique amino acid sequence. Primary protein structure always involves more than just the numbers and kinds of amino acids present; it also involves the *order of attachment* of the amino acids to each other through peptide bonds.

Insulin, the hormone that regulates blood-glucose levels, was the first protein for which primary structure was determined; the "sequencing" of its 51 amino acids was completed in 1953, after eight years of work by the British biochemist Frederick Sanger (Figure 9.3). Today, primary structures are known for many thousands of proteins, and the sequencing procedures involve automated methods that require relatively short periods of time (days). Figure 9.4 shows the primary structure of myoglobin, a protein involved in oxygen transport in muscles; it contains 153 amino acids assembled in the particular, definite order shown in this diagram.

The primary structure of a specific protein is always the same regardless of where the protein is found within an organism. The structures of certain proteins are even similar among different species of animals. For example, the primary structures of insulin in cows, pigs, sheep, and horses are very similar both to each other and to human insulin. Until recently, this similarity was particularly important for diabetics who required supplemental injections of insulin. The focus on relevancy feature Chemical Connections 9-A on the next page considers several old and new aspects of insulin chemistry.

Hulton Archives/Getty Images

Figure 9.3 The British biochemist Frederick Sanger (1918–) determined the primary structure of the protein hormone *insulin* in 1953. His work is a landmark in biochemistry because it showed for the first time that a protein has a precisely defined amino acid sequence. Sanger was awarded the Nobel Prize in chemistry in 1958 for this work. Later, in 1980, he was awarded a second Nobel Prize in chemistry, this time for work that involved the sequencing of units in nucleic acids (Chapter 11).

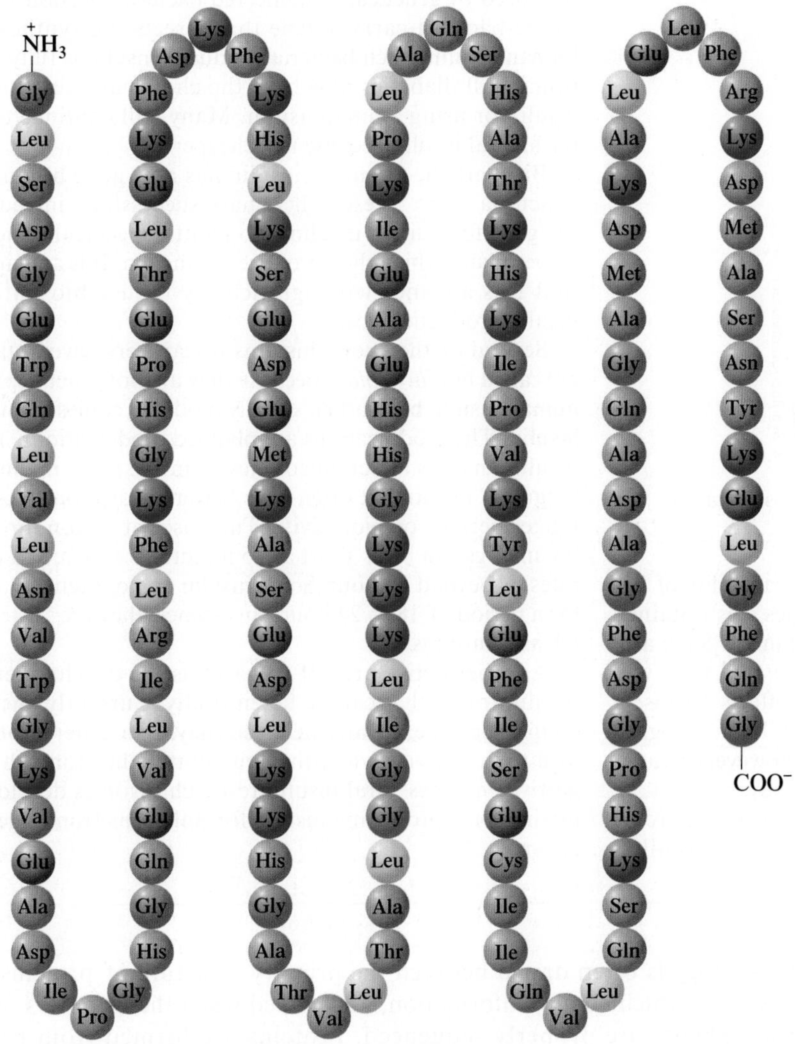

Figure 9.4 The primary structure of human myoglobin. This diagram gives only the sequence of the amino acids present and conveys no information about the actual three-dimensional shape of the protein. The "wavy" pattern for the 153 amino acid sequence was chosen to minimize the space used to present the needed information. The actual shape of the protein is determined by secondary and tertiary levels of protein structure, levels yet to be discussed.

CHEMICAL CONNECTIONS 9-A

"Substitutes" for Human Insulin

Insulin is a 51-amino-acid-containing protein hormone whose structure is given later in the chapter (Section 9.12). Its function within the human body primarily involves regulation of blood-glucose levels. It assists the entry of blood glucose into cells by interacting with receptors on cell membranes (Section 8.10). It also helps facilitate the conversion of glucose to the storage polysaccharide glycogen (Section 7.15) when blood-glucose levels become too high and facilitates the reverse process (conversion of glycogen back to glucose) when blood-glucose levels become too low.

Insufficient insulin production or an inability to use insulin produced (insulin resistance) result in the condition *diabetes mellitus*. Treatment of this condition often involves giving a person insulin via subcutaneous injection (see accompanying photo).

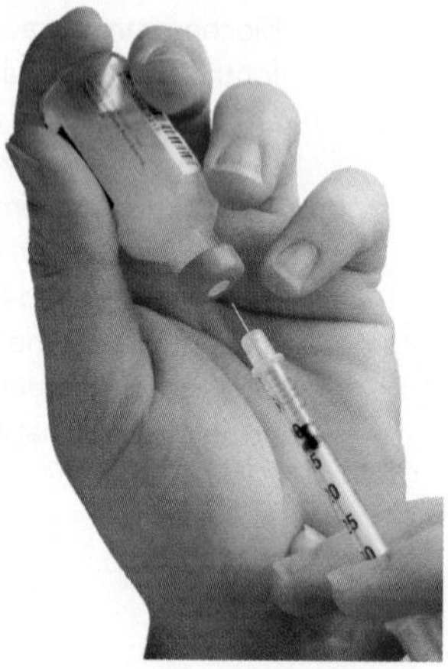

Currently, diabetics who require insulin must obtain it through injection rather than orally.

For many years, because of the limited availability of human insulin, most insulin used by diabetics was obtained from the pancreases of slaughter-house animals. Such animal insulin, obtained primarily from cows and pigs, was used by most diabetics without serious side effects because it is structurally very similar to human insulin. Immunological reactions gradually do increase over time, however, because the animal insulin is foreign to the human body.

In terms of primary structure, human insulin, porcine (pig) insulin, and bovine (cow) insulin are very similar. Porcine and human insulin match at 50 of the 51 amino acid positions, and bovine and human insulin match at 48 of the 51 amino acid positions (see the table that follows and Figure 9.10).

	Chain A			Chain B
Species	**#8**	**#9**	**#10**	**#30**
human	Thr	Ser	Ile	Thr
pig (porcine)	Thr	Ser	Ile	Ala
cow (bovine)	Ala	Ser	Val	Ala

The dependency of diabetics on animal insulin is now on the decline because of the availability of human insulin produced by genetically engineered bacteria (Section 11.14). These bacteria carry a gene that directs the synthesis of human insulin. Such bacteria-produced insulin is fully functional. All diabetics now have the choice of using human insulin or using animal insulin. Many still continue to use the animal insulin because it is cheaper.

Biosynthetic "human" insulin has now gone beyond the "bacteria stage." Researchers have successfully introduced the gene for human insulin into plants, specifically the safflower plant, which then produce the insulin. It is anticipated that this advance will significantly reduce biosynthetic insulin production costs.

Several of the "new" insulins researchers have obtained are called *insulin analogs* because they are not exact copies of human insulin but, rather, slightly modified copies of human insulin. The modifications are planned modifications. These insulin analogs affect glucose levels in the same manner as "regular" insulin but often have better absorption rates and longer periods of bioactivity. The onset of action for insulin analogs can be as short as 15 minutes compared to usual rates of beyond an hour. Some insulin analogs remain active for a period of 18 to 24 hours; in essence, they are "extended release" insulins.

Another active area of research is the development of insulin forms that can be taken orally. Currently, insulins cannot be taken orally because they, like other proteins, lose their activity when they encounter the stomach's digestive enzymes. Oral insulin research involves developing methods for protecting insulin formulations from digestive enzymes.

An analogy is often drawn between the primary structure of proteins and words. Words, which convey information, are formed when the 26 letters of the English alphabet are properly sequenced. Proteins are formed from proper sequences of the 20 standard amino acids. Just as the proper sequence of letters in a word is necessary for it to make sense, the proper sequence of amino acids is necessary to make biochemically active protein. Furthermore, the letters that form a word are written from left to right, as are amino acids in protein formulas. As any dictionary of the English language will document, a tremendous variety of words can be formed by different letter sequences. Imagine the number of amino acid sequences possible for a large protein. There are 1.55×10^{66} sequences possible for the 51 amino acids found in insulin! From these possibilities, the body reliably produces only *one,* illustrating the remarkable precision of life processes. From the simplest bacterium to the human brain cell, only those amino acid sequences needed by the cell are produced. The fascinating process of protein biosynthesis and the way in which genes in DNA direct this process will be discussed in Chapter 11.

The various amino acids present in a protein, whose order is the primary structure of the protein, are linked to each other by peptide linkages. The peptide bonds are part of the "backbone" of the protein. The structural characteristics of a protein backbone are the same as those of a peptide backbone (Section 9.7); relative to backbone structure, a protein is simply an "extra long" peptide. A representative segment of a protein backbone is as follows:

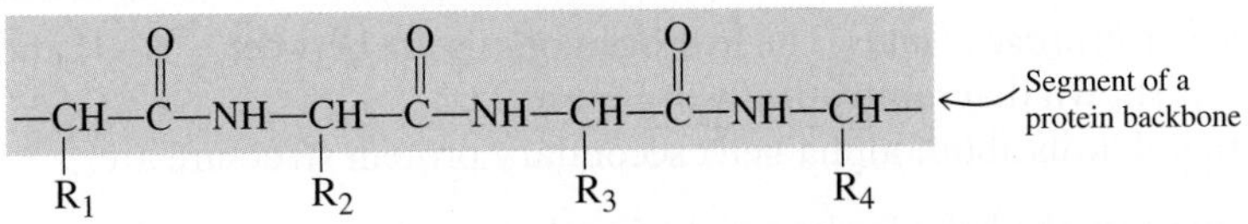

Attached to the backbone, at the CH locations, are the various amino acid R groups.

The carbon and nitrogen atoms of a protein backbone are arranged in a "zigzag" manner. This zigzag pattern arises from the geometric characteristics of peptide bonds. Important points concerning peptide bond geometry are:

1. The peptide linkages are essentially planar. This means that for two amino acids linked through a peptide linkage, six atoms lie in the same plane: the α-carbon atom and the C═O group from the first amino acid and the N—H group and the α-carbon atom from the second amino acid.

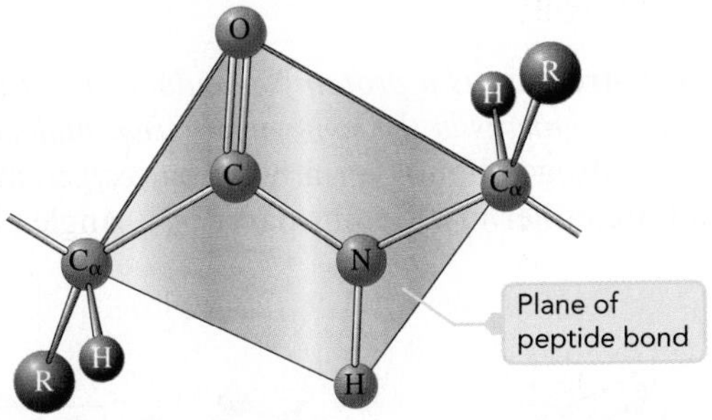

2. The planar peptide linkage structure has considerable rigidity, which means that rotation of groups about the C—N bond is hindered, and *cis–trans* isomerism is possible about this bond. The *trans* isomer orientation is the preferred orientation, as shown in the preceding diagram. The O atom of the C═O group and the H atom of the N—H group are positioned *trans* to each other.

The net effect of "peptide bond planarity" is the zigzag arrangement, previously mentioned, of atoms within a protein backbone, which is shown in the following diagram.

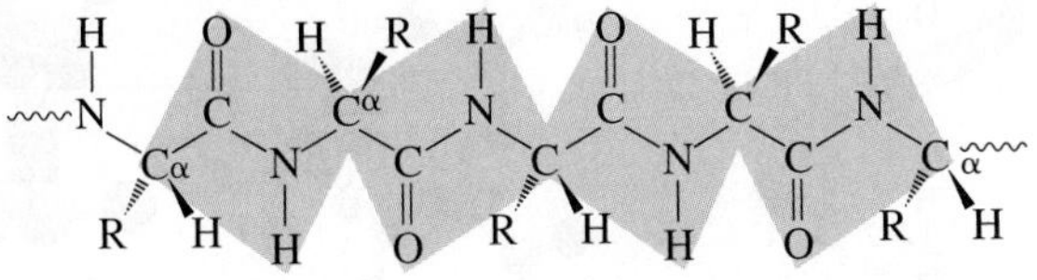

9.11 Secondary Structure of Proteins

Secondary protein structure *is the arrangement in space adopted by the backbone portion of a protein.* The two most common types of secondary structure are the *alpha helix* (α helix) and the *beta pleated sheet* (β pleated sheet). The type of interaction responsible for both of these types of secondary structure is hydrogen

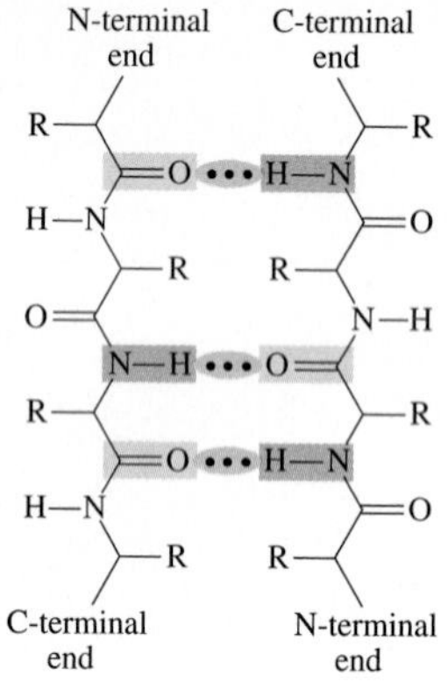

Figure 9.5 The hydrogen bonding between the carbonyl oxygen atom of one peptide linkage and the amide hydrogen atom of another peptide linkage.

The hydrogen bonding present in an α helix is *intra*molecular. In a β pleated sheet, the hydrogen bonding can be *inter*molecular (between two different chains) or *intra*molecular (a single chain folding back on itself).

bonding between a carbonyl oxygen atom of a peptide linkage and the hydrogen atom of an amino group of another peptide linkage farther along the protein backbone. Figure 9.5 shows the hydrogen bonding possibilities that exist between carbonyl oxygen atoms and amino hydrogen atoms associated with different peptide linkages in a protein backbone. The protein backbone segments involved in hydrogen bonding can be two segments from different backbones (as shown in Figure 9.5) or two segments of the same backbone that has folded back upon itself. Both of these situations are considered in further detail in this section.

The Alpha Helix

An **alpha helix structure** *is a protein secondary structure in which a single protein chain adopts a shape that resembles a coiled spring (helix), with the coil configuration maintained by hydrogen bonds.* The hydrogen bonds are between >N—H and >C=O groups as is shown diagrammatically in Figure 9.6.

Further details about alpha helix secondary protein structure are:

1. The twist of the helix forms a right-handed, or clockwise, spiral.
2. The hydrogen bonds between C=O and N—H entities are orientated parallel to the axis of the helix (Figure 9.6b).
3. A given hydrogen bond involves a C=O group of one amino acid and a N—H group of another amino acid located four amino acid residues further along the spiral (Figure 9.6b). This is because one turn of the spiral includes 3.6 amino acid residues.
4. All of the amino acid R groups extend outward from the spiral (Figure 9.6d). There is not enough room for the R groups within the spiral.

The Beta Pleated Sheet

A **beta pleated sheet structure** *is a protein secondary structure in which two fully extended protein chain segments in the same or different molecules are held together by hydrogen bonds.* Hydrogen bonds form between oxygen and hydrogen peptide linkage atoms that are either in different parts of a single chain that folds back

Figure 9.6 Four representations of the α helix protein secondary structure.

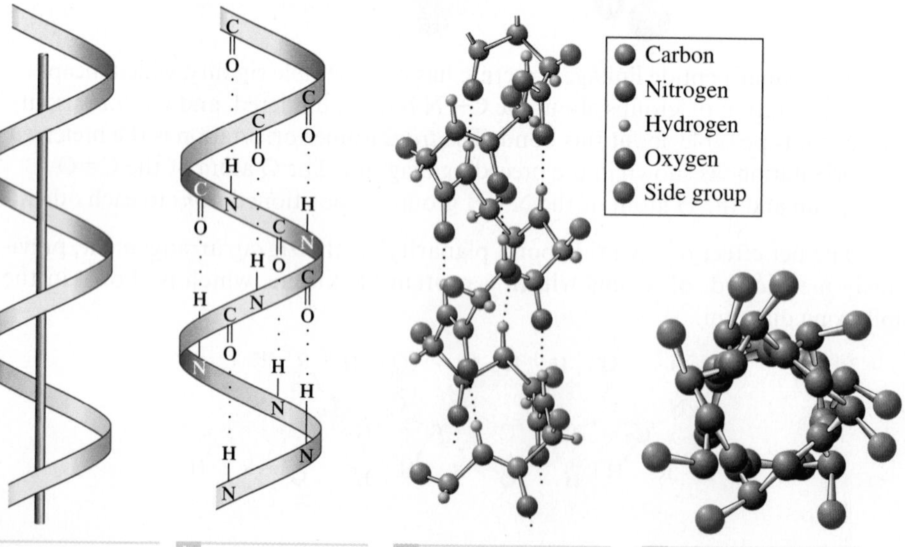

a Arrangement of protein backbone with no detail shown.

b Backbone arrangement with hydrogen-bonding interactions shown.

c Backbone atomic detail shown, as well as hydrogen-bonding interactions.

d Top view of an α helix showing that amino acid side chains (R groups) point away from the long axis of the helix.

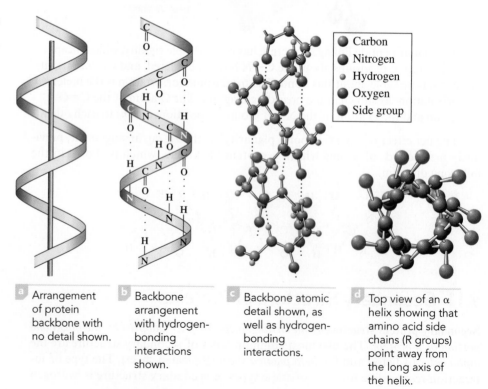

on itself (intrachain bonds) or between atoms in different peptide chains in those proteins that contain more than one chain (interchain bonds). In molecules where the β pleated sheet involves a single molecule, several U-turns in the protein chain arrangement are needed in order to form the structure.

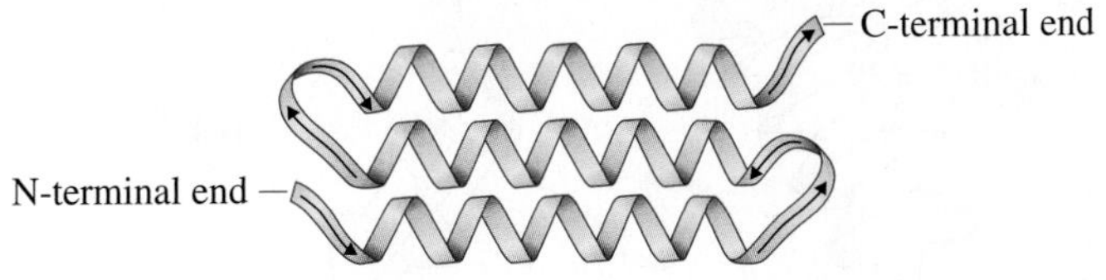

This "U-turn structure" is the most frequently encountered type of β pleated sheet structure.

Figure 9.7a shows a representation of the β pleated sheet structure that occurs when portions of two different peptide chains are aligned parallel to each other (interchain bonds). The term *pleated sheet* arises from the repeated zigzag pattern in the structure (Figure 9.7b).

The β pleated sheet is found extensively in the protein of silk. Because such proteins are already fully extended, silk fibers cannot be stretched. When wool, which has an α helix structure, becomes wet, it stretches as hydrogen bonds of the helix are broken. The wool returns to its original shape as it dries. Wet stretched wool, dried under tension, maintains its stretched length because it has assumed a β pleated sheet configuration.

Further features of the β pleated sheet secondary protein structure are:

1. The hydrogen bonds between C=O and N—H entities lie in the plane of the sheet (Figure 9.7a).
2. The amino acid R groups are found above and below the plane of the sheet and within a given backbone segment alternating between the top and bottom positions (Figure 9.7b).

Unstructured Segments

Very few proteins have entirely α helix or β pleated sheet structures. Instead, only certain portions of the molecules of most proteins are in these conformations. It is also possible to have both α helix and β pleated sheet structures within the same protein (see Figure 9.8). Helical structure and pleated sheet structure are found only in the portions of a protein where the amino acid R groups present are relatively small; large R groups tend to disrupt both of these types of secondary structure.

The portions of a protein that have neither α helix nor β pleated sheet structure are called *unstructured segments*. This designation is somewhat of a misnomer because all molecules of a given protein exhibit *identical* unstructured segments. An **unstructured protein segment** *is a protein secondary structure that is neither an α helix nor a β pleated sheet.*

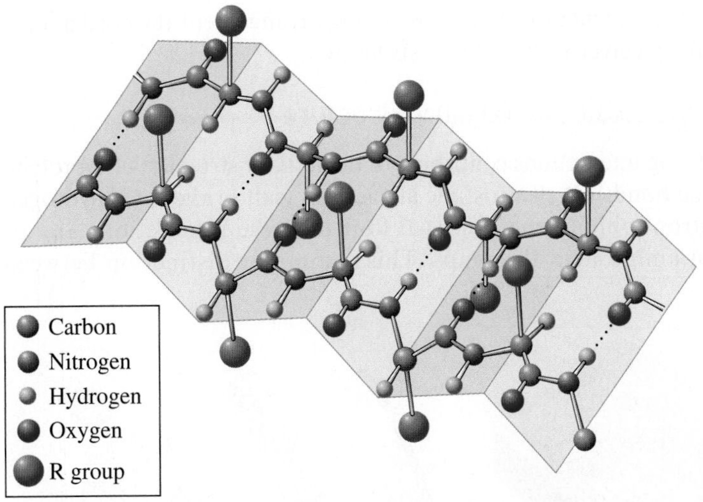

a A representation emphasizing the hydrogen bonds between protein chains.

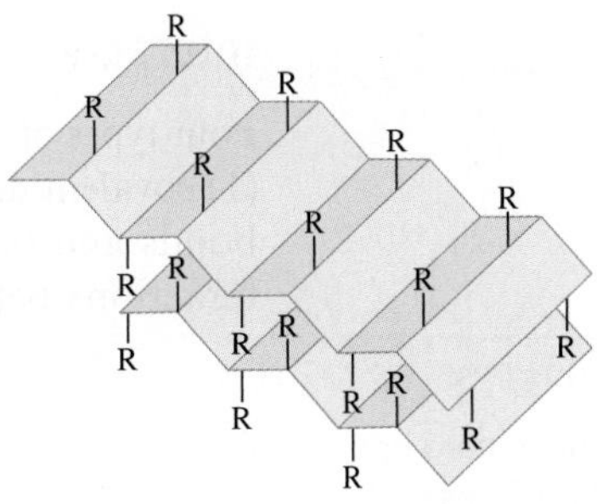

b A representation emphasizing the pleats and the location of the R groups.

Figure 9.7 Two representations of the β pleated sheet protein structure.

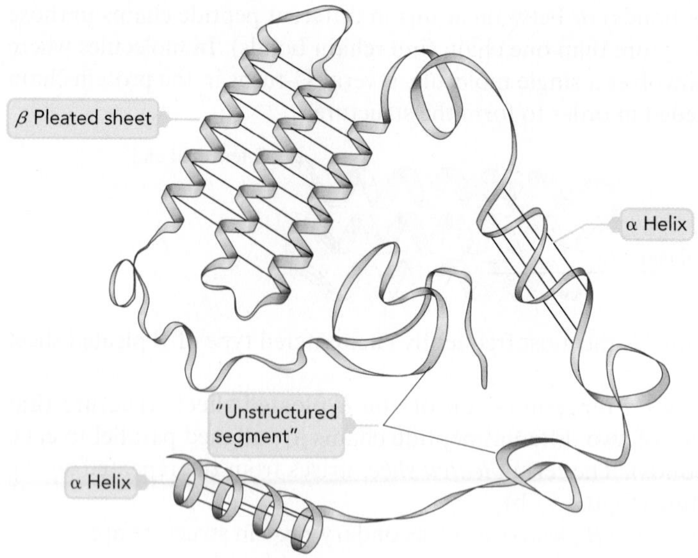

Figure 9.8 The secondary structure of a single protein often shows areas of α helix and β pleated sheet configurations, as well as areas of "unstructure."

An active area of protein research involves learning more about the biochemical functions of the unstructured portions of proteins. A growing number of researchers now believe that some "unstructure" is essential to the functioning of many proteins. It confers flexibility to proteins, thereby allowing them to interact with several different substances, an important mechanism for rapid response to changing cellular conditions. Often unstructured regions of a protein have the flexibility to bind to several different protein partners, allowing the same amino acid sequence to multitask. The act of binding to another protein does bring added structure to the unstructured portion of the protein, but the added structure is lost when the binding interaction ceases.

9.12 Tertiary Structure of Proteins

Tertiary protein structure *is the overall three-dimensional shape of a protein that results from the interactions between amino acid side chains (R groups) that are widely separated from each other within a peptide chain.*

A good analogy for the relationships among the primary, secondary, and tertiary structures of a protein is that of a telephone cord (Figure 9.9). The primary structure is the long, straight cord. The coiling of the cord into a helical arrangement gives the secondary structure. The supercoiling arrangement the cord adopts after you hang up the receiver is the tertiary structure.

Interactions Responsible for Tertiary Structure

Four types of stabilizing interactions contribute to the tertiary structure of a protein: (1) covalent disulfide bonds (2) electrostatic attractions (salt bridges) (3) hydrogen bonds and (4) hydrophobic attractions. All four of these interactions are interactions between amino acid R groups. This is a major distinction between

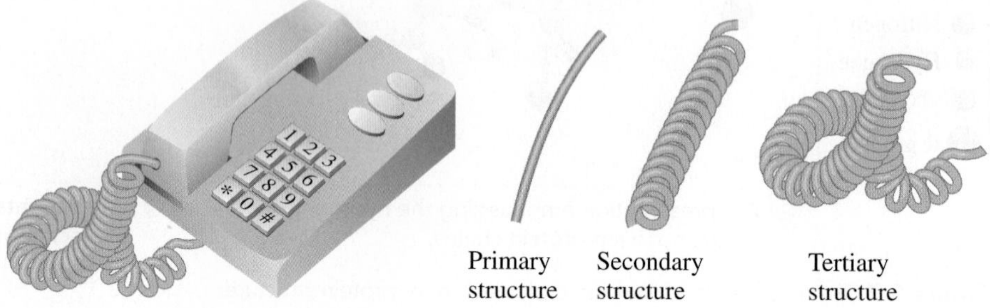

Figure 9.9 A telephone cord has three levels of structure. These structural levels are a good analogy for the first three levels of protein structure.

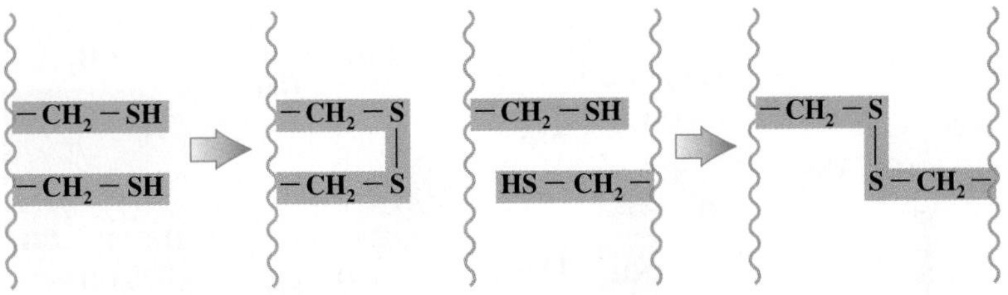

a Between two —SH groups on the same chain.

b Between two —SH groups on different chains.

Figure 9.10 Disulfide bonds involving cysteine residues can form in two different ways.

tertiary-structure interactions and secondary-structure interactions. Tertiary-structure interactions involve the R groups of amino acids; secondary-structure interactions involve the peptide linkages between amino acid residues.

Disulfide bonds, the strongest of the tertiary-structure interactions, result from the —SH groups of two cysteine residues reacting with each other to form a *covalent* disulfide bond (Section 9.6). This type of interaction is the only one of the four tertiary-structure interactions that involves a covalent bond. Disulfide bond formation may involve two cysteine units in the same peptide chain (an intramolecular disulfide bond; Figure 9.10a) or two cysteine units in different chains (an intermolecular disulfide bond; Figure 9.10b). Figure 9.11 gives the structure of the protein hormone insulin, a protein that has two peptide chains and a total of 51 amino acid residues; both inter- and intramolecular disulfide bonds are present in its structure.

Cysteine is the only α-amino acid that contains a sulfhydryl group (—SH).

Electrostatic interactions, also called *salt bridges,* always involve the interaction between an acidic side chain (R group) and a basic side chain (R group). The side chains of acidic and basic amino acids, at the appropriate pH, carry charges, with the acidic side chain being negatively charged and the basic side chain being positively charged. Such side chain charges occur when a —COOH group becomes a —COO$^-$ group and when a —NH_2 group becomes a —$\overset{+}{N}H_3$ group. The interaction that occurs between the two types of side chains is a positive–negative ion–ion attraction. Figure 9.12b shows an electrostatic interaction.

Hydrogen bonds can occur between amino acids with polar R groups. A variety of polar side chains can be involved, especially those that possess the following functional groups:

$$-OH \qquad -NH_2 \qquad -\overset{O}{\overset{\|}{C}}-OH \qquad -\overset{O}{\overset{\|}{C}}-NH_2$$

Hydrogen bonds are relatively weak and are easily disrupted by changes in pH and temperature. Figure 9.12c shows the hydrogen-bonding interactions between the R groups of glutamine and serine.

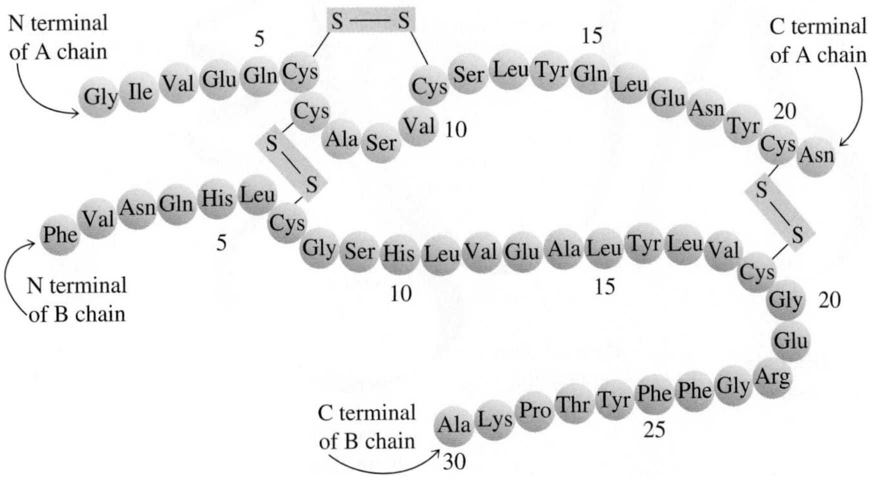

Figure 9.11 Human insulin, a small two-chain protein, has both intrachain and interchain disulfide linkages as part of its tertiary structure.

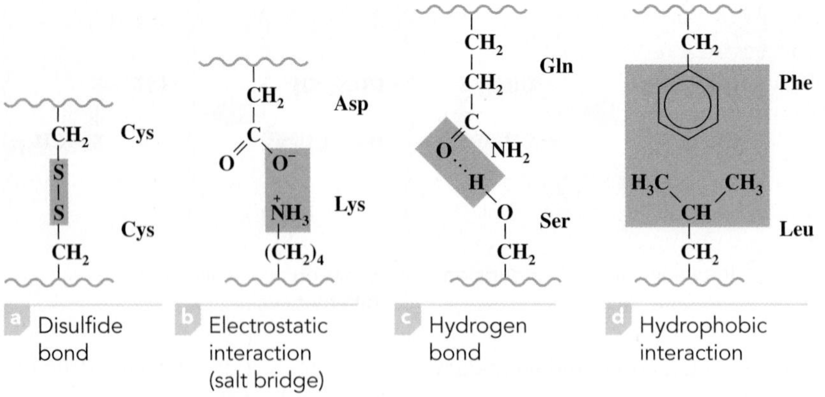

Figure 9.12 The four types of stabilizing interactions between amino acid R groups that contribute to the tertiary structure of a protein.

Hydrophobic interactions result when two nonpolar side chains are close to each other. In aqueous solution, many proteins have their polar R groups pointing outward, toward the aqueous solvent (which is also polar), and their nonpolar R groups pointing inward (away from the polar water molecules). The nonpolar R groups then interact with each other. The attractive forces are London forces resulting from the momentary uneven distribution of electrons within the side chains. Hydrophobic interactions are common between phenyl rings and alkyl side chains. Although hydrophobic interactions are weaker than hydrogen bonds or electrostatic interactions, they are a significant force in some proteins because there are so many of them; their cumulative effect can be greater in magnitude than the effects of hydrogen bonding. Figure 9.12d shows the hydrophobic interactions between the R groups of phenylalanine and leucine.

Example 9.4 considers the mechanics involved in identifying and drawing structural representations for the various stabilizing interactions associated with protein tertiary structure. Figure 9.13 shows all four of the stabilizing interactions

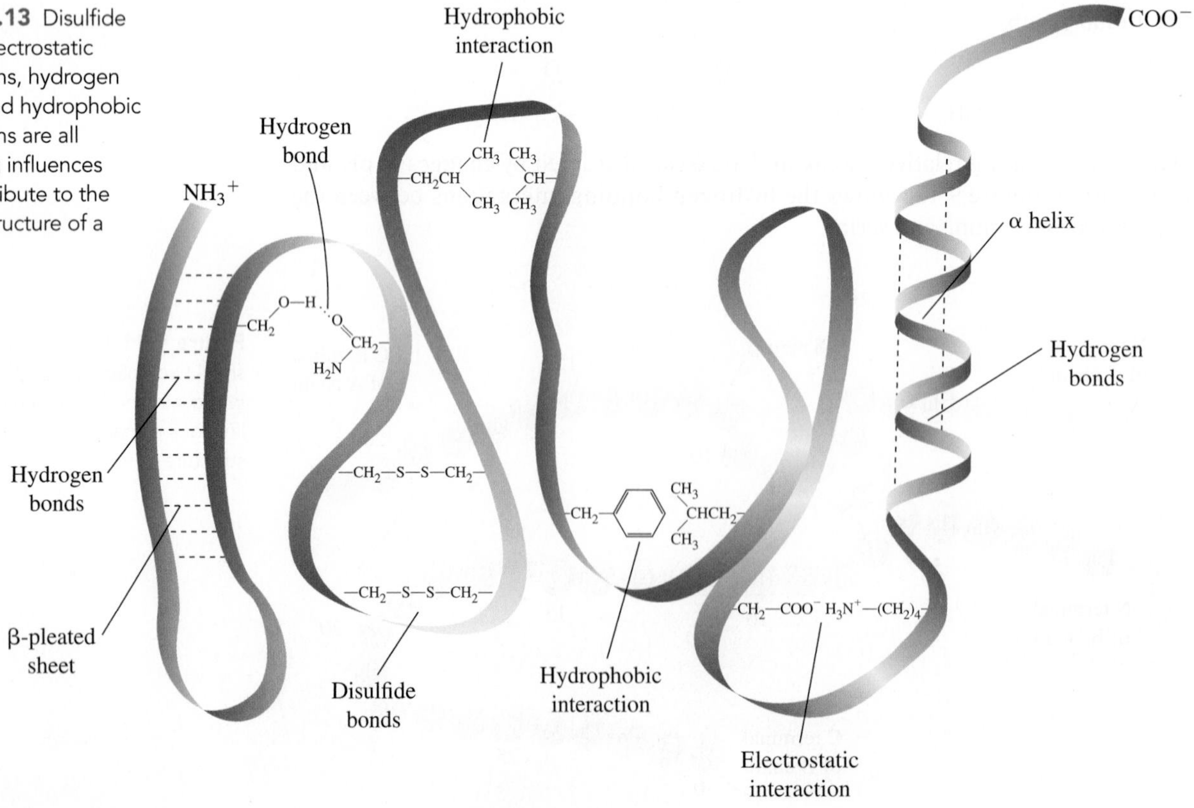

Figure 9.13 Disulfide bonds, electrostatic interactions, hydrogen bonds, and hydrophobic interactions are all stabilizing influences that contribute to the tertiary structure of a protein.

that contribute to tertiary protein structure in the context of a single peptide chain (a monomeric protein; Section 9.9).

In 1959, a protein tertiary structure was determined for the first time. The determination involved myoglobin, a conjugated protein (Section 9.9) whose function is oxygen storage in muscle tissue. Figure 9.14 shows myoglobin's tertiary structure. It involves a single peptide chain of 153 amino acids with numerous α helix segments within the chain. The structure also contains a prosthetic heme group, an iron-containing group with the ability to bind molecular oxygen.

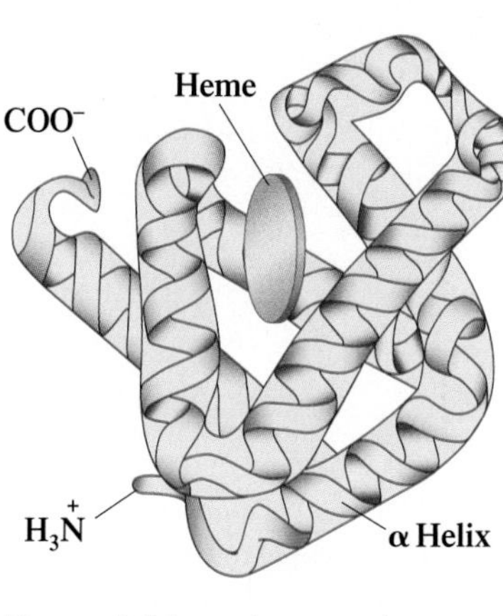

Figure 9.14 A schematic diagram showing the tertiary structure of the single-chain protein myoglobin.

EXAMPLE 9.4 Drawing Structural Representations for Amino Acid Side-Chain Interactions Associated with Protein Tertiary Structure

Identify the type of noncovalent interaction that occurs between the side chains of the following amino acids, and show the interaction using structural representations for the side chains.

a. Serine and asparagine
b. Glutamic acid and lysine

Solution

a. Both serine and asparagine are polar neutral amino acids (see Table 9.1). The side chains of such amino acids interact through *hydrogen bonding.* A structural representation for this hydrogen bonding interaction is:

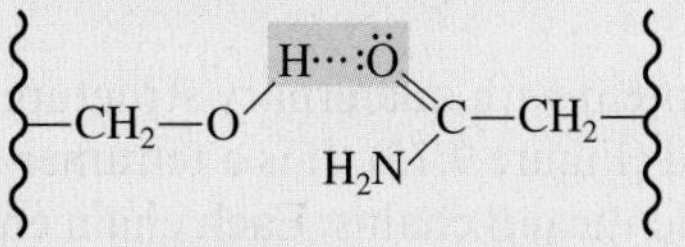

Ser Asn

b. Glutamic acid and lysine have, respectively, acidic and basic side chains (see Table 9.1). Such side chains carry a charge, in solution, as the result of proton transfer. The interaction between the negatively charged acidic side chain and the positively charged basic side chain is an *electrostatic interaction.* A structural representation for this electrostatic interaction is:

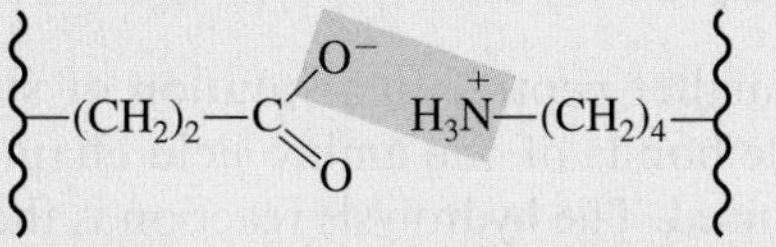

Glu Lys

Practice Exercise 9.4

Identify the type of noncovalent interaction that occurs between the side chains of the following amino acids, and show the interaction using structural representations for the side chains.

a. Threonine and asparagine
b. Aspartic acid and arginine

Answers:

a. Hydrogen bonding;

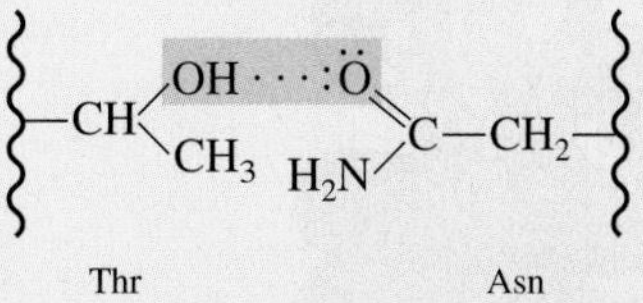

Thr Asn

b. Electrostatic interaction;

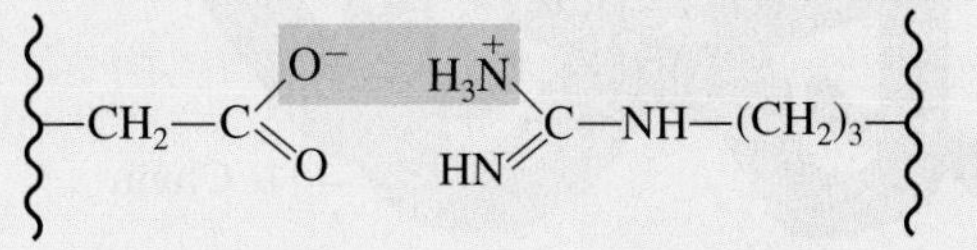

Asp Arg

A comparison of Figure 9.14 (myoglobin's tertiary structure) with Figure 9.4 (myoglobin's primary structure) shows how different the perspectives of primary and tertiary structure are for a protein.

9.13 Quaternary Structure of Proteins

Quaternary structure is the highest level of protein organization. It is found only in multimeric proteins (Section 9.9). Such proteins have structures involving two or more peptide chains that are independent of each other—that is, are not covalently bonded to each other. **Quaternary protein structure** *is the organization among the various peptide chains in a multimeric protein.*

Most multimeric proteins contain an even number of subunits (two subunits = a dimer, four subunits = a tetramer, and so on). The subunits are held together mainly by hydrophobic interactions between amino acid R groups.

The noncovalent interactions that contribute to tertiary structure (electrostatic interactions, hydrogen bonds, and hydrophobic interactions) are also responsible for the maintenance of quaternary structure. The noncovalent interactions that contribute to quaternary structure are, however, more easily disrupted. For example, only small changes in cellular conditions can cause a tetrameric protein to fall apart, dissociating into dimers or perhaps four separate subunits, with a resulting temporary loss of protein activity. As original cellular conditions are restored, the tertiary structure automatically re-forms, and normal protein function is restored.

An example of a protein with quaternary structure is hemoglobin, the oxygen-carrying protein in blood (Figure 9.15). It is a tetramer in which there are two identical α chains and two identical β chains. Each chain enfolds a heme group, the site where oxygen binds to the protein.

The Chemistry at a Glance feature on the next page reviews basic concepts about all four levels of protein structure.

9.14 Protein Hydrolysis

When a protein or smaller peptide in a solution of strong acid or strong base is heated, the peptide bonds of the amino acid chain are hydrolyzed and free amino acids are produced. The hydrolysis reaction is the reverse of the formation reaction for a peptide bond. Amine and carboxylic acid functional groups are regenerated.

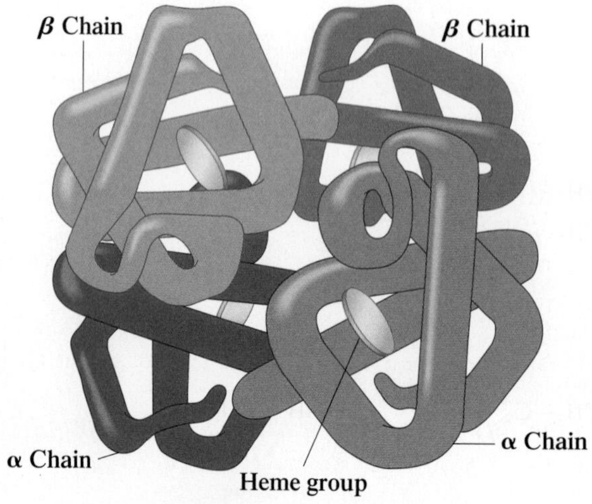

Figure 9.15 A schematic diagram showing the tertiary and quaternary structure of the oxygen-carrying protein hemoglobin.

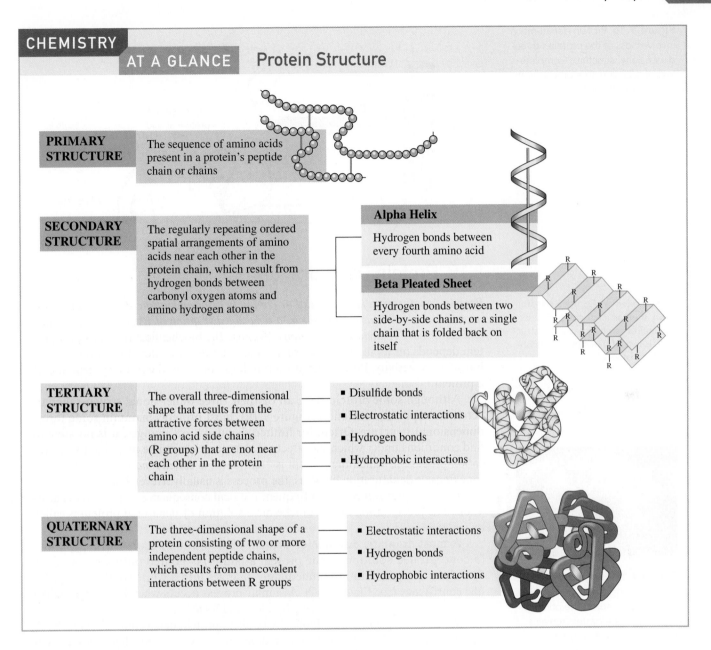

CHEMISTRY AT A GLANCE Protein Structure

PRIMARY STRUCTURE

The sequence of amino acids present in a protein's peptide chain or chains

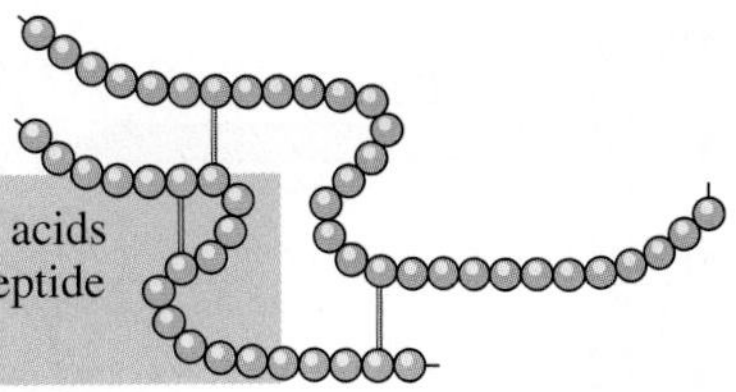

SECONDARY STRUCTURE

The regularly repeating ordered spatial arrangements of amino acids near each other in the protein chain, which result from hydrogen bonds between carbonyl oxygen atoms and amino hydrogen atoms

Alpha Helix

Hydrogen bonds between every fourth amino acid

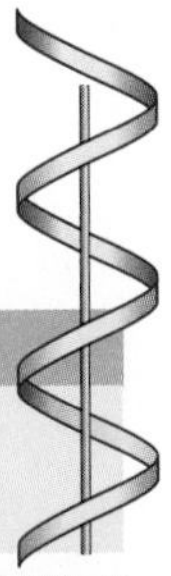

Beta Pleated Sheet

Hydrogen bonds between two side-by-side chains, or a single chain that is folded back on itself

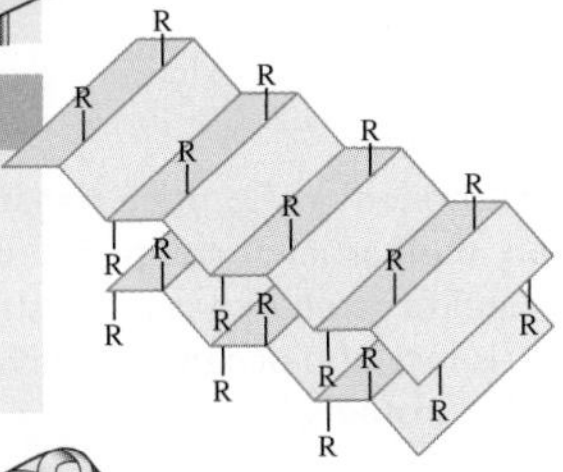

TERTIARY STRUCTURE

The overall three-dimensional shape that results from the attractive forces between amino acid side chains (R groups) that are not near each other in the protein chain

- Disulfide bonds
- Electrostatic interactions
- Hydrogen bonds
- Hydrophobic interactions

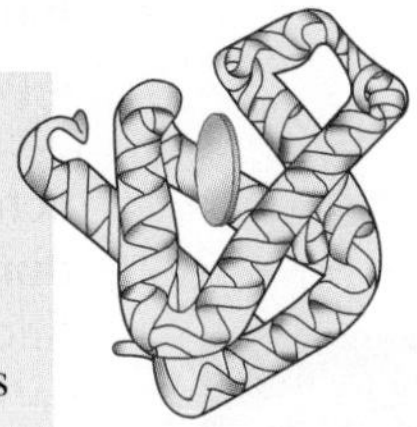

QUATERNARY STRUCTURE

The three-dimensional shape of a protein consisting of two or more independent peptide chains, which results from noncovalent interactions between R groups

- Electrostatic interactions
- Hydrogen bonds
- Hydrophobic interactions

For example, the complete hydrolysis of the tripeptide Ala–Gly–Cys under acidic conditions produces one unit each of the amino acids alanine, glycine, and cysteine. The equation for the hydrolysis is

$$\underset{\text{Ala–Gly–Cys}}{H_3\overset{+}{N}-CH(CH_3)-C(=O)-NH-CH_2-C(=O)-NH-CH(CH_2-SH)-COOH} \xrightarrow[\text{heat}]{H_2O,\ H^+} \underset{\text{Ala}}{H_3\overset{+}{N}-CH(CH_3)-COOH} + \underset{\text{Gly}}{H_3\overset{+}{N}-CH_2-COOH} + \underset{\text{Cys}}{H_3\overset{+}{N}-CH(CH_2-SH)-COOH}$$

Note that the product amino acids in this reaction are written in positive-ion form because of the acidic reaction conditions.

Protein digestion (Section 15.1) is simply enzyme-catalyzed hydrolysis of ingested protein. The free amino acids produced from this process are absorbed through the intestinal wall into the bloodstream and transported to the liver. Here they become the raw materials for the synthesis of new protein. Also, the hydrolysis of cellular proteins to amino acids is an ongoing process, as the body resynthesizes needed molecules and tissue.

Protein hydrolysis produces free amino acids. This process is the reverse of protein synthesis, where free amino acids are combined.

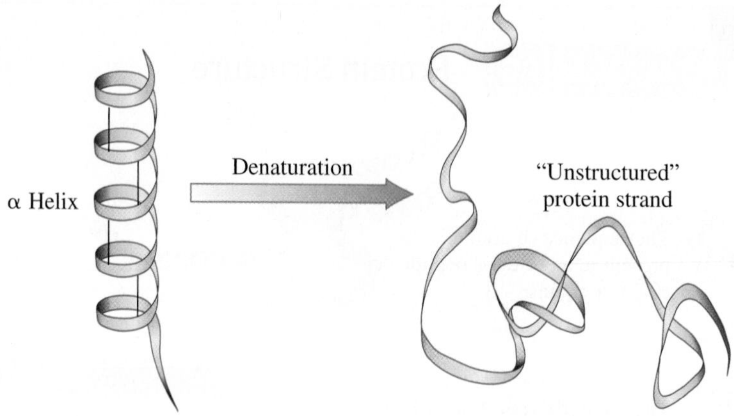

Figure 9.16 Protein denaturation involves loss of the protein's three-dimensional structure. Complete loss of such structure produces an "unstructured" protein strand.

9.15 Protein Denaturation

A consequence of protein denaturation, the partial or complete loss of a protein's three-dimensional structure, is loss of biochemical activity for the protein.

Protein denaturation *is the partial or complete disorganization of a protein's characteristic three-dimensional shape as a result of disruption of its secondary, tertiary, and quaternary structural interactions.* Because the biochemical function of a protein depends on its three-dimensional shape, the result of denaturation is loss of biochemical activity. Protein denaturation does not affect the primary structure of a protein.

Although some proteins lose all of their three-dimensional structural characteristics upon denaturation (Figure 9.16), most proteins maintain some three-dimensional structure. Often, for limited denaturation changes, it is possible to find conditions under which the effects of denaturation can be reversed; this restoration process, in which the protein is "refolded," is called *renaturation.* However, for extensive denaturation changes, the process is usually irreversible.

Loss of water solubility is a frequent physical consequence of protein denaturation. The precipitation out of biochemical solution of denatured protein is called *coagulation.*

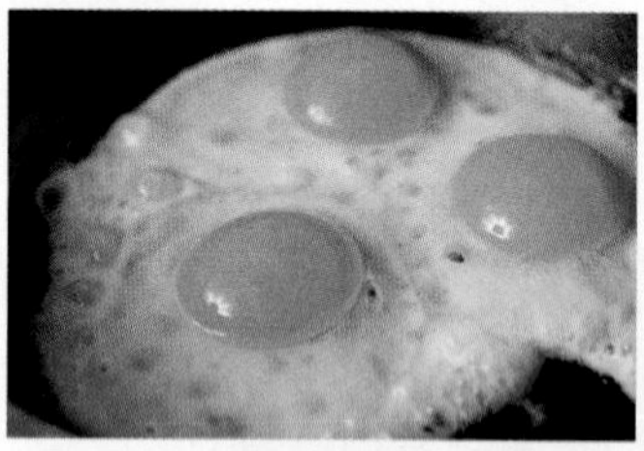

© Erica Stone/Peter Arnold, Inc./Photolibrary

Figure 9.17 Heat denatures the protein in egg white, producing a white jellylike solid. The primary structure of the protein remains intact, but all higher levels of protein structure are disrupted.

A most dramatic example of protein denaturation occurs when egg white (a concentrated solution of the protein albumin) is poured onto a hot surface. The clear albumin solution immediately changes into a white solid with a jelly-like consistency (see Figure 9.17). A similar process occurs when hamburger juices encounter a hot surface. A brown jelly-like solid forms.

When protein-containing foods are cooked, protein denaturation occurs. Such "cooked" protein is more easily digested because it is easier for digestive enzymes to "work on" denatured (unraveled) protein. Cooking foods also kills microorganisms through protein denaturation. For example, ham and bacon can harbor parasites that cause trichinosis. Cooking the ham or bacon denatures parasite protein.

In surgery, heat is often used to seal small blood vessels. This process is called *cauterization.* Small wounds can also be sealed by cauterization. Heat-induced denaturation is used in sterilizing surgical instruments and in canning foods; bacteria are destroyed when the heat denatures their protein.

Rosenfield Images, Ltd./SPL/Photo Researchers RM

Figure 9.18 Storage room for cheese; during storage, cheese "matures" as bacteria and enzymes ferment the cheese, giving it a stronger flavor.

The body temperature of a patient with fever may rise to 102°F, 103°F, or even 104°F without serious consequences. A temperature above 106°F (41°C) is extremely dangerous, as the body's enzymes begin to be inactivated at this level. Enzymes, which function as catalysts for almost all body reactions, are protein. Inactivation of enzymes through denaturation can have lethal effects on body chemistry.

The effect of ultraviolet radiation from the sun, an ionizing radiation, is similar to that of heat. Denatured skin proteins cause most of the problems associated with sunburn.

A curdy precipitate of casein, the principal protein in milk, is formed in the stomach when the hydrochloric acid of gastric juice denatures the casein. The curdling of milk that takes place when milk sours or cheese is made (Figure 9.18) results from the presence of lactic acid, a by-product of bacterial growth. Yogurt is

Table 9.5 Selected Physical and Chemical Denaturing Agents

Denaturing Agent	Mode of Action
heat	disrupts hydrogen bonds by making molecules vibrate too violently; produces coagulation, as in the frying of an egg
microwave radiation	causes violent vibrations of molecules that disrupt hydrogen bonds
ultraviolet radiation	operates very similarly to the action of heat (e.g., sunburning)
violent whipping or shaking	causes molecules in globular shapes to extend to longer lengths, which then entangle (e.g., beating egg white into meringue)
detergent	affects R-group interactions
organic solvents (e.g., ethanol, 2-propanol, acetone)	interferes with R-group interactions because these solvents also can form hydrogen bonds; quickly denatures proteins in bacteria, killing them (e.g., the disinfectant action of 70% ethanol)
strong acids and bases	disrupts hydrogen bonds and salt bridges; prolonged action leads to actual hydrolysis of peptide bonds
salts of heavy metals (e.g., salts of Hg^{2+}, Ag^{+}, Pb^{2+})	metal ions combine with —SH groups and form poisonous salts
reducing agents	reduces disulfide linkages to produce —SH groups

Disulfide bridges, which involve covalent bonds, impart considerable resistance to denaturation because they are much stronger than the noncovalent interactions otherwise present.

prepared by growing lactic-acid-producing bacteria in skim milk. The coagulated denatured protein gives yogurt its semi-solid consistency.

Serious eye damage can result from eye tissue contact with acids or bases, when irreversibly denatured and coagulated protein causes a clouded cornea. This reaction is part of the basis for the rule that students wear protective eyewear in the chemistry laboratory.

Alcohols are an important type of denaturing agent. Denaturation of bacterial protein takes place when isopropyl or ethyl alcohol is used as a disinfectant—hence the common practice of swabbing the skin with alcohol before giving an injection. Interestingly, pure isopropyl or ethyl alcohol is less effective than the commonly used 70% alcohol solution. Pure alcohol quickly denatures and coagulates the bacterial surface, thereby forming an effective barrier to further penetration by the alcohol. The 70% solution denatures more slowly and allows complete penetration to be achieved before coagulation of the surface proteins takes place.

The chemical details associated with the process of giving someone a "hair permanent," which involves protein denaturation through the use of reducing agents, are considered in the focus on relevancy feature Chemical Connections 9-B on the next page.

Table 9.5 is a listing of selected physical and chemical agents that cause protein denaturation. The effectiveness of a given denaturing agent depends on the type of protein upon which it is acting.

If fresh pineapple is added to gelatin desserts, the gelatin will not gel. Fresh pineapple contains an enzyme that facilitates the hydrolysis of gelatin peptide linkages. Canned pineapple can be added to gelatin without preventing it from gelling. The heat used in processing the canned pineapple has denatured the hydrolysis enzymes.

9.16 Protein Classification Based on Shape

Based on molecular shape, which is determined primarily by tertiary and quaternary structural features, there are three main types of proteins: *fibrous, globular,* and *membrane.* A **fibrous protein** *is a protein whose molecules have an elongated shape with one dimension much longer than the others.* Fibrous proteins tend to have simple, regular, linear structures. There is a tendency for such proteins to aggregate together to form macromolecular structures. A **globular protein** *is a protein whose*

CHEMICAL CONNECTIONS 9-B

Denaturation and Human Hair

The process used in waving hair—that is, in a hair permanent—involves reversible denaturation. Hair is protein in which many disulfide (—S—S—) linkages occur as part of its tertiary structure; 16%–18% of hair is the amino acid cysteine. It is these disulfide linkages that give hair protein its overall shape. When a permanent is administered, hair is first treated with a reducing agent (ammonium thioglycolate) that breaks the disulfide linkages in the hair, producing two sulfhydryl (—SH) groups:

$$\text{Disulfide bridges} \xrightarrow[\text{agents}]{\text{Reducing}} \text{sulfhydryl groups}$$

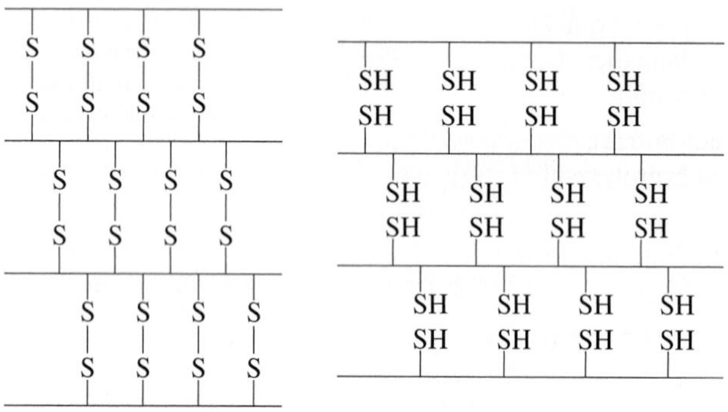

The "reduced" hair, whose tertiary structure has been disrupted, is then wound on curlers (see accompanying photo) to give it a new configuration.

Finally, the reduced and rearranged hair is treated with an oxidizing agent (potassium bromate) to form disulfide linkages at new locations within the hair:

$$\text{Sulfhydryl groups} \xrightarrow[\text{agents}]{\text{Oxidizing}} \text{re-formed disulfide bridges}$$

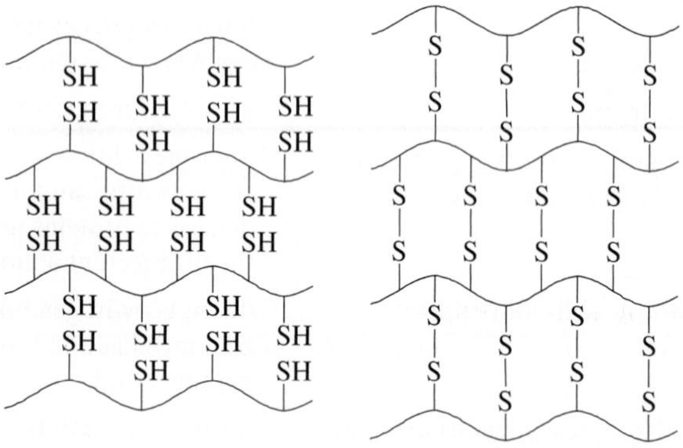

The new shape and curl of the hair are maintained by the newly formed disulfide bonds and the resulting new tertiary structure accompanying their formation. Of course, as new hair grows in, the "permanent" process has to be repeated.

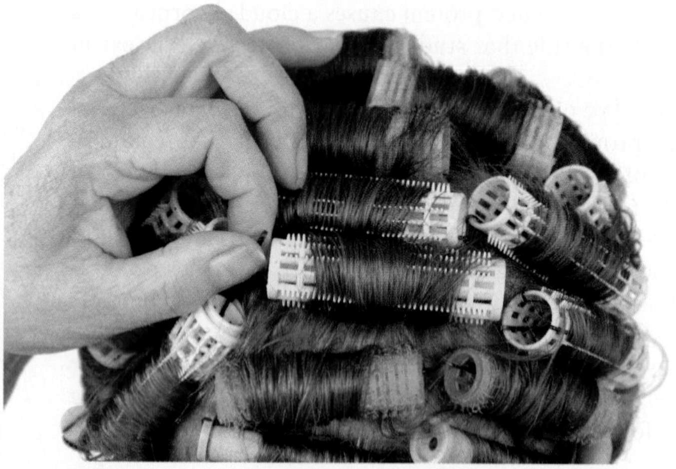

© Zenpix/Dreamstime.com

Winding hair on curlers and then treating it with an oxidizing agent, as part of receiving a hair permanent, is the stage inwhich disulfide linkages that had been previously broken are reformed.

molecules have peptide chains that are folded into spherical or globular shapes. The folding in such proteins is such that most of the amino acids with hydrophobic side chains (nonpolar R groups) are in the interior of the molecule and most of the hydrophilic side chains (polar R groups) are on the outside of the molecule. Generally, globular proteins are water-soluble substances. A **membrane protein** *is a protein that is found associated with a membrane system of a cell.* Membrane protein structure is somewhat opposite that of globular proteins, with most of the

Table 9.6 Some Common Fibrous and Globular Proteins

Name	Occurrence and Function
Fibrous proteins (insoluble)	
keratins	found in wool, feathers, hooves, silk, and fingernails
collagens	found in tendons, bone, and other connective tissue
elastins	found in blood vessels and ligaments
myosins	found in muscle tissue
fibrin	found in blood clots
Globular proteins (soluble)	
insulin	regulatory hormone for controlling glucose metabolism
myoglobin	involved in oxygen storage in muscles
hemoglobin	involved in oxygen transport in blood
transferrin	involved in iron transport in blood
immunoglobulins	involved in immune system responses

hydrophobic amino acid side chains oriented outward. Thus, such proteins tend to be water-insoluble and they usually have fewer hydrophobic amino acids than globular proteins.

Further discussion of proteins in this section is limited to fibrous and globular proteins, as membrane proteins were considered in the previous chapter in relation to cell membrane structure (Section 8.10).

Globular proteins denature more readily than fibrous proteins because of weaker secondary and tertiary attractive forces.

Table 9.6 gives examples of selected fibrous and globular proteins. Comparison of fibrous and globular proteins in terms of general properties shows the following differences.

1. Fibrous proteins are generally water-insoluble, whereas globular proteins dissolve in water. This enables globular proteins to travel through the blood and other body fluids to sites where their activity is needed.
2. Fibrous proteins usually have a single type of secondary structure, whereas globular proteins often contain several types of secondary structure.
3. Fibrous proteins generally have structural functions that provide support and external protection, whereas globular proteins are involved in metabolic chemistry, performing functions such as catalysis, transport, and regulation.
4. The number of different kinds of globular protein far exceeds the number of different kinds of fibrous protein. However, because the most abundant proteins in the human body are fibrous proteins rather than globular proteins, the total mass of fibrous proteins present exceeds the total mass of globular proteins present.

The characteristics of two fibrous proteins (α keratin and collagen) and two globular proteins (hemoglobin and myoglobin) as representatives of their types are now examined.

PhotoDisc/Getty Images

Figure 9.19 The tail feathers of a peacock contain the fibrous protein α keratin.

α Keratin

The fibrous protein *α keratin* is particularly abundant in nature, where it is found in protective coatings for organisms. It is the major protein constituent of hair, feathers (Figure 9.19), wool, fingernails and toenails, claws, scales, horns, turtle shells, quills, and hooves.

The structure of a typical α keratin, that of hair, is depicted in Figure 9.20. The individual molecules are almost wholly α helical (Figure 9.20a). Pairs of these helices twine about one another to produce a coiled coil (Figure 9.20b). In hair, two of the coiled coils then further twist together to form a protofilament (Figure 9.20c). Protofilaments then coil together in groups of four to form microfilaments (Figure 9.20d), which become the "core" unit in the structure of the α keratin of hair. These microfilaments in turn coil at even higher levels. This coiling at higher

Natural silk (silkworm silk) and spider silk (spider webs) are made of *fibroin,* a fibrous protein that exists mainly in a beta pleated sheet form. The great strength and toughness of silk fibers, which exceed those of many synthetic fibers, is related to the *close* stacking of the beta sheets. A high percentage of the amino acid residues (primary structure) in silk are either glycine (R = H) or alanine (R = CH_3). It is the smallness of these two R groups that makes the close stacking possible.

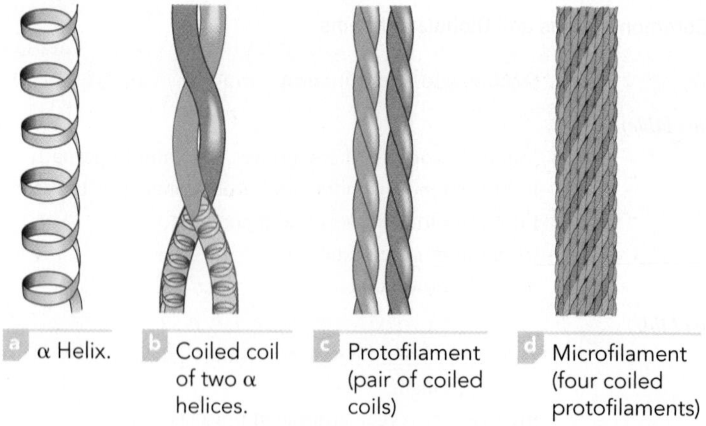

Figure 9.20 The coiled-coil structure of the fibrous protein α keratin.

and higher levels is what produces the strength associated with α-keratin-containing proteins. All levels of coiling organization are stabilized by attractive forces of the types previously considered in the discussion of generalized secondary and tertiary protein structure (Sections 9.11 and 9.12). Particularly important are *inter*coil disulfide bridges that form between cysteine residues.

Introduction of disulfide bridges within the several levels of coiling structure determines the "hardness" of an α keratin. "Hard" keratins, such as those found in horns and nails, have considerably more disulfide bridges than their softer counterparts found in hair, wool, and feathers.

Collagen

Collagen, pronounced "KAHL-uh-jen," is the most abundant protein in the human body.

Collagen, the most abundant of all proteins in humans (30% of total body protein), is a major structural material in tendons, ligaments, blood vessels, and skin; it is also the organic component of bones and teeth. Table 9.7 gives the collagen content of selected body tissues. The predominant structural feature within collagen molecules is a *triple helix* formed when three chains of amino acids wrap around each other to give a ropelike arrangement of polypeptide chains (see Figure 9.21).

Table 9.7 The Collagen Content of Selected Body Tissues

Tissue	Collagen (% dry mass)
Achilles tendon	86
aorta	12–24
bone (mineral-free)	88
cartilage	46–63
cornea	68
ligament	17
skin	72

The rich content of the amino acid proline (up to 20%) in collagen is one reason why it has a triple-helix conformation rather than the simpler α helix structure (Section 9.10). Proline amino acid residues do not fit into regular α helices because of the cyclic nature of the side chain present and its accompanying different "geometry."

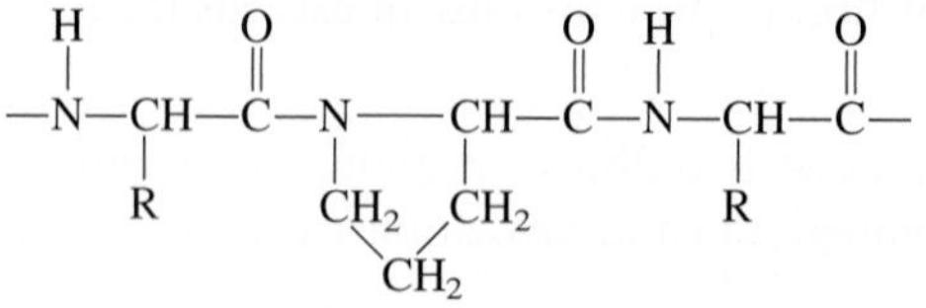

Portion of a collagen chain

Collagen molecules (triple helices) are very long, thin, and rigid. Many such molecules, lined up alongside each other, combine to make collagen fibrils. Cross-linking between helices gives the fibrils extra strength. The greater the number

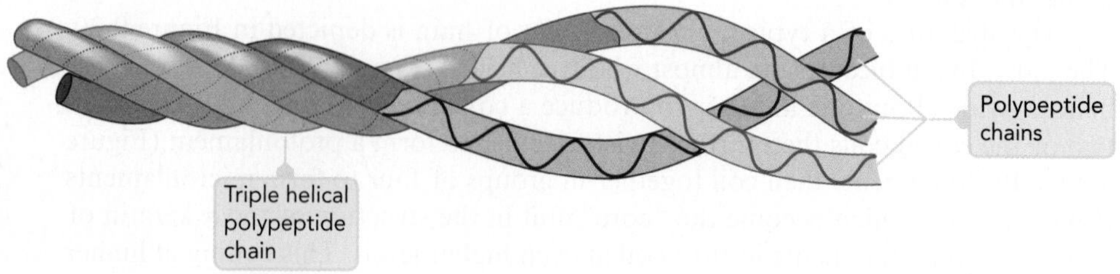

Figure 9.21 A schematic diagram emphasizing how three helical polypeptide chains intertwine to form a triple helix. The chains are partially unwound and cut away to show their structure.

of cross links, the more rigid the fibril is. The stiffening of skin and other tissues associated with aging is thought to result, at least in part, from an increasing amount of cross-linking between collagen molecules. The process of tanning, which converts animal hides to leather, involves increasing the degree of cross-linking.

Figure 9.22 shows an electron micrograph of collagen fibers.

Prof. P.M. Motta & E. Vizza/Photo Researchers

Figure 9.22 Electron micrograph of collagen fibers.

Hemoglobin

The globular protein *hemoglobin* transports oxygen from the lungs to tissue. Its tertiary structure was shown in Figure 9.14. It is a tetramer (four peptide chains) with each subunit also containing a *heme group,* the entity that binds oxygen. With four heme groups present, a hemoglobin molecule can transport four oxygen molecules at the same time.

The structure of a heme group is

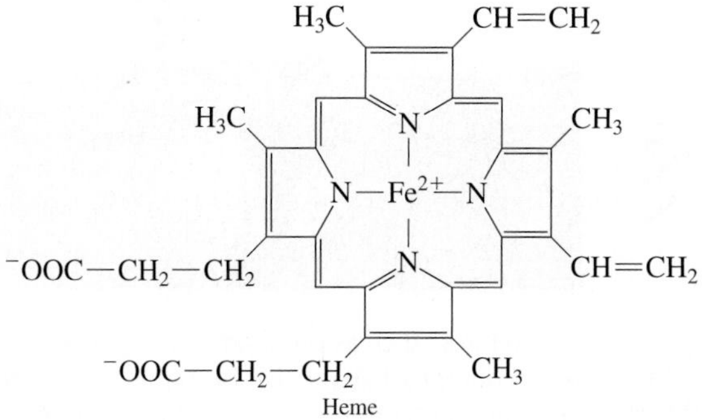

Heme

It is the iron atom at the center of the heme molecule that actually interacts with the O_2.

The hemoglobin of a fetus is slightly different in structure from adult hemoglobin. Called *fetal hemoglobin,* this hemoglobin has a greater affinity for oxygen than the mother's hemoglobin. This ensures a steady flow of oxygen to the fetus. Shortly after birth, a baby's body ceases to produce fetal hemoglobin, and its production of "adult" hemoglobin begins.

Myoglobin

The globular protein *myoglobin* functions as an oxygen storage molecule in muscles. Its tertiary structure was shown in Figure 9.13. Myoglobin is a monomer, whereas hemoglobin is a tetramer. That is, myoglobin consists of a single peptide chain and a heme unit, and hemoglobin has four peptide chains and four heme units. Thus only one O_2 molecule can be carried by a myoglobin molecule. The tertiary structure of the single peptide chain of myoglobin is almost identical to the tertiary structure of each of the subunits of hemoglobin.

Myoglobin has a higher affinity for oxygen than does hemoglobin. Thus the transfer of oxygen from hemoglobin to myoglobin occurs readily. Oxygen stored in myoglobin molecules serves as a reserve oxygen source for working muscles when their demand for oxygen exceeds that which can be supplied by hemoglobin.

The function of hemoglobin is oxygen *transfer,* and the function of myoglobin is oxygen *storage.*

The focus on relevancy feature Chemical Connections 9-C on the next page considers how the amount of myoglobin present in animal muscle tissue is related to the color of the meats (chicken, turkey, fish, etc.) that humans eat.

9.17 Protein Classification Based on Function

Proteins play crucial roles in almost all biochemical processes. The diversity of functions exhibited by proteins far exceeds that of other major types of biochemical molecules. The functional versatility of proteins stems from (1) their ability to bind small molecules specifically and strongly to themselves (2) their ability to bind other proteins, often other like proteins, to form fiber-like structures and (3) their ability to bind to, and often become integrated into, cell membranes.

The following list includes several major categories of proteins based on function. The order should not be taken as an indication of relative functional importance. The list should also be considered a selective list, with a number of functions not included because of space considerations.

CHEMICAL CONNECTIONS 9-C

Protein Structure and the Color of Meat

The meat that humans eat is composed primarily of muscle tissue. The major proteins present in such muscle tissue are *myosin* and *actin,* which lie in alternating layers and which slide past each other during muscle contraction. Contraction is temporarily maintained through interactions between these two types of proteins.

Structurally, myosin consists of a rodlike coil of two alpha helices (fibrous protein) with two globular protein heads. It is the "head portions" of myosin that interact with the actin.

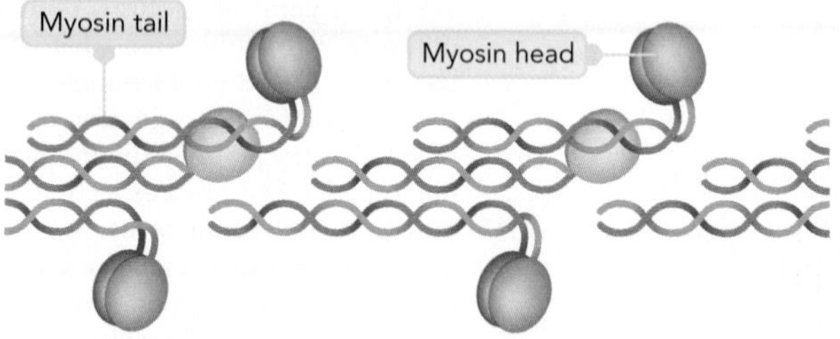

Structurally, actin has the appearance of two filaments spiraling about one another (see diagram below). Each circle in this structural diagram represents a monomeric unit of actin (called globular actin). The monomeric actin units associate to form a long polymer (called fibrous actin). Each identical monomeric actin unit is a globular protein containing many amino acid residues.

The chemical process associated with muscle contraction (interaction between myosin and actin) requires molecular oxygen. The oxygen storage protein *myoglobin* (Section 9.16) is the oxygen source. The amount of myoglobin present in a muscle is determined by how the muscle is used. Heavily used muscles require larger amounts of myoglobin than infrequently used muscles require.

The amount of myoglobin present in muscle tissue is a major determiner of the color of the muscle tissue. Myoglobin molecules have a red color when oxygenated and a purple color when deoxygenated. Thus, heavily worked muscles have a darker color than infrequently used muscles.

The different colors of meat reflect the concentration of myoglobin in the muscle tissue. In turkeys and chickens, which walk around a lot but rarely fly, the leg meat is dark, the breast meat is white. On the other hand, game birds that do fly a lot have dark breast meat (see accompanying photo). In general, game animals (which use all of their muscles regularly) tend to have darker meat than domesticated animals.

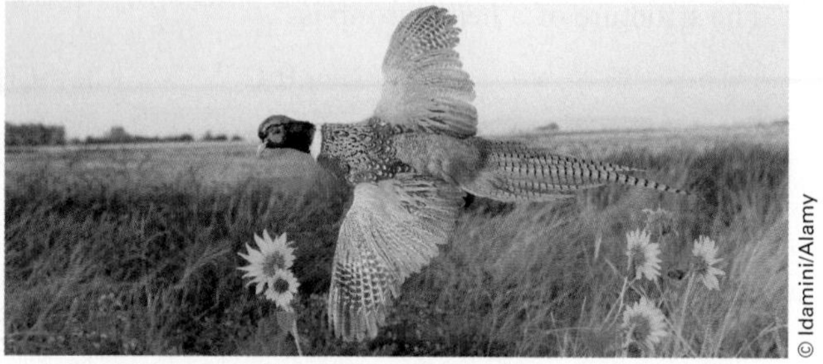

Game birds (such as pheasants, wild geese, and wild ducks), which often fly, have darker breast meat than domestic birds (such as chickens and turkeys), which rarely fly. This difference is related to the amount of myoglobin present in muscle tissue.

All land animals and birds need to support their own weight. Fish, on the other hand, are supported by water as they swim, which reduces the need for myoglobin oxygen support. Hence fish tend to have lighter flesh. Fish that spend most of their time lying at the bottom of a body of water have the lightest (whitest) flesh of all. Salmon flesh contains additional pigments that give it its characteristic "orange-pink" color.

Meat, when cooked, turns brown as the result of changes in myoglobin structure caused by the heat; the iron atom in the heme unit of myoglobin (Section 9.16) becomes oxidized. When meat is heavily salted with preservatives (NaCl, $NaNO_2$, or the like), as in the preparation of ham, the myoglobin picks up nitrite ions, and its color changes to pink.

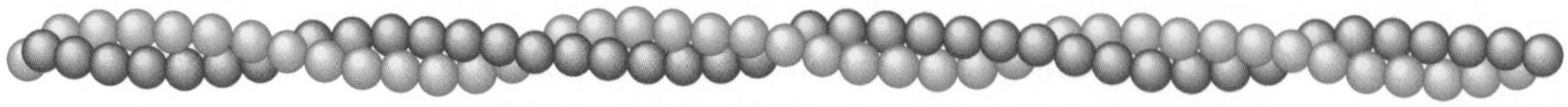

1. **Catalytic proteins.** Proteins are probably best known for their role as catalysts. Proteins with the role of biochemical catalyst are called *enzymes.* An entire chapter in this text, Chapter 10, is devoted to this most important role for proteins. Enzymes participate in almost all of the metabolic reactions that occur in cells. The chemistry of human genetics, discussed in detail in Chapter 11, is very dependent on the presence of enzymes.
2. **Defense proteins.** These proteins, also called *immunoglobulins* or *antibodies* (Section 9.18), are central to the functioning of the body's immune system. They bind to foreign substances, such as bacteria and viruses, to help combat invasion of the body by foreign particles.
3. **Transport proteins.** These proteins bind to particular small biomolecules and transport them to other locations in the body and then release the small molecules as needed at the destination location. The most well-known example of

a transport protein is *hemoglobin* (Section 9.13), which carries oxygen from the lungs to other organs and tissues. Another transport protein is *transferrin,* which carries iron from the liver to the bone marrow. *High-* and *low-density lipoproteins* (Section 9.18) are carriers of cholesterol in the bloodstream.

4. **Messenger proteins.** These proteins transmit signals to coordinate biochemical processes between different cells, tissues, and organs. A number of hormones (Section 7.12) that regulate body processes are messenger proteins, including *insulin* and *glucagon* (Section 13.9). *Human growth hormone* is another example of a messenger protein.
5. **Contractile proteins.** These proteins are necessary for all forms of movement. Muscles are composed of filament-like contractile proteins that, in response to nerve stimuli, undergo conformation changes that involve contraction and extension. *Actin* and *myosin* (Section 9.16) are examples of such proteins. Human reproduction depends on the movement of sperm. Sperm can "swim" because of long flagella made up of contractile proteins.
6. **Structural proteins.** These proteins confer stiffness and rigidity to otherwise fluid-like biochemical systems. *Collagen* (Section 9.16) is a component of cartilage, and *α keratin* (Section 9.16) gives mechanical strength as well as protective covering to hair, fingernails, feathers, hooves, and some animal shells.
7. **Transmembrane proteins.** These proteins, which span a cell membrane (Section 7.10), help control the movement of small molecules and ions through the cell membrane. Many such proteins have channels through which molecules can enter and exit a cell. Such protein channels are very selective, often allowing passage of just one type of molecule or ion.
8. **Storage proteins.** These proteins bind (and store) small molecules for future use. During degradation of hemoglobin (Section 15.7) the iron atoms present are released and become part of *ferritin,* an iron-storage protein, which saves the iron for use in the biosynthesis of new hemoglobin molecules. *Myoglobin* (Section 9.16) is an oxygen-storage protein present in muscle; the oxygen so stored is a reserve oxygen source for working muscle.
9. **Regulatory proteins.** These proteins are often found "embedded" in the exterior surface of cell membranes. They act as sites at which messenger molecules, including messenger proteins such as insulin, can bind and thereby initiate the effect that the messenger "carries." Regulatory proteins are often the molecules that bind to enzymes (catalytic proteins), thereby turning them "on" and "off" and thus controlling enzymatic action (Section 10.8).
10. **Nutrient proteins.** These proteins are particularly important in the early stages of life, from embryo to infant. *Casein,* found in milk, and *ovalbumin,* found in egg white, are two examples of such proteins. The role of milk in nature is to nourish and provide immunological protection for mammalian young. Three-fourths of the protein in milk is casein. More than 50% of the protein in egg white is *ovalbumin.*
11. **Buffer proteins.** These proteins are part of the system by which the acid-base balance within body fluids is maintained. Within the blood, the protein *hemoglobin* has a buffering role in addition to being an oxygen carrier. Transmembrane proteins regulate the movement of ions in and out of cells, ensuring that ion concentrations are those needed for correct acidity/alkalinity.
12. **Fluid-balance proteins.** These proteins help maintain fluid balance between blood and surrounding tissue. Two well-known fluid-balance proteins, found in the capillary beds of the circulatory system, are *albumin* and *globulin.* When increased blood pressure generated by a pumping heart forces water and nutrients out of the capillaries, these proteins remain behind (since they are too big to cross cellular membranes). As their concentration increases (due to less fluid being present), osmotic pressure "forces" draw water back into the capillaries, which is necessary for fluid balance to be maintained.

Though not exhaustive, this list of protein functions is long. Its length draws attention to the diversity of functions for proteins within the human body, as well as to the diversity of protein types.

9.18 Glycoproteins

A glycoprotein is a protein that contains carbohydrates or carbohydrate derivatives in addition to amino acids (Section 7.18). The carbohydrate content of glycoproteins is variable (from a few percent up to 85%), but it is fixed for any specific glycoprotein.

Glycoproteins include a number of very important substances; two of these, collagen and immunoglobulins, are considered in this section. Many of the proteins in cell membranes (lipid bilayers; see Section 8.10) are actually glycoproteins. The blood group markers of the ABO system (see Chemical Connections 7-A on page 285) are also glycoproteins in which the carbohydrate content can reach 85%.

Collagen

Nonstandard amino acids consist of amino acid residues that have been chemically modified after their incorporation into a protein (as is the case with 4-hydroxyproline and 5-hydroxylysine) and amino acids that occur in living organisms but are not found in proteins.

The fibrous protein collagen, whose structure was first considered in Section 9.16, qualifies as a *glycoprotein* because carbohydrate units are present in its structure. This structural feature of collagen, not considered previously, involves the presence of the *nonstandard* amino acids 4-hydroxyproline (5%) and 5-hydroxylysine (1%)—derivatives of the standard amino acids proline and lysine (Table 9.1).

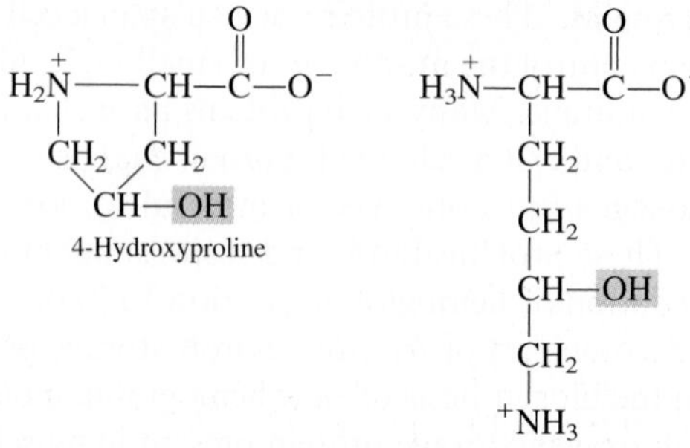

4-Hydroxyproline

5-Hydroxylysine

A primary biochemical function of vitamin C involves the hydroxylation of proline and lysine during collagen formation. These hydroxylation processes require the enzymes *proline hydroxylase* and *lysine hydroxylase.* These enzymes can function only in the presence of vitamin C.

The presence of carbohydrate units (mostly glucose, galactose, and their disaccharides) attached by glycosidic linkages (Section 7.13) to collagen at its 5-hydroxylysine residues causes collagen to be classified as a *glycoprotein.* The function of the carbohydrate groups in collagen is related to cross-linking; they direct the assembly of collagen triple helices into more complex aggregations called *collagen fibrils.*

When collagen is boiled in water, under basic conditions, it is converted to the water-soluble protein gelatin. This process involves both denaturation (Section 9.15) and hydrolysis (Section 9.14). Heat acts as a denaturant, causing rupture of the hydrogen bonds supporting collagen's triple-helix structure. Regions in the amino acid chains where proline and hydroxyproline concentrations are high are particularly susceptible to hydrolysis, which breaks up the polypeptide chains. Meats become more tender when cooked because of the conversion of some collagen to gelatin. Tougher cuts of meat (more cross-linking), such as stew meat, need longer cooking times.

Immunoglobulins

Immunoglobulins are among the most important and interesting of the soluble proteins in the human body. An **immunoglobulin** *is a glycoprotein produced by an organism as a protective response to the invasion of microorganisms or foreign molecules.* Different classes of immunoglobulins, identified by differing carbohydrate content and molecular mass, exist.

Immunoglobulins serve as *antibodies* to combat invasion of the body by *antigens.* An **antigen** *is a foreign substance, such as a bacterium or virus, that invades the human body.* An **antibody** *is a biochemical molecule that counteracts a specific antigen.* The immune system of the human body has the capability to produce immunoglobulins that respond to several million different antigens.

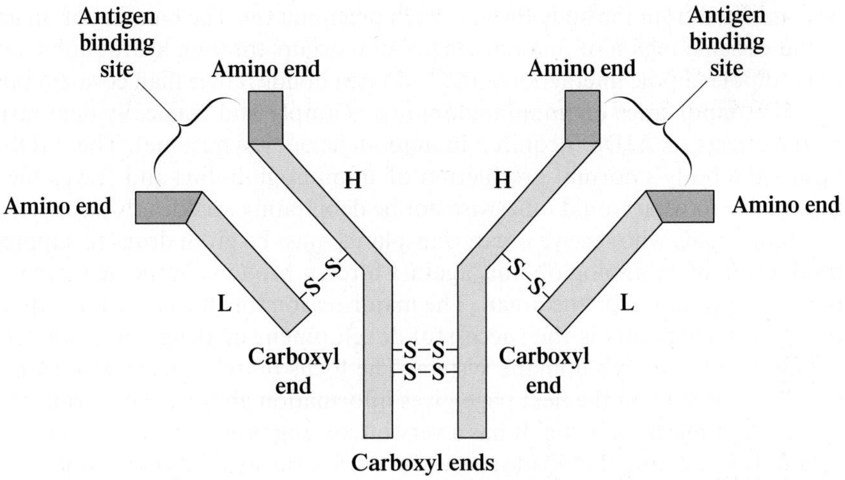

Figure 9.23 This schematic diagram shows the structure of an immunoglobulin. Two heavy (H) polypeptide chains and two light (L) polypeptide chains are cross-linked by disulfide bridges. The purple areas are the constant amino acid regions, and the areas shown in red are the variable amino acid regions of each chain. Carbohydrate molecules attached to the heavy chains aid in determining the destinations of immunoglobulins in the tissues.

All types of immunoglobulin molecules have much the same basic structure, which includes the following features:

1. Four polypeptide chains are present: two identical heavy (H) chains and two identical light (L) chains.
2. The H chains, which usually contain 400–500 amino acid residues, are approximately twice as long as the L chains.
3. Both the H and L chains have constant and variable regions. The constant regions have the same amino acid sequence from immunoglobulin to immunoglobulin, and the variable regions have a different amino acid sequence in each immunoglobulin.
4. The carbohydrate content of various immunoglobulins varies from 1% to 12% by mass.
5. The secondary and tertiary structures are similar for all immunoglobulins. They involve a Y-shaped conformation (Figure 9.23) with disulfide linkages between H and L chains stabilizing the structure.

The interaction of an immunoglobulin molecule with an antigen occurs at the "tips" (upper-most part) of the Y structure. These tips are the variable-composition region of the immunoglobulin structure. It is here that the antigen binds specifically, and it is here that the amino acid sequence differs from one immunoglobulin to another.

Each immunoglobulin has two identical active sites and can thus bind to two molecules of the antigen it is "designed for." The action of many such immunoglobulins of a given type in concert with each other creates an *antigen–antibody complex* that precipitates from solution (Figure 9.24). Eventually, an invading antigen can

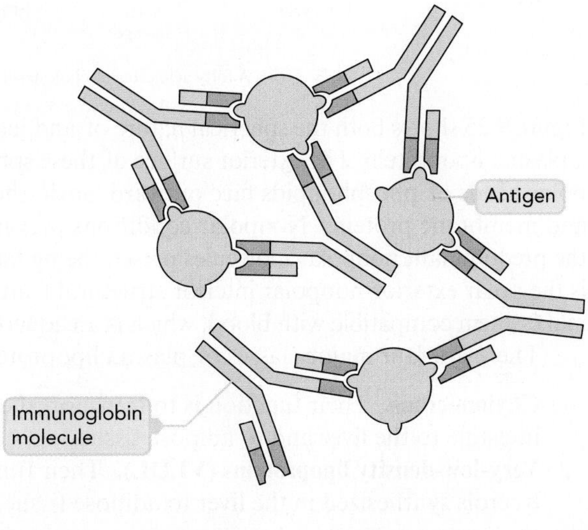

Figure 9.24 In this immunoglobulin–antigen complex, note that more than one immunoglobulin molecule can attach itself to a given antigen. Also, any given immunoglobulin has only two sites where an antigen can bind.

be eliminated from the body through such precipitation. The bonding of an antigen to the variable region of an immunoglobulin occurs through hydrophobic interactions, dipole–dipole interactions, and hydrogen bonds rather than covalent bonds.

The importance of immunoglobulins is amply and tragically demonstrated by the effects of AIDS (acquired immunodeficiency syndrome). The AIDS virus upsets the body's normal production of immunoglobulins and leaves the body susceptible to what would otherwise not be debilitating and deadly infections.

Individuals who receive organ transplants must be given drugs to suppress the production of immunoglobulins against foreign proteins in the new organ, thus preventing rejection of the organ. The major reason for the increasing importance of organ transplants is the successful development of drugs that can properly manipulate the body's immune system. The focus on relevancy feature Chemical Connections 9-D on the next page gives information about cyclosporine, the best known "antirejection" drug. It has a very interesting origin, structure, and mode of action. It is the drug that made successful organ transplantation possible.

Many reasons exist for a mother to breast-feed a newborn infant. One of the most important involves the transmission of immunoglobulins. During the first few days of lactation, immunoglobulins from the mother's blood are passed to the newborn via "milk," which gives the infant protection from the pathogens to which the mother has developed immunity. The focus on relevancy feature Chemical Connections 9-E on page 405 considers further the protein chemistry associated with "mother's milk."

9.19 Lipoproteins

A **lipoprotein** *is a conjugated protein that contains lipids in addition to amino acids.* The major function of such proteins is to help suspend lipids and transport them through the bloodstream. Lipids, in general, are insoluble in blood (an aqueous medium) because of their nonpolar nature (Section 8.1).

A **plasma lipoprotein** *is a lipoprotein that is involved in the transport system for lipids in the bloodstream.* These proteins have a spherical structure that involves a central core of lipid material (triacylglycerols and cholesterol esters) surrounded by a shell (membrane structure) of phospholipids, cholesterol, and proteins. In the blood, cholesterol exists primarily in the form of cholesterol esters formed from the esterification of cholesterol's hydroxyl group with a fatty acid.

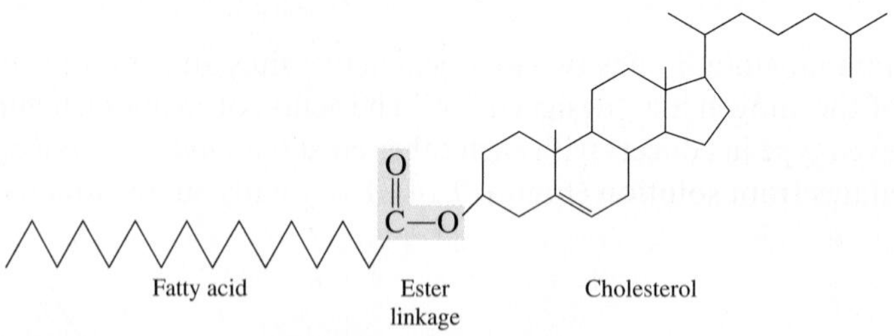

A fatty acid ester of cholesterol

Figure 9.25 shows both the spherical nature of and individual components present in a plasma lipoprotein. The exterior surface of these spherical lipoproteins is polar; the polar heads of phospholipids face outward, as do the polar portions of cholesterol and membrane proteins. Nonpolar conditions prevail within these structures, with the predominant nonpolar molecules present being fatty acid esters of cholesterol. It is the polar exterior-nonpolar interior structural feature of plasma lipoproteins that makes them compatible with blood, which is an aqueous-based (polar) medium.

There are four major classes of plasma lipoproteins:

1. **Chylomicrons.** Their function is to transport dietary triacylglycerols from the intestine to the liver and to adipose tissue.
2. **Very-low-density lipoproteins (VLDL).** Their function is to transport triacylglycerols synthesized in the liver to adipose tissue.

CHEMICAL CONNECTIONS 9-D

Cyclosporine: An Antirejection Drug

The survival rate for patients undergoing human organ transplants such as heart, liver, or kidney replacements was relatively low until the mid-1980s when it dramatically increased, an increase that has since been maintained. Many patients who received transplanted organs 25 years ago are leading normal lives today (see accompanying photo). Increased transplant success is directly related to the development of new drugs to prevent transplant rejection by a patient's own immune system. The key new drug introduced, which remains an important drug today, was the immunosuppressive agent (antirejection drug) cyclosporine. Cyclosporine is a compound that was initially isolated from a soil fungus; it is a fungal peptide.

Structurally, cyclosporine is a cyclic, 11-amino-acid-residue peptide (see accompanying structural diagram). The structure is simple, but yet complex.

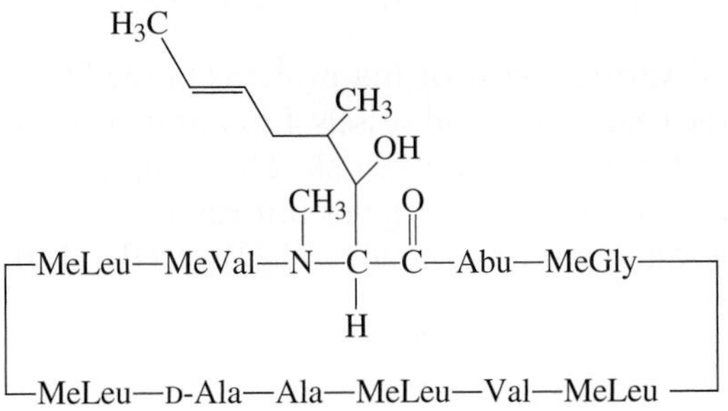

Cyclosporin

Structural characteristics of cyclosporine include:

1. Nine of the eleven amino acid residues present are those of standard amino acids with short hydrocarbon side chains—Gly, Ala, Val, and Leu. These are the simplest of the standard amino acids with respect to side chains.
2. The tenth amino acid residue present, shown as Abu in the structural diagram, also has a simple side chain; it is an ethyl group. The designation Abu for this nonstandard amino acid comes from the IUPAC name for the acid, which is 2-aminobutyric acid.
3. The eleventh amino acid residue present, which is the key to cyclosporine's pharmacological activity, had never been previously encountered in a natural product. Its side chain is a 7-carbon branched, unsaturated, hydroxylated entity. This side chain is shown in detail in the structural diagram for cyclosporine.
4. Seven of the amino acid residues have their nitrogen atoms methylated; that is, a methyl group has replaced the nitrogen's hydrogen atom. Methylation is denoted in the structural diagram using the notation MeVal, MeLeu, etc.
5. An additional rarely encountered structural feature is present. One amino acid residue has a D-configuration rather than the normal L-configuration for amino acids. One of the Ala residues present is D-Ala. Interestly, next to it in the structure is an L-Ala residue

The fat solubility of cyclosporine allows it to cross cell membranes readily and to be widely distributed in the body. It is administered either intravenously or orally. Because of its low water solubility, the drug is supplied in olive oil for oral administration. Cyclosporine has a narrow therapeutic index. When the blood concentration of cyclosporine is too low, inadequate immunosuppression occurs. On the other hand, a high cyclosporine concentration can lead to kidney problems.

Before the cyclosporine-era, available antirejection drugs weakened a patient's *entire* immune system, with the result that patients could not survive severe infections. Cyclosporine suppresses only part of the body's immune system (T-cells), the part associated with rejection of transplanted tissue. The bulk of the patient's immune system continues to function normally and remains able to fight general infections.

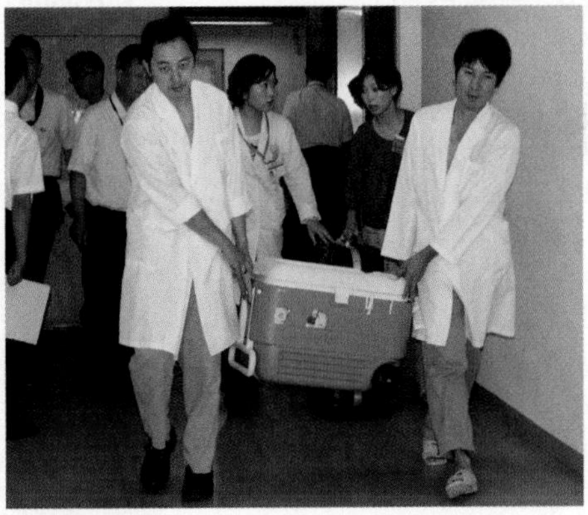

Kyodo/Landov

A donated organ (in ice) being carried to an operating room where a donor recipient awaits its arrival.

3. **Low-density lipoproteins (LDL).** Their function is to transport cholesterol synthesized in the liver to cells throughout the body.
4. **High-density lipoproteins (HDL).** Their function is to collect excess cholesterol from body tissues and transport it back to the liver for degradation to bile acids.

The density of a lipoprotein is related to the fractions of protein and lipid material present. The greater the amount of protein in the lipoprotein, the higher the density. Figure 9.26 characterizes the major plasma lipoprotein types in terms of density as well as lipid–protein composition.

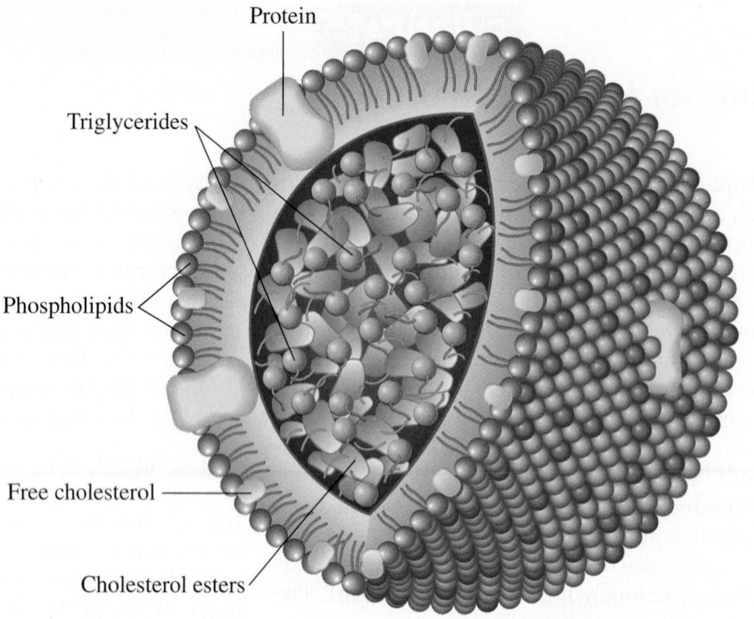

Figure 9.25 A model for the structure of a plasma lipoprotein.

The presence or absence of various types of lipoproteins in the blood have implications for the health of the heart and blood vessels. Lipoprotein levels in the blood are now used as an indicator of heart disease risk. The focus on relevancy feature Chemical Connections 9-F on page 406 gives information concerning the chemical interrelationships among the substances VLDL, LDL, HDL, and cholesterol in blood chemistry.

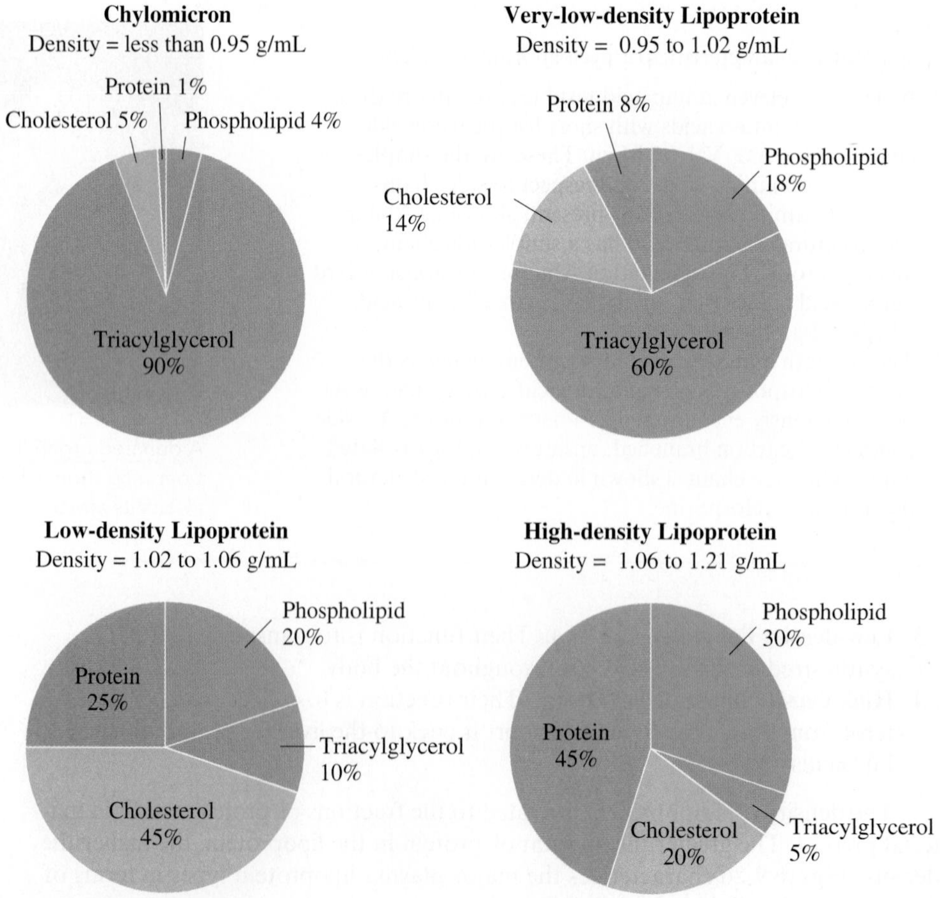

Figure 9.26 Densities and relative amounts of proteins, phospholipids, cholesterol, and triacylglycerols present in the four major classes of plasma lipoproteins.

CHEMICAL CONNECTIONS 9-E

Colostrum: Immunoglobulins and Much More

Two types of milk are produced by a lactating mother as she feeds a newborn baby. They are:

1. **Colostrum:** During the first two to three days of lactation, a mother's breasts produce a *foremilk* called colostrum. This substance, secreted in low volume amounts, is a thick yellow secretion that is rich in maternal antibodies (immunoglobulins) and carbohydrates and low in fat.
2. **Mature milk:** Within a week, mature milk has replaced colostrum production. Mature milk is much whiter in color than colostrum and has a watery appearance. It contains increased amounts of fat, carbohydrates, and calories, but less protein, particularly immunoglobulins. Some immunologically active substances are, however, still present. The high water content of mature milk is sufficient to meet an infant's hydration needs when he or she is exclusively breast-fed.

The composition of mature human milk and cow's milk differ significantly, as is shown in the accompanying table.

Comparison of Mature Human Milk and Cow's Milk (g/100 mL)

Nutrient	Human Milk	Cow's Milk
Total protein	1.1	3.1
Total fat	4.5	3.8
Total lactose	7.1	4.7

Lactose is the principle carbohydrate in milk (Section 7.13). Human milk obtained by a nursing infant (see accompanying photo) contains 7%–8% lactose, almost double the 4%–5% lactose found in cow's milk.

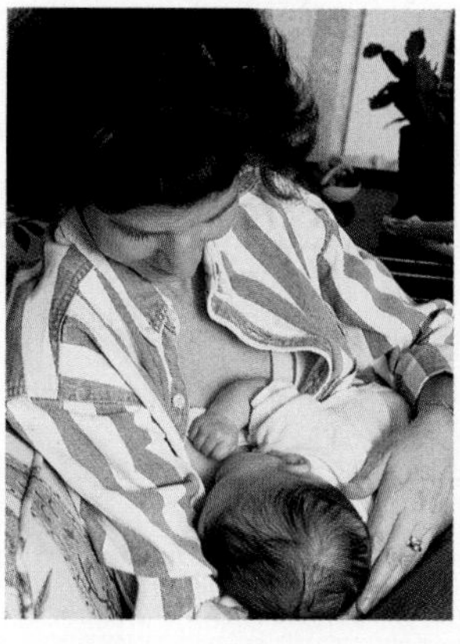

Human milk contains more lactose and fat, but less protein, than does cow's milk.

Many reasons exist for a mother to breast-feed a newborn. One of the most important is the immunoglobulins present in colostrum and, to a lesser extent, in mature milk. These immunoglobulins come from the mother's bloodstream and represent protection from those substances to which the mother has developed immunity. These substances in the mother's environment are precisely the entities the infant needs protection from. Infant formula is usually nutritionally equivalent to breast milk, but it is definitely not immunologically equivalent.

It is particularly important that the mother's immunological protection be transmitted to the infant's body during the first week of life. It is at this time, and no other time in life, that immunoglobulins and other whole proteins can pass unaltered through the infant's immature intestinal tract directly into the bloodstream.

A major antibody present in colostrum is secretory immunoglobulin A (IgA). IgA gives the infant protection in the locations where attack from germs is most likely to occur, namely, the mucous membranes in the throat, lungs, and intestines. Prior to birth, the baby received the benefit of immunoglobulin G (IgG) through the placenta. IgG is present in the baby's circulatory system. It is supplied to the baby via the placenta.

Newborns have *very small* digestive systems, and colostrum delivers its ingredients in a very concentrated low-volume form. Colostrum volumes obtained by breast-feeding are small—teaspoon amounts rather than several-ounce amounts. This is adequate for a newborn. The stomach capacity of a newborn is 0.20 oz.–0.25 oz. (one teaspoon) at day one, 0.75 oz.–1.0 oz. (one tablespoon) at day three, and 1.50 oz.–2.0 oz. (the volume of a ping-pong ball) at day seven. The small-volume amounts of colostrum produced are the right amounts needed for the infant's initial feedings.

Proteins other than those associated with immunological responses are also present in colostrum. Two of these additional proteins are *lactalbumin* and *lactoferrin*. These additional proteins are mostly synthesized within breast tissue, whereas immunoproteins come from the mother's bloodstream.

Lactalbumin is an easily digested protein that is less stressful on the infant's immature digestive system than are other proteins such as casein (the principle protein in cow's milk). Lactoferrin increases the rate of iron absorption by the infant. Due to its presence, more iron is absorbed from human milk than from infant formula or cow's milk even though human milk contains less iron. Lactoferrin also offers protection against bacteria and other pathogens by binding the iron, thus making it unavailable to these agents, which require it for growth.

Colostrum also helps the infant's developing digestive system in other ways. It has a mild laxative effect, which encourages the passing of the baby's first stool. It also reduces the permeability of the intestinal tract by giving it a protective coating that prevents foreign substances from penetrating it and entering the bloodstream.

CHEMICAL CONNECTIONS 9-F

Lipoproteins and Heart Disease Risk

In the United States, heart disease is the number-one health problem for both men and women. There are many risk factors associated with heart disease, some of which are controllable and some of which are not. You can't change your age, race, or family history (genetics). But you can control (manage) factors such as your weight, whether or not you smoke, and your cholesterol level.

Relative to high cholesterol as a risk factor for heart disease, studies now indicate that focusing just on the numerical value of cholesterol present in the blood is an oversimplification of the problem. Consideration of the "form" in which the cholesterol is present is as important as the total amount. Much of the research concerning the "form" of cholesterol present relates to the lipoproteins LDL and HDL present in the blood.

Both LDLs and HDLs are involved in cholesterol transport (Section 9.19). LDLs carry approximately 80% of the cholesterol and HDLs the remainder. Of significance, LDLs and HDLs carry cholesterol for different purposes. LDLs carry cholesterol to cells for their use, whereas HDLs carry excess cholesterol away from cells to the liver for processing and excretion from the body.

Studies show that LDL levels correlate *directly* with heart disease, whereas HDL levels correlate *inversely* with heart disease risk. Thus HDL is sometimes referred to as "good" cholesterol (*H*DL = *H*ealthy) and LDL as "bad" cholesterol (*L*DL = *L*ess healthy).

The goal of dietary measures to slow the advance of atherosclerosis is to reduce LDL cholesterol levels. Reduction in the dietary intake of saturated fat appears to be a key action (Section 8.5).

High HDL levels are desirable because they give the body an efficient means of removing excess cholesterol. Low HDL levels can result in excess cholesterol depositing within the circulatory system.

In general, women have higher HDL levels than men—an average of 55 mg per 100 mL of blood serum versus 45 mg per 100 mL. This may explain in part why proportionately fewer women have heart attacks than men. Nonsmokers have uniformly higher HDL levels than smokers. Exercise on a regular basis tends to increase HDL levels. This discovery has increased the popularity of walking and running exercise

Moderate exercise done on a regular basis tends to increase HDL levels.

(see accompanying photo). Genetics also plays a role in establishing HDL as well as other lipoprotein concentrations in the blood.

A person's *total* blood cholesterol level does not necessarily correlate with that individual's real risk for heart and blood vessel disease. A better measure is the *cholesterol ratio,* which is defined as

$$\text{Cholesterol ratio} = \frac{\text{total cholesterol}}{\text{HDL cholesterol}}$$

For example, if a person's total cholesterol is 200 and his or her HDL is 45, then the cholesterol ratio would be 4.4. According to the accompanying guidelines for interpreting cholesterol ratio values, this indicates an average risk for heart disease.

What Cholesterol Ratio Means

Ratio	Heart Disease Risk
6.0	high
5.0	above average
4.5	average
4.0	below average
3.0	low

Concepts to Remember

OWL Sign in at www.cengage.com/owl to view tutorials and simulations, develop problem-solving skills, and complete online homework assigned by your professor.

Protein. A protein is a polymer in which the monomer units are amino acids (Section 9.1).

α-Amino acid. An α-amino acid is an amino acid in which the amino group and the carboxyl group are both attached to the α-carbon atom (Section 9.2).

Standard amino acid. A standard amino acid is one of the 20 α-amino acids that are normally present in protein (Section 9.2).

Amino acid classifications. Amino acids are classified as nonpolar, polar neutral, polar basic, or polar acidic depending on the nature of the side chain (R group) present (Section 9.2).

Essential amino acid. A standard amino acid needed for protein synthesis that must be obtained from dietary sources because the human body cannot synthesize it in adequate amounts from other substances (Section 9.3).

Chirality of amino acids. Amino acids found in proteins are always left-handed (L isomer) (Section 9.4).

Zwitterion. A zwitterion is a molecule that has a positive charge on one atom and a negative charge on another atom. In neutral solution and in the solid state, amino acids exist as zwitterions. For amino acids in solution, the isoelectric point is the pH at which the amino acid exists primarily in its zwitterion form (Section 9.5).

Disulfide bond formation. The amino acid cysteine readily dimerizes; the —SH groups of two cysteine molecules interact to form a covalent disulfide bond (Section 9.6).

Peptide bond. A peptide bond is an amide bond involving the carboxyl group of one amino acid and the amino group of another amino acid. In a protein, the amino acids are linked to each other through peptide bonds (Section 9.7).

Biochemically important peptides. Numerous small peptides are biochemically active. Their functions include hormonal action, neurotransmission functions, and antioxidant activity (Section 9.8).

General characteristics of proteins. Proteins are peptides with at least 40 amino acid residues. A single peptide chain is present in a monomeric protein, and two or more peptide chains are present in a multimeric protein. A simple protein contains only one or more peptide chains. A conjugated protein contains one or more additional chemical components, called prosthetic groups, in addition to peptide chains (Section 9.9).

Primary protein structure. The primary structure of a protein is the sequence of amino acids present in the peptide chain or chains of the protein (Section 9.10).

Secondary protein structure. The secondary structure of a protein is the arrangement in space of the backbone portion of the protein. The two major types of protein secondary structure are the α helix and the β pleated sheet (Section 9.11).

Tertiary protein structure. The tertiary structure of a protein is the overall three-dimensional shape that results from the attractive forces among amino acid side chains (R groups) (Section 9.12).

Quaternary protein structure. The quaternary structure of a protein involves the associations among the peptide chains present in a multimeric protein (Section 9.13).

Protein hydrolysis. Protein hydrolysis is a chemical reaction in which peptide bonds within a protein are broken through reaction with water. Complete hydrolysis produces free amino acids (Section 9.14).

Protein denaturation. Protein denaturation is the partial or complete disorganization of a protein's characteristic three-dimensional shape as a result of disruption of its secondary, tertiary, and quaternary structural interactions (Section 9.15).

Fibrous and globular proteins. Fibrous proteins are generally insoluble in water and have a long, thin, fibrous shape. α Keratin and collagen are important fibrous proteins. Globular proteins are generally soluble in water and have a roughly spherical or globular overall shape. Hemoglobin and myoglobin are important globular proteins (Section 9.16).

Glycoproteins. Glycoproteins are conjugated proteins that contain carbohydrates or carbohydrate derivatives in addition to amino acids. Collagen and immunoglobulins are important glycoproteins (Section 9.18).

Lipoproteins. Lipoproteins are conjugated proteins that are composed of both lipids and amino acids. Lipoproteins are classified on the basis of their density (Section 9.19).

Exercises and Problems

OWL Interactive versions of these problems may be assigned in OWL.

Exercises and problems are arranged in matched pairs with the two members of a pair addressing the same concept(s). The answer to the odd-numbered member of a pair is given at the back of the book. Problems denoted with a ▲ involve concepts found not only in the section under consideration but also concepts found in one or more earlier sections of the chapter. Problems denoted with a ● cover concepts found in a Chemical Connections feature box.

Characteristics of Proteins (Section 9.1)

9.1 What is the general name for the building blocks (monomers) from which a protein (polymer) is made?

9.2 What element is always present in proteins that is seldom present in carbohydrates and lipids?

9.3 What percent of a cell's overall mass is accounted for by proteins?

9.4 Approximately how many different proteins are present in a typical human cell?

Amino Acid Structural Characteristics (Section 9.2)

9.5 Which of the following structures represent α-amino acids?

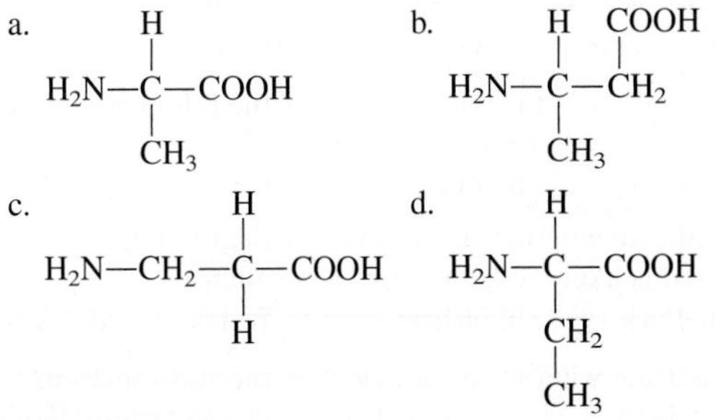

9.6 What is the significance of the prefix α in the designation α-amino acid?

9.7 What is the major structural difference among the various standard amino acids?

9.8 On the basis of polarity, what are the four types of side chains found in the standard amino acids?

9.9 With the help of Table 9.1, determine which of the standard amino acids have a side chain with the following characteristics.

a. Contains an aromatic group
b. Contains the element sulfur
c. Contains a carboxyl group
d. Contains a hydroxyl group

9.10 With the help of Table 9.1, determine which of the standard amino acids have a side chain with the following characteristics.
a. Contains only carbon and hydrogen
b. Contains an amino group
c. Contains an amide group
d. Contains more than four carbon atoms

9.11 What is the distinguishing characteristic of a polar basic amino acid?

9.12 What is the distinguishing characteristic of a polar acidic amino acid?

9.13 In what way is the structure of the amino acid proline different from that of the other 19 standard amino acids?

9.14 Which two of the standard amino acids are constitutional isomers?

9.15 What amino acids do these abbreviations stand for?
a. Ala b. Leu c. Met d. Trp

9.16 What amino acids do these abbreviations stand for?
a. Asp b. Cys c. Phe d. Val

9.17 Which four standard amino acids have three-letter abbreviations that are not the first three letters of their common names?

9.18 What are the three-letter abbreviations for the three polar basic amino acids?

9.19 Classify each of the following amino acids as nonpolar, polar neutral, polar acidic, or polar basic.
a. Asn b. Glu c. Pro d. Ser

9.20 Classify each of the following amino acids as nonpolar, polar neutral, polar acidic, or polar basic.
a. Gly b. Thr c. Tyr d. His

Essential Amino Acids (Section 9.3)

9.21 Indicate whether or not each of the following amino acids is classified as an *essential* amino acid.
a. Valine b. Isoleucine c. Trp d. Pro

9.22 Indicate whether or not each of the following amino acids is classified as an *essential* amino acid.
a. Lysine b. Glycine c. Glu d. Met

9.23 Indicate whether or not each of the following types of food is likely to be a *complete* protein.
a. Egg b. Oats c. Corn d. Soy

9.24 Indicate whether or not each of the following types of food is likely to be a *complete* protein.
a. Pork b. Wheat c. Beans d. Gelatin

9.25 Indicate whether or not each of the substances in Problem 9.23 would likely contain one or more *limiting* amino acids.

9.26 Indicate whether or not each of the substances in Problem 9.24 would likely contain one or more *limiting* amino acids.

9.27 Indicate whether or not the designation *complementary proteins* applies to each of the following pairs of dietary substances.
a. Soy and rice b. Egg and milk
c. Beef and oats d. Rice and corn

9.28 Indicate whether or not the designation *complementary proteins* applies to each of the following pairs of dietary substances.
a. Gelatin and wheat b. Oats and corn
c. Salmon and trout d. Beans and rice

9.29 Identify the *limiting* amino acid(s) present in each of the following protein sources. If none are present, indicate that such is the case.
a. Wheat b. Beans c. Soy d. Peas

9.30 Identify the *limiting* amino acid(s) present in each of the following protein sources. If none are present, indicate that such is the case.
a. Oats b. Rice c. Corn d. Gelatin

Chirality and Amino Acids (Section 9.4)

9.31 To which family of mirror-image isomers do nearly all naturally occurring amino acids belong?

9.32 In what way is the structure of glycine different from that of the other 19 common amino acids?

9.33 Draw Fischer projection formulas for the following amino acids.
a. L-Serine b. D-Serine
c. D-Alanine d. L-Leucine

9.34 Draw Fischer projection formulas for the following amino acids.
a. L-Cysteine b. D-Cysteine
c. D-Alanine d. L-Valine

9.35 Draw a structural formula for the amino acid threonine, and circle each of the four different groups attached to the chiral center present in the amino acid.

9.36 Draw a structural formula for the amino acid isoleucine, and circle each of the four different groups attached to the chiral center present in the amino acid.

▲9.37 Answer the following questions about the amino acid whose Fischer projection formula is

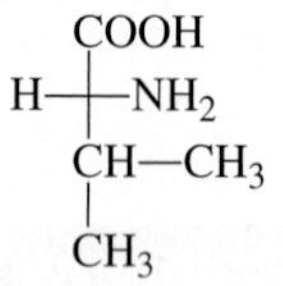

a. Is it a D-amino acid or an L-amino acid?
b. Is it a nonpolar or polar amino acid?
c. Is it an essential or nonessential amino acid?
d. Is it a standard or nonstandard amino acid?

▲**9.38** Answer the following questions about the amino acid whose Fischer projection formula is

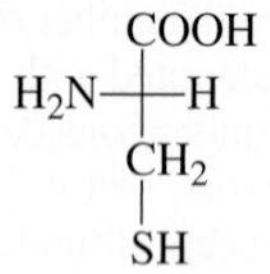

a. Is it a D-amino acid or an L-amino acid?
b. Is it a nonpolar or polar amino acid?
c. Is it an essential or nonessential amino acid?
d. Is it a standard or nonstandard amino acid?

Acid–Base Properties of Amino Acids (Section 9.5)

9.39 At room temperature, amino acids are solids with relatively high decomposition points. Explain why.

9.40 At room temperature, most amino acids are not very soluble in water. Explain why.

9.41 Draw the zwitterion structure for each of the following amino acids.
a. Leucine b. Isoleucine
c. Cysteine d. Glycine

9.42 Draw the zwitterion structure for each of the following amino acids.
a. Serine b. Methionine
c. Threonine d. Phenylalanine

9.43 Draw the structure of serine at each of the following pH values.
a. 5.68 b. 1.0 c. 1.0 d. 3.0

9.44 Draw the structure of glycine at each of the following pH values.
a. 5.97 b. 2.0 c. 2.0 d. 11.0

9.45 Explain what is meant by the term *isoelectric point.*

9.46 Most amino acids have isoelectric points between 5.0 and 6.0, but the isoelectric point of lysine is 9.7. Explain why lysine has such a high value for its isoelectric point.

9.47 Glutamic acid exists in two low-pH forms instead of the usual one. Explain why.

9.48 Arginine exists in two high-pH forms instead of the usual one. Explain why.

9.49 In a low-pH aqueous solution, indicate whether each of the following amino acids has (1) a net positive charge of +2 (2) a net positive charge of +1 (3) a net charge of zero (4) a net negative charge of −1 or (5) a net negative charge of −2.
a. Valine b. Lysine c. Aspartic acid d. Serine

9.50 In a high-pH aqueous solution, indicate whether each of the following amino acids has (1) a net positive charge of +2 (2) a net positive charge of +1 (3) a net charge of zero (4) a net negative charge of −1 or (5) a net negative charge of −2.
a. Arginine b. Glutamic acid
c. Glutamine d. Alanine

▲**9.51** Write an ionic equation that shows how a valine zwitterion in aqueous solution acts as a buffer when
a. OH^- ion is added b. H_3O^+ ion is added

▲**9.52** Write an ionic equation that shows how an alanine zwitterion in aqueous solution acts as a buffer when
a. OH^- ion is added b. H_3O^+ ion is added

Cysteine and Disulfide Bonds (Section 9.6)

9.53 When two cysteine molecules dimerize, what happens to the R groups present?

9.54 What chemical reaction involving the cysteine molecule produces a disulfide bond?

Peptides (Section 9.7)

9.55 What two functional groups are involved in the formation of a peptide bond?

9.56 Write a generalized structural representation for a peptide bond.

9.57 Write a generalized structural representation for the N-terminal end of a peptide.

9.58 Write a generalized structural representation for the C-terminal end of a peptide.

9.59 Write out the full structure of the tripeptide Val–Phe–Cys.

9.60 Write out the full structure of the tripeptide Glu–Ala–Leu.

9.61 Identify the amino acids contained in each of the following tripeptides.

a.
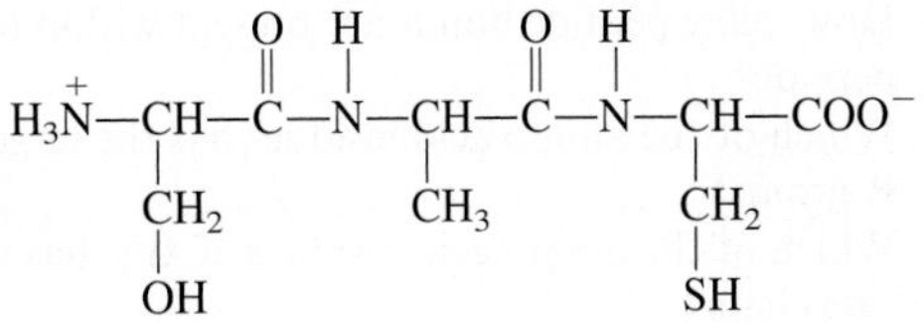

b.
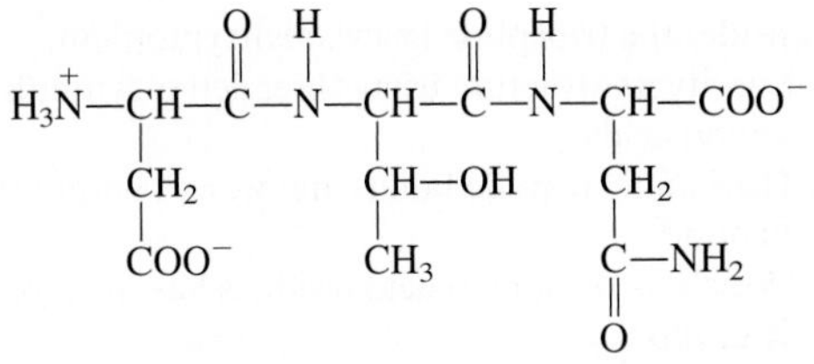

9.62 Identify the amino acids contained in each of the following tripeptides.

a.
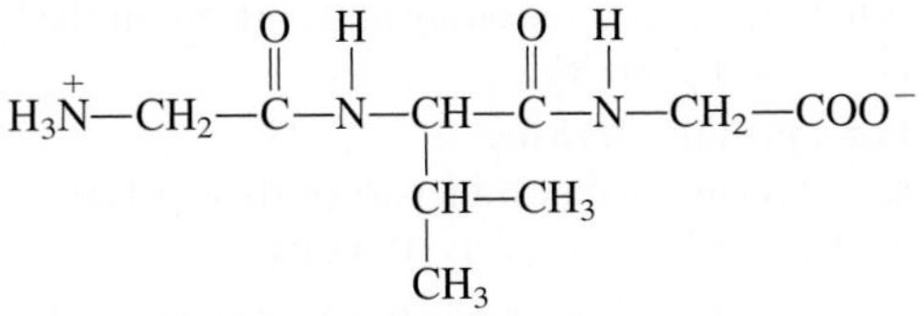

b.
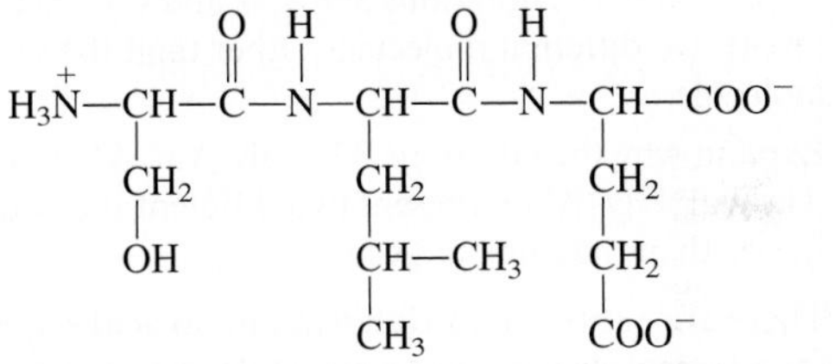

9.63 How many peptide bonds are present in each of the molecules in Problem 9.61?

9.64 How many peptide bonds are present in each of the molecules in Problem 9.62?

9.65 With the help of Table 9.1, assign an IUPAC name to each of the following small peptides.
a. Ser–Cys b. Gly–Ala–Val
c. Tyr–Asp–Gln d. Leu–Lys–Trp–Met

9.66 With the help of Table 9.1, assign an IUPAC name to each of the following small peptides.
a. Cys–Ser b. Val–Ala–Gly
c. Tyr–Gln–Asp d. Phe–Met–Tyr–Asn

▲**9.67** Consider the tripeptide Ala–Val–Gly.
a. Which amino acid is at the N-terminal end of the peptide?
b. Which amino acid is at the C-terminal end of the peptide?
c. Which of the amino acids present, if any, are essential amino acids?
d. Which of the amino acids present, if any, are polar neutral amino acids?

▲**9.68** Consider the tripeptide Ser–Arg–Ile.
a. Which amino acid is at the N-terminal end of the peptide?
b. Which amino acid is at the C-terminal end of the peptide?
c. Which of the amino acids present, if any, are essential amino acids?
d. Which of the amino acids present, if any, are nonpolar amino acids?

▲**9.69** Consider the tripeptide tyrosylleucylisoleucine.
a. Specify its structure using three-letter symbols for the amino acids.

b. How many peptide bonds are present within the peptide?
c. Which of the amino acid residues has the largest R group?
d. Which of the amino acid residues, if any, has an acidic side chain?

▲**9.70** Consider the tripeptide leucylvalyltryptophan.
a. Specify its structure using three-letter symbols for the amino acids.
b. How many peptide bonds are present within the peptide?
c. Which of the amino acid residues has the largest R group?
d. Which of the amino acid residues, if any, has a basic side chain?

9.71 What are the two repeating units present in the "backbone" of a peptide?

9.72 For a peptide, describe
a. the regularly repeating part of its structure.
b. the variable part of its structure.

9.73 Explain why the notations Ser–Cys and Cys–Ser represent two different molecules rather than the same molecule.

9.74 Explain why the notations Ala–Gly–Val–Ala and Ala–Val–Gly–Ala represent two different molecules rather than the same molecule.

9.75 There are a total of six different amino acid sequences for a tripeptide containing one molecule each of serine, valine, and glycine. Using three-letter abbreviations for the amino acids, draw the six possible sequences of amino acids.

9.76 There are a total of six different amino acid sequences for a tetrapeptide containing two molecules each of serine and valine. Using three-letter abbreviations for the amino acids, draw the six possible sequences of amino acids.

Biochemically Important Small Peptides (Section 9.8)

9.77 Compare the structures of the protein hormones oxytocin and vasopressin in terms of
a. what they have in common.
b. how they differ.

9.78 Compare the protein hormones oxytocin and vasopressin in terms of their biochemical functions.

9.79 Compare the binding-site locations in the brain for enkephalins and the prescription painkillers morphine and codeine.

9.80 Compare the structures of the peptide neurotransmitters Met-enkephalin and Leu-enkephalin in terms of
a. what they have in common.
b. how they differ.

9.81 What is the unusual structural feature present in the molecule glutathione?

9.82 What is the major biochemical function of glutathione?

General Structural Characteristics of Proteins (Section 9.9)

9.83 What is the major difference between a monomeric protein and a multimeric protein?

9.84 What is the major difference between a simple protein and a conjugated protein?

9.85 Indicate whether each of the following statements about proteins is true or false.
a. Two or more peptide chains are always present in a multimeric protein.
b. A simple protein contains only one type of amino acid.
c. A conjugated protein can also be a monomeric protein.
d. The prosthetic group(s) present in a glycoprotein are carbohydrate groups.

9.86 Indicate whether each of the following statements about proteins is true or false.
a. Conjugated proteins always have only one peptide chain.
b. All peptide chains in a multimeric protein must be identical to each other.
c. A simple protein can also be a multimeric protein.
d. Both monomeric proteins and multimeric proteins can contain prosthetic groups.

Primary Protein Structure (Section 9.10)

9.87 What is meant by the *primary structure* of a protein?

9.88 Two proteins with the same amino acid composition do not have to have the same primary structure. Explain why.

9.89 What type of bond is responsible for the primary structure of a protein?

9.90 A segment of a protein contains two alanine and two glycine amino acid residues. How many different primary structures are possible for this segment of the protein?

▲9.91 Draw a segment of the "backbone" of a protein that is long enough to attach three R groups (amino acid side chains).

▲**9.92** Draw a segment of the "backbone" of a protein that is long enough for three peptide linkages to be present.

•9.93 (Chemical Connections 9-A) Indicate whether each of the following statements relating to the protein insulin is true or false.
a. Biosynthetic insulin produced using genetically modified bacteria has the same primary structure as human insulin.
b. The primary structures of porcine and human insulin match at all except three amino acid positions.
c. The primary biochemical function of human insulin is regulation of blood-glucose levels.
d. Orally administered insulin undergoes digestion in the stomach.

•**9.94** (Chemical Connections 9-A) Indicate whether each of the following statements relating to the protein insulin is true or false.
a. Insulin was the first protein for which the primary structure was obtained.
b. For many years, the source of the insulin used by diabetics was slaughter-house animals.
c. Because of its differing primary structure bovine insulin, is not compatible with the human body.
d. Insulin production using "bioengineered" plants is now possible.

Secondary Protein Structure (Section 9.11)

9.95 What are the two common types of secondary protein structure?

9.96 Hydrogen bonding between which functional groups stabilizes protein secondary structure arrangements?

9.97 Indicate whether each of the following statements concerning secondary protein structure is true or false.
a. Hydrogen bonds present in an α-helix structure always involve two adjacent amino acids.
b. Both α-helix and β-pleated sheet structures can be present in the same protein.
c. In a β-pleated sheet structure, the hydrogen bonding is always between different protein chains.
d. In an α-helix, all of the amino acid R groups lie inside the helix.

9.98 Indicate whether each of the following statements concerning secondary protein structure is true or false.
a. In a β-pleated sheet structure, the hydrogen bonds lie in the plane of the sheet.
b. The coil configuration in an α-helix structure is maintained by covalent bonds.
c. In a β-pleated sheet structure, the R groups are all oriented below the plane of the sheet.
d. Hydrogen bonds present in both the α-helix structure and the β-pleated sheet structure are always interactions between C═O and N—H groups.

9.99 What is meant by the statement that a portion of the secondary structure of a protein is "unstructured"?

9.100 Why is the phrase "unstructured segment of a protein" somewhat of a misnomer?

Tertiary Protein Structure (Section 9.12)

9.101 State the four types of attractive forces that give rise to tertiary protein structure.

9.102 What is the difference between the types of hydrogen bonding that occur in secondary and tertiary protein structure?

9.103 Give the type of amino acid R group that is involved in each of the following interactions that contribute to tertiary protein structure.
a. Hydrophobic interaction
b. Hydrogen bond
c. Disulfide bond
d. Electrostatic interaction

9.104 Classify each of the interactions listed in Problem 9.103 as a covalent bond or as a noncovalent interaction.

9.105 With the help of Table 9.1, specify the nature of each of the following tertiary-structure interactions, using the choices hydrophobic, electrostatic, hydrogen bond, and disulfide bond.
a. Phenylalanine and leucine
b. Arginine and glutamic acid
c. Two cysteines
d. Serine and tyrosine

9.106 With the help of Table 9.1, specify the nature of each of the following tertiary-structure interactions using the choices hydrophobic, electrostatic, hydrogen bond, and disulfide bond.
a. Lysine and aspartic acid
b. Threonine and tyrosine
c. Alanine and valine
d. Leucine and isoleucine

▲9.107 Indicate whether each of the following statements applies to *primary protein structure, secondary protein structure,* or *tertiary protein structure.* More than one type of structure may apply in a given situation.
a. Salt bridges between amino acids with acidic and basic side chains are present.
b. Hydrogen bonds are present.
c. A segment of a protein chain folds back upon itself.
d. The C-terminal amino acid in the protein chain is specified.

▲9.108 Indicate whether each of the following statements applies to *primary protein structure, secondary protein structure,* or *tertiary protein structure.* More than one type of structure may apply in a given situation.
a. Peptide linkages are present in the protein chain.
b. Disulfide bonds between cysteine amino acids are present.
c. Hydrogen bonds between C═O and N—H groups are present.
d. The sequential order of amino acids in a protein chain is given.

Quaternary Protein Structure (Section 9.13)

9.109 What is meant by the *quaternary structure* of a protein?

9.110 Not all proteins have quaternary structure. Explain why this is so.

9.111 Compare the types of noncovalent interactions that contribute to tertiary protein structure with those that contribute to quaternary protein structure.

9.112 Quaternary protein structure is more easily disrupted than tertiary protein structure. Explain why this is so.

Protein Hydrolysis (Section 9.14)

9.113 Will hydrolysis of the dipeptides Ala–Val and Val–Ala yield the same products? Explain your answer.

9.114 A shampoo bottle lists "partially hydrolyzed protein" as one of its ingredients. What is the difference between partially hydrolyzed protein and completely hydrolyzed protein?

9.115 Drugs that are proteins, such as insulin, must always be injected rather than taken by mouth. Explain why.

9.116 Which structural levels of a protein are affected by hydrolysis?

9.117 How many different di- and tripeptides could be present in a solution of partially hydrolyzed Ala–Gly–Ser–Tyr?

9.118 How many different di- and tripeptides could be present in a solution of partially hydrolyzed Ala–Gly–Ala–Gly?

9.119 Identify the primary structure of a hexapeptide containing six different amino acids if the following smaller peptides are among the partial-hydrolysis products: Ala–Gly, His–Val–Arg, Ala–Gly–Met, and Gly–Met–His.

9.120 Identify the primary structure of a hexapeptide containing five different amino acids if the following smaller peptides are among the partial-hydrolysis products: Gly–Cys, Ala–Ser, Ala–Gly, and Cys–Val–Ala.

▲9.121 Draw structural formulas for the products obtained when the tripeptide Ala–Gly–Ser undergoes complete hydrolysis under
a. acidic conditions
b. basic conditions

▲9.122 Draw structural formulas for the products obtained when the tripeptide Thr–Leu–Ile undergoes complete hydrolysis under
a. acidic conditions
b. basic conditions

Protein Denaturation (Section 9.15)

9.123 Which structural levels of a protein are affected by denaturation?

9.124 Why is complete hydrolysis of a protein not also protein denaturation?

9.125 In what way is the protein in a cooked egg the same as that in a raw egg?

9.126 Why is 70% ethanol rather than pure ethanol preferred for use as an antiseptic agent?

•9.127 (Chemical Connections 9-C) Indicate whether each of the following statements concerning hair permanents ("perms") is true or false.
a. The protein in hair contains large amounts of the amino acid cysteine.
b. The "perm" process involves reversible protein denaturation.
c. The reducing agent used in giving a "perm" facilitates the re-formation of disulfide linkages that had previously been broken.
d. At the curler stage of a "perm," the tertiary structure of hair protein is completely intact.

•9.128 (Chemical Connections 9-C) Indicate whether each of the following statements concerning hair permanents ("perms") is true or false.
a. The protein in hair contains many more disulfide linkages than do most other proteins.
b. In the "perm" process, S—S bonds are broken and then re-formed in the exact locations where they previously were.
c. The oxidizing agents used in giving a "perm" facilitates the re-formation of peptide linkages that had been previously broken.
d. As part of the "perm" process, disulfide linkages are converted to sulfhydryl groups.

Protein Classification Based on Shape (Section 9.16)

9.129 Contrast fibrous and globular proteins in terms of
a. solubility characteristics in water.
b. general biochemical function.

9.130 Contrast fibrous and globular proteins in terms of
a. general secondary structure.
b. relative abundance within the human body.

9.131 Classify each of the following proteins as a globular protein or a fibrous protein.
a. α-Keratin b. Collagen
c. Hemoglobin d. Myoglobin

9.132 What is the major biochemical function of each of the following proteins?
a. α-Keratin b. Collagen
c. Hemoglobin d. Myoglobin

•9.133 (Chemical Connections 9-B) Indicate whether each of the following statements concerning protein structure and the color of meat is true or false.
a. The amount of myosin present in muscle tissue determines its color.
b. The O_2 source needed for muscle contraction is the protein actin.
c. Myoglobin is purple when oxygenated and red when deoxygenated.
d. The major proteins present in muscle tissue (meat) are myosin and myoglobin.

•9.134 (Chemical Connections 9-B) Indicate whether each of the following statements concerning protein structure and the color of meat is true or false.
a. Heavily used muscles have higher concentrations of myoglobin than do lightly used muscles.
b. Myosin proteins have two globular protein "heads."
c. Muscle contraction involves interactions between the proteins myosin and actin.
d. Meat, when cooked, turns brown because iron atoms in myoglobin are oxidized.

Protein Classification Based on Function (Section 9.17)

9.135 Using the list in Section 9.17, characterize each of the following proteins in terms of its function classification.
a. Actin b. Myoglobin
c. Transferrin d. Insulin

9.136 Using the list in Section 9.17, characterize each of the following proteins in terms of its function classification.
a. Collagen b. Hemoglobin
c. Ferritin d. Casein

Glycoproteins (Section 9.18)

9.137 What two nonstandard amino acids are present in collagen?

9.138 Where are the carbohydrate units located in collagen?

9.139 What is the function of the carbohydrate groups present in collagen?

9.140 What is the role of vitamin C in the biosynthesis of collagen?

9.141 What is the difference between an antigen and an antibody?

9.142 What is an immunoglobulin?

9.143 Describe the structural features of a typical immunoglobulin molecule.

9.144 Describe the process by which blood immunoglobulins help protect the body from invading bacteria and viruses.

•9.145 (Chemical Connections 9-D) Indicate whether each of the following statements relating to the antirejection drug cyclosporin is true or false.
a. Cyclosporin is a peptide in which all amino acids present have D-configurations rather than L-configurations.
b. Cyclosporin's effectiveness is related to it affecting only part of the body's immune system.
c. Cyclosporin's original source was a soil fungus.
d. Most of the amino acid residues present in cyclosporin's peptide structure have "simple" side chains.

•9.146 (Chemical Connections 9-D) Indicate whether each of the following statements relating to the antirejection drug cyclosporin is true or false.
a. One of the amino acid residues present in cyclosporin is unique to this natural product.
b. Many of cyclosporin's amino acid residues have their nitrogen atoms methylated.
c. Cyclosporin is a peptide in which three nonstandard amino acids are present.
d. The pharmacological activity of cyclosporin involves the side chain of the amino acid Abu.

●9.147 (Chemical Connections 9-E) Indicate whether each of the following statements relating to human breast milk is true or false.
a. Human milk contains almost twice as much lactose as cow's milk.
b. The immunoglobulins present in human milk come from the mother's bloodstream.
c. An important immunoglobulin present in colostrum is IgG.
d. Colostrum production in breast tissue stops after about six weeks of nursing.

●**9.148** (Chemical Connections 9-E) Indicate whether each of the following statements relating to human breast milk is true or false.
a. Infant formula is usually, but not always, immunologically equivalent to human breast milk.
b. Most immunologically active substances present in colostrum are synthesized in breast tissue.
c. Colostrum contains the easily digested protein lactoalbumin.
d. Immunoglobulins and other whole proteins can pass unaltered through a newborn's intestinal tract into the bloodstream.

Lipoproteins (Section 9.19)

9.149 Describe the general overall structure of a plasma lipoprotein.

9.150 In what chemical form does cholesterol usually exist in the bloodstream?

9.151 What are the four major classes of plasma lipoproteins?

9.152 What do the designations VLDL, LDL, and HDL stand for?

9.153 What factor determines the density of a plasma lipoprotein?

9.154 As the lipid content of a plasma lipoprotein increases, does its density increase or decrease?

9.155 What is the biochemical function of the following?
a. Chylomicrons
b. Low-density lipoproteins

9.156 What is the biochemical function of the following?
a. Very-low-density lipoproteins
b. High-density lipoproteins

●9.157 (Chemical Connections 9-F) Indicate whether each of the following statements concerning cholesterol-carrying substances in the bloodstream is true or false.
a. The form in which cholesterol is present in the blood is as important as the amount of cholesterol present.
b. In general, women have higher HDL levels than men.
c. LDLs carry excess cholesterol to the liver for destruction.
d. 80% of blood cholesterol is associated with HDLs.

●**9.158** (Chemical Connections 9-F) Indicate whether each of the following statements concerning cholesterol-carrying substances in the bloodstream is true or false.
a. HDLs carry cholesterol to cells for their use.
b. LDLs correlate inversely with heart disease risk.
c. Exercise tends to reduce HDL levels.
d. The cholesterol ratio is defined as the ratio of HDL to LDL.

10

Enzymes and Vitamins

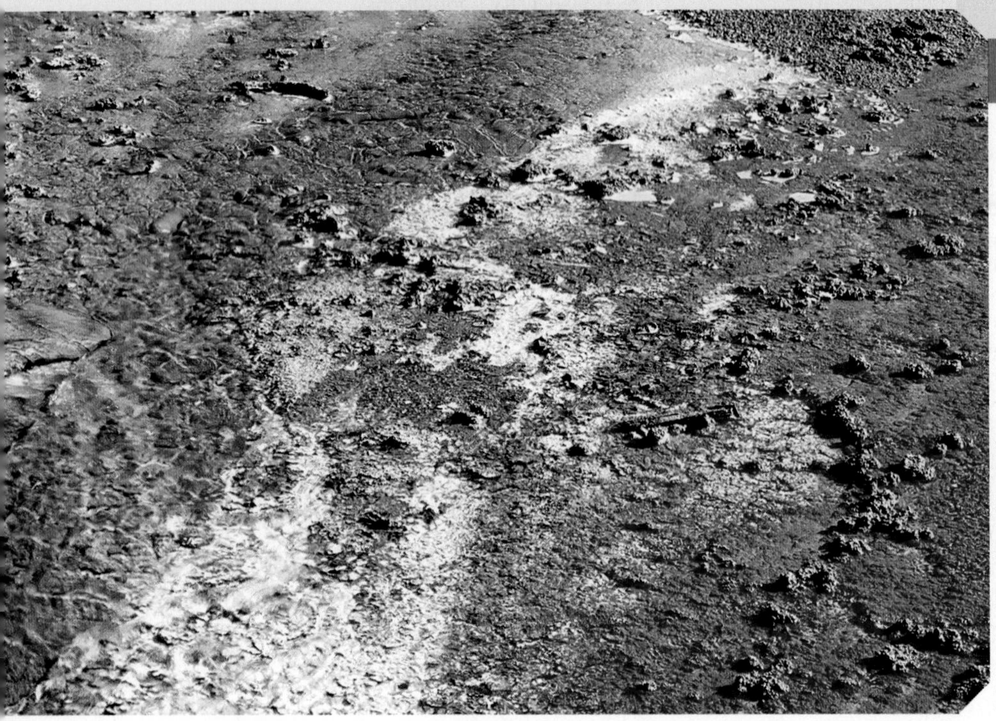

© Ed Reschke/Peter Arnold/Photo Library

The microbial life present in hot springs and thermal vents possesses enzymes that are specially adapted to higher temperature conditions.

CHAPTER OUTLINE

Two topics constitute the subject matter for this chapter: enzymes and vitamins. Most enzymes are specialized proteins that function as biochemical catalysts. Without enzymes, most chemical reactions in biochemical systems would occur too slowly to produce adequate amounts of the substances needed for cells to function properly. In some cases, needed reactions would not occur at all without the presence of enzymes.

Vitamins are dietary organic compounds required in very small quantities for normal cellular function. Many enzymes have vitamins as part of their structures, the presence of which is an absolute necessity for the enzymes to carry out their catalytic function. Hence the topic of vitamins is considered in conjunction with the topic of enzymes.

10.1 General Characteristics of Enzymes

An **enzyme** *is a compound, usually a protein, that acts as a catalyst for a biochemical reaction.* Each cell in the human body contains thousands of different enzymes because almost every reaction in a cell requires its own specific enzyme. Enzymes cause cellular reactions to occur millions of times faster than corresponding uncatalyzed reactions. As catalysts, enzymes are not consumed during the reaction but merely help the reaction occur more rapidly.

The word *enzyme* comes from the Greek words *en,* which means "in," and *zyme,* which means "yeast." Long before their chemical nature was understood, yeast enzymes were used in the production of bread and alcoholic beverages. The action of yeast on sugars produces the carbon dioxide gas that causes bread to rise (Figure 10.1). Fermentation of sugars in fruit juices using the same yeast enzymes produces alcoholic beverages.

Most enzymes are globular proteins (Section 9.16). Some are simple proteins, consisting entirely of amino acid chains. Others are conjugated proteins, containing additional chemical components (Section 10.2). Until the 1980s, it was thought that *all* enzymes were proteins. A few enzymes are now known that are made of ribonucleic acids (RNA; Section 11.7) and that catalyze cellular reactions involving nucleic acids. In this chapter, only enzymes that are proteins are considered.

Figure 10.1 Bread dough rises as the result of carbon dioxide production resulting from the action of yeast enzymes on sugars present in the dough.

Enzymes undergo all the reactions of proteins, including *denaturation* (Section 9.15). Slight alterations in pH or temperature affect enzyme activity dramatically. Good cooks realize that overheating yeast kills the action of the yeast. A person suffering from a high fever (greater than 106°F) runs the risk of having cellular enzymes denatured. The biochemist must exercise extreme caution in handling enzymes to avoid the loss of their activity. Even vigorous shaking of an enzyme solution can destroy enzyme activity.

Enzymes differ from nonbiochemical (laboratory) catalysts not only in size, being much larger, but also in that their activity is usually regulated by other substances present in the cell in which they are found. Most laboratory catalysts need to be removed from a reaction mixture to stop their catalytic action; this is not so with enzymes. In some cases, if a certain chemical is needed in the cell, the enzyme responsible for its production is activated by other cellular components. When a sufficient quantity has been produced, the enzyme is then deactivated. In other situations, the cell may produce more or less enzyme as required. Because different enzymes are required for nearly all cellular reactions, certain necessary reactions can be accelerated or decelerated without affecting the rest of the cellular chemistry.

Enzymes, the most efficient catalysts known, increase the rates of biochemical reactions by factors of up to 10^{20} over uncatalyzed reactions. Nonenzymatic catalysts, on the other hand, typically enhance the rate of a reaction by factors of 10^2 to 10^4.

10.2 Enzyme Structure

Enzymes can be divided into two general structural classes: simple enzymes and conjugated enzymes. A **simple enzyme** *is an enzyme composed only of protein (amino acid chains).* A **conjugated enzyme** *is an enzyme that has a nonprotein part in addition to a protein part.* By itself, neither the protein part nor the nonprotein portion of a conjugated enzyme has catalytic properties. An **apoenzyme** *is the protein part of a conjugated enzyme.* A **cofactor** *is the nonprotein part of a conjugated enzyme.* The term *holoenzyme* is often used to designate a biologically active combined apoenzyme-cofactor entity. A **holoenzyme** *is the biochemically active conjugated enzyme produced from an apoenzyme and a cofactor.*

$$\text{Apoenzyme} + \text{cofactor} = \text{holoenzyme}$$

Why do apoenzymes need cofactors? Cofactors provide additional chemically reactive functional groups besides those present in the amino acid side chains of apoenzymes.

A cofactor is generally either a small organic molecule or an inorganic ion (usually a metal ion). A **coenzyme** *is a small organic molecule that serves as a cofactor in a conjugated enzyme.* Many vitamins (see Section 10.12) have coenzyme functions in the human body.

Typical inorganic ion cofactors include Zn^{2+}, Mg^{2+}, Mn^{2+}, and Fe^{2+}. The nonmetallic Cl^- ion occasionally acts as a cofactor. Dietary minerals are an important source of inorganic ion cofactors.

10.3 Nomenclature and Classification of Enzymes

Enzymes are most commonly named by using a system that attempts to provide information about the *function* (rather than the structure) of the enzyme. The type of reaction catalyzed and the *substrate* identity are focal points for the nomenclature. A **substrate** *is the reactant in an enzyme-catalyzed reaction.* The substrate is the substance upon which the enzyme "acts."

Three important aspects of the enzyme-naming process are the following:

1. The suffix *-ase* identifies a substance as an enzyme. Thus ur*ase,* sucr*ase,* and lip*ase* are all enzyme designations. The suffix *-in* is still found in the names of some of the first enzymes studied, many of which are digestive enzymes. Such names include *trypsin, chymotrypsin,* and *pepsin.*
2. The type of reaction catalyzed by an enzyme is often noted with a prefix. An *oxidase* enzyme catalyzes an oxidation reaction, and a *hydrolase* enzyme catalyzes a hydrolysis reaction.
3. The identity of the substrate is often noted in addition to the type of reaction. Enzyme names of this type include *glucose oxidase, pyruvate carboxylase,* and *succinate dehydrogenase.* Infrequently, the substrate but not the reaction type is given, as in the names *urease* and *lactase.* In such names, the reaction involved is hydrolysis; *urease* catalyzes the hydrolysis of urea, *lactase* the hydrolysis of lactose.

EXAMPLE 10.1 Predicting Enzyme Function from an Enzyme's Name

Predict the function of the following enzymes.

a. Cellulase **b.** Sucrase
c. L-Amino acid oxidase **d.** Aspartate aminotransferase

Solution

a. Cellulase catalyzes the hydrolysis of cellulose.
b. Sucrase catalyzes the hydrolysis of the disaccharide sucrose.
c. L-Amino acid oxidase catalyzes the oxidation of L-amino acids.
d. Aspartate aminotransferase catalyzes the transfer of an amino group from aspartate to a different molecule.

Practice Exercise 10.1

Predict the function of the following enzymes.

a. Maltase
b. Lactate dehydrogenase
c. Fructose oxidase
d. Maleate isomerase

Answers: **a.** Hydrolysis of maltose; **b.** Removal of hydrogen from lactate ion; **c.** Oxidation of fructose; **d.** Rearrangement (isomerization) of maleate ion

Recall from Section 3.9 that for organic redox reactions, the following two operational rules are used instead of oxidation numbers to characterize oxidation and reduction processes:

1. An *organic oxidation reaction* is an oxidation that *increases* the number of C—O bonds and/or *decreases* the number of C—H bonds.
2. An *organic reduction reaction* is a reduction that *decreases* the number of C—O bonds and/or *increases* the number of C—H bonds.

Enzymes are grouped into six major classes on the basis of the types of reactions they catalyze.

1. An **oxidoreductase** *is an enzyme that catalyzes an oxidation–reduction reaction.* Because oxidation and reduction are not independent processes but linked processes that must occur together, an oxidoreductase requires a coenzyme that is oxidized or reduced as the substrate is reduced or oxidized. *Lactate dehydrogenase* is an oxidoreductase that removes hydrogen atoms from a molecule.

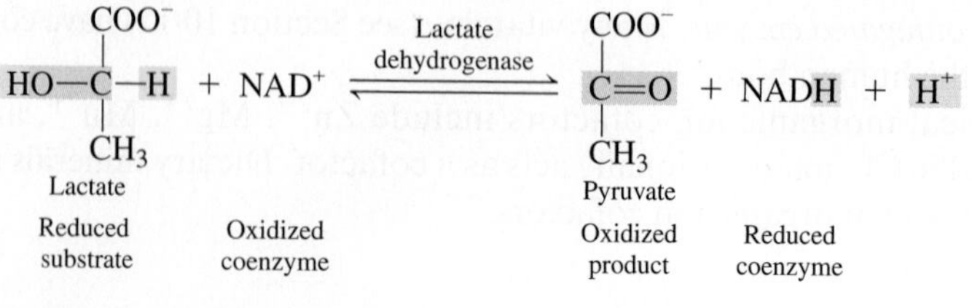

2. A **transferase** *is an enzyme that catalyzes the transfer of a functional group from one molecule to another.* Two major subtypes of transferases are *transaminases* and *kinases.* A transaminase catalyzes the transfer of an amino group from one molecule to another. Kinases, which play a major role in metabolic energy-production reactions (see Section 13.2), catalyze the transfer of a phosphate group from adenosine triphosphate (ATP) to give adenosine diphosphate (ADP) and a phosphorylated product (a product containing an additional phosphate group).

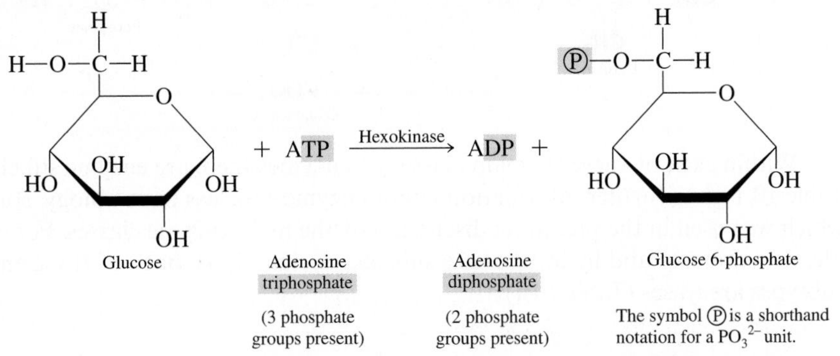

3. A **hydrolase** *is an enzyme that catalyzes a hydrolysis reaction in which the addition of a water molecule to a bond causes the bond to break.* Hydrolysis reactions are central to the process of digestion. *Carbohydrases* effect the breaking of glycosidic bonds in oligo- and polysaccharides, *proteases* effect the breaking of peptide linkages in proteins, and *lipases* effect the breaking of ester linkages in triacylglycerols.

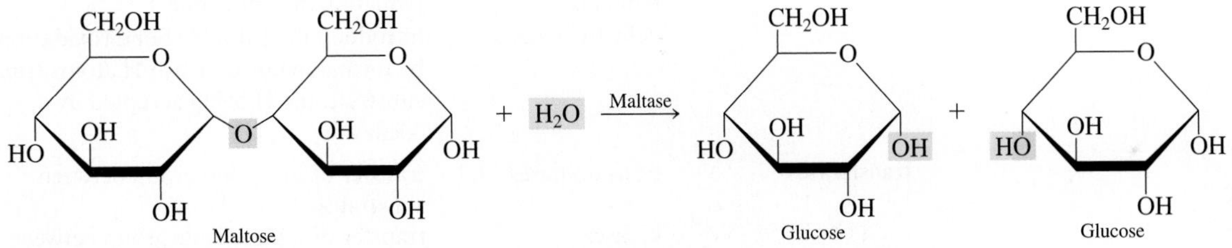

4. A **lyase** *is an enzyme that catalyzes the addition of a group to a double bond or the removal of a group to form a double bond in a manner that does not involve hydrolysis or oxidation.* A *dehydratase* effects the removal of the components of water from a double bond and a *hydratase* effects the addition of the components of water to a double bond.

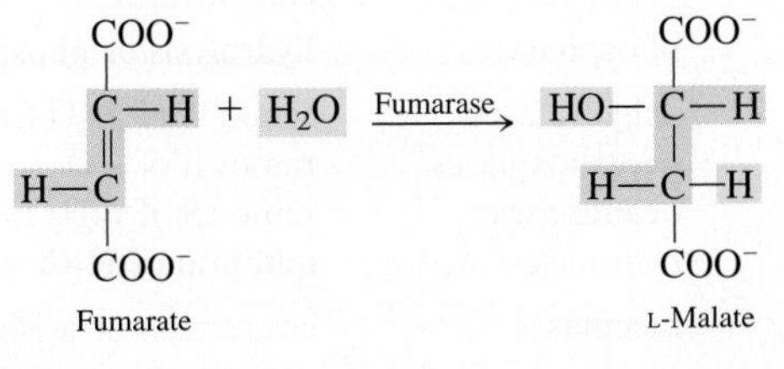

5. An **isomerase** *is an enzyme that catalyzes the isomerization (rearrangement of atoms) of a substrate in a reaction, converting it into a molecule isomeric with itself.* There is only one reactant and one product in reactions where isomerases are operative.

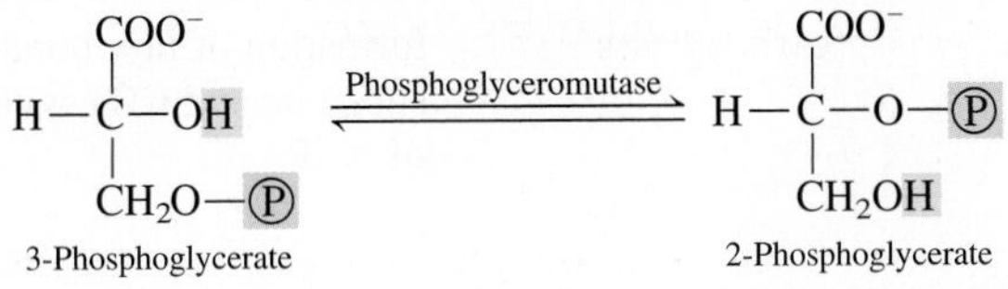

6. A **ligase** *is an enzyme that catalyzes the bonding together of two molecules into one with the participation of ATP.* ATP involvement is required because such reactions are generally energetically unfavorable and they require the simultaneous input of energy obtained by a hydrolysis reaction in which ATP is converted to ADP (such energy release is considered in Section 12.3).

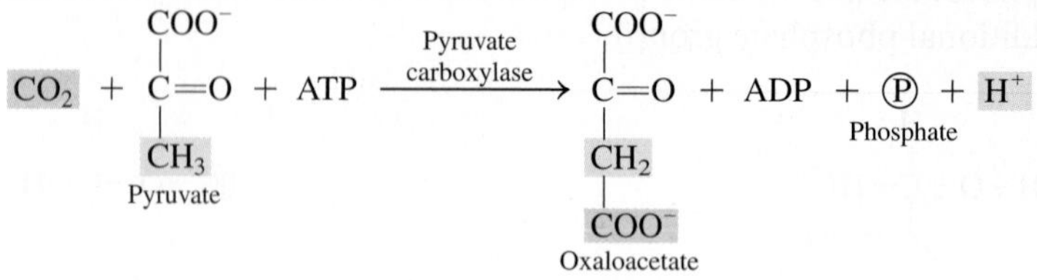

Within each of these six main classes of enzymes there are enzyme subclasses. Table 10.1 gives further information about enzyme subclass terminology, some of which was used in the preceding discussion of the main enzyme classes. For example, dehydratases and hydratases are subclass designations; both of these enzyme subtypes are lyases (Table 10.1).

Table 10.1 Main Classes and Subclasses of Enzymes

Main Classes	Selected Subclasses	Type of Reaction Catalyzed
oxidoreductases	oxidases	oxidation of a substrate
	reductases	reduction of a substrate
	dehydrogenases	introduction of double bond (oxidation) by formal removal of two H atoms from substrate, the H being accepted by a coenzyme
transferases	transaminases	transfer of an amino group between substrates
	kinases	transfer of a phosphate group between substrates
hydrolases	lipases	hydrolysis of ester linkages in lipids
	proteases	hydrolysis of amide linkages in proteins
	nucleases	hydrolysis of sugar–phosphate ester bonds in nucleic acids
	carbohydrases	hydrolysis of glycosidic bonds in carbohydrates
	phosphatases	hydrolysis of phosphate–ester bonds
lyases	dehydratases	removal of H_2O from a substrate
	decarboxylases	removal of CO_2 from a substrate
	deaminases	removal of NH_3 from a substrate
	hydratases	addition of H_2O to a substrate
isomerases	racemases	conversion of D isomer to L isomer, or vice versa
	mutases	transfer of a functional group from one position to another in the same molecule
ligases	synthetases	formation of new bond between two substrates, with participation of ATP
	carboxylases	formation of new bond between a substrate and CO_2, with participation of ATP

EXAMPLE 10.2 Classifying Enzymes by the Type of Chemical Reaction They Catalyze

To what main enzyme class do the enzymes that catalyze the following chemical reactions belong?

a.

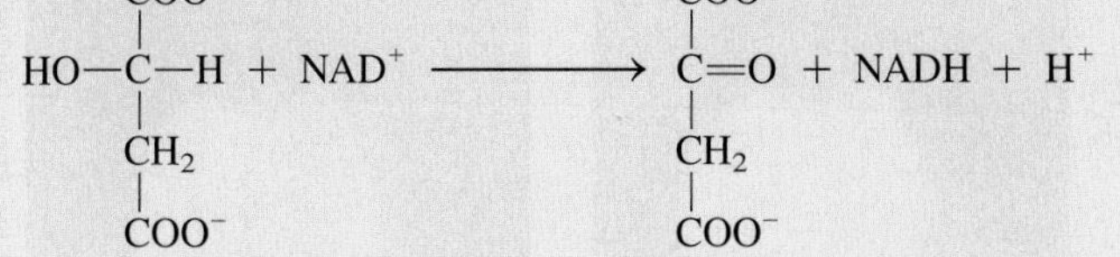

b.

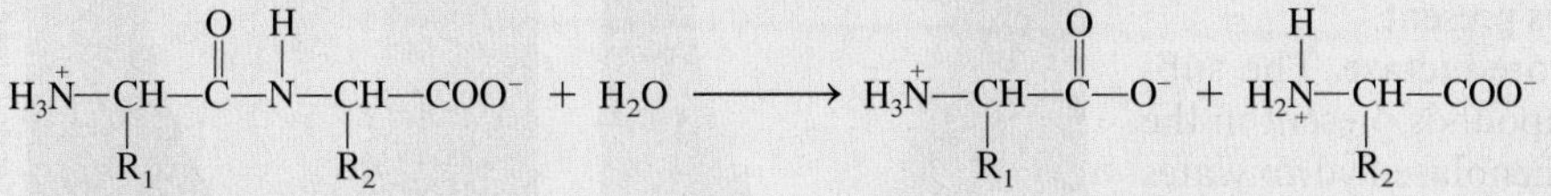

Solution

a. In this reaction, two H atoms are removed from the substrate and a carbon–oxygen double bond is formed. Loss of two hydrogen atoms by a molecule is an indication of oxidation. This is an oxidation–reduction reaction, and the enzyme needed to effect the change will be an *oxidoreductase.*

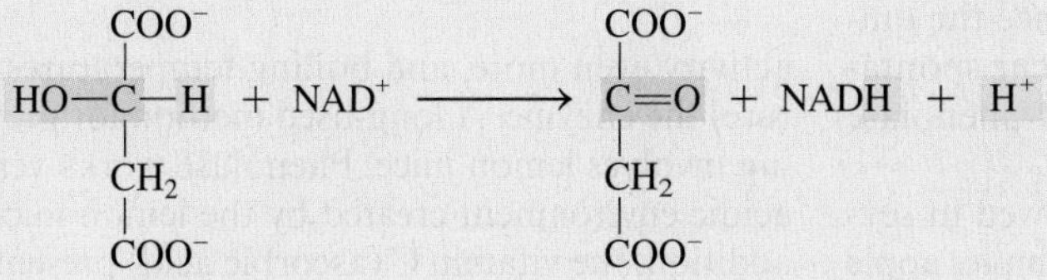

Malate Oxaloacetate

b. In this reaction, the components of a molecule of H_2O are added to the substrate (a dipeptide) with the resulting breaking of the peptide bond to produce two amino acids. This is an example of a hydrolysis reaction. An enzyme that effects such a change is called a *hydrolase.*

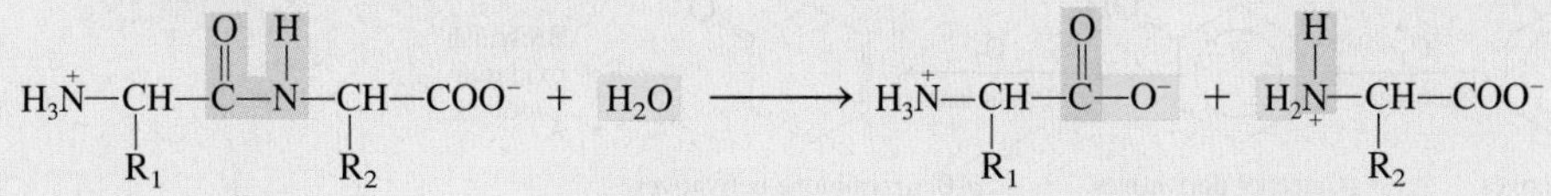

Dipeptide Amino acid Amino acid

Practice Exercise 10.2

To what main enzyme class do the enzymes that catalyze the following chemical reactions belong?

a.

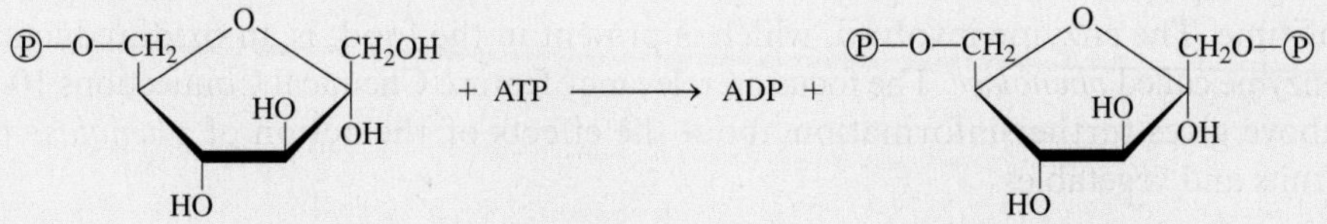

Fructose 6-phosphate Fructose 1,6-bisphosphate

b.

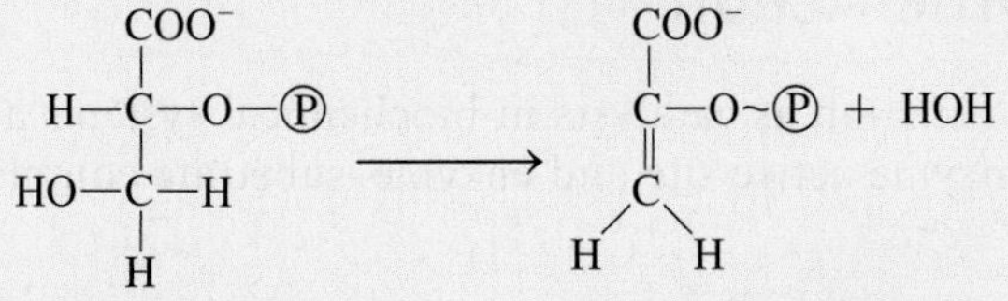

2-Phosphoglycerate Phosphoenolpyruvate

Answers: **a.** Transferase; **b.** Lyase

CHEMICAL CONNECTIONS 10-A

Enzymatic Browning: Discoloration of Fruits and Vegetables

Everyone is familiar with the way fruits such as apples, pears, peaches, apricots, and bananas, and vegetables such as potatoes, quickly turn brown when their tissue is exposed to oxygen. Such oxygen exposure occurs when the food is sliced or bitten into or when it has sustained bruises, cuts, or other injury to the peel. This "browning reaction" is related to the work of an enzyme called phenolase (or polyphenoloxidase), a conjugated enzyme in which copper is present.

Phenolase is classified as an oxidoreductase. The substrates for phenolase are phenolic compounds present in the tissues of the fruits and vegetables. Phenolase hydroxylates phenol derivatives to catechol (Section 3.13) derivatives and then oxidizes the catechol derivatives to *o*-benzoquinone derivatives (see the chemical equations below for the structural relationships). The *o*-benzoquinone derivatives then enter into a number of other reactions, which produce the "unwanted" brown discolorations. *o*-Benzoquinone derivative formation is enzyme- and oxygen-dependent. Once the quinones have formed, the subsequent reactions occur spontaneously and no longer depend on the presence of phenolase or oxygen.

Enzymatic browning can be prevented or slowed in several ways. Immersing the "injured" food (for example, apple slices) in cold water slows the browning process. The lower temperature decreases enzyme activity, and the water limits the enzyme's access to oxygen. Refrigeration slows enzyme activity even more, and boiling temperatures destroy (denature) the enzyme. A long-used method for preventing browning involves lemon juice. Phenolase works very slowly in the acidic environment created by the lemon juice's presence. In addition, the vitamin C (ascorbic acid) present in lemon juice functions as an antioxidant. It is more easily oxidized than the phenolic-derived compounds, and its oxidation products are colorless.

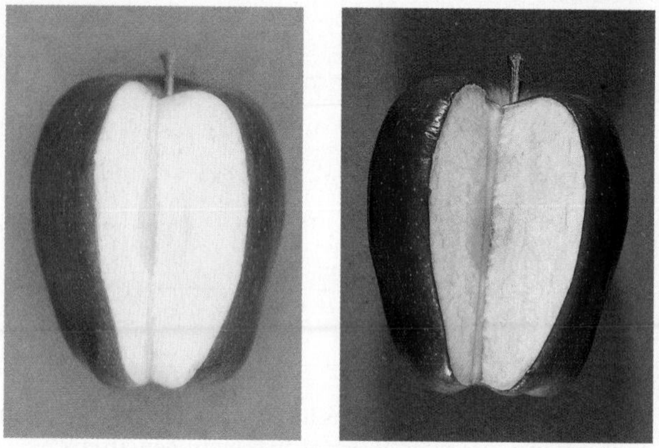

At left, a freshly cut apple. Brownish oxidation products form in a few minutes (at right).

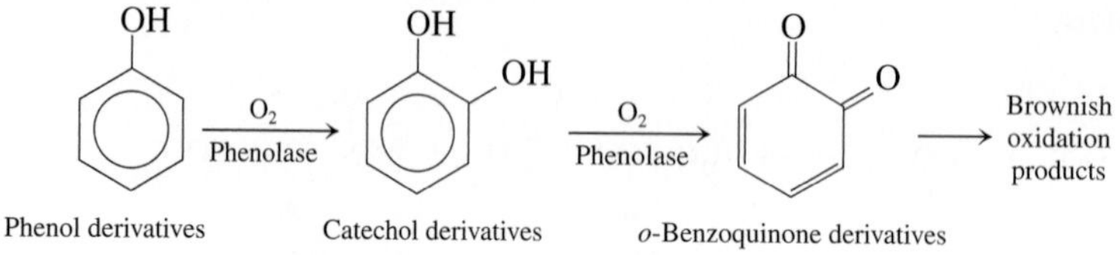

A commonly observed enzyme-influenced phenomenon that occurs outside the human body is the discoloration (browning) that occurs when freshly cut fruit (apples, pears, etc.) and vegetables (potatoes) are exposed to air for a short period of time. The enzyme involved, which is present in the food, is an oxidoreductase enzyme called *phenolase*. The focus on relevancy feature Chemical Connections 10-A above gives further information about the effects of the action of *phenolase* on fruits and vegetables.

10.4 Models of Enzyme Action

Explanations of *how* enzymes function as catalysts in biochemical systems are based on the concepts of an enzyme active site and enzyme–substrate complex formation.

Enzyme Active Site

Studies show that only a small portion of an enzyme molecule called the active site participates in the interaction with a substrate or substrates during a reaction. The **active site** *is the relatively small part of an enzyme's structure that is actually involved in catalysis.*

The active site in an enzyme is a three-dimensional entity formed by groups that come from different parts of the protein chain(s); these groups are brought together by the folding and bending (secondary and tertiary structure; Sections 9.11 and 9.12) of the protein. The active site is usually a "crevicelike" location in the enzyme (Figure 10.2).

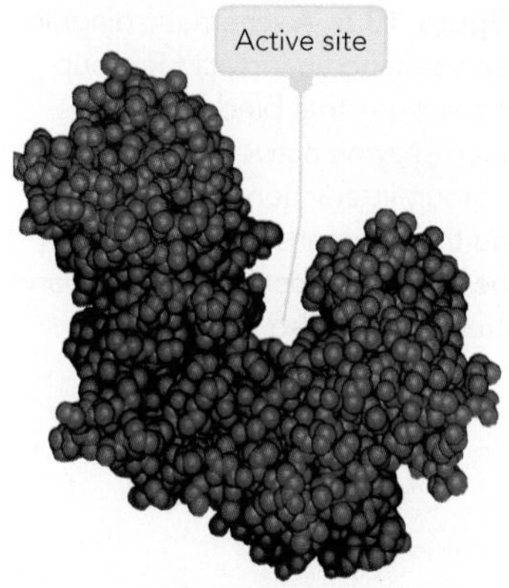

Figure 10.2 The active site of an enzyme is usually a crevicelike region formed as a result of the protein's secondary and tertiary structural characteristics.

Enzyme–Substrate Complex

Catalysts offer an alternative pathway with lower activation energy through which a reaction can occur (Section 9.6). In enzyme-controlled reactions, this alternative pathway involves the formation of an enzyme–substrate complex as an intermediate species in the reaction. An **enzyme–substrate complex** *is the intermediate reaction species that is formed when a substrate binds to the active site of an enzyme.* Within the enzyme–substrate complex, the substrate encounters more favorable reaction conditions than if it were free. The result is faster formation of product.

Lock-and-Key Model

To account for the highly specific way an enzyme recognizes a substrate and binds it to the active site, researchers have proposed several models. The simplest of these models is the lock-and-key model.

In the lock-and-key model, the active site in the enzyme has a fixed, rigid geometrical conformation. Only substrates with a complementary geometry can be accommodated at such a site, much as a lock accepts only certain keys. Figure 10.3 illustrates the lock-and-key concept of substrate–enzyme interaction.

The lock-and-key model is more than just a "shape fit." In addition, there are weak binding forces (R group interactions) between parts.

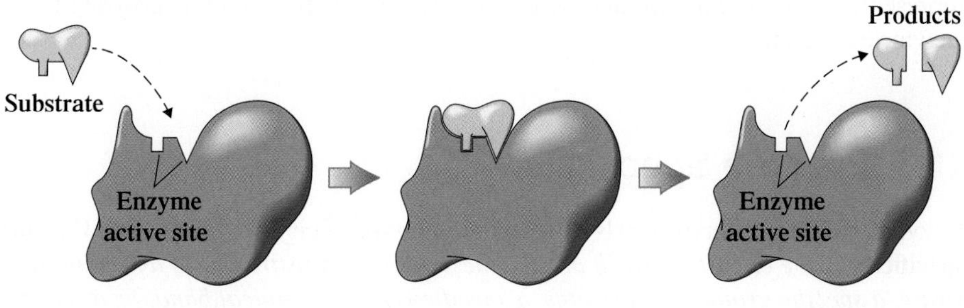

Figure 10.3 The lock-and-key model for enzyme activity. Only a substrate whose shape and chemical nature are complementary to those of the active site can interact with the enzyme.

Induced-Fit Model

The lock-and-key model explains the action of numerous enzymes. It is, however, too restrictive for the action of many other enzymes. Experimental evidence indicates that many enzymes have flexibility in their shapes. They are not rigid and static; there is constant change in their shape. The induced-fit model is used for this type of situation.

The induced-fit model allows for small changes in the shape or geometry of the active site of an enzyme to accommodate a substrate. A good analogy is the changes that occur in the shape of a glove when a hand is inserted into it. The induced fit is a result of the enzyme's flexibility; it adapts to accept the incoming substrate. This model, illustrated in Figure 10.4, is a more thorough explanation

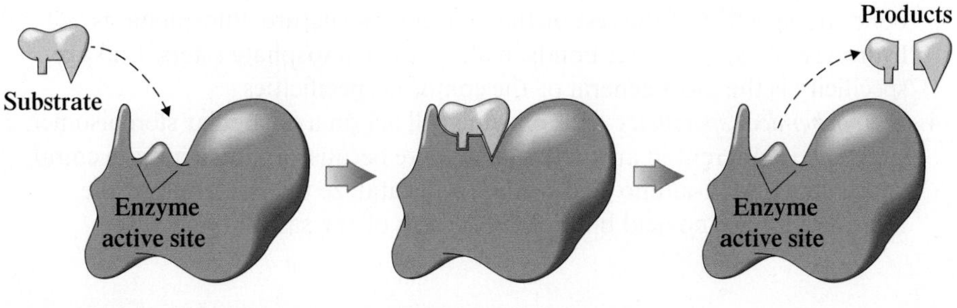

Figure 10.4 The induced-fit model for enzyme activity. The enzyme active site, although not exactly complementary in shape to that of the substrate, is flexible enough that it can adapt to the shape of the substrate.

Figure 10.5 A schematic diagram representing amino acid R group interactions that bind a substrate to an enzyme active site. The R group interactions that maintain the three-dimensional structure of the enzyme (secondary and tertiary structure) are also shown.

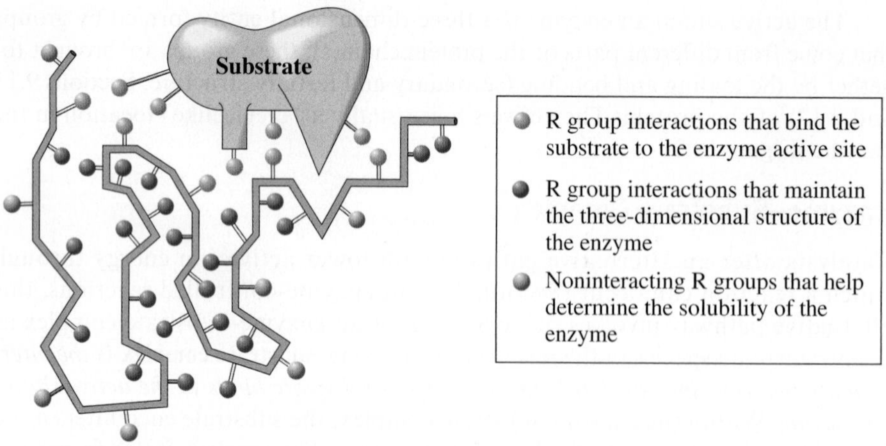

for the active-site properties of an enzyme because it includes the specificity of the lock-and-key model coupled with the flexibility of the enzyme protein.

The forces that draw the substrate into the active site are many of the same forces that maintain tertiary structure in the folding of peptide chains. Electrostatic interactions, hydrogen bonds, and hydrophobic interactions all help attract and bind substrate molecules. For example, a protonated (positively charged) amino group in a substrate could be attracted and held at the active site by a negatively charged aspartate or glutamate residue. Alternatively, cofactors such as positively charged metal ions often help bind substrate molecules. Figure 10.5 is a schematic representation of the amino acid R group interactions that bind a substrate to an enzyme active site.

10.5 Enzyme Specificity

Enzymes exhibit different levels of selectivity, or specificity, for substrates. **Enzyme specificity** *is the extent to which an enzyme's activity is restricted to a specific substrate, a specific group of substrates, a specific type of chemical bond, or a specific type of chemical reaction.* The degree of enzyme specificity is determined by the active site. Some active sites accommodate only one particular compound, whereas others can accommodate a "family" of closely related compounds. Types of enzyme specificity include:

1. *Absolute specificity*—the enzyme will catalyze only one reaction. This most restrictive of all specificities is not common. *Catalase* is an enzyme with absolute specificity. It catalyzes the conversion of hydrogen peroxide (H_2O_2) to O_2 and H_2O. Hydrogen peroxide is the only substrate it will accept.
2. *Group specificity*—the enzyme will act only on molecules that have a specific functional group, such as hydroxyl, amino, or phosphate groups. Carboxypeptidase is group-specific; it cleaves amino acids, one at a time, from the carboxyl end of a peptide chain.
3. *Linkage specificity*—the enzyme will act on a particular type of chemical bond, irrespective of the rest of the molecular structure. Phosphatases hydrolyze phosphate-ester bonds in all types of phosphate esters. Linkage specificity is the most general of the common specificities.
4. *Stereochemical specificity*—the enzyme will act on a particular stereoisomer. Chirality is inherent in an enzyme active site because amino acids are chiral compounds. An L-amino acid oxidase will catalyze the oxidation of the L-form of an amino acid but not the D-form of the same amino acid.

10.6 Factors That Affect Enzyme Activity

Enzyme activity *is a measure of the rate at which an enzyme converts substrate to products in a biochemical reaction.* Four factors affect enzyme activity: temperature, pH, substrate concentration, and enzyme concentration.

Temperature

Temperature is a measure of the kinetic energy (energy of motion) of molecules. Higher temperatures mean molecules are moving faster and colliding more frequently. This concept applies to collisions between substrate molecules and enzymes. As the temperature of an enzymatically catalyzed reaction increases, so does the rate (velocity) of the reaction.

However, when the temperature increases beyond a certain point, the increased energy begins to cause disruptions in the tertiary structure of the enzyme; denaturation is occurring. Change in tertiary structure at the active site impedes catalytic action, and the enzyme activity quickly decreases as the temperature climbs past this point (Figure 10.6). The temperature that produces maximum activity for an enzyme is known as the optimum temperature for that enzyme. **Optimum temperature** *is the temperature at which an enzyme exhibits maximum activity.*

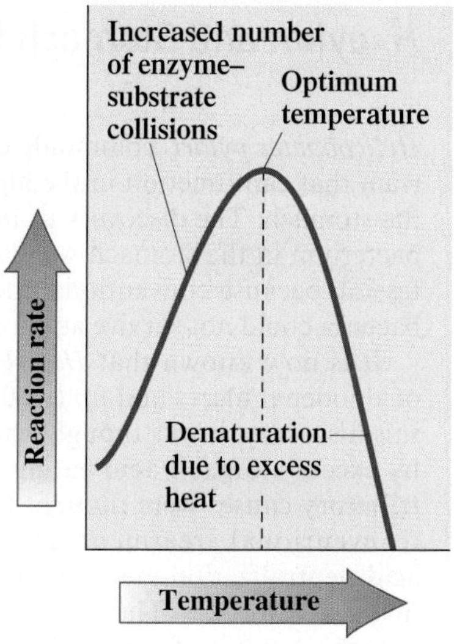

Figure 10.6 A graph showing the effect of temperature on the rate of an enzymatic reaction.

For human enzymes, the optimum temperature is around 37°C, normal body temperature. A person who has a fever where body core temperature exceeds 40°C can be in a life-threatening situation because such a temperature is sufficient to initiate enzyme denaturation. The loss of function of critical enzymes, particularly those of the central nervous system, can result in dysfunction sufficient to cause death.

The "destroying" effect of temperature on bacterial enzymes is used in a hospital setting to sterilize medical instruments and laundry. In high-temperature, high-pressure vessels called *autoclaves,* super-heated steam is used to produce a temperature sufficient to denature bacterial enzymes.

pH

The pH of an enzyme's environment can affect its activity. This is not surprising because the *charge* on acidic and basic amino acids (Section 9.2) located at the active site depends on pH. Small changes in pH (less than one unit) can result in enzyme denaturation (Section 9.16) and subsequent loss of catalytic activity.

Most enzymes exhibit maximum activity over a very narrow pH range. Only within this narrow pH range do the enzyme's amino acids exist in properly charged forms (Section 9.4). **Optimum pH** *is the pH at which an enzyme exhibits maximum activity.* Figure 10.7 shows the effect of pH on an enzyme's activity. Biochemical buffers help maintain the optimum pH for an enzyme.

Each enzyme has a characteristic optimum pH, which usually falls within the physiological pH range of 7.0–7.5. Notable exceptions to this generalization are the digestive enzymes pepsin and trypsin. Pepsin, which is active in the stomach, functions best at a pH of 2.0. On the other hand, trypsin, which operates in the small intestine, functions best at a pH of 8.0. The amino acid sequences present in pepsin and trypsin are those needed such that the R groups present can maintain protein tertiary structure (Section 9.11) at low (2.0) and high (8.0) pH values, respectively.

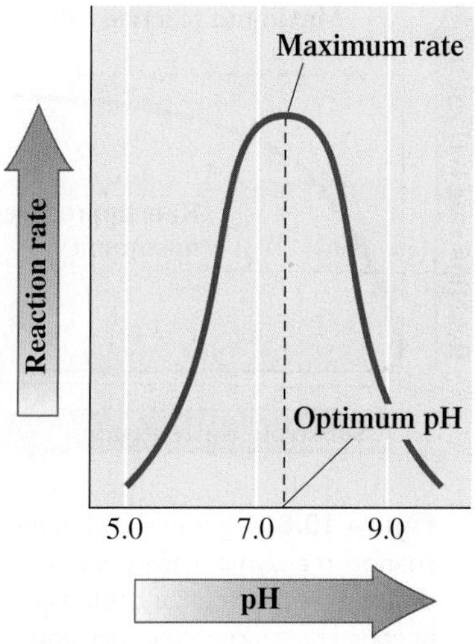

Figure 10.7 A graph showing the effect of pH on the rate of an enzymatic reaction.

A variation from normal pH can also affect substrates, causing either protonation or deprotonation of groups on the substrate. The interaction between the altered substrate and the enzyme active site may be less efficient than normal—or even impossible.

The reason that pickles and other pickled foods do not readily undergo spoilage is due to the acidic conditions associated with their preparation. These acidic conditions significantly reduce the enzymatic activity of any microorganisms present.

An interesting bacterial adaptation to an acidic environment occurs within the human stomach. Bacteria are now known to be the cause of most stomach

CHEMICAL CONNECTIONS 10-B

H. pylori and Stomach Ulcers

Helicobacter pylori, commonly called *H. pylori,* is a bacterium that can function in the highly acidic environment of the stomach. The discovery in 1982 of the existence of this bacterium in the stomach was startling to the medical profession because conventional thought at the time was that bacteria could not survive at the stomach's pH of about 1.4.

It is now known that *H. pylori* causes more than 90% of duodenal ulcers and up to 80% of gastric ulcers. Before this discovery, it was thought that most ulcers were caused by excess stomach acid eating the stomach lining. Contributory causes were thought to be spicy food and stress. Conventional treatment involved acid-suppression or acid-neutralization medications. Now, treatment regimens involve antibiotics. The medical profession was slow to accept the concept of a bacterial cause for most ulcers, and it was not until the mid-1990s that antibiotic treatment became common.

How the enzymes present in the *H. pylori* bacterium can function in the acidic environment of the stomach (where they should be denatured) is now known. Present on the surface of the bacterium is the enzyme *urease,* an enzyme that converts urea to the basic substance ammonia. The ammonia then neutralizes acid present in its immediate vicinity; a protective barrier is thus created. The *urease* itself is protected from denaturation by its complex quaternary structure.

H. pylori causes ulcers by weakening the protective mucous coating of the stomach and duodenum, which allows acid to get through to the sensitive lining beneath. Both the acid and the bacteria irritate the lining and cause a sore—the ulcer. Ultimately the *H. pylori* themselves burrow into the lining to an acid-safe area within the lining.

Approximately two-thirds of the world's population is infected with *H. pylori.* In the United States 30% of the adult population is infected, with the infection most prevalent among older adults. About 20% of people under the age of 40 and half of those over 60 have it. Only one out of every six people infected with *H. pylori* ever suffer symptoms related to ulcers. Why *H. pylori* does not cause ulcers in every infected person is not known.

H. pylori bacteria are most likely spread from person to person through fecal–oral or oral–oral routes. Possible environmental sources include contaminated water sources. The infection is more common in crowded living conditions with poor sanitation. In countries with poor sanitation, 90% of the adult population can be infected.

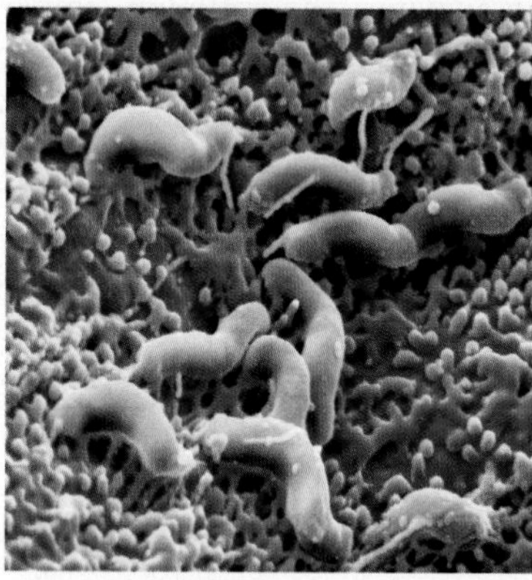

H. pylori bacteria.

ulcers. How these ulcer-causing bacteria survive the highly acidic conditions of the human stomach is considered in the focus on relevancy feature Chemical Connections 10-B above.

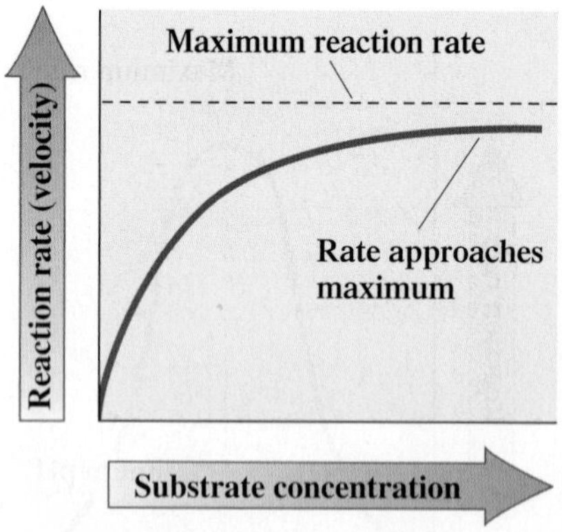

Figure 10.8 A graph showing the change in enzyme activity with a change in substrate concentration at constant temperature, pH, and enzyme concentration. Enzyme activity remains constant after a certain substrate concentration is reached.

Substrate Concentration

When the concentration of an enzyme is kept constant and the concentration of substrate is increased, the enzyme activity pattern shown in Figure 10.8 is obtained. This activity pattern is called a *saturation curve.* Enzyme activity increases up to a certain substrate concentration and thereafter remains constant.

What limits enzymatic activity to a certain maximum value? As substrate concentration increases, the point is eventually reached where enzyme capabilities are used to their maximum extent. The rate remains constant from this point on (Figure 10.8). Each substrate must occupy an enzyme active site for a finite amount of time, and the products must leave the site before the cycle can be repeated. When each enzyme molecule is working at full capacity, the incoming substrate molecules must "wait their turn" for an empty active site. At this point, the enzyme is said to be under saturation conditions.

The rate at which an enzyme accepts substrate molecules and releases product molecules at substrate saturation is given by its turnover number. An enzyme's **turnover number** *is the number of substrate molecules transformed per minute by one*

Table 10.2 Turnover Numbers for Selected Enzymes

Enzyme	Turnover Number (per minute)	Reaction Catalyzed
carbonic anhydrase	36,000,000	$CO_2 + H_2O \rightleftharpoons H_2CO_3$
catalase	5,600,000	$2H_2O_2 \rightleftharpoons 2H_2O + O_2$
cholinesterase	1,500,000	hydrolysis of acetylcholine
penicillinase	120,000	hydrolysis of penicillin
lactate dehydrogenase	60,000	conversion of pyruvate to lactate
DNA polymerase I	900	addition of nucleotides to DNA chains

molecule of enzyme under optimum conditions of temperature, pH, and saturation. Table 10.2 gives turnover numbers for selected enzymes. Some enzymes have a much faster mode of operation than others.

Enzyme Concentration

Because enzymes are not consumed in the reactions they catalyze, the cell usually keeps the number of enzymes low compared with the number of substrate molecules. This is efficient; the cell avoids paying the energy costs of synthesizing and maintaining a large work force of enzyme molecules. Thus, in general, the concentration of substrate in a reaction is much higher than that of the enzyme.

If the amount of substrate present is kept constant and the enzyme concentration is increased, the reaction rate increases because more substrate molecules can be accommodated in a given amount of time. A plot of enzyme activity versus enzyme concentration, at a constant substrate concentration that is high relative to enzyme concentration, is shown in Figure 10.9. The greater the enzyme concentration, the greater the reaction rate.

The Chemistry at a Glance feature on the next page reviews what has been presented about enzyme activity.

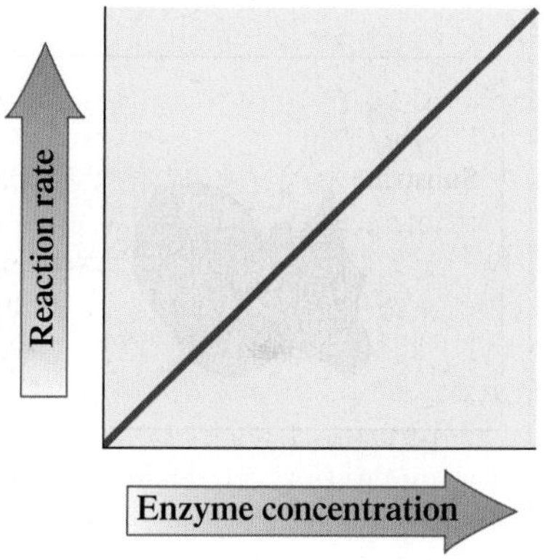

Figure 10.9 A graph showing the change in reaction rate with a change in enzyme concentration for an enzymatic reaction. Temperature, pH, and substrate concentration are constant. The substrate concentration is high relative to enzyme concentration.

EXAMPLE 10.3 Determining How Enzyme Activity Is Affected by Various Changes

Describe the effect that each of the following changes would have on the rate of a biochemical reaction that involves the substrate urea and the liver enzyme urease.

a. Increasing the urea concentration
b. Increasing the urease concentration
c. Increasing the temperature from its optimum value to a value 10°C higher than this value
d. Decreasing the pH by one unit from its optimum value

Solution

a. The enzyme activity rate will *increase* until all of the enzyme molecules are engaged with urea substrate.
b. The enzyme activity rate will *increase* until all of the urea molecules are engaged with urease enzymes.
c. At temperatures higher than the optimum temperature, enzyme activity will *decrease* from that at the optimum temperature.
d. At pH values lower than the optimum pH value, enzyme activity will *decrease* from that at the optimum pH.

(continued)

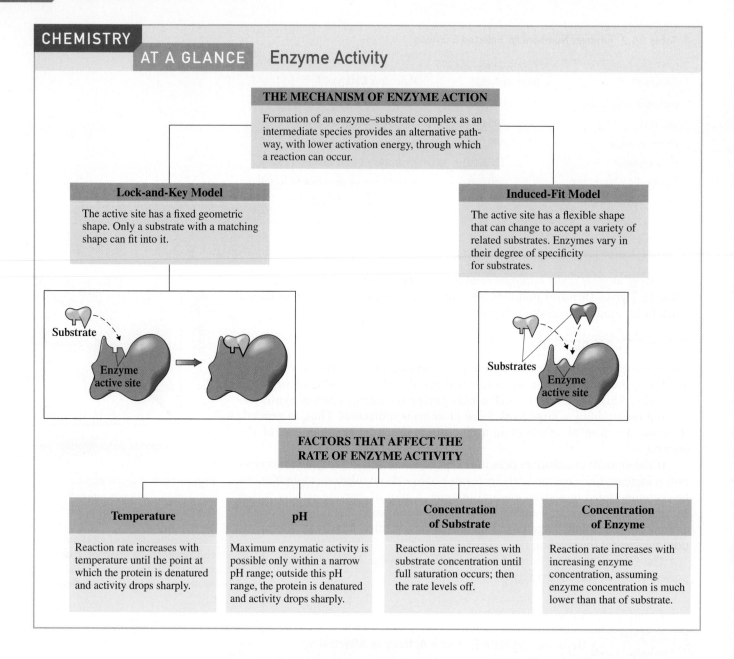

CHEMISTRY AT A GLANCE Enzyme Activity

THE MECHANISM OF ENZYME ACTION

Formation of an enzyme–substrate complex as an intermediate species provides an alternative pathway, with lower activation energy, through which a reaction can occur.

Lock-and-Key Model

The active site has a fixed geometric shape. Only a substrate with a matching shape can fit into it.

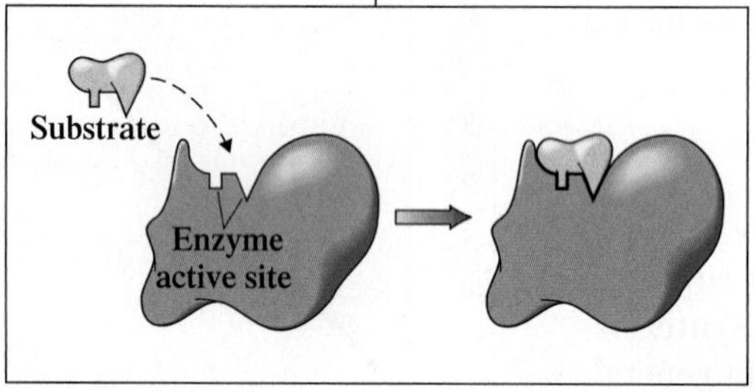

Induced-Fit Model

The active site has a flexible shape that can change to accept a variety of related substrates. Enzymes vary in their degree of specificity for substrates.

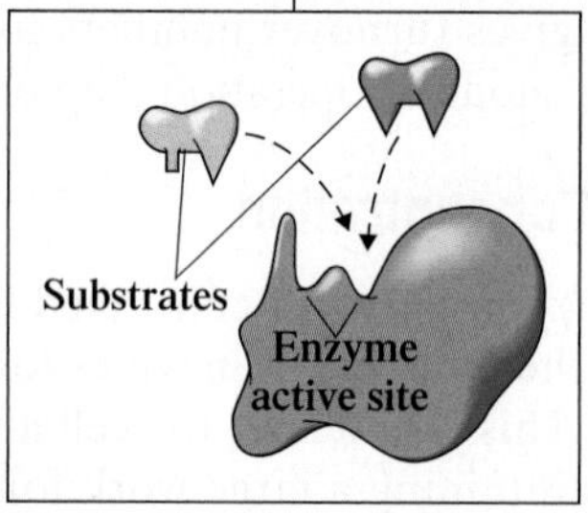

FACTORS THAT AFFECT THE RATE OF ENZYME ACTIVITY

Temperature

Reaction rate increases with temperature until the point at which the protein is denatured and activity drops sharply.

pH

Maximum enzymatic activity is possible only within a narrow pH range; outside this pH range, the protein is denatured and activity drops sharply.

Concentration of Substrate

Reaction rate increases with substrate concentration until full saturation occurs; then the rate levels off.

Concentration of Enzyme

Reaction rate increases with increasing enzyme concentration, assuming enzyme concentration is much lower than that of substrate.

Practice Exercise 10.3

Describe the effect that each of the following changes would have on the rate of a biochemical reaction that involves the substrate sucrose and the intestinal enzyme sucrase.

a. Decreasing the sucrase concentration
b. Increasing the sucrose concentration
c. Lowering the temperature to 10°C
d. Increasing the pH by one unit from its optimum value

Answers: **a.** Decrease rate; **b.** Increase rate; **c.** Decrease rate; **d.** Decrease rate

10.7 Extremozymes

An **extremophile** *is a microorganism that thrives in extreme environments, environments in which humans and most other forms of life could not survive.* Extremophile environments include the hydrothermal areas of Yellowstone National

Park (Figure 10.10) and hydrothermal vents on the ocean floor where temperatures and pressures can be extremely high. The ability of extremophiles to survive under such harsh conditions is related to the amino acid sequences present in the enzymes (proteins) of extremophiles. These sequences are stable under the extraordinary conditions present.

During the 1980s and 1990s research carried out on extremophiles found at many different locations resulted in the identification of numerous diverse types of such entities. Types identified include *acidophiles* (optimal growth at pH levels of 3.0 or below), *alkaliphiles* (optimal growth at pH levels of 9.0 or above), *halophiles* (a salinity that exceeds 0.2 M NaCl needed for growth), *hypothermophiles* (a temperature between 80°C and 122°C needed to thrive), *piezophiles* (a high hydrostatic pressure needed for growth), and *cryophiles* (a temperature of 15°C or lower needed for growth).

The enzymes present in extremophiles are called *extremozymes.* An **extremozyme** *is a microbial enzyme active at conditions that would inactivate human enzymes as well as enzymes present in other types of higher organisms.*

The study of extremozymes is an area of high interest and active research for industrial chemists. Enzymes are heavily used in industrial processes, a context in which they offer advantages similar to those associated with enzyme function in biochemical reactions within living cells. Because industrial processes usually require higher temperatures and pressures than do physiological processes, extremozymes have characteristics that have been found to be useful. The enzymes present in some detergent formulations, which must function in hot water, are the result of research associated with high-temperature microbial enzymes. Similarly, cold-water-wash laundry detergents contain enzymes originally characterized in cold-environment microbial organisms.

The development of commercially useful enzymes using extremophile sources involves the following general approach:

1. Samples containing the extremophile are gathered from the extreme environment where it is found.
2. DNA material is extracted from the extremophile and processed.
3. Macroscopic amounts of the DNA are produced using the polymerase chain reaction (Section 11.15).
4. The macroscopic amount of DNA is analyzed to identify the genes present (Section 11.8) that are involved in extremozyme production.
5. Genetic engineering techniques (Section 11.14) are used to insert the extremozyme gene into bacteria, which then produce the extremozyme.
6. The process is then commercialized.

Commercially produced enzymes are used in the petroleum industry during oil-well drilling operations. A thick mixture of enzymes, guar gum, sand, and water is forced into the bore hole as the drill works its way through rock. At the appropriate time, explosives are used to crack the rock and the enzyme-containing mixture enters the cracks produced. The enzymes present convert the gum mixture into a freely flowing liquid, through hydrolysis of the gum. The freely flowing liquid carries the oil and gas out of the rock. Because of the high temperature generated from the drilling operations and explosive use, the enzymes used must be stable at high temperatures; such enzymes must thus be extremozymes.

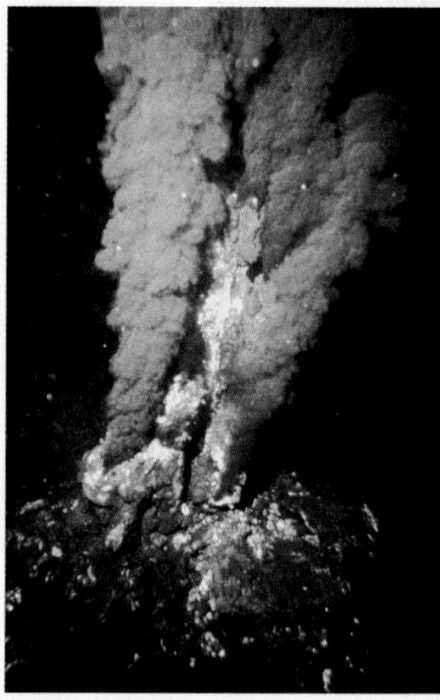

Science Source/Photo Researchers

Figure 10.10 Extremophiles that live in the harsh environment of deep-ocean thermal vents possess enzymes that are adapted to such harsh conditions.

The upper temperature limit for life now stands at 121°C as the result of the discovery, in 2004, of a new "heat-loving" microbe. The microbe was found in a water sample from a hydrothermal vent deep in the Northeast Pacific Ocean. Its method of respiration involves reduction of Fe(III) to Fe(II) to produce energy.

10.8 Enzyme Inhibition

The rates of enzyme-catalyzed reactions can be *decreased* by a group of substances called inhibitors. An **enzyme inhibitor** *is a substance that slows or stops the normal catalytic function of an enzyme by binding to it.* In this section, three modes by which inhibition takes place are considered: reversible competitive inhibition, reversible noncompetitive inhibition, and irreversible inhibition.

The treatment for methanol poisoning involves giving a patient intravenous ethanol (Section 3.5). This action is based on the principle of competitive enzyme inhibition. The same enzyme, *alcohol dehydrogenase*, detoxifies both methanol and ethanol. Ethanol has 10 times the affinity for the enzyme than methanol has. Keeping the enzyme busy with ethanol as the substrate gives the body time to excrete the methanol before it is oxidized to the potentially deadly formaldehyde (Section 3.5).

Reversible Competitive Inhibition

In Section 10.5 it was noted that enzymes are quite specific about the molecules they accept at their active sites. Molecular shape and charge distribution are key determining factors in whether an enzyme accepts a molecule. A **competitive enzyme inhibitor** *is a molecule that sufficiently resembles an enzyme substrate in shape and charge distribution that it can compete with the substrate for occupancy of the enzyme's active site.*

When a competitive inhibitor binds to an enzyme active site, the inhibitor remains unchanged (no reaction occurs), but its physical presence at the site prevents a normal substrate molecule from occupying the site. The result is a decrease in enzyme activity.

The formation of an enzyme–competitive inhibitor complex is a reversible process because it is maintained by weak interactions (hydrogen bonds, etc.). With time (a fraction of a second), the complex breaks up. The empty active site is then available for a new occupant. Substrate and inhibitor again compete for the empty active site. Thus the active site of an enzyme binds either inhibitor or normal substrate on a random basis. If inhibitor concentration is greater than substrate concentration, the inhibitor dominates the occupancy process. The reverse is also true. Competitive inhibition can be reduced by simply increasing the concentration of the substrate.

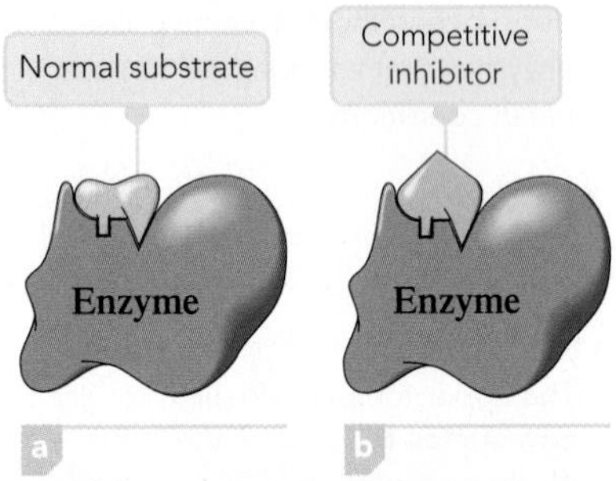

Figure 10.11 A comparison of an enzyme with a substrate at its active site (a) and an enzyme with a competitive inhibitor at its active site (b).

Figure 10.11 compares the binding of a normal substrate and that of a competitive inhibitor at an enzyme's active site. Note that the portions of these two molecules that bind to the active site have the same shape, but that the two molecules differ in *overall* shape. It is because of this overall difference in shape that the substrate reacts at the active site but the inhibitor does not.

Numerous drugs act by means of competitive inhibition. For example, antihistamines are competitive inhibitors of histidine decarboxylation, the enzymatic reaction that converts histidine to histamine. Histamine causes the usual allergy and cold symptoms: watery eyes and runny nose.

Reversible Noncompetitive Inhibition

A **noncompetitive enzyme inhibitor** *is a molecule that decreases enzyme activity by binding to a site on an enzyme other than the active site.* The substrate can still occupy the active site, but the presence of the inhibitor causes a change in the structure of the enzyme sufficient to prevent the catalytic groups at the active site from properly effecting their catalyzing action. Figure 10.12 contrasts the processes of reversible competitive inhibition and reversible noncompetitive inhibition.

Figure 10.12 The difference between a reversible competitive inhibitor and a reversible noncompetitive inhibitor.

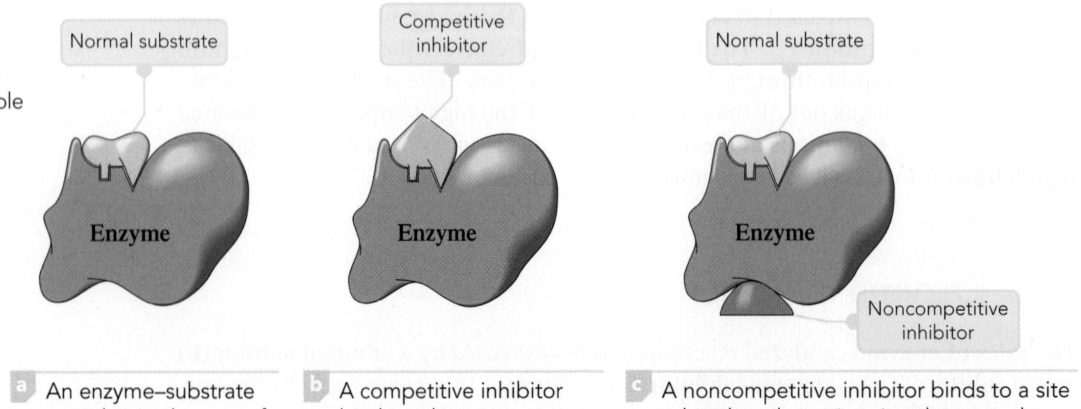

a An enzyme–substrate complex in absence of an inhibitor.

b A competitive inhibitor binds to the active site, which prevents the normal substrate from binding to the site.

c A noncompetitive inhibitor binds to a site other than the active site; the normal substrate still binds to the active site but the enzyme cannot catalyze the reaction due to the presence of the inhibitor.

Unlike the situation in competitive inhibition, increasing the concentration of substrate does not completely overcome the inhibitory effect in this case. However, lowering the concentration of a noncompetitive inhibitor sufficiently does free up many enzymes, which then return to normal activity.

Examples of noncompetitive inhibitors include the heavy metal ions Pb^{2+}, Ag^{+}, and Hg^{2+}. The binding sites for these ions are sulfhydryl (—SH) groups located away from the active site. Metal sulfide linkages are formed, an effect that disrupts secondary and tertiary structure.

Irreversible Inhibition

An **irreversible enzyme inhibitor** *is a molecule that inactivates enzymes by forming a strong covalent bond to an amino acid side-chain group at the enzyme's active site.* In general, such inhibitors do *not* have structures similar to that of the enzyme's normal substrate. The inhibitor–active site bond is sufficiently strong that addition of excess substrate does not reverse the inhibition process. Thus the enzyme is permanently deactivated. The actions of chemical warfare agents (nerve gases) and organophosphate insecticides are based on irreversible inhibition.

The Chemistry at a Glance feature on the next page summarizes what has been considered concerning enzyme inhibition.

EXAMPLE 10.4 Identifying the Type of Enzyme Inhibition from Inhibitor Characteristics

Identify the type of enzyme inhibition each of the following inhibitor characteristics is associated with.

a. An inhibitor that decreases enzyme activity by binding to a site on the enzyme other than the active site
b. An inhibitor that inactivates enzymes by forming a strong covalent bond at the enzyme active site

Solution

a. Inhibitor binding at a nonactive site location is a characteristic of a *reversible noncompetitive inhibitor.*
b. Covalent bond formation at the active site, with amino acid residues located there, is a characteristic of an *irreversible inhibitor*.

Practice Exercise 10.4

Identify the type of enzyme inhibition each of the following inhibitor characteristics is associated with.

a. An inhibitor that has a shape and charge distribution similar to that of the enzyme's normal substrate
b. An inhibitor whose effect can be reduced by simply increasing the concentration of normal substrate present

Answers: **a.** Reversible competitive inhibitor; **b.** Reversible competitive inhibitor

10.9 Regulation of Enzyme Activity

Regulation of enzyme activity within a cell is a necessity for many reasons. Illustrative of this need are the following two situations, both of which involve the concept of energy conservation.

1. A cell that continually produces large amounts of an enzyme for which substrate concentration is always very low is wasting energy. The production of the enzyme needs to be "turned off."
2. A product of an enzyme-catalyzed reaction that is present in plentiful (more than needed) amounts in a cell is a waste of energy if the enzyme continues to catalyze the reaction that produces the product. The enzyme needs to be "turned off."

CHEMISTRY AT A GLANCE Enzyme Inhibition

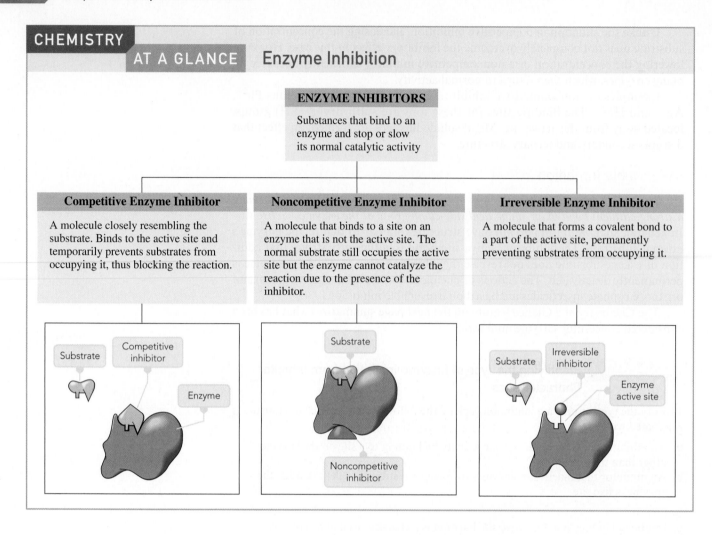

ENZYME INHIBITORS

Substances that bind to an enzyme and stop or slow its normal catalytic activity

Competitive Enzyme Inhibitor

A molecule closely resembling the substrate. Binds to the active site and temporarily prevents substrates from occupying it, thus blocking the reaction.

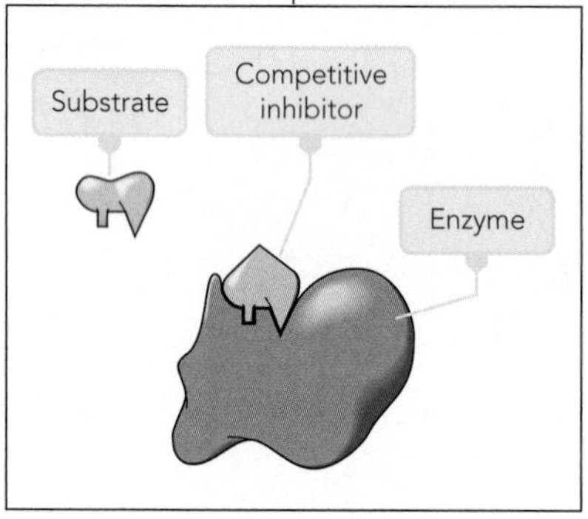

Noncompetitive Enzyme Inhibitor

A molecule that binds to a site on an enzyme that is not the active site. The normal substrate still occupies the active site but the enzyme cannot catalyze the reaction due to the presence of the inhibitor.

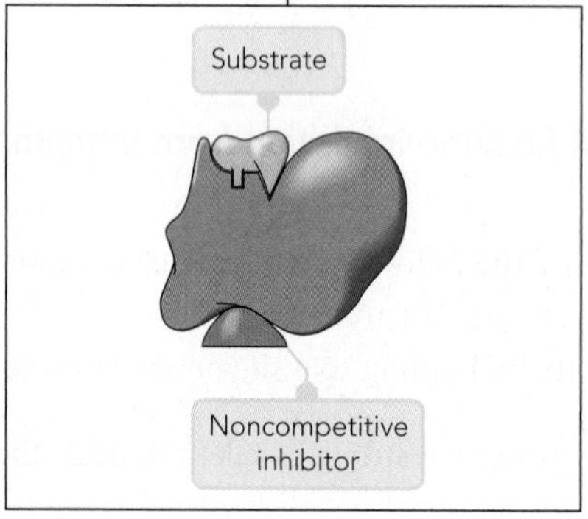

Irreversible Enzyme Inhibitor

A molecule that forms a covalent bond to a part of the active site, permanently preventing substrates from occupying it.

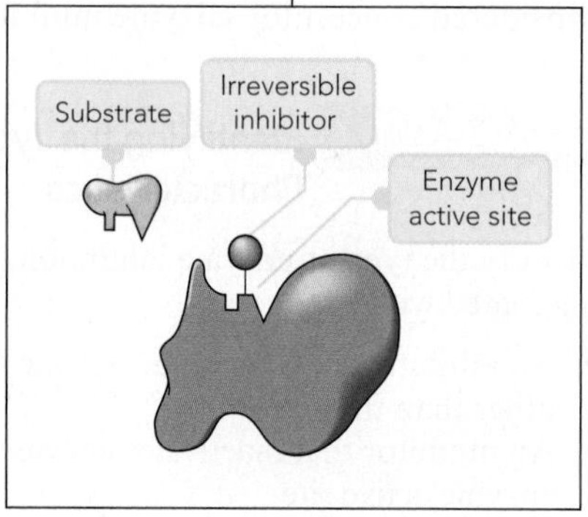

Many mechanisms exist by which enzymes within a cell can be "turned on" and "turned off." In this section, three such mechanisms are considered: (1) feedback control associated with allosteric enzymes (2) proteolytic enzymes and zymogens and (3) covalent modification.

Allosteric Enzymes

Many, but not all, of the enzymes responsible for regulating cellular processes are *allosteric enzymes.* Characteristics of allosteric enzymes are as follows:

1. All allosteric enzymes have quaternary structure; that is, they are composed of two or more protein chains.
2. All allosteric enzymes have two kinds of binding sites: those for substrate and those for regulators.
3. Active and regulatory binding sites are distinct from each other in both location and shape. Often the regulatory site is on one protein chain and the active site is on another.
4. Binding of a molecule at the regulatory site causes changes in the overall three-dimensional structure of the enzyme, including structural changes at the active site.

The term *allosteric* comes from the Greek *allo*, which means "other," and *stereos*, which means "site or space."

Thus an **allosteric enzyme** *is an enzyme with two or more protein chains (quaternary structure) and two kinds of binding sites (substrate and regulator).*

Substances that bind at regulatory sites of allosteric enzymes are called *regulators.* The binding of a *positive regulator* increases enzyme activity; the shape of the active site is changed such that it can more readily accept substrate. The binding of a *negative regulator* (a noncompetitive inhibitor; Section 10.8) decreases enzyme

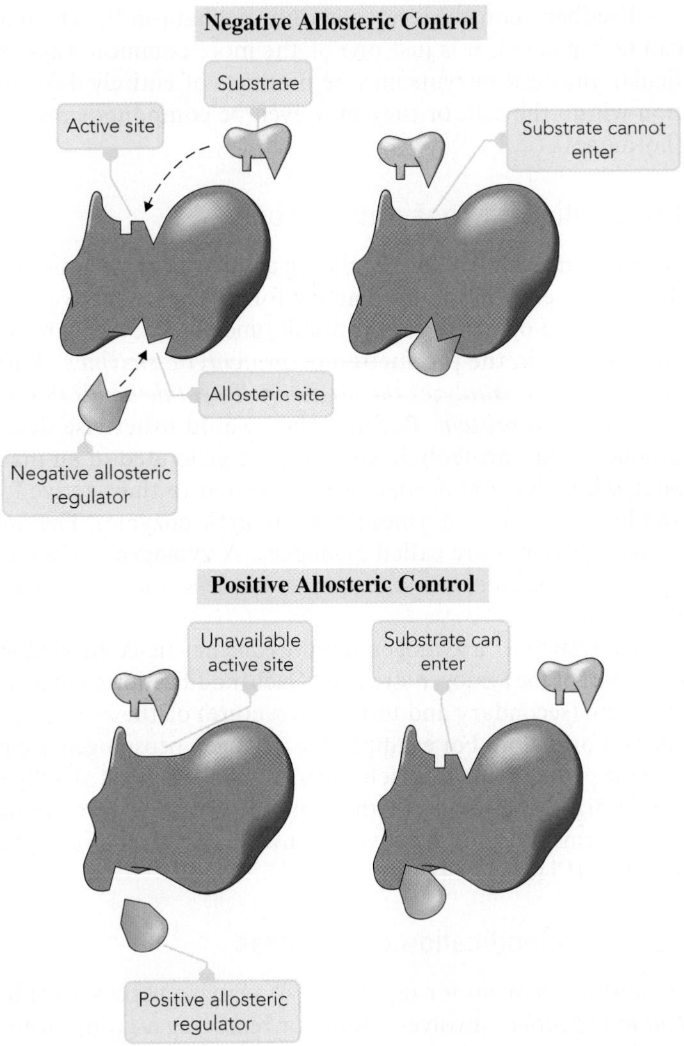

Figure 10.13 The differing effects of positive and negative regulators on an allosteric enzyme.

activity; changes to the active site are such that substrate is less readily accepted. Figure 10.13 contrasts the different effects on an allosteric enzyme that positive and negative regulators produce.

Some regulators of allosteric enzyme function are inhibitors (negative regulators), and some increase enzyme activity (positive regulators).

Feedback Control

One of the mechanisms by which allosteric enzyme activity is regulated is feedback control. **Feedback control** *is a process in which activation or inhibition of the first reaction in a reaction sequence is controlled by a product of the reaction sequence.*

As illustrative of the feedback control mechanism, consider a biochemical process within a cell that occurs in several steps, each step catalyzed by a different enzyme.

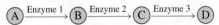

$$\text{(A)} \xrightarrow{\text{Enzyme 1}} \text{(B)} \xrightarrow{\text{Enzyme 2}} \text{(C)} \xrightarrow{\text{Enzyme 3}} \text{(D)}$$

The product of each step is the substrate for the next enzyme.

Most biochemical processes within cells take place in several steps rather than in a single step. A different enzyme is required for each step of the process.

What will happen in this reaction series if the final product (D) is a negative regulator of the first enzyme (enzyme 1)? At low concentrations of D, the reaction sequence proceeds rapidly. At higher concentrations of D, the activity of enzyme 1 becomes inhibited (by feedback), and eventually the activity stops. At the stopping point, there is sufficient D present in the cell to meet its needs. Later, when the concentration of D decreases through use in other cell reactions, the activity of enzyme 1 increases and more D is produced.

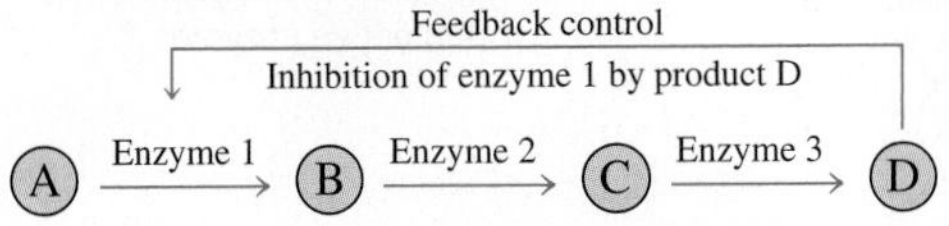

Feedback control is operative in devices that maintain a constant temperature, such as a furnace or an oven. If you set the control thermostat at 68°F, the furnace or oven produces heat until that temperature is reached, and then, through electronic feedback, the furnace or oven is shut off.

The general term *allosteric control* is often used to describe a process in which a regulatory molecule that binds at one site in an enzyme influences substrate binding at the active site in the enzyme.

Feedback control is not the only mechanism by which an allosteric enzyme can be regulated; it is just one of the more common ways. Regulators of a particular allosteric enzyme may be products of entirely different pathways of reaction within the cell, or they may even be compounds produced outside the cell (hormones).

Proteolytic Enzymes and Zymogens

A second mechanism for regulating cellular enzyme activity is based on the production of enzymes in an inactive form. These inactive enzyme precursors are then "turned on" at the appropriate time. Such a mechanism for control is often encountered in the production of *proteolytic enzymes.* A **proteolytic enzyme** *is an enzyme that catalyzes the breaking of peptide bonds that maintain the primary structure of a protein.* Because they would otherwise destroy the tissues that produce them, proteolytic enzymes are generated in an inactive form and then later, when they are needed, are converted to their active form. Most digestive and blood-clotting enzymes are proteolytic enzymes. The inactive forms of proteolytic enzymes are called *zymogens.* A **zymogen** *is the inactive precursor of a proteolytic enzyme.* (An alternative, but less often used, name for a zymogen is *proenzyme.*)

The names of zymogens can be recognized by the suffix *-ogen* or the prefix *pre-* or *pro-*.

Activation of a zymogen requires an enzyme-controlled reaction that removes some part of the zymogen structure. Such modification changes the three-dimensional structure (secondary and tertiary structure) of the zymogen, which affects active site conformation. For example, the zymogen pepsinogen is converted to the active enzyme pepsin in the stomach, where it then functions as a digestive enzyme. Pepsin would digest the tissues of the stomach wall if it were prematurely generated in active form. Pepsinogen activation involves removal of a peptide fragment from its structure (Figure 10.14).

Covalent Modification of Enzymes

A third mechanism for regulation of enzyme activity within a cell, called *covalent modification,* involves adding or removing a group from an enzyme through the forming or breaking of a covalent bond. **Covalent modification** *is a process in which enzyme activity is altered by covalently modifying the structure of the enzyme through attachment of a chemical group to or removal of a chemical group from a particular amino acid within the enzyme's structure.*

The most commonly encountered type of covalent modification involves the processes by which a phosphate group is added to or removed from an enzyme. The source of the added phosphate group is often an ATP molecule. The process of

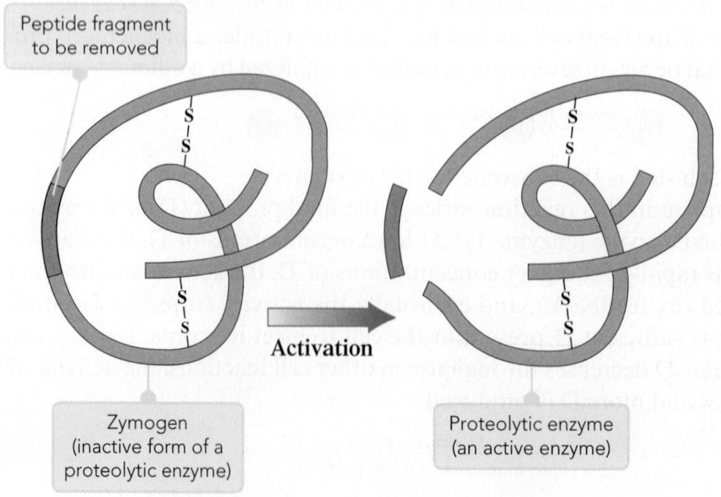

Figure 10.14 Conversion of a zymogen (the inactive form of a proteolytic enzyme) to a proteolytic enzyme (the active form of the enzyme) often involves removal of a peptide chain segment from the zymogen structure.

addition of the phosphate group to the enzyme is called *phosphorylation,* and the removal of the phosphate group from the enzyme is called *dephosphorylation.* This phosphorylation/dephosphorylation process is the off/on or on/off switch for the enzyme. For some enzymes, the active ("turned-on" form) is the phosphorylated version of the enzyme; however, for other enzymes, it is the dephosphorylated version that is active.

The preceding covalent modification processes are governed by other enzymes. *Protein kinases* effect the addition of phosphate groups, and *phosphatases* catalyze removal of the phosphate groups. Usually, the phosphate group is added to (or removed from) the R group of a serine, tyrosine, or threonine amino acid residue present in the protein (enzyme). The R groups of these three amino acids have a common structural feature, the presence of a free —OH group (see Table 10.1). The hydroxyl group is the site where phosphorylation or dephosphorylation occurs.

Glycogen phosphorylase, an enzyme involved in the breakdown of glycogen to glucose (Section 13.5), is activated by the addition of a phosphate group. *Glycogen synthase,* an enzyme involved in the synthesis of glycogen (Section 13.5), is deactivated by phosphorylation.

10.10 Prescription Drugs That Inhibit Enzyme Activity

Several common types of prescription drugs have modes of action that involve enzyme inhibition. Included among them are ACE inhibitors, sulfa drugs, and penicillins. ACE inhibitors are used to treat high blood pressure conditions as well as several heart conditions. Sulfa drugs and penicillins are two well-known families of antibiotics. Discussion of how these three types of medications function in the human body, in terms of enzyme effects, is illustrative of how enzyme inhibition can be used in a positive manner in treating human disease.

ACE Inhibitors

The acronym ACE stands for *angiotensin-converting enzyme.* Angiotensin is an octapeptide hormone (Section 9.8) involved in blood pressure regulation. It increases blood pressure by narrowing blood vessels. Until needed, angiotensin is present in the body in an inactive form as the zymogen (Section 10.9) angiotensinogen, which is a decapeptide. ACE converts the inactive decapeptide zymogen to the active octapeptide form (angiotensin) by cleaving two amino acids from the zymogen structure.

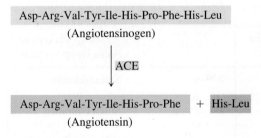

Asp-Arg-Val-Tyr-Ile-His-Pro-Phe-His-Leu
(Angiotensinogen)

↓ ACE

Asp-Arg-Val-Tyr-Ile-His-Pro-Phe + His-Leu
(Angiotensin)

ACE inhibitor medications block the action of ACE in converting angiotensinogen to angiotensin. The effect of this is a lower blood pressure than if the zymogen activation had occurred.

Representative of ACE inhibitors in use today is the compound lisinopril, a compound heavily prescribed for treatment of moderately elevated blood pressure conditions. Structurally, lisinopril is a 1-carboxy-3-phenylpropyl derivative of the dipeptide Lys-Pro.

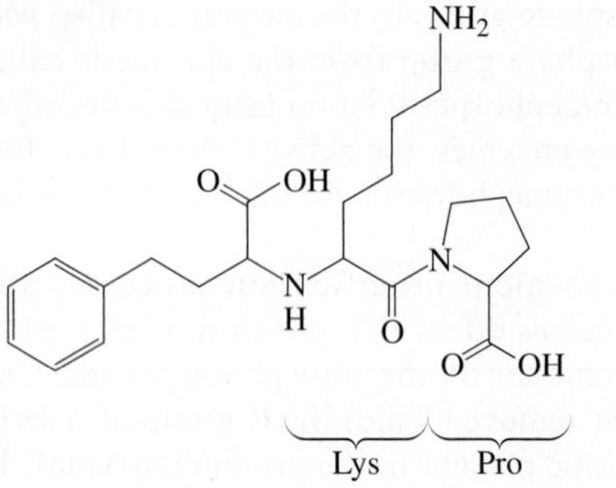

Unlike most other ACE inhibitors, lisinopril is not metabolized (broken down) by liver enzymes; it is excreted unchanged in the urine. Interestingly, lisinopril, as well as several other ACE inhibitors, was first obtained from snake venom, the venom of the jararaca (a Brazilian pit viper).

Sulfa Drugs

The 1932 discovery of the antibacterial activity of the compound sulfanilamide by the German bacteriologist Gerhard Domagk (1895–1964) led to the characterization of a whole family of sulfanilamide derivatives collectively called sulfa drugs—the first "antibiotics" in the medical field.

An **antibiotic** *is a substance that kills bacteria or inhibits their growth.* Antibiotics exert their action selectively on bacteria and do not affect the normal metabolism of the host organism. Antibiotics usually inhibit specific enzymes essential to the life processes of the bacteria.

Sulfanilamide inhibits bacterial growth because it is structurally similar to PABA (*p*-aminobenzoic acid).

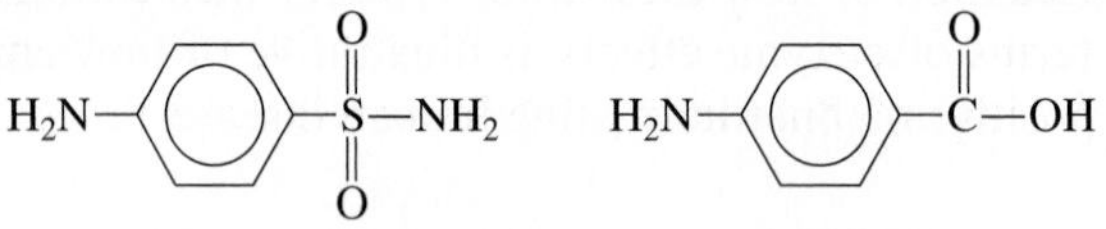

Sulfanilamide

p-Aminobenzoic acid (PABA)

Many bacteria need PABA in order to produce an important coenzyme, folic acid. Sulfanilamide acts as a competitive inhibitor to enzymes in the biosynthetic pathway for converting PABA into folic acid in these bacteria. Folic acid deficiency retards growth of the bacteria and can eventually kill them.

Sulfa drugs selectively inhibit only bacteria metabolism and growth because humans absorb folic acid from their diet and thus do not use PABA for its synthesis. A few of the most common sulfa drugs and their structures are shown in Figure 10.15.

Figure 10.15 Structures of selected sulfa drugs in use today as antibiotics.

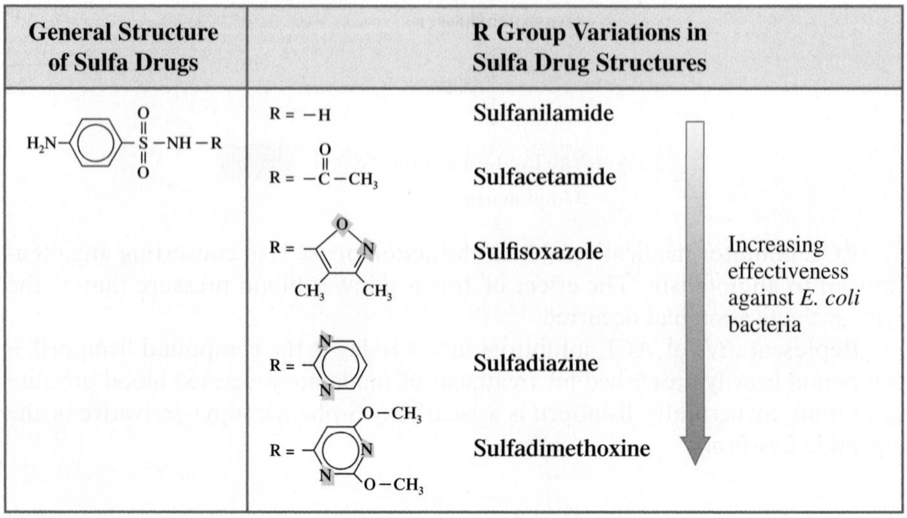

Penicillins

Penicillin, one of the most widely used antibiotics, was accidentally discovered by Alexander Fleming in 1928 while he was working with cultures of an infectious staphylococcus bacterium. A decade later, the scientists Howard Flory and Ernst Chain isolated penicillin in pure form and proved its effectiveness as an antibiotic.

Several naturally occurring penicillins have now been isolated, and numerous derivatives of these substances have been synthetically produced. All have structures containing a four-membered β-lactam ring (Section 6.12) fused with a five-membered thiazolidine ring (Figure 10.16). As with sulfa drugs, derivatives of the basic structure differ from each other in the identity of a particular R group.

Penicillins inhibit *transpeptidase,* an enzyme that catalyzes the formation of peptide cross links between polysaccharide strands in bacterial cell walls. These cross links strengthen cell walls. A strong cell wall is necessary to protect the bacterium from lysis (breaking open). By inhibiting transpeptidase, penicillin prevents the formation of a strong cell wall. Any osmotic or mechanical shock then causes lysis, killing the bacterium.

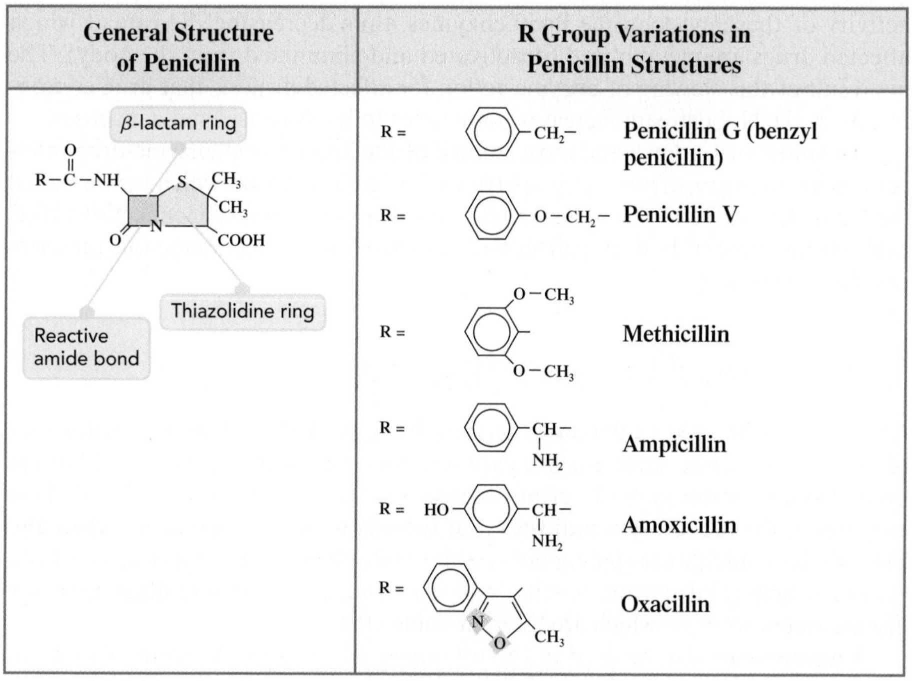

Figure 10.16 Structures of selected penicillins in use today as antibiotics.

Penicillin's unique action depends on two aspects of enzyme deactivation that have been discussed before: structural similarity to the enzyme's natural substrate and irreversible inhibition. Penicillin is *highly specific* in binding to the active site of transpeptidase. In this sense, it acts as a very selective competitive inhibitor. However, unlike a normal competitive inhibitor, once bound to the active site, the β-lactam ring opens as the highly reactive amide bond forms a covalent bond to a critical serine residue required for normal catalytic action. The result is an irreversibly inhibited transpeptidase enzyme (Figure 10.17).

Some bacteria produce the enzyme *penicillinase,* which protects them from penicillin. Penicillinase selectively binds penicillin and catalyzes the opening of the β-lactam ring before penicillin can form a covalent bond to the enzyme. Once the ring is opened, the penicillin is no longer capable of inactivating transpeptidase.

Certain semi-synthetic penicillins such as methicillin and amoxicillin have been produced that are resistant to penicillinase activity and are thus clinically important.

Penicillin does not usually interfere with normal metabolism in humans because of its highly selective binding to bacterial transpeptidase. This selectivity makes penicillin an extremely useful antibiotic.

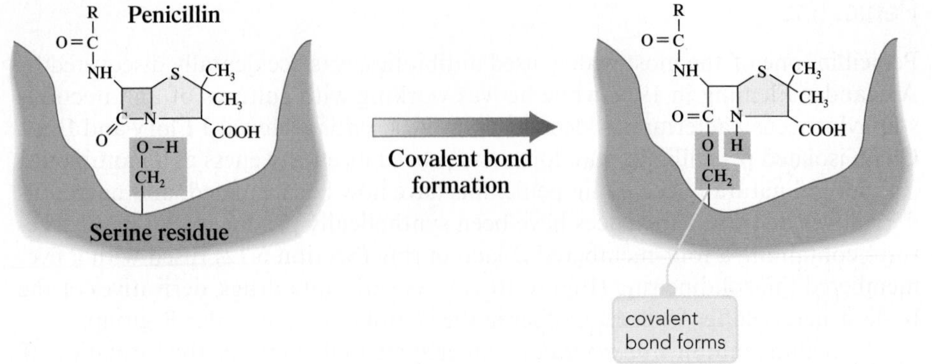

Figure 10.17 The selective binding of penicillin to the active site of transpeptidase. Subsequent irreversible inhibition through formation of a covalent bond to a serine residue permanently blocks the active site.

Food-Enzyme Interactions That Affect Prescription Medications

Liver enzymes known as *cytochrome P450s* are involved in the processes by which many prescription medications are metabolized in the body. It is now known that several foods and herbal remedies contain compounds that decrease (inhibit) the activity of these cytochrome P450 enzymes, thus decreasing the rate at which affected drugs are metabolized (deactivated and eliminated from the body). The net result of this slowing of enzyme action for affected drugs is that drug concentration in the bloodstream increases, sometimes to levels considered dangerous.

The most well-known and most studied of the known food-enzyme-drug interactions involves grapefruit and grapefruit juice, an interaction which is now called the "grapefruit effect." The focus on relevancy feature Chemical Connections 10-C explores the topic of how grapefruit and grapefruit juice affect numerous prescription medications.

10.11 Medical Uses of Enzymes

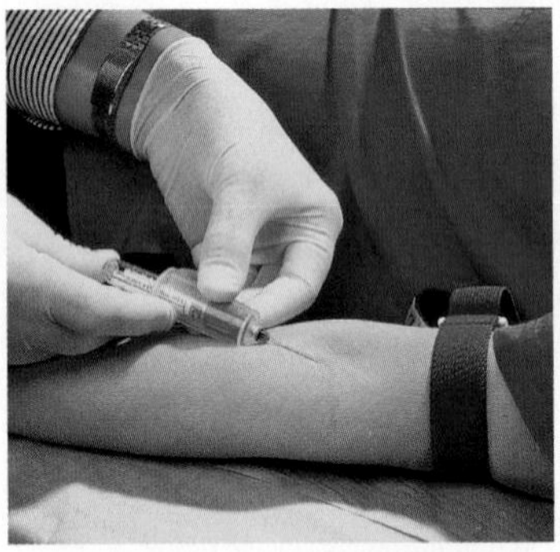

Saturn Stills/SPL/Photo Researchers

Figure 10.18 Drawing of a blood sample. Determination of enzyme concentrations in blood provides important information about the "state" of various organs within the human body.

Enzymes can be used to diagnose certain diseases. Although blood serum contains many enzymes, some enzymes are not normally found in the blood but are produced only inside cells of certain organs and tissues. The appearance of these enzymes in the blood often indicates that there is tissue damage in an organ and that cellular contents are spilling out (leaking) into the bloodstream (Figure 10.18). Assays of abnormal enzyme activity in blood serum can be used to diagnose many disease states, some of which are listed in Table 10.3.

Enzymes can also be used in the treatment of diseases. A recent advance in treating heart attacks is the use of tissue plasminogen activator (TPA), which activates the enzyme plasminogen. When so activated, this enzyme dissolves blood clots in the heart and often provides immediate relief.

Table 10.3 Selected Blood Enzyme Assays Used in Diagnostic Medicine

Enzyme	Condition Indicated by Abnormal Level
lactate dehydrogenase (LDH)	heart disease, liver disease
creatine phosphokinase (CPK)	heart disease
aspartate transaminase (AST)	heart disease, liver disease, muscle damage
alanine transaminase (ALT)	heart disease, liver disease, muscle damage
gamma-glutamyl transpeptidase (GGTP)	heart disease, liver disease
alkaline phosphatase (ALP)	bone disease, liver disease

CHEMICAL CONNECTIONS 10-C

Enzymes, Prescription Medications, and the "Grapefruit Effect"

Grapefruit and grapefruit juice are excellent sources of healthful compounds such as vitamin C and lycopene. The vitamin C content of half a grapefruit can be up to 80% of the recommended daily amount of this vitamin. The pink and red colors of grapefruit are due to the carotenoid lycopene (Section 2.7). Among the common dietary carotenoids, lycopene has one of the highest antioxidant activities (Section 3.14). Antioxidant activity is important in controlling oxygen free radicals, which are substances that can cause significant cellular damage.

Grapefruit and grapefruit juice also contain compounds that can slow down the metabolism of several common prescription medications by interacting with (inhibiting) the enzymes needed for their metabolism—liver enzymes called *cytochrome P450s*. This enzyme inhibition, known as the "grapefruit effect," leads to increased concentrations of the affected drugs in the bloodstream, which increases the risk of potentially serious side effects from the drugs.

More than 50 medications of diverse types are now known to be affected by the "grapefruit effect." A partial listing of affected medications is as follows.

Grapefruit and grapefruit juice contain several healthful compounds (vitamin C, lycopene, etc.), as well as a compound that, through enzyme interaction, can adversely affect the breakdown of prescription medications within the body.

Drug name	Type of drug
Amiodarone (Cordarone)	A drug used to treat and prevent abnormal heart rhythms (arrhythmias)
Buspirone (BuSpar), sertraline (Zoloft)	Antidepressants
Carbamazepine (Carbatrol, Tegretol)	An anti-seizure medication
Cyclosporine (Neoral, Sandimmune), tacrolimus (Prograf)	Immunosuppressant drugs
Felodipine (Plendil), nifedipine (Procardia), nimodipine (Nimotop), nisoldipine (Sular)	Calcium-channel blockers used to treat high blood pressure
Saquinavir	An HIV medication
Simvastatin (Zocor), lovastatin (Mevacor), atorvastatin (Lipitor)	Statins used to treat high cholesterol

This grapefruit effect is significant enough that the United States Food and Drug Administration (FDA) now requires that all new medications be tested for problems arising from grapefruit interactions before the medication is approved for use. Drugs whose metabolism is affected by grapefruit and grapefruit juice must carry warnings in their patient information sheets.

The actual compounds responsible for the grapefruit effect have been identified. They are the compound bergamottin and one of its metabolites (a dihydroxyderivative of bergamottin). Both compounds have a fused three-ring structure containing ether and ester functional groups to which an unsaturated carbon chain is attached.

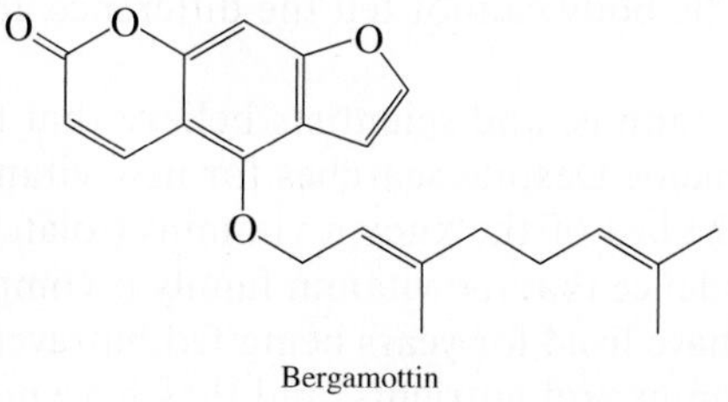

Bergamottin

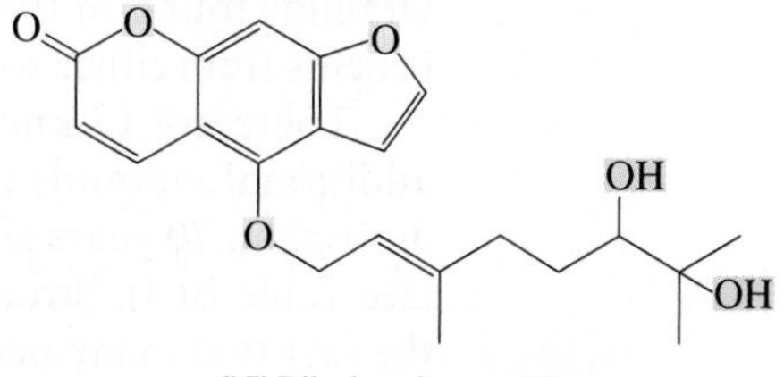

6′,7′-Dihydroxybergamottin

(continued)

Bergamottin and its metabolite exert their effect by a process called *suicide inactivation:* The cytochrome P450 enzyme transforms them into highly reactive intermediates, which then react with (inhibit) the enzyme. The inhibiting interaction involves modification of a glutamine residue (an amino acid residue; Section 9.2) on the enzyme's surface; it is not the enzyme's active site that is affected.

Normally, the grapefruit effect is thought of as a negative interaction. There is, however, a potentially positive side to this effect. Several research studies underway are exploring the possibility of using bergamottin, its dihydroxy derivative, and synthetic analogs to increase the oral availability of drugs that would otherwise have limited use because they are metabolized too rapidly. In such a case, the dose required to achieve a required effective bloodstream concentration of the drug would be lowered, thus increasing the drug's usefulness.

There are also potential solutions in the offering for "grapefruit lovers" whose medications prevent them from consuming a desired food. Several research projects are underway whose goal is to remove bergamottin from grapefruit juice. One approach involves genetic engineering (Section 11.15) of grapefruit trees so that they produce fruit that does not contain bergamottin. Another approach seeks methods to remove bergamottin from grapefruit juice after the fact. This latter approach is a parallel idea to that which led to decaffeinated coffee; de-bergamottined grapefruit juice could someday be available for consumption.

Another medical use for enzymes is in clinical laboratory chemical analysis. For example, no simple direct test for the measurement of urea in the blood is available. However, if the urea in the blood is converted to ammonia via the enzyme *urease,* the ammonia produced, which is easily measured, becomes an indicator of urea. This *blood urea nitrogen (BUN) test* is a common clinical laboratory procedure. High urea levels in the blood indicate kidney malfunction.

10.12 General Characteristics of Vitamins

This section and those that follow deal with vitamins. Vitamins are considered in conjunction with enzymes because many enzymes contain vitamins as part of their structure. Recall from Section 10.2 that conjugated enzymes have a protein part (apoenzyme) and a nonprotein part (cofactor). Vitamins, in many cases, are cofactors in conjugated enzymes.

The spelling of the term *vitamin* was originally *vitamine,* a word derived from the Latin *vita,* meaning "life," and from the fact that these substances were all thought to contain the *amine* functional group. When this supposition was found to be false, the final *e* was dropped from *vitamine,* and the term *vitamin* came into use. Some vitamins contain amine functional groups, but others do not.

A **vitamin** *is an organic compound, essential in small amounts for the proper functioning of the human body, that must be obtained from dietary sources because the body cannot synthesize it.*

Vitamins differ from the major classes of nutrients in foods (carbohydrates, lipids, and proteins) in the amount required; for vitamins, it is *micro*gram to *milli*gram quantities per day compared with 50–200 grams per day for the major food nutrients categories. To better understand the small amount of vitamins needed by the human body, consider the recommended daily allowance (RDA) of vitamin B_{12}, which is 2.0 micrograms per day for an adult. Just 1.0 gram of this vitamin could theoretically supply the daily needs of 500,000 people.

A well-balanced diet usually meets all the body's vitamin requirements. However, supplemental vitamins are often required for women during pregnancy and for people recovering from certain illnesses.

One of the most common myths associated with the nutritional aspects of vitamins is that vitamins from natural sources are superior to synthetic vitamins. In truth, synthetic vitamins, manufactured in the laboratory, are identical to the vitamins found in foods. The body cannot tell the difference and gets the same benefits from either source.

There are 13 known vitamins, and scientists believe that the discovery of additional vitamins is unlikely. Despite searches for new vitamins, it has been more than 70 years since the last of the known vitamins (folate) was discovered (see Table 10.4). Strong evidence that the vitamin family is complete comes from the fact that many people have lived for years being fed, intravenously, solutions containing the known vitamins and nutrients, and they have not developed any known vitamin deficiency disease.

Table 10.4 The Discovery Dates of the 13 Known Vitamins. (Years can vary in different tabulations depending on the definition of "discovery.")

Year of Discovery	Vitamin
1910	Thiamin
1913	Vitamin A
1920	Vitamin C
1920	Vitamin D
1920	Riboflavin
1922	Vitamin E
1926	Vitamin B_{12}
1929	Vitamin K
1931	Pantothenic acid
1931	Biotin
1934	Vitamin B_6
1936	Niacin
1941	Folate

Solubility characteristics divide the vitamins into two major classes: the water-soluble vitamins and the fat (lipid)-soluble vitamins. There are nine water-soluble vitamins and four fat-soluble vitamins. Table 10.5 lists the vitamins in each of these solubility categories. Water-soluble vitamins must be constantly replenished in the body because they are rapidly eliminated from the body in the urine. They are carried in the bloodstream, are needed in frequent, small doses, and are unlikely to be toxic except when taken in unusually large doses. The fat-soluble vitamins are found dissolved in lipid materials. They are, in general, carried in the blood by protein carriers, are stored in fat tissues, are needed in periodic doses, and are more likely to be toxic when consumed in excess of need.

An important difference exists, in terms of function, between water-soluble and fat-soluble vitamins. Water-soluble vitamins function as coenzymes for a number of important biochemical reactions in humans, animals, and microorganisms. Fat-soluble vitamins generally do not function as coenzymes in humans and animals and are rarely utilized in any manner by microorganisms. Other differences between the two categories of vitamins are summarized in Table 10.6. A few exceptions occur, but the differences shown in this table are generally valid.

Table 10.5 The Water-Soluble and Fat-Soluble Vitamins

Water-Soluble Vitamins	Fat-Soluble Vitamins
Vitamin C	Vitamin A
Thiamin	Vitamin D
Riboflavin	Vitamin E
Niacin	Vitamin K
Pantothenic acid	
Vitamin B_6	
Biotin	
Folate	
Vitamin B_{12}	

Table 10.6 The General Properties of Water-Soluble Vitamins and Fat-Soluble Vitamins

	Water-Soluble Vitamins (B vitamins and vitamin C)	Fat-Soluble Vitamins (vitamins A, D, E, and K)
absorption	directly into the blood	first enter into the lymph system
transport	travel without carriers	many require protein carriers
storage	circulate in the water-filled parts of the body	found in the cells associated with fat
excretion	kidneys remove excess in urine	tend to remain in fat-storage sites
toxicity	not likely to reach toxic levels when consumed from supplements	likely to reach toxic levels when consumed from supplements
dosage frequency	needed in frequent doses	needed in periodic doses
relationship to coenzymes	function as coenzymes	do not function as coenzymes

10.13 Water-Soluble Vitamins: Vitamin C

Vitamin C, which has the simplest structure of the 13 vitamins, exists in two active forms in the human body: an oxidized form and a reduced form.

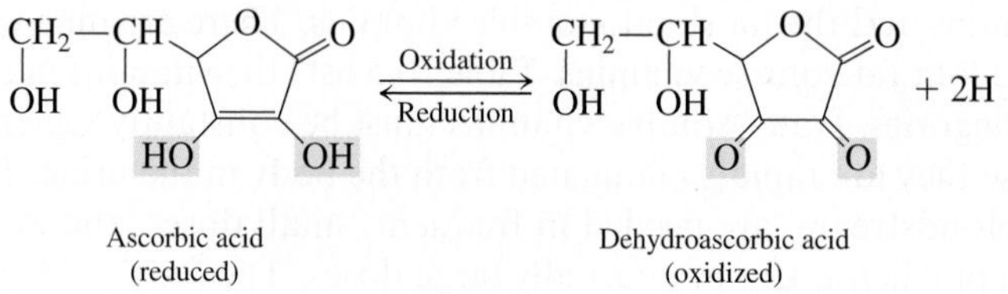

Vitamin C, the best known of all vitamins, was the first to be structurally characterized (1933), and the first to be synthesized in the laboratory (1933). Laboratory production of vitamin C, which exceeds 80 million pounds per year, is greater than the combined production of all the other vitamins. In addition to its use as a vitamin supplement, synthetic vitamin C is used as a food additive (preservative), a flour additive, and an animal feed additive.

Why is vitamin C called ascorbic *acid* when there is no carboxyl group (acid group) present in its structure? Vitamin C is a cyclic ester in which the carbon 1 carboxyl group has reacted with a carbon 4 hydroxyl group, forming the ring structure.

Other naturally occurring dietary antioxidants include glutathione (Section 9.8), vitamin E (Section 10.15), beta-carotene (Section 10.15), and flavonoids (Section 12.11).

Humans, monkeys, apes, and guinea pigs are among the relatively few species that require dietary sources of vitamin C. Other species synthesize vitamin C from carbohydrates. Vitamin C's biosynthesis involves L-gulonic acid, an acid derivative of the monosaccharide L-gulose (see Figure 7.13). L-Gulonic acid is changed by the enzyme *lactonase* into a cyclic ester (lactone, Section 5.11); ring closure involves carbons 1 and 4. An *oxidase* then introduces a double bond into the ring, producing L-ascorbic acid.

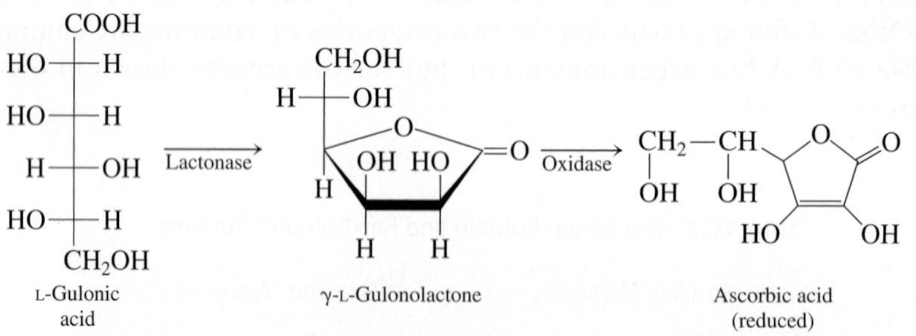

The four —OH groups present in vitamin C's reduced form are suggestive of its biosynthetic monosaccharide (*polyhydroxy* aldehyde) origins. Its chemical name, L-ascorbic acid, correctly indicates that vitamin C is a weak acid. Although no carboxyl group is present, the carbon 3 hydroxyl group hydrogen atom exhibits acidic behavior as a result of its attachment to an unsaturated carbon atom.

The most completely characterized role of vitamin C is its function as a cosubstrate in the formation of the structural protein collagen (Section 9.16), which makes up much of the skin, ligaments, and tendons and also serves as the matrix on which bone and teeth are formed. Specifically, biosynthesis of the amino acids hydroxyproline and hydroxylysine (important in binding collagen fibers together) from proline and lysine requires the presence of both vitamin C and iron.

Iron serves as a cofactor in the reaction, and vitamin C maintains iron in the oxidation state that allows it to function. In this role, vitamin C is functioning as a *specific* antioxidant.

Vitamin C also functions as a *general* antioxidant (Section 3.14) for water-soluble substances in the blood and other body fluids. Its antioxidant properties are also beneficial for several other vitamins. The active form of vitamin E is regenerated by vitamin C, and it also helps keep the active form of folate (a B vitamin) in its reduced state. Because of its antioxidant properties, vitamin C is often added to foods as a preservative.

Vitamin C is also involved in the metabolism of several amino acids that end up being converted to the hormones norepinephrine and thyroxine. The adrenal glands contain a higher concentration of vitamin C than any other organ in the body.

An intake of 100 mg/day of vitamin C saturates all body tissues with the compound. After the tissues are saturated, all additional vitamin C is excreted. The RDA for vitamin C varies from country to country. It is 30 mg/day in Great Britain, 60 mg/day in the United States and Canada, and 75 mg/day in Germany. A variety of fruits and vegetables have a relatively high vitamin C content (Figure 10.19).

Figure 10.19 Rows of cabbage plants. Although many people think citrus fruits (50 mg per 100 g) are the best source of vitamin C, peppers (128 mg per 100 g), cauliflower (70 mg per 100 g), strawberries (60 mg per 100 g), and spinach or cabbage (60 mg per 100 g) are all richer in vitamin C.

10.14 Water-Soluble Vitamins: The B Vitamins

There are nine water-soluble vitamins (Table 10.5). One is called vitamin C (Section 10.13). The other eight are grouped together and are called B vitamins. This grouping is based on biochemical function. All of the B vitamins serve as *precursors for enzyme cofactors.* Vitamin C is not an enzyme cofactor precursor.

Nomenclature for B Vitamins

Perhaps the most confusing aspect of the chemistry of B vitamins is their nomenclature. Nomenclature in use today is a combination of three different nomenclature systems that have evolved over time. Early on in vitamin chemistry research, it was thought that only two vitamins existed—a fat-soluble one (designated as vitamin A) and a water-soluble one (designated as vitamin B). When further research showed that there were several fat-soluble vitamins and even more water-soluble vitamins, naming continued in an alphabetical manner (vitamin A, B, C, D, E, F, G, etc.). With the discovery of the similarity in function (coenzyme precursors) of the water-soluble vitamins (except vitamin C), the coenzyme vitamins were all renamed as numbered B vitamins (B_1, B_2, B_3, etc.). Later, the numbered B vitamin system was replaced with a system that gave common names to the B vitamins. The currently *preferred* names for the B vitamins (alternative names in parentheses) are

1. Thiamin (vitamin B_1)
2. Riboflavin (vitamin B_2)
3. Niacin (nicotinic acid, nicotinamide, vitamin B_3)
4. Pantothenic acid (vitamin B_5)
5. Vitamin B_6 (pyridoxine, pyridoxal, pyridoxamine)
6. Biotin (vitamin B_7)
7. Folate (folic acid, vitamin B_9)
8. Vitamin B_{12} (cobalamin)

Pronunciation guidelines for the standard names of the B vitamins:

THIGH-a-min
RYE-boh-flay-vin
NIGH-a-sin
PAN-toe-THEN-ick acid
PEER-a-DOX-all
BYE-oat-in
FOAL-ate
CO-ball-a-min

An additional nomenclatural complication, which shows up indirectly in the preceding listing, is that in early research several substances were mistakenly characterized as vitamins. Note that no vitamin B_4 or vitamin B_8 entry is found in the previous listing. The substance originally identified as vitamin B_4 was later found to be adenine, a DNA metabolite (Section 11.3), and vitamin B_8 was found to be adenylic acid, another DNA metabolite. Thus "vitamins" B_4 and B_8 dropped out of the system.

The following two examples illustrate the changing nature of B vitamin nomenclature:

1. Vitamin G became vitamin B2, which became riboflavin.
2. Vitamin H became vitamin B7, which became biotin.

Structural Characteristics of the B Vitamins

The structural form(s) in which B vitamins are found in food is (are) not the form(s) in which they are used in the human body. As mentioned previously, B vitamins serve as *precursors* for enzyme cofactors. The "active form" for vitamins in the body is their enzyme cofactor form, to which they are converted once they are obtained from food through digestion of the food. As structural aspects of the various B vitamins are now considered, emphasis is given to the chemical modifications that occur as the "free" vitamins" are converted to enzyme cofactor form(s), that is, to their active forms in the human body.

Thiamin (Vitamin B_1)

The structures for the "free" and coenzyme form of thiamin are as follows:

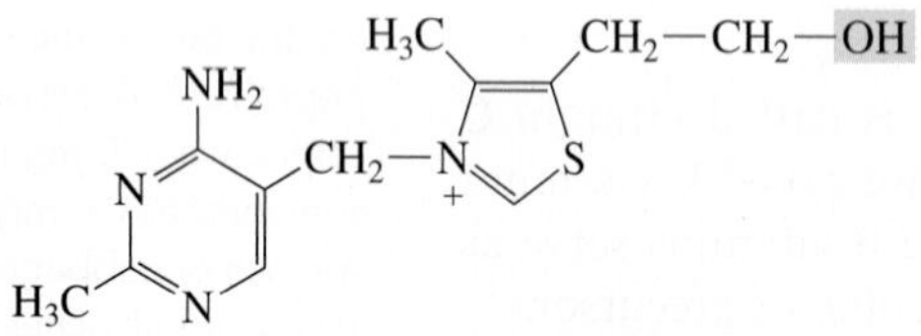

Thiamin (vitamin B_1)

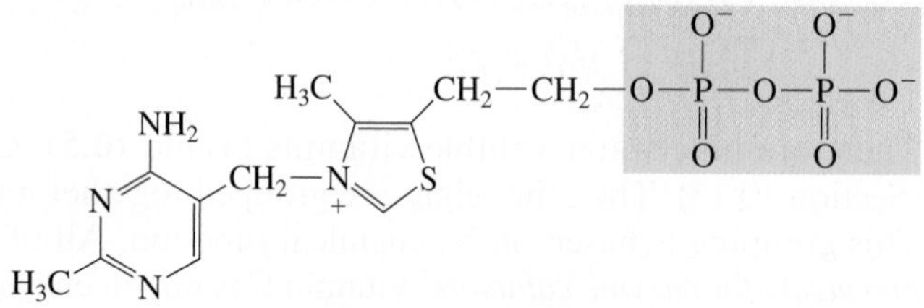

Thiamin pyrophosphate (TPP)

"Free" thiamin's structure consists of a central carbon atom to which is attached a six-membered heterocylic amine and a five-membered thiazole (sulfur-nitrogen) ring system. The name *thiamin* comes from "thio," which means "sulfur" and "amine" which refers to the numerous amine groups present.

The coenzyme form of thiamin is called thiamin pyrophosphate (TPP), a molecule in which a diphosphate group has been attached to the side chain. The coenzyme TPP functions in the decarboxylation of α-keto acids (Section 12.6).

Riboflavin (Vitamin B_2)

Riboflavin's structure involves three fused six-membered rings (two of which contain nitrogen) with the monosaccharide ribose (Section 7.9) attached to the middle ring.

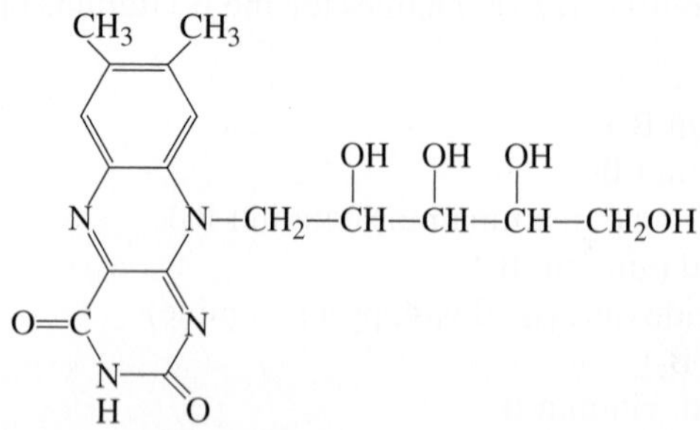

Riboflavin

Riboflavin was once called the "yellow vitamin" because of its color. Its name comes from its color (flavin means "yellow" in Latin) and its ribose component.

Two important riboflavin-based coenzymes exist: flavin adenine dinucleotide (FAD) and flavin mononucleotide (FMN). (Detailed structural information for FAD is given in Section 12.3 and for FMN in Section 12.7.) Both coenzymes are involved with oxidation-reduction reactions in which hydrogen atoms are transferred from one molecule to another.

Niacin (Vitamin B_3)

Niacin occurs in food in two different, but similar, forms: nicotinic acid and nicotinamide.

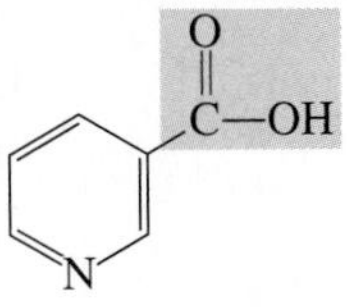

Nicotinic acid

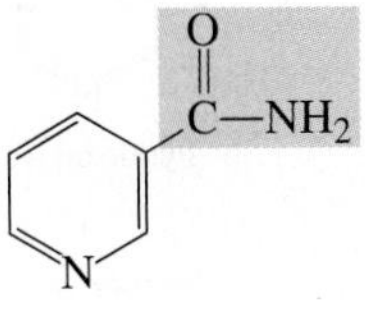

Nicotinamide

Both forms convert to the same coenzymes, two of which will be encountered repeatedly in Chapters 12–15: nicotinamide adenine dinucleotide (NAD^+) and nicotinamide adenine dinucleotide phosphate ($NADP^+$). Detailed structural information for these two coenzymes is found, respectively, in Sections 12.3 and 13.8. Both coenzymes are involved with oxidation-reduction reactions in which hydrogen atoms are transferred from one molecule to another.

The nicotinic acid form of niacin was first described in 1873, long before the concept of vitamins was known. It was prepared by oxidizing nicotine using nitric acid; hence the name nicotinic acid. When the biological significance of nicotinic acid was realized, the name niacin was coined to disassociate this vitamin from the name nicotine and to avoid the perception that niacin-rich foods contain nicotine or that cigarettes contain vitamins. The name *niacin* is derived in the following manner.

nicotinic acid + vitamin

Pantothenic Acid (Vitamin B_5)

The name pantothenic acid is based on the Greek word "pantothen," which means "from everywhere." This vitamin is found in almost every plant and animal tissue. The structure of pantothenic acid is

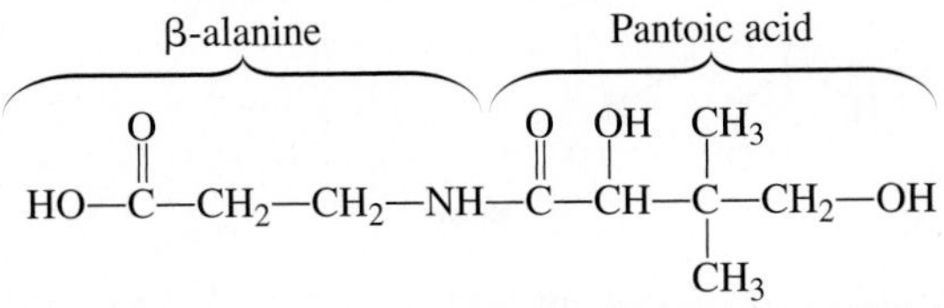

Pantothenic acid

This structure can be envisioned as an amide (Section 6.16) formed from the reaction of β-alanine (an amino acid; Section 9.2) and pantoic acid (2,4-dihydroxy-3,3-dimethylbutanoic acid), a carboxylic acid (Section 5.4).

Coenzyme A (CoA), one of the most used of all vitamin B coenzymes, contains pantothenic acid as part of its structure. Coenzyme A is required in the metabolism of carbohydrates, lipids, and proteins, where it is involved in the transfer of acetyl groups (Section 12.3) between molecules. Structural details for coenzyme A are given in Section 12.3. Another pantothenic acid-containing coenzyme is acyl carrier protein (ACP), which may be regarded as a "giant coenzyme A molecule." ACP is important in the biosynthesis of fatty acids (Section 14.7).

Vitamin B_6 (Pyridoxine, Pyridoxal, and Pyridoxamine)

Vitamin B_6 is a collective term for three related compounds: pyridoxine (found in foods of plant origin) and pyridoxal and pyridoxamine (found in foods of animal origin). The coenzyme forms of these three compounds, which contain an added phosphate group, are related to each other in the same manner that the "free" forms are related to each other, as shown in the following structures.

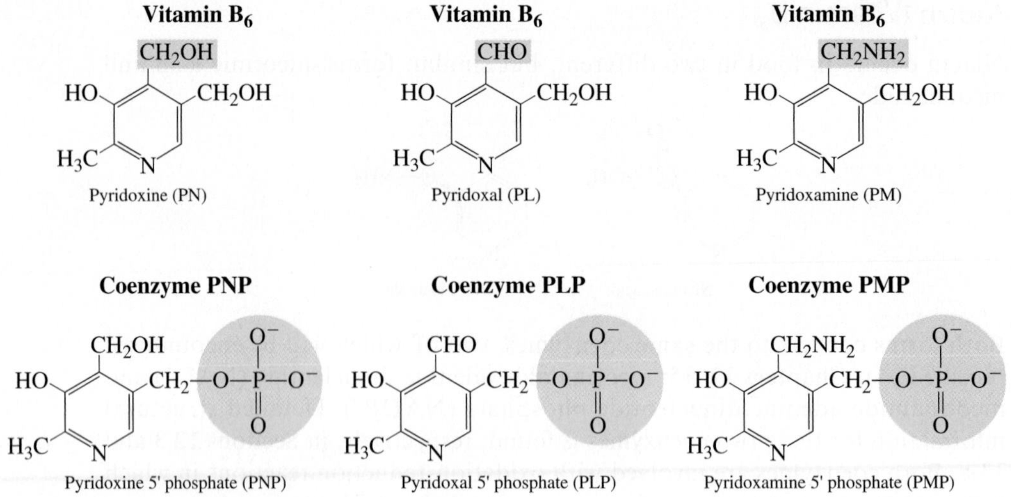

Vitamin B_6 coenzymes participate in reactions where amino groups are transferred between molecules. Such transfer occurs repeatedly when protein molecules are metabolized.

Biotin (Vitamin B_7)

Biotin is unique among the B vitamins in that it can be obtained both from dietary intake and also via biotin-producing bacteria (micro*biota*, hence the name *biot*in) present in the human large intestine. Structurally, biotin is a fused two-ring system with one ring containing sulfur and the other ring containing nitrogen. Attached to the sulfur-containing ring is a pentanoic acid residue.

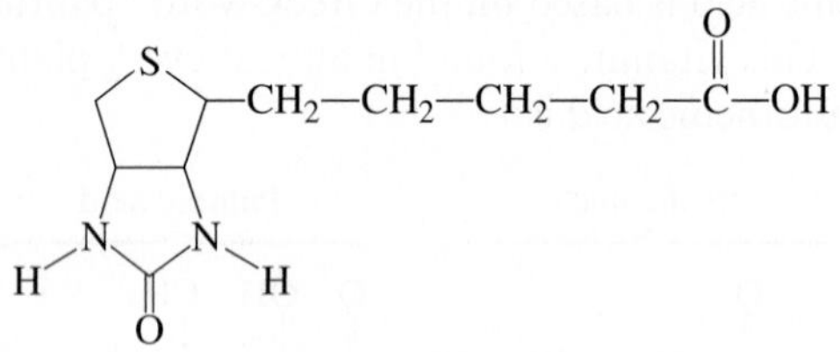

Biotin

"Free" biotin is biologically active. The coenzyme form of biotin is formed by the carboxyl group of biotin's pentanoic acid attachment forming an amide linkage with a residue of the amino acid lysine present at the enzyme's active site.

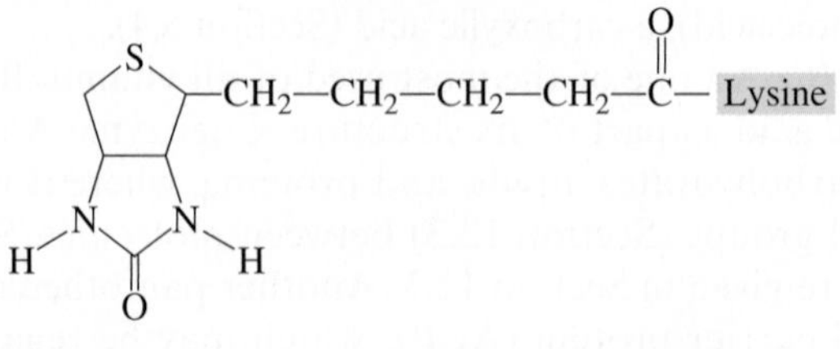

As a coenzyme, biotin is a carrier for CO_2; it has a specific site (a nitrogen atom) where a CO_2 molecule can become attached.

Folate (Vitamin B_9)

Several forms of folate are found in foods. All of them have structures that consist of three parts: (1) a nitrogen-containing double-ring system (pteridine) (2) para-aminobenzoic acid (PABA) and (3) one or more residues of the amino acid glutamate. When only one glutamate residue is present, as is shown in the following structure, the folate is called *folic acid.*

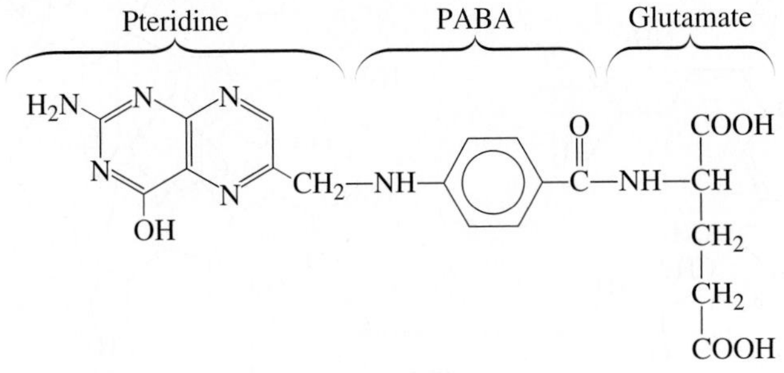

Folic acid

In food, about 90% of the folate molecules have three or more glutamate residues present; such molecules are called *polyglutamates*.

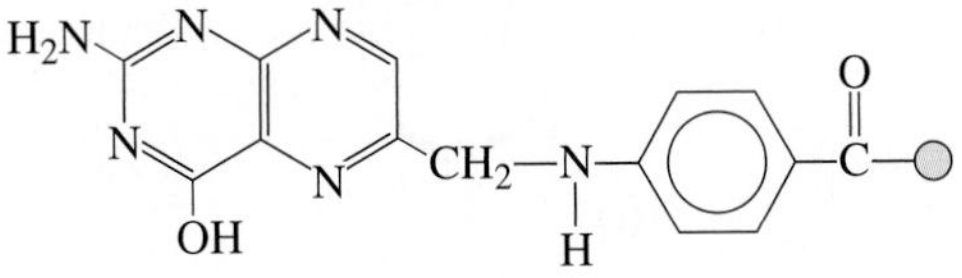

Folic acid

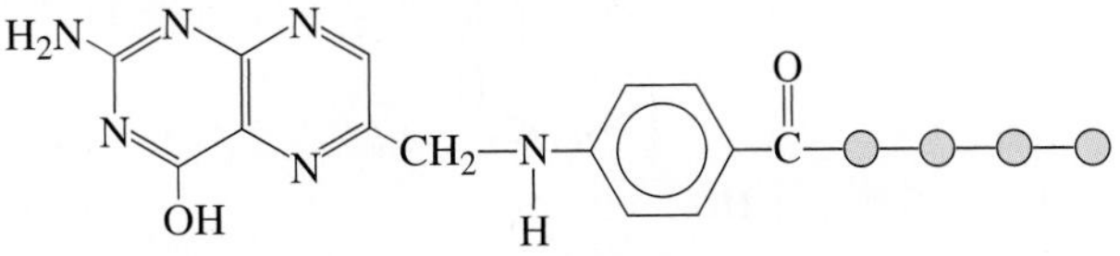

a polyglutamate

The active coenzyme form of folate, which is known as tetrahydrofolate (THF), has only one glutamate, and four hydrogen atoms have been added to the double-ring nitrogen system.

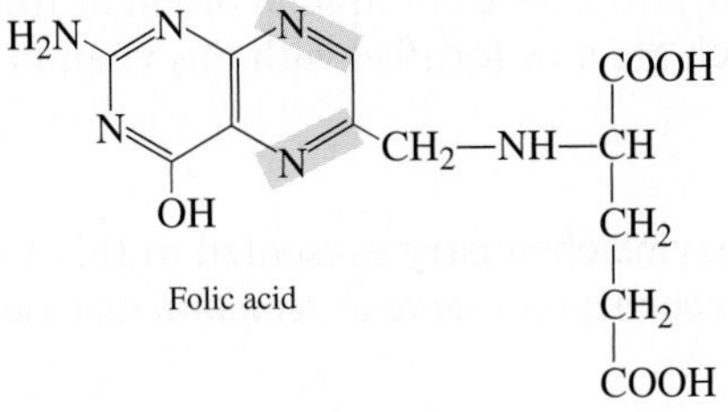

Folic acid

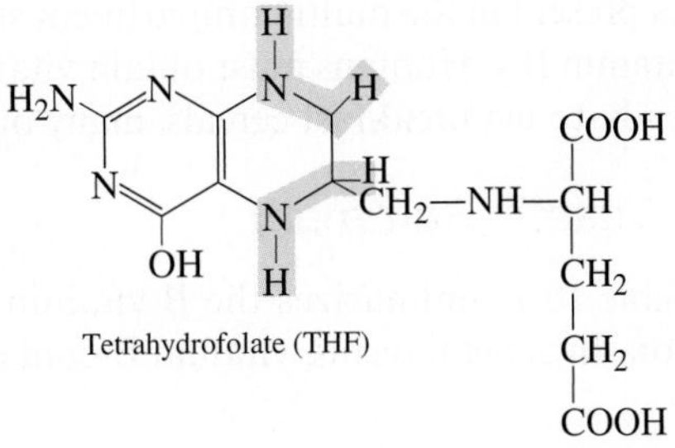

Tetrahydrofolate (THF)

THF is needed in methylation reactions, reactions in which one or more methyl groups are transferred from one molecule to another.

The name *folate* comes from the Latin word "folium," which means "leaf." Dark green leafy vegetables are the best natural source for folate. Legislation that dates back to the 1940s requires that all grain products that cross state lines be enriched in thiamin, riboflavin, and niacin. Folate was added to the legislated enrichment list in 1996 when research showed that folate was essential in the prevention of certain birth defects.

Vitamin B_{12} (Cobalamin)

The name *cobalamin* comes from the fact that an atom of the metal cobalt and numerous amine groups are present in the structure of vitamin B_{12}, which is by far the most complex of all vitamin structures. Vitamin B_{12} is unique in that it is the only vitamin that contains a metal atom. As shown in the following two structures, "free" vitamin B_{12} and coenzyme vitamin B_{12} differ only in one attachment to the cobalt atom; the free form is *cyano*cobalamin, and the coenzyme form is *methyl*cobalamin.

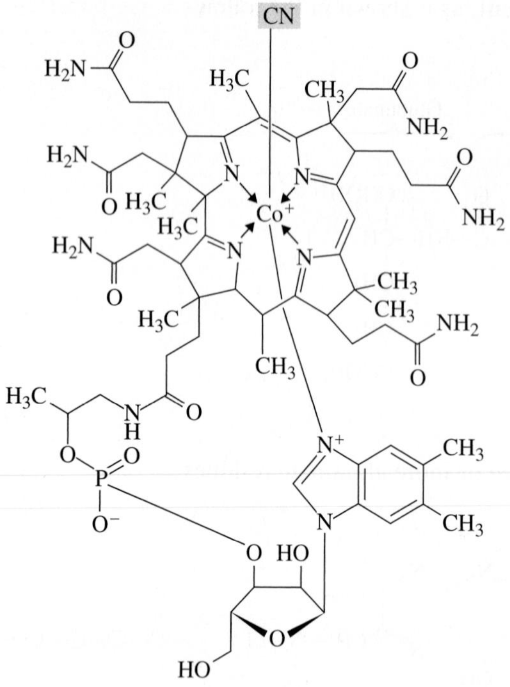

Cyanocobalamin

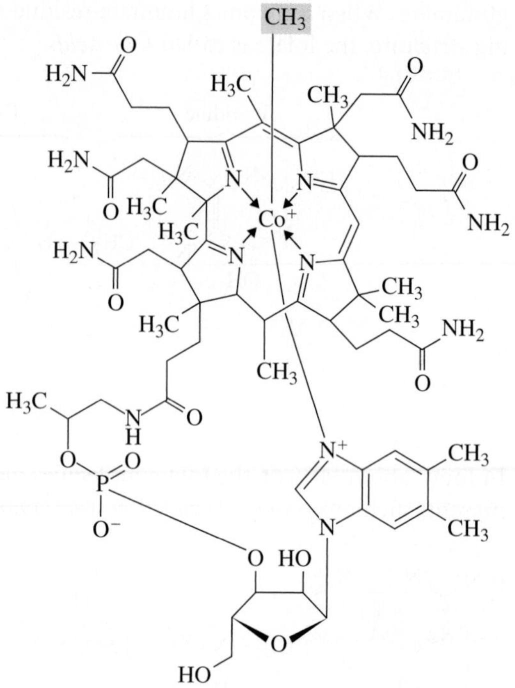

Methylcobalamin

Functionally, B_{12} coenzymes participate in the transfer of alkyl groups and hydrogen atoms from one molecule to another.

Vitamin B_{12} is also unique among vitamins in that only microorganisms can produce it; it cannot be made by plants, animals, birds, or humans. Grazing animals acquire vitamin B_{12} by ingesting some soil during the grazing process. Bacteria present in the multi-compartment stomachs of cows and sheep can also produce vitamin B_{12}. Humans must obtain vitamin B_{12} from foods of animal origin or from ready-to-eat breakfast cereals, many of which are now fortified with this vitamin.

B Vitamin Summary

Table 10.7 summarizes the B vitamin coenzyme chemistry presented in this section. In general terms, vitamin B-containing coenzymes serve as *temporary* carriers

Table 10.7 Selected Important Coenzymes in Which B Vitamins Are Present

B Vitamin	Coenzymes	Groups Transferred
thiamin	thiamin pyrophosphate (TPP)	aldehydes
riboflavin	flavin mononucleotide (FMN) flavin adenine dinucleotide (FAD)	hydrogen atoms
niacin	nicotinamide adenine dinucleotide (NAD^+) nicotinamide adenine dinucleotide phosphate ($NADP^+$)	hydrogen atoms
pantothenic acid	coenzyme A (CoA)	acyl groups
vitamin B_6	pyridoxal-5-phosphate (PLP) pyridoxine-5′-phophate (PNP) pyridoxamine-5′-phosphate (PMP)	amino groups
biotin	biotin	carbon dioxide (carboxyl group)
folate	tetrahydrofolate (THF)	one-carbon groups other than CO_2
vitamin B_{12}	methylcobalamin	methyl groups, hydrogen atoms

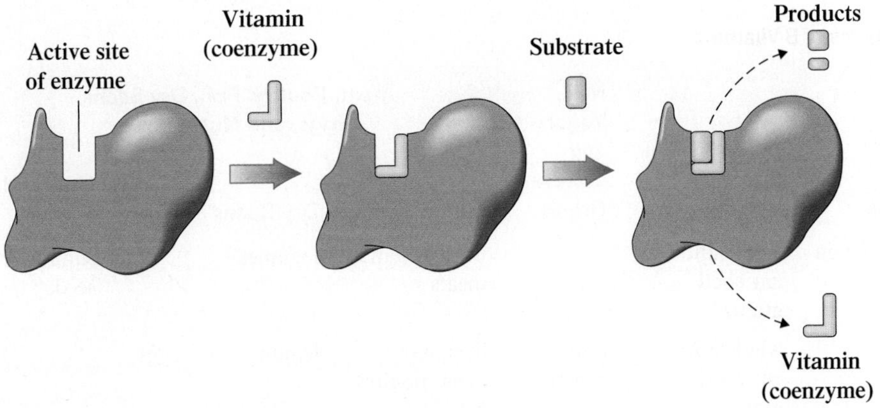

Figure 10.20 Many enzymes require a vitamin-based coenzyme in order to be active. After the catalytic action, the vitamin is released and can be reused.

of atoms or functional groups in redox and group transfer reactions associated with the metabolism of ingested food in order to obtain energy from the food. The details of metabolic reactions as they occur in the human body are the subject matter of Chapters 12–15.

In their function as coenzymes, B vitamins usually do not remain permanently bonded to the apoenzyme (Section 10.2) that they are associated with. This means that they can be repeatedly used by various enzymes. This reuse (recycling) diminishes the need for large amounts of the B vitamins in biochemical systems. Figure 10.20 shows diagrammatically how a vitamin is used and then released in an enzyme-catalyzed reaction.

An ample supply of the B vitamins can be obtained from normal dietary intake as long as a variety of foods are consumed. A certain food may be a better source of a particular B vitamin than others; however, there are multiple sources for each of the B vitamins, as Table 10.8 shows. Note from Table 10.8 that fruits, in general, are very poor sources of B vitamins and that only certain vegetables are good B vitamin sources.

10.15 Fat-Soluble Vitamins

The four fat-soluble vitamins are designated using the letters A, D, E, and K. Many of the functions of the fat-soluble vitamins involve processes that occur in cell membranes. The structures of the fat-soluble vitamins are more hydrocarbon-like, with fewer functional groups than the water-soluble vitamins. Their structures as a whole are nonpolar, which enhances their solubility in cell membranes.

Lipids are substances that are not soluble in water but are soluble in nonpolar substances such as fat (Section 8.1). Thus the four fat-soluble vitamins are a type of lipid. They could have been, but were not, discussed in Chapter 8.

Vitamin A

Normal dietary intake provides a person with both *preformed* and *precursor forms* (provitamin forms) of vitamin A. Preformed vitamin A forms are called *retinoids*. The retinoids include retin*al*, retin*ol*, and retin*oic acid*.

Beta-carotene is a deep yellow (almost orange) compound. If a plant food is white or colorless, it possesses little or no vitamin A activity. Potatoes, pasta, and rice are foods in this category.

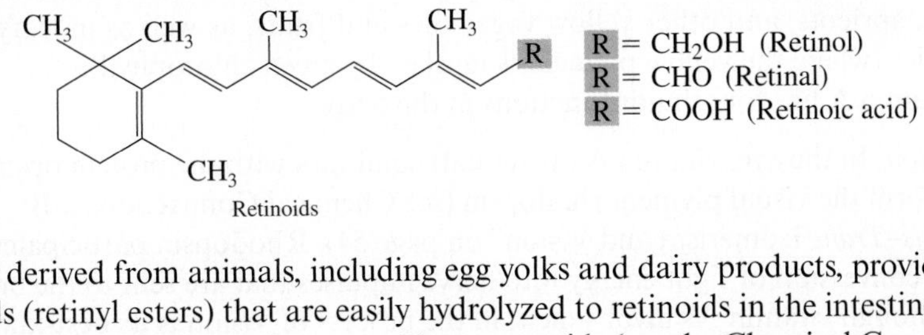

Retinoids

Foods derived from animals, including egg yolks and dairy products, provide compounds (retinyl esters) that are easily hydrolyzed to retinoids in the intestine.

Foods derived from plants provide carotenoids (see the Chemical Connections 2-C "Carotenoids: A Source of Color" on page 57), which serve as precursor

The retinoids are terpenes (Section 2.7) in which four isoprene units are present. Beta-carotene has an eight-unit terpene structure.

Table 10.8 A Summary of Dietary Sources of B Vitamins

Vitamin	Vegetable Group	Fruit Group	Bread, Cereal, Rice, and Pasta Group	Milk, Yogurt, and Cheese Group	Meat, Poultry, Fish, Dry Beans, Eggs, and Nuts Group			
					Meats	Dry Beans	Eggs	Nuts and Seeds
thiamin		watermelon	whole and enriched grains		pork, organ meats	legumes		sunflower seeds
riboflavin	mushrooms, asparagus, broccoli, leafy greens		whole and enriched grains	milk, cheeses	liver, red meat, poultry, fish	legumes	eggs	
niacin	mushrooms, asparagus, potato		whole and enriched grains, wheat bran		tuna, chicken, beef, turkey	legumes, peanuts		sunflower seeds
pantothenic acid	mushrooms, broccoli, avocados		whole grains		meat	legumes	egg yolk	
vitamin B_6	broccoli, spinach, potato, squash	bananas, watermelon	whole wheat, brown rice		chicken, fish, pork, organ meats	soybeans		sunflower seeds
biotin			fortified cereals	yogurt	liver (muscle meats are poor sources)	soybeans	egg yolk	nuts
folate	mushrooms, leafy greens, broccoli, asparagus, corn	oranges	fortified grains		organ meats (muscle meats are poor sources)	legumes		sunflower seeds, nuts
vitamin B_{12}				milk products	beef, poultry, fish, shellfish		egg yolk	

Beta-carotene cleavage does not always occur in the "middle" of the molecule, so only one molecule of vitamin A is produced. Furthermore, not all beta-carotene is converted to vitamin A, and its absorption is not as efficient as that of vitamin A itself. It is estimated that 6 mg of beta-carotene is needed to produce 1 mg of retinol. Unconverted beta-carotene serves as an antioxidant (see Section 2.7), a role independent of its conversion to vitamin A.

forms of vitamin A. The major carotenoid with vitamin A activity is beta-carotene (β-carotene), which can be cleaved to yield two molecules of vitamin A.

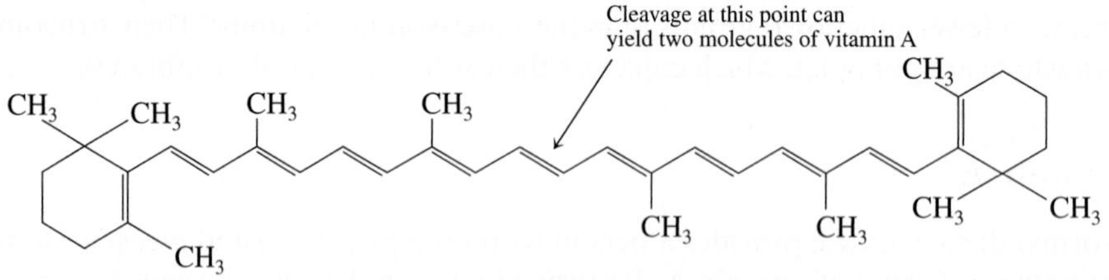

Beta-carotene, a precursor for vitamin A

Beta-carotene is a yellow to red-orange pigment plentiful in carrots, squash, cantaloupe, apricots, and other yellow vegetables and fruits, as well as in leafy green vegetables (where the yellow pigment is masked by green chlorophyll).

Vitamin A has four major functions in the body.

1. *Vision.* In the eye, vitamin A (as retinal) combines with the protein opsin to form the visual pigment rhodopsin (see Chemical Connections 2-B "*Cis–Trans* Isomerism and Vision" on page 54). Rhodopsin participates in the conversion of light energy into nerve impulses that are sent to the brain. Although vitamin A's involvement in the process of vision is its best-known function ("Eat your carrots and you'll see better"), only 0.1% of the body's vitamin A is found in the eyes.

2. *Regulating Cell Differentiation.* Cell differentiation is the process whereby immature cells change in structure and function to become specialized cells. For example, some immature bone marrow cells differentiate into white blood cells and others into red blood cells. In the cellular differentiation process, vitamin A (as retinoic acid) binds to protein receptors; these vitamin A–protein receptor complexes then bind to regulatory regions of DNA molecules.
3. *Maintenance of the Health of Epithelial Tissues.* Epithelial tissue covers outer body surfaces in addition to lining internal cavities and tubes. It includes skin and the linings of the mouth, stomach, lungs, vagina, and bladder. Lack of vitamin A (as retinoic acid) causes such surfaces to become drier and harder than normal. Vitamin A's role here is related to cellular differentiation involving mucus-secreting cells.
4. *Reproduction and Growth.* In men, vitamin A participates in sperm development. In women, normal fetal development during pregnancy requires vitamin A. In both cases, it is the retinoic acid form of vitamin A that is needed. Again vitamin A's role is related to cellular differentiation processes.

Melissa Grimes-Guy/Photo Researchers

Figure 10.21 The quantity of vitamin D synthesized by exposure of the skin to sunlight (ultraviolet radiation) varies with latitude, the length of exposure time, and skin pigmentation. (Darker-skinned people synthesize less vitamin D because the pigmentation filters out ultraviolet light.)

Vitamin D

The two most important members of the vitamin D family of molecules are vitamin D_3 (cholecalciferol) and vitamin D_2 (ergocalciferol). Vitamin D_3 is produced in the skin of humans and animals by the action of sunlight (ultraviolet light) on its precursor molecule, the cholesterol derivative 7-dehydrocholesterol (a normal metabolite of cholesterol found in the skin) (Figure 10.21). Absorption of light energy induces breakage of the 9, 10 carbon bond; a spontaneous isomerization (shifting of double bonds) then occurs.

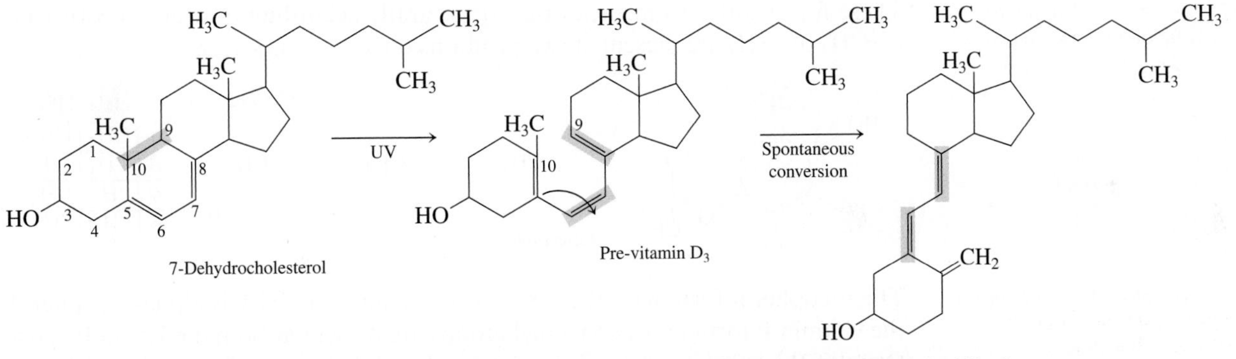

Vitamin D_2 (ergocalciferol) differs from vitamin D_3 only in the side-chain structure. It is produced from the plant sterol ergosterol through the action of light.

Vitamin D_3 (cholecalciferol) is sometimes called the "sunshine vitamin" because of its synthesis in the skin by sunlight irradiation.

The common names for Vitamin D_2 and Vitamin D_3 are pronounced as follows:

er-go-cal-CIF-er-ol (Vitamin D_2)
cho-le-cal-CIF-er-ol (Vitamin D_3)

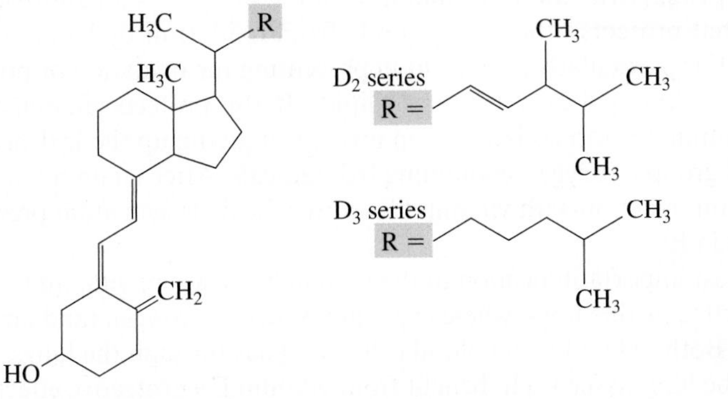

Vitamin D

Both the cholecalciferol and the ergocalciferol forms of vitamin D must undergo two further hydroxylation steps before the vitamin D becomes fully functional. The first step, which occurs in the liver, adds a —OH group to carbon 25. The second step, which occurs in the kidneys, adds a —OH group to carbon 1.

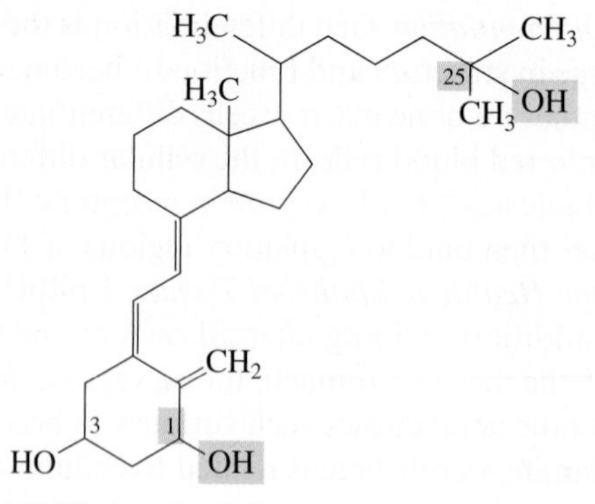

1,25-Dihydroxyvitamin D_3

Milk is enriched in vitamin D by exposure to ultraviolet light. Cholesterol in milk is converted to cholecalciferol (vitamin D) by ultraviolet light.

Only a few foods, including liver, fatty fish (such as salmon), and egg yolks, are good natural sources of vitamin D. Such vitamin D is vitamin D_3. Foods fortified with vitamin D include milk and margarine. The rest of the body's vitamin D supplies are made within the body (skin) with the help of sunlight.

When it comes to strong bones, calcium won't do you a lot of good unless you are also getting enough vitamin D. In one study, women consuming 500 IU of vitamin D a day had a 37% lower risk of hip fracture than women consuming only 140 IU daily of vitamin D. Some researchers now recommend a standard of 800–1000 IU per day instead of the long-established standard recommendation for vitamin D of 400 IU per day.

The principal function of vitamin D is to maintain normal blood levels of calcium ion and phosphate ion so that bones can absorb these ions. Vitamin D stimulates absorption of these ions from the gastrointestinal tract and aids in their retention by the kidneys. Vitamin D triggers the deposition of calcium salts into the organic matrix of bones by activating the biosynthesis of calcium-binding proteins.

Vitamin E

There are four forms of vitamin E: alpha-, beta-, delta-, and gamma-tocopherol. These forms differ from each other structurally according to which substituents ($-CH_3$ or $-H$) are present at two positions on an aromatic ring.

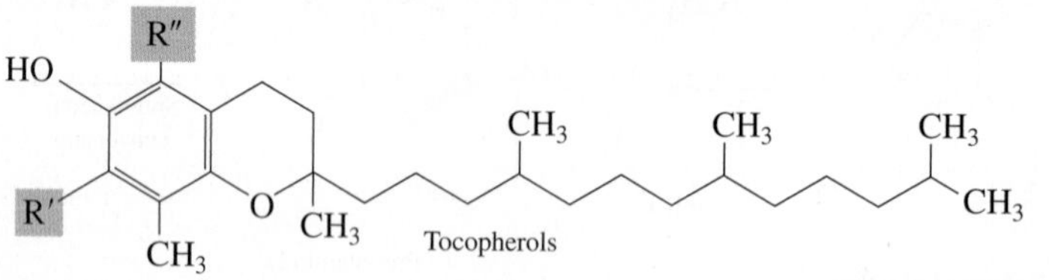

Tocopherols

	R′	R″
α	CH_3	CH_3
β	CH_3	H
γ	H	CH_3
δ	H	H

The word *tocopherol* is pronounced "tuh-KOFF-er-ol."

The tocopherol form with the greatest biochemical activity is alpha-tocopherol, the vitamin E form in which methyl groups are present at both the R and R′ positions on the aromatic ring. Gamma-tocopherol is the main form of vitamin E in vitamin-E rich foods.

Vitamin E is unique among the vitamins in that antioxidant activity is its *principal* biochemical role.

Plant oils (margarine, salad dressings, and shortenings), green and leafy vegetables, and whole-grain products are sources of vitamin E.

The primary function of vitamin E in the body is as an antioxidant—a compound that protects other compounds from oxidation by being oxidized itself. Vitamin E is particularly important in preventing the oxidation of polyunsaturated fatty acids (Section 8.2) in membrane lipids. It also protects vitamin A from oxidation. Vitamin E's antioxidant action involves it giving up the hydrogen present on its $-OH$ group to oxygen-containing free radicals. After vitamin E is "spent" as an antioxidant, reaction with vitamin C restores the hydrogen atom previously lost by the vitamin E.

A most important location in the human body where vitamin E exerts its antioxidant effect is the lungs, where exposure of cells to oxygen (and air pollutants) is greatest. Both red and white blood cells that pass through the lungs, as well as the cells of the lung tissue itself, benefit from vitamin E's protective effect.

Infants, particularly premature infants, do not have a lot of vitamin E, which is passed from the mother to the infant only in the last weeks of pregnancy. Often, premature infants require oxygen supplementation for the purpose of controlling respiratory distress. In such situations, vitamin E is administered to the infant along with oxygen to give antioxidant protection.

Vitamin E has also been found to be involved in the conversion of arachidonic acid (20:4) to prostaglandins (Section 8.13).

Vitamin K

Like the other fat-soluble vitamins, vitamin K has more than one form. Structurally, all forms have a methylated napthoquinone structure to which a long side chain of carbon atoms is attached. The various forms differ structurally in the length and degree of unsaturation of the side chain.

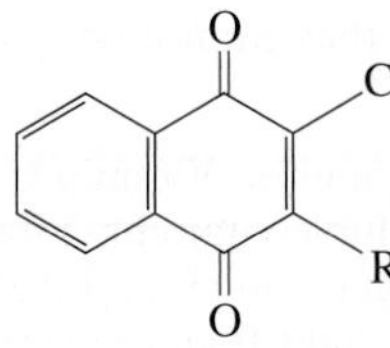

Vitamin K

Vitamin K_1, also called phylloquinone, has a side chain that is predominantly saturated; only one carbon–carbon double bond is present. It is a substance found in plants. Vitamin K_2 has several forms, called menaquinones, with the various forms differing in the length of the side chain. Menaquinone side chains have several carbon–carbon double bonds, in contrast to the one carbon–carbon double bond present in phylloquinone. Vitamin K_2 is found in animals and humans and can be synthesized by bacteria, including those found in the human intestinal tract.

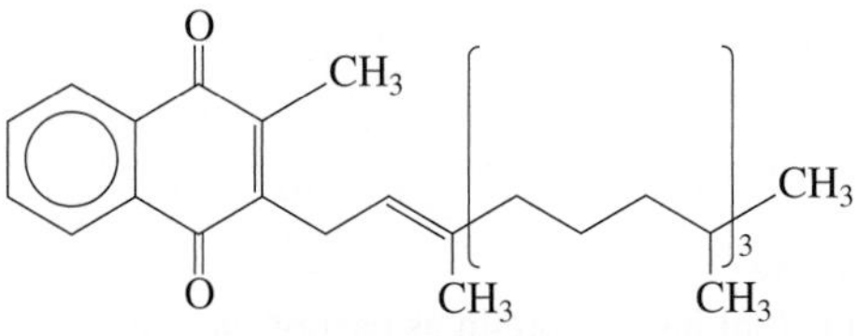

Vitamin K_1 (phylloquinone)

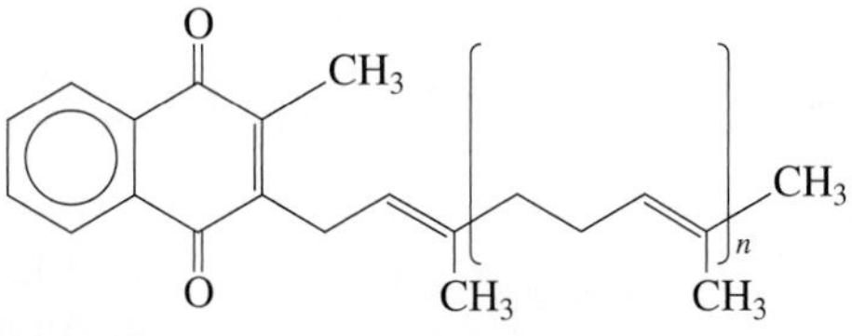

(where n may be 1 to 13 but is mostly 7 to 9)

Vitamin K_2 (menaquinone)

Typically, about half of the human body's vitamin K is synthesized by intestinal bacteria and half comes from the diet. Menaquinones are the form of vitamin K found in vitamin K supplements. Only leafy green vegetables such as spinach and cabbage are particularly rich in vitamin K. Other vegetables such as peas and tomatoes, as well as animal tissues including liver, contain lesser amounts.

Vitamin K is essential to the blood-clotting process. More than a dozen different proteins and the mineral calcium are involved in the formation of a blood clot. Vitamin K is essential for the formation of prothrombin and at least five other proteins involved in the regulation of blood clotting. Vitamin K is sometimes given to presurgical patients to ensure adequate prothrombin levels and to prevent hemorrhaging.

Vitamin K is also required for the biosynthesis of several other proteins found in the plasma, bone, and kidney.

All of the fat-soluble vitamins share a common structural feature; they all have terpene-like structures. That is, they are all made up of five-carbon isoprene units (Section 2.5). No common structural pattern exists for the water-soluble vitamins. On the other hand, the water-soluble vitamins have *functional* uniformity, whereas the fat-soluble vitamins have diverse functions.

Concepts to Remember

OWL Sign in at **www.cengage.com/owl** to view tutorials and simulations, develop problem-solving skills, and complete online homework assigned by your professor.

Enzymes. Enzymes are highly specialized protein molecules that act as biochemical catalysts. Enzymes have common names that provide information about their function rather than their structure. The suffix *-ase* is characteristic of most enzyme names (Section 10.1).

Enzyme structure. Simple enzymes are composed only of protein (amino acids). Conjugated enzymes have a nonprotein portion (cofactor) in addition to a protein portion (apoenzyme). Cofactors may be small organic molecules (coenzymes) or inorganic ions (Section 10.2).

Enzyme classification. There are six classes of enzymes based on function: oxidoreductases, transferases, hydrolases, lyases, isomerases, and ligases (Section 10.3).

Enzyme active site. An enzyme active site is the relatively small part of the enzyme that is actually involved in catalysis. It is where substrate binds to the enzyme (Section 10.4).

Lock-and-key model of enzyme activity. The active site in an enzyme has a fixed, rigid geometrical conformation. Only substrates with a complementary geometry can be accommodated at the active site (Section 10.4).

Induced-fit model of enzyme activity. The active site in an enzyme can undergo small changes in geometry in order to accommodate a series of related substrates (Section 10.4).

Enzyme activity. Enzyme activity is a measure of the rate at which an enzyme converts substrate to products. Four factors that affect enzyme activity are temperature, pH, substrate concentration, and enzyme concentration (Section 10.6).

Enzyme inhibition. An enzyme inhibitor slows or stops the normal catalytic function of an enzyme by binding to it. Three modes of inhibition are reversible competitive inhibition, reversible noncompetitive inhibition, and irreversible inhibition (Section 10.7).

Allosteric enzyme. An allosteric enzyme is an enzyme with two or more protein chains and two kinds of binding sites (for substrate and regulator) (Section 10.8).

Zymogen. A zymogen is an inactive precursor of a proteolytic enzyme; the zymogen is activated by a chemical reaction that removes part of its structure (Section 10.8).

Covalent modification. Covalent modification is a cellular process for regulation of enzyme activity in which the structure of an enzyme is modified through formation of, or breaking of, a covalent bond. The most commonly encountered type of covalent modification involves a phosphate group being added to, or removed from, an enzyme (Section 10.8).

Vitamins. A vitamin is an organic compound necessary in small amounts for the normal growth of humans and some animals. Vitamins must be obtained from dietary sources because they cannot be synthesized in the body (Section 10.11).

Water-soluble vitamins. Vitamin C and the eight B vitamins are the water-soluble vitamins. Vitamin C is essential for the proper formation of bones and teeth and is also an important antioxidant. All eight B vitamins function as coenzymes (Section 10.12).

Fat-soluble vitamins. The four fat-soluble vitamins are vitamins A, D, E, and K. The best-known function of vitamin A is its role in vision. Vitamin D is essential for the proper use of calcium and phosphorus to form bones and teeth. The primary function of vitamin E is as an antioxidant. Vitamin K is essential in the regulation of blood clotting (Section 10.13).

Exercises and Problems

OWL Interactive versions of these problems may be assigned in OWL.

Exercises and problems are arranged in matched pairs with the two members of a pair addressing the same concept(s). The answer to the odd-numbered member of a pair is given at the back of the book. Problems denoted with a ▲ involve concepts found not only in the section under consideration but also concepts found in one or more earlier sections of the chapter. Problems denoted with a ● cover concepts found in a Chemical Connections feature box.

Importance of Enzymes (Section 10.1)

10.1 What is the general role of enzymes in the human body?

10.2 Why does the body need so many different enzymes?

10.3 List two ways in which enzymes differ from inorganic laboratory catalysts.

10.4 Occasionally the "delicate" nature of enzymes is referred to. Explain why this adjective is appropriate.

Enzyme Structure (Section 10.2)

10.5 Indicate whether each of the following phrases describes a simple or a conjugated enzyme.
- a. An enzyme that has both a protein and a nonprotein portion
- b. An enzyme that requires Mg^{2+} ion for activity
- c. An enzyme in which only amino acids are present
- d. An enzyme in which a cofactor is present

10.6 Indicate whether each of the following phrases describes a simple or a conjugated enzyme.
- a. An enzyme that contains a carbohydrate portion
- b. An enzyme that contains only protein
- c. A holoenzyme
- d. An enzyme that has a vitamin as part of its structure

10.7 What is the difference between a cofactor and a coenzyme?

10.8 All coenzymes are cofactors, but not all cofactors are coenzymes. Explain this statement.

10.9 Why are cofactors present in most enzymes?

10.10 What is the difference between an apoenzyme and a holoenzyme?

Enzyme Nomenclature (Section 10.3)

10.11 Which of the following substances are enzymes?
- a. Sucrase
- b. Galactose
- c. Trypsin
- d. Xylulose reductase

10.12 Which of the following substances are enzymes?
- a. Sucrose
- b. Pepsin
- c. Glutamine synthetase
- d. Cellulase

10.13 Predict the function of each of the following enzymes.
- a. Pyruvate carboxylase
- b. Alcohol dehydrogenase
- c. L-Amino acid reductase
- d. Maltase

10.14 Predict the function of each of the following enzymes.
- a. Cytochrome oxidase
- b. *Cis–trans* isomerase
- c. Succinate dehydrogenase
- d. Lactase

10.15 Suggest a name for an enzyme that catalyzes each of the following reactions.
- a. Hydrolysis of sucrose
- b. Decarboxylation of pyruvate
- c. Isomerization of glucose
- d. Removal of hydrogen from lactate

10.16 Suggest a name for an enzyme that catalyzes each of the following reactions.
a. Hydrolysis of lactose b. Oxidation of nitrite
c. Decarboxylation of citrate d. Reduction of oxalate

10.17 Give the name of the substrate on which each of the following enzymes acts.
a. Pyruvate carboxylase b. Galactase
c. Alcohol dehydrogenase d. L-Amino acid reductase

10.18 Give the name of the substrate on which each of the following enzymes acts.
a. Cytochrome oxidase b. Lactase
c. Succinate dehydrogenase d. Tyrosine kinase

10.19 To which of the six major classes of enzymes does each of the following belong?
a. Mutase b. Dehydratase
c. Carboxylase d. Kinase

10.20 To which of the six major classes of enzymes does each of the following belong?
a. Protease b. Racemase
c. Dehydrogenase d. Synthetase

10.21 To which of the six major classes of enzymes does the enzyme that catalyzes each of the following reactions belong?
a. A *cis* double bond is converted to a *trans* double bond.
b. An alcohol is dehydrated to form a compound with a double bond.
c. An amino group is transferred from one substrate to another.
d. An ester linkage is hydrolyzed.

10.22 To which of the six major classes of enzymes does the enzyme that catalyzes each of the following reactions belong?
a. An L isomer is converted to a D isomer.
b. A phosphate group is transferred from one substrate to another.
c. An amide linkage is hydrolyzed.
d. Hydrolysis of a carbohydrate to monosaccharides occurs.

10.23 Identify the enzyme needed in each of the following reactions as an isomerase, a decarboxylase, a dehydrogenase, a lipase, or a phosphatase.

a.
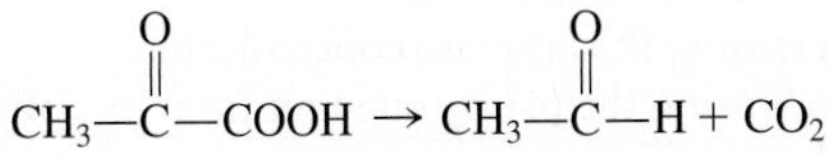

b.
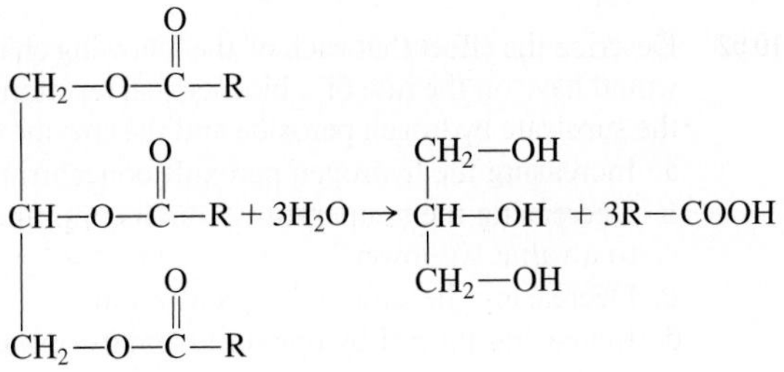

c.
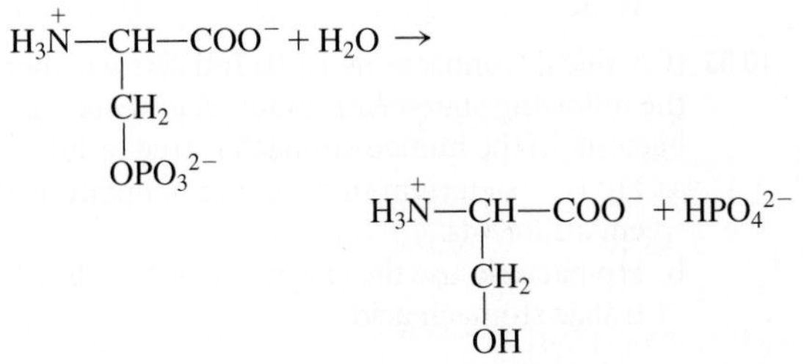

d.
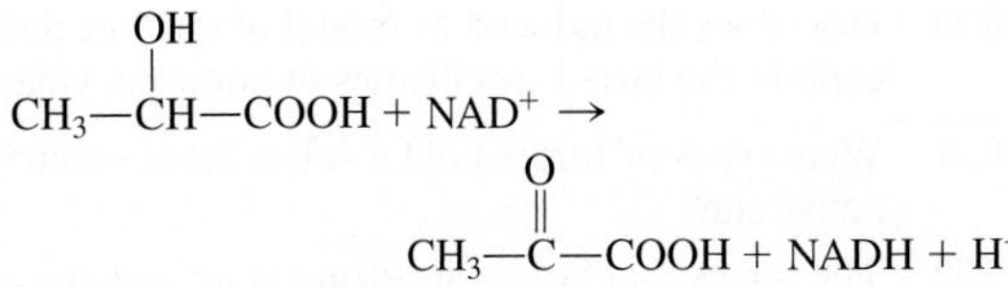

10.24 Identify the enzyme needed in each of the following reactions as an isomerase, a decarboxylase, a dehydrogenase, a protease, or a phosphatase.

a.
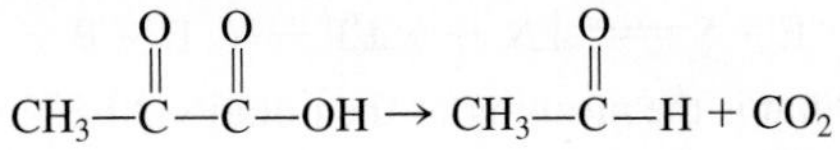

b.
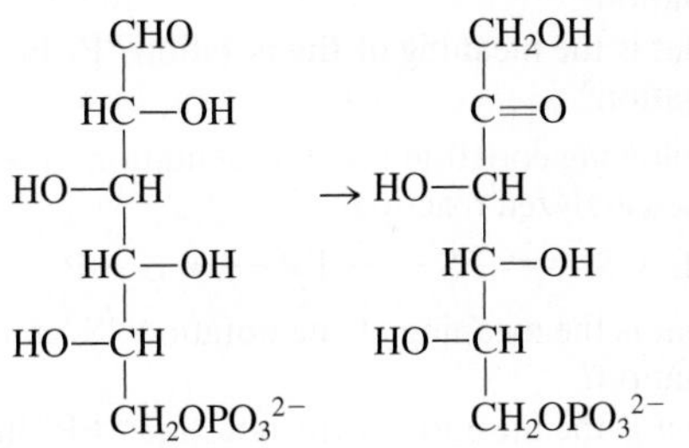

c.
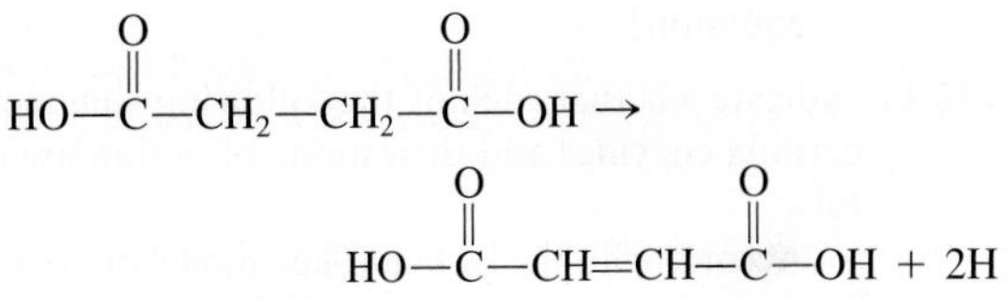

d.
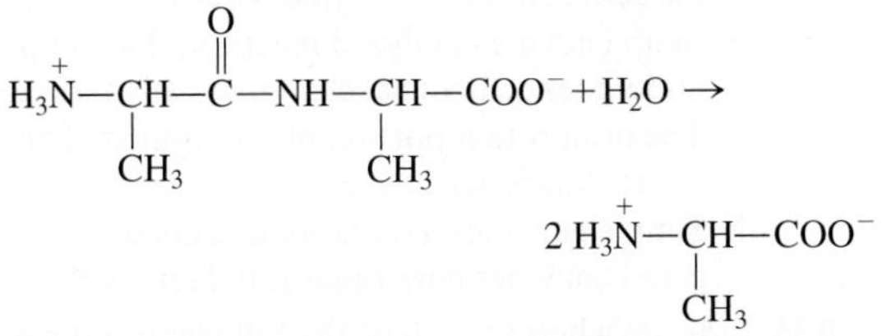

•**10.25** (Chemical Connections 10-A) Indicate whether each of the following statements about "enzymatic browning" is true or false.
a. The enzyme involved in the discoloration of cut fruits and vegetables is called benzoquinonase.
b. Enzymatic browning can be "slowed" by immersing affected food in cold water.
c. Catechol derivatives are substrates for the enzyme phenolase.
d. *o*-Hydroquinone derivative formation is both enzyme- and oxygen-dependent.

•**10.26** (Chemical Connections 10-A) Indicate whether each of the following statements about "enzymatic browning" is true or false.
a. The enzyme involved in the discoloration of cut fruits and vegetables contains the metal copper.
b. Enzymatic browning can be "slowed" by putting affected food in the refrigerator.
c. Phenol derivatives are substrates for the enzyme phenolase.
d. Catechol derivative formation is both enzyme- and oxygen-dependent.

Models of Enzyme Action (Section 10.4)

10.27 What is an enzyme active site?

10.28 What is an enzyme–substrate complex?

10.29 How does the lock-and-key model of enzyme action explain the highly specific way some enzymes select a substrate?

10.30 How does the induced-fit model of enzyme action explain the broad specificities of some enzymes?

10.31 What types of forces hold a substrate at an enzyme active site?

10.32 The forces that hold a substrate at an enzyme active site are not covalent bonds. Explain why not.

10.33 The following equation is a representation of an enzyme-catalyzed reaction.

$$E + S \rightleftharpoons ES \longrightarrow EP \longrightarrow E + P$$

a. What is the meaning of the notation "ES" in the equation?
b. What is the meaning of the notation "P" in the equation?

10.34 The following equation is a representation of an enzyme-catalyzed reaction.

$$E + S \rightleftharpoons ES \longrightarrow EP \longrightarrow E + P$$

a. What is the meaning of the notation "S" in the equation?
b. What is the meaning of the notation "EP" in the equation?

▲10.35 Indicate whether each of the following statements concerning enzymes and their mode of action are true or false.
a. According to the lock-and-key model of enzyme action, the active site of an enzyme is flexible in shape.
b. In an enzyme-catalyzed reaction, the compound that undergoes a chemical change is called the substrate.
c. The nonprotein portion of a conjugated enzyme is the enzyme's active site.
d. Simple enzymes have inorganic cofactors, and conjugated enzymes have organic cofactors.

▲10.36 Indicate whether each of the following statements concerning enzymes and their mode of action are true or false.
a. According to the induced-fit model of enzyme action, a substrate must be changed into a cofactor before a chemical reaction occurs.
b. The active site of an enzyme is the location within the enzyme where catalysis takes place.
c. The protein portion of a conjugated enzyme can accommodate several substrates at the same time.
d. Simple enzymes do not have active sites, whereas conjugated enzymes do have active sites.

Enzyme Specificity (Section 10.5)

10.37 Define the following terms dealing with enzyme specificity.
a. Absolute specificity b. Linkage specificity

10.38 Define the following terms dealing with enzyme specificity.
a. Group specificity b. Stereochemical specificity

10.39 Which enzyme in each of the following pairs would be more limited in its catalytic scope?
a. An enzyme that exhibits absolute specificity or an enzyme that exhibits group specificity
b. An enzyme that exhibits stereochemical specificity or an enzyme that exhibits linkage specificity

10.40 Which enzyme in each of the following pairs would be more limited in its catalytic scope?
a. An enzyme that exhibits linkage specificity or an enzyme that exhibits absolute specificity
b. An enzyme that exhibits group specificity or an enzyme that exhibits stereochemical specificity

▲10.41 What type of specificity (absolute, group, linkage, or stereochemical) is associated with each of the following enzymes?
a. Sucrase b. A lipase
c. A decarboxylase d. L-glutamate oxidase

▲10.42 What type of specificity (absolute, group, linkage, or stereochemical) is associated with each of the following enzymes?
a. A deaminase b. A phosphatase
c. Maltase d. L-lactate dehydrogenase

Factors That Affect Enzyme Activity (Section 10.6)

10.43 Temperature affects enzymatic reaction rates in two ways. An increase in temperature can accelerate the rate of a reaction or it can stop the reaction. Explain each of these effects.

10.44 Define the optimum temperature for an enzyme.

10.45 Explain why all enzymes do not possess the same optimum pH.

10.46 Why does an enzyme lose activity when the pH is drastically changed from the optimum pH?

10.47 Draw a graph that shows the effect of increasing substrate concentration on the rate of an enzyme-catalyzed reaction (at constant temperature, pH, and enzyme concentration).

10.48 Draw a graph that shows the effect of increasing enzyme concentration on the rate of an enzyme-catalyzed reaction (at constant temperature, pH, and substrate concentration).

10.49 In an enzyme-catalyzed reaction, all of the enzyme active sites are saturated by substrate molecules at a certain substrate concentration. What happens to the rate of the reaction when the substrate concentration is doubled?

10.50 What is an enzyme turnover number?

10.51 Describe the effect that each of the following changes would have on the rate of a biochemical reaction involving the substrate arginine and the enzyme arginase.
a. Decreasing the arginine concentration
b. Increasing the temperature from its optimum value to a value 10° higher
c. Increasing the arginase concentration
d. Decreasing the pH by one unit from its optimum value

10.52 Describe the effect that each of the following changes would have on the rate of a biochemical reaction involving the substrate hydrogen peroxide and the enzyme catalase.
a. Increasing the hydrogen peroxide concentration
b. Decreasing the temperature from its optimum value to a value 10° lower
c. Decreasing the catalase concentration
d. Increasing the pH by one unit from its optimum value

•10.53 (Chemical Connections 10-B) Indicate whether each of the following statements about *Helicobacter pylori* (Hp) bacteria in the human stomach is true or false.
a. Hp is a bacterium that can live in highly acidic environments.
b. Hp bacteria use the enzyme urease to directly neutralize stomach acid.

c. Hp bacteria irritate the stomach lining by interfering with the protective mucous coating present.
d. Poor sanitation conditions correlate well with Hp infection rates.

•10.54 (Chemical Connections 10-B) Indicate whether each of the following statements about *Helicobacter pylori* (Hp) bacteria in the human stomach is true or false.
a. The enzyme urease is present on the exterior surface of Hp bacteria.
b. Hp bacteria cause ulcers through generation of ammonia.
c. The majority of people have Hp bacteria present in their stomach.
d. Ulcer conditions occur within weeks of infection with Hp bacteria.

Extremozymes (Section 10.7)

10.55 What is an extremophile?

10.56 What are two common environmental settings where extremophiles are found?

10.57 What environmental parameter defines the following types of extremophiles?
a. Alkaliphile b. Halophile

10.58 What environmental parameter defines the following types of extremophiles?
a. Acidophile b. Piezophile

10.59 What type(s) of extremophiles are used in commercial laundry detergent formulations?

10.60 What type(s) of extremophiles are used in oil well drilling operations?

Enzyme Inhibition (Section 10.8)

10.61 In competitive inhibition, can both the inhibitor and the substrate bind to an enzyme at the same time? Explain your answer.

10.62 Compare the sites where competitive and noncompetitive inhibitors bind to enzymes.

10.63 Indicate whether each of the following statements describes a reversible competitive inhibitor, a reversible noncompetitive inhibitor, or an irreversible inhibitor. More than one answer may apply.
a. Both inhibitor and substrate bind at the active site on a random basis.
b. The inhibitor effect cannot be reversed by the addition of more substrate.
c. Inhibitor structure does not have to resemble substrate structure.
d. The inhibitor and substrate can bind to the enzyme simultaneously.

10.64 Indicate whether each of the following statements describes a reversible competitive inhibitor, a reversible noncompetitive inhibitor, or an irreversible inhibitor. More than one answer may apply.
a. It bonds covalently to the enzyme active site.
b. The inhibitor effect can be reversed by the addition of more substrate.
c. Inhibitor structure must be somewhat similar to that of the substrate.
d. The inhibitor and substrate cannot bind to the enzyme simultaneously.

Regulation of Enzyme Activity (Section 10.9)

10.65 What is an allosteric enzyme?

10.66 What is a regulator molecule?

10.67 What is feedback control?

10.68 What is the difference between positive and negative feedback to an allosteric enzyme?

10.69 What is the general relationship between zymogens and proteolytic enzymes?

10.70 What, if any, is the difference in meaning between the terms *zymogen* and *proenzyme*?

10.71 Why are proteolytic enzymes always produced in an inactive form?

10.72 What is the mechanism by which most zymogens are activated?

10.73 What is covalent modification?

10.74 What are the two most commonly encountered covalent modification processes?

10.75 What is the most common source of the phosphate group involved in phosphorylation?

10.76 Is the phosphorylated version of an enzyme the "turned-on" or "turned-off" form of the enzyme?

10.77 What is the general name for enzymes that effect the phosphorylation of another enzyme?

10.78 What is the general name for enzymes that effect the dephosphorylation of another enzyme?

▲10.79 Explain the difference in meaning, if any, between the following pairs of "enzyme terms."
a. Apoenzyme and proenzyme
b. Simple enzyme and allosteric enzyme

▲10.80 Explain the difference in meaning, if any, between the following pairs of "enzyme terms."
a. Coenzyme and zymogen
b. Extremozyme and proteolytic enzyme

Prescription Drugs That Inhibit Enzyme Activity (Section 10.10)

10.81 What does the acronym ACE stand for?

10.82 Characterize the compound angiotensin in terms of general structure and biochemical function.

10.83 What is the structural relationship between angiotensinogen and angiotensin?

10.84 Contrast the compounds angiotensinogen and angiotensin in terms of enzyme activity.

10.85 By what mechanism do sulfa drugs kill bacteria?

10.86 By what mechanism do penicillins kill bacteria?

10.87 Why is penicillin toxic to bacteria but not to higher organisms?

10.88 What amino acid in transpeptidase forms a covalent bond to penicillin?

•10.89 (Chemical Connections 10-C) Indicate whether each of the following statements relating to the "grapefruit effect" are true or false.
a. Lycopene is the compound responsible for the "grapefruit effect."
b. Bergamottin is the prescription drug most affected by the "grapefruit effect."

c. The "grapefruit effect" causes the concentrations of affected prescription drugs in the bloodstream to be lower than they should be.
d. Grapefruit and grapefruit juice contain cytochrome P450 enzymes.

•10.90 (Chemical Connections 10-C) Indicate whether each of the following statements relating to the "grapefruit effect" are true or false.
a. Bergamottin is the compound responsible for the "grapefruit effect."
b. Cytochrome P450 enzymes are involved in the process called the "grapefruit effect."
c. Lipitor, one of the most commonly used prescription medications, is affected by the "grapefruit effect."
d. The "grapefruit effect" increases the rate of metabolism of affected drugs.

Medical Uses of Enzymes (Section 10.11)

10.91 What does the acronym TPA stand for, and how is TPA used in therapeutic medicine?

10.92 What does the acronym BUN stand for, and how is this test used in clinical laboratory analysis?

10.93 What enzyme is denoted by each of the following acronyms?
a. LDH b. AST

10.94 What enzyme is denoted by each of the following acronyms?
a. CPK b. ALT

10.95 What is the medical diagnostic value associated with the presence of each of the following enzymes in the bloodstream?
a. CPK b. ALT

10.96 What is the medical diagnostic value associated with the presence of each of the following enzymes in the bloodstream?
a. LDH b. AST

General Characteristics of Vitamins (Section 10.12)

10.97 What is a vitamin?

10.98 List a way in which vitamins differ from carbohydrates, fats, and proteins (the major classes of nutrients).

10.99 Indicate whether each of the following is a fat-soluble or a water-soluble vitamin.
a. Vitamin K
b. Vitamin B_{12}
c. Vitamin C
d. Thiamin

10.100 Indicate whether each of the following is a fat-soluble or a water-soluble vitamin.
a. Vitamin A
b. Vitamin B_6
c. Vitamin E
d. Riboflavin

10.101 Indicate whether each of the vitamins in Problem 10.99 would be likely or unlikely to be toxic when consumed in excess.

10.102 Indicate whether each of the vitamins in Problem 10.100 would be likely or unlikely to be toxic when consumed in excess.

Water-Soluble Vitamins: Vitamin C (Section 10.13)

10.103 What are the two most completely characterized roles of vitamin C in the human body?

10.104 Structurally, how do the oxidized and reduced forms of vitamin C differ?

10.105 Vitamin C is biosynthesized in a two-step process. What is the reactant and product in each of these biosynthetic steps?

10.106 Two enzymes, lactonase and an oxidase, are involved in the biosynthesis of vitamin C. What is the substrate for each of these enzymes?

Water-Soluble Vitamins: The B Vitamins (Section 10.14)

10.107 What is the most characterized role of the B vitamins in the human body?

10.108 Only eight of the nine water-soluble vitamins are called B vitamins. What characteristic "separates" the ninth water-soluble vitamin from the other eight?

10.109 Indicate whether each of the following B vitamin names is the *preferred name* or an *alternative name* for the vitamin.
a. Vitamin B_1 b. Vitamin B_6
c. Cobalamin d. Niacin

10.110 Indicate whether each of the following B vitamin names is the *preferred name* or an *alternative name* for the vitamin.
a. Vitamin B_5 b. Vitamin B_{12}
c. Biotin d. Pyridoxal

10.111 Which of the B vitamins has a structure that fits each of the following characterizations? For a given characterization, more than one B vitamin may qualify.
a. Contains the element sulfur
b. Contains a fused two-ring component
c. Exists in two or more different structural forms
d. Contains a monosaccharide component

10.112 Which of the B vitamins has a structure that fits each of the following structural characterizations? For a given characterization, more than one B vitamin may qualify.
a. Contains the element nitrogen
b. Contains a fused three-ring component
c. Exists in three or more different structural forms
d. Contains an amino acid component

10.113 What is the difference structurally between the two forms of niacin?

10.114 What is the difference structurally between the three forms of vitamin B_6?

10.115 Name the B vitamin precursor for each of the following coenzymes.
a. PLP b. TPP c. CoA d. $NADP^+$

10.116 Name the B vitamin precursor for each of the following coenzymes.
a. NAD^+ b. FMN c. FAD d. THF

10.117 The coenzyme forms of B vitamins are involved in the transfer of an atom or functional group from one molecule to another in metabolic reactions. Identify the atom or group transferred by coenzymes containing the following B vitamins.
a. Thiamin b. Folate
c. Pantothenic acid d. Vitamin B_{12}

10.118 The coenzyme forms of B vitamins are involved in the transfer of an atom or functional group from one molecule to another in metabolic reactions. Identify the atom or group transferred by coenzymes containing the following B vitamins.
a. Niacin
b. Biotin
c. Riboflavin
d. Vitamin B_6

Fat-Soluble Vitamins (Section 10.15)

10.119 Describe the structural differences among the three retinoid forms of vitamin A.

10.120 What is the relationship between the plant pigment beta-carotene and vitamin A?

10.121 What is *cell differentiation,* and how does vitamin A participate in this process?

10.122 List four major functions of vitamin A in the human body.

10.123 How do vitamin D_2 and vitamin D_3 differ in structure?

10.124 In terms of their source, how do vitamin D_2 and vitamin D_3 differ?

10.125 What is the principal function of vitamin D in the human body?

10.126 Why is vitamin D often called the "sunshine" vitamin?

10.127 Which form of tocopherol (vitamin E) exhibits the greatest biochemical activity?

10.128 How do the various forms of tocopherol differ in structure?

10.129 What is the principal function of vitamin E in the human body?

10.130 Why is vitamin E often given to premature infants who are on oxygen therapy?

10.131 How do vitamin K_1 and vitamin K_2 differ in structure?

10.132 In terms of their source, how do vitamin K_1 and vitamin K_2 differ?

10.133 How are menaquinones, phylloquinones, and vitamin K related?

10.134 What is the principal function of vitamin K in the human body?

•10.135 Which vitamin or vitamins has (have) each of the following functions?
a. Is a water-soluble antioxidant
b. Is a fat-soluble antioxidant
c. Is involved in the process of vision
d. Is involved in the formation of collagen

•10.136 Which vitamin or vitamins has (have) each of the following functions?
a. Is involved in prostaglandin synthesis
b. Is a coenzyme precursor
c. Is involved in calcium deposition in bone
d. Is involved in cell differentiation

•10.137 Which of the 13 vitamins has a structure that fits each of the following characterizations? For a given characterization, more than one vitamin may qualify.
a. Does not contain nitrogen
b. Contains a metal atom
c. Has a saturated or an unsaturated hydrocarbon chain component
d. Contains one or more carboxyl (—COOH) functional groups

•10.138 Which of the 13 vitamins has a structure that fits each of the following characterizations? For a given characterization, more than one vitamin may qualify.
a. Contains sulfur
b. Does not have a cyclic component
c. Contains one or more alcohol (—OH) functional groups
d. Contains both fused and nonfused ring systems

Nucleic Acids

11

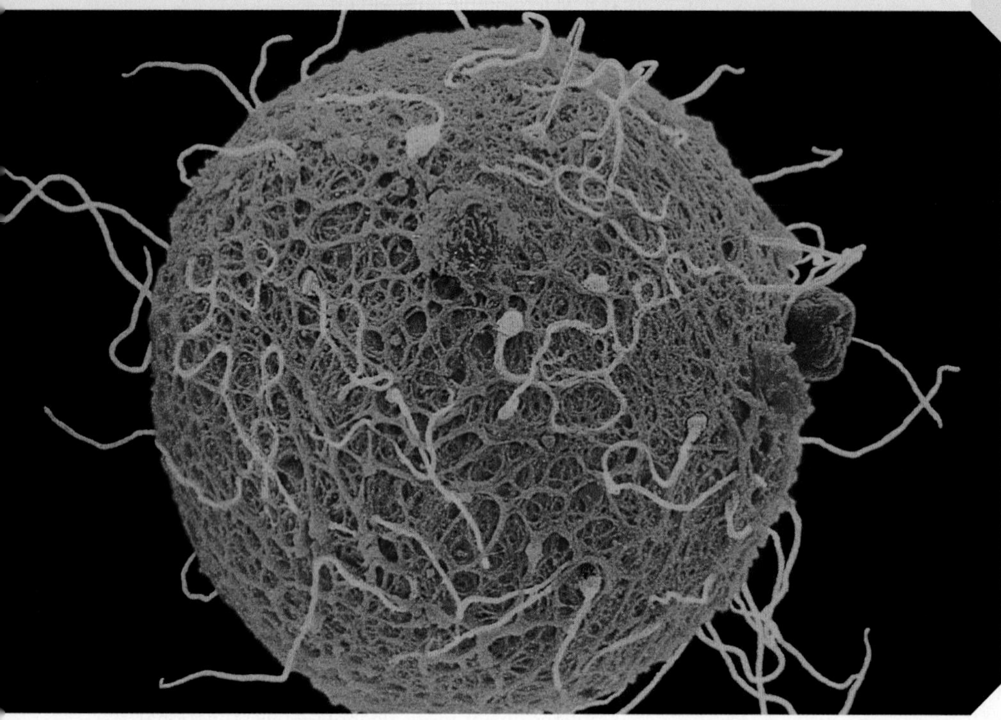

Dr. Nikos/James Burns/Phototake

Human egg and sperm.

CHAPTER OUTLINE

A most remarkable property of living cells is their ability to produce exact replicas of themselves. Furthermore, cells contain all the instructions needed for making the complete organism of which they are a part. The molecules within a cell that are responsible for these amazing capabilities are nucleic acids.

The Swiss physiologist Friedrich Miescher (1844–1895) discovered nucleic acids in 1869 while studying the nuclei of white blood cells. The fact that they were initially found in cell nuclei and are acidic accounts for the name *nucleic acid.* Although it is now known that nucleic acids are found throughout a cell, not just in the nucleus, the name is still used for such materials.

11.1 Types of Nucleic Acids

Two types of nucleic acids are found within cells of higher organisms: *deoxyribonucleic acid* (DNA) and *ribonucleic acid* (RNA). Nearly all the DNA is found within the cell nucleus. Its primary function is the storage and transfer of genetic information. This information is used (indirectly) to control many functions of a living cell. In addition, DNA is passed from existing cells to new cells during cell division. RNA occurs in all parts of a cell. It functions primarily in synthesis of proteins, the molecules that carry out essential cellular functions. The structural distinctions between DNA and RNA molecules are considered in Section 11.4.

All nucleic acid molecules are unbranched polymers. A **nucleic acid** *is an unbranched polymer in which the monomer units are nucleotides.* Thus the starting point for a discussion of nucleic acids is an understanding of the structures and chemical properties of nucleotides.

It was not until 1944, 75 years after the discovery of nucleic acids, that scientists obtained the first evidence that these molecules are responsible for the storage and transfer of genetic information.

11.2 Nucleotide Building Blocks

A **nucleotide** *is a three-subunit molecule in which a pentose sugar is bonded to both a phosphate group and a nitrogen-containing heterocyclic base.* With a three-subunit structure, nucleotides are more complex monomers than the monosaccharides of polysaccharides (Section 7.8) and the amino acids of proteins (Section 9.2). A block structural diagram for a nucleotide is

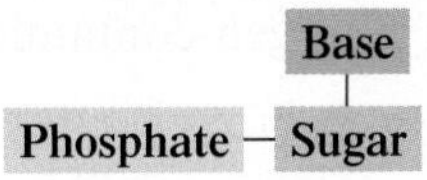

Proteins are polypeptides, many carbohydrates are polysaccharides, and nucleic acids are polynucleotides.

Pentose Sugars

The sugar unit of a nucleotide is either the pentose *ribose* or the pentose *2′-deoxyribose.*

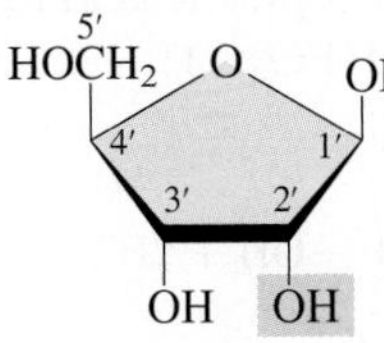

β-D-Ribose

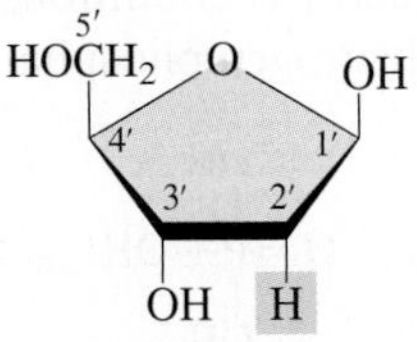

β-D-2′-Deoxyribose

Structurally, the only difference between these two sugars occurs at carbon 2′. The —OH group present on this carbon in ribose becomes a —H atom in 2′-deoxyribose. (The prefix *deoxy-* means "without oxygen.")

RNA and DNA differ in the identity of the sugar unit in their nucleotides. In RNA, the sugar unit is *r*ibose—hence the *R* in RNA. In DNA, the sugar unit is 2′-*d*eoxyribose—hence the *D* in DNA.

The systems for numbering the atoms in the pentose and nitrogen-containing base subunits of a nucleotide are important and will be used extensively in later sections of this chapter. The convention is that

1. Pentose ring atoms are designated with *primed* numbers.
2. Nitrogen-containing base ring atoms are designated with *unprimed* numbers.

Nitrogen-Containing Heterocyclic Bases

Five nitrogen-containing heterocyclic bases are nucleotide components. Three of them are derivatives of pyrimidine (Section 6.9), a monocyclic base with a six-membered ring, and two are derivatives of purine (Section 6.9), a bicyclic base with fused five- and six-membered rings.

Pyrimidine and purine do not themselves occur naturally; numerous derivatives of these two compounds, however, are naturally occurring substances.

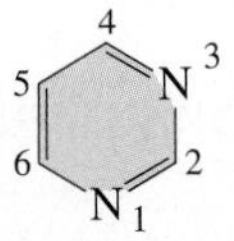
Pyrimidine

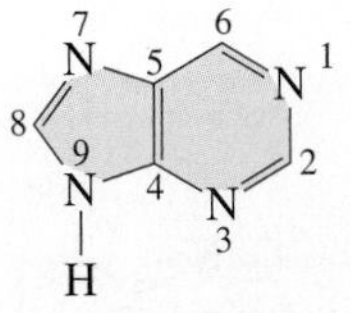
Purine

A pyrimidine derivative that was encountered previously is the B vitamin thiamine (see Section 10.14).

Both of these heterocyclic compounds are bases because they contain amine functional groups (secondary or tertiary), and amine functional groups exhibit basic behavior (proton acceptors; Section 6.6).

The three pyrimidine derivatives found in nucleotides are thymine (T), cytosine (C), and uracil (U).

Caffeine, the most widely used nonprescription central nervous system stimulant, is the 1,3,7-trimethyl-2,6-dioxo derivative of purine (Section 6.9).

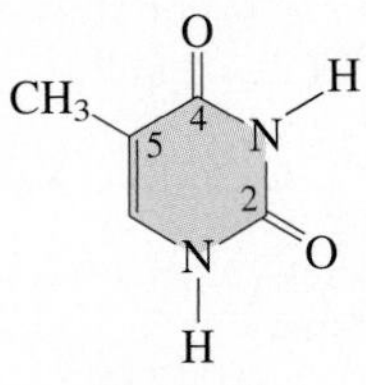
Thymine (T)

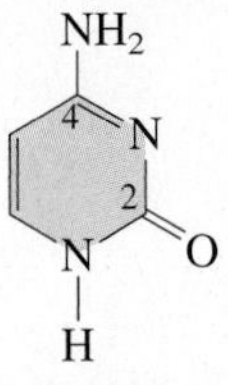
Cytosine (C)

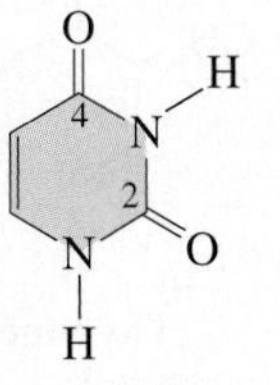
Uracil (U)

Thymine is the 5-methyl-2,4-dioxo derivative, cytosine the 4-amino-2-oxo derivative, and uracil the 2,4-dioxo derivative of pyrimidine.

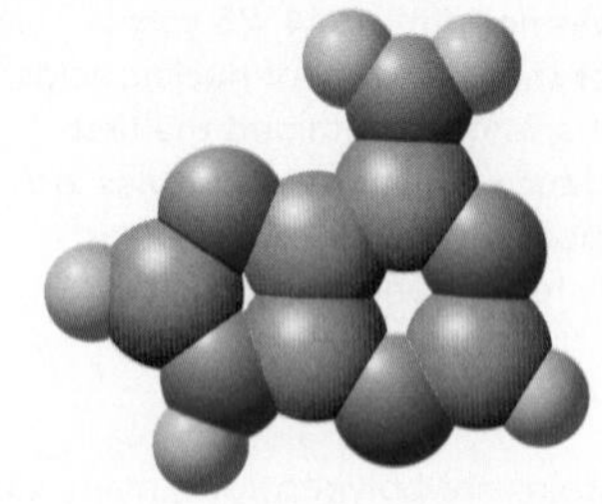

Figure 11.1 Space-filling model of the molecule adenine, a nitrogen-containing heterocyclic base present in both DNA and RNA.

The two purine derivatives found in nucleotides are adenine (A) and guanine (G).

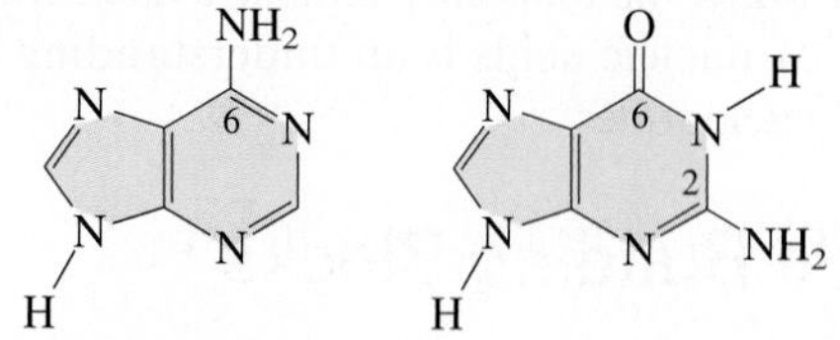

Adenine (A) Guanine (G)

Adenine is the 6-amino derivative of purine, and guanine is the 2-amino-6-oxo purine derivative. A space-filling model for adenine is shown in Figure 11.1.

Adenine, guanine, and cytosine are found in both DNA and RNA. Uracil is found only in RNA, and thymine usually occurs only in DNA. Figure 11.2 summarizes the occurrences of nitrogen-containing heterocyclic bases in nucleic acids.

Phosphate

Phosphate, the third component of a nucleotide, is derived from phosphoric acid (H_3PO_4). Under cellular pH conditions, the phosphoric acid loses two of its hydrogen atoms to give a hydrogen phosphate ion (HPO_4^{2-}).

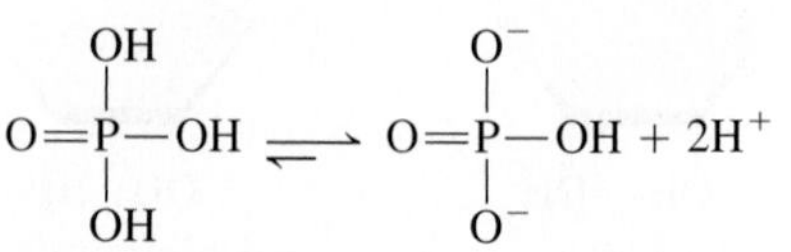

Phosphoric acid Hydrogen phosphate ion

11.3 Nucleotide Formation

The formation of a nucleotide from a sugar, a base, and a phosphate can be visualized as a two-step process.

Nucleoside = Sugar + Base
Nucleotide = Nucleoside + Phosphate

1. First, the pentose sugar and nitrogen-containing base react to form a two-subunit entity called a *nucleoside* (not *nucleotide*, *s* versus *t*).
2. The nucleoside reacts with a phosphate group to form the three-subunit entity called a *nucleotide*. It is nucleotides that become the building blocks for nucleic acids.

Figure 11.2 Two purine bases and three pyrimidine bases are found in the nucleotides present in nucleic acids.

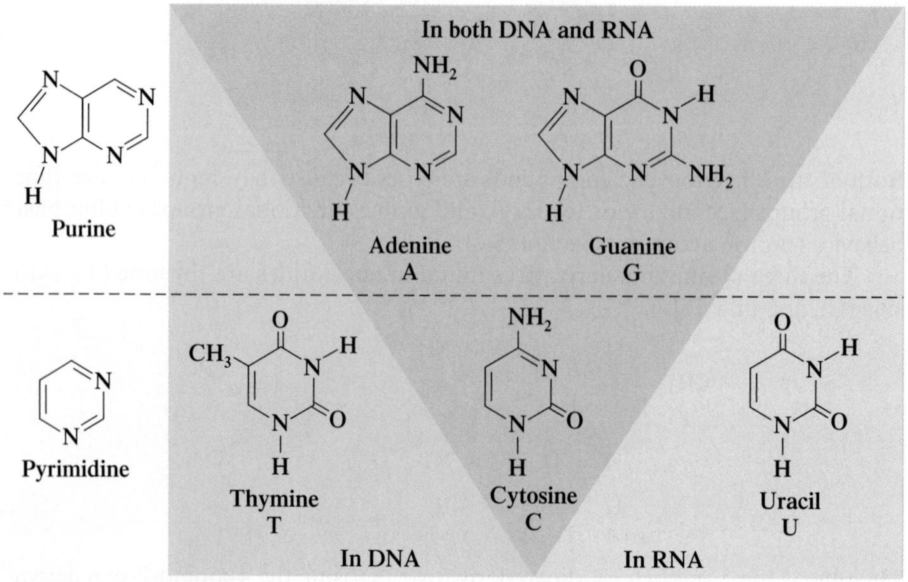

To remember which two of the five nucleotide bases are the purine derivatives (fused rings), use the phrase "pure silver" and substitute the chemical symbol for silver, which is Ag.

pure Ag
purine A and G

Nucleoside Formation

A **nucleoside** *is a two-subunit molecule in which a pentose sugar is bonded to a nitrogen-containing heterocyclic base.* The following structural equation is representative of nucleoside formation.

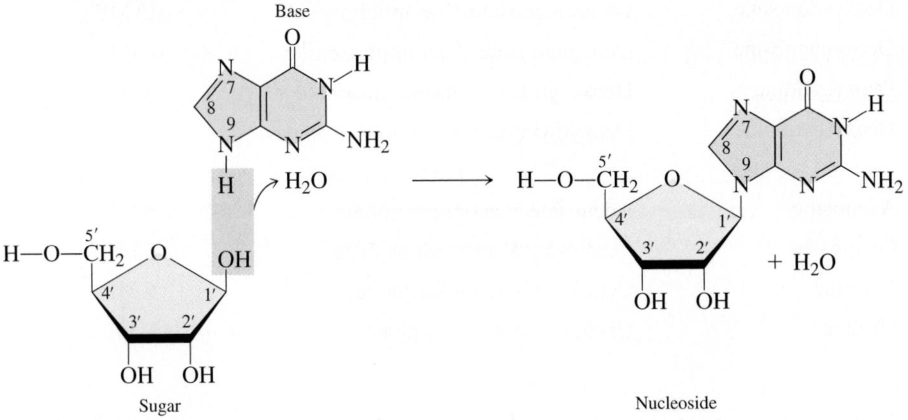

Important characteristics of the nucleoside formation process of combining two molecules into one are:

1. The base is always attached to C-1′ of the sugar (the anomeric carbon atom (Section 7.10)), which is always in a β-configuration. For purine bases, attachment is through N-9; for pyrimidine bases, N-1 is involved. The bond connecting the sugar and base is a β-N-glycosidic linkage (Section 7.3).
2. A molecule of water is formed as the two molecules bond together; a condensation reaction occurs.

Eight nucleosides are associated with nucleic acid chemistry—four involve ribose (RNA nucleosides) and four involve deoxyribose (DNA nucleosides). The eight combinations are:

RNA Nucleosides	DNA Nucleosides
ribose-adenine	deoxyribose-adenine
ribose-cytosine	deoxyribose-cytosine
ribose-guanine	deoxyribose-guanine
ribose-uracil	deoxyribose-thymine

Nucleosides are named as derivatives of the base that they contain; the base's name is modified using a suffix.

1. For pyrimidine bases, the suffix *-idine* is used (cytidine, thymidine, uridine).
2. For purine bases, the suffix *-osine* is used (adenosine, guanosine).
3. The prefix *-deoxy* is used to indicate that the sugar present is deoxyribose. No prefix is used when the sugar present is ribose.

Using these rules, the nucleoside containing ribose and adenine is called *adenosine,* and the nucleoside containing deoxyribose and thymine is called *deoxythymidine.*

Nucleotide Formation

Addition of a phosphate group to a nucleoside produces a nucleotide. The following structural equation is representative of nucleotide formation.

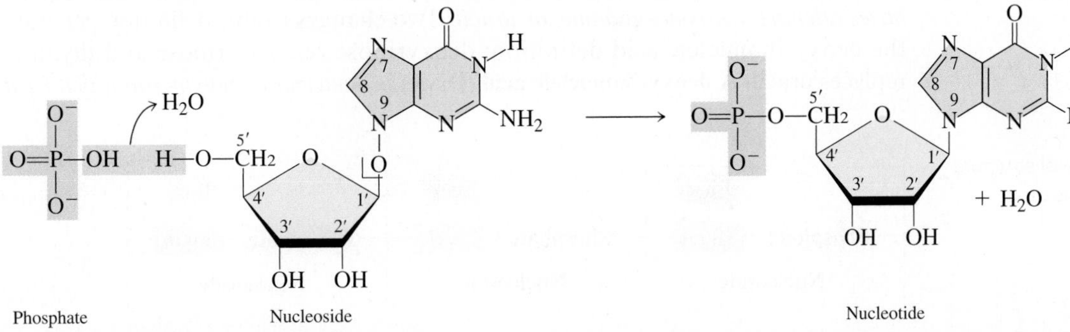

Table 11.1 Information Concerning the Eight Nucleotides That Are Building Blocks for DNA and RNA

Base	Abbreviation	Nucleoside	Nucleotide	Abbreviation
DNA				
Adenine	A	Deoxyadenosine	Deoxyadenosine 5′-monophosphate	dAMP
Guanine	G	Deoxyguanosine	Deoxguanosine 5′-monophosphate	dGMP
Cytosine	C	Deoxycytidine	Deoxycytidine 5′-monophosphate	dCMP
Thymine	T	Deoxythymidine	Deoxythymidine 5′-monophosphate	dTMP
RNA				
Adenine	A	Adenosine	Adenosine 5′-monophosphate	AMP
Guanine	G	Guanosine	Guanosine 5′-monophosphate	GMP
Cytosine	C	Cytidine	Cytidine 5′-monophosphate	CMP
Uracil	U	Uridine	Uridine 5′-monophosphate	UMP

Important characteristics of the nucleotide formation process of adding a phosphate group to a nucleoside are the following:

1. The phosphate group is attached to the sugar at the C-5′ position through a phosphate-ester linkage.
2. As with nucleoside formation, a molecule of water is produced in nucleotide formation. Thus, overall, two molecules of water are produced in combining a sugar, base, and phosphate into a nucleotide.

Nucleotides are named by appending the term *5′-monophosphate* to the name of the nucleoside from which they are derived. Addition of a phosphate group to the nucleoside adenosine produces the nucleotide adenosine 5′-monophosphate.

Abbreviations for nucleotides exist, which are used in a manner similar to that for amino acids (Section 9.2). The abbreviations use the one-letter symbols for the base (*A, C, G, T,* and *U*), *MP* for monophosphate, and a lowercase *d* at the start of the abbreviation when deoxyribose is the sugar. The abbreviation for adenosine 5′-monophosphate is *AMP* and that for deoxyadenosine 5′-monophosphate is *dAMP.* Table 11.1 summarizes information presented in this section about nucleosides and nucleotides.

11.4 Primary Nucleic Acid Structure

Nucleotides are related to nucleic acids in the same way that amino acids are related to proteins.

Nucleic acids are polymers in which the repeating units, the monomers, are nucleotides (Section 11.2). The nucleotide units within a nucleic acid molecule are linked to each other through sugar–phosphate bonds. The resulting molecular structure (Figure 11.3) involves a chain of alternating sugar and phosphate groups with a base group protruding from the chain at regular intervals.

In Section 11.1, the two general types of nucleic acids—ribonucleic acids and deoxyribonucleic acids—were mentioned, but their definitions were not given. Definitions are now in order. A **ribonucleic acid (RNA)** *is a nucleotide polymer in which each of the monomers contains ribose, a phosphate group, and one of the heterocyclic bases adenine, cytosine, guanine, or uracil.* Two changes to this definition generate the deoxyribonucleic acid definition; deoxyribose replaces ribose and thymine replaces uracil. A **deoxyribonucleic acid (DNA)** *is a nucleotide polymer in which each*

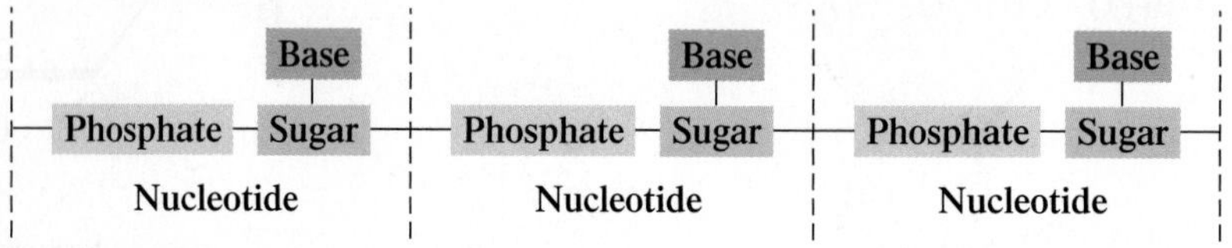

Figure 11.3 The general structure of a nucleic acid in terms of nucleotide subunits.

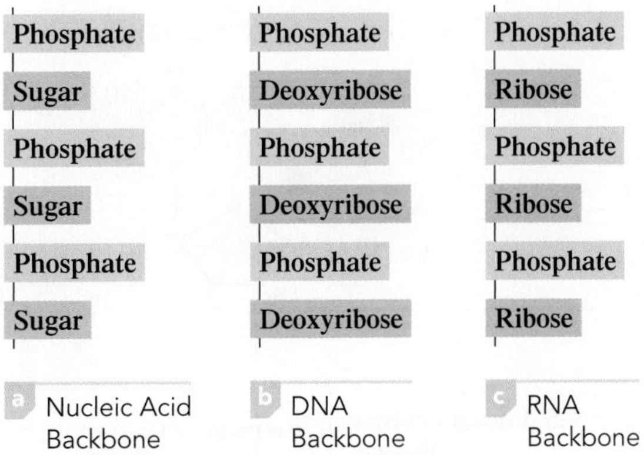

Figure 11.4 (a) The generalized backbone structure of a nucleic acid. (b) The specific backbone structure for a deoxyribonucleic acid (DNA). (c) The specific backbone structure for a ribonucleic acid (RNA).

of the monomers contains deoxyribose, a phosphate group, and one of the heterocyclic bases adenine, cytosine, guanine, or thymine.

The alternating sugar–phosphate chain in a nucleic acid structure is often called the *nucleic acid backbone.* This backbone is constant throughout the entire nucleic acid structure. For DNA molecules, the backbone consists of alternating phosphate and *deoxyribose* sugar units; for RNA molecules, the backbone consists of alternating phosphate and *ribose* sugar units. Figure 11.4 contrasts the generalized backbone structure for a nucleic acid with the specific backbone structures of DNAs and RNAs.

The backbone of a nucleic acid structure is always an alternating sequence of phosphate and sugar groups. The sugar is ribose in RNA and deoxyribose in DNA.

The variable portion of nucleic acid structure is the sequence of bases attached to the sugar units of the backbone. The sequence of these base side chains distinguishes various DNAs from each other and various RNAs from each other. Only four types of bases are found in any given nucleic acid structure. This situation is much simpler than that for proteins, where 20 side-chain entities (amino acids) are available (Section 9.2). In both RNA and DNA, adenine, guanine, and cytosine are encountered as side-chain components; thymine is found mainly in DNA, and uracil is found only in RNA (Figure 11.2).

Primary nucleic acid structure *is the sequence in which nucleotides are linked together in a nucleic acid.* Because the sugar–phosphate backbone of a given nucleic acid does not vary, the primary structure of the nucleic acid depends only on the sequence of bases present. Further information about nucleic acid structure can be obtained by considering the detailed four-nucleotide segment of a DNA molecule shown in Figure 11.5.

Just as the order of amino acid side chains determines the primary structure of a protein (Section 9.10), the order of nucleotide bases determines the primary structure of a nucleic acid.

The following list describes some important points about nucleic acid structure that are illustrated in Figure 11.5:

1. Each nonterminal phosphate group of the sugar–phosphate backbone is bonded to two sugar molecules through a *3′,5′-phosphodiester linkage.* There is a phosphoester bond to the 5′ carbon of one sugar unit and a phosphoester bond to the 3′ carbon of the other sugar.
2. A nucleotide chain has *directionality.* One end of the nucleotide chain, the *5′ end,* normally carries a free phosphate group attached to the 5′ carbon atom. The other end of the nucleotide chain, the *3′ end,* normally has a free hydroxyl group attached to the 3′ carbon atom. By convention, the sequence of bases of a nucleic acid strand is read from the 5′ end to the 3′ end.
3. Each nonterminal phosphate group in the backbone of a nucleic acid carries a −1 charge. The parent phosphoric acid molecule from which the phosphate was derived originally had three —OH groups (Section 11.2). Two of these become involved in the 3′,5′-phosphodiester linkage. The remaining —OH group is free to exhibit acidic behavior—that is, to produce a H^+ ion.

For both nucleic acids and proteins, a distinction is made between the two ends of the polymer chain. For nucleic acids, there is a 5′ end and a 3′ end; for proteins, there is an N-terminal end and a C-terminal end (Section 9.7).

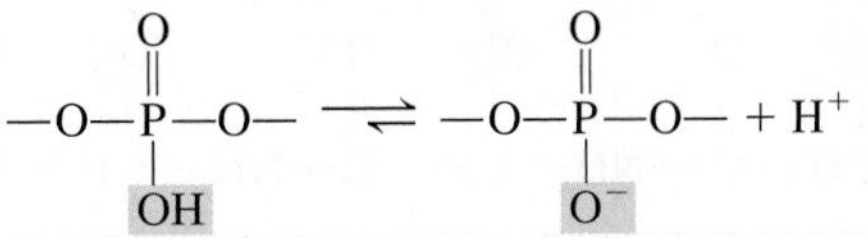

This behavior by the many phosphate groups in a nucleic acid backbone gives nucleic acids their acidic properties.

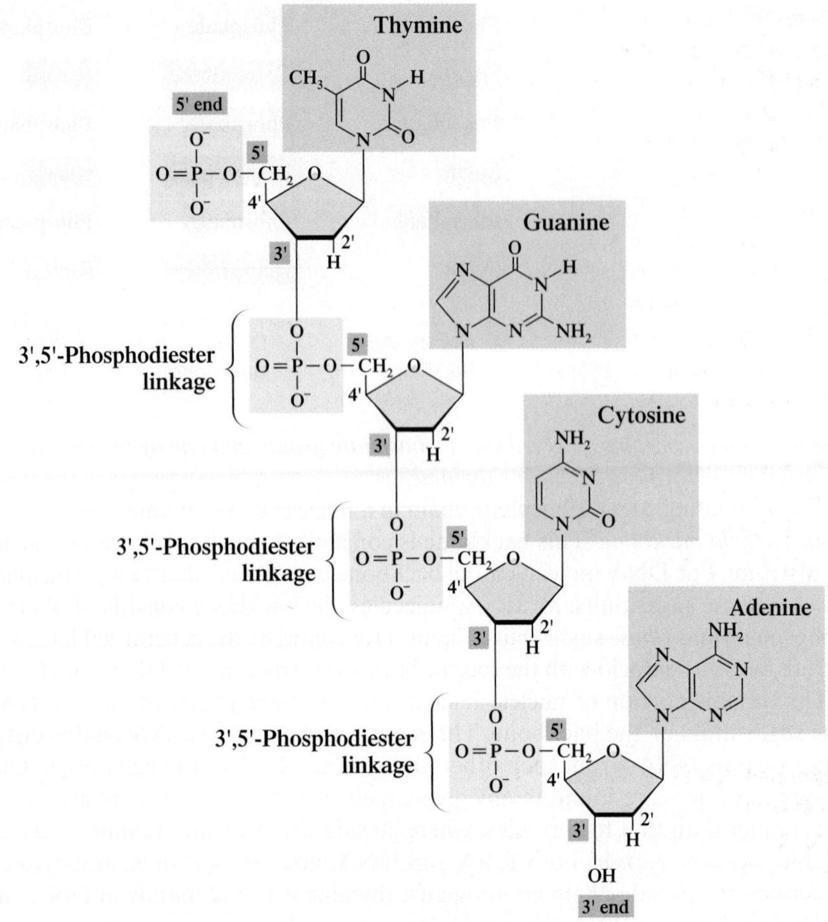

Figure 11.5 A four-nucleotide-long segment of DNA. (The choice of bases was arbitrary.)

Specifying the primary structure for a nucleic acid is done by listing nucleotide base components (using their one-letter abbreviations) in sequential order starting with the base at the 5′ end of the nucleotide strand. The primary structure for the four-nucleotide DNA segment shown in Figure 11.5 is

5′ T–G–C–A 3′

Three parallels between primary nucleic acid structure and primary protein structure (Section 9.10) are worth noting:

1. DNAs, RNAs, and proteins all have backbones that do not vary in structure (see Figure 11.6).
2. The sequence of attachments to the backbones (nitrogen bases in nucleic acids and amino acid R groups in proteins) distinguishes one

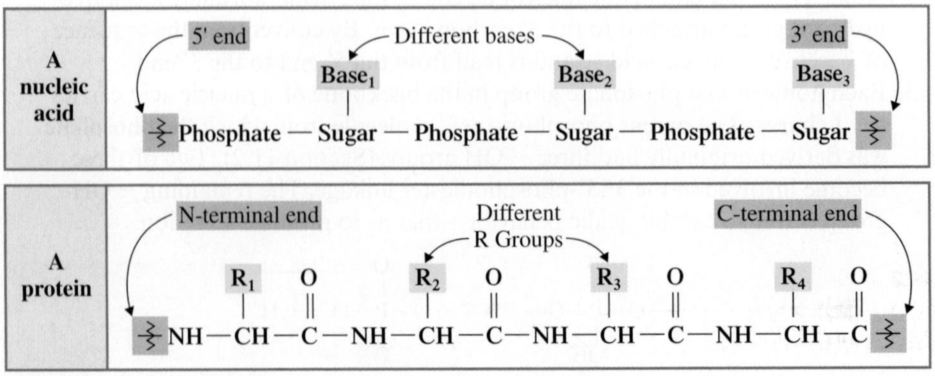

Figure 11.6 A comparison of the general primary structures of nucleic acids and proteins.

CHEMISTRY AT A GLANCE Nucleic Acid Structure

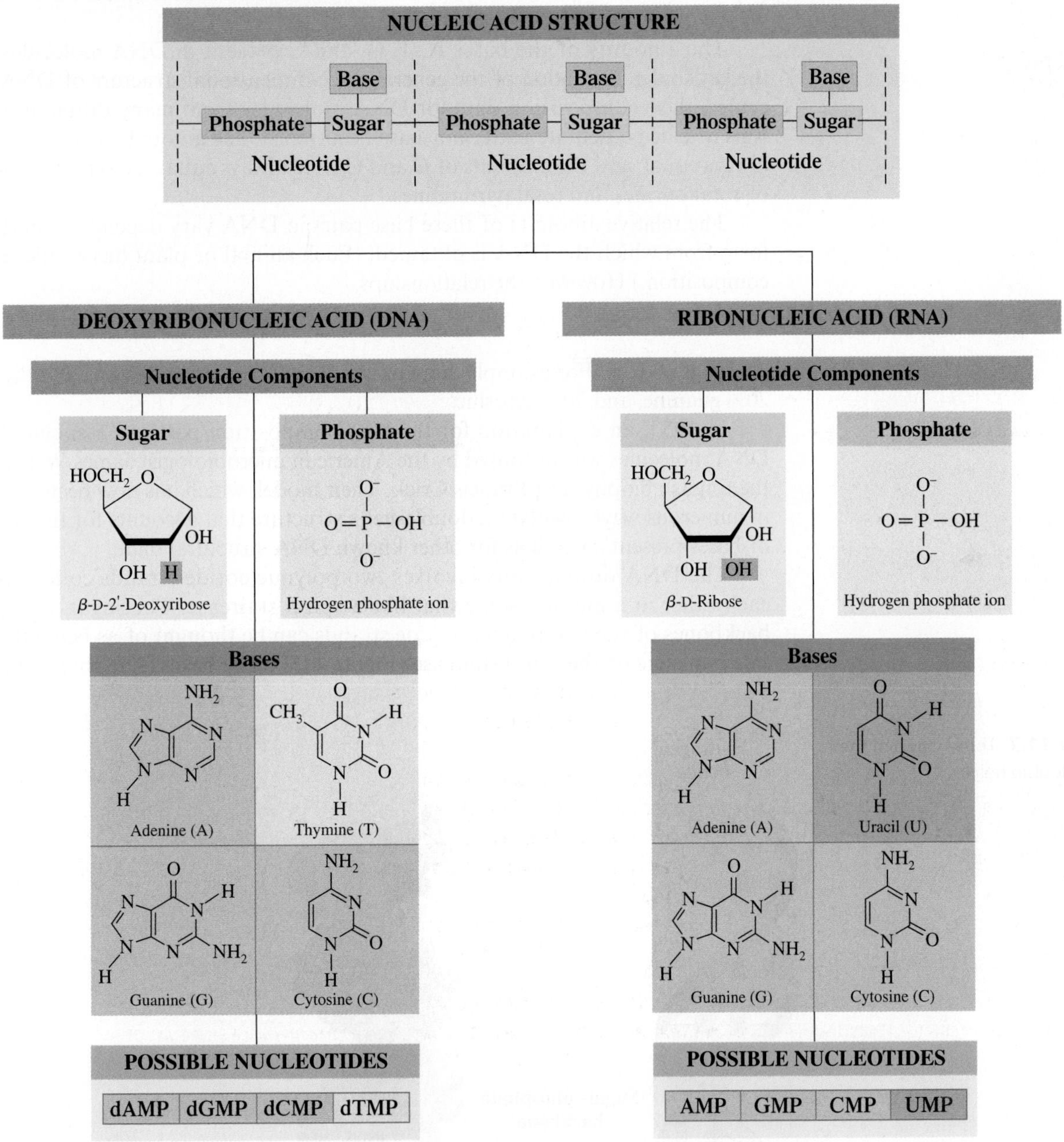

DNA from another, one RNA from another, and one protein from another (Figure 11.6).

3. Both nucleic acid polymer chains and protein polymer chains have directionality; for nucleic acids, there is a 5′ end and a 3′ end, and for proteins, there is an N-terminal end and a C-terminal end.

The Chemistry at a Glance feature above summarizes important concepts relative to the makeup of the nucleotide building blocks (monomers) present in polymeric DNA and RNA molecules.

11.5 The DNA Double Helix

Like proteins, nucleic acids have secondary, or three-dimensional, structure as well as primary structure. The secondary structures of DNAs and RNAs differ, so they will be discussed separately.

The amounts of the bases A, T, G, and C present in DNA molecules were the key to determination of the general three-dimensional structure of DNA molecules. Base composition data for DNA molecules from many different organisms revealed a definite pattern of base occurrence. The amounts of A and T were always equal, and the amounts of C and G were always equal, as were the amounts of total purines and total pyrimidines.

The relative amounts of these base pairs in DNA vary depending on the life form from which the DNA is obtained. (Each animal or plant has a unique base composition.) However, the relationships

$$\%A = \%T \qquad \text{and} \qquad \%C = \%G$$

always hold true. For example, human DNA contains 30% adenine, 30% thymine, 20% guanine, and 20% cytosine.

In 1953, an explanation for the base composition patterns associated with DNA molecules was proposed by the American microbiologist James Watson and the English biophysicist Francis Crick. Their model, which has now been validated in numerous ways, involves a double-helix structure that accounts for the equality of bases present, as well as for other known DNA structural data.

The DNA double helix involves two polynucleotide strands coiled around each other in a manner somewhat like a spiral staircase. The sugar–phosphate backbones of the two polynucleotide strands can be thought of as being the outside banisters of the spiral staircase (Figure 11.7). The bases (side chains) of each

Figure 11.7 Three views of the DNA double helix.

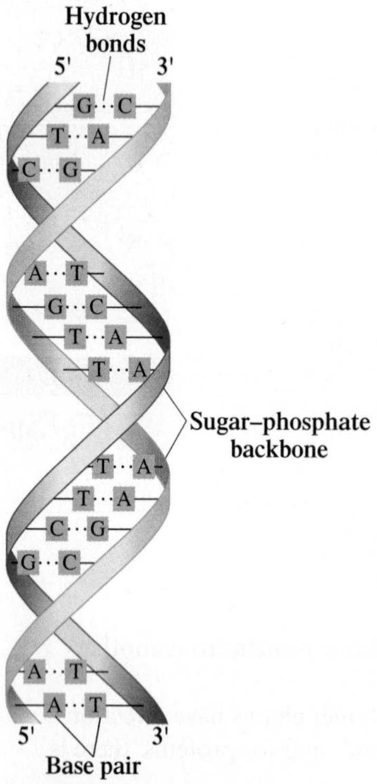

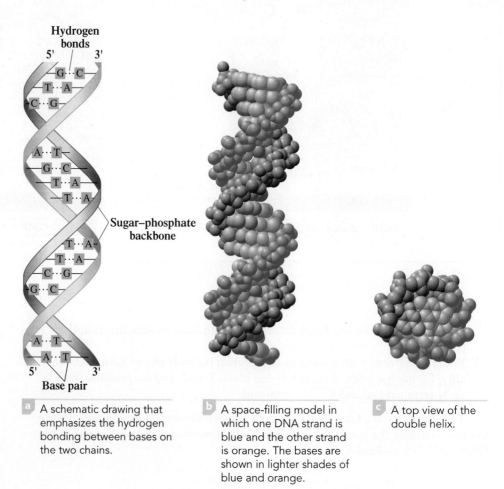

a A schematic drawing that emphasizes the hydrogen bonding between bases on the two chains.

b A space-filling model in which one DNA strand is blue and the other strand is orange. The bases are shown in lighter shades of blue and orange.

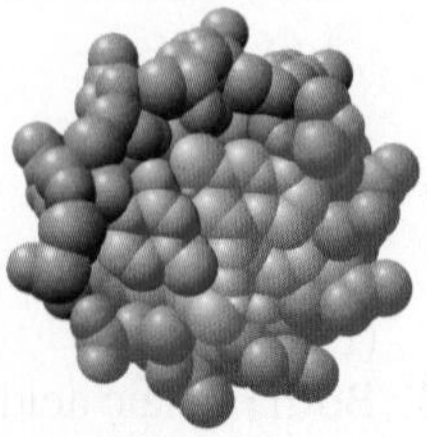

c A top view of the double helix.

backbone extend inward toward the bases of the other strand. The two strands are connected by *hydrogen bonds* between their bases. Additionally, the two strands of the double helix are *antiparallel*—that is, they run in opposite directions. One strand runs in the 5′-to-3′ direction, and the other is oriented in the 3′-to-5′ direction.

Base Pairing

A physical restriction, the size of the interior of the DNA double helix, limits the base pairs that can hydrogen-bond to one another. Only pairs involving one small base (a pyrimidine) and one large base (a purine) correctly "fit" within the helix interior. There is not enough room for two large purine bases to fit opposite each other (they overlap), and two small pyrimidine bases are too far apart to hydrogen-bond to one another effectively. Of the four possible purine–pyrimidine combinations (A–T, A–C, G–T, and G–C), hydrogen-bonding possibilities are *most favorable* for the A–T and G–C pairings, and these two combinations are the *only two* that normally occur in DNA. Figure 11.8 shows the specific hydrogen-bonding interactions for the four possible purine–pyrimidine base-pairing combinations.

The pairing of A with T and that of G with C are said to be *complementary.* A and T are complementary bases, as are G and C. **Complementary bases** *are pairs of bases in a nucleic acid structure that can hydrogen-bond to each other.* The fact that complementary base pairing occurs in DNA molecules explains, very simply, why the amounts of the bases A and T present are always equal, as are the amounts of G and C.

The two strands of DNA in a double helix are *not identical*—they are complementary. **Complementary DNA strands** *are strands of DNA in a double helix with base pairing such that each base is located opposite its complementary base.* Wherever G occurs in one strand, there is a C in the other strand; wherever T occurs in one

The α-helix secondary structure of proteins (Section 9.11) involves *one* polypeptide chain; the double-helix secondary structure of DNA involves *two* polynucleotide chains. In the α-helix of proteins, the R groups are on the *outside* of the helix; in the double helix of DNA, the bases are on the *inside* of the double helix.

The *antiparallel* nature of the two polynucleotide chains in the DNA double helix means that there is a 5′ end and a 3′ end at both ends of the double helix.

The two strands of DNA in a double helix are complementary. This means that if you know the order of bases in one strand, you can predict the order of bases in the other strand.

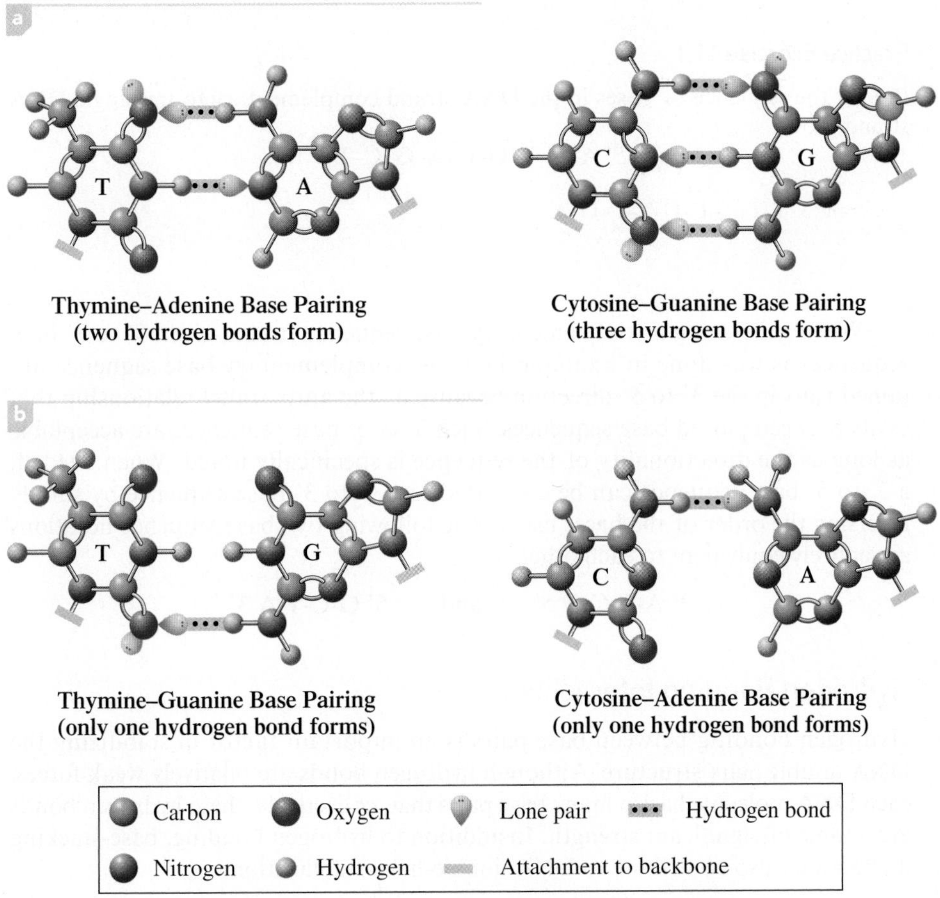

Figure 11.8 Hydrogen-bonding possibilities are more favorable when A–T and G–C base pairing occurs than when A–C and G–T base pairing occurs. (a) Two and three hydrogen bonds can form, respectively, between A–T and G–C base pairs. These combinations are present in DNA molecules. (b) Only one hydrogen bond can form between G–T and A–C base pairs. These combinations are not present in DNA molecules.

A mnemonic device for recalling base-pairing combinations in DNA involves listing the base abbreviations in alphabetical order. Then the first and last bases pair, and so do the middle two bases.

DNA: A C G T

Another way to remember these base-pairing combinations is to note that AT spells a word and that C and G look very much alike.

strand, there is an A in the other strand. An important ramification of this complementary relationship is that knowing the base sequence of one strand of DNA enables prediction of the base sequence of the complementary strand.

The base sequence of a single strand of a DNA molecule segment is always written in the direction from the 5′ end to the 3′ end of the segment.

5′ A–A–G–C–T–A–G–C–T–T–A–C–T 3′

If the end designations for a base sequence (5′ and 3′) are not specified for a sequence of bases, it is assumed that the sequence starts with the 5′ end base. In the base sequence

A–C–G–T–T–C

It is assumed that A is the 5′ end base.

EXAMPLE 11.1 Predicting Base Sequence in a Complementary DNA Strand

Predict the sequence of bases in the DNA strand that is complementary to the single DNA strand shown.

5′ C–G–A–A–T–C–C–T–A 3′

Solution

Because only A forms a complementary base pair with T, and only G with C, the complementary strand is as follows:

Given: 5′ C–G–A–A–T–C–C–T–A 3′

Complementary strand: 3′ G–C–T–T–A–G–G–A–T 5′

Note the reversal of the numbering of the ends of the complementary strand compared to the given strand. This is due to the antiparallel nature of the two strands in a DNA double helix.

Practice Exercise 11.1

Predict the sequence of bases in the DNA strand complementary to the single DNA strand shown.

5′ A–A–T–G–C–A–G–C–T 3′

Answer: 3′ T–T–A–C–G–T–C–G–A 5′

When generating a complementary base sequence from a given 5′ to 3′ base sequence, as was done in Example 11.1, the complementary base sequence obtained runs in the 3′ to 5′ direction because of the antiparallel relationship that exists between paired base sequences. Such 3′ to 5′ base sequences are acceptable as long as the directionality of the sequence is specifically noted. When needed, a 3′ to 5′ base sequence can be converted to a 5′ to 3′ base sequence by simply reversing the order of the bases listed. The following two base sequence notations are entirely equivalent to each other.

3′ A-T-C-G 5′ and 5′ G-C-T-A 3′

Hydrogen Bonding Interactions

Hydrogen bonding is responsible for the secondary structure (double helix) of DNA. Hydrogen bonding is also responsible for secondary structure in proteins (Section 9.11).

Hydrogen bonding between base pairs is an important factor in stabilizing the DNA double helix structure. Although hydrogen bonds are relatively weak forces, each DNA molecule has so many base pairs that, collectively, these hydrogen bonds are a force of significant strength. In addition to hydrogen bonding, base-stacking interactions also contribute to DNA double-helix stabilization.

Base-Stacking Interactions

The bases in a DNA double helix are positioned with the planes of their rings parallel (like a stack of coins). Stacking interactions involving a given base and the parallel bases directly above and below it also contribute to the stabilization of the DNA double helix. These stacking interactions are as important in their stabilization effects as is the hydrogen bonding associated with base pairing—perhaps even more important. Purine and pyrimidine bases are hydrophobic in nature, so their stacking interactions are those associated with hydrophobic molecules—mainly London forces. The concept of hydrophobic interactions has been encountered twice previously. Hydrophobic interactions involving the nonpolar tails of membrane lipids contribute to the structural stability of cell membranes (Section 8.10), and hydrophobic interactions involving nonpolar R groups of amino acids contribute to protein tertiary structure stability (Section 9.12).

Use of the Term "DNA Molecule"

The term *DNA molecule* is actually a misnomer, even though general usage of the term is common in news reports, in textbooks, and even in the vocabulary of scientists. It is technically a misnomer for two reasons.

1. Cellular solutions have pH values such that the phosphate groups present in the DNA backbone structure are negatively charged. This means DNA is actually a multicharged ionic species rather than a neutral molecule.
2. The two strands of DNA in a double-helix structure are not held together by covalent bonds but, rather, by hydrogen bonds, which are noncovalent interactions. Thus, double-helix DNA is an entity that involves two intertwined ionic species rather than a single molecule.

Despite these considerations, usage of the term *DNA molecule* is accepted by most scientists and is used through the remainder of this textbook.

11.6 Replication of DNA Molecules

DNA molecules are the carriers of genetic information within a cell; that is, they are the molecules of heredity. Each time a cell divides, an exact copy of the DNA of the parent cell is needed for the new daughter cell. The process by which new DNA molecules are generated is DNA replication. **DNA replication** *is the biochemical process by which DNA molecules produce exact duplicates of themselves.* The key concept in understanding DNA replication is the base pairing associated with the DNA double helix.

DNA Replication Overview

In DNA replication, the two strands of the DNA double helix are regarded as a pair of *templates,* or patterns. During replication, the strands separate. Each can then act as a template for the synthesis of a new, complementary strand. The result is two daughter DNA molecules with base sequences identical to those of the parent double helix. Details of this replication are as follows.

Under the influence of the enzyme *DNA helicase,* the DNA double helix unwinds, and the hydrogen bonds between complementary bases are broken. This unwinding process, as shown in Figure 11.9, is somewhat like opening a zipper. The point at which the DNA double helix is unwinding, which is constantly changing (moving), is called the *replication fork* (Figure 11.9).

The bases of the separated strands are no longer connected by hydrogen bonds. They can pair with *free* individual nucleotides present in the cell's nucleus. As shown in Figure 11.9, the base pairing always involves C pairing with G and A pairing with T. The pairing process occurs one nucleotide at a time. After a free nucleotide has formed hydrogen bonds with a base of the old strand (the template),

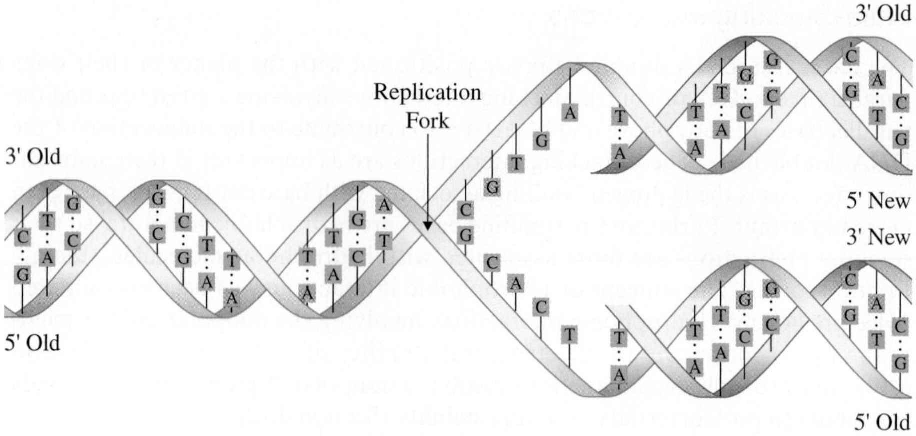

Figure 11.9 In DNA replication, the two strands of the DNA double helix unwind, with the separated strands serving as templates for the formation of new DNA strands. Free nucleotides pair with the complementary bases on the separated strands of DNA. This process ultimately results in the complete replication of the DNA molecule.

the enzyme *DNA polymerase* verifies that the base pairing is correct and then catalyzes the formation of a new phosphodiester linkage between the nucleotide and the growing strand (represented by the darker blue ribbons in Figure 11.9). The *DNA polymerase* then slides down the strand to the next unpaired base of the template, and the same process is repeated.

The net result of DNA replication is the production of two *daughter DNA molecules,* both of which are identical to the one *parent DNA molecule* from which they were formed. Figure 11.10 gives a "close-up" view of the relationships between parent DNA and the daughter DNA produced from it. Note that each daughter DNA molecule contains one strand from the parent DNA and one strand that is newly formed.

The Replication Process in Finer Detail

Though simple in principle, the DNA replication process has many intricacies.

1. The enzyme *DNA polymerase* can operate on a forming DNA daughter strand only in the 5′-to-3′ direction. Because the two strands of parent DNA run in opposite directions (one is 5′ to 3′ and the other 3′ to 5′; Section 11.4), only one strand can grow continuously in the 5′-to-3′ direction. The other strand must be formed in short segments, called *Okazaki fragments* (after their discoverer, Reiji Okazaki), as the DNA unwinds (Figure 11.11). The breaks or gaps in this daughter strand are called *nicks.* To complete the formation of this strand, the Okazaki fragments are connected by action of the enzyme *DNA ligase.* The strand that grows continuously is called the *leading strand,* and the strand that is synthesized in small segments is called the *lagging strand.*

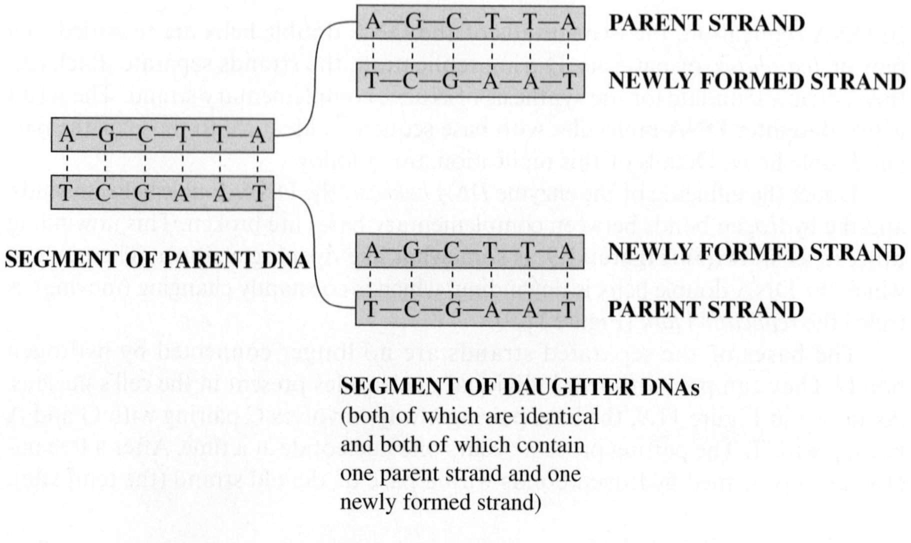

Figure 11.10 In DNA replication, two daughter DNA molecules are produced from one parent DNA molecule, with each daughter DNA molecule containing one parent DNA strand and one newly formed DNA strand.

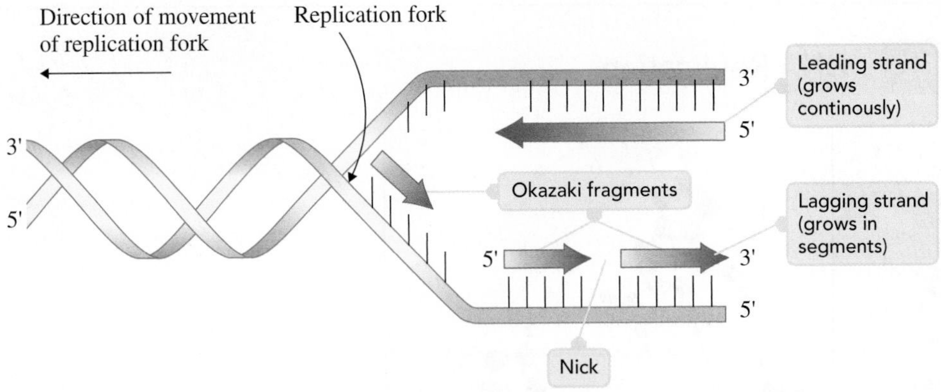

Figure 11.11 Because the enzyme *DNA polymerase* can act only in the 5′-to-3′ direction, one strand (top) grows continuously in the direction of the unwinding, and the other strand grows in segments in the opposite direction. The segments in this latter chain are then connected by a different enzyme, *DNA ligase.*

2. The process of DNA unwinding does not have to begin at an end of the DNA molecule. It may occur at any location within the molecule. Indeed, studies show that unwinding usually occurs at several interior locations simultaneously and that DNA replication is bidirectional for these locations; that is, it proceeds in both directions from the unwinding sites. As shown in Figure 11.12, the result of this multiple-site replication process is the formation of "bubbles" of newly synthesized DNA. The bubbles grow larger and eventually coalesce, giving rise to two complete daughter DNAs. Multiple-site replication enables large DNA molecules to be replicated rapidly.

Based on mode of action, several types of anticancer drugs exist. One large group of such drugs are the *antimetabolites,* drugs which are DNA-replication inhibitors. The focus on relevancy feature Chemical Connections 11-A on page 473 discusses several substances that find use as DNA-replication inhibitors.

The Chemistry at a Glance feature on the next page summarizes the steps, as discussed in this section, that occur in the process of DNA replication.

Chromosomes

Once the DNA within a cell has been replicated, it interacts with specific proteins in the cell called *histones* to form structural units that provide the most stable arrangement for the long DNA molecules. These histone–DNA complexes are called *chromosomes.* A **chromosome** *is an individual DNA molecule bound to a group of proteins.* Typically, a chromosome is about 15% by mass DNA and 85% by mass protein.

Chromosomes are *nucleoproteins.* They are a combination of nucleic acid (DNA) and various proteins.

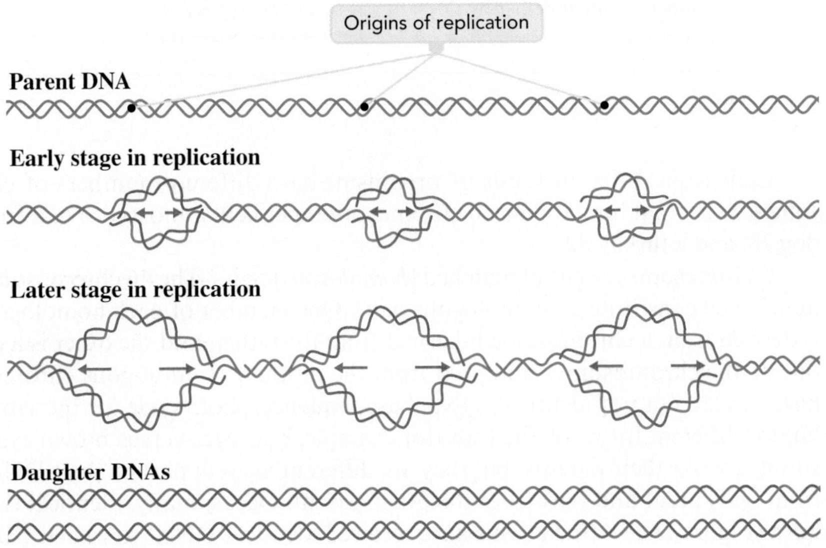

Figure 11.12 DNA replication usually occurs at multiple sites within a molecule, and the replication is bidirectional from these sites.

CHEMISTRY AT A GLANCE DNA Replication

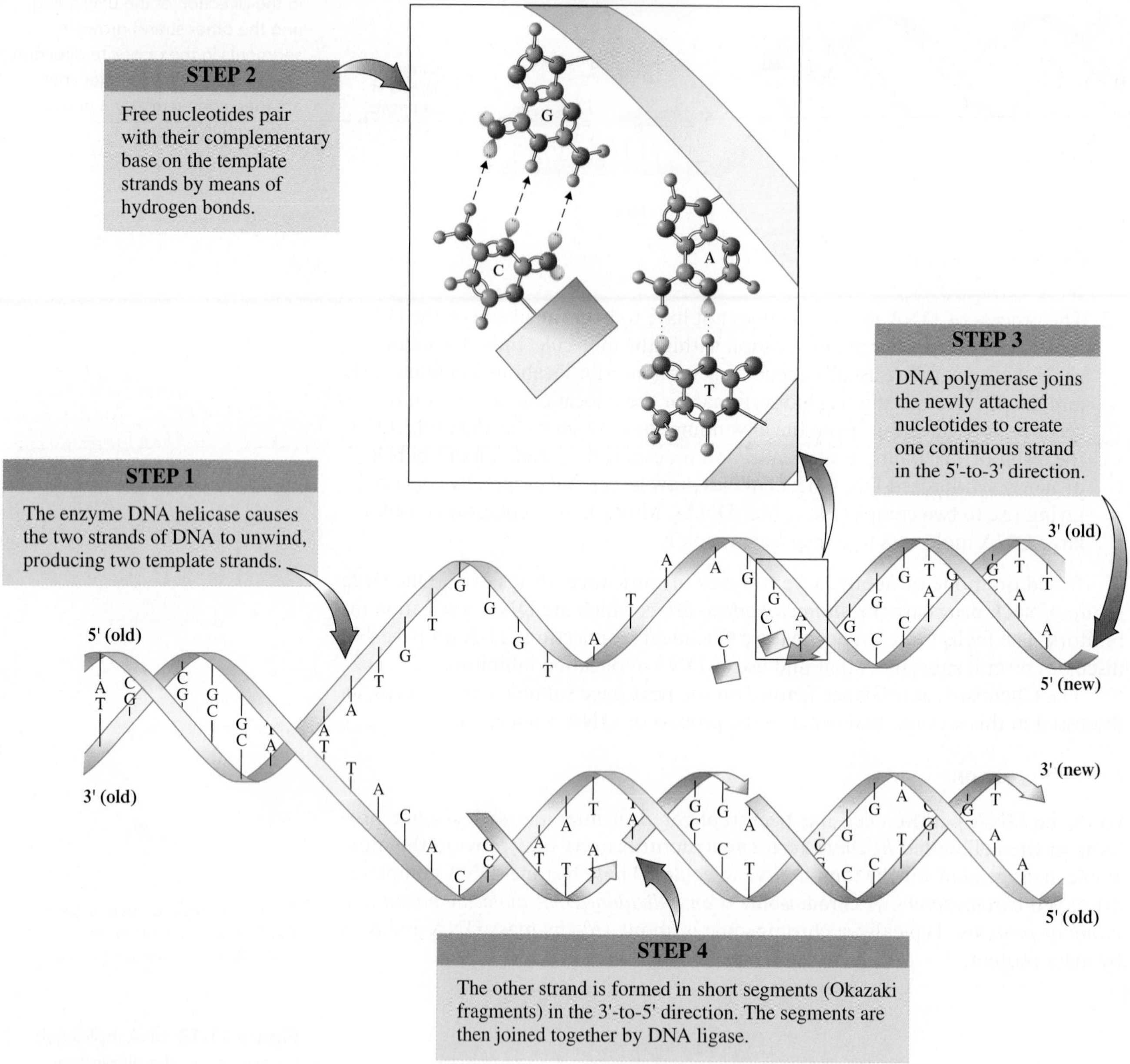
STEP 2
Free nucleotides pair with their complementary base on the template strands by means of hydrogen bonds.
STEP 3
DNA polymerase joins the newly attached nucleotides to create one continuous strand in the 5'-to-3' direction.
STEP 1
The enzyme DNA helicase causes the two strands of DNA to unwind, producing two template strands.
5' (old)
3' (old)
3' (old)
5' (new)
3' (new)
5' (old)
STEP 4
The other strand is formed in short segments (Okazaki fragments) in the 3'-to-5' direction. The segments are then joined together by DNA ligase.

© Erica Stone/Peter Arnold, Inc./Photolibrary

Figure 11.13 Identical twins share identical physical characteristics because they received identical DNA from their parents.

Cells from different kinds of organisms have different numbers of chromosomes. A normal human has 46 chromosomes per cell, a mosquito 6, a frog 26, a dog 78, and a turkey 82.

Chromosomes occur in matched (*homologous*) pairs. The 46 chromosomes of a human cell constitute 23 homologous pairs. One member of each homologous pair is derived from a chromosome inherited from the father, and the other is a copy of one of the chromosomes inherited from the mother. Homologous chromosomes have similar, but not identical, DNA base sequences; both code for the same traits but for different forms of the trait (for example, blue eyes versus brown eyes). Offspring are like their parents, but they are different as well; part of their DNA came from one parent and part from the other parent. Occasionally, identical twins are born (Figure 11.13). Such twins have received identical DNA from their parents.

CHEMICAL CONNECTIONS 11-A

Antimetabolites: Anticancer Drugs That Inhibit DNA Synthesis

Cancer is a disease characterized by rapid uncontrolled cell division. Rapid cell division necessitates the synthesis of large amounts of DNA, as DNA must be present in each new cell produced. Numerous anticancer drugs are now available that block DNA synthesis and therefore decrease the rate at which new cancer cells are produced.

Antimetabolites are a class of anticancer drugs that interfere with DNA replication because their structures are similar to molecules required for normal DNA replication. The structural similarity is close enough that enzymes can be "tricked" into using the drug rather than the real substrate needed. This "trickery" shuts down DNA synthesis, which causes cells to die.

Four examples of commonly used antimetabolites and the molecules they "mimic" are as follows:

1. 6-Mercaptopurine (6-MP): 6-MP structurally resembles adenine, one of the four nitrogen-containing bases present in all DNA molecules. Synthesis of adenine-containing nucleotides is inhibited when 6-MP is present; nonfunctional DNA results when 6-MP, rather than adenine, is incorporated into a nucleotide.

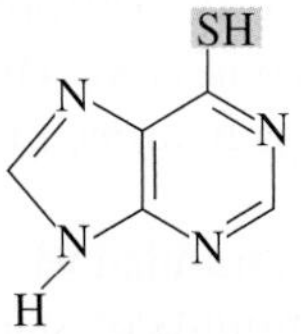

6-Mercaptopurine
(a modified adenine)

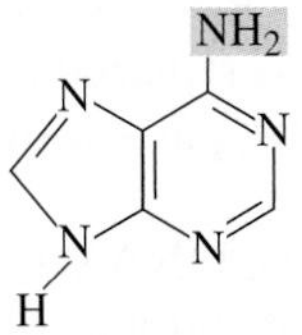

Adenine

2. Thioguanine: As was the case with 6-MP, the close structural resemblance between thioguanine and guanine leads to the incorporation of thioguanine, rather than guanine, into nucleotides. Nonfunctional DNA is the result.

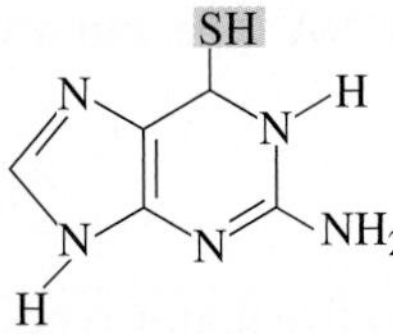

Thioguanine
(a modified guanine)

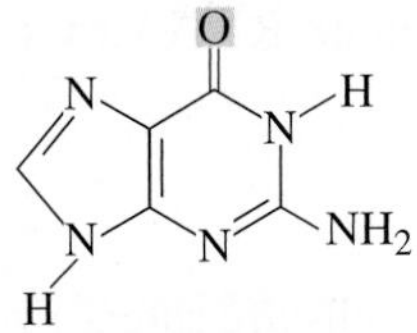

Guanine

3. 5-Fluorouracil: Uracil is a base found in RNA rather than DNA. However, its structure is close enough to that of thymine (which is methyluracil) that it can pass for thymine. It thus inhibits the synthesis of active thymine-containing nucleotides needed for DNA synthesis.

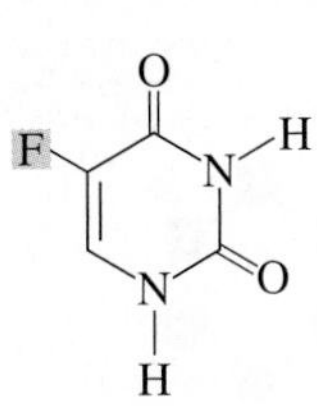

5-Fluorouracil
(a modified thymine)

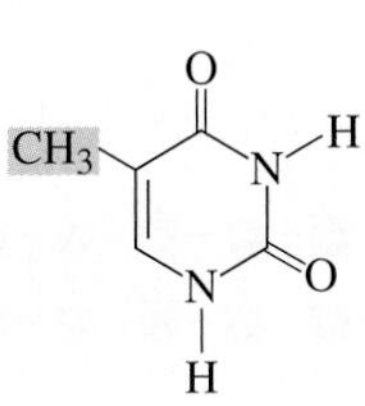

Thymine

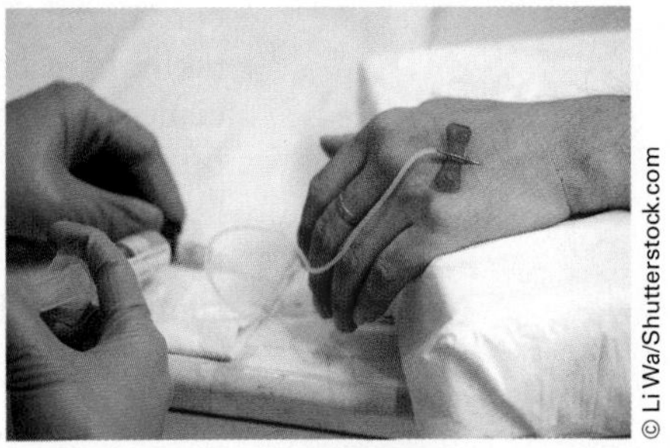

A cancer patient undergoing chemotherapy to inhibit DNA synthesis.

4. Methotrexate: The previous three antimetabolites were all close structural analogs of bases found in DNA nucleotides. Some antimetabolites are not base analogs. Methotrexate is one of these non-base analogs. It is a structural analog of folic acid (folate), which is one of the B vitamins (Section 10.14). A derivative of folic acid is needed in one of the early steps of nucleotide synthesis. Methotrexate inhibits the conversion of folic acid to this needed derivative, which shuts down DNA synthesis.

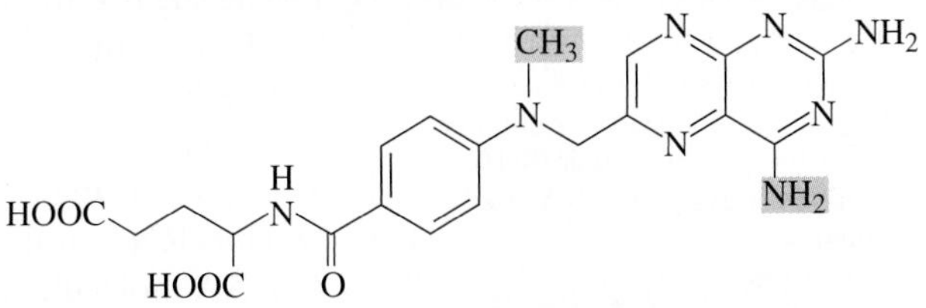

Methotrexate

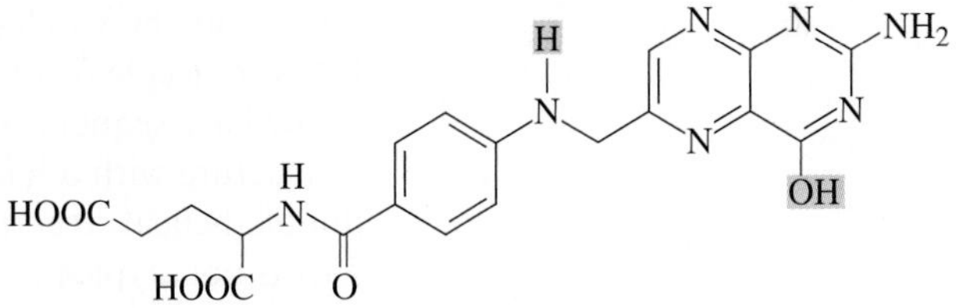

Folic acid

Note that all four of these antimetabolites are synthetically produced molecules rather than naturally occurring ones. Many different synthetically modified purine and pyrimidine derivatives are now available for use in studies that involve DNA.

While some types of cancer respond very well to chemotherapy using antimetabolite drugs (leukemia is one such cancer), other types of cancer do not respond well to such treatment. Two other general types of anticancer drugs are available for use in these treatment situations. They are (1) DNA-damaging agents and (2) cell-division inhibitors. Cisplatin is a well-known DNA damaging agent, and the natural products vincristine (obtained from the periwinkle plant) and taxol (obtained from yew tree bark) are cell-division inhibitors.

Cells most affected by anticancer drugs are those undergoing rapid cell division (the cancer cells). However, normal cells are also affected, to a lesser extent, by these drugs. Eventually, the normal cells are affected to such a degree that use of the drugs must be discontinued, at least for a period of time.

11.7 Overview of Protein Synthesis

In the previous section it was shown how the replication of DNA makes it possible for a new cell to contain the same genetic information as its parent cell. How the genetic information contained in a cell is expressed in cell operation will now be considered. This leads to the topic of protein synthesis. The synthesis of proteins (skin, hair, enzymes, hormones, and so on) is under the direction of DNA molecules. It is this role of DNA that establishes the similarities between parent and offspring that are regarded as hereditary characteristics.

The overall process of protein synthesis is divided into two phases. The first phase is called *transcription* and the second *translation.* The following diagram summarizes the relationship between transcription and translation.

$$\boxed{\text{DNA}} \xrightarrow{\text{Transcription}} \boxed{\text{RNA}} \xrightarrow{\text{Translation}} \boxed{\text{protein}}$$

Before discussing the details of transcription and translation, more information about RNA molecules is needed. They are involved in transcription, as the end products, and translation, as the starting materials. Particularly important are the differences between RNA and DNA and among various types of RNA molecules.

11.8 Ribonucleic Acids

Four major differences exist between RNA molecules and DNA molecules:

The bases thymine (T) and uracil (U) have similar structures. Thymine is a methyluracil (Section 11.2). The hydrogen-bonding patterns (Figure 11.7) for the A–U base pair (RNA) and the A–T base pair (DNA) are identical.

1. The sugar unit in the backbone of RNA is ribose; it is deoxyribose in DNA.
2. The base thymine found in DNA is replaced by uracil in RNA (Figure 11.2). In RNA, uracil, instead of thymine, pairs with (forms hydrogen bonds with) adenine.
3. RNA is a single-stranded molecule; DNA is double-stranded (double helix). Thus RNA, unlike DNA, does not contain equal amounts of specific bases.
4. RNA molecules are much smaller than DNA molecules, ranging from 75 nucleotides to a few thousand nucleotides.

Note that the single-stranded nature of RNA does not prevent *portions* of an RNA molecule from folding back upon itself and forming double-helical regions. If the base sequences along two portions of an RNA strand are complementary, a structure with a hairpin loop results, as shown in Figure 11.14. The amount of double-helical structure present in an RNA varies with RNA type, but a value of 50% is not atypical.

Types of RNA Molecules

RNA molecules found in human cells are categorized into five major types, distinguished by their function. These five RNA types are heterogeneous nuclear RNA (hnRNA), messenger RNA (mRNA), small nuclear RNA (snRNA), ribosomal RNA (rRNA), and transfer RNA (tRNA).

Heterogeneous nuclear RNA (hnRNA) also goes by the name *primary transcript RNA* (ptRNA).

Heterogeneous nuclear RNA (hnRNA) *is RNA formed directly by DNA transcription.* Post-transcription processing converts the heterogeneous nuclear RNA to messenger RNA.

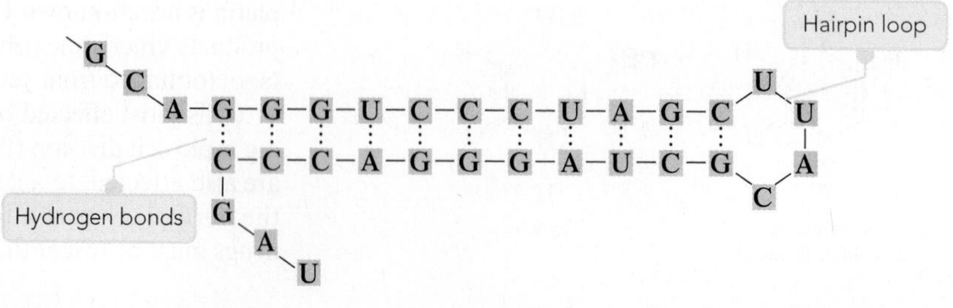

Figure 11.14 A hairpin loop is produced when single-stranded RNA doubles back on itself and complementary base pairing occurs.

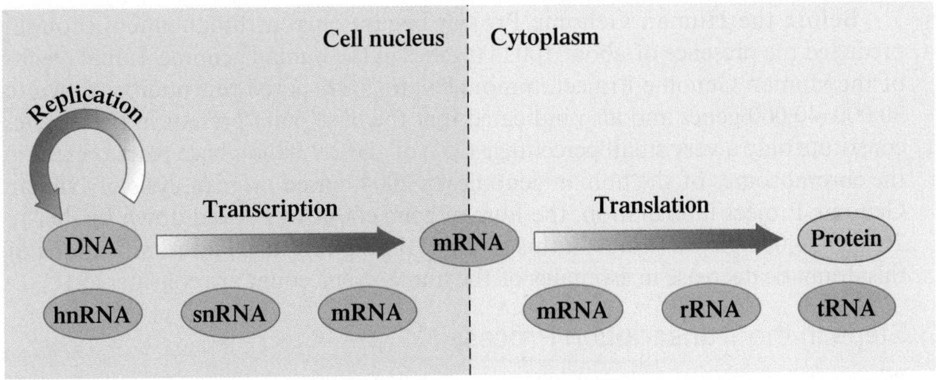

Figure 11.15 An overview of types of RNA in terms of cellular locations where they are encountered and processes in which they are involved.

Messenger RNA (mRNA) *is RNA that carries instructions for protein synthesis (genetic information) to the sites for protein synthesis.* The molecular mass of messenger RNA varies with the length of the protein whose synthesis it will direct.

Small nuclear RNA (snRNA) *is RNA that facilitates the conversion of heterogeneous nuclear RNA to messenger RNA.* It contains from 100 to 200 nucleotides.

Ribosomal RNA (rRNA) *is RNA that combines with specific proteins to form ribosomes, the physical sites for protein synthesis.* Ribosomes have molecular masses on the order of 3 million amu. The rRNA present in ribosomes has no informational function.

Transfer RNA (tRNA) *is RNA that delivers amino acids to the sites for protein synthesis.* Transfer RNAs are the smallest of the RNAs, possessing only 75–90 nucleotide units.

At a nondetail level, a cell consists of a nucleus and an extranuclear region called the cytoplasm. The process of DNA transcription occurs in the nucleus, as does the processing of hnRNA to mRNA. [DNA replication (Section 11.5) also occurs in the nucleus.] The mRNA formed in the nucleus travels to the cytoplasm where translation (protein synthesis) occurs. Figure 11.15 summarizes the transcription and translation processes in terms of the types of RNA involved and the cellular locations where the processes occur.

The most abundant type of RNA in a cell is ribosomal RNA (75% to 80% by mass). Transfer RNA constitutes 10%–15% of cellular RNA; messenger RNA and its precursor, heterogeneous nuclear RNA, make up 5%–10% of RNA material in a cell.

An additional role for RNA, besides its major involvement in protein synthesis, has been recently discovered by scientists. It also plays a part in the process of blood coagulation near a wound. RNA that is released from damaged cells associated with the wound helps to activate two enzymes needed for the blood coagulation process.

A detailed look at cellular structure is found in Section 12.2.

11.9 Transcription: RNA Synthesis

Transcription *is the process by which DNA directs the synthesis of hnRNA/mRNA molecules that carry the coded information needed for protein synthesis.* Messenger RNA production via transcription is actually a "two-step" process in which an hnRNA molecule is initially produced and then is "edited" to yield the desired mRNA molecule. The mRNA molecule so produced then functions as the carrier of the information needed to direct protein synthesis.

Within a strand of a DNA molecule are instructions for the synthesis of numerous hnRNA/mRNA molecules. During transcription, a DNA molecule unwinds, under enzyme influence, at the particular location where the appropriate base sequence is found for the hnRNA/mRNA of concern, and the "exposed" base sequence is transcribed. A short segment of a DNA strand so transcribed, which contains instructions for the formation of a particular hnRNA/mRNA, is called a *gene.* A **gene** *is a segment of a DNA strand that contains the base sequence for the production of a specific hnRNA/mRNA molecule.*

In humans, most genes are composed of 1000–3500 nucleotide units. Hundreds of genes can exist along a DNA strand. Obtaining information concerning the total number of genes and the total number of nucleotide base pairs present in human DNA has been an area of intense research activity for the last two decades. The central activity in this research was the *Human Genome Project,* a decade-long internationally based research project to determine the location and base sequence of each of the genes in the human *genome.* A **genome** *is all of the genetic material (the total DNA) contained in the chromosomes of an organism.*

The overall reaction of the scientific community to the results of the Human Genome Project is aptly summarized by the expression "There are things we didn't know that we didn't know."

Before the Human Genome Project began, current biochemical thought predicted the presence of about 100,000 genes in the human genome. Initial results of the Human Genome Project, announced in 2001, pared this number down to 30,000–40,000 genes and also indicated that the base pairs present in these genes constitute only a very small percentage (2%) of the 2.9 billion base pairs present in the chromosomes of the human genome. In 2004, based on reanalysis of Human Genome Project information, the human gene count was pared down further to 20,000–25,000 genes. (Later in this section, the significance and ramifications of this dramatic decrease in estimates of the human gene count are considered.)

Steps in the Transcription Process

The mechanics of transcription are in many ways similar to those of DNA replication. Four steps are involved:

In DNA–RNA base pairing, the complementary base pairs are

DNA	RNA
A	U
G	C
C	G
T	A

RNA molecules contain the base U instead of the base T.

1. A *portion* of the DNA double helix unwinds, exposing some bases (a gene). The unwinding process is governed by the enzyme *RNA polymerase* rather than by *DNA helicase* (replication enzyme).
2. Free *ribo*nucleotides, one nucleotide at a time, align along *one* of the exposed strands of DNA bases, the *template* strand, forming new base pairs. In this process, U rather than T aligns with A in the base-pairing process. Only about 10 base pairs of the DNA template strand are exposed at a time. Because ribonucleotides rather than deoxyribonucleotides are involved in the base pairing, ribose, rather than deoxyribose, becomes incorporated into the new nucleic acid backbone.
3. *RNA polymerase* is involved in the linkage of ribonucleotides, one by one, to the growing hnRNA molecule.
4. Transcription ends when the *RNA polymerase* enzyme encounters a sequence of bases that is "read" as a stop signal. The newly formed hnRNA molecule and the *RNA polymerase* enzyme are released, and the DNA then rewinds to re-form the original double helix.

The strand of DNA used for hnRNA/mRNA synthesis is called the *template strand.* It is copied proceeding in the 3′ to 5′ direction. The other DNA strand (the non-template strand) is called the *informational strand.* The informational strand, although not involved in RNA synthesis, gives the base sequence present in the hnRNA strand being synthesized (with the exception of U replacing T).

Figure 11.16 shows the overall process of transcription of DNA to form hnRNA.

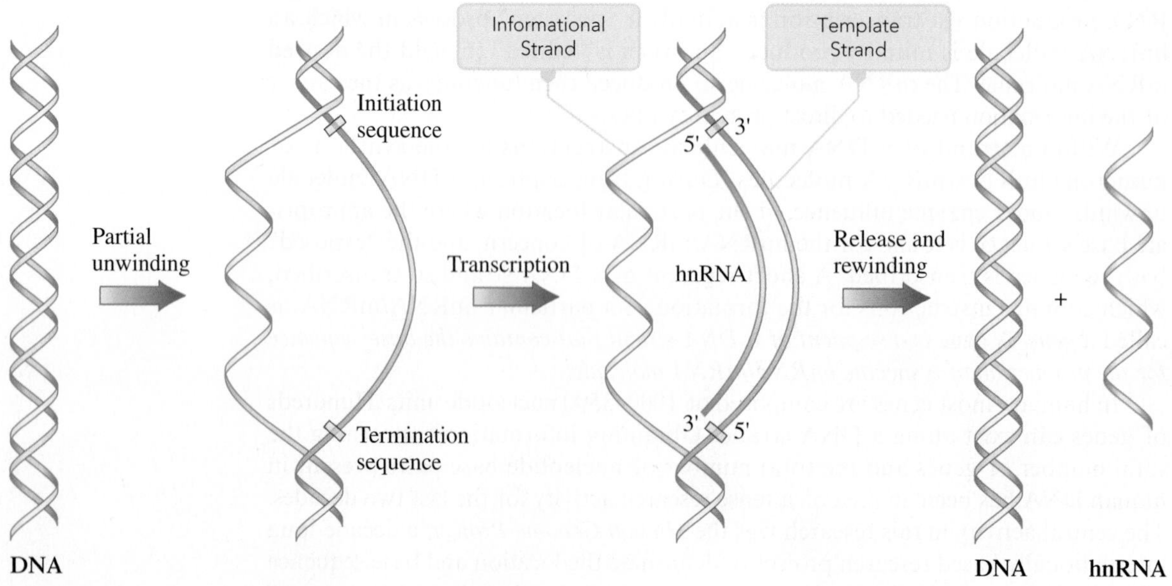

Figure 11.16 The transcription of DNA to form hnRNA involves an unwinding of a portion of the DNA double helix. Only one strand of the DNA is copied during transcription.

EXAMPLE 11.2 Base Pairing Associated with the Transcription Process

From the base sequence 5′ A–T–G–C–C–A 3′ in a DNA template strand, determine the base sequence in the hnRNA synthesized from the DNA template strand.

Solution

An RNA molecule cannot contain the base T. The base U is present instead. Therefore, U–A base pairing will occur instead of T–A base pairing. The other base-pairing combination, G–C, remains the same. The base sequence of the hnRNA strand is obtained using the following relationships:

DNA template strand: 5′ A-T-G-C-C-A 3′

complementary base pairing

DNA informational strand: 3′ T-A-C-G-G-T 5′

T bases are replaced with U bases

hnRNA strand: 3′ U-A-C-G-G-U 5′

Note that the direction of the hnRNA strand is antiparallel to that of the DNA template. This will always be the case during transcription.

It is standard procedure, when writing and reading base sequences for nucleic acids (both DNAs and RNAs), always to specify base sequence in the 5′ → 3′ direction unless otherwise directed. Thus

3′ U–A–C–G–G–U 5′ becomes 5′ U–G–G–C–A–U 3′

Practice Exercise 11.2

From the base sequence 5′ T–A–A–C–C–T 3′ in a DNA template strand, determine the base sequence in the hnRNA synthesized from the DNA template strand.

Answer: 3′ A–U–U–G–G–A 5′, which becomes 5′ A–G–G–U–U–A 3′

The base sequence in both DNA and RNA is most often specified in the 5′ → 3′ direction.

Post-Transcription Processing: Formation of mRNA

The RNA produced from a gene through transcription is hnRNA, the precursor for mRNA. The conversion of hnRNA to mRNA involves *post-transcription processing* of the hnRNA. In this processing, certain portions of the hnRNA are deleted and the retained parts are then spliced together. This process leads to the concepts of *exons* and *introns.*

It is now known that not all bases in a gene convey genetic information. Instead, a gene is *segmented;* it has portions called *exons* that contain genetic information and portions called *introns* that do not convey genetic information.

An **exon** *is a gene segment that conveys (codes for) genetic information. Ex*ons are DNA segments that help *ex*press a genetic message. An **intron** *is a gene segment that does not convey (code for) genetic information. Int*rons are DNA segments that *int*errupt a genetic message. A gene consists of alternating exon and intron segments (Figure 11.17).

Both the exons and the introns of a gene are transcribed during production of hnRNA. The hnRNA is then "edited," under enzyme direction, to remove the

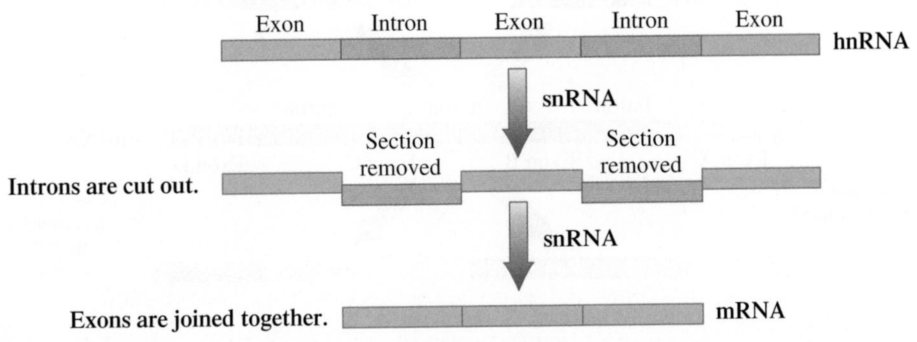

Figure 11.17 Heterogeneous nuclear RNA contains both exons and introns. Messenger RNA is heterogeneous nuclear RNA from which the introns have been excised.

introns, and the remaining exons are joined together to form a shortened RNA strand that carries the genetic information of the transcribed gene. The removal of the introns and joining together of the exons takes place simultaneously in a single process. The "edited" RNA so produced is the messenger RNA (mRNA) that serves as a blueprint for protein assembly. Much is yet to be learned about introns and why they are present in genes; investigating their function is an active area of biochemical research.

Splicing *is the process of removing introns from an hnRNA molecule and joining the remaining exons together to form an mRNA molecule.* The splicing process involves snRNA molecules, the most recent of the RNA types to be discovered. This type of RNA is never found "free" in a cell. An snRNA molecule is always found complexed with proteins in particles called *small nuclear ribonucleoprotein particles,* which are usually called *snRNPs* (pronounced "snurps"). A **small nuclear ribonucleoprotein particle** *is a complex formed from an snRNA molecule and several proteins.* "Snurps" always further collect together into larger complexes called *spliceosomes.* A **spliceosome** *is a large assembly of snRNA molecules and proteins involved in the conversion of hnRNA molecules to mRNA molecules.*

Humans have a huge proteome (an estimated 150,000 unique proteins) and a relatively small genome (20,000–25,000 unique genes). The "machinery" that bridges the genome-proteome "gap" is spliceosomes. It is spliceosomes that give humans their chemical complexity.

Alternative Splicing

Prior to the announcement of the Human Genome Project's results, biochemistry had largely embraced the "one-protein-one-gene" concept. It was generally assumed that each type of protein had "its own" gene that carried the instructions for its synthesis. This is no longer plausible because the estimated number of different proteins present in the human body now significantly exceeds the estimated number of genes.

The concept of *alternative splicing* bridges the gap between the larger estimated number of proteins and the now-smaller estimated number of genes. **Alternative splicing** *is a process by which several different proteins that are variations of a basic structural motif can be produced from a single gene.* In alternative splicing, an hnRNA molecule with multiple exons present is spliced in several different ways. Figure 11.18 shows the four alternative splicing patterns that can occur when an hnRNA contains four exons, two of which are *alternative exons.*

Spliceosome research lags behind that of most other biochemical frontiers because spliceosomes are difficult entities to study for two reasons. First, cells do not produce large amounts of these substances. They are scarce, making up less than 1% of the dry weight of a cell. This contrasts with the 25% of a cell's dry weight that is due to similarly-sized ribosomes (the sites for protein synthesis; Section 11.12). Second, bacteria, which are widely used in biochemical research, do not possess a spliceosome. Thus, research must always be carried out using multicellular organisms which is a more difficult endeavor.

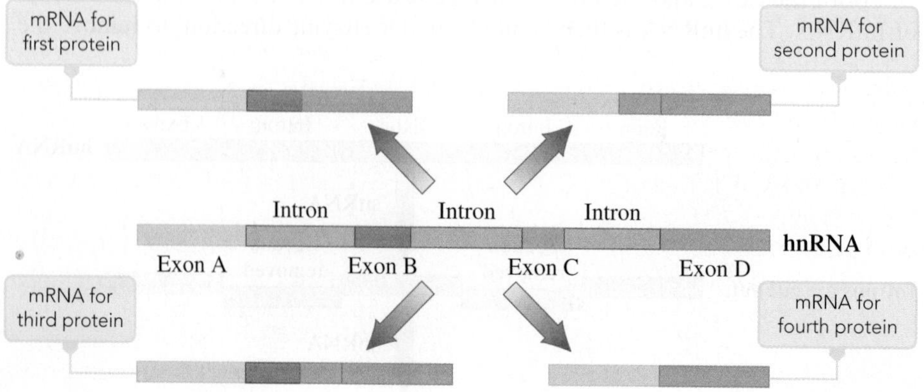

Figure 11.18 An hnRNA molecule containing four exons, two of which (B and C) are alternative exons, can be spliced in four different ways, producing four different proteins. Proteins can be produced with neither, either, or both of the alternative exons present.

The Human Transcriptome

As biochemists were mapping the human genome, they anticipated that they were close to unlocking many secrets of the human body and that the results of the project would provide a list of human genes for which the function of each could be quickly investigated. The results of the project, however, only complicated the situation.

Results indicate that the total number of genes present in a genome is not as important in understanding human cell behavior as was previously thought, whereas mRNA transcripts obtained from the genes are *more* important in understanding human cell behavior than was previously thought. Because of alternative splicing, different cell types interpret the information encoded a DNA molecule differently and so produce a different number of mRNA molecules and ultimately a different number of proteins. For each cell type, the number of mRNA transcripts generated varies in response to complex signals within a cell and between cells.

Research now shows that the information-bearing sections of DNA within a gene can be spliced together an average of eight different ways. There could turn out to be around 150,000–200,000 relevant mRNA molecules as compared to 20,000–25,000 genes within the human genome. Collectively, the total number of mRNA molecules for an organism is known as its *transcriptome.* A **transcriptome** *is all of the mRNA molecules that can be generated from the genetic material in a genome.* A transcriptome differs from a genome in that it acknowledges the biochemical complexity created by splice variants obtained from hnRNA. Transcriptome research is now a developing biochemical frontier.

Completion of the Human Genome Project is not the end to high-profile cooperative international study of the human genome. A new multi-year, perhaps multidecade, project called ENCODE (encyclopedia of DNA elements) is in its initial stages. The focus of this new project is identification of functional elements within the genome. Functional elements not only include base sequences that code for proteins, but also regulatory sequences that control DNA transcription, and sequences that control the packaging of the genome.

11.10 The Genetic Code

The nucleotide (base) sequence of an mRNA molecule is the informational part of such a molecule. This base sequence in a given mRNA determines the amino acid sequence for the protein synthesized under that mRNA's direction.

How can the base sequence of an mRNA molecule (which involves only *four* different bases—A, C, G, and U) encode enough information to direct proper sequencing of *20* amino acids in proteins? If each base encoded for a particular standard amino acid, then only four amino acids would be specified out of the 20 needed for protein synthesis, a clearly inadequate number. If two-base sequences were used to code amino acids, then there would be $4^2 = 16$ possible combinations, so 16 amino acids could be represented uniquely. This is still an inadequate number. If three-base sequences were used to code for amino acids, there would be $4^3 = 64$ possible combinations, which is more than enough combinations for uniquely specifying each of the 20 standard amino acids found in proteins.

Research has verified that sequences of three nucleotides in mRNA molecules specify the amino acids that go into synthesis of a protein. Such three-nucleotide sequences are called codons. A **codon** *is a three-nucleotide sequence in an mRNA molecule that codes for a specific amino acid.*

Which amino acid is specified by which codon? (There are 64 codons to choose from.) Researchers deciphered codon–amino acid relationships by adding different *synthetic* mRNA molecules (whose base sequences were known) to cell extracts and then determining the structure of any newly formed protein. After many such experiments, researchers finally matched all 64 possible codons with their functions in protein synthesis. It was found that 61 of the 64 codons formed by various combinations of the bases A, C, G, and U were related to specific amino acids; the other three combinations were termination codons ("stop" signals) for protein synthesis. Collectively, these relationships between three-nucleotide sequences in mRNA and amino acid identities are known as the genetic code. The **genetic code** *is the assignment of the 64 mRNA codons to specific amino acids (or stop signals).* The determination of this code during the early 1960s is one of the most remarkable twentieth-century scientific achievements. The 1968 Nobel Prize in chemistry

Table 11.2 The Genetic Code

The code is composed of 64 three-nucleotide sequences (codons), which can be read from the table. The left-hand column indicates the nucleotide base found in the first (5′) position of the codon. The nucleotide in the second (middle) position of the codon is given by the base listing at the top of the table. The right-hand column indicates the nucleotide found in the third (3′) position. Thus the codon ACG encodes for the amino acid Thr, and the codon GGG encodes for the amino acid Gly.

First letter (5′ end)	Second letter: U	Second letter: C	Second letter: A	Second letter: G	Third letter (3′ end)
U	UUU Phenylalanine Phe	UCU Serine Ser	UAU Tyrosine Tyr	UGU Cysteine Cys	U
U	UUC Phenylalanine Phe	UCC Serine Ser	UAC Tyrosine Tyr	UGC Cysteine Cys	C
U	UUA Leucine Leu	UCA Serine Ser	UAA Stop codon	UGA Stop codon	A
U	UUG Leucine Leu	UCG Serine Ser	UAG Stop codon	UGG Tryptophan Trp	G
C	CUU Leucine Leu	CCU Proline Pro	CAU Histidine His	CGU Arginine Arg	U
C	CUC Leucine Leu	CCC Proline Pro	CAC Histidine His	CGC Arginine Arg	C
C	CUA Leucine Leu	CCA Proline Pro	CAA Glutamine Gln	CGA Arginine Arg	A
C	CUG Leucine Leu	CCG Proline Pro	CAG Glutamine Gln	CGG Arginine Arg	G
A	AUU Isoleucine Ile	ACU Threonine Thr	AAU Asparagine Asn	AGU Serine Ser	U
A	AUC Isoleucine Ile	ACC Threonine Thr	AAC Asparagine Asn	AGC Serine Ser	C
A	AUA Isoleucine Ile	ACA Threonine Thr	AAA Lysine Lys	AGA Arginine Arg	A
A	AUG Methionine Met / Initiation codon	ACG Threonine Thr	AAG Lysine Lys	AGG Arginine Arg	G
G	GUU Valine Val	GCU Alanine Ala	GAU Aspartic acid Asp	GGU Glycine Gly	U
G	GUC Valine Val	GCC Alanine Ala	GAC Aspartic acid Asp	GGC Glycine Gly	C
G	GUA Valine Val	GCA Alanine Ala	GAA Glutamic acid Glu	GGA Glycine Gly	A
G	GUG Valine Val	GCG Alanine Ala	GAG Glutamic acid Glu	GGG Glycine Gly	G

was awarded to Marshall Nirenberg and Har Gobind Khorana for their work in illuminating how mRNA encodes for proteins.

The genetic code is given in Table 11.2. Examination of this table indicates that the genetic code has several remarkable features:

1. *The genetic code is highly degenerate; that is, many amino acids are designated by more than one codon.* Three amino acids (Arg, Leu, and Ser) are represented by six codons. Two or more codons exist for all other amino acids except Met and Trp, which have only a single codon. Codons that specify the same amino acid are called *synonyms.*
2. *There is a pattern to the arrangement of synonyms in the genetic code table.* All synonyms for an amino acid fall within a single box in Table 11.2, unless there are more than four synonyms, where two boxes are needed. The significance of the "single box" pattern is that with synonyms, the first two bases of the codon are the same—they differ only in the third base. For example, the four synonyms for the amino acid proline (Pro) are CCU, CCC, CCA, and CCG.
3. *The genetic code is almost universal.* Although Table 11.2 does not show this feature, studies of many organisms indicate that with minor exceptions, the code is the same in all of them. The same codon specifies the same amino acid whether the cell is a bacterial cell, a corn plant cell, or a human cell.
4. *An initiation codon exists.* The existence of "stop" codons (UAG, UAA, and UGA) suggests the existence of "start" codons. There is one initiation codon. Besides coding for the amino acid methionine, the codon AUG functions as an initiator of protein synthesis when it occurs as the first codon in an amino acid sequence.

There is a rough correlation between the number of codons for a particular amino acid and that amino acid's frequency of occurrence in proteins. For example, the two amino acids that have a single codon, Met and Trp, are two of the least common amino acids in proteins.

EXAMPLE 11.3 Using the Genetic Code and mRNA Codons to Predict Amino Acid Sequences

Using the genetic code in Table 11.2, determine the sequence of amino acids encoded by the mRNA codon sequence

5′ GCC–AUG–GUA–AAA–UGC–GAC–CCA 3′

Solution

Matching the codons with the amino acids, using Table 11.2, yields

mRNA: 5′ GCC–AUG–GUA–AAA–UGC–GAC–CCA 3′
Peptide: Ala–Met–Val–Lys–Cys–Asp–Pro

Practice Exercise 11.3

Using the genetic code in Table 11.2, determine the sequence of amino acids encoded by the mRNA codon sequence

5′ CAU–CCU–CAC–ACU–GUU–UGU–UGG 3′

Answer: His–Pro–His–Thr–Val–Cys–Trp

EXAMPLE 11.4 Relating Exons and Introns to hnRNA and mRNA Structures

Sections A, C, and E of the following base sequence section of a DNA template strand are exons, and sections B and D are introns.

DNA 5′ ATT – CGT – TGT – TTT – CCC – AGT – GCC 3′
A B C D E

a. What is the structure of the hnRNA transcribed from this template?
b. What is the structure of the mRNA obtained by splicing the hnRNA?

Solution

a. The base sequence in the hnRNA will be complementary to that of the template DNA, except that U is used in the RNA instead of T. The hnRNA will have a directionality antiparallel to that of the DNA sequence.

hnRNA 3′ UAA – GCA – ACA – AAA – GGG – UCA – CGG 5′

Rewriting this base sequence so that it reads from the 5′ end to the 3′ end (which is standard notation) gives

hnRNA 5′ GGC – ACU – GGG – AAA – ACA – ACG – AAU 3′

Note that in reversing the directionality from 3′-to-5′ to 5′-to-3′, the sequence of bases in a codon is also reversed; for example, GAC becomes CAG.

b. In the splicing process, introns are removed and the exons combined to give the mRNA.

mRNA 5′ GGC – AAA – ACA – AAU 3′

Introns and exons are actually never as short as those given in this simplified example.

Practice Exercise 11.4

Sections A, C, and E of the following base sequence section of a DNA template strand are exons, and sections B and D are introns.

DNA 5′ CGC – CGT – AGT – TGG – CCC – GGA – GGA 3′
A B C D E

a. What is the structure of the hnRNA transcribed from this template?
b. What is the structure of the mRNA obtained by splicing the hnRNA?

Answers: **a.** 5′ UCC–UCC–GGG–CCA–ACU–ACG–GCG 3′ **b.** 5′ UCC–UCC–CCA–GCG 3′

EXAMPLE 11.5 Relating Protein Amino Acid Sequence to the Directionality of an mRNA Segment

The structure of an mRNA segment obtained from a DNA template strand is

mRNA 3′ AUU – CCG – UAC – GAC 5′

What polypeptide amino acid sequence will be synthesized using this mRNA?

Solution

The directionality of an mRNA segment obtained from template DNA is 3′-to-5′ because the two segments must be antiparallel to each other (Section 11.5). The codons in an mRNA must be read in the 5′-to-3′ direction to correctly use genetic code relationships to determine the sequence of amino acids in the peptide. Rewriting the given mRNA with reversed directionality (5′-to-3′ direction) gives

mRNA 5′ CAG – CAU – GCC – UUA 3′

Note that in reversing the directionality from 3′-to-5′ to 5′-to-3′, the sequence of bases in a codon is also reversed; for example, GAC becomes CAG.

Using the genetic code relationships between codon and amino acid (Table 11.2) shows that this mRNA codon sequence codes for the amino acid sequence

N-end Gln–His–Ala–Leu C-end

Codons written from the 5′ to 3′ end in an mRNA give amino acids that correspond to a peptide written from the N-terminal end to the C-terminal end.

Practice Exercise 11.5

The structure of an mRNA segment obtained from a DNA template strand is

mRNA 3′ ACG – AGC – CCU – CUU 5′

What polypeptide amino acid sequence will be synthesized using this mRNA?

Answer: N-end Phe–Ser–Arg–Ala C-end

11.11 Anticodons and tRNA Molecules

The amino acids used in protein synthesis do not directly interact with the codons of an mRNA molecule. Instead, tRNA molecules function as intermediaries that deliver amino acids to the mRNA. At least one type of tRNA molecule exists for each of the 20 amino acids found in proteins.

All tRNA molecules have the same general shape, and this shape is crucial to how they function. Figure 11.19a shows the general *two-dimensional* "cloverleaf" shape of a tRNA molecule, a shape produced by the molecule's folding and twisting into regions of parallel strands and regions of hairpin loops. (The actual three-dimensional shape of a tRNA molecule involves considerable additional twisting of the "cloverleaf" shape—Figure 11.19b.)

Two features of the tRNA structure are of particular importance:

1. The 3′ end of the open part of the cloverleaf structure is where an amino acid becomes *covalently* bonded to the tRNA molecule through an ester bond. Each of the different tRNA molecules is specifically recognized by an *amino-acyl tRNA synthetase* enzyme. These enzymes also recognize the one kind of amino acid that "belongs" with the particular tRNA and facilitates its bonding to the tRNA (see Figure 11.20).
2. The loop *opposite* the open end of the cloverleaf is the site for a sequence of three bases called an anticodon. An **anticodon** *is a three-nucleotide sequence on a tRNA molecule that is complementary to a codon on an mRNA molecule.*

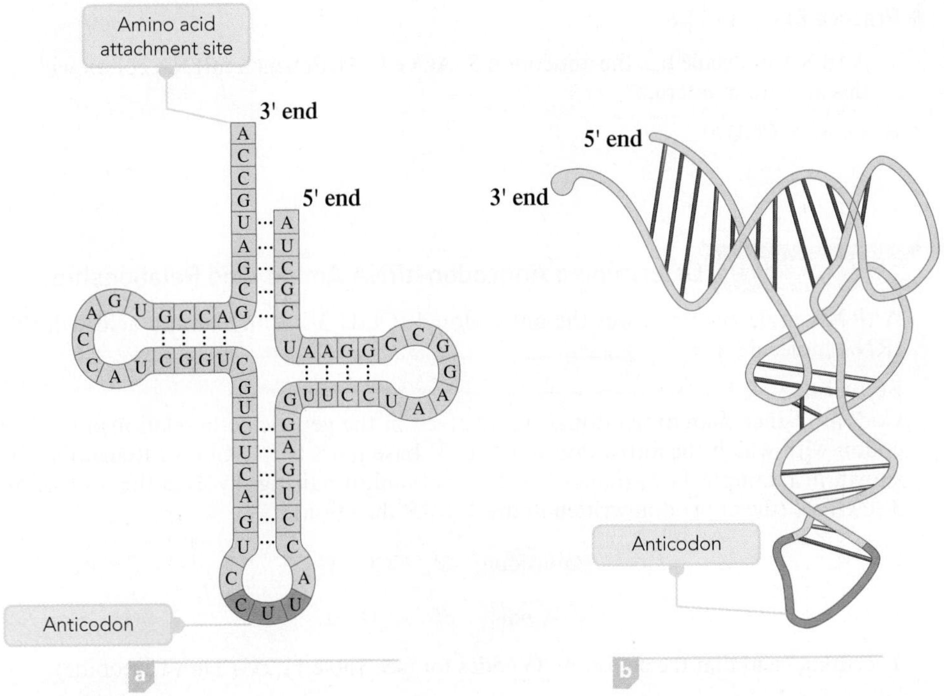

Figure 11.19 A tRNA molecule. The amino acid attachment site is at the open end of the cloverleaf (the 3′ end), and the anticodon is located in the hairpin loop opposite the open end.

The interaction between the anticodon of the tRNA and the codon of the mRNA leads to the proper placement of an amino acid into a growing peptide chain during protein synthesis. This interaction, which involves complementary base pairing, is shown in Figure 11.21.

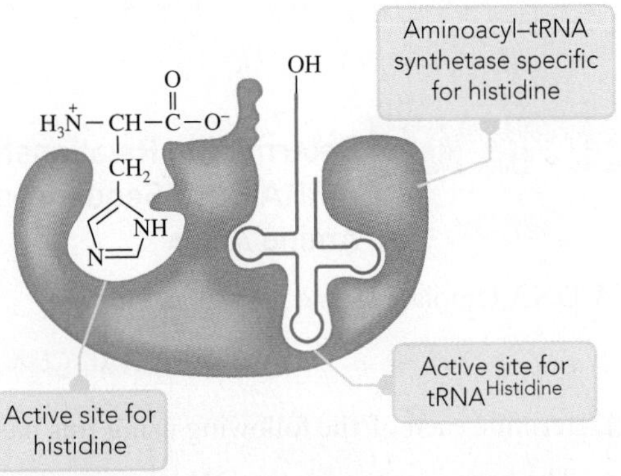

Figure 11.20 An aminoacyl–tRNA synthetase has an active site for tRNA and a binding site for the particular amino acid that is to be attached to that tRNA.

EXAMPLE 11.6 Determining Codon–Anticodon Relationships

A tRNA molecule has the anticodon 5′ AAG 3′. With which mRNA codon will this anticodon interact?

Solution

The interaction between a codon and an anticodon has directionality (antiparallel) considerations, that is,

3′	Anticodon	5′
	⋮ ⋮ ⋮	
5′	Codon	3′

The given anticodon, with reversed directionality, is

3′ GAA 5′

and its codon interaction, which involves complementary base pairing, is

Anticodon	3′	GAA	5′
		⋮ ⋮ ⋮	
Codon	5′	CUU	3′

The codon is, thus, 5′ CUU 3′.

(continued)

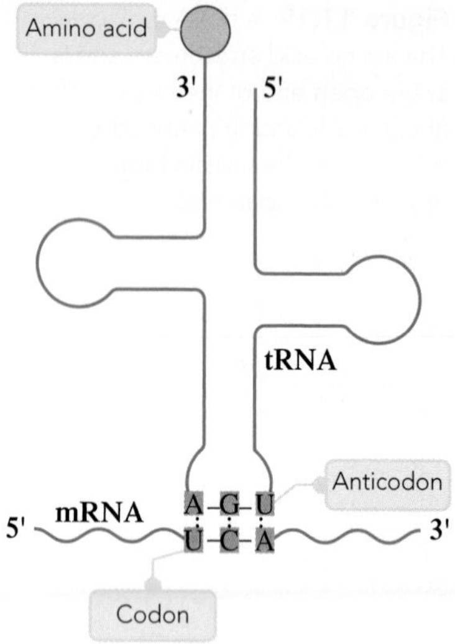

Figure 11.21 The interaction between anticodon (tRNA) and codon (mRNA), which involves complementary base pairing, governs the proper placement of amino acids in a protein.

Practice Exercise 11.6

A tRNA molecule has the anticodon 5′ ACG 3′. With which mRNA codon will this anticodon interact?

Answer: 5′ CGU 3′

EXAMPLE 11.7 Determining Anticodon–tRNA Amino Acid Relationships

A tRNA molecule possesses the anticodon 5′ CGU 3′. Which amino acid will this tRNA molecule carry?

Solution

Codons, rather than anticodons, are involved in the genetic code relationships. The codon with which the anticodon 5′ CGU 3′ base pairs is 5′ ACG 3′. Remember, as shown in Example 11.6, that the codon–anticodon pairing involves the anticodon 3′ UGC 5′ (the anticodon written in the 3′-to-5′ direction).

Anticodon	3′	UGC	5′
		⋮ ⋮ ⋮	
Codon	5′	ACG	3′

The amino acid that the codon ACG codes for (see Table 11.2) is Thr (Threonine).

Practice Exercise 11.7

A tRNA molecule possesses the anticodon 5′ UGA 3′. Which amino acid will this tRNA molecule carry?

Answer: Ser (Serine)

EXAMPLE 11.8 Determining Relationships Among DNA Base Sequences, mRNA Base Sequences, Codons, Anticodons, and Amino Acids

A DNA template strand is read as follows:

3′ ACG AAC CGA TAG GGA 5′

Determine each of the following using this base sequence.

a. The base sequence in the DNA informational strand
b. The codons present in the mRNA transcribed from the DNA template strand (assuming that no introns are present)
c. The tRNA anticodons that interact with the mRNA codons
d. The amino acids that the tRNA molecules carry
e. The structure of the peptide formed from the overall translation process

Solution

a. The DNA informational strand has bases complementary to those in the DNA template strand.

Template strand 3′ ACG AAC CGA TAG GGA 5′

Informational strand 5′ TGC TTG GCT ATC CCT 3′

b. The mRNA bases will be the same as those in the informational strand, except that U has replaced T. The codons are derived from the mRNA base sequence.

Informational strand 5′ TGC TTG GCT ATC CCT 3′

mRNA 5′ UGC UUG GCU AUC CCU 3′

Codons 5′ UGC UUG GCU AUC CCU 3′

c. The anticodon bases will be complementary bases of the codon bases.

Codons 5′ UGC UUG GCU AUC CCU 3′

Anticodons 3′ ACG AAC CGA UAG GGA 5′

Rewriting the anticodons in the 5′ to 3′ direction gives

Anticodons 5′ AGG GAU AGC CAA GCA 3′

d. The amino acids that the tRNAs carry are determined by codons rather than anticodons. The genetic code is used to make the connection between codons and amino acids.

Codons 5′ UGC UUG GCU AUC CCU 3′

Amino acids Cys Leu Ala Ile Pro

e. The peptide structure parallels the amino acid sequence obtained from the codon sequence.

N-terminal end Cys-Leu-Ala-Ile-Pro C-terminal end

Practice Exercise 11.8

Determining relationships among DNA base sequences, mRNA base sequences, codons, anticodons, and amino acids

A DNA template strand is read as follows:

3′ GCA AAA CAA ATA GTG 5′

Determine each of the following using this information.

a. The base sequence in the DNA informational strand
b. The codons present in the mRNA transcribed from the DNA template strand (assuming that no introns are present)
c. The tRNA anticodons that interact with the mRNA codons
d. The amino acids that the tRNA molecules carry
e. The structure of the peptide formed from the overall translation process

Answers: a. 5′ CGT TTT GTT TAT CAC 3′ b. 5′ CGU UUU GUU UAU CAC 3′

c. 5′ AUG AUA AAC AAA ACG 3′ d. Arg Phe Val Tyr His

e. Arg-Phe-Val-Tyr-His

11.12 Translation: Protein Synthesis

Translation *is the process by which mRNA codons are deciphered and a particular protein molecule is synthesized.* The substances needed for the translation phase of protein synthesis are mRNA molecules, tRNA molecules, amino acids, ribosomes, and a number of different enzymes. A **ribosome** *is an rRNA–protein complex that serves as the site for the translation phase of protein synthesis.*

The number of ribosomes present in a cell for higher organisms varies from hundreds of thousands to even a few million. Recent research concerning ribosome structure suggests the following for such structures:

1. They contain four rRNA molecules and about 80 proteins that are packed into two rRNA-protein subunits, one small subunit, and one large subunit (Figure 11.22).
2. Each subunit contains approximately 65% rRNA and 35% protein by mass.
3. A ribosome's active site, the location where proteins are synthesized by one-at-a-time addition of amino acids to a growing peptide chain, is located in the large ribosomal subunit.
4. The active site is mostly rRNA, with only one of the ribosome's many protein components being present.

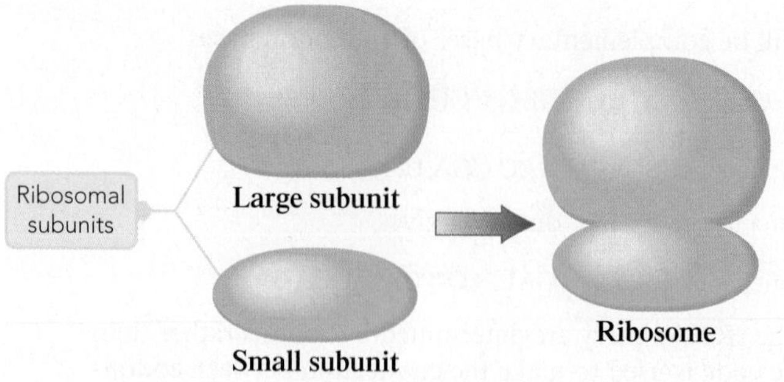

Figure 11.22 Ribosomes, which contain both rRNA and protein, have structures that contain two subunits. One subunit is much larger than the other.

5. Because rRNA is so predominant at the active site, the ribosome is thought to be an RNA enzyme (Section 10.1), that is, a *ribozyme.*
6. The mRNA involved in the translation phase of protein synthesis binds to the small subunit of the ribosome.

There are five general steps to the translation process: (1) activation of tRNA (2) initiation (3) elongation (4) termination and (5) post-translational processing.

Activation of tRNA

There are two steps involved in tRNA activation. First, an amino acid interacts with an activator molecule (ATP; Section 12.3) to form a highly energetic complex. This complex then reacts with the appropriate tRNA molecule to produce an *activated tRNA molecule,* a tRNA molecule that has an amino acid covalently bonded to it at its 3′ end through an ester linkage.

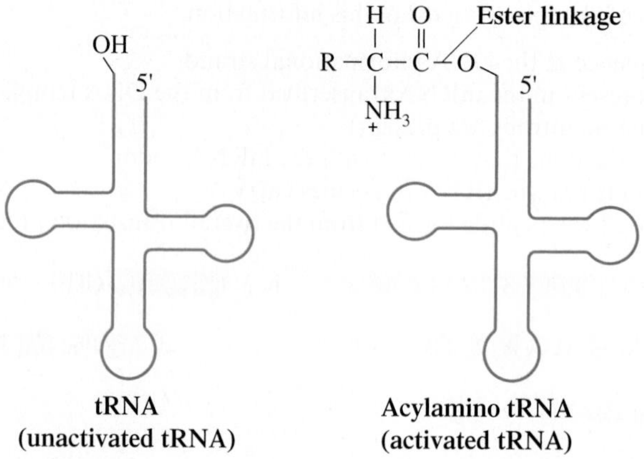

Initiation

The initiation of protein synthesis in human cells begins when mRNA attaches itself to the surface of a small ribosomal subunit such that its first codon, which is always the initiating codon AUG, occupies a site called the P site (peptidyl site) (Figure 11.23a). An activated tRNA molecule with an anticodon complementary to

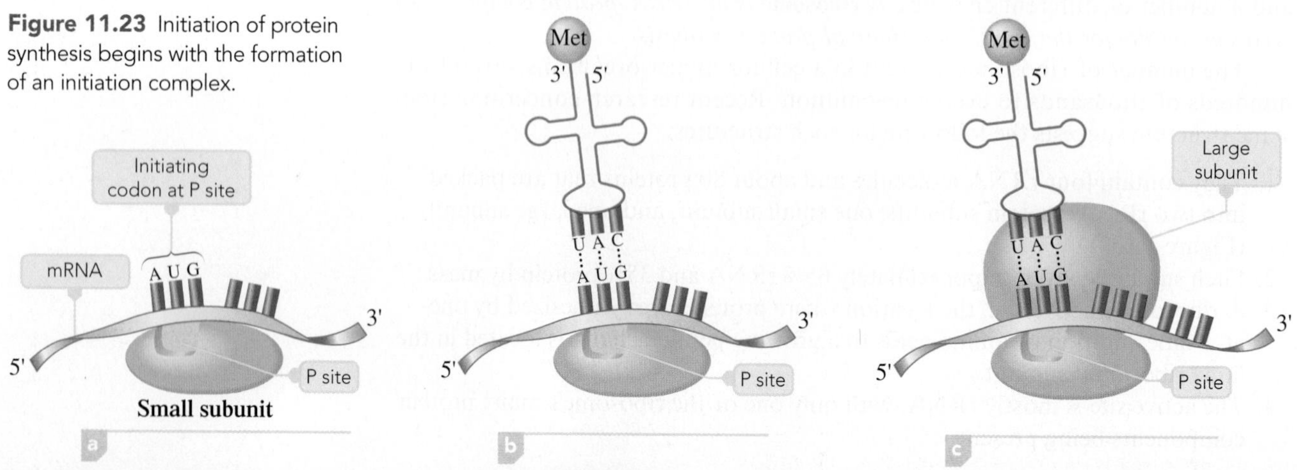

Figure 11.23 Initiation of protein synthesis begins with the formation of an initiation complex.

the codon AUG attaches itself, through complementary base pairing, to the AUG codon (Figure 11.23b). The resulting complex then interacts with a large ribosomal subunit to complete the formation of an initiation complex (Figure 11.23c). (Since the initiating codon AUG codes for the amino acid methionine, the first amino acid in a developing human protein chain will always be methionine.)

Elongation

Next to the P site in an mRNA–ribosome complex is a second binding site called the A site (aminoacyl site) (Figure 11.24a). At this second site the next mRNA codon is exposed, and a tRNA with the appropriate anticodon binds to it (Figure 11.24b). With amino acids in place at both the P and the A sites, the enzyme *peptidyl transferase* effects the linking of the P site amino acid to the A site amino acid to form a dipeptide. Such peptide bond formation leaves the tRNA at the P site empty and the tRNA at the A site bearing the dipeptide (Figure 11.24c).

In elongation, the polypeptide chain grows one amino acid at a time.

The empty tRNA at the P site now leaves that site and is free to pick up another molecule of its specific amino acid. Simultaneously with the release of tRNA from the P site, the ribosome shifts along the mRNA. This shift puts the newly

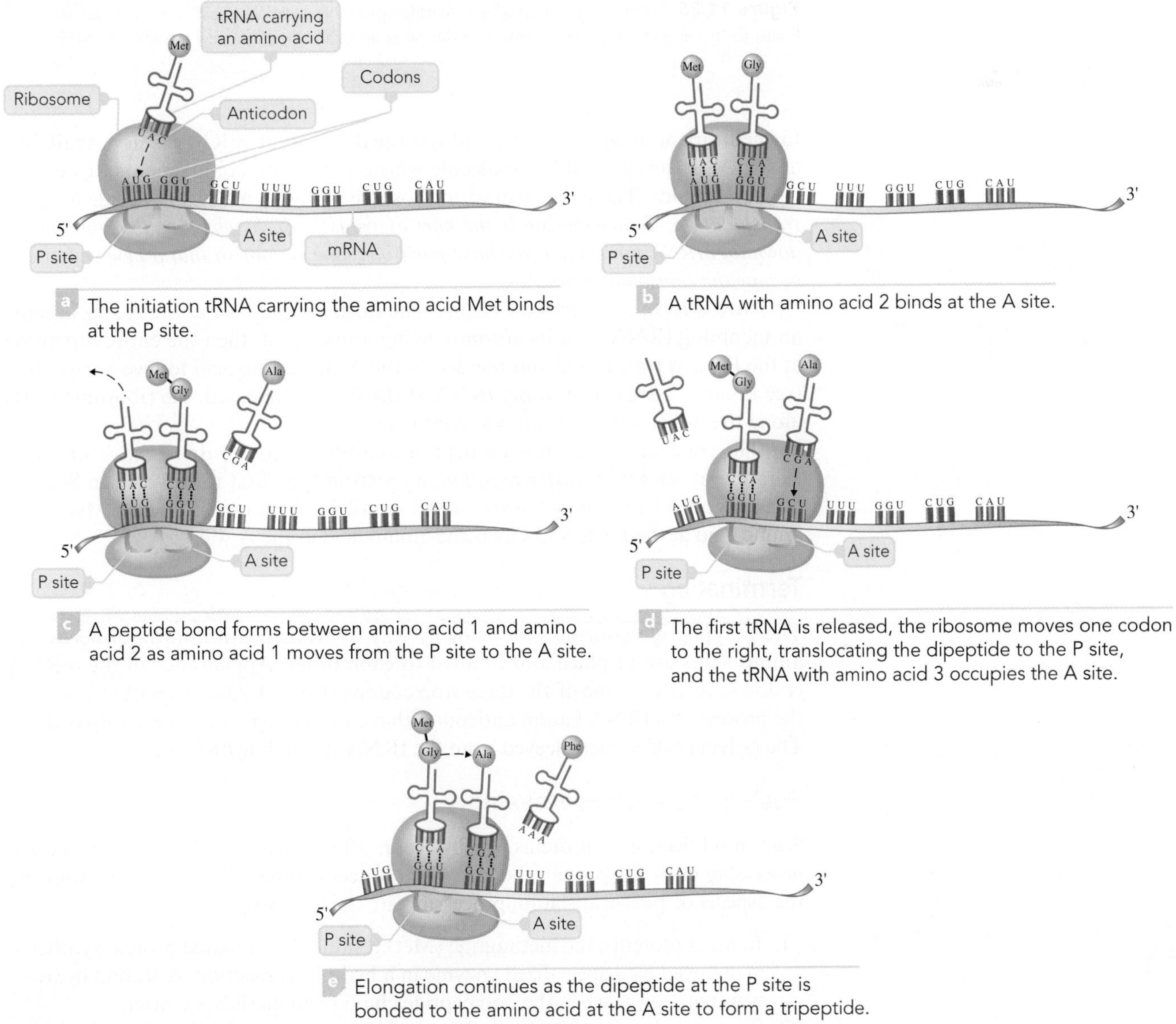

a The initiation tRNA carrying the amino acid Met binds at the P site.

b A tRNA with amino acid 2 binds at the A site.

c A peptide bond forms between amino acid 1 and amino acid 2 as amino acid 1 moves from the P site to the A site.

d The first tRNA is released, the ribosome moves one codon to the right, translocating the dipeptide to the P site, and the tRNA with amino acid 3 occupies the A site.

e Elongation continues as the dipeptide at the P site is bonded to the amino acid at the A site to form a tripeptide.

Figure 11.24 The process of translation that occurs during protein synthesis. The anticodons of tRNA molecules are paired with the codons of an mRNA molecule to bring the appropriate amino acids into sequence for protein formation.

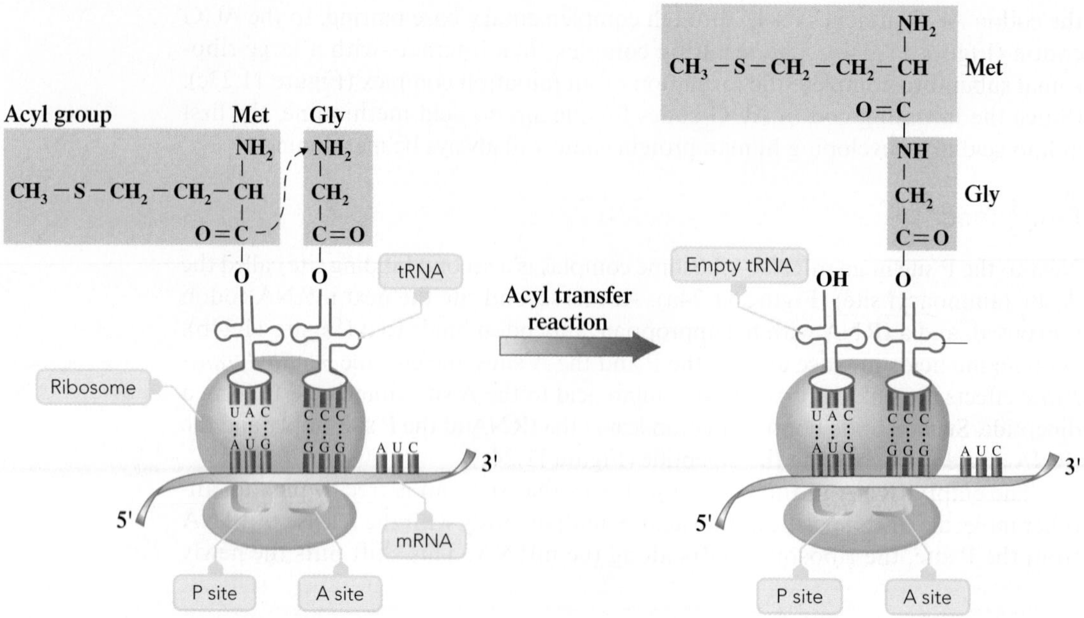

Figure 11.25 The transfer of an amino acid (or growing peptide chain) from the ribosomal P site to the ribosomal A site during translation is an example of an acyl transfer reaction.

The process of *translocation* occurs at approximately 50-millisecond intervals, that is, 20 times per second.

formed dipeptide at the P site, and the third codon of mRNA is now available, at site A, to accept a tRNA molecule whose anticodon complements this codon (Figure 11.24d). The movement of a ribosome along an mRNA molecule is called *translocation.* **Translocation** *is the part of translation in which a ribosome moves down an mRNA molecule three base positions (one codon) so that a new codon can occupy the ribosomal A site.*

Now a repetitious process begins. The third codon, now at the A site, accepts an incoming tRNA with its accompanying amino acid; then the entire dipeptide at the P site is transferred and bonded to the A site amino acid to give a tripeptide (see Figure 11.24e). The empty tRNA at the P site is released, the ribosome shifts along the mRNA, and the process continues.

The transfer of the growing peptide chain from the P site to the A site is an example of an *acyl transfer reaction,* a reaction type first introduced in Section 5.19. Figure 11.25 shows the structural detail for such transfer when Met is the amino acid at the P site and Gly is the amino acid at the A site.

Termination

The polypeptide continues to grow by way of translocation until all necessary amino acids are in place and bonded to each other. Appearance in the mRNA codon sequence of one of the three stop codons (UAA, UAG, or UGA) terminates the process. No tRNA has an anticodon that can base-pair with these stop codons. The polypeptide is then cleaved from the tRNA through hydrolysis.

Post-Translation Processing

Some modification of proteins usually occurs after translation. This *post-translation processing* gives the protein the final form it needs to be fully functional. Some of the aspects of post-translation processing are the following:

1. In most proteins, the methionine (Met) residue that initiated protein synthesis is removed by a specialized enzyme in a hydrolysis reaction. A second hydrolysis reaction releases the polypeptide chain from its tRNA carrier.
2. Some covalent modification of a protein can occur, such as the formation of disulfide bridges between cysteine residues (Section 9.6).

CHEMISTRY AT A GLANCE Protein Synthesis: Transcription and Translation

TRANSCRIPTION PHASE

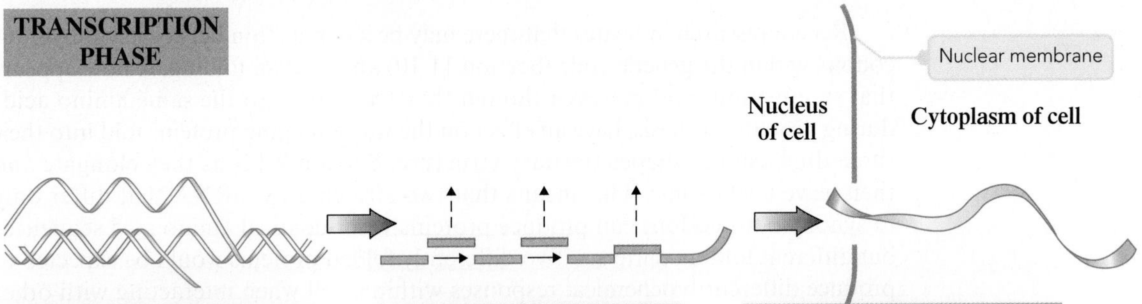

Step 1: Formation of hnRNA

DNA in the nucleus partially unwinds to allow a strand of hnRNA to be made.

Step 2: Formation of mRNA

Introns are removed from the hnRNA strand.

Step 3: mRNA Enters the Cytoplasm

The mRNA leaves the nucleus and enters the cytoplasm.

TRANSLATION PHASE

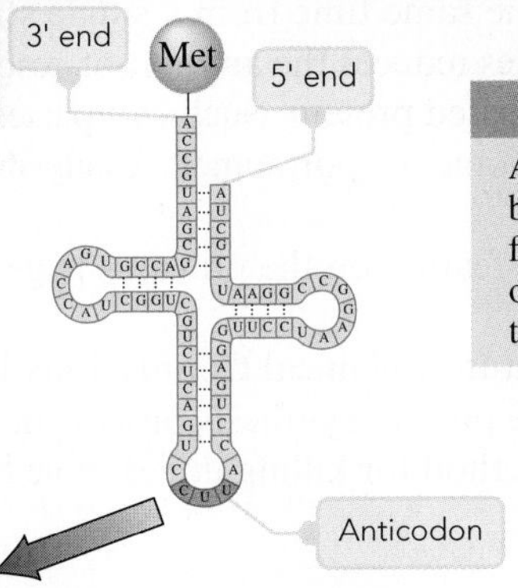

Step 1: Activation of tRNA

An amino acid interacts with ATP to become highly energized. It then forms a covalent bond with the 3' end of a tRNA molecule. Amino acid–tRNA pairing is governed by enzymes.

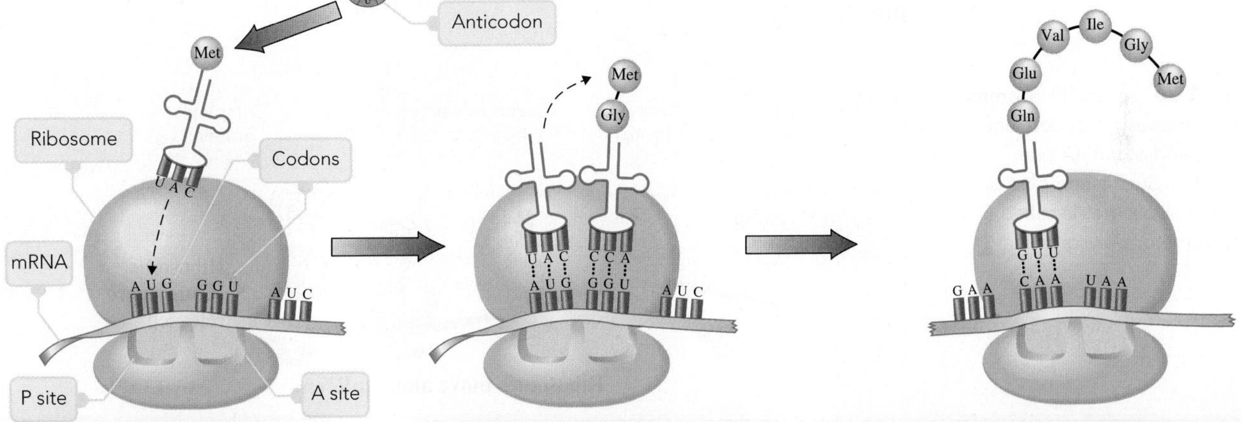

Step 2: Initiation

The mRNA attaches to a ribosome so that the first codon (AUG) is at the P site. A tRNA carrying methionine attaches to the first codon.

Step 3: Elongation

Another tRNA with the second amino acid binds at the A site. The methionine transfers from the P site to the A site. The ribosome shifts to the next codon, making its A site available for the tRNA carrying the third amino acid.

Steps 4 and 5: Termination and Post-Translation Processing

The polypeptide chain continues to lengthen until a stop codon appears on the mRNA. The new protein is cleaved from the last tRNA.

During post-translation processing, cleavage of Met (the initiation codon) usually occurs. S—S bonds between Cys units also can form.

3. Completion of the folding of polypeptides into their active conformations occurs. Protein folding actually begins as the polypeptide chain is elongated on the ribosome. For proteins with quaternary structure (Section 9.13), the various components are assembled together.

Recent research indicates that there may be a connection between synonymous codons within the genetic code (Section 11.10) and protein folding. It now appears that synonymous codons, even though they translate into the same amino acids during protein synthesis, have an effect on the way emerging proteins fold into their three-dimensional shapes (tertiary structure; Section 9.12) as they elongate and then leave a ribosome. This means that two stretches of mRNA that differ only in synonymous codons can produce proteins with identical amino acid sequences but different folding patterns. Two differently folded proteins would be expected to produce different biochemical responses within a cell when interacting with other substances; there is now some evidence that this is the case.

Efficiency of mRNA Utilization

Many ribosomes can move simultaneously along a single mRNA molecule (Figure 11.26). In this highly efficient arrangement, many identical protein chains can be synthesized almost at the same time from a single strand of mRNA. This multiple use of mRNA molecules reduces the amount of resources and energy that the cell expends to synthesize needed protein. Such complexes of several ribosomes and mRNA are called polyribosomes or polysomes. A **polyribosome** *is a complex of mRNA and several ribosomes.*

The Chemistry at a Glance feature on the previous page summarizes the steps in protein synthesis.

The focus on relevancy feature Chemical Connections 11-B on the next page discusses how disruption of the protein synthesis process in bacteria, through use of antibiotics, in an effective method for killing undesirable bacteria present in the human body.

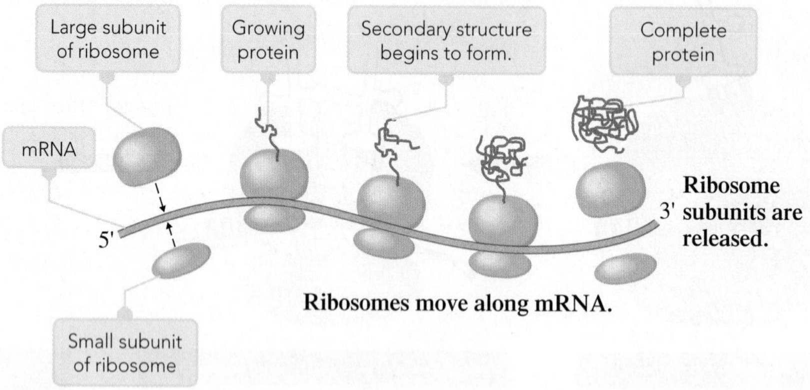

Figure 11.26 Several ribosomes can simultaneously proceed along a single strand of mRNA one after another. Such a complex of mRNA and ribosomes is called a polyribosome or polysome.

11.13 Mutations

A **mutation** *is an error in base sequence in a gene that is reproduced during DNA replication.* Such errors alter the genetic information that is passed on during transcription. The altered information can cause changes in amino acid sequence during protein synthesis. Sometimes, such changes have a profound effect on an organism.

A **mutagen** *is a substance or agent that causes a change in the structure of a gene.* Radiation and chemical agents are two important types of mutagens. Radiation, in the form of ultraviolet light, X-rays, radioactivity, and cosmic rays, has the potential to be mutagenic. Ultraviolet light from the sun is the

CHEMICAL CONNECTIONS 11-B

Antibiotic Protein Synthesis Inhibitors

Protein synthesis proceeds in the same general way in all forms of life. The synthesis details are, however, less complex for organisms such as bacteria.

Comparison of protein synthesis in bacteria with that in human cells shows the following similarities and differences:

1. In both cases, ribosomes are the sites for protein synthesis. Human ribosomes are much larger than bacterial ribosomes.
2. In bacteria, the initiator codon is N-formylmethionine. In human cells, it is methionine.
3. In bacteria, mRNA translation begins while the mRNA is still being transcribed from DNA. In human cells, mRNA translation begins only after transcription.
4. Bacteria/mRNAs undergo very little processing after being transcribed and are very short-lived. In some cases, degradation of mRNA begins before completion of transcription. In human cells, transcribed mRNA (hnRNA) undergoes considerable processing before mature RNA is formed.

The differences between bacterial and human protein synthesis are sufficient to make bacterial protein synthesis an ideal target for antibacterial chemotherapy, that is, use of antibiotics to kill bacteria. Many antibiotics now in use are bacterial protein synthesis inhibitors (see accompanying photo).

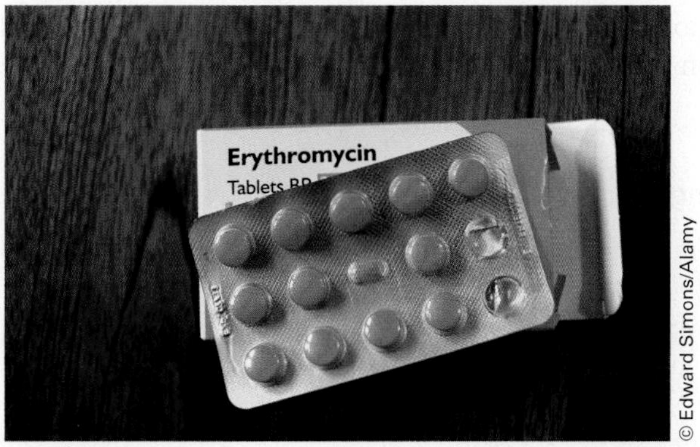

© Edward Simons/Alamy

Erythromycin and similar drugs inhibit protein synthesis in bacteria.

These inhibitors disrupt bacterial protein synthesis but do not affect human cell protein synthesis. This is because the inhibitors target the bacterial ribosome, which is much smaller than human ribosomes. Human ribosomes are unsuitable targets for the antibiotics because of their larger size.

Many, but not all, of the protein synthesis inhibitor antibiotics now in use have names that end in "mycin." The mode of action of some of these inhibitors is as follows:

1. **Erythromycin:** binds to the larger bacterial ribosome subunit, blocking the exit of a growing peptide chain.
2. **Terramycin:** blocks the A-site location on the ribosome, preventing the attachment of amino-acid carrying tRNAs.
3. **Streptomycin** (see accompanying structural diagram): binds to the smaller bacterial ribosome subunit causing a shape change, which in turn causes a misreading of mRNA information.

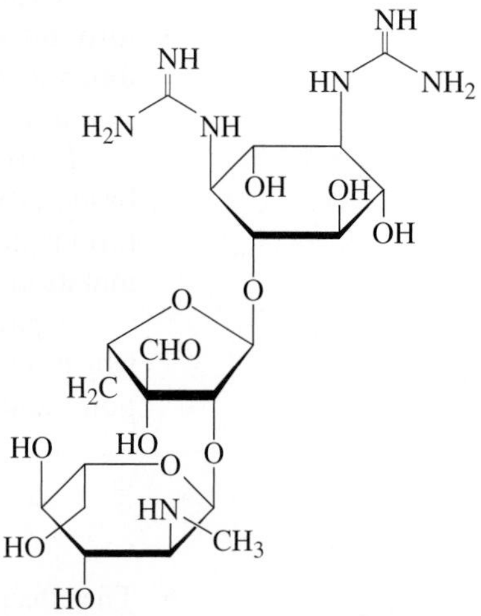

Streptomycin
(a trisaccharide derivative)

4. **Neomycin:** binds to the smaller bacterial ribosome subunit in a manner similar to streptomycin.
5. **Chloramphenicol:** binds to the ribosome and interferes with the formation of peptide bonds between amino acids.

The well-known penicillin family of antibiotics are not protein synthesis inhibitors but, rather, enzyme inhibitors (Section 10.9).

In a reversal of the process in which human intervention (drugs) disrupt bacterial protein synthesis, bacterial intervention (diseases) can block human protein synthesis. Several protein synthesis toxins produced by microorganisms are known that shut down human protein synthesis with fatal results. Diphtheria is caused by a protein synthesis inhibitor produced by a bacterium that grows in the upper respiratory tract of an infected person. This inhibitor affects the chain elongation phase of protein synthesis. Diphtheria was a major cause of death in children before the advent of effective immunization procedures.

radiation that causes sunburn and can induce changes in the DNA of skin cells. Sustained exposure to ultraviolet light can lead to serious problems such as skin cancer.

Chemical agents can also have mutagenic effects. Nitrous acid (HNO_2) is a mutagen that causes deamination of heterocyclic nitrogen bases. For example, HNO_2 can convert cytosine to uracil.

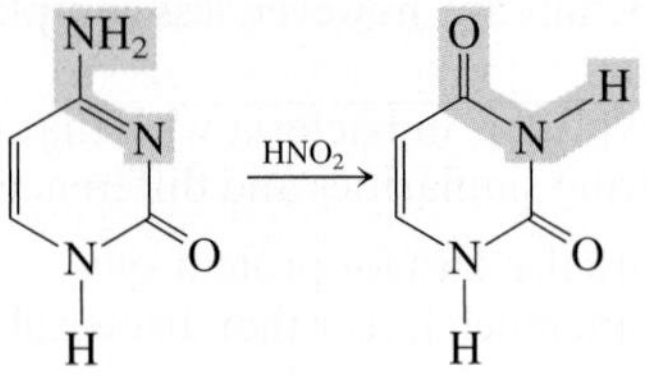

Cytosine Uracil

Deamination of a cytosine that was part of an mRNA codon would change the codon; for example, CGG would become UGG.

A variety of chemicals—including nitrites, nitrates, and nitrosamines—can form nitrous acid in the body. The use of nitrates and nitrites as preservatives in foods such as bologna and hot dogs is a cause of concern because of their conversion to nitrous acid in the body and subsequent possible damage to DNA.

Fortunately, the body has *repair enzymes* that recognize and replace altered bases. Normally, the vast majority of altered DNA bases are repaired, and mutations are avoided. Occasionally, however, the damage is not repaired, and the mutation persists.

A particularly common type of mutation is a *point* mutation, a mutation in which one nucleotide is substituted for another. The following in an example of a point mutation.

Original DNA	T A (G) C A C	
		C has replaced G
Point Mutated DNA	T A (C) C A C	

The effect of a point mutation can vary from no effect to a change in primary protein structure to termination of protein synthesis, as is illustrated in Example 11.9.

EXAMPLE 11.9 Predicting the Effect of a Point Mutation

Predict the change that occurs in amino acid identity when each of the following point mutations occur on 3′ to 5′ DNA base segments.

a. GAG is point mutated to GAA. **b.** AAA is point mutated to AAT.
c. ATA is point mutated to ATT.

Solution

The same analysis applies to each of the three parts of this example.

1. Determine the DNA informational strand base sequence that is complementary to the given DNA template strand base sequence.
2. Determine the mRNA base sequence using the informational strand base sequence. They will be the same except that the RNA base U has replaced the DNA base T.
3. Using the genetic code, determine the amino acid that is specified by the mRNA base sequence (codon).

a. Unmutated base sequence:	GAG	→	CTC	→	CUC	→	Leu
	template strand		informational strand		mRNA		amino acid
Mutated base sequence:	GAA	→	CTT	→	CUU	→	Leu
	template strand		informational strand		mRNA		amino acid

This is a *silent* mutation. Amino acid identity is not affected, as both mRNA codons code for the same amino acid.

b. Unmutated base sequence:

AAA	→	TTT	→	UUU	→	Phe
template strand		informational strand		mRNA		amino acid

Mutated base sequence:

AAT	→	TTA	→	UUA	→	Leu
template strand		informational strand		mRNA		amino acid

The point mutation has produced a codon that codes for a different amino acid.

c. Unmutated base sequence:

ATA	→	TAT	→	UAU	→	Tyr
template strand		informational strand		mRNA		amino acid

Mutated base sequence:

ATT	→	TAA	→	UAA	→	stop codon
template strand		informational strand		mRNA		amino acid

The point mutation has produced a stop codon, resulting in termination of protein synthesis.

Practice Exercise 11.9

Predict the change that occurs in amino acid identity when each of the following point mutations occur on 3′ to 5′ DNA base segments.

a. CTA is point mutated to CTT.
b. CGA is point mutated to GGA.
c. GTC is point mutated to ATC.

Answers: **a.** Asp becomes Glu. **b.** Ala becomes Pro. **c.** Gln becomes a stop codon.

11.14 Nucleic Acids and Viruses

Viruses are very small disease-causing agents that are considered the lowest order of life. Indeed, their structure is so simple that some scientists do not consider them truly alive because they are unable to reproduce in the absence of other organisms. Figure 11.27 shows an electron microscope image of an influenza virus.

A **virus** *is a small particle that contains DNA or RNA (but not both) surrounded by a coat of protein and that cannot reproduce without the aid of a host cell.* Viruses do not possess the nucleotides, enzymes, amino acids, and other molecules necessary to replicate their nucleic acid or to synthesize proteins. To reproduce, viruses must invade the cells of another organism and cause these host cells to carry out the reproduction of the virus. Such an invasion disrupts the normal operation of cells, causing diseases within the host organism. The only function of a virus is reproduction; viruses do not generate energy.

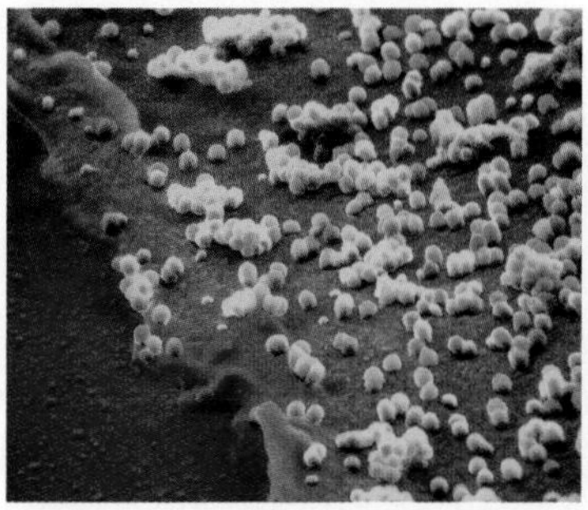

NIBSC/SPL/Photo Researchers

Figure 11.27 An electron microscope image of an influenza virus.

There is no known form of life that is not subject to attack by viruses. Viruses attack bacteria, plants, animals, and humans. Many human diseases are of viral origin. Among them are the common cold, mumps, measles, smallpox, rabies, influenza, infectious mononucleosis, hepatitis, and AIDS.

Viruses most often attach themselves to the outside of specific cells in a host organism. An enzyme within the protein overcoat of the virus catalyzes the breakdown of the cell membrane, opening a hole in the membrane. The virus then injects its DNA or RNA into the cell. Once inside, this nucleic acid material is mistaken by the host cell for its own, whereupon that cell begins to translate and/or transcribe the viral nucleic acid. When all the virus components have been synthesized by the host cell, they assemble automatically to form many new virus particles. Within 20 to 30 minutes after a single molecule of viral nucleic acid enters the host cell, hundreds of new virus particles have formed. So many are formed that they eventually burst the host cell and are free to infect other cells.

Viral infections are more difficult to treat than bacterial infections because viruses, unlike bacteria, replicate inside cells. It is difficult to design drugs that prevent the replication of the virus that do not also affect the normal activities of the host cells.

If a virus contains DNA, the host cell replicates the viral DNA in a manner similar to the way it replicates its own DNA. The newly produced viral DNA then proceeds to make the proteins needed for the production of protein coats for additional viruses.

An RNA-containing virus is called a *retrovirus.* Once inside a host, such viruses first make viral DNA. This *reverse* synthesis is governed by the enzyme *reverse transcriptase.* The template is the viral RNA rather than DNA. The viral DNA so produced then produces additional viral DNA and the proteins necessary for the protein coats.

The AIDS (acquired immunodeficiency syndrome) virus is an example of a retrovirus. This virus has an affinity for a specific type of white blood cell called a *helper T cell,* which is an important part of the body's immune system. When helper T cells are unable to perform their normal functions as a result of such viral infection, the body becomes more susceptible to infection and disease.

A **vaccine** *is a preparation containing an inactive or weakened form of a virus or bacterium.* The antibodies produced by the body against these specially modified viruses or bacteria effectively act against the naturally occurring active forms as well. Thanks to vaccination programs, many diseases, such as polio and mumps (caused by RNA-containing viruses) and smallpox and yellow fever (caused by DNA-containing viruses), are now seldom encountered.

11.15 Recombinant DNA and Genetic Engineering

Ever-increasing knowledge about DNA molecules and how they function under various chemical conditions has opened the door to an increasingly important field of technology known by several names, including *genetic engineering, genetic modification, bioengineering,* and *biotechnology.* The term *genetic engineering* will be used in this text to refer to this developing field.

Genetic engineering *is the process whereby an organism is intentionally changed at the molecular (DNA) level so that it exhibits different traits.* The first organisms to be genetically engineered were bacteria in 1973 and mice in 1974. Insulin-producing bacteria were commercialized in 1982, and genetically modified food crops have been available since 1994. Genetically modified forms of foods and fibers now dominate several major crops in the United States, as is shown by the data in Figure 11.28.

For the plant crops listed in Figure 11.28, the most common genetic modification involves introduction of a herbicide tolerance trait. A gene is inserted that allows the crops to be sprayed with the weed killer glyphosate (also known as Round-up) without harm to the plants. These crops also frequently have an insect

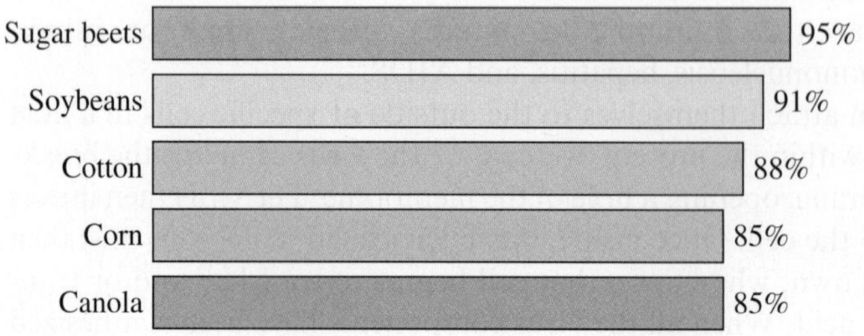

Figure 11.28 Percentage of major crops in the United States that were genetically engineered in 2009.

resistant trait that is obtained from the presence of a gene obtained from the soil bacteria *Bacillus thuringiensis.* This gene's presence causes the plants to produce their own pesticide. Plantings of insect resistant corn and cotton are common and research continues on tomato plants that carry this gene.

In 2011, the United States Department of Agriculture (USDA) gave approval for planting corn that is genetically modified to produce the enzyme *α-amylase,* an enzyme that rapidly breaks down starch into glucose (Section 7.6). For corn destined for ethanol production, the presence of this *amylase* improves the economics of the production process.

Two additional examples, still in the developmental stage, of the direction that genetic engineering can take are:

1. Genetically engineered tomato plants with a longer shelf life now exist. In this case, gene insertion is not needed. Instead, a gene naturally present in the tomato plant that is involved in the ripening process is "turned off" (deactivated). This slows, but does not completely stop, the ripening process, resulting in less overripening, softening, and overall deterioration.
2. A strawberry gene is introduced into a mustard plant variety that is very susceptible to attack by spider mites. The gene produces a chemical attractant for predator mites that eat the spidermites, thus solving a "mitey" problem.

Genetic engineering has been alluded to in several sections in previous chapters of the text, although no details were given about the process. Mentions of this subject are found in the following locations:

Section 8.5 in the omega-3 fatty acid discussion

Section 8.6 in the trans-fatty acid discussion

Section 9.3 in the complete protein discussion

Section 9.10 in the insulin discussion

Section 10.8 in the extremozyme discussion

Section 10.11 in the "grapefruit effect" discussion

Bacteria are now routinely used as "protein factories." These genetically engineered bacteria, which contain genes for human proteins, can produce large quantities of designated proteins because of their rapid reproduction rate. Human proteins in short supply that are produced in this manner include insulin (see Chemical Connections 9-A on page 382) and human growth hormone. Table 11.3 gives additional examples of human proteins used in therapeutic medicine that have become available through the use of genetic engineering technology.

Principles and Procedures of Genetic Engineering

Genetic engineering procedures involve a type of DNA called *recombinant DNA.* **Recombinant DNA (rDNA)** *is DNA that contains genetic material from two different organisms.* The procedures used in obtaining recombinant DNA are as follows.

The notation rDNA is often used to designate recombinant DNA.

Table 11.3 Selected Human Proteins Produced Using Recombinant DNA Technology and Their Uses

Protein	Treatment
insulin	diabetes
erythropoietin (EPO)	anemia
human growth hormone (HGH)	stimulate growth
interleukins	stimulate immune system
interferons	leukemia and other cancers
lung surfactant protein	respiratory distress
serum albumin	plasma supplement
tumor necrosis factor (TNF)	cancers
tissue plasminogen activator (TPA)	heart attacks
epidermal growth factor	healing of wounds and burns
fibroblast growth factor	ulcers

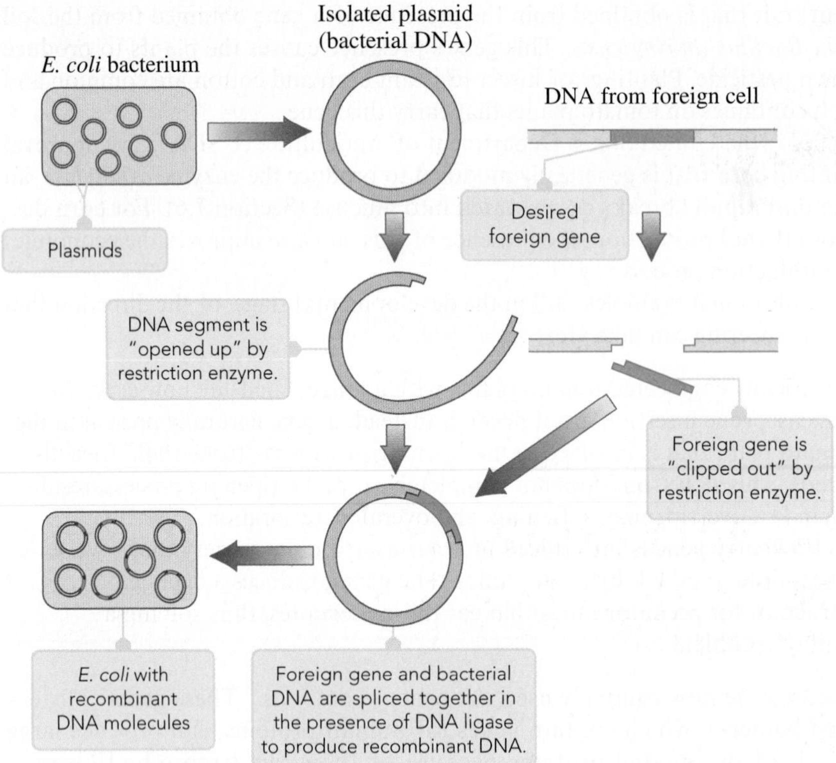

Figure 11.29 Recombinant DNA is made by inserting a gene obtained from DNA of one organism into the DNA from another kind of organism.

The bacterium *E. coli,* which is found in the intestinal tract of humans and animals, is the organism most often used in recombinant DNA experiments. Yeast cells are also used, with increasing frequency, in this research.

In addition to their chromosomal DNA, *E. coli* (and other bacteria) contain DNA in the form of small, circular, double-stranded molecules called *plasmids.* These plasmids, which carry only a few genes, replicate independently of the chromosome. Also, they are transferred relatively easily from one cell to another. Plasmids from *E. coli* are used in recombinant DNA work.

The procedure used to obtain *E. coli* cells that contain recombinant DNA involves the following steps (Figure 11.29):

Step 1: ***Cell membrane dissolution.*** E. coli *cells of a specific strain are placed in a solution that dissolves cell membranes, thus releasing the contents of the cells.*

Step 2: ***Isolation of plasmid fraction.*** *The released cell components are separated into fractions, one fraction being the plasmids. The isolated plasmid fraction is the material used in further steps.*

Step 3: ***Cleavage of plasmid DNA.*** *A special enzyme, called a* restriction enzyme, *is used to cleave the double-stranded DNA of a circular plasmid. The result is a linear (noncircular) DNA molecule.*

Step 4: ***Gene removal from another organism.*** *The same* restriction enzyme *is then used to remove a desired gene from a chromosome of another organism.*

Step 5: ***Gene–plasmid splicing.*** *The gene (from Step 4) and the opened plasmid (from Step 3) are mixed in the presence of the enzyme* DNA ligase, *which splices the two together. This splicing, which attaches one end of the gene to one end of the opened plasmid and attaches the other end of the gene to the other end of the plasmid, results in an altered circular plasmid (the recombinant DNA).*

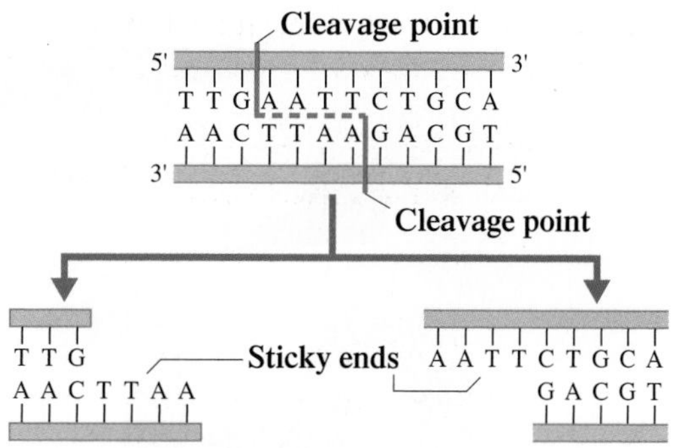

Figure 11.30 Cleavage pattern resulting from the use of a restriction enzyme that cleaves DNA between G and A bases in the 5′-to-3′ direction in the sequence G—A—A—T—T—C. The double-helix structure is not cut straight across.

Step 6: ***Uptake of recombinant DNA.*** *The altered plasmids (recombinant DNA) are placed in a live* E. coli *culture, where they are taken up by the* E. coli *bacteria. The* E. coli *culture into which the plasmids are placed need not be identical to that from which the plasmids were originally obtained.*

Note in Step 3 that the conversion of a circular plasmid into a linear DNA molecule requires a restriction enzyme. A **restriction enzyme** *is an enzyme that recognizes specific base sequences in DNA and cleaves the DNA in a predictable manner at these sequences.* The discovery of restriction enzymes made genetic engineering possible.

Restriction enzymes occur naturally in numerous types of bacterial cells. Their function is to protect the bacteria from invasion by foreign DNA by catalyzing the cleavage of the invading DNA. The term *restriction* relates to such enzymes placing a "restriction" on the type of DNA allowed into the bacterial cells.

To understand how a restriction enzyme works, consider one that cleaves DNA between G and A bases in the 5′-to-3′ direction in the sequence G–A–A–T–T–C. This enzyme will cleave the double-helix structure of a DNA molecule in the manner shown in Figure 11.30.

Note that the double helix is not cut straight across; the individual strands are cut at different points, giving a staircase cut. (Both cuts must be between G and A in the 5′-to-3′ direction.) This staircase cut leaves unpaired bases on each cut strand. These ends with unpaired bases are called "sticky ends" because they are ready to "stick to" (pair up with) a complementary section of DNA if they can find one.

If the same restriction enzyme used to cut a plasmid is also used to cut a gene from another DNA molecule, the sticky ends of the gene will be complementary to those of the plasmid. This enables the plasmid and gene to combine readily, forming a new, modified plasmid molecule. This modified plasmid molecule is called recombinant DNA. In addition to the newly spliced gene, the recombinant DNA plasmid contains all of the genes and characteristics of the original plasmid. Figure 11.31 shows diagrammatically the match between sticky ends that occurs when plasmid and gene combine.

Step 6 involves inserting the recombinant DNA (modified plasmids) back into *E. coli* cells. The process is called transformation. **Transformation** *is the process of incorporating recombinant DNA into a host cell.*

The transformed cells then reproduce, resulting in large numbers of identical cells called clones. **Clones** *are cells with identical DNA that have descended from a single cell.* Within a few hours, a single genetically altered bacterial cell can give rise to thousands of clones. Each clone has the capacity to synthesize the protein directed by the foreign gene it carries.

Researchers are not limited to selection of naturally occurring genes for transforming bacteria. Chemists have developed nonenzymatic methods of linking nucleotides together such that they can construct artificial genes of any sequence

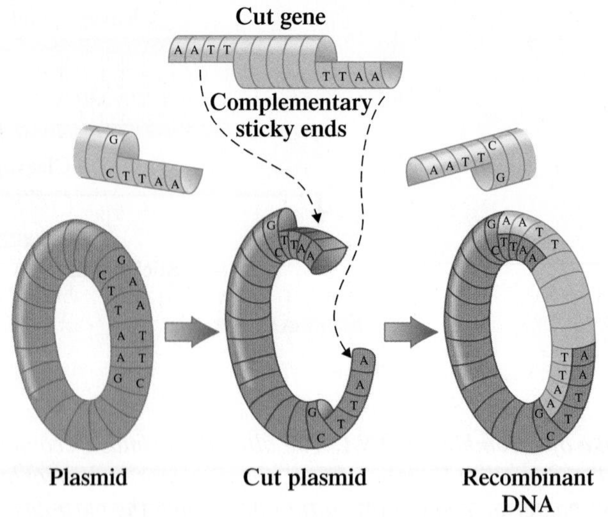

Figure 11.31 The "sticky ends" of the cut plasmid and the cut gene are complementary and combine to form recombinant DNA.

they desire. In fact, benchtop instruments are now available that can be programmed by a microprocessor to synthesize any DNA base sequence *automatically.* The operator merely enters a sequence of desired bases, starts the instrument, and returns later to obtain the product. Such flexibility in manufacturing DNA has opened many doors, accelerated the pace of recombinant DNA research, and redefined the term *designer genes*!

11.16 The Polymerase Chain Reaction

PCR temperature conditions are higher than those in the human body. This is possible because the DNA polymerase used was isolated from an organism that lives in the "hot pots" of Yellowstone National Park at temperatures of 70°C–75°C.

The **polymerase chain reaction (PCR)** *is a method for rapidly producing multiple copies of a DNA nucleotide sequence.* Billions of copies of a specific DNA sequence (gene) can be produced in a few hours via this reaction. The PCR is easy to carry out, requiring only a few chemicals, a container, and a source of heat. (In actuality, the PCR process is now completely automated.)

By means of the PCR process, DNA that is available only in very small quantities can be amplified to quantities large enough to analyze. The PCR process, devised in 1983, has become a valuable tool for diagnosing diseases and detecting pathogens in the body. It is now used in the prenatal diagnosis of a number of genetic disorders, including muscular dystrophy and cystic fibrosis, and in the identification of bacterial pathogens. It is also the definitive way to detect the AIDS virus.

After n cycles of the PCR process, the amount of DNA will have increased 2^n times.

2^{10} is approximately 1000.
2^{20} is approximately 1,000,000.

Twenty-five cycles of the PCR can be carried out in an hour in a process that is fully automated.

The PCR process has also proved useful in certain types of forensic investigations. A DNA sample may be obtained from a single drop of blood or semen or a single strand of hair at a crime scene and amplified by the PCR process. A forensic chemist can then compare the amplified samples with DNA samples taken from suspects. Work with DNA in the forensic area is often referred to as *DNA fingerprinting.*

DNA polymerase, an enzyme present in all living organisms, is a key substance in the PCR process. It can attach additional nucleotides to a short starter nucleotide chain, called a *primer,* when the primer is bound to a complementary strand of DNA that functions as a template. The original DNA is heated to separate its strands, and then primers, DNA polymerase, and deoxyribonucleotides are added so that the *DNA polymerase* can replicate the original strand. The process is repeated until, in a short time, millions of copies of the original DNA have been made.

Figure 11.32 shows diagrammatically, in very simplified terms, the basic steps in the PCR process.

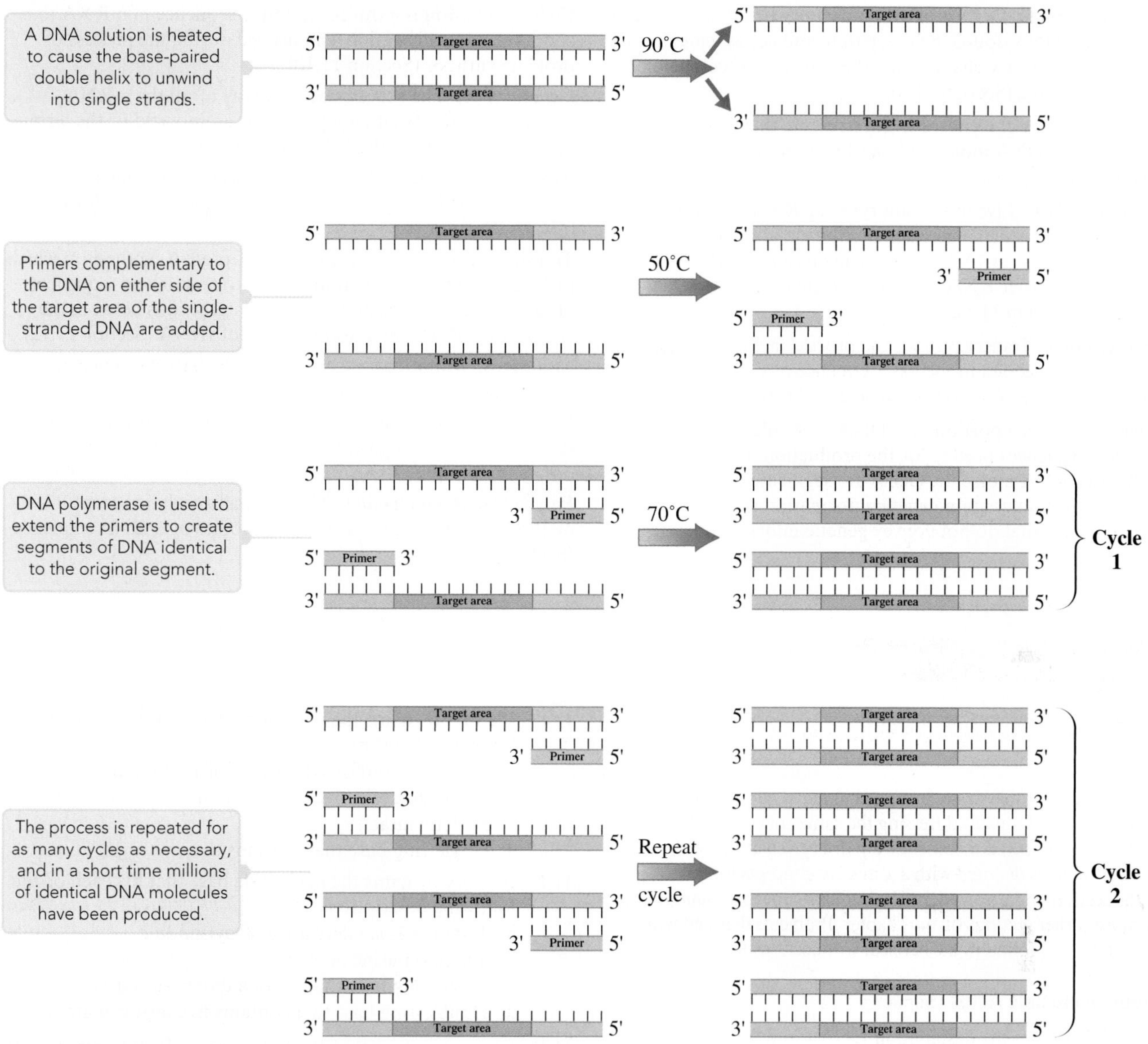
A DNA solution is heated to cause the base-paired double helix to unwind into single strands.
Primers complementary to the DNA on either side of the target area of the single-stranded DNA are added.
DNA polymerase is used to extend the primers to create segments of DNA identical to the original segment.
The process is repeated for as many cycles as necessary, and in a short time millions of identical DNA molecules have been produced.
Target area
Primer
5'
3'
90°C
50°C
70°C
Repeat cycle
Cycle 1
Cycle 2

Figure 11.32 The basic steps, in simplified terms, of the polymerase chain reaction process. Each cycle of the polymerase chain reaction doubles the number of copies of the target DNA sequence.

Concepts to Remember

OWL Sign in at **www.cengage.com/owl** to view tutorials and simulations, develop problem-solving skills, and complete online homework assigned by your professor.

Nucleic acids. Nucleic acids are polymeric molecules in which the repeating units are nucleotides. Cells contain two kinds of nucleic acids—deoxyribonucleic acids (DNA) and ribonucleic acids (RNA). The major biochemical functions of DNA and RNA are, respectively, transfer of genetic information and synthesis of proteins (Section 11.1).

Nucleic acid building blocks. Three types of subunits are present in a nucleic acid. They are: (1) a pentose sugar (ribose or deoxyribose) (2) a nitrogen-containing base (either a purine or pyrimidine derivative) and (3) a phosphate group. The nitrogen-containing bases are of five types: adenine (A), guanine (G), cytosine (C), thymine (T), and uracil (U) (Section 11.2).

Nucleosides and nucleotides. A nucleoside is a compound formed from a pentose sugar and a purine or pyrimidine base derivative. A nucleotide is a nucleoside to which a phosphate group has been added (bonded to the sugar). Nucleotides are the monomers for nucleic acid polymers (Section 11.3).

Primary nucleic acid structure. The "backbone" of a nucleic acid molecule is a constant alternating sequence of sugar and phosphate groups. Each sugar unit has a nitrogen-containing base attached to it (Section 11.4).

Complementary bases. Complementary bases are specific pairs of bases in nucleic acid structures that hydrogen-bond to each other (Section 11.5).

Secondary DNA structure. A DNA molecule exists as two polynucleotide chains coiled around each other in a double-helix arrangement. The double helix is held together by hydrogen bonding between complementary pairs of bases. Only two base-pairing combinations occur: A with T, and C with G (Section 11.5).

DNA replication. DNA replication occurs when the two strands of a parent DNA double helix separate and act as templates for the synthesis of new chains using the principle of complementary base pairing (Section 11.6).

Chromosome. A chromosome is a structure that consists of an individual DNA molecule bound to a group of proteins (Section 11.6).

RNA molecules. Five important types of RNA molecules, distinguished by their function, are ribosomal RNA (rRNA), messenger RNA (mRNA), heterogeneous nuclear RNA (hnRNA), transfer RNA (tRNA), and small nuclear RNA (snRNA) (Section 11.8).

Transcription. Transcription is the process in which the genetic information encoded in the base sequence of DNA is copied into hnRNA/mRNA molecules (Section 11.9).

Gene. A gene is a portion of a DNA molecule that contains the base sequences needed for the production of a specific hnRNA/mRNA molecule. Genes are segmented, with portions called exons that contain genetic information and portions called introns that do not convey genetic information (Section 11.9).

Codon. A codon is a three-nucleotide sequence in mRNA that codes for a specific amino acid needed during the process of protein synthesis (Section 11.10).

Genetic code. The genetic code consists of all the mRNA codons that specify either a particular amino acid or the termination of protein synthesis (Section 11.10).

Anticodon. An anticodon is a three-nucleotide sequence in tRNA that binds to a complementary sequence (a codon) in mRNA (Section 11.11).

Translation. Translation is the stage of protein synthesis in which the codons in mRNA are translated into amino acid sequences of new proteins. Translation involves interactions between the codons of mRNA and the anticodons of tRNA (Section 11.12).

Mutations. Mutations are changes in the base sequence in DNA molecules (Section 11.13).

Recombinant DNA. Recombinant DNA molecules are synthesized by splicing a segment of DNA, usually a gene, from one organism into the DNA of another organism (Section 11.15).

Polymerase chain reaction. The polymerase chain reaction is a method for rapidly producing many copies of a DNA sequence (Section 11.16).

Exercises and Problems

OWL Interactive versions of these problems may be assigned in OWL.

Exercises and problems are arranged in matched pairs with the two members of a pair addressing the same concept(s). The answer to the odd-numbered member of a pair is given at the back of the book. Problems denoted with a ▲ involve concepts found not only in the section under consideration but also concepts found in one or more earlier sections of the chapter. Problems denoted with a ● cover concepts found in a Chemical Connections feature box.

Types of Nucleic Acids (Section 11.1)

11.1 What does the designation DNA stand for?

11.2 What does the designation RNA stand for?

11.3 What is the primary function within cells of RNA?

11.4 What is the primary function within cells of DNA?

11.5 Within human cells, where is the DNA located?

11.6 Within human cells, where is the RNA located?

11.7 What is the general name for the building blocks (monomers) from which an RNA molecule is made?

11.8 What is the general name for the building blocks (monomers) from which a DNA molecule is made?

Nucleotide Building Blocks (Section 11.2)

11.9 What is the structural difference between the pentose sugars ribose and 2-deoxyribose?

11.10 What are the names of the pentose sugars present, respectively, in DNA and RNA molecules?

11.11 Characterize each of the following nitrogen-containing bases as a purine derivative or a pyrimidine derivative.
a. Thymine b. Cytosine
c. Adenine d. Guanine

11.12 Characterize each of the following nitrogen-containing bases as a "single-ring" base or a "double-ring" base.
a. Thymine b. Cytosine c. Guanine d. Adenine

11.13 Identify by name the nucleotide base that fits each of the following descriptions.
a. 5-methyl-2,4-dioxo derivative of pyrimidine
b. 2,4-dioxo derivative of pyrimidine
c. 5-Methyluracil
d. Single-ring base that contains three nitrogen atoms

11.14 Identify by name the nucleotide base that fits each of the following descriptions.
a. 4-amino-2-oxo derivative of pyrimidine
b. 6-amino purine derivative
c. Single-ring base that is not a dioxo derivative
d. Double-ring base that contains five nitrogen atoms

11.15 Characterize each of the following nucleotide subunits as (1) present in both DNA and RNA (2) present in DNA but not in RNA or (3) present in RNA but not in DNA.
a. Phosphate b. 2-deoxyribose c. Uracil d. Adenine

11.16 Characterize each of the following nucleotide subunits as (1) present in both DNA and RNA (2) present in DNA but not in RNA or (3) present in RNA but not in DNA.
a. Guanine b. Ribose c. Thymine d. Cytosine

11.17 How many different choices are there for each of the following subunits in the specified type of nucleotide?
a. Pentose sugar subunit in DNA nucleotides
b. Nitrogen-containing base subunit in RNA nucleotides
c. Phosphate subunit in DNA nucleotides

11.18 How many different choices are there for each of the following subunits in the specified type of nucleotide?
a. Pentose sugar subunit in RNA nucleotides
b. Nitrogen-containing base subunit in DNA nucleotides
c. Phosphate subunit in RNA nucleotides

Nucleotide Formation (Section 11.3)

11.19 How many subunits are present within a nucleoside?

11.20 What are the general names of the subunits present within a nucleoside?

11.21 Which carbon atom (by number) and which nitrogen atom (by number) are involved in the bond that joins nucleoside components together?

11.22 What type of bond, by name, joins nucleoside components together?

11.23 What is the name of each of the following nucleosides?
a. Ribose-adenine b. Ribose-cytosine
c. Deoxyribose-thymine d. Deoxyribose-guanine

11.24 What is the name of each of the following nucleosides?
a. Ribose-uracil b. Ribose-guanine
c. Deoxyribose-adenine d. Deoxyribose-cytosine

11.25 How many subunits are present within a nucleotide?

11.26 What are the general names of the subunits present within a nucleotide?

11.27 What is the composition, in terms of specific subunits present, of the four types of nucleotides found in RNA molecules?

11.28 What is the composition, in terms of specific subunits present, of the four types of nucleotides found in DNA molecules?

11.29 What is the name of the type of linkage that connects the
a. sugar and base in a nucleotide
b. sugar and phosphate in a nucleotide

11.30 What carbon atom, by number, is involved in the
a. sugar-base bond in a nucleotide
b. sugar-phosphate bond in a nucleotide

11.31 What nitrogen-containing base and what sugar are present in each of the following nucleotides?
a. AMP b. dGMP c. dTMP d. UMP

11.32 What nitrogen-containing base and what sugar are present in each of the following nucleotides?
a. GMP b. dAMP c. CMP d. dCMP

11.33 What is the name of each of the nucleotides in Problem 11.31?

11.34 What is the name of each of the nucleotides in Problem 11.32?

11.35 Indicate whether each of the nucleotides in Problem 11.31 is (1) present in both DNA and RNA (2) present in DNA but not in RNA or (3) present in RNA but not in DNA.

11.36 Indicate whether each of the nucleotides in Problem 11.32 is (1) present in both DNA and RNA (2) present in DNA but not in RNA or (3) present in RNA but not in DNA.

11.37 Characterize as true or false each of the following statements about the given nucleotide.

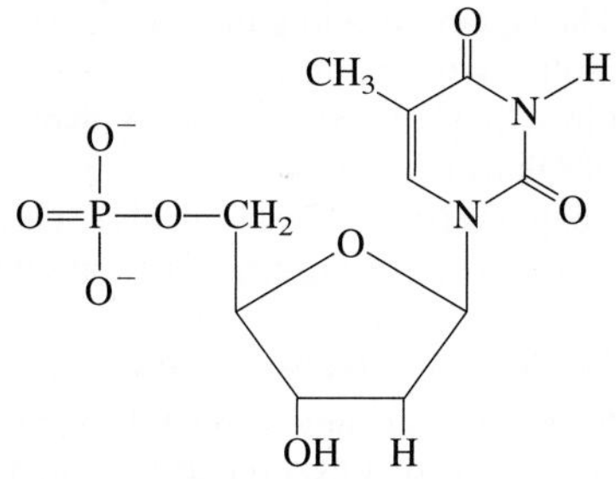

a. The nitrogen-containing base is a purine derivative.
b. The phosphate group is attached to the sugar unit at carbon 3′.
c. The sugar unit is ribose.
d. The nucleotide could be a component of both DNA and RNA.

11.38 Characterize as true or false each of the following statements about the given nucleotide.

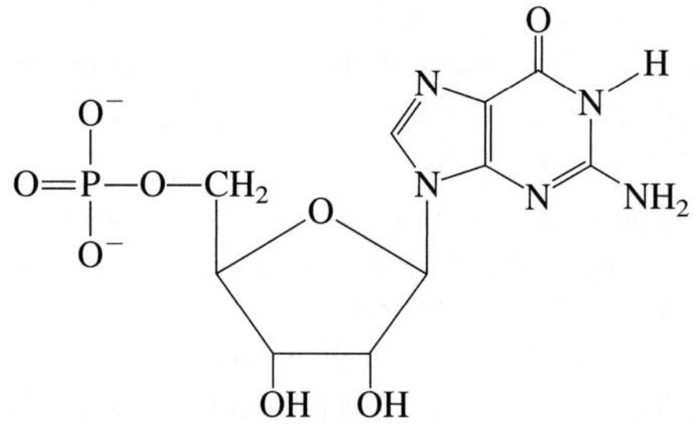

a. The sugar unit is 2-deoxyribose.
b. The sugar unit is attached to the nitrogen-containing base at nitrogen 3.
c. The nitrogen-containing base is a pyrimidine derivative.
d. The nucleotide could be a component of both DNA and RNA.

11.39 Draw the structures of the three products produced when the nucleotide in Problem 11.37 undergoes hydrolysis.

11.40 Draw the structures of the three products produced when the nucleotide in Problem 11.38 undergoes hydrolysis.

▲11.41 Indicate whether each of the following is (1) a nucleoside (2) a nucleotide or (3) neither a nucleoside nor a nucleotide.
a. Adenosine b. Adenine
c. dAMP d. Adenosine 5′-monophosphate

▲11.42 Indicate whether each of the following is (1) a nucleoside (2) a nucleotide or (3) neither a nucleoside nor a nucleotide.
a. Cytosine b. Cytidine
c. CMP d. Deoxycytidine 5′-monophosphate

Primary Nucleic Acid Structure (Section 11.4)

11.43 What are the two repeating subunits present in the *backbone* portion of a nucleic acid?

11.44 To which type of subunit in a nucleic acid *backbone* are the nitrogen-containing bases attached?

11.45 What distinguishes various DNA molecules from each other?

11.46 What distinguishes various RNA molecules from each other?

11.47 What is the difference between a nucleic acid's 3′ end and its 5′ end?

11.48 In the lengthening of a polynucleotide chain, which type of nucleotide subunit would bond to the 3′ end of the polynucleotide chain?

11.49 What are the nucleotide subunits that participate in a nucleic acid 3′,5′-phosphodiester linkage?

11.50 How many 3′,5′-phosphodiester linkages are present in a tetranucleotide segment of a nucleic acid?

11.51 Draw the structure of the dinucleotide product obtained by combining the nucleotides of Problems 11.37 and 11.38 such that the Problem 11.37 nucleotide is the 5′ end of the dinucleotide.

11.52 Draw the structure of the dinucleotide product obtained by combining the nucleotides of Problems 11.37 and 11.38 such that the Problem 11.38 nucleotide is the 3′ end of the dinucleotide.

The DNA Double Helix (Section 11.5)

11.53 Describe the DNA double helix in terms of
a. general shape.
b. what is on the outside of the helix and what is within the interior of the helix.

11.54 Describe the DNA double helix in terms of
a. the directionality of the two polynucleotide chains present.
b. a comparison of the total number of nitrogen-containing bases present in each of the two polynucleotide chains.

11.55 The base content of a particular DNA molecule is 36% thymine. What is the percentage of each of the following bases in the molecule?
a. Adenine b. Guanine c. Cytosine

11.56 The base content of a particular DNA molecule is 24% guanine. What is the percentage of each of the following bases in the molecule?
a. Adenine b. Cytosine c. Thymine

11.57 In terms of hydrogen bonding, a G–C base pair is more stable than an A–T base pair. Explain why this is so.

11.58 What structural consideration prevents the following bases from forming complementary base pairs?
a. A and G b. T and C

11.59 What is the relationship between the total number of purine bases (A and G) and the total number of pyrimidine bases (C and T) present in a DNA double helix?

11.60 The base composition for one of the strands of a DNA double helix is 19% A, 34% C, 28% G, and 19% T. What is the percent base composition for the other strand of the DNA double helix?

11.61 Identify the 3′ and 5′ ends of the DNA base sequence TAGCC.

11.62 The two-base DNA sequences TA and AT represent different dinucleotides. Explain why this is so.

11.63 Convert each of the following 3′ to 5′ DNA base sequences to 5′ to 3′ DNA base sequences.
a. 3′ ATCG 5′ b. 3′ AATA 5′
c. 3′ CACA 5′ d. 3′ CAAC 5′

11.64 Convert each of the following 3′ to 5′ DNA base sequences to 5′ to 3′ DNA base sequences.
a. 3′ CGTA 5′ b. 3′ CGCG 5′
c. 3′ ATTA 5′ d. 3′ CTAA 5′

11.65 Using the concept of complementary base pairing, write the complementary DNA strands, with their 5′ and 3′ ends labeled, for each of the following DNA base sequences.
a. 5′ ACGTAT 3′ b. 5′ TTACCG 3′
c. 3′ GCATAA 5′ d. AACTGG

11.66 Using the concept of complementary base pairing, write the complementary DNA strands, with their 5′ and 3′ ends labeled, for each of the following DNA base sequences.
a. 5′ CCGGTA 3′ b. 5′ CACAGA 3′
c. 3′ TTTAGA 5′ d. CATTAC

11.67 How many total hydrogen bonds would exist between the DNA strand 5′ AGTCCTCA 3′ and its complementary strand?

11.68 How many total hydrogen bonds would exist between the DNA strand 5′ CCTAGGAT 3′ and its complementary strand?

▲11.69 For the DNA base sequences 5′ TTGCATC 3′ and 3′ TTGCATC 5′, how many of the following are present, respectively?
a. Nucleotides b. Purine bases
c. Phosphate units d. Ribose units

▲11.70 For the DNA base sequences 5′ CAAGTAGT 3′ and 3′ CAAGTAGT 5′, how many of the following are present, respectively?
a. Nucleotides b. Pyrimidine bases
c. Ribose units d. Deoxyribose units

Replication of DNA Molecules (Section 11.6)

11.71 What is the function of the enzyme *DNA helicase* in the DNA replication process?

11.72 What are two functions of the enzyme *DNA polymerase* in the DNA replication process?

11.73 What is the base sequence, specified in the 5′ to 3′ direction, for a segment of newly formed DNA if it was formed using the following template DNA segments?
a. 3′ AATGC 5′ b. 5′ AATGC 3′
c. 3′ GCAGC 5′ d. 5′ GCAGC 3′

11.74 What is the base sequence, specified in the 5′ to 3′ direction, for a segment of newly formed DNA if it was formed using the following template DNA segments?
a. 3′ GGCAA 5′ b. 5′ GTCCG 3′
c. 3′ ACGTA 5′ d. 5′ ACGTA 3′

11.75 In the replication of a DNA molecule, two daughter molecules, Q and R, are formed. The following base sequence is part of the newly formed strand in daughter molecule Q.

5′ ACTTAG 3′

Indicate the corresponding base sequence in
a. the newly formed strand in daughter molecule R.
b. the "parent" strand in daughter molecule Q.
c. the "parent" strand in daughter molecule R.

11.76 In the replication of a DNA molecule, two daughter molecules, S and T, are formed. The following base sequence is part of the "parent" strand in daughter molecule S.

5′ TTCAGAG 3′

Indicate the corresponding base sequence in
a. the newly formed strand in daughter molecule T.
b. the newly formed strand in daughter molecule S.
c. the "parent" strand in daughter molecule T.

11.77 How does the synthesis of a daughter DNA strand growing toward a replication fork differ from the synthesis of a daughter DNA strand growing away from a replication fork?

11.78 In the context of DNA replication, what is the difference between a "leading DNA strand" and a "lagging DNA strand"?

11.79 Indicate whether each of the following statements relating to aspects of DNA replication is true or false.
a. The lagging strand grows in the same direction as the replication fork moves.
b. Growth of the leading strand involves the production of Okazaki fragments.
c. Lagging strands always involve "daughter" DNA segments, and leading strands always involve "parent" DNA segments.
d. Leading strands always grow in the 5′ to 3′ direction.

11.80 Indicate whether each of the following statements relating to aspects of DNA replication is true or false.
a. Leading strands always involve newly synthesized DNA, and lagging strands always involve template DNA.
b. The concept of "nicks" is associated with the growth of a lagging strand.
c. The leading strand involves template DNA.
d. The lagging strand always grows in a continuous manner.

11.81 What is a chromosome?

11.82 Chromosomes are nucleoproteins. Explain.

▲11.83 Suppose that 28% of the nucleotides in a DNA molecule are deoxythymidine 5′ monophosphate, and that during DNA replication the percentage amounts of *available* nucleotide bases are 22% A, 22% C, 28% G, and 28% T. Which base would be depleted first in the replication process?

▲11.84 Suppose that 30% of the nucleotides in a DNA molecule are deoxyguanosine 5′ monophosphate, and that during DNA replication the percentage amounts of *available* nucleotide bases are 20% A, 20% C, 30% G, and 30% T. Which base would be depleted first in the replication process?

•11.85 (Chemical Connections 11-A) Indicate whether each of the following statements concerning antimetabolites used in chemotherapy is true or false.
a. Most metabolites were initially obtained as natural products from molds and fungi.
b. The MP in the designation 6-MP stands for monophosphate.
c. Methotrexate "mimics" the DNA base guanosine.
d. 5-Fluorouracil "mimics" the DNA base uracil.

•11.86 (Chemical Connections 11-A) Indicate whether each of the following statements concerning antimetabolites used in chemotherapy is true or false.
a. Antimetabolites function as cell-division inhibitors.
b. Folic acid is an antimetabolite whose structure contains three ring systems.
c. Both thioguanine and 6-MP differ from the DNA base they "mimic" in that a —SH group has replaced a —NH_2 group.
d. Antimetabolites are always derivatives of DNA bases.

Overview of Protein Synthesis (Section 11.7)

11.87 What type of molecule is the starting material for the transcription phase of protein synthesis?

11.88 What type of molecule is the end product for the transcription phase of protein synthesis?

11.89 What type of molecule is the starting material for the translation phase of protein synthesis?

11.90 What type of molecule is the end product for the translation phase of protein synthesis?

RNA Molecules (Section 11.8)

11.91 What are the four major differences between RNA molecules and DNA molecules?

11.92 What are the names and abbreviations for the five major types of RNA molecules?

11.93 State whether each of the following phrases applies to hnRNA, mRNA, tRNA, rRNA, or snRNA.
a. Material from which messenger RNA is made
b. Delivers amino acids to protein synthesis sites
c. Smallest of the RNAs in terms of nucleotide units present
d. Also goes by the designation ptRNA

11.94 State whether each of the following phrases applies to hnRNA, mRNA, tRNA, rRNA, or snRNA.
a. Associated with a series of proteins in a complex structure
b. Contains genetic information needed for protein synthesis
c. Most abundant type of RNA in a cell
d. Involved in the editing of hnRNA molecules

11.95 For each of the following types of RNA, indicate whether the predominant cellular location for the RNA is the nuclear region, the extranuclear region, or both the nuclear and the extranuclear regions.
a. hnRNA b. tRNA c. rRNA d. mRNA

11.96 Indicate whether each of the following processes occurs in the nuclear or the extranuclear region of a cell.
a. DNA transcription
b. Processing of hnRNA to mRNA
c. mRNA translation (protein synthesis)
d. DNA replication

Transcription: RNA Synthesis (Section 11.9)

11.97 What serves as a template in the process of *transcription*?

11.98 What is the initial product of the *transcription* process?

11.99 What are two functions of the enzyme *RNA polymerase* in the transcription process?

11.100 What is a *gene*?

11.101 What are the complementary base pairs in DNA–RNA interactions?

11.102 In DNA–DNA interactions there are two complementary base pairs, and in DNA–RNA interactions there are three complementary base pairs. Explain.

11.103 Which of the following characterizations applies to each of the base-pairing situations below (1) involves two DNA strands (2) involves a DNA strand and an RNA strand or (3) could involve either two DNA strands or a DNA strand and an RNA strand?

a. A G T
⋮ ⋮ ⋮
U C A

b. A C T
⋮ ⋮ ⋮
T G A

c. A G U
⋮ ⋮ ⋮
T C A

d. C G C
⋮ ⋮ ⋮
G C G

11.104 Which of the following characterizations applies to each of the base-pairing situations below (1) involves two DNA strands (2) involves a DNA strand and an RNA strand or (3) could involve either two DNA strands or a DNA strand and an RNA strand?

a. A A T
⋮ ⋮ ⋮
T T A

b. A A T
⋮ ⋮ ⋮
U U A

c. C G U
⋮ ⋮ ⋮
G C A

d. C C G
⋮ ⋮ ⋮
G G C

11.105 For each of the following DNA template strands, determine (1) the base sequence of the DNA informational strand and (2) the base sequence of the hnRNA synthesized using the DNA template strand.
a. 3′ TACGGC 5′ b. 3′ CCATTA 5′
c. 3′ ACATGG 5′ d. 3′ ACGTAC 5′

11.106 For each of the following DNA template strands, determine (1) the base sequence of the DNA informational strand and (2) the base sequence of the hnRNA synthesized using the DNA template strand.
a. 3′ TTCGTA 5′ b. 3′ CCGAAT 5′
c. 3′ CATCAT 5′ d. 3′ TGCTGC 5′

11.107 From what DNA template strand was each of the following hnRNA base sequences transcribed?

a. 5′ CCUUAA 3′ b. 5′ ACGUAC 3′
c. 5′ ACGACG 3′ d. 5′ UACCAU 3′

11.108 From what DNA template strand was each of the following hnRNA base sequences transcribed?
a. 5′ CUAGGC 3′ b. 5′ UACUGG 3′
c. 5′ GACGAC 3′ d. 5′ UAAUAU 3′

11.109 What is the base sequence of the complementary strand of the DNA from which the hnRNA segments in Problem 11.107 were synthesized?

11.110 What is the base sequence of the complementary strand of the DNA from which the hnRNA segments in Problem 11.108 were synthesized?

11.111 What is the relationship between an exon and a gene?

11.112 What is the relationship between an intron and a gene?

11.113 What mRNA base sequence would be obtained from the following portion of a hnRNA molecule?

| exon | intron | exon |
3′ UUAC–AACG–GCAU 5′

11.114 What mRNA base sequence would be obtained from the following portion of a hnRNA molecule?

| exon | intron | exon |
3′ UCAG–UAGC–UUCA 5′

11.115 What mRNA base sequence would be obtained from the following portion of a gene?

| exon | intron | exon |
5′ TCAG–TAGC–TTCA 3′

11.116 What mRNA base sequence would be obtained from the following portion of a gene?

| intron | exon | intron |
5′ TTAC–AACG–GCAT 3′

11.117 In the process of splicing, which type of RNA
a. undergoes the splicing?
b. is present in the spliceosomes?

11.118 What is the difference between snRNA and snRNPs?

11.119 What is *alternative splicing*?

11.120 How many different mRNAs can be produced from an hnRNA that contains three exons, one of which is an "alternative" exon?

The Genetic Code (Section 11.10)

11.121 What is a codon?

11.122 On what type of RNA molecule are codons found?

11.123 Using the information in Table 11.2, determine what amino acid is coded for by each of the following codons.
a. CUU b. AAU c. AGU d. GGG

11.124 Using the information in Table 11.2, determine what amino acid is coded for by each of the following codons.
a. GUA b. CCC c. CAC d. CCA

11.125 Using the information in Table 11.2, determine the synonyms, if any, of each of the codons in Problem 11.123.

11.126 Using the information in Table 11.2, determine the synonyms, if any, of each of the codons in Problem 11.124.

11.127 Explain why the base sequence ATC could not be a codon.

11.128 Explain why the base sequence AGAC could not be a codon.

11.129 Predict the sequence of amino acids coded by the mRNA sequence

5′ AUG–AAA–GAA–GAC–CUA 3′

11.130 Predict the sequence of amino acids coded by the mRNA sequence

5′ GGA–GGC–ACA–UGG–GAA 3′

11.131 Predict the sequence of amino acids coded by the mRNA sequence

3′ AUG–AAA–GAA–GAC–CUA 5′

11.132 Predict the sequence of amino acids coded by the mRNA sequence

3′ GGA–GGC–ACA–UGG–GAA 5′

▲**11.133** Determine each of the following items using the hnRNA nucleotide sequence

| exon | intron | exon |
5′ UCCG–CCAU–UAACA 3′

a. The sequence of bases on the DNA template strand from which the hnRNA was synthesized
b. The sequence of bases on the DNA informational strand used to synthesize the hnRNA
c. The sequence of bases in the mRNA produced from the hnRNA
d. The sequence of amino acids in the peptide synthesized from the mRNA

▲**11.134** Determine each of the following items using the hnRNA nucleotide sequence

| exon | intron | exon |
5′ AGGC–GGAU–UCACA 3′

a. The sequence of bases on the DNA template strand from which the hnRNA was synthesized
b. The sequence of bases on the DNA informational strand used to synthesize the hnRNA
c. The sequence of bases in the mRNA produced from the hnRNA
d. The sequence of amino acids in the peptide synthesized from the mRNA

Anticodons and tRNA Molecules (Section 11.11)

11.135 Describe the general structure of a tRNA molecule.

11.136 Where is the anticodon site on a tRNA molecule?

11.137 By what type of bond is an amino acid attached to a tRNA molecule?

11.138 What principle governs the codon–anticodon interaction that leads to proper placement of amino acids in proteins?

11.139 What is the anticodon that would interact with each of the following codons?
a. AGA b. CGU c. UUU d. CAA

11.140 What is the anticodon that would interact with each of the following codons?
a. CCU b. GUA c. AUC d. GCA

11.141 Which amino acid will a tRNA molecule be carrying if its anticodon is the following?
a. CCC b. CAG c. UGC d. GAG

11.142 Which amino acid will a tRNA molecule be carrying if its anticodon is the following?
a. GGG b. GAC c. AUA d. CGA

11.143 What are the possible codons that a tRNA molecule could react with if it is carrying each of the following amino acids? (Note that a given tRNA will be specific for one or more of the possible codons.)
a. Serine b. Leucine c. Isoleucine d. Glycine

11.144 What are the possible codons that a tRNA molecule could interact with if it is carrying each of the following

amino acids? (Note that a given tRNA will be specific for one or more of the possible codons.)
a. Valine b. Cysteine c. Alanine d. Tyrosine

11.145 Identify the amino acid associated with each of the following entities.
a. 5′ ACG 3′ codon
b. 3′ ACG 5′ codon
c. 5′ ACG 3′ anticodon
d. 3′ ACG 5′ anticodon

11.146 Identify the amino acid associated with each of the following entities.
a. 5′ CCG 3′ codon
b. 3′ CCG 5′ codon
c. 5′ CCG 3′ anticodon
d. 3′ CCG 5′ anticodon

▲11.147 The following is the base sequence of an exon portion of a template strand of a DNA molecule:

5′ TCCGCATTAACA 3′

a. What is the base sequence of the hnRNA strand synthesized from the DNA template strand?
b. What is the base sequence of the mRNA strand synthesized from the hnRNA strand?
c. What codons are present in the mRNA strand produced from the DNA template strand?
d. What tRNA molecule anticodons are needed to interact with the codons present in the mRNA strand produced from the template DNA strand?

▲11.148 The following is a base sequence for an exon portion of a template strand of a DNA molecule:

5′ TAACGCTTCACG 3′

a. What is the base sequence of the hnRNA strand synthesized from the DNA template strand?
b. What is the base sequence of the mRNA strand synthesized from the hnRNA strand?
c. What codons are present in the mRNA strand produced from the DNA template strand?
d. What tRNA molecule anticodons are needed to interact with the codons present in the mRNA strand produced from the template DNA strand?

Translation: Protein Synthesis (Section 11.12)

11.149 Indicate whether each of the following statements about ribosomes is true or false.
a. Ribosomes have two subunits that are approximately the same size.
b. The active site of a ribosome is predominantly RNA rather than protein.
c. The mRNA involved in protein synthesis binds to the small subunit of a ribosome.
d. A ribosome functions as an enzyme in protein synthesis.

11.150 Indicate whether each of the following statements about ribosomes is true or false.
a. The active site of a ribosome is located in its smaller subunit.
b. A ribosome is a tRNA-protein complex.
c. Ribosome subunits are approximately 75% protein and 25% RNA.
d. Human cells contain from 10–20 ribosomes.

11.151 What are the two steps involved in the activation of a tRNA molecule?

11.152 Why is the first amino acid in a developing human protein always the amino acid Met?

11.153 In the elongation phase of translation, at which site in the ribosome does new peptide bond formation actually take place?

11.154 What two changes occur at a ribosome during protein synthesis immediately after peptide bond formation?

11.155 What types of events occur during post-translation processing of a protein?

11.156 At what stage of protein formation does folding of the protein into its active conformation occur?

11.157 Write a possible mRNA base sequence that would lead to the production of the following pentapeptide. (There is more than one correct answer.)

Gly–Ala–Cys–Val–Tyr

11.158 Write a possible mRNA base sequence that would lead to the production of the following pentapeptide. (There is more than one correct answer.)

Lys–Met–Thr–His–Phe

11.159 Consider the translation of the following mRNA base sequence during protein synthesis.

5′ CAA CGA AAG 3′

a. What are the codons associated with this base sequence?
b. What tRNA anticodons are compatible with this base sequence?
c. What are the amino acids specified by this base sequence?

11.160 Consider the translation of the following mRNA base sequence during protein synthesis.

5′ CGC CGU UAC 3′

a. What are the codons associated with this base sequence?
b. What tRNA anticodons are compatible with this base sequence?
c. What are the amino acids specified by this base sequence?

▲11.161 Which of these RNA types—(1) mRNA (2) hnRNA (3) rRNA or (4) tRNA—is most closely associated with each of the following terms or descriptions?
a. Codon b. Intron
c. Amino acid carrier d. Interacts with ribosomes

▲11.162 Which of these RNA types—(1) mRNA (2) hnRNA (3) rRNA or (4) tRNA—is most closely associated with each of the following terms or descriptions?
a. Exon
b. Anticodon
c. Ribosome structural component
d. Interacts with spliceosomes

●11.163 (Chemical Connections 11-B) Indicate whether each of the following statements concerning bacterial protein synthesis and antimetabolite cancer drugs is true or false.
a. Bacterial protein synthesis does not require the use of ribosomes.
b. Human ribosomes are unsuitable targets for antimetabolites because they are too large.
c. Structurally, the antimetabolite streptomycin is a trisaccharide derivative.
d. Terramycin disrupts bacterial protein synthesis by reacting with all alanine amino acids present.

•**11.164** (Chemical Connections 11-B) Indicate whether each of the following statements concerning bacterial protein synthesis and antimetabolite cancer drugs is true or false.
a. In bacterial protein synthesis, translation begins before transcription has been completed.
b. All antimetabolites have names that end in -mycin.
c. Streptomycin and neomycin disrupt protein synthesis by causing a ribosomal shape change.
d. Diphtheria is caused by a bacteria-generated protein toxin that disrupts human protein synthesis.

Mutations (Section 11.13)

11.165 Consider the following mRNA base sequence

5′ CUU CAG 3′

a. What dipeptide is coded for by this mRNA?
b. What dipeptide is formed if a mutation converts CUU to CUC?
c. What dipeptide is formed if a mutation converts CAG to AAG?
d. What dipeptide is formed if a mutation converts CUU to CUC and CAG to AAG?

11.166 Consider the following mRNA base sequence

5′ ACC CAC 3′

a. What dipeptide is coded for by this mRNA?
b. What dipeptide is formed if a mutation converts CAC to AAC?
c. What dipeptide is formed if a mutation converts ACC to ACU?
d. What dipeptide is formed if a mutation converts CAC to AAC and ACC to ACU?

▲11.167 Consider the following DNA base sequence

3′ TTA ATA 5′

a. What dipeptide is formed from the transcription and translation of this DNA segment?
b. What dipeptide is formed if a DNA mutation converts ATA to ATG?
c. What dipeptide is formed if a DNA mutation converts ATA to AGA?
d. What dipeptide is formed if a DNA mutation converts TTA to TTT?

▲**11.168** Consider the following DNA base sequence

3′ TAT CGG 5′

a. What dipeptide is formed from the transcription and translation of this DNA segment?
b. What dipeptide is formed if a DNA mutation converts CGG to CGT?
c. What dipeptide is formed if a DNA mutation converts CGG to AGG?
d. What dipeptide is formed if a DNA mutation converts TAT to TTT?

Viruses and Vaccines (Section 11.14)

11.169 Describe the general structure of a virus.

11.170 What is the only function of a virus?

11.171 What is the most common method by which viruses invade cells?

11.172 Why must a virus infect another organism in order to reproduce?

Recombinant DNA and Genetic Engineering (Section 11.15)

11.173 How does recombinant DNA differ from normal DNA?

11.174 Give two reasons why bacterial cells are used for recombinant DNA procedures.

11.175 What role do plasmids play in recombinant DNA procedures?

11.176 Describe what occurs when a particular restriction enzyme operates on a segment of double-stranded DNA.

11.177 Describe what happens during transformation.

11.178 How are plasmids obtained from *E. coli* bacteria?

11.179 A particular restriction enzyme will cleave DNA between A and A in the sequence AAGCTT in the 5′-to-3′ direction. Draw a diagram showing the structural details of the "sticky ends" that result from cleavage of the following DNA segment.

5′ — 3′
C C A A G C T T G
G G T T C G A A C
3′ — 5′

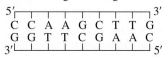

11.180 A particular restriction enzyme will cleave DNA between A and A in the sequence AAGCTT in the 5′-to-3′ direction. Draw a diagram showing the structural details of the "sticky ends" that result from cleavage of the following DNA segment.

5′ — 3′
G G A A G C T T A
C C T T C G A A T
3′ — 5′

▲11.181 Which of the following processes—(1) transcription phase of protein synthesis (2) translation phase of protein synthesis (3) replication of DNA and (4) formation of recombinant DNA—is associated with each of the events below?
a. Complete unwinding of a DNA molecule occurs.
b. Partial unwinding of a DNA molecule occurs.
c. An mRNA-ribosome complex is formed.
d. The process of transformation occurs.

▲**11.182** Which of the following processes—(1) transcription phase of protein synthesis (2) translation phase of protein synthesis (3) replication of DNA and (4) formation of recombinant DNA—is associated with each of the events below?
a. Amino acid-tRNA molecules are formed.
b. Anticodon-codon base pairing occurs.
c. Restriction enzymes are used.
d. Okazaki fragments are formed.

Polymerase Chain Reaction (Section 11.16)

11.183 What is the function of the polymerase chain reaction?

11.184 What is the function of the enzyme *DNA polymerase* in the PCR process?

11.185 What is a *primer,* and what is its function in the PCR process?

11.186 What are the four types of substances needed to carry out the PCR process?

Biochemical Energy Production

12

The energy consumed by these scarlet ibises in flight is generated by numerous sequences of biochemical reactions that occur within their bodies.

CHAPTER OUTLINE

OWL

Sign in to OWL at **www.cengage.com/owl** to view tutorials and simulations, develop problem-solving skills, and complete online homework assigned by your professor.

This chapter is the first of four dealing with the chemical reactions that occur in a living organism. In this first chapter, those molecules that are repeatedly encountered in biological reactions are considered as well as those reactions that are common to the processing of carbohydrates, lipids, and proteins. The three following chapters consider the reactions associated uniquely with carbohydrate, lipid, and protein processing, respectively.

12.1 Metabolism

Metabolism *is the sum total of all the biochemical reactions that take place in a living organism.* Human metabolism is quite remarkable. An average human adult whose weight remains the same for 40 years processes about 6 tons of solid food and 10,000 gallons of water, during which time the composition of the body is essentially constant. Just as gasoline is put into a car to make it go or a kitchen appliance is plugged in to make it run, the human body needs a source of energy to make it function. Even the simplest living cell is continually carrying on energy-demanding processes such as protein synthesis, DNA replication, RNA transcription, and membrane transport.

Metabolic reactions fall into one of two subtypes: catabolism and anabolism. **Catabolism** *is all metabolic reactions in which large biochemical molecules are broken down to smaller ones.* Catabolic reactions usually release energy.

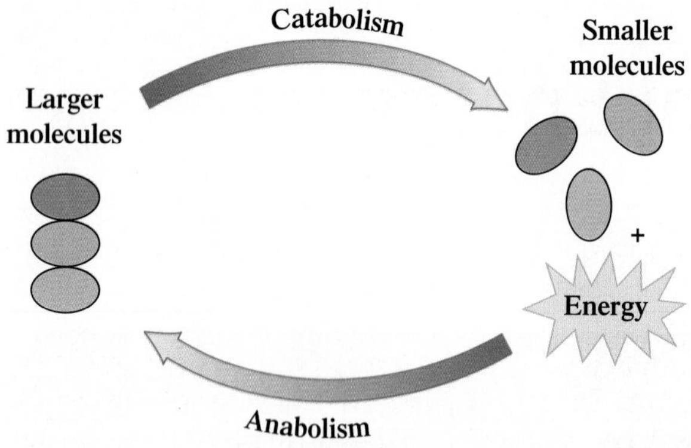

Figure 12.1 The processes of catabolism and anabolism are opposite in nature. The first usually produces energy, and the second usually consumes energy.

Catabolism is pronounced ca-TAB-o-lism, and *anabolism* is pronounced an-ABB-o-lism. *Catabolic* is pronounced CAT-a-bol-ic, and *anabolic* is pronounced AN-a-bol-ic.

The reactions involved in the oxidation of glucose are catabolic. **Anabolism** *is all metabolic reactions in which small biochemical molecules are joined together to form larger ones.* Anabolic reactions usually require energy in order to proceed. The synthesis of proteins from amino acids is an anabolic process. Figure 12.1 contrasts catabolic and anabolic processes.

The metabolic reactions that occur in a cell are usually organized into sequences called *metabolic pathways.* A **metabolic pathway** *is a series of consecutive biochemical reactions used to convert a starting material into an end product.* Such pathways may be *linear,* in which a series of reactions generates a final product, or *cyclic,* in which a series of reactions regenerates the first reactant.

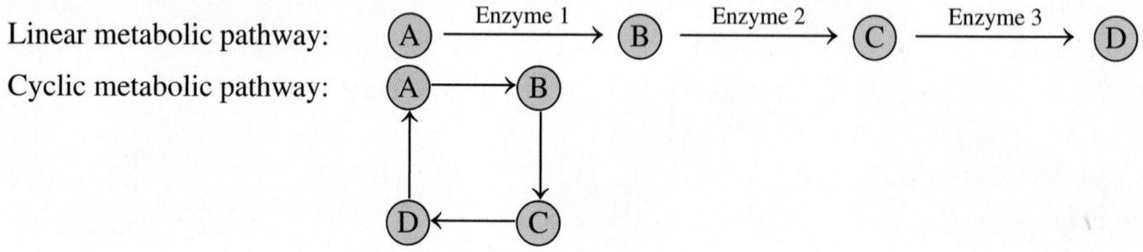

The major metabolic pathways for all life forms are similar. This enables scientists to study metabolic reactions in simpler life forms and use the results to help understand the corresponding metabolic reactions in more complex organisms, including humans.

EXAMPLE 12.1 Distinguishing Between Anabolic and Catabolic Processes

Classify each of the following chemical processes as anabolic or catabolic.

a. Synthesis of a polysaccharide from monosaccharides
b. Hydrolysis of a pentasaccharide to monosaccharides
c. Formation of a nucleotide from phosphate, nitrogenous base, and pentose sugar
d. Hydrolysis of a triacylglycerol to glycerol and fatty acids

Solution

a. *Anabolic.* A large molecule is formed from many small molecules.
b. *Catabolic.* Five monosaccharides are produced from a larger molecule.
c. *Anabolic.* Three subunits are combined to give a larger unit.
d. *Catabolic.* A four-subunit molecule is broken down into four smaller molecules.

Practice Exercise 12.1

Classify each of the following chemical processes as anabolic or catabolic.

a. Synthesis of a protein from amino acids
b. Formation of a triacylglycerol from glycerol and fatty acids
c. Hydrolysis of a polysaccharide to monosaccharides
d. Formation of a nucleic acid from nucleotides

Answers: **a.** Anabolic; **b.** Anabolic; **c.** Catabolic; **d.** Anabolic

12.2 Metabolism and Cell Structure

Knowledge of the major structural features of a cell is a prerequisite to understanding *where* metabolic reactions take place.

Cells are of two types: prokaryotic and eukaryotic. *Prokaryotic cells* have no nucleus and are found only in bacteria. The DNA that governs the reproduction of prokaryotic cells is usually a single circular molecule found near the center of the cell in a region called the *nucleoid.* A **eukaryotic cell** *is a cell in which the DNA is found in a membrane-enclosed nucleus.* Cells of this type, which are found in all higher organisms, are about 1000 times larger than bacterial cells. The remainder of this section focuses on eukaryotic cells, the type present in humans. Figure 12.2 shows the general internal structure of a eukaryotic cell. Note the key components shown: the outer membrane, nucleus, cytosol, ribosomes, lysosomes, and mitochondria.

The term *eukaryotic,* pronounced you-KAHR-ee-ah-tic, is from the Greek *eu,* meaning "true," and *karyon,* meaning "nucleus." The term *prokaryotic,* which contains the Greek *pro,* meaning "before," literally means "before the nucleus."

The **cytoplasm** *is the water-based material of a eukaryotic cell that lies between the nucleus and the outer membrane of the cell.* Within the cytoplasm are several kinds of small structures called *organelles.* An **organelle** *is a minute structure within the cytoplasm of a cell that carries out a specific cellular function.* The organelles are surrounded by the *cytosol.* The **cytosol** *is the water-based fluid part of the cytoplasm of a cell.*

Eukaryotic and prokaryotic cells differ in that the former contain a well-defined nucleus, set off from the rest of the cell by a membrane.

Three important types of organelles are ribosomes, lysosomes, and mitochondria. Ribosomes were encountered in the last chapter; they are the sites where protein synthesis occurs (Section 11.12). A **lysosome** *is an organelle that contains hydrolytic enzymes needed for cellular rebuilding, repair, and degradation.* Some lysosome enzymes hydrolyze proteins to amino acids; others hydrolyze polysaccharides to monosaccharides. Bacteria and viruses "trapped" by the body's immune system (Section 9.18) are degraded and destroyed by enzymes from lysosomes.

A protective mechanism exists to prevent lysosome enzymes from destroying the cell in which they are found if they should be accidently released (via membrane rupture or leakage). The optimum pH (Section 10.6) for lysosome enzyme activity is 4.8. The cytoplasmic pH of 7.0–7.3 renders such enzymes inactive.

A **mitochondrion** *is an organelle that is responsible for the generation of most of the energy for a cell.* Much of the discussion of this chapter deals with the energy-producing chemical reactions that occur within mitochondria.

Mitochondria are sausage-shaped organelles containing both an *outer membrane* and a *multifolded inner membrane* (see Figure 12.3). The outer membrane, which is about 50% lipid and 50% protein, is freely permeable to small molecules. The inner membrane, which is about 20% lipid and 80% protein, is highly impermeable to most substances. The nonpermeable nature of the inner membrane divides a mitochondrion into two separate compartments—an interior region

Mitochondria, pronounced my-toe-KON-dree-ah, is plural. The singular form of the term is *mitochondrion.* The threadlike shape of the inner membrane of the mitochondria is responsible for this organelle's name; *mitos* is Greek for "thread," and *chondrion* is Greek for "granule."

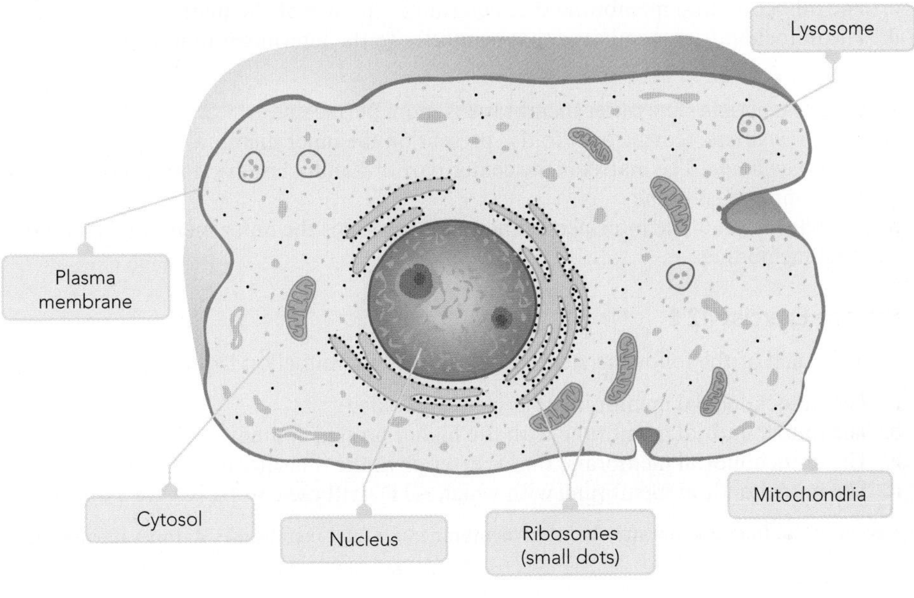

Figure 12.2 A schematic representation of a eukaryotic cell with selected internal components identified.

Figure 12.3

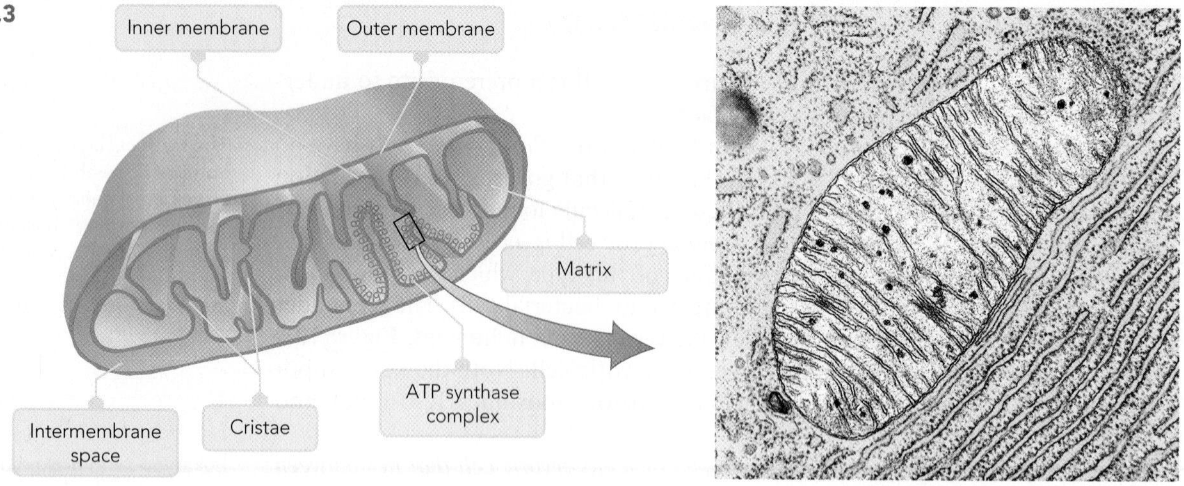

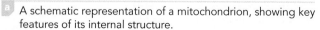

a A schematic representation of a mitochondrion, showing key features of its internal structure.

b An electron micrograph showing the ATP synthase knobs extending into the matrix.

called the *matrix* and the region between the inner and outer membranes, called the *intermembrane space.* The folds of the inner membrane that protrude into the matrix are called *cristae.*

The invention of high-resolution electron microscopes allowed researchers to see the interior structure of the mitochondrion more clearly and led to the discovery, in 1962, of small spherical knobs attached to the cristae called *ATP synthase complexes.* As their name implies, these relatively small knobs, which are located on the matrix side of the inner membrane, are responsible for ATP synthesis, and their association with the inner membrane is critically important for this task. More will be said about ATP in the next section.

EXAMPLE 12.2 Recognizing Structural Characteristics of a Mitochondrion

Identify each of the following structural features of a mitochondrion.

a. The more permeable of the two mitochondrial membranes
b. The mitochondrial membrane that has cristae
c. The mitochondrial membrane that determines the size of the matrix
d. The mitochondrial membrane that is interior to the intermembrane space

Solution

a. *Outer membrane.* The outer membrane is more permeable.
b. *Inner membrane.* Cristae are "folds" present on the inner membrane.
c. *Inner membrane.* The matrix is the center part of a mitochondrion and is surrounded by the inner membrane.
d. *Inner membrane.* The intermembrane space separates the outer membrane from the inner membrane.

Practice Exercise 12.2

Identify each of the following structural features of a mitochondrion.

a. The mitochondrial membrane that is highly folded.
b. The mitochondrial membrane with the higher protein content.
c. The mitochondrial membrane that is exterior to the intermembrane space.
d. The mitochondrial membrane with which ATP synthase complexes are associated.

Answers: **a.** Inner membrane; **b.** Inner membrane; **c.** Outer membrane; **d.** Inner membrane

12.3 Important Nucleotide-Containing Compounds in Metabolic Pathways

As a prelude to an overview presentation (Section 12.5) of the metabolic processes by which our food is converted to energy, several compounds that repeatedly function as key intermediates in these metabolic pathways will be considered. Knowing about these compounds will make it easier to understand the details of metabolic pathways. The compounds to be discussed all have *nucleotides* (Section 11.2) as part of their structures.

Nucleotides, besides being the monomer units from which nucleic acids are made, are also present in several *nonpolymeric* molecules that are important in energy production in living things.

Adenosine Phosphates (ATP, ADP, and AMP)

Several adenosine phosphates exist. Of importance in metabolism are adenosine *mono*phosphate (AMP), adenosine *di*phosphate (ADP), and adenosine *tri*phosphate (ATP). AMP is one of the nucleotides present in RNA molecules (Section 11.2). ADP and ATP differ structurally from AMP only in the number of phosphate groups present. Block structural diagrams for these three adenosine phosphates follow.

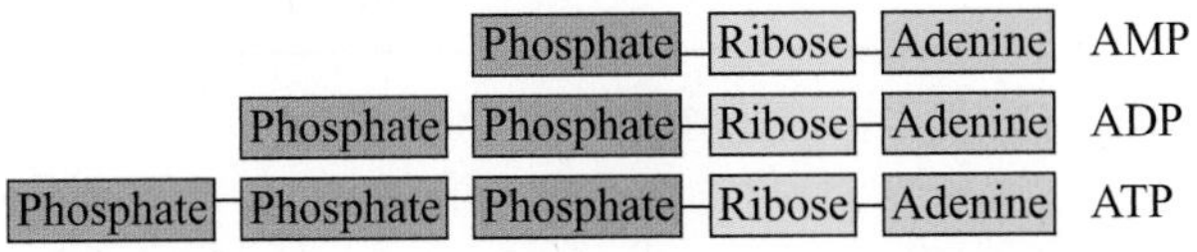

Figure 12.4 shows actual structural formulas for these three adenosine phosphates. Highlighted within the structural formulas are the bonds that the phosphate groups participate in. The phosphate-ribose bond is a *phosphoester bond,* and the phosphate-phosphate bonds are *phosphoanhydride bonds.* The word *anhydride* present in the term *phosphoanhydride* refers to the production of (loss of) a molecule of water when two phosphate groups bond to each other. A **phosphoanhydride bond** *is the chemical bond formed when two phosphate groups react with each other and a water molecule is produced.*

In Chapter 5 when inorganic phosphate esters were considered (Section 5.19), the term *phosphoryl group* was introduced. A phosphoryl group, which has the formula PO_3^{2-}, is the functional group derived from a phosphate ion when the latter becomes part of another molecule. As shown in Figure 12.4, ATP contains three phosphoryl groups, ADP two phosphoryl groups, and AMP one phosphoryl group.

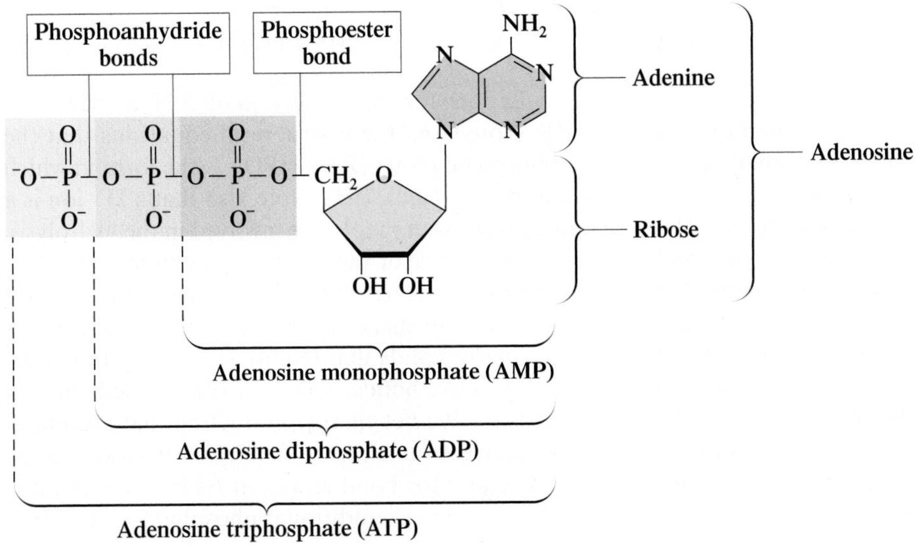

Figure 12.4 Structures of the various phosphate forms of adenosine.

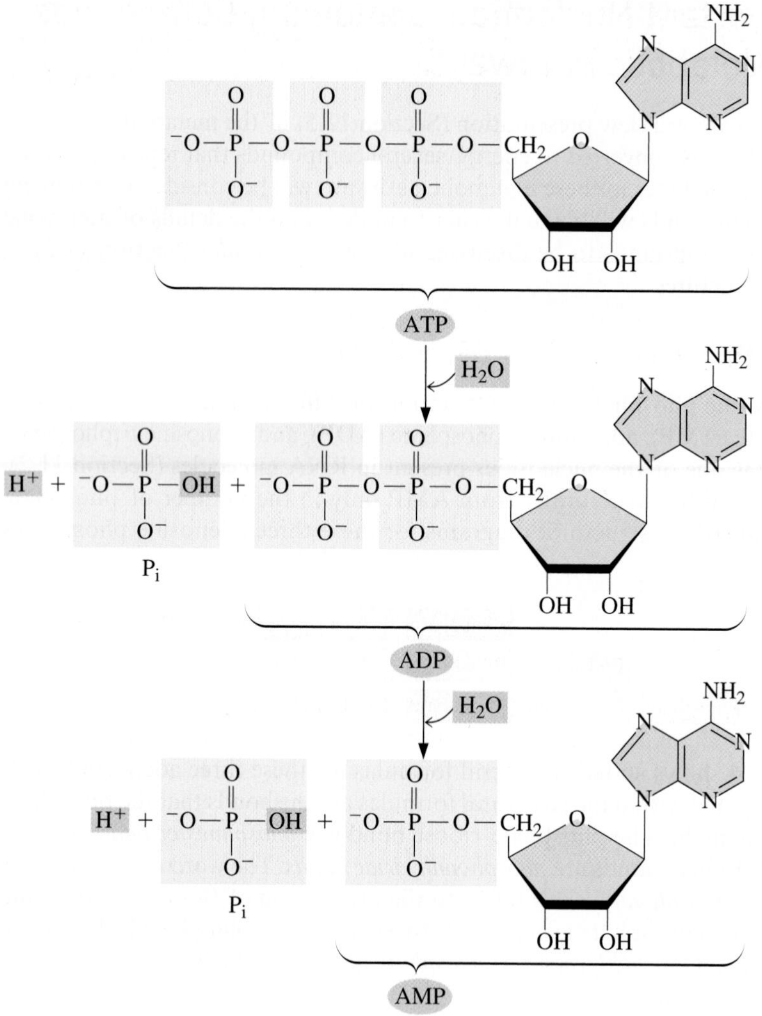

Figure 12.5 Structural equations for the hydrolysis of ATP to ADP and the hydrolysis of ADP to AMP.

ATP and ADP molecules readily undergo hydrolysis reactions in which phosphate groups (P_i, inorganic phosphate) are released.

$$\text{ATP} + H_2O \longrightarrow \text{ADP} + P_i + H^+ + \text{energy}$$
$$\text{ADP} + H_2O \longrightarrow \text{AMP} + P_i + H^+ + \text{energy}$$
$$\text{ATP} + 2H_2O \longrightarrow \text{AMP} + 2P_i + 2H^+ + \text{energy}$$

Figure 12.5 gives structural equations for the conversion of ATP to ADP and the conversion of ADP to AMP. Note from these structural equations that the chemical formula for a released phosphate group (P_i) is HPO_4^{2-} (the form in which phosphate ion exists in solution at physiological pH). Note also that a H^+ ion is a product of the hydrolysis; its source is the water molecule involved in the hydrolysis. The reacting water molecule is also the source of the OH unit present in P_i.

In metabolic pathways in which they are involved, the adenosine phosphates continually change back and forth among the various forms:

$$\text{ATP} \rightleftharpoons \text{ADP} \rightleftharpoons \text{AMP}$$

The preceding hydrolysis reactions are energy-producing reactions that are used to drive cellular processes that require energy input. The phosphoanhydride bonds in ATP and ADP are *very reactive* bonds that require less energy than normal to break. The presence of such reactive bonds, which are often called *strained bonds* (see Section 12.5), is the basis for the net energy production that accompanies hydrolysis. Greater-than-normal electron–electron repulsive forces at specific locations within a molecule are the cause for bond strain; in ATP and ADP, it is the highly electronegative oxygen atoms in the additional phosphate groups that cause the increased repulsive strain.

A typical cellular reaction in which ATP functions as both a source of a phosphate group and a source of energy is the conversion of glucose to glucose-6-phosphate, a reaction that is the first step in the process of glycolysis (Section 13.2).

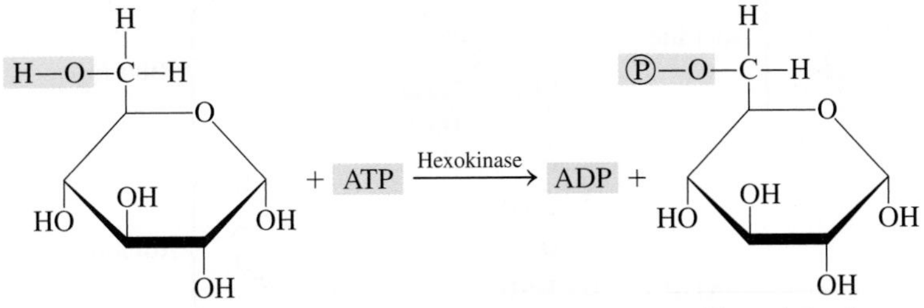

Glucose Glucose 6-phosphate

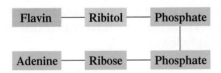

The symbol Ⓟ is a shorthand notation for a PO_3^{2-} unit.

ATP is not the only nucleotide triphosphate present in cells, although it is the most prevalent. The other nitrogen-containing bases associated with nucleotides (Section 11.2) are also present in triphosphate form. Uridine triphosphate (UTP) is involved in carbohydrate metabolism, guanosine triphosphate (GTP) participates in protein and carbohydrate metabolism, and cytidine triphosphate (CTP) is involved in lipid metabolism.

Flavin Adenine Dinucleotide (FAD, $FADH_2$)

Flavin adenine dinucleotide (FAD) is a coenzyme (Section 10.2) required in numerous metabolic redox reactions. Structurally, FAD can be visualized as containing either three subunits or six subunits. A block diagram of FAD from the three-subunit viewpoint is

Flavin — **Ribitol** — **ADP**

The flavin and ribitol subunits in this structure together constitute the B vitamin riboflavin (Section 10.14). The coenzyme FAD is thus one of the biochemically active forms of riboflavin (Section 10.14); the activating factor is the ADP subunit.

The block diagram for FAD from the six-subunit viewpoint is

Flavin — **Ribitol** — **Phosphate**
|
Adenine — **Ribose** — **Phosphate**

This block diagram shows the basis for the name *f*lavin *a*denine *d*inucleotide. Ribitol is a reduced form of ribose; a —CH_2OH group is present in place of the —CHO group (Section 7.12).

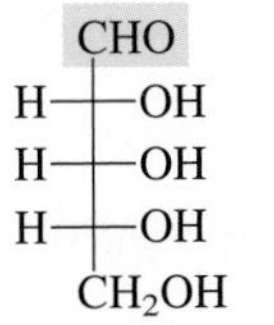

D-Ribose

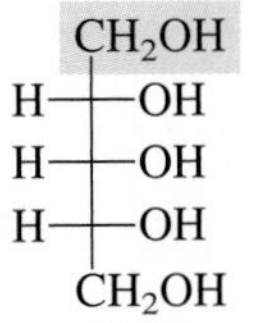

D-Ribitol

The complete structural formula of FAD is given in Figure 12.6a.

In examining what happens to coenzymes such as FAD when they participate in *redox* reactions, the definitions for oxidation and reduction used are those that relate to hydrogen atom change. These definitions, first given in Section 3.9, are:

1. Oxidation involves hydrogen atom loss.
2. Reduction involves hydrogen atom gain.

Also, in order to better keep track of changes that occur within a coenzyme as a result of oxidation or reduction, it is convenient to consider hydrogen atoms

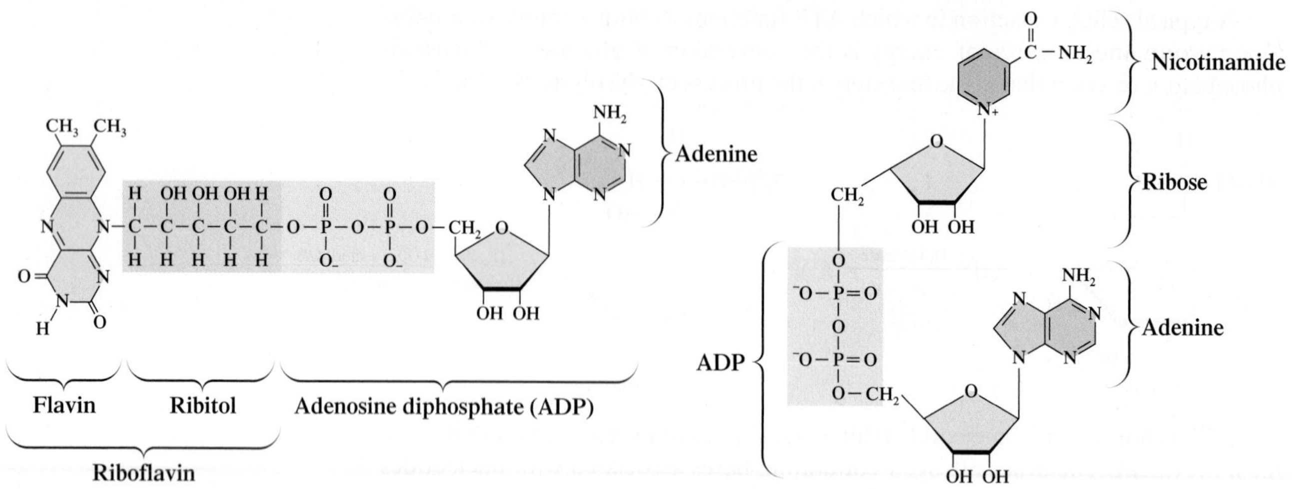

a Flavin adenine dinucleotide (FAD)

b Nicotinamide adenine dinucleotide (NAD+)

Figure 12.6 Structural formulas of the molecules flavin adenine dinucleotide, FAD (a) and nicotinamide adenine dinucleotide, NAD^+ (b).

participating in redox reactions as being composed of a proton (H^+) and an electron (e^-). As a result of this consideration, when two hydrogens participate as reactants in a redox process, they are often shown in the structural equation for the process as $2H^+ + 2e^-$ rather than as 2H.

Flavin adenine dinucleotide has two forms—an oxidized form and a reduced form. The notation FAD denotes the oxidized form, the structure of which is given in Figure 12.6a. The reduced form, denoted by the notation $FADH_2$, contains two more H atoms than the oxidized form, which is consistent with the process of reduction involving hydrogen atom gain.

The active portion of flavin adenine dinucleotide in redox reactions is the flavin subunit of the molecule. It is this portion of the molecule that undergoes change (gains hydrogen atoms) when the oxidized form (FAD) is converted to the reduced form ($FADH_2$). The actual structural change that the flavin subunit undergoes upon reduction is as follows.

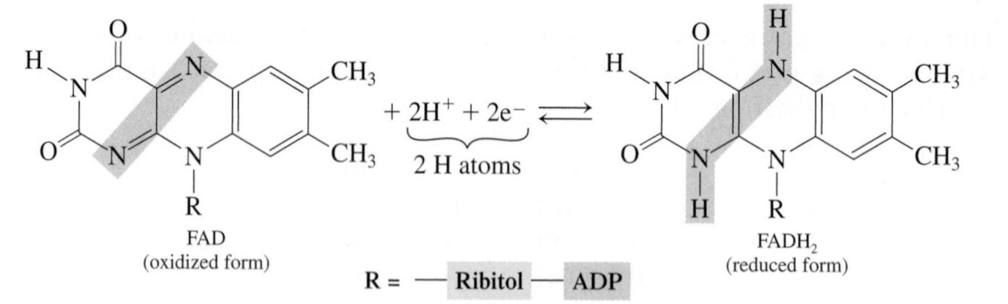

A typical cellular reaction in which FAD serves as the oxidizing agent involves a $—CH_2—CH_2—$ portion of a substrate being oxidized to produce a carbon–carbon double bond.

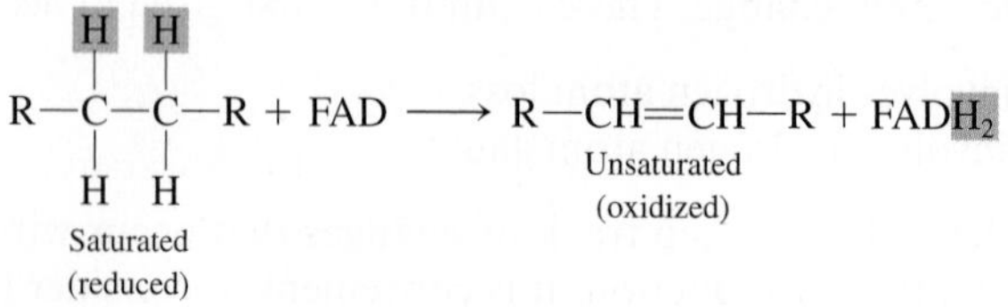

In metabolic pathways in which it is involved, flavin adenine dinucleotide continually changes back and forth between its oxidized form and its reduced form.

$$2H^+ + 2e^- + FAD \rightleftharpoons FADH_2$$

The summary equation relating the oxidized and reduced forms of flavin adenine dinucleotide is usually written as

$$\underbrace{2H^+ + 2e^-}_{\text{2 H atoms}} + \text{FAD} \rightleftharpoons \text{FADH}_2$$

Nicotinamide Adenine Dinucleotide (NAD^+, NADH)

Several parallels exist between the characteristics of nicotinamide adenine dinucleotide and those of FAD/$FADH_2$. Both have coenzyme functions in metabolic redox pathways, both have a B vitamin as a structural component, and both can be represented structurally by using a three-subunit and six-subunit formulation, and both have an oxidized and a reduced form. The notation for the oxidized form of nicotinamide adenine dinucleotide is NAD^+ and that for the reduced form is NADH. The B vitamin present in NAD^+/NADH is nicotinamide (Section 10.14).

The three-subunit block diagram for the structure of NAD^+ is

Nicotinamide — Ribose — ADP

The six-subunit block diagram, which emphasizes the dinucleotide nature of the coenzyme, as well as the origin of its name, is

Nicotinamide — Ribose — Phosphate
|
Adenine — Ribose — Phosphate

Examination of the detailed structure of NAD^+ (Figure 12.6b) reveals the basis for the positive electrical charge. The + sign refers to the positive charge on the nitrogen atom in the nicotinamide component of the structure; this nitrogen atom has four bonds instead of the usual three (Section 6.6).

The active portion of NAD^+ in metabolic redox reactions is the nicotinamide subunit of the molecule, the six-membered ring that contains the positively charged nitrogen atom. When this ring is reduced by reacting with two hydrogen atoms ($2H^+ + 2e^-$), the ring gains one H^+ ion and two electrons and one H^+ ion is left over (which enters cellular solution). In this way, NAD^+ is reduced to NADH. Note that there is no positively charged nitrogen atom in NADH because of the second electron added.

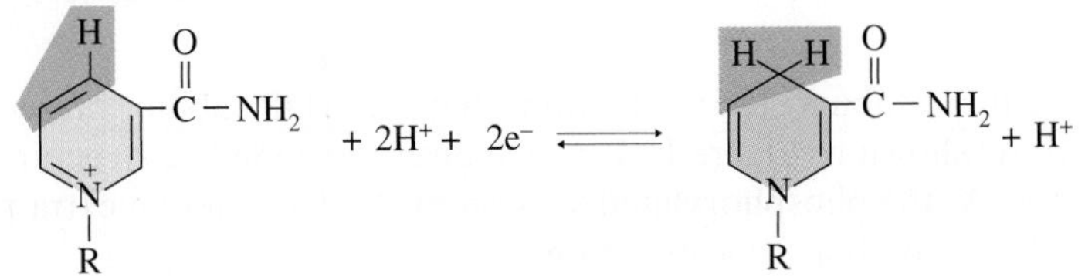

NAD^+ (oxidized form) — NADH (reduced form)

R = — Ribose — ADP

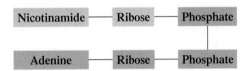

A typical cellular reaction in which NAD^+ serves as the oxidizing agent is the oxidation of a secondary alcohol to give a ketone.

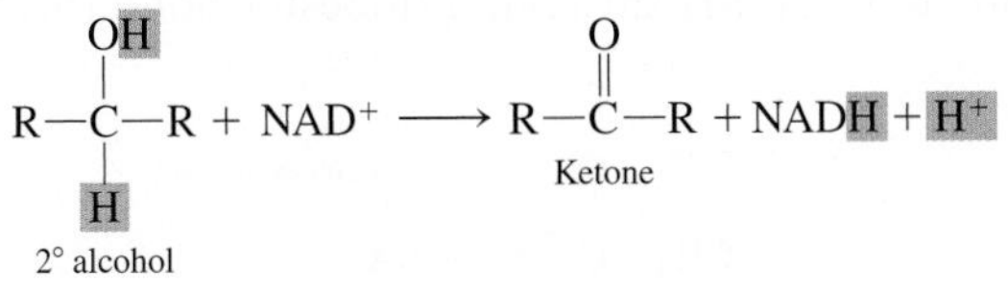

In metabolic pathways, nicotinamide adenine dinucleotide continually changes back and forth between its oxidized form and its reduced form.

$$2H^+ + 2e^- + NAD^+ \rightleftharpoons NADH + H^+$$

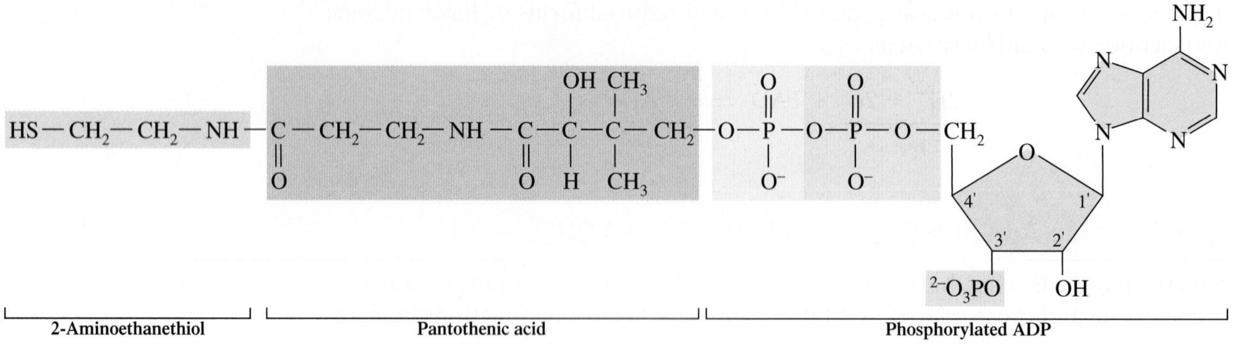

Figure 12.7 Structural formula for coenzyme A (CoA–SH).

In this reaction, one hydrogen atom of the alcohol substrate is directly transferred to NAD^+, whereas the other appears in solution as H^+ ion. Both electrons lost by the alcohol go to the nicotinamide ring in NADH. (Two electrons are required, rather than one, because of the original positive charge on NAD^+.) Thus the summary equation relating the oxidized and reduced forms of nicotinamide adenine dinucleotide is written as

$$\underbrace{2H^+ + 2e^-}_{\text{2 H atoms}} + NAD^+ \rightleftharpoons NADH + H^+$$

Coenzyme A (CoA–SH)

Coenzyme A differs from the coenzymes $FAD/FADH_2$ and NAD^+/NADH because it is not involved in oxidation/reduction processes. Its basic function is that of an acetyl group carrier. Coenzyme A is similar to $FAD/FADH_2$ and NAD^+/NADH in that they are all vitamin B-based coenzymes. Coenzyme A contains the B vitamin pantothenic acid as one of its subunits.

Another important coenzyme in metabolic pathways is coenzyme A, a derivative of the B vitamin pantothenic acid (Section 10.14). The three-subunit and six-subunit block diagrams for coenzyme A are

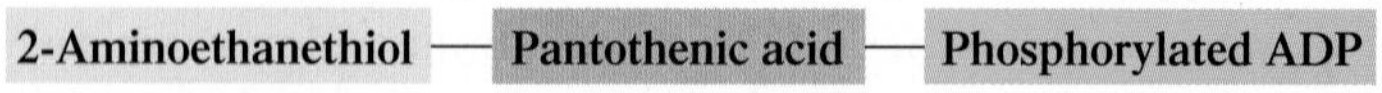

and

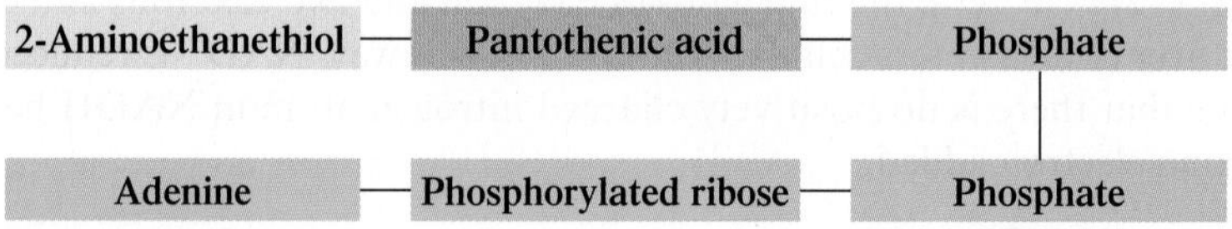

In metabolic pathways, coenzyme A is continually changing back and forth between its CoA form and its acetyl CoA form.

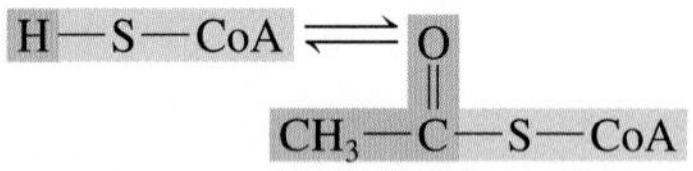

Note, in the three-subunit block diagram, that the ADP subunit present is phosphorylated. As shown in Figure 12.7, which gives the complete structural formula for coenzyme A, the phosphorylated version of ADP carries an extra phosphate group attached to carbon 3′ of its ribose.

The active portion of coenzyme A is the sulfhydryl group (—SH group; Section 3.20) in the ethanethiol subunit of the coenzyme. For this reason, the abbreviation CoA–SH is used for coenzyme A.

Think of the letter A in the name *coenzyme A* as reflecting a general metabolic function of this substance; it is the transfer of *acetyl* groups in metabolic pathways. An **acetyl group** *is the portion of an acetic acid molecule* (CH_3—COOH) *that remains after the —OH group is removed from the carboxyl carbon atom.* An acetyl group bonds to CoA–SH through a thioester bond (Section 5.16) to give acetyl CoA.

An acetyl group, which can be considered to be derived from acetic acid, has the structure

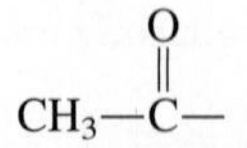

Acetyl group

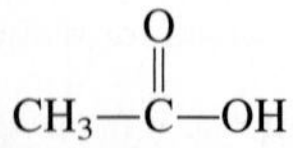

Acetic acid

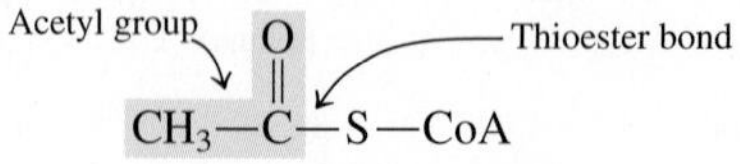

Acetyl CoA

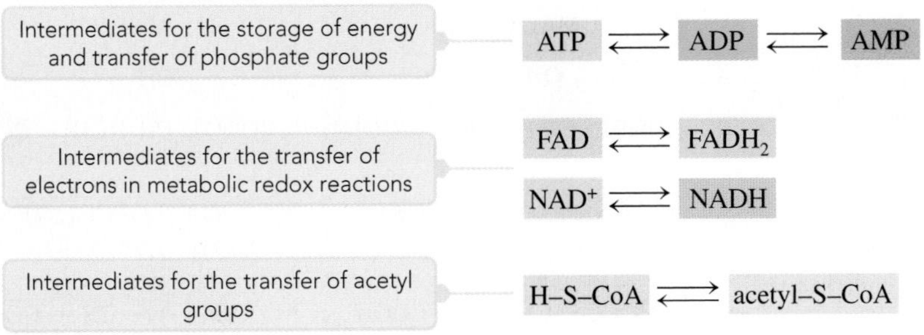

Figure 12.8 Classification of metabolic intermediate compounds in terms of function.

Classification of Metabolic Intermediate Compounds

The metabolic intermediate compounds considered in this section can be classified into three groups based on function. The classifications are:

1. Intermediates for the storage of energy and transfer of phosphate groups
2. Intermediates for the transfer of electrons in metabolic redox reactions
3. Intermediates for the transfer of acetyl groups

Figure 12.8 shows the category assignment for the intermediates previously considered.

EXAMPLE 12.3 Recognizing Relationships Among Metabolic Intermediate Compounds

Give the abbreviated formula for the following metabolic intermediate compounds.

a. The intermediate produced when FAD is reduced
b. The intermediate produced when $FADH_2$ is oxidized
c. The intermediate produced when ATP loses two phosphoryl groups
d. The intermediate produced when acetyl–S–CoA transfers an acetyl group

Solution

a. In FAD reduction, two hydrogen ions and two electrons are acquired by the FAD to produce *$FADH_2$*.
b. In $FADH_2$ oxidation, two hydrogen ions and two electrons are released by the $FADH_2$ to produce *FAD*. $FADH_2$ oxidation and FAD reduction (Part a) are reverse processes.
c. Loss of one phosphoryl group by ATP as P_i produces ADP. Loss of two phosphoryl groups by ATP produces *AMP*.
d. Release of the acetyl group from acetyl CoA (acetyl–S–CoA) produces *coenzyme A (H–S–CoA)* itself.

Practice Exercise 12.3

Give the abbreviated formula for the following metabolic intermediate compounds.

a. The intermediate produced when NADH is oxidized
b. The intermediate produced when NAD^+ is reduced
c. The intermediate produced when a phosphate group is added to AMP
d. The intermediate produced when CoA–S–H bonds to an acetyl group

Answers: **a.** NAD^+; **b.** NADH; **c.** ADP; **d.** Acetyl–S–CoA

12.4 Important Carboxylate Ions in Metabolic Pathways

In the previous section, it was noted that knowing about several nucleotide-containing compounds that function as key intermediates in metabolic pathways makes it easier to understand the yet-to-come details of metabolic processes. In a like manner, knowing about several structurally related polyfunctional carboxylate ions (Section 5.8) will facilitate a better understanding of the details of metabolic processes.

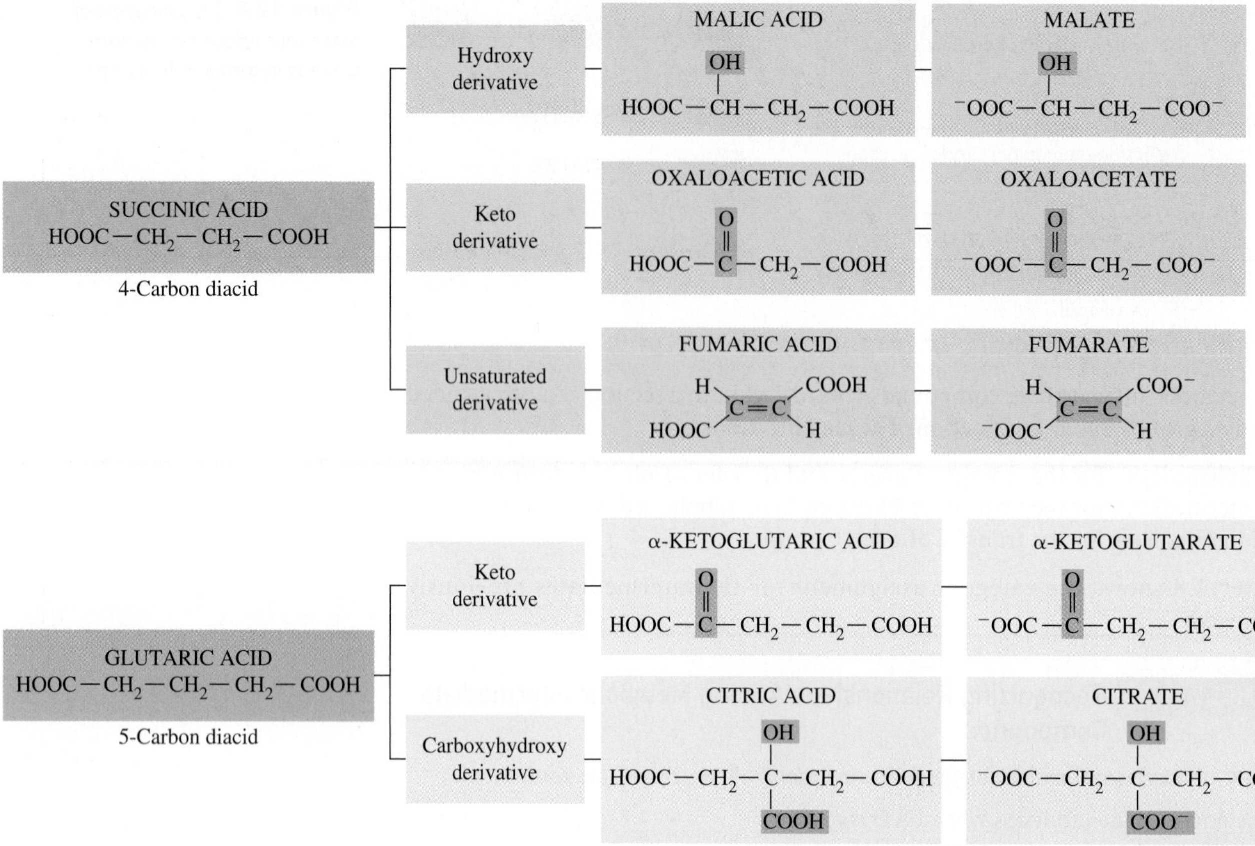

Figure 12.9 Structural formulas for polyfunctional carboxylate ions that serve as substrates for frequently encountered metabolic reactions.

Five polyfunctional carboxylate ions serve as substrates for enzymes in the metabolic reactions to be considered later in this chapter. As shown in Figure 12.9, these five carboxylate ions have structures related to just two simple carboxylic acids—succinic acid (the four-carbon dicarboxylic acid) and glutaric acid (the five-carbon dicarboxylic acid). These two "parent acids" were previously considered in Section 5.3.

The following structural relationships among these five carboxylate ions are of importance.

1. The three carboxylate ions derived from succinic acid all contain four carbon atoms, and all have a charge of -2.
2. The positioning of the double bond in the fumarate ion is such that *cis-trans* isomerism is possible. It is the *trans* isomer that is involved in metabolic reactions.
3. The two carboxylate ions derived from glutaric acid differ in both the number of carbon atoms present and ion charge. The α-ketoglutarate ion is a five-carbon species that possesses a -2 charge. The citrate ion is a six-carbon species that possesses a -3 charge. These differences stem from the citrate ion having three carboxyl groups rather than the two present in α-ketoglutarate.

12.5 High-Energy Phosphate Compounds

In the previous two sections, it was noted that knowing about several key intermediate compounds in metabolic reactions makes it easier to understand the yet-to-come details of metabolic processes. In like manner, knowing about a particular type of bond present in certain phosphate-containing metabolic intermediates makes the details of metabolic processes easier to understand.

Several phosphate-containing compounds found in metabolic pathways are known as high-energy compounds. A **high-energy compound** *is a compound that has* *a greater free energy of hydrolysis than that of a typical compound.* High-energy compounds differ from other compounds in that they contain one or more *very reactive* bonds, often called *strained bonds.* The energy required to break these strained bonds during hydrolysis is less than that generally required to break a chemical bond. Consequently, the balance between the energy needed to break bonds in the reactants and that released by bond formation in the products is such that more than the typical amount of free energy is released during the hydrolysis reaction.

In the definition for a high-energy compound, the term *free energy* rather than simply *energy* was used. Free energy is the amount of energy released by a chemical reaction that is actually available for further use at a given temperature and pressure. In reality, the energy released in a chemical reaction is divisible into two parts. One part, lost as heat, is not available for further use. The other part, the free energy, is available for further use; in cells, it can be used to "drive" reactions that require energy.

Greater-than-normal electron–electron repulsive forces at specific locations within a molecule are the cause of bond strain. Highly electronegative atoms and/or highly charged atoms occurring together in a molecule cause increased repulsive forces and thus increase bond strain.

Futher details concerning bond strain as it is related to phosphate-containing organic molecules involved in metabolic pathways are now considered. The parent molecule for phosphate groups is phosphoric acid, H_3PO_4, a weak triprotic inorganic acid. This acid exists in aqueous solution in several forms, the dominant form at cellular pH being HPO_4^{2-} ion.

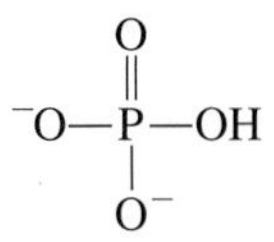

In a chemical reaction, the energy balance between bond breaking among reactants (energy input) and new bond formation among products (energy release) determines whether there is a net loss or a net gain of energy.

Diphosphate and triphosphate ions can also exist in cellular fluids.

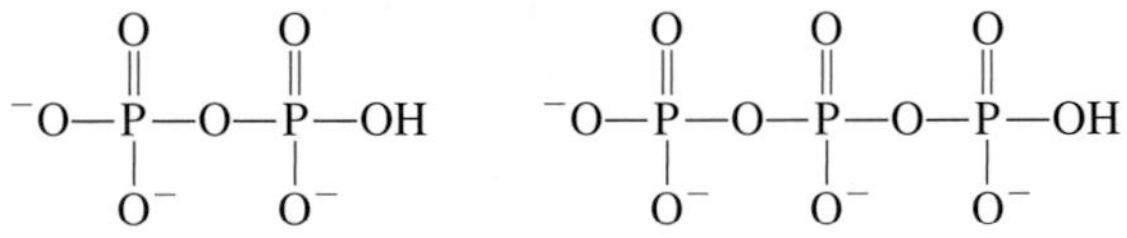

Note the presence in these three phosphate structures of highly electronegative oxygen atoms, many of which bear negative charges. The factors that can produce bond strain are present when phosphates (mono-, di-, and tri-) are bonded to certain organic molecules.

Table 12.1 gives the structures of commonly encountered phosphate-containing compounds, as well as a numerical parameter—the free energy of hydrolysis—that can be considered a measure of the extent of bond strain in the molecules. The more negative the free energy of hydrolysis, the greater the bond strain. A free-energy release greater than 6.0 kcal/mole is generally considered indicative of bond strain. In Table 12.1, strained bonds within the molecules are noted with a squiggle (~), a notation often employed to denote strained bonds.

The designation *high-energy compound* does not mean that a compound is different from other compounds in terms of bonding. High-energy compounds obey the normal rules for chemical bonding. The only difference between such compounds and other compounds is the presence of one or more *strained bonds.* The breaking of such bonds requires lower-than-normal amounts of energy.

12.6 An Overview of Biochemical Energy Production

The energy needed to run the human body is obtained from ingested food through a multistep process that involves several different catabolic pathways. There are four general stages in the biochemical energy production process, and numerous reactions are associated with each stage.

Stage 1: The first stage, **digestion**, begins in the mouth (saliva contains starch-digesting enzymes), continues in the stomach (gastric juices), and is completed in the small intestine (the majority of digestive enzymes and bile salts). The end products of digestion—glucose and other monosaccharides from carbohydrates, amino acids from proteins, and fatty acids and glycerol from fats and oils—are small enough to pass across intestinal membranes and into the blood, where they are transported to the body's cells.

The first stage of biochemical energy production, digestion, is not considered part of metabolism because it is extracellular. Metabolic processes are intracellular.

Stage 2: The second stage, **acetyl group formation**, involves numerous reactions, some of which occur in the cytosol of cells and some in cellular mitochondria. The small molecules from digestion are further oxidized during this stage. Primary products include two-carbon acetyl units (which become attached to coenzyme A to give acetyl CoA) and the reduced coenzyme NADH.

The $—PO_3^{2-}$ group as part of a larger organic phosphate molecule is referred to as a *phosphoryl group.*

Table 12.1 Free Energies of Hydrolysis of Common Phosphate-Containing Metabolic Compounds

Type	Example	Free Energy of Hydrolysis (kcal/mole)
enol phosphates 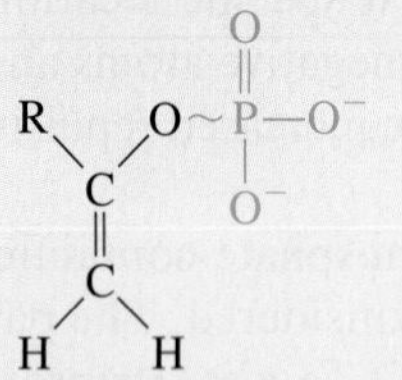	phosphoenolpyruvate	−14.8
acyl phosphates 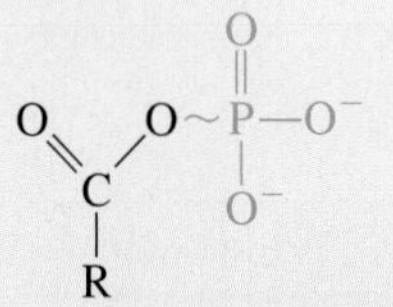	1,3-bisphosphoglycerate acetyl phosphate	−11.8 −11.3
guanidine phosphates 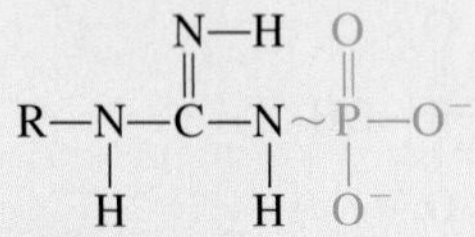	creatine phosphate arginine phosphate	−10.3 −9.1
triphosphates 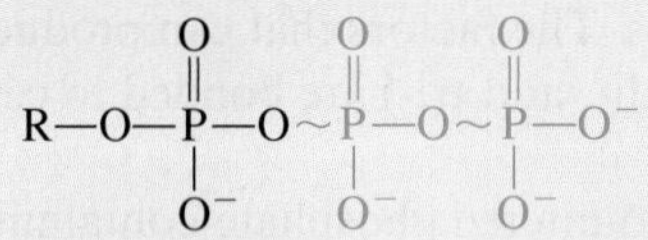	$ATP \longrightarrow AMP + PP_i$* $ATP \longrightarrow ADP + P_i$*	−7.7 −7.5
diphosphates 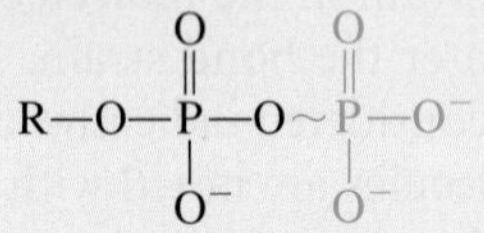	$PP_i \longrightarrow 2P_i$ $ADP \longrightarrow AMP + P_i$	−7.8 −7.5
sugar phosphates 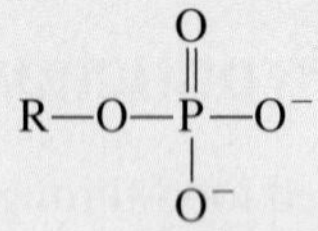	glucose 1-phosphate fructose 6-phosphate $AMP \longrightarrow$ adenosine $+ P_i$ glucose 6-phosphate glycerol 3-phosphate	−5.0 −3.8 −3.4 −3.3 −2.2

*The notation P_i is used as a general designation for any free monophosphate species present in cellular fluid. Free diphosphate ions are designated as PP_i ("i" stands for *inorganic*).

Stage 3: The third stage, the **citric acid cycle**, occurs inside mitochondria. Here acetyl groups are oxidized to produce CO_2 and energy. Some of the energy released by these reactions is lost as heat, and some is carried by the reduced coenzymes NADH and $FADH_2$ to the fourth stage. The CO_2 that is exhaled as part of the breathing process comes primarily from this stage.

Stage 4: The fourth stage, the **electron transport chain and oxidative phosphorylation**, also occurs inside mitochondria. NADH and $FADH_2$ supply the "fuel" (hydrogen ions and electrons) needed for the production of ATP molecules, the primary energy carriers in metabolic pathways. Molecular O_2, inhaled via breathing, is converted to H_2O in this stage.

CHEMISTRY AT A GLANCE — Simplified Summary of the Four Stages of Biochemical Energy Production

STAGE 1

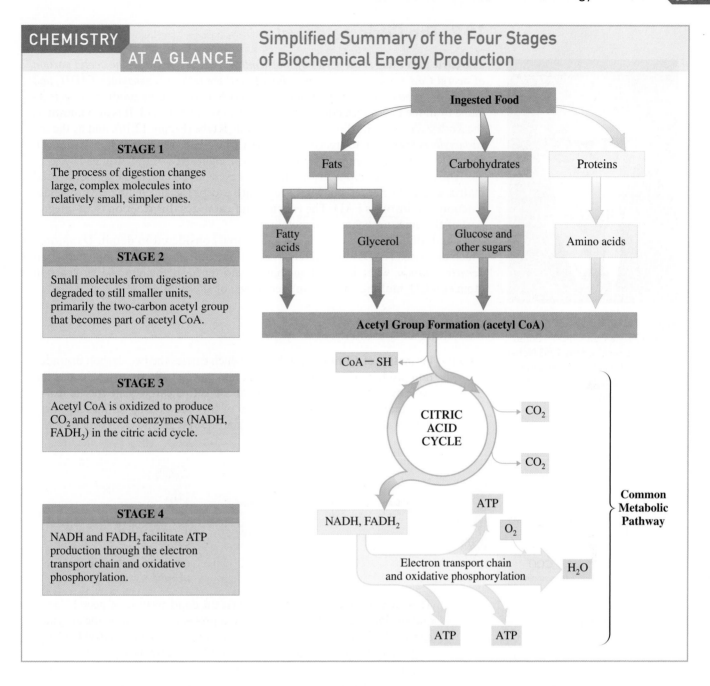

The process of digestion changes large, complex molecules into relatively small, simpler ones.

STAGE 2

Small molecules from digestion are degraded to still smaller units, primarily the two-carbon acetyl group that becomes part of acetyl CoA.

STAGE 3

Acetyl CoA is oxidized to produce CO_2 and reduced coenzymes (NADH, $FADH_2$) in the citric acid cycle.

STAGE 4

NADH and $FADH_2$ facilitate ATP production through the electron transport chain and oxidative phosphorylation.

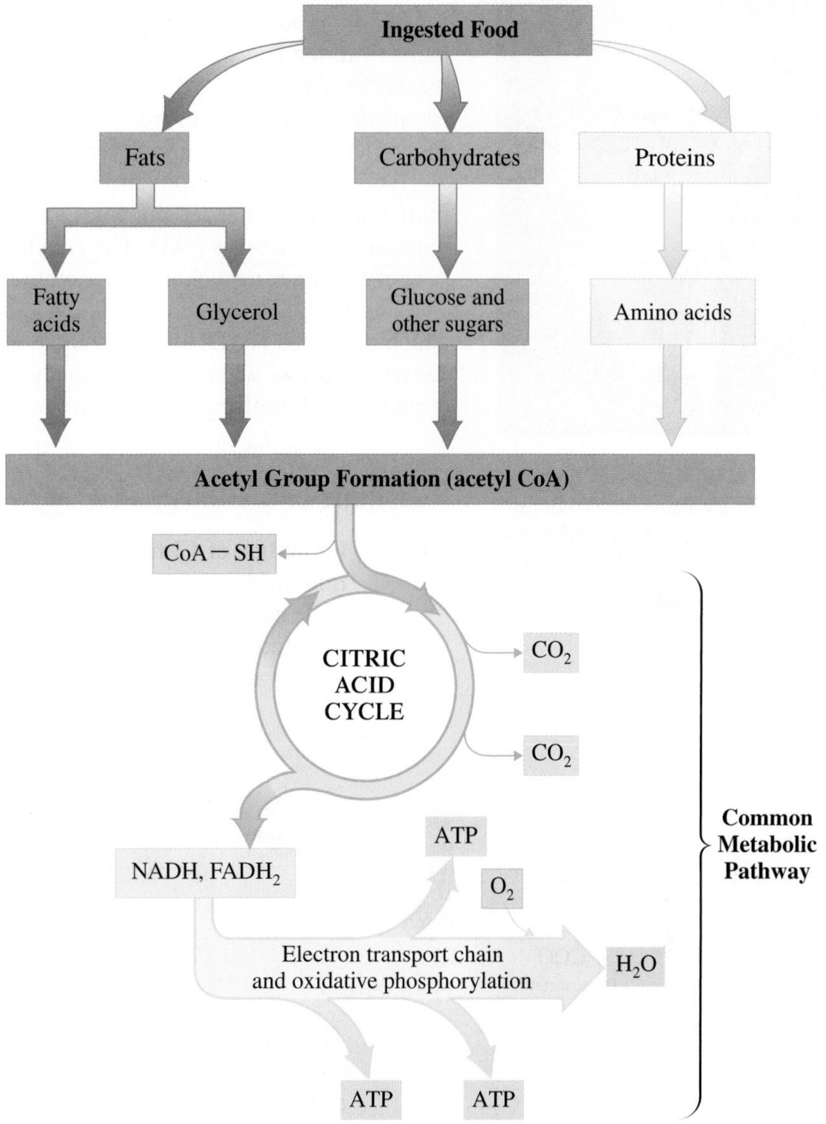

The reactions in stages 3 and 4 are the same for all types of foods (carbohydrates, fats, proteins). These reactions constitute the common metabolic pathway. The **common metabolic pathway** *is the sum total of the biochemical reactions of the citric acid cycle, the electron transport chain, and oxidative phosphorylation.* The remainder of this chapter deals with the common metabolic pathway. The reactions of stages 1 and 2 of biochemical energy production differ for different types of foodstuffs. They are discussed in Chapters 13–15, which cover the metabolism of carbohydrates, fats (lipids), and proteins, respectively.

The Chemistry at a Glance feature above summarizes the four general stages in the process of production of biochemical energy from ingested food. This diagram is a *very simplified* version of the "energy generation" process that occurs in the human body, as will become clear from the discussions presented in later sections of this chapter, which give further details of the process.

12.7 The Citric Acid Cycle

Hulton Archive/Getty Images

Figure 12.10 Hans Adolf Krebs (1900–1981), a German-born British biochemist, received the 1953 Nobel Prize in medicine for establishing the relationships among the different compounds in the cycle that carries his name, the Krebs cycle.

The **citric acid cycle** *is the series of biochemical reactions in which the acetyl portion of acetyl CoA is oxidized to carbon dioxide and the reduced coenzymes $FADH_2$ and NADH are produced.* This cycle, stage 3 of biochemical energy production, gets its name from the first intermediate product in the cycle, citric acid. It is also known as the *Krebs cycle,* after its discoverer Hans Adolf Krebs (Figure 12.10), and as the *tricarboxylic acid cycle,* in reference to the three carboxylate groups present in citric acid. Figure 12.11 lists the compounds produced in all eight steps of the citric acid cycle.

The chemical reactions of the citric acid cycle take place in the mitochondrial matrix where the needed enzymes are found, except the succinate dehydrogenase reaction that involves FAD. The enzyme that catalyzes this reaction is an integral part of the inner mitochondrial membrane.

The individual steps of the cycle are now considered in detail. *Oxidation,* which produces NADH or $FADH_2$, is encountered in four of the eight steps, and *decarboxylation,* wherein a carbon chain is shortened by the removal of a carbon atom as a CO_2 molecule, is encountered in two of the eight steps.

Reactions of the Citric Acid Cycle

Step 1: **Formation of Citrate.** Acetyl CoA, which carries the two-carbon degradation product of carbohydrates, fats, and proteins (Section 12.6), enters the cycle by combining with the four-carbon keto dicarboxylate species oxaloacetate. This results in the transfer of the acetyl group from coenzyme A to oxaloacetate, producing the C_6 citrate species and free coenzyme A.

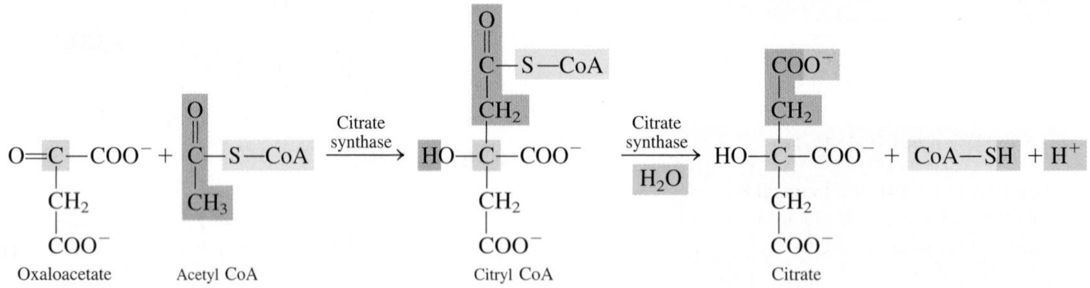

The formation of citryl CoA involves addition of acetyl CoA to the carbon–oxygen double bond. A hydrogen atom of the acetyl —CH_3 group adds to the oxygen atom of the double bond, and the remainder of the acetyl CoA adds to the carbon atom of the double bond (Section 4.10).

There are two parts to the reaction: (1) the condensation of acetyl CoA and oxaloacetate to form citryl CoA, a process catalyzed by the enzyme *citrate synthase* and (2) hydrolysis of the thioester bond in citryl CoA to produce CoA—SH and citrate, also catalyzed by the enzyme *citrate synthase.*

At cellular pH, citric acid is actually present as citrate ion. Despite this, the name of the cycle is the citric acid cycle, which references the molecular, rather than ionic, form of the substance.

Step 2: **Formation of Isocitrate.** Citrate is converted to its less symmetrical isomer isocitrate in an isomerization process that involves a dehydration followed by a hydration, both catalyzed by the enzyme *aconitase.* The net result of these reactions is that the —OH group from citrate is moved to a different carbon atom.

A *synthase* is an enzyme that makes a new covalent bond during a reaction without the direct involvement of an ATP molecule.

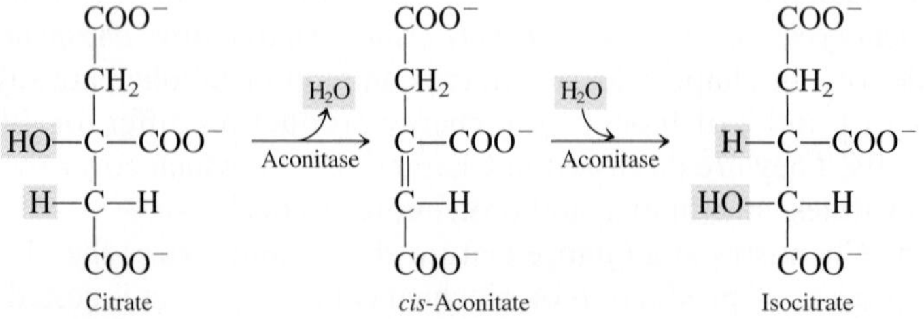

All acids found in the citric acid cycle exist as carboxylate ions (Section 5.7) at cellular pH.

Citrate is an achiral compound (Section 7.4), and isocitrate is a chiral compound with two chiral centers (four stereoisomers possible). *Aconitase* produces only one of the four stereoisomers isocitrate.

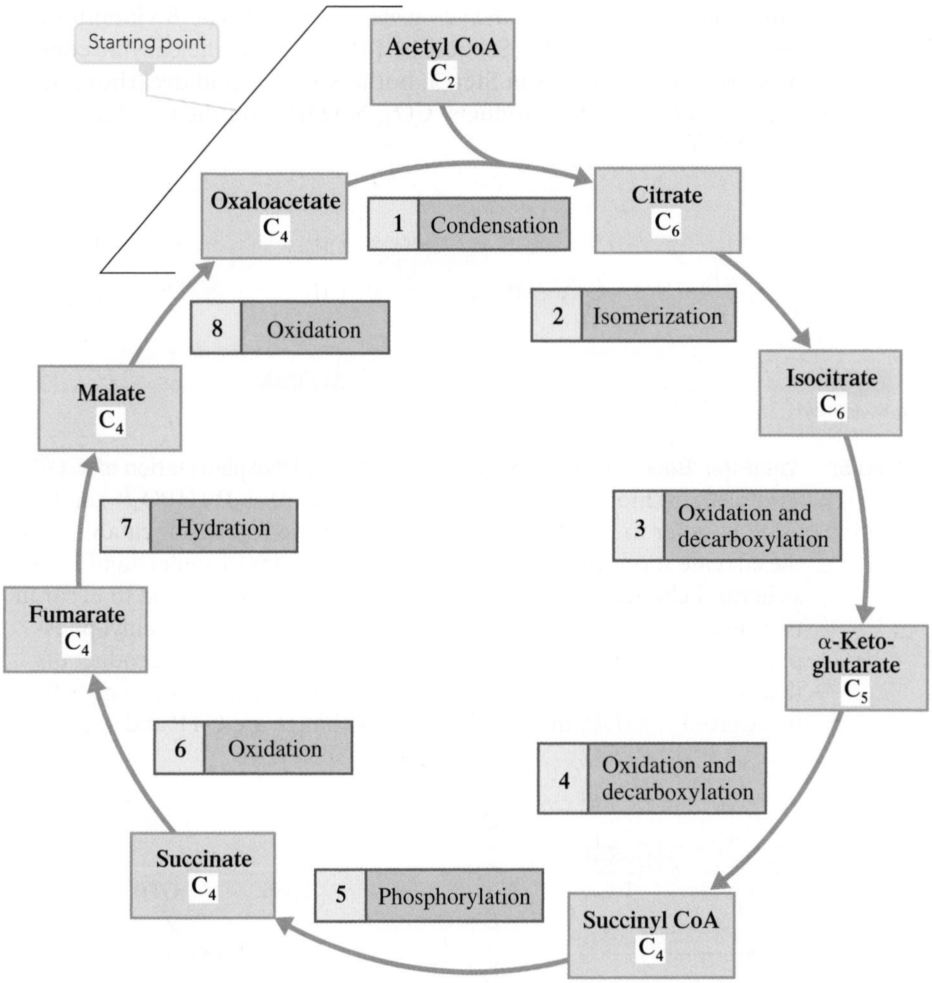

Figure 12.11 The citric acid cycle. Details of the numbered steps are given in the text.

Citrate is a tertiary alcohol and isocitrate a secondary alcohol. Tertiary alcohols are not readily oxidized; secondary alcohols are easier to oxidize (Section 3.9). The next step in the cycle involves oxidation.

Step 3: **Oxidation of Isocitrate and Formation of CO_2.** This step involves oxidation–reduction (the first of four redox reactions in the citric acid cycle) and decarboxylation. The reactants are a NAD^+ molecule and isocitrate. The reaction, catalyzed by *isocitrate dehydrogenase,* is complex: (1) Isocitrate is oxidized to a ketone (oxalosuccinate) by NAD^+, releasing two hydrogens. (2) One hydrogen and two electrons are transferred to NAD^+ to form NADH; the remaining hydrogen ion (H^+) is released. (3) The oxalosuccinate remains bound to the enzyme and undergoes decarboxylation (loses CO_2), which produces the C_5 α-ketoglutarate (a keto dicarboxylate species).

The key happenings in Step 3 are the following:

1. The hydroxyl group of isocitrate is oxidized to a ketone group.
2. NAD^+ is converted to its reduced form, NADH.
3. A carboxyl group from the original oxaloacetate is removed as CO_2.

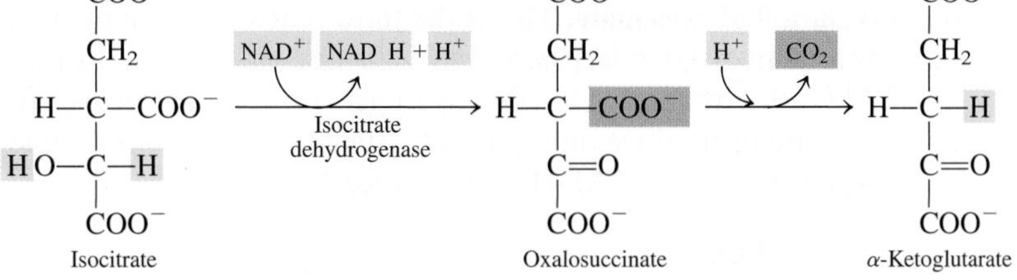

This step yields the first molecules of CO_2 and NADH in the cycle.

Step 4: **Oxidation of α-Ketoglutarate and Formation of CO_2.** This second redox reaction of the cycle involves one molecule each of NAD^+, CoA—SH, and α-ketoglutarate. The catalyst is a three-enzyme system

The CO_2 molecules produced in Steps 3 and 4 of the citric acid cycle are the CO_2 molecules exhaled in the process of respiration.

The key happenings in Step 4 are the following:

1. A second NAD^+ is converted to its reduced form, NADH.
2. A second carboxyl group is removed as CO_2.
3. Coenzyme A reacts with the decarboxylation product succinate to produce succinyl CoA, a compound with a high-energy thioester bond. This is the second involvement of a coenzyme A molecule in the cycle, the other instance occurring in Step 1.

called the *α-ketoglutarate dehydrogenase complex*. The B vitamin thiamin, in the form of TPP (Section 10.14), is part of the enzyme complex, as is Mg^{2+} ion. As in Step 3, both oxidation and decarboxylation occur. There are three products: CO_2, NADH, and the C_4 species succinyl CoA.

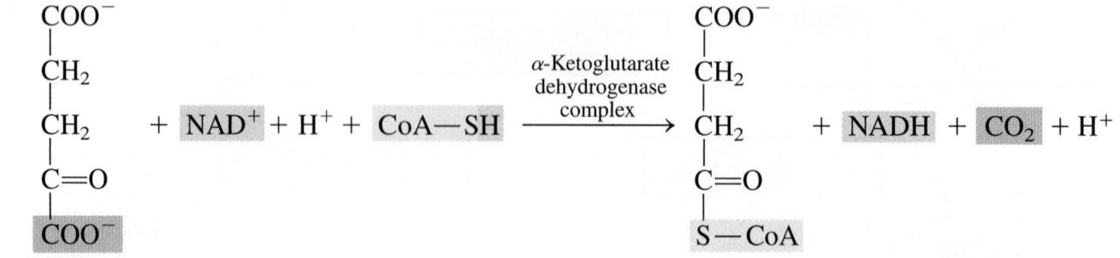

α-Ketoglutarate   Succinyl CoA

Step 5: **Thioester Bond Cleavage in Succinyl CoA and Phosphorylation of GDP.** Two reactant molecules are involved in this step—a P_i (HPO_4^{2-}) and a GDP (similar to ADP; Section 12.3). The entire reaction is catalyzed by the enzyme *succinyl-CoA synthetase*. For purposes of understanding the structural changes that occur, the reaction can be considered to occur in two steps. In the first step, succinyl CoA is converted to succinyl phosphate (a high-energy phosphate compound); CoA–SH is a product of this change. The phosphoryl group present in succinyl phosphate is then transferred to GDP; the products of this change are GTP and succinate.

The enzyme needed in Step 1 is called a *synthase*, and the enzyme for Step 5 is called a *synthatase*. The difference between a *synthase* and a *synthatase* is that the latter uses energy from the breaking of a high-energy phosphate bond, whereas the former does not require such energy.

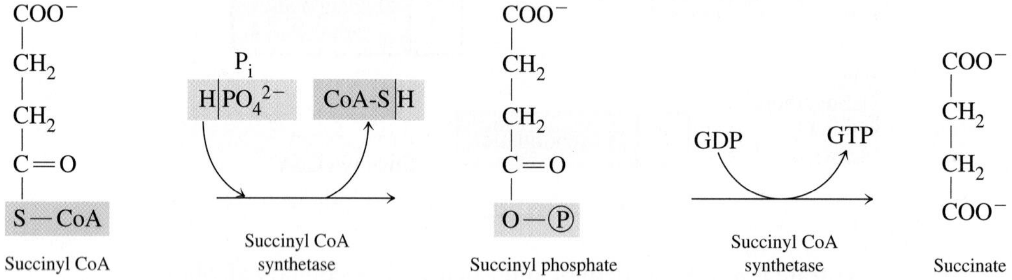

Succinyl CoA   Succinyl phosphate   Succinate

Thinking of the two steps as occurring concurrently gives the following energy analysis: When broken, the high-energy thioester bond in succinyl CoA releases energy, which is trapped by formation of GTP. The function of the GTP produced is similar to that of ATP, which is to store energy in the form of a high-energy phosphate bond (Section 12.5).

Steps 6 through 8 of the citric acid cycle involve a sequence of functional group changes that have been encountered several times in the organic sections of the text. The reaction sequence is

$$\text{Alkane} \xrightarrow[\text{(dehydrogenation)}]{\text{① Oxidation}} \text{alkene} \xrightarrow{\text{② Hydration}} \text{secondary alcohol} \xrightarrow[\text{(dehydrogenation)}]{\text{③ Oxidation}} \text{ketone}$$

Step 6: **Oxidation of Succinate.** This is the third redox reaction of the cycle. The enzyme involved is *succinate dehydrogenase,* and the oxidizing agent is FAD rather than NAD^+. Two hydrogen atoms are removed from the succinate to produce fumarate, a C_4 species with a *trans* double bond. FAD is reduced to $FADH_2$ in the process.

Fumarate, with its *trans* double bond, is an essential metabolic intermediate in both plants and animals. Its isomer, with a *cis* double bond, is called maleate, and it is toxic and irritating to tissues. *Succinate dehydrogenase* produces only the *trans* isomer of this unsaturated diacid.

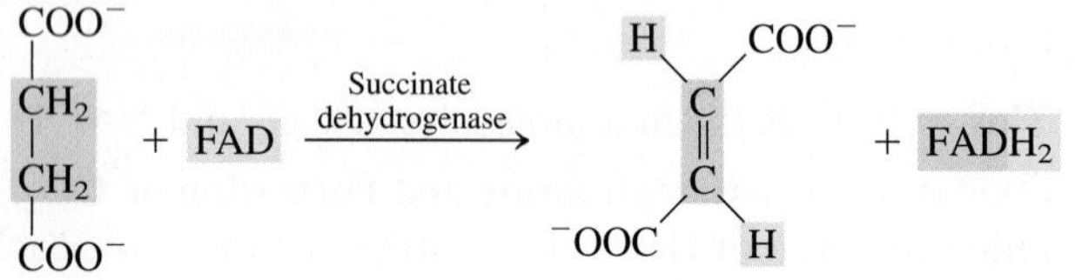

Succinate   Fumarate

Step 7: **Hydration of Fumarate.** The enzyme *fumarase* catalyzes the addition of water to the double bond of fumarate. The enzyme is stereospecific, so only the L isomer of the product malate is produced.

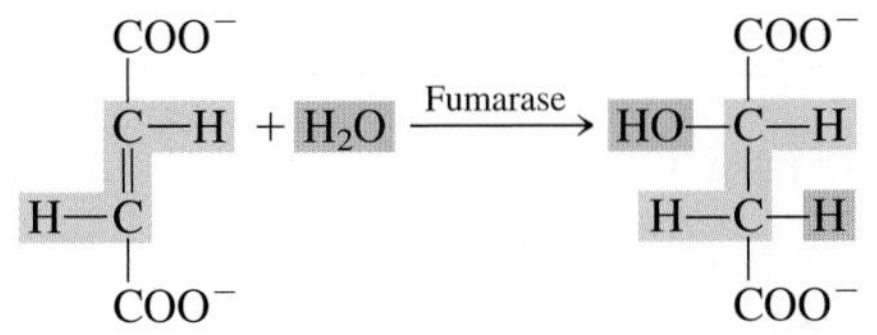

Fumarate L-Malate

Step 8: **Oxidation of L-Malate to Regenerate Oxaloacetate.** In the fourth oxidation–reduction reaction of the cycle, a molecule of NAD^+ reacts with malate, picking up two hydrogen atoms with their associated energy to form NADH + H^+. The needed enzyme is *malate dehydrogenase.* The product of this reaction is regenerated oxaloacetate, which can combine with another molecule of acetyl CoA (Step 1), and the cycle can begin again.

Step 8 is the second step of the cycle in which oxidation of a secondary alcohol occurs. Such an oxidation also occurred in Step 3.

The product from Step 8 is the starting material for Step 1. Thus the cycle can repeat itself provided that an additional acetyl CoA is also available for reaction in Step 1.

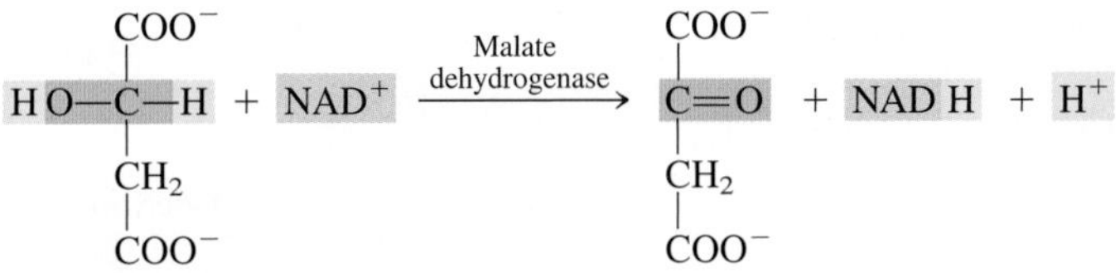

L-Malate Oxaloacetate

Summary of the Citric Acid Cycle

An overall summary equation for the citric acid cycle is obtained by adding together the individual reactions of the cycle:

$$\text{Acetyl CoA} + 3\text{NAD}^+ + \text{FAD} + \text{GDP} + \text{HPO}_4^{2-}(\text{P}_i) + 2\text{H}_2\text{O} \longrightarrow 2\text{CO}_2 + \text{CoA—SH} + 3\text{NADH} + 3\text{H}^+ + \text{FADH}_2 + \text{GTP}$$

Important features of the cycle include the following:

1. The "fuel" for the cycle is acetyl CoA, obtained from the breakdown of carbohydrates, fats, and proteins.
2. Four of the cycle reactions involve oxidation and reduction. The oxidizing agent is either NAD^+ (three times) or FAD (once). The operation of the cycle depends on the availability of these oxidizing agents.
3. In redox reactions, NAD^+ is the oxidizing agent when a carbon–oxygen double bond is formed; FAD is the oxidizing agent when a carbon–carbon double bond is formed.
4. The three NADH and one $FADH_2$ that are formed during the cycle carry electrons and H^+ to the electron transport chain (Section 12.8) through which ATP is synthesized.
5. Two carbon atoms enter the cycle as the acetyl unit of acetyl CoA, and two carbon atoms leave the cycle as two molecules of CO_2. The carbon atoms that enter and leave are not the same ones. The carbon atoms that leave during one turn of the cycle are carbon atoms that entered during the previous turn of the cycle.
6. Four B vitamins are necessary for the proper functioning of the cycle: riboflavin (in both FAD and the α-ketoglutarate dehydrogenase complex), nicotinamide (in NAD^+), pantothenic acid (in CoA—SH), and thiamine (in the α-ketoglutarate dehydrogenase complex).
7. One high-energy GTP molecule is produced by phosphorylation.

The eight B vitamins and their structures were discussed in Section 10.14.

The Chemistry at a Glance feature on the next page gives a detailed diagrammatic summary of the reactions that occur in the citric acid cycle.

CHEMISTRY AT A GLANCE Summary of the Reactions of the Citric Acid Cycle

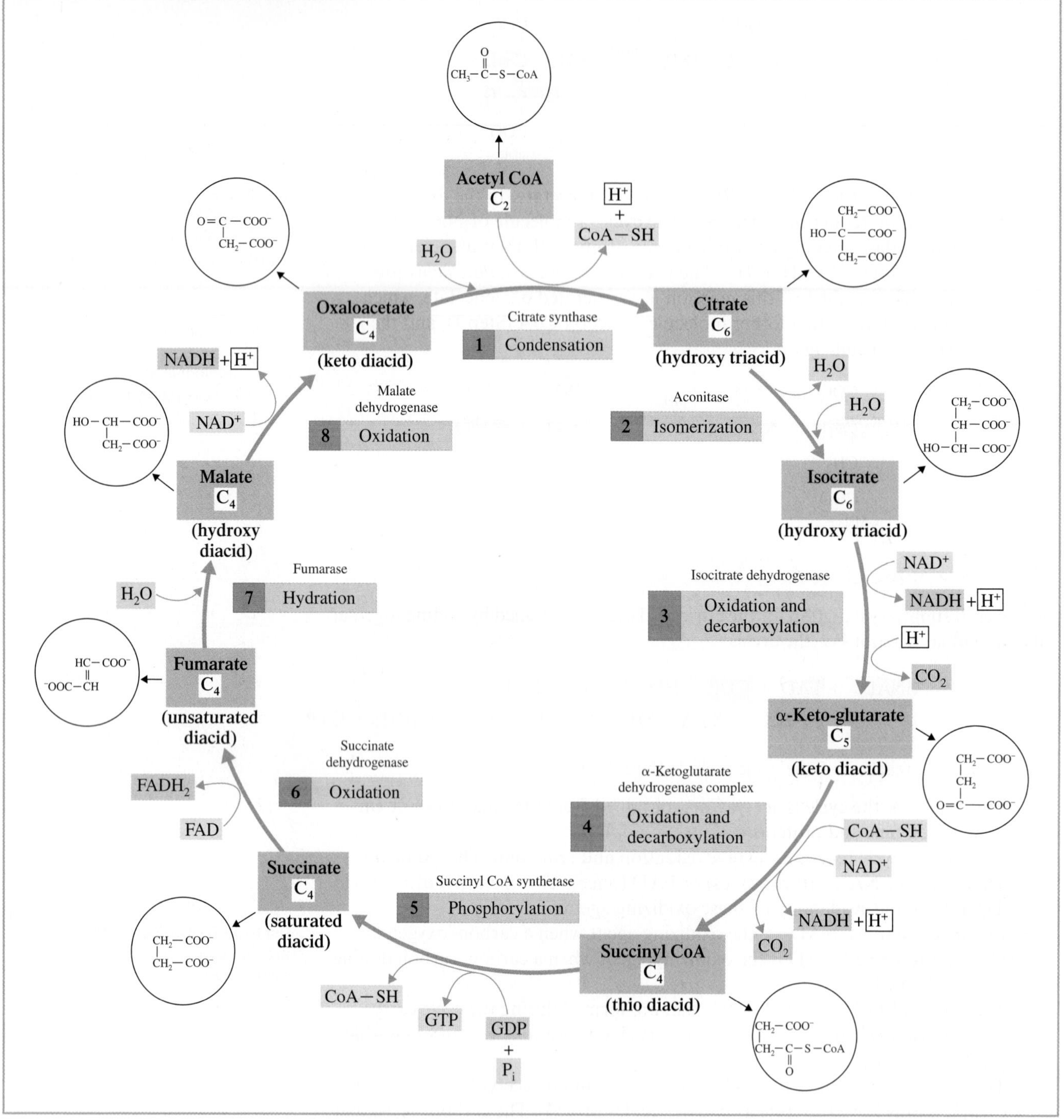

EXAMPLE 12.4 Recognizing the Reactants and Products of Various Steps in the Citric Acid Cycle

When one acetyl CoA is processed through the citric acid cycle, how many times does each of the following events occur?

a. A secondary alcohol group is oxidized to a ketone group.
b. A NADH molecule is produced.
c. A decarboxylation reaction occurs.
d. A C_6 molecule is produced.

Solution

a. *Two.* Both isocitrate (Step 3) and malate (Step 8) are secondary alcohols that are oxidized by NAD^+ to ketones.
b. *Three.* The product of the use of NAD^+ as an oxidizing agent is NADH. Oxidation using NAD^+ occurs in Steps 3, 4, and 8.
c. *Two.* In Step 3, isocitrate is decarboxylated, and in Step 4 α-ketoglutarate is decarboxylated. In each case, the product is a CO_2 molecule.
d. *Two.* Citrate and isocitrate are C_6 molecules. Citrate is produced in Step 1, and isocitrate is produced in Step 2.

Practice Exercise 12.4

When one acetyl CoA is processed through the citric acid cycle, how many times does each of the following events occur?

a. A FAD molecule is a reactant.
b. A CoA–S–H molecule is produced.
c. A dehydrogenase enzyme is needed for the reaction to occur.
d. A C_5 molecule is produced.

Answers: **a.** One (Step 6); **b.** Two (Steps 1 and 5); **c.** Four (Steps 3, 4, 6, and 8); **d.** One (Step 3)

Regulation of the Citric Acid Cycle

The rate at which the citric acid cycle operates is controlled by the body's need for energy (ATP). When the body's ATP supply is high, the ATP present inhibits the activity of citrate synthase, the enzyme in Step 1 of the cycle. When energy is being used at a high rate, a state of low ATP and high ADP concentrations, the ADP activates citrate synthase and the cycle speeds up. A similar control mechanism exists at Step 3, which involves isocitrate dehydrogenase; here NADH acts as an inhibitor and ADP as an activator.

12.8 The Electron Transport Chain

The NADH and $FADH_2$ produced in the citric acid cycle pass to the electron transport chain. The **electron transport chain** *is a series of biochemical reactions in which electrons and hydrogen ions from NADH and $FADH_2$ are passed to intermediate carriers and then ultimately react with molecular oxygen to produce water.* NADH and $FADH_2$ are oxidized in this process.

The *electron transport chain* is also frequently called the *respiratory chain.*

$$NADH + H^+ \longrightarrow NAD^+ + 2H^+ + 2e^-$$
$$FADH_2 \longrightarrow FAD + 2H^+ + 2e^-$$

Water is formed when the electrons and hydrogen ions that originate from these reactions react with molecular oxygen.

$$O_2 + 4e^- + 4H^+ \longrightarrow 2H_2O$$

The oxygen involved in the water formation associated with the electron transport chain is the oxygen that is inhaled during the human breathing process.

The electrons that pass through the various steps of the electron transport chain (ETC) lose some energy with each transfer along the chain. Some of this "lost" energy is used to make ATP from ADP (oxidative phosphorylation), as is shown in Section 12.9.

The enzymes and electron carriers needed for the ETC are located along the inner mitochondrial membrane. Within this membrane are four distinct protein complexes, each containing some of the molecules needed for the ETC process to occur. These four protein complexes, which are tightly bound to the membrane, are

Complex I:	NADH–coenzyme Q reductase
Complex II:	Succinate–coenzyme Q reductase
Complex III:	Coenzyme Q–cytochrome c reductase
Complex IV:	Cytochrome c oxidase

Two electron carriers, coenzyme Q and cytochrome c, which are not tightly associated with any of the four complexes, serve as mobile electron carriers that shuttle electrons between the various complexes.

The discussion of the individual reactions that occur in the ETC is divided into four parts, each part dealing with the reactions associated with one of the four protein complexes.

Complex I: NADH–Coenzyme Q Reductase

NADH, from the citric acid cycle, is the source for the electrons that are processed through complex I, the largest of the four protein complexes. Complex I contains more than 40 subunits, including the B vitamin-containing flavin mononucleotide (FMN) and several iron–sulfur proteins (FeSP). The net result of electron movement through complex I is the transfer of electrons from NADH to coenzyme Q (CoQ), a result implied by the name of complex I: *NADH–coenzyme Q reductase.* The actual electron transfer process is not, however, a single-step direct transfer of electrons from NADH to CoQ; several intermediate carriers are involved.

The first electron transfer step that occurs in complex I involves the interaction of NADH with flavin mononucleotide (FMN). The NADH is oxidized to NAD^+ (which can again participate in the citric acid cycle) as it passes two hydrogen ions and two electrons to FMN, which is reduced to $FMNH_2$.

The $FMN/FMNH_2$ pair is the third biochemical situation we have encountered in which a flavin molecule is present. The other two are the $FAD/FADH_2$ pair and the B vitamin riboflavin. FMN differs from FAD in not having an adenine nucleotide. Both FMN and FAD are synthesized within the body from riboflavin.

$$NADH + H^+ \xrightarrow{\text{Oxidation}} NAD^+ + 2H^+ + 2e^-$$

$$2H^+ + 2e^- + FMN \xrightarrow{\text{Reduction}} FMNH_2$$

NADH supplies both electrons and one of the H^+ ions that are transferred; the other H^+ ion comes from the matrix solution. The actual changes that occur within the structure of FMN as it accepts the two electrons and two H^+ ions are shown in Figure 12.12a.

The next steps involve transfer of electrons from the reduced $FMNH_2$ through a series of iron/sulfur proteins (FeSPs). The iron present in these FeSPs is Fe^{3+}, which is reduced to Fe^{2+}. The two H atoms of $FMNH_2$ are released to solution as two H^+ ions. Two FeSP molecules are needed to accommodate the two electrons released by $FMNH_2$ because an Fe^{3+}/Fe^{2+} reduction involves only one electron.

Figure 12.12 Structural characteristics of the electron carriers flavin mononucleotide and coenzyme Q.

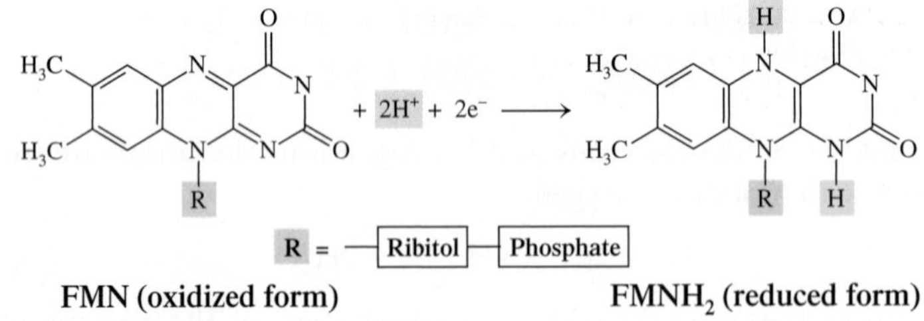

a The oxidized form (FMN) and reduced form ($FMNH_2$) of the electron carrier flavin mononucleotide.

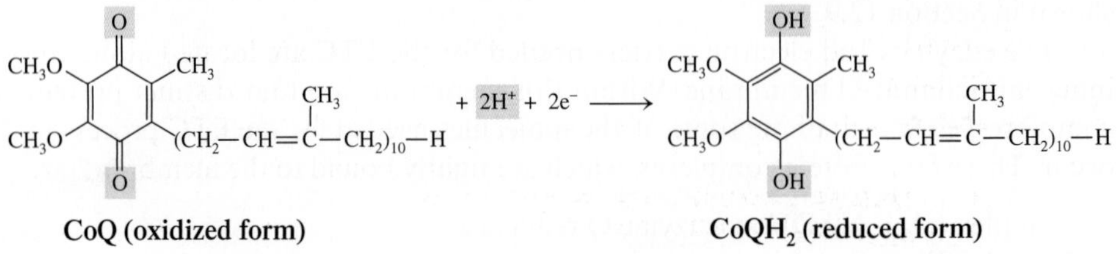

b The oxidized form (CoQ) and reduced form ($CoQH_2$) of the electron carrier coenzyme Q.

$$\text{FMNH}_2 \xrightarrow{\text{Oxidation}} \text{FMN} + 2\text{H}^+ + 2e^-$$

$$2e^- + 2\text{Fe(III)SP} \xrightarrow{\text{Reduction}} 2\text{Fe(II)SP}$$

In the final complex I reaction, Fe(II)SP is reconverted into Fe(III)SP as each of two Fe(II)SP units passes an electron to CoQ, changing it from its oxidized form (CoQ) to its reduced form (CoQH_2).

$$2\text{Fe(II)SP} \xrightarrow{\text{Oxidation}} 2\text{Fe(III)SP} + 2e^-$$

$$2e^- + 2\text{H}^+ + \text{CoQ} \xrightarrow{\text{Reduction}} \text{CoQH}_2$$

Coenzyme Q, in both its oxidized and reduced forms, is lipid soluble and can move laterally within the mitochondrial membrane. Its function is to shuttle its newly acquired electrons to complex III, where it becomes the initial substrate for reactions at this complex.

The Q in the designation coenzyme Q comes from the name quinone. Structurally, coenzyme Q is a quinone derivative. In its most common form, coenzyme Q has a long carbon chain containing 10 isoprene units (Section 2.7) attached to its quinone unit. The actual changes that occur within the structure of CoQ as it accepts the two electrons and the two H^+ ions involve the quinone part of its structure, as is shown in Figure 12.12b. The two H^+ ions that CoQ picks up in forming CoQH_2 come from solution.

Quinone is the common name for the cyclic ketone (Section 4.3) with the structure

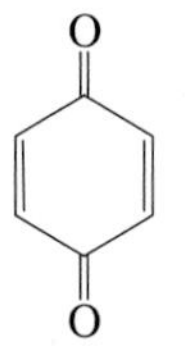

The IUPAC name for this compound is o-benzoquinone. Quinone derivatives were previously encountered in Chemical Connections 10-A on page 420.

Complex II: Succinate–Coenzyme Q Reductase

Complex II, which is much smaller than complex I, contains only four subunits, including two FeSPs. This complex is used to process the FADH_2 that is generated in the citric acid cycle when succinate is converted to fumarate. (Thus the use of the term *succinate* in the name of complex II.)

CoQ is associated with the operations in complex II in a manner similar to its actions in complex I. It is the final recipient of the electrons from FADH_2, with iron–sulfur proteins serving as intermediaries.

Thus complexes I and II produce a common product, the reduced form of coenzyme Q (CoQH_2). As was the case with complex I, the reduced CoQH_2 shuttles electrons to complex III.

$$\text{FADH}_2 \xrightarrow{\text{Oxidation}} \text{FAD} + 2\text{H}^+ + 2e^-$$

$$2e^- + 2\text{Fe(III)SP} \xrightarrow{\text{Reduction}} 2\text{Fe(II)SP}$$

$$2\text{Fe(II)SP} \xrightarrow{\text{Oxidation}} 2\text{Fe(III)SP} + 2e^-$$

$$2e^- + 2\text{H}^+ + \text{CoQ} \xrightarrow{\text{Reduction}} \text{CoQH}_2$$

Figure 12.13 summarizes the electron transport chain reactions associated with complexes I and II. In Figure 12.13a the net process is shown with only starting and end products shown. In Figure 12.13b individual reaction detail is shown. Note the general pattern that is developing for the electron carriers. They are reduced (accept electrons) in one step and then regenerated (oxidized; lose electrons) in the next step so that they can again participate in electron transport chain reactions.

Complex III: Coenzyme Q–Cytochrome c Reductase

Complex III contains 11 different subunits. Electron carriers present include several iron–sulfur proteins as well as several cytochromes. A **cytochrome** *is a heme-containing protein in which reversible oxidation and reduction of an iron atom occur.*

All H^+ ions required for the reactions of NADH, CoQ, and O_2 in the ETC come from the matrix side of the inner mitochondrial membrane.

Figure 12.13 An overview of electron movement through complexes I and II of the electron transport chain.

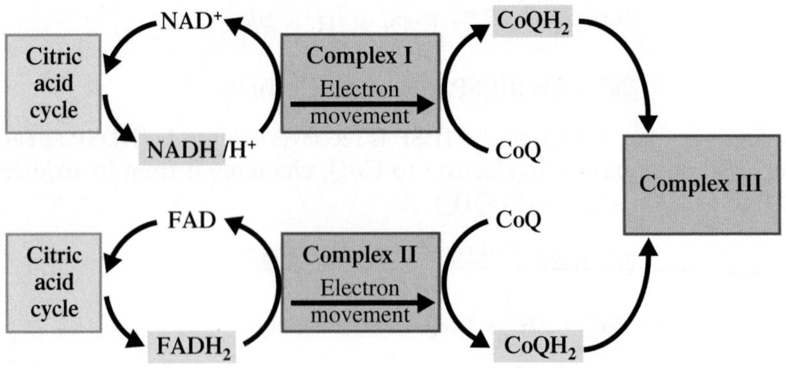

a $CoQH_2$ carries electrons from both complexes I and II to complex III.

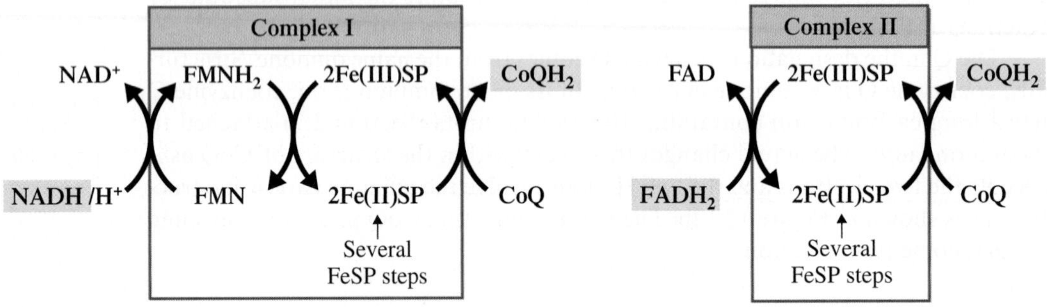

b NADH is the substrate for complex I, and $FADH_2$ is the substrate for complex II. $CoQH_2$ is the common product from both electron transfer processes.

In cytochromes, the iron of the heme is involved in redox reactions in which the iron changes back and forth between the +2 and +3 oxidation states.

In cytochromes, the heme present is bound to protein in such a way as to prevent the heme from combining with oxygen as it does when it is present in hemoglobin.

Iron/sulfur protein (FeSP) is a *non-heme iron protein.* Most proteins of this type contain sulfur, as is the case with FeSP. Often the iron is bound to the sulfur atom in the amino acid cysteine.

A feature that all steps in the ETC share is that as each electron carrier passes electrons along the chain, it becomes reoxidized and thus able to accept more electrons.

Heme, a compound also present in hemoglobin and myoglobin (Section 9.16), has the structure

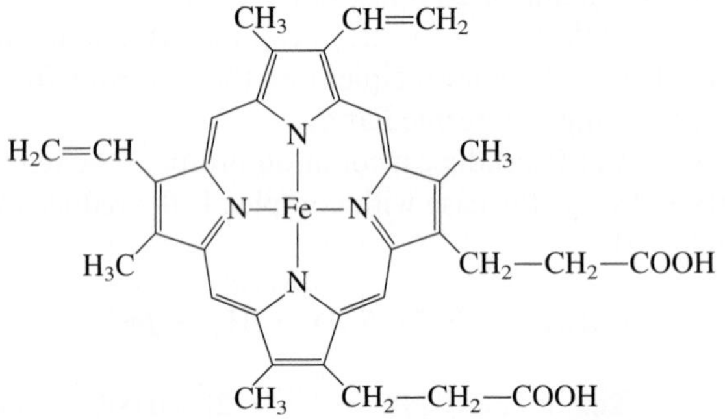

Heme-containing proteins function similarly to FeSP; iron changes back and forth between the +3 and +2 oxidation states.

Various cytochromes, abbreviated cyt a, cyt b, cyt c, and so on, differ from each other in (1) their protein constituents (2) the manner in which the heme is bound to the protein and (3) attachments to the heme ring. Again, because the Fe^{3+}/Fe^{2+} system involves only a one-electron change, two cytochrome molecules are needed to move two electrons along the chain.

The initial substrate for complex III is $CoQH_2$ molecules carrying the electrons that have been processed through complex I (from NADH) and also those processed through complex II (from $FADH_2$). The electron transfer process proceeds from $CoQH_2$ to an FeSP, then to cyt b, then to another FeSP, then to cyt c_1, and finally to cyt c. Cyt c can move laterally in the intermembrane space; it delivers its electrons to complex IV. Cyt c is the only one of the cytochromes that is water soluble.

The initial oxidation–reduction reaction at complex III is between $CoQH_2$ and an iron–sulfur protein (FeSP).

$$CoQH_2 \xrightarrow{\text{Oxidation}} CoQ + 2e^- + 2H^+$$

$$2e^- + 2Fe(III)SP \xrightarrow{\text{Reduction}} 2Fe(II)SP$$

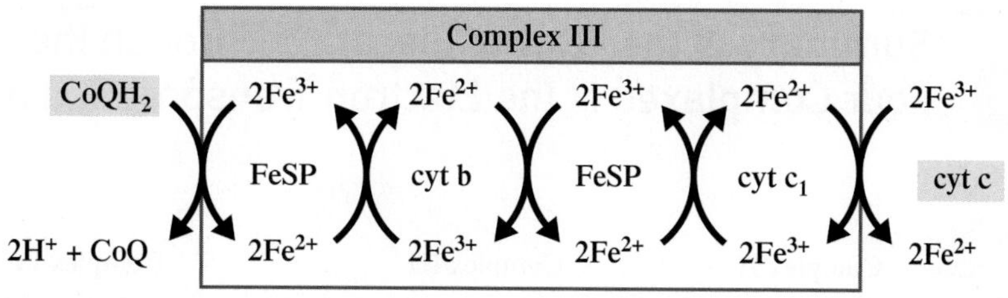

Figure 12.14 Electron movement through complex III is initiated by the electron carrier $CoQH_2$. In several steps, the electrons are passed to cyt c.

The H^+ ions produced from the oxidation of $CoQH_2$ go into cellular solution. All further redox reactions at complex III involve only electrons, which are conveyed further down the enzyme complex chain. Figure 12.14 shows diagrammatically the electron transfer steps associated with complex III.

Complex IV: Cytochrome c Oxidase

Complex IV contains 13 subunits, including two cytochromes. The electron movement flows from cyt c (carrying electrons from complex III) to cyt a to cyt a_3. In the final step of electron transfer, the electrons from cyt a_3 and hydrogen ions from cellular solution combine with oxygen (O_2) to form water.

$$O_2 + 4H^+ + 4e^- \longrightarrow 2H_2O$$

It is estimated that 95% of the oxygen used by cells serves as the final electron acceptor for the ETC.

The two cytochromes present in cytochrome c oxidase (a and a_3) differ from previously encountered cytochromes in that each has a copper atom associated with it in addition to its iron center. The copper atom sites participate in the electron transfer process as do the iron atom sites, with the copper atoms going back and forth between the reduced Cu^+ state and the oxidized Cu^{2+} state. Figure 12.15 shows the electron transfer sequence through these copper and iron sites.

The Chemistry at a Glance feature on the next page is a schematic diagram summarizing the flow of electrons through the four complexes of the electron transport chain.

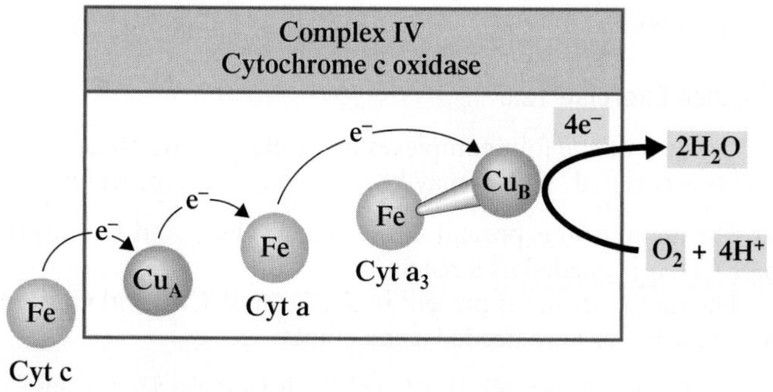

Figure 12.15 The electron transfer pathway through complex IV (cytochrome c oxidase). Electrons pass through both copper and iron centers and in the last step interact with molecular O_2. Reduction of one O_2 molecule requires the passage of four electrons through complex IV, one at a time.

EXAMPLE 12.5 Recognizing Relationships Among Electron Carriers and Enzyme Complexes in the Electron Transport Chain

With which of the four complexes in the electron transport chain is each of the following events associated? (There may be more than one correct answer in a given situation.)

a. Iron–sulfur proteins (FeSPs) are needed as reactants.
b. The mobile electron carrier CoQ serves as a "shuttle molecule."
c. Molecular O_2 is needed as a reactant.
d. FAD is a product.

(continued)

CHEMISTRY AT A GLANCE

Summary of the Flow of Electrons Through the Four Complexes of the Electron Transport Chain

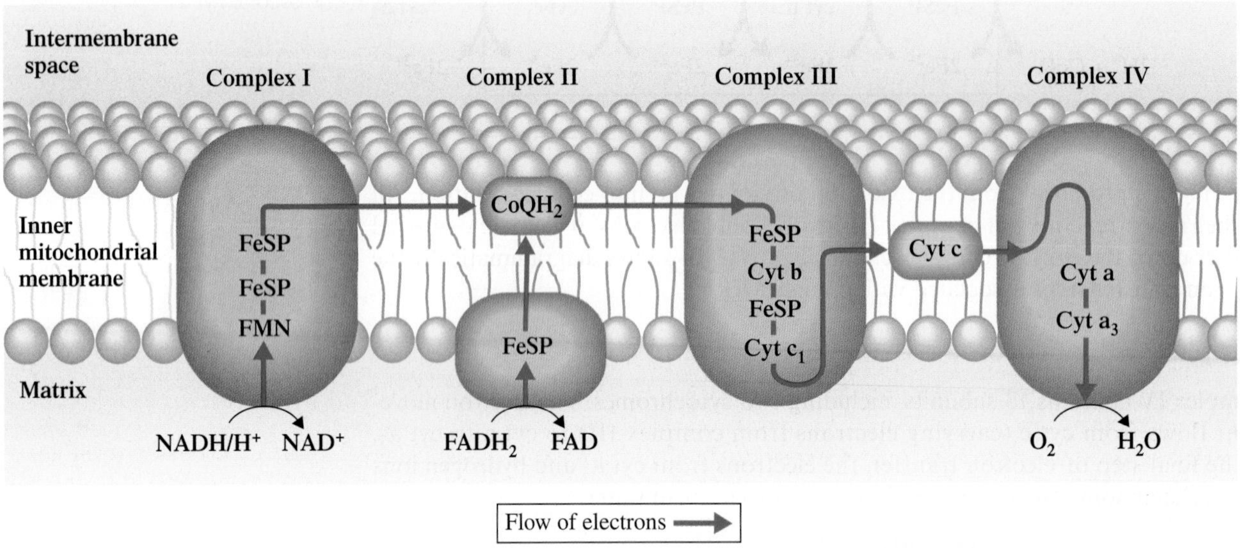

Both hydrogen and electrons from NADH and $FADH_2$ participate in the reactions involving enzyme complexes I and II. Following the formation of $CoQH_2$, hydrogen ions no longer directly participate in enzyme complex reactions; that is, they are not passed further down the enzyme complex chain. Instead, they become part of the cellular solution, where they participate in several reactions including the process by which ATP is synthesized (Section 12.8).

Solution

a. *Complexes I, II, and III.* Iron-sulfur proteins accept electrons from NADH (complex I), $FADH_2$ (complex II), and $CoQH_2$ (complex III).

b. *Complexes I, II, and III.* $CoQH_2$ functions as a shuttle molecule between complex I and complex III and also between complex II and complex III.

c. *Complex IV.* The final electron acceptor in the ETC is molecular O_2. It combines with electrons and H^+ ions to produce H_2O.

d. *Complex II.* $FADH_2$ is converted to FAD at complex II.

Practice Exercise 12.5

With which of the four complexes in the electron transport chain is each of the following events associated? (There may be more than one correct answer in a given situation.)

a. The metal iron is present in the form of Fe^{2+} and Fe^{3+} ions.

b. $FADH_2$ is needed as a reactant.

c. The metal copper is present in the form of Cu^+ and Cu^{2+} ions.

d. Cytochromes are needed as reactants.

Answers: **a.** Complexes I, II, III, and IV; **b.** Complex II; **c.** Complex IV; **d.** Complexes III and IV

12.9 Oxidative Phosphorylation

Oxidative phosphorylation *is the biochemical process by which ATP is synthesized from ADP as a result of the transfer of electrons and hydrogen ions from NADH or $FADH_2$ to O_2 through the electron carriers involved in the electron transport chain.* Oxidative phosphorylation is conceptually simple but mechanistically complex. Determining the details of oxidative phosphorylation has been—and still is—one of the most challenging research areas in biochemistry.

One concept central to the oxidative phosphorylation process is that of *coupled* reactions. **Coupled reactions** *are pairs of biochemical reactions that occur concurrently*

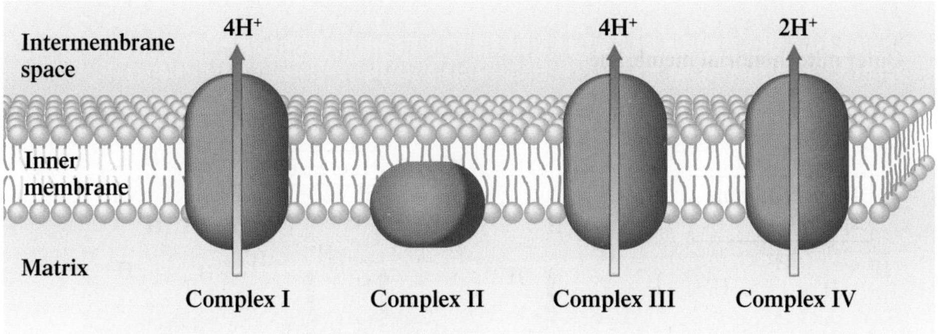

Figure 12.16 A second function for protein complexes I, III, and IV involved in the electron transport chain is that of proton pump. For every two electrons passed through the ETC, 10 H^+ ions are transferred from the mitochondrial matrix to the intermembrane space through these complexes.

in which energy released by one reaction is used in the other reaction. Oxidative phosphorylation and the oxidation reactions of the electron transport chain are coupled systems.

The interdependence (coupling) of ATP synthesis with the reactions of the ETC is related to the movement of protons (H^+ ions) across the inner mitochondrial membrane. Three of the four protein complexes involved in the ETC chain (I, III, and IV) have a second function besides electron transfer down the chain. They also serve as "proton pumps," transferring protons from the matrix side of the inner mitochondrial membrane to the intermembrane space (Figure 12.16).

Some of the H^+ ions crossing the inner mitochondrial membrane come from the reduced electron carriers, and some come from the matrix; the details of how the H^+ ions cross the inner mitochondrial membrane are not fully understood.

For every two electrons passed through the ETC, four protons cross the inner mitochondrial membrane through complex I, four through complex III, and two more through complex IV. This proton flow causes a buildup of H^+ ions (protons) in the intermembrane space; this high concentration of protons becomes the basis for ATP synthesis (Figure 12.15).

The "proton flow" explanation for ATP–ETC coupling is formally called chemiosmotic coupling. **Chemiosmotic coupling** *is an explanation for the coupling of ATP synthesis with electron transport chain reactions that requires a proton gradient across the inner mitochondrial membrane.* The main concepts in this explanation for coupling follow.

1. The result of the pumping of protons from the mitochondrial matrix across the inner mitochondrial membrane is a higher concentration of protons in the intermembrane space than in the matrix. This concentration difference constitutes an *electrochemical (proton) gradient.* A chemical gradient exists whenever a substance has a higher concentration in one region than in another. Because the proton has an electrical charge (H^+ ion), an electrical gradient also exists. Potential energy is always associated with an electrochemical gradient.
2. A spontaneous flow of protons from the region of high concentration to the region of low concentration occurs because of the electrochemical gradient. This proton flow is not through the membrane itself (it is not permeable to H^+ ions) but rather through enzyme complexes called *ATP synthases* located on the inner mitochondrial membrane (Section 12.2). This proton flow through the ATP synthases "powers" the synthesis of ATP. ATP synthases are thus the *coupling factors* that link the processes of oxidative phosphorylation and the electron transport chain.
3. ATP synthase has two subunits, the F_0 and F_1 subunits (Figure 12.17). The F_0 part of the synthase is the channel for proton flow, whereas the formation of ATP takes place in the F_1 subunit. As protons return to the mitochondrial matrix through the F_0 subunit, the potential energy associated with the electrochemical gradient is released and used in the F_1 subunit for the synthesis of ATP.

$$H^+ + \text{ADP} \quad P_i \xrightarrow{\text{ATP synthase}} \text{ATP} + H_2O$$

Oxidative phosphorylation is not the only process by which ATP is produced in cells. A second process, *substrate phosphorylation* (Section 13.2), can also be an ATP source. However, the amount of ATP produced by this second process is much less than that produced by oxidative phosphorylation.

The difference in H^+ ion concentration between the two sides of the inner mitochondrial membrane causes a pH difference of about 1.4 units. A pH difference of 1.4 units means that the intermembrane space, the more acidic region, has 25 times more protons than the matrix.

Some of the energy released at each of the protein complexes I, III, and IV is consumed in the movement of H^+ ions across the inner membrane from the matrix into the intermembrane space. Movement of ions from a region of lower concentration (the matrix) to one of higher concentration (the intermembrane space) requires the expenditure of energy because it opposes the natural tendency, as exhibited in the process of osmosis, to equalize concentrations.

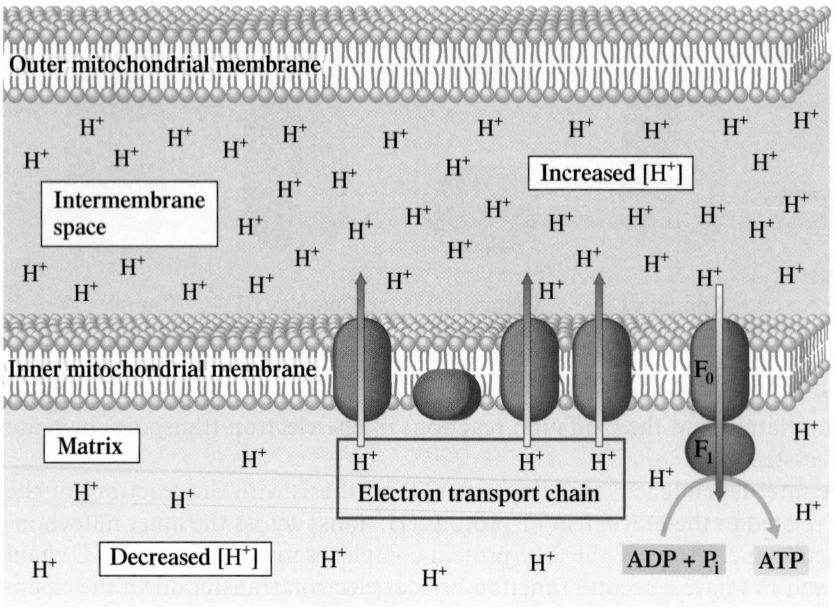

Figure 12.17 Formation of ATP accompanies the flow of protons from the intermembrane space back into the mitochondrial matrix. The proton flow results from an electrochemical gradient across the inner mitochondrial membrane.

The Chemistry at a Glance feature below brings together into one diagram the three processes that constitute the common metabolic pathway: the citric acid cycle, the electron transport chain, and oxidative phosphorylation. These three processes operate together. Discussing them separately, as we have done, is a matter of convenience only.

The enzymes involved in the operation of the ETC can be inhibited by other substances just as other enzymes can. The gas hydrogen cyanide (HCN) exerts its deadly effect by inhibiting the ETC enzyme cytochrome c oxidase. The focus on relevancy feature Chemical Connections 12-A on the next page explores further the topic of the biochemical effects of cyanide ion and hydrogen cyanide on the human body.

CHEMISTRY AT A GLANCE Summary of the Common Metabolic Pathway

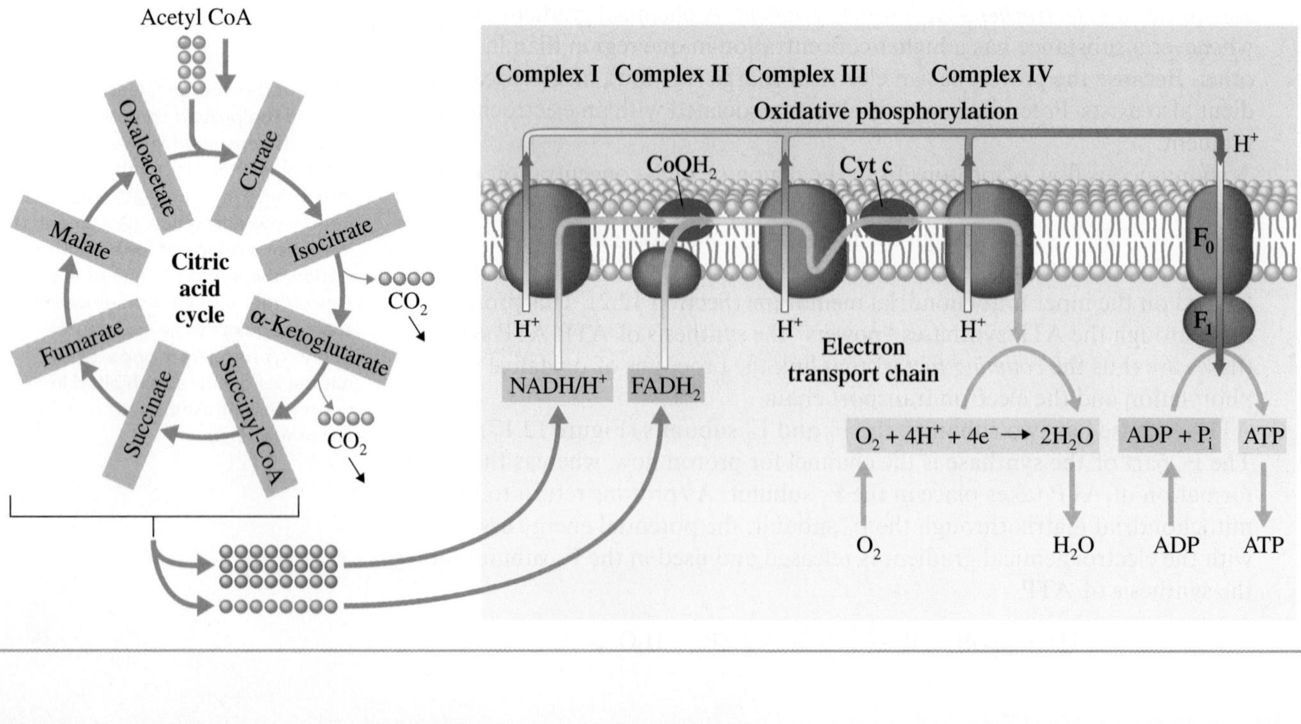

CHEMICAL CONNECTIONS 12-A

Cyanide Poisoning

Inhalation of hydrogen cyanide gas (HCN) or ingestion of solid potassium cyanide (KCN) rapidly inhibits the electron transport chain in all tissues, making cyanide one of the most potent and rapidly acting poisons known. The attack point for the cyanide ion (CN^-) is cytochrome c oxidase, the last of the four protein complexes in the electron transport chain. Cyanide inactivates this complex by bonding itself to the Fe^{3+} in the complex's heme portions. As a result, Fe^{3+} is unable to transfer electrons to oxygen, blocking the cell's use of oxygen. Death results from tissue asphyxiation, particularly of the central nervous system. Cyanide also binds to the heme group in hemoglobin, blocking oxygen transport in the bloodstream.

One treatment for cyanide poisoning is to administer various nitrites, NO_2^-, which oxidize the iron atoms of hemoglobin to Fe^{3+}. This form of hemoglobin helps draw CN^- back into the bloodstream, where it can be converted to thiocyanate (SCN^-) by thiosulfate ($S_2O_3^{2-}$), which is administered along with the nitrite (see the accompanying figure).

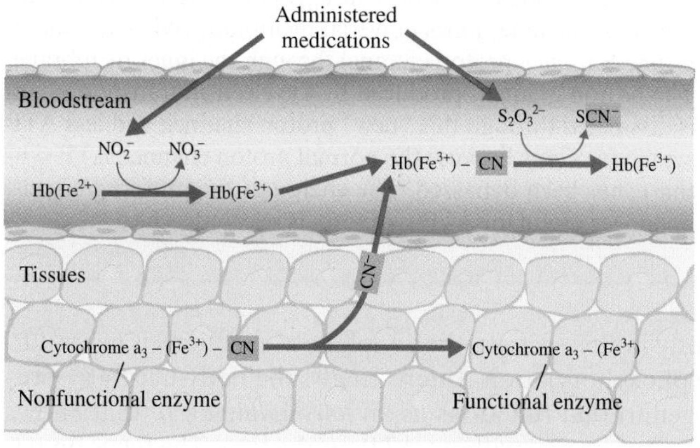

Several plant-produced compounds are known that contain cyano (CN) groups. These compounds, called *cyanogenic glycosides,* can produce toxic levels of HCN when they are enzymatically hydrolyzed. The best known of the cyanogenic glycosides is *amygdalin,* a substance found in the pits of apricots, peaches, and plums (see accompanying photo).

Apricot pits are the most common source of amygdalin.

The structure of amygdalin is below in the abbreviated equation for its hydrolysis.

The disaccharide part of amygdalin's structure involves two D-glucose units joined via a β-(1→6) glycosidic linkage (Section 7.13).

Amygdalin, produced primarily in Mexico and sold under the name *laetrile,* was once heavily promoted as a substance useful in treating cancer. Studies have now established that laetrile has little or no effect in treating cancer. The HCN produced when ingested laetrile is hydrolyzed affects all cells rather than selectively targeting cancer cells; the side effects produced closely resemble those associated with chronic HCN exposure: headache, vomiting, and in some cases coma and death.

The United States Federal Drug Administration (FDA) now seeks jail sentences for vendors who sell laetrile within the United States for cancer treatment. In scientific literature, the use of laetrile as an anticancer agent has been described as the most sophisticated and most remunerative example of medical quackery in medical history.

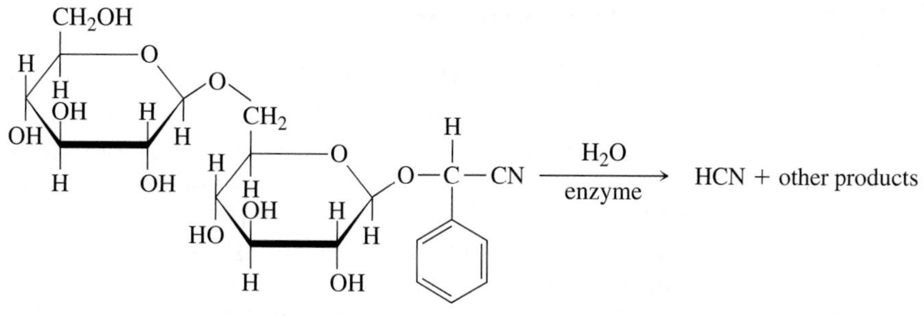

Amygdalin

CHEMICAL CONNECTIONS 12-B

Brown Fat, Newborn Babies, and Hibernating Animals

Ordinarily, metabolic processes generate enough heat to maintain normal body temperature. In certain cases, however, including newborn infants and hibernating animals, normal metabolism is not sufficient to meet the body's heat requirements. In these cases, a supplemental method of heat generation, which involves *brown fat tissue,* occurs.

Brown fat tissue, as the name implies, is darker in color than ordinary fat tissue, which is white. Brown fat is specialized for heat production. It contains many more blood vessels and mitochondria than white fat. (The increased number of mitochondria gives brown fat its color.)

Another difference between the two types of fat is that the mitochondria in brown fat cells contain a protein called *thermogenin,* which functions as an *uncoupling agent.* This protein "uncouples" the ATP production associated with the electron transport chain. The ETC reactions still take place, but the energy that would ordinarily be used for ATP synthesis is simply released as heat.

Brown fat tissue is of major importance for newborn infants. Newborns are immediately faced with a temperature regulation problem. They leave an environment of constant 37°C temperature and enter a much colder environment (25°C). A supply of *active* brown fat, present at birth, helps the baby adapt to the cooler environment.

Very limited amounts of brown fat are present in most adults. However, stores of brown fat increase in adults who are regularly exposed to cold environs. Thus the production of brown fat is one of the body's mechanisms for adaptation to cold.

Hibernating bears rely on brown fat tissue to help meet their bodies' heat requirements.

Thermogenin, the uncoupling agent in brown fat, is a protein bound to the inner mitochondrial membrane. When activated, it functions as a proton channel through the inner membrane. The proton gradient produced by the electron transport chain is dissipated through this "new" proton channel, and less ATP synthesis occurs because the normal proton channel, ATP synthase, has been bypassed. The energy of the proton gradient, no longer useful for ATP synthesis, is released as heat.

As indicated previously in this section, the coupling of reactions is central to the process of oxidative phosphorylation. Interestingly, the body naturally produces a protein (thermogenin) that functions as an *uncoupling agent* that allows the electron transport chain to proceed without ATP production. The focus on relevancy feature Chemical Connections 12-B above addresses the topic of *when* and *why* the uncoupling agent thermogenin is needed by the human body.

Without oxygen, the biochemical systems of the human body quickly shut down and death occurs. Why? Without oxygen as the final electron acceptor in the ETC, the ETC chain shuts down and ATP production stops. Without ATP to power life's processes (Chapters 13–15), these processes stop.

12.10 ATP Production for the Common Metabolic Pathway

For each mole of NADH oxidized in the ETC, 2.5 moles of ATP are formed. $FADH_2$, which does not enter the ETC at its start, produces only 1.5 moles of ATP per mole of $FADH_2$ oxidized. $FADH_2$'s entrance point into the chain, complex II, is beyond the first "proton-pumping" site, complex I. Hence fewer ATP molecules are produced from $FADH_2$ than from NADH.

Biochemistry textbooks published before the mid-1990s make the following statements:

1 NADH produces 3 ATP in the ETC.

1 $FADH_2$ produces 2 ATP in the ETC.

As more has been learned about the electron transport chain and oxidative phosphorylation, these numbers have had to be reduced. The overall conversion process is more complex than was originally thought, and not as much ATP is produced.

The energy yield, in terms of ATP production, can now be totaled for the common metabolic pathway (Section 12.6). Every acetyl CoA entering the citric acid cycle (CAC) produces three NADH, one $FADH_2$, and one GTP (which is equivalent in energy to ATP; Section 12.6). Thus 10 molecules of ATP are produced for each acetyl CoA catabolized.

$$\begin{aligned} 3\text{ NADH} &\longrightarrow 7.5\text{ ATP} \\ 1\text{ FADH}_2 &\longrightarrow 1.5\text{ ATP} \\ 1\text{ GTP} &\longrightarrow \underline{1\quad\text{ ATP}} \\ & \qquad 10\quad\text{ ATP} \end{aligned}$$

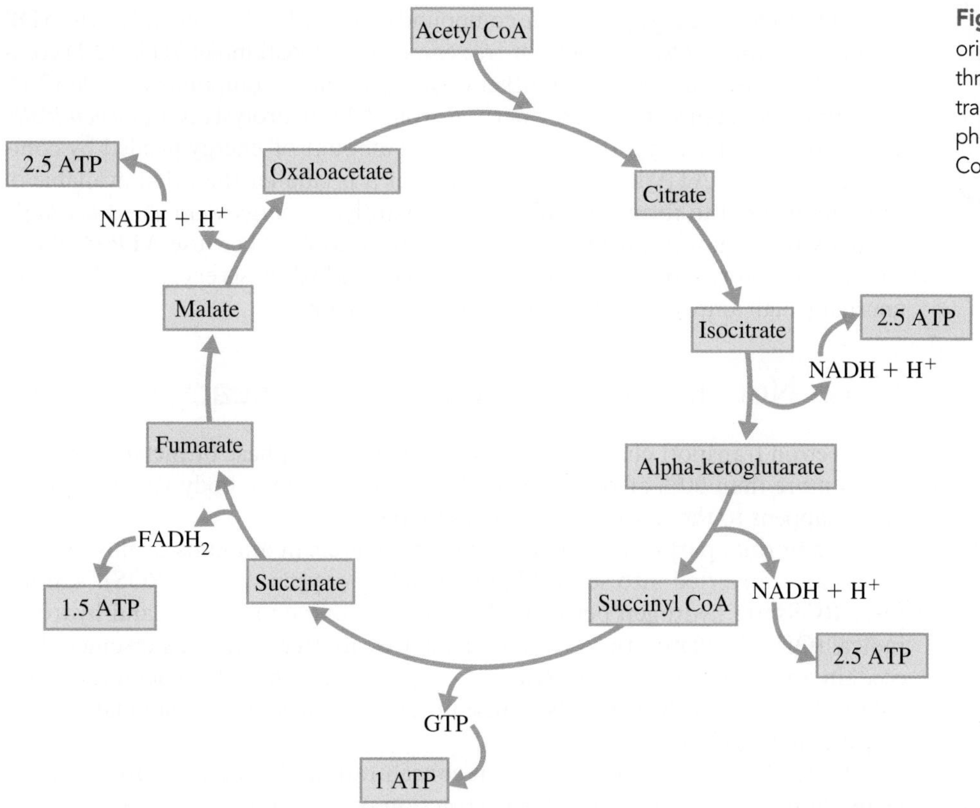

Figure 12.18 Citric acid cycle origins for the ATP produced through operation of the electron transfer chain and oxidative phosphorylation per one acetyl CoA entering the citric acid cycle.

Figure 12.18 relates the preceding 10 ATP production figures to individual steps in the citric acid cycle in which NADH, $FADH_2$, and GTP are produced.

12.11 The Importance of ATP

The cycling of ATP and ADP in metabolic processes is the principal medium for energy exchange in biochemical processes. The conversion

$$\text{ATP} \longrightarrow \text{ADP} + P_i$$

powers life processes (the biosynthesis of essential compounds, muscle contraction, nutrient transport, and so on). The conversion

$$P_i + \text{ADP} \longrightarrow \text{ATP}$$

which occurs in food catabolism cycles, regenerates the ATP expended in cell operation. Figure 12.19 summarizes the ATP–ADP cycling process.

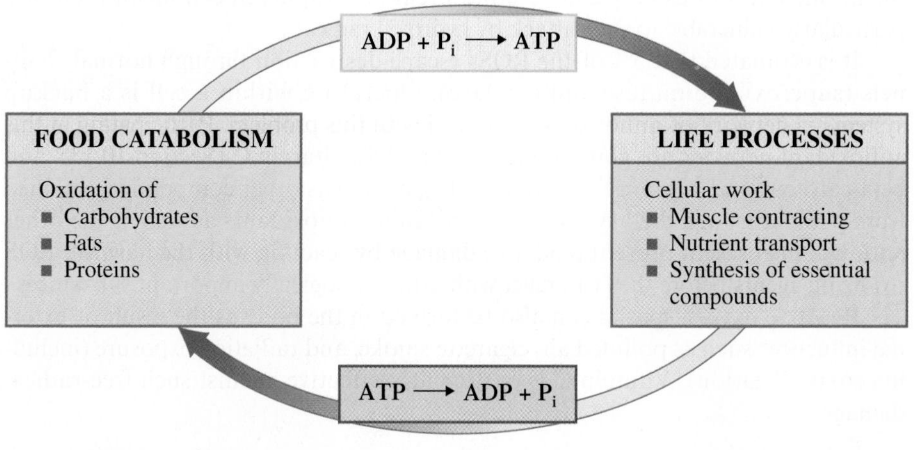

Figure 12.19 The interconversion of ATP and ADP is the principal medium for energy exchange in biochemical processes.

ATP is a high-energy phosphate compound (Section 12.5). Its hydrolysis to ADP produces an *intermediate* amount of free energy (−7.5 kcal/mole; Table 12.1) compared with hydrolysis energies for other organophosphate compounds (Table 12.1). Of major importance, the energy derived from ATP hydrolysis is a *biochemically useful* amount of energy. It is larger than the amount of energy needed by compounds to which ATP donates energy, and yet it is smaller than that available in compounds used to form ATP. If the ATP hydrolysis energy were *unusually high,* the body would not be able to convert ADP back to ATP because ATP synthesis requires an energy input equal to or greater than the hydrolysis energy, and such an unusually high amount of energy would not be available.

ATP molecules in cells have a high turnover rate. Normally, a given ATP molecule in a cell does not last more than a minute before it is converted to ADP. The concentration of ATP in a cell varies from 0.5 to 2.5 milligrams per milliliter of cellular fluid.

12.12 Non-ETC Oxygen-Consuming Reactions

The electron transport chain/oxidative phosphorylation phase of metabolism consumes more than 90% of the oxygen taken into the human body via respiration. What happens to the remainder of the inspired O_2?

As a normal part of metabolic chemistry, significant amounts of this remaining O_2 are converted into several highly reactive oxygen species (ROS). Among these ROSs are hydrogen peroxide (H_2O_2), superoxide ion (O_2^-), and hydroxyl radical (OH). The latter two of these substances are free radicals, substances that contain an unpaired electron. Reactive oxygen species have beneficial functions within the body, but they can also cause problems if they are not eliminated when they are no longer needed.

Commercially, hydrogen peroxide (H_2O_2) is used as a bleaching agent. In the human body, H_2O_2 is the indirect cause of dark hair turning gray and then often white as a person ages. This natural bleaching effect is caused by the presence of excess H_2O_2 within cells. The H_2O_2 buildup occurs because of a decline in the production of the enzyme *catalase*, the enzyme that converts H_2O_2 to H_2O. The H_2O_2 oxidizes amino acid residues (methioine) in proteins associated with hair pigment production, limiting their ability to function properly; hence hair with less pigmentation results.

White blood cells have a significant concentration of superoxide free radicals. Here, these free radicals aid in the destruction of invading bacteria and viruses. Their formation reaction is

$$2O_2 + \text{NADPH} \longrightarrow 2O_2^- + \text{NADP}^+ + H^+$$

(NADP is a phosphorylated version of the coenzyme NADH; see Section 13.8.)

Superoxide ion that is not needed is eliminated from cells in a two-step process governed by the enzymes *superoxide dismutase* and *catalase,* two of the most rapidly working enzymes known (see Table 10.2). In the first step, superoxide ion is converted to hydrogen peroxide, which is then, in the second step, converted to H_2O.

$$2O_2^- + 2H^+ \xrightarrow[\text{dismutase}]{\text{superoxide}} H_2O_2 + O_2$$

$$2H_2O_2 \xrightarrow{\text{catalase}} 2H_2O + O_2$$

Immediate destruction of the hydrogen peroxide produced in the first of these two steps is critical, because if it persisted, then unwanted production of hydroxyl radical would occur via hydrogen peroxide's reaction with superoxide ion.

$$H_2O_2 + O_2^- + H^+ \longrightarrow H_2O + O_2 + OH$$

Hydroxyl radicals quickly react with other substances by taking an electron from them. Such action usually causes bond breaking. Lipids in cell membranes are particularly vulnerable to such attack by hydroxyl radicals.

It is estimated that 5% of the ROSs escape destruction through normal channels (superoxide dimutase and catalase). Operating within a cell is a backup system—a network of antioxidants—to deal with this problem. Participating in this antioxidant network are glutathione (Section 9.8), vitamin C (Section 10.13), and beta-carotene and vitamin E (Section 10.15), as well as other compounds obtained from plants through dietary intake. The vitamin antioxidants as well as the other antioxidants present prevent oxidative damage by reacting with the harmful ROS oxidizing agents before they can react with other biologically important substances.

Antioxidant molecules provide electrons to convert free radicals and other ROSs into less-reactive substances.

Reactive oxygen species can also be formed in the body as the result of external influences such as polluted air, cigarette smoke, and radiation exposure (including solar radiation). Vitamin C is particularly effective against such free-radical damage.

CHEMICAL CONNECTIONS 12-C

Flavonoids: An Important Class of Dietary Antioxidants

Numerous studies indicate that diets high in fruits and vegetables are associated with a healthy lifestyle. One reason for this is that fruits and vegetables contain compounds called *phytochemicals.* Phytochemicals are compounds found in plants that have biochemical activity in the human body even though they have no nutritional value. The functions that phytochemicals perform in the human body include antioxidant activity, cancer inhibition, cholesterol regulation, and anti-inflammatory activity.

Each fruit and vegetable is a unique package of phytochemicals, so consuming a wide variety of fruits and vegetables provides the body with the broadest spectrum of benefits. In such a situation, many phytochemicals are consumed in *small* amounts. This approach is much safer than taking supplemental doses of particular phytochemicals; in larger doses, some phytochemicals are toxic.

A major group of phytochemicals are the *flavonoids,* of which more than 4000 individual compounds are known. All flavonoids are antioxidants (Section 3.14), but some are stronger antioxidants than others, depending on their molecular structure. About 50 flavonoids are present in foods and in beverages derived from plants (tea leaves, grapes, oranges, and so on).

The core *flavonoid* structure is

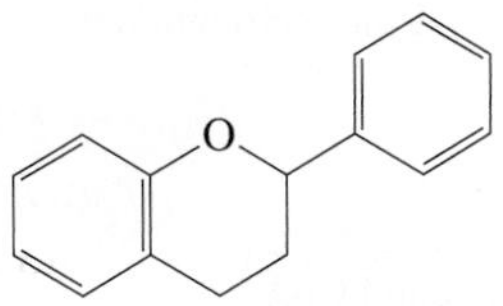

Both aromatic and cyclic ether ring systems are present. Of particular importance as antioxidants in foods are flavonoids known as *flavones* and *flavonols,* flavonoids whose core structures are enhanced by the presence of ketone and/or hydroxyl groups and a double bond in the oxygen-containing ring system.

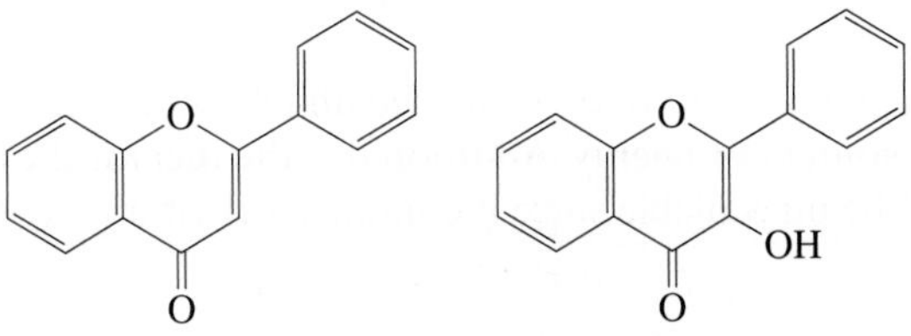

The formation of flavone and flavonol compounds normally depends on the action of light, so the highest concentration of these compounds generally occurs in leaves or in the skins of fruits, whereas only traces are found in parts of plants that grow below the ground. The common onion is, however, a well-known exception to this generalization.

Apples are the fruit that contains the greatest amount of the antioxidant quercetin; the skin (peel) contains the majority of the quercetin.

The most widespread flavonoid in food is the flavonol *quercetin.*

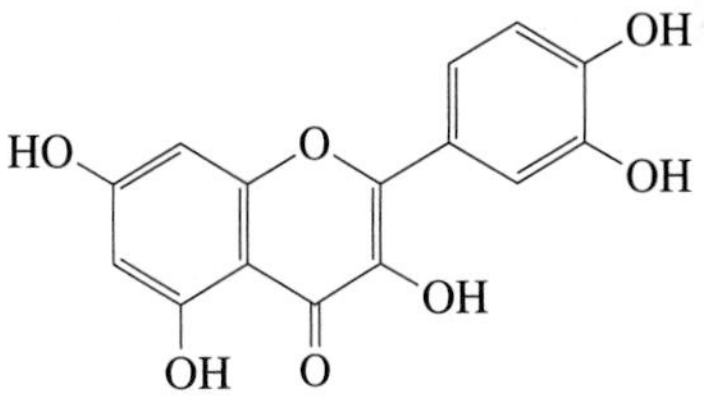

It is predominant in fruits, vegetables, and the leaves of various vegetables. In fruits, apples contain the highest amounts of quercetin, the majority of it being found in the outer tissues (skin, peel). A small peeled apple contains about 5.7 mg of the antioxidant vitamin C. But the same amount of apple *with the skin* contains flavonoids and other phytochemicals that have the effect of 1500 mg of vitamin C. Onions are also major dietary sources of quercetin.

In addition to their antioxidant benefits, flavonoids may also help fight bacterial infections. Recent studies indicate that flavonoids can stop the growth of some strains of drug-resistant bacteria.

Phytochemicals are biologically active compounds found in plants. A major group of phytochemicals are the *flavonoids,* substances that have high antioxidant activity. The focus on relevancy feature Chemical Connections 12-C above gives further information about dietary flavonoids found in fruits and vegetables.

12.13 B Vitamins and the Common Metabolic Pathway

Structurally modified B vitamins function as coenzymes in metabolic pathways (Section 10.14). Now that the reactions of the common metabolic pathway have been considered, it is useful to formalize, in a summary fashion, B vitamin involvement in the citric acid cycle and the electron transport chain. As is shown in Figure 12.20, four vitamins have involvement in these metabolic reactions.

1. Niacin—as NAD^+ and NADH
2. Riboflavin—as FAD, $FADH_2$, and FMN
3. Thiamin—as TPP
4. Pantothenic acid—as CoA

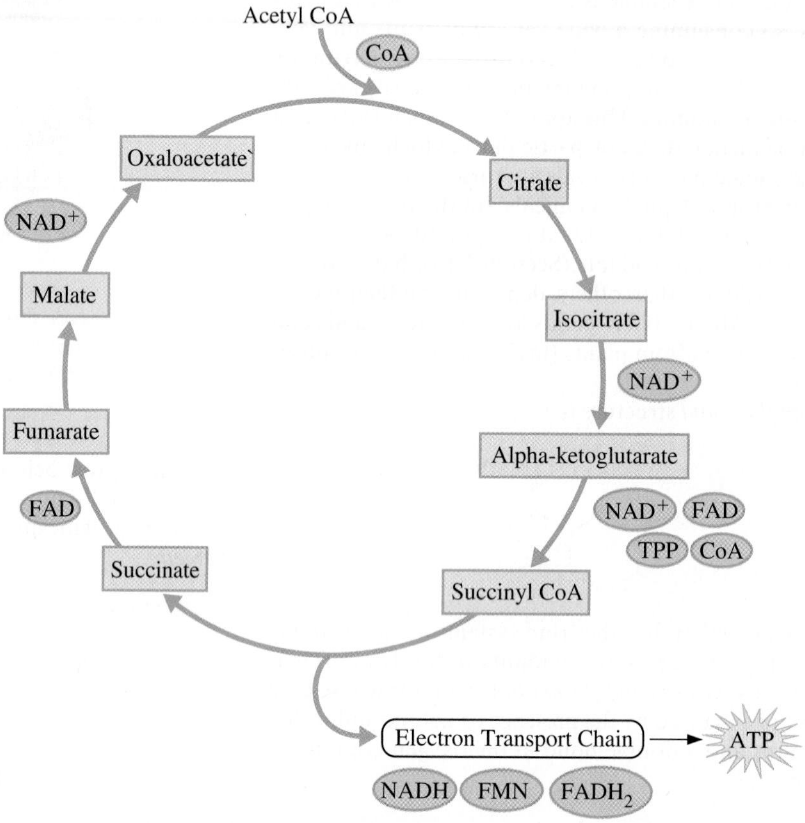

Figure 12.20 B vitamin participation in chemical reactions associated with the common metabolic pathway.

Without these B vitamins, the human body would be unable to utilize carbohydrates, fats, and proteins as sources of energy. Additionally, the fuel for the citric acid cycle, acetyl CoA, would be unavailable since it contains a B vitamin (pantothenic acid).

Concepts to Remember

OWL Sign in at www.cengage.com/owl to view tutorials and simulations, develop problem-solving skills, and complete online homework assigned by your professor.

Metabolism. Metabolism is the sum total of all the biochemical reactions that take place in a living organism. Metabolism consists of catabolism and anabolism. Catabolic biochemical reactions involve the breakdown of large molecules into smaller fragments. Anabolic biochemical reactions synthesize large molecules from smaller ones (Section 12.1).

Mitochondria. Mitochondria are membrane-enclosed subcellular structures that are the site of energy production in the form of ATP molecules. Enzymes for both the citric acid cycle and the electron transport chain are housed in the mitochondria (Section 12.2).

Metabolic coenzymes. Three important coenzymes involved in metabolic reactions are NAD^+, FAD, and CoA. NAD^+ and FAD are oxidizing agents that participate in the oxidation reactions of the citric acid cycle. They transport hydrogen atoms and electrons from the citric acid cycle to the electron transport chain. CoA interacts with acetyl groups produced from food degradation to form acetyl CoA. Acetyl CoA is the "fuel" for the citric acid cycle (Section 12.3).

Metabolic carboxylate ions. Five important carboxylate ions involved as substrates in metabolic reactions are malate, oxaloacetate, fumarate, α-ketoglutarate, and citrate. The first three of these carboxylate ions are polyfunctional derivatives of succinic acid, the four-carbon dicarboxylic acid, and the latter two are polyfunctional derivatives of glutaric acid, the five-carbon dicarboxylic acid (Section 12.4).

High-energy compounds. A high-energy compound liberates a larger-than-normal amount of free energy upon hydrolysis because structural features in the molecule contribute to repulsive strain in one or more bonds. Most high-energy biochemical molecules contain phosphate groups (Section 12.5).

Common metabolic pathway. The common metabolic pathway includes the reactions of the citric acid cycle and those of the electron transport chain and oxidative phosphorylation. The degradation products from all types of foods (carbohydrates, fats, and proteins) participate in the reactions of the common metabolic pathway (Section 12.6).

Citric acid cycle. The citric acid cycle is a cyclic series of eight reactions that oxidize the acetyl portion of acetyl CoA, resulting in the production of two molecules of CO_2. The complete oxidation of one acetyl group produces three molecules of NADH, one of $FADH_2$, and one of GTP besides the CO_2 (Section 12.7).

Electron transport chain. The electron transport chain is a series of reactions that passes electrons from NADH and $FADH_2$ to molecular oxygen. Each electron carrier that participates in the chain has an increasing affinity for electrons. Upon accepting the electrons and hydrogen ions, the O_2 is reduced to H_2O (Section 12.8).

Oxidative phosphorylation. Oxidative phosphorylation is the biochemical process by which ATP is synthesized from ADP as the result of a proton gradient across the inner mitochondrial membrane. Oxidative phosphorylation is coupled to the reactions of the electron transport chain (Section 12.9).

Chemiosmotic coupling. Chemiosmotic coupling explains how the energy needed for ATP synthesis is obtained. Synthesis takes place because of a flow of protons across the inner mitochondrial membrane (Section 12.10).

Importance of ATP. ATP is the link between energy production and energy use in cells. The conversion of ATP to ADP powers life processes, and the conversion of ADP back to ATP regenerates the energy expended in cell operation (Section 12.11).

Exercises and Problems

OWL Interactive versions of these problems may be assigned in OWL.

Exercises and problems are arranged in matched pairs with the two members of a pair addressing the same concept(s). The answer to the odd-numbered member of a pair is given at the back of the book. Problems denoted with a ▲ involve concepts found not only in the section under consideration but also concepts found in one or more earlier sections of the chapter. Problems denoted with a ● cover concepts found in a Chemical Connections feature box.

Metabolism (Section 12.1)

12.1 Classify anabolism and catabolism as synthetic or degradative processes.

12.2 Classify anabolism and catabolism as energy-producing or energy-consuming processes.

12.3 What is a metabolic pathway?

12.4 What is the difference between a linear and a cyclic metabolic pathway?

12.5 What general characteristics are associated with a catabolic pathway?

12.6 What general characteristics are associated with an anabolic pathway?

Cell Structure (Section 12.2)

12.7 List several differences between prokaryotic cells and eukaryotic cells.

12.8 What kinds of organisms have prokaryotic cells and what kinds have eukaryotic cells?

12.9 What is an organelle?

12.10 What is the general function of each of the following types of organelles?
a. Ribosome b. Lysosome c. Mitochondrion

12.11 In a mitochondrion, what separates the matrix from the intermembrane space?

12.12 In what major way do the inner and outer mitochondrial membranes differ?

12.13 What is the intermembrane space of a mitochondrion?

12.14 Where are ATP synthase complexes located in a mitochondrion?

Nucleotide-Containing Compounds in Metabolic Pathways (Section 12.3)

12.15 What does each letter in ATP stand for?

12.16 What does each letter in ADP stand for?

12.17 Draw a block diagram structure for ATP.

12.18 Draw a block diagram structure for ADP.

12.19 What is the structural difference between ATP and AMP?

12.20 What is the structural difference between ADP and AMP?

12.21 What is the structural difference between ATP and GTP?

12.22 What is the structural difference between ATP and CTP?

12.23 In a biochemical context, what is the chemical formula for a free phosphate group (P_i)?

12.24 In a biochemical context, what is the chemical formula for a phosphoryl group?

12.25 Write a generalized chemical equation, containing acronyms, for the hydrolysis of ATP to ADP.

12.26 Write a generalized chemical equation, containing acronyms, for the hydrolysis of ATP to AMP.

12.27 What is the fate of the three atoms present in a water molecule when it participates in the hydrolysis of ATP to ADP?

12.28 What is the fate of the three atoms present in a water molecule when it participates in the hydrolysis of ADP to AMP?

12.29 What does each letter in FAD stand for?

12.30 What does each letter in NAD^+ stand for?

12.31 Draw a block diagram structure for FAD based on the presence of an ADP core (three-block diagram).

12.32 Draw a block diagram structure for FAD based on the presence of two nucleotides (six-block diagram).

12.33 Draw a block diagram structure for NAD^+ based on the presence of two nucleotides (six-block diagram).

12.34 Draw a block diagram structure for NAD^+ based on the presence of an ADP core (three-block diagram).

12.35 Which part of an NAD^+ molecule is the active participant in redox reactions?

12.36 Which part of an FAD molecule is the active participant in redox reactions?

12.37 Give the letter designation for
a. the reduced form of FAD.
b. the oxidized form of NADH.

12.38 Give the letter designation for
a. the oxidized form of $FADH_2$.
b. the reduced form of NAD^+.

12.39 Name the vitamin B molecule that is part of the structure of
a. NAD^+ b. FAD

12.40 Indicate whether or not the vitamin B portion of the following molecules is the "active" portion of the molecule in redox processes.
a. NAD^+ b. FAD

12.41 What identical structural subunits do the following pairs of molecules have in common?
a. NAD^+ and NADH
b. NAD^+ and $FADH_2$

12.42 What identical structural subunits do the following pairs of molecules have in common?
a. FAD and $FADH_2$
b. FAD and NADH

12.43 Draw the three-block diagram structure for coenzyme A.

12.44 Which part of a coenzyme A molecule is the active participant in an acetyl transfer reaction?

12.45 Classify each of the following substances as an *oxidizing agent,* a *reducing agent,* or as *neither an oxidizing agent nor a reducing agent.*
a. $FADH_2$ b. ADP c. NAD^+

12.46 Classify each of the following substances as an *oxidizing agent,* a *reducing agent,* or as *neither an oxidizing agent nor a reducing agent.*
a. ATP b. NADH c. FAD

12.47 Name the B vitamin that is part of the structure of each of the following molecules.
a. CoA-SH b. NAD^+ c. $FADH_2$

12.48 Name the B vitamin that is part of the structure of each of the following molecules.
a. NADH b. FAD c. acetyl-CoA

12.49 Which of the substances ATP, CoA-SH, FAD, and NAD^+ contains the following subunits within their structure? More than one substance may apply in a given situation.
a. Contains two ribose subunits
b. Contains two phosphate subunits
c. Contains one adenine subunit
d. Contains four different kinds of subunits

12.50 Which of the substances ATP, CoA-SH, FAD, and NAD^+ contains the following subunits within their structure? More than one substance may apply in a given situation.
a. Contains one ribose subunit
b. Contains one phosphorylated ribose subunit
c. Contains both a ribitol and a ribose subunit
d. Contains five different kinds of subunits

Carboxylate Ions in Metabolic Pathways (Section 12.4)

12.51 Draw structural formulas for each of the following substances.
a. Oxaloacetic acid b. Oxaloacetate ion
c. Citric acid d. Citrate ion

12.52 Draw structural formulas for each of the following substances.
a. Fumaric acid b. Fumarate ion
c. Malic acid d. Malate ion

12.53 Which of the substances *malate, oxaloacetate, fumarate, α-ketoglutarate,* and *citrate* have each of the following characteristics? More than one substance may apply in a given situation.
a. Contains four carbon atoms
b. Contains a keto functional group
c. Has a charge of −2
d. Is a succinic acid derivative

12.54 Which of the substances *malate, oxaloacetate, fumarate, α-ketoglutarate,* and *citrate* have each of the following characteristics? More than one substance may apply in a given situation.
a. Contains six carbon atoms
b. Contains a hydroxyl functional group
c. Has a charge of −3
d. Is a glutaric acid derivative

High-Energy Phosphate Compounds (Section 12.5)

12.55 What is a high-energy compound?

12.56 What factors contribute to a strained bond in high-energy phosphate compounds?

12.57 What does the designation P_i denote?

12.58 What does the designation PP_i denote?

12.59 With the help of Table 12.1, determine which compound in each of the following pairs of phosphate-containing compounds releases more free energy upon hydrolysis.
a. ATP and phosphoenolpyruvate
b. Creatine phosphate and ADP
c. Glucose 1-phosphate and 1,3-bisphosphoglycerate
d. AMP and glycerol 3-phosphate

12.60 With the help of Table 12.1, determine which compound in each of the following pairs of phosphate-containing compounds releases more free energy upon hydrolysis.
a. ATP and creatine phosphate
b. Glucose 1-phosphate and glucose 6-phosphate
c. ADP and AMP
d. Phosphoenolpyruvate and PP_i

Biochemical Energy Production (Section 12.6)

12.61 Describe the four general stages of the process by which biochemical energy is obtained from food.

12.62 Of the four general stages of biochemical energy production from food, which are part of the common metabolic pathway?

The Citric Acid Cycle (Section 12.7)

12.63 What are two other names for the citric acid cycle?

12.64 What is the basis for the name *citric acid cycle*?

12.65 What is the "fuel" for the citric acid cycle?

12.66 What are the products of the citric acid cycle?

12.67 Consider the reactions that occur during *one turn* of the citric acid cycle in answering each of the following questions.
a. How many CO_2 molecules are formed?
b. How many molecules of $FADH_2$ are formed?
c. How many times is a secondary alcohol oxidized?
d. How many times does water add to a carbon–carbon double bond?

12.68 Consider the reactions that occur during *one turn* of the citric acid cycle in answering each of the following questions.
a. How many molecules of NADH are formed?
b. How many GTP molecules are formed?
c. How many decarboxylation reactions occur?
d. How many oxidation–reduction reactions occur?

12.69 There are eight steps in the citric acid cycle. List those steps, by number, that involve
a. oxidation.
b. isomerization.
c. hydration.

12.70 There are eight steps in the citric acid cycle. List those steps, by number, that involve
a. oxidation and decarboxylation.
b. phosphorylation.
c. condensation.

12.71 There are four C_4 dicarboxylic acid species in the citric acid cycle. What are their names and structures?

12.72 There are two keto carboxylic acid species in the citric acid cycle. What are their names and structures?

12.73 What type of reaction occurs in the citric acid cycle whereby a C_6 compound is converted to a C_5 compound?

12.74 What type of reaction occurs in the citric acid cycle whereby a C_5 compound is converted to a C_4 compound?

12.75 Characterize, in terms of number of carbon atoms present, each of the following citric acid cycle changes as (1) a C_6 to C_6 change (2) a C_6 to C_5 change (3) a C_5 to C_4 change or (4) a C_4 to C_4 change.
a. Citrate to isocitrate
b. Succinate to fumarate
c. Malate to oxaloacetate
d. Isocitrate to α-ketoglutarate

12.76 Characterize, in terms of number of carbon atoms present, each of the following citric acid cycle changes as (1) a C_4 to C_4 change (2) a C_4 to C_6 change (3) a C_5 to C_4 change or (4) a C_6 to C_5 change.
a. Oxaloacetate to citrate
b. α-Ketoglutarate to succinyl CoA
c. Fumarate to malate
d. Succinyl CoA to succinate

12.77 Identify the oxidized coenzyme (NAD^+ or FAD) that participates in each of the following citric acid cycle reactions.
a. Isocitrate $\longrightarrow$ α-ketoglutarate
b. Succinate $\longrightarrow$ fumarate

12.78 Identify the oxidized coenzyme (NAD^+ or FAD) that participates in each of the following citric acid cycle reactions.
a. Malate $\longrightarrow$ oxaloacetate
b. α-Ketoglutarate $\longrightarrow$ succinyl CoA

12.79 List the two citric acid cycle intermediates involved in the reaction governed by each of the following enzymes. List the reactant first.
a. Isocitrate dehydrogenase
b. Fumarase
c. Malate dehydrogenase
d. Aconitase

12.80 List the two citric acid cycle intermediates involved in the reaction governed by each of the following enzymes. List the reactant first.
a. α-Ketoglutarate dehydrogenase
b. Succinate dehydrogenase
c. Citrate synthase
d. Succinyl CoA synthetase

The Electron Transport Chain (Section 12.8)

12.81 By what other name is the electron transport chain known?

12.82 Give a one-sentence summary of what occurs during the reactions known as the electron transport chain.

12.83 What is the final electron acceptor of the electron transport chain?

12.84 Which substances generated in the citric acid cycle participate in the electron transport chain?

12.85 Give the abbreviation for each of the following electron carriers.
a. The oxidized form of flavin mononucleotide
b. The reduced form of coenzyme Q

12.86 Give the abbreviation for each of the following electron carriers.
a. The reduced form of flavin mononucleotide
b. The oxidized form of coenzyme Q

12.87 Indicate whether each of the following electron carriers is in its oxidized form or its reduced form.
a. Fe(III)SP
b. Cyt b (Fe^{3+})
c. NADH
d. FAD

12.88 Indicate whether each of the following electron carriers is in its oxidized form or its reduced form.
a. $FMNH_2$
b. Fe(II)SP
c. Cyt c_1 (Fe^{2+})
d. NAD^+

12.89 Indicate whether each of the following changes represents oxidation or reduction.
a. $CoQH_2$ $\longrightarrow$ CoQ
b. NAD^+ $\longrightarrow$ NADH
c. Cyt c (Fe^{2+}) $\longrightarrow$ cyt c (Fe^{3+})
d. Cyt b (Fe^{3+}) $\longrightarrow$ cyt b (Fe^{2+})

12.90 Indicate whether each of the following changes represents oxidation or reduction.
a. $FADH_2$ $\longrightarrow$ FAD
b. FMN $\longrightarrow$ $FMNH_2$
c. Fe(III)SP $\longrightarrow$ Fe(II)SP
d. Cyt c_1 (Fe^{3+}) $\longrightarrow$ cyt c_1 (Fe^{2+})

12.91 With which of the protein complexes (I, II, III, and IV) of the ETC is each of the following electron carriers associated? More than one answer may apply in a given situation.
a. NADH
b. CoQ
c. Cyt b
d. Cyt a

12.92 With which of the protein complexes (I, II, III, and IV) of the ETC is each of the following electron carriers associated? More than one answer may apply in a given situation.
a. $FADH_2$ b. FeSP c. Cyt c d. Cyt c_1

12.93 Which electron carrier shuttles electrons between protein complexes I and III?

12.94 Which electron carrier shuttles electrons between protein complexes II and III?

12.95 How many electrons does the electron carrier between complexes II and III carry per "trip"?

12.96 How many electrons does the electron carrier between complexes III and IV carry per "trip"?

12.97 Fill in the missing substances in the following electron transport chain reaction sequences.

a.
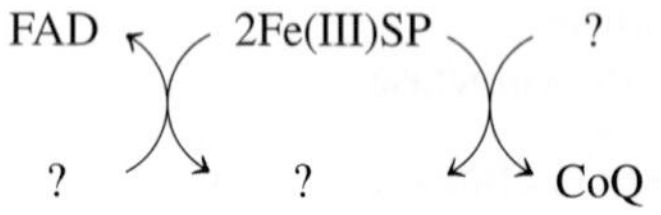

b.
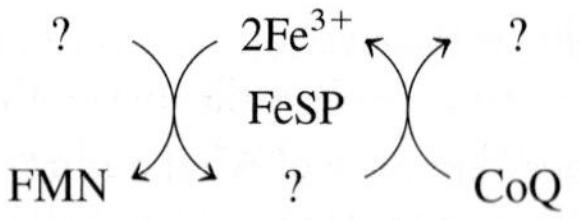

12.98 Fill in the missing substances in the following electron transport chain reaction sequences.

a.
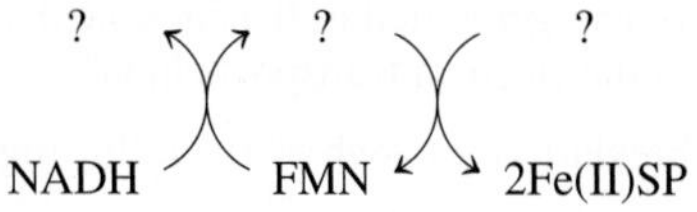

b.
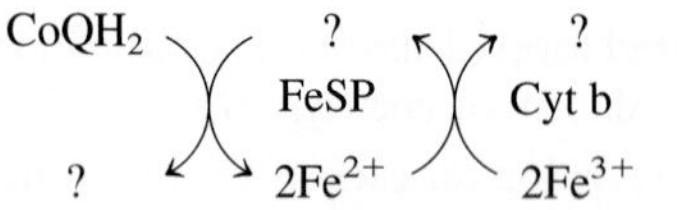

▲12.99 Classify each of the following substances as (1) a reactant in the CAC (2) a reactant in the ETC or (3) a reactant in both the CAC and the ETC.
a. NADH b. O_2
c. Fumarate d. Cytochrome a

▲12.100 Classify each of the following substances as (1) a product in the CAC (2) a product in the ETC or (3) a product in both the CAC and the ETC.
a. FAD b. CO_2
c. Malate d. $FMNH_2$

▲12.101 Put the following substances in the correct order in which they are first encountered in the common metabolic pathway: succinate, FeSP, CO_2, $FADH_2$.

▲12.102 Put the following substances in the correct order in which they are first encountered in the common metabolic pathway: $CoQH_2$, GTP, O_2, isocitrate.

▲12.103 Indicate whether each of the following substances is (1) oxidized but not reduced (2) reduced but not oxidized (3) both oxidized and reduced or (4) neither oxidized nor reduced in the common metabolic pathway.
a. Oxaloacetate b. NAD^+
c. $FADH_2$ d. FeSP

▲12.104 Indicate whether each of the following substances is (1) oxidized but not reduced (2) reduced but not oxidized (3) both oxidized and reduced or (4) neither oxidized nor reduced in the common metabolic pathway.
a. Malate b. NADH c. FAD d. Cyt b

Oxidative Phosphorylation (Section 12.9)

12.105 What is oxidative phosphorylation?

12.106 What are coupled reactions?

12.107 The coupling of ATP synthesis with the reactions of the ETC is related to the movement of what chemical species across the inner mitochrondial membrane?

12.108 At what protein complex location(s) in the electron transport chain does proton pumping occur?

12.109 At what mitochondrial location does H^+ ion buildup occur as the result of proton pumping?

12.110 How many protons cross the inner mitochondrial membrane for every two electrons that are passed through the electron transport chain?

12.111 What is the name of the enzyme that catalyzes ATP production during oxidative phosphorylation?

12.112 What is the location of the enzyme that uses stored energy in a proton gradient to drive the reaction that produces ATP?

12.113 In oxidative phosphorylation, what is oxidized and what is phosphorylated?

12.114 In oxidative phosphorylation, what are the identities of the
a. three reactants b. two products

•12.115 (Chemical Connections 12-A) Indicate whether each of the following statements relating to the effects of hydrogen cyanide on ETC operation is true or false.
a. Hydrogen cyanide exerts its toxic effects by inactivating the first enzyme complex in the ETC chain.
b. In treating persons suffering from hydrogen cyanide poisoning, NO_2^- ions are used to oxidize $Hb(Fe^{2+})$ to $Hb(Fe^{3+})$.
c. Amygdalin and laetrile are two names for the same substance.
d. The flesh of apricots and peaches contains amygdalin.

•12.116 (Chemical Connections 12-A) Indicate whether each of the following statements relating to the effects of hydrogen cyanide on ETC operation is true or false.
a. Hydrogen cyanide exerts its toxic effects by bonding to copper atoms present in the fourth enzyme complex in the ETC chain.
b. In treating persons suffering from hydrogen cyanide poisoning, $S_2O_3^{2-}$ ions are used to convert CN^- ions to SCN^- ions.
c. Amygdalin contains two glucose residues as part of its structure.
d. Hydrogen cyanide is one of the products produced from enzymatic hydrolysis of amygdalin.

•12.117 (Chemical Connections 12-B) Indicate whether each of the following statements relating to brown fat production and its use in the human body is true or false.
a. Brown fat present in newborn infants helps them fight off infections.
b. Brown fat cells contain a protein called thermogenin, which functions as an ETC uncoupling agent.
c. Activated brown fat functions as a channel through which ATP can cross the inner mitochondrial membrane.
d. Hibernating animals rely on brown fat tissue as a supplemental source for body heat production.

•**12.118** (Chemical Connections 12-B) Indicate whether each of the following statements relating to brown fat production and its use in the human body is true or false.
a. Adults have more brown fat tissue than do young children.
b. Brown fat production is a mechanism by which the human body adapts to cold environs.
c. ATP production cannot be disassociated from the operation of the ETC chain.
d. The protein thermogenin present in brown fat tissue significantly increases ATP production associated with the ETC.

ATP Production (Section 12.10)

12.119 How many ATP molecules are formed for
a. each NADH molecule that enters the ETC?
b. each $FADH_2$ molecule that enters the ETC?

12.120 NADH and $FADH_2$ molecules do not yield the same number of ATP molecules. Explain why.

12.121 In which step(s) of the CAC are ATP-precursor-molecules generated that are equivalent to
a. 2.5 units of ATP b. 1.5 units of ATP
c. 1.0 unit of ATP

12.122 In which step(s) of the CAC are ATP-precursor-molecules generated that
a. enter the ETC at enzyme complex I
b. enter the ETC at enzyme complex II
c. do not enter the ETC

The Importance of ATP (Section 12.11)

12.123 How does the free-energy release associated with the hydrolysis of ATP to ADP compare with the hydrolysis energies of other high-energy phosphate compounds?

12.124 What would the biochemical consequences be if the free-energy release associated with the hydrolysis of ATP to ADP had an unusually high value?

12.125 Indicate whether each of the following processes would be expected to involve the conversion of ATP to ADP or the conversion of ADP to ATP.
a. Heart muscle contraction
b. Transport of nutrients to various locations in the body

12.126 Indicate whether each of the following processes would be expected to involve the conversion of ATP to ADP or the conversion of ADP to ATP.
a. Degradation of dietary carbohydrates
b. Synthesis of protein from amino acids

Non-ETC Oxygen-Consuming Reactions (Section 12.12)

12.127 What does the designation ROS stand for?

12.128 Give the chemical formula for each of the following.
a. Superoxide ion b. Hydroxyl radical

12.129 Give the chemical equation for the reaction by which
a. superoxide ion is generated within cells.
b. superoxide ion is converted to hydrogen peroxide within cells.

12.130 Give the chemical equation for the reaction by which
a. hydrogen peroxide is converted to desirable products within cells.
b. hydrogen peroxide is converted to an undesirable product within cells.

•12.131 (Chemical Connections 12-C) Indicate whether each of the following statements concerning phytochemicals called flavonoids is true or false.
a. Flavonoids are a group of phytochemicals that deactivate the B vitamin riboflavin.
b. The core part of a flavonoid structure contains both aromatic and cyclic ether ring systems.
c. The formation of flavonoids found in food is usually dependent on the action of sunlight.
d. The major sources of the flavonoid quercetin are apples and onions.

•**12.132** (Chemical Connections 12-C) Indicate whether each of the following statements concerning phytochemicals called flavonoids is true or false.
a. Flavonoids are a group of phytochemicals that have antioxidant properties.
b. The fleshy parts of fruits and vegetables have a higher flavonoid content than does their skin.
c. Structurally, the flavonoid quercetin is a polyhydroxy compound.
d. The most abundant flavonoid in food is the compound quercetin.

B Vitamins and the Common Metabolic Pathway (Section 12.13)

12.133 Indicate whether or not each of the following B vitamins is needed for the proper functioning of the CAC.
a. Thiamin b. Biotin
c. Niacin d. Vitamin B_6

12.134 Indicate whether or not each of the following B vitamins is needed for the proper functioning of the CAC.
a. Riboflavin b. Pantothenic acid
c. Folate d. Vitamin B_{12}

12.135 Indicate whether or not each of the following B vitamins is needed for the proper functioning of the ETC.
a. Thiamin b. Riboflavin
c. Folate d. Vitamin B_{12}

12.136 Indicate whether or not each of the following B vitamins is needed for the proper functioning of the ETC.
a. Niacin b. Biotin
c. Pantothenic acid d. Vitamin B_6

Carbohydrate Metabolism

13

© John Kropewnicki/Shutterstock.com

Carbohydrates are the major energy source for human beings.

CHAPTER OUTLINE

In this chapter the relationship between carbohydrate metabolism and energy production in cells is explored. The molecule glucose is the focal point of carbohydrate metabolism. Commonly called blood sugar, glucose is supplied to the body via the circulatory system and, after being absorbed by a cell, can be either oxidized to yield energy or stored as glycogen for future use. When sufficient oxygen is present, glucose is totally oxidized to CO_2 and H_2O. However, in the absence of oxygen, glucose is only partially oxidized to lactic acid. Besides supplying energy needs, glucose and other six-carbon sugars can be converted into a variety of different sugars (C_3, C_4, C_5, and C_7) needed for biosynthesis. Some of the oxidative steps in carbohydrate metabolism also produce NADH and NADPH, sources of reductive power in cells.

13.1 Digestion and Absorption of Carbohydrates

Digestion *is the biochemical process by which food molecules, through hydrolysis, are broken down into simpler chemical units that can be used by cells for their metabolic needs.* Digestion is the first stage in the processing of food products.

The digestion of carbohydrates begins in the mouth, where the enzyme *salivary α-amylase* catalyzes the hydrolysis of α-glycosidic linkages (Section 7.13) in starch from plants and glycogen from meats to produce smaller polysaccharides and the disaccharide maltose.

Only a small amount of carbohydrate digestion occurs in the mouth because food is swallowed so quickly. Although the food mass remains longer in the stomach, very little further carbohydrate digestion occurs there either, because *salivary*

OWL

Sign in to OWL at **www.cengage.com/owl** to view tutorials and simulations, develop problem-solving skills, and complete online homework assigned by your professor.

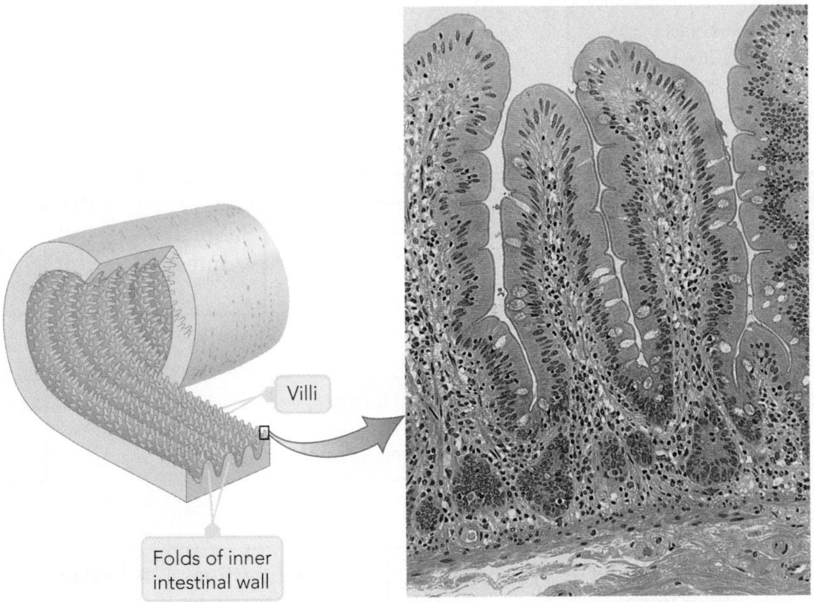

Figure 13.1 A section of the small intestine, showing its folds and the villi that cover the inner surface of the folds. Villi greatly increase the inner intestinal surface area.

α-amylase is inactivated by the acidic environment of the stomach, and the stomach's own secretions do not contain any carbohydrate-digesting enzymes.

The primary site for carbohydrate digestion is within the small intestine, where α-amylase, this time secreted by the pancreas, again begins to function. The *pancreatic α-amylase* breaks down polysaccharide chains into shorter and shorter segments until the disaccharide maltose (two glucose units; Section 7.13) and glucose itself are the dominant species.

The final step in carbohydrate digestion occurs on the outer membranes of intestinal mucosal cells, where the enzymes that convert disaccharides to monosaccharides are located. The important disaccharidase enzymes are *maltase, sucrase,* and *lactase.* These enzymes convert, respectively, maltose to two glucose units, sucrose to one glucose and one fructose unit, and lactose to one glucose and one galactose unit (Section 7.13). (The disaccharides sucrose and lactose present in food are not digested until they reach this point.)

The three major breakdown products from carbohydrate digestion are thus glucose, galactose, and fructose. These monosaccharides are absorbed into the bloodstream through the intestinal wall. The folds of the intestinal wall are lined with fingerlike projections called *villi,* which are rich in blood capillaries (Figure 13.1). Absorption is by *active transport* (Section 8.10), which, unlike passive transport, is an energy-requiring process. In this case, ATP is needed. Protein carriers mediate the passage of the monosaccharides through cell membranes. Figure 13.2 summarizes the different phases in the digestive process for carbohydrates.

After their absorption into the bloodstream, monosaccharides are transported to the liver, where fructose and galactose are rapidly converted into compounds that are metabolized by the same pathway as glucose. Thus the central focus of carbohydrate metabolism is the pathway by which glucose is further processed, a pathway called *glycolysis* (Section 13.2)—a series of ten reactions, each of which involves a different enzyme.

Salivary α-amylase is a constituent of saliva, the fluid secreted by the salivary glands. Saliva is 99% water plus small amounts of several inorganic ions and organic molecules. Saliva secretion can be triggered by the taste, smell, sight, and even thought of food. Average saliva output is about 1.5 L per day.

EXAMPLE 13.1 Determining the Sites Where Various Aspects of Carbohydrate Digestion Occur

Based on the information in Figure 13.2, determine the location within the human body where each of the following aspects of carbohydrate digestion occurs.

a. The enzyme sucrase is active.
b. Hydrolysis reactions converting polysaccharides to disaccharides occur.
c. First site where breaking of glycosidic linkages occurs.
d. The monosaccharides glucose, fructose, and galactose are produced.

(continued)

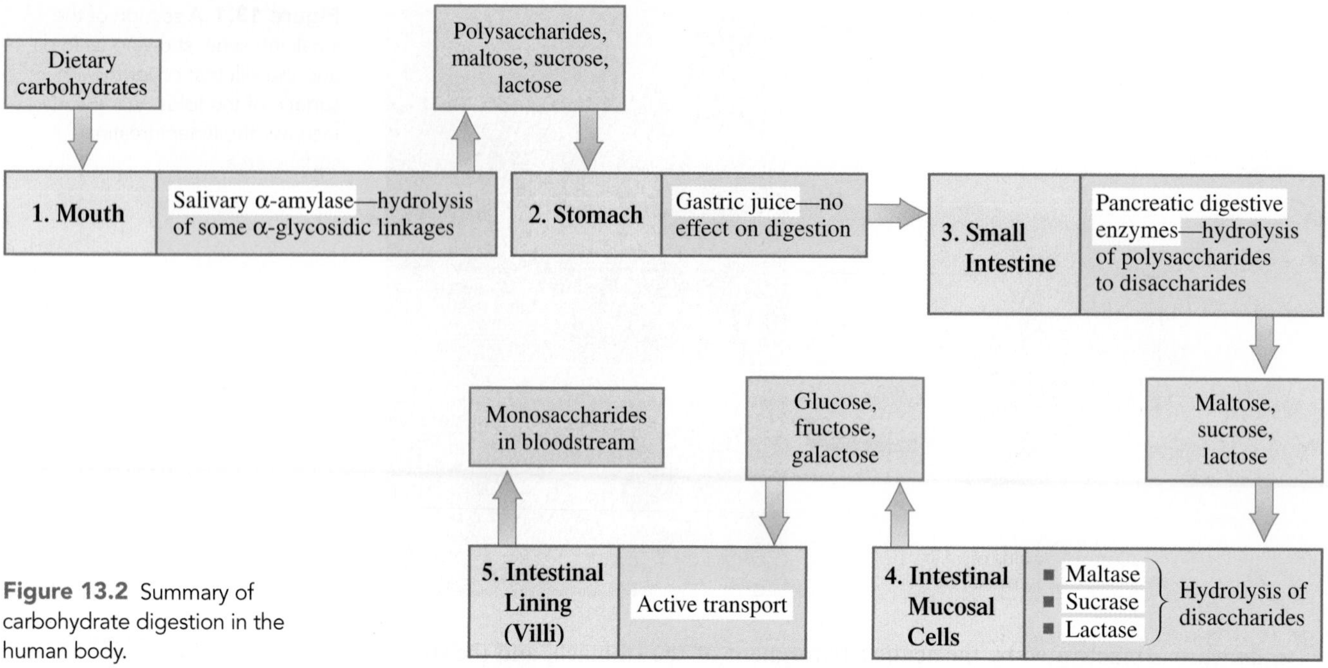

Figure 13.2 Summary of carbohydrate digestion in the human body.

Solution

a. *Intestinal mucosal cells* are the sites where hydrolysis of disaccharides, effected by the enzymes maltase, sucrase, and lactase, occurs.

b. The *small intestine* is the location where pancreatic digestive enzymes convert polysaccharides to disaccharides.

c. Digestion of carbohydrates begins in the *mouth,* where salivary enzymes convert some polysaccharides to smaller polysaccharides through breakage of glycosidic linkages.

d. *Intestinal mucosal cells* are the sites where disaccharides, through hydrolysis, are converted to the monosaccharides glucose, fructose, and galactose.

Practice Exercise 13.1

Based on the information in Figure 13.2, determine the location within the human body where each of the following aspects of carbohydrate digestion occurs.

a. Pancreatic enzymes are active.

b. Hydrolysis reactions converting disaccharides to monosaccharides occur.

c. Monosaccharides enter the bloodstream.

d. The primary site for carbohydrate digestion is located here.

Answers: **a.** Small intestine; **b.** Intestinal mucosal cells; **c.** Intestinal lining villi; **d.** Small intestine

13.2 Glycolysis

The term *glycolysis,* pronounced "gligh-KOLL-ih-sis," comes from the Greek *glyco,* meaning "sweet," and *lysis,* meaning "breakdown."

Pyruvate, pronounced "PIE-roo-vate," is the carboxylate ion (Section 5.7) produced when pyruvic acid (a three-carbon keto acid) loses its acidic hydrogen atom.

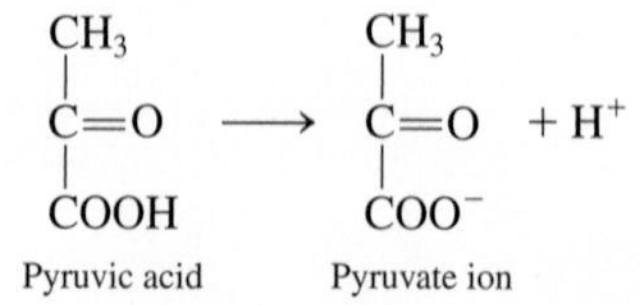

Glycolysis *is the metabolic pathway by which glucose (a C_6 molecule) is converted into two molecules of pyruvate (a C_3 molecule), chemical energy in the form of ATP is produced, and NADH-reduced coenzymes are produced.* It is a linear rather than cyclic pathway (see Section 12.1) that functions in almost all cells.

The conversion of glucose to pyruvate is an oxidation process in which no molecular oxygen is utilized. The oxidizing agent is the coenzyme NAD^+. Metabolic pathways in which molecular oxygen is not a participant are called *anaerobic* pathways. Pathways that require molecular oxygen are called *aerobic* pathways. Glycolysis is an anaerobic pathway.

Glycolysis is a ten-step process (compared to the eight steps of the citric acid cycle; Section 12.7) in which every step is enzyme-catalyzed. Figure 13.3 gives an

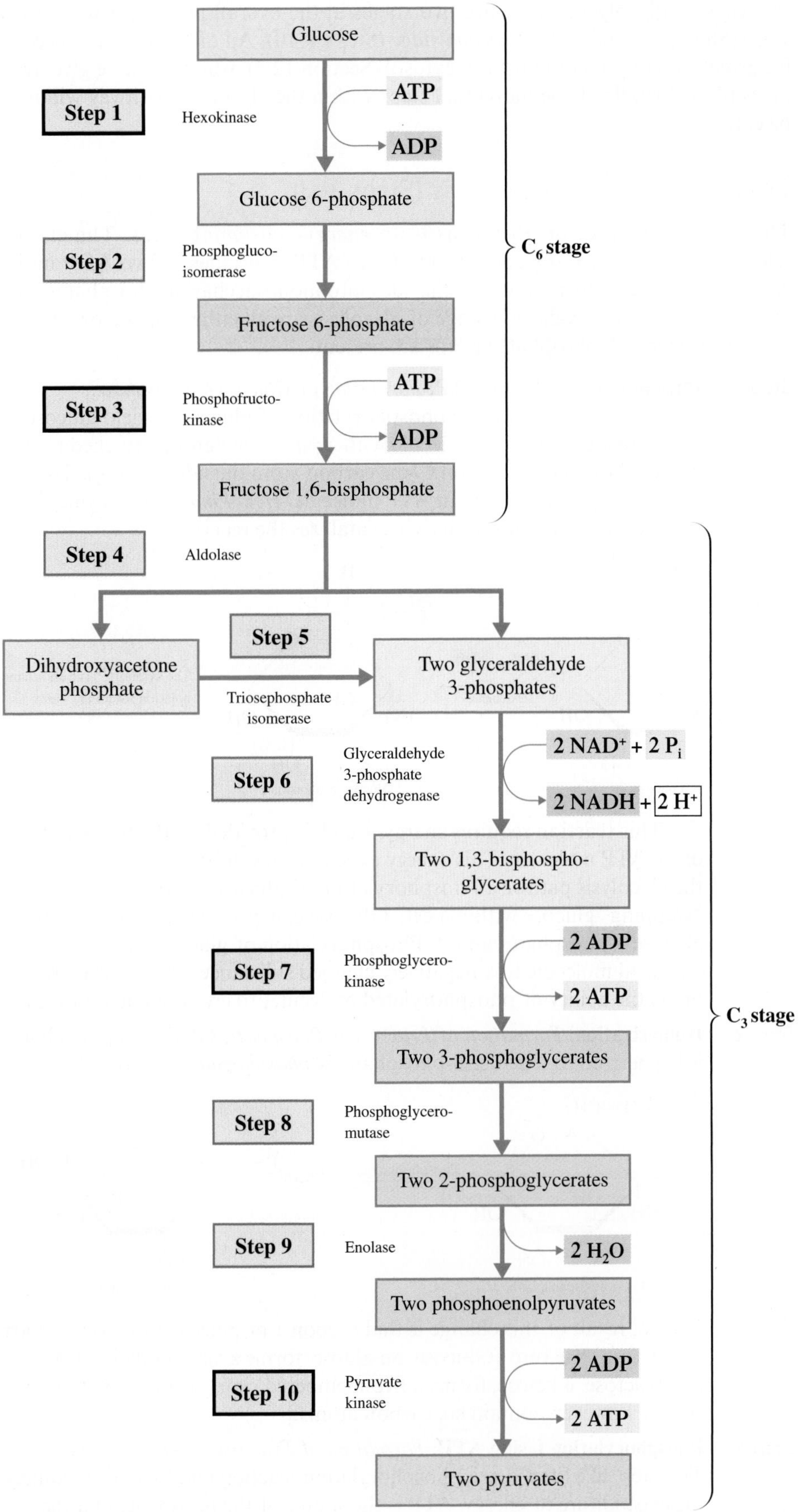

Figure 13.3 An overview of glycolysis.

Anaerobic is pronounced "AN-air-ROE-bic." *Aerobic* is pronounced "air-ROE-bic."

Glycolysis is also called the *Embden–Meyerhof pathway* after the German chemists Gustav Embden (1874–1933) and Otto Meyerhof (1884–1951), who discovered many of the details of the pathway in the early 1930s.

overview of glycolysis. There are two stages in the overall process, a *six-carbon stage* (Steps 1–3) and a *three-carbon stage* (Steps 4–10). All of the enzymes needed for glycolysis are present in the cell cytosol (Section 12.2), which is where glycolysis takes place. Details of the individual steps within the glycolysis pathway will now be considered.

Six-Carbon Stage of Glycolysis (Steps 1–3)

The six-carbon stage of glycolysis is an *energy-consuming stage.* The energy release associated with the conversion of two ATP molecules to two ADP molecules is used to transform monosaccharides into monosaccharide phosphates. The intermediates of the six-carbon stage of glycolysis are all either *glucose* or *fructose* derivatives in which phosphate groups are present.

Step 1: **Phosphorylation Using ATP:** *Formation of Glucose 6-Phosphate.* Glycolysis begins with the phosphorylation of glucose to yield glucose 6-phosphate, a glucose molecule with a phosphate group attached to the hydroxyl oxygen on carbon 6 (the carbon atom outside the ring). The phosphate group is from an ATP molecule. *Hexokinase,* an enzyme that requires Mg^{2+} ion for its activity, catalyzes the reaction.

A *kinase* is an enzyme that catalyzes the transfer of a phosphoryl group (PO_3^{2-}) from ATP (or some other high-energy phosphate compound) to a substrate (Section 10.3).

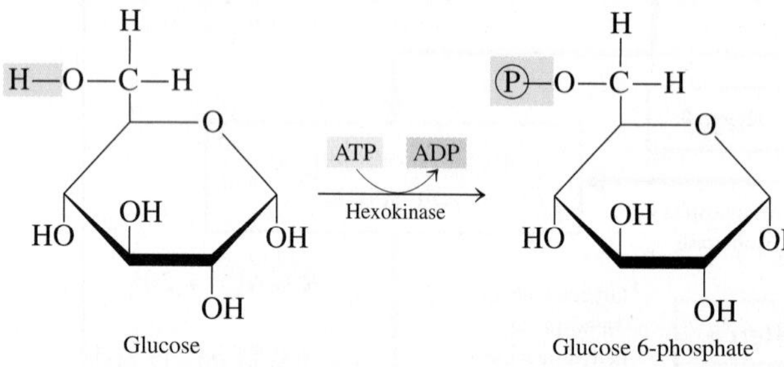

The symbol Ⓟ is a shorthand notation for a PO_3^{2-} unit.

Step 1 of glycolysis is the first of two steps in which an ATP molecule is converted to an ADP with the energy released used to effect a phosphorylation. The other ATP-phosphorylation reaction occurs in Step 3.

This reaction requires energy, which is provided by the breakdown of an ATP molecule. This energy expenditure will be recouped later in the glycolysis pathway. Phosphorylation of glucose provides a way of "trapping" glucose within a cell. Glucose can cross cell membranes, but glucose 6-phosphate cannot. Phosphorylation of glucose changes it from a neutral molecule to a negatively charged substance; charge severely limits the ability of phosphorylated molecules to cross cell membranes.

Step 2: **Isomerization:** *Formation of Fructose 6-Phosphate.* Glucose 6-phosphate is isomerized to fructose 6-phosphate by *phosphoglucoisomerase.*

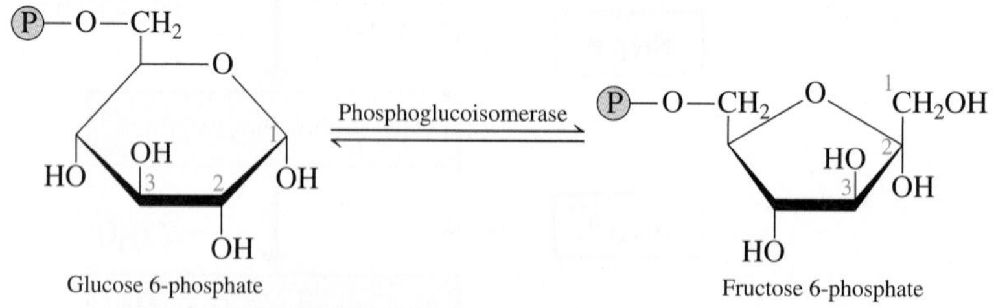

The net result of this change is that carbon 1 of glucose is no longer part of the ring structure. [Glucose, an aldose, forms a six-membered ring, and fructose, a ketose, forms a five-membered ring (Section 7.10); both sugars, however, contain six carbon atoms.]

Step 3 of glycolysis commits the original glucose molecule to the glycolysis pathway. Glucose 6-phosphate (Step 1) and fructose 6-phosphate (Step 2) can enter other metabolic pathways, but fructose 1,6-bisphosphate can enter only glycolysis.

Step 3: **Phosphorylation Using ATP:** *Formation of Fructose 1,6-Bisphosphate.* This step, like Step 1, is a phosphorylation reaction and therefore requires the expenditure of energy. ATP is the source of the phosphate and the energy. The enzyme involved, *phosphofructokinase,* is another enzyme that requires Mg^{2+} ion for its activity. The fructose molecule now contains two phosphate groups.

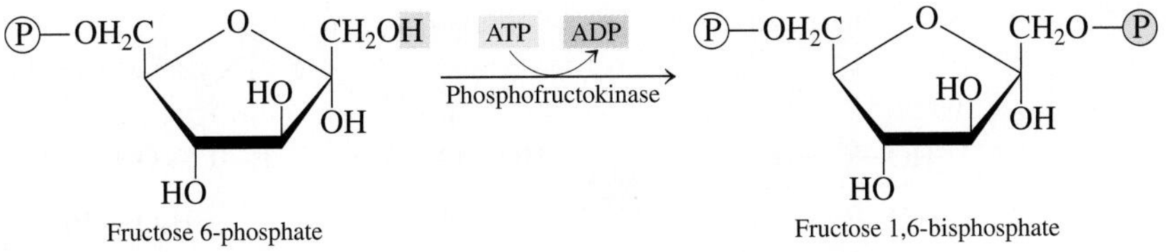

Step 3 of glycolysis, like Step 1, is a step in which phosphorylation is effected as an ATP molecule is converted to an ADP molecule.

The term *bisphosphate* is used instead of *diphosphate* to indicate the two phosphates are on different carbon atoms in fructose and not connected to each other.

Three-Carbon Stage of Glycolysis (Steps 4–10)

The three-carbon stage of glycolysis is an *energy-generating stage* rather than an energy-consuming stage. All of the intermediates in this stage are C_3-phosphates, two of which are high-energy phosphate species (Section 12.5). Loss of a phosphate from these high-energy species effects the conversion of ADP molecules to ATP molecules.

The C_3 intermediates in this stage of glycolysis are all phosphorylated derivatives of *dihydroxyacetone, glyceraldehyde, glycerate,* or *pyruvate,* which in turn are derivatives of either glycerol or acetone. Figure 13.4 shows the structural relationships among these molecules.

Step 4: **Cleavage:** *Formation of Two Triose Phosphates.* In this step, the reacting C_6 species is split into two C_3 (triose) species. Because fructose 1,6-bisphosphate, the molecule being split, is unsymmetrical, the two trioses produced are not identical. One product is dihydroxyacetone phosphate, and the other is glyceraldehyde 3-phosphate. *Aldolase* is the enzyme that catalyzes this reaction. A better understanding of the structural relationships between reactant and products is obtained if the fructose 1,6-bisphosphate is written in its open-chain form (Section 7.10) rather than in its cyclic form.

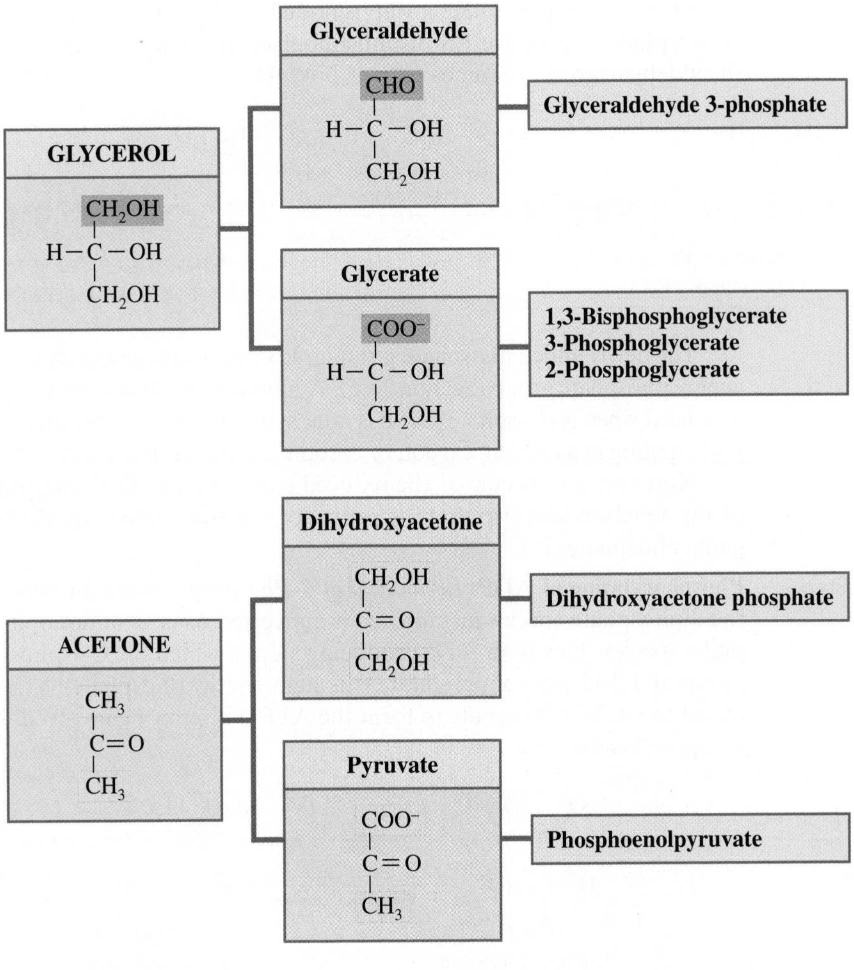

Figure 13.4 Structural relationships among glycerol and acetone and the C_3 intermediates in the process of glycolysis.

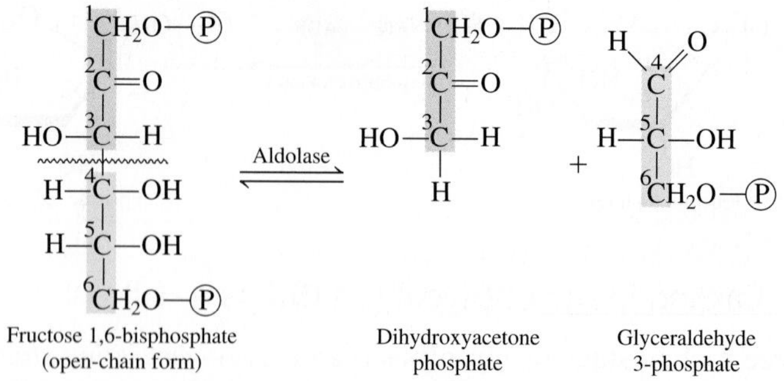

Step 5: **Isomerization:** *Formation of Glyceraldehyde 3-Phosphate.* Only one of the two trioses produced in Step 4, glyceraldehyde 3-phosphate, is a glycolysis intermediate. Dihydroxyacetone phosphate, the other triose, can, however, be readily converted into glyceraldehyde 3-phosphate. Dihydroxyacetone phosphate (a ketose) and glyceraldehyde 3-phosphate (an aldose) are isomers, and the isomerization process from ketose to aldose is catalyzed by the enzyme *triosephosphate isomerase*.

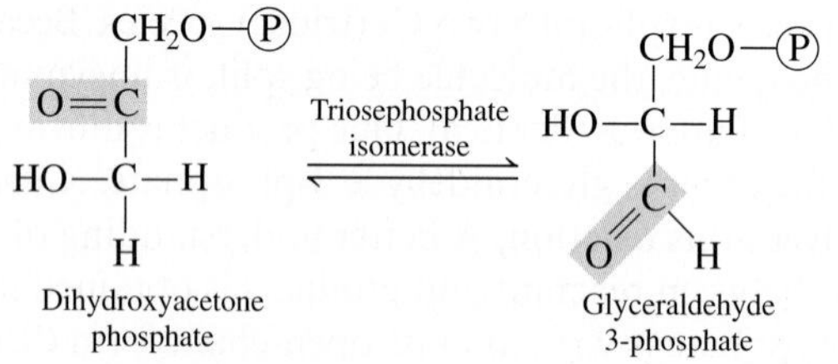

Step 6: **Oxidation and Phosphorylation Using P_i:** *Formation of 1,3-Bisphosphoglycerate*. In a reaction catalyzed by *glyceraldehyde 3-phosphate dehydrogenase,* a phosphate group is added to glyceraldehyde 3-phosphate to produce 1,3-bisphosphoglycerate. The hydrogen of the aldehyde group becomes part of NADH.

Step 6 is the first of two glycolysis steps in which a high-energy phosphate compound, that is, an "energy-rich" compound, is formed. The other high-energy phosphate compound is formed in Step 9. In the step that follows each of these two situations, energy from the high-energy phosphate compound is used to convert ADP to ATP.

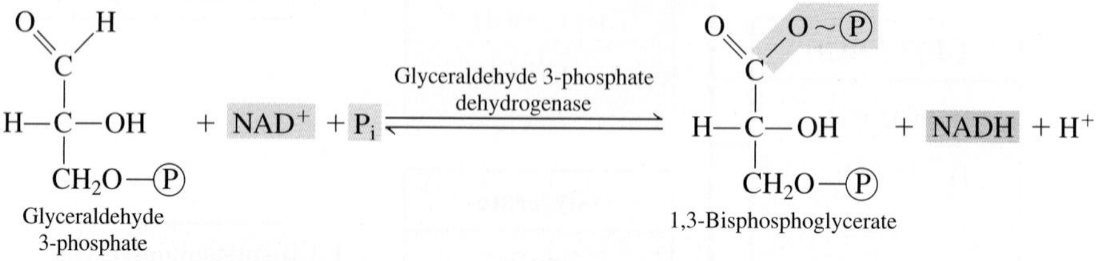

Keep in mind that from Step 6 onward, two molecules of each of the C_3 compounds take part in every reaction for each original C_6 glucose molecule.

The newly added phosphate group in 1,3-bisphosphoglycerate is a high-energy phosphate group (Section 12.5). A high-energy phosphate group is produced when a phosphate group is attached to a carbon atom that is also participating in a carbon–carbon or carbon–oxygen double bond.

Note that a molecule of the reduced coenzyme NADH is a product of this reaction and also that the source of the added phosphate is inorganic phosphate (P_i).

Step 7: **Phosphorylation of ADP:** *Formation of 3-Phosphoglycerate.* In this step, the diphosphate species just formed is converted back to a monophosphate species. This is an ATP-producing step in which the C-1 phosphate group of 1,3-bisphosphoglycerate (the high-energy phosphate) is transferred to an ADP molecule to form the ATP. The enzyme involved is *phosphoglycerokinase.*

Step 7 is the first of two steps in which ATP is formed from ADP. This process, called *substrate-level phosphorylation,* always involves a high-energy phosphate compound. This same process also occurs in Step 10.

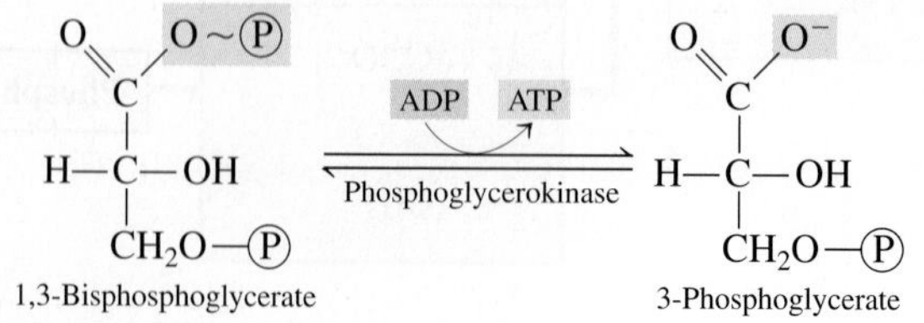

Remember that two ATP molecules are produced for each original glucose molecule because both C_3 molecules produced from the glucose react.

ATP production in this step involves substrate-level phosphorylation. **Substrate-level phosphorylation** *is the biochemical process whereby ATP is produced from ADP through direct transfer of a high-energy phosphoryl group from a reaction substrate to ADP.* Substrate-level phosphorylation differs from *oxidative phosphorylation* (Section 12.9) in that the latter involves production of ATP from ADP using P_i (free phosphate) and the energy "harvested" from the oxidation-reduction reactions of the electron transport chain.

Step 8: **Isomerization:** *Formation of 2-Phosphoglycerate.* In this isomerization step, the phosphate group of 3-phosphoglycerate is moved from carbon 3 to carbon 2. The enzyme *phosphoglyceromutase* catalyzes the exchange of the phosphate group between the two carbons.

A *mutase* is an enzyme that effects the shift of a functional group from one position to another within a molecule (Section 10.2).

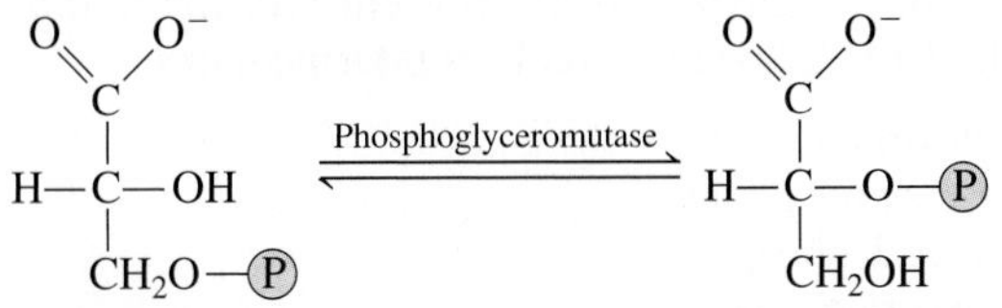

Step 9: **Dehydration:** *Formation of Phosphoenolpyruvate.* This is an alcohol dehydration reaction that proceeds with the enzyme *enolase,* another Mg^{2+}-requiring enzyme. The result is another compound containing a high-energy phosphate group; the phosphate group is attached to a carbon atom that is involved in a carbon–carbon double bond.

An *enol* (from *ene* + *ol*), as in phos*phoenol*pyruvate, is a compound in which an —OH group is attached to a carbon atom involved in a carbon–carbon double bond. Note that in phosphoenolpyruvate, the —OH group has been phosphorylated.

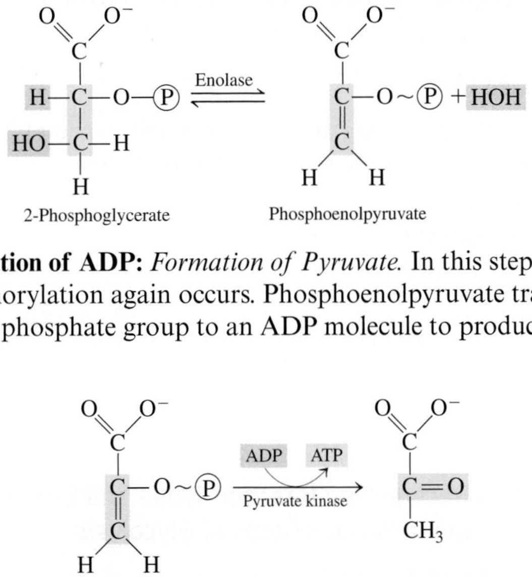

Step 9 is the second of two glycolytic steps in which a high-energy phosphate compound, that is, "energy-rich" compound, is formed; the other step was Step 6. In the next step, energy from the high-energy phosphate compound will be used to convert ADP and ATP.

Step 10: **Phosphorylation of ADP:** *Formation of Pyruvate.* In this step, substrate-level phosphorylation again occurs. Phosphoenolpyruvate transfers its high-energy phosphate group to an ADP molecule to produce ATP and pyruvate.

Step 10 is the second of two steps in which ATP is formed from ADP. This same process also occurred in Step 7.

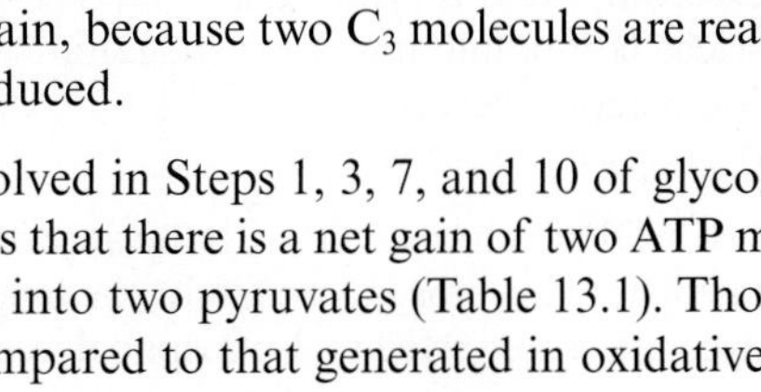

The enzyme involved, *pyruvate kinase,* requires both Mg^{2+} and K^+ ions for its activity. Again, because two C_3 molecules are reacting, two ATP molecules are produced.

ATP molecules are involved in Steps 1, 3, 7, and 10 of glycolysis. Considering these steps collectively shows that there is a net gain of two ATP molecules for every glucose molecule converted into two pyruvates (Table 13.1). Though useful, this is a small amount of ATP compared to that generated in oxidative phosphorylation (Section 12.9).

The net overall equation for the process of glycolysis is

$$\text{Glucose} + 2\text{NAD}^+ + 2\text{ADP} + 2\text{P}_i \longrightarrow 2\text{ pyruvate} + 2\text{NADH} + 2\text{ATP} + 2\text{H}^+ + 2\text{H}_2\text{O}$$

Table 13.1 ATP Production and Consumption During Glycolysis

Step	Reaction	ATP Change per Glucose
1	Glucose → glucose 6-phosphate	−1
3	Fructose 6-phosphate → fructose 1,6-bisphosphate	−1
7	2(1,3-Bisphosphoglycerate → 3-phosphoglycerate)	+2
10	2(Phosphoenolpyruvate → pyruvate)	+2
		Net +2

EXAMPLE 13.2 Recognizing Structural Characteristics of Glycolysis Intermediates

Specify the number of carbon atoms present and the number of phosphate groups present in each of the following glycolysis intermediates.

a. Glucose 6-phosphate
b. Fructose 1,6-bisphosphate
c. Glyceraldehyde 3-phosphate
d. 3-Phosphoglycerate

Solution

a. Glucose is a hexose. It is a *C_6 molecule* that carries *one phosphate group,* attached to carbon 6.
b. Fructose is also a hexose. It is a *C_6 molecule* that carries *two phosphate groups,* one attached to carbon 1 and the other attached to carbon 6.
c. Glyceraldehyde is a *C_3 molecule* with *one phosphate group,* attached to carbon 3.
d. Glycerate, like glyceraldehyde, is a *C_3 molecule* with *one phosphate group,* attached to carbon 3.

Practice Exercise 13.2

Specify the number of carbon atoms present and the number of phosphate groups present in each of the following glycolysis intermediates.

a. Fructose 6-phosphate
b. 1,3-Bisphosphoglycerate
c. 2-Phosphoglycerate
d. Phosphoenolpyruvate

Answers: **a.** C_6 with one phosphate; **b.** C_3 with two phosphates; **c.** C_3 with one phosphate; **d.** C_3 with one phosphate

EXAMPLE 13.3 Recognizing Reaction Types and Events That Occur in the Various Steps of Glycolysis

Indicate at what step in the glycolysis pathway each of the following events occurs.

a. First phosphorylation of ADP occurs.
b. First "energy-rich" compound is produced.
c. Second "energy-rich" compound undergoes reaction.
d. First isomerization reaction occurs.

Solution

a. Phosphorylation of ADP produces ATP with the added phosphate coming from a glycolysis intermediate. This process occurs for the first time in *Step 7,* where 1,3-bisphosphoglycerate is converted to 3-phosphoglycerate. It occurs a second time in Step 10.
b. In carbohydrate metabolism, an "energy-rich" compound is a high-energy phosphate. Two such compounds are produced during glycolysis, the first in *Step 6* and the second in Step 9. The Step 6 compound is 1,3-bisphosphoglycerate.

c. The second "energy-rich" compound is phosphoenolpyruvate, which is produced in Step 9 and undergoes reaction in *Step 10.*
d. There are three isomerization reactions in glycolysis. They occur in Steps 2, 5, and 8. In the first occurrence, *Step 2*, glucose 6-phosphate is converted to fructose 6-phosphate. Glucose and fructose are isomeric hexose (Section 7.8) molecules.

Practice Exercise 13.3

Indicate at what step in the glycolysis pathway each of the following events occurs.

a. Second formation of ATP occurs.
b. Second "energy-rich" compound is produced.
c. ATP is converted to ADP for the second time.
d. A dehydration reaction occurs.

Answers: **a.** Step 10; **b.** Step 9; **c.** Step 3; **d.** Step 9

EXAMPLE 13.4 Relating Enzyme Names and Functions to Steps in the Process of Glycolysis

Relate the names and functions of the following glycolytic enzymes to steps in the process of glycolysis.

a. Phosphofructokinase
b. Phosphoglyceromutase
c. Triosephosphate isomerase
d. Enolase

Solution

a. A kinase is an enzyme that is involved with phosphate group transfer. The phosphofructo portion of the enzyme name refers to a fructose phosphate. Transfer of a phosphate group to a fructose phosphate occurs in *Step 3.* Fructose 6-phosphate is converted to fructose 1,6-bisphosphate.
b. A mutase is an enzyme that shifts the position of a functional group within a molecule. Such a functional group shift occurs in *Step 8*, where 3-phosphoglycerate is isomerized to 2-phosphoglycerate. The phosphoglycero portion of the enzyme name indicates that a glycerate phosphate is the substrate for the enzyme.
c. A triosephosphate isomerase effects the isomerization of a triose. Such a process occurs in *Step 5*, where dihydroxyacetone phosphate (a ketone) is converted to glyceraldehyde 3-phosphate (an aldehyde). Ketones and aldehydes with the same number of carbon atoms are often isomers.
d. The only enol species in glycolysis is the compound phosphoenolpyruvate produced in *Step 9* through a dehydration reaction that introduces a carbon–carbon double bond in the molecule. An enolase effects such a change.

Practice Exercise 13.4

Relate the names and functions of the following glycolytic enzymes to steps in the process of glycolysis.

a. Hexokinase
b. Pyruvate kinase
c. Glyceraldehyde 3-phosphate dehydrogenase
d. Phosphoglycerokinase

Answers: **a.** Step 1; **b.** Step 10; **c.** Step 6; **d.** Step 7

Entry of Galactose and Fructose into Glycolysis

The breakdown products from carbohydrate digestion are glucose, galactose, and fructose (Section 13.1). Both galactose and fructose are converted, in the liver, to intermediates that enter into the glycolysis pathway.

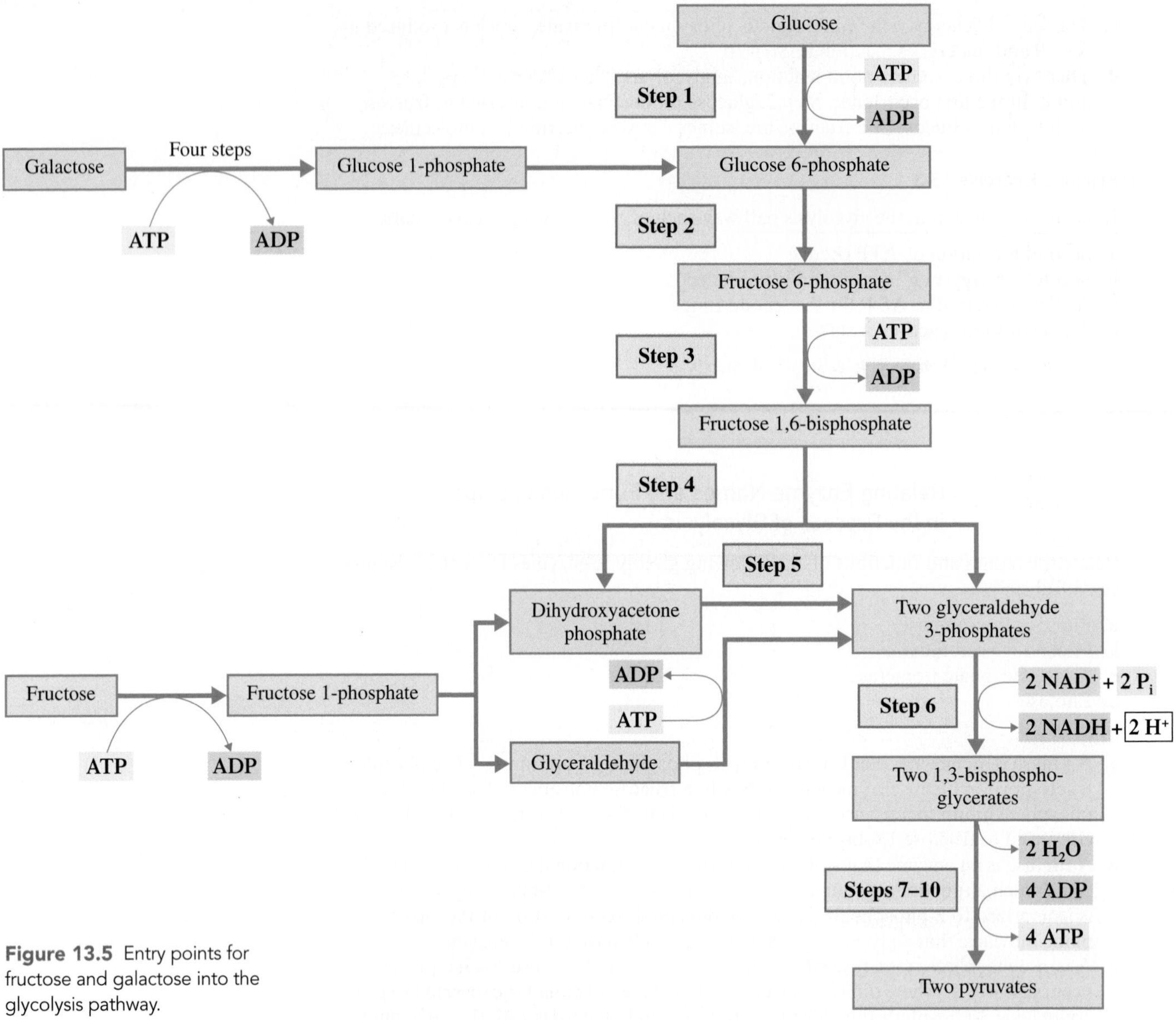

Figure 13.5 Entry points for fructose and galactose into the glycolysis pathway.

The entry of galactose into the glycolytic pathway begins with its conversion to glucose 1-phosphate (a four-step sequence), which is then converted to glucose 6-phosphate, a glycolysis intermediate (Figure 13.5).

The entry of fructose into the glycolytic pathway involves phosphorylation by ATP to produce fructose 1-phosphate, which is then split into two trioses—glyceraldehyde and dihydroxyacetone phosphate. Dihydroxyacetone phosphate enters glycolysis directly; glyceraldehyde must be phosphorylated by ATP to glyceraldehyde 3-phosphate before it enters the pathway (Figure 13.5).

Regulation of Glycolysis

Glycolysis, like all metabolic pathways, must have control mechanisms associated with it. In glycolysis, the control points are Steps 1, 3, and 10 (see Figure 13.3).

Step 1, the conversion of glucose to glucose 6-phosphate, involves the enzyme *hexokinase.* This particular enzyme is inhibited by glucose 6-phosphate, the substance produced by its action (feedback inhibition; Section 10.9).

At Step 3, where fructose 6-phosphate is converted to fructose 1,6-bisphosphate by the enzyme *phosphofructokinase,* high concentrations of ATP and citrate inhibit enzyme activity. A high ATP concentration, which is characteristic of a state of low energy consumption, thus stops glycolysis at the fructose 6-phosphate stage. This stoppage also causes increases in glucose 6-phosphate stores because glucose 6-phosphate is in equilibrium with fructose 6-phosphate.

The third control point involves the last step of glycolysis, the conversion of phosphoenolpyruvate to pyruvate. *Pyruvate kinase,* the enzyme needed at this point, is inhibited by high ATP concentrations. Both pyruvate kinase (Step 10) and phosphofructokinase (Step 3) are allosteric enzymes (Section 10.8).

13.3 Fates of Pyruvate

The production of pyruvate from glucose (glycolysis) occurs in a similar manner in most cells. In contrast, the fate of the pyruvate so produced varies with cellular conditions and the nature of the organism. Three common fates for pyruvate, all of importance, exist. They are conversion to acetyl CoA, conversion to lactate, and conversion to ethanol. As is shown in Figure 13.6, acetyl CoA formation requires aerobic (oxygen-rich) conditions, lactate and ethanol formation occur under anaerobic (oxygen-deficient) conditions, and ethanol formation is limited to some microorganisms.

An additional fate for pyruvate is conversion to oxaloacetate. This fate for pyruvate, which occurs during the process called *gluconeogenesis,* is discussed in Section 13.6.

A key concept in considering these fates of pyruvate is the need for a continuous supply of NAD^+ for glycolysis. As glucose is oxidized to pyruvate in glycolysis, NAD^+ is reduced to NADH.

$$\text{Glucose} + 2\text{NAD}^+ \xrightarrow[2\text{ADP} + 2\text{P}_i \;\to\; 2\text{ATP}]{} 2\text{ pyruvate} + 2\text{NADH} + 2\text{H}^+$$

It is significant that each of the three pathways for processing pyruvate, which will now be discussed, have a provision for regenerating NAD^+ from NADH so that glycolysis can continue.

Oxidation to Acetyl CoA

Under *aerobic* (oxygen-rich) conditions, pyruvate is oxidized to acetyl CoA. Pyruvate formed in the cytosol through glycolysis crosses the two mitochondrial membranes and enters the mitochondrial matrix, where the oxidation takes place. The overall reaction, in simplified terms, is

Pyruvate, the end product of glycolysis, can leave the cytosol, cross the two mitochondrial membranes, and enter the mitochondrial matrix. None of the glycolysis intermediates can do this. The glycolysis intermediates are all phosphorylated substances; pyruvate is not phosphorylated.

$$\underset{\text{Pyruvate}}{CH_3{-}\overset{\overset{\displaystyle O}{\|}}{C}{-}COO^-} + \text{CoA{-}SH} + \text{NAD}^+ \xrightarrow[\text{complex}]{\text{Pyruvate dehydrogenase}} \underset{\text{Acetyl CoA}}{CH_3{-}\overset{\overset{\displaystyle O}{\|}}{C}{-}S{-}\text{CoA}} + \text{NADH} + CO_2$$

This reaction, which involves both oxidation and decarboxylation (CO_2 is produced), is far more complex than the simple stoichiometry of the equation suggests. The enzyme complex involved contains three different enzymes, each with

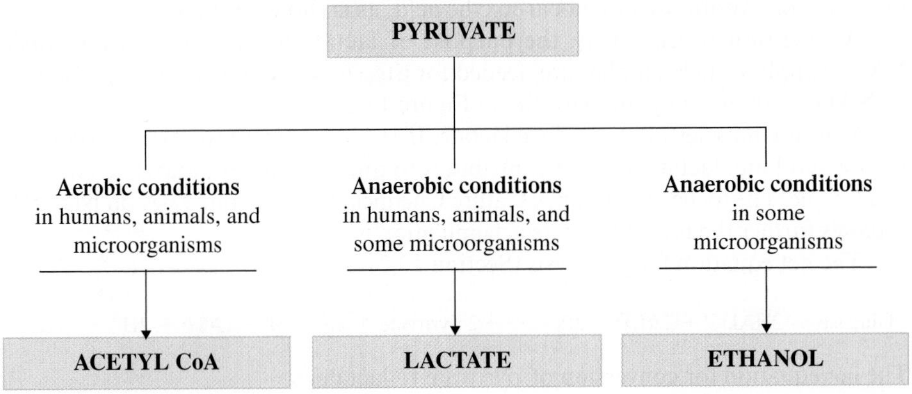

Figure 13.6 The three common fates of pyruvate generated by glycolysis.

numerous subunits. The overall reaction process involves four separate steps and requires NAD^+, CoA—SH, FAD, and two other coenzymes (lipoic acid and thiamin pyrophosphate, the latter derived from the B vitamin thiamin).

Not all acetyl CoA produced from pyruvate enters the citric acid cycle. Particularly when high levels of acetyl CoA are produced (from excess ingestion of dietary carbohydrates), some acetyl CoA is used as the starting material for the production of the fatty acids needed for fat (triacylglycerol) formation (Section 14.7).

Most acetyl CoA molecules produced from pyruvate enter the citric acid cycle. Citric acid cycle operations change more NAD^+ to its reduced form, NADH. The NADH from glycolysis, from the conversion of pyruvate to acetyl CoA, and from the citric acid cycle enters the electron transport chain directly (Section 12.8) or indirectly (Section 13.4). In the ETC, electrons from NADH are transferred to O_2, and the NADH is changed back to NAD^+. The NAD^+ needed for glycolysis, pyruvate–acetyl CoA conversion, and the citric acid cycle is regenerated.

The net overall reaction for processing one glucose molecule to two molecules of acetyl CoA is

$$\text{Glucose} + 2\text{ADP} + 2\text{P}_\text{i} + 4\text{NAD}^+ + 2\text{CoA—SH} \longrightarrow 2\text{ acetyl CoA} + 2\text{CO}_2 + 2\text{ATP} + 4\text{NADH} + 4\text{H}^+ + 2\text{H}_2\text{O}$$

Fermentation Processes

When the body becomes oxygen deficient (anaerobic conditions), such as during strenuous exercise, the electron transport chain process slows down because its last step is dependent on oxygen. The result of this "slowing down" is a buildup in NADH concentration (it is not being consumed as fast) and a decreased amount of available NAD^+ (it is not being produced as fast). Decreased NAD^+ concentration then negatively affects the rate of glycolysis. An alternative method for conversion of NADH to NAD^+—a method that does not require oxygen—is needed if glycolysis is to continue, it being the only available source of *new* ATP under these conditions.

Fermentation processes solve this problem. **Fermentation** *is a biochemical process by which NADH is oxidized to* NAD^+ *without the need for oxygen.* Two fermentation processes—lactate fermentation and ethanol fermentation—are now considered.

Lactate Fermentation

Lactate fermentation *is the enzymatic anaerobic reduction of pyruvate to lactate.* The equation for lactate formation from pyruvate is

$$\underset{\text{Pyruvate}}{\text{CH}_3\text{—}\overset{\displaystyle \text{O}}{\overset{\|}{\text{C}}}\text{—COO}^-} + \text{NADH} + \text{H}^+ \xrightarrow[\text{dehydrogenase}]{\text{Lactate}} \underset{\text{Lactate}}{\text{CH}_3\text{—}\overset{\displaystyle \text{OH}}{\overset{|}{\text{CH}}}\text{—COO}^-} + \text{NAD}^+$$

The sole purpose of this process is the conversion of NADH to NAD^+. The lactate so formed is converted back to pyruvate when aerobic conditions are again established in a cell (Section 13.6).

The chemical structure of lactate closely resembles that of pyruvate and also glycerate. Lactate, pyruvate, and glycerate are all derivatives of propionic acid, the three-carbon unsaturated monocarboxylic acid, as is shown in Figure 13.7.

As mentioned previously, the purpose of lactate formation is to replenish NAD^+ supplies; such supplies are needed for Step 6 of glycolysis. This "recycling" of NAD^+ is shown diagrammatically in Figure 13.8.

Working muscles often produce lactate. If strenuous work (or exercise) continues for too long, lactate buildup contributes to muscle soreness, muscle cramping, and fatigue. The focus on relevancy feature Chemical Connections 13-A on page 560 explores further the topic of lactate accumulation in muscles.

The net equation for glycolysis (Section 13.2) is

$$\text{Glucose} + 2\text{NAD}^+ + 2\text{ADP} + 2\text{P}_\text{i} \longrightarrow 2\text{ pyruvate} + 2\text{NADH} + 2\text{ATP} + 2\text{H}^+ + 2\text{H}_2\text{O}$$

The net equation for conversion of pyruvate to lactate is

$$\text{Pyruvate} + \text{NADH} + \text{H}^+ \longrightarrow \text{lactate} + \text{NAD}^+$$

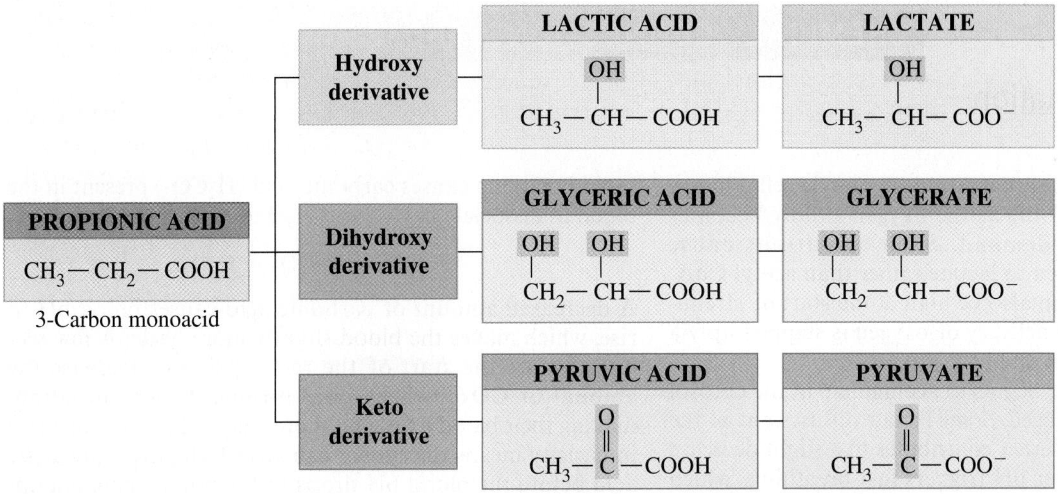

Figure 13.7 Structural relationships among the carboxylate ions lactate, pyruvate, and glycerate.

Adding these two equations together, and taking into account that one glucose produces two pyruvates and therefore two lactates, yields the following equation for the overall conversion of glucose to lactate.

Red blood cells have no mitochondria and therefore always form lactate as the end product of glycolysis.

$$\text{Glucose} + \boxed{2\text{ADP}} + 2\text{P}_i \longrightarrow 2\text{ lactate} + \boxed{2\text{ATP}} + 2\text{H}_2\text{O}$$

There is a net gain of two ATP for the conversion of glucose to two lactates.

Note that NADH and NAD^+ do not appear in this equation, even though the process cannot proceed without them. The NADH generated during glycolysis (Step 6) is consumed in the conversion of pyruvate to lactate. Thus there is no net oxidation–reduction in the conversion of glucose to lactate.

Ethanol Fermentation

Under anaerobic conditions, several simple organisms, including yeast, possess the ability to regenerate NAD^+ through ethanol, rather than lactate, production. Such a process is called ethanol fermentation. **Ethanol fermentation** *is the enzymatic anaerobic conversion of pyruvate to ethanol and carbon dioxide.* Ethanol fermentation involving yeast causes bread and related products to rise as a result of CO_2

With bread and other related products obtained using yeast, the ethanol produced by fermentation evaporates during baking.

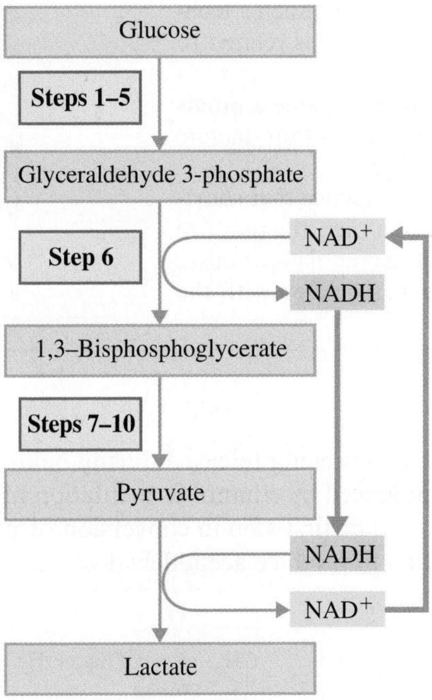

Figure 13.8 Anaerobic lactate formation allows for "recycling" of NAD^+, providing the NAD^+ needed for Step 6 of glycolysis.

CHEMICAL CONNECTIONS 13-A

Lactate Accumulation

During strenuous exercise, conditions in muscle cells change from aerobic to anaerobic as the oxygen supply becomes inadequate to meet demand. Such conditions cause pyruvate to be converted to lactate rather than acetyl CoA. (Lactate production can also be high at the start of strenuous exercise before the delivery of oxygen is stepped up via an increased respiration rate.)

The resulting lactate begins to accumulate in the cytosol of cells where it is produced. Some lactate diffuses out of the cells into the blood, where it contributes to a slight decrease in blood pH. This lower pH triggers fast breathing, which helps supply more oxygen to the cells.

Lactate accumulation and pH change are the cause of muscle pain and cramping during prolonged, strenuous exercise. As a result of such cramping, muscles may be stiff and sore the next day. Regular, hard exercise increases the efficiency with which oxygen is delivered to the body. Thus athletes can function longer than nonathletes under aerobic conditions without lactate production.

Recent research indicates that pH change in muscle cells (H^+ accumulation) may be more important than lactate accumulation as a cause of muscle pain. Hydrogen ions are produced when NAD^+ is reduced to NADH when glucose (as well as fats) is used by the body as a source of energy.

$$NAD^+ + 2H^+ + 2e^- \longrightarrow NADH + H^+$$

Lactate production consumes hydrogen ions as the reverse of the preceding reaction occurs. During strenuous exercise, lactate production (H^+ ion consumption) may not be fast enough to keep up with H^+ ion production.

Lactate accumulation can also occur in heart muscle if it experiences decreased oxygen supply (from artery blockage). The heart muscle experiences cramps and stops beating (cardiac arrest). Massage of heart muscle often reduces such cramps, just as it does for skeletal muscle, and it is sometimes possible to start the heart beating again by using such a technique. The pain associated with a heart attack is related to lactate and H^+ accumulation.

Hunters are usually aware that meat from game animals that have been run to exhaustion usually tastes sour; lactate accumulation is the reason for this problem.

Lactate formation is also relevant to a practice that short-distance sprinters often use just prior to a race, the practice of hyperventilation. Rapid breathing (hyperventilation) raises slightly the pH of blood. The CO_2 loss associated with the rapid breathing causes carbonic acid (H_2CO_3) present in the blood to dissociate in CO_2 and H_2O to replace the lost CO_2.

$$H_2CO_3 \rightleftharpoons CO_2 + H_2O$$

A decreased amount of carbonic acid causes blood pH to rise, which makes the blood slightly more basic. A few seconds before the start of the race, sprinters decrease the amount of CO_2 in their lungs through hyperventilation, making their blood a little bit more basic. This slight increase in basicity means the runner can absorb slightly more lactic acid before the blood pH drops to the point where cramping becomes a problem. Having such an advantage for only a few seconds in a short race can be helpful.

In diagnostic medicine, lactate levels in blood can often be used to determine the severity of a patient's condition. Higher than normal lactate levels are a sign of impaired oxygen delivery to tissue. Conditions that can cause higher lactate levels include lung disease and congestive heart failure.

Premature infants with underdeveloped lungs are often given increased amounts of oxygen to minimize lactate accumulation. They are also often given bicarbonate (HCO_3^{2-}) solution to counteract the acidity change in the blood that accompanies lactate buildup.

© Dan McCoy/Rainbow

Strenuous muscular activity can result in lactate accumulation.

bubbles being released during baking. Beer, wine, and other alcoholic drinks are produced by ethanol fermentation of the sugars in grain and fruit products.

The first step in conversion of pyruvate to ethanol is a decarboxylation reaction to produce acetaldehyde.

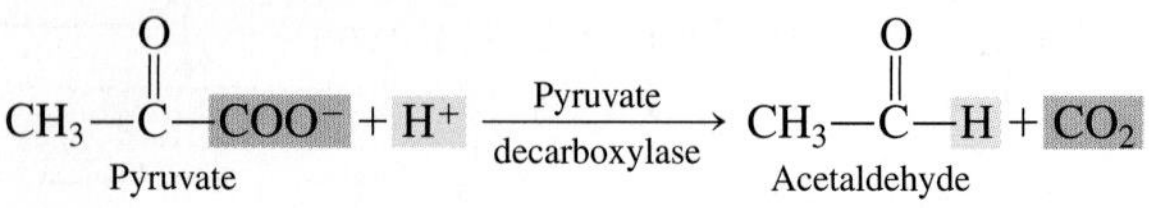

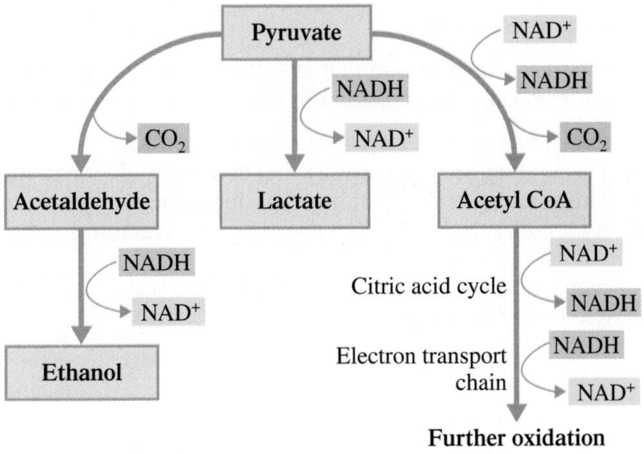

Figure 13.9 All three of the common fates of pyruvate from glycolysis provide for the regeneration of NAD^+ from NADH.

The second step involves acetaldehyde reduction to produce ethanol.

$$\underset{\text{Acetaldehyde}}{CH_3-\overset{\overset{\displaystyle O}{\|}}{C}-H} + NADH + H^+ \xrightarrow[\text{dehydrogenase}]{\text{Alcohol}} \underset{\text{Ethanol}}{CH_3-\overset{\overset{\displaystyle OH}{|}}{\underset{\underset{\displaystyle H}{|}}{C}}-H} + NAD^+$$

The overall equation for the conversion of pyruvate to ethanol (the sum of the two steps) is

$$\underset{\text{Pyruvate}}{CH_3-\overset{\overset{\displaystyle O}{\|}}{C}-COO^-} + 2H^+ + NADH \xrightarrow{\text{Two steps}} \underset{\text{Ethanol}}{CH_3-CH_2-OH} + NAD^+ + CO_2$$

An overall reaction for the production of ethanol from glucose is obtained by combining the reaction for the conversion of pyruvate with the net reaction for glycolysis (Section 13.2).

$$\text{Glucose} + 2ADP + 2P_i \longrightarrow 2\ \text{ethanol} + 2CO_2 + 2ATP + 2H_2O$$

Again note that NADH and NAD^+ do not appear in the final equation; they are both generated and consumed.

Figure 13.9 summarizes the relationship between the fates of pyruvate and the regeneration of NAD^+ from NADH.

EXAMPLE 13.5 Identifying Characteristics of the Three Common Pathways by Which Pyruvate Is Converted to Other Products

Which of the three common metabolic pathways for pyruvate is compatible with each of the following characterizations concerning the reactions that pyruvate undergoes?

a. CO_2 is not a product for this pathway.
b. A C_3 molecule is a product for this pathway.
c. NAD^+ is needed as an oxidizing agent for this pathway.
d. A C_2 molecule is a product under anaerobic reaction conditions for this pathway.

Solution

a. CO_2 is a product for both acetyl CoA formation and ethanol fermentation. In both of these processes, the C_3 pyruvate is converted to a C_2 species and CO_2. No CO_2 production occurs during *lactate fermentation*, as C_3 pyruvate is converted to C_3 lactate.
b. The only pathway that produces a C_3 molecule is *lactate fermentation.* In the other two pathways, a C_2 molecule and CO_2 are produced.
c. NAD^+, as an oxidizing agent, is needed in pathways that function under aerobic conditions. NADH, as a reducing agent, is needed in pathways that function under anaerobic conditions. The only pathway that involves aerobic conditions is *acetyl CoA formation*.
d. *Alcohol fermentation,* an anaerobic process, produces ethanol, a C_2 molecule. The other anaerobic process, lactate fermentation, produces the C_3 molecule lactate.

(continued)

Practice Exercise 13.5

Which of the three common metabolic pathways for pyruvate is compatible with each of the following characterizations concerning the reactions that pyruvate undergoes?

a. Acetaldehyde is an intermediate in this pathway.
b. This is an anaerobic pathway that does not function in humans.
c. This is an anaerobic pathway that does function in humans.
d. A C_2 molecule is a product under aerobic reaction conditions for this pathway.

Answers: **a.** Ethanol fermentation; **b.** Ethanol fermentation; **c.** Lactate fermentation; **d.** Acetyl CoA formation

13.4 ATP Production for the Complete Oxidation of Glucose

Using assembled energy production figures for glycolysis, oxidation of pyruvate to acetyl CoA, the citric acid cycle, and the electron transport chain—with one added piece of information—gives the ATP yield for the *complete* oxidation of one molecule of glucose.

The new piece of information involves the NADH produced during Step 6 of glycolysis. This NADH, produced in the cytosol, cannot *directly* participate in the electron transport chain because mitochondria are impermeable to NADH (and NAD^+). A transport system shuttles the electrons from NADH, but not NADH itself, across the membrane. This shuttle involves dihydroxyacetone phosphate (a glycolysis intermediate) and glycerol 3-phosphate.

The first step in the shuttle is the cytosolic reduction of dihydroxyacetone phosphate by NADH to produce glycerol 3-phosphate and NAD^+ (Figure 13.10). Glycerol 3-phosphate then crosses the outer mitochondrial membrane, where it is reoxidized to dihydroxyacetone phosphate. The oxidizing agent is FAD rather than NAD^+. The regenerated dihydroxyacetone phosphate diffuses out of the mitochondrion and returns to the cytosol for participation in another "turn" of the shuttle. The $FADH_2$ coproduced in the mitochondrial reaction

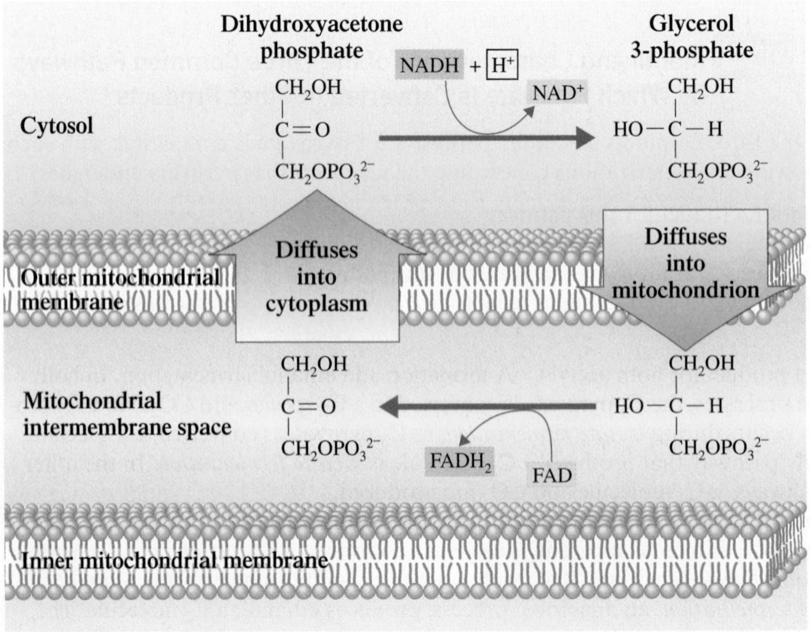

Figure 13.10 The dihydroxyacetone phosphate–glycerol 3-phosphate shuttle.

Table 13.2 Production of ATP from the Complete Oxidation of One Glucose Molecule in a Skeletal Muscle Cell

Reaction	Comments	Yield of ATP
Glycolysis		
glucose → glucose 6-phosphate	consumes 1 ATP	−1
glucose 6-phosphate → fructose 1,6-bisphosphate	consumes 1 ATP	−1
2(glyceraldehyde 3-phosphate → 1,3-bisphosphoglycerate)	each produces 1 cytosolic NADH	—
2(1,3-bisphosphoglycerate → 3-phosphoglycerate)	each produces 1 ATP	+2
2(phosphoenolpyruvate → pyruvate)	each produces 1 ATP	+2
Oxidation of Pyruvate		
2(pyruvate → acetyl CoA + CO_2)	each produces 1 NADH	—
Citric Acid Cycle		
2(isocitrate → α-ketoglutarate + CO_2)	each produces 1 NADH	—
2(α-ketoglutarate → succinyl CoA + CO_2)	each produces 1 NADH	—
2(succinyl CoA → succinate)	each produces 1 GTP	+2
2(succinate → fumarate)	each produces 1 $FADH_2$	—
2(malate → oxaloacetate)	each produces 1 NADH	—
Electron Transport Chain and Oxidative Phosphorylation		
2 cytosolic NADH formed in glycolysis	each produces 1.5 ATP	+3
2 NADH formed in the oxidation of pyruvate	each produces 2.5 ATP	+5
2 $FADH_2$ formed in the citric acid cycle	each produces 1.5 ATP	+3
6 NADH formed in the citric acid cycle	each produces 2.5 ATP	+15
Net Production of ATP		+30

can participate in the electron transport chain reactions. The net reaction of this shuttle process is

$$\underset{\text{(cytosolic)}}{\text{NADH}} + \text{H}^+ + \underset{\text{(mitochondrial)}}{\text{FAD}} \longrightarrow \underset{\text{(cytosolic)}}{\text{NAD}^+} + \underset{\text{(mitochondrial)}}{\text{FADH}_2}$$

The consequence of this reaction is that only 1.5 rather than 2.5 molecules of ATP are formed for each cytosolic NADH, because $FADH_2$ yields one less ATP than does NADH in the electron transport chain.

Table 13.2 shows ATP production for the complete oxidation of a molecule of glucose. The final number is 30 ATP, 26 of which come from the oxidative phosphorylation associated with the electron transport chain. This total of 30 ATP for complete oxidation contrasts markedly with a total of 2 ATP for oxidation of glucose to lactate and 2 ATP for oxidation of glucose to ethanol. Neither of these latter processes involves the citric acid cycle or the electron transport chain. Thus the aerobic oxidation of glucose is 15 times more efficient in the production of ATP than the anaerobic lactate and ethanol processes.

The production of 30 ATP molecules per glucose (Table 13.2) is for those cells where the dihydroxyacetone phosphate–glycerol 3-phosphate shuttle operates (skeletal muscle and nerve cells). In certain other cells, particularly heart and liver cells, a more complex shuttle system called the malate–aspartate shuttle functions. In this shuttle, 2.5 ATP molecules result from 1 cytosolic NADH, which changes the total ATP production to 32 molecules per glucose.

The net overall reaction for the *complete* metabolism (oxidation) of a glucose molecule is the simple equation

$$\text{Glucose} + 6\text{O}_2 + 30\text{ADP} + 30\text{P}_i \longrightarrow 6\text{CO}_2 + 6\text{H}_2\text{O} + 30\ \text{ATP}$$

Note that substances such as NADH, NAD^+, and $FADH_2$ are not part of this equation. Why? They cancel out—that is, they are consumed in one step (reactant) and regenerated in another step (product). Note also what the net equation does not acknowledge: the many dozens of reactions that are needed to generate the 30 molecules of ATP.

13.5 Glycogen Synthesis and Degradation

Glycogen, a branched polymeric form of glucose (Section 7.15), is the storage form of carbohydrates in humans and animals. It is found primarily in muscle and liver tissue. In muscles, it is the source of glucose needed for glycolysis. In the liver, it is the source of glucose needed to maintain normal glucose levels in the blood.

Glycogenesis

Glycogenesis *is the metabolic pathway by which glycogen is synthesized from glucose 6-phosphate.* Glycogenesis involves three reactions (steps).

Step 1: **Isomerization:** *Formation of Glucose 1-phosphate.* The starting material for this step is not glucose itself but, rather, glucose 6-phosphate (available from the first step of glycolysis). The enzyme *phosphoglucomutase* effects the change from a 6-phosphate to a 1-phosphate.

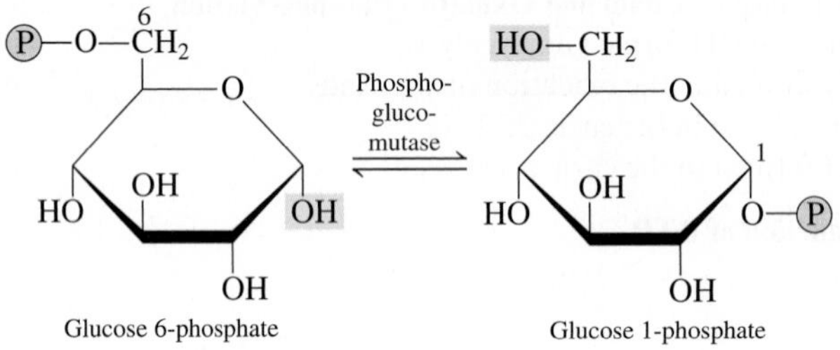

Step 2: **Activation:** *Formation of UDP-glucose.* Glucose 1-phosphate from Step 1 must be activated before it can be added to a growing glycogen chain. The activator is the high-energy compound UTP (uridine triphosphate). A UMP is transferred to glucose 1-phosphate, and the resulting PP_i is hydrolyzed to $2P_i$.

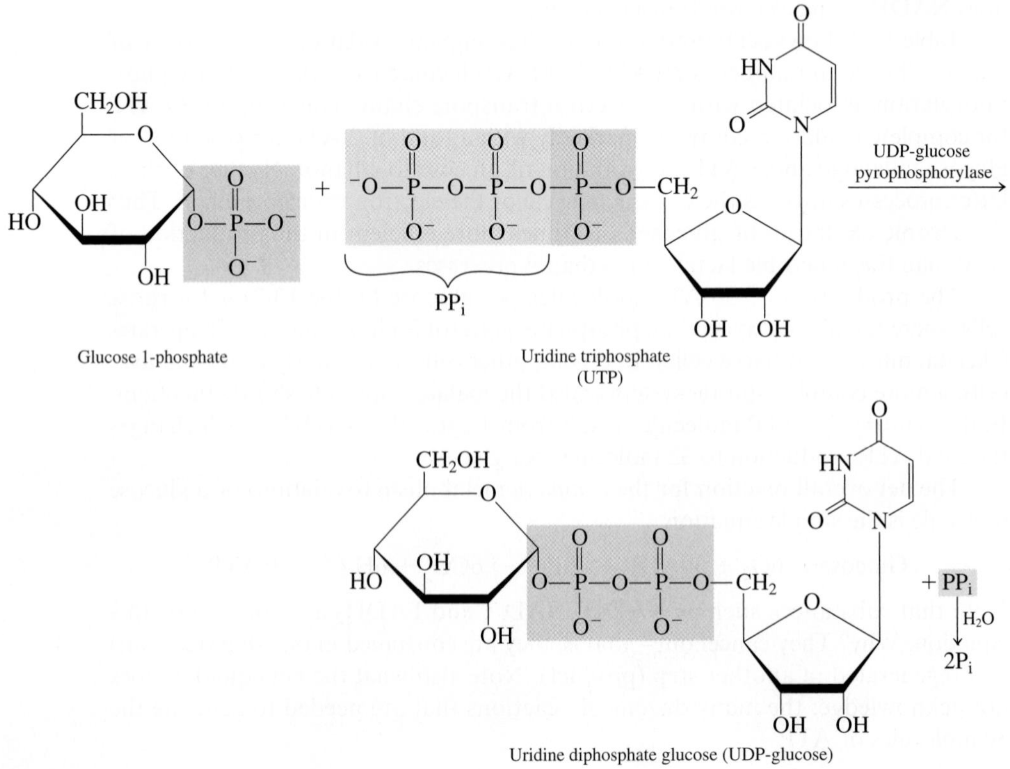

Step 3: **Linkage to Chain:** *Glucose Transfer to a Glycogen Chain.* The glucose unit of UDP-glucose is then attached to the end of a glycogen chain.

$$\underset{\text{}}{\text{UDP-glucose}} + \underset{\text{Glycogen chain}}{(\text{glucose})_n} \xrightarrow[\text{}]{\text{Glycogen synthase}} \underset{\text{Glycogen with an additional glucose unit}}{(\text{glucose})_{n+1}} + \text{UDP}$$

Glucose-UDP is the activated carrier of glucose in glycogen synthesis (glycogenesis).

In a subsequent reaction, the UDP produced in Step 3 is converted back to UTP, which can then react with another glucose 1-phosphate (Step 2). The conversion reaction requires ATP.

$$\text{UDP} + \text{ATP} \longrightarrow \text{UTP} + \text{ADP}$$

Adding a single glucose unit to a growing glycogen chain requires the investment of two ATP molecules: one in the formation of glucose 6-phosphate and one in the regeneration of UTP.

Glycogenolysis

Glycogenolysis *is the metabolic pathway by which glucose 6-phosphate is produced from glycogen.* This process is not simply the reverse of glycogen synthesis (glycogenesis) because it does not require UTP or UDP molecules. Glycogenolysis is a two-step process rather than a three-step process.

The chain cleavage reaction that occurs in Step 1 that releases a glucose residue is called *phosphorolysis* by analogy with the process of *hydrolysis,* which also involves breaking a molecule into two parts. Here, it is P_i rather than H_2O that participates in the bond breaking. A phosphoryl group is added to the released glucose residue.

Step1: ***Phosphorolysis:*** *Formation of Glucose-1-phosphate.* The enzyme glycogen phosphorylase effects the removal of an end glucose unit from a glycogen molecule as glucose 1-phosphate.

$$\underset{\text{Glycogen}}{(\text{Glucose})_n} + P_i \xrightarrow{\text{Glycogen Phosphorylase}} \underset{\text{Glycogen with one fewer glucose unit}}{(\text{glucose})_{n-1}} + \text{glucose 1-phosphate}$$

A *phosphorylase* is an enzyme that catalyzes the cleavage of a bond by P_i (in contrast to a hydrolase, which effects bond cleavage by water), such as removal of a glucose unit from glycogen to give glucose 1-phosphate.

Step 2: ***Isomerization:*** *Formation of Glucose 6-phosphate.* The enzyme phosphoglucomutase catalyzes the isomerization process whereby the phosphate group of glucose 1-phosphate is moved to the carbon 6 position.

$$\text{Glucose 1-phosphate} \xrightleftharpoons{\text{Phosphogluco-mutase}} \text{glucose 6-phosphate}$$

This process is the reverse of the first step of glycogenesis.

A *phosphatase* is an enzyme that effects the removal of a phosphate group (P_i) from a molecule, such as converting glucose 6-phosphate to glucose, with H_2O as the attacking species.

Neither ATP nor any of the other nucleotide triphosphates is needed in the reactions of glycogenolysis.

In muscle and brain cells, an immediate need for energy is the stimulus that initiates glycogenolysis. The glucose 6-phosphate that is produced directly enters the glycolysis pathway at Step 2 (Figure 13.3), and its multistep conversion to pyruvate begins. A low level of glucose is the stimulus that initiates glycogenolysis in liver cells. Here, the glucose 6-phosphate produced must be converted to free glucose before it can enter the bloodstream, as glucose 6-phosphate cannot cross cell membranes (Section 13.2). This change is effected by the enzyme *glucose 6-phosphatase,* an enzyme found in liver cells but not in muscle cells or brain cells.

The fact that glycogen synthesis (glycogenesis) and glycogen degradation (glycogenolysis) are not totally reverse processes has significance. In fact, it is almost always the case in biochemistry that "opposite" biosynthetic and degradative pathways differ in some steps. This allows for separate control of the pathways.

$$\text{Glucose 6-phosphate} + H_2O \xrightarrow{\text{Glucose 6-phosphatase}} \text{glucose} + P_i$$

Because muscle and brain cells lack glucose 6-phosphatase, they cannot form free glucose from glucose 6-phosphate. Thus muscle and brain cells can use glucose 6-phosphate from glycogen for energy production only. The liver, however, with this enzyme present, has the capacity to use glucose 6-phosphate obtained from glycogen to supply additional glucose to the blood.

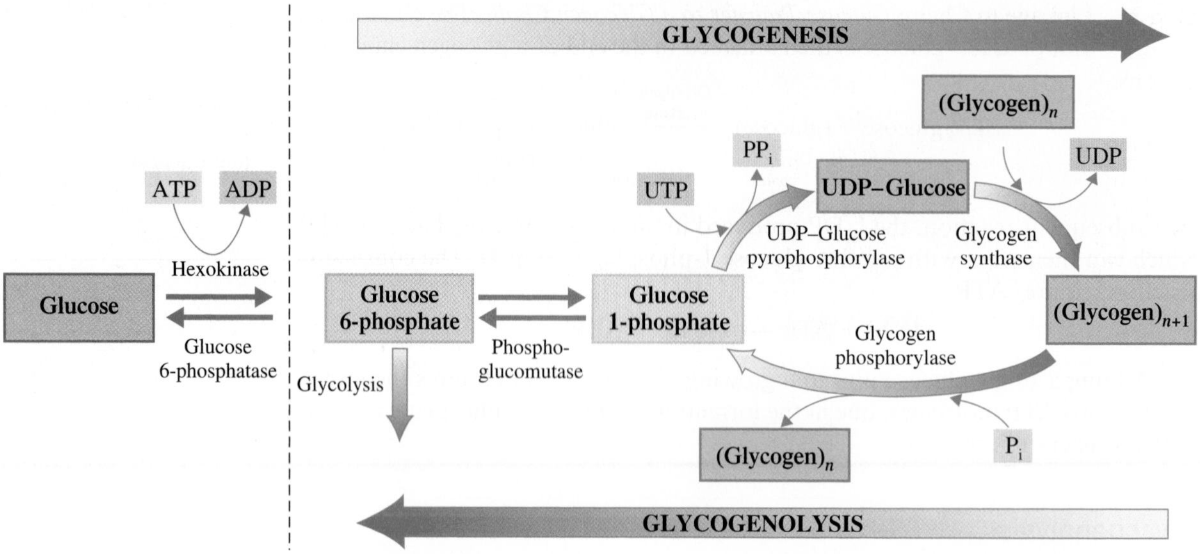

Figure 13.11 The processes of glycogenesis and glycogenolysis contrasted. The intermediate UDP—glucose is part of glycogenesis but not of glycogenolysis.

Figure 13.11 contrasts the "opposite" processes of glycogenesis and glycogenolysis. Both processes involve the glycolysis intermediate glucose 1-phosphate. UDP–glucose is unique to glycogenesis.

When glycogen rather than free glucose is the starting material for glycolysis, the net gain in ATP is three molecules rather than two for each glucose processed. Glucose from glycogen enters glycolysis at Step 2, as glucose 6-phosphate, and thus bypasses ATP-consuming Step 1. Thus glycogen is a more effective energy source than is free glucose. Remember, however, that it costs the equivalent of two ATP to incorporate a glucose molecule into glycogen (glycogenesis).

13.6 Gluconeogenesis

Gluconeogenesis *is the metabolic pathway by which glucose is synthesized from noncarbohydrate materials.* Glycogen stores in muscle and liver tissue are depleted within 12–18 hours of fasting or in even less time from heavy work or strenuous exercise. Without gluconeogenesis, the brain, which is dependent on glucose as a fuel, would have problems functioning if food intake were restricted for even one day.

The noncarbohydrate starting materials for gluconeogenesis are lactate (from hard-working muscles and from red blood cells), glycerol (from triacylglycerol hydrolysis), and certain amino acids (from dietary protein hydrolysis or from muscle protein during starvation). About 90% of gluconeogenesis takes place in the liver. Hence gluconeogenesis helps to maintain normal blood-glucose levels in times of inadequate dietary carbohydrate intake (such as between meals).

The processes of gluconeogenesis (pyruvate to glucose) and glycolysis (glucose to pyruvate) are not exact opposites. The most obvious difference

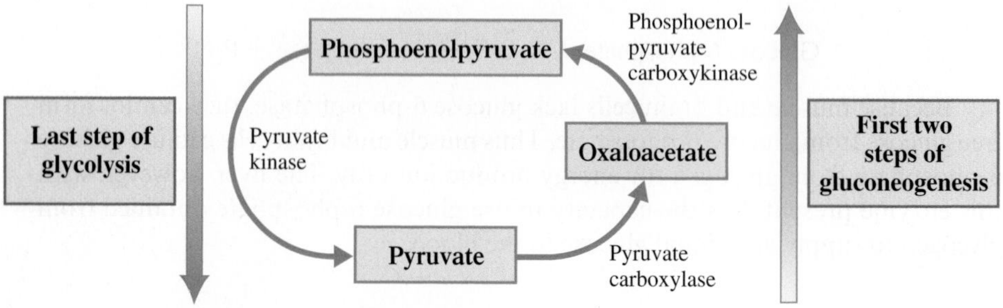

Figure 13.12 The "opposite" processes of gluconeogenesis (pyruvate to glucose) and glycolysis (glucose to pyruvate) are not exact opposites. The reversal of the last step of glycolysis requires two steps in gluconeogenesis. Therefore, gluconeogenesis has 11 steps, whereas glycolysis has only 10 steps.

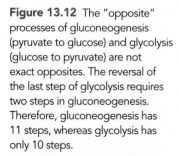

between these two processes is that 12 compounds are involved in gluconeogenesis and only 11 in glycolysis. Why the difference? The last step of glycolysis is the conversion of the high-energy compound phosphoenolpyruvate to pyruvate. The reverse of this process, which is the beginning of gluconeogenesis, cannot be accomplished in a single step because of the large energy difference between the two compounds and the slow rate of the reaction. Instead, a two-step process by way of oxaloacetate is required to effect the change, and this adds an extra compound to the gluconeogenesis pathway (Figure 13.12). Both an ATP molecule and a GTP molecule are needed to drive this two-step process.

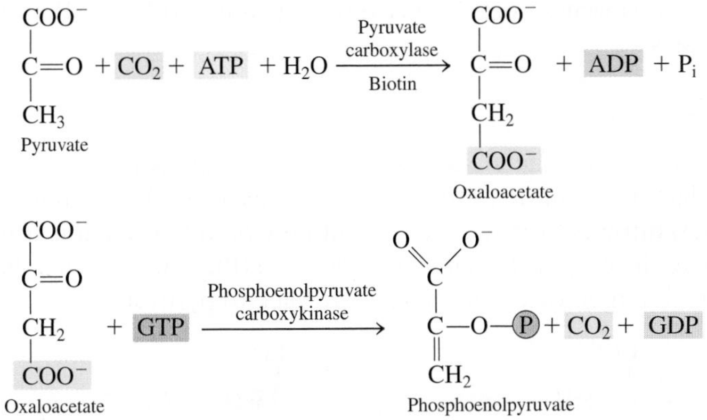

When carbohydrate intake is high (glucose is plentiful), little use of the gluconeogenesis metabolic pathway occurs. On the other hand, gluconeogenesis becomes an important pathway for a person on a low carbohydrate diet (glucose is available in only limited amounts). Through gluconeogenesis, fat and even proteins can be converted to glucose.

The oxaloacetate intermediate in this two-step process provides a connection to the citric acid cycle. In the first step of this cycle, oxaloacetate combines with acetyl CoA. If energy rather than glucose is needed, then oxaloacetate can go directly into the citric acid cycle.

As is shown in Figure 13.13, there are two other locations where gluconeogenesis and glycolysis differ. In Steps 9 and 11 of gluconeogenesis (Steps 1 and 3

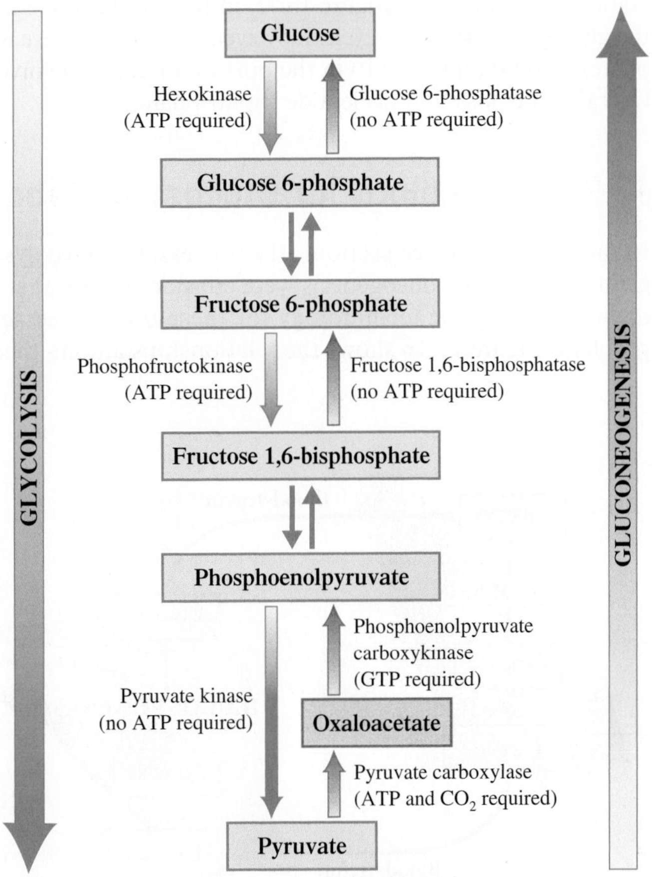

Figure 13.13 The pathway for gluconeogenesis is similar, but not identical, to the pathway for glycolysis.

of glycolysis), the reactant–product combinations match between pathways. However, different enzymes are required for the forward and reverse processes. The new enzymes for gluconeogenesis are *fructose 1,6-bisphosphatase* and *glucose 6-phosphatase.*

Glycolysis has a net production of 2 ATP (Section 13.2). Gluconeogenesis has a net expenditure of 4 ATP and 2 GTP, which is equivalent to the expenditure of 6 ATP.

The overall net reaction for gluconeogenesis is

$$2\ \text{Pyruvate} + 4\text{ATP} + 2\text{GTP} + 2\text{NADH} + 2\text{H}_2\text{O} \longrightarrow \text{glucose} + 4\text{ADP} + 2\text{GDP} + 6\text{P}_i + 2\text{NAD}^+$$

Thus to reconvert pyruvate to glucose requires the expenditure of 4 ATP and 2 GTP. Whenever gluconeogenesis occurs, it is at the expense of other ATP-producing metabolic processes.

The Cori Cycle

The Cori cycle is named in honor of Gerty Radnitz Cori (1896–1957) and Carl Cori (1896–1984), the husband-and-wife team who discovered it. They were awarded a Nobel Prize in 1947, the third husband-and-wife team to be so recognized. Marie and Pierre Curie were the first, Irene and Frederic Joliot-Curie the second.

Gluconeogenesis using lactate as a source of pyruvate is particularly important because of lactate formation during strenuous exercise. The lactate so produced (Section 13.3) diffuses from muscle cells into the blood, where it is transported to the liver. Here the enzyme lactate dehydrogenase (the same enzyme that catalyzes lactate formation in muscle) converts lactate back to pyruvate.

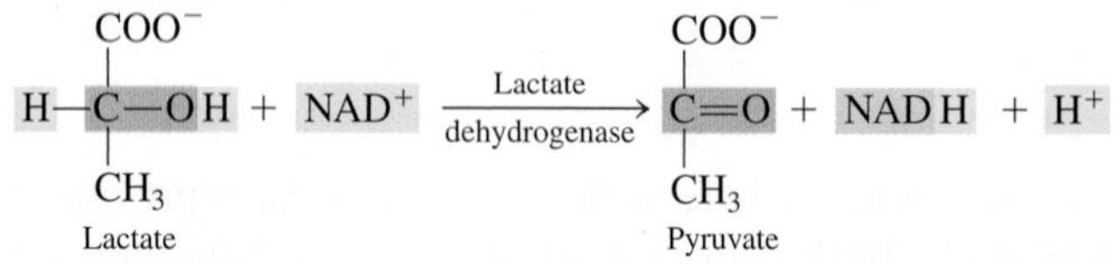

The newly formed pyruvate is then converted via gluconeogenesis to glucose, which enters the bloodstream and goes to the muscles. This cyclic process, which is called the Cori cycle, is diagrammed in Figure 13.14. The **Cori cycle** *is a cyclic biochemical process in which glucose is converted to lactate in muscle tissue, the lactate is reconverted to glucose in the liver, and the glucose is returned to the muscle tissue.*

A net loss of nucleotide triphosphates (ATP, GTP) accompanies Cori cycle operation. As is shown in Figure 13.15, in the muscle, there is a net gain of 2 ATP in the glycolytic part of the cycle. However, in the liver, there is a net loss of 6 triphosphates (4 ATP and 2 GTP) in the portion of the cycle involving gluconeogenesis. Overall, a net loss of 4 nucleotide triphosphates occurs.

13.7 Terminology for Glucose Metabolic Pathways

In the preceding three sections, the processes of glycolysis, glycogenesis, glycogenolysis, and gluconeogenesis were considered. Because of their like-sounding names, keeping the terminology for these four processes "straight" is often a problem. Figure 13.16 shows the relationships among these processes. Note that

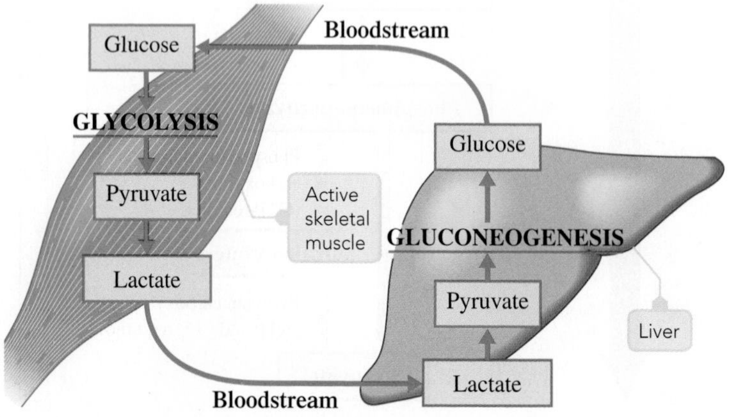

Figure 13.14 The Cori cycle. Lactate, formed from glucose under anaerobic conditions in muscle cells, is transferred to the liver, where it is reconverted to glucose, which is then transferred back to the muscle cells.

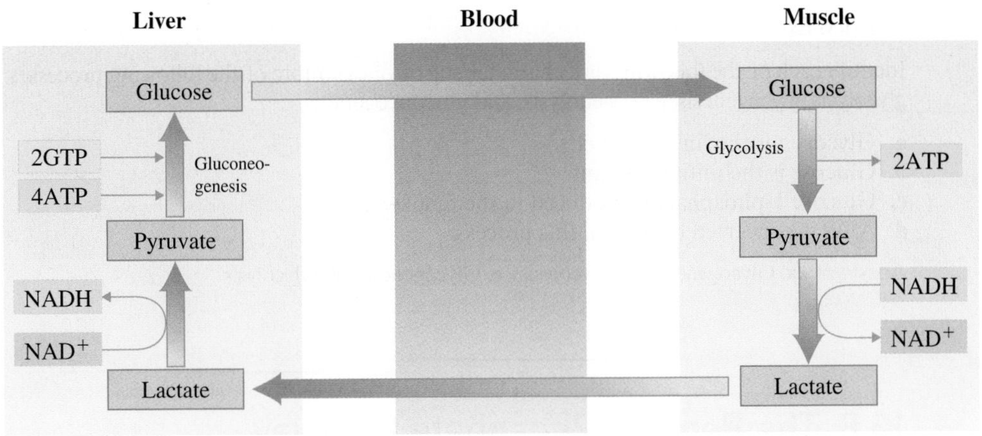

Figure 13.15 Nucleotide triphosphate change (gain or loss) associated with the two parts of the Cori cycle.

the glycogen degradation pathways (left side of Figure 13.16) have names ending in *-lysis,* which means "breakdown." The pathways associated with glycogen synthesis (right side of Figure 13.16) have names ending in *-genesis,* which means "making."

EXAMPLE 13.6 Recognizing Characteristics of Glycolysis, Glycogenesis, Glycogenolysis, and Gluconeogenesis

Identify each of the following as a characteristic of one or more of the following processes: glycolysis, glycogenesis, glycogenolysis, and gluconeogenesis.

a. Glucose 6-phosphate is the initial reactant.
b. Glucose is the final product.
c. Glucose 6-phosphate is produced in the first step.
d. UTP is involved in this process.

Solution

a. *Glycogenesis.* In this three-step process, glucose 6-phosphate units are added to a growing glycogen molecule.
b. *Gluconeogenesis.* This 11-step process converts pyruvate into glucose.
c. *Glycolysis.* The first step of glycolysis is the production of glucose 6-phosphate from glucose. The glucose 6-phosphate produced can then be further processed in the glycolysis pathway or processed to glycogen using the glycogenesis pathway.
d. *Glycogenesis.* The second step of glycogenesis is the conversion of glucose 1-phosphate to UDP-glucose.

(continued)

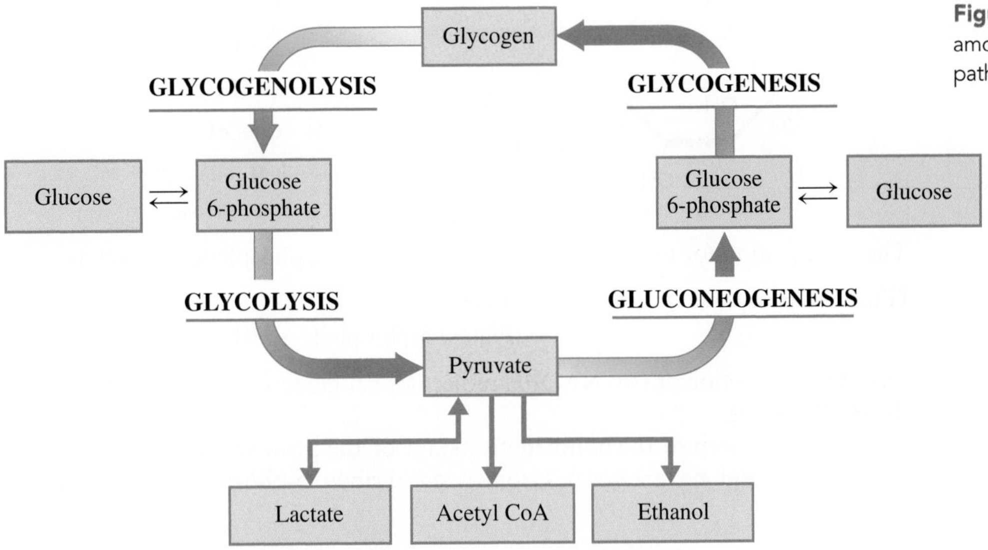

Figure 13.16 The relationships among four common metabolic pathways that involve glucose.

Practice Exercise 13.6

Identify each of the following as a characteristic of one or more of the following processes: glycolysis, glycogenesis, glycogenolysis, and gluconeogenesis.

a. Glycogen is the final product.
b. Glucose is the initial reactant.
c. Glucose 1-phosphate is produced in the first step.
d. ADP is converted to ATP in this process.

Answers: **a.** Glycogenesis; **b.** Glycolysis; **c.** Glycogenesis; **d.** Glycolysis

13.8 The Pentose Phosphate Pathway

Glycolysis is not the only pathway by which glucose may be degraded. Depending on the type of cell, various amounts of glucose are degraded by the pentose phosphate pathway, a pathway whose main focus is *not* subsequent ATP production as is the case for glycolysis. Major functions of this alternative pathway are (1) synthesis of the coenzyme NADPH needed in lipid biosynthesis (Section 14.7) and (2) production of ribose 5-phosphate, a pentose derivative needed for the synthesis of nucleic acids and many coenzymes. The **pentose phosphate pathway** *is the metabolic pathway by which glucose is used to produce NADPH, ribose 5-phosphate (a pentose phosphate), and numerous other sugar phosphates.* The operation of the pentose phosphate pathway is significant in cells that produce lipids: fatty tissue, the liver, mammary glands, and the adrenal cortex (an active producer of steroid lipids).

NADPH, the coenzyme produced in the pentose phosphate pathway, is the reduced form of $NADP^+$ (nicotinamide adenine dinucleotide phosphate). Structurally, $NADP^+$/NADPH is a phosphorylated version of NAD^+/NADH (Figure 13.17).

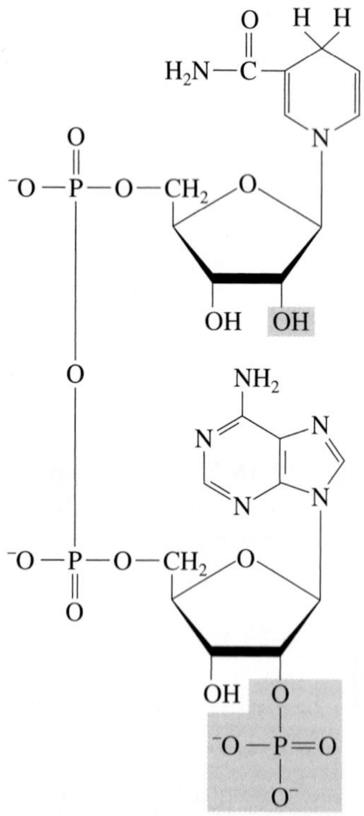

Figure 13.17 The structure of NADPH. The phosphate group shown in color is the structural feature that distinguishes NADPH from NADH.

The nonphosphorylated and phosphorylated versions of this coenzyme have significantly different functions. The nonphosphorylated version is involved, mainly in its oxidized form (NAD^+), in the reactions of the common metabolic pathway (Section 12.6). The phosphorylated version is involved, mainly in its reduced form (NADPH), in biosynthetic reactions of lipids and nucleic acids.

There are two stages within the pentose phosphate pathway—an oxidative stage and a nonoxidative stage. The oxidative stage, which occurs first, involves three steps through which glucose 6-phosphate is converted to ribulose 5-phosphate and CO_2.

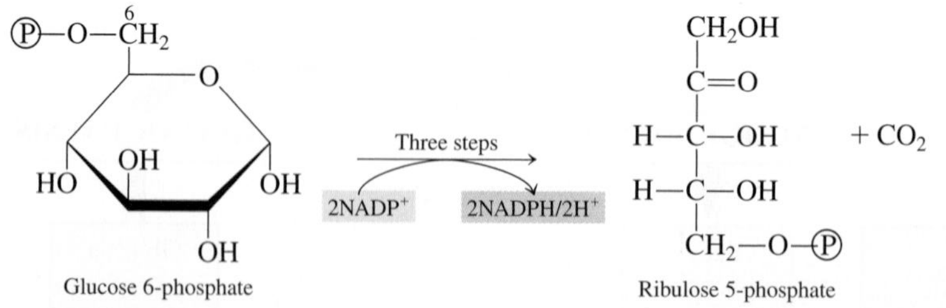

The net equation for the oxidative stage of the pentose phosphate pathway is

$$\text{Glucose 6-phosphate} + 2NADP^+ + H_2O \longrightarrow \text{ribulose 5-phosphate} + CO_2 + 2NADPH + 2H^+$$

Note the production of two NADPH molecules per glucose 6-phosphate processed during this stage.

In the first step of the nonoxidative stage of the pentose phosphate pathway, ribulose 5-phosphate (a ketose) is isomerized to ribose 5-phosphate (an aldose).

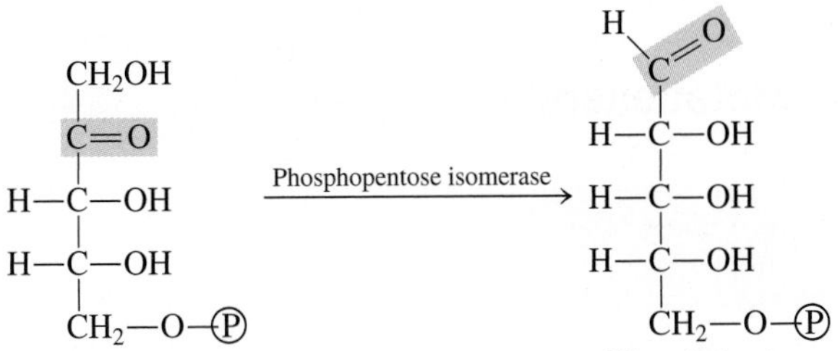

Ribulose 5-phosphate → Ribose 5-phosphate

The pentose ribose is a component of ATP, GTP, UTP, CoA, NAD^+/NADH, FAD/$FADH_2$, and RNA. Further steps in the nonoxidative stage contain provision for the conversion of ribose 5-phosphate to numerous other sugar phosphates. Ultimately, glyceraldehyde 3-phosphate and fructose 6-phosphate (both glycolysis intermediates) are formed. The overall net reaction for the pentose phosphate pathway is

$$3 \text{ Glucose 6-phosphate} + 6\text{NADP}^+ + 3\text{H}_2\text{O} \longrightarrow$$
$$2 \text{ fructose 6-phosphate} + 3\text{CO}_2 + \text{glyceraldehyde 3-phosphate} + 6\text{NADPH} + 6\text{H}^+$$

The pentose phosphate pathway, with its many intermediates, helps meet cellular needs in numerous ways:

1. When ATP demand is high, the pathway continues to its end products, which enter glycolysis.
2. When NADPH demand is high, intermediates are recycled to glucose 6-phosphate (the start of the pathway), and further NADPH is produced.
3. When ribose 5-phosphate demand is high, for nucleic acid and coenzyme production, most of the nonoxidative stage is nonfunctional, leaving ribose 5-phosphate as a major product.

The Chemistry at a Glance feature on the next page shows how the pentose phosphate pathway is related to the other major pathways of glucose metabolism that have been considered.

Glutathione (Section 9.8) is the main antioxidant used by the body to keep hemoglobin in its reduced form. NADPH is needed for the regeneration of depleted glutathione. An insufficient supply of NADPH, and the ensuing lack of regeneration of glutathione, leads to the destruction of hemoglobin-containing red blood cells; severe anemia can result.

13.9 Hormonal Control of Carbohydrate Metabolism

A second major method for regulating carbohydrate metabolism, besides enzyme inhibition by metabolites (Section 13.2), is hormonal control. Among others, three hormones—insulin, glucagon, and epinephrine—affect carbohydrate metabolism.

Insulin

Insulin, a 51-amino-acid protein hormone whose structure was considered in Section 9.12, is produced by the beta cells of the pancreas. Insulin promotes the uptake and utilization of glucose by cells. Thus its function is to lower blood-glucose levels. It is also involved in lipid metabolism.

The release of insulin is triggered by *high* blood-glucose levels. The mechanism for insulin action involves insulin binding to protein receptors on the outer surfaces of cells, which facilitates entry of glucose into the cells. Insulin also produces an increase in the rates of glycogenesis, glycolysis, and fatty acid synthesis.

Insulin is at the core of the metabolic disorder known as diabetes; either the body does not produce enough insulin or body cells do not respond properly to the insulin that is produced. The focus on relevancy feature Chemical Connections 13-B on page 573 examines further the diabetic condition that afflicts an ever increasing portion of the human population.

CHEMISTRY AT A GLANCE Glucose Metabolism

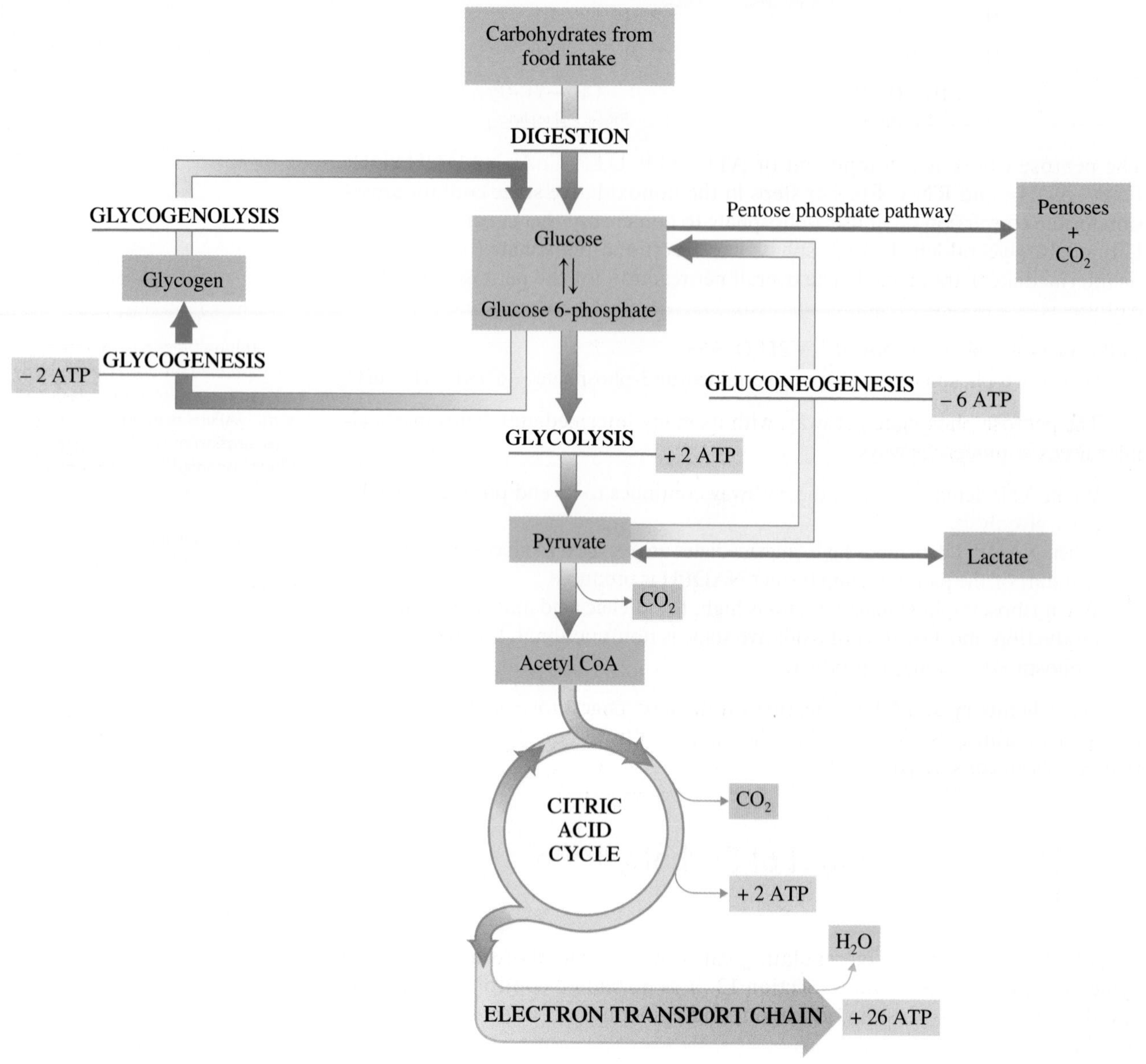

Glucagon

Glucagon is a polypeptide hormone (29 amino acids) produced in the pancreas by alpha cells. It is released when blood-glucose levels are *low.* Its principal function is to increase blood-glucose concentrations by speeding up the conversion of glycogen to glucose (glycogenolysis) and gluconeogenesis in the liver. Thus glucagon's effects are opposite those of insulin.

Epinephrine

Epinephrine (Section 6.10), also called adrenaline, is released by the adrenal glands in response to anger, fear, or excitement. Its function is similar to that of glucagon—stimulation of glycogenolysis, the release of glucose from glycogen. Its primary target is muscle cells, where energy is needed for quick action. It also functions in lipid metabolism.

CHEMICAL CONNECTIONS 13-B

Diabetes Mellitus

Diabetes mellitus, which is usually simply referred to as *diabetes,* is a metabolic disorder characterized by elevated levels of glucose in the blood. Classic symptoms associated with an uncontrolled diabetic condition are frequent urination, increased thirst, and increased hunger. These symptoms are the basis for the name *diabetes mellitus*, which originates from the Greek words "diabetes," meaning "siphon," and "mellitus," meaning "sweet." In the second century A.D., the Greek physician Aretus the Cappadocian named this condition; he observed that some people had a condition in which the body acts like a siphon—taking water in at one end and discharging it at the other—and that the urine produced was sweet to the taste. The name *diabetes mellitus* can roughly be translated as "sweet urine."

As of 2010, it is estimated that 26 million Americans (about 1 in 12) are diabetic. Diagnosis of diabetes is based on measurement of fasting blood-glucose levels. A fasting blood-glucose level greater than 126 mg/dL is considered a positive test, and a level less than 100 mg/dL is considered a negative test. Readings between 100 mg/dL and 126 mg/dL indicate a prediabetic condition; the blood-glucose level is higher than it should be but not high enough to be classified as diabetic. Prediabetic conditions are found in 15% of Americans.

There are two major forms of diabetes mellitus: type I (insulin-dependent) and type II (non-insulin-dependent).

Type 1 diabetes is the result of inadequate insulin production by the beta cells of the pancreas. Control of this condition involves insulin injections and special dietary programs. A risk associated with the insulin injections is that too much insulin can produce severe hypoglycemia (insulin shock); blackout or a coma can result. Treatment involves a quick infusion of glucose. Diabetics often carry candy bars (quick glucose sources) for use if they feel any of the symptoms that signal the onset of insulin shock.

Type 2 diabetes results from insulin resistance, a condition in which cells fail to use insulin properly. Bodily insulin production may be normal, but the cells do not respond to it normally. Treatment involves use of medications that decrease glucose production and/or increase insulin levels, as well as a carefully regulated diet to decrease obesity if the latter is a problem. More efficient use of undamaged insulin receptors occurs at increased insulin levels.

About 10% of all cases of diabetes are type I. The more common non–insulin-dependent type II diabetes occurs in the other 90% of cases. The effects of both types of diabetes are the same—inadequate glucose uptake by cells. The result is blood-glucose levels much higher than normal (hyperglycemia). With an inadequate glucose intake, cells must resort to other procedures for energy production, procedures that involve the breakdown of fats and protein.

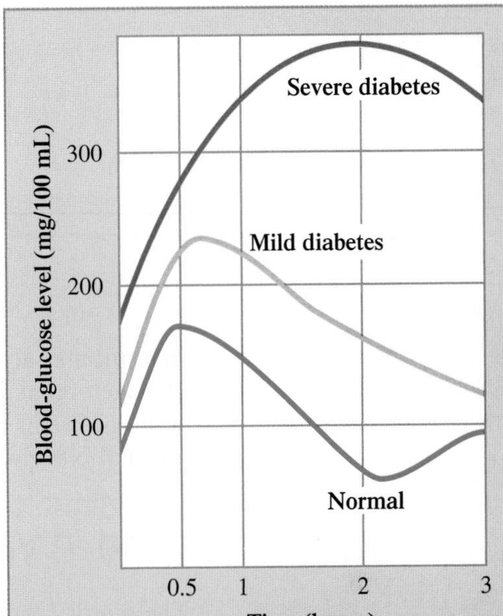

Typical Tolerance Curves for Glucose

The above graph contrasts blood-glucose levels for diabetic and nondiabetic individuals in the context of a two-hour oral glucose-tolerance test. A person must fast for eight hours prior to testing. A blood sample is taken at the beginning of the test, a 50-g glucose beverage is consumed, and a second blood sample is taken two hours later.

Most people with diabetes take oral medication rather than insulin, and the proportion who do so is increasing. Oral medication use rose from 60% in 1997 to 77% in 2007. One of the most used oral anti-diabetic drugs is the compound metformin. Structurally, metformin is a noncyclic organic compound that contains more nitrogen atoms (five) than carbon atoms (four).

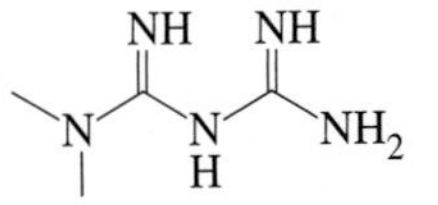

Metformin does not increase how much insulin the pancreas makes; instead it acts on the liver, decreasing the amount of glucose it produces. An average person with type 2 diabetes has a gluconeogenesis rate that is three times the normal rate. Metformin slows down the production of glucose via gluconeogenesis (Section 13.6).

Epinephrine acts by binding to a receptor site on the outside of the cell membrane, stimulating the enzyme *adenyl cyclase* to begin production of a *second messenger,* cyclic AMP (cAMP) from ATP. The cAMP is released in the cell interior, where, in a series of reactions, it activates *glycogen phosphorylase,* the enzyme

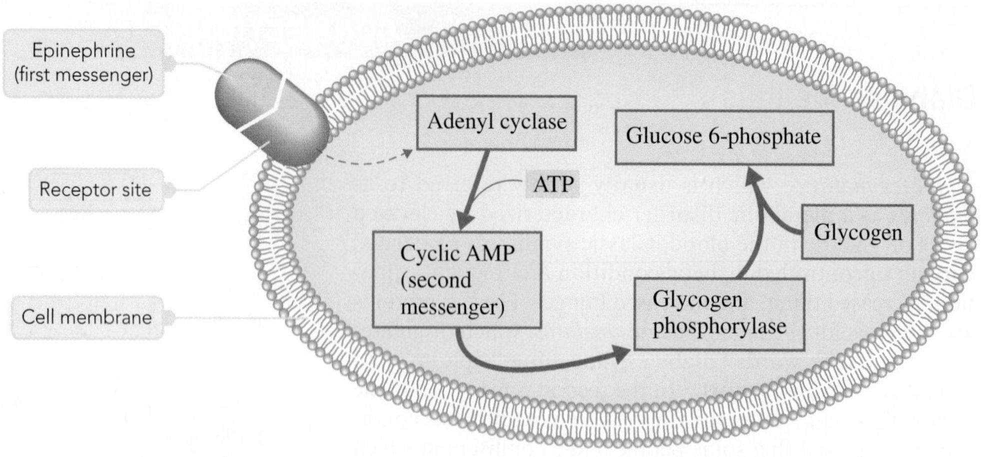

Figure 13.18 The series of events by which the hormone epinephrine stimulates glucose production.

that initiates glycogenolysis. The glucose 6-phosphate that is produced from the glycogen breakdown provides a source of quick energy. Figure 13.18 shows the series of events initiated by the release of the hormone epinephrine. Cyclic AMP also inhibits glycogenesis, thus preventing glycogen production at the same time.

13.10 B Vitamins and Carbohydrate Metabolism

The major function of B vitamins is that of coenzymes in metabolic reactions (Section 10.14). Now that the reactions involved in carbohydrate metabolism have been considered, it is informative to consider, in summary fashion, B vitamin involvement in the processes of glycolysis, gluconeogenesis, glycogenesis, glycogenolysis, and conversion of pyruvate to acetyl CoA and lactate. As is shown in Figure 13.19, six of the eight B vitamins are involved in carbohydrate metabolism.

Four of the six B vitamins involved in carbohydrate metabolism are the same four that are involved in the common metabolic pathway (see Figure 11.20). These four vitamins are niacin (as NAD^+, NADH), riboflavin (as FAD), thiamin (as TPP), and pantothenic acid (as CoA).

The two newly involved B vitamins are biotin and vitamin B_6:

1. Biotin involvement occurs in the enzyme pyruvate carboxylate, the enzyme needed to convert pyruvate to oxaloacetate (the new first step in gluconeogenesis).
2. Vitamin B_6 in the form of PLP is involved in glycogenosis (Section 10.14).

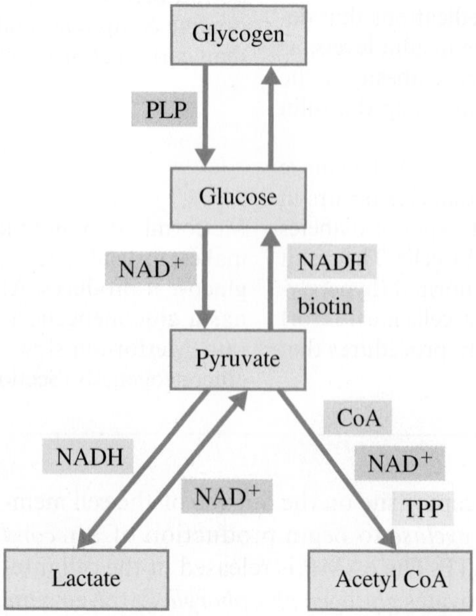

Figure 13.19 B vitamin participation in chemical reactions associated with carbohydrate metabolism.

Concepts to Remember

OWL Sign in at www.cengage.com/owl to view tutorials and simulations, develop problem-solving skills, and complete online homework assigned by your professor.

Glycolysis. Glycolysis, a series of ten reactions that occur in the cytosol, is a process in which one glucose molecule is converted into two molecules of pyruvate. A net gain of two molecules of ATP and two molecules of NADH results from the metabolizing of glucose to pyruvate (Section 13.2).

Fates of pyruvate. With respect to energy-yielding metabolism, the pyruvate produced by glycolysis can be converted to acetyl CoA under aerobic conditions or to lactate under anaerobic conditions. Some microorganisms convert pyruvate to ethanol, an anaerobic process (Section 13.3).

Glycogenesis. Glycogenesis is the process whereby excess glucose 6-phosphate is converted into glycogen. The glycogen is stored in the liver and in muscle tissue (Section 13.5).

Glycogenolysis. Glycogenolysis is the breakdown of glycogen into glucose 6-phosphate. This process occurs when muscles need energy and when the liver is restoring a low blood-sugar level to normal (Section 13.5).

Gluconeogenesis. Gluconeogenesis is the formation of glucose from pyruvate, lactate, and certain other substances. This process takes place in the liver when glycogen supplies are being depleted and when carbohydrate intake is low (Section 13.6).

Cori cycle. The Cori cycle is the cyclic process involving the transport of lactate from muscle tissue to the liver, the resynthesis of glucose by gluconeogenesis, and the return of glucose to muscle tissue (Section 13.6).

Pentose phosphate pathway. The pentose phosphate pathway metabolizes glucose to produce ribose (a pentose), NADPH, and other sugars needed for biosynthesis (Section 13.8).

Carbohydrate metabolism and hormones. Insulin decreases blood-glucose levels by promoting the uptake of glucose by cells. Glucagon increases blood-glucose levels by promoting the conversion of glycogen to glucose. Epinephrine stimulates the release of glucose from glycogen in muscle cells (Section 13.9).

Exercises and Problems

OWL Interactive versions of these problems may be assigned in OWL.

Exercises and problems are arranged in matched pairs with the two members of a pair addressing the same concept(s). The answer to the odd-numbered member of a pair is given at the back of the book. Problems denoted with a ▲ involve concepts found not only in the section under consideration but also concepts found in one or more earlier sections of the chapter. Problems denoted with a ● cover concepts found in a Chemical Connections feature box.

Carbohydrate Digestion (Section 13.1)

13.1 Where does carbohydrate digestion begin in the body, and what is the name of the enzyme involved in this initial digestive process?

13.2 Very little digestion of carbohydrates occurs in the stomach. Why?

13.3 What is the primary site for carbohydrate digestion, and what organ produces the enzymes that are active at this location?

13.4 Where does the final step in carbohydrate digestion take place, and in what form are carbohydrates as they enter this final step?

13.5 Where does the digestion of sucrose begin, and what is the reaction that occurs?

13.6 Where does the digestion of lactose begin, and what is the reaction that occurs?

13.7 Identify the three major monosaccharides produced by digestion of carbohydrates.

13.8 The various stages of carbohydrate digestion all involve the same general type of reaction. What is this reaction type?

Glycolysis (Section 13.2)

13.9 What is the starting material of glycolysis?

13.10 What is the end product of glycolysis?

13.11 What coenzyme functions as the oxidizing agent in glycolysis?

13.12 What is meant by the statement that glycolysis is an anaerobic pathway?

13.13 What is the first step of glycolysis, and why is it important in retaining glucose inside the cell?

13.14 Step 3 of glycolysis is the commitment step. Explain.

13.15 What two C_3 fragments are formed by the splitting of a fructose 1,6-bisphosphate molecule?

13.16 In one step of the glycolysis pathway, a C_6 chain is broken into two C_3 fragments, only one of which can be further degraded. What happens to the other C_3 fragment?

13.17 How many pyruvate molecules are produced per glucose molecule during glycolysis?

13.18 How many molecules of ATP and NADH are produced per glucose molecule during glycolysis?

13.19 How many steps in the glycolysis pathway produce ATP?

13.20 How many steps in the glycolysis pathway consume ATP?

13.21 Of the 10 steps of glycolysis, which ones involve phosphorylation?

13.22 Of the 10 steps of glycolysis, which ones involve oxidation?

13.23 Where in a cell does glycolysis occur?

13.24 Do the reactions of glycolysis and the citric acid cycle occur at the same location in a cell? Explain.

13.25 Replace the question mark in each of the following word equations with the name of a substance.

a. Glucose + ATP $\xrightarrow{\text{Hexokinase}}$? + ADP

b. ? $\xrightarrow{\text{Enolase}}$ phosphoenolpyruvate + water

c. 3-Phosphoglycerate $\xrightarrow{?}$ 2-phosphoglycerate

d. 1,3-Bisphosphoglycerate + ? $\xrightarrow[\text{kinase}]{\text{Phosphoglycero-}}$ 3-phosphoglycerate + ATP

13.26 Replace the question mark in each of the following word equations with the name of a substance.

a. Glucose 6-phosphate $\xrightarrow[\text{isomerase}]{\text{Phosphogluco-}}$?

b. ? $\xrightarrow{\text{Aldolase}}$ dihydroxyacetone phosphate + glyceraldehyde 3-phosphate

c. Phosphoenolpyruvate + ? $\xrightarrow[\text{kinase}]{\text{Pyruvate}}$ pyruvate + ATP

d. Dihydroxyacetone phosphate $\xrightarrow{?}$ glyceraldehyde 3-phosphate

13.27 In which step of glycolysis does each of the following occur?
a. Second substrate-level phosphorylation reaction
b. First ATP-consuming reaction
c. Third isomerization reaction
d. Use of NAD^+ as an oxidizing agent

13.28 In which step of glycolysis does each of the following occur?
a. First energy-producing reaction
b. First ATP-producing reaction
c. A dehydration reaction
d. First isomerization reaction

13.29 What is the net ATP production when each of the following molecules is processed through the glycolysis pathway?
a. One glucose molecule
b. One sucrose molecule

13.30 What is the net ATP production when each of the following molecules is processed through the glycolysis pathway?
a. One lactose molecule
b. One maltose molecule

13.31 Draw structural formulas for each of the following pairs of molecules.
a. Pyruvic acid and pyruvate
b. Dihydroxyacetone and dihydroxyacetone phosphate
c. Fructose 6-phosphate and fructose 1,6-bisphosphate
d. Glyceric acid and glyceraldehyde

13.32 Draw structural formulas for each of the following pairs of molecules.
a. Glyceric acid and glycerate
b. Glycerate and pyruvate
c. Glucose 6-phosphate and fructose 6-phosphate
d. Dihydroxyacetone and glyceric acid

13.33 Number the carbon atoms of fructose 1,6-bisphosphate 1 through 6, and show the location of each carbon in the two trioses produced during Step 4 of glycolysis.

13.34 Number the carbon atoms of glucose 1 through 6, and show the location of each carbon in the two molecules of pyruvate produced by glycolysis.

▲13.35 Indicate whether each of the following molecules or reaction types is associated with (1) carbohydrate digestion or (2) the glycolysis metabolic pathway.
a. Hexokinase b. Lactase
c. Hydrolysis d. Dehydration

▲13.36 Indicate whether each of the following molecules is associated with (1) carbohydrate digestion or (2) the glycolysis metabolic pathway.
a. α-Amylase b. Aldolase
c. Fructose d. Fructose 6-phosphate

Fates of Pyruvate (Section 13.3)

13.37 What are the three common possible fates for pyruvate produced from glycolysis?

13.38 Compare the fates of pyruvate in the body under aerobic and anaerobic conditions.

13.39 What is the overall reaction equation for the conversion of pyruvate to acetyl CoA?

13.40 What is the overall reaction equation for the conversion of pyruvate to lactate?

13.41 Explain how lactate fermentation allows glycolysis to continue under anaerobic conditions.

13.42 How is the ethanol fermentation in yeast similar to lactate fermentation in skeletal muscle?

13.43 In ethanol fermentation, a C_3 pyruvate molecule is changed to a C_2 ethanol molecule. What is the fate of the third pyruvate carbon?

13.44 What are the structural differences between pyruvate and lactate ions?

13.45 What is the net reaction for the conversion of one glucose molecule to two lactate molecules?

13.46 What is the net reaction for the conversion of one glucose molecule to two ethanol molecules?

13.47 With which of the possible fates of pyruvate—acetyl CoA, lactate, and ethanol—is each of the following reaction characterizations associated? More than one answer (fate) may apply in a given situation.
a. CO_2 is produced.
b. NADH is a reactant.
c. NAD^+ is a reactant.
d. The end product is a C_3 molecule.

13.48 With which of the possible fates of pyruvate—acetyl CoA, lactate, and ethanol—is each of the following reaction characterizations associated? More than one answer (fate) may apply in a given situation.
a. Acetaldehyde is an intermediate.
b. NADH is a product.
c. NAD^+ is a product.
d. The end product is a C_2 molecule.

▲13.49 Indicate whether each of the following compounds is associated with (1) glycolysis (2) pyruvate oxidation (3) lactate fermentation or (4) ethanol fermentation. There may be more than one correct response in a given situation.
a. CO_2 b. Acetyl CoA
c. ATP d. NADH

▲13.50 Indicate whether each of the following compounds is associated with (1) glycolysis (2) pyruvate oxidation (3) lactate fermentation or (4) ethanol fermentation. There may be more than one correct response in a given situation.
a. Pyruvate b. CoA c. ADP d. NAD^+

•13.51 (Chemical Connections 13-A) Indicate whether each of the following statements concerning "lactate accumulation" is true or false.
a. In the conversion of pyruvate to lactate, NAD^+ is a reactant.
b. The pH of blood increases when lactate diffuses from cells into the bloodstream.
c. Some sprinters use hyperventilation just prior to a race to increase lactate concentration in the bloodstream.
d. Higher than normal lactate concentration in the blood of a person at rest is a sign of impaired oxygen delivery.

•13.52 (Chemical Connections 13-A) Indicate whether each of the following statements concerning "lactate accumulation" is true or false.
a. The conversion of pyruvate to lactate is an anaerobic process.
b. More rapid breathing results when lactate diffusion from cells into the bloodstream occurs.
c. Hyperventilation causes lactate to move from the bloodstream to cells, causing the blood to become more basic.
d. Premature infants are often given oxygen to minimize lactate production.

Complete Oxidation of Glucose (Section 13.4)

13.53 How does the fact that cytosolic $NADH/H^+$ cannot cross the mitochondrial membranes affect ATP production from cytosolic $NADH/H^+$?

13.54 What is the net reaction for the shuttle mechanism involving glycerol 3-phosphate by which NADH electrons are shuttled across the mitochondrial membrane?

13.55 Contrast, in terms of ATP production, the oxidation of glucose to CO_2 and H_2O with the oxidation of glucose to pyruvate.

13.56 Contrast, in terms of ATP production, the oxidation of glucose to CO_2 and H_2O with the oxidation of glucose to ethanol.

13.57 How many of the 30 ATP molecules produced from the complete oxidation of 1 glucose molecule are produced during glycolysis?

13.58 How many of the 30 ATP molecules produced from the complete oxidation of 1 glucose molecule are produced during the oxidation of pyruvate to acetyl CoA?

Glycogen Metabolism (Section 13.5)

13.59 Compare the meanings of the terms *glycogenesis* and *glycogenolysis.*

13.60 Where is most of the body's glycogen stored?

13.61 Glucose 1-phosphate is the product of the first step of glycogenesis. What is the reactant?

13.62 Glucose 1-phosphate is the product of the first step of glycogenolysis. What are the reactants?

13.63 What is the source of the PP_i produced during the second step of glycogenesis?

13.64 What is the function of the PP_i produced during the second step of glycogenesis?

13.65 How is ATP involved in glycogenesis?

13.66 How many ATP molecules are needed to attach a single glucose molecule to a growing glycogen chain?

13.67 Which step of glycogenolysis is the reverse of Step 1 of glycogenesis?

13.68 What reaction determines whether glucose 6-phosphate formed by glycogenolysis can leave a cell?

13.69 What is the difference between the processing of glucose 6-phosphate in liver cells and that in muscle cells?

13.70 The liver, but not the brain or muscle cells, has the capacity to supply free glucose to the blood. Explain.

13.71 In what form does glycogen enter the glycolysis pathway?

13.72 Explain why one more ATP is produced when glucose is obtained from glycogen than when it is obtained directly from the blood.

▲13.73 Indicate whether each of the following compounds is associated with (1) glycolysis (2) glycogenesis or (3) glycogenolysis. There may be more than one correct response in a given situation.
a. Glucose 6-phosphate b. UTP
c. Phosphoglucomutase d. Pyruvate

▲13.74 Indicate whether each of the following compounds is associated with (1) glycolysis (2) glycogenesis or (3) glycogenolysis. There may be more than one correct response in a given situation.
a. Glucose 1-phosphate b. Glucose
c. Phosphoglucoisomerase d. UDP-glucose

Gluconeogenesis (Section 13.6)

13.75 What organ is primarily responsible for gluconeogenesis?

13.76 What is the physiological function of gluconeogenesis?

13.77 How does gluconeogenesis get around the three irreversible steps of glycolysis?

13.78 Although gluconeogenesis and glycolysis are "reverse" processes, there are 11 steps in gluconeogenesis and only 10 steps in glycolysis. Explain.

13.79 What intermediate in gluconeogenesis is also an intermediate in the citric acid cycle?

13.80 What are the sources of high-energy bonds in gluconeogenesis?

13.81 What is the fate of lactate formed by muscular activity?

13.82 What is the physiological function of the Cori cycle?

▲13.83 Indicate whether each of the following enzymes is involved in (1) glycolysis but not gluconeogenesis (2) gluconeogenesis but not glycolysis or (3) both glycolysis and gluconeogenesis.
a. Hexokinase b. Phosphofructokinase
c. Pyruvate carboxylase d. Phosphoglyceromutase

▲13.84 Indicate whether each of the following enzymes is involved in (1) glycolysis but not gluconeogenesis (2) gluconeogenesis but not glycolysis or (3) both glycolysis and gluconeogenesis.
a. Glucose 6-phosphatase
b. Fructose 1,6-bisphosphatase
c. Pyruvate kinase
d. Phosphoglucoisomerase

▲13.85 Indicate whether each of the following compounds is involved in (1) glycolysis but not gluconeogenesis (2) gluconeogenesis but not glycolysis or (3) both glycolysis and gluconeogenesis.
a. 2-Phosphoglycerate
b. Phosphoenolpyruvate
c. Fructose 1,6-bisphosphate
d. Glucose 6-phosphate

▲13.86 Indicate whether each of the following compounds is involved in (1) glycolysis but not gluconeogenesis (2) gluconeogenesis but not glycolysis or (3) both glycolysis and gluconeogenesis.
a. 3-Phosphoglycerate
b. Oxaloacetate
c. Fructose 6-phosphate
d. Dihydroxyacetone phosphate

▲13.87 How many different "C_3 molecules" are involved in
a. the Cori cycle
b. gluconeogenesis
c. glycolysis
d. lactate fermentation

▲13.88 How many different "C_6 molecules" are involved in
a. the Cori cycle
b. gluconeogenesis
c. glycolysis
d. ethanol fermentation

Terminology for Glucose Metabolic Pathways (Section 13.7)

13.89 Indicate in which of the four processes *glycolysis, glycogenesis, glycogenolysis,* and *gluconeogenesis* each of the following compounds is encountered. There may be more than one correct answer for a given compound.
a. Glucose 6-phosphate
b. Dihydroxyacetone phosphate
c. Oxaloacetate
d. UDP-glucose

13.90 Indicate in which of the four processes *glycolysis, glycogenesis, glycogenolysis,* and *gluconeogenesis* each of the following compounds is encountered. There may be more than one correct answer for a given compound.
a. Glucose 1-phosphate
b. Glycogen
c. Pyruvate
d. Fructose 6-phosphate

13.91 Indicate in which of the four processes *glycolysis, glycogenesis, glycogenolysis,* and *gluconeogenesis* each of the following situations is encountered. There may be more than one correct answer for each.
a. NAD^+ is consumed.
b. ATP is produced.
c. UDP is involved.
d. ADP is consumed.

13.92 Indicate in which of the four processes *glycolysis, glycogenesis, glycogenolysis,* and *gluconeogenesis* each of the following situations is encountered. There may be more than one correct answer for each.
a. NADH is consumed.
b. ATP is consumed.
c. CO_2 is involved.
d. H_2O is a product.

13.93 Indicate in which of the four processes *glycolysis, glycogenesis, glycogenolysis,* and *gluconeogenesis* each of the following enzymes is needed. There may be more than one correct answer for each.
a. Fructose 1,6-bisphosphatase
b. Pyruvate kinase
c. Glycogen synthase
d. Phosphoglucomutase

13.94 Indicate in which of the four processes *glycolysis, glycogenesis, glycogenolysis,* and *gluconeogenesis* each of the following enzymes is needed. There may be more than one correct answer for each.
a. Pyruvate carboxylase
b. Glycogen phosphorylase
c. Hexokinase
d. Glucose 6-phosphatase

▲13.95 What is the net gain or net loss in triphosphates (ATP, UTP, etc.) in each of the following metabolic processes?
a. Gluconeogenesis
b. Glycogenesis
c. Glycogen glucose unit to pyruvate
d. Cori cycle

▲13.96 What is the net gain or net loss in triphosphates (ATP, UTP, etc.) in each of the following metabolic processes?
a. Glycolysis
b. Glycogenolysis
c. Glycogen glucose unit to glucose
d. Glycogen glucose unit to lactate

The Pentose Phosphate Pathway (Section 13.8)

13.97 What is the starting material for the pentose phosphate pathway?

13.98 What are two major functions of the pentose phosphate pathway?

13.99 How do the biochemical functions of NADH and NADPH differ?

13.100 How do the structures of NADH and NADPH differ?

13.101 Write a general equation for the oxidative stage of the pentose phosphate pathway.

13.102 Write a general equation for the entire pentose phosphate pathway.

13.103 What compound contains the carbon atom lost from glucose (a hexose) in its conversion to ribose (a pentose)?

13.104 How many molecules of NADPH are produced per glucose 6-phosphate in the pentose phosphate pathway?

▲13.105 Indicate whether each of the following compounds is involved in (1) the pentose phosphate pathway (2) the Cori cycle (3) glycolysis or (4) lactate fermentation. More than one response may be correct in a given situation.
a. Lactate
b. NAD^+
c. Glucose 6-phosphate
d. Ribose 5-phosphate

▲13.106 Indicate whether each of the following compounds is involved in (1) the pentose phosphate pathway (2) the Cori cycle (3) glycolysis or (4) lactate fermentation. More than one response may be correct in a given situation.
a. Pyruvate
b. $NADP^+$
c. Glucose
d. Fructose 6-phosphate

Control of Carbohydrate Metabolism (Section 13.9)

13.107 What effect does insulin have on glycogen metabolism?

13.108 What effect does insulin have on blood-glucose levels?

13.109 What effect does glucagon have on blood-glucose levels?

13.110 What effect does glucagon have on glycogen metabolism?

13.111 What organ is the source of insulin?

13.112 What organ is the source of glucagon?

13.113 The hormone epinephrine generates a "second messenger." Explain.

13.114 What is the relationship between cAMP and the hormone epinephrine?

13.115 Compare the target tissues for glucagon and epinephrine.

13.116 Compare the biological functions of glucagon and epinephrine.

•13.117 (Chemical Connections 13-B) Indicate whether each of the following statements concerning diabetes mellitus is true or false.
a. A fasting blood-glucose level of 131 g/dL is interpreted as a "positive" test for diabetes.
b. Type 1 diabetes is associated with insulin resistance.
c. The anti-diabetic drug metformin increases insulin production.
d. Structurally, the drug metformin contains more N atoms than C atoms.

•13.118 (Chemical Connections 13-B) Indicate whether each of the following statements concerning diabetes mellitus is true or false.
a. A fasting blood-glucose level of 115 g/dL is interpreted as a "negative" test for diabetes.
b. Type 2 diabetes is associated with too much insulin production.
c. The anti-diabetic drug metformin decreases the rate at which gluconeogenesis occurs.
d. The chemical formula for the anti-diabetic drug metformin is $C_4N_5H_5$.

B Vitamins and Carbohydrate Metabolism (Section 13.10)

13.119 For the B vitamins below, indicate which of the following processes each vitamin is involved in as a cofactor: (1) glycolysis (2) gluconeogenesis (3) lactate fermentation or (4) glycogenolysis. For a given vitamin, there may be more than one correct response, or only one or none of the responses may apply.
a. Thiamin
b. Riboflavin
c. Pantothenic acid
d. Vitamin B_6

13.120 For the B vitamins below, indicate which of the following processes each vitamin is involved in as a cofactor: (1) glycolysis (2) gluconeogenesis (3) lactate fermentation or (4) glycogenolysis. For a given vitamin, there may be more than one correct response, or only one or none of the responses may apply.
a. Niacin
b. Biotin
c. Folate
d. Vitamin B_{12}

Lipid Metabolism

14

© Huguet Pierre/Peter Arnold, Inc.

The fat stored in a camel's hump serves not only as a source of energy but also as a source of water. Water is one of the products of fat (triacylglycerol) metabolism.

CHAPTER OUTLINE

Certain classes of lipids play an extremely important role in cellular metabolism because they represent an energy-rich "fuel" that can be stored in large amounts in adipose (fat) tissue. Between one-third and one-half of the calories present in the diet of the average U.S. resident are supplied by lipids. Furthermore, excess energy derived from carbohydrates and proteins beyond normal daily needs is stored in lipid molecules (in adipose tissue), to be mobilized later and used when needed.

14.1 Digestion and Absorption of Lipids

Because 98% of total *dietary* lipids are triacylglycerols (fats and oils; Section 8.4), this chapter focuses on triacylglycerol metabolism. Like all lipids, triacylglycerols (TAGs) are insoluble in water. Hence water-based salivary enzymes in the mouth have little effect on them. The *major* change that TAGs undergo in the stomach is physical rather than chemical. The churning action of the stomach breaks up triacylglycerol materials into small globules, or droplets, which float as a layer above the other components of swallowed food. The resulting material is called *chyme.* **Chyme** *is a thick semi-liquid material made up of partially digested food and gastric secretions (hydrochloric acid and several enzymes).*

High-fat foods remain in the stomach longer than low-fat foods. The conversion of high-fat materials into chyme takes longer than the breakup of low-fat

The saliva of infants contains a lipase that can hydrolyze TAGs, so digestion begins in the mouth for nursing infants. Because mother's milk is already a lipid-in-water emulsion, emulsifications by stomach churning is a much less important factor in an infant's processing of fat. Mother's milk also contains a lipase that supplements the action of the salivary lipases the infant itself produces. After weaning, infants cease to produce salivary lipases.

OWL

Sign in to OWL at **www.cengage.com/owl** to view tutorials and simulations, develop problem-solving skills, and complete online homework assigned by your professor.

materials. This is why a high-fat meal causes a person to feel "full" for a longer period of time.

Chyme is pronounced "kyme" (rhymes with *dime*).

Lipid digestion also begins in the stomach. Under the action of *gastric lipase* enzymes, hydrolysis of TAGs occurs. Normally, about 10% of TAGs undergo hydrolysis in the stomach, but regular consumption of a high-fat diet can induce the production of higher levels of gastric lipases.

The arrival of chyme from the stomach triggers in the small intestine, through the action of the hormone *cholecystokinin,* the release of bile stored in the gallbladder. The bile (Section 8.11), which contains no enzymes, acts as an emulsifier (Section 8.11). Colloid particle formation through bile emulsification "solubilizes" the triacylglycerol globules, and digestion of the TAGs resumes. The major enzymes involved at this point are the *pancreatic lipases,* which hydrolyze ester linkages between the glycerol and fatty acid units of the TAGs. *Complete* hydrolysis does not usually occur; only two of the three fatty acid units are liberated, producing a monoacylglycerol and two free fatty acids. Occasionally, enzymes remove all three fatty acid units, leaving a free glycerol molecule.

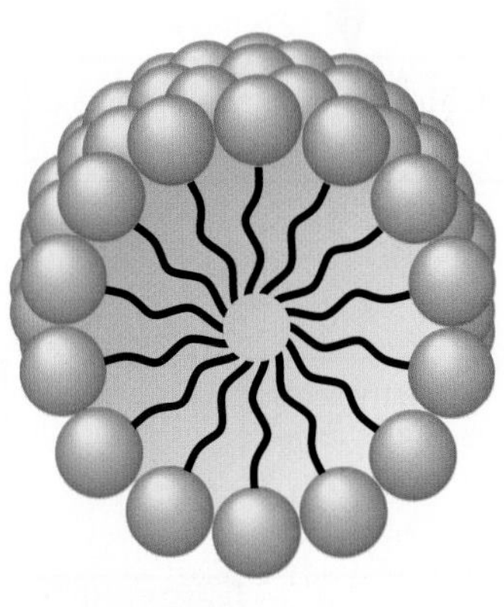

Figure 14.1 In a fatty acid micelle, the hydrophobic chains of the fatty acids and monoacylglycerols are in the interior of the micelle.

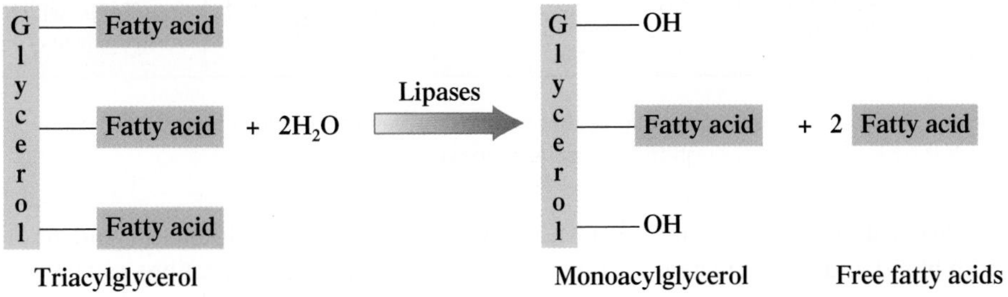

When freed of the triacylglycerol molecules they "transport" during digestion, bile acids are mostly recycled. Small amounts are excreted.

With the help of bile, the free fatty acids and monoacylglycerols produced from hydrolysis are combined into tiny spherical droplets called micelles (Section 8.6). A **fatty acid micelle** *is a micelle in which fatty acids and/or monoacylglycerols and some bile are present.* Fatty acid micelles are very small compared to the original triacylglycerol globules, which contain thousands of triacylglycerol molecules. Figure 14.1 shows a cross-section of the three-dimensional structure of a fatty acid micelle.

Chylomicron is pronounced "kye-lo-MY-cron."

Micelles, containing free fatty acid and monoacylglycerol components, are small enough to be readily absorbed through the membranes of intestinal cells. Within the intestinal cells, a "repackaging" occurs in which the free fatty acids and monoacylglycerols are reassembled into triacylglycerols. The newly formed triacylglycerols are then combined with membrane lipids (phospholipids and cholesterol) and water-soluble proteins to produce a type of lipoprotein (Section 9.19) called a *chylomicron* (Figure 14.2). A **chylomicron** *is a lipoprotein that transports triacylglycerols from intestinal cells, via the lymphatic system, to the bloodstream.* Triacylglycerols constitute 95% of the core lipids present in a chylomicron.

Chylomicrons are too large to pass through capillary walls directly into the bloodstream. Consequently, delivery of the chylomicrons to the bloodstream is accomplished through the body's lymphatic system. Chylomicrons enter the lymphatic system through tiny lymphatic vessels in the intestinal lining. They enter the bloodstream through the thoracic duct (a large lymphatic vessel just below the collarbone), where the fluid of the lymphatic system flows into a vein, joining the bloodstream.

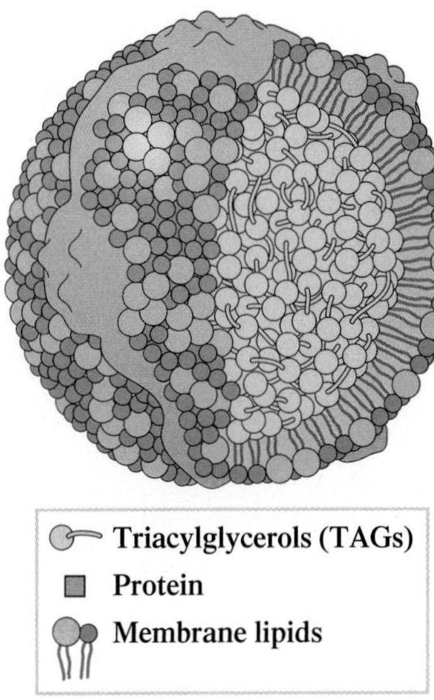

Figure 14.2 A three-dimensional model of a chylomicron, a type of lipoprotein. Chylomicrons are the form in which TAGs are delivered to the bloodstream via the lymphatic system.

Once the chylomicrons reach the bloodstream, the TAGs they carry are again hydrolyzed to produce glycerol and free fatty acids. TAG release from chylomicrons and their ensuing hydrolysis is mediated by *lipoprotein lipases.* These enzymes are located on the lining of blood vessels in muscle and other tissues that use fatty acids for fuel and in fat synthesis. The fatty acid and glycerol hydrolysis products from TAG hydrolysis are absorbed by the cells of the body and are either broken down to acetyl CoA for energy or stored as lipids (they are again repackaged as TAGs). Figure 14.3 summarizes the events that must occur before triacylglycerols can reach the bloodstream through the digestive process.

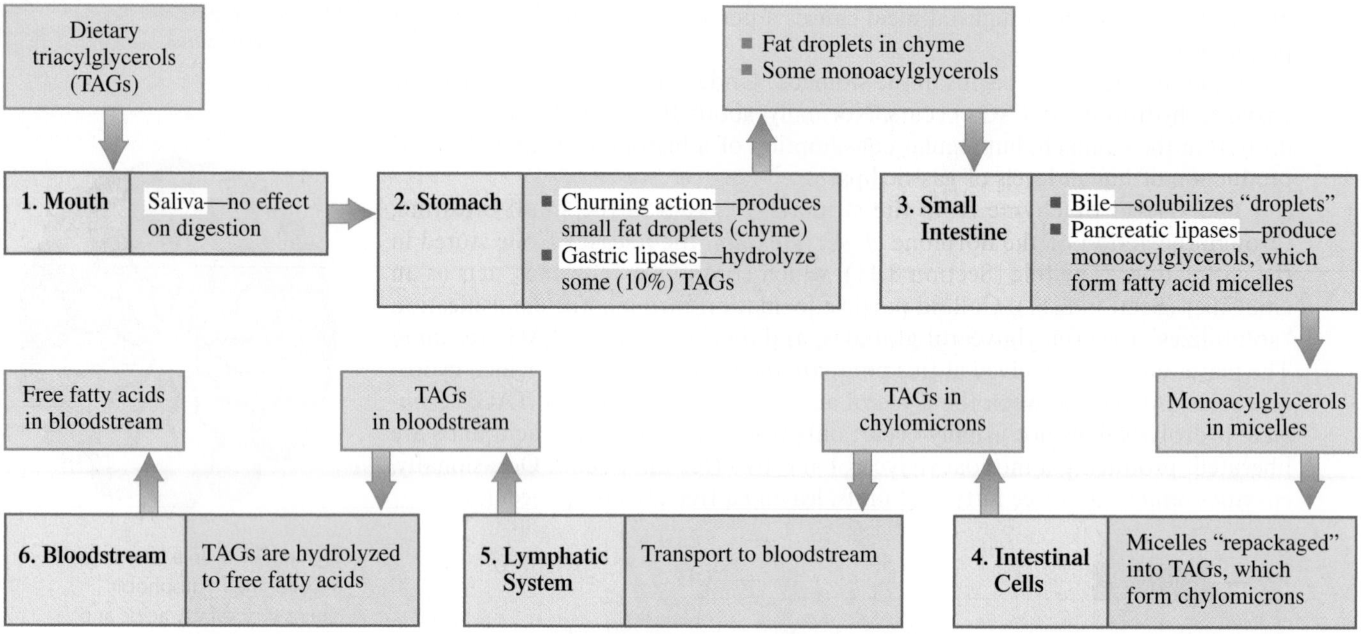

Figure 14.3 A summary of the events that must occur before triacylglycerols (TAGs) can reach the bloodstream through the digestive process.

Soon after a meal heavily laden with TAGs is ingested, the chylomicron content of both blood and lymph increases dramatically. Chylomicron concentrations usually begin to rise within 2 hours after a meal, reach a peak in 4–6 hours, and then drop rather rapidly to a normal level as they move into adipose cells (Section 14.2) or into the liver.

EXAMPLE 14.1 Determining the Sites Where Various Aspects of Lipid Digestion Occur

Based on the information in Figure 14.3, determine the location within the human body where each of the following aspects of lipid digestion occurs.

a. Interaction with bile occurs.
b. Monoacylglycerols are produced.
c. Chyme is produced.
d. Gastric lipases are active.

Solution

a. The *small intestine* is the location where bile released from the gallbladder interacts with the fat droplets present in chyme. The bile functions as an emulsifying agent.
b. Pancreatic lipases present in the *small intestine* convert triacylglycerols to monoacylglycerols.
c. The churning action of the *stomach* breaks triacylglycerol materials into small droplets that float on top of other ingested food materials; the resulting mixture is called chyme.
d. Gastric lipases present in the *stomach* begin the triacylglycerol hydrolysis process; about 10% of triacylglycerols undergo hydrolysis in the stomach.

Practice Exercise 14.1

Based on the information in Figure 14.3, determine the location within the human body where each of the following aspects of lipid digestion occurs.

a. Pancreatic lipases are active.
b. Fatty acid micelles are produced.
c. Chylomicrons are produced.
d. Monoacylglycerols are converted back to triacylglycerols.

Answers: **a.** Small intestine; **b.** Small intestine; **c.** Intestinal cells; **d.** Intestinal cells

14.2 Triacylglycerol Storage and Mobilization

Most cells in the body have limited capability for storage of TAGs. However, this activity is the major function of specialized cells called adipocytes, found in adipose tissue. An **adipocyte** *is a triacylglycerol-storing cell.* **Adipose tissue** *is tissue that contains large numbers of adipocyte cells.*

Adipose tissue is located primarily directly beneath the skin (subcutaneous), particularly in the abdominal region, and in areas around vital organs. Besides its function as a storage location for the chemical energy inherent in TAGs, subcutaneous adipose tissue also serves as an insulator against excessive heat loss to the environment and provides organs with protection against physical shock.

Adipose cells are among the largest cells in the body. They differ from other cells in that most of the cytoplasm has been replaced with a large triacylglycerol droplet (Figure 14.4). This droplet accounts for nearly the entire volume of the cell. As newly formed TAGs are imported into an adipose cell, they form small droplets at the periphery of the cell that later merge with the large central droplet.

Use of the TAGs stored in adipose tissue for energy production is triggered by several hormones, including epinephrine and glucagon. Hormonal interaction with adipose cell membrane receptors stimulates production of cAMP from ATP inside the adipose cell. In a series of enzymatic reactions, the cAMP activates *hormone-sensitive lipase (HSL)* through phosphorylation. HSL is the lipase needed for triacylglycerol hydrolysis, a prerequisite for fatty acids to enter the bloodstream from an adipose cell. This cAMP activation process is illustrated in Figure 14.5.

The overall process of tapping the body's triacylglycerol energy reserves (adipose tissue) for energy is called triacylglycerol mobilization. **Triacylglycerol mobilization** *is the hydrolysis of triacylglycerols stored in adipose tissue, followed by release into the bloodstream of the fatty acids and glycerol so produced.* Triacylglycerol mobilization is an ongoing process. On the average, about 10% of the TAGs in adipose tissue are replaced daily by new triacylglycerol molecules.

Triacylglycerol energy reserves (fat reserves) are the human body's major source of stored energy. Energy reserves associated with protein, glycogen, and glucose are small to very small when compared to fat reserves. Table 14.1 shows relative amounts of stored energy associated with the various types of energy reserves present in the human body.

Dietary TAGs deposited in adipose tissue have undergone hydrolysis two times (to form free fatty acids and/or monoacylglycerols) and are repackaged twice (to re-form TAGs) in reaching that state. They undergo hydrolysis for a third time when *triacylglycerol mobilization* occurs.

Adipose tissue is the only tissue in which *free* TAGs occur in appreciable amounts. In other types of cells and in the bloodstream, TAGs are part of lipoprotein particles.

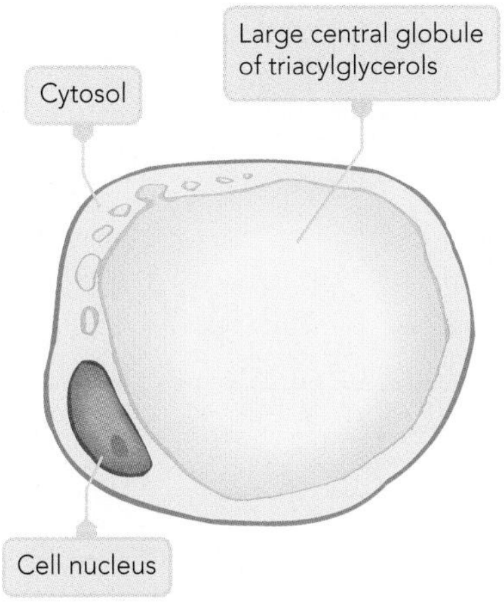

Figure 14.4 Structural characteristics of an adipose cell.

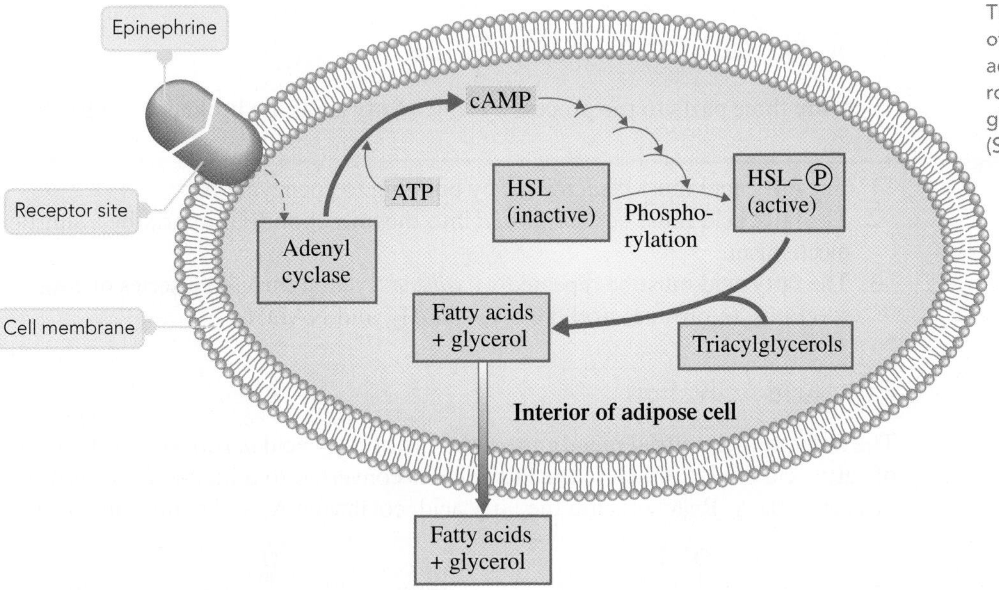

Figure 14.5 Hydrolysis of stored triacylglycerols in adipose tissue is triggered by hormones that stimulate cAMP production within adipose cells.

The use of cAMP in the activation of hormone-sensitive lipase in adipose cells is similar to cAMP's role in the activation of the glycogenolysis process (Section 13.9).

Triacylglycerol reserves would enable the average person to survive starvation for about 30 days, given sufficient water. Glycogen reserves (stored glucose) would be depleted within 1 day.

Table 14.1 Stored Energy Reserves of Various Types for a 150-lb (70-kg) Person

Type of Energy Reserve	Amount of Stored Energy	Percent of Total Stored Energy
triacylglycerol	135,000 kcal	84.3%
protein	24,000 kcal	15%
glycogen	720 kcal	0.45%
blood glucose	80 kcal	0.05%

14.3 Glycerol Metabolism

During triacylglycerol mobilization, one molecule of glycerol is produced for each triacylglycerol completely hydrolyzed. Glycerol metabolism primarily involves processes considered in the previous chapter. After entering the bloodstream, glycerol travels to the liver or kidneys, where it is converted, in a two-step process, to dihydroxyacetone phosphate.

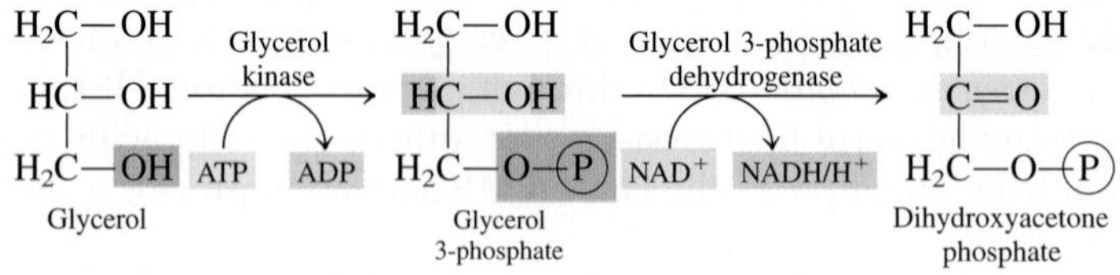

The first step involves phosphorylation of a primary hydroxyl group of the glycerol. In the second step, glycerol's secondary alcohol group (C-2) is oxidized to a ketone.

The following equation represents the overall reaction for the metabolism of glycerol.

$$\text{Glycerol} + \text{ATP} + \text{NAD}^+ \longrightarrow \text{Dihydroxyacetone phosphate} + \text{ADP} + \text{NADH} + \text{H}^+$$

Dihydroxyacetone phosphate is an intermediate in both glycolysis (Section 13.2) and gluconeogenesis (Section 13.6). It can be converted to pyruvate, then acetyl CoA, and finally carbon dioxide, or it can be used to form glucose. Dihydroxyacetone phosphate formation from glycerol represents the first of several situations we will consider wherein carbohydrate and lipid metabolism are connected.

14.4 Oxidation of Fatty Acids

The stored TAGs in adipose tissue supply approximately 60% of the body's energy needs when the body is in a resting state.

There are three parts to the process by which fatty acids are broken down to obtain energy.

1. The fatty acid must be *activated* by bonding to coenzyme A.
2. The fatty acid must be *transported* into the mitochondrial matrix by a shuttle mechanism.
3. The fatty acid must be repeatedly *oxidized,* cycling through a series of four reactions, to produce acetyl CoA, $FADH_2$, and NADH.

Fatty Acid Activation

The outer mitochondrial membrane is the site of fatty acid *activation,* the first stage of fatty acid oxidation. Here the fatty acid is converted to a high-energy derivative of coenzyme A. Reactants are the fatty acid, coenzyme A, and a molecule of ATP.

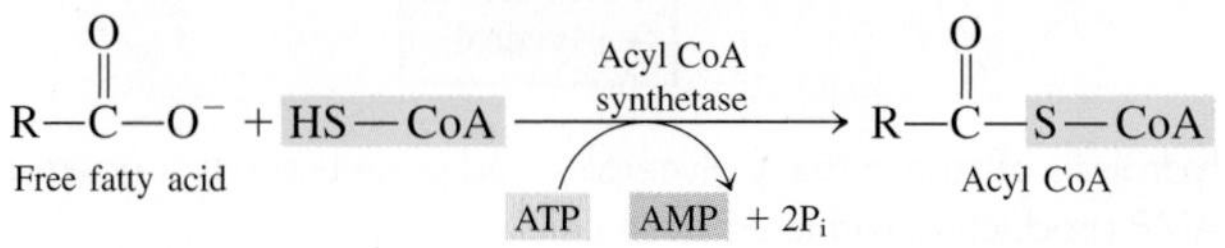

This reaction requires the expenditure of two high-energy phosphate bonds from a single ATP molecule; the ATP is converted to AMP rather than ADP, and the resulting pyrophosphate (PP_i) is hydrolyzed to $2P_i$.

The activated fatty acid–CoA molecule is called *acyl* CoA. The difference between the designations *acyl* CoA and *acetyl* CoA is that *acyl* refers to a random-length fatty acid carbon chain that is covalently bonded to coenzyme A, whereas *acetyl* refers to a two-carbon chain covalently bonded to coenzyme A.

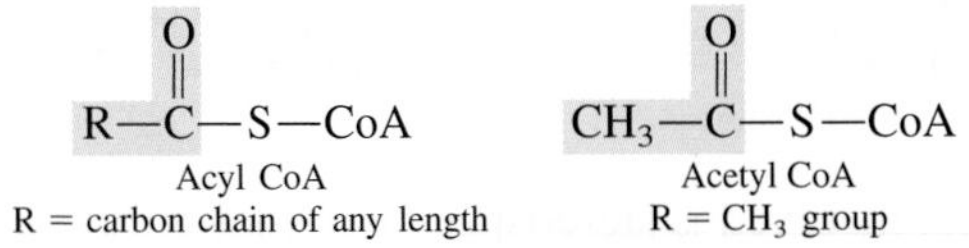

Recall, from Section 5.1, that *acyl* is a generic term for

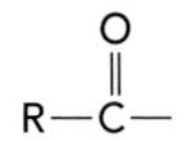

which is the species formed when the carboxyl —OH is removed from a carboxylic acid. The R group can involve a carbon chain of any length.

Fatty Acid Transport

Acyl CoA is too large to pass through the inner mitochondrial membrane to the mitochondrial matrix, where the enzymes needed for fatty acid oxidation are located. A shuttle mechanism involving the molecule carnitine effects the entry of acyl CoA into the matrix (Figure 14.6). The acyl group is transferred to a carnitine molecule, which carries it through the membrane. The acyl group is then transferred from the carnitine back to a CoA molecule.

Reactions of the β-Oxidation Pathway

In the mitochondrial matrix, a sequence of four reactions *repeatedly* cleaves two-carbon units from the carboxyl end of the acyl CoA molecule. This repetitive four-reaction sequence is called the *β-oxidation pathway* because the second carbon from the carboxyl end of the chain, the beta carbon, is the carbon atom that is oxidized. The **β-oxidation pathway** *is a repetitive series of four biochemical reactions that degrades acyl CoA to acetyl CoA by removing two carbon atoms at a time, with $FADH_2$ and NADH also being produced.* Each repetition of the four-reaction sequence generates an acetyl CoA molecule and an acyl CoA molecule that has two fewer carbon atoms.

The reaction sequence dehydrogenation–hydration–dehydrogenation in the β-oxidation pathway has a parallel in Steps 6–8 of the citric acid cycle (Section 12.7), where succinate is dehydrogenated to fumarate, which is hydrated to malate, which is dehydrogenated to oxaloacetate.

For a *saturated* fatty acid, the β-oxidation pathway involves the following functional group changes at the β carbon and the following reaction types.

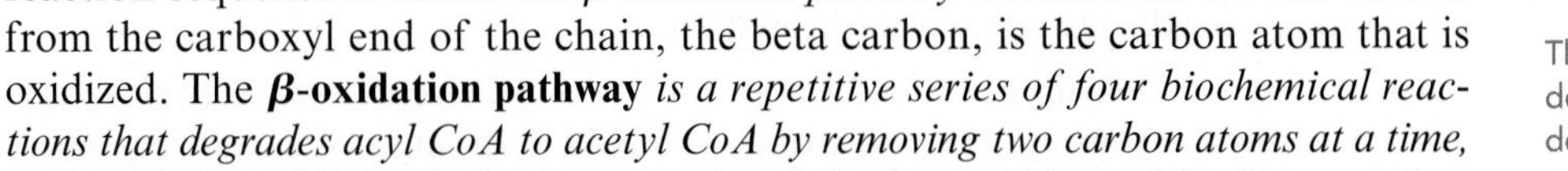

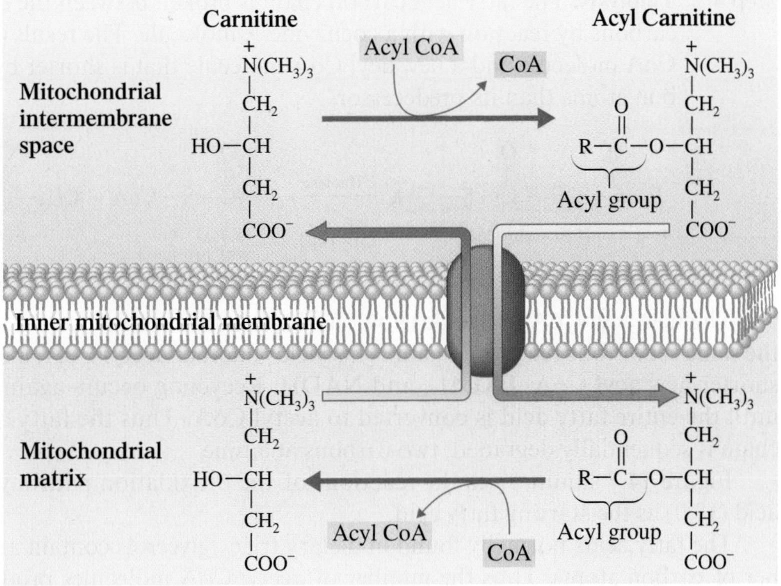

Figure 14.6 Fatty acids are transported across the inner mitochondrial membrane in the form of acyl carnitine.

Details about Steps 1–4 of the β-oxidation pathway follow.

Step 1: **First Dehydrogenation.** Hydrogen atoms are removed from the α and β carbons, creating a double bond between these two carbon atoms. FAD is the oxidizing agent, and a $FADH_2$ molecule is a product.

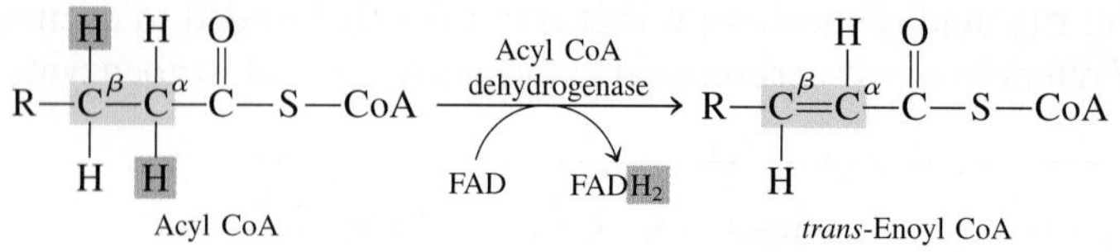

The enzyme involved is stereospecific in that only *trans* double bonds are produced.

Step 2: **Hydration.** A molecule of water is added across the *trans* double bond, producing a secondary alcohol at the β-carbon position. Again, the enzyme involved is stereospecific in that only the L-hydroxy isomer is produced from the *trans* double bond.

The loss of hydrogen atoms (oxidation) occurs in both Steps 1 and 3. The oxidizing agents differ because the type of double bond formed differs. The body uses FAD as the oxidizing agent when C═C double bond formation occurs and uses NAD^+ as the oxidizing agent when C═O double bond formation occurs.

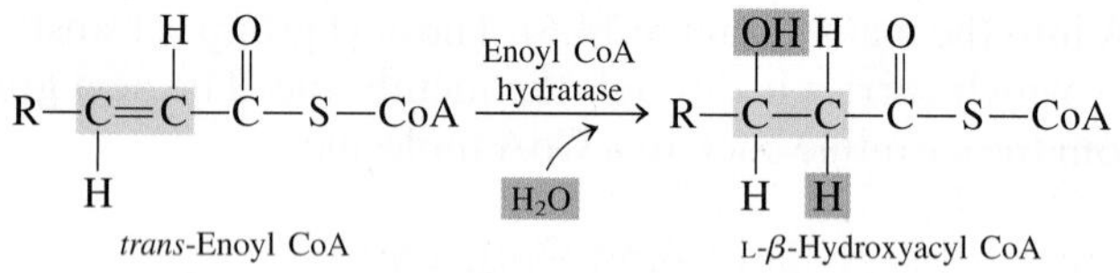

The enzyme involved in this hydration will also hydrate a *cis* double bond, but the product then is the D isomer. D isomer formation is of importance when considering how unsaturated fatty acids are oxidized, a topic discussed later in this section.

Step 3: **Second Dehydrogenation.** Removal of two hydrogen atoms converts the β-hydroxy group to a keto group, with NAD^+ serving as the oxidizing agent. The required enzyme exhibits absolute stereospecificity for the L isomer.

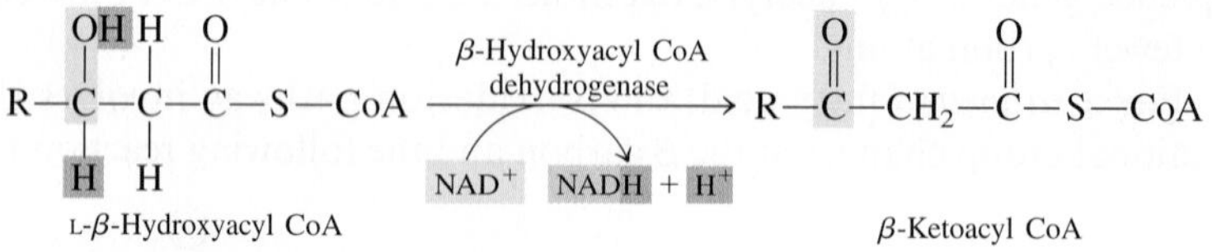

It is now apparent why the name for this series of reactions is the β-oxidation pathway. The β-carbon atom has been oxidized from a —CH_2— group to a ketone group.

Step 4: **Thiolysis.** The fatty acid carbon chain is broken between the α and β carbons by reaction with a coenzyme A molecule. The result is an acetyl CoA molecule and a new acyl CoA molecule that is shorter by two carbon atoms than its predecessor.

The chain cleavage reaction that occurs in Step 4 is called *thiolysis* by analogy with the process of *hydrolysis*, which also involves breaking a molecule into two parts. It is the thiol group of coenzyme A that undergoes reaction.

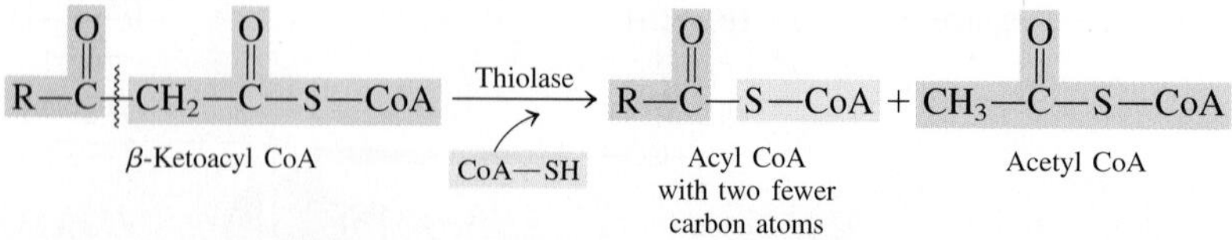

The new acyl CoA molecule (now shorter by two carbons) is *recycled* through the same set of four reactions again. This yields another acetyl CoA, a two-carbon-shorter new acyl CoA, $FADH_2$, and NADH. Recycling occurs again and again, until the entire fatty acid is converted to acetyl CoA. Thus the fatty acid carbon chain is sequentially degraded, two carbons at a time.

Figure 14.7 summarizes the reactions of the β-oxidation pathway for stearic acid (18:0) as the starting fatty acid.

The fatty acids normally found in dietary triacylglycerols contain an *even* number of carbon atoms. Thus the number of acetyl CoA molecules produced in the

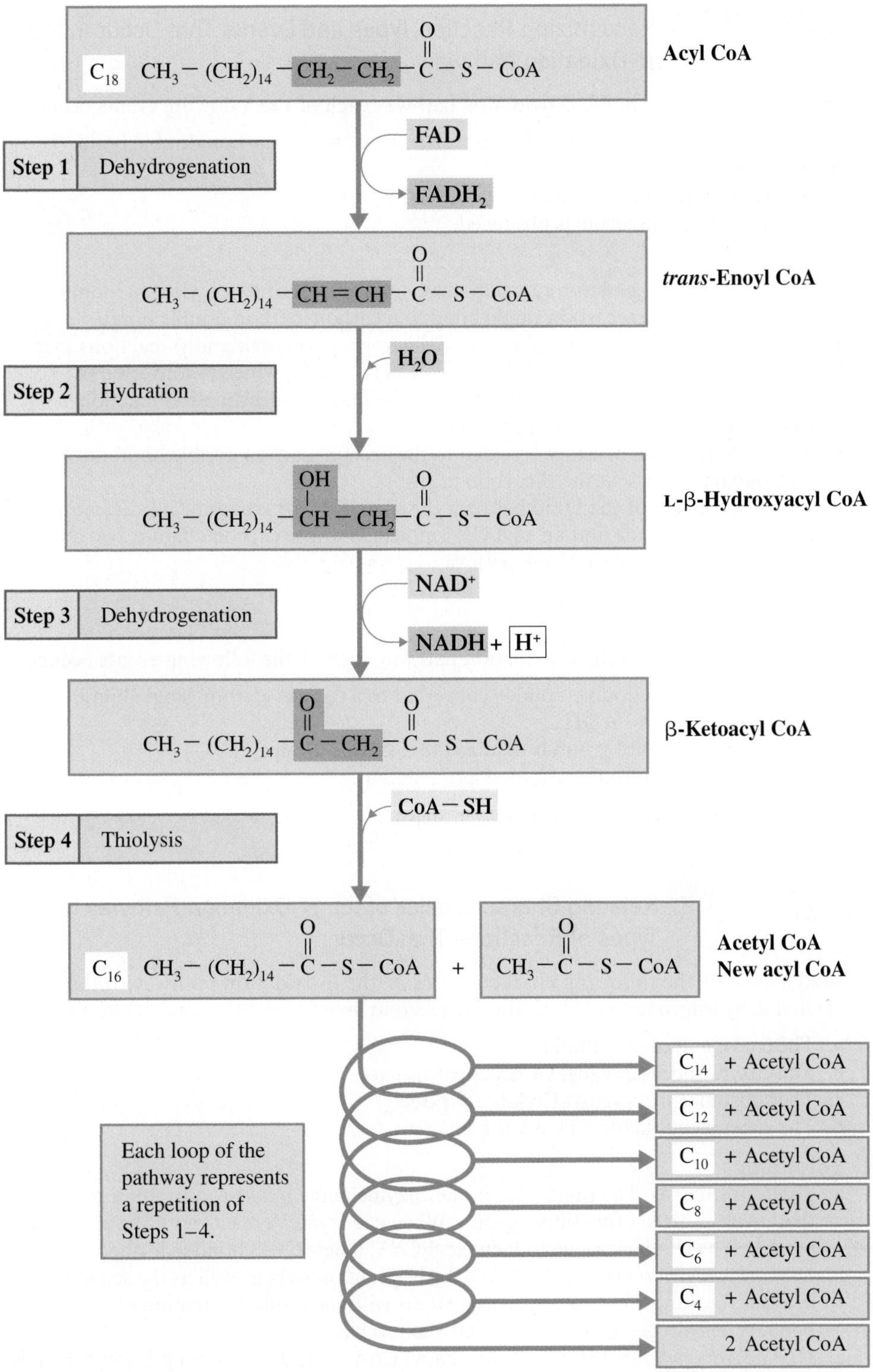

Figure 14.7 Reactions of the *β*-oxidation pathway for an 18:0 fatty acid (stearic acid).

β-oxidation pathway is equal to half the number of carbon atoms in the fatty acid. The number of *repetitions* of the *β*-oxidation pathway that are needed to produce the acetyl CoA is always one less than the number of acetyl CoA molecules produced because the last repetition produces two acetyl CoA molecules as a C_4 unit splits into two C_2 units.

$$C_{18} \text{ fatty acid} \longrightarrow 9 \text{ acetyl CoA (8 repetitive sequences)}$$
$$C_{14} \text{ fatty acid} \longrightarrow 7 \text{ acetyl CoA (6 repetitive sequences)}$$

This sequence of reactions is called the *β*-oxidation *pathway* rather than the *β*-oxidation *cycle* because a different product results from each repetition.

EXAMPLE 14.2 Recognizing Reaction Types and Events That Occur in the β-Oxidation Pathway

Indicate at what step in the β-oxidation pathway each of the following events occurs.

a. A carbon–carbon single bond is converted to a carbon–carbon double bond.
b. NAD^+ is reduced to NADH.
c. A hydration reaction occurs.
d. An acetyl CoA molecule is produced.

Solution

a. *Step 1.* A dehydrogenation reaction involving removal of two hydrogen atoms changes the carbon–carbon single bond to a carbon–carbon double bond.
b. *Step 3.* This step is the second of two dehydrogenation (oxidation) reactions that occur. The oxidizing agent for this oxidation is NAD^+, which is converted to NADH. Note that the oxidizing agent for the first dehydrogenation reaction (Step 1) is not NAD^+ but rather FAD.
c. *Step 2.* A molecule of water is added to the carbon–carbon double bond producing a secondary alcohol at the β-carbon position.
d. *Step 4.* Breakage of the bond between the α- and β-carbon atoms produces an acetyl CoA molecule and an acyl CoA molecule whose carbon chain is two atoms shorter than at the start of the pathway.

Practice Exercise 14.2

Indicate at what step in the β-oxidation pathway each of the following events occurs.

a. A carbon–carbon double bond is converted to a carbon–carbon single bond.
b. FAD is reduced to $FADH_2$.
c. A secondary alcohol group is oxidized to a ketone group.
d. A coenzyme A molecule is needed as a reactant.

Answers: **a.** Step 2; **b.** Step 1; **c.** Step 3; **d.** Step 4

EXAMPLE 14.3 Relating Characteristics of the β-Oxidation Pathway to Types of Reactions That Occur

Match each of the following characteristics of the β-oxidation pathway to the terms (1) first dehydrogenation (2) hydration (3) second dehydrogenation and (4) thiolysis.

a. The enzyme needed is thiolase.
b. The enzyme needed is acyl CoA dehydrogenase.
c. The substance *trans*-enoyl CoA is a product.
d. The substance β-ketoacyl CoA is a reactant.

Solution

a. *Thiolysis* (Step 4). The prefix *thio* in the enzyme name *thiolase* indicates a reaction that involves an SH functional group. When coenzyme A reacts with the acyl group produced by chain cleavage to form acyl CoA, a new C—S bond is formed.
b. *First dehydrogenation* (Step 1). This dehydrogenation step, as well as the second one, involves the removal of two hydrogen atoms with the resulting creation of a double bond. A dehydrogenase enzyme effects such a change.
c. *First dehydrogenation* (Step 1). *Trans*-enoyl CoA is produced in Step 1 and becomes the reactant for Step 2.
d. *Thiolysis* (Step 4). β-ketoacyl CoA is produced in Step 3 and becomes the reactant for Step 4.

Practice Exercise 14.3

Match each of the following characteristics of the β-oxidation pathway to the terms (1) first dehydrogenation (2) hydration (3) second dehydrogenation and (4) thiolysis.

a. The enzyme enoyl CoA hydratase is needed.
b. A stereospecific enzyme that produces *trans*-carbon–carbon double bonds is needed.
c. The substance acyl CoA is a reactant.
d. The substance acyl CoA is a product.

Answers: **a.** Hydration; **b.** First dehydrogenation; **c.** First dehydrogenation; **d.** Thiolysis

Unsaturated Fatty Acids

Unsaturated fatty acids are common components of dietary triacylglycerols. Their oxidation through the β-oxidation pathway requires two additional enzymes besides those needed for oxidation of saturated fatty acids. These two—an epimerase that can change a D configuration to an L configuration and a *cis–trans* isomerase—are needed for two reasons. First, the double bonds in naturally occurring unsaturated fatty acids are nearly always *cis* double bonds, which yield on hydration a D-hydroxy product rather than the L-hydroxy product needed for Step 3 of the pathway. The epimerase enzyme effects a configuration change from the D form to the L form.

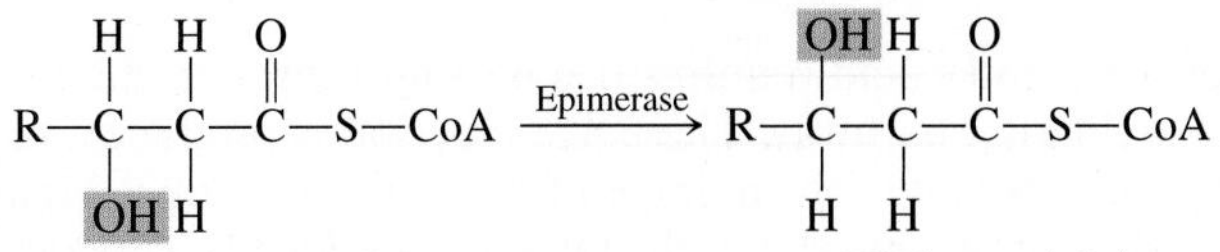

D-β-Hydroxyacyl CoA   L-β-Hydroxyacyl CoA

The oxidation of an unsaturated fatty acid does not generate as many ATPs as does the oxidation of an unsaturated fatty acid with the same number of carbon atoms. Step 1 of the β-oxidation parthway, which introduces a carbon–carbon double bond into the fatty acid, can be skipped since the double bond is already there. Skipping Step 1 means less $FADH_2$ is produced.

Second, the double bonds in naturally occurring unsaturated fatty acids often occupy odd-numbered positions (Section 8.2). The hydratase in Step 2 of the pathway can effect hydration of only an even-numbered double bond. The *cis–trans* isomerase produces a *trans*-(2,3) double bond from a *cis*-(3,4) double bond.

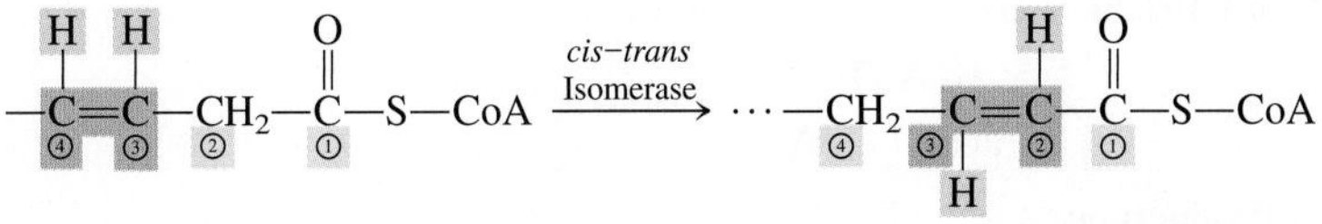

cis-(3,4)   *trans*-(2,3)

The Step 2 hydratase can then work on the *trans*-(2,3) double bond in the normal fashion.

14.5 ATP Production from Fatty Acid Oxidation

How does the total energy output from fatty acid oxidation compare to that of glucose oxidation? Calculation of ATP production for the oxidation of a specific fatty acid molecule, stearic acid (18:0), and comparison of the result with that from glucose is now considered.

Figure 14.7 shows that for each four-reaction sequence except the last one, one $FADH_2$ molecule, one NADH molecule, and one acetyl CoA molecule are produced. In the final four-reaction sequence, two acetyl CoA molecules are produced in addition to the $FADH_2$ and NADH molecules.

Eight repetitions of the β-oxidation pathway are required for the oxidation of stearic acid, an 18-carbon acid. These eight repetitions of the pathway produce 9 acetyl CoA molecules, 8 $FADH_2$ molecules, and 8 NADH molecules. Further processing of these products through the common metabolic pathway (citric acid cycle, electron transport chain, and oxidative phosphorylation) leads to ATP production as follows:

$$9 \text{ acetyl CoA} \times \frac{10 \text{ ATP}}{1 \text{ acetyl CoA}} = 90 \text{ ATP}$$

$$8 \text{ FADH}_2 \times \frac{1.5 \text{ ATP}}{1 \text{ FADH}_2} = 12 \text{ ATP}$$

$$8 \text{ NADH} \times \frac{2.5 \text{ ATP}}{1 \text{ NADH}} = \underline{20 \text{ ATP}}$$

$$122 \text{ ATP}$$

The conversion factors used in this calculation were first presented in Section 12.10.

This *gross* production of 122 ATP must be decreased by the ATP needed to activate the fatty acid before it enters the β-oxidation pathway. The activation consumes two high-energy phosphate bonds of an ATP molecule. For accounting purposes, this is equivalent to hydrolyzing 2 ATP molecules to ADP. Thus the *net* ATP production from oxidation of stearic acid is 120 ATP (122 minus 2).

EXAMPLE 14.4 Calculating Net ATP Production from the Complete Oxidation of a Fatty Acid

What is the net ATP production for the complete oxidation of lauric acid, the C_{12} saturated fatty acid, to CO_2 and H_2O?

Solution

The total ATP production for oxidation of a fatty acid is the sum of the ATP produced from the metabolic products $FADH_2$, NADH, and acetyl CoA. In obtaining the *net* ATP production, it must be remembered that the activation of a fatty acid molecule prior to its oxidation requires ATP consumption equivalent to two ATP molecules (Section 14.4).

Oxidation of a C_{12} fatty acid requires five passages through the four-reaction sequence of the β-oxidation pathway. These five repetitions produce a total of 6 acetyl CoA, 5 $FADH_2$, and 5 NADH. The number of acetyl CoA produced is always equal to the number of carbon atoms present in the fatty acid divided by 2, which is 12/2 = 6 in this case. The number of $FADH_2$ and NADH produced will always be one less than the number of acetyl CoA produced.

Further processing of these metabolic products through the citric acid cycle, electron transport chain, and oxidative phosphorylation leads to ATP production as follows:

Fatty Acid Activation: −2 ATP

Acetyl CoA Production:

$$6 \text{ acetyl CoA} \times \frac{10 \text{ ATP}}{1 \text{ acetyl CoA}} \qquad +60 \text{ ATP}$$

$FADH_2$ Production:

$$5\ FADH_2 \times \frac{1.5 \text{ ATP}}{1\ FADH_2} \qquad +7.5 \text{ ATP}$$

NADH Production:

$$5 \text{ FADH} \times \frac{2.5 \text{ ATP}}{1 \text{ NADH}} \qquad +12.5 \text{ ATP}$$

Net ATP Production: +78 ATP

The ATP equivalents for acetyl CoA, $FADH_2$, and NADH used in this calculation were first presented in Section 12.10.

Practice Exercise 14.4

What is the net ATP production for the complete oxidation of palmitic acid, the C_{16} saturated fatty acid, to CO_2 and H_2O?

Answer: 106 ATP

The comparison between complete fatty acid oxidation and complete glucose oxidation (Section 13.4) shows that a C_{18} stearic acid molecule produces four times as much ATP as a glucose molecule.

$$\begin{aligned} 1 \text{ glucose} &\longrightarrow 30 \text{ ATP} \\ 1 \text{ stearic acid} &\longrightarrow 120 \text{ ATP} \end{aligned}$$

Taking into account the fact that glucose has only 6 carbon atoms and stearic acid has 18 carbon atoms still shows more ATP production from the fatty acid.

$$\begin{aligned} 3 \text{ glucose (18 C)} &\longrightarrow 90 \text{ ATP} \\ 1 \text{ stearic acid (18 C)} &\longrightarrow 120 \text{ ATP} \end{aligned}$$

Thus, on the basis of equal numbers of carbon atoms, lipids are 33% more efficient than carbohydrates as energy-storage systems.

On an equal-mass basis, fatty acids produce 2.5 times as much energy per gram as carbohydrates (glucose); this is shown by the following calculation involving 1.00 gram of stearic acid and 1.00 gram of glucose.

$$1.00\ \text{g stearic acid} \times \left(\frac{1\ \text{mole stearic acid}}{284\ \text{g stearic acid}}\right) \times \left(\frac{120\ \text{moles ATP}}{1\ \text{mole stearic acid}}\right) = 0.423\ \text{mole ATP}$$

$$1.00\ \text{g glucose} \times \left(\frac{1\ \text{mole glucose}}{180\ \text{g glucose}}\right) \times \left(\frac{30\ \text{moles ATP}}{1\ \text{mole glucose}}\right) = 0.167\ \text{mole ATP}$$

The fact that fatty acids (stearic acid) yield 2.5 times as much energy per gram as carbohydrates (glucose) means that, in terms of calories consumed, the former "do 2.5 times as much damage" to a person on a diet. On the other hand, fatty acids are much better energy-storing molecules than is glucose; they can store more than twice as much energy per gram than glucose.

In dietary considerations, nutritionists say that 1 gram of carbohydrate equals 4 kcal and that 1 gram of fat equals 9 kcal. We now know the basis for these numbers. The value of 9 kcal for fat takes into account the fact that not all fatty acids present in fat contain 18 carbon atoms (the basis for the preceding calculations) and also the fact that fats contain glycerol, which produces ATP when degraded.

Is the preferred fuel for "running" the human body fatty acids, which yield 2.5 times as much energy per gram as glucose, or is it glucose? In a normally functioning human body, certain organs use both fuels, others prefer glucose, and still others prefer fatty acids. Some generalizations about "fuel" use are:

1. Skeletal muscle uses glucose (from glycogen) when in an active state. In a resting state, it uses fatty acids.
2. Cardiac muscle depends first on fatty acids and secondarily on ketone bodies (Section 14.6), glucose, and lactate.
3. The liver uses fatty acids as the preferred fuel.
4. Brain function is maintained by glucose and ketone bodies (Section 14.6). Fatty acids cannot cross the blood–brain barrier and thus are unavailable.

The focus on relevancy feature Chemical Connections 14-A on the next page contrasts how fuel use (carbohydrate versus fatty acid) changes within the human body over time during an exercise workout and also contrasts the fuel use mix associated with high- and low-intensity workouts.

14.6 Ketone Bodies

Ordinarily, when there is adequate balance between lipid and carbohydrate metabolism, most of the acetyl CoA produced from the β-oxidation pathway is further processed through the citric acid cycle. The first step of the citric acid cycle (Section 12.7) involves the reaction between oxaloacetate and acetyl CoA. Sufficient oxaloacetate must be present for the acetyl CoA to react with. Oxaloacetate concentration depends on pyruvate produced from glycolysis (Section 13.2); pyruvate can be converted to oxaloacetate by *pyruvate carboxylase* (Section 13.6).

Certain body conditions upset the lipid–carbohydrate balance required for acetyl CoA generated by fatty acids to be processed by the citric acid cycle. These conditions include (1) dietary intake high in fat and low in carbohydrates (2) diabetic conditions in which the body cannot adequately process glucose even though it is present and (3) *prolonged* fasting conditions, including starvation, where glycogen supplies are exhausted. Under these conditions, the problem of inadequate oxaloacetate supplies arises, which is compounded by the body's using oxaloacetate that is present to produce glucose through gluconeogenesis (Section 13.6).

CHEMICAL CONNECTIONS 14-A

High-Intensity Versus Low-Intensity Workouts

In a resting state, the human body burns more fat than carbohydrate. The fuel consumed is about one-third carbohydrate and two-thirds fat.

Information about fuel consumption ratios is obtainable from respiratory gas measurements, specifically from the respiratory exchange ratio (RER). The RER is the ratio of carbon dioxide to oxygen inhaled divided by the ratio of carbon dioxide to oxygen exhaled. For 100% fat burning, the RER would be 0.7; for 100% carbohydrate burning, the RER would be 1.0.

When a person at rest begins exercising, his or her body suddenly needs energy at a greater rate—more fuel and more oxygen are needed. It takes 0.7 L of oxygen to burn 1 gram of carbohydrate and 1.0 L of oxygen to burn 1 gram of fat. At the onset of exercise, the body is immediately short of oxygen. Also, there is a time delay in triacylglycerol mobilization. Triacylglycerols have to be broken down to fatty acids, which have to be attached to protein carriers before they can be carried in the bloodstream to working muscles. At their destination, they must be released from the carriers and then undergo energy-producing reactions. By contrast, glycogen is already present in muscle cells, and it can release glucose 6-phosphate as an instant fuel.

Adam G. Sylvester/Photo Researchers

The initial stages of exercise are fueled primarily by glucose; in later stages, triacylglycerols become the primary fuel.

Consequently, the initial stages of exercise are fueled primarily by glucose—it requires less oxygen and can even be burned anaerobically (to lactate). During the first few minutes of exercise, up to 80% of the fuel used comes from glycogen.

With time, increased breathing rates increase oxygen supplies to muscles, and triacylglycerol use increases. Continued activity for three-quarters of an hour achieves a 50–50 balance of triacylglycerol and glucose use. Beyond an hour, triacylglycerol use may be as high as 80%.

Suppose a person is exercising at a moderate rate and decides to speed up. Immediately, body fuel and oxygen needs are increased. The response is increased use of glycogen supplies.

The accompanying table compares exercise on a stationary cycle at 45% and 70% of maximum oxygen uptake sufficient to burn 300 calories.

	Low-Intensity Exercise	High-Intensity Exercise
percent of maximum oxygen uptake	45%	70%
time required to burn 300 calories	48 min	30 min
calories obtained from fat	133 cal	65 cal
percent of calories from fat	44%	22%
rate of fat burning per minute	2.8 cal/min	2.1 cal/min

What happens when oxaloacetate supplies are too low for all acetyl CoA present to be processed through the citric acid cycle? The excess acetyl CoA is diverted to the formation of ketone bodies. A **ketone body** *is one of three substances (acetoacetate, β-hydroxybutyrate, and acetone) produced from acetyl CoA when an excess of acetyl CoA from fatty acid degradation accumulates because of triacylglycerol–carbohydrate metabolic imbalances.* The structural formulas for the three ketone bodies, two of which are C_4 molecules and the other a C_3 molecule, are

Ketone bodies are produced when the amount of acetyl CoA is excessive compared with the amount of oxaloacetate available to react with it (Step 1 of the citric acid cycle).

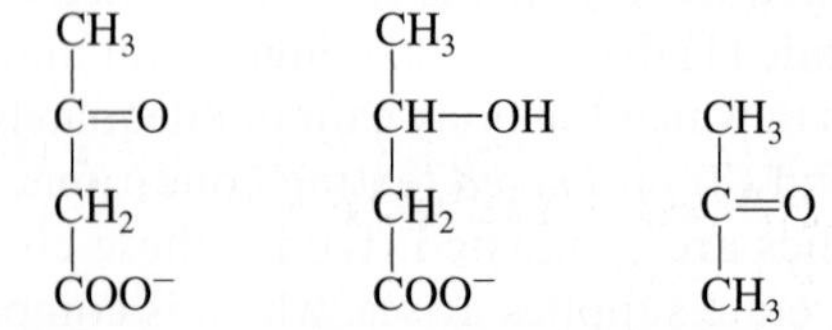

Acetoacetate — C_4 ketoacid

β-Hydroxybutyrate — C_4 hydroxyacid

Acetone — C_3 ketone

Chemically, these three structures are closely related. The relationships are most easily seen if the focus starts with the molecule acetoacetate.

1. Reduction of the ketone group present in acetoacetate to a secondary alcohol produces β-hydroxybutyrate. Such a reduction process was initially considered in Section 4.10.

$$
\begin{array}{ccc}
CH_3 & & CH_3 \\
| & \text{Reducing} & | \\
C=O & \xrightarrow{\text{agent}} & CH-OH \\
| & \text{NADH/H}^+ \quad \text{NAD}^+ & | \\
CH_2 & & CH_2 \\
| & & | \\
COO^- & & COO^- \\
\text{Acetoacetate} & & \beta\text{-Hydroxybutyrate}
\end{array}
$$

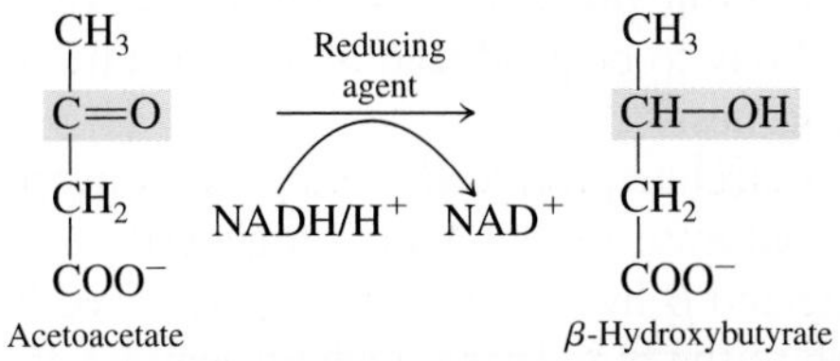

Structurally, there is no ketone functional group present in β-hydroxybutyrate. Despite it not being a ketone, it is still called a ketone body because of its structural relationship to acetoacetate.

2. Decarboxylation of acetoacetate produces acetone.

$$
\begin{array}{ccc}
CH_3 & \text{Decarboxy-} & CH_3 \\
| & \text{lation} & | \\
C=O & \xrightarrow{\quad} & C=O \;+\; CO_2 \\
| & H^+ & | \\
CH_2 & & CH_3 \\
| & & \text{Acetone} \\
COO^- & & \\
\text{Acetoacetate} & &
\end{array}
$$

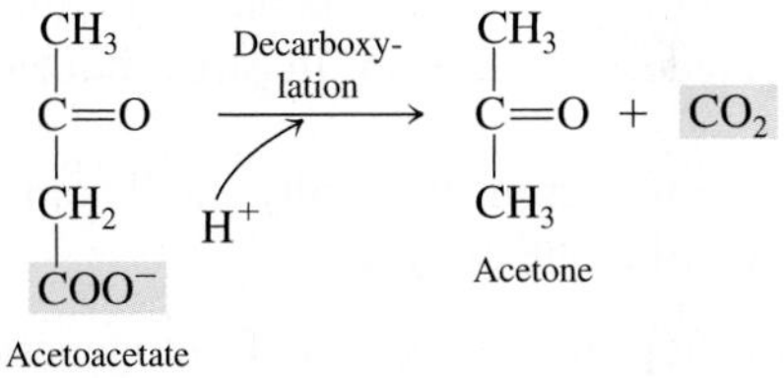

For a number of years, ketone bodies were thought of as degradation products that had little physiological significance. It is now known that ketone bodies can serve as sources of energy for various tissues and are very important energy sources in heart muscle and the renal cortex. Even the brain, which requires glucose, can adapt to obtain a portion of its energy from ketone bodies in dieting situations that involve a properly constructed low-carbohydrate diet.

EXAMPLE 14.5 Recognizing Structural Characteristics of Ketone Bodies

For each of the following structural characterizations for ketone bodies, identify the ketone body to which it applies. There may be more than one correct answer for a given characterization.

a. It is a C_4 molecule.
b. It is a ketoacid.
c. It can be produced by reduction of acetoacetate.
d. Its structure contains a ketone functional group.

Solution

a. *Acetoacetate* and *β-hydroxybutyrate.* Acetoacetate is a C_4 ketoacid, and β-hydroxybutyrate is a C_4 hydroxyacid.
b. *Acetoacetate.* Both acetoacetate and β-hydroxybutyrate are acids; the first is a ketoacid, and the second is a hydroxyacid.
c. *β-Hydroxybutyrate.* Reduction of the ketone functional group in acetoacetate to a secondary alcohol group produces the molecule β-hydroxybutyrate.
d. *Acetoacetate* and *acetone.* Acetoacetate is a ketoacid, and acetone is a simple ketone.

Practice Exercise 14.5

For each of the following structural characterizations for ketone bodies, identify the ketone body to which it applies. There may be more than one correct answer for a given characterization.

a. It is a C_3 molecule.
b. It can be produced by decarboxylation of acetoacetate.
c. Its structure contains a carboxyl functional group.
d. It is classified as a ketone body, but it is not a ketone.

Answers: **a.** Acetone; **b.** Acetone; **c.** Acetoacetate and β-hydroxybutyrate; **d.** β-Hydroxybutyrate

Ketogenesis

Ketogenesis *is the metabolic pathway by which ketone bodies are synthesized from acetyl CoA.* Items to consider about this process prior to looking at the actual steps in this four-step process are:

1. The primary site for the process is liver mitochondria.
2. The first ketone body to be produced is acetoacetate. This production occurs in Step 3 of ketogenesis.
3. Some of the acetoacetate produced in Step 3 is converted to the second ketone body, β-hydroxybutyrate, in Step 4 of ketogenesis.
4. The acetoacetate and β-hydroxybutyrate synthesized by ketogenesis in the liver are released to the bloodstream where acetone, the third ketone body, is produced.
5. Acetoacetate is somewhat unstable and can spontaneously or enzymatically lose its carboxyl group to form acetone. Thus the ketone body acetone is not actually a product of the metabolic pathway ketogenesis.
6. The ketone body acetone present in the bloodstream is a volatile substance that is mainly excreted by exhalation. Its sweet odor is detectible in the breath of a diabetic.
7. The amount of acetone present is usually small compared to the concentrations of the other two ketone bodies.

Even when ketogenic conditions are not present in the human body, the liver produces a *small amount* of ketone bodies.

The actual reaction steps in the process of ketogenesis are:

Step 1: **First condensation.** Two acetyl CoA molecules combine to produce acetoacetyl CoA, a reversal of the last step of the β-oxidation pathway (Section 14.4) via a condensation reaction.

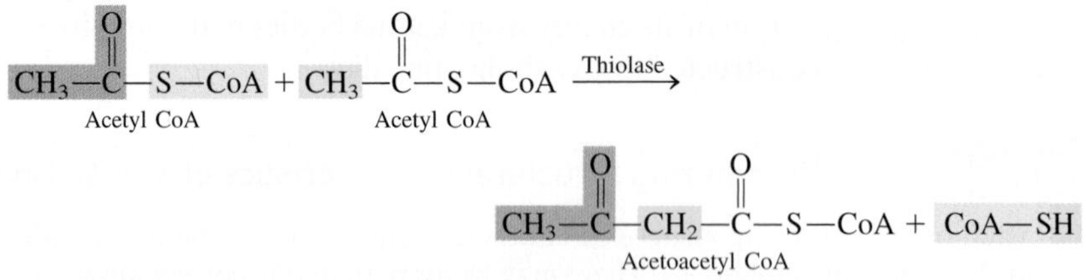

Step 2: **Second condensation.** Acetoacetyl CoA reacts with a third acetyl CoA and water to produce 3-hydroxy-3-methylglutaryl CoA (HMG-CoA) and CoA—SH.

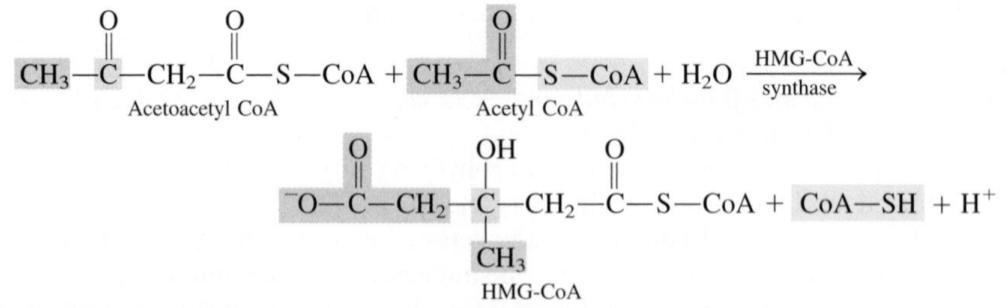

Step 3: **Chain cleavage.** HMG-CoA is cleaved to acetyl CoA and acetoacetate.

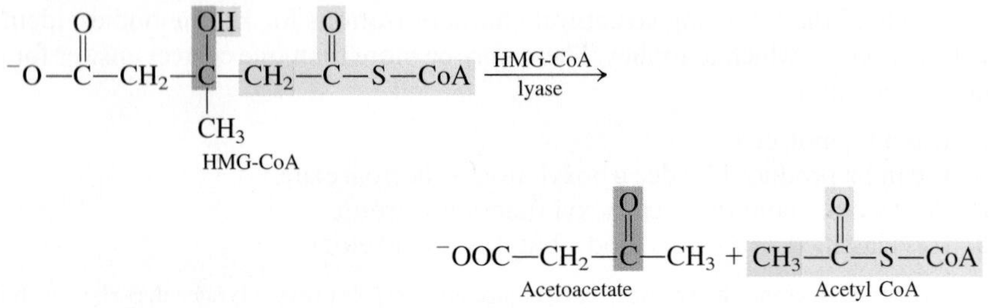

Step 4: **Hydrogenation.** Acetoacetate is reduced to *β*-hydroxybutyrate. The reducing agent is NADH.

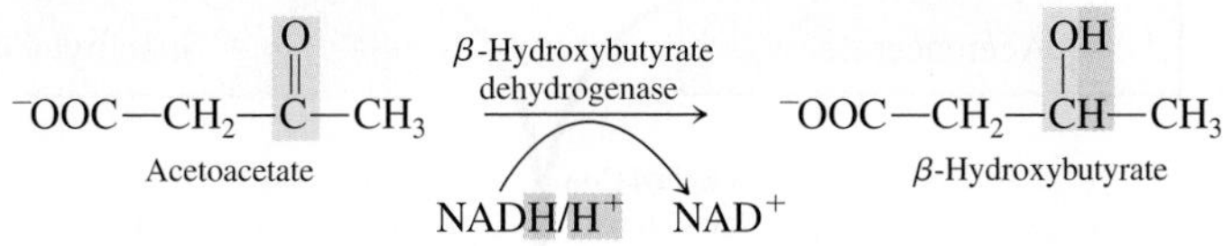

The four chemical reactions associated with ketogenesis are summarized diagrammatically in Figure 14.8.

Energy Production from Acetoacetate

The ketone bodies *β*-hydroxybutyrate and acetoacetate are connected by a reversible reaction. Using a different enzyme than used in ketogenesis, *β*-hydroxybutyrate can be converted back to acetoacetate when cellular energy needs require it.

For acetoacetate to be used as a fuel—in heart muscle, for example—it must first be activated. Acetoacetate is activated by transfer of a CoA group from succinyl CoA (a citric acid cycle intermediate). The resulting acetoacetyl CoA is then cleaved to give two acetyl CoA molecules that can enter the citric acid cycle (see Figure 14.9). In effect, acetoacetate is a water-soluble, transportable form of acetyl units.

Heart muscle and the renal cortex use acetoacetate in preference to glucose. The brain adapts to the utilization of acetoacetate with starvation or diabetes. 75% of the fuel needs of the brain are obtained from acetoacetate during prolonged starvation.

Ketosis

Under normal metabolic conditions (an appropriate glucose–fatty acid balance), the concentration of ketone bodies in the blood is very low—about 1 mg/100 mL. Abnormal metabolic conditions, such as those mentioned at the start of this section, produce elevated blood ketone levels, levels 50–100 times greater than normal. Excess accumulation of ketone bodies in blood (20 mg/100 mL) is called *ketonemia.* At a level of 70 mg/100 mL, the renal threshold is exceeded and ketone

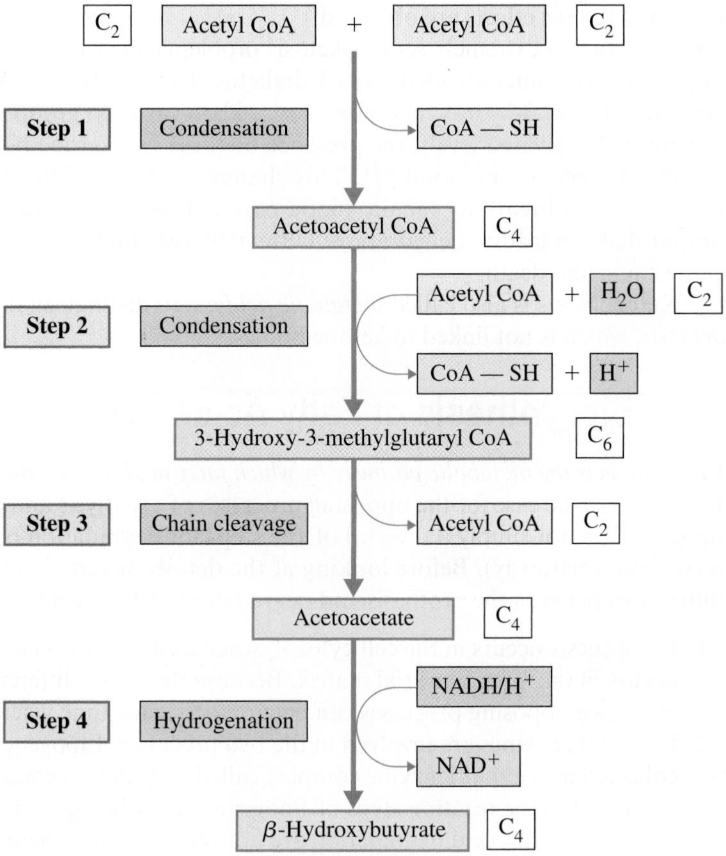

Figure 14.8 Ketogenesis involves the production of ketone bodies from acetyl CoA.

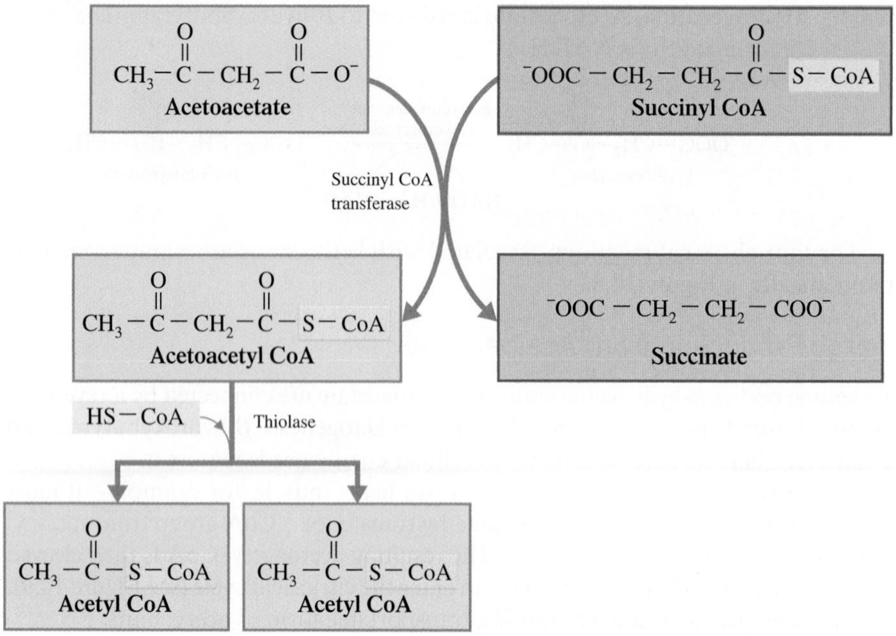

Figure 14.9 The pathway for utilization of acetoacetate as a fuel. The required succinyl CoA comes from the citric acid cycle.

bodies are excreted in the urine, a condition called *ketonuria.* The overall accumulation of ketone bodies in the blood and urine is called *ketosis.* Ketosis is often detectable by the smell of acetone on a person's breath; acetone is very volatile and is excreted through the lungs.

For the vast majority of persons following a low-carbohydrate diet, the effects of ketosis appear to be harmless or nearly so. The symptoms of the *mild* ketosis that occurs as the result of such dieting include headache, dry mouth, and sometimes acetone-smelling breath. Typically, the same is true for fasting situations; only mild ketosis effects are observed.

A serious to extremely serious ketosis problem called *ketoacidosis* can develop in persons with uncontrolled Type 1 diabetes. Two of the three ketone bodies—acetoacetate and β-hydroxybutyrate—are acids; a carboxyl group is present in their structure. At elevated levels, the presence of these two ketone bodies can cause a significant decrease in blood pH. This change in blood acidity, if left untreated, leads to heavy breathing (acidic blood carries less oxygen) and increased urine output that can lead to dehydration. Ultimately the condition of ketoacidosis can cause coma and death.

Ketoacidosis is also called *metabolic acidosis* to distinguish it from *respiratory acidosis*, which is not linked to ketone bodies.

14.7 Biosynthesis of Fatty Acids: Lipogenesis

Lipogenesis *is the metabolic pathway by which fatty acids are synthesized from acetyl CoA.* As was the case for the opposing processes of glycolysis and gluconeogenesis, lipogenesis is not simply a reversal of the steps for degradation of fatty acids (the β-oxidation pathway). Before looking at the details of fatty acid synthesis, some differences between the synthesis and degradation of fatty acids are considered.

1. Lipogenesis occurs in the cell cytosol, whereas degradation of fatty acids occurs in the mitochondrial matrix. Because they have different reaction sites, these two opposing processes can occur at the same time when necessary.
2. Different enzymes are involved in the two processes. Lipogenesis enzymes are collected into a multienzyme complex called *fatty acid synthase.* This enzyme complex ties the reaction steps of lipogenesis closely together. The enzymes involved in fatty acid degradation are not physically associated, so the reaction steps are independent.

3. Intermediates of the two processes are covalently bonded to different carriers. The carrier for fatty acid degradation intermediates is CoA. Lipogenesis intermediates are bonded to ACP (acyl carrier protein).
4. Fatty acid synthesis is dependent on the reducing agent NADPH. Fatty acid degradation is dependent on the oxidizing agents FAD and NAD^+.
5. Fatty acids are built up two carbons at a time during synthesis and are broken down two carbons at a time during degradation. The source of the two carbon units differs between the two processes. In lipogenesis, acetyl CoA is used to form malonyl ACP, which becomes the carrier of the two carbon units. CoA derivatives are involved in all steps of fatty acid degradation.

In general, fatty acid biosynthesis (lipogenesis) occurs any time dietary intake provides more nutrients than are needed for energy requirements. The primary lipogenesis sites are the liver, adipose tissue, and mammary glands. The mammary glands show increased synthetic activity during periods of lactation.

The Citrate–Malate Shuttle System

Acetyl CoA is the starting material for lipogenesis. Because acetyl CoA is generated in mitochondria and lipogenesis occurs in the cytosol, the acetyl CoA must first be transported to the cytosol. It exits the mitochondria through a transport system that involves citrate ion.

The outer mitochondrial matrix is freely permeable to acetyl CoA, as well as many other substances such as citrate, malate, and pyruvate. The inner mitochondrial membrane, however, is not permeable to acetyl CoA. An indirect shuttle system involving citrate solves this problem.

This shuttle system, which is diagrammed in Figure 14.10, functions as follows. Mitochondrial acetyl CoA reacts with oxaloacetate (the first step of the citric acid cycle; Section 12.6) to produce citrate, which is then transported through the inner mitochondrial membrane by a citrate transporter (a membrane protein structure).

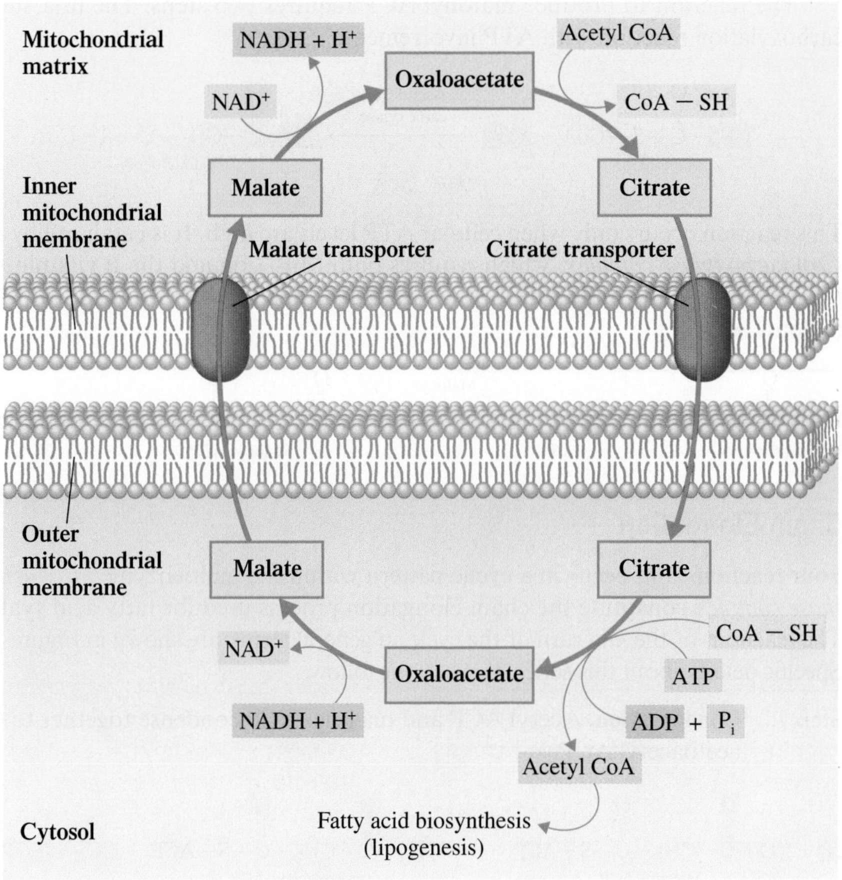

Figure 14.10 The citrate–malate shuttle system for transferring acetyl CoA from a mitochondrion to the cytosol.

Once in the cytosol, the citrate undergoes the reverse reaction to its formation to regenerate the acetyl CoA and oxaloacetate, with ATP involved in the process. The acetyl CoA so generated becomes the "fuel" for lipogenesis; the oxaloacetate so generated reacts further to produce malate, in an NADH dependent change. The malate reenters the mitochondrial matrix through a malate transporter and is then converted to oxaloacetate, which can then react with another acetyl CoA molecule to form citrate and the shuttle process repeats itself. Additional particulars about the shuttle system are found in Figure 14.10.

Compared to the carnitine shuttle system for long-chain fatty acid groups (acyl groups; Figure 14.6), the citrate–malate shuttle system is more complex. However, the intermediates in the shuttle have been encountered before, in glycolysis and the citric acid cycle.

ACP Complex Formation

Studies show that all intermediates in fatty acid biosynthesis (lipogenesis) are bound to acyl carrier proteins (ACP—SH) rather than coenzyme A (CoA—SH). This applies even to the C_2 acetyl group. An acyl carrier protein can be regarded as a "giant coenzyme A molecule." Involved in the ACP structure are the 2-ethanethiol and pantothenic acid components present in CoA—SH (Section 12.3), which are attached to a polypeptide chain containing 77 amino acid residues.

Two simple ACP complexes are needed to start the lipogenesis process. They are acetyl ACP, a C_2—ACP, and malonyl ACP, a C_3—ACP. Additional malonyl ACP molecules are needed as the lipogenesis process proceeds.

Cytosolic acetyl CoA is the starting material for the production of both of these simple ACP complexes. Acetyl ACP is produced by the direct reaction of acetyl CoA with an ACP molecule.

The parent compound for the malonyl group is malonic acid, the C_3 dicarboxylic acid

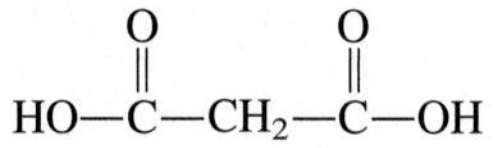

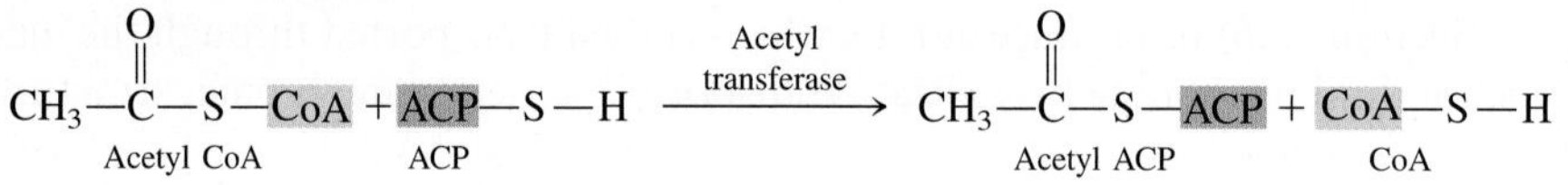

The reaction to produce malonyl ACP requires two steps. The first step is a carboxylation reaction with ATP involvement.

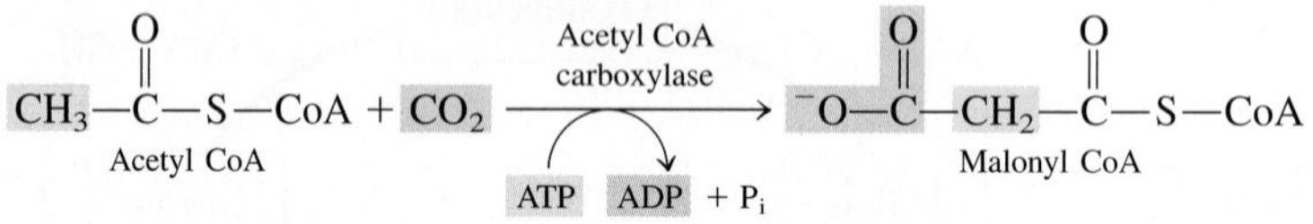

This reaction occurs only when cellular ATP levels are high. It is catalyzed by *acetyl CoA carboxylase complex,* which requires both Mn^{2+} ion and the B vitamin biotin for its activity. The malonyl CoA so produced then reacts with ACP to produce malonyl ACP.

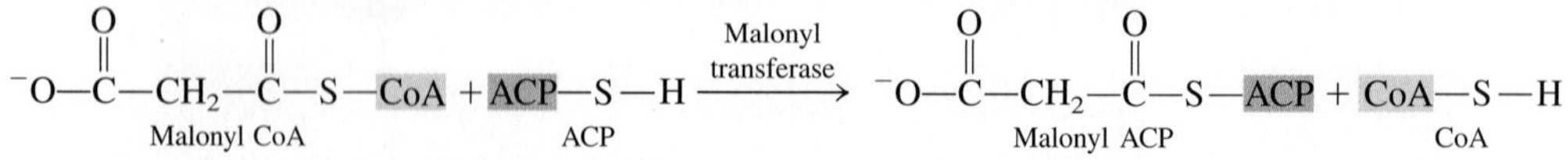

Chain Elongation

Four reactions that occur in a cyclic pattern within the multienzyme *fatty acid synthase complex* constitute the chain elongation process used for fatty acid synthesis. The reactions of the *first* turn of the cycle, in general terms, are shown in Figure 14.11. Specific details about this series of reactions follow.

Step 1: **Condensation.** Acetyl ACP and malonyl ACP condense together to form acetoacetyl ACP.

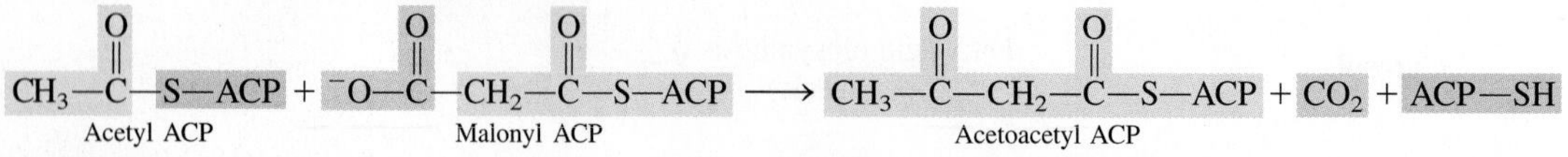

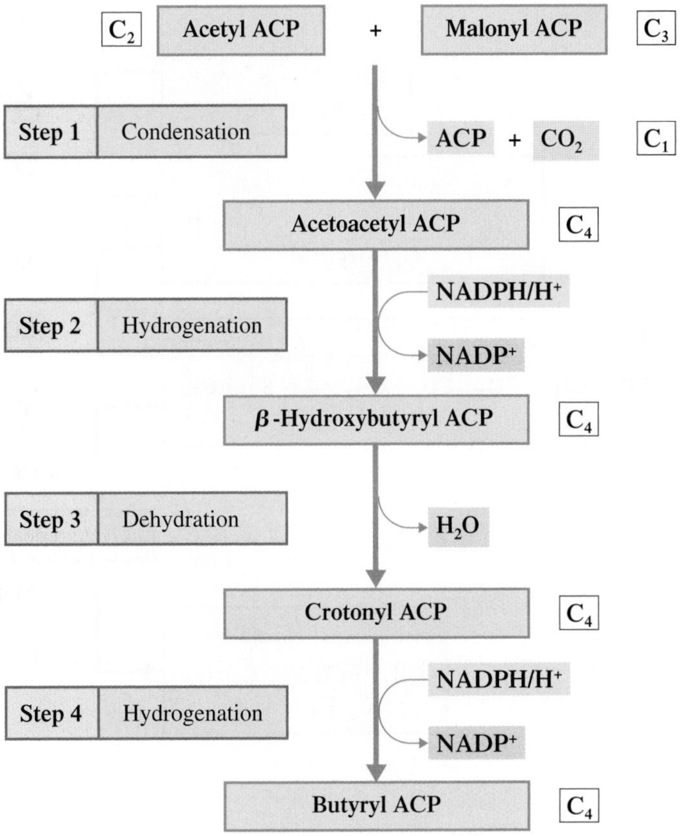

Figure 14.11 In the first "turn" of the fatty acid biosynthetic pathway, acetyl ACP is converted to butyryl ACP. In the next cycle (not shown), the butyryl ACP reacts with another malonyl ACP to produce a 6-carbon acid. Continued cycles produce acids with 8, 10, 12, 14, and 16 carbon atoms.

Note that a C_2 species (acetyl) and a C_3 species (malonyl) react to produce a C_4 species (acetoacetyl) rather than a C_5 species. One carbon atom leaves the reaction in the form of a CO_2 molecule.

Steps 2 through 4 involve a sequence of functional group changes that have been encountered twice before—in fatty acid degradation (Section 14.4) and in the citric acid cycle (Section 12.7). This time, however, the changes occur in the reverse sequence to that previously encountered. The functional group changes are

$$\textbf{Ketone} \xrightarrow[\text{(Reduction)}]{\text{② Hydrogenation}} \textbf{secondary alcohol} \xrightarrow{\text{③ Dehydration}} \textbf{alkene} \xrightarrow[\text{(Reduction)}]{\text{④ Hydrogenation}} \textbf{alkane}$$

Step 2: **First hydrogenation.** The keto group of the acetoacetyl complex, which involves the β-carbon atom, is reduced to the corresponding alcohol by NADPH.

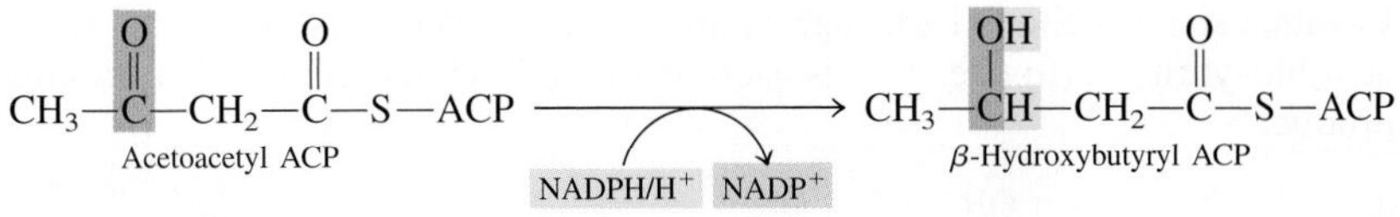

Step 3: **Dehydration.** The alcohol produced in Step 2 is dehydrated to introduce a double bond into the molecule (between the α and β carbons).

Steps 2, 3, and 4 of fatty acid biosynthesis accomplish the reverse of Steps 3, 2, and 1 of the β-oxidation pathway.

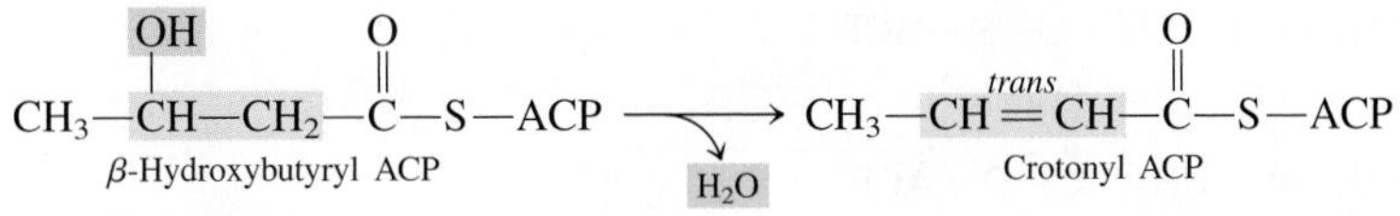

Step 4: **Second hydrogenation.** The double bond introduced in Step 3 is converted to a single bond through hydrogenation. As in Step 2, NADPH is the reducing agent.

Figure 14.12 The sequence of cycles needed to produce a C_{16} fatty acid from acetyl ACP. Each loop represents one cycle.

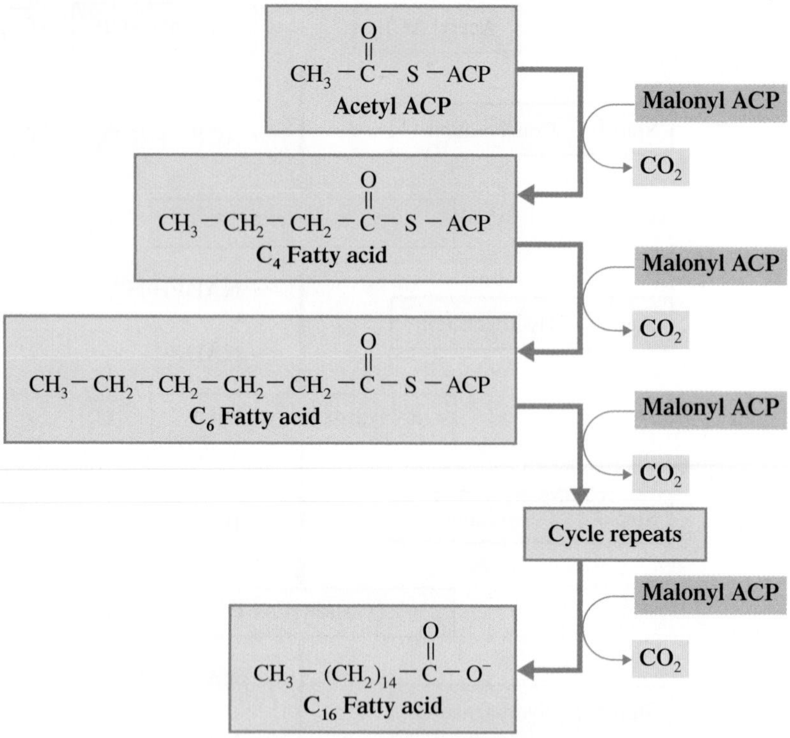

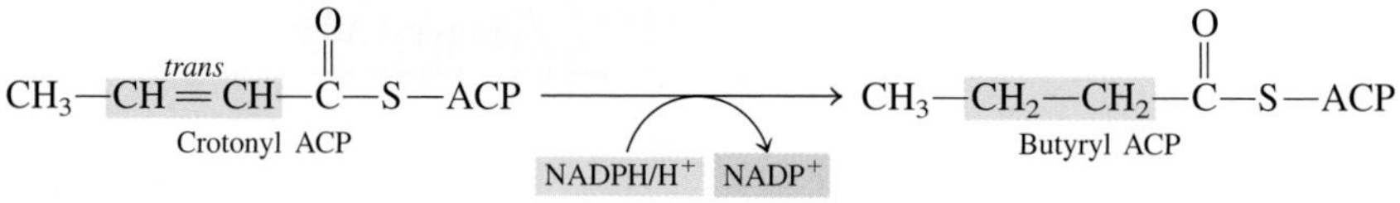

Among enzymes, the multienzyme complex *fatty acid synthase* is considered to be a "monster" or "gargantuan" enzyme in terms of size. Recent research relating to a fungal fatty acid synthase indicates that the enzyme has two reaction chambers with three full sets of active sites, so six fatty acids can be synthesized simultaneously. ACP shuttles intermediates around the reaction chamber, and the fatty acid chain length increases by two carbons with each cycle.

Further cycles of the preceding four-step process convert the four-carbon acyl group to a six-carbon acyl group, then to an eight-carbon acyl group, and so on (Figure 14.12). Elongation of the acyl group chain through this procedure, which is tied to the fatty acid synthase complex, stops upon formation of the C_{16} acyl group (palmitic acid). Different enzyme systems and different cellular locations are required for elongation of the chain beyond C_{16} and for introduction of double bonds into the acyl group (unsaturated fatty acids).

A relatively large input of energy is needed to biosynthesize a fatty acid molecule, as can be seen from the data in Table 14.2, which gives a net summary of the reactants and products involved in the synthesis of one molecule of palmitic acid, the 16:0 fatty acid.

Table 14.2 Reactants and Products in the Biosynthesis of One Molecule of Palmitic Acid, the 16:0 Fatty Acid

Reactants	Products
8 acetyl CoA	1 palmitate
7 ATP	8 CoA
14 NADPH	7 ADP
6 H^+	7 P_i
	14 $NADP^+$
	6 H_2O

EXAMPLE 14.6 Characterizing Intermediates as to When They Are Produced in the Lipogenesis Process

In which step (of Steps 1 through 4) and in which cycle (first or second turn) of fatty acid biosynthesis (lipogenesis) is each of the following compounds encountered as a product?

a.

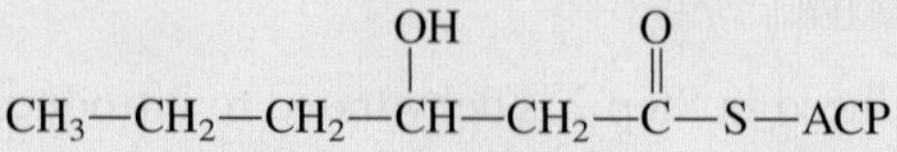

b.

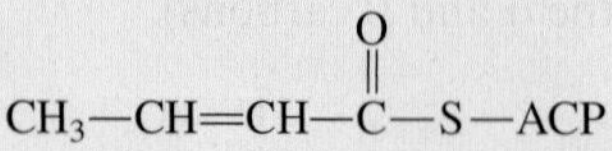

c.

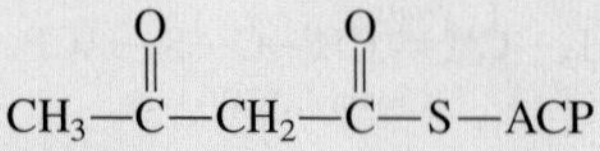

d.

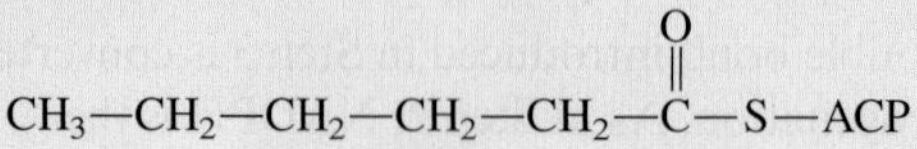

Solution

All intermediates produced during the first turn of the cycle will have a C_4 carbon chain, those produced during the second turn a C_6 carbon chain, those during the third turn a C_8 carbon chain, and so on.

Determination of which step in a turn is involved is based on the functional group present in the carbon chain. Step 1 produces a keto acid chain, Step 2 a hydroxy acid chain, Step 3 an unsaturated acid chain, and Step 4 a saturated acid chain.

a. The characterization for this molecule's carbon chain is C_6 hydroxyacid. The hydroxyacid designation indicates *Step 2,* and the C_6 chain length indicates *second turn.*
b. The carbon chain is a C_4 unsaturated acid; this indicates *Step 3* and *first turn.*
c. The carbon chain is a C_4 keto acid; this indicates *Step 1* and *first turn.*
d. The carbon chain is a C_6 saturated acid; this indicates *Step 4* and *second turn.*

Practice Exercise 14.6

In which step (of Steps 1 through 4) and in which cycle (first or second turn) of fatty acid biosynthesis (lipogenesis) is each of the following compounds encountered as a product?

a.
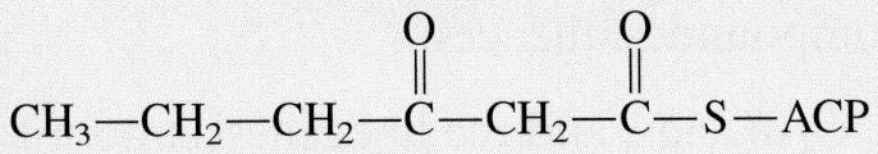

b.
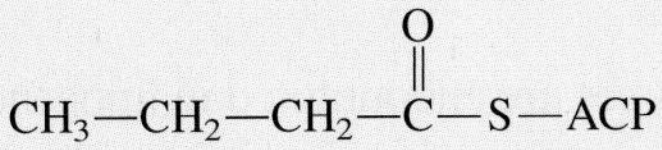

c.
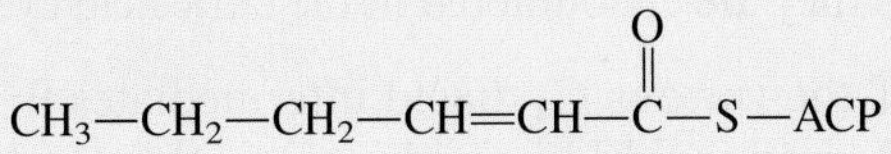

d.
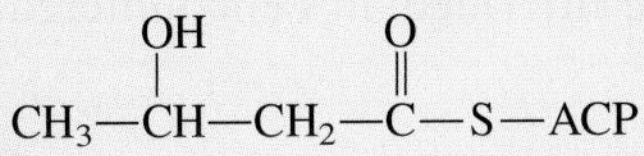

Answers: **a.** Step 1, second turn; **b.** Step 4, first turn; **c.** Step 3, second turn; **d.** Step 2, first turn

Unsaturated Fatty Acid Biosynthesis

Production of unsaturated fatty acids (insertion of double bonds) requires molecular oxygen (O_2). In an oxidation step, hydrogen is removed and combined with the O_2 to form water.

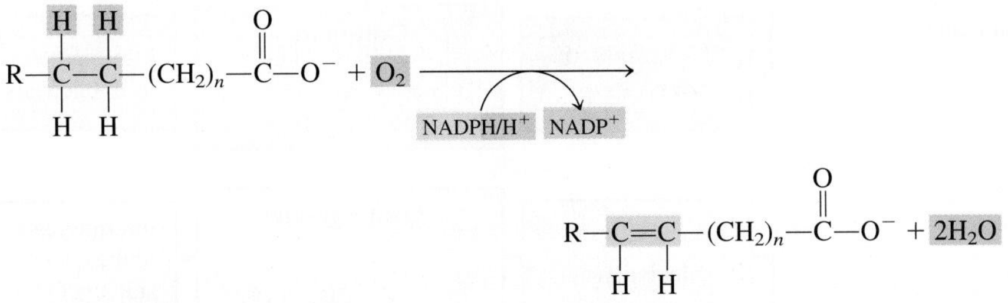

In humans and animals, enzymes can introduce double bonds only between C-4 and C-5 and between C-9 and C-10. Thus the important unsaturated fatty acids linoleic (C_{18} with C-9 and C-12 double bonds) and linolenic (C_{18} with C-9, C-12, and C-15 double bonds) cannot be biosynthesized. They must be obtained from the diet. (Plants have the enzymes necessary to synthesize these acids.) Acids such as linoleic and linolenic, which cannot be synthesized by the body but are necessary for its proper functioning, are called *essential fatty acids* (Section 8.2).

Lipogenesis can be used to convert glucose to fatty acids via acetyl CoA. The reverse process, conversion of fatty acids to glucose, is not possible within the human body. Fatty acids can be broken down to acetyl CoA, but there is no enzyme present for the conversion of acetyl CoA to pyruvate or oxaloacetate, starting materials for gluconeogenesis (Section 13.6). Plants and some bacteria do possess the needed enzymes and thus can convert fatty acids to carbohydrates.

14.8 Relationships Between Lipogenesis and Citric Acid Cycle Intermediates

The intermediates in the last four steps of the citric acid cycle are all C_4 molecules (Section 12.7). In the first cycle of the four repetitive reactions in lipogenesis, all of the carbon chains attached to ACP are C_4 chains. Several relationships exist between these two sets of C_4 entities.

The last four intermediates of the citric acid cycle bear the following relationship to each other.

Saturated C_4 diacid → unsaturated C_4 diacid → hydroxy C_4 diacid → keto C_4 diacid

The intermediate C_4 carbon chains of lipogenesis bear the following relationship to each other.

Keto C_4 monoacid → hydroxy C_4 monoacid → unsaturated C_4 monoacid → saturated C_4 monoacid

Note two important contrasts in these compound sequences:

1. The citric acid intermediates involve C_4 *di*acids, and the lipogenesis intermediates involve C_4 *mono*acids.
2. The order in which the various acid derivative types are encountered in lipogenesis is the reverse of the order in which they are encountered in the citric acid cycle.

Figure 14.13 contrasts the structures of the various C_4 diacid intermediates from the citric acid cycle with the various C_4 monoacid intermediates encountered in lipogenesis.

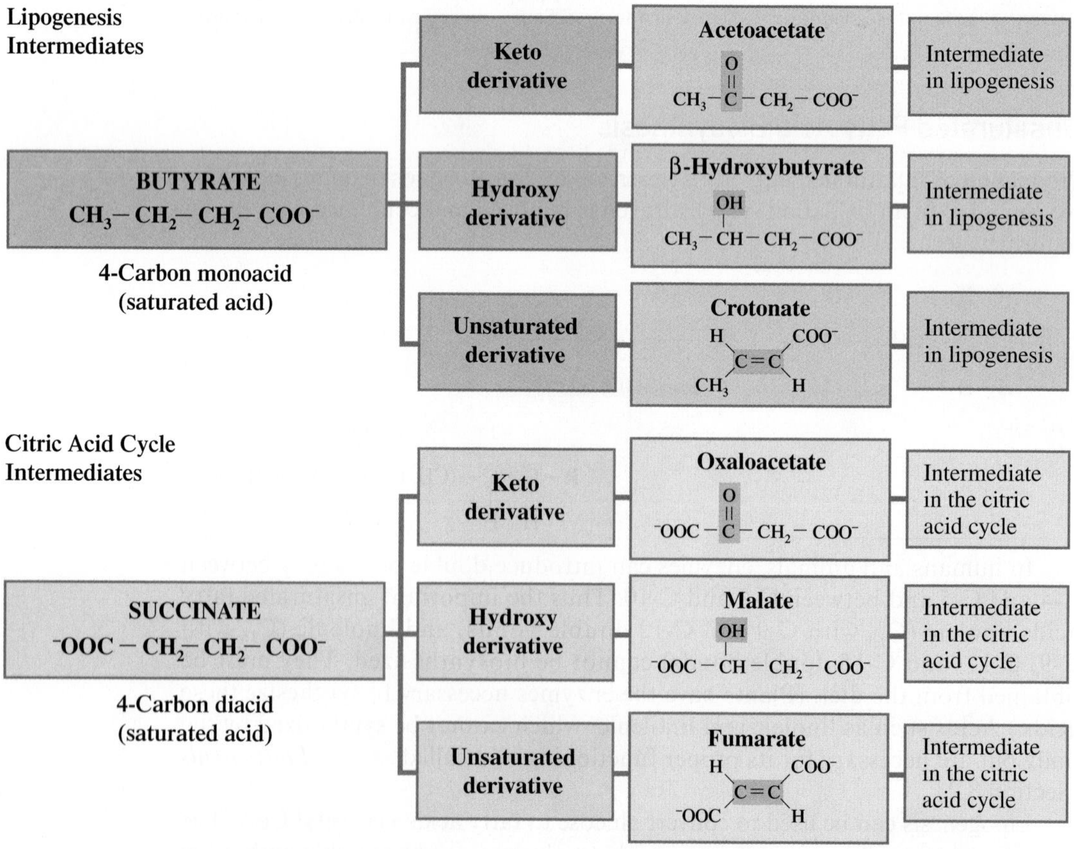

Figure 14.13 Structural relationships between C_4 citric acid cycle intermediates and C_4 lipogenesis intermediates. C_4 citric acid cycle intermediates are diacid derivatives, and C_4 lipogenesis intermediates are monoacid derivatives. The same derivative types are present in both series of intermediates.

14.9 Fate of Fatty Acid Generated Acetyl CoA

What happens to the acetyl-CoA obtained from fatty acid oxidation? Several different options are available for its use. These options include the following:

1. The acetyl-CoA can be further processed through the common metabolic pathway (CAC, ETC, and oxidative phosphorylation) to obtain ATP. Section 14.5 considered how much ATP is obtained from the oxidation of selected fatty acids.
2. The acetyl-CoA can undergo conversion to ketone bodies (Section 14.6). Such ketone bodies can be reconverted, when needed, back to acetyl-CoA, which can then be processed for ATP production.
3. The acetyl-CoA can be stored in the body in the form of triacylglycerols (Section 8.4) by reconverting via lipogenesis the acetyl-CoA to fatty acids (Section 14.7) that are then converted to triacylglycerols.
4. The acetyl-CoA can be used as the starting material for the production of lipids other than fatty acids that the body needs. A specific example of such "other lipid" production is the biosynthesis of cholesterol. As was previously discussed (Section 8.9), cholesterol is a necessary component of cell membranes. It is also the precursor for bile salts, sex hormones, and adrenal hormones (Sections 8.11 and 8.12).

   The biosynthesis of cholesterol, a C_{27} molecule, occurs primarily in the liver. Biosynthesis of cholesterol consumes 18 molecules of acetyl CoA and involves at least 27 separate enzymatic steps. An overview of cholesterol biosynthesis, which shows key intermediate compounds, is given in Figure 14.14.

   The rate-determining step in the formation of cholesterol occurs early in its biosynthesis. It involves the formation of the C_6 intermediate mevalonate; in a multistep sequence, three acetyl CoA molecules are condensed together to produce this C_6 species. Extensive research has been carried out concerning this "slow step," with the goal being prevention of its occurrence as a means of decreasing the body's internal production of cholesterol in order to decrease blood cholesterol levels. The focus on relevancy feature Chemical Connections 14-B on the next page discusses the success obtained in achieving this goal.
5. It is also important to note what *cannot* happen to the acetyl CoA obtained from fatty acid oxidation. In humans and animals, it *cannot* be used for the *net* synthesis of glucose.

Pyruvate or oxaloacetate is the needed starting material for glucose production via gluconeogenesis (Section 13.6). Humans and animals do not possess the enzymes needed to convert acetyl CoA to pyruvate. (By contrast, plants do contain the two additional enzymes needed.) Pyruvate can be converted to acetyl CoA (Section 13.3) in a reaction sequence that is *irreversible*. Acetyl CoA cannot be converted to pyruvate.

In Step 1 of gluconeogenesis, pyruvate is converted to oxaloacetate (Section 13.6). Oxaloacetate is also produced in the last reaction of the citric acid cycle. However, use of such oxaloacetate in glucose production cannot lead to a *net* increase in glucose. In each round of the citric acid cycle, two carbon atoms enter

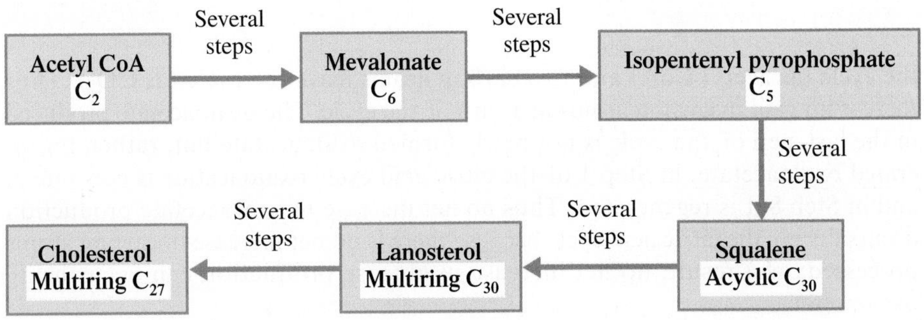

Figure 14.14 An overview of the biosynthetic pathway for cholesterol synthesis.

CHEMICAL CONNECTIONS 14-B

Statins: Drugs That Lower Plasma Levels of Cholesterol

More than half of all deaths in the United States are directly or indirectly related to heart disease, in particular to atherosclerosis. Atherosclerosis results from the buildup of plaque (fatty deposits) on the inner walls of arteries. Cholesterol, obtained from low-density-lipoproteins (LDL) that circulate in blood plasma, is also a major component of plaque.

Because most of the cholesterol in the human body is synthesized in the liver, from acetyl CoA, much research has focused on finding ways to inhibit its biosynthesis. The rate-determining step in cholesterol biosynthesis involves the conversion of 3-hydroxy-3-methylglutaryl CoA (HMG-CoA) to mevalonate, a process catalyzed by the enzyme HMG-CoA reductase.

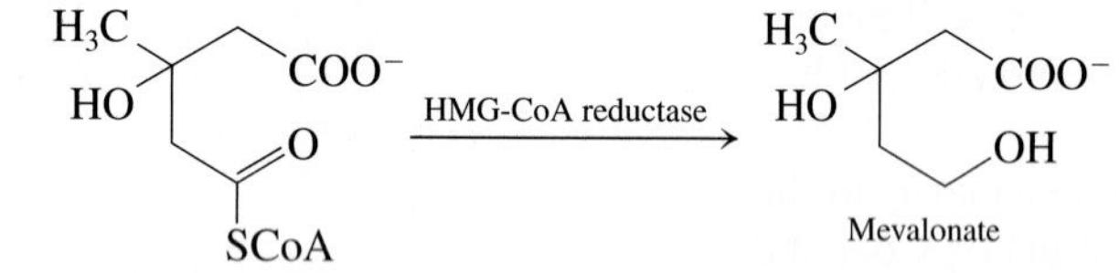

3-Hydroxy-3-methylglutaryl-CoA (HMG-CoA)

In 1976, as the result of screening more than 8000 strains of microorganisms, a compound now called *mevastatin*—a potent inhibitor of HMG-CoA reductase—was isolated from culture broths of a fungus. Soon thereafter, a second, more active compound called *lovastatin* was isolated.

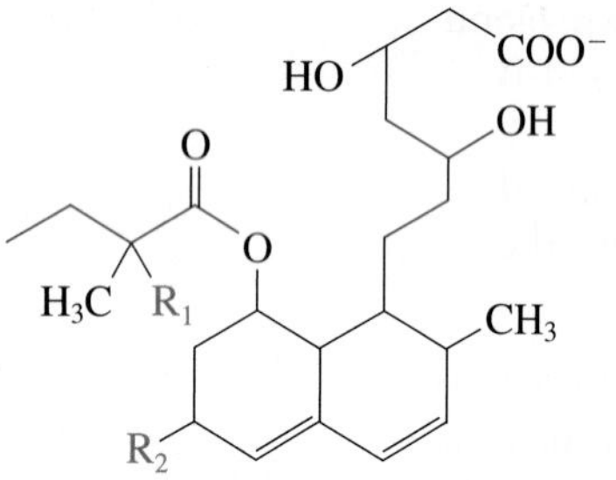

$R_1 = R_2 = H$, mevastatin
$R_1 = H$, $R_2 = CH_3$, lovastatin (Mevacor)
$R_1 = R_2 = CH_3$, simvastatin (Zocor)

These "statins" are very effective in lowering plasma concentrations of LDL by functioning as competitive inhibitors of HMG-CoA reductase.

After years of testing, the statins are now available as prescription drugs for lowering blood cholesterol levels. Clinical studies indicate that use of these drugs lowers the incidence of heart disease in individuals with mildly elevated blood cholesterol levels. A later-generation statin with a ring structure distinctly different from that of earlier statins—atorvastatin (Lipitor)—became the most prescribed medication in the United States starting in the year 2000. Note the structural resemblance between part of the structure of Lipitor and that of mevalonate.

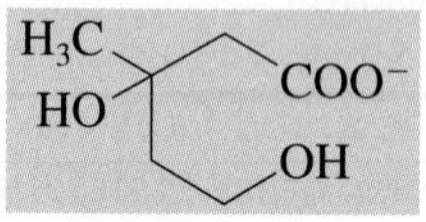

Mevalonate

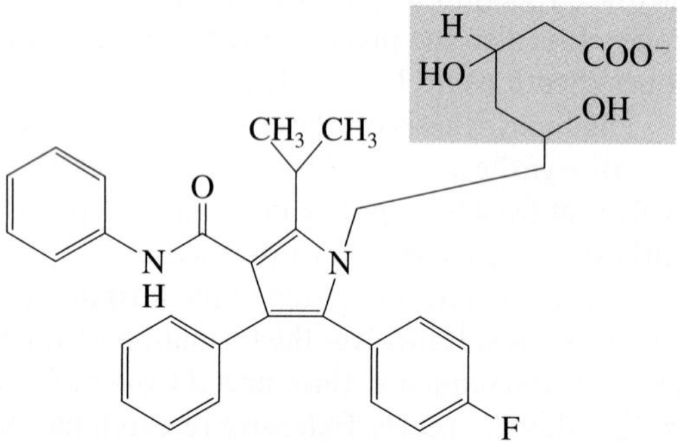

Atorvastatin (Lipitor)

Recent research studies have unexpectedly shown that the cholesterol-lowering statins have two added benefits.

Laboratory studies with animals indicate that statins prompt growth of cells to build new bone, replacing bone that has been leached away by osteoporosis ("brittle-bone disease"). A retrospective study of osteoporosis patients who also took statins shows evidence that their bones became more dense than did bones of osteoporosis patients who did not take the drugs.

Statins have also been shown to function as antiinflammatory agents that counteract the effects of a common virus, cytomegalovirus, which is now believed to contribute to the development of coronary heart disease. Researchers believe that by age 65, more than 70% of all people have been exposed to this virus. The virus, along with other infecting agents in blood, may actually trigger the inflammation mechanism for heart disease.

the cycle (as acetyl CoA) and two carbon atoms leave the cycle (as CO_2). Thus there is no gain in carbon atoms in a turn of the cycle. The oxaloacetate produced in the last step of the cycle is not newly formed oxaloacetate but, rather, regenerated oxaloacetate; in Step 1 of the citric acid cycle oxaloacetate is consumed, and in Step 8 it is regenerated. Thus no net increase in oxaloacetate production occurs during the citric acid cycle because there is no net increase in carbon atoms processed; as a result, no net increase in glucose production can occur using oxaloacetate.

CHEMISTRY AT A GLANCE

Interrelationships Between Carbohydrate and Lipid Metabolism

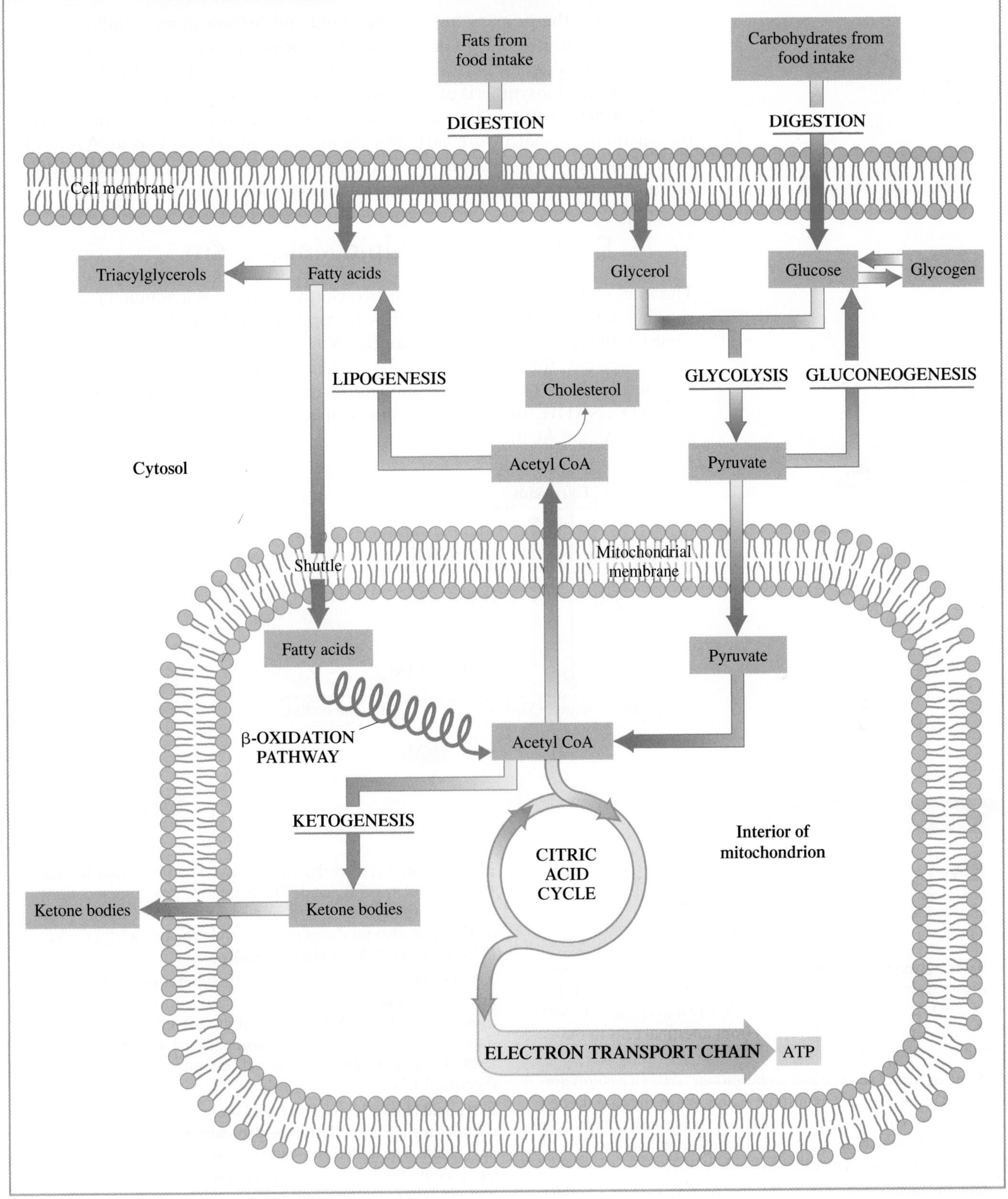

14.10 Relationships Between Lipid and Carbohydrate Metabolism

Acetyl CoA is the primary link between lipid and carbohydrate metabolism. As shown in the Chemistry at a Glance feature on the previous page, acetyl CoA is the degradation product for glucose, glycerol, and fatty acids, and it is also the starting material for the biosynthesis of fatty acids, cholesterol, and ketone bodies.

This Chemistry at a Glance summary diagram also depicts *where* within a cell the reactions associated with carbohydrate and lipid metabolism occur. As shown in the diagram, for both types of metabolism some of the reactions occur within the mitochondrion and others occur within the cytosol.

14.11 B Vitamins and Lipid Metabolism

The final section in each of the last two chapters includes a summary diagram showing how B vitamins, as coenzymes, participate in the metabolic reactions discussed in the chapter. This pattern continues in this chapter.

Collectively, for the processes of β-oxidation, ketogenesis, and lipogenesis, as is shown in Figure 14.15, four of the eight B vitamins are needed. These are: niacin (as NAD^+, NADH, and NADPH), riboflavin (as FAD), pantothenic acid (as CoA, acetyl-CoA, and ACP), and biotin.

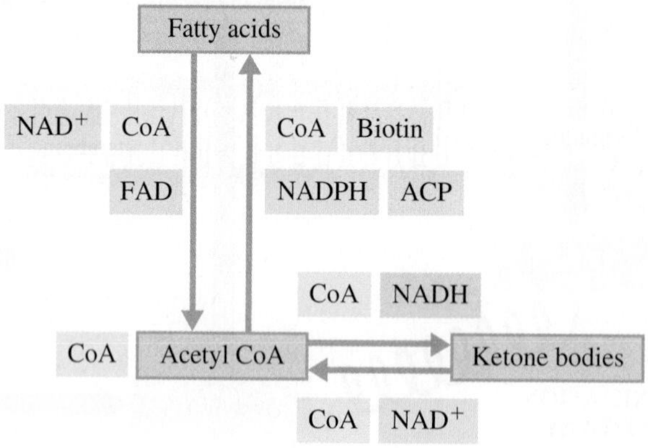

Figure 14.15 B vitamin participation, as coenzymes, in chemical reactions associated with lipid metabolism.

Concepts to Remember

OWL Sign in at **www.cengage.com/owl** to view tutorials and simulations, develop problem-solving skills, and complete online homework assigned by your professor.

Triacylglycerol digestion and absorption. Triacylglycerols are digested (hydrolyzed) in the intestine and then reassembled after passage into the intestinal wall. Chylomicrons transport the reassembled triacylglycerols from intestinal cells to the bloodstream (Section 14.1).

Triacylglycerol storage and mobilization. Triacylglycerols are stored as fat droplets in adipose tissue. When they are needed for energy, enzyme-controlled hydrolysis reactions liberate the fatty acids, which then enter the bloodstream and travel to tissues where they are utilized (Section 14.2).

Glycerol metabolism. Glycerol is first phosphorylated and then oxidized to dihydroxyacetone phosphate, a glycolysis pathway intermediate. Through glycolysis and the common metabolic pathway, the glycerol can be converted to CO_2 and H_2O (Section 14.3).

Fatty acid degradation. Fatty acid degradation is accomplished through the β-oxidation pathway. The degradation process involves removal of carbon atoms, two at a time, from the carboxyl end of the fatty acid. There are four repeating reactions that accompany the removal of each two-carbon unit. A turn of the cycle also produces one molecule each of acetyl CoA, NADH, and $FADH_2$ (Section 14.4).

Ketone bodies. Acetoacetate, β-hydroxybutyrate, and acetone are known as ketone bodies. Synthesis occurs mainly in the liver from acetyl CoA as a result of excessive fatty acid degradation. During starvation and in unchecked diabetes, the level of ketone bodies in the blood becomes very high (Section 14.6).

Fatty acid biosynthesis. Fatty acid biosynthesis, lipogenesis, occurs through the addition of two-carbon units to a growing acyl chain. The added two-carbon units come from malonyl CoA. A multienzyme complex, an acyl carrier protein (ACP), and NADPH are important parts of the biosynthetic process (Section 14.7).

Biosynthesis of cholesterol. Cholesterol is biosynthesized from acetyl CoA in a complex series of reactions in which isoprene units are key intermediates. Cholesterol is the precursor for the various classes of steroid hormones (Section 14.9).

Exercises and Problems

OWL Interactive versions of these problems may be assigned in OWL.

Exercises and problems are arranged in matched pairs with the two members of a pair addressing the same concept(s). The answer to the odd-numbered member of a pair is given at the back of the book. Problems denoted with a ▲ involve concepts found not only in the section under consideration but also concepts found in one or more earlier sections of the chapter. Problems denoted with a ● cover concepts found in a Chemical Connections feature box.

Digestion and Absorption of Lipids (Section 14.1)

14.1 What percent of dietary lipids are triacylglycerols?

14.2 What are the solubility characteristics of triacylglycerols?

14.3 What effect do salivary enzymes have on triacylglycerols?

14.4 What effect do stomach fluids have on triacylglycerols?

14.5 The process of lipid digestion occurs primarily at two sites within the human body.
a. What are the identities of these two sites?
b. What is the relative amount of TAG digestion that occurs at each site?
c. What type of digestive enzyme functions at each site?

14.6 Why does ingestion of lipids make one feel "full" for a long time?

14.7 What function does bile serve in lipid digestion?

14.8 What are the major products of triacylglycerol digestion?

14.9 *Complete* hydrolysis of triacylglycerols during digestion is unusual. Explain.

14.10 What is a fatty acid micelle?

14.11 What happens to the products of triacylglycerol digestion after they pass through the intestinal wall?

14.12 What is a chylomicron, and what is its function?

Triacylglycerol Storage and Mobilization (Section 14.2)

14.13 What is the distinctive structural feature of *adipocytes*?

14.14 What is the major metabolic function of adipose tissue?

14.15 What is triacylglycerol mobilization?

14.16 What situation signals the need for mobilization of triacylglycerols from adipose tissue?

14.17 What role does cAMP play in triacylglycerol mobilization?

14.18 Triacylglycerols in adipose tissue do not enter the bloodstream as triacylglycerols. Explain.

Glycerol Metabolism (Section 14.3)

14.19 In which step of glycerol metabolism does each of the following events occur?
a. ATP is consumed.
b. A phosphorylation reaction occurs.
c. Dihydroxyacetone phosphate is produced.
d. Glycerol kinase is active.

14.20 In which step of glycerol metabolism does each of the following events occur?
a. The oxidizing agent NAD^+ is needed.
b. ADP is produced.
c. Glycerol 3-phosphate is a reactant.
d. A secondary alcohol group is oxidized.

14.21 How many ATP molecules are expended in the conversion of glycerol to a glycolysis intermediate?

14.22 What are the two fates of glycerol after it has been converted to a glycolysis intermediate?

Oxidation of Fatty Acids (Section 14.4)

14.23 Indicate whether each of the following statements concerning the fatty acid activation process that occurs prior to β-oxidation is true or false.
a. CoA—SH is a reactant.
b. A molecule of ADP is produced.
c. The site for the process is the outer mitochondrial membrane.
d. Two ATP molecules are consumed.

14.24 Indicate whether each of the following statements concerning the fatty acid activation process that occurs prior to β-oxidation is true or false.
a. Acetyl CoA is a product.
b. A phosphorylation enzyme is needed.
c. A molecule of AMP is produced.
d. The activation product is acyl CoA.

14.25 Indicate whether each of the following aspects of the carnitine shuttle system associated with the process of β-oxidation occurs in the mitochondrial matrix or in the mitochondrial intermembrane space.
a. Acyl CoA is a reactant.
b. Carnitine enters the inner mitochondrial membrane.
c. Carnitine is converted to acyl carnitine.
d. Free CoA is a reactant.

14.26 Indicate whether each of the following aspects of the carnitine shuttle system associated with the process of β-oxidation occurs in the mitochondrial matrix or in the mitochondrial intermembrane space.
a. Acyl CoA is a product.
b. Acyl carnitine enters the inner mitochondrial membrane.
c. Acyl carnitine is converted to carnitine.
d. Coenzyme A is a product.

14.27 Only one molecule of ATP is needed to activate fatty acids before oxidation occurs, yet in "bookkeeping" on this energy expenditure it is counted as the loss of two ATP molecules. Explain why.

14.28 What is the difference between an acetyl CoA molecule and an acyl CoA molecule?

14.29 Explain what functional group change occurs, during one turn of the β-oxidation pathway in
a. Step 1.
b. Step 2.
c. Step 3.

14.30 For one turn of the β-oxidation pathway, arrange the following β-carbon functional groups in the order in which they are encountered: secondary alcohol, ketone, alkane, and alkene.

14.31 What is the configuration of the unsaturated enoyl CoA formed by dehydrogenation during a turn of the β-oxidation pathway?

14.32 What is the configuration of the β-hydroxyacyl CoA formed by hydration during a turn of the β-oxidation pathway?

14.33 Draw structural formulas for the following compounds, all of which are associated with the fatty acid β-oxidation pathway.

a. C_6 acyl CoA b. C_4 hydroxyacyl CoA
c. C_8 enoyl CoA d. C_{10} ketoacyl CoA

14.34 Draw structural formulas for the following compounds, all of which are associated with the fatty acid β-oxidation pathway.

a. C_4 acyl CoA b. C_8 hydroxyacyl CoA
c. C_{10} enoyl CoA d. C_6 ketoacyl CoA

14.35 What type of compound—acyl CoA, enoyl CoA, hydroxyacyl CoA, or ketoacyl CoA—is involved in each of the following aspects of one "loop" of the β-oxidation pathway?

a. Reactant in step 1 b. Product in step 2
c. Product in step 3 d. Reactant in step 4

14.36 What type of compound—acyl CoA, enoyl CoA, hydroxyacyl CoA, or ketoacyl CoA—is involved in each of the following aspects of one "loop" of the β-oxidation pathway?

a. Product in step 1 b. Reactant in step 2
c. Reactant in step 3 d. Product in step 4

14.37 In the context of fatty acid oxidation to acetyl CoA, indicate whether each of the following substances is involved in (1) fatty acid activation (2) fatty acid transport or (3) β-oxidation pathway. More than one choice may be correct in a given situation.

a. AMP b. FAD c. acyl CoA d. H_2O

14.38 In the context of fatty acid oxidation to acetyl CoA, indicate whether each of the following substances is involved in (1) fatty acid activation (2) fatty acid transport or (3) β-oxidation pathway. More than one choice may be correct in a given situation.

a. acyl carnitine b. enoyl CoA
c. CoA d. NAD^+

14.39 In which step (of Steps 1 through 4) and in which turn (second or third) of the β-oxidation pathway is each of the following compounds encountered as a reactant if the fatty acid to be degraded is decanoic acid?

a. $CH_3-(CH_2)_4-CH(OH)-CH_2-C(=O)-S-CoA$

b. $CH_3-(CH_2)_2-CH{=}CH-C(=O)-S-CoA$

c. $CH_3-(CH_2)_2-C(=O)-CH_2-C(=O)-S-CoA$

d. $CH_3-(CH_2)_6-C(=O)-S-CoA$

14.40 In which step (of Steps 1 through 4) and in which turn (second or third) of the β-oxidation pathway is each of the following compounds encountered as a reactant if the fatty acid to be degraded is decanoic acid?

a. $CH_3-(CH_2)_4-C(=O)-CH_2-C(=O)-S-CoA$

b. $CH_3-(CH_2)_4-C(=O)-S-CoA$

c. $CH_3-(CH_2)_4-CH{=}CH-C(=O)-S-CoA$

d. $CH_3-(CH_2)_2-CH(OH)-CH_2-C(=O)-S-CoA$

14.41 Which compound(s) in Problem 14.39 undergo(es) a dehydrogenation reaction during a turn of the β-oxidation pathway?

14.42 Which compound(s) in Problem 14.40 undergo(es) a thiolysis reaction during a turn of the β-oxidation pathway?

14.43 How many turns of the β-oxidation pathway would be needed to degrade each of the following fatty acids to acetyl CoA?

a. 16:0 fatty acid b. 12:0 fatty acid

14.44 How many turns of the β-oxidation pathway would be needed to degrade each of the following fatty acids to acetyl CoA?

a. 20:0 fatty acid b. 10:0 fatty acid

14.45 The degradation of *cis*-3-hexenoic acid, a 6:1 acid, requires one more step than the degradation of hexanoic acid, a 6:0 acid. Describe the nature of the extra step.

14.46 The degradation of *cis*-4-hexenoic acid, a 6:1 acid, requires one more step than the degradation of hexanoic acid, a 6:0 acid. Describe the nature of this extra step.

▲14.47 Indicate whether each of the following substances is involved in (1) glycerol metabolism to dihydroxyacetone phosphate (2) fatty acid metabolism to acetyl CoA or (3) both glycerol metabolism and fatty acid metabolism. More than one choice may be correct in a given situation.

a. NAD^+ b. ADP
c. Kinase d. Ketoacyl CoA

▲**14.48** Indicate whether each of the following substances is involved in (1) glycerol metabolism to dihydroxyacetone phosphate (2) fatty acid metabolism to acetyl CoA or (3) both glycerol metabolism and fatty acid metabolism. More than one choice may be correct in a given situation.

a. Dehydrogenase b. AMP
c. Acyl CoA d. Glycerol 3-phosphate

ATP Production from Fatty Acid Oxidation (Section 14.5)

14.49 Identify the major fuel for skeletal muscle in

a. an active state.
b. a resting state.

14.50 Explain why fatty acids cannot serve as fuel for the brain.

14.51 Consider the conversion of a C_{10} saturated acid entirely to acetyl CoA.

a. How many turns of the β-oxidation pathway are required?
b. What is the yield of acetyl CoA?
c. What is the yield of NADH?
d. What is the yield of $FADH_2$?
e. How many high-energy ATP bonds are consumed?

14.52 Consider the conversion of a C_{14} saturated acid entirely to acetyl CoA.

a. How many turns of the β-oxidation pathway are required?
b. What is the yield of acetyl CoA?
c. What is the yield of NADH?
d. What is the yield of $FADH_2$?
e. How many high-energy ATP bonds are consumed?

14.53 What is the net ATP production for the complete oxidation to CO_2 and H_2O of the fatty acid in Problem 14.51?

14.54 What is the net ATP production for the complete oxidation to CO_2 and H_2O of the fatty acid in Problem 14.52?

14.55 Which yield more $FADH_2$, saturated or unsaturated fatty acids? Explain.

14.56 Which yield more NADH, saturated or unsaturated fatty acids? Explain.

14.57 Compare the energy released when 1 g of carbohydrate and 1 g of lipid are completely degraded in the body.

14.58 Compare the net ATP produced from 1 molecule of glucose and 1 molecule of hexanoic acid when they are completely degraded in the body.

•14.59 (Chemical Connections 14-A) Indicate whether each of the following statements concerning high- and low-intensity exercise is true or false.
a. In a resting state, the human body burns more fat than carbohydrates.
b. Both O_2 and CO_2 measurements are needed in order to determine a RER.
c. Initial stages of exercise are fueled primarily by glucose.
d. Beyond an hour of exercise, a 50-50 balance between triacylglycerol and glucose use as fuel occurs.

•14.60 (Chemical Connections 14-A) Indicate whether each of the following statements concerning high- and low-intensity exercise is true or false.
a. The acronym RER stands for *respiratory energy ratio*.
b. The O_2 requirement for burning 1 gram of carbohydrate is greater than that for burning 1 gram of fat.
c. During the first few minutes of exercise, glycogen stores are the body's main energy source.
d. Low-intensity exercise burns fat at a greater rate than does high-intensity exercise.

Ketone Bodies and Ketogenesis (Section 14.6)

14.61 What three body conditions are conducive to ketone body formation?

14.62 Why does a deficiency of carbohydrates in the diet lead to ketone body formation?

14.63 What is the relationship between oxaloacetate concentration and ketone body formation?

14.64 What is the relationship between pyruvate concentration and ketone body formation?

14.65 Draw the structures of the three compounds classified as ketone bodies.

14.66 Two of the three ketone bodies can be formed from the third one. Write equations for the formation of these two compounds.

14.67 Identify the ketone body (or bodies) to which each of the following statements applies.
a. Its structure contains a ketone functional group.
b. Its structure contains a hydroxyl functional group.
c. It is produced in Step 3 of ketogenesis.
d. It is produced from the decomposition of acetoacetate.

14.68 Identify the ketone body (or bodies) to which each of the following statements applies.
a. Its structure contains two methyl groups.
b. Its structure contains a carboxyl functional group.
c. It is produced in Step 4 of ketogenesis.
d. It is produced from the reduction of acetoacetate.

14.69 In which step of ketogenesis is each of the following changes effected?
a. $C_2 + C_2 \longrightarrow C_4$
b. $C_4 \longrightarrow C_4$
c. $C_6 \longrightarrow C_4 + C_2$
d. $C_4 + C_2 \longrightarrow C_6$

14.70 In the context of ketogenesis, classify each of the reactions in Problem 14.69 as (1) a condensation (2) a chain cleavage (3) a hydrogenation or (4) an oxidation.

14.71 In the context of ketogenesis, indicate whether each of the following molecules is a C_2, C_4, C_6, or C_8 species.
a. Acetoacetyl CoA
b. 3-Hydroxy-3-methylglutaryl CoA
c. Acetoacetate
d. β-Hydroxybutyrate

14.72 In the context of ketogenesis reaction steps, indicate whether each of the molecules in Problem 14.71 is (1) a reactant but not a product (2) a product but not a reactant or (3) a reactant in one step and a product in another step.

14.73 Before acetoacetate can be used as a "fuel," it must first be activated. What are the names of the reactants and products in the activation reaction?

14.74 What are the names of the reactants and products in the reaction by which "usable fuel" is produced from the activated form of acetoacetate?

14.75 What is ketosis?

14.76 Severe ketosis situations produce acidosis. Explain.

▲14.77 In the context of lipid metabolism, indicate whether each of the following compounds is associated with (1) the β-oxidation pathway, (2) ketogenesis, or (3) both the β-oxidation pathway and ketogenesis.
a. Acyl CoA
b. Enoyl CoA
c. Acetyl CoA
d. β-Hydroxybutyrate

▲14.78 In the context of lipid metabolism, indicate whether each of the following compounds is associated with (1) the β-oxidation pathway, (2) ketogenesis, or (3) both the β-oxidation pathway and ketogenesis.
a. Ketoacyl CoA
b. Acetoacetyl CoA
c. Acetoacetate
d. Hydroxyacyl CoA

▲14.79 Indicate whether each of the following is a characteristic of (1) the β-oxidation pathway but not ketogenesis, (2) ketogenesis but not the β-oxidation pathway, (3) both the β-oxidation pathway and ketogenesis, or (4) neither the β-oxidation pathway nor ketogenesis.
a. Two different condensation reactions occur.
b. The process has four distinct steps.
c. Thiolysis occurs.
d. A hydrogenation reaction occurs.

▲14.80 Indicate whether each of the following is a characteristic of (1) the β-oxidation pathway but not ketogenesis, (2) ketogenesis but not the β-oxidation pathway, (3) both the β-oxidation pathway and ketogenesis, or (4) neither the β-oxidation pathway nor ketogenesis.
a. Two different dehydrogenation reactions occur.
b. A molecule is broken into two parts.
c. A hydrolysis reaction occurs.
d. A hydration reaction occurs.

Lipogenesis (Section 14.7)

14.81 Compare the intracellular locations where fatty acid biosynthesis and fatty acid degradation take place.

14.82 How does the structure of fatty acid synthase differ from that of the enzymes that degrade fatty acids?

14.83 Coenzyme A plays an important role in fatty acid degradation. What is its counterpart in fatty acid biosynthesis, and how does its structure differ from that of coenzyme A?

14.84 What does the designation ACP stand for?

14.85 Indicate whether each of the following aspects of the citrate–malate shuttle system associated with the process of lipogenesis occurs in the mitochondrial matrix or in the cytosol.
a. Citrate is produced from oxaloacetate and acetyl CoA.
b. ATP is converted to ADP.
c. Acetyl CoA and oxaloacetate are produced from citrate.
d. NADH is used as a reducing agent.

14.86 Indicate whether each of the following aspects of the citrate–malate shuttle system associated with the process of lipogenesis occurs in the mitochondrial matrix or in the cytosol.
a. Citrate enters the inner mitochondrial membrane.
b. Malate is produced from oxaloacetate.
c. NAD^+ is used to convert malate to oxaloacetate.
d. CoA—SH is generated.

14.87 Indicate whether each of the following lipogenesis events associated with ACP complex formation applies to (1) acetyl CoA (2) acetyl ACP (3) malonyl CoA or (4) malonyl ACP.
a. A carboxylation occurs in its production.
b. The enzyme acetyl transferase is needed in its production.
c. The B vitamin biotin is involved in its production.
d. Acetyl CoA and ACP are the reactants in its production.

14.88 Indicate whether each of the following lipogenesis events associated with ACP complex formation applies to (1) acetyl CoA (2) acetyl ACP (3) malonyl CoA or (4) malonyl ACP.
a. Acetyl CoA and CO_2 are the reactants in its production.
b. The enzyme malonyl transferase is needed in its production.
c. ATP is consumed in its production.
d. Malonyl CoA and ACP are the reactants in its production.

14.89 For the first cycle of the lipogenesis pathway, identify the step to which each of the following statements apply.
a. First hydrogenation reaction occurs.
b. A dehydration reaction occurs.
c. Acetoacetyl ACP is a product.
d. Crotonyl ACP is a reactant.

14.90 For the first cycle of the lipogenesis pathway, identify the step to which each of the following statements apply.
a. Second hydrogenation reaction occurs.
b. A condensation reaction occurs.
c. Acetoacetyl ACP is a reactant.
d. Crotonyl ACP is a product.

14.91 What is the name of the intermediate compound generated in the first cycle of the lipogenesis pathway that has each of the following characteristics?
a. Has a carbon chain that is derived from a C_4 keto monoacid
b. Has a carbon chain that is derived from a C_4 unsaturated monoacid
c. Produced by a dehydration reaction
d. Produced in a reaction that also produces CO_2

14.92 What is the name of the intermediate compound generated in the first cycle of the lipogenesis pathway that has each of the following characteristics?
a. Has a carbon chain that is derived from a C_4 hydroxy monoacid
b. Has a carbon chain that is derived from a C_4 saturated monoacid
c. Undergoes a dehydration reaction
d. Produced by a condensation reaction

14.93 Indicate whether each of the following intermediate compounds generated in the first or second cycle of the lipogenesis pathway is produced by (1) a dehydration reaction (2) a hydrogenation reaction or (3) a condensation reaction.
a. $CH_3{-}C({=}O){-}CH_2{-}C({=}O){-}S{-}ACP$
b. $CH_3{-}CH_2{-}CH_2{-}CH(OH){-}CH_2{-}C({=}O){-}S{-}ACP$
c. $CH_3{-}CH_2{-}CH_2{-}CH{=}CH{-}C({=}O){-}S{-}ACP$
d. $CH_3{-}CH_2{-}CH_2{-}C({=}O){-}S{-}ACP$

14.94 Indicate whether each of the following intermediate compounds generated in the first or second cycle of the lipogenesis pathway is produced by (1) a dehydration reaction (2) a hydrogenation reaction or (3) a condensation reaction.
a. $CH_3{-}CH_2{-}CH_2{-}CH_2{-}CH_2{-}C({=}O){-}S{-}ACP$
b. $CH_3{-}CH{=}CH{-}C({=}O){-}S{-}ACP$
c. $CH_3{-}CH(OH){-}CH_2{-}C({=}O){-}S{-}ACP$
d. $CH_3{-}CH_2{-}CH_2{-}C({=}O){-}CH_2{-}C({=}O){-}S{-}ACP$

14.95 What role does molecular oxygen, O_2, play in fatty acid biosynthesis?

14.96 What is the characteristic structural feature of an essential fatty acid?

14.97 Consider the biosynthesis of a C_{14} saturated fatty acid from acetyl CoA molecules.
a. How many turns of the fatty acid biosynthetic pathway are needed?
b. How many molecules of malonyl ACP must be formed?
c. How many high-energy ATP bonds are consumed?
d. How many NADPH molecules are needed?

14.98 Consider the biosynthesis of a C_{16} saturated fatty acid from acetyl CoA molecules.
a. How many turns of the fatty acid biosynthetic pathway are needed?
b. How many molecules of malonyl ACP must be formed?
c. How many high-energy ATP bonds are consumed?
d. How many NADPH molecules are needed?

▲14.99 Indicate whether each of the following is a characteristic of (1) the β-oxidation pathway (2) ketogenesis or (3) the chain elongation phase of lipogenesis. More than one answer may be correct in a given situation.
a. The process has four distinct reaction steps.
b. Two different hydrogenation reactions occur.
c. Two different dehydrogenation reactions occur.
d. A thiolysis reaction occurs.

▲**14.100** Indicate whether each of the following is a characteristic of (1) the β-oxidation pathway (2) ketogenesis or (3) the chain elongation phase of lipogenesis. More than one answer may be correct in a given situation.
a. Acetyl CoA is a reactant.
b. Acyl CoA is a reactant.
c. Two different condensation reactions occur.
d. A dehydration reaction occurs.

▲14.101 Indicate whether each of the following is a characteristic of (1) the β-oxidation pathway (2) ketogenesis or (3) lipogenesis. More than one answer may be correct in a given situation.
a. The carnitine shuttle system is used.
b. Malonyl ACP is a reactant.
c. CO_2 is a product.
d. Molecular O_2 is sometimes needed.

▲**14.102** Indicate whether each of the following is a characteristic of (1) the β-oxidation pathway (2) ketogenesis or (3) lipogenesis. More than one answer may be correct in a given situation.
a. The citrate shuttle system is used.
b. Acetyl ACP is a reactant.
c. H_2O is a product.
d. H_2O is a reactant.

▲14.103 In which of the processes (1) glycerol metabolism to dihydroxyacetone phosphate (2) β-oxidation pathway (3) ketogenesis and (4) lipogenesis is each of the following molecules encountered? There may be more than one correct response in a given situation.
a. Enoyl CoA
b. FAD
c. β–Hydroxybutyrate
d. Glycerol 3-phosphate

▲**14.104** In which of the processes (1) glycerol metabolism to dihydroxyacetone phosphate (2) β-oxidation pathway (3) ketogenesis and (4) lipogenesis is each of the following molecules encountered? There may be more than one correct response in a given situation.
a. HMG-CoA
b. NADPH
c. Malonyl ACP
d. Acetoacetyl CoA

Relationships Between Lipogenesis and Citric Acid Cycle Intermediates (Section 14.8)

14.105 What is the citric acid cycle diacid intermediate counterpart for each of the following lipogenesis C_4—ACP monoacid intermediates?
a. Butyrate
b. Acetoacetate
c. β-Hydroxybutyrate
d. Crotonate

14.106 What is the lipogenesis C_4—ACP monoacid intermediate counterpart for each of the following citric acid cycle C_4 diacid intermediates?
a. Succinate
b. Malate
c. Oxaloacetate
d. Fumarate

14.107 Identify each of the following C_4 species as a (1) hydroxy acid (2) keto acid (3) saturated acid or (4) unsaturated acid.
a. Crotonate
b. Oxaloacetate
c. Acetoacetate
d. Malate

14.108 Identify each of the following C_4 species as a monocarboxylic acid or dicarboxylic acid.
a. Succinate
b. Butyrate
c. β-Hydroxybutyrate
d. Fumarate

Fate of Fatty Acid Generated Acetyl CoA (Section 14.9)

14.109 Indicate whether or not each of the following substances can be produced from acetyl CoA using the choices (1) can be produced in a one-step process (2) can be produced in a multistep process or (3) cannot be produced from acetyl CoA.
a. Cholesterol
b. Acetoacetyl CoA
c. Malonyl CoA
d. Pyruvate

14.110 Indicate whether or not each of the following substances can be produced from acetyl CoA using the choices (1) can be produced in a one-step process (2) can be produced in a multistep process, or (3) cannot be produced from acetyl CoA.
a. HMG-CoA
b. Acetyl ACP
c. Mevalonate
d. Acetoacetate

14.111 Explain why pyruvate can be used to produce glucose, but acetyl CoA cannot be used for this purpose.

14.112 Explain why the *net* production of glucose using oxaloacetate obtained from acetyl CoA as the starting material is zero.

•14.113 (Chemical Connections 14-B) Indicate whether each of the following statements relating to the control of the biosynthesis of cholesterol in the human body is true or false.
a. Mevalonate is a reactant in the rate-determining step for the biosynthesis of cholesterol.
b. The mode of action of the statin family of anticholesterol medications is competitive enzyme inhibition.
c. Part of the structure of the statin Lipitor resembles that of mevalonate.
d. Cholesterol-lowering statins also have antiinflammatory properties.

•**14.114** (Chemical Connections 14-B) Indicate whether each of the following statements relating to the control of the biosynthesis of cholesterol in the human body is true or false.
a. HMG-CoA is the product in the rate-determining step for the biosynthesis of cholesterol.
b. Lipitor inhibits the formation of the cholesterol precursor HMG-CoA.
c. The original source for the first statin-type drugs was a fungus extract.
d. Cholesterol-lowering statins have the side effect of decreasing bone density.

Relationships Between Lipid and Carbohydrate Metabolism (Section 14.10)

14.115 Indicate whether each of the following metabolic processes occurs within mitochondria of a cell or in a cell's cytosol.
a. Ketogenesis
b. Glycolysis
c. Citric acid cycle
d. β-Oxidation pathway

14.116 Indicate whether each of the following metabolic processes occurs within mitochondria of a cell or in a cell's cytosol.
a. Electron transport chain
b. Lipogenesis
c. Glycerol metabolism
d. Gluconeogenesis

B Vitamins and Lipid Metabolism (Section 14.11)

14.117 Indicate whether each of the following B vitamins is involved (as a cofactor) in (1) β-oxidation pathway (2) ketogenesis (3) lipogenesis or (4) conversion of ketone bodies to acetyl CoA. There may be more than one correct response or no correct response for a particular B vitamin.
a. Niacin b. Thiamin
c. Pantothenic acid d. Folate

14.118 Indicate whether each of the following B vitamins is involved (as a cofactor) in (1) β-oxidation pathway (2) ketogenesis (3) lipogenesis or (4) conversion of ketone bodies to acetyl CoA. There may be more than one correct response or no correct response for a particular B vitamin.
a. Biotin b. Vitamin B_6
c. Vitamin B_{12} d. Riboflavin

Protein Metabolism

15

Marevision Marevision/Photo Library

Fish, such as the Atlantic salmon, and other aquatic species process (eliminate) the nitrogen from protein in a manner different from that which occurs in human beings.

CHAPTER OUTLINE

From an energy production standpoint, proteins supply only a small portion of the body's needs. With a normal diet, carbohydrates and fats supply 90% of the body's energy, and only 10% comes from proteins. However, despite its minor role in energy production, protein metabolism plays an important role in maintaining good health. The amino acids obtained from proteins are needed for both protein synthesis and synthesis of other nitrogen-containing compounds in the cell. In this chapter, protein digestion, the oxidative degradation of amino acids, and amino acid biosynthesis are examined.

15.1 Protein Digestion and Absorption

Protein digestion begins in the stomach rather than in the mouth because saliva contains no enzymes that affect proteins. Both protein denaturation (Section 9.15) and protein hydrolysis (Section 9.14) occur in the stomach. The partially digested protein (large polypeptides) passes from the stomach into the small intestine, where digestion is completed (Figure 15.1).

Dietary protein entering the stomach effects the release of the hormone *gastrin* by stomach mucosa cells. Gastrin's presence causes both hydrochloric acid and pepsinogen secretion. Hydrochloric acid has three major functions within the stomach: (1) its antiseptic properties kill most bacteria (2) its denaturing action (Section 9.15) "unwinds" globular proteins, making peptide

A very small number of people are unable to synthesize enough stomach acid, and these individuals must ingest capsules of dilute hydrochloric acid with every meal.

OWL

Sign in to OWL at **www.cengage.com/owl** to view tutorials and simulations, develop problem-solving skills, and complete online homework assigned by your professor.

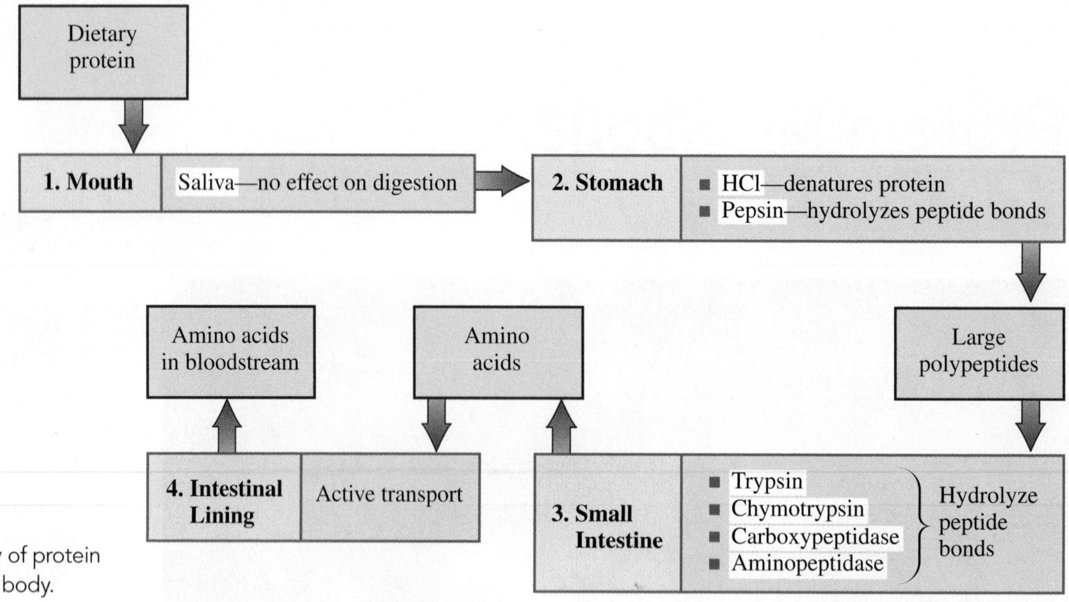

Figure 15.1 Summary of protein digestion in the human body.

bonds more accessible to digestive enzymes and (3) its acidity leads to the activation of pepsinogen, which is the inactive form of the digestive enzyme *pepsin*; because of the hydrochloric acid present, gastric juice has a pH between 1.5 and 2.0. Pepsin effects the hydrolysis of approximately 10% of peptide bonds present in proteins, producing a variety of polypeptides.

Passage, in small batches, of the stomach's acidic protein contents into the small intestine stimulates production of the hormone *secretin*, which in turn stimulates pancreatic production of bicarbonate ion, HCO_3^-, whose function is neutralization of gastric hydrochloric acid. With such neutralization, the partially digested protein mix becomes slightly basic, with a pH value between 7.0 and 8.0. This basic environment allows for the production of the pancreatic digestive enzymes *trypsin, chymotrypsin, and carboxypeptidase*, all of which attack peptide bonds. *Aminopeptidase,* secreted by intestinal mucosal cells, also attacks peptide bonds. Pepsin, trypsin, chymotrypsin, carboxypeptidase, and aminopeptidase are all examples of *proteolytic* enzymes (Section 10.9). Enzymes of this type are produced in inactive forms called *zymogens* that are activated at their site of action (Section 10.9).

The passage of polypeptide chains and small proteins across the intestinal wall is uncommon in adults. In infants, however, such transport allows the passage of antibodies (proteins) in colostral milk from a mother to a nursing infant to build up immunologic protection in the infant.

The net result of protein digestion is the release of the protein's constituent amino acids. Absorption of these "free" amino acids through the intestinal wall requires active transport with the expenditure of energy (Section 8.10). Different transport systems exist for the various kinds of amino acids. After passing through the intestinal wall, the free amino acids enter the bloodstream, which distributes them throughout the body.

EXAMPLE 15.1 Determining the Sites Where Various Aspects of Protein Digestion Occur

Based on the information in Figure 15.1, determine the location within the human body where each of the following aspects of protein digestion occurs.

a. The enzyme trypsin is active.
b. Large polypeptides are produced.
c. Protein is denatured by HCl.
d. Active transport moves amino acids into the bloodstream.

Solution

a. Trypsin is one of several proteolytic enzymes found in the *small intestine* that hydrolyze peptide bonds.
b. Large polypeptides are the product of enzymatic action that occurs in the *stomach.*
c. Protein denaturation, effected by HCl (stomach acid), occurs in the *stomach.*
d. Active transport processes effect the passage of amino acids through the *intestinal lining* into the bloodstream.

Practice Exercise 15.1

Based on the information in Figure 15.1, determine the location within the human body where each of the following aspects of protein digestion occurs.

a. The enzyme aminopeptidase is active.
b. Peptide bonds are hydrolyzed under the action of pepsin.
c. Individual amino acids are produced.
d. The enzyme carboxypeptidase is active.

Answers: **a.** Small intestine; **b.** Stomach; **c.** Small intestine; **d.** Small intestine

15.2 Amino Acid Utilization

Amino acids produced from the digestion of proteins enter the amino acid pool of the body. The **amino acid pool** *is the total supply of free amino acids available for use in the human body.* Dietary protein is one of three sources that contributes amino acids to the amino acid pool. The other two sources are *protein turnover* and *biosynthesis* of amino acids in the liver.

Within the human body, proteins are continually being degraded (hydrolyzed) to amino acids and resynthesized. Disease, injury, and "wear and tear" are all causes of degradation. The degradation–resynthesis process is called protein turnover. **Protein turnover** *is the repetitive process in which proteins are degraded and resynthesized within the human body.*

Biosynthesis of amino acids by the liver also supplies the amino acid pool with amino acids. However, only the *nonessential* amino acids (Sections 9.3 and 15.6) can be produced in this manner.

In a healthy adult, the amount of nitrogen taken into the body each day (dietary proteins) equals the amount of nitrogen excreted from the body. Such a person is said to be in a state of nitrogen balance. **Nitrogen balance** *is the state that results when the amount of nitrogen taken into the human body as protein equals the amount of nitrogen excreted from the body in waste materials.*

Two types of nitrogen imbalance can occur. When protein degradation exceeds protein synthesis, the amount of nitrogen in the urine exceeds the amount of nitrogen ingested (dietary protein). This condition of *negative nitrogen balance* accompanies a state of "tissue wasting," because more tissue proteins are being catabolized than are being replaced by protein synthesis. Protein-poor diets, starvation, and wasting illnesses produce a negative nitrogen balance.

A *positive nitrogen balance* (nitrogen intake exceeds nitrogen output) indicates that the rate of protein anabolism (synthesis) exceeds that of protein catabolism. This state indicates that large amounts of tissue are being synthesized, such as during growth, pregnancy, and convalescence from an emaciating illness.

Although the overall nitrogen balance in the body often varies, the relative concentrations of amino acids within the amino acid pool remain essentially constant. No specialized storage forms for amino acids exist in the body, as is the case for glucose (glycogen) and fatty acids (triacylglycerols). Therefore, the body needs a relatively constant source of amino acids to maintain normal metabolism. During negative nitrogen balance, the body must resort to degradation of proteins that were synthesized for other functions.

The amino acids from the body's amino acid pool are used in four different ways.

1. *Protein synthesis.* It is estimated that about 75% of the free amino acids in a healthy, well-nourished adult go into protein synthesis. Proteins are continually needed to replace old tissue (protein turnover) and also to build new tissue (growth). The subject of protein synthesis was considered in Section 11.12.
2. *Synthesis of nonprotein nitrogen-containing compounds.* Amino acids are regularly withdrawn from the amino acid pool for the synthesis of nonprotein nitrogen-containing compounds. Such molecules include the purines and

The amino acid pool is not a specific location within the body where free amino acids "congregate." Rather, these free amino acids are present throughout the body, accumulating in the blood and within cells where they are available when needed for further use, which in most instances is for protein synthesis.

The rate of protein turnover varies from a few minutes to several hours. Proteins with short turnover rates include many enzymes and regulatory hormones. In a healthy adult, about 2% of the body's protein is broken down and resynthesized every day.

Higher plants and certain microorganisms are capable of synthesizing all the protein amino acids from carbon dioxide, water, and inorganic salts.

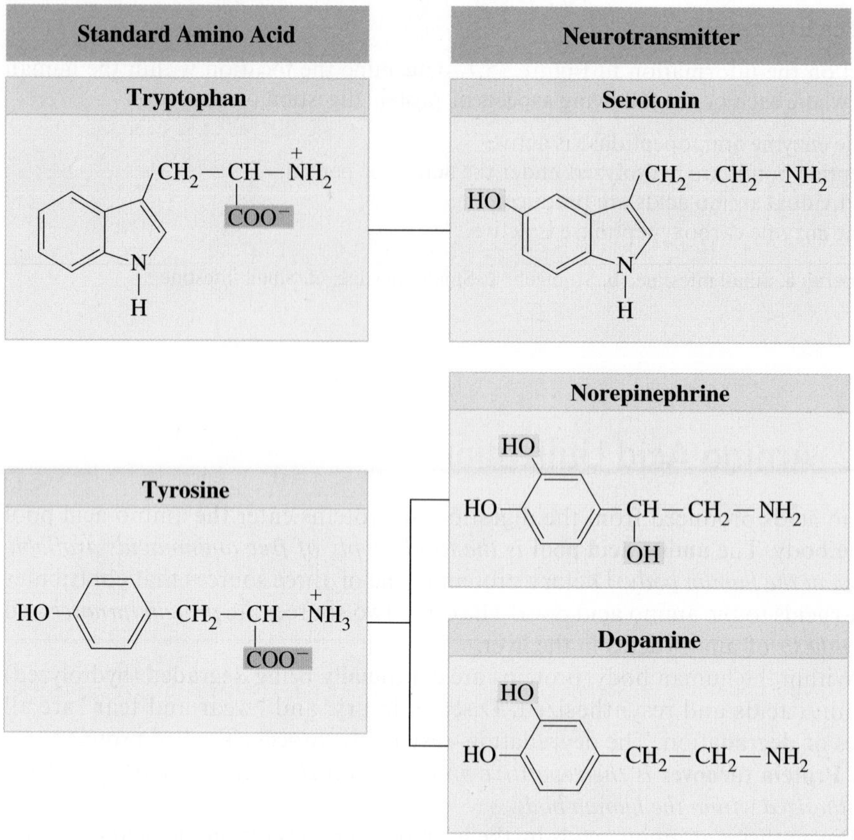

Figure 15.2 Structural relationships between the neurotransmitters serotonin, norepinephrine, and dopamine and their amino acid precursors.

pyrimidines of nucleic acids (A, C, G, T, and U; Section 11.2), the heme of hemoglobin (Section 15.7), the choline and ethanolamine of phosphoglycerides (Section 8.7), hormones such as adrenaline (Section 6.10), and neurotransmitters such as norepinephrine, dopamine, and serotonin (Section 6.10). The neurotransmitter serotonin is synthesized from the amino acid tryptophan, and the neurotransmitters norepinephrine and dopamine are synthesized from the amino acid tyrosine. Figure 15.2 shows the close structural resemblance between these two amino acids and the neurotransmitters synthesized from them.

There are approximately 100 grams of free amino acids present in the amino acid pool. Glutamine is the most abundant, followed closely by alanine, glycine, and valine. The essential amino acids constitute approximately 10 grams of the pool.

3. *Synthesis of nonessential amino acids.* When required, the body draws on the amino acid pool for raw materials for the production of *nonessential* amino acids that are in short supply. The "roadblock" preventing the synthesis of the *essential* amino acids is not lack of nitrogen but lack of a correct carbon skeleton upon which enzymes can work. In general, the essential amino acids contain carbon chains or aromatic rings not present in other amino acids or the intermediates of carbohydrate or lipid metabolism. Table 15.1 lists the essential amino acids and the nonessential amino acids with the precursors needed to form the latter.
4. *Production of energy.* Because excess amino acids cannot be stored for later use, the body's response is to degrade them. The degradation process is complex because each of the 20 standard amino acids has a different degradation pathway.

In all the degradation pathways, the amino nitrogen atom is removed and ultimately is excreted from the body as urea. The remaining carbon skeleton is then converted to pyruvate, acetyl CoA, or a citric acid cycle intermediate, depending on its makeup, with the resulting energy production or energy storage. Figure 15.3 shows the various pathways available for the further use of amino acid catabolism products. Subsequent sections of this chapter give further details about these processes.

Table 15.1 Essential and Nonessential Amino Acids

Nutritionally Essential Amino Acids	Nutritionally Nonessential Amino Acids	
	Amino Acid	Precursor
histidine	alanine	pyruvate
isoleucine	arginine	glutamate
leucine	asparagine	aspartate
lysine	aspartic acid	oxaloacetate
methionine	cysteine	serine
phenylalanine	glutamic acid	α-ketoglutarate
threonine	glutamine	glutamate
tryptophan	glycine	serine
valine	proline	glutamate
	serine	3-phosphoglycerate
	tyrosine	phenylalanine

15.3 Transamination and Oxidative Deamination

Degradation of an amino acid has two stages: (1) the removal of the α-amino group and (2) the degradation of the remaining carbon skeleton. In this section and the next, what happens to the amino group is considered; in Section 15.5, the fate of the carbon skeleton is considered.

Two different types of biochemical reactions are needed for the removal of amino groups from amino acids. They are *transamination reactions* and *oxidative deamination reactions*, both of which are considered in this section.

Transamination Reactions

Transamination reactions always involve two amino acids (one as a reactant and one as a product) and two keto acids (one as a reactant and one as a product). Two keto/amino acid pairs are involved, with the members of a pair having a common carbon-chain base.

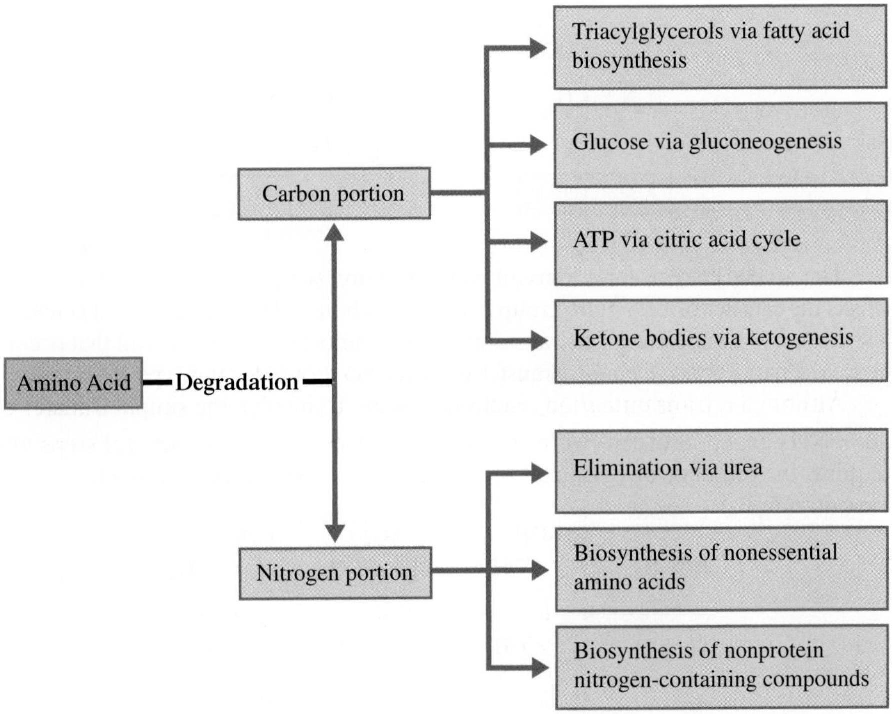

Figure 15.3 Possible fates for amino acid degradation products.

Figure 15.4 Key keto/amino acid pairs encountered in transamination reactions.

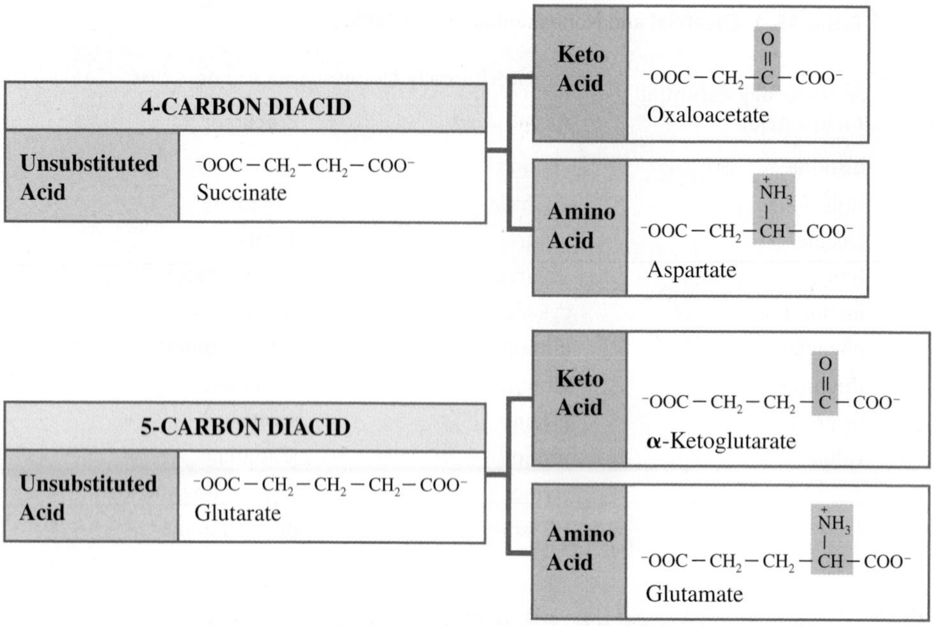

The two keto acids in Figure 15.4, oxaloacetate and α-ketoglutarate, have been previously encountered; they are intermediates in the citric acid cycle.

The two most encountered keto/amino acid pairs in transamination are α-ketoglutarate/glutamate and oxaloacetate/aspartate. Their structural formulas, as well as those of their "parent" acids, are shown in Figure 15.4.

A **transamination reaction** *is a biochemical reaction that involves the interchange of the amino group of an α-amino acid with the keto group of an α-keto acid.* A generalized structural equation for a transamination reaction is as follows:

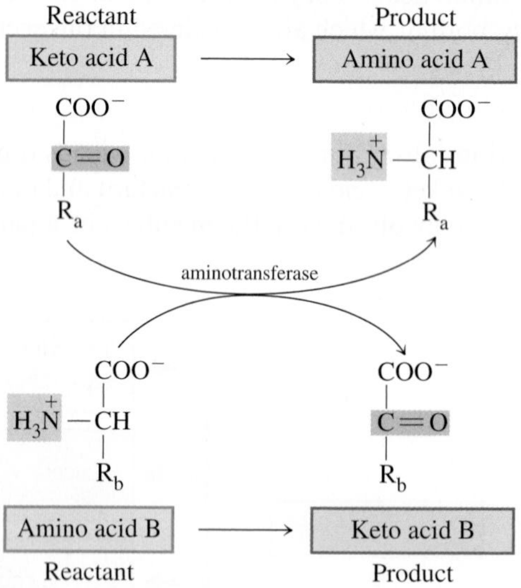

The needed enzyme for a transamination reaction is an *aminotransferase*. Its action effects the transfer of the amino group from one carbon skeleton to another. There is no loss or gain of amino groups in transamination. Amino group transfer is all that occurs; hence the name *transamination* (transfer of an amino group) for this type of reaction.

Although a transamination reaction appears to involve the simple transfer of an $—\overset{+}{N}H_3$ group between two molecules, the reaction involves several steps and requires the presence of pyridoxal phosphate, a coenzyme produced from pyridoxine (vitamin B_6).

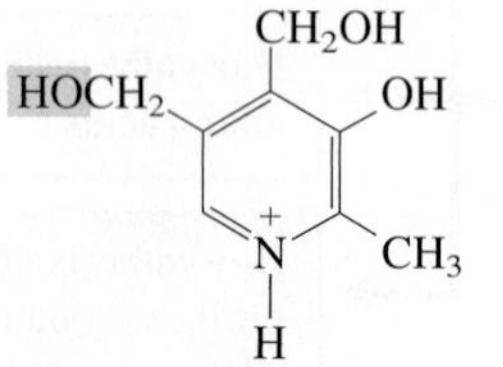

Pyridoxine (vitamin B_6)

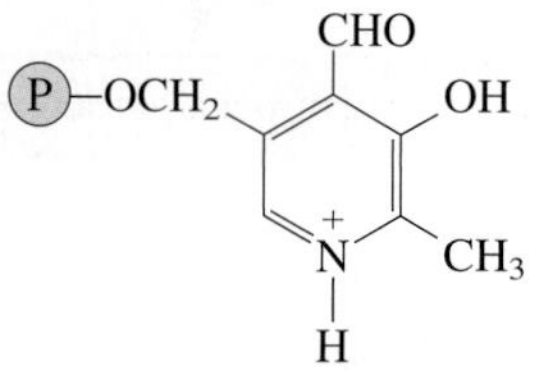

Pyridoxal phosphate (coenzyme)

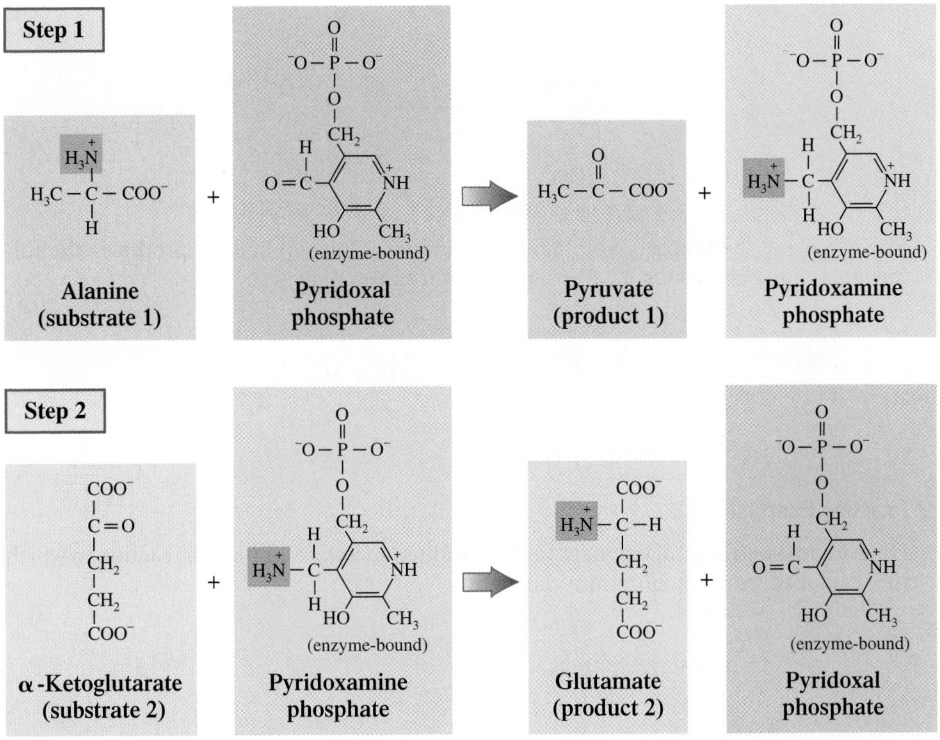

Figure 15.5 The role of pyridoxal phosphate in the process of transamination.

This coenzyme is an integral part of the transamination process. The amino group of the amino acid is transferred first to the coenzyme and then from the coenzyme to the α-keto acid. Figure 15.5 shows the role of this coenzyme in the transamination process, where alanine is the amino acid and α-ketoglutarate is the α-keto acid.

Transamination reactions are reversible and can easily go in either direction, depending on the reactant concentrations. This reversibility is the basis for regulation of amino acid concentrations in the body.

EXAMPLE 15.2 Determining the Products in a Transamination Reaction When Given the Reactants

Determine the structural formulas for the products in a transamination reaction in which the reactants are aspartate and pyruvate.

Aspartate (amino acid) Pyruvate (keto acid)

Solution

The interchange of functional groups associated with transamination is such that the new keto acid produced will have the same carbon skeleton as the reacting amino acid, and the new amino acid produced will have the same carbon skeleton as the reacting keto acid.

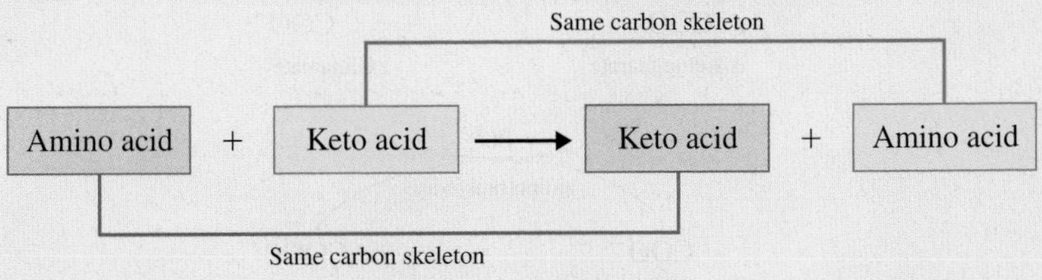

For aspartate, the reacting amino acid, the functional group change produces the following keto acid, which has the same carbon skeleton as the aspartate.

(continued)

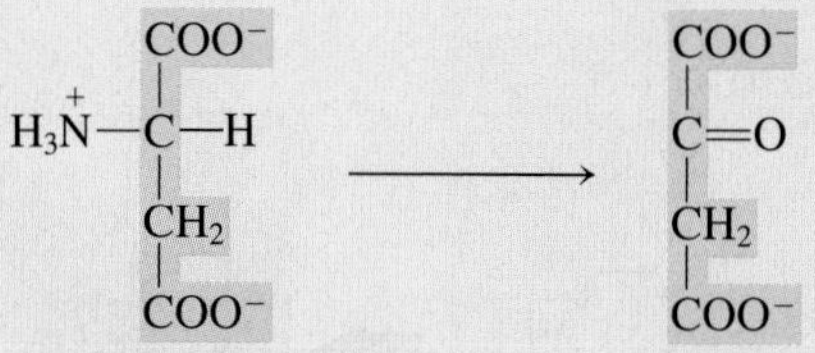

Reacting amino acid Product keto acid

For pyruvate, the reacting keto acid, the functional group change produces the following amino acid, which has the same carbon skeleton as pyruvate.

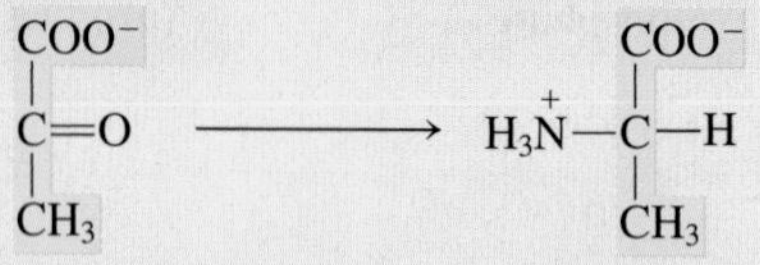

Reacting keto acid Product amino acid

Practice Exercise 15.2

Determine the structural formulas for the products in a transamination reaction in which the reactants are α-ketoglutarate and alanine.

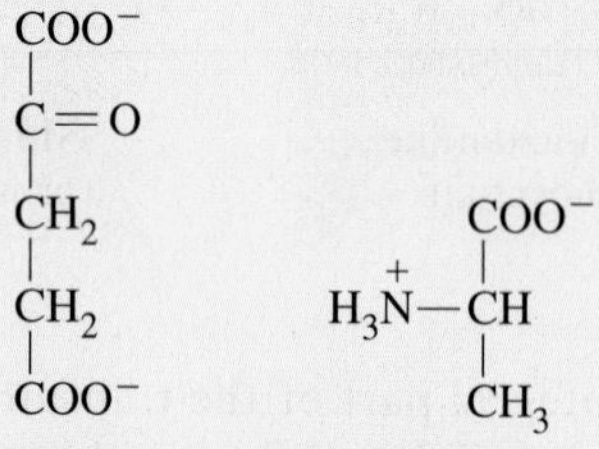

α-Ketoglutarate Alanine

Answer:

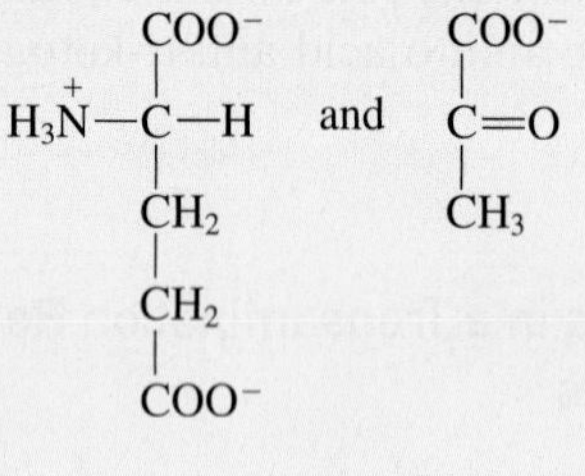

Glutamate Production Via Transamination

The most important transamination reaction in protein metabolism, in terms of frequency of occurrence, is the one in which the keto acid α-ketoglutarate is the amino group acceptor. Glutamate is the amino acid produced when α-ketoglutarate accepts the amino group.

Various aminotransferases are differentiated from each other, namewise, by including as part of the name the amino group donor that participates in the transamination. *Aspartate aminotransferase* is involved in the transfer of an amino group from the amino acid aspartate, and *alanine aminotransferase* transfers an amino group from the amino acid alanine.

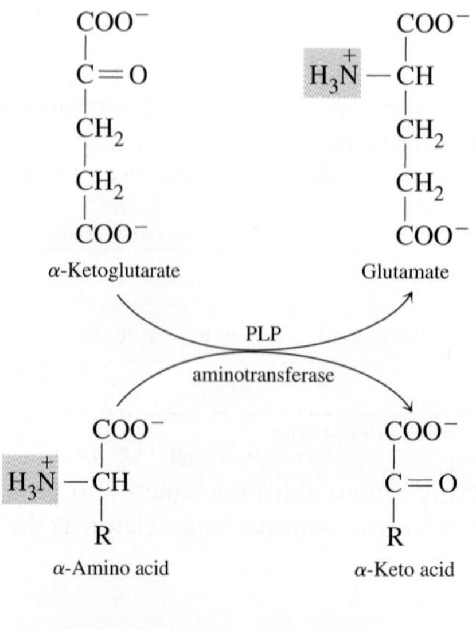

There are at least 50 aminotransferases associated with transamination reactions. They differ in which amino acid substrates they accept. Aminotransferases are highly specific relative to which keto acid substrates they accept. Most aminotransferases accept only α-ketoglutarate and to a lesser extent oxaloacetate. Thus glutamate (from α-ketoglutarate) and aspartate (from oxaloacetate) are the two amino acids produced in transamination reactions.

The high specificity of aminotransferases for α-ketoglutarate as the keto acid substrate has an important ramification that reveals a major purpose for transamination reactions in protein metabolism. Through transamination involving α-ketoglutarate, the amino groups from many different amino acids are collected into one type of molecule, the amino acid glutamate. The glutamate then functions as an amino group donor for further processing of the amino groups, with the ultimate fate of the amino groups being elimination from the body in the form of urea. The "nitrogen removal problem" has been reduced from dealing with twenty amino acids to dealing with one amino acid.

The concentration of aminotransferases in blood is used to diagnose liver and heart disorders. Liver damage releases *alanine aminotransferase* (ALT) into the blood. *Aspartate aminotransferase* (AST) is abundant in heart muscle, and increased blood levels of this enzyme indicate that heart muscle damage (myocardial infarction) has occurred.

There are two pathways for further processing of the many glutamate molecules produced via transamination. One pathway is a second transamination reaction that produces the amino acid aspartate, and the other pathway involves production of ammonium ion (NH_4^+) via *oxidative deamination*. Both the ammonium ion and the aspartate, which now become the "amino group carriers," are participants in the urea cycle (Section 15.4) through which, as the name implies, urea is produced. Further details concerning these two pathways—a second transamination reaction and oxidative deamination—are as follows.

Aspartate Production Via Transamination

Transamination in which glutamate is the reacting amino acid and oxaloacetate is the reacting keto acid produces aspartate as the new amino acid.

Both pathways for the further processing of glutamate—transamination and oxidative deamination—produce the keto acid α-ketoglutarate as one of the products. The N-containing product differs, being aspartate in the first case and ammonium ion in the second case.

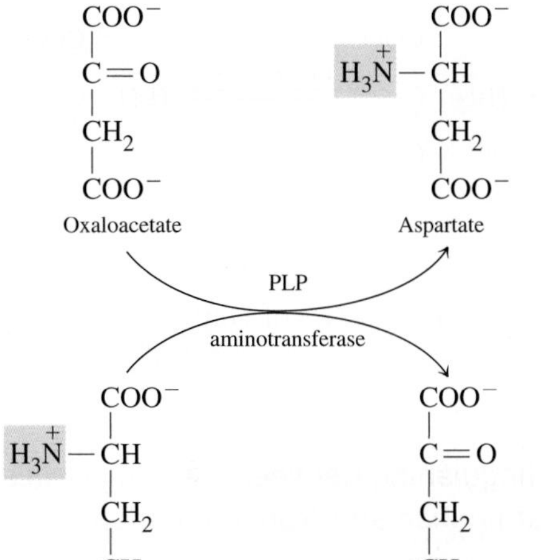

Note that the second product in this reaction is α-ketoglutarate. The "regenerated" α-ketoglutarate becomes available for re-participation in the transamination reactions whereby glutamate was initially produced. Note also that aspartate is now a carrier of nitrogen atoms that are to be eliminated from the body as urea.

Ammonium Ion Production Via Oxidative Deamination

Not all glutamates undergo transanimation to produce aspartate. A significant number instead undergo *oxidative deamination.* An **oxidative deamination reaction** *is a biochemical reaction in which an α-amino acid is converted into an α-keto acid with release of an ammonium ion.* Oxidative deamination occurs primarily in liver and kidney mitochondria.

Oxidative deamination of glutamate requires the enzyme *glutamate dehydrogenase.* This enzyme is unusual in that it is the only known enzyme that can function with either $NADP^+$ or NAD^+ as a coenzyme. With NAD^+ as the coenzyme, the reaction is:

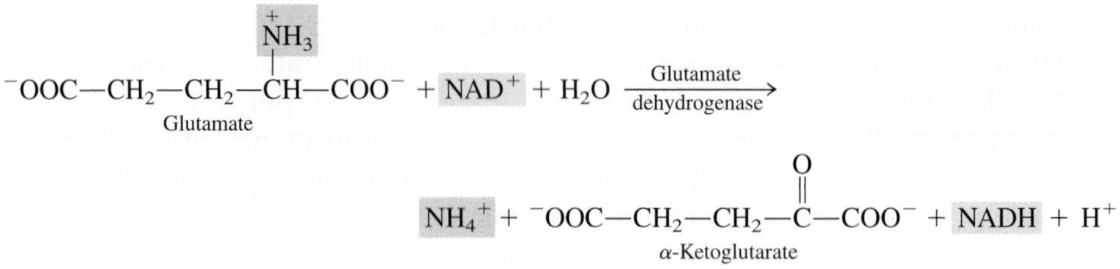

Note that α-ketoglutarate is a product of this process. It can be reused in the first series of transamination reactions. The NADH and H^+ formed can participate in the electron transport chain and oxidative phosphorylation to produce ATP molecules (Sections 12.8 and 12.9).

The net equation for combined action of an aminotransferase (transamination) and glutamate dehydrogenase (oxidative deamination) is:

$$\alpha\text{-Amino acid} + NAD^+ + H_2O \longrightarrow \alpha\text{-keto acid} + NH_4^+ + NADH + H^+$$

The toxicity of ammonium ion is related to the oxidative deamination reaction by which it is formed, the conversion of glutamate to α-ketoglutarate. This reaction, which is an equilibrium situation, is shifted to the glutamate side by increased ammonium ion levels. This shift decreases α-ketoglutarate levels significantly, which affects the citric acid cycle, of which α-ketoglutarate is an intermediate. Cellular ATP production drops, and the lack of ATP causes central nervous system problems.

The NH_4^+ so produced, a toxic substance if left to accumulate in the body, is then converted to urea in the urea cycle (Section 15.4).

Two amino acids, serine and threonine, exhibit different behavior from the other amino acids. They undergo *direct deamination* by a dehydration–hydration process rather than *oxidative deamination.* This different behavior results from the presence of a side-chain β-hydroxyl group, a feature unique to these two acids. The direct deamination reaction for serine is:

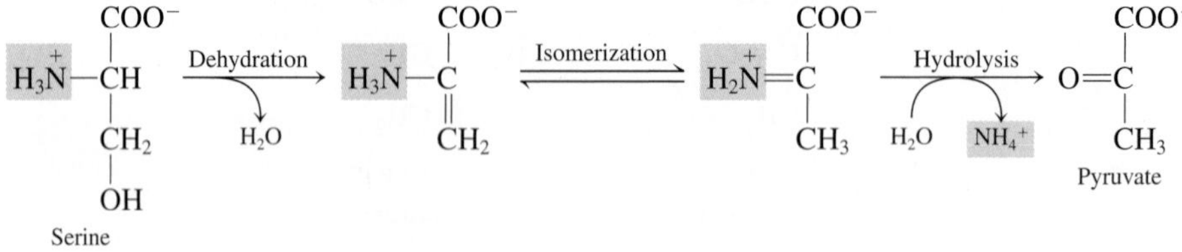

Threonine goes through a similar series of steps.

EXAMPLE 15.3 Distinguishing Between Characteristics of Oxidative Deamination and Transamination

Indicate whether each of the following reaction characteristics is associated with the process of *transamination* or with the process of *oxidative deamination.*

a. The enzyme glutamate dehydrogenase is active.
b. The coenzyme pyridoxal phosphate is needed.
c. An amino acid is converted into a keto acid with release of ammonium ion.
d. One of the reactants is an amino acid, and one of the products is an amino acid.

Solution

a. *Oxidative deamination.* The substrate for oxidative deamination is glutamate, and the process is an oxidation reaction (removal of H atoms) with NAD^+ serving as the oxidizing agent.
b. *Transamination.* The coenzyme pyridoxal phosphate, a derivative of vitamin B_6, is an intermediate in the amino group transfer process.
c. *Oxidative deamination.* Ammonium ion production is a characteristic of oxidative deamination; no ammonium ion is produced during transamination.
d. *Transamination.* In transamination, there are two reactants (an amino acid and a keto acid) and two products (a new amino acid and a new keto acid).

Practice Exercise 15.3

Indicate whether each of the following reaction characteristics is associated with the process of *transamination* or with the process of *oxidative deamination*.

a. One of the reactants is a keto acid, and one of the products is a keto acid.
b. Enzymes with a specificity toward α-ketoglutarate are often active.
c. NAD^+ is used as an oxidizing agent.
d. An aminotransferase enzyme is active.

Answers: **a.** Transamination; **b.** Transamination; **c.** Oxidative deamination; **d.** Transamination

15.4 The Urea Cycle

From a nitrogen standpoint, the net effect of transamination and deamination reactions (Section 15.3) is production of ammonium ions and aspartate molecules. Both of these nitrogen-carrying entities are processed further in the urea cycle. Ammonium ions enter the cycle indirectly, being first incorporated into another molecule (carbamoyl phosphate) that then enters the cycle's first step. Aspartate molecules enter the cycle directly in the cycle's second step.

The **urea cycle** *is a cyclic biochemical pathway in which urea is produced, for excretion, using ammonium ions and aspartate molecules as nitrogen sources.* The urea, produced in the liver, is transported in the blood to the kidneys and eliminated from the body in urine.

In the pure state, urea is a white solid with a melting point of 133°C. Its structure is

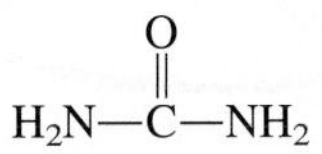

Urea is very soluble in water (1 g per 1 mL), is odorless and colorless, and has a salty taste. (Urea does not contribute to the odor or color of urine.) With normal metabolism, an adult excretes about 30 g of urea daily in urine, although the exact amount varies with the protein content of the diet.

Three amino acids are involved as intermediates in the operation of the urea cycle. These acids are arginine, ornithine, and citrulline, the latter two of which are nonstandard amino acids—that is, amino acids not found in protein. Structurally, all three of these amino acids have the same carbon chain.

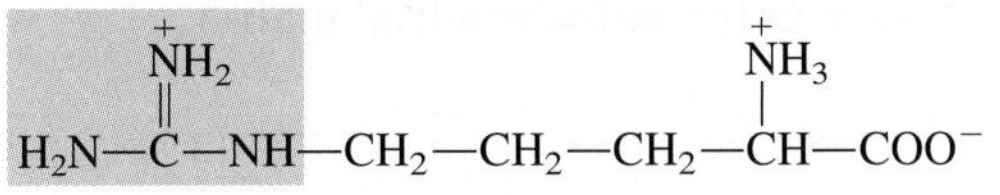

Arginine
(standard amino acid)

Arginine is the most nitrogen-rich of the standard amino acids. It contains four nitrogen atoms.

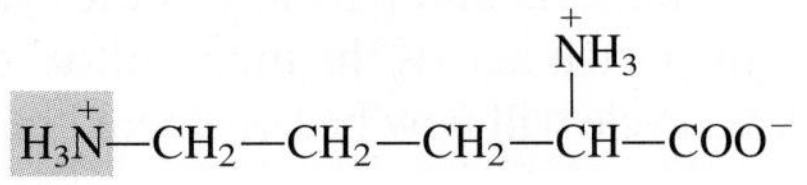

Ornithine
(nonstandard amino acid)

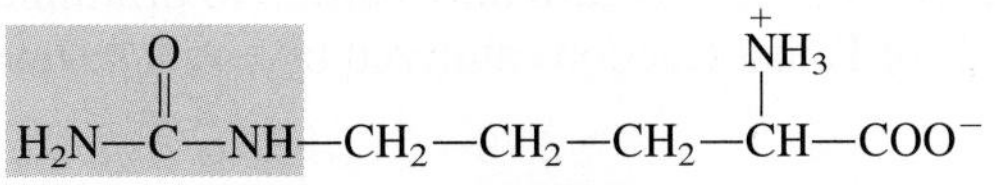

Citrulline
(nonstandard amino acid)

Carbamoyl Phosphate

The "fuel" for the urea cycle is the compound *carbamoyl phosphate.* This fuel is formed from ammonium ion (from oxidative deamination; Section 15.3), carbon dioxide (from the citric acid cycle), water, and two ATP molecules. The formation equation for carbamoyl phosphate is

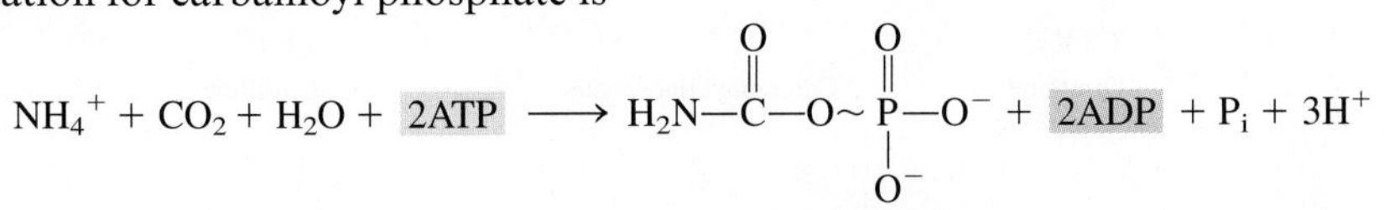

Carbamoyl phosphate

The functional group attached to the phosphate in carbamoyl phosphate is the simple amide functional group

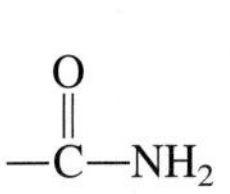

The term *carbamoyl* is the *prefix* that denotes an amide group. Most often, amide groups are named by using the *suffix* system, in which case the suffix is *amide.*

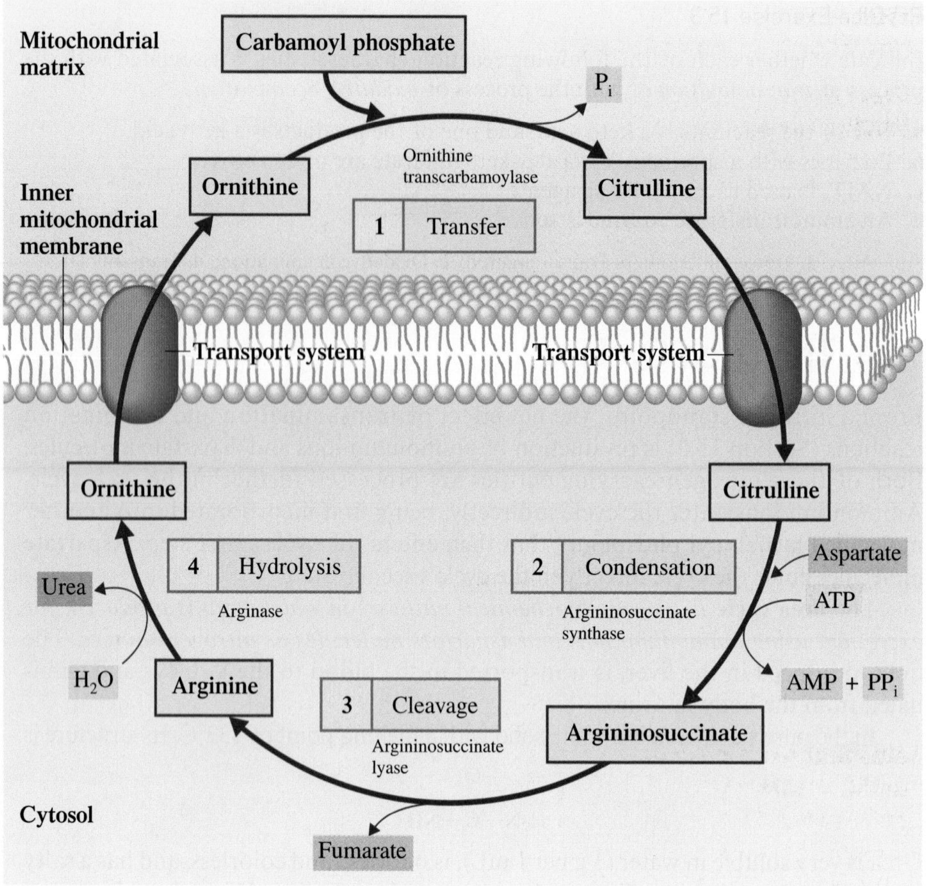

Figure 15.6 The four-step urea cycle in which carbamoyl phosphate is converted to urea.

Note that two ATP molecules are expended in the formation of one carbamoyl phosphate molecule and that carbamoyl phosphate contains a high-energy phosphate bond. The carbamoyl phosphate formation reaction, like the reactions of the citric acid cycle, takes place in the mitochondrial matrix.

Steps of the Urea Cycle

Figure 15.6 shows the four-step urea cycle in outline form. Note that the urea cycle occurs partially in the mitochondria and partially in the cytosol and that ornithine and citrulline must be transported across the inner mitochondrial membrane. The individual steps of the urea cycle will now be considered in detail.

Ornithine has a role similar to that of oxaloacetate in the citric acid cycle (Section 12.6), accepting an entering group at the start of each turn of the cycle.

Step 1: **Carbamoyl group transfer.** Carbamoyl phosphate, the entering "fuel" for the urea cycle, transfers its carbamoyl group to ornithine to form citrulline, with release of P_i in a reaction catalyzed by *ornithine transcarbamoylase.*

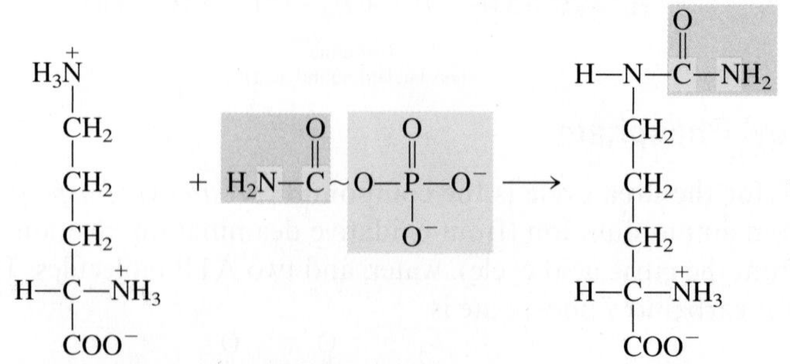

Ornithine   Carbamoyl phosphate   Citrulline

The breaking of the high-energy phosphate bond in carbamoyl phosphate drives the transfer process. With the carbamoyl transfer, the first of the two nitrogen atoms and the carbon atom needed for the formation of urea have been introduced into the cycle.

Step 2: **Citrulline–aspartate condensation.** In contrast to Step 1, which occurred in the mitochondrial matrix, Steps 2, 3, and 4 of the cycle occur in the cytosol. Citrulline, from Step 1, is transported into the cytosol where a condensation reaction between it and aspartate occurs. Aspartate is a second molecule of "fuel" entering the cycle, having been previously generated from glutamate by transamination (Section 15.3). The condensation, which produces argininosuccinate, is catalyzed by *argininosuccinate synthase;* the reaction is driven by the expenditure of ATP.

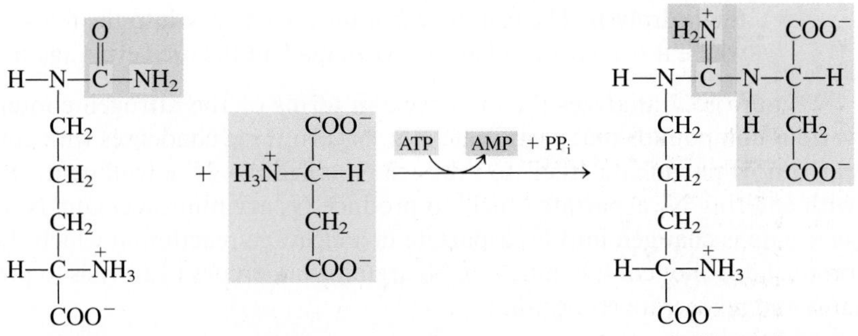

Citrulline   Aspartate   Argininosuccinate

With this reaction, the second of the two nitrogen atoms that will be part of the end-product urea has been introduced into the cycle. One nitrogen atom comes from carbamoyl phosphate, the other from aspartate.

Recall that the original source of both nitrogens is glutamate, the collecting agent for amino acid nitrogen atoms (Section 15.3). The flow of the nitrogen can be shown by these reactions:

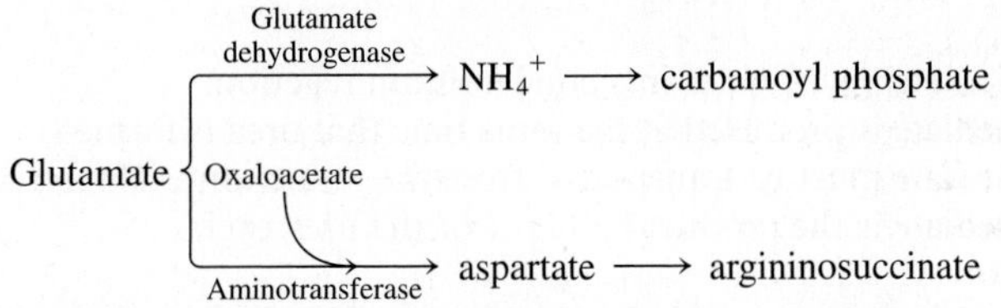

Step 3: **Argininosuccinate cleavage.** The enzyme *argininosuccinate lyase* catalyzes the cleavage of argininosuccinate into arginine, a standard amino acid, and fumarate, a citric acid cycle intermediate. The significance of fumarate production will be considered shortly.

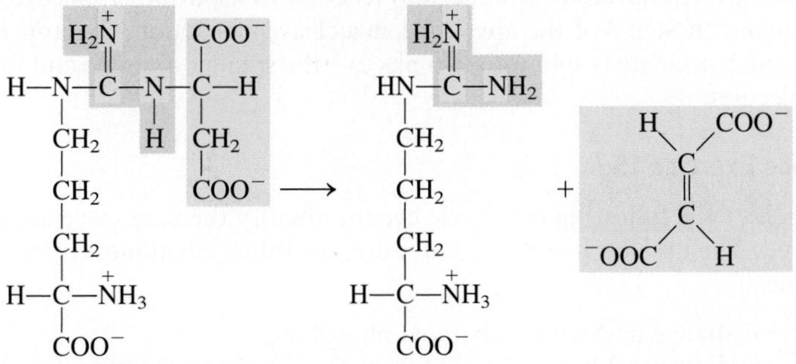

Argininosuccinate   Arginine   Fumarate

Step 4: **Urea from arginine hydrolysis.** Hydrolysis of arginine produces urea and regenerates ornithine, one of the cycle's starting materials. The enzyme involved is *arginase.*

The standard amino acid lysine and the nonstandard amino acid ornithine, both basic amino acids, have closely related structures.

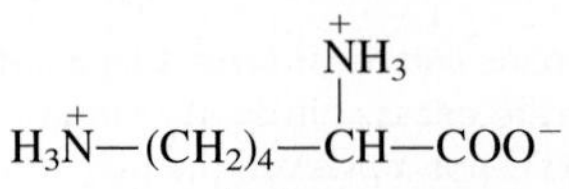

Lysine

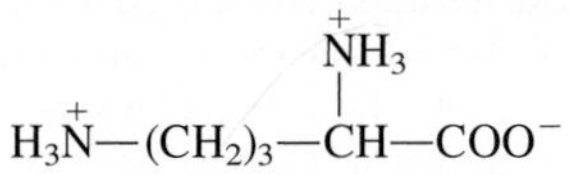

Ornithine

Lysine has one more CH_2 group than does ornithine.

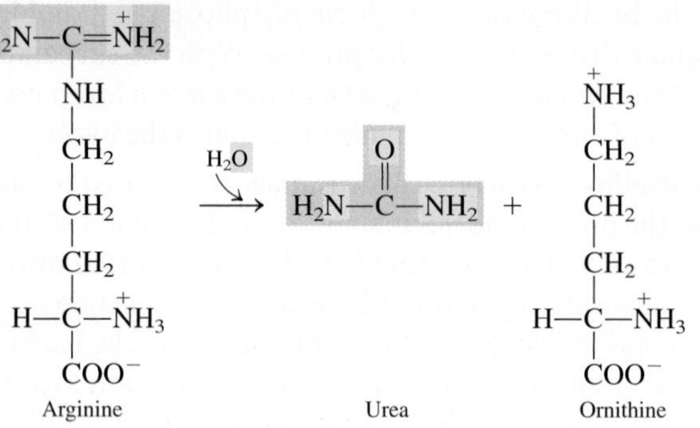

The oxygen atom present in the urea comes from the water involved in the hydrolysis. The ornithine is transported back into the mitochondria, where it becomes available to participate in the urea cycle again.

Humans and most terrestrial animals excrete excess nitrogen as urea. Urea is not, however, the only biochemical means for disposing of excess nitrogen. Aquatic species (bacteria and fish) release ammonia directly into the surrounding water. Birds, terrestrial reptiles, and many insects secrete nitrogen as uric acid; it is the familar white solid in bird droppings. The structure of uric acid, a compound with a purine ring system (Section 6.9), is

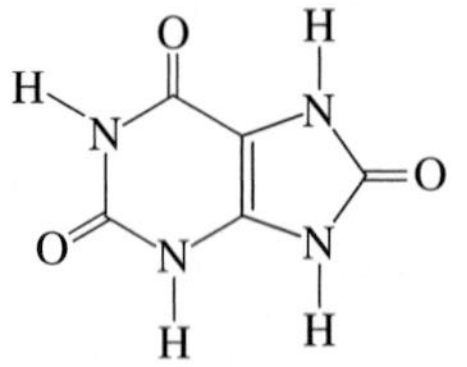

Uric acid

Figure 15.7 analyzes the urea cycle in terms of the nitrogen content of the various compounds that participate in it. N_2 ornithine condenses with entering N_1 carbamoyl phosphate "fuel" to produce N_3 citrulline. N_3 citrulline then interacts with entering N_1 aspartate "fuel" to produce N_4 argininosuccinate. N_4 argininosuccinate is changed into N_4 aspartate in a cleavage reaction in which the second product does not contain nitrogen. N_4 arginine undergoes hydrolysis to produce N_2 urea and regenerate N_2 ornithine.

EXAMPLE 15.4 Relating Urea Cycle Intermediates to Events That Occur in the Urea Cycle

For each of the following urea cycle events, identify the urea cycle intermediate that is involved. The urea cycle intermediates are ornithine, citrulline, argininosuccinate, and arginine.

a. Intermediate participates in a condensation reaction.
b. Intermediate is produced at the same time that urea is formed.
c. Intermediate must be transported from the mitochondrial matrix to the cytosol.
d. Intermediate is the product in Step 3 of the urea cycle.

Solution

a. *Citrulline.* The reaction between citrulline and aspartate in Step 2 of the urea cycle is a condensation reaction.
b. *Ornithine.* In Step 4 of the urea cycle, arginine undergoes hydrolysis to produce ornithine and urea.
c. *Citrulline.* Citrulline is formed in the mitochondrial matrix, transported across the inner mitochondrial membrane, and reacts with aspartate in the cytosol.
d. *Arginine.* In Step 3 of the urea cycle, in a cleavage reaction, the large molecule argininosuccinate is split into two pieces—the smaller arginine and fumarate molecules.

Practice Exercise 15.4

For each of the following urea cycle events, identify the urea cycle intermediate that is involved. The urea cycle intermediates are ornithine, citrulline, argininosuccinate, and arginine.

a. Intermediate reacts with carbamoyl phosphate.
b. Intermediate must be transported from the cytosol to the mitochondrial matrix.
c. Intermediate is a reactant in a hydrolysis reaction.
d. Intermediate is the reactant in Step 2 of the urea cycle.

Answers: **a.** Ornithine; **b.** Ornithine; **c.** Arginine; **d.** Citrulline

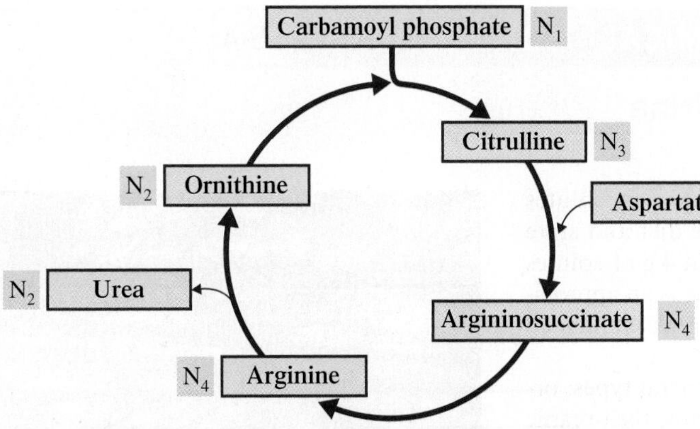

Figure 15.7 The nitrogen content of the various compounds that participate in the urea cycle.

Urea Cycle Net Reaction

The net reaction for urea formation, in which all of the urea cycle intermediates cancel out of the equation, is:

$$NH_4^+ + CO_2 + 3ATP + 2H_2O + \text{aspartate} \longrightarrow$$
$$\text{urea} + 2ADP + AMP + PP_i + 2P_i + \text{fumarate}$$

Carbamoyl phosphate is not shown as a reactant in this net equation; precursor molecules needed for its formation (NH_4^+, CO_2, ATP, H_2O) have been substituted for it.

The equivalent of a total of four ATP molecules is expended in the production of one urea molecule. Two ATP molecules are consumed in the production of carbamoyl phosphate, and the equivalent of two ATP molecules is consumed in Step 2 of the urea cycle, where an ATP is hydrolyzed to AMP and PP_i and the PP_i is then further hydrolyzed to two P_i.

Urea formed via the urea cycle is excreted in urine. Urea is one of several molecules eliminated from the body through urine. The focus on relevancy feature Chemical Connections 15-A on the next page considers some of the additional compounds found in urine.

The Chemistry at a Glance feature on page 629 combines the reactions of Section 15.3 (transamination and oxidative deamination) and those of the urea cycle (Section 15.4) into a single diagram, which serves as a useful summary of important reactions associated with the nitrogen portion of protein metabolism.

Sulfur-containing amino acids (cysteine and methionine) contain both sulfur and nitrogen. The nitrogen-containing group is lost through transamination and oxidative deamination and processed to urea. The sulfur-containing group is processed to sulfur dioxide (SO_2), which is then oxidized to sulfate (SO_4^{2-}). The SO_4^{2-} ion, the negative ion from sulfuric acid, is eliminated in urine.

Linkage Between the Urea and Citric Acid Cycles

The net equation for urea formation shows fumarate, a citric acid cycle intermediate, as a product. This fumarate enters the citric acid cycle, where it is converted to malate and then to oxaloacetate, which can then be converted to aspartate through transamination. The aspartate then re-enters the urea cycle at Step 2 (Figure 15.8).

Besides undergoing transamination, the oxaloacetate produced from fumarate of the urea cycle can be (1) converted to glucose via gluconeogenesis (2) condensed with acetyl CoA to form citrate or (3) converted to pyruvate.

Urea Cycle Intermediates and NO Production

In the mid-1980s, it was discovered that two of the urea cycle intermediates, arginine and citrulline, are involved in an important biochemical reaction that is completely independent of the urea cycle. This newly discovered biochemical role is participation in the production of the biochemical messenger molecule nitric oxide, NO. The focus on relevancy feature Chemical Connections 15-B on page 630 explores this discovery further.

CHEMICAL CONNECTIONS 15-A

The Chemical Composition of Urine

Urine is a dilute aqueous solution containing many solutes whose concentrations are dependent on the diet and state of health of the individual. On average, about 4 g of solutes are present in a 100-g urine sample; thus urine is an approximately 4%-by-mass aqueous solution of materials eliminated from the body.

The solutes present in urine are of two general types: organic compounds and inorganic ions. Generally, the organic compounds are more abundant because of the dominance of urea, as shown in the following composition data.

Major Constituents of Urine (for a 1400-mL specimen obtained over a 24-hour period)

Organic Constituents		Inorganic Constituents	
urea	25.0 g	chloride (Cl^-)	6.3 g
creatinine	1.5 g	sodium (Na^+)	3.0 g
amino acids	0.8 g	potassium (K^+)	1.7 g
uric acid	0.7 g	sulfate (SO_4^{2-})	1.4 g
		dihydrogen phosphate ($H_2PO_4^-$)	1.2 g
		ammonium (NH_4^+)	0.8 g
		calcium (Ca^{2+})	0.2 g
		magnesium (Mg^{2+})	0.2 g

Urea, the solute present in the greatest quantity in urine, is odorless and colorless in solution (Section 15.4). (The pale yellow color of urine is due to small amounts of urobilin and related compounds, as discussed in Section 15.7.) Urea is the principal nitrogen-containing end product of protein metabolism.

The amount of urea in urine is significantly affected by dietary intake. Large high-protein-content meals (see accompanying photo) often provide protein amounts in excess of the body's needs, and the excess protein cannot be stored. Processing of the nitrogen content of the excess protein load increases the urea concentration in urine.

Creatinine (not to be confused with creatine, to be discussed shortly) is the second-most abundant organic product in urea. The structure of this nitrogen-containing compound is:

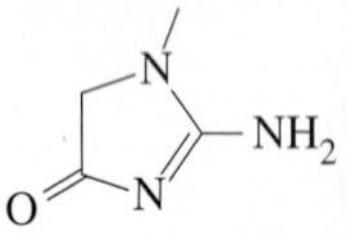

Creatinine and creatine are related molecules. Creatinine is produced from the metabolism (degradation) of creatine. Approximately 2% of the body's creatine is converted to creatinine each day.

Urea concentrations in urine increase after ingestion of large amounts of dietary protein as the nitrogen content of excess protein is metabolized to urea.

Creatine, in the form of creatine phosphate, is a high-energy phosphate (Section 12.5) compound naturally present in skeletal muscle. Its function is to supply phosphate groups to ADP (produced from spent ATP due to muscle activity) to regenerate ATP for further use:

$$\text{Creatine phosphate} + \text{ADP} \rightleftharpoons \text{Creatine} + \text{ATP}$$

The most abundant inorganic constituent of urine is chloride ion. Its primary source is dietary table salt (NaCl). Correspondingly, the second most abundant ion present is sodium ion, the positive ion in table salt. The sulfate ion present in urine comes primarily from the metabolism of sulfur-containing amino acids. Ammonium ions come primarily from the hydrolysis of urea.

Urine is normally slightly acidic, having an average pH value of 6.6. However, the pH range is wide—from 4.5 to 8.0. Fruits and vegetables in the diet tend to raise urine pH, and high-protein foods tend to lower urine pH.

A normal adult excretes 1000–1500 mL of urine daily. Actual urine volume depends on liquid intake and weather. During hot weather, urine volume decreases as a result of increased water loss through perspiration.

15.5 Amino Acid Carbon Skeletons

The removal of the amino group of an amino acid by transamination/oxidative deamination (Section 15.3) produces an α-keto acid that contains the carbon skeleton from the amino acid. Because each of the 20 amino acids has a different carbon skeleton, each amino acid has a different degradation pathway for its carbon

CHEMISTRY AT A GLANCE Metabolic Reactions That Involve Nitrogen-Containing Compounds

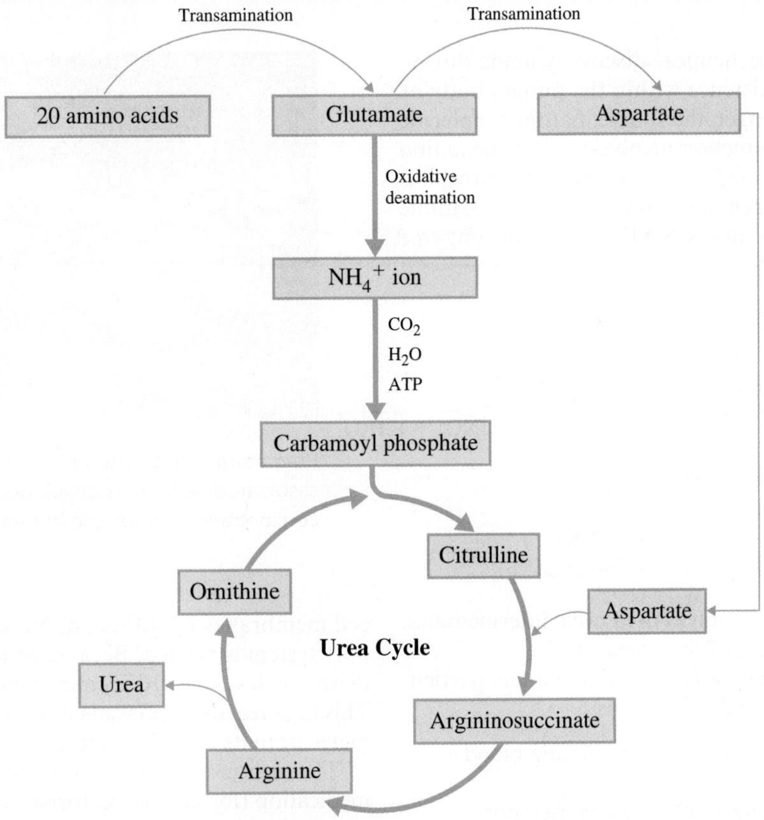

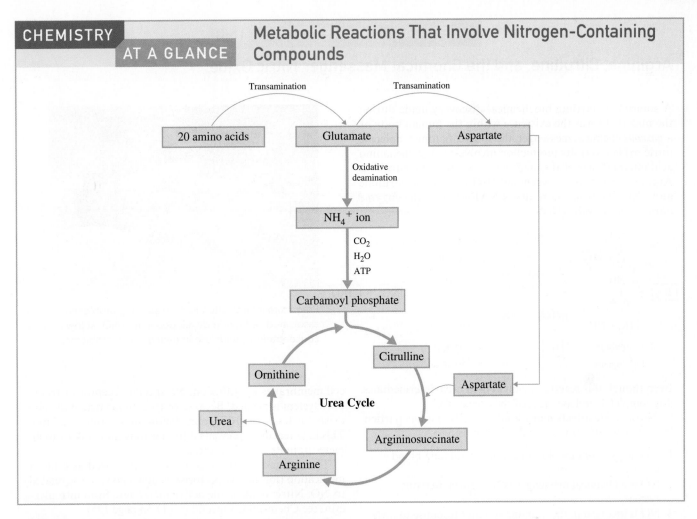

skeleton. For alanine and serine, the degradation requires a single step. For most carbon arrangements, however, multistep sequences are required. The details of the various degradation pathways are not considered in this text. Of importance, however, are the products obtained from the various degradations. The 20 degradation pathways do not produce 20 different products. Rather, the pathways converge

The 20 standard amino acids are degraded by 20 different pathways that converge to produce just 7 products (metabolic products).

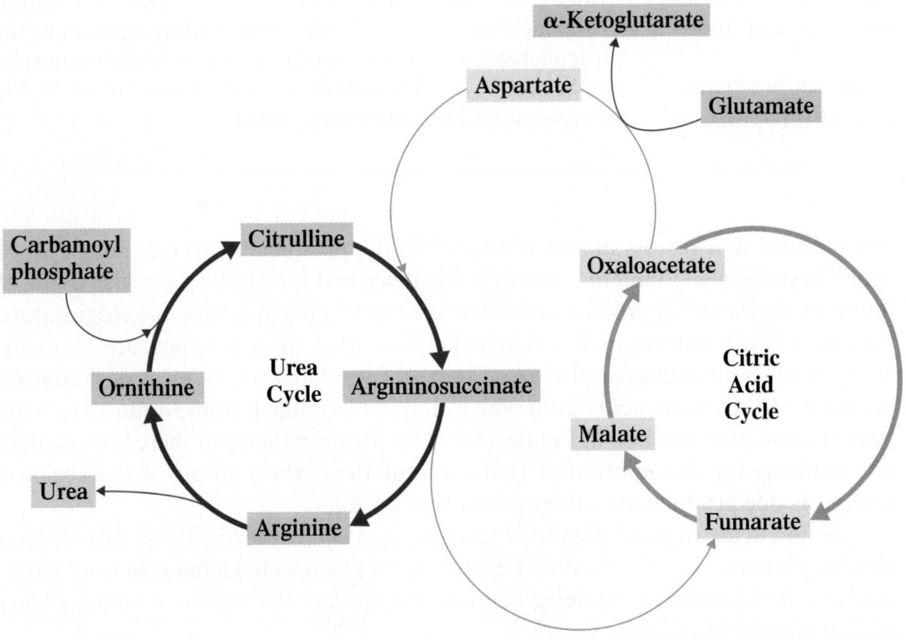

Figure 15.8 Fumarate from the urea cycle enters the citric acid cycle, and aspartate produced from oxaloacetate of the citric acid cycle enters the urea cycle.

CHEMICAL CONNECTIONS 15-B

Arginine, Citrulline, and the Chemical Messenger Nitric Oxide

A somewhat startling biochemical discovery made during the mid-1980s was the existence within the human body of a *gaseous* chemical messenger, the simple diatomic molecule nitric oxide (NO). Its production involves two of the amino acid intermediates of the urea cycle—arginine and citrulline. Arginine reacts with oxygen and H_2O to produce citrulline and NO. The reaction requires NADPH and the enzyme *nitric oxide synthase* (NOS).

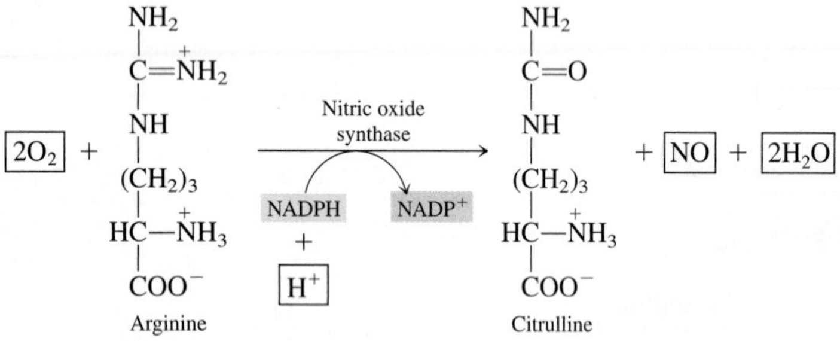

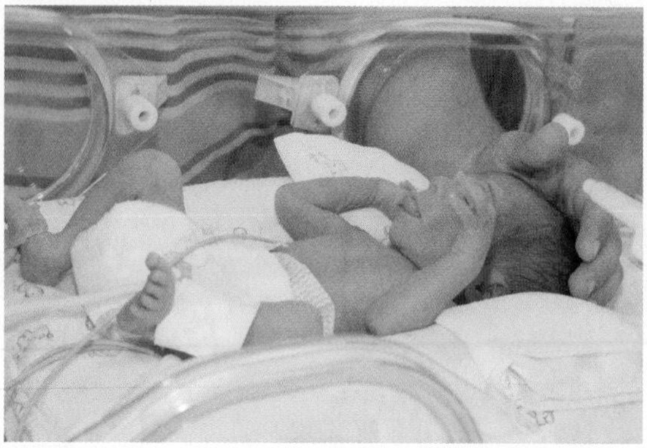

Premature infants often have respiratory problems associated with underdeveloped lungs. NO, at low concentrations, finds use in treating such problems.

Even though this reaction involves urea cycle intermediates, it is completely independent of the urea cycle.

Nitric oxide affects many kinds of cells and has particularly striking effects in the following areas:

1. NO helps maintain blood pressure by dilating blood vessels.
2. NO is a chemical messenger in the central nervous system.
3. NO is involved in the immune system's response to invasion by foreign organisms or materials.
4. NO is found in the brain and may be a major biochemical component of long-term memory.

At low levels NO in NO/O_2 blends finds use in newborn intensive care units in treating pulmonary hypertension (high blood pressure in the lungs) in premature infants. NO action reduces blood pressure by dilating arteries. Its use can save the life of an infant at risk for pulmonary vascular diseases (see accompanying photo).

In humans, nitric oxide is the first known biochemical messenger compound that is a gas. It can easily pass through cell membranes by diffusion. No specific receptor or transport system is needed. Because of its extreme reactivity, NO exists for less than 10 seconds before undergoing reaction. This high reactivity prevents it from getting more than 1 millimeter from its site of synthesis.

The action of nitroglycerin, when it is used as a heart medication (for angina pectoris), is now known to be related to NO. Nitric oxide is the active metabolite from nitroglycerin (see Chemical Connections 5-D on page 195).

Before the discovery of nitric oxide's role as a biochemical messenger, this gas was mainly regarded as a noxious atmospheric gas found in cigarette smoke and smog, as a destroyer of ozone, and as a precursor of acid rain. The contrast between nitric oxide's role in environmental pollution and its function in the human body as a chemical messenger is indeed startling.

Once NO has delivered its message, it must be "deactivated" so as not to interfere with subsequent NO signals or the lack thereof. The body's deactivating agents are the proteins hemoglobin (in blood) and myoglobin (in muscle) (see Section 9.13). The action of these proteins result in NO being converted to nitrate ion (NO_3^-).

in a manner that results in just seven metabolic products (intermediates), all of which have been encountered in previously discussed metabolic pathways. Four of the seven degradation products are citric acid cycle intermediates: α-ketoglutarate, succinyl CoA, fumarate, and oxaloacetate. The other three products are pyruvate, acetyl CoA, and acetoacetyl CoA. Figure 15.9 relates these seven degradation products to the amino acids from which they are obtained. Some amino acids appear in more than one box in Figure 15.9. This means either that there is more than one pathway for degradation or that some of the carbon atoms of the skeleton emerge as one product and others as another product.

Only amino acids that can replenish, either directly or indirectly, oxaloacetate supplies are glucogenic, that is, can be used to produce glucose through gluconeogenesis.

Amino acids that are degraded to citric acid cycle intermediates can serve as glucose precursors and are called glucogenic. A **glucogenic amino acid** *is an amino acid that has a carbon-containing degradation product that can be used to produce glucose via gluconeogenesis.*

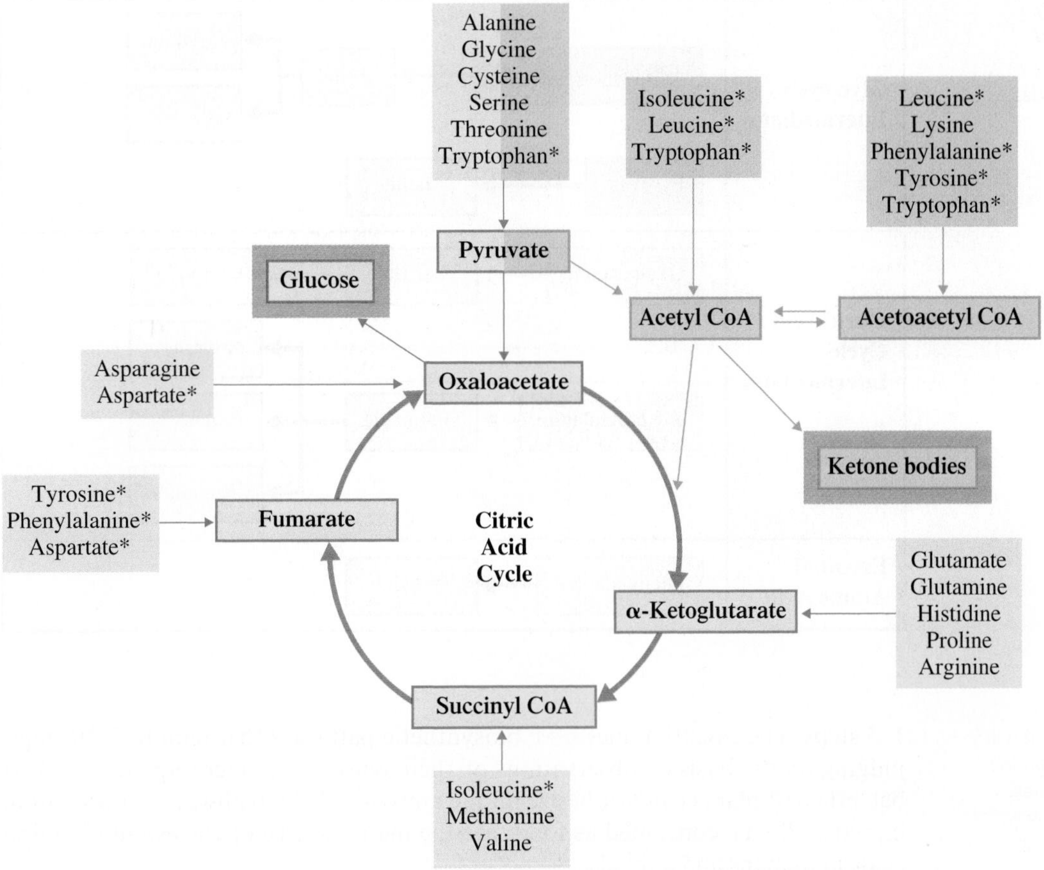

Figure 15.9 Fates of the carbon skeletons of amino acids. Glucogenic amino acids are shaded blue, and ketogenic amino acids are shaded green. Some amino acids (marked with an asterisk) have more than one degradation pathway and are thus present more than once in the diagram.

Amino acids that are degraded to acetyl CoA or acetoacetyl CoA can contribute to the formation of fatty acids or ketone bodies and are called ketogenic. A **ketogenic amino acid** *is an amino acid that has a carbon-containing degradation product that can be used to produce ketone bodies.* Even though acetyl CoA can enter the citric acid cycle, there can be no *net* production of glucose from it. Acetyl groups are C_2 species, and such species only maintain the carbon count in the cycle because two CO_2 molecules exit the cycle (Section 12.6). Thus amino acids that are degraded to acetyl CoA (or acetoacetyl CoA) are not glucogenic.

Amino acids that are degraded to pyruvate can be either glucogenic or ketogenic. Pyruvate can be metabolized to either oxaloacetate (glucogenic) or acetyl CoA (ketogenic).

Only two amino acids are purely ketogenic: leucine and lysine. Nine amino acids are both glucogenic and ketogenic: those degraded to pyruvate (see Figure 15.9), as well as tyrosine, phenylalanine, and isoleucine (which have two degradation products). The remaining nine amino acids are purely glucogenic.

The existence of glucogenicity and ketogenicity for amino acids points out that ATP production (common metabolic pathway) is not the only fate for amino acid degradation products. They can also be converted to glucose, ketone bodies, or fatty acids (via acetyl CoA).

15.6 Amino Acid Biosynthesis

The classification of amino acids as essential or nonessential for humans (Section 9.3) roughly parallels the number of steps in their biosynthetic pathways and the energy required for their synthesis. The nonessential amino acids can be made in

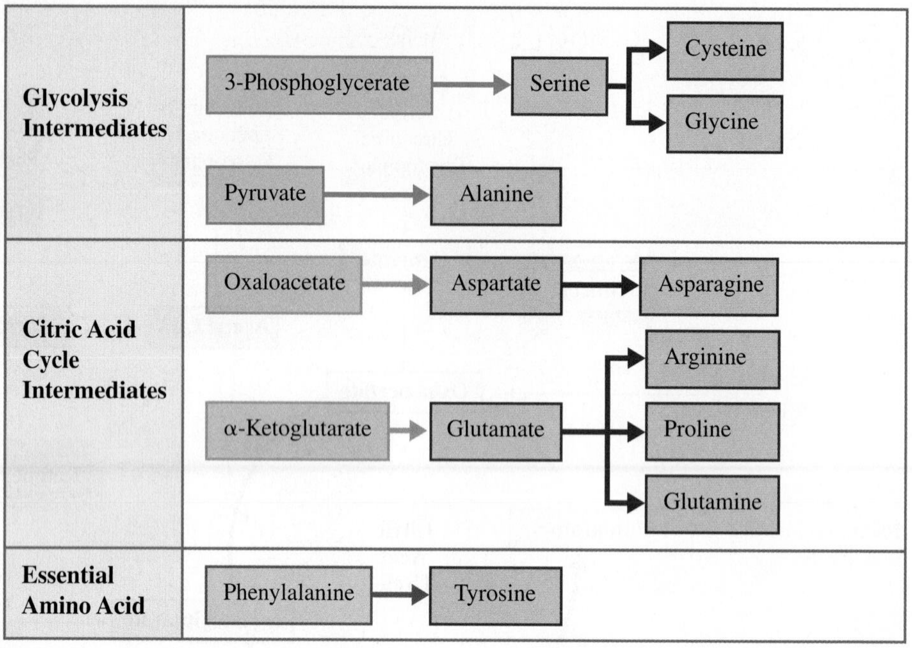

Figure 15.10 A summary of the starting materials for the biosynthesis of the 11 nonessential amino acids.

There is considerable variation in biosynthetic pathways for amino acids among different species. By contrast, the basic pathways of carbohydrate and lipid metabolism are almost universal.

1–3 steps. The essential ones have biosynthetic pathways that require 7–10 steps, judging on the basis of observations of their synthesis in microorganisms. Most bacteria and plants can synthesize all the amino acids by pathways not present in humans. Plants, consumed as food, are the major source of the essential amino acids in humans and animals.

The starting materials for the biosynthesis of the 11 nonessential amino acids are the glycolysis intermediates 3-phosphoglycerate and pyruvate and the citric acid cycle intermediates oxaloacetate and α-ketoglutarate (Figure 15.10).

Three of the nonessential amino acids—alanine, aspartate, and glutamate—are biosynthesized by transamination (Section 15.3) of the appropriate α-keto acid starting material.

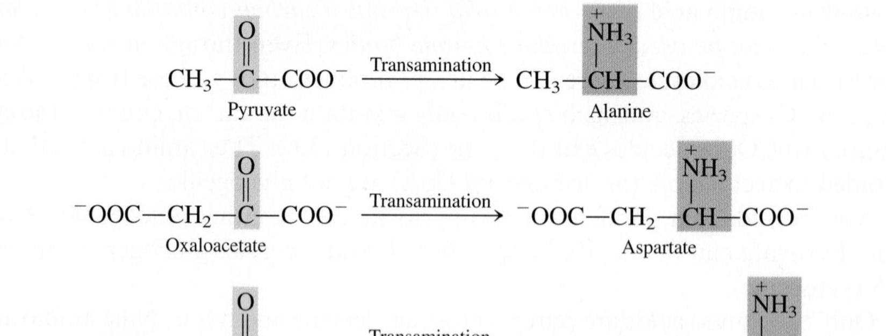

PKU is characterized by elevated blood levels of phenylalanine and phenylpyruvate. The physical consequence of PKU is damage to *developing* brain cells. In children up to six years old, PKU leads to retarded mental development. The major defense against PKU is mandatory screening of newborns to identify the one in every 20,000 who is afflicted and then restricting those children's dietary phenylalanine intake to that needed for protein synthesis until they are six years old. After that age, brain cells are not so susceptible to the toxic effect of phenylpyruvate.

The nonessential amino acid tyrosine is obtained from the essential amino acid phenylalanine in a one-step oxidation that involves molecular O_2, NADPH, and the enzyme *phenylalanine hydroxylase.* Lack of this enzyme causes the metabolic disease phenylketonuria (PKU).

15.7 Hemoglobin Catabolism

Red blood cells are highly specialized cells whose primary function is to deliver oxygen to, and remove carbon dioxide from, body tissues. Mature red blood cells have no nucleus or DNA. Instead, they are filled with the red pigment hemoglobin.

Red blood cell formation occurs in the bone marrow, and about 200 billion new red blood cells are formed daily. The life span of a red blood cell is about four months.

The oxygen-carrying ability of red blood cells is due to the protein hemoglobin present in such cells (Figure 15.11). Hemoglobin is a conjugated protein (Section 9.9); the protein portion is called *globin,* and the prosthetic group (nonprotein portion) is *heme.* Heme contains four pyrrole groups (Section 6.9) joined together with an iron atom in the center.

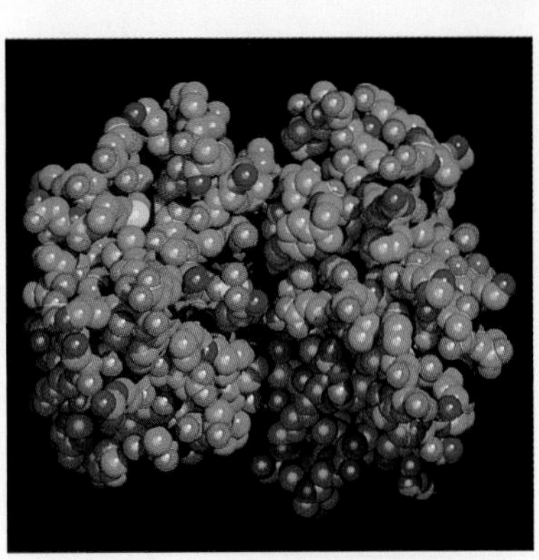

Figure 15.11 A molecular model of the protein hemoglobin.

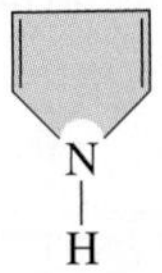

Pyrrole

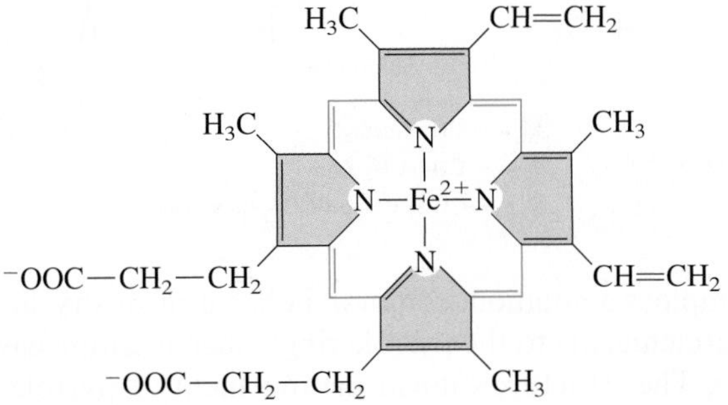

Heme

It is the iron atom in heme that interacts with O_2, forming a reversible complex with it. This complexation increases the amount of O_2 that the blood can carry by a factor of 80 over that which simply "dissolves" in the blood.

Old red blood cells are broken down in the spleen (primary site) and liver (secondary site). Part of this process is degradation of hemoglobin. The globin protein is hydrolyzed to amino acids, which become part of the amino acid pool (Section 15.2). The iron atom of heme becomes part of *ferritin,* an iron-storage protein, which saves the iron for use in the biosynthesis of new hemoglobin molecules. The tetrapyrrole carbon arrangement of heme is degraded to *bile pigments* that are eliminated in feces and to a lesser extent in urine.

The tetrapyrrole heme ring is the only component of hemoglobin that is not reused by the body.

Degradation of heme begins with a ring-opening reaction in which a single carbon atom is removed. The product is called *biliverdin.*

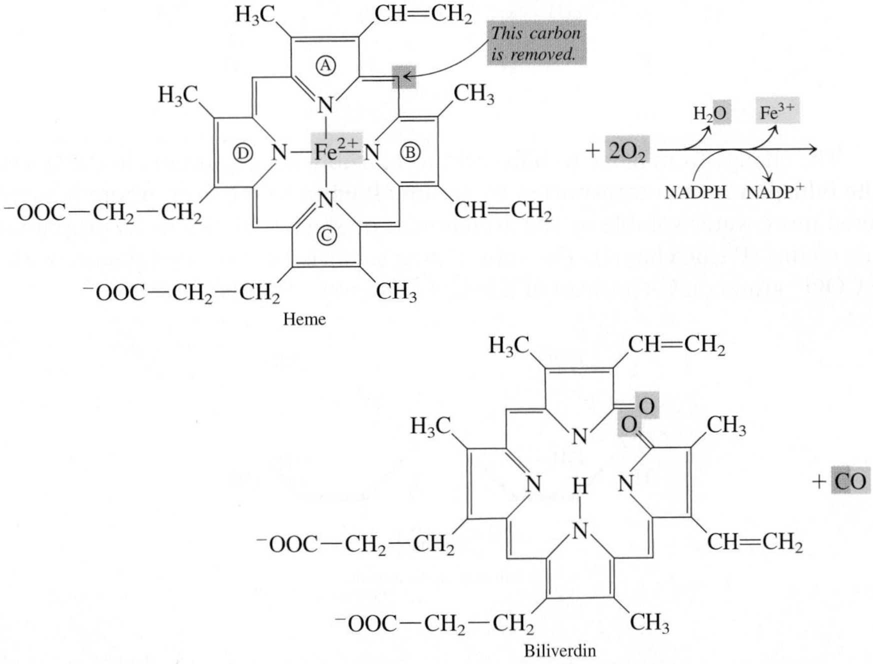

The level of carbon monoxide produced in the first step of hemoglobin degradation is sufficient to complex 1% of the oxygen-binding sites of the blood's hemoglobin.

This reaction has several important characteristics. (1) Molecular oxygen, O_2, is required as a reactant. (2) Ring opening releases the iron atom to be incorporated into ferritin. (3) The product containing the excised carbon atom is *carbon monoxide* (a substance toxic to the human body). The carbon monoxide so produced reacts with functioning hemoglobin, forming a CO–hemoglobin complex; this decreases the oxygen-carrying ability of the blood. CO–hemoglobin complexes are very stable; CO release to the lungs is a slow process.

An alternative rendering of the structure of biliverdin is:

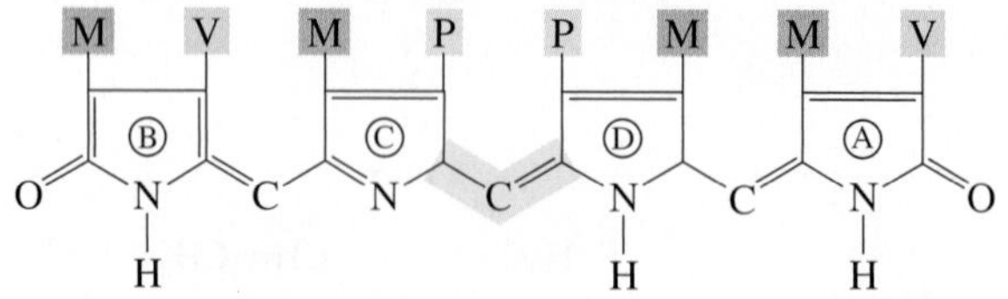

Biliverdin

M = $-CH_3$ (methyl)
V = $-CH{=}CH_2$ (vinyl)
P = $-CH_2-CH_2-COO^-$ (propionate)

This structure employs a notation, common in heme chemistry, in which letters are used to denote attachments to the pyrrole rings; such notation easily distinguishes the attachments. The structure's linear arrangement of pyrrole rings also saves space compared to the heme-like representation of the rings. However, the linear structure incorrectly implies that the arrangement of the pyrrole rings that results from the ring opening is linear (straight-line); rather, the pyrrole rings actually have a hemi-like arrangement.

In the second step of heme degradation, biliverdin is converted to bilirubin. This change involves reduction of the central methylene bridge of biliverdin.

In 2002, it was discovered that bilirubin has antioxidant properties. It protects against peroxyl radicals (Section 12.12) by being oxidized back to biliverdin. Its antioxidant properties are significantly better than those of glutathione (Section 9.8), the molecule believed for 80 years to be the most important cellular antioxidant.

Bilirubin is found only in low concentrations in cells but in higher concentrations in blood. This new research suggests that bilirubin is probably the major antioxidant protector for cell membranes, while glutathione protects components inside cells.

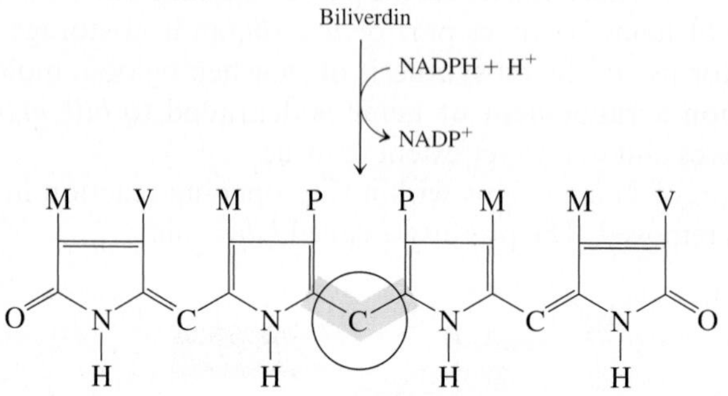

Bilirubin

The change from heme to biliverdin to bilirubin usually occurs in the spleen. The bilirubin is then transported by serum albumin to the liver, where it is rendered more water-soluble by the attachment of sugar residues to its propionate side chains (P side chains). The solubilizing sugar is *glucuronate* (glucose with a $-COO^-$ group on C-6 instead of a $-CH_2OH$ group; Section 7.12).

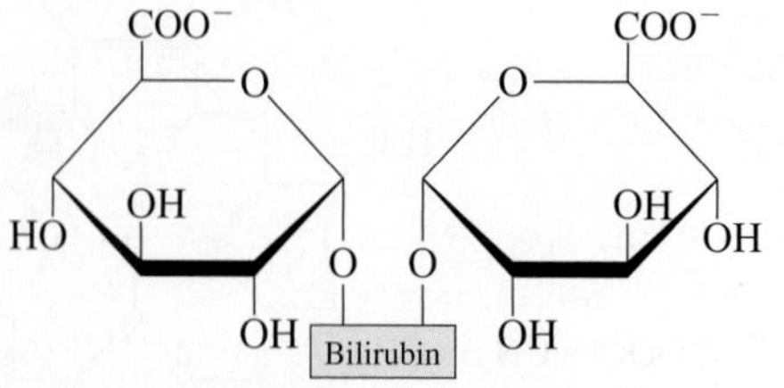

Bilirubin diglucuronide

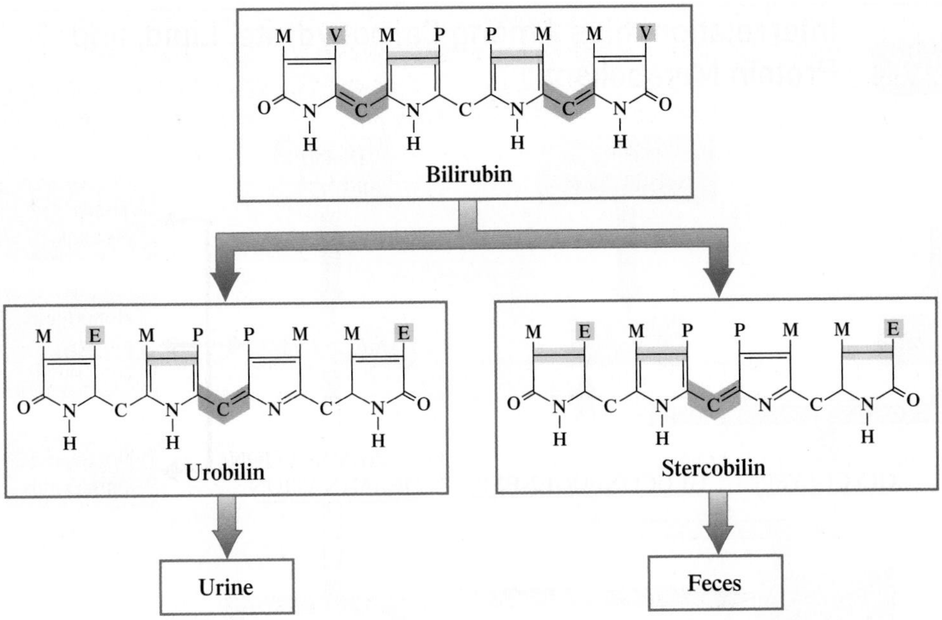

Figure 15.12 Stercobilin and urobilin have structures closely resembling that of bilirubin. Changes include reduction of vinyl (V) groups to ethyl (E) groups and reduction of —CH_2— bridges.

The solubilized bilirubin is excreted from the liver in bile, which flows into the small intestine. Here the bilirubin diglucuronide is changed, in a multistep process, to either stercobilin for excretion in feces or urobilin for excretion in urine. Both stercobilin and urobilin still have tetrapyrrole structures (Figure 15.12). Intestinal bacteria are primarily responsible for the changes that produce stercobilin and urobilin.

The first part of the names *biliverdin* and *bilirubin* and the last part of the names *stercobilin* and *urobilin* all come from the Latin *bilis,* which means "bile." As for the other parts of the names:

1. Latin *virdis* means "green"; biliverdin = "green bile."
2. Latin *rubin* means "red"; bilirubin = "red bile."
3. Latin *urina* means "urine"; urobilin = "urine bile."
4. Latin *sterco* means "dung"; stercobilin = "dung bile."

Bile Pigments

The tetrapyrrole degradation products obtained from heme are known as bile *pigments* because they are secreted with the bile (Section 8.11), and most of them are highly colored. A **bile pigment** *is a colored tetrapyrrole degradation product present in bile.* Biliverdin and bilirubin are, respectively, green and reddish-orange in color. Stercobilin has a brownish hue and is the compound that gives feces their characteristic color. Urobilin is the pigment that gives urine its characteristic yellow color. Normally, the body excretes 1–2 mg of bile pigments in urine daily and 250–350 mg of bile pigments in feces daily.

The word *jaundice* comes from the French *jaune,* which means "yellow."

When the body is functioning properly, the degradation of heme in the spleen to bilirubin and the removal of bilirubin from the blood by the liver balance each other.

Jaundice is the condition that occurs when this balance is upset such that bilirubin concentrations in the blood become higher than normal. The skin and the white of the eyes acquire a yellowish tint because of the excess bilirubin in the blood. Jaundice can occur as a result of liver diseases, such as infectious hepatitis and cirrhosis, that decrease the liver's ability to process bilirubin; from spleen malfunction, in which heme is degraded more rapidly than it can be absorbed by the liver; and from gallbladder malfunction, usually from an obstruction of the bile duct.

A mild form of jaundice is common among premature infants because of underdeveloped liver function. Treatment involves the use of white or ultraviolet light, which breaks the bilirubin down to simpler compounds that are more easily excreted.

The local coloration associated with a deep bruise is also related to the pigmentation associated with heme, biliverdin, and bilirubin. The changing color of the bruise as it heals reflects the dominant degradation product present as the tissue repairs itself.

CHEMISTRY AT A GLANCE

Interrelationships Among Carbohydrate, Lipid, and Protein Metabolism

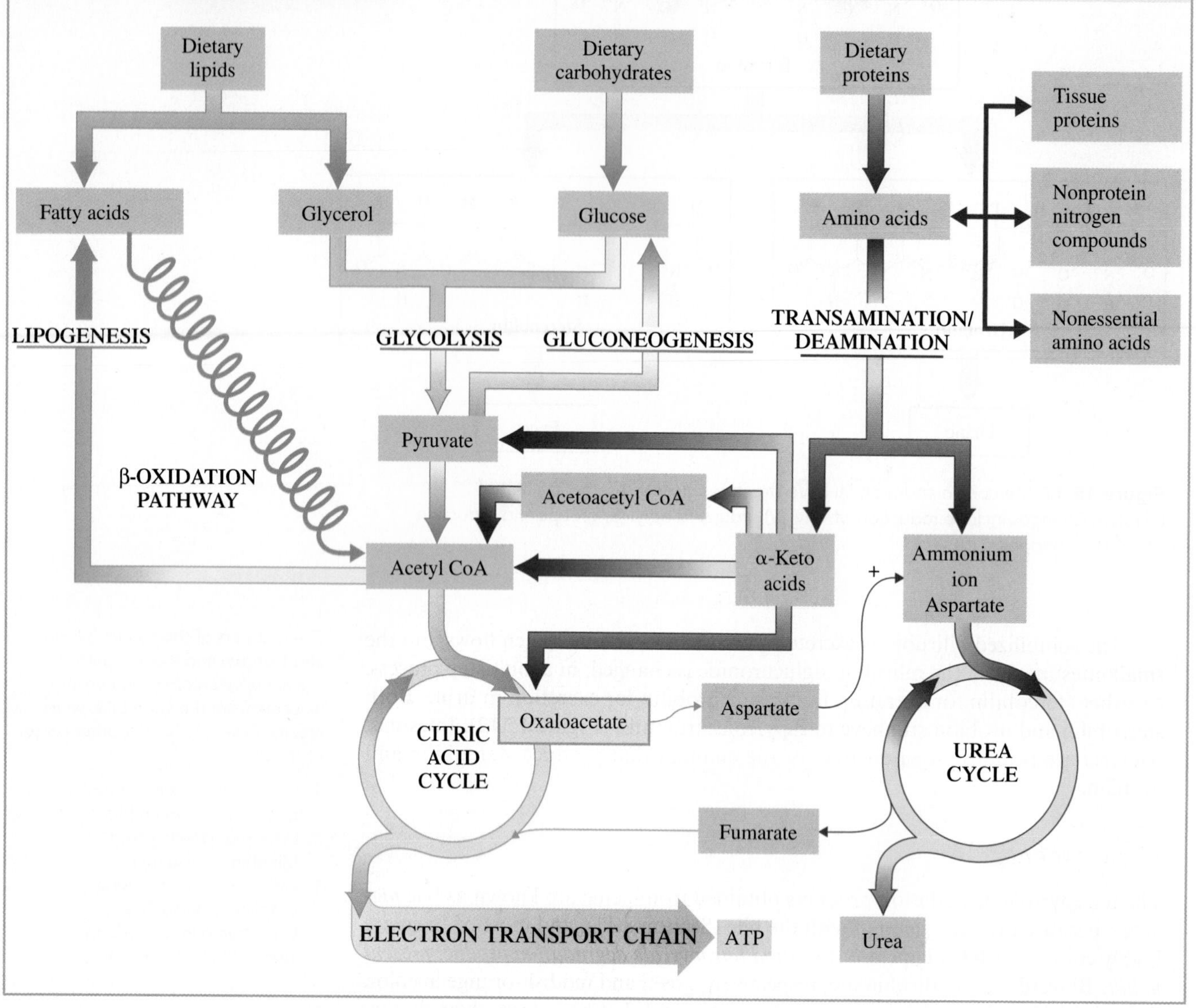

15.8 Interrelationships Among Metabolic Pathways

In this chapter and the previous two chapters, metabolic pathways of carbohydrates, lipids, and proteins have been considered. These pathways are not independent of each other but, rather, are integrally linked, as shown in the Chemistry at a Glance feature above. The numerous connections among pathways mean that a change in one pathway can affect many other pathways.

A good illustration of the interrelationships among pathways emerges from comparing the processes of eating (feasting), not eating for a short period (fasting), and not eating for a prolonged period (starvation). Figure 15.13 shows how the body responds to each of these situations.

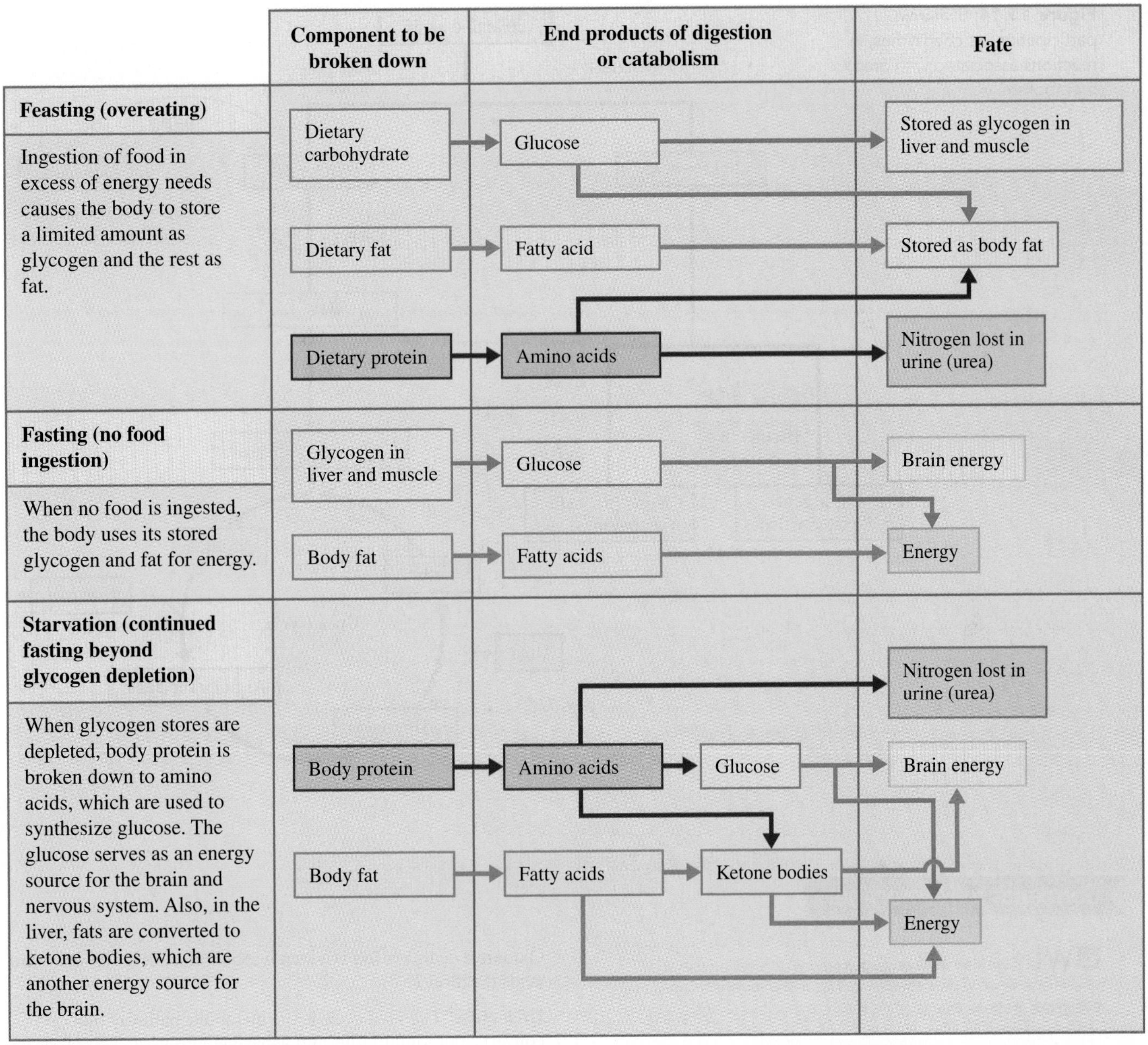

Figure 15.13 The human body's response to feasting, to fasting, and to starvation.

15.9 B Vitamins and Protein Metabolism

The final section in each of the last three chapters contains a summary diagram showing how B vitamins, as cofactors, participate in the metabolic reactions discussed in the chapter. That pattern continues in this chapter.

Transamination reactions are dependent on the cofactor PLP, which involves vitamin B_6. Oxidation deamination requires use of NAD^+, which involves niacin, as an oxidizing agent.

The details for the degradation of the 20 amino acid carbon skeletons were not included in the text; the final degradation products were, however, given. All eight B vitamins—including vitamin B_{12}, the least used B vitamin in terms of cofactor function—are needed as cofactors at least once in obtaining these degradation products. Vitamin B_{12} is needed in the formation of the degradation product succinyl CoA.

Figure 15.14 summarizes B vitamin requirements associated with the aspects of protein metabolism discussed in this chapter.

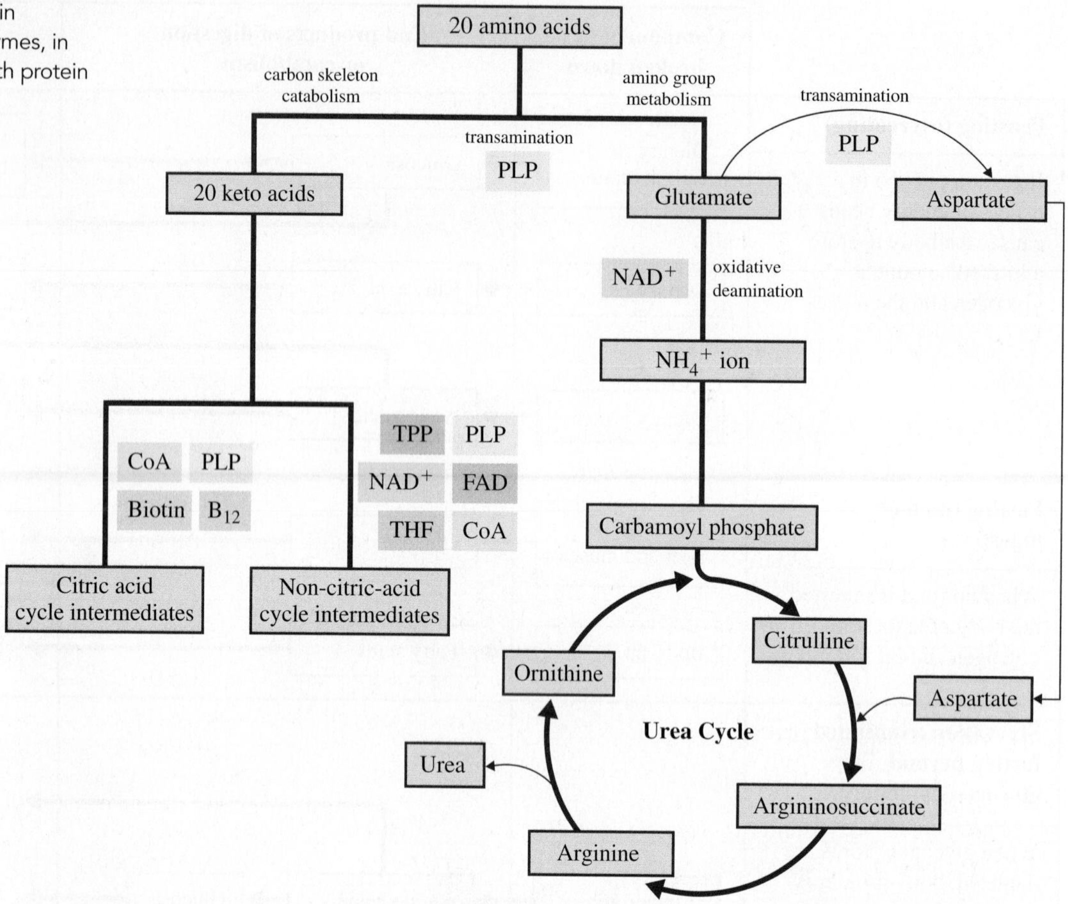

Figure 15.14 B vitamin participation, as coenzymes, in reactions associated with protein metabolism.

Concepts to Remember

OWL Sign in at **www.cengage.com/owl** to view tutorials and simulations, develop problem-solving skills, and complete online homework assigned by your professor.

Protein digestion and absorption. Digestion of proteins involves the hydrolysis of the peptide bonds that link amino acids to each other. This process begins in the stomach and is completed in the small intestine. The amino acids released by digestion are absorbed through the intestinal wall into the bloodstream (Section 15.1).

Amino acid pool. The amino acid pool within cells consists of varying amounts of each of the 20 standard amino acids found in proteins (Section 15.2).

Amino acid utilization. Amino acids from the amino acid pool are used for protein synthesis, synthesis of nonprotein nitrogen compounds, synthesis of nonessential amino acids, and energy production (Section 15.2).

Transamination. A transamination reaction is an enzyme-catalyzed transfer of an amino group from an α-amino acid to an α-keto acid. Transamination is a step in obtaining energy from amino acids (Section 15.3).

Oxidative deamination. An oxidative deamination reaction is a reaction in which an α-amino acid is converted into an α-keto acid, accompanied by the release of a free ammonium ion. Oxidative deamination is a step in obtaining energy from amino acids (Section 15.3).

Urea cycle. The urea cycle is the metabolic pathway that converts ammonium ions and aspartate into urea. This cycle processes the ammonium ions in the form of carbamoyl phosphate, a compound formed from CO_2, NH_4^+, ATP, and H_2O (Section 15.4).

Amino acid carbon skeletons. Amino acid carbon skeletons (keto acids) are classified as glucogenic or ketogenic on the basis of their catabolic pathways. Glucogenic amino acids are degraded to intermediates of the citric acid cycle and can be used for glucose synthesis. Ketogenic amino acids are degraded into acetoacetyl CoA or acetyl CoA and can be used to make ketone bodies (Section 15.5).

Amino acid biosynthesis. Amino acid biosynthesis is the process in which the body synthesizes amino acids from intermediates of the glycolysis pathway and the citric acid cycle. Eleven amino acids can be synthesized by the body. The other nine amino acids, called essential amino acids, must be obtained from the diet (Section 15.6).

Hemoglobin catabolism. Hemoglobin from red blood cells undergoes a stepwise degradation to biliverdin, to bilirubin, and then to bile pigments that are excreted from the body (Section 15.7).

Exercises and Problems

OWL Interactive versions of these problems may be assigned in OWL.

Exercises and problems are arranged in matched pairs with the two members of a pair addressing the same concept(s). The answer to the odd-numbered member of a pair is given at the back of the book. Problems denoted with a ▲ involve concepts found not only in the section under consideration but also concepts found in one or more earlier sections of the chapter. Problems denoted with a ● cover concepts found in a Chemical Connections feature box.

Protein Digestion and Absorption (Section 15.1)

15.1 The first step in protein digestion is denaturation. Where does denaturation occur in the body, and what is the denaturant?

15.2 What is the first digestive enzyme that protein encounters, and where does this encounter take place?

15.3 What is the relationship between pepsinogen and pepsin?

15.4 What is the relationship between trypsinogen and trypsin?

15.5 Contrast gastric juice and pancreatic juice in terms of pH.

15.6 Contrast gastric juice and pancreatic juice in terms of enzymes present.

15.7 Absorption of amino acids through the intestinal wall requires a transport system. Explain.

15.8 The passage of small polypeptides through the intestinal wall is particularly important in infants. Explain.

15.9 In the context of protein digestion, indicate whether each of the following substances is associated with the *stomach* or the *small intestine*.

a. Gastrin b. Trypsin
c. HCl d. Carboxypeptidase

15.10 In the context of protein digestion, indicate whether each of the following substances is associated with the *stomach* or the *small intestine*.

a. Aminopeptidase b. Pepsin
c. HCO_3^- d. Secretin

15.11 Indicate whether each of the substances in Problem 15.9 is (1) a hormone (2) a digestive enzyme or (3) neither a hormone nor a digestive enzyme.

15.12 Indicate whether each of the substances in Problem 15.10 is (1) a hormone (2) a digestive enzyme or (3) neither a hormone nor a digestive enzyme.

Amino Acid Utilization (Section 15.2)

15.13 What is the amino acid pool?

15.14 What are the three major sources of amino acids for the amino acid pool?

15.15 What is protein turnover?

15.16 The protein turnover rate is not the same for all proteins. Explain.

15.17 What is the difference between a positive nitrogen balance and a negative nitrogen balance?

15.18 What happens to the nitrogen balance during a period of fasting?

15.19 What happens to the nitrogen balance when the diet is lacking in one of the essential amino acids?

15.20 What happens to the nitrogen balance of a pregnant woman?

15.21 What four types of processes draw amino acids out of the amino acid pool?

15.22 What percent of amino acid utilization from the amino acid pool is for protein synthesis?

15.23 Classify each of the following amino acids as essential or nonessential.

a. Lysine b. Arginine
c. Serine d. Tryptophan

15.24 Classify each of the following amino acids as essential or nonessential.

a. Proline b. Asparagine
c. Glutamic acid d. Tyrosine

15.25 Which of the amino acids in Problem 15.23 can be biosynthesized in the human body?

15.26 Which of the amino acids in Problem 15.24 can be biosynthesized in the human body?

Transamination and Oxidative Deamination (Section 15.3)

15.27 Indicate whether each of the following structural characterizations applies to *oxaloacetate, aspartate, α-ketoglutarate*, or *glutamate*. More than one response may be correct in a given situation, or no correct response is possible.

a. Derivative of succinate
b. A keto acid
c. A C_5 molecule
d. Contains two carboxyl groups

15.28 Indicate whether each of the following structural characterizations applies to *oxaloacetate, aspartate, α-ketoglutarate*, or *glutamate*. More than one response may be correct in a given situation, or no correct response is possible.

a. A C_4 molecule
b. Derivative of glutarate
c. An amino acid
d. Contains two amino groups

15.29 In general terms, what are the two reactants in a transamination reaction?

15.30 In general terms, what are the two products in a transamination reaction?

15.31 Write structural equations for the transamination reactions that involve the following pairs of reactants. Use Table 9.1 as the source for any needed amino acid structural information.

a. Serine and oxaloacetate
b. Alanine and oxaloacetate
c. Glycine and α-ketoglutarate
d. Threonine and α-ketoglutarate

15.32 Write structural equations for the transamination reactions that involve the following pairs of reactants. Use Table 9.1 as the source for any needed amino acid structural information.

a. Threonine and oxaloacetate
b. Glycine and oxaloacetate
c. Alanine and α-ketoglutarate
d. Serine and α-ketoglutarate

15.33 What are the names of the enzymes needed to effect the reactions in Problem 15.31?

15.34 What are the names of the enzymes needed to effect the reactions in Problem 15.32?

15.35 What are the two α-keto acids that are most often reactants in transamination reactions?

15.36 What are the two α-keto acids that are most often substrates for aminotransferases?

15.37 Via transamination reactions, the amino groups from a variety of amino acids are collected into a single molecule, the amino acid glutamate. What is the underlying cause for this situation?

15.38 What are the two major pathways for further processing of the glutamate molecules produced via transamination?

15.39 What is the function of pyridoxal phosphate in transamination processes?

15.40 Which one of the B vitamins is important in the process of transamination?

15.41 Describe the process of oxidative deamination.

15.42 What coenzyme is required for an oxidative deamination reaction?

15.43 How does oxidative deamination differ from transamination?

15.44 What do the processes of oxidative deamination and transamination have in common?

15.45 Draw the structure of the α-keto acid produced from the oxidative deamination of each of the following amino acids.

a. Glutamate b. Cysteine
c. Alanine d. Phenylalanine

15.46 Draw the structure of the α-keto acid produced from the oxidative deamination of each of the following amino acids.

a. Glycine b. Leucine
c. Aspartate d. Tyrosine

15.47 The following α-keto acid can be used as a substitute for a particular *essential* amino acid in the diet. Explain how this is possible, and draw the structure of the essential amino acid.

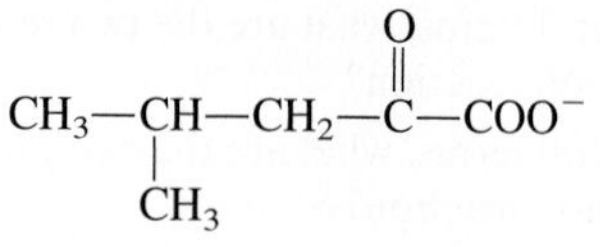

15.48 The following α-keto acid can be used as a substitute for a particular *essential* amino acid in the diet. Explain how this is possible, and draw the structure of the essential amino acid.

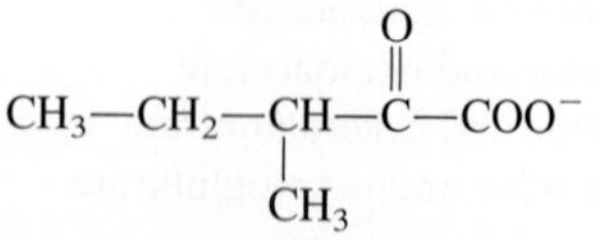

15.49 Give the *name* of the compound produced from each reactant or the reactant needed to produce each product using transamination.

a. Oxaloacetate ⟶ ? b. ? ⟶ α-ketoglutarate
c. Alanine ⟶ ? d. ? ⟶ glutamate

15.50 Give the *name* of the compound produced from each reactant or the reactant needed to produce each product using transamination.

a. Pyruvate ⟶ ? b. ? ⟶ oxaloacetate
c. Aspartate ⟶ ? d. ? ⟶ alanine

15.51 Characterize each of the following as a possible reactant, product, or enzyme involved in (1) transamination (2) oxidative deamination or (3) both transamination and oxidative deamination.

a. α-Ketoglutarate b. Glutamate
c. Glutamate dehydrogenase d. NH_4^+

15.52 Characterize each of the following as a possible reactant, product, or enzyme involved in (1) transamination (2) oxidative deamination or (3) both transamination and oxidative deamination.

a. Oxaloacetate b. Aspartate
c. Glutamate aminotransferase d. H_2O

The Urea Cycle (Section 15.4)

15.53 Draw the chemical structure of urea.

15.54 What are some of the physical characteristics of urea?

15.55 In what chemical form do ammonium ions enter the urea cycle?

15.56 What are the chemical reactants for the formation of carbamoyl phosphate?

15.57 What is a carbamoyl group?

15.58 Draw the structure of the molecule carbamoyl phosphate.

15.59 What are the names of the two *standard* amino acids that participate in the urea cycle?

15.60 What are the names of the two *nonstandard* amino acids that participate in the urea cycle?

15.61 Name the compound that enters the urea cycle by combining with

a. ornithine. b. citrulline.

15.62 Identify the first reaction of the urea cycle that occurs in the

a. mitochondrial matrix. b. cytosol.

15.63 What are the names of the two fuels for the urea cycle, and in which step in the cycle is each of the fuels encountered?

15.64 If the urea cycle were named in the same way as the citric acid cycle, what would the cycle's name be?

15.65 Characterize each of the following "urea cycle compounds" in terms of its nitrogen content (N_1, N_2, N_3, or N_4).

a. Ornithine b. Citrulline
c. Aspartate d. Argininosuccinate

15.66 Characterize each of the following "urea cycle compounds" in terms of its nitrogen content (N_1, N_2, N_3, or N_4).

a. Carbamoyl phosphate b. Ammonium ion
c. Aspartate d. Urea

15.67 In each of the following pairs of compounds associated with the urea cycle, specify which one is encountered first in the cycle.

a. Citrulline and arginine
b. Ornithine and aspartate
c. Argininosuccinate and fumarate
d. Carbamoyl phosphate and citrulline

15.68 In each of the following pairs of compounds associated with the urea cycle, specify which one is encountered first in the cycle.

a. Carbamoyl phosphate and fumarate
b. Argininosuccinate and arginine

c. Ornithine and aspartate
d. Citrulline and ATP

15.69 In which step of the urea cycle does each of the following events occur?
a. Aspartate enters the cycle.
b. A condensation reaction occurs.
c. Ornithine is a product.
d. The reaction $N_4 \longrightarrow N_2 + N_2$ occurs.

15.70 In which step of the urea cycle does each of the following events occur?
a. Citrulline is a product.
b. Carbamoyl phosphate enters the cycle.
c. A hydrolysis reaction occurs.
d. The reaction $N_3 + N_1 \longrightarrow N_4$ occurs.

15.71 How much energy, in terms of ATP, is expended in the synthesis of a molecule of urea?

15.72 What are the sources of the carbon atom and the two nitrogen atoms in urea?

15.73 What is the fate of the fumarate formed in the urea cycle?

15.74 Explain how the urea cycle is linked to the citric acid cycle.

▲15.75 Indicate whether each of the following compounds is associated with (1) transamination (2) oxidative deamination or (3) the urea cycle. More than one response may be correct in a given situation.
a. Oxaloacetate b. Arginine c. H_2O d. ATP

▲**15.76** Indicate whether each of the following compounds is directly associated with (1) transamination (2) oxidative deamination or (3) the urea cycle. More than one response may be correct in a given situation.
a. Ornithine b. NH_4^+ c. NAO^+ d. Aspartate

▲15.77 Classify each of the following nitrogen-containing entities as an N_1, N_2, N_3, or N_4 species.
a. Carbamoyl phosphate b. Glutamate
c. Urea d. Citrulline

▲**15.78** Classify each of the following nitrogen-containing entities as an N_1, N_2, N_3, or N_4 species.
a. Ornithine b. Ammonium ion
c. Aspartate d. Arginine

•15.79 (Chemical Connections 15-A) Indicate whether each of the following statements concerning the chemical composition of urine is true or false.
a. The total mass of organic solutes in urine exceeds that of inorganic solutes.
b. The two most abundant solutes in urine, in terms of mass, are sodium and chloride ions.
c. Urea is the substance that gives urine its pale yellow color.
d. The function of creatine in the body is as a source of phosphate for ADP molecules.

•**15.80** (Chemical Connections 15-A) Indicate whether each of the following statements concerning the chemical composition of urine is true or false.
a. The concentration of total solutes in urine is approximately 8% by mass.
b. The two most abundant solutes in urine, in terms of mass, are urea and creatinine.
c. Creatinine's chemical structure closely resembles that of urea.
d. Sulfate ion present in urine comes from the metabolism of sulfur-containing amino acids.

•15.81 (Chemical Connections 15-B) Indicate whether each of the following statements relating to the chemical messenger NO is true or false.
a. NO is a gaseous chemical messenger that is a urea cycle intermediate.
b. The amino acid arginine is a coproduct in the biochemical production of NO.
c. NO helps regulate blood pressure by dilating blood vessels.
d. NO is a very stable molecule that exerts its effect in the body over an extended period of time.

•**15.82** (Chemical Connections 15-B) Indicate whether each of the following statements relating to the chemical messenger NO is true or false.
a. The N-containing source from which NO is produced is the amino acid citrulline.
b. Both O_2 and H_2O are needed as reactants for the biosynthesis of NO.
c. NO participates in the body's immune response to foreign organisms.
d. Deactivation of NO molecules occurs when they interact with either hemoglobin or myoglobin.

Amino Acid Carbon Skeletons (Section 15.5)

15.83 What are the four possible degradation products of the carbon skeletons of amino acids that are citric acid cycle intermediates?

15.84 What are the three possible degradation products of the carbon skeletons of amino acids that are not citric acid cycle intermediates?

15.85 With the help of Figure 15.9, write the name of the compound (or compounds) to which each of the following amino acid carbon skeletons is metabolized.
a. Leucine b. Isoleucine
c. Aspartate d. Arginine

15.86 With the help of Figure 15.9, write the name of the compound (or compounds) to which each of the following amino acid carbon skeletons is metabolized.
a. Serine b. Tyrosine
c. Tryptophan d. Histidine

15.87 With the help of Figure 15.9, classify each of the amino acids in Problem 15.85 as (1) ketogenic (2) glucogenic or (3) both ketogenic and glucogenic.

15.88 With the help of Figure 15.9, classify each of the amino acids in Problem 15.86 as (1) ketogenic (2) glucogenic or (3) both ketogenic and glucogenic.

15.89 What degradation characteristics do all purely glucogenic amino acids share?

15.90 What degradation characteristics do all purely ketogenic amino acids share?

Amino Acid Biosynthesis (Section 15.6)

15.91 What compound is a major source of amino groups in amino acid biosynthesis?

15.92 How does transamination play a role in both catabolism and anabolism of amino acids?

15.93 What are the five starting materials for the biosynthesis of the 11 nonessential amino acids?

15.94 What is a major difference between the biosynthetic pathways for the essential and the nonessential amino acids?

Hemoglobin Catabolism (Section 15.7)

15.95 What happens to the globin produced from the breakdown of hemoglobin?

15.96 What happens to the iron (Fe^{2+}) produced from the breakdown of hemoglobin?

15.97 What are the structural differences between heme and biliverdin?

15.98 What are the structural differences between biliverdin and bilirubin?

15.99 Arrange the following substances in the order in which they are encountered during the catabolism of heme: bilirubin, urobilin, biliverdin, and bilirubin diglucuronide.

15.100 Carbon monoxide is a byproduct of the degradation of heme. At what point in the degradation process is it formed, and what happens to it once it is formed?

15.101 Which bile pigment is responsible for the yellow color of urine?

15.102 Which bile pigment is responsible for the brownish-red color of feces?

15.103 What chemical condition is responsible for jaundice?

15.104 What physical conditions cause jaundice?

15.105 Which of the heme degradation products (1) bilirubin (2) biliverdin (3) stercobilin and (4) urobilin is associated with each of the following heme degradation characterizations?
a. CO is produced at the same time as this substance.
b. This substance is associated with the condition called jaundice.
c. Molecular O_2 is a reactant in the production of this substance.
d. This bile pigment has a brownish color.

15.106 Which of the heme degradation products (1) bilirubin (2) biliverdin (3) stercobilin and (4) urobilin is associated with each of the following heme degradation characterizations?
a. "Ring-opening" occurs in the production of this substance.
b. A carbon–carbon double bond is changed to a carbon–carbon single bond in the production of this substance.
c. This substance is rendered more water-soluble by use of a glucose derivative.
d. This bile pigment has a yellowish color.

▲15.107 In which of the processes (1) urea cycle (2) heme degradation and (3) oxidative deamination would each of the following molecules be encountered?
a. Citrulline
b. Glutamate
c. Bilirubin
d. Ammonium ion

▲15.108 In which of the processes (1) urea cycle (2) heme degradation and (3) oxidative deamination would each of the following molecules be encountered?
a. Carbon monoxide
b. Stercobilin
c. Biliverdin
d. Ornithine

Interrelationships Among Metabolic Pathways (Section 15.8)

15.109 Briefly explain how the carbon atoms from amino acids can end up in ketone bodies.

15.110 Briefly explain how the carbon atoms from amino acids can end up in glucose.

15.111 How are the amino acids from protein "processed" when they are present in amounts that exceed the body's needs?

15.112 How are the amino acids from protein "processed" when an individual is in a state of starvation?

B Vitamins and Protein Metabolism (Section 15.9)

15.113 Indicate whether each of the following B vitamins is involved as a cofactor in the processes of (1) transamination (2) oxidative deamination (3) urea cycle (4) carbon skeleton degradation to CAC intermediates or (5) carbon skeleton degradation to non-CAC intermediates. For a given vitamin, more than one response may be correct.
a. Niacin
b. Folate
c. Biotin
d. Vitamin B_6

15.114 Indicate whether each of the following B vitamins is involved as a cofactor in the processes of (1) transamination (2) oxidative deamination (3) urea cycle (4) carbon skeleton degradation to CAC intermediates or (5) carbon skeleton degradation to non-CAC intermediates. For a given vitamin, more than one response may be correct.
a. Riboflavin
b. Thiamin
c. Pantothenic acid
d. Vitamin B_{12}

Answers to Selected Exercises

Chapter 1 **1.1** (a) false (b) false (c) true (d) true **1.3** (a) meets (b) does not meet (c) does not meet (d) does not meet **1.5** Hydrocarbons contain C and H, and hydrocarbon derivatives contain at least one additional element besides C and H. **1.7** All bonds are single bonds in a saturated hydrocarbon, and at least one carbon–carbon multiple bond is present in an unsaturated hydrocarbon. **1.9** (a) saturated (b) unsaturated (c) unsaturated (d) unsaturated **1.11** (a) 18 (b) 4 (c) 13 (d) 22 **1.13** (a) true (b) true (c) true (d) false **1.15** (a) $CH_3—CH_2—CH_2—CH_3$ (b) $CH_3—CH_2—CH_2—CH_2—CH_2—CH_2—CH_3$ **1.17** (a) $CH_3—(CH_2)_2—CH_3$ (b) $CH_3—(CH_2)_5—CH_3$ **1.19** (a) C—C—C—C (b) C—C—C—C—C—C—C

1.21 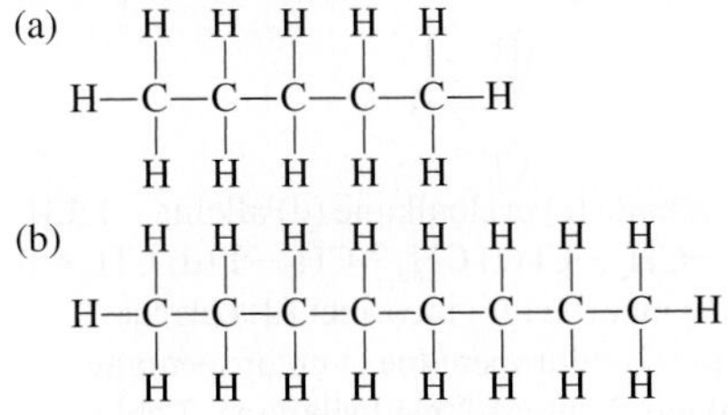

(c) $CH_3—(CH_2)_8—CH_3$ (d) C_6H_{14}

1.23 (a) 14 (b) 5 (c) 4 (d) 19 **1.25** the same molecular formula **1.27** (a) the same (b) different (c) different (d) different **1.29** a continuous chain of carbon atoms versus a continuous chain of carbon atoms to which one or more branches of carbon atoms are attached **1.31** (a) 2 (b) 5 (c) 18 (d) 75 **1.33** one **1.35** (a) different compounds that are not constitutional isomers (b) different compounds that are constitutional isomers (c) different conformations of the same molecule (d) different compounds that are constitutional isomers

1.37

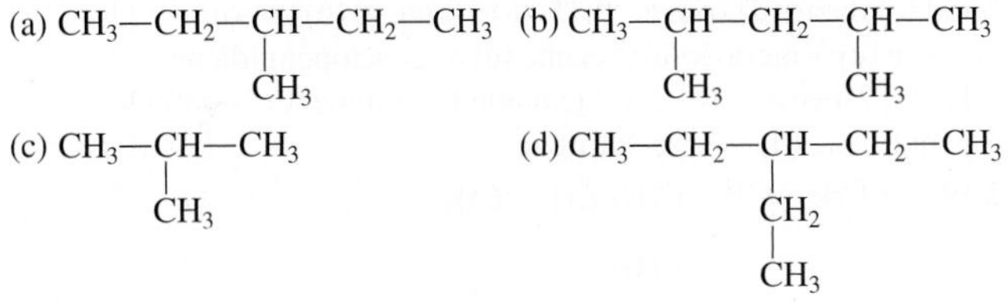

1.39

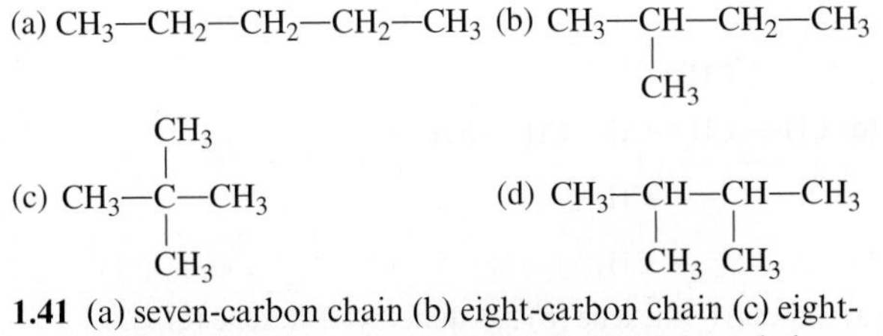

1.41 (a) seven-carbon chain (b) eight-carbon chain (c) eight-carbon chain (d) seven-carbon chain **1.43** (a) 3-methylpentane (b) 2-methylhexane (c) 2-methylhexane (d) 2,4-dimethylhexane **1.45** (a) 2,3,5-trimethylhexane (b) 2,2,4-trimethylpentane (c) 3-ethyl-3-methylpentane (d) 3-ethyl-3-methylhexane

1.47 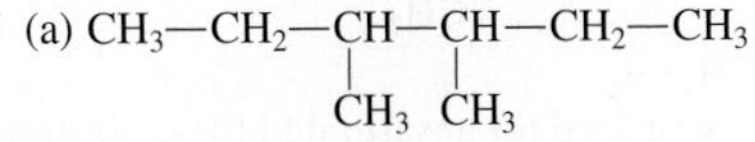

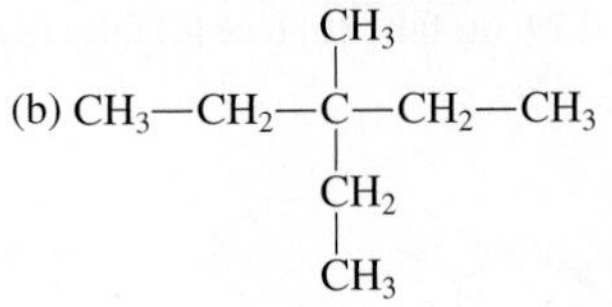

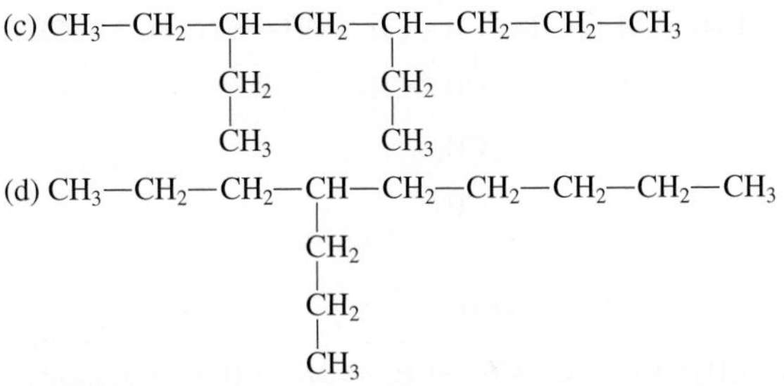

1.49 (a) 2, 2 (b) 2, 2 (c) 2, 2 (d) 1, 1 **1.51** (a) not based on longest carbon chain; 2,2-dimethylbutane (b) carbon chain numbered from wrong end; 2,2,3-trimethylbutane (c) carbon chain numbered from wrong end and alkyl groups not listed alphabetically; 3-ethyl-4-methylhexane (d) like alkyl groups listed separately; 2,4-dimethylhexane **1.53** (a) 1 (b) 3 (c) 7 (d) 6

1.55 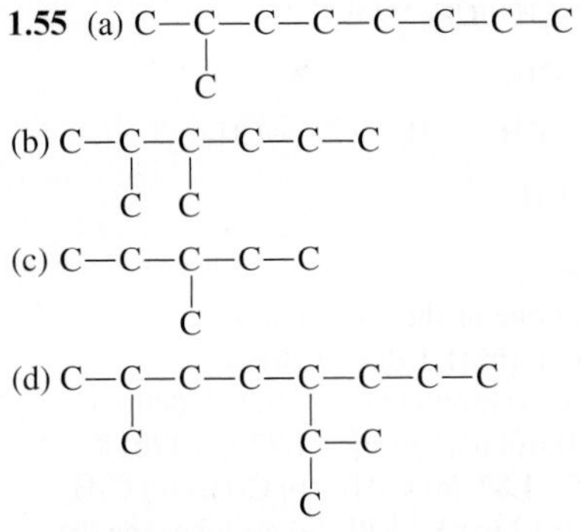

1.57 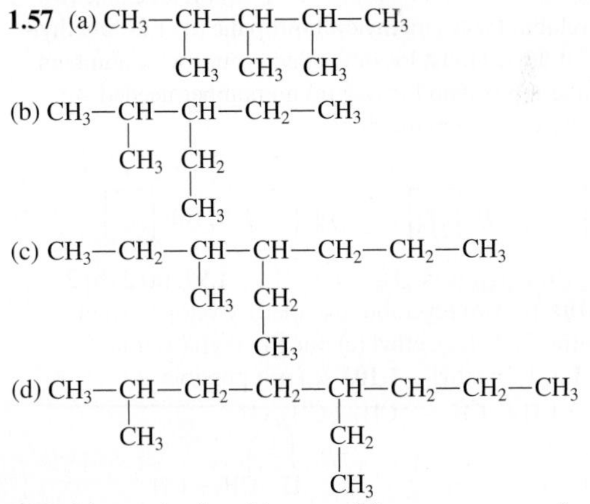

1.59 (a) constitutional isomers (b) same compound

1.61 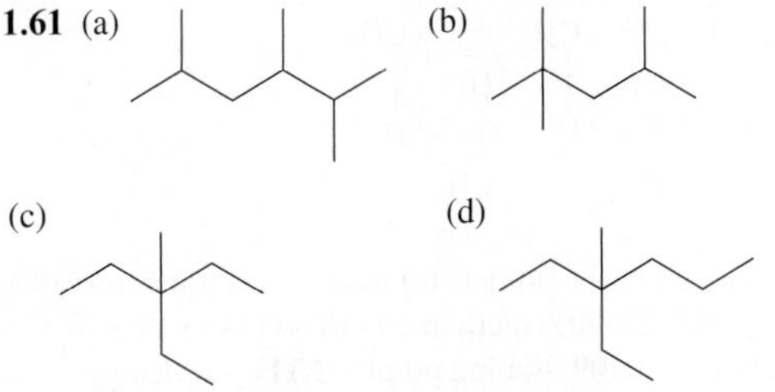

1.63 (a) 2-methyloctane (b) 2,3-dimethylhexane (c) 3-methylpentane (d) 5-isopropyl-2-methyloctane **1.65** (a) C_8H_{18} (b) C_9H_{20} (c) $C_{10}H_{22}$ (d) $C_{11}H_{24}$ **1.67** (a) 5, 1, 3, 0 (b) 5, 1, 1, 1 (c) 4, 3, 0, 1 (d) 4, 4, 0, 1 **1.69** (a) 2 (b) 0 (c) 1 (d) 1

1.71 (a) $CH_3—CH_2—CH_2—CH_2—CH_2—CH_2—CH_3$

(b) 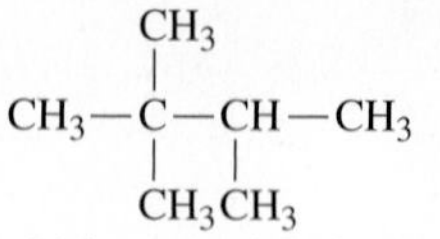

1.73 (a) isopropyl (b) isobutyl (c) isopropyl (d) *sec*-butyl

1.75

(a) 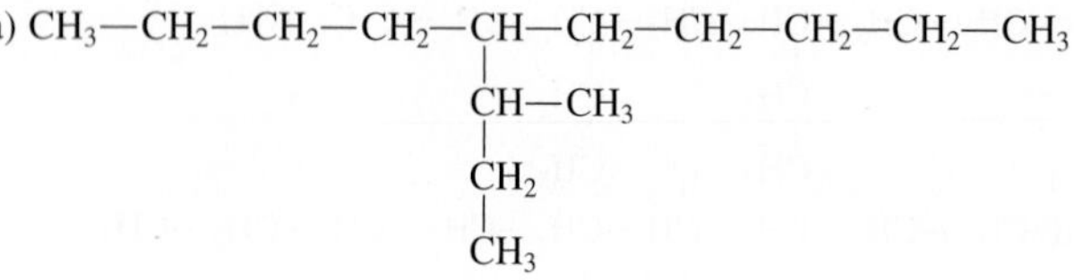

(b) 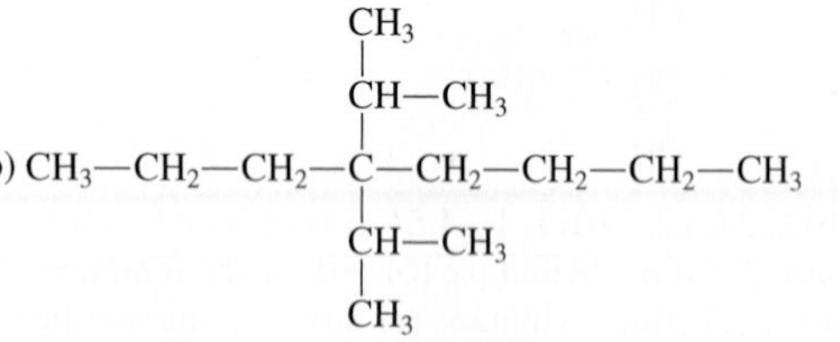

(c) 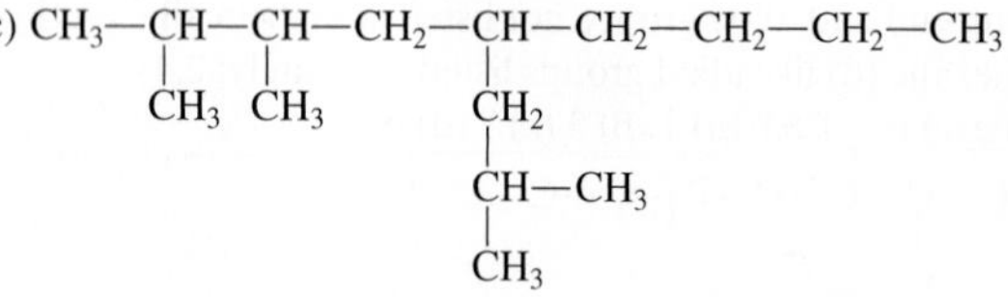

(d) 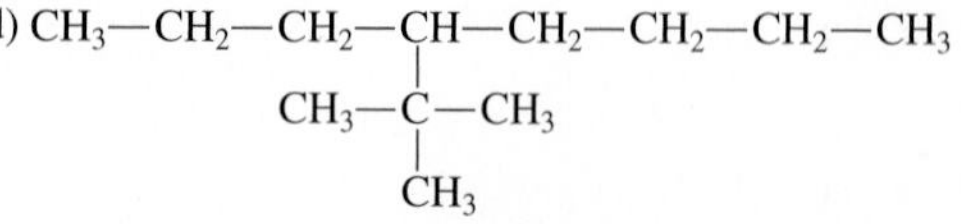

1.77 (a) 3 or 4 (b) 3 or 4 (c) none of them (d) 3 or 4
1.79 (a) (2-methylbutyl) group (b) (1,1-dimethylpropyl) group **1.81** (a) (1-methylethyl) group (b) (1,1-dimethylethyl) group (c) *sec*-butyl group (d) isobutyl group **1.83** (a) 4 (b) 8 **1.85** (a) 16 (b) 6 (c) 5 (d) 15 **1.87** (a) C_6H_{12} (b) C_6H_{12} (c) C_4H_8 (d) C_7H_{14} **1.89** (a) 6 (b) 2 (c) 2 (d) 3 **1.91** (a) cyclohexane (b) 1,2-dimethylcyclobutane (c) methylcyclopropane (d) 1,2-dimethylcyclopentane **1.93** (a) must locate methyl groups with numbers (b) wrong numbering system for ring (c) no number needed (d) wrong numbering system for ring

1.95 (a) 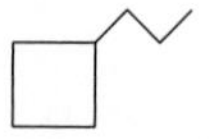 (b) 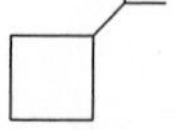 (c) 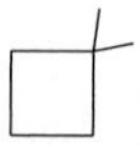 (d) 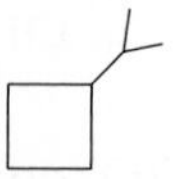

1.97 (a) C_8H_{16} (b) C_8H_{18} (c) C_8H_{16} (d) C_8H_{18} **1.99** (a) 2 (b) 2 (c) 3 (d) 1 **1.101** (a) two (cyclobutane, methylcyclopropane) (b) three: dimethyl (1,1; 1,2); ethyl (c) one (methyl) (d) four: dimethyl (1,1; 1,2; 1,3); ethyl **1.103** (a) not possible

(b) 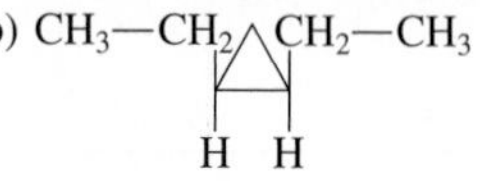 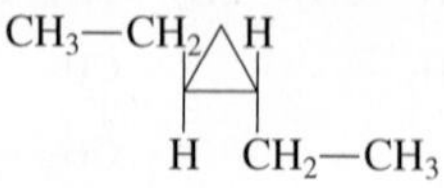

cis *trans*

(c) not possible (d) 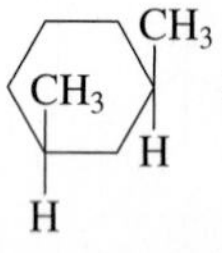 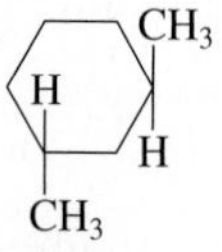

cis *trans*

1.105 (a) not isomers (b) not isomers (c) isomers (d) not constitutional isomers **1.107** 50–90% methane, 1–10% ethane, up to 8% propane and butanes **1.109** boiling point **1.111** (a) octane (b) cyclopentane (c) pentane (d) cyclopentane **1.113** (a) different states (b) same state (c) same state (d) same state **1.115** (a) liquid (b) less dense (c) insoluble (d) flammable **1.117** (a) true (b) true (c) true (d) false **1.119** (a) CO_2 and H_2O (b) CO_2 and H_2O (c) CO_2 and H_2O (d) CO_2 and H_2O **1.121** CH_3Br, CH_2Br_2, $CHBr_3$, CBr_4

1.123
(a) 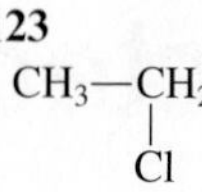

(b) 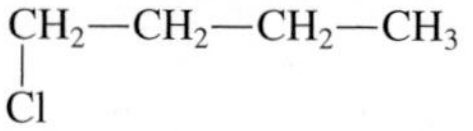 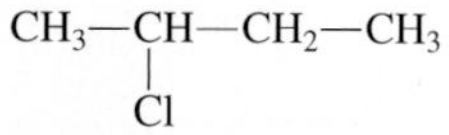

(c) 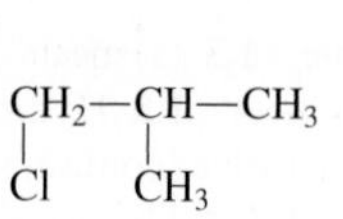 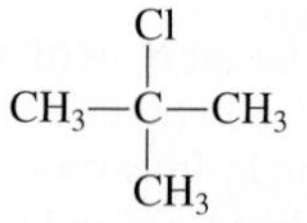 (d) 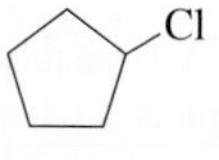

1.125 (a) iodomethane, methyl iodide (b) 1-chloropropane, propyl chloride (c) 2-fluorobutane, *sec*-butyl fluoride (d) chlorocyclobutane, cyclobutyl chloride
1.127 (a) 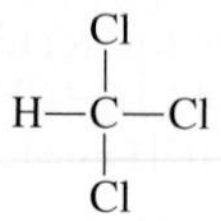 (b) 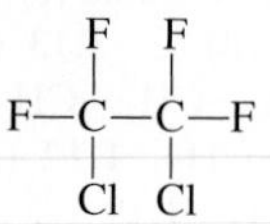

(c) 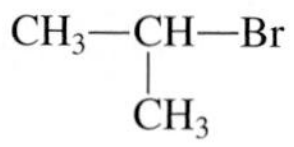 (d) 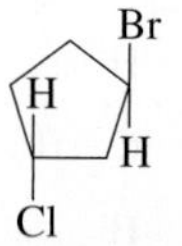

1.129 (a) alkane (b) alkane (c) cycloalkane (d) alkane **1.131** (a) $CH_3—I$ (b) $CH_3—CH_2—Cl$ (c) $CH_3—CH_2—I$ (d) $CH_3—Br$
1.133 (a) incorrect (b) incorrect (c) incorrect (d) correct
1.135 1-chloropentane, 2-chloropentane, 3-chloropentane, 1-chloro-2-methylbutane, 2-chloro-2-methylbutane, 2-chloro-3-methylbutane, 1-chloro-3-methylbutane, 1-chloro-2,2-dimethylpropane **1.137** (a) true (b) false (c) false (d) true

Chapter 2

2.1 one or more carbon–carbon multiple bonds are present **2.3** carbon–carbon double bond **2.5** they are very similar **2.7** (a) unsaturated, alkene with one double bond (b) unsaturated, alkene with one double bond (c) unsaturated, diene (d) unsaturated, triene **2.9** (a) C_4H_{10} (b) C_5H_{10} (c) C_5H_8 (d) C_7H_{10} **2.11** (a) C_nH_{2n-2} (b) C_nH_{2n-2} (c) C_nH_{2n-2} (d) C_nH_{2n-6}
2.13 (a) alkene with one double bond (b) alkene with one double bond (c) diene (d) triene **2.15** (a) 2-butene (b) 2,4-dimethyl-2-pentene (c) 3-methylcyclohexane (d) 1,3-cyclopentadiene
2.17 (a) 2-pentene (b) 2,3,3-trimethyl-1-butene (c) 2-methyl-1,4-pentadiene (d) 1,3,5-hexatriene
2.19 (a) 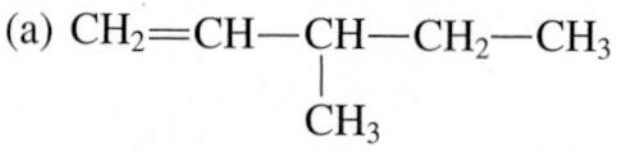

(b) 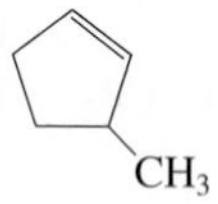 (c) $CH_2{=}CH—CH{=}CH_2$

(d) 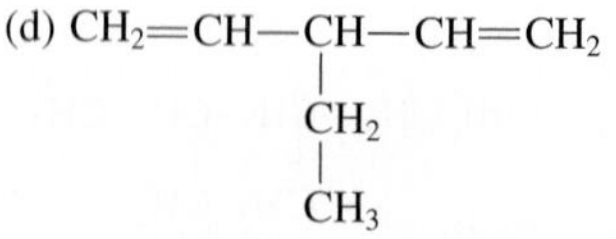

2.21 (a) 3-methyl-3-hexene (b) 2,3-dimethyl-2-hexene (c) 1,3-cyclopentadiene (d) 4,5-dimethylcyclohexene
2.23 (a) $CH_2{=}CH_2$ (b) 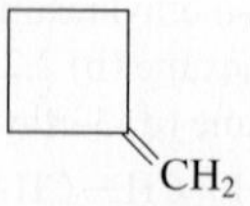 (c) $CH_2{=}CH—Br$

(d) $CH_2{=}CH—CH_2—I$
2.25 (a) saturated (b) unsaturated (c) unsaturated (d) unsaturated
2.27 (a) 14 (b) 12 (c) 6 (d) 12 **2.29** (a) false (b) true (c) false (d) false
2.31 (a)  (b) 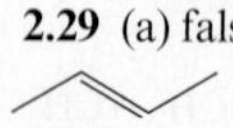

(c) 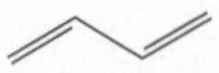 (d) 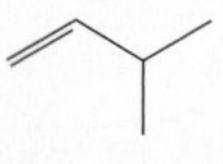

2.33 (a) 8 (b) 8 (c) 8 (d)10 **2.35** (a) C_8H_{16} (b) C_8H_{16} (c) C_8H_{14} (d) $C_{10}H_{18}$ **2.37** (a) 3-octene (b) 3-octene (c) 1,3-octadiene (d) 3,7-dimethyl-1,5-octadiene **2.39** (a) positional (b) skeletal (c) skeletal (d) positional **2.41** (a) 2 (b) 4 (c) 3 (d) zero

2.43

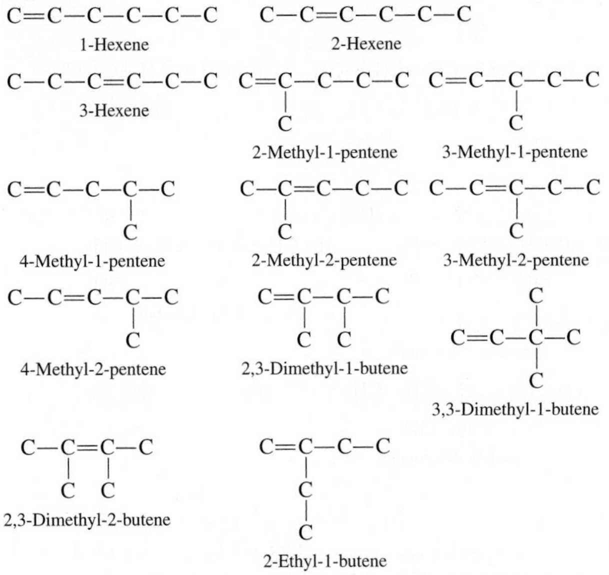

2.45 C=C−C−C C−C=C−C 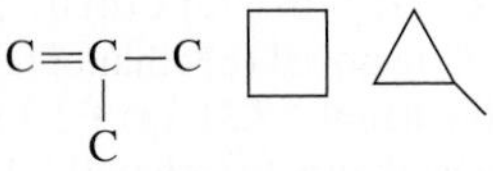

2.47 (a) no (b) no

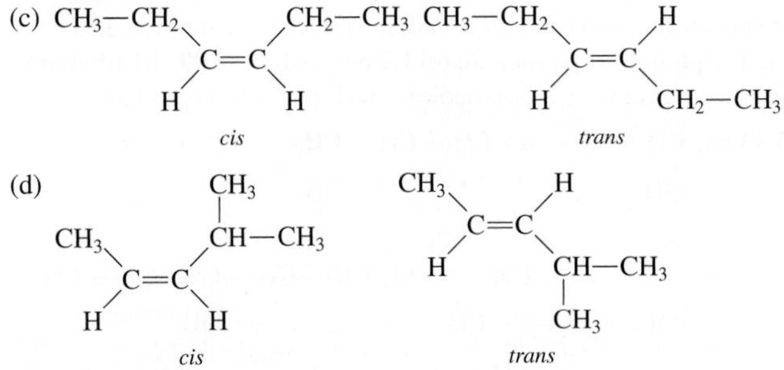

2.49 (a) *cis*-2-pentene (b) *trans*-1-bromo-2-iodoethene (c) tetrafluoroethene (d) 2-methyl-2-butene

2.51

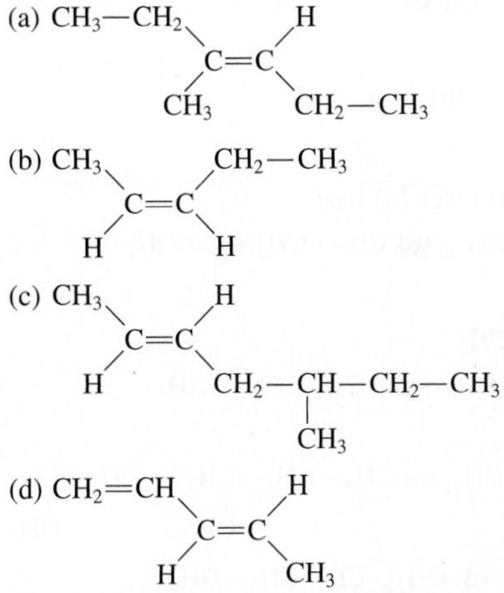

2.53 (a) no (b) no (c) no (d) yes **2.55** (a) false (b) false (c) true (d) true **2.57** molecule used to transmit a message to other members of the same species **2.59** isopentyl unit **2.61** Splitting of a β-carotene molecule produces two vitamin A molecules **2.63** (a) false (b) true (c) false (d) true **2.65** (a) gas (b) liquid (c) liquid (d) liquid **2.67** (a) yes (b) no (c) yes (d) no

2.69

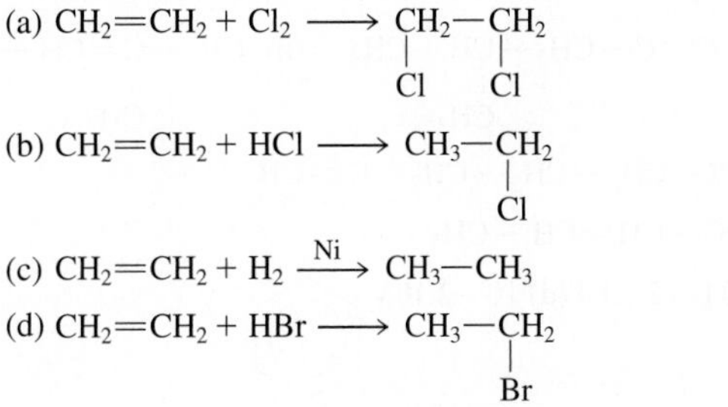

2.71

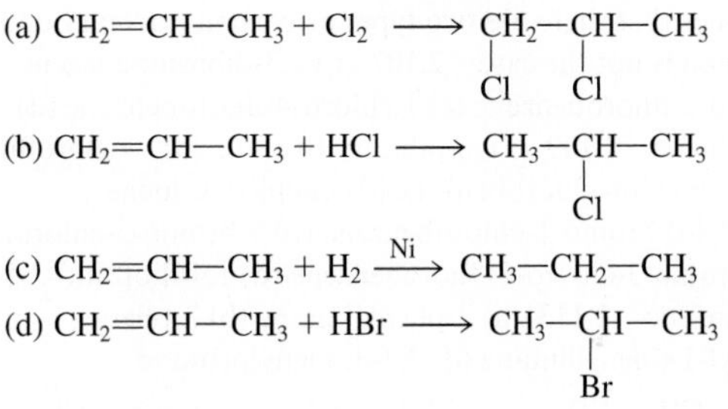

2.73

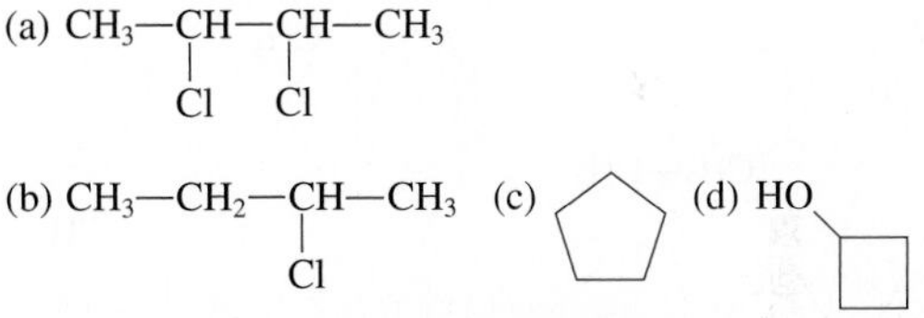

2.75 (a) Br_2 (b) H_2 + Ni catalyst (c) HCl (d) H_2O + H_2SO_4 catalyst **2.77** (a) 2 (b) 2 (c) 2 (d) 3 **2.79** a large molecule formed by the repetitive bonding together of many smaller molecules **2.81** a polymer in which the monomers add together to give the polymer as the only product **2.83** (a) $CF_2{=}CF_2$

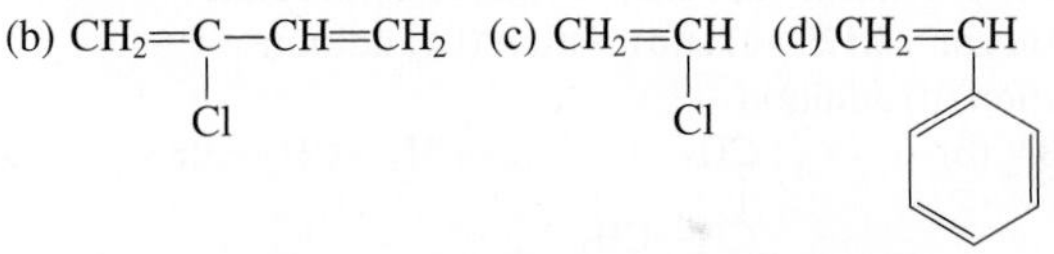

2.85

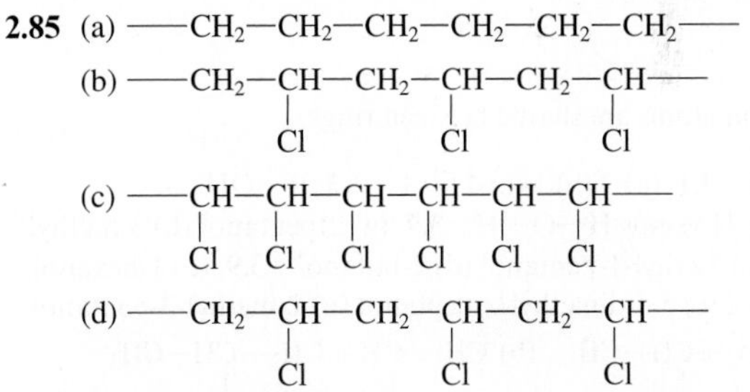

2.87 (a) C_nH_{2n-6} **2.89** (a) 1-hexyne (b) 4-methyl-2-pentyne (c) 2,2-dimethyl-3-heptyne (d) 1-butyne

2.91 C≡C−C−C−C (1-pentyne)
C−C≡C−C−C (2-pentyne)
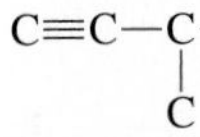 (3-methyl-1-butyne)

2.93 because of the linearity (180° angles) about an alkyne's carbon–carbon triple bond **2.95** Their physical properties are very similar.

2.97

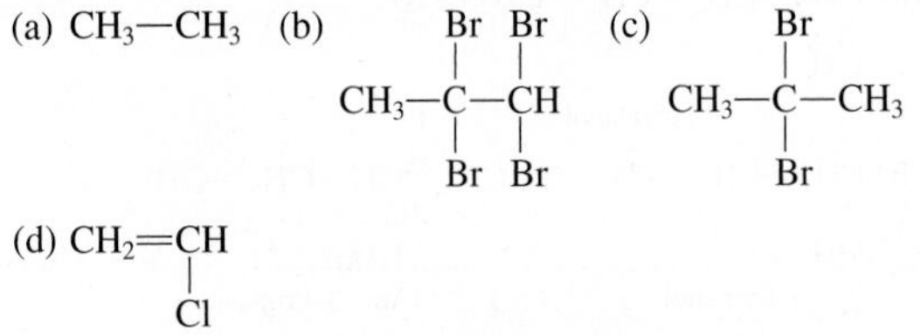

2.99

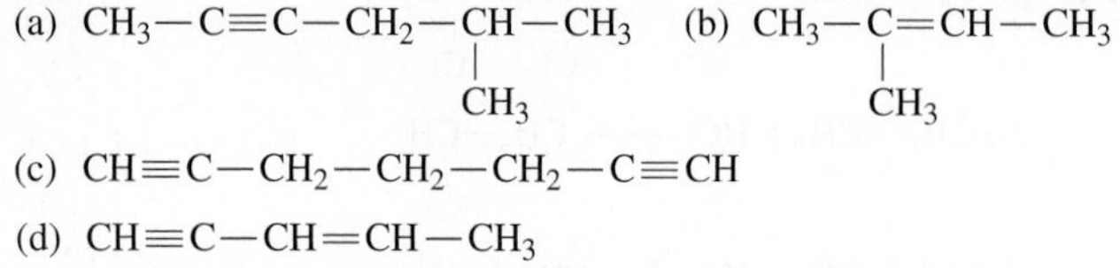

2.101 (a) 3 (b) 2 (c) 4 (d) 10 **2.103**

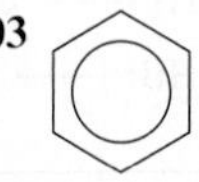

2.105 Implies that there are two types of carbon–carbon bonds present which is not the case **2.107** (a) 1,3-dibromobenzene (b) 1-chloro-2-fluorobenzene (c) 1-chloro-4-fluorobenzene (d) 3-chlorotoluene **2.109** (a) *m*-dibromobenzene (b) *o*-chlorofluorobenzene (c) *p*-chlorofluorobenzene (d) *m*-chlorotoluene **2.111** (a) 2,4-dibromo-1-chlorobenzene (b) 3-bromo-5-chlorotoluene (c) 1-bromo-3-chloro-2-fluorobenzene (d) 1,4-dibromo-2,5-dichlorobenzene **2.113** (a) 2-phenylbutane (b) 3-phenyl-1-butene (c) 3-methyl-1-phenylbutane (d) 2,4-diphenylpentane

2.115

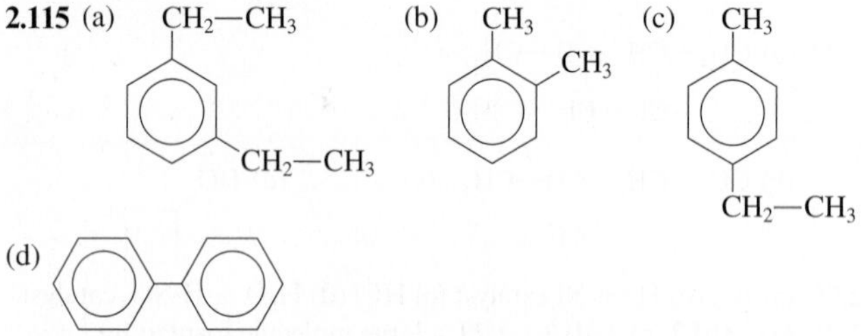

2.117 1,2,3-trimethylbenzene, 1,2,4-trimethylbenzene, 1,3,5-trimethylbenzene, 2-ethyltoluene, 3-ethyltoluene, 4-ethyltoluene, propylbenzene, isopropylbenzene **2.119** (a) all four (b) cyclohexene (c) cyclohexane, benzene (d) benzene **2.121** liquid state **2.123** petroleum **2.125** (a) substitution (b) addition (c) substitution (d) addition
2.127 (a) Br_2 (b)

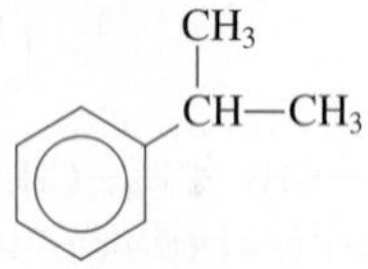

(c) $CH_3—CH_2—Br$

2.129 carbon atoms are shared between rings

Chapter 3

3.1 (a) 2 (b) 1 (c) 4 (d) 1 **3.3** R—OH **3.5** R—O—H versus H—O—H **3.7** (a) 2-pentanol (b) 3-methyl-2-butanol (c) 2-ethyl-1-pentanol (d) 2-butanol **3.9** (a) 1-hexanol (b) 3-hexanol (c) 5,6-dimethyl-2-heptanol (d) 2-methyl-3-pentanol
3.11

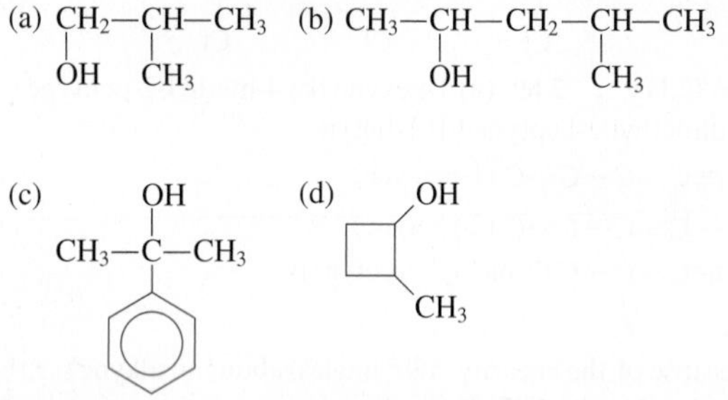

3.13

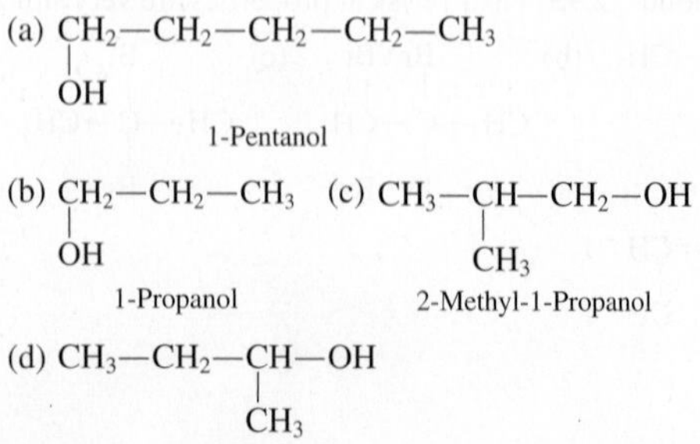

1-Pentanol; 1-Propanol; 2-Methyl-1-Propanol; 2-Butanol

3.15 (a) 1,2-propanediol (b) 1,4-pentanediol (c) 1,3-pentanediol (d) 3-methyl-1,2,4-butanetriol **3.17** (a) cyclohexanol (b) *trans*-3-chlorocyclohexanol (c) *cis*-2-methylcyclohexanol (d) 1-methylcyclobutanol
3.19

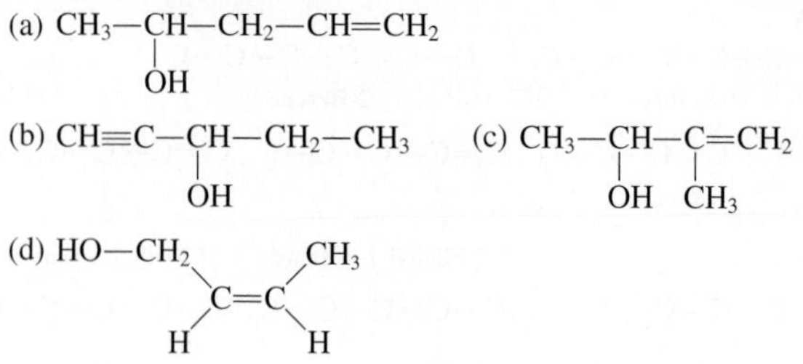

3.21

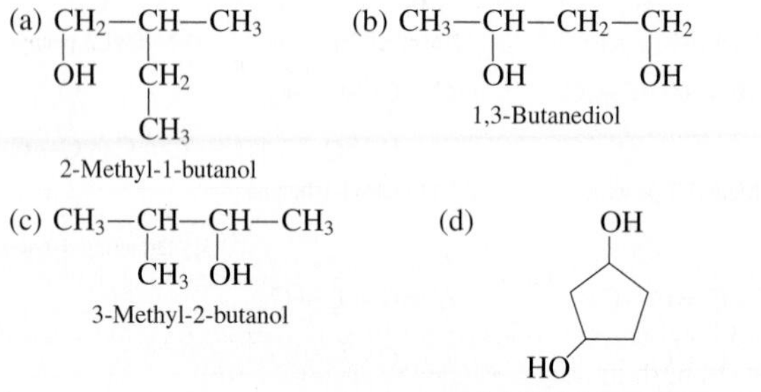

2-Methyl-1-butanol; 1,3-Butanediol; 3-Methyl-2-butanol; 1,3-Cyclopentanediol

3.23 (a) no (b) yes (c) yes (d) yes **3.25** (a) 4 (b) 8 (c) 8 (d) 4 **3.27** (a) 4 (b) 3 (c) 1 (d) 0 **3.29** (a) ethanol with all traces of H_2O removed (b) ethanol (c) 70% solution of isopropyl alcohol (d) ethanol **3.31** (a) 1,2,3-propanetriol (b) 1,2-propanediol (c) methanol (d) ethanol **3.33** Methanol fires, but not gasoline fires, can be put out using water. **3.35** Alcohol molecules can hydrogen-bond to each other; alkane molecules cannot. **3.37** (a) 1-heptanol (b) 1-propanol (c) 1,2-ethanediol **3.39** (a) 1-butanol (b) 1-pentanol (c) 1,2-butanediol **3.41** (a) 3 (b) 3 (c) 3 (d) 3
3.43

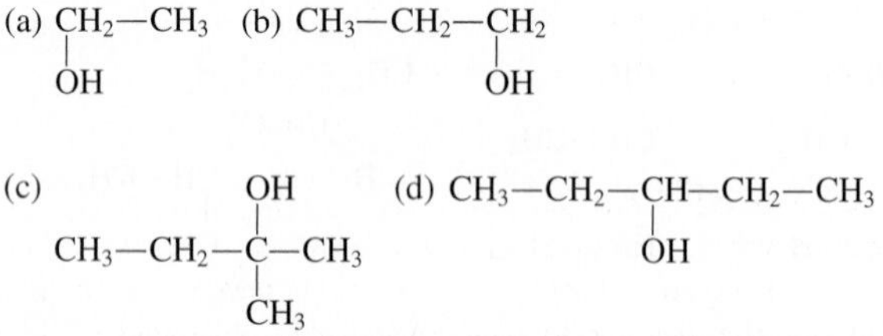

3.45 (a) 2° (b) 2° (c) 1° (d) 2° **3.47** (a) 2° (b) 2° (c) 3° (d) 1° **3.49** (a) 1° (b) 2° (c) 1° (d) 3°

3.51

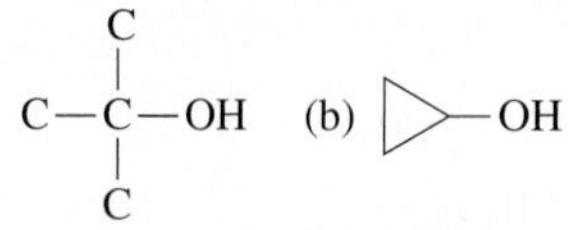

3.53 (a) true (b) true (c) true (d) true
3.55

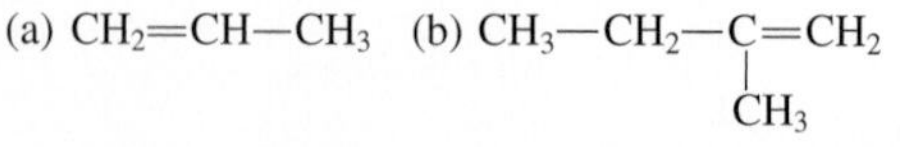

(c) $CH_3—CH{=}CH_2$
(d) $CH_3—CH_2—CH_2—O—CH_2—CH_2—CH_3$

3.57

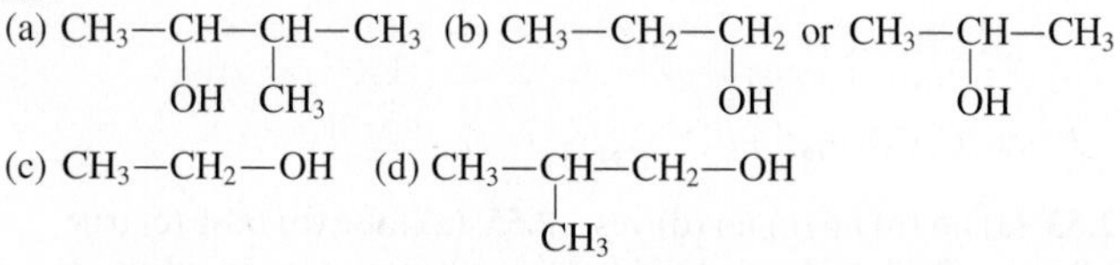

3.59

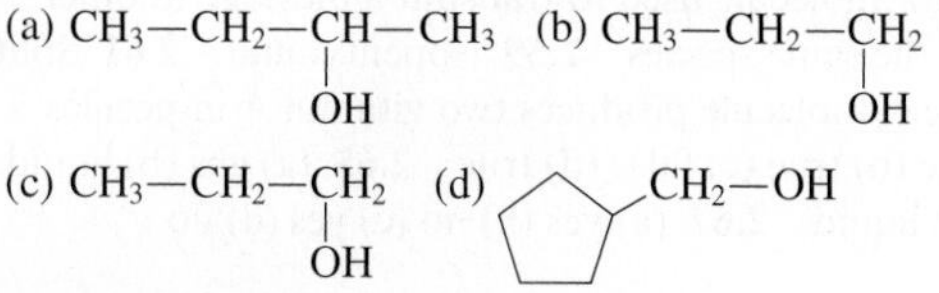

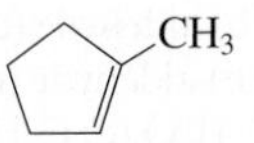

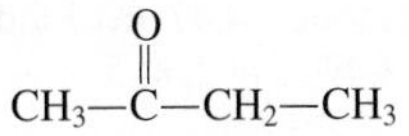

3.61 (a) CH₃—CH₂—CH₂—Cl (b)

(c)

$$CH_3-\overset{\overset{\displaystyle O}{\|}}{C}-CH_2-CH_3$$

(d) CH₃—CH₂—O—CH₂—CH₃

3.63 1-pentanol **3.65** white solid, water soluble, hydrocarbon insoluble, does not absorb oxygen **3.67** Phenols require the —OH groups to be attached directly to the benzene ring. **3.69** (a) 3-ethylphenol (b) 2-chlorophenol (c) *o*-cresol (d) hydroquinone

3.71 (a)

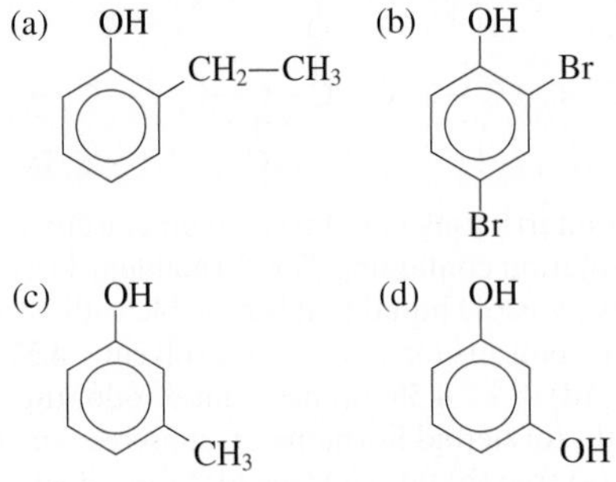

(b)

(c)

(d)

3.73 low-melting solids or oily liquids **3.75** (a) both are flammable (b) both undergo halogenation

3.77

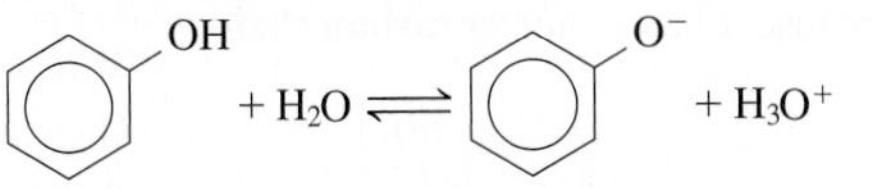

3.79 An antiseptic kills microorganisms on living tissue; a disinfectant kills microorganisms on inanimate objects. **3.81** BHA has methoxy and *tert*-butyl groups; BHT has a methyl group and two *tert*-butyl groups. **3.83** (a) true (b) false (c) true (d) true **3.85** (a) yes (b) no (c) yes (d) yes **3.87** (a) ether (b) alcohol (c) alcohol (d) ether

3.89 (a)

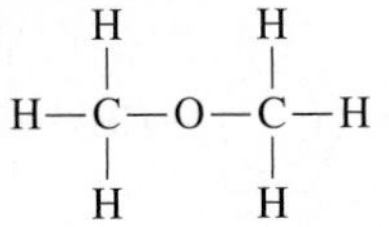

$$H-\overset{\overset{\displaystyle H}{|}}{\underset{\underset{\displaystyle H}{|}}{C}}-O-\overset{\overset{\displaystyle H}{|}}{\underset{\underset{\displaystyle H}{|}}{C}}-H$$

(b) CH₃—O—CH₃

(c) C—O—C (d) C₂H₆O **3.91** (a) 1-ethoxypropane (b) 2-methoxypropane (c) methoxybenzene (d) cyclohexoxycyclohexane **3.93** (a) ethyl propyl ether (b) isopropyl methyl ether (c) methyl phenyl ether (d) dicyclohexyl ether **3.95** (a) 1-methoxypentane (b) 1-ethoxy-2-methylpropane (c) 2-ethoxybutane (d) 2-methoxybutane

3.97

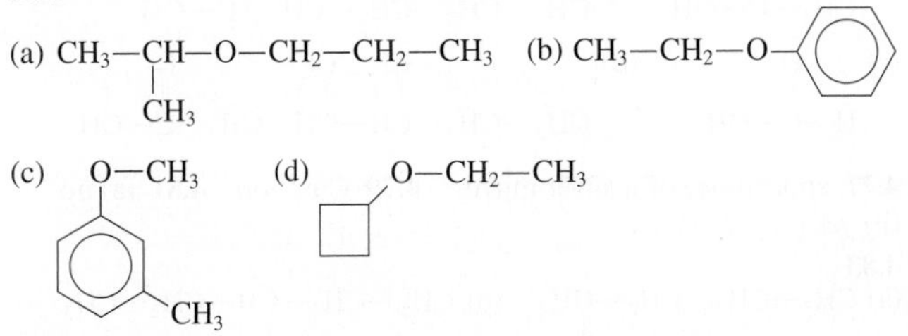

(a)

$$CH_3-\underset{\underset{\displaystyle CH_3}{|}}{CH}-O-CH_2-CH_2-CH_3$$

(b) CH₃—CH₂—O—

(c)

(d)

3.99 (a) ethoxyethane (b) 2-methoxy-2-methylpropane (c) methoxymethanol (d) 2-methylanisole **3.101** (a) false (b) false (c) true (d) true **3.103** (a) no (b) no (c) yes (d) no **3.105** butyl methyl ether, *sec*-butyl methyl ether, isobutyl methyl ether, *tert*-butyl methyl ether, ethyl isopropyl ether **3.107** (a) 1 (b) 6 (c) 2 (d) 4

3.109 (a) CH₃—O—CH₂—CH₂—CH₃,

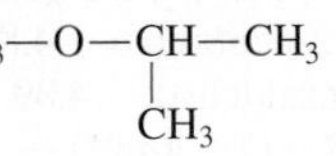

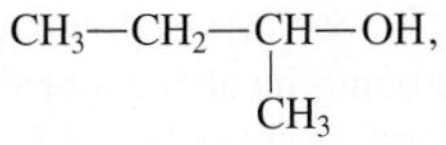

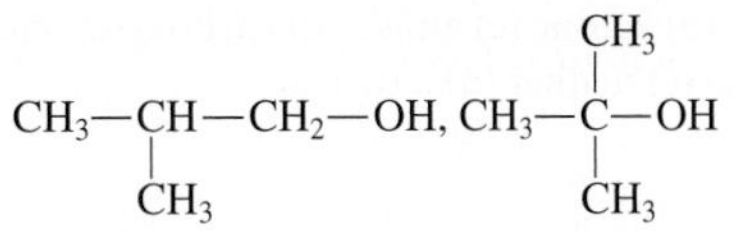

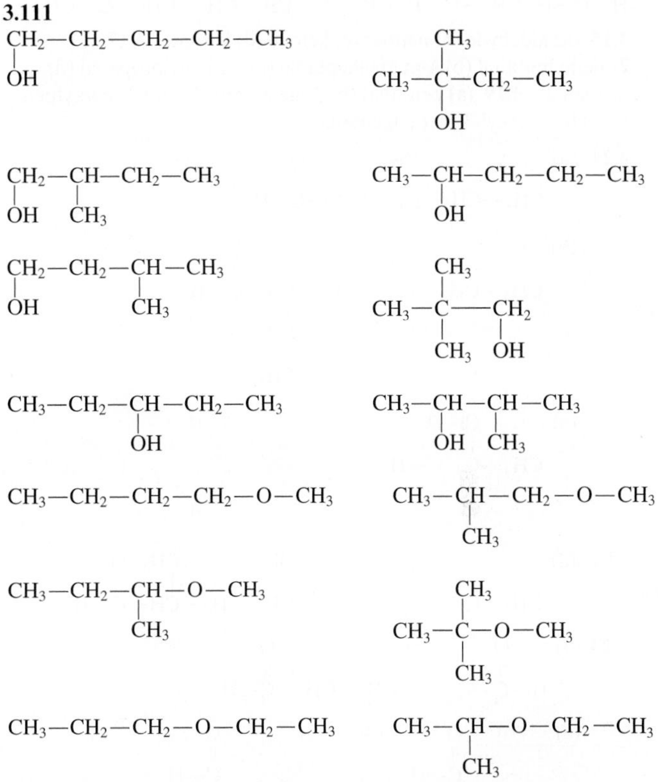

$$CH_3-O-\underset{\underset{\displaystyle CH_3}{|}}{CH}-CH_3$$

CH₃—CH₂—O—CH₂—CH₃

(b) CH₃—CH₂—CH₂—CH₂—OH,

$$CH_3-CH_2-\underset{\underset{\displaystyle CH_3}{|}}{CH}-OH,$$

$$CH_3-\underset{\underset{\displaystyle CH_3}{|}}{CH}-CH_2-OH, \quad CH_3-\overset{\overset{\displaystyle CH_3}{|}}{\underset{\underset{\displaystyle CH_3}{|}}{C}}-OH$$

3.111

$$\underset{\underset{\displaystyle OH}{|}}{CH_2}-CH_2-CH_2-CH_2-CH_3$$

$$CH_3-\overset{\overset{\displaystyle CH_3}{|}}{\underset{\underset{\displaystyle OH}{|}}{C}}-CH_2-CH_3$$

$$\underset{\underset{\displaystyle OH}{|}}{CH_2}-\underset{\underset{\displaystyle CH_3}{|}}{CH}-CH_2-CH_3$$

$$CH_3-\underset{\underset{\displaystyle OH}{|}}{CH}-CH_2-CH_2-CH_3$$

$$\underset{\underset{\displaystyle OH}{|}}{CH_2}-CH_2-\underset{\underset{\displaystyle CH_3}{|}}{CH}-CH_3$$

$$CH_3-\overset{\overset{\displaystyle CH_3}{|}}{\underset{\underset{\displaystyle CH_3}{|}}{C}}-\underset{\underset{\displaystyle OH}{|}}{CH_2}$$

$$CH_3-CH_2-\underset{\underset{\displaystyle OH}{|}}{CH}-CH_2-CH_3$$

$$CH_3-\underset{\underset{\displaystyle OH}{|}}{CH}-\underset{\underset{\displaystyle CH_3}{|}}{CH}-CH_3$$

CH₃—CH₂—CH₂—CH₂—O—CH₃

$$CH_3-\underset{\underset{\displaystyle CH_3}{|}}{CH}-CH_2-O-CH_3$$

$$CH_3-CH_2-\underset{\underset{\displaystyle CH_3}{|}}{CH}-O-CH_3$$

$$CH_3-\overset{\overset{\displaystyle CH_3}{|}}{\underset{\underset{\displaystyle CH_3}{|}}{C}}-O-CH_3$$

CH₃—CH₂—CH₂—O—CH₂—CH₃

$$CH_3-\underset{\underset{\displaystyle CH_3}{|}}{CH}-O-CH_2-CH_3$$

3.113 Dimethyl ether molecules cannot hydrogen-bond to each other; ethanol molecules can. **3.115** flammability and peroxide formation **3.117** No oxygen–hydrogen bonds are present. **3.119** (a) noncyclic ether (b) noncyclic ether (c) cyclic ether (d) nonether **3.121** (a) false (b) true (c) true (d) true **3.123** R—S—H versus R—O—H

3.125

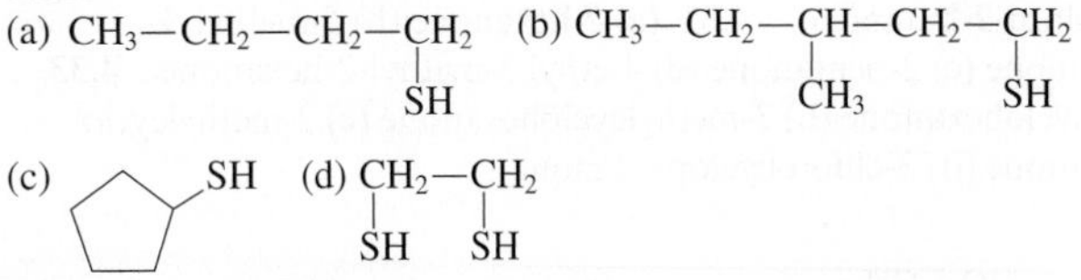

(a)

$$CH_3-CH_2-CH_2-\underset{\underset{\displaystyle SH}{|}}{CH_2}$$

(b)

$$CH_3-CH_2-\underset{\underset{\displaystyle CH_3}{|}}{CH}-CH_2-\underset{\underset{\displaystyle SH}{|}}{CH_2}$$

(c)

(d)

$$\underset{\underset{\displaystyle SH}{|}}{CH_2}-\underset{\underset{\displaystyle SH}{|}}{CH_2}$$

3.127 (a) methyl mercaptan (b) propyl mercaptan (c) *sec*-butyl mercaptan (d) isobutyl mercaptan **3.129** Alcohol oxidation produces aldehydes and ketones; thioalcohol oxidation produces disulfides. **3.131** (a) 1-pentanethiol (b) 1-pentanol (c) 3-methyl-2-butanethiol (d) 3-methyl-2-butanol **3.133** R—S—R versus R—O—R **3.135** (a) methylthioethane; ethyl methyl sulfide (b) 2-methylthiopropane; isopropyl methyl sulfide (c) methyl-thiocyclohexane; cyclohexyl methyl sulfide (d) 3-(methylthio)-1-propene; allyl methyl sulfide **3.137** (a) disulfide (b) peroxide (c) diol (d) ether, thioether, sulfide **3.139** (a) true (b) true (c) true (d) true

Chapter 4 **4.1** (a) yes (b) no (c) yes (d) yes **4.3** similarity; both have bonds involving four shared electrons; difference; C=O is polar, C=C is not polar **4.5** 120° **4.7** (a) yes (b) yes (c) no (d) yes **4.9** (a) aldehyde (b) ketone (c) amide (d) carboxylic acid **4.11** (a) neither (b) ketone (c) neither (d) aldehyde

4.13

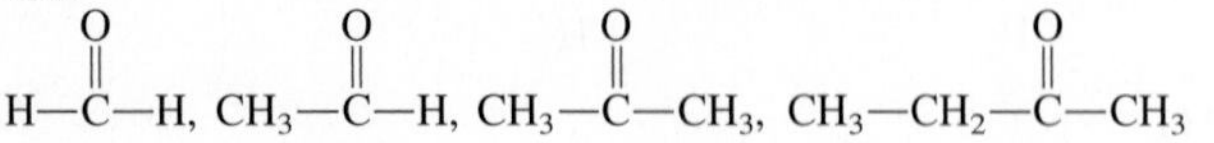

4.15 (a) aldehyde (b) neither (c) ketone (d) ketone **4.17** (a) 2-methylbutanal (b) 4-methylheptanal (c) 3-phenylpropanal (d) propanal **4.19** (a) pentanal (b) 3-methylbutanal (c) 3-methylpentanal (d) 2-ethyl-3-methylpentanal

4.21

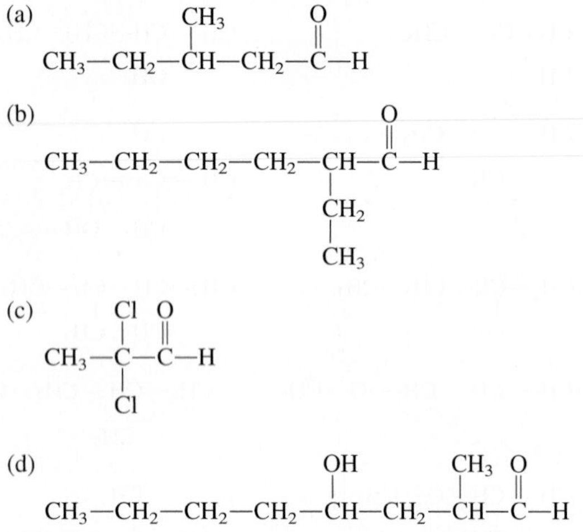

4.23

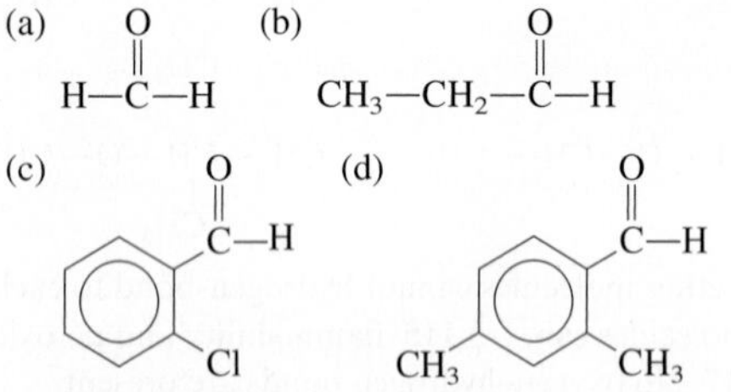

4.25 (a) propionaldehyde (b) propionaldehyde (c) dichloroacetaldehyde (d) *o*-chlorobenzaldehyde **4.27** (a) aldehyde (b) alkene, aldehyde (c) alcohol (d) alcohol, aldehyde **4.29** (a) 2-butanone (b) 2,4,5-trimethyl-3-hexanone (c) 6-methyl-3-heptanone (d) 1,1-dichloro-2-butanone **4.31** (a) 2-hexanone (b) 5-methyl-3-hexanone (c) 2-pentanone (d) 4-ethyl-3-methyl-2-hexanone **4.33** (a) cyclohexanone (b) 3-methylcyclohexanone (c) 2-methylcyclohexanone (d) 3-chlorocyclopentanone

4.35

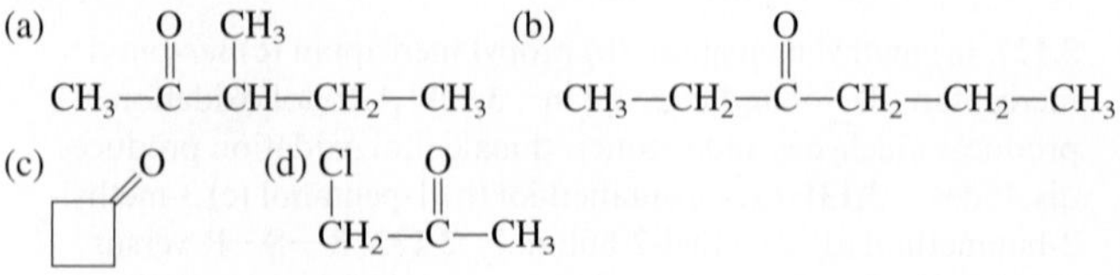

4.37

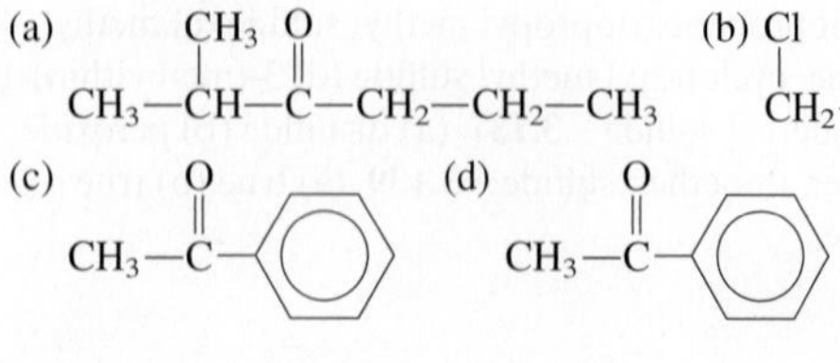

4.39 (a) ketone (b) aldehyde (c) ketone, aldehyde (d) alkene, ketone **4.41** (a) ketone (b) aldehyde (c) aldehyde (d) aldehyde **4.43** (a) $C_4H_{10}O$ (b) C_4H_8O (c) C_4H_8O (d) $C_4H_6O_2$ **4.45** (a) heptanal (b) 2-heptanone, 3-heptanone, 4-heptanone **4.47** (a) 1 aldehyde, no ketones (b) 1 aldehyde, 1 ketone **4.49** $x = 2, 4, 5$

4.51

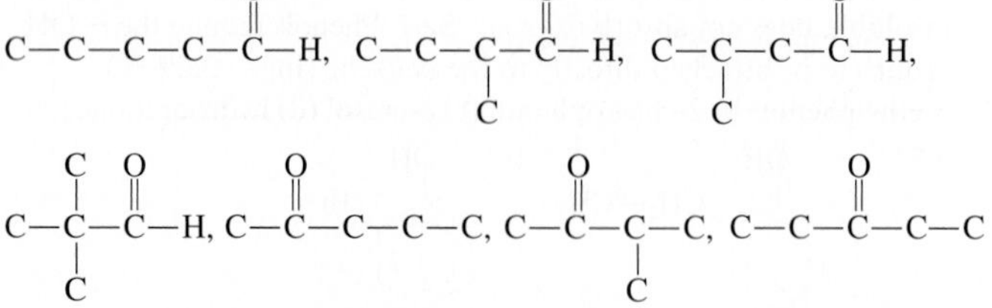

4.53 formaldehyde is an irritating gas at room temperature; formalin is an aqueous solution containing 37% formaldehyde by mass **4.55** a colorless, volatile liquid that is miscible with both water and nonpolar solvents; its main use is as a solvent **4.57** (a) 1, 0 (b) 0, 1 (c) 0, 3 (d) 0, 2 **4.59** (a) flavorant or odorant (b) flavorant or odorant (c) steroid hormone (d) sunscreen or suntanning agent **4.61** (a) true (b) false (c) true (d) false **4.63** (a) gas (b) liquid (c) liquid (d) liquid **4.65** Dipole–dipole attractions between molecules raise the boiling point. **4.67** 2 **4.69** ethanal, because it has a shorter carbon chain

4.71

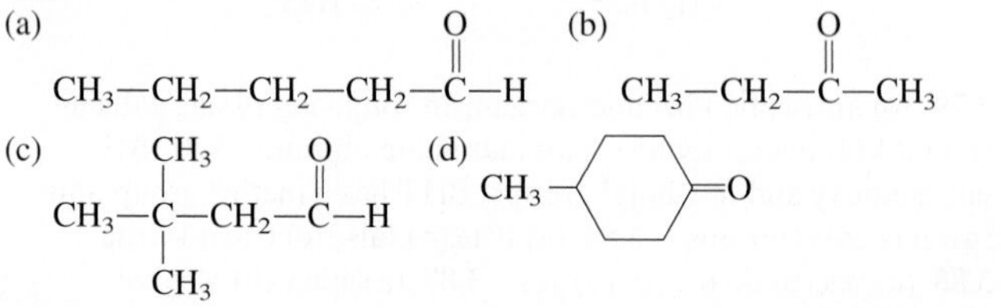

4.73

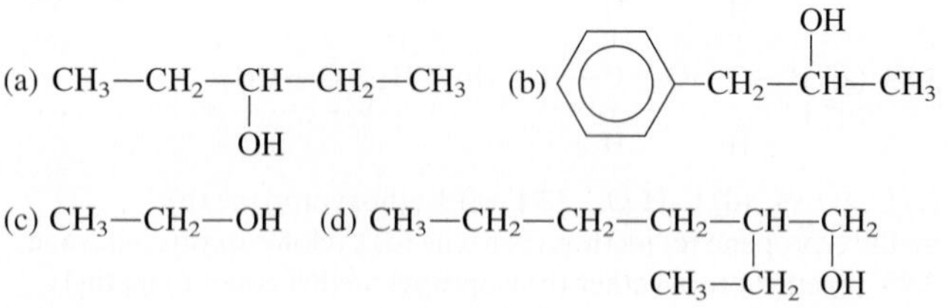

4.75

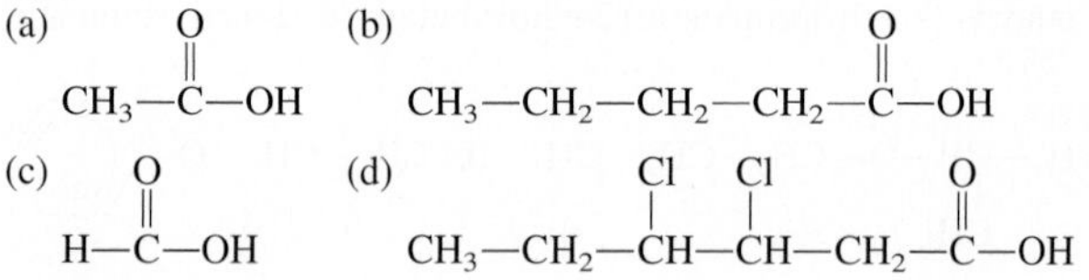

4.77 appearance of a silver mirror **4.79** Cu^{2+} ion **4.81** (a) no (b) yes (c) yes (d) no

4.83

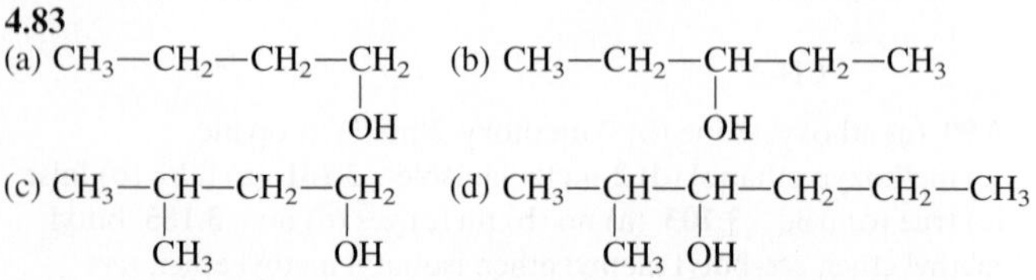

4.85 (a) pentanal, 2-pentanol (b) pentanal (c) pentanal (d) pentanal, 2-pentanone **4.87** (a) ethanal (b) butanone (c) propanal (d) benzaldehyde **4.89** (a) false (b) true (c) true (d) false **4.91** R—O— and H— **4.93** (a) no (b) yes (c) yes (d) no

4.95

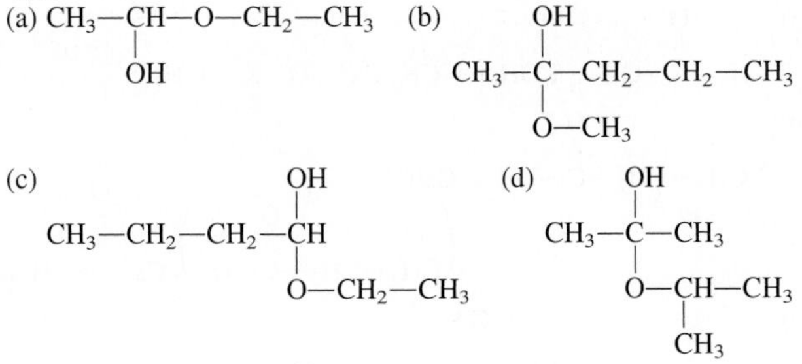

4.97

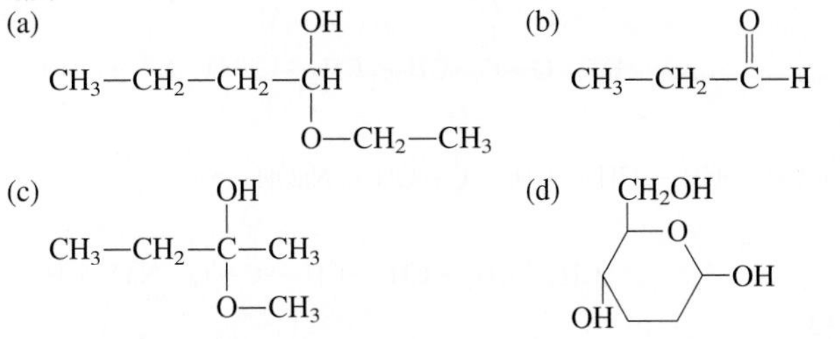

4.99 (a) yes (b) yes (c) no (d) yes

4.101

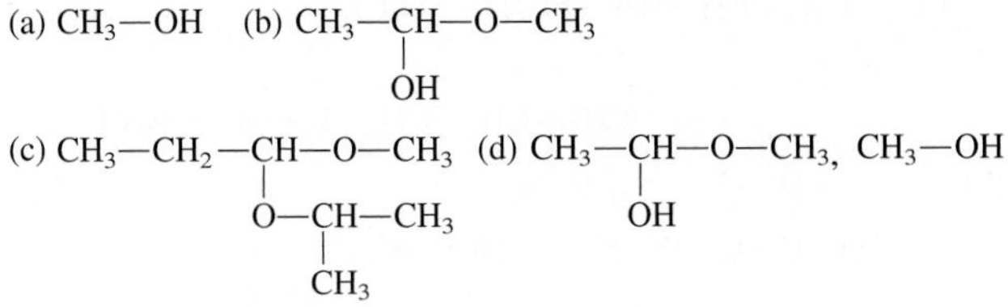

4.103

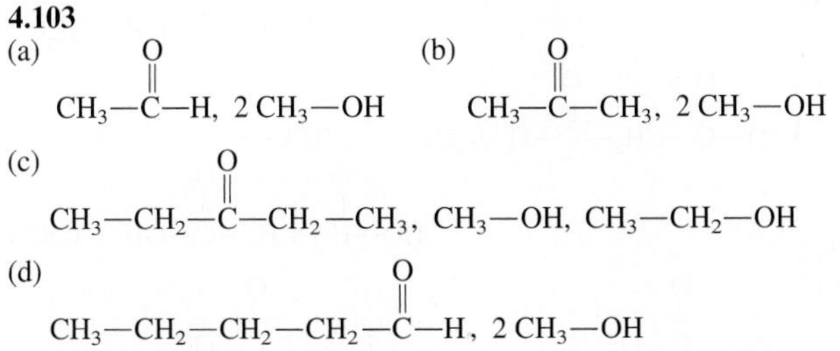

4.105 (a) dimethyl acetal of ethanal (b) dimethyl acetal of propanone (c) ethyl methyl acetal of 3-pentanone (d) dimethyl acetal of pentanal **4.107** (a) methanol, propanal (b) methanol, propanal (c) methanol, 2-methylpropanal (d) methanol, ethanol, propanal **4.109** (a) $CH_3—CH_2—CH_2—CH_2—CH_2—OH$

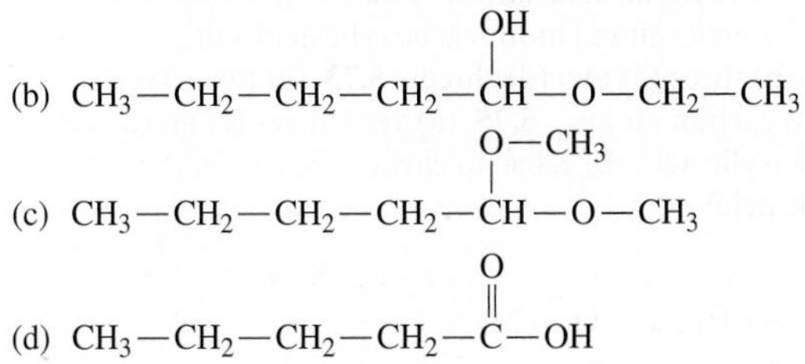

4.111

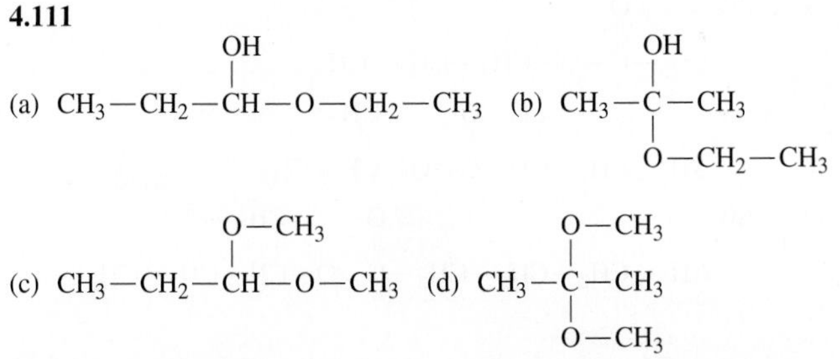

4.113 monomers are connected in a three-dimensional cross-linked network **4.115** mono-, di-, and tri-substituted phenols **4.117** thiocarbonyl compound

4.119

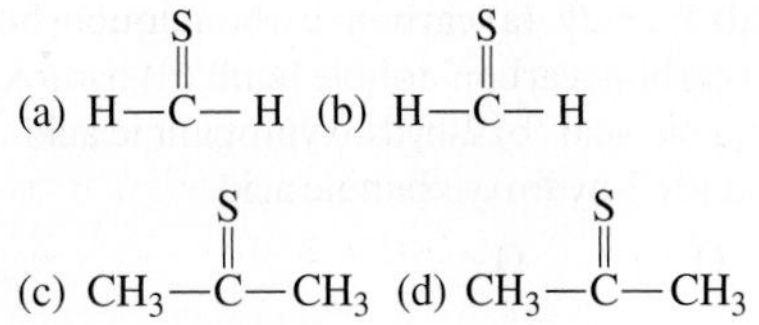

4.121 (a) true (b) true (c) true (d) false

Chapter 5 **5.1** (a) yes (b) no (c) no (d) yes **5.3** (a) yes (b) no (c) no (d) yes

5.5

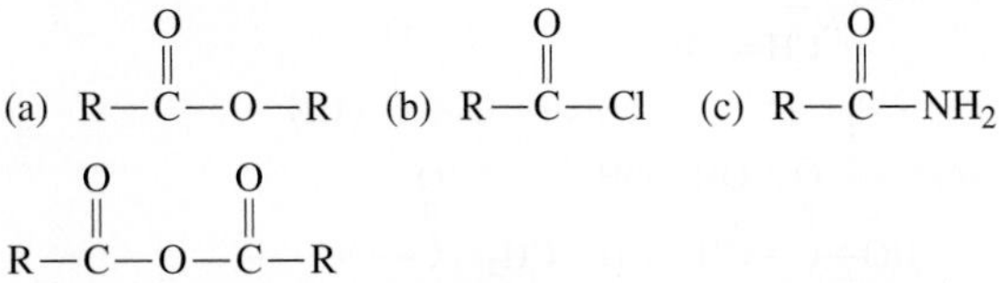

5.7 (a) polar (b) nonpolar (c) polar (d) polar **5.9** (a) acyl (b) carbonyl (c) acyl (d) acyl **5.11** (a) carbonyl, acyl (b) carboxyl, carbonyl, acyl (c) carbonyl, acyl (d) carbonyl, acyl **5.13** (a) butanoic acid (b) 4-bromopentanoic acid (c) 3-methylpentanoic acid (d) chloroethanoic acid **5.15** (a) hexanoic acid (b) 3-methylpentanoic acid (c) 2,3-dimethylbutanoic acid (d) 4,5-dimethylhexanoic acid

5.17

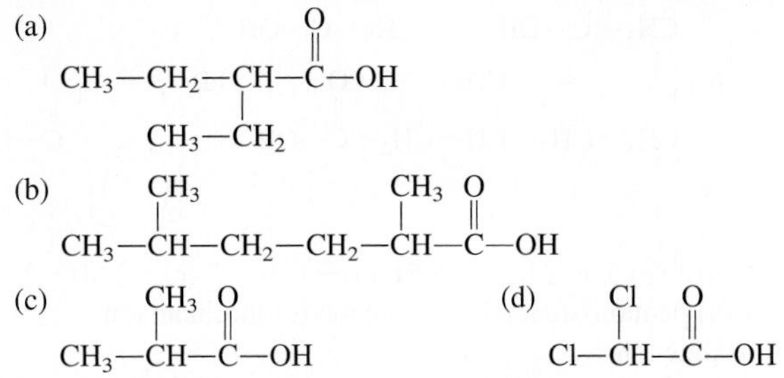

5.19 (a) propanedioic acid (b) 3-methylpentanedioic acid (c) 2-chlorobenzoic acid (d) 2-chlorocyclopentanecarboxylic acid

5.21

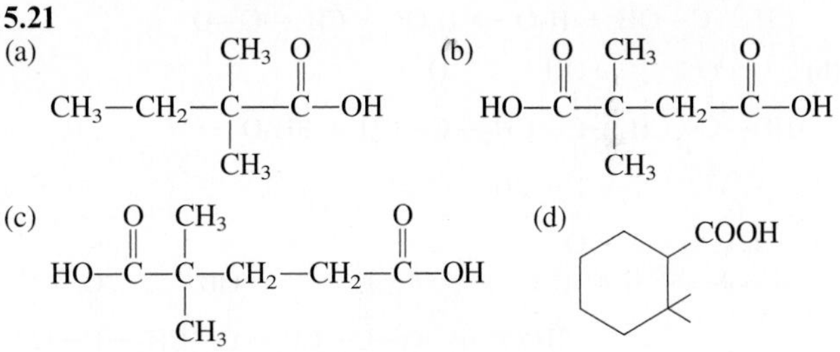

5.23

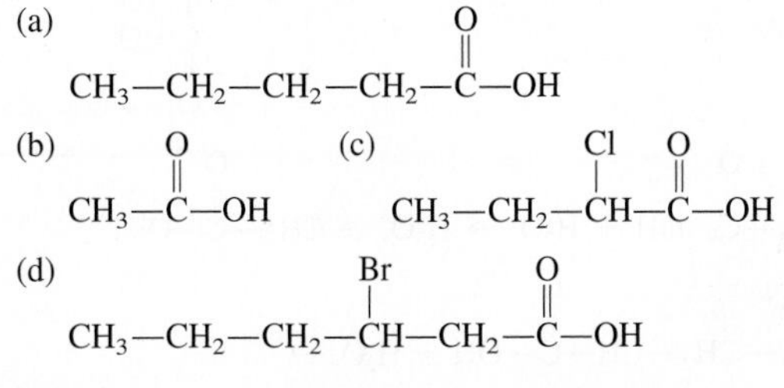

5.25

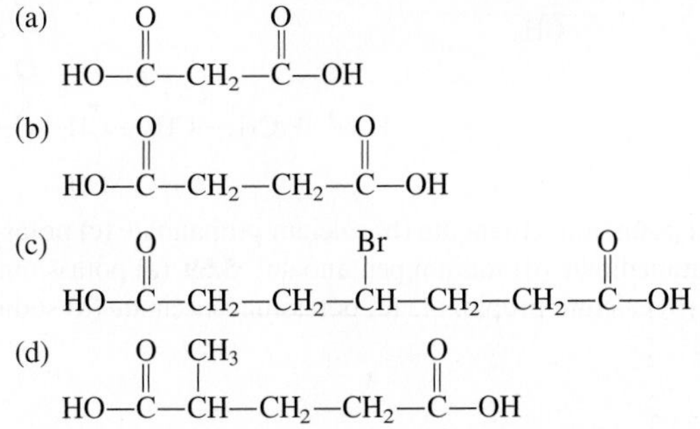

5.27 (a) 3 (b) 1 (c) 2 (d) 1 **5.29** (a) carbon–carbon double bond (b) hydroxyl group (c) carbon–carbon double bond (d) hydroxyl group **5.31** (a) propenoic acid (b) 2-hydroxypropanoic acid (c) *cis*-butenedioic acid (d) 2-hydroxyethanoic acid

5.33 (a)

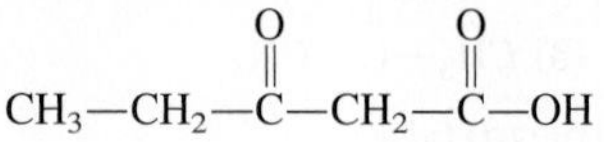

(b)

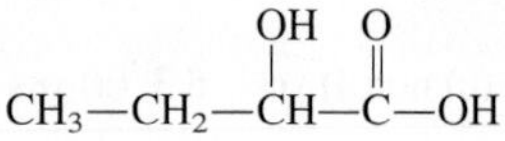

(c)

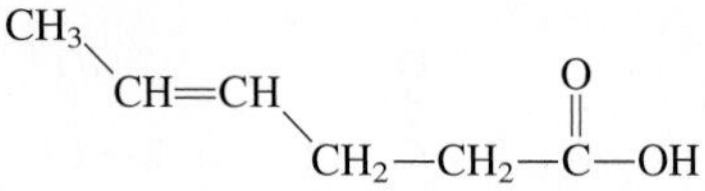

(d)

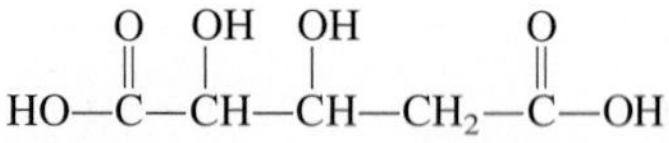

5.35 (a) unsaturated, monocarboxylic (b) unsaturated, dicarboxylic (c) dicarboxylic (d) monocarboxylic **5.37** (a) true (b) false (c) false (d) true **5.39** (a) true (b) true (c) true (d) false **5.41** (a) liquid (b) solid (c) liquid (d) solid **5.43** (a) 2 (b) 5 **5.45** (a) solid (b) solid (c) liquid (d) solid

5.47 (a)

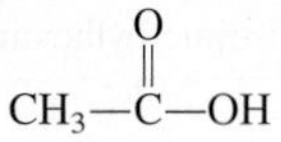

(b)

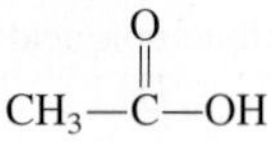

(c)

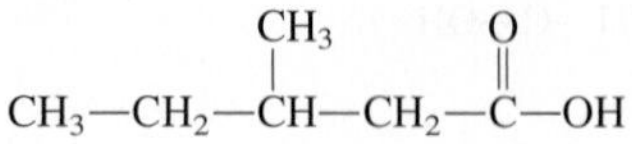

(d)

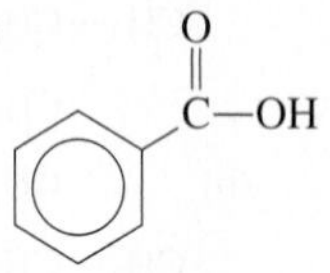

5.49 (a) 1 (b) 3 (c) 2 (d) 2 **5.51** (a) −1 (b) −3 (c) −2 (d) −2 **5.53** (a) pentanoate ion (b) citrate ion (c) succinate ion (d) oxalate ion

5.55

(a)

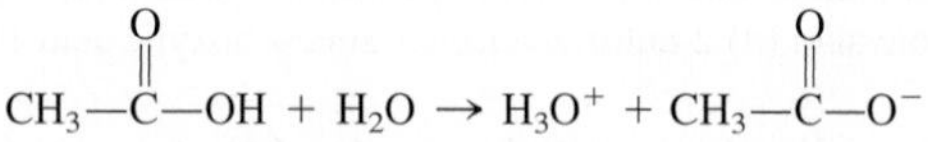

(b)

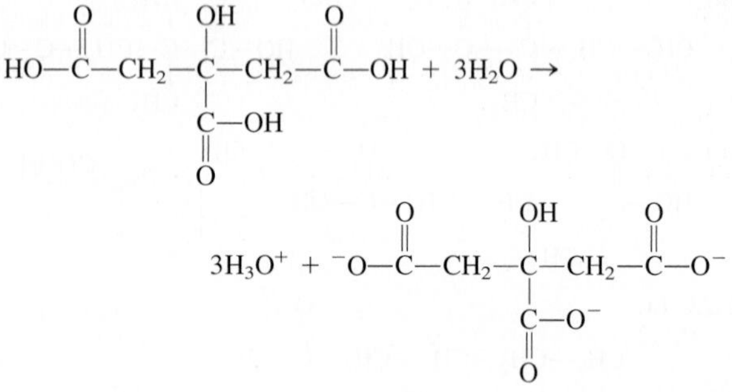

(c)

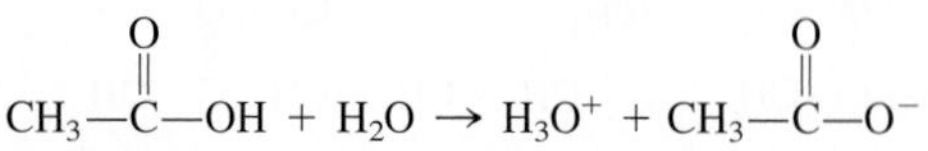

(d)

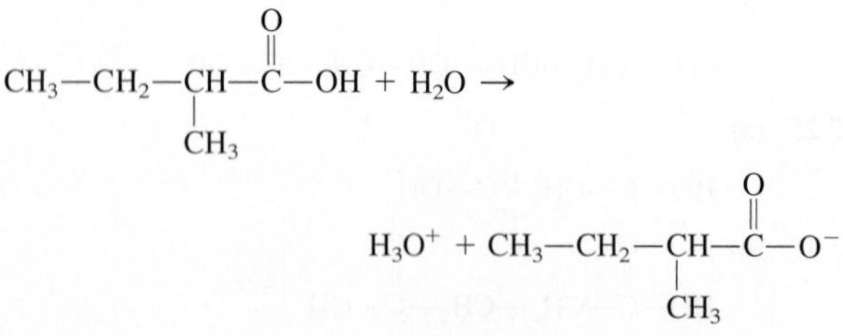

5.57 (a) potassium ethanoate (b) calcium propanoate (c) potassium butanedioate (d) sodium pentanoate **5.59** (a) potassium acetate (b) calcium propionate (c) potassium succinate (d) sodium valerate

5.61

(a)

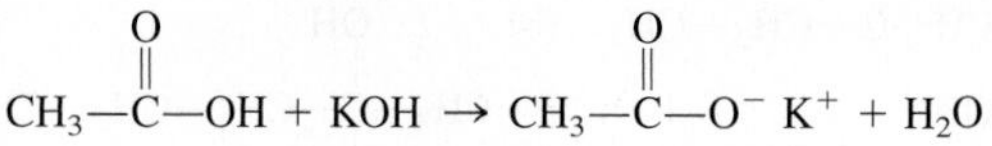

(b)

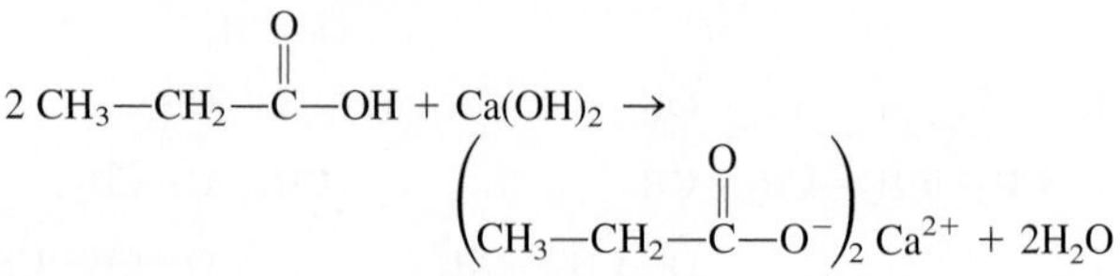

(c)

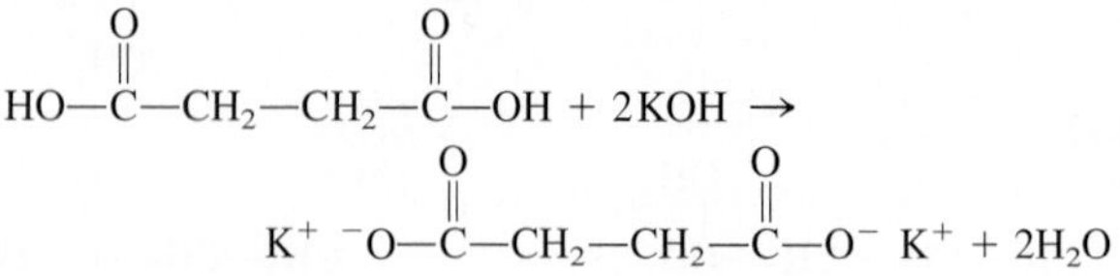

(d)

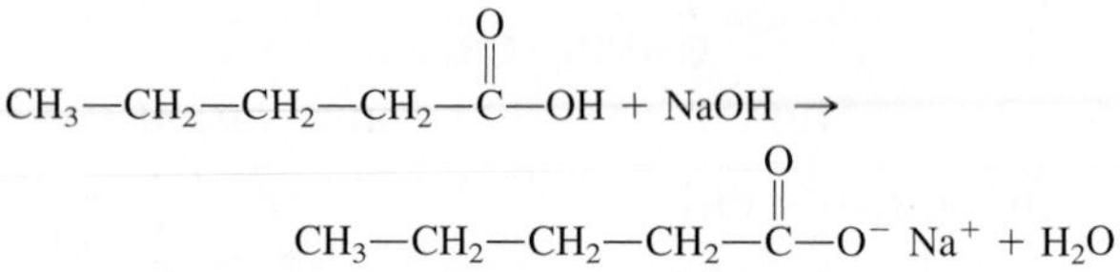

5.63

(a)

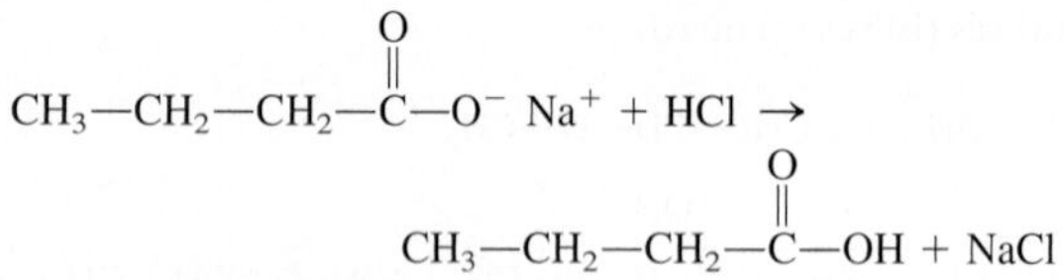

(b)

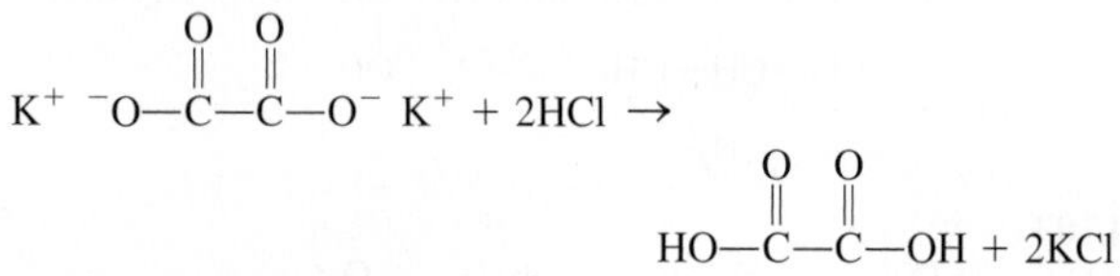

(c)

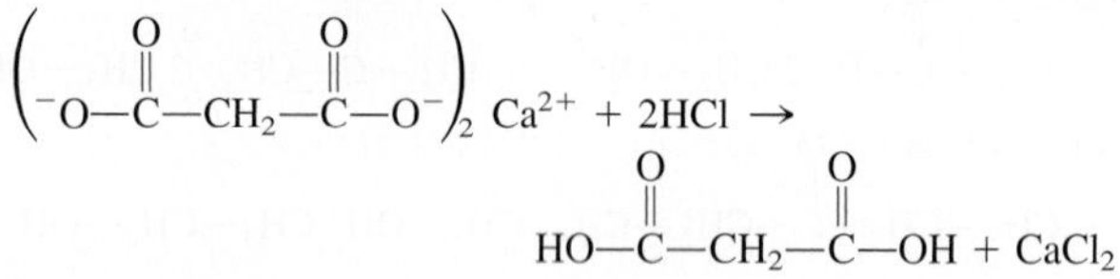

(d)

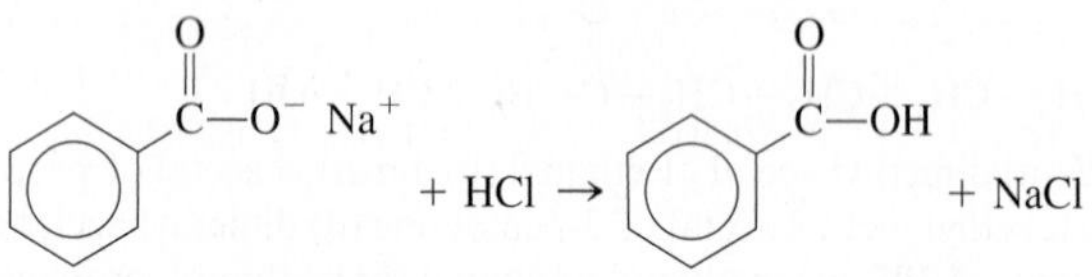

5.65 a compound used as a food preservative **5.67** (a) benzoic acid (b) sorbic acid (c) sorbic acid (d) propionic acid **5.69** (a) monocarboxylic acid salt (b) dicarboxylic acid salt (c) dicarboxylic acid salt (d) monocarboxylic acid salt **5.71** (a) two (b) three (c) two (d) three **5.73** (a) two oxygen atoms (b) two carbon atoms **5.75** (a) yes (b) yes (c) no (d) yes **5.77** (a) carboxylic acid (b) ester (c) carboxylic acid salt (d) carboxylic acid

5.79 (a)

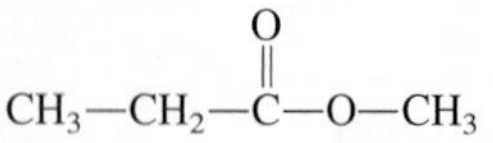

(b)

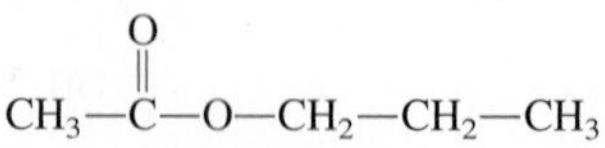

(c)

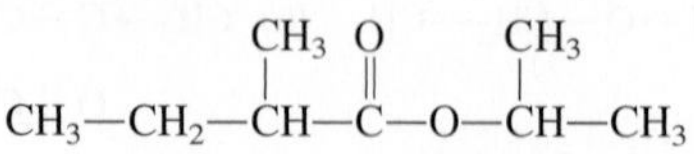

(d)

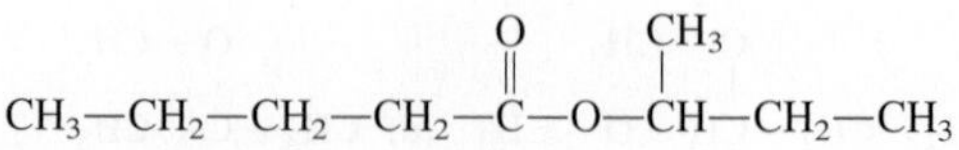

5.81
(a) CH$_3$—CH$_2$—C(=O)—OH; CH$_3$—CH$_2$—OH

(b) CH$_3$—CH$_2$—CH$_2$—C(=O)—OH; CH$_3$—OH

(c), (d)

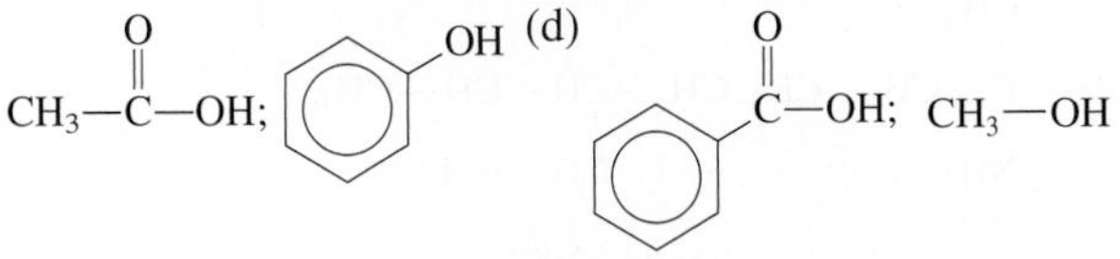

5.83 a cyclic ester; intermolecular esterification of a hydroxy-acid **5.85** (a) methyl propanoate (b) methyl methanoate (c) propyl ethanoate (d) isopropyl propanoate **5.87** (a) methyl propionate (b) methyl formate (c) propyl acetate (d) isopropyl propionate **5.89** (a) ethyl butanoate (b) propyl pentanoate (c) methyl 3-methylpentanoate (d) ethyl propanoate

5.91
(a) H—C(=O)—O—CH$_3$

(b)

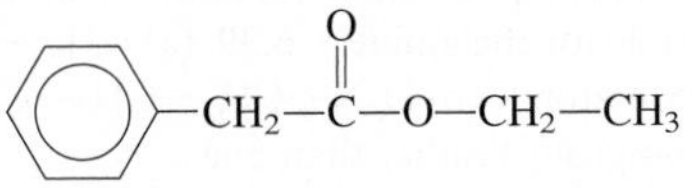

(c) CH$_3$—C(=O)—O—CH(CH$_3$)—CH$_3$ (d) CH$_3$—C(=O)—O—CH$_2$—CH(Br)—CH$_3$

5.93 (a) ethyl ethanoate (b) methyl ethanoate (c) ethyl butanoate (d) 1-methylpropyl hexanoate **5.95** (a) ester (b) carboxylic acid salt (c) ester (d) ester **5.97** (a) five (b) four (c) five (d) six **5.99** acid part is the same; alcohol part is methyl (apple) versus ethyl (pineapple) **5.101** the —OH group of salicylic acid has been esterified (methyl ester) in aspirin **5.103** Different species are involved (plant and animal). **5.105** (a) false (b) true (c) true (d) true **5.107** pentanoic acid, 2-methylbutanoic acid, 3-methylbutanoic acid, 2,2-dimethylpropanoic acid **5.109** methyl pentanoate, methyl 2-methylbutanoate, methyl 3-methylbutanoate methyl 2,2-dimethylpropanoate **5.111** nine (methyl butanoate, methyl 2-methylpropanoate, ethyl propanoate, propyl ethanoate, isopropyl ethanoate, butyl methanoate, *sec*-butyl methanoate, isobutyl methanoate, *tert*-butyl methanoate)

5.113
CH$_3$—CH$_2$—C(=O)—OH, CH$_3$—C(=O)—O—CH$_3$, H—C(=O)—O—CH$_2$—CH$_3$

5.115 $C_nH_{2n}O_2$ **5.117** No oxygen–hydrogen bonds are present. **5.119** There is no hydrogen bonding between ester molecules.

5.121
(a) CH$_3$—CH$_2$—C(=O)—OH; CH$_3$—CH$_2$—OH

(b)

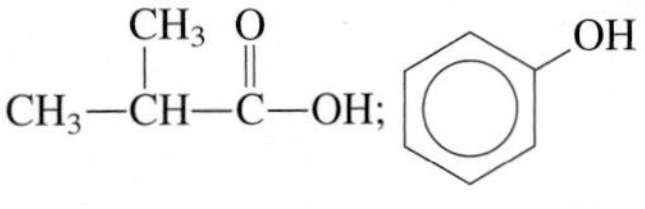

(c) CH$_3$—CH$_2$—CH$_2$—C(=O)—OH; CH$_3$—OH

(d)

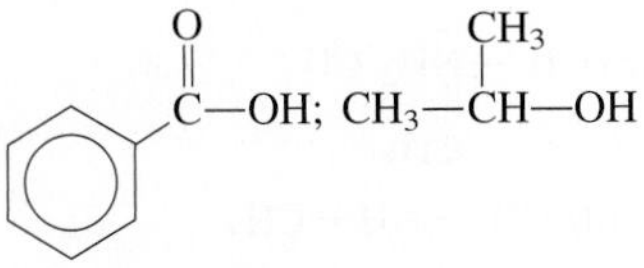

5.123 (a) CH$_3$—CH$_2$—C(=O)—O$^-$ Na$^+$; CH$_3$—CH$_2$—OH

(b)

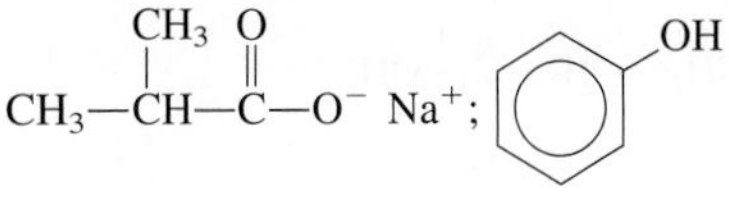

(c) CH$_3$—CH$_2$—CH$_2$—C(=O)—O$^-$ Na$^+$; CH$_3$—OH

(d)

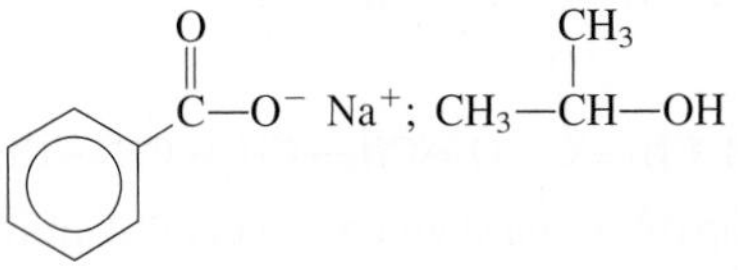

5.125 (a) CH$_3$—CH(CH$_3$)—C(=O)—OH; CH$_3$—CH$_2$—OH
(b) CH$_3$—CH(CH$_3$)—C(=O)—O$^-$ Na$^+$; CH$_3$—CH$_2$—OH
(c) H—C(=O)—OH; CH$_3$—CH$_2$—CH$_2$—CH$_2$—OH
(d) CH$_3$—C(=O)—O$^-$ Na$^+$; CH$_3$—CH(CH$_3$)—CH(CH$_3$)—CH$_2$—OH

5.127 (a) carboxylic acid salt (b) ester (c) carboxylic acid, alcohol (d) carboxylic acid salt, alcohol

5.129
(a) CH$_3$—C(=O)—S—CH$_2$—CH$_3$

(b) CH$_3$—(CH$_2$)$_8$—C(=O)—S—CH$_3$ (c)

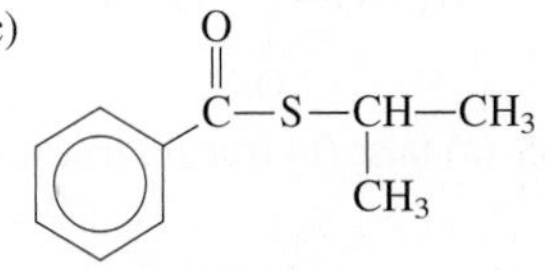

(d) H—C(=O)—S—CH$_2$—CH$_2$—CH$_3$

5.131 (a) CH$_3$—C(=O)—S—CH$_3$ (b) same as part a.
(c) H—C(=O)—S—CH$_2$—CH$_3$ (d) same as part c.

5.133 acetic acid and coenzyme A

5.135 —C(=O)—C(=O)—O—(CH$_2$)$_3$—O—C(=O)—C(=O)—O—(CH$_2$)$_3$—O—

5.137
HO—C(=O)—CH$_2$—CH$_2$—C(=O)—OH; HO—CH$_2$—CH$_2$—CH$_2$—OH

5.139 ethylene glycol and terephthalic acid **5.141** specialty packaging, orthopedic devices, and controlled drug-release formulations

5.143 (a) CH$_3$—CH$_2$—C(=O)—Cl (b) CH$_3$—CH—CH$_2$—C(=O)—Cl
(c) CH$_3$—CH$_2$—CH$_2$—C(=O)—O—C(=O)—CH$_2$—CH$_2$—CH$_3$
(d) CH$_3$—CH$_2$—CH$_2$—C(=O)—O—C(=O)—CH$_3$

5.145 (a) ethanoic propanoic anhydride (b) pentanoyl chloride (c) 2,3-dimethylbutanoyl chloride (d) methanoic propanoic anhydride
5.147 (a)
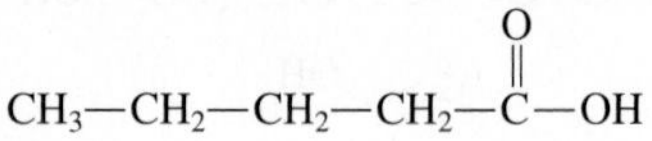
(b)
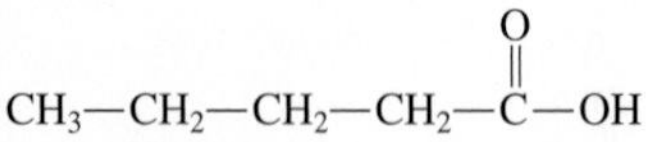
5.149 (a)
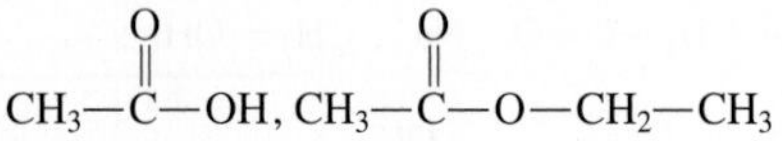
(b)
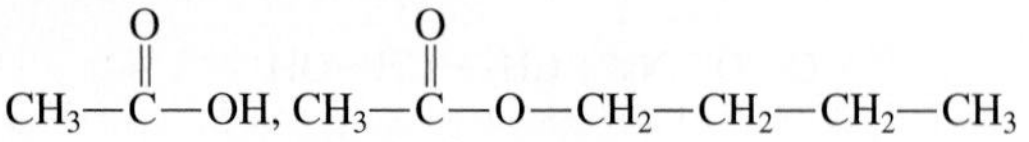
5.151 (a) propanoyl group (b) butanoyl group (c) butanoyl group (d) ethanoyl group **5.153** an ester and HCl
5.155
(a)
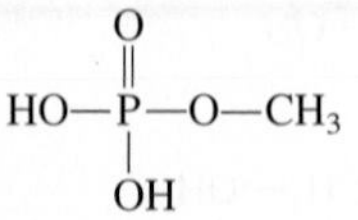
(b)
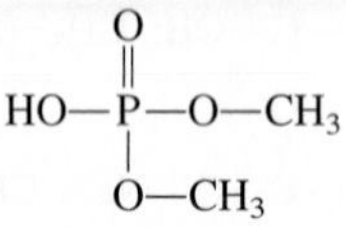
(c)
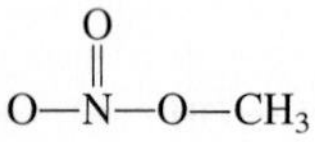
(d)
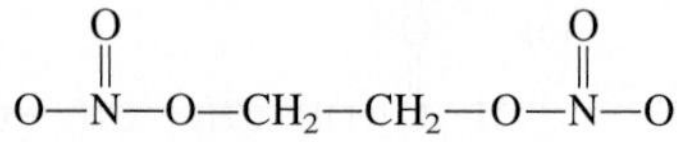
5.157 H_3PO_4 is a triprotic acid, and H_2SO_4 is a diprotic acid.
5.159
(a)
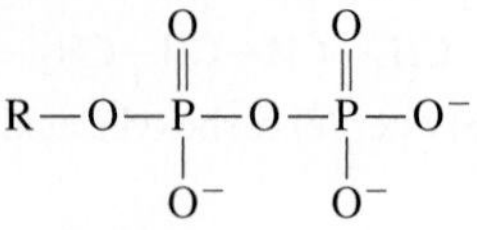
(b)
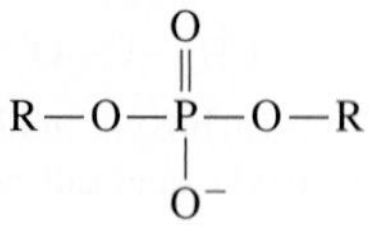
5.161 $—PO_3^{2-}$ group
5.163
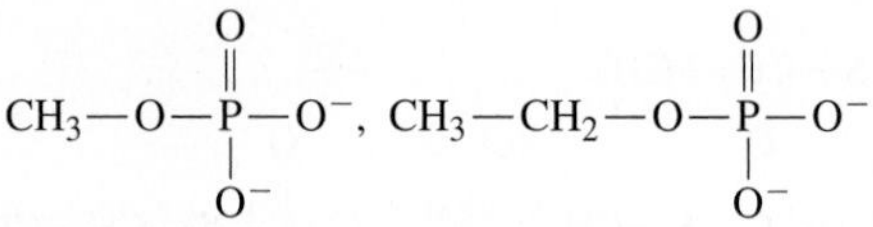
5.165 (a) false (b) true (c) true (d) true

Chapter 6 **6.1** 3 (N), 2 (O), and 4 (C)
6.3 (a) R—NH_2 (b) R—NH—R (c)
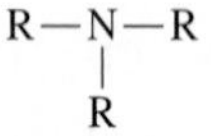
6.5 (a) 1 (b) 2 (c) 3 **6.7** (a) yes (b) yes (c) no (d) yes
6.9 (a) 1° (b) 1° (c) 2° (d) 3° **6.11** (a) 2° (b) 3° (c) 3° (d) 1°
6.13 (a) ethylmethylamine (b) propylamine (c) diethylmethylamine (d) isopropylmethylamine **6.15** (a) 3-pentanamine (b) 2-methyl-3-pentanamine (c) *N*-methyl-3-pentanamine (d) 2,3-butanediamine
6.17 (a) propylamine, 1-propanamine (b) diethylmethylamine, *N*-ethyl-*N*-methylethanamine (c) methylpropylamine, *N*-methyl-1-propanamine (d) (*sec*-butyl)methylamine, *N*-methyl-2-butanamine
6.19 (a) 2-bromoaniline (b) *N*-isopropylaniline (c) *N*-ethyl-*N*-methylaniline (d) *N*-methyl-*N*-phenylaniline
6.21 (a)
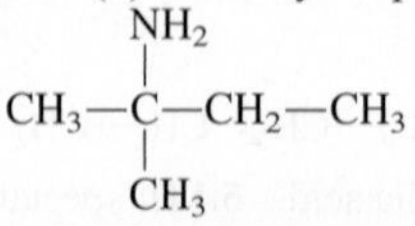
(b) $H_2N—CH_2—CH_2—CH_2—CH_2—CH_2—CH_2—NH_2$
(c)
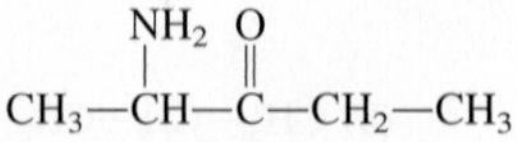
(d)
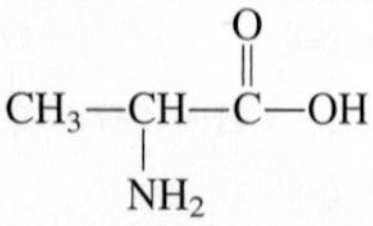
6.23 (a) 2° amine (b) 1° amine (c) 2° amine (d) 1° amine
6.25 (a) $C_5H_{13}N$ (b) C_3H_9N (c) $C_{12}H_{11}N$ (d) $C_8H_{17}N$

6.27
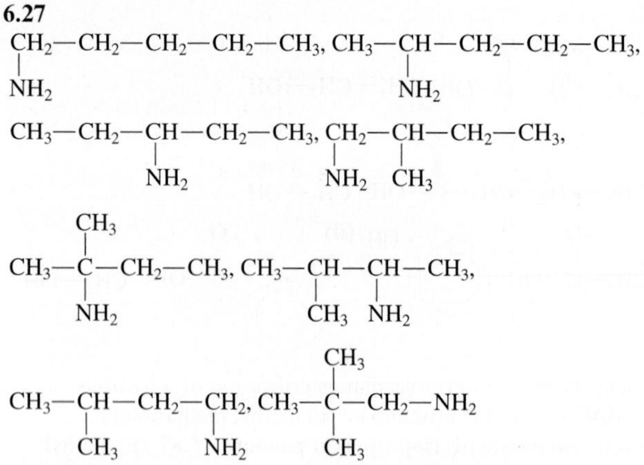
6.29 dimethylpropylamine, isopropyldimethylamine, diethylmethylamine **6.31** 1-propanamine, 2-propanamine, *N*-methylethanamine, *N,N*-dimethylmethanamine **6.33** (a) liquid (b) gas (c) gas (d) liquid **6.35** (a) 3 (b) 3 **6.37** Hydrogen bonding is possible for the amine. **6.39** (a) $CH_3—CH_2—NH_2$; it has fewer carbon atoms. (b) $H_2N—CH_2—CH_2—CH_2—NH_2$; it has two amino groups rather than one.
6.41 (a) $CH_3—CH_2—\overset{+}{N}H_3$ (b) OH^-
(c)
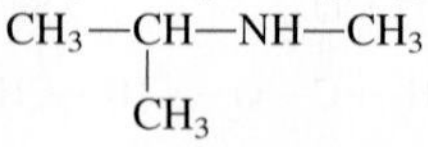
(d) $CH_3—CH_2—\overset{+}{N}H_2—CH_2—CH_3$; OH^-
6.43 (a) dimethylammonium ion (b) triethylammonium ion (c) *N,N*-diethylanilinium ion (d) *N*-isopropylanilinium ion
6.45 (a) $CH_3—NH—CH_3$
(b)
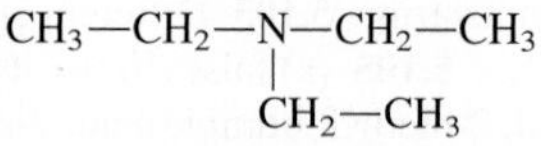
(c)
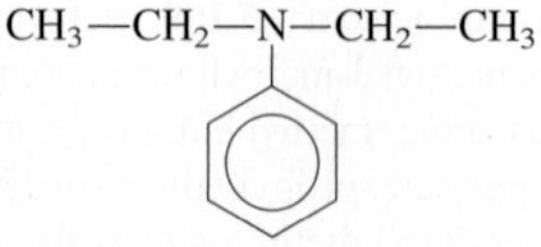
(d)
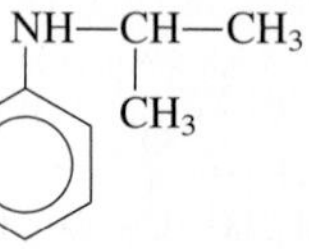
6.47
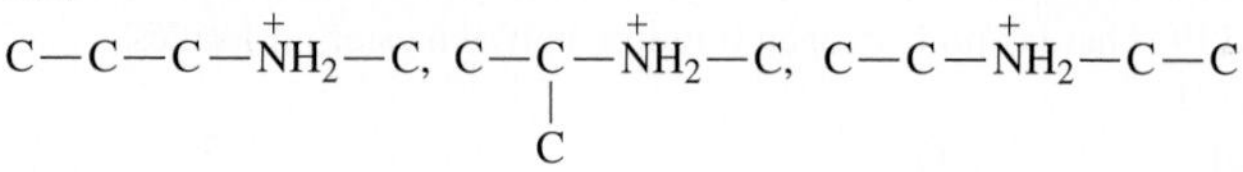
6.49 (a) $CH_3—CH_2—\overset{+}{N}H_3$ Cl^-
(b)
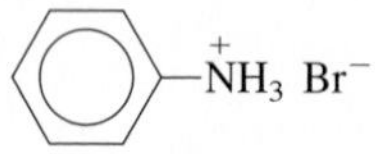
(c)
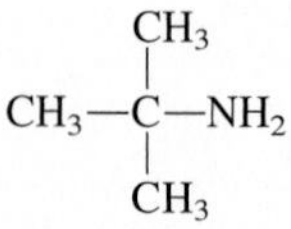
(d) HCl
6.51 (a)
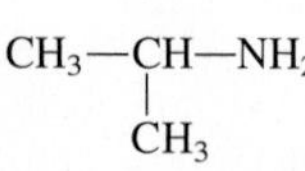
(b)
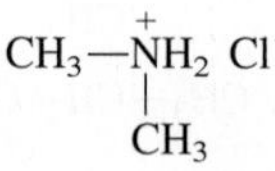
(c)
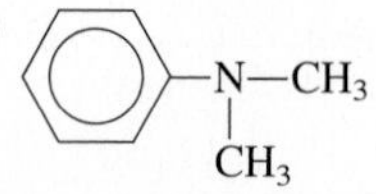
(d) $CH_3—NH—CH_3$

6.53 (a) propylammonium chloride (b) methylpropylammonium chloride (c) ethyldimethylammonium bromide (d) *N,N*-dimethylanilinium bromide **6.55** (a) 1-propanamine (b) *N*-methyl-1-propanamine (c) *N,N*-dimethylethanamine (d) *N,N*-dimethylphenylamine **6.57** to increase water solubility **6.59** (a) free amine, free base, deprotonated base (b) free amine, free base, deprotonated base (c) protonated base (d) protonated base

6.61 (a) $CH_3—CH_2—CH_2—NH_2$, NaCl, H_2O

(b) 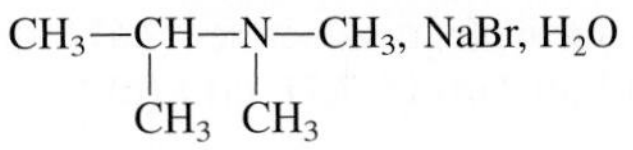

(c) $CH_3—CH_2—NH—CH_2—CH_3$, NaCl, H_2O

(d) 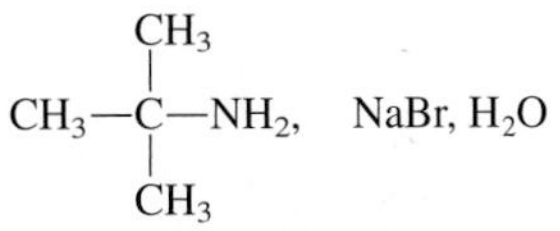

6.63 ethylmethylamine and propyl chloride, ethylpropylamine and methyl chloride, methylpropylamine and ethyl chloride

6.65 (a) 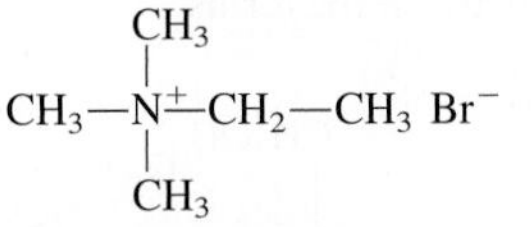

(b) 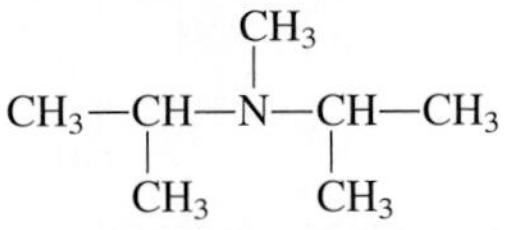

(c) 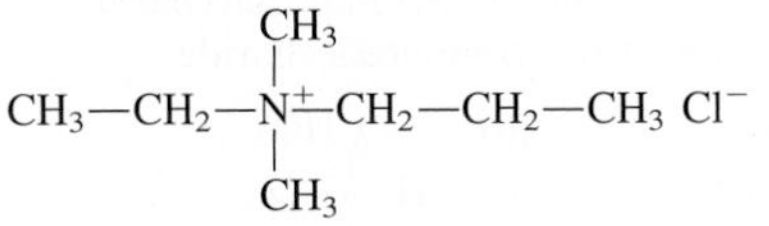

(d) $CH_3—CH_2—NH—CH_2—CH_3$

6.67 (a) amine salt (b) quaternary ammonium salt (c) amine salt (d) quaternary ammonium salt **6.69** (a) yes (b) no (c) yes (d) no **6.71** (a) trimethylammonium bromide (b) tetramethylammonium chloride (c) ethylmethylammonium bromide (d) diethyldimethylammonium chloride **6.73** (a) amine salt (b) quaternary ammonium salt (c) 3° amine (d) 1° amine **6.75** (a) yes (b) yes (c) no (d) yes **6.77** (a) saturated (b) unsaturated (c) unsaturated (d) unsaturated, fused **6.79** no difference **6.81** (a) purine (b) pyrrole (c) imidazole (d) indole **6.83** (a) false (b) true (c) true (d) true **6.85** (a) true (b) true (c) true (d) false **6.87** (a) true (b) true (c) false (d) false **6.89** (a) yes (b) yes (c) no (d) yes **6.91** (a) yes (b) yes (c) yes (d) yes **6.93** (a) true (b) true (c) false (d) false **6.95** (a) hydroxy, methoxy (b) two hydroxy (c) methoxy, keto (d) methoxy, keto, hydroxy **6.97** (a) true (b) true (c) true (d) false

6.99 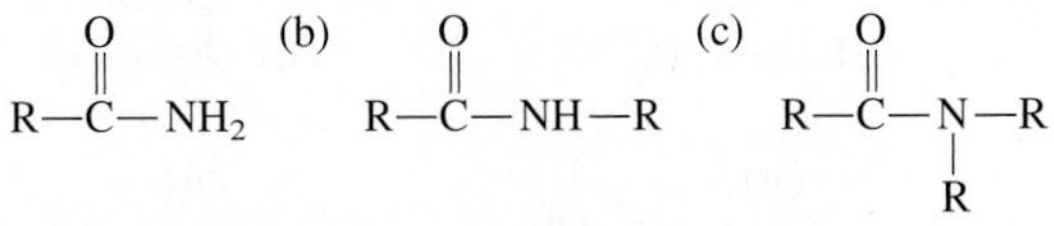

6.101 (a) 1° amide (b) 1° amide (c) all three (d) 3° amide **6.103** (a) yes (b) yes (c) no (d) yes **6.105** (a) monosubstituted (b) disubstituted (c) unsubstituted (d) monosubstituted **6.107** (a) secondary amide (b) tertiary amide (c) primary amide (d) secondary amide **6.109** (a) amide (b) amine (c) amine (d) amine and amide **6.111** (a) *N*-ethylethanamide (b) *N,N*-dimethylpropanamide (c) butanamide (d) 2-chloropropanamide **6.113** (a) *N*-ethylacetamide (b) *N,N*-dimethylpropionamide (c) butyramide (d) 2-chloropropionamide **6.115** (a) propanamide (b) *N*-methylpropanamide (c) 3,5-dimethylhexanamide (d) *N,N*-dimethylbutanamide

6.117 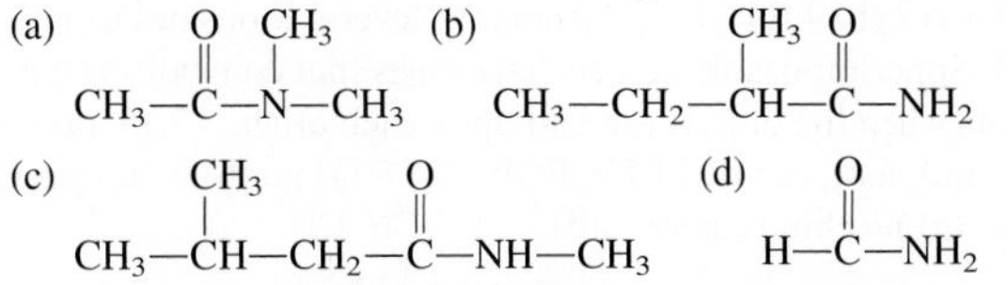

6.119 (a) hexanamide (b) 1-hexanamine (c) 5-aminopentanamide (d) 2,5-dimethyl-1,6-hexanediamine **6.121** (a) true (b) false (c) true (d) true **6.123** inhibition of insect olfactory receptors **6.125** (a) true (b) true (c) true (d) true **6.127** draws N lone pair of electrons closer to the N atom **6.129** (a) no, an amide (b) yes, an amine (c) no, an amide (d) yes, an amine **6.131** An electronegativity effect induced by the carbonyl oxygen atom makes the lone pair of electrons on the nitrogen atom unavailable. **6.133** (a) 4 (b) 4

6.135 (a) $CH_3—NH_2$ (b) 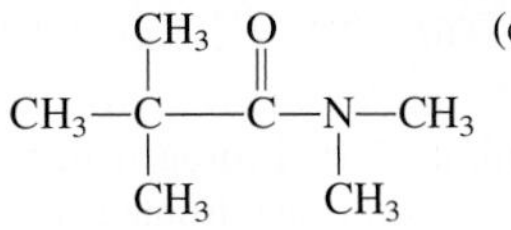 (c) NH_3

(d) 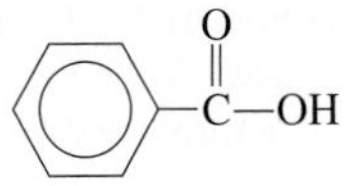

6.137 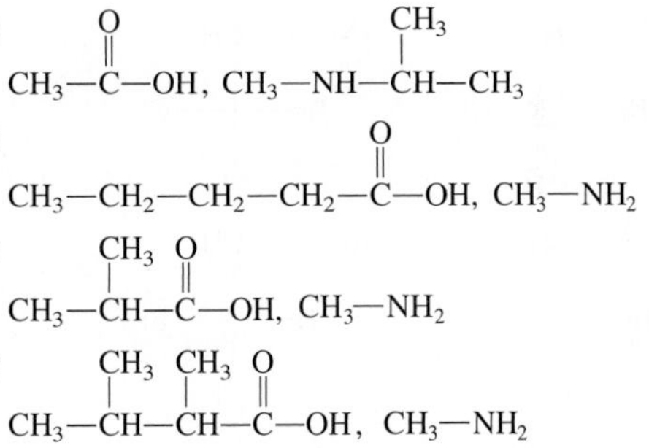

6.139 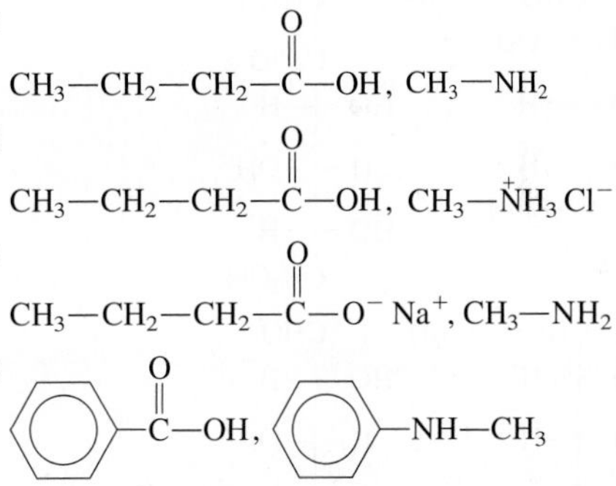

6.141 (a) amine salt (b) carboxylic acid, amine salt (c) carboxylic acid salt, amine (d) amine **6.143** diacid and diamine

6.145 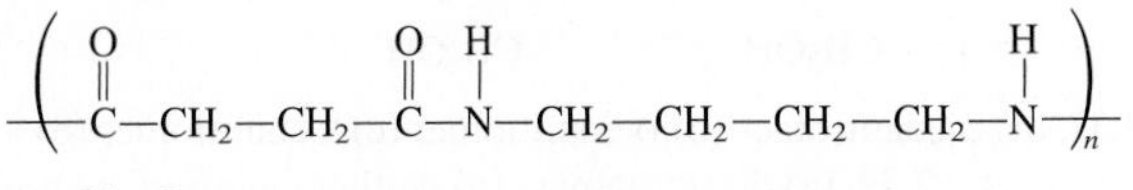

6.147 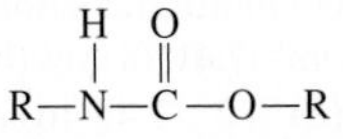

Chapter 7 **7.1** (a) study of the chemical substances in living organisms and the interactions of these substances with each other (b) chemical substance found within a living organism **7.3** proteins, lipids, carbohydrates, and nucleic acids

7.5 $CO_2 + H_2O + \text{solar energy} \xrightarrow[\text{Plant enzymes}]{\text{Chlorophyll}} \text{carbohydrates} + O_2$

7.7 serve as structural elements, provide energy reserves **7.9** Carbohydrates are polyhydroxy aldehydes, polyhydroxy ketones, or compounds that yield such substances upon hydrolysis.

7.11 (a) 2 (b) 4 (c) 3 to 10 (d) many (several thousand usually) **7.13** Superimposable objects have parts that coincide exactly at all points when the objects are laid upon each other. **7.15** (a) drill bit (b) hand, foot, ear (c) PEEP, POP **7.17** (a) no (b) no (c) yes (d) yes **7.19** (a) no chiral center (b)

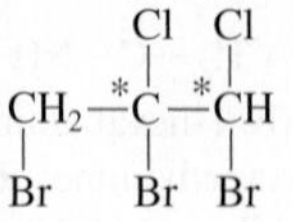

(c)

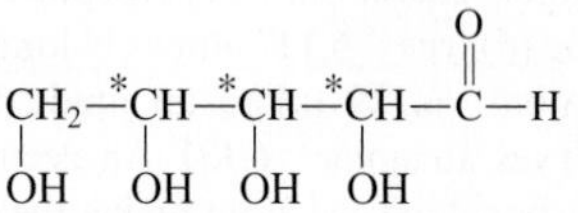

(d)

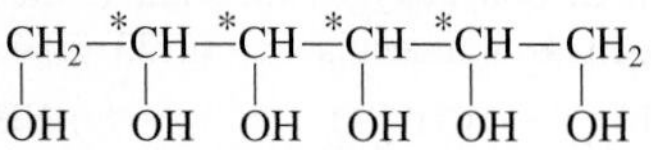

7.21 (a) zero (b) two (c) zero (d) zero **7.23** (a) achiral (b) chiral (c) chiral (d) chiral **7.25** hydrogen, methyl, ethyl, propyl or isopropyl **7.27** Constitutional isomers have a different connectivity of atoms. Stereoisomers have the same connectivity of atoms with different arrangements of the atoms in space. **7.29** the presence of a chiral center and the presence of "structural rigidity" **7.31** (a) true (b) true (c) false (d) false

7.33 (a)

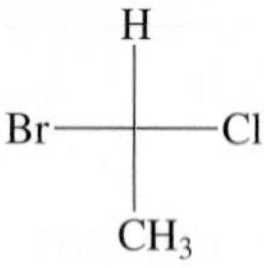

(b)

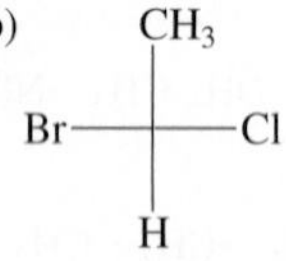

(c)

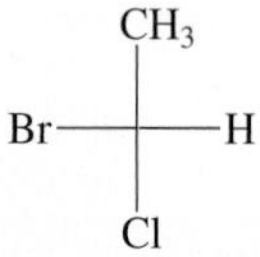

(d)

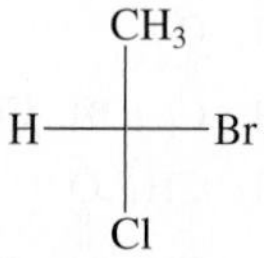

7.35 (a)

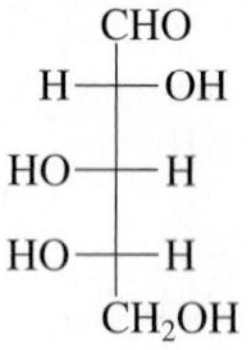

(b)

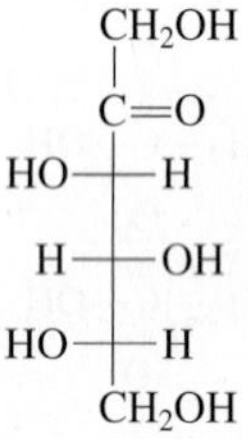

(c)

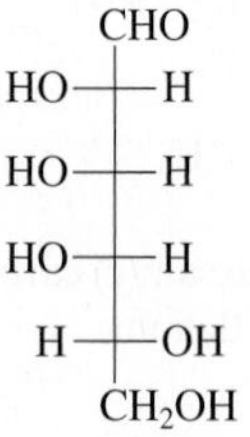

(d)

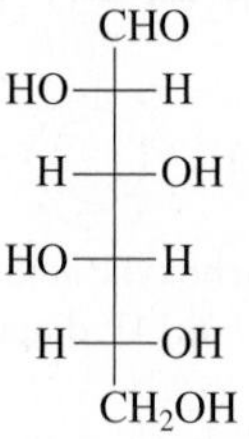

7.37 (a) D enantiomer (b) D enantiomer (c) L enantiomer (d) L enantiomer **7.39** (a) diastereomers (b) neither enantiomers nor diastereomers (c) enantiomers (d) diastereomers **7.41** (a) yes (b) no (c) no (d) yes **7.43** (a) 3, 3 (b) 2, 2 (c) 4, 4 (d) 2, 2 **7.45** (d) effect on plane-polarized light **7.47** (a) same (b) different (c) same (d) different **7.49** (a) no (b) yes (c) no (d) no **7.51** (a) aldose (b) ketose (c) ketose (d) ketose **7.53** (a) aldohexose (b) ketohexose (c) ketotriose (d) ketotetrose **7.55** (a) D-galactose (b) D-psicose (c) dihydroxyacetone (d) L-erythrulose **7.57** (a) 4 (b) 3 (c) 0 (d) 1 **7.59** (a) carbon 4 (b) carbons 1 and 2 (c) carbons 1 and 2 (d) carbon 2 **7.61** (a) aldoses, hexoses, aldohexoses (b) hexoses (c) hexoses (d) aldoses

7.63
(a)

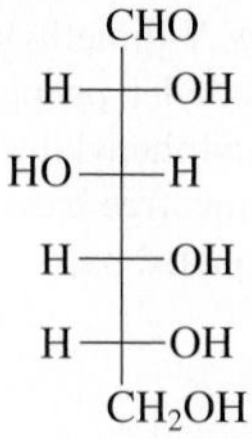

(b)

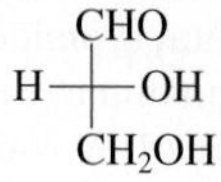

(c)

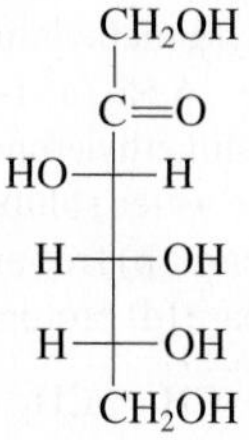

(d)

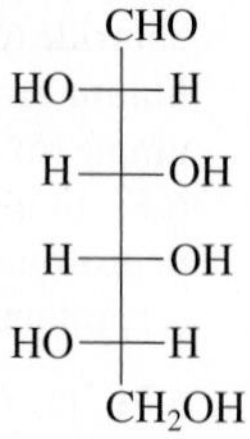

7.65 (a) D-fructose (b) D-glucose (c) D-galactose **7.67** (a) 6 (b) 5 (c) 5 (d) 4 **7.69** (a) 1 (b) 1 (c) 2 (d) 1 **7.71** (a) 1 (b) 1

7.73

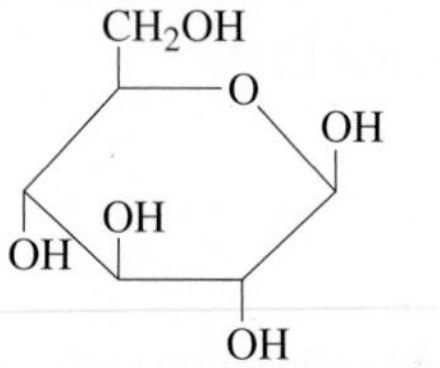

7.75 alpha-anomer **7.77** The cyclic and noncyclic forms interconvert; an equilibrium exists between the forms.

7.79 (a)

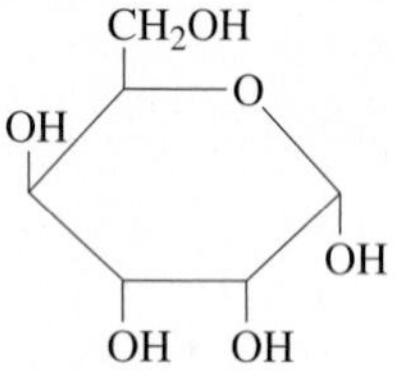

(b)

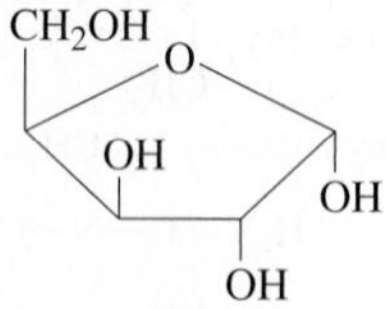

7.81 (a) α-D-monosaccharide (b) α-D-monosaccharide (c) β-D-monosaccharide (d) α-D-monosaccharide

7.83 (a)

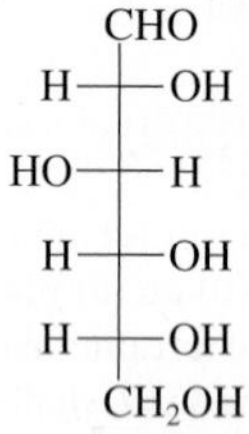

(b)

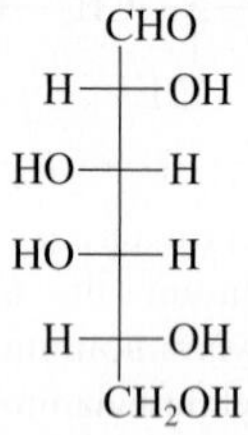

(c)

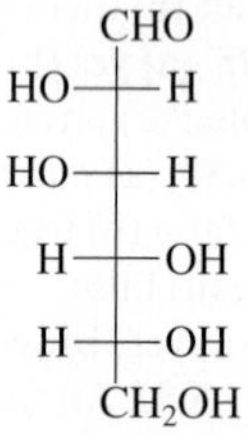

(d)

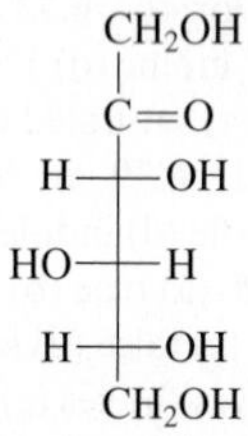

7.85 (a) α-D-glucose (b) α-D-galactose (c) β-D-mannose (d) α-D-sorbose

7.87 (a)

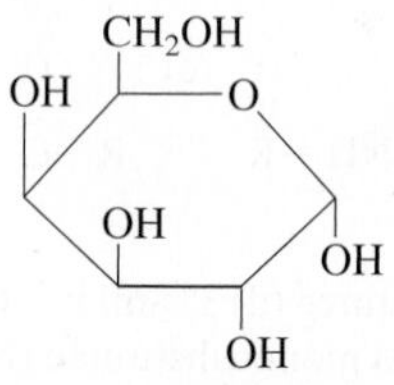

(b)

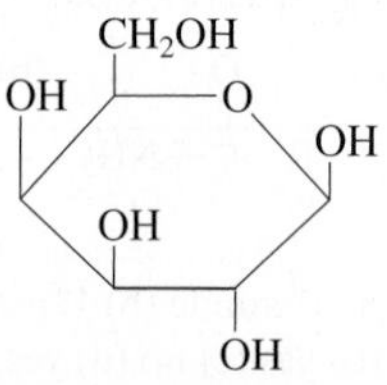

(c)

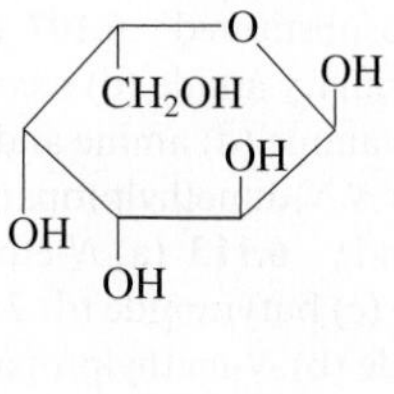

(d)

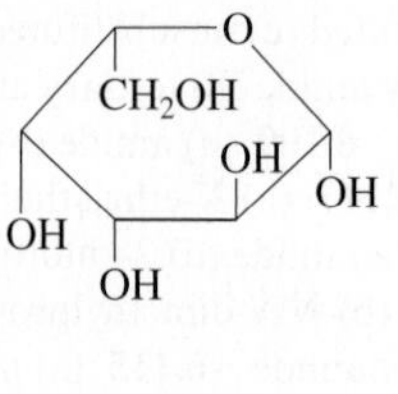

7.89 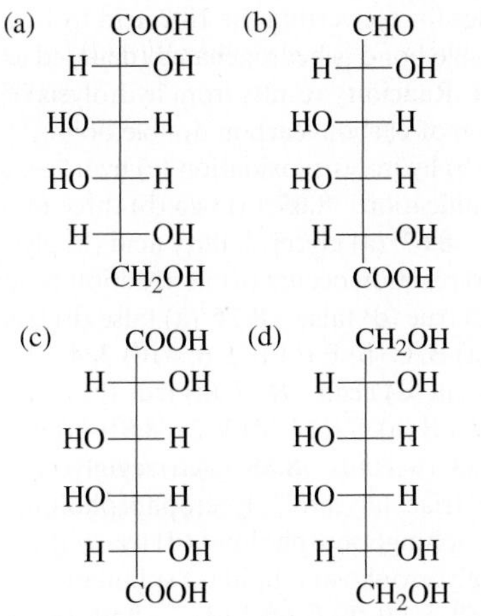

7.91 (a) acidic sugar (b) acidic sugar (c) acidic sugar (d) sugar alcohol **7.93** (a) galactonic acid (b) galacturonic acid (c) galactaric acid (d) galactitol **7.95** (a) reducing sugar (b) reducing sugar (c) reducing sugar (d) reducing sugar **7.97** (a) yes (b) yes (c) yes (d) yes **7.99** (a) alpha (b) beta (c) alpha (d) beta **7.101** (a) methyl alcohol (b) ethyl alcohol (c) ethyl alcohol (d) methyl alcohol **7.103** (a) methyl-α-D-alloside (b) ethyl-β-D-altroside (c) ethyl-α-D-fructoside (d) methyl-β-D-glucoside

7.105

(a) 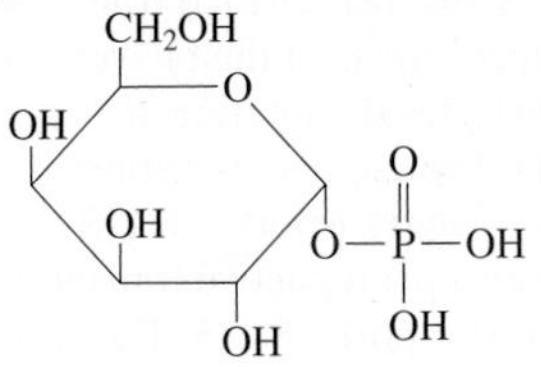 (b) 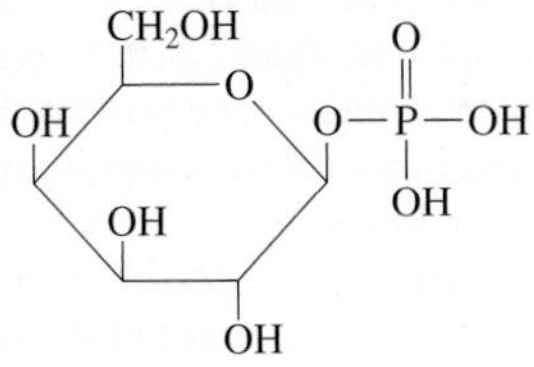

7.107 (a) 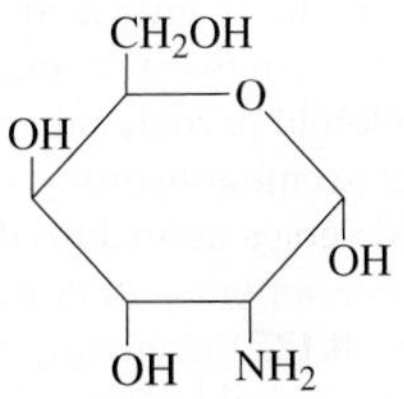 (b) 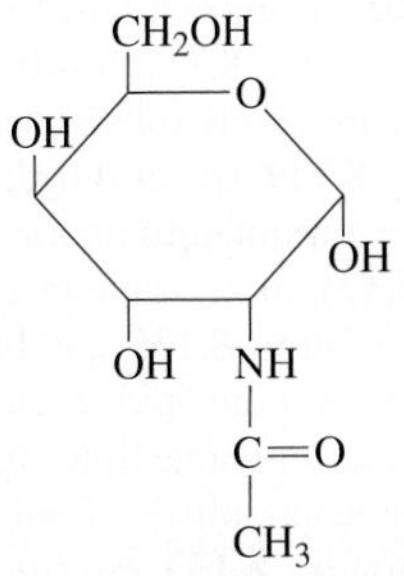

7.109 (a) maltose, cellobiose (b) maltose (c) lactose (d) lactose, sucrose **7.111** (a) no (b) yes (c) yes (d) yes **7.113** (a) one acetal and one hemiacetal (b) one acetal and one hemiacetal (c) one acetal and one hemiacetal (d) one acetal and one hemiacetal **7.115** (a) alpha (b) beta (c) alpha (d) beta **7.117** (a) reducing sugar (b) reducing sugar (c) reducing sugar (d) reducing sugar

7.119 (a) 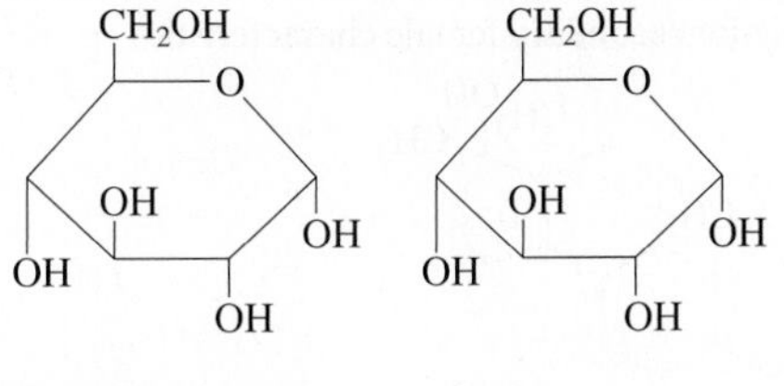

(b) 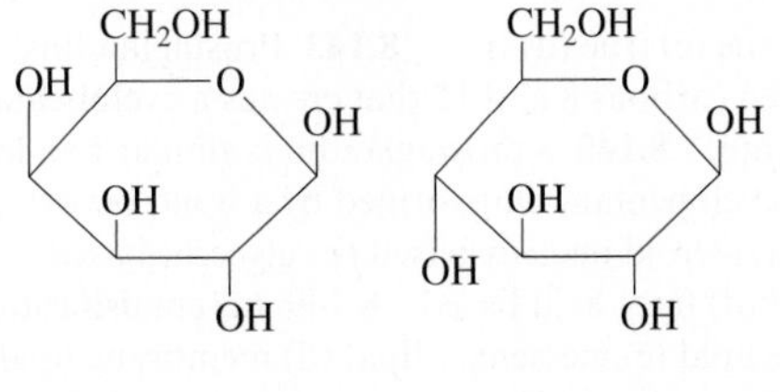

(c) 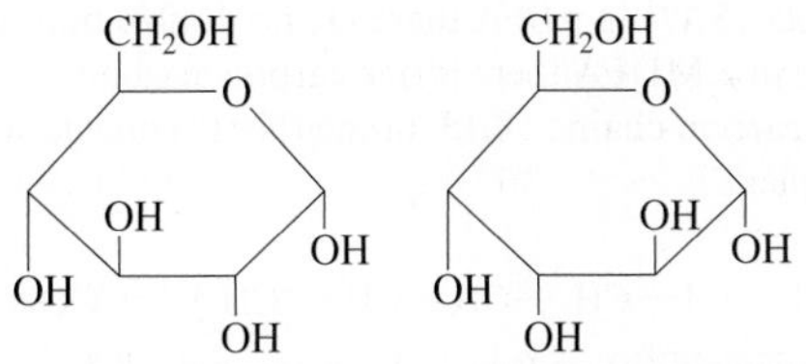

(d) 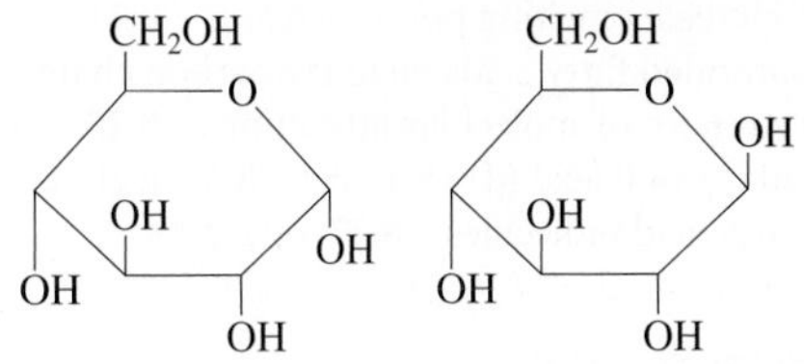

7.121 (a) $\alpha(1\rightarrow6)$ (b) $\beta(1\rightarrow4)$ (c) $\alpha(1\rightarrow4)$ (d) $\alpha(1\rightarrow4)$

7.123 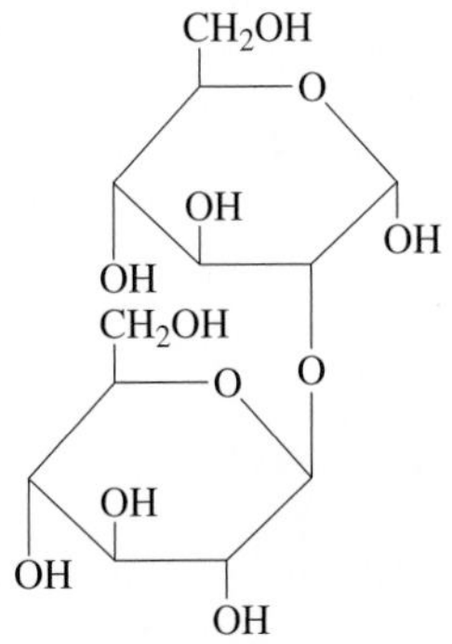

7.125 (a) monosaccharide, reducing sugar, anomers, aldohexose (b) disaccharide (c) monosaccharide, reducing sugar, enantiomers (d) reducing sugar **7.127** (a) true (b) false (c) false (d) false **7.129** (a) false (b) false (c) true (d) false **7.131** (a) true (b) false (c) false (d) false **7.133** (a) 3 (b) 3 (c) 2 (d) 2 **7.135** (a) no (b) no (c) yes (d) yes **7.137** (a) $\alpha(1\rightarrow4)$ (b) none (c) $\alpha(1\rightarrow6)$, $\alpha,\beta(1\rightarrow2)$ (d) none **7.139** (a) false (b) true (c) true (d) true **7.141** they are two names for the same thing **7.143** less than 100 monomer units up to a million monomer units **7.145** (a) correct (b) incorrect (c) incorrect (d) correct **7.147** (a) amylopectin is more abundant (b) amylopectin has the longer polymer chain (c) amylopectin has two types of glycosidic linkages and amylose one type of glycosidic linkage (d) same for both **7.149** (a) correct (b) incorrect (c) incorrect (d) correct **7.151** (a) to neither (b) to cellulose only (c) to neither (d) to both **7.153** (a) amylopectin, amylose, glycogen (b) amylose, cellulose, chitin (c) amylose, cellulose, chitin (d) chitin **7.155** (a) true (b) false (c) true (d) true **7.157** (a) homopolysaccharide (b) homopolysaccharide, branched polysaccharide (c) homopolysaccharide, unbranched polysaccharide (d) homopolysaccharide, unbranched polysaccharide **7.159** (a) storage (b) non (c) acidic (d) structural **7.161** (a) amylose, cellulose, chitin, heparin (b) amylopectin, glycogen (c) hyaluronic acid (d) amylose, heparin **7.163** Simple carbohydrates are the mono- and disaccharides, and complex carbohydrates are the polysaccharides. **7.165** carbohydrates that provide energy but few other nutrients **7.167** (a) false (b) false (c) true (d) true **7.169** a lipid molecule that has a carbohydrate molecule covalently bonded to it **7.171** interaction between the carbohydrate unit of one cell and a protein imbedded in the cell membrane of another cell

Chapter 8 **8.1** All lipids are insoluble or only sparingly soluble in water. **8.3** (a) insoluble (b) soluble (c) insoluble (d) soluble **8.5** energy-storage lipids, membrane lipids, emulsification lipids, messenger lipids, and protective-coating lipids **8.7** (a) long-chain (b) short-chain (c) long-chain (d) medium-chain **8.9** (a) saturated (b) polyunsaturated (c) polyunsaturated

(d) monounsaturated **8.11** In a SFA there are no double bonds in the carbon chain; in a MUFA there is one carbon–carbon double bond in the carbon chain. **8.13** (a) neither (b) omega-3 (c) omega-3 (d) neither

8.15

$$CH_3-(CH_2)_4-CH=CH-CH_2-CH=CH-(CH_2)_7-COOH$$

8.17 (a) tetradecanoic acid (b) cis-9-hexadecenoic acid **8.19** as carbon chain length increases melting point increases **8.21** *cis* double bonds in unsaturated fatty acids bend the carbon chain, which decreases the strength of molecular attractions **8.23** (a) 18:1 acid (b) 18:3 acid (c) 14:0 acid (d) 18:1 acid **8.25** a glycerol molecule and three fatty acid molecules **8.27** one, ester

8.29

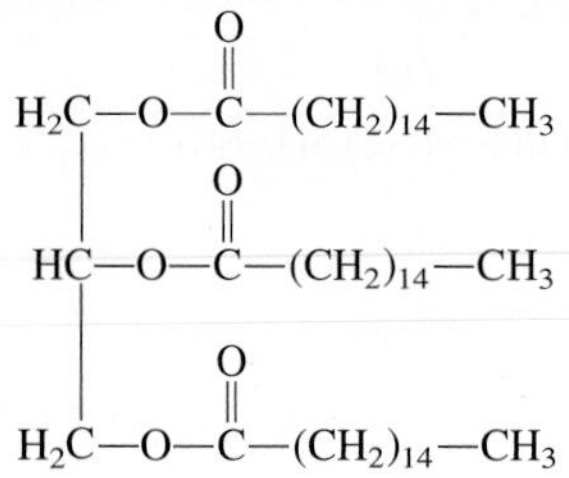

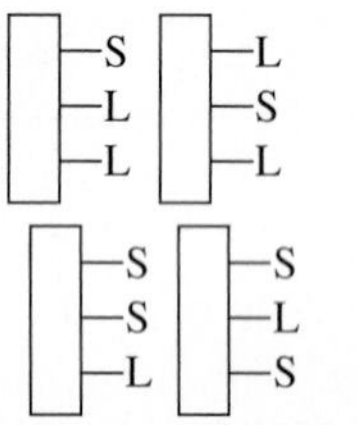

$$H_2C-O-\overset{\overset{O}{\|}}{C}-(CH_2)_{14}-CH_3$$
$$HC-O-\overset{\overset{O}{\|}}{C}-(CH_2)_{14}-CH_3$$
$$H_2C-O-\overset{\overset{O}{\|}}{C}-(CH_2)_{14}-CH_3$$

8.31

S	L
L	S
L	L

S	S
S	L
L	S

8.33 (a) palmitic, myristic, oleic (b) oleic, palmitic, palmitoleic
8.35 (a) no difference (b) A triacylglycerol may be a solid or a liquid; a fat is a triacylglycerol that is a solid. (c) A triacylglycerol can have fatty acid residues that are all the same, or two or more different kinds may be present. In a mixed triacylglycerol, two or more different fatty acid residues must be present. (d) A fat is a triacylglycerol that is a solid; an oil is a triacylglycerol that is a liquid.
8.37 a. (a) 3 (b) 0 (c) 2 (d) 0 b. (a) 3 (b) 0 (c) 1 (d) 0
8.39 (a) not correct (b) not correct **8.41** (a) correct (b) not correct **8.43** (a) nonessential fatty acid (b) essential fatty acid (c) nonessential fatty acid (d) nonessential fatty acid **8.45** a. (a) 0 (b) 0 (c) 1 (d) 1 b. (a) 0 (b) 0 (c) 2 (d) 3 **8.47** (a) false (b) false (c) true (d) true **8.49** (a) true (b) true (c) false (d) false

8.51

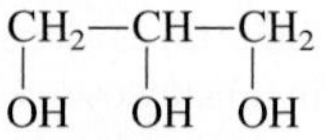

$$CH_2-CH-CH_2$$
$$\ \ |\quad\ \ |\quad\ \ |$$
$$OH\ \ OH\ \ OH$$

$$CH_3-(CH_2)_{12}-COOH$$
$$CH_3-(CH_2)_{14}-COOH$$
$$CH_3-(CH_2)_7-CH=CH-(CH_2)_7-COOH$$

8.53

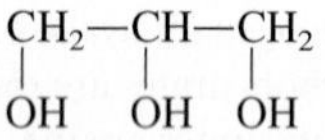

$$CH_2-CH-CH_2$$
$$\ \ |\quad\ \ |\quad\ \ |$$
$$OH\ \ OH\ \ OH$$

$$CH_3-(CH_2)_{12}-COO^-\ Na^+$$
$$CH_3-(CH_2)_{14}-COO^-\ Na^+$$
$$CH_3-(CH_2)_7-CH=CH-(CH_2)_7-COO^-\ Na^+$$

8.55 Carbon–carbon double bond(s) must be present. **8.57** six

8.59 (a)

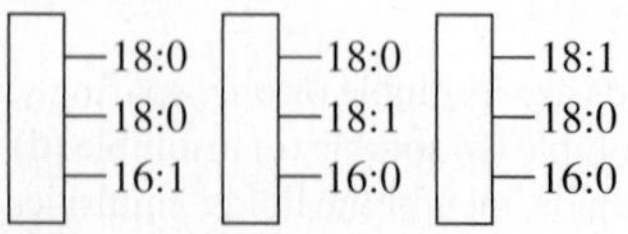

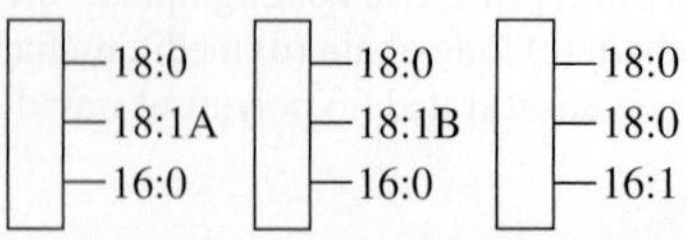

18:0	18:0	18:1
18:0	18:1	18:0
16:1	16:0	16:0

(b)

18:0	18:0	18:0
18:1A	18:1B	18:0
16:0	16:0	16:1

There are two possibilities for converting the 18:2 acid to 18:1 acid depending on which double bond is hydrogenated (denoted as 18:1A and 18:1B). **8.61** Rancidity results from hydrolysis of ester linkages and oxidation of carbon–carbon double bonds.
8.63 (a) hydrogenation (b) hydrolysis, oxidation (c) hydrogenation (d) hydrolysis, saponification **8.65** (a) two (b) three or four (c) two (d) three or four **8.67** (a) glycerol, fatty acid (b) glycerol, fatty acid salt (c) no reaction occurs (d) no reaction occurs
8.69 (a) false (b) false (c) true (d) false **8.71** (a) false (b) true (c) true (d) true **8.73** (a) B, C (b) E (c) 1, 2, 3, 4 (d) 3, 4
8.75 (a) tail (b) tail (c) head (d) head **8.77** (a) true (b) true (c) false (d) true **8.79** (a) B (b) C (c) 1 (d) 1, 2 **8.81** (a) yes (b) yes (c) no (d) no **8.83** two tails **8.85** (a) triacylglycerol, sphingophospholipid (b) triacylglycerol, glycerophospholipid (c) glycerophospholipid, sphingophospholipid (d) triacylglycerol
8.87 (a) none apply (b) glycerophospholipid (c) sphingophospholipid (d) none apply **8.89** (a) B (b) C (c) 1 (d) 2 **8.91** (a) (1) (b) (2) (c) (3) (d) (3) **8.93** (a) glycerophospholipid, sphingophospholipid, sphingoglycolipid (b) glycerophospholipid, sphingophospholipid, sphingoglycolipid (c) sphingophospholipid, sphingoglycolipid (d) glycerophospholipid, sphingophospholipid, sphingoglycolipid **8.95** (a) true (b) false (c) true (d) false
8.97 (a) 3 (b) 0 (c) 1 (d) 2 **8.99** Swiss cheese, fish fillet, chicken, liver **8.101** "Good cholesterol" is that present in HDLs, and "bad cholesterol" is that present in LDLs.
8.103 head and two tail structure **8.105** (a) false (b) true (c) false (d) false **8.107** creates "open" areas in the bilayer
8.109 a membrane protein that penetrates the interior of the lipid bilayer (cell membrane) **8.111** Protein help is required in facilitated transport but not in passive transport. **8.113** (a) active transport (b) facilitated transport (c) active transport (d) passive transport and facilitated transport **8.115** The first membrane has more fluid-like character; more fatty acid unsaturation is present. **8.117** a substance that can disperse and stabilize water-insoluble substances as colloidal particles in an aqueous solution **8.119** tri- or dihydroxy versus monohydroxy; oxidized side chain amidified to an amino acid versus nonoxidized side chain **8.121** amino acid glycine versus amino acid taurine
8.123 bile fluid **8.125** gall bladder **8.127** (a) energy-storage lipid (b) membrane lipid (c) energy-storage lipid (d) membrane lipid **8.129** (a) saponifiable lipid (b) saponifiable lipid (c) saponifiable lipid (d) nonsaponifiable lipid **8.131** sex hormones, adrenocortical hormones **8.133** estradiol has an —OH on carbon 3, while testosterone has a ketone group at this location; testosterone has an extra —CH₃ group at carbon 10 **8.135** (a) adrenocorticoid hormone (b) sex hormone (c) adrenocorticoid hormone (d) sex hormone **8.137** (a) control Na^+/K^+ ion balance in cells and body fluids (b) responsible for secondary male characteristics (c) controls glucose metabolism and is an anti-inflammatory agent (d) responsible for secondary female characteristics

8.139

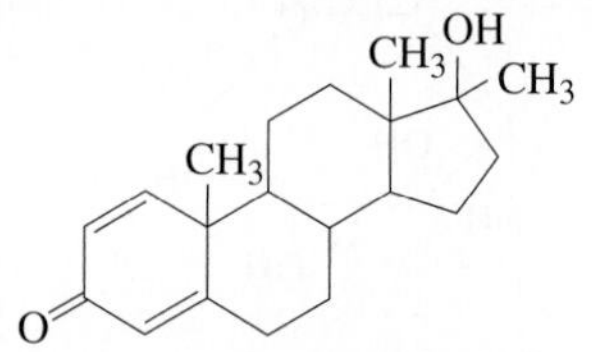

8.141 (a) true (b) true (c) true (d) true **8.143** Prostaglandins have a bond between carbons 8 and 12 that creates a cyclopentane ring structural feature. **8.145** a prostaglandin is similar to a leukotriene but has a cyclopentane ring formed by a bond between C8 and C12 **8.147** (a) steroid nucleus-based (b) glycerol-based (c) fatty acid-based (d) fatty-acid based **8.149** (a) emulsification lipid (b) membrane lipid (c) messenger lipid (d) membrane lipid
8.151 (a) false (b) false (c) true (d) true

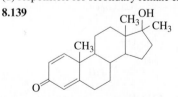

8.153 | long-chain alchohol | — | long-chain fatty acid | **8.155** mixture of esters involving a long-chain fatty acid and a long-chain alcohol versus a long-chain alkane mixture **8.157** (a) bile acid (b) biological wax, sphingophospholipid (c) biological wax, sphingophospholipid (d) sphingophospholipid **8.159** (a) yes (b) no (c) yes (d) yes **8.161** (a) ester linkage (b) ester linkage (c) none apply (d) none apply **8.163** (a) yes (b) no (c) no (d) no **8.165** (a) 0 (b) 3 (c) 0 (d) 1

Chapter 9 **9.1** amino acids **9.3** 15% by mass **9.5** (a) yes (b) no (c) no (d) yes **9.7** the identity of the R group (side chain) **9.9** (a) phenylalanine, tyrosine, tryptophan (b) methionine, cysteine (c) aspartic acid, glutamic acid (d) serine, threonine, tyrosine **9.11** An amino group is part of the side chain. **9.13** The side chain covalently bonds to the amino acid's amino group, producing a cyclic structure. **9.15** (a) alanine (b) leucine (c) methionine (d) tryptophan **9.17** asparagine, glutamine, isoleucine, tryptophan **9.19** (a) polar neutral (b) polar acidic (c) nonpolar (d) polar neutral **9.21** (a) yes (b) yes (c) yes (d) no **9.23** (a) yes (b) no (c) no (d) yes **9.25** (a) no (b) yes (c) yes (d) no **9.27** (a) no (b) no (c) no (d) no **9.29** (a) lysine (b) methionine, tryptophan (c) none (d) methionine **9.31** L family

9.33

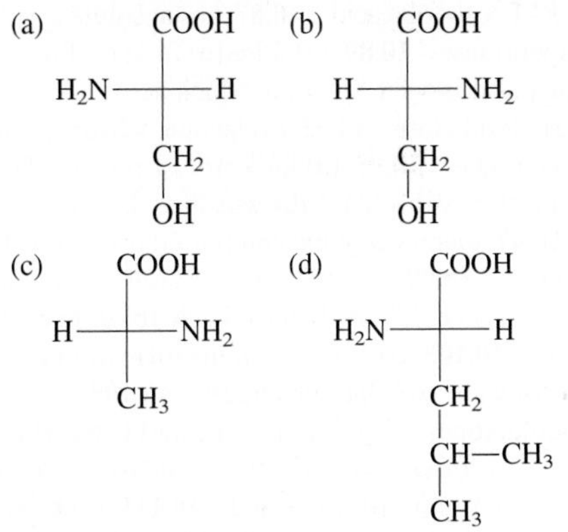

9.35

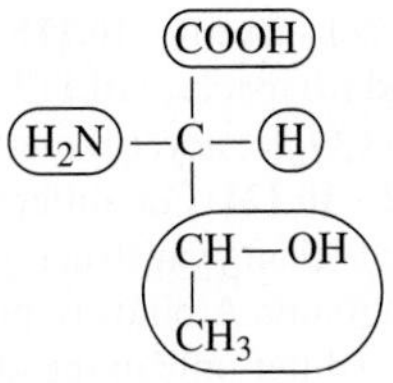

9.37 (a) D- (b) nonpolar (c) essential (d) standard
9.39 They exist as zwitterions.

9.41

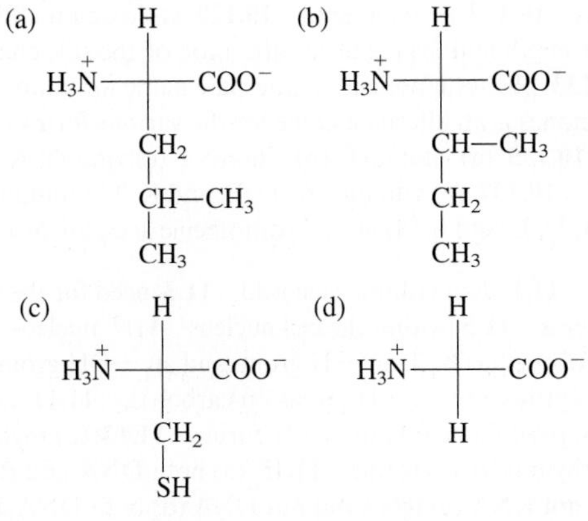

9.43

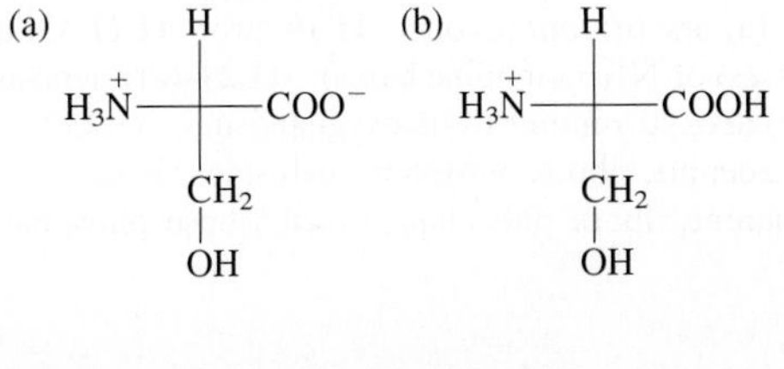

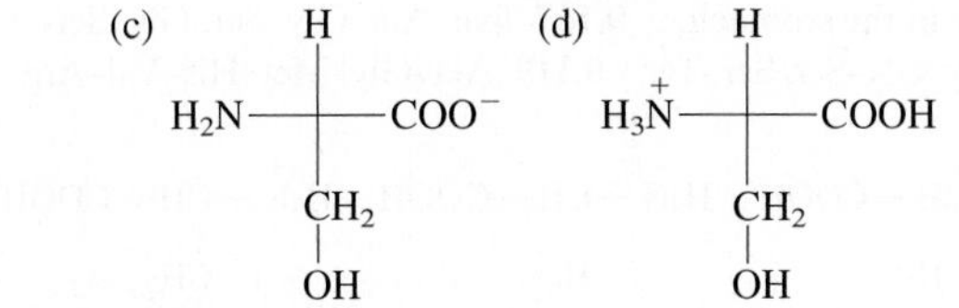

9.45 the pH at which zwitterion concentration in a solution is maximized **9.47** Two —COOH groups are present, which deprotonate at different pH values. **9.49** (a) +1 (b) +2 (c) +1 (d) +1

9.51

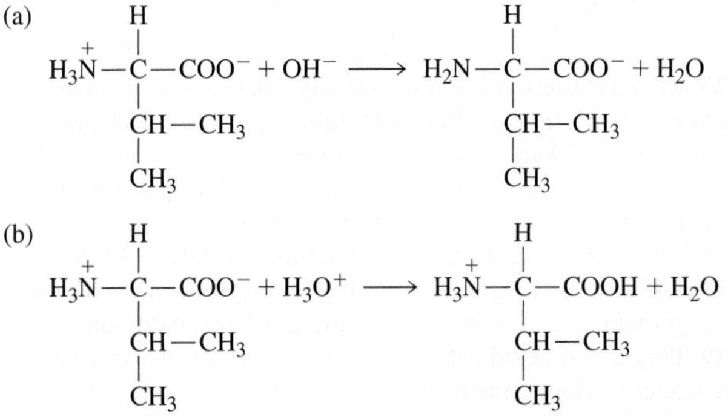

9.53 They react with each other to produce a covalent disulfide bond. **9.55** —COOH and —NH_2 **9.57** $H_3\overset{+}{N}$—CH—

9.59

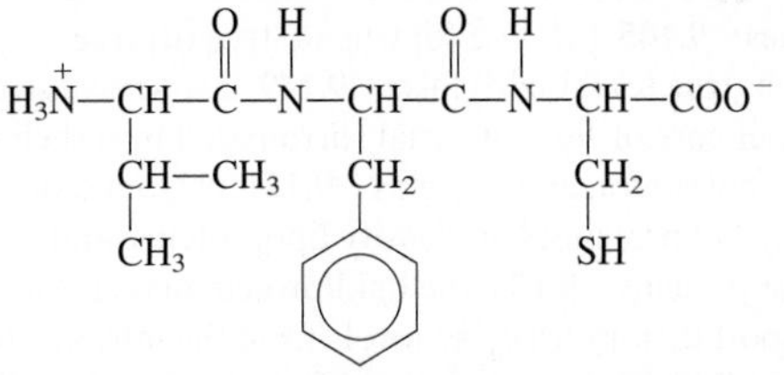

9.61 (a) Ser–Ala–Cys (b) Asp–Thr–Asn **9.63** two in each **9.65** (a) serylcysteine (b) glycylalanylvaline (c) tyrosylaspartylglutamine (d) leucyllysyltryptophylmethionine **9.67** (a) Ala (b) Gly (c) Val (d) none **9.69** (a) Tyr-Leu-Ile (b) two (c) Tyr (d) none **9.71** peptide bonds and α-carbon —CH groups **9.73** Ser is the N-terminal end of Ser–Cys and Cys is the N-terminal end of Cys–Ser. **9.75** Ser–Val–Gly, Val–Ser–Gly, Gly–Ser–Val, Ser–Gly–Val, Val–Gly–Ser, Gly–Val–Ser **9.77** (a) Both are nonapeptides with six of the residues held in the form of a loop by a disulfide bond. (b) They differ in the identity of the amino acid present at two positions in the nonapeptide. **9.79** They bind at the same sites. **9.81** Glu is bonded to Cys through the side-chain carboxyl group rather than through the α-carbon carboxyl group. **9.83** Monomeric proteins contain a single peptide chain and multimeric proteins have two or more peptide chains. **9.85** (a) true (b) false (c) true (d) true **9.87** the order in which the amino acids are bonded to each other **9.89** peptide bond

9.91

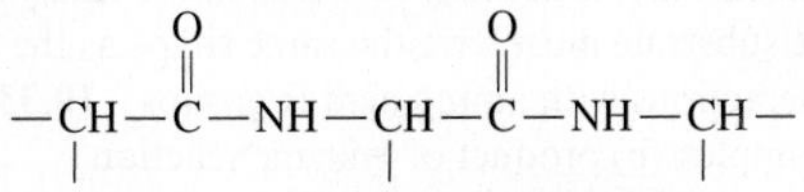

9.93 (a) true (b) false (c) true (d) true **9.95** α helix, β pleated sheet **9.97** (a) false (b) true (c) false (d) false **9.99** an arrangement other than α-helix or β-pleated sheet **9.101** disulfide bonds, electrostatic interactions, hydrogen bonds, and hydrophobic interactions **9.103** (a) nonpolar (b) polar neutral R groups (c) —SH groups (d) acidic and basic R groups **9.105** (a) hydrophobic (b) electrostatic (c) disulfide bond (d) hydrogen bonding **9.107** (a) tertiary (b) secondary, tertiary (c) secondary (d) primary **9.109** the organization among the various peptide chains in a multimeric protein **9.111** They are the same. **9.113** Yes, both Ala and Val are products in each case. **9.115** Drug hydrolysis would occur in the stomach. **9.117** five: Ala–Gly–Ser, Gly–Ser–Tyr, Ala–Gly, Gly–Ser, Ser–Tyr **9.119** Ala–Gly–Met–His–Val–Arg **9.121**

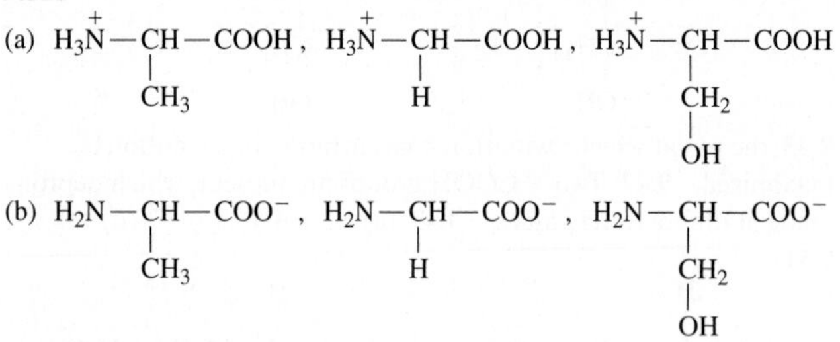

9.123 secondary, tertiary, and quaternary **9.125** same primary structure **9.127** (a) true (b) true (c) false (d) false **9.129** (a) fibrous: generally water-insoluble; globular: generally water-soluble (b) fibrous: support and external protection; globular: involvement in metabolic reactions **9.131** (a) fibrous (b) fibrous (c) globular (d) globular **9.133** (a) false (b) false (c) false (d) false **9.135** (a) contractile protein (b) storage protein (c) transport protein (d) messenger protein **9.137** 4-hydroxyproline and 5-hydroxylysine **9.139** They are involved with cross-linking. **9.141** An antigen is a substance foreign to the human body, and an antibody is a substance that defends against an invading antigen. **9.143** four polypeptide chains that have constant and variable amino acid regions; two chains are longer than the other two; 1%–12% carbohydrates present; long and short chains are connected through disulfide linkages **9.145** (a) false (b) true (c) true (d) true **9.147** (a) true (b) true (c) false (d) false **9.149** a spherical structure with an inner core of lipid material surrounded by a shell of phospholipids, cholesterol, and proteins **9.151** chylomicrons, very-low-density lipoproteins, low-density lipoproteins, and high-density lipoproteins **9.153** the lipid/protein mass ratio **9.155** (a) transport dietary triacylglycerols from the intestine to various locations (b) transport cholesterol from the liver to cells throughout the body **9.157** (a) true (b) true (c) false (d) false

Chapter 10

10.1 catalyst **10.3** larger molecular size, activity regulated by other substances **10.5** (a) conjugated (b) conjugated (c) simple (d) conjugated **10.7** A coenzyme is a cofactor that is an organic substance. A cofactor can be an inorganic or an organic substance. **10.9** to provide additional functional groups **10.11** (a) yes (b) no (c) yes (d) yes **10.13** (a) add a carboxylate group to pyruvate (b) remove H_2 from an alcohol (c) reduce an L-amino acid (d) hydrolyze maltose **10.15** (a) sucrase (or sucrose hydrolase) (b) pyruvate decarboxylase (c) glucose isomerase (d) lactate dehydrogenase **10.17** (a) pyruvate (b) galactose (c) an alcohol (d) an L-amino acid **10.19** (a) isomerase (b) lyase (c) ligase (d) transferase **10.21** (a) isomerase (b) lyase (c) transferase (d) hydrolase **10.23** (a) decarboxylase (b) lipase (c) phosphatase (d) dehydrogenase **10.25** (a) false (b) true (c) true (d) true **10.27** the portion of an enzyme actually involved in the catalysis process **10.29** The substrate must have the same shape as the active site. **10.31** interactions with amino acid R groups **10.33** (a) enzyme-substrate complex (b) product of enzyme reaction **10.35** (a) false (b) true (c) false (d) false **10.37** (a) accepts only one substrate (b) accepts substrate with a particular type of bond **10.39** (a) absolute (b) stereochemical **10.41** (a) absolute (b) linkage (c) group (d) stereochemical **10.43** Rate increases until enzyme denaturation occurs. **10.45** Enzymes vary in the number of acidic and basic amino acids present.
10.47

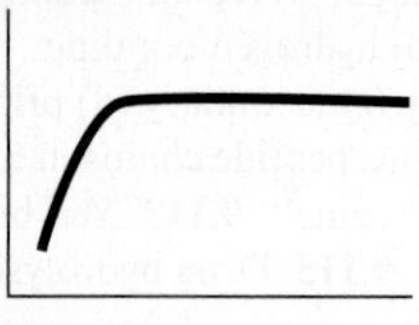

10.49 nothing; the rate remains constant **10.51** (a) decreased rate (b) decreased rate (c) increased rate (d) decreased rate **10.53** (a) true (b) false (c) true (d) true **10.55** a microorganism that thrives in extreme environments **10.57** (a) pH of 9.0 or above (b) salinity that exceeds 0.2 M **10.59** those that can withstand a hot-water environment and/or cold-water environment **10.61** no; only one molecule may occupy the active site at a given time **10.63** (a) reversible competitive (b) reversible noncompetitive, irreversible (c) reversible noncompetitive, irreversible (d) reversible noncompetitive **10.65** enzyme that has quaternary structure and more than one binding site **10.67** The product of a subsequent reaction in a series of reactions inhibits a prior reaction. **10.69** A zymogen is an inactive precursor for a proteolytic enzyme. **10.71** so that they will not destroy the tissues that produce them **10.73** process in which enzyme activity is altered by covalently modifying the structure of the enzyme **10.75** ATP molecules **10.77** protein kinases **10.79** (a) apoenzyme—protein portion of a conjugated enzyme; proenzyme—inactive precursor of an enzyme (b) simple enzyme—contains only protein; allosteric enzyme—contains two or more protein chains and two binding sites **10.81** angiotensin-converting enzyme **10.83** decapeptide versus octapeptide, with both having a common 8-amino-acid sequence **10.85** competitive inhibition of the conversion of PABA to folic acid **10.87** has absolute specificity for bacterial transpeptidase **10.89** (a) false (b) false (c) false (d) false **10.91** tissue plasminogen activator, which activates an enzyme that dissolves blood clots **10.93** (a) lactate dehydrogenase (b) aspartate transaminase **10.95** (a) indicator of possible heart disease (b) indicator of possible heart disease, liver disease, and muscle damage **10.97** dietary organic compound needed by the body in trace amounts **10.99** (a) fat-soluble (b) water-soluble (c) water-soluble (d) water-soluble **10.101** (a) likely (b) unlikely (c) unlikely (d) unlikely **10.103** cosubstrate in the formation of collagen; antioxidant for water-soluble substances **10.105.** L-gulonic acid and γ-L-gulonolactone; γ-L-gulonolactone and L-ascorbic acid **10.107** precursor for enzyme cofactors **10.109** (a) alternative (b) preferred (c) alternative (d) preferred **10.111** (a) thiamin, biotin (b) folate, biotin (c) niacin, vitamin B_6, folate (d) riboflavin **10.113** carboxyl substituent versus amide constituent **10.115** (a) vitamin B_6 (b) thiamin (c) pantothenic acid (d) niacin **10.117** (a) aldehyde (b) one-carbon group other than CO_2 (c) acyl (d) methyl, hydrogen **10.119** alcohol, aldehyde, acid **10.121** Cell differentiation is the process whereby immature cells change in structure and function to become specialized cells. Vitamin A binds to protein receptors in the process. **10.123** They differ only in the identity of the side chain present. **10.125** to maintain normal blood levels of calcium and phosphate ion so that bones can absorb these minerals **10.127** α-tocopherol **10.129** antioxidant effect **10.131** in the length and degree of unsaturation of the side chain present **10.133** phylloquinone is an alternate name for vitamin K_1 and menaquinones is an alternate name for the various forms of vitamin K_2 **10.135** (a) vitamin C (b) vitamin E (c) vitamin A (d) vitamin C **10.137** (a) vitamins A, D, E, and K (b) vitamin B_{12} (c) vitamins A, D, E, and K (d) niacin, pantothenic acid, folate, biotin

Chapter 11

11.1 deoxyribonucleic acid **11.3** need for the synthesis of proteins **11.5** within the cell nucleus **11.7** nucleotides **11.9** Ribose has both an —H group and an —OH group on carbon 2; deoxyribose has 2 —H atoms on carbon 2. **11.11** (a) pyrimidine (b) pyrimidine (c) purine (d) purine **11.13** (a) thymine (b) uracil (c) thymine (d) cytosine **11.15** (a) both DNA and RNA (b) DNA but not RNA (c) RNA but not DNA (d) both DNA and RNA **11.17** (a) one (b) four (c) one **11.19** two **11.21** C1 and N9 (purine bases) or N1(pyrimidine bases) **11.23** (a) adenosine (b) cytidine (c) deoxythymidine (d) deoxyguanosine **11.25** three **11.27** adenine, ribose, phosphate; cytosine, ribose, phosphate; guanine, ribose, phosphate; uracil, ribose, phosphate

11.29 (a) β-*N*-glycosidic linkage (b) phosphate-ester linkage **11.31** (a) adenine, ribose (b) guanine, deoxyribose (c) thymine, deoxyribose (d) uracil, ribose **11.33** (a) adenosine 5′-monophosphate (b) deoxyguanosine 5′-monophosphate (c) deoxythymidine 5′-monophosphate (d) uridine 5′-monophosphate **11.35** (a) RNA but not DNA (b) DNA but not RNA (c) DNA but not RNA (d) RNA but not DNA **11.37** (a) false (b) false (c) false (d) false

11.39

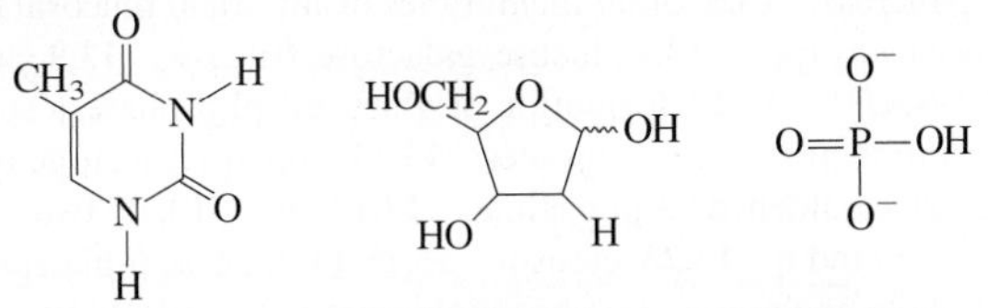

11.41 (a) nucleoside (b) neither nucleoside nor nucleotide (c) nucleotide (d) nucleotide **11.43** a pentose sugar and a phosphate **11.45** base sequence **11.47** 5′ end has a phosphate group attached to the 5′ carbon; 3′ end has a hydroxyl group attached to the 3′ carbon **11.49** a phosphate group and two pentose sugars

11.51

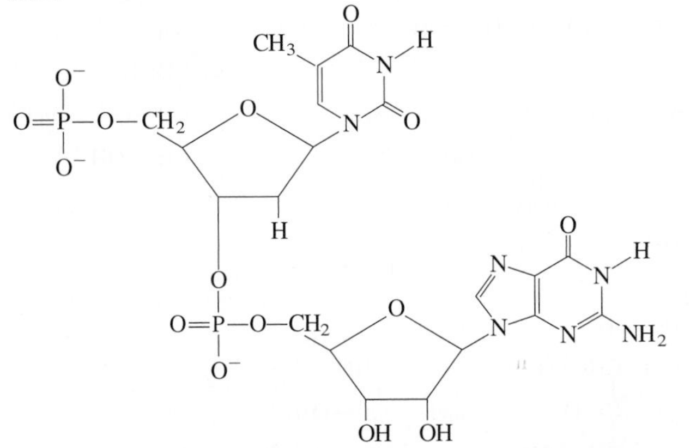

11.53 (a) two polynucleotide chains coiled around each other in a helical fashion (b) The nucleic acid backbones are the outside, and the nitrogen-containing bases are on the inside. **11.55** (a) 36% (b) 14% (c) 14% **11.57** A G–C pairing involves 3 hydrogen bonds, and an A–T pairing involves 2 hydrogen bonds. **11.59** They are the same. **11.61** 5′ TAGCC 3′ **11.63** (a) 5′ GCTA 3′ (b) 5′ ATAA 3′ (c) 5′ ACAC 3′ (d) 5′ CAAC 3′ **11.65** (a) 3′ TGCATA 5′ (b) 3′ AATGGC 5′ (c) 5′ CGTATT 3′ (d) 3′ TTGACC 5′ **11.67** 20 hydrogen bonds **11.69** (a) 7,7 (b) 2, 2 (c) 7, 7 (d) 0, 0 **11.71** catalyzes the unwinding of the double helix structure **11.73** (a) 5′ TTACG 3′ (b) 5′ GCATT 3′ (c) 5′ CGTCG 3′ (d) 5′ GCTGC 3′ **11.75** (a) 3′ TGAATC 5′ (b) 3′ TGAATC 5′ (c) 5′ ACTTAG 3′ **11.77** strand growing toward fork is synthesized continuously; strand growing away from fork is synthesized in segments **11.79** (a) false (b) false (c) false (d) true **11.81** a DNA molecule bound to a group of small proteins **11.83** A **11.85** (a) false (b) false (c) false (d) false **11.87** DNA **11.89** RNA **11.91** (1) RNA contains ribose instead of deoxyribose, (2) RNA contains the base U instead of T, (3) RNA is single-stranded rather than double-stranded, and (4) RNA has a lower molecular mass. **11.93** (a) hnRNA (b) tRNA (c) tRNA (d) hnRNA **11.95** (a) nuclear region (b) extranuclear region (c) extranuclear region (d) both nuclear and extranuclear regions **11.97** a strand of DNA **11.99** causes a DNA helix to unwind; links aligned ribonucleotides together **11.101** T–A, A–U, G–C, C–G **11.103** (a) DNA-RNA (b) DNA-DNA (c) DNA-RNA (d) DNA-DNA or DNA-RNA **11.105** (a) 5′ ATGCCG 3′ and 5′ AUGCCG 3′ (b) 5′ GGTAAT 3′ and 5′ GGUAAU 3′ (c) 5′ TGTACC 3′ and 5′ UGUACC 3′ (d) 5′ TGCATG 3′ and 5′ UGCAUG 3′ **11.107** (a) 3′ GGAATT 5′ (b) 3′ TGCATG 5′ (c) 3′ TGCTGC 5′ (d) 3′ ATGGTA 5′ **11.109** (a) 5′ CCTTAA 3′ (b) 5′ ACGTAC 3′ (c) 5′ ACGACG 3′ (d) 5′ TACC AT 3′ **11.111** Exons convey genetic information whereas introns do not. **11.113** 3′ UUACGCAU 5′ **11.115** 3′ AGUCAAGU 5′ **11.117** (a) hnRNA (b) snRNA **11.119** a mechanism by which a number of proteins that are variations of a basic structural motif can be produced from a single gene **11.121** A three-nucleotide sequence in mRNA that codes for a specific amino acid **11.123** (a) Leu (b) Asn (c) Ser (d) Gly **11.125** (a) CUC, CUA, CUG, UUA, UUG (b) AAC (c) AGC, UCU, UCC, UCA, UCG (d) GGU, GGC, GGA **11.127** The base T cannot be present in a codon. **11.129** Met–Lys–Glu–Asp–Leu **11.131** Ile-Gln-Lys-Lys-Val **11.133** (a) 3′ AGGC-GGTA-ATTGT 5′ (b) 5′ TCCG-CCAT-TAACA 3′(c) 5′ UCCG-UAACA 3′ (d) Ser-Val-Thr **11.135** A cloverleaf shape with three hairpin loops and one open side. **11.137** covalent bond (ester) **11.139** anticodons written in the 3′-to-5′ direction: (a) UCU (b) ACG (c) AAA (d) UUG **11.141** (a) Gly (b) Leu (c) Ala (d) Leu **11.143** (a) UCU, UCC, UCA, UCG, AGU, AGC (b) UUA, UUG, CUU, CUC, CUA, CUG (c) AUU, AUC, AUA (d) GGU, GGC, GGA, GGG **11.145** (a) Thr (b) Ala (c) Arg (d) Cys **11.147** (a) 3′ AGGC-GUAAUUGU 5′ (b) 3′ AGGCGUAAUUGU 5′ (c) UGU UAA UGC GGA (d) ACA AUU ACG CCU **11.149** (a) false (b) true (c) true (d) true **11.151** an amino acid reacts with ATP, the resulting complex reacts with tRNA **11.153** A site **11.155** initial Met residue is removed; covalent modification occurs if needed; completion of folding of protein occurs **11.157** Gly: GGU, GGC, GGA or GGG; Ala: GCU, GCC, GCA or GCG; Cys: UGU or UGC; Val: GUU, GUC, GUA or GUG; Tyr: UAU or UAC **11.159** (a) CAA CGA AAG (b) GUU GCU UUC (c) Gln-Arg-Lys **11.161** (a) mRNA (b) hnRNA (c) tRNA (d) mRNA **11.163** (a) false (b) true (c) true (d) false **11.165** (a) Leu-Gln (b) Leu-Gln (c) Leu-Lys (d) Leu-Lys **11.167** (a) Asn-Tyr (b) Asn-Tyr (c) Asn-Ser (d) Lys-Tyr **11.169** a DNA or an RNA molecule with a protein coating **11.171** (1) attaches itself to cell membrane, (2) opens a hole in the membrane, and (3) injects itself into the cell **11.173** contains a "foreign" gene **11.175** host for a "foreign" gene **11.177** Recombinant DNA is incorporated into a host cell.

11.179

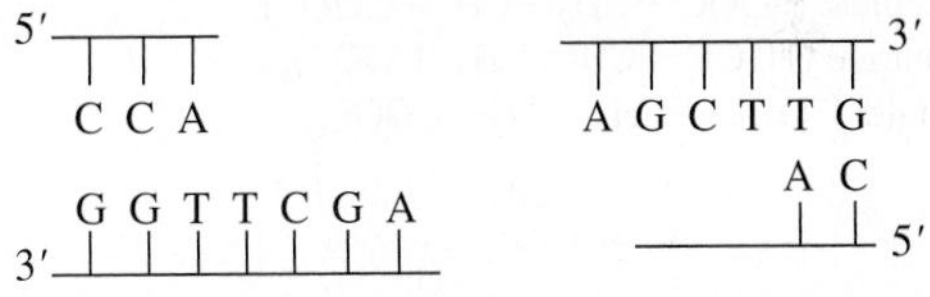

11.181 (a) replication (b) transcription (c) translation (d) recombinant DNA **11.183** to produce many copies of a specific DNA sequence in a relatively short time **11.185** a short nucleotide chain bound to the template DNA strand to which new nucleotides can be attached

Chapter 12

12.1 anabolism—synthetic; catabolism—degradative **12.3** a series of consecutive biochemical reactions **12.5** Large molecules are broken down to smaller ones; energy is released. **12.7** Prokaryotic cells have no nucleus, and the DNA is usually a single circular molecule. Eukaryotic cells have their DNA in a membrane-enclosed nucleus. **12.9** An organelle is a small structure within the cell cytosol that carries out a specific cellular function. **12.11** inner membrane **12.13** region between inner and outer membranes **12.15** adenosine triphosphate

12.17 [phosphate]–[phosphate]–[phosphate]–[ribose]–[adenine]

12.19 three phosphates versus one phosphate **12.21** adenine versus guanine **12.23** HPO_4^{2-} **12.25** ATP + $H_2O \rightarrow$ ADP + P_i + H^+ + energy **12.27** OH becomes part of P_i; H becomes H^+ ion. **12.29** flavin adenine dinucleotide **12.31** [flavin]–[ribitol]–[ADP]

12.33 [nicotinamide]—[ribose]—[phosphate]
|
[adenine]—[ribose]—[phosphate]

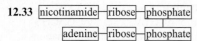

12.35 nicotinamide subunit **12.37** (a) $FADH_2$ (b) NAD^+ **12.39** (a) nicotinamide (b) riboflavin **12.41** (a) two riboses, two phosphates, adenine (b) two phosphates, ribose, adenine

12.43 [2-aminoethanethiol]—[pantothenic acid]—[phosphorylated ADP]

12.45 (a) reducing agent (b) neither (c) oxidizing agent **12.47** (a) pantothenic acid (b) niacin (c) riboflavin **12.49** (a) NAD^+ (b) CoA-SH, FAD, NAD^+ (c) all four (d) CoA-SH, FAD, NAD^+

12.51

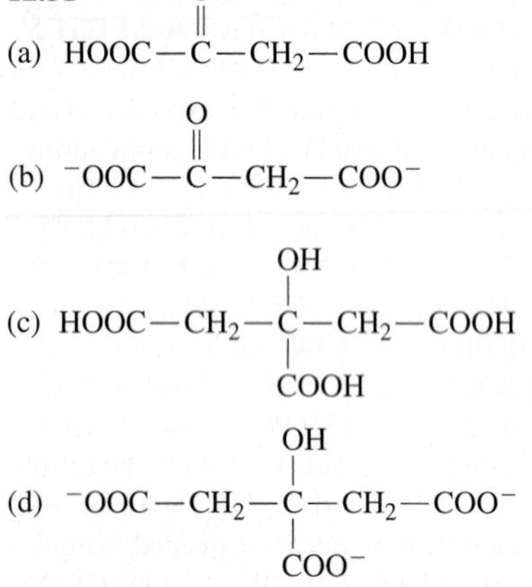

12.53 (a) malate, oxaloacetate, fumarate (b) oxaloacetate, α-ketoglutarate (c) malate, oxaloacetate, fumarate, α-ketoglutarate (d) malate, oxaloacetate, fumarate **12.55** a compound with a greater free energy of hydrolysis than is typical for a compound **12.57** free monophosphate species **12.59** (a) phosphoenolpyruvate (b) creatine phosphate (c) 1,3-bisphosphoglycerate (d) AMP **12.61** (1) digestion, (2) acetyl group formation, (3) citric acid cycle, (4) electron transport chain and oxidative phosphorylation **12.63** tricarboxylic acid cycle, Krebs cycle **12.65** acetyl CoA **12.67** (a) 2 (b) 1 (c) 2 (Steps 3, 8) (d) 2 (Steps 2, 7) **12.69** (a) Steps 3, 4, 6, 8 (b) Step 2 (c) Step 7 **12.71** succinate ($^-OOC-CH_2-CH_2-COO^-$); fumarate ($^-OOC-CH=CH-COO^-$); malate ($^-OOC-CH_2-CH(OH)-COO^-$); oxaloacetate ($^-OOC-CH_2-C(=O)-COO^-$) **12.73** oxidation and decarboxylation **12.75** (a) C_6 to C_6 (b) C_4 to C_4 (c) C_4 to C_4 (d) C_6 to C_5 **12.77** (a) NAD^+ (b) FAD **12.79** (a) isocitrate, α-ketoglutarate (b) fumarate, malate (c) malate, oxaloacetate (d) citrate, isocitrate **12.81** respiratory chain **12.83** O_2 **12.85** (a) FMN (b) $CoQH_2$ **12.87** (a) oxidized (b) oxidized (c) reduced (d) oxidized **12.89** (a) oxidation (b) reduction (c) oxidation (d) reduction **12.91** (a) I (b) I, II, III (c) III (d) IV **12.93** $CoQH_2$ **12.95** two **12.97** (a) $FADH_2$, 2 Fe(II)SP, $CoQH_2$ (b) $FMNH_2$, 2 Fe^{2+}, $CoQH_2$ **12.99** (a) ETC (b) ETC (c) CAC (d) ETC **12.101** CO_2, succinate, $FADH_2$, FeSP **12.103** (a) neither (b) reduced (c) oxidized (d) both **12.105** ATP synthesis from ADP and Pi using energy from the electron transport chain **12.107** protons (H^+ ions) **12.109** intermembrane space **12.111** ATP synthase **12.113** Coenzymes are oxidized; ADP is phosphorylated. **12.115** (a) false (b) true (c) true (d) false **12.117** (a) false (b) true (c) false (d) true **12.119** (a) 2.5 molecules ATP (b) 1.5 molecules ATP **12.121** (a) 3,4, and 8 (b) 6 (c) 5 **12.123** It is an intermediate amount of free energy, higher than some reactions and lower than others. **12.125** (a) ATP to ADP (b) ATP to ADP **12.127** reactive oxygen species

12.129 (a) $2O_2 + NADPH \longrightarrow 2O_2^- + NADP^+ + H^+$
(b) $2O_2^- + 2H^+ \longrightarrow H_2O_2 + O_2$

12.131 (a) false (b) true (c) true (d) true **12.133** (a) yes (b) no (c) yes (d) no **12.135** (a) no (b) yes (c) no (d) no

Chapter 13 **13.1** mouth, salivary α-amylase **13.3** small intestine, pancreas **13.5** outer membranes of intestinal mucosal cells, sucrose hydrolysis **13.7** glucose, galactose, fructose **13.9** glucose **13.11** NAD^+ **13.13** formation of glucose 6-phosphate, a species that cannot cross cell membranes **13.15** dihydroxyacetone phosphate, glyceraldehyde 3-phosphate **13.17** two **13.19** two **13.21** Steps 1, 3, and 6 **13.23** cytosol **13.25** (a) glucose 6-phosphate (b) 2-phosphoglycerate (c) phosphoglyceromutase (d) ADP **13.27** (a) Step 10 (b) Step 1 (c) Step 8 (d) Step 6 **13.29** (a) +2 (b) +4 **13.31**

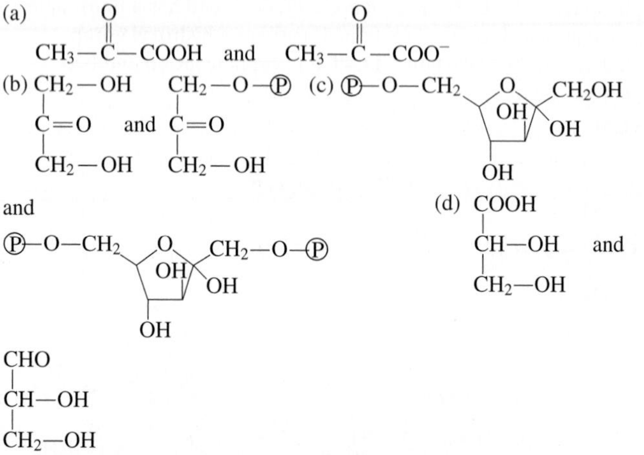

13.33

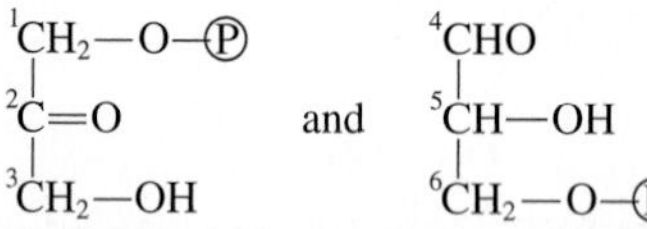

13.35 (a) glycolysis (b) digestion (c) digestion (d) glycolysis **13.37** acetyl CoA, lactate, ethanol **13.39** pyruvate + CoA + $NAD^+ \rightarrow$ acetyl CoA + NADH + CO_2 **13.41** NADH is oxidized to NAD^+, a substance needed for glycolysis. **13.43** CO_2 **13.45** glucose + 2ADP + $2P_i \rightarrow$ 2lactate + 2ATP + $2H_2O$ **13.47** (a) acetyl CoA, ethanol (b) lactate, ethanol (c) acetyl CoA (d) lactate **13.49** (a) pyruvate oxidation, ethanol fermentation (b) pyruvate oxidation (c) glycolysis (d) all four **13.51** (a) false (b) false (c) false (d) true **13.53** decreases ATP production by 2 **13.55** 30 ATP versus 2 ATP **13.57** two **13.59** Glycogenesis converts glucose 6-phosphate to glycogen and glycogenolysis is the reverse process. **13.61** glucose 6-phosphate **13.63** UTP **13.65** UDP + ATP → UTP + ADP **13.67** Step 2 **13.69** Glucose 6-phosphate is converted to glucose in liver cells and used as is in muscle cells. **13.71** as glucose 6-phosphate **13.73** (a) all three (b) glycogenesis (c) glycogenesis (d) glycolysis **13.75** the liver **13.77** two-step pathway for Step 10; different enzymes for Steps 1 and 3 **13.79** oxaloacetate **13.81** goes to the liver, where it is converted to glucose **13.83** (a) glycolysis but not gluconeogenesis (b) glycolysis but not gluconeogenesis (c) gluconeogenesis but not glycolysis (d) both glycolysis and gluconeogenesis **13.85** (a) both glycolysis and gluconeogenesis (b) both glycolysis and gluconeogenesis (c) both glycolysis and gluconeogenesis (d) both glycolysis and gluconeogenesis **13.87** (a) eight (b) seven (c) seven (d) two **13.89** (a) all four processes (b) glycolysis and gluconeogenesis (c) gluconeogenesis (d) glycogenesis **13.91** (a) glycolysis (b) glycolysis (c) glycogenesis (d) glycolysis **13.93** (a) gluconeogenesis (b) glycolysis (c) glycogenesis (d) glycogenesis and glycogenolysis **13.95** (a) loss of six (b) loss of two (c) gain of three (d) loss of four **13.97** glucose 6-phosphate **13.99** NADPH is consumed in its reduced form; NADH is consumed in its oxidized form (NAD^+). **13.101** Glucose 6-phospate + 2 $NADP^+$ + $H_2O \rightarrow$ ribulose 5-phosphate + CO_2 + 2NADPH + $2H^+$ **13.103** CO_2 **13.105** (a) Cori cycle, lactate fermentation (b) Cori cycle, lactate fermentation, glycolysis (c) pentose phosphate pathway, Cori cycle, glycolysis (d) pentose phosphate pathway **13.107** increases rate of glycogen synthesis **13.109** increases blood glucose levels **13.111** pancreas **13.113** Epinephrine attaches to cell membrane and stimulates the production of cAMP, which activates glycogen phosphorylase. **13.115** glucagon (liver cells) and epinephrine (muscle cells) **13.117** (a) true (b) false (c) false (d) true **13.119** (a) none (b) none (c) none (d) glycogenolysis

Chapter 14 **14.1** 98% **14.3** no effect **14.5** (a) stomach and small intestine (b) stomach (10%) and small intestine (90%) (c) stomach (gastric lipases), small intestine (pancreatic lipases) **14.7** acts as an emulsifier **14.9** monoacylglycerols are the major product **14.11** reassembled into triacylglycerols; converted to chylomicrons **14.13** They have a large storage capacity for triacylglycerols. **14.15** hydrolysis of triacylglycerols in adipose tissue; entry of hydrolysis products into bloodstream **14.17** activates hormone-sensitive lipase **14.19** (a) Step 1 (b) Step 1 (c) Step 2 (d) Step 1 **14.21** one **14.23** (a) true (b) false (c) true 2 (d) false **14.25** (a) intermembrane space (b) matrix (c) intermembrane space (d) matrix **14.27** the ATP is converted to AMP rather than ADP **14.29** (a) alkane to alkene (b) alkene to 2° alcohol (c) 2° alcohol to ketone **14.31** *trans* isomer

14.33

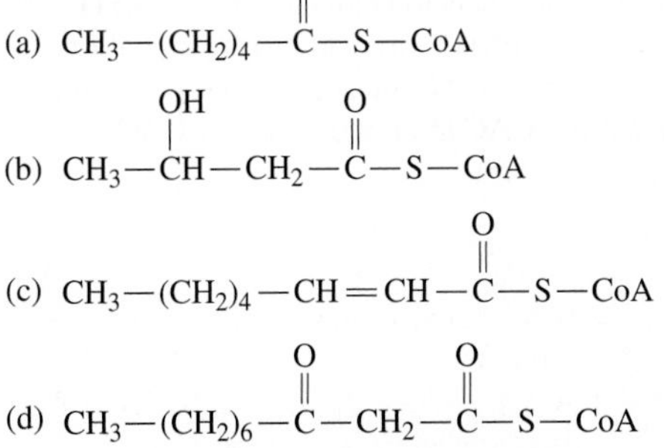

14.35 (a) acyl CoA (b) hydroxyacyl CoA (c) ketoacyl CoA (d) ketoacyl CoA **14.37** (a) activation (b) oxidation (c) all three (d) oxidation **14.39** (a) Step 3, turn 2 (b) Step 2, turn 3 (c) Step 4, turn 3 (d) Step 1, turn 2 **14.41** compounds a and d **14.43** (a) 7 turns (b) 5 turns **14.45** A *cis–trans* isomerase converts a *cis* bond to a *trans* bond **14.47** (a) both (b) glycerol metabolism (c) glycerol metabolism (d) fatty acid metabolism **14.49** (a) glucose (b) fatty acids **14.51** (a) 4 turns (b) 5 acetyl CoA (c) 4 NADH (d) 4 $FADH_2$ (e) 2 high-energy bonds **14.53** 64 ATP **14.55** unsaturated fatty acid produces less $FADH_2$ **14.57** 4 kcal versus 9 kcal **14.59** (a) true (b) true (c) true (d) false **14.61** (1) dietary intakes high in fat and low in carbohydrates, (2) inadequate processing of glucose present, and (3) prolonged fasting **14.63** Ketone body formation occurs when oxaloacetate concentrations are low.

14.65

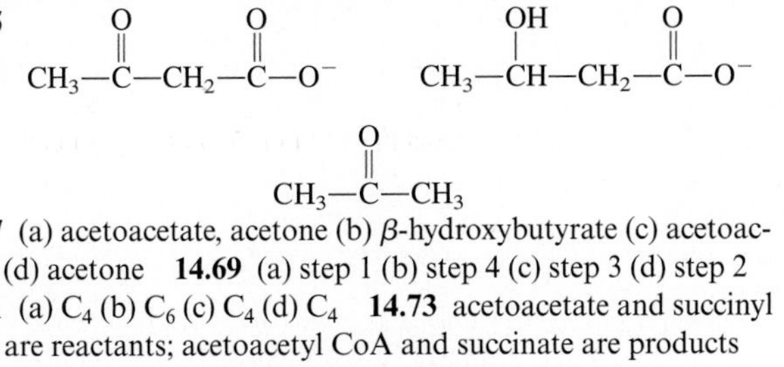

14.67 (a) acetoacetate, acetone (b) β-hydroxybutyrate (c) acetoacetate (d) acetone **14.69** (a) step 1 (b) step 4 (c) step 3 (d) step 2 **14.71** (a) C_4 (b) C_6 (c) C_4 (d) C_4 **14.73** acetoacetate and succinyl CoA are reactants; acetoacetyl CoA and succinate are products **14.75** accumulation of ketone bodies in blood and urine **14.77** (a) β-oxidation (b) β-oxidation (c) both (d) ketogenesis **14.79** (a) ketogenesis only (b) both (c) β-oxidation only (d) ketogenesis **14.81** cytosol versus mitochondrial matrix **14.83** acyl carrier protein; polypeptide chain replaces phosphorylated AMP **14.85** (a) matrix (b) cytosol (c) cytosol (d) cytosol **14.87** (a) malonyl CoA (b) acetyl ACP (c) malonyl CoA (d) acetyl ACP **14.89** (a) Step 2 (b) Step 3 (c) Step 1 (d) Step 4 **14.91** (a) acetoacetyl ACP (b) crotonyl ACP (c) crotonyl ACP (d) acetoacetyl ACP **14.93** (a) condensation (b) hydrogenation (c) dehydration (d) hydrogenation **14.95** needed to convert saturated fatty acids to unsaturated fatty acids **14.97** (a) 6 turns (b) 6 malonyl ACP (c) 6 ATP bonds (d) 12 NADPH **14.99** (a) all three (b) lipogenesis (c) β-oxidation (d) β-oxidation **14.101** (a) β-oxidation (b) lipogenesis (c) lipogenesis (d) lipogenesis **14.103** (a) β-oxidation (b) β-oxidation (c) ketogenesis (d) glycerol metabolism **14.105** (a) succinate (b) oxaloacetate (c) malate (d) fumarate **14.107** (a) unsaturated acid (b) ketoacid (c) ketoacid (d) hydroxyacid **14.109** (a) produced in multi-step process (b) produced in one step (c) produced in one step (d) cannot be produced **14.111** The human body lacks the enzymes needed to convert acetyl CoA to pyruvate. **14.113** (a) false (b) true (c) true (d) true **14.115** (a) mitochondria (b) cytosol (c) mitochondria (d) mitochondria **14.117** (a) all four (b) no correct response (c) all four (d) no correct response

Chapter 15 **15.1** Denaturation occurs in the stomach with gastric juice as the denaturant. **15.3** Pepsinogen is the inactive precursor of pepsin. **15.5** Gastric juice is acidic (1.5–2.0 pH) and pancreatic juice is basic (7.0–8.0 pH). **15.7** Membrane protein molecules facilitate the passage of amino acids through the intestinal wall. **15.9** (a) stomach (b) small intestine (c) stomach (d) small intestine **15.11** (a) hormone (b) digestive enzyme (c) neither (d) digestive enzyme **15.13** total supply of free amino acids available for use **15.15** cyclic process of protein degradation and resynthesis **15.17** A positive nitrogen balance has nitrogen intake exceeding nitrogen output; a negative nitrogen balance has nitrogen output exceeding nitrogen intake. **15.19** negative balance; proteins are degraded to get the needed amino acids **15.21** protein synthesis; synthesis of nonprotein nitrogen-containing compounds; nonessential amino acid synthesis; energy production **15.23** (a) essential (b) nonessential (c) nonessential (d) essential **15.25** b and c **15.27** (a) oxaloacetate, aspartate (b) oxaloacetate, α-ketoglutarate (c) α-ketoglutarate, glutamate (d) all four **15.29** an amino acid and an α-keto acid

15.31

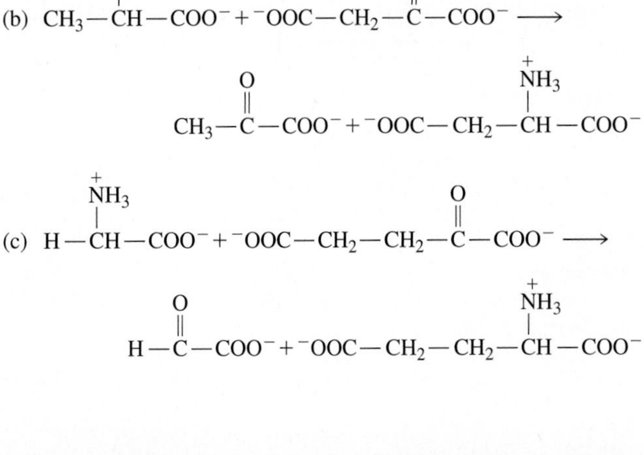

(d)

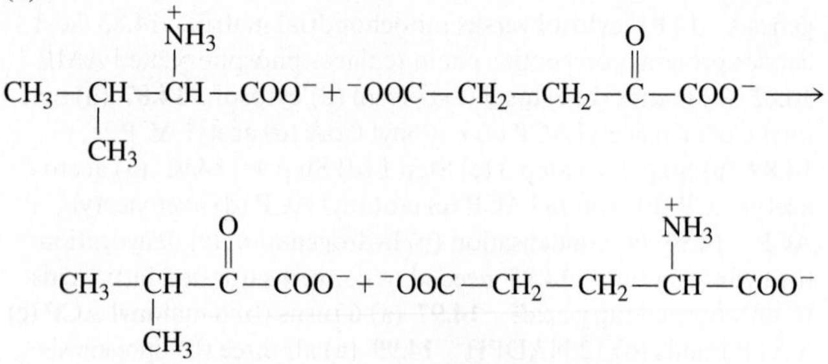

15.33 (a) serine aminotransferase (b) alanine aminotransferase (c) glycine aminotransferase (d) threonine aminotransferase **15.35** pyruvate, α-ketoglutarate **15.37** the specificity of aminotransferases for α-ketoglutarate **15.39** coenzyme that participates in the amino group transfer **15.41** conversion of an amino acid into a keto acid with the release of ammonium ion **15.43** Oxidative deamination produces ammonium ion, and transamination produces an amino acid.
15.45 (a)

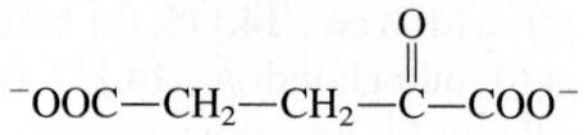

(b)

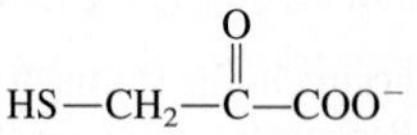

(c)

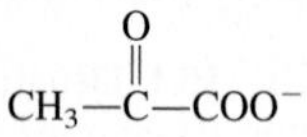

(d)

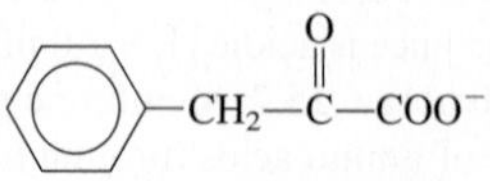

15.47 Transamination of the α-keto acid produces the amino acid.

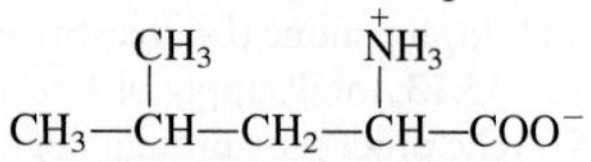

15.49 (a) aspartate (b) glutamate (c) pyruvate (d) α-ketoglutarate **15.51** (a) both (b) both (c) oxidative deamination (d) oxidative deamination

15.53

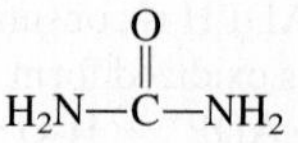

15.55 carbamoyl phosphate **15.57** an amide group,

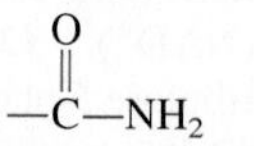

15.59 aspartate and arginine **15.61** (a) carbamoyl phosphate (b) aspartate **15.63** carbamoyl phosphate (step 1) and aspartate (step 2) **15.65** (a) N_2 (b) N_3 (c) N_1 (d) N_4 **15.67** (a) citrulline (b) ornithine (c) argininosuccinate (d) carbamoyl phosphate **15.69** (a) step 2 (b) step 2 (c) step 4 (d) step 4 **15.71** equivalent of four ATP molecules **15.73** goes to the citric acid cycle where it is converted to oxaloacetate, which is then converted to aspartate **15.75** (a) transamination (b) urea cycle (c) oxidative deamination, urea cycle (d) urea cycle **15.77** (a) N_1 (b) N_1 (c) N_2 (d) N_3 **15.79** (a) true (b) false (c) false (d) true **15.81** (a) false (b) false (c) true (d) false **15.83** α-ketoglutarate, succinyl CoA, fumarate, oxaloacetate **15.85** (a) acetoacetyl CoA and acetyl CoA (b) succinyl CoA and acetyl CoA (c) fumarate and oxaloacetate (d) α-ketoglutarate **15.87** (a) ketogenic (b) both (c) glucogenic (d) glucogenic **15.89** Degradation products can be used to make glucose. **15.91** glutamate **15.93** pyruvate, α-ketoglutarate, 3-phosphoglycerate, oxaloacetate, and phenylalanine **15.95** hydrolyzed to amino acids **15.97** In biliverdin the heme ring has been opened and one carbon atom has been lost (as CO). **15.99** biliverdin, bilirubin, bilirubin diglucuronide, urobilin **15.101** urobilin **15.103** excess bilirubin **15.105** (a) biliverdin (b) bilirubin (c) biliverdin (d) stercobilin **15.107** (a) urea cycle (b) oxidative deamination (c) heme degradation (d) oxidative deamination **15.109** Amino acid carbon skeletons are degraded to acetyl CoA or acetoacetyl CoA; ketogenesis converts these degradation products to ketone bodies. **15.111** converted to body fat stores **15.113** (a) oxidative deamination, non-CAC intermediates (b) non-CAC intermediates (c) CAC intermediates (d) transamination, CAC intermediates, non-CAC intermediates

Index/Glossary